MW01284231

BIG IDEAS MATH®

Geometry

A Common Core Curriculum

Ron Larson and Laurie Boswell

Erie, Pennsylvania
BigIdeasLearning.com

Big Ideas Learning, LLC
1762 Norcross Road
Erie, PA 16510-3838
USA

For product information and customer support, contact Big Ideas Learning
at **1-877-552-7766** or visit us at ***BigIdeasLearning.com***.

Cover Image
Image of metal sculpture by Vladimir Bulatov

Printed in the U.S.A.

ISBN 13: 978-1-60840-839-9
ISBN 10: 1-60840-839-6

7 8 9 10 WEB 18 17 16 15

Authors

Ron Larson, Ph.D., is well known as the lead author of a comprehensive program for mathematics that spans middle school, high school, and college courses. He holds the distinction of Professor Emeritus from Penn State Erie, The Behrend College, where he taught for nearly 40 years. He received his Ph.D. in mathematics from the University of Colorado. Dr. Larson's numerous professional activities keep him actively involved in the mathematics education community and allow him to fully understand the needs of students, teachers, supervisors, and administrators.

Ron Larson

Laurie Boswell, Ed.D., is the Head of School and a mathematics teacher at the Riverside School in Lyndonville, Vermont. Dr. Boswell is a recipient of the Presidential Award for Excellence in Mathematics Teaching and has taught mathematics to students at all levels, from elementary through college. Dr. Boswell was a Tandy Technology Scholar and served on the NCTM Board of Directors from 2002 to 2005. She currently serves on the board of NCSM and is a popular national speaker.

Laurie Boswell

Dr. Ron Larson and **Dr. Laurie Boswell** began writing together in 1992. Since that time, they have authored over two dozen textbooks. In their collaboration, Ron is primarily responsible for the student edition while Laurie is primarily responsible for the teaching edition.

For the Student

Welcome to *Big Ideas Math Geometry*. From start to finish, this program was designed with you, the learner, in mind.

As you work through the chapters in your Geometry course, you will be encouraged to think and to make conjectures while you persevere through challenging problems and exercises. You will make errors—and that is ok! Learning and understanding occur when you make errors and push through mental roadblocks to comprehend and solve new and challenging problems.

In this program, you will also be required to explain your thinking and your analysis of diverse problems and exercises. You will master content through engaging explorations that will provide deeper understanding, concise stepped-out examples, and rich thought-provoking exercises. Being actively involved in learning will help you develop mathematical reasoning and use it to solve math problems and work through other everyday challenges.

We wish you the best of luck as you explore Geometry. We are excited to be a part of your preparation for the challenges you will face in the remainder of your high school career and beyond.

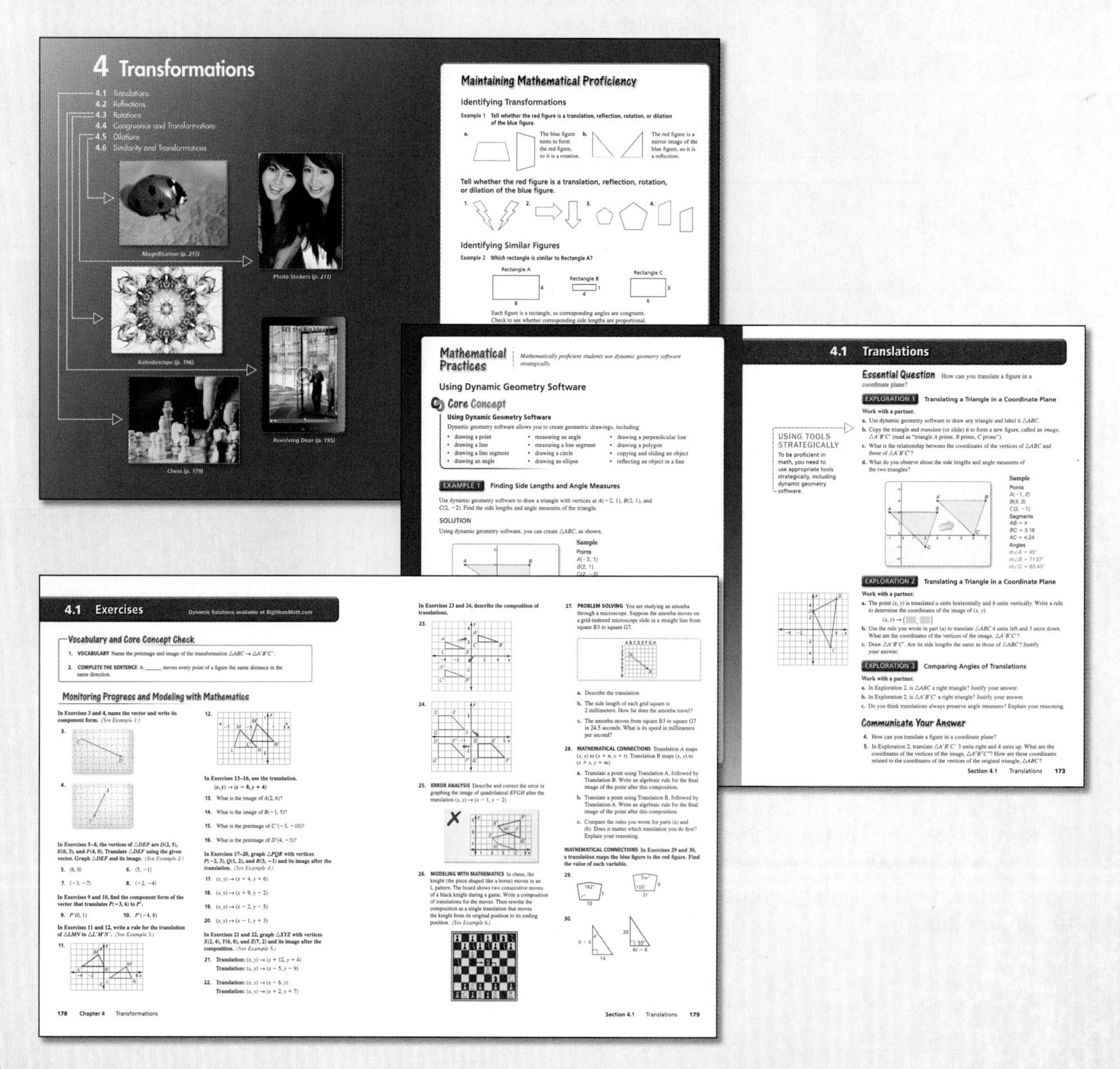

4 Transformations

4.1 Translations
4.2 Reflections
4.3 Rotations
4.4 Congruence and Transformations
4.5 Dilations
4.6 Similarity and Transformations

Magnification (p. 213)
Photo Stickers (p. 211)
Kaleidoscope (p. 196)
Revolving Door (p. 195)
Chess (p. 179)

Maintaining Mathematical Proficiency

Identifying Transformations

Tell whether the red figure is a translation, reflection, rotation, or dilation of the blue figure.

Identifying Similar Figures

Mathematical Practices

Using Dynamic Geometry Software

Core Concept

4.1 Translations

Essential Question How can you translate a figure in a coordinate plane?

Communicate Your Answer

4.1 Exercises

Vocabulary and Core Concept Check

Monitoring Progress and Modeling with Mathematics

Big Ideas Math High School Research

Big Ideas Math Algebra 1, *Geometry*, and *Algebra 2* is a research-based program providing a rigorous, focused, and coherent curriculum for high school students. Ron Larson and Laurie Boswell utilized their expertise as well as the body of knowledge collected by additional expert mathematicians and researchers to develop each course.

The pedagogical approach to this program follows the best practices outlined in the most prominent and widely-accepted educational research and standards, including:

Achieve, ACT, and The College Board

Adding It Up: Helping Children Learn Mathematics
National Research Council ©2001

Common Core State Standards for Mathematics
National Governors Association Center for Best Practices and the Council of Chief State School Officers ©2010

Curriculum Focal Points and the *Principles and Standards for School Mathematics* ©2000
National Council of Teachers of Mathematics (NCTM)

Project Based Learning
The Buck Institute

Rigor/Relevance Framework™
International Center for Leadership in Education

Universal Design for Learning Guidelines
CAST ©2011

Big Ideas Math would like to express our gratitude to the mathematics education and instruction experts who served as consultants during the writing of *Big Ideas Math Algebra 1*, *Geometry*, and *Algebra 2*. Their input was an invaluable asset during the development of this program.

Kristen Karbon
Curriculum and Assessment Coordinator
Troy School District
Troy, Michigan

Jean Carwin
Math Specialist/TOSA
Snohomish School District
Snohomish, Washington

Carolyn Briles
Performance Tasks Consultant
Mathematics Teacher, Loudoun County Public Schools
Leesburg, Virginia

Bonnie Spence
Differentiated Instruction Consultant
Mathematics Lecturer, The University of Montana
Missoula, Montana

Connie Schrock, Ph.D.
Performance Tasks Consultant
Mathematics Professor, Emporia State University
Emporia, Kansas

We would also like to thank all of our reviewers who took the time to provide feedback during the final development phases. For a complete list of the *Big Ideas Math* program reviewers, please visit *www.BigIdeasLearning.com*.

Common Core State Standards for Mathematical Practice

Make sense of problems and persevere in solving them.

- *Essential Questions* help students focus on core concepts as they analyze and work through each *Exploration*.
- Section opening *Explorations* allow students to struggle with new mathematical concepts and explain their reasoning in the *Communicate Your Answer* questions.

Reason abstractly and quantitatively.

- *Reasoning*, *Critical Thinking*, *Abstract Reasoning*, and *Problem Solving* exercises challenge students to apply their acquired knowledge and reasoning skills to solve each problem.
- *Thought Provoking* exercises test the reasoning skills of students as they analyze and interpret perplexing scenarios.

Construct viable arguments and critique the reasoning of others.

- Students must justify their responses to each *Essential Question* in the *Communicate Your Answer* questions at the end of each *Exploration* set.
- Students are asked to construct arguments and critique the reasoning of others in specialized exercises, including *Making an Argument*, *How Do You See It?*, *Drawing Conclusions*, *Reasoning*, *Error Analysis*, *Problem Solving*, and *Writing*.

Model with mathematics.

- Real-life scenarios are utilized in *Explorations*, *Examples*, *Exercises*, and *Assessments* so students have opportunities to apply the mathematical concepts they have learned to realistic situations.
- *Modeling with Mathematics* exercises allow students to interpret a problem in the context of a real-life situation, often utilizing tables, graphs, visual representations, and formulas.

Use appropriate tools strategically.

- Students are provided opportunities for selecting and utilizing the appropriate mathematical tool in *Using Tools* exercises. Students work with graphing calculators, dynamic geometry software, models, and more.
- A variety of tool papers and manipulatives are available for students to use in problems as strategically appropriate.

Attend to precision.

- *Vocabulary and Core Concept Check* exercises require students to use clear, precise mathematical language in their solutions and explanations.
- The many opportunities for cooperative learning in this program, including working with partners for each *Exploration*, support precise, explicit mathematical communication.

Look for and make use of structure.

- *Using Structure* exercises provide students with the opportunity to explore patterns and structure in mathematics.
- *Proof* exercises require students to understand and apply the structure of geometric theorems to solve each problem.

Look for and express regularity in repeated reasoning.

- Students are continually encouraged to evaluate the reasonableness of their solutions and their steps in the problem-solving process.
- Stepped-out *Examples* encourage students to maintain oversight of their problem-solving process and pay attention to the relevant details in each step.

Go to *BigIdeasLearning.com* for more information on the Common Core State Standards for Mathematical Practice.

Common Core State Standards for Mathematical Content for Geometry

Chapter Coverage for Standards

1 **2** **3** **4** **5** **6** **7** 8 9 **10** **11** 12

Conceptual Category Geometry

- Congruence

1 2 3 **4** **5** 6 **7** **8** **9** 10 11 12

Conceptual Category Geometry

- Similarity, Right Triangles, and Trigonometry

1 2 3 4 5 **6** 7 8 9 **10** **11** 12

Conceptual Category Geometry

- Circles

1 2 **3** 4 **5** 6 7 **8** 9 **10** 11 12

Conceptual Category Geometry

- Expressing Geometric Properties with Equations

1 2 3 4 5 6 7 8 9 10 **11** 12

Conceptual Category Geometry

- Geometric Measurement and Dimension

1 2 3 4 5 6 7 8 9 10 11 **12**

Conceptual Category Statistics and Probability

- Probability

Go to *BigIdeasLearning.com* for more information on the Common Core State Standards for Mathematical Content.

1 Basics of Geometry

See the Big Idea

Learn how bridges are designed using compression and tension.

2 Reasoning and Proofs

See the Big Idea

Tigers and humans display obvious differences between males and females. Use logic to determine whether other mammals do.

3 Parallel and Perpendicular Lines

See the Big Idea

Discover why parallel lines and reference points are so important to builders.

Transformations 4

See the Big Idea
Investigate the rotational symmetry of revolving doors and discover why this technology allows skyscrapers to be built.

5 Congruent Triangles

See the Big Idea
Learn how to use triangle congruence in a model hang glider challenge.

Relationships Within Triangles

See the Big Idea
Discover why triangles are used in building for strength.

7 Quadrilaterals and Other Polygons

See the Big Idea
Explore what the refractive index, reflected light, and light dispersion have to do with diamonds.

Similarity 8

See the Big Idea
Discover how many different ways you can scale a model.

Right Triangles and Trigonometry

See the Big Idea
Test the accuracy of two measurement methods and discover which one prevails.

10 Circles

See the Big Idea
Utilize trilateration to find the epicenters of historical earthquakes and discover where they lie on known fault lines.

11 Circumference, Area, and Volume

See the Big Idea

Analyze the population density in various parts of Los Angeles—as viewed from an observation point high in the hills of Santa Monica.

Probability 12

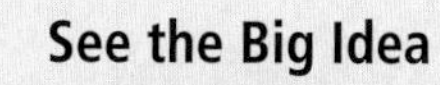

See the Big Idea
Learn about caring for trees at an arboretum.

How to Use Your Math Book

Get ready for each chapter by **Maintaining Mathematical Proficiency** and reviewing the **Mathematical Practices**. Begin each section by working through the **EXPLORATIONS** to **Communicate Your Answer** to the **Essential Question**. Each **Lesson** will explain **What You Will Learn** through **EXAMPLES**, **Core Concepts**, and **Core Vocabulary**.

Answer the **Monitoring Progress** questions as you work through each lesson. Look for STUDY TIPS, COMMON ERRORS, and suggestions for looking at a problem ANOTHER WAY throughout the lessons. We will also provide you with guidance for accurate mathematical READING and concept details you should REMEMBER.

Sharpen your newly acquired skills with **Exercises** at the end of every section. Halfway through each chapter you will be asked **What Did You Learn?** and you can use the Mid-Chapter **Quiz** to check your progress. You can also use the **Chapter Review** and **Chapter Test** to review and assess yourself after you have completed a chapter.

Apply what you learned in each chapter to a **Performance Task** and build your confidence for taking standardized tests with each chapter's **Cumulative Assessment**.

For extra practice in any chapter, use your *Online Resources*, *Skills Review Handbook*, or your *Student Journal*.

1 Basics of Geometry

Alamillo Bridge *(p. 53)*

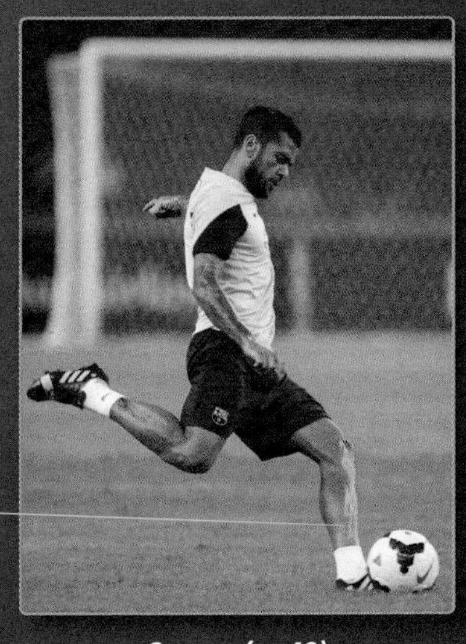

Soccer *(p. 49)*

Shed *(p. 33)*

Skateboard *(p. 20)*

Sulfur Hexafluoride *(p. 7)*

Maintaining Mathematical Proficiency

Finding Absolute Value

Example 1 **Simplify $|-7 - 1|$.**

$|-7 - 1| = |-7 + (-1)|$ Add the opposite of 1.

$= |-8|$ Add.

$= 8$ Find the absolute value.

▶ $|-7 - 1| = 8$

Simplify the expression.

1. $|8 - 12|$ **2.** $|-6 - 5|$ **3.** $|4 + (-9)|$

4. $|13 + (-4)|$ **5.** $|6 - (-2)|$ **6.** $|5 - (-1)|$

7. $|-8 - (-7)|$ **8.** $|8 - 13|$ **9.** $|-14 - 3|$

Finding the Area of a Triangle

Example 2 **Find the area of the triangle.**

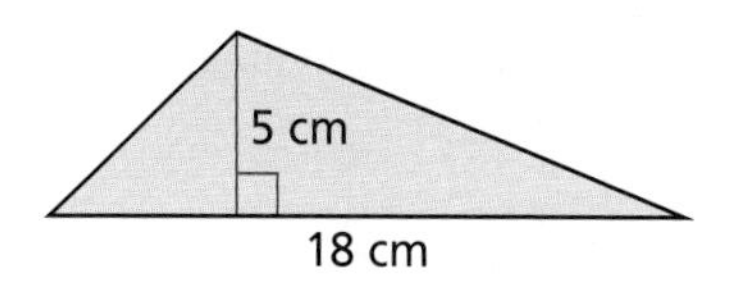

$A = \frac{1}{2}bh$ Write the formula for area of a triangle.

$= \frac{1}{2}(18)(5)$ Substitute 18 for b and 5 for h.

$= \frac{1}{2}(90)$ Multiply 18 and 5.

$= 45$ Multiply $\frac{1}{2}$ and 90.

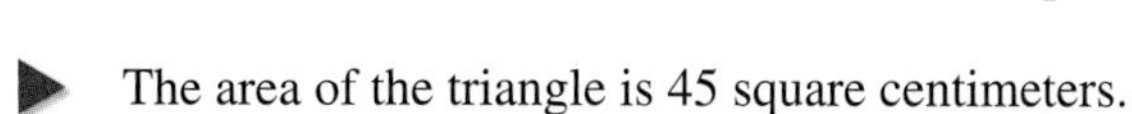

▶ The area of the triangle is 45 square centimeters.

Find the area of the triangle.

10.

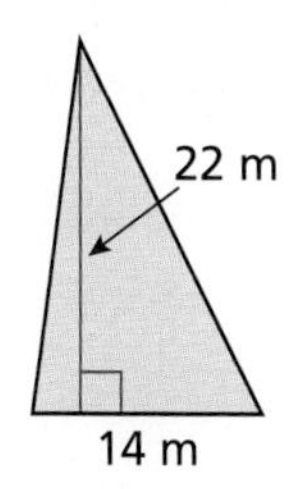

11.

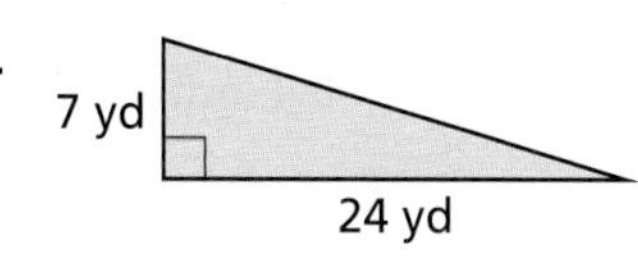

12.

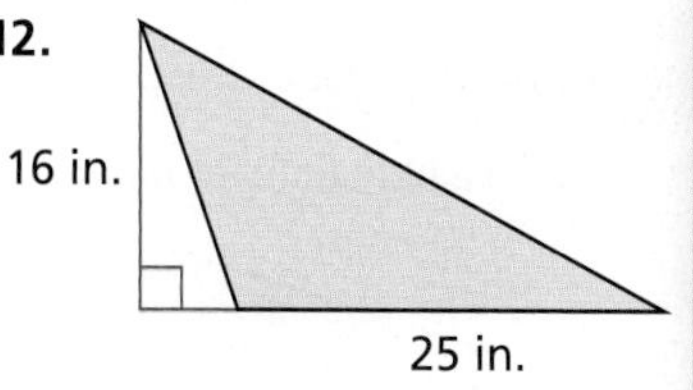

13. ABSTRACT REASONING Describe the possible values for x and y when $|x - y| > 0$. What does it mean when $|x - y| = 0$? Can $|x - y| < 0$? Explain your reasoning.

Mathematical Practices

Mathematically proficient students carefully specify units of measure.

Specifying Units of Measure

Customary Units of Length

1 foot = 12 inches
1 yard = 3 feet
1 mile = 5280 feet = 1760 yards

Metric Units of Length

1 centimeter = 10 millimeters
1 meter = 1000 millimeters
1 kilometer = 1000 meters

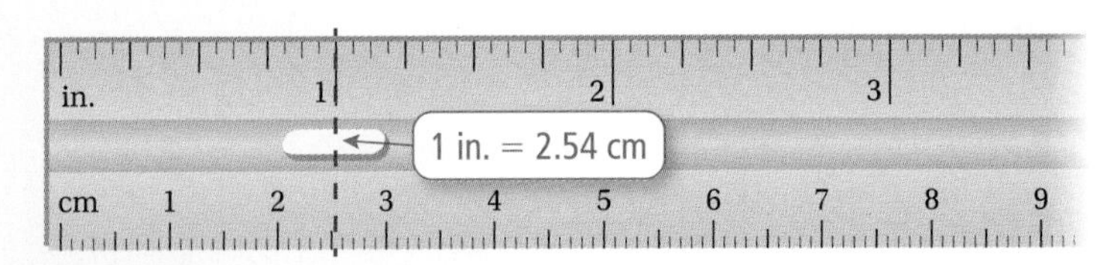

EXAMPLE 1 Converting Units of Measure

Find the area of the rectangle in square centimeters. Round your answer to the nearest hundredth.

2 in.

6 in.

SOLUTION

Use the formula for the area of a rectangle. Convert the units of length from customary units to metric units.

Area = (Length)(Width)	Formula for area of a rectangle
= (6 in.)(2 in.)	Substitute given length and width.
$= \left[(6\ \not{\text{in.}})\left(\frac{2.54\text{ cm}}{1\ \not{\text{in.}}}\right)\right]\left[(2\ \not{\text{in.}})\left(\frac{2.54\text{ cm}}{1\ \not{\text{in.}}}\right)\right]$	Multiply each dimension by the conversion factor.
= (15.24 cm)(5.08 cm)	Multiply.
$\approx 77.42\text{ cm}^2$	Multiply and round to the nearest hundredth.

▶ The area of the rectangle is about 77.42 square centimeters.

Monitoring Progress

Find the area of the polygon using the specified units. Round your answer to the nearest hundredth.

1. triangle (square inches)

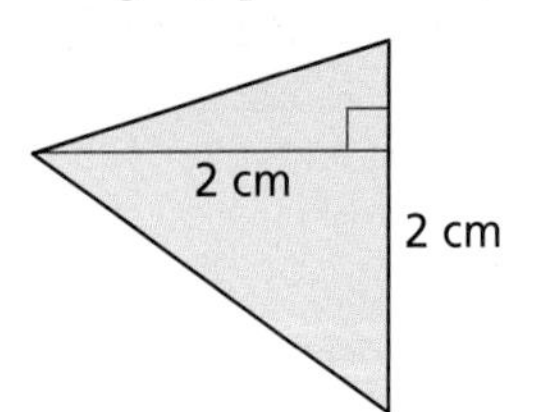

2. parallelogram (square centimeters)

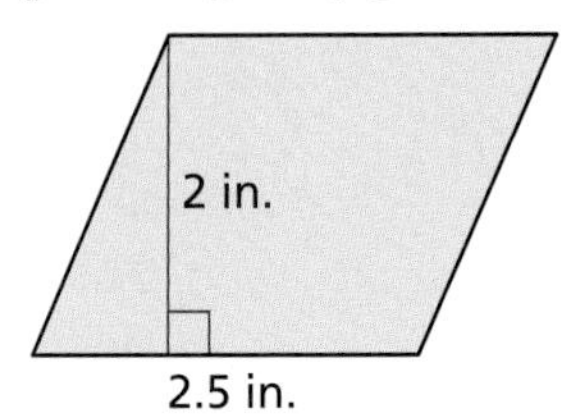

3. The distance between two cities is 120 miles. What is the distance in kilometers? Round your answer to the nearest whole number.

1.1 Points, Lines, and Planes

Essential Question How can you use dynamic geometry software to visualize geometric concepts?

EXPLORATION 1 Using Dynamic Geometry Software

Work with a partner. Use dynamic geometry software to draw several points. Also, draw some lines, line segments, and rays. What is the difference between a line, a line segment, and a ray?

Sample

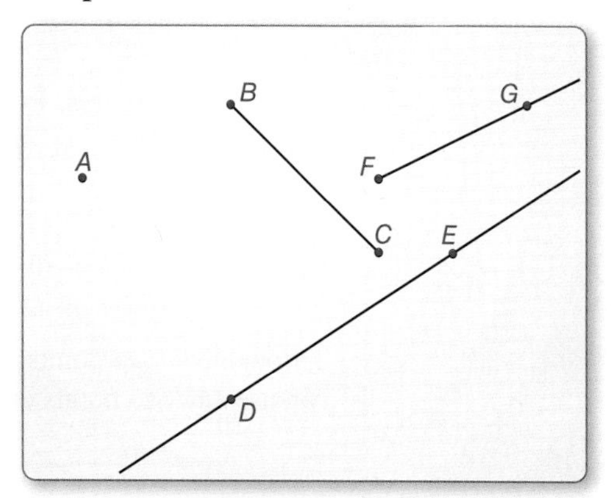

EXPLORATION 2 Intersections of Lines and Planes

Work with a partner.

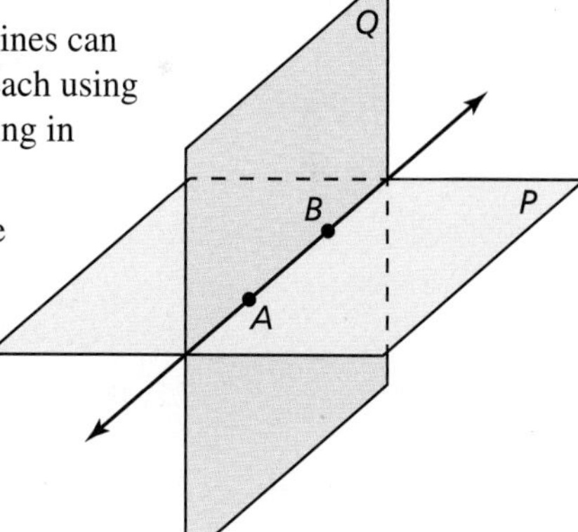

a. Describe and sketch the ways in which two lines can intersect or not intersect. Give examples of each using the lines formed by the walls, floor, and ceiling in your classroom.

b. Describe and sketch the ways in which a line and a plane can intersect or not intersect. Give examples of each using the walls, floor, and ceiling in your classroom.

c. Describe and sketch the ways in which two planes can intersect or not intersect. Give examples of each using the walls, floor, and ceiling in your classroom.

UNDERSTANDING MATHEMATICAL TERMS

To be proficient in math, you need to understand definitions and previously established results. An appropriate tool, such as a software package, can sometimes help.

EXPLORATION 3 Exploring Dynamic Geometry Software

Work with a partner. Use dynamic geometry software to explore geometry. Use the software to find a term or concept that is unfamiliar to you. Then use the capabilities of the software to determine the meaning of the term or concept.

Communicate Your Answer

4. How can you use dynamic geometry software to visualize geometric concepts?

1.1 Lesson

Core Vocabulary

undefined terms, *p. 4*
point, *p. 4*
line, *p. 4*
plane, *p. 4*
collinear points, *p. 4*
coplanar points, *p. 4*
defined terms, *p. 5*
line segment, or segment, *p. 5*
endpoints, *p. 5*
ray, *p. 5*
opposite rays, *p. 5*
intersection, *p. 6*

What You Will Learn

- Name points, lines, and planes.
- Name segments and rays.
- Sketch intersections of lines and planes.
- Solve real-life problems involving lines and planes.

Using Undefined Terms

In geometry, the words *point*, *line*, and *plane* are **undefined terms**. These words do not have formal definitions, but there is agreement about what they mean.

Core Concept

Undefined Terms: Point, Line, and Plane

Point A **point** has no dimension. A dot represents a point.

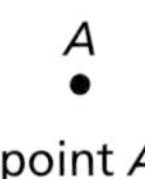

point A

Line A **line** has one dimension. It is represented by a line with two arrowheads, but it extends without end.

Through any two points, there is exactly one line. You can use any two points on a line to name it.

line ℓ, line AB ($\overleftrightarrow{AB}$), or line BA ($\overleftrightarrow{BA}$)

Plane A **plane** has two dimensions. It is represented by a shape that looks like a floor or a wall, but it extends without end.

Through any three points not on the same line, there is exactly one plane. You can use three points that are not all on the same line to name a plane.

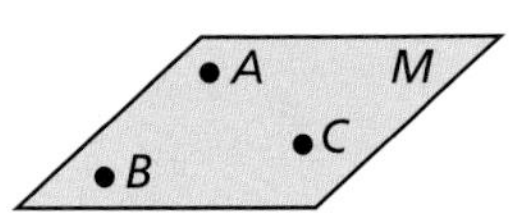

plane M, or plane ABC

Collinear points are points that lie on the same line. **Coplanar points** are points that lie in the same plane.

EXAMPLE 1 Naming Points, Lines, and Planes

a. Give two other names for $\overleftrightarrow{PQ}$ and plane R.

b. Name three points that are collinear. Name four points that are coplanar.

SOLUTION

a. Other names for $\overleftrightarrow{PQ}$ are $\overleftrightarrow{QP}$ and line n. Other names for plane R are plane SVT and plane PTV.

b. Points S, P, and T lie on the same line, so they are collinear. Points S, P, T, and V lie in the same plane, so they are coplanar.

Monitoring Progress

Help in English and Spanish at *BigIdeasMath.com*

1. Use the diagram in Example 1. Give two other names for $\overleftrightarrow{ST}$. Name a point that is *not* coplanar with points Q, S, and T.

Using Defined Terms

In geometry, terms that can be described using known words such as *point* or *line* are called **defined terms**.

Core Concept

Defined Terms: Segment and Ray

The definitions below use line AB (written as $\overleftrightarrow{AB}$) and points A and B.

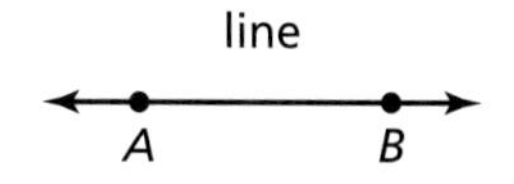

segment

endpoint endpoint

A B

Segment The **line segment** AB, or **segment** AB, (written as $\overline{AB}$) consists of the **endpoints** A and B and all points on $\overleftrightarrow{AB}$ that are between A and B. Note that $\overline{AB}$ can also be named $\overline{BA}$.

Ray The **ray** AB (written as $\overrightarrow{AB}$) consists of the endpoint A and all points on $\overleftrightarrow{AB}$ that lie on the same side of A as B.

Note that $\overrightarrow{AB}$ and $\overrightarrow{BA}$ are different rays.

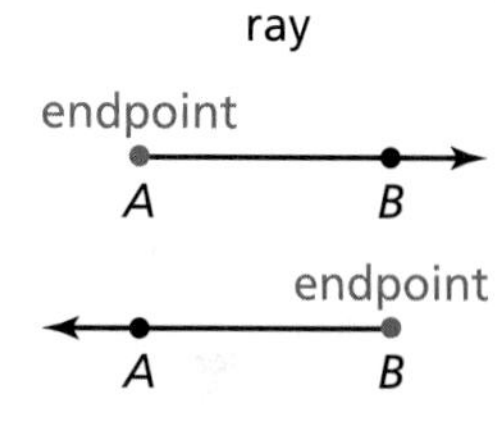

Opposite Rays If point C lies on $\overleftrightarrow{AB}$ between A and B, then $\overrightarrow{CA}$ and $\overrightarrow{CB}$ are **opposite rays**.

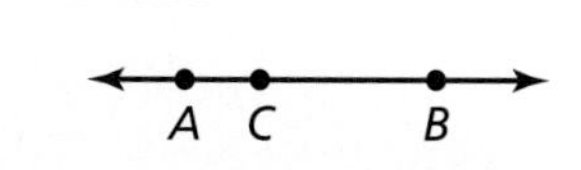

Segments and rays are collinear when they lie on the same line. So, opposite rays are collinear. Lines, segments, and rays are coplanar when they lie in the same plane.

EXAMPLE 2 Naming Segments, Rays, and Opposite Rays

a. Give another name for $\overline{GH}$.

b. Name all rays with endpoint J. Which of these rays are opposite rays?

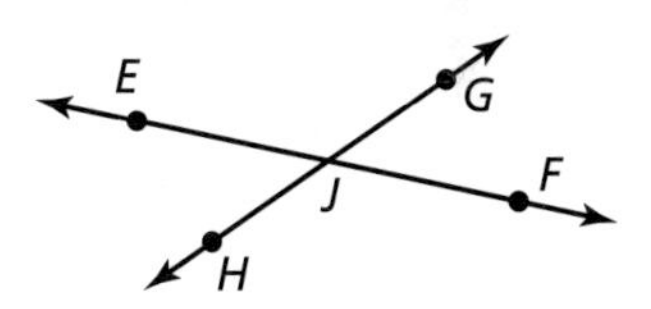

COMMON ERROR

In Example 2, $\overrightarrow{JG}$ and $\overrightarrow{JF}$ have a common endpoint, but they are not collinear. So, they are *not* opposite rays.

SOLUTION

a. Another name for $\overline{GH}$ is $\overline{HG}$.

b. The rays with endpoint J are $\overrightarrow{JE}$, $\overrightarrow{JG}$, $\overrightarrow{JF}$, and $\overrightarrow{JH}$. The pairs of opposite rays with endpoint J are $\overrightarrow{JE}$ and $\overrightarrow{JF}$, and $\overrightarrow{JG}$ and $\overrightarrow{JH}$.

Monitoring Progress

Help in English and Spanish at *BigIdeasMath.com*

Use the diagram.

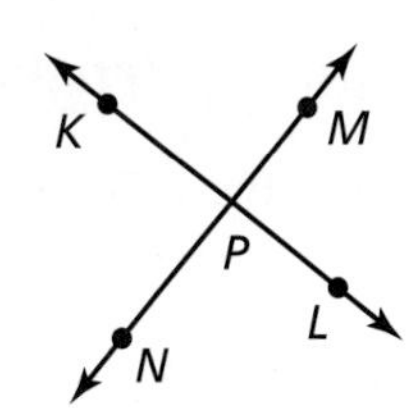

2. Give another name for $\overline{KL}$.

3. Are $\overrightarrow{KP}$ and $\overrightarrow{PK}$ the same ray? Are $\overrightarrow{NP}$ and $\overrightarrow{NM}$ the same ray? Explain.

Sketching Intersections

Two or more geometric figures *intersect* when they have one or more points in common. The **intersection** of the figures is the set of points the figures have in common. Some examples of intersections are shown below.

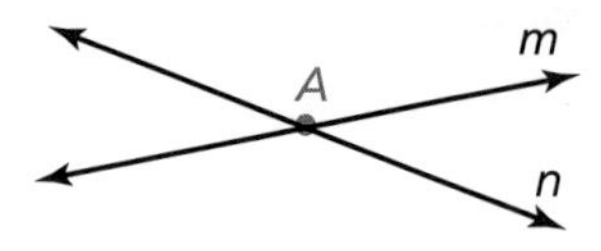

The intersection of two different lines is a point.

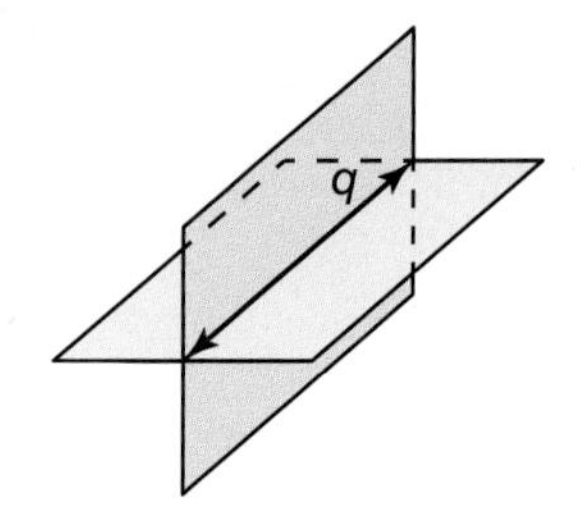

The intersection of two different planes is a line.

EXAMPLE 3 Sketching Intersections of Lines and Planes

a. Sketch a plane and a line that is in the plane.

b. Sketch a plane and a line that does not intersect the plane.

c. Sketch a plane and a line that intersects the plane at a point.

SOLUTION

a.

b.

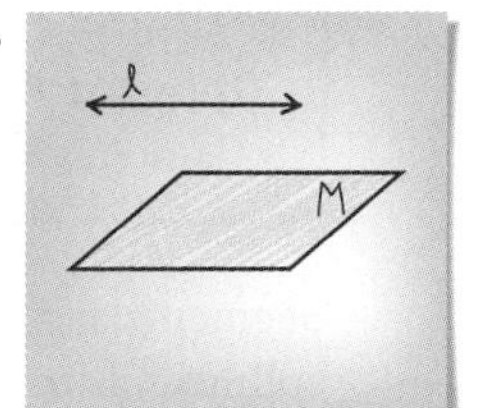

c.

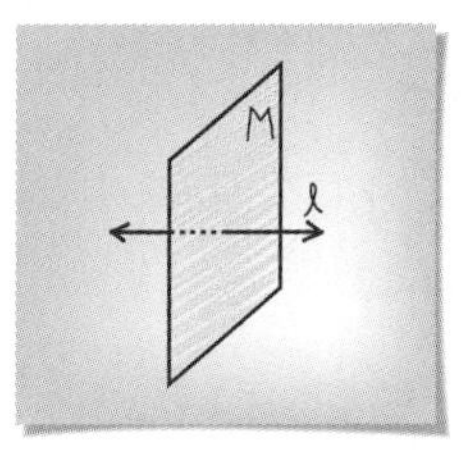

EXAMPLE 4 Sketching Intersections of Planes

Sketch two planes that intersect in a line.

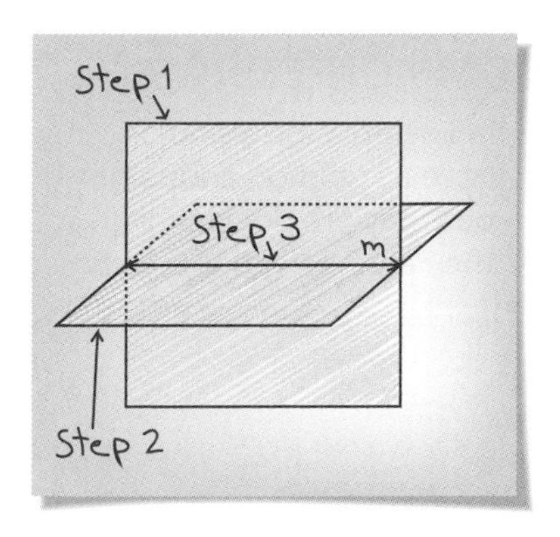

SOLUTION

Step 1 Draw a vertical plane. Shade the plane.

Step 2 Draw a second plane that is horizontal. Shade this plane a different color. Use dashed lines to show where one plane is hidden.

Step 3 Draw the line of intersection.

Monitoring Progress  Help in English and Spanish at *BigIdeasMath.com*

4. Sketch two different lines that intersect a plane at the same point.

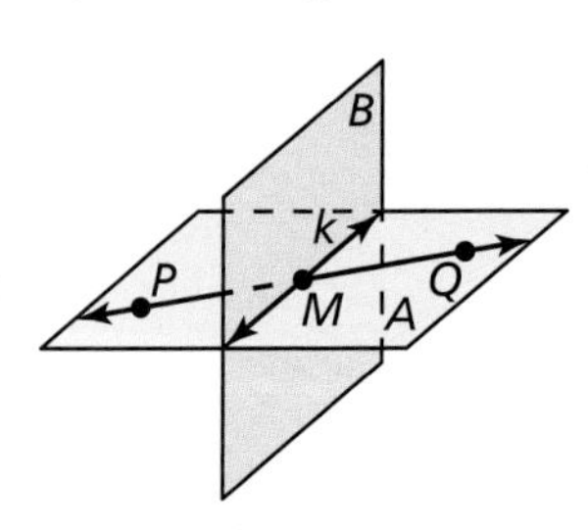

Use the diagram.

5. Name the intersection of $\overleftrightarrow{PQ}$ and line k.

6. Name the intersection of plane A and plane B.

7. Name the intersection of line k and plane A.

Solving Real-Life Problems

EXAMPLE 5 Modeling with Mathematics

The diagram shows a molecule of sulfur hexafluoride, the most potent greenhouse gas in the world. Name two different planes that contain line r.

Electric utilities use sulfur hexafluoride as an insulator. Leaks in electrical equipment contribute to the release of sulfur hexafluoride into the atmosphere.

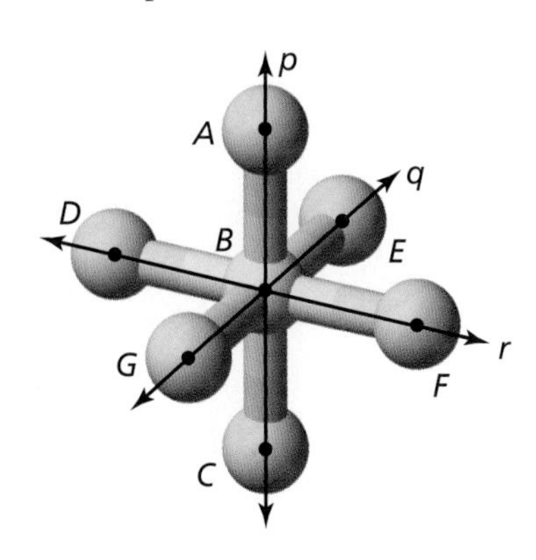

SOLUTION

1. **Understand the Problem** In the diagram, you are given three lines, p, q, and r, that intersect at point B. You need to name two different planes that contain line r.
2. **Make a Plan** The planes should contain two points on line r and one point not on line r.
3. **Solve the Problem** Points D and F are on line r. Point E does not lie on line r. So, plane DEF contains line r. Another point that does not lie on line r is C. So, plane CDF contains line r.

   Note that you cannot form a plane through points D, B, and F. By definition, three points that do not lie on the same line form a plane. Points D, B, and F are collinear, so they do *not* form a plane.
4. **Look Back** The question asks for two *different* planes. You need to check whether plane DEF and plane CDF are two unique planes or the same plane named differently. Because point C does not lie on plane DEF, plane DEF and plane CDF are different planes.

Monitoring Progress  Help in English and Spanish at *BigIdeasMath.com*

Use the diagram that shows a molecule of phosphorus pentachloride.

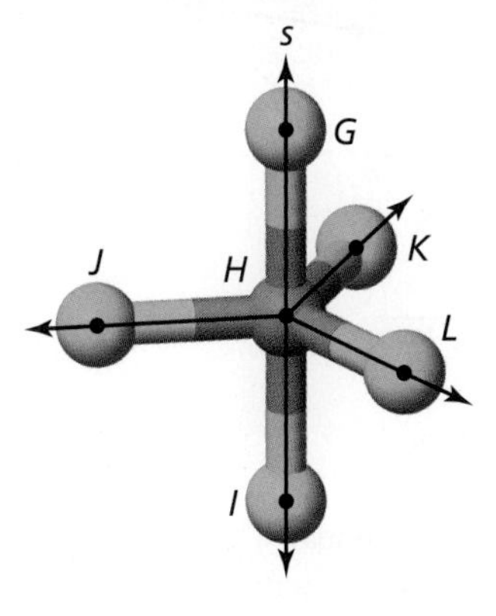

8. Name two different planes that contain line s.
9. Name three different planes that contain point K.
10. Name two different planes that contain $\overrightarrow{HJ}$.

1.1 Exercises

Dynamic Solutions available at *BigIdeasMath.com*

Vocabulary and Core Concept Check

1. **WRITING** Compare collinear points and coplanar points.

2. **WHICH ONE DOESN'T BELONG?** Which term does *not* belong with the other three? Explain your reasoning.

$\overline{AB}$ | plane CDE | $\overleftrightarrow{FG}$ | $\overrightarrow{HI}$

Monitoring Progress and Modeling with Mathematics

In Exercises 3–6, use the diagram.

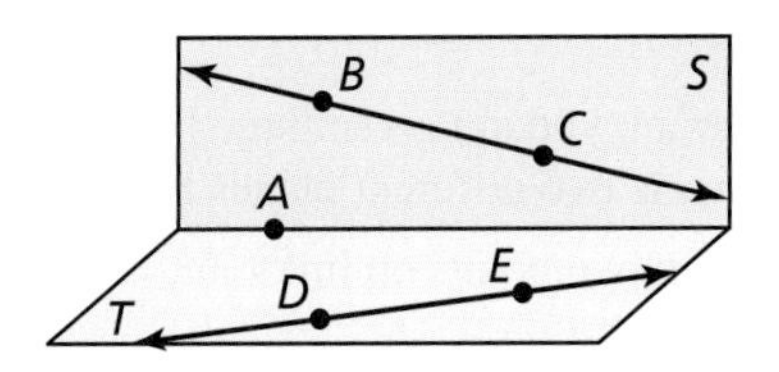

3. Name four points.

4. Name two lines.

5. Name the plane that contains points A, B, and C.

6. Name the plane that contains points A, D, and E.

In Exercises 7–10, use the diagram. *(See Example 1.)*

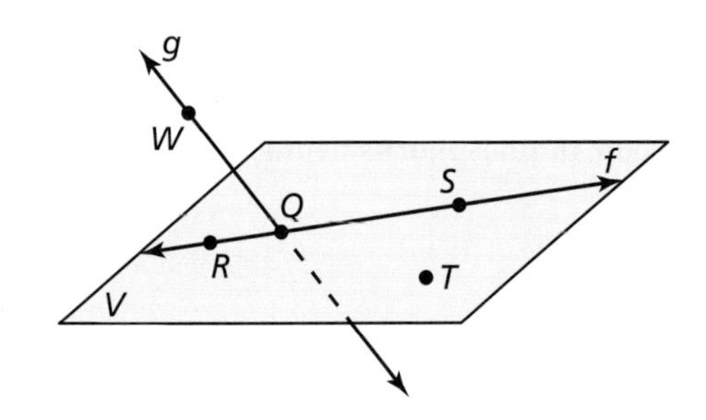

7. Give two other names for $\overleftrightarrow{WQ}$.

8. Give another name for plane V.

9. Name three points that are collinear. Then name a fourth point that is not collinear with these three points.

10. Name a point that is not coplanar with R, S, and T.

In Exercises 11–16, use the diagram. *(See Example 2.)*

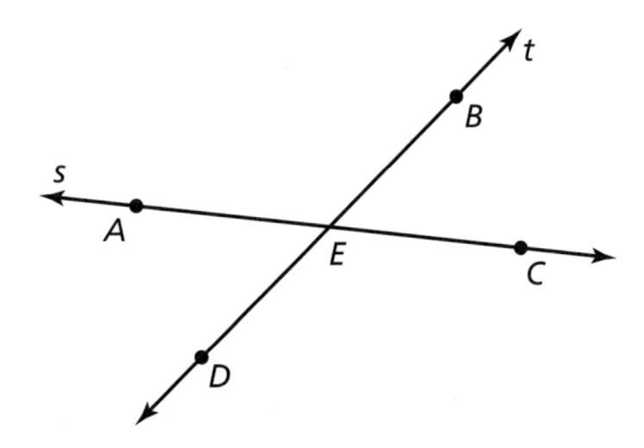

11. What is another name for $\overline{BD}$?

12. What is another name for $\overline{AC}$?

13. What is another name for ray $\overrightarrow{AE}$?

14. Name all rays with endpoint E.

15. Name two pairs of opposite rays.

16. Name one pair of rays that are not opposite rays.

In Exercises 17–24, sketch the figure described. *(See Examples 3 and 4.)*

17. plane P and line ℓ intersecting at one point

18. plane K and line m intersecting at all points on line m

19. $\overrightarrow{AB}$ and $\overleftrightarrow{AC}$

20. $\overrightarrow{MN}$ and $\overrightarrow{NX}$

21. plane M and $\overrightarrow{NB}$ intersecting at B

22. plane M and $\overrightarrow{NB}$ intersecting at A

23. plane A and plane B not intersecting

24. plane C and plane D intersecting at $\overleftrightarrow{XY}$

ERROR ANALYSIS **In Exercises 25 and 26, describe and correct the error in naming opposite rays in the diagram.**

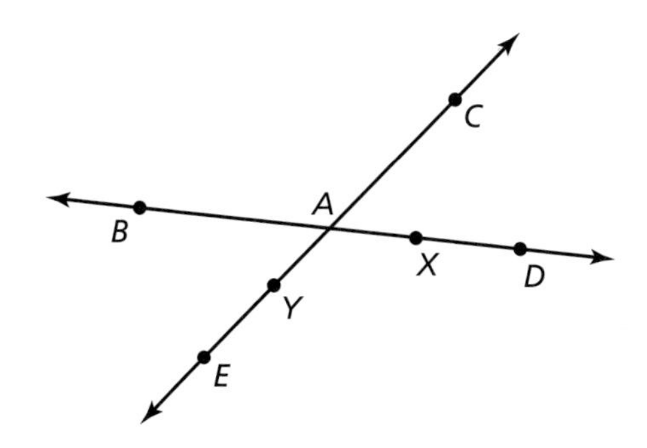

25. ✗ $\overrightarrow{AD}$ and $\overrightarrow{AC}$ are opposite rays.

26. ✗ $\overline{YC}$ and $\overline{YE}$ are opposite rays.

In Exercises 27–34, use the diagram.

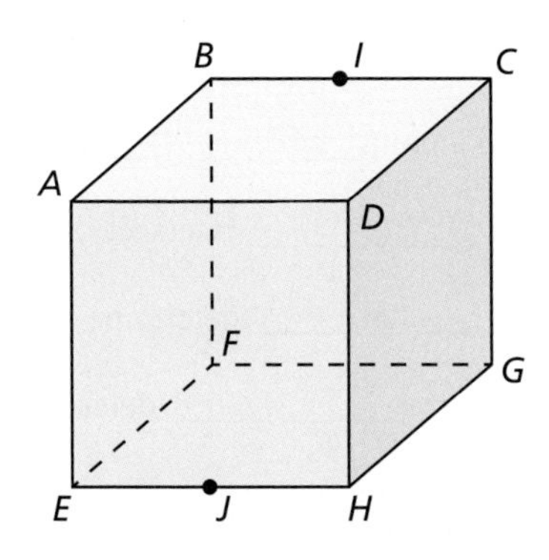

27. Name a point that is collinear with points *E* and *H*.

28. Name a point that is collinear with points *B* and *I*.

29. Name a point that is not collinear with points *E* and *H*.

30. Name a point that is not collinear with points *B* and *I*.

31. Name a point that is coplanar with points *D*, *A*, and *B*.

32. Name a point that is coplanar with points *C*, *G*, and *F*.

33. Name the intersection of plane *AEH* and plane *FBE*.

34. Name the intersection of plane *BGF* and plane *HDG*.

In Exercises 35–38, name the geometric term modeled by the object.

35.

36.

37. **38.**

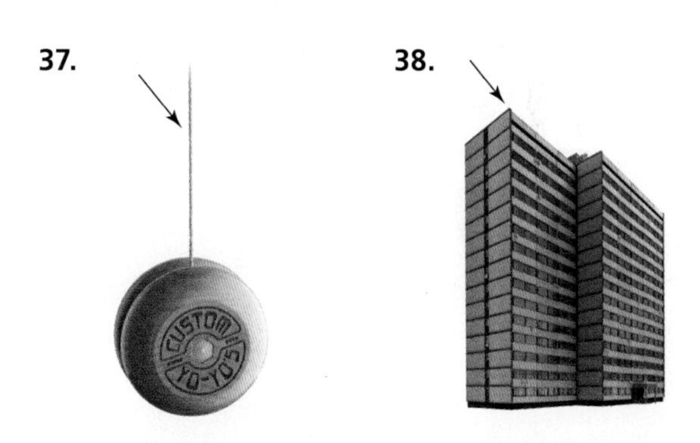

In Exercises 39–44, use the diagram to name all the points that are not coplanar with the given points.

39. *N*, *K*, and *L*

40. *P*, *Q*, and *N*

41. *P*, *Q*, and *R*

42. *R*, *K*, and *N*

43. *P*, *S*, and *K*

44. *Q*, *K*, and *L*

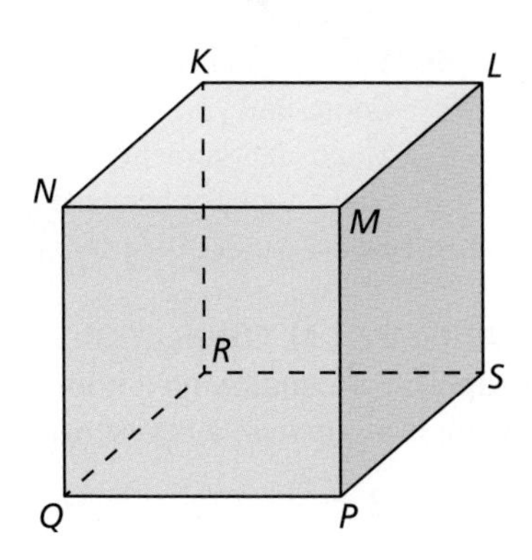

45. **CRITICAL THINKING** Given two points on a line and a third point not on the line, is it possible to draw a plane that includes the line and the third point? Explain your reasoning.

46. **CRITICAL THINKING** Is it possible for one point to be in two different planes? Explain your reasoning.

47. **REASONING** Explain why a four-legged chair may rock from side to side even if the floor is level. Would a three-legged chair on the same level floor rock from side to side? Why or why not?

48. **THOUGHT PROVOKING** You are designing the living room of an apartment. Counting the floor, walls, and ceiling, you want the design to contain at least eight different planes. Draw a diagram of your design. Label each plane in your design.

49. **LOOKING FOR STRUCTURE** Two coplanar intersecting lines will always intersect at one point. What is the greatest number of intersection points that exist if you draw four coplanar lines? Explain.

50. **HOW DO YOU SEE IT?** You and your friend walk in opposite directions, forming opposite rays. You were originally on the corner of Apple Avenue and Cherry Court.

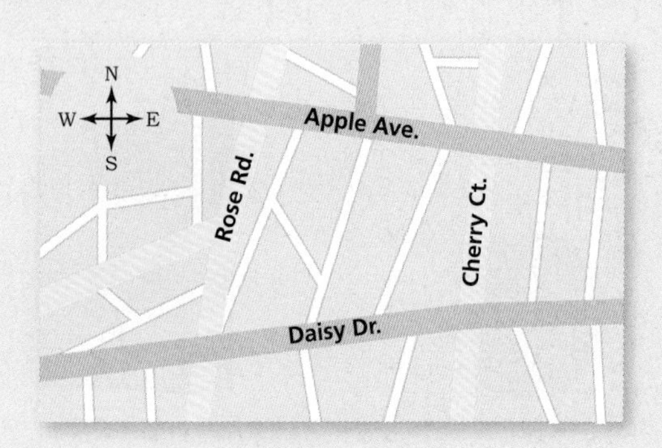

a. Name two possibilities of the road and direction you and your friend may have traveled.

b. Your friend claims he went north on Cherry Court, and you went east on Apple Avenue. Make an argument as to why you know this could not have happened.

MATHEMATICAL CONNECTIONS **In Exercises 51–54, graph the inequality on a number line. Tell whether the graph is a *segment*, a *ray* or *rays*, a *point*, or a *line*.**

51. $x \leq 3$

52. $-7 \leq x \leq 4$

53. $x \geq 5$ or $x \leq -2$

54. $|x| \leq 0$

55. **MODELING WITH MATHEMATICS** Use the diagram.

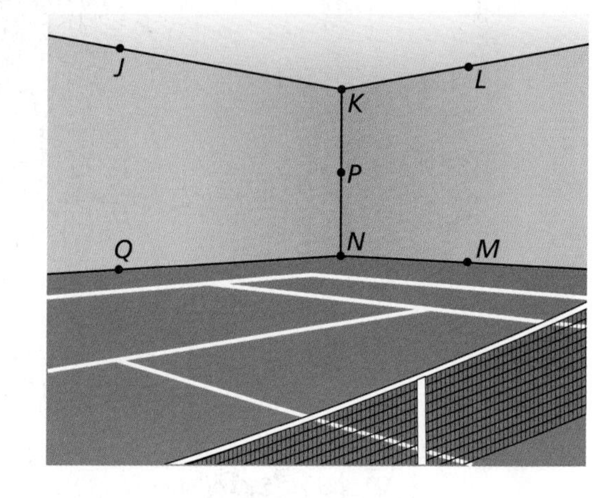

a. Name two points that are collinear with P.

b. Name two planes that contain J.

c. Name all the points that are in more than one plane.

CRITICAL THINKING **In Exercises 56–63, complete the statement with *always*, *sometimes*, or *never*. Explain your reasoning.**

56. A line ____________ has endpoints.

57. A line and a point ____________ intersect.

58. A plane and a point ____________ intersect.

59. Two planes ____________ intersect in a line.

60. Two points ____________ determine a line.

61. Any three points ____________ determine a plane.

62. Any three points not on the same line ____________ determine a plane.

63. Two lines that are not parallel ____________ intersect.

64. **ABSTRACT REASONING** Is it possible for three planes to never intersect? intersect in one line? intersect in one point? Sketch the possible situations.

Maintaining Mathematical Proficiency

Reviewing what you learned in previous grades and lessons

Find the absolute value. *(Skills Review Handbook)*

65. $|6 + 2|$

66. $|3 - 9|$

67. $|-8 - 2|$

68. $|7 - 11|$

Solve the equation. *(Skills Review Handbook)*

69. $18 + x = 43$

70. $36 + x = 20$

71. $x - 15 = 7$

72. $x - 23 = 19$

1.2 Measuring and Constructing Segments

Essential Question How can you measure and construct a line segment?

EXPLORATION 1 Measuring Line Segments Using Nonstandard Units

Work with a partner.

a. Draw a line segment that has a length of 6 inches.

b. Use a standard-sized paper clip to measure the length of the line segment. Explain how you measured the line segment in "paper clips."

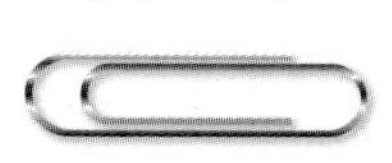

c. Write conversion factors from paper clips to inches and vice versa.

1 paper clip = ___ in.

1 in. = ___ paper clip

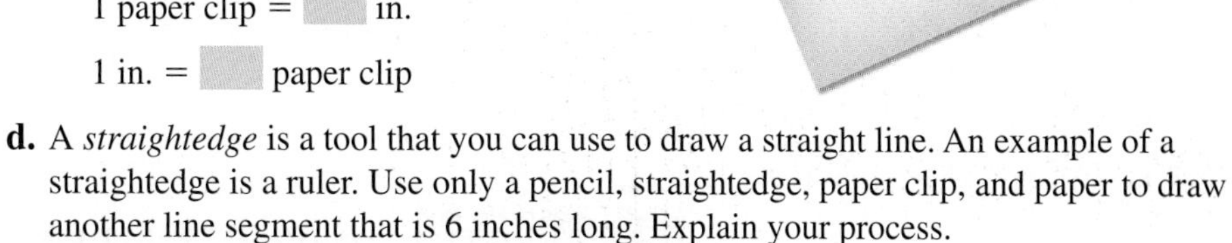

d. A *straightedge* is a tool that you can use to draw a straight line. An example of a straightedge is a ruler. Use only a pencil, straightedge, paper clip, and paper to draw another line segment that is 6 inches long. Explain your process.

MAKING SENSE OF PROBLEMS

To be proficient in math, you need to explain to yourself the meaning of a problem and look for entry points to its solution.

EXPLORATION 2 Measuring Line Segments Using Nonstandard Units

Work with a partner.

a. Fold a 3-inch by 5-inch index card on one of its diagonals.

b. Use the Pythagorean Theorem to algebraically determine the length of the diagonal in inches. Use a ruler to check your answer.

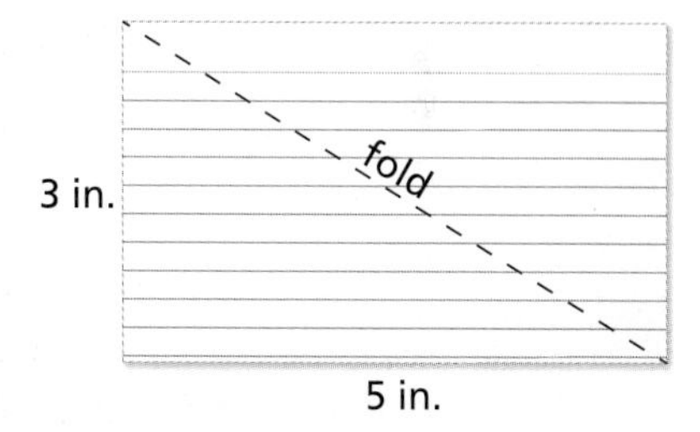

c. Measure the length and width of the index card in paper clips.

d. Use the Pythagorean Theorem to algebraically determine the length of the diagonal in paper clips. Then check your answer by measuring the length of the diagonal in paper clips. Does the Pythagorean Theorem work for any unit of measure? Justify your answer.

EXPLORATION 3 Measuring Heights Using Nonstandard Units

Work with a partner. Consider a unit of length that is equal to the length of the diagonal you found in Exploration 2. Call this length "1 diag." How tall are you in diags? Explain how you obtained your answer.

Communicate Your Answer

4. How can you measure and construct a line segment?

1.2 Lesson

Core Vocabulary

postulate, *p. 12*
axiom, *p. 12*
coordinate, *p. 12*
distance, *p. 12*
construction, *p. 13*
congruent segments, *p. 13*
between, *p. 14*

What You Will Learn

- Use the Ruler Postulate.
- Copy segments and compare segments for congruence.
- Use the Segment Addition Postulate.

Using the Ruler Postulate

In geometry, a rule that is accepted without proof is called a **postulate** or an **axiom**. A rule that can be proved is called a *theorem*, as you will see later. Postulate 1.1 shows how to find the distance between two points on a line.

Postulate

Postulate 1.1 Ruler Postulate

The points on a line can be matched one to one with the real numbers. The real number that corresponds to a point is the **coordinate** of the point.

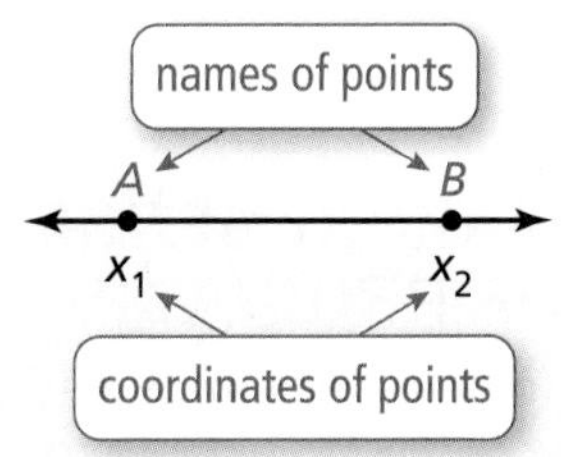

The **distance** between points A and B, written as AB, is the absolute value of the difference of the coordinates of A and B.

A AB B
x_1 x_2

$AB = |x_2 - x_1|$

EXAMPLE 1 Using the Ruler Postulate

Measure the length of $\overline{ST}$ to the nearest tenth of a centimeter.

SOLUTION

Align one mark of a metric ruler with S. Then estimate the coordinate of T. For example, when you align S with 2, T appears to align with 5.4.

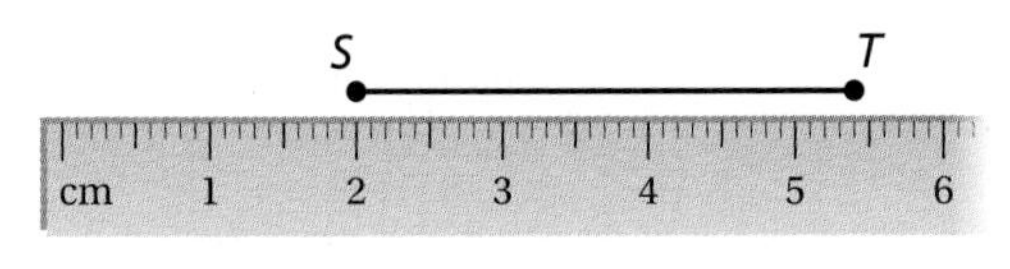

$ST = |5.4 - 2| = 3.4$ Ruler Postulate

▶ So, the length of $\overline{ST}$ is about 3.4 centimeters.

Monitoring Progress

Help in English and Spanish at *BigIdeasMath.com*

Use a ruler to measure the length of the segment to the nearest $\frac{1}{8}$ inch.

1. 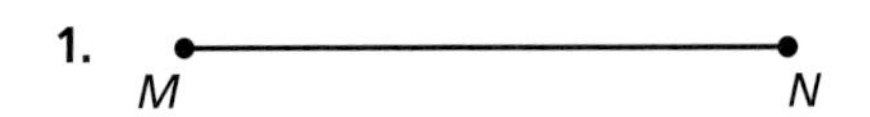

2. 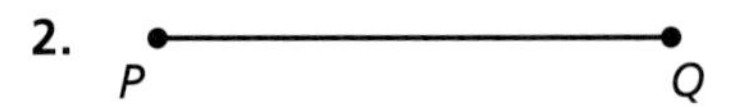

3. U ——— V

4. 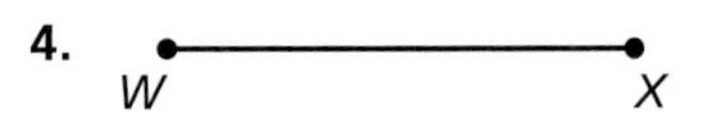

Constructing and Comparing Congruent Segments

A **construction** is a geometric drawing that uses a limited set of tools, usually a *compass* and *straightedge*.

CONSTRUCTION **Copying a Segment**

Use a compass and straightedge to construct a line segment that has the same length as $\overline{AB}$.

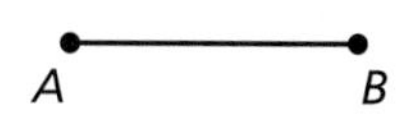

SOLUTION

Step 1

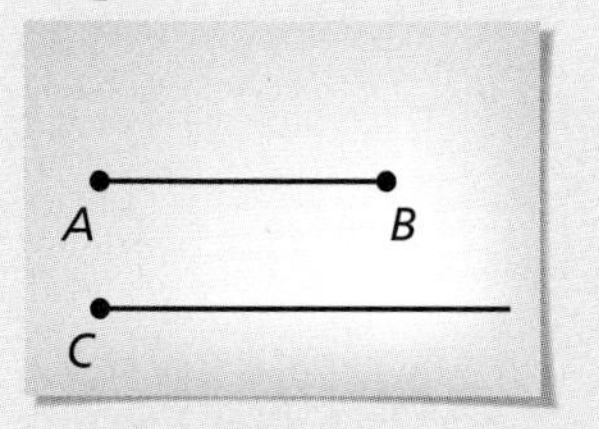

Draw a segment Use a straightedge to draw a segment longer than $\overline{AB}$. Label point C on the new segment.

Step 2

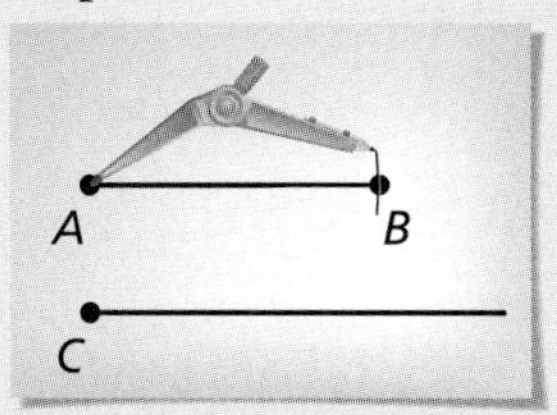

Measure length Set your compass at the length of $\overline{AB}$.

Step 3

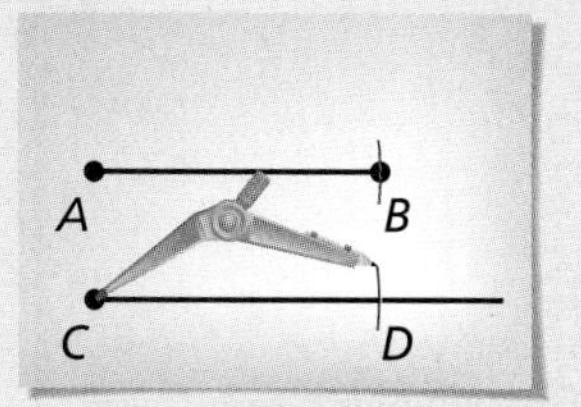

Copy length Place the compass at C. Mark point D on the new segment. So, $\overline{CD}$ has the same length as $\overline{AB}$.

Core Concept

Congruent Segments

Line segments that have the same length are called **congruent segments.** You can say "the length of $\overline{AB}$ is equal to the length of $\overline{CD}$," or you can say "$\overline{AB}$ *is congruent to* $\overline{CD}$." The symbol $\cong$ means "is congruent to."

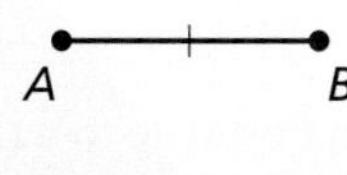

Lengths are equal.

$AB = CD$

"is equal to"

Segments are congruent.

$\overline{AB} \cong \overline{CD}$

"is congruent to"

READING

In the diagram, the red tick marks indicate $\overline{AB} \cong \overline{CD}$. When there is more than one pair of congruent segments, use multiple tick marks.

EXAMPLE 2 **Comparing Segments for Congruence**

Plot $J(-3, 4)$, $K(2, 4)$, $L(1, 3)$, and $M(1, -2)$ in a coordinate plane. Then determine whether $\overline{JK}$ and $\overline{LM}$ are congruent.

SOLUTION

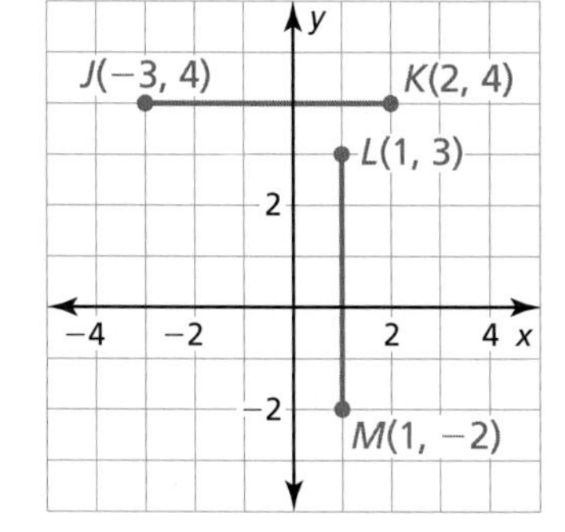

Plot the points, as shown. To find the length of a horizontal segment, find the absolute value of the difference of the x-coordinates of the endpoints.

$JK = |2 - (-3)| = 5$ Ruler Postulate

To find the length of a vertical segment, find the absolute value of the difference of the y-coordinates of the endpoints.

$LM = |-2 - 3| = 5$ Ruler Postulate

▶ $\overline{JK}$ and $\overline{LM}$ have the same length. So, $\overline{JK} \cong \overline{LM}$.

Monitoring Progress

Help in English and Spanish at *BigIdeasMath.com*

5. Plot $A(-2, 4)$, $B(3, 4)$, $C(0, 2)$, and $D(0, -2)$ in a coordinate plane. Then determine whether $\overline{AB}$ and $\overline{CD}$ are congruent.

Using the Segment Addition Postulate

When three points are collinear, you can say that one point is **between** the other two.

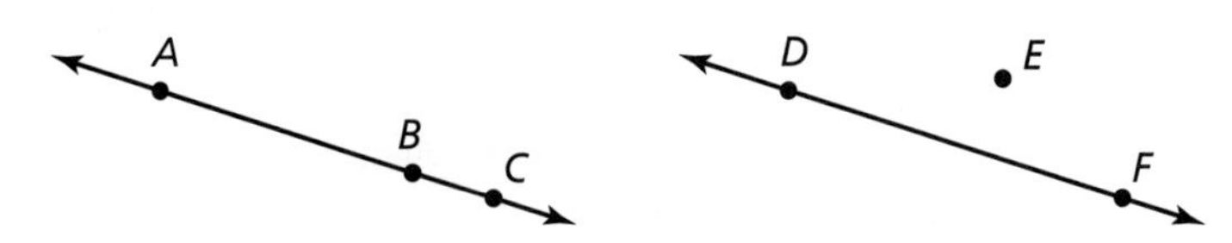

Point B is between points A and C.

Point E is not between points D and F.

Postulate

Postulate 1.2 Segment Addition Postulate

If B is between A and C, then $AB + BC = AC$.

If $AB + BC = AC$, then B is between A and C.

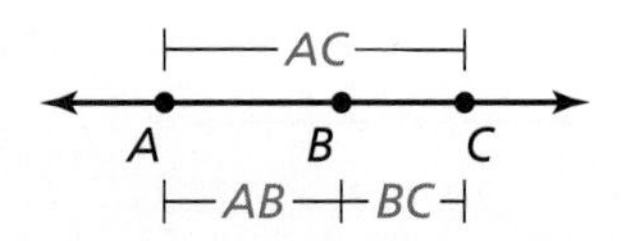

EXAMPLE 3 **Using the Segment Addition Postulate**

a. Find DF.

b. Find GH.

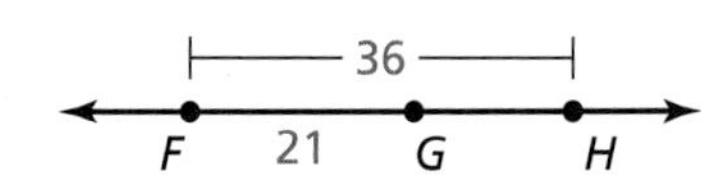

SOLUTION

a. Use the Segment Addition Postulate to write an equation. Then solve the equation to find DF.

$DF = DE + EF$	Segment Addition Postulate
$DF = 23 + 35$	Substitute 23 for DE and 35 for EF.
$DF = 58$	Add.

b. Use the Segment Addition Postulate to write an equation. Then solve the equation to find GH.

$FH = FG + GH$	Segment Addition Postulate
$36 = 21 + GH$	Substitute 36 for FH and 21 for FG.
$15 = GH$	Subtract 21 from each side.

Monitoring Progress

Help in English and Spanish at *BigIdeasMath.com*

Use the diagram at the right.

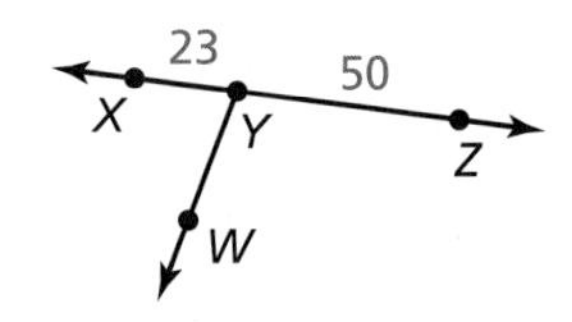

6. Use the Segment Addition Postulate to find XZ.

7. In the diagram, $WY = 30$. Can you use the Segment Addition Postulate to find the distance between points W and Z? Explain your reasoning.

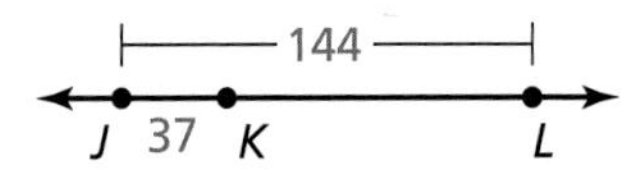

8. Use the diagram at the left to find KL.

EXAMPLE 4 Using the Segment Addition Postulate

The cities shown on the map lie approximately in a straight line. Find the distance from Tulsa, Oklahoma, to St. Louis, Missouri.

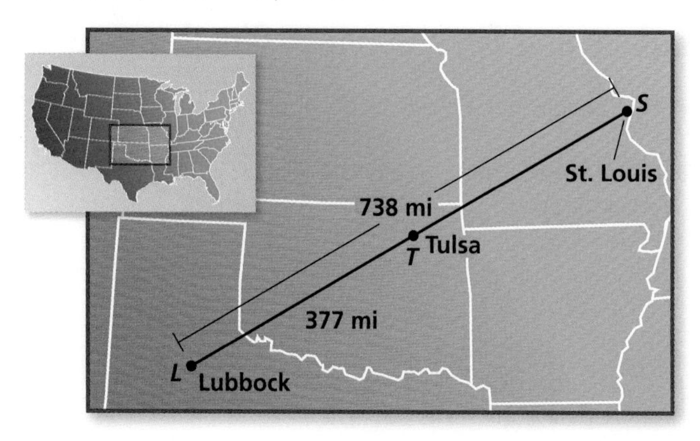

SOLUTION

1. **Understand the Problem** You are given the distance from Lubbock to St. Louis and the distance from Lubbock to Tulsa. You need to find the distance from Tulsa to St. Louis.
2. **Make a Plan** Use the Segment Addition Postulate to find the distance from Tulsa to St. Louis.
3. **Solve the Problem** Use the Segment Addition Postulate to write an equation. Then solve the equation to find TS.

$LS = LT + TS$	Segment Addition Postulate
$738 = 377 + TS$	Substitute 738 for LS and 377 for LT.
$361 = TS$	Subtract 377 from each side.

▶ So, the distance from Tulsa to St. Louis is about 361 miles.

4. **Look Back** Does the answer make sense in the context of the problem? The distance from Lubbock to St. Louis is 738 miles. By the Segment Addition Postulate, the distance from Lubbock to Tulsa plus the distance from Tulsa to St. Louis should equal 738 miles.

$377 + 361 = 738$ ✓

Monitoring Progress

Help in English and Spanish at *BigIdeasMath.com*

9. The cities shown on the map lie approximately in a straight line. Find the distance from Albuquerque, New Mexico, to Provo, Utah.

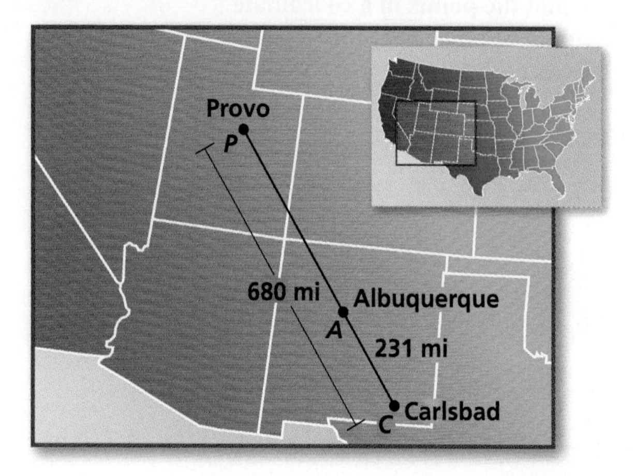

1.2 Exercises

Dynamic Solutions available at *BigIdeasMath.com*

Vocabulary and Core Concept Check

1. **WRITING** Explain how $\overline{XY}$ and XY are different.

2. **DIFFERENT WORDS, SAME QUESTION** Which is different? Find "both" answers.

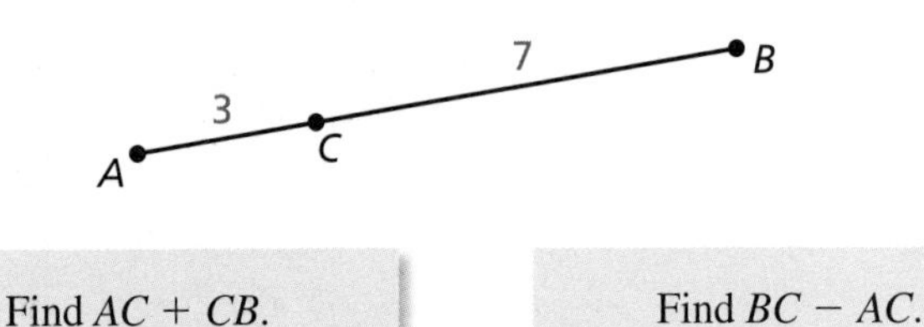

Find $AC + CB$.	Find $BC - AC$.
Find AB.	Find $CA + BC$.

Monitoring Progress and Modeling with Mathematics

In Exercises 3–6, use a ruler to measure the length of the segment to the nearest tenth of a centimeter. *(See Example 1.)*

3. 

4.

5.

6.

CONSTRUCTION In Exercises 7 and 8, use a compass and straightedge to construct a copy of the segment.

7. Copy the segment in Exercise 3.

8. Copy the segment in Exercise 4.

In Exercises 9–14, plot the points in a coordinate plane. Then determine whether $\overline{AB}$ and $\overline{CD}$ are congruent. *(See Example 2.)*

9. $A(-4, 5)$, $B(-4, 8)$, $C(2, -3)$, $D(2, 0)$

10. $A(6, -1)$, $B(1, -1)$, $C(2, -3)$, $D(4, -3)$

11. $A(8, 3)$, $B(-1, 3)$, $C(5, 10)$, $D(5, 3)$

12. $A(6, -8)$, $B(6, 1)$, $C(7, -2)$, $D(-2, -2)$

13. $A(-5, 6)$, $B(-5, -1)$, $C(-4, 3)$, $D(3, 3)$

14. $A(10, -4)$, $B(3, -4)$, $C(-1, 2)$, $D(-1, 5)$

In Exercises 15–22, find FH. *(See Example 3.)*

15. 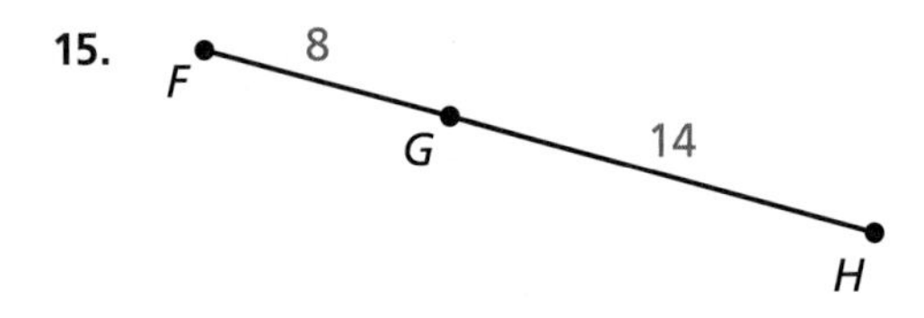

16. 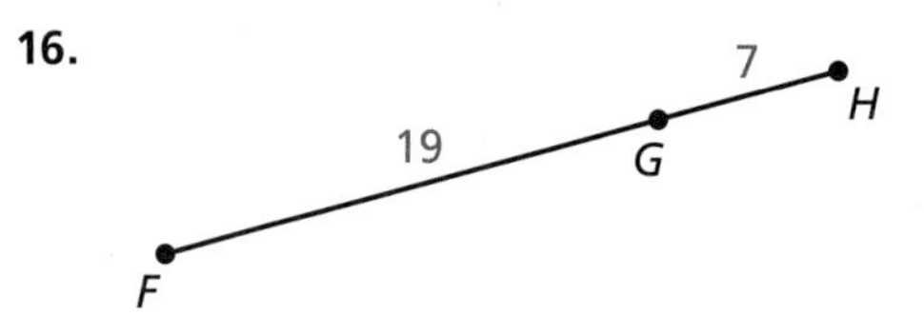

17. 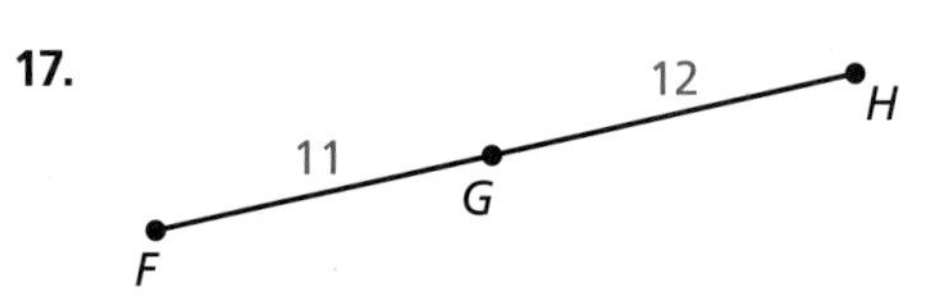

18. 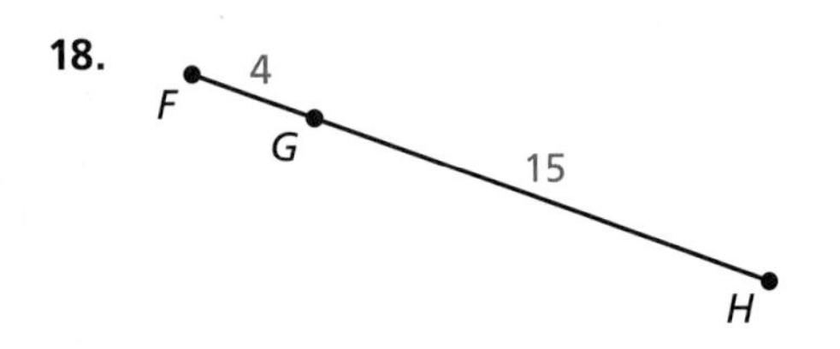

19. 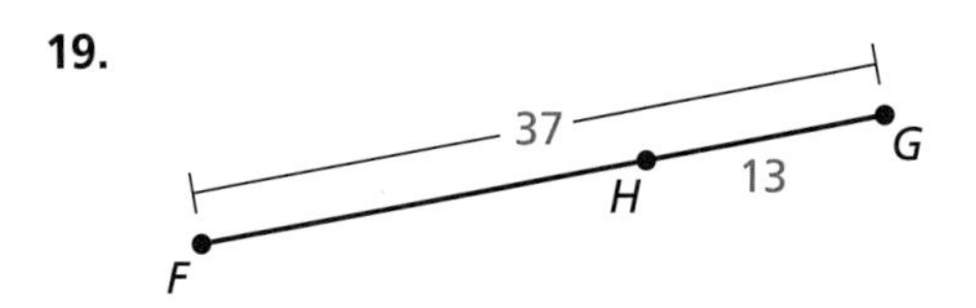

20. 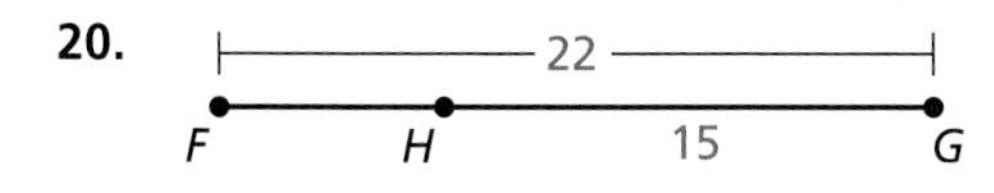

21. 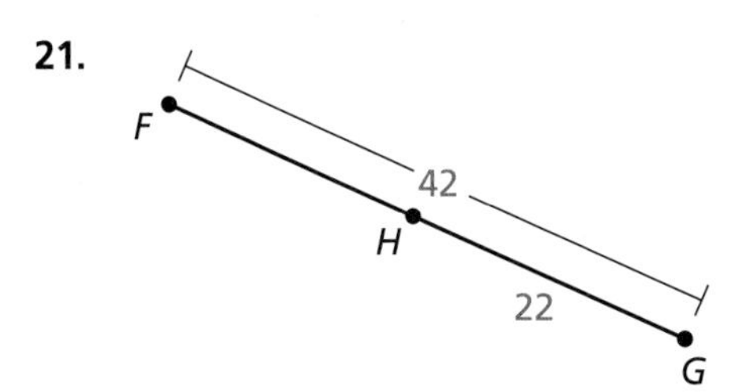

22. 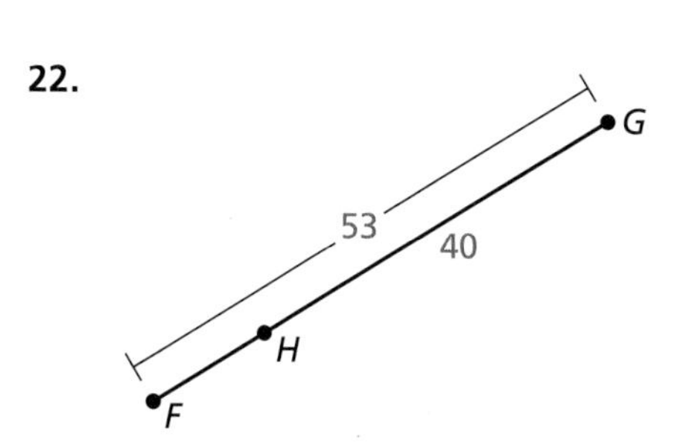

ERROR ANALYSIS In Exercises 23 and 24, describe and correct the error in finding the length of $\overline{AB}$.

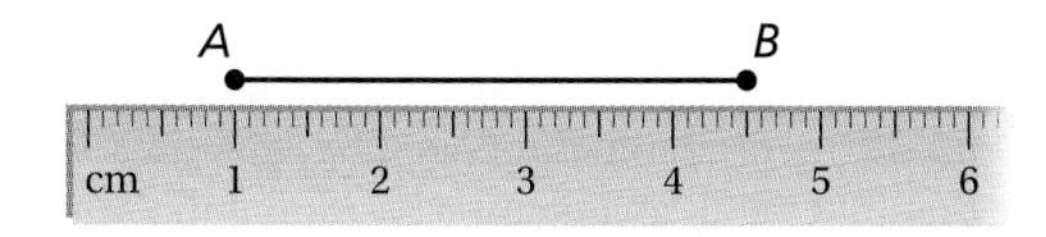

23. 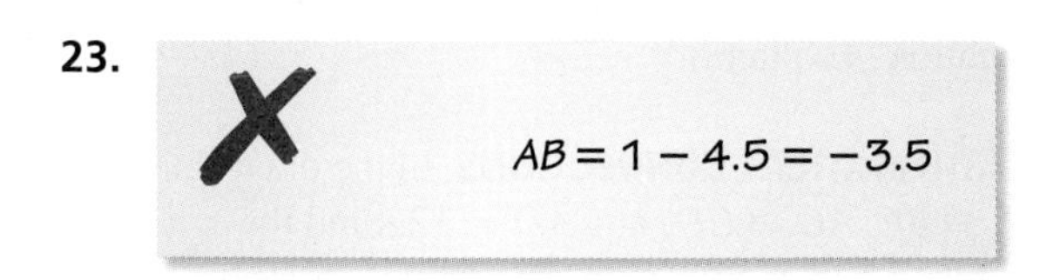

24. ✗ $AB = |1 + 4.5| = 5.5$

25. ATTENDING TO PRECISION The diagram shows an insect called a walking stick. Use the ruler to estimate the length of the abdomen and the length of the thorax to the nearest $\frac{1}{4}$ inch. How much longer is the walking stick's abdomen than its thorax? How many times longer is its abdomen than its thorax?

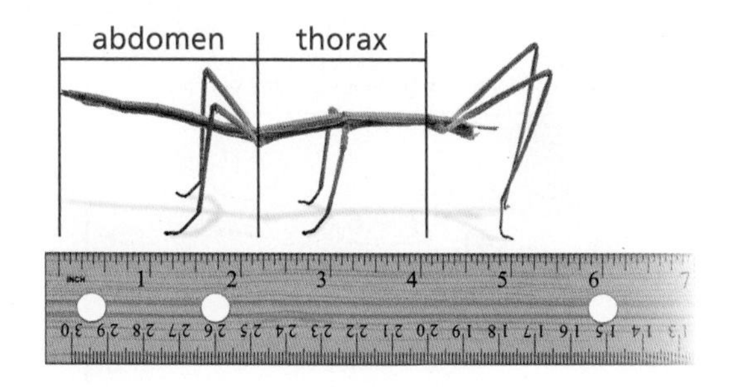

26. MODELING WITH MATHEMATICS In 2003, a remote-controlled model airplane became the first ever to fly nonstop across the Atlantic Ocean. The map shows the airplane's position at three different points during its flight. Point A represents Cape Spear, Newfoundland, point B represents the approximate position after 1 day, and point C represents Mannin Bay, Ireland. The airplane left from Cape Spear and landed in Mannin Bay. *(See Example 4.)*

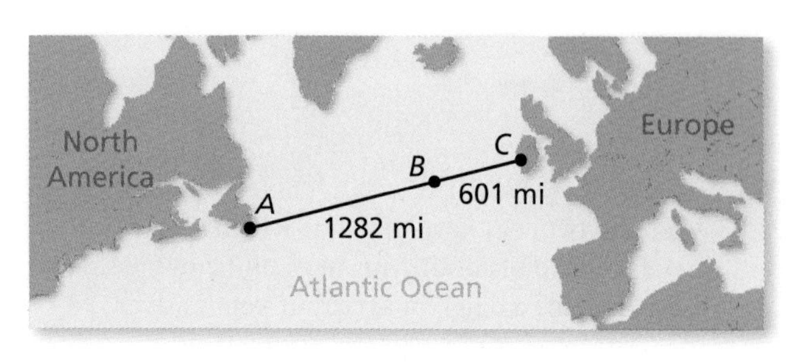

a. Find the total distance the model airplane flew.

b. The model airplane's flight lasted nearly 38 hours. Estimate the airplane's average speed in miles per hour.

27. USING STRUCTURE Determine whether the statements are true or false. Explain your reasoning.

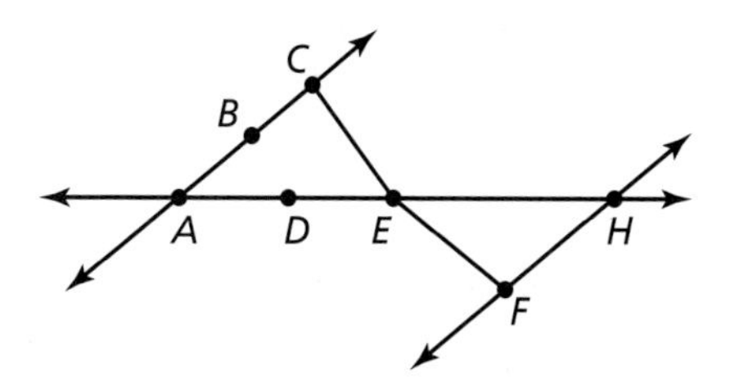

a. B is between A and C.

b. C is between B and E.

c. D is between A and H.

d. E is between C and F.

28. MATHEMATICAL CONNECTIONS Write an expression for the length of the segment.

a. $\overline{AC}$

A $x + 2$ B $7x - 3$ C

b. $\overline{QR}$

$13y + 25$

P $8y + 5$ Q R

29. **MATHEMATICAL CONNECTIONS** Point S is between points R and T on $\overline{RT}$. Use the information to write an equation in terms of x. Then solve the equation and find RS, ST, and RT.

a. $RS = 2x + 10$
$ST = x - 4$
$RT = 21$

b. $RS = 3x - 16$
$ST = 4x - 8$
$RT = 60$

c. $RS = 2x - 8$
$ST = 11$
$RT = x + 10$

d. $RS = 4x - 9$
$ST = 19$
$RT = 8x - 14$

30. **THOUGHT PROVOKING** Is it possible to design a table where no two legs have the same length? Assume that the endpoints of the legs must all lie in the same plane. Include a diagram as part of your answer.

31. **MODELING WITH MATHEMATICS** You have to walk from Room 103 to Room 117.

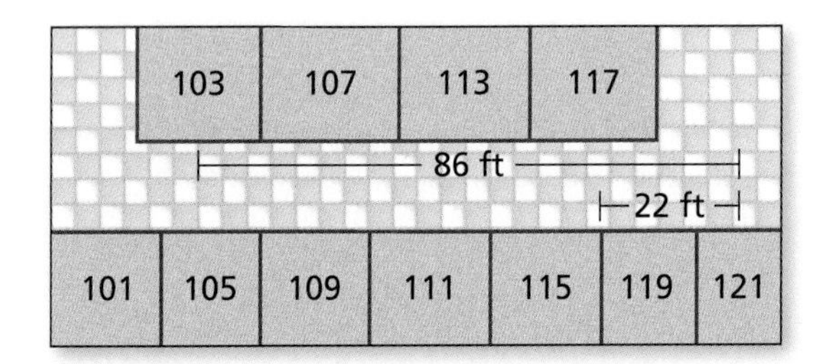

a. How many feet do you travel from Room 103 to Room 117?

b. You can walk 4.4 feet per second. How many minutes will it take you to get to Room 117?

c. Why might it take you longer than the time in part (b)?

32. **MAKING AN ARGUMENT** Your friend and your cousin discuss measuring with a ruler. Your friend says that you must always line up objects at the zero on a ruler. Your cousin says it does not matter. Decide who is correct and explain your reasoning.

33. **REASONING** You travel from City X to City Y. You know that the round-trip distance is 647 miles. City Z, a city you pass on the way, is 27 miles from City X. Find the distance from City Z to City Y. Justify your answer.

34. **HOW DO YOU SEE IT?** The bar graph shows the win-loss record for a lacrosse team over a period of three years. Explain how you can apply the Ruler Postulate (Post. 1.1) and the Segment Addition Postulate (Post. 1.2) when interpreting a stacked bar graph like the one shown.

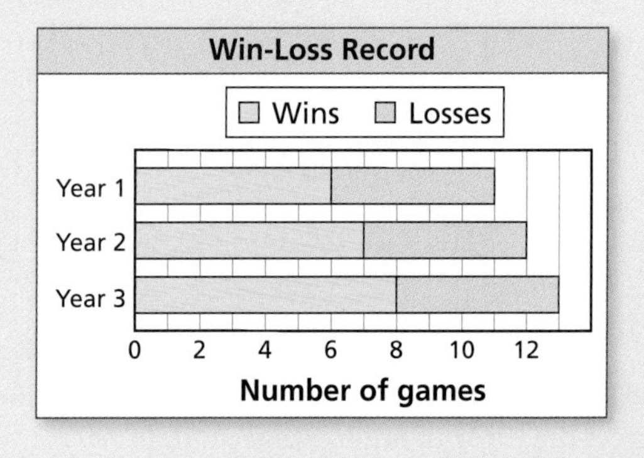

35. **ABSTRACT REASONING** The points (a, b) and (c, b) form a segment, and the points (d, e) and (d, f) form a segment. Create an equation assuming the segments are congruent. Are there any letters not used in the equation? Explain.

36. **MATHEMATICAL CONNECTIONS** In the diagram, $\overline{AB} \cong \overline{BC}, \overline{AC} \cong \overline{CD}$, and $AD = 12$. Find the lengths of all segments in the diagram. Suppose you choose one of the segments at random. What is the probability that the measure of the segment is greater than 3? Explain your reasoning.

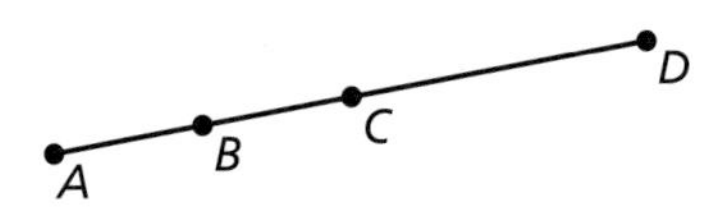

37. **CRITICAL THINKING** Is it possible to use the Segment Addition Postulate (Post. 1.2) to show $FB > CB$ or that $AC > DB$? Explain your reasoning.

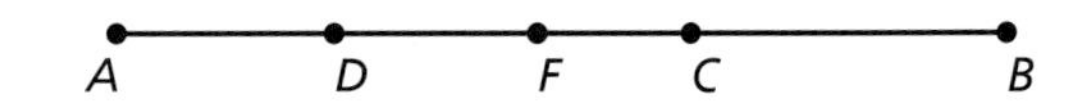

Maintaining Mathematical Proficiency

Reviewing what you learned in previous grades and lessons

Simplify. *(Skills Review Handbook)*

38. $\dfrac{-4 + 6}{2}$

39. $\sqrt{20 + 5}$

40. $\sqrt{25 + 9}$

41. $\dfrac{7 + 6}{2}$

Solve the equation. *(Skills Review Handbook)*

42. $5x + 7 = 9x - 17$

43. $\dfrac{3 + y}{2} = 6$

44. $\dfrac{-5 + x}{2} = -9$

45. $-6x - 13 = -x - 23$

1.3 Using Midpoint and Distance Formulas

Essential Question How can you find the midpoint and length of a line segment in a coordinate plane?

EXPLORATION 1 Finding the Midpoint of a Line Segment

Work with a partner. Use centimeter graph paper.

a. Graph $\overline{AB}$, where the points A and B are as shown.

b. Explain how to *bisect* $\overline{AB}$, that is, to divide $\overline{AB}$ into two congruent line segments. Then bisect $\overline{AB}$ and use the result to find the *midpoint* M of $\overline{AB}$.

c. What are the coordinates of the midpoint M?

d. Compare the x-coordinates of A, B, and M. Compare the y-coordinates of A, B, and M. How are the coordinates of the midpoint M related to the coordinates of A and B?

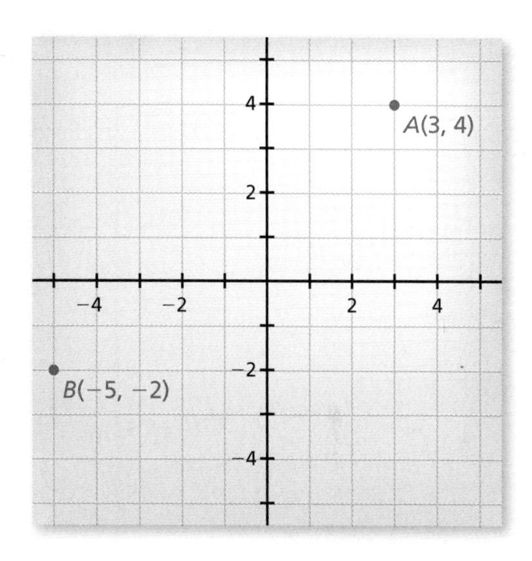

EXPLORATION 2 Finding the Length of a Line Segment

Work with a partner. Use centimeter graph paper.

a. Add point C to your graph as shown.

b. Use the Pythagorean Theorem to find the length of $\overline{AB}$.

c. Use a centimeter ruler to verify the length you found in part (b).

d. Use the Pythagorean Theorem and point M from Exploration 1 to find the lengths of $\overline{AM}$ and $\overline{MB}$. What can you conclude?

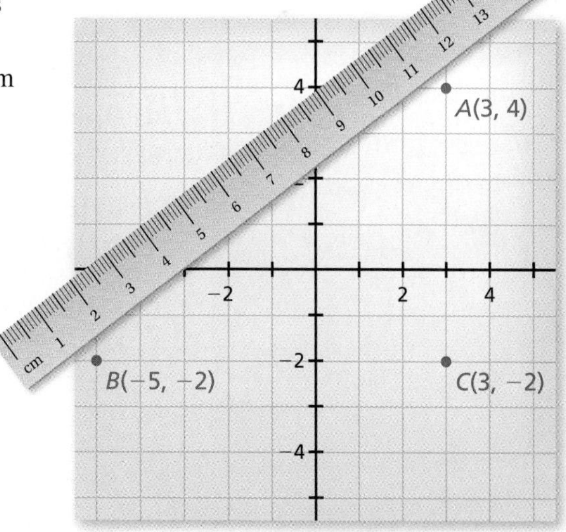

MAKING SENSE OF PROBLEMS

To be proficient in math, you need to check your answers and continually ask yourself, "Does this make sense?"

Communicate Your Answer

3. How can you find the midpoint and length of a line segment in a coordinate plane?

4. Find the coordinates of the midpoint M and the length of the line segment whose endpoints are given.

a. $D(-10, -4)$, $E(14, 6)$ **b.** $F(-4, 8)$, $G(9, 0)$

1.3 Lesson

What You Will Learn

- Find segment lengths using midpoints and segment bisectors.
- Use the Midpoint Formula.
- Use the Distance Formula.

Core Vocabulary

midpoint, *p. 20*
segment bisector, *p. 20*

Midpoints and Segment Bisectors

Core Concept

Midpoints and Segment Bisectors

The **midpoint** of a segment is the point that divides the segment into two congruent segments.

M is the midpoint of $\overline{AB}$.
So, $\overline{AM} \cong \overline{MB}$ and $AM = MB$.

READING

The word *bisect* means "to cut into two equal parts."

A **segment bisector** is a point, ray, line, line segment, or plane that intersects the segment at its midpoint. A midpoint or a segment bisector *bisects* a segment.

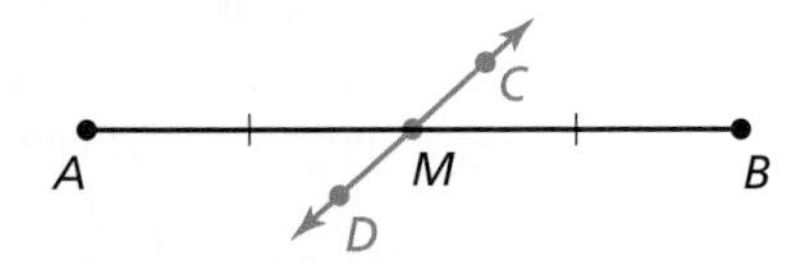

$\overleftrightarrow{CD}$ is a segment bisector of $\overline{AB}$.
So, $\overline{AM} \cong \overline{MB}$ and $AM = MB$.

EXAMPLE 1 Finding Segment Lengths

In the skateboard design, $\overline{VW}$ bisects $\overline{XY}$ at point T, and $XT = 39.9$ cm. Find XY.

SOLUTION

Point T is the midpoint of $\overline{XY}$. So, $XT = TY = 39.9$ cm.

$XY = XT + TY$	Segment Addition Postulate (Postulate 1.2)
$= 39.9 + 39.9$	Substitute.
$= 79.8$	Add.

▶ So, the length of $\overline{XY}$ is 79.8 centimeters.

Monitoring Progress

Help in English and Spanish at *BigIdeasMath.com*

Identify the segment bisector of $\overline{PQ}$. Then find PQ.

1.

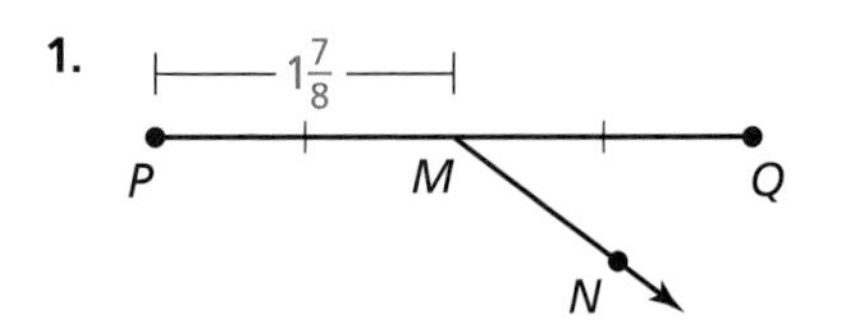

2.

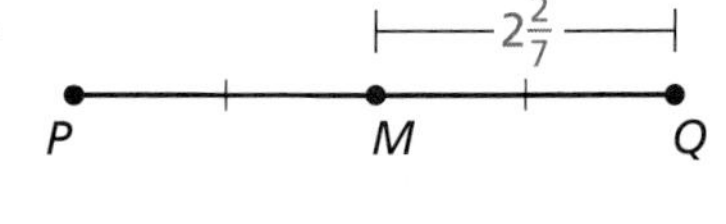

EXAMPLE 2 Using Algebra with Segment Lengths

Point M is the midpoint of $\overline{VW}$. Find the length of $\overline{VM}$.

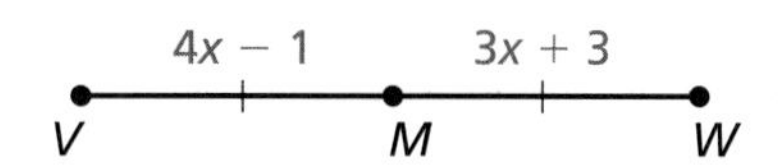

SOLUTION

Step 1 Write and solve an equation. Use the fact that $VM = MW$.

$VM = MW$	Write the equation.
$4x - 1 = 3x + 3$	Substitute.
$x - 1 = 3$	Subtract $3x$ from each side.
$x = 4$	Add 1 to each side.

Step 2 Evaluate the expression for VM when $x = 4$.

$VM = 4x - 1 = 4(4) - 1 = 15$

▶ So, the length of $\overline{VM}$ is 15.

Check

Because $VM = MW$, the length of $\overline{MW}$ should be 15.

$MW = 3x + 3 = 3(4) + 3 = 15$ ✓

Monitoring Progress

Help in English and Spanish at *BigIdeasMath.com*

3. Identify the segment bisector of $\overline{PQ}$. Then find MQ.

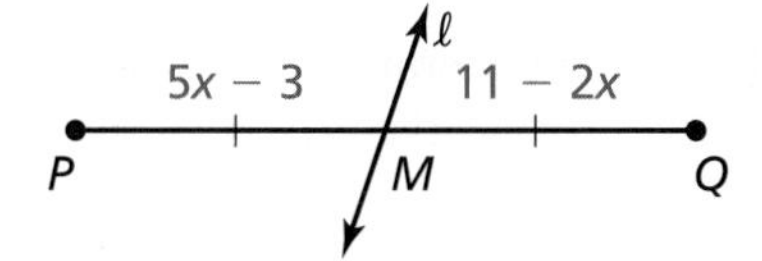

4. Identify the segment bisector of $\overline{RS}$. Then find RS.

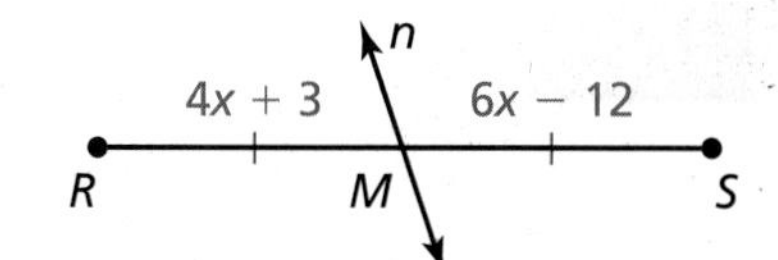

CONSTRUCTION Bisecting a Segment

Construct a segment bisector of $\overline{AB}$ by paper folding. Then find the midpoint M of $\overline{AB}$.

SOLUTION

Step 1

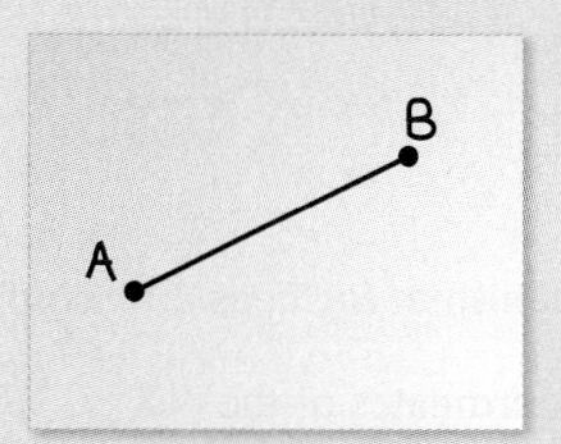

Draw the segment
Draw $\overline{AB}$ on a piece of paper.

Step 2

Fold the paper
Fold the paper so that B is on top of A.

Step 3

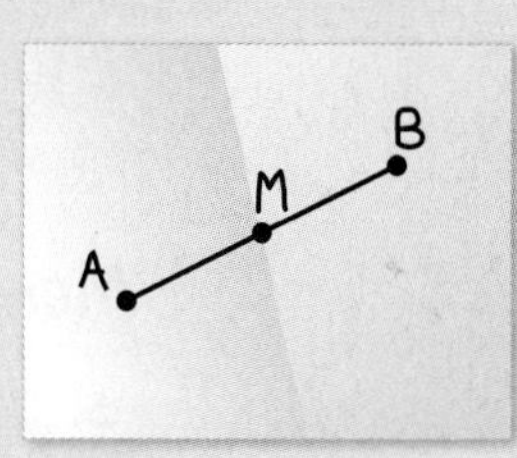

Label the midpoint
Label point M. Compare AM, MB, and AB.

$AM = MB = \frac{1}{2}AB$

Using the Midpoint Formula

You can use the coordinates of the endpoints of a segment to find the coordinates of the midpoint.

The Midpoint Formula

The coordinates of the midpoint of a segment are the averages of the x-coordinates and of the y-coordinates of the endpoints.

If $A(x_1, y_1)$ and $B(x_2, y_2)$ are points in a coordinate plane, then the midpoint M of $\overline{AB}$ has coordinates

$$\left(\frac{x_1 + x_2}{2}, \frac{y_1 + y_2}{2}\right).$$

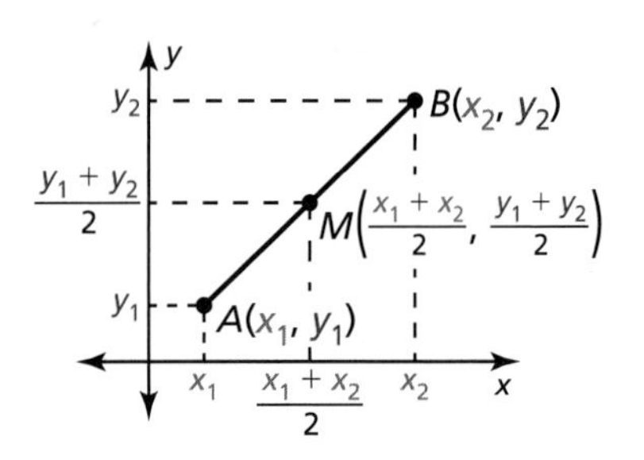

EXAMPLE 3 Using the Midpoint Formula

a. The endpoints of $\overline{RS}$ are $R(1, -3)$ and $S(4, 2)$. Find the coordinates of the midpoint M.

b. The midpoint of $\overline{JK}$ is $M(2, 1)$. One endpoint is $J(1, 4)$. Find the coordinates of endpoint K.

SOLUTION

a. Use the Midpoint Formula.

$$M\left(\frac{1 + 4}{2}, \frac{-3 + 2}{2}\right) = M\left(\frac{5}{2}, -\frac{1}{2}\right)$$

▶ The coordinates of the midpoint M are $\left(\frac{5}{2}, -\frac{1}{2}\right)$.

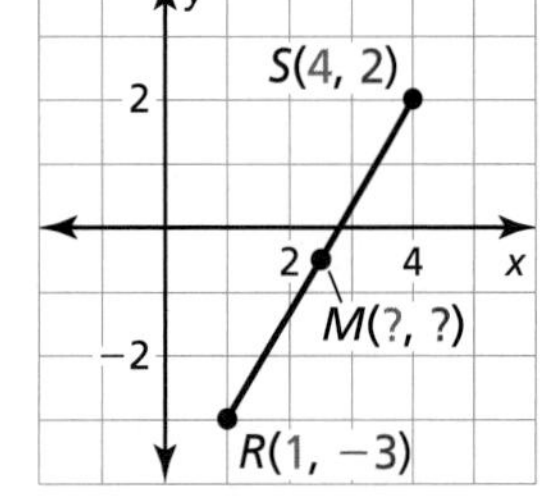

b. Let (x, y) be the coordinates of endpoint K. Use the Midpoint Formula.

Step 1 Find x.

$$\frac{1 + x}{2} = 2$$
$$1 + x = 4$$
$$x = 3$$

Step 2 Find y.

$$\frac{4 + y}{2} = 1$$
$$4 + y = 2$$
$$y = -2$$

▶ The coordinates of endpoint K are $(3, -2)$.

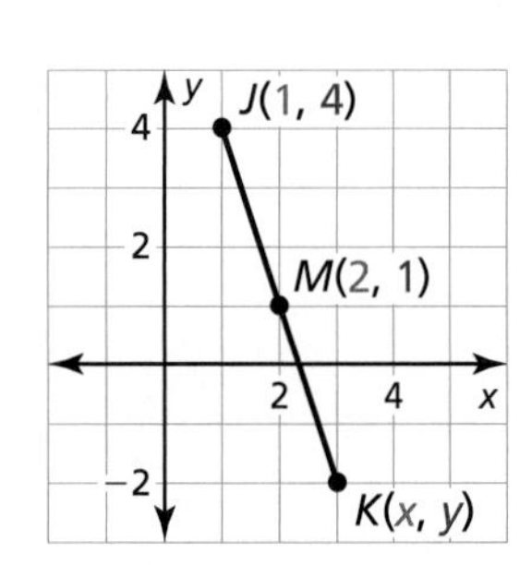

Monitoring Progress  Help in English and Spanish at *BigIdeasMath.com*

5. The endpoints of $\overline{AB}$ are $A(1, 2)$ and $B(7, 8)$. Find the coordinates of the midpoint M.

6. The endpoints of $\overline{CD}$ are $C(-4, 3)$ and $D(-6, 5)$. Find the coordinates of the midpoint M.

7. The midpoint of $\overline{TU}$ is $M(2, 4)$. One endpoint is $T(1, 1)$. Find the coordinates of endpoint U.

8. The midpoint of $\overline{VW}$ is $M(-1, -2)$. One endpoint is $W(4, 4)$. Find the coordinates of endpoint V.

Using the Distance Formula

You can use the Distance Formula to find the distance between two points in a coordinate plane.

READING

The red mark at the corner of the triangle that makes a right angle indicates a right triangle.

Core Concept

The Distance Formula

If $A(x_1, y_1)$ and $B(x_2, y_2)$ are points in a coordinate plane, then the distance between A and B is

$$AB = \sqrt{(x_2 - x_1)^2 + (y_2 - y_1)^2}.$$

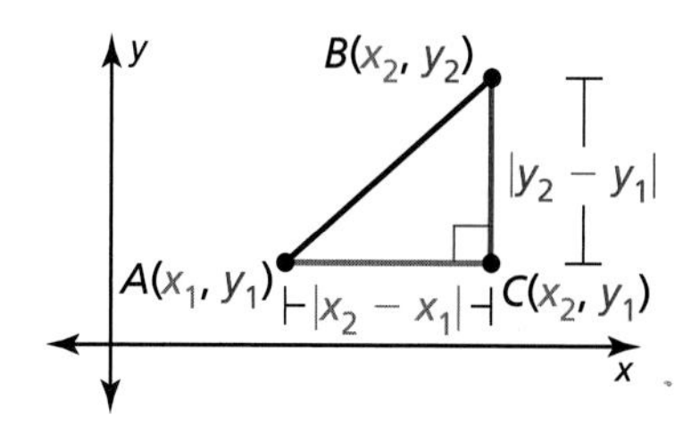

The Distance Formula is related to the *Pythagorean Theorem*, which you will see again when you work with right triangles.

Distance Formula

$(AB)^2 = (x_2 - x_1)^2 + (y_2 - y_1)^2$

Pythagorean Theorem

$c^2 = a^2 + b^2$

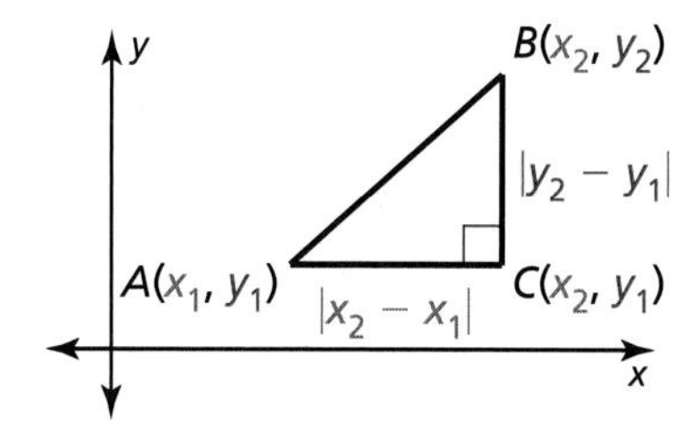

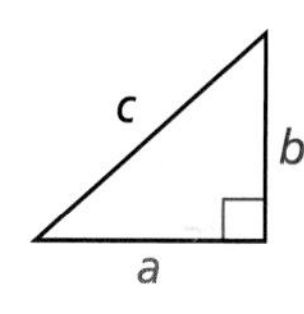

EXAMPLE 4 Using the Distance Formula

Your school is 4 miles east and 1 mile south of your apartment. A recycling center, where your class is going on a field trip, is 2 miles east and 3 miles north of your apartment. Estimate the distance between the recycling center and your school.

SOLUTION

You can model the situation using a coordinate plane with your apartment at the origin (0, 0). The coordinates of the recycling center and the school are $R(2, 3)$ and $S(4, -1)$, respectively. Use the Distance Formula. Let $(x_1, y_1) = (2, 3)$ and $(x_2, y_2) = (4, -1)$.

$RS = \sqrt{(x_2 - x_1)^2 + (y_2 - y_1)^2}$	Distance Formula
$= \sqrt{(4 - 2)^2 + (-1 - 3)^2}$	Substitute.
$= \sqrt{2^2 + (-4)^2}$	Subtract.
$= \sqrt{4 + 16}$	Evaluate powers.
$= \sqrt{20}$	Add.
≈ 4.5	Use a calculator.

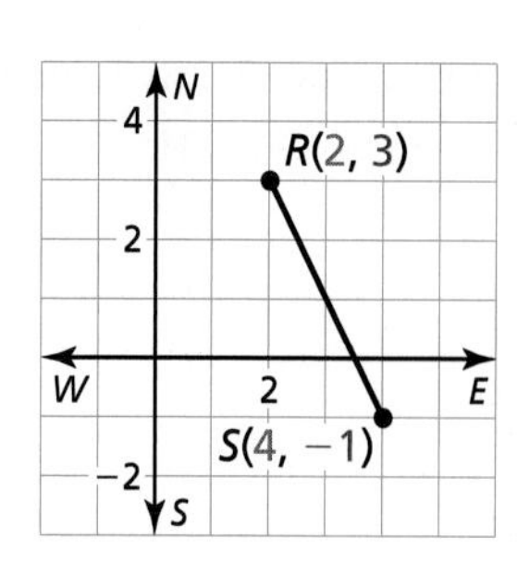

READING

The symbol ≈ means "is approximately equal to."

▶ So, the distance between the recycling center and your school is about 4.5 miles.

Monitoring Progress

Help in English and Spanish at *BigIdeasMath.com*

9. In Example 4, a park is 3 miles east and 4 miles south of your apartment. Find the distance between the park and your school.

1.3 Exercises

Dynamic Solutions available at *BigIdeasMath.com*

Vocabulary and Core Concept Check

1. **VOCABULARY** If a point, ray, line, line segment, or plane intersects a segment at its midpoint, then what does it do to the segment?

2. **COMPLETE THE SENTENCE** To find the length of $\overline{AB}$, with endpoints $A(-7, 5)$ and $B(4, -6)$, you can use the _____________.

Monitoring Progress and Modeling with Mathematics

In Exercises 3–6, identify the segment bisector of $\overline{RS}$. Then find RS. *(See Example 1.)*

3.

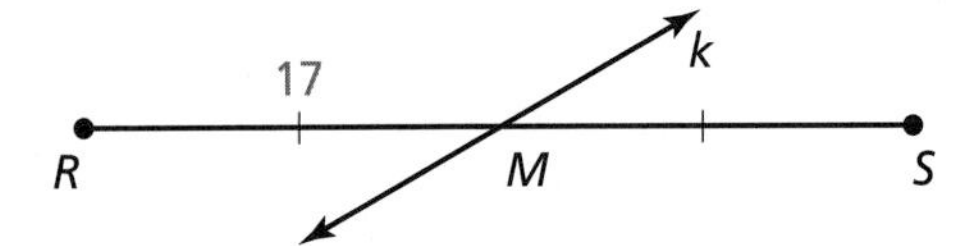

4.

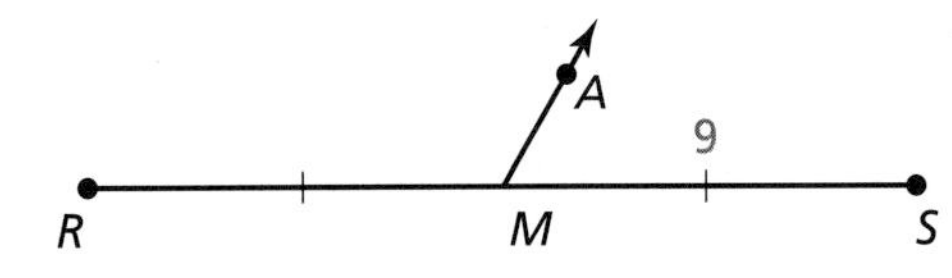

5.

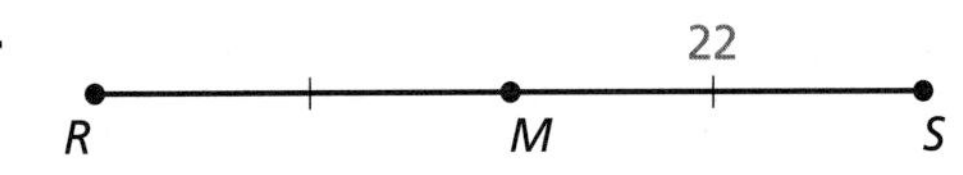

6.

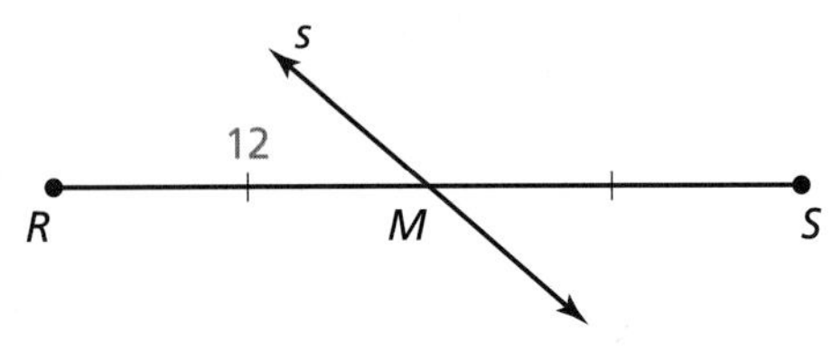

In Exercises 7 and 8, identify the segment bisector of $\overline{JK}$. Then find JM. *(See Example 2.)*

7.

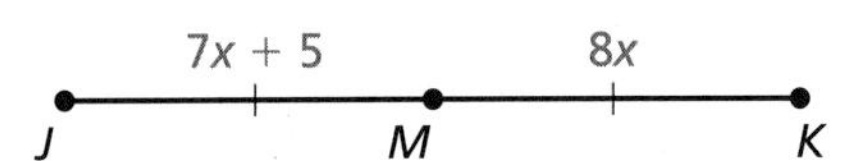

8.

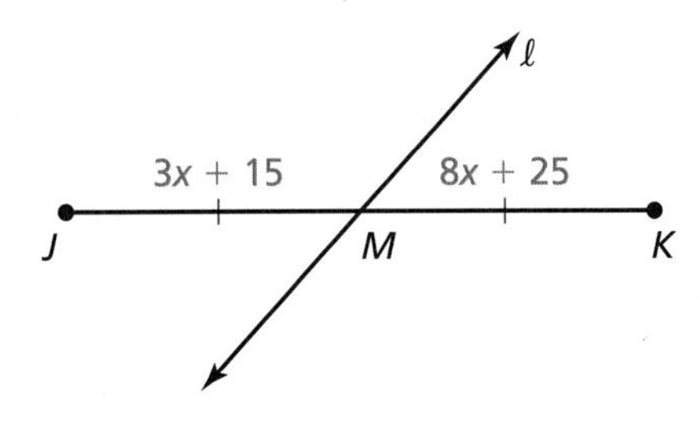

In Exercises 9 and 10, identify the segment bisector of $\overline{XY}$. Then find XY. *(See Example 2.)*

9.

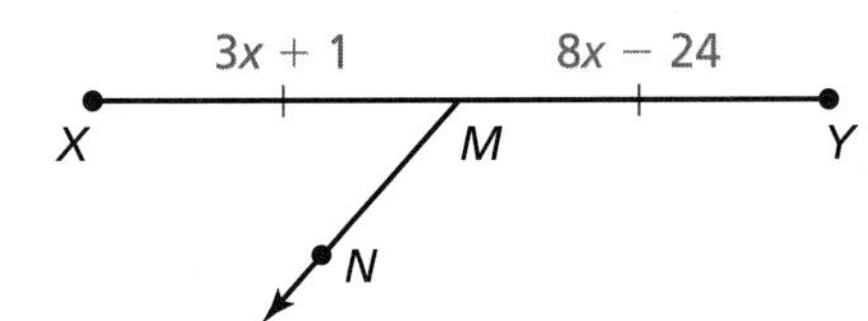

10.

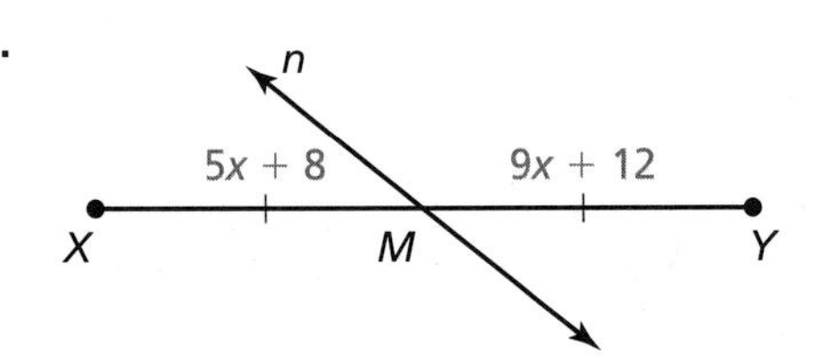

CONSTRUCTION In Exercises 11–14, copy the segment and construct a segment bisector by paper folding. Then label the midpoint M.

11. A B

12.

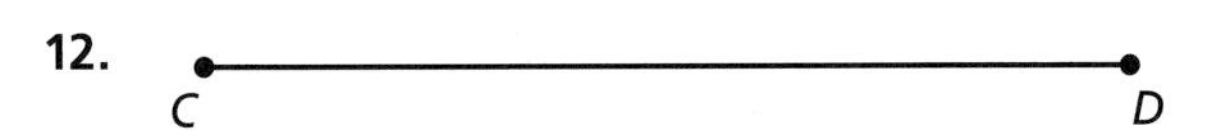

13.

14. G H

In Exercises 15–18, the endpoints of $\overline{CD}$ are given. Find the coordinates of the midpoint M. *(See Example 3.)*

15. $C(3, -5)$ and $D(7, 9)$

16. $C(-4, 7)$ and $D(0, -3)$

17. $C(-2, 0)$ and $D(4, 9)$

18. $C(-8, -6)$ and $D(-4, 10)$

In Exercises 19–22, the midpoint M and one endpoint of $\overline{GH}$ are given. Find the coordinates of the other endpoint. *(See Example 3.)*

19. $G(5, -6)$ and $M(4, 3)$

20. $H(-3, 7)$ and $M(-2, 5)$

21. $H(-2, 9)$ and $M(8, 0)$

22. $G(-4, 1)$ and $M\left(-\frac{13}{2}, -6\right)$

In Exercises 23–30, find the distance between the two points. *(See Example 4.)*

23. $A(13, 2)$ and $B(7, 10)$

24. $C(-6, 5)$ and $D(-3, 1)$

25. $E(3, 7)$ and $F(6, 5)$

26. $G(-5, 4)$ and $H(2, 6)$

27. $J(-8, 0)$ and $K(1, 4)$

28. $L(7, -1)$ and $M(-2, 4)$

29. $R(0, 1)$ and $S(6, 3.5)$

30. $T(13, 1.6)$ and $V(5.4, 3.7)$

ERROR ANALYSIS In Exercises 31 and 32, describe and correct the error in finding the distance between $A(6, 2)$ and $B(1, -4)$.

31. ✗

$$\begin{aligned} AB &= (6-1)^2 + [2-(-4)]^2 \\ &= 5^2 + 6^2 \\ &= 25 + 36 \\ &= 61 \end{aligned}$$

32. ✗

$$\begin{aligned} AB &= \sqrt{(6-2)^2 + [1-(-4)]^2} \\ &= \sqrt{4^2 + 5^2} \\ &= \sqrt{16 + 25} \\ &= \sqrt{41} \\ &\approx 6.4 \end{aligned}$$

COMPARING SEGMENTS In Exercises 33 and 34, the endpoints of two segments are given. Find each segment length. Tell whether the segments are congruent. If they are not congruent, state which segment length is greater.

33. $\overline{AB}$: $A(0, 2)$, $B(-3, 8)$ and $\overline{CD}$: $C(-2, 2)$, $D(0, -4)$

34. $\overline{EF}$: $E(1, 4)$, $F(5, 1)$ and $\overline{GH}$: $G(-3, 1)$, $H(1, 6)$

35. WRITING Your friend is having trouble understanding the Midpoint Formula.

a. Explain how to find the midpoint when given the two endpoints in your own words.

b. Explain how to find the other endpoint when given one endpoint and the midpoint in your own words.

36. PROBLEM SOLVING In baseball, the strike zone is the region a baseball needs to pass through for the umpire to declare it a strike when the batter does not swing. The top of the strike zone is a horizontal plane passing through the midpoint of the top of the batter's shoulders and the top of the uniform pants when the player is in a batting stance. Find the height of T. (*Note:* All heights are in inches.)

37. MODELING WITH MATHEMATICS The figure shows the position of three players during part of a water polo match. Player A throws the ball to Player B, who then throws the ball to Player C.

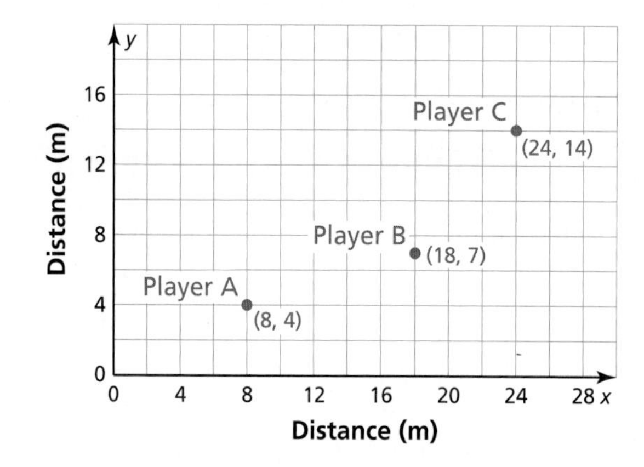

a. How far did Player A throw the ball? Player B?

b. How far would Player A have to throw the ball to throw it directly to Player C?

38. MODELING WITH MATHEMATICS Your school is 20 blocks east and 12 blocks south of your house. The mall is 10 blocks north and 7 blocks west of your house. You plan on going to the mall right after school. Find the distance between your school and the mall assuming there is a road directly connecting the school and the mall. One block is 0.1 mile.

39. PROBLEM SOLVING A path goes around a triangular park, as shown.

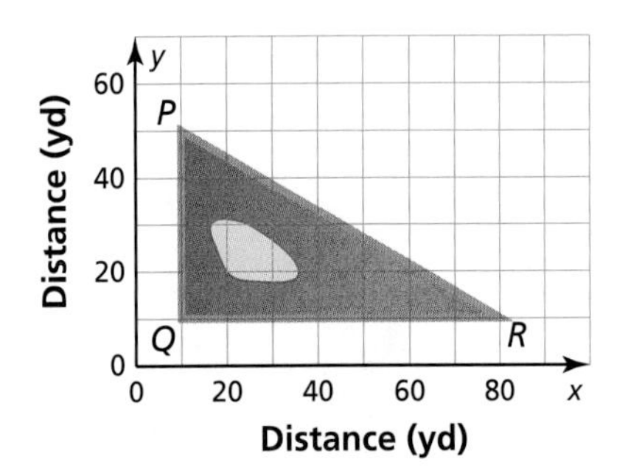

a. Find the distance around the park to the nearest yard.

b. A new path and a bridge are constructed from point Q to the midpoint M of $\overline{PR}$. Find QM to the nearest yard.

c. A man jogs from P to Q to M to R to Q and back to P at an average speed of 150 yards per minute. About how many minutes does it take? Explain your reasoning.

40. MAKING AN ARGUMENT Your friend claims there is an easier way to find the length of a segment than the Distance Formula when the x-coordinates of the endpoints are equal. He claims all you have to do is subtract the y-coordinates. Do you agree with his statement? Explain your reasoning.

41. MATHEMATICAL CONNECTIONS Two points are located at (a, c) and (b, c). Find the midpoint and the distance between the two points.

42. HOW DO YOU SEE IT? $\overline{AB}$ contains midpoint M and points C and D, as shown. Compare the lengths. If you cannot draw a conclusion, write *impossible to tell*. Explain your reasoning.

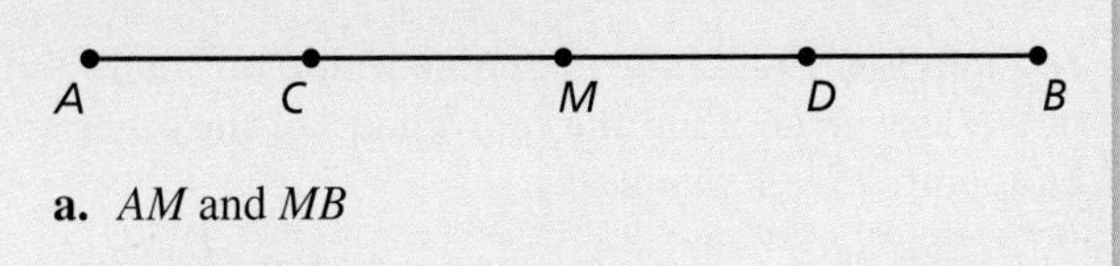

a. AM and MB

b. AC and MB

c. MC and MD

d. MB and DB

43. ABSTRACT REASONING Use the diagram in Exercise 42. The points on $\overline{AB}$ represent locations you pass on your commute to work. You travel from your home at location A to location M before realizing that you left your lunch at home. You could turn around to get your lunch and then continue to work at location B. Or you could go to work and go to location D for lunch today. You want to choose the option that involves the least distance you must travel. Which option should you choose? Explain your reasoning.

44. THOUGHT PROVOKING Describe three ways to divide a rectangle into two congruent regions. Do the regions have to be triangles? Use a diagram to support your answer.

45. ANALYZING RELATIONSHIPS The length of $\overline{XY}$ is 24 centimeters. The midpoint of $\overline{XY}$ is M, and C is on $\overline{XM}$ so that XC is $\frac{2}{3}$ of XM. Point D is on $\overline{MY}$ so that MD is $\frac{3}{4}$ of MY. What is the length of $\overline{CD}$?

Maintaining Mathematical Proficiency

Reviewing what you learned in previous grades and lessons

Find the perimeter and area of the figure. *(Skills Review Handbook)*

46.

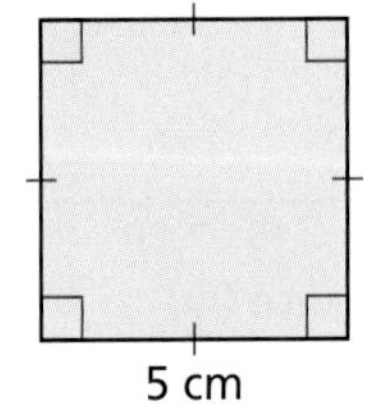

47.

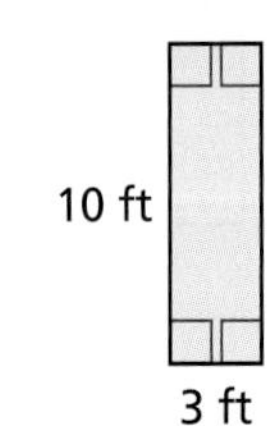

48.

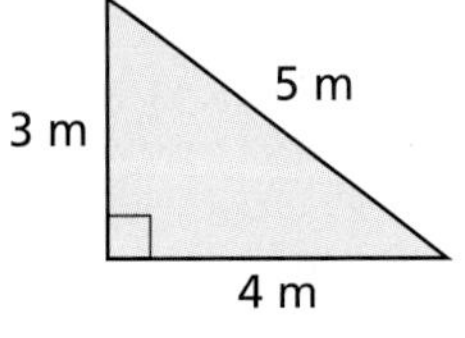

49.

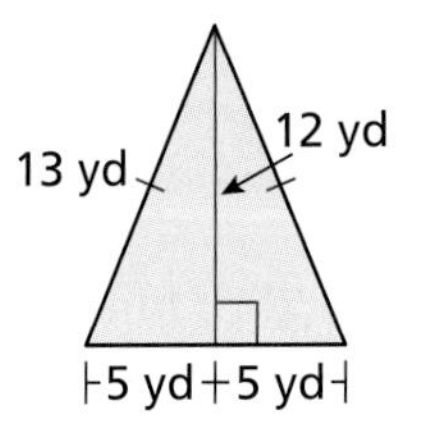

Solve the inequality. Graph the solution. *(Skills Review Handbook)*

50. $a + 18 < 7$

51. $y - 5 \geq 8$

52. $-3x > 24$

53. $\frac{z}{4} \leq 12$

1.1–1.3 What Did You Learn?

Core Vocabulary

undefined terms, *p. 4*
point, *p. 4*
line, *p. 4*
plane, *p. 4*
collinear points, *p. 4*
coplanar points, *p. 4*
defined terms, *p. 5*
line segment, or segment, *p. 5*
endpoints, *p. 5*
ray, *p. 5*
opposite rays, *p. 5*
intersection, *p. 6*
postulate, *p. 12*
axiom, *p. 12*
coordinate, *p. 12*
distance, *p. 12*
construction, *p. 13*
congruent segments, *p. 13*
between, *p. 14*
midpoint, *p. 20*
segment bisector, *p. 20*

Core Concepts

Section 1.1

Undefined Terms: Point, Line, and Plane, *p. 4*
Defined Terms: Segment and Ray, *p. 5*
Intersections of Lines and Planes, *p. 6*

Section 1.2

Postulate 1.1 Ruler Postulate, *p. 12*
Congruent Segments, *p. 13*
Postulate 1.2 Segment Addition Postulate, *p. 14*

Section 1.3

Midpoints and Segment Bisectors, *p. 20*
The Midpoint Formula, *p. 22*
The Distance Formula, *p. 23*

Mathematical Practices

1. Sketch an example of the situation described in Exercise 49 on page 10 in a coordinate plane. Label your figure.
2. Explain how you arrived at your answer for Exercise 35 on page 18.
3. What assumptions did you make when solving Exercise 43 on page 26?

Study Skills

Keeping Your Mind Focused

- Keep a notebook just for vocabulary, formulas, and core concepts.
- Review this notebook before completing homework and before tests.

1.1–1.3 Quiz

Use the diagram. *(Section 1.1)*

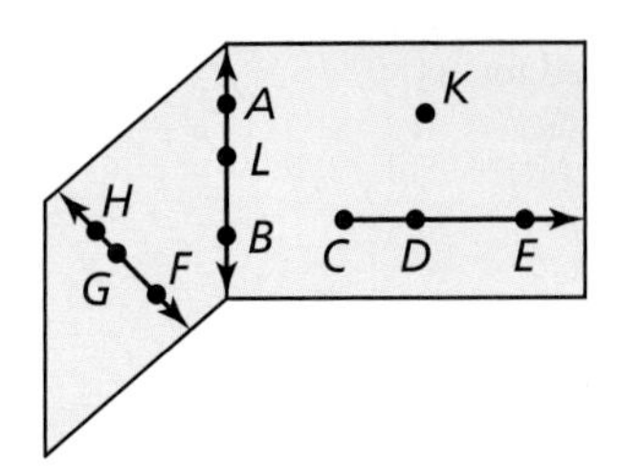

1. Name four points.
2. Name three collinear points.
3. Name two lines.
4. Name three coplanar points.
5. Name the plane that is shaded green.
6. Give two names for the plane that is shaded blue.
7. Name three line segments.
8. Name three rays.

Sketch the figure described. *(Section 1.1)*

9. $\overrightarrow{QR}$ and $\overleftrightarrow{QS}$
10. plane P intersecting $\overleftrightarrow{YZ}$ at Z

Plot the points in a coordinate plane. Then determine whether $\overline{AB}$ and $\overline{CD}$ are congruent. *(Section 1.2)*

11. $A(-3, 3)$, $B(1, 3)$, $C(3, 2)$, $D(3, -2)$
12. $A(-8, 7)$, $B(1, 7)$, $C(-3, -6)$, $D(5, -6)$

Find AC. *(Section 1.2)*

13. 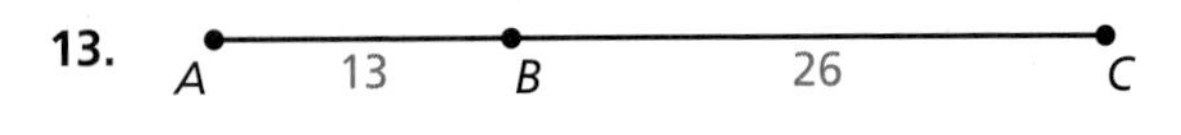

14. 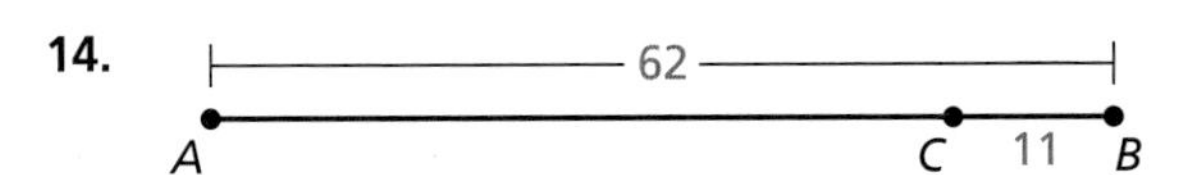

Find the coordinates of the midpoint M and the distance between the two points. *(Section 1.3)*

15. $J(4, 3)$ and $K(2, -3)$
16. $L(-4, 5)$ and $N(5, -3)$
17. $P(-6, -1)$ and $Q(1, 2)$

18. Identify the segment bisector of $\overline{RS}$. Then find RS. *(Section 1.3)*

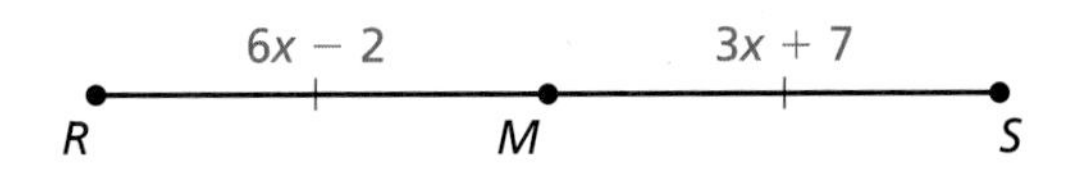

19. The midpoint of $\overline{JK}$ is $M(0, 1)$. One endpoint is $J(-6, 3)$. Find the coordinates of endpoint K. *(Section 1.3)*

20. Your mom asks you to run some errands on your way home from school. She wants you to stop at the post office and the grocery store, which are both on the same straight road between your school and your house. The distance from your school to the post office is 376 yards, the distance from the post office to your house is 929 yards, and the distance from the grocery store to your house is 513 yards. *(Section 1.2)*

    a. Where should you stop first?

    b. What is the distance from the post office to the grocery store?

    c. What is the distance from your school to your house?

    d. You walk at a speed of 75 yards per minute. How long does it take you to walk straight home from school? Explain your answer.

21. The figure shows a coordinate plane on a baseball field. The distance from home plate to first base is 90 feet. The pitching mound is the midpoint between home plate and second base. Find the distance from home plate to second base. Find the distance between home plate and the pitching mound. Explain how you found your answers. *(Section 1.3)*

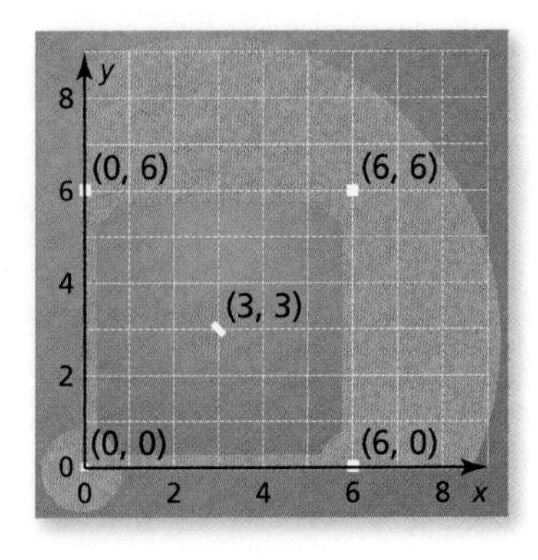

1.4 Perimeter and Area in the Coordinate Plane

Essential Question

How can you find the perimeter and area of a polygon in a coordinate plane?

EXPLORATION 1 Finding the Perimeter and Area of a Quadrilateral

Work with a partner.

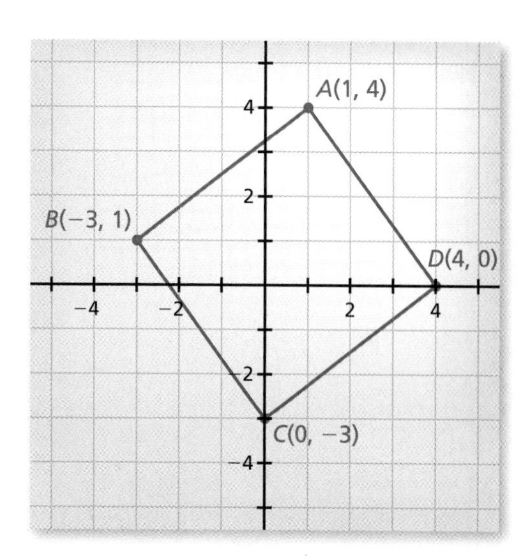

a. On a piece of centimeter graph paper, draw quadrilateral $ABCD$ in a coordinate plane. Label the points $A(1, 4)$, $B(-3, 1)$, $C(0, -3)$, and $D(4, 0)$.

b. Find the perimeter of quadrilateral $ABCD$.

c. Are adjacent sides of quadrilateral $ABCD$ perpendicular to each other? How can you tell?

d. What is the definition of a square? Is quadrilateral $ABCD$ a square? Justify your answer. Find the area of quadrilateral $ABCD$.

LOOKING FOR STRUCTURE

To be proficient in math, you need to visualize single objects as being composed of more than one object.

EXPLORATION 2 Finding the Area of a Polygon

Work with a partner.

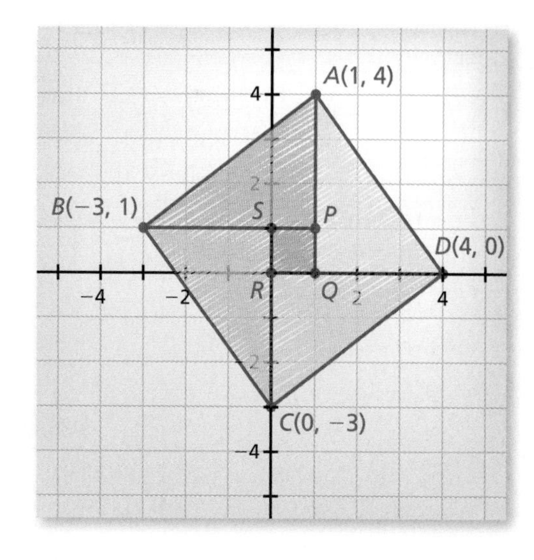

a. Partition quadrilateral $ABCD$ into four right triangles and one square, as shown. Find the coordinates of the vertices for the five smaller polygons.

b. Find the areas of the five smaller polygons.

Area of Triangle BPA:

Area of Triangle AQD:

Area of Triangle DRC:

Area of Triangle CSB:

Area of Square $PQRS$:

c. Is the sum of the areas of the five smaller polygons equal to the area of quadrilateral $ABCD$? Justify your answer.

Communicate Your Answer

3. How can you find the perimeter and area of a polygon in a coordinate plane?

4. Repeat Exploration 1 for quadrilateral $EFGH$, where the coordinates of the vertices are $E(-3, 6)$, $F(-7, 3)$, $G(-1, -5)$, and $H(3, -2)$.

1.4 Lesson

Core Vocabulary

Previous
polygon
side
vertex
n-gon
convex
concave

What You Will Learn

- Classify polygons.
- Find perimeters and areas of polygons in the coordinate plane.

Classifying Polygons

Core Concept

Polygons

In geometry, a figure that lies in a plane is called a plane figure. Recall that a *polygon* is a closed plane figure formed by three or more line segments called *sides*. Each side intersects exactly two sides, one at each *vertex*, so that no two sides with a common vertex are collinear. You can name a polygon by listing the vertices in consecutive order.

side $\overline{BC}$, vertex D

polygon $ABCDE$

The number of sides determines the name of a polygon, as shown in the table.

You can also name a polygon using the term *n*-gon, where *n* is the number of sides. For instance, a 14-gon is a polygon with 14 sides.

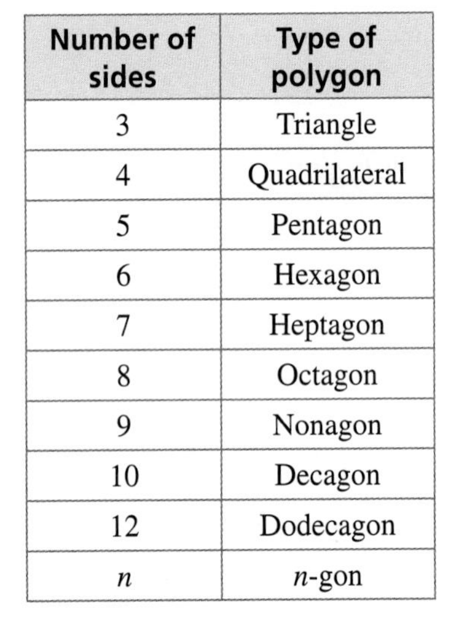

Number of sides	Type of polygon
3	Triangle
4	Quadrilateral
5	Pentagon
6	Hexagon
7	Heptagon
8	Octagon
9	Nonagon
10	Decagon
12	Dodecagon
n	*n*-gon

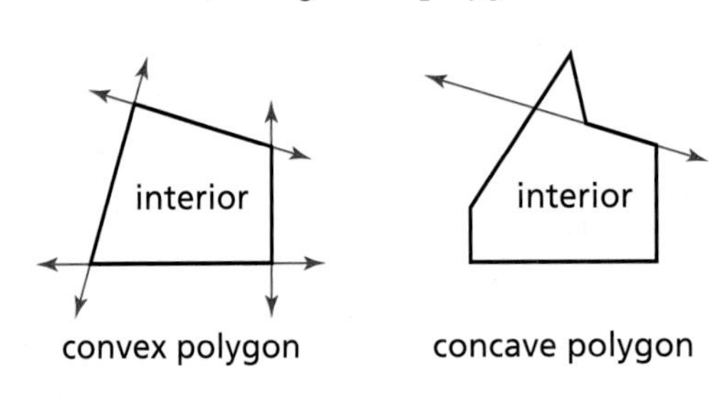

A polygon is *convex* when no line that contains a side of the polygon contains a point in the interior of the polygon. A polygon that is not convex is *concave*.

EXAMPLE 1 Classifying Polygons

Classify each polygon by the number of sides. Tell whether it is *convex* or *concave*.

a.

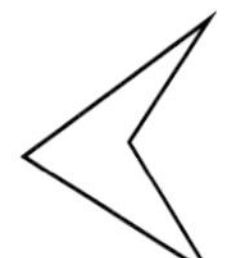

b.

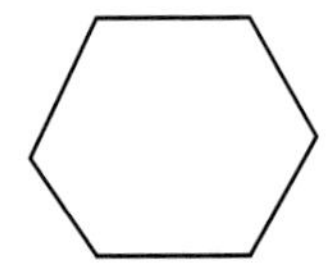

SOLUTION

a. The polygon has four sides. So, it is a quadrilateral. The polygon is concave.

b. The polygon has six sides. So, it is a hexagon. The polygon is convex.

Monitoring Progress

Help in English and Spanish at *BigIdeasMath.com*

Classify the polygon by the number of sides. Tell whether it is *convex* or *concave*.

1.

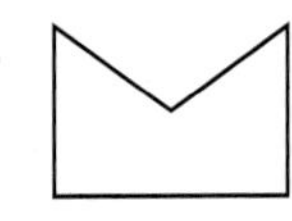

2.

Finding Perimeter and Area in the Coordinate Plane

You can use the formulas given below and the Distance Formula to find the perimeters and areas of polygons in the coordinate plane.

> **REMEMBER**
> Perimeter has linear units, such as feet or meters. Area has square units, such as square feet or square meters.

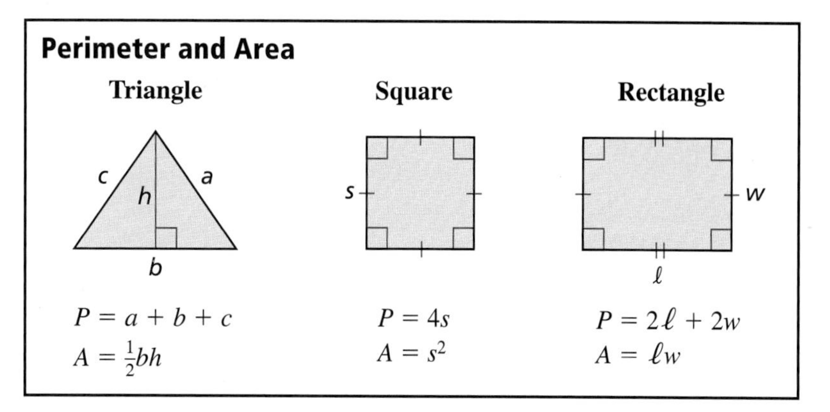

EXAMPLE 2 Finding Perimeter in the Coordinate Plane

> **READING**
> You can read the notation $\triangle ABC$ as "triangle *A B C*."

Find the perimeter of $\triangle ABC$ with vertices $A(-2, 3)$, $B(3, -3)$, and $C(-2, -3)$.

SOLUTION

Step 1 Draw the triangle in a coordinate plane. Then find the length of each side.

Side $\overline{AB}$

$AB = \sqrt{(x_2 - x_1)^2 + (y_2 - y_1)^2}$	Distance Formula
$= \sqrt{[3 - (-2)]^2 + (-3 - 3)^2}$	Substitute.
$= \sqrt{5^2 + (-6)^2}$	Subtract.
$= \sqrt{61}$	Simplify.
≈ 7.81	Use a calculator.

Side $\overline{BC}$

$BC = |-2 - 3| = 5$ — Ruler Postulate (Postulate 1.1)

Side $\overline{CA}$

$CA = |3 - (-3)| = 6$ — Ruler Postulate (Postulate 1.1)

Step 2 Find the sum of the side lengths.

$$AB + BC + CA \approx 7.81 + 5 + 6 = 18.81$$

▶ So, the perimeter of $\triangle ABC$ is about 18.81 units.

Monitoring Progress  Help in English and Spanish at *BigIdeasMath.com*

Find the perimeter of the polygon with the given vertices.

3. $D(-3, 2)$, $E(4, 2)$, $F(4, -3)$
4. $G(-3, 2)$, $H(2, 2)$, $J(-1, -3)$
5. $K(-1, 1)$, $L(4, 1)$, $M(2, -2)$, $N(-3, -2)$
6. $Q(-4, -1)$, $R(1, 4)$, $S(4, 1)$, $T(-1, -4)$

EXAMPLE 3 Finding Area in the Coordinate Plane

Find the area of $\triangle DEF$ with vertices $D(1, 3)$, $E(4, -3)$, and $F(-4, -3)$.

SOLUTION

Step 1 Draw the triangle in a coordinate plane by plotting the vertices and connecting them.

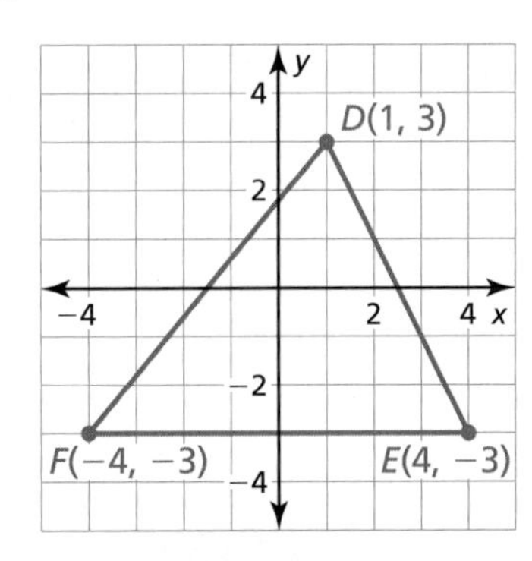

Step 2 Find the lengths of the base and height.

Base

The base is $\overline{FE}$. Use the Ruler Postulate (Postulate 1.1) to find the length of $\overline{FE}$.

$FE = \lvert 4 - (-4) \rvert$	Ruler Postulate (Postulate 1.1)
$= \lvert 8 \rvert$	Subtract.
$= 8$	Simplify.

So, the length of the base is 8 units.

Height

The height is the distance from point D to line segment $\overline{FE}$. By counting grid lines, you can determine that the height is 6 units.

Step 3 Substitute the values for the base and height into the formula for the area of a triangle.

$A = \frac{1}{2}bh$	Write the formula for area of a triangle.
$= \frac{1}{2}(8)(6)$	Substitute.
$= 24$	Multiply.

▶ So, the area of $\triangle DEF$ is 24 square units.

Monitoring Progress  Help in English and Spanish at *BigIdeasMath.com*

Find the area of the polygon with the given vertices.

7. $G(2, 2)$, $H(3, -1)$, $J(-2, -1)$

8. $N(-1, 1)$, $P(2, 1)$, $Q(2, -2)$, $R(-1, -2)$

9. $F(-2, 3)$, $G(1, 3)$, $H(1, -1)$, $J(-2, -1)$

10. $K(-3, 3)$, $L(3, 3)$, $M(3, -1)$, $N(-3, -1)$

EXAMPLE 4 Modeling with Mathematics

You are building a shed in your backyard. The diagram shows the four vertices of the shed. Each unit in the coordinate plane represents 1 foot. Find the area of the floor of the shed.

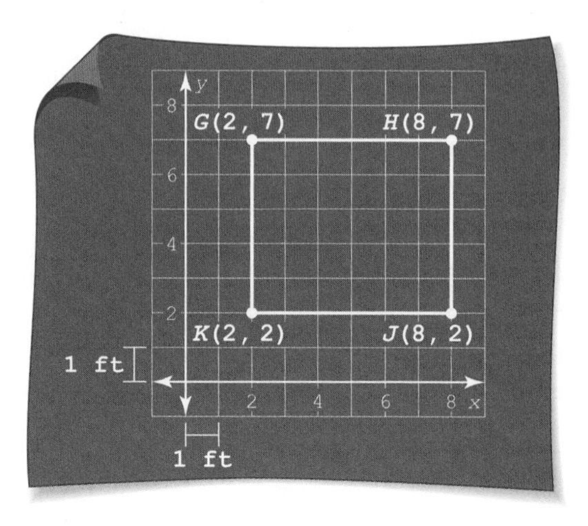

SOLUTION

1. **Understand the Problem** You are given the coordinates of a shed. You need to find the area of the floor of the shed.

2. **Make a Plan** The shed is rectangular, so use the coordinates to find the length and width of the shed. Then use a formula to find the area.

3. **Solve the Problem**

   Step 1 Find the length and width.

   Length $GH = |8 - 2| = 6$ Ruler Postulate (Postulate 1.1)

   Width $GK = |7 - 2| = 5$ Ruler Postulate (Postulate 1.1)

   The shed has a length of 6 feet and a width of 5 feet.

   Step 2 Substitute the values for the length and width into the formula for the area of a rectangle.

   $A = \ell w$ Write the formula for area of a rectangle.

   $= (6)(5)$ Substitute.

   $= 30$ Multiply.

   ▶ So, the area of the floor of the shed is 30 square feet.

4. **Look Back** Make sure your answer makes sense in the context of the problem. Because you are finding an area, your answer should be in square units. An answer of 30 square feet makes sense in the context of the problem. ✓

Monitoring Progress  Help in English and Spanish at *BigIdeasMath.com*

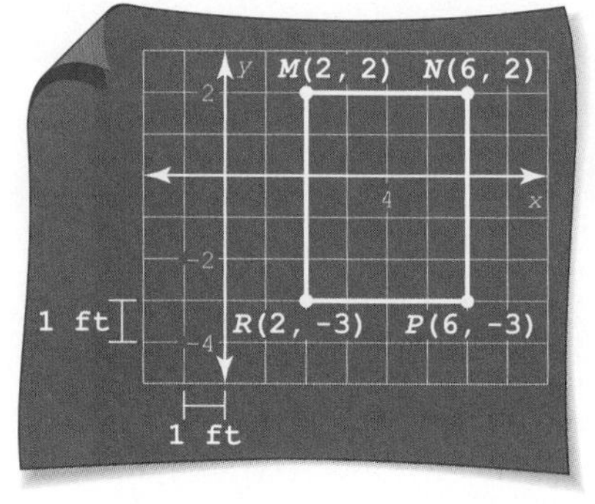

11. You are building a patio in your school's courtyard. In the diagram at the left, the coordinates represent the four vertices of the patio. Each unit in the coordinate plane represents 1 foot. Find the area of the patio.

1.4 Exercises

Dynamic Solutions available at *BigIdeasMath.com*

Vocabulary and Core Concept Check

1. **COMPLETE THE SENTENCE** The perimeter of a square with side length s is $P =$ ____.

2. **WRITING** What formulas can you use to find the area of a triangle in a coordinate plane?

Monitoring Progress and Modeling with Mathematics

In Exercises 3–6, classify the polygon by the number of sides. Tell whether it is *convex* or *concave*. *(See Example 1.)*

3. 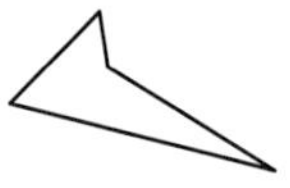

4. 

5. 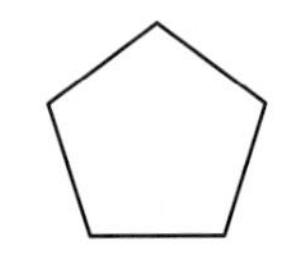

6. 

In Exercises 7–12, find the perimeter of the polygon with the given vertices. *(See Example 2.)*

7. $G(2, 4), H(2, -3), J(-2, -3), K(-2, 4)$

8. $Q(-3, 2), R(1, 2), S(1, -2), T(-3, -2)$

9. $U(-2, 4), V(3, 4), W(3, -4)$

10. $X(-1, 3), Y(3, 0), Z(-1, -2)$

11. 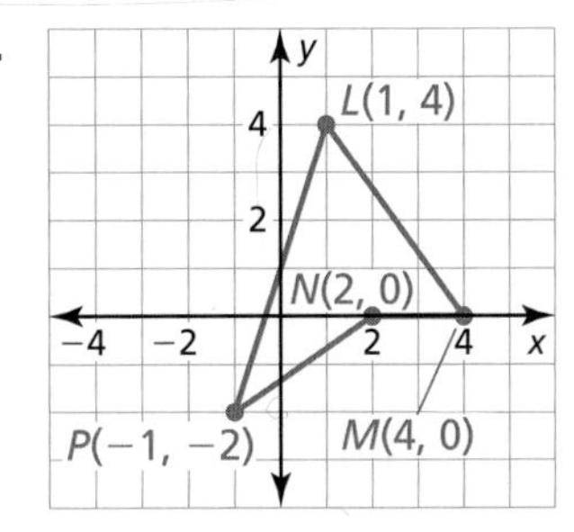

12. 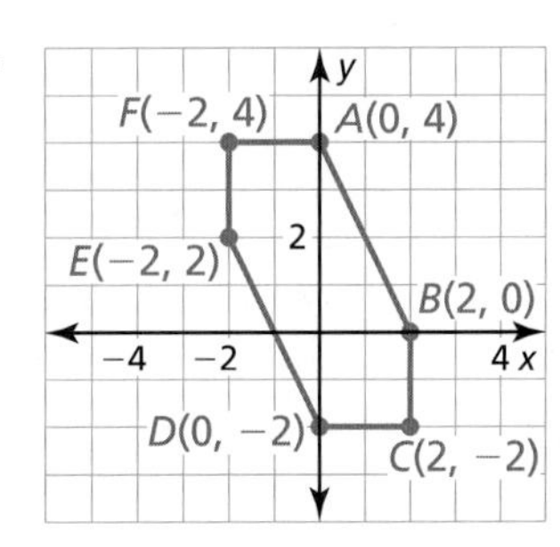

In Exercises 13–16, find the area of the polygon with the given vertices. *(See Example 3.)*

13. $E(3, 1), F(3, -2), G(-2, -2)$

14. $J(-3, 4), K(4, 4), L(3, -3)$

15. $W(0, 0), X(0, 3), Y(-3, 3), Z(-3, 0)$

16. $N(-2, 1), P(3, 1), Q(3, -1), R(-2, -1)$

In Exercises 17–24, use the diagram.

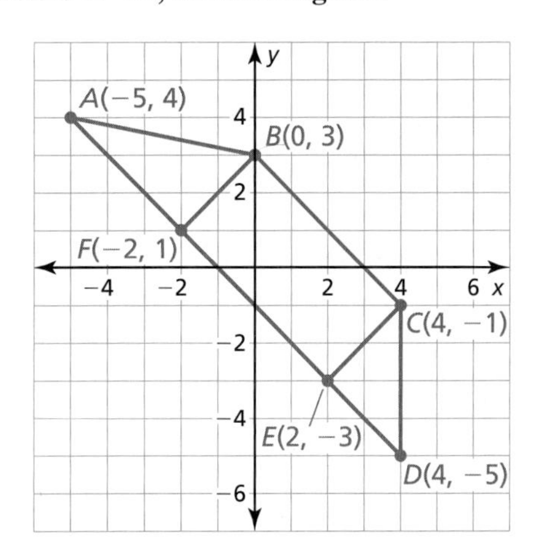

17. Find the perimeter of $\triangle CDE$.

18. Find the perimeter of rectangle $BCEF$.

19. Find the perimeter of $\triangle ABF$.

20. Find the perimeter of quadrilateral $ABCD$.

21. Find the area of $\triangle CDE$.

22. Find the area of rectangle $BCEF$.

23. Find the area of $\triangle ABF$.

24. Find the area of quadrilateral $ABCD$.

ERROR ANALYSIS In Exercises 25 and 26, describe and correct the error in finding the perimeter or area of the polygon.

25.

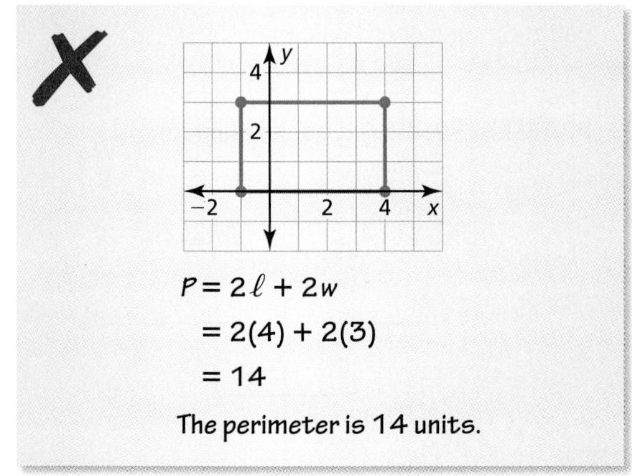

26.

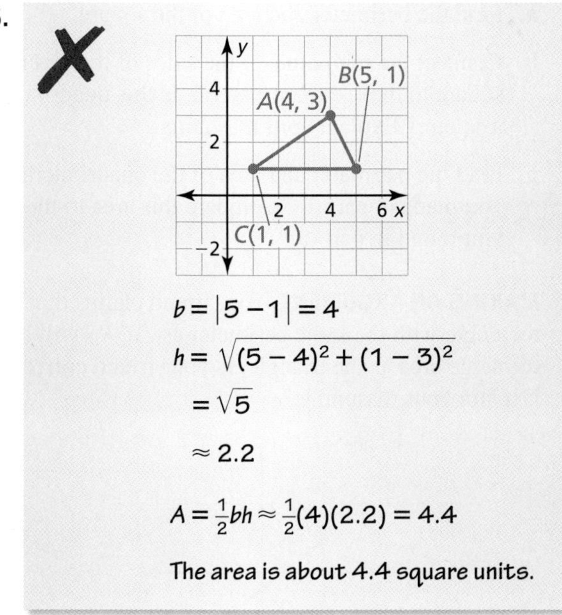

In Exercises 27 and 28, use the diagram.

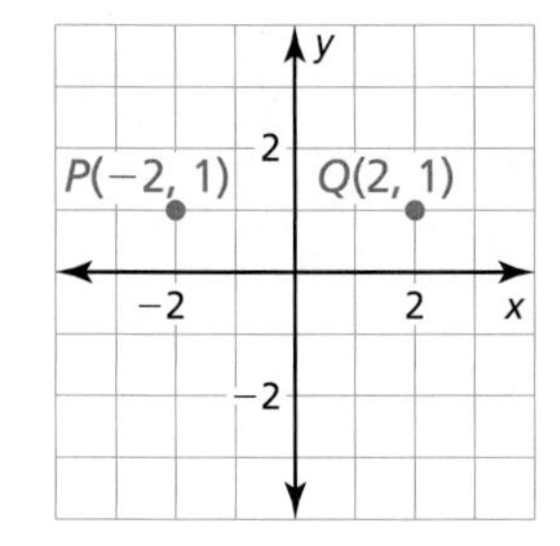

27. Determine which point is the remaining vertex of a triangle with an area of 4 square units.

Ⓐ $R(2, 0)$

Ⓑ $S(-2, -1)$

Ⓒ $T(-1, 0)$

Ⓓ $U(2, -2)$

28. Determine which points are the remaining vertices of a rectangle with a perimeter of 14 units.

Ⓐ $A(2, -2)$ and $B(2, -1)$

Ⓑ $C(-2, -2)$ and $D(-2, 2)$

Ⓒ $E(-2, -2)$ and $F(2, -2)$

Ⓓ $G(2, 0)$ and $H(-2, 0)$

29. USING STRUCTURE Use the diagram.

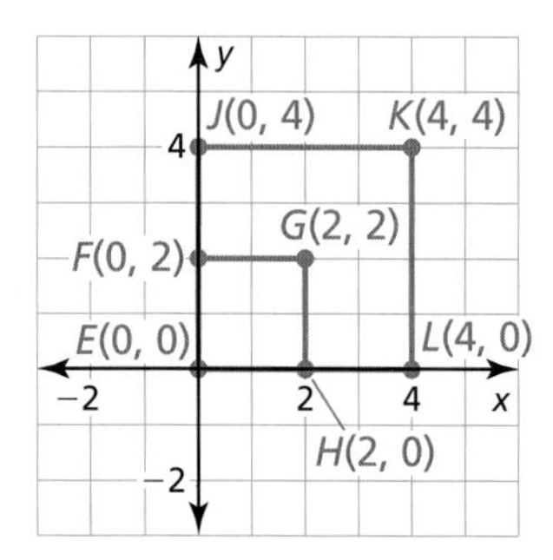

a. Find the areas of square *EFGH* and square *EJKL*. What happens to the area when the perimeter of square *EFGH* is doubled?

b. Is this true for every square? Explain.

30. MODELING WITH MATHEMATICS You are growing zucchini plants in your garden. In the figure, the entire garden is rectangle *QRST*. Each unit in the coordinate plane represents 1 foot. *(See Example 4.)*

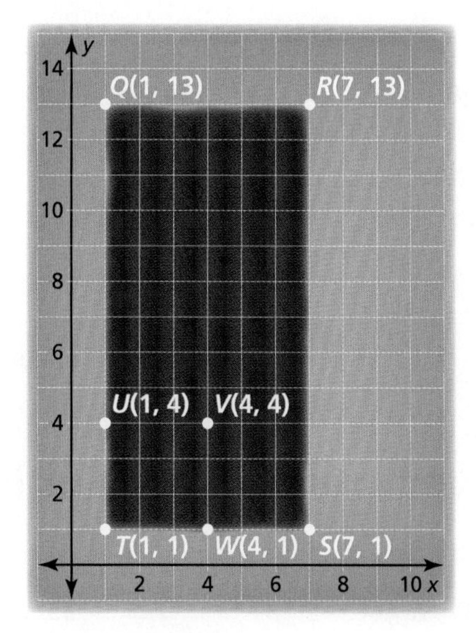

a. Find the area of the garden.

b. Zucchini plants require 9 square feet around each plant. How many zucchini plants can you plant?

c. You decide to use square *TUVW* to grow lettuce. You can plant four heads of lettuce per square foot. How many of each vegetable can you plant? Explain.

31. MODELING WITH MATHEMATICS You are going for a hike in the woods. You hike to a waterfall that is 4 miles east of where you left your car. You then hike to a lookout point that is 2 miles north of your car. From the lookout point, you return to your car.

a. Map out your route in a coordinate plane with your car at the origin. Let each unit in the coordinate plane represent 1 mile. Assume you travel along straight paths.

b. How far do you travel during the entire hike?

c. When you leave the waterfall, you decide to hike to an old wishing well before going to the lookout point. The wishing well is 3 miles north and 2 miles west of the lookout point. How far do you travel during the entire hike?

32. HOW DO YOU SEE IT? Without performing any calculations, determine whether the triangle or the rectangle has a greater area. Which one has a greater perimeter? Explain your reasoning.

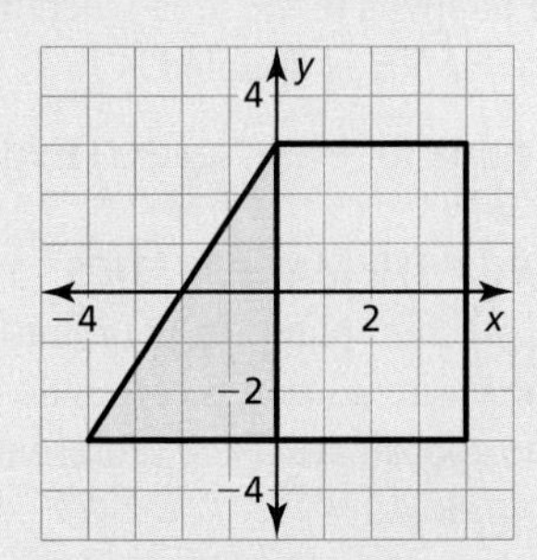

33. MATHEMATICAL CONNECTIONS The lines $y_1 = 2x - 6$, $y_2 = -3x + 4$, and $y_3 = -\frac{1}{2}x + 4$ are the sides of a right triangle.

a. Use slopes to determine which sides are perpendicular.

b. Find the vertices of the triangle.

c. Find the perimeter and area of the triangle.

34. THOUGHT PROVOKING Your bedroom has an area of 350 square feet. You are remodeling to include an attached bathroom that has an area of 150 square feet. Draw a diagram of the remodeled bedroom and bathroom in a coordinate plane.

35. PROBLEM SOLVING Use the diagram.

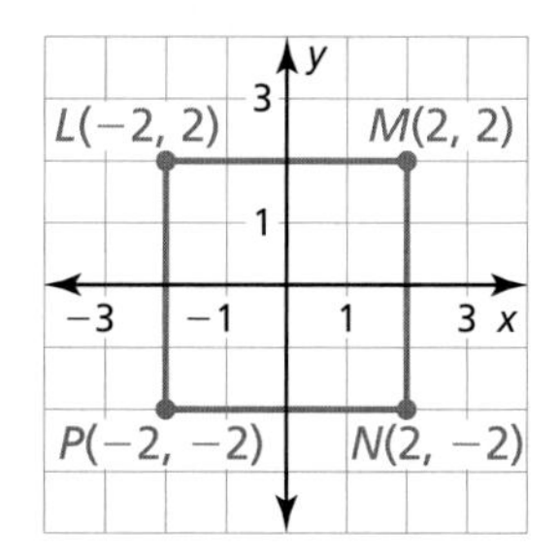

a. Find the perimeter and area of the square.

b. Connect the midpoints of the sides of the given square to make a quadrilateral. Is this quadrilateral a square? Explain your reasoning.

c. Find the perimeter and area of the quadrilateral you made in part (b). Compare this area to the area you found in part (a).

36. MAKING AN ARGUMENT Your friend claims that a rectangle with the same perimeter as $\triangle QRS$ will have the same area as the triangle. Is your friend correct? Explain your reasoning.

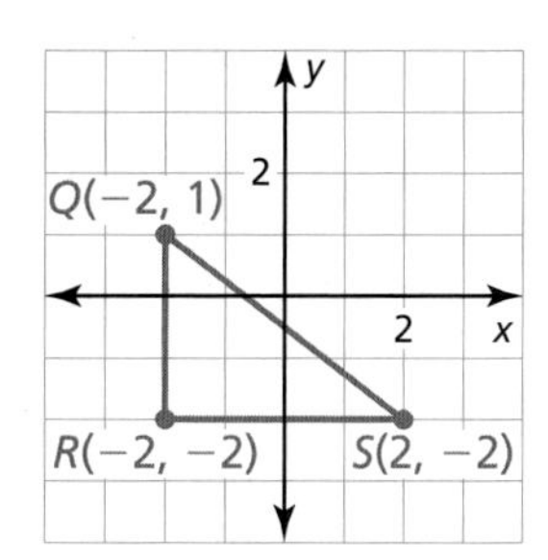

37. REASONING Triangle ABC has a perimeter of 12 units. The vertices of the triangle are $A(x, 2)$, $B(2, -2)$, and $C(-1, 2)$. Find the value of x.

Maintaining Mathematical Proficiency

Reviewing what you learned in previous grades and lessons

Solve the equation. *(Skills Review Handbook)*

38. $3x - 7 = 2$

39. $5x + 9 = 4$

40. $x + 4 = x - 12$

41. $4x - 9 = 3x + 5$

42. $11 - 2x = 5x - 3$

43. $\frac{x + 1}{2} = 4x - 3$

44. Use a compass and straightedge to construct a copy of the line segment. *(Section 1.2)*

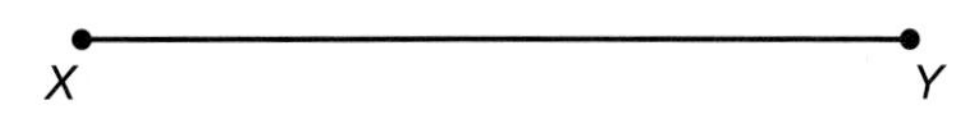

1.5 Measuring and Constructing Angles

Essential Question How can you measure and classify an angle?

EXPLORATION 1 Measuring and Classifying Angles

Work with a partner. Find the degree measure of each of the following angles. Classify each angle as acute, right, or obtuse.

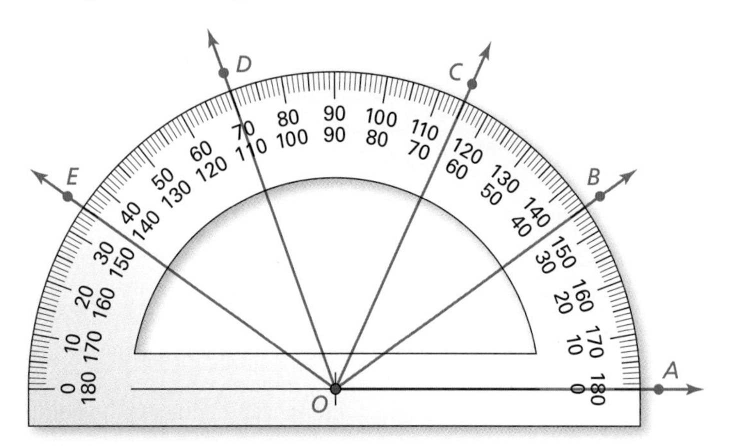

a. $\angle AOB$ **b.** $\angle AOC$ **c.** $\angle BOC$ **d.** $\angle BOE$
e. $\angle COE$ **f.** $\angle COD$ **g.** $\angle BOD$ **h.** $\angle AOE$

EXPLORATION 2 Drawing a Regular Polygon

Work with a partner.

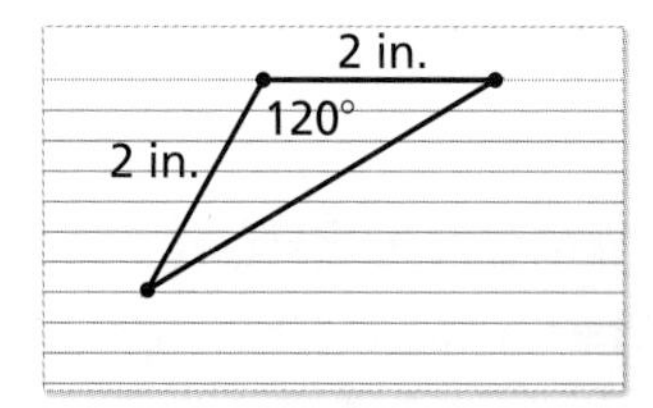

a. Use a ruler and protractor to draw the triangular pattern shown at the right.

b. Cut out the pattern and use it to draw three regular hexagons, as shown below.

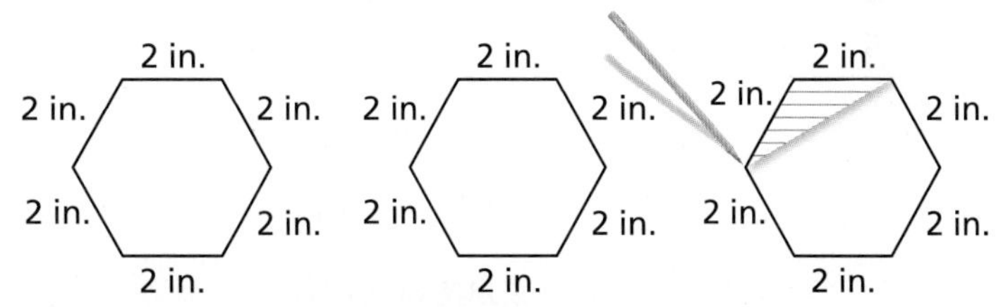

ATTENDING TO PRECISION

To be proficient in math, you need to calculate and measure accurately and efficiently.

c. The sum of the angle measures of a polygon with n sides is equal to $180(n - 2)°$. Do the angle measures of your hexagons agree with this rule? Explain.

d. Partition your hexagons into smaller polygons, as shown below. For each hexagon, find the sum of the angle measures of the smaller polygons. Does each sum equal the sum of the angle measures of a hexagon? Explain.

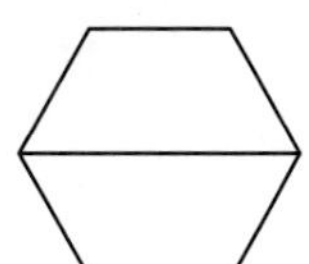
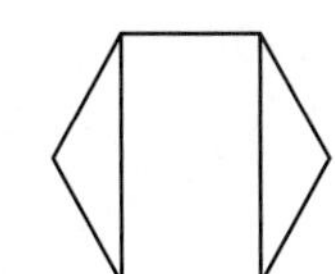

Communicate Your Answer

3. How can you measure and classify an angle?

1.5 Lesson

Core Vocabulary

angle, *p. 38*
vertex, *p. 38*
sides of an angle, *p. 38*
interior of an angle, *p. 38*
exterior of an angle, *p. 38*
measure of an angle, *p. 39*
acute angle, *p. 39*
right angle, *p. 39*
obtuse angle, *p. 39*
straight angle, *p. 39*
congruent angles, *p. 40*
angle bisector, *p. 42*

Previous
protractor
degrees

What You Will Learn

- Name angles.
- Measure and classify angles.
- Identify congruent angles.
- Use the Angle Addition Postulate to find angle measures.
- Bisect angles.

Naming Angles

An **angle** is a set of points consisting of two different rays that have the same endpoint, called the **vertex**. The rays are the **sides** of the angle.

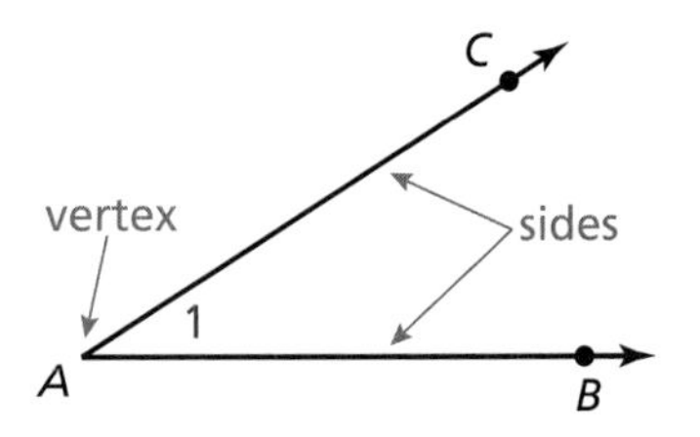

You can name an angle in several different ways.

- Use its vertex, such as $\angle A$.
- Use a point on each ray and the vertex, such as $\angle BAC$ or $\angle CAB$.
- Use a number, such as $\angle 1$.

The region that contains all the points between the sides of the angle is the **interior of the angle**. The region that contains all the points outside the angle is the **exterior of the angle**.

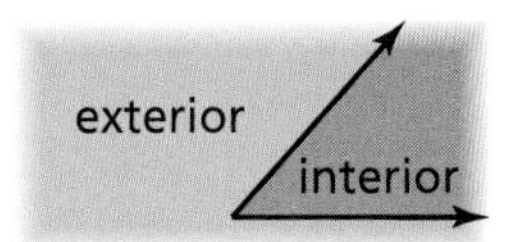

EXAMPLE 1 Naming Angles

A lighthouse keeper measures the angles formed by the lighthouse at point M and three boats. Name three angles shown in the diagram.

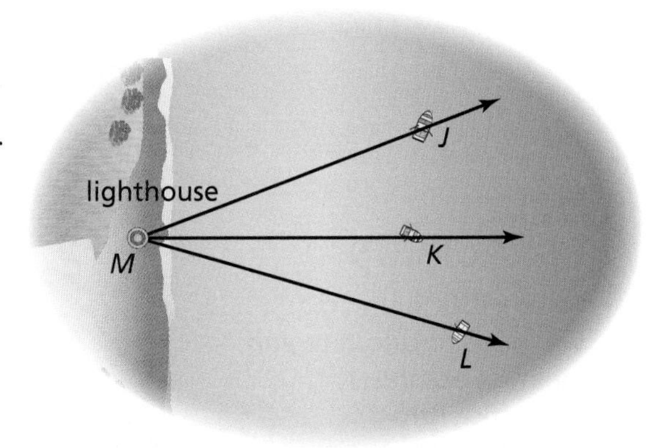

SOLUTION

$\angle JMK$ or $\angle KMJ$

$\angle KML$ or $\angle LMK$

$\angle JML$ or $\angle LMJ$

COMMON ERROR

When a point is the vertex of more than one angle, you cannot use the vertex alone to name the angle.

Monitoring Progress

Help in English and Spanish at *BigIdeasMath.com*

Write three names for the angle.

1.

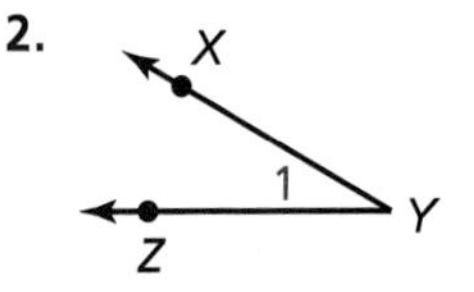

2.

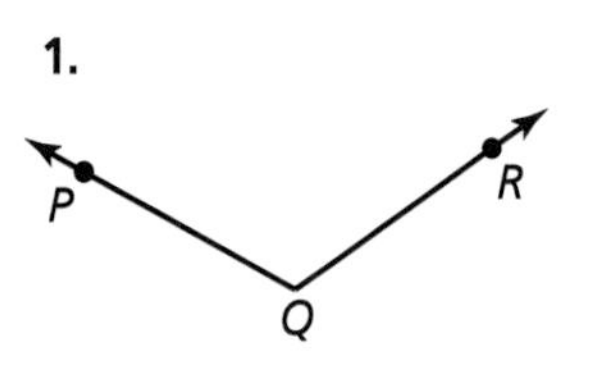

3.

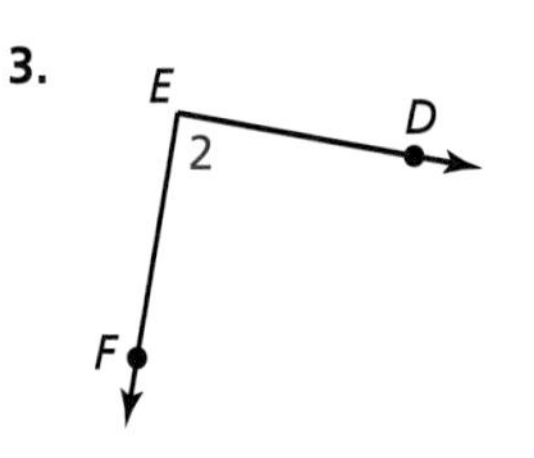

Measuring and Classifying Angles

A protractor helps you approximate the *measure* of an angle. The measure is usually given in *degrees*.

COMMON ERROR

Most protractors have an inner and an outer scale. When measuring, make sure you are using the correct scale.

Postulate

Postulate 1.3 Protractor Postulate

Consider $\overleftrightarrow{OB}$ and a point A on one side of $\overleftrightarrow{OB}$. The rays of the form $\overrightarrow{OA}$ can be matched one to one with the real numbers from 0 to 180.

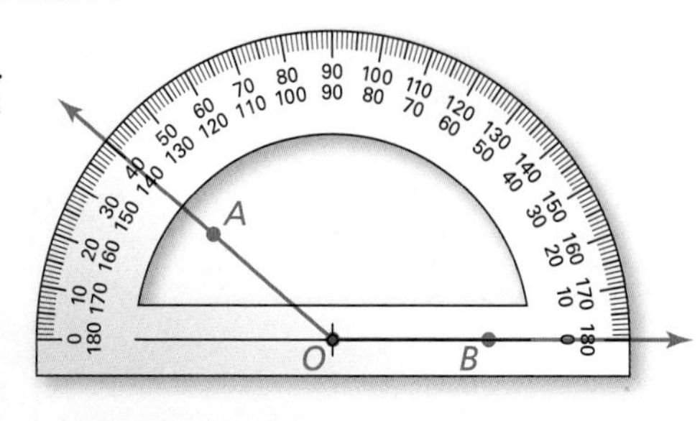

The **measure** of $\angle AOB$, which can be written as $m\angle AOB$, is equal to the absolute value of the difference between the real numbers matched with $\overrightarrow{OA}$ and $\overrightarrow{OB}$ on a protractor.

You can classify angles according to their measures.

Core Concept

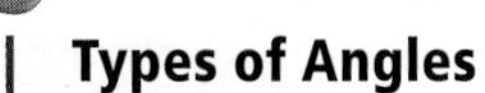

Types of Angles

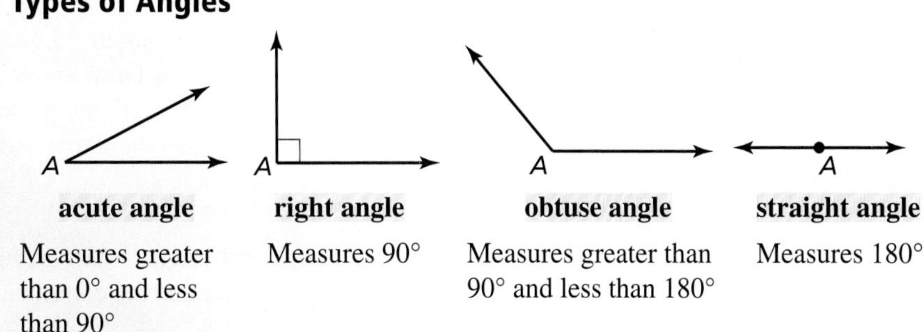

acute angle	**right angle**	**obtuse angle**	**straight angle**
Measures greater than 0° and less than 90°	Measures 90°	Measures greater than 90° and less than 180°	Measures 180°

EXAMPLE 2 Measuring and Classifying Angles

Find the measure of each angle. Then classify each angle.

a. $\angle GHK$ **b.** $\angle JHL$ **c.** $\angle LHK$

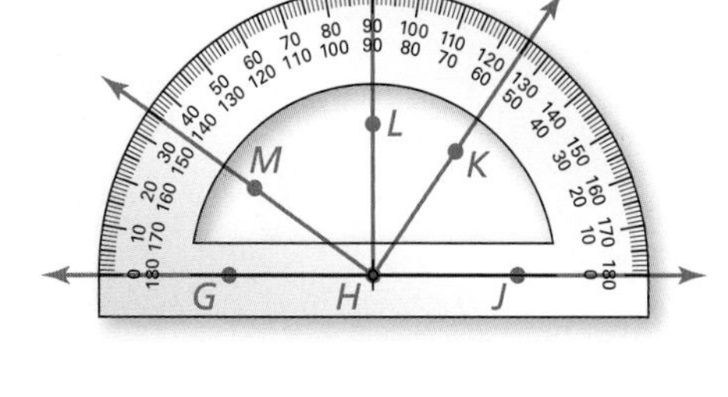

SOLUTION

a. $\overrightarrow{HG}$ lines up with 0° on the outer scale of the protractor. $\overrightarrow{HK}$ passes through 125° on the outer scale. So, $m\angle GHK = 125°$. It is an *obtuse* angle.

b. $\overrightarrow{HJ}$ lines up with 0° on the inner scale of the protractor. $\overrightarrow{HL}$ passes through 90°. So, $m\angle JHL = 90°$. It is a *right* angle.

c. $\overrightarrow{HL}$ passes through 90°. $\overrightarrow{HK}$ passes through 55° on the inner scale. So, $m\angle LHK = |90 - 55| = 35°$. It is an *acute* angle.

Monitoring Progress  Help in English and Spanish at *BigIdeasMath.com*

Use the diagram in Example 2 to find the angle measure. Then classify the angle.

4. $\angle JHM$ **5.** $\angle MHK$ **6.** $\angle MHL$

Identifying Congruent Angles

You can use a compass and straightedge to construct an angle that has the same measure as a given angle.

CONSTRUCTION Copying an Angle

Use a compass and straightedge to construct an angle that has the same measure as $\angle A$. In this construction, the *center* of an arc is the point where the compass point rests. The *radius* of an arc is the distance from the center of the arc to a point on the arc drawn by the compass.

SOLUTION

Step 1

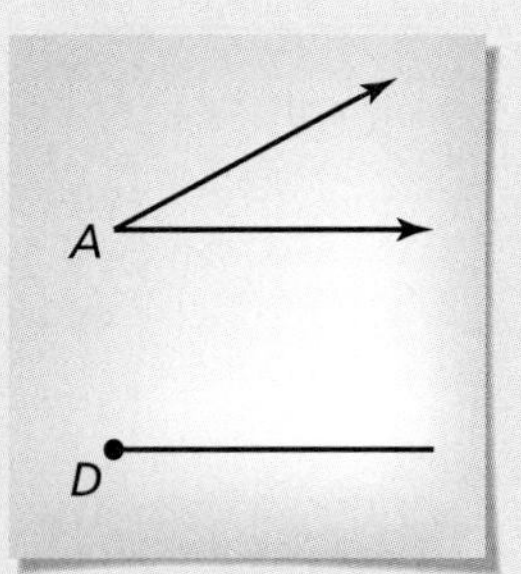

Draw a segment Draw an angle such as $\angle A$, as shown. Then draw a segment. Label a point D on the segment.

Step 2

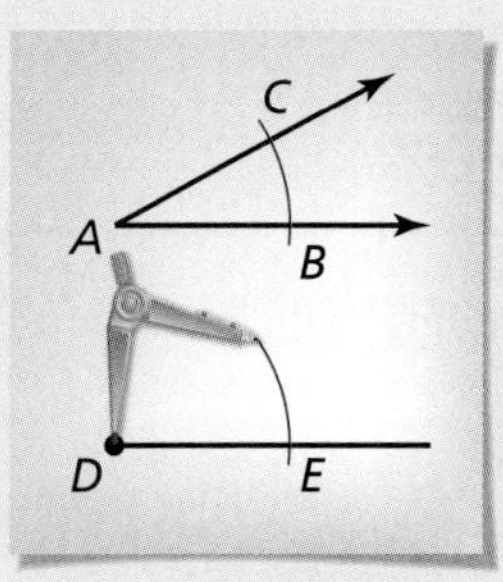

Draw arcs Draw an arc with center A. Using the same radius, draw an arc with center D.

Step 3

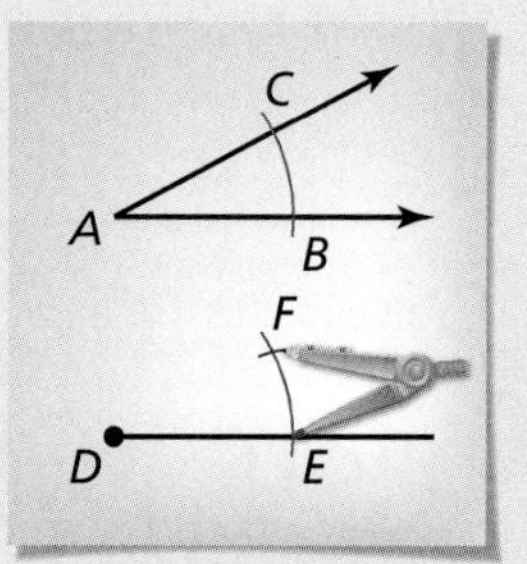

Draw an arc Label B, C, and E. Draw an arc with radius BC and center E. Label the intersection F.

Step 4

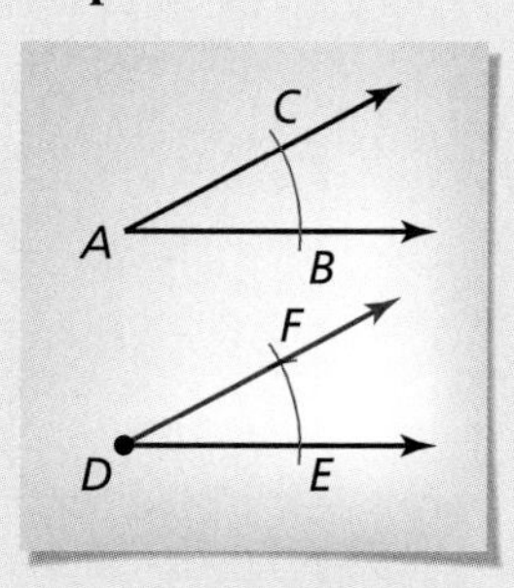

Draw a ray Draw $\overrightarrow{DF}$. $\angle EDF \cong \angle BAC$.

Two angles are **congruent angles** when they have the same measure. In the construction above, $\angle A$ and $\angle D$ are congruent angles. So,

$m\angle A = m\angle D$ — The measure of angle A is *equal to* the measure of angle D.

and

$\angle A \cong \angle D$. — Angle A is *congruent* to angle D.

EXAMPLE 3 Identifying Congruent Angles

a. Identify the congruent angles labeled in the quilt design.

b. $m\angle ADC = 140°$. What is $m\angle EFG$?

SOLUTION

a. There are two pairs of congruent angles:

$\angle ABC \cong \angle FGH$ and $\angle ADC \cong \angle EFG$.

b. Because $\angle ADC \cong \angle EFG$, $m\angle ADC = m\angle EFG$.

So, $m\angle EFG = 140°$.

READING

In diagrams, matching arcs indicate congruent angles. When there is more than one pair of congruent angles, use multiple arcs.

Monitoring Progress

Help in English and Spanish at *BigIdeasMath.com*

7. Without measuring, is $\angle DAB \cong \angle FEH$ in Example 3? Explain your reasoning. Use a protractor to verify your answer.

Using the Angle Addition Postulate

Postulate

Postulate 1.4 Angle Addition Postulate

Words If P is in the interior of $\angle RST$, then the measure of $\angle RST$ is equal to the sum of the measures of $\angle RSP$ and $\angle PST$.

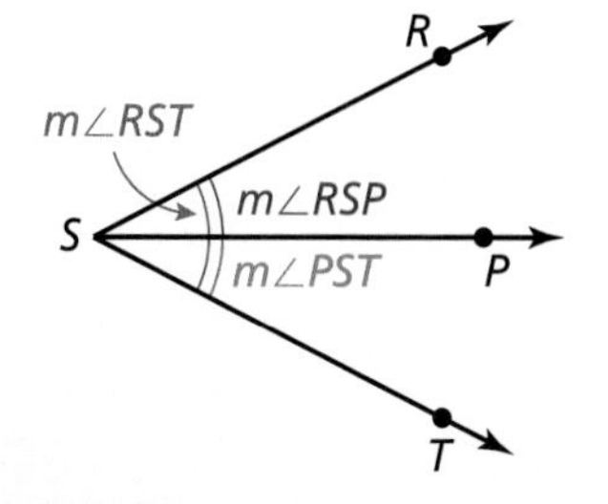

Symbols If P is in the interior of $\angle RST$, then

$$m\angle RST = m\angle RSP + m\angle PST.$$

EXAMPLE 4 **Finding Angle Measures**

Given that $m\angle LKN = 145°$, find $m\angle LKM$ and $m\angle MKN$.

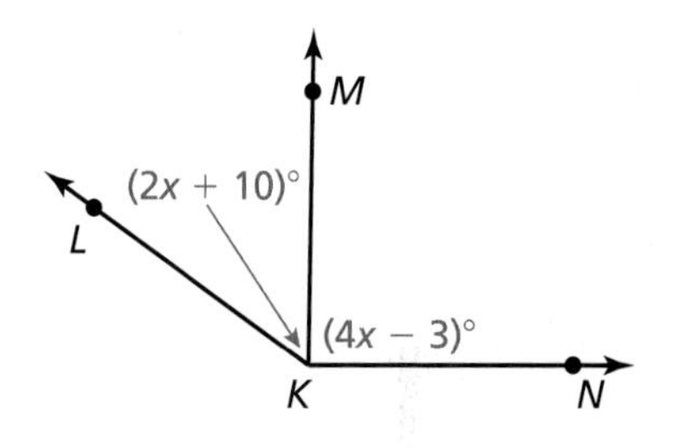

SOLUTION

Step 1 Write and solve an equation to find the value of x.

$m\angle LKN = m\angle LKM + m\angle MKN$	Angle Addition Postulate
$145° = (2x + 10)° + (4x - 3)°$	Substitute angle measures.
$145 = 6x + 7$	Combine like terms.
$138 = 6x$	Subtract 7 from each side.
$23 = x$	Divide each side by 6.

Step 2 Evaluate the given expressions when $x = 23$.

$$m\angle LKM = (2x + 10)° = (2 \cdot 23 + 10)° = 56°$$

$$m\angle MKN = (4x - 3)° = (4 \cdot 23 - 3)° = 89°$$

▶ So, $m\angle LKM = 56°$, and $m\angle MKN = 89°$.

Monitoring Progress

Help in English and Spanish at *BigIdeasMath.com*

Find the indicated angle measures.

8. Given that $\angle KLM$ is a straight angle, find $m\angle KLN$ and $m\angle NLM$.

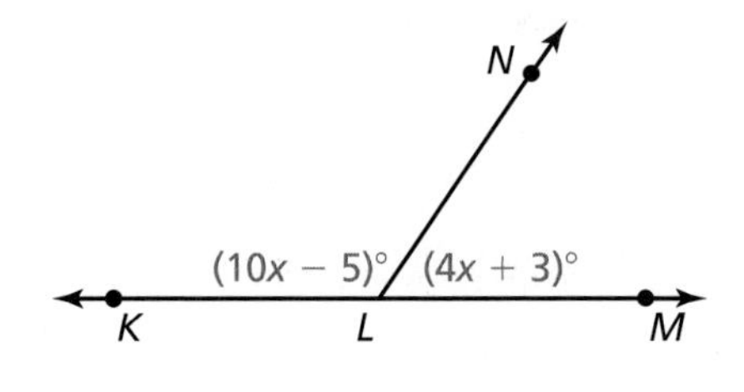

9. Given that $\angle EFG$ is a right angle, find $m\angle EFH$ and $m\angle HFG$.

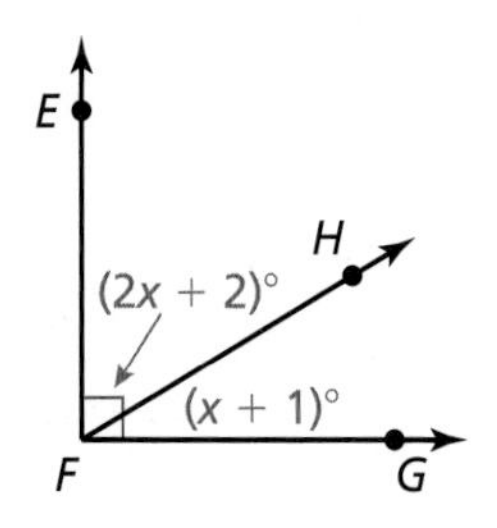

Bisecting Angles

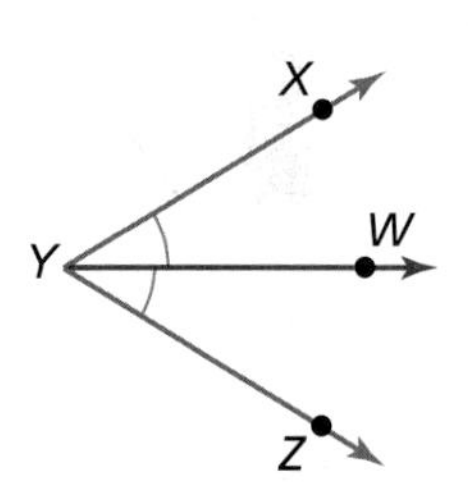

An **angle bisector** is a ray that divides an angle into two angles that are congruent. In the figure, $\overrightarrow{YW}$ bisects $\angle XYZ$, so $\angle XYW \cong \angle ZYW$.

You can use a compass and straightedge to bisect an angle.

CONSTRUCTION Bisecting an Angle

Construct an angle bisector of $\angle A$ with a compass and straightedge.

SOLUTION

Step 1

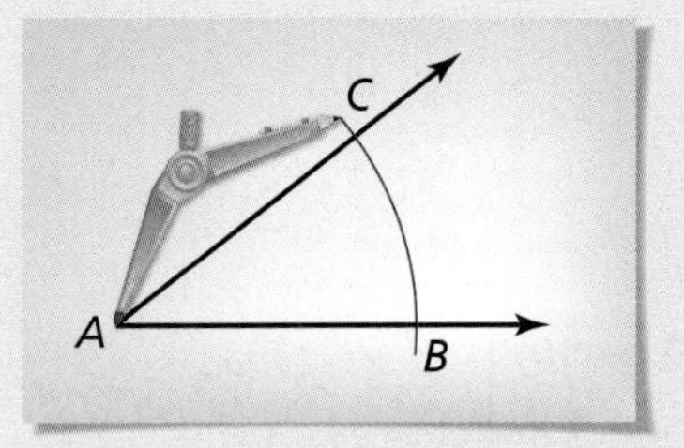

Draw an arc Draw an angle such as $\angle A$, as shown. Place the compass at A. Draw an arc that intersects both sides of the angle. Label the intersections B and C.

Step 2

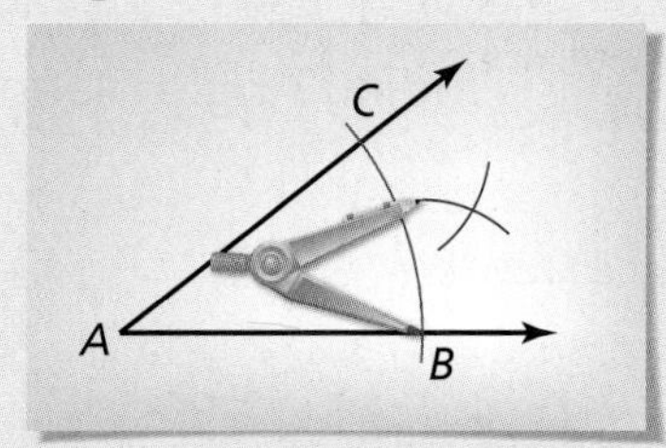

Draw arcs Place the compass at C. Draw an arc. Then place the compass point at B. Using the same radius, draw another arc.

Step 3

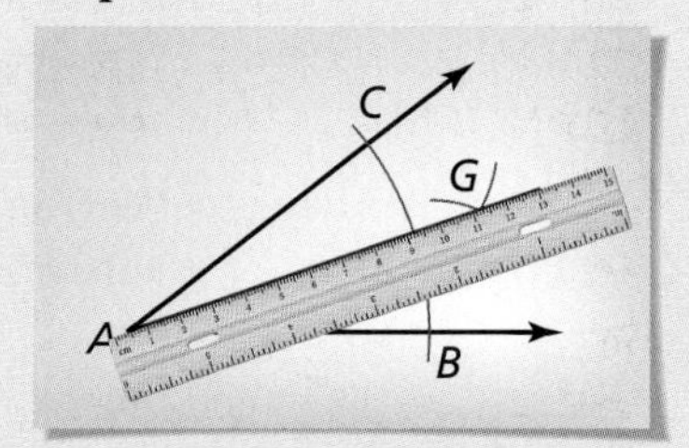

Draw a ray Label the intersection G. Use a straightedge to draw a ray through A and G.

$\overrightarrow{AG}$ bisects $\angle A$.

EXAMPLE 5 Using a Bisector to Find Angle Measures

$\overrightarrow{QS}$ bisects $\angle PQR$, and $m\angle PQS = 24°$. Find $m\angle PQR$.

SOLUTION

Step 1 Draw a diagram.

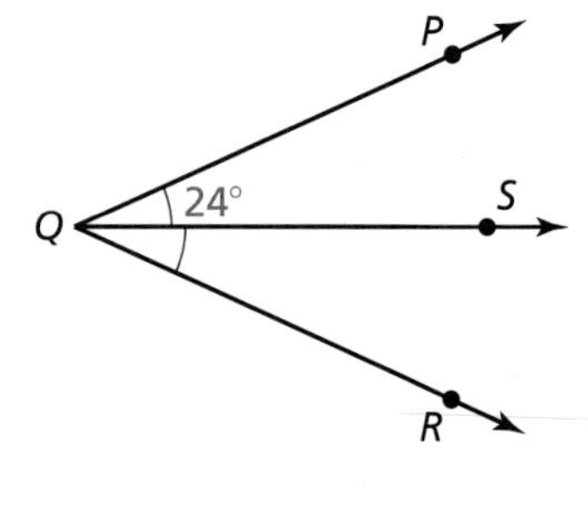

Step 2 Because $\overrightarrow{QS}$ bisects $\angle PQR$, $m\angle PQS = m\angle RQS$. So, $m\angle RQS = 24°$. Use the Angle Addition Postulate to find $m\angle PQR$.

$m\angle PQR = m\angle PQS + m\angle RQS$	Angle Addition Postulate
$= 24° + 24°$	Substitute angle measures.
$= 48°$	Add.

▶ So, $m\angle PQR = 48°$.

Monitoring Progress  Help in English and Spanish at *BigIdeasMath.com*

10. Angle MNP is a straight angle, and $\overrightarrow{NQ}$ bisects $\angle MNP$. Draw $\angle MNP$ and $\overrightarrow{NQ}$. Use arcs to mark the congruent angles in your diagram. Find the angle measures of these congruent angles.

1.5 Exercises

Dynamic Solutions available at *BigIdeasMath.com*

Vocabulary and Core Concept Check

1. **COMPLETE THE SENTENCE** Two angles are ___________ angles when they have the same measure.

2. **WHICH ONE DOESN'T BELONG?** Which angle name does *not* belong with the other three? Explain your reasoning.

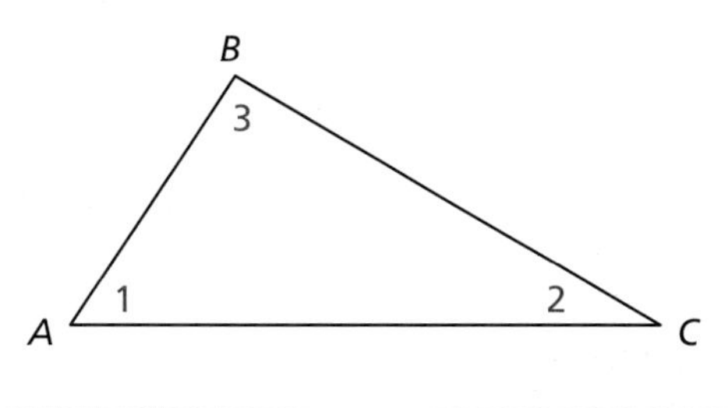

$\angle BCA$	$\angle BAC$	$\angle 1$	$\angle CAB$

Monitoring Progress and Modeling with Mathematics

In Exercises 3–6, write three names for the angle. *(See Example 1.)*

3.

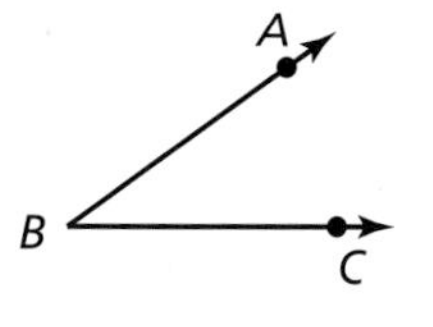

4.

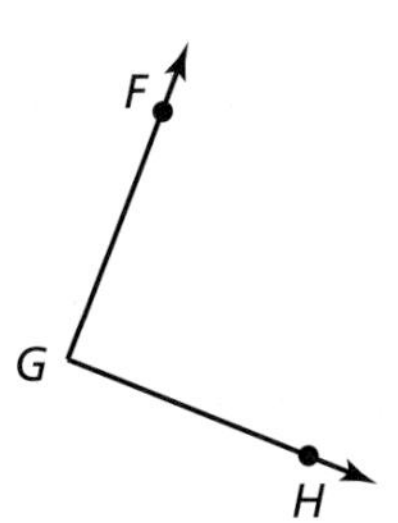

5.

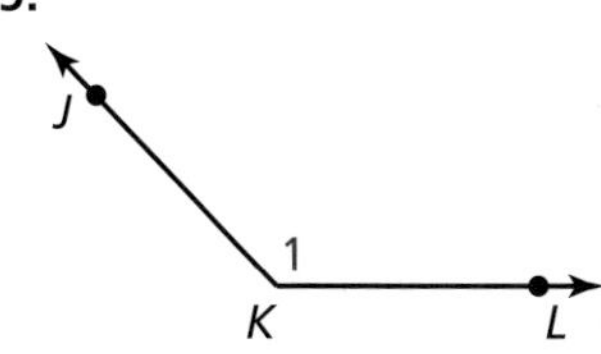

6.

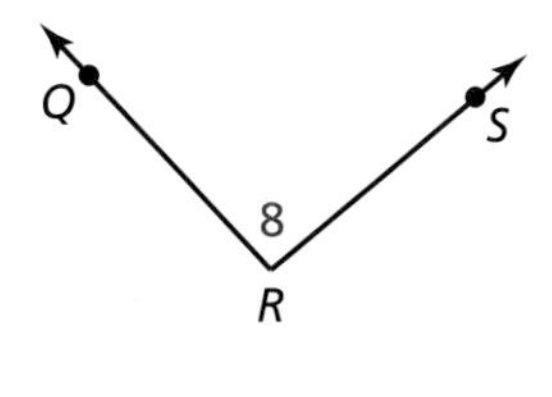

In Exercises 7 and 8, name three different angles in the diagram. *(See Example 1.)*

7.

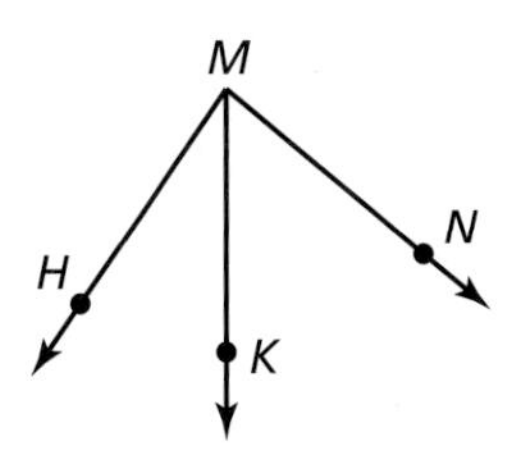

8.

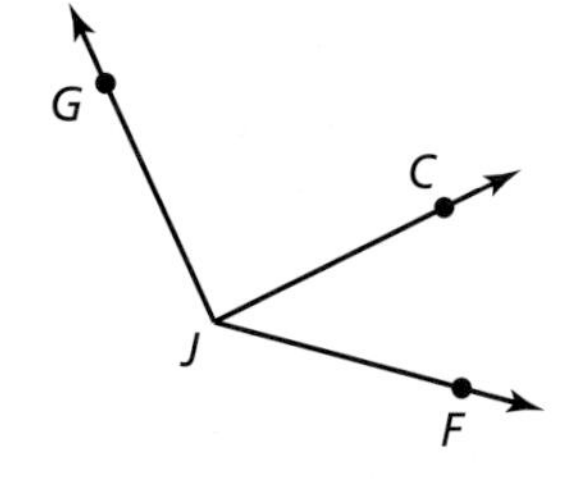

In Exercises 9–12, find the angle measure. Then classify the angle. *(See Example 2.)*

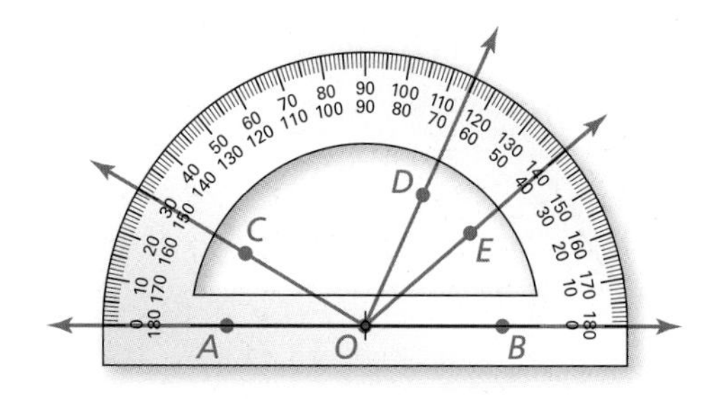

9. $m\angle AOC$

10. $m\angle BOD$

11. $m\angle COD$

12. $m\angle EOD$

ERROR ANALYSIS In Exercises 13 and 14, describe and correct the error in finding the angle measure. Use the diagram from Exercises 9–12.

13.

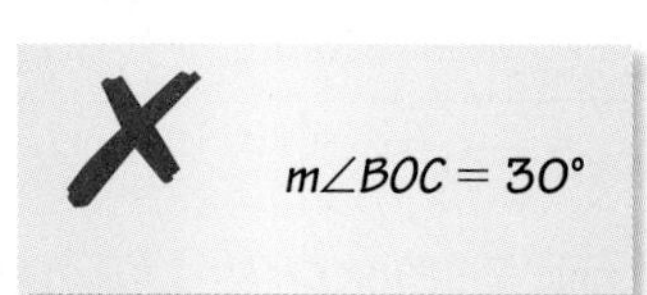

14.

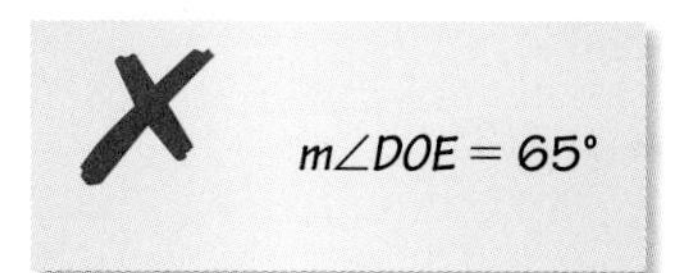

CONSTRUCTION In Exercises 15 and 16, use a compass and straightedge to copy the angle.

15. **16.**

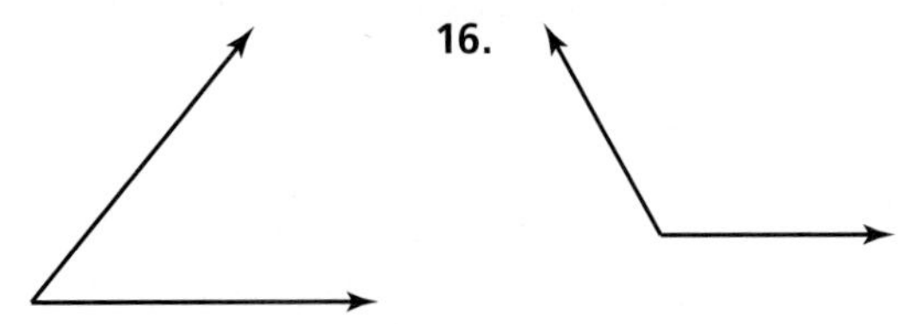

In Exercises 17–20, $m\angle AED = 34°$ and $m\angle EAD = 112°$. *(See Example 3.)*

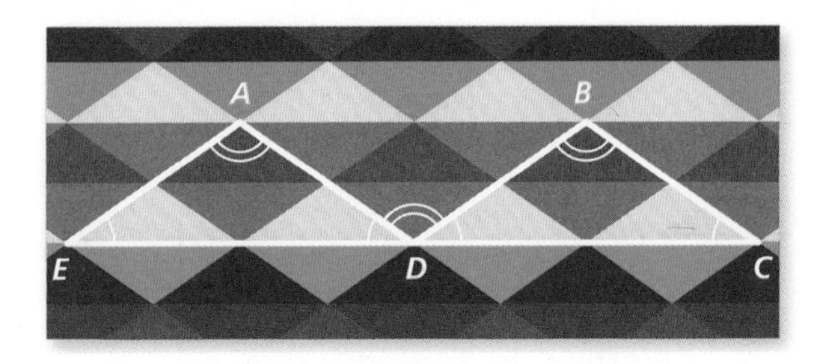

17. Identify the angles congruent to $\angle AED$.

18. Identify the angles congruent to $\angle EAD$.

19. Find $m\angle BDC$.

20. Find $m\angle ADB$.

In Exercises 21–24, find the indicated angle measure.

21. Find $m\angle ABC$. **22.** Find $m\angle LMN$.

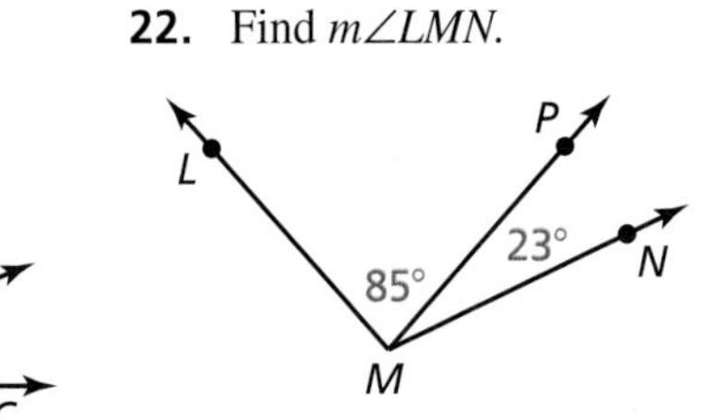

23. $m\angle RST = 114°$. Find $m\angle RSV$.

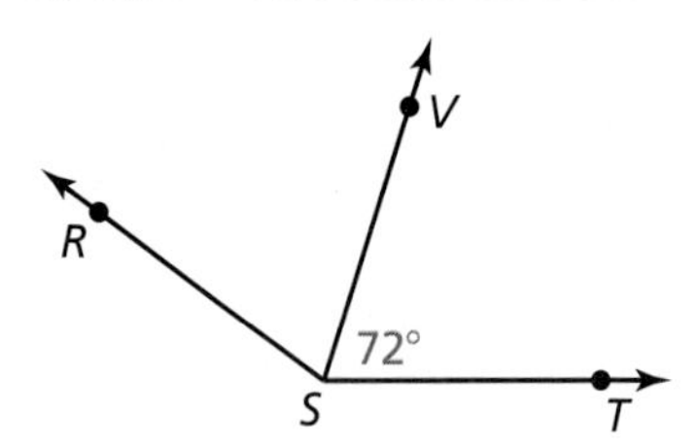

24. $\angle GHK$ is a straight angle. Find $m\angle LHK$.

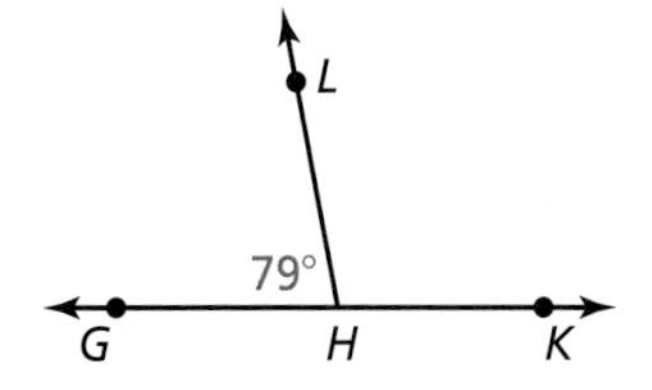

In Exercises 25–30, find the indicated angle measures. *(See Example 4.)*

25. $m\angle ABC = 95°$. Find $m\angle ABD$ and $m\angle DBC$.

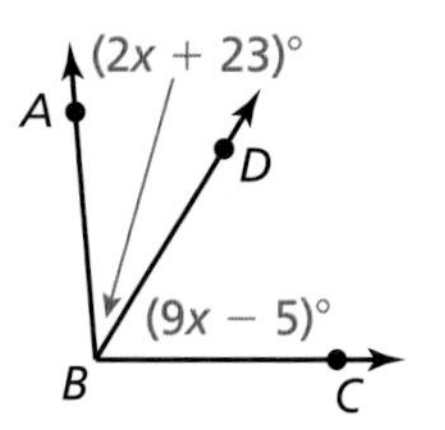

26. $m\angle XYZ = 117°$. Find $m\angle XYW$ and $m\angle WYZ$.

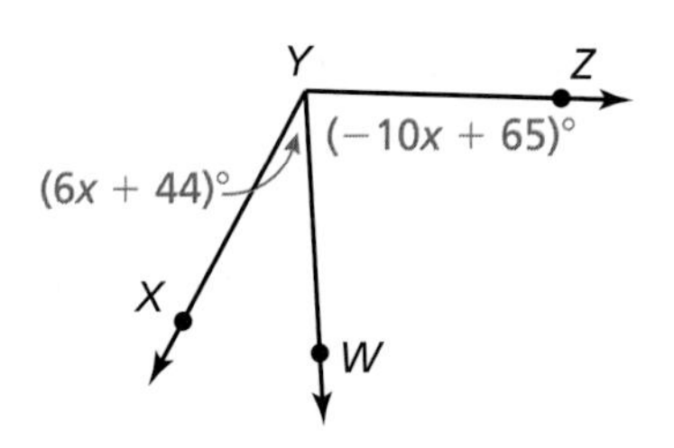

27. $\angle LMN$ is a straight angle. Find $m\angle LMP$ and $m\angle NMP$.

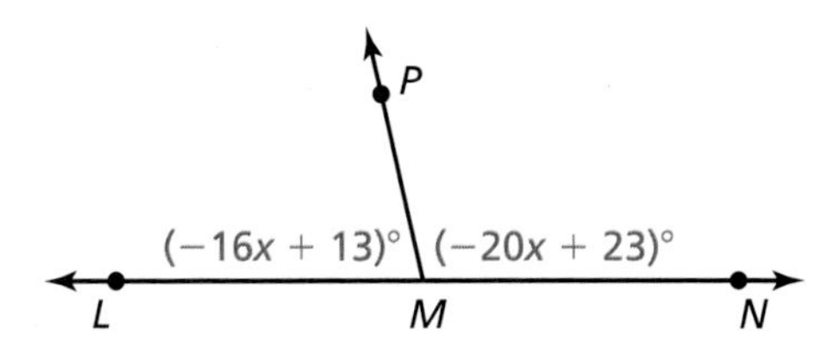

28. $\angle ABC$ is a straight angle. Find $m\angle ABX$ and $m\angle CBX$.

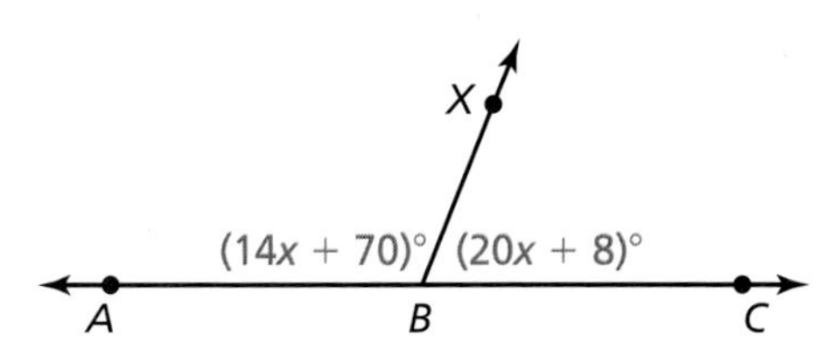

29. Find $m\angle RSQ$ and $m\angle TSQ$.

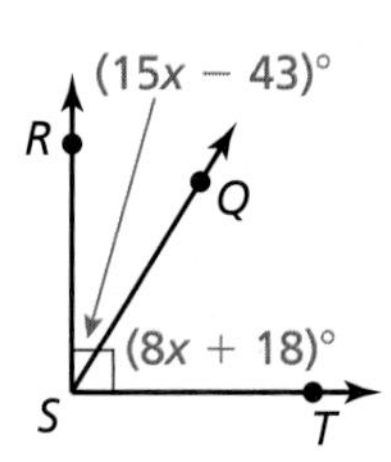

30. Find $m\angle DEH$ and $m\angle FEH$.

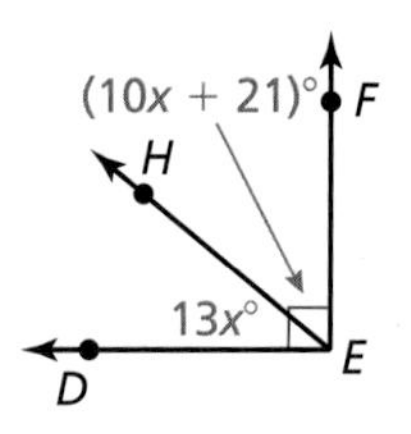

CONSTRUCTION In Exercises 31 and 32, copy the angle. Then construct the angle bisector with a compass and straightedge.

31. **32.**

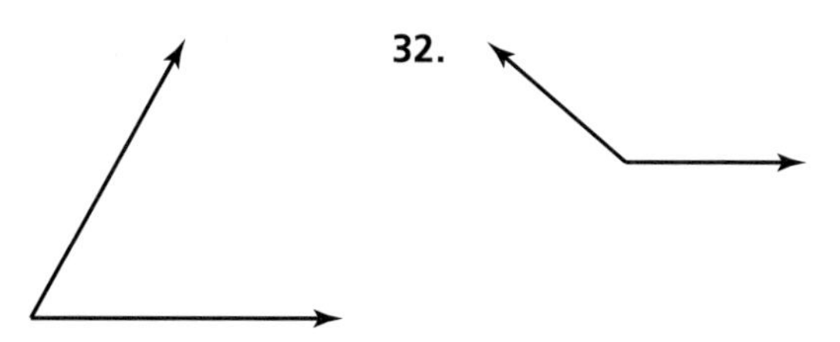

In Exercises 33–36, $\overrightarrow{QS}$ bisects $\angle PQR$. Use the diagram and the given angle measure to find the indicated angle measures. *(See Example 5.)*

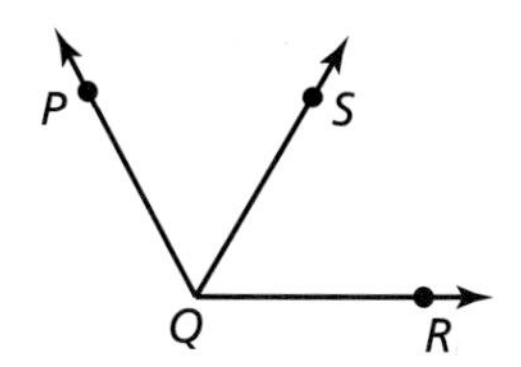

33. $m\angle PQS = 63°$. Find $m\angle RQS$ and $m\angle PQR$.

34. $m\angle RQS = 71°$. Find $m\angle PQS$ and $m\angle PQR$.

35. $m\angle PQR = 124°$. Find $m\angle PQS$ and $m\angle RQS$.

36. $m\angle PQR = 119°$. Find $m\angle PQS$ and $m\angle RQS$.

In Exercises 37–40, $\overrightarrow{BD}$ bisects $\angle ABC$. Find $m\angle ABD$, $m\angle CBD$, and $m\angle ABC$.

37.

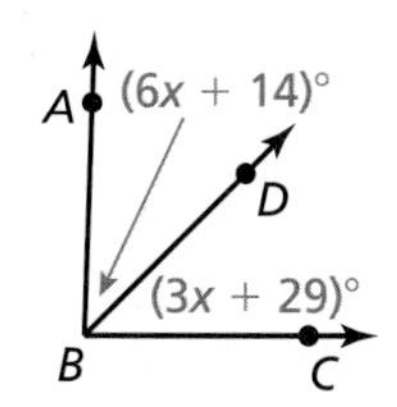

38.

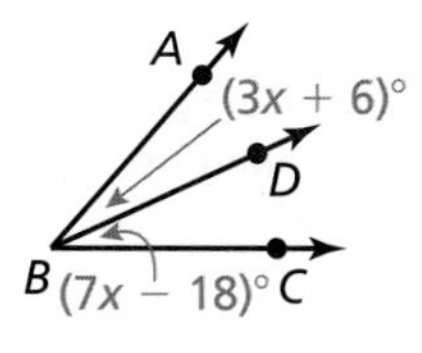

39.

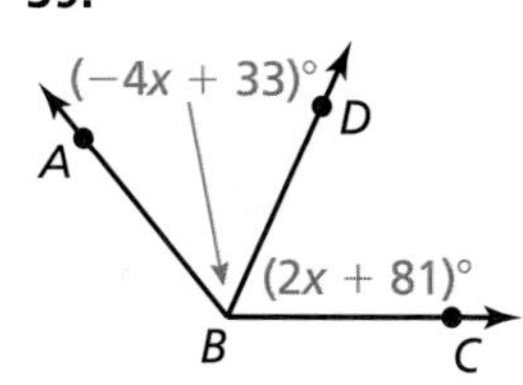

40.

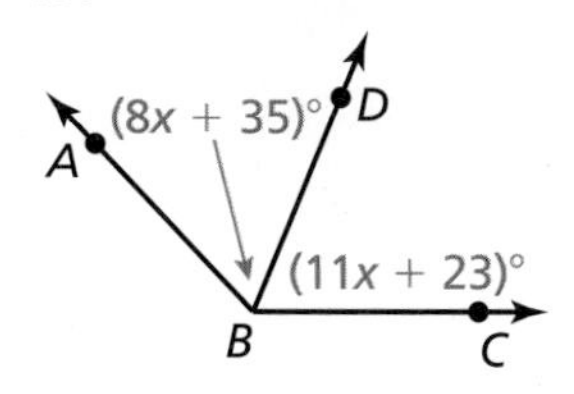

41. WRITING Explain how to find $m\angle ABD$ when you are given $m\angle ABC$ and $m\angle CBD$.

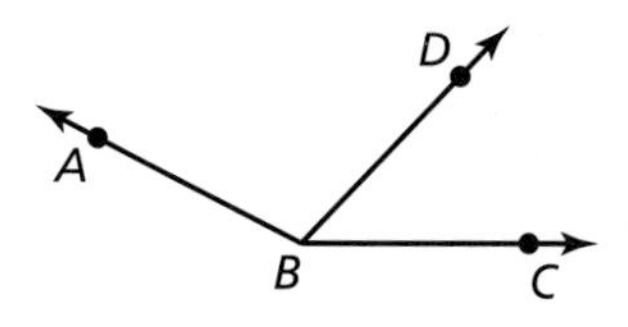

42. ANALYZING RELATIONSHIPS The map shows the intersections of three roads. Malcom Way intersects Sydney Street at an angle of 162°. Park Road intersects Sydney Street at an angle of 87°. Find the angle at which Malcom Way intersects Park Road.

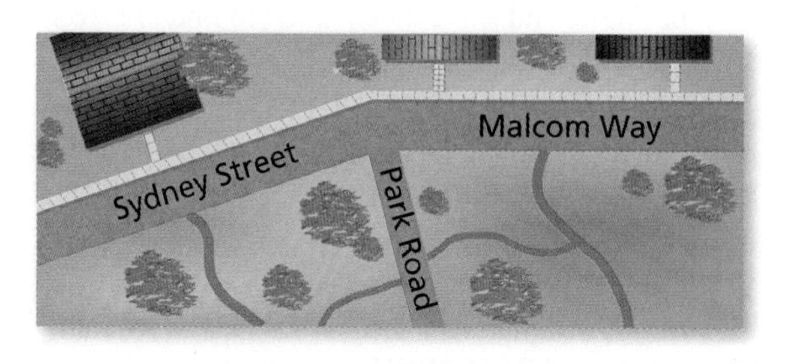

43. ANALYZING RELATIONSHIPS In the sculpture shown in the photograph, the measure of $\angle LMN$ is 76° and the measure of $\angle PMN$ is 36°. What is the measure of $\angle LMP$?

USING STRUCTURE In Exercises 44–46, use the diagram of the roof truss.

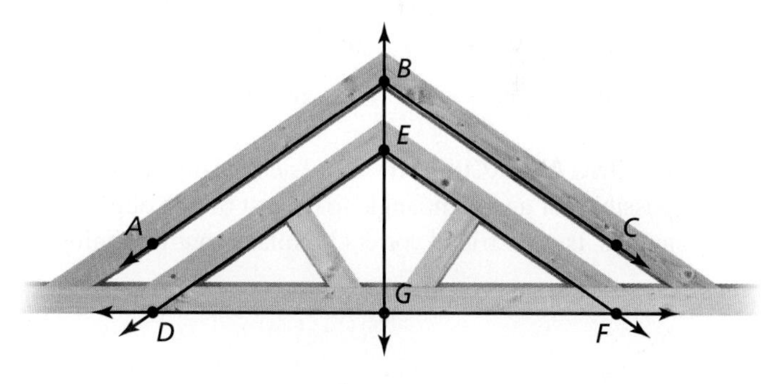

44. In the roof truss, $\overrightarrow{BG}$ bisects $\angle ABC$ and $\angle DEF$, $m\angle ABC = 112°$, and $\angle ABC \cong \angle DEF$. Find the measure of each angle.

a. $m\angle DEF$ **b.** $m\angle ABG$

c. $m\angle CBG$ **d.** $m\angle DEG$

45. In the roof truss, $\angle DGF$ is a straight angle and $\overrightarrow{GB}$ bisects $\angle DGF$. Find $m\angle DGE$ and $m\angle FGE$.

46. Name an example of each of the four types of angles according to their measures in the diagram.

47. MATHEMATICAL CONNECTIONS In $\angle ABC$, $\overrightarrow{BX}$ is in the interior of the angle, $m\angle ABX$ is 12 more than 4 times $m\angle CBX$, and $m\angle ABC = 92°$.

a. Draw a diagram to represent the situation.

b. Write and solve an equation to find $m\angle ABX$ and $m\angle CBX$.

48. THOUGHT PROVOKING The angle between the minute hand and the hour hand of a clock is 90°. What time is it? Justify your answer.

49. ABSTRACT REASONING Classify the angles that result from bisecting each type of angle.

a. acute angle
b. right angle
c. obtuse angle
d. straight angle

50. ABSTRACT REASONING Classify the angles that result from drawing a ray in the interior of each type of angle. Include all possibilities and explain your reasoning.

a. acute angle
b. right angle
c. obtuse angle
d. straight angle

51. CRITICAL THINKING The ray from the origin through (4, 0) forms one side of an angle. Use the numbers below as x- and y-coordinates to create each type of angle in a coordinate plane.

−2	−1	0	1	2

a. acute angle
b. right angle
c. obtuse angle
d. straight angle

52. MAKING AN ARGUMENT Your friend claims it is possible for a straight angle to consist of two obtuse angles. Is your friend correct? Explain your reasoning.

53. CRITICAL THINKING Two acute angles are added together. What type(s) of angle(s) do they form? Explain your reasoning.

54. HOW DO YOU SEE IT? Use the diagram.

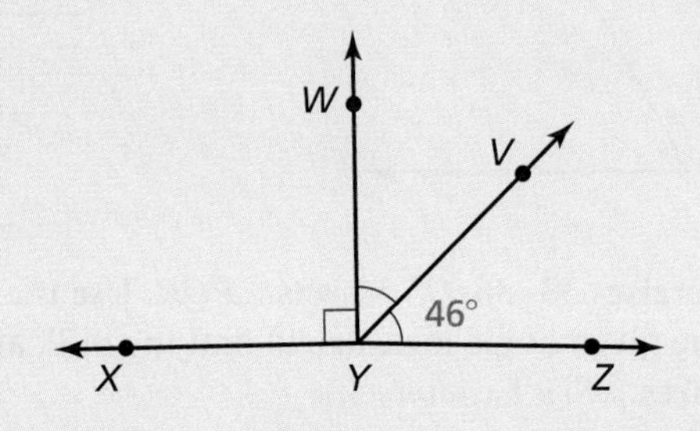

a. Is it possible for $\angle XYZ$ to be a straight angle? Explain your reasoning.

b. What can you change in the diagram so that $\angle XYZ$ is a straight angle?

55. WRITING Explain the process of bisecting an angle in your own words. Compare it to bisecting a segment.

56. ANALYZING RELATIONSHIPS $\overrightarrow{SQ}$ bisects $\angle RST$, $\overrightarrow{SP}$ bisects $\angle RSQ$, and $\overrightarrow{SV}$ bisects $\angle RSP$. The measure of $\angle VSP$ is 17°. Find $m\angle TSQ$. Explain.

57. ABSTRACT REASONING A bubble level is a tool used to determine whether a surface is horizontal, like the top of a picture frame. If the bubble is not exactly in the middle when the level is placed on the surface, then the surface is not horizontal. What is the most realistic type of angle formed by the level and a horizontal line when the bubble is not in the middle? Explain your reasoning.

Maintaining Mathematical Proficiency

Reviewing what you learned in previous grades and lessons

Solve the equation. *(Skills Review Handbook)*

58. $x + 67 = 180$

59. $x + 58 = 90$

60. $16 + x = 90$

61. $109 + x = 180$

62. $(6x + 7) + (13x + 21) = 180$

63. $(3x + 15) + (4x - 9) = 90$

64. $(11x - 25) + (24x + 10) = 90$

65. $(14x - 18) + (5x + 8) = 180$

1.6 Describing Pairs of Angles

Essential Question

How can you describe angle pair relationships and use these descriptions to find angle measures?

EXPLORATION 1 Finding Angle Measures

Work with a partner. The five-pointed star has a regular pentagon at its center.

a. What do you notice about the following angle pairs?

$x°$ and $y°$

$y°$ and $z°$

$x°$ and $z°$

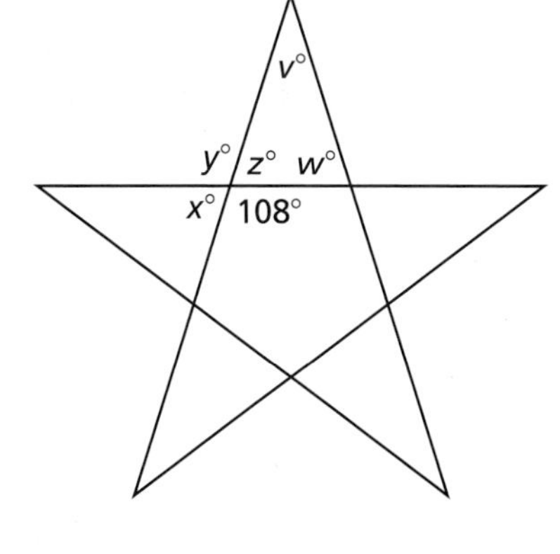

b. Find the values of the indicated variables. Do not use a protractor to measure the angles.

$x =$

$y =$

$z =$

$w =$

$v =$

Explain how you obtained each answer.

EXPLORATION 2 Finding Angle Measures

Work with a partner. A square is divided by its diagonals into four triangles.

a. What do you notice about the following angle pairs?

$a°$ and $b°$

$c°$ and $d°$

$c°$ and $e°$

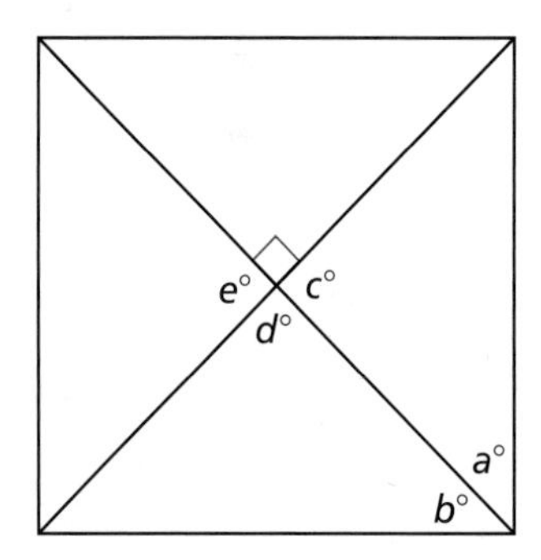

b. Find the values of the indicated variables. Do not use a protractor to measure the angles.

$c =$ 

$d =$

$e =$

Explain how you obtained each answer.

ATTENDING TO PRECISION

To be proficient in math, you need to communicate precisely with others.

Communicate Your Answer

3. How can you describe angle pair relationships and use these descriptions to find angle measures?

4. What do you notice about the angle measures of complementary angles, supplementary angles, and vertical angles?

1.6 Lesson

Core Vocabulary

complementary angles, *p. 48*
supplementary angles, *p. 48*
adjacent angles, *p. 48*
linear pair, *p. 50*
vertical angles, *p. 50*

Previous
vertex
sides of an angle
interior of an angle
opposite rays

What You Will Learn

- Identify complementary and supplementary angles.
- Identify linear pairs and vertical angles.

Using Complementary and Supplementary Angles

Pairs of angles can have special relationships. The measurements of the angles or the positions of the angles in the pair determine the relationship.

Core Concept

Complementary and Supplementary Angles

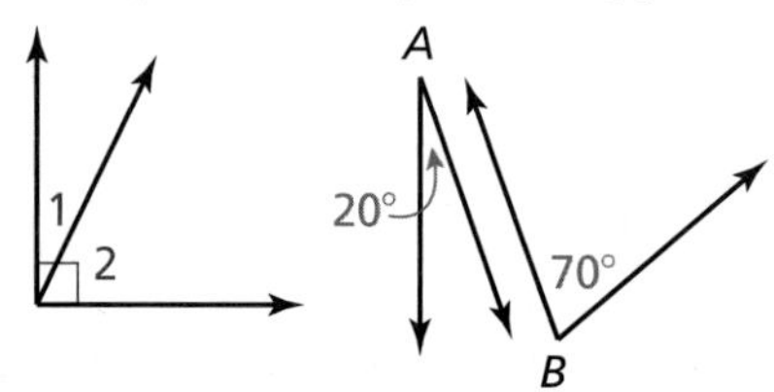

$\angle 1$ and $\angle 2$ $\angle A$ and $\angle B$

complementary angles

Two positive angles whose measures have a sum of 90°. Each angle is the *complement* of the other.

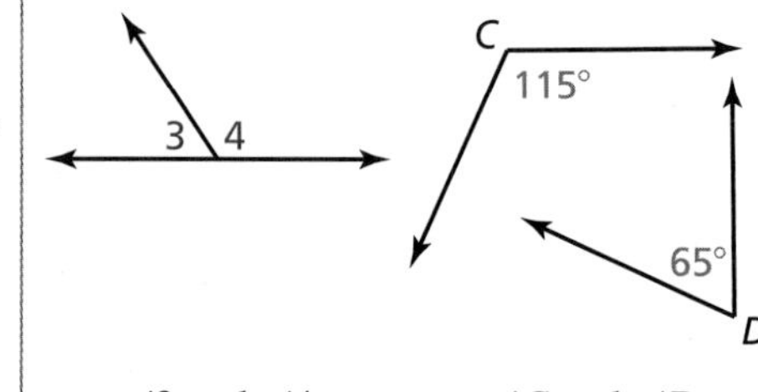

$\angle 3$ and $\angle 4$ $\angle C$ and $\angle D$

supplementary angles

Two positive angles whose measures have a sum of 180°. Each angle is the *supplement* of the other.

Adjacent Angles

Complementary angles and supplementary angles can be *adjacent angles* or *nonadjacent angles*. **Adjacent angles** are two angles that share a common vertex and side, but have no common interior points.

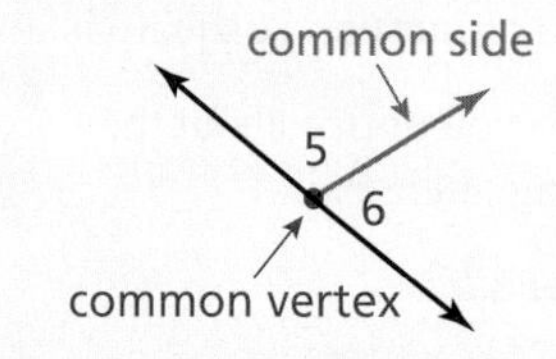

$\angle 5$ and $\angle 6$ are adjacent angles.

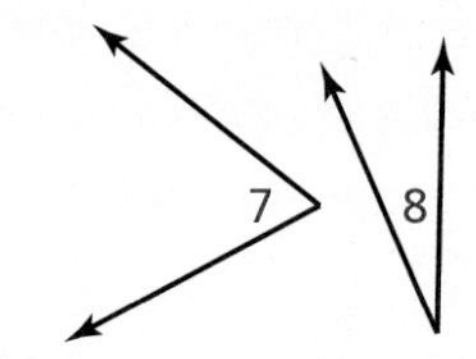

$\angle 7$ and $\angle 8$ are nonadjacent angles.

EXAMPLE 1 Identifying Pairs of Angles

In the figure, name a pair of complementary angles, a pair of supplementary angles, and a pair of adjacent angles.

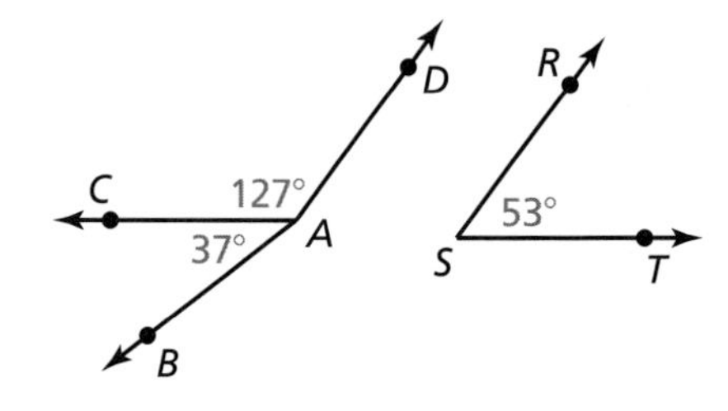

SOLUTION

Because 37° + 53° = 90°, $\angle BAC$ and $\angle RST$ are complementary angles.

Because 127° + 53° = 180°, $\angle CAD$ and $\angle RST$ are supplementary angles.

Because $\angle BAC$ and $\angle CAD$ share a common vertex and side, they are adjacent angles.

COMMON ERROR

In Example 1, $\angle DAC$ and $\angle DAB$ share a common vertex and a common side. But they also share common interior points. So, they are *not* adjacent angles.

EXAMPLE 2 Finding Angle Measures

a. $\angle 1$ is a complement of $\angle 2$, and $m\angle 1 = 62°$. Find $m\angle 2$.

b. $\angle 3$ is a supplement of $\angle 4$, and $m\angle 4 = 47°$. Find $m\angle 3$.

COMMON ERROR

Do not confuse angle names with angle measures.

SOLUTION

a. Draw a diagram with complementary adjacent angles to illustrate the relationship.

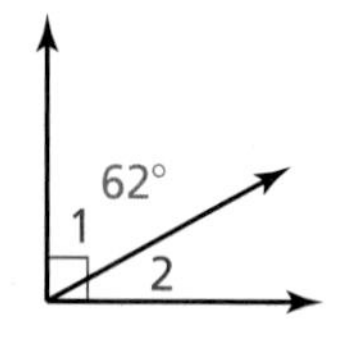

$$m\angle 2 = 90° - m\angle 1 = 90° - 62° = 28°$$

b. Draw a diagram with supplementary adjacent angles to illustrate the relationship.

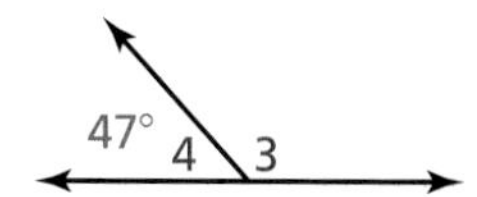

$$m\angle 3 = 180° - m\angle 4 = 180° - 47° = 133°$$

Monitoring Progress  Help in English and Spanish at *BigIdeasMath.com*

In Exercises 1 and 2, use the figure.

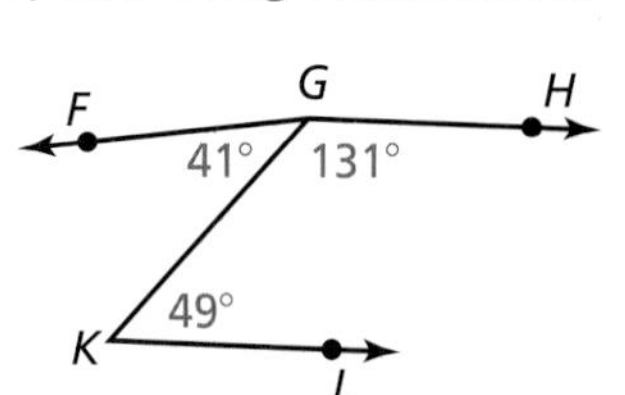

1. Name a pair of complementary angles, a pair of supplementary angles, and a pair of adjacent angles.
2. Are $\angle KGH$ and $\angle LKG$ adjacent angles? Are $\angle FGK$ and $\angle FGH$ adjacent angles? Explain.
3. $\angle 1$ is a complement of $\angle 2$, and $m\angle 2 = 5°$. Find $m\angle 1$.
4. $\angle 3$ is a supplement of $\angle 4$, and $m\angle 3 = 148°$. Find $m\angle 4$.

EXAMPLE 3 Real-Life Application

When viewed from the side, the frame of a ball-return net forms a pair of supplementary angles with the ground. Find $m\angle BCE$ and $m\angle ECD$.

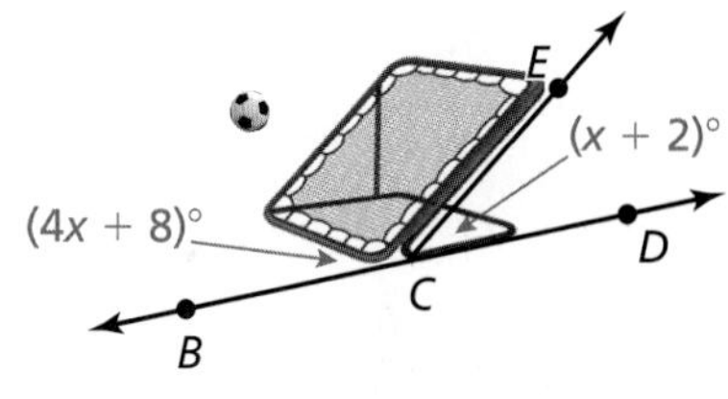

SOLUTION

Step 1 Use the fact that the sum of the measures of supplementary angles is 180°.

$m\angle BCE + m\angle ECD = 180°$	Write an equation.
$(4x + 8)° + (x + 2)° = 180°$	Substitute angle measures.
$5x + 10 = 180$	Combine like terms.
$x = 34$	Solve for x.

Step 2 Evaluate the given expressions when $x = 34$.

$$m\angle BCE = (4x + 8)° = (4 \cdot 34 + 8)° = 144°$$

$$m\angle ECD = (x + 2)° = (34 + 2)° = 36°$$

▶ So, $m\angle BCE = 144°$ and $m\angle ECD = 36°$.

Monitoring Progress  Help in English and Spanish at *BigIdeasMath.com*

5. $\angle LMN$ and $\angle PQR$ are complementary angles. Find the measures of the angles when $m\angle LMN = (4x - 2)°$ and $m\angle PQR = (9x + 1)°$.

Using Other Angle Pairs

Linear Pairs and Vertical Angles

Two adjacent angles are a **linear pair** when their noncommon sides are opposite rays. The angles in a linear pair are supplementary angles.

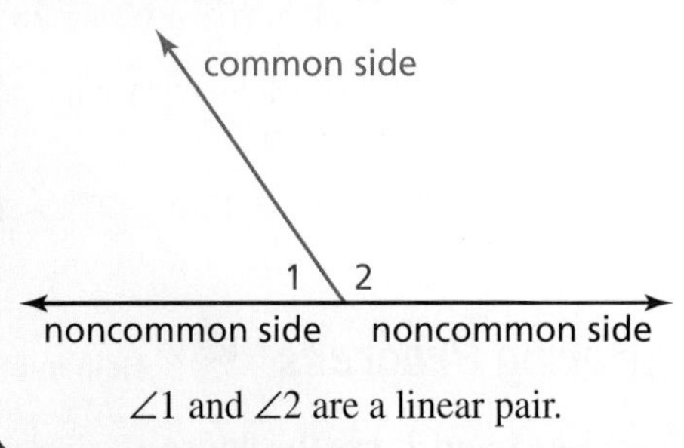

$\angle 1$ and $\angle 2$ are a linear pair.

Two angles are **vertical angles** when their sides form two pairs of opposite rays.

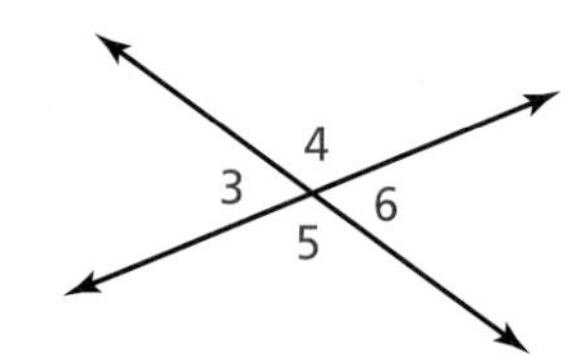

$\angle 3$ and $\angle 6$ are vertical angles.

$\angle 4$ and $\angle 5$ are vertical angles.

EXAMPLE 4 Identifying Angle Pairs

Identify all the linear pairs and all the vertical angles in the figure.

COMMON ERROR

In Example 4, one side of $\angle 1$ and one side of $\angle 3$ are opposite rays. But the angles are not a linear pair because they are *nonadjacent*.

SOLUTION

To find vertical angles, look for angles formed by intersecting lines.

▶ $\angle 1$ and $\angle 5$ are vertical angles.

To find linear pairs, look for adjacent angles whose noncommon sides are opposite rays.

▶ $\angle 1$ and $\angle 4$ are a linear pair. $\angle 4$ and $\angle 5$ are also a linear pair.

EXAMPLE 5 Finding Angle Measures in a Linear Pair

Two angles form a linear pair. The measure of one angle is five times the measure of the other angle. Find the measure of each angle.

SOLUTION

Step 1 Draw a diagram. Let x° be the measure of one angle. The measure of the other angle is $5x^\circ$.

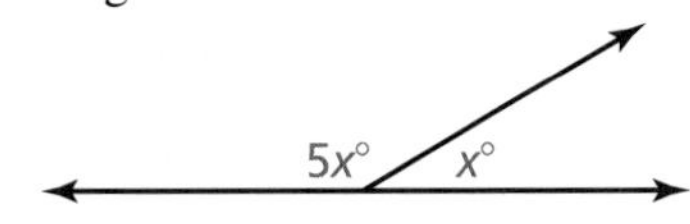

Step 2 Use the fact that the angles of a linear pair are supplementary to write an equation.

$x^\circ + 5x^\circ = 180^\circ$ Write an equation.

$6x = 180$ Combine like terms.

$x = 30$ Divide each side by 6.

▶ The measures of the angles are 30° and $5(30^\circ) = 150^\circ$.

Monitoring Progress  Help in English and Spanish at *BigIdeasMath.com*

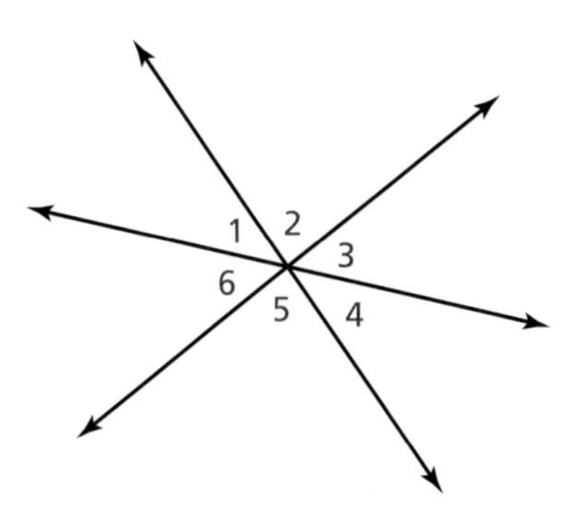

6. Do any of the numbered angles in the figure form a linear pair? Which angles are vertical angles? Explain your reasoning.

7. The measure of an angle is twice the measure of its complement. Find the measure of each angle.

8. Two angles form a linear pair. The measure of one angle is $1\frac{1}{2}$ times the measure of the other angle. Find the measure of each angle.

Concept Summary

Interpreting a Diagram

There are some things you can conclude from a diagram, and some you cannot. For example, here are some things that you ***can* conclude** from the diagram below.

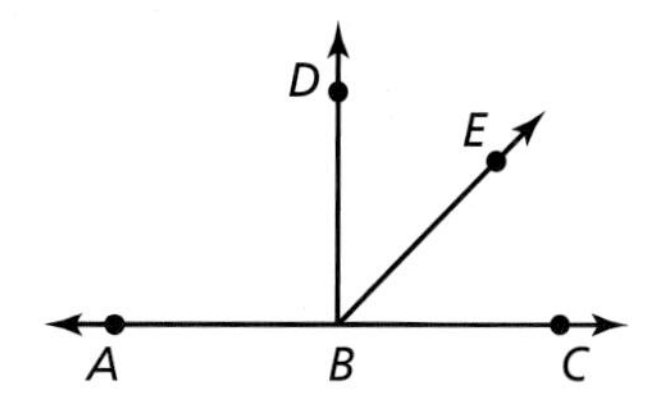

YOU CAN CONCLUDE

- All points shown are coplanar.
- Points A, B, and C are collinear, and B is between A and C.
- $\overleftrightarrow{AC}$, $\overrightarrow{BD}$, and $\overrightarrow{BE}$ intersect at point B.
- $\angle DBE$ and $\angle EBC$ are adjacent angles, and $\angle ABC$ is a straight angle.
- Point E lies in the interior of $\angle DBC$.

Here are some things you ***cannot* conclude** from the diagram above.

YOU CANNOT CONCLUDE

- $\overline{AB} \cong \overline{BC}$.
- $\angle DBE \cong \angle EBC$.
- $\angle ABD$ is a right angle.

To make such conclusions, the following information must be given.

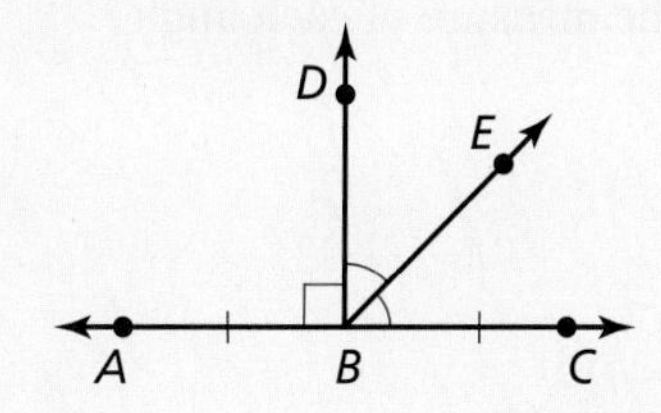

1.6 Exercises

Dynamic Solutions available at *BigIdeasMath.com*

Vocabulary and Core Concept Check

1. **WRITING** Explain what is different between adjacent angles and vertical angles.

2. **WHICH ONE DOESN'T BELONG?** Which one does *not* belong with the other three? Explain your reasoning.

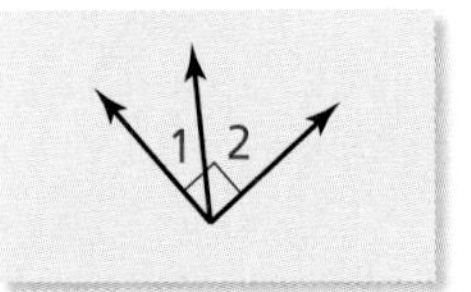

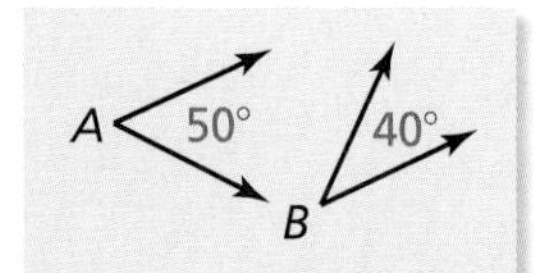

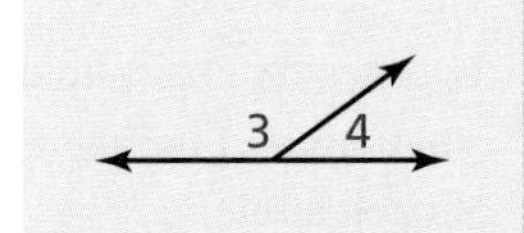

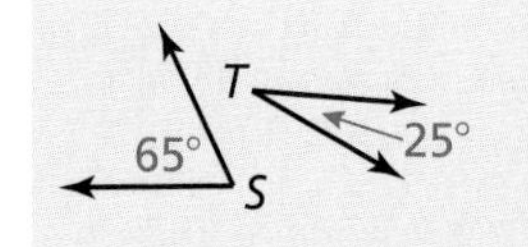

Monitoring Progress and Modeling with Mathematics

In Exercises 3–6, use the figure. *(See Example 1.)*

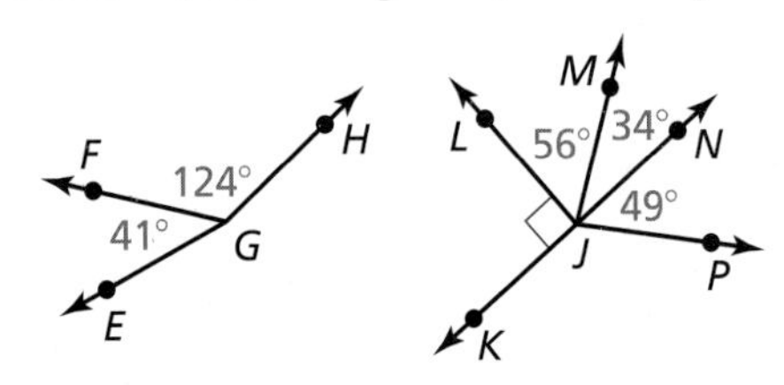

3. Name a pair of adjacent complementary angles.

4. Name a pair of adjacent supplementary angles.

5. Name a pair of nonadjacent complementary angles.

6. Name a pair of nonadjacent supplementary angles.

In Exercises 7–10, find the angle measure. *(See Example 2.)*

7. $\angle 1$ is a complement of $\angle 2$, and $m\angle 1 = 23^\circ$. Find $m\angle 2$.

8. $\angle 3$ is a complement of $\angle 4$, and $m\angle 3 = 46^\circ$. Find $m\angle 4$.

9. $\angle 5$ is a supplement of $\angle 6$, and $m\angle 5 = 78^\circ$. Find $m\angle 6$.

10. $\angle 7$ is a supplement of $\angle 8$, and $m\angle 7 = 109^\circ$. Find $m\angle 8$.

In Exercises 11–14, find the measure of each angle. *(See Example 3.)*

11.

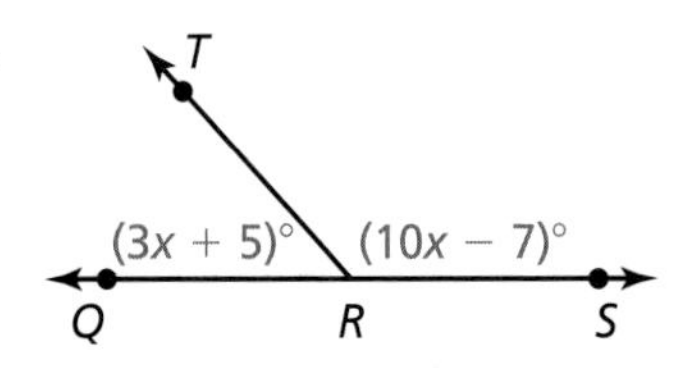

12.

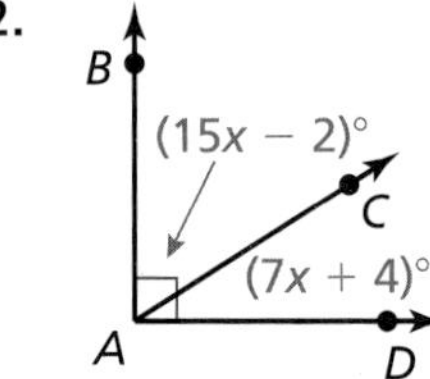

13. $\angle UVW$ and $\angle XYZ$ are complementary angles, $m\angle UVW = (x - 10)^\circ$, and $m\angle XYZ = (4x - 10)^\circ$.

14. $\angle EFG$ and $\angle LMN$ are supplementary angles, $m\angle EFG = (3x + 17)^\circ$, and $m\angle LMN = \left(\frac{1}{2}x - 5\right)^\circ$.

In Exercises 15–18, use the figure. *(See Example 4.)*

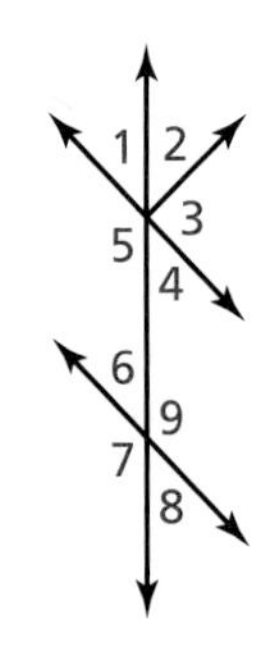

15. Identify the linear pair(s) that include $\angle 1$.

16. Identify the linear pair(s) that include $\angle 7$.

17. Are $\angle 6$ and $\angle 8$ vertical angles? Explain your reasoning.

18. Are $\angle 2$ and $\angle 5$ vertical angles? Explain your reasoning.

In Exercises 19–22, find the measure of each angle. *(See Example 5.)*

19. Two angles form a linear pair. The measure of one angle is twice the measure of the other angle.

20. Two angles form a linear pair. The measure of one angle is $\frac{1}{3}$ the measure of the other angle.

21. The measure of an angle is nine times the measure of its complement.

22. The measure of an angle is $\frac{1}{4}$ the measure of its complement.

ERROR ANALYSIS In Exercises 23 and 24, describe and correct the error in identifying pairs of angles in the figure.

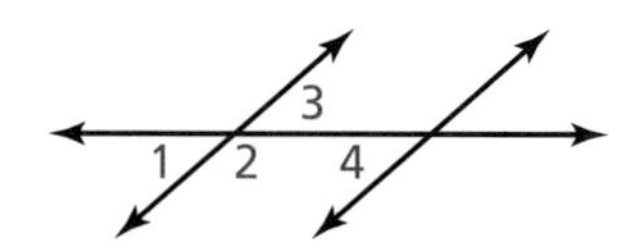

23.  $\angle 2$ and $\angle 4$ are adjacent.

24. 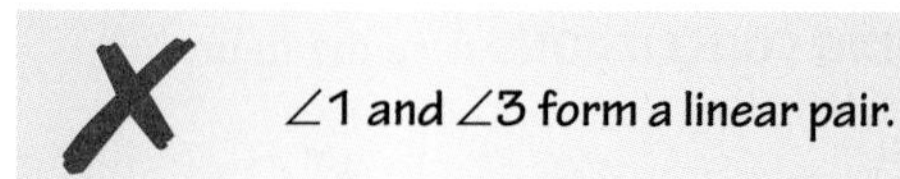

In Exercises 25 and 26, the picture shows the Alamillo Bridge in Seville, Spain. In the picture, $m\angle 1 = 58°$ and $m\angle 2 = 24°$.

25. Find the measure of the supplement of $\angle 1$.

26. Find the measure of the supplement of $\angle 2$.

27. **PROBLEM SOLVING** The arm of a crossing gate moves 42° from a vertical position. How many more degrees does the arm have to move so that it is horizontal?

A. 42° **B.** 138°

C. 48° **D.** 90°

28. **REASONING** The foul lines of a baseball field intersect at home plate to form a right angle. A batter hits a fair ball such that the path of the baseball forms an angle of 27° with the third base foul line. What is the measure of the angle between the first base foul line and the path of the baseball?

29. **CONSTRUCTION** Construct a linear pair where one angle measure is 115°.

30. **CONSTRUCTION** Construct a pair of adjacent angles that have angle measures of 45° and 97°.

31. **PROBLEM SOLVING** $m\angle U = 2x°$, and $m\angle V = 4m\angle U$. Which value of x makes $\angle U$ and $\angle V$ complements of each other?

A. 25 **B.** 9 **C.** 36 **D.** 18

MATHEMATICAL CONNECTIONS In Exercises 32–35, write and solve an algebraic equation to find the measure of each angle based on the given description.

32. The measure of an angle is 6° less than the measure of its complement.

33. The measure of an angle is 12° more than twice the measure of its complement.

34. The measure of one angle is 3° more than $\frac{1}{2}$ the measure of its supplement.

35. Two angles form a linear pair. The measure of one angle is 15° less than $\frac{2}{3}$ the measure of the other angle.

CRITICAL THINKING In Exercises 36–41, tell whether the statement is *always*, *sometimes*, or *never* true. Explain your reasoning.

36. Complementary angles are adjacent.

37. Angles in a linear pair are supplements of each other.

38. Vertical angles are adjacent.

39. Vertical angles are supplements of each other.

40. If an angle is acute, then its complement is greater than its supplement.

41. If two complementary angles are congruent, then the measure of each angle is 45°.

42. **WRITING** Explain why the supplement of an acute angle must be obtuse.

43. **WRITING** Explain why an obtuse angle does not have a complement.

44. THOUGHT PROVOKING Sketch an intersection of roads. Identify any supplementary, complementary, or vertical angles.

45. ATTENDING TO PRECISION Use the figure.

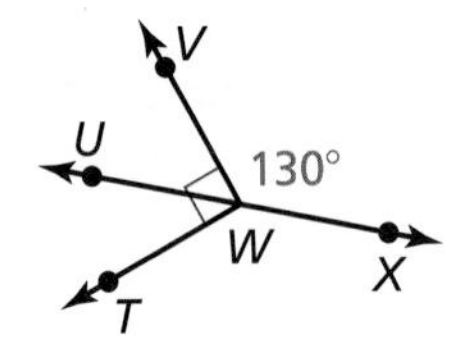

a. Find $m\angle UWV$, $m\angle TWU$, and $m\angle TWX$.

b. You write the measures of $\angle TWU$, $\angle TWX$, $\angle UWV$, and $\angle VWX$ on separate pieces of paper and place the pieces of paper in a box. Then you pick two pieces of paper out of the box at random. What is the probability that the angle measures you choose are supplementary? Explain your reasoning.

46. HOW DO YOU SEE IT? Tell whether you can conclude that each statement is true based on the figure. Explain your reasoning.

a. $\overline{CA} \cong \overline{AF}$.

b. Points C, A, and F are collinear.

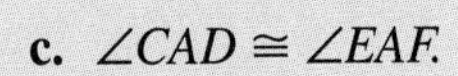

c. $\angle CAD \cong \angle EAF$.

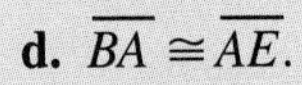

d. $\overline{BA} \cong \overline{AE}$.

e. $\overleftrightarrow{CF}$, $\overleftrightarrow{BE}$, and $\overleftrightarrow{AD}$ intersect at point A.

f. $\angle BAC$ and $\angle CAD$ are complementary angles.

g. $\angle DAE$ is a right angle.

47. REASONING $\angle KJL$ and $\angle LJM$ are complements, and $\angle MJN$ and $\angle LJM$ are complements. Can you show that $\angle KJL \cong \angle MJN$? Explain your reasoning.

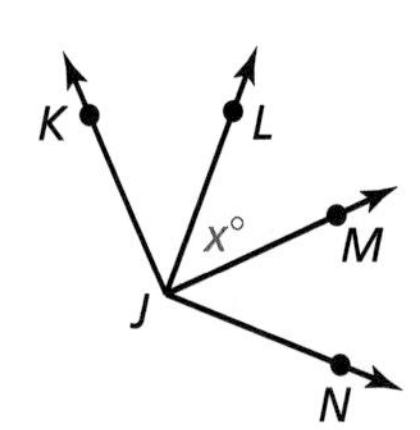

48. MAKING AN ARGUMENT Light from a flashlight strikes a mirror and is reflected so that the angle of reflection is congruent to the angle of incidence. Your classmate claims that $\angle QPR$ is congruent to $\angle TPU$ regardless of the measure of $\angle RPS$. Is your classmate correct? Explain your reasoning.

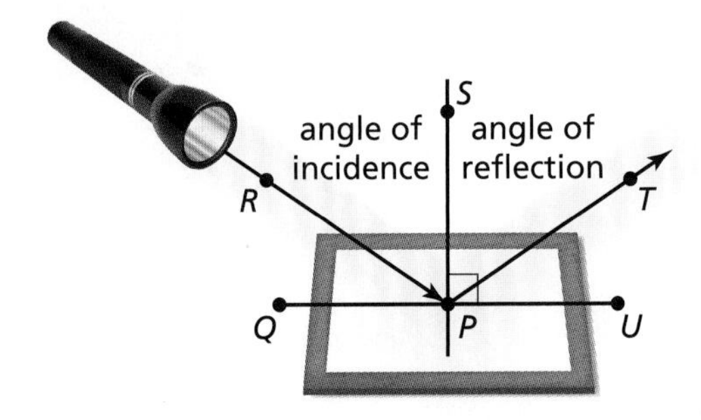

49. DRAWING CONCLUSIONS Use the figure.

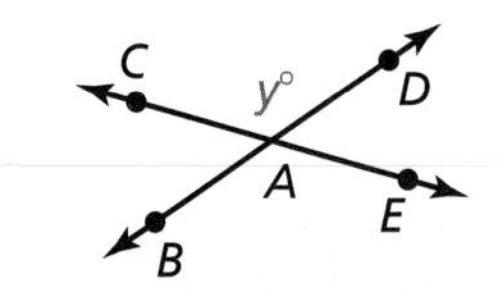

a. Write expressions for the measures of $\angle BAE$, $\angle DAE$, and $\angle CAB$.

b. What do you notice about the measures of vertical angles? Explain your reasoning.

50. MATHEMATICAL CONNECTIONS Let $m\angle 1 = x°$, $m\angle 2 = y_1°$, and $m\angle 3 = y_2°$. $\angle 2$ is the complement of $\angle 1$, and $\angle 3$ is the supplement of $\angle 1$.

a. Write equations for y_1 as a function of x and for y_2 as a function of x. What is the domain of each function? Explain.

b. Graph each function and describe its range.

51. MATHEMATICAL CONNECTIONS The sum of the measures of two complementary angles is 74° greater than the difference of their measures. Find the measure of each angle. Explain how you found the angle measures.

Maintaining Mathematical Proficiency

Reviewing what you learned in previous grades and lessons

Determine whether the statement is *always*, *sometimes*, or *never* true. Explain your reasoning. *(Skills Review Handbook)*

52. An integer is a whole number.

53. An integer is an irrational number.

54. An irrational number is a real number.

55. A whole number is negative.

56. A rational number is an integer.

57. A natural number is an integer.

58. A whole number is a rational number.

59. An irrational number is negative.

1.4–1.6 What Did You Learn?

Core Vocabulary

angle, *p. 38*
vertex, *p. 38*
sides of an angle, *p. 38*
interior of an angle, *p. 38*
exterior of an angle, *p. 38*
measure of an angle, *p. 39*
acute angle, *p. 39*
right angle, *p. 39*
obtuse angle, *p. 39*
straight angle, *p. 39*
congruent angles, *p. 40*
angle bisector, *p. 42*
complementary angles, *p. 48*
supplementary angles, *p. 48*
adjacent angles, *p. 48*
linear pair, *p. 50*
vertical angles, *p. 50*

Core Concepts

Section 1.4

Classifying Polygons, *p. 30*
Finding Perimeter and Area in the Coordinate Plane, *p. 31*

Section 1.5

Postulate 1.3 Protractor Postulate, *p. 39*
Types of Angles, *p. 39*
Postulate 1.4 Angle Addition Postulate, *p. 41*
Bisecting Angles, *p. 42*

Section 1.6

Complementary and Supplementary Angles, *p. 48*
Adjacent Angles, *p. 48*
Linear Pairs and Vertical Angles, *p. 50*
Interpreting a Diagram, *p. 51*

Mathematical Practices

1. How could you explain your answers to Exercise 33 on page 36 to a friend who is unable to hear?
2. What tool(s) could you use to verify your answers to Exercises 25–30 on page 44?
3. Your friend says that the angles in Exercise 28 on page 53 are supplementary angles. Explain why you agree or disagree.

Performance Task

Comfortable Horse Stalls

The plan for a new barn includes standard, rectangular horse stalls. The architect is sure that this will provide the most comfort for your horse because it is the greatest area for the stall. Is that correct? How can you investigate to find out?

To explore the answers to this question and more, go to ***BigIdeasMath.com***.

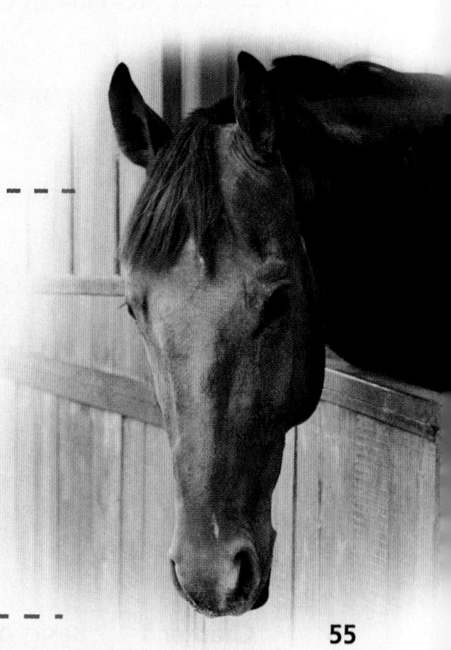

1 Chapter Review

Dynamic Solutions available at *BigIdeasMath.com*

1.1 Points, Lines, and Planes (pp. 3–10)

Use the diagram at the right. Give another name for plane *P*. Then name a line in the plane, a ray, a line intersecting the plane, and three collinear points.

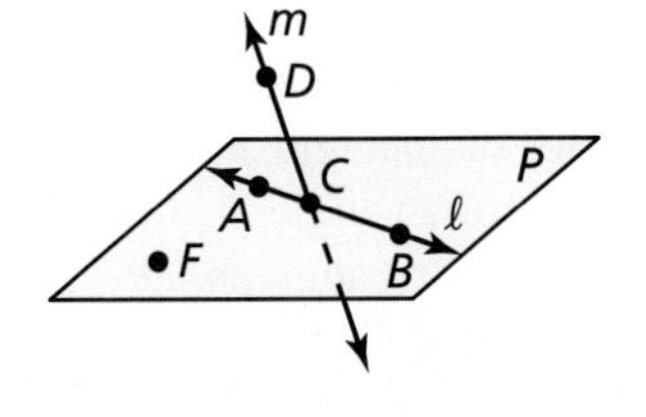

You can find another name for plane *P* by using any three points in the plane that are not on the same line. So, another name for plane *P* is plane *FAB*.

A line in the plane is $\overleftrightarrow{AB}$, a ray is $\overrightarrow{CB}$, a line intersecting the plane is $\overleftrightarrow{CD}$, and three collinear points are *A*, *C*, and *B*.

Use the diagram.

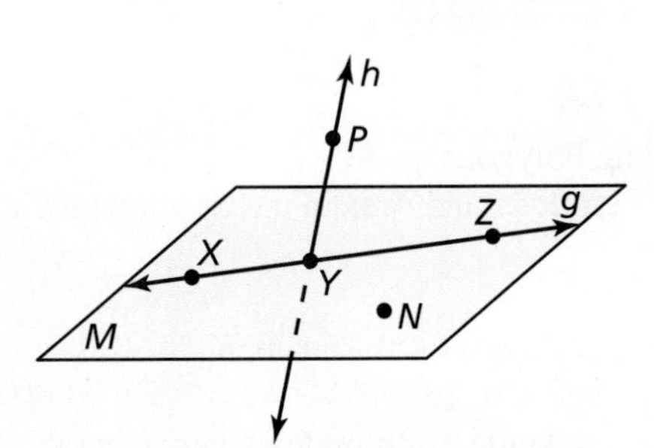

1. Give another name for plane *M*.
2. Name a line in the plane.
3. Name a line intersecting the plane.
4. Name two rays.
5. Name a pair of opposite rays.
6. Name a point not in plane *M*.

1.2 Measuring and Constructing Segments (pp. 11–18)

a. Find *AC*.

$$\begin{aligned} AC &= AB + BC && \text{Segment Addition Postulate (Postulate 1.2)} \\ &= 12 + 25 && \text{Substitute 12 for } AB \text{ and 25 for } BC. \\ &= 37 && \text{Add.} \end{aligned}$$

▶ So, $AC = 37$.

b. Find *EF*.

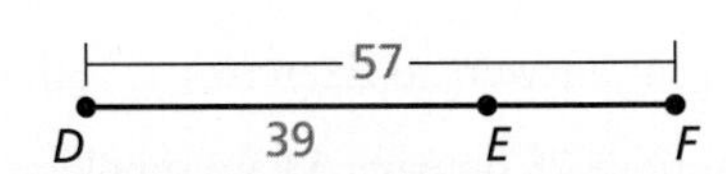

$$\begin{aligned} DF &= DE + EF && \text{Segment Addition Postulate (Postulate 1.2)} \\ 57 &= 39 + EF && \text{Substitute 57 for } DF \text{ and 39 for } DE. \\ 18 &= EF && \text{Subtract 39 from each side.} \end{aligned}$$

▶ So, $EF = 18$.

Find *XZ*.

7. 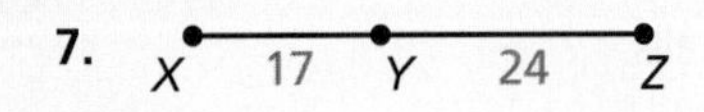

8. 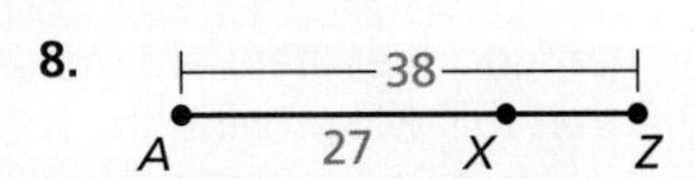

9. Plot $A(8, -4)$, $B(3, -4)$, $C(7, 1)$, and $D(7, -3)$ in a coordinate plane. Then determine whether $\overline{AB}$ and $\overline{CD}$ are congruent.

1.3 Using Midpoint and Distance Formulas (pp. 19–26)

The endpoints of $\overline{AB}$ are $A(6, -1)$ and $B(3, 5)$. Find the coordinates of the midpoint M. Then find the distance between points A and B.

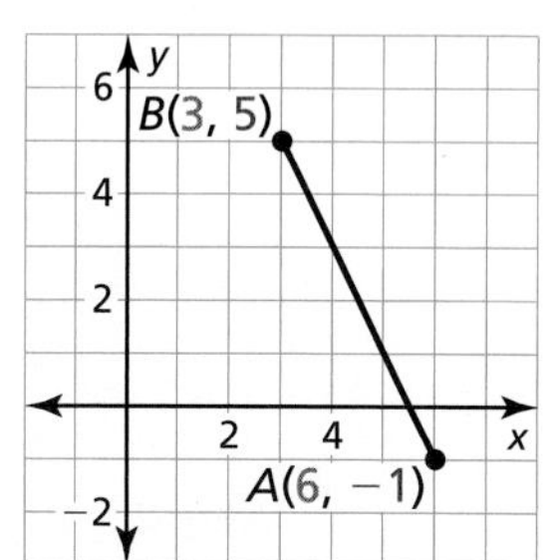

Use the Midpoint Formula.

$$M\left(\frac{6+3}{2}, \frac{-1+5}{2}\right) = M\left(\frac{9}{2}, 2\right)$$

Use the Distance Formula.

$AB = \sqrt{(x_2 - x_1)^2 + (y_2 - y_1)^2}$	Distance Formula
$= \sqrt{(3-6)^2 + [5-(-1)]^2}$	Substitute.
$= \sqrt{(-3)^2 + 6^2}$	Subtract.
$= \sqrt{9 + 36}$	Evaluate powers.
$= \sqrt{45}$	Add.
≈ 6.7 units	Use a calculator.

▶ So, the midpoint is $M\left(\frac{9}{2}, 2\right)$, and the distance is about 6.7 units.

Find the coordinates of the midpoint M. Then find the distance between points S and T.

10. $S(-2, 4)$ and $T(3, 9)$

11. $S(6, -3)$ and $T(7, -2)$

12. The midpoint of $\overline{JK}$ is $M(6, 3)$. One endpoint is $J(14, 9)$. Find the coordinates of endpoint K.

13. Point M is the midpoint of $\overline{AB}$ where $AM = 3x + 8$ and $MB = 6x - 4$. Find AB.

1.4 Perimeter and Area in the Coordinate Plane (pp. 29–36)

Find the perimeter and area of rectangle $ABCD$ with vertices $A(-3, 4)$, $B(6, 4)$, $C(6, -1)$, and $D(-3, -1)$.

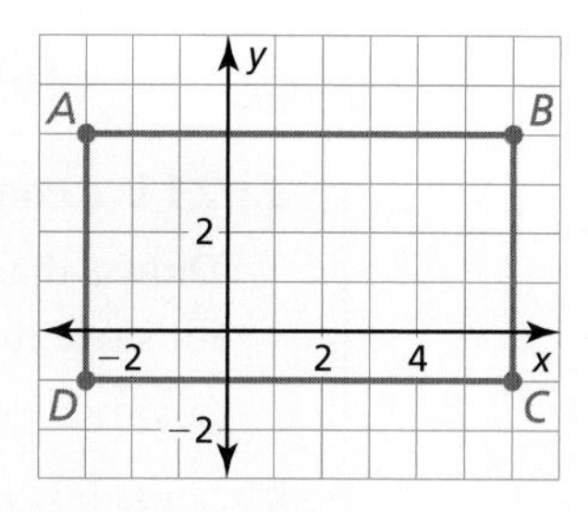

Draw the rectangle in a coordinate plane. Then find the length and width using the Ruler Postulate (Postulate 1.1).

Length $AB = |-3 - 6| = 9$

Width $BC = |4 - (-1)| = 5$

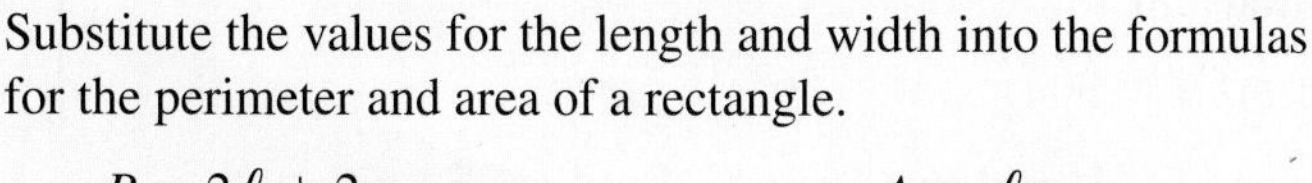

Substitute the values for the length and width into the formulas for the perimeter and area of a rectangle.

$P = 2\ell + 2w$	$A = \ell w$
$= 2(9) + 2(5)$	$= (9)(5)$
$= 18 + 10$	$= 45$
$= 28$	

▶ So, the perimeter is 28 units, and the area is 45 square units.

Find the perimeter and area of the polygon with the given vertices.

14. $W(5, -1)$, $X(5, 6)$, $Y(2, -1)$, $Z(2, 6)$

15. $E(6, -2)$, $F(6, 5)$, $G(-1, 5)$

1.5 Measuring and Constructing Angles (pp. 37–46)

Given that $m\angle DEF = 87°$, find $m\angle DEG$ and $m\angle GEF$.

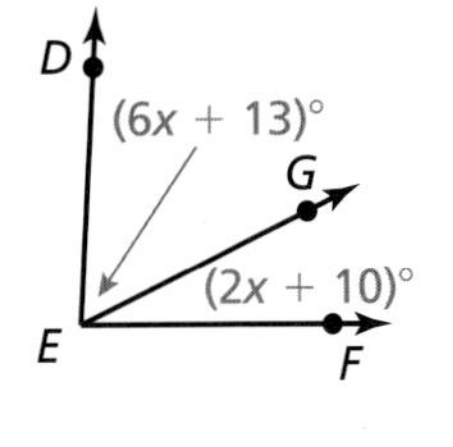

Step 1 Write and solve an equation to find the value of x.

$m\angle DEF = m\angle DEG + m\angle GEF$	Angle Addition Postulate (Post. 1.4)
$87° = (6x + 13)° + (2x + 10)°$	Substitute angle measures.
$87 = 8x + 23$	Combine like terms.
$64 = 8x$	Subtract 23 from each side.
$8 = x$	Divide each side by 8.

Step 2 Evaluate the given expressions when $x = 8$.

$m\angle DEG = (6x + 13)° = (6 \cdot 8 + 13)° = 61°$

$m\angle GEF = (2x + 10)° = (2 \cdot 8 + 10)° = 26°$

▶ So, $m\angle DEG = 61°$, and $m\angle GEF = 26°$.

Find $m\angle ABD$ and $m\angle CBD$.

16. $m\angle ABC = 77°$

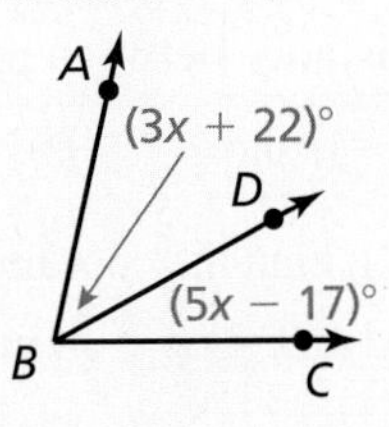

17. $m\angle ABC = 111°$

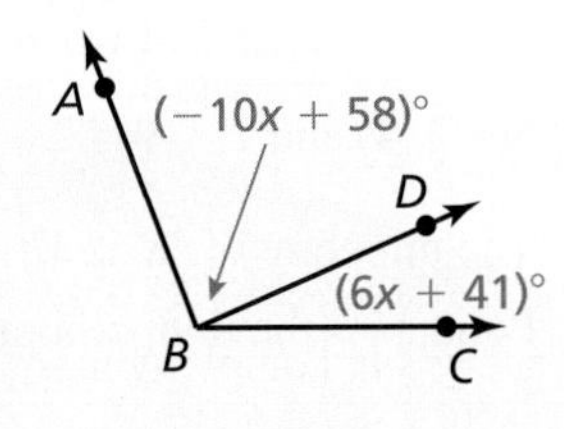

18. Find the measure of the angle using a protractor.

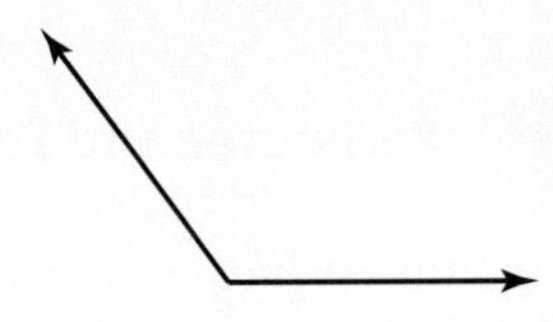

1.6 Describing Pairs of Angles (pp. 47–54)

a. $\angle 1$ is a complement of $\angle 2$, and $m\angle 1 = 54°$. Find $m\angle 2$.

Draw a diagram with complementary adjacent angles to illustrate the relationship.

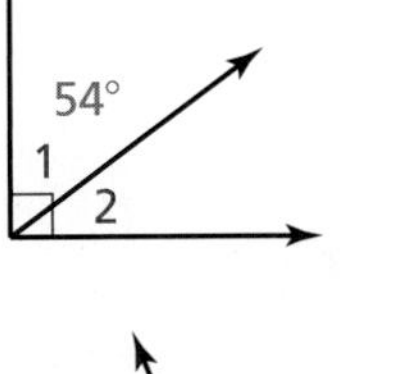

$m\angle 2 = 90° - m\angle 1 = 90° - 54° = 36°$

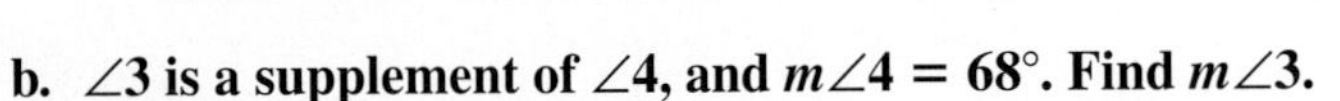

b. $\angle 3$ is a supplement of $\angle 4$, and $m\angle 4 = 68°$. Find $m\angle 3$.

Draw a diagram with supplementary adjacent angles to illustrate the relationship.

68°
4
3

$m\angle 3 = 180° - m\angle 4 = 180° - 68° = 112°$

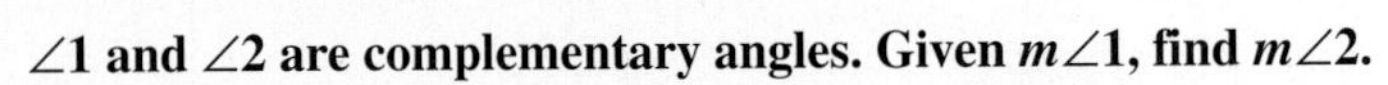

$\angle 1$ and $\angle 2$ are complementary angles. Given $m\angle 1$, find $m\angle 2$.

19. $m\angle 1 = 12°$

20. $m\angle 1 = 83°$

$\angle 3$ and $\angle 4$ are supplementary angles. Given $m\angle 3$, find $m\angle 4$.

21. $m\angle 3 = 116°$

22. $m\angle 3 = 56°$

1 Chapter Test

Find the length of $\overline{QS}$. Explain how you found your answer.

1. 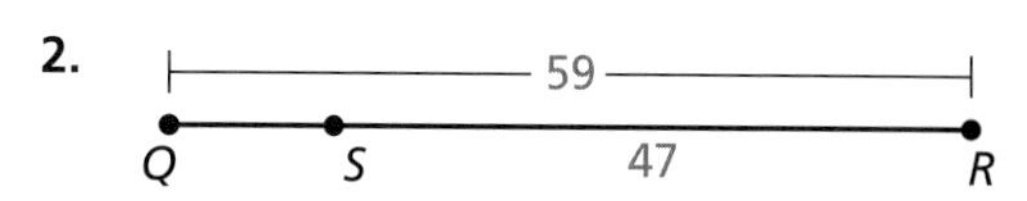

2.

Find the coordinates of the midpoint M. Then find the distance between the two points.

3. $A(-4, -8)$ and $B(-1, 4)$

4. $C(-1, 7)$ and $D(-8, -3)$

5. The midpoint of $\overline{EF}$ is $M(1, -1)$. One endpoint is $E(-3, 2)$. Find the coordinates of endpoint F.

Use the diagram to decide whether the statement is true or false.

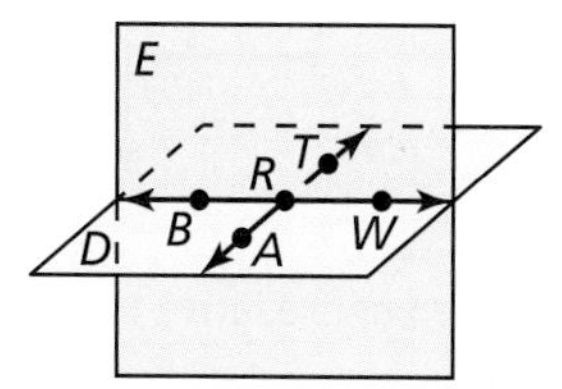

6. Points A, R, and B are collinear.

7. $\overleftrightarrow{BW}$ and $\overleftrightarrow{AT}$ are lines.

8. $\overrightarrow{BR}$ and $\overrightarrow{RT}$ are opposite rays.

9. Plane D could also be named plane ART.

Find the perimeter and area of the polygon with the given vertices. Explain how you found your answer.

10. $P(-3, 4)$, $Q(1, 4)$, $R(-3, -2)$, $S(3, -2)$

11. $J(-1, 3)$, $K(5, 3)$, $L(2, -2)$

12. In the diagram, $\angle AFE$ is a straight angle and $\angle CFE$ is a right angle. Identify all supplementary and complementary angles. Explain. Then find $m\angle DFE$, $m\angle BFC$, and $m\angle BFE$.

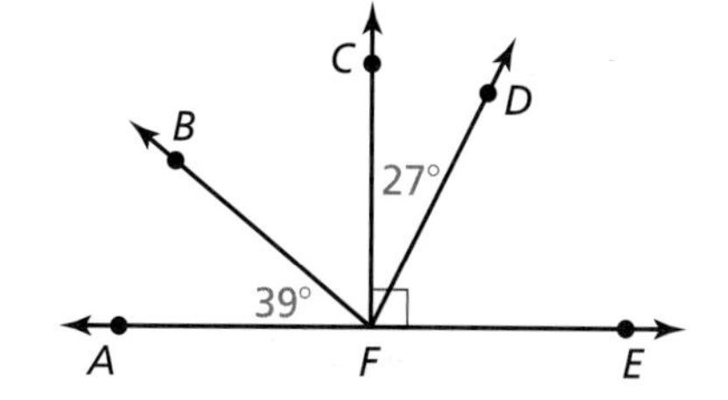

13. Use the clock at the left.

 a. What is the measure of the acute angle created when the clock is at 10:00?

 b. What is the measure of the obtuse angle created when the clock is at 5:00?

 c. Find a time where the hour and minute hands create a straight angle.

14. Sketch a figure that contains a plane and two lines that intersect the plane at one point.

15. Your parents decide they would like to install a rectangular swimming pool in the backyard. There is a 15-foot by 20-foot rectangular area available. Your parents request a 3-foot edge around each side of the pool. Draw a diagram of this situation in a coordinate plane. What is the perimeter and area of the largest swimming pool that will fit?

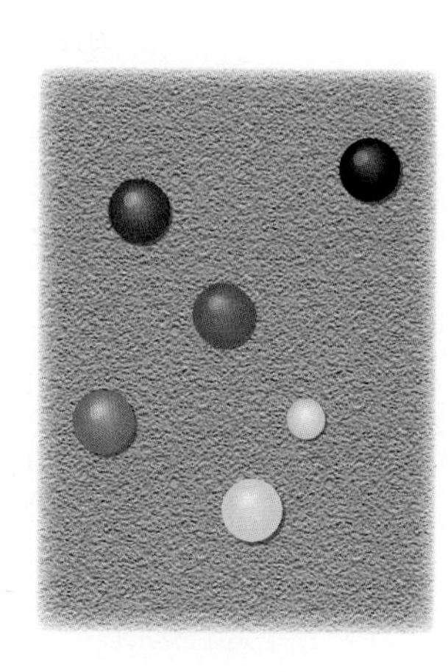

16. The picture shows the arrangement of balls in a game of boccie. The object of the game is to throw your ball closest to the small, white ball, which is called the *pallino*. The green ball is the midpoint between the red ball and the pallino. The distance between the green ball and the red ball is 10 inches. The distance between the yellow ball and the pallino is 8 inches. Which ball is closer to the pallino, the green ball or the yellow ball? Explain.

1 Cumulative Assessment

1. Use the diagram to determine which segments, if any, are congruent. List all congruent segments.

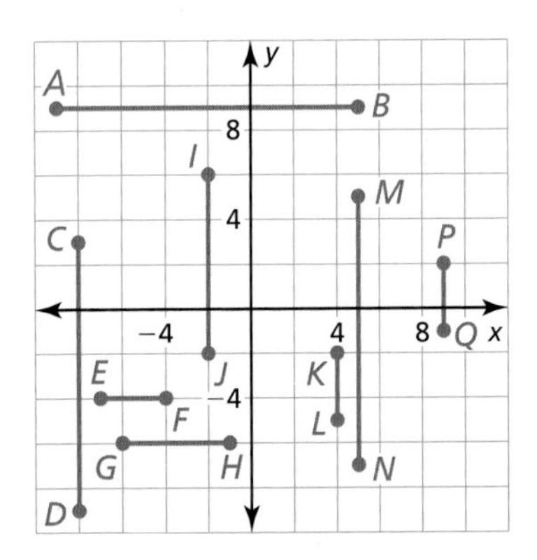

2. Order the terms so that each consecutive term builds off the previous term.

plane	segment	line	point	ray

3. The endpoints of a line segment are $(-6, 13)$ and $(11, 5)$. Which choice shows the correct midpoint and distance between these two points?

Ⓐ $\left(\frac{5}{2}, 4\right)$; 18.8 units

Ⓑ $\left(\frac{5}{2}, 9\right)$; 18.8 units

Ⓒ $\left(\frac{5}{2}, 4\right)$; 9.4 units

Ⓓ $\left(\frac{5}{2}, 9\right)$; 9.4 units

4. Find the perimeter and area of the figure shown.

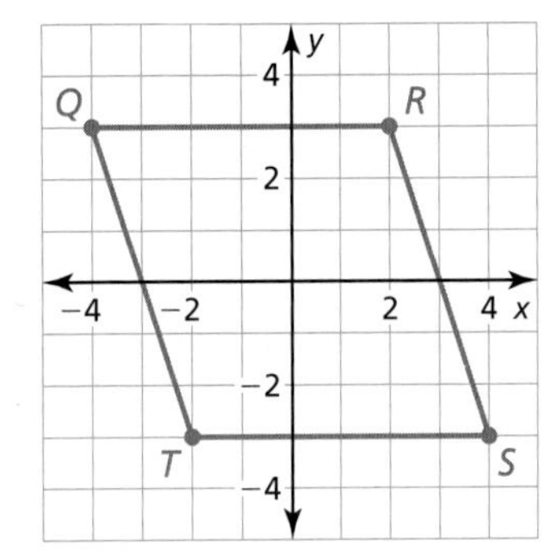

5. Plot the points $W(-1, 1)$, $X(5, 1)$, $Y(5, -2)$, and $Z(-1, -2)$ in a coordinate plane. What type of polygon do the points form? Your friend claims that you could use this figure to represent a basketball court with an area of 4050 square feet and a perimeter of 270 feet. Do you support your friend's claim? Explain.

6. Use the steps in the construction to explain how you know that $\overrightarrow{AG}$ is the angle bisector of $\angle CAB$.

Step 1

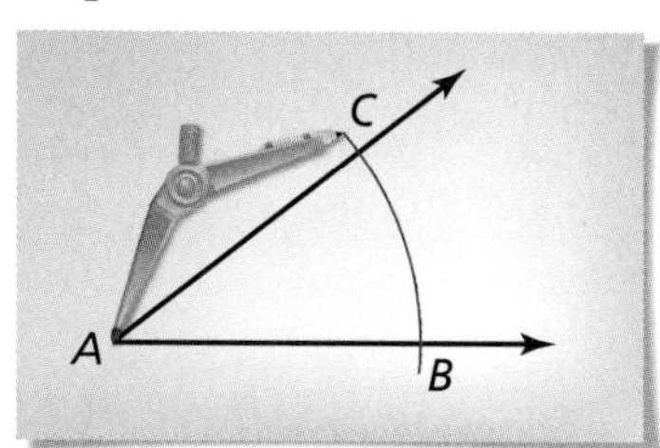

Step 2

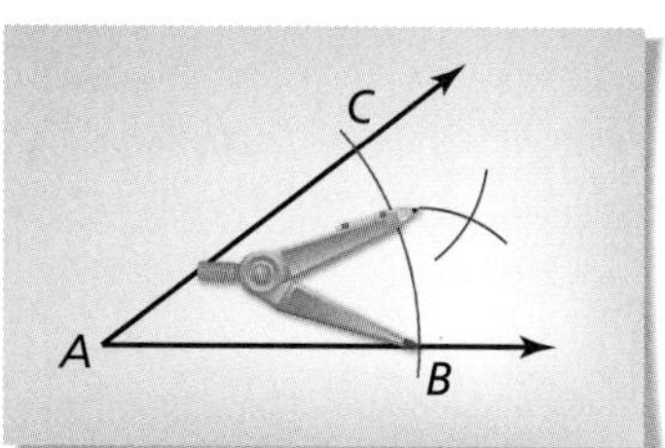

Step 3

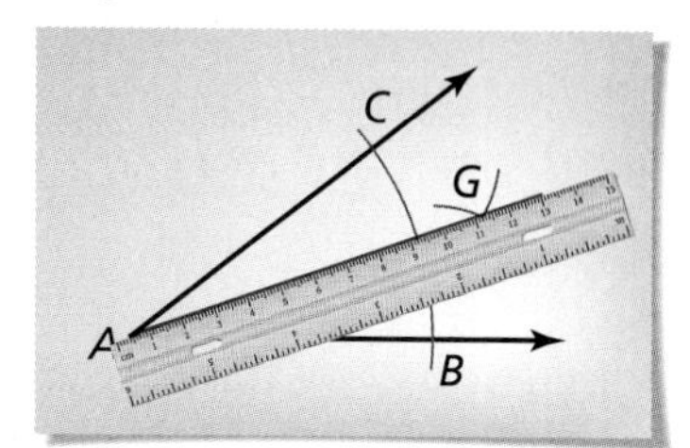

7. The picture shows an aerial view of a city. Use the streets highlighted in red to identify all congruent angles. Assume all streets are straight angles.

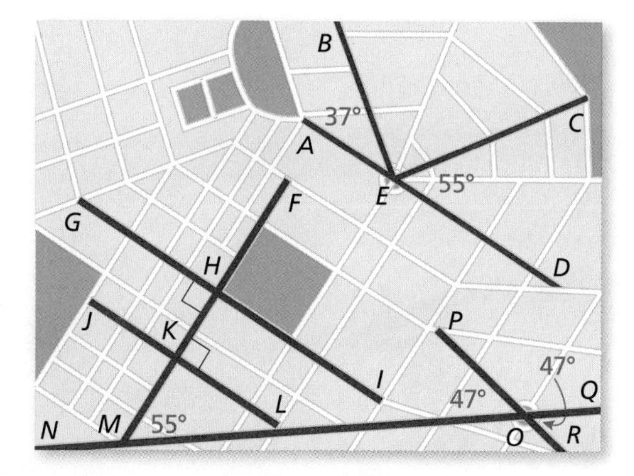

8. Three roads come to an intersection point that the people in your town call Five Corners, as shown in the figure.

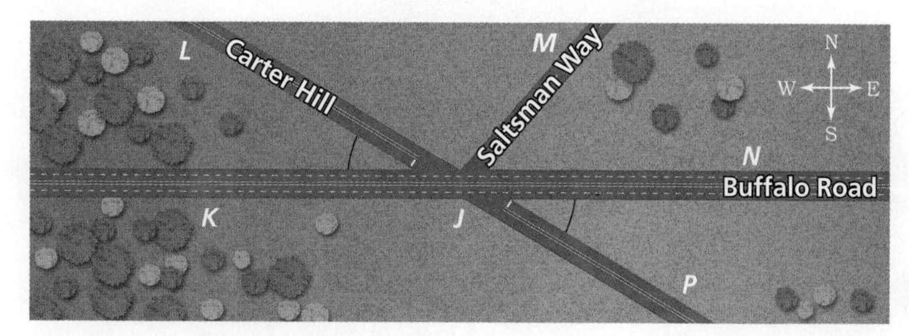

a. Identify all vertical angles.

b. Identify all linear pairs.

c. You are traveling east on Buffalo Road and decide to turn left onto Carter Hill. Name the angle of the turn you made.

$\angle KJL$	$\angle KJM$	$\angle KJN$	$\angle KJP$	$\angle LJM$
$\angle LJN$	$\angle LJP$	$\angle MJN$	$\angle MJP$	$\angle NJP$

2 Reasoning and Proofs

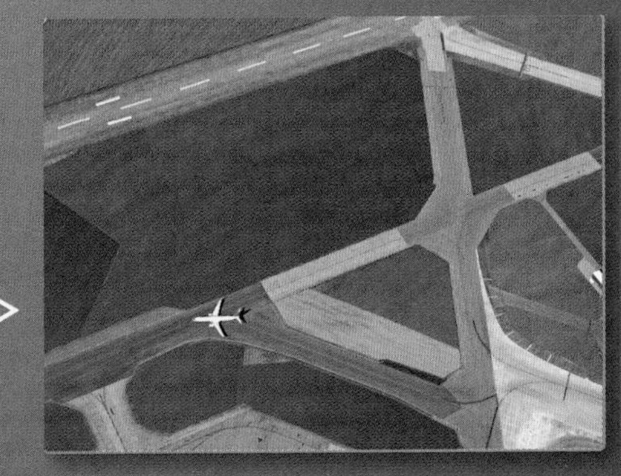

Airport Runway *(p. 108)*

Sculpture *(p. 104)*

City Street *(p. 95)*

Tiger *(p. 81)*

Guitar *(p. 67)*

Maintaining Mathematical Proficiency

Finding the *n*th Term of an Arithmetic Sequence

Example 1 **Write an equation for the *n*th term of the arithmetic sequence 2, 5, 8, 11, ... Then find a_{20}.**

The first term is 2, and the common difference is 3.

$a_n = a_1 + (n - 1)d$	Equation for an arithmetic sequence
$a_n = 2 + (n - 1)3$	Substitute 2 for a_1 and 3 for d.
$a_n = 3n - 1$	Simplify.

Use the equation to find the 20th term.

$a_n = 3n - 1$	Write the equation.
$a_{20} = 3(20) - 1$	Substitute 20 for n.
$= 59$	Simplify.

▶ The 20th term of the arithmetic sequence is 59.

Write an equation for the *n*th term of the arithmetic sequence. Then find a_{50}.

1. 3, 9, 15, 21, . . .

2. −29, −12, 5, 22, . . .

3. 2.8, 3.4, 4.0, 4.6, . . .

4. $\frac{1}{3}, \frac{1}{2}, \frac{2}{3}, \frac{5}{6}, \ldots$

5. 26, 22, 18, 14, . . .

6. 8, 2, −4, −10, . . .

Rewriting Literal Equations

Example 2 **Solve the literal equation $3x + 6y = 24$ for y.**

$3x + 6y = 24$	Write the equation.
$3x - 3x + 6y = 24 - 3x$	Subtract $3x$ from each side.
$6y = 24 - 3x$	Simplify.
$\frac{6y}{6} = \frac{24 - 3x}{6}$	Divide each side by 6.
$y = 4 - \frac{1}{2}x$	Simplify.

▶ The rewritten literal equation is $y = 4 - \frac{1}{2}x$.

Solve the literal equation for *x*.

7. $2y - 2x = 10$

8. $20y + 5x = 15$

9. $4y - 5 = 4x + 7$

10. $y = 8x - x$

11. $y = 4x + zx + 6$

12. $z = 2x + 6xy$

13. **ABSTRACT REASONING** Can you use the equation for an arithmetic sequence to write an equation for the sequence 3, 9, 27, 81, . . . ? Explain your reasoning.

Dynamic Solutions available at *BigIdeasMath.com*

Mathematical Practices

Mathematically proficient students distinguish correct reasoning from flawed reasoning.

Using Correct Reasoning

Core Concept

Deductive Reasoning

When you use *deductive reasoning*, you start with two or more true statements and *deduce* or *infer* the truth of another statement. Here is an example.

1. Premise: If a polygon is a triangle, then the sum of its angle measures is 180°.
2. Premise: Polygon ABC is a triangle.
3. Conclusion: The sum of the angle measures of polygon ABC is 180°.

This pattern for deductive reasoning is called a *syllogism*.

EXAMPLE 1 Recognizing Flawed Reasoning

The syllogisms below represent common types of *flawed reasoning*. Explain why each conclusion is not valid.

a. When it rains, the ground gets wet.
The ground is wet.
Therefore, it must have rained.

b. If $\triangle ABC$ is equilateral, then it is isosceles.
$\triangle ABC$ is not equilateral.
Therefore, it must not be isosceles.

c. All squares are polygons.
All trapezoids are quadrilaterals.
Therefore, all squares are quadrilaterals.

d. No triangles are quadrilaterals.
Some quadrilaterals are not squares.
Therefore, some squares are not triangles.

SOLUTION

a. The ground may be wet for another reason.

b. A triangle can be isosceles but not equilateral.

c. All squares are quadrilaterals, but not because all trapezoids are quadrilaterals.

d. No squares are triangles.

Monitoring Progress

Decide whether the syllogism represents correct or flawed reasoning. If flawed, explain why the conclusion is not valid.

1. All triangles are polygons.
Figure ABC is a triangle.
Therefore, figure ABC is a polygon.

2. No trapezoids are rectangles.
Some rectangles are not squares.
Therefore, some squares are not trapezoids.

3. If polygon $ABCD$ is a square, then it is a rectangle.
Polygon $ABCD$ is a rectangle.
Therefore, polygon $ABCD$ is a square.

4. If polygon $ABCD$ is a square, then it is a rectangle.
Polygon $ABCD$ is not a square.
Therefore, polygon $ABCD$ is not a rectangle.

2.1 Conditional Statements

Essential Question

When is a conditional statement true or false?

A *conditional statement*, symbolized by $p \rightarrow q$, can be written as an "if-then statement" in which p is the *hypothesis* and q is the *conclusion*. Here is an example.

If a polygon is a triangle, then the sum of its angle measures is 180°.

hypothesis, p — conclusion, q

EXPLORATION 1 Determining Whether a Statement Is True or False

Work with a partner. A hypothesis can either be true or false. The same is true of a conclusion. For a conditional statement to be true, the hypothesis and conclusion do not necessarily both have to be true. Determine whether each conditional statement is true or false. Justify your answer.

a. If yesterday was Wednesday, then today is Thursday.

b. If an angle is acute, then it has a measure of 30°.

c. If a month has 30 days, then it is June.

d. If an even number is not divisible by 2, then 9 is a perfect cube.

EXPLORATION 2 Determining Whether a Statement Is True or False

Work with a partner. Use the points in the coordinate plane to determine whether each statement is true or false. Justify your answer.

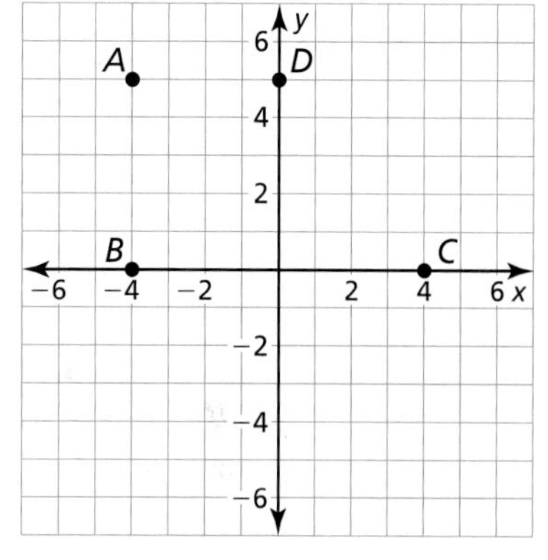

a. $\triangle ABC$ is a right triangle.

b. $\triangle BDC$ is an equilateral triangle.

c. $\triangle BDC$ is an isosceles triangle.

d. Quadrilateral $ABCD$ is a trapezoid.

e. Quadrilateral $ABCD$ is a parallelogram.

CONSTRUCTING VIABLE ARGUMENTS

To be proficient in math, you need to distinguish correct logic or reasoning from that which is flawed.

EXPLORATION 3 Determining Whether a Statement Is True or False

Work with a partner. Determine whether each conditional statement is true or false. Justify your answer.

a. If $\triangle ADC$ is a right triangle, then the Pythagorean Theorem is valid for $\triangle ADC$.

b. If $\angle A$ and $\angle B$ are complementary, then the sum of their measures is 180°.

c. If figure $ABCD$ is a quadrilateral, then the sum of its angle measures is 180°.

d. If points A, B, and C are collinear, then they lie on the same line.

e. If $\overleftrightarrow{AB}$ and $\overleftrightarrow{BD}$ intersect at a point, then they form two pairs of vertical angles.

Communicate Your Answer

4. When is a conditional statement true or false?

5. Write one true conditional statement and one false conditional statement that are different from those given in Exploration 3. Justify your answer.

2.1 Lesson

Core Vocabulary

conditional statement, *p. 66*
if-then form, *p. 66*
hypothesis, *p. 66*
conclusion, *p. 66*
negation, *p. 66*
converse, *p. 67*
inverse, *p. 67*
contrapositive, *p. 67*
equivalent statements, *p. 67*
perpendicular lines, *p. 68*
biconditional statement, *p. 69*
truth value, *p. 70*
truth table, *p. 70*

What You Will Learn

- Write conditional statements.
- Use definitions written as conditional statements.
- Write biconditional statements.
- Make truth tables.

Writing Conditional Statements

Core Concept

Conditional Statement

A **conditional statement** is a logical statement that has two parts, a *hypothesis* p and a *conclusion* q. When a conditional statement is written in **if-then form**, the "if" part contains the **hypothesis** and the "then" part contains the **conclusion**.

Words If p, then q. **Symbols** $p \rightarrow q$ (read as "p implies q")

EXAMPLE 1 Rewriting a Statement in If-Then Form

Use red to identify the hypothesis and blue to identify the conclusion. Then rewrite the conditional statement in if-then form.

a. All birds have feathers.

b. You are in Texas if you are in Houston.

SOLUTION

a. All birds have feathers.

- If an animal is a bird, then it has feathers.

b. You are in Texas if you are in Houston.

- If you are in Houston, then you are in Texas.

Monitoring Progress

Help in English and Spanish at *BigIdeasMath.com*

Use red to identify the hypothesis and blue to identify the conclusion. Then rewrite the conditional statement in if-then form.

1. All 30° angles are acute angles.

2. $2x + 7 = 1$, because $x = -3$.

Core Concept

Negation

The **negation** of a statement is the *opposite* of the original statement. To write the negation of a statement p, you write the symbol for negation ($\sim$) before the letter. So, "not p" is written $\sim p$.

Words not p **Symbols** $\sim p$

EXAMPLE 2 Writing a Negation

Write the negation of each statement.

a. The ball is red.

b. The cat is *not* black.

SOLUTION

a. The ball is *not* red.

b. The cat is black.

Core Concept

Related Conditionals

Consider the conditional statement below.

Words If p, then q. **Symbols** $p \rightarrow q$

Converse To write the **converse** of a conditional statement, exchange the hypothesis and the conclusion.

Words If q, then p. **Symbols** $q \rightarrow p$

Inverse To write the **inverse** of a conditional statement, negate both the hypothesis and the conclusion.

Words If not p, then not q. **Symbols** $\sim p \rightarrow \sim q$

Contrapositive To write the **contrapositive** of a conditional statement, first write the converse. Then negate both the hypothesis and the conclusion.

Words If not q, then not p. **Symbols** $\sim q \rightarrow \sim p$

A conditional statement and its contrapositive are either both true or both false. Similarly, the converse and inverse of a conditional statement are either both true or both false. In general, when two statements are both true or both false, they are called **equivalent statements**.

COMMON ERROR

Just because a conditional statement and its contrapositive are both true does not mean that its converse and inverse are both false. The converse and inverse could also both be true.

EXAMPLE 3 Writing Related Conditional Statements

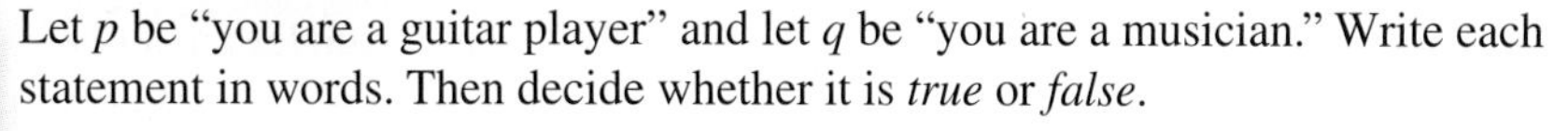

Let p be "you are a guitar player" and let q be "you are a musician." Write each statement in words. Then decide whether it is *true* or *false*.

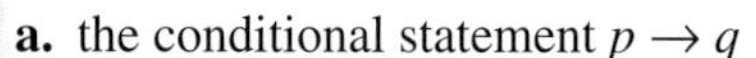

a. the conditional statement $p \rightarrow q$

b. the converse $q \rightarrow p$

c. the inverse $\sim p \rightarrow \sim q$

d. the contrapositive $\sim q \rightarrow \sim p$

SOLUTION

a. Conditional: If you are a guitar player, then you are a musician.
true; Guitar players are musicians.

b. Converse: If you are a musician, then you are a guitar player.
false; Not all musicians play the guitar.

c. Inverse: If you are not a guitar player, then you are not a musician.
false; Even if you do not play a guitar, you can still be a musician.

d. Contrapositive: If you are not a musician, then you are not a guitar player.
true; A person who is not a musician cannot be a guitar player.

Monitoring Progress

Help in English and Spanish at *BigIdeasMath.com*

In Exercises 3 and 4, write the negation of the statement.

3. The shirt is green.

4. The shoes are *not* red.

5. Repeat Example 3. Let p be "the stars are visible" and let q be "it is night."

Using Definitions

You can write a definition as a conditional statement in if-then form or as its converse. Both the conditional statement and its converse are true for definitions. For example, consider the definition of *perpendicular lines*.

If two lines intersect to form a right angle, then they are **perpendicular lines**.

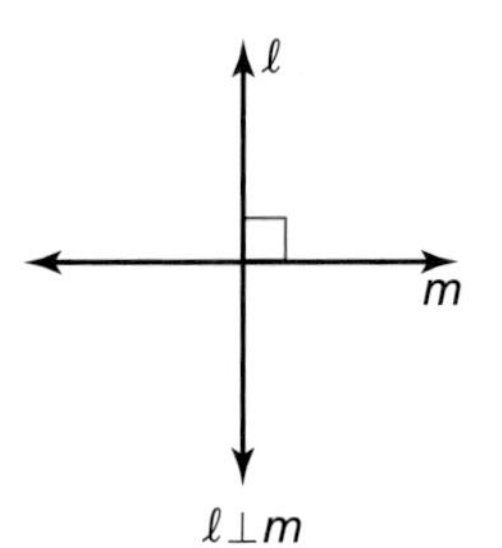

You can also write the definition using the converse: If two lines are perpendicular lines, then they intersect to form a right angle.

You can write "line ℓ is perpendicular to line m" as $\ell \perp m$.

EXAMPLE 4 Using Definitions

Decide whether each statement about the diagram is true. Explain your answer using the definitions you have learned.

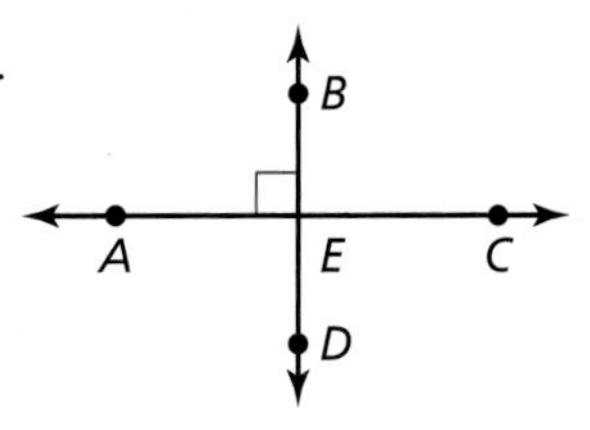

a. $\overleftrightarrow{AC} \perp \overleftrightarrow{BD}$

b. $\angle AEB$ and $\angle CEB$ are a linear pair.

c. $\overrightarrow{EA}$ and $\overrightarrow{EB}$ are opposite rays.

SOLUTION

a. This statement is *true*. The right angle symbol in the diagram indicates that the lines intersect to form a right angle. So, you can say the lines are perpendicular.

b. This statement is *true*. By definition, if the noncommon sides of adjacent angles are opposite rays, then the angles are a linear pair. Because $\overrightarrow{EA}$ and $\overrightarrow{EC}$ are opposite rays, $\angle AEB$ and $\angle CEB$ are a linear pair.

c. This statement is *false*. Point E does not lie on the same line as A and B, so the rays are not opposite rays.

Monitoring Progress

Help in English and Spanish at *BigIdeasMath.com*

Use the diagram. Decide whether the statement is true. Explain your answer using the definitions you have learned.

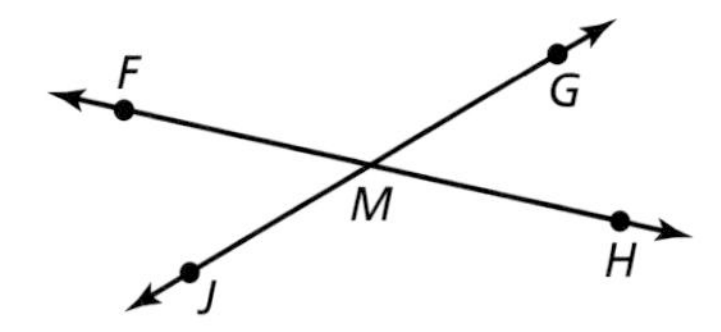

6. $\angle JMF$ and $\angle FMG$ are supplementary.

7. Point M is the midpoint of $\overline{FH}$.

8. $\angle JMF$ and $\angle HMG$ are vertical angles.

9. $\overleftrightarrow{FH} \perp \overleftrightarrow{JG}$

Writing Biconditional Statements

Core Concept

Biconditional Statement

When a conditional statement and its converse are both true, you can write them as a single *biconditional statement*. A **biconditional statement** is a statement that contains the phrase "if and only if."

Words p if and only if q **Symbols** $p \leftrightarrow q$

Any definition can be written as a biconditional statement.

EXAMPLE 5 Writing a Biconditional Statement

Rewrite the definition of perpendicular lines as a single biconditional statement.

Definition If two lines intersect to form a right angle, then they are perpendicular lines.

SOLUTION

Let p be "two lines intersect to form a right angle" and let q be "they are perpendicular lines."
Use red to identify p and blue to identify q.
Write the definition $p \rightarrow q$.

Definition If two lines intersect to form a right angle, then they are perpendicular lines.

Write the converse $q \rightarrow p$.

Converse If two lines are perpendicular lines, then they intersect to form a right angle.

s
t
$s \perp t$

Use the definition and its converse to write the biconditional statement $p \leftrightarrow q$.

▶ **Biconditional** Two lines intersect to form a right angle if and only if they are perpendicular lines.

Monitoring Progress

Help in English and Spanish at *BigIdeasMath.com*

10. Rewrite the definition of a right angle as a single biconditional statement.

Definition If an angle is a right angle, then its measure is 90°.

11. Rewrite the definition of congruent segments as a single biconditional statement.

Definition If two line segments have the same length, then they are congruent segments.

12. Rewrite the statements as a single biconditional statement.

If Mary is in theater class, then she will be in the fall play. If Mary is in the fall play, then she must be taking theater class.

13. Rewrite the statements as a single biconditional statement.

If you can run for President, then you are at least 35 years old. If you are at least 35 years old, then you can run for President.

Making Truth Tables

The **truth value** of a statement is either true (T) or false (F). You can determine the conditions under which a conditional statement is true by using a **truth table**. The truth table below shows the truth values for hypothesis p and conclusion q.

Conditional		
p	q	$p \rightarrow q$
T	T	T
T	F	F
F	T	T
F	F	T

The conditional statement $p \rightarrow q$ is only false when a true hypothesis produces a false conclusion.

Two statements are *logically equivalent* when they have the same truth table.

EXAMPLE 6 Making a Truth Table

Use the truth table above to make truth tables for the converse, inverse, and contrapositive of a conditional statement $p \rightarrow q$.

SOLUTION

The truth tables for the converse and the inverse are shown below. Notice that the converse and the inverse are logically equivalent because they have the same truth table.

Converse		
p	q	$q \rightarrow p$
T	T	T
T	F	T
F	T	F
F	F	T

Inverse				
p	q	$\sim p$	$\sim q$	$\sim p \rightarrow \sim q$
T	T	F	F	T
T	F	F	T	T
F	T	T	F	F
F	F	T	T	T

The truth table for the contrapositive is shown below. Notice that a conditional statement and its contrapositive are logically equivalent because they have the same truth table.

Contrapositive				
p	q	$\sim q$	$\sim p$	$\sim q \rightarrow \sim p$
T	T	F	F	T
T	F	T	F	F
F	T	F	T	T
F	F	T	T	T

Monitoring Progress  Help in English and Spanish at *BigIdeasMath.com*

14. Make a truth table for the conditional statement $p \rightarrow \sim q$.

15. Make a truth table for the conditional statement $\sim(p \rightarrow q)$.

2.1 Exercises

Dynamic Solutions available at *BigIdeasMath.com*

Vocabulary and Core Concept Check

1. **VOCABULARY** What type of statements are either both true or both false?

2. **WHICH ONE DOESN'T BELONG?** Which statement does *not* belong with the other three? Explain your reasoning.

If today is Tuesday, then tomorrow is Wednesday.	If it is Independence Day, then it is July.
If an angle is acute, then its measure is less than 90°.	If you are an athlete, then you play soccer.

Monitoring Progress and Modeling with Mathematics

In Exercises 3–6, copy the conditional statement. Underline the hypothesis and circle the conclusion.

3. If a polygon is a pentagon, then it has five sides.

4. If two lines form vertical angles, then they intersect.

5. If you run, then you are fast.

6. If you like math, then you like science.

In Exercises 7–12, rewrite the conditional statement in if-then form. *(See Example 1.)*

7. $9x + 5 = 23$, because $x = 2$.

8. Today is Friday, and tomorrow is the weekend.

9. You are in a band, and you play the drums.

10. Two right angles are supplementary angles.

11. Only people who are registered are allowed to vote.

12. The measures of complementary angles sum to 90°.

In Exercises 13–16, write the negation of the statement. *(See Example 2.)*

13. The sky is blue.

14. The lake is cold.

15. The ball is *not* pink.

16. The dog is *not* a Lab.

In Exercises 17–24, write the conditional statement $p \rightarrow q$, the converse $q \rightarrow p$, the inverse $\sim p \rightarrow \sim q$, and the contrapositive $\sim q \rightarrow \sim p$ in words. Then decide whether each statement is true or false. *(See Example 3.)*

17. Let p be "two angles are supplementary" and let q be "the measures of the angles sum to 180°."

18. Let p be "you are in math class" and let q be "you are in Geometry."

19. Let p be "you do your math homework" and let q be "you will do well on the test."

20. Let p be "you are not an only child" and let q be "you have a sibling."

21. Let p be "it does not snow" and let q be "I will run outside."

22. Let p be "the Sun is out" and let q be "it is daytime."

23. Let p be "$3x - 7 = 20$" and let q be "$x = 9$."

24. Let p be "it is Valentine's Day" and let q be "it is February."

In Exercises 25–28, decide whether the statement about the diagram is true. Explain your answer using the definitions you have learned. *(See Example 4.)*

25. $m\angle ABC = 90°$

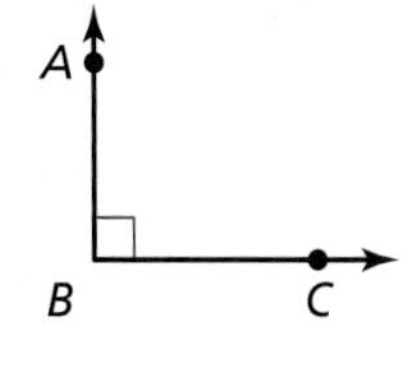

26. $\overleftrightarrow{PQ} \perp \overleftrightarrow{ST}$

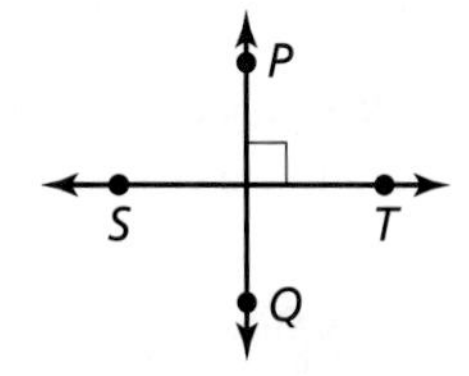

27. $m\angle 2 + m\angle 3 = 180°$

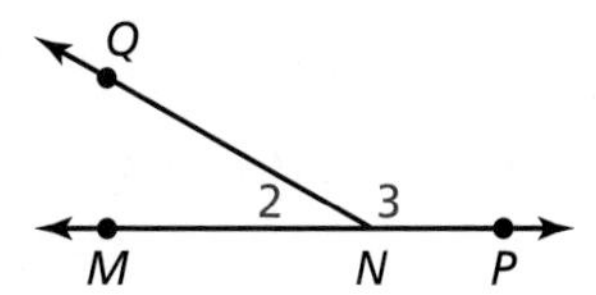

28. M is the midpoint of $\overline{AB}$.

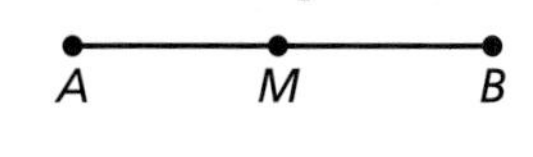

In Exercises 29–32, rewrite the definition of the term as a biconditional statement. *(See Example 5.)*

29. The *midpoint* of a segment is the point that divides the segment into two congruent segments.

30. Two angles are *vertical angles* when their sides form two pairs of opposite rays.

31. *Adjacent angles* are two angles that share a common vertex and side but have no common interior points.

32. Two angles are *supplementary angles* when the sum of their measures is 180°.

In Exercises 33–36, rewrite the statements as a single biconditional statement. *(See Example 5.)*

33. If a polygon has three sides, then it is a triangle.
    If a polygon is a triangle, then it has three sides.

34. If a polygon has four sides, then it is a quadrilateral.
    If a polygon is a quadrilateral, then it has four sides.

35. If an angle is a right angle, then it measures 90°.
    If an angle measures 90°, then it is a right angle.

36. If an angle is obtuse, then it has a measure between 90° and 180°.
    If an angle has a measure between 90° and 180°, then it is obtuse.

37. **ERROR ANALYSIS** Describe and correct the error in rewriting the conditional statement in if-then form.

Conditional statement
All high school students take four English courses.

If-then form
If a high school student takes four courses, then all four are English courses.

38. **ERROR ANALYSIS** Describe and correct the error in writing the converse of the conditional statement.

Conditional statement
If it is raining, then I will bring an umbrella.

Converse
If it is not raining, then I will not bring an umbrella.

In Exercises 39–44, create a truth table for the logical statement. *(See Example 6.)*

39. $\sim p \rightarrow q$

40. $\sim q \rightarrow p$

41. $\sim(\sim p \rightarrow \sim q)$

42. $\sim(p \rightarrow \sim q)$

43. $q \rightarrow \sim p$

44. $\sim(q \rightarrow p)$

45. **USING STRUCTURE** The statements below describe three ways that rocks are formed.

Igneous rock is formed from the cooling of molten rock.

Sedimentary rock is formed from pieces of other rocks.

Metamorphic rock is formed by changing temperature, pressure, or chemistry.

a. Write each statement in if-then form.

b. Write the converse of each of the statements in part (a). Is the converse of each statement true? Explain your reasoning.

c. Write a true if-then statement about rocks that is different from the ones in parts (a) and (b). Is the converse of your statement true or false? Explain your reasoning.

46. **MAKING AN ARGUMENT** Your friend claims the statement "If I bought a shirt, then I went to the mall" can be written as a true biconditional statement. Your sister says you cannot write it as a biconditional. Who is correct? Explain your reasoning.

47. **REASONING** You are told that the contrapositive of a statement is true. Will that help you determine whether the statement can be written as a true biconditional statement? Explain your reasoning.

48. PROBLEM SOLVING Use the conditional statement to identify the if-then statement as the converse, inverse, or contrapositive of the conditional statement. Then use the symbols to represent both statements.

Conditional statement
If I rode my bike to school, then I did not walk to school.

If-then statement
If I did not ride my bike to school, then I walked to school.

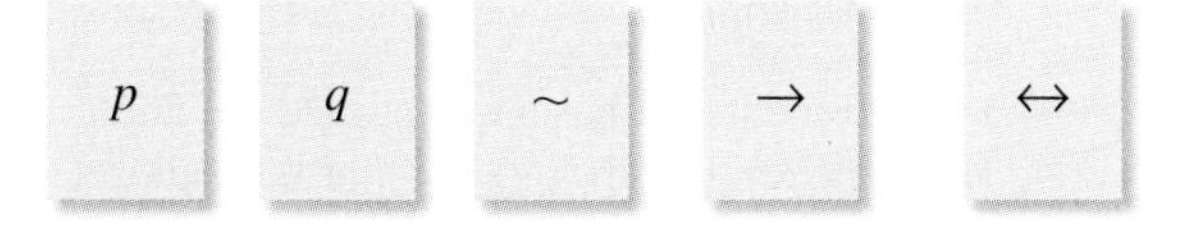

USING STRUCTURE In Exercises 49–52, rewrite the conditional statement in if-then form. Then underline the hypothesis and circle the conclusion.

49.

If you tell the truth, you don't have to remember anything.
Mark Twain

50.

You have to expect things of yourself before you can do them.
Michael Jordan

51.

If one is lucky, a solitary fantasy can totally transform one million realities.
Maya Angelou

52.

Whoever is happy will make others happy too.
Anne Frank

53. MATHEMATICAL CONNECTIONS Can the statement "If $x^2 - 10 = x + 2$, then $x = 4$" be combined with its converse to form a true biconditional statement?

54. CRITICAL THINKING The largest natural arch in the United States is Landscape Arch, located in Thompson, Utah. It spans 290 feet.

a. Use the information to write at least two true conditional statements.

b. Which type of related conditional statement must also be true? Write the related conditional statements.

c. What are the other two types of related conditional statements? Write the related conditional statements. Then determine their truth values. Explain your reasoning.

55. REASONING Which statement has the same meaning as the given statement?

Given statement
You can watch a movie after you do your homework.

Ⓐ If you do your homework, then you can watch a movie afterward.

Ⓑ If you do not do your homework, then you can watch a movie afterward.

Ⓒ If you cannot watch a movie afterward, then do your homework.

Ⓓ If you can watch a movie afterward, then do not do your homework.

56. THOUGHT PROVOKING Write three conditional statements, where one is always true, one is always false, and one depends on the person interpreting the statement.

57. CRITICAL THINKING One example of a conditional statement involving dates is "If today is August 31, then tomorrow is September 1." Write a conditional statement using dates from two different months so that the truth value depends on when the statement is read.

58. HOW DO YOU SEE IT? The Venn diagram represents all the musicians at a high school. Write three conditional statements in if-then form describing the relationships between the various groups of musicians.

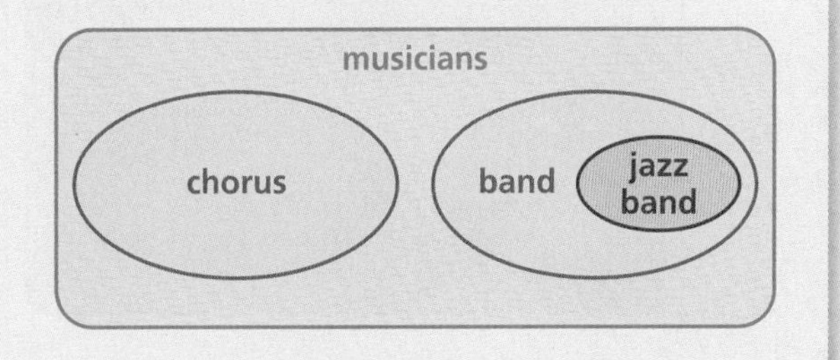

59. MULTIPLE REPRESENTATIONS Create a Venn diagram representing each conditional statement. Write the converse of each conditional statement. Then determine whether each conditional statement and its converse are true or false. Explain your reasoning.

a. If you go to the zoo to see a lion, then you will see a cat.

b. If you play a sport, then you wear a helmet.

c. If this month has 31 days, then it is not February.

60. DRAWING CONCLUSIONS You measure the heights of your classmates to get a data set.

a. Tell whether this statement is true: If x and y are the least and greatest values in your data set, then the mean of the data is between x and y.

b. Write the converse of the statement in part (a). Is the converse true? Explain your reasoning.

c. Copy and complete the statement below using *mean*, *median*, or *mode* to make a conditional statement that is true for any data set. Explain your reasoning.

If a data set has a mean, median, and a mode, then the _______ of the data set will always be a data value.

61. WRITING Write a conditional statement that is true, but its converse is false.

62. CRITICAL THINKING Write a series of if-then statements that allow you to find the measure of each angle, given that $m\angle 1 = 90°$. Use the definition of linear pairs.

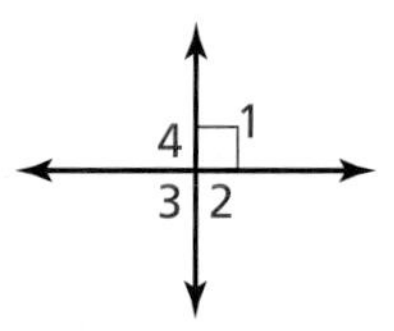

63. WRITING Advertising slogans such as "Buy these shoes! They will make you a better athlete!" often imply conditional statements. Find an advertisement or write your own slogan. Then write it as a conditional statement.

Maintaining Mathematical Proficiency

Reviewing what you learned in previous grades and lessons

Find the pattern. Then draw the next two figures in the sequence. *(Skills Review Handbook)*

64.

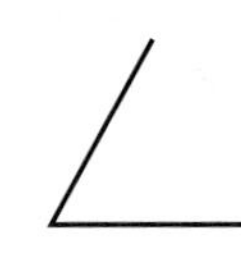
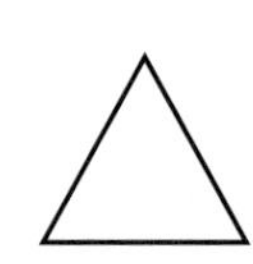
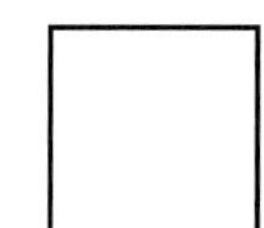

65.

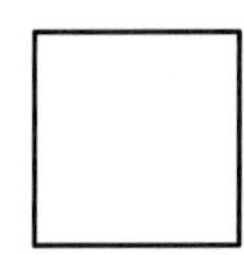
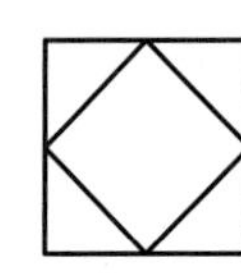
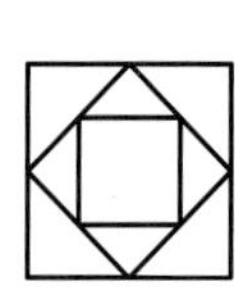

Find the pattern. Then write the next two numbers. *(Skills Review Handbook)*

66. 1, 3, 5, 7, . . .

67. 12, 23, 34, 45, . . .

68. $2, \frac{4}{3}, \frac{8}{9}, \frac{16}{27}, \ldots$

69. 1, 4, 9, 16, . . .

2.2 Inductive and Deductive Reasoning

Essential Question

How can you use reasoning to solve problems?

A **conjecture** is an unproven statement based on observations.

EXPLORATION 1 Writing a Conjecture

Work with a partner. Write a conjecture about the pattern. Then use your conjecture to draw the 10th object in the pattern.

1 2 3 4 5 6 7

a.

b.

c.

CONSTRUCTING VIABLE ARGUMENTS

To be proficient in math, you need to justify your conclusions and communicate them to others.

EXPLORATION 2 Using a Venn Diagram

Work with a partner. Use the Venn diagram to determine whether the statement is true or false. Justify your answer. Assume that no region of the Venn diagram is empty.

Property A

Property C

Property B

a. If an item has Property B, then it has Property A.

b. If an item has Property A, then it has Property B.

c. If an item has Property A, then it has Property C.

d. Some items that have Property A do not have Property B.

e. If an item has Property C, then it does not have Property B.

f. Some items have both Properties A and C.

g. Some items have both Properties B and C.

EXPLORATION 3 Reasoning and Venn Diagrams

Work with a partner. Draw a Venn diagram that shows the relationship between different types of quadrilaterals: squares, rectangles, parallelograms, trapezoids, rhombuses, and kites. Then write several conditional statements that are shown in your diagram, such as "If a quadrilateral is a square, then it is a rectangle."

Communicate Your Answer

4. How can you use reasoning to solve problems?

5. Give an example of how you used reasoning to solve a real-life problem.

2.2 Lesson

Core Vocabulary

conjecture, *p. 76*
inductive reasoning, *p. 76*
counterexample, *p. 77*
deductive reasoning, *p. 78*

What You Will Learn

- Use inductive reasoning.
- Use deductive reasoning.

Using Inductive Reasoning

Core Concept

Inductive Reasoning

A **conjecture** is an unproven statement that is based on observations. You use **inductive reasoning** when you find a pattern in specific cases and then write a conjecture for the general case.

EXAMPLE 1 Describing a Visual Pattern

Describe how to sketch the fourth figure in the pattern. Then sketch the fourth figure.

Figure 1

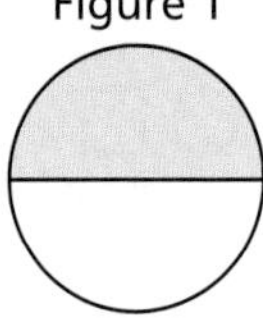

Figure 2

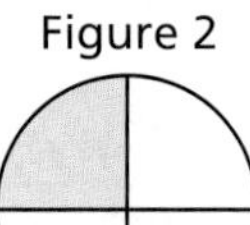

Figure 3

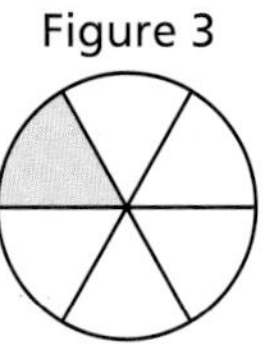

SOLUTION

Each circle is divided into twice as many equal regions as the figure number. Sketch the fourth figure by dividing a circle into eighths. Shade the section just above the horizontal segment at the left.

Figure 4

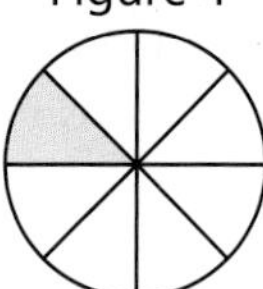

Monitoring Progress

Help in English and Spanish at *BigIdeasMath.com*

1. Sketch the fifth figure in the pattern in Example 1.

Sketch the next figure in the pattern.

2.

3.

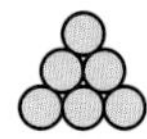

EXAMPLE 2 Making and Testing a Conjecture

Numbers such as 3, 4, and 5 are called *consecutive integers*. Make and test a conjecture about the sum of any three consecutive integers.

SOLUTION

Step 1 Find a pattern using a few groups of small numbers.

$3 + 4 + 5 = 12 = 4 \cdot 3$ $\quad$ $7 + 8 + 9 = 24 = 8 \cdot 3$

$10 + 11 + 12 = 33 = 11 \cdot 3$ $\quad$ $16 + 17 + 18 = 51 = 17 \cdot 3$

Step 2 Make a conjecture.

Conjecture The sum of any three consecutive integers is three times the second number.

Step 3 Test your conjecture using other numbers. For example, test that it works with the groups $-1, 0, 1$ and 100, 101, 102.

$-1 + 0 + 1 = 0 = 0 \cdot 3$ ✓

$100 + 101 + 102 = 303 = 101 \cdot 3$ ✓

Core Concept

Counterexample

To show that a conjecture is true, you must show that it is true for all cases. You can show that a conjecture is false, however, by finding just one *counterexample*. A **counterexample** is a specific case for which the conjecture is false.

EXAMPLE 3 Finding a Counterexample

A student makes the following conjecture about the sum of two numbers. Find a counterexample to disprove the student's conjecture.

Conjecture The sum of two numbers is always more than the greater number.

SOLUTION

To find a counterexample, you need to find a sum that is less than the greater number.

$-2 + (-3) = -5$

$-5 \not> -2$

▶ Because a counterexample exists, the conjecture is false.

Monitoring Progress

Help in English and Spanish at *BigIdeasMath.com*

4. Make and test a conjecture about the sign of the product of any three negative integers.

5. Make and test a conjecture about the sum of any five consecutive integers.

Find a counterexample to show that the conjecture is false.

6. The value of x^2 is always greater than the value of x.

7. The sum of two numbers is always greater than their difference.

Using Deductive Reasoning

Deductive Reasoning

Deductive reasoning uses facts, definitions, accepted properties, and the laws of logic to form a logical argument. This is different from *inductive reasoning*, which uses specific examples and patterns to form a conjecture.

Laws of Logic

Law of Detachment

If the hypothesis of a true conditional statement is true, then the conclusion is also true.

Law of Syllogism

If hypothesis p, then conclusion q. } If these statements are true,

If hypothesis q, then conclusion r.

If hypothesis p, then conclusion r. ← then this statement is true.

EXAMPLE 4 Using the Law of Detachment

If two segments have the same length, then they are congruent. You know that $BC = XY$. Using the Law of Detachment, what statement can you make?

SOLUTION

Because $BC = XY$ satisfies the hypothesis of a true conditional statement, the conclusion is also true.

▶ So, $\overline{BC} \cong \overline{XY}$.

EXAMPLE 5 Using the Law of Syllogism

If possible, use the Law of Syllogism to write a new conditional statement that follows from the pair of true statements.

a. If $x^2 > 25$, then $x^2 > 20$.
If $x > 5$, then $x^2 > 25$.

b. If a polygon is regular, then all angles in the interior of the polygon are congruent.
If a polygon is regular, then all its sides are congruent.

SOLUTION

a. Notice that the conclusion of the second statement is the hypothesis of the first statement. The order in which the statements are given does not affect whether you can use the Law of Syllogism. So, you can write the following new statement.

▶ If $x > 5$, then $x^2 > 20$.

b. Neither statement's conclusion is the same as the other statement's hypothesis.

▶ You cannot use the Law of Syllogism to write a new conditional statement.

EXAMPLE 6 Using Inductive and Deductive Reasoning

What conclusion can you make about the product of an even integer and any other integer?

SOLUTION

> **MAKING SENSE OF PROBLEMS**
> In geometry, you will frequently use inductive reasoning to make conjectures. You will also use deductive reasoning to show that conjectures are true or false. You will need to know which type of reasoning to use.

Step 1 Look for a pattern in several examples. Use inductive reasoning to make a conjecture.

$(-2)(2) = -4$ $\quad (-1)(2) = -2$ $\quad 2(2) = 4$ $\quad 3(2) = 6$

$(-2)(-4) = 8$ $\quad (-1)(-4) = 4$ $\quad 2(-4) = -8$ $\quad 3(-4) = -12$

Conjecture Even integer • Any integer = Even integer

Step 2 Let n and m each be any integer. Use deductive reasoning to show that the conjecture is true.

$2n$ is an even integer because any integer multiplied by 2 is even.

$2nm$ represents the product of an even integer $2n$ and any integer m.

$2nm$ is the product of 2 and an integer nm. So, $2nm$ is an even integer.

▶ The product of an even integer and any integer is an even integer.

EXAMPLE 7 Comparing Inductive and Deductive Reasoning

Decide whether inductive reasoning or deductive reasoning is used to reach the conclusion. Explain your reasoning.

a. Each time Monica kicks a ball up in the air, it returns to the ground. So, the next time Monica kicks a ball up in the air, it will return to the ground.

b. All reptiles are cold-blooded. Parrots are not cold-blooded. Sue's pet parrot is not a reptile.

SOLUTION

a. Inductive reasoning, because a pattern is used to reach the conclusion.

b. Deductive reasoning, because facts about animals and the laws of logic are used to reach the conclusion.

Monitoring Progress

Help in English and Spanish at *BigIdeasMath.com*

8. If $90° < m\angle R < 180°$, then $\angle R$ is obtuse. The measure of $\angle R$ is 155°. Using the Law of Detachment, what statement can you make?

9. Use the Law of Syllogism to write a new conditional statement that follows from the pair of true statements.

If you get an A on your math test, then you can go to the movies.
If you go to the movies, then you can watch your favorite actor.

10. Use inductive reasoning to make a conjecture about the sum of a number and itself. Then use deductive reasoning to show that the conjecture is true.

11. Decide whether inductive reasoning or deductive reasoning is used to reach the conclusion. Explain your reasoning.

All multiples of 8 are divisible by 4.
64 is a multiple of 8.
So, 64 is divisible by 4.

2.2 Exercises

Dynamic Solutions available at *BigIdeasMath.com*

Vocabulary and Core Concept Check

1. **VOCABULARY** How does the prefix "counter-" help you understand the term counterexample?

2. **WRITING** Explain the difference between inductive reasoning and deductive reasoning.

Monitoring Progress and Modeling with Mathematics

In Exercises 3–8, describe the pattern. Then write or draw the next two numbers, letters, or figures. *(See Example 1.)*

3. $1, -2, 3, -4, 5, \ldots$

4. $0, 2, 6, 12, 20, \ldots$

5. Z, Y, X, W, V, . . .

6. J, F, M, A, M, . . .

7. 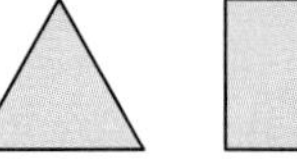  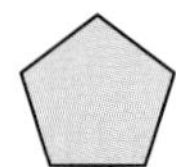

8. 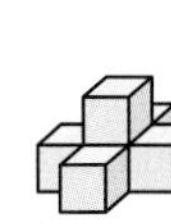 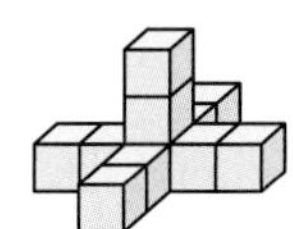

In Exercises 9–12, make and test a conjecture about the given quantity. *(See Example 2.)*

9. the product of any two even integers

10. the sum of an even integer and an odd integer

11. the quotient of a number and its reciprocal

12. the quotient of two negative integers

In Exercises 13–16, find a counterexample to show that the conjecture is false. *(See Example 3.)*

13. The product of two positive numbers is always greater than either number.

14. If n is a nonzero integer, then $\frac{n+1}{n}$ is always greater than 1.

15. If two angles are supplements of each other, then one of the angles must be acute.

16. A line s divides $\overline{MN}$ into two line segments. So, the line s is a segment bisector of $\overline{MN}$.

In Exercises 17–20, use the Law of Detachment to determine what you can conclude from the given information, if possible. *(See Example 4.)*

17. If you pass the final, then you pass the class. You passed the final.

18. If your parents let you borrow the car, then you will go to the movies with your friend. You will go to the movies with your friend.

19. If a quadrilateral is a square, then it has four right angles. Quadrilateral $QRST$ has four right angles.

20. If a point divides a line segment into two congruent line segments, then the point is a midpoint. Point P divides $\overline{LH}$ into two congruent line segments.

In Exercises 21–24, use the Law of Syllogism to write a new conditional statement that follows from the pair of true statements, if possible. *(See Example 5.)*

21. If $x < -2$, then $|x| > 2$. If $x > 2$, then $|x| > 2$.

22. If $a = 3$, then $5a = 15$. If $\frac{1}{2}a = 1\frac{1}{2}$, then $a = 3$.

23. If a figure is a rhombus, then the figure is a parallelogram. If a figure is a parallelogram, then the figure has two pairs of opposite sides that are parallel.

24. If a figure is a square, then the figure has four congruent sides. If a figure is a square, then the figure has four right angles.

In Exercises 25–28, state the law of logic that is illustrated.

25. If you do your homework, then you can watch TV. If you watch TV, then you can watch your favorite show.

    If you do your homework, then you can watch your favorite show.

26. If you miss practice the day before a game, then you will not be a starting player in the game.

You miss practice on Tuesday. You will not start the game Wednesday.

27. If $x > 12$, then $x + 9 > 20$. The value of x is 14.

So, $x + 9 > 20$.

28. If $\angle 1$ and $\angle 2$ are vertical angles, then $\angle 1 \cong \angle 2$.
If $\angle 1 \cong \angle 2$, then $m\angle 1 = m\angle 2$.

If $\angle 1$ and $\angle 2$ are vertical angles, then $m\angle 1 = m\angle 2$.

In Exercises 29 and 30, use inductive reasoning to make a conjecture about the given quantity. Then use deductive reasoning to show that the conjecture is true. *(See Example 6.)*

29. the sum of two odd integers

30. the product of two odd integers

In Exercises 31–34, decide whether inductive reasoning or deductive reasoning is used to reach the conclusion. Explain your reasoning. *(See Example 7.)*

31. Each time your mom goes to the store, she buys milk. So, the next time your mom goes to the store, she will buy milk.

32. Rational numbers can be written as fractions. Irrational numbers cannot be written as fractions. So, $\frac{1}{2}$ is a rational number.

33. All men are mortal. Mozart is a man, so Mozart is mortal.

34. Each time you clean your room, you are allowed to go out with your friends. So, the next time you clean your room, you will be allowed to go out with your friends.

ERROR ANALYSIS In Exercises 35 and 36, describe and correct the error in interpreting the statement.

35. If a figure is a rectangle, then the figure has four sides. A trapezoid has four sides.

Using the Law of Detachment, you can conclude that a trapezoid is a rectangle.

36. Each day, you get to school before your friend.

Using deductive reasoning, you can conclude that you will arrive at school before your friend tomorrow.

37. REASONING The table shows the average weights of several subspecies of tigers. What conjecture can you make about the relation between the weights of female tigers and the weights of male tigers? Explain your reasoning.

	Weight of female (pounds)	Weight of male (pounds)
Amur	370	660
Bengal	300	480
South China	240	330
Sumatran	200	270
Indo-Chinese	250	400

38. HOW DO YOU SEE IT? Determine whether you can make each conjecture from the graph. Explain your reasoning.

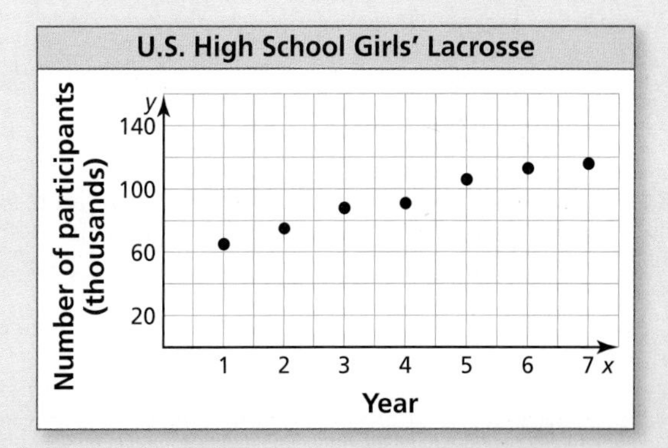

a. More girls will participate in high school lacrosse in Year 8 than those who participated in Year 7.

b. The number of girls participating in high school lacrosse will exceed the number of boys participating in high school lacrosse in Year 9.

39. MATHEMATICAL CONNECTIONS Use inductive reasoning to write a formula for the sum of the first n positive even integers.

40. FINDING A PATTERN The following are the first nine *Fibonacci numbers*.

$$1, 1, 2, 3, 5, 8, 13, 21, 34, \ldots$$

a. Make a conjecture about each of the Fibonacci numbers after the first two.

b. Write the next three numbers in the pattern.

c. Research to find a real-world example of this pattern.

41. **MAKING AN ARGUMENT** Which argument is correct? Explain your reasoning.

Argument 1: If two angles measure 30° and 60°, then the angles are complementary. $\angle 1$ and $\angle 2$ are complementary. So, $m\angle 1 = 30°$ and $m\angle 2 = 60°$.

Argument 2: If two angles measure 30° and 60°, then the angles are complementary. The measure of $\angle 1$ is 30° and the measure of $\angle 2$ is 60°. So, $\angle 1$ and $\angle 2$ are complementary.

42. **THOUGHT PROVOKING** The first two terms of a sequence are $\frac{1}{4}$ and $\frac{1}{2}$. Describe three different possible patterns for the sequence. List the first five terms for each sequence.

43. **MATHEMATICAL CONNECTIONS** Use the table to make a conjecture about the relationship between x and y. Then write an equation for y in terms of x. Use the equation to test your conjecture for other values of x.

x	0	1	2	3	4
y	2	5	8	11	14

44. **REASONING** Use the pattern below. Each figure is made of squares that are 1 unit by 1 unit.

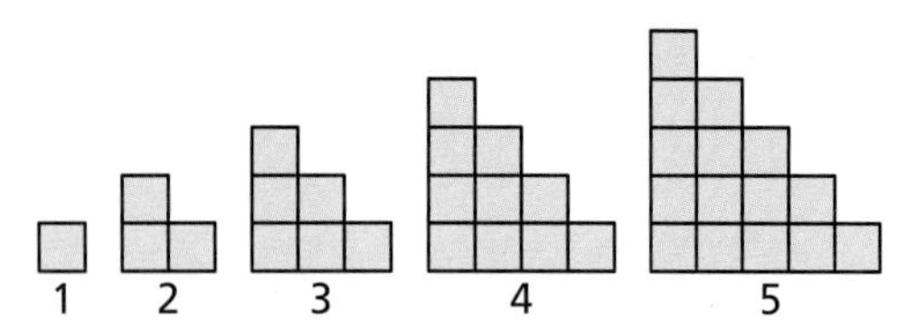

a. Find the perimeter of each figure. Describe the pattern of the perimeters.

b. Predict the perimeter of the 20th figure.

45. **DRAWING CONCLUSIONS** Decide whether each conclusion is valid. Explain your reasoning.

- Yellowstone is a national park in Wyoming.
- You and your friend went camping at Yellowstone National Park.
- When you go camping, you go canoeing.
- If you go on a hike, your friend goes with you.
- You go on a hike.
- There is a 3-mile-long trail near your campsite.

a. You went camping in Wyoming.

b. Your friend went canoeing.

c. Your friend went on a hike.

d. You and your friend went on a hike on a 3-mile-long trail.

46. **CRITICAL THINKING** Geologists use the Mohs' scale to determine a mineral's hardness. Using the scale, a mineral with a higher rating will leave a scratch on a mineral with a lower rating. Testing a mineral's hardness can help identify the mineral.

Mineral	Talc	Gypsum	Calcite	Fluorite
Mohs' rating	1	2	3	4

a. The four minerals are randomly labeled A, B, C, and D. Mineral A is scratched by Mineral B. Mineral C is scratched by all three of the other minerals. What can you conclude? Explain your reasoning.

b. What additional test(s) can you use to identify *all* the minerals in part (a)?

Maintaining Mathematical Proficiency

Reviewing what you learned in previous grades and lessons

Determine which postulate is illustrated by the statement. *(Section 1.2 and Section 1.5)*

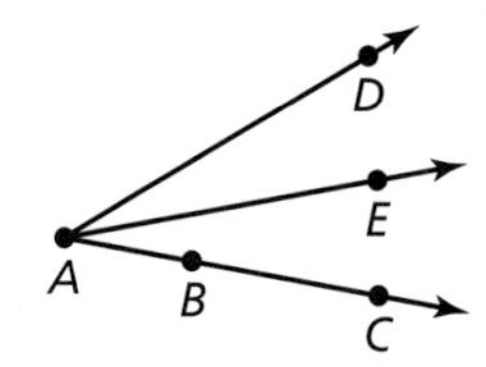

47. $AB + BC = AC$

48. $m\angle DAC = m\angle DAE + m\angle EAB$

49. AD is the absolute value of the difference of the coordinates of A and D.

50. $m\angle DAC$ is equal to the absolute value of the difference between the real numbers matched with $\overrightarrow{AD}$ and $\overrightarrow{AC}$ on a protractor.

2.3 Postulates and Diagrams

Essential Question

In a diagram, what can be assumed and what needs to be labeled?

EXPLORATION 1 Looking at a Diagram

Work with a partner. On a piece of paper, draw two perpendicular lines. Label them $\overleftrightarrow{AB}$ and $\overleftrightarrow{CD}$. Look at the diagram from different angles. Do the lines appear perpendicular regardless of the angle at which you look at them? Describe *all* the angles at which you can look at the lines and have them appear perpendicular.

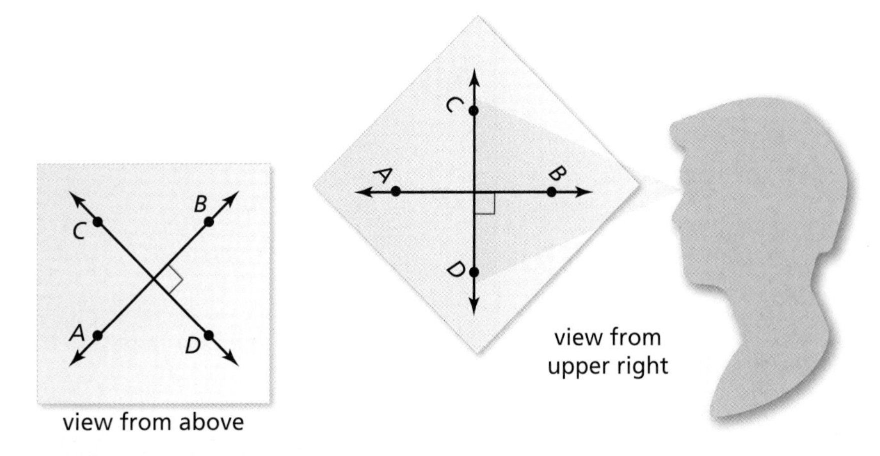

ATTENDING TO PRECISION

To be proficient in math, you need to state the meanings of the symbols you choose.

EXPLORATION 2 Interpreting a Diagram

Work with a partner. When you draw a diagram, you are communicating with others. It is important that you include sufficient information in the diagram. Use the diagram to determine which of the following statements you can assume to be true. Explain your reasoning.

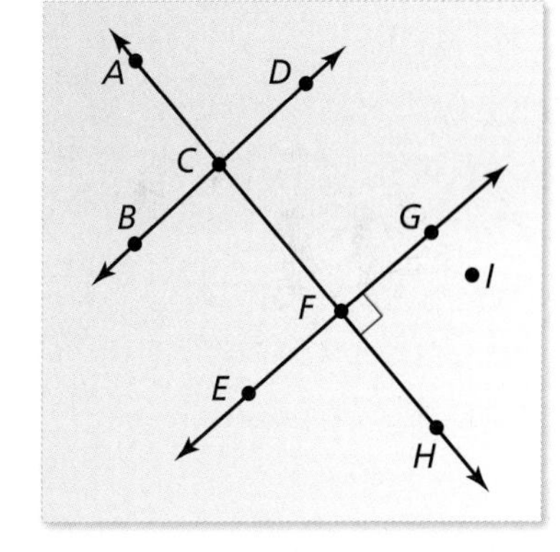

a. All the points shown are coplanar.

b. Points D, G, and I are collinear.

c. Points A, C, and H are collinear.

d. $\overleftrightarrow{EG}$ and $\overleftrightarrow{AH}$ are perpendicular.

e. $\angle BCA$ and $\angle ACD$ are a linear pair.

f. $\overleftrightarrow{AF}$ and $\overleftrightarrow{BD}$ are perpendicular.

g. $\overleftrightarrow{EG}$ and $\overleftrightarrow{BD}$ are parallel.

h. $\overleftrightarrow{AF}$ and $\overleftrightarrow{BD}$ are coplanar.

i. $\overleftrightarrow{EG}$ and $\overleftrightarrow{BD}$ do not intersect.

j. $\overleftrightarrow{AF}$ and $\overleftrightarrow{BD}$ intersect.

k. $\overleftrightarrow{EG}$ and $\overleftrightarrow{BD}$ are perpendicular.

l. $\angle ACD$ and $\angle BCF$ are vertical angles.

m. $\overleftrightarrow{AC}$ and $\overleftrightarrow{FH}$ are the same line.

Communicate Your Answer

3. In a diagram, what can be assumed and what needs to be labeled?

4. Use the diagram in Exploration 2 to write two statements you can assume to be true and two statements you cannot assume to be true. Your statements should be different from those given in Exploration 2. Explain your reasoning.

2.3 Lesson

Core Vocabulary

line perpendicular to a plane, *p. 86*

Previous
postulate
point
line
plane

What You Will Learn

- Identify postulates using diagrams.
- Sketch and interpret diagrams.

Identifying Postulates

Here are seven more postulates involving points, lines, and planes.

Postulates

Point, Line, and Plane Postulates

Postulate | **Example**

2.1 Two Point Postulate

Through any two points, there exists exactly one line.

2.2 Line-Point Postulate

A line contains at least two points.

Through points A and B, there is exactly one line ℓ. Line ℓ contains at least two points.

2.3 Line Intersection Postulate

If two lines intersect, then their intersection is exactly one point.

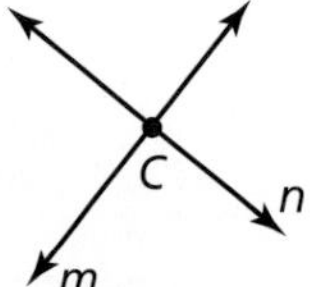

The intersection of line m and line n is point C.

2.4 Three Point Postulate

Through any three noncollinear points, there exists exactly one plane.

2.5 Plane-Point Postulate

A plane contains at least three noncollinear points.

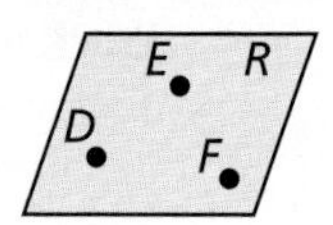

Through points D, E, and F, there is exactly one plane, plane R. Plane R contains at least three noncollinear points.

2.6 Plane-Line Postulate

If two points lie in a plane, then the line containing them lies in the plane.

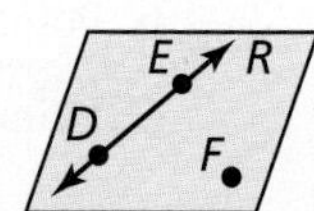

Points D and E lie in plane R, so $\overleftrightarrow{DE}$ lies in plane R.

2.7 Plane Intersection Postulate

If two planes intersect, then their intersection is a line.

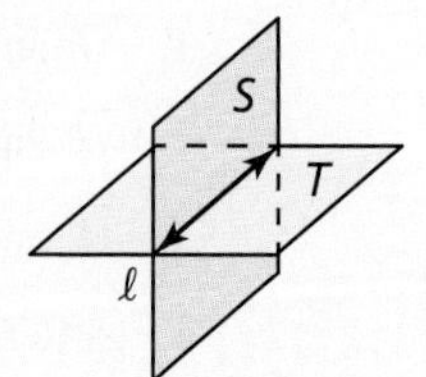

The intersection of plane S and plane T is line ℓ.

EXAMPLE 1 Identifying a Postulate Using a Diagram

State the postulate illustrated by the diagram.

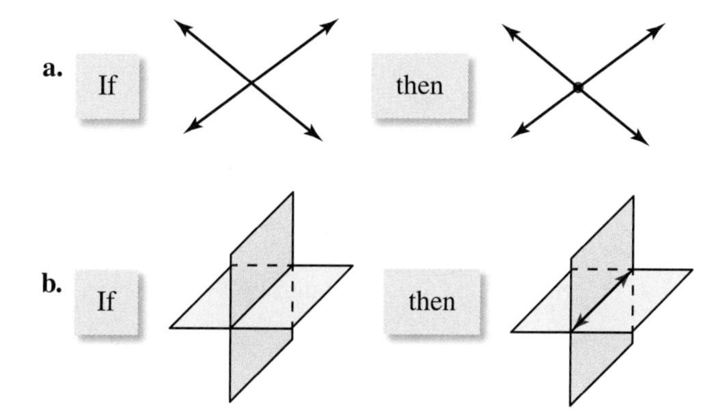

SOLUTION

a. Line Intersection Postulate If two lines intersect, then their intersection is exactly one point.

b. Plane Intersection Postulate If two planes intersect, then their intersection is a line.

EXAMPLE 2 Identifying Postulates from a Diagram

Use the diagram to write examples of the Plane-Point Postulate and the Plane-Line Postulate.

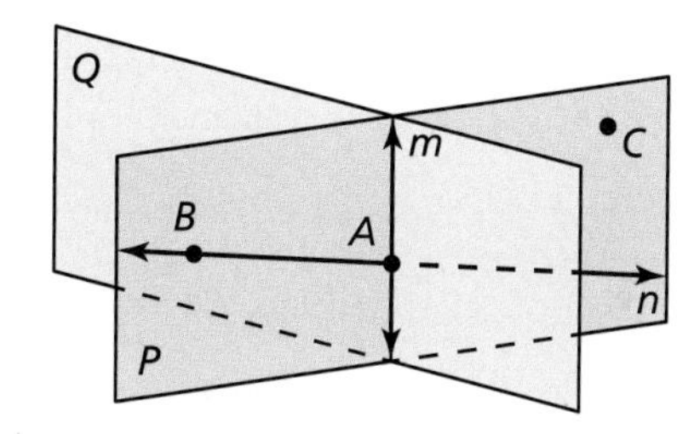

SOLUTION

Plane-Point Postulate Plane P contains at least three noncollinear points, A, B, and C.

Plane-Line Postulate Point A and point B lie in plane P. So, line n containing points A and B also lies in plane P.

Monitoring Progress

Help in English and Spanish at *BigIdeasMath.com*

1. Use the diagram in Example 2. Which postulate allows you to say that the intersection of plane P and plane Q is a line?
2. Use the diagram in Example 2 to write an example of the postulate.
   - **a.** Two Point Postulate
   - **b.** Line-Point Postulate
   - **c.** Line Intersection Postulate

Sketching and Interpreting Diagrams

EXAMPLE 3 Sketching a Diagram

Sketch a diagram showing $\overleftrightarrow{TV}$ intersecting $\overline{PQ}$ at point W, so that $\overline{TW} \cong \overline{WV}$.

SOLUTION

Step 1 Draw $\overleftrightarrow{TV}$ and label points T and V.

Step 2 Draw point W at the midpoint of $\overline{TV}$. Mark the congruent segments.

Step 3 Draw $\overline{PQ}$ through W.

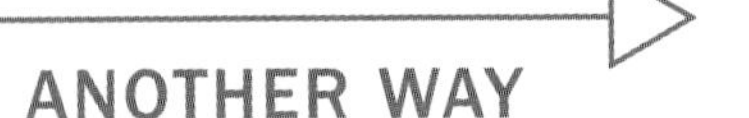

ANOTHER WAY

In Example 3, there are many ways you can sketch the diagram. Another way is shown below.

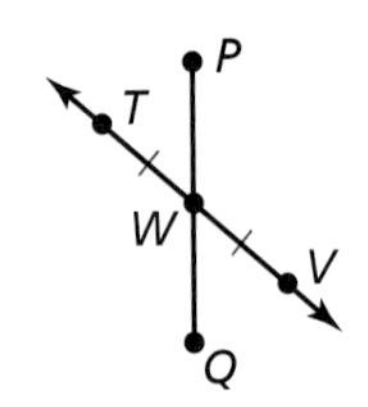

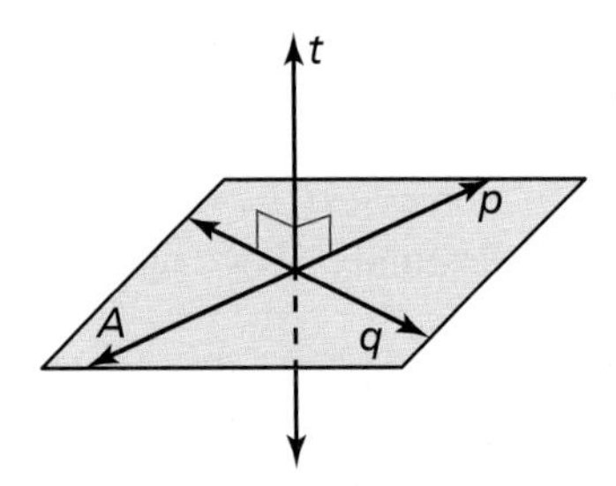

A line is a **line perpendicular to a plane** if and only if the line intersects the plane in a point and is perpendicular to every line in the plane that intersects it at that point.

In a diagram, a line perpendicular to a plane must be marked with a right angle symbol, as shown.

EXAMPLE 4 Interpreting a Diagram

Which of the following statements *cannot* be assumed from the diagram?

Points A, B, and F are collinear.

Points E, B, and D are collinear.

$\overleftrightarrow{AB} \perp$ plane S

$\overleftrightarrow{CD} \perp$ plane T

$\overleftrightarrow{AF}$ intersects $\overleftrightarrow{BC}$ at point B.

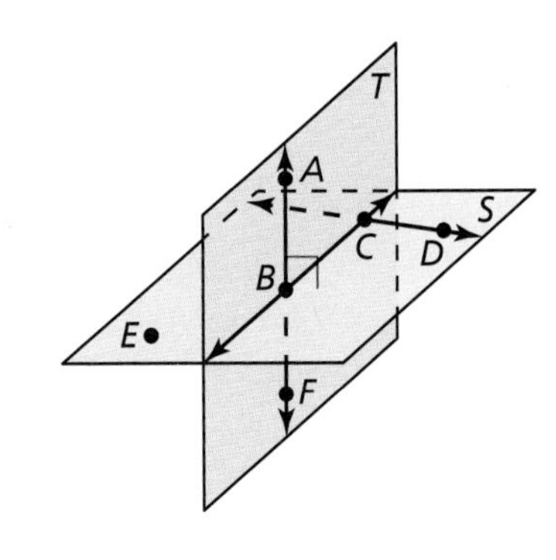

SOLUTION

No drawn line connects points E, B, and D. So, you cannot assume they are collinear. With no right angle marked, you cannot assume $\overleftrightarrow{CD} \perp$ plane T.

Monitoring Progress  Help in English and Spanish at *BigIdeasMath.com*

Refer back to Example 3.

3. If the given information states that $\overline{PW}$ and $\overline{QW}$ are congruent, how can you indicate that in the diagram?

4. Name a pair of supplementary angles in the diagram. Explain.

Use the diagram in Example 4.

5. Can you assume that plane S intersects plane T at $\overleftrightarrow{BC}$?

6. Explain how you know that $\overleftrightarrow{AB} \perp \overleftrightarrow{BC}$.

2.3 Exercises

Dynamic Solutions available at *BigIdeasMath.com*

Vocabulary and Core Concept Check

1. **COMPLETE THE SENTENCE** Through any __________ noncollinear points, there exists exactly one plane.

2. **WRITING** Explain why you need at least three noncollinear points to determine a plane.

Monitoring Progress and Modeling with Mathematics

In Exercises 3 and 4, state the postulate illustrated by the diagram. *(See Example 1.)*

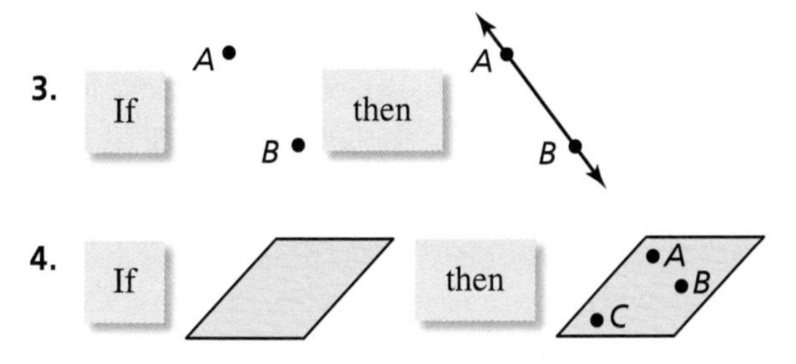

In Exercises 5–8, use the diagram to write an example of the postulate. *(See Example 2.)*

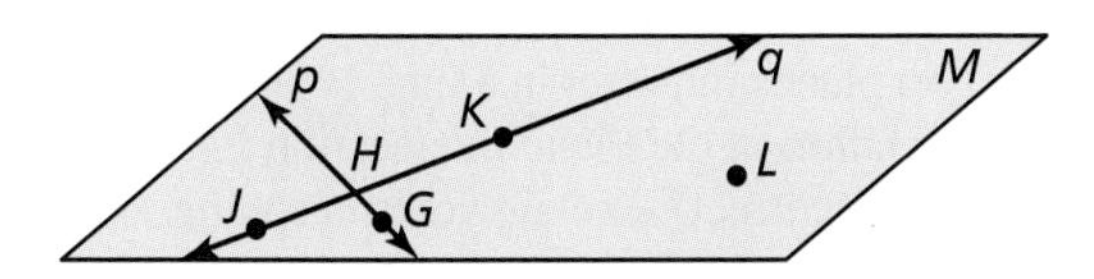

5. Line-Point Postulate (Postulate 2.2)

6. Line Intersection Postulate (Postulate 2.3)

7. Three Point Postulate (Postulate 2.4)

8. Plane-Line Postulate (Postulate 2.6)

In Exercises 9–12, sketch a diagram of the description. *(See Example 3.)*

9. plane P and line m intersecting plane P at a 90° angle

10. $\overline{XY}$ in plane P, $\overline{XY}$ bisected by point A, and point C not on $\overline{XY}$

11. $\overline{XY}$ intersecting $\overline{WV}$ at point A, so that $XA = VA$

12. $\overline{AB}$, $\overline{CD}$, and $\overline{EF}$ are all in plane P, and point X is the midpoint of all three segments.

In Exercises 13–20, use the diagram to determine whether you can assume the statement. *(See Example 4.)*

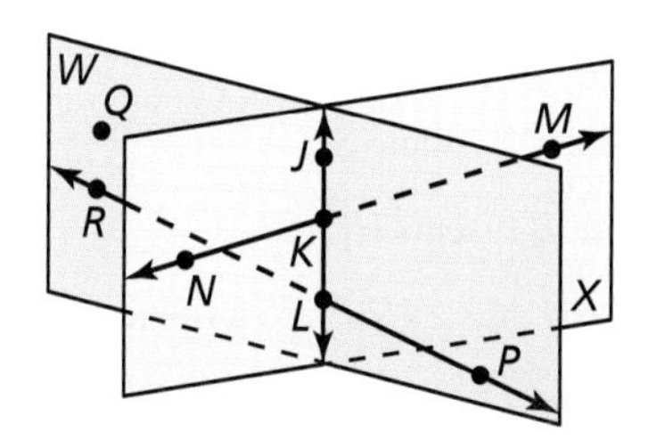

13. Planes W and X intersect at $\overleftrightarrow{KL}$.

14. Points K, L, M, and N are coplanar.

15. Points Q, J, and M are collinear.

16. $\overleftrightarrow{MN}$ and $\overleftrightarrow{RP}$ intersect.

17. $\overleftrightarrow{JK}$ lies in plane X.

18. $\angle PLK$ is a right angle.

19. $\angle NKL$ and $\angle JKM$ are vertical angles.

20. $\angle NKJ$ and $\angle JKM$ are supplementary angles.

ERROR ANALYSIS In Exercises 21 and 22, describe and correct the error in the statement made about the diagram.

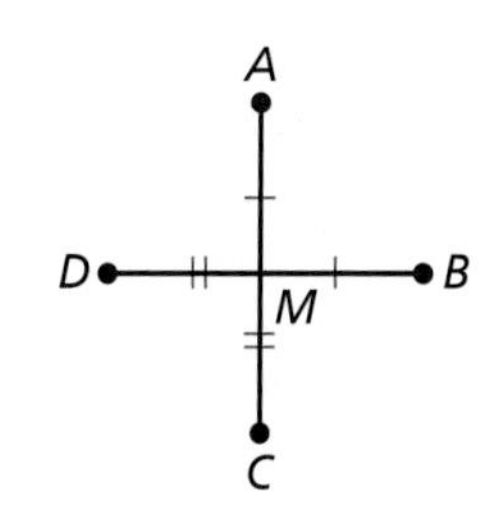

21.

M is the midpoint of $\overline{AC}$ and $\overline{BD}$.

22.

$\overline{AC}$ intersects $\overline{BD}$ at a 90° angle, so $\overline{AC} \perp \overline{BD}$.

23. ATTENDING TO PRECISION Select all the statements about the diagram that you *cannot* conclude.

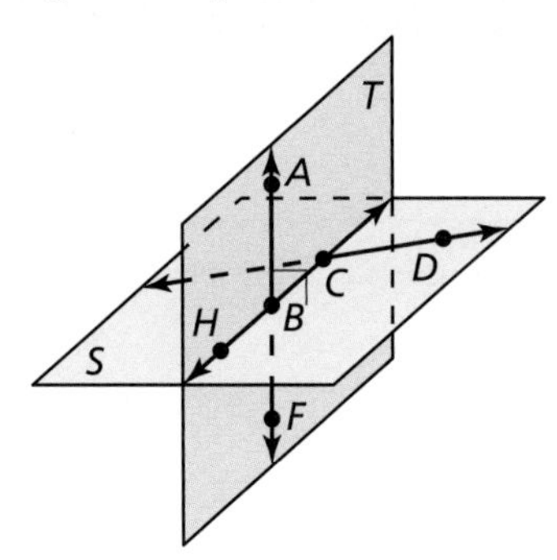

Ⓐ A, B, and C are coplanar.

Ⓑ Plane T intersects plane S in $\overleftrightarrow{BC}$.

Ⓒ $\overleftrightarrow{AB}$ intersects $\overleftrightarrow{CD}$.

Ⓓ H, F, and D are coplanar.

Ⓔ Plane $T \perp$ plane S.

Ⓕ Point B bisects $\overline{HC}$.

Ⓖ $\angle ABH$ and $\angle HBF$ are a linear pair.

Ⓗ $\overleftrightarrow{AF} \perp \overleftrightarrow{CD}$.

24. HOW DO YOU SEE IT? Use the diagram of line m and point C. Make a conjecture about how many planes can be drawn so that line m and point C lie in the same plane. Use postulates to justify your conjecture.

25. MATHEMATICAL CONNECTIONS One way to graph a linear equation is to plot two points whose coordinates satisfy the equation and then connect them with a line. Which postulate guarantees this process works for any linear equation?

26. MATHEMATICAL CONNECTIONS A way to solve a system of two linear equations that intersect is to graph the lines and find the coordinates of their intersection. Which postulate guarantees this process works for any two linear equations?

In Exercises 27 and 28, (a) rewrite the postulate in if-then form. Then (b) write the converse, inverse, and contrapositive and state which ones are true.

27. Two Point Postulate (Postulate 2.1)

28. Plane-Point Postulate (Postulate 2.5)

29. REASONING Choose the correct symbol to go between the statements.

number of points to determine a line ▢ number of points to determine a plane

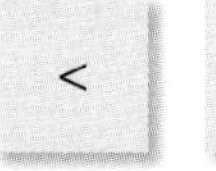

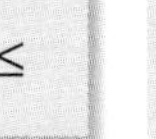

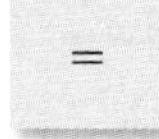

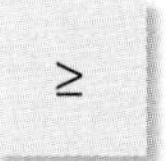

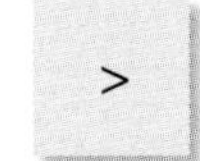

30. CRITICAL THINKING If two lines intersect, then they intersect in exactly one point by the Line Intersection Postulate (Postulate 2.3). Do the two lines have to be in the same plane? Draw a picture to support your answer. Then explain your reasoning.

31. MAKING AN ARGUMENT Your friend claims that even though two planes intersect in a line, it is possible for three planes to intersect in a point. Is your friend correct? Explain your reasoning.

32. MAKING AN ARGUMENT Your friend claims that by the Plane Intersection Postulate (Post. 2.7), any two planes intersect in a line. Is your friend's interpretation of the Plane Intersection Postulate (Post. 2.7) correct? Explain your reasoning.

33. ABSTRACT REASONING Points E, F, and G all lie in plane P and in plane Q. What must be true about points E, F, and G so that planes P and Q are different planes? What must be true about points E, F, and G to force planes P and Q to be the same plane? Make sketches to support your answers.

34. THOUGHT PROVOKING The postulates in this book represent Euclidean geometry. In spherical geometry, all points are points on the surface of a sphere. A line is a circle on the sphere whose diameter is equal to the diameter of the sphere. A plane is the surface of the sphere. Find a postulate on page 84 that is not true in spherical geometry. Explain your reasoning.

Maintaining Mathematical Proficiency

Reviewing what you learned in previous grades and lessons

Solve the equation. Tell which algebraic property of equality you used. *(Skills Review Handbook)*

35. $t - 6 = -4$ **36.** $3x = 21$ **37.** $9 + x = 13$ **38.** $\frac{x}{7} = 5$

2.1–2.3 What Did You Learn?

Core Vocabulary

conditional statement, *p. 66*
if-then form, *p. 66*
hypothesis, *p. 66*
conclusion, *p. 66*
negation, *p. 66*
converse, *p. 67*
inverse, *p. 67*
contrapositive, *p. 67*
equivalent statements, *p. 67*
perpendicular lines, *p. 68*
biconditional statement, *p. 69*
truth value, *p. 70*
truth table, *p. 70*
conjecture, *p. 76*
inductive reasoning, *p. 76*
counterexample, *p. 77*
deductive reasoning, *p. 78*
line perpendicular to a plane, *p. 86*

Core Concepts

Section 2.1

Conditional Statement, *p. 66*
Negation, *p. 66*
Related Conditionals, *p. 67*
Biconditional Statement, *p. 69*
Making a Truth Table, *p. 70*

Section 2.2

Inductive Reasoning, *p. 76*
Counterexample, *p. 77*
Deductive Reasoning, *p. 78*
Laws of Logic, *p. 78*

Section 2.3

Postulates 2.1–2.7 Point, Line, and Plane Postulates, *p. 84*
Identifying Postulates, *p. 85*
Sketching and Interpreting Diagrams, *p. 86*

Mathematical Practices

1. Provide a counterexample for each *false* conditional statement in Exercises 17–24 on page 71. (You do not need to consider the converse, inverse, and contrapositive statements.)
2. Create a truth table for each of your answers to Exercise 59 on page 74.
3. For Exercise 32 on page 88, write a question you would ask your friend about his or her interpretation.

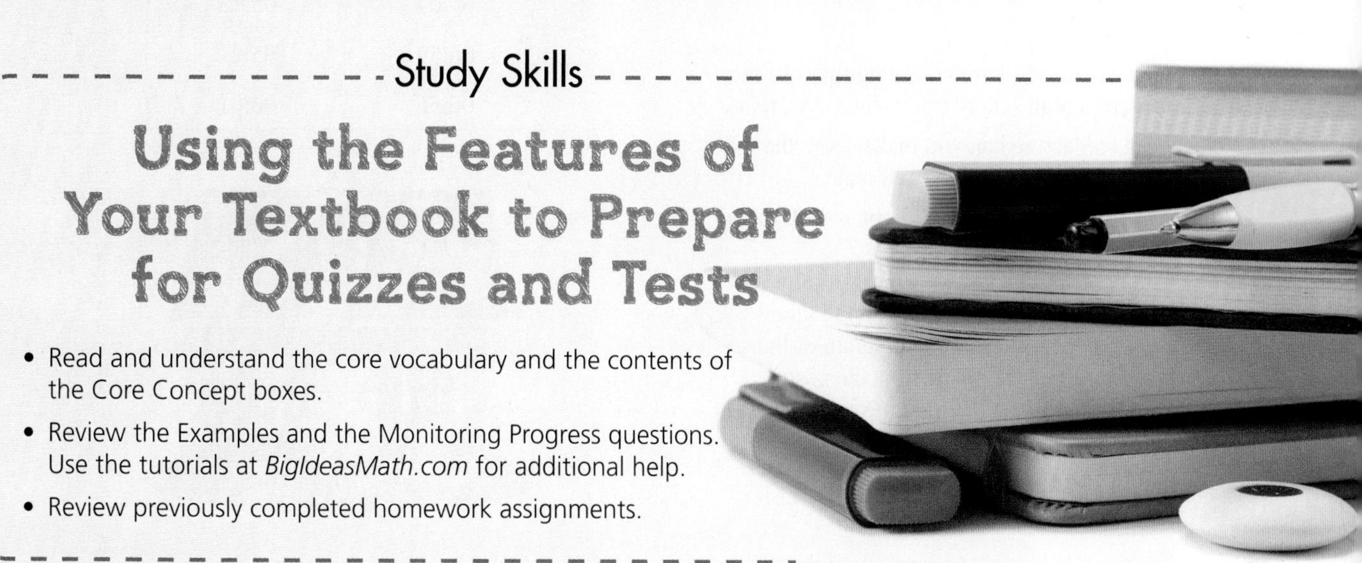

Study Skills

Using the Features of Your Textbook to Prepare for Quizzes and Tests

- Read and understand the core vocabulary and the contents of the Core Concept boxes.
- Review the Examples and the Monitoring Progress questions. Use the tutorials at *BigIdeasMath.com* for additional help.
- Review previously completed homework assignments.

2.1–2.3 Quiz

Rewrite the conditional statement in if-then form. Then write the converse, inverse, and contrapositive of the conditional statement. Decide whether each statement is true or false. *(Section 2.1)*

1. An angle measure of 167° is an obtuse angle.
2. You are in a physics class, so you always have homework.
3. I will take my driving test, so I will get my driver's license.

Find a counterexample to show that the conjecture is false. *(Section 2.2)*

4. The sum of a positive number and a negative number is always positive.
5. If a figure has four sides, then it is a rectangle.

Use inductive reasoning to make a conjecture about the given quantity. Then use deductive reasoning to show that the conjecture is true. *(Section 2.2)*

6. the sum of two negative integers
7. the difference of two even integers

Use the diagram to determine whether you can assume the statement. *(Section 2.3)*

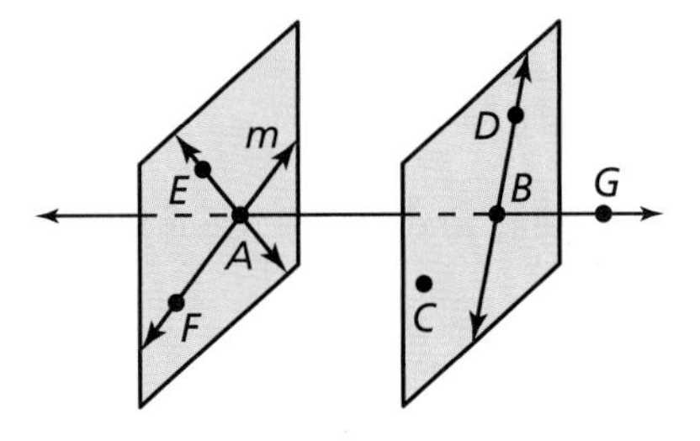

8. Points D, B, and C are coplanar.
9. Plane EAF is parallel to plane DBC.
10. Line m intersects line $\overleftrightarrow{AB}$ at point A.
11. Line $\overleftrightarrow{DC}$ lies in plane DBC.
12. $m\angle DBG = 90°$

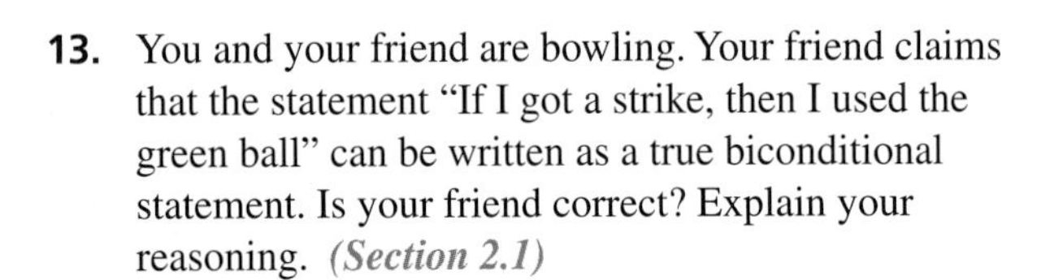

13. You and your friend are bowling. Your friend claims that the statement "If I got a strike, then I used the green ball" can be written as a true biconditional statement. Is your friend correct? Explain your reasoning. *(Section 2.1)*

Females	Males
06:43	05:41
07:22	06:07
07:04	05:13
06:39	05:21
06:56	06:01

14. The table shows the 1-mile running times of the members of a high school track team. *(Section 2.2)*
    a. What conjecture can you make about the running times of females and males?
    b. What type of reasoning did you use? Explain.

15. List five of the seven Point, Line, and Plane Postulates on page 84 that the diagram of the house demonstrates. Explain how the postulate is demonstrated in the diagram. *(Section 2.3)*

2.4 Algebraic Reasoning

Essential Question How can algebraic properties help you solve an equation?

EXPLORATION 1 Justifying Steps in a Solution

Work with a partner. In previous courses, you studied different properties, such as the properties of equality and the Distributive, Commutative, and Associative Properties. Write the property that justifies each of the following solution steps.

Algebraic Step	Justification
$2(x + 3) - 5 = 5x + 4$	Write given equation.
$2x + 6 - 5 = 5x + 4$	
$2x + 1 = 5x + 4$	
$2x - 2x + 1 = 5x - 2x + 4$	
$1 = 3x + 4$	
$1 - 4 = 3x + 4 - 4$	
$-3 = 3x$	
$\frac{-3}{3} = \frac{3x}{3}$	
$-1 = x$	
$x = -1$	

LOOKING FOR STRUCTURE

To be proficient in math, you need to look closely to discern a pattern or structure.

EXPLORATION 2 Stating Algebraic Properties

Work with a partner. The symbols ♦ and • represent addition and multiplication (not necessarily in that order). Determine which symbol represents which operation. Justify your answer. Then state each algebraic property being illustrated.

Example of Property	Name of Property
5 ♦ 6 = 6 ♦ 5	
5 • 6 = 6 • 5	
4 ♦ (5 ♦ 6) = (4 ♦ 5) ♦ 6	
4 • (5 • 6) = (4 • 5) • 6	
0 ♦ 5 = 0	
0 • 5 = 5	
1 ♦ 5 = 5	
4 ♦ (5 • 6) = 4 ♦ 5 • 4 ♦ 6	

Communicate Your Answer

3. How can algebraic properties help you solve an equation?
4. Solve $3(x + 1) - 1 = -13$. Justify each step.

2.4 Lesson

Core Vocabulary

Previous
equation
solve an equation
formula

What You Will Learn

- Use Algebraic Properties of Equality to justify the steps in solving an equation.
- Use the Distributive Property to justify the steps in solving an equation.
- Use properties of equality involving segment lengths and angle measures.

Using Algebraic Properties of Equality

When you *solve an equation*, you use properties of real numbers. Segment lengths and angle measures are real numbers, so you can also use these properties to write logical arguments about geometric figures.

Core Concept

Algebraic Properties of Equality

Let a, b, and c be real numbers.

Addition Property of Equality If $a = b$, then $a + c = b + c$.

Subtraction Property of Equality If $a = b$, then $a - c = b - c$.

Multiplication Property of Equality If $a = b$, then $a \cdot c = b \cdot c, c \neq 0$.

Division Property of Equality If $a = b$, then $\frac{a}{c} = \frac{b}{c}, c \neq 0$.

Substitution Property of Equality If $a = b$, then a can be substituted for b (or b for a) in any equation or expression.

EXAMPLE 1 Justifying Steps

Solve $3x + 2 = 23 - 4x$. Justify each step.

REMEMBER
Inverse operations "undo" each other. Addition and subtraction are inverse operations. Multiplication and division are inverse operations.

SOLUTION

Equation	Explanation	Reason
$3x + 2 = 23 - 4x$	Write the equation.	Given
$3x + 2 + 4x = 23 - 4x + 4x$	Add $4x$ to each side.	Addition Property of Equality
$7x + 2 = 23$	Combine like terms.	Simplify.
$7x + 2 - 2 = 23 - 2$	Subtract 2 from each side.	Subtraction Property of Equality
$7x = 21$	Combine constant terms.	Simplify.
$x = 3$	Divide each side by 7.	Division Property of Equality

- The solution is $x = 3$.

Monitoring Progress  Help in English and Spanish at *BigIdeasMath.com*

Solve the equation. Justify each step.

1. $6x - 11 = -35$ **2.** $-2p - 9 = 10p - 17$ **3.** $39 - 5z = -1 + 5z$

Using the Distributive Property

Distributive Property

Let a, b, and c be real numbers.

Sum $a(b + c) = ab + ac$ **Difference** $a(b - c) = ab - ac$

EXAMPLE 2 Using the Distributive Property

Solve $-5(7w + 8) = 30$. Justify each step.

SOLUTION

Equation	Explanation	Reason
$-5(7w + 8) = 30$	Write the equation.	Given
$-35w - 40 = 30$	Multiply.	Distributive Property
$-35w = 70$	Add 40 to each side.	Addition Property of Equality
$w = -2$	Divide each side by -35.	Division Property of Equality

▶ The solution is $w = -2$.

EXAMPLE 3 Solving a Real-Life Problem

You get a raise at your part-time job. To write your raise as a percent, use the formula $p(r + 1) = n$, where p is your previous wage, r is the percent increase (as a decimal), and n is your new wage. Solve the formula for r. What is your raise written as a percent when your hourly wage increases from \$7.25 to \$7.54 per hour?

SOLUTION

Step 1 Solve for r in the formula $p(r + 1) = n$.

Equation	Explanation	Reason
$p(r + 1) = n$	Write the equation.	Given
$pr + p = n$	Multiply.	Distributive Property
$pr = n - p$	Subtract p from each side.	Subtraction Property of Equality
$r = \frac{n - p}{p}$	Divide each side by p.	Division Property of Equality

Step 2 Evaluate $r = \frac{n - p}{p}$ when $n = 7.54$ and $p = 7.25$.

$$r = \frac{n - p}{p} = \frac{7.54 - 7.25}{7.25} = \frac{0.29}{7.25} = 0.04$$

REMEMBER

When evaluating expressions, use the order of operations.

▶ Your raise is 4%.

Monitoring Progress

Help in English and Spanish at *BigIdeasMath.com*

Solve the equation. Justify each step.

4. $3(3x + 14) = -3$

5. $4 = -10b + 6(2 - b)$

6. Solve the formula $A = \frac{1}{2}bh$ for b. Justify each step. Then find the base of a triangle whose area is 952 square feet and whose height is 56 feet.

Using Other Properties of Equality

The following properties of equality are true for all real numbers. Segment lengths and angle measures are real numbers, so these properties of equality are true for all segment lengths and angle measures.

Core Concept

Reflexive, Symmetric, and Transitive Properties of Equality

	Real Numbers	Segment Lengths	Angle Measures
Reflexive Property	$a = a$	$AB = AB$	$m\angle A = m\angle A$
Symmetric Property	If $a = b$, then $b = a$.	If $AB = CD$, then $CD = AB$.	If $m\angle A = m\angle B$, then $m\angle B = m\angle A$.
Transitive Property	If $a = b$ and $b = c$, then $a = c$.	If $AB = CD$ and $CD = EF$, then $AB = EF$.	If $m\angle A = m\angle B$ and $m\angle B = m\angle C$, then $m\angle A = m\angle C$.

EXAMPLE 4 Using Properties of Equality with Angle Measures

You reflect the beam of a spotlight off a mirror lying flat on a stage, as shown. Determine whether $m\angle DBA = m\angle EBC$.

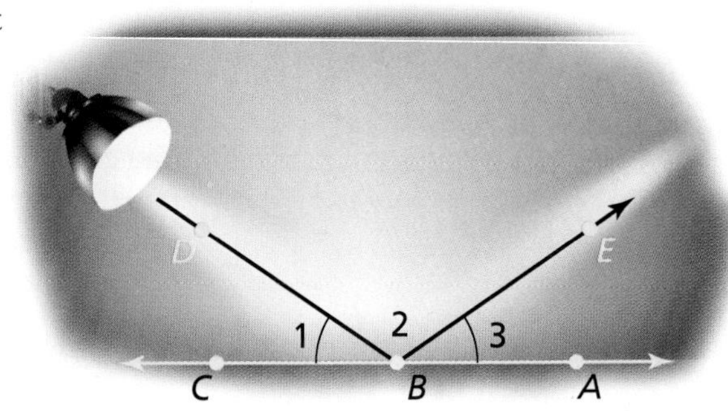

SOLUTION

Equation	Explanation	Reason
$m\angle 1 = m\angle 3$	Marked in diagram.	Given
$m\angle DBA = m\angle 3 + m\angle 2$	Add measures of adjacent angles.	Angle Addition Postulate (Post. 1.4)
$m\angle DBA = m\angle 1 + m\angle 2$	Substitute $m\angle 1$ for $m\angle 3$.	Substitution Property of Equality
$m\angle 1 + m\angle 2 = m\angle EBC$	Add measures of adjacent angles.	Angle Addition Postulate (Post. 1.4)
$m\angle DBA = m\angle EBC$	Both measures are equal to the sum $m\angle 1 + m\angle 2$.	Transitive Property of Equality

Monitoring Progress

Help in English and Spanish at *BigIdeasMath.com*

Name the property of equality that the statement illustrates.

7. If $m\angle 6 = m\angle 7$, then $m\angle 7 = m\angle 6$.

8. $34° = 34°$

9. $m\angle 1 = m\angle 2$ and $m\angle 2 = m\angle 5$. So, $m\angle 1 = m\angle 5$.

EXAMPLE 5 Modeling with Mathematics

A park, a shoe store, a pizza shop, and a movie theater are located in order on a city street. The distance between the park and the shoe store is the same as the distance between the pizza shop and the movie theater. Show that the distance between the park and the pizza shop is the same as the distance between the shoe store and the movie theater.

SOLUTION

1. **Understand the Problem** You know that the locations lie in order and that the distance between two of the locations (park and shoe store) is the same as the distance between the other two locations (pizza shop and movie theater). You need to show that two of the other distances are the same.

2. **Make a Plan** Draw and label a diagram to represent the situation.

park shoe store pizza shop movie theater

Modify your diagram by letting the points *P*, *S*, *Z*, and *M* represent the park, the shoe store, the pizza shop, and the movie theater, respectively. Show any mathematical relationships.

Use the Segment Addition Postulate (Postulate 1.2) to show that $PZ = SM$.

3. **Solve the Problem**

Equation	Explanation	Reason
$PS = ZM$	Marked in diagram.	Given
$PZ = PS + SZ$	Add lengths of adjacent segments.	Segment Addition Postulate (Post. 1.2)
$SM = SZ + ZM$	Add lengths of adjacent segments.	Segment Addition Postulate (Post. 1.2)
$PS + SZ = ZM + SZ$	Add SZ to each side of $PS = ZM$.	Addition Property of Equality
$PZ = SM$	Substitute PZ for $PS + SZ$ and SM for $ZM + SZ$.	Substitution Property of Equality

4. **Look Back** Reread the problem. Make sure your diagram is drawn precisely using the given information. Check the steps in your solution.

Monitoring Progress

Help in English and Spanish at *BigIdeasMath.com*

Name the property of equality that the statement illustrates.

10. If $JK = KL$ and $KL = 16$, then $JK = 16$.

11. $PQ = ST$, so $ST = PQ$.

12. $ZY = ZY$

13. In Example 5, a hot dog stand is located halfway between the shoe store and the pizza shop, at point *H*. Show that $PH = HM$.

2.4 Exercises

Dynamic Solutions available at *BigIdeasMath.com*

Vocabulary and Core Concept Check

1. **VOCABULARY** The statement "The measure of an angle is equal to itself" is true because of what property?

2. **DIFFERENT WORDS, SAME QUESTION** Which is different? Find both answers.

   What property justifies the following statement?

If $c = d$, then $d = c$.	If $JK = LM$, then $LM = JK$.
If $e = f$ and $f = g$, then $e = g$.	If $m\angle R = m\angle S$, then $m\angle S = m\angle R$.

Monitoring Progress and Modeling with Mathematics

In Exercises 3 and 4, write the property that justifies each step.

3. $3x - 12 = 7x + 8$ Given

   $-4x - 12 = 8$ ________

   $-4x = 20$ ________

   $x = -5$ ________

4. $5(x - 1) = 4x + 13$ Given

   $5x - 5 = 4x + 13$ ________

   $x - 5 = 13$ ________

   $x = 18$ ________

In Exercises 5–14, solve the equation. Justify each step. *(See Examples 1 and 2.)*

5. $5x - 10 = -40$
6. $6x + 17 = -7$
7. $2x - 8 = 6x - 20$
8. $4x + 9 = 16 - 3x$
9. $5(3x - 20) = -10$
10. $3(2x + 11) = 9$
11. $2(-x - 5) = 12$
12. $44 - 2(3x + 4) = -18x$
13. $4(5x - 9) = -2(x + 7)$
14. $3(4x + 7) = 5(3x + 3)$

In Exercises 15–20, solve the equation for *y*. Justify each step. *(See Example 3.)*

15. $5x + y = 18$
16. $-4x + 2y = 8$
17. $2y + 0.5x = 16$
18. $\frac{1}{2}x - \frac{3}{4}y = -2$
19. $12 - 3y = 30x + 6$
20. $3x + 7 = -7 + 9y$

In Exercises 21–24, solve the equation for the given variable. Justify each step. *(See Example 3.)*

21. $C = 2\pi r$; r
22. $I = Prt$; P
23. $S = 180(n - 2)$; n
24. $S = 2\pi r^2 + 2\pi rh$; h

In Exercises 25–32, name the property of equality that the statement illustrates.

25. If $x = y$, then $3x = 3y$.
26. If $AM = MB$, then $AM + 5 = MB + 5$.
27. $x = x$
28. If $x = y$, then $y = x$.
29. $m\angle Z = m\angle Z$
30. If $m\angle A = 29°$ and $m\angle B = 29°$, then $m\angle A = m\angle B$.
31. If $AB = LM$, then $LM = AB$.
32. If $BC = XY$ and $XY = 8$, then $BC = 8$.

In Exercises 33–40, use the property to copy and complete the statement.

33. Substitution Property of Equality:
If $AB = 20$, then $AB + CD =$ ____.

34. Symmetric Property of Equality:
If $m\angle 1 = m\angle 2$, then ____.

35. Addition Property of Equality:
If $AB = CD$, then $AB + EF =$ ____.

36. Multiplication Property of Equality:
If $AB = CD$, then $5 \cdot AB =$ ____.

37. Subtraction Property of Equality:
If $LM = XY$, then $LM - GH =$ ____.

38. Distributive Property:
If $5(x + 8) = 2$, then ____ + ____ = 2.

39. Transitive Property of Equality:
If $m\angle 1 = m\angle 2$ and $m\angle 2 = m\angle 3$, then ____.

40. Reflexive Property of Equality:
$m\angle ABC =$ ____.

ERROR ANALYSIS In Exercises 41 and 42, describe and correct the error in solving the equation.

41.

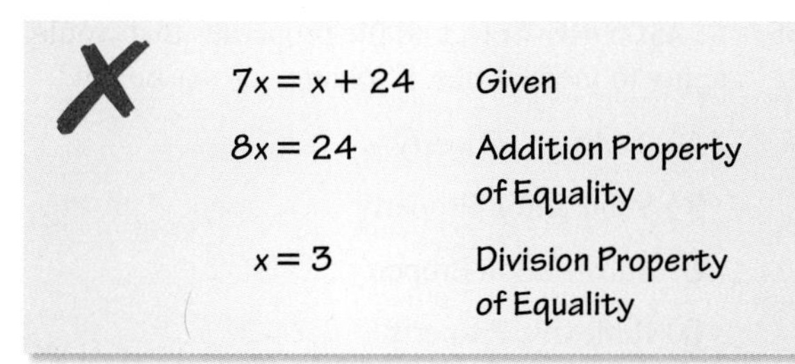

42.

$6x + 14 = 32$	Given
$6x = 18$	Division Property of Equality
$x = 3$	Simplify.

43. REWRITING A FORMULA The formula for the perimeter P of a rectangle is $P = 2\ell + 2w$, where ℓ is the length and w is the width. Solve the formula for ℓ. Justify each step. Then find the length of a rectangular lawn with a perimeter of 32 meters and a width of 5 meters.

44. REWRITING A FORMULA The formula for the area A of a trapezoid is $A = \frac{1}{2}h(b_1 + b_2)$, where h is the height and b_1 and b_2 are the lengths of the two bases. Solve the formula for b_1. Justify each step. Then find the length of one of the bases of the trapezoid when the area of the trapezoid is 91 square meters, the height is 7 meters, and the length of the other base is 20 meters.

45. ANALYZING RELATIONSHIPS In the diagram, $m\angle ABD = m\angle CBE$. Show that $m\angle 1 = m\angle 3$. *(See Example 4.)*

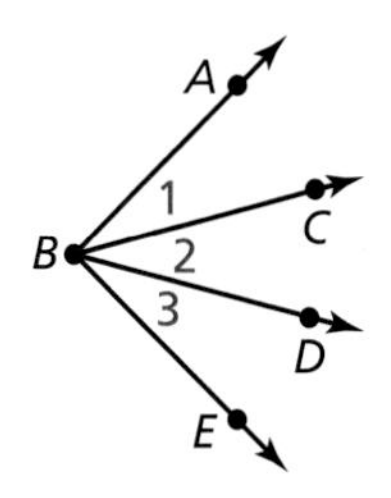

46. ANALYZING RELATIONSHIPS In the diagram, $AC = BD$. Show that $AB = CD$. *(See Example 5.)*

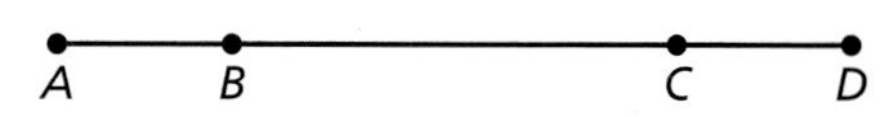

47. ANALYZING RELATIONSHIPS Copy and complete the table to show that $m\angle 2 = m\angle 3$.

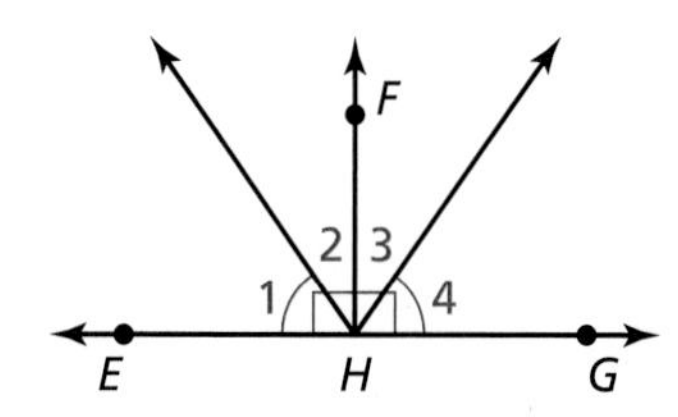

Equation	Reason
$m\angle 1 = m\angle 4$, $m\angle EHF = 90°$, $m\angle GHF = 90°$	Given
$m\angle EHF = m\angle GHF$	
$m\angle EHF = m\angle 1 + m\angle 2$ $m\angle GHF = m\angle 3 + m\angle 4$	
$m\angle 1 + m\angle 2 = m\angle 3 + m\angle 4$	
	Substitution Property of Equality
$m\angle 2 = m\angle 3$	

48. WRITING Compare the Reflexive Property of Equality with the Symmetric Property of Equality. How are the properties similar? How are they different?

REASONING In Exercises 49 and 50, show that the perimeter of $\triangle ABC$ is equal to the perimeter of $\triangle ADC$.

49.

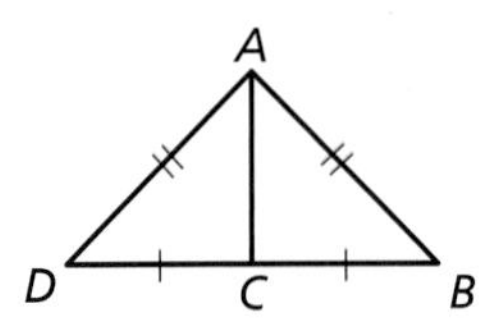

50.

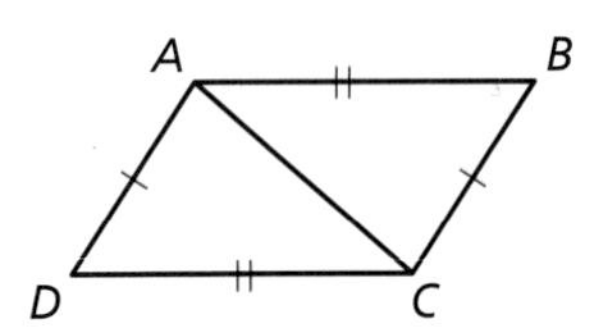

51. MATHEMATICAL CONNECTIONS In the figure, $\overline{ZY} \cong \overline{XW}$, $ZX = 5x + 17$, $YW = 10 - 2x$, and $YX = 3$. Find ZY and XW.

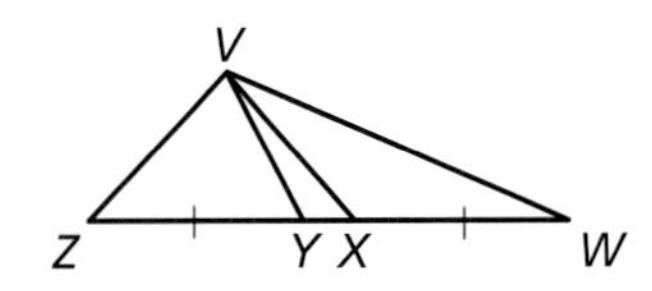

52. HOW DO YOU SEE IT? The bar graph shows the number of hours each employee works at a grocery store. Give an example of the Reflexive, Symmetric, and Transitive Properties of Equality.

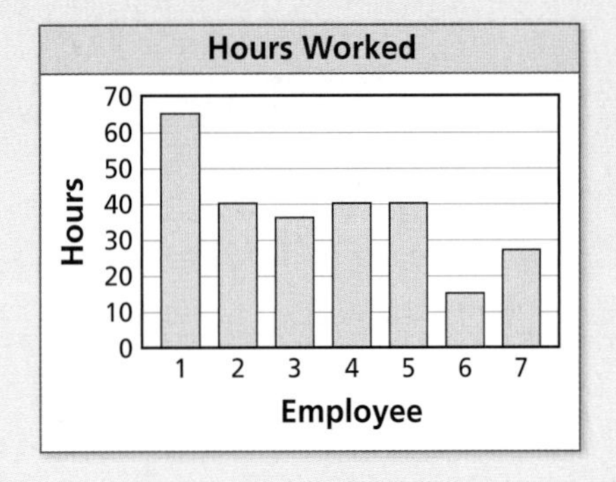

53. ATTENDING TO PRECISION Which of the following statements illustrate the Symmetric Property of Equality? Select all that apply.

Ⓐ If $AC = RS$, then $RS = AC$.

Ⓑ If $x = 9$, then $9 = x$.

Ⓒ If $AD = BC$, then $DA = CB$.

Ⓓ $AB = BA$

Ⓔ If $AB = LM$ and $LM = RT$, then $AB = RT$.

Ⓕ If $XY = EF$, then $FE = XY$.

54. THOUGHT PROVOKING Write examples from your everyday life to help you remember the Reflexive, Symmetric, and Transitive Properties of Equality. Justify your answers.

55. MULTIPLE REPRESENTATIONS The formula to convert a temperature in degrees Fahrenheit (°F) to degrees Celsius (°C) is $C = \frac{5}{9}(F - 32)$.

a. Solve the formula for F. Justify each step.

b. Make a table that shows the conversion to Fahrenheit for each temperature: 0°C, 20°C, 32°C, and 41°C.

c. Use your table to graph the temperature in degrees Fahrenheit as a function of the temperature in degrees Celsius. Is this a linear function?

56. REASONING Select all the properties that would also apply to inequalities. Explain your reasoning.

Ⓐ Addition Property

Ⓑ Subtraction Property

Ⓒ Substitution Property

Ⓓ Reflexive Property

Ⓔ Symmetric Property

Ⓕ Transitive Property

Maintaining Mathematical Proficiency

Reviewing what you learned in previous grades and lessons

Name the definition, property, or postulate that is represented by each diagram. *(Section 1.2, Section 1.3, and Section 1.5)*

57.

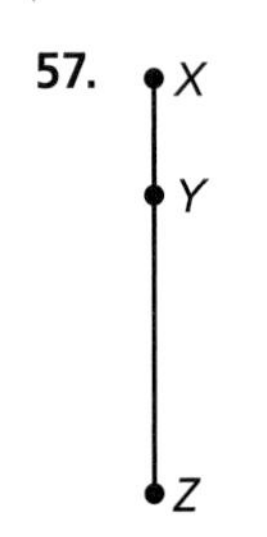

$XY + YZ = XZ$

58.

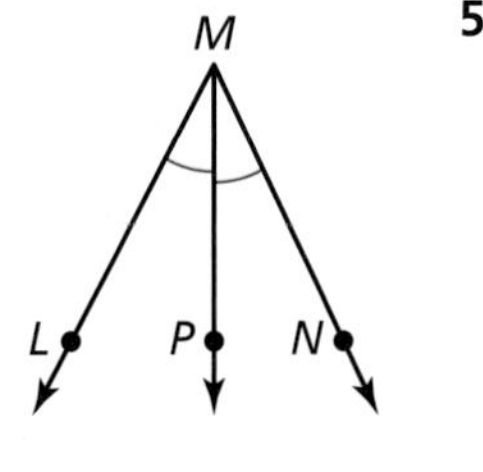

59.

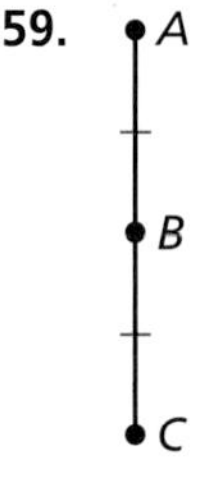

60.

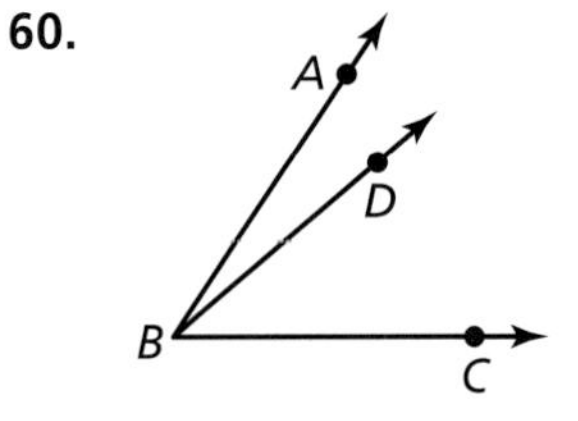

$m\angle ABD + m\angle DBC = m\angle ABC$

2.5 Proving Statements about Segments and Angles

Essential Question

How can you prove a mathematical statement?

A **proof** is a logical argument that uses deductive reasoning to show that a statement is true.

EXPLORATION 1 Writing Reasons in a Proof

Work with a partner. Four steps of a proof are shown. Write the reasons for each statement.

Given $AC = AB + AB$

Prove $AB = BC$

A B C

STATEMENTS	REASONS
1. $AC = AB + AB$	**1.** Given
2. $AB + BC = AC$	**2.**
3. $AB + AB = AB + BC$	**3.**
4. $AB = BC$	**4.**

REASONING ABSTRACTLY

To be proficient in math, you need to know and be able to use algebraic properties.

EXPLORATION 2 Writing Steps in a Proof

Work with a partner. Six steps of a proof are shown. Complete the statements that correspond to each reason.

Given $m\angle 1 = m\angle 3$

Prove $m\angle EBA = m\angle CBD$

E D C A 1 2 3 B

STATEMENTS	REASONS
1.	**1.** Given
2. $m\angle EBA = m\angle 2 + m\angle 3$	**2.** Angle Addition Postulate (Post.1.4)
3. $m\angle EBA = m\angle 2 + m\angle 1$	**3.** Substitution Property of Equality
4. $m\angle EBA =$	**4.** Commutative Property of Addition
5. $m\angle 1 + m\angle 2 =$	**5.** Angle Addition Postulate (Post.1.4)
6.	**6.** Transitive Property of Equality

Communicate Your Answer

3. How can you prove a mathematical statement?

4. Use the given information and the figure to write a proof for the statement.

Given B is the midpoint of $\overline{AC}$.
C is the midpoint of $\overline{BD}$.

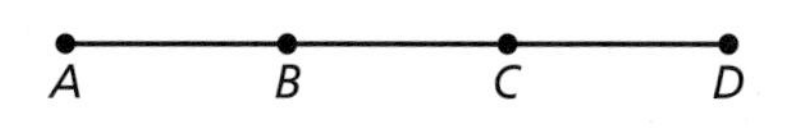

Prove $AB = CD$

2.5 Lesson

Core Vocabulary

proof, *p. 100*
two-column proof, *p. 100*
theorem, *p. 101*

What You Will Learn

- Write two-column proofs.
- Name and prove properties of congruence.

Writing Two-Column Proofs

A **proof** is a logical argument that uses deductive reasoning to show that a statement is true. There are several formats for proofs. A **two-column proof** has numbered statements and corresponding reasons that show an argument in a logical order.

In a two-column proof, each statement in the left-hand column is either given information or the result of applying a known property or fact to statements already made. Each reason in the right-hand column is the explanation for the corresponding statement.

EXAMPLE 1 Writing a Two-Column Proof

Write a two-column proof for the situation in Example 4 from the Section 2.4 lesson.

Given $m\angle 1 = m\angle 3$

Prove $m\angle DBA = m\angle EBC$

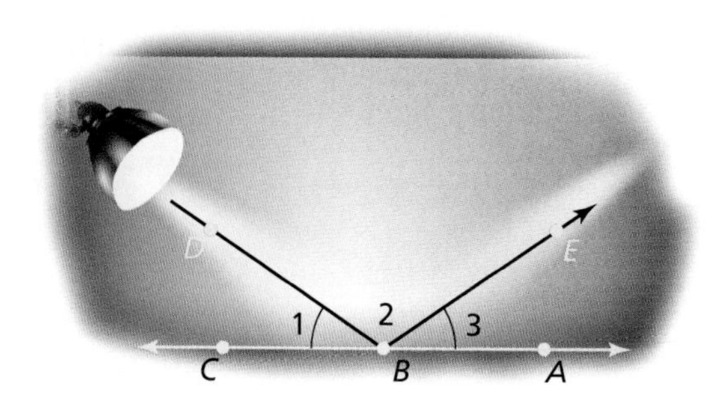

STATEMENTS	REASONS
1. $m\angle 1 = m\angle 3$	**1.** Given
2. $m\angle DBA = m\angle 3 + m\angle 2$	**2.** Angle Addition Postulate (Post.1.4)
3. $m\angle DBA = m\angle 1 + m\angle 2$	**3.** Substitution Property of Equality
4. $m\angle 1 + m\angle 2 = m\angle EBC$	**4.** Angle Addition Postulate (Post.1.4)
5. $m\angle DBA = m\angle EBC$	**5.** Transitive Property of Equality

Monitoring Progress  Help in English and Spanish at *BigIdeasMath.com*

1. Six steps of a two-column proof are shown. Copy and complete the proof.

Given T is the midpoint of $\overline{SU}$.

S ——— 7x ——— T ——— 3x + 20 ——— U

Prove $x = 5$

STATEMENTS	REASONS
1. T is the midpoint of $\overline{SU}$.	**1.** ________________
2. $\overline{ST} \cong \overline{TU}$	**2.** Definition of midpoint
3. $ST = TU$	**3.** Definition of congruent segments
4. $7x = 3x + 20$	**4.** ________________
5. ________________	**5.** Subtraction Property of Equality
6. $x = 5$	**6.** ________________

Using Properties of Congruence

The reasons used in a proof can include definitions, properties, postulates, and *theorems*. A **theorem** is a statement that can be proven. Once you have proven a theorem, you can use the theorem as a reason in other proofs.

Theorems

Theorem 2.1 Properties of Segment Congruence

Segment congruence is reflexive, symmetric, and transitive.

Reflexive For any segment AB, $\overline{AB} \cong \overline{AB}$.

Symmetric If $\overline{AB} \cong \overline{CD}$, then $\overline{CD} \cong \overline{AB}$.

Transitive If $\overline{AB} \cong \overline{CD}$ and $\overline{CD} \cong \overline{EF}$, then $\overline{AB} \cong \overline{EF}$.

Proofs Ex. 11, p. 103; Example 3, p. 101; Chapter Review 2.5 Example, p. 118

Theorem 2.2 Properties of Angle Congruence

Angle congruence is reflexive, symmetric, and transitive.

Reflexive For any angle A, $\angle A \cong \angle A$.

Symmetric If $\angle A \cong \angle B$, then $\angle B \cong \angle A$.

Transitive If $\angle A \cong \angle B$ and $\angle B \cong \angle C$, then $\angle A \cong \angle C$.

Proofs Ex. 25, p. 118; 2.5 Concept Summary, p. 102; Ex. 12, p. 103

EXAMPLE 2 Naming Properties of Congruence

Name the property that the statement illustrates.

a. If $\angle T \cong \angle V$ and $\angle V \cong \angle R$, then $\angle T \cong \angle R$.

b. If $\overline{JL} \cong \overline{YZ}$, then $\overline{YZ} \cong \overline{JL}$.

SOLUTION

a. Transitive Property of Angle Congruence

b. Symmetric Property of Segment Congruence

In this lesson, most of the proofs involve showing that congruence and equality are equivalent. You may find that what you are asked to prove seems to be obviously true. It is important to practice writing these proofs to help you prepare for writing more-complicated proofs in later chapters.

STUDY TIP

When writing a proof, organize your reasoning by copying or drawing a diagram for the situation described. Then identify the **Given** and **Prove** statements.

EXAMPLE 3 Proving a Symmetric Property of Congruence

Write a two-column proof for the Symmetric Property of Segment Congruence.

Given $\overline{LM} \cong \overline{NP}$

Prove $\overline{NP} \cong \overline{LM}$

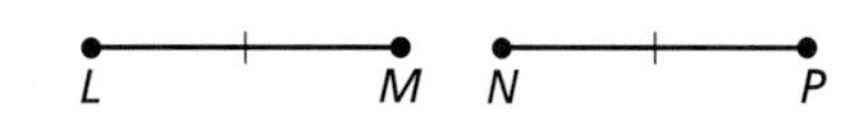

STATEMENTS	REASONS
1. $\overline{LM} \cong \overline{NP}$	**1.** Given
2. $LM = NP$	**2.** Definition of congruent segments
3. $NP = LM$	**3.** Symmetric Property of Equality
4. $\overline{NP} \cong \overline{LM}$	**4.** Definition of congruent segments

EXAMPLE 4 Writing a Two-Column Proof

Prove this property of midpoints: If you know that M is the midpoint of $\overline{AB}$, prove that AB is two times AM and AM is one-half AB.

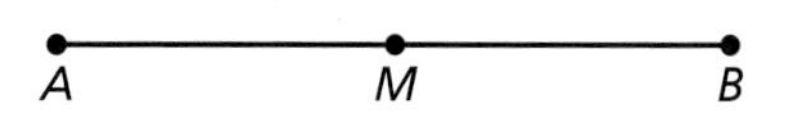

Given M is the midpoint of $\overline{AB}$.

Prove $AB = 2AM$, $AM = \frac{1}{2}AB$

STATEMENTS	REASONS
1. M is the midpoint of $\overline{AB}$.	**1.** Given
2. $\overline{AM} \cong \overline{MB}$	**2.** Definition of midpoint
3. $AM = MB$	**3.** Definition of congruent segments
4. $AM + MB = AB$	**4.** Segment Addition Postulate (Post. 1.2)
5. $AM + AM = AB$	**5.** Substitution Property of Equality
6. $2AM = AB$	**6.** Distributive Property
7. $AM = \frac{1}{2}AB$	**7.** Division Property of Equality

Monitoring Progress

Help in English and Spanish at *BigIdeasMath.com*

Name the property that the statement illustrates.

2. $\overline{GH} \cong \overline{GH}$

3. If $\angle K \cong \angle P$, then $\angle P \cong \angle K$.

4. Look back at Example 4. What would be different if you were proving that $AB = 2 \cdot MB$ and that $MB = \frac{1}{2}AB$ instead?

Concept Summary

Writing a Two-Column Proof

In a proof, you make one statement at a time until you reach the conclusion. Because you make statements based on facts, you are using deductive reasoning. Usually the first statement-and-reason pair you write is given information.

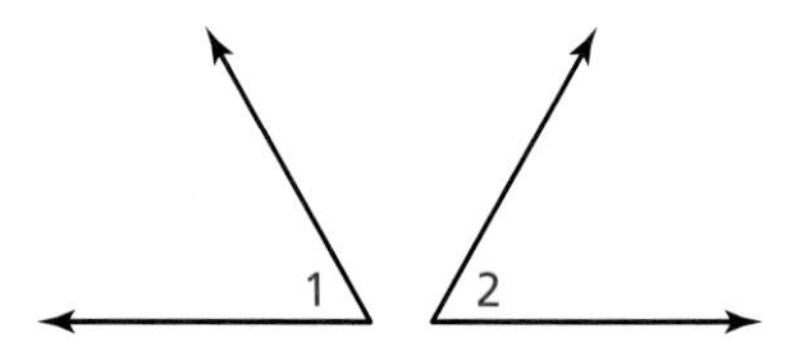

Copy or draw diagrams and label given information to help develop proofs. Do not mark or label the information in the Prove statement on the diagram.

Proof of the Symmetric Property of Angle Congruence

Given $\angle 1 \cong \angle 2$

Prove $\angle 2 \cong \angle 1$

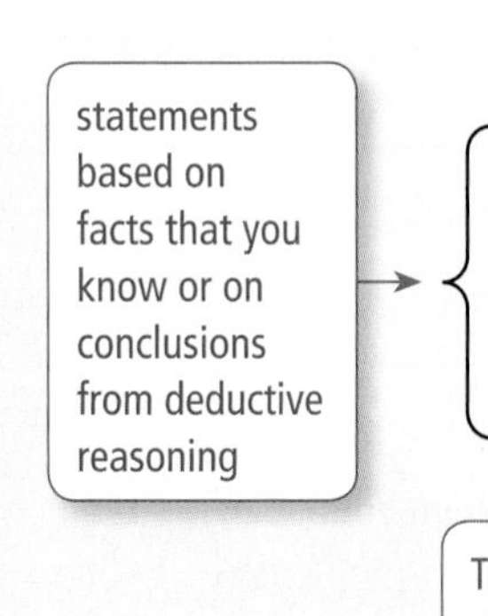

STATEMENTS	REASONS
1. $\angle 1 \cong \angle 2$	**1.** Given
2. $m\angle 1 = m\angle 2$	**2.** Definition of congruent angles
3. $m\angle 2 = m\angle 1$	**3.** Symmetric Property of Equality
4. $\angle 2 \cong \angle 1$	**4.** Definition of congruent angles

definitions, postulates, or proven theorems that allow you to state the corresponding statement

The number of statements will vary.

Remember to give a reason for the last statement.

2.5 Exercises

Dynamic Solutions available at *BigIdeasMath.com*

Vocabulary and Core Concept Check

1. **WRITING** How is a theorem different from a postulate?

2. **COMPLETE THE SENTENCE** In a two-column proof, each ______ is on the left and each _____ is on the right.

Monitoring Progress and Modeling with Mathematics

In Exercises 3 and 4, copy and complete the proof. *(See Example 1.)*

3. **Given** $PQ = RS$

   Prove $PR = QS$

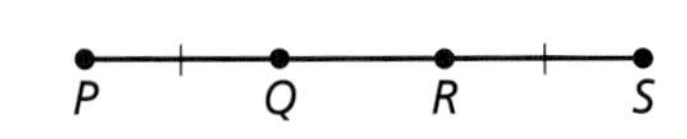

STATEMENTS	REASONS
1. $PQ = RS$	1. ______________
2. $PQ + QR = RS + QR$	2. ______________
3. ______________	3. Segment Addition Postulate (Post. 1.2)
4. $RS + QR = QS$	4. Segment Addition Postulate (Post. 1.2)
5. $PR = QS$	5. ______________

4. **Given** $\angle 1$ is a complement of $\angle 2$. $\angle 2 \cong \angle 3$

   Prove $\angle 1$ is a complement of $\angle 3$.

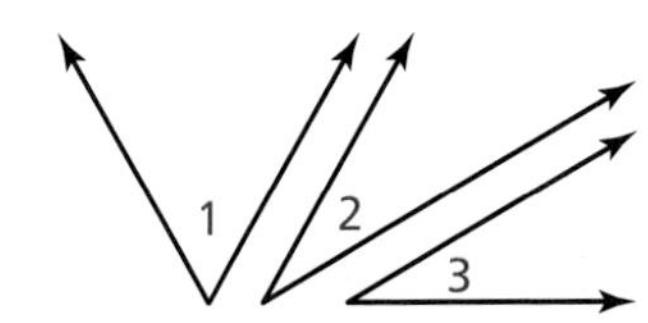

STATEMENTS	REASONS
1. $\angle 1$ is a complement of $\angle 2$.	1. Given
2. $\angle 2 \cong \angle 3$	2. ______________
3. $m\angle 1 + m\angle 2 = 90°$	3. ______________
4. $m\angle 2 = m\angle 3$	4. Definition of congruent angles
5. ______________	5. Substitution Property of Equality
6. $\angle 1$ is a complement of $\angle 3$.	6. ______________

In Exercises 5–10, name the property that the statement illustrates. *(See Example 2.)*

5. If $\overline{PQ} \cong \overline{ST}$ and $\overline{ST} \cong \overline{UV}$, then $\overline{PQ} \cong \overline{UV}$.

6. $\angle F \cong \angle F$

7. If $\angle G \cong \angle H$, then $\angle H \cong \angle G$.

8. $\overline{DE} \cong \overline{DE}$

9. If $\overline{XY} \cong \overline{UV}$, then $\overline{UV} \cong \overline{XY}$.

10. If $\angle L \cong \angle M$ and $\angle M \cong \angle N$, then $\angle L \cong \angle N$.

PROOF In Exercises 11 and 12, write a two-column proof for the property. *(See Example 3.)*

11. Reflexive Property of Segment Congruence (Thm. 2.1)

12. Transitive Property of Angle Congruence (Thm. 2.2)

PROOF In Exercises 13 and 14, write a two-column proof. *(See Example 4.)*

13. **Given** $\angle GFH \cong \angle GHF$

    Prove $\angle EFG$ and $\angle GHF$ are supplementary.

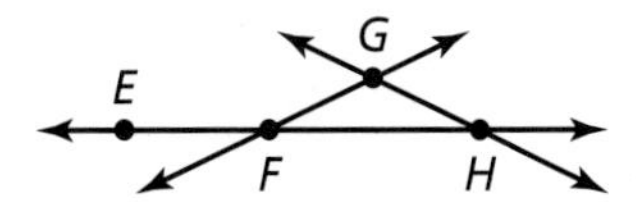

14. **Given** $\overline{AB} \cong \overline{FG}$, $\overleftrightarrow{BF}$ bisects $\overline{AC}$ and $\overline{DG}$.

    Prove $\overline{BC} \cong \overline{DF}$

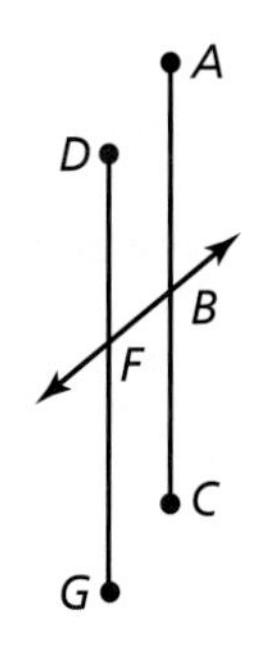

15. ERROR ANALYSIS In the diagram, $\overline{MN} \cong \overline{LQ}$ and $\overline{LQ} \cong \overline{PN}$. Describe and correct the error in the reasoning.

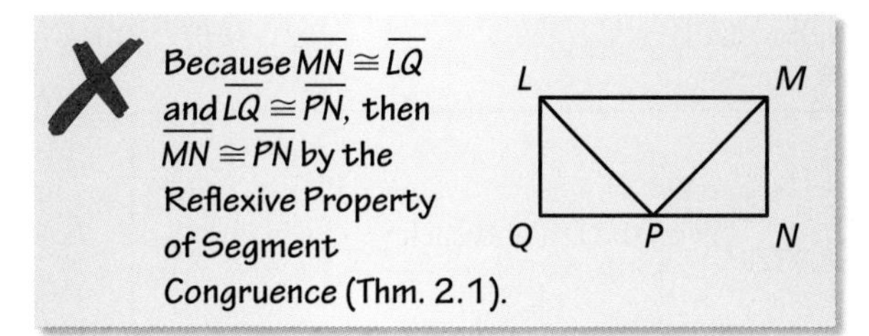

16. MODELING WITH MATHEMATICS The distance from the restaurant to the shoe store is the same as the distance from the café to the florist. The distance from the shoe store to the movie theater is the same as the distance from the movie theater to the cafe, and from the florist to the dry cleaners.

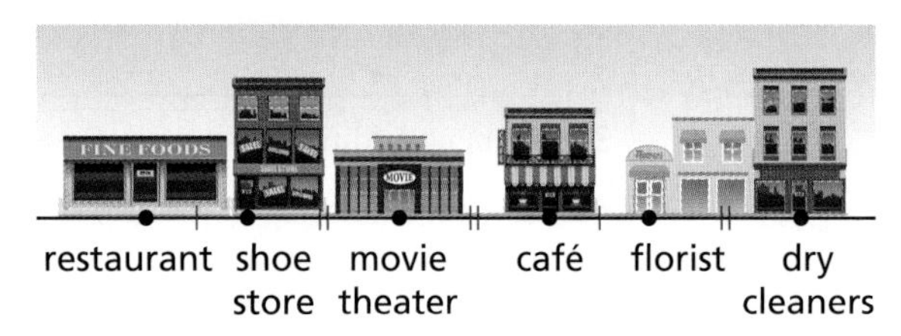

Use the steps below to prove that the distance from the restaurant to the movie theater is the same as the distance from the café to the dry cleaners.

a. State what is given and what is to be proven for the situation.

b. Write a two-column proof.

17. REASONING In the sculpture shown, $\angle 1 \cong \angle 2$ and $\angle 2 \cong \angle 3$. Classify the triangle and justify your answer.

18. MAKING AN ARGUMENT In the figure, $\overline{SR} \cong \overline{CB}$ and $\overline{AC} \cong \overline{QR}$. Your friend claims that, because of this, $\overline{CB} \cong \overline{AC}$ by the Transitive Property of Segment Congruence (Thm. 2.1). Is your friend correct? Explain your reasoning.

A Q
C B S R

19. WRITING Explain why you do not use inductive reasoning when writing a proof.

20. HOW DO YOU SEE IT? Use the figure to write Given and Prove statements for each conclusion.

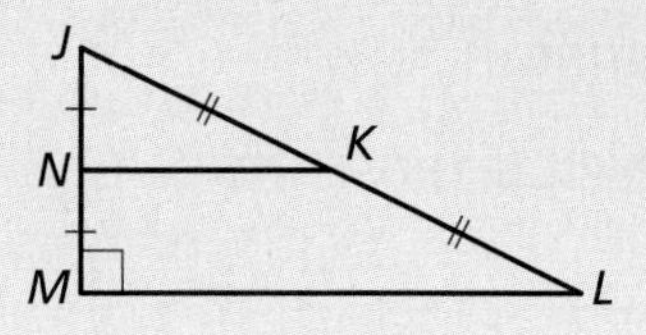

a. The acute angles of a right triangle are complementary.

b. A segment connecting the midpoints of two sides of a triangle is half as long as the third side.

21. REASONING Fold two corners of a piece of paper so their edges match, as shown.

1 1 2 2

a. What do you notice about the angle formed at the top of the page by the folds?

b. Write a two-column proof to show that the angle measure is always the same no matter how you make the folds.

22. THOUGHT PROVOKING The distance from Springfield to Lakewood City is equal to the distance from Springfield to Bettsville. Janisburg is 50 miles farther from Springfield than Bettsville. Moon Valley is 50 miles farther from Springfield than Lakewood City is. Use line segments to draw a diagram that represents this situation.

23. MATHEMATICAL CONNECTIONS Solve for x using the given information. Justify each step.

Given $\overline{QR} \cong \overline{PQ}$, $\overline{RS} \cong \overline{PQ}$

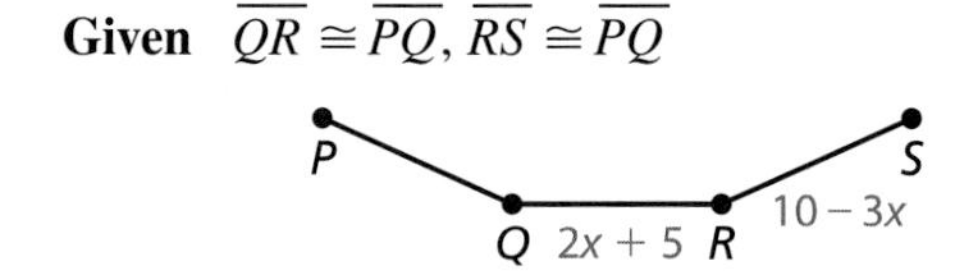

Maintaining Mathematical Proficiency

Reviewing what you learned in previous grades and lessons

Use the figure. *(Section 1.6)*

24. $\angle 1$ is a complement of $\angle 4$, and $m\angle 1 = 33°$. Find $m\angle 4$.

25. $\angle 3$ is a supplement of $\angle 2$, and $m\angle 2 = 147°$. Find $m\angle 3$.

1 2 3 4

26. Name a pair of vertical angles.

2.6 Proving Geometric Relationships

Essential Question

How can you use a flowchart to prove a mathematical statement?

EXPLORATION 1 Matching Reasons in a Flowchart Proof

Work with a partner. Match each reason with the correct step in the flowchart.

Given $AC = AB + AB$

Prove $AB = BC$

MODELING WITH MATHEMATICS

To be proficient in math, you need to map relationships using such tools as diagrams, two-way tables, graphs, flowcharts, and formulas.

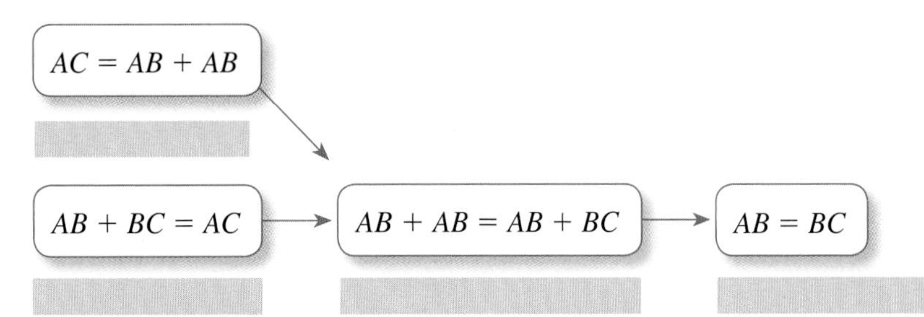

A. Segment Addition Postulate (Post. 1.2) **B.** Given

C. Transitive Property of Equality **D.** Subtraction Property of Equality

EXPLORATION 2 Matching Reasons in a Flowchart Proof

Work with a partner. Match each reason with the correct step in the flowchart.

Given $m\angle 1 = m\angle 3$

Prove $m\angle EBA = m\angle CBD$

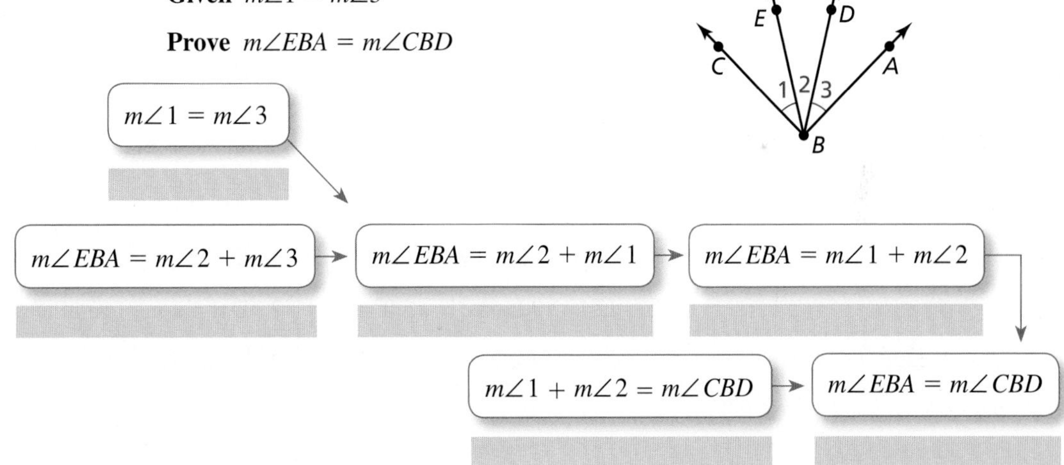

A. Angle Addition Postulate (Post. 1.4) **B.** Transitive Property of Equality

C. Substitution Property of Equality **D.** Angle Addition Postulate (Post. 1.4)

E. Given **F.** Commutative Property of Addition

Communicate Your Answer

3. How can you use a flowchart to prove a mathematical statement?

4. Compare the flowchart proofs above with the two-column proofs in the Section 2.5 Explorations. Explain the advantages and disadvantages of each.

2.6 Lesson

What You Will Learn

- Write flowchart proofs to prove geometric relationships.
- Write paragraph proofs to prove geometric relationships.

Core Vocabulary

flowchart proof, or flow proof, *p. 106*
paragraph proof, *p. 108*

Writing Flowchart Proofs

Another proof format is a **flowchart proof**, or **flow proof**, which uses boxes and arrows to show the flow of a logical argument. Each reason is below the statement it justifies. A flowchart proof of the *Right Angles Congruence Theorem* is shown in Example 1. This theorem is useful when writing proofs involving right angles.

Theorem

Theorem 2.3 Right Angles Congruence Theorem

All right angles are congruent.

Proof Example 1, p. 106

STUDY TIP

When you prove a theorem, write the hypothesis of the theorem as the **Given** statement. The conclusion is what you must **Prove**.

EXAMPLE 1 Proving the Right Angles Congruence Theorem

Use the given flowchart proof to write a two-column proof of the Right Angles Congruence Theorem.

Given $\angle 1$ and $\angle 2$ are right angles.

Prove $\angle 1 \cong \angle 2$

Flowchart Proof

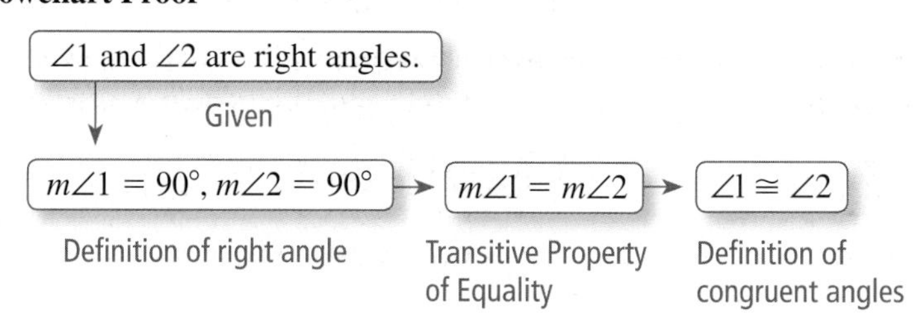

∠1 and ∠2 are right angles.
Given
m∠1 = 90°, m∠2 = 90°
Definition of right angle
m∠1 = m∠2
Transitive Property of Equality
∠1 ≅ ∠2
Definition of congruent angles

Two-Column Proof

STATEMENTS	REASONS
1. $\angle 1$ and $\angle 2$ are right angles.	**1.** Given
2. $m\angle 1 = 90^\circ, m\angle 2 = 90^\circ$	**2.** Definition of right angle
3. $m\angle 1 = m\angle 2$	**3.** Transitive Property of Equality
4. $\angle 1 \cong \angle 2$	**4.** Definition of congruent angles

Monitoring Progress

Help in English and Spanish at *BigIdeasMath.com*

1. Copy and complete the flowchart proof. Then write a two-column proof.

Given $\overline{AB} \perp \overline{BC}, \overline{DC} \perp \overline{BC}$

Prove $\angle B \cong \angle C$

$\overline{AB} \perp \overline{BC}, \overline{DC} \perp \overline{BC}$ (Given) → [ ] (Definition of $\perp$ lines) → $\angle B \cong \angle C$ ([ ])

Theorems

Theorem 2.4 Congruent Supplements Theorem

If two angles are supplementary to the same angle (or to congruent angles), then they are congruent.

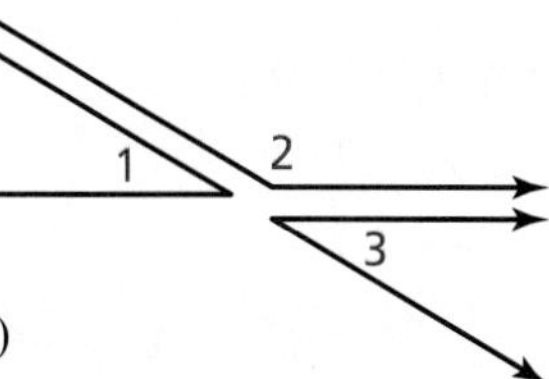

If $\angle 1$ and $\angle 2$ are supplementary and $\angle 3$ and $\angle 2$ are supplementary, then $\angle 1 \cong \angle 3$.

Proof Example 2, p. 107 (case 1); Ex. 20, p. 113 (case 2)

Theorem 2.5 Congruent Complements Theorem

If two angles are complementary to the same angle (or to congruent angles), then they are congruent.

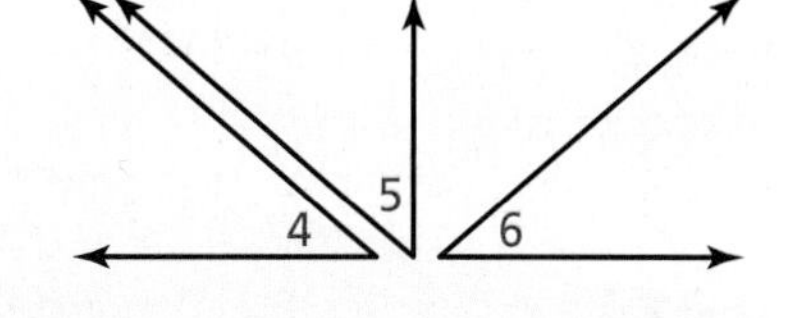

If $\angle 4$ and $\angle 5$ are complementary and $\angle 6$ and $\angle 5$ are complementary, then $\angle 4 \cong \angle 6$.

Proof Ex. 19, p. 112 (case 1); Ex. 22, p. 113 (case 2)

To prove the Congruent Supplements Theorem, you must prove two cases: one with angles supplementary to the same angle and one with angles supplementary to congruent angles. The proof of the Congruent Complements Theorem also requires two cases.

EXAMPLE 2 Proving a Case of Congruent Supplements Theorem

Use the given two-column proof to write a flowchart proof that proves that two angles supplementary to the same angle are congruent.

Given $\angle 1$ and $\angle 2$ are supplementary.
$\angle 3$ and $\angle 2$ are supplementary.

Prove $\angle 1 \cong \angle 3$

Two-Column Proof

STATEMENTS	REASONS
1. $\angle 1$ and $\angle 2$ are supplementary. $\angle 3$ and $\angle 2$ are supplementary.	**1.** Given
2. $m\angle 1 + m\angle 2 = 180°$, $m\angle 3 + m\angle 2 = 180°$	**2.** Definition of supplementary angles
3. $m\angle 1 + m\angle 2 = m\angle 3 + m\angle 2$	**3.** Transitive Property of Equality
4. $m\angle 1 = m\angle 3$	**4.** Subtraction Property of Equality
5. $\angle 1 \cong \angle 3$	**5.** Definition of congruent angles

Flowchart Proof

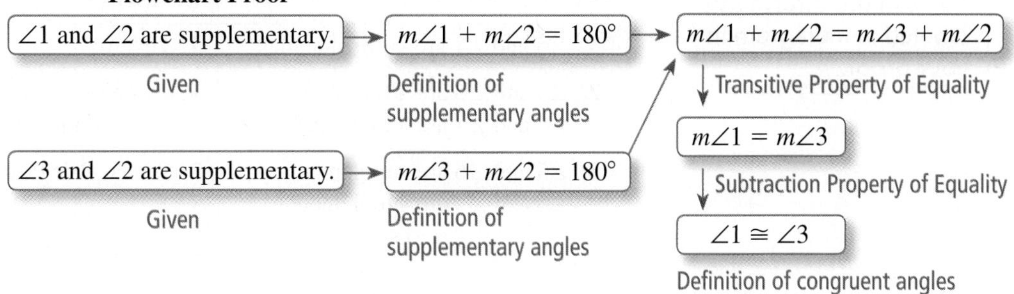

Writing Paragraph Proofs

Another proof format is a **paragraph proof**, which presents the statements and reasons of a proof as sentences in a paragraph. It uses words to explain the logical flow of the argument.

Two intersecting lines form pairs of vertical angles and linear pairs. The *Linear Pair Postulate* formally states the relationship between linear pairs. You can use this postulate to prove the *Vertical Angles Congruence Theorem*.

Postulate and Theorem

Postulate 2.8 Linear Pair Postulate

If two angles form a linear pair, then they are supplementary.

$\angle 1$ and $\angle 2$ form a linear pair, so $\angle 1$ and $\angle 2$ are supplementary and $m\angle 1 + m\angle 2 = 180°$.

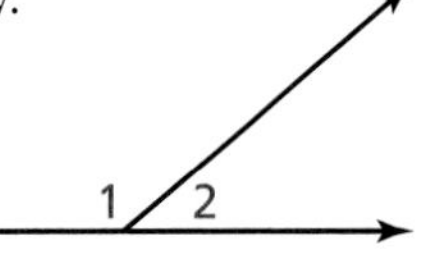

Theorem 2.6 Vertical Angles Congruence Theorem

Vertical angles are congruent.

1 2 3 4

$\angle 1 \cong \angle 3$, $\angle 2 \cong \angle 4$

Proof Example 3, p. 108

EXAMPLE 3 Proving the Vertical Angles Congruence Theorem

Use the given paragraph proof to write a two-column proof of the Vertical Angles Congruence Theorem.

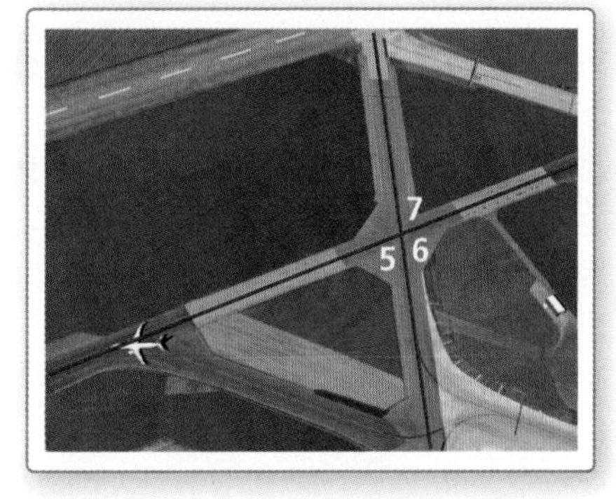

Given $\angle 5$ and $\angle 7$ are vertical angles.

Prove $\angle 5 \cong \angle 7$

STUDY TIP

In paragraph proofs, *transitional words* such as *so*, *then*, and *therefore* help make the logic clear.

Paragraph Proof

$\angle 5$ and $\angle 7$ are vertical angles formed by intersecting lines. As shown in the diagram, $\angle 5$ and $\angle 6$ are a linear pair, and $\angle 6$ and $\angle 7$ are a linear pair. Then, by the Linear Pair Postulate, $\angle 5$ and $\angle 6$ are supplementary and $\angle 6$ and $\angle 7$ are supplementary. So, by the Congruent Supplements Theorem, $\angle 5 \cong \angle 7$.

JUSTIFYING STEPS

You can use information labeled in a diagram in your proof.

Two-Column Proof

STATEMENTS	REASONS
1. $\angle 5$ and $\angle 7$ are vertical angles.	**1.** Given
2. $\angle 5$ and $\angle 6$ are a linear pair. $\angle 6$ and $\angle 7$ are a linear pair.	**2.** Definition of linear pair, as shown in the diagram
3. $\angle 5$ and $\angle 6$ are supplementary. $\angle 6$ and $\angle 7$ are supplementary.	**3.** Linear Pair Postulate
4. $\angle 5 \cong \angle 7$	**4.** Congruent Supplements Theorem

Monitoring Progress  Help in English and Spanish at *BigIdeasMath.com*

2. Copy and complete the two-column proof. Then write a flowchart proof.

Given $AB = DE, BC = CD$

Prove $\overline{AC} \cong \overline{CE}$

STATEMENTS	REASONS
1. $AB = DE, BC = CD$	**1.** Given
2. $AB + BC = BC + DE$	**2.** Addition Property of Equality
3. ______________	**3.** Substitution Property of Equality
4. $AB + BC = AC, CD + DE = CE$	**4.** ______________
5. ______________	**5.** Substitution Property of Equality
6. $\overline{AC} \cong \overline{CE}$	**6.** ______________

3. Rewrite the two-column proof in Example 3 without using the Congruent Supplements Theorem. How many steps do you save by using the theorem?

EXAMPLE 4 Using Angle Relationships

Find the value of x.

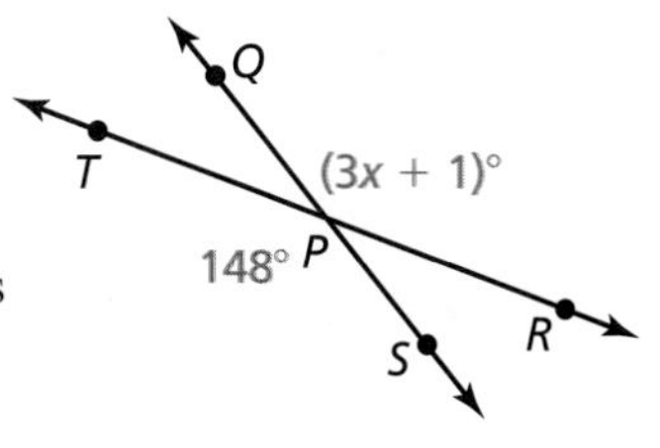

SOLUTION

$\angle TPS$ and $\angle QPR$ are vertical angles. By the Vertical Angles Congruence Theorem, the angles are congruent. Use this fact to write and solve an equation.

$m\angle TPS = m\angle QPR$	Definition of congruent angles
$148° = (3x + 1)°$	Substitute angle measures.
$147 = 3x$	Subtract 1 from each side.
$49 = x$	Divide each side by 3.

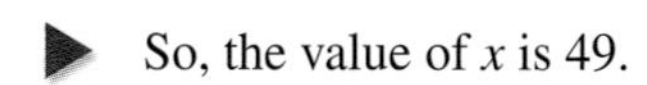

▶ So, the value of x is 49.

Monitoring Progress  Help in English and Spanish at *BigIdeasMath.com*

Use the diagram and the given angle measure to find the other three angle measures.

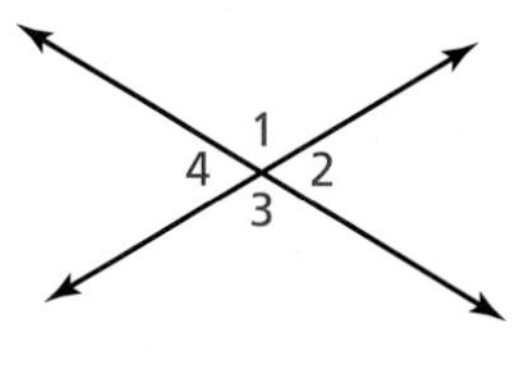

4. $m\angle 1 = 117°$

5. $m\angle 2 = 59°$

6. $m\angle 4 = 88°$

7. Find the value of w.

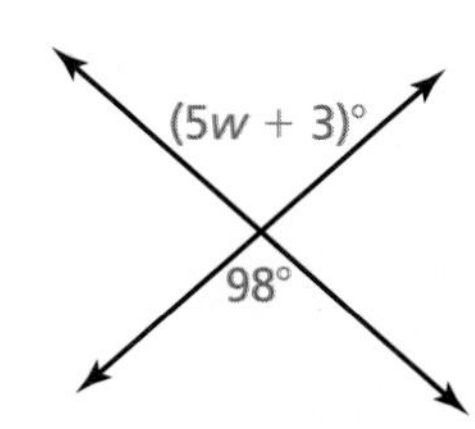

EXAMPLE 5 Using the Vertical Angles Congruence Theorem

Write a paragraph proof.

Given $\angle 1 \cong \angle 4$

Prove $\angle 2 \cong \angle 3$

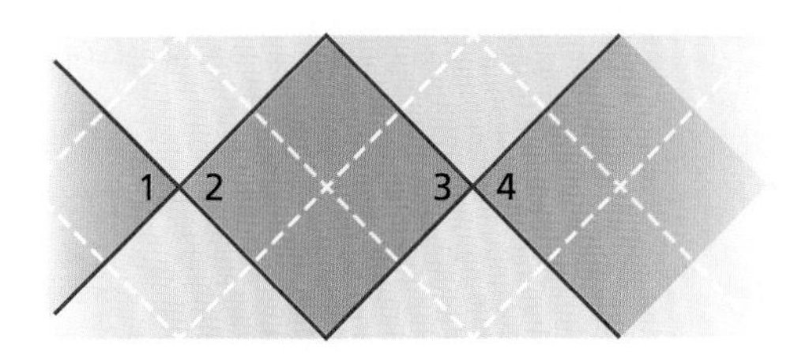

Paragraph Proof

$\angle 1$ and $\angle 4$ are congruent. By the Vertical Angles Congruence Theorem, $\angle 1 \cong \angle 2$ and $\angle 3 \cong \angle 4$. By the Transitive Property of Angle Congruence (Theorem 2.2), $\angle 2 \cong \angle 4$. Using the Transitive Property of Angle Congruence (Theorem 2.2) once more, $\angle 2 \cong \angle 3$.

Monitoring Progress

Help in English and Spanish at *BigIdeasMath.com*

8. Write a paragraph proof.

Given $\angle 1$ is a right angle.

Prove $\angle 2$ is a right angle.

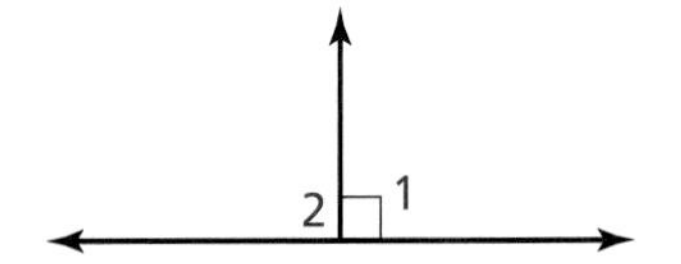

Concept Summary

Types of Proofs

Symmetric Property of Angle Congruence (Theorem 2.2)

Given $\angle 1 \cong \angle 2$

Prove $\angle 2 \cong \angle 1$

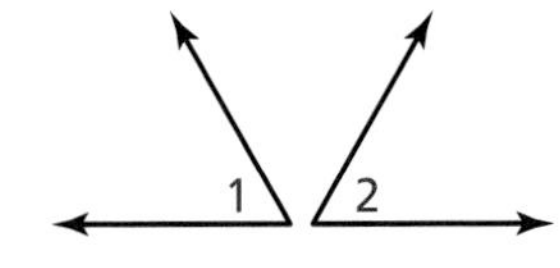

Two-Column Proof

STATEMENTS	REASONS
1. $\angle 1 \cong \angle 2$	**1.** Given
2. $m\angle 1 = m\angle 2$	**2.** Definition of congruent angles
3. $m\angle 2 = m\angle 1$	**3.** Symmetric Property of Equality
4. $\angle 2 \cong \angle 1$	**4.** Definition of congruent angles

Flowchart Proof

$\angle 1 \cong \angle 2$ → $m\angle 1 = m\angle 2$ → $m\angle 2 = m\angle 1$ → $\angle 2 \cong \angle 1$

Given | Definition of congruent angles | Symmetric Property of Equality | Definition of congruent angles

Paragraph Proof

$\angle 1$ is congruent to $\angle 2$. By the definition of congruent angles, the measure of $\angle 1$ is equal to the measure of $\angle 2$. The measure of $\angle 2$ is equal to the measure of $\angle 1$ by the Symmetric Property of Equality. Then by the definition of congruent angles, $\angle 2$ is congruent to $\angle 1$.

2.6 Exercises

Dynamic Solutions available at *BigIdeasMath.com*

Vocabulary and Core Concept Check

1. **WRITING** Explain why all right angles are congruent.

2. **VOCABULARY** What are the two types of angles that are formed by intersecting lines?

Monitoring Progress and Modeling with Mathematics

In Exercises 3–6, identify the pair(s) of congruent angles in the figures. Explain how you know they are congruent. *(See Examples 1, 2, and 3.)*

3.

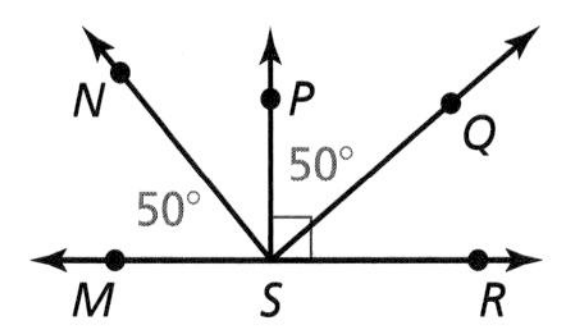

4.

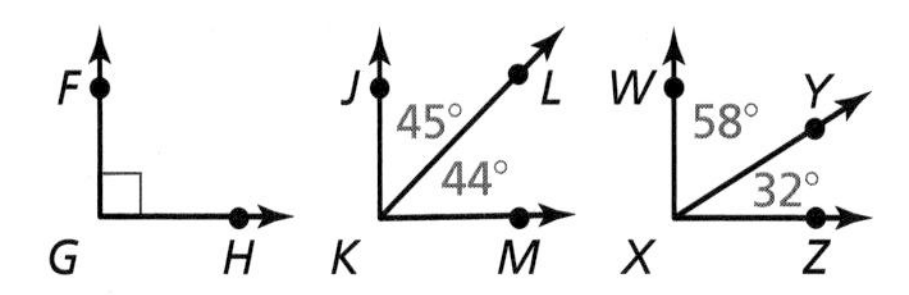

5.

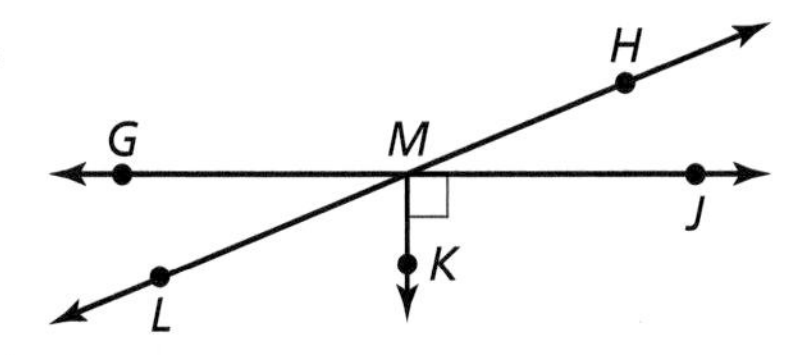

6. $\angle ABC$ is supplementary to $\angle CBD$.
$\angle CBD$ is supplementary to $\angle DEF$.

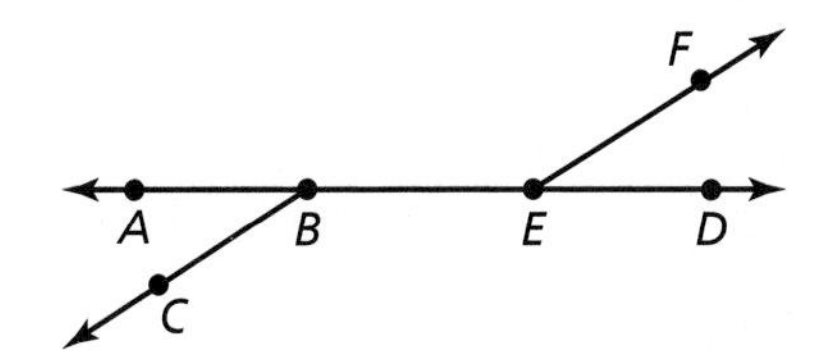

In Exercises 7–10, use the diagram and the given angle measure to find the other three measures. *(See Example 3.)*

7. $m\angle 1 = 143°$

8. $m\angle 3 = 159°$

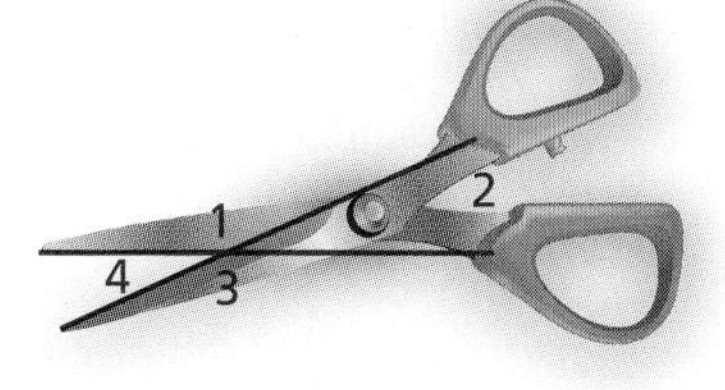

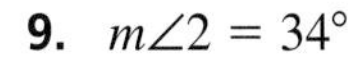

9. $m\angle 2 = 34°$

10. $m\angle 4 = 29°$

In Exercises 11–14, find the values of x and y. *(See Example 4.)*

11.

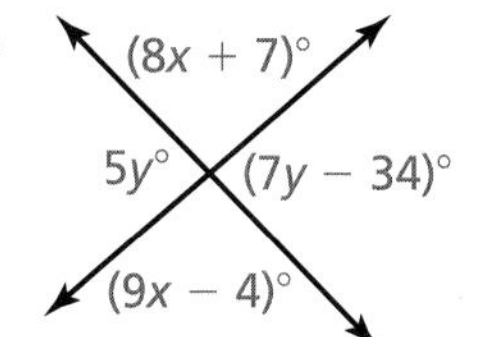

12.

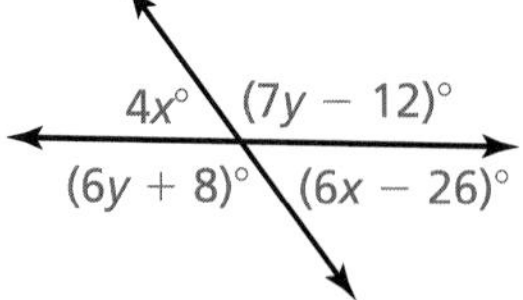

13.

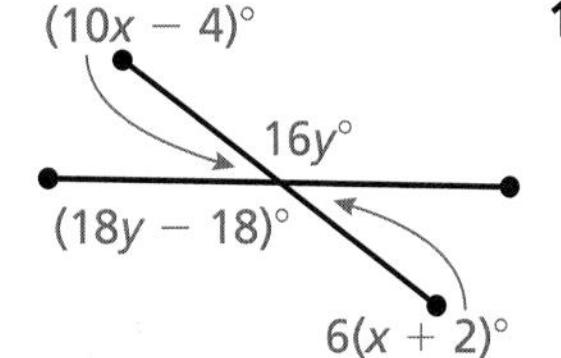

14.

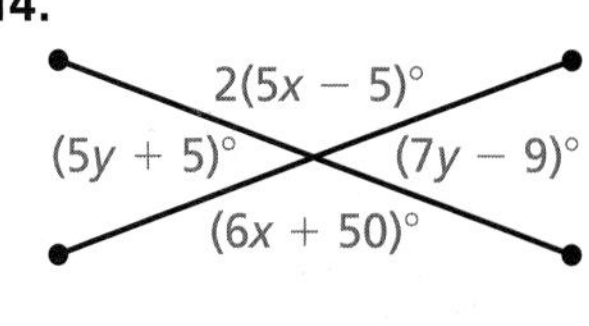

ERROR ANALYSIS In Exercises 15 and 16, describe and correct the error in using the diagram to find the value of x.

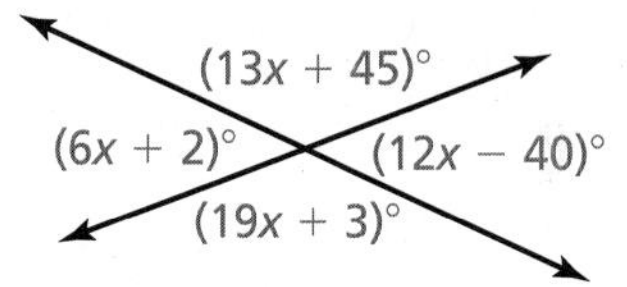

15.

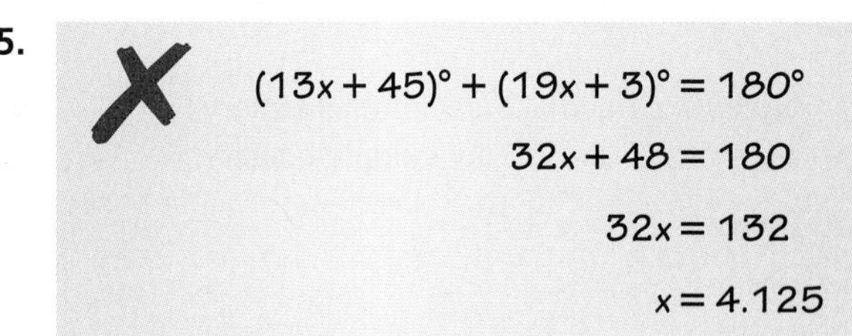

16.

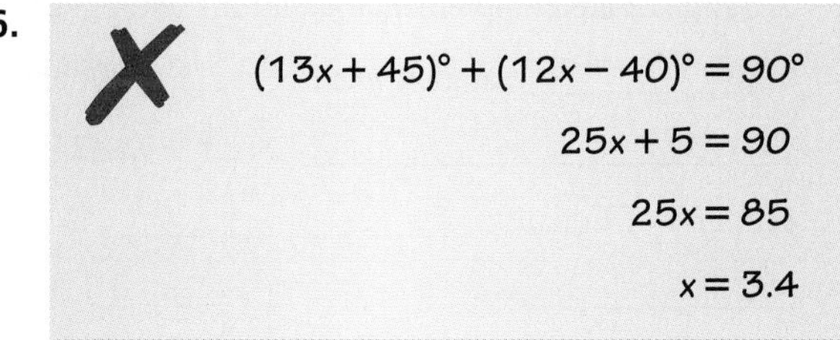

17. PROOF Copy and complete the flowchart proof. Then write a two-column proof. *(See Example 1.)*

Given $\angle 1 \cong \angle 3$

Prove $\angle 2 \cong \angle 4$

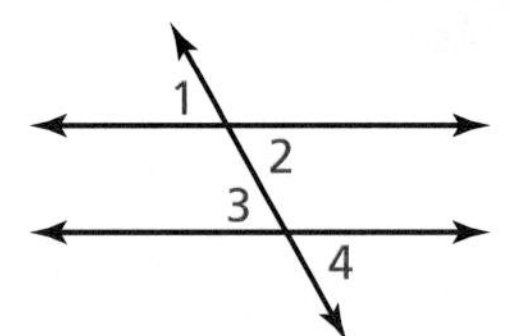

$\angle 1 \cong \angle 3$ → $\angle 1 \cong \angle 2, \angle 3 \cong \angle 4$ → $\angle 2 \cong \angle 3$ → $\angle 2 \cong \angle 4$

Given | Vertical Angles Congruence Theorem (Theorem 2.6) | ______ | ______

18. PROOF Copy and complete the two-column proof. Then write a flowchart proof. *(See Example 2.)*

Given $\angle ABD$ is a right angle.
$\angle CBE$ is a right angle.

Prove $\angle ABC \cong \angle DBE$

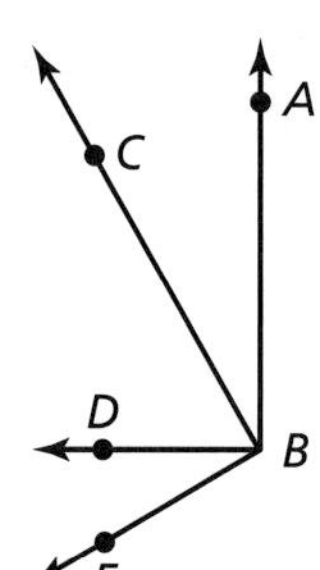

STATEMENTS	REASONS
1. $\angle ABD$ is a right angle. $\angle CBE$ is a right angle.	**1.** ______________________
2. $\angle ABC$ and $\angle CBD$ are complementary.	**2.** Definition of complementary angles
3. $\angle DBE$ and $\angle CBD$ are complementary.	**3.** ______________________
4. $\angle ABC \cong \angle DBE$	**4.** ______________________

19. PROVING A THEOREM Copy and complete the paragraph proof for the Congruent Complements Theorem (Theorem 2.5). Then write a two-column proof. *(See Example 3.)*

Given $\angle 1$ and $\angle 2$ are complementary.
$\angle 1$ and $\angle 3$ are complementary.

Prove $\angle 2 \cong \angle 3$

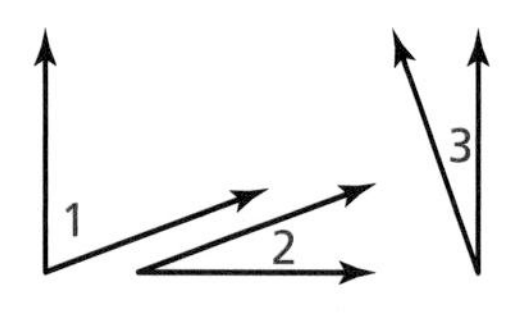

$\angle 1$ and $\angle 2$ are complementary, and $\angle 1$ and $\angle 3$ are complementary. By the definition of ____________ angles, $m\angle 1 + m\angle 2 = 90°$ and ____________ $= 90°$. By the ____________________________, $m\angle 1 + m\angle 2 = m\angle 1 + m\angle 3$. By the Subtraction Property of Equality, ____________________. So, $\angle 2 \cong \angle 3$ by the definition of ______________.

20. PROVING A THEOREM Copy and complete the two-column proof for the Congruent Supplement Theorem (Theorem 2.4). Then write a paragraph proof. *(See Example 5.)*

Given $\angle 1$ and $\angle 2$ are supplementary.
$\angle 3$ and $\angle 4$ are supplementary.
$\angle 1 \cong \angle 4$

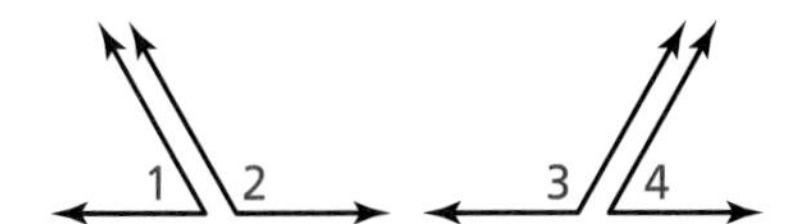

Prove $\angle 2 \cong \angle 3$

STATEMENTS	REASONS
1. $\angle 1$ and $\angle 2$ are supplementary. $\angle 3$ and $\angle 4$ are supplementary. $\angle 1 \cong \angle 4$	**1.** Given
2. $m\angle 1 + m\angle 2 = 180°$, $m\angle 3 + m\angle 4 = 180°$	**2.** ________________
3. ________ $= m\angle 3 + m\angle 4$	**3.** Transitive Property of Equality
4. $m\angle 1 = m\angle 4$	**4.** Definition of congruent angles
5. $m\angle 1 + m\angle 2 =$ ________	**5.** Substitution Property of Equality
6. $m\angle 2 = m\angle 3$	**6.** ________________
7. ________________	**7.** ________________

PROOF In Exercises 21–24, write a proof using any format.

21. Given $\angle QRS$ and $\angle PSR$ are supplementary.

Prove $\angle QRL \cong \angle PSR$

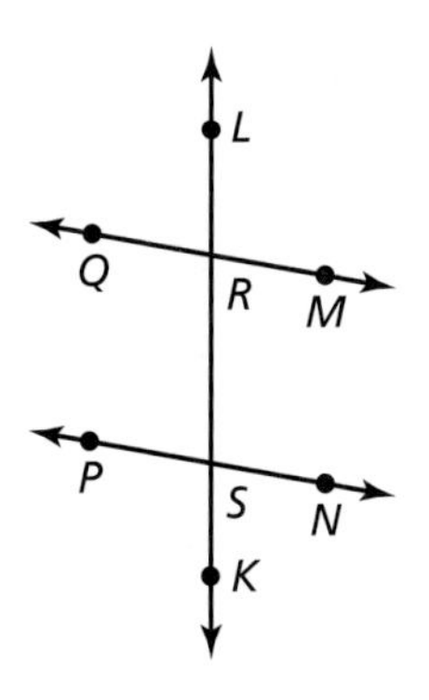

22. Given $\angle 1$ and $\angle 3$ are complementary.
$\angle 2$ and $\angle 4$ are complementary.

Prove $\angle 1 \cong \angle 4$

23. Given $\angle AEB \cong \angle DEC$

Prove $\angle AEC \cong \angle DEB$

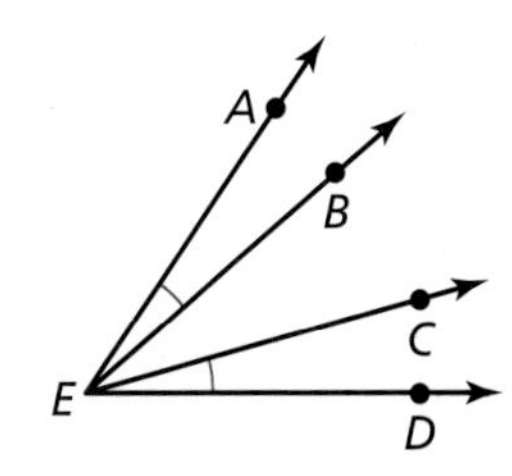

24. Given $\overline{JK} \perp \overline{JM}$, $\overline{KL} \perp \overline{ML}$,
$\angle J \cong \angle M$, $\angle K \cong \angle L$

Prove $\overline{JM} \perp \overline{ML}$ and $\overline{JK} \perp \overline{KL}$

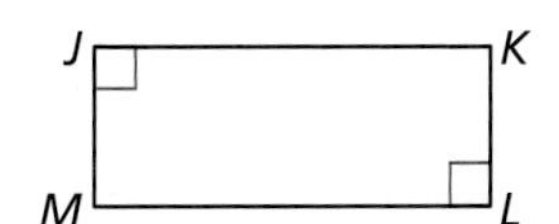

25. MAKING AN ARGUMENT You overhear your friend discussing the diagram shown with a classmate. Your classmate claims $\angle 1 \cong \angle 4$ because they are vertical angles. Your friend claims they are not congruent because he can tell by looking at the diagram. Who is correct? Support your answer with definitions or theorems.

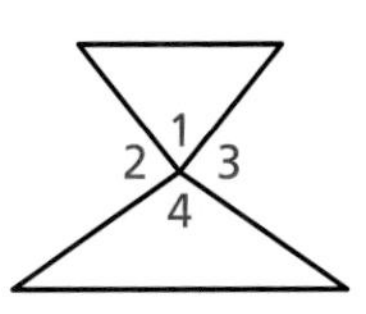

26. THOUGHT PROVOKING Draw three lines all intersecting at the same point. Explain how you can give two of the angle measures so that you can find the remaining four angle measures.

27. CRITICAL THINKING Is the converse of the Linear Pair Postulate (Postulate 2.8) true? If so, write a biconditional statement. Explain your reasoning.

28. WRITING How can you save time writing proofs?

29. MATHEMATICAL CONNECTIONS Find the measure of each angle in the diagram.

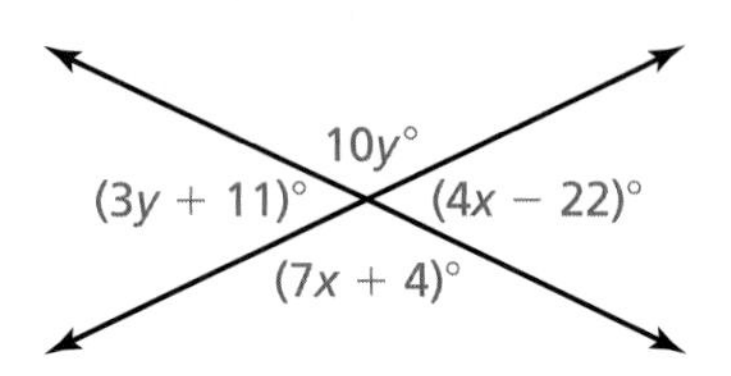

30. HOW DO YOU SEE IT? Use the student's two-column proof.

Given $\angle 1 \cong \angle 2$
$\angle 1$ and $\angle 2$ are supplementary.

Prove ________________

STATEMENTS	REASONS
1. $\angle 1 \cong \angle 2$ $\angle 1$ and $\angle 2$ are supplementary.	**1.** Given
2. $m\angle 1 = m\angle 2$	**2.** Definition of congruent angles
3. $m\angle 1 + m\angle 2 = 180°$	**3.** Definition of supplementary angles
4. $m\angle 1 + m\angle 1 = 180°$	**4.** Substitution Property of Equality
5. $2m\angle 1 = 180°$	**5.** Simplify.
6. $m\angle 1 = 90°$	**6.** Division Property of Equality
7. $m\angle 2 = 90°$	**7.** Transitive Property of Equality
8. ________________	**8.** ________________

a. What is the student trying to prove?

b. Your friend claims that the last line of the proof should be $\angle 1 \cong \angle 2$, because the measures of the angles are both 90°. Is your friend correct? Explain.

Maintaining Mathematical Proficiency

Reviewing what you learned in previous grades and lessons

Use the cube. *(Section 1.1)*

31. Name three collinear points.

32. Name the intersection of plane *ABF* and plane *EHG*.

33. Name two planes containing $\overline{BC}$.

34. Name three planes containing point *D*.

35. Name three points that are not collinear.

36. Name two planes containing point *J*.

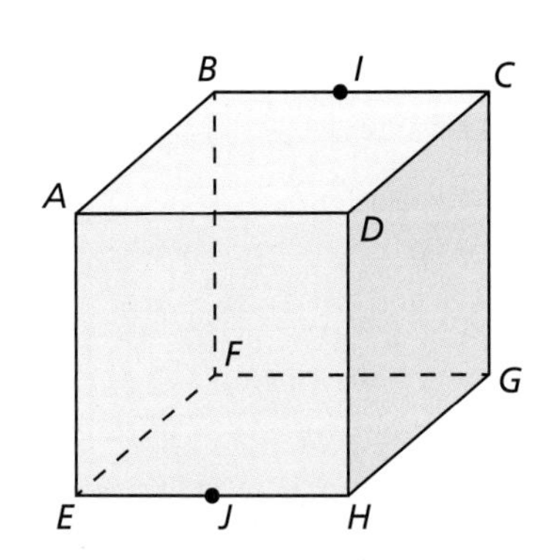

2.4–2.6 What Did You Learn?

Core Vocabulary

proof, *p. 100*
two-column proof, *p. 100*
theorem, *p. 101*
flowchart proof, or flow proof, *p. 106*
paragraph proof, *p. 108*

Core Concepts

Section 2.4

Algebraic Properties of Equality, *p. 92*
Distributive Property, *p. 93*
Reflexive, Symmetric, and Transitive Properties of Equality, *p. 94*

Section 2.5

Writing Two-Column Proofs, *p. 100*
Theorem 2.1 Properties of Segment Congruence Theorem, *p. 101*
Theorem 2.2 Properties of Angle Congruence Theorem, *p. 101*

Section 2.6

Writing Flowchart Proofs, *p. 106*
Theorem 2.3 Right Angles Congruence Theorem, *p. 106*
Theorem 2.4 Congruent Supplements Theorem, *p. 107*
Theorem 2.5 Congruent Complements Theorem, *p. 107*
Writing Paragraph Proofs, *p. 108*
Postulate 2.8 Linear Pair Postulate, *p. 108*
Theorem 2.6 Vertical Angles Congruence Theorem, *p. 108*

Mathematical Practices

1. Explain the purpose of justifying each step in Exercises 5–14 on page 96.
2. Create a diagram to model each statement in Exercises 5–10 on page 103.
3. Explain why you would not be able to prove the statement in Exercise 21 on page 113 if you were not provided with the given information or able to use any postulates or theorems.

Performance Task

Induction and the Next Dimension

Before you took Geometry, you could find the midpoint of a segment on a number line (a one-dimensional system). In Chapter 1, you learned how to find the midpoint of a segment in a coordinate plane (a two-dimensional system). How would you find the midpoint of a segment in a three-dimensional system?

To explore the answers to this question and more, go to ***BigIdeasMath.com***.

2 Chapter Review

Dynamic Solutions available at *BigIdeasMath.com*

2.1 Conditional Statements (pp. 65–74)

Write the if-then form, the converse, the inverse, the contrapositive, and the biconditional of the conditional statement "A leap year is a year with 366 days."

If-then form: If it is a leap year, then it is a year with 366 days.

Converse: If it is a year with 366 days, then it is a leap year.

Inverse: If it is not a leap year, then it is not a year with 366 days.

Contrapositive: If it is not a year with 366 days, then it is not a leap year.

Biconditional: It is a leap year if and only if it is a year with 366 days.

Write the if-then form, the converse, the inverse, the contrapositive, and the biconditional of the conditional statement.

1. Two lines intersect in a point.
2. $4x + 9 = 21$ because $x = 3$.
3. Supplementary angles sum to 180°.
4. Right angles are 90°.

2.2 Inductive and Deductive Reasoning (pp. 75–82)

What conclusion can you make about the sum of any two even integers?

Step 1 Look for a pattern in several examples. Use inductive reasoning to make a conjecture.

$2 + 4 = 6$ $\quad$ $6 + 10 = 16$ $\quad$ $12 + 16 = 28$

$-2 + 4 = 2$ $\quad$ $6 + (-10) = -4$ $\quad$ $-12 + (-16) = -28$

Conjecture Even integer + Even integer = Even integer

Step 2 Let n and m each be any integer. Use deductive reasoning to show that the conjecture is true.

$2n$ and $2m$ are even integers because any integer multiplied by 2 is even.

$2n + 2m$ represents the sum of two even integers.

$2n + 2m = 2(n + m)$ by the Distributive Property.

$2(n + m)$ is the product of 2 and an integer $(n + m)$.
So, $2(n + m)$ is an even integer.

▶ The sum of any two even integers is an even integer.

5. What conclusion can you make about the difference of any two odd integers?
6. What conclusion can you make about the product of an even and an odd integer?
7. Use the Law of Detachment to make a valid conclusion.
If an angle is a right angle, then the angle measures 90°. $\angle B$ is a right angle.
8. Use the Law of Syllogism to write a new conditional statement that follows from the pair of true statements: If $x = 3$, then $2x = 6$. If $4x = 12$, then $x = 3$.

2.3 Postulates and Diagrams (pp. 83–88)

Use the diagram to make three statements that can be concluded and three statements that *cannot* be concluded. Justify your answers.

You can conclude:

1. Points A, B, and C are coplanar because they lie in plane M.
2. $\overleftrightarrow{FG}$ lies in plane P by the Plane-Line Postulate (Post. 2.6).
3. $\overleftrightarrow{CD}$ and $\overleftrightarrow{FH}$ intersect at point H by the Line Intersection Postulate (Post. 2.3).

You *cannot* conclude:

1. $\overleftrightarrow{CD} \perp$ to plane P because no right angle is marked.
2. Points A, F, and G are coplanar because point A lies in plane M and point G lies in plane P.
3. Points G, D, and J are collinear because no drawn line connects the points.

Use the diagram at the right to determine whether you can assume the statement.

9. Points A, B, C, and E are coplanar.

10. $\overleftrightarrow{HC} \perp \overleftrightarrow{GE}$

11. Points F, B, and G are collinear.

12. $\overleftrightarrow{AB} \parallel \overleftrightarrow{GE}$

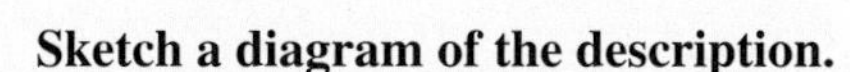

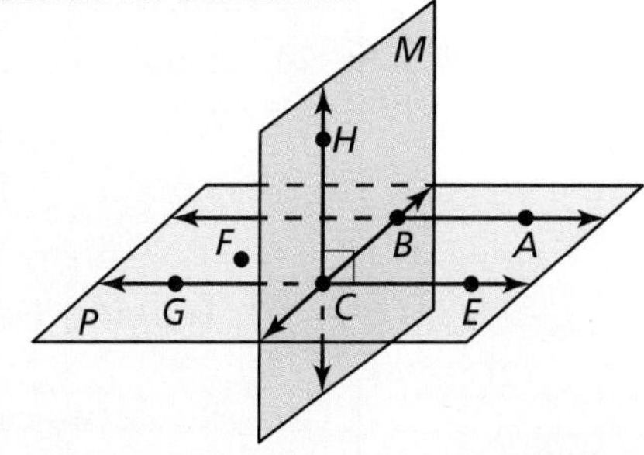

Sketch a diagram of the description.

13. $\angle ABC$, an acute angle, is bisected by $\overrightarrow{BE}$.

14. $\angle CDE$, a straight angle, is bisected by $\overleftrightarrow{DK}$.

15. Plane P and plane R intersect perpendicularly in $\overleftrightarrow{XY}$. $\overline{ZW}$ lies in plane P.

2.4 Algebraic Reasoning (pp. 91–98)

Solve $2(2x + 9) = -10$. Justify each step.

Equation	Explanation	Reason
$2(2x + 9) = -10$	Write the equation.	Given
$4x + 18 = -10$	Multiply.	Distributive Property
$4x = -28$	Subtract 18 from each side.	Subtraction Property of Equality
$x = -7$	Divide each side by 4.	Division Property of Equality

▶ The solution is $x = -7$.

Solve the equation. Justify each step.

16. $-9x - 21 = -20x - 87$

17. $15x + 22 = 7x + 62$

18. $3(2x + 9) = 30$

19. $5x + 2(2x - 23) = -154$

Name the property of equality that the statement illustrates.

20. If $LM = RS$ and $RS = 25$, then $LM = 25$.

21. $AM = AM$

2.5 Proving Statements about Segments and Angles (pp. 99–104)

Write a two-column proof for the Transitive Property of Segment Congruence (Theorem 2.1).

Given $\overline{AB} \cong \overline{CD}, \overline{CD} \cong \overline{EF}$

Prove $\overline{AB} \cong \overline{EF}$

STATEMENTS	REASONS
1. $\overline{AB} \cong \overline{CD}, \overline{CD} \cong \overline{EF}$	**1.** Given
2. $AB = CD,\ CD = EF$	**2.** Definition of congruent segments
3. $AB = EF$	**3.** Transitive Property of Equality
4. $\overline{AB} \cong \overline{EF}$	**4.** Definition of congruent segments

Name the property that the statement illustrates.

22. If $\angle DEF \cong \angle JKL$, then $\angle JKL \cong \angle DEF$.

23. $\angle C \cong \angle C$

24. If $MN = PQ$ and $PQ = RS$, then $MN = RS$.

25. Write a two-column proof for the Reflexive Property of Angle Congruence (Thm. 2.2).

2.6 Proving Geometric Relationships (pp. 105–114)

Rewrite the two-column proof into a paragraph proof.

Given $\angle 2 \cong \angle 3$

Prove $\angle 3 \cong \angle 6$

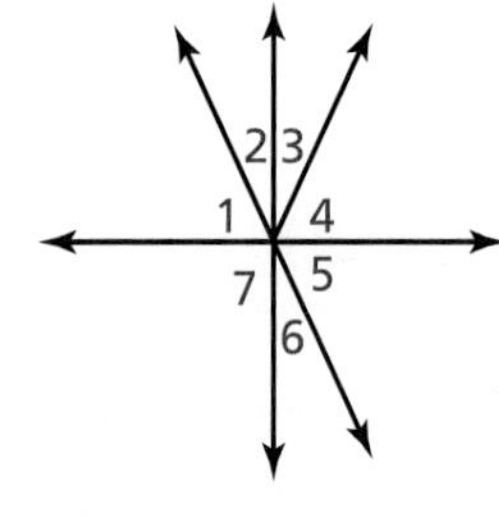

Two-Column Proof

STATEMENTS	REASONS
1. $\angle 2 \cong \angle 3$	**1.** Given
2. $\angle 2 \cong \angle 6$	**2.** Vertical Angles Congruence Theorem (Thm. 2.6)
3. $\angle 3 \cong \angle 6$	**3.** Transitive Property of Angle Congruence (Thm. 2.2)

Paragraph Proof

$\angle 2$ and $\angle 3$ are congruent. By the Vertical Angles Congruence Theorem (Theorem 2.6), $\angle 2 \cong \angle 6$. So, by the Transitive Property of Angle Congruence (Theorem 2.2), $\angle 3 \cong \angle 6$.

26. Write a proof using any format.

Given $\angle 3$ and $\angle 2$ are complementary.
$m\angle 1 + m\angle 2 = 90°$

Prove $\angle 3 \cong \angle 1$

2 Chapter Test

Use the diagram to determine whether you can assume the statement. Explain your reasoning.

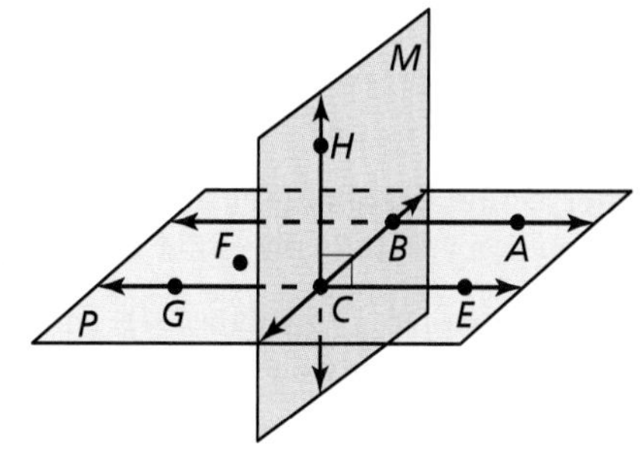

1. $\overleftrightarrow{AB} \perp$ plane M
2. Points F, G, and A are coplanar.
3. Points E, C, and G are collinear.
4. Planes M and P intersect at $\overleftrightarrow{BC}$.
5. $\overleftrightarrow{FA}$ lies in plane P.
6. $\overleftrightarrow{FG}$ intersects $\overleftrightarrow{AB}$ at point B.

Solve the equation. Justify each step.

7. $9x + 31 = -23 + 3x$
8. $26 + 2(3x + 11) = -18$
9. $3(7x - 9) - 19x = -15$

Write the if-then form, the converse, the inverse, the contrapositive, and the biconditional of the conditional statement.

10. Two planes intersect at a line.
11. A relation that pairs each input with exactly one output is a function.

Use inductive reasoning to make a conjecture about the given quantity. Then use deductive reasoning to show that the conjecture is true.

12. the sum of three odd integers
13. the product of three even integers
14. Give an example of two statements for which the Law of Detachment does not apply.
15. The formula for the area A of a triangle is $A = \frac{1}{2}bh$, where b is the base and h is the height. Solve the formula for h and justify each step. Then find the height of a standard yield sign when the area is 558 square inches and each side is 36 inches.

16. You visit the zoo and notice the following.
    - The elephants, giraffes, lions, tigers, and zebras are located along a straight walkway.
    - The giraffes are halfway between the elephants and the lions.
    - The tigers are halfway between the lions and the zebras.
    - The lions are halfway between the giraffes and the tigers.

    Draw and label a diagram that represents this information. Then prove that the distance between the elephants and the giraffes is equal to the distance between the tigers and the zebras. Use any proof format.

17. Write a proof using any format.

    Given $\angle 2 \cong \angle 3$

    $\overrightarrow{TV}$ bisects $\angle UTW$.

    Prove $\angle 1 \cong \angle 3$

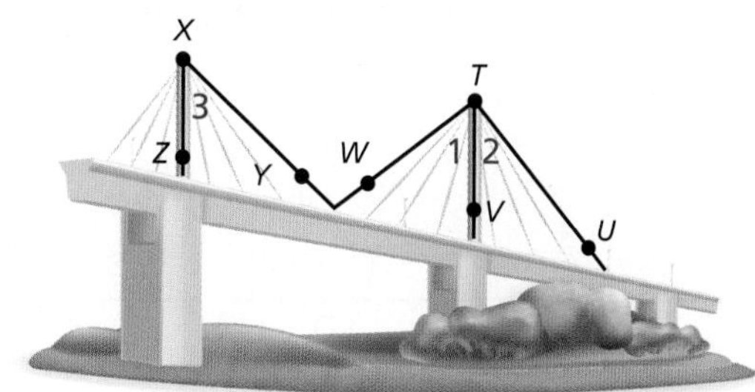

2 Cumulative Assessment

1. Use the diagram to write an example of each postulate.

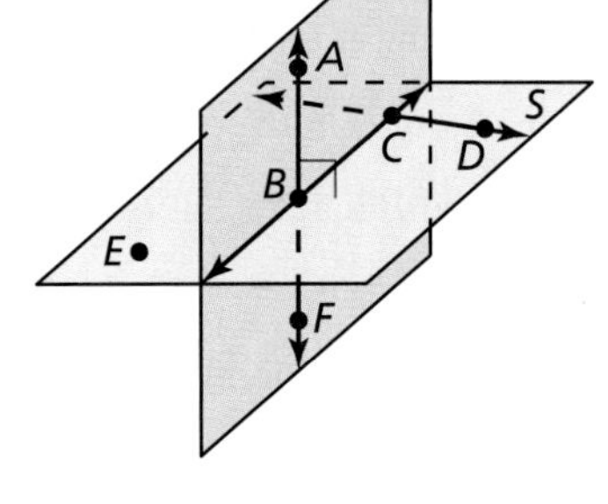

a. **Two Point Postulate (Postulate 2.1)** Through any two points, there exists exactly one line.

b. **Line Intersection Postulate (Postulate 2.3)** If two lines intersect, then their intersection is exactly one point.

c. **Three Point Postulate (Postulate 2.4)** Through any three noncollinear points, there exists exactly one plane.

d. **Plane-Line Postulate (Postulate 2.6)** If two points lie in a plane, then the line containing them lies in the plane.

e. **Plane Intersection Postulate (Postulate 2.7)** If two planes intersect, then their intersection is a line.

2. Enter the reasons in the correct positions to complete the two-column proof.

Given $\overline{AX} \cong \overline{DX}, \overline{XB} \cong \overline{XC}$
Prove $\overline{AC} \cong \overline{BD}$

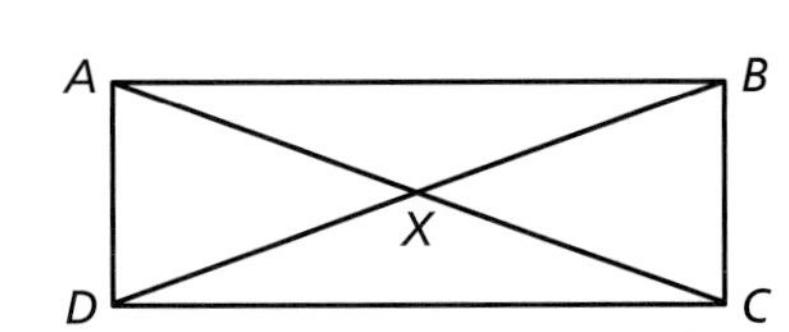

STATEMENTS	REASONS
1. $\overline{AX} \cong \overline{DX}$	1. Given
2. $AX = DX$	2. ________
3. $\overline{XB} \cong \overline{XC}$	3. Given
4. $XB = XC$	4. ________
5. $AX + XC = AC$	5. ________
6. $DX + XB = DB$	6. ________
7. $AC = DX + XB$	7. ________
8. $AC = BD$	8. ________
9. $\overline{AC} \cong \overline{BD}$	9. ________

Segment Addition Postulate (Postulate 1.2)

Definition of congruent segments

Substitution Property of Equality

3. Classify each related conditional statement, based on the conditional statement "If I study, then I will pass the final exam."

a. I will pass the final exam if and only if I study.

b. If I do not study, then I will not pass the final exam.

c. If I pass the final exam, then I studied.

d. If I do not pass the final exam, then I did not study.

4. List all segment bisectors given $x = 3$.

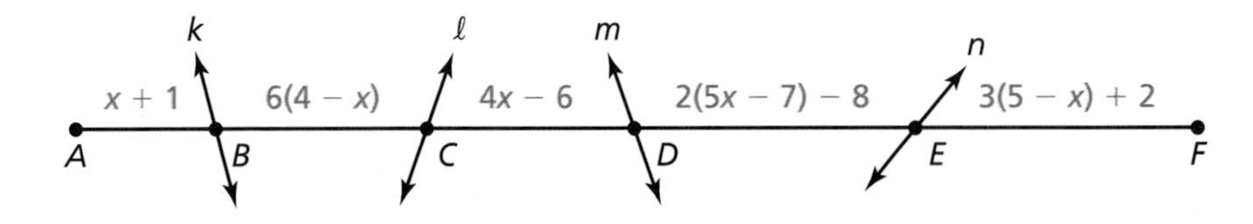

5. You are given $m\angle FHE = m\angle BHG = m\angle AHF = 90°$. Choose the symbol that makes each statement true. State which theorem or postulate, if any, supports your answer.

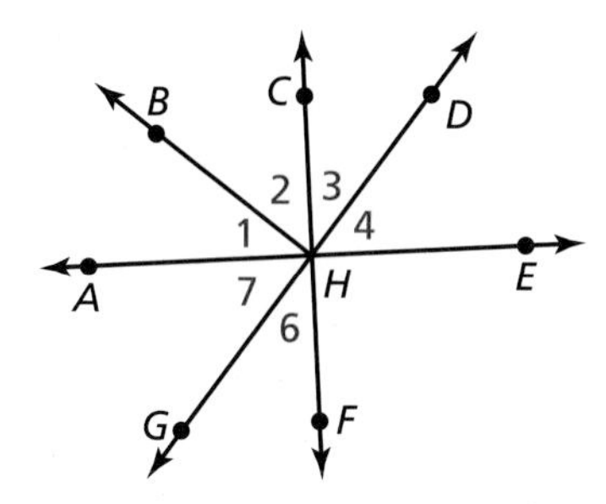

a. $\angle 3$ ___ $\angle 6$

b. $m\angle 4$ ___ $m\angle 7$

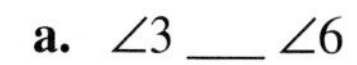

c. $m\angle FHE$ ___ $m\angle AHG$

d. $m\angle AHG + m\angle GHE$ ___ $180°$

$=$ $\quad \cong$ $\quad \neq$

6. Find the distance between each pair of points. Then order each line segment from longest to shortest.

a. $A(-6, 1), B(-1, 6)$

b. $C(-5, 8), D(5, 8)$

c. $E(2, 7), F(4, -2)$

d. $G(7, 3), H(7, -1)$

e. $J(-4, -2), K(1, -5)$

f. $L(3, -8), M(7, -5)$

7. The proof shows that $\angle MRL$ is congruent to $\angle NSR$. Select all other angles that are also congruent to $\angle NSR$.

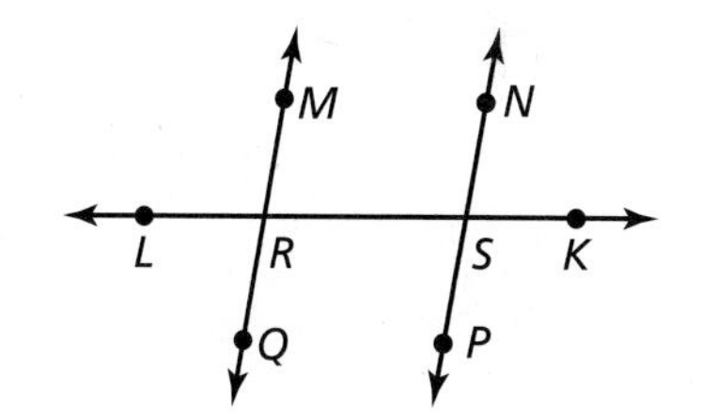

Given $\angle MRS$ and $\angle NSR$ are supplementary.

Prove $\angle MRL \cong \angle NSR$

STATEMENTS	REASONS
1. $\angle MRS$ and $\angle NSR$ are supplementary.	1. Given
2. $\angle MRL$ and $\angle MRS$ are a linear pair.	2. Definition of linear pair, as shown in the diagram
3. $\angle MRL$ and $\angle MRS$ are supplementary.	3. Linear Pair Postulate (Postulate 2.8)
4. $\angle MRL \cong \angle NSR$	4. Congruent Supplements Theorem (Theorem 2.4)

$\angle PSK$ $\quad \angle KSN$ $\quad \angle PSR$ $\quad \angle QRS$ $\quad \angle QRL$

8. Your teacher assigns your class a homework problem that asks you to prove the Vertical Angles Congruence Theorem (Theorem 2.6) using the picture and information given at the right. Your friend claims that this can be proved without using the Linear Pair Postulate (Postulate 2.8). Is your friend correct? Explain your reasoning.

Given $\angle 1$ and $\angle 3$ are vertical angles.

Prove $\angle 1 \cong \angle 3$

3 Parallel and Perpendicular Lines

Bike Path *(p. 161)*

Crosswalk *(p. 154)*

Kiteboarding *(p. 143)*

Gymnastics *(p. 130)*

Tree House *(p. 130)*

Maintaining Mathematical Proficiency

Finding the Slope of a Line

Example 1 **Find the slope of the line shown.**

Let $(x_1, y_1) = (-2, -2)$ and $(x_2, y_2) = (1, 0)$.

$\text{slope} = \dfrac{y_2 - y_1}{x_2 - x_1}$ — Write formula for slope.

$= \dfrac{0 - (-2)}{1 - (-2)}$ — Substitute.

$= \dfrac{2}{3}$ — Simplify.

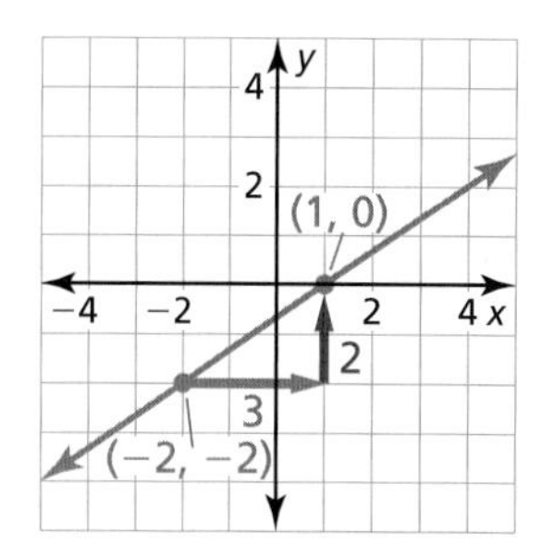

Find the slope of the line.

1.

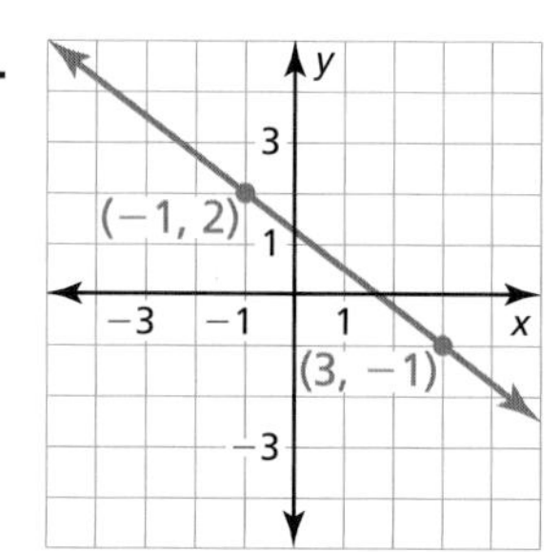

2.

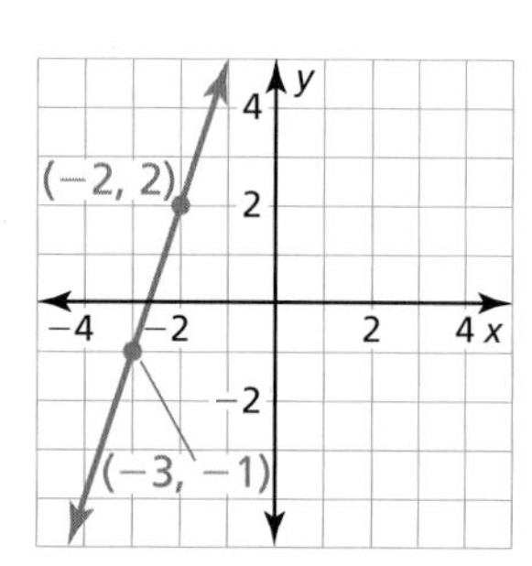

3.

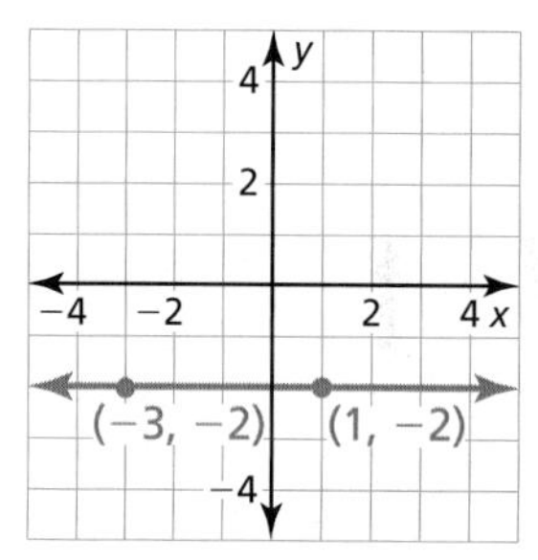

Writing Equations of Lines

Example 2 **Write an equation of the line that passes through the point (−4, 5) and has a slope of $\frac{3}{4}$.**

$y = mx + b$ — Write the slope-intercept form.

$5 = \frac{3}{4}(-4) + b$ — Substitute $\frac{3}{4}$ for m, −4 for x, and 5 for y.

$5 = -3 + b$ — Simplify.

$8 = b$ — Solve for b.

▶ So, an equation is $y = \frac{3}{4}x + 8$.

Write an equation of the line that passes through the given point and has the given slope.

4. $(6, 1);\ m = -3$

5. $(-3, 8);\ m = -2$

6. $(-1, 5);\ m = 4$

7. $(2, -4);\ m = \frac{1}{2}$

8. $(-8, -5);\ m = -\frac{1}{4}$

9. $(0, 9);\ m = \frac{2}{3}$

10. **ABSTRACT REASONING** Why does a horizontal line have a slope of 0, but a vertical line has an undefined slope?

Mathematical Practices

Mathematically proficient students use technological tools to explore concepts.

Characteristics of Lines in a Coordinate Plane

Core Concept

Lines in a Coordinate Plane

1. In a coordinate plane, two lines are *parallel* if and only if they are both vertical lines or they both have the same slope.
2. In a coordinate plane, two lines are *perpendicular* if and only if one is vertical and the other is horizontal or the slopes of the lines are negative reciprocals of each other.
3. In a coordinate plane, two lines are *coincident* if and only if their equations are equivalent.

EXAMPLE 1 Classifying Pairs of Lines

Here are some examples of pairs of lines in a coordinate plane.

a. $2x + y = 2$
$x - y = 4$

These lines are not parallel or perpendicular. They intersect at $(2, -2)$.

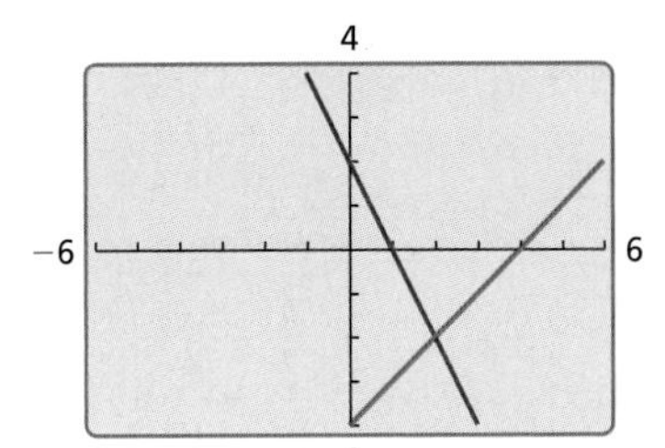

b. $2x + y = 2$
$4x + 2y = 4$

These lines are coincident because their equations are equivalent.

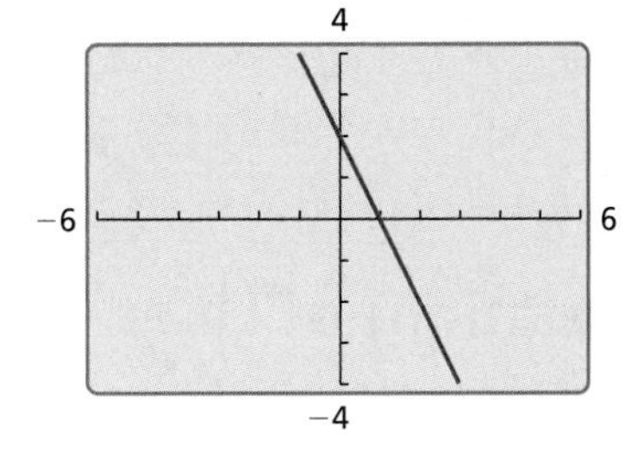

c. $2x + y = 2$
$2x + y = 4$

These lines are parallel. Each line has a slope of $m = -2$.

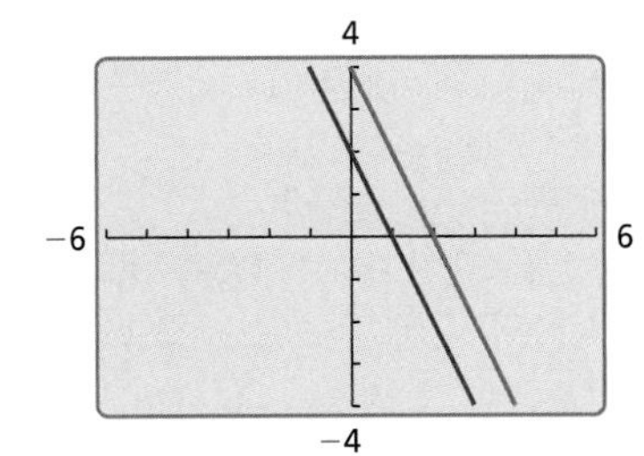

d. $2x + y = 2$
$x - 2y = 4$

These lines are perpendicular. They have slopes of $m_1 = -2$ and $m_2 = \frac{1}{2}$.

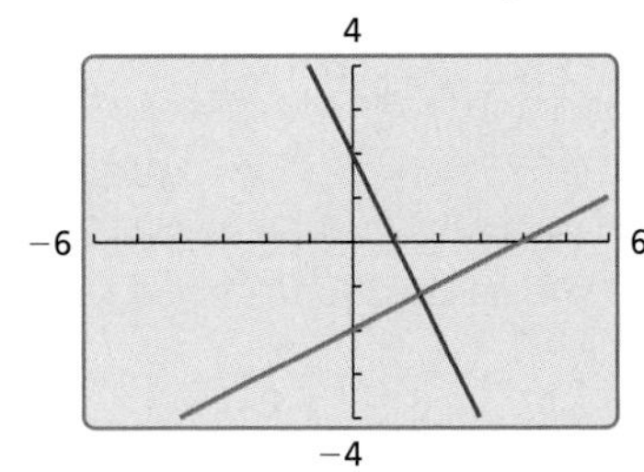

Monitoring Progress

Use a graphing calculator to graph the pair of lines. Use a square viewing window. Classify the lines as parallel, perpendicular, coincident, or nonperpendicular intersecting lines. Justify your answer.

1. $x + 2y = 2$
$2x - y = 4$

2. $x + 2y = 2$
$2x + 4y = 4$

3. $x + 2y = 2$
$x + 2y = -2$

4. $x + 2y = 2$
$x - y = -4$

3.1 Pairs of Lines and Angles

Essential Question

What does it mean when two lines are parallel, intersecting, coincident, or skew?

EXPLORATION 1 Points of Intersection

Work with a partner. Write the number of points of intersection of each pair of coplanar lines.

a. parallel lines

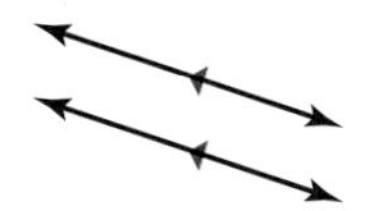

b. intersecting lines

c. coincident lines

EXPLORATION 2 Classifying Pairs of Lines

Work with a partner. The figure shows a *right rectangular prism.* All its angles are right angles. Classify each of the following pairs of lines as *parallel*, *intersecting*, *coincident*, or *skew*. Justify your answers. (Two lines are **skew lines** when they do not intersect and are not coplanar.)

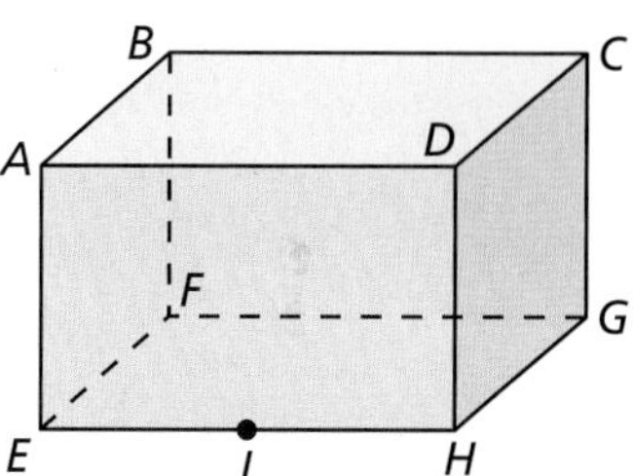

Pair of Lines	Classification	Reason
a. $\overleftrightarrow{AB}$ and $\overleftrightarrow{BC}$		
b. $\overleftrightarrow{AD}$ and $\overleftrightarrow{BC}$		
c. $\overleftrightarrow{EI}$ and $\overleftrightarrow{IH}$		
d. $\overleftrightarrow{BF}$ and $\overleftrightarrow{EH}$		
e. $\overleftrightarrow{EF}$ and $\overleftrightarrow{CG}$		
f. $\overleftrightarrow{AB}$ and $\overleftrightarrow{GH}$		

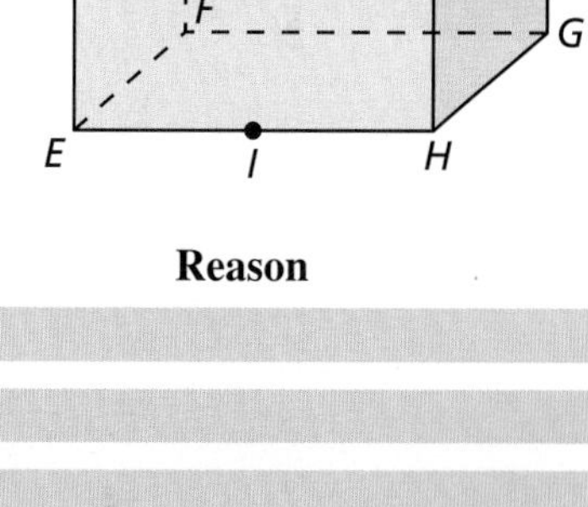

CONSTRUCTING VIABLE ARGUMENTS

To be proficient in math, you need to understand and use stated assumptions, definitions, and previously established results.

EXPLORATION 3 Identifying Pairs of Angles

Work with a partner. In the figure, two parallel lines are intersected by a third line called a *transversal.*

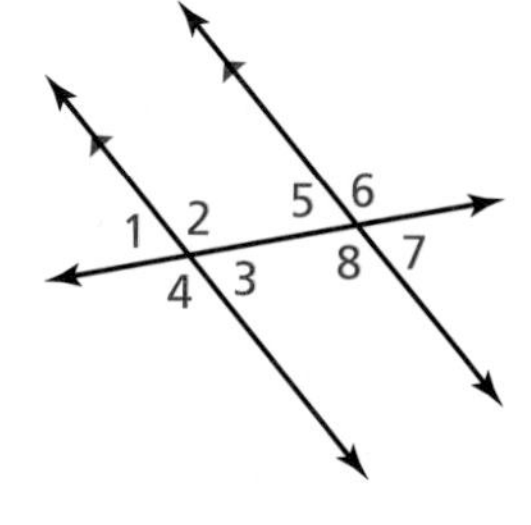

a. Identify all the pairs of vertical angles. Explain your reasoning.

b. Identify all the linear pairs of angles. Explain your reasoning.

Communicate Your Answer

4. What does it mean when two lines are parallel, intersecting, coincident, or skew?

5. In Exploration 2, find three more pairs of lines that are different from those given. Classify the pairs of lines as *parallel*, *intersecting*, *coincident*, or *skew*. Justify your answers.

3.1 Lesson

Core Vocabulary

parallel lines, *p. 126*
skew lines, *p. 126*
parallel planes, *p. 126*
transversal, *p. 128*
corresponding angles, *p. 128*
alternate interior angles, *p. 128*
alternate exterior angles, *p. 128*
consecutive interior angles, *p. 128*

Previous
perpendicular lines

What You Will Learn

- Identify lines and planes.
- Identify parallel and perpendicular lines.
- Identify pairs of angles formed by transversals.

Identifying Lines and Planes

Core Concept

Parallel Lines, Skew Lines, and Parallel Planes

Two lines that do not intersect are either *parallel lines* or *skew lines*. Two lines are **parallel lines** when they do not intersect and are coplanar. Two lines are **skew lines** when they do not intersect and are not coplanar. Also, two planes that do not intersect are **parallel planes**.

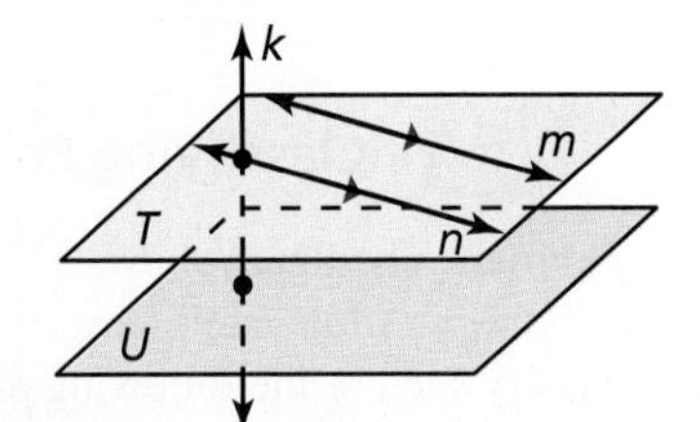

Lines m and n are parallel lines ($m \parallel n$).

Lines m and k are skew lines.

Planes T and U are parallel planes ($T \parallel U$).

Lines k and n are intersecting lines, and there is a plane (not shown) containing them.

Small directed arrows, as shown in red on lines m and n above, are used to show that lines are parallel. The symbol $\parallel$ means "is parallel to," as in $m \parallel n$.

Segments and rays are parallel when they lie in parallel lines. A line is parallel to a plane when the line is in a plane parallel to the given plane. In the diagram above, line n is parallel to plane U.

EXAMPLE 1 Identifying Lines and Planes

Think of each segment in the figure as part of a line. Which line(s) or plane(s) appear to fit the description?

a. line(s) parallel to $\overleftrightarrow{CD}$ and containing point A

b. line(s) skew to $\overleftrightarrow{CD}$ and containing point A

c. line(s) perpendicular to $\overleftrightarrow{CD}$ and containing point A

d. plane(s) parallel to plane EFG and containing point A

REMEMBER
Recall that if two lines intersect to form a right angle, then they are perpendicular lines.

SOLUTION

a. $\overleftrightarrow{AB}$, $\overleftrightarrow{HG}$, and $\overleftrightarrow{EF}$ all appear parallel to $\overleftrightarrow{CD}$, but only $\overleftrightarrow{AB}$ contains point A.

b. Both $\overleftrightarrow{AG}$ and $\overleftrightarrow{AH}$ appear skew to $\overleftrightarrow{CD}$ and contain point A.

c. $\overleftrightarrow{BC}$, $\overleftrightarrow{AD}$, $\overleftrightarrow{DE}$, and $\overleftrightarrow{FC}$ all appear perpendicular to $\overleftrightarrow{CD}$, but only $\overleftrightarrow{AD}$ contains point A.

d. Plane ABC appears parallel to plane EFG and contains point A.

Monitoring Progress  Help in English and Spanish at *BigIdeasMath.com*

1. Look at the diagram in Example 1. Name the line(s) through point F that appear skew to $\overleftrightarrow{EH}$.

Identifying Parallel and Perpendicular Lines

Two distinct lines in the same plane either are parallel, like line ℓ and line n, or intersect in a point, like line j and line n.

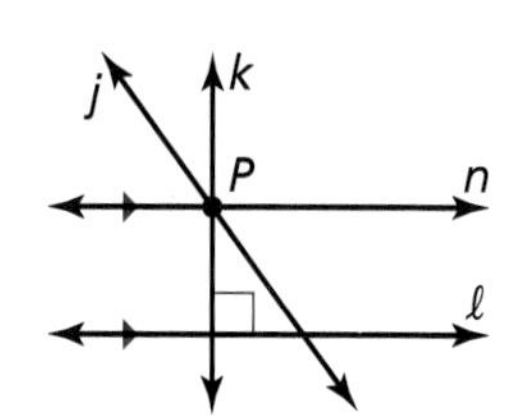

Through a point not on a line, there are infinitely many lines. Exactly one of these lines is parallel to the given line, and exactly one of them is perpendicular to the given line. For example, line k is the line through point P perpendicular to line ℓ, and line n is the line through point P parallel to line ℓ.

Postulates

Postulate 3.1 Parallel Postulate

If there is a line and a point not on the line, then there is exactly one line through the point parallel to the given line.

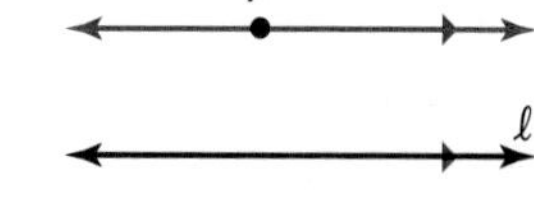

There is exactly one line through P parallel to ℓ.

Postulate 3.2 Perpendicular Postulate

If there is a line and a point not on the line, then there is exactly one line through the point perpendicular to the given line.

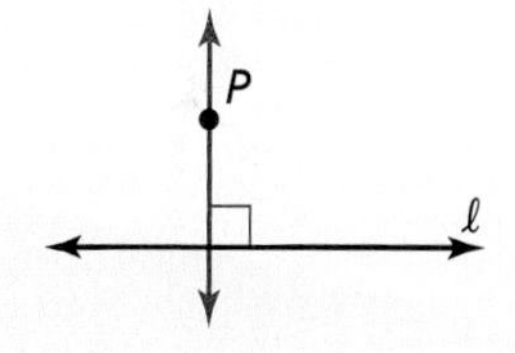

There is exactly one line through P perpendicular to ℓ.

EXAMPLE 2 Identifying Parallel and Perpendicular Lines

The given line markings show how the roads in a town are related to one another.

a. Name a pair of parallel lines.

b. Name a pair of perpendicular lines.

c. Is $\overleftrightarrow{FE} \parallel \overleftrightarrow{AC}$? Explain.

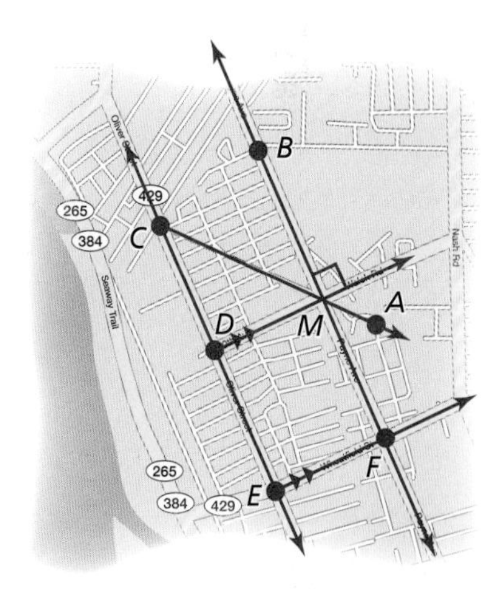

SOLUTION

a. $\overleftrightarrow{MD} \parallel \overleftrightarrow{FE}$

b. $\overleftrightarrow{MD} \perp \overleftrightarrow{BF}$

c. $\overleftrightarrow{FE}$ is not parallel to $\overleftrightarrow{AC}$, because $\overleftrightarrow{MD}$ is parallel to $\overleftrightarrow{FE}$, and by the Parallel Postulate, there is exactly one line parallel to $\overleftrightarrow{FE}$ through M.

Monitoring Progress  Help in English and Spanish at *BigIdeasMath.com*

2. In Example 2, can you use the Perpendicular Postulate to show that $\overleftrightarrow{AC}$ is *not* perpendicular to $\overleftrightarrow{BF}$? Explain why or why not.

Identifying Pairs of Angles

A **transversal** is a line that intersects two or more coplanar lines at different points.

Core Concept

Angles Formed by Transversals

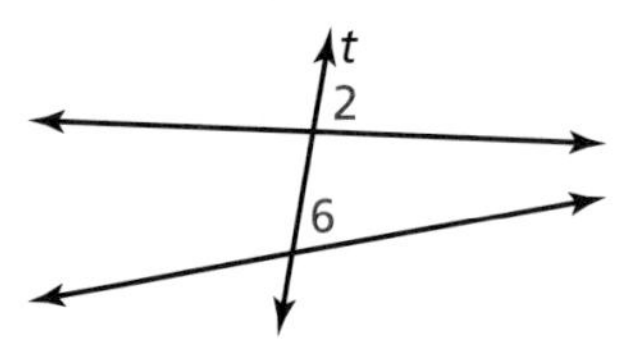

Two angles are **corresponding angles** when they have corresponding positions. For example, $\angle 2$ and $\angle 6$ are above the lines and to the right of the transversal t.

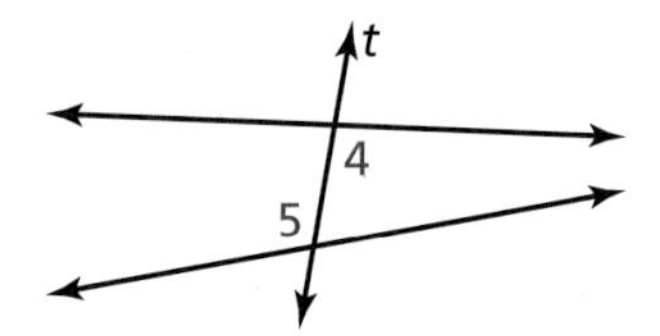

Two angles are **alternate interior angles** when they lie between the two lines and on opposite sides of the transversal t.

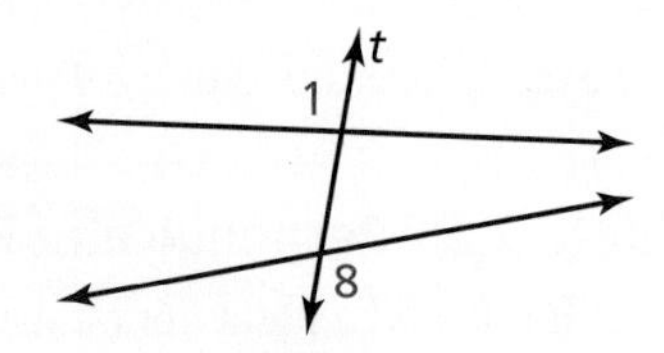

Two angles are **alternate exterior angles** when they lie outside the two lines and on opposite sides of the transversal t.

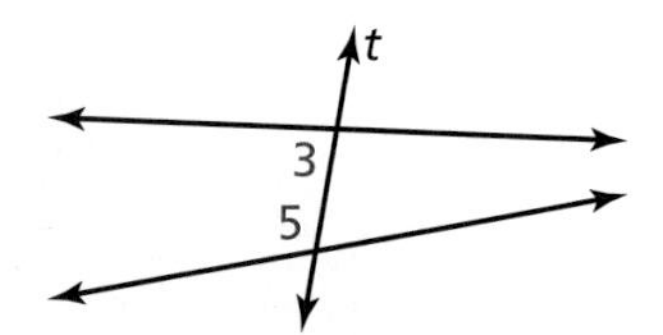

Two angles are **consecutive interior angles** when they lie between the two lines and on the same side of the transversal t.

EXAMPLE 3 Identifying Pairs of Angles

Identify all pairs of angles of the given type.

a. corresponding
b. alternate interior
c. alternate exterior
d. consecutive interior

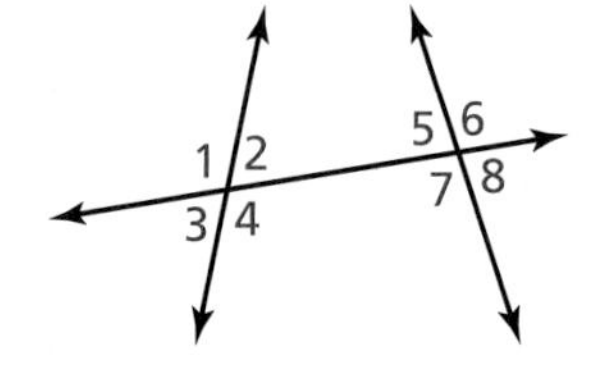

SOLUTION

a. $\angle 1$ and $\angle 5$
$\angle 2$ and $\angle 6$
$\angle 3$ and $\angle 7$
$\angle 4$ and $\angle 8$

b. $\angle 2$ and $\angle 7$
$\angle 4$ and $\angle 5$

c. $\angle 1$ and $\angle 8$
$\angle 3$ and $\angle 6$

d. $\angle 2$ and $\angle 5$
$\angle 4$ and $\angle 7$

Monitoring Progress  Help in English and Spanish at *BigIdeasMath.com*

Classify the pair of numbered angles.

3. 1 5

4.
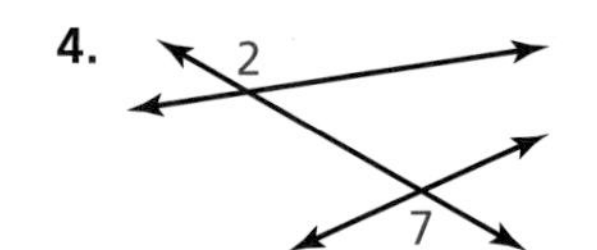

5. 5 4

3.1 Exercises

Dynamic Solutions available at *BigIdeasMath.com*

Vocabulary and Core Concept Check

1. **COMPLETE THE SENTENCE** Two lines that do not intersect and are also not parallel are ________ lines.

2. **WHICH ONE DOESN'T BELONG?** Which angle pair does *not* belong with the other three? Explain your reasoning.

$\angle 2$ and $\angle 3$	$\angle 4$ and $\angle 5$
$\angle 1$ and $\angle 8$	$\angle 2$ and $\angle 7$

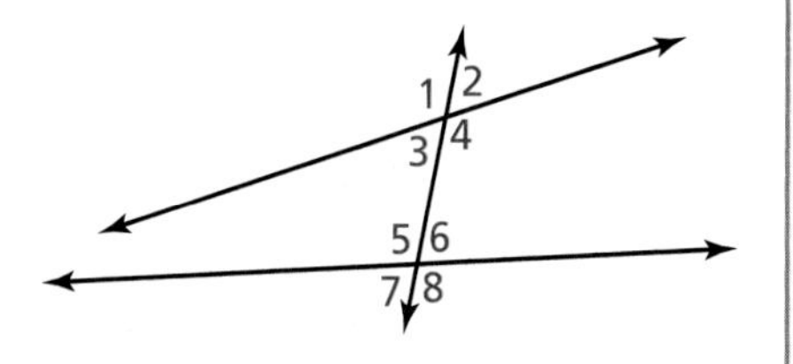

Monitoring Progress and Modeling with Mathematics

In Exercises 3–6, think of each segment in the diagram as part of a line. All the angles are right angles. Which line(s) or plane(s) contain point *B* and appear to fit the description? *(See Example 1.)*

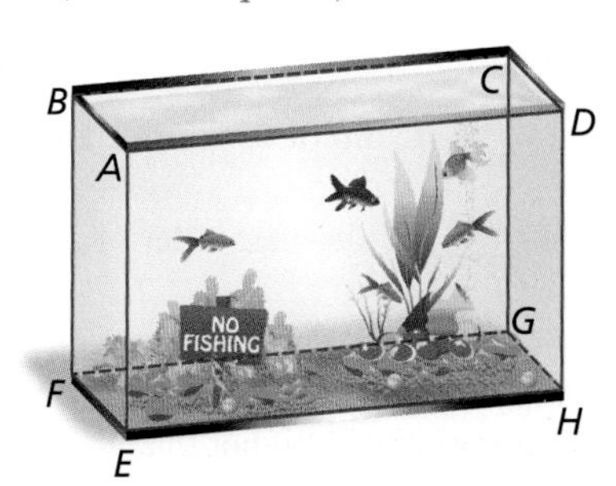

3. line(s) parallel to $\overleftrightarrow{CD}$

4. line(s) perpendicular to $\overleftrightarrow{CD}$

5. line(s) skew to $\overleftrightarrow{CD}$

6. plane(s) parallel to plane *CDH*

In Exercises 7–10, use the diagram. *(See Example 2.)*

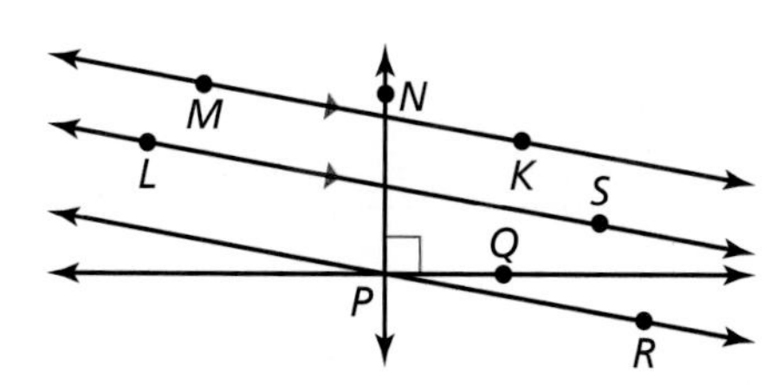

7. Name a pair of parallel lines.

8. Name a pair of perpendicular lines.

9. Is $\overleftrightarrow{PN} \parallel \overleftrightarrow{KM}$? Explain.

10. Is $\overleftrightarrow{PR} \perp \overleftrightarrow{NP}$? Explain.

In Exercises 11–14, identify all pairs of angles of the given type. *(See Example 3.)*

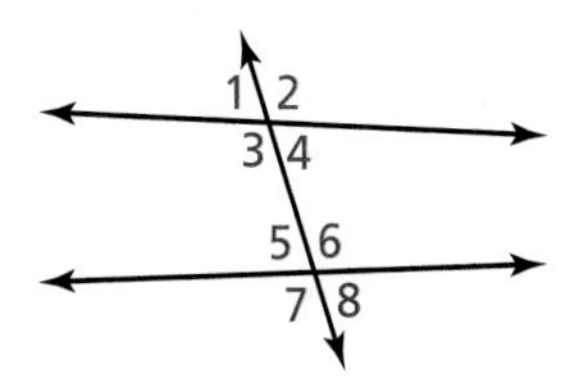

11. corresponding

12. alternate interior

13. alternate exterior

14. consecutive interior

USING STRUCTURE In Exercises 15–18, classify the angle pair as *corresponding*, *alternate interior*, *alternate exterior*, or *consecutive interior* angles.

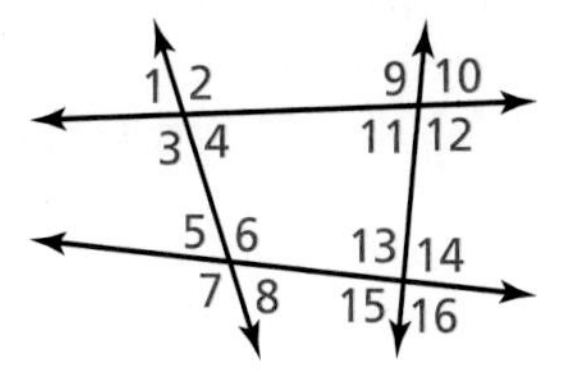

15. $\angle 5$ and $\angle 1$

16. $\angle 11$ and $\angle 13$

17. $\angle 6$ and $\angle 13$

18. $\angle 2$ and $\angle 11$

ERROR ANALYSIS **In Exercises 19 and 20, describe and correct the error in the conditional statement about lines.**

19. 

If two lines do not intersect, then they are parallel.

20. 

If there is a line and a point not on the line, then there is exactly one line through the point that intersects the given line.

21. **MODELING WITH MATHEMATICS** Use the photo to decide whether the statement is true or false. Explain your reasoning.

a. The plane containing the floor of the tree house is parallel to the ground.

b. The lines containing the railings of the staircase, such as $\overleftrightarrow{AB}$, are skew to all lines in the plane containing the ground.

c. All the lines containing the balusters, such as $\overleftrightarrow{CD}$, are perpendicular to the plane containing the floor of the tree house.

22. **THOUGHT PROVOKING** If two lines are intersected by a third line, is the third line necessarily a transversal? Justify your answer with a diagram.

23. **MATHEMATICAL CONNECTIONS** Two lines are cut by a transversal. Is it possible for all eight angles formed to have the same measure? Explain your reasoning.

24. **HOW DO YOU SEE IT?** Think of each segment in the figure as part of a line.

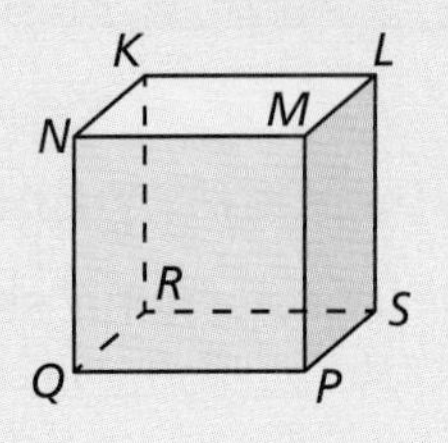

a. Which lines are parallel to $\overleftrightarrow{NQ}$?

b. Which lines intersect $\overleftrightarrow{NQ}$?

c. Which lines are skew to $\overleftrightarrow{NQ}$?

d. Should you have named all the lines on the cube in parts (a)–(c) except $\overleftrightarrow{NQ}$? Explain.

In Exercises 25–28, copy and complete the statement. List all possible correct answers.

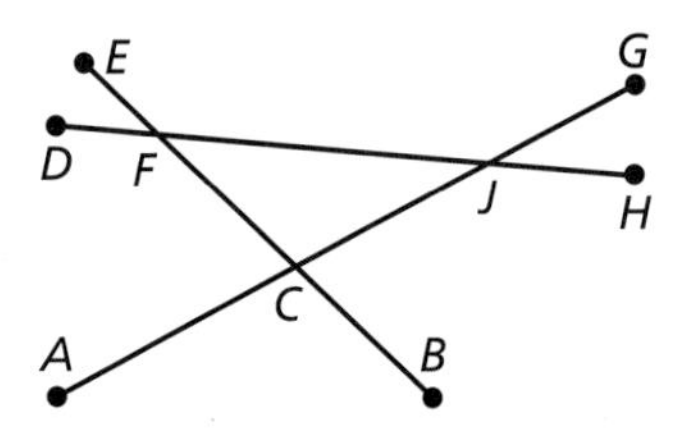

25. $\angle BCG$ and ____ are corresponding angles.

26. $\angle BCG$ and ____ are consecutive interior angles.

27. $\angle FCJ$ and ____ are alternate interior angles.

28. $\angle FCA$ and ____ are alternate exterior angles.

29. **MAKING AN ARGUMENT** Your friend claims the uneven parallel bars in gymnastics are not really parallel. She says one is higher than the other, so they cannot be in the same plane. Is she correct? Explain.

Maintaining Mathematical Proficiency

Reviewing what you learned in previous grades and lessons

Use the diagram to find the measures of all the angles. *(Section 2.6)*

30. $m\angle 1 = 76^\circ$

31. $m\angle 2 = 159^\circ$

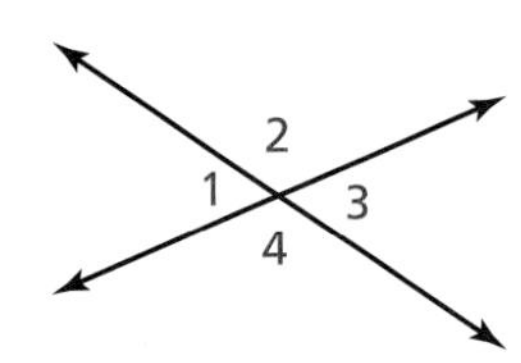

3.2 Parallel Lines and Transversals

Essential Question

When two parallel lines are cut by a transversal, which of the resulting pairs of angles are congruent?

EXPLORATION 1 Exploring Parallel Lines

Work with a partner.
Use dynamic geometry software to draw two parallel lines. Draw a third line that intersects both parallel lines. Find the measures of the eight angles that are formed. What can you conclude?

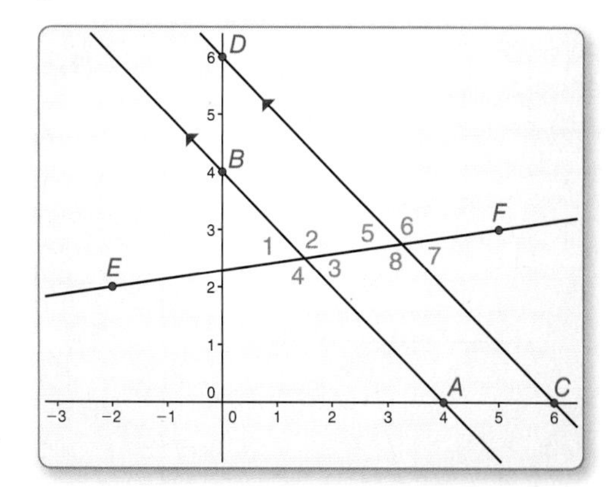

ATTENDING TO PRECISION

To be proficient in math, you need to communicate precisely with others.

EXPLORATION 2 Writing Conjectures

Work with a partner. Use the results of Exploration 1 to write conjectures about the following pairs of angles formed by two parallel lines and a transversal.

a. corresponding angles

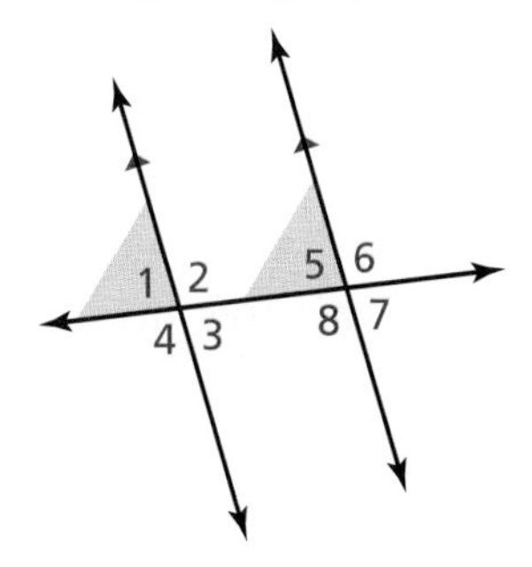

b. alternate interior angles

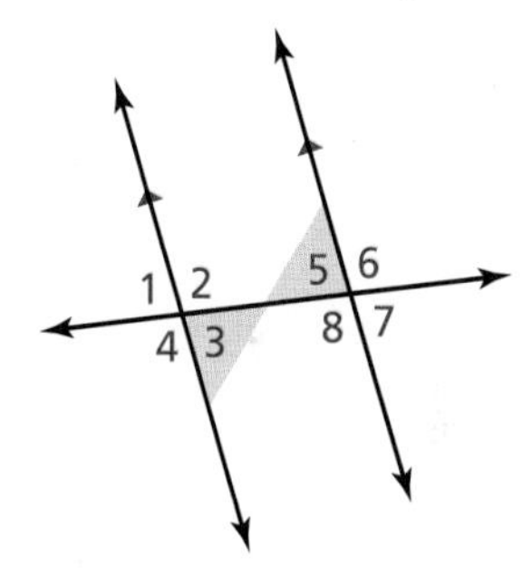

c. alternate exterior angles

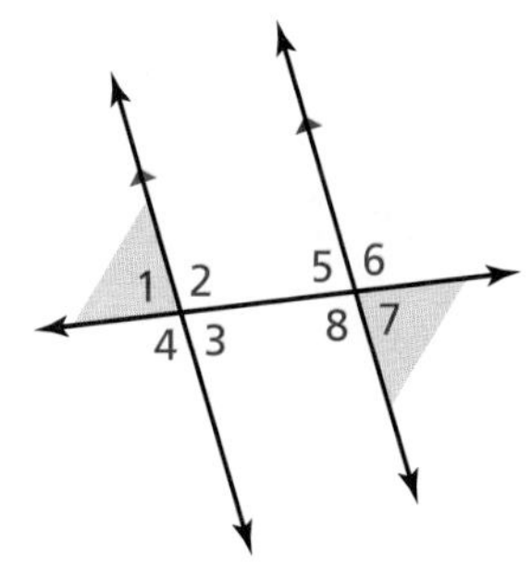

d. consecutive interior angles

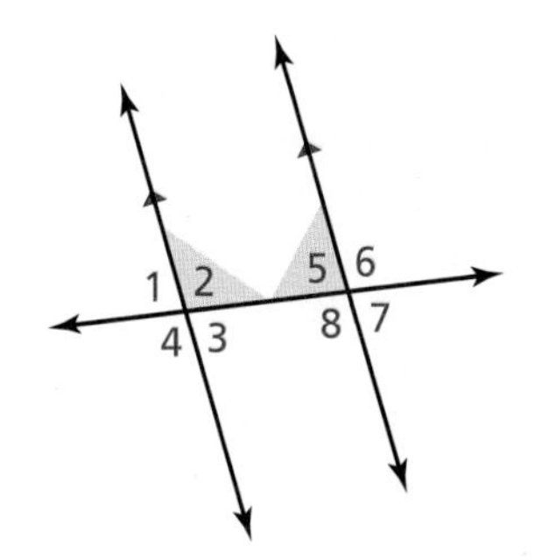

Communicate Your Answer

3. When two parallel lines are cut by a transversal, which of the resulting pairs of angles are congruent?

4. In Exploration 2, $m\angle 1 = 80°$. Find the other angle measures.

3.2 Lesson

What You Will Learn

- Use properties of parallel lines.
- Prove theorems about parallel lines.
- Solve real-life problems.

Core Vocabulary

Previous
corresponding angles
parallel lines
supplementary angles
vertical angles

Using Properties of Parallel Lines

Theorems

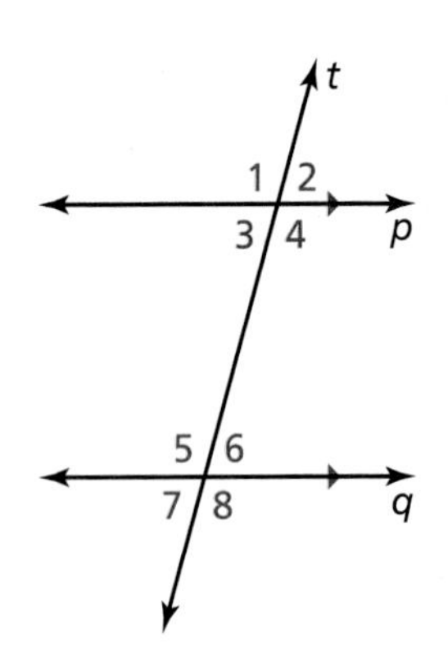

Theorem 3.1 Corresponding Angles Theorem

If two parallel lines are cut by a transversal, then the pairs of corresponding angles are congruent.

Examples In the diagram at the left, $\angle 2 \cong \angle 6$ and $\angle 3 \cong \angle 7$.

Proof Ex. 36, p. 180

Theorem 3.2 Alternate Interior Angles Theorem

If two parallel lines are cut by a transversal, then the pairs of alternate interior angles are congruent.

Examples In the diagram at the left, $\angle 3 \cong \angle 6$ and $\angle 4 \cong \angle 5$.

Proof Example 4, p. 134

Theorem 3.3 Alternate Exterior Angles Theorem

If two parallel lines are cut by a transversal, then the pairs of alternate exterior angles are congruent.

Examples In the diagram at the left, $\angle 1 \cong \angle 8$ and $\angle 2 \cong \angle 7$.

Proof Ex. 15, p. 136

Theorem 3.4 Consecutive Interior Angles Theorem

If two parallel lines are cut by a transversal, then the pairs of consecutive interior angles are supplementary.

Examples In the diagram at the left, $\angle 3$ and $\angle 5$ are supplementary, and $\angle 4$ and $\angle 6$ are supplementary.

Proof Ex. 16, p. 136

ANOTHER WAY

There are many ways to solve Example 1. Another way is to use the Corresponding Angles Theorem to find $m\angle 5$ and then use the Vertical Angles Congruence Theorem (Theorem 2.6) to find $m\angle 4$ and $m\angle 8$.

EXAMPLE 1 Identifying Angles

The measures of three of the numbered angles are 120°. Identify the angles. Explain your reasoning.

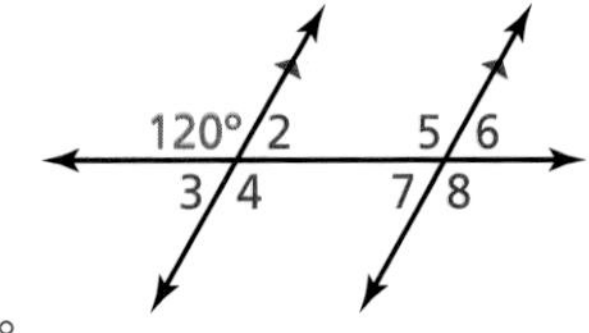

SOLUTION

By the Alternate Exterior Angles Theorem, $m\angle 8 = 120°$.

$\angle 5$ and $\angle 8$ are vertical angles. Using the Vertical Angles Congruence Theorem (Theorem 2.6), $m\angle 5 = 120°$.

$\angle 5$ and $\angle 4$ are alternate interior angles. By the Alternate Interior Angles Theorem, $\angle 4 = 120°$.

▶ So, the three angles that each have a measure of 120° are $\angle 4$, $\angle 5$, and $\angle 8$.

EXAMPLE 2 Using Properties of Parallel Lines

Find the value of x.

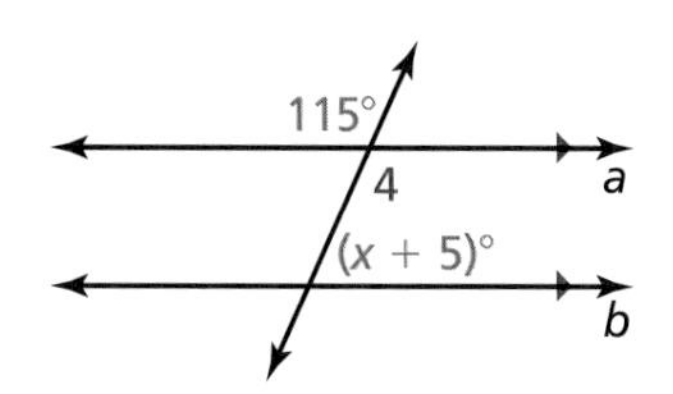

SOLUTION

By the Vertical Angles Congruence Theorem (Theorem 2.6), $m\angle 4 = 115°$. Lines a and b are parallel, so you can use the theorems about parallel lines.

$m\angle 4 + (x + 5)° = 180°$	Consecutive Interior Angles Theorem
$115° + (x + 5)° = 180°$	Substitute 115° for $m\angle 4$.
$x + 120 = 180$	Combine like terms.
$x = 60$	Subtract 120 from each side.

▶ So, the value of x is 60.

Check

$115° + (x + 5)° = 180°$

$115 + (60 + 5) \stackrel{?}{=} 180$

$180 = 180$ ✓

EXAMPLE 3 Using Properties of Parallel Lines

Find the value of x.

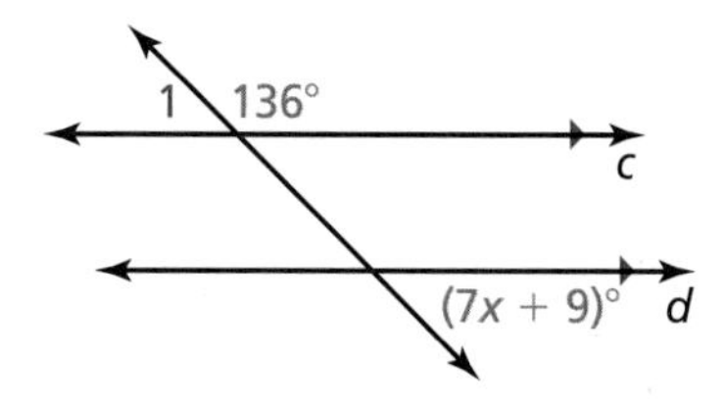

SOLUTION

By the Linear Pair Postulate (Postulate 2.8), $m\angle 1 = 180° - 136° = 44°$. Lines c and d are parallel, so you can use the theorems about parallel lines.

$m\angle 1 = (7x + 9)°$	Alternate Exterior Angles Theorem
$44° = (7x + 9)°$	Substitute 44° for $m\angle 1$.
$35 = 7x$	Subtract 9 from each side.
$5 = x$	Divide each side by 7.

▶ So, the value of x is 5.

Check

$44° = (7x + 9)°$

$44 \stackrel{?}{=} 7(5) + 9$

$44 = 44$ ✓

Monitoring Progress

Help in English and Spanish at *BigIdeasMath.com*

Use the diagram.

1. Given $m\angle 1 = 105°$, find $m\angle 4$, $m\angle 5$, and $m\angle 8$. Tell which theorem you use in each case.
2. Given $m\angle 3 = 68°$ and $m\angle 8 = (2x + 4)°$, what is the value of x? Show your steps.

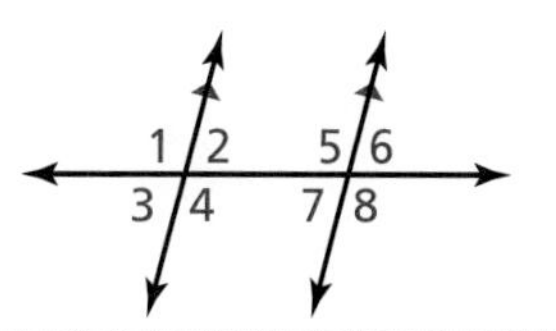

Proving Theorems about Parallel Lines

EXAMPLE 4 Proving the Alternate Interior Angles Theorem

Prove that if two parallel lines are cut by a transversal, then the pairs of alternate interior angles are congruent.

STUDY TIP

Before you write a proof, identify the **Given** and **Prove** statements for the situation described or for any diagram you draw.

SOLUTION

Draw a diagram. Label a pair of alternate interior angles as $\angle 1$ and $\angle 2$. You are looking for an angle that is related to both $\angle 1$ and $\angle 2$. Notice that one angle is a vertical angle with $\angle 2$ and a corresponding angle with $\angle 1$. Label it $\angle 3$.

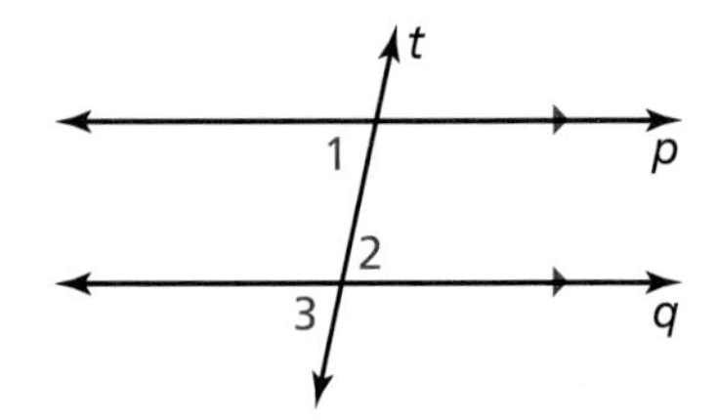

Given $p \parallel q$

Prove $\angle 1 \cong \angle 2$

STATEMENTS	REASONS
1. $p \parallel q$	**1.** Given
2. $\angle 1 \cong \angle 3$	**2.** Corresponding Angles Theorem
3. $\angle 3 \cong \angle 2$	**3.** Vertical Angles Congruence Theorem (Theorem 2.6)
4. $\angle 1 \cong \angle 2$	**4.** Transitive Property of Congruence (Theorem 2.2)

Monitoring Progress  Help in English and Spanish at *BigIdeasMath.com*

3. In the proof in Example 4, if you use the third statement before the second statement, could you still prove the theorem? Explain.

Solving Real-Life Problems

EXAMPLE 5 Solving a Real-life Problem

When sunlight enters a drop of rain, different colors of light leave the drop at different angles. This process is what makes a rainbow. For violet light, $m\angle 2 = 40°$. What is $m\angle 1$? How do you know?

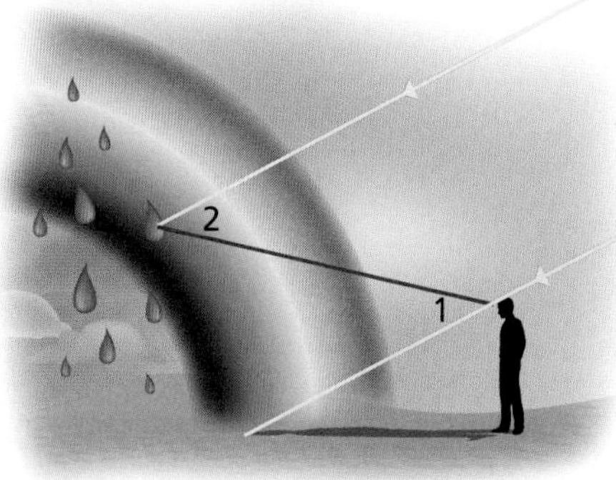

SOLUTION

Because the Sun's rays are parallel, $\angle 1$ and $\angle 2$ are alternate interior angles. By the Alternate Interior Angles Theorem, $\angle 1 \cong \angle 2$.

▶ So, by the definition of congruent angles, $m\angle 1 = m\angle 2 = 40°$.

Monitoring Progress  Help in English and Spanish at *BigIdeasMath.com*

4. WHAT IF? In Example 5, yellow light leaves a drop at an angle of $m\angle 2 = 41°$. What is $m\angle 1$? How do you know?

3.2 Exercises

Dynamic Solutions available at *BigIdeasMath.com*

Vocabulary and Core Concept Check

1. **WRITING** How are the Alternate Interior Angles Theorem (Theorem 3.2) and the Alternate Exterior Angles Theorem (Theorem 3.3) alike? How are they different?

2. **WHICH ONE DOESN'T BELONG?** Which pair of angle measures does *not* belong with the other three? Explain.

$m\angle 1$ and $m\angle 3$	$m\angle 2$ and $m\angle 4$
$m\angle 2$ and $m\angle 3$	$m\angle 1$ and $m\angle 5$

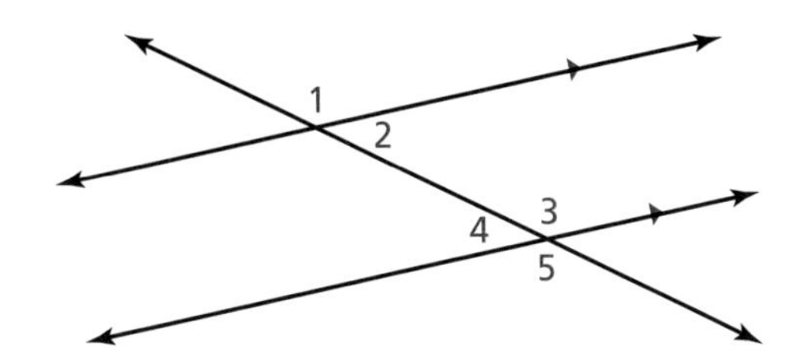

Monitoring Progress and Modeling with Mathematics

In Exercises 3–6, find $m\angle 1$ and $m\angle 2$. Tell which theorem you use in each case. *(See Example 1.)*

3.
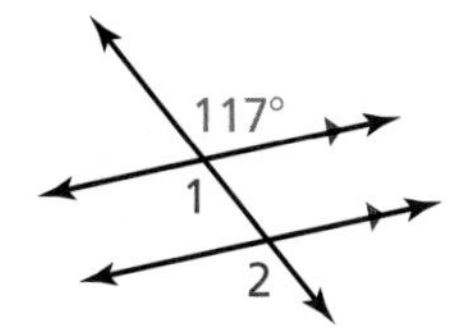

4.
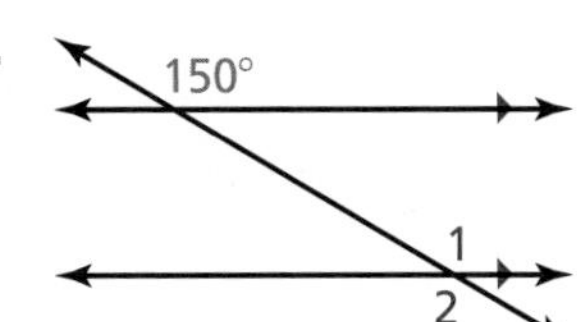

5.
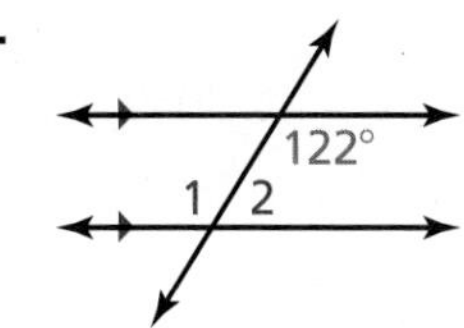

6.
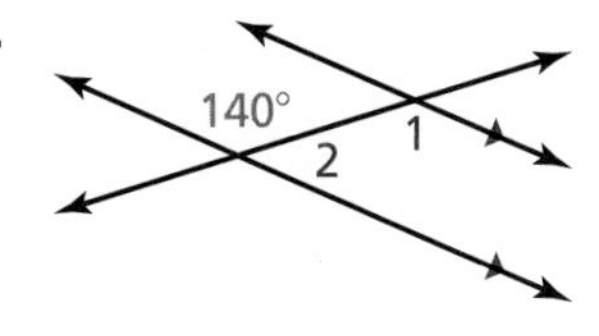

In Exercises 7–10, find the value of x. Show your steps. *(See Examples 2 and 3.)*

7.
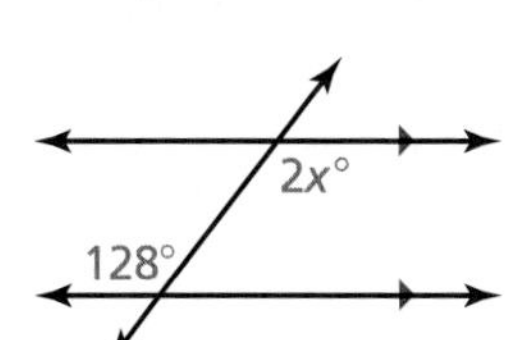

8.
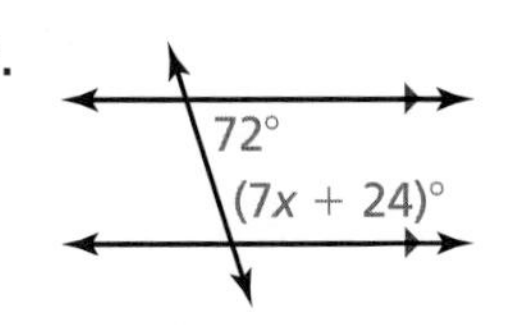

9.
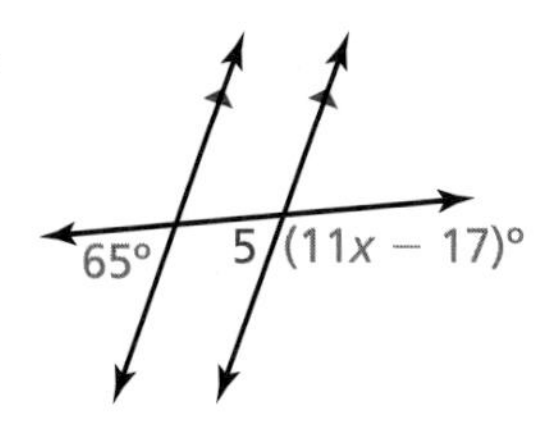

10.
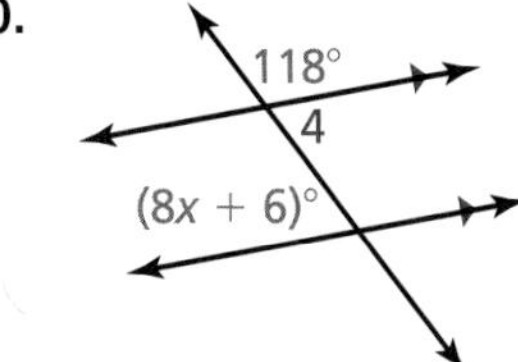

In Exercises 11 and 12, find $m\angle 1$, $m\angle 2$, and $m\angle 3$. Explain your reasoning.

11.
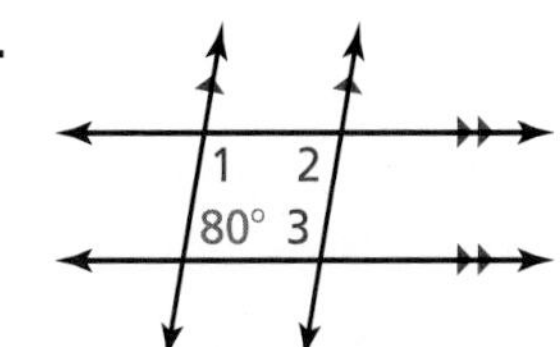

12.
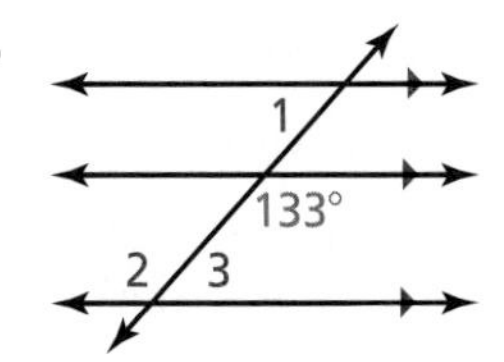

13. **ERROR ANALYSIS** Describe and correct the error in the student's reasoning.

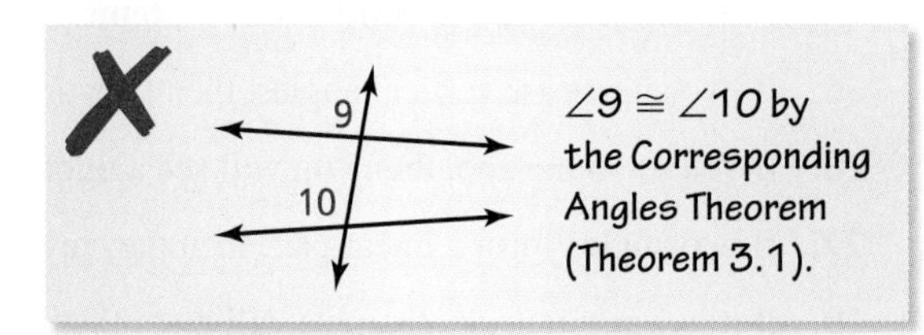

14. HOW DO YOU SEE IT? Use the diagram.

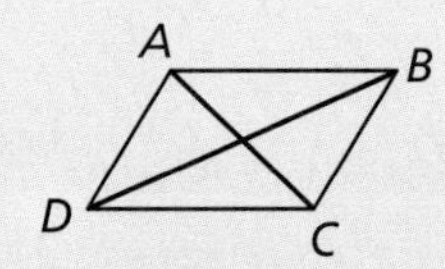

a. Name two pairs of congruent angles when $\overline{AD}$ and $\overline{BC}$ are parallel. Explain your reasoning.

b. Name two pairs of supplementary angles when $\overline{AB}$ and $\overline{DC}$ are parallel. Explain your reasoning.

19. CRITICAL THINKING Is it possible for consecutive interior angles to be congruent? Explain.

20. THOUGHT PROVOKING The postulates and theorems in this book represent Euclidean geometry. In spherical geometry, all points are points on the surface of a sphere. A line is a circle on the sphere whose diameter is equal to the diameter of the sphere. In spherical geometry, is it possible that a transversal intersects two parallel lines? Explain your reasoning.

PROVING A THEOREM In Exercises 15 and 16, prove the theorem. *(See Example 4.)*

15. Alternate Exterior Angles Theorem (Thm. 3.3)

16. Consecutive Interior Angles Theorem (Thm. 3.4)

17. PROBLEM SOLVING A group of campers tie up their food between two parallel trees, as shown. The rope is pulled taut, forming a straight line. Find $m\angle 2$. Explain your reasoning. *(See Example 5.)*

18. DRAWING CONCLUSIONS You are designing a box like the one shown.

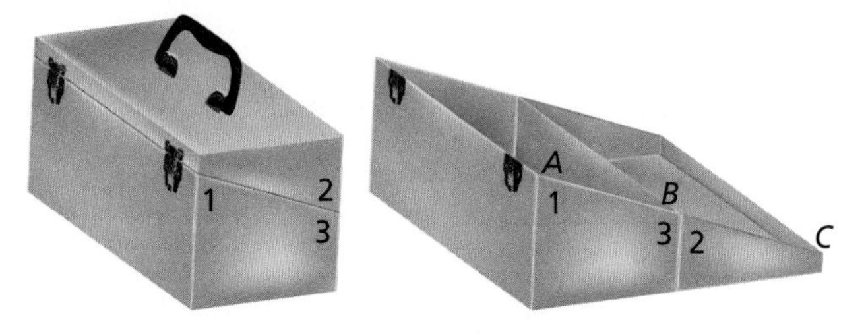

a. The measure of $\angle 1$ is 70°. Find $m\angle 2$ and $m\angle 3$.

b. Explain why $\angle ABC$ is a straight angle.

c. If $m\angle 1$ is 60°, will $\angle ABC$ still be a straight angle? Will the opening of the box be *more steep* or *less steep*? Explain.

MATHEMATICAL CONNECTIONS In Exercises 21 and 22, write and solve a system of linear equations to find the values of *x* and *y*.

21.

(14x − 10)° 2y° 5x°

22.

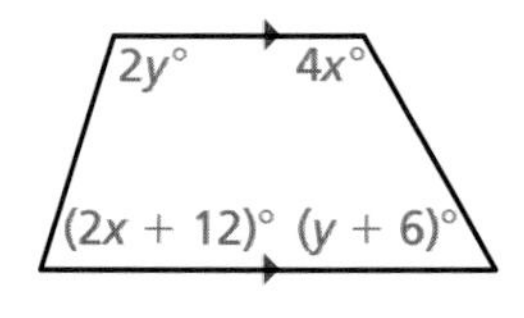

23. MAKING AN ARGUMENT During a game of pool, your friend claims to be able to make the shot shown in the diagram by hitting the cue ball so that $m\angle 1 = 25°$. Is your friend correct? Explain your reasoning.

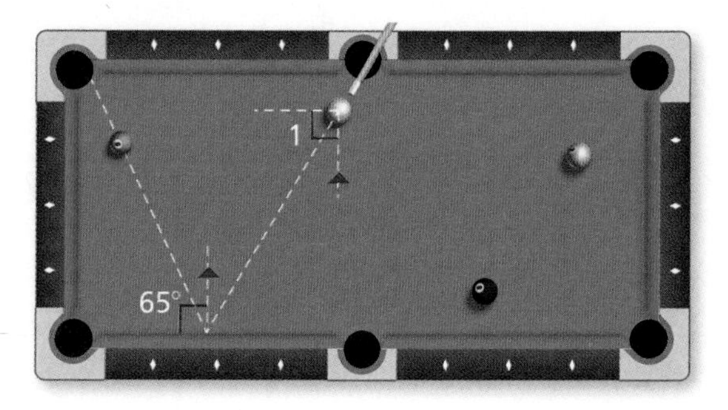

24. REASONING In the diagram, $\angle 4 \cong \angle 5$ and $\overline{SE}$ bisects $\angle RSF$. Find $m\angle 1$. Explain your reasoning.

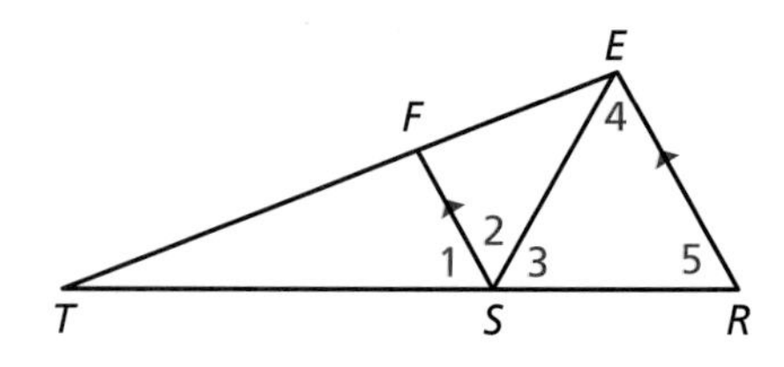

Maintaining Mathematical Proficiency

Reviewing what you learned in previous grades and lessons

Write the converse of the conditional statement. Decide whether it is true or false. *(Section 2.1)*

25. If two angles are vertical angles, then they are congruent.

26. If you go to the zoo, then you will see a tiger.

27. If two angles form a linear pair, then they are supplementary.

28. If it is warm outside, then we will go to the park.

3.3 Proofs with Parallel Lines

Essential Question For which of the theorems involving parallel lines and transversals is the converse true?

EXPLORATION 1 Exploring Converses

Work with a partner. Write the converse of each conditional statement. Draw a diagram to represent the converse. Determine whether the converse is true. Justify your conclusion.

> **CONSTRUCTING VIABLE ARGUMENTS**
> To be proficient in math, you need to make conjectures and build a logical progression of statements to explore the truth of your conjectures.

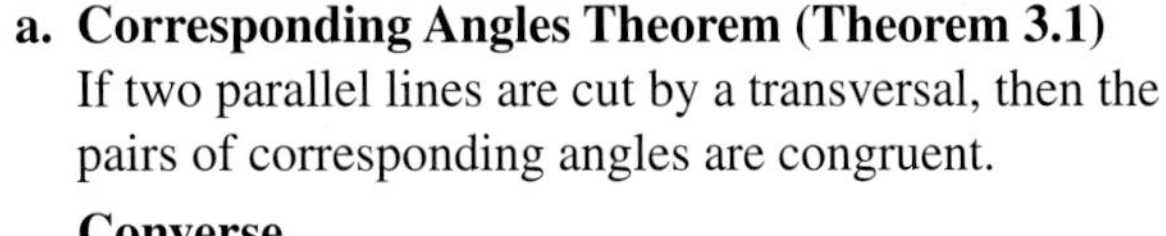

a. Corresponding Angles Theorem (Theorem 3.1)
If two parallel lines are cut by a transversal, then the pairs of corresponding angles are congruent.

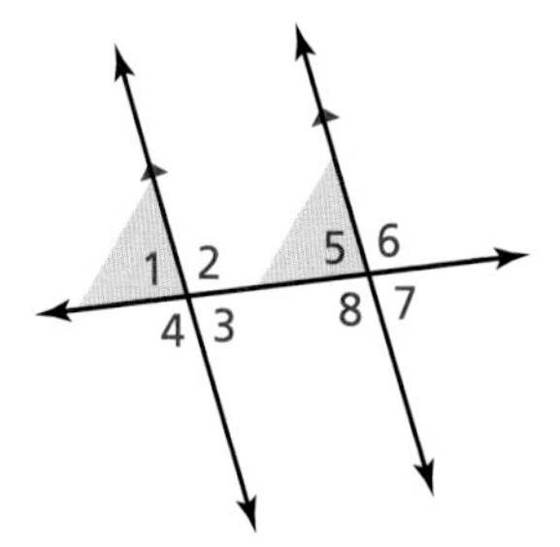

Converse

b. Alternate Interior Angles Theorem (Theorem 3.2)
If two parallel lines are cut by a transversal, then the pairs of alternate interior angles are congruent.

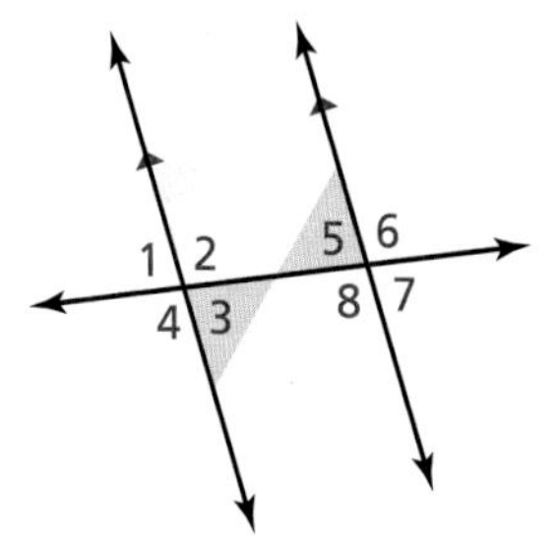

Converse

c. Alternate Exterior Angles Theorem (Theorem 3.3)
If two parallel lines are cut by a transversal, then the pairs of alternate exterior angles are congruent.

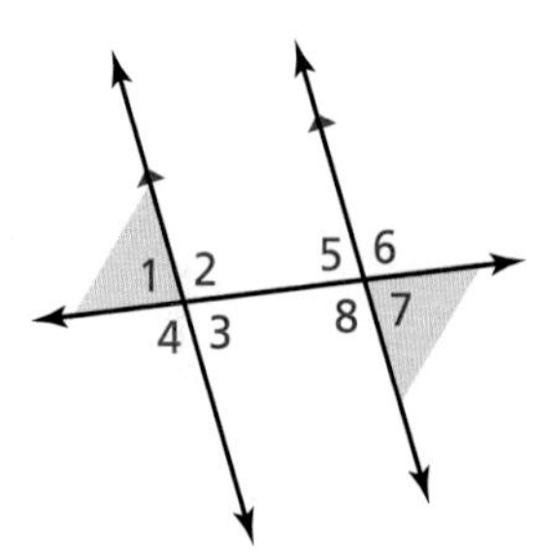

Converse

d. Consecutive Interior Angles Theorem (Theorem 3.4)
If two parallel lines are cut by a transversal, then the pairs of consecutive interior angles are supplementary.

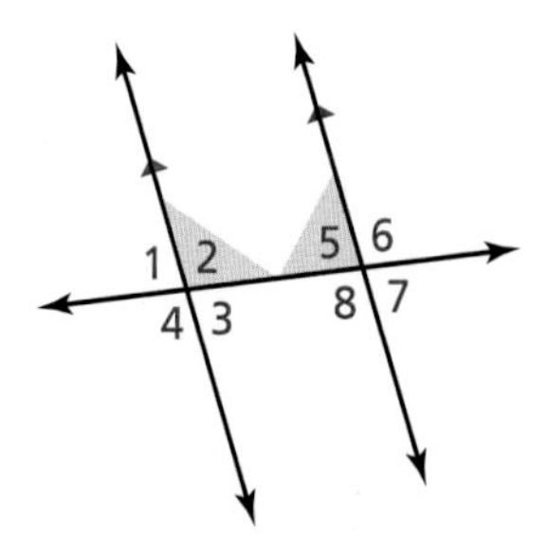

Converse

Communicate Your Answer

2. For which of the theorems involving parallel lines and transversals is the converse true?

3. In Exploration 1, explain how you would prove any of the theorems that you found to be true.

3.3 Lesson

Core Vocabulary

Previous
converse
parallel lines
transversal
corresponding angles
congruent
alternate interior angles
alternate exterior angles
consecutive interior angles

What You Will Learn

- Use the Corresponding Angles Converse.
- Construct parallel lines.
- Prove theorems about parallel lines.
- Use the Transitive Property of Parallel Lines.

Using the Corresponding Angles Converse

Theorem 3.5 below is the converse of the Corresponding Angles Theorem (Theorem 3.1). Similarly, the other theorems about angles formed when parallel lines are cut by a transversal have true converses. Remember that the converse of a true conditional statement is not necessarily true, so you must prove each converse of a theorem.

Theorem

Theorem 3.5 Corresponding Angles Converse

If two lines are cut by a transversal so the corresponding angles are congruent, then the lines are parallel.

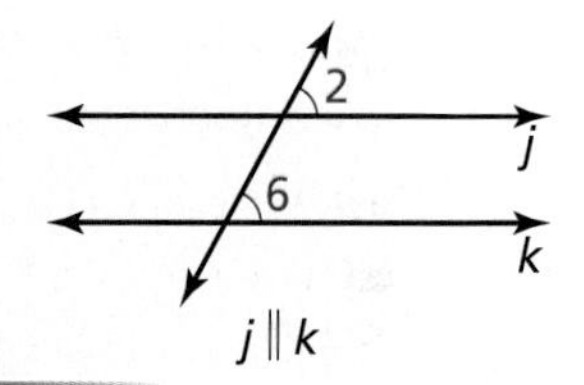

$j \parallel k$

Proof Ex. 36, p. 180

EXAMPLE 1 Using the Corresponding Angles Converse

Find the value of x that makes $m \parallel n$.

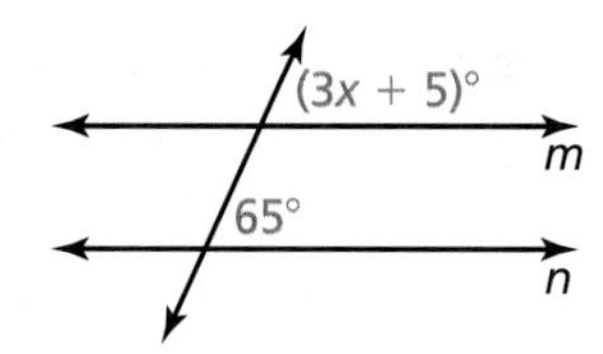

SOLUTION

Lines m and n are parallel when the marked corresponding angles are congruent.

$(3x + 5)° = 65°$	Use the Corresponding Angles Converse to write an equation.
$3x = 60$	Subtract 5 from each side.
$x = 20$	Divide each side by 3.

- So, lines m and n are parallel when $x = 20$.

Monitoring Progress  Help in English and Spanish at *BigIdeasMath.com*

1. Is there enough information in the diagram to conclude that $m \parallel n$? Explain.

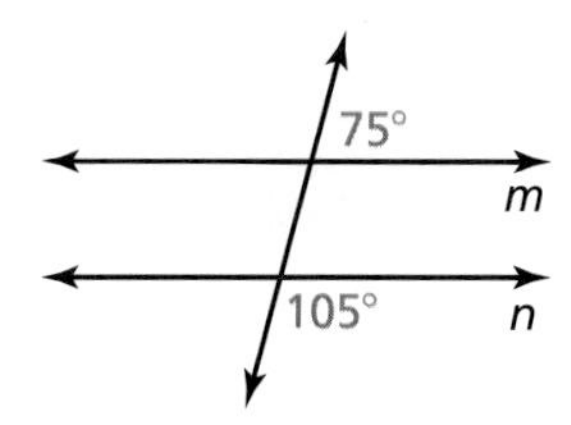

2. Explain why the Corresponding Angles Converse is the converse of the Corresponding Angles Theorem (Theorem 3.1).

Constructing Parallel Lines

The Corresponding Angles Converse justifies the construction of parallel lines, as shown below.

CONSTRUCTION **Constructing Parallel Lines**

Use a compass and straightedge to construct a line through point P that is parallel to line m.

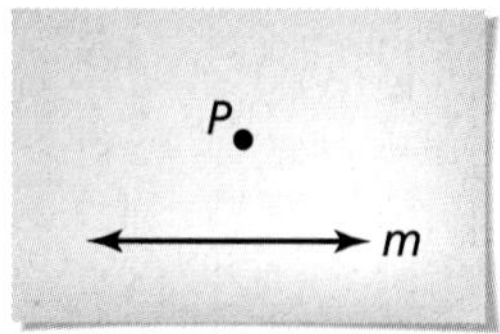

SOLUTION

Step 1

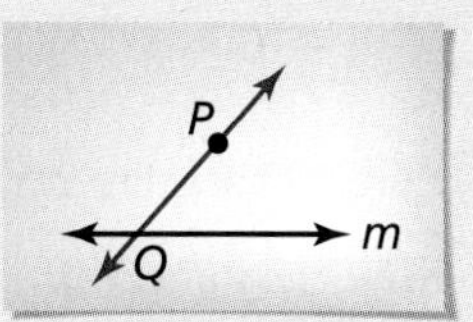

Draw a point and line Start by drawing point P and line m. Choose a point Q anywhere on line m and draw $\overleftrightarrow{QP}$.

Step 2

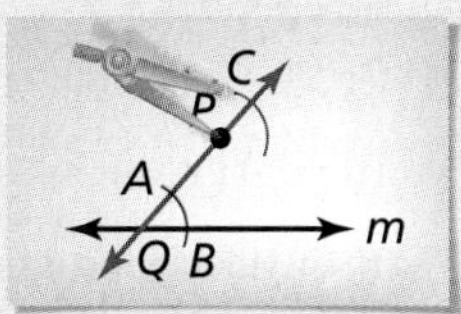

Draw arcs Draw an arc with center Q that crosses $\overleftrightarrow{QP}$ and line m. Label points A and B. Using the same compass setting, draw an arc with center P. Label point C.

Step 3

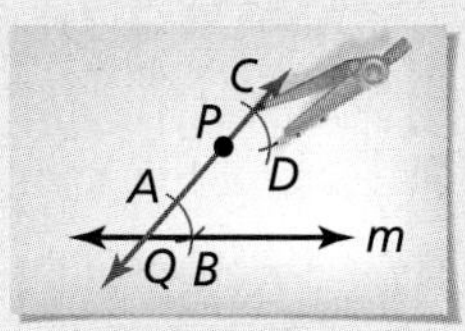

Copy angle Draw an arc with radius AB and center A. Using the same compass setting, draw an arc with center C. Label the intersection D.

Step 4

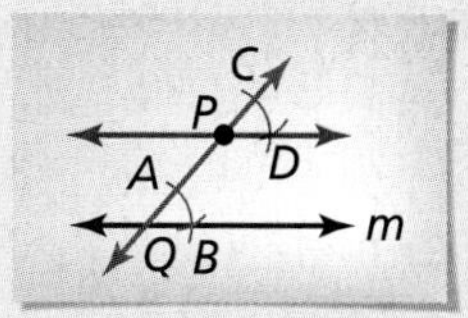

Draw parallel lines Draw $\overleftrightarrow{PD}$. This line is parallel to line m.

Theorems

Theorem 3.6 Alternate Interior Angles Converse

If two lines are cut by a transversal so the alternate interior angles are congruent, then the lines are parallel.

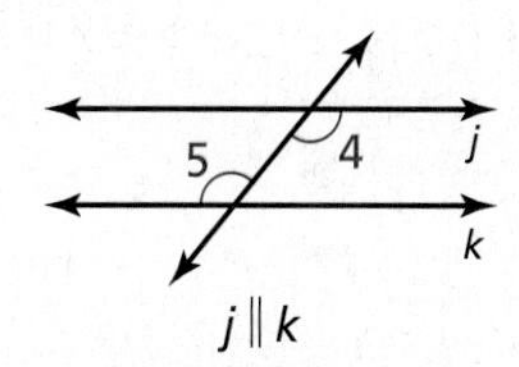

$j \parallel k$

Proof Example 2, p. 140

Theorem 3.7 Alternate Exterior Angles Converse

If two lines are cut by a transversal so the alternate exterior angles are congruent, then the lines are parallel.

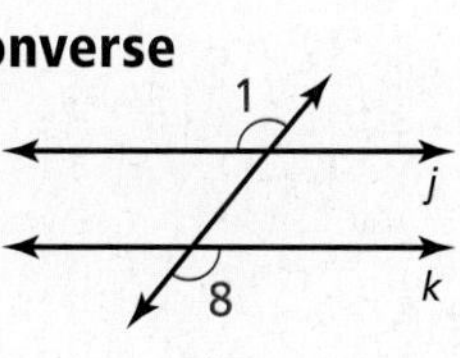

$j \parallel k$

Proof Ex. 11, p. 142

Theorem 3.8 Consecutive Interior Angles Converse

If two lines are cut by a transversal so the consecutive interior angles are supplementary, then the lines are parallel.

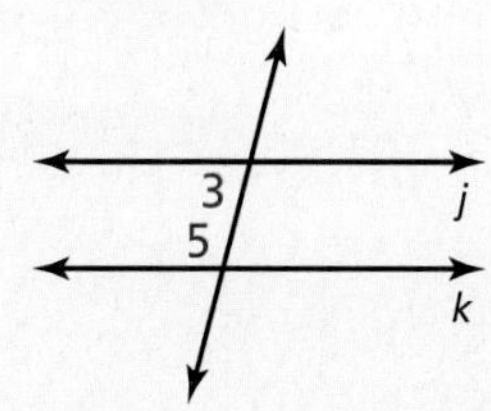

If $\angle 3$ and $\angle 5$ are supplementary, then $j \parallel k$.

Proof Ex. 12, p. 142

Proving Theorems about Parallel Lines

EXAMPLE 2 Proving the Alternate Interior Angles Converse

Prove that if two lines are cut by a transversal so the alternate interior angles are congruent, then the lines are parallel.

SOLUTION

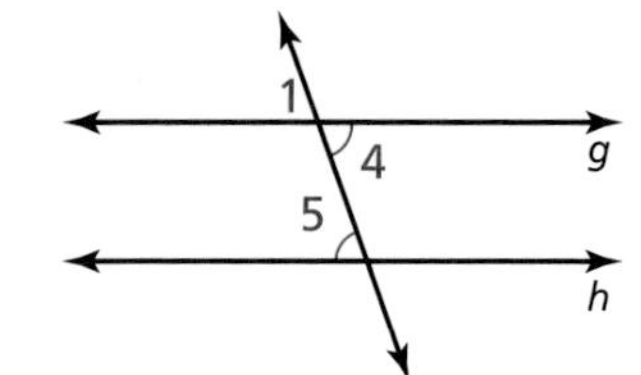

Given $\angle 4 \cong \angle 5$

Prove $g \parallel h$

STATEMENTS	REASONS
1. $\angle 4 \cong \angle 5$	**1.** Given
2. $\angle 1 \cong \angle 4$	**2.** Vertical Angles Congruence Theorem (Theorem 2.6)
3. $\angle 1 \cong \angle 5$	**3.** Transitive Property of Congruence (Theorem 2.2)
4. $g \parallel h$	**4.** Corresponding Angles Converse

EXAMPLE 3 Determining Whether Lines Are Parallel

In the diagram, $r \parallel s$ and $\angle 1$ is congruent to $\angle 3$. Prove $p \parallel q$.

SOLUTION

Look at the diagram to make a plan. The diagram suggests that you look at angles 1, 2, and 3. Also, you may find it helpful to focus on one pair of lines and one transversal at a time.

Plan for Proof **a.** Look at $\angle 1$ and $\angle 2$. $\angle 1 \cong \angle 2$ because $r \parallel s$.

b. Look at $\angle 2$ and $\angle 3$. If $\angle 2 \cong \angle 3$, then $p \parallel q$.

Plan for Action **a.** It is given that $r \parallel s$, so by the Corresponding Angles Theorem (Theorem 3.1), $\angle 1 \cong \angle 2$.

b. It is also given that $\angle 1 \cong \angle 3$. Then $\angle 2 \cong \angle 3$ by the Transitive Property of Congruence (Theorem 2.2).

▶ So, by the Alternate Interior Angles Converse, $p \parallel q$.

Monitoring Progress  Help in English and Spanish at *BigIdeasMath.com*

3. If you use the diagram below to prove the Alternate Exterior Angles Converse, what **Given** and **Prove** statements would you use?

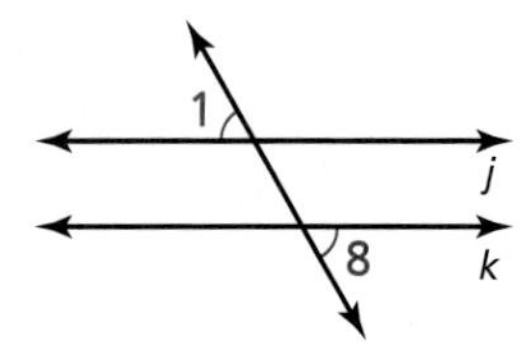

4. Copy and complete the following paragraph proof of the Alternate Interior Angles Converse using the diagram in Example 2.

It is given that $\angle 4 \cong \angle 5$. By the ______, $\angle 1 \cong \angle 4$. Then by the Transitive Property of Congruence (Theorem 2.2), ______. So, by the ______, $g \parallel h$.

Using the Transitive Property of Parallel Lines

Theorem

Theorem 3.9 Transitive Property of Parallel Lines

If two lines are parallel to the same line, then they are parallel to each other.

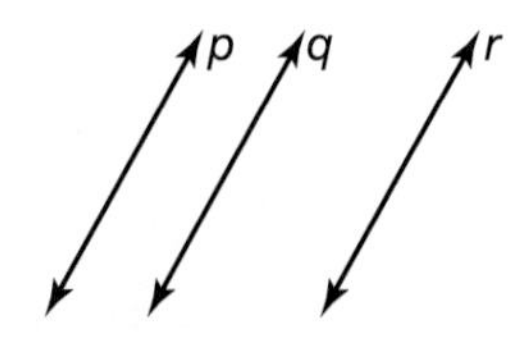

Proof Ex. 39, p. 144; Ex. 48, p. 162

If $p \parallel q$ and $q \parallel r$, then $p \parallel r$.

EXAMPLE 4 **Using the Transitive Property of Parallel Lines**

The flag of the United States has 13 alternating red and white stripes. Each stripe is parallel to the stripe immediately below it. Explain why the top stripe is parallel to the bottom stripe.

SOLUTION

You can name the stripes from top to bottom as $s_1, s_2, s_3, \ldots, s_{13}$. Each stripe is parallel to the one immediately below it, so $s_1 \parallel s_2$, $s_2 \parallel s_3$, and so on. Then $s_1 \parallel s_3$ by the Transitive Property of Parallel Lines. Similarly, because $s_3 \parallel s_4$, it follows that $s_1 \parallel s_4$. By continuing this reasoning, $s_1 \parallel s_{13}$.

So, the top stripe is parallel to the bottom stripe.

Monitoring Progress

Help in English and Spanish at *BigIdeasMath.com*

5. Each step is parallel to the step immediately above it. The bottom step is parallel to the ground. Explain why the top step is parallel to the ground.

6. In the diagram below, $p \parallel q$ and $q \parallel r$. Find $m\angle 8$. Explain your reasoning.

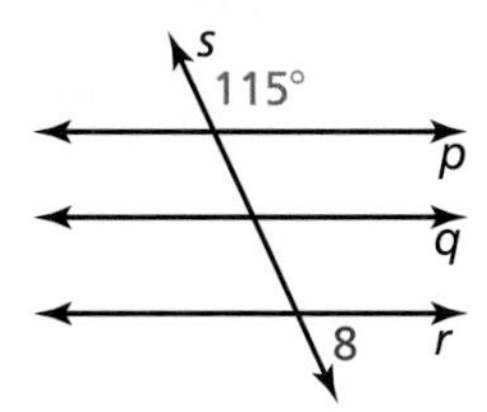

3.3 Exercises

Dynamic Solutions available at *BigIdeasMath.com*

Vocabulary and Core Concept Check

1. **VOCABULARY** Two lines are cut by a transversal. Which angle pairs must be congruent for the lines to be parallel?

2. **WRITING** Use the theorems from Section 3.2 and the converses of those theorems in this section to write three biconditional statements about parallel lines and transversals.

Monitoring Progress and Modeling with Mathematics

In Exercises 3–8, find the value of x that makes $m \parallel n$. Explain your reasoning. *(See Example 1.)*

3. 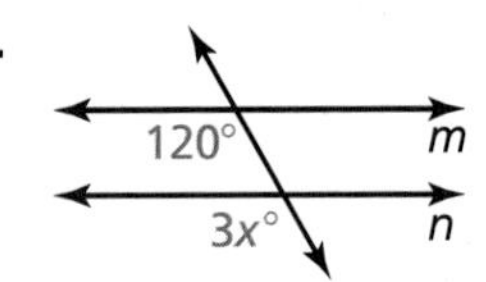

4. 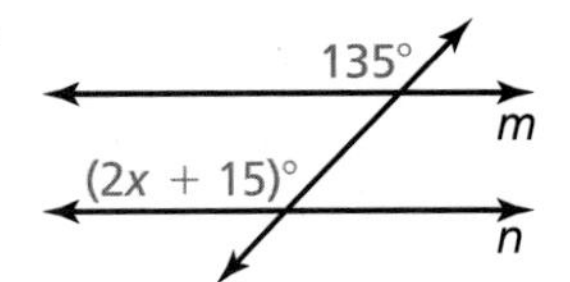

5. 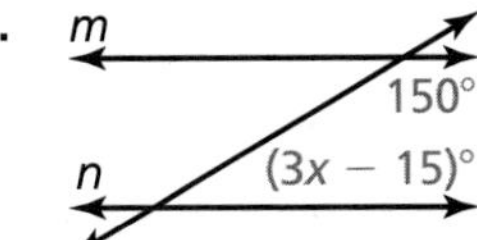

6. 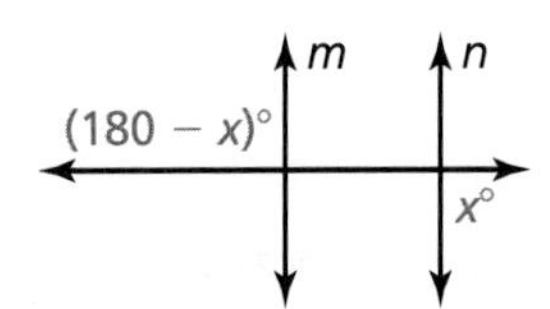

7. 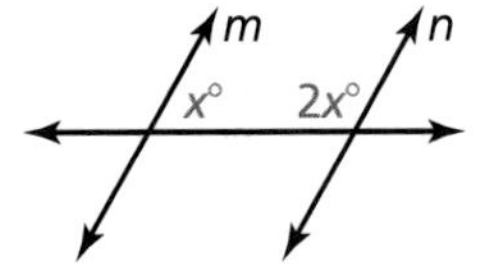

8. 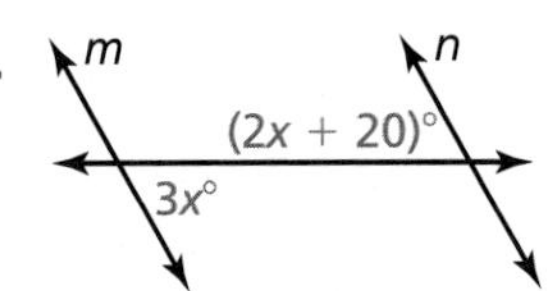

In Exercises 9 and 10, use a compass and straightedge to construct a line through point P that is parallel to line m.

9. 10. 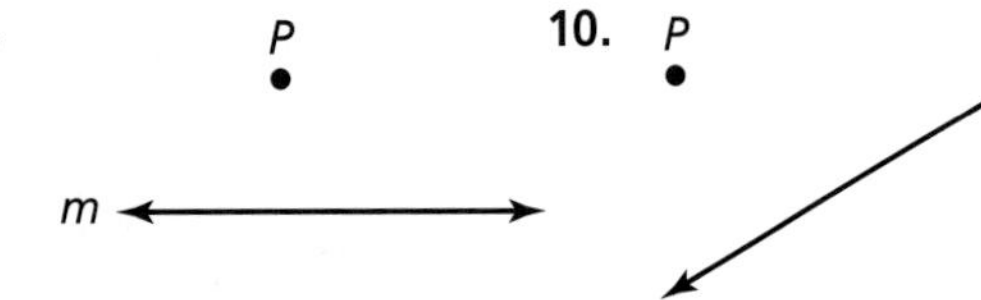

PROVING A THEOREM In Exercises 11 and 12, prove the theorem. *(See Example 2.)*

11. Alternate Exterior Angles Converse (Theorem 3.7)

12. Consecutive Interior Angles Converse (Theorem 3.8)

In Exercises 13–18, decide whether there is enough information to prove that $m \parallel n$. If so, state the theorem you would use. *(See Example 3.)*

13. 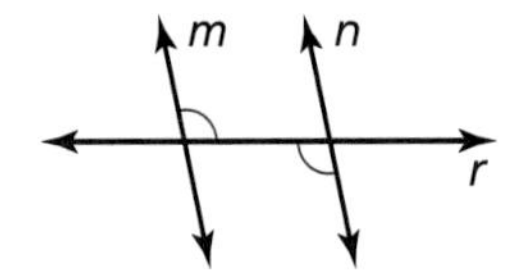

14. 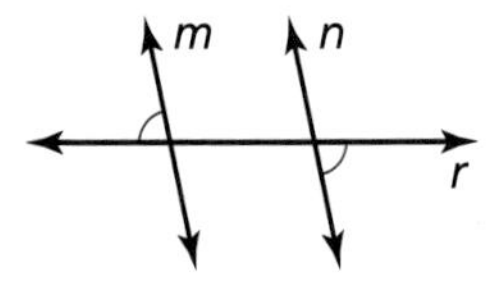

15. 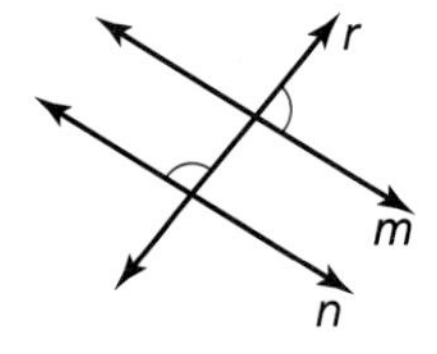

16. 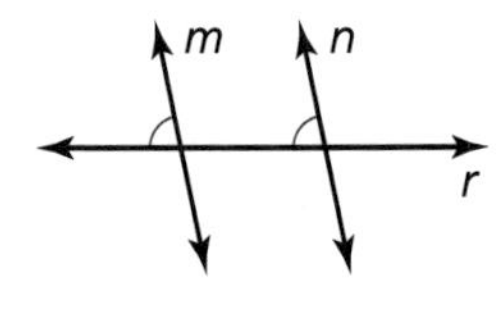

17. 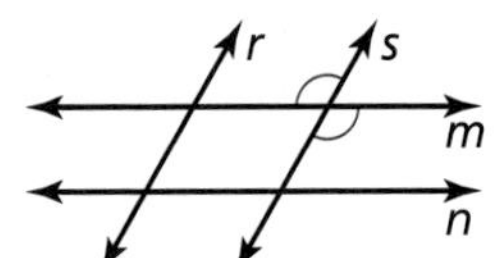

18. 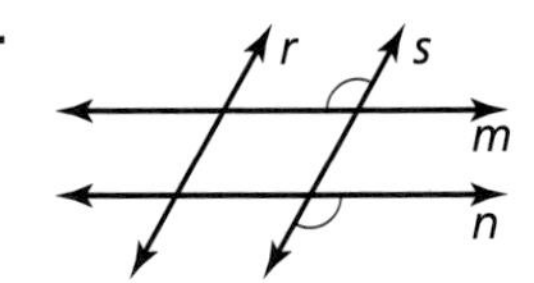

ERROR ANALYSIS In Exercises 19 and 20, describe and correct the error in the reasoning.

19. 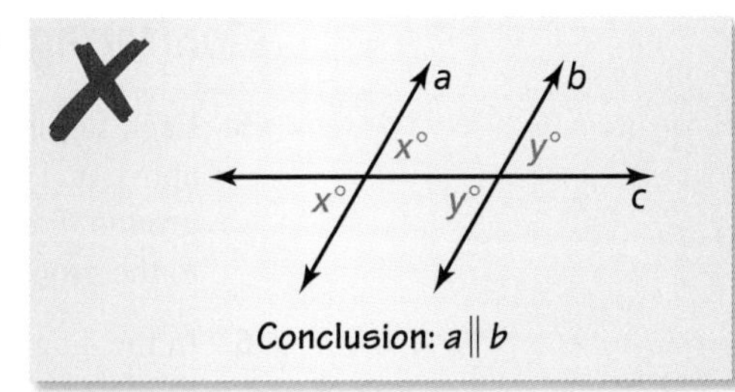

20.

2

1

a

b

Conclusion: $a \parallel b$

In Exercises 21–24, are $\overleftrightarrow{AC}$ and $\overleftrightarrow{DF}$ parallel? Explain your reasoning.

21.

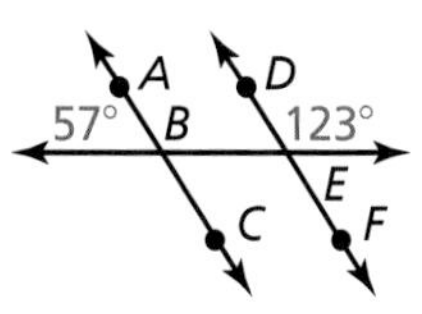

22.

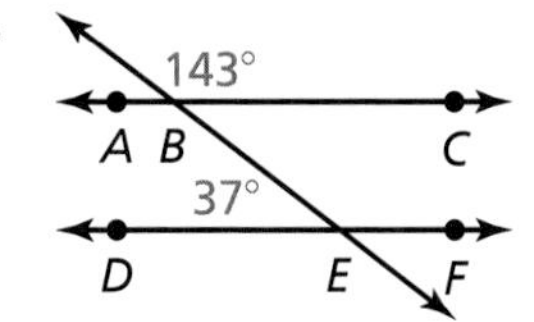

23.

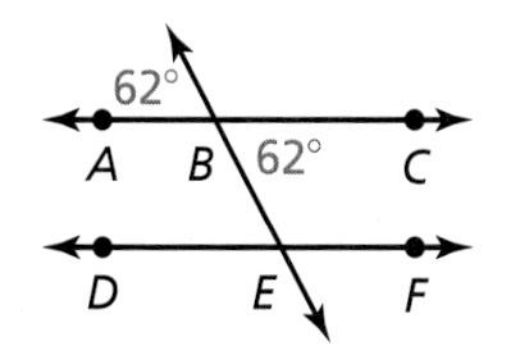

24.

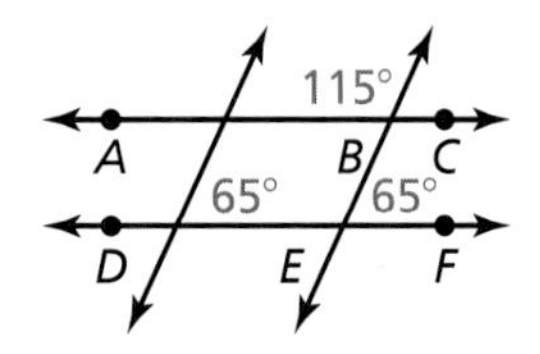

25. ANALYZING RELATIONSHIPS The map shows part of Denver, Colorado. Use the markings on the map. Are the numbered streets parallel to one another? Explain your reasoning. *(See Example 4.)*

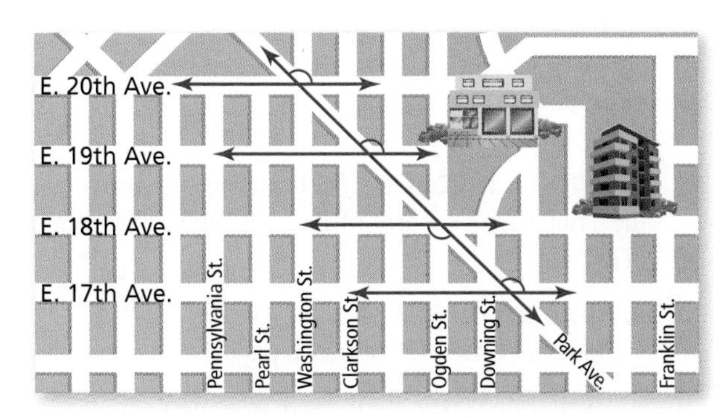

26. ANALYZING RELATIONSHIPS Each rung of the ladder is parallel to the rung directly above it. Explain why the top rung is parallel to the bottom rung.

27. MODELING WITH MATHEMATICS The diagram of the control bar of the kite shows the angles formed between the control bar and the kite lines. How do you know that n is parallel to m?

28. REASONING Use the diagram. Which rays are parallel? Which rays are not parallel? Explain your reasoning.

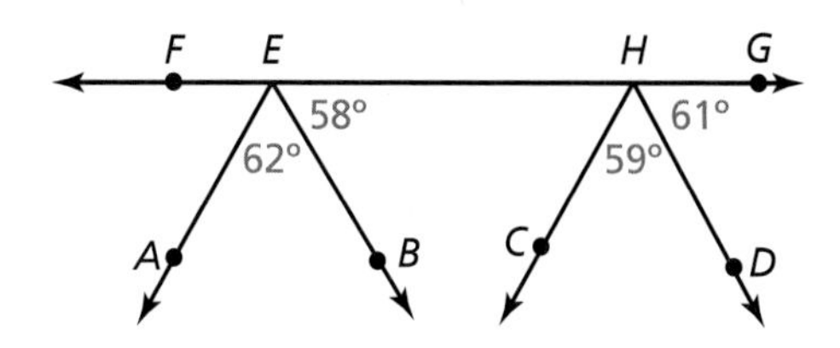

29. ATTENDING TO PRECISION Use the diagram. Which theorems allow you to conclude that $m \parallel n$? Select all that apply. Explain your reasoning.

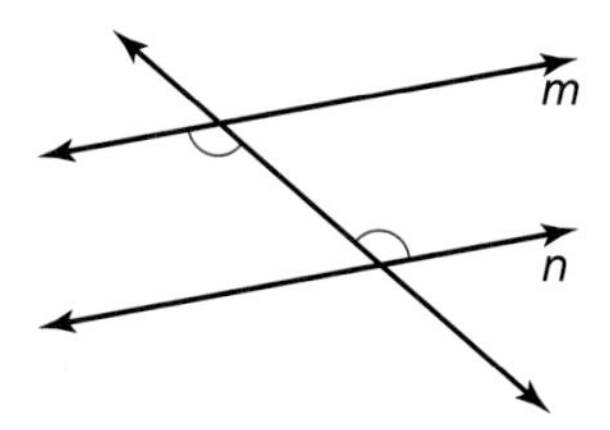

Ⓐ Corresponding Angles Converse (Thm. 3.5)

Ⓑ Alternate Interior Angles Converse (Thm. 3.6)

Ⓒ Alternate Exterior Angles Converse (Thm. 3.7)

Ⓓ Consecutive Interior Angles Converse (Thm. 3.8)

30. MODELING WITH MATHEMATICS One way to build stairs is to attach triangular blocks to an angled support, as shown. The sides of the angled support are parallel. If the support makes a 32° angle with the floor, what must $m\angle 1$ be so the top of the step will be parallel to the floor? Explain your reasoning.

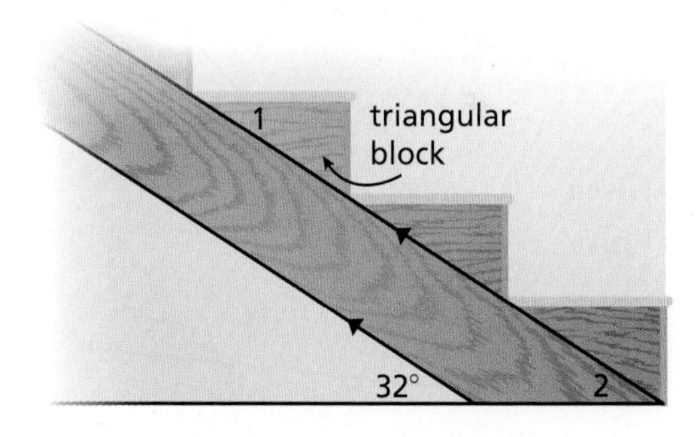

31. ABSTRACT REASONING In the diagram, how many angles must be given to determine whether $j \parallel k$? Give four examples that would allow you to conclude that $j \parallel k$ using the theorems from this lesson.

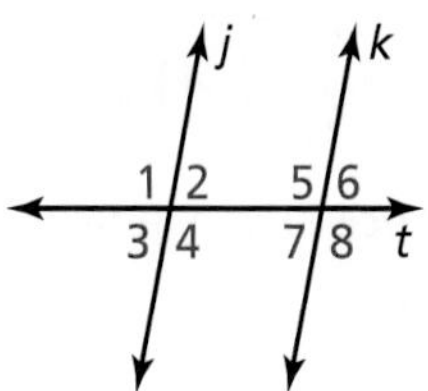

32. THOUGHT PROVOKING Draw a diagram of at least two lines cut by at least one transversal. Mark your diagram so that it cannot be proven that any lines are parallel. Then explain how your diagram would need to change in order to prove that lines are parallel.

PROOF In Exercises 33–36, write a proof.

33. Given $m\angle 1 = 115^\circ$, $m\angle 2 = 65^\circ$

Prove $m \parallel n$

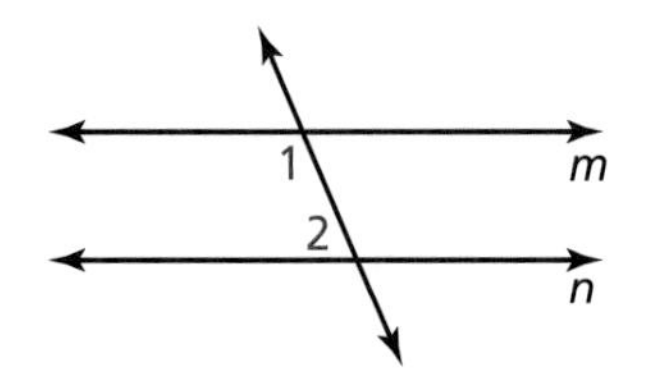

34. Given $\angle 1$ and $\angle 3$ are supplementary.

Prove $m \parallel n$

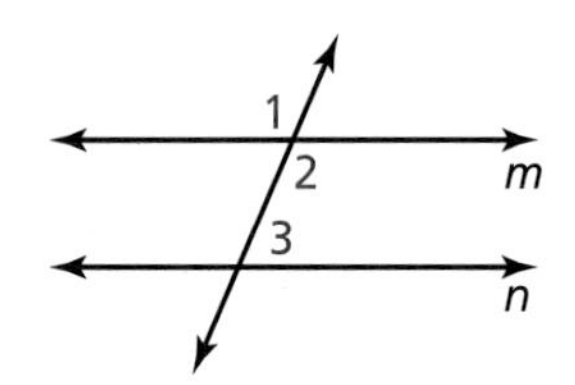

35. Given $\angle 1 \cong \angle 2$, $\angle 3 \cong \angle 4$

Prove $\overline{AB} \parallel \overline{CD}$

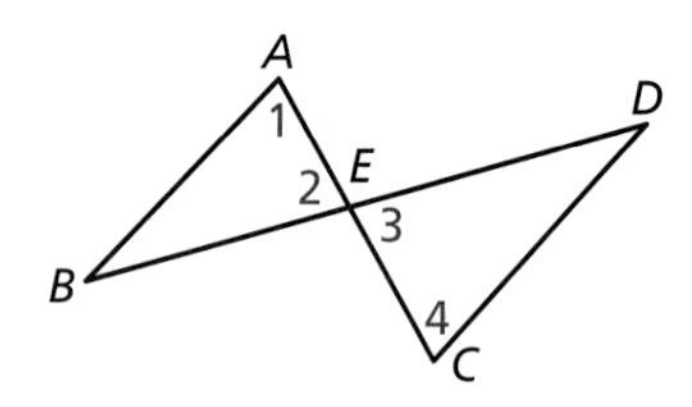

36. Given $a \parallel b$, $\angle 2 \cong \angle 3$

Prove $c \parallel d$

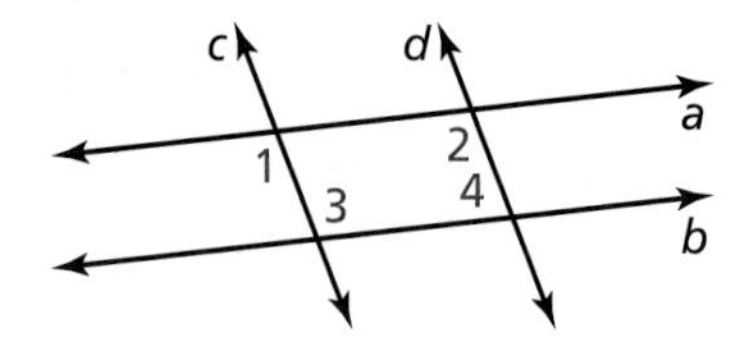

37. MAKING AN ARGUMENT Your classmate decided that $\overleftrightarrow{AD} \parallel \overleftrightarrow{BC}$ based on the diagram. Is your classmate correct? Explain your reasoning.

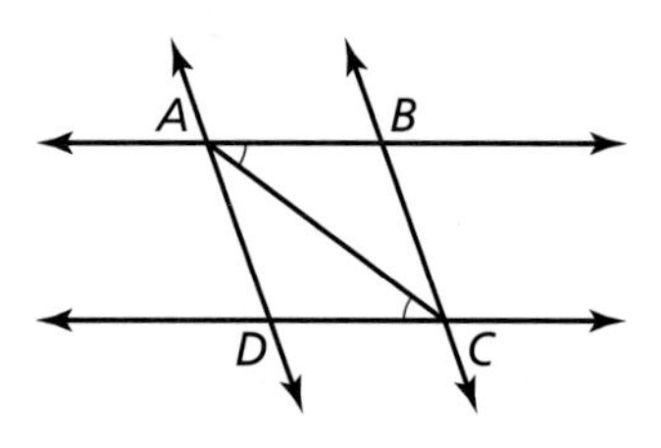

38. HOW DO YOU SEE IT? Are the markings on the diagram enough to conclude that any lines are parallel? If so, which ones? If not, what other information is needed?

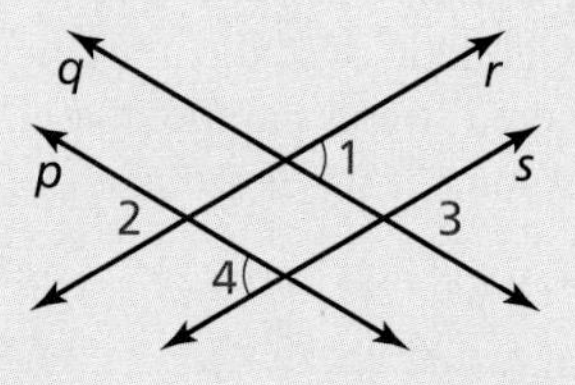

39. PROVING A THEOREM Use these steps to prove the Transitive Property of Parallel Lines Theorem (Theorem 3.9).

a. Copy the diagram with the Transitive Property of Parallel Lines Theorem on page 141.

b. Write the **Given** and **Prove** statements.

c. Use the properties of angles formed by parallel lines cut by a transversal to prove the theorem.

40. MATHEMATICAL CONNECTIONS Use the diagram.

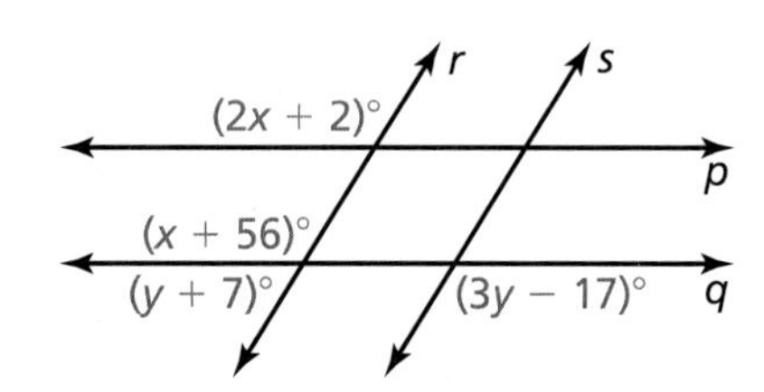

a. Find the value of x that makes $p \parallel q$.

b. Find the value of y that makes $r \parallel s$.

c. Can r be parallel to s and can p be parallel to q at the same time? Explain your reasoning.

Maintaining Mathematical Proficiency

Reviewing what you learned in previous grades and lessons

Use the Distance Formula to find the distance between the two points. *(Section 1.3)*

41. $(1, 3)$ and $(-2, 9)$

42. $(-3, 7)$ and $(8, -6)$

43. $(5, -4)$ and $(0, 8)$

44. $(13, 1)$ and $(9, -4)$

3.1–3.3 What Did You Learn?

Core Vocabulary

parallel lines, *p. 126*
skew lines, *p. 126*
parallel planes, *p. 126*
transversal, *p. 128*
corresponding angles, *p. 128*
alternate interior angles, *p. 128*
alternate exterior angles, *p. 128*
consecutive interior angles, *p. 128*

Core Concepts

Section 3.1

Parallel Lines, Skew Lines, and Parallel Planes, *p. 126*
Postulate 3.1 Parallel Postulate, *p. 127*
Postulate 3.2 Perpendicular Postulate, *p. 127*
Angles Formed by Transversals, *p. 128*

Section 3.2

Theorem 3.1 Corresponding Angles Theorem, *p. 132*
Theorem 3.2 Alternate Interior Angles Theorem, *p. 132*
Theorem 3.3 Alternate Exterior Angles Theorem, *p. 132*
Theorem 3.4 Consecutive Interior Angles Theorem, *p. 132*

Section 3.3

Theorem 3.5 Corresponding Angles Converse, *p. 138*
Theorem 3.6 Alternate Interior Angles Converse, *p. 139*
Theorem 3.7 Alternate Exterior Angles Converse, *p. 139*
Theorem 3.8 Consecutive Interior Angles Converse, *p. 139*
Theorem 3.9 Transitive Property of Parallel Lines, *p. 141*

Mathematical Practices

1. Draw the portion of the diagram that you used to answer Exercise 26 on page 130.
2. In Exercise 40 on page 144, explain how you started solving the problem and why you started that way.

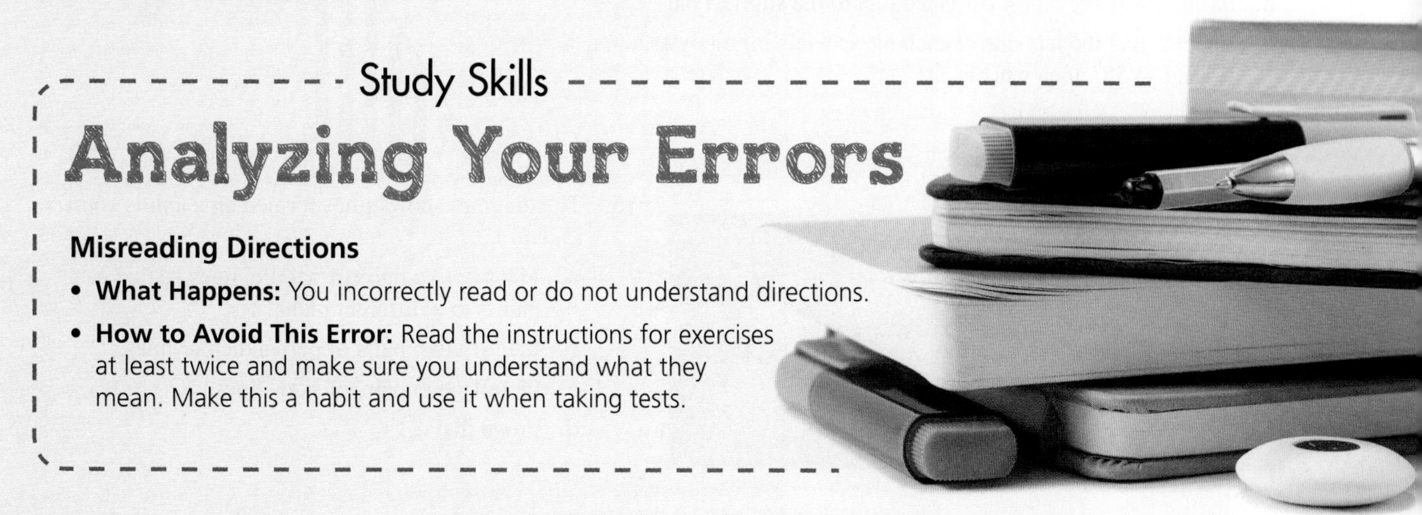

Study Skills

Analyzing Your Errors

Misreading Directions

- **What Happens:** You incorrectly read or do not understand directions.
- **How to Avoid This Error:** Read the instructions for exercises at least twice and make sure you understand what they mean. Make this a habit and use it when taking tests.

3.1–3.3 Quiz

Think of each segment in the diagram as part of a line. Which line(s) or plane(s) contain point G and appear to fit the description? *(Section 3.1)*

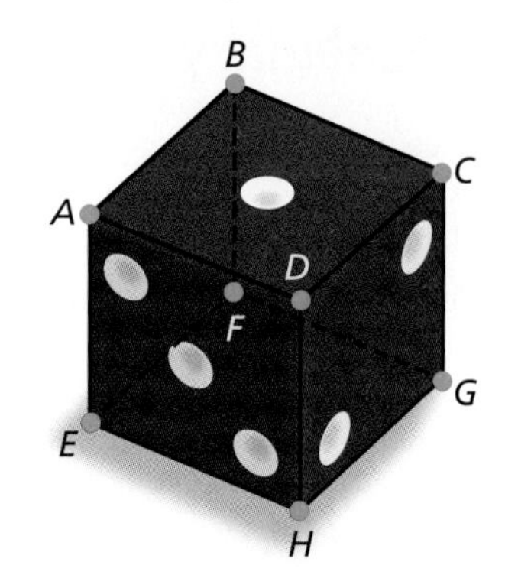

1. line(s) parallel to $\overleftrightarrow{EF}$

2. line(s) perpendicular to $\overleftrightarrow{EF}$

3. line(s) skew to $\overleftrightarrow{EF}$

4. plane(s) parallel to plane ADE

Identify all pairs of angles of the given type. *(Section 3.1)*

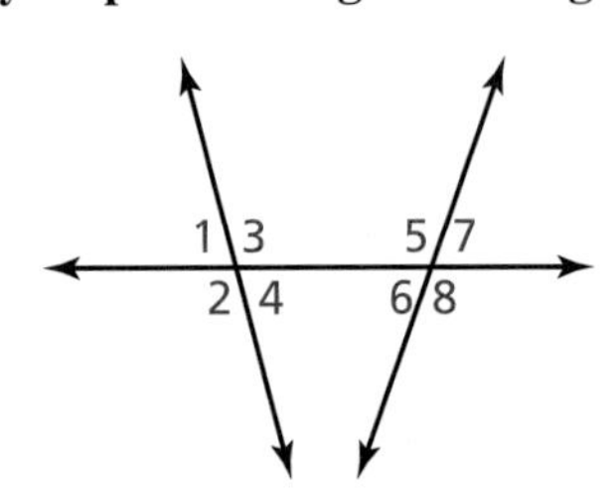

5. consecutive interior

6. alternate interior

7. corresponding

8. alternate exterior

Find $m\angle 1$ and $m\angle 2$. Tell which theorem you use in each case. *(Section 3.2)*

9.

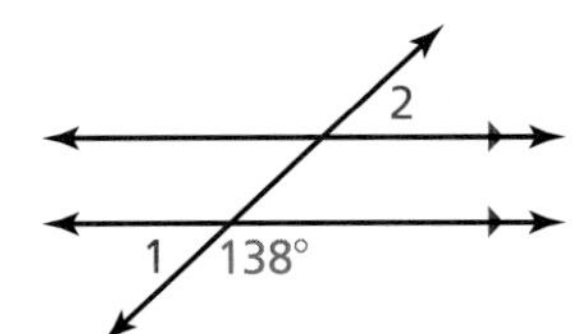

10.

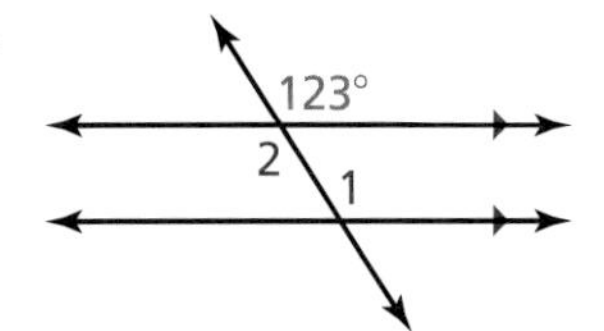

11.

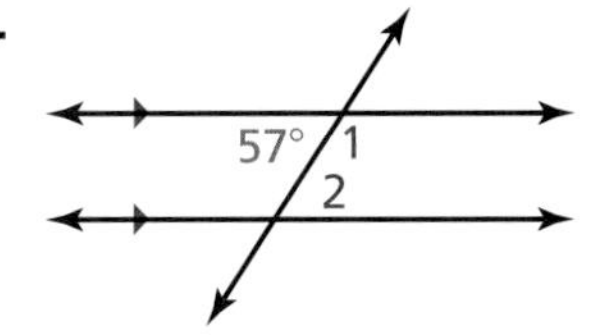

Decide whether there is enough information to prove that $m \parallel n$. If so, state the theorem you would use. *(Section 3.3)*

12.

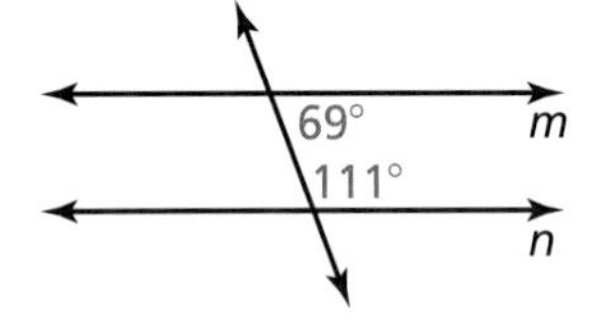

13.

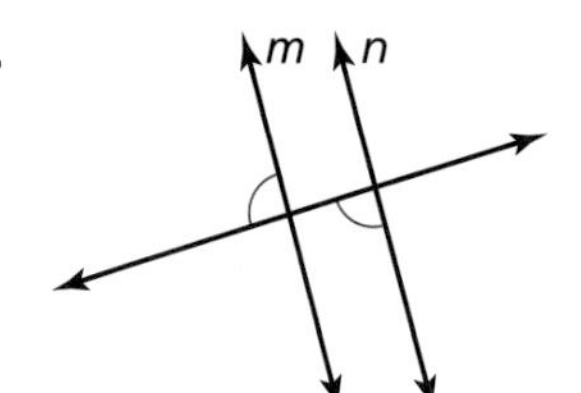

14.

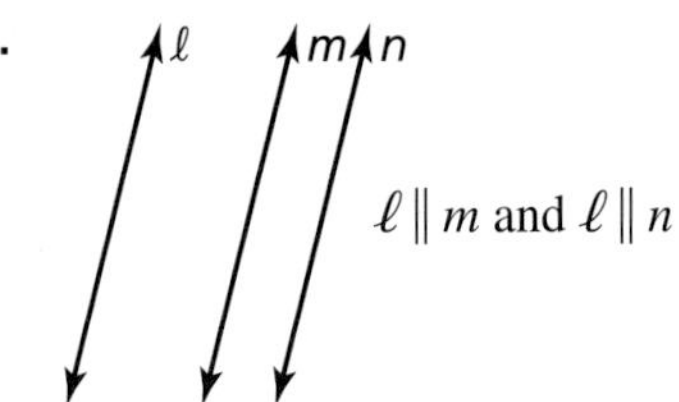

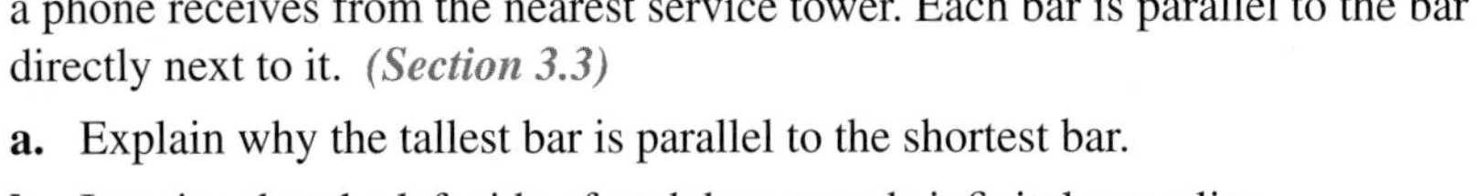

15. Cellular phones use bars like the ones shown to indicate how much signal strength a phone receives from the nearest service tower. Each bar is parallel to the bar directly next to it. *(Section 3.3)*

a. Explain why the tallest bar is parallel to the shortest bar.

b. Imagine that the left side of each bar extends infinitely as a line. If $m\angle 1 = 58°$, then what is $m\angle 2$?

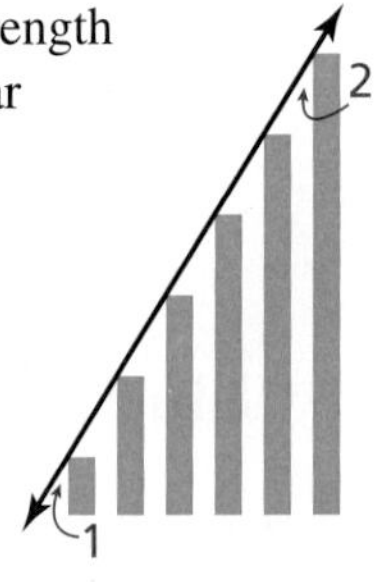

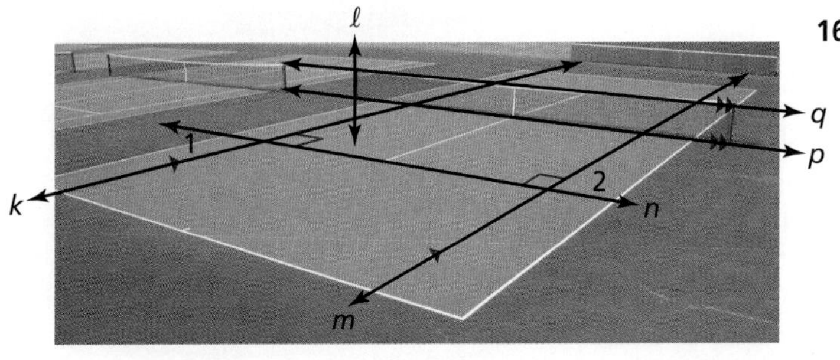

16. The diagram shows lines formed on a tennis court. *(Section 3.1 and Section 3.3)*

a. Identify two pairs of parallel lines so that each pair is in a different plane.

b. Identify two pairs of perpendicular lines.

c. Identify two pairs of skew lines.

d. Prove that $\angle 1 \cong \angle 2$.

3.4 Proofs with Perpendicular Lines

Essential Question

What conjectures can you make about perpendicular lines?

EXPLORATION 1 Writing Conjectures

Work with a partner. Fold a piece of paper in half twice. Label points on the two creases, as shown.

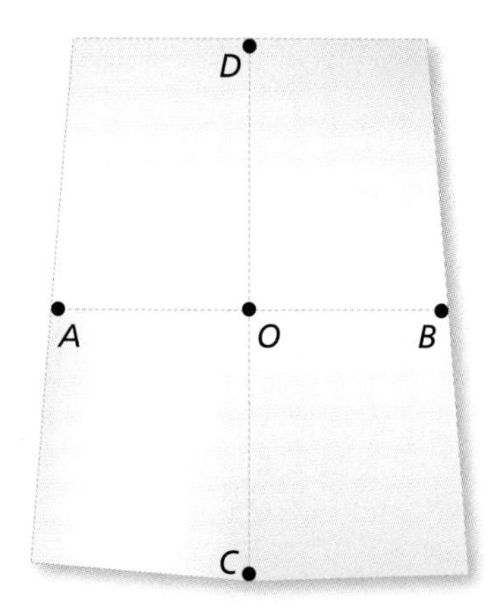

a. Write a conjecture about $\overline{AB}$ and $\overline{CD}$. Justify your conjecture.

b. Write a conjecture about $\overline{AO}$ and $\overline{OB}$. Justify your conjecture.

EXPLORATION 2 Exploring a Segment Bisector

Work with a partner. Fold and crease a piece of paper, as shown. Label the ends of the crease as *A* and *B*.

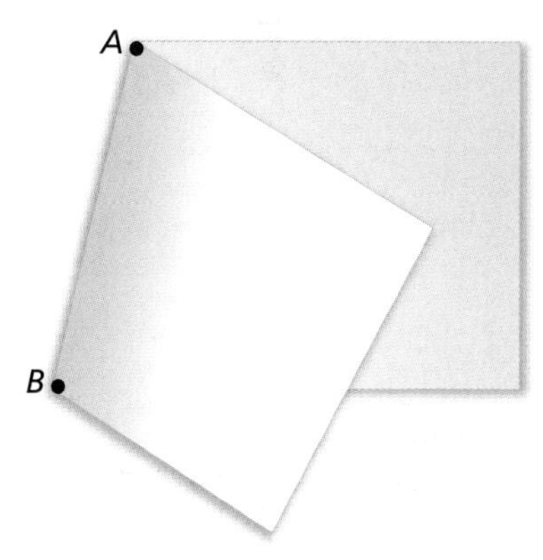

a. Fold the paper again so that point *A* coincides with point *B*. Crease the paper on that fold.

b. Unfold the paper and examine the four angles formed by the two creases. What can you conclude about the four angles?

EXPLORATION 3 Writing a Conjecture

Work with a partner.

a. Draw $\overline{AB}$, as shown.

b. Draw an arc with center *A* on each side of $\overline{AB}$. Using the same compass setting, draw an arc with center *B* on each side of $\overline{AB}$. Label the intersections of the arcs *C* and *D*.

c. Draw $\overline{CD}$. Label its intersection with $\overline{AB}$ as *O*. Write a conjecture about the resulting diagram. Justify your conjecture.

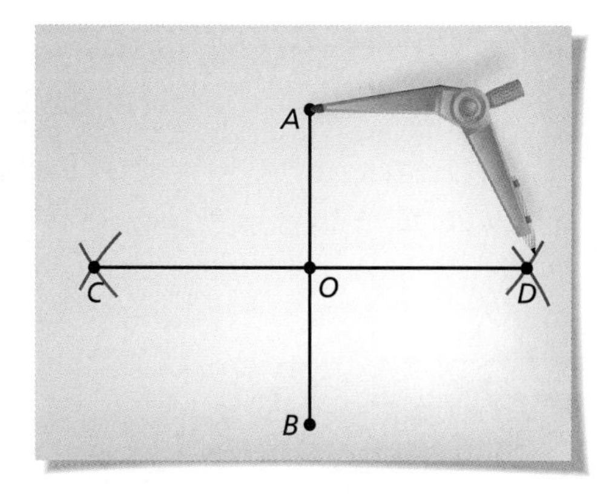

CONSTRUCTING VIABLE ARGUMENTS

To be proficient in math, you need to make conjectures and build a logical progression of statements to explore the truth of your conjectures.

Communicate Your Answer

4. What conjectures can you make about perpendicular lines?

5. In Exploration 3, find *AO* and *OB* when $AB = 4$ units.

3.4 Lesson

Core Vocabulary

distance from a point to a line, *p. 148*
perpendicular bisector, *p. 149*

What You Will Learn

- Find the distance from a point to a line.
- Construct perpendicular lines.
- Prove theorems about perpendicular lines.
- Solve real-life problems involving perpendicular lines.

Finding the Distance from a Point to a Line

The **distance from a point to a line** is the length of the perpendicular segment from the point to the line. This perpendicular segment is the shortest distance between the point and the line. For example, the distance between point A and line k is AB.

distance from a point to a line

EXAMPLE 1 Finding the Distance from a Point to a Line

Find the distance from point A to $\overleftrightarrow{BD}$.

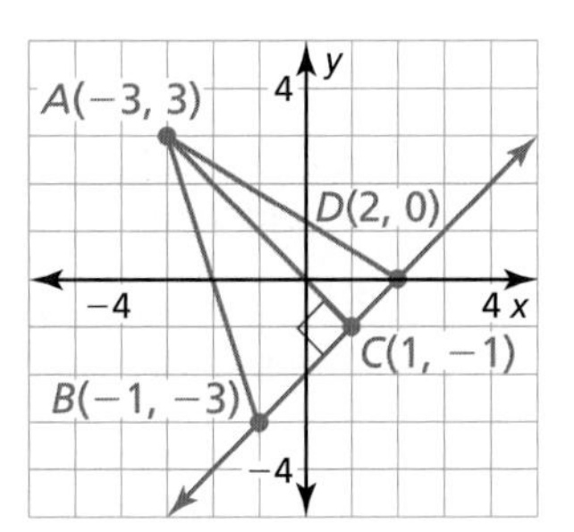

REMEMBER

Recall that if $A(x_1, y_1)$ and $C(x_2, y_2)$ are points in a coordinate plane, then the distance between A and C is

$AC = \sqrt{(x_2 - x_1)^2 + (y_2 - y_1)^2}$.

SOLUTION

Because $\overline{AC} \perp \overleftrightarrow{BD}$, the distance from point A to $\overleftrightarrow{BD}$ is AC. Use the Distance Formula.

$$AC = \sqrt{(-3 - 1)^2 + [3 - (-1)]^2} = \sqrt{(-4)^2 + 4^2} = \sqrt{32} \approx 5.7$$

- So, the distance from point A to $\overleftrightarrow{BD}$ is about 5.7 units.

Monitoring Progress  Help in English and Spanish at *BigIdeasMath.com*

1. Find the distance from point E to $\overleftrightarrow{FH}$.

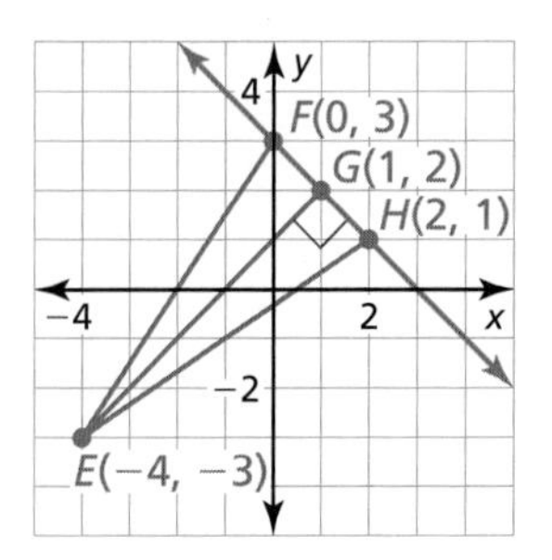

Constructing Perpendicular Lines

CONSTRUCTION **Constructing a Perpendicular Line**

Use a compass and straightedge to construct a line perpendicular to line m through point P, which is not on line m.

SOLUTION

Step 1

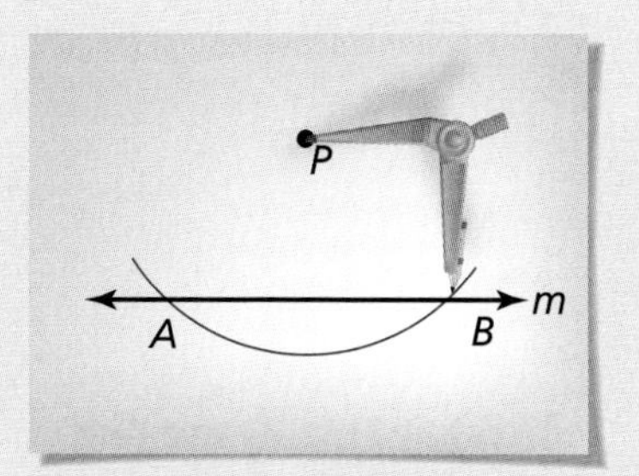

Draw arc with center P Place the compass at point P and draw an arc that intersects the line twice. Label the intersections A and B.

Step 2

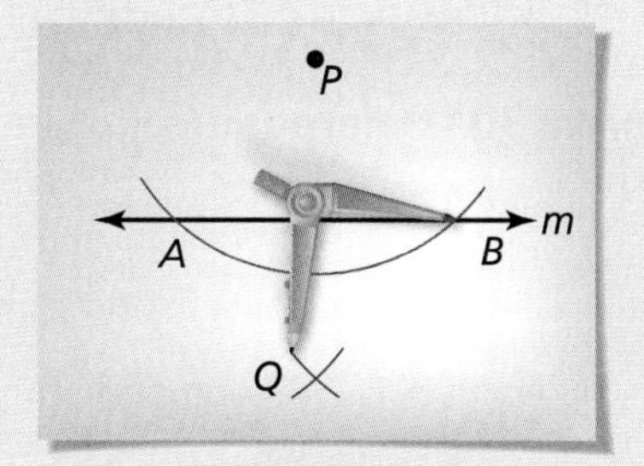

Draw intersecting arcs Draw an arc with center A. Using the same radius, draw an arc with center B. Label the intersection of the arcs Q.

Step 3

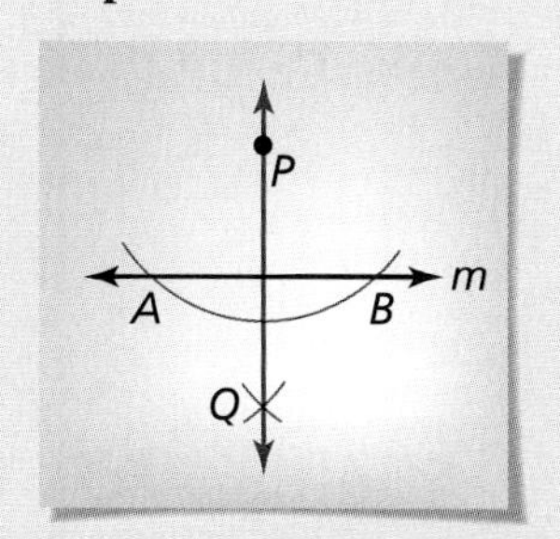

Draw perpendicular line Draw $\overleftrightarrow{PQ}$. This line is perpendicular to line m.

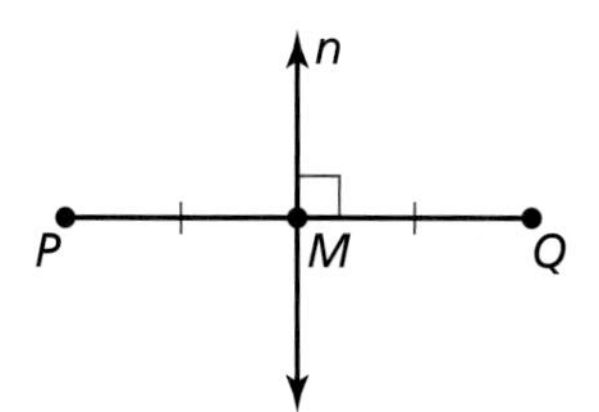

The **perpendicular bisector** of a line segment $\overline{PQ}$ is the line n with the following two properties.

- $n \perp \overline{PQ}$
- n passes through the midpoint M of $\overline{PQ}$.

CONSTRUCTION **Constructing a Perpendicular Bisector**

Use a compass and straightedge to construct the perpendicular bisector of $\overline{AB}$.

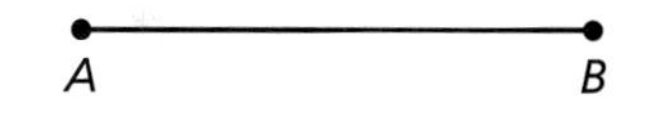

SOLUTION

Step 1

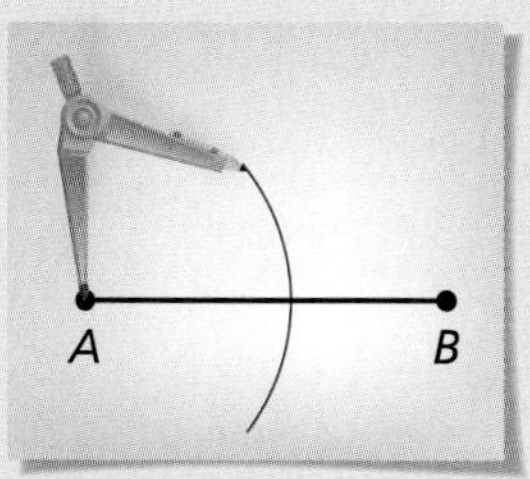

Draw an arc Place the compass at A. Use a compass setting that is greater than half the length of $\overline{AB}$. Draw an arc.

Step 2

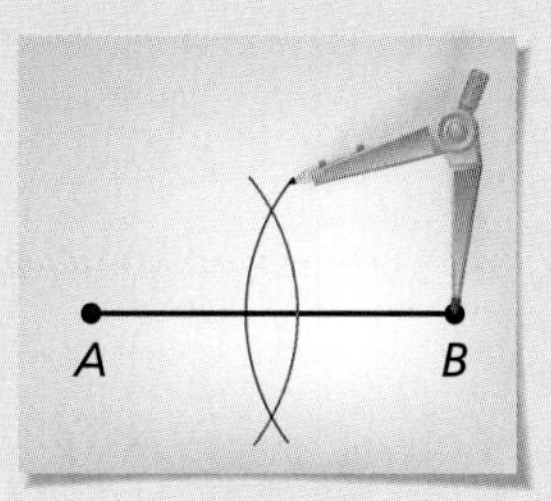

Draw a second arc Keep the same compass setting. Place the compass at B. Draw an arc. It should intersect the other arc at two points.

Step 3

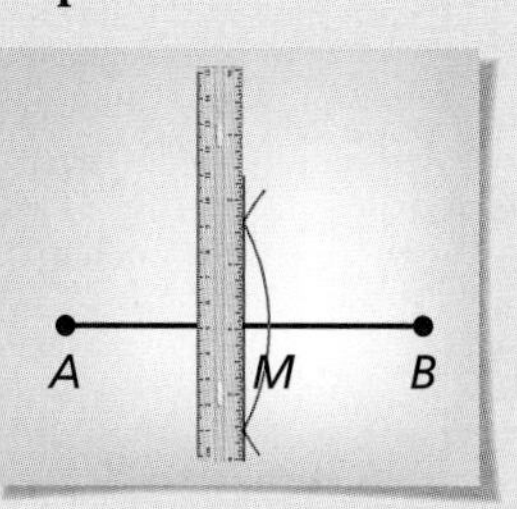

Bisect segment Draw a line through the two points of intersection. This line is the perpendicular bisector of $\overline{AB}$. It passes through M, the midpoint of $\overline{AB}$. So, $AM = MB$.

Proving Theorems about Perpendicular Lines

Theorems

Theorem 3.10 Linear Pair Perpendicular Theorem

If two lines intersect to form a linear pair of congruent angles, then the lines are perpendicular.

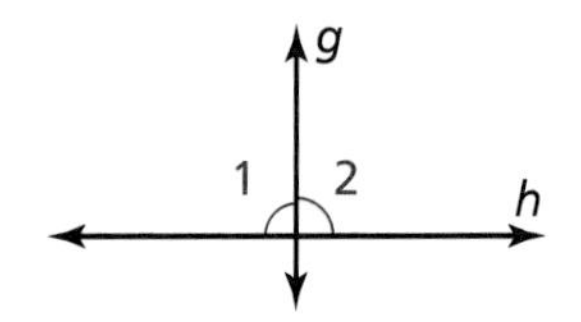

If $\angle 1 \cong \angle 2$, then $g \perp h$.

Proof Ex. 13, p. 153

Theorem 3.11 Perpendicular Transversal Theorem

In a plane, if a transversal is perpendicular to one of two parallel lines, then it is perpendicular to the other line.

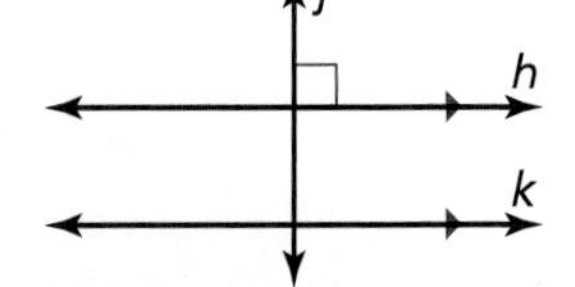

If $h \parallel k$ and $j \perp h$, then $j \perp k$.

Proof Example 2, p. 150; Question 2, p. 150

Theorem 3.12 Lines Perpendicular to a Transversal Theorem

In a plane, if two lines are perpendicular to the same line, then they are parallel to each other.

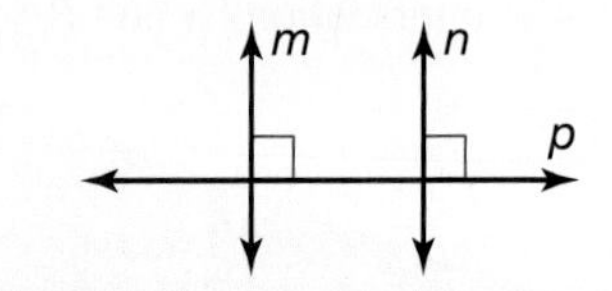

If $m \perp p$ and $n \perp p$, then $m \parallel n$.

Proof Ex. 14, p. 153; Ex. 47, p. 162

EXAMPLE 2 Proving the Perpendicular Transversal Theorem

Use the diagram to prove the Perpendicular Transversal Theorem.

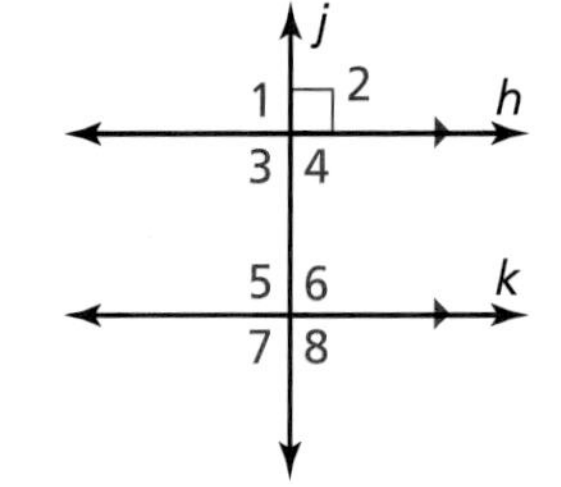

SOLUTION

Given $h \parallel k, j \perp h$

Prove $j \perp k$

STATEMENTS	REASONS
1. $h \parallel k, j \perp h$	**1.** Given
2. $m\angle 2 = 90°$	**2.** Definition of perpendicular lines
3. $\angle 2 \cong \angle 6$	**3.** Corresponding Angles Theorem (Theorem 3.1)
4. $m\angle 2 = m\angle 6$	**4.** Definition of congruent angles
5. $m\angle 6 = 90°$	**5.** Transitive Property of Equality
6. $j \perp k$	**6.** Definition of perpendicular lines

Monitoring Progress  Help in English and Spanish at *BigIdeasMath.com*

2. Prove the Perpendicular Transversal Theorem using the diagram in Example 2 and the Alternate Exterior Angles Theorem (Theorem 3.3).

Solving Real-Life Problems

EXAMPLE 3 Proving Lines Are Parallel

The photo shows the layout of a neighborhood. Determine which lines, if any, must be parallel in the diagram. Explain your reasoning.

SOLUTION

Lines p and q are both perpendicular to s, so by the Lines Perpendicular to a Transversal Theorem, $p \parallel q$. Also, lines s and t are both perpendicular to q, so by the Lines Perpendicular to a Transversal Theorem, $s \parallel t$.

▶ So, from the diagram you can conclude $p \parallel q$ and $s \parallel t$.

Monitoring Progress  Help in English and Spanish at *BigIdeasMath.com*

Use the lines marked in the photo.

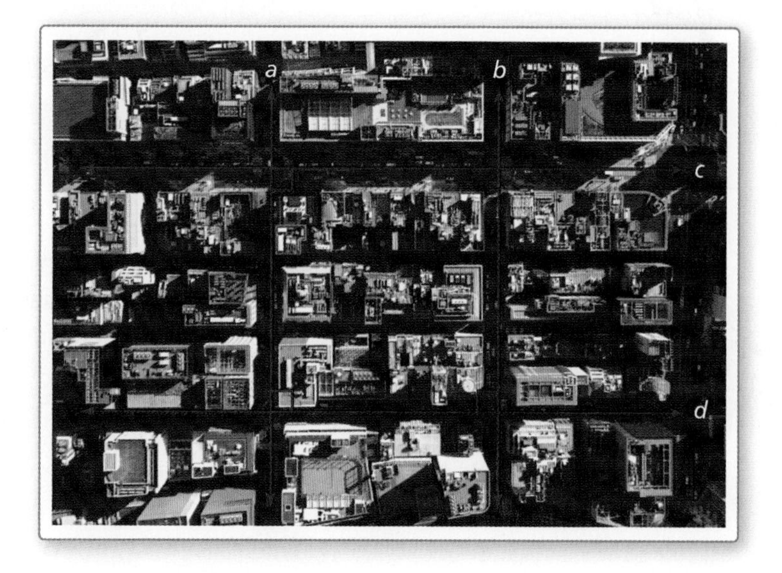

3. Is $b \parallel a$? Explain your reasoning.

4. Is $b \perp c$? Explain your reasoning.

3.4 Exercises

Dynamic Solutions available at *BigIdeasMath.com*

Vocabulary and Core Concept Check

1. **COMPLETE THE SENTENCE** The perpendicular bisector of a segment is the line that passes through the ________ of the segment at a ________ angle.

2. **DIFFERENT WORDS, SAME QUESTION** Which is different? Find "both" answers.

Find the distance from point X to line $\overleftrightarrow{WZ}$.

Find XZ.

Find the length of $\overline{XY}$.

Find the distance from line ℓ to point X.

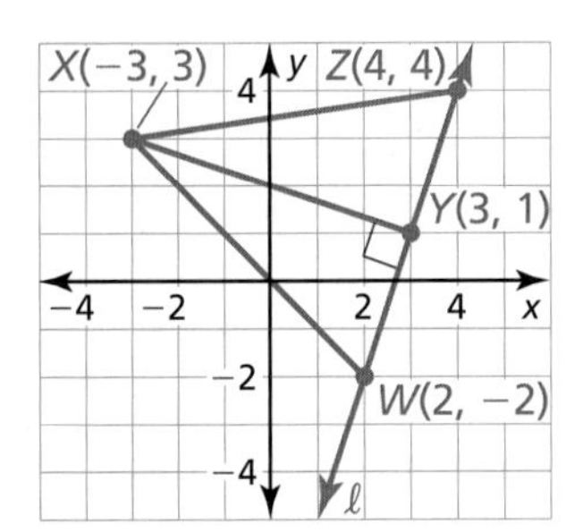

Monitoring Progress and Modeling with Mathematics

In Exercises 3 and 4, find the distance from point A to $\overleftrightarrow{XZ}$. *(See Example 1.)*

3.

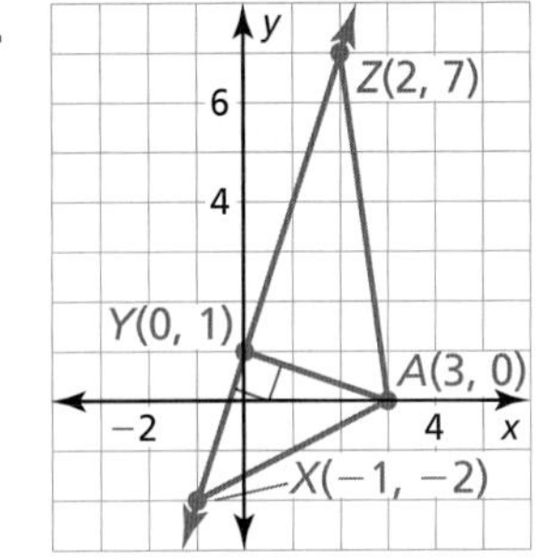

4.

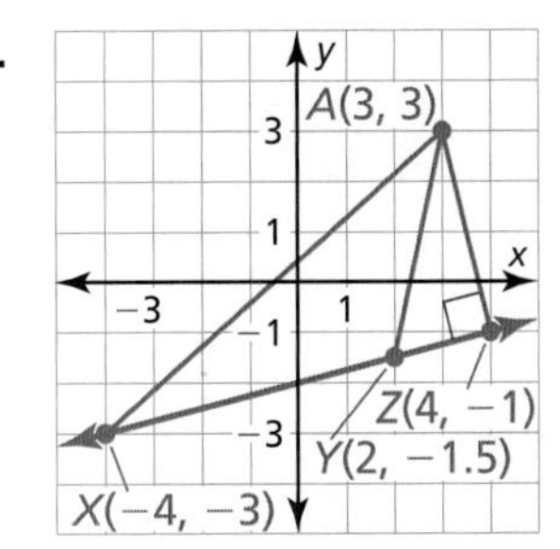

CONSTRUCTION In Exercises 5–8, trace line m and point P. Then use a compass and straightedge to construct a line perpendicular to line m through point P.

5.

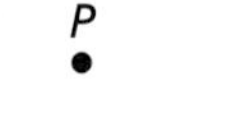

6.

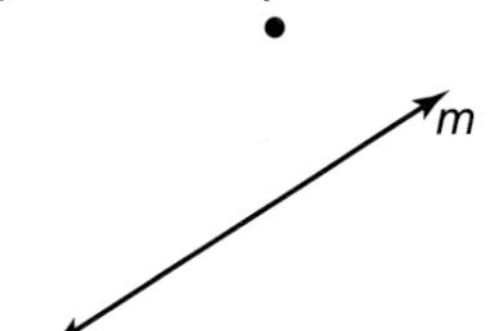

7.

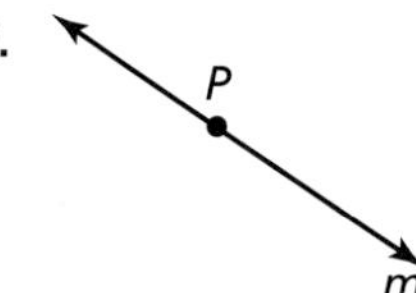

8.

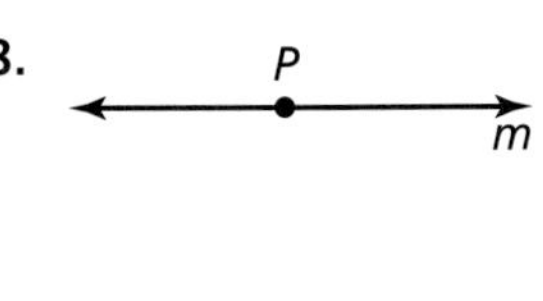

CONSTRUCTION In Exercises 9 and 10, trace $\overline{AB}$. Then use a compass and straightedge to construct the perpendicular bisector of $\overline{AB}$.

9.

10.

A B

ERROR ANALYSIS In Exercises 11 and 12, describe and correct the error in the statement about the diagram.

11.

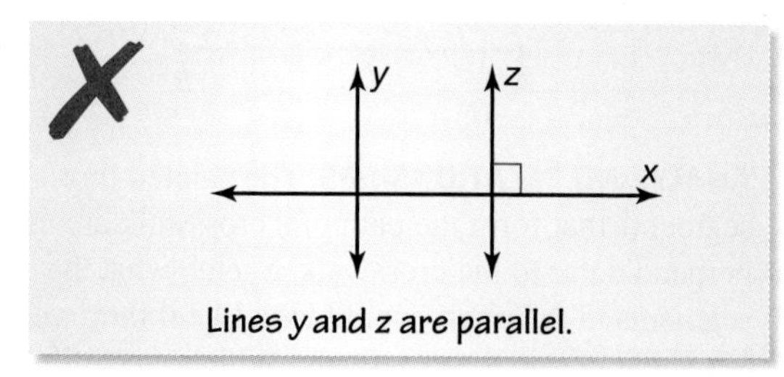

12.

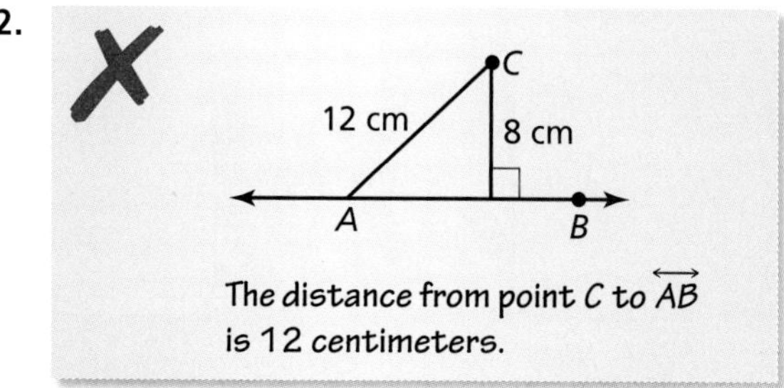

PROVING A THEOREM In Exercises 13 and 14, prove the theorem. *(See Example 2.)*

13. Linear Pair Perpendicular Theorem (Thm. 3.10)

14. Lines Perpendicular to a Transversal Theorem (Thm. 3.12)

PROOF In Exercises 15 and 16, use the diagram to write a proof of the statement.

15. If two intersecting lines are perpendicular, then they intersect to form four right angles.

Given $a \perp b$

Prove $\angle 1$, $\angle 2$, $\angle 3$, and $\angle 4$ are right angles.

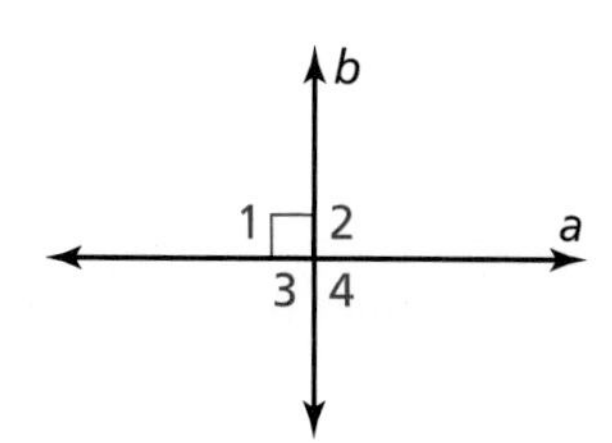

16. If two sides of two adjacent acute angles are perpendicular, then the angles are complementary.

Given $\overrightarrow{BA} \perp \overrightarrow{BC}$

Prove $\angle 1$ and $\angle 2$ are complementary.

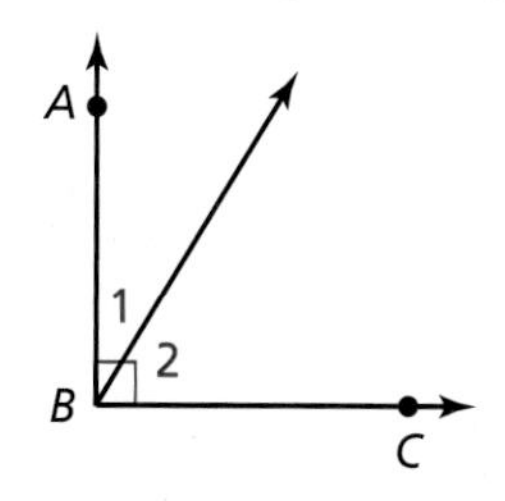

In Exercises 17–22, determine which lines, if any, must be parallel. Explain your reasoning. *(See Example 3.)*

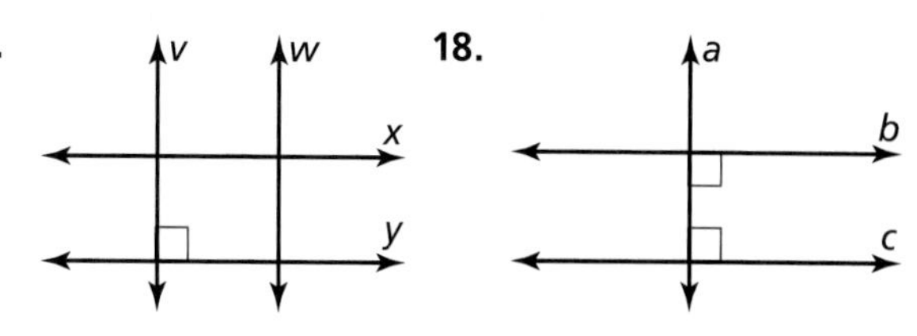

19.

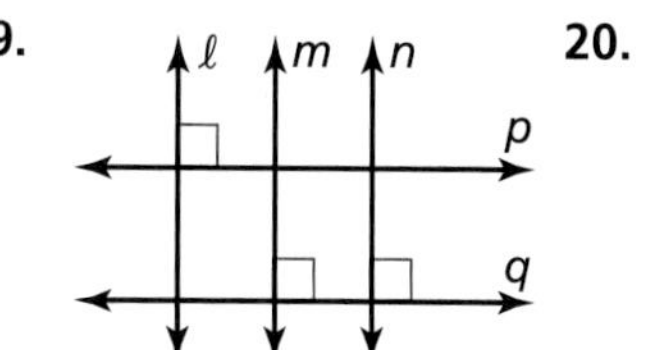

20.

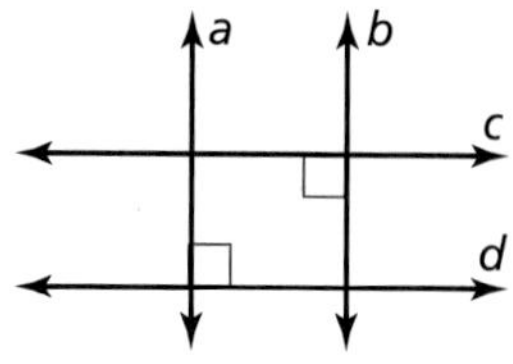

21.

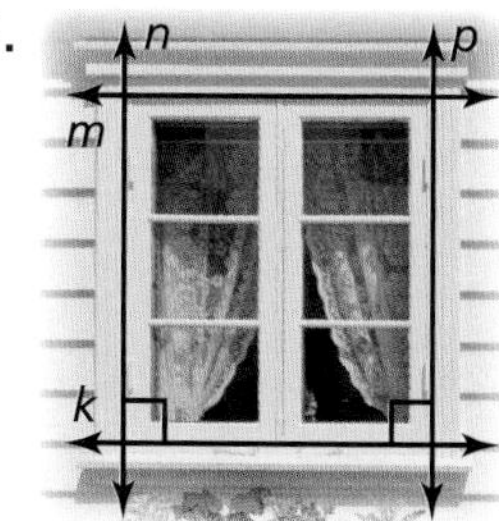

22.

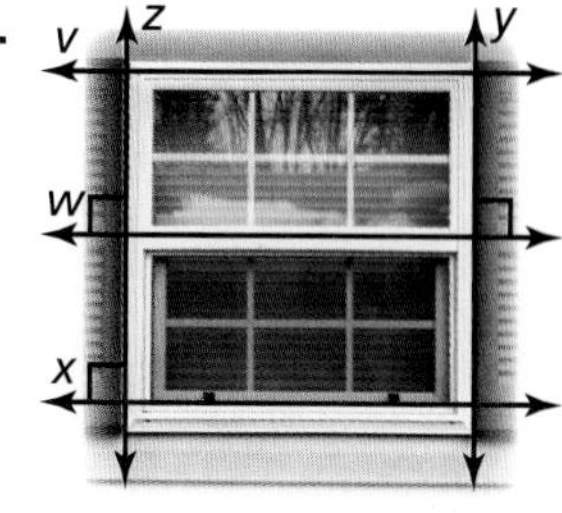

23. USING STRUCTURE Find all the unknown angle measures in the diagram. Justify your answer for each angle measure.

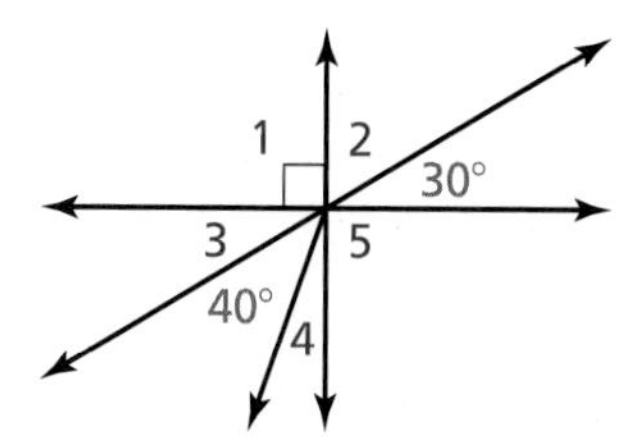

24. MAKING AN ARGUMENT Your friend claims that because you can find the distance from a point to a line, you should be able to find the distance between any two lines. Is your friend correct? Explain your reasoning.

25. MATHEMATICAL CONNECTIONS Find the value of x when $a \perp b$ and $b \parallel c$.

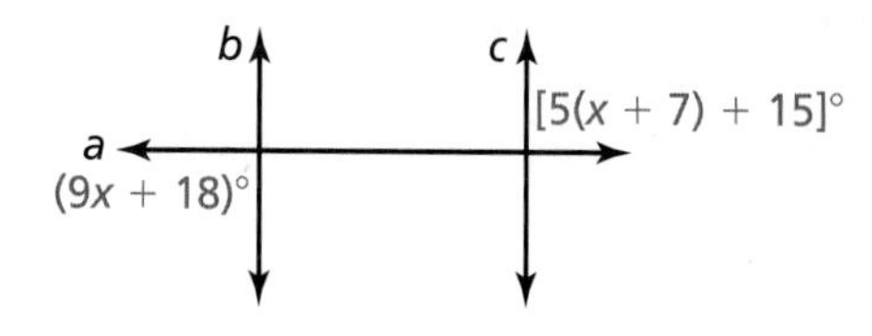

26. HOW DO YOU SEE IT? You are trying to cross a stream from point A. Which point should you jump to in order to jump the shortest distance? Explain your reasoning.

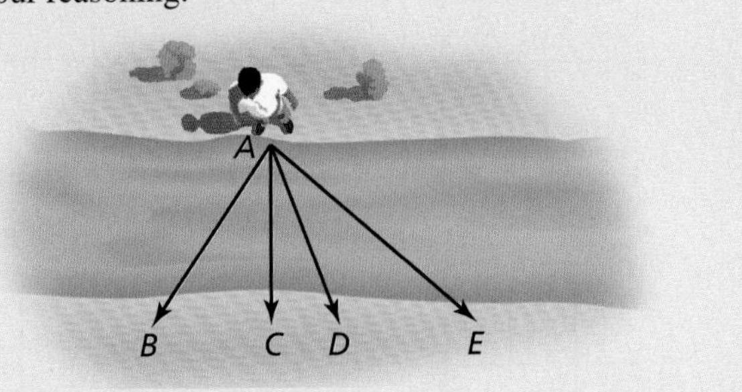

27. ATTENDING TO PRECISION In which of the following diagrams is $\overline{AC} \parallel \overline{BD}$ and $\overline{AC} \perp \overline{CD}$? Select all that apply.

Ⓐ 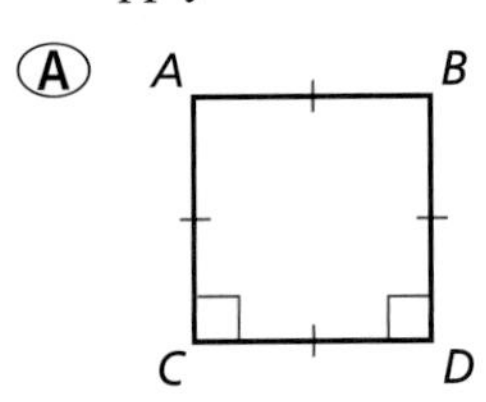

Ⓑ 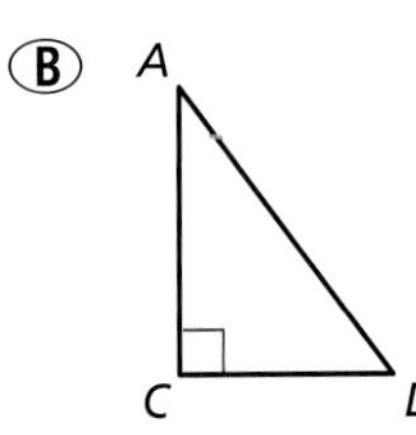

Ⓒ 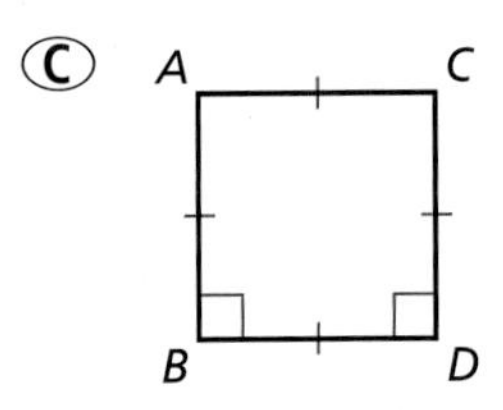

Ⓓ 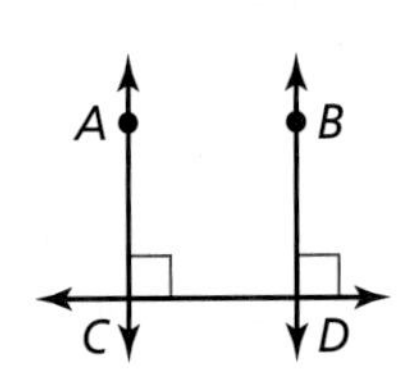

Ⓔ 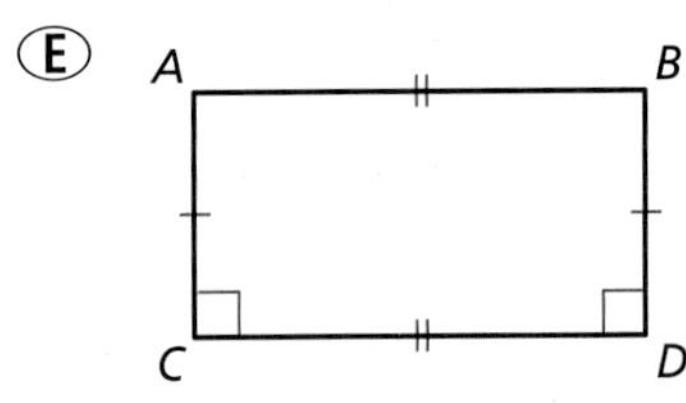

28. THOUGHT PROVOKING The postulates and theorems in this book represent Euclidean geometry. In spherical geometry, all points are points on the surface of a sphere. A line is a circle on the sphere whose diameter is equal to the diameter of the sphere. In spherical geometry, how many right angles are formed by two perpendicular lines? Justify your answer.

29. CONSTRUCTION Construct a square of side length AB.

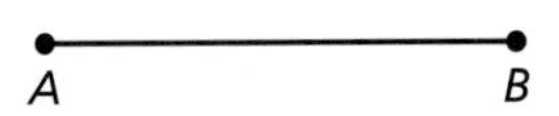

30. ANALYZING RELATIONSHIPS The painted line segments that form the path of a crosswalk are usually perpendicular to the crosswalk. Sketch what the segments in the photo would look like if they were perpendicular to the crosswalk. Which type of line segment requires less paint? Explain your reasoning.

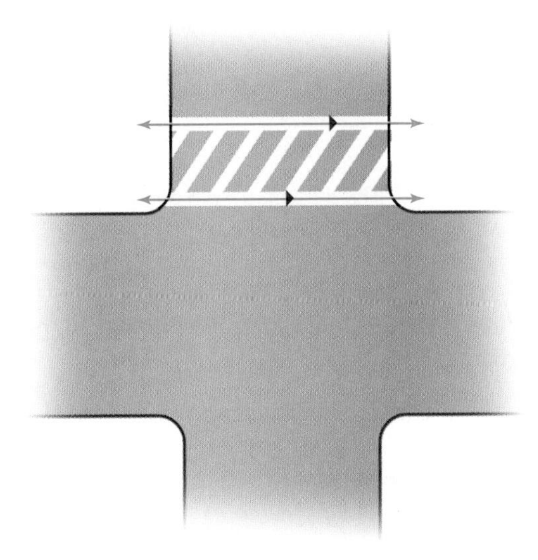

31. ABSTRACT REASONING Two lines, a and b, are perpendicular to line c. Line d is parallel to line c. The distance between lines a and b is x meters. The distance between lines c and d is y meters. What shape is formed by the intersections of the four lines?

32. MATHEMATICAL CONNECTIONS Find the distance between the lines with the equations $y = \frac{3}{2}x + 4$ and $-3x + 2y = -1$.

33. WRITING Describe how you would find the distance from a point to a plane. Can you find the distance from a line to a plane? Explain your reasoning.

Maintaining Mathematical Proficiency

Reviewing what you learned in previous grades and lessons

Simplify the ratio. *(Skills Review Handbook)*

34. $\dfrac{6-(-4)}{8-3}$ **35.** $\dfrac{3-5}{4-1}$ **36.** $\dfrac{8-(-3)}{7-(-2)}$ **37.** $\dfrac{13-4}{2-(-1)}$

Identify the slope and the *y*-intercept of the line. *(Skills Review Handbook)*

38. $y = 3x + 9$ **39.** $y = -\frac{1}{2}x + 7$ **40.** $y = \frac{1}{6}x - 8$ **41.** $y = -8x - 6$

3.5 Equations of Parallel and Perpendicular Lines

Essential Question

How can you write an equation of a line that is parallel or perpendicular to a given line and passes through a given point?

EXPLORATION 1 Writing Equations of Parallel and Perpendicular Lines

Work with a partner. Write an equation of the line that is parallel or perpendicular to the given line and passes through the given point. Use a graphing calculator to verify your answer. What is the relationship between the slopes?

a.

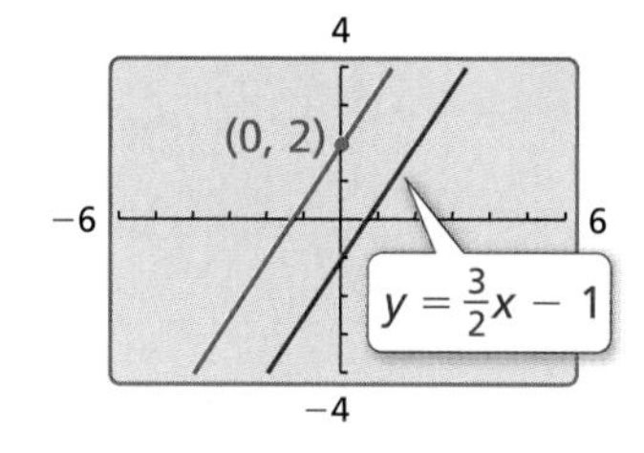

b.

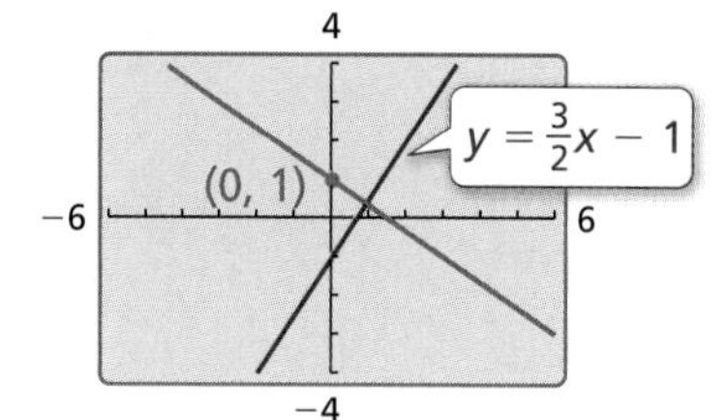

c.

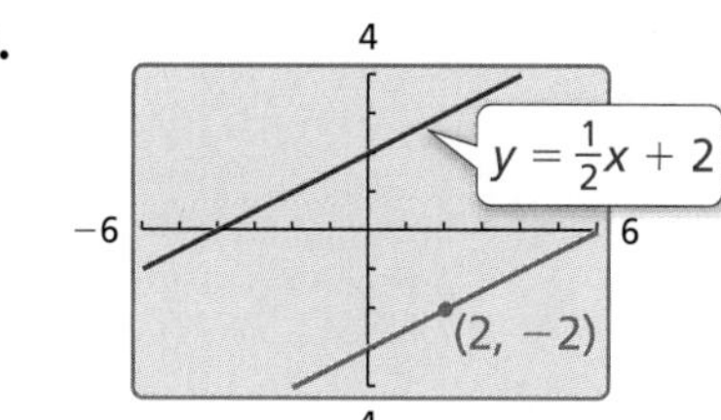

d.

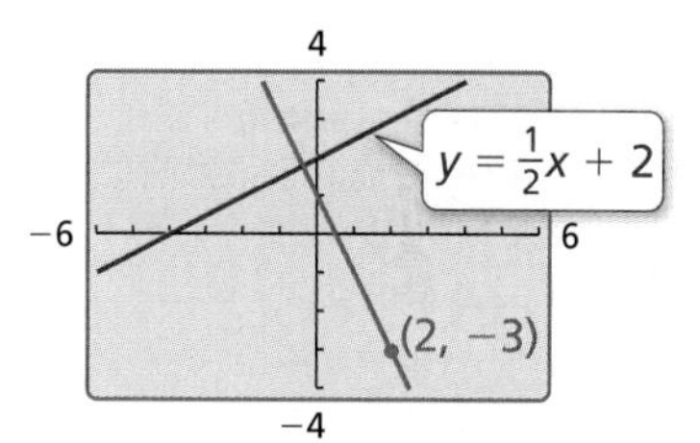

e.

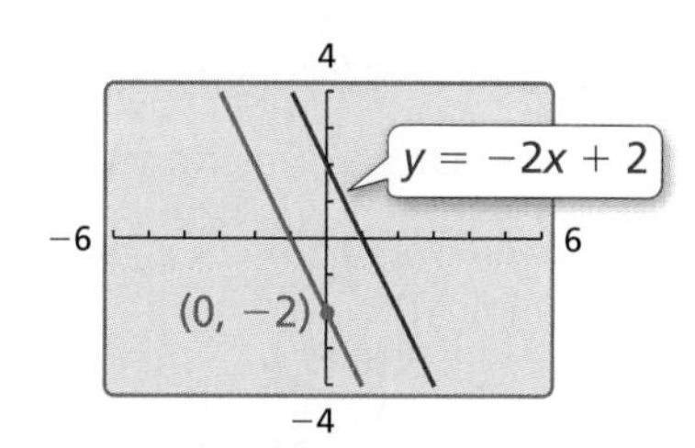

f.

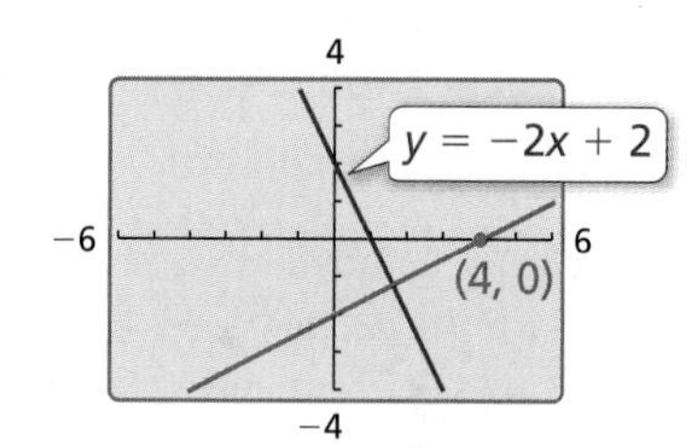

EXPLORATION 2 Writing Equations of Parallel and Perpendicular Lines

Work with a partner. Write the equations of the parallel or perpendicular lines. Use a graphing calculator to verify your answers.

a.

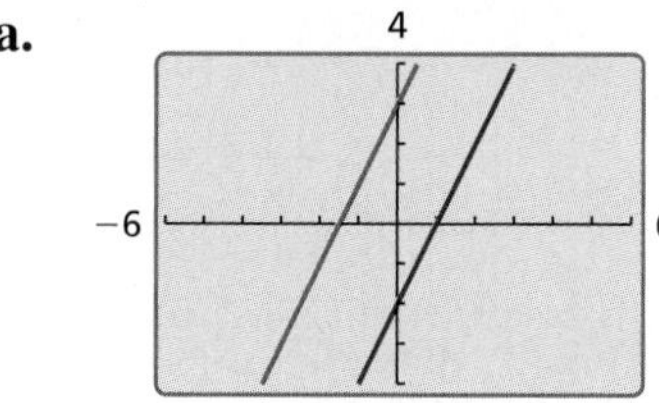

b.

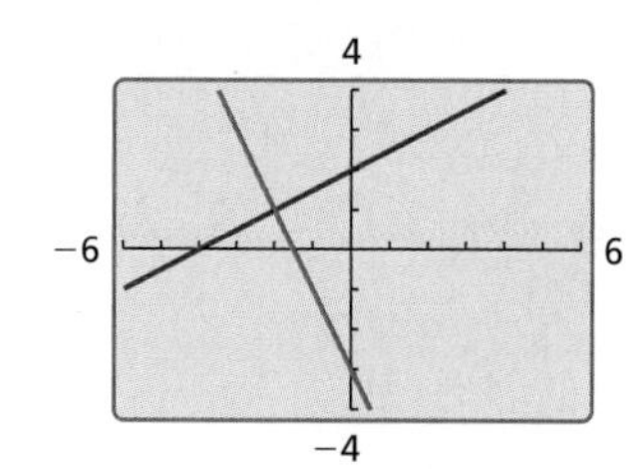

MODELING WITH MATHEMATICS

To be proficient in math, you need to analyze relationships mathematically to draw conclusions.

Communicate Your Answer

3. How can you write an equation of a line that is parallel or perpendicular to a given line and passes through a given point?

4. Write an equation of the line that is (a) parallel and (b) perpendicular to the line $y = 3x + 2$ and passes through the point $(1, -2)$.

3.5 Lesson

Core Vocabulary

directed line segment, *p. 156*

Previous
slope
slope-intercept form
y-intercept

What You Will Learn

- Use slope to partition directed line segments.
- Identify parallel and perpendicular lines.
- Write equations of parallel and perpendicular lines.
- Use slope to find the distance from a point to a line.

Partitioning a Directed Line Segment

A **directed line segment** AB is a segment that represents moving from point A to point B. The following example shows how to use slope to find a point on a directed line segment that partitions the segment in a given ratio.

EXAMPLE 1 Partitioning a Directed Line Segment

Find the coordinates of point P along the directed line segment AB so that the ratio of AP to PB is 3 to 2.

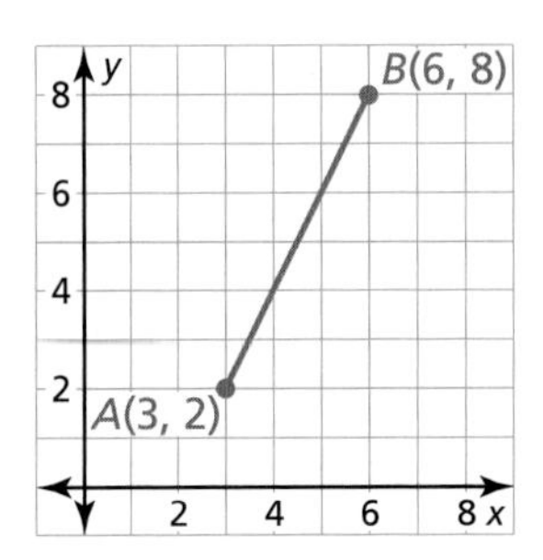

REMEMBER

Recall that the slope of a line or line segment through two points, (x_1, y_1) and (x_2, y_2), is defined as follows.

$$m = \frac{y_2 - y_1}{x_2 - x_1} = \frac{\text{change in } y}{\text{change in } x} = \frac{\text{rise}}{\text{run}}$$

You can choose either of the two points to be (x_1, y_1).

SOLUTION

In order to divide the segment in the ratio 3 to 2, think of dividing, or *partitioning*, the segment into $3 + 2$, or 5 congruent pieces.

Point P is the point that is $\frac{3}{5}$ of the way from point A to point B.

Find the rise and run from point A to point B. Leave the slope in terms of rise and run and do not simplify.

$$\text{slope of } \overline{AB}\text{: } m = \frac{8-2}{6-3} = \frac{6}{3} = \frac{\text{rise}}{\text{run}}$$

To find the coordinates of point P, add $\frac{3}{5}$ of the run to the x-coordinate of A, and add $\frac{3}{5}$ of the rise to the y-coordinate of A.

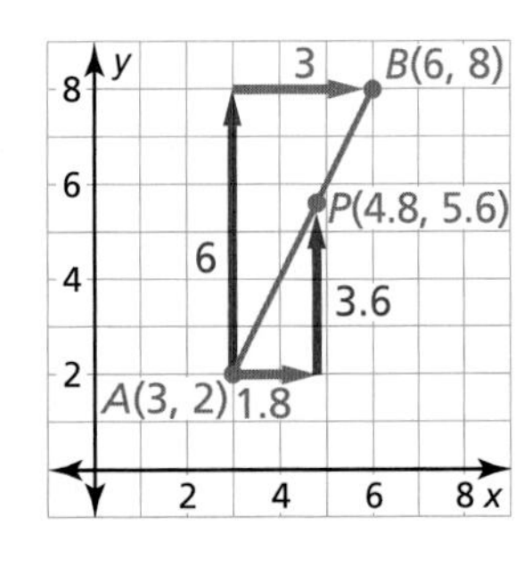

$$\text{run: } \frac{3}{5} \text{ of } 3 = \frac{3}{5} \cdot 3 = 1.8$$

$$\text{rise: } \frac{3}{5} \text{ of } 6 = \frac{3}{5} \cdot 6 = 3.6$$

- So, the coordinates of P are

$$(3 + 1.8, 2 + 3.6) = (4.8, 5.6).$$

The ratio of AP to PB is 3 to 2.

Monitoring Progress  Help in English and Spanish at *BigIdeasMath.com*

Find the coordinates of point *P* along the directed line segment *AB* so that *AP* to *PB* is the given ratio.

1. $A(1, 3)$, $B(8, 4)$; 4 to 1

2. $A(-2, 1)$, $B(4, 5)$; 3 to 7

Identifying Parallel and Perpendicular Lines

In the coordinate plane, the x-axis and the y-axis are perpendicular. Horizontal lines are parallel to the x-axis, and vertical lines are parallel to the y-axis.

Theorems

Theorem 3.13 Slopes of Parallel Lines

In a coordinate plane, two distinct nonvertical lines are parallel if and only if they have the same slope.

Any two vertical lines are parallel.

Proof p. 439; Ex. 41, p. 444

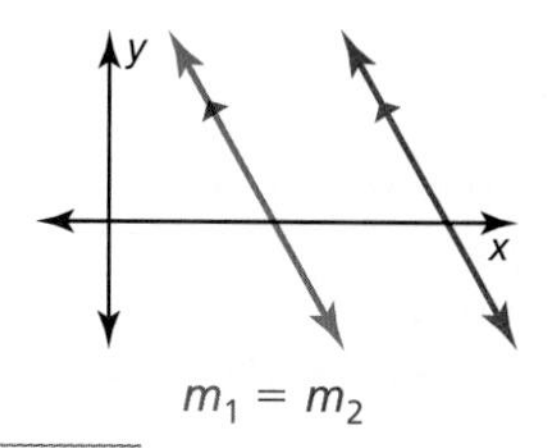

$m_1 = m_2$

READING

If the product of two numbers is -1, then the numbers are called *negative reciprocals*.

Theorem 3.14 Slopes of Perpendicular Lines

In a coordinate plane, two nonvertical lines are perpendicular if and only if the product of their slopes is -1.

Horizontal lines are perpendicular to vertical lines.

Proof p. 440; Ex. 42, p. 444

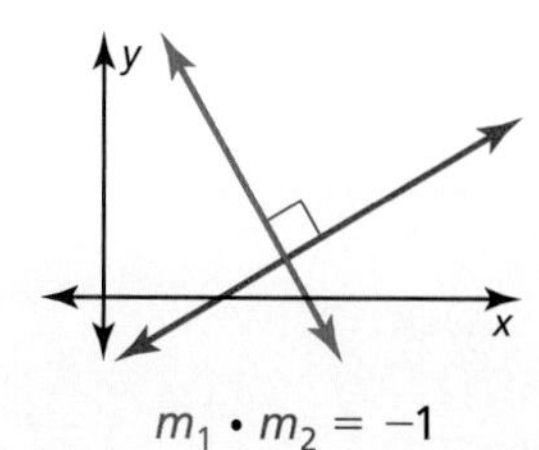

$m_1 \cdot m_2 = -1$

EXAMPLE 2 **Identifying Parallel and Perpendicular Lines**

Determine which of the lines are parallel and which of the lines are perpendicular.

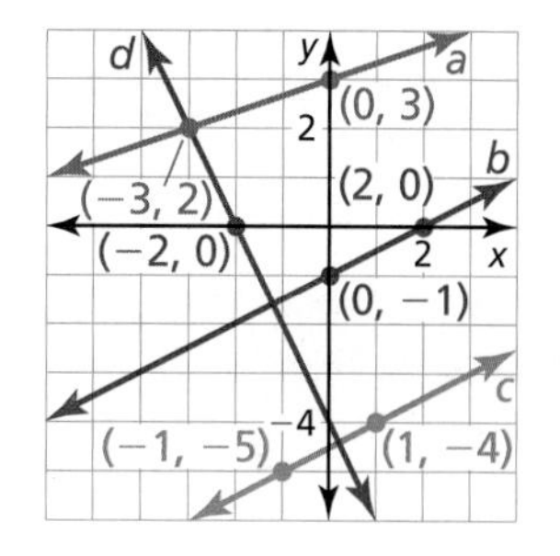

SOLUTION

Find the slope of each line.

$$\text{Line } a\text{: } m = \frac{3 - 2}{0 - (-3)} = \frac{1}{3}$$

$$\text{Line } b\text{: } m = \frac{0 - (-1)}{2 - 0} = \frac{1}{2}$$

$$\text{Line } c\text{: } m = \frac{-4 - (-5)}{1 - (-1)} = \frac{1}{2}$$

$$\text{Line } d\text{: } m = \frac{2 - 0}{-3 - (-2)} = -2$$

Because lines b and c have the same slope, lines b and c are parallel. Because $\frac{1}{2}(-2) = -1$, lines b and d are perpendicular and lines c and d are perpendicular.

Monitoring Progress

Help in English and Spanish at *BigIdeasMath.com*

3. Determine which of the lines are parallel and which of the lines are perpendicular.

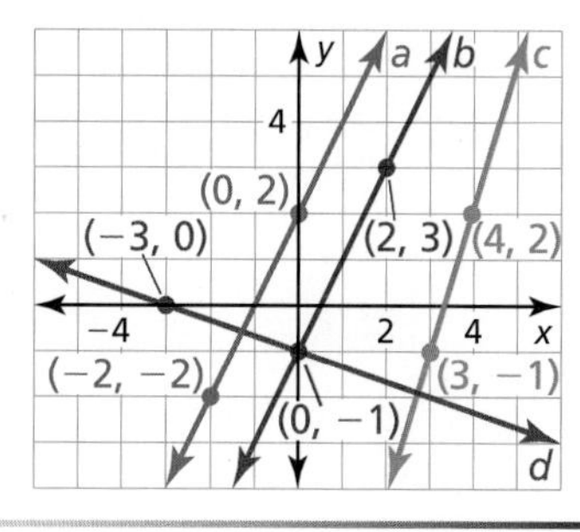

Writing Equations of Parallel and Perpendicular Lines

You can apply the Slopes of Parallel Lines Theorem and the Slopes of Perpendicular Lines Theorem to write equations of parallel and perpendicular lines.

EXAMPLE 3 Writing an Equation of a Parallel Line

Write an equation of the line passing through the point $(-1, 1)$ that is parallel to the line $y = 2x - 3$.

REMEMBER

The linear equation $y = 2x - 3$ is written in slope-intercept form $y = mx + b$, where m is the slope and b is the y-intercept.

SOLUTION

Step 1 Find the slope m of the parallel line. The line $y = 2x - 3$ has a slope of 2. By the Slopes of Parallel Lines Theorem, a line parallel to this line also has a slope of 2. So, $m = 2$.

Step 2 Find the y-intercept b by using $m = 2$ and $(x, y) = (-1, 1)$.

$y = mx + b$	Use slope-intercept form.
$1 = 2(-1) + b$	Substitute for m, x, and y.
$3 = b$	Solve for b.

▶ Because $m = 2$ and $b = 3$, an equation of the line is $y = 2x + 3$. Use a graph to check that the line $y = 2x - 3$ is parallel to the line $y = 2x + 3$.

Check

$y = 2x + 3$

$(-1, 1)$

$y = 2x - 3$

EXAMPLE 4 Writing an Equation of a Perpendicular Line

Write an equation of the line passing through the point $(2, 3)$ that is perpendicular to the line $2x + y = 2$.

SOLUTION

Step 1 Find the slope m of the perpendicular line. The line $2x + y = 2$, or $y = -2x + 2$, has a slope of -2. Use the Slopes of Perpendicular Lines Theorem.

$-2 \cdot m = -1$	The product of the slopes of $\perp$ lines is -1.
$m = \frac{1}{2}$	Divide each side by -2.

Step 2 Find the y-intercept b by using $m = \frac{1}{2}$ and $(x, y) = (2, 3)$.

$y = mx + b$	Use slope-intercept form.
$3 = \frac{1}{2}(2) + b$	Substitute for m, x, and y.
$2 = b$	Solve for b.

▶ Because $m = \frac{1}{2}$ and $b = 2$, an equation of the line is $y = \frac{1}{2}x + 2$. Check that the lines are perpendicular by graphing their equations and using a protractor to measure one of the angles formed by their intersection.

Check

$y = \frac{1}{2}x + 2$

$y = -2x + 2$

$(2, 3)$

Monitoring Progress  Help in English and Spanish at *BigIdeasMath.com*

4. Write an equation of the line that passes through the point $(1, 5)$ and is (a) parallel to the line $y = 3x - 5$ and (b) perpendicular to the line $y = 3x - 5$.

5. How do you know that the lines $x = 4$ and $y = 2$ are perpendicular?

Finding the Distance from a Point to a Line

Recall that the distance from a point to a line is the length of the perpendicular segment from the point to the line.

EXAMPLE 5 Finding the Distance from a Point to a Line

Find the distance from the point (1, 0) to the line $y = -x + 3$.

SOLUTION

Step 1 Find an equation of the line perpendicular to the line $y = -x + 3$ that passes through the point (1, 0).

First, find the slope m of the perpendicular line. The line $y = -x + 3$ has a slope of -1. Use the Slopes of Perpendicular Lines Theorem.

$-1 \cdot m = -1$ The product of the slopes of $\perp$ lines is -1.

$m = 1$ Divide each side by -1.

Then find the y-intercept b by using $m = 1$ and $(x, y) = (1, 0)$.

$y = mx + b$ Use slope-intercept form.

$0 = 1(1) + b$ Substitute for x, y, and m.

$-1 = b$ Solve for b.

Because $m = 1$ and $b = -1$, an equation of the line is $y = x - 1$.

REMEMBER

Recall that the solution of a system of two linear equations in two variables gives the coordinates of the point of intersection of the graphs of the equations.

There are two special cases when the lines have the same slope.

- When the system has no solution, the lines are parallel.
- When the system has infinitely many solutions, the lines coincide.

Step 2 Use the two equations to write and solve a system of equations to find the point where the two lines intersect.

$y = -x + 3$ Equation 1

$y = x - 1$ Equation 2

Substitute $-x + 3$ for y in Equation 2.

$y = x - 1$ Equation 2

$-x + 3 = x - 1$ Substitute $-x + 3$ for y.

$x = 2$ Solve for x.

Substitute 2 for x in Equation 1 and solve for y.

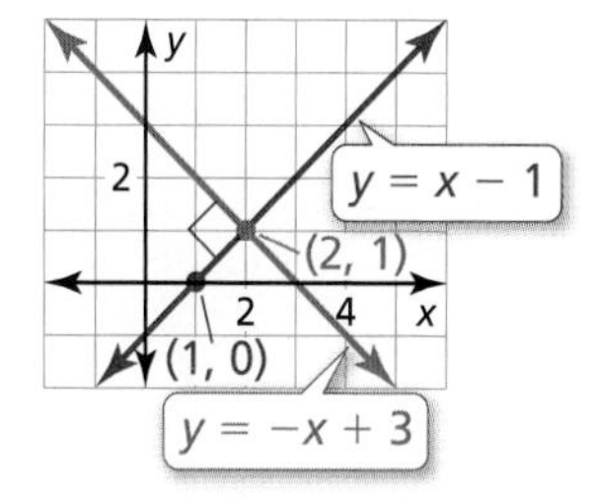

$y = -x + 3$ Equation 1

$y = -2 + 3$ Substitute 2 for x.

$y = 1$ Simplify.

So, the perpendicular lines intersect at (2, 1).

Step 3 Use the Distance Formula to find the distance from (1, 0) to (2, 1).

$$\text{distance} = \sqrt{(1-2)^2 + (0-1)^2} = \sqrt{(-1)^2 + (-1)^2} = \sqrt{2} \approx 1.4$$

▶ So, the distance from the point (1, 0) to the line $y = -x + 3$ is about 1.4 units.

Monitoring Progress  Help in English and Spanish at *BigIdeasMath.com*

6. Find the distance from the point (6, 4) to the line $y = x + 4$.

7. Find the distance from the point $(-1, 6)$ to the line $y = -2x$.

3.5 Exercises

Dynamic Solutions available at *BigIdeasMath.com*

Vocabulary and Core Concept Check

1. **COMPLETE THE SENTENCE** A ________ line segment AB is a segment that represents moving from point A to point B.

2. **WRITING** How are the slopes of perpendicular lines related?

Monitoring Progress and Modeling with Mathematics

In Exercises 3–6, find the coordinates of point P along the directed line segment AB so that AP to PB is the given ratio. *(See Example 1.)*

3. $A(8, 0)$, $B(3, -2)$; 1 to 4

4. $A(-2, -4)$, $B(6, 1)$; 3 to 2

5. $A(1, 6)$, $B(-2, -3)$; 5 to 1

6. $A(-3, 2)$, $B(5, -4)$; 2 to 6

In Exercises 7 and 8, determine which of the lines are parallel and which of the lines are perpendicular. *(See Example 2.)*

7.

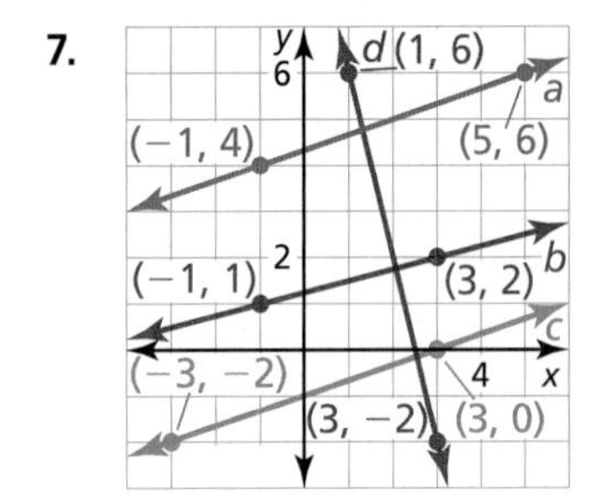

8.

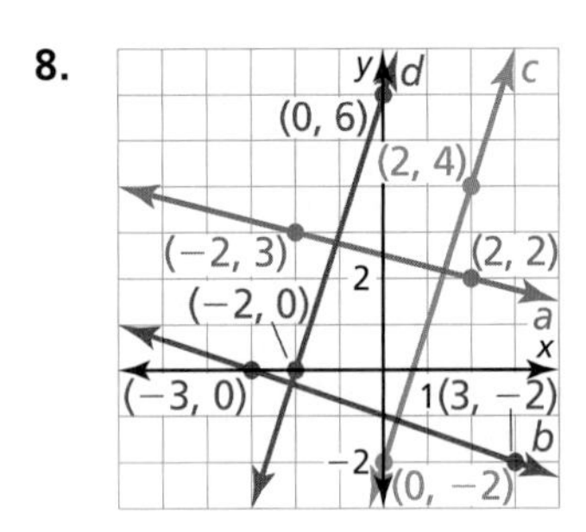

In Exercises 9–12, tell whether the lines through the given points are *parallel*, *perpendicular*, or *neither*. Justify your answer.

9. **Line 1:** $(1, 0)$, $(7, 4)$
   Line 2: $(7, 0)$, $(3, 6)$

10. **Line 1:** $(-3, 1)$, $(-7, -2)$
    Line 2: $(2, -1)$, $(8, 4)$

11. **Line 1:** $(-9, 3)$, $(-5, 7)$
    Line 2: $(-11, 6)$, $(-7, 2)$

12. **Line 1:** $(10, 5)$, $(-8, 9)$
    Line 2: $(2, -4)$, $(11, -6)$

In Exercises 13–16, write an equation of the line passing through point P that is parallel to the given line. Graph the equations of the lines to check that they are parallel. *(See Example 3.)*

13. $P(0, -1)$, $y = -2x + 3$

14. $P(3, 8)$, $y = \frac{1}{5}(x + 4)$

15. $P(-2, 6)$, $x = -5$

16. $P(4, 0)$, $-x + 2y = 12$

In Exercises 17–20, write an equation of the line passing through point P that is perpendicular to the given line. Graph the equations of the lines to check that they are perpendicular. *(See Example 4.)*

17. $P(0, 0)$, $y = -9x - 1$

18. $P(4, -6)$, $y = -3$

19. $P(2, 3)$, $y - 4 = -2(x + 3)$

20. $P(-8, 0)$, $3x - 5y = 6$

In Exercises 21–24, find the distance from point A to the given line. *(See Example 5.)*

21. $A(-1, 7)$, $y = 3x$

22. $A(-9, -3)$, $y = x - 6$

23. $A(15, -21)$, $5x + 2y = 4$

24. $A\left(-\frac{1}{4}, 5\right)$, $-x + 2y = 14$

25. **ERROR ANALYSIS** Describe and correct the error in determining whether the lines are parallel, perpendicular, or neither.

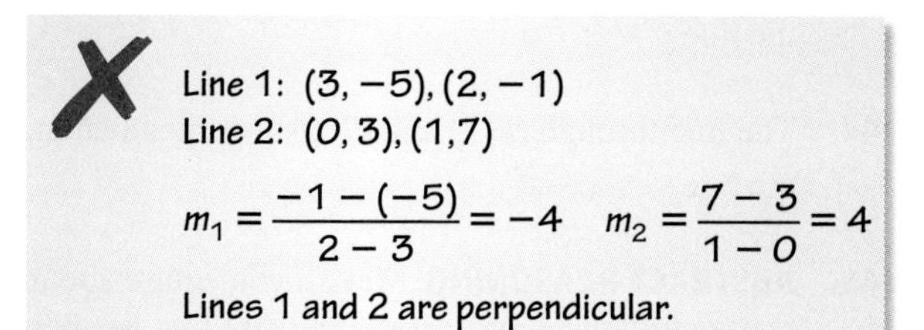

26. **ERROR ANALYSIS** Describe and correct the error in writing an equation of the line that passes through the point (3, 4) and is parallel to the line $y = 2x + 1$.

$y = 2x + 1, (3, 4)$

$4 = m(3) + 1$

$1 = m$

The line $y = x + 1$ is parallel to the line $y = 2x + 1$.

In Exercises 27–30, find the midpoint of $\overline{PQ}$. Then write an equation of the line that passes through the midpoint and is perpendicular to $\overline{PQ}$. This line is called the *perpendicular bisector*.

27. $P(-4, 3), Q(4, -1)$

28. $P(-5, -5), Q(3, 3)$

29. $P(0, 2), Q(6, -2)$

30. $P(-7, 0), Q(1, 8)$

31. **MODELING WITH MATHEMATICS** Your school lies directly between your house and the movie theater. The distance from your house to the school is one-fourth of the distance from the school to the movie theater. What point on the graph represents your school?

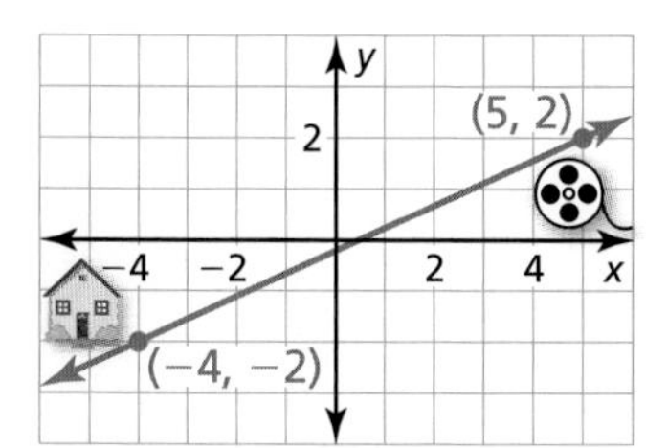

32. **REASONING** Is quadrilateral $QRST$ a parallelogram? Explain your reasoning.

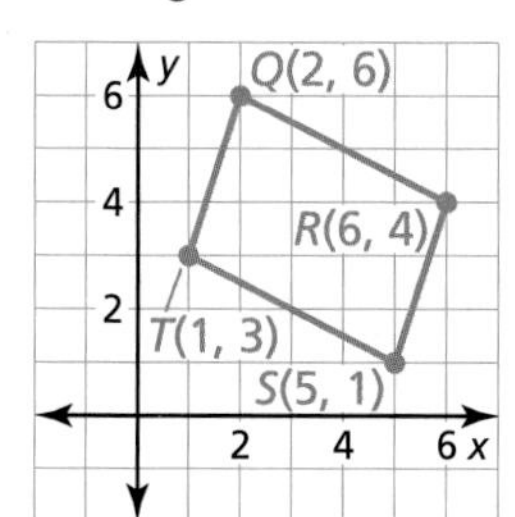

33. **REASONING** A triangle has vertices $L(0, 6)$, $M(5, 8)$, and $N(4, -1)$. Is the triangle a right triangle? Explain your reasoning.

34. **MODELING WITH MATHEMATICS** A new road is being constructed parallel to the train tracks through point V. An equation of the line representing the train tracks is $y = 2x$. Find an equation of the line representing the new road.

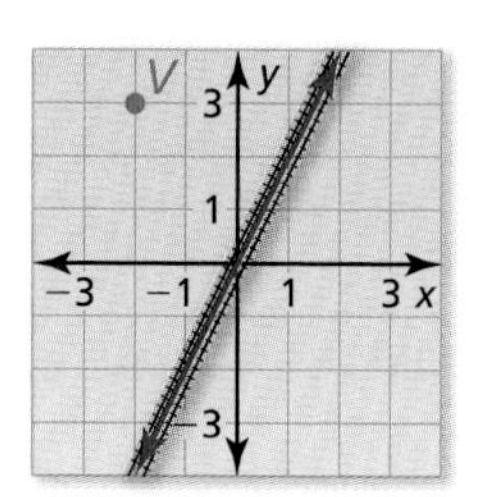

35. **MODELING WITH MATHEMATICS** A bike path is being constructed perpendicular to Washington Boulevard through point $P(2, 2)$. An equation of the line representing Washington Boulevard is $y = -\frac{2}{3}x$. Find an equation of the line representing the bike path.

36. **PROBLEM SOLVING** A gazebo is being built near a nature trail. An equation of the line representing the nature trail is $y = \frac{1}{3}x - 4$. Each unit in the coordinate plane corresponds to 10 feet. Approximately how far is the gazebo from the nature trail?

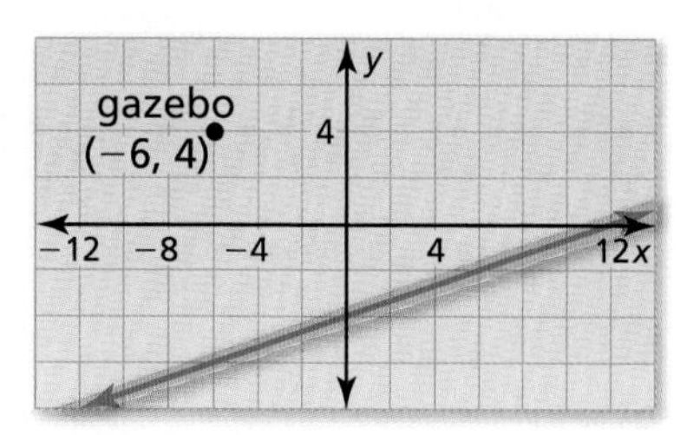

37. **CRITICAL THINKING** The slope of line ℓ is greater than 0 and less than 1. Write an inequality for the slope of a line perpendicular to ℓ. Explain your reasoning.

38. **HOW DO YOU SEE IT?** Determine whether quadrilateral *JKLM* is a square. Explain your reasoning.

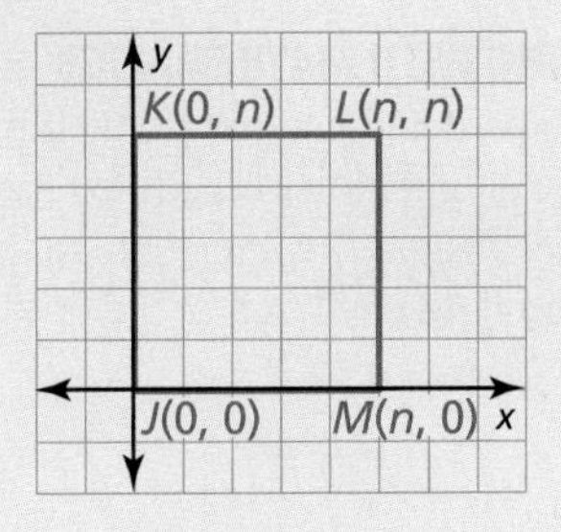

39. **CRITICAL THINKING** Suppose point *P* divides the directed line segment *XY* so that the ratio of *XP* to *PY* is 3 to 5. Describe the point that divides the directed line segment *YX* so that the ratio of *YP* to *PX* is 5 to 3.

40. **MAKING AN ARGUMENT** Your classmate claims that no two nonvertical parallel lines can have the same *y*-intercept. Is your classmate correct? Explain.

41. **MATHEMATICAL CONNECTIONS** Solve each system of equations algebraically. Make a conjecture about what the solution(s) can tell you about whether the lines intersect, are parallel, or are the same line.

 a. $y = 4x + 9$
 $4x - y = 1$

 b. $3y + 4x = 16$
 $2x - y = 18$

 c. $y = -5x + 6$
 $10x + 2y = 12$

42. **THOUGHT PROVOKING** Find a formula for the distance from the point (x_0, y_0) to the line $ax + by = 0$. Verify your formula using a point and a line.

MATHEMATICAL CONNECTIONS In Exercises 43 and 44, find a value for *k* based on the given description.

43. The line through $(-1, k)$ and $(-7, -2)$ is parallel to the line $y = x + 1$.

44. The line through $(k, 2)$ and $(7, 0)$ is perpendicular to the line $y = x - \frac{28}{5}$.

45. **ABSTRACT REASONING** Make a conjecture about how to find the coordinates of a point that lies beyond point *B* along $\overrightarrow{AB}$. Use an example to support your conjecture.

46. **PROBLEM SOLVING** What is the distance between the lines $y = 2x$ and $y = 2x + 5$? Verify your answer.

PROVING A THEOREM In Exercises 47 and 48, use the slopes of lines to write a paragraph proof of the theorem.

47. Lines Perpendicular to a Transversal Theorem (Theorem 3.12): In a plane, if two lines are perpendicular to the same line, then they are parallel to each other.

48. Transitive Property of Parallel Lines Theorem (Theorem 3.9): If two lines are parallel to the same line, then they are parallel to each other.

49. **PROOF** Prove the statement: If two lines are vertical, then they are parallel.

50. **PROOF** Prove the statement: If two lines are horizontal, then they are parallel.

51. **PROOF** Prove that horizontal lines are perpendicular to vertical lines.

Maintaining Mathematical Proficiency

Reviewing what you learned in previous grades and lessons

Plot the point in a coordinate plane. *(Skills Review Handbook)*

52. $A(3, 6)$

53. $B(0, -4)$

54. $C(5, 0)$

55. $D(-1, -2)$

Copy and complete the table. *(Skills Review Handbook)*

56.

x	-2	-1	0	1	2
$y = x + 9$					

57.

x	-2	-1	0	1	2
$y = x - \frac{3}{4}$					

3.4–3.5 What Did You Learn?

Core Vocabulary

distance from a point to a line, *p. 148*
perpendicular bisector, *p. 149*
directed line segment, *p. 156*

Core Concepts

Section 3.4

Finding the Distance from a Point to a Line, *p. 148*
Constructing Perpendicular Lines, *p. 149*
Theorem 3.10 Linear Pair Perpendicular Theorem, *p. 150*
Theorem 3.11 Perpendicular Transversal Theorem, *p. 150*
Theorem 3.12 Lines Perpendicular to a Transversal Theorem, *p. 150*

Section 3.5

Partitioning a Directed Line Segment, *p. 156*
Theorem 3.13 Slopes of Parallel Lines, *p. 157*
Theorem 3.14 Slopes of Perpendicular Lines, *p. 157*
Writing Equations of Parallel and Perpendicular Lines, *p. 158*
Finding the Distance from a Point to a Line, *p. 159*

Mathematical Practices

1. Compare the effectiveness of the argument in Exercise 24 on page 153 with the argument "You can find the distance between any two parallel lines." What flaw(s) exist in the argument(s)? Does either argument use correct reasoning? Explain.
2. Look back at your construction of a square in Exercise 29 on page 154. How would your construction change if you were to construct a rectangle?
3. In Exercise 31 on page 161, a classmate tells you that your answer is incorrect because you should have divided the segment into four congruent pieces. Respond to your classmate's argument by justifying your original answer.

Performance Task

Navajo Rugs

Navajo rugs use mathematical properties to enhance their beauty. How can you describe these creative works of art with geometry? What properties of lines can you see and use to describe the patterns?

To explore the answers to this question and more, go to ***BigIdeasMath.com***.

3 Chapter Review

Dynamic Solutions available at *BigIdeasMath.com*

3.1 Pairs of Lines and Angles (pp. 125–130)

Think of each segment in the figure as part of a line.

a. Which line(s) appear perpendicular to $\overleftrightarrow{AB}$?

▶ $\overleftrightarrow{BD}$, $\overleftrightarrow{AC}$, $\overleftrightarrow{BH}$, and $\overleftrightarrow{AG}$ appear perpendicular to $\overleftrightarrow{AB}$.

b. Which line(s) appear parallel to $\overleftrightarrow{AB}$?

▶ $\overleftrightarrow{CD}$, $\overleftrightarrow{GH}$, and $\overleftrightarrow{EF}$ appear parallel to $\overleftrightarrow{AB}$.

c. Which line(s) appear skew to $\overleftrightarrow{AB}$?

▶ $\overleftrightarrow{CF}$, $\overleftrightarrow{CE}$, $\overleftrightarrow{DF}$, $\overleftrightarrow{FH}$, and $\overleftrightarrow{EG}$ appear skew to $\overleftrightarrow{AB}$.

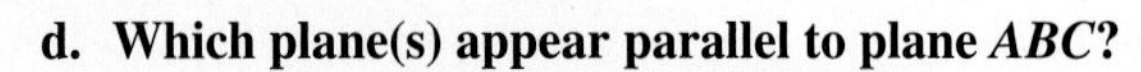

d. Which plane(s) appear parallel to plane *ABC*?

▶ Plane *EFG* appears parallel to plane *ABC*.

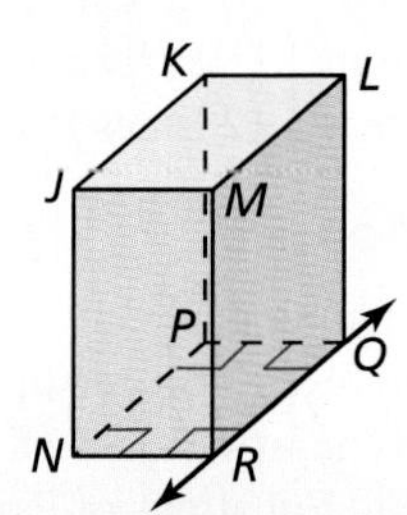

Think of each segment in the figure as part of a line. Which line(s) or plane(s) appear to fit the description?

1. line(s) perpendicular to $\overleftrightarrow{QR}$
2. line(s) parallel to $\overleftrightarrow{QR}$
3. line(s) skew to $\overleftrightarrow{QR}$
4. plane(s) parallel to plane *LMQ*

3.2 Parallel Lines and Transversals (pp. 131–136)

Find the value of *x*.

By the Vertical Angles Congruence Theorem (Theorem 2.6), $m\angle 6 = 50°$.

$(x - 5)° + m\angle 6 = 180°$	Consecutive Interior Angles Theorem (Thm. 3.4)
$(x - 5)° + 50° = 180°$	Substitute 50° for $m\angle 6$.
$x + 45 = 180$	Combine like terms.
$x = 135$	Subtract 45 from each side.

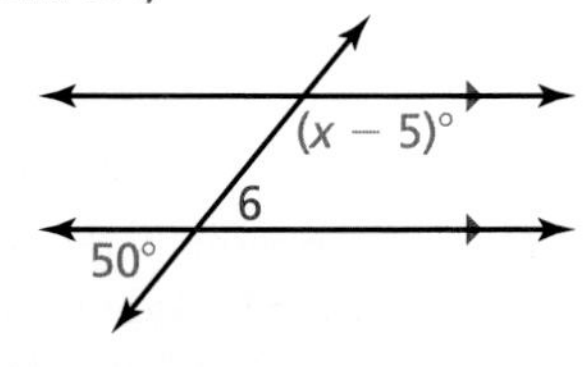

▶ So, the value of *x* is 135.

Find the values of *x* and *y*.

5.

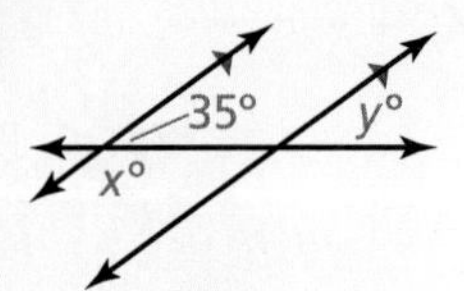

6.

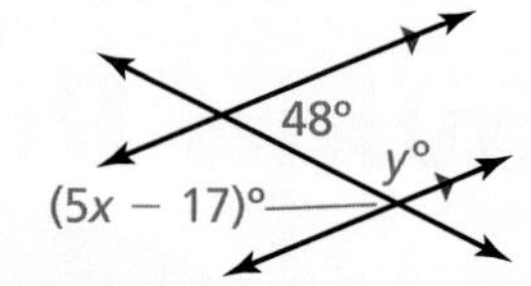

7.

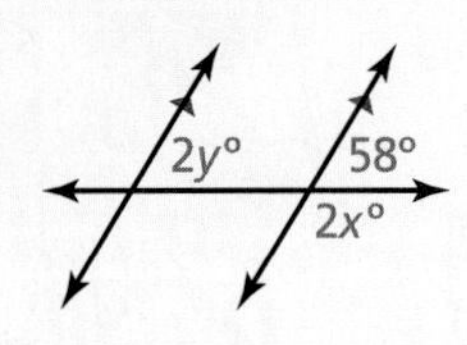

8.

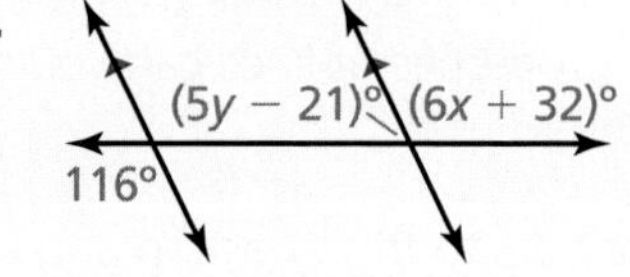

3.3 Proofs with Parallel Lines (pp. 137–144)

Find the value of x that makes $m \parallel n$.

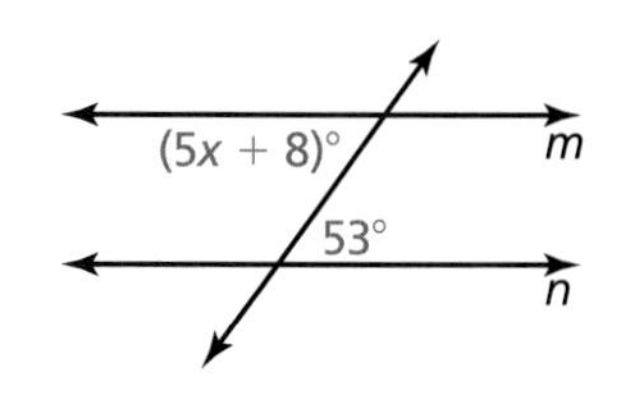

By the Alternate Interior Angles Converse (Theorem 3.6), $m \parallel n$ when the marked angles are congruent.

$$(5x + 8)° = 53°$$

$$5x = 45$$

$$x = 9$$

▶ The lines m and n are parallel when $x = 9$.

Find the value of x that makes $m \parallel n$.

9. $x°$, $73°$, m, n

10. $147°$, $(x + 14)°$, m, n

11. $(2x + 20)°$, $3x°$, m, n

12. $(7x - 11)°$, $(4x + 58)°$, m, n

3.4 Proofs with Perpendicular Lines (pp. 147–154)

Determine which lines, if any, must be parallel. Explain your reasoning.

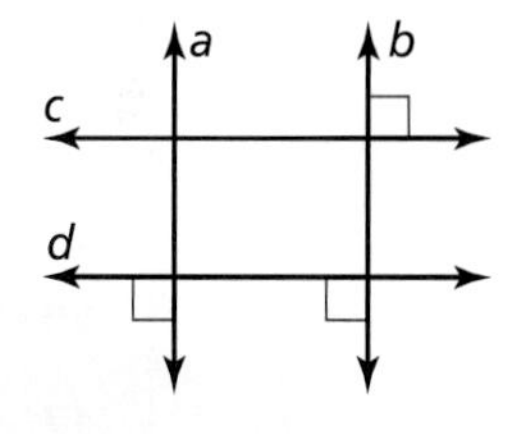

Lines a and b are both perpendicular to d, so by the Lines Perpendicular to a Transversal Theorem (Theorem 3.12), $a \parallel b$.

Also, lines c and d are both perpendicular to b, so by the Lines Perpendicular to a Transversal Theorem (Theorem 3.12), $c \parallel d$.

Determine which lines, if any, must be parallel. Explain your reasoning.

13.

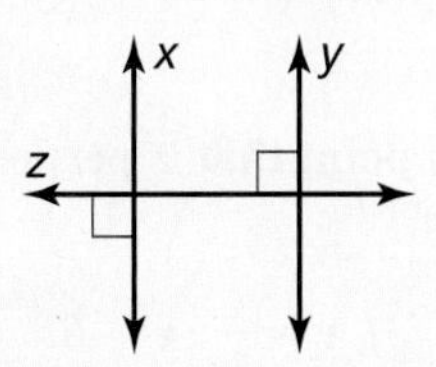

14.

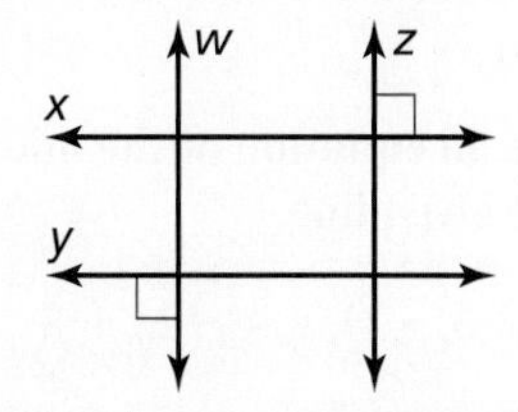

15.

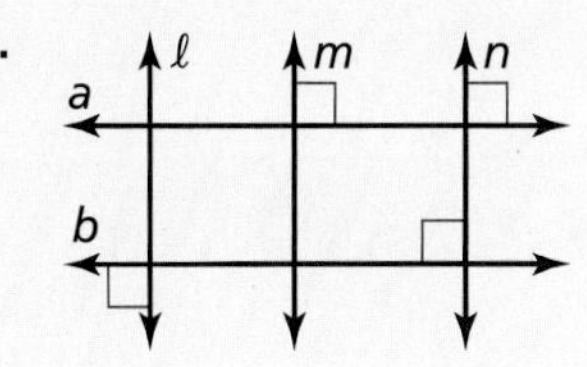

16.

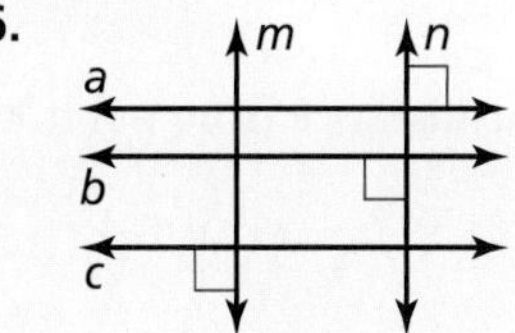

3.5 Equations of Parallel and Perpendicular Lines (pp. 155–162)

a. Write an equation of the line passing through the point $(-2, 4)$ that is parallel to the line $y = 5x - 7$.

Step 1 Find the slope m of the parallel line. The line $y = 5x - 7$ has a slope of 5. By the Slopes of Parallel Lines Theorem (Theorem 3.13), a line parallel to this line also has a slope of 5. So, $m = 5$.

Step 2 Find the y-intercept b by using $m = 5$ and $(x, y) = (-2, 4)$.

$y = mx + b$ — Use slope-intercept form.

$4 = 5(-2) + b$ — Substitute for m, x, and y.

$14 = b$ — Solve for b.

▶ Because $m = 5$ and $b = 14$, an equation of the line is $y = 5x + 14$.

b. Write an equation of the line passing through the point $(6, 1)$ that is perpendicular to the line $3x + y = 9$.

Step 1 Find the slope m of the perpendicular line. The line $3x + y = 9$, or $y = -3x + 9$, has a slope of -3. Use the Slopes of Perpendicular Lines Theorem (Theorem 3.14).

$-3 \cdot m = -1$ — The product of the slopes of $\perp$ lines is -1.

$m = \frac{1}{3}$ — Divide each side by -3.

Step 2 Find the y-intercept b by using $m = \frac{1}{3}$ and $(x, y) = (6, 1)$.

$y = mx + b$ — Use slope-intercept form.

$1 = \frac{1}{3}(6) + b$ — Substitute for m, x, and y.

$-1 = b$ — Solve for b.

▶ Because $m = \frac{1}{3}$ and $b = -1$, an equation of the line is $y = \frac{1}{3}x - 1$.

Write an equation of the line passing through the given point that is parallel to the given line.

17. $A(3, -4), y = -x + 8$

18. $A(-6, 5), y = \frac{1}{2}x - 7$

19. $A(2, 0), y = 3x - 5$

20. $A(3, -1), y = \frac{1}{3}x + 10$

Write an equation of the line passing through the given point that is perpendicular to the given line.

21. $A(6, -1), y = -2x + 8$

22. $A(0, 3), y = -\frac{1}{2}x - 6$

23. $A(8, 2), y = 4x - 7$

24. $A(-1, 5), y = \frac{1}{7}x + 4$

Find the distance from point A to the given line.

25. $A(2, -1), y = -x + 4$

26. $A(-2, 3), y = \frac{1}{2}x + 1$

3 Chapter Test

Find the values of x and y. State which theorem(s) you used.

1.

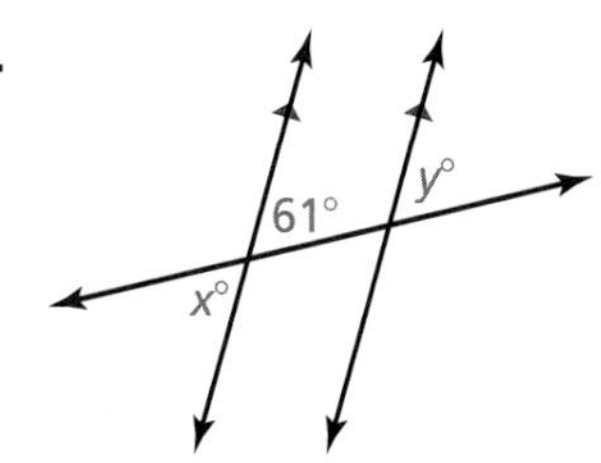

2.

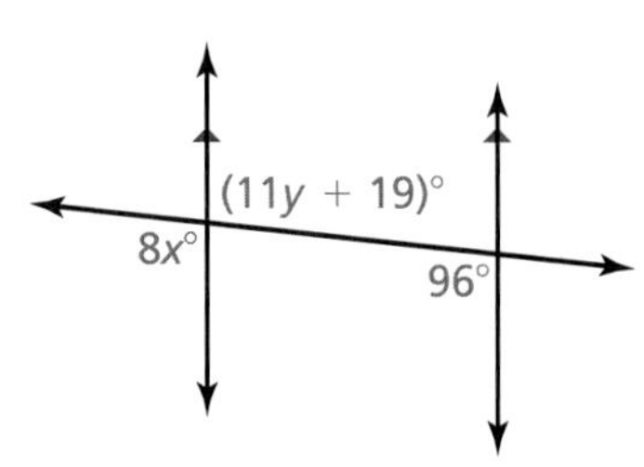

3.

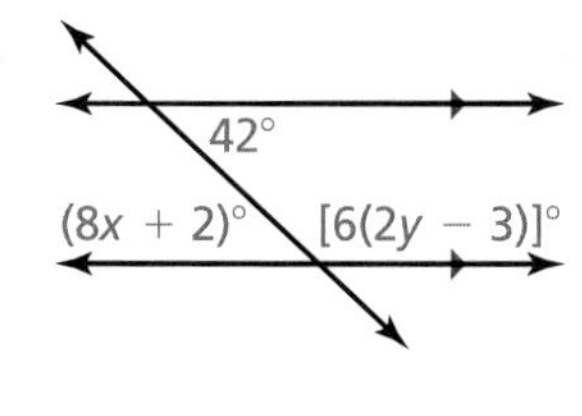

Find the distance from point A to the given line.

4. $A(3, 4)$, $y = -x$

5. $A(-3, 7)$, $y = \frac{1}{3}x - 2$

Find the value of x that makes $m \parallel n$.

6.

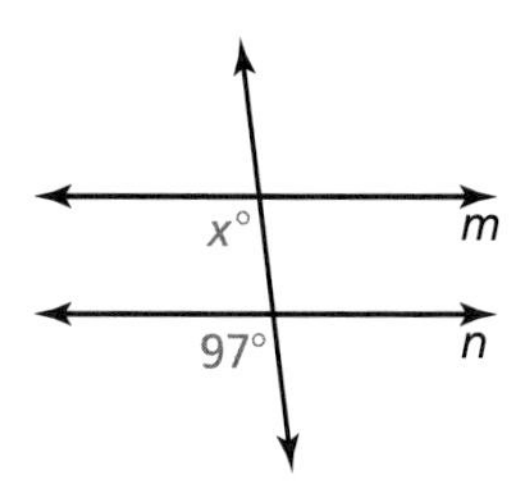

7.

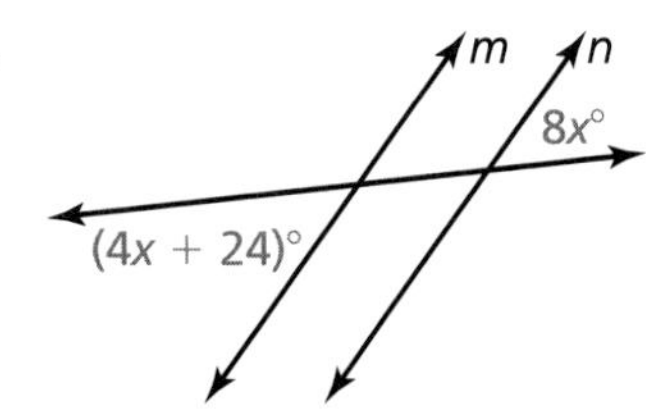

8.

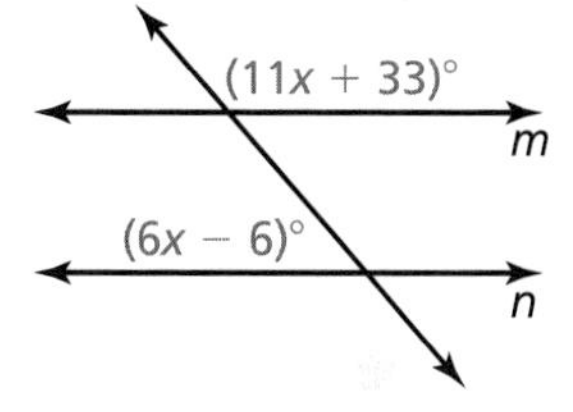

Write an equation of the line that passes through the given point and is (a) parallel to and (b) perpendicular to the given line.

9. $(-5, 2)$, $y = 2x - 3$

10. $(-1, -9)$, $y = -\frac{1}{3}x + 4$

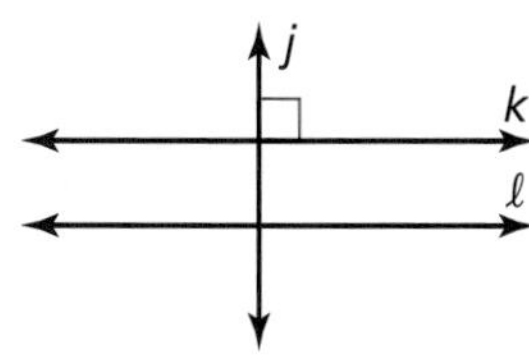

11. A student says, "Because $j \perp k$, $j \perp \ell$." What missing information is the student assuming from the diagram? Which theorem is the student trying to use?

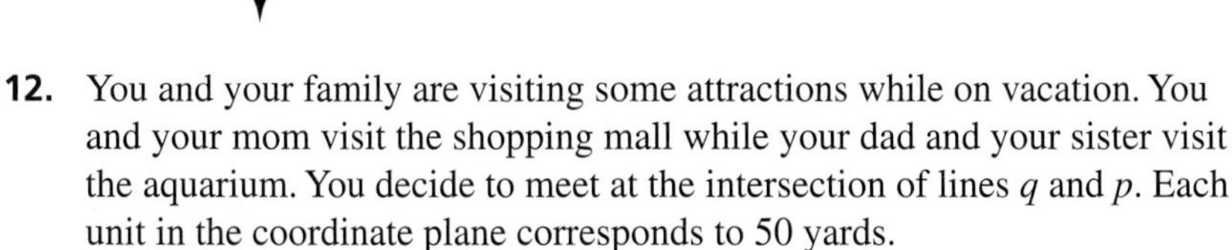

12. You and your family are visiting some attractions while on vacation. You and your mom visit the shopping mall while your dad and your sister visit the aquarium. You decide to meet at the intersection of lines q and p. Each unit in the coordinate plane corresponds to 50 yards.

a. Find an equation of line q.

b. Find an equation of line p.

c. What are the coordinates of the meeting point?

d. What is the distance from the meeting point to the subway?

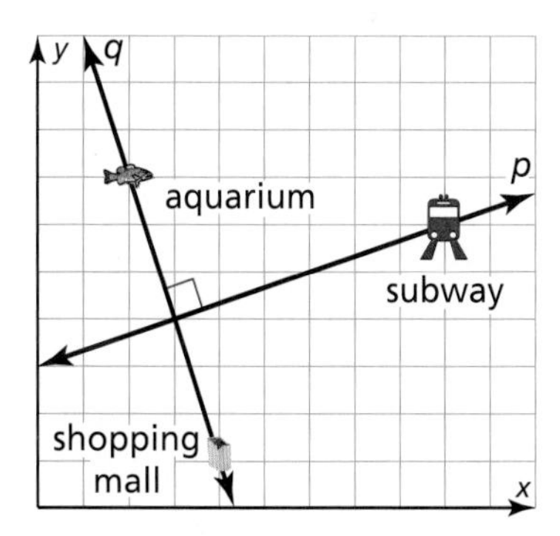

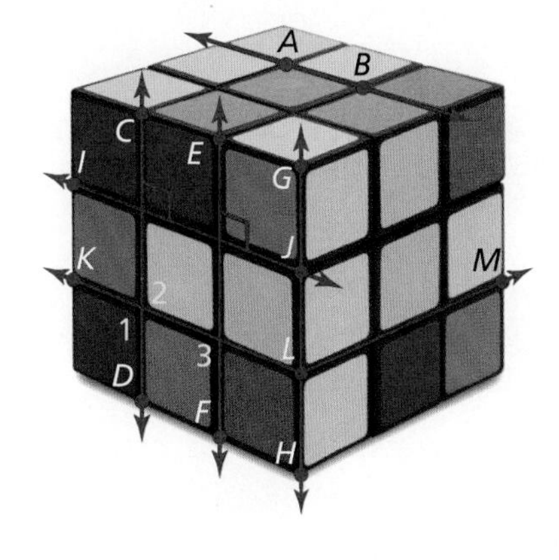

13. Identify an example on the puzzle cube of each description. Explain your reasoning.

a. a pair of skew lines

b. a pair of perpendicular lines

c. a pair of parallel lines

d. a pair of congruent corresponding angles

e. a pair of congruent alternate interior angles

3 Cumulative Assessment

1. Use the steps in the construction to explain how you know that $\overleftrightarrow{CD}$ is the perpendicular bisector of $\overline{AB}$.

Step 1

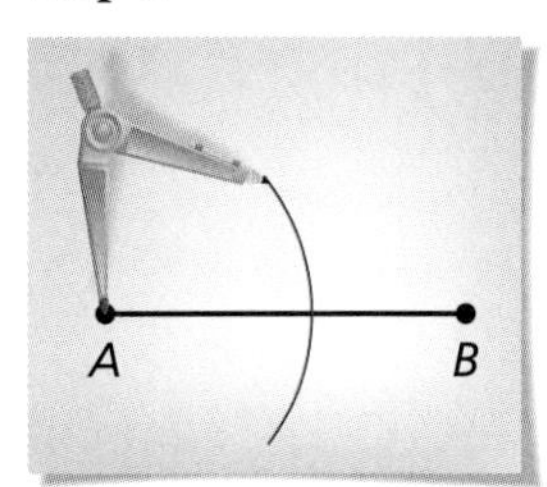

Step 2

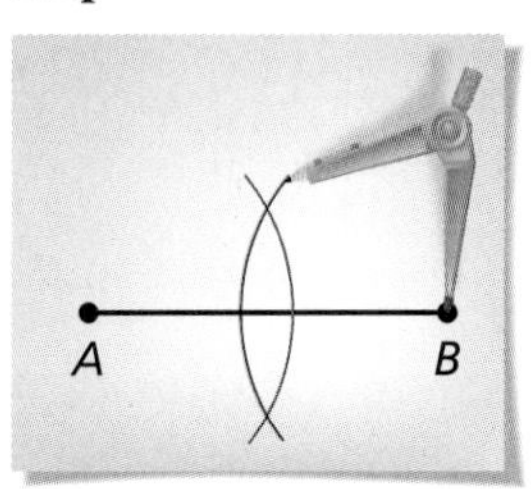

Step 3

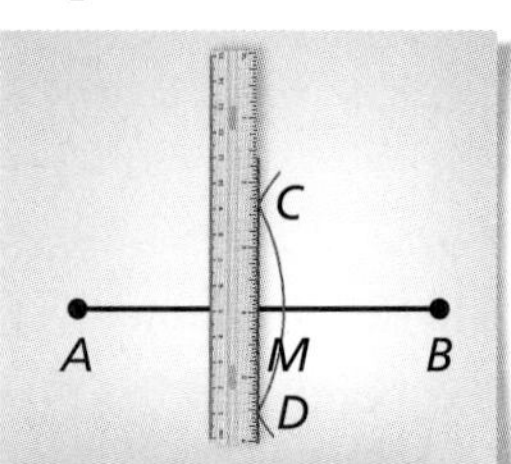

2. The equation of a line is $x + 2y = 10$.

a. Use the numbers and symbols to create the equation of a line in slope-intercept form that passes through the point $(4, -5)$ and is parallel to the given line.

b. Use the numbers and symbols to create the equation of a line in slope-intercept form that passes through the point $(2, -1)$ and is perpendicular to the given line.

y	x	$=$	$+$	$-$	-9	-2	-1
$-\frac{1}{2}$	$\frac{1}{2}$	1	$\frac{3}{2}$	2	3	4	5

3. Classify each pair of angles whose measurements are given.

a.

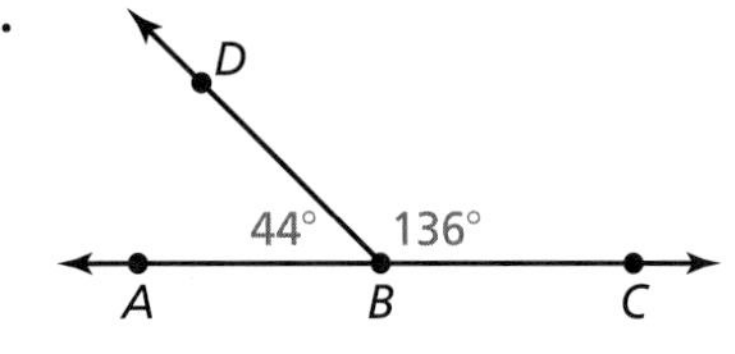

b.

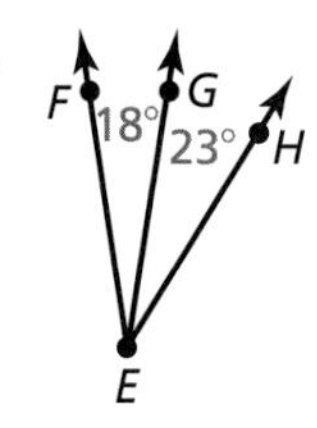

c.

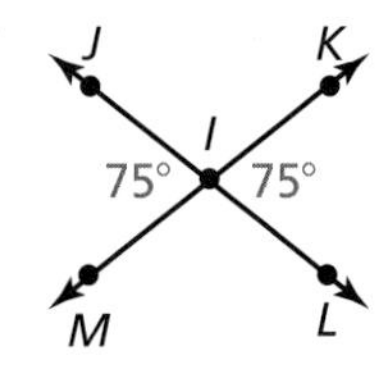

d.

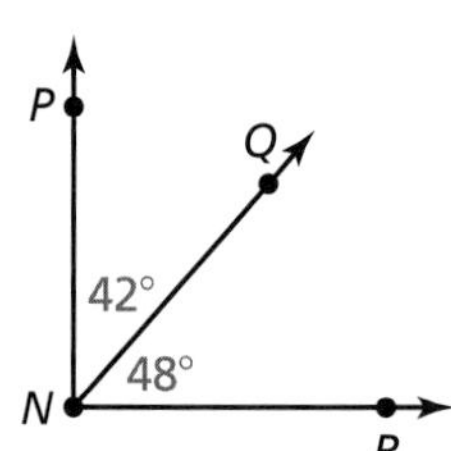

4. Your school is installing new turf on the football field. A coordinate plane has been superimposed on a diagram of the football field where 1 unit = 20 feet.

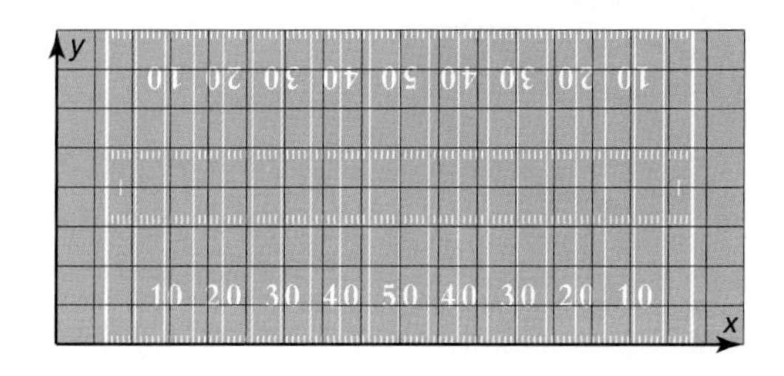

a. What is the length of the field?

b. What is the perimeter of the field?

c. Turf costs \$2.69 per square foot. Your school has a \$150,000 budget. Does the school have enough money to purchase new turf for the entire field?

5. Enter a statement or reason in each blank to complete the two-column proof.

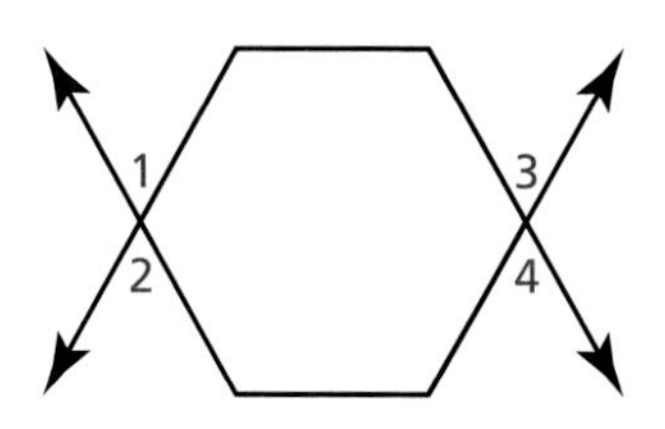

Given $\angle 1 \cong \angle 3$

Prove $\angle 2 \cong \angle 4$

STATEMENTS	REASONS
1. $\angle 1 \cong \angle 3$	1. Given
2. $\angle 1 \cong \angle 2$	2. ________________
3. $\angle 2 \cong \angle 3$	3. ________________
4. ____________	4. Vertical Angles Congruence Theorem (Thm. 2.6)
5. $\angle 2 \cong \angle 4$	5. ________________

6. Your friend claims that lines m and n are parallel. Do you support your friend's claim? Explain your reasoning.

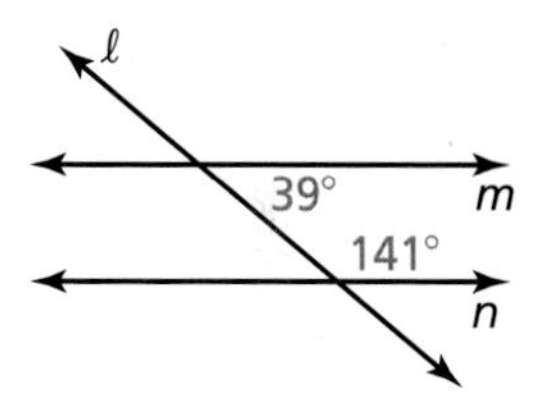

7. Which of the following is true when $\overleftrightarrow{AB}$ and $\overleftrightarrow{CD}$ are skew?

Ⓐ $\overleftrightarrow{AB}$ and $\overleftrightarrow{CD}$ are parallel.

Ⓑ $\overleftrightarrow{AB}$ and $\overleftrightarrow{CD}$ intersect.

Ⓒ $\overleftrightarrow{AB}$ and $\overleftrightarrow{CD}$ are perpendicular.

Ⓓ A, B, and C are noncollinear.

8. Select the angle that makes the statement true.

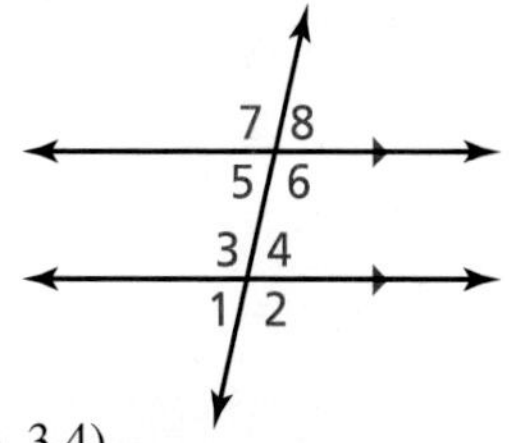

a. $\angle 4 \cong$ ___ by the Alternate Interior Angles Theorem (Thm. 3.2).

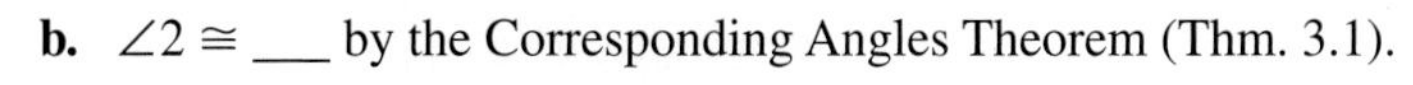

b. $\angle 2 \cong$ ___ by the Corresponding Angles Theorem (Thm. 3.1).

c. $\angle 1 \cong$ ___ by the Alternate Exterior Angles Theorem (Thm. 3.3).

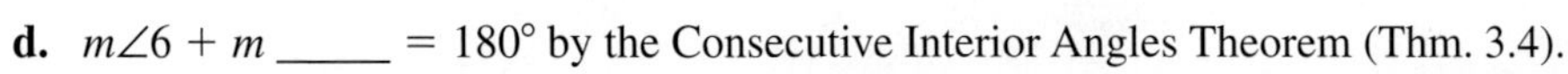

d. $m\angle 6 + m$ _____ $= 180°$ by the Consecutive Interior Angles Theorem (Thm. 3.4).

9. You and your friend walk to school together every day. You meet at the halfway point between your houses first and then walk to school. Each unit in the coordinate plane corresponds to 50 yards.

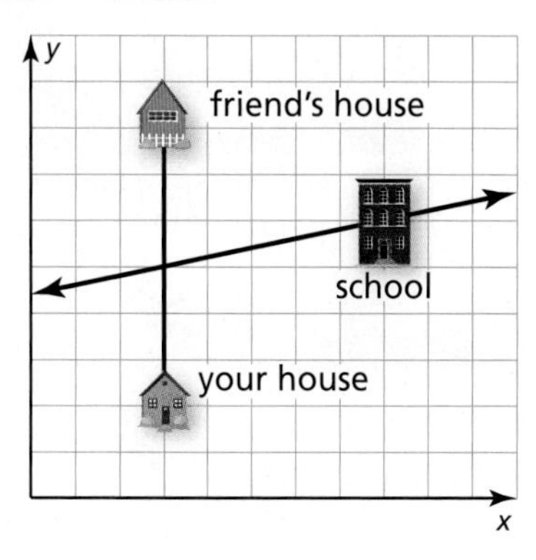

a. What are the coordinates of the midpoint of the line segment joining the two houses?

b. What is the distance that the two of you walk together?

4 Transformations

Magnification *(p. 213)*

Photo Stickers *(p. 211)*

Kaleidoscope *(p. 196)*

Revolving Door *(p. 195)*

Chess *(p. 179)*

Maintaining Mathematical Proficiency

Identifying Transformations

Example 1 **Tell whether the red figure is a translation, reflection, rotation, or dilation of the blue figure.**

a.

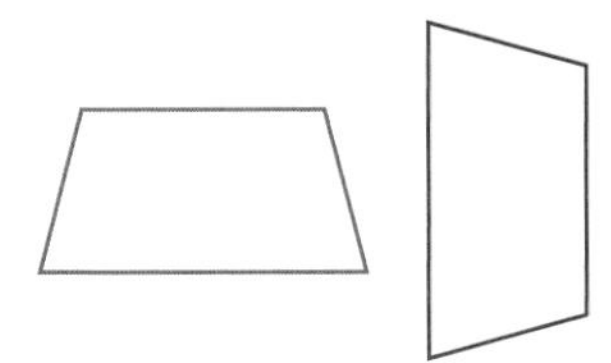

The blue figure turns to form the red figure, so it is a rotation.

b.

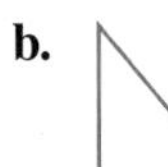

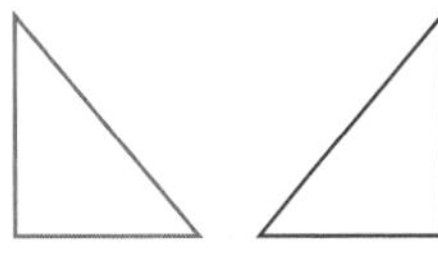

The red figure is a mirror image of the blue figure, so it is a reflection.

Tell whether the red figure is a translation, reflection, rotation, or dilation of the blue figure.

1.

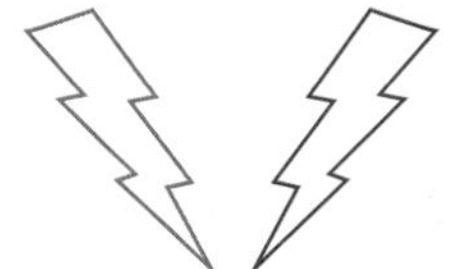

2.

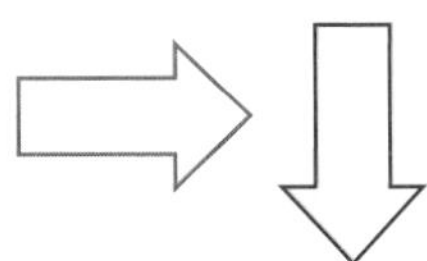

3.

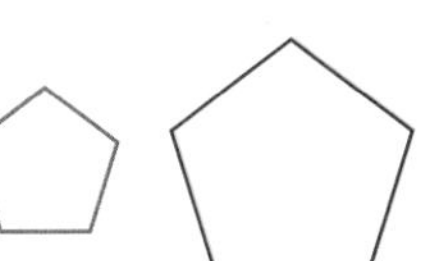

4.

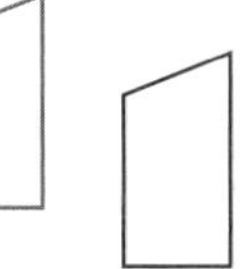

Identifying Similar Figures

Example 2 **Which rectangle is similar to Rectangle A?**

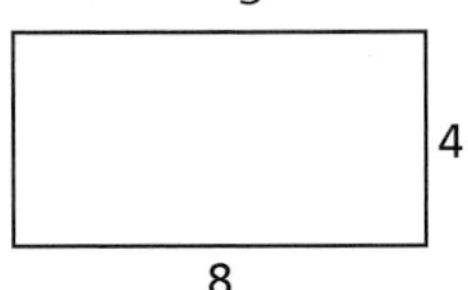

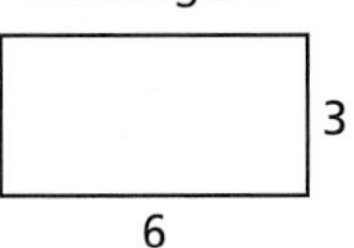

Each figure is a rectangle, so corresponding angles are congruent. Check to see whether corresponding side lengths are proportional.

Rectangle A and Rectangle B

$$\frac{\text{Length of A}}{\text{Length of B}} = \frac{8}{4} = 2 \qquad \frac{\text{Width of A}}{\text{Width of B}} = \frac{4}{1} = 4$$

not proportional

Rectangle A and Rectangle C

$$\frac{\text{Length of A}}{\text{Length of C}} = \frac{8}{6} = \frac{4}{3} \qquad \frac{\text{Width of A}}{\text{Width of C}} = \frac{4}{3}$$

proportional

▶ So, Rectangle C is similar to Rectangle A.

Tell whether the two figures are similar. Explain your reasoning.

5.

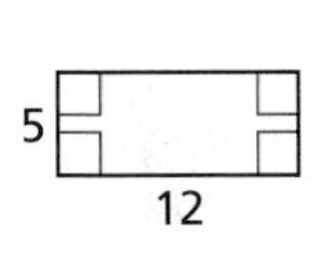

6.

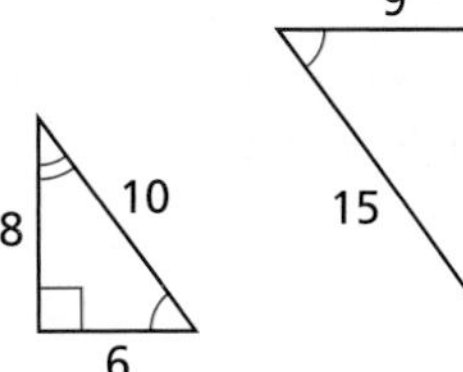

7.

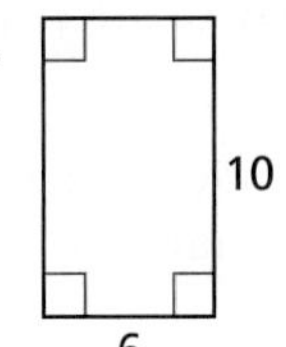

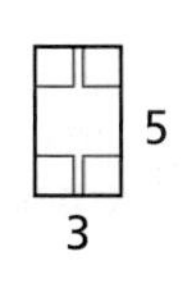

8. **ABSTRACT REASONING** Can you draw two squares that are not similar? Explain your reasoning.

Mathematical Practices

Mathematically proficient students use dynamic geometry software strategically.

Using Dynamic Geometry Software

Core Concept

Using Dynamic Geometry Software

Dynamic geometry software allows you to create geometric drawings, including:

- drawing a point
- drawing a line
- drawing a line segment
- drawing an angle
- measuring an angle
- measuring a line segment
- drawing a circle
- drawing an ellipse
- drawing a perpendicular line
- drawing a polygon
- copying and sliding an object
- reflecting an object in a line

EXAMPLE 1 Finding Side Lengths and Angle Measures

Use dynamic geometry software to draw a triangle with vertices at $A(-2, 1)$, $B(2, 1)$, and $C(2, -2)$. Find the side lengths and angle measures of the triangle.

SOLUTION

Using dynamic geometry software, you can create $\triangle ABC$, as shown.

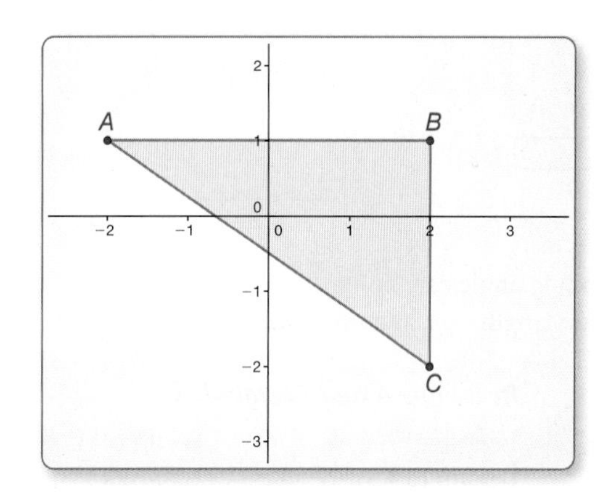

Sample

Points
$A(-2, 1)$
$B(2, 1)$
$C(2, -2)$
Segments
$AB = 4$
$BC = 3$
$AC = 5$
Angles
$m\angle A = 36.87°$
$m\angle B = 90°$
$m\angle C = 53.13°$

▶ From the display, the side lengths are $AB = 4$ units, $BC = 3$ units, and $AC = 5$ units. The angle measures, rounded to two decimal places, are $m\angle A \approx 36.87°$, $m\angle B = 90°$, and $m\angle C \approx 53.13°$.

Monitoring Progress

Use dynamic geometry software to draw the polygon with the given vertices. Use the software to find the side lengths and angle measures of the polygon. Round your answers to the nearest hundredth.

1. $A(0, 2)$, $B(3, -1)$, $C(4, 3)$
2. $A(-2, 1)$, $B(-2, -1)$, $C(3, 2)$
3. $A(1, 1)$, $B(-3, 1)$, $C(-3, -2)$, $D(1, -2)$
4. $A(1, 1)$, $B(-3, 1)$, $C(-2, -2)$, $D(2, -2)$
5. $A(-3, 0)$, $B(0, 3)$, $C(3, 0)$, $D(0, -3)$
6. $A(0, 0)$, $B(4, 0)$, $C(1, 1)$, $D(0, 3)$

4.1 Translations

Essential Question How can you translate a figure in a coordinate plane?

EXPLORATION 1 Translating a Triangle in a Coordinate Plane

Work with a partner.

a. Use dynamic geometry software to draw any triangle and label it $\triangle ABC$.

b. Copy the triangle and *translate* (or slide) it to form a new figure, called an *image*, $\triangle A'B'C'$ (read as "triangle *A* prime, *B* prime, *C* prime").

c. What is the relationship between the coordinates of the vertices of $\triangle ABC$ and those of $\triangle A'B'C'$?

d. What do you observe about the side lengths and angle measures of the two triangles?

USING TOOLS STRATEGICALLY

To be proficient in math, you need to use appropriate tools strategically, including dynamic geometry software.

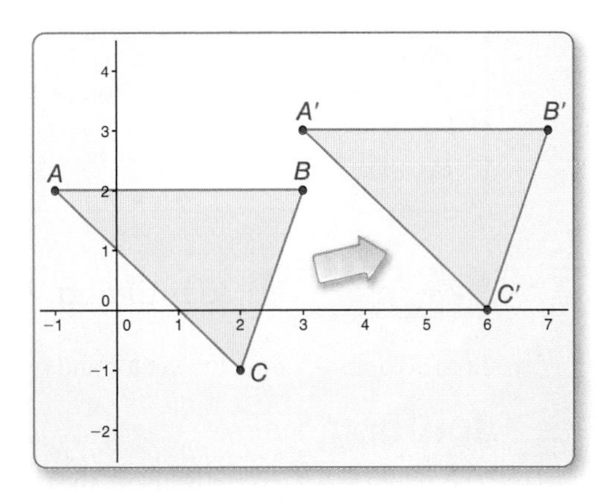

Sample

Points
$A(-1, 2)$
$B(3, 2)$
$C(2, -1)$
Segments
$AB = 4$
$BC = 3.16$
$AC = 4.24$
Angles
$m\angle A = 45°$
$m\angle B = 71.57°$
$m\angle C = 63.43°$

EXPLORATION 2 Translating a Triangle in a Coordinate Plane

Work with a partner.

a. The point (x, y) is translated *a* units horizontally and *b* units vertically. Write a rule to determine the coordinates of the image of (x, y).

$$(x, y) \rightarrow (\square, \square)$$

b. Use the rule you wrote in part (a) to translate $\triangle ABC$ 4 units left and 3 units down. What are the coordinates of the vertices of the image, $\triangle A'B'C'$?

c. Draw $\triangle A'B'C'$. Are its side lengths the same as those of $\triangle ABC$? Justify your answer.

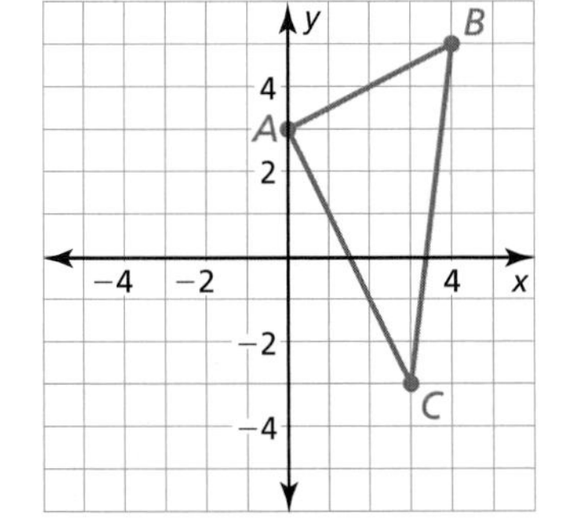

EXPLORATION 3 Comparing Angles of Translations

Work with a partner.

a. In Exploration 2, is $\triangle ABC$ a right triangle? Justify your answer.

b. In Exploration 2, is $\triangle A'B'C'$ a right triangle? Justify your answer.

c. Do you think translations always preserve angle measures? Explain your reasoning.

Communicate Your Answer

4. How can you translate a figure in a coordinate plane?

5. In Exploration 2, translate $\triangle A'B'C'$ 3 units right and 4 units up. What are the coordinates of the vertices of the image, $\triangle A''B''C''$? How are these coordinates related to the coordinates of the vertices of the original triangle, $\triangle ABC$?

4.1 Lesson

Core Vocabulary

vector, *p. 174*
initial point, *p. 174*
terminal point, *p. 174*
horizontal component, *p. 174*
vertical component, *p. 174*
component form, *p. 174*
transformation, *p. 174*
image, *p. 174*
preimage, *p. 174*
translation, *p. 174*
rigid motion, *p. 176*
composition of transformations, *p. 176*

What You Will Learn

- Perform translations.
- Perform compositions.
- Solve real-life problems involving compositions.

Performing Translations

A **vector** is a quantity that has both direction and *magnitude*, or size, and is represented in the coordinate plane by an arrow drawn from one point to another.

Core Concept

Vectors

The diagram shows a vector. The **initial point**, or starting point, of the vector is P, and the **terminal point**, or ending point, is Q. The vector is named $\overrightarrow{PQ}$, which is read as "vector PQ." The **horizontal component** of $\overrightarrow{PQ}$ is 5, and the **vertical component** is 3. The **component form** of a vector combines the horizontal and vertical components. So, the component form of $\overrightarrow{PQ}$ is $\langle 5, 3\rangle$.

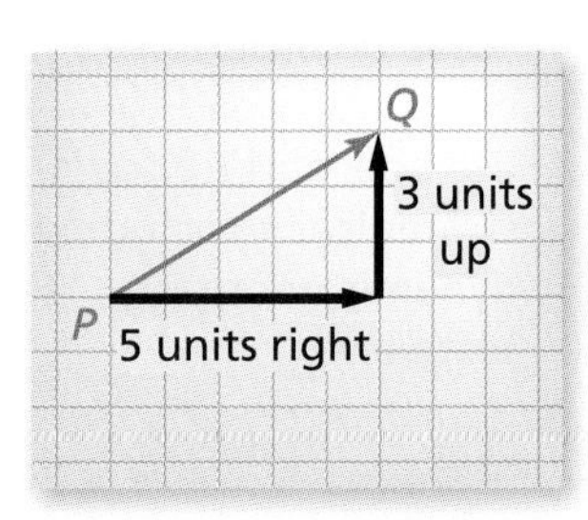

EXAMPLE 1 **Identifying Vector Components**

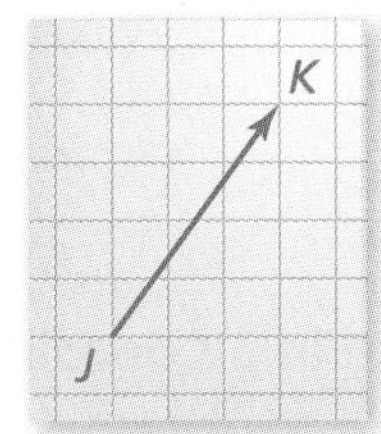

In the diagram, name the vector and write its component form.

SOLUTION

The vector is $\overrightarrow{JK}$. To move from the initial point J to the terminal point K, you move 3 units right and 4 units up. So, the component form is $\langle 3, 4\rangle$.

A **transformation** is a function that moves or changes a figure in some way to produce a new figure called an **image**. Another name for the original figure is the **preimage**. The points on the preimage are the inputs for the transformation, and the points on the image are the outputs.

Core Concept

Translations

A **translation** moves every point of a figure the same distance in the same direction. More specifically, a translation *maps*, or moves, the points P and Q of a plane figure along a vector $\langle a, b\rangle$ to the points P' and Q', so that one of the following statements is true.

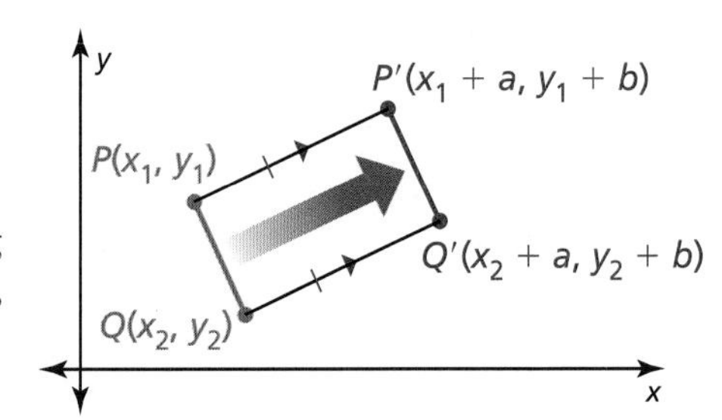

- $PP' = QQ'$ and $\overline{PP'} \parallel \overline{QQ'}$, or
- $PP' = QQ'$ and $\overline{PP'}$ and $\overline{QQ'}$ are collinear.

STUDY TIP

You can use *prime notation* to name an image. For example, if the preimage is point P, then its image is point P', read as "point P *prime*."

Translations map lines to parallel lines and segments to parallel segments. For instance, in the figure above, $\overline{PQ} \parallel \overline{P'Q'}$.

EXAMPLE 2 Translating a Figure Using a Vector

The vertices of $\triangle ABC$ are $A(0, 3)$, $B(2, 4)$, and $C(1, 0)$. Translate $\triangle ABC$ using the vector $\langle 5, -1 \rangle$.

SOLUTION

First, graph $\triangle ABC$. Use $\langle 5, -1 \rangle$ to move each vertex 5 units right and 1 unit down. Label the image vertices. Draw $\triangle A'B'C'$. Notice that the vectors drawn from preimage vertices to image vertices are parallel.

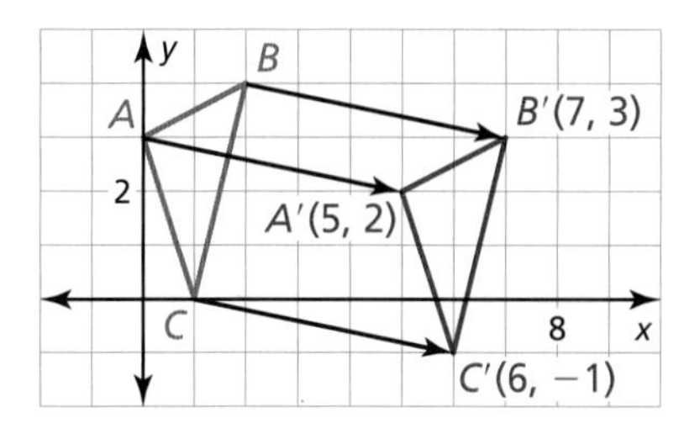

You can also express a translation along the vector $\langle a, b \rangle$ using a rule, which has the notation $(x, y) \rightarrow (x + a, y + b)$.

EXAMPLE 3 Writing a Translation Rule

Write a rule for the translation of $\triangle ABC$ to $\triangle A'B'C'$.

SOLUTION

To go from A to A', you move 4 units left and 1 unit up, so you move along the vector $\langle -4, 1 \rangle$.

▶ So, a rule for the translation is $(x, y) \rightarrow (x - 4, y + 1)$.

EXAMPLE 4 Translating a Figure in the Coordinate Plane

Graph quadrilateral $ABCD$ with vertices $A(-1, 2)$, $B(-1, 5)$, $C(4, 6)$, and $D(4, 2)$ and its image after the translation $(x, y) \rightarrow (x + 3, y - 1)$.

SOLUTION

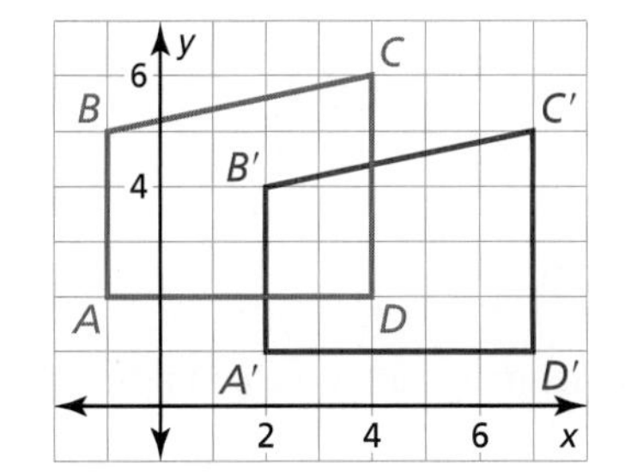

Graph quadrilateral $ABCD$. To find the coordinates of the vertices of the image, add 3 to the x-coordinates and subtract 1 from the y-coordinates of the vertices of the preimage. Then graph the image, as shown at the left.

$$(x, y) \rightarrow (x + 3, y - 1)$$

$$A(-1, 2) \rightarrow A'(2, 1)$$
$$B(-1, 5) \rightarrow B'(2, 4)$$
$$C(4, 6) \rightarrow C'(7, 5)$$
$$D(4, 2) \rightarrow D'(7, 1)$$

Monitoring Progress  Help in English and Spanish at *BigIdeasMath.com*

1. Name the vector and write its component form.
2. The vertices of $\triangle LMN$ are $L(2, 2)$, $M(5, 3)$, and $N(9, 1)$. Translate $\triangle LMN$ using the vector $\langle -2, 6 \rangle$.
3. In Example 3, write a rule to translate $\triangle A'B'C'$ back to $\triangle ABC$.
4. Graph $\triangle RST$ with vertices $R(2, 2)$, $S(5, 2)$, and $T(3, 5)$ and its image after the translation $(x, y) \rightarrow (x + 1, y + 2)$.

Performing Compositions

A **rigid motion** is a transformation that preserves length and angle measure. Another name for a rigid motion is an *isometry*. A rigid motion maps lines to lines, rays to rays, and segments to segments.

Postulate

Postulate 4.1 Translation Postulate

A translation is a rigid motion.

Because a translation is a rigid motion, and a rigid motion preserves length and angle measure, the following statements are true for the translation shown.

- $DE = D'E'$, $EF = E'F'$, $FD = F'D'$
- $m\angle D = m\angle D'$, $m\angle E = m\angle E'$, $m\angle F = m\angle F'$

When two or more transformations are combined to form a single transformation, the result is a **composition of transformations**.

Theorem

Theorem 4.1 Composition Theorem

The composition of two (or more) rigid motions is a rigid motion.

Proof Ex. 35, p. 180

The theorem above is important because it states that no matter how many rigid motions you perform, lengths and angle measures will be preserved in the final image. For instance, the composition of two or more translations is a translation, as shown.

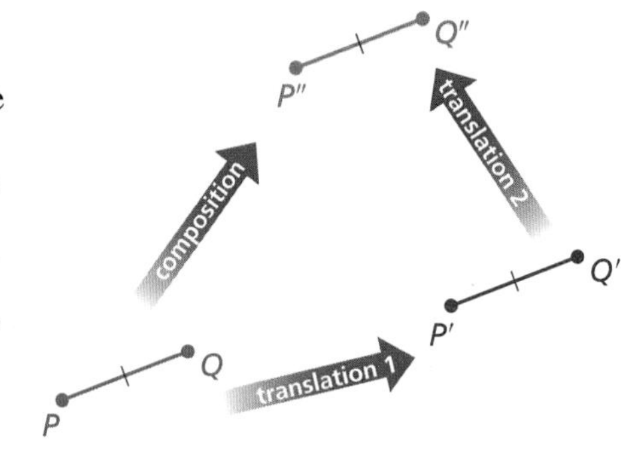

EXAMPLE 5 **Performing a Composition**

Graph $\overline{RS}$ with endpoints $R(-8, 5)$ and $S(-6, 8)$ and its image after the composition.

Translation: $(x, y) \rightarrow (x + 5, y - 2)$

Translation: $(x, y) \rightarrow (x - 4, y - 2)$

SOLUTION

Step 1 Graph $\overline{RS}$.

Step 2 Translate $\overline{RS}$ 5 units right and 2 units down. $\overline{R'S'}$ has endpoints $R'(-3, 3)$ and $S'(-1, 6)$.

Step 3 Translate $\overline{R'S'}$ 4 units left and 2 units down. $\overline{R''S''}$ has endpoints $R''(-7, 1)$ and $S''(-5, 4)$.

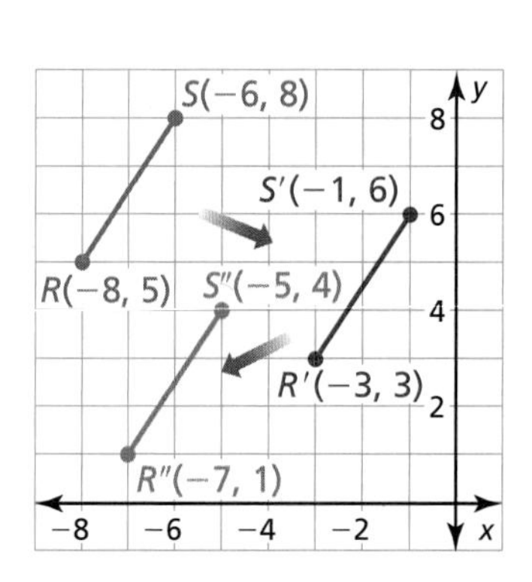

Solving Real-Life Problems

EXAMPLE 6 Modeling with Mathematics

You are designing a favicon for a golf website. In an image-editing program, you move the red rectangle 2 units left and 3 units down. Then you move the red rectangle 1 unit right and 1 unit up. Rewrite the composition as a single translation.

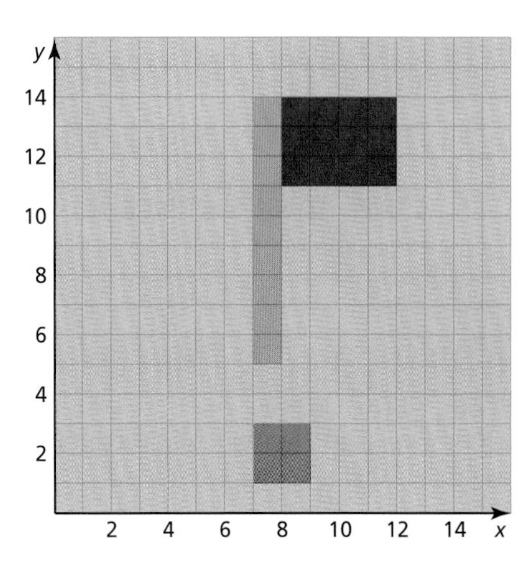

SOLUTION

1. **Understand the Problem** You are given two translations. You need to rewrite the result of the composition of the two translations as a single translation.
2. **Make a Plan** You can choose an arbitrary point (x, y) in the red rectangle and determine the horizontal and vertical shift in the coordinates of the point after both translations. This tells you how much you need to shift each coordinate to map the original figure to the final image.
3. **Solve the Problem** Let $A(x, y)$ be an arbitrary point in the red rectangle. After the first translation, the coordinates of its image are

   $A'(x - 2, y - 3)$.

   The second translation maps $A'(x - 2, y - 3)$ to

   $A''(x - 2 + 1, y - 3 + 1) = A''(x - 1, y - 2)$.

   The composition of translations uses the original point (x, y) as the input and returns the point $(x - 1, y - 2)$ as the output.

    So, the single translation rule for the composition is $(x, y) \rightarrow (x - 1, y - 2)$.

4. **Look Back** Check that the rule is correct by testing a point. For instance, (10, 12) is a point in the red rectangle. Apply the two translations to (10, 12).

   $(10, 12) \rightarrow (8, 9) \rightarrow (9, 10)$

   Does the final result match the rule you found in Step 3?

   $(10, 12) \rightarrow (10 - 1, 12 - 2) = (9, 10)$ ✓

Monitoring Progress

Help in English and Spanish at *BigIdeasMath.com*

5. Graph $\overline{TU}$ with endpoints $T(1, 2)$ and $U(4, 6)$ and its image after the composition.

   Translation: $(x, y) \rightarrow (x - 2, y - 3)$
   Translation: $(x, y) \rightarrow (x - 4, y + 5)$

6. Graph $\overline{VW}$ with endpoints $V(-6, -4)$ and $W(-3, 1)$ and its image after the composition.

   Translation: $(x, y) \rightarrow (x + 3, y + 1)$
   Translation: $(x, y) \rightarrow (x - 6, y - 4)$

7. In Example 6, you move the gray square 2 units right and 3 units up. Then you move the gray square 1 unit left and 1 unit down. Rewrite the composition as a single transformation.

4.1 Exercises

Dynamic Solutions available at *BigIdeasMath.com*

Vocabulary and Core Concept Check

1. **VOCABULARY** Name the preimage and image of the transformation $\triangle ABC \rightarrow \triangle A'B'C'$.

2. **COMPLETE THE SENTENCE** A ______ moves every point of a figure the same distance in the same direction.

Monitoring Progress and Modeling with Mathematics

In Exercises 3 and 4, name the vector and write its component form. *(See Example 1.)*

3.

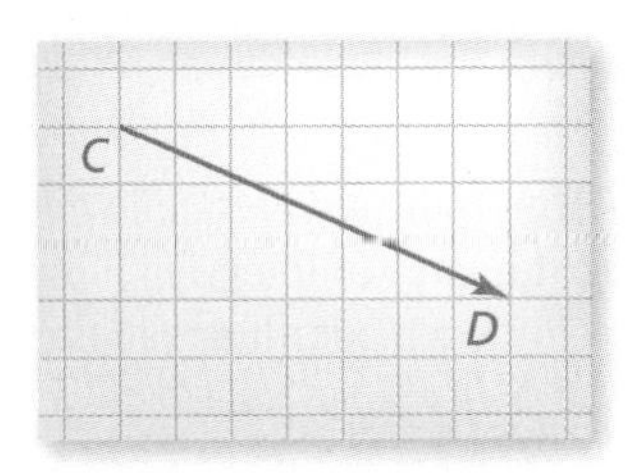

4.

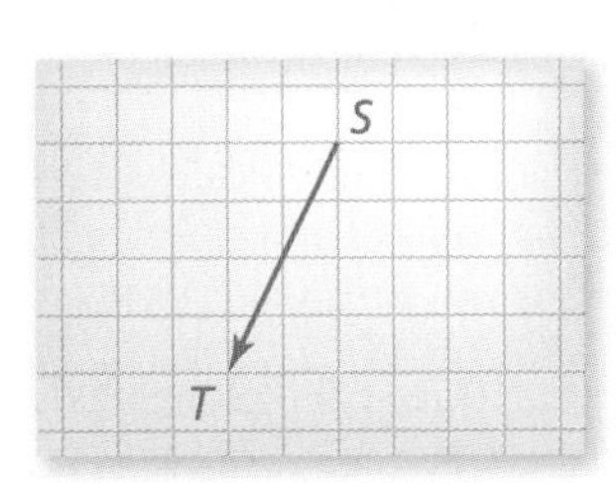

In Exercises 5–8, the vertices of $\triangle DEF$ are $D(2, 5)$, $E(6, 3)$, and $F(4, 0)$. Translate $\triangle DEF$ using the given vector. Graph $\triangle DEF$ and its image. *(See Example 2.)*

5. $\langle 6, 0 \rangle$

6. $\langle 5, -1 \rangle$

7. $\langle -3, -7 \rangle$

8. $\langle -2, -4 \rangle$

In Exercises 9 and 10, find the component form of the vector that translates $P(-3, 6)$ to P'.

9. $P'(0, 1)$

10. $P'(-4, 8)$

In Exercises 11 and 12, write a rule for the translation of $\triangle LMN$ to $\triangle L'M'N'$. *(See Example 3.)*

11.

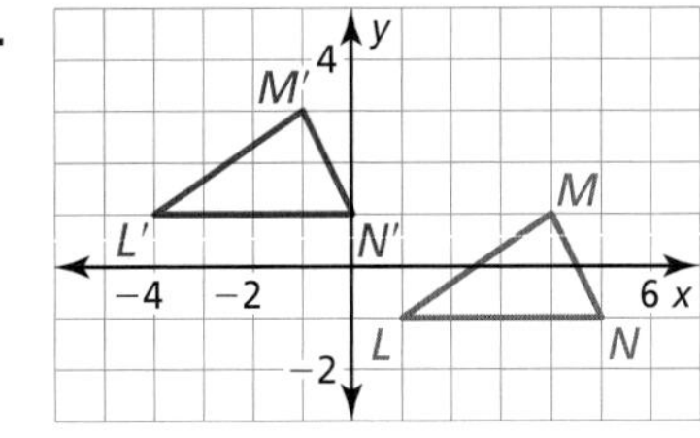

12.

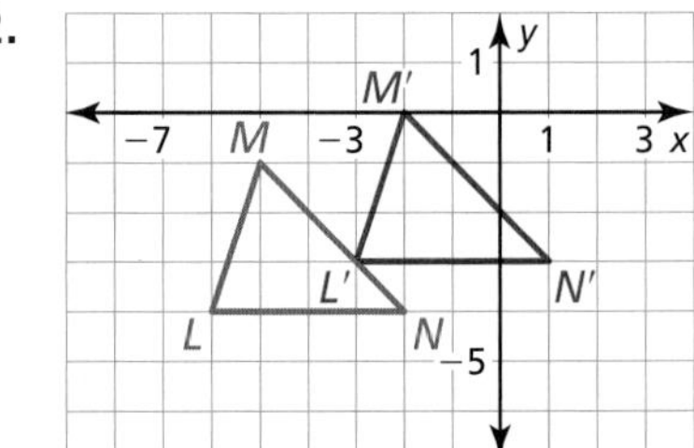

In Exercises 13–16, use the translation.

$$(x, y) \rightarrow (x - 8, y + 4)$$

13. What is the image of $A(2, 6)$?

14. What is the image of $B(-1, 5)$?

15. What is the preimage of $C'(-3, -10)$?

16. What is the preimage of $D'(4, -3)$?

In Exercises 17–20, graph $\triangle PQR$ with vertices $P(-2, 3)$, $Q(1, 2)$, and $R(3, -1)$ and its image after the translation. *(See Example 4.)*

17. $(x, y) \rightarrow (x + 4, y + 6)$

18. $(x, y) \rightarrow (x + 9, y - 2)$

19. $(x, y) \rightarrow (x - 2, y - 5)$

20. $(x, y) \rightarrow (x - 1, y + 3)$

In Exercises 21 and 22, graph $\triangle XYZ$ with vertices $X(2, 4)$, $Y(6, 0)$, and $Z(7, 2)$ and its image after the composition. *(See Example 5.)*

21. **Translation:** $(x, y) \rightarrow (x + 12, y + 4)$
    Translation: $(x, y) \rightarrow (x - 5, y - 9)$

22. **Translation:** $(x, y) \rightarrow (x - 6, y)$
    Translation: $(x, y) \rightarrow (x + 2, y + 7)$

In Exercises 23 and 24, describe the composition of translations.

23.

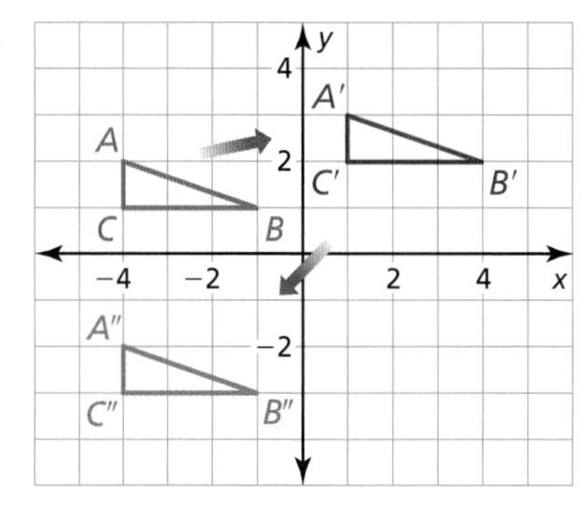

24.

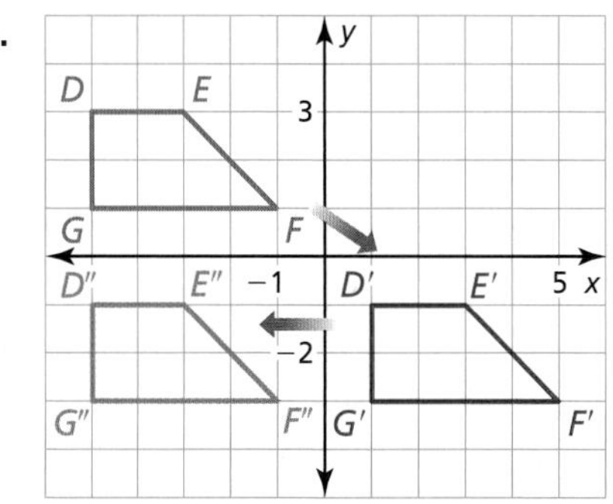

25. ERROR ANALYSIS Describe and correct the error in graphing the image of quadrilateral $EFGH$ after the translation $(x, y) \rightarrow (x - 1, y - 2)$.

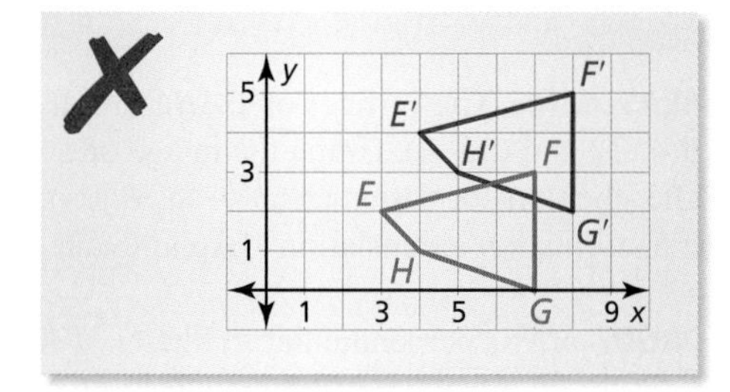

26. MODELING WITH MATHEMATICS In chess, the knight (the piece shaped like a horse) moves in an L pattern. The board shows two consecutive moves of a black knight during a game. Write a composition of translations for the moves. Then rewrite the composition as a single translation that moves the knight from its original position to its ending position. *(See Example 6.)*

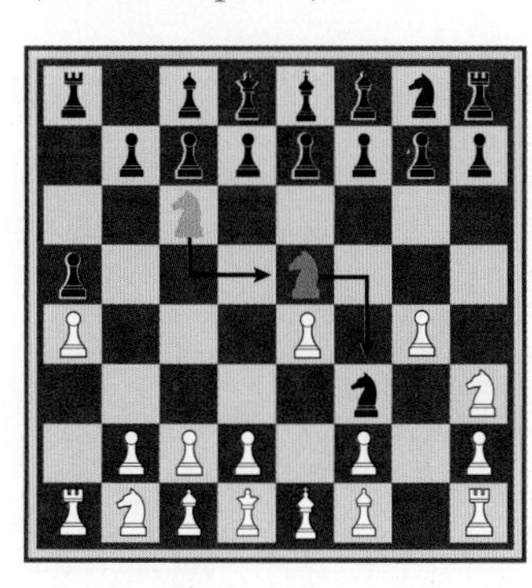

27. PROBLEM SOLVING You are studying an amoeba through a microscope. Suppose the amoeba moves on a grid-indexed microscope slide in a straight line from square B3 to square G7.

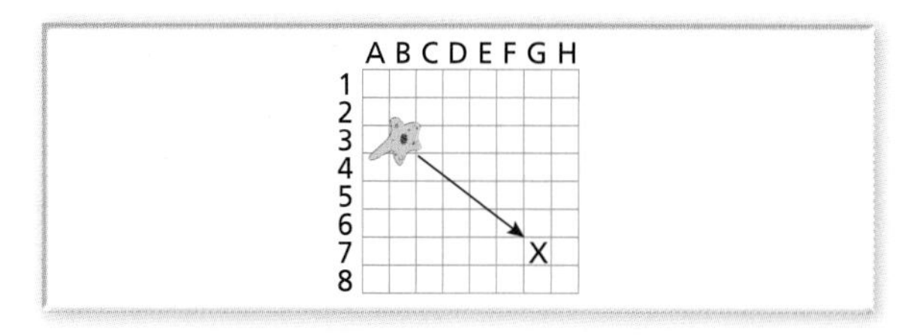

a. Describe the translation.

b. The side length of each grid square is 2 millimeters. How far does the amoeba travel?

c. The amoeba moves from square B3 to square G7 in 24.5 seconds. What is its speed in millimeters per second?

28. MATHEMATICAL CONNECTIONS Translation A maps (x, y) to $(x + n, y + t)$. Translation B maps (x, y) to $(x + s, y + m)$.

a. Translate a point using Translation A, followed by Translation B. Write an algebraic rule for the final image of the point after this composition.

b. Translate a point using Translation B, followed by Translation A. Write an algebraic rule for the final image of the point after this composition.

c. Compare the rules you wrote for parts (a) and (b). Does it matter which translation you do first? Explain your reasoning.

MATHEMATICAL CONNECTIONS In Exercises 29 and 30, a translation maps the blue figure to the red figure. Find the value of each variable.

29.

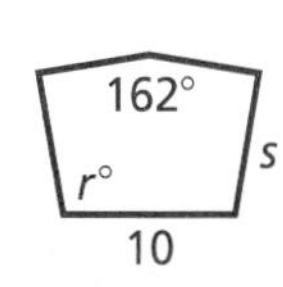

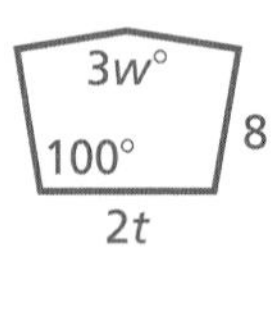

30.

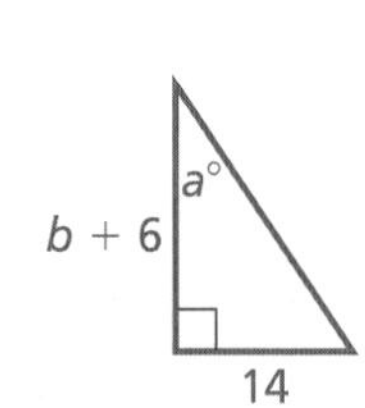

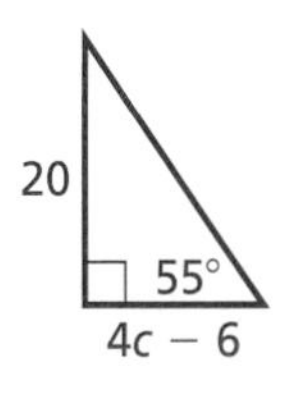

31. USING STRUCTURE Quadrilateral $DEFG$ has vertices $D(-1, 2)$, $E(-2, 0)$, $F(-1, -1)$, and $G(1, 3)$. A translation maps quadrilateral $DEFG$ to quadrilateral $D'E'F'G'$. The image of D is $D'(-2, -2)$. What are the coordinates of E', F', and G'?

32. HOW DO YOU SEE IT? Which two figures represent a translation? Describe the translation.

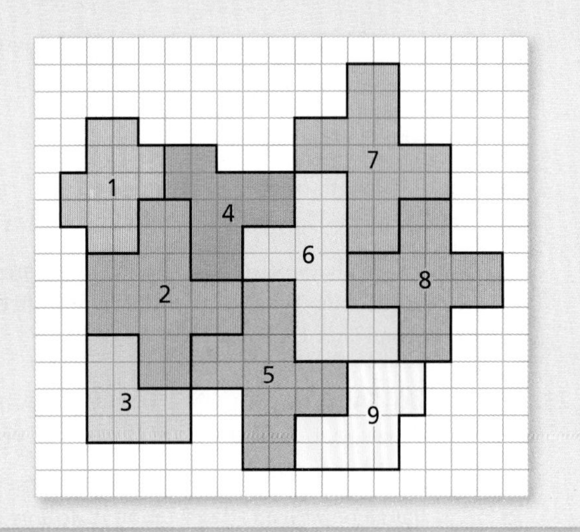

33. REASONING The translation $(x, y) \rightarrow (x + m, y + n)$ maps $\overline{PQ}$ to $\overline{P'Q'}$. Write a rule for the translation of $\overline{P'Q'}$ to $\overline{PQ}$. Explain your reasoning.

34. DRAWING CONCLUSIONS The vertices of a rectangle are $Q(2, -3)$, $R(2, 4)$, $S(5, 4)$, and $T(5, -3)$.

a. Translate rectangle $QRST$ 3 units left and 3 units down to produce rectangle $Q'R'S'T'$. Find the area of rectangle $QRST$ and the area of rectangle $Q'R'S'T'$.

b. Compare the areas. Make a conjecture about the areas of a preimage and its image after a translation.

35. PROVING A THEOREM Prove the Composition Theorem (Theorem 4.1).

36. PROVING A THEOREM Use properties of translations to prove each theorem.

a. Corresponding Angles Theorem (Theorem 3.1)

b. Corresponding Angles Converse (Theorem 3.5)

37. WRITING Explain how to use translations to draw a rectangular prism.

38. MATHEMATICAL CONNECTIONS The vector $\overrightarrow{PQ} = \langle 4, 1 \rangle$ describes the translation of $A(-1, w)$ onto $A'(2x + 1, 4)$ and $B(8y - 1, 1)$ onto $B'(3, 3z)$. Find the values of w, x, y, and z.

39. MAKING AN ARGUMENT A translation maps $\overline{GH}$ to $\overline{G'H'}$. Your friend claims that if you draw segments connecting G to G' and H to H', then the resulting quadrilateral is a parallelogram. Is your friend correct? Explain your reasoning.

40. THOUGHT PROVOKING You are a graphic designer for a company that manufactures floor tiles. Design a floor tile in a coordinate plane. Then use translations to show how the tiles cover an entire floor. Describe the translations that map the original tile to four other tiles.

41. REASONING The vertices of $\triangle ABC$ are $A(2, 2)$, $B(4, 2)$, and $C(3, 4)$. Graph the image of $\triangle ABC$ after the transformation $(x, y) \rightarrow (x + y, y)$. Is this transformation a translation? Explain your reasoning.

42. PROOF $\overline{MN}$ is perpendicular to line ℓ. $\overline{M'N'}$ is the translation of $\overline{MN}$ 2 units to the left. Prove that $\overline{M'N'}$ is perpendicular to ℓ.

Maintaining Mathematical Proficiency

Reviewing what you learned in previous grades and lessons

Tell whether the figure can be folded in half so that one side matches the other. *(Skills Review Handbook)*

43. 

44. 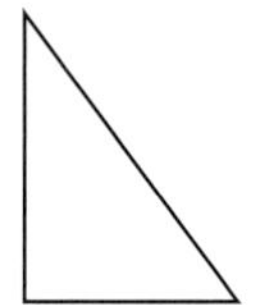

45. 

46. 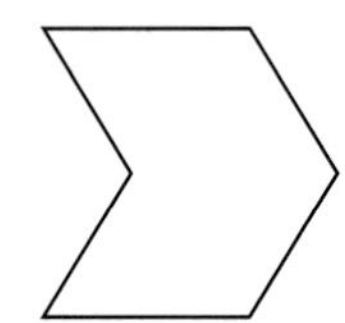

Simplify the expression. *(Skills Review Handbook)*

47. $-(-x)$

48. $-(x + 3)$

49. $x - (12 - 5x)$

50. $x - (-2x + 4)$

4.2 Reflections

Essential Question How can you reflect a figure in a coordinate plane?

EXPLORATION 1 **Reflecting a Triangle Using a Reflective Device**

Work with a partner. Use a straightedge to draw any triangle on paper. Label it $\triangle ABC$.

a. Use the straightedge to draw a line that does not pass through the triangle. Label it m.

b. Place a reflective device on line m.

c. Use the reflective device to plot the images of the vertices of $\triangle ABC$. Label the images of vertices A, B, and C as A', B', and C', respectively.

d. Use a straightedge to draw $\triangle A'B'C'$ by connecting the vertices.

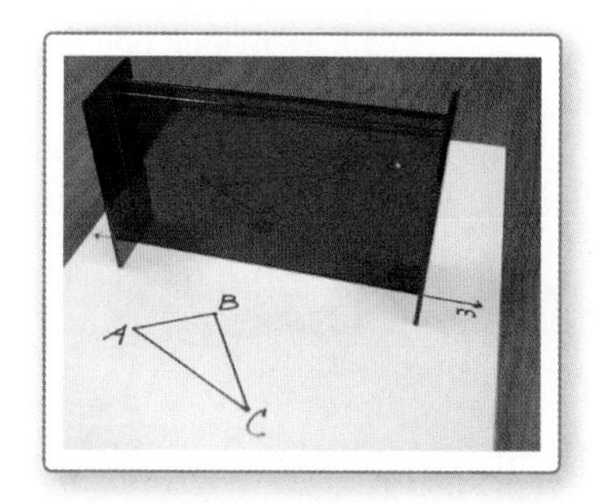

LOOKING FOR STRUCTURE

To be proficient in math, you need to look closely to discern a pattern or structure.

EXPLORATION 2 **Reflecting a Triangle in a Coordinate Plane**

Work with a partner. Use dynamic geometry software to draw any triangle and label it $\triangle ABC$.

a. *Reflect* $\triangle ABC$ in the y-axis to form $\triangle A'B'C'$.

b. What is the relationship between the coordinates of the vertices of $\triangle ABC$ and those of $\triangle A'B'C'$?

c. What do you observe about the side lengths and angle measures of the two triangles?

d. *Reflect* $\triangle ABC$ in the x-axis to form $\triangle A'B'C'$. Then repeat parts (b) and (c).

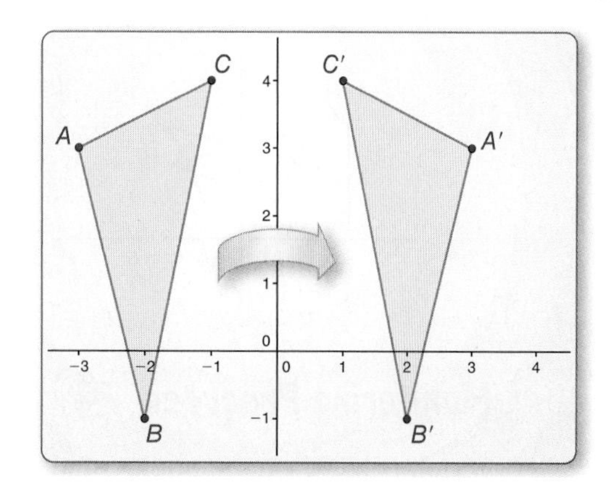

Sample

Points
$A(-3, 3)$
$B(-2, -1)$
$C(-1, 4)$
Segments
$AB = 4.12$
$BC = 5.10$
$AC = 2.24$
Angles
$m\angle A = 102.53°$
$m\angle B = 25.35°$
$m\angle C = 52.13°$

Communicate Your Answer

3. How can you reflect a figure in a coordinate plane?

4.2 Lesson

Core Vocabulary

reflection, *p. 182*
line of reflection, *p. 182*
glide reflection, *p. 184*
line symmetry, *p. 185*
line of symmetry, *p. 185*

What You Will Learn

- Perform reflections.
- Perform glide reflections.
- Identify lines of symmetry.
- Solve real-life problems involving reflections.

Performing Reflections

Core Concept

Reflections

A **reflection** is a transformation that uses a line like a mirror to reflect a figure. The mirror line is called the **line of reflection**.

A reflection in a line m maps every point P in the plane to a point P', so that for each point one of the following properties is true.

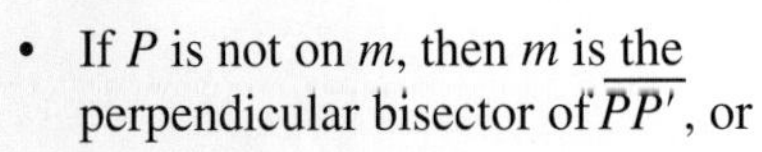

- If P is not on m, then m is the perpendicular bisector of $\overline{PP'}$, or

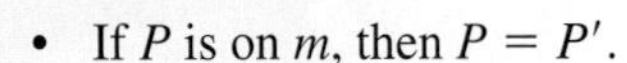

- If P is on m, then $P = P'$.

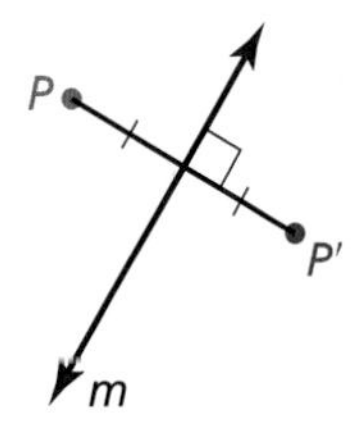

point P not on m

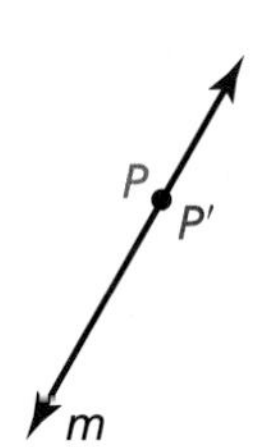

point P on m

EXAMPLE 1 Reflecting in Horizontal and Vertical Lines

Graph $\triangle ABC$ with vertices $A(1, 3)$, $B(5, 2)$, and $C(2, 1)$ and its image after the reflection described.

a. In the line n: $x = 3$

b. In the line m: $y = 1$

SOLUTION

a. Point A is 2 units left of line n, so its reflection A' is 2 units right of line n at $(5, 3)$. Also, B' is 2 units left of line n at $(1, 2)$, and C' is 1 unit right of line n at $(4, 1)$.

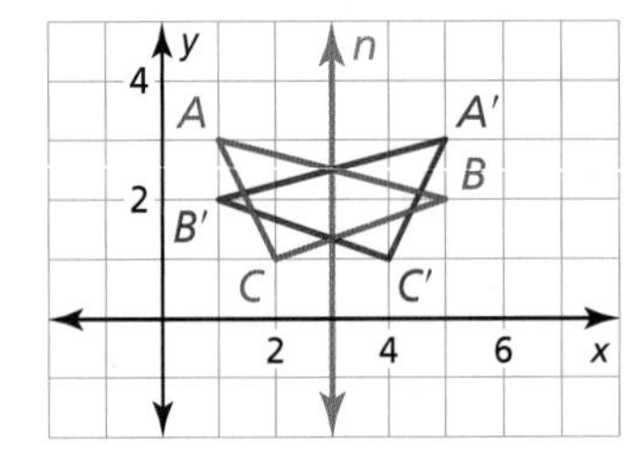

b. Point A is 2 units above line m, so A' is 2 units below line m at $(1, -1)$. Also, B' is 1 unit below line m at $(5, 0)$. Because point C is on line m, you know that $C = C'$.

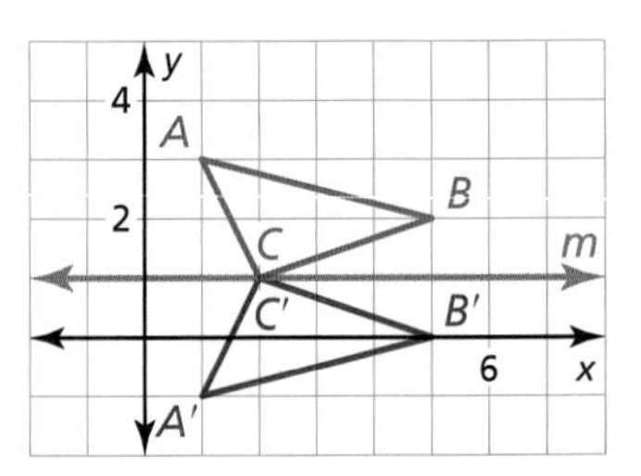

Monitoring Progress  Help in English and Spanish at *BigIdeasMath.com*

Graph $\triangle ABC$ from Example 1 and its image after a reflection in the given line.

1. $x = 4$

2. $x = -3$

3. $y = 2$

4. $y = -1$

EXAMPLE 2 Reflecting in the Line $y = x$

Graph $\overline{FG}$ with endpoints $F(-1, 2)$ and $G(1, 2)$ and its image after a reflection in the line $y = x$.

REMEMBER

The product of the slopes of perpendicular lines is -1.

SOLUTION

The slope of $y = x$ is 1. The segment from F to its image, $\overline{FF'}$, is perpendicular to the line of reflection $y = x$, so the slope of $\overline{FF'}$ will be -1 (because $1(-1) = -1$). From F, move 1.5 units right and 1.5 units down to $y = x$. From that point, move 1.5 units right and 1.5 units down to locate $F'(2, -1)$.

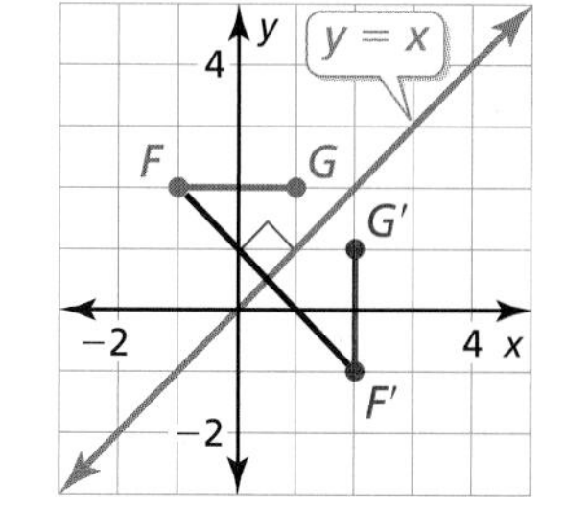

The slope of $\overline{GG'}$ will also be -1. From G, move 0.5 unit right and 0.5 unit down to $y = x$. Then move 0.5 unit right and 0.5 unit down to locate $G'(2, 1)$.

You can use coordinate rules to find the images of points reflected in four special lines.

Core Concept

Coordinate Rules for Reflections

- If (a, b) is reflected in the x-axis, then its image is the point $(a, -b)$.
- If (a, b) is reflected in the y-axis, then its image is the point $(-a, b)$.
- If (a, b) is reflected in the line $y = x$, then its image is the point (b, a).
- If (a, b) is reflected in the line $y = -x$, then its image is the point $(-b, -a)$.

EXAMPLE 3 Reflecting in the Line $y = -x$

Graph $\overline{FG}$ from Example 2 and its image after a reflection in the line $y = -x$.

SOLUTION

Use the coordinate rule for reflecting in the line $y = -x$ to find the coordinates of the endpoints of the image. Then graph $\overline{FG}$ and its image.

$$\mathbf{(a, b) \rightarrow (-b, -a)}$$

$$F(-1, 2) \rightarrow F'(-2, 1)$$

$$G(1, 2) \rightarrow G'(-2, -1)$$

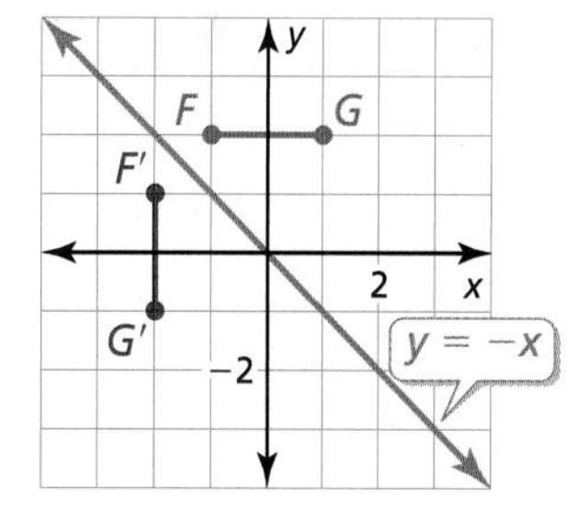

Monitoring Progress  Help in English and Spanish at *BigIdeasMath.com*

The vertices of $\triangle JKL$ are $J(1, 3)$, $K(4, 4)$, and $L(3, 1)$.

5. Graph $\triangle JKL$ and its image after a reflection in the x-axis.

6. Graph $\triangle JKL$ and its image after a reflection in the y-axis.

7. Graph $\triangle JKL$ and its image after a reflection in the line $y = x$.

8. Graph $\triangle JKL$ and its image after a reflection in the line $y = -x$.

9. In Example 3, verify that $\overline{FF'}$ is perpendicular to $y = -x$.

Performing Glide Reflections

Postulate

Postulate 4.2 Reflection Postulate

A reflection is a rigid motion.

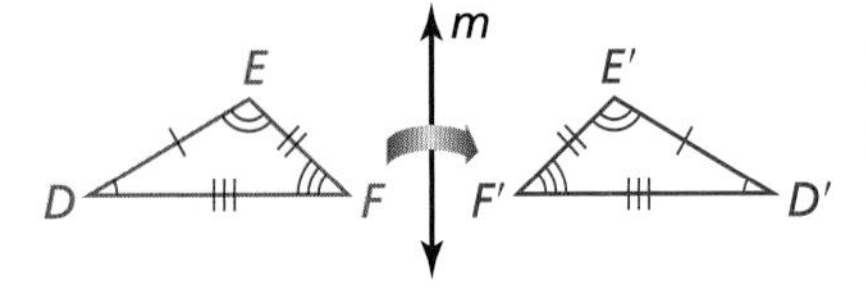

Because a reflection is a rigid motion, and a rigid motion preserves length and angle measure, the following statements are true for the reflection shown.

- $DE = D'E'$, $EF = E'F'$, $FD = F'D'$
- $m\angle D = m\angle D'$, $m\angle E = m\angle E'$, $m\angle F = m\angle F'$

Because a reflection is a rigid motion, the Composition Theorem (Theorem 4.1) guarantees that any composition of reflections and translations is a rigid motion.

STUDY TIP

The line of reflection must be parallel to the direction of the translation to be a glide reflection.

A **glide reflection** is a transformation involving a translation followed by a reflection in which every point P is mapped to a point P'' by the following steps.

Step 1 First, a translation maps P to P'.

Step 2 Then, a reflection in a line k parallel to the direction of the translation maps P' to P''.

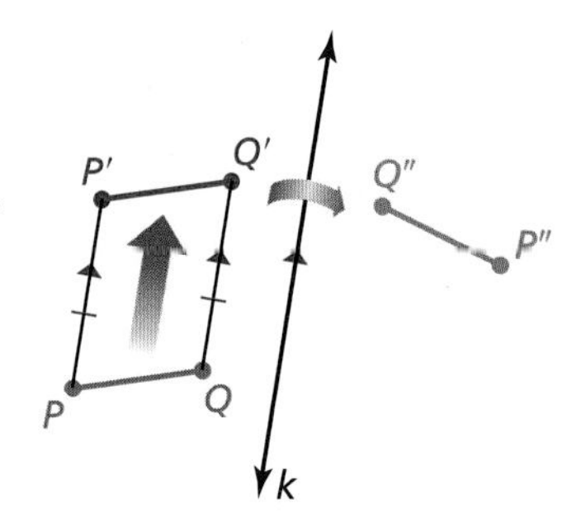

EXAMPLE 4 Performing a Glide Reflection

Graph $\triangle ABC$ with vertices $A(3, 2)$, $B(6, 3)$, and $C(7, 1)$ and its image after the glide reflection.

Translation: $(x, y) \rightarrow (x - 12, y)$

Reflection: in the x-axis

SOLUTION

Begin by graphing $\triangle ABC$. Then graph $\triangle A'B'C'$ after a translation 12 units left. Finally, graph $\triangle A''B''C''$ after a reflection in the x-axis.

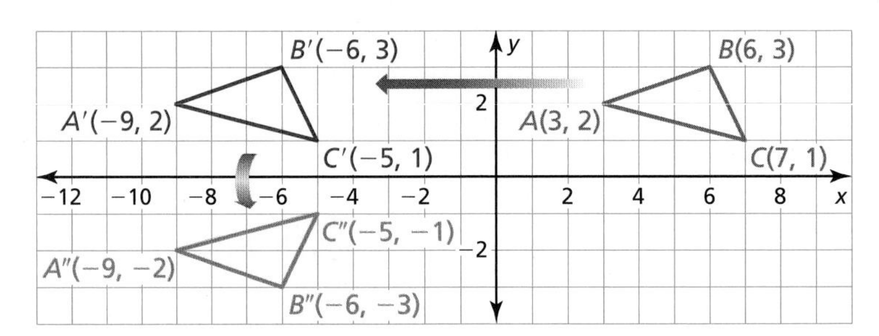

Monitoring Progress  Help in English and Spanish at *BigIdeasMath.com*

10. WHAT IF? In Example 4, $\triangle ABC$ is translated 4 units down and then reflected in the y-axis. Graph $\triangle ABC$ and its image after the glide reflection.

11. In Example 4, describe a glide reflection from $\triangle A''B''C''$ to $\triangle ABC$.

Identifying Lines of Symmetry

A figure in the plane has **line symmetry** when the figure can be mapped onto itself by a reflection in a line. This line of reflection is a **line of symmetry**, such as line m at the left. A figure can have more than one line of symmetry.

EXAMPLE 5 Identifying Lines of Symmetry

How many lines of symmetry does each hexagon have?

a. 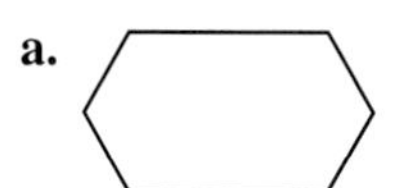 **b.** 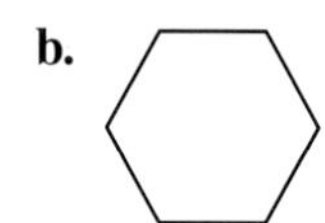 **c.** 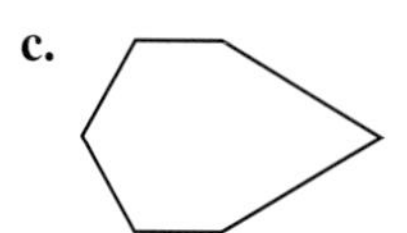

SOLUTION

a. 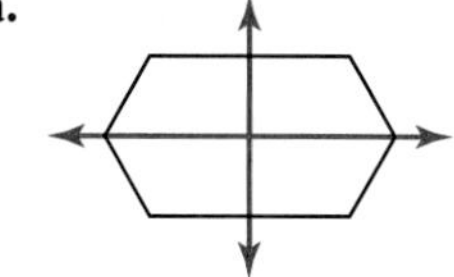 **b.** 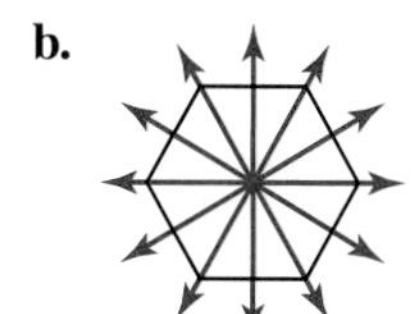 **c.** 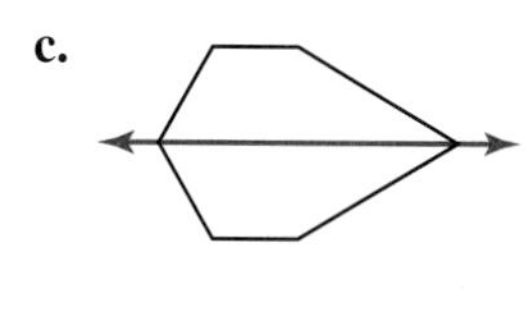

Monitoring Progress  Help in English and Spanish at *BigIdeasMath.com*

Determine the number of lines of symmetry for the figure.

12. 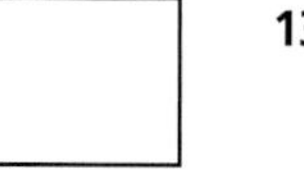

13. 

14. 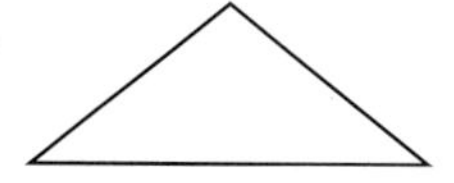

15. Draw a hexagon with no lines of symmetry.

Solving Real-Life Problems

EXAMPLE 6 Finding a Minimum Distance

You are going to buy books. Your friend is going to buy CDs. Where should you park to minimize the distance you both will walk?

SOLUTION

Reflect B in line m to obtain B'. Then draw $\overline{AB'}$. Label the intersection of $\overline{AB'}$ and m as C. Because AB' is the shortest distance between A and B' and $BC = B'C$, park at point C to minimize the combined distance, $AC + BC$, you both have to walk.

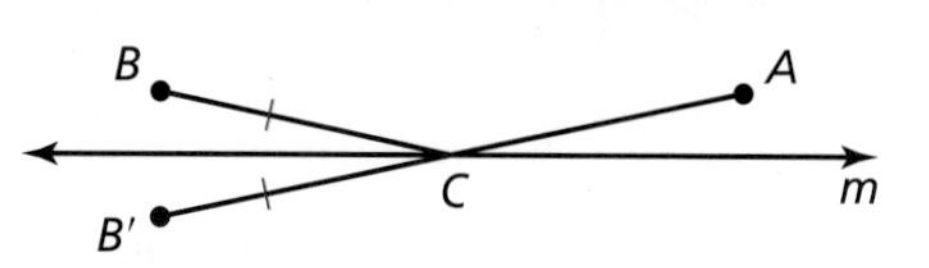

Monitoring Progress  Help in English and Spanish at *BigIdeasMath.com*

16. Look back at Example 6. Answer the question by using a reflection of point A instead of point B.

4.2 Exercises

Dynamic Solutions available at *BigIdeasMath.com*

Vocabulary and Core Concept Check

1. **VOCABULARY** A glide reflection is a combination of which two transformations?

2. **WHICH ONE DOESN'T BELONG?** Which transformation does *not* belong with the other three? Explain your reasoning.

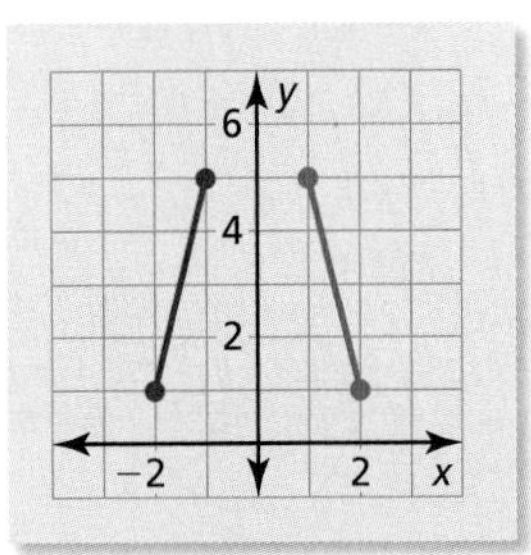

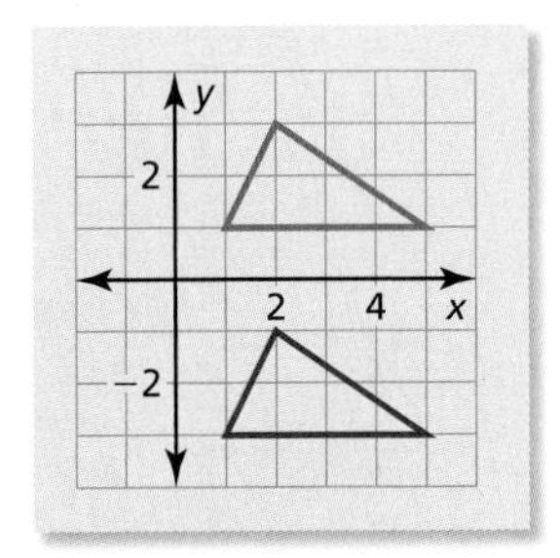

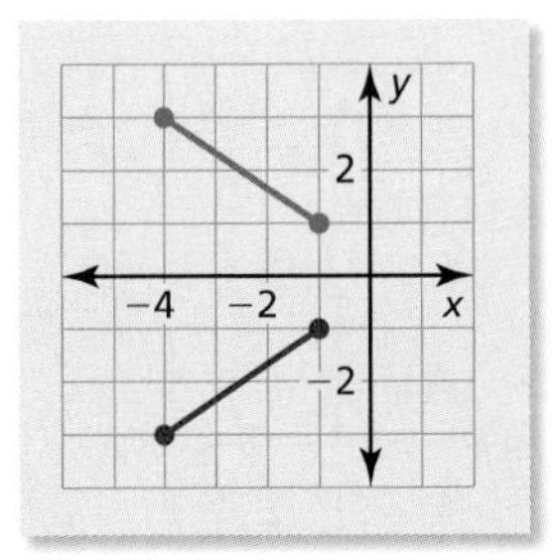

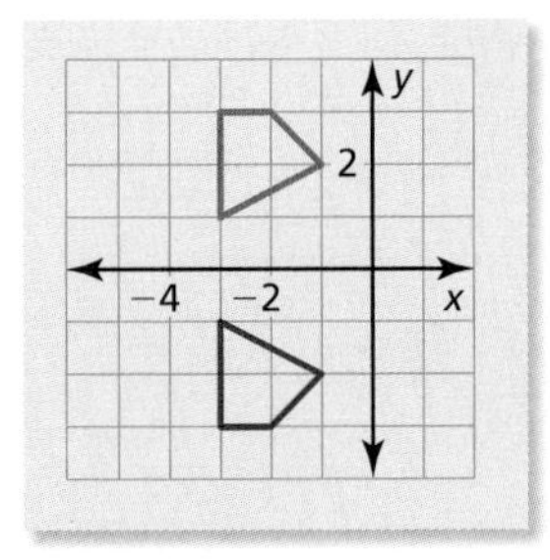

Monitoring Progress and Modeling with Mathematics

In Exercises 3–6, determine whether the coordinate plane shows a reflection in the *x-axis*, *y-axis*, or *neither*.

3.

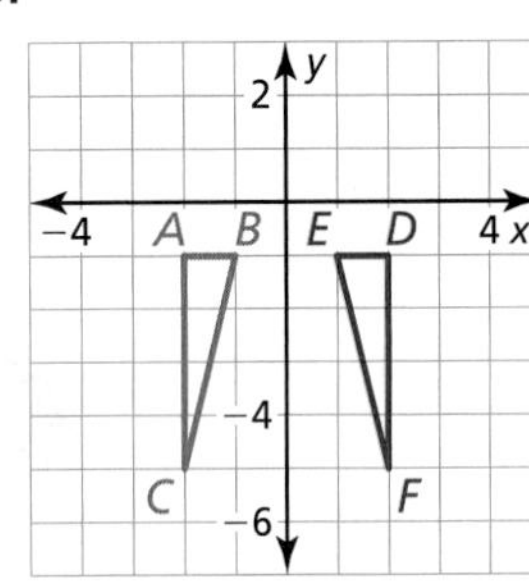

4.

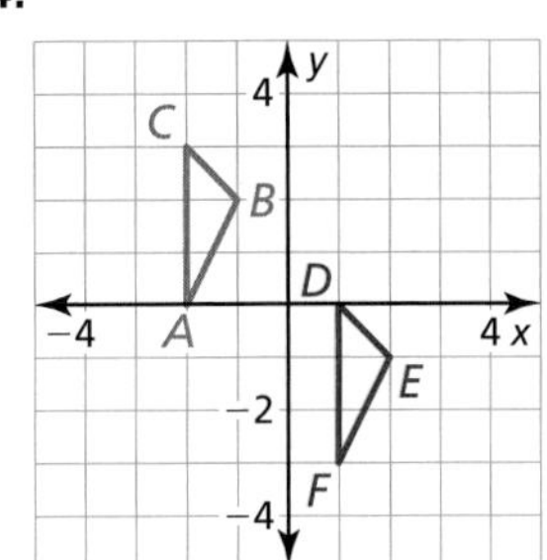

5.

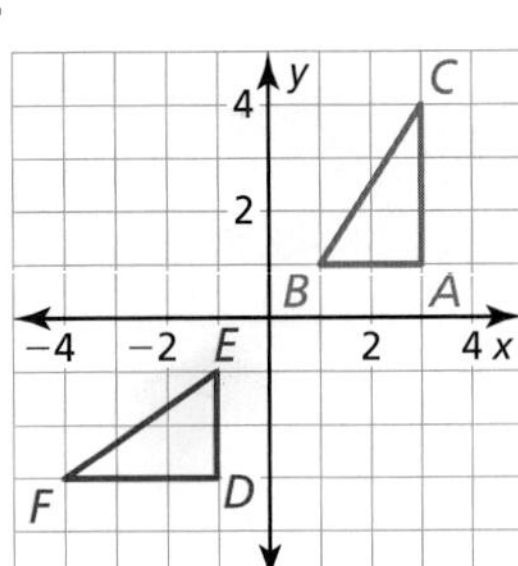

6.

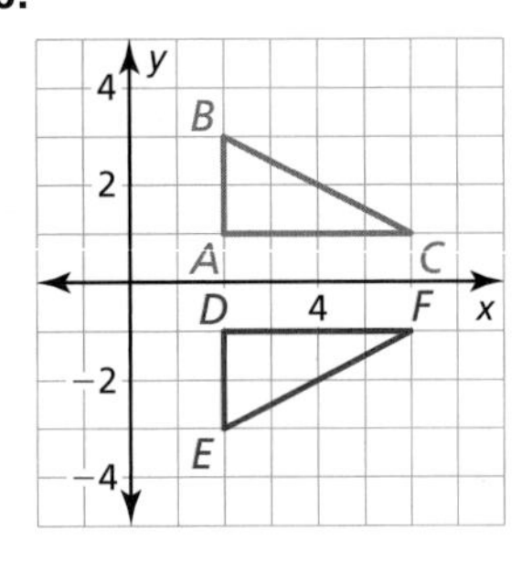

In Exercises 7–12, graph $\triangle JKL$ and its image after a reflection in the given line. *(See Example 1.)*

7. $J(2, -4)$, $K(3, 7)$, $L(6, -1)$; x-axis

8. $J(5, 3)$, $K(1, -2)$, $L(-3, 4)$; y-axis

9. $J(2, -1)$, $K(4, -5)$, $L(3, 1)$; $x = -1$

10. $J(1, -1)$, $K(3, 0)$, $L(0, -4)$; $x = 2$

11. $J(2, 4)$, $K(-4, -2)$, $L(-1, 0)$; $y = 1$

12. $J(3, -5)$, $K(4, -1)$, $L(0, -3)$; $y = -3$

In Exercises 13–16, graph the polygon and its image after a reflection in the given line. *(See Examples 2 and 3.)*

13. $y = x$

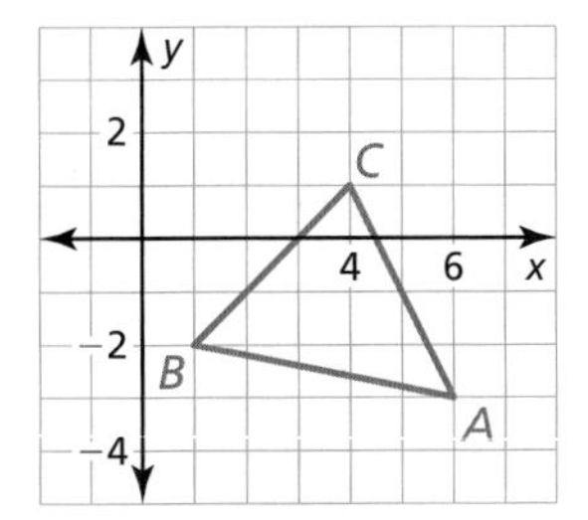

14. $y = x$

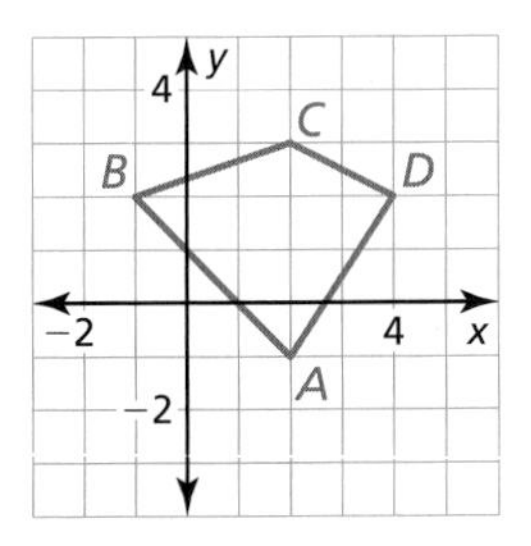

15. $y = -x$

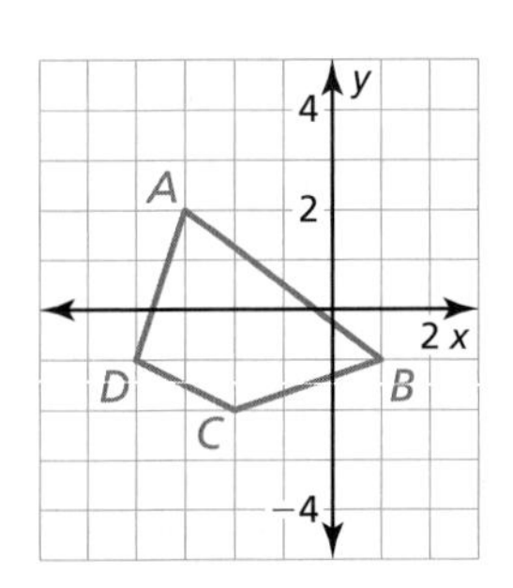

16. $y = -x$

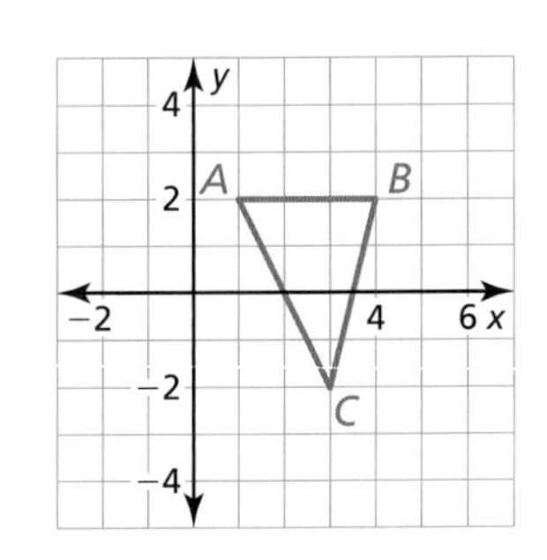

In Exercises 17–20, graph $\triangle RST$ with vertices $R(4, 1)$, $S(7, 3)$, and $T(6, 4)$ and its image after the glide reflection. *(See Example 4.)*

17. **Translation:** $(x, y) \rightarrow (x, y - 1)$
Reflection: in the y-axis

18. **Translation:** $(x, y) \rightarrow (x - 3, y)$
Reflection: in the line $y = -1$

19. **Translation:** $(x, y) \rightarrow (x, y + 4)$
Reflection: in the line $x = 3$

20. **Translation:** $(x, y) \rightarrow (x + 2, y + 2)$
Reflection: in the line $y = x$

In Exercises 21–24, determine the number of lines of symmetry for the figure. *(See Example 5.)*

21.

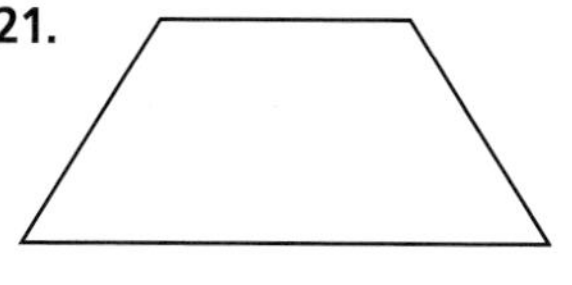

22.

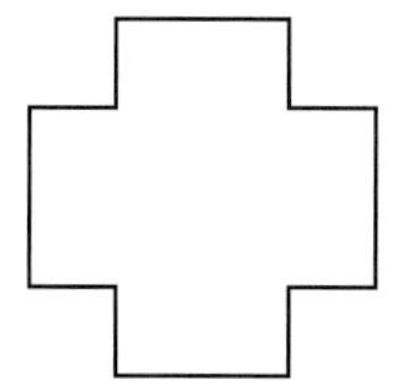

23.

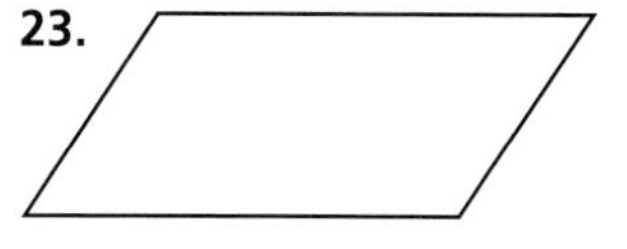

24.

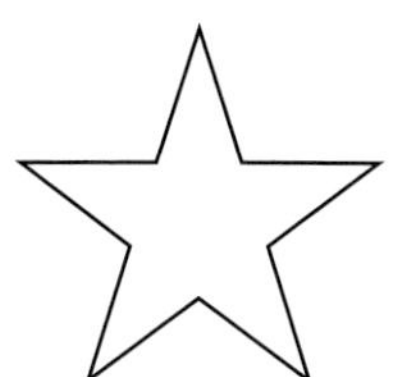

25. **USING STRUCTURE** Identify the line symmetry (if any) of each word.

a. LOOK

b. MOM

c. OX

d. DAD

26. **ERROR ANALYSIS** Describe and correct the error in describing the transformation.

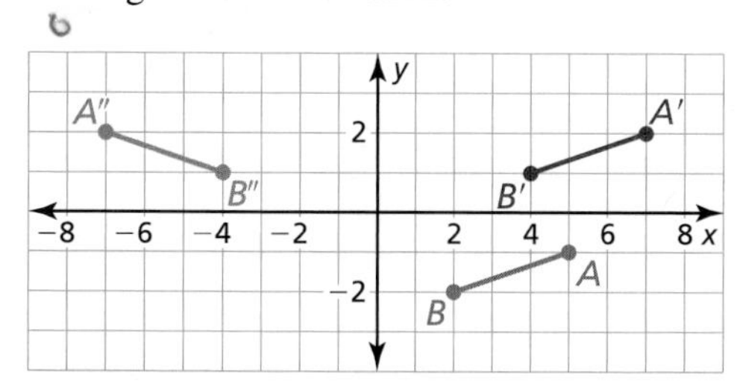

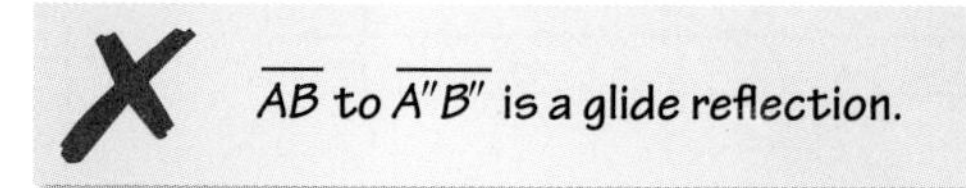

27. **MODELING WITH MATHEMATICS** You park at some point K on line n. You deliver a pizza to House H, go back to your car, and deliver a pizza to House J. Assuming that you can cut across both lawns, how can you determine the parking location K that minimizes the distance $HK + KJ$? *(See Example 6.)*

28. **ATTENDING TO PRECISION** Use the numbers and symbols to create the glide reflection resulting in the image shown.

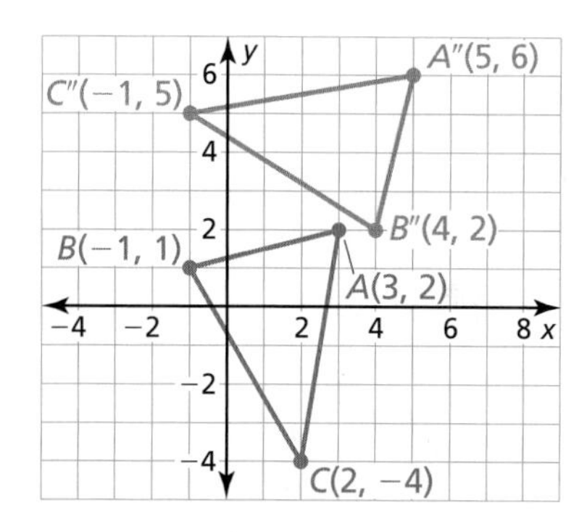

Translation: $(x, y) \rightarrow (\ \square\ , \ \square\ )$
Reflection: in $y = x$

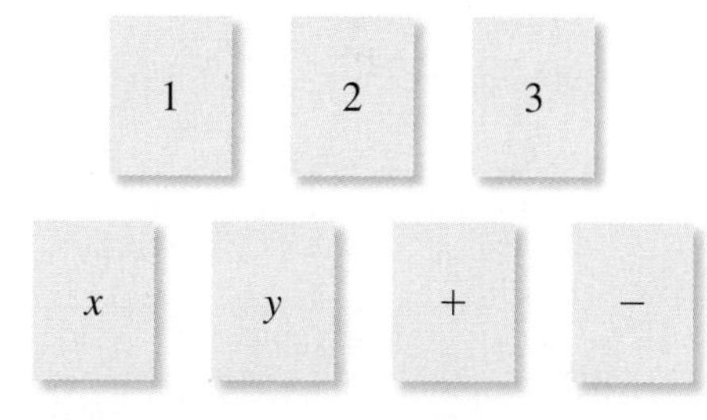

In Exercises 29–32, find point C on the x-axis so $AC + BC$ is a minimum.

29. $A(1, 4)$, $B(6, 1)$

30. $A(4, -5)$, $B(12, 3)$

31. $A(-8, 4)$, $B(-1, 3)$

32. $A(-1, 7)$, $B(5, -4)$

33. **MATHEMATICAL CONNECTIONS** The line $y = 3x + 2$ is reflected in the line $y = -1$. What is the equation of the image?

34. HOW DO YOU SEE IT? Use Figure A.

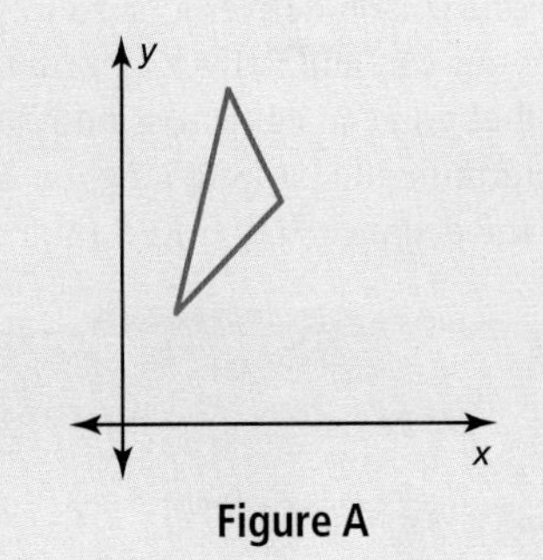

Figure A

Figure 1

Figure 2

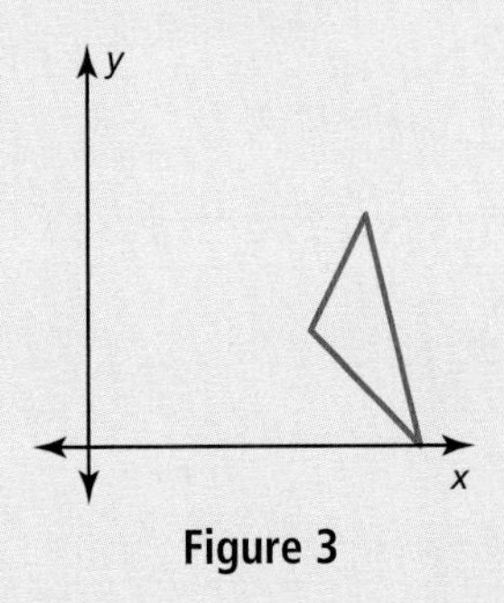

Figure 3

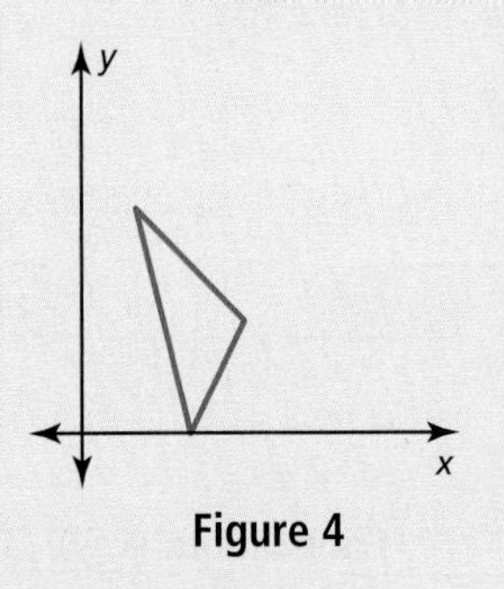

Figure 4

a. Which figure is a reflection of Figure A in the line $x = a$? Explain.

b. Which figure is a reflection of Figure A in the line $y = b$? Explain.

c. Which figure is a reflection of Figure A in the line $y = x$? Explain.

d. Is there a figure that represents a glide reflection? Explain your reasoning.

35. CONSTRUCTION Follow these steps to construct a reflection of $\triangle ABC$ in line m. Use a compass and straightedge.

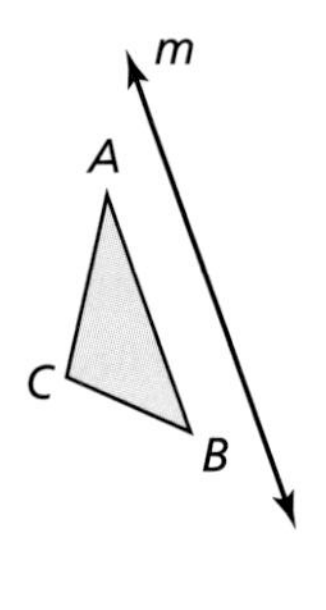

Step 1 Draw $\triangle ABC$ and line m.

Step 2 Use one compass setting to find two points that are equidistant from A on line m. Use the same compass setting to find a point on the other side of m that is the same distance from these two points. Label that point as A'.

Step 3 Repeat Step 2 to find points B' and C'. Draw $\triangle A'B'C'$.

36. USING TOOLS Use a reflective device to verify your construction in Exercise 35.

37. MATHEMATICAL CONNECTIONS Reflect $\triangle MNQ$ in the line $y = -2x$.

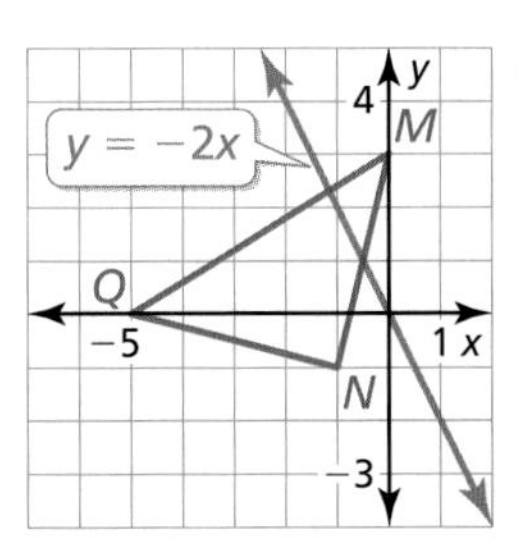

38. THOUGHT PROVOKING Is the composition of a translation and a reflection commutative? (In other words, do you obtain the same image regardless of the order in which you perform the transformations?) Justify your answer.

39. MATHEMATICAL CONNECTIONS Point $B'(1, 4)$ is the image of $B(3, 2)$ after a reflection in line c. Write an equation for line c.

Maintaining Mathematical Proficiency

Reviewing what you learned in previous grades and lessons

Use the diagram to find the angle measure. *(Section 1.5)*

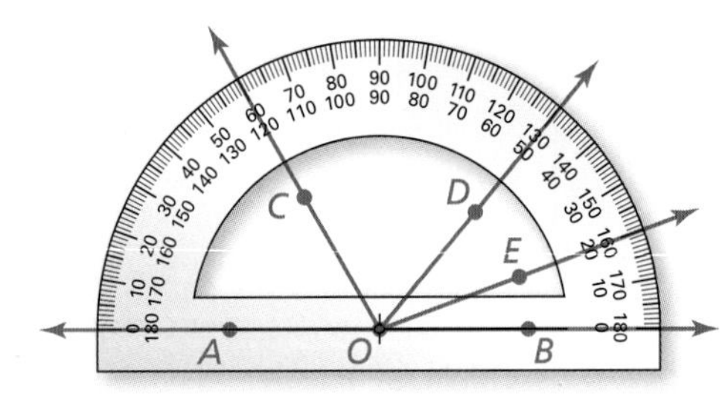

40. $m\angle AOC$

41. $m\angle AOD$

42. $m\angle BOE$

43. $m\angle AOE$

44. $m\angle COD$

45. $m\angle EOD$

46. $m\angle COE$

47. $m\angle AOB$

48. $m\angle COB$

49. $m\angle BOD$

4.3 Rotations

Essential Question How can you rotate a figure in a coordinate plane?

EXPLORATION 1 Rotating a Triangle in a Coordinate Plane

Work with a partner.

a. Use dynamic geometry software to draw any triangle and label it $\triangle ABC$.

b. *Rotate* the triangle 90° counterclockwise about the origin to form $\triangle A'B'C'$.

c. What is the relationship between the coordinates of the vertices of $\triangle ABC$ and those of $\triangle A'B'C'$?

d. What do you observe about the side lengths and angle measures of the two triangles?

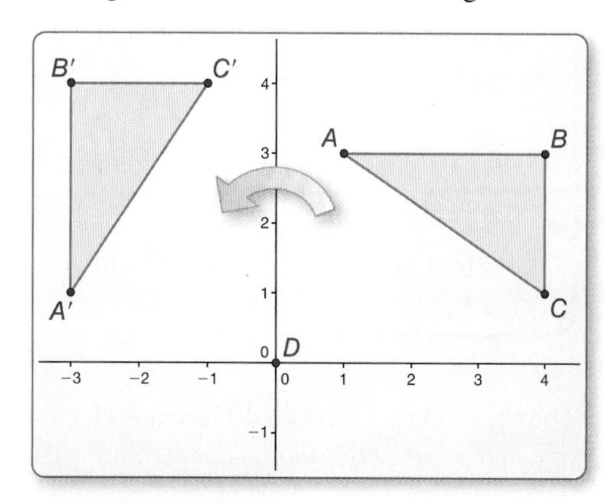

Sample

Points
$A(1, 3)$
$B(4, 3)$
$C(4, 1)$
$D(0, 0)$
Segments
$AB = 3$
$BC = 2$
$AC = 3.61$
Angles
$m\angle A = 33.69°$
$m\angle B = 90°$
$m\angle C = 56.31°$

CONSTRUCTING VIABLE ARGUMENTS

To be proficient in math, you need to use previously established results in constructing arguments.

EXPLORATION 2 Rotating a Triangle in a Coordinate Plane

Work with a partner.

a. The point (x, y) is rotated 90° counterclockwise about the origin. Write a rule to determine the coordinates of the image of (x, y).

b. Use the rule you wrote in part (a) to rotate $\triangle ABC$ 90° counterclockwise about the origin. What are the coordinates of the vertices of the image, $\triangle A'B'C'$?

c. Draw $\triangle A'B'C'$. Are its side lengths the same as those of $\triangle ABC$? Justify your answer.

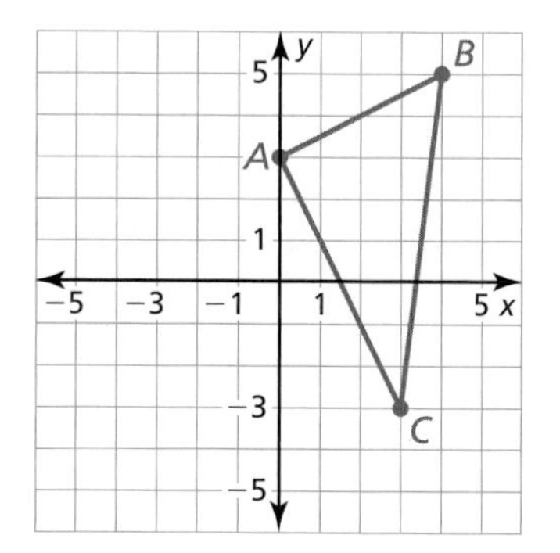

EXPLORATION 3 Rotating a Triangle in a Coordinate Plane

Work with a partner.

a. The point (x, y) is rotated 180° counterclockwise about the origin. Write a rule to determine the coordinates of the image of (x, y). Explain how you found the rule.

b. Use the rule you wrote in part (a) to rotate $\triangle ABC$ (from Exploration 2) 180° counterclockwise about the origin. What are the coordinates of the vertices of the image, $\triangle A'B'C'$?

Communicate Your Answer

4. How can you rotate a figure in a coordinate plane?

5. In Exploration 3, rotate $\triangle A'B'C'$ 180° counterclockwise about the origin. What are the coordinates of the vertices of the image, $\triangle A''B''C''$? How are these coordinates related to the coordinates of the vertices of the original triangle, $\triangle ABC$?

4.3 Lesson

Core Vocabulary

rotation, *p. 190*
center of rotation, *p. 190*
angle of rotation, *p. 190*
rotational symmetry, *p. 193*
center of symmetry, *p. 193*

What You Will Learn

- Perform rotations.
- Perform compositions with rotations.
- Identify rotational symmetry.

Performing Rotations

Core Concept

Rotations

A **rotation** is a transformation in which a figure is turned about a fixed point called the **center of rotation**. Rays drawn from the center of rotation to a point and its image form the **angle of rotation**.

A rotation about a point P through an angle of $x°$ maps every point Q in the plane to a point Q' so that one of the following properties is true.

- If Q is not the center of rotation P, then $QP = Q'P$ and $m\angle QPQ' = x°$, or
- If Q is the center of rotation P, then $Q = Q'$.

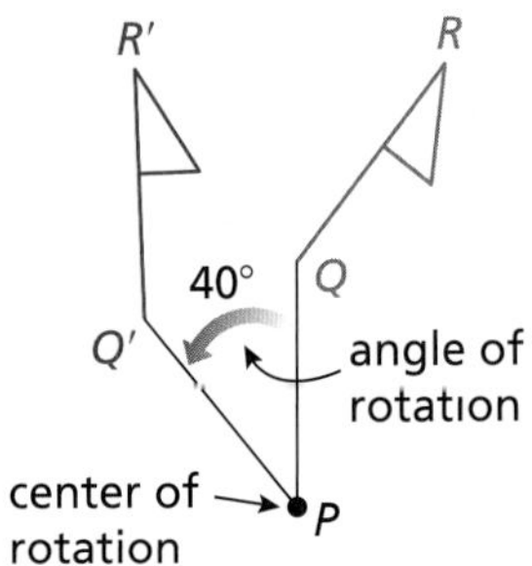

Direction of rotation

clockwise

counterclockwise

The figure above shows a 40° counterclockwise rotation. Rotations can be *clockwise* or *counterclockwise*. In this chapter, all rotations are counterclockwise unless otherwise noted.

EXAMPLE 1 Drawing a Rotation

Draw a 120° rotation of $\triangle ABC$ about point P.

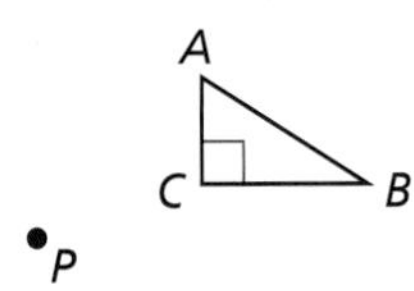

SOLUTION

Step 1 Draw a segment from P to A.

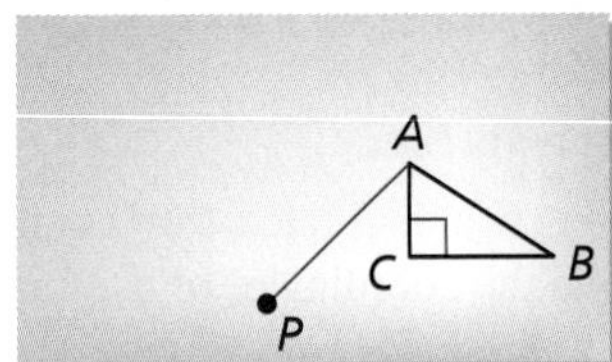

Step 2 Draw a ray to form a 120° angle with $\overline{PA}$.

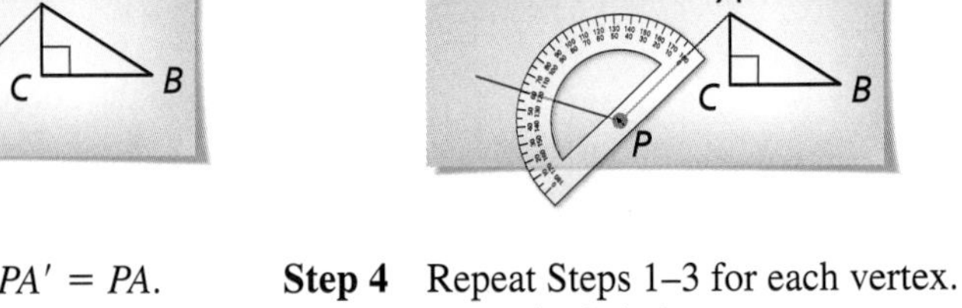

Step 3 Draw A' so that $PA' = PA$.

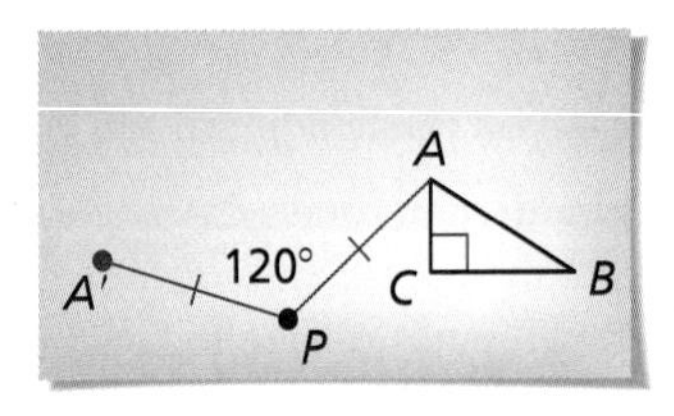

Step 4 Repeat Steps 1–3 for each vertex. Draw $\triangle A'B'C'$.

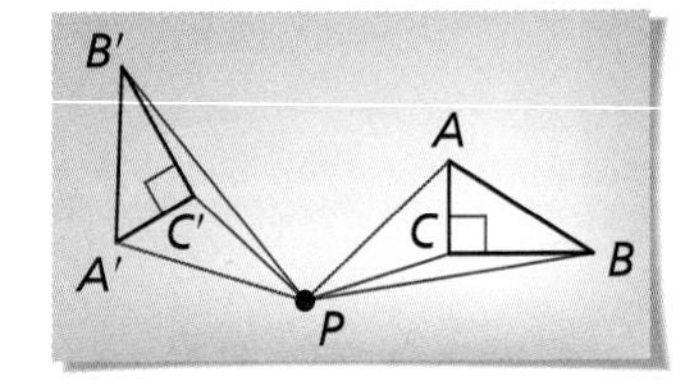

USING ROTATIONS

You can rotate a figure more than 360°. The effect, however, is the same as rotating the figure by the angle minus 360°.

You can rotate a figure more than 180°. The diagram shows rotations of point A 130°, 220°, and 310° about the origin. Notice that point A and its images all lie on the same circle. A rotation of 360° maps a figure onto itself.

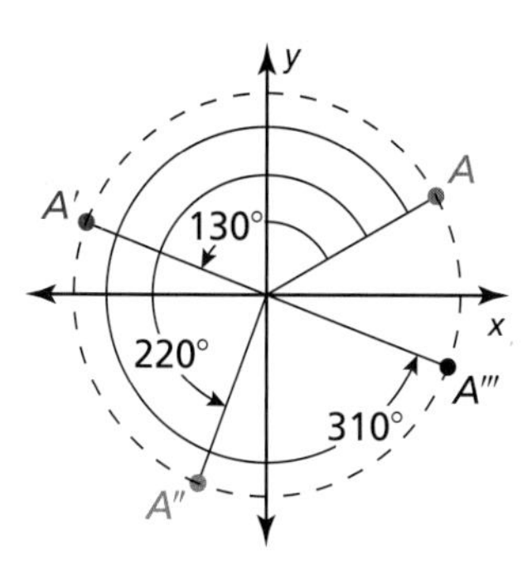

You can use coordinate rules to find the coordinates of a point after a rotation of 90°, 180°, or 270° about the origin.

Core Concept

Coordinate Rules for Rotations about the Origin

When a point (a, b) is rotated counterclockwise about the origin, the following are true.

- For a rotation of 90°, $(a, b) \rightarrow (-b, a)$.
- For a rotation of 180°, $(a, b) \rightarrow (-a, -b)$.
- For a rotation of 270°, $(a, b) \rightarrow (b, -a)$.

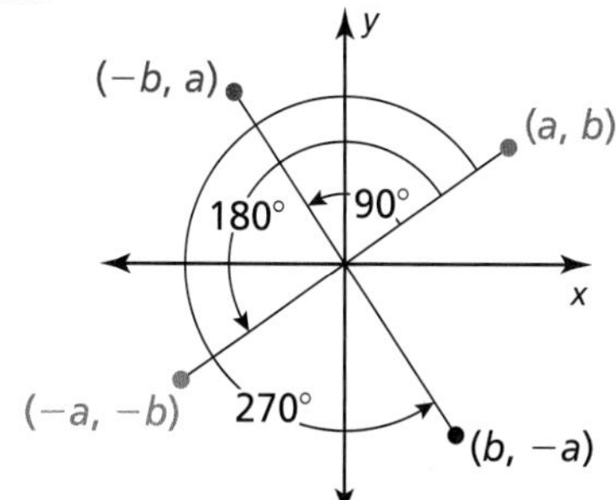

EXAMPLE 2 Rotating a Figure in the Coordinate Plane

Graph quadrilateral $RSTU$ with vertices $R(3, 1)$, $S(5, 1)$, $T(5, -3)$, and $U(2, -1)$ and its image after a 270° rotation about the origin.

SOLUTION

Use the coordinate rule for a 270° rotation to find the coordinates of the vertices of the image. Then graph quadrilateral $RSTU$ and its image.

$$(a, b) \rightarrow (b, -a)$$

$$R(3, 1) \rightarrow R'(1, -3)$$

$$S(5, 1) \rightarrow S'(1, -5)$$

$$T(5, -3) \rightarrow T'(-3, -5)$$

$$U(2, -1) \rightarrow U'(-1, -2)$$

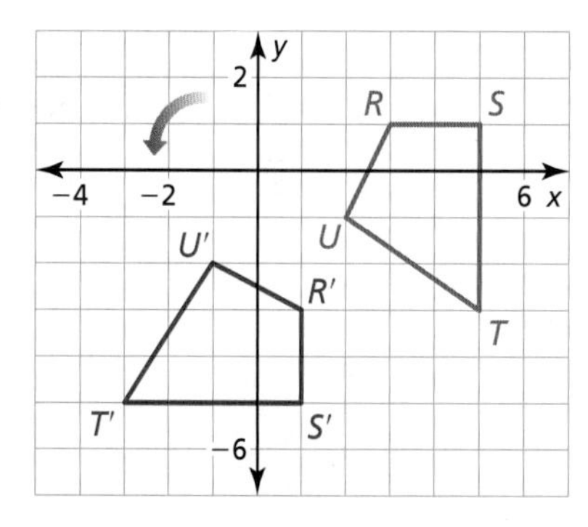

Monitoring Progress

Help in English and Spanish at *BigIdeasMath.com*

1. Trace $\triangle DEF$ and point P. Then draw a 50° rotation of $\triangle DEF$ about point P.

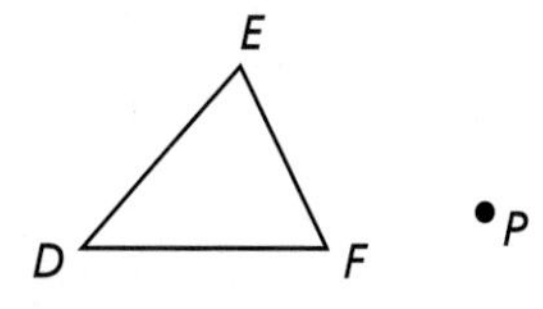

2. Graph $\triangle JKL$ with vertices $J(3, 0)$, $K(4, 3)$, and $L(6, 0)$ and its image after a 90° rotation about the origin.

Performing Compositions with Rotations

Postulate

Postulate 4.3 Rotation Postulate

A rotation is a rigid motion.

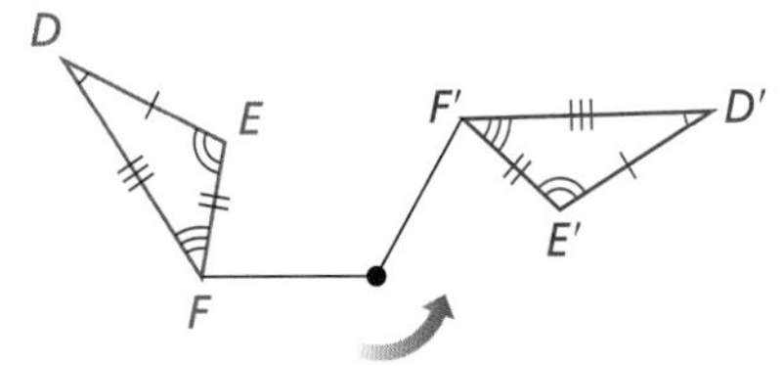

Because a rotation is a rigid motion, and a rigid motion preserves length and angle measure, the following statements are true for the rotation shown.

- $DE = D'E'$, $EF = E'F'$, $FD = F'D'$
- $m\angle D = m\angle D'$, $m\angle E = m\angle E'$, $m\angle F = m\angle F'$

Because a rotation is a rigid motion, the Composition Theorem (Theorem 4.1) guarantees that compositions of rotations and other rigid motions, such as translations and reflections, are rigid motions.

EXAMPLE 3 Performing a Composition

Graph $\overline{RS}$ with endpoints $R(1, -3)$ and $S(2, -6)$ and its image after the composition.

Reflection: in the y-axis

Rotation: $90°$ about the origin

COMMON ERROR

Unless you are told otherwise, perform the transformations in the order given.

SOLUTION

Step 1 Graph $\overline{RS}$.

Step 2 Reflect $\overline{RS}$ in the y-axis. $\overline{R'S'}$ has endpoints $R'(-1, -3)$ and $S'(-2, -6)$.

Step 3 Rotate $\overline{R'S'}$ $90°$ about the origin. $\overline{R''S''}$ has endpoints $R''(3, -1)$ and $S''(6, -2)$.

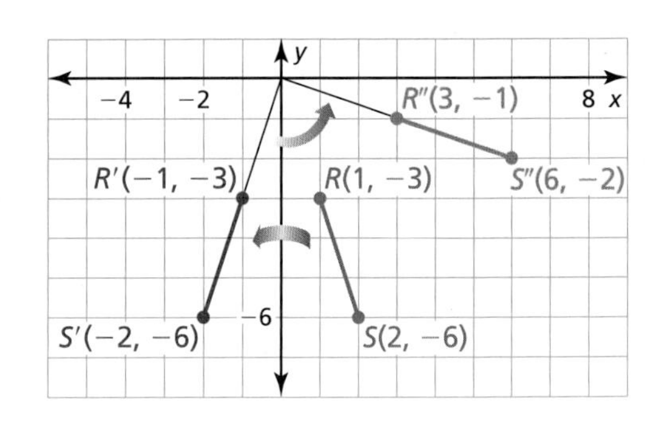

Monitoring Progress

Help in English and Spanish at *BigIdeasMath.com*

3. Graph $\overline{RS}$ from Example 3. Perform the rotation first, followed by the reflection. Does the order of the transformations matter? Explain.

4. **WHAT IF?** In Example 3, $\overline{RS}$ is reflected in the x-axis and rotated $180°$ about the origin. Graph $\overline{RS}$ and its image after the composition.

5. Graph $\overline{AB}$ with endpoints $A(-4, 4)$ and $B(-1, 7)$ and its image after the composition.

   Translation: $(x, y) \rightarrow (x - 2, y - 1)$

   Rotation: $90°$ about the origin

6. Graph $\triangle TUV$ with vertices $T(1, 2)$, $U(3, 5)$, and $V(6, 3)$ and its image after the composition.

   Rotation: $180°$ about the origin

   Reflection: in the x-axis

Identifying Rotational Symmetry

A figure in the plane has **rotational symmetry** when the figure can be mapped onto itself by a rotation of 180° or less about the center of the figure. This point is the **center of symmetry**. Note that the rotation can be either clockwise or counterclockwise.

For example, the figure below has rotational symmetry, because a rotation of either 90° or 180° maps the figure onto itself (although a rotation of 45° does not).

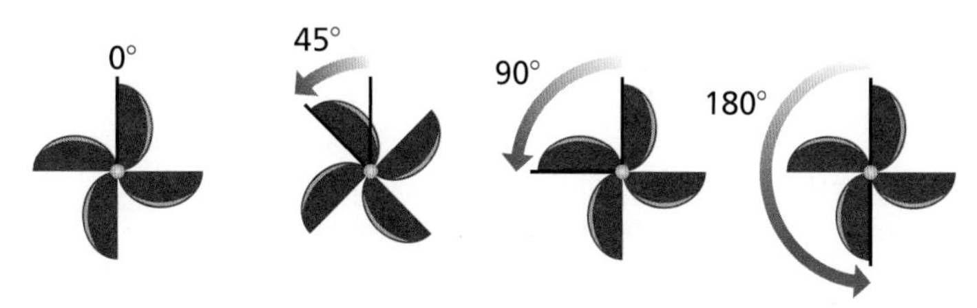

The figure above also has *point symmetry*, which is 180° rotational symmetry.

EXAMPLE 4 Identifying Rotational Symmetry

Does the figure have rotational symmetry? If so, describe any rotations that map the figure onto itself.

a. parallelogram

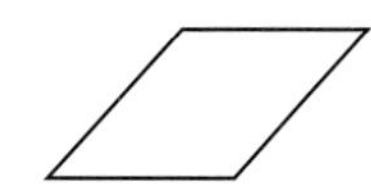

b. regular octagon

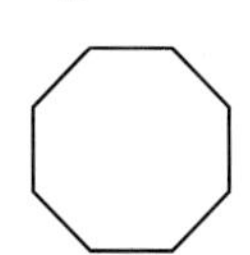

c. trapezoid

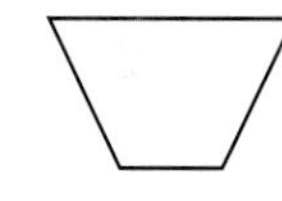

SOLUTION

a. The parallelogram has rotational symmetry. The center is the intersection of the diagonals. A 180° rotation about the center maps the parallelogram onto itself.

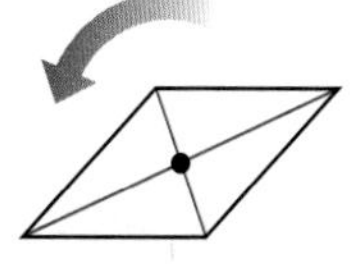

b. The regular octagon has rotational symmetry. The center is the intersection of the diagonals. Rotations of 45°, 90°, 135°, or 180° about the center all map the octagon onto itself.

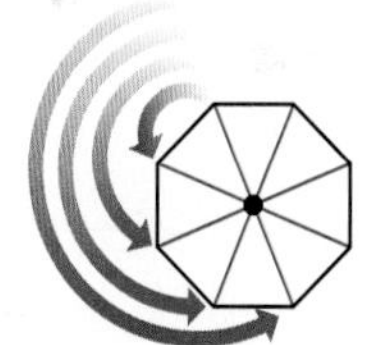

c. The trapezoid does not have rotational symmetry because no rotation of 180° or less maps the trapezoid onto itself.

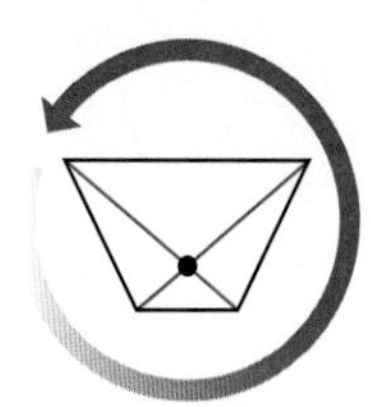

Monitoring Progress

Help in English and Spanish at *BigIdeasMath.com*

Determine whether the figure has rotational symmetry. If so, describe any rotations that map the figure onto itself.

7. rhombus

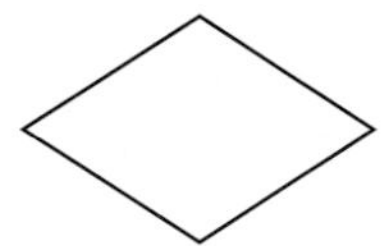

8. octagon

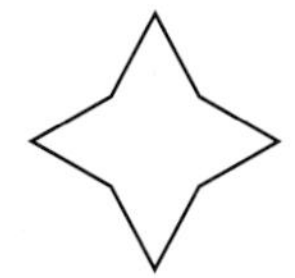

9. right triangle

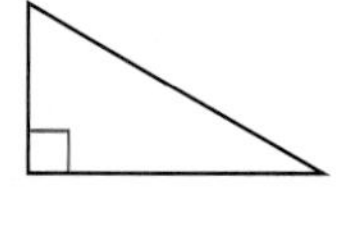

4.3 Exercises

Dynamic Solutions available at *BigIdeasMath.com*

Vocabulary and Core Concept Check

1. **COMPLETE THE SENTENCE** When a point (a, b) is rotated counterclockwise about the origin, $(a, b) \rightarrow (b, -a)$ is the result of a rotation of ______.

2. **DIFFERENT WORDS, SAME QUESTION** Which is different? Find "both" answers.

What are the coordinates of the vertices of the image after a 90° counterclockwise rotation about the origin?

What are the coordinates of the vertices of the image after a 270° clockwise rotation about the origin?

What are the coordinates of the vertices of the image after turning the figure 90° to the left about the origin?

What are the coordinates of the vertices of the image after a 270° counterclockwise rotation about the origin?

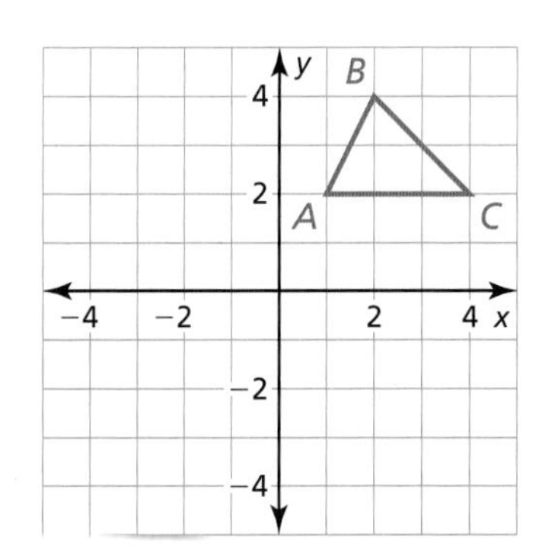

Monitoring Progress and Modeling with Mathematics

In Exercises 3–6, trace the polygon and point *P*. Then draw a rotation of the polygon about point *P* using the given number of degrees. *(See Example 1.)*

3. 30°

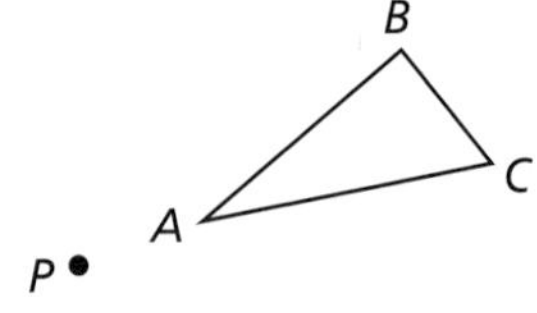

4. 80°

5. 150°

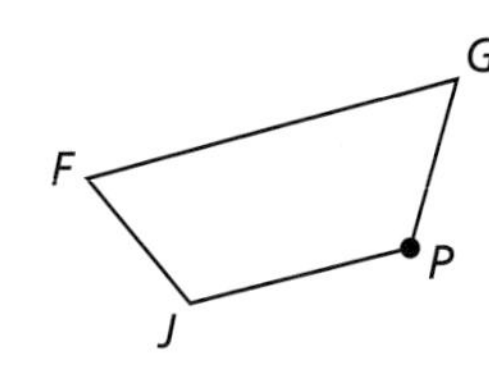

6. 130°

In Exercises 7–10, graph the polygon and its image after a rotation of the given number of degrees about the origin. *(See Example 2.)*

7. 90°

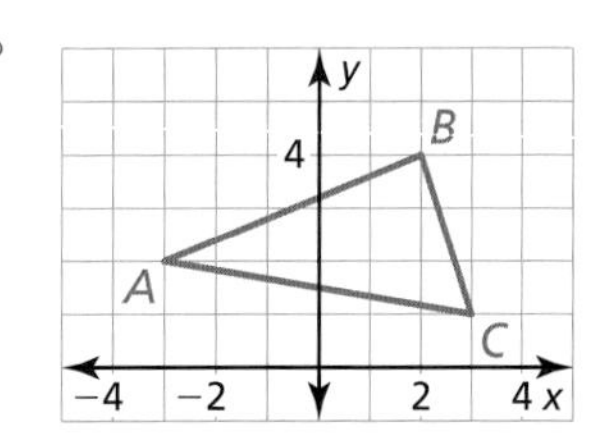

8. 180°

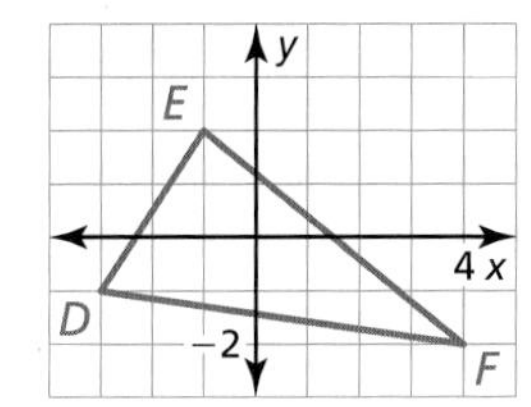

9. 180°

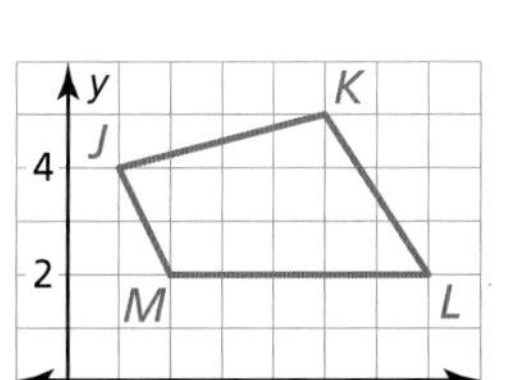

10. 270°

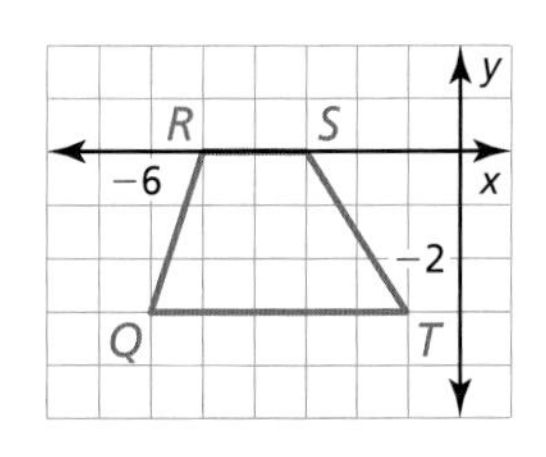

In Exercises 11–14, graph $\overline{XY}$ with endpoints $X(-3, 1)$ and $Y(4, -5)$ and its image after the composition. *(See Example 3.)*

11. **Translation:** $(x, y) \rightarrow (x, y + 2)$
Rotation: 90° about the origin

12. **Rotation:** 180° about the origin
Translation: $(x, y) \rightarrow (x - 1, y + 1)$

13. **Rotation:** 270° about the origin
Reflection: in the y-axis

14. **Reflection:** in the line $y = x$
Rotation: 180° about the origin

In Exercises 15 and 16, graph $\triangle LMN$ with vertices $L(1, 6)$, $M(-2, 4)$, and $N(3, 2)$ and its image after the composition. *(See Example 3.)*

15. **Rotation:** 90° about the origin
Translation: $(x, y) \rightarrow (x - 3, y + 2)$

16. **Reflection:** in the x-axis
Rotation: 270° about the origin

In Exercises 17–20, determine whether the figure has rotational symmetry. If so, describe any rotations that map the figure onto itself. *(See Example 4.)*

17.

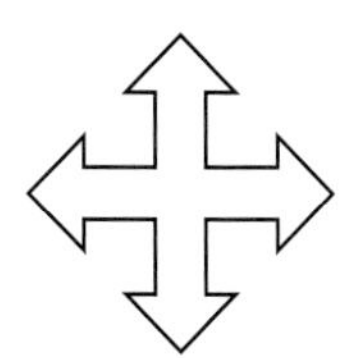

18.

19.

20.

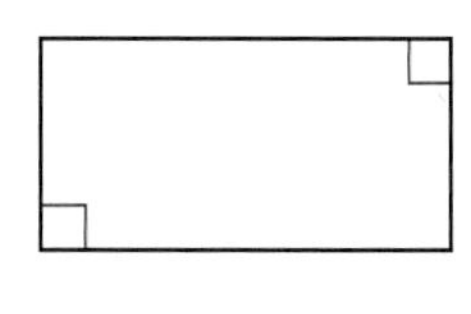

REPEATED REASONING **In Exercises 21–24, select the angles of rotational symmetry for the regular polygon. Select all that apply.**

Ⓐ 30° Ⓑ 45° Ⓒ 60° Ⓓ 72°
Ⓔ 90° Ⓕ 120° Ⓖ 144° Ⓗ 180°

21.

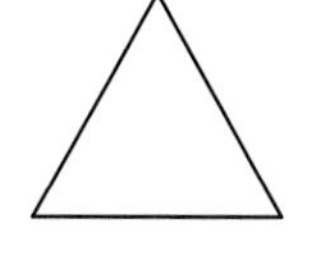

22.

23.

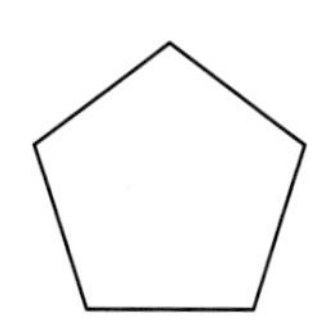

24.

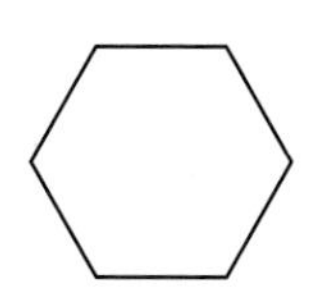

ERROR ANALYSIS **In Exercises 25 and 26, the endpoints of $\overline{CD}$ are $C(-1, 1)$ and $D(2, 3)$. Describe and correct the error in finding the coordinates of the vertices of the image after a rotation of 270° about the origin.**

25.

$C(-1, 1) \rightarrow C'(-1, -1)$
$D(2, 3) \rightarrow D'(2, -3)$

26.

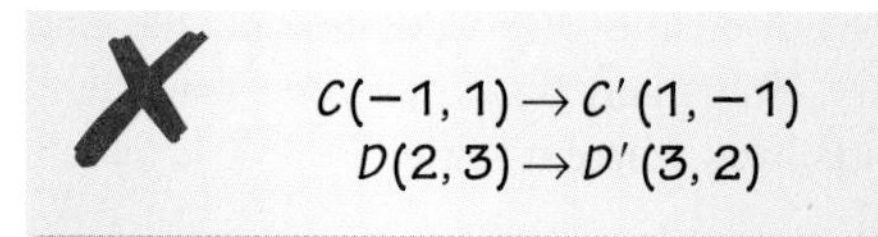

$C(-1, 1) \rightarrow C'(1, -1)$
$D(2, 3) \rightarrow D'(3, 2)$

27. **CONSTRUCTION** Follow these steps to construct a rotation of $\triangle ABC$ by angle D around a point O. Use a compass and straightedge.

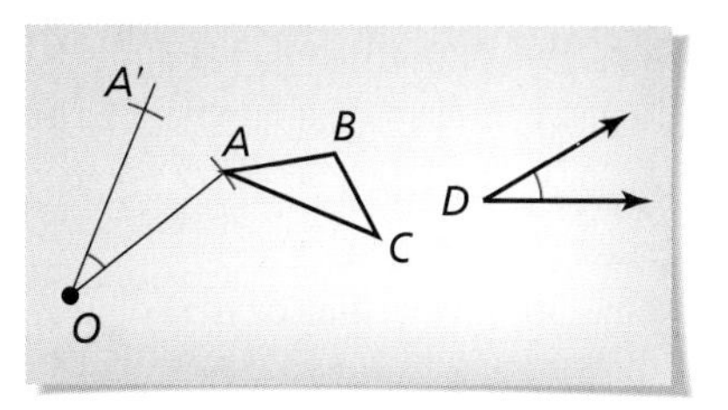

Step 1 Draw $\triangle ABC$, $\angle D$, and O, the center of rotation.

Step 2 Draw $\overline{OA}$. Use the construction for copying an angle to copy $\angle D$ at O, as shown. Then use distance OA and center O to find A'.

Step 3 Repeat Step 2 to find points B' and C'. Draw $\triangle A'B'C'$.

28. **REASONING** You enter the revolving door at a hotel.

a. You rotate the door 180°. What does this mean in the context of the situation? Explain.

b. You rotate the door 360°. What does this mean in the context of the situation? Explain.

29. **MATHEMATICAL CONNECTIONS** Use the graph of $y = 2x - 3$.

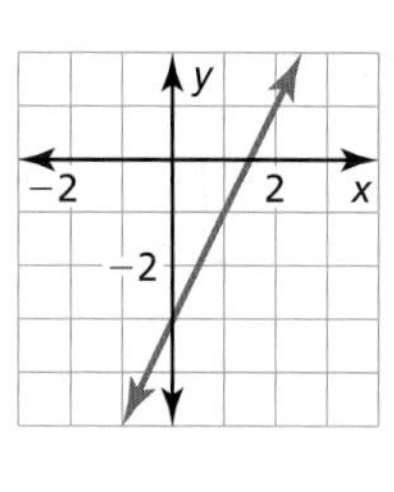

a. Rotate the line 90°, 180°, 270°, and 360° about the origin. Write the equation of the line for each image. Describe the relationship between the equation of the preimage and the equation of each image.

b. Do you think that the relationships you described in part (a) are true for any line? Explain your reasoning.

30. **MAKING AN ARGUMENT** Your friend claims that rotating a figure by 180° is the same as reflecting a figure in the y-axis and then reflecting it in the x-axis. Is your friend correct? Explain your reasoning.

31. **DRAWING CONCLUSIONS** A figure only has point symmetry. How many times can you rotate the figure before it is back where it started?

32. **ANALYZING RELATIONSHIPS** Is it possible for a figure to have 90° rotational symmetry but not 180° rotational symmetry? Explain your reasoning.

33. **ANALYZING RELATIONSHIPS** Is it possible for a figure to have 180° rotational symmetry but not 90° rotational symmetry? Explain your reasoning.

34. **THOUGHT PROVOKING** Can rotations of 90°, 180°, 270°, and 360° be written as the composition of two reflections? Justify your answer.

35. **USING AN EQUATION** Inside a kaleidoscope, two mirrors are placed next to each other to form a V. The angle between the mirrors determines the number of lines of symmetry in the image. Use the formula $n(m\angle 1) = 180°$ to find the measure of $\angle 1$, the angle between the mirrors, for the number n of lines of symmetry.

mirror

1

black glass

a. b.

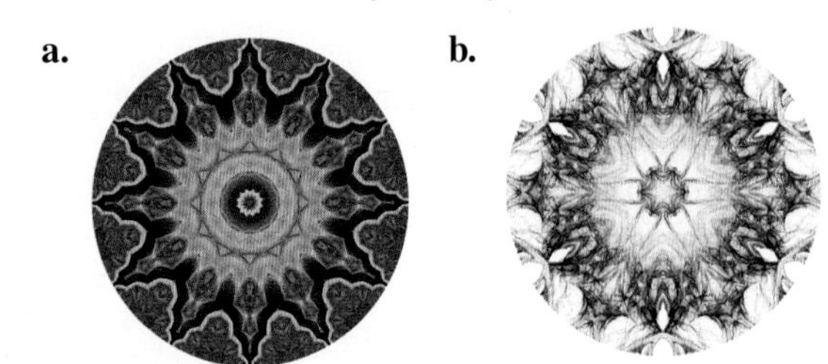

36. **REASONING** Use the coordinate rules for counterclockwise rotations about the origin to write coordinate rules for clockwise rotations of 90°, 180°, or 270° about the origin.

37. **USING STRUCTURE** $\triangle XYZ$ has vertices $X(2, 5)$, $Y(3, 1)$, and $Z(0, 2)$. Rotate $\triangle XYZ$ 90° about the point $P(-2, -1)$.

38. **HOW DO YOU SEE IT?** You are finishing the puzzle. The remaining two pieces both have rotational symmetry.

a. Describe the rotational symmetry of Piece 1 and of Piece 2.

b. You pick up Piece 1. How many different ways can it fit in the puzzle?

c. Before putting Piece 1 into the puzzle, you connect it to Piece 2. Now how many ways can it fit in the puzzle? Explain.

39. **USING STRUCTURE** A polar coordinate system locates a point in a plane by its distance from the origin O and by the measure of an angle with its vertex at the origin. For example, the point $A(2, 30°)$ is 2 units from the origin and $m\angle XOA = 30°$. What are the polar coordinates of the image of point A after a 90° rotation? a 180° rotation? a 270° rotation? Explain.

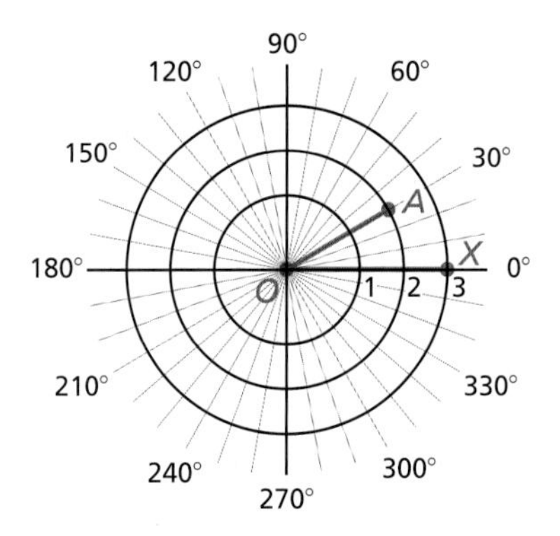

Maintaining Mathematical Proficiency

Reviewing what you learned in previous grades and lessons

The figures are congruent. Name the corresponding angles and the corresponding sides. *(Skills Review Handbook)*

40.

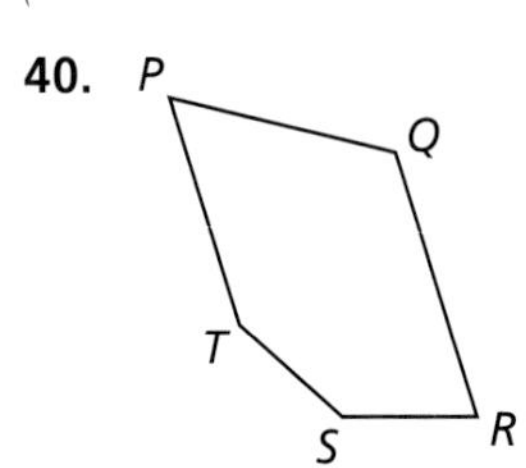

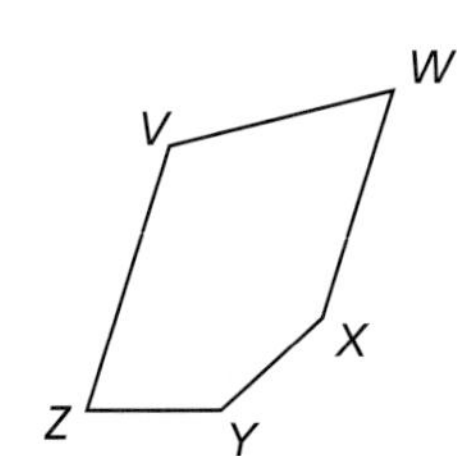

41.

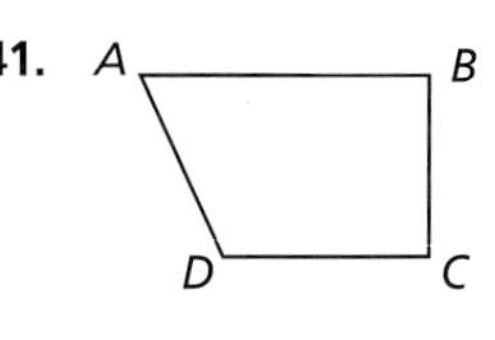

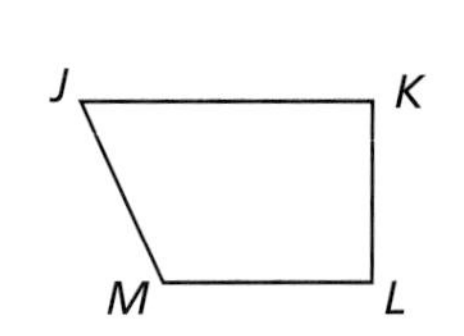

4.1–4.3 What Did You Learn?

Core Vocabulary

vector, *p. 174*
initial point, *p. 174*
terminal point, *p. 174*
horizontal component, *p. 174*
vertical component, *p. 174*
component form, *p. 174*
transformation, *p. 174*
image, *p. 174*
preimage, *p. 174*
translation, *p. 174*
rigid motion, *p. 176*
composition of transformations, *p. 176*
reflection, *p. 182*
line of reflection, *p. 182*
glide reflection, *p. 184*
line symmetry, *p. 185*
line of symmetry, *p. 185*
rotation, *p. 190*
center of rotation, *p. 190*
angle of rotation, *p. 190*
rotational symmetry, *p. 193*
center of symmetry, *p. 193*

Core Concepts

Section 4.1

Vectors, *p. 174*
Translations, *p. 174*
Postulate 4.1 Translation Postulate, *p. 176*
Theorem 4.1 Composition Theorem, *p. 176*

Section 4.2

Reflections, *p. 182*
Coordinate Rules for Reflections, *p. 183*
Postulate 4.2 Reflection Postulate, *p. 184*
Line Symmetry, *p. 185*

Section 4.3

Rotations, *p. 190*
Coordinate Rules for Rotations about the Origin, *p. 191*
Postulate 4.3 Rotation Postulate, *p. 192*
Rotational Symmetry, *p. 193*

Mathematical Practices

1. How could you determine whether your results make sense in Exercise 26 on page 179?
2. State the meaning of the numbers and symbols you chose in Exercise 28 on page 187.
3. Describe the steps you would take to arrive at the answer to Exercise 29 part (a) on page 195.

Study Skills

Keeping a Positive Attitude

Ever feel frustrated or overwhelmed by math? You're not alone. Just take a deep breath and assess the situation. Try to find a productive study environment, review your notes and examples in the textbook, and ask your teacher or peers for help.

4.1–4.3 Quiz

Graph quadrilateral $ABCD$ with vertices $A(-4, 1)$, $B(-3, 3)$, $C(0, 1)$, and $D(-2, 0)$ and its image after the translation. *(Section 4.1)*

1. $(x, y) \rightarrow (x + 4, y - 2)$

2. $(x, y) \rightarrow (x - 1, y - 5)$

3. $(x, y) \rightarrow (x + 3, y + 6)$

Graph the polygon with the given vertices and its image after a reflection in the given line. *(Section 4.2)*

4. $A(-5, 6)$, $B(-7, 8)$, $C(-3, 11)$; x-axis

5. $D(-5, -1)$, $E(-2, 1)$, $F(-1, -3)$; $y = x$

6. $J(-1, 4)$, $K(2, 5)$, $L(5, 2)$, $M(4, -1)$; $x = 3$

7. $P(2, -4)$, $Q(6, -1)$, $R(9, -4)$, $S(6, -6)$; $y = -2$

Graph $\triangle ABC$ with vertices $A(2, -1)$, $B(5, 2)$, and $C(8, -2)$ and its image after the glide reflection. *(Section 4.2)*

8. Translation: $(x, y) \rightarrow (x, y + 6)$
Reflection: in the y-axis

9. Translation: $(x, y) \rightarrow (x - 9, y)$
Reflection: in the line $y = 1$

Determine the number of lines of symmetry for the figure. *(Section 4.2)*

10.

11.

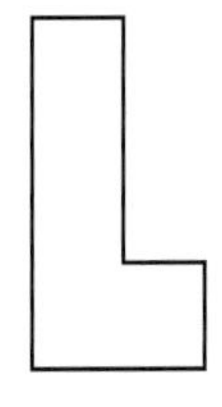

12.

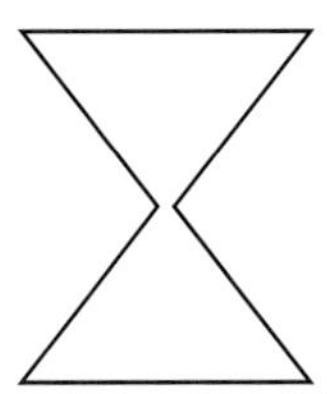

13.

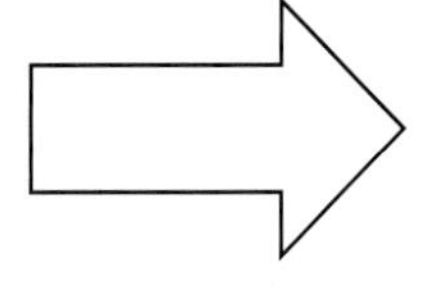

Graph the polygon and its image after a rotation of the given number of degrees about the origin. *(Section 4.3)*

14. 90°

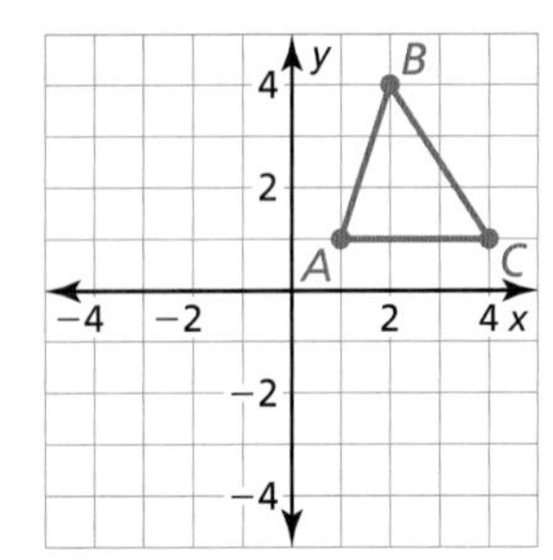

15. 270°

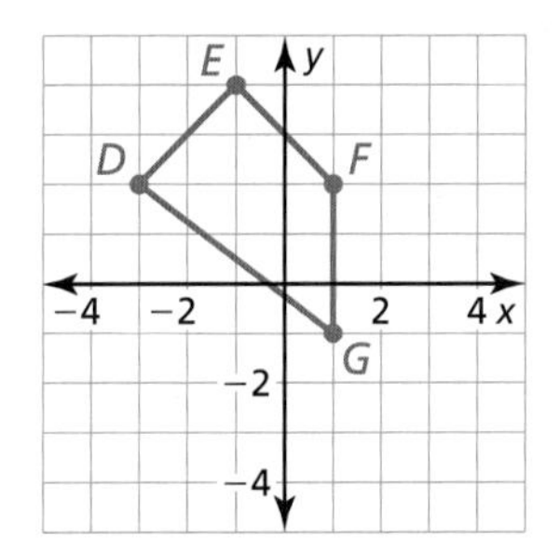

16. 180°

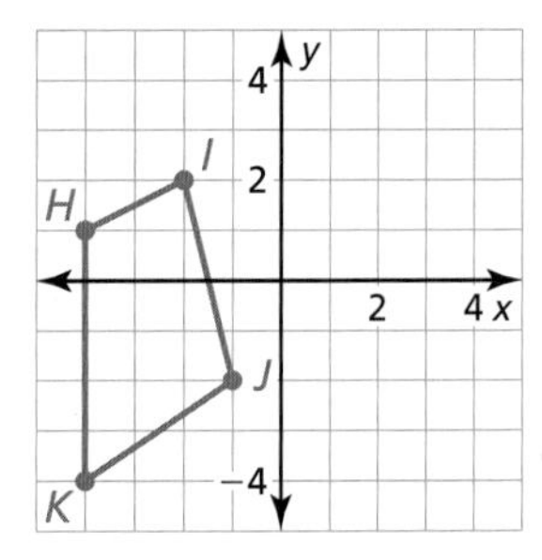

Graph $\triangle LMN$ with vertices $L(-3, -2)$, $M(-1, 1)$, and $N(2, -3)$ and its image after the composition. *(Sections 4.1–4.3)*

17. Translation: $(x, y) \rightarrow (x - 4, y + 3)$
Rotation: 180° about the origin

18. Rotation: 90° about the origin
Reflection: in the y-axis

19. The figure shows a game in which the object is to create solid rows using the pieces given. Using only translations and rotations, describe the transformations for each piece at the top that will form two solid rows at the bottom. *(Section 4.1 and Section 4.3)*

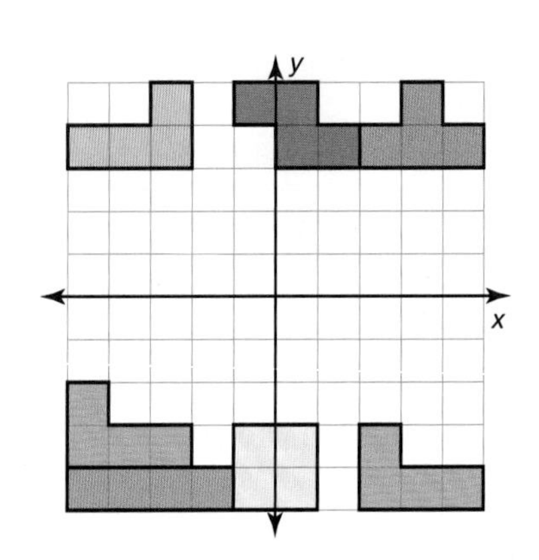

4.4 Congruence and Transformations

Essential Question What conjectures can you make about a figure reflected in two lines?

EXPLORATION 1 Reflections in Parallel Lines

Work with a partner. Use dynamic geometry software to draw any scalene triangle and label it $\triangle ABC$.

Sample

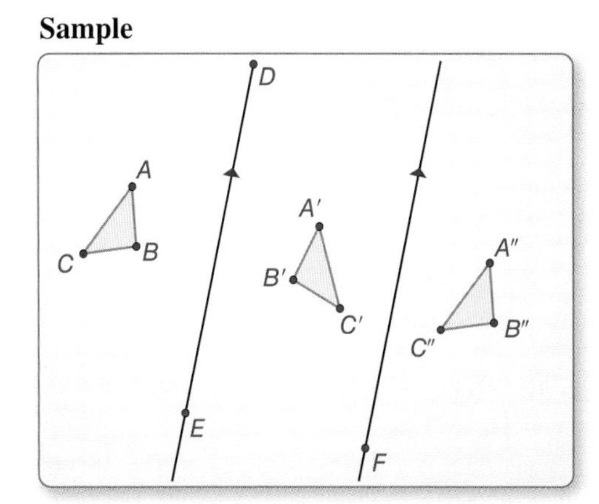

a. Draw any line $\overleftrightarrow{DE}$. Reflect $\triangle ABC$ in $\overleftrightarrow{DE}$ to form $\triangle A'B'C'$.

b. Draw a line parallel to $\overleftrightarrow{DE}$. Reflect $\triangle A'B'C'$ in the new line to form $\triangle A''B''C''$.

c. Draw the line through point A that is perpendicular to $\overleftrightarrow{DE}$. What do you notice?

d. Find the distance between points A and A''. Find the distance between the two parallel lines. What do you notice?

e. Hide $\triangle A'B'C'$. Is there a single transformation that maps $\triangle ABC$ to $\triangle A''B''C''$? Explain.

f. Make conjectures based on your answers in parts (c)–(e). Test your conjectures by changing $\triangle ABC$ and the parallel lines.

CONSTRUCTING VIABLE ARGUMENTS

To be proficient in math, you need to make conjectures and justify your conclusions.

EXPLORATION 2 Reflections in Intersecting Lines

Work with a partner. Use dynamic geometry software to draw any scalene triangle and label it $\triangle ABC$.

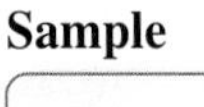

Sample

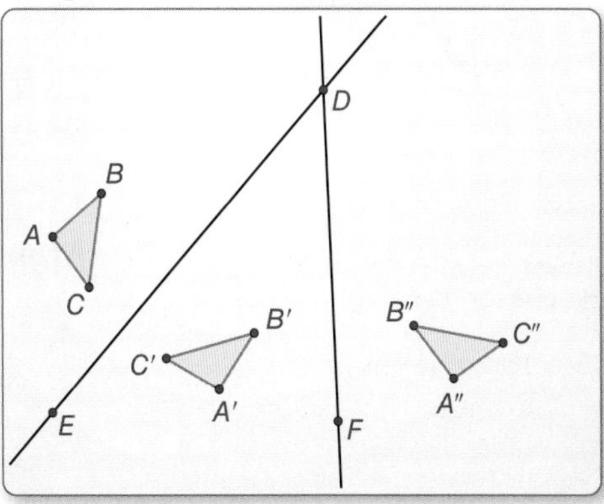

a. Draw any line $\overleftrightarrow{DE}$. Reflect $\triangle ABC$ in $\overleftrightarrow{DE}$ to form $\triangle A'B'C'$.

b. Draw any line $\overleftrightarrow{DF}$ so that angle EDF is less than or equal to 90°. Reflect $\triangle A'B'C'$ in $\overleftrightarrow{DF}$ to form $\triangle A''B''C''$.

c. Find the measure of $\angle EDF$. Rotate $\triangle ABC$ counterclockwise about point D using an angle twice the measure of $\angle EDF$.

d. Make a conjecture about a figure reflected in two intersecting lines. Test your conjecture by changing $\triangle ABC$ and the lines.

Communicate Your Answer

3. What conjectures can you make about a figure reflected in two lines?

4. Point Q is reflected in two parallel lines, $\overleftrightarrow{GH}$ and $\overleftrightarrow{JK}$, to form Q' and Q''. The distance from $\overleftrightarrow{GH}$ to $\overleftrightarrow{JK}$ is 3.2 inches. What is the distance QQ''?

4.4 Lesson

Core Vocabulary

congruent figures, *p. 200*
congruence transformation, *p. 201*

What You Will Learn

- Identify congruent figures.
- Describe congruence transformations.
- Use theorems about congruence transformations.

Identifying Congruent Figures

Two geometric figures are **congruent figures** if and only if there is a rigid motion or a composition of rigid motions that maps one of the figures onto the other. Congruent figures have the same size and shape.

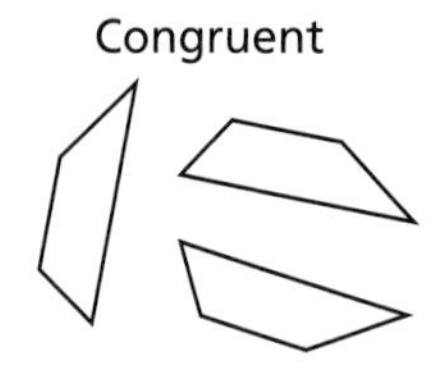

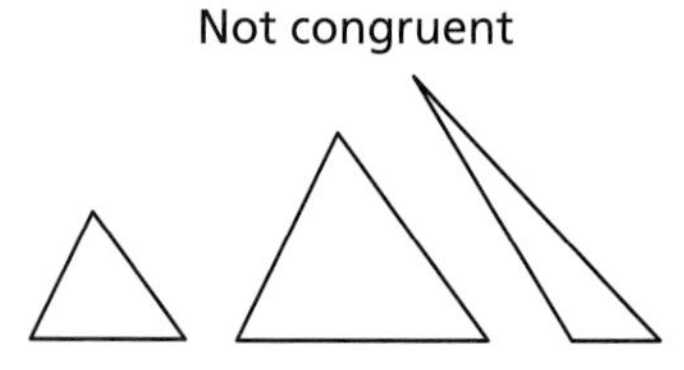

You can identify congruent figures in the coordinate plane by identifying the rigid motion or composition of rigid motions that maps one of the figures onto the other. Recall from Postulates 4.1–4.3 and Theorem 4.1 that translations, reflections, rotations, and compositions of these transformations are rigid motions.

EXAMPLE 1 Identifying Congruent Figures

Identify any congruent figures in the coordinate plane. Explain.

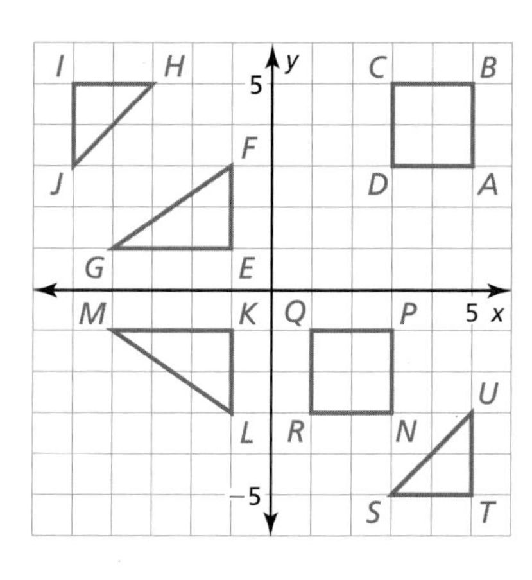

SOLUTION

Square *NPQR* is a translation of square *ABCD* 2 units left and 6 units down. So, square *ABCD* and square *NPQR* are congruent.

$\triangle KLM$ is a reflection of $\triangle EFG$ in the x-axis. So, $\triangle EFG$ and $\triangle KLM$ are congruent.

$\triangle STU$ is a 180° rotation of $\triangle HIJ$. So, $\triangle HIJ$ and $\triangle STU$ are congruent.

Monitoring Progress  Help in English and Spanish at *BigIdeasMath.com*

1. Identify any congruent figures in the coordinate plane. Explain.

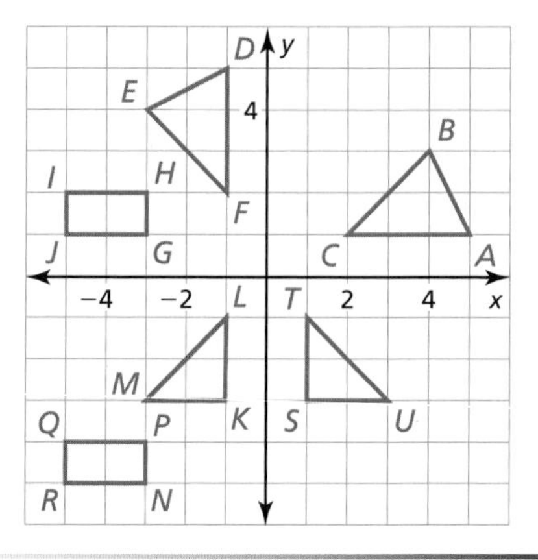

Congruence Transformations

Another name for a rigid motion or a combination of rigid motions is a **congruence transformation** because the preimage and image are congruent. The terms "rigid motion" and "congruence transformation" are interchangeable.

READING

You can read the notation ▱*ABCD* as "parallelogram *A, B, C, D*."

EXAMPLE 2 **Describing a Congruence Transformation**

Describe a congruence transformation that maps ▱*ABCD* to ▱*EFGH*.

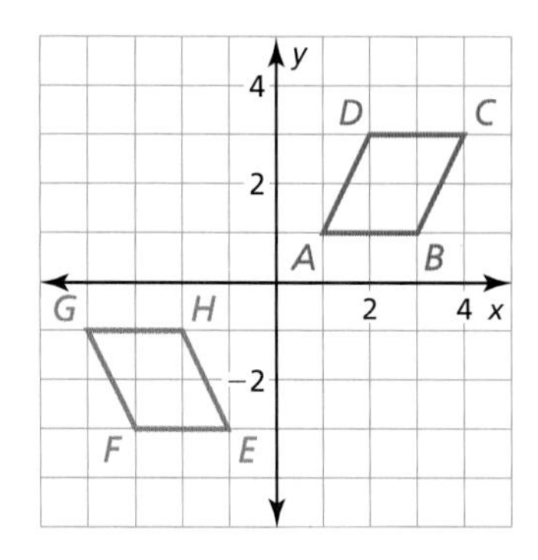

SOLUTION

The two vertical sides of ▱*ABCD* rise from left to right, and the two vertical sides of ▱*EFGH* fall from left to right. If you reflect ▱*ABCD* in the *y*-axis, as shown, then the image, ▱*A′B′C′D′*, will have the same orientation as ▱*EFGH*.

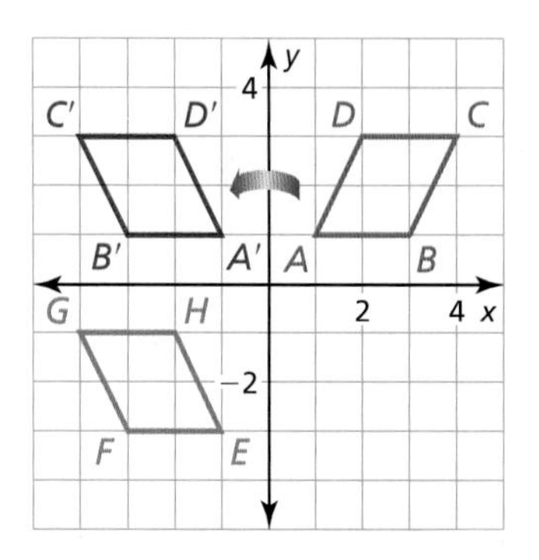

Then you can map ▱*A′B′C′D′* to ▱*EFGH* using a translation of 4 units down.

▶ So, a congruence transformation that maps ▱*ABCD* to ▱*EFGH* is a reflection in the *y*-axis followed by a translation of 4 units down.

Monitoring Progress Help in English and Spanish at *BigIdeasMath.com*

2. In Example 2, describe another congruence transformation that maps ▱*ABCD* to ▱*EFGH*.

3. Describe a congruence transformation that maps $\triangle JKL$ to $\triangle MNP$.

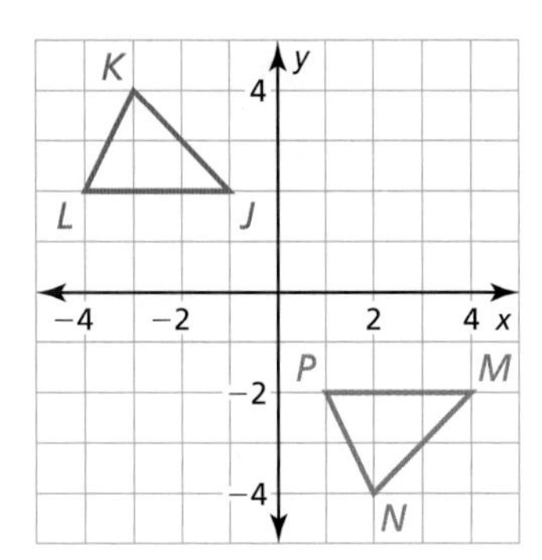

Using Theorems about Congruence Transformations

Compositions of two reflections result in either a translation or a rotation. A composition of two reflections in parallel lines results in a translation, as described in the following theorem.

Theorem

Theorem 4.2 Reflections in Parallel Lines Theorem

If lines k and m are parallel, then a reflection in line k followed by a reflection in line m is the same as a translation.

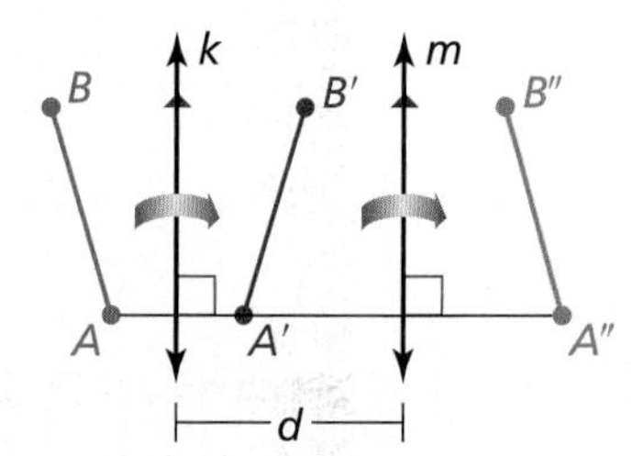

If A'' is the image of A, then

1. $\overline{AA''}$ is perpendicular to k and m, and
2. $AA'' = 2d$, where d is the distance between k and m.

Proof Ex. 31, p. 206

EXAMPLE 3 Using the Reflections in Parallel Lines Theorem

In the diagram, a reflection in line k maps $\overline{GH}$ to $\overline{G'H'}$. A reflection in line m maps $\overline{G'H'}$ to $\overline{G''H''}$. Also, $HB = 9$ and $DH'' = 4$.

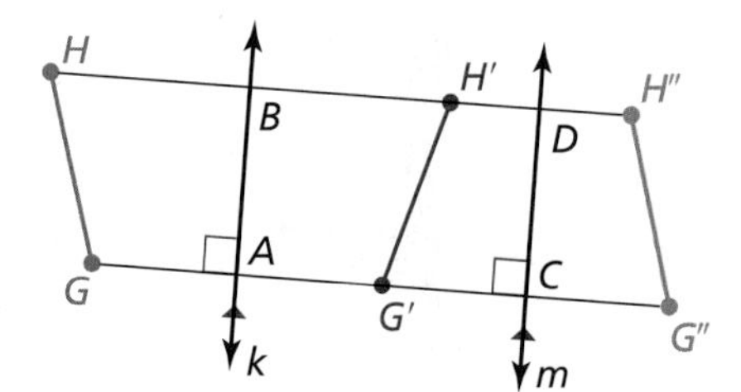

a. Name any segments congruent to each segment: $\overline{GH}$, $\overline{HB}$, and $\overline{GA}$.

b. Does $AC = BD$? Explain.

c. What is the length of $\overline{GG''}$?

SOLUTION

a. $\overline{GH} \cong \overline{G'H'}$, and $\overline{GH} \cong \overline{G''H''}$. $\overline{HB} \cong \overline{H'B}$. $\overline{GA} \cong \overline{G'A}$.

b. Yes, $AC = BD$ because $\overline{GG''}$ and $\overline{HH''}$ are perpendicular to both k and m. So, $\overline{BD}$ and $\overline{AC}$ are opposite sides of a rectangle.

c. By the properties of reflections, $H'B = 9$ and $H'D = 4$. The Reflections in Parallel Lines Theorem implies that $GG'' = HH'' = 2 \cdot BD$, so the length of $\overline{GG''}$ is $2(9 + 4) = 26$ units.

Monitoring Progress  Help in English and Spanish at *BigIdeasMath.com*

Use the figure. The distance between line k and line m is 1.6 centimeters.

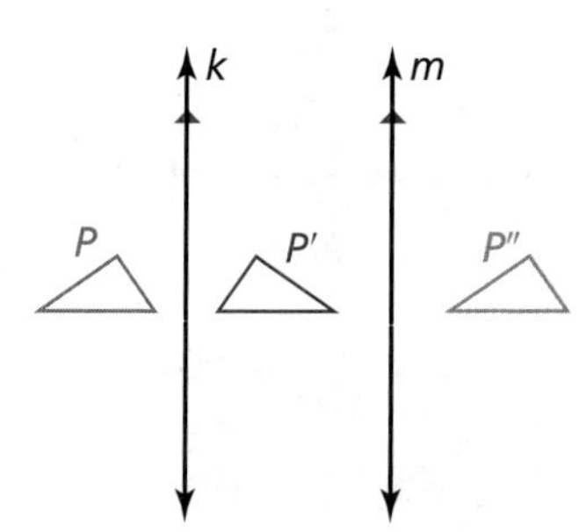

4. The preimage is reflected in line k, then in line m. Describe a single transformation that maps the blue figure to the green figure.
5. What is the relationship between $\overline{PP'}$ and line k? Explain.
6. What is the distance between P and P''?

A composition of two reflections in intersecting lines results in a rotation, as described in the following theorem.

Theorem

Theorem 4.3 Reflections in Intersecting Lines Theorem

If lines k and m intersect at point P, then a reflection in line k followed by a reflection in line m is the same as a rotation about point P.

The angle of rotation is $2x°$, where $x°$ is the measure of the acute or right angle formed by lines k and m.

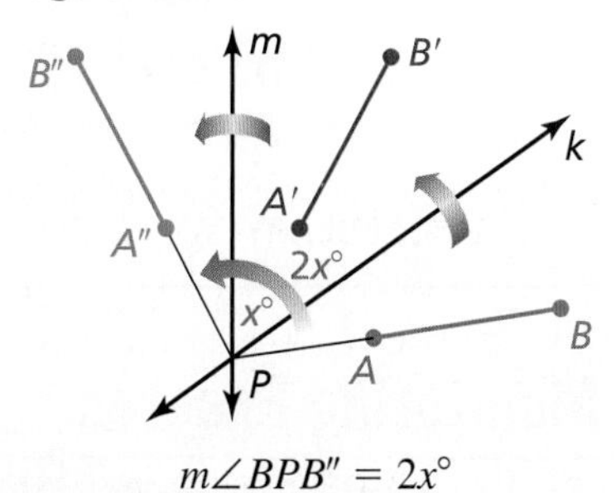

$m\angle BPB'' = 2x°$

Proof Ex. 31, p. 250

EXAMPLE 4 Using the Reflections in Intersecting Lines Theorem

In the diagram, the figure is reflected in line k. The image is then reflected in line m. Describe a single transformation that maps F to F''.

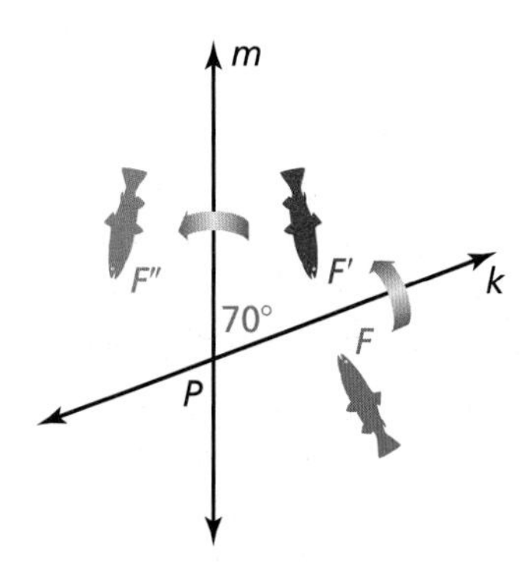

SOLUTION

By the Reflections in Intersecting Lines Theorem, a reflection in line k followed by a reflection in line m is the same as a rotation about point P. The measure of the acute angle formed between lines k and m is 70°. So, by the Reflections in Intersecting Lines Theorem, the angle of rotation is $2(70°) = 140°$. A single transformation that maps F to F'' is a 140° rotation about point P.

▶ You can check that this is correct by tracing lines k and m and point F, then rotating the point 140°.

Monitoring Progress  Help in English and Spanish at *BigIdeasMath.com*

7. In the diagram, the preimage is reflected in line k, then in line m. Describe a single transformation that maps the blue figure onto the green figure.

8. A rotation of 76° maps C to C'. To map C to C' using two reflections, what is the measure of the angle formed by the intersecting lines of reflection?

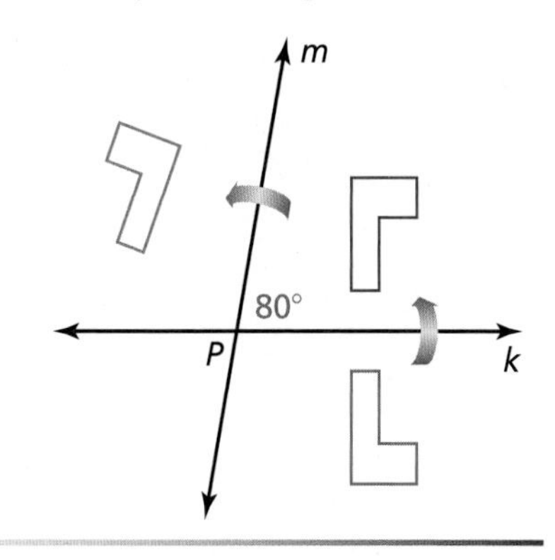

4.4 Exercises

Dynamic Solutions available at *BigIdeasMath.com*

Vocabulary and Core Concept Check

1. **COMPLETE THE SENTENCE** Two geometric figures are _________ if and only if there is a rigid motion or a composition of rigid motions that moves one of the figures onto the other.

2. **VOCABULARY** Why is the term *congruence transformation* used to refer to a rigid motion?

Monitoring Progress and Modeling with Mathematics

In Exercises 3 and 4, identify any congruent figures in the coordinate plane. Explain. *(See Example 1.)*

3.

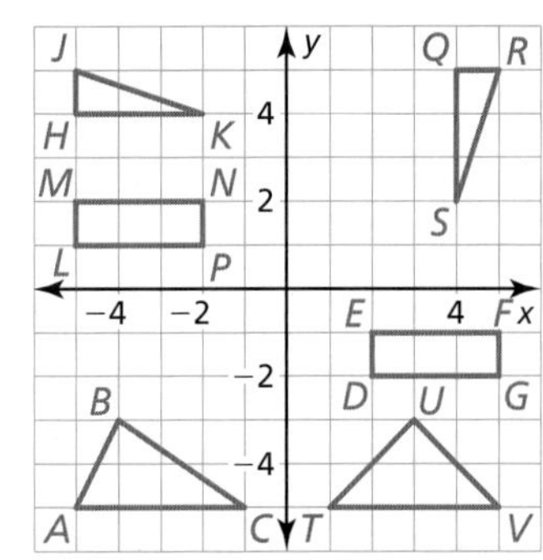

4.

In Exercises 5 and 6, describe a congruence transformation that maps the blue preimage to the green image. *(See Example 2.)*

5.

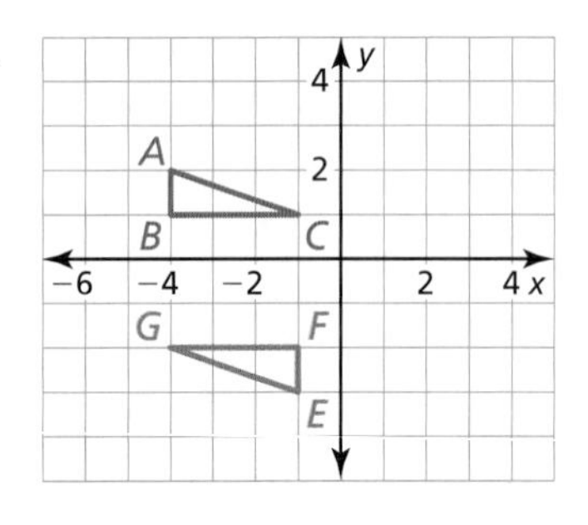

6.

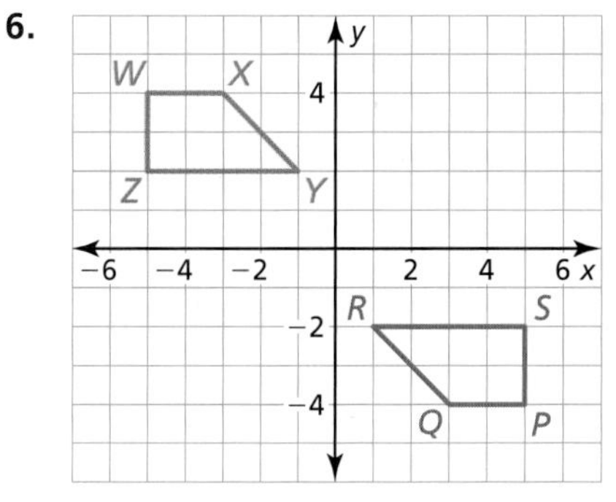

In Exercises 7–10, determine whether the polygons with the given vertices are congruent. Use transformations to explain your reasoning.

7. $Q(2, 4)$, $R(5, 4)$, $S(4, 1)$ and $T(6, 4)$, $U(9, 4)$, $V(8, 1)$

8. $W(-3, 1)$, $X(2, 1)$, $Y(4, -4)$, $Z(-5, -4)$ and $C(-1, -3)$, $D(-1, 2)$, $E(4, 4)$, $F(4, -5)$

9. $J(1, 1)$, $K(3, 2)$, $L(4, 1)$ and $M(6, 1)$, $N(5, 2)$, $P(2, 1)$

10. $A(0, 0)$, $B(1, 2)$, $C(4, 2)$, $D(3, 0)$ and $E(0, -5)$, $F(-1, -3)$, $G(-4, -3)$, $H(-3, -5)$

In Exercises 11–14, $k \parallel m$, $\triangle ABC$ is reflected in line k, and $\triangle A'B'C'$ is reflected in line m. *(See Example 3.)*

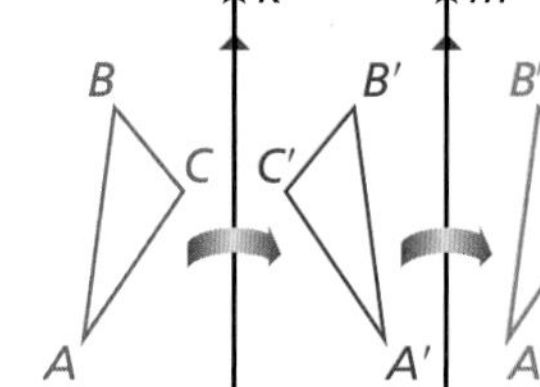

11. A translation maps $\triangle ABC$ onto which triangle?

12. Which lines are perpendicular to $\overline{AA''}$?

13. If the distance between k and m is 2.6 inches, what is the length of $\overline{CC''}$?

14. Is the distance from B' to m the same as the distance from B'' to m? Explain.

In Exercises 15 and 16, find the angle of rotation that maps A onto A''. *(See Example 4.)*

15.

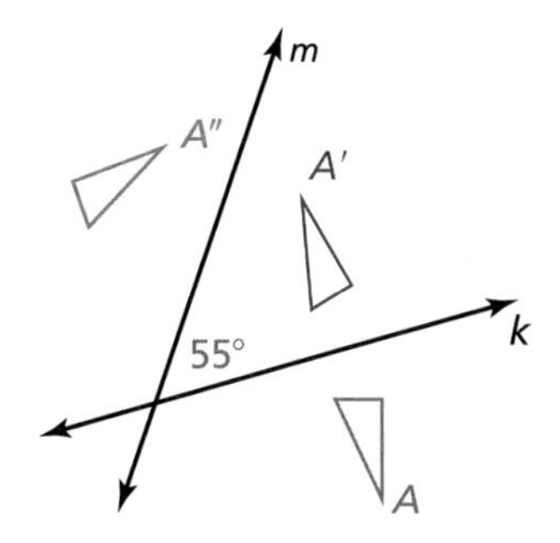

16.

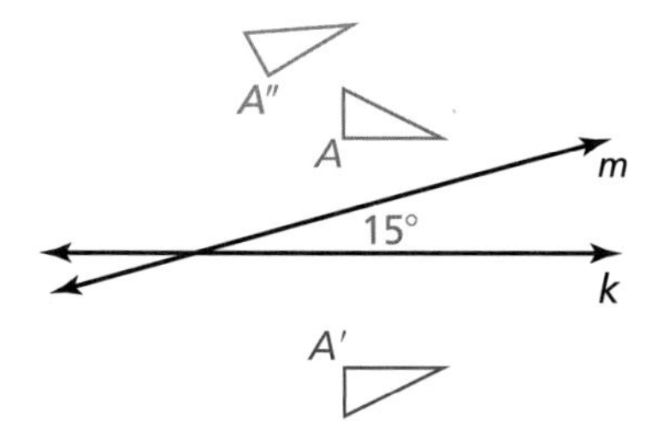

17. ERROR ANALYSIS Describe and correct the error in describing the congruence transformation.

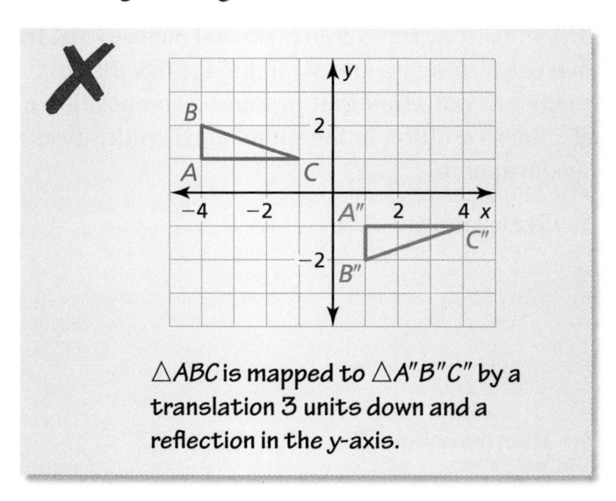

18. ERROR ANALYSIS Describe and correct the error in using the Reflections in Intersecting Lines Theorem (Theorem 4.3).

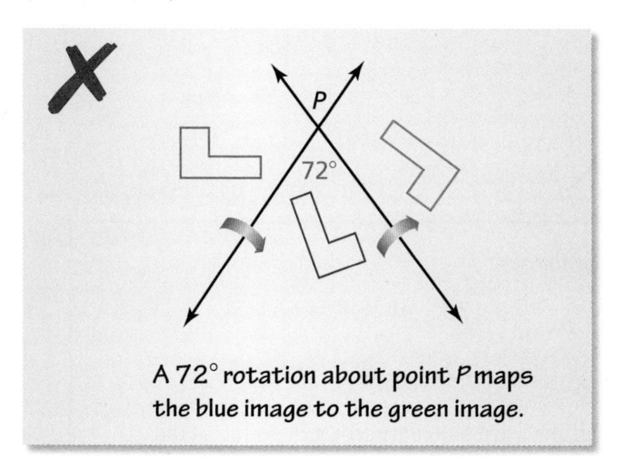

P
72°
A 72° rotation about point P maps the blue image to the green image.

In Exercises 19–22, find the measure of the acute or right angle formed by intersecting lines so that C can be mapped to C' using two reflections.

19. A rotation of 84° maps C to C'.

20. A rotation of 24° maps C to C'.

21. The rotation $(x, y) \rightarrow (-x, -y)$ maps C to C'.

22. The rotation $(x, y) \rightarrow (y, -x)$ maps C to C'.

23. REASONING Use the Reflections in Parallel Lines Theorem (Theorem 4.2) to explain how you can make a glide reflection using three reflections. How are the lines of reflection related?

24. DRAWING CONCLUSIONS The pattern shown is called a *tessellation*.

a. What transformations did the artist use when creating this tessellation?

b. Are the individual figures in the tessellation congruent? Explain your reasoning.

CRITICAL THINKING In Exercises 25–28, tell whether the statement is *always*, *sometimes*, or *never* true. Explain your reasoning.

25. A congruence transformation changes the size of a figure.

26. If two figures are congruent, then there is a rigid motion or a composition of rigid motions that maps one figure onto the other.

27. The composition of two reflections results in the same image as a rotation.

28. A translation results in the same image as the composition of two reflections.

29. REASONING During a presentation, a marketing representative uses a projector so everyone in the auditorium can view the advertisement. Is this projection a congruence transformation? Explain your reasoning.

30. **HOW DO YOU SEE IT?** What type of congruence transformation can be used to verify each statement about the stained glass window?

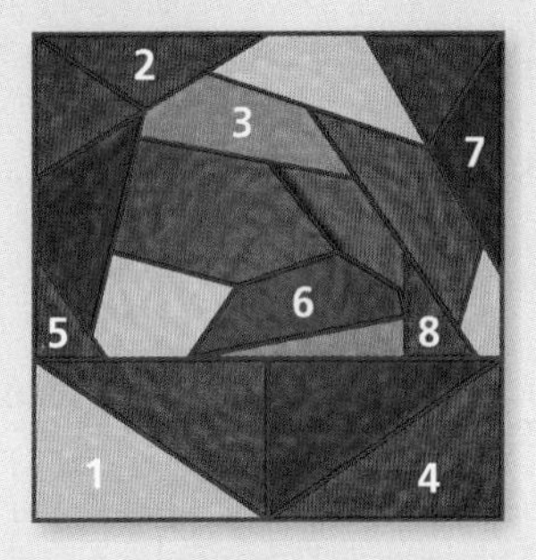

a. Triangle 5 is congruent to Triangle 8.

b. Triangle 1 is congruent to Triangle 4.

c. Triangle 2 is congruent to Triangle 7.

d. Pentagon 3 is congruent to Pentagon 6.

31. **PROVING A THEOREM** Prove the Reflections in Parallel Lines Theorem (Theorem 4.2).

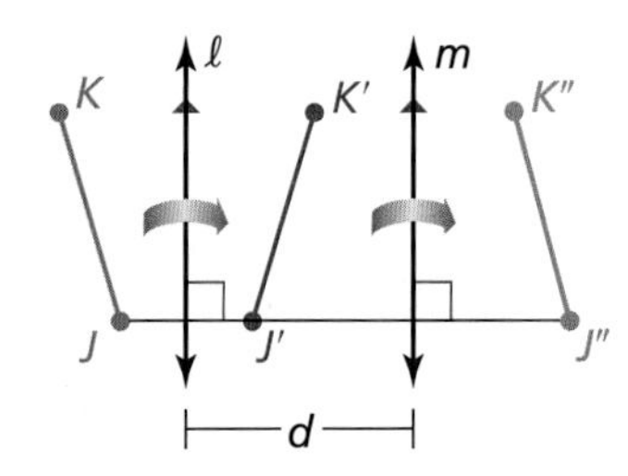

Given A reflection in line ℓ maps $\overline{JK}$ to $\overline{J'K'}$, a reflection in line m maps $\overline{J'K'}$ to $\overline{J''K''}$, and $\ell \parallel m$.

Prove **a.** $\overline{KK''}$ is perpendicular to ℓ and m.

b. $KK'' = 2d$, where d is the distance between ℓ and m.

32. **THOUGHT PROVOKING** A *tessellation* is the covering of a plane with congruent figures so that there are no gaps or overlaps (see Exercise 24). Draw a tessellation that involves two or more types of transformations. Describe the transformations that are used to create the tessellation.

33. **MAKING AN ARGUMENT** $\overline{PQ}$, with endpoints $P(1, 3)$ and $Q(3, 2)$, is reflected in the y-axis. The image $\overline{P'Q'}$ is then reflected in the x-axis to produce the image $\overline{P''Q''}$. One classmate says that $\overline{PQ}$ is mapped to $\overline{P''Q''}$ by the translation $(x, y) \rightarrow (x - 4, y - 5)$. Another classmate says that $\overline{PQ}$ is mapped to $\overline{P''Q''}$ by a $(2 \cdot 90)°$, or $180°$, rotation about the origin. Which classmate is correct? Explain your reasoning.

34. **CRITICAL THINKING** Does the order of reflections for a composition of two reflections in parallel lines matter? For example, is reflecting $\triangle XYZ$ in line ℓ and then its image in line m the same as reflecting $\triangle XYZ$ in line m and then its image in line ℓ?

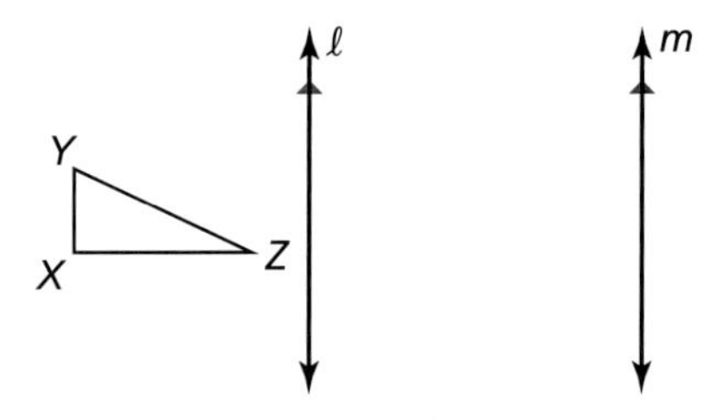

CONSTRUCTION In Exercises 35 and 36, copy the figure. Then use a compass and straightedge to construct two lines of reflection that produce a composition of reflections resulting in the same image as the given transformation.

35. **Translation:** $\triangle ABC \rightarrow \triangle A''B''C''$

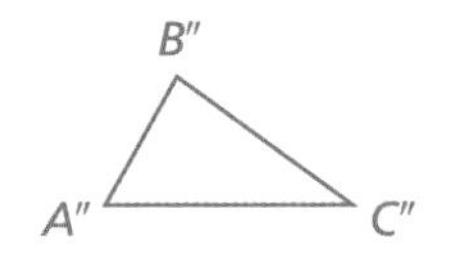

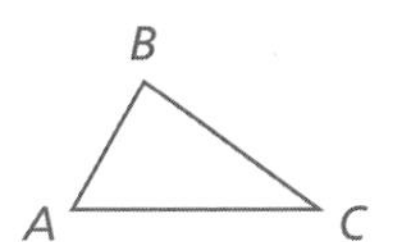

36. **Rotation about *P*:** $\triangle XYZ \rightarrow \triangle X''Y''Z''$

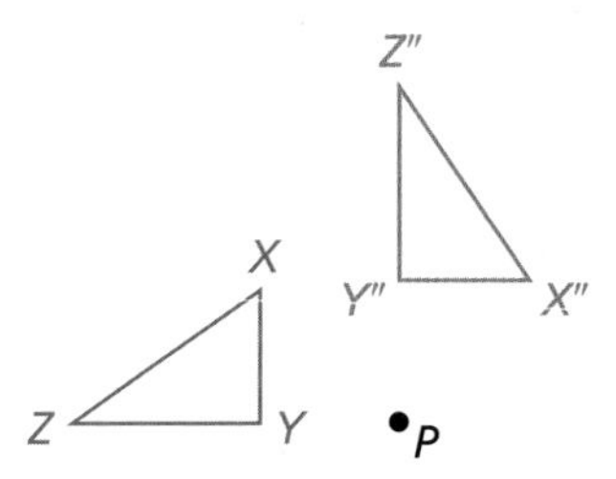

Maintaining Mathematical Proficiency

Reviewing what you learned in previous grades and lessons

Solve the equation. Check your solution. *(Skills Review Handbook)*

37. $5x + 16 = -3x$

38. $12 + 6m = 2m$

39. $4b + 8 = 6b - 4$

40. $7w - 9 = 13 - 4w$

41. $7(2n + 11) = 4n$

42. $-2(8 - y) = -6y$

43. Last year, the track team's yard sale earned \$500. This year, the yard sale earned \$625. What is the percent of increase? *(Skills Review Handbook)*

4.5 Dilations

Essential Question What does it mean to dilate a figure?

EXPLORATION 1 Dilating a Triangle in a Coordinate Plane

Work with a partner. Use dynamic geometry software to draw any triangle and label it $\triangle ABC$.

a. *Dilate* $\triangle ABC$ using a *scale factor* of 2 and a *center of dilation* at the origin to form $\triangle A'B'C'$. Compare the coordinates, side lengths, and angle measures of $\triangle ABC$ and $\triangle A'B'C'$.

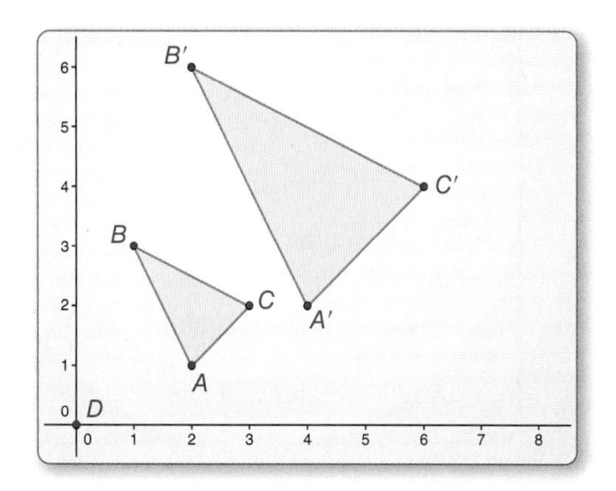

Sample

Points

$A(2, 1)$

$B(1, 3)$

$C(3, 2)$

Segments

$AB = 2.24$

$BC = 2.24$

$AC = 1.41$

Angles

$m\angle A = 71.57°$

$m\angle B = 36.87°$

$m\angle C = 71.57°$

LOOKING FOR STRUCTURE

To be proficient in math, you need to look closely to discern a pattern or structure.

b. Repeat part (a) using a *scale factor* of $\frac{1}{2}$.

c. What do the results of parts (a) and (b) suggest about the coordinates, side lengths, and angle measures of the image of $\triangle ABC$ after a dilation with a scale factor of k?

EXPLORATION 2 Dilating Lines in a Coordinate Plane

Work with a partner. Use dynamic geometry software to draw $\overleftrightarrow{AB}$ that passes through the origin and $\overleftrightarrow{AC}$ that does not pass through the origin.

a. *Dilate* $\overleftrightarrow{AB}$ using a *scale factor* of 3 and a *center of dilation* at the origin. Describe the image.

b. *Dilate* $\overleftrightarrow{AC}$ using a *scale factor* of 3 and a *center of dilation* at the origin. Describe the image.

c. Repeat parts (a) and (b) using a scale factor of $\frac{1}{4}$.

d. What do you notice about dilations of lines passing through the center of dilation and dilations of lines not passing through the center of dilation?

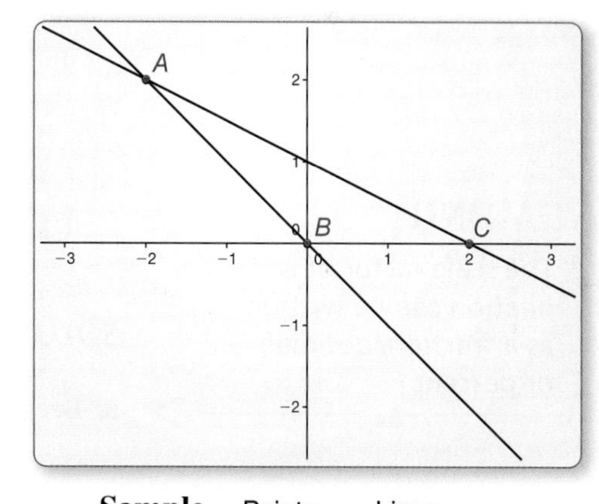

Sample	Points	Lines
	$A(-2, 2)$	$x + y = 0$
	$B(0, 0)$	$x + 2y = 2$
	$C(2, 0)$	

Communicate Your Answer

3. What does it mean to dilate a figure?

4. Repeat Exploration 1 using a center of dilation at a point other than the origin.

4.5 Lesson

Core Vocabulary

dilation, *p. 208*
center of dilation, *p. 208*
scale factor, *p. 208*
enlargement, *p. 208*
reduction, *p. 208*

What You Will Learn

- Identify and perform dilations.
- Solve real-life problems involving scale factors and dilations.

Identifying and Performing Dilations

Core Concept

Dilations

A **dilation** is a transformation in which a figure is enlarged or reduced with respect to a fixed point C called the **center of dilation** and a **scale factor** k, which is the ratio of the lengths of the corresponding sides of the image and the preimage.

A dilation with center of dilation C and scale factor k maps every point P in a figure to a point P' so that the following are true.

- If P is the center point C, then $P = P'$.
- If P is not the center point C, then the image point P' lies on $\overrightarrow{CP}$. The scale factor k is a positive number such that $k = \dfrac{CP'}{CP}$.
- Angle measures are preserved.

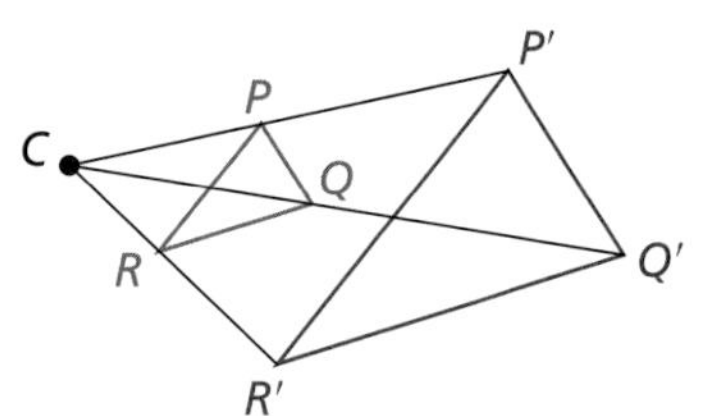

A dilation does not change any line that passes through the center of dilation. A dilation maps a line that does not pass through the center of dilation to a parallel line. In the figure above, $\overleftrightarrow{PR} \parallel \overleftrightarrow{P'R'}$, $\overleftrightarrow{PQ} \parallel \overleftrightarrow{P'Q'}$, and $\overleftrightarrow{QR} \parallel \overleftrightarrow{Q'R'}$.

When the scale factor $k > 1$, a dilation is an **enlargement**. When $0 < k < 1$, a dilation is a **reduction**.

EXAMPLE 1 Identifying Dilations

Find the scale factor of the dilation. Then tell whether the dilation is a *reduction* or an *enlargement*.

a.

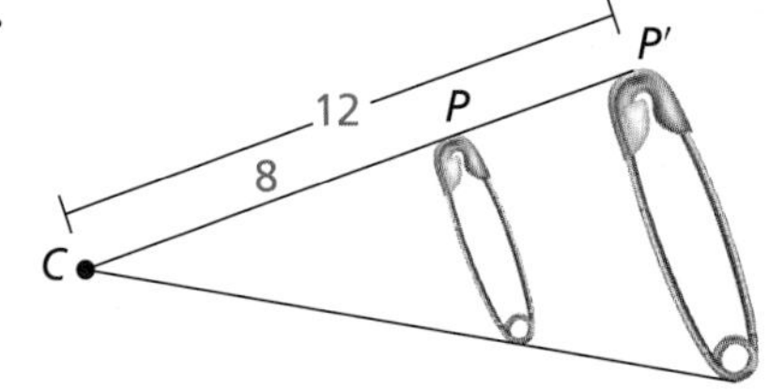

b.

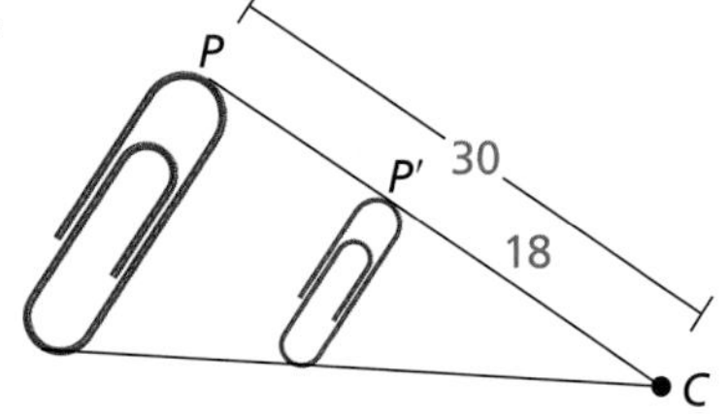

READING

The scale factor of a dilation can be written as a fraction, decimal, or percent.

SOLUTION

a. Because $\dfrac{CP'}{CP} = \dfrac{12}{8}$, the scale factor is $k = \dfrac{3}{2}$. So, the dilation is an enlargement.

b. Because $\dfrac{CP'}{CP} = \dfrac{18}{30}$, the scale factor is $k = \dfrac{3}{5}$. So, the dilation is a reduction.

Monitoring Progress  Help in English and Spanish at *BigIdeasMath.com*

1. In a dilation, $CP' = 3$ and $CP = 12$. Find the scale factor. Then tell whether the dilation is a *reduction* or an *enlargement*.

Core Concept

Coordinate Rule for Dilations

If $P(x, y)$ is the preimage of a point, then its image after a dilation centered at the origin $(0, 0)$ with scale factor k is the point $P'(kx, ky)$.

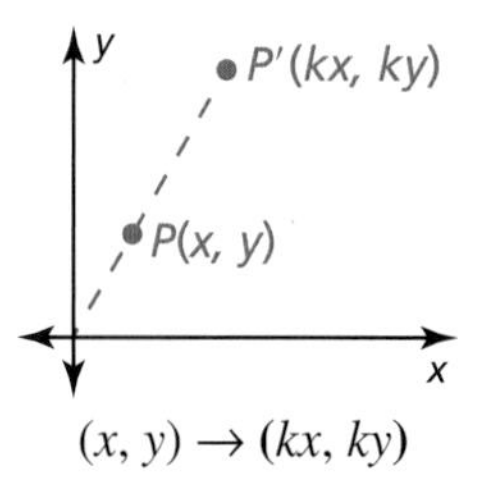

$(x, y) \rightarrow (kx, ky)$

READING DIAGRAMS

In this chapter, for all of the dilations in the coordinate plane, the center of dilation is the origin unless otherwise noted.

EXAMPLE 2 Dilating a Figure in the Coordinate Plane

Graph $\triangle ABC$ with vertices $A(2, 1)$, $B(4, 1)$, and $C(4, -1)$ and its image after a dilation with a scale factor of 2.

SOLUTION

Use the coordinate rule for a dilation with $k = 2$ to find the coordinates of the vertices of the image. Then graph $\triangle ABC$ and its image.

$$(x, y) \rightarrow (2x, 2y)$$

$$A(2, 1) \rightarrow A'(4, 2)$$

$$B(4, 1) \rightarrow B'(8, 2)$$

$$C(4, -1) \rightarrow C'(8, -2)$$

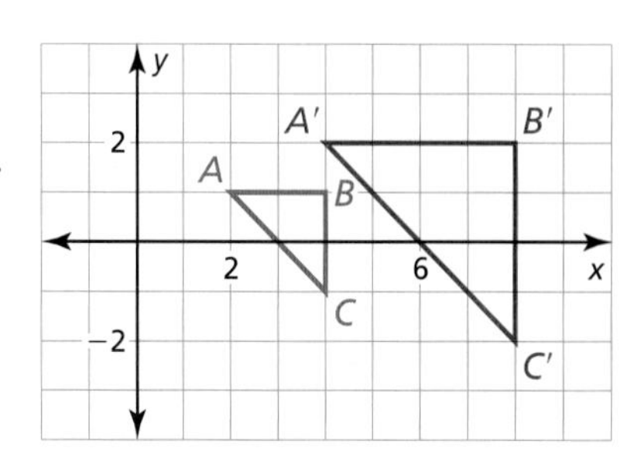

Notice the relationships between the lengths and slopes of the sides of the triangles in Example 2. Each side length of $\triangle A'B'C'$ is longer than its corresponding side by the scale factor. The corresponding sides are parallel because their slopes are the same.

EXAMPLE 3 Dilating a Figure in the Coordinate Plane

Graph quadrilateral $KLMN$ with vertices $K(-3, 6)$, $L(0, 6)$, $M(3, 3)$, and $N(-3, -3)$ and its image after a dilation with a scale factor of $\frac{1}{3}$.

SOLUTION

Use the coordinate rule for a dilation with $k = \frac{1}{3}$ to find the coordinates of the vertices of the image. Then graph quadrilateral $KLMN$ and its image.

$$(x, y) \rightarrow \left(\tfrac{1}{3}x, \tfrac{1}{3}y\right)$$

$$K(-3, 6) \rightarrow K'(-1, 2)$$

$$L(0, 6) \rightarrow L'(0, 2)$$

$$M(3, 3) \rightarrow M'(1, 1)$$

$$N(-3, -3) \rightarrow N'(-1, -1)$$

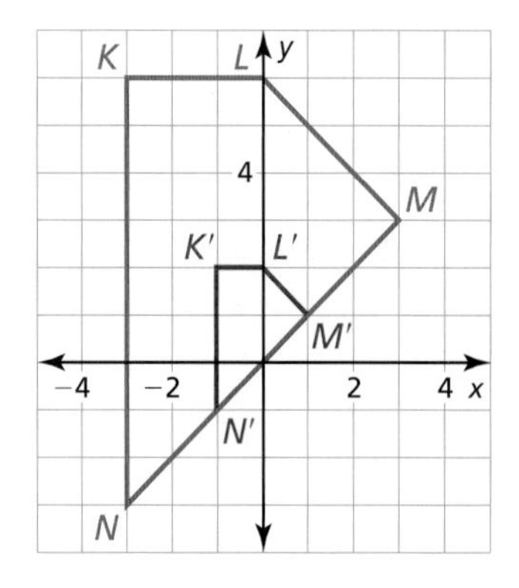

Monitoring Progress

Help in English and Spanish at *BigIdeasMath.com*

Graph $\triangle PQR$ and its image after a dilation with scale factor k.

2. $P(-2, -1)$, $Q(-1, 0)$, $R(0, -1)$; $k = 4$

3. $P(5, -5)$, $Q(10, -5)$, $R(10, 5)$; $k = 0.4$

CONSTRUCTION Constructing a Dilation

Use a compass and straightedge to construct a dilation of $\triangle PQR$ with a scale factor of 2. Use a point C outside the triangle as the center of dilation.

SOLUTION

Step 1

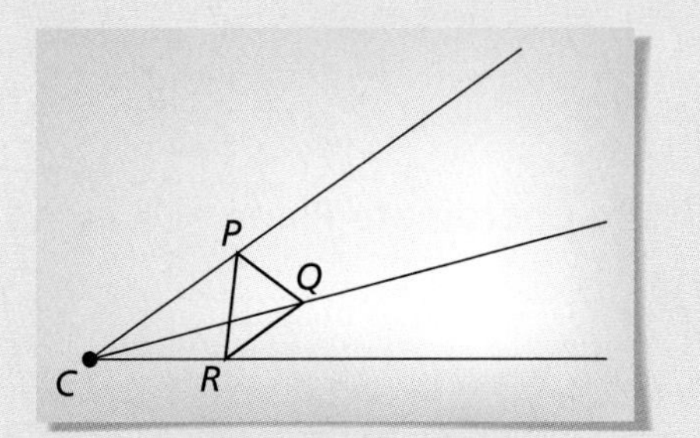

Draw a triangle Draw $\triangle PQR$ and choose the center of the dilation C outside the triangle. Draw rays from C through the vertices of the triangle.

Step 2

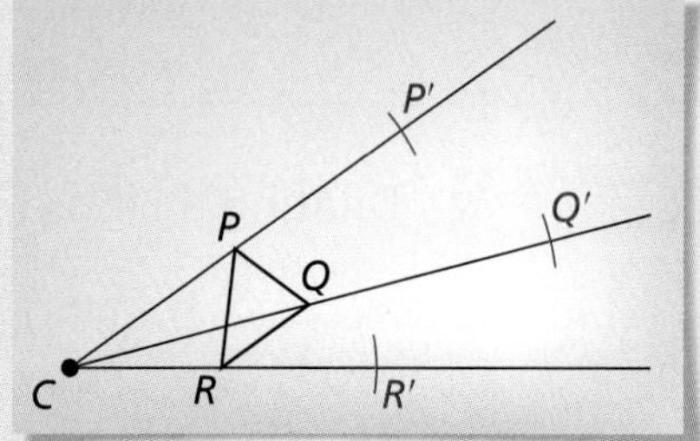

Use a compass Use a compass to locate P' on $\overrightarrow{CP}$ so that $CP' = 2(CP)$. Locate Q' and R' using the same method.

Step 3

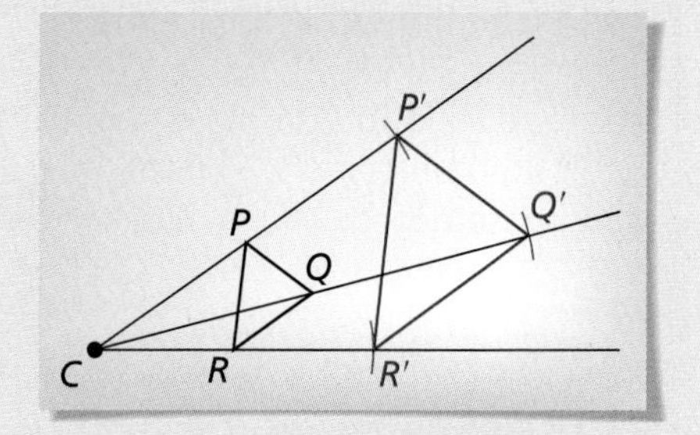

Connect points Connect points P', Q', and R' to form $\triangle P'Q'R'$.

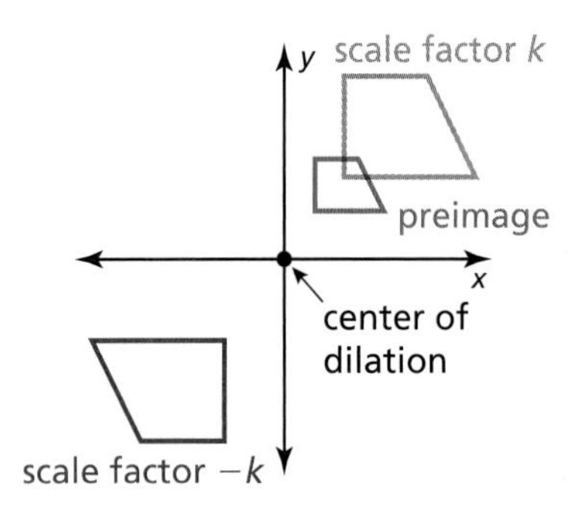

In the coordinate plane, you can have scale factors that are negative numbers. When this occurs, the figure rotates 180°. So, when $k > 0$, a dilation with a scale factor of $-k$ is the same as the composition of a dilation with a scale factor of k followed by a rotation of 180° about the center of dilation. Using the coordinate rules for a dilation and a rotation of 180°, you can think of the notation as

$$(x, y) \rightarrow (kx, ky) \rightarrow (-kx, -ky).$$

EXAMPLE 4 Using a Negative Scale Factor

Graph $\triangle FGH$ with vertices $F(-4, -2)$, $G(-2, 4)$, and $H(-2, -2)$ and its image after a dilation with a scale factor of $-\frac{1}{2}$.

SOLUTION

Use the coordinate rule for a dilation with $k = -\frac{1}{2}$ to find the coordinates of the vertices of the image. Then graph $\triangle FGH$ and its image.

$$(x, y) \rightarrow \left(-\tfrac{1}{2}x, -\tfrac{1}{2}y\right)$$

$F(-4, -2) \rightarrow F'(2, 1)$

$G(-2, 4) \rightarrow G'(1, -2)$

$H(-2, -2) \rightarrow H'(1, 1)$

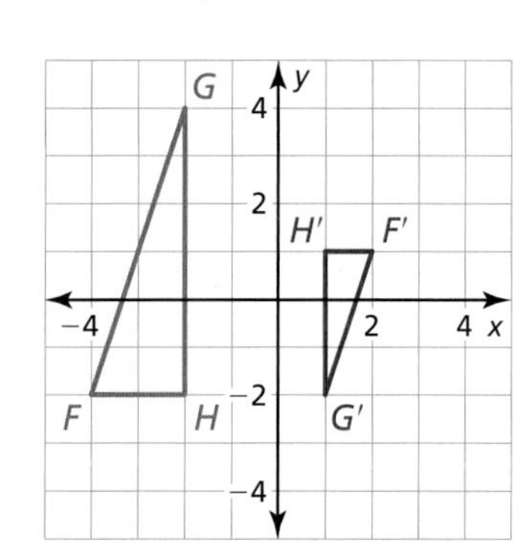

Monitoring Progress  Help in English and Spanish at *BigIdeasMath.com*

4. Graph $\triangle PQR$ with vertices $P(1, 2)$, $Q(3, 1)$, and $R(1, -3)$ and its image after a dilation with a scale factor of -2.

5. Suppose a figure containing the origin is dilated. Explain why the corresponding point in the image of the figure is also the origin.

Solving Real-Life Problems

EXAMPLE 5 Finding a Scale Factor

You are making your own photo stickers. Your photo is 4 inches by 4 inches. The image on the stickers is 1.1 inches by 1.1 inches. What is the scale factor of this dilation?

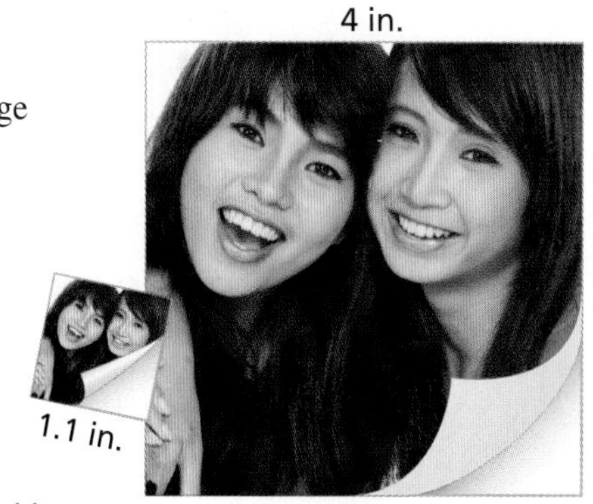

READING

Scale factors are written so that the units in the numerator and denominator divide out.

SOLUTION

The scale factor is the ratio of a side length of the sticker image to a side length of the original photo, or $\frac{1.1 \text{ in.}}{4 \text{ in.}}$.

▶ So, in simplest form, the scale factor is $\frac{11}{40}$.

EXAMPLE 6 Finding the Length of an Image

You are using a magnifying glass that shows the image of an object that is six times the object's actual size. Determine the length of the image of the spider seen through the magnifying glass.

SOLUTION

$$\frac{\text{image length}}{\text{actual length}} = k$$

$$\frac{x}{1.5} = 6$$

$$x = 9$$

▶ So, the image length through the magnifying glass is 9 centimeters.

Monitoring Progress  Help in English and Spanish at *BigIdeasMath.com*

6. An optometrist dilates the pupils of a patient's eyes to get a better look at the back of the eyes. A pupil dilates from 4.5 millimeters to 8 millimeters. What is the scale factor of this dilation?
7. The image of a spider seen through the magnifying glass in Example 6 is shown at the left. Find the actual length of the spider.

When a transformation, such as a dilation, changes the shape or size of a figure, the transformation is *nonrigid*. In addition to dilations, there are many possible nonrigid transformations. Two examples are shown below. It is important to pay close attention to whether a nonrigid transformation preserves lengths and angle measures.

Horizontal Stretch

A

C B B′

Vertical Stretch

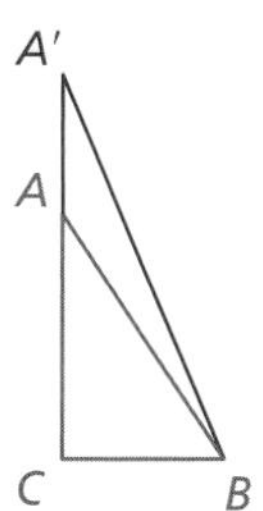

4.5 Exercises

Dynamic Solutions available at *BigIdeasMath.com*

Vocabulary and Core Concept Check

1. **COMPLETE THE SENTENCE** If $P(x, y)$ is the preimage of a point, then its image after a dilation centered at the origin $(0, 0)$ with scale factor k is the point __________.

2. **WHICH ONE DOESN'T BELONG?** Which scale factor does *not* belong with the other three? Explain your reasoning.

$\frac{5}{4}$	60%	115%	2

Monitoring Progress and Modeling with Mathematics

In Exercises 3–6, find the scale factor of the dilation. Then tell whether the dilation is a *reduction* or an *enlargement*. (See Example 1.)

3.

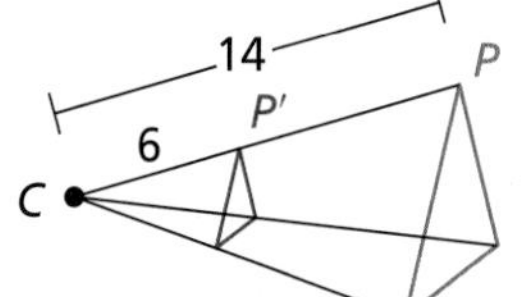

4.

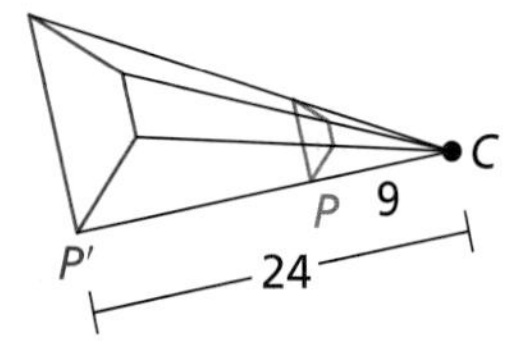

5.

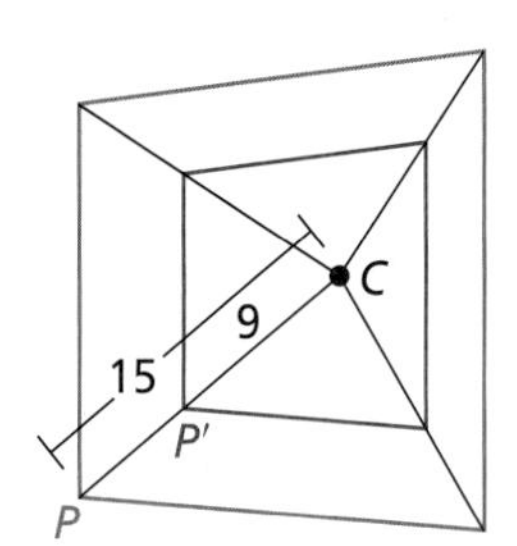

6.

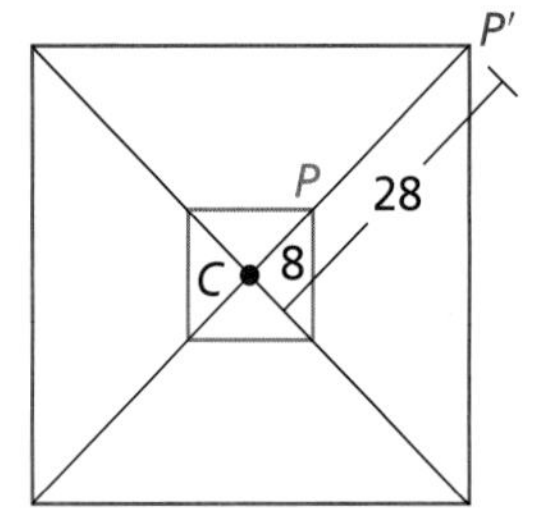

CONSTRUCTION In Exercises 7–10, copy the diagram. Then use a compass and straightedge to construct a dilation of $\triangle LMN$ with the given center and scale factor k.

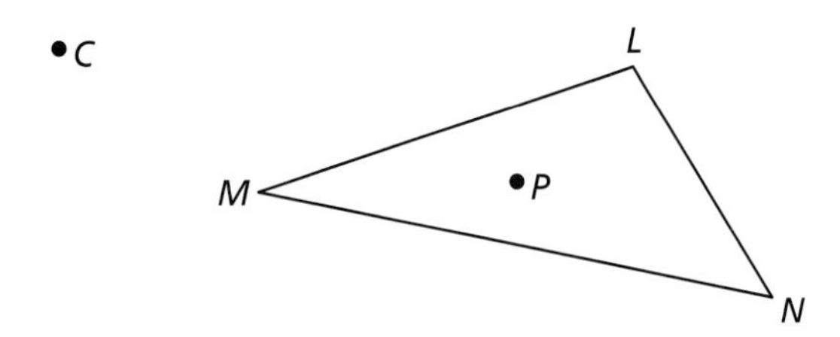

7. Center C, $k = 2$

8. Center P, $k = 3$

9. Center M, $k = \frac{1}{2}$

10. Center C, $k = 25\%$

CONSTRUCTION In Exercises 11–14, copy the diagram. Then use a compass and straightedge to construct a dilation of quadrilateral $RSTU$ with the given center and scale factor k.

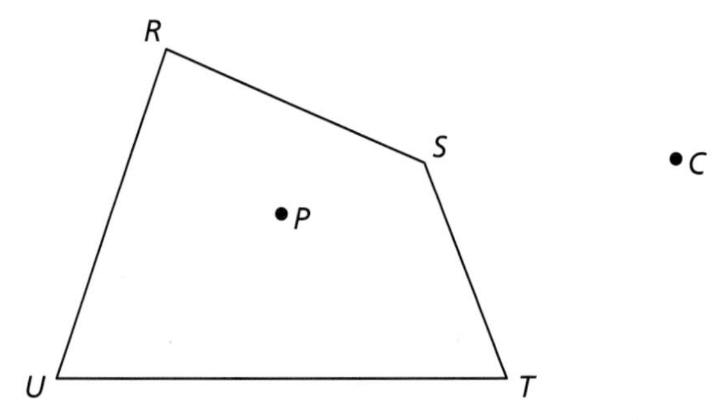

11. Center C, $k = 3$

12. Center P, $k = \frac{1}{3}$

13. Center R, $k = 0.25$

14. Center C, $k = 75\%$

In Exercises 15–18, graph the polygon and its image after a dilation with scale factor k. (See Examples 2 and 3.)

15. $X(6, -1)$, $Y(-2, -4)$, $Z(1, 2)$; $k = 3$

16. $A(0, 5)$, $B(-10, -5)$, $C(5, -5)$; $k = 120\%$

17. $T(9, -3)$, $U(6, 0)$, $V(3, 9)$, $W(0, 0)$; $k = \frac{2}{3}$

18. $J(4, 0)$, $K(-8, 4)$, $L(0, -4)$, $M(12, -8)$; $k = 0.25$

In Exercises 19–22, graph the polygon and its image after a dilation with scale factor k. (See Example 4.)

19. $B(-5, -10)$, $C(-10, 15)$, $D(0, 5)$; $k = -\frac{1}{5}$

20. $L(0, 0)$, $M(-4, 1)$, $N(-3, -6)$; $k = -3$

21. $R(-7, -1)$, $S(2, 5)$, $T(-2, -3)$, $U(-3, -3)$; $k = -4$

22. $W(8, -2)$, $X(6, 0)$, $Y(-6, 4)$, $Z(-2, 2)$; $k = -0.5$

ERROR ANALYSIS In Exercises 23 and 24, describe and correct the error in finding the scale factor of the dilation.

23.

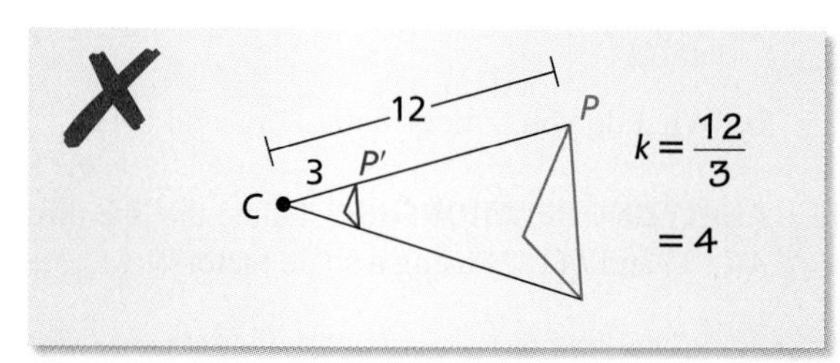

24.

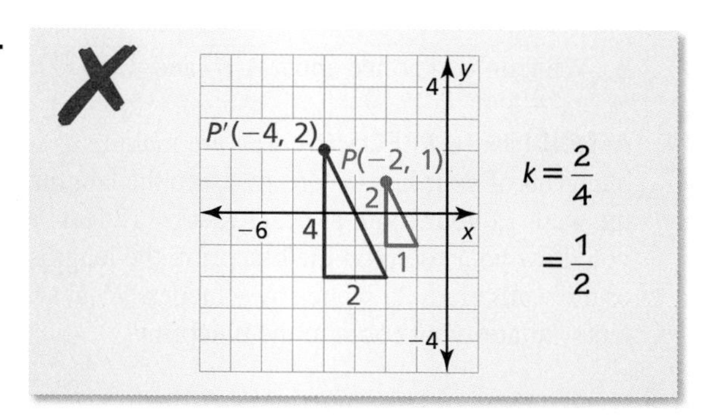

In Exercises 25–28, the red figure is the image of the blue figure after a dilation with center *C*. Find the scale factor of the dilation. Then find the value of the variable.

25.

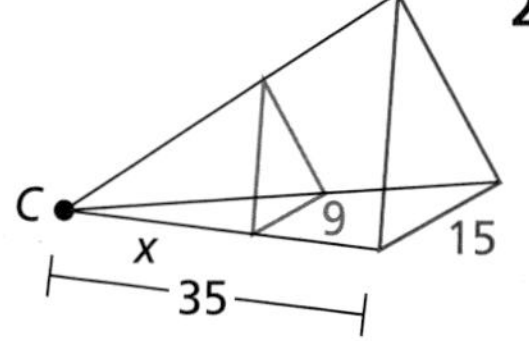

26.

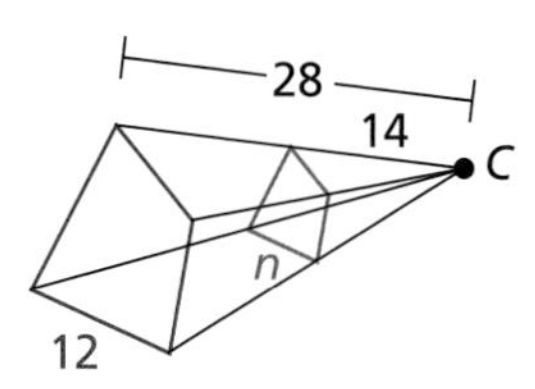

27.

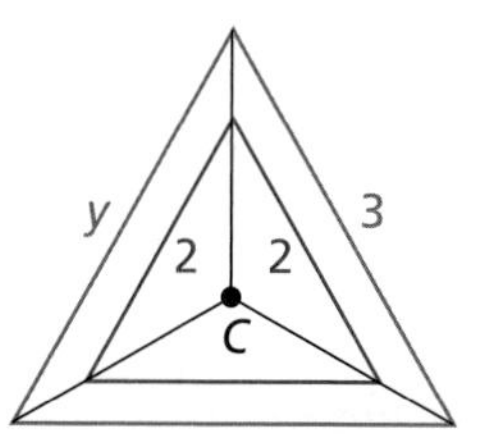

28.

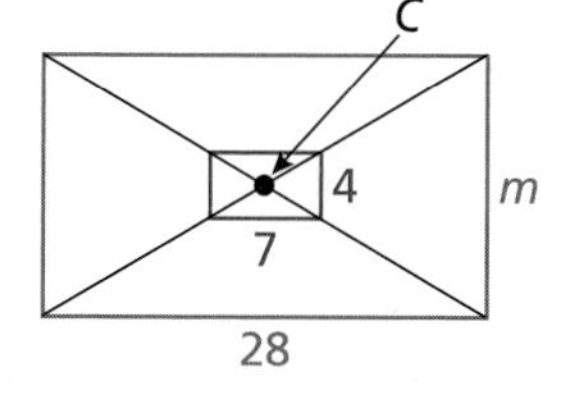

29. FINDING A SCALE FACTOR You receive wallet-sized photos of your school picture. The photo is 2.5 inches by 3.5 inches. You decide to dilate the photo to 5 inches by 7 inches at the store. What is the scale factor of this dilation? *(See Example 5.)*

30. FINDING A SCALE FACTOR Your visually impaired friend asked you to enlarge your notes from class so he can study. You took notes on 8.5-inch by 11-inch paper. The enlarged copy has a smaller side with a length of 10 inches. What is the scale factor of this dilation? *(See Example 5.)*

In Exercises 31–34, you are using a magnifying glass. Use the length of the insect and the magnification level to determine the length of the image seen through the magnifying glass. *(See Example 6.)*

31. emperor moth
Magnification: 5×

32. ladybug
Magnification: 10×

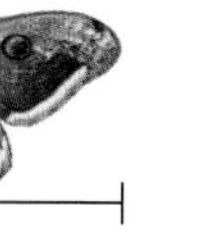

33. dragonfly
Magnification: 20×

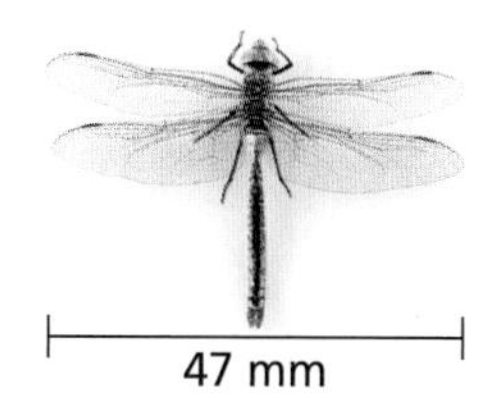

34. carpenter ant
Magnification: 15×

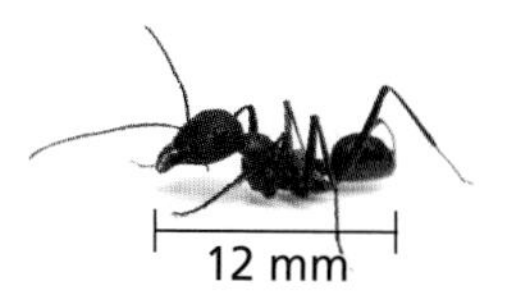

35. ANALYZING RELATIONSHIPS Use the given actual and magnified lengths to determine which of the following insects were looked at using the same magnifying glass. Explain your reasoning.

grasshopper
Actual: 2 in.
Magnified: 15 in.

black beetle
Actual: 0.6 in.
Magnified: 4.2 in.

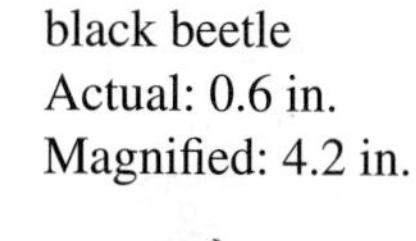

honeybee
Actual: $\frac{5}{8}$ in.
Magnified: $\frac{75}{16}$ in.

monarch butterfly
Actual: 3.9 in.
Magnified: 29.25 in.

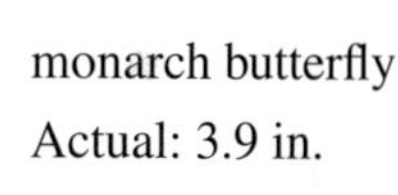

36. THOUGHT PROVOKING Draw $\triangle ABC$ and $\triangle A'B'C'$ so that $\triangle A'B'C'$ is a dilation of $\triangle ABC$. Find the center of dilation and explain how you found it.

37. REASONING Your friend prints a 4-inch by 6-inch photo for you from the school dance. All you have is an 8-inch by 10-inch frame. Can you dilate the photo to fit the frame? Explain your reasoning.

38. HOW DO YOU SEE IT? Point C is the center of dilation of the images. The scale factor is $\frac{1}{3}$. Which figure is the original figure? Which figure is the dilated figure? Explain your reasoning.

39. MATHEMATICAL CONNECTIONS The larger triangle is a dilation of the smaller triangle. Find the values of x and y.

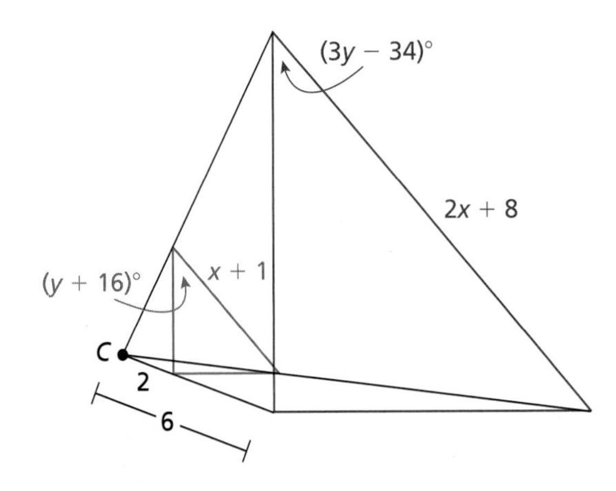

40. WRITING Explain why a scale factor of 2 is the same as 200%.

In Exercises 41–44, determine whether the dilated figure or the original figure is closer to the center of dilation. Use the given location of the center of dilation and scale factor *k*.

41. Center of dilation: inside the figure; $k = 3$

42. Center of dilation: inside the figure; $k = \frac{1}{2}$

43. Center of dilation: outside the figure; $k = 120\%$

44. Center of dilation: outside the figure; $k = 0.1$

45. ANALYZING RELATIONSHIPS Dilate the line through $O(0, 0)$ and $A(1, 2)$ using a scale factor of 2.

a. What do you notice about the lengths of $\overline{O'A'}$ and $\overline{OA}$?

b. What do you notice about $\overleftrightarrow{O'A'}$ and $\overleftrightarrow{OA}$?

46. ANALYZING RELATIONSHIPS Dilate the line through $A(0, 1)$ and $B(1, 2)$ using a scale factor of $\frac{1}{2}$.

a. What do you notice about the lengths of $\overline{A'B'}$ and $\overline{AB}$?

b. What do you notice about $\overleftrightarrow{A'B'}$ and $\overleftrightarrow{AB}$?

47. ATTENDING TO PRECISION You are making a blueprint of your house. You measure the lengths of the walls of your room to be 11 feet by 12 feet. When you draw your room on the blueprint, the lengths of the walls are 8.25 inches by 9 inches. What scale factor dilates your room to the blueprint?

48. MAKING AN ARGUMENT Your friend claims that dilating a figure by 1 is the same as dilating a figure by -1 because the original figure will not be enlarged or reduced. Is your friend correct? Explain your reasoning.

49. USING STRUCTURE Rectangle $WXYZ$ has vertices $W(-3, -1)$, $X(-3, 3)$, $Y(5, 3)$, and $Z(5, -1)$.

a. Find the perimeter and area of the rectangle.

b. Dilate the rectangle using a scale factor of 3. Find the perimeter and area of the dilated rectangle. Compare with the original rectangle. What do you notice?

c. Repeat part (b) using a scale factor of $\frac{1}{4}$.

d. Make a conjecture for how the perimeter and area change when a figure is dilated.

50. REASONING You put a reduction of a page on the original page. Explain why there is a point that is in the same place on both pages.

51. REASONING $\triangle ABC$ has vertices $A(4, 2)$, $B(4, 6)$, and $C(7, 2)$. Find the coordinates of the vertices of the image after a dilation with center $(4, 0)$ and a scale factor of 2.

Maintaining Mathematical Proficiency

Reviewing what you learned in previous grades and lessons

The vertices of $\triangle ABC$ are $A(2, -1)$, $B(0, 4)$, and $C(-3, 5)$. Find the coordinates of the vertices of the image after the translation. *(Section 4.1)*

52. $(x, y) \rightarrow (x, y - 4)$	**53.** $(x, y) \rightarrow (x - 1, y + 3)$	**54.** $(x, y) \rightarrow (x + 3, y - 1)$
55. $(x, y) \rightarrow (x - 2, y)$	**56.** $(x, y) \rightarrow (x + 1, y - 2)$	**57.** $(x, y) \rightarrow (x - 3, y + 1)$

4.6 Similarity and Transformations

Essential Question When a figure is translated, reflected, rotated, or dilated in the plane, is the image always similar to the original figure?

Two figures are *similar figures* when they have the same shape but not necessarily the same size.

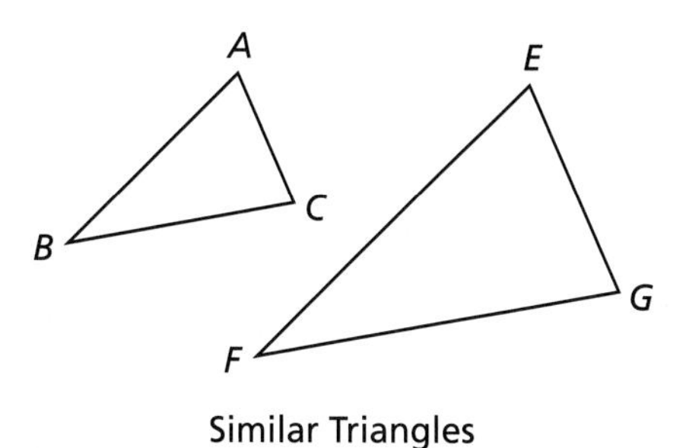

Similar Triangles

ATTENDING TO PRECISION

To be proficient in math, you need to use clear definitions in discussions with others and in your own reasoning.

EXPLORATION 1 Dilations and Similarity

Work with a partner.

a. Use dynamic geometry software to draw any triangle and label it $\triangle ABC$.

b. Dilate the triangle using a scale factor of 3. Is the image similar to the original triangle? Justify your answer.

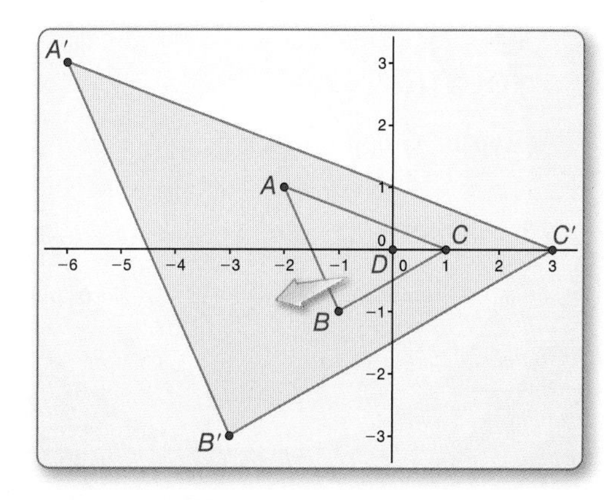

Sample

Points
$A(-2, 1)$
$B(-1, -1)$
$C(1, 0)$
$D(0, 0)$
Segments
$AB = 2.24$
$BC = 2.24$
$AC = 3.16$
Angles
$m\angle A = 45°$
$m\angle B = 90°$
$m\angle C = 45°$

EXPLORATION 2 Rigid Motions and Similarity

Work with a partner.

a. Use dynamic geometry software to draw any triangle.

b. Copy the triangle and translate it 3 units left and 4 units up. Is the image similar to the original triangle? Justify your answer.

c. Reflect the triangle in the y-axis. Is the image similar to the original triangle? Justify your answer.

d. Rotate the original triangle 90° counterclockwise about the origin. Is the image similar to the original triangle? Justify your answer.

Communicate Your Answer

3. When a figure is translated, reflected, rotated, or dilated in the plane, is the image always similar to the original figure? Explain your reasoning.

4. A figure undergoes a composition of transformations, which includes translations, reflections, rotations, and dilations. Is the image similar to the original figure? Explain your reasoning.

4.6 Lesson

Core Vocabulary

similarity transformation, *p. 216*
similar figures, *p. 216*

What You Will Learn

- Perform similarity transformations.
- Describe similarity transformations.
- Prove that figures are similar.

Performing Similarity Transformations

A dilation is a transformation that preserves shape but not size. So, a dilation is a nonrigid motion. A **similarity transformation** is a dilation or a composition of rigid motions and dilations. Two geometric figures are **similar figures** if and only if there is a similarity transformation that maps one of the figures onto the other. Similar figures have the same shape but not necessarily the same size.

Congruence transformations preserve length and angle measure. When the scale factor of the dilation(s) is not equal to 1 or -1, similarity transformations preserve angle measure only.

EXAMPLE 1 Performing a Similarity Transformation

Graph $\triangle ABC$ with vertices $A(-4, 1)$, $B(-2, 2)$, and $C(-2, 1)$ and its image after the similarity transformation.

Translation: $(x, y) \rightarrow (x + 5, y + 1)$

Dilation: $(x, y) \rightarrow (2x, 2y)$

SOLUTION

Step 1 Graph $\triangle ABC$.

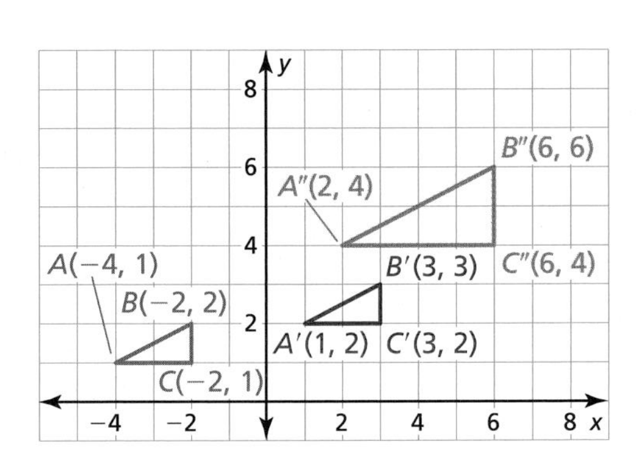

Step 2 Translate $\triangle ABC$ 5 units right and 1 unit up. $\triangle A'B'C'$ has vertices $A'(1, 2)$, $B'(3, 3)$, and $C'(3, 2)$.

Step 3 Dilate $\triangle A'B'C'$ using a scale factor of 2. $\triangle A''B''C''$ has vertices $A''(2, 4)$, $B''(6, 6)$, and $C''(6, 4)$.

Monitoring Progress  Help in English and Spanish at *BigIdeasMath.com*

1. Graph $\overline{CD}$ with endpoints $C(-2, 2)$ and $D(2, 2)$ and its image after the similarity transformation.

   Rotation: 90° about the origin

   Dilation: $(x, y) \rightarrow \left(\frac{1}{2}x, \frac{1}{2}y\right)$

2. Graph $\triangle FGH$ with vertices $F(1, 2)$, $G(4, 4)$, and $H(2, 0)$ and its image after the similarity transformation.

   Reflection: in the x-axis

   Dilation: $(x, y) \rightarrow (1.5x, 1.5y)$

Describing Similarity Transformations

EXAMPLE 2 Describing a Similarity Transformation

Describe a similarity transformation that maps trapezoid *PQRS* to trapezoid *WXYZ*.

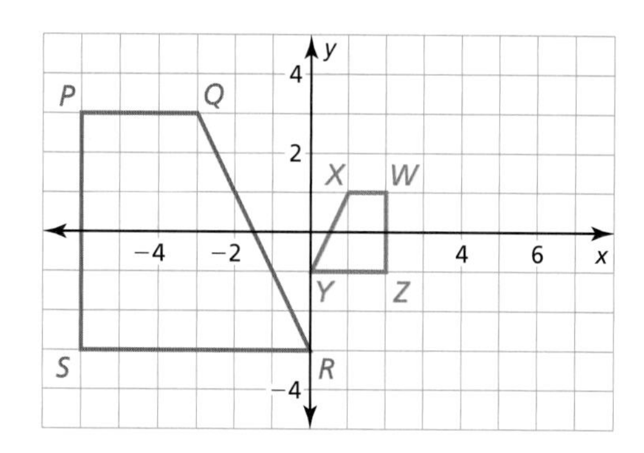

SOLUTION

$\overline{QR}$ falls from left to right, and $\overline{XY}$ rises from left to right. If you reflect trapezoid *PQRS* in the *y*-axis as shown, then the image, trapezoid $P'Q'R'S'$, will have the same orientation as trapezoid *WXYZ*.

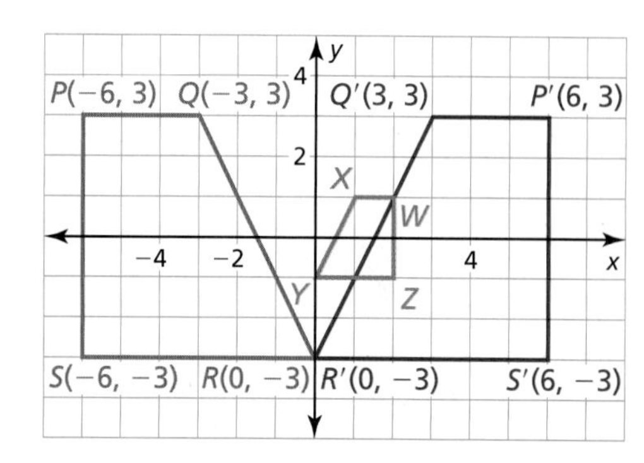

Trapezoid *WXYZ* appears to be about one-third as large as trapezoid $P'Q'R'S'$. Dilate trapezoid $P'Q'R'S'$ using a scale factor of $\frac{1}{3}$.

$$(x, y) \rightarrow \left(\tfrac{1}{3}x, \tfrac{1}{3}y\right)$$

$$P'(6, 3) \rightarrow P''(2, 1)$$

$$Q'(3, 3) \rightarrow Q''(1, 1)$$

$$R'(0, -3) \rightarrow R''(0, -1)$$

$$S'(6, -3) \rightarrow S''(2, -1)$$

The vertices of trapezoid $P''Q''R''S''$ match the vertices of trapezoid *WXYZ*.

▶ So, a similarity transformation that maps trapezoid *PQRS* to trapezoid *WXYZ* is a reflection in the *y*-axis followed by a dilation with a scale factor of $\frac{1}{3}$.

Monitoring Progress  Help in English and Spanish at *BigIdeasMath.com*

3. In Example 2, describe another similarity transformation that maps trapezoid *PQRS* to trapezoid *WXYZ*.

4. Describe a similarity transformation that maps quadrilateral *DEFG* to quadrilateral *STUV*.

Proving Figures Are Similar

To prove that two figures are similar, you must prove that a similarity transformation maps one of the figures onto the other.

EXAMPLE 3 Proving That Two Squares Are Similar

Prove that square $ABCD$ is similar to square $EFGH$.

Given Square $ABCD$ with side length r, square $EFGH$ with side length s, $\overline{AD} \parallel \overline{EH}$

Prove Square $ABCD$ is similar to square $EFGH$.

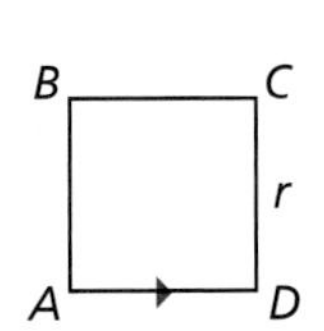

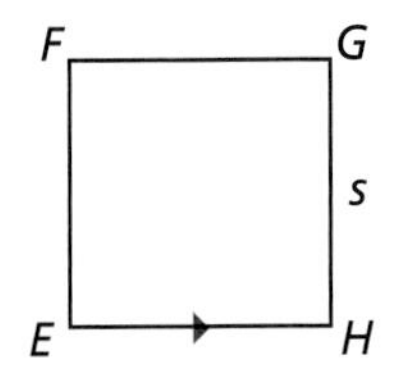

SOLUTION

Translate square $ABCD$ so that point A maps to point E. Because translations map segments to parallel segments and $\overline{AD} \parallel \overline{EH}$, the image of $\overline{AD}$ lies on $\overline{EH}$.

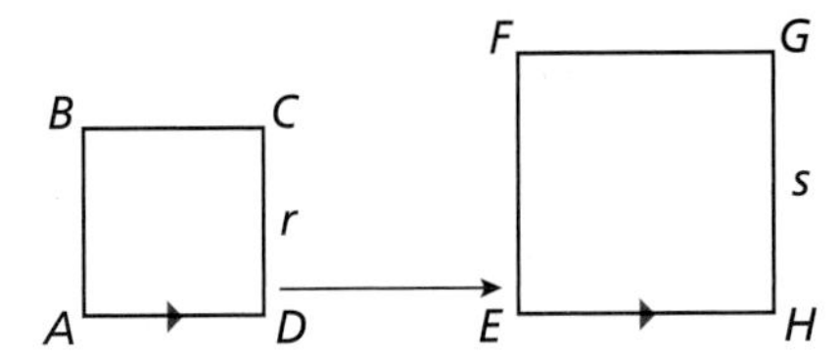

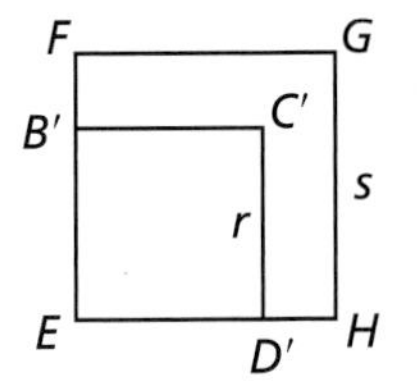

Because translations preserve length and angle measure, the image of $ABCD$, $EB'C'D'$, is a square with side length r. Because all the interior angles of a square are right angles, $\angle B'ED' \cong \angle FEH$. When $\overrightarrow{ED'}$ coincides with $\overrightarrow{EH}$, $\overrightarrow{EB'}$ coincides with $\overrightarrow{EF}$. So, $\overline{EB'}$ lies on $\overline{EF}$. Next, dilate square $EB'C'D'$ using center of dilation E. Choose the scale factor to be the ratio of the side lengths of $EFGH$ and $EB'C'D'$, which is $\frac{s}{r}$.

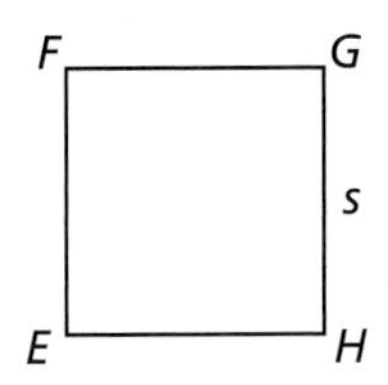

This dilation maps $\overline{ED'}$ to $\overline{EH}$ and $\overline{EB'}$ to $\overline{EF}$ because the images of $\overline{ED'}$ and $\overline{EB'}$ have side length $\frac{s}{r}(r) = s$ and the segments $\overline{ED'}$ and $\overline{EB'}$ lie on lines passing through the center of dilation. So, the dilation maps B' to F and D' to H. The image of C' lies $\frac{s}{r}(r) = s$ units to the right of the image of B' and $\frac{s}{r}(r) = s$ units above the image of D'. So, the image of C' is G.

▶ A similarity transformation maps square $ABCD$ to square $EFGH$. So, square $ABCD$ is similar to square $EFGH$.

Monitoring Progress

Help in English and Spanish at *BigIdeasMath.com*

5. Prove that $\triangle JKL$ is similar to $\triangle MNP$.

Given Right isosceles $\triangle JKL$ with leg length t, right isosceles $\triangle MNP$ with leg length v, $\overline{LJ} \parallel \overline{PM}$

Prove $\triangle JKL$ is similar to $\triangle MNP$.

4.6 Exercises

Dynamic Solutions available at *BigIdeasMath.com*

Vocabulary and Core Concept Check

1. **VOCABULARY** What is the difference between *similar figures* and *congruent figures*?

2. **COMPLETE THE SENTENCE** A transformation that produces a similar figure, such as a dilation, is called a _________.

Monitoring Progress and Modeling with Mathematics

In Exercises 3–6, graph $\triangle FGH$ with vertices $F(-2, 2)$, $G(-2, -4)$, and $H(-4, -4)$ and its image after the similarity transformation. *(See Example 1.)*

3. **Translation:** $(x, y) \rightarrow (x + 3, y + 1)$
   Dilation: $(x, y) \rightarrow (2x, 2y)$

4. **Dilation:** $(x, y) \rightarrow \left(\frac{1}{2}x, \frac{1}{2}y\right)$
   Reflection: in the y-axis

5. **Rotation:** 90° about the origin
   Dilation: $(x, y) \rightarrow (3x, 3y)$

6. **Dilation:** $(x, y) \rightarrow \left(\frac{3}{4}x, \frac{3}{4}y\right)$
   Reflection: in the x-axis

In Exercises 7 and 8, describe a similarity transformation that maps the blue preimage to the green image. *(See Example 2.)*

7.

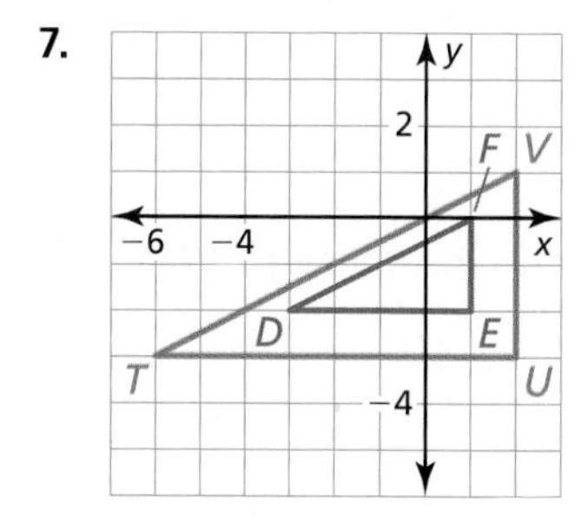

8.

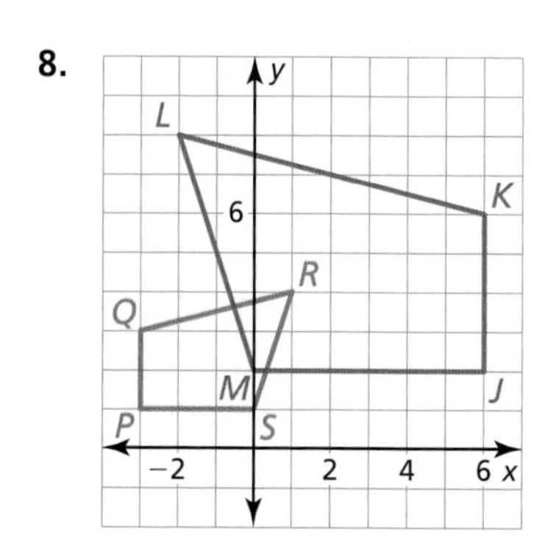

In Exercises 9–12, determine whether the polygons with the given vertices are similar. Use transformations to explain your reasoning.

9. $A(6, 0)$, $B(9, 6)$, $C(12, 6)$ and $D(0, 3)$, $E(1, 5)$, $F(2, 5)$

10. $Q(-1, 0)$, $R(-2, 2)$, $S(1, 3)$, $T(2, 1)$ and $W(0, 2)$, $X(4, 4)$, $Y(6, -2)$, $Z(2, -4)$

11. $G(-2, 3)$, $H(4, 3)$, $I(4, 0)$ and $J(1, 0)$, $K(6, -2)$, $L(1, -2)$

12. $D(-4, 3)$, $E(-2, 3)$, $F(-1, 1)$, $G(-4, 1)$ and $L(1, -1)$, $M(3, -1)$, $N(6, -3)$, $P(1, -3)$

In Exercises 13 and 14, prove that the figures are similar. *(See Example 3.)*

13. **Given** Right isosceles $\triangle ABC$ with leg length j, right isosceles $\triangle RST$ with leg length k, $\overline{CA} \parallel \overline{RT}$
    Prove $\triangle ABC$ is similar to $\triangle RST$.

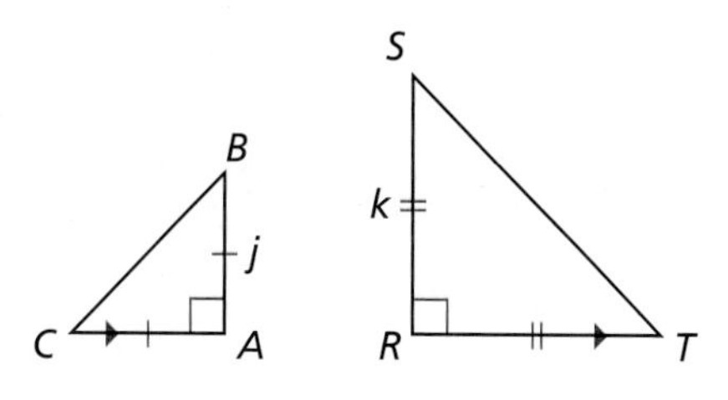

14. **Given** Rectangle $JKLM$ with side lengths x and y, rectangle $QRST$ with side lengths $2x$ and $2y$
    Prove Rectangle $JKLM$ is similar to rectangle $QRST$.

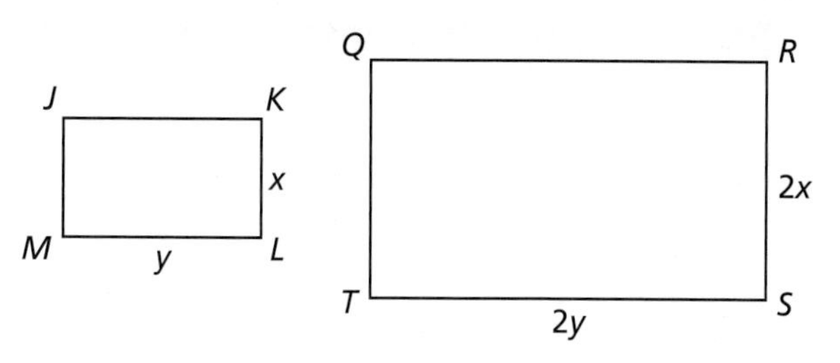

15. **MODELING WITH MATHEMATICS** Determine whether the regular-sized stop sign and the stop sign sticker are similar. Use transformations to explain your reasoning.

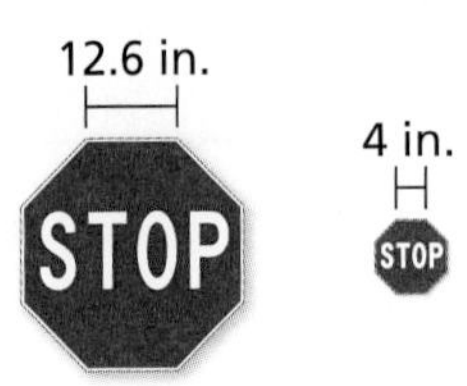

16. **ERROR ANALYSIS** Describe and correct the error in comparing the figures.

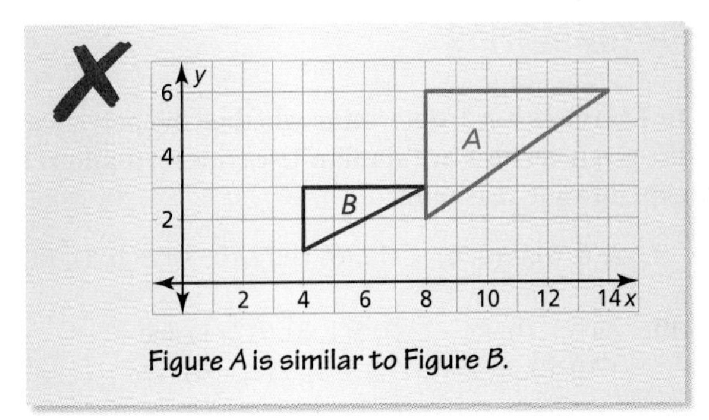

Figure *A* is similar to Figure *B*.

17. **MAKING AN ARGUMENT** A member of the homecoming decorating committee gives a printing company a banner that is 3 inches by 14 inches to enlarge. The committee member claims the banner she receives is distorted. Do you think the printing company distorted the image she gave it? Explain.

18. **HOW DO YOU SEE IT?** Determine whether each pair of figures is similar. Explain your reasoning.

a. **b.**

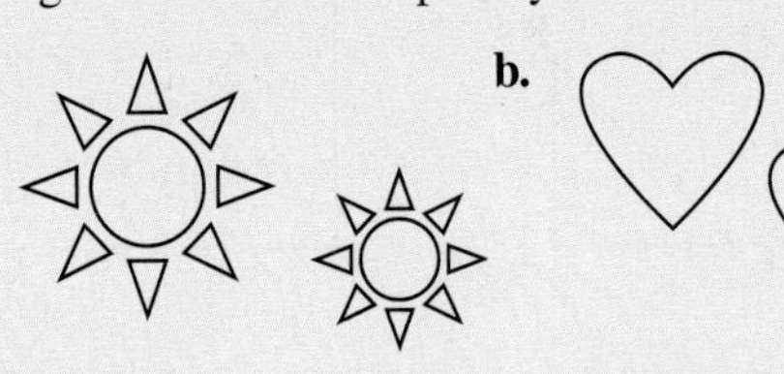

19. **ANALYZING RELATIONSHIPS** Graph a polygon in a coordinate plane. Use a similarity transformation involving a dilation (where k is a whole number) and a translation to graph a second polygon. Then describe a similarity transformation that maps the second polygon onto the first.

20. **THOUGHT PROVOKING** Is the composition of a rotation and a dilation commutative? (In other words, do you obtain the same image regardless of the order in which you perform the transformations?) Justify your answer.

21. **MATHEMATICAL CONNECTIONS** Quadrilateral $JKLM$ is mapped to quadrilateral $J'K'L'M'$ using the dilation $(x, y) \rightarrow \left(\frac{3}{2}x, \frac{3}{2}y\right)$. Then quadrilateral $J'K'L'M'$ is mapped to quadrilateral $J''K''L''M''$ using the translation $(x, y) \rightarrow (x + 3, y - 4)$. The vertices of quadrilateral $J'K'L'M'$ are $J(-12, 0)$, $K(-12, 18)$, $L(-6, 18)$, and $M(-6, 0)$. Find the coordinates of the vertices of quadrilateral $JKLM$ and quadrilateral $J''K''L''M''$. Are quadrilateral $JKLM$ and quadrilateral $J''K''L''M''$ similar? Explain.

22. **REPEATED REASONING** Use the diagram.

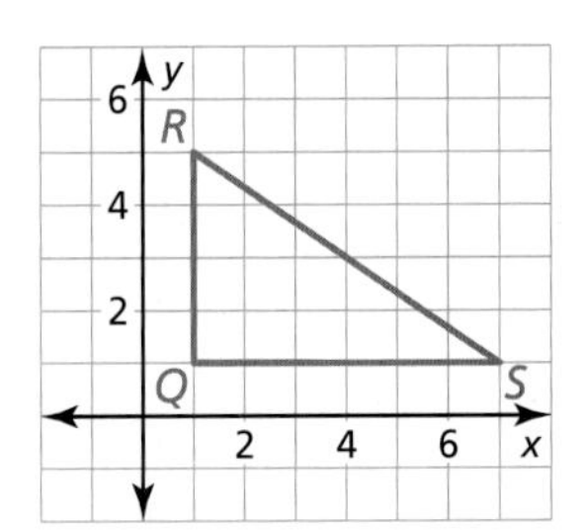

a. Connect the midpoints of the sides of $\triangle QRS$ to make another triangle. Is this triangle similar to $\triangle QRS$? Use transformations to support your answer.

b. Repeat part (a) for two other triangles. What conjecture can you make?

Maintaining Mathematical Proficiency

Reviewing what you learned in previous grades and lessons

Classify the angle as *acute*, *obtuse*, *right*, or *straight*. *(Section 1.5)*

23.

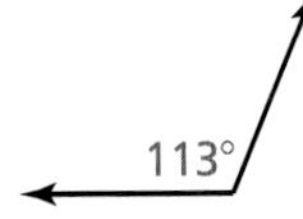

24.

25.

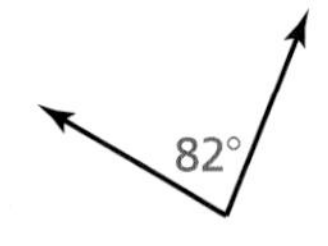

26.

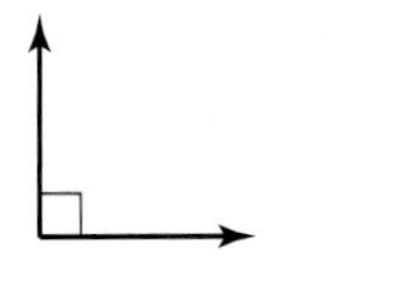

4.4–4.6 What Did You Learn?

Core Vocabulary

congruent figures, *p. 200*
congruence transformation, *p. 201*
dilation, *p. 208*
center of dilation, *p. 208*
scale factor, *p. 208*
enlargement, *p. 208*
reduction, *p. 208*
similarity transformation, *p. 216*
similar figures, *p. 216*

Core Concepts

Section 4.4

Identifying Congruent Figures, *p. 200*
Describing a Congruence Transformation, *p. 201*
Theorem 4.2 Reflections in Parallel Lines Theorem, *p. 202*
Theorem 4.3 Reflections in Intersecting Lines Theorem, *p. 203*

Section 4.5

Dilations and Scale Factor, *p. 208*
Coordinate Rule for Dilations, *p. 209*
Negative Scale Factors, *p. 210*

Section 4.6

Similarity Transformations, *p. 216*

Mathematical Practices

1. Revisit Exercise 31 on page 206. Try to recall the process you used to reach the solution. Did you have to change course at all? If so, how did you approach the situation?
2. Describe a real-life situation that can be modeled by Exercise 28 on page 213.

Performance Task

The Magic of Optics

Look at yourself in a shiny spoon. What happened to your reflection? Can you describe this mathematically? Now turn the spoon over and look at your reflection on the back of the spoon. What happened? Why?

To explore the answers to these questions and more, go to ***BigIdeasMath.com***.

4 Chapter Review

Dynamic Solutions available at *BigIdeasMath.com*

4.1 Translations (pp. 173–180)

Graph quadrilateral $ABCD$ with vertices $A(1, -2)$, $B(3, -1)$, $C(0, 3)$, and $D(-4, 1)$ and its image after the translation $(x, y) \rightarrow (x + 2, y - 2)$.

Graph quadrilateral $ABCD$. To find the coordinates of the vertices of the image, add 2 to the x-coordinates and subtract 2 from the y-coordinates of the vertices of the preimage. Then graph the image.

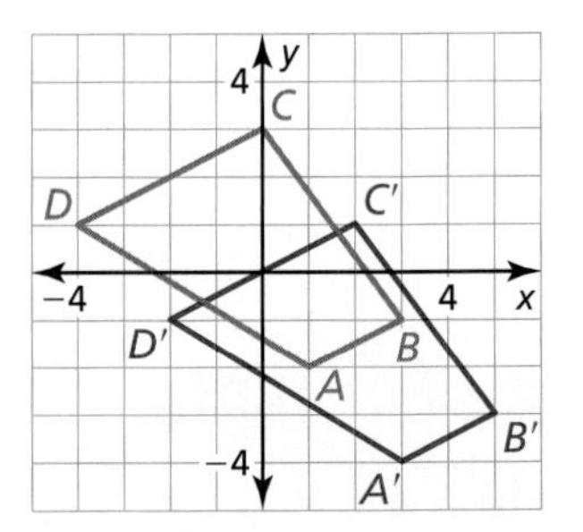

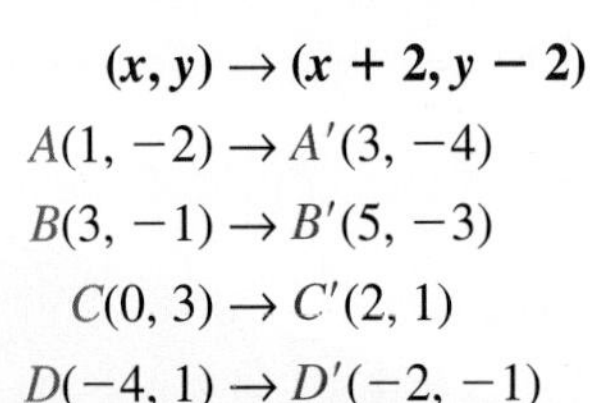

$$(x, y) \rightarrow (x + 2, y - 2)$$

$A(1, -2) \rightarrow A'(3, -4)$

$B(3, -1) \rightarrow B'(5, -3)$

$C(0, 3) \rightarrow C'(2, 1)$

$D(-4, 1) \rightarrow D'(-2, -1)$

Graph $\triangle XYZ$ with vertices $X(2, 3)$, $Y(-3, 2)$, and $Z(-4, -3)$ and its image after the translation.

1. $(x, y) \rightarrow (x, y + 2)$

2. $(x, y) \rightarrow (x - 3, y)$

3. $(x, y) \rightarrow (x + 3, y - 1)$

4. $(x, y) \rightarrow (x + 4, y + 1)$

Graph $\triangle PQR$ with vertices $P(0, -4)$, $Q(1, 3)$, and $R(2, -5)$ and its image after the composition.

5. **Translation:** $(x, y) \rightarrow (x + 1, y + 2)$
Translation: $(x, y) \rightarrow (x - 4, y + 1)$

6. **Translation:** $(x, y) \rightarrow (x, y + 3)$
Translation: $(x, y) \rightarrow (x - 1, y + 1)$

4.2 Reflections (pp. 181–188)

Graph $\triangle ABC$ with vertices $A(1, -1)$, $B(3, 2)$, and $C(4, -4)$ and its image after a reflection in the line $y = x$.

Graph $\triangle ABC$ and the line $y = x$. Then use the coordinate rule for reflecting in the line $y = x$ to find the coordinates of the vertices of the image.

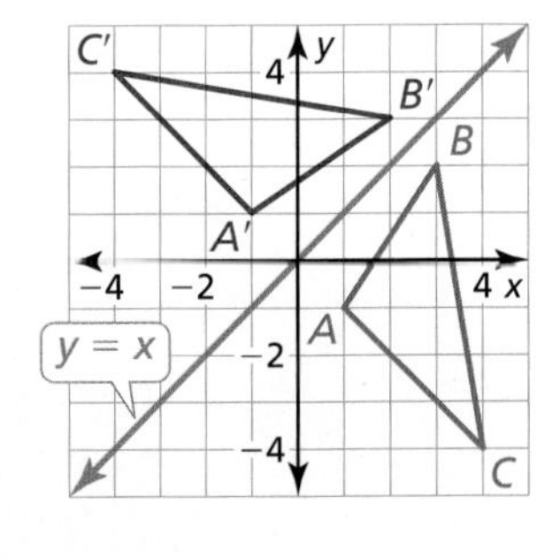

$$(a, b) \rightarrow (b, a)$$

$A(1, -1) \rightarrow A'(-1, 1)$

$B(3, 2) \rightarrow B'(2, 3)$

$C(4, -4) \rightarrow C'(-4, 4)$

Graph the polygon and its image after a reflection in the given line.

7. $x = 4$

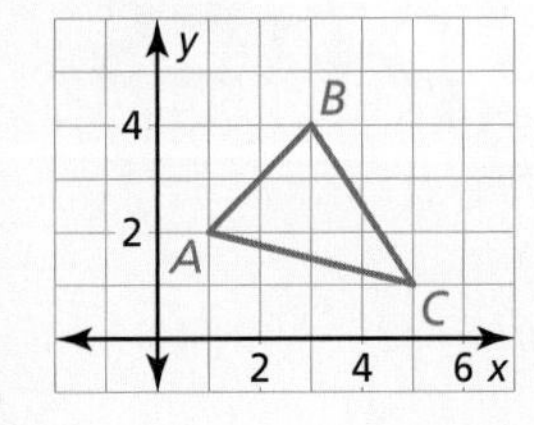

8. $y = 3$

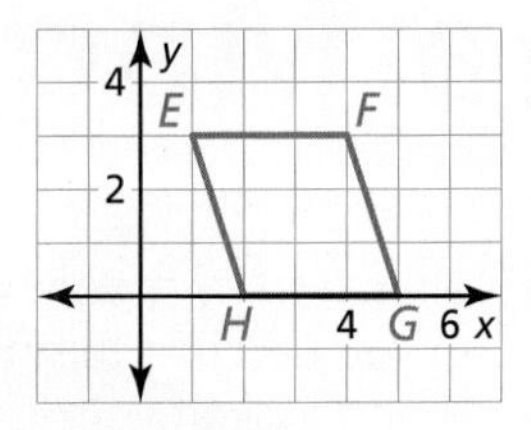

9. How many lines of symmetry does the figure have?

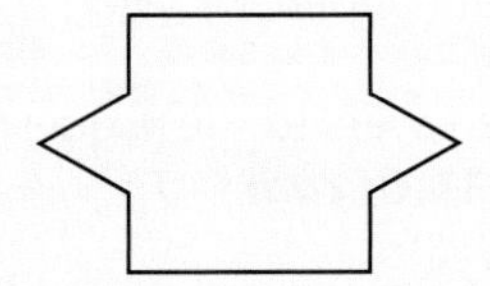

4.3 Rotations (pp. 189–196)

Graph $\triangle LMN$ with vertices $L(1, -1)$, $M(2, 3)$, and $N(4, 0)$ and its image after a 270° rotation about the origin.

Use the coordinate rule for a 270° rotation to find the coordinates of the vertices of the image. Then graph $\triangle LMN$ and its image.

$$(a, b) \rightarrow (b, -a)$$
$$L(1, -1) \rightarrow L'(-1, -1)$$
$$M(2, 3) \rightarrow M'(3, -2)$$
$$N(4, 0) \rightarrow N'(0, -4)$$

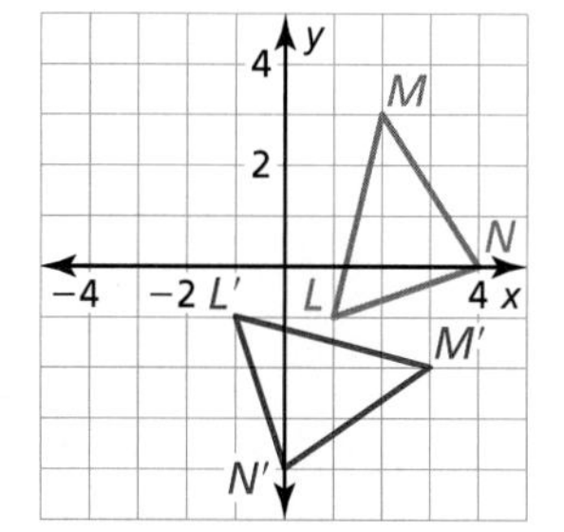

Graph the polygon with the given vertices and its image after a rotation of the given number of degrees about the origin.

10. $A(-3, -1)$, $B(2, 2)$, $C(3, -3)$; 90°

11. $W(-2, -1)$, $X(-1, 3)$, $Y(3, 3)$, $Z(3, -3)$; 180°

12. Graph $\overline{XY}$ with endpoints $X(5, -2)$ and $Y(3, -3)$ and its image after a reflection in the x-axis and then a rotation of 270° about the origin.

Determine whether the figure has rotational symmetry. If so, describe any rotations that map the figure onto itself.

13.

14.

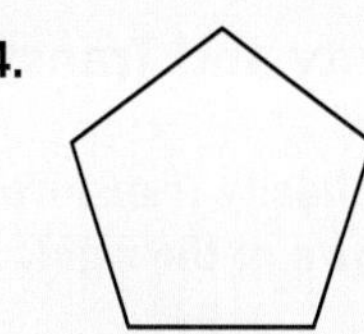

4.4 Congruence and Transformations (pp. 199–206)

Describe a congruence transformation that maps quadrilateral $ABCD$ to quadrilateral $WXYZ$, as shown at the right.

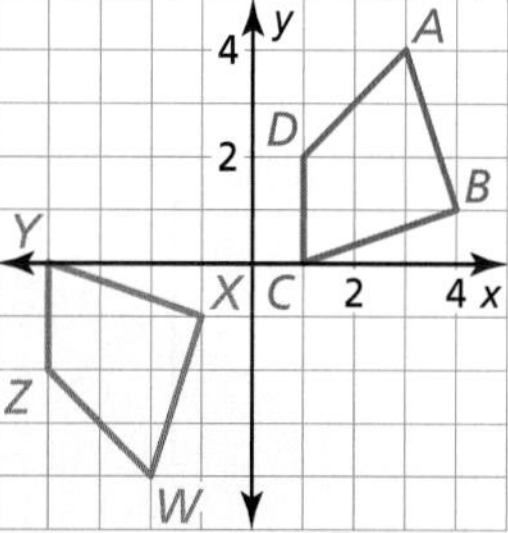

$\overline{AB}$ falls from left to right, and $\overline{WX}$ rises from left to right. If you reflect quadrilateral $ABCD$ in the x-axis as shown at the bottom right, then the image, quadrilateral $A'B'C'D'$, will have the same orientation as quadrilateral $WXYZ$. Then you can map quadrilateral $A'B'C'D'$ to quadrilateral $WXYZ$ using a translation of 5 units left.

▶ So, a congruence transformation that maps quadrilateral $ABCD$ to quadrilateral $WXYZ$ is a reflection in the x-axis followed by a translation of 5 units left.

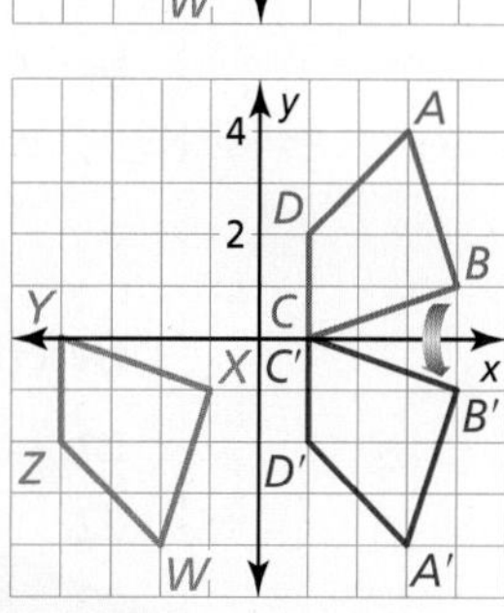

Describe a congruence transformation that maps $\triangle DEF$ to $\triangle JKL$.

15. $D(2, -1)$, $E(4, 1)$, $F(1, 2)$ and $J(-2, -4)$, $K(-4, -2)$, $L(-1, -1)$

16. $D(-3, -4)$, $E(-5, -1)$, $F(-1, 1)$ and $J(1, 4)$, $K(-1, 1)$, $L(3, -1)$

17. Which transformation is the same as reflecting an object in two parallel lines? in two intersecting lines?

4.5 Dilations (pp. 207–214)

Graph trapezoid *ABCD* with vertices $A(1, 1)$, $B(1, 3)$, $C(3, 2)$, and $D(3, 1)$ and its image after a dilation with a scale factor of 2.

Use the coordinate rule for a dilation with $k = 2$ to find the coordinates of the vertices of the image. Then graph trapezoid *ABCD* and its image.

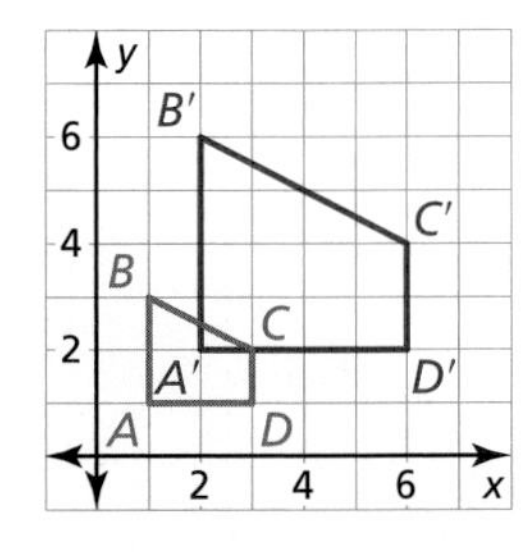

$$(x, y) \rightarrow (2x, 2y)$$

$A(1, 1) \rightarrow A'(2, 2)$

$B(1, 3) \rightarrow B'(2, 6)$

$C(3, 2) \rightarrow C'(6, 4)$

$D(3, 1) \rightarrow D'(6, 2)$

Graph the triangle and its image after a dilation with scale factor *k*.

18. $P(2, 2)$, $Q(4, 4)$, $R(8, 2)$; $k = \frac{1}{2}$

19. $X(-3, 2)$, $Y(2, 3)$, $Z(1, -1)$; $k = -3$

20. You are using a magnifying glass that shows the image of an object that is eight times the object's actual size. The image length is 15.2 centimeters. Find the actual length of the object.

4.6 Similarity and Transformations (pp. 215–220)

Describe a similarity transformation that maps $\triangle FGH$ to $\triangle LMN$, as shown at the right.

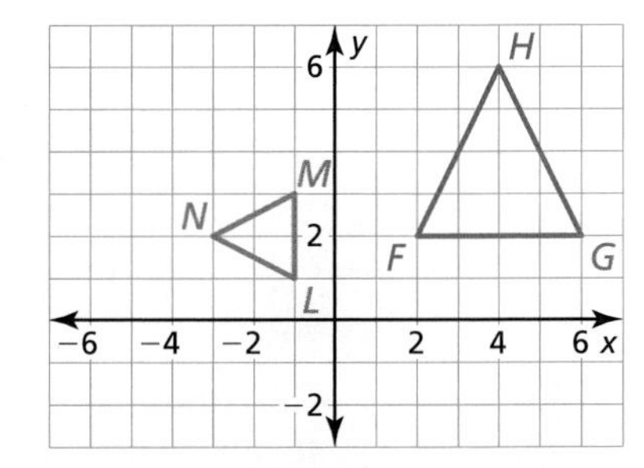

$\overline{FG}$ is horizontal, and $\overline{LM}$ is vertical. If you rotate $\triangle FGH$ 90° about the origin as shown at the bottom right, then the image, $\triangle F'G'H'$, will have the same orientation as $\triangle LMN$. $\triangle LMN$ appears to be half as large as $\triangle F'G'H'$. Dilate $\triangle F'G'H'$ using a scale factor of $\frac{1}{2}$.

$$(x, y) \rightarrow \left(\tfrac{1}{2}x, \tfrac{1}{2}y\right)$$

$F'(-2, 2) \rightarrow F''(-1, 1)$

$G'(-2, 6) \rightarrow G''(-1, 3)$

$H'(-6, 4) \rightarrow H''(-3, 2)$

The vertices of $\triangle F''G''H''$ match the vertices of $\triangle LMN$.

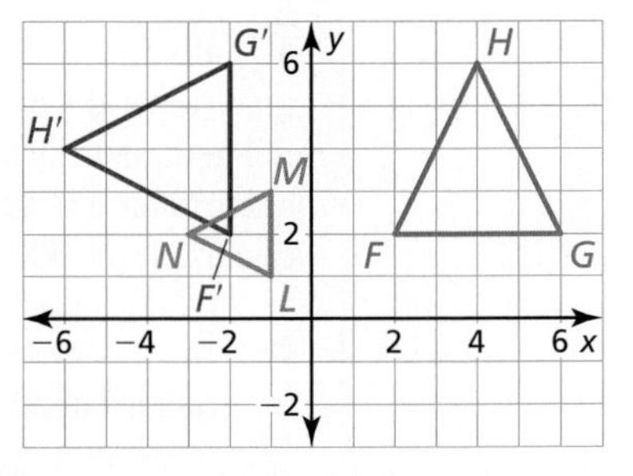

▶ So, a similarity transformation that maps $\triangle FGH$ to $\triangle LMN$ is a rotation of 90° about the origin followed by a dilation with a scale factor of $\frac{1}{2}$.

Describe a similarity transformation that maps $\triangle ABC$ to $\triangle RST$.

21. $A(1, 0)$, $B(-2, -1)$, $C(-1, -2)$ and $R(-3, 0)$, $S(6, -3)$, $T(3, -6)$

22. $A(6, 4)$, $B(-2, 0)$, $C(-4, 2)$ and $R(2, 3)$, $S(0, -1)$, $T(1, -2)$

23. $A(3, -2)$, $B(0, 4)$, $C(-1, -3)$ and $R(-4, -6)$, $S(8, 0)$, $T(-6, 2)$

4 Chapter Test

Graph $\triangle RST$ with vertices $R(-4, 1)$, $S(-2, 2)$, and $T(3, -2)$ and its image after the translation.

1. $(x, y) \rightarrow (x - 4, y + 1)$

2. $(x, y) \rightarrow (x + 2, y - 2)$

Graph the polygon with the given vertices and its image after a rotation of the given number of degrees about the origin.

3. $D(-1, -1)$, $E(-3, 2)$, $F(1, 4)$; 270°

4. $J(-1, 1)$, $K(3, 3)$, $L(4, -3)$, $M(0, -2)$; 90°

Determine whether the polygons with the given vertices are congruent or similar. Use transformations to explain your reasoning.

5. $Q(2, 4)$, $R(5, 4)$, $S(6, 2)$, $T(1, 2)$ and $W(6, -12)$, $X(15, -12)$, $Y(18, -6)$, $Z(3, -6)$

6. $A(-6, 6)$, $B(-6, 2)$, $C(-2, -4)$ and $D(9, 7)$, $E(5, 7)$, $F(-1, 3)$

Determine whether the object has line symmetry and whether it has rotational symmetry. Identify all lines of symmetry and angles of rotation that map the figure onto itself.

7.

8.

9.

10. Draw a diagram using a coordinate plane, two parallel lines, and a parallelogram that demonstrates the Reflections in Parallel Lines Theorem (Theorem 4.2).

11. A rectangle with vertices $W(-2, 4)$, $X(2, 4)$, $Y(2, 2)$, and $Z(-2, 2)$ is reflected in the y-axis. Your friend says that the image, rectangle $W'X'Y'Z'$, is exactly the same as the preimage. Is your friend correct? Explain your reasoning.

12. Write a composition of transformations that maps $\triangle ABC$ onto $\triangle CDB$ in the tesselation shown. Is the composition a congruence transformation? Explain your reasoning.

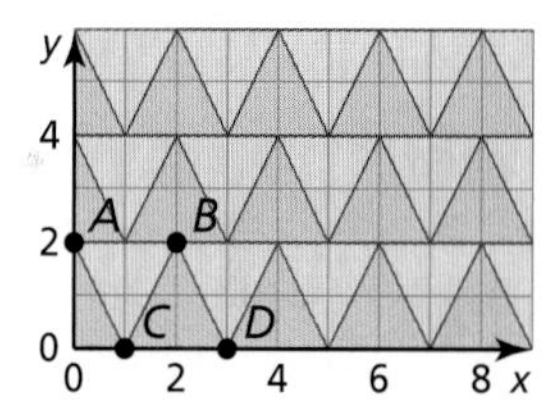

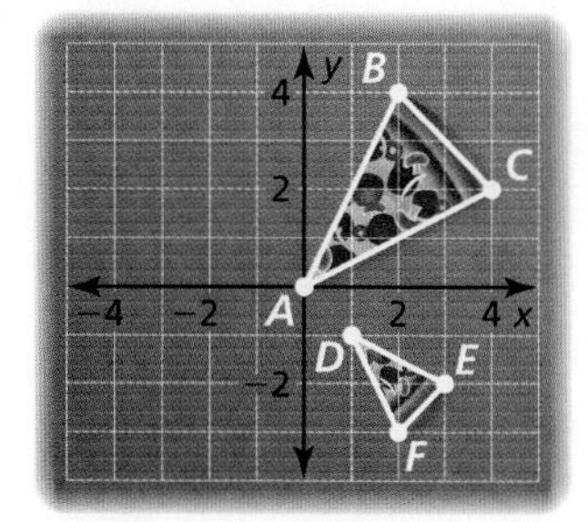

13. There is one slice of a large pizza and one slice of a small pizza in the box.

a. Describe a similarity transformation that maps pizza slice ABC to pizza slice DEF.

b. What is one possible scale factor for a medium slice of pizza? Explain your reasoning. (Use a dilation on the large slice of pizza.)

14. The original photograph shown is 4 inches by 6 inches.

a. What transfomations can you use to produce the new photograph?

b. You dilate the original photograph by a scale factor of $\frac{1}{2}$. What are the dimensions of the new photograph?

c. You have a frame that holds photos that are 8.5 inches by 11 inches. Can you dilate the original photograph to fit the frame? Explain your reasoning.

4 Cumulative Assessment

1. Which composition of transformations maps $\triangle ABC$ to $\triangle DEF$?

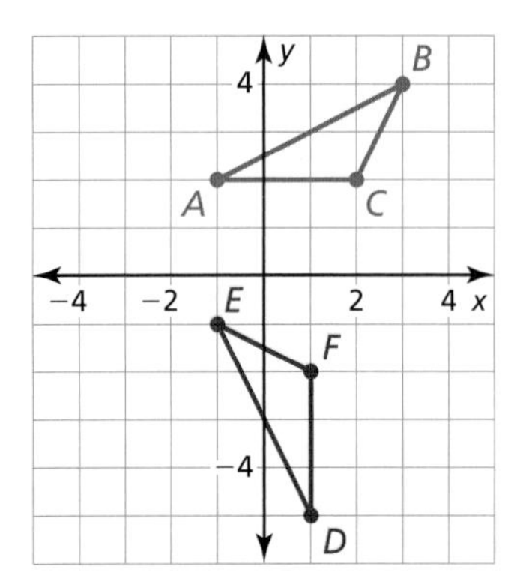

Ⓐ **Rotation:** 90° counterclockwise about the origin
Translation: $(x, y) \rightarrow (x + 4, y - 3)$

Ⓑ **Translation:** $(x, y) \rightarrow (x - 4, y - 3)$
Rotation: 90° counterclockwise about the origin

Ⓒ **Translation:** $(x, y) \rightarrow (x + 4, y - 3)$
Rotation: 90° counterclockwise about the origin

Ⓓ **Rotation:** 90° counterclockwise about the origin
Translation: $(x, y) \rightarrow (x - 4, y - 3)$

2. Use the diagrams to describe the steps you would take to construct a line perpendicular to line m through point P, which is not on line m.

Step 1

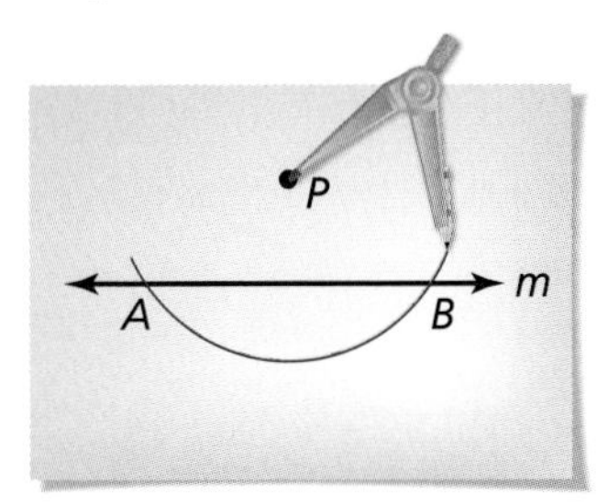

Step 2

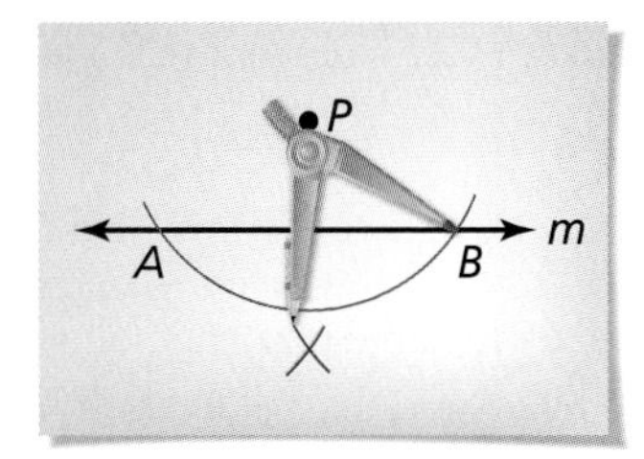

Step 3

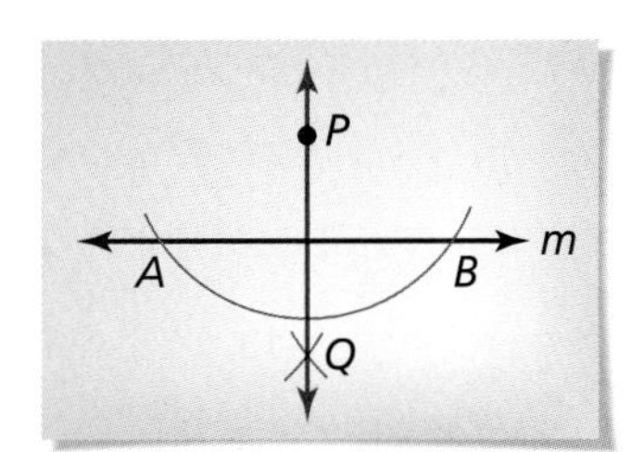

3. Your friend claims that she can find the perimeter of the school crossing sign without using the Distance Formula. Do you support your friend's claim? Explain your reasoning.

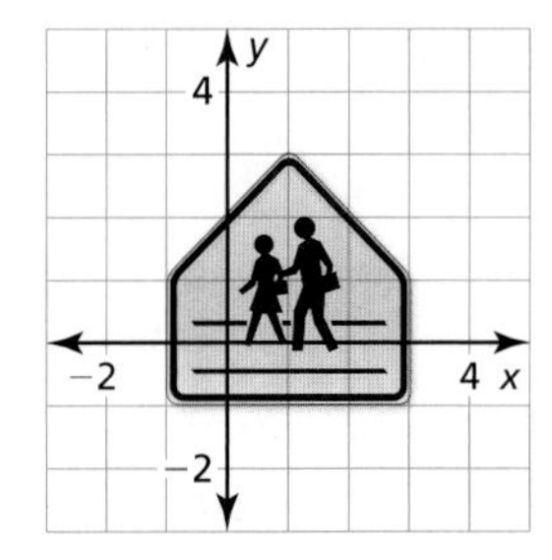

4. Graph the directed line segment ST with endpoints $S(-3, -2)$ and $T(4, 5)$. Then find the coordinates of point P along the directed line segment ST so that the ratio of SP to PT is 3 to 4.

5. The graph shows quadrilateral $WXYZ$ and quadrilateral $ABCD$.

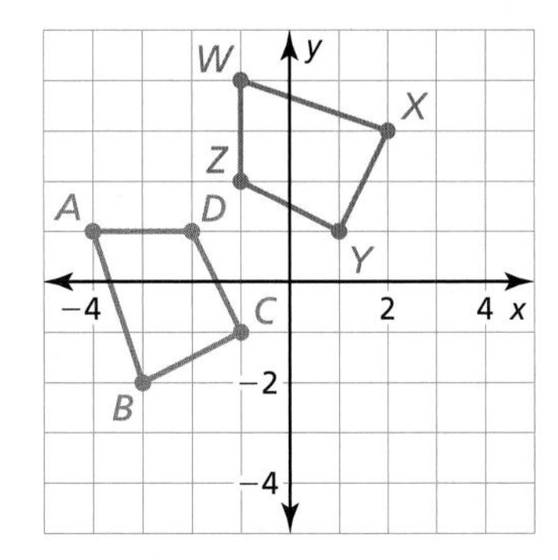

a. Write a composition of transformations that maps quadrilateral $WXYZ$ to quadrilateral $ABCD$.

b. Are the quadrilaterals congruent? Explain your reasoning.

6. Which equation represents the line passing through the point $(-6, 3)$ that is parallel to the line $y = -\frac{1}{3}x - 5$?

Ⓐ $y = 3x + 21$

Ⓑ $y = -\frac{1}{3}x - 5$

Ⓒ $y = 3x - 15$

Ⓓ $y = -\frac{1}{3}x + 1$

7. Which scale factor(s) would create a dilation of $\overline{AB}$ that is shorter than $\overline{AB}$? Select all that apply.

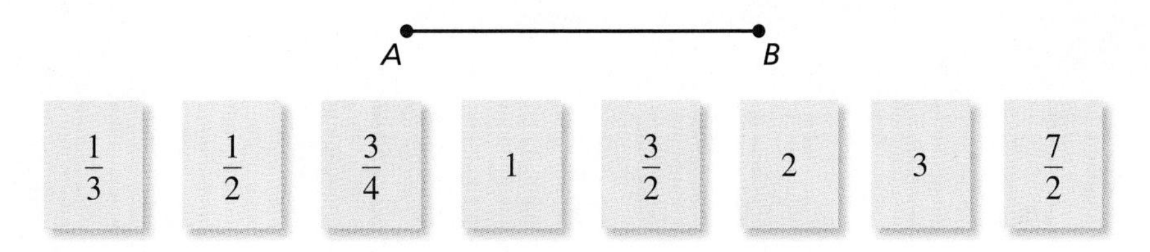

8. List one possible set of coordinates of the vertices of quadrilateral $ABCD$ for each description.

a. A reflection in the y-axis maps quadrilateral $ABCD$ onto itself.

b. A reflection in the x-axis maps quadrilateral $ABCD$ onto itself.

c. A rotation of 90° about the origin maps quadrilateral $ABCD$ onto itself.

d. A rotation of 180° about the origin maps quadrilateral $ABCD$ onto itself.

5 Congruent Triangles

Hang Glider ***(p. 278)***

Lifeguard Tower ***(p. 255)***

Barn ***(p. 248)***

Home Decor ***(p. 241)***

Painting ***(p. 235)***

Maintaining Mathematical Proficiency

Using the Midpoint and Distance Formulas

Example 1 **The endpoints of $\overline{AB}$ are $A(-2, 3)$ and $B(4, 7)$. Find the coordinates of the midpoint M.**

Use the Midpoint Formula.

$$M\left(\frac{-2+4}{2}, \frac{3+7}{2}\right) = M\left(\frac{2}{2}, \frac{10}{2}\right)$$

$$= M(1, 5)$$

▶ The coordinates of the midpoint M are (1, 5).

Example 2 **Find the distance between $C(0, -5)$ and $D(3, 2)$.**

$CD = \sqrt{(x_2 - x_1)^2 + (y_2 - y_1)^2}$	Distance Formula
$= \sqrt{(3-0)^2 + [2-(-5)]^2}$	Substitute.
$= \sqrt{3^2 + 7^2}$	Subtract.
$= \sqrt{9 + 49}$	Evaluate powers.
$= \sqrt{58}$	Add.
≈ 7.6	Use a calculator.

▶ The distance between $C(0, -5)$ and $D(3, 2)$ is about 7.6.

Find the coordinates of the midpoint *M* of the segment with the given endpoints. Then find the distance between the two points.

1. $P(-4, 1)$ and $Q(0, 7)$ **2.** $G(3, 6)$ and $H(9, -2)$ **3.** $U(-1, -2)$ and $V(8, 0)$

Solving Equations with Variables on Both Sides

Example 3 **Solve $2 - 5x = -3x$.**

$2 - 5x = -3x$	Write the equation.
$\underline{+5x} \quad \underline{+5x}$	Add $5x$ to each side.
$2 = 2x$	Simplify.
$\frac{2}{2} = \frac{2x}{2}$	Divide each side by 2.
$1 = x$	Simplify.

▶ The solution is $x = 1$.

Solve the equation.

4. $7x + 12 = 3x$ **5.** $14 - 6t = t$ **6.** $5p + 10 = 8p + 1$

7. $w + 13 = 11w - 7$ **8.** $4x + 1 = 3 - 2x$ **9.** $z - 2 = 4 + 9z$

10. **ABSTRACT REASONING** Is it possible to find the length of a segment in a coordinate plane without using the Distance Formula? Explain your reasoning.

Mathematical Practices

Mathematically proficient students understand and use given definitions.

Definitions, Postulates, and Theorems

Core Concept

Definitions and Biconditional Statements

A definition is always an "if and only if" statement. Here is an example.

Definition: Two geometric figures are *congruent figures* if and only if there is a rigid motion or a composition of rigid motions that maps one of the figures onto the other.

Because this is a definition, it is a biconditional statement. It implies the following two conditional statements.

1. If two geometric figures are congruent figures, then there is a rigid motion or a composition of rigid motions that maps one of the figures onto the other.
2. If there is a rigid motion or a composition of rigid motions that maps one geometric figure onto another, then the two geometric figures are congruent figures.

Definitions, postulates, and theorems are the building blocks of geometry. In two-column proofs, the statements in the *reason* column are almost always definitions, postulates, or theorems.

EXAMPLE 1 Identifying Definitions, Postulates, and Theorems

Classify each statement as a definition, a postulate, or a theorem.

a. If two lines are cut by a transversal so that alternate interior angles are congruent, then the lines are parallel.

b. If two coplanar lines have no point of intersection, then the lines are parallel.

c. If there is a line and a point not on the line, then there is exactly one line through the point parallel to the given line.

SOLUTION

a. This is a theorem. It is the Alternate Interior Angles Converse Theorem (Theorem 3.6) studied in Section 3.3.

b. This is the definition of parallel lines.

c. This is a postulate. It is the Parallel Postulate (Postulate 3.1) studied in Section 3.1. In Euclidean geometry, it is assumed, not proved, to be true.

Monitoring Progress

Classify each statement as a definition, a postulate, or a theorem. Explain your reasoning.

1. In a coordinate plane, two nonvertical lines are perpendicular if and only if the product of their slopes is -1.
2. If two lines intersect to form a linear pair of congruent angles, then the lines are perpendicular.
3. If two lines intersect to form a right angle, then the lines are perpendicular.
4. Through any two points, there exists exactly one line.

5.1 Angles of Triangles

Essential Question How are the angle measures of a triangle related?

EXPLORATION 1 Writing a Conjecture

Work with a partner.

a. Use dynamic geometry software to draw any triangle and label it $\triangle ABC$.

b. Find the measures of the interior angles of the triangle.

c. Find the sum of the interior angle measures.

d. Repeat parts (a)–(c) with several other triangles. Then write a conjecture about the sum of the measures of the interior angles of a triangle.

CONSTRUCTING VIABLE ARGUMENTS

To be proficient in math, you need to reason inductively about data and write conjectures.

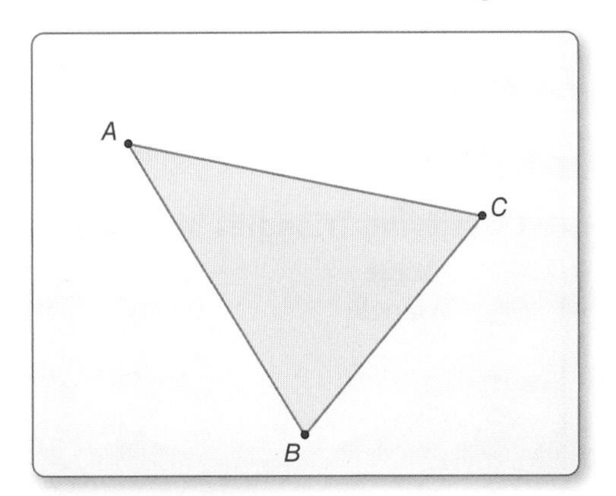

Sample

Angles

$m\angle A = 43.67°$

$m\angle B = 81.87°$

$m\angle C = 54.46°$

EXPLORATION 2 Writing a Conjecture

Work with a partner.

a. Use dynamic geometry software to draw any triangle and label it $\triangle ABC$.

b. Draw an exterior angle at any vertex and find its measure.

c. Find the measures of the two nonadjacent interior angles of the triangle.

d. Find the sum of the measures of the two nonadjacent interior angles. Compare this sum to the measure of the exterior angle.

e. Repeat parts (a)–(d) with several other triangles. Then write a conjecture that compares the measure of an exterior angle with the sum of the measures of the two nonadjacent interior angles.

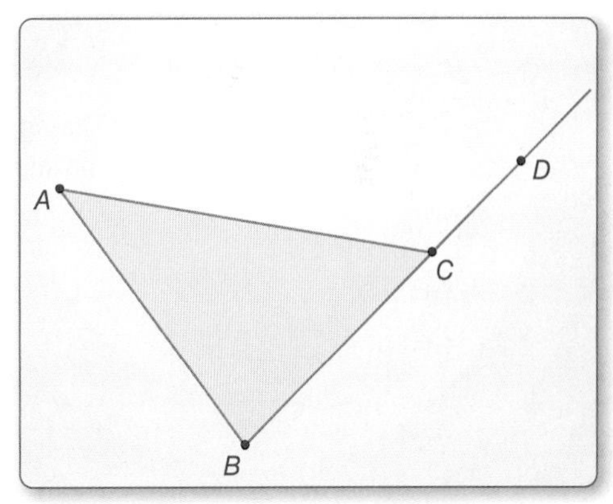

Sample

Angles

$m\angle A = 43.67°$

$m\angle B = 81.87°$

$m\angle ACD = 125.54°$

Communicate Your Answer

3. How are the angle measures of a triangle related?

4. An exterior angle of a triangle measures 32°. What do you know about the measures of the interior angles? Explain your reasoning.

5.1 Lesson

Core Vocabulary

interior angles, *p. 233*
exterior angles, *p. 233*
corollary to a theorem, *p. 235*

Previous
triangle

What You Will Learn

- Classify triangles by sides and angles.
- Find interior and exterior angle measures of triangles.

Classifying Triangles by Sides and by Angles

Recall that a *triangle* is a polygon with three sides. You can classify triangles by sides and by angles, as shown below.

READING

Notice that an equilateral triangle is also isosceles. An equiangular triangle is also acute.

Core Concept

Classifying Triangles by Sides

Scalene Triangle

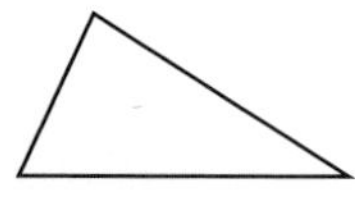

no congruent sides

Isosceles Triangle

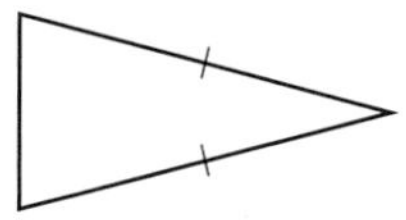

at least 2 congruent sides

Equilateral Triangle

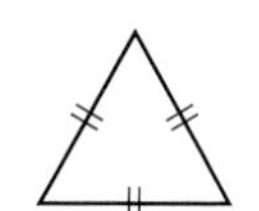

3 congruent sides

Classifying Triangles by Angles

Acute Triangle

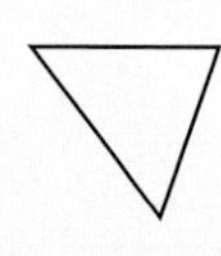

3 acute angles

Right Triangle

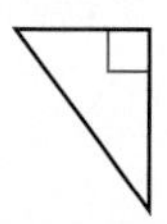

1 right angle

Obtuse Triangle

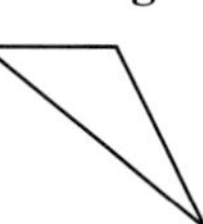

1 obtuse angle

Equiangular Triangle

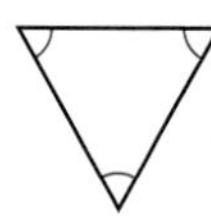

3 congruent angles

EXAMPLE 1 Classifying Triangles by Sides and by Angles

Classify the triangular shape of the support beams in the diagram by its sides and by measuring its angles.

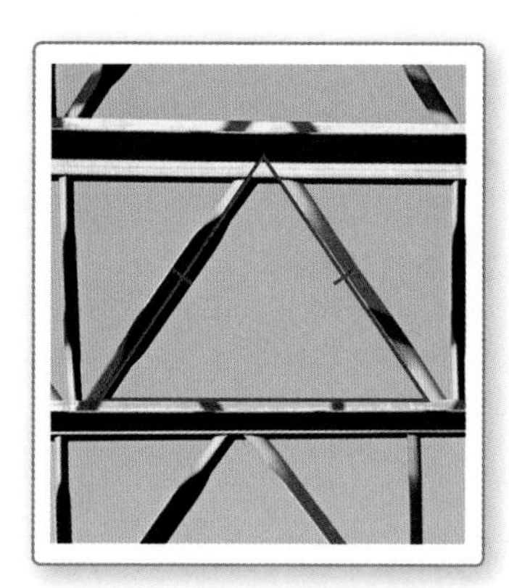

SOLUTION

The triangle has a pair of congruent sides, so it is isosceles. By measuring, the angles are 55°, 55°, and 70°.

- So, it is an acute isosceles triangle.

Monitoring Progress

Help in English and Spanish at *BigIdeasMath.com*

1. Draw an obtuse isosceles triangle and an acute scalene triangle.

EXAMPLE 2 Classifying a Triangle in the Coordinate Plane

Classify $\triangle OPQ$ by its sides. Then determine whether it is a right triangle.

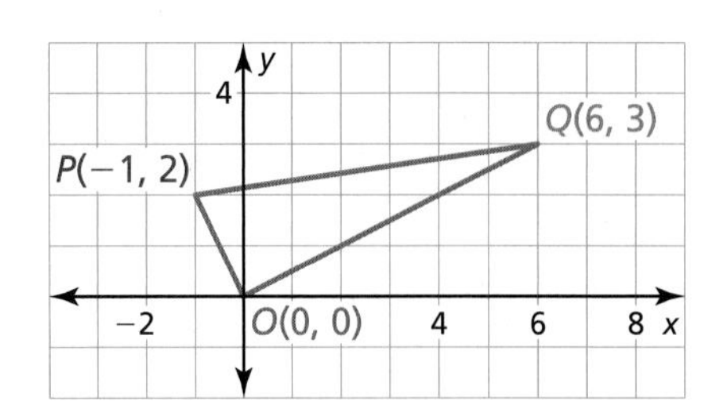

SOLUTION

Step 1 Use the Distance Formula to find the side lengths.

$$OP = \sqrt{(x_2 - x_1)^2 + (y_2 - y_1)^2} = \sqrt{(-1 - 0)^2 + (2 - 0)^2} = \sqrt{5} \approx 2.2$$

$$OQ = \sqrt{(x_2 - x_1)^2 + (y_2 - y_1)^2} = \sqrt{(6 - 0)^2 + (3 - 0)^2} = \sqrt{45} \approx 6.7$$

$$PQ = \sqrt{(x_2 - x_1)^2 + (y_2 - y_1)^2} = \sqrt{[6 - (-1)]^2 + (3 - 2)^2} = \sqrt{50} \approx 7.1$$

Because no sides are congruent, $\triangle OPQ$ is a scalene triangle.

Step 2 Check for right angles. The slope of $\overline{OP}$ is $\frac{2 - 0}{-1 - 0} = -2$. The slope of $\overline{OQ}$ is $\frac{3 - 0}{6 - 0} = \frac{1}{2}$. The product of the slopes is $-2\left(\frac{1}{2}\right) = -1$. So, $\overline{OP} \perp \overline{OQ}$ and $\angle POQ$ is a right angle.

▶ So, $\triangle OPQ$ is a right scalene triangle.

Monitoring Progress

Help in English and Spanish at *BigIdeasMath.com*

2. $\triangle ABC$ has vertices $A(0, 0)$, $B(3, 3)$, and $C(-3, 3)$. Classify the triangle by its sides. Then determine whether it is a right triangle.

Finding Angle Measures of Triangles

When the sides of a polygon are extended, other angles are formed. The original angles are the **interior angles**. The angles that form linear pairs with the interior angles are the **exterior angles**.

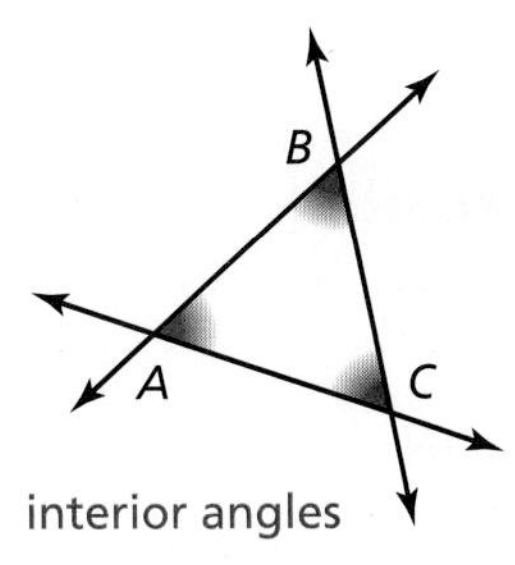

interior angles

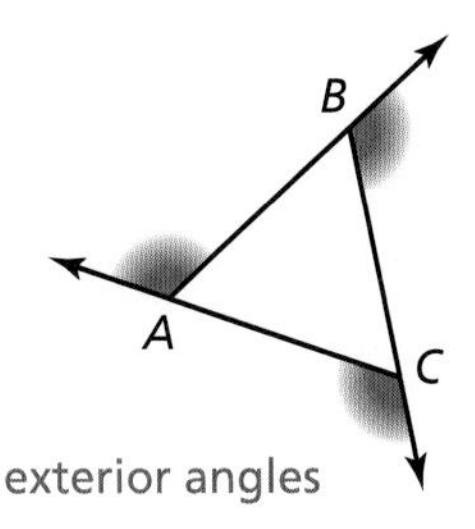

exterior angles

Theorem

Theorem 5.1 Triangle Sum Theorem

The sum of the measures of the interior angles of a triangle is 180°.

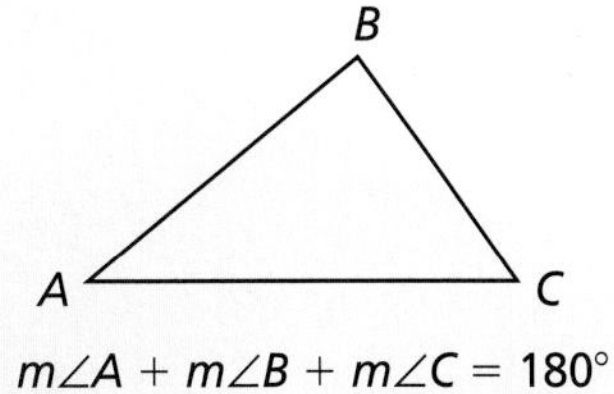

$m\angle A + m\angle B + m\angle C = 180°$

Proof p. 234; Ex. 53, p. 238

To prove certain theorems, you may need to add a line, a segment, or a ray to a given diagram. An *auxiliary* line is used in the proof of the Triangle Sum Theorem.

PROOF **Triangle Sum Theorem**

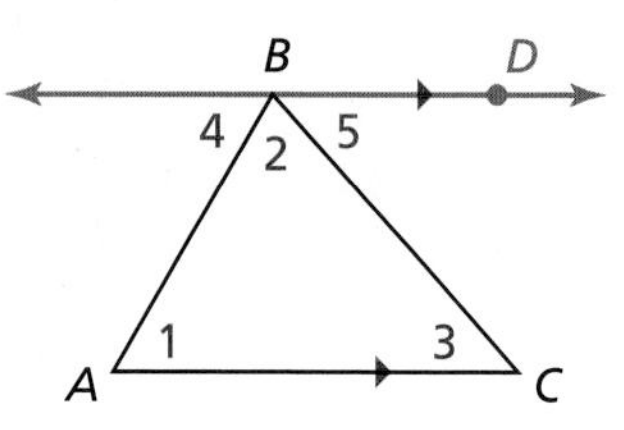

Given $\triangle ABC$

Prove $m\angle 1 + m\angle 2 + m\angle 3 = 180°$

Plan for Proof

a. Draw an auxiliary line through B that is parallel to $\overline{AC}$.

b. Show that $m\angle 4 + m\angle 2 + m\angle 5 = 180°$, $\angle 1 \cong \angle 4$, and $\angle 3 \cong \angle 5$.

c. By substitution, $m\angle 1 + m\angle 2 + m\angle 3 = 180°$.

Plan in Action

	STATEMENTS	REASONS
a.	**1.** Draw $\overleftrightarrow{BD}$ parallel to $\overline{AC}$.	**1.** Parallel Postulate (Post. 3.1)
b.	**2.** $m\angle 4 + m\angle 2 + m\angle 5 = 180°$	**2.** Angle Addition Postulate (Post. 1.4) and definition of straight angle
	3. $\angle 1 \cong \angle 4$, $\angle 3 \cong \angle 5$	**3.** Alternate Interior Angles Theorem (Thm. 3.2)
	4. $m\angle 1 = m\angle 4$, $m\angle 3 = m\angle 5$	**4.** Definition of congruent angles
c.	**5.** $m\angle 1 + m\angle 2 + m\angle 3 = 180°$	**5.** Substitution Property of Equality

Theorem

Theorem 5.2 Exterior Angle Theorem

The measure of an exterior angle of a triangle is equal to the sum of the measures of the two nonadjacent interior angles.

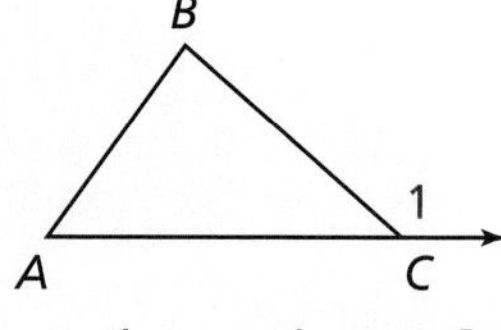

$m\angle 1 = m\angle A + m\angle B$

Proof Ex. 42, p. 237

EXAMPLE 3 **Finding an Angle Measure**

Find $m\angle JKM$.

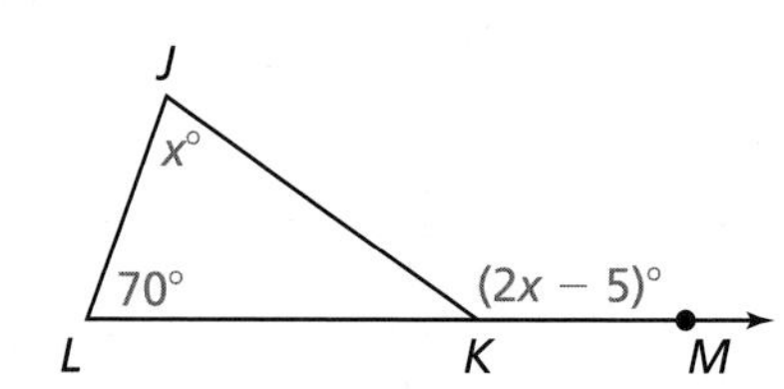

SOLUTION

Step 1 Write and solve an equation to find the value of x.

$(2x - 5)° = 70° + x°$ Apply the Exterior Angle Theorem.

$x = 75$ Solve for x.

Step 2 Substitute 75 for x in $2x - 5$ to find $m\angle JKM$.

$2x - 5 = 2 \cdot 75 - 5 = 145$

▶ So, the measure of $\angle JKM$ is 145°.

A **corollary to a theorem** is a statement that can be proved easily using the theorem. The corollary below follows from the Triangle Sum Theorem.

Corollary

Corollary 5.1 Corollary to the Triangle Sum Theorem

The acute angles of a right triangle are complementary.

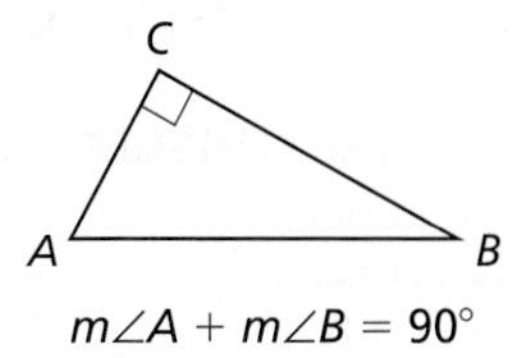

$m\angle A + m\angle B = 90°$

Proof Ex. 41, p. 237

EXAMPLE 4 Modeling with Mathematics

In the painting, the red triangle is a right triangle. The measure of one acute angle in the triangle is twice the measure of the other. Find the measure of each acute angle.

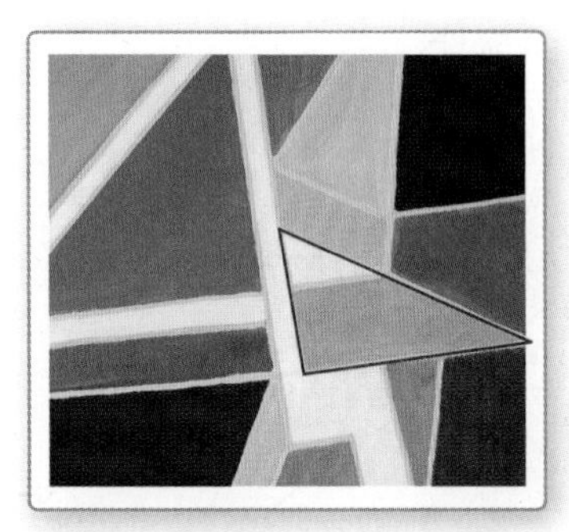

SOLUTION

1. **Understand the Problem** You are given a right triangle and the relationship between the two acute angles in the triangle. You need to find the measure of each acute angle.

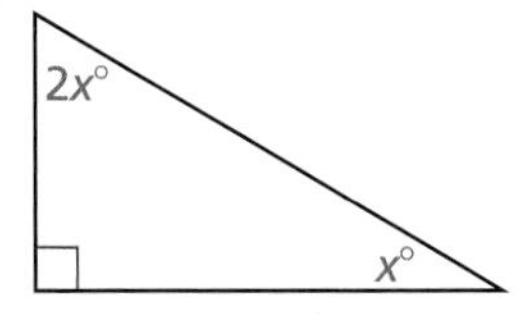

2. **Make a Plan** First, sketch a diagram of the situation. You can use the Corollary to the Triangle Sum Theorem and the given relationship between the two acute angles to write and solve an equation to find the measure of each acute angle.
3. **Solve the Problem** Let the measure of the smaller acute angle be $x°$. Then the measure of the larger acute angle is $2x°$. The Corollary to the Triangle Sum Theorem states that the acute angles of a right triangle are complementary.

   Use the corollary to set up and solve an equation.

   $x° + 2x° = 90°$ — Corollary to the Triangle Sum Theorem

   $x = 30$ — Solve for x.

   ▶ So, the measures of the acute angles are 30° and 2(30°) = 60°.

4. **Look Back** Add the two angles and check that their sum satisfies the Corollary to the Triangle Sum Theorem.

   $30° + 60° = 90°$ 

Monitoring Progress

Help in English and Spanish at *BigIdeasMath.com*

3. Find the measure of $\angle 1$.

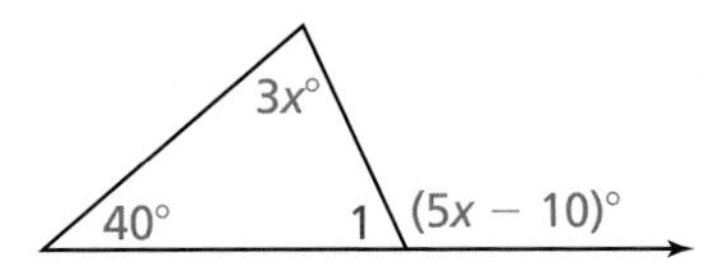

4. Find the measure of each acute angle.

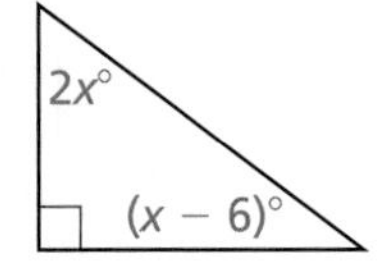

5.1 Exercises

Dynamic Solutions available at *BigIdeasMath.com*

Vocabulary and Core Concept Check

1. **WRITING** Can a right triangle also be obtuse? Explain your reasoning.

2. **COMPLETE THE SENTENCE** The measure of an exterior angle of a triangle is equal to the sum of the measures of the two ________________ interior angles.

Monitoring Progress and Modeling with Mathematics

In Exercises 3–6, classify the triangle by its sides and by measuring its angles. *(See Example 1.)*

3. 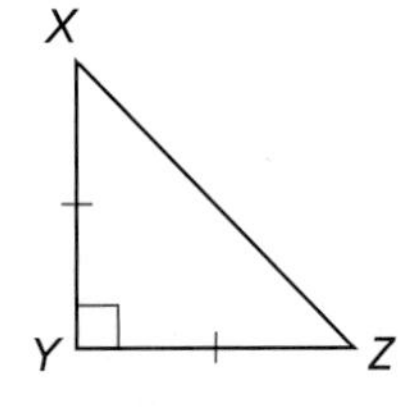

4. 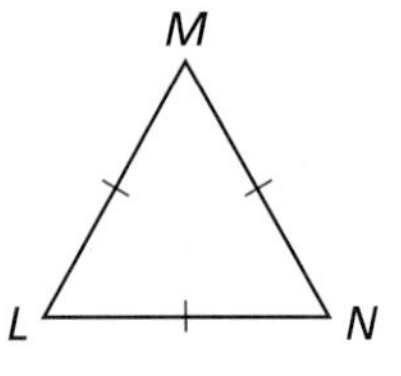

5. 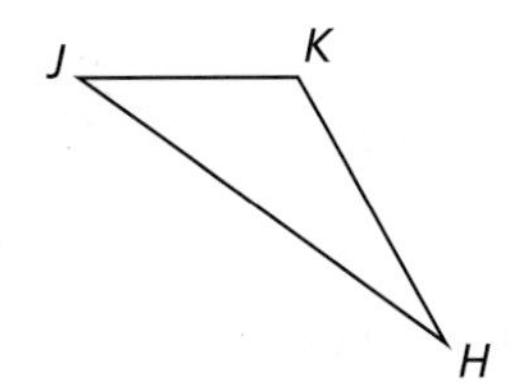

6. 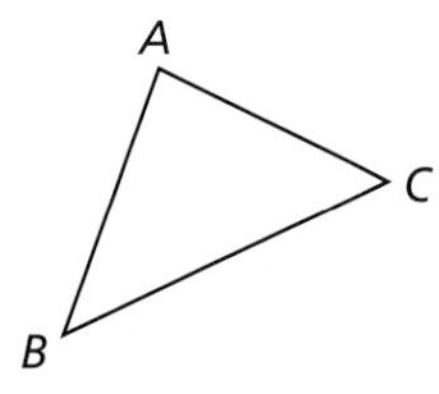

In Exercises 7–10, classify $\triangle ABC$ by its sides. Then determine whether it is a right triangle. *(See Example 2.)*

7. $A(2, 3)$, $B(6, 3)$, $C(2, 7)$

8. $A(3, 3)$, $B(6, 9)$, $C(6, -3)$

9. $A(1, 9)$, $B(4, 8)$, $C(2, 5)$

10. $A(-2, 3)$, $B(0, -3)$, $C(3, -2)$

In Exercises 11–14, find $m\angle 1$. Then classify the triangle by its angles.

11. 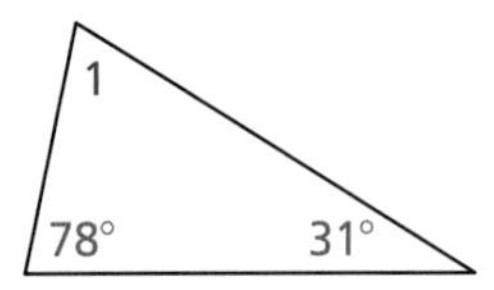

12. 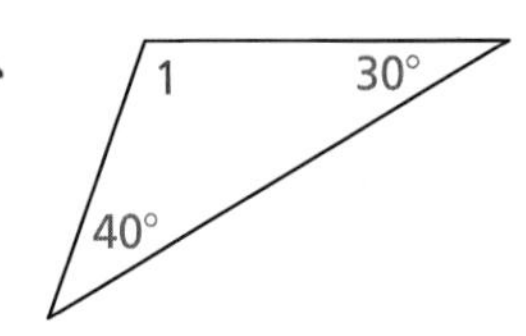

13. 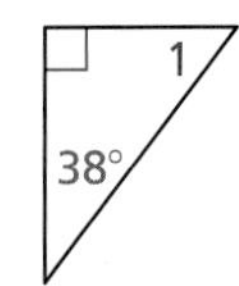

14. 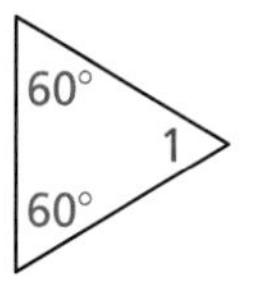

In Exercises 15–18, find the measure of the exterior angle. *(See Example 3.)*

15. 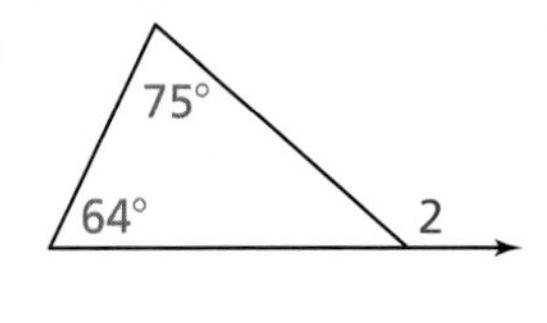

16. 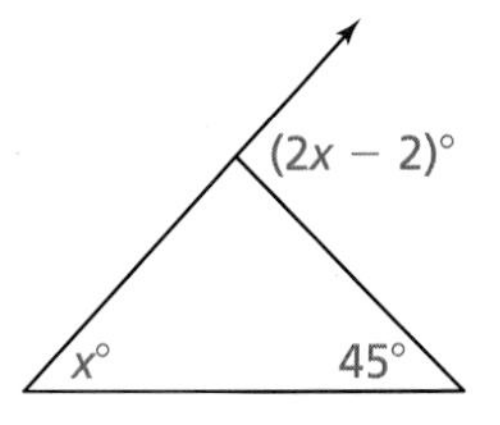

17. 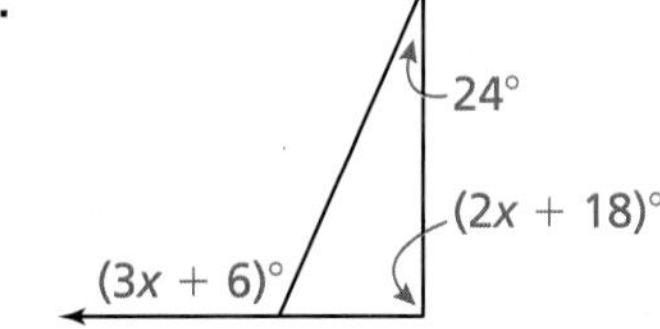

18. 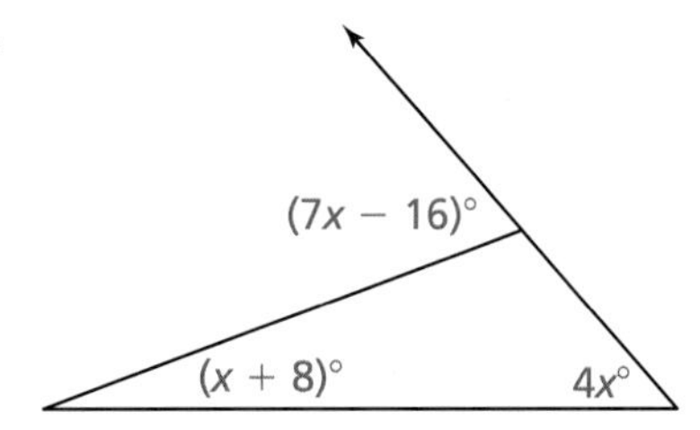

In Exercises 19–22, find the measure of each acute angle. *(See Example 4.)*

19. 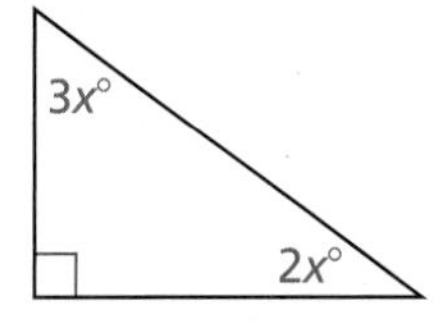

20. 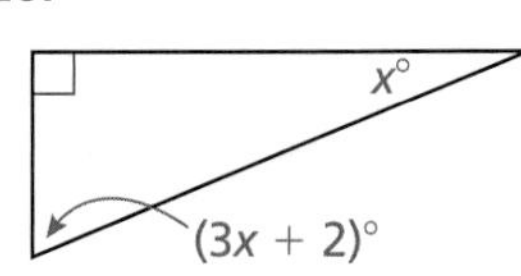

21. 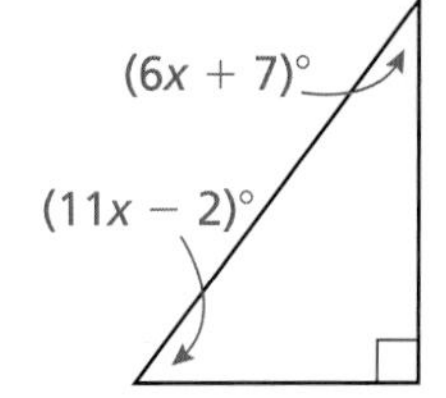

22. 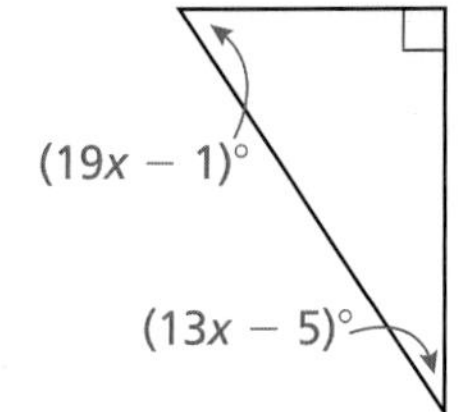

In Exercises 23–26, find the measure of each acute angle in the right triangle. *(See Example 4.)*

23. The measure of one acute angle is 5 times the measure of the other acute angle.

24. The measure of one acute angle is 8 times the measure of the other acute angle.

25. The measure of one acute angle is 3 times the sum of the measure of the other acute angle and 8.

26. The measure of one acute angle is twice the difference of the measure of the other acute angle and 12.

ERROR ANALYSIS In Exercises 27 and 28, describe and correct the error in finding $m\angle 1$.

27.

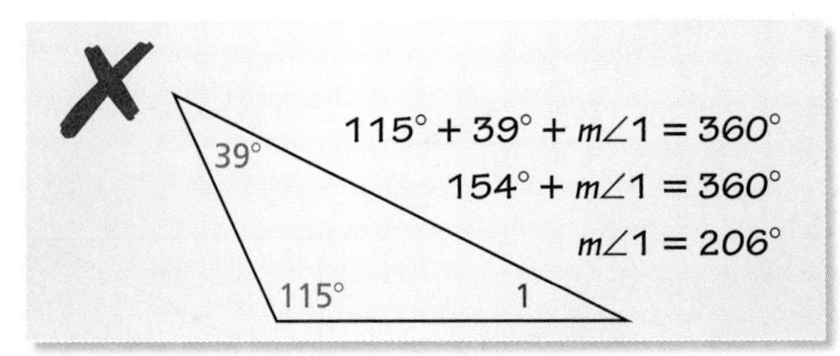

28.

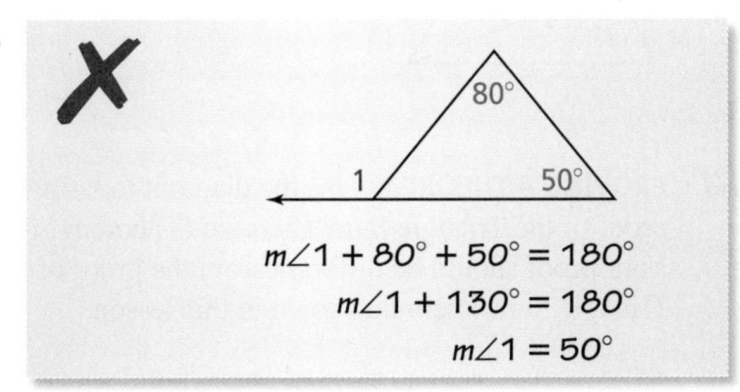

In Exercises 29–36, find the measure of the numbered angle.

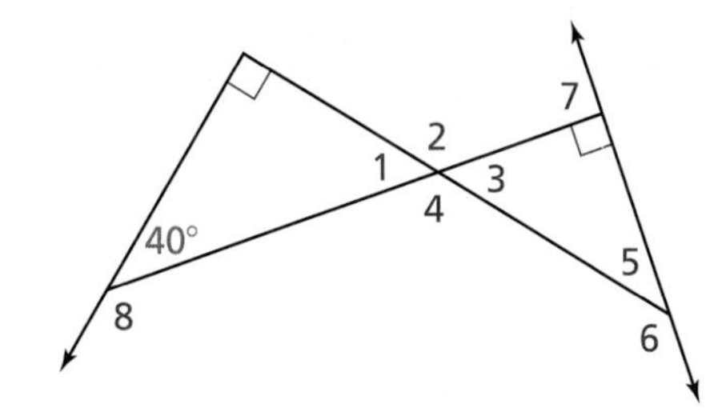

29. $\angle 1$ **30.** $\angle 2$

31. $\angle 3$ **32.** $\angle 4$

33. $\angle 5$ **34.** $\angle 6$

35. $\angle 7$ **36.** $\angle 8$

37. USING TOOLS Three people are standing on a stage. The distances between the three people are shown in the diagram. Classify the triangle by its sides and by measuring its angles.

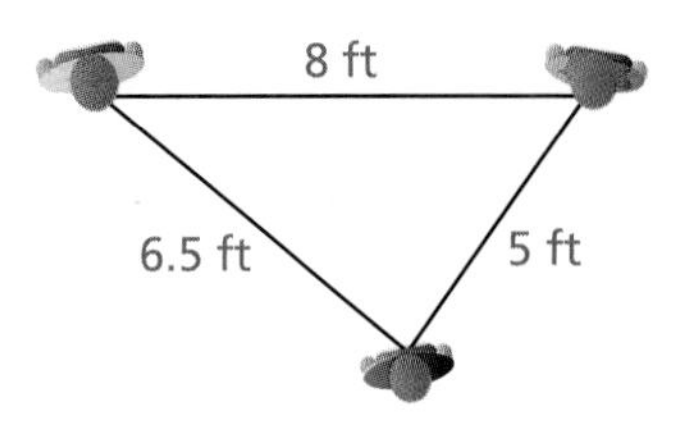

38. USING STRUCTURE Which of the following sets of angle measures could form a triangle? Select all that apply.

Ⓐ 100°, 50°, 40° Ⓑ 96°, 74°, 10°

Ⓒ 165°, 113°, 82° Ⓓ 101°, 41°, 38°

Ⓔ 90°, 45°, 45° Ⓕ 84°, 62°, 34°

39. MODELING WITH MATHEMATICS You are bending a strip of metal into an isosceles triangle for a sculpture. The strip of metal is 20 inches long. The first bend is made 6 inches from one end. Describe two ways you could complete the triangle.

40. THOUGHT PROVOKING Find and draw an object (or part of an object) that can be modeled by a triangle and an exterior angle. Describe the relationship between the interior angles of the triangle and the exterior angle in terms of the object.

41. PROVING A COROLLARY Prove the Corollary to the Triangle Sum Theorem (Corollary 5.1).

Given $\triangle ABC$ is a right triangle.

Prove $\angle A$ and $\angle B$ are complementary.

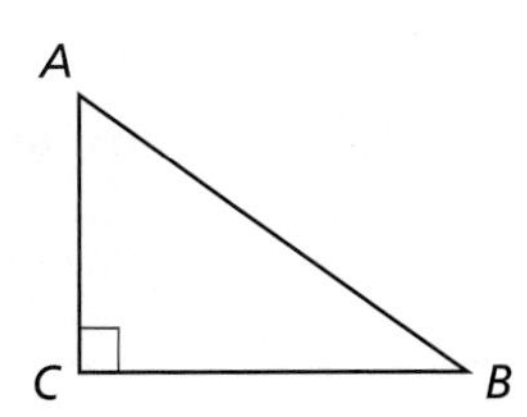

42. PROVING A THEOREM Prove the Exterior Angle Theorem (Theorem 5.2).

Given $\triangle ABC$, exterior $\angle BCD$

Prove $m\angle A + m\angle B = m\angle BCD$

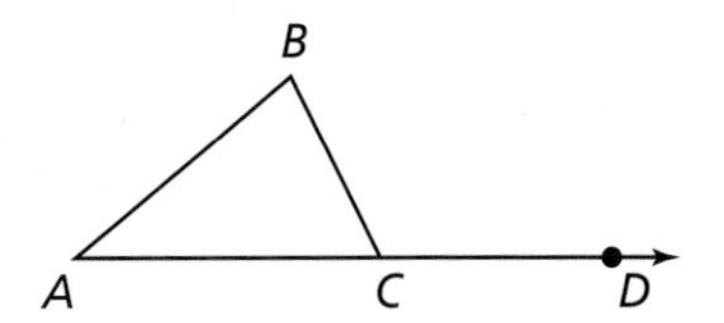

43. CRITICAL THINKING Is it possible to draw an obtuse isosceles triangle? obtuse equilateral triangle? If so, provide examples. If not, explain why it is not possible.

44. CRITICAL THINKING Is it possible to draw a right isosceles triangle? right equilateral triangle? If so, provide an example. If not, explain why it is not possible.

45. MATHEMATICAL CONNECTIONS $\triangle ABC$ is isosceles, $AB = x$, and $BC = 2x - 4$.

a. Find two possible values for x when the perimeter of $\triangle ABC$ is 32.

b. How many possible values are there for x when the perimeter of $\triangle ABC$ is 12?

46. HOW DO YOU SEE IT? Classify the triangles, in as many ways as possible, without finding any measurements.

a.

b.

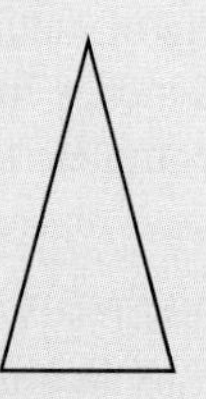

c.

d.

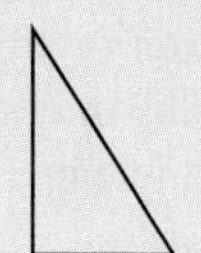

47. ANALYZING RELATIONSHIPS Which of the following could represent the measures of an exterior angle and two interior angles of a triangle? Select all that apply.

Ⓐ 100°, 62°, 38°

Ⓑ 81°, 57°, 24°

Ⓒ 119°, 68°, 49°

Ⓓ 95°, 85°, 28°

Ⓔ 92°, 78°, 68°

Ⓕ 149°, 101°, 48°

48. MAKING AN ARGUMENT Your friend claims the measure of an exterior angle will always be greater than the sum of the nonadjacent interior angle measures. Is your friend correct? Explain your reasoning.

MATHEMATICAL CONNECTIONS In Exercises 49–52, find the values of x and y.

49.

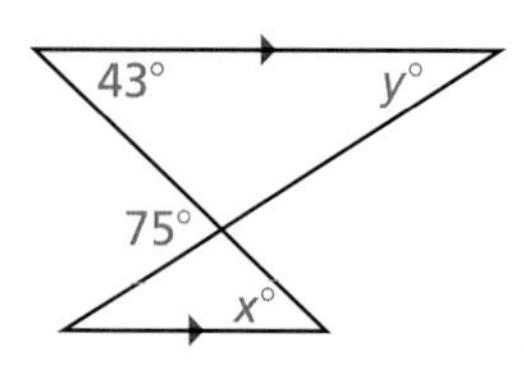

50.

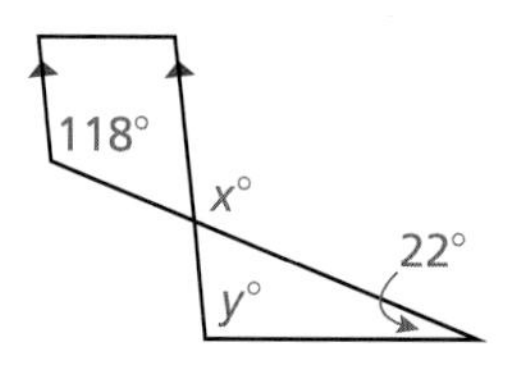

51.

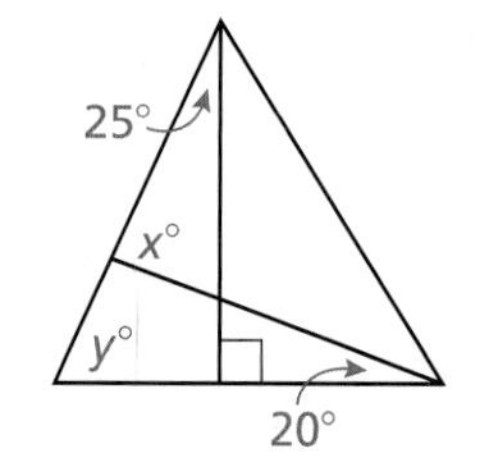

52.

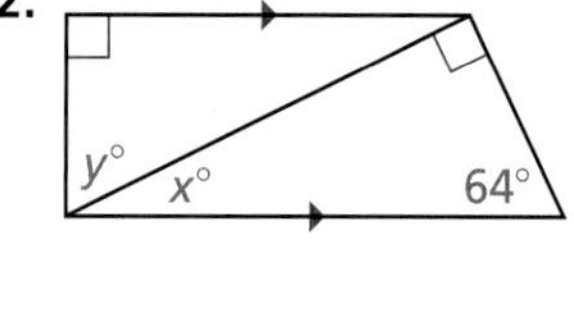

53. PROVING A THEOREM Use the diagram to write a proof of the Triangle Sum Theorem (Theorem 5.1). Your proof should be different from the proof of the Triangle Sum Theorem shown in this lesson.

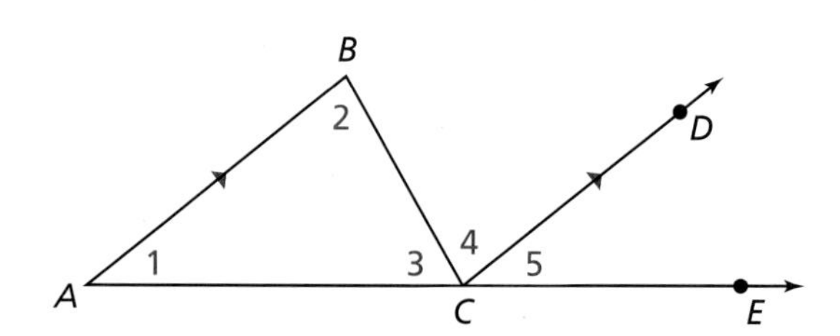

Maintaining Mathematical Proficiency

Reviewing what you learned in previous grades and lessons

Use the diagram to find the measure of the segment or angle. *(Section 1.2 and Section 1.5)*

54. $m\angle KHL$

55. $m\angle ABC$

56. GH

57. BC

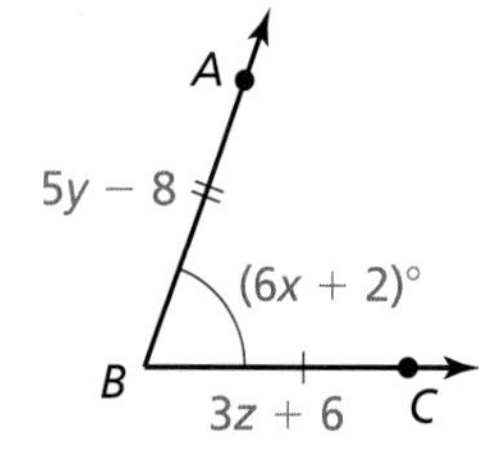

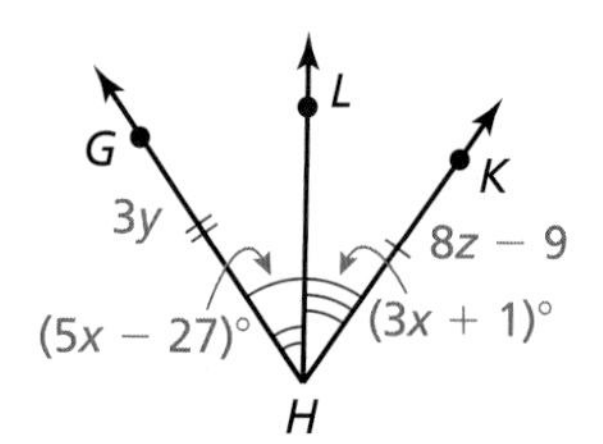

5.2 Congruent Polygons

Essential Question Given two congruent triangles, how can you use rigid motions to map one triangle to the other triangle?

EXPLORATION 1 **Describing Rigid Motions**

Work with a partner. Of the four transformations you studied in Chapter 4, which are rigid motions? Under a rigid motion, why is the image of a triangle always congruent to the original triangle? Explain your reasoning.

Translation

Reflection

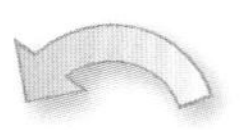
Rotation

Dilation

LOOKING FOR STRUCTURE

To be proficient in math, you need to look closely to discern a pattern or structure.

EXPLORATION 2 **Finding a Composition of Rigid Motions**

Work with a partner. Describe a composition of rigid motions that maps $\triangle ABC$ to $\triangle DEF$. Use dynamic geometry software to verify your answer.

a. $\triangle ABC \cong \triangle DEF$

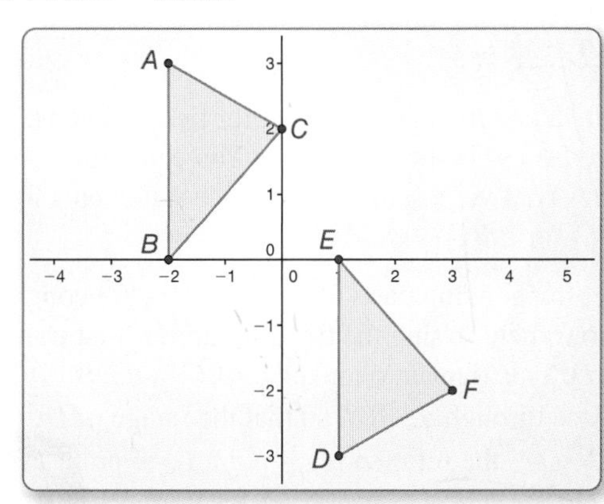

b. $\triangle ABC \cong \triangle DEF$

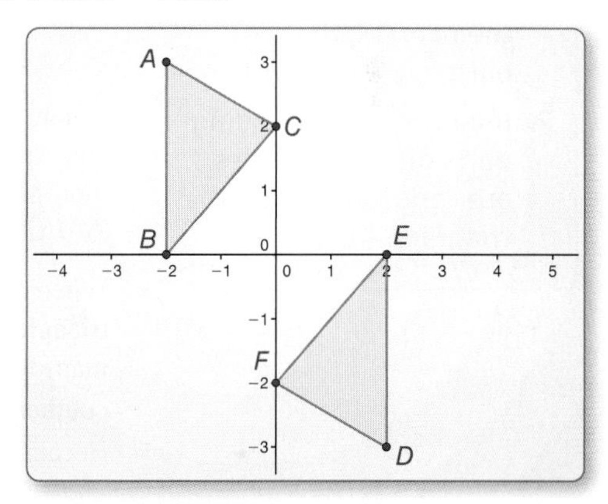

c. $\triangle ABC \cong \triangle DEF$

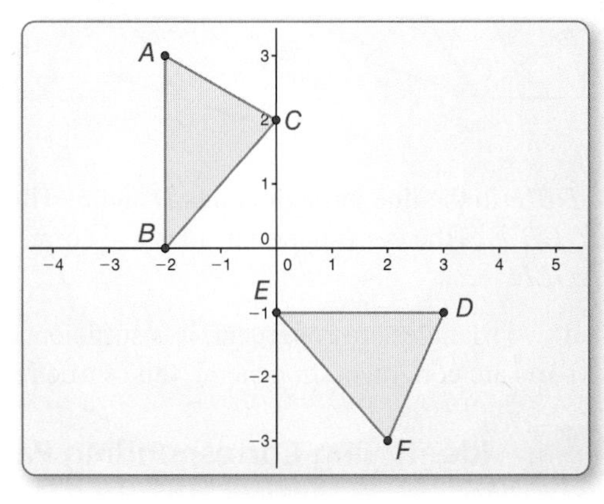

d. $\triangle ABC \cong \triangle DEF$

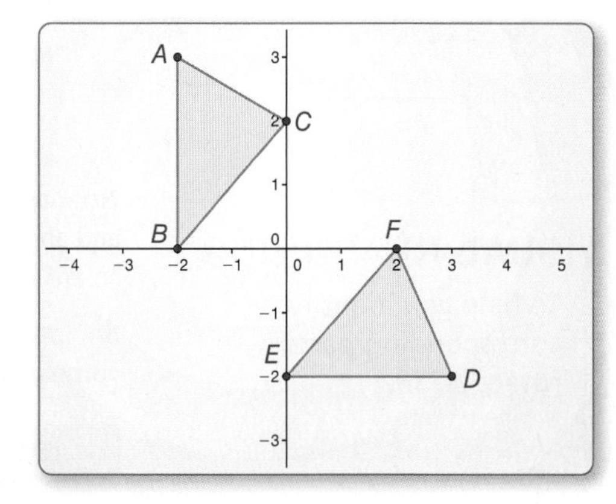

Communicate Your Answer

3. Given two congruent triangles, how can you use rigid motions to map one triangle to the other triangle?

4. The vertices of $\triangle ABC$ are $A(1, 1)$, $B(3, 2)$, and $C(4, 4)$. The vertices of $\triangle DEF$ are $D(2, -1)$, $E(0, 0)$, and $F(-1, 2)$. Describe a composition of rigid motions that maps $\triangle ABC$ to $\triangle DEF$.

5.2 Lesson

Core Vocabulary

corresponding parts, *p. 240*

Previous

congruent figures

What You Will Learn

- Identify and use corresponding parts.
- Use the Third Angles Theorem.

Identifying and Using Corresponding Parts

Recall that two geometric figures are congruent if and only if a rigid motion or a composition of rigid motions maps one of the figures onto the other. A rigid motion maps each part of a figure to a **corresponding part** of its image. Because rigid motions preserve length and angle measure, corresponding parts of congruent figures are congruent. In congruent polygons, this means that the *corresponding sides* and the *corresponding angles* are congruent.

When $\triangle DEF$ is the image of $\triangle ABC$ after a rigid motion or a composition of rigid motions, you can write congruence statements for the corresponding angles and corresponding sides.

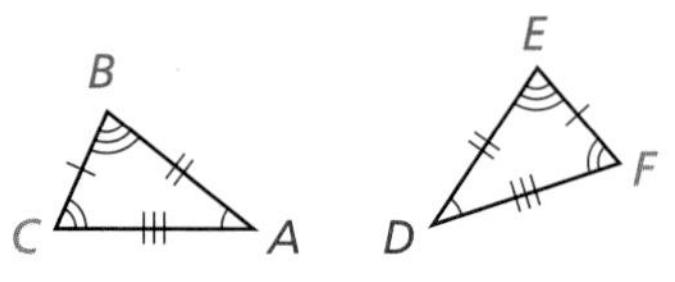

Corresponding angles

$\angle A \cong \angle D$, $\angle B \cong \angle E$, $\angle C \cong \angle F$

Corresponding sides

$\overline{AB} \cong \overline{DE}$, $\overline{BC} \cong \overline{EF}$, $\overline{AC} \cong \overline{DF}$

STUDY TIP

Notice that both of the following statements are true.

1. If two triangles are congruent, then all their corresponding parts are congruent.
2. If all the corresponding parts of two triangles are congruent, then the triangles are congruent.

When you write a congruence statement for two polygons, always list the corresponding vertices in the same order. You can write congruence statements in more than one way. Two possible congruence statements for the triangles above are $\triangle ABC \cong \triangle DEF$ or $\triangle BCA \cong \triangle EFD$.

When all the corresponding parts of two triangles are congruent, you can show that the triangles are congruent. Using the triangles above, first translate $\triangle ABC$ so that point A maps to point D. This translation maps $\triangle ABC$ to $\triangle DB'C'$. Next, rotate $\triangle DB'C'$ counterclockwise through $\angle C'DF$ so that the image of $\overrightarrow{DC'}$ coincides with $\overrightarrow{DF}$. Because $\overline{DC'} \cong \overline{DF}$, the rotation maps point C' to point F. So, this rotation maps $\triangle DB'C'$ to $\triangle DB''F$.

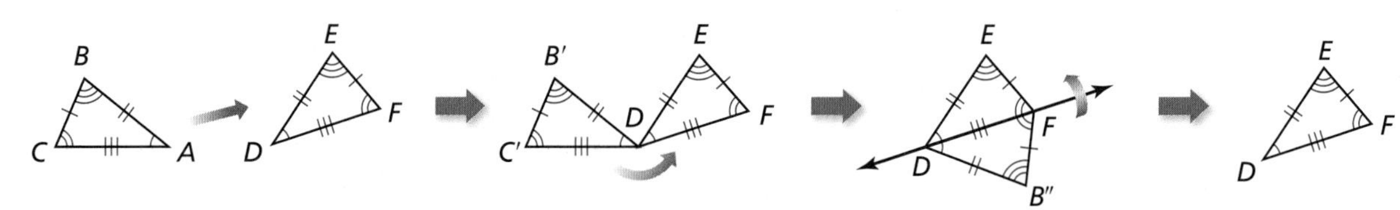

Now, reflect $\triangle DB''F$ in the line through points D and F. This reflection maps the sides and angles of $\triangle DB''F$ to the corresponding sides and corresponding angles of $\triangle DEF$, so $\triangle ABC \cong \triangle DEF$.

So, to show that two triangles are congruent, it is sufficient to show that their corresponding parts are congruent. In general, this is true for all polygons.

VISUAL REASONING

To help you identify corresponding parts, rotate $\triangle TSR$.

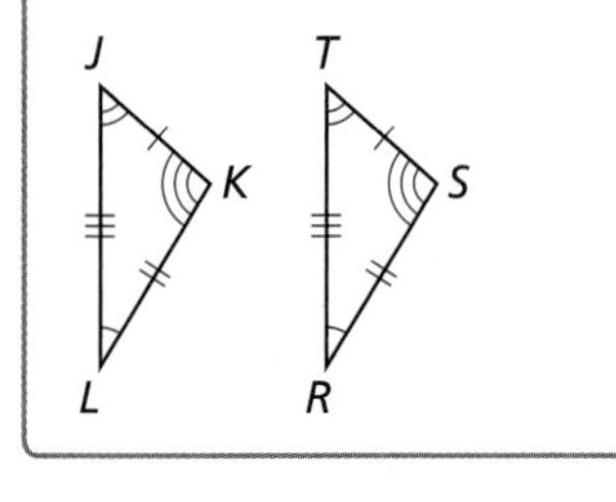

EXAMPLE 1 Identifying Corresponding Parts

Write a congruence statement for the triangles. Identify all pairs of congruent corresponding parts.

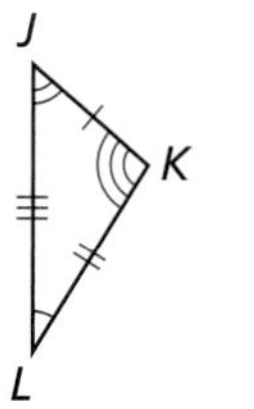

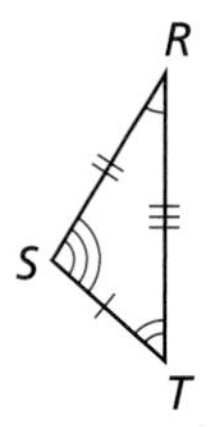

SOLUTION

The diagram indicates that $\triangle JKL \cong \triangle TSR$.

Corresponding angles $\angle J \cong \angle T$, $\angle K \cong \angle S$, $\angle L \cong \angle R$

Corresponding sides $\overline{JK} \cong \overline{TS}$, $\overline{KL} \cong \overline{SR}$, $\overline{LJ} \cong \overline{RT}$

EXAMPLE 2 Using Properties of Congruent Figures

In the diagram, $DEFG \cong SPQR$.

a. Find the value of x.

b. Find the value of y.

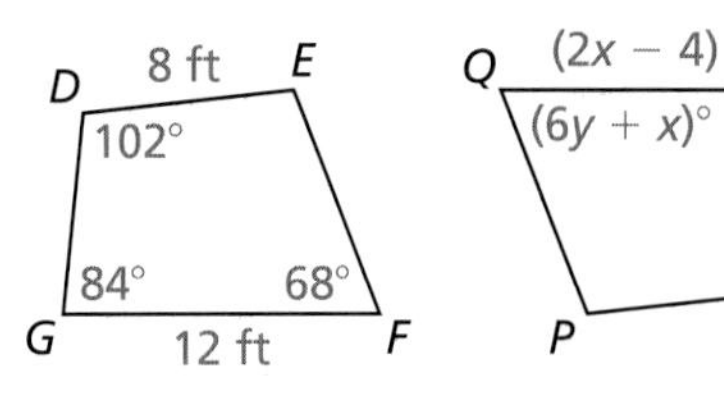

SOLUTION

a. You know that $\overline{FG} \cong \overline{QR}$.

$$FG = QR$$
$$12 = 2x - 4$$
$$16 = 2x$$
$$8 = x$$

b. You know that $\angle F \cong \angle Q$.

$$m\angle F = m\angle Q$$
$$68^\circ = (6y + x)^\circ$$
$$68 = 6y + 8$$
$$10 = y$$

EXAMPLE 3 Showing That Figures Are Congruent

You divide the wall into orange and blue sections along $\overline{JK}$. Will the sections of the wall be the same size and shape? Explain.

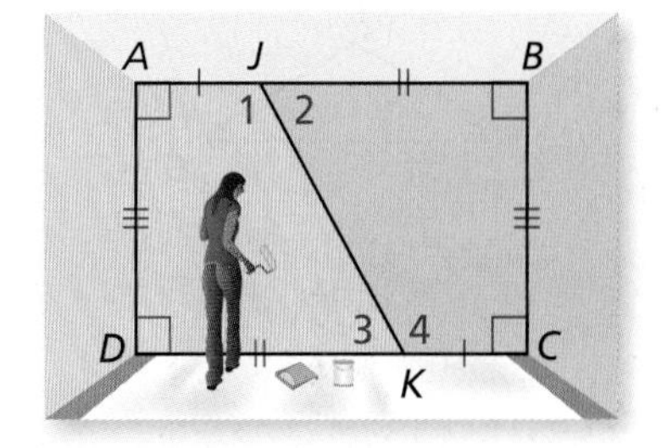

SOLUTION

From the diagram, $\angle A \cong \angle C$ and $\angle D \cong \angle B$ because all right angles are congruent. Also, by the Lines Perpendicular to a Transversal Theorem (Thm. 3.12), $\overline{AB} \parallel \overline{DC}$. Then $\angle 1 \cong \angle 4$ and $\angle 2 \cong \angle 3$ by the Alternate Interior Angles Theorem (Thm. 3.2). So, all pairs of corresponding angles are congruent. The diagram shows $\overline{AJ} \cong \overline{CK}$, $\overline{KD} \cong \overline{JB}$, and $\overline{DA} \cong \overline{BC}$. By the Reflexive Property of Congruence (Thm. 2.1), $\overline{JK} \cong \overline{KJ}$. So, all pairs of corresponding sides are congruent. Because all corresponding parts are congruent, $AJKD \cong CKJB$.

▶ Yes, the two sections will be the same size and shape.

Monitoring Progress  Help in English and Spanish at *BigIdeasMath.com*

In the diagram, $ABGH \cong CDEF$.

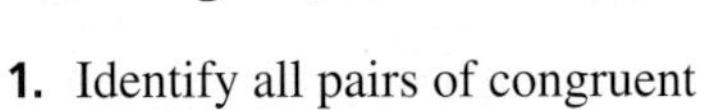

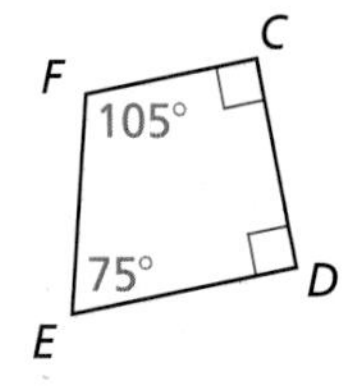

1. Identify all pairs of congruent corresponding parts.
2. Find the value of x.
3. In the diagram at the left, show that $\triangle PTS \cong \triangle RTQ$.

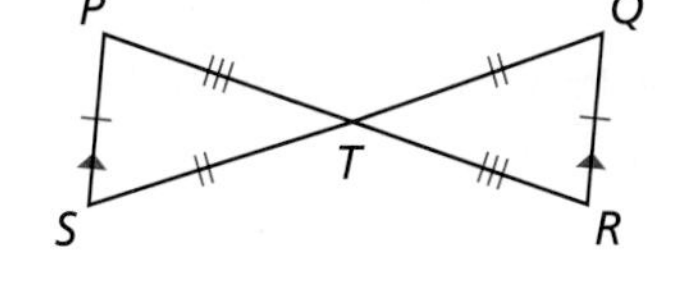

Theorem

Theorem 5.3 Properties of Triangle Congruence

Triangle congruence is reflexive, symmetric, and transitive.

Reflexive For any triangle ABC, $\triangle ABC \cong \triangle ABC$.

Symmetric If $\triangle ABC \cong \triangle DEF$, then $\triangle DEF \cong \triangle ABC$.

Transitive If $\triangle ABC \cong \triangle DEF$ and $\triangle DEF \cong \triangle JKL$, then $\triangle ABC \cong \triangle JKL$.

Proof BigIdeasMath.com

STUDY TIP

The properties of congruence that are true for segments and angles are also true for triangles.

Using the Third Angles Theorem

Theorem

Theorem 5.4 Third Angles Theorem

If two angles of one triangle are congruent to two angles of another triangle, then the third angles are also congruent.

Proof Ex. 19, p. 244

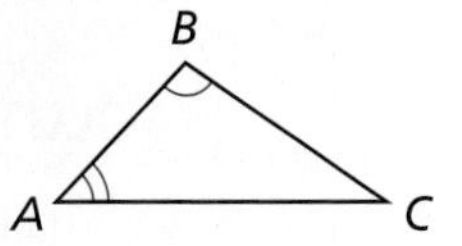

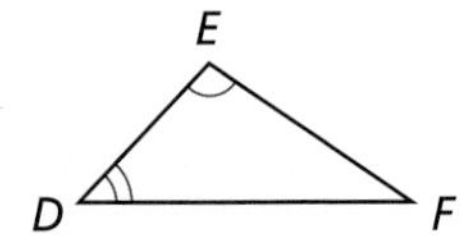

If $\angle A \cong \angle D$ and $\angle B \cong \angle E$, then $\angle C \cong \angle F$.

EXAMPLE 4 Using the Third Angles Theorem

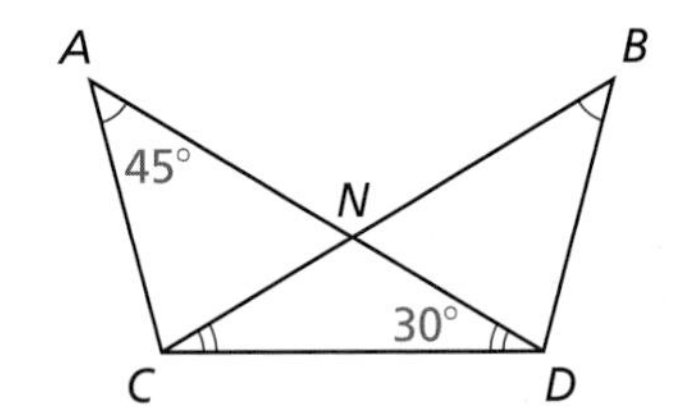

Find $m\angle BDC$.

SOLUTION

$\angle A \cong \angle B$ and $\angle ADC \cong \angle BCD$, so by the Third Angles Theorem, $\angle ACD \cong \angle BDC$. By the Triangle Sum Theorem (Theorem 5.1), $m\angle ACD = 180° - 45° - 30° = 105°$.

▶ So, $m\angle BDC = m\angle ACD = 105°$ by the definition of congruent angles.

EXAMPLE 5 Proving That Triangles Are Congruent

Use the information in the figure to prove that $\triangle ACD \cong \triangle CAB$.

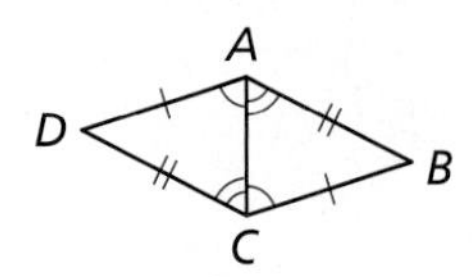

SOLUTION

Given $\overline{AD} \cong \overline{CB}, \overline{DC} \cong \overline{BA}, \angle ACD \cong \angle CAB, \angle CAD \cong \angle ACB$

Prove $\triangle ACD \cong \triangle CAB$

Plan for Proof
a. Use the Reflexive Property of Congruence (Thm. 2.1) to show that $\overline{AC} \cong \overline{CA}$.
b. Use the Third Angles Theorem to show that $\angle B \cong \angle D$.

Plan in Action

	STATEMENTS	REASONS
	1. $\overline{AD} \cong \overline{CB}, \overline{DC} \cong \overline{BA}$	**1.** Given
a.	**2.** $\overline{AC} \cong \overline{CA}$	**2.** Reflexive Property of Congruence (Theorem 2.1)
	3. $\angle ACD \cong \angle CAB$, $\angle CAD \cong \angle ACB$	**3.** Given
b.	**4.** $\angle B \cong \angle D$	**4.** Third Angles Theorem
	5. $\triangle ACD \cong \triangle CAB$	**5.** All corresponding parts are congruent.

Monitoring Progress

Help in English and Spanish at *BigIdeasMath.com*

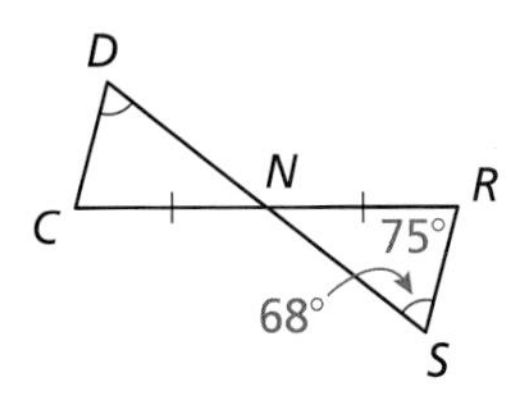

Use the diagram.

4. Find $m\angle DCN$.

5. What additional information is needed to conclude that $\triangle NDC \cong \triangle NSR$?

5.2 Exercises

Dynamic Solutions available at *BigIdeasMath.com*

Vocabulary and Core Concept Check

1. **WRITING** Based on this lesson, what information do you need to prove that two triangles are congruent? Explain your reasoning.

2. **DIFFERENT WORDS, SAME QUESTION** Which is different? Find "both" answers.

Is $\triangle JKL \cong \triangle RST$?	Is $\triangle KJL \cong \triangle SRT$?
Is $\triangle JLK \cong \triangle STR$?	Is $\triangle LKJ \cong \triangle TSR$?

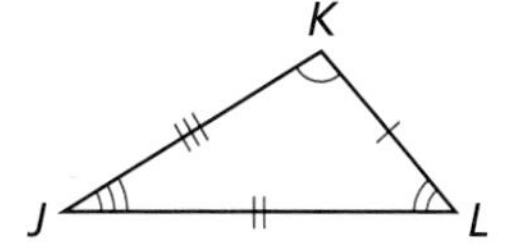

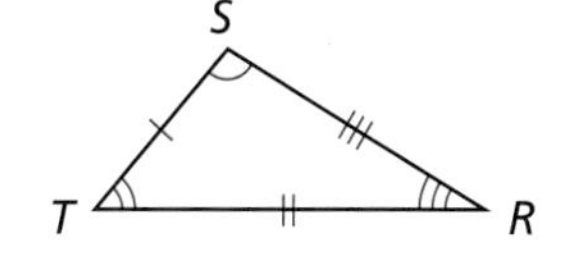

Monitoring Progress and Modeling with Mathematics

In Exercises 3 and 4, identify all pairs of congruent corresponding parts. Then write another congruence statement for the polygons. *(See Example 1.)*

3. $\triangle ABC \cong \triangle DEF$

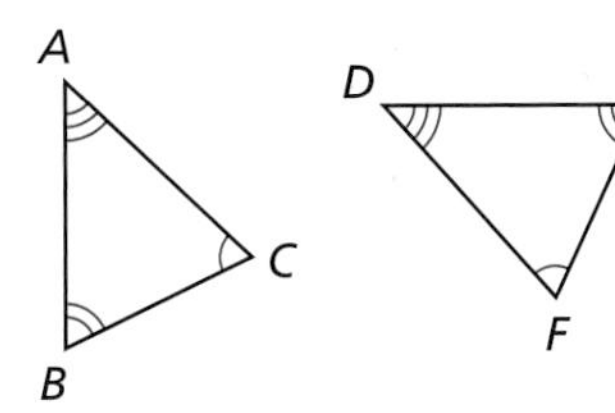

4. $GHJK \cong QRST$

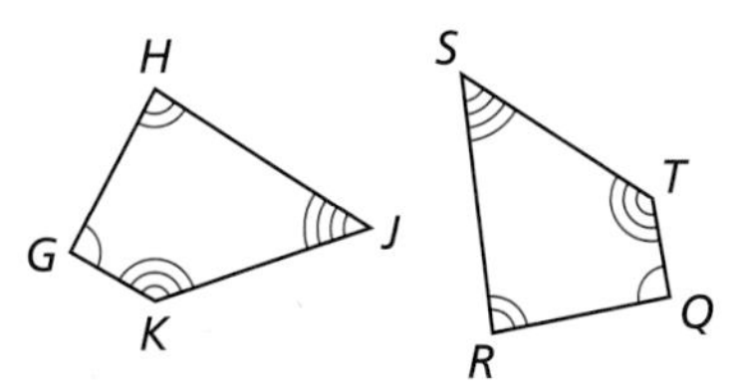

In Exercises 5–8, $\triangle XYZ \cong \triangle MNL$. Copy and complete the statement.

5. $m\angle Y =$ ______

6. $m\angle M =$ ______

7. $m\angle Z =$ ______

8. $XY =$ ______

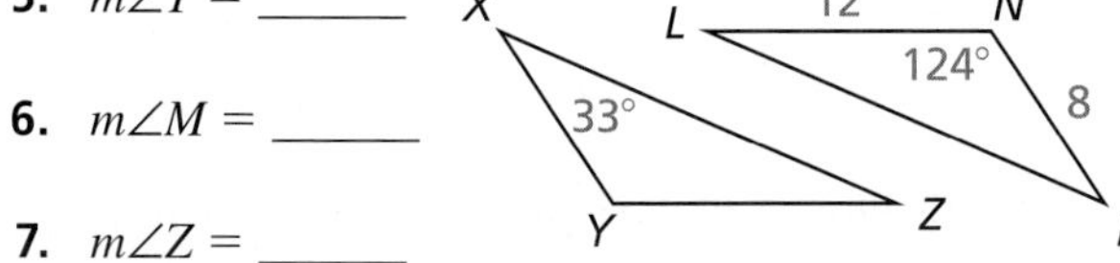

In Exercises 9 and 10, find the values of *x* and *y*. *(See Example 2.)*

9. $ABCD \cong EFGH$

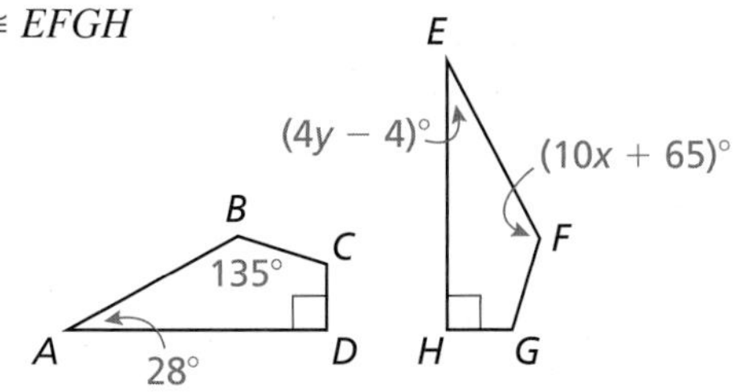

10. $\triangle MNP \cong \triangle TUS$

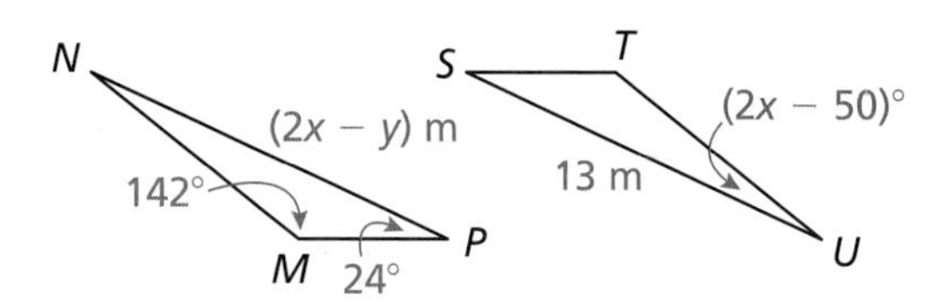

In Exercises 11 and 12, show that the polygons are congruent. Explain your reasoning. *(See Example 3.)*

11.

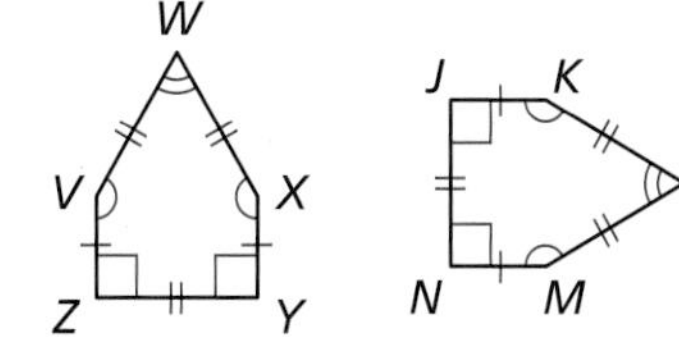

12.

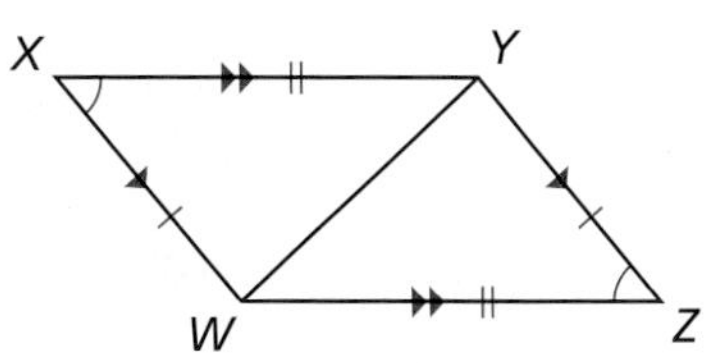

In Exercises 13 and 14, find $m\angle 1$. *(See Example 4.)*

13.

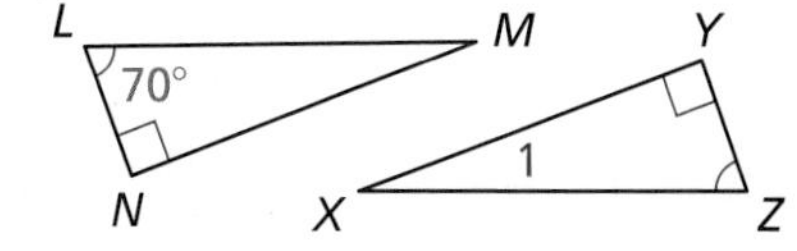

14.

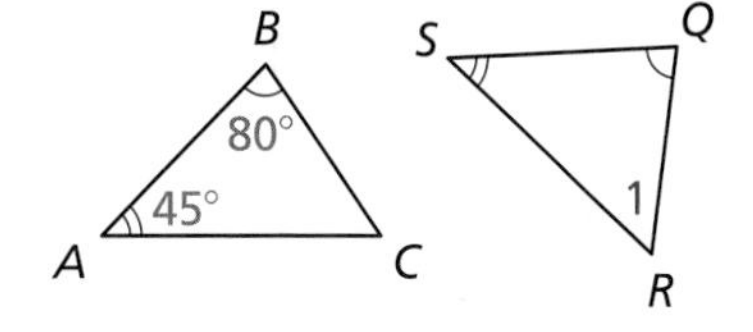

15. PROOF Triangular postage stamps, like the ones shown, are highly valued by stamp collectors. Prove that $\triangle AEB \cong \triangle CED$. *(See Example 5.)*

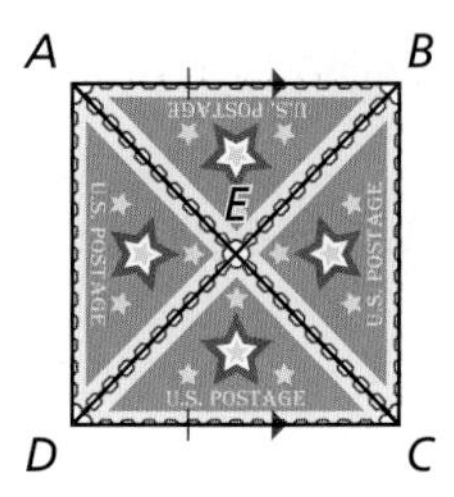

Given $\overline{AB} \parallel \overline{DC}$, $\overline{AB} \cong \overline{DC}$, E is the midpoint of $\overline{AC}$ and $\overline{BD}$.

Prove $\triangle AEB \cong \triangle CED$

16. PROOF Use the information in the figure to prove that $\triangle ABG \cong \triangle DCF$.

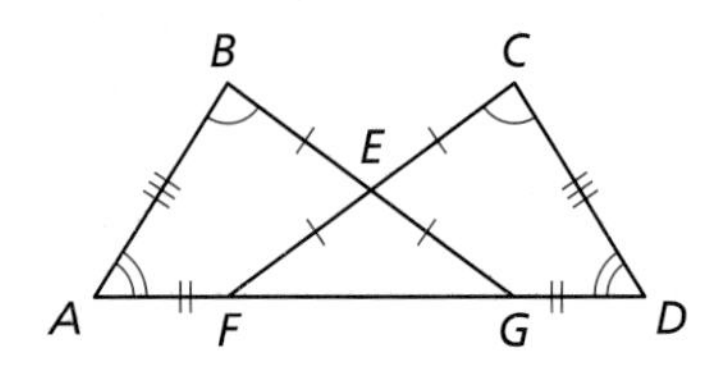

ERROR ANALYSIS In Exercises 17 and 18, describe and correct the error.

17.

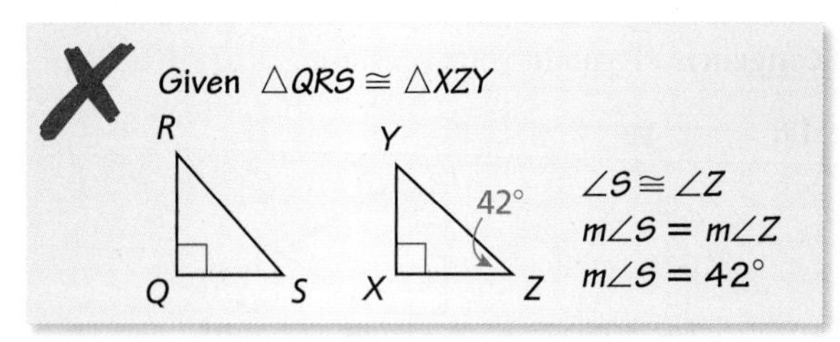

18.

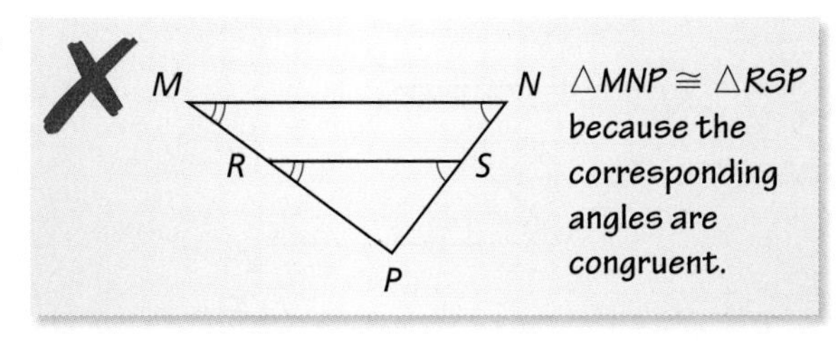

19. PROVING A THEOREM Prove the Third Angles Theorem (Theorem 5.4) by using the Triangle Sum Theorem (Theorem 5.1).

20. THOUGHT PROVOKING Draw a triangle. Copy the triangle multiple times to create a rug design made of congruent triangles. Which property guarantees that all the triangles are congruent?

21. REASONING $\triangle JKL$ is congruent to $\triangle XYZ$. Identify all pairs of congruent corresponding parts.

22. HOW DO YOU SEE IT? In the diagram, $ABEF \cong CDEF$.

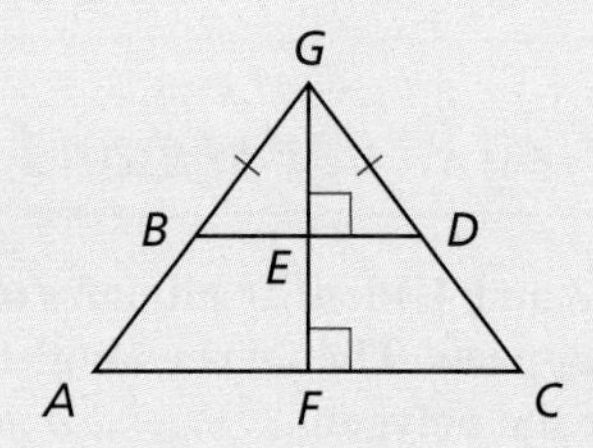

a. Explain how you know that $\overline{BE} \cong \overline{DE}$ and $\angle ABE \cong \angle CDE$.

b. Explain how you know that $\angle GBE \cong \angle GDE$.

c. Explain how you know that $\angle GEB \cong \angle GED$.

d. Do you have enough information to prove that $\triangle BEG \cong \triangle DEG$? Explain.

MATHEMATICAL CONNECTIONS In Exercises 23 and 24, use the given information to write and solve a system of linear equations to find the values of *x* and *y*.

23. $\triangle LMN \cong \triangle PQR$, $m\angle L = 40°$, $m\angle M = 90°$, $m\angle P = (17x - y)°$, $m\angle R = (2x + 4y)°$

24. $\triangle STU \cong \triangle XYZ$, $m\angle T = 28°$, $m\angle U = (4x + y)°$, $m\angle X = 130°$, $m\angle Y = (8x - 6y)°$

25. PROOF Prove that the criteria for congruent triangles in this lesson is equivalent to the definition of congruence in terms of rigid motions.

Maintaining Mathematical Proficiency

Reviewing what you learned in previous grades and lessons

What can you conclude from the diagram? *(Section 1.6)*

26.

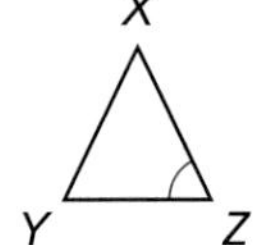

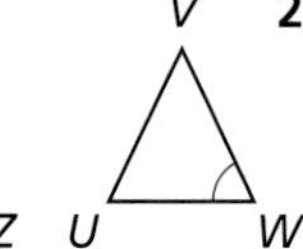

27.

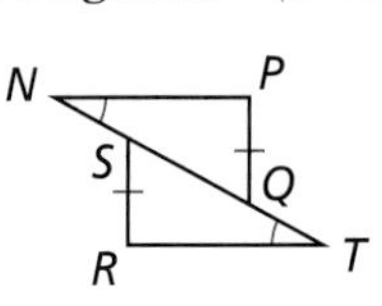

28.

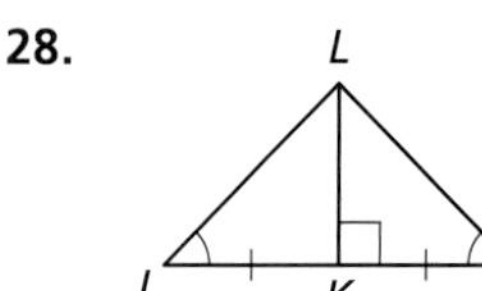

29.

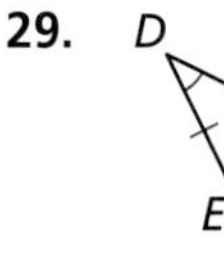

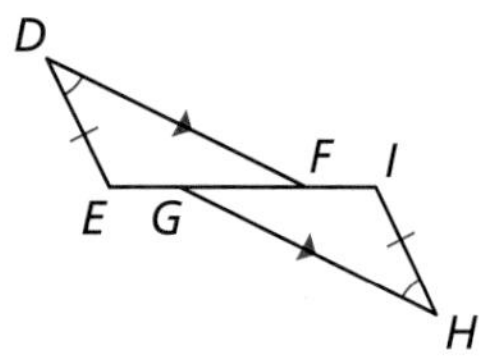

5.3 Proving Triangle Congruence by SAS

Essential Question

What can you conclude about two triangles when you know that two pairs of corresponding sides and the corresponding included angles are congruent?

EXPLORATION 1 Drawing Triangles

Work with a partner. Use dynamic geometry software.

a. Construct circles with radii of 2 units and 3 units centered at the origin. Construct a 40° angle with its vertex at the origin. Label the vertex A.

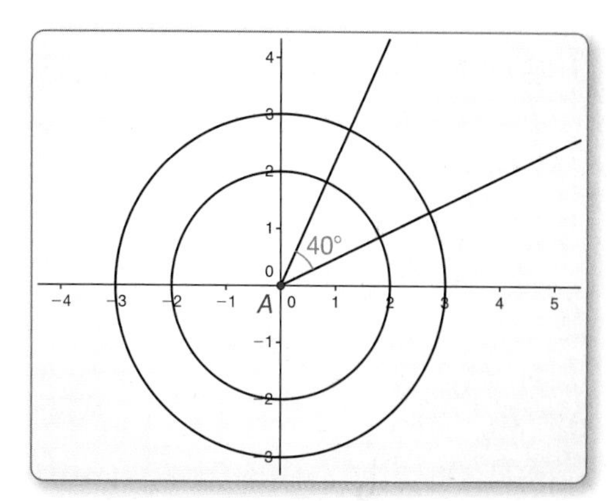

b. Locate the point where one ray of the angle intersects the smaller circle and label this point B. Locate the point where the other ray of the angle intersects the larger circle and label this point C. Then draw $\triangle ABC$.

USING TOOLS STRATEGICALLY

To be proficient in math, you need to use technology to help visualize the results of varying assumptions, explore consequences, and compare predictions with data.

c. Find BC, $m\angle B$, and $m\angle C$.

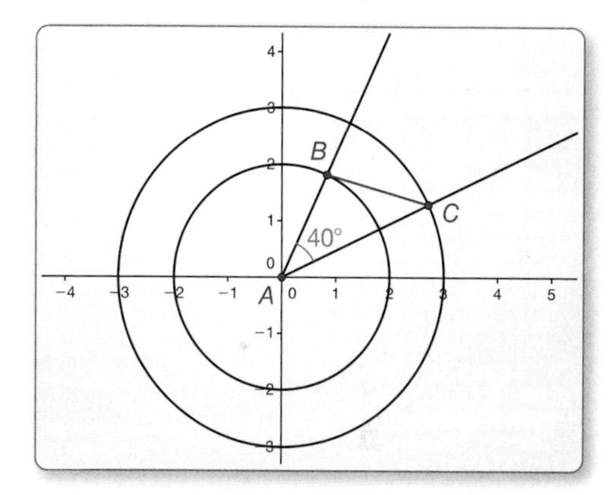

d. Repeat parts (a)–(c) several times, redrawing the angle in different positions. Keep track of your results by copying and completing the table below. What can you conclude?

	A	B	C	AB	AC	BC	m∠A	m∠B	m∠C
1.	(0, 0)			2	3		40°		
2.	(0, 0)			2	3		40°		
3.	(0, 0)			2	3		40°		
4.	(0, 0)			2	3		40°		
5.	(0, 0)			2	3		40°		

Communicate Your Answer

2. What can you conclude about two triangles when you know that two pairs of corresponding sides and the corresponding included angles are congruent?

3. How would you prove your conclusion in Exploration 1(d)?

5.3 Lesson

Core Vocabulary

Previous
congruent figures
rigid motion

What You Will Learn

- Use the Side-Angle-Side (SAS) Congruence Theorem.
- Solve real-life problems.

Using the Side-Angle-Side Congruence Theorem

Theorem

Theorem 5.5 Side-Angle-Side (SAS) Congruence Theorem

If two sides and the included angle of one triangle are congruent to two sides and the included angle of a second triangle, then the two triangles are congruent.

If $\overline{AB} \cong \overline{DE}$, $\angle A \cong \angle D$, and $\overline{AC} \cong \overline{DF}$, then $\triangle ABC \cong \triangle DEF$.

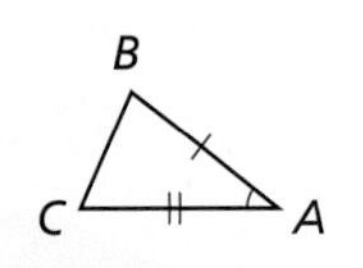

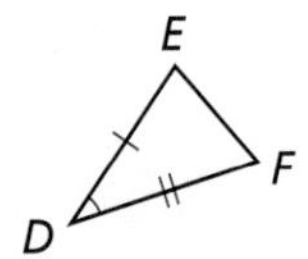

Proof p. 246

STUDY TIP

The *included angle* of two sides of a triangle is the angle formed by the two sides.

PROOF Side-Angle-Side (SAS) Congruence Theorem

Given $\overline{AB} \cong \overline{DE}$, $\angle A \cong \angle D$, $\overline{AC} \cong \overline{DF}$

Prove $\triangle ABC \cong \triangle DEF$

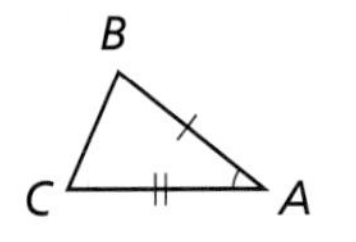

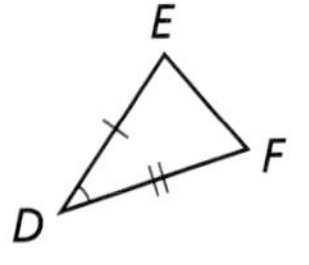

First, translate $\triangle ABC$ so that point A maps to point D, as shown below.

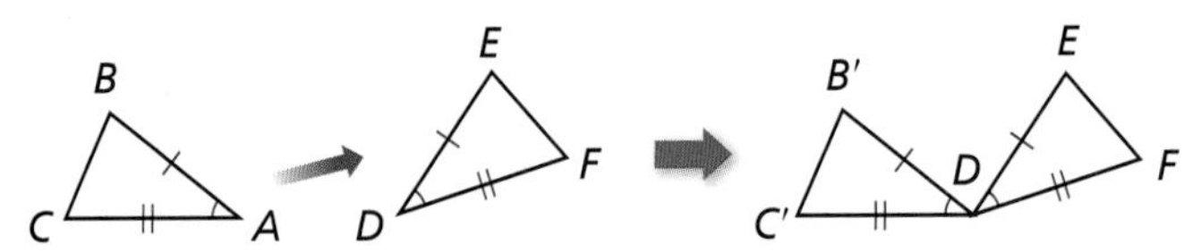

This translation maps $\triangle ABC$ to $\triangle DB'C'$. Next, rotate $\triangle DB'C'$ counterclockwise through $\angle C'DF$ so that the image of $\overrightarrow{DC'}$ coincides with $\overrightarrow{DF}$, as shown below.

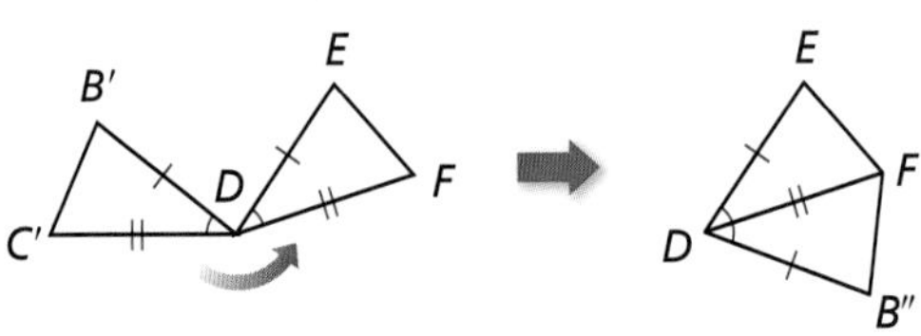

Because $\overline{DC'} \cong \overline{DF}$, the rotation maps point C' to point F. So, this rotation maps $\triangle DB'C'$ to $\triangle DB''F$. Now, reflect $\triangle DB''F$ in the line through points D and F, as shown below.

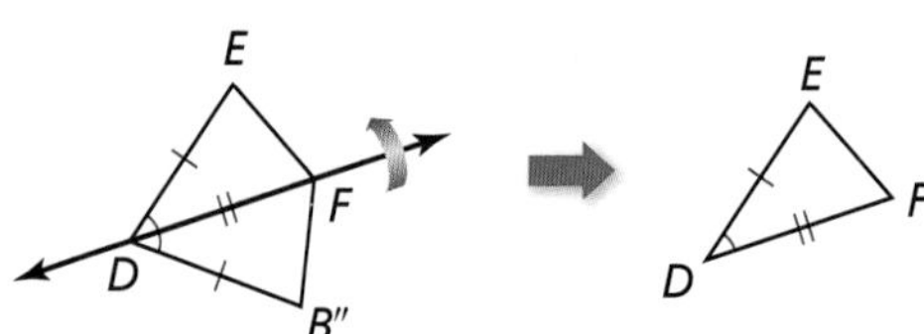

Because points D and F lie on $\overleftrightarrow{DF}$, this reflection maps them onto themselves. Because a reflection preserves angle measure and $\angle B''DF \cong \angle EDF$, the reflection maps $\overrightarrow{DB''}$ to $\overrightarrow{DE}$. Because $\overline{DB''} \cong \overline{DE}$, the reflection maps point B'' to point E. So, this reflection maps $\triangle DB''F$ to $\triangle DEF$.

Because you can map $\triangle ABC$ to $\triangle DEF$ using a composition of rigid motions, $\triangle ABC \cong \triangle DEF$.

EXAMPLE 1 Using the SAS Congruence Theorem

Write a proof.

Given $\overline{BC} \cong \overline{DA}$, $\overline{BC} \parallel \overline{AD}$

Prove $\triangle ABC \cong \triangle CDA$

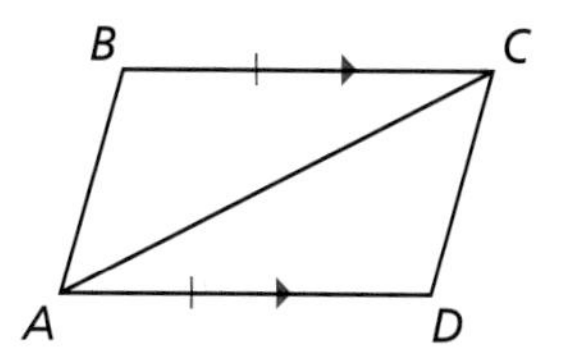

STUDY TIP

Make your proof easier to read by identifying the steps where you show congruent sides (S) and angles (A).

SOLUTION

STATEMENTS	REASONS
S **1.** $\overline{BC} \cong \overline{DA}$	**1.** Given
2. $\overline{BC} \parallel \overline{AD}$	**2.** Given
A **3.** $\angle BCA \cong \angle DAC$	**3.** Alternate Interior Angles Theorem (Thm. 3.2)
S **4.** $\overline{AC} \cong \overline{CA}$	**4.** Reflexive Property of Congruence (Thm. 2.1)
5. $\triangle ABC \cong \triangle CDA$	**5.** SAS Congruence Theorem

EXAMPLE 2 Using SAS and Properties of Shapes

In the diagram, $\overline{QS}$ and $\overline{RP}$ pass through the center M of the circle. What can you conclude about $\triangle MRS$ and $\triangle MPQ$?

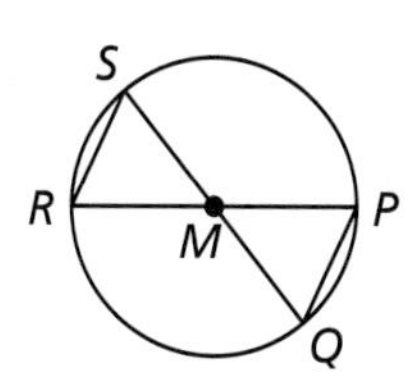

SOLUTION

Because they are vertical angles, $\angle PMQ \cong \angle RMS$. All points on a circle are the same distance from the center, so $\overline{MP}$, $\overline{MQ}$, $\overline{MR}$, and $\overline{MS}$ are all congruent.

▶ So, $\triangle MRS$ and $\triangle MPQ$ are congruent by the SAS Congruence Theorem.

Monitoring Progress

Help in English and Spanish at *BigIdeasMath.com*

In the diagram, *ABCD* is a square with four congruent sides and four right angles. *R*, *S*, *T*, and *U* are the midpoints of the sides of *ABCD*. Also, $\overline{RT} \perp \overline{SU}$ and $\overline{SV} \cong \overline{VU}$.

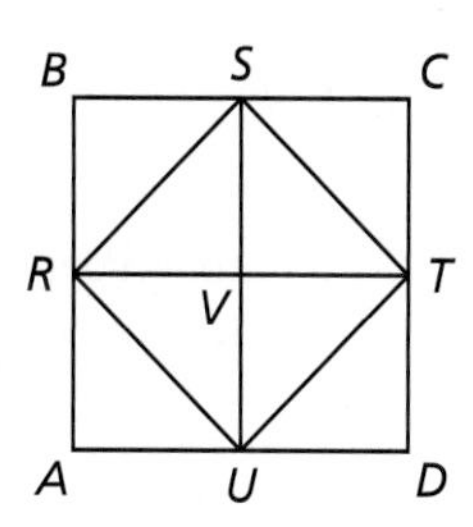

1. Prove that $\triangle SVR \cong \triangle UVR$.

2. Prove that $\triangle BSR \cong \triangle DUT$.

CONSTRUCTION Copying a Triangle Using SAS

Construct a triangle that is congruent to $\triangle ABC$ using the SAS Congruence Theorem. Use a compass and straightedge.

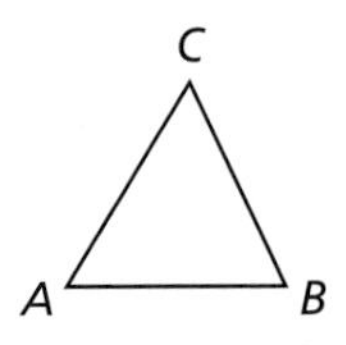

SOLUTION

Step 1

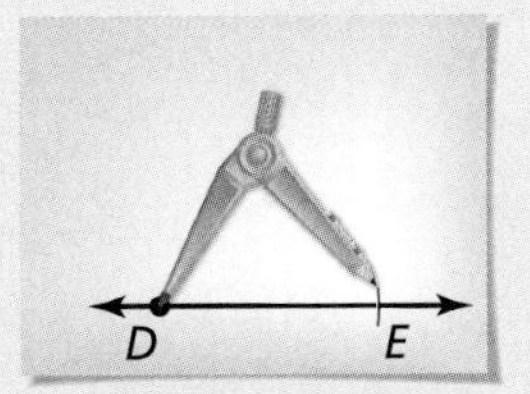

Construct a side
Construct $\overline{DE}$ so that it is congruent to $\overline{AB}$.

Step 2

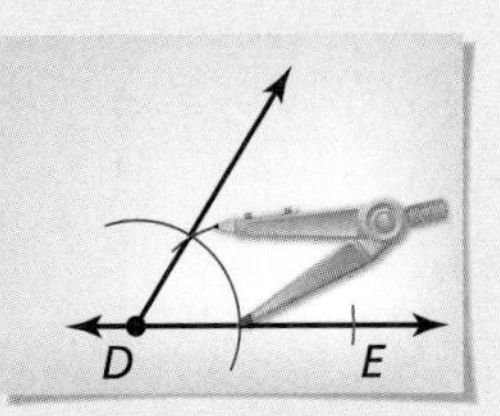

Construct an angle
Construct $\angle D$ with vertex D and side $\overrightarrow{DE}$ so that it is congruent to $\angle A$.

Step 3

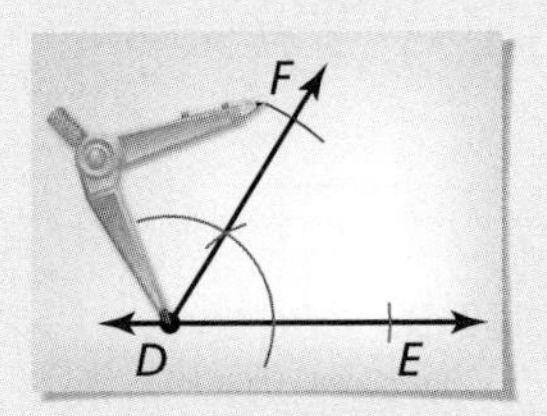

Construct a side
Construct $\overline{DF}$ so that it is congruent to $\overline{AC}$.

Step 4

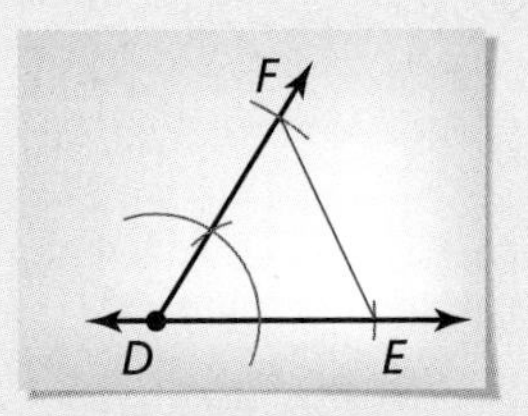

Draw a triangle
Draw $\triangle DEF$. By the SAS Congruence Theorem, $\triangle ABC \cong \triangle DEF$.

Solving Real-Life Problems

EXAMPLE 3 Solving a Real-Life Problem

You are making a canvas sign to hang on the triangular portion of the barn wall shown in the picture. You think you can use two identical triangular sheets of canvas. You know that $\overline{RP} \perp \overline{QS}$ and $\overline{PQ} \cong \overline{PS}$. Use the SAS Congruence Theorem to show that $\triangle PQR \cong \triangle PSR$.

SOLUTION

You are given that $\overline{PQ} \cong \overline{PS}$. By the Reflexive Property of Congruence (Theorem 2.1), $\overline{RP} \cong \overline{RP}$. By the definition of perpendicular lines, both $\angle RPQ$ and $\angle RPS$ are right angles, so they are congruent. So, two pairs of sides and their included angles are congruent.

▶ $\triangle PQR$ and $\triangle PSR$ are congruent by the SAS Congruence Theorem.

Monitoring Progress

Help in English and Spanish at *BigIdeasMath.com*

3. You are designing the window shown in the photo. You want to make $\triangle DRA$ congruent to $\triangle DRG$. You design the window so that $\overline{DA} \cong \overline{DG}$ and $\angle ADR \cong \angle GDR$. Use the SAS Congruence Theorem to prove $\triangle DRA \cong \triangle DRG$.

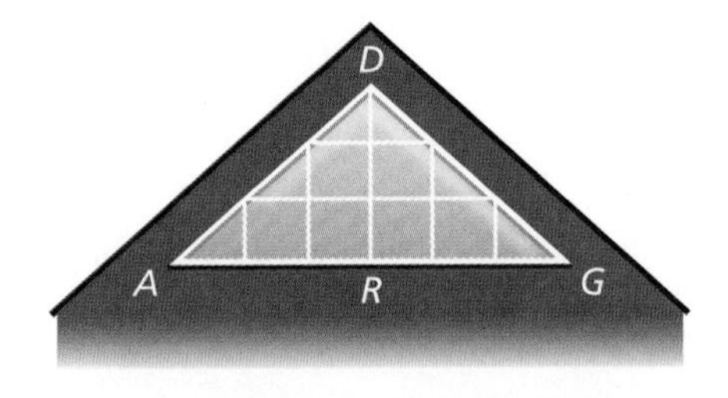

5.3 Exercises

Dynamic Solutions available at *BigIdeasMath.com*

Vocabulary and Core Concept Check

1. **WRITING** What is an included angle?

2. **COMPLETE THE SENTENCE** If two sides and the included angle of one triangle are congruent to two sides and the included angle of a second triangle, then __________.

Monitoring Progress and Modeling with Mathematics

In Exercises 3–8, name the included angle between the pair of sides given.

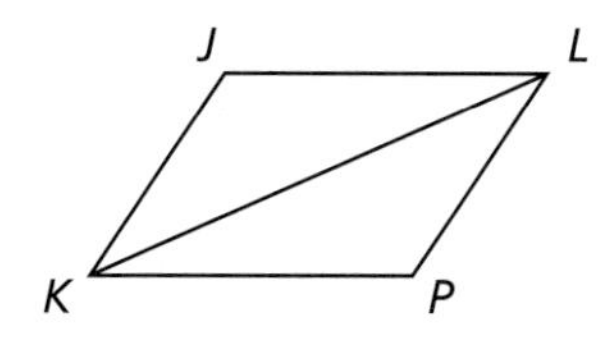

3. $\overline{JK}$ and $\overline{KL}$
4. $\overline{PK}$ and $\overline{LK}$
5. $\overline{LP}$ and $\overline{LK}$
6. $\overline{JL}$ and $\overline{JK}$
7. $\overline{KL}$ and $\overline{JL}$
8. $\overline{KP}$ and $\overline{PL}$

In Exercises 9–14, decide whether enough information is given to prove that the triangles are congruent using the SAS Congruence Theorem (Theorem 5.5). Explain.

9. $\triangle ABD$, $\triangle CDB$

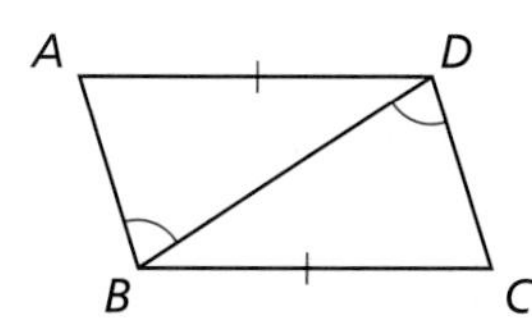

10. $\triangle LMN$, $\triangle NQP$

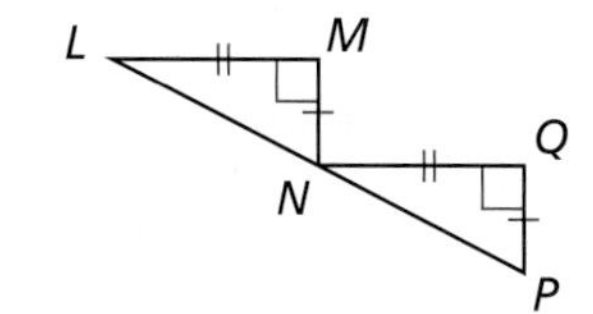

11. $\triangle YXZ$, $\triangle WXZ$

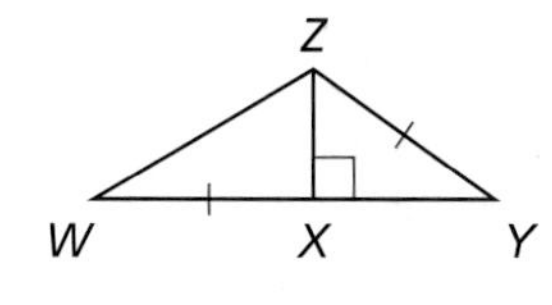

12. $\triangle QRV$, $\triangle TSU$

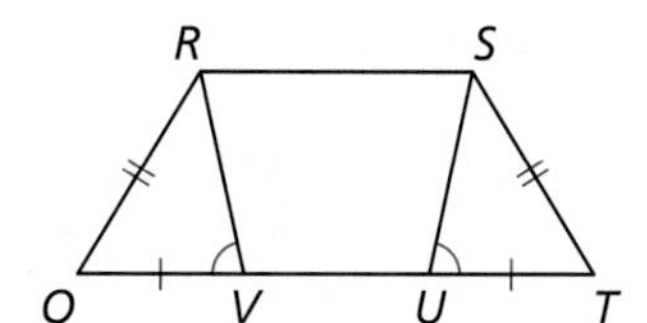

13. $\triangle EFH$, $\triangle GHF$

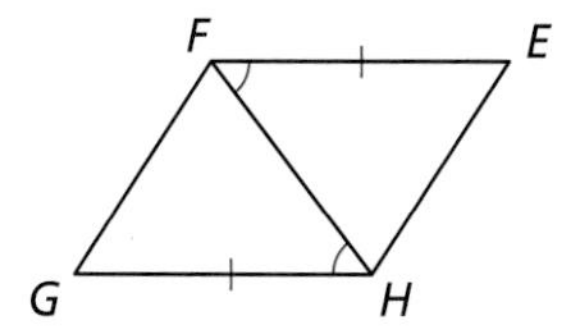

14. $\triangle KLM$, $\triangle MNK$

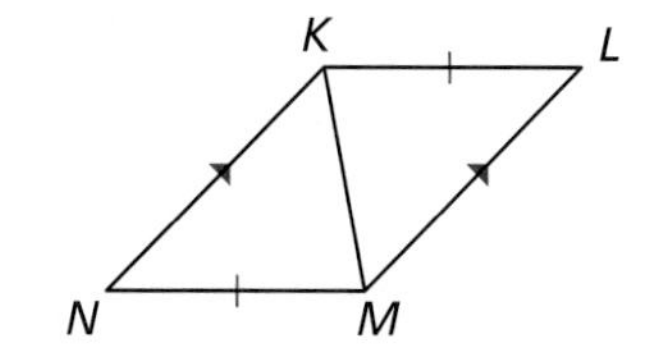

In Exercises 15–18, write a proof. *(See Example 1.)*

15. **Given** $\overline{PQ}$ bisects $\angle SPT$, $\overline{SP} \cong \overline{TP}$

    Prove $\triangle SPQ \cong \triangle TPQ$

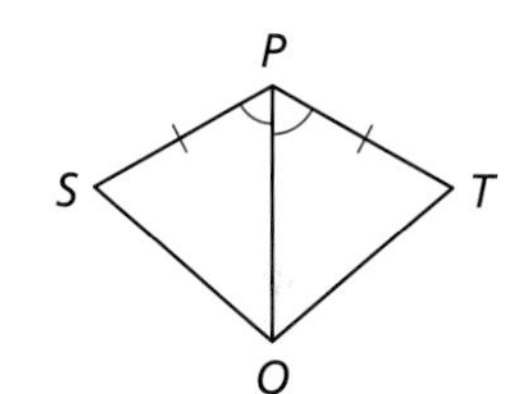

16. **Given** $\overline{AB} \cong \overline{CD}$, $\overline{AB} \parallel \overline{CD}$

    Prove $\triangle ABC \cong \triangle CDA$

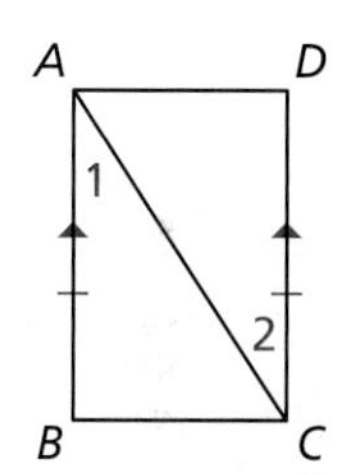

17. **Given** C is the midpoint of $\overline{AE}$ and $\overline{BD}$.

    Prove $\triangle ABC \cong \triangle EDC$

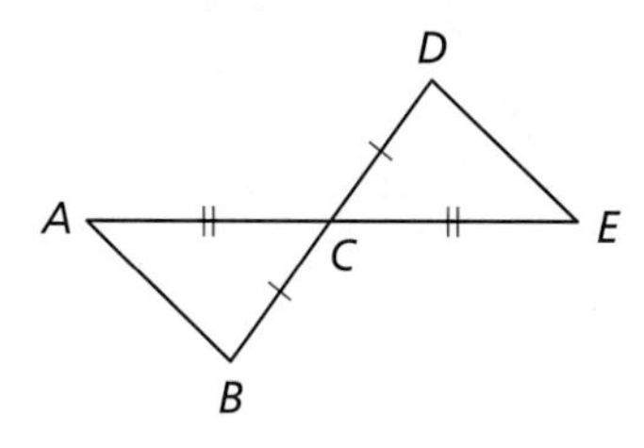

18. **Given** $\overline{PT} \cong \overline{RT}$, $\overline{QT} \cong \overline{ST}$

    Prove $\triangle PQT \cong \triangle RST$

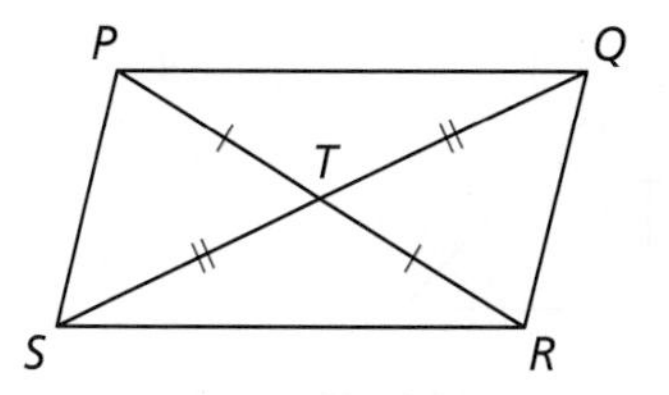

In Exercises 19–22, use the given information to name two triangles that are congruent. Explain your reasoning. *(See Example 2.)*

19. $\angle SRT \cong \angle URT$, and R is the center of the circle.

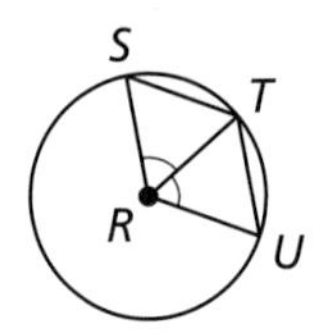

20. $ABCD$ is a square with four congruent sides and four congruent angles.

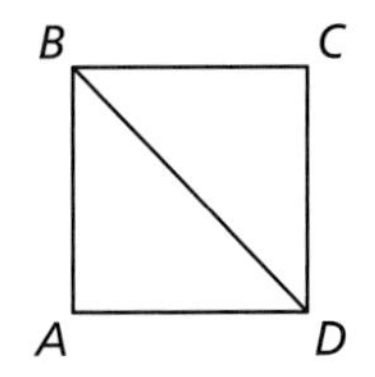

21. $RSTUV$ is a regular pentagon.

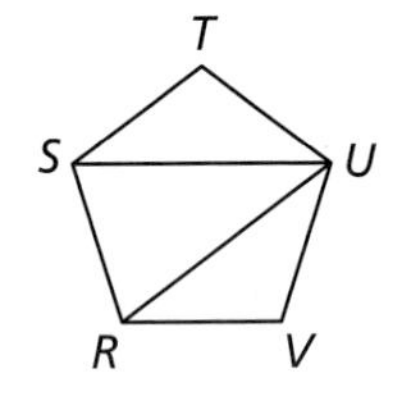

22. $\overline{MK} \perp \overline{MN}$, $\overline{KL} \perp \overline{NL}$, and M and L are centers of circles.

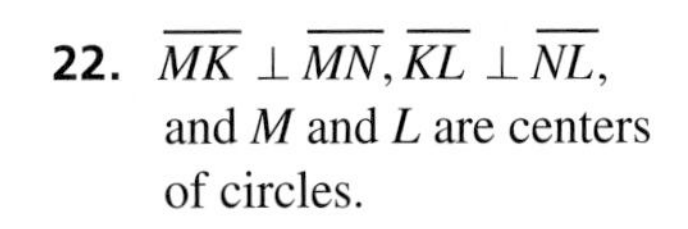

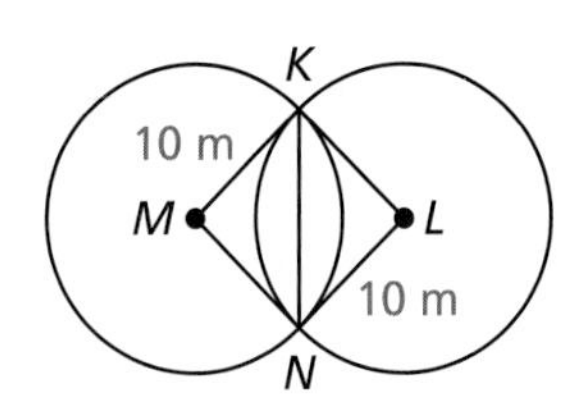

CONSTRUCTION In Exercises 23 and 24, construct a triangle that is congruent to $\triangle ABC$ using the SAS Congruence Theorem (Theorem 5.5).

23. **24.**

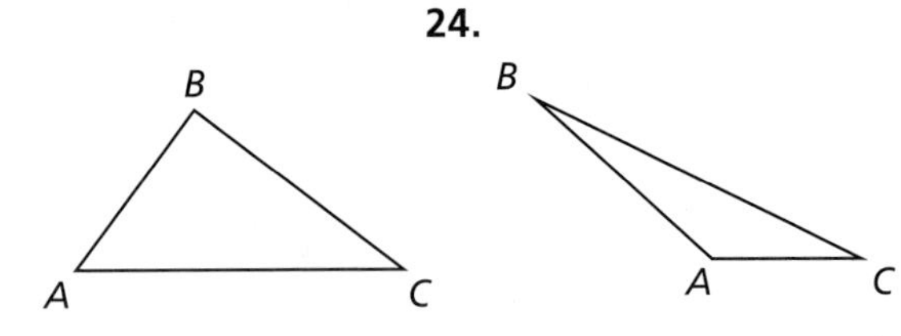

25. ERROR ANALYSIS Describe and correct the error in finding the value of x.

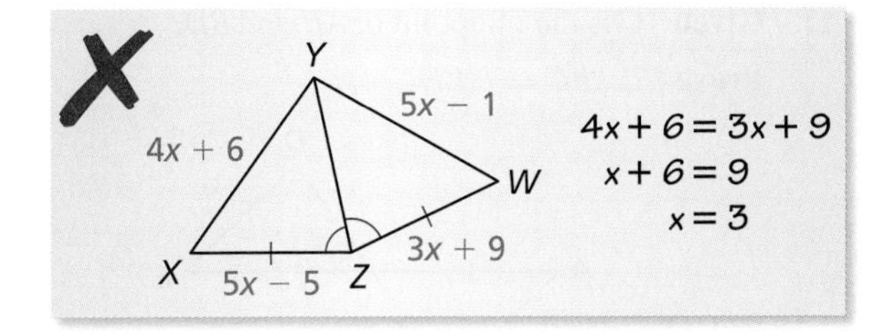

26. HOW DO YOU SEE IT? What additional information do you need to prove that $\triangle ABC \cong \triangle DBC$?

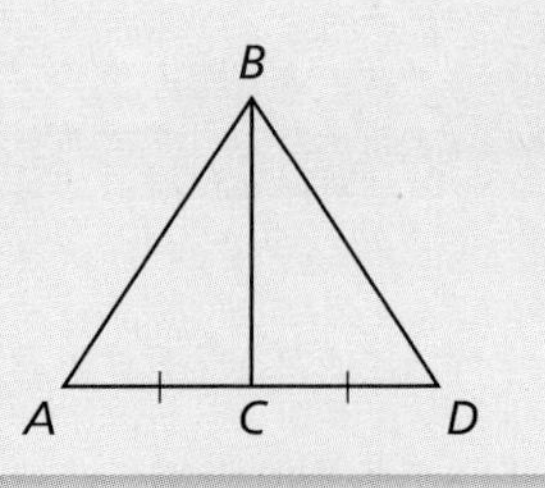

27. PROOF The Navajo rug is made of isosceles triangles. You know $\angle B \cong \angle D$. Use the SAS Congruence Theorem (Theorem 5.5) to show that $\triangle ABC \cong \triangle CDE$. *(See Example 3.)*

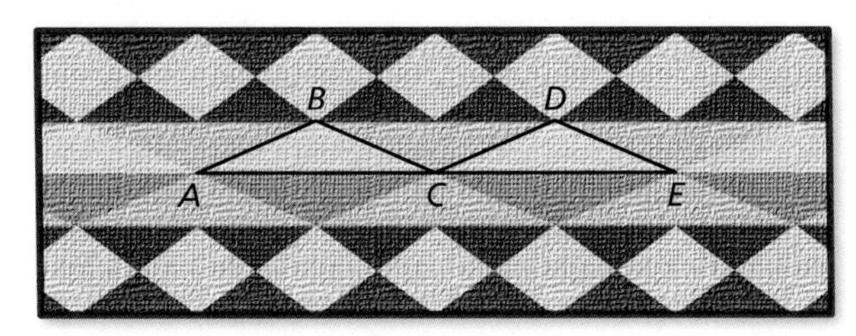

28. THOUGHT PROVOKING There are six possible subsets of three sides or angles of a triangle: SSS, SAS, SSA, AAA, ASA, and AAS. Which of these correspond to congruence theorems? For those that do not, give a counterexample.

29. MATHEMATICAL CONNECTIONS Prove that $\triangle ABC \cong \triangle DEC$. Then find the values of x and y.

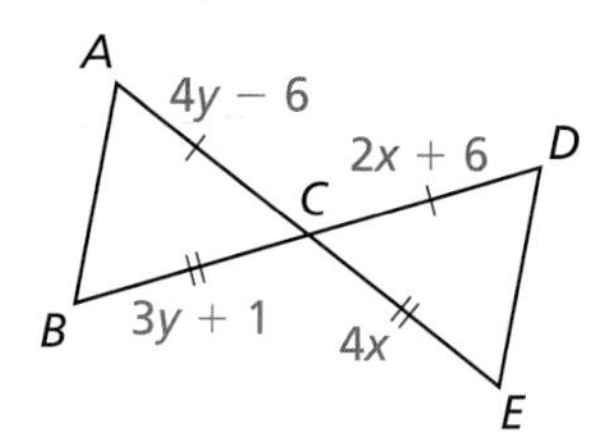

30. MAKING AN ARGUMENT Your friend claims it is possible to construct a triangle congruent to $\triangle ABC$ by first constructing $\overline{AB}$ and $\overline{AC}$, and then copying $\angle C$. Is your friend correct? Explain your reasoning.

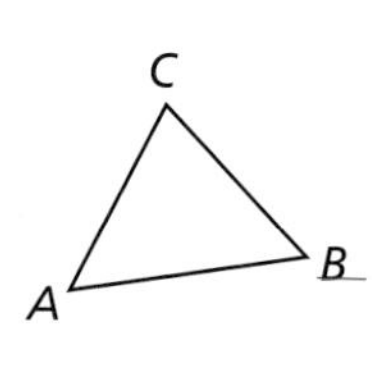

31. PROVING A THEOREM Prove the Reflections in Intersecting Lines Theorem (Theorem 4.3).

Maintaining Mathematical Proficiency

Reviewing what you learned in previous grades and lessons

Classify the triangle by its sides and by measuring its angles. *(Section 5.1)*

32.

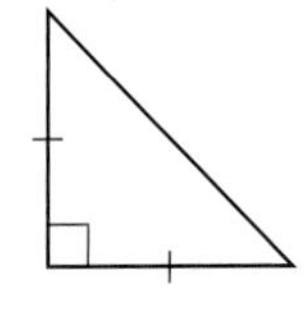

33.

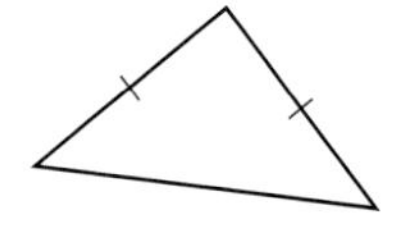

34.

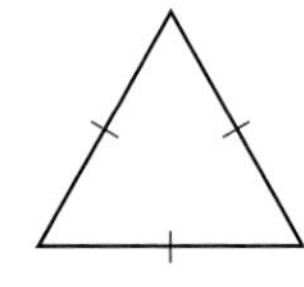

35.

5.4 Equilateral and Isosceles Triangles

Essential Question What conjectures can you make about the side lengths and angle measures of an isosceles triangle?

EXPLORATION 1 Writing a Conjecture about Isosceles Triangles

Work with a partner. Use dynamic geometry software.

a. Construct a circle with a radius of 3 units centered at the origin.

b. Construct $\triangle ABC$ so that B and C are on the circle and A is at the origin.

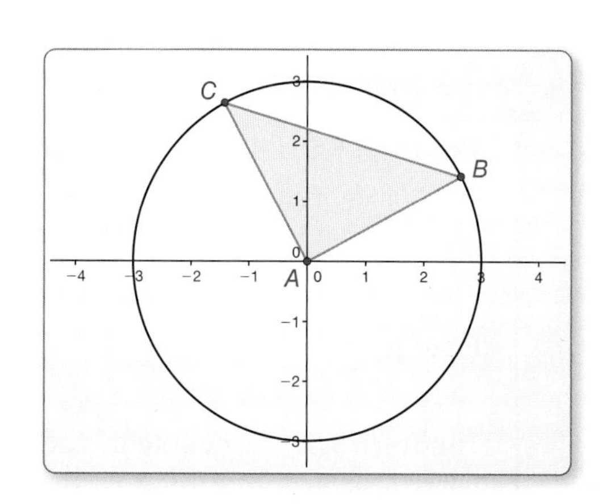

Sample

Points
$A(0, 0)$
$B(2.64, 1.42)$
$C(-1.42, 2.64)$
Segments
$AB = 3$
$AC = 3$
$BC = 4.24$
Angles
$m\angle A = 90°$
$m\angle B = 45°$
$m\angle C = 45°$

c. Recall that a triangle is *isosceles* if it has at least two congruent sides. Explain why $\triangle ABC$ is an isosceles triangle.

d. What do you observe about the angles of $\triangle ABC$?

e. Repeat parts (a)–(d) with several other isosceles triangles using circles of different radii. Keep track of your observations by copying and completing the table below. Then write a conjecture about the angle measures of an isosceles triangle.

CONSTRUCTING VIABLE ARGUMENTS

To be proficient in math, you need to make conjectures and build a logical progression of statements to explore the truth of your conjectures.

		A	*B*	*C*	*AB*	*AC*	*BC*	$m\angle A$	$m\angle B$	$m\angle C$
Sample	**1.**	(0, 0)	(2.64, 1.42)	(−1.42, 2.64)	3	3	4.24	90°	45°	45°
	2.	(0, 0)								
	3.	(0, 0)								
	4.	(0, 0)								
	5.	(0, 0)								

f. Write the converse of the conjecture you wrote in part (e). Is the converse true?

Communicate Your Answer

2. What conjectures can you make about the side lengths and angle measures of an isosceles triangle?

3. How would you prove your conclusion in Exploration 1(e)? in Exploration 1(f)?

5.4 Lesson

Core Vocabulary

legs, *p. 252*
vertex angle, *p. 252*
base, *p. 252*
base angles, *p. 252*

What You Will Learn

- Use the Base Angles Theorem.
- Use isosceles and equilateral triangles.

Using the Base Angles Theorem

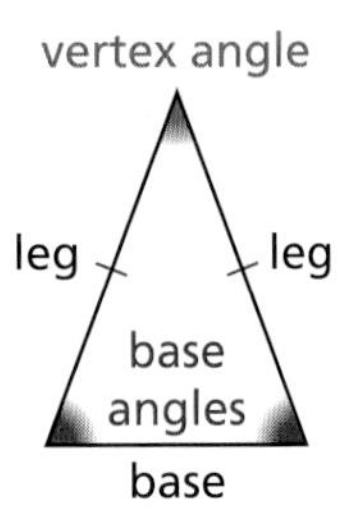

A triangle is isosceles when it has at least two congruent sides. When an isosceles triangle has exactly two congruent sides, these two sides are the **legs**. The angle formed by the legs is the **vertex angle**. The third side is the **base** of the isosceles triangle. The two angles adjacent to the base are called **base angles**.

Theorems

Theorem 5.6 Base Angles Theorem

If two sides of a triangle are congruent, then the angles opposite them are congruent.

If $\overline{AB} \cong \overline{AC}$, then $\angle B \cong \angle C$.

Proof p. 252

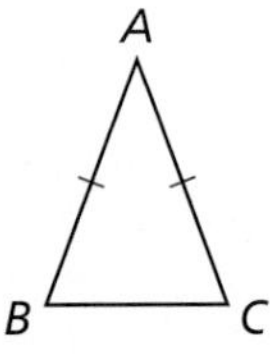

Theorem 5.7 Converse of the Base Angles Theorem

If two angles of a triangle are congruent, then the sides opposite them are congruent.

If $\angle B \cong \angle C$, then $\overline{AB} \cong \overline{AC}$.

Proof Ex. 27, p. 275

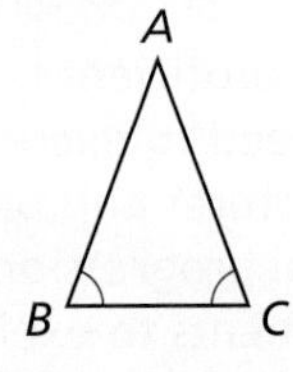

PROOF Base Angles Theorem

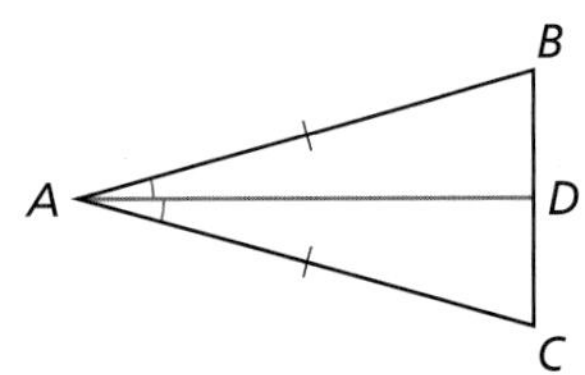

Given $\overline{AB} \cong \overline{AC}$

Prove $\angle B \cong \angle C$

Plan for Proof

a. Draw $\overline{AD}$ so that it bisects $\angle CAB$.

b. Use the SAS Congruence Theorem to show that $\triangle ADB \cong \triangle ADC$.

c. Use properties of congruent triangles to show that $\angle B \cong \angle C$.

Plan in Action	STATEMENTS	REASONS
a.	**1.** Draw $\overline{AD}$, the angle bisector of $\angle CAB$.	**1.** Construction of angle bisector
	2. $\angle CAD \cong \angle BAD$	**2.** Definition of angle bisector
	3. $\overline{AB} \cong \overline{AC}$	**3.** Given
	4. $\overline{DA} \cong \overline{DA}$	**4.** Reflexive Property of Congruence (Thm. 2.1)
b.	**5.** $\triangle ADB \cong \triangle ADC$	**5.** SAS Congruence Theorem (Thm. 5.5)
c.	**6.** $\angle B \cong \angle C$	**6.** Corresponding parts of congruent triangles are congruent.

EXAMPLE 1 Using the Base Angles Theorem

In $\triangle DEF$, $\overline{DE} \cong \overline{DF}$. Name two congruent angles.

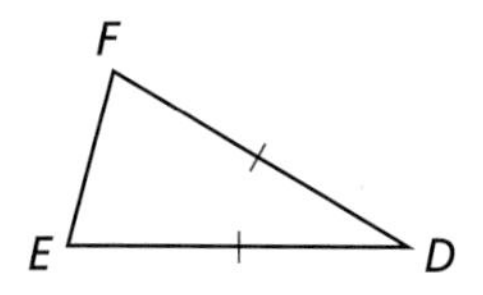

SOLUTION

▶ $\overline{DE} \cong \overline{DF}$, so by the Base Angles Theorem, $\angle E \cong \angle F$.

Monitoring Progress  Help in English and Spanish at *BigIdeasMath.com*

Copy and complete the statement.

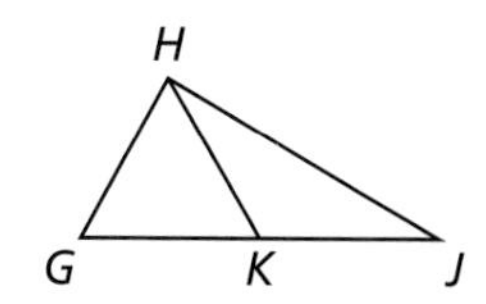

1. If $\overline{HG} \cong \overline{HK}$, then $\angle$____ $\cong \angle$____.
2. If $\angle KHJ \cong \angle KJH$, then ____ $\cong$ ____.

Recall that an equilateral triangle has three congruent sides.

Corollaries

READING

The corollaries state that a triangle is *equilateral* if and only if it is *equiangular.*

Corollary 5.2 Corollary to the Base Angles Theorem

If a triangle is equilateral, then it is equiangular.

Proof Ex. 37, p. 258; Ex. 10, p. 353

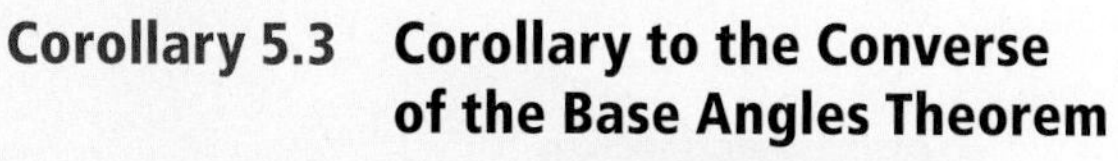

Corollary 5.3 Corollary to the Converse of the Base Angles Theorem

If a triangle is equiangular, then it is equilateral.

Proof Ex. 39, p. 258

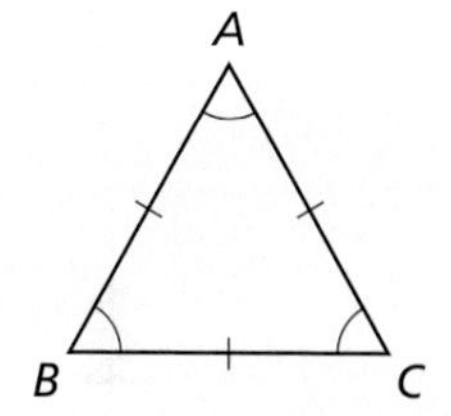

EXAMPLE 2 Finding Measures in a Triangle

Find the measures of $\angle P$, $\angle Q$, and $\angle R$.

SOLUTION

The diagram shows that $\triangle PQR$ is equilateral. So, by the Corollary to the Base Angles Theorem, $\triangle PQR$ is equiangular. So, $m\angle P = m\angle Q = m\angle R$.

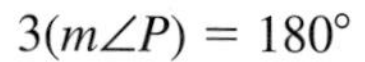

$3(m\angle P) = 180^\circ$	Triangle Sum Theorem (Theorem 5.1)
$m\angle P = 60^\circ$	Divide each side by 3.

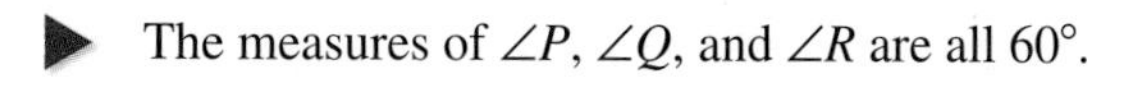

▶ The measures of $\angle P$, $\angle Q$, and $\angle R$ are all 60°.

Monitoring Progress  Help in English and Spanish at *BigIdeasMath.com*

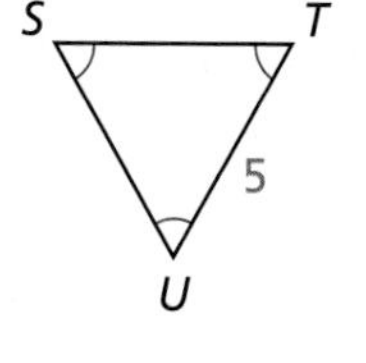

3. Find the length of $\overline{ST}$ for the triangle at the left.

Using Isosceles and Equilateral Triangles

CONSTRUCTION Constructing an Equilateral Triangle

Construct an equilateral triangle that has side lengths congruent to $\overline{AB}$. Use a compass and straightedge.

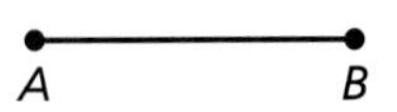

SOLUTION

Step 1

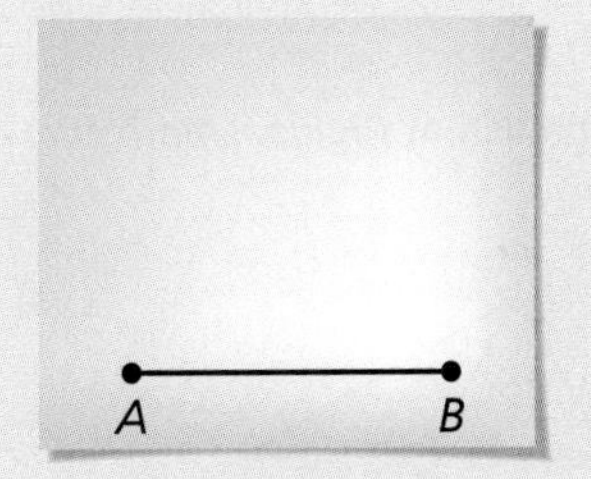

Copy a segment Copy $\overline{AB}$.

Step 2

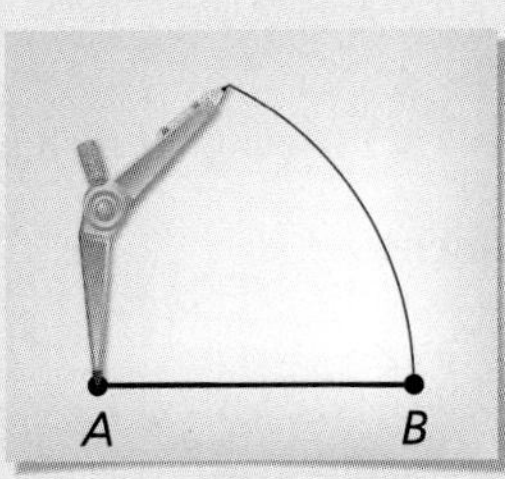

Draw an arc Draw an arc with center A and radius AB.

Step 3

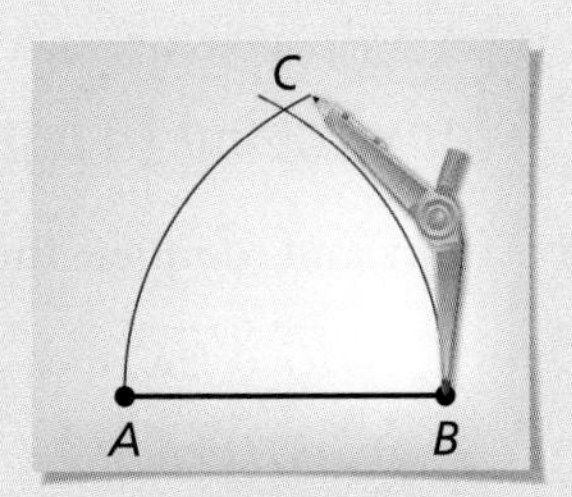

Draw an arc Draw an arc with center B and radius AB. Label the intersection of the arcs from Steps 2 and 3 as C.

Step 4

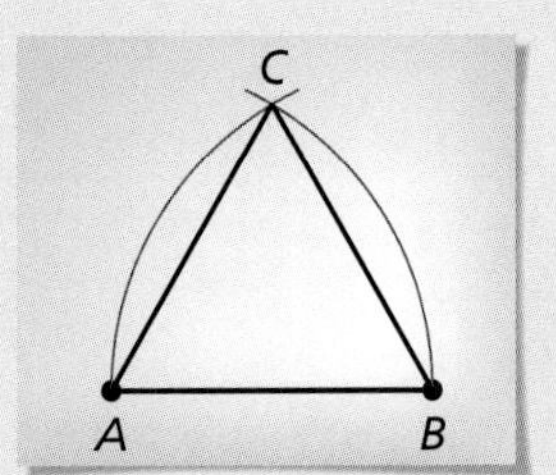

Draw a triangle Draw $\triangle ABC$. Because $\overline{AB}$ and $\overline{AC}$ are radii of the same circle, $\overline{AB} \cong \overline{AC}$. Because $\overline{AB}$ and $\overline{BC}$ are radii of the same circle, $\overline{AB} \cong \overline{BC}$. By the Transitive Property of Congruence (Theorem 2.1), $\overline{AC} \cong \overline{BC}$. So, $\triangle ABC$ is equilateral.

EXAMPLE 3 Using Isosceles and Equilateral Triangles

Find the values of x and y in the diagram.

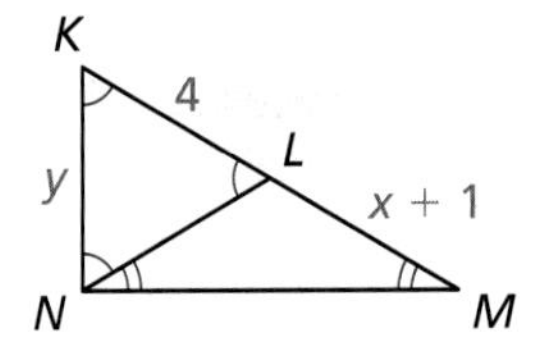

COMMON ERROR

You cannot use N to refer to $\angle LNM$ because three angles have N as their vertex.

SOLUTION

Step 1 Find the value of y. Because $\triangle KLN$ is equiangular, it is also equilateral and $\overline{KN} \cong \overline{KL}$. So, $y = 4$.

Step 2 Find the value of x. Because $\angle LNM \cong \angle LMN$, $\overline{LN} \cong \overline{LM}$, and $\triangle LMN$ is isosceles. You also know that $LN = 4$ because $\triangle KLN$ is equilateral.

$LN = LM$	Definition of congruent segments
$4 = x + 1$	Substitute 4 for LN and $x + 1$ for LM.
$3 = x$	Subtract 1 from each side.

EXAMPLE 4 Solving a Multi-Step Problem

In the lifeguard tower, $\overline{PS} \cong \overline{QR}$ and $\angle QPS \cong \angle PQR$.

a. Explain how to prove that $\triangle QPS \cong \triangle PQR$.

b. Explain why $\triangle PQT$ is isosceles.

COMMON ERROR

When you redraw the triangles so that they do not overlap, be careful to copy all given information and labels correctly.

SOLUTION

a. Draw and label $\triangle QPS$ and $\triangle PQR$ so that they do not overlap. You can see that $\overline{PQ} \cong \overline{QP}$, $\overline{PS} \cong \overline{QR}$, and $\angle QPS \cong \angle PQR$. So, by the SAS Congruence Theorem (Theorem 5.5), $\triangle QPS \cong \triangle PQR$.

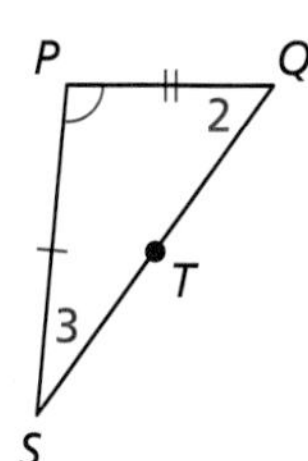

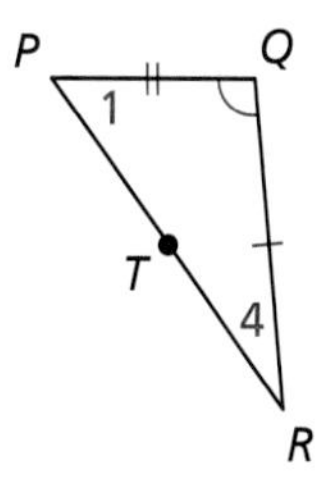

b. From part (a), you know that $\angle 1 \cong \angle 2$ because corresponding parts of congruent triangles are congruent. By the Converse of the Base Angles Theorem, $\overline{PT} \cong \overline{QT}$, and $\triangle PQT$ is isosceles.

Monitoring Progress

Help in English and Spanish at *BigIdeasMath.com*

4. Find the values of x and y in the diagram.

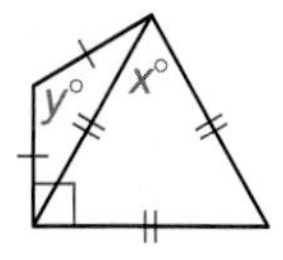

5. In Example 4, show that $\triangle PTS \cong \triangle QTR$.

5.4 Exercises

Dynamic Solutions available at *BigIdeasMath.com*

Vocabulary and Core Concept Check

1. **VOCABULARY** Describe how to identify the *vertex angle* of an isosceles triangle.

2. **WRITING** What is the relationship between the base angles of an isosceles triangle? Explain.

Monitoring Progress and Modeling with Mathematics

In Exercises 3–6, copy and complete the statement. State which theorem you used. *(See Example 1.)*

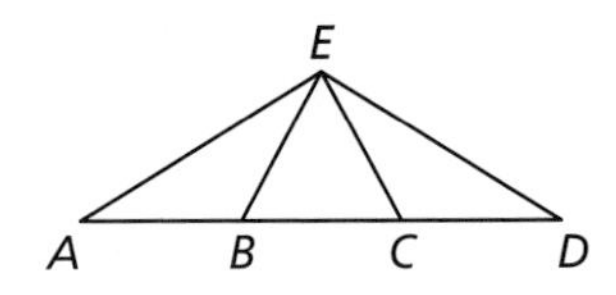

3. If $\overline{AE} \cong \overline{DE}$, then $\angle$___ $\cong \angle$___.

4. If $\overline{AB} \cong \overline{EB}$, then $\angle$___ $\cong \angle$___.

5. If $\angle D \cong \angle CED$, then ___ $\cong$ ___.

6. If $\angle EBC \cong \angle ECB$, then ___ $\cong$ ___.

In Exercises 7–10, find the value of x. *(See Example 2.)*

7.

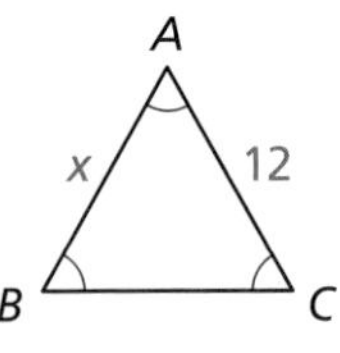

8.

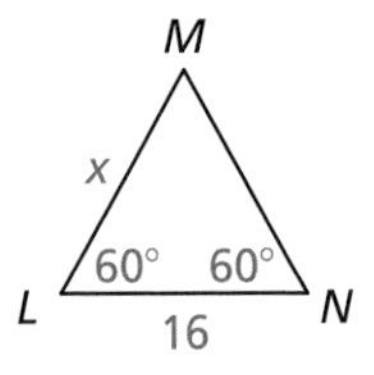

9.

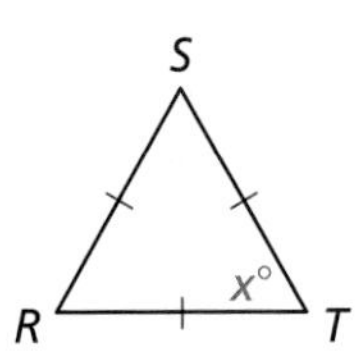

10.

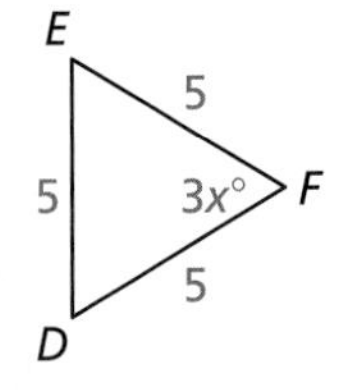

11. **MODELING WITH MATHEMATICS** The dimensions of a sports pennant are given in the diagram. Find the values of x and y.

12. **MODELING WITH MATHEMATICS** A logo in an advertisement is an equilateral triangle with a side length of 7 centimeters. Sketch the logo and give the measure of each side.

In Exercises 13–16, find the values of x and y. *(See Example 3.)*

13.

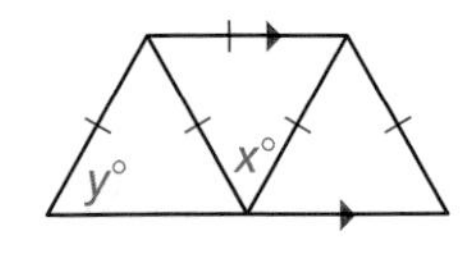

14.

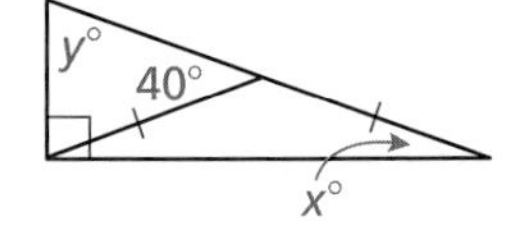

15.

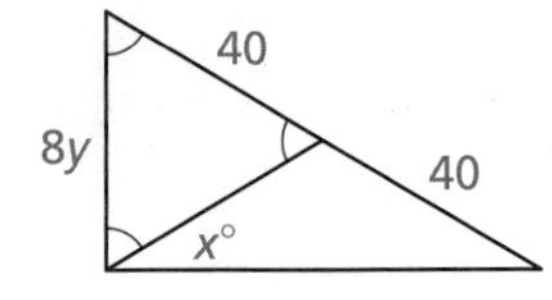

16.

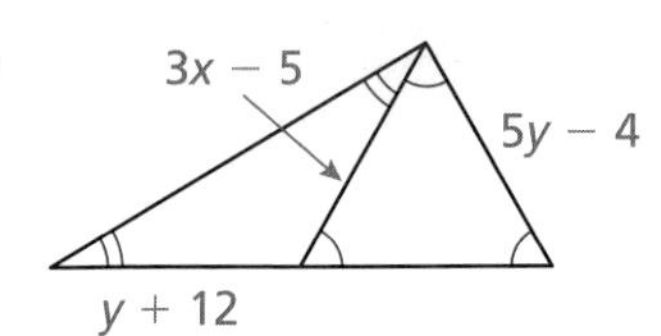

CONSTRUCTION In Exercises 17 and 18, construct an equilateral triangle whose sides are the given length.

17. 3 inches

18. 1.25 inches

19. **ERROR ANALYSIS** Describe and correct the error in finding the length of $\overline{BC}$.

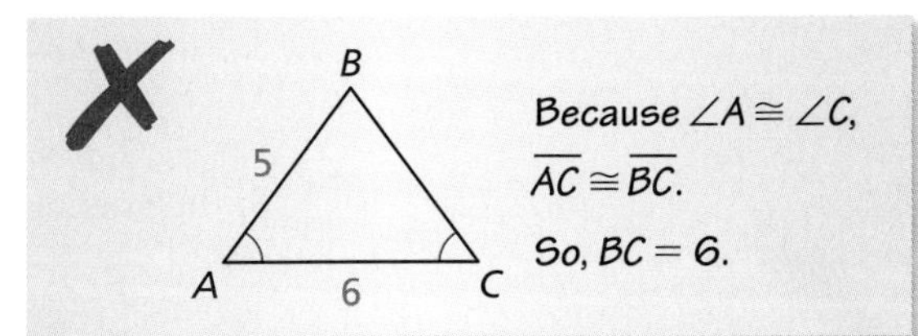

20. PROBLEM SOLVING The diagram represents part of the exterior of the Bow Tower in Calgary, Alberta, Canada. In the diagram, $\triangle ABD$ and $\triangle CBD$ are congruent equilateral triangles. *(See Example 4.)*

a. Explain why $\triangle ABC$ is isosceles.

b. Explain why $\angle BAE \cong \angle BCE$.

c. Show that $\triangle ABE$ and $\triangle CBE$ are congruent.

d. Find the measure of $\angle BAE$.

21. FINDING A PATTERN In the pattern shown, each small triangle is an equilateral triangle with an area of 1 square unit.

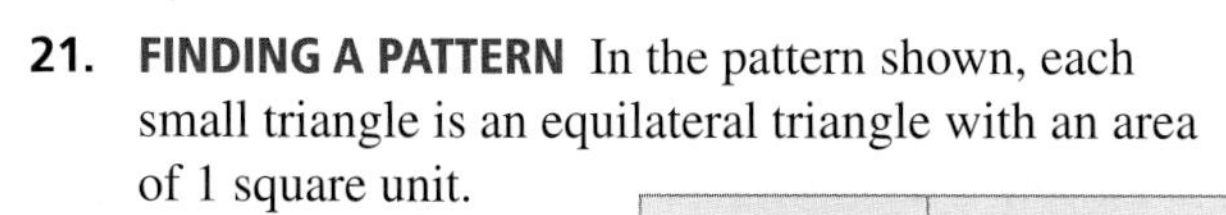

Triangle	Area
△	1 square unit

a. Explain how you know that any triangle made out of equilateral triangles is equilateral.

b. Find the areas of the first four triangles in the pattern.

c. Describe any patterns in the areas. Predict the area of the seventh triangle in the pattern. Explain your reasoning.

22. REASONING The base of isosceles $\triangle XYZ$ is $\overline{YZ}$. What can you prove? Select all that apply.

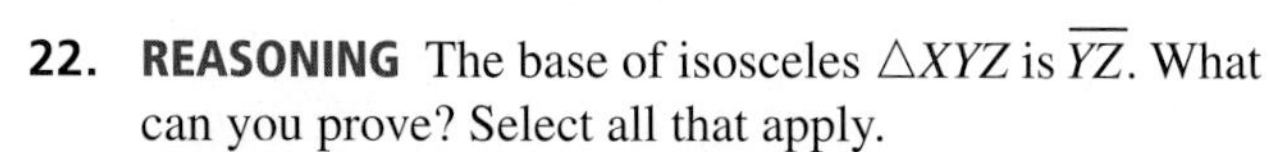

Ⓐ $\overline{XY} \cong \overline{XZ}$　　Ⓑ $\angle X \cong \angle Y$

Ⓒ $\angle Y \cong \angle Z$　　Ⓓ $\overline{YZ} \cong \overline{ZX}$

In Exercises 23 and 24, find the perimeter of the triangle.

23.

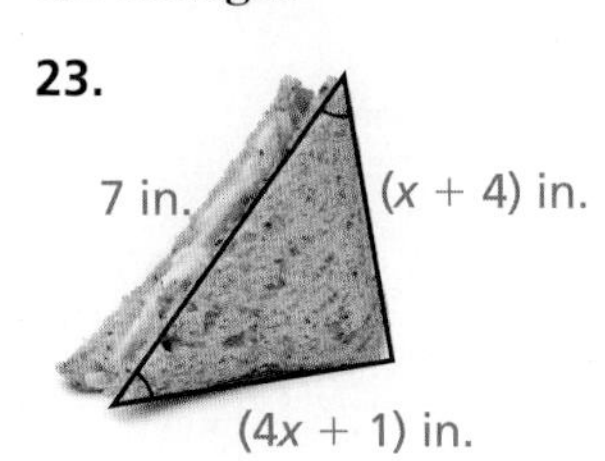

24.

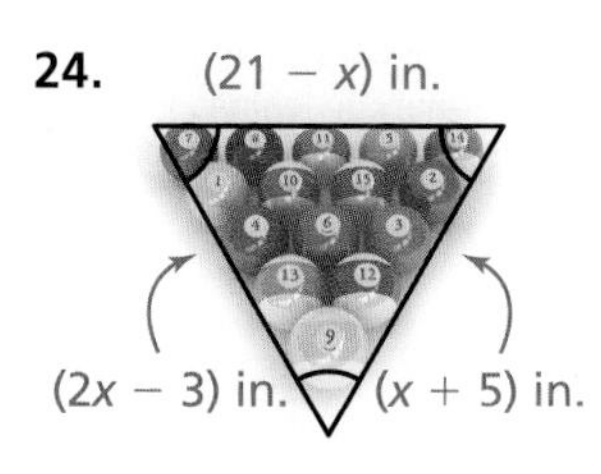

MODELING WITH MATHEMATICS In Exercises 25–28, use the diagram based on the color wheel. The 12 triangles in the diagram are isosceles triangles with congruent vertex angles.

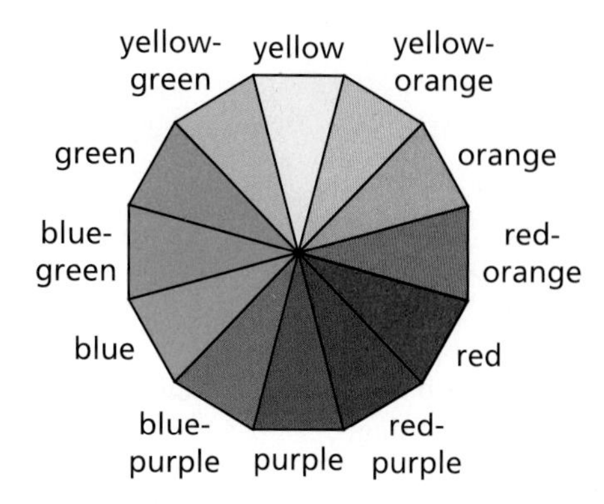

25. Complementary colors lie directly opposite each other on the color wheel. Explain how you know that the yellow triangle is congruent to the purple triangle.

26. The measure of the vertex angle of the yellow triangle is 30°. Find the measures of the base angles.

27. Trace the color wheel. Then form a triangle whose vertices are the midpoints of the bases of the red, yellow, and blue triangles. (These colors are the *primary colors*.) What type of triangle is this?

28. Other triangles can be formed on the color wheel that are congruent to the triangle in Exercise 27. The colors on the vertices of these triangles are called *triads*. What are the possible triads?

29. CRITICAL THINKING Are isosceles triangles always acute triangles? Explain your reasoning.

30. CRITICAL THINKING Is it possible for an equilateral triangle to have an angle measure other than 60°? Explain your reasoning.

31. MATHEMATICAL CONNECTIONS The lengths of the sides of a triangle are $3t$, $5t - 12$, and $t + 20$. Find the values of t that make the triangle isosceles. Explain your reasoning.

32. MATHEMATICAL CONNECTIONS The measure of an exterior angle of an isosceles triangle is $x°$. Write expressions representing the possible angle measures of the triangle in terms of x.

33. WRITING Explain why the measure of the vertex angle of an isosceles triangle must be an even number of degrees when the measures of all the angles of the triangle are whole numbers.

34. **PROBLEM SOLVING** The triangular faces of the peaks on a roof are congruent isosceles triangles with vertex angles U and V.

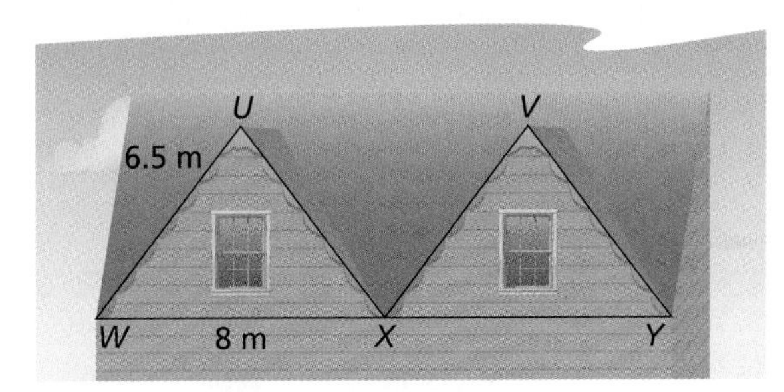

a. Name two angles congruent to $\angle WUX$. Explain your reasoning.

b. Find the distance between points U and V.

35. **PROBLEM SOLVING** A boat is traveling parallel to the shore along $\overrightarrow{RT}$. When the boat is at point R, the captain measures the angle to the lighthouse as 35°. After the boat has traveled 2.1 miles, the captain measures the angle to the lighthouse to be 70°.

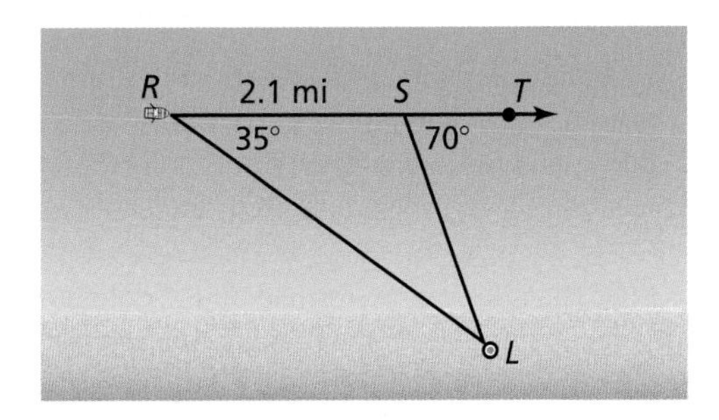

a. Find SL. Explain your reasoning.

b. Explain how to find the distance between the boat and the shoreline.

36. **THOUGHT PROVOKING** The postulates and theorems in this book represent Euclidean geometry. In spherical geometry, all points are points on the surface of a sphere. A line is a circle on the sphere whose diameter is equal to the diameter of the sphere. In spherical geometry, do all equiangular triangles have the same angle measures? Justify your answer.

37. **PROVING A COROLLARY** Prove that the Corollary to the Base Angles Theorem (Corollary 5.2) follows from the Base Angles Theorem (Theorem 5.6).

38. **HOW DO YOU SEE IT?** You are designing fabric purses to sell at the school fair.

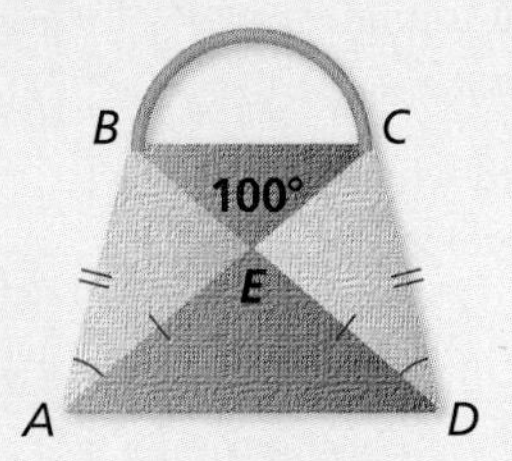

a. Explain why $\triangle ABE \cong \triangle DCE$.

b. Name the isosceles triangles in the purse.

c. Name three angles that are congruent to $\angle EAD$.

39. **PROVING A COROLLARY** Prove that the Corollary to the Converse of the Base Angles Theorem (Corollary 5.3) follows from the Converse of the Base Angles Theorem (Theorem 5.7).

40. **MAKING AN ARGUMENT** The coordinates of two points are $T(0, 6)$ and $U(6, 0)$. Your friend claims that points T, U, and V will always be the vertices of an isosceles triangle when V is any point on the line $y = x$. Is your friend correct? Explain your reasoning.

41. **PROOF** Use the diagram to prove that $\triangle DEF$ is equilateral.

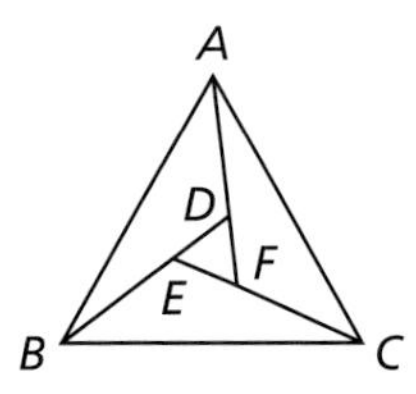

Given $\triangle ABC$ is equilateral.
$\angle CAD \cong \angle ABE \cong \angle BCF$

Prove $\triangle DEF$ is equilateral.

Maintaining Mathematical Proficiency

Reviewing what you learned in previous grades and lessons

Use the given property to complete the statement. *(Section 2.5)*

42. Reflexive Property of Congruence (Theorem 2.1): ____ $\cong \overline{SE}$

43. Symmetric Property of Congruence (Theorem 2.1): If ____ $\cong$ ____, then $\overline{RS} \cong \overline{JK}$.

44. Transitive Property of Congruence (Theorem 2.1): If $\overline{EF} \cong \overline{PQ}$, and $\overline{PQ} \cong \overline{UV}$, then ____ $\cong$ ____.

5.1–5.4 What Did You Learn?

Core Vocabulary

interior angles, *p. 233*
exterior angles, *p. 233*
corollary to a theorem, *p. 235*
corresponding parts, *p. 240*
legs (of an isosceles triangle), *p. 252*
vertex angle (of an isosceles triangle), *p. 252*
base (of an isosceles triangle), *p. 252*
base angles (of an isosceles triangle), *p. 252*

Core Concepts

Classifying Triangles by Sides, *p. 232*
Classifying Triangles by Angles, *p. 232*
Theorem 5.1 Triangle Sum Theorem, *p. 233*
Theorem 5.2 Exterior Angle Theorem, *p. 234*
Corollary 5.1 Corollary to the Triangle Sum Theorem, *p. 235*
Identifying and Using Corresponding Parts, *p. 240*
Theorem 5.3 Properties of Triangle Congruence, *p. 241*
Theorem 5.4 Third Angles Theorem, *p. 242*
Theorem 5.5 Side-Angle-Side (SAS) Congruence Theorem, *p. 246*
Theorem 5.6 Base Angles Theorem, *p. 252*
Theorem 5.7 Converse of the Base Angles Theorem, *p. 252*
Corollary 5.2 Corollary to the Base Angles Theorem, *p. 253*
Corollary 5.3 Corollary to the Converse of the Base Angles Theorem, *p. 253*

Mathematical Practices

1. In Exercise 37 on page 237, what are you given? What relationships are present? What is your goal?
2. Explain the relationships present in Exercise 23 on page 244.
3. Describe at least three different patterns created using triangles for the picture in Exercise 20 on page 257.

Study Skills

Visual Learners

Draw a picture of a word problem.

- Draw a picture of a word problem before starting to solve the problem. You do not have to be an artist.
- When making a review card for a word problem, include a picture. This will help you recall the information while taking a test.
- Make sure your notes are visually neat for easy recall.

5.1–5.4 Quiz

Find the measure of the exterior angle. *(Section 5.1)*

1.

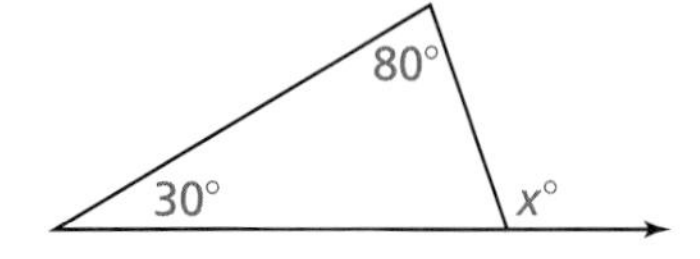

2.

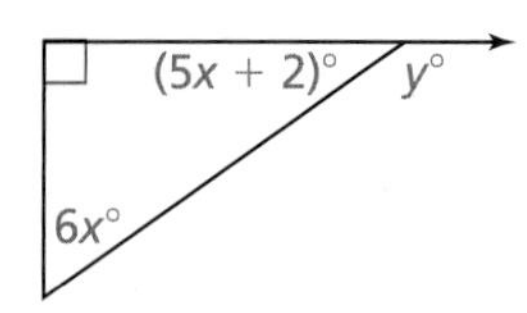

3.

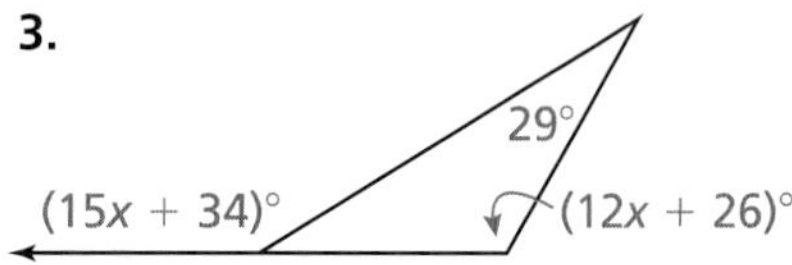

Identify all pairs of congruent corresponding parts. Then write another congruence statement for the polygons. *(Section 5.2)*

4. $\triangle ABC \cong \triangle DEF$

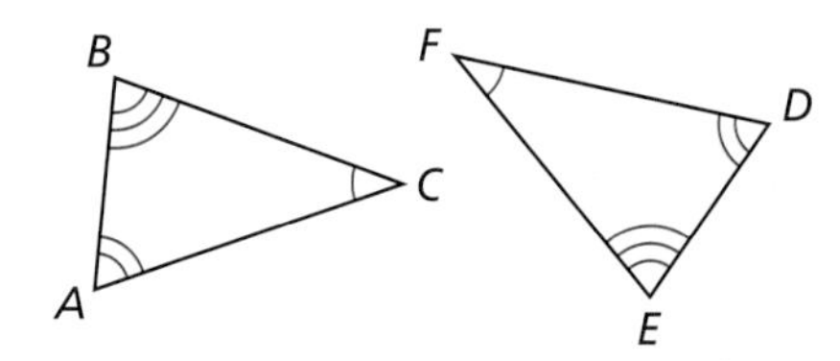

5. $QRST \cong WXYZ$

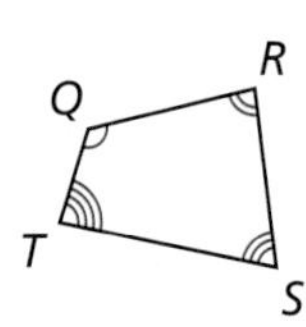

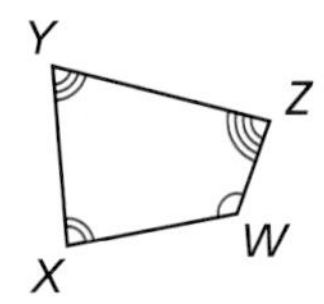

Decide whether enough information is given to prove that the triangles are congruent using the SAS Congruence Theorem (Thm. 5.5). If so, write a proof. If not, explain why. *(Section 5.3)*

6. $\triangle CAD, \triangle CBD$

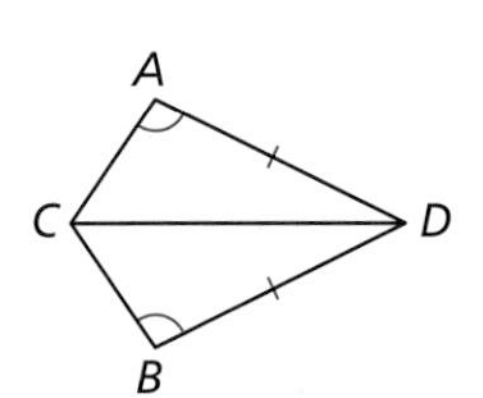

7. $\triangle GHF, \triangle KHJ$

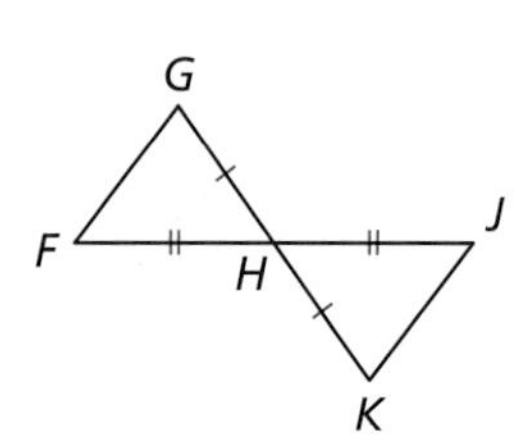

8. $\triangle LMP, \triangle NMP$

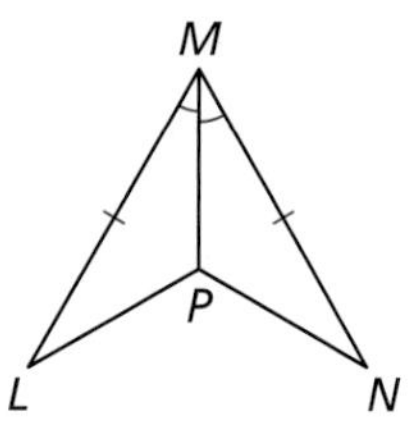

Copy and complete the statement. State which theorem you used. *(Section 5.4)*

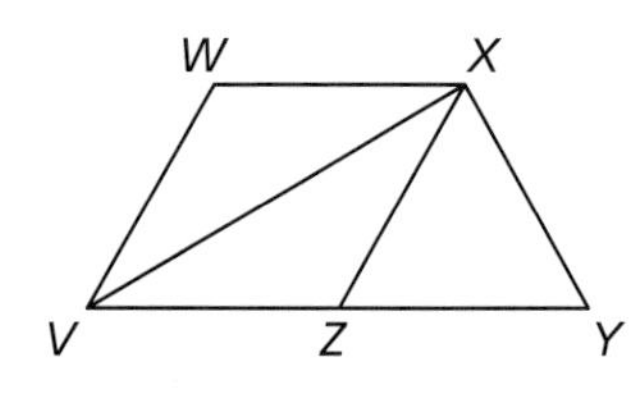

9. If $\overline{VW} \cong \overline{WX}$, then $\angle$___ $\cong \angle$___.

10. If $\overline{XZ} \cong \overline{XY}$, then $\angle$___ $\cong \angle$___.

11. If $\angle ZVX \cong \angle ZXV$, then ___ $\cong$ ___.

12. If $\angle XYZ \cong \angle ZXY$, then ___ $\cong$ ___.

Find the values of *x* and *y*. *(Section 5.2 and Section 5.4)*

13. $\triangle DEF \cong \triangle QRS$

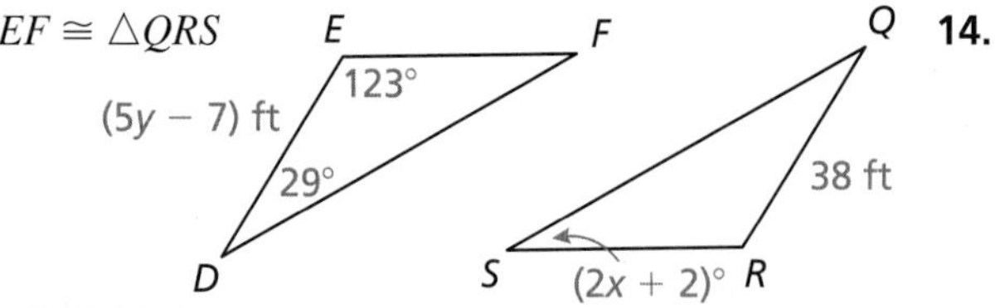

14.

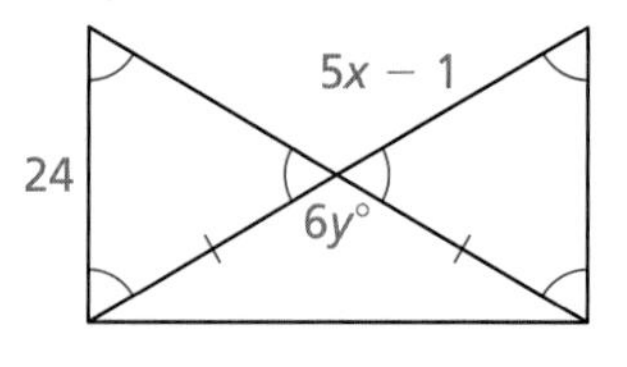

15. In a right triangle, the measure of one acute angle is 4 times the difference of the measure of the other acute angle and 5. Find the measure of each acute angle in the triangle. *(Section 5.1)*

16. The figure shows a stained glass window. *(Section 5.1 and Section 5.3)*

a. Classify triangles 1–4 by their angles.

b. Classify triangles 4–6 by their sides.

c. Is there enough information given to prove that $\triangle 7 \cong \triangle 8$? If so, label the vertices and write a proof. If not, determine what additional information is needed.

5.5 Proving Triangle Congruence by SSS

Essential Question What can you conclude about two triangles when you know the corresponding sides are congruent?

EXPLORATION 1 Drawing Triangles

Work with a partner. Use dynamic geometry software.

a. Construct circles with radii of 2 units and 3 units centered at the origin. Label the origin A. Then draw $\overline{BC}$ of length 4 units.

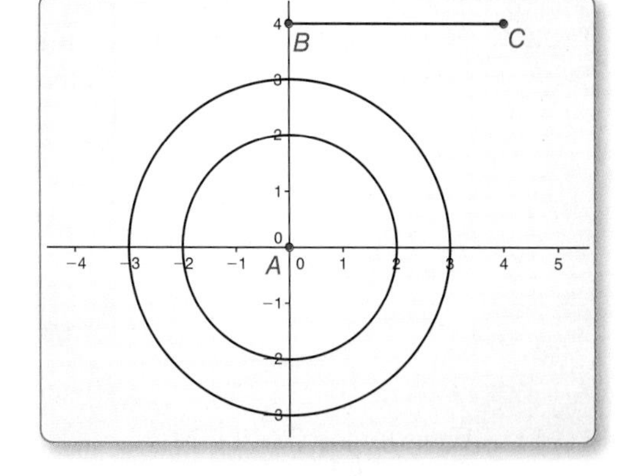

b. Move $\overline{BC}$ so that B is on the smaller circle and C is on the larger circle. Then draw $\triangle ABC$.

c. Explain why the side lengths of $\triangle ABC$ are 2, 3, and 4 units.

d. Find $m\angle A$, $m\angle B$, and $m\angle C$.

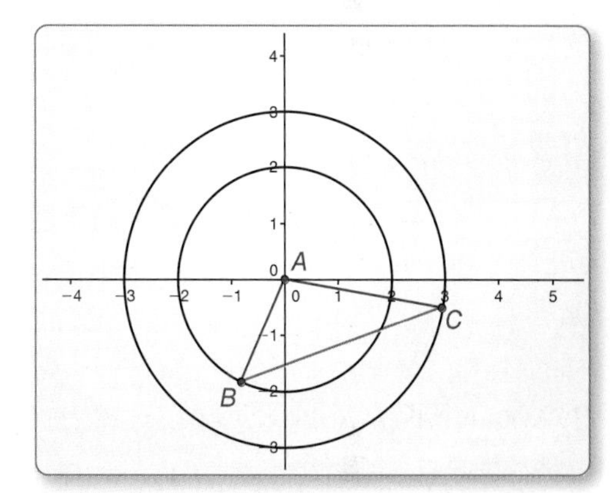

e. Repeat parts (b) and (d) several times, moving $\overline{BC}$ to different locations. Keep track of your results by copying and completing the table below. What can you conclude?

USING TOOLS STRATEGICALLY

To be proficient in math, you need to use technology to help visualize the results of varying assumptions, explore consequences, and compare predictions with data.

	A	B	C	AB	AC	BC	$m\angle A$	$m\angle B$	$m\angle C$
1.	(0, 0)			2	3	4			
2.	(0, 0)			2	3	4			
3.	(0, 0)			2	3	4			
4.	(0, 0)			2	3	4			
5.	(0, 0)			2	3	4			

Communicate Your Answer

2. What can you conclude about two triangles when you know the corresponding sides are congruent?

3. How would you prove your conclusion in Exploration 1(e)?

5.5 Lesson

Core Vocabulary

legs, *p. 264*
hypotenuse, *p. 264*

Previous
congruent figures
rigid motion

What You Will Learn

- Use the Side-Side-Side (SSS) Congruence Theorem.
- Use the Hypotenuse-Leg (HL) Congruence Theorem.

Using the Side-Side-Side Congruence Theorem

Theorem

Theorem 5.8 Side-Side-Side (SSS) Congruence Theorem

If three sides of one triangle are congruent to three sides of a second triangle, then the two triangles are congruent.

If $\overline{AB} \cong \overline{DE}$, $\overline{BC} \cong \overline{EF}$, and $\overline{AC} \cong \overline{DF}$, then $\triangle ABC \cong \triangle DEF$.

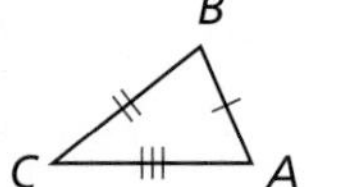

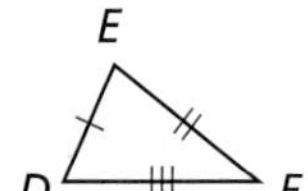

PROOF Side-Side-Side (SSS) Congruence Theorem

Given $\overline{AB} \cong \overline{DE}$, $\overline{BC} \cong \overline{EF}$, $\overline{AC} \cong \overline{DF}$

Prove $\triangle ABC \cong \triangle DEF$

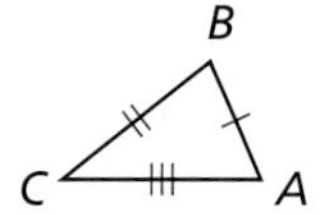

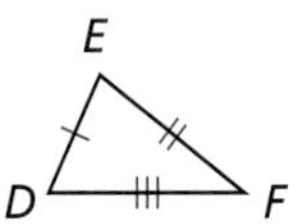

First, translate $\triangle ABC$ so that point A maps to point D, as shown below.

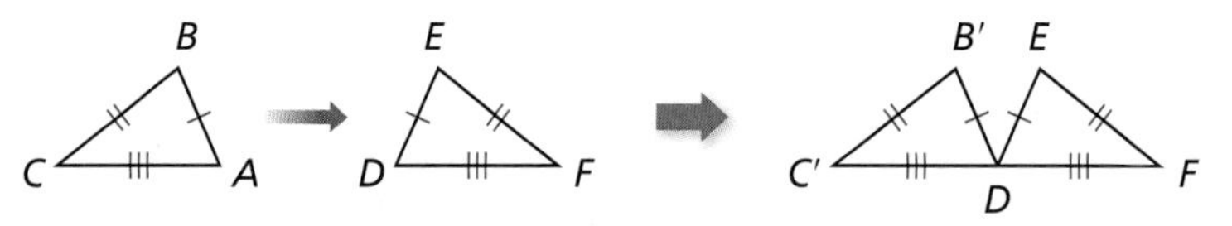

This translation maps $\triangle ABC$ to $\triangle DB'C'$. Next, rotate $\triangle DB'C'$ counterclockwise through $\angle C'DF$ so that the image of $\overrightarrow{DC'}$ coincides with $\overrightarrow{DF}$, as shown below.

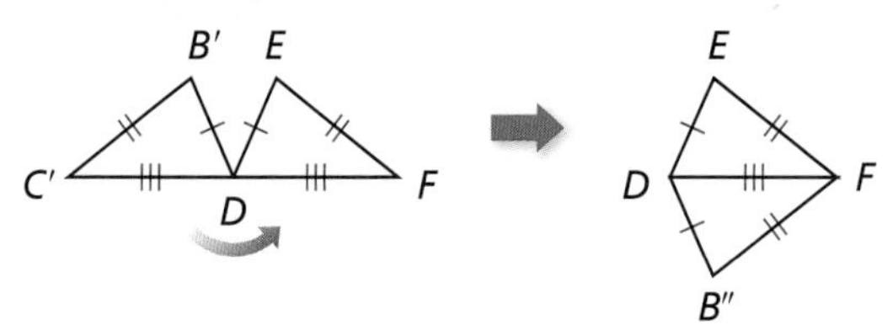

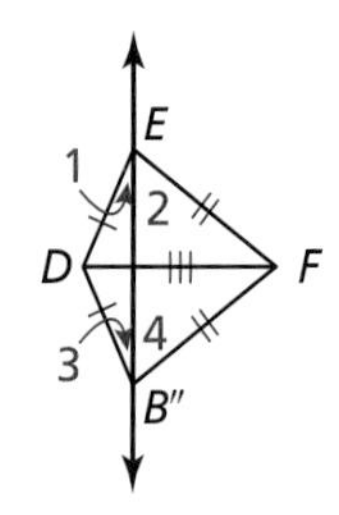

Because $\overline{DC'} \cong \overline{DF}$, the rotation maps point C' to point F. So, this rotation maps $\triangle DB'C'$ to $\triangle DB''F$. Draw an auxiliary line through points E and B''. This line creates $\angle 1$, $\angle 2$, $\angle 3$, and $\angle 4$, as shown at the left.

Because $\overline{DE} \cong \overline{DB''}$, $\triangle DEB''$ is an isosceles triangle. Because $\overline{FE} \cong \overline{FB''}$, $\triangle FEB''$ is an isosceles triangle. By the Base Angles Theorem (Thm. 5.6), $\angle 1 \cong \angle 3$ and $\angle 2 \cong \angle 4$. By the definition of congruence, $m\angle 1 = m\angle 3$ and $m\angle 2 = m\angle 4$. By construction, $m\angle DEF = m\angle 1 + m\angle 2$ and $m\angle DB''F = m\angle 3 + m\angle 4$. You can now use the Substitution Property of Equality to show $m\angle DEF = m\angle DB''F$.

$m\angle DEF = m\angle 1 + m\angle 2$	Angle Addition Postulate (Postulate 1.4)
$= m\angle 3 + m\angle 4$	Substitute $m\angle 3$ for $m\angle 1$ and $m\angle 4$ for $m\angle 2$.
$= m\angle DB''F$	Angle Addition Postulate (Postulate 1.4)

By the definition of congruence, $\angle DEF \cong \angle DB''F$. So, two pairs of sides and their included angles are congruent. By the SAS Congruence Theorem (Thm. 5.5), $\triangle DB''F \cong \triangle DEF$. So, a composition of rigid motions maps $\triangle DB''F$ to $\triangle DEF$. Because a composition of rigid motions maps $\triangle ABC$ to $\triangle DB''F$ and a composition of rigid motions maps $\triangle DB''F$ to $\triangle DEF$, a composition of rigid motions maps $\triangle ABC$ to $\triangle DEF$. So, $\triangle ABC \cong \triangle DEF$.

EXAMPLE 1 Using the SSS Congruence Theorem

Write a proof.

Given $\overline{KL} \cong \overline{NL}$, $\overline{KM} \cong \overline{NM}$

Prove $\triangle KLM \cong \triangle NLM$

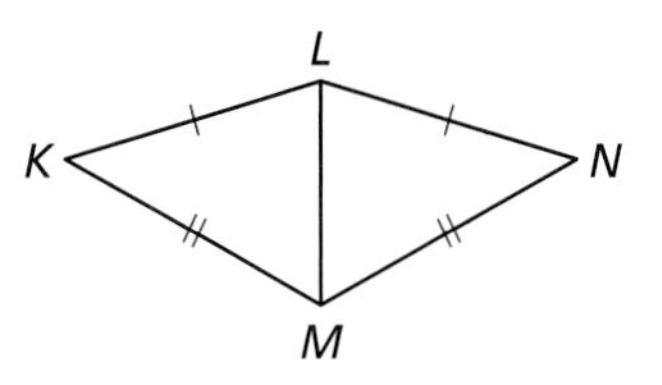

SOLUTION

STATEMENTS	REASONS
S **1.** $\overline{KL} \cong \overline{NL}$	**1.** Given
S **2.** $\overline{KM} \cong \overline{NM}$	**2.** Given
S **3.** $\overline{LM} \cong \overline{LM}$	**3.** Reflexive Property of Congruence (Thm. 2.1)
4. $\triangle KLM \cong \triangle NLM$	**4.** SSS Congruence Theorem

Monitoring Progress

Help in English and Spanish at *BigIdeasMath.com*

Decide whether the congruence statement is true. Explain your reasoning.

1. $\triangle DFG \cong \triangle HJK$

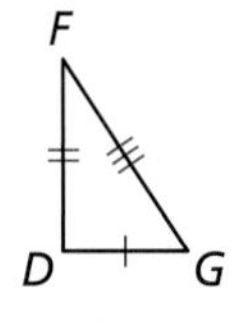

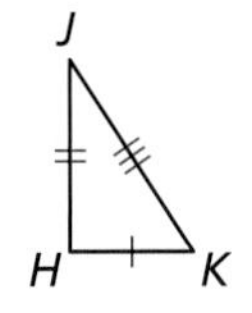

2. $\triangle ACB \cong \triangle CAD$

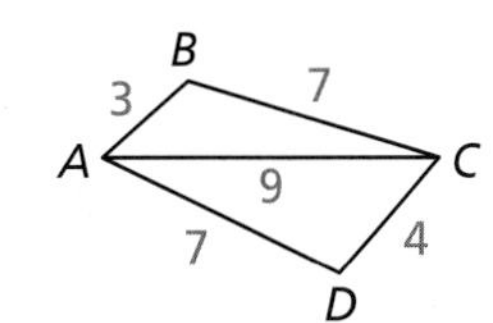

3. $\triangle QPT \cong \triangle RST$

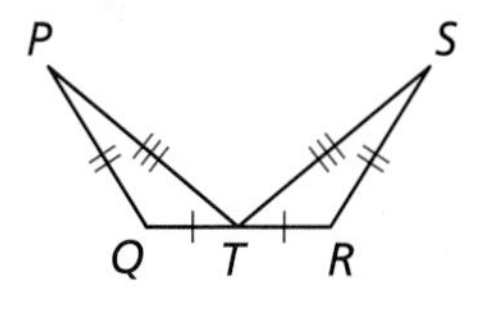

EXAMPLE 2 Solving a Real-Life Problem

Explain why the bench with the diagonal support is stable, while the one without the support can collapse.

SOLUTION

The bench with the diagonal support forms triangles with fixed side lengths. By the SSS Congruence Theorem, these triangles cannot change shape, so the bench is stable. The bench without the diagonal support is not stable because there are many possible quadrilaterals with the given side lengths.

Monitoring Progress

Help in English and Spanish at *BigIdeasMath.com*

Determine whether the figure is stable. Explain your reasoning.

4.

5.

6.

CONSTRUCTION **Copying a Triangle Using SSS**

Construct a triangle that is congruent to $\triangle ABC$ using the SSS Congruence Theorem. Use a compass and straightedge.

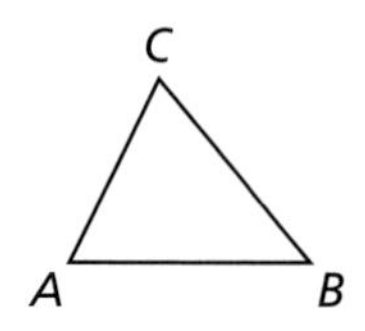

SOLUTION

Step 1

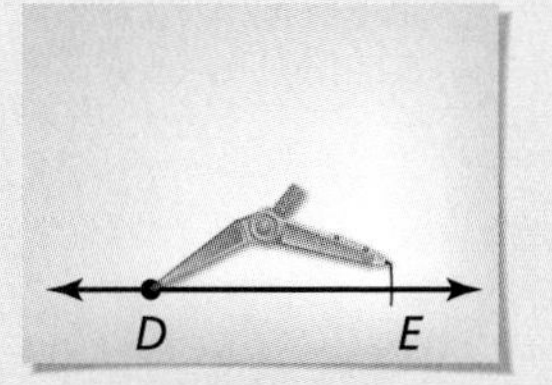

Construct a side
Construct $\overline{DE}$ so that it is congruent to $\overline{AB}$.

Step 2

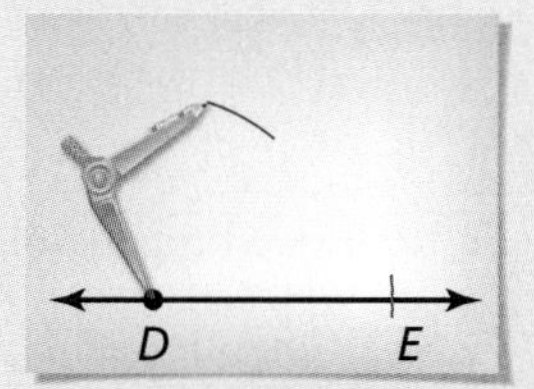

Draw an arc
Open your compass to the length AC. Use this length to draw an arc with center D.

Step 3

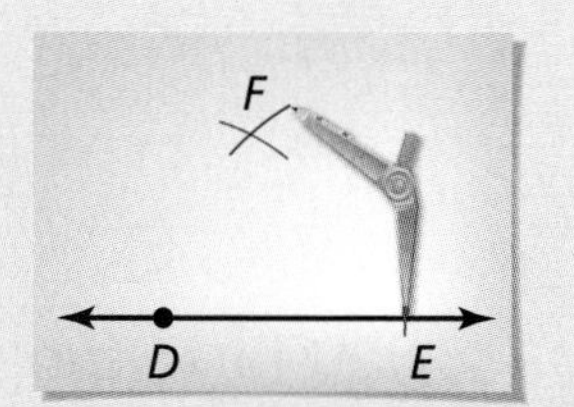

Draw an arc
Draw an arc with radius BC and center E that intersects the arc from Step 2. Label the intersection point F.

Step 4

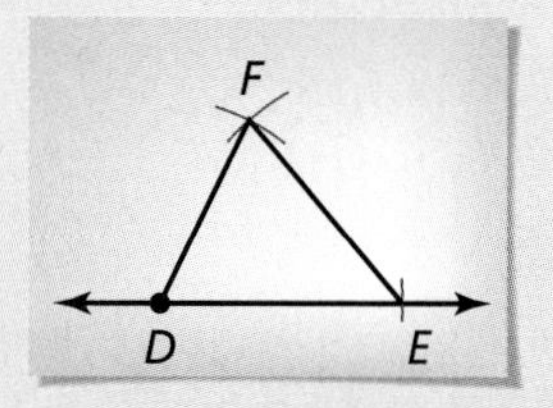

Draw a triangle
Draw $\triangle DEF$. By the SSS Congruence Theorem, $\triangle ABC \cong \triangle DEF$.

Using the Hypotenuse-Leg Congruence Theorem

You know that SAS and SSS are valid methods for proving that triangles are congruent. What about SSA?

In general, SSA is *not* a valid method for proving that triangles are congruent. In the triangles below, two pairs of sides and a pair of angles not included between them are congruent, but the triangles are not congruent.

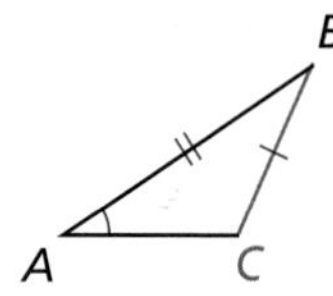

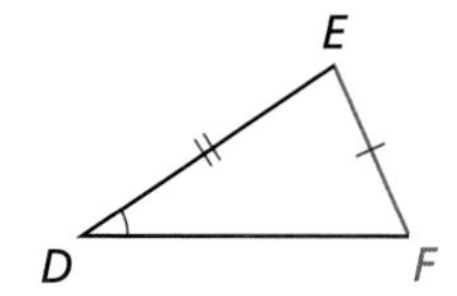

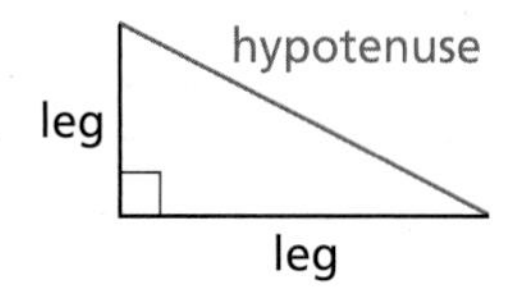

While SSA is not valid in general, there is a special case for right triangles.

In a right triangle, the sides adjacent to the right angle are called the **legs**. The side opposite the right angle is called the **hypotenuse** of the right triangle.

Theorem

Theorem 5.9 Hypotenuse-Leg (HL) Congruence Theorem

If the hypotenuse and a leg of a right triangle are congruent to the hypotenuse and a leg of a second right triangle, then the two triangles are congruent.

If $\overline{AB} \cong \overline{DE}$, $\overline{AC} \cong \overline{DF}$, and $m\angle C = m\angle F = 90°$, then $\triangle ABC \cong \triangle DEF$.

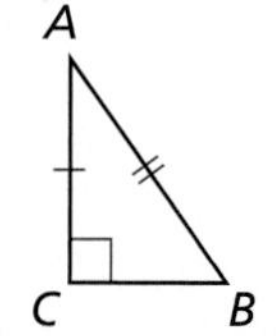

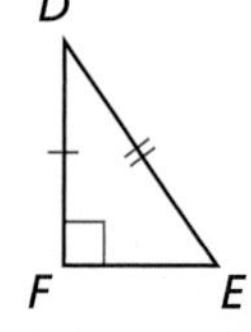

Proof Ex. 38, p. 470; *BigIdeasMath.com*

EXAMPLE 3 Using the Hypotenuse-Leg Congruence Theorem

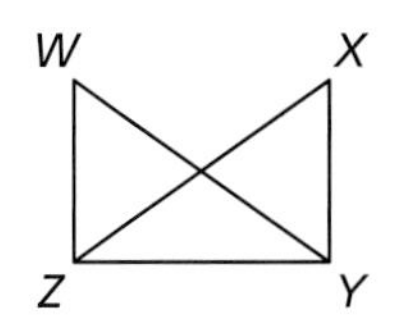

Write a proof.

Given $\overline{WY} \cong \overline{XZ}$, $\overline{WZ} \perp \overline{ZY}$, $\overline{XY} \perp \overline{ZY}$

Prove $\triangle WYZ \cong \triangle XZY$

SOLUTION

Redraw the triangles so they are side by side with corresponding parts in the same position. Mark the given information in the diagram.

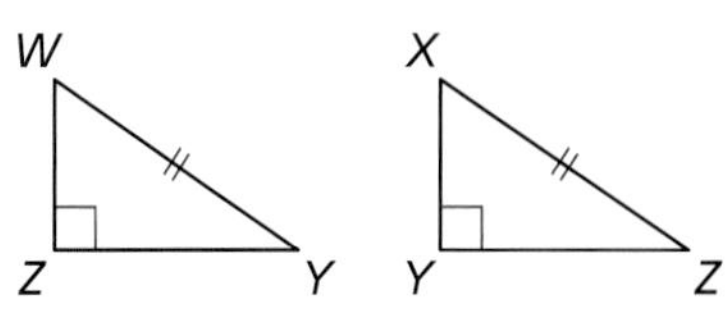

STUDY TIP

If you have trouble matching vertices to letters when you separate the overlapping triangles, leave the triangles in their original orientations.

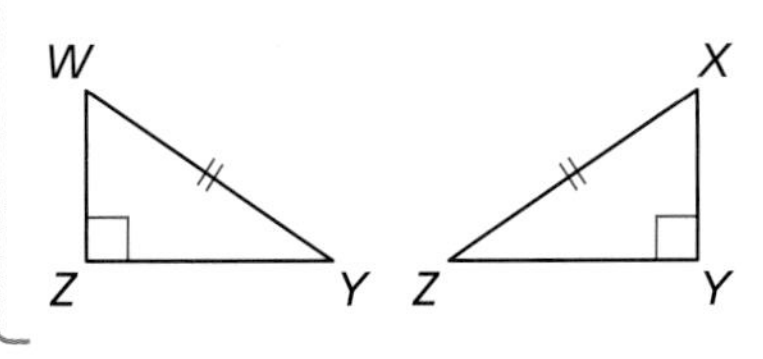

STATEMENTS	REASONS
H **1.** $\overline{WY} \cong \overline{XZ}$	**1.** Given
2. $\overline{WZ} \perp \overline{ZY}, \overline{XY} \perp \overline{ZY}$	**2.** Given
3. $\angle Z$ and $\angle Y$ are right angles.	**3.** Definition of $\perp$ lines
4. $\triangle WYZ$ and $\triangle XZY$ are right triangles.	**4.** Definition of a right triangle
L **5.** $\overline{ZY} \cong \overline{YZ}$	**5.** Reflexive Property of Congruence (Thm. 2.1)
6. $\triangle WYZ \cong \triangle XZY$	**6.** HL Congruence Theorem

EXAMPLE 4 Using the Hypotenuse-Leg Congruence Theorem

The television antenna is perpendicular to the plane containing points B, C, D, and E. Each of the cables running from the top of the antenna to B, C, and D has the same length. Prove that $\triangle AEB$, $\triangle AEC$, and $\triangle AED$ are congruent.

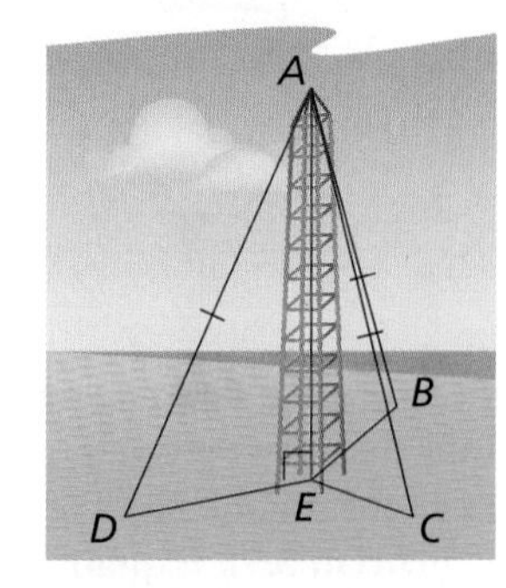

Given $\overline{AE} \perp \overline{EB}$, $\overline{AE} \perp \overline{EC}$, $\overline{AE} \perp \overline{ED}$, $\overline{AB} \cong \overline{AC} \cong \overline{AD}$

Prove $\triangle AEB \cong \triangle AEC \cong \triangle AED$

SOLUTION

You are given that $\overline{AE} \perp \overline{EB}$ and $\overline{AE} \perp \overline{EC}$. So, $\angle AEB$ and $\angle AEC$ are right angles by the definition of perpendicular lines. By definition, $\triangle AEB$ and $\triangle AEC$ are right triangles. You are given that the hypotenuses of these two triangles, $\overline{AB}$ and $\overline{AC}$, are congruent. Also, $\overline{AE}$ is a leg for both triangles, and $\overline{AE} \cong \overline{AE}$ by the Reflexive Property of Congruence (Thm. 2.1). So, by the Hypotenuse-Leg Congruence Theorem, $\triangle AEB \cong \triangle AEC$. You can use similar reasoning to prove that $\triangle AEC \cong \triangle AED$.

▶ So, by the Transitive Property of Triangle Congruence (Thm. 5.3), $\triangle AEB \cong \triangle AEC \cong \triangle AED$.

Monitoring Progress

Help in English and Spanish at *BigIdeasMath.com*

Use the diagram.

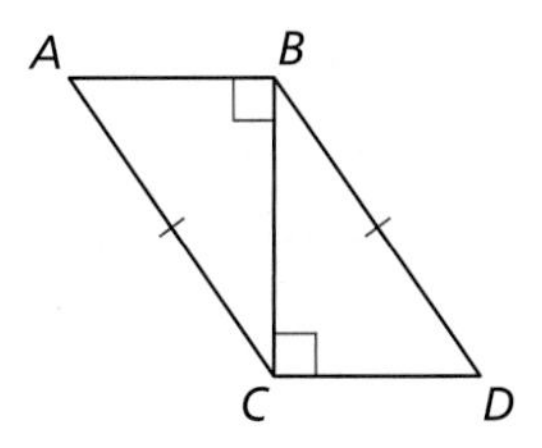

7. Redraw $\triangle ABC$ and $\triangle DCB$ side by side with corresponding parts in the same position.

8. Use the information in the diagram to prove that $\triangle ABC \cong \triangle DCB$.

5.5 Exercises

Dynamic Solutions available at *BigIdeasMath.com*

Vocabulary and Core Concept Check

1. **COMPLETE THE SENTENCE** The side opposite the right angle is called the __________ of the right triangle.

2. **WHICH ONE DOESN'T BELONG?** Which triangle's legs do *not* belong with the other three? Explain your reasoning.

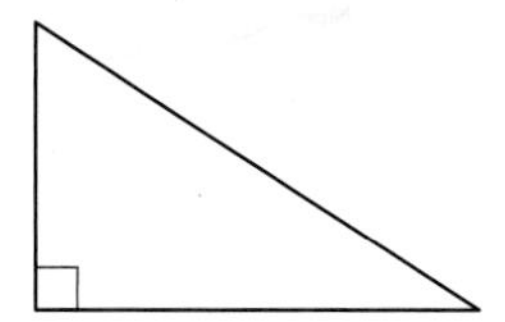

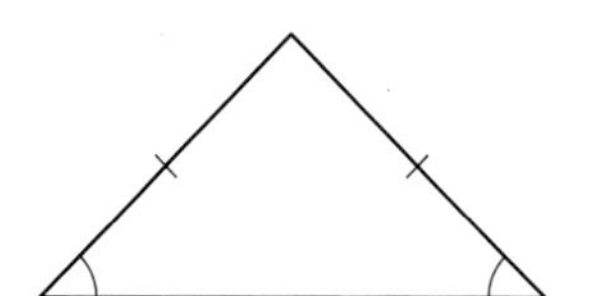

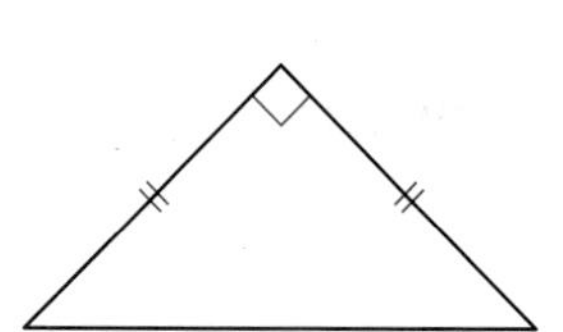

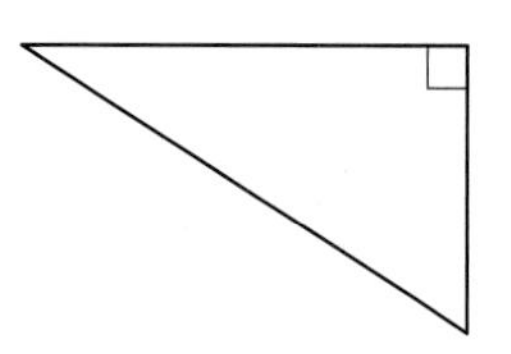

Monitoring Progress and Modeling with Mathematics

In Exercises 3 and 4, decide whether enough information is given to prove that the triangles are congruent using the SSS Congruence Theorem (Theorem 5.8). Explain.

3. $\triangle ABC$, $\triangle DBE$

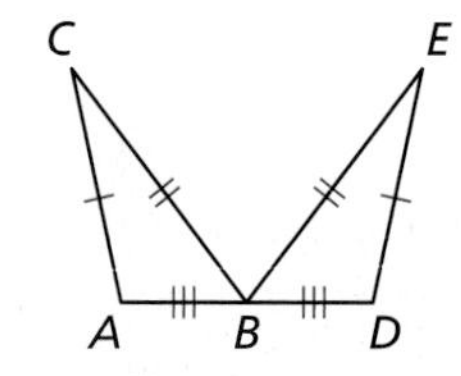

4. $\triangle PQS$, $\triangle RQS$

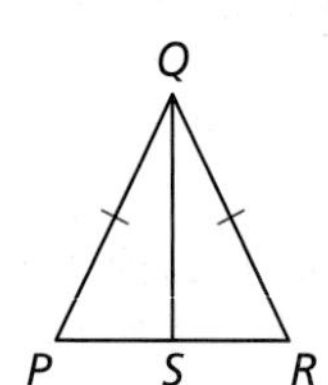

In Exercises 5 and 6, decide whether enough information is given to prove that the triangles are congruent using the HL Congruence Theorem (Theorem 5.9). Explain.

5. $\triangle ABC$, $\triangle FED$

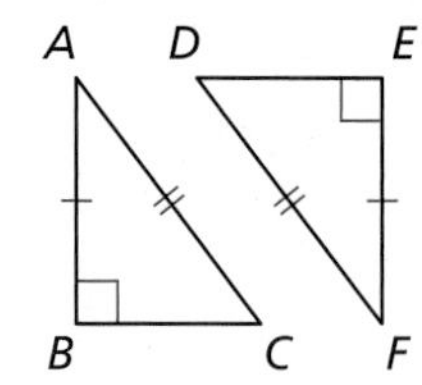

6. $\triangle PQT$, $\triangle SRT$

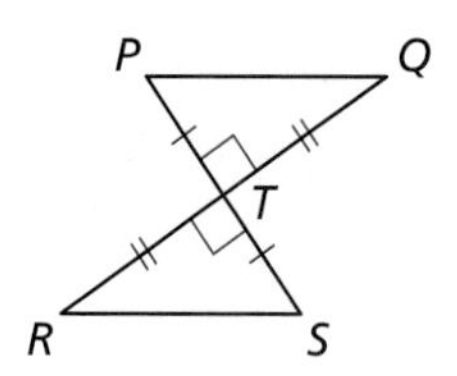

In Exercises 7–10, decide whether the congruence statement is true. Explain your reasoning. *(See Example 1.)*

7. $\triangle RST \cong \triangle TQP$

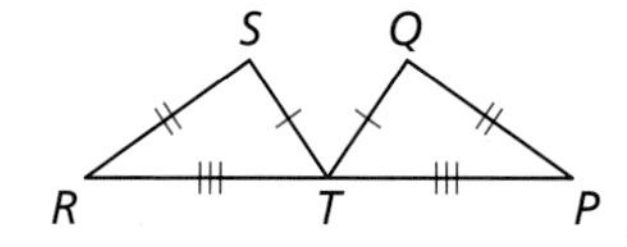

8. $\triangle ABD \cong \triangle CDB$

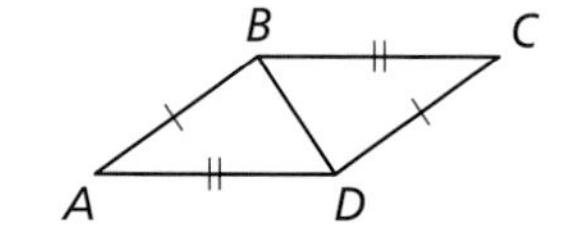

9. $\triangle DEF \cong \triangle DGF$

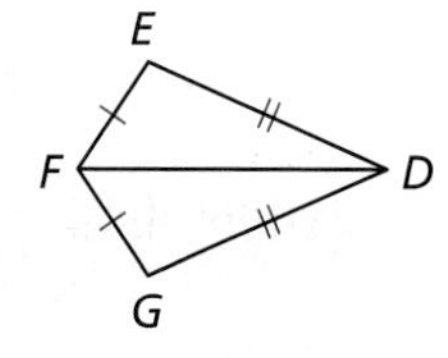

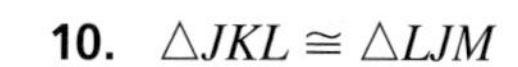

10. $\triangle JKL \cong \triangle LJM$

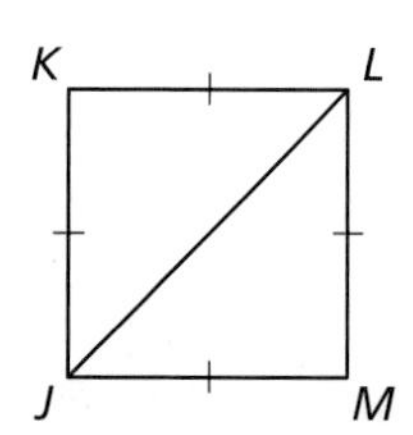

In Exercises 11 and 12, determine whether the figure is stable. Explain your reasoning. *(See Example 2.)*

11.

12.

In Exercises 13 and 14, redraw the triangles so they are side by side with corresponding parts in the same position. Then write a proof. *(See Example 3.)*

13. **Given** $\overline{AC} \cong \overline{BD}$, $\overline{AB} \perp \overline{AD}$, $\overline{CD} \perp \overline{AD}$

Prove $\triangle BAD \cong \triangle CDA$

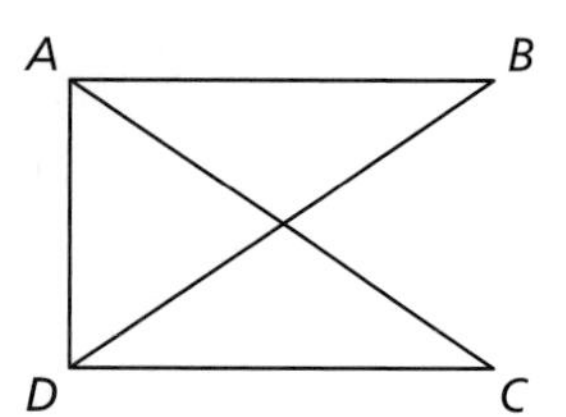

14. **Given** G is the midpoint of $\overline{EH}$, $\overline{FG} \cong \overline{GI}$, $\angle E$ and $\angle H$ are right angles.

Prove $\triangle EFG \cong \triangle HIG$

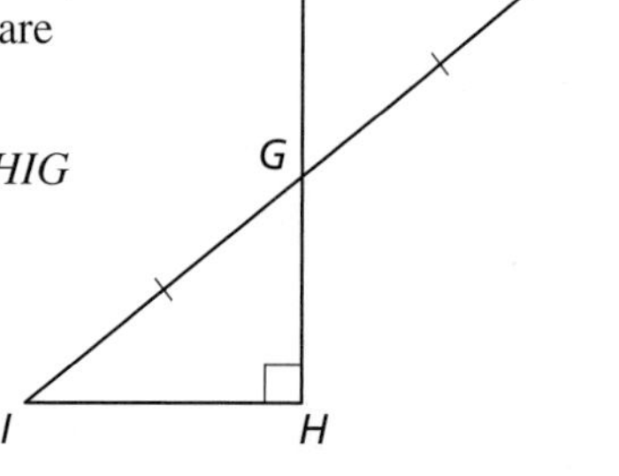

In Exercises 15 and 16, write a proof.

15. Given $\overline{LM} \cong \overline{JK}$, $\overline{MJ} \cong \overline{KL}$

Prove $\triangle LMJ \cong \triangle JKL$

16. Given $\overline{WX} \cong \overline{VZ}$, $\overline{WY} \cong \overline{VY}$, $\overline{YZ} \cong \overline{YX}$

Prove $\triangle VWX \cong \triangle WVZ$

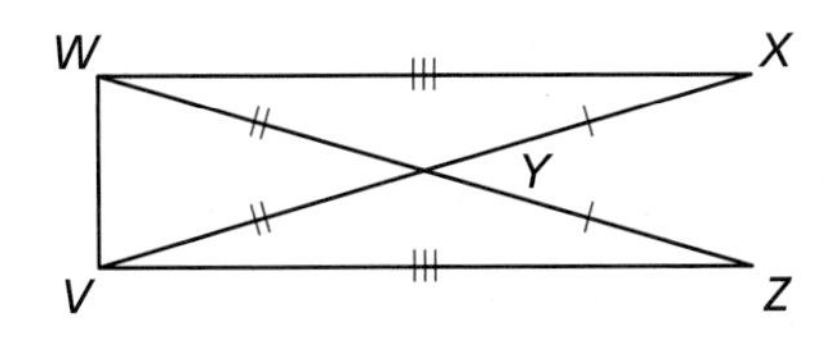

CONSTRUCTION In Exercises 17 and 18, construct a triangle that is congruent to $\triangle QRS$ using the SSS Congruence Theorem (Theorem 5.8).

17.

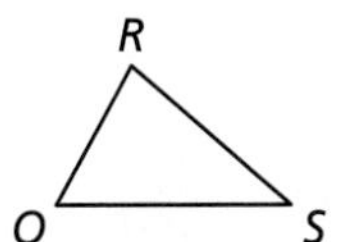

18.

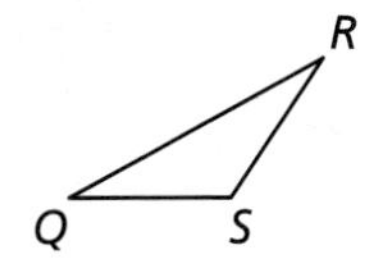

19. ERROR ANALYSIS Describe and correct the error in identifying congruent triangles.

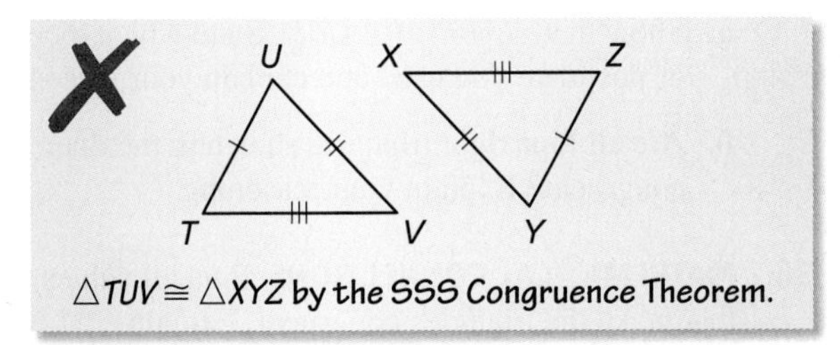

20. ERROR ANALYSIS Describe and correct the error in determining the value of x that makes the triangles congruent.

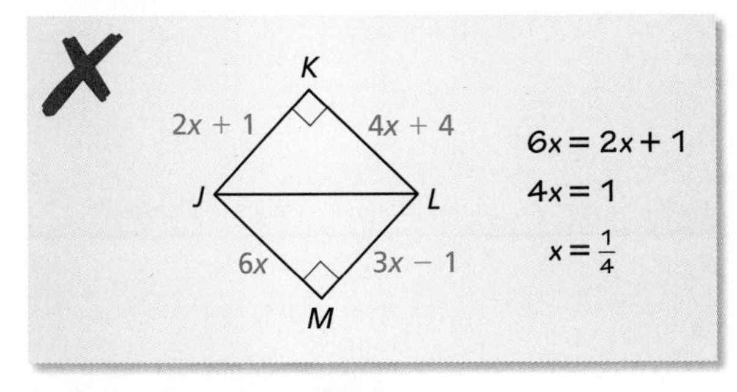

21. MAKING AN ARGUMENT Your friend claims that in order to use the SSS Congruence Theorem (Theorem 5.8) to prove that two triangles are congruent, both triangles must be equilateral triangles. Is your friend correct? Explain your reasoning.

22. MODELING WITH MATHEMATICS The distances between consecutive bases on a softball field are the same. The distance from home plate to second base is the same as the distance from first base to third base. The angles created at each base are 90°. Prove $\triangle HFS \cong \triangle FST \cong \triangle STH$. *(See Example 4.)*

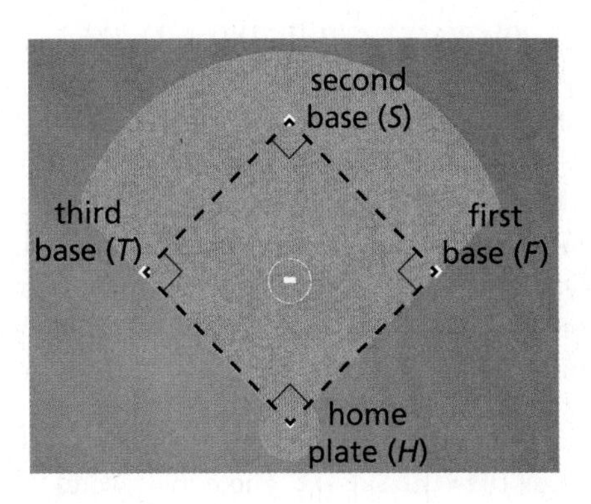

23. REASONING To support a tree, you attach wires from the trunk of the tree to stakes in the ground, as shown in the diagram.

a. What additional information do you need to use the HL Congruence Theorem (Theorem 5.9) to prove that $\triangle JKL \cong \triangle MKL$?

b. Suppose K is the midpoint of JM. Name a theorem you could use to prove that $\triangle JKL \cong \triangle MKL$. Explain your reasoning.

24. REASONING Use the photo of the Navajo rug, where $\overline{BC} \cong \overline{DE}$ and $\overline{AC} \cong \overline{CE}$.

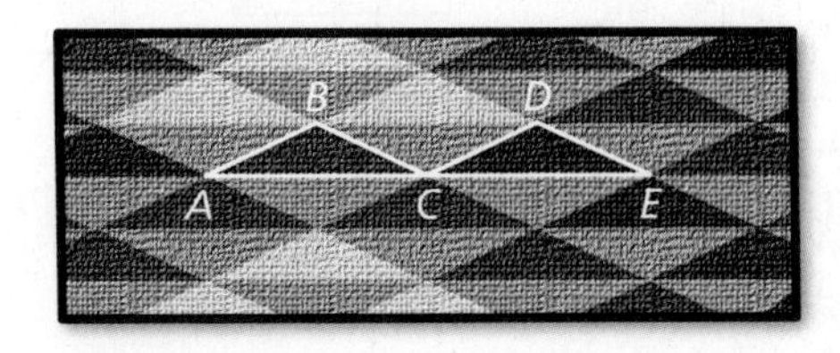

a. What additional information do you need to use the SSS Congruence Theorem (Theorem 5.8) to prove that $\triangle ABC \cong \triangle CDE$?

b. What additional information do you need to use the HL Congruence Theorem (Theorem 5.9) to prove that $\triangle ABC \cong \triangle CDE$?

In Exercises 25–28, use the given coordinates to determine whether $\triangle ABC \cong \triangle DEF$.

25. $A(-2, -2), B(4, -2), C(4, 6), D(5, 7), E(5, 1), F(13, 1)$

26. $A(-2, 1), B(3, -3), C(7, 5), D(3, 6), E(8, 2), F(10, 11)$

27. $A(0, 0), B(6, 5), C(9, 0), D(0, -1), E(6, -6), F(9, -1)$

28. $A(-5, 7), B(-5, 2), C(0, 2), D(0, 6), E(0, 1), F(4, 1)$

29. **CRITICAL THINKING** You notice two triangles in the tile floor of a hotel lobby. You want to determine whether the triangles are congruent, but you only have a piece of string. Can you determine whether the triangles are congruent? Explain.

30. **HOW DO YOU SEE IT?** There are several theorems you can use to show that the triangles in the "square" pattern are congruent. Name two of them.

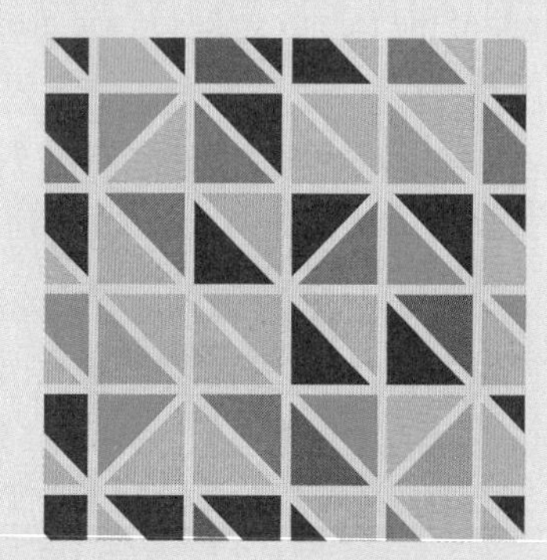

31. **MAKING AN ARGUMENT** Your cousin says that $\triangle JKL$ is congruent to $\triangle LMJ$ by the SSS Congruence Theorem (Thm. 5.8). Your friend says that $\triangle JKL$ is congruent to $\triangle LMJ$ by the HL Congruence Theorem (Thm. 5.9). Who is correct? Explain your reasoning.

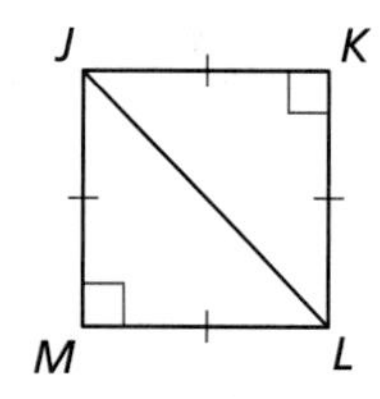

32. **THOUGHT PROVOKING** The postulates and theorems in this book represent Euclidean geometry. In spherical geometry, all points are points on the surface of a sphere. A line is a circle on the sphere whose diameter is equal to the diameter of the sphere. In spherical geometry, do you think that two triangles are congruent if their corresponding sides are congruent? Justify your answer.

USING TOOLS In Exercises 33 and 34, use the given information to sketch $\triangle LMN$ and $\triangle STU$. Mark the triangles with the given information.

33. $\overline{LM} \perp \overline{MN}, \overline{ST} \perp \overline{TU}, \overline{LM} \cong \overline{NM} \cong \overline{UT} \cong \overline{ST}$

34. $\overline{LM} \perp \overline{MN}, \overline{ST} \perp \overline{TU}, \overline{LM} \cong \overline{ST}, \overline{LN} \cong \overline{SU}$

35. **CRITICAL THINKING** The diagram shows the light created by two spotlights. Both spotlights are the same distance from the stage.

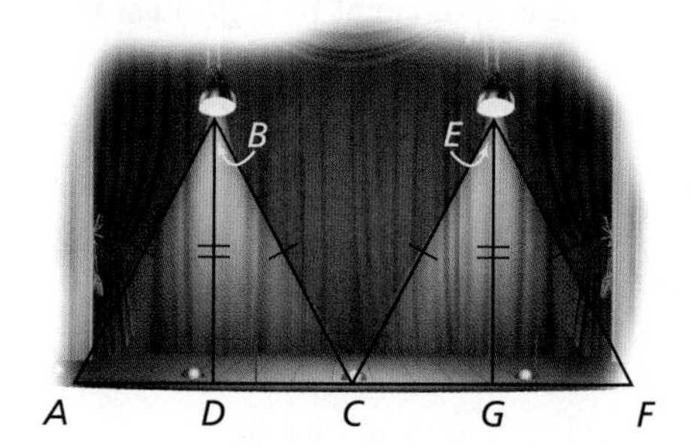

a. Show that $\triangle ABD \cong \triangle CBD$. State which theorem or postulate you used and explain your reasoning.

b. Are all four right triangles shown in the diagram congruent? Explain your reasoning.

36. **MATHEMATICAL CONNECTIONS** Find all values of x that make the triangles congruent. Explain.

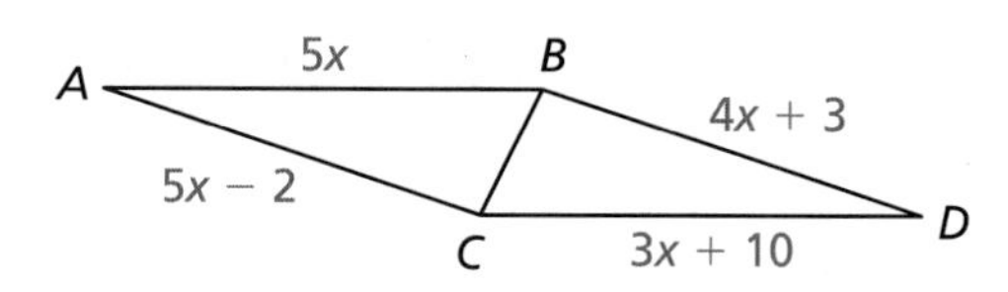

Maintaining Mathematical Proficiency

Reviewing what you learned in previous grades and lessons

Use the congruent triangles. *(Section 5.2)*

37. Name the segment in $\triangle DEF$ that is congruent to $\overline{AC}$.

38. Name the segment in $\triangle ABC$ that is congruent to $\overline{EF}$.

39. Name the angle in $\triangle DEF$ that is congruent to $\angle B$.

40. Name the angle in $\triangle ABC$ that is congruent to $\angle F$.

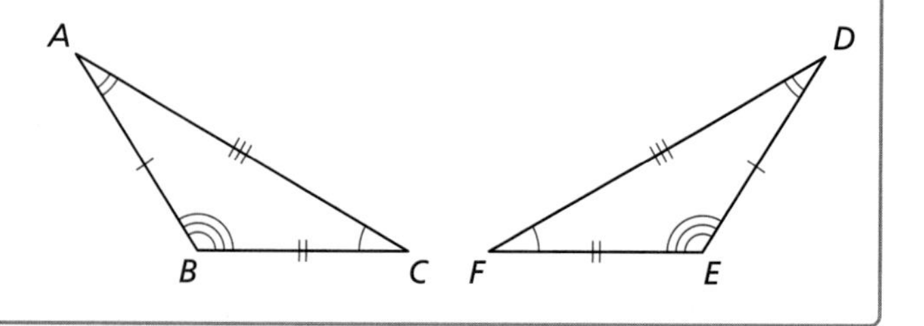

5.6 Proving Triangle Congruence by ASA and AAS

Essential Question What information is sufficient to determine whether two triangles are congruent?

EXPLORATION 1 Determining Whether SSA Is Sufficient

Work with a partner.

a. Use dynamic geometry software to construct $\triangle ABC$. Construct the triangle so that vertex B is at the origin, $\overline{AB}$ has a length of 3 units, and $\overline{BC}$ has a length of 2 units.

b. Construct a circle with a radius of 2 units centered at the origin. Locate point D where the circle intersects $\overline{AC}$. Draw $\overline{BD}$.

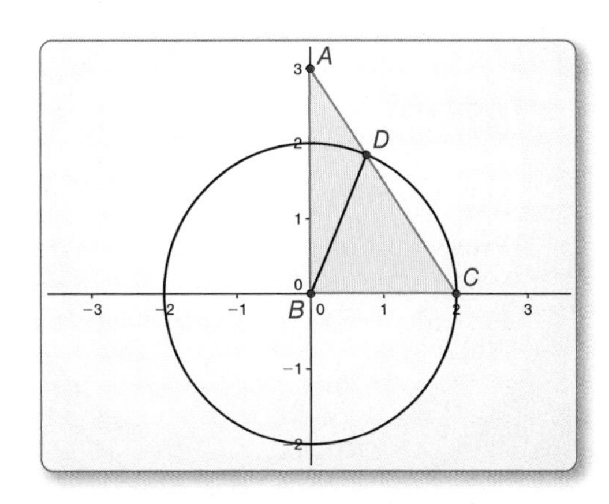

Sample

Points
$A(0, 3)$
$B(0, 0)$
$C(2, 0)$
$D(0.77, 1.85)$
Segments
$AB = 3$
$AC = 3.61$
$BC = 2$
$AD = 1.38$
Angle
$m\angle A = 33.69°$

c. $\triangle ABC$ and $\triangle ABD$ have two congruent sides and a nonincluded congruent angle. Name them.

d. Is $\triangle ABC \cong \triangle ABD$? Explain your reasoning.

e. Is SSA sufficient to determine whether two triangles are congruent? Explain your reasoning.

CONSTRUCTING VIABLE ARGUMENTS

To be proficient in math, you need to recognize and use counterexamples.

EXPLORATION 2 Determining Valid Congruence Theorems

Work with a partner. Use dynamic geometry software to determine which of the following are valid triangle congruence theorems. For those that are not valid, write a counterexample. Explain your reasoning.

Possible Congruence Theorem	Valid or not valid?
SSS	
SSA	
SAS	
AAS	
ASA	
AAA	

Communicate Your Answer

3. What information is sufficient to determine whether two triangles are congruent?

4. Is it possible to show that two triangles are congruent using more than one congruence theorem? If so, give an example.

5.6 Lesson

Core Vocabulary

Previous
congruent figures
rigid motion

What You Will Learn

- Use the ASA and AAS Congruence Theorems.

Using the ASA and AAS Congruence Theorems

Theorem

Theorem 5.10 Angle-Side-Angle (ASA) Congruence Theorem

If two angles and the included side of one triangle are congruent to two angles and the included side of a second triangle, then the two triangles are congruent.

If $\angle A \cong \angle D$, $\overline{AC} \cong \overline{DF}$, and $\angle C \cong \angle F$, then $\triangle ABC \cong \triangle DEF$.

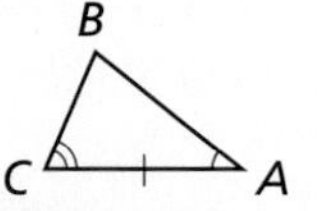

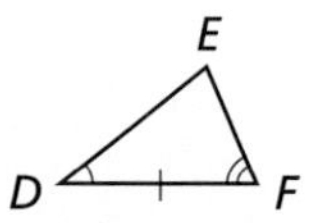

Proof p. 270

PROOF Angle-Side-Angle (ASA) Congruence Theorem

Given $\angle A \cong \angle D$, $\overline{AC} \cong \overline{DF}$, $\angle C \cong \angle F$

Prove $\triangle ABC \cong \triangle DEF$

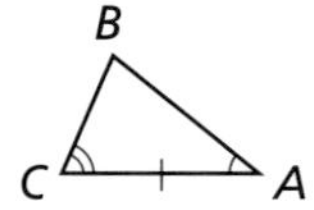

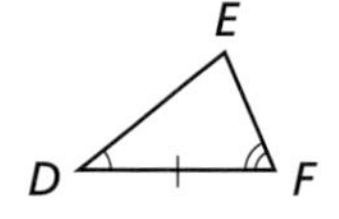

First, translate $\triangle ABC$ so that point A maps to point D, as shown below.

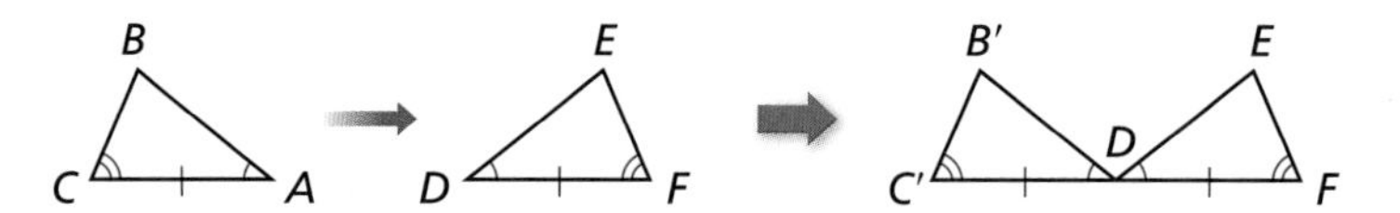

This translation maps $\triangle ABC$ to $\triangle DB'C'$. Next, rotate $\triangle DB'C'$ counterclockwise through $\angle C'DF$ so that the image of $\overrightarrow{DC'}$ coincides with $\overrightarrow{DF}$, as shown below.

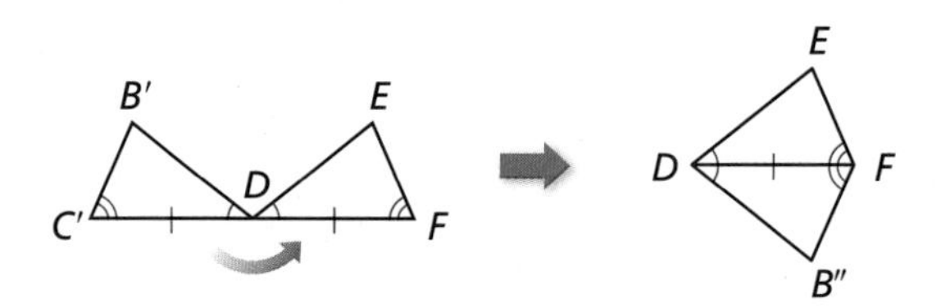

Because $\overline{DC'} \cong \overline{DF}$, the rotation maps point C' to point F. So, this rotation maps $\triangle DB'C'$ to $\triangle DB''F$. Now, reflect $\triangle DB''F$ in the line through points D and F, as shown below.

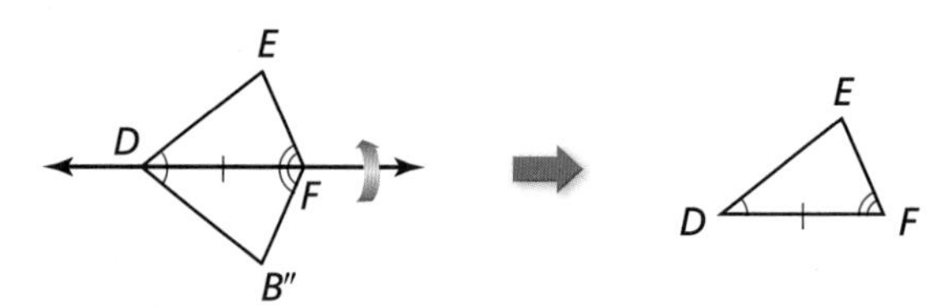

Because points D and F lie on $\overleftrightarrow{DF}$, this reflection maps them onto themselves. Because a reflection preserves angle measure and $\angle B''DF \cong \angle EDF$, the reflection maps $\overrightarrow{DB''}$ to $\overrightarrow{DE}$. Similarly, because $\angle B''FD \cong \angle EFD$, the reflection maps $\overrightarrow{FB''}$ to $\overrightarrow{FE}$. The image of B'' lies on $\overrightarrow{DE}$ and $\overrightarrow{FE}$. Because $\overrightarrow{DE}$ and $\overrightarrow{FE}$ only have point E in common, the image of B'' must be E. So, this reflection maps $\triangle DB''F$ to $\triangle DEF$.

Because you can map $\triangle ABC$ to $\triangle DEF$ using a composition of rigid motions, $\triangle ABC \cong \triangle DEF$.

Theorem

Theorem 5.11 Angle-Angle-Side (AAS) Congruence Theorem

If two angles and a non-included side of one triangle are congruent to two angles and the corresponding non-included side of a second triangle, then the two triangles are congruent.

If $\angle A \cong \angle D$, $\angle C \cong \angle F$, and $\overline{BC} \cong \overline{EF}$, then $\triangle ABC \cong \triangle DEF$.

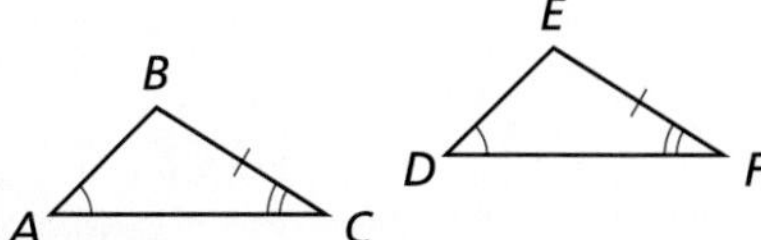

Proof p. 271

PROOF Angle-Angle-Side (AAS) Congruence Theorem

Given $\angle A \cong \angle D$,
$\angle C \cong \angle F$,
$\overline{BC} \cong \overline{EF}$

Prove $\triangle ABC \cong \triangle DEF$

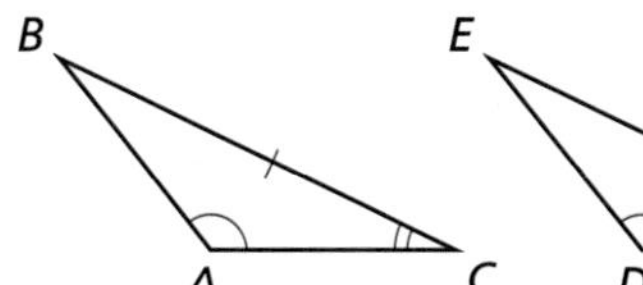

You are given $\angle A \cong \angle D$ and $\angle C \cong \angle F$. By the Third Angles Theorem (Theorem 5.4), $\angle B \cong \angle E$. You are given $\overline{BC} \cong \overline{EF}$. So, two pairs of angles and their included sides are congruent. By the ASA Congruence Theorem, $\triangle ABC \cong \triangle DEF$.

EXAMPLE 1 Identifying Congruent Triangles

Can the triangles be proven congruent with the information given in the diagram? If so, state the theorem you would use.

a. 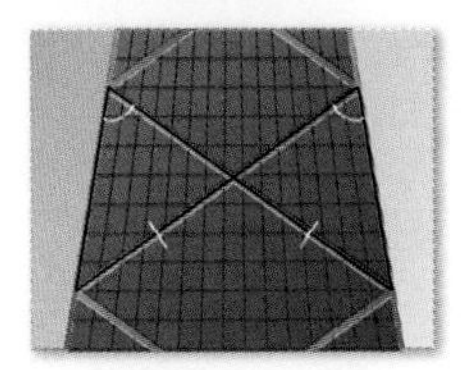

b. 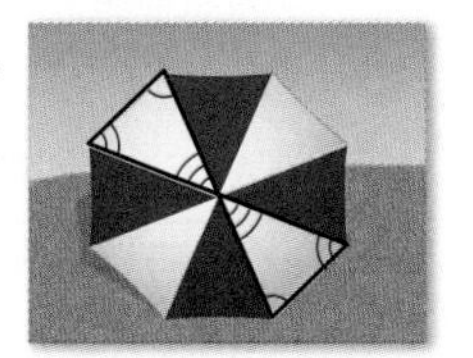

c. 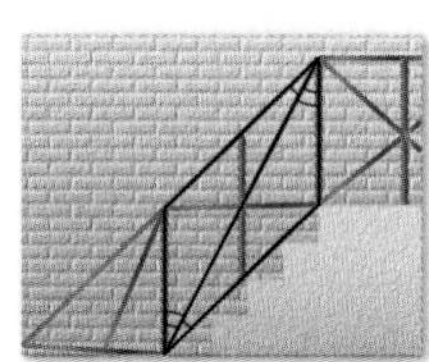

COMMON ERROR

You need at least one pair of congruent corresponding sides to prove two triangles are congruent.

SOLUTION

a. The vertical angles are congruent, so two pairs of angles and a pair of non-included sides are congruent. The triangles are congruent by the AAS Congruence Theorem.

b. There is not enough information to prove the triangles are congruent, because no sides are known to be congruent.

c. Two pairs of angles and their included sides are congruent. The triangles are congruent by the ASA Congruence Theorem.

Monitoring Progress

Help in English and Spanish at *BigIdeasMath.com*

1. Can the triangles be proven congruent with the information given in the diagram? If so, state the theorem you would use.

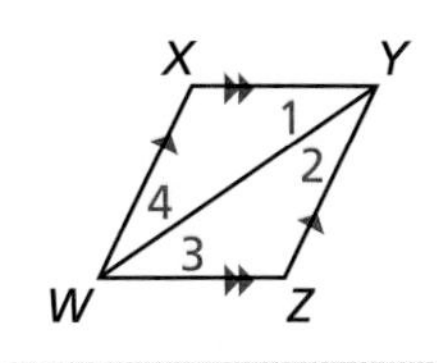

CONSTRUCTION Copying a Triangle Using ASA

Construct a triangle that is congruent to $\triangle ABC$ using the ASA Congruence Theorem. Use a compass and straightedge.

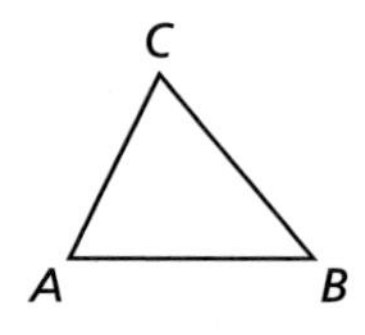

SOLUTION

Step 1

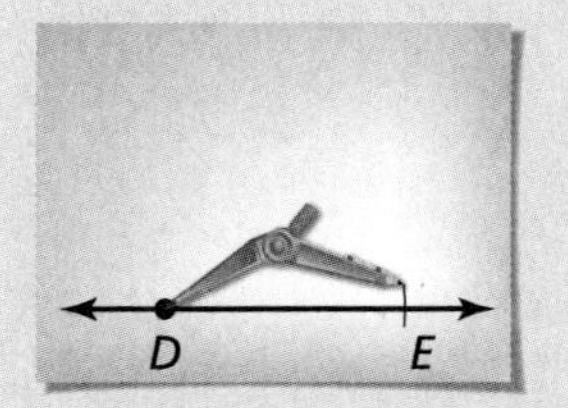

Construct a side Construct $\overline{DE}$ so that it is congruent to $\overline{AB}$.

Step 2

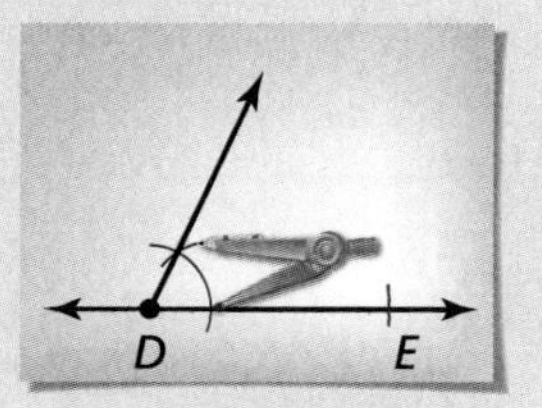

Construct an angle Construct $\angle D$ with vertex D and side $\overrightarrow{DE}$ so that it is congruent to $\angle A$.

Step 3

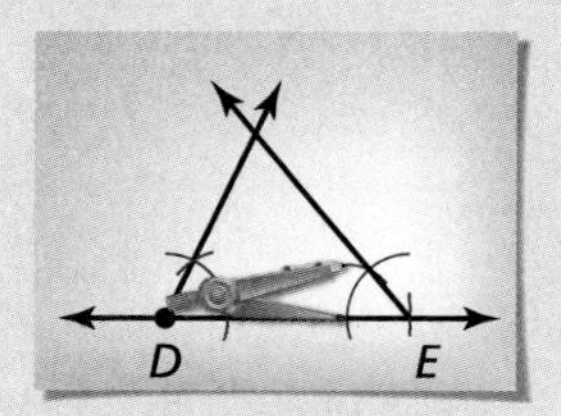

Construct an angle Construct $\angle E$ with vertex E and side $\overrightarrow{ED}$ so that it is congruent to $\angle B$.

Step 4

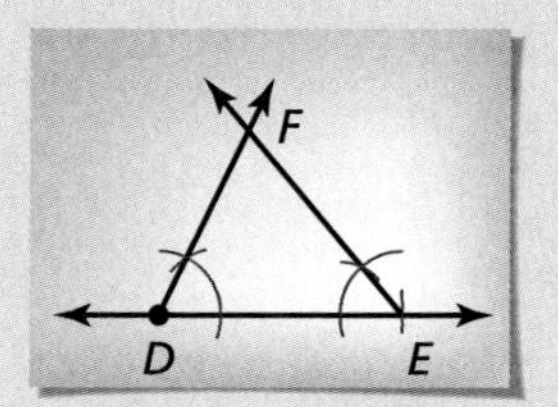

Label a point Label the intersection of the sides of $\angle D$ and $\angle E$ that you constructed in Steps 2 and 3 as F. By the ASA Congruence Theorem, $\triangle ABC \cong \triangle DEF$.

EXAMPLE 2 Using the ASA Congruence Theorem

Write a proof.

Given $\overline{AD} \parallel \overline{EC}$, $\overline{BD} \cong \overline{BC}$

Prove $\triangle ABD \cong \triangle EBC$

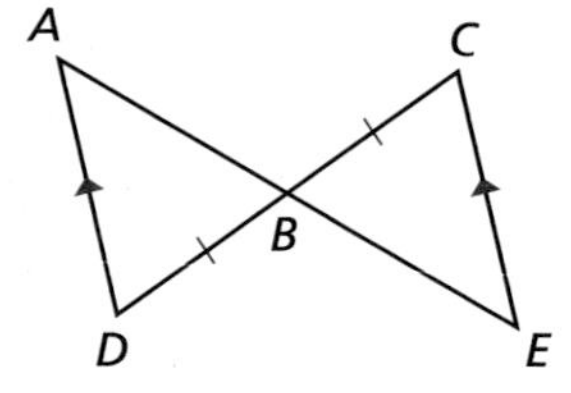

SOLUTION

STATEMENTS	REASONS
1. $\overline{AD} \parallel \overline{EC}$	1. Given
A 2. $\angle D \cong \angle C$	2. Alternate Interior Angles Theorem (Thm. 3.2)
S 3. $\overline{BD} \cong \overline{BC}$	3. Given
A 4. $\angle ABD \cong \angle EBC$	4. Vertical Angles Congruence Theorem (Thm 2.6)
5. $\triangle ABD \cong \triangle EBC$	5. ASA Congruence Theorem

Monitoring Progress  Help in English and Spanish at *BigIdeasMath.com*

2. In the diagram, $\overline{AB} \perp \overline{AD}$, $\overline{DE} \perp \overline{AD}$, and $\overline{AC} \cong \overline{DC}$. Prove $\triangle ABC \cong \triangle DEC$.

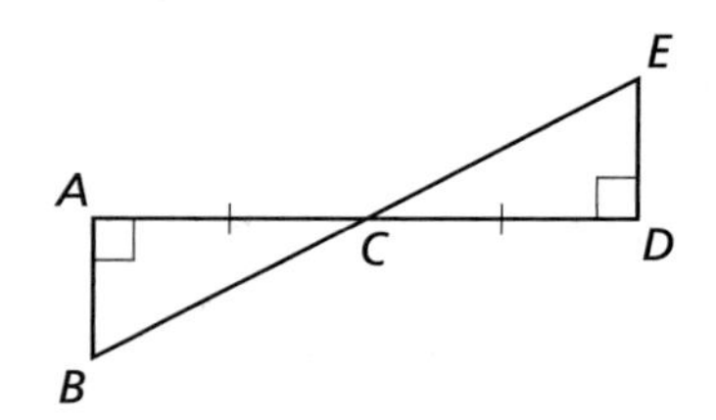

EXAMPLE 3 Using the AAS Congruence Theorem

Write a proof.

Given $\overline{HF} \parallel \overline{GK}$, $\angle F$ and $\angle K$ are right angles.

Prove $\triangle HFG \cong \triangle GKH$

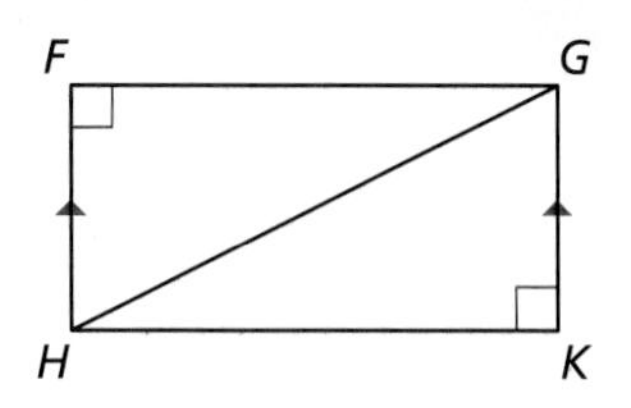

SOLUTION

STATEMENTS	REASONS
1. $\overline{HF} \parallel \overline{GK}$	1. Given
A 2. $\angle GHF \cong \angle HGK$	2. Alternate Interior Angles Theorem (Theorem 3.2)
3. $\angle F$ and $\angle K$ are right angles.	3. Given
A 4. $\angle F \cong \angle K$	4. Right Angles Congruence Theorem (Theorem 2.3)
S 5. $\overline{HG} \cong \overline{GH}$	5. Reflexive Property of Congruence (Theorem 2.1)
6. $\triangle HFG \cong \angle GKH$	6. AAS Congruence Theorem

Monitoring Progress

Help in English and Spanish at *BigIdeasMath.com*

3. In the diagram, $\angle S \cong \angle U$ and $\overline{RS} \cong \overline{VU}$. Prove $\triangle RST \cong \triangle VUT$.

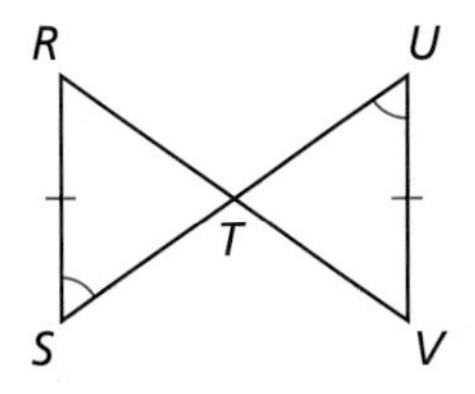

Concept Summary

Triangle Congruence Theorems

You have learned five methods for proving that triangles are congruent.

SAS	SSS	HL (right △s only)	ASA	AAS
Two sides and the included angle are congruent.	All three sides are congruent.	The hypotenuse and one of the legs are congruent.	Two angles and the included side are congruent.	Two angles and a non-included side are congruent.

In the Exercises, you will prove three additional theorems about the congruence of right triangles: **H**ypotenuse-**A**ngle, **L**eg-**L**eg, and **A**ngle-**L**eg.

5.6 Exercises

Dynamic Solutions available at *BigIdeasMath.com*

Vocabulary and Core Concept Check

1. **WRITING** How are the AAS Congruence Theorem (Theorem 5.11) and the ASA Congruence Theorem (Theorem 5.10) similar? How are they different?

2. **WRITING** You know that a pair of triangles has two pairs of congruent corresponding angles. What other information do you need to show that the triangles are congruent?

Monitoring Progress and Modeling with Mathematics

In Exercises 3–6, decide whether enough information is given to prove that the triangles are congruent. If so, state the theorem you would use. *(See Example 1.)*

3. $\triangle ABC$, $\triangle QRS$

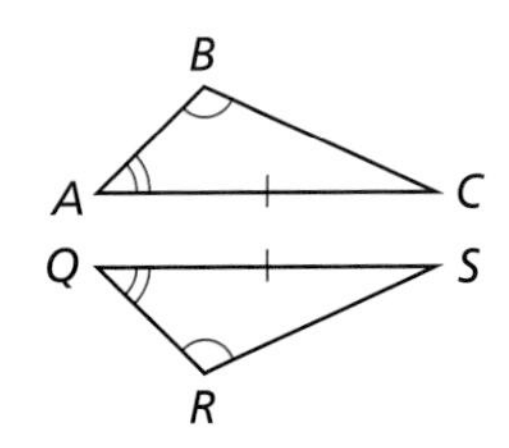

4. $\triangle ABC$, $\triangle DBC$

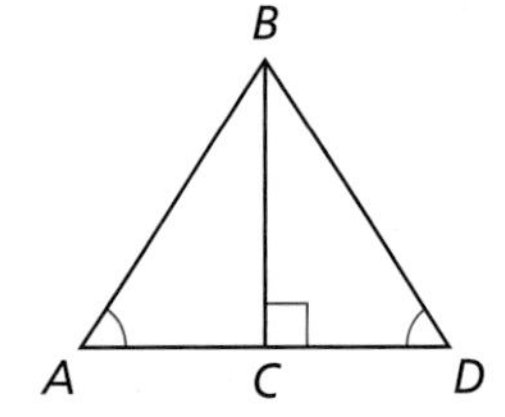

5. $\triangle XYZ$, $\triangle JKL$

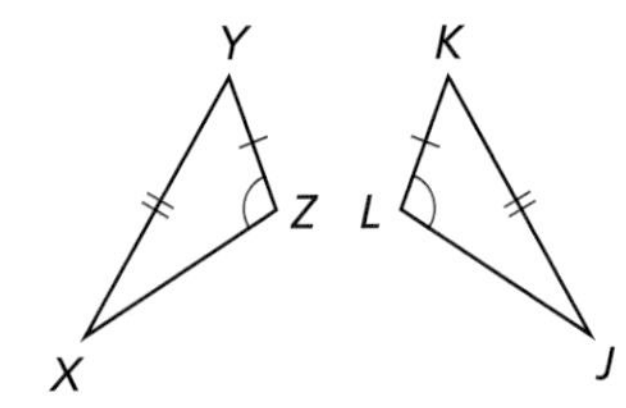

6. $\triangle RSV$, $\triangle UTV$

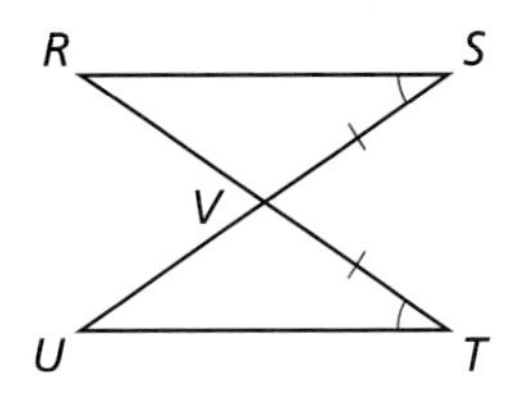

In Exercises 7 and 8, state the third congruence statement that is needed to prove that $\triangle FGH \cong \triangle LMN$ using the given theorem.

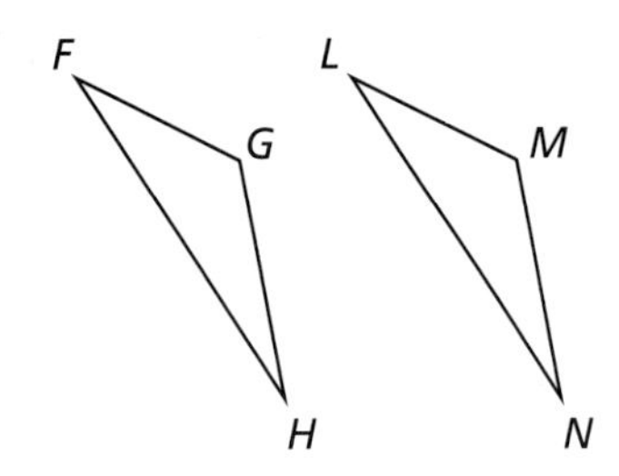

7. **Given** $\overline{GH} \cong \overline{MN}$, $\angle G \cong \angle M$, ___ $\cong$ ___

   Use the AAS Congruence Theorem (Thm. 5.11).

8. **Given** $\overline{FG} \cong \overline{LM}$, $\angle G \cong \angle M$, ___ $\cong$ ___

   Use the ASA Congruence Theorem (Thm. 5.10).

In Exercises 9–12, decide whether you can use the given information to prove that $\triangle ABC \cong \triangle DEF$. Explain your reasoning.

9. $\angle A \cong \angle D$, $\angle C \cong \angle F$, $\overline{AC} \cong \overline{DF}$

10. $\angle C \cong \angle F$, $\overline{AB} \cong \overline{DE}$, $\overline{BC} \cong \overline{EF}$

11. $\angle B \cong \angle E$, $\angle C \cong \angle F$, $\overline{AC} \cong \overline{DE}$

12. $\angle A \cong \angle D$, $\angle B \cong \angle E$, $\overline{BC} \cong \overline{EF}$

CONSTRUCTION **In Exercises 13 and 14, construct a triangle that is congruent to the given triangle using the ASA Congruence Theorem (Theorem 5.10). Use a compass and straightedge.**

13.

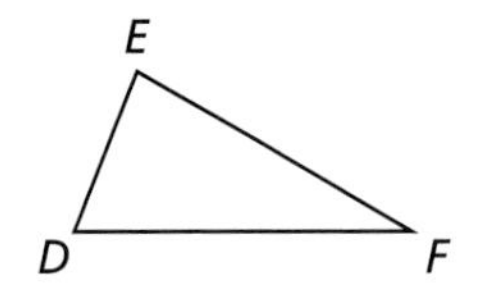

14.

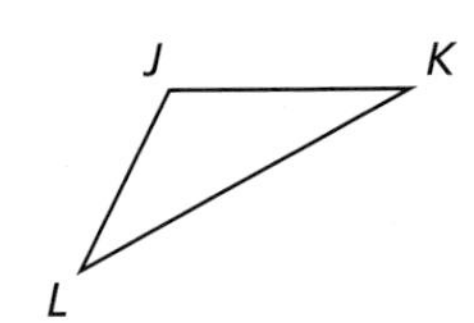

ERROR ANALYSIS **In Exercises 15 and 16, describe and correct the error.**

15.

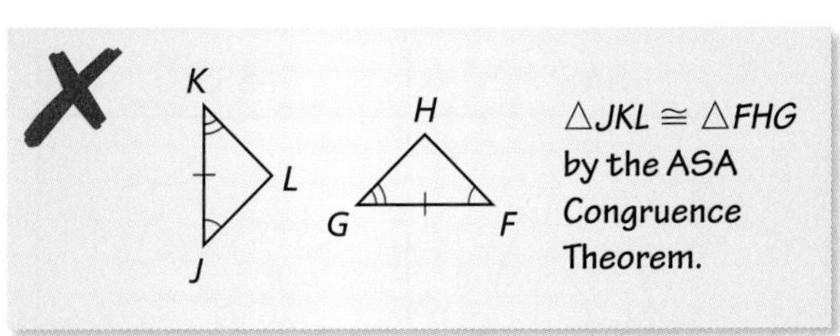

16.

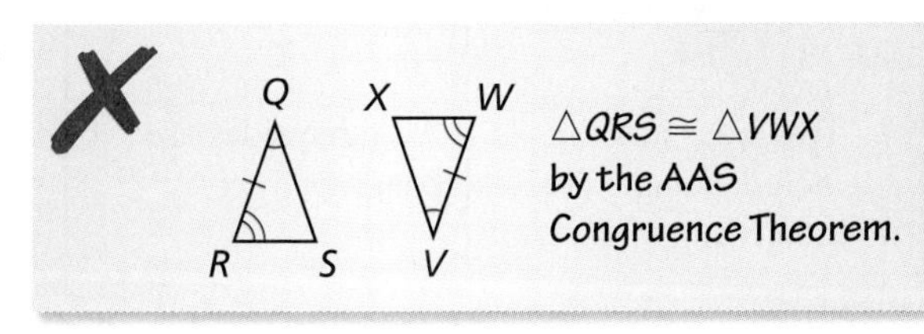

PROOF **In Exercises 17 and 18, prove that the triangles are congruent using the ASA Congruence Theorem (Theorem 5.10).** *(See Example 2.)*

17. Given M is the midpoint of $\overline{NL}$.
$\overline{NL} \perp \overline{NQ}$, $\overline{NL} \perp \overline{MP}$, $\overline{QM} \parallel \overline{PL}$

Prove $\triangle NQM \cong \triangle MPL$

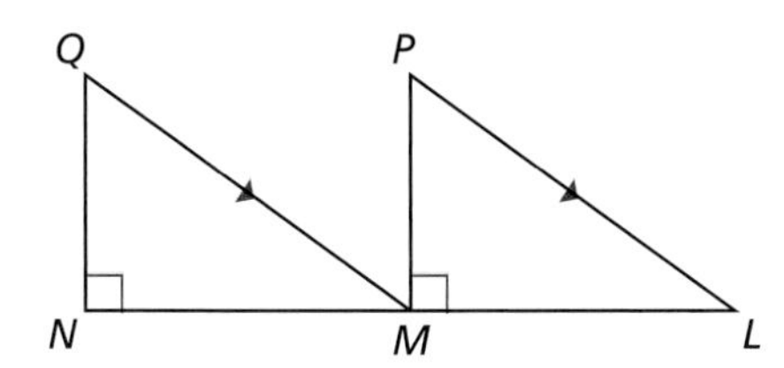

18. Given $\overline{AJ} \cong \overline{KC}$, $\angle BJK \cong \angle BKJ$, $\angle A \cong \angle C$

Prove $\triangle ABK \cong \triangle CBJ$

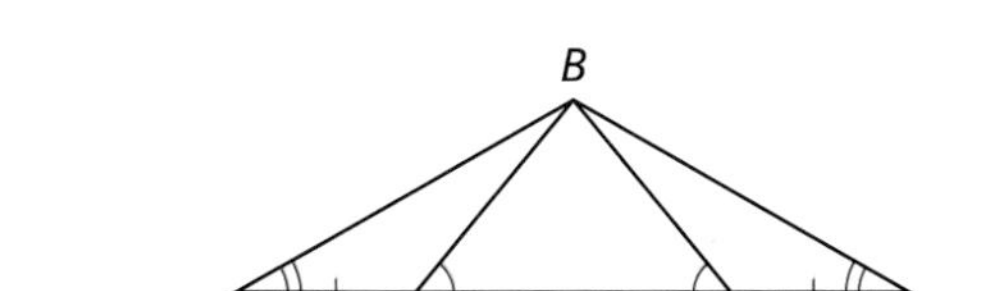

PROOF **In Exercises 19 and 20, prove that the triangles are congruent using the AAS Congruence Theorem (Theorem 5.11).** *(See Example 3.)*

19. Given $\overline{VW} \cong \overline{UW}$, $\angle X \cong \angle Z$

Prove $\triangle XWV \cong \triangle ZWU$

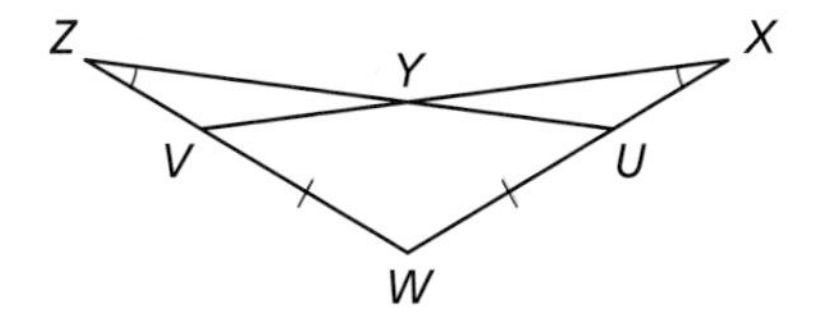

20. Given $\angle NKM \cong \angle LMK$, $\angle L \cong \angle N$

Prove $\triangle NMK \cong \triangle LKM$

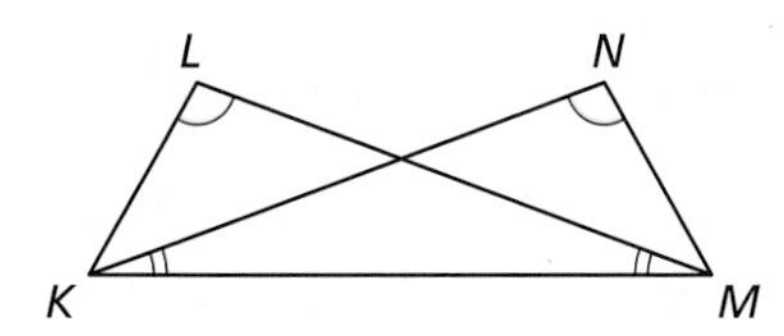

PROOF **In Exercises 21–23, write a paragraph proof for the theorem about right triangles.**

21. Hypotenuse-Angle (HA) Congruence Theorem If an angle and the hypotenuse of a right triangle are congruent to an angle and the hypotenuse of a second right triangle, then the triangles are congruent.

22. Leg-Leg (LL) Congruence Theorem If the legs of a right triangle are congruent to the legs of a second right triangle, then the triangles are congruent.

23. Angle-Leg (AL) Congruence Theorem If an angle and a leg of a right triangle are congruent to an angle and a leg of a second right triangle, then the triangles are congruent.

24. REASONING What additional information do you need to prove $\triangle JKL \cong \triangle MNL$ by the ASA Congruence Theorem (Theorem 5.10)?

Ⓐ $\overline{KM} \cong \overline{KJ}$

Ⓑ $\overline{KH} \cong \overline{NH}$

Ⓒ $\angle M \cong \angle J$

Ⓓ $\angle LKJ \cong \angle LNM$

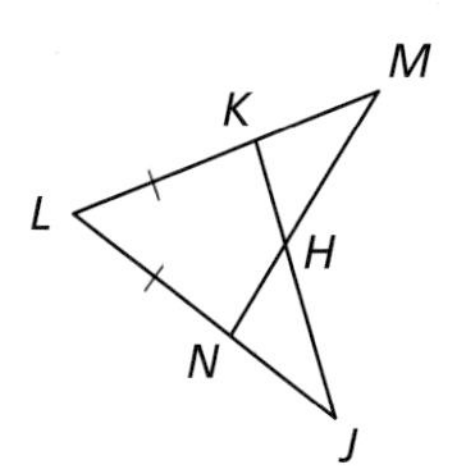

25. MATHEMATICAL CONNECTIONS This toy contains $\triangle ABC$ and $\triangle DBC$. Can you conclude that $\triangle ABC \cong \triangle DBC$ from the given angle measures? Explain.

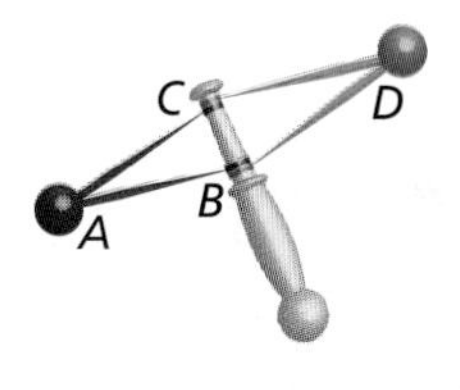

$m\angle ABC = (8x - 32)°$

$m\angle DBC = (4y - 24)°$

$m\angle BCA = (5x + 10)°$

$m\angle BCD = (3y + 2)°$

$m\angle CAB = (2x - 8)°$

$m\angle CDB = (y - 6)°$

26. REASONING Which of the following congruence statements are true? Select all that apply.

Ⓐ $\overline{TU} \cong \overline{UV}$

Ⓑ $\triangle STV \cong \triangle XVW$

Ⓒ $\triangle TVS \cong \triangle VWU$

Ⓓ $\triangle VST \cong \triangle VUW$

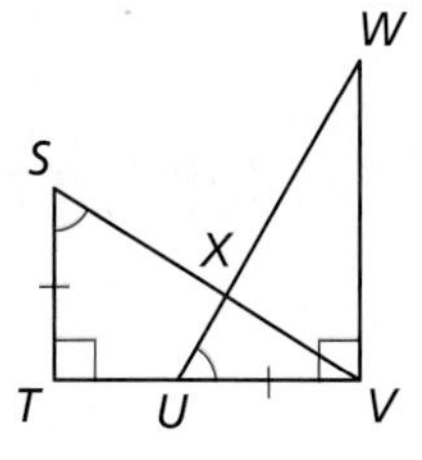

27. PROVING A THEOREM Prove the Converse of the Base Angles Theorem (Theorem 5.7). (*Hint:* Draw an auxiliary line inside the triangle.)

28. MAKING AN ARGUMENT Your friend claims to be able to rewrite any proof that uses the AAS Congruence Theorem (Thm. 5.11) as a proof that uses the ASA Congruence Theorem (Thm. 5.10). Is this possible? Explain your reasoning.

29. MODELING WITH MATHEMATICS When a light ray from an object meets a mirror, it is reflected back to your eye. For example, in the diagram, a light ray from point C is reflected at point D and travels back to point A. The *law of reflection* states that the angle of incidence, $\angle CDB$, is congruent to the angle of reflection, $\angle ADB$.

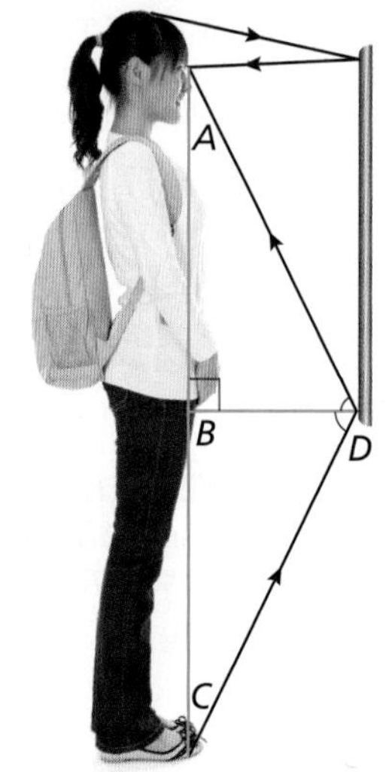

a. Prove that $\triangle ABD$ is congruent to $\triangle CBD$.

Given $\angle CDB \cong \angle ADB$, $\overline{DB} \perp \overline{AC}$

Prove $\triangle ABD \cong \triangle CBD$

b. Verify that $\triangle ACD$ is isosceles.

c. Does moving away from the mirror have any effect on the amount of his or her reflection a person sees? Explain.

30. HOW DO YOU SEE IT? Name as many pairs of congruent triangles as you can from the diagram. Explain how you know that each pair of triangles is congruent.

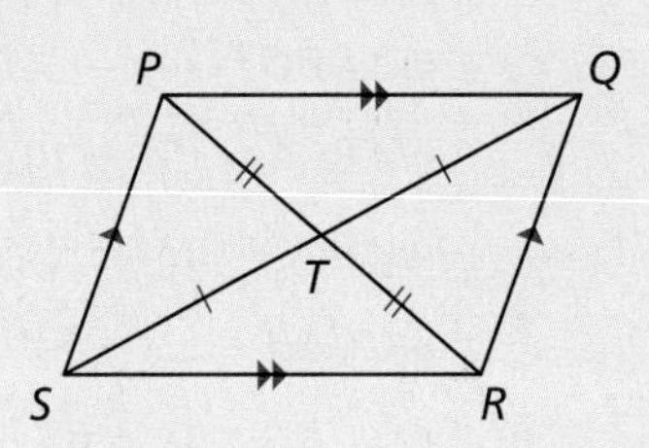

31. CONSTRUCTION Construct a triangle. Show that there is no AAA congruence rule by constructing a second triangle that has the same angle measures but is not congruent.

32. THOUGHT PROVOKING Graph theory is a branch of mathematics that studies vertices and the way they are connected. In graph theory, two polygons are *isomorphic* if there is a one-to-one mapping from one polygon's vertices to the other polygon's vertices that preserves adjacent vertices. In graph theory, are any two triangles isomorphic? Explain your reasoning.

33. MATHEMATICAL CONNECTIONS Six statements are given about $\triangle TUV$ and $\triangle XYZ$.

$\overline{TU} \cong \overline{XY}$ $\quad$ $\overline{UV} \cong \overline{YZ}$ $\quad$ $\overline{TV} \cong \overline{XZ}$

$\angle T \cong \angle X$ $\quad$ $\angle U \cong \angle Y$ $\quad$ $\angle V \cong \angle Z$

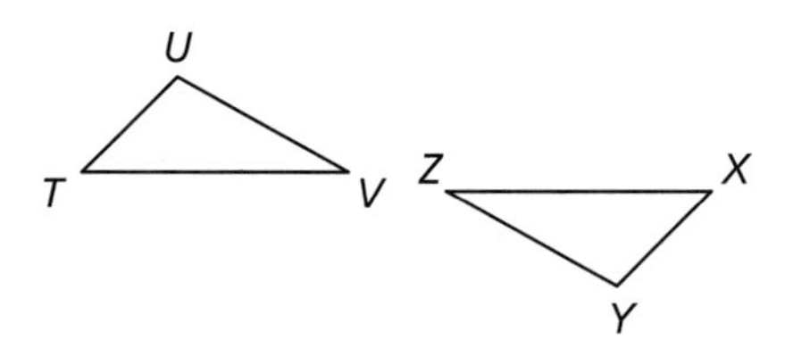

a. List all combinations of three given statements that would provide enough information to prove that $\triangle TUV$ is congruent to $\triangle XYZ$.

b. You choose three statements at random. What is the probability that the statements you choose provide enough information to prove that the triangles are congruent?

Maintaining Mathematical Proficiency

Reviewing what you learned in previous grades and lessons

Find the coordinates of the midpoint of the line segment with the given endpoints. *(Section 1.3)*

34. $C(1, 0)$ and $D(5, 4)$ **35.** $J(-2, 3)$ and $K(4, -1)$ **36.** $R(-5, -7)$ and $S(2, -4)$

Copy the angle using a compass and straightedge. *(Section 1.5)*

37.

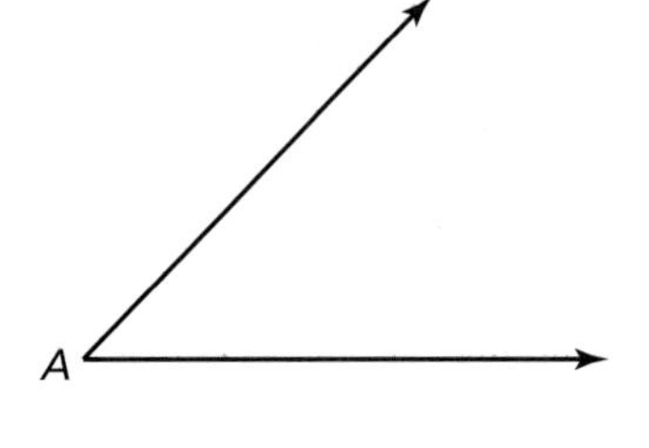

38.

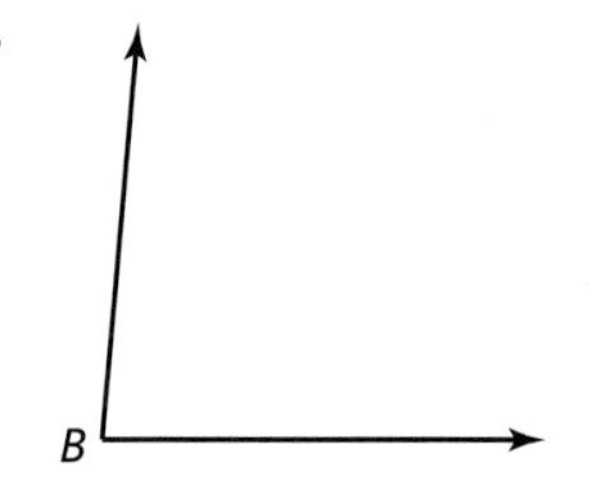

5.7 Using Congruent Triangles

Essential Question How can you use congruent triangles to make an indirect measurement?

CRITIQUING THE REASONING OF OTHERS

To be proficient in math, you need to listen to or read the arguments of others, decide whether they make sense, and ask useful questions to clarify or improve the arguments.

EXPLORATION 1 **Measuring the Width of a River**

Work with a partner. The figure shows how a surveyor can measure the width of a river by making measurements on only one side of the river.

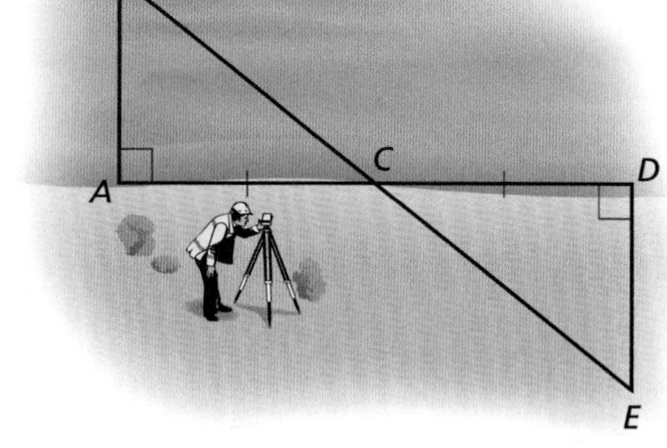

a. Study the figure. Then explain how the surveyor can find the width of the river.

b. Write a proof to verify that the method you described in part (a) is valid.

Given $\angle A$ is a right angle, $\angle D$ is a right angle, $\overline{AC} \cong \overline{CD}$

c. Exchange proofs with your partner and discuss the reasoning used.

EXPLORATION 2 **Measuring the Width of a River**

Work with a partner. It was reported that one of Napoleon's officers estimated the width of a river as follows. The officer stood on the bank of the river and lowered the visor on his cap until the farthest thing visible was the edge of the bank on the other side. He then turned and noted the point on his side that was in line with the tip of his visor and his eye. The officer then paced the distance to this point and concluded that distance was the width of the river.

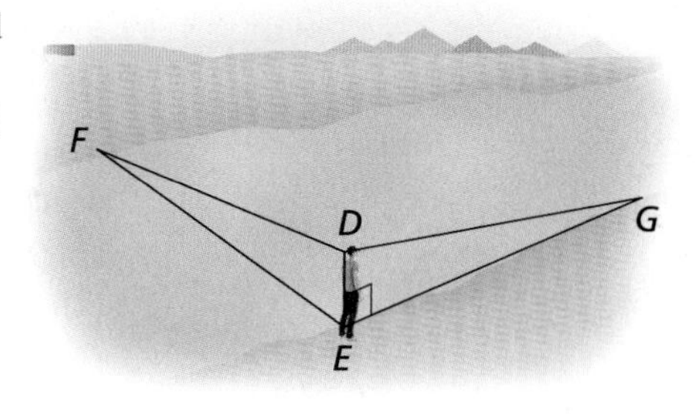

a. Study the figure. Then explain how the officer concluded that the width of the river is EG.

b. Write a proof to verify that the conclusion the officer made is correct.

Given $\angle DEG$ is a right angle, $\angle DEF$ is a right angle, $\angle EDG \cong \angle EDF$

c. Exchange proofs with your partner and discuss the reasoning used.

Communicate Your Answer

3. How can you use congruent triangles to make an indirect measurement?

4. Why do you think the types of measurements described in Explorations 1 and 2 are called *indirect* measurements?

5.7 Lesson

Core Vocabulary

Previous
congruent figures
corresponding parts
construction

What You Will Learn

- Use congruent triangles.
- Prove constructions.

Using Congruent Triangles

Congruent triangles have congruent corresponding parts. So, if you can prove that two triangles are congruent, then you know that their corresponding parts must be congruent as well.

EXAMPLE 1 Using Congruent Triangles

Explain how you can use the given information to prove that the hang glider parts are congruent.

Given $\angle 1 \cong \angle 2$, $\angle RTQ \cong \angle RTS$

Prove $\overline{QT} \cong \overline{ST}$

SOLUTION

If you can show that $\triangle QRT \cong \triangle SRT$, then you will know that $\overline{QT} \cong \overline{ST}$. First, copy the diagram and mark the given information. Then mark the information that you can deduce. In this case, $\angle RQT$ and $\angle RST$ are supplementary to congruent angles, so $\angle RQT \cong \angle RST$. Also, $\overline{RT} \cong \overline{RT}$ by the Reflexive Property of Congruence (Theorem 2.1).

Mark given information.

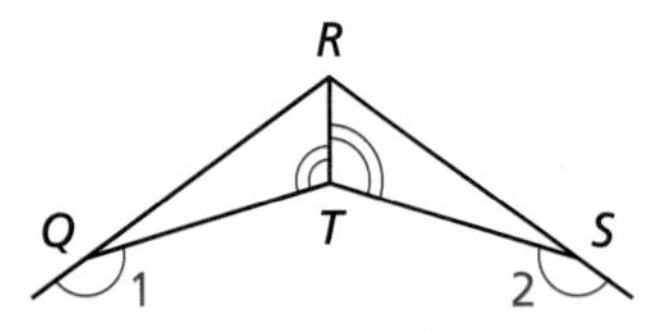

Mark deduced information.

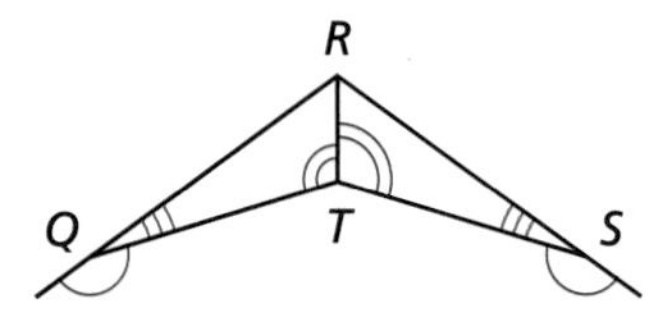

Two angle pairs and a non-included side are congruent, so by the AAS Congruence Theorem (Theorem 5.11), $\triangle QRT \cong \triangle SRT$.

- Because corresponding parts of congruent triangles are congruent, $\overline{QT} \cong \overline{ST}$.

Monitoring Progress  Help in English and Spanish at *BigIdeasMath.com*

1. Explain how you can prove that $\angle A \cong \angle C$.

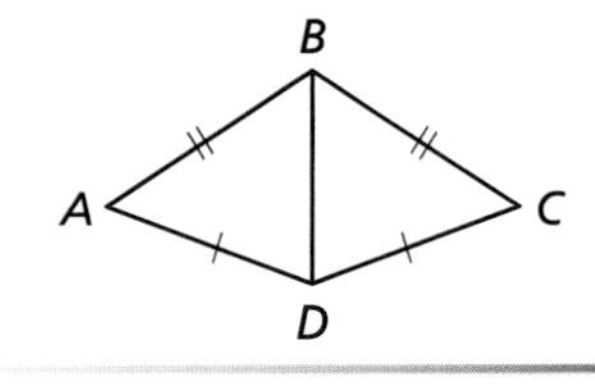

EXAMPLE 2 Using Congruent Triangles for Measurement

MAKING SENSE OF PROBLEMS

When you cannot easily measure a length directly, you can make conclusions about the length *indirectly*, usually by calculations based on known lengths.

Use the following method to find the distance across a river, from point N to point P.

- Place a stake at K on the near side so that $\overline{NK} \perp \overline{NP}$.
- Find M, the midpoint of $\overline{NK}$.
- Locate the point L so that $\overline{NK} \perp \overline{KL}$ and L, P, and M are collinear.

Explain how this plan allows you to find the distance.

SOLUTION

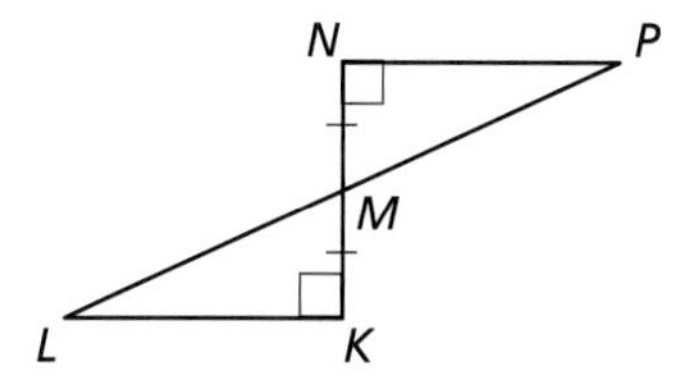

Because $\overline{NK} \perp \overline{NP}$ and $\overline{NK} \perp \overline{KL}$, $\angle N$ and $\angle K$ are congruent right angles. Because M is the midpoint of $\overline{NK}$, $\overline{NM} \cong \overline{KM}$. The vertical angles $\angle KML$ and $\angle NMP$ are congruent. So, $\triangle MLK \cong \triangle MPN$ by the ASA Congruence Theorem (Theorem 5.10). Then because corresponding parts of congruent triangles are congruent, $\overline{KL} \cong \overline{NP}$. So, you can find the distance NP across the river by measuring $\overline{KL}$.

EXAMPLE 3 Planning a Proof Involving Pairs of Triangles

Use the given information to write a plan for proof.

Given $\angle 1 \cong \angle 2$, $\angle 3 \cong \angle 4$

Prove $\triangle BCE \cong \triangle DCE$

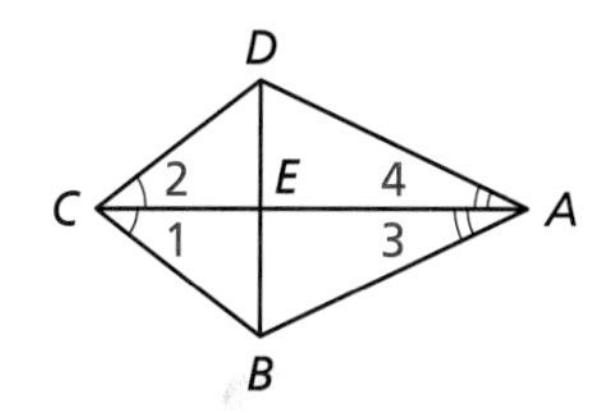

SOLUTION

In $\triangle BCE$ and $\triangle DCE$, you know that $\angle 1 \cong \angle 2$ and $\overline{CE} \cong \overline{CE}$. If you can show that $\overline{CB} \cong \overline{CD}$, then you can use the SAS Congruence Theorem (Theorem 5.5).

To prove that $\overline{CB} \cong \overline{CD}$, you can first prove that $\triangle CBA \cong \triangle CDA$. You are given $\angle 1 \cong \angle 2$ and $\angle 3 \cong \angle 4$. $\overline{CA} \cong \overline{CA}$ by the Reflexive Property of Congruence (Theorem 2.1). You can use the ASA Congruence Theorem (Theorem 5.10) to prove that $\triangle CBA \cong \triangle CDA$.

▶ **Plan for Proof** Use the ASA Congruence Theorem (Theorem 5.10) to prove that $\triangle CBA \cong \triangle CDA$. Then state that $\overline{CB} \cong \overline{CD}$. Use the SAS Congruence Theorem (Theorem 5.5) to prove that $\triangle BCE \cong \triangle DCE$.

Monitoring Progress

Help in English and Spanish at *BigIdeasMath.com*

2. In Example 2, does it matter how far from point N you place a stake at point K? Explain.

3. Write a plan to prove that $\triangle PTU \cong \triangle UQP$.

P Q R S T U

Proving Constructions

Recall that you can use a compass and a straightedge to copy an angle. The construction is shown below. You can use congruent triangles to prove that this construction is valid.

Step 1

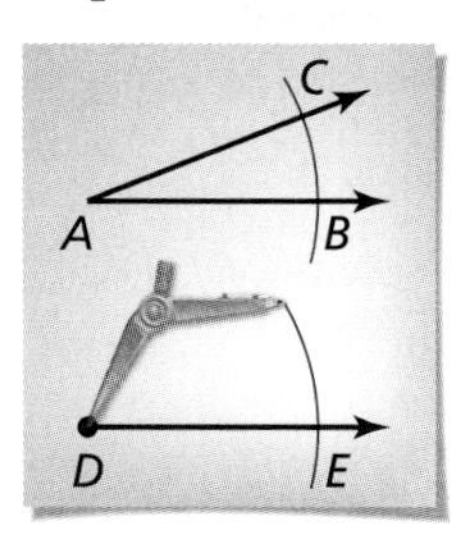

Draw a segment and arcs
To copy $\angle A$, draw a segment with initial point D. Draw an arc with center A. Using the same radius, draw an arc with center D. Label points B, C, and E.

Step 2

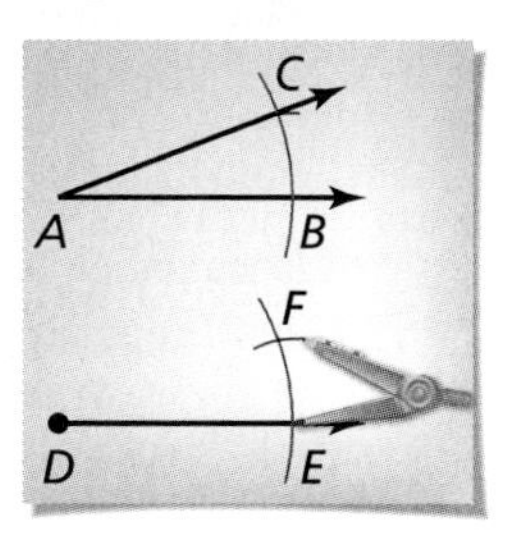

Draw an arc
Draw an arc with radius BC and center E. Label the intersection F.

Step 3

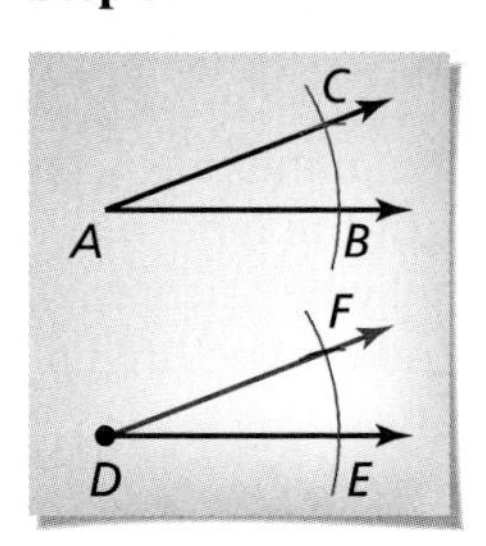

Draw a ray
Draw $\overrightarrow{DF}$. In Example 4, you will prove that $\angle D \cong \angle A$.

EXAMPLE 4 Proving a Construction

Write a proof to verify that the construction for copying an angle is valid.

SOLUTION

Add $\overline{BC}$ and $\overline{EF}$ to the diagram. In the construction, one compass setting determines $\overline{AB}, \overline{DE}, \overline{AC}$, and $\overline{DF}$, and another compass setting determines $\overline{BC}$ and $\overline{EF}$. So, you can assume the following as given statements.

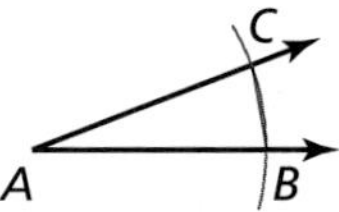

Given $\overline{AB} \cong \overline{DE}$, $\overline{AC} \cong \overline{DF}$, $\overline{BC} \cong \overline{EF}$

Prove $\angle D \cong \angle A$

Plan for Proof Show that $\triangle DEF \cong \triangle ABC$, so you can conclude that the corresponding parts $\angle D$ and $\angle A$ are congruent.

Plan in Action

STATEMENTS	REASONS
1. $\overline{AB} \cong \overline{DE}, \overline{AC} \cong \overline{DF}, \overline{BC} \cong \overline{EF}$	**1.** Given
2. $\triangle DEF \cong \triangle ABC$	**2.** SSS Congruence Theorem (Theorem 5.8)
3. $\angle D \cong \angle A$	**3.** Corresponding parts of congruent triangles are congruent.

Monitoring Progress

Help in English and Spanish at *BigIdeasMath.com*

4. Use the construction of an angle bisector on page 42. What segments can you assume are congruent?

5.7 Exercises

Dynamic Solutions available at *BigIdeasMath.com*

Vocabulary and Core Concept Check

1. **COMPLETE THE SENTENCE** ______________ parts of congruent triangles are congruent.

2. **WRITING** Describe a situation in which you might choose to use indirect measurement with congruent triangles to find a measure rather than measuring directly.

Monitoring Progress and Modeling with Mathematics

In Exercises 3–8, explain how to prove that the statement is true. *(See Example 1.)*

3. $\angle A \cong \angle D$

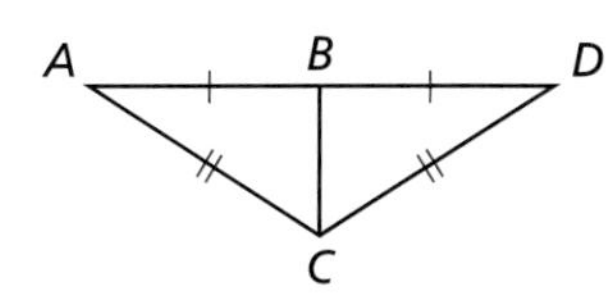

4. $\angle Q \cong \angle T$

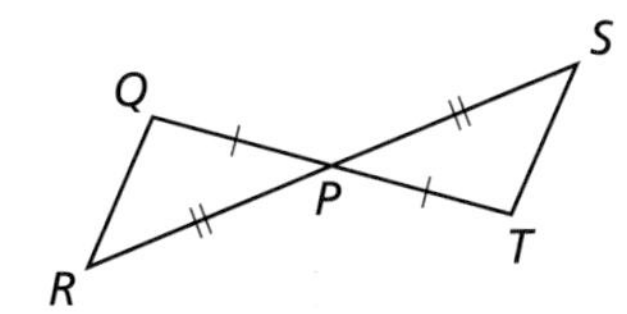

5. $\overline{JM} \cong \overline{LM}$

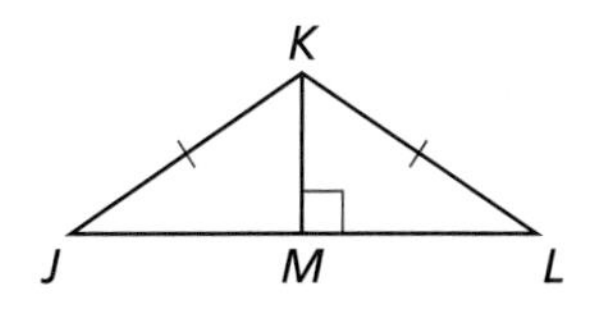

6. $\overline{AC} \cong \overline{DB}$

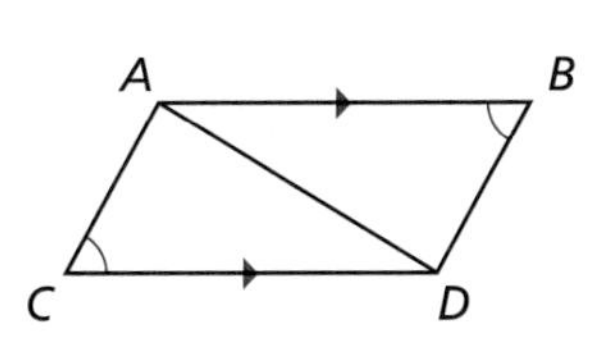

7. $\overline{GK} \cong \overline{HJ}$

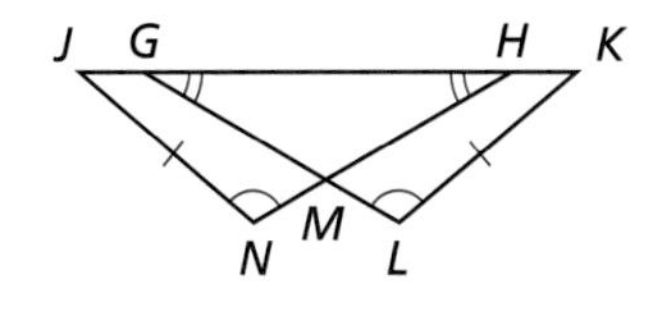

8. $\overline{QW} \cong \overline{VT}$

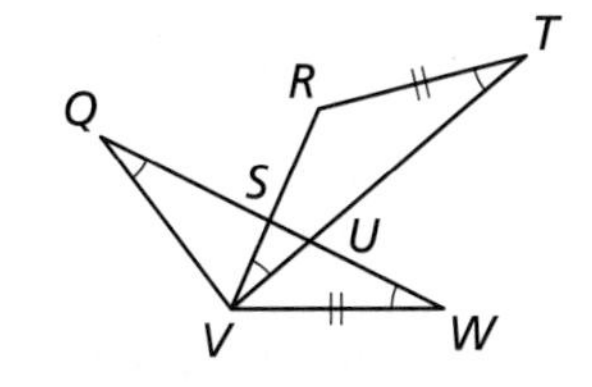

In Exercises 9–12, write a plan to prove that $\angle 1 \cong \angle 2$. *(See Example 3.)*

9.

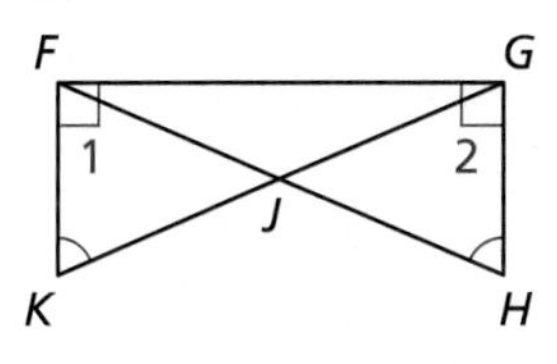

10.

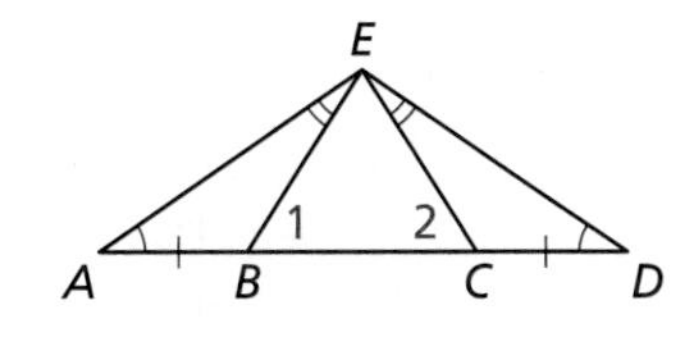

11.

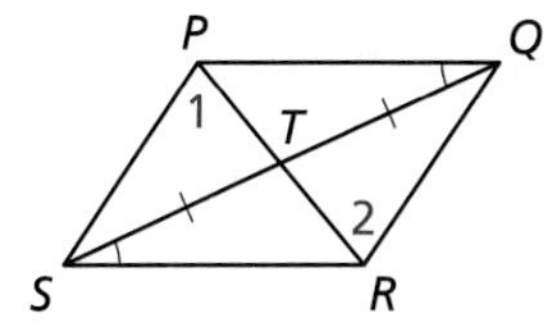

12.

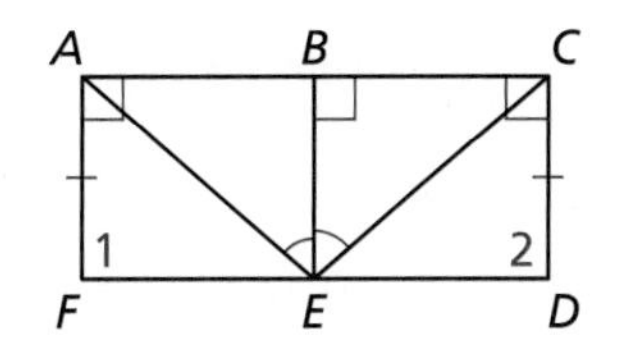

In Exercises 13 and 14, write a proof to verify that the construction is valid. *(See Example 4.)*

13. Line perpendicular to a line through a point not on the line

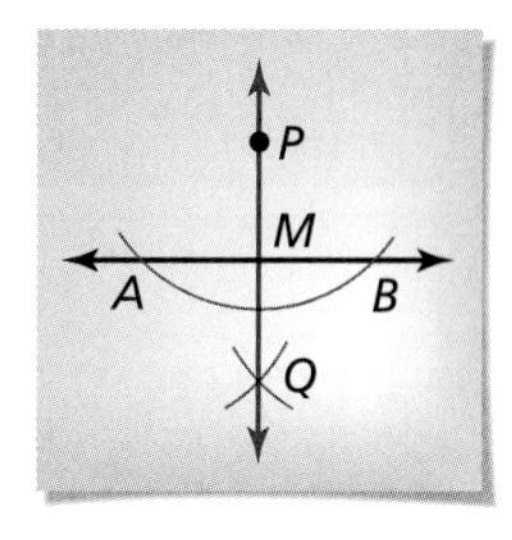

Plan for Proof Show that $\triangle APQ \cong \triangle BPQ$ by the SSS Congruence Theorem (Theorem 5.8). Then show that $\triangle APM \cong \triangle BPM$ using the SAS Congruence Theorem (Theorem 5.5). Use corresponding parts of congruent triangles to show that $\angle AMP$ and $\angle BMP$ are right angles.

14. Line perpendicular to a line through a point on the line

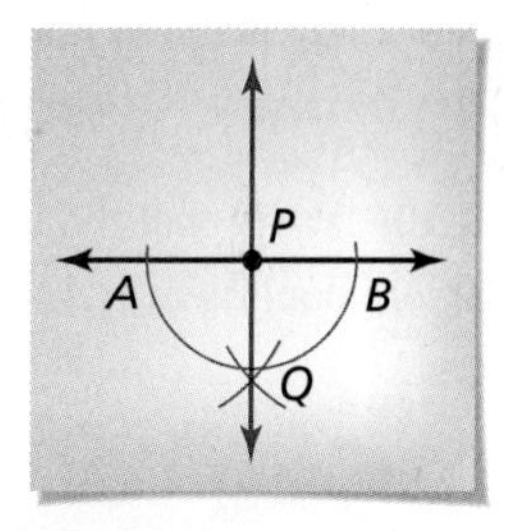

Plan for Proof Show that $\triangle APQ \cong \triangle BPQ$ by the SSS Congruence Theorem (Theorem 5.8). Use corresponding parts of congruent triangles to show that $\angle QPA$ and $\angle QPB$ are right angles.

In Exercises 15 and 16, use the information given in the diagram to write a proof.

15. **Prove** $\overline{FL} \cong \overline{HN}$

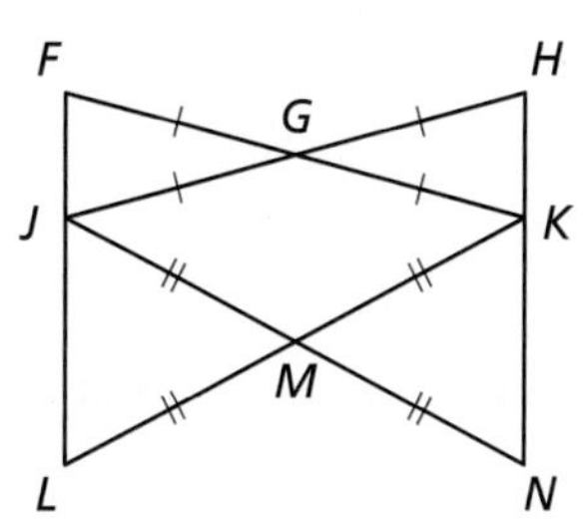

16. Prove $\triangle PUX \cong \triangle QSY$

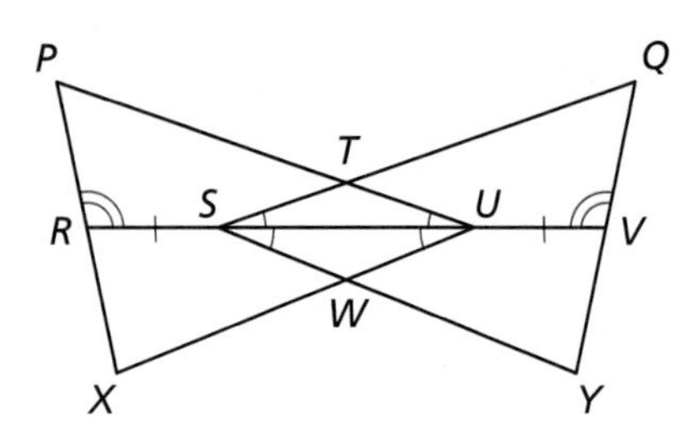

17. MODELING WITH MATHEMATICS Explain how to find the distance across the canyon. *(See Example 2.)*

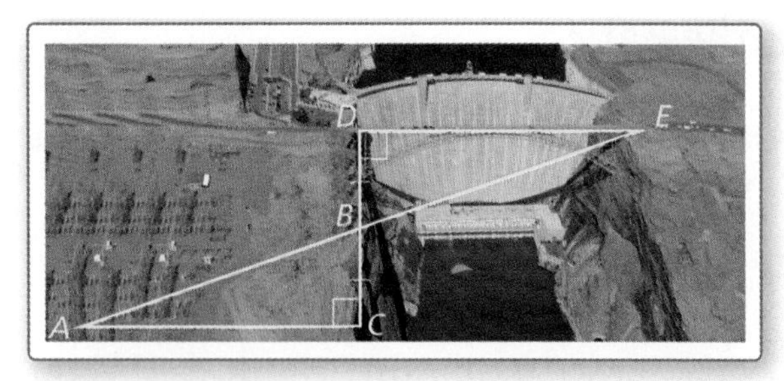

18. HOW DO YOU SEE IT? Use the tangram puzzle.

a. Which triangle(s) have an area that is twice the area of the purple triangle?

b. How many times greater is the area of the orange triangle than the area of the purple triangle?

19. PROOF Prove that the green triangles in the Jamaican flag are congruent if $\overline{AD} \parallel \overline{BC}$ and E is the midpoint of $\overline{AC}$.

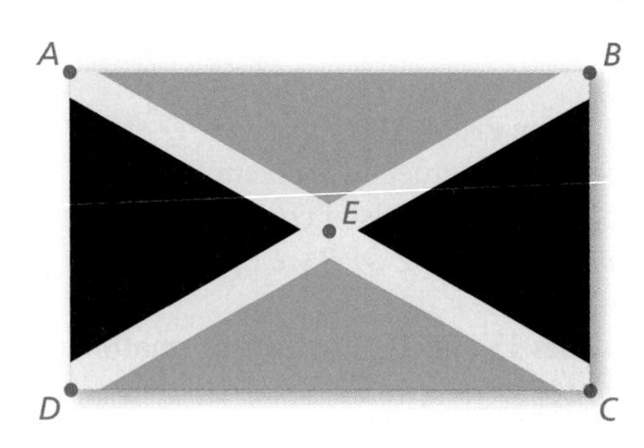

20. THOUGHT PROVOKING The Bermuda Triangle is a region in the Atlantic Ocean in which many ships and planes have mysteriously disappeared. The vertices are Miami, San Juan, and Bermuda. Use the Internet or some other resource to find the side lengths, the perimeter, and the area of this triangle (in miles). Then create a congruent triangle on land using cities as vertices.

21. MAKING AN ARGUMENT Your friend claims that $\triangle WZY$ can be proven congruent to $\triangle YXW$ using the HL Congruence Theorem (Thm. 5.9). Is your friend correct? Explain your reasoning.

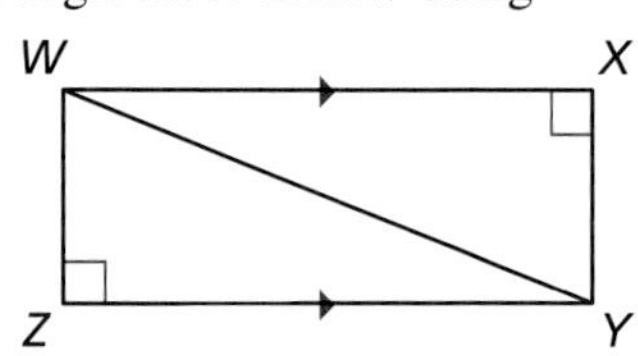

22. CRITICAL THINKING Determine whether each conditional statement is true or false. If the statement is false, rewrite it as a true statement using the converse, inverse, or contrapositive.

a. If two triangles have the same perimeter, then they are congruent.

b. If two triangles are congruent, then they have the same area.

23. ATTENDING TO PRECISION Which triangles are congruent to $\triangle ABC$? Select all that apply.

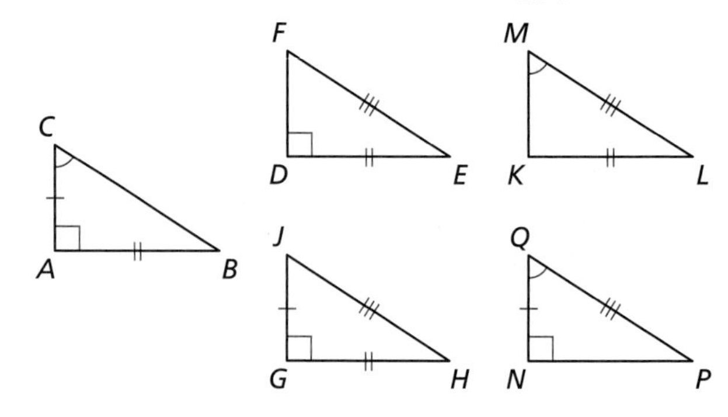

Maintaining Mathematical Proficiency

Reviewing what you learned in previous grades and lessons

Find the perimeter of the polygon with the given vertices. *(Section 1.4)*

24. $A(-1, 1), B(4, 1), C(4, -2), D(-1, -2)$

25. $J(-5, 3), K(-2, 1), L(3, 4)$

5.8 Coordinate Proofs

Essential Question How can you use a coordinate plane to write a proof?

EXPLORATION 1 Writing a Coordinate Proof

Work with a partner.

a. Use dynamic geometry software to draw $\overline{AB}$ with endpoints $A(0, 0)$ and $B(6, 0)$.

b. Draw the vertical line $x = 3$.

c. Draw $\triangle ABC$ so that C lies on the line $x = 3$.

d. Use your drawing to prove that $\triangle ABC$ is an isosceles triangle.

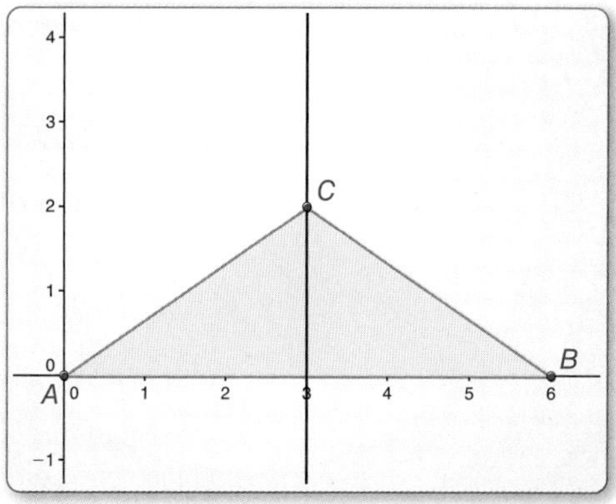

Sample

Points

$A(0, 0)$

$B(6, 0)$

$C(3, y)$

Segments

$AB = 6$

Line

$x = 3$

EXPLORATION 2 Writing a Coordinate Proof

Work with a partner.

a. Use dynamic geometry software to draw $\overline{AB}$ with endpoints $A(0, 0)$ and $B(6, 0)$.

b. Draw the vertical line $x = 3$.

c. Plot the point $C(3, 3)$ and draw $\triangle ABC$. Then use your drawing to prove that $\triangle ABC$ is an isosceles right triangle.

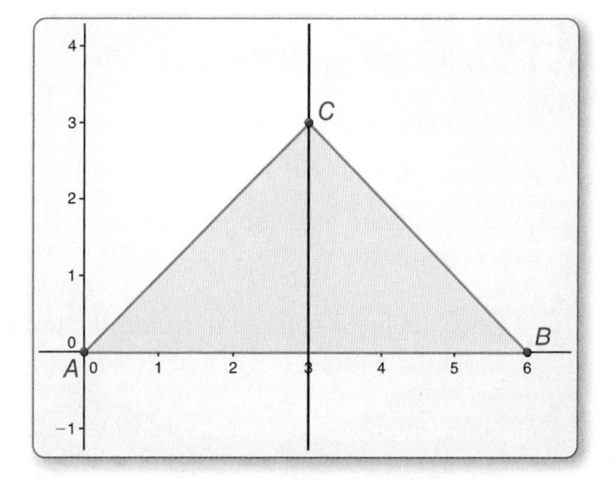

Sample

Points

$A(0, 0)$

$B(6, 0)$

$C(3, 3)$

Segments

$AB = 6$

$BC = 4.24$

$AC = 4.24$

Line

$x = 3$

d. Change the coordinates of C so that C lies below the x-axis and $\triangle ABC$ is an isosceles right triangle.

e. Write a coordinate proof to show that if C lies on the line $x = 3$ and $\triangle ABC$ is an isosceles right triangle, then C must be the point $(3, 3)$ or the point found in part (d).

CRITIQUING THE REASONING OF OTHERS

To be proficient in math, you need to understand and use stated assumptions, definitions, and previously established results.

Communicate Your Answer

3. How can you use a coordinate plane to write a proof?

4. Write a coordinate proof to prove that $\triangle ABC$ with vertices $A(0, 0)$, $B(6, 0)$, and $C(3, 3\sqrt{3})$ is an equilateral triangle.

5.8 Lesson

Core Vocabulary

coordinate proof, *p. 284*

What You Will Learn

- Place figures in a coordinate plane.
- Write coordinate proofs.

Placing Figures in a Coordinate Plane

A **coordinate proof** involves placing geometric figures in a coordinate plane. When you use variables to represent the coordinates of a figure in a coordinate proof, the results are true for all figures of that type.

EXAMPLE 1 Placing a Figure in a Coordinate Plane

Place each figure in a coordinate plane in a way that is convenient for finding side lengths. Assign coordinates to each vertex.

a. a rectangle

b. a scalene triangle

SOLUTION

It is easy to find lengths of horizontal and vertical segments and distances from (0, 0), so place one vertex at the origin and one or more sides on an axis.

a. Let h represent the length and k represent the width.

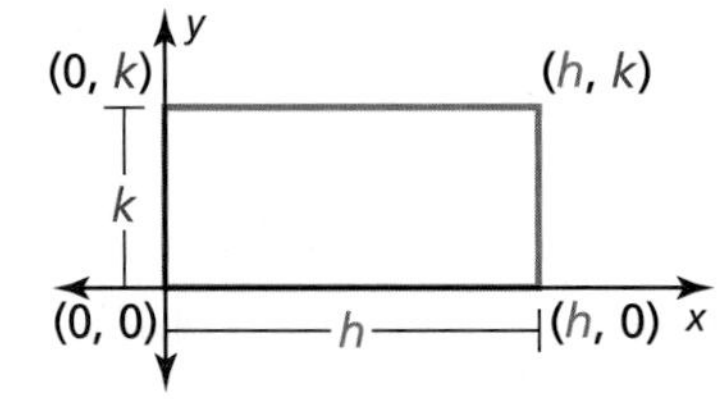

b. Notice that you need to use three different variables.

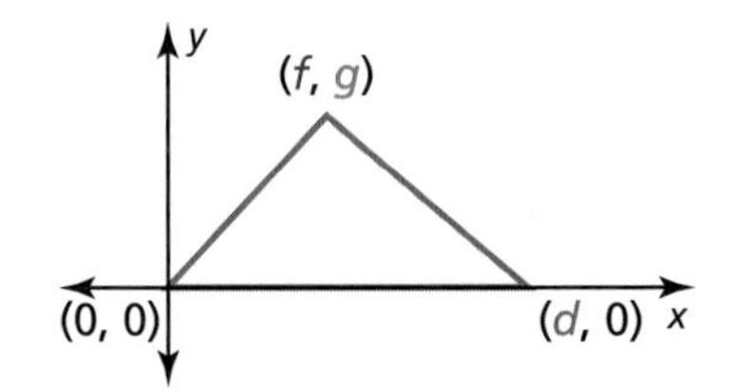

Monitoring Progress

Help in English and Spanish at *BigIdeasMath.com*

1. Show another way to place the rectangle in Example 1 part (a) that is convenient for finding side lengths. Assign new coordinates.
2. A square has vertices (0, 0), $(m, 0)$, and $(0, m)$. Find the fourth vertex.

Once a figure is placed in a coordinate plane, you may be able to prove statements about the figure.

EXAMPLE 2 Writing a Plan for a Coordinate Proof

Write a plan to prove that $\overrightarrow{SO}$ bisects $\angle PSR$.

Given Coordinates of vertices of $\triangle POS$ and $\triangle ROS$

Prove $\overrightarrow{SO}$ bisects $\angle PSR$.

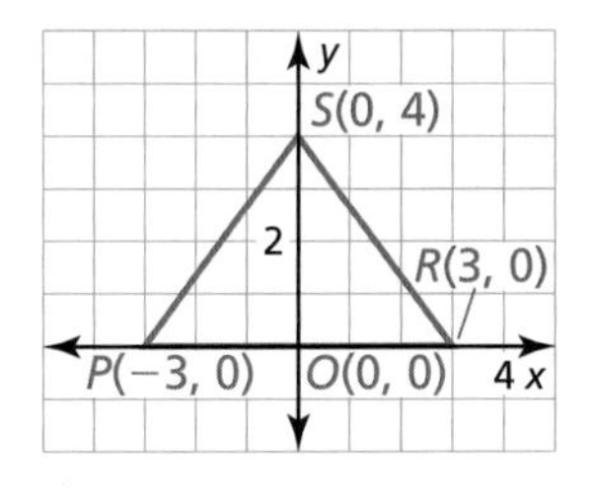

SOLUTION

Plan for Proof Use the Distance Formula to find the side lengths of $\triangle POS$ and $\triangle ROS$. Then use the SSS Congruence Theorem (Theorem 5.8) to show that $\triangle POS \cong \triangle ROS$. Finally, use the fact that corresponding parts of congruent triangles are congruent to conclude that $\angle PSO \cong \angle RSO$, which implies that $\overrightarrow{SO}$ bisects $\angle PSR$.

Monitoring Progress  Help in English and Spanish at *BigIdeasMath.com*

3. Write a plan for the proof.

Given $\overrightarrow{GJ}$ bisects $\angle OGH$.

Prove $\triangle GJO \cong \triangle GJH$

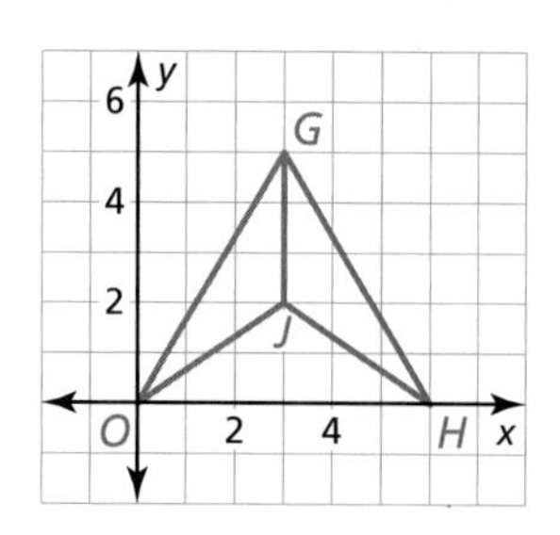

The coordinate proof in Example 2 applies to a specific triangle. When you want to prove a statement about a more general set of figures, it is helpful to use variables as coordinates.

For instance, you can use variable coordinates to duplicate the proof in Example 2. Once this is done, you can conclude that $\overrightarrow{SO}$ bisects $\angle PSR$ for any triangle whose coordinates fit the given pattern.

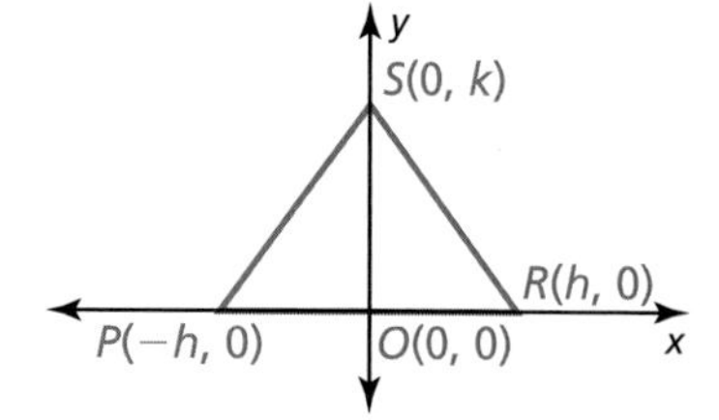

EXAMPLE 3 Applying Variable Coordinates

Place an isosceles right triangle in a coordinate plane. Then find the length of the hypotenuse and the coordinates of its midpoint M.

SOLUTION

Place $\triangle PQO$ with the right angle at the origin. Let the length of the legs be k. Then the vertices are located at $P(0, k)$, $Q(k, 0)$, and $O(0, 0)$.

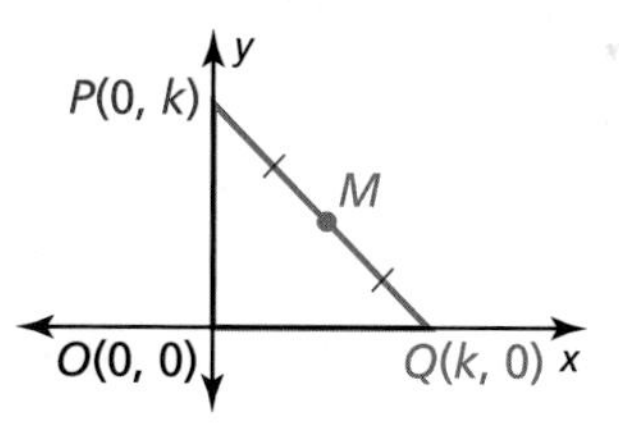

Use the Distance Formula to find PQ, the length of the hypotenuse.

$$PQ = \sqrt{(k-0)^2 + (0-k)^2} = \sqrt{k^2 + (-k)^2} = \sqrt{k^2 + k^2} = \sqrt{2k^2} = k\sqrt{2}$$

Use the Midpoint Formula to find the midpoint M of the hypotenuse.

$$M\left(\frac{0+k}{2}, \frac{k+0}{2}\right) = M\left(\frac{k}{2}, \frac{k}{2}\right)$$

▶ So, the length of the hypotenuse is $k\sqrt{2}$ and the midpoint of the hypotenuse is $\left(\frac{k}{2}, \frac{k}{2}\right)$.

FINDING AN ENTRY POINT

Another way to solve Example 3 is to place a triangle with point C at $(0, h)$ on the y-axis and hypotenuse $\overline{AB}$ on the x-axis. To make $\angle ACB$ a right angle, position A and B so that legs $\overline{CA}$ and $\overline{CB}$ have slopes of 1 and -1, respectively.

Slope is 1. Slope is −1. $C(0, h)$ $A(-h, 0)$ $B(h, 0)$ x y

Length of hypotenuse $= 2h$

$$M\left(\frac{-h+h}{2}, \frac{0+0}{2}\right) = M(0, 0)$$

Monitoring Progress  Help in English and Spanish at *BigIdeasMath.com*

4. Graph the points $O(0, 0)$, $H(m, n)$, and $J(m, 0)$. Is $\triangle OHJ$ a right triangle? Find the side lengths and the coordinates of the midpoint of each side.

Writing Coordinate Proofs

EXAMPLE 4 Writing a Coordinate Proof

Write a coordinate proof.

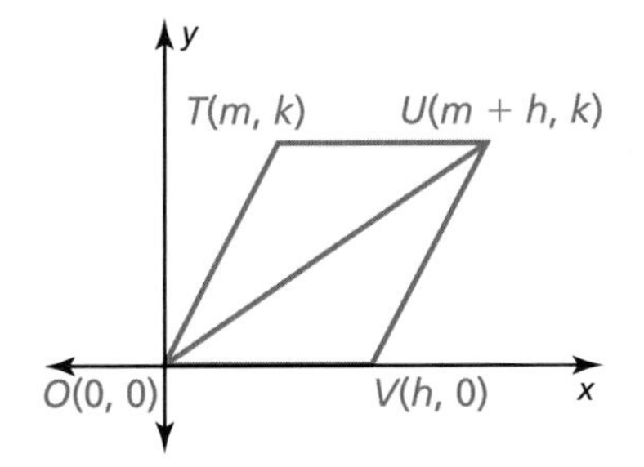

Given Coordinates of vertices of quadrilateral $OTUV$

Prove $\triangle OTU \cong \triangle UVO$

SOLUTION

Segments $\overline{OV}$ and $\overline{UT}$ have the same length.

$$OV = |h - 0| = h$$

$$UT = |(m + h) - m| = h$$

Horizontal segments $\overline{UT}$ and $\overline{OV}$ each have a slope of 0, which implies that they are parallel. Segment $\overline{OU}$ intersects $\overline{UT}$ and $\overline{OV}$ to form congruent alternate interior angles, $\angle TUO$ and $\angle VOU$. By the Reflexive Property of Congruence (Theorem 2.1), $\overline{OU} \cong \overline{OU}$.

▶ So, you can apply the SAS Congruence Theorem (Theorem 5.5) to conclude that $\triangle OTU \cong \triangle UVO$.

EXAMPLE 5 Writing a Coordinate Proof

You buy a tall, three-legged plant stand. When you place a plant on the stand, the stand appears to be unstable under the weight of the plant. The diagram at the right shows a coordinate plane superimposed on one pair of the plant stand's legs. The legs are extended to form $\triangle OBC$. Prove that $\triangle OBC$ is a scalene triangle. Explain why the plant stand may be unstable.

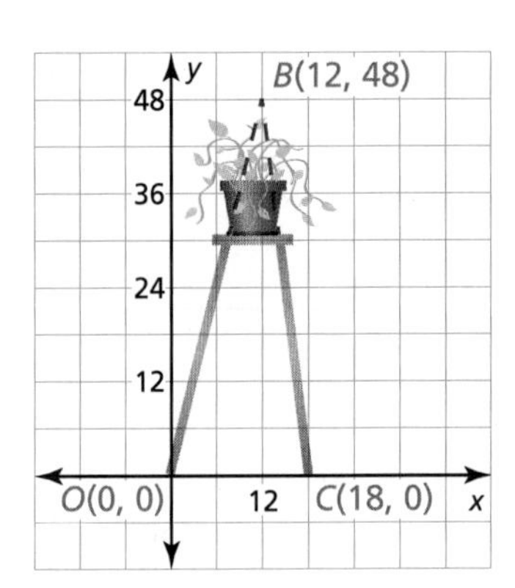

SOLUTION

First, find the side lengths of $\triangle OBC$.

$$OB = \sqrt{(48 - 0)^2 + (12 - 0)^2} = \sqrt{2448} \approx 49.5$$

$$BC = \sqrt{(18 - 12)^2 + (0 - 48)^2} = \sqrt{2340} \approx 48.4$$

$$OC = |18 - 0| = 18$$

▶ Because $\triangle OBC$ has no congruent sides, $\triangle OBC$ is a scalene triangle by definition. The plant stand may be unstable because $\overline{OB}$ is longer than $\overline{BC}$, so the plant stand is leaning to the right.

Monitoring Progress  Help in English and Spanish at *BigIdeasMath.com*

5. Write a coordinate proof.

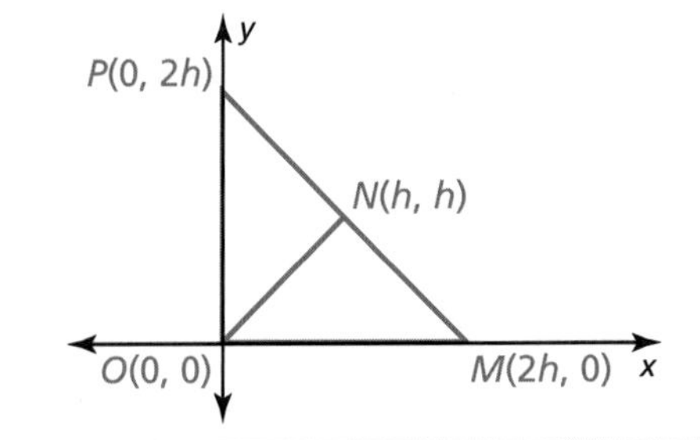

Given Coordinates of vertices of $\triangle NPO$ and $\triangle NMO$

Prove $\triangle NPO \cong \triangle NMO$

5.8 Exercises

Dynamic Solutions available at *BigIdeasMath.com*

Vocabulary and Core Concept Check

1. **VOCABULARY** How is a *coordinate proof* different from other types of proofs you have studied? How is it the same?

2. **WRITING** Explain why it is convenient to place a right triangle on the grid as shown when writing a coordinate proof.

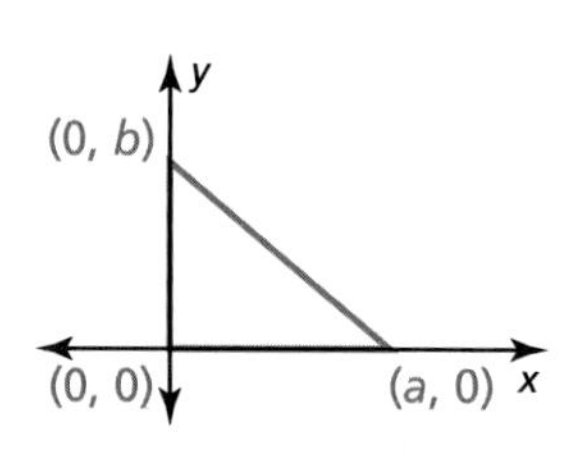

Monitoring Progress and Modeling with Mathematics

In Exercises 3–6, place the figure in a coordinate plane in a convenient way. Assign coordinates to each vertex. Explain the advantages of your placement. *(See Example 1.)*

3. a right triangle with leg lengths of 3 units and 2 units

4. a square with a side length of 3 units

5. an isosceles right triangle with leg length p

6. a scalene triangle with one side length of $2m$

In Exercises 7 and 8, write a plan for the proof. *(See Example 2.)*

7. **Given** Coordinates of vertices of $\triangle OPM$ and $\triangle ONM$
   Prove $\triangle OPM$ and $\triangle ONM$ are isosceles triangles.

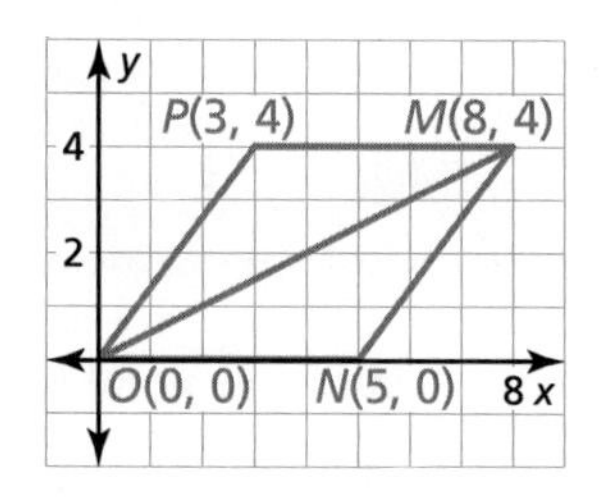

8. **Given** G is the midpoint of $\overline{HF}$.
   Prove $\triangle GHJ \cong \triangle GFO$

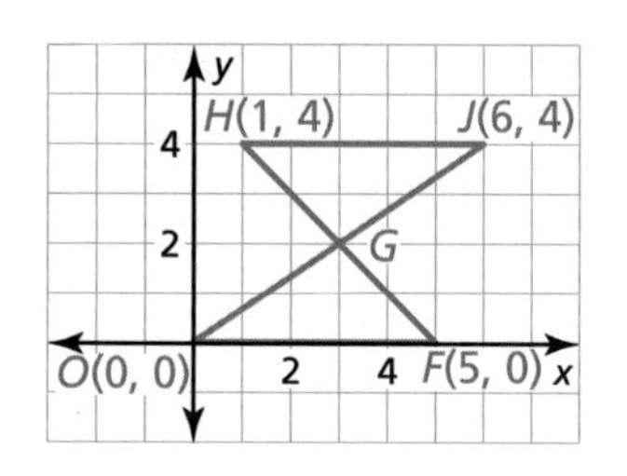

In Exercises 9–12, place the figure in a coordinate plane and find the indicated length.

9. a right triangle with leg lengths of 7 and 9 units; Find the length of the hypotenuse.

10. an isosceles triangle with a base length of 60 units and a height of 50 units; Find the length of one of the legs.

11. a rectangle with a length of 5 units and a width of 4 units; Find the length of the diagonal.

12. a square with side length n; Find the length of the diagonal.

In Exercises 13 and 14, graph the triangle with the given vertices. Find the length and the slope of each side of the triangle. Then find the coordinates of the midpoint of each side. Is the triangle a right triangle? isosceles? Explain. (Assume all variables are positive and $m \neq n$.) *(See Example 3.)*

13. $A(0, 0)$, $B(h, h)$, $C(2h, 0)$

14. $D(0, n)$, $E(m, n)$, $F(m, 0)$

In Exercises 15 and 16, find the coordinates of any unlabeled vertices. Then find the indicated length(s).

15. Find ON and MN.

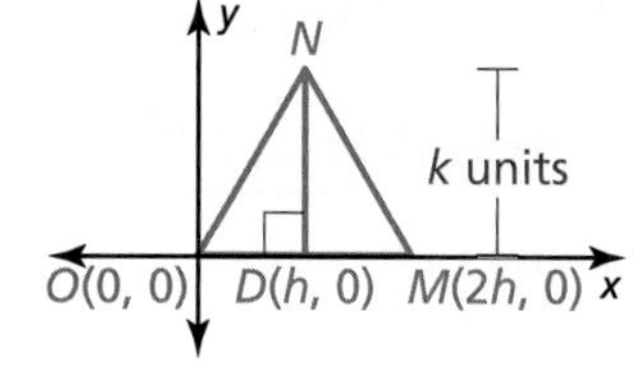

16. Find OT.

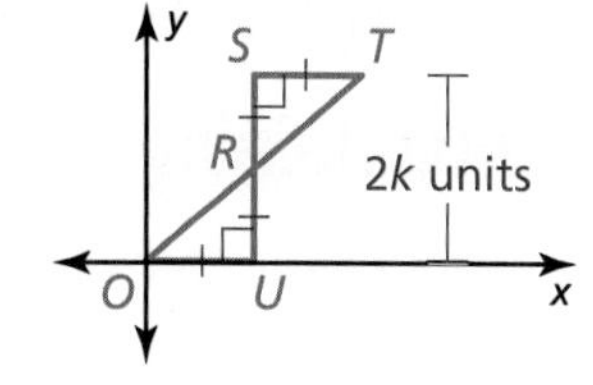

PROOF In Exercises 17 and 18, write a coordinate proof. *(See Example 4.)*

17. Given Coordinates of vertices of $\triangle DEC$ and $\triangle BOC$

Prove $\triangle DEC \cong \triangle BOC$

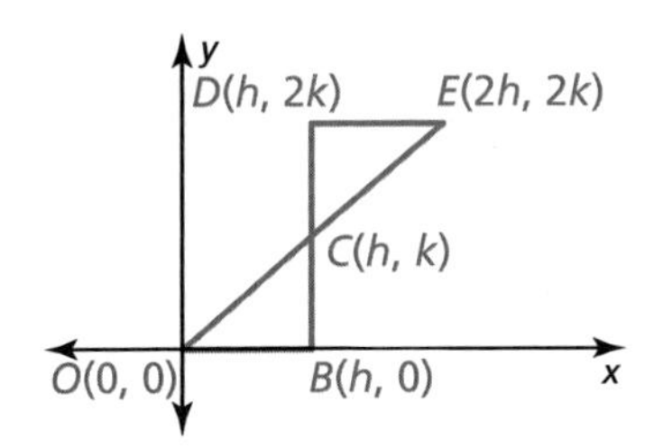

18. Given Coordinates of $\triangle DEA$, H is the midpoint of $\overline{DA}$, G is the midpoint of $\overline{EA}$.

Prove $\overline{DG} \cong \overline{EH}$

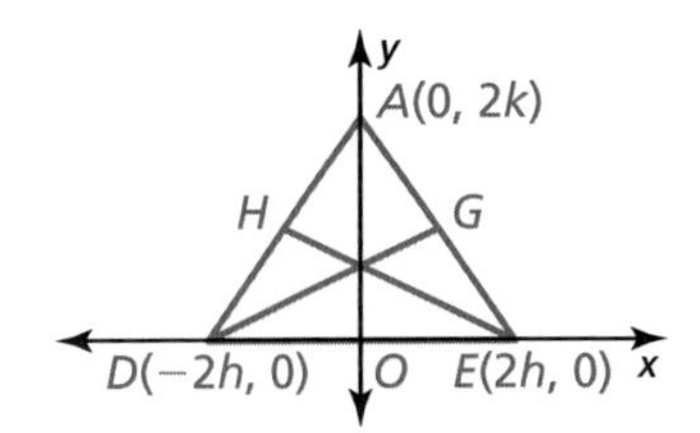

19. MODELING WITH MATHEMATICS You and your cousin are camping in the woods. You hike to a point that is 500 meters east and 1200 meters north of the campsite. Your cousin hikes to a point that is 1000 meters east of the campsite. Use a coordinate proof to prove that the triangle formed by your position, your cousin's position, and the campsite is isosceles. *(See Example 5.)*

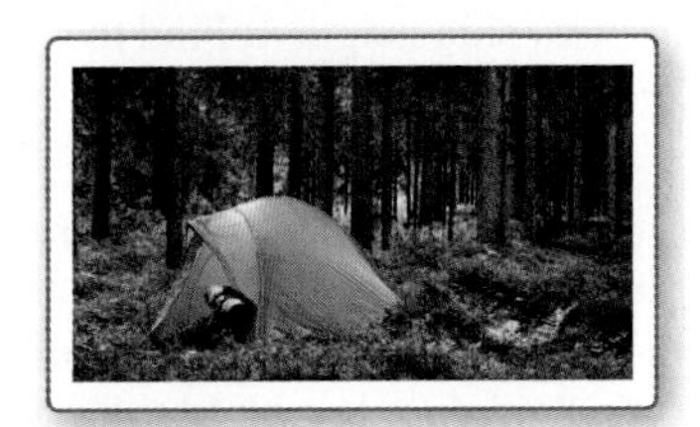

20. MAKING AN ARGUMENT Two friends see a drawing of quadrilateral $PQRS$ with vertices $P(0, 2)$, $Q(3, -4)$, $R(1, -5)$, and $S(-2, 1)$. One friend says the quadrilateral is a parallelogram but not a rectangle. The other friend says the quadrilateral is a rectangle. Which friend is correct? Use a coordinate proof to support your answer.

21. MATHEMATICAL CONNECTIONS Write an algebraic expression for the coordinates of each endpoint of a line segment whose midpoint is the origin.

22. REASONING The vertices of a parallelogram are $(w, 0)$, $(0, v)$, $(-w, 0)$, and $(0, -v)$. What is the midpoint of the side in Quadrant III?

Ⓐ $\left(\frac{w}{2}, \frac{v}{2}\right)$ Ⓑ $\left(-\frac{w}{2}, -\frac{v}{2}\right)$

Ⓒ $\left(-\frac{w}{2}, \frac{v}{2}\right)$ Ⓓ $\left(\frac{w}{2}, -\frac{v}{2}\right)$

23. REASONING A rectangle with a length of $3h$ and a width of k has a vertex at $(-h, k)$. Which point cannot be a vertex of the rectangle?

Ⓐ (h, k) Ⓑ $(-h, 0)$

Ⓒ $(2h, 0)$ Ⓓ $(2h, k)$

24. THOUGHT PROVOKING Choose one of the theorems you have encountered up to this point that you think would be easier to prove with a coordinate proof than with another type of proof. Explain your reasoning. Then write a coordinate proof.

25. CRITICAL THINKING The coordinates of a triangle are $(5d, -5d)$, $(0, -5d)$, and $(5d, 0)$. How should the coordinates be changed to make a coordinate proof easier to complete?

26. HOW DO YOU SEE IT? Without performing any calculations, how do you know that the diagonals of square $TUVW$ are perpendicular to each other? How can you use a similar diagram to show that the diagonals of any square are perpendicular to each other?

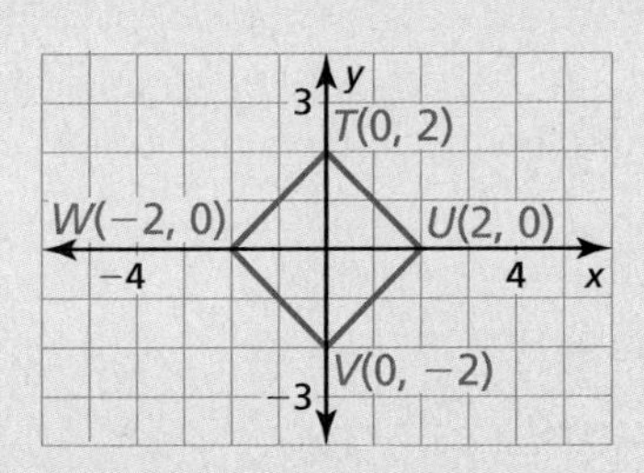

27. PROOF Write a coordinate proof for each statement.

a. The midpoint of the hypotenuse of a right triangle is the same distance from each vertex of the triangle.

b. Any two congruent right isosceles triangles can be combined to form a single isosceles triangle.

Maintaining Mathematical Proficiency

Reviewing what you learned in previous grades and lessons

$\overrightarrow{YW}$ bisects $\angle XYZ$ such that $m\angle XYW = (3x - 7)^\circ$ and $m\angle WYZ = (2x + 1)^\circ$. *(Section 1.5)*

28. Find the value of x.

29. Find $m\angle XYZ$.

5.5–5.8 What Did You Learn?

Core Vocabulary

legs (of a right triangle), *p. 264*
hypotenuse (of a right triangle), *p. 264*
coordinate proof, *p. 284*

Core Concepts

Theorem 5.8 Side-Side-Side (SSS) Congruence Theorem, *p. 262*
Theorem 5.9 Hypotenuse-Leg (HL) Congruence Theorem, *p. 264*
Theorem 5.10 Angle-Side-Angle (ASA) Congruence Theorem, *p. 270*
Theorem 5.11 Angle-Angle-Side (AAS) Congruence Theorem, *p. 271*
Using Congruent Triangles, *p. 278*
Proving Constructions, *p. 280*
Placing Figures in a Coordinate Plane, *p. 284*
Writing Coordinate Proofs, *p. 286*

Mathematical Practices

1. Write a simpler problem that is similar to Exercise 22 on page 267. Describe how to use the simpler problem to gain insight into the solution of the more complicated problem in Exercise 22.
2. Make a conjecture about the meaning of your solutions to Exercises 21–23 on page 275.
3. Identify at least two external resources that you could use to help you solve Exercise 20 on page 282.

Performance Task

Creating the Logo

Congruent triangles are often used to create company logos. Why are they used and what are the properties that make them attractive? Following the required constraints, create your new logo and justify how your shape contains the required properties.

To explore the answers to these questions and more, go to ***BigIdeasMath.com***.

5 Chapter Review

Dynamic Solutions available at *BigIdeasMath.com*

5.1 Angles of Triangles *(pp. 231–238)*

Classify the triangle by its sides and by measuring its angles.

The triangle does not have any congruent sides, so it is scalene. The measure of $\angle B$ is 117°, so the triangle is obtuse.

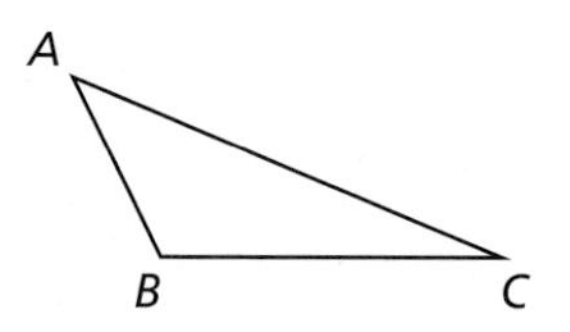

▶ The triangle is an obtuse scalene triangle.

1. Classify the triangle at the right by its sides and by measuring its angles.

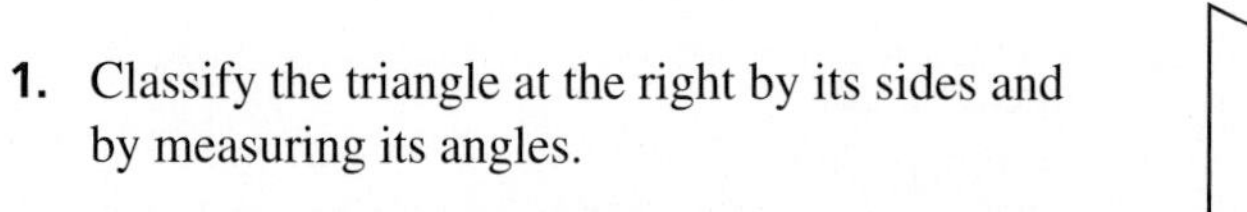

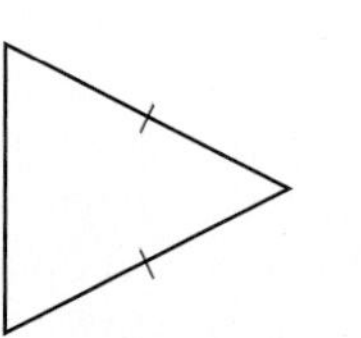

Find the measure of the exterior angle.

2.

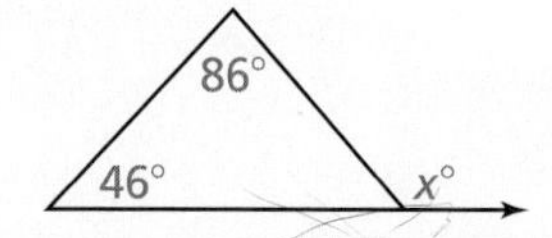

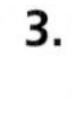

3.

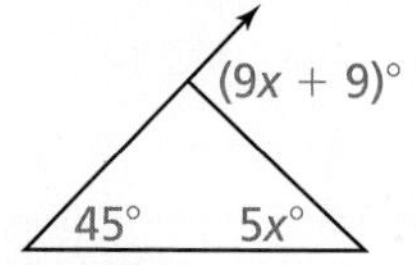

Find the measure of each acute angle.

4.

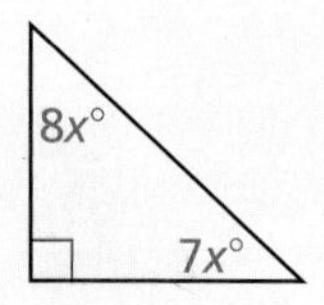

5.

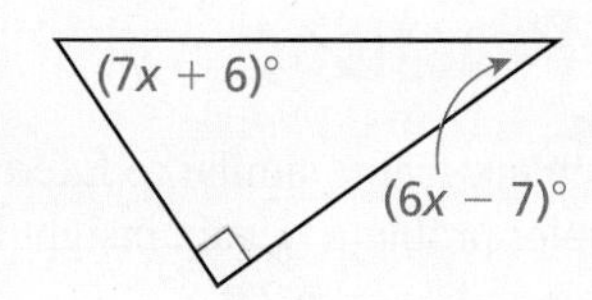

5.2 Congruent Polygons *(pp. 239–244)*

Write a congruence statement for the triangles. Identify all pairs of congruent corresponding parts.

The diagram indicates that $\triangle ABC \cong \triangle FED$.

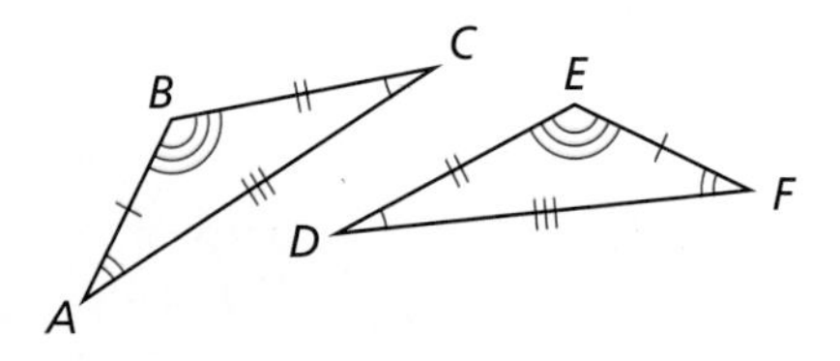

Corresponding angles $\angle A \cong \angle F$, $\angle B \cong \angle E$, $\angle C \cong \angle D$

Corresponding sides $\overline{AB} \cong \overline{FE}$, $\overline{BC} \cong \overline{ED}$, $\overline{AC} \cong \overline{FD}$

6. In the diagram, $GHJK \cong LMNP$. Identify all pairs of congruent corresponding parts. Then write another congruence statement for the quadrilaterals.

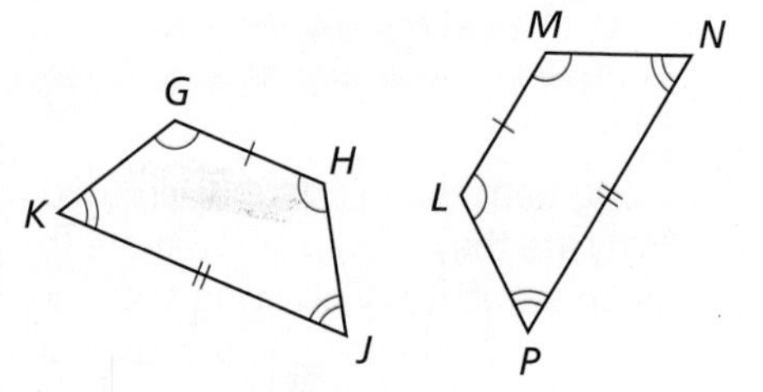

7. Find $m\angle V$.

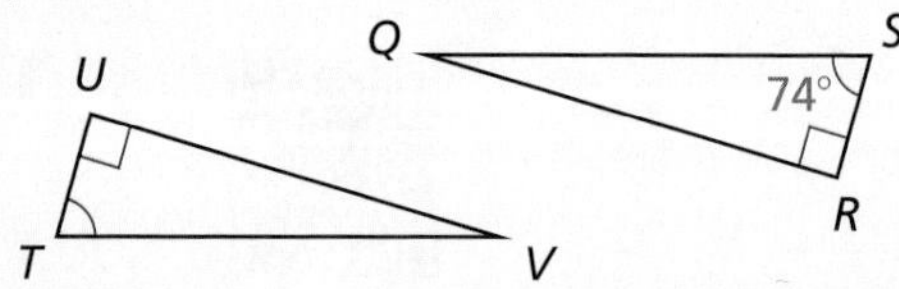

5.3 Proving Triangle Congruence by SAS (pp. 245–250)

Write a proof.

Given $\overline{AC} \cong \overline{EC}, \overline{BC} \cong \overline{DC}$

Prove $\triangle ABC \cong \triangle EDC$

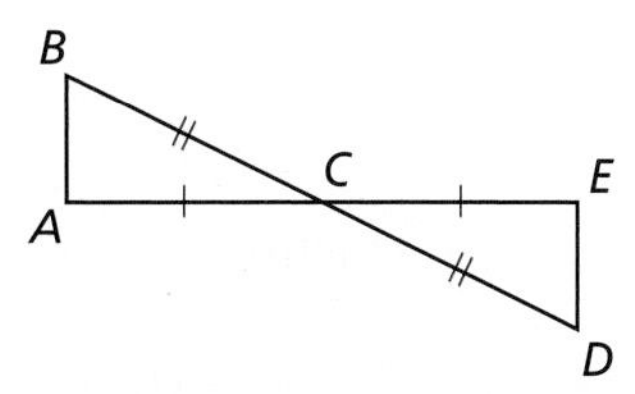

STATEMENTS	REASONS
1. $\overline{AC} \cong \overline{EC}$	**1.** Given
2. $\overline{BC} \cong \overline{DC}$	**2.** Given
3. $\angle ACB \cong \angle ECD$	**3.** Vertical Angles Congruence Theorem (Theorem 2.6)
4. $\triangle ABC \cong \triangle EDC$	**4.** SAS Congruence Theorem (Theorem 5.5)

Decide whether enough information is given to prove that $\triangle WXZ \cong \triangle YZX$ using the SAS Congruence Theorem (Theorem 5.5). If so, write a proof. If not, explain why.

8.

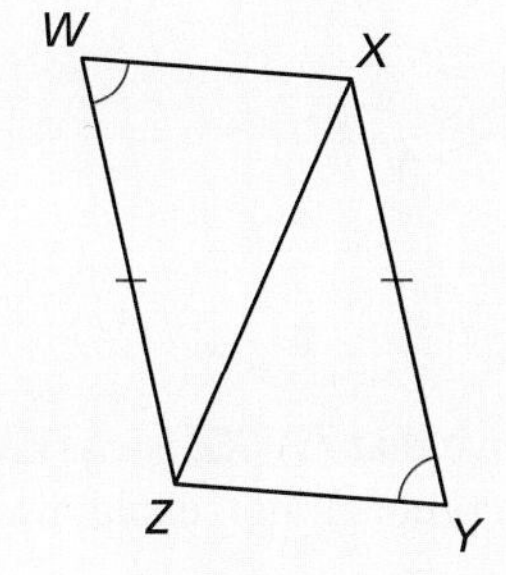

9.

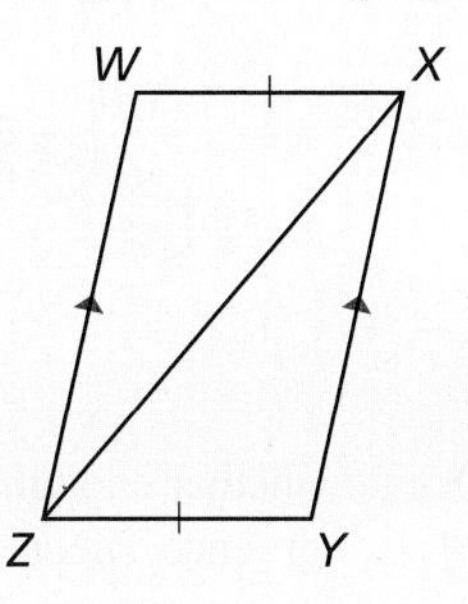

5.4 Equilateral and Isosceles Triangles (pp. 251–258)

In $\triangle LMN$, $\overline{LM} \cong \overline{LN}$. Name two congruent angles.

▶ $\overline{LM} \cong \overline{LN}$, so by the Base Angles Theorem (Theorem 5.6), $\angle M \cong \angle N$.

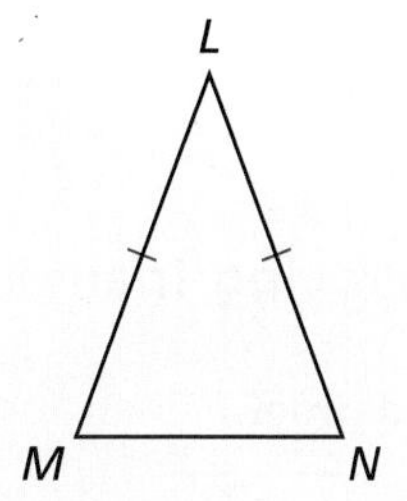

Copy and complete the statement.

10. If $\overline{QP} \cong \overline{QR}$, then $\angle$__ $\cong \angle$__.

11. If $\angle TRV \cong \angle TVR$, then ___ $\cong$ ___.

12. If $\overline{RQ} \cong \overline{RS}$, then $\angle$__ $\cong \angle$__.

13. If $\angle SRV \cong \angle SVR$, then ___ $\cong$ ___.

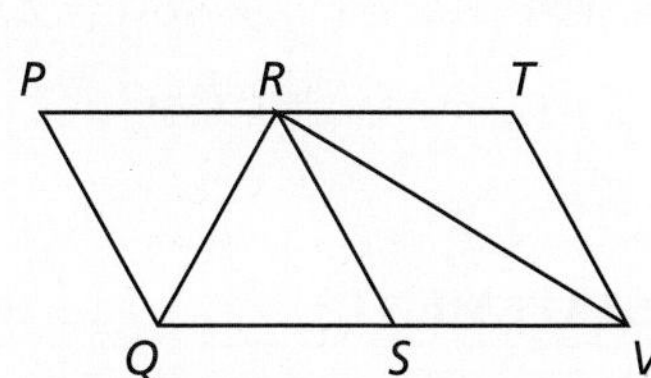

14. Find the values of x and y in the diagram.

5.5 Proving Triangle Congruence by SSS (pp. 261–268)

Write a proof.

Given $\overline{AD} \cong \overline{CB}, \overline{AB} \cong \overline{CD}$

Prove $\triangle ABD \cong \triangle CDB$

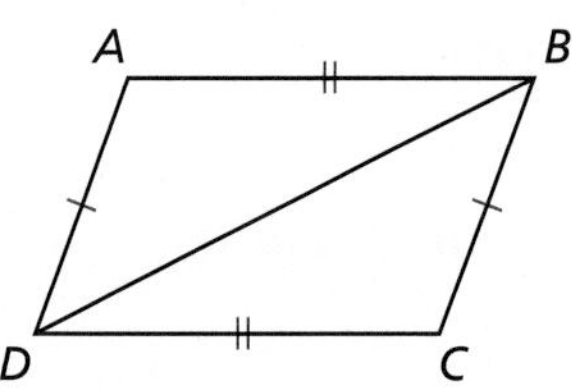

STATEMENTS	REASONS
1. $\overline{AD} \cong \overline{CB}$	**1.** Given
2. $\overline{AB} \cong \overline{CD}$	**2.** Given
3. $\overline{BD} \cong \overline{DB}$	**3.** Reflexive Property of Congruence (Theorem 2.1)
4. $\triangle ABD \cong \triangle CDB$	**4.** SSS Congruence Theorem (Theorem 5.8)

15. Decide whether enough information is given to prove that $\triangle LMP \cong \triangle NPM$ using the SSS Congruence Theorem (Thm. 5.8). If so, write a proof. If not, explain why.

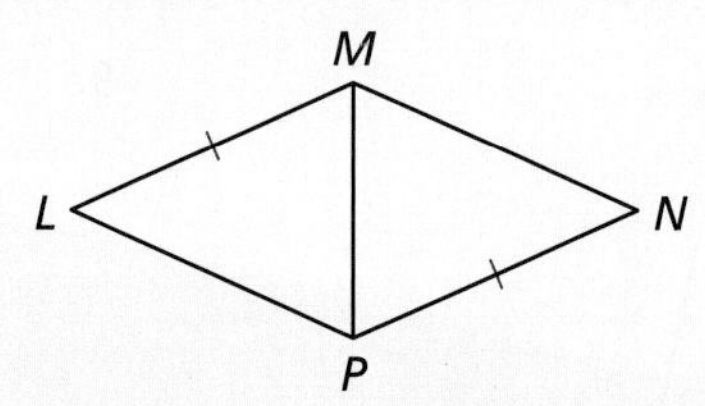

16. Decide whether enough information is given to prove that $\triangle WXZ \cong \triangle YZX$ using the HL Congruence Theorem (Thm. 5.9). If so, write a proof. If not, explain why.

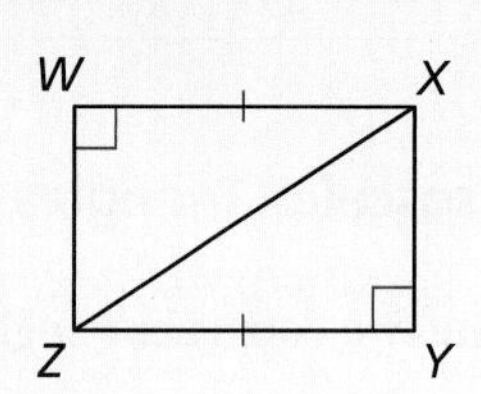

5.6 Proving Triangle Congruence by ASA and AAS (pp. 269–276)

Write a proof.

Given $\overline{AB} \cong \overline{DE}$, $\angle ABC \cong \angle DEC$

Prove $\triangle ABC \cong \triangle DEC$

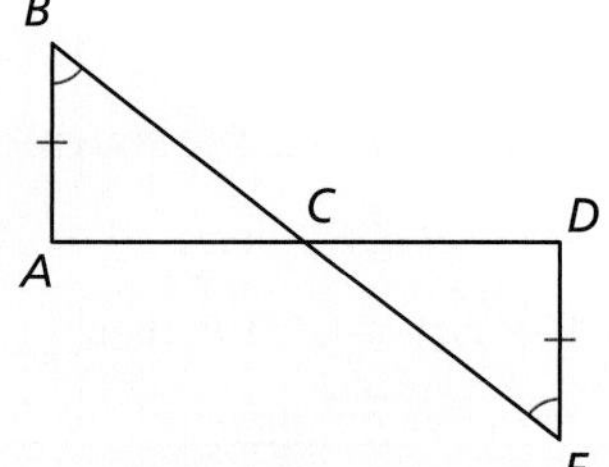

STATEMENTS	REASONS
1. $\overline{AB} \cong \overline{DE}$	**1.** Given
2. $\angle ABC \cong \angle DEC$	**2.** Given
3. $\angle ACB \cong \angle DCE$	**3.** Vertical Angles Congruence Theorem (Thm. 2.6)
4. $\triangle ABC \cong \triangle DEC$	**4.** AAS Congruence Theorem (Thm. 5.11)

Decide whether enough information is given to prove that the triangles are congruent using the AAS Congruence Theorem (Thm. 5.11). If so, write a proof. If not, explain why.

17. $\triangle EFG, \triangle HJK$

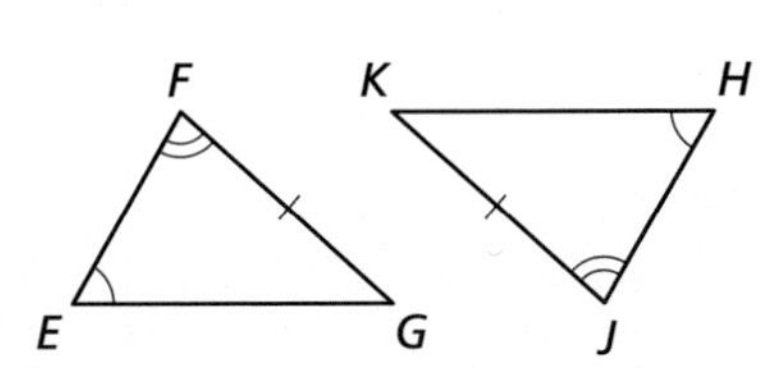

18. $\triangle TUV, \triangle QRS$

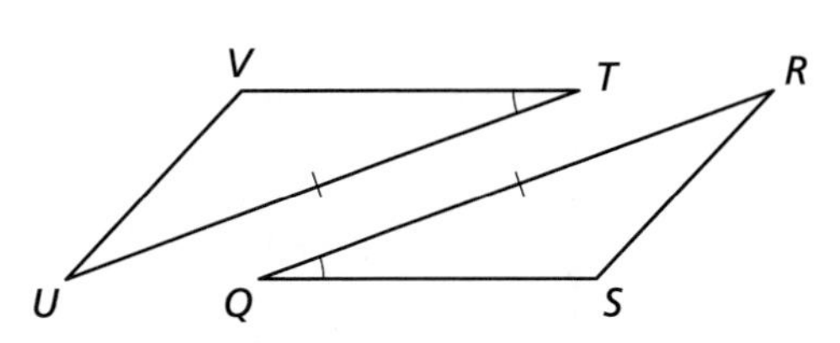

Decide whether enough information is given to prove that the triangles are congruent using the ASA Congruence Theorem (Thm. 5.10). If so, write a proof. If not, explain why.

19. $\triangle LPN, \triangle LMN$

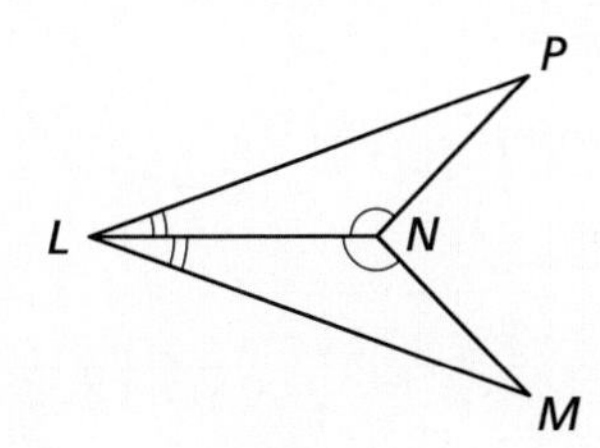

20. $\triangle WXZ, \triangle YZX$

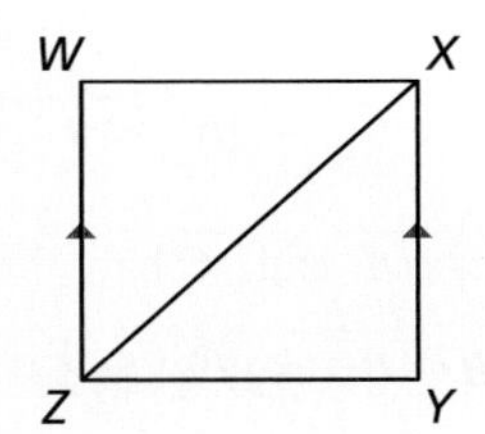

5.7 Using Congruent Triangles *(pp. 277–282)*

Explain how you can prove that $\angle A \cong \angle D$.

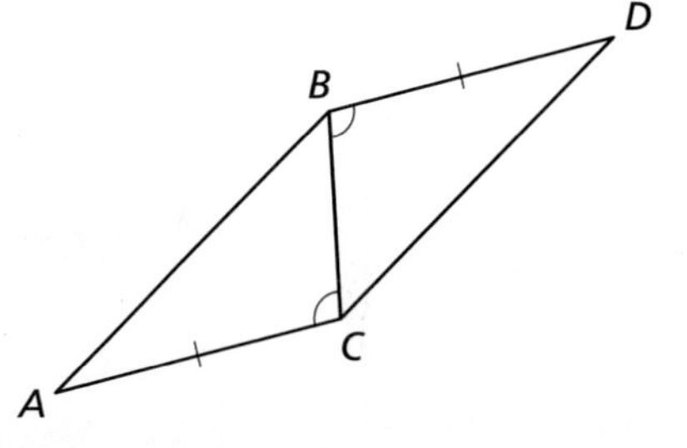

If you can show that $\triangle ABC \cong \triangle DCB$, then you will know that $\angle A \cong \angle D$. You are given $\overline{AC} \cong \overline{DB}$ and $\angle ACB \cong \angle DBC$. You know that $\overline{BC} \cong \overline{CB}$ by the Reflexive Property of Congruence (Thm. 2.1). Two pairs of sides and their included angles are congruent, so by the SAS Congruence Theorem (Thm. 5.5), $\triangle ABC \cong \triangle DCB$.

▶ Because corresponding parts of congruent triangles are congruent, $\angle A \cong \angle D$.

21. Explain how to prove that $\angle K \cong \angle N$.

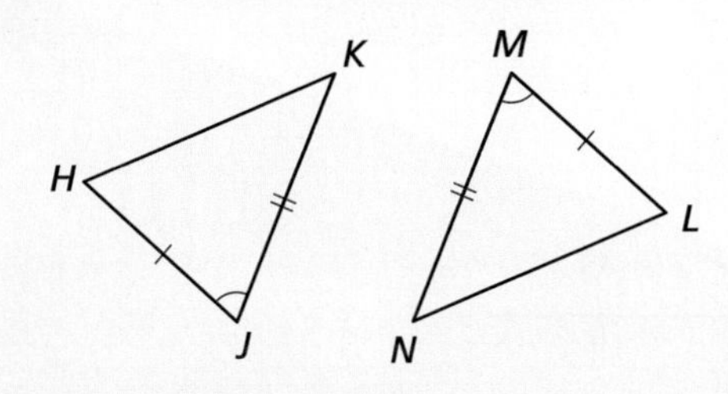

22. Write a plan to prove that $\angle 1 \cong \angle 2$.

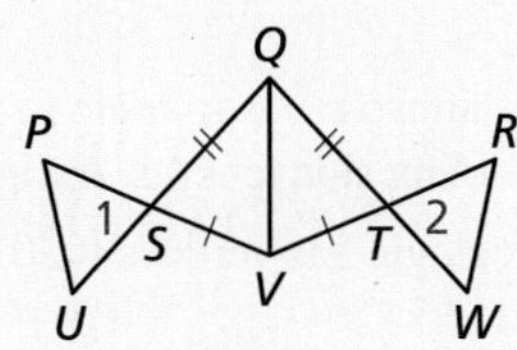

5.8 Coordinate Proofs (pp. 283–288)

Write a coordinate proof.

Given Coordinates of vertices of $\triangle ODB$ and $\triangle BDC$

Prove $\triangle ODB \cong \triangle BDC$

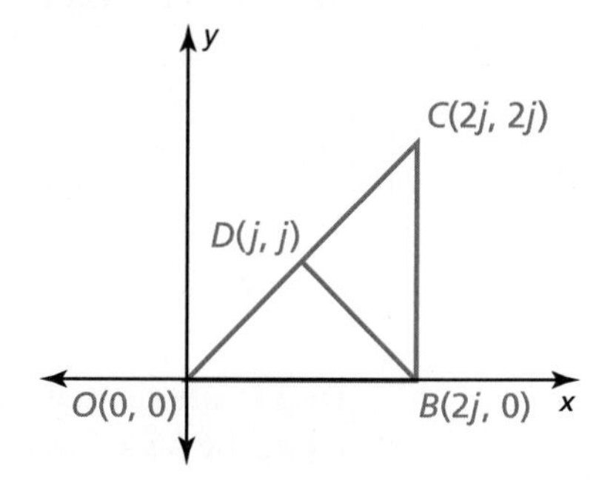

Segments $\overline{OD}$ and $\overline{BD}$ have the same length.

$$OD = \sqrt{(j-0)^2 + (j-0)^2} = \sqrt{j^2 + j^2} = \sqrt{2j^2} = j\sqrt{2}$$

$$BD = \sqrt{(j-2j)^2 + (j-0)^2} = \sqrt{(-j)^2 + j^2} = \sqrt{2j^2} = j\sqrt{2}$$

Segments $\overline{DB}$ and $\overline{DC}$ have the same length.

$$DB = BD = j\sqrt{2}$$

$$DC = \sqrt{(2j-j)^2 + (2j-j)^2} = \sqrt{j^2 + j^2} = \sqrt{2j^2} = j\sqrt{2}$$

Segments $\overline{OB}$ and $\overline{BC}$ have the same length.

$$OB = |2j - 0| = 2j$$

$$BC = |2j - 0| = 2j$$

▶ So, you can apply the SSS Congruence Theorem (Theorem 5.8) to conclude that $\triangle ODB \cong \triangle BDC$.

23. Write a coordinate proof.

Given Coordinates of vertices of quadrilateral $OPQR$

Prove $\triangle OPQ \cong \triangle QRO$

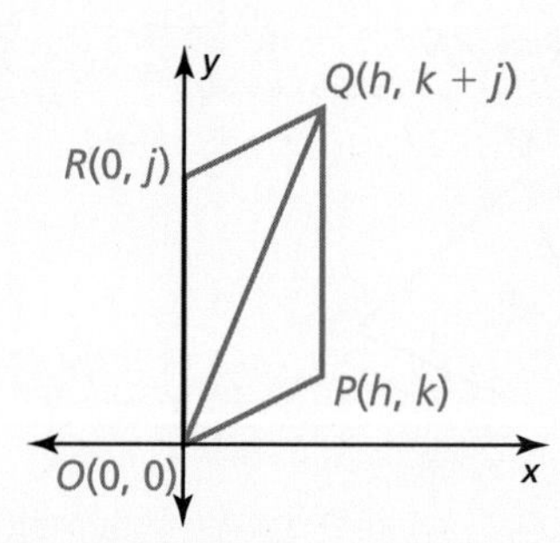

24. Place an isosceles triangle in a coordinate plane in a way that is convenient for finding side lengths. Assign coordinates to each vertex.

25. A rectangle has vertices (0, 0), (2*k*, 0), and (0, *k*). Find the fourth vertex.

5 Chapter Test

Write a proof.

1. **Given** $\overline{CA} \cong \overline{CB} \cong \overline{CD} \cong \overline{CE}$
Prove $\triangle ABC \cong \triangle EDC$

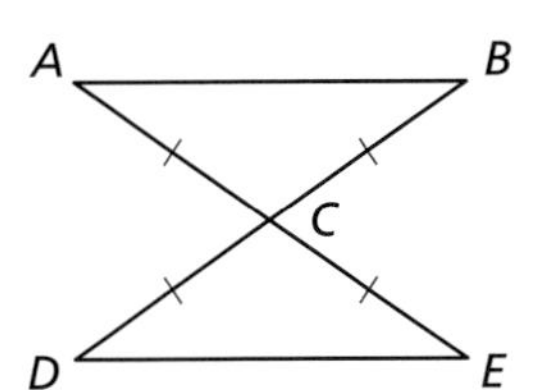

2. **Given** $\overline{JK} \parallel \overline{ML}$, $\overline{MJ} \parallel \overline{KL}$
Prove $\triangle MJK \cong \triangle KLM$

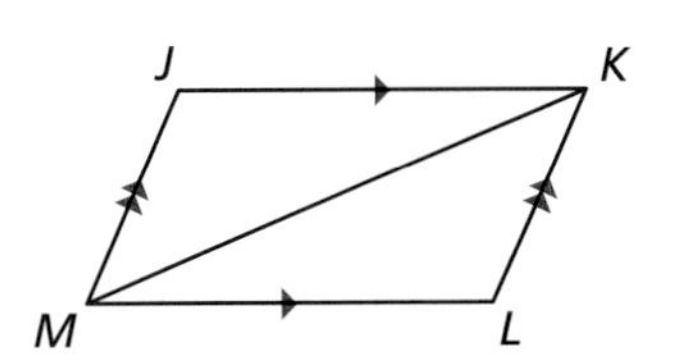

3. **Given** $\overline{QR} \cong \overline{RS}$, $\angle P \cong \angle T$
Prove $\triangle SRP \cong \triangle QRT$

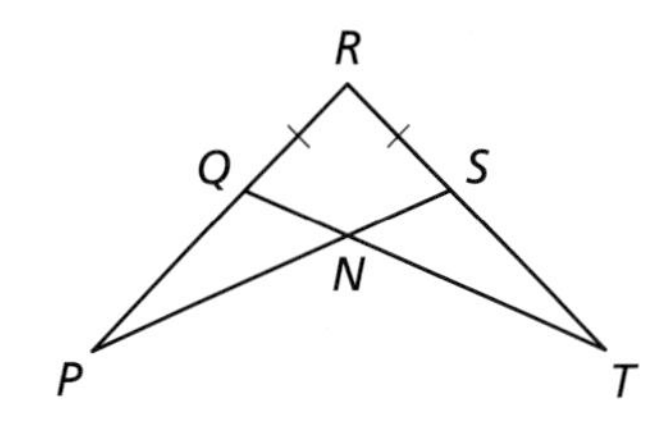

4. Find the measure of each acute angle in the figure at the right.

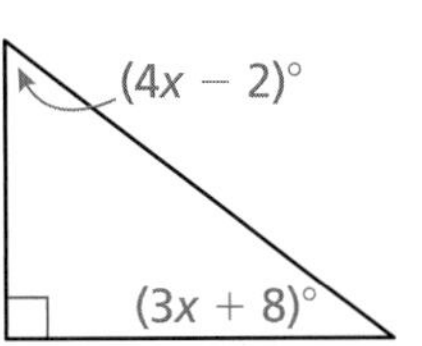

5. Is it possible to draw an equilateral triangle that is not equiangular? If so, provide an example. If not, explain why.

6. Can you use the Third Angles Theorem (Theorem 5.4) to prove that two triangles are congruent? Explain your reasoning.

Write a plan to prove that $\angle 1 \cong \angle 2$.

7.

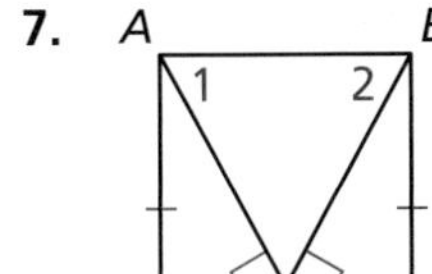
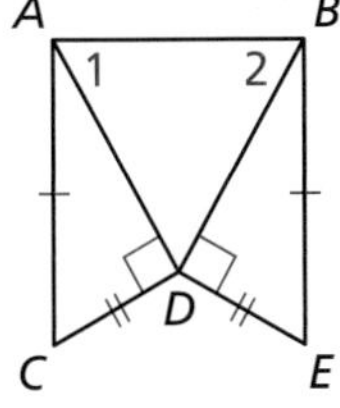

8.

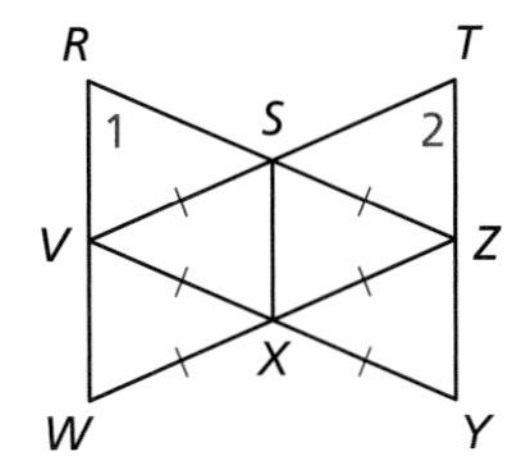

9. Is there more than one theorem that could be used to prove that $\triangle ABD \cong \triangle CDB$? If so, list all possible theorems.

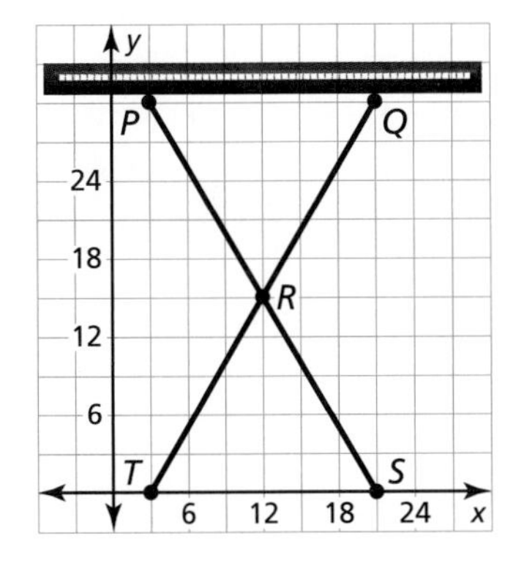

10. Write a coordinate proof to show that the triangles created by the keyboard stand are congruent.

11. The picture shows the Pyramid of Cestius, which is located in Rome, Italy. The measure of the base for the triangle shown is 100 Roman feet. The measures of the other two sides of the triangle are both 144 Roman feet.

a. Classify the triangle shown by its sides.

b. The measure of $\angle 3$ is 40°. What are the measures of $\angle 1$ and $\angle 2$? Explain your reasoning.

5 Cumulative Assessment

1. Your friend claims that the Exterior Angle Theorem (Theorem 5.2) can be used to prove the Triangle Sum Theorem (Theorem 5.1). Is your friend correct? Explain your reasoning.

2. Use the steps in the construction to explain how you know that the line through point P is parallel to line m.

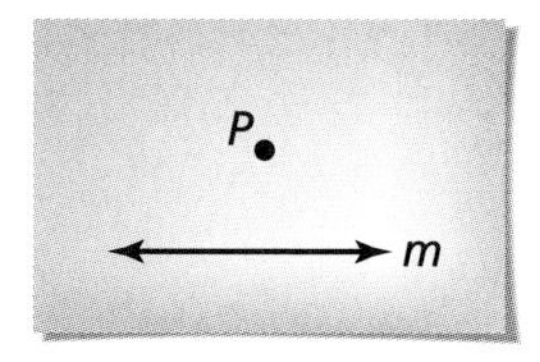

Step 1

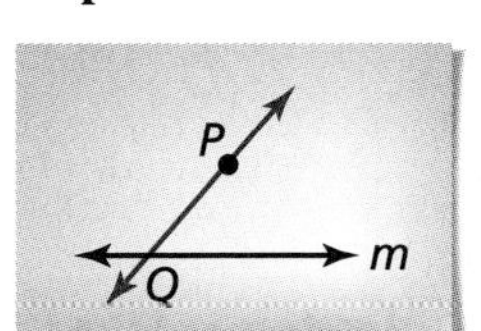

Step 2

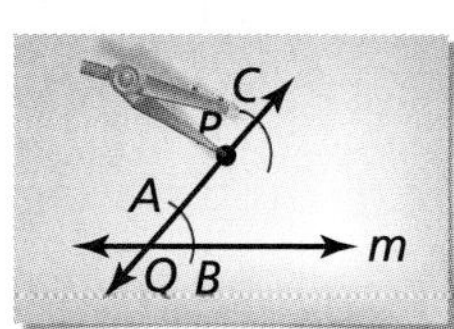

Step 3

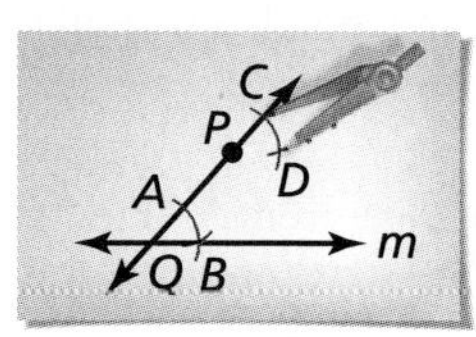

Step 4

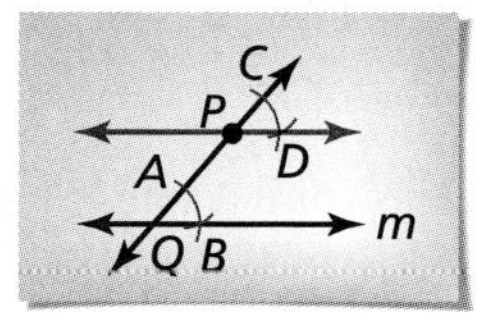

3. The coordinate plane shows $\triangle JKL$ and $\triangle XYZ$.

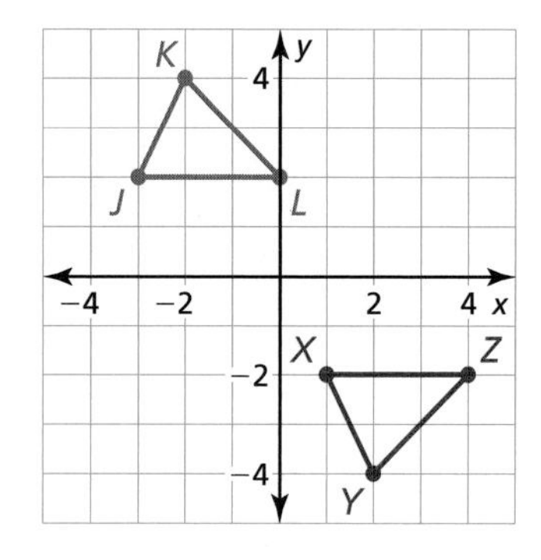

   a. Write a composition of transformations that maps $\triangle JKL$ to $\triangle XYZ$.

   b. Is the composition a congruence transformation? If so, identify all congruent corresponding parts.

4. The directed line segment RS is shown. Point Q is located along $\overline{RS}$ so that the ratio of RQ to QS is 2 to 3. What are the coordinates of point Q?

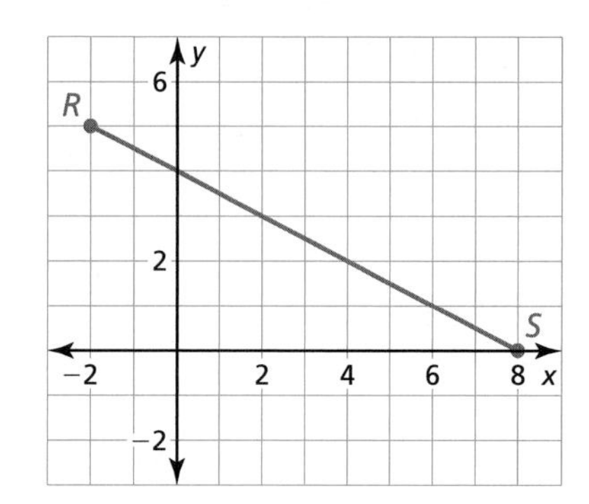

Ⓐ $Q(1.2, 3)$ Ⓑ $Q(4, 2)$ Ⓒ $Q(2, 3)$ Ⓓ $Q(-6, 7)$

5. The coordinate plane shows $\triangle ABC$ and $\triangle DEF$.

   a. Prove $\triangle ABC \cong \triangle DEF$ using the given information.

   b. Describe the composition of rigid motions that maps $\triangle ABC$ to $\triangle DEF$.

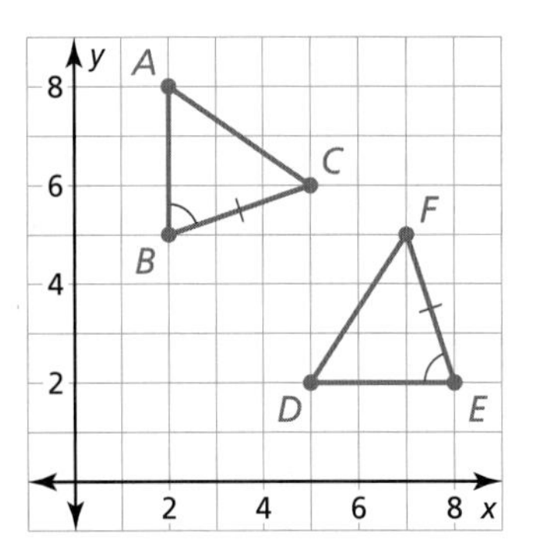

6. The vertices of a quadrilateral are $W(0, 0)$, $X(-1, 3)$, $Y(2, 7)$, and $Z(4, 2)$. Your friend claims that point W will not change after dilating quadrilateral $WXYZ$ by a scale factor of 2. Is your friend correct? Explain your reasoning.

7. Which figure(s) have rotational symmetry? Select all that apply.

Ⓐ 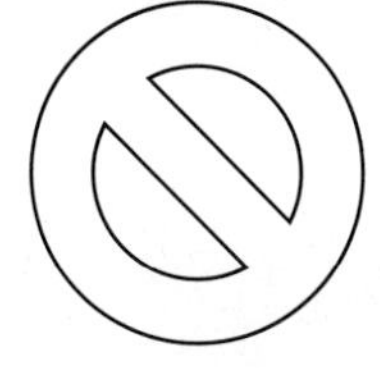

Ⓑ 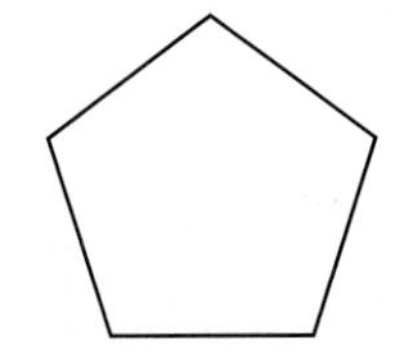

Ⓒ 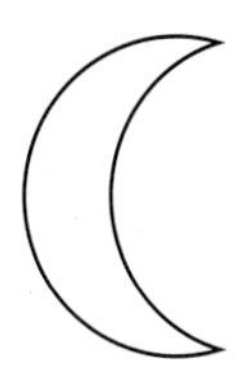

Ⓓ 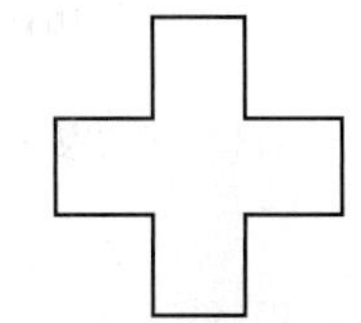

8. Write a coordinate proof.

   Given Coordinates of vertices of quadrilateral $ABCD$

   Prove Quadrilateral $ABCD$ is a rectangle.

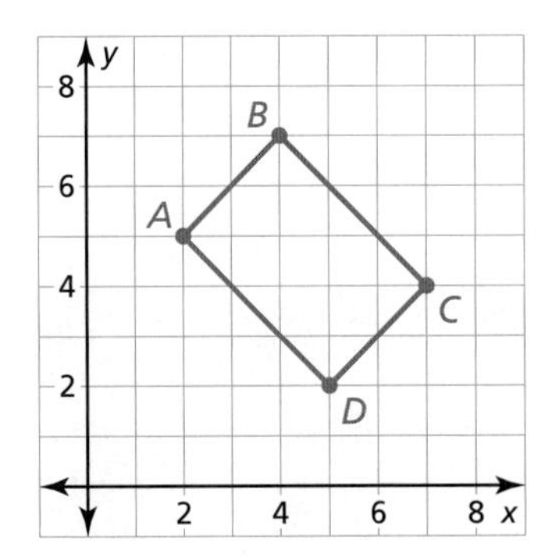

9. Write a proof to verify that the construction of the equilateral triangle shown below is valid.

Step 1

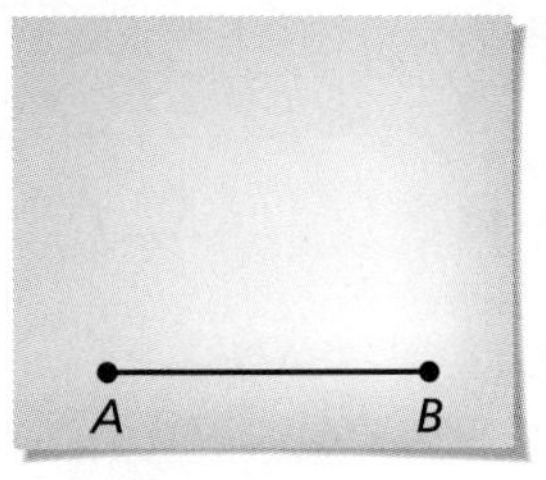

Step 2

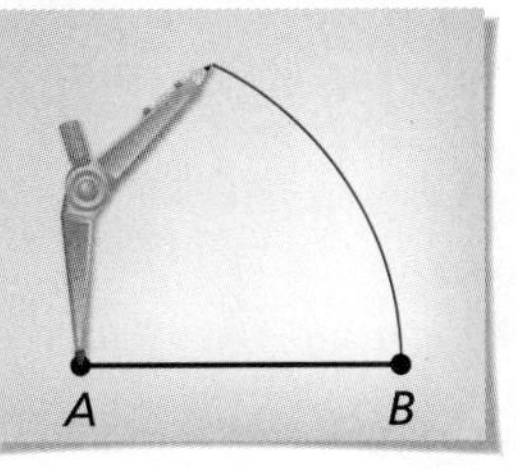

Step 3

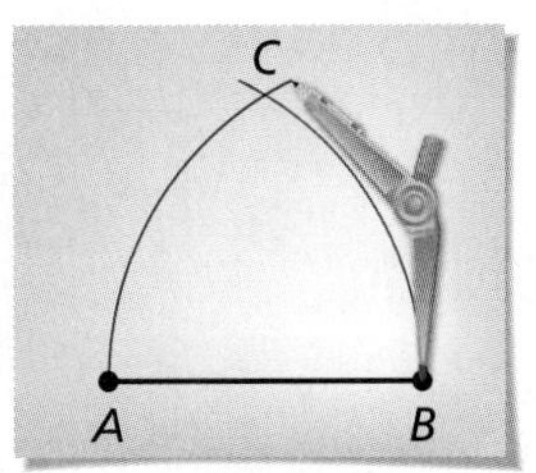

Step 4

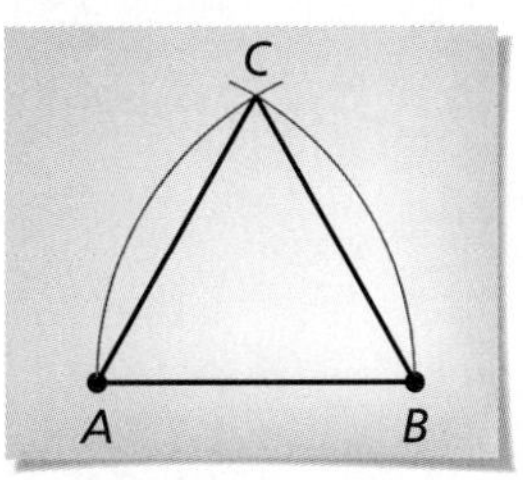

6 Relationships Within Triangles

Biking *(p. 346)*

Montana *(p. 341)*

Roof Truss *(p. 331)*

Windmill *(p. 318)*

Bridge *(p. 303)*

Maintaining Mathematical Proficiency

Writing an Equation of a Perpendicular Line

Example 1 **Write the equation of a line passing through the point $(-2, 0)$ that is perpendicular to the line $y = 2x + 8$.**

Step 1 Find the slope m of the perpendicular line. The line $y = 2x + 8$ has a slope of 2. Use the Slopes of Perpendicular Lines Theorem (Theorem 3.14).

$2 \cdot m = -1$ — The product of the slopes of $\perp$ lines is -1.

$m = -\frac{1}{2}$ — Divide each side by 2.

Step 2 Find the y-intercept b by using $m = -\frac{1}{2}$ and $(x, y) = (-2, 0)$.

$y = mx + b$ — Use the slope-intercept form.

$0 = -\frac{1}{2}(-2) + b$ — Substitute for m, x, and y.

$-1 = b$ — Solve for b.

▶ Because $m = -\frac{1}{2}$ and $b = -1$, an equation of the line is $y = -\frac{1}{2}x - 1$.

Write an equation of the line passing through point *P* that is perpendicular to the given line.

1. $P(3, 1)$, $y = \frac{1}{3}x - 5$ **2.** $P(4, -3)$, $y = -x - 5$ **3.** $P(-1, -2)$, $y = -4x + 13$

Writing Compound Inequalities

Example 2 **Write each sentence as an inequality.**

a. A number x is greater than or equal to -1 and less than 6.

A number x is greater than or equal to -1 and less than 6.

$x \geq -1$ *and* $x < 6$

▶ An inequality is $-1 \leq x < 6$.

b. A number y is at most 4 or at least 9.

A number y is at most 4 or at least 9.

$y \leq 4$ *or* $y \geq 9$

▶ An inequality is $y \leq 4$ *or* $y \geq 9$.

Write the sentence as an inequality.

4. A number w is at least -3 and no more than 8.

5. A number m is more than 0 and less than 11.

6. A number s is less than or equal to 5 or greater than 2.

7. A number d is fewer than 12 or no less than -7.

8. **ABSTRACT REASONING** Is it possible for the solution of a compound inequality to be all real numbers? Explain your reasoning.

Mathematical Practices

Mathematically proficient students use technological tools to explore concepts.

Lines, Rays, and Segments in Triangles

Core Concept

Lines, Rays, and Segments in Triangles

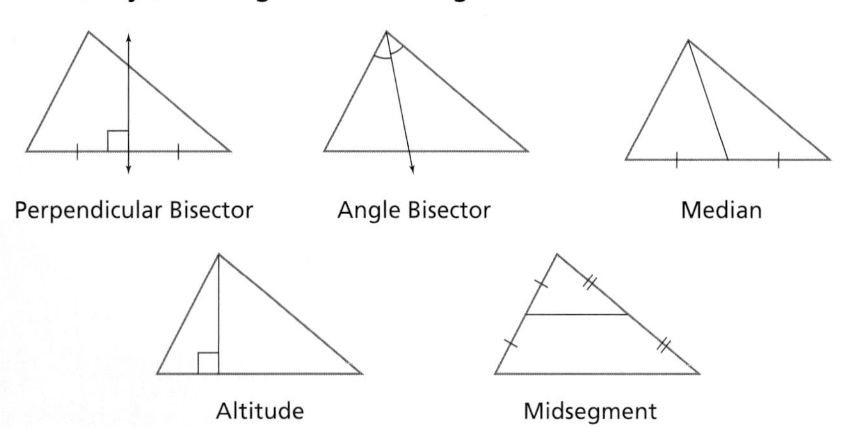

EXAMPLE 1 **Drawing a Perpendicular Bisector**

Use dynamic geometry software to construct the perpendicular bisector of one of the sides of the triangle with vertices $A(-1, 2)$, $B(5, 4)$, and $C(4, -1)$. Find the lengths of the two segments of the bisected side.

SOLUTION

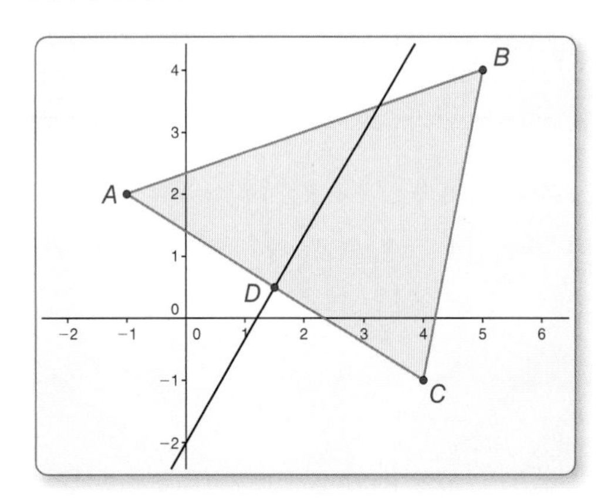

Sample
Points
$A(-1, 2)$
$B(5, 4)$
$C(4, -1)$
Line
$-5x + 3y = -6$
Segments
$AD = 2.92$
$CD = 2.92$

▶ The two segments of the bisected side have the same length, $AD = CD = 2.92$ units.

Monitoring Progress

Refer to the figures at the top of the page to describe each type of line, ray, or segment in a triangle.

1. perpendicular bisector **2.** angle bisector **3.** median

4. altitude **5.** midsegment

6.1 Perpendicular and Angle Bisectors

Essential Question What conjectures can you make about a point on the perpendicular bisector of a segment and a point on the bisector of an angle?

EXPLORATION 1 Points on a Perpendicular Bisector

Work with a partner. Use dynamic geometry software.

a. Draw any segment and label it $\overline{AB}$. Construct the perpendicular bisector of $\overline{AB}$.

b. Label a point C that is on the perpendicular bisector of $\overline{AB}$ but is not on $\overline{AB}$.

c. Draw $\overline{CA}$ and $\overline{CB}$ and find their lengths. Then move point C to other locations on the perpendicular bisector and note the lengths of $\overline{CA}$ and $\overline{CB}$.

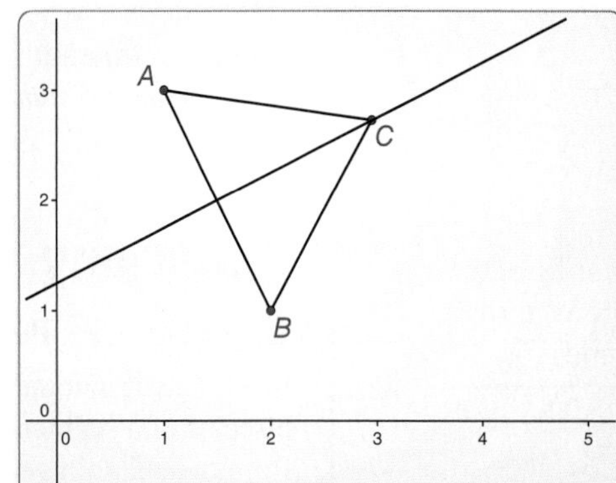

Sample
Points
$A(1, 3)$
$B(2, 1)$
$C(2.95, 2.73)$
Segments
$AB = 2.24$
$CA = ?$
$CB = ?$
Line
$-x + 2y = 2.5$

d. Repeat parts (a)–(c) with other segments. Describe any relationship(s) you notice.

USING TOOLS STRATEGICALLY

To be proficient in math, you need to visualize the results of varying assumptions, explore consequences, and compare predictions with data.

EXPLORATION 2 Points on an Angle Bisector

Work with a partner. Use dynamic geometry software.

a. Draw two rays $\overrightarrow{AB}$ and $\overrightarrow{AC}$ to form $\angle BAC$. Construct the bisector of $\angle BAC$.

b. Label a point D on the bisector of $\angle BAC$.

c. Construct and find the lengths of the perpendicular segments from D to the sides of $\angle BAC$. Move point D along the angle bisector and note how the lengths change.

d. Repeat parts (a)–(c) with other angles. Describe any relationship(s) you notice.

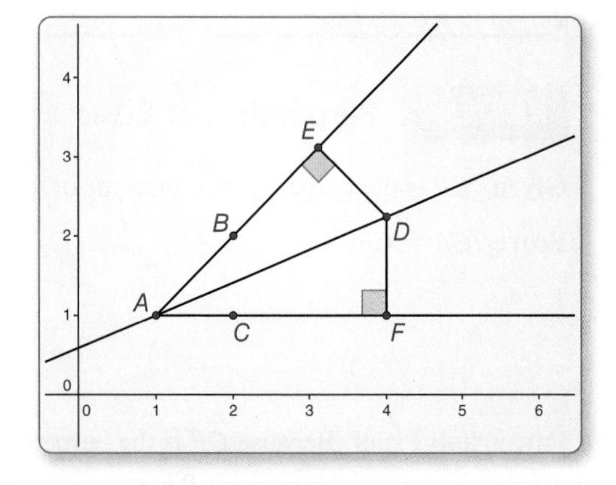

Sample
Points
$A(1, 1)$
$B(2, 2)$
$C(2, 1)$
$D(4, 2.24)$
Rays
$AB = -x + y = 0$
$AC = y = 1$
Line
$-0.38x + 0.92y = 0.54$

Communicate Your Answer

3. What conjectures can you make about a point on the perpendicular bisector of a segment and a point on the bisector of an angle?

4. In Exploration 2, what is the distance from point D to $\overrightarrow{AB}$ when the distance from D to $\overrightarrow{AC}$ is 5 units? Justify your answer.

6.1 Lesson

Core Vocabulary

equidistant, *p. 302*

Previous
perpendicular bisector
angle bisector

What You Will Learn

- Use perpendicular bisectors to find measures.
- Use angle bisectors to find measures and distance relationships.
- Write equations for perpendicular bisectors.

Using Perpendicular Bisectors

In Section 3.4, you learned that a *perpendicular bisector* of a line segment is the line that is perpendicular to the segment at its midpoint.

A point is **equidistant** from two figures when the point is the *same distance* from each figure.

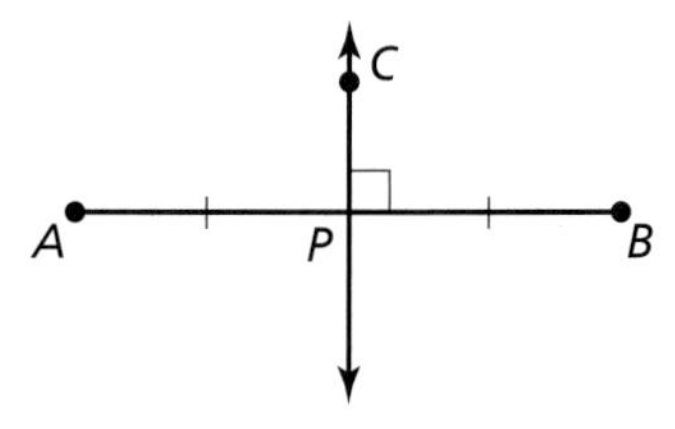

$\overleftrightarrow{CP}$ is a $\perp$ bisector of $\overline{AB}$.

STUDY TIP

A perpendicular bisector can be a segment, a ray, a line, or a plane.

Theorems

Theorem 6.1 Perpendicular Bisector Theorem

In a plane, if a point lies on the perpendicular bisector of a segment, then it is equidistant from the endpoints of the segment.

If $\overleftrightarrow{CP}$ is the $\perp$ bisector of $\overline{AB}$, then $CA = CB$.

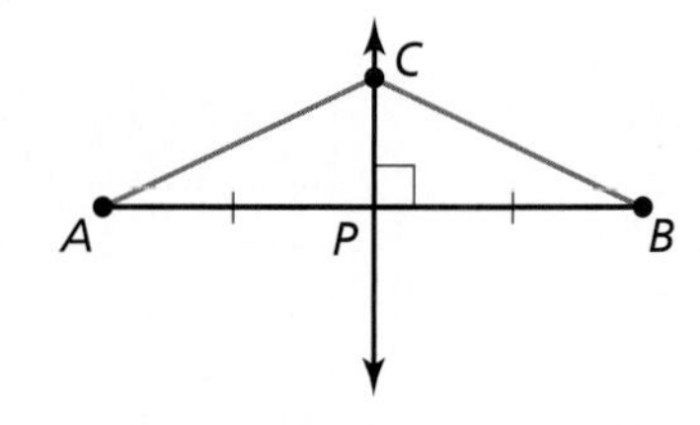

Proof p. 302

Theorem 6.2 Converse of the Perpendicular Bisector Theorem

In a plane, if a point is equidistant from the endpoints of a segment, then it lies on the perpendicular bisector of the segment.

If $DA = DB$, then point D lies on the $\perp$ bisector of $\overline{AB}$.

Proof Ex. 32, p. 308

PROOF Perpendicular Bisector Theorem

Given $\overleftrightarrow{CP}$ is the perpendicular bisector of $\overline{AB}$.

Prove $CA = CB$

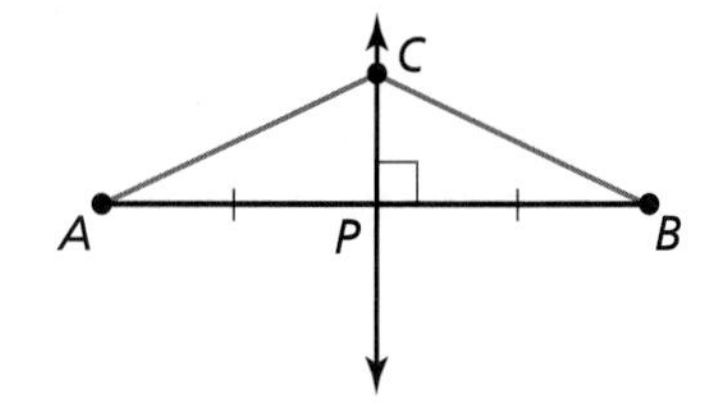

Paragraph Proof Because $\overleftrightarrow{CP}$ is the perpendicular bisector of $\overline{AB}$, $\overleftrightarrow{CP}$ is perpendicular to $\overline{AB}$ and point P is the midpoint of $\overline{AB}$. By the definition of midpoint, $AP = BP$, and by the definition of perpendicular lines, $m\angle CPA = m\angle CPB = 90°$. Then by the definition of segment congruence, $\overline{AP} \cong \overline{BP}$, and by the definition of angle congruence, $\angle CPA \cong \angle CPB$. By the Reflexive Property of Congruence (Theorem 2.1), $\overline{CP} \cong \overline{CP}$. So, $\triangle CPA \cong \triangle CPB$ by the SAS Congruence Theorem (Theorem 5.5), and $\overline{CA} \cong \overline{CB}$ because corresponding parts of congruent triangles are congruent. So, $CA = CB$ by the definition of segment congruence.

EXAMPLE 1 Using the Perpendicular Bisector Theorems

Find each measure.

a. *RS*

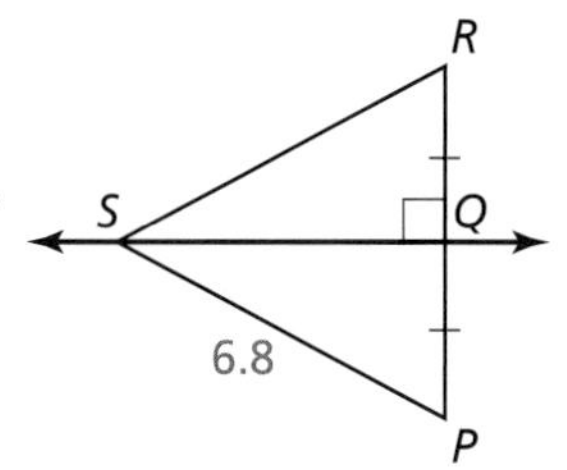

From the figure, $\overleftrightarrow{SQ}$ is the perpendicular bisector of $\overline{PR}$. By the Perpendicular Bisector Theorem, $PS = RS$.

▶ So, $RS = PS = 6.8$.

b. *EG*

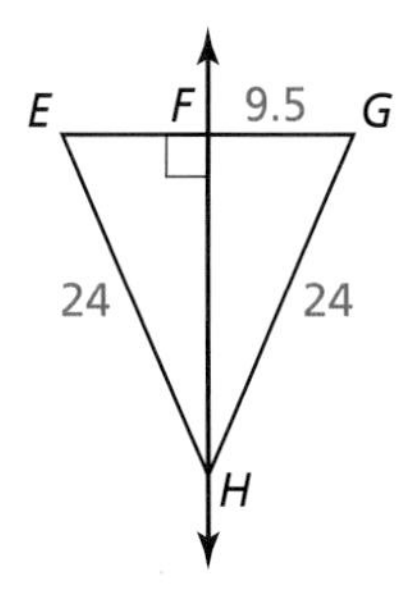

Because $EH = GH$ and $\overleftrightarrow{HF} \perp \overline{EG}$, $\overleftrightarrow{HF}$ is the perpendicular bisector of $\overline{EG}$ by the Converse of the Perpendicular Bisector Theorem. By the definition of segment bisector, $EG = 2GF$.

▶ So, $EG = 2(9.5) = 19$.

c. *AD*

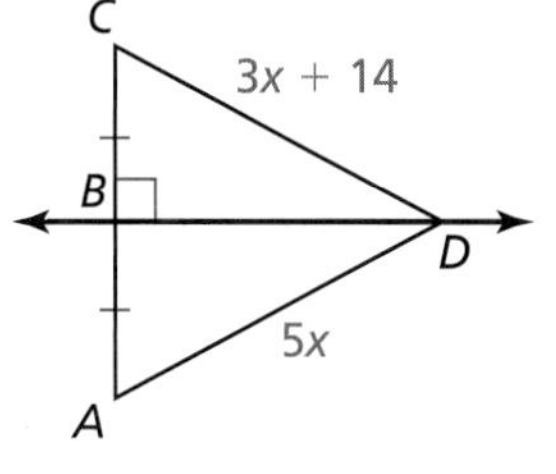

From the figure, $\overleftrightarrow{BD}$ is the perpendicular bisector of $\overline{AC}$.

$AD = CD$	Perpendicular Bisector Theorem
$5x = 3x + 14$	Substitute.
$x = 7$	Solve for x.

▶ So, $AD = 5x = 5(7) = 35$.

EXAMPLE 2 Solving a Real-Life Problem

Is there enough information in the diagram to conclude that point N lies on the perpendicular bisector of $\overline{KM}$?

SOLUTION

It is given that $\overline{KL} \cong \overline{ML}$. So, $\overline{LN}$ is a segment bisector of $\overline{KM}$. You do not know whether $\overline{LN}$ is perpendicular to $\overline{KM}$ because it is not indicated in the diagram.

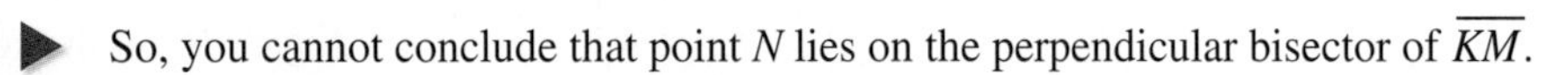

▶ So, you cannot conclude that point N lies on the perpendicular bisector of $\overline{KM}$.

Monitoring Progress  Help in English and Spanish at *BigIdeasMath.com*

Use the diagram and the given information to find the indicated measure.

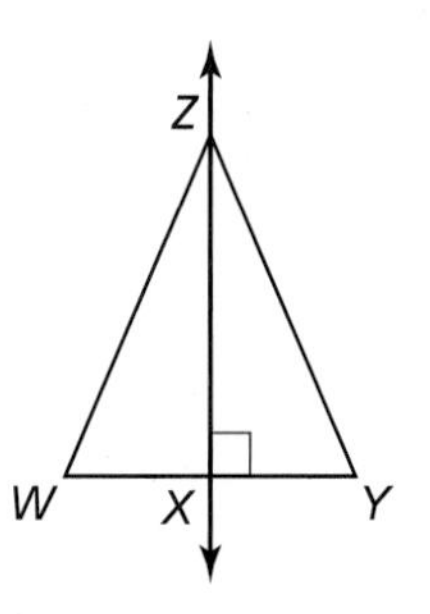

1. $\overleftrightarrow{ZX}$ is the perpendicular bisector of $\overline{WY}$, and $YZ = 13.75$. Find WZ.

2. $\overleftrightarrow{ZX}$ is the perpendicular bisector of $\overline{WY}$, $WZ = 4n - 13$, and $YZ = n + 17$. Find YZ.

3. Find WX when $WZ = 20.5$, $WY = 14.8$, and $YZ = 20.5$.

Using Angle Bisectors

In Section 1.5, you learned that an *angle bisector* is a ray that divides an angle into two congruent adjacent angles. You also know that the *distance from a point to a line* is the length of the perpendicular segment from the point to the line. So, in the figure, $\overrightarrow{AD}$ is the bisector of $\angle BAC$, and the distance from point D to $\overrightarrow{AB}$ is DB, where $\overline{DB} \perp \overrightarrow{AB}$.

Theorems

Theorem 6.3 Angle Bisector Theorem

If a point lies on the bisector of an angle, then it is equidistant from the two sides of the angle.

If $\overrightarrow{AD}$ bisects $\angle BAC$ and $\overline{DB} \perp \overrightarrow{AB}$ and $\overline{DC} \perp \overrightarrow{AC}$, then $DB = DC$.

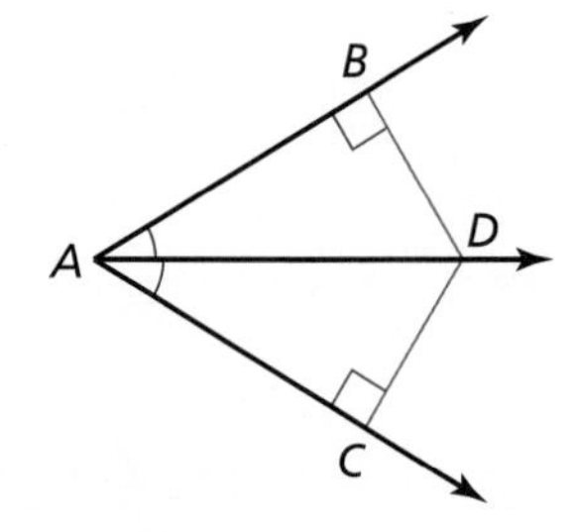

Proof Ex. 33(a), p. 308

Theorem 6.4 Converse of the Angle Bisector Theorem

If a point is in the interior of an angle and is equidistant from the two sides of the angle, then it lies on the bisector of the angle.

If $\overline{DB} \perp \overrightarrow{AB}$ and $\overline{DC} \perp \overrightarrow{AC}$ and $DB = DC$, then $\overrightarrow{AD}$ bisects $\angle BAC$.

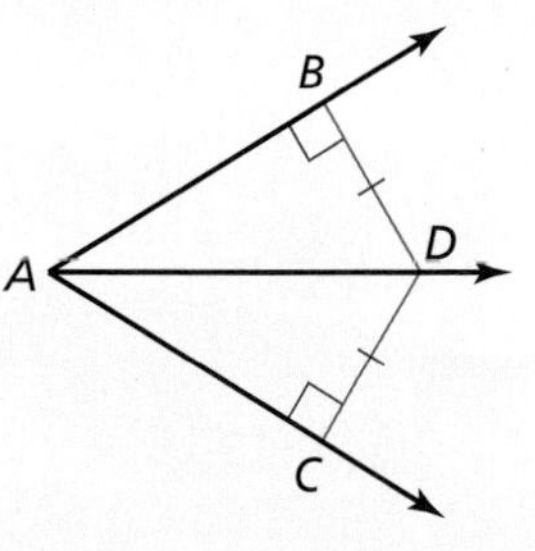

Proof Ex. 33(b), p. 308

EXAMPLE 3 Using the Angle Bisector Theorems

Find each measure.

a. $m\angle GFJ$

Because $\overline{JG} \perp \overrightarrow{FG}$ and $\overline{JH} \perp \overrightarrow{FH}$ and $JG = JH = 7$, $\overrightarrow{FJ}$ bisects $\angle GFH$ by the Converse of the Angle Bisector Theorem.

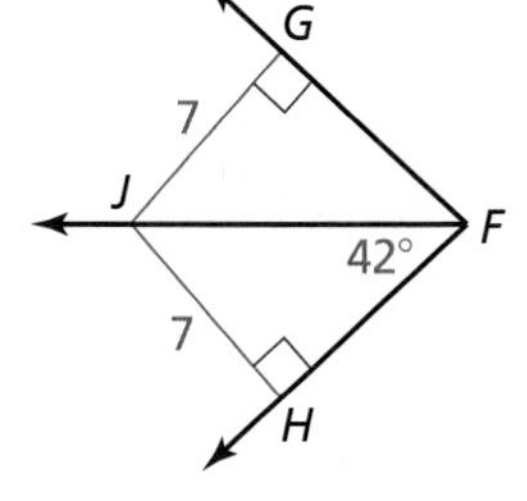

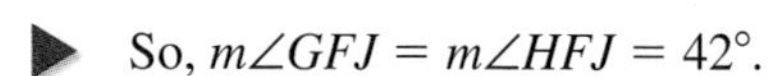
▶ So, $m\angle GFJ = m\angle HFJ = 42°$.

b. RS

$PS = RS$	Angle Bisector Theorem
$5x = 6x - 5$	Substitute.
$5 = x$	Solve for x.

▶ So, $RS = 6x - 5 = 6(5) - 5 = 25$.

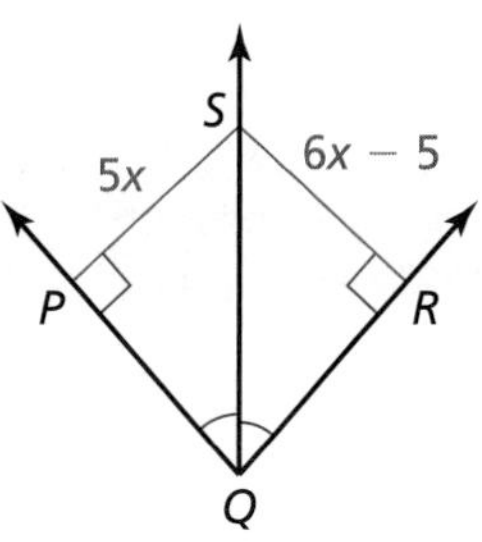

Monitoring Progress  Help in English and Spanish at *BigIdeasMath.com*

Use the diagram and the given information to find the indicated measure.

4. $\overrightarrow{BD}$ bisects $\angle ABC$, and $DC = 6.9$. Find DA.

5. $\overrightarrow{BD}$ bisects $\angle ABC$, $AD = 3z + 7$, and $CD = 2z + 11$. Find CD.

6. Find $m\angle ABC$ when $AD = 3.2$, $CD = 3.2$, and $m\angle DBC = 39°$.

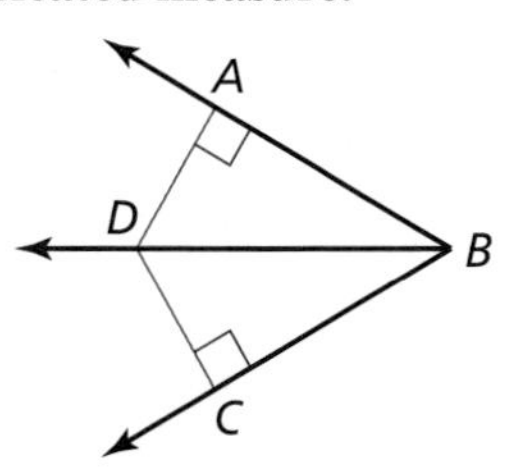

EXAMPLE 4 Solving a Real-Life Problem

A soccer goalie's position relative to the ball and goalposts forms congruent angles, as shown. Will the goalie have to move farther to block a shot toward the right goalpost R or the left goalpost L?

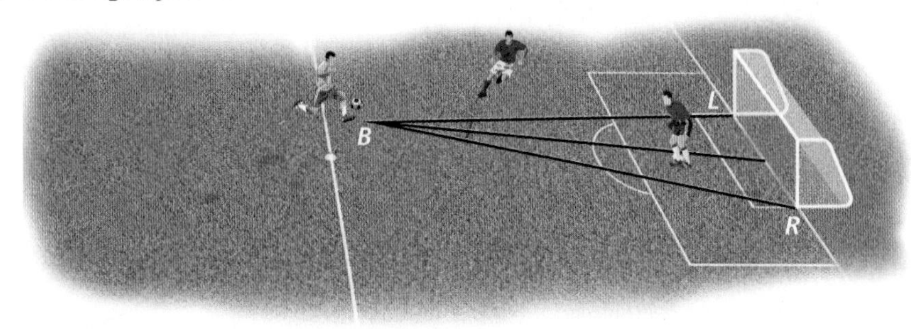

SOLUTION

The congruent angles tell you that the goalie is on the bisector of $\angle LBR$. By the Angle Bisector Theorem, the goalie is equidistant from $\overrightarrow{BR}$ and $\overrightarrow{BL}$.

▶ So, the goalie must move the same distance to block either shot.

Writing Equations for Perpendicular Bisectors

EXAMPLE 5 Writing an Equation for a Bisector

Write an equation of the perpendicular bisector of the segment with endpoints $P(-2, 3)$ and $Q(4, 1)$.

SOLUTION

Step 1 Graph $\overline{PQ}$. By definition, the perpendicular bisector of $\overline{PQ}$ is perpendicular to $\overline{PQ}$ at its midpoint.

Step 2 Find the midpoint M of $\overline{PQ}$.

$$M\left(\frac{-2+4}{2}, \frac{3+1}{2}\right) = M\left(\frac{2}{2}, \frac{4}{2}\right) = M(1, 2)$$

Step 3 Find the slope of the perpendicular bisector.

$$\text{slope of } \overline{PQ} = \frac{1-3}{4-(-2)} = \frac{-2}{6} = -\frac{1}{3}$$

Because the slopes of perpendicular lines are negative reciprocals, the slope of the perpendicular bisector is 3.

Step 4 Write an equation. The bisector of $\overline{PQ}$ has slope 3 and passes through $(1, 2)$.

$y = mx + b$	Use slope-intercept form.
$2 = 3(1) + b$	Substitute for m, x, and y.
$-1 = b$	Solve for b.

▶ So, an equation of the perpendicular bisector of $\overline{PQ}$ is $y = 3x - 1$.

Monitoring Progress  Help in English and Spanish at *BigIdeasMath.com*

7. Do you have enough information to conclude that $\overrightarrow{QS}$ bisects $\angle PQR$? Explain.

8. Write an equation of the perpendicular bisector of the segment with endpoints $(-1, -5)$ and $(3, -1)$.

6.1 Exercises

Dynamic Solutions available at *BigIdeasMath.com*

Vocabulary and Core Concept Check

1. **COMPLETE THE SENTENCE** Point C is in the interior of $\angle DEF$. If $\angle DEC$ and $\angle CEF$ are congruent, then $\overrightarrow{EC}$ is the ________ of $\angle DEF$.

2. **DIFFERENT WORDS, SAME QUESTION** Which is different? Find "both" answers.

Is point B the same distance from both X and Z?

Is point B equidistant from X and Z?

Is point B collinear with X and Z?

Is point B on the perpendicular bisector of $\overline{XZ}$?

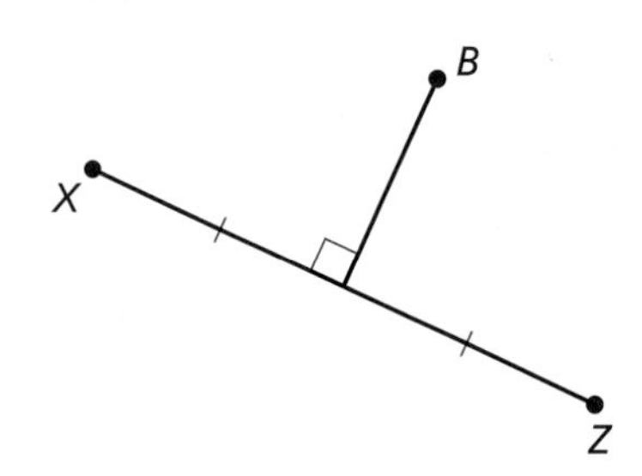

Monitoring Progress and Modeling with Mathematics

In Exercises 3–6, find the indicated measure. Explain your reasoning. *(See Example 1.)*

3. GH

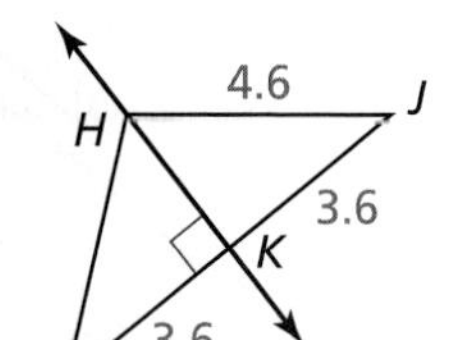

4. QR

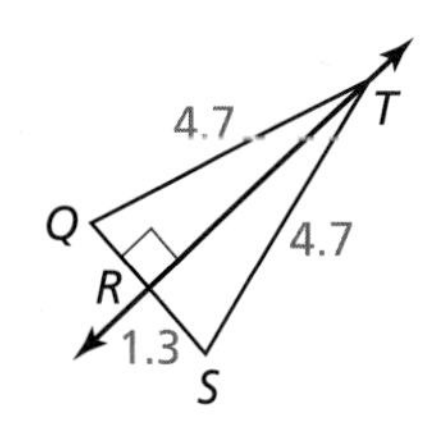

5. AB

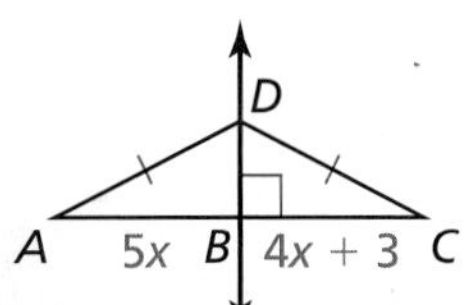

6. UW

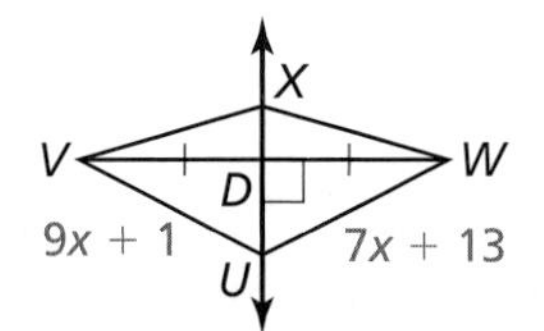

In Exercises 7–10, tell whether the information in the diagram allows you to conclude that point P lies on the perpendicular bisector of $\overline{LM}$. Explain your reasoning. *(See Example 2.)*

7.

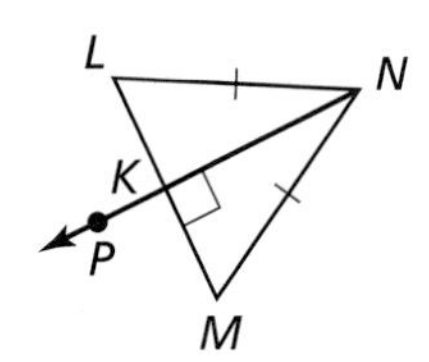

8.

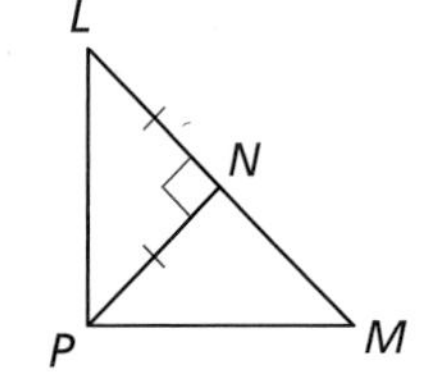

9.

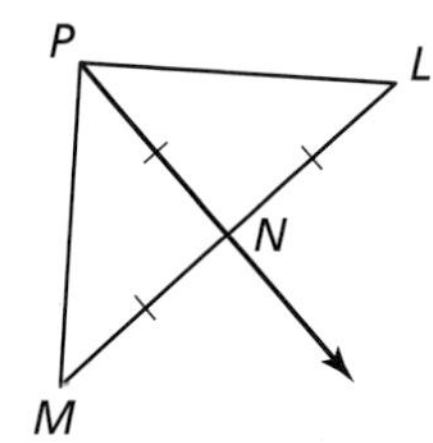

10.

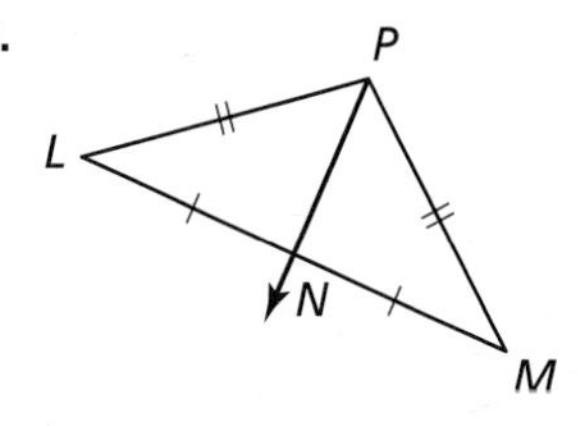

In Exercises 11–14, find the indicated measure. Explain your reasoning. *(See Example 3.)*

11. $m\angle ABD$

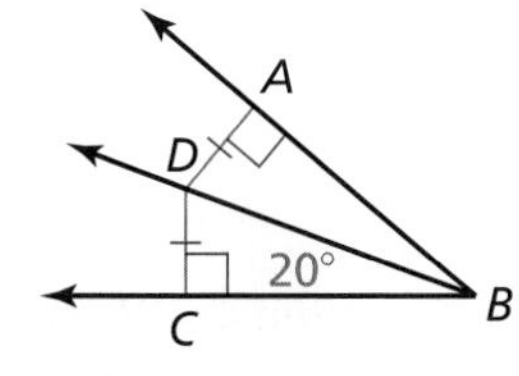

12. PS

13. $m\angle KJL$

14. FG

In Exercises 15 and 16, tell whether the information in the diagram allows you to conclude that $\overrightarrow{EH}$ bisects $\angle FEG$. Explain your reasoning. *(See Example 4.)*

15.

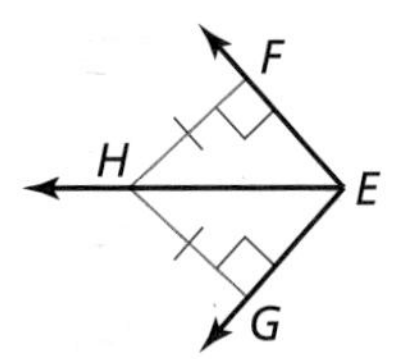

16.

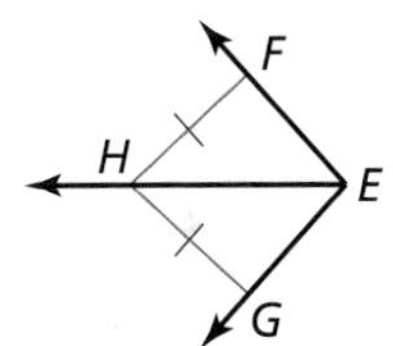

In Exercises 17 and 18, tell whether the information in the diagram allows you to conclude that $DB = DC$. Explain your reasoning.

17.

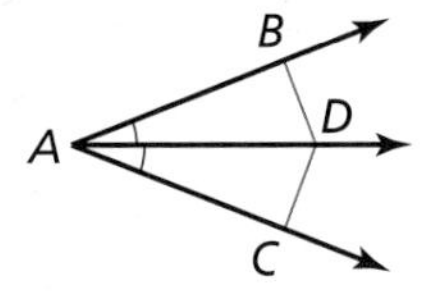

18.

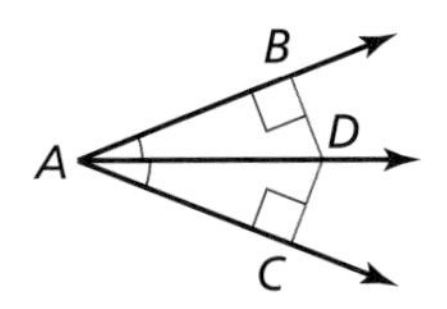

In Exercises 19–22, write an equation of the perpendicular bisector of the segment with the given endpoints. *(See Example 5.)*

19. $M(1, 5)$, $N(7, -1)$

20. $Q(-2, 0)$, $R(6, 12)$

21. $U(-3, 4)$, $V(9, 8)$

22. $Y(10, -7)$, $Z(-4, 1)$

ERROR ANALYSIS In Exercises 23 and 24, describe and correct the error in the student's reasoning.

23.

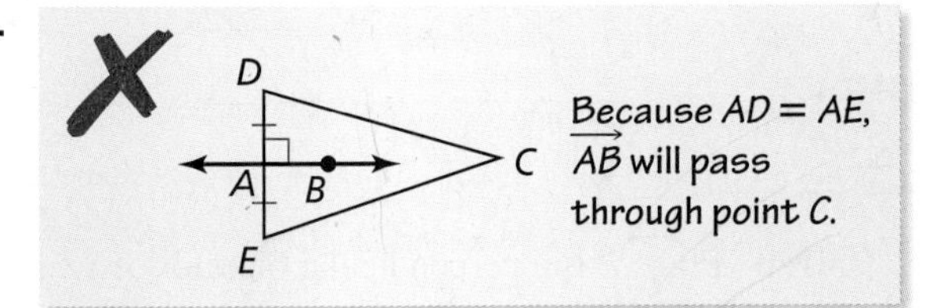

24.

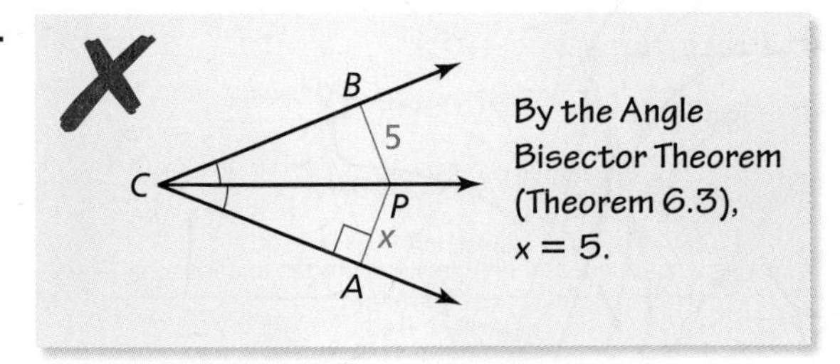

25. MODELING MATHEMATICS In the photo, the road is perpendicular to the support beam and $\overline{AB} \cong \overline{CB}$. Which theorem allows you to conclude that $\overline{AD} \cong \overline{CD}$?

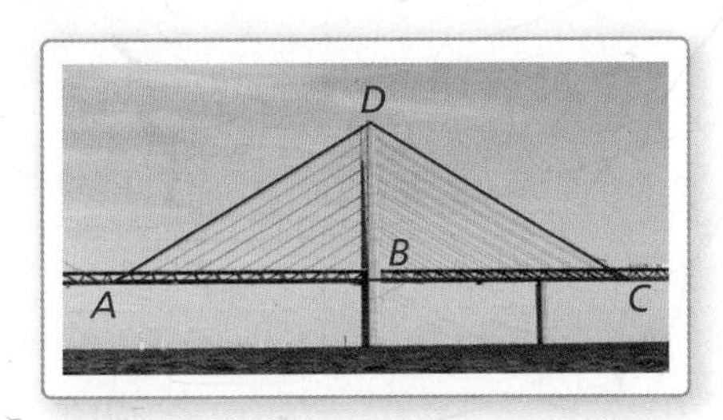

26. MODELING WITH MATHEMATICS The diagram shows the position of the goalie and the puck during a hockey game. The goalie is at point G, and the puck is at point P.

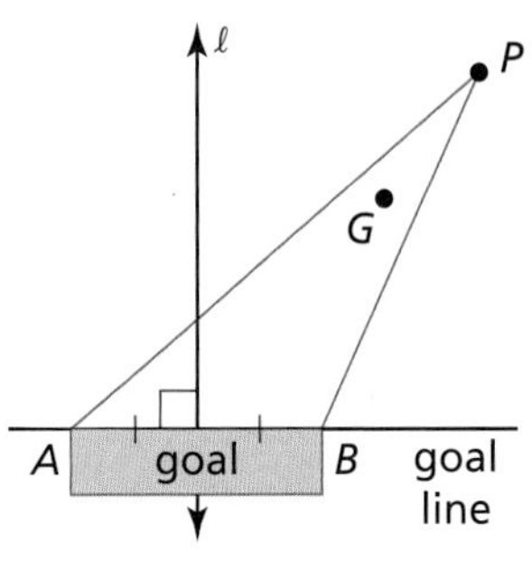

a. What should be the relationship between $\overrightarrow{PG}$ and $\angle APB$ to give the goalie equal distances to travel on each side of $\overrightarrow{PG}$?

b. How does $m\angle APB$ change as the puck gets closer to the goal? Does this change make it easier or more difficult for the goalie to defend the goal? Explain your reasoning.

27. CONSTRUCTION Use a compass and straightedge to construct a copy of $\overline{XY}$. Construct a perpendicular bisector and plot a point Z on the bisector so that the distance between point Z and $\overline{XY}$ is 3 centimeters. Measure $\overline{XZ}$ and $\overline{YZ}$. Which theorem does this construction demonstrate?

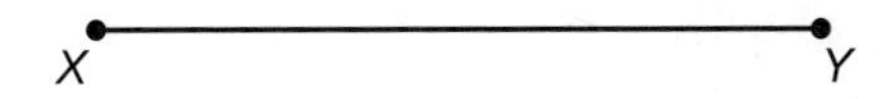

28. WRITING Explain how the Converse of the Perpendicular Bisector Theorem (Theorem 6.2) is related to the construction of a perpendicular bisector.

29. REASONING What is the value of x in the diagram?

Ⓐ 13

Ⓑ 18

Ⓒ 33

Ⓓ not enough information

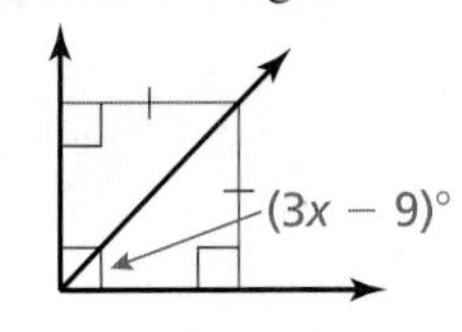

30. REASONING Which point lies on the perpendicular bisector of the segment with endpoints $M(7, 5)$ and $N(-1, 5)$?

Ⓐ $(2, 0)$

Ⓑ $(3, 9)$

Ⓒ $(4, 1)$

Ⓓ $(1, 3)$

31. MAKING AN ARGUMENT Your friend says it is impossible for an angle bisector of a triangle to be the same line as the perpendicular bisector of the opposite side. Is your friend correct? Explain your reasoning.

32. PROVING A THEOREM Prove the Converse of the Perpendicular Bisector Theorem (Thm. 6.2). (*Hint:* Construct a line through point C perpendicular to $\overline{AB}$ at point P.)

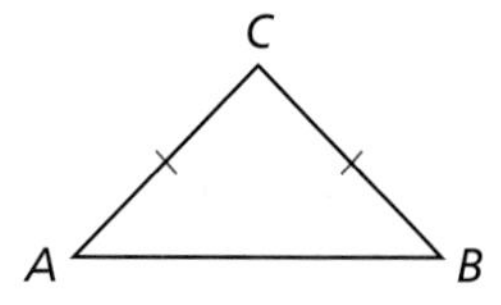

Given $CA = CB$

Prove Point C lies on the perpendicular bisector of $\overline{AB}$.

33. PROVING A THEOREM Use a congruence theorem to prove each theorem.

a. Angle Bisector Theorem (Thm. 6.3)

b. Converse of the Angle Bisector Theorem (Thm. 6.4)

34. HOW DO YOU SEE IT? The figure shows a map of a city. The city is arranged so each block north to south is the same length and each block east to west is the same length.

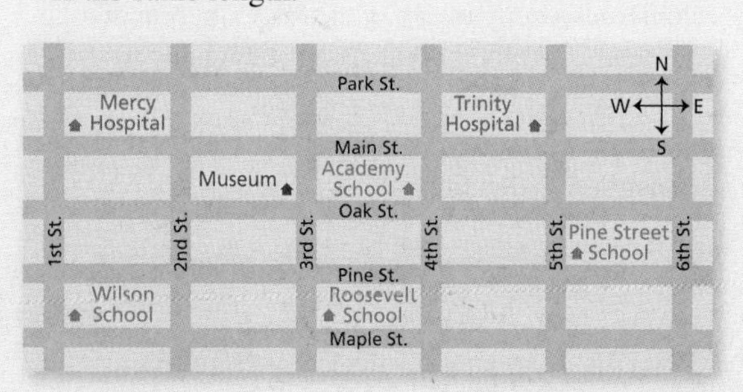

a. Which school is approximately equidistant from both hospitals? Explain your reasoning.

b. Is the museum approximately equidistant from Wilson School and Roosevelt School? Explain your reasoning.

35. MATHEMATICAL CONNECTIONS Write an equation whose graph consists of all the points in the given quadrants that are equidistant from the x- and y-axes.

a. I and III **b.** II and IV **c.** I and II

36. THOUGHT PROVOKING The postulates and theorems in this book represent Euclidean geometry. In spherical geometry, all points are on the surface of a sphere. A line is a circle on the sphere whose diameter is equal to the diameter of the sphere. In spherical geometry, is it possible for two lines to be perpendicular but not bisect each other? Explain your reasoning.

37. PROOF Use the information in the diagram to prove that $\overline{AB} \cong \overline{CB}$ if and only if points D, E, and B are collinear.

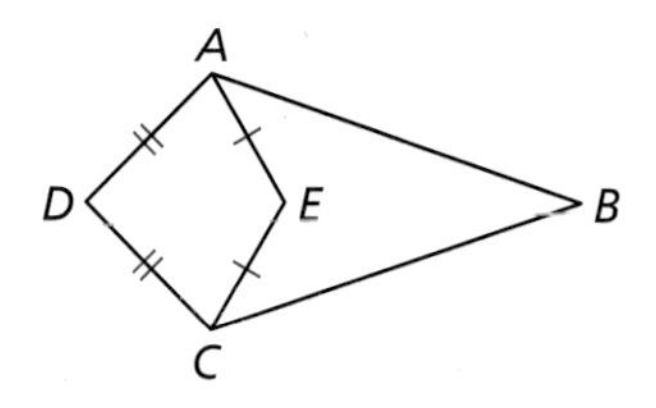

38. PROOF Prove the statements in parts (a)–(c).

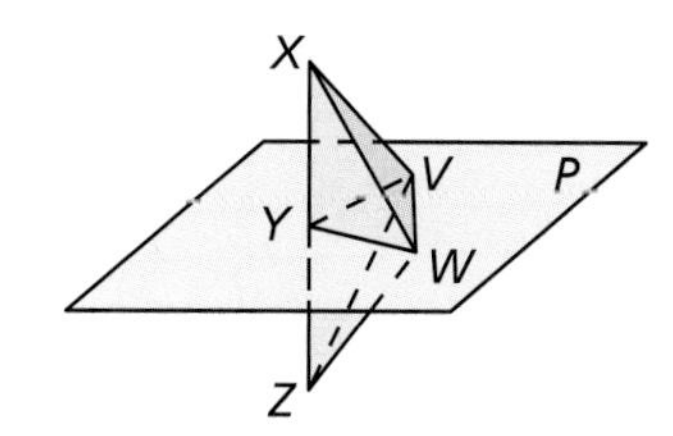

Given Plane P is a perpendicular bisector of $\overline{XZ}$ at point Y.

Prove **a.** $\overline{XW} \cong \overline{ZW}$

b. $\overline{XV} \cong \overline{ZV}$

c. $\angle VXW \cong \angle VZW$

Maintaining Mathematical Proficiency

Reviewing what you learned in previous grades and lessons

Classify the triangle by its sides. *(Section 5.1)*

39.

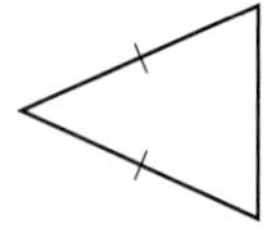

40.

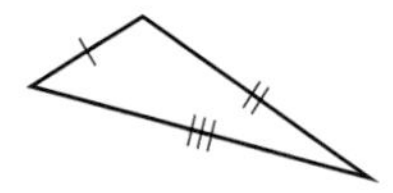

41.

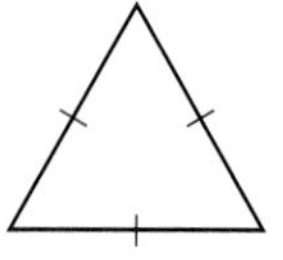

Classify the triangle by its angles. *(Section 5.1)*

42.

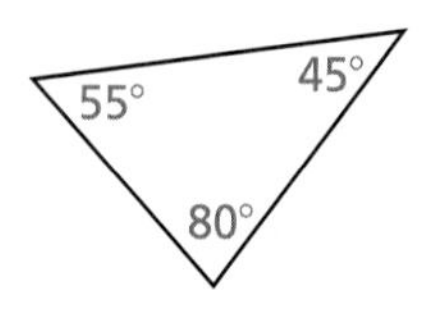

43.

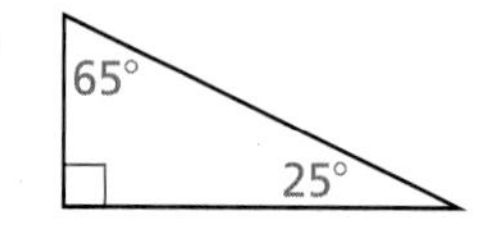

44.

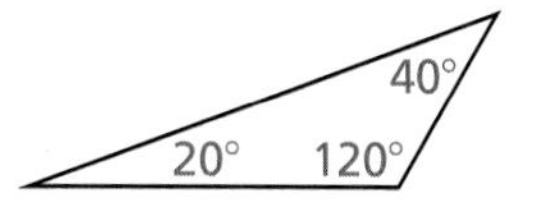

6.2 Bisectors of Triangles

Essential Question What conjectures can you make about the perpendicular bisectors and the angle bisectors of a triangle?

EXPLORATION 1 Properties of the Perpendicular Bisectors of a Triangle

Work with a partner. Use dynamic geometry software. Draw any $\triangle ABC$.

a. Construct the perpendicular bisectors of all three sides of $\triangle ABC$. Then drag the vertices to change $\triangle ABC$. What do you notice about the perpendicular bisectors?

b. Label a point D at the intersection of the perpendicular bisectors.

c. Draw the circle with center D through vertex A of $\triangle ABC$. Then drag the vertices to change $\triangle ABC$. What do you notice?

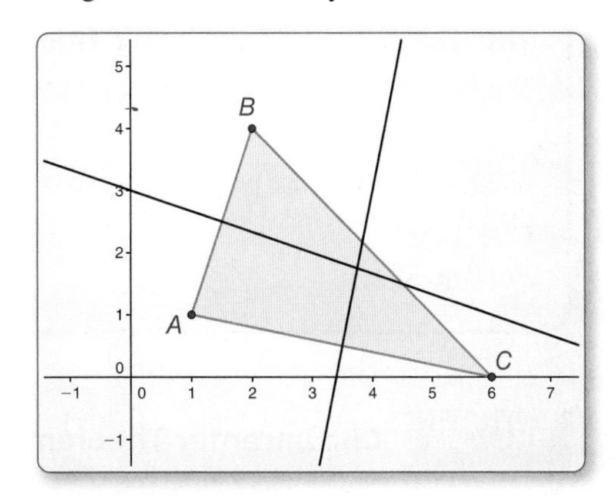

Sample
Points
$A(1, 1)$
$B(2, 4)$
$C(6, 0)$
Segments
$BC = 5.66$
$AC = 5.10$
$AB = 3.16$
Lines
$x + 3y = 9$
$-5x + y = -17$

LOOKING FOR STRUCTURE

To be proficient in math, you need to see complicated things as single objects or as being composed of several objects.

EXPLORATION 2 Properties of the Angle Bisectors of a Triangle

Work with a partner. Use dynamic geometry software. Draw any $\triangle ABC$.

a. Construct the angle bisectors of all three angles of $\triangle ABC$. Then drag the vertices to change $\triangle ABC$. What do you notice about the angle bisectors?

b. Label a point D at the intersection of the angle bisectors.

c. Find the distance between D and $\overline{AB}$. Draw the circle with center D and this distance as a radius. Then drag the vertices to change $\triangle ABC$. What do you notice?

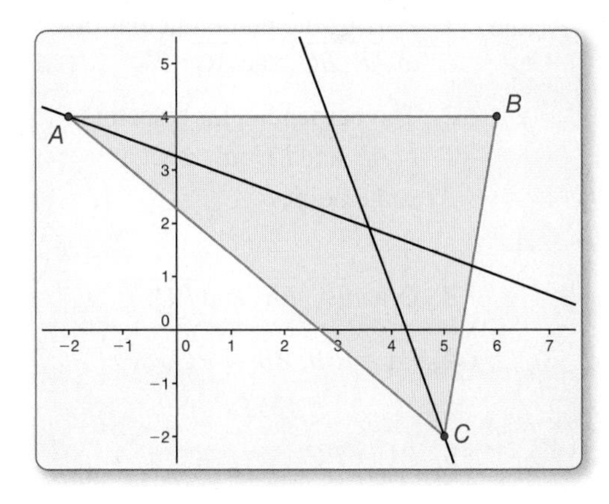

Sample
Points
$A(-2, 4)$
$B(6, 4)$
$C(5, -2)$
Segments
$BC = 6.08$
$AC = 9.22$
$AB = 8$
Lines
$0.35x + 0.94y = 3.06$
$-0.94x - 0.34y = -4.02$

Communicate Your Answer

3. What conjectures can you make about the perpendicular bisectors and the angle bisectors of a triangle?

6.2 Lesson

Core Vocabulary

concurrent, *p. 310*
point of concurrency, *p. 310*
circumcenter, *p. 310*
incenter, *p. 313*

Previous
perpendicular bisector
angle bisector

What You Will Learn

- Use and find the circumcenter of a triangle.
- Use and find the incenter of a triangle.

Using the Circumcenter of a Triangle

When three or more lines, rays, or segments intersect in the same point, they are called **concurrent** lines, rays, or segments. The point of intersection of the lines, rays, or segments is called the **point of concurrency**.

In a triangle, the three perpendicular bisectors are concurrent. The point of concurrency is the **circumcenter** of the triangle.

Theorems

Theorem 6.5 Circumcenter Theorem

The circumcenter of a triangle is equidistant from the vertices of the triangle.

If $\overline{PD}$, $\overline{PE}$, and $\overline{PF}$ are perpendicular bisectors, then $PA = PB = PC$.

Proof p. 310

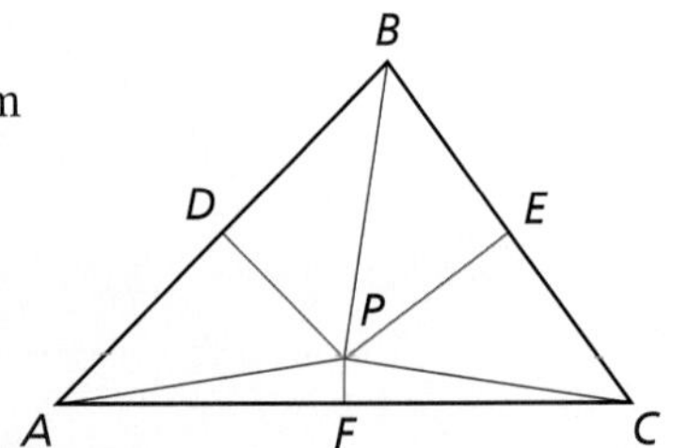

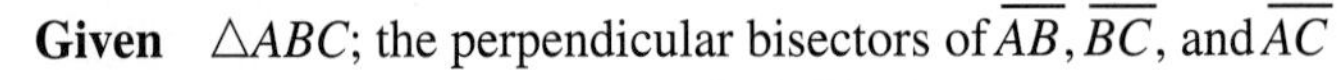

PROOF Circumcenter Theorem

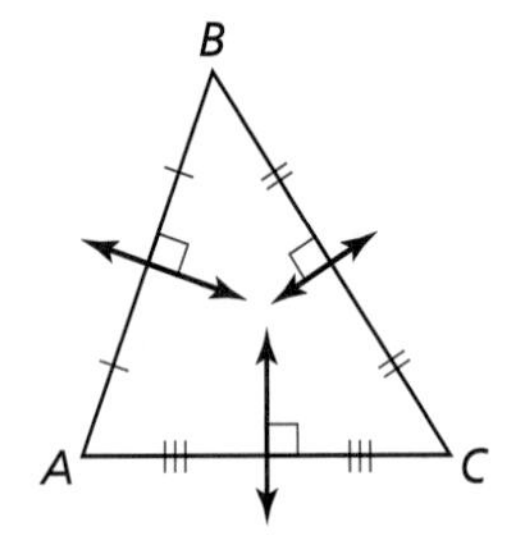

Given $\triangle ABC$; the perpendicular bisectors of $\overline{AB}$, $\overline{BC}$, and $\overline{AC}$

Prove The perpendicular bisectors intersect in a point; that point is equidistant from A, B, and C.

Plan for Proof Show that P, the point of intersection of the perpendicular bisectors of $\overline{AB}$ and $\overline{BC}$, also lies on the perpendicular bisector of $\overline{AC}$. Then show that point P is equidistant from the vertices of the triangle.

Plan in Action

STATEMENTS	REASONS
1. $\triangle ABC$; the perpendicular bisectors of $\overline{AB}$, $\overline{BC}$, and $\overline{AC}$	**1.** Given
2. The perpendicular bisectors of $\overline{AB}$ and $\overline{BC}$ intersect at some point P.	**2.** Because the sides of a triangle cannot be parallel, these perpendicular bisectors must intersect in some point. Call it P.
3. Draw $\overline{PA}$, $\overline{PB}$, and $\overline{PC}$.	**3.** Two Point Postulate (Post. 2.1)
4. $PA = PB$, $PB = PC$	**4.** Perpendicular Bisector Theorem (Thm. 6.1)
5. $PA = PC$	**5.** Transitive Property of Equality
6. P is on the perpendicular bisector of $\overline{AC}$.	**6.** Converse of the Perpendicular Bisector Theorem (Thm. 6.2)
7. $PA = PB = PC$. So, P is equidistant from the vertices of the triangle.	**7.** From the results of Steps 4 and 5 and the definition of equidistant

STUDY TIP

Use diagrams like the one below to help visualize your proof.

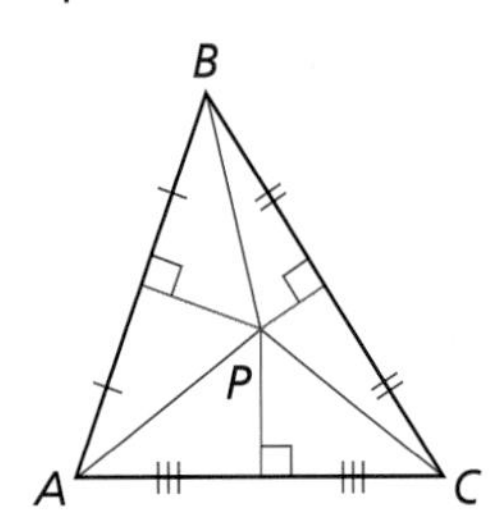

EXAMPLE 1 Solving a Real-Life Problem

Three snack carts sell frozen yogurt from points A, B, and C outside a city. Each of the three carts is the same distance from the frozen yogurt distributor.

Find the location of the distributor.

SOLUTION

The distributor is equidistant from the three snack carts. The Circumcenter Theorem shows that you can find a point equidistant from three points by using the perpendicular bisectors of the triangle formed by those points.

Copy the positions of points A, B, and C and connect the points to draw $\triangle ABC$. Then use a ruler and protractor to draw the three perpendicular bisectors of $\triangle ABC$. The circumcenter D is the location of the distributor.

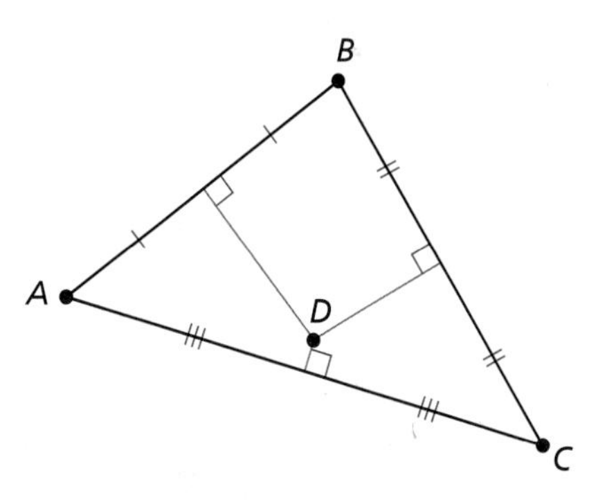

Monitoring Progress  Help in English and Spanish at *BigIdeasMath.com*

1. Three snack carts sell hot pretzels from points A, B, and E. What is the location of the pretzel distributor if it is equidistant from the three carts? Sketch the triangle and show the location.

B •

A • • E

READING

The prefix *circum-* means "around" or "about," as in *circumference* (distance around a circle).

The circumcenter P is equidistant from the three vertices, so P is the center of a circle that passes through all three vertices. As shown below, the location of P depends on the type of triangle. The circle with center P is said to be *circumscribed* about the triangle.

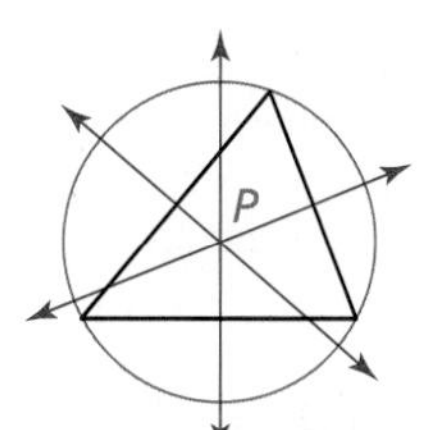

Acute triangle
P is inside triangle.

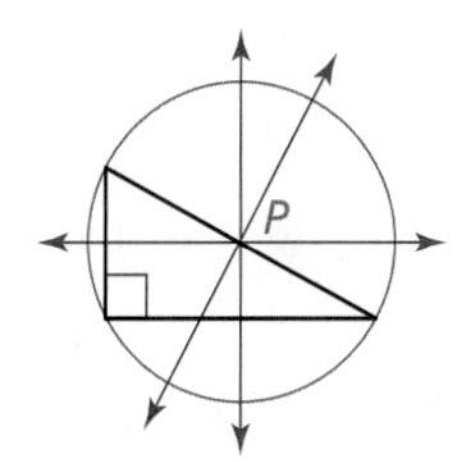

Right triangle
P is on triangle.

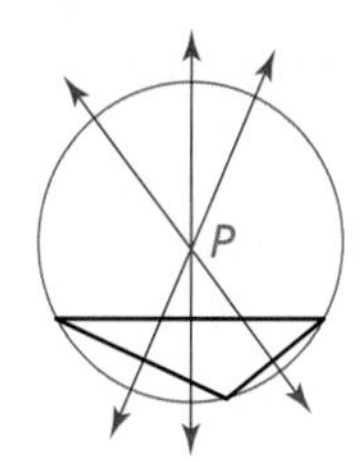

Obtuse triangle
P is outside triangle.

CONSTRUCTION Circumscribing a Circle About a Triangle

Use a compass and straightedge to construct a circle that is circumscribed about $\triangle ABC$.

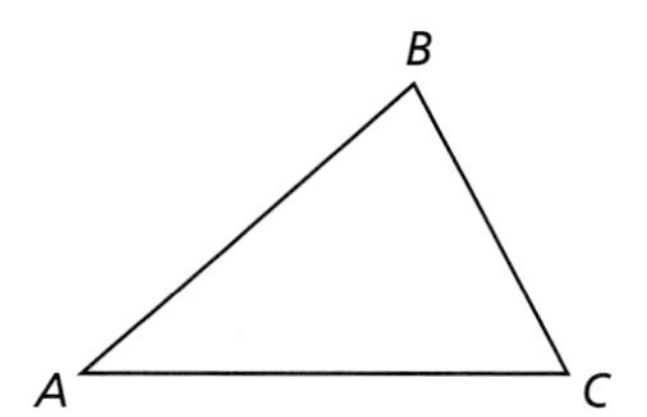

SOLUTION

Step 1

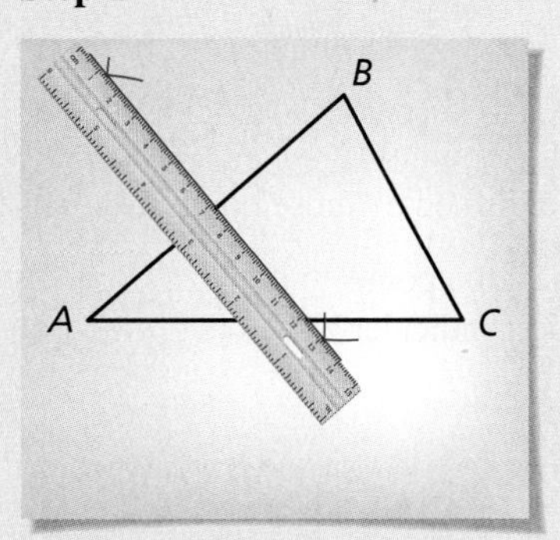

Draw a bisector Draw the perpendicular bisector of $\overline{AB}$.

Step 2

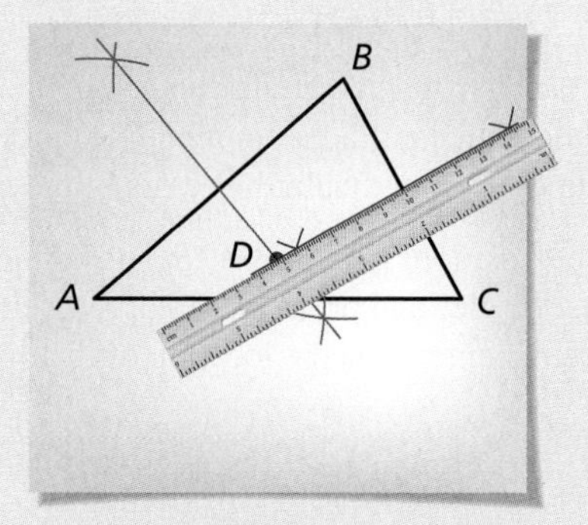

Draw a bisector Draw the perpendicular bisector of $\overline{BC}$. Label the intersection of the bisectors D. This is the circumcenter.

Step 3

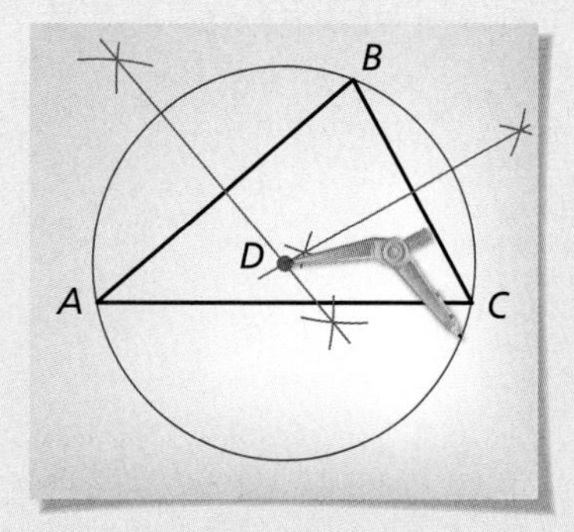

Draw a circle Place the compass at D. Set the width by using any vertex of the triangle. This is the radius of the *circumcircle*. Draw the circle. It should pass through all three vertices A, B, and C.

STUDY TIP

Note that you only need to find the equations for *two* perpendicular bisectors. You can use the perpendicular bisector of the third side to verify your result.

EXAMPLE 2 Finding the Circumcenter of a Triangle

Find the coordinates of the circumcenter of $\triangle ABC$ with vertices $A(0, 3)$, $B(0, -1)$, and $C(6, -1)$.

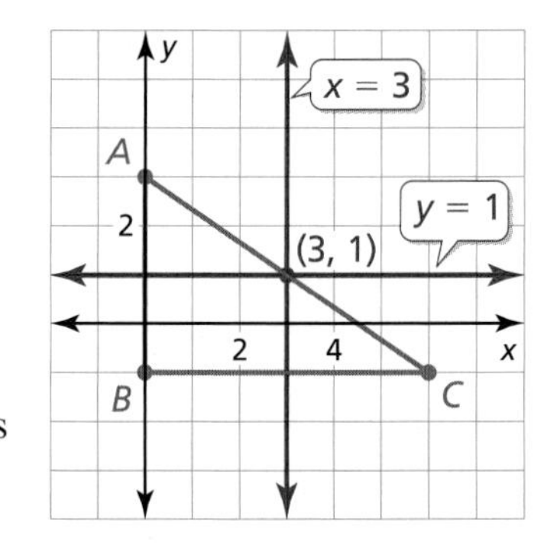

SOLUTION

Step 1 Graph $\triangle ABC$.

Step 2 Find equations for two perpendicular bisectors. Use the Slopes of Perpendicular Lines Theorem (Theorem 3.14), which states that horizontal lines are perpendicular to vertical lines.

The midpoint of $\overline{AB}$ is $(0, 1)$. The line through $(0, 1)$ that is perpendicular to $\overline{AB}$ is $y = 1$.

The midpoint of $\overline{BC}$ is $(3, -1)$. The line through $(3, -1)$ that is perpendicular to $\overline{BC}$ is $x = 3$.

Step 3 Find the point where $x = 3$ and $y = 1$ intersect. They intersect at $(3, 1)$.

▶ So, the coordinates of the circumcenter are $(3, 1)$.

MAKING SENSE OF PROBLEMS

Because $\triangle ABC$ is a right triangle, the circumcenter lies on the triangle.

Monitoring Progress

Help in English and Spanish at *BigIdeasMath.com*

Find the coordinates of the circumcenter of the triangle with the given vertices.

2. $R(-2, 5)$, $S(-6, 5)$, $T(-2, -1)$

3. $W(-1, 4)$, $X(1, 4)$, $Y(1, -6)$

Using the Incenter of a Triangle

Just as a triangle has three perpendicular bisectors, it also has three angle bisectors. The angle bisectors of a triangle are also concurrent. This point of concurrency is the **incenter** of the triangle. For any triangle, the incenter always lies inside the triangle.

Theorem

Theorem 6.6 Incenter Theorem

The incenter of a triangle is equidistant from the sides of the triangle.

If $\overline{AP}$, $\overline{BP}$, and $\overline{CP}$ are angle bisectors of $\triangle ABC$, then $PD = PE = PF$.

Proof Ex. 38, p. 317

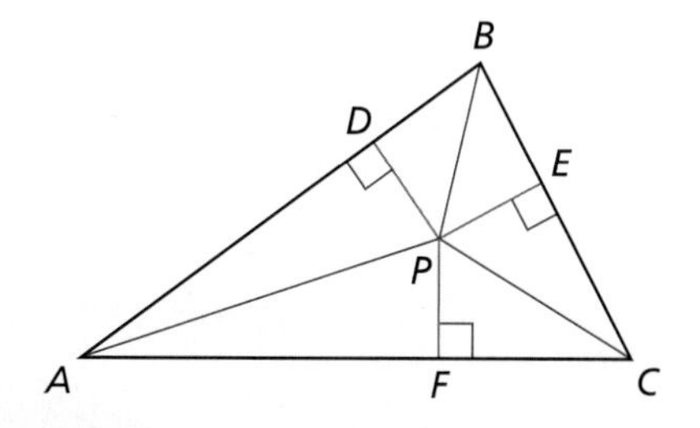

EXAMPLE 3 **Using the Incenter of a Triangle**

In the figure shown, $ND = 5x - 1$ and $NE = 2x + 11$.

a. Find NF.

b. Can NG be equal to 18? Explain your reasoning.

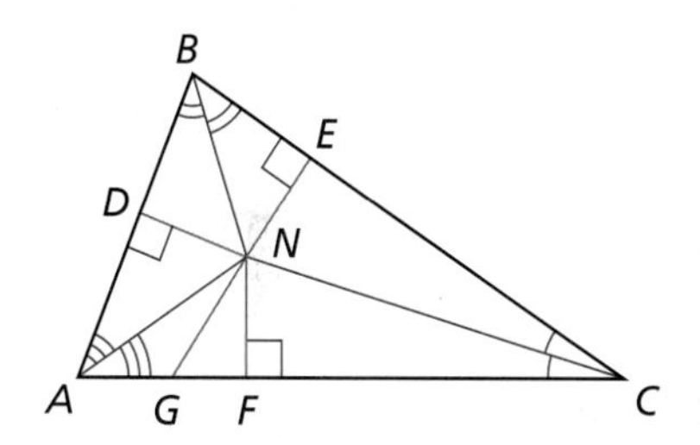

SOLUTION

a. N is the incenter of $\triangle ABC$ because it is the point of concurrency of the three angle bisectors. So, by the Incenter Theorem, $ND = NE = NF$.

Step 1 Solve for x.

$ND = NE$	Incenter Theorem
$5x - 1 = 2x + 11$	Substitute.
$x = 4$	Solve for x.

Step 2 Find ND (or NE).

$$ND = 5x - 1 = 5(4) - 1 = 19$$

▶ So, because $ND = NF$, $NF = 19$.

b. Recall that the shortest distance between a point and a line is a perpendicular segment. In this case, the perpendicular segment is $\overline{NF}$, which has a length of 19. Because $18 < 19$, NG cannot be equal to 18.

Monitoring Progress

Help in English and Spanish at *BigIdeasMath.com*

4. In the figure shown, $QM = 3x + 8$ and $QN = 7x + 2$. Find QP.

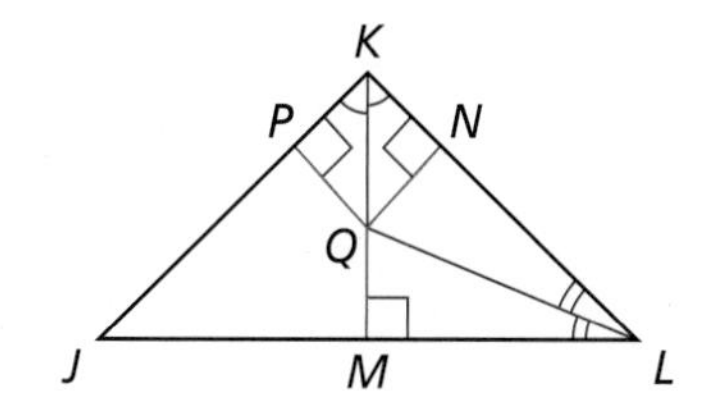

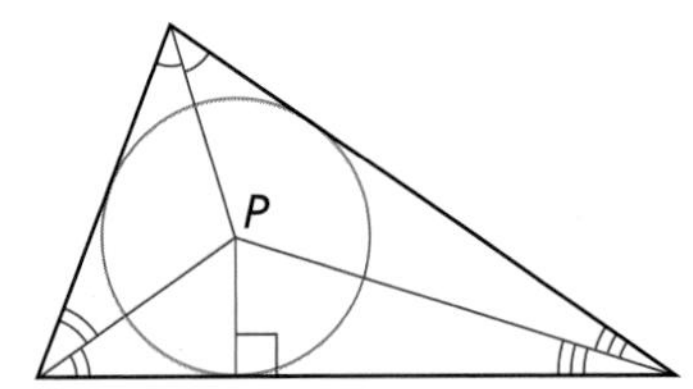

Because the incenter P is equidistant from the three sides of the triangle, a circle drawn using P as the center and the distance to one side of the triangle as the radius will just touch the other two sides of the triangle. The circle is said to be *inscribed* within the triangle.

CONSTRUCTION Inscribing a Circle Within a Triangle

Use a compass and straightedge to construct a circle that is inscribed within $\triangle ABC$.

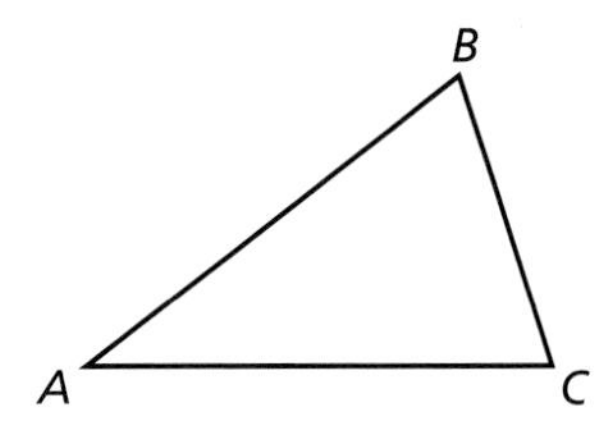

SOLUTION

Step 1

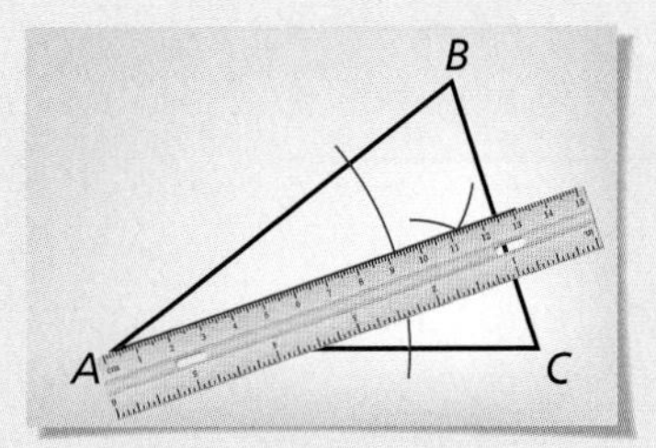

Draw a bisector Draw the angle bisector of $\angle A$.

Step 2

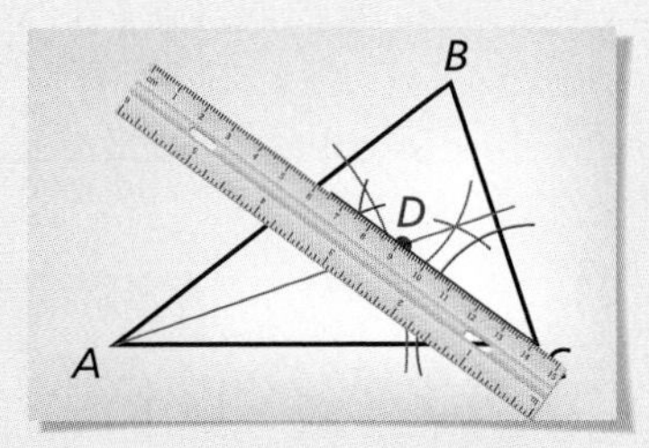

Draw a bisector Draw the angle bisector of $\angle C$. Label the intersection of the bisectors D. This is the incenter.

Step 3

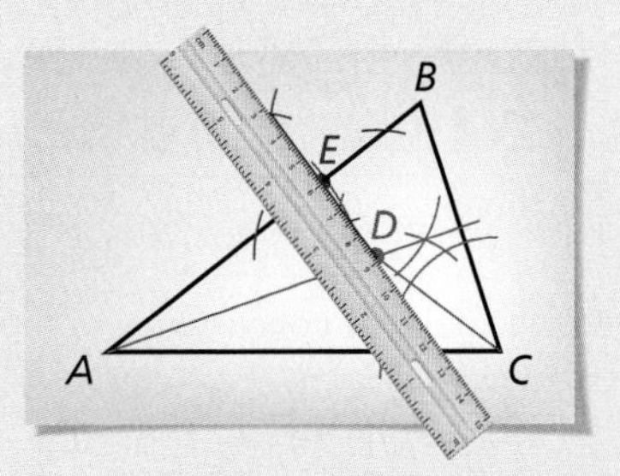

Draw a perpendicular line Draw the perpendicular line from D to $\overline{AB}$. Label the point where it intersects $\overline{AB}$ as E.

Step 4

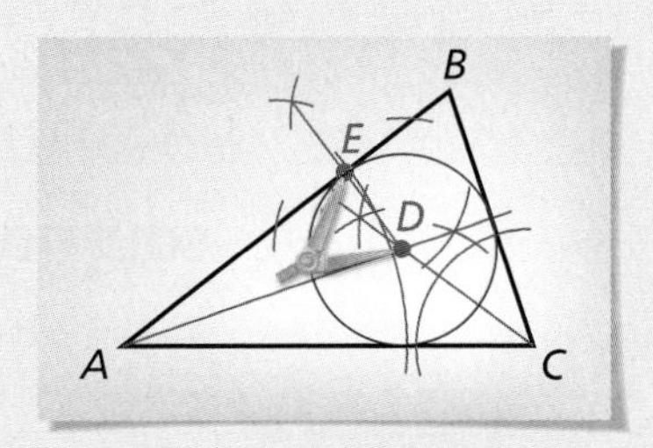

Draw a circle Place the compass at D. Set the width to E. This is the radius of the *incircle*. Draw the circle. It should touch each side of the triangle.

EXAMPLE 4 Solving a Real-Life Problem

A city wants to place a lamppost on the boulevard shown so that the lamppost is the same distance from all three streets. Should the location of the lamppost be at the *circumcenter* or *incenter* of the triangular boulevard? Explain.

SOLUTION

Because the shape of the boulevard is an obtuse triangle, its circumcenter lies outside the triangle. So, the location of the lamppost cannot be at the circumcenter. The city wants the lamppost to be the same distance from all three streets. By the Incenter Theorem, the incenter of a triangle is equidistant from the sides of a triangle.

▶ So, the location of the lamppost should be at the incenter of the boulevard.

ATTENDING TO PRECISION

Pay close attention to how a problem is stated. The city wants the lamppost to be the *same distance* from the three streets, not from where the streets intersect.

Monitoring Progress  Help in English and Spanish at *BigIdeasMath.com*

5. Draw a sketch to show the location L of the lamppost in Example 4.

6.2 Exercises

Dynamic Solutions available at *BigIdeasMath.com*

Vocabulary and Core Concept Check

1. **VOCABULARY** When three or more lines, rays, or segments intersect in the same point, they are called ___________ lines, rays, or segments.

2. **WHICH ONE DOESN'T BELONG?** Which triangle does *not* belong with the other three? Explain your reasoning.

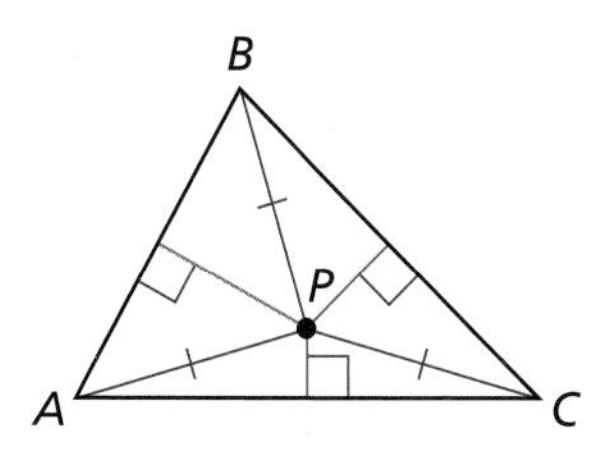

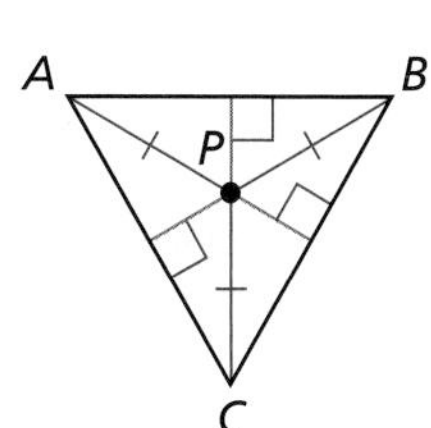

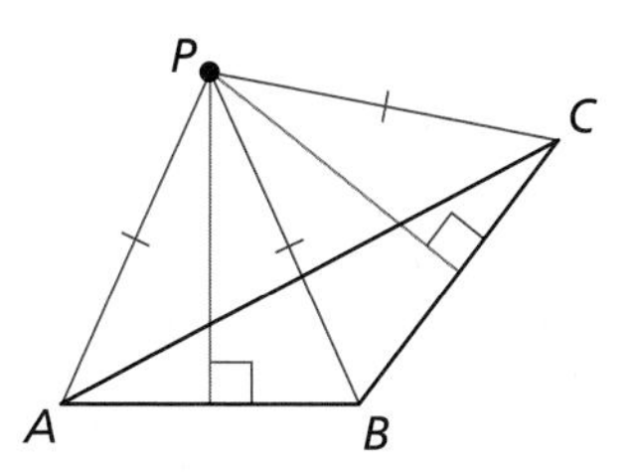

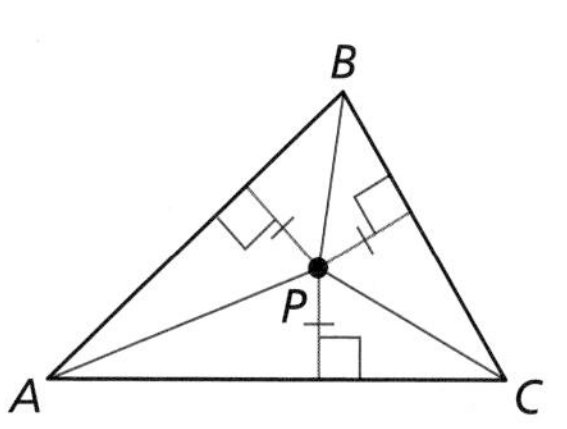

Monitoring Progress and Modeling with Mathematics

In Exercises 3 and 4, the perpendicular bisectors of $\triangle ABC$ intersect at point G and are shown in blue. Find the indicated measure.

3. Find BG.

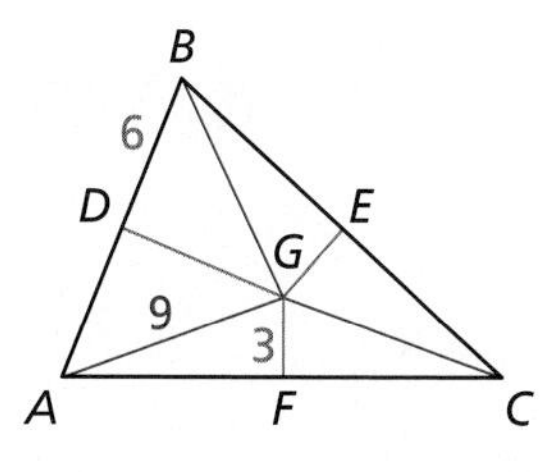

4. Find GA.

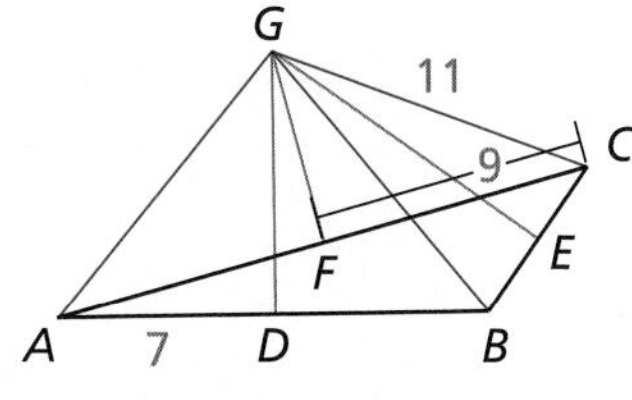

In Exercises 5 and 6, the angle bisectors of $\triangle XYZ$ intersect at point P and are shown in red. Find the indicated measure.

5. Find PB.

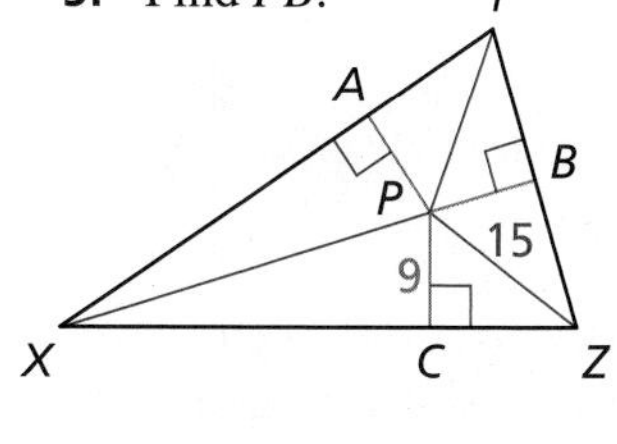

6. Find HP.

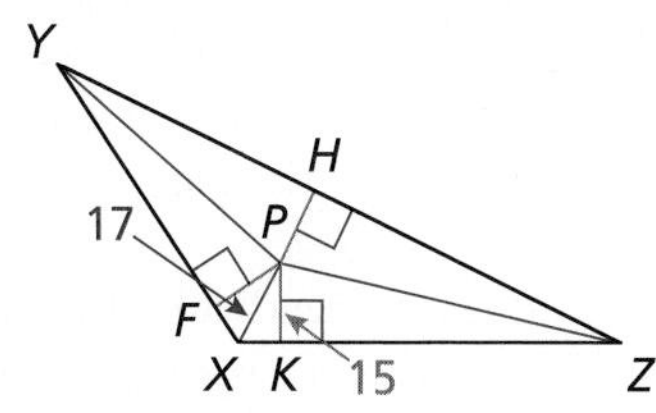

In Exercises 7–10, find the coordinates of the circumcenter of the triangle with the given vertices. *(See Example 2.)*

7. $A(2, 6)$, $B(8, 6)$, $C(8, 10)$

8. $D(-7, -1)$, $E(-1, -1)$, $F(-7, -9)$

9. $H(-10, 7)$, $J(-6, 3)$, $K(-2, 3)$

10. $L(3, -6)$, $M(5, -3)$, $N(8, -6)$

In Exercises 11–14, N is the incenter of $\triangle ABC$. Use the given information to find the indicated measure. *(See Example 3.)*

11. $ND = 6x - 2$
    $NE = 3x + 7$
    Find NF.

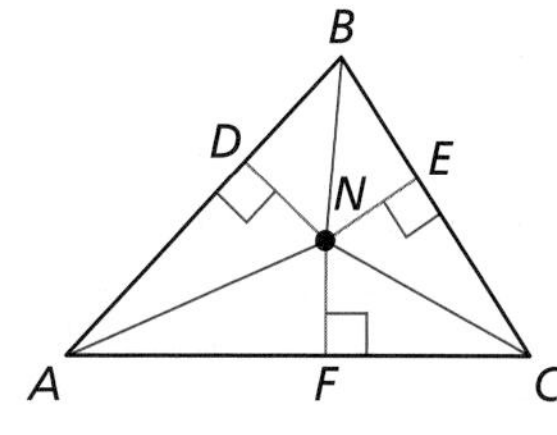

12. $NG = x + 3$
    $NH = 2x - 3$
    Find NJ.

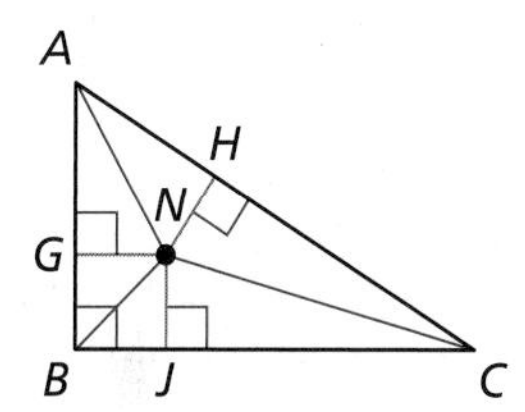

13. $NK = 2x - 2$
    $NL = -x + 10$
    Find NM.

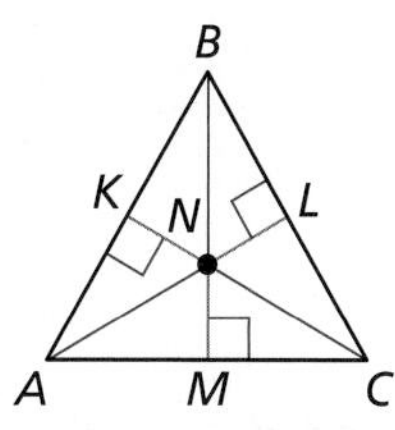

14. $NQ = 2x$
    $NR = 3x - 2$
    Find NS.

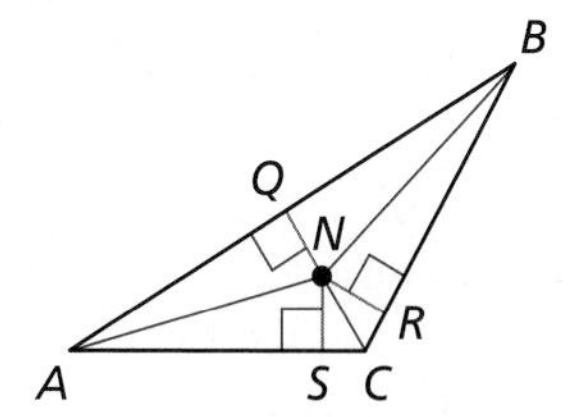

15. P is the circumcenter of $\triangle XYZ$. Use the given information to find PZ.

    $PX = 3x + 2$
    $PY = 4x - 8$

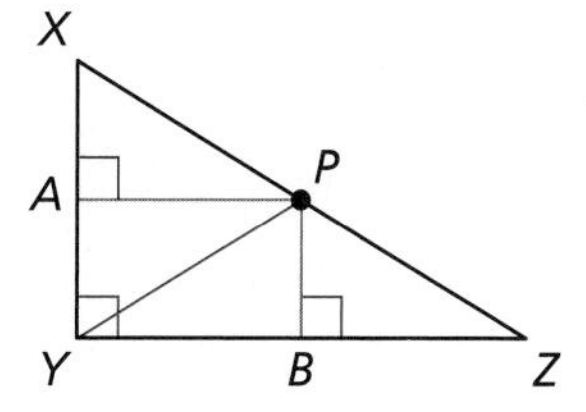

16. P is the circumcenter of $\triangle XYZ$. Use the given information to find PY.

$PX = 4x + 3$

$PZ = 6x - 11$

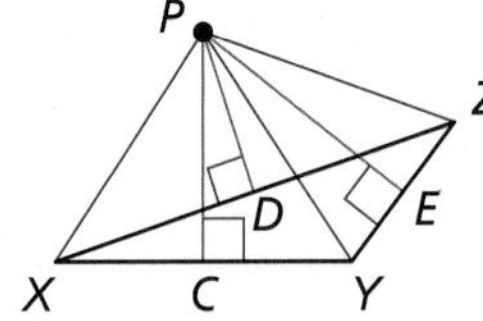

CONSTRUCTION In Exercises 17–20, draw a triangle of the given type. Find the circumcenter. Then construct the circumscribed circle.

17. right

18. obtuse

19. acute isosceles

20. equilateral

CONSTRUCTION In Exercises 21–24, copy the triangle with the given angle measures. Find the incenter. Then construct the inscribed circle.

21.

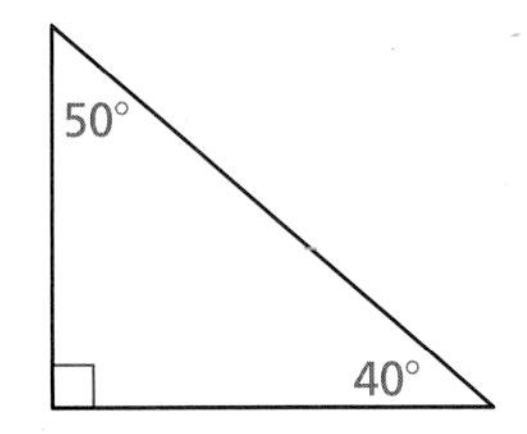

22.

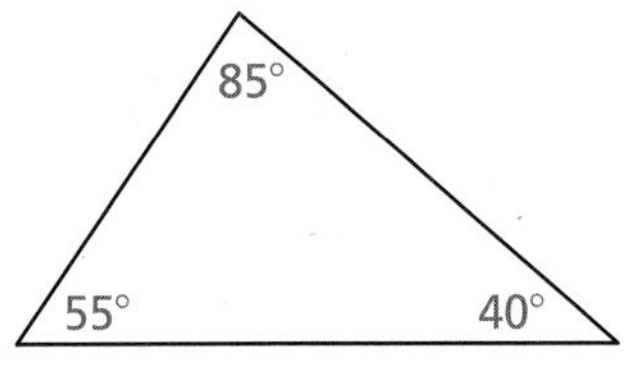

23.

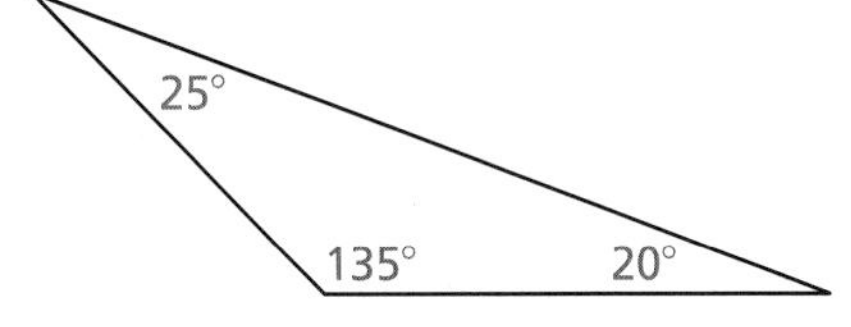

24.

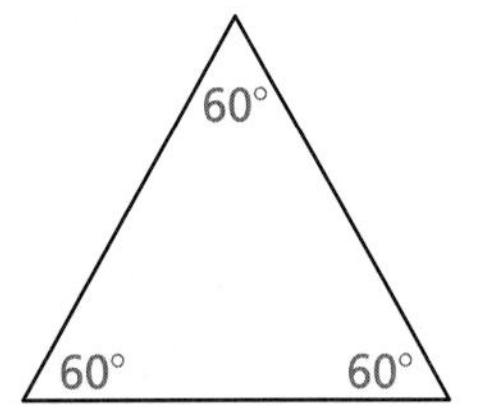

ERROR ANALYSIS In Exercises 25 and 26, describe and correct the error in identifying equal distances inside the triangle.

25.

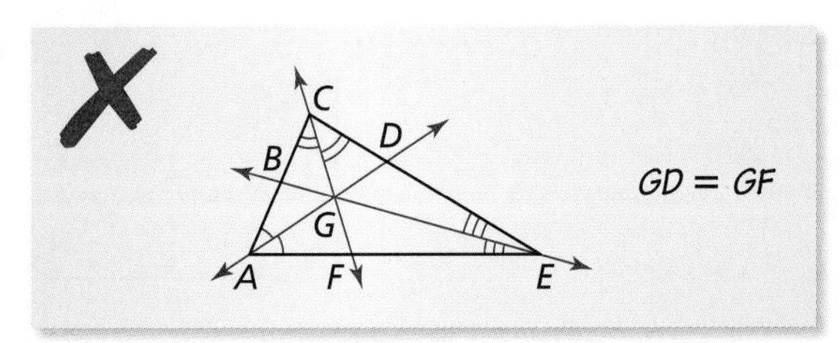

26.

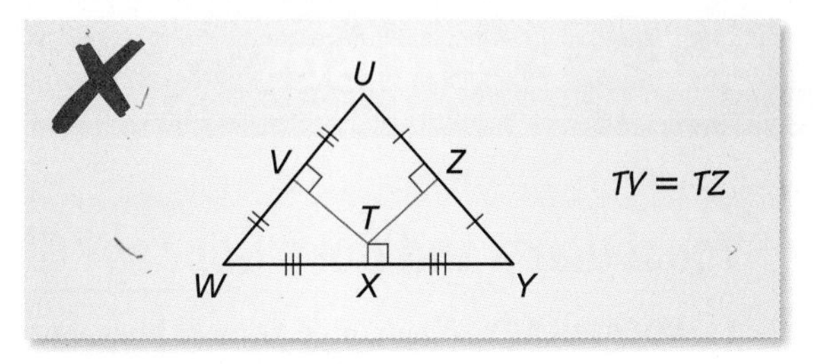

27. **MODELING WITH MATHEMATICS** You and two friends plan to meet to walk your dogs together. You want the meeting place to be the same distance from each person's house. Explain how you can use the diagram to locate the meeting place. *(See Example 1.)*

friend's house

28. **MODELING WITH MATHEMATICS** You are placing a fountain in a triangular koi pond. You want the fountain to be the same distance from each edge of the pond. Where should you place the fountain? Explain your reasoning. Use a sketch to support your answer. *(See Example 4.)*

CRITICAL THINKING In Exercises 29–32, complete the statement with *always*, *sometimes*, or *never*. Explain your reasoning.

29. The circumcenter of a scalene triangle is _________ inside the triangle.

30. If the perpendicular bisector of one side of a triangle intersects the opposite vertex, then the triangle is _________ isosceles.

31. The perpendicular bisectors of a triangle intersect at a point that is _________ equidistant from the midpoints of the sides of the triangle.

32. The angle bisectors of a triangle intersect at a point that is _________ equidistant from the sides of the triangle.

CRITICAL THINKING In Exercises 33 and 34, find the coordinates of the circumcenter of the triangle with the given vertices.

33. $A(2, 5)$, $B(6, 6)$, $C(12, 3)$

34. $D(-9, -5)$, $E(-5, -9)$, $F(-2, -2)$

MATHEMATICAL CONNECTIONS In Exercises 35 and 36, find the value of x that makes N the incenter of the triangle.

35.

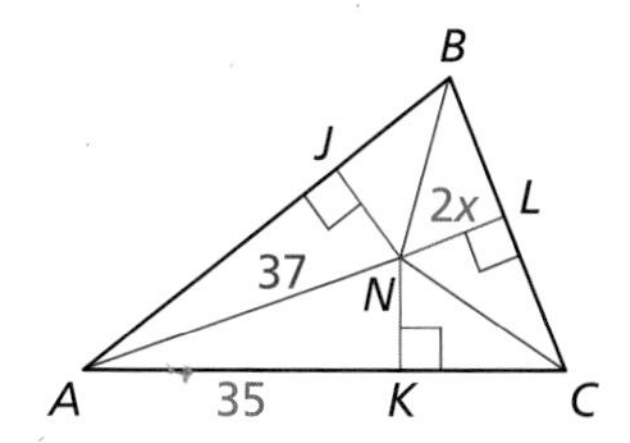

36.

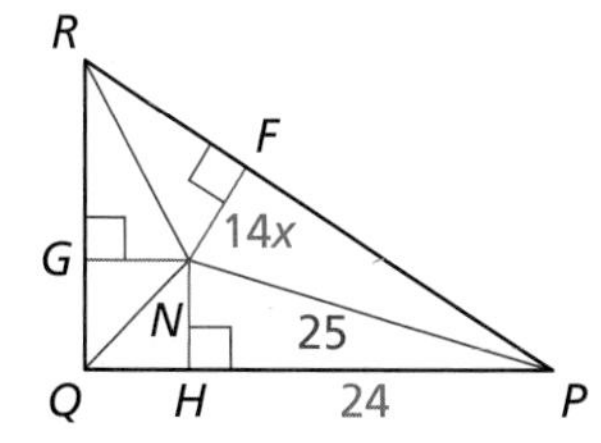

37. PROOF Where is the circumcenter located in any right triangle? Write a coordinate proof of this result.

38. PROVING A THEOREM Write a proof of the Incenter Theorem (Theorem 6.6).

Given $\triangle ABC$, $\overline{AD}$ bisects $\angle CAB$, $\overline{BD}$ bisects $\angle CBA$, $\overline{DE} \perp \overline{AB}$, $\overline{DF} \perp \overline{BC}$, and $\overline{DG} \perp \overline{CA}$

Prove The angle bisectors intersect at D, which is equidistant from $\overline{AB}$, $\overline{BC}$, and $\overline{CA}$.

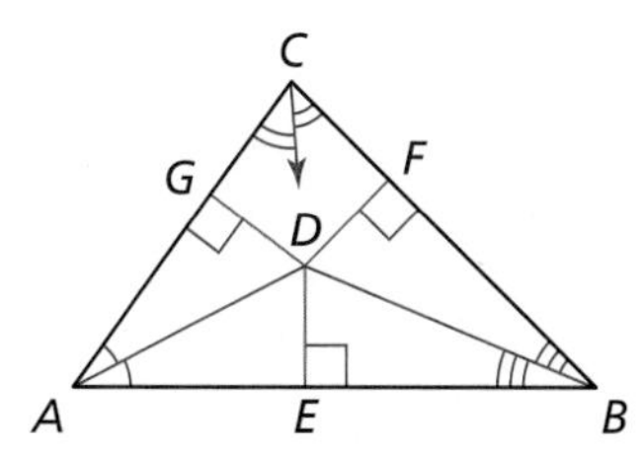

39. WRITING Explain the difference between the circumcenter and the incenter of a triangle.

40. REASONING Is the incenter of a triangle ever located outside the triangle? Explain your reasoning.

41. MODELING WITH MATHEMATICS You are installing a circular pool in the triangular courtyard shown. You want to have the largest pool possible on the site without extending into the walkway.

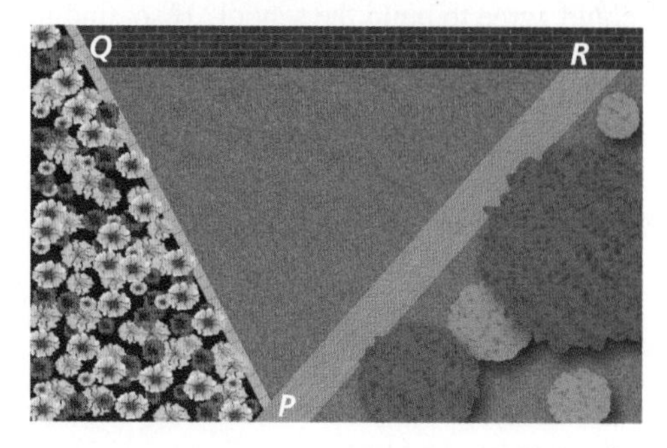

a. Copy the triangle and show how to install the pool so that it just touches each edge. Then explain how you can be sure that you could not fit a larger pool on the site.

b. You want to have the largest pool possible while leaving at least 1 foot of space around the pool. Would the center of the pool be in the same position as in part (a)? Justify your answer.

42. MODELING WITH MATHEMATICS Archaeologists find three stones. They believe that the stones were once part of a circle of stones with a community fire pit at its center. They mark the locations of stones A, B, and C on a graph, where distances are measured in feet.

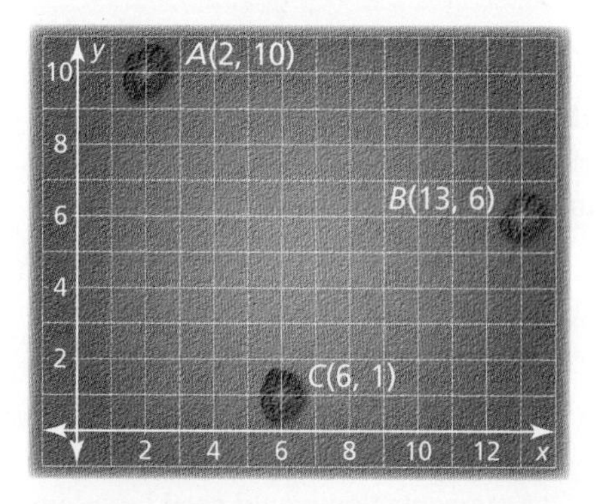

a. Explain how archaeologists can use a sketch to estimate the center of the circle of stones.

b. Copy the diagram and find the approximate coordinates of the point at which the archaeologists should look for the fire pit.

43. REASONING Point P is inside $\triangle ABC$ and is equidistant from points A and B. On which of the following segments must P be located?

Ⓐ $\overline{AB}$

Ⓑ the perpendicular bisector of $\overline{AB}$

Ⓒ $\overline{AC}$

Ⓓ the perpendicular bisector of $\overline{AC}$

44. CRITICAL THINKING A high school is being built for the four towns shown on the map. Each town agrees that the school should be an equal distance from each of the four towns. Is there a single point where they could agree to build the school? If so, find it. If not, explain why not. Justify your answer with a diagram.

45. MAKING AN ARGUMENT Your friend says that the circumcenter of an equilateral triangle is also the incenter of the triangle. Is your friend correct? Explain your reasoning.

46. HOW DO YOU SEE IT? The arms of the windmill are the angle bisectors of the red triangle. What point of concurrency is the point that connects the three arms?

47. ABSTRACT REASONING You are asked to draw a triangle and all its perpendicular bisectors and angle bisectors.

a. For which type of triangle would you need the fewest segments? What is the minimum number of segments you would need? Explain.

b. For which type of triangle would you need the most segments? What is the maximum number of segments you would need? Explain.

48. THOUGHT PROVOKING The diagram shows an official hockey rink used by the National Hockey League. Create a triangle using hockey players as vertices in which the center circle is inscribed in the triangle. The center dot should be the incenter of your triangle. Sketch a drawing of the locations of your hockey players. Then label the actual lengths of the sides and the angle measures in your triangle.

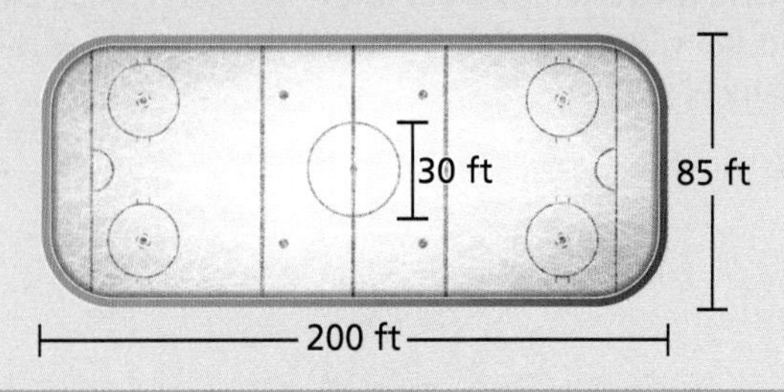

COMPARING METHODS In Exercises 49 and 50, state whether you would use *perpendicular bisectors* or *angle bisectors*. Then solve the problem.

49. You need to cut the largest circle possible from an isosceles triangle made of paper whose sides are 8 inches, 12 inches, and 12 inches. Find the radius of the circle.

50. On a map of a camp, you need to create a circular walking path that connects the pool at (10, 20), the nature center at (16, 2), and the tennis court at (2, 4). Find the coordinates of the center of the circle and the radius of the circle.

51. CRITICAL THINKING Point D is the incenter of $\triangle ABC$. Write an expression for the length x in terms of the three side lengths AB, AC, and BC.

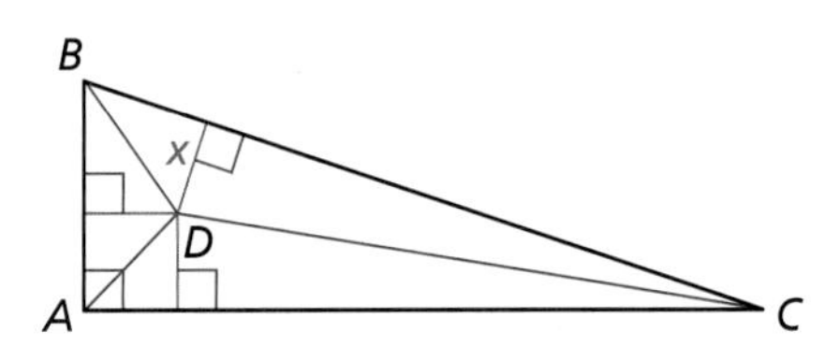

Maintaining Mathematical Proficiency

Reviewing what you learned in previous grades and lessons

The endpoints of $\overline{AB}$ are given. Find the coordinates of the midpoint M. Then find AB. *(Section 1.3)*

52. $A(-3, 5), B(3, 5)$

53. $A(2, -1), B(10, 7)$

54. $A(-5, 1), B(4, -5)$

55. $A(-7, 5), B(5, 9)$

Write an equation of the line passing through point P that is perpendicular to the given line. Graph the equations of the lines to check that they are perpendicular. *(Section 3.5)*

56. $P(2, 8), y = 2x + 1$

57. $P(6, -3), y = -5$

58. $P(-8, -6), 2x + 3y = 18$

59. $P(-4, 1), y + 3 = -4(x + 3)$

6.3 Medians and Altitudes of Triangles

Essential Question What conjectures can you make about the medians and altitudes of a triangle?

EXPLORATION 1 Finding Properties of the Medians of a Triangle

Work with a partner. Use dynamic geometry software. Draw any $\triangle ABC$.

a. Plot the midpoint of $\overline{BC}$ and label it D. Draw $\overline{AD}$, which is a *median* of $\triangle ABC$. Construct the medians to the other two sides of $\triangle ABC$.

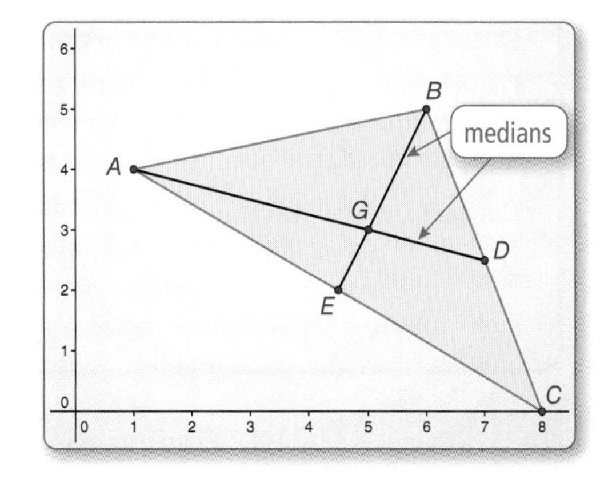

Sample
Points
$A(1, 4)$
$B(6, 5)$
$C(8, 0)$
$D(7, 2.5)$
$E(4.5, 2)$
$G(5, 3)$

b. What do you notice about the medians? Drag the vertices to change $\triangle ABC$. Use your observations to write a conjecture about the medians of a triangle.

c. In the figure above, point G divides each median into a shorter segment and a longer segment. Find the ratio of the length of each longer segment to the length of the whole median. Is this ratio always the same? Justify your answer.

EXPLORATION 2 Finding Properties of the Altitudes of a Triangle

Work with a partner. Use dynamic geometry software. Draw any $\triangle ABC$.

a. Construct the perpendicular segment from vertex A to $\overline{BC}$. Label the endpoint D. $\overline{AD}$ is an *altitude* of $\triangle ABC$.

b. Construct the altitudes to the other two sides of $\triangle ABC$. What do you notice?

c. Write a conjecture about the altitudes of a triangle. Test your conjecture by dragging the vertices to change $\triangle ABC$.

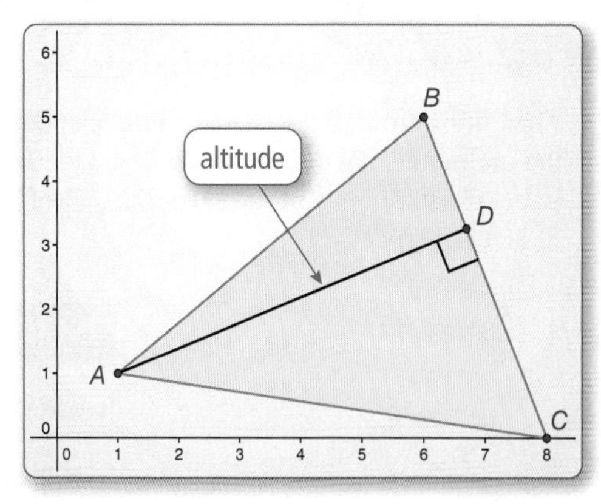

LOOKING FOR STRUCTURE

To be proficient in math, you need to look closely to discern a pattern or structure.

Communicate Your Answer

3. What conjectures can you make about the medians and altitudes of a triangle?

4. The length of median $\overline{RU}$ in $\triangle RST$ is 3 inches. The point of concurrency of the three medians of $\triangle RST$ divides $\overline{RU}$ into two segments. What are the lengths of these two segments?

6.3 Lesson

Core Vocabulary

median of a triangle, *p. 320*
centroid, *p. 320*
altitude of a triangle, *p. 321*
orthocenter, *p. 321*

Previous
midpoint
concurrent
point of concurrency

What You Will Learn

- Use medians and find the centroids of triangles.
- Use altitudes and find the orthocenters of triangles.

Using the Median of a Triangle

A **median of a triangle** is a segment from a vertex to the midpoint of the opposite side. The three medians of a triangle are concurrent. The point of concurrency, called the **centroid**, is inside the triangle.

Theorem

Theorem 6.7 Centroid Theorem

The centroid of a triangle is two-thirds of the distance from each vertex to the midpoint of the opposite side.

The medians of $\triangle ABC$ meet at point P, and $AP = \frac{2}{3}AE$, $BP = \frac{2}{3}BF$, and $CP = \frac{2}{3}CD$.

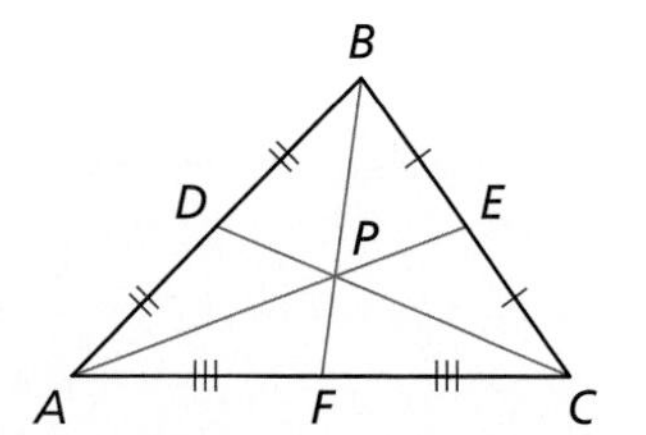

Proof BigIdeasMath.com

CONSTRUCTION Finding the Centroid of a Triangle

Use a compass and straightedge to construct the medians of $\triangle ABC$.

SOLUTION

Step 1

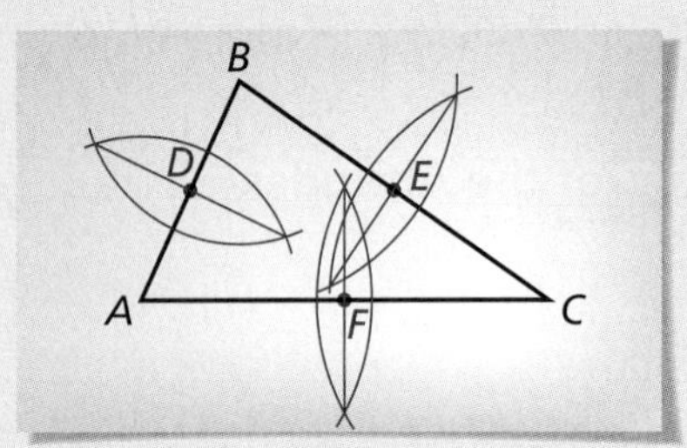

Find midpoints Draw $\triangle ABC$. Find the midpoints of $\overline{AB}$, $\overline{BC}$, and $\overline{AC}$. Label the midpoints of the sides D, E, and F, respectively.

Step 2

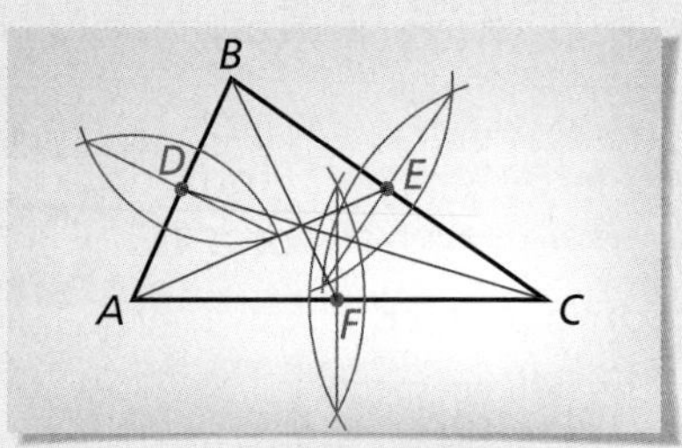

Draw medians Draw $\overline{AE}$, $\overline{BF}$, and $\overline{CD}$. These are the three medians of $\triangle ABC$.

Step 3

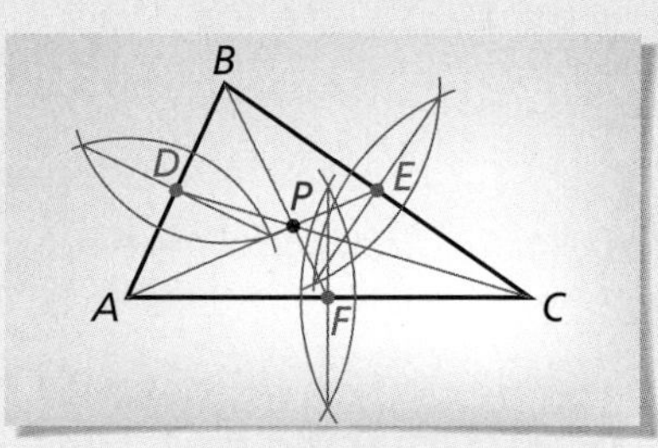

Label a point Label the point where $\overline{AE}$, $\overline{BF}$, and $\overline{CD}$ intersect as P. This is the centroid.

EXAMPLE 1 Using the Centroid of a Triangle

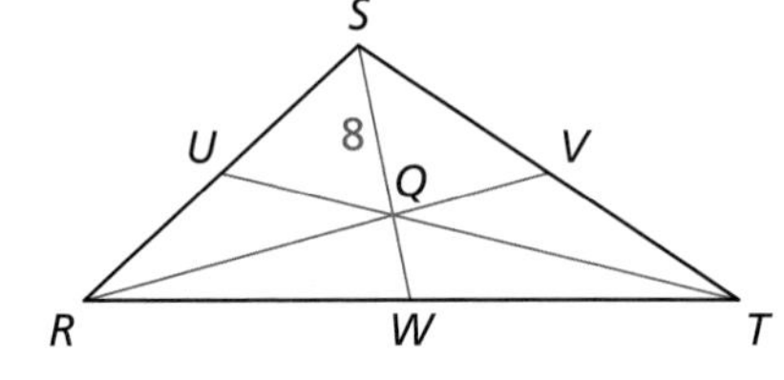

In $\triangle RST$, point Q is the centroid, and $SQ = 8$. Find QW and SW.

SOLUTION

$SQ = \frac{2}{3}SW$	Centroid Theorem
$8 = \frac{2}{3}SW$	Substitute 8 for SQ.
$12 = SW$	Multiply each side by the reciprocal, $\frac{3}{2}$.

Then $QW = SW - SQ = 12 - 8 = 4$.

- So, $QW = 4$ and $SW = 12$.

FINDING AN ENTRY POINT

The median $\overline{SV}$ is chosen in Example 2 because it is easier to find a distance on a vertical segment.

JUSTIFYING CONCLUSIONS

You can check your result by using a different median to find the centroid.

READING

In the area formula for a triangle, $A = \frac{1}{2}bh$, you can use the length of any side for the base b. The height h is the length of the altitude to that side from the opposite vertex.

EXAMPLE 2 Finding the Centroid of a Triangle

Find the coordinates of the centroid of $\triangle RST$ with vertices $R(2, 1)$, $S(5, 8)$, and $T(8, 3)$.

SOLUTION

Step 1 Graph $\triangle RST$.

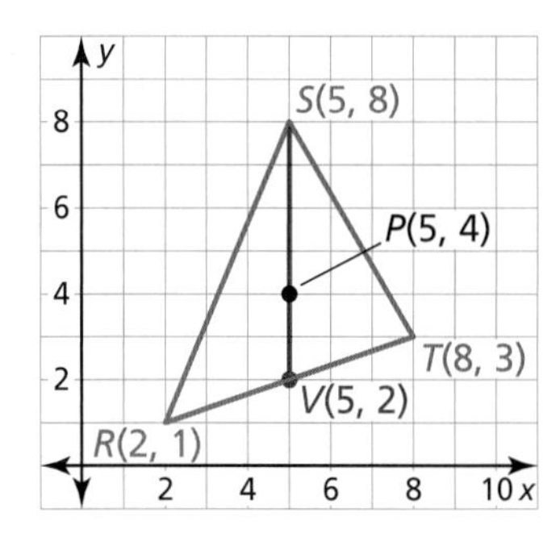

Step 2 Use the Midpoint Formula to find the midpoint V of $\overline{RT}$ and sketch median $\overline{SV}$.

$$V\left(\frac{2+8}{2}, \frac{1+3}{2}\right) = (5, 2)$$

Step 3 Find the centroid. It is two-thirds of the distance from each vertex to the midpoint of the opposite side.

The distance from vertex $S(5, 8)$ to $V(5, 2)$ is $8 - 2 = 6$ units.
So, the centroid is $\frac{2}{3}(6) = 4$ units down from vertex S on $\overline{SV}$.

▶ So, the coordinates of the centroid P are $(5, 8 - 4)$, or $(5, 4)$.

Monitoring Progress  Help in English and Spanish at *BigIdeasMath.com*

There are three paths through a triangular park. Each path goes from the midpoint of one edge to the opposite corner. The paths meet at point *P*.

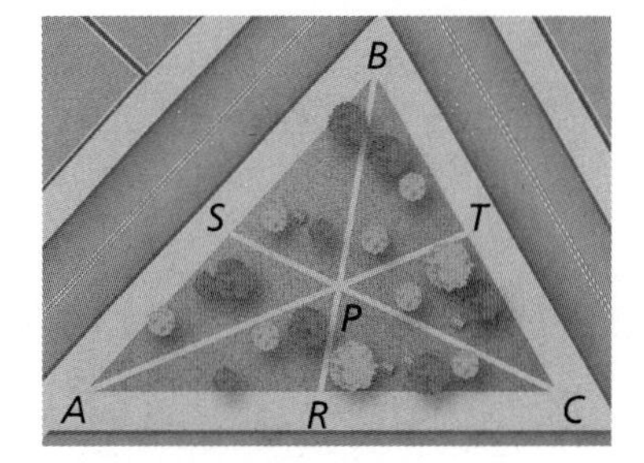

1. Find PS and PC when $SC = 2100$ feet.
2. Find TC and BC when $BT = 1000$ feet.
3. Find PA and TA when $PT = 800$ feet.

Find the coordinates of the centroid of the triangle with the given vertices.

4. $F(2, 5)$, $G(4, 9)$, $H(6, 1)$
5. $X(-3, 3)$, $Y(1, 5)$, $Z(-1, -2)$

Using the Altitude of a Triangle

An **altitude of a triangle** is the perpendicular segment from a vertex to the opposite side or to the line that contains the opposite side.

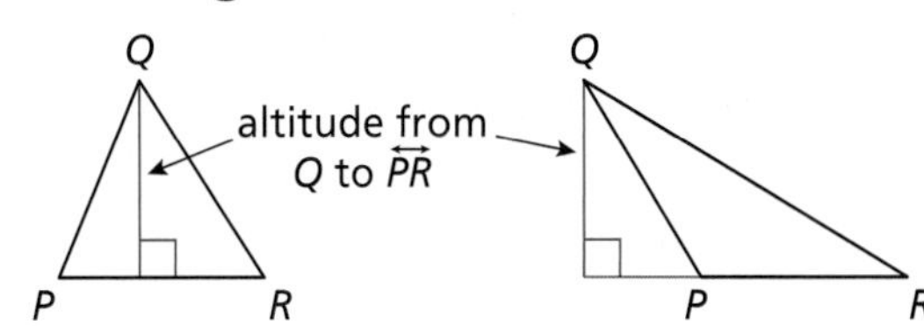

Core Concept

Orthocenter

The lines containing the altitudes of a triangle are concurrent. This point of concurrency is the **orthocenter** of the triangle.

The lines containing $\overline{AF}$, $\overline{BD}$, and $\overline{CE}$ meet at the orthocenter G of $\triangle ABC$.

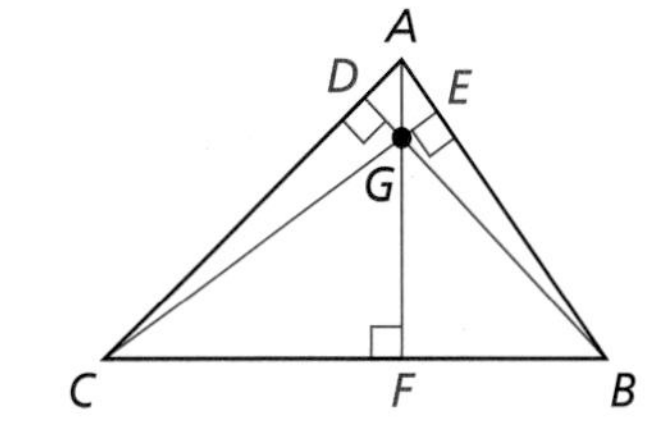

As shown below, the location of the orthocenter P of a triangle depends on the type of triangle.

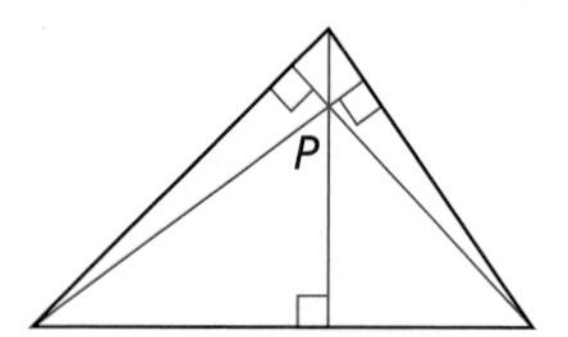

Acute triangle
P is inside triangle.

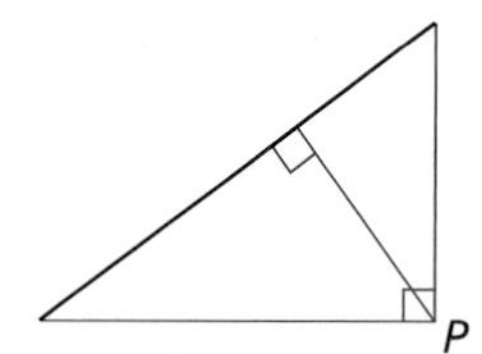

Right triangle
P is on triangle.

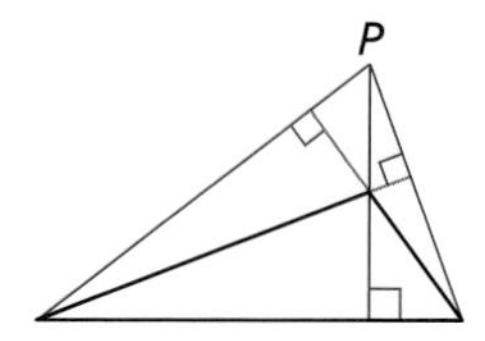

Obtuse triangle
P is outside triangle.

READING

The altitudes are shown in red. Notice that in the right triangle, the legs are also altitudes. The altitudes of the obtuse triangle are extended to find the orthocenter.

EXAMPLE 3 Finding the Orthocenter of a Triangle

Find the coordinates of the orthocenter of $\triangle XYZ$ with vertices $X(-5, -1)$, $Y(-2, 4)$, and $Z(3, -1)$.

SOLUTION

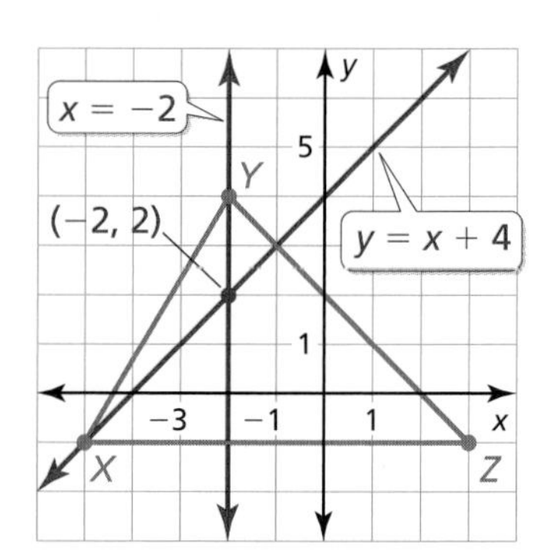

Step 1 Graph $\triangle XYZ$.

Step 2 Find an equation of the line that contains the altitude from Y to $\overline{XZ}$. Because $\overline{XZ}$ is horizontal, the altitude is vertical. The line that contains the altitude passes through $Y(-2, 4)$. So, the equation of the line is $x = -2$.

Step 3 Find an equation of the line that contains the altitude from X to $\overline{YZ}$.

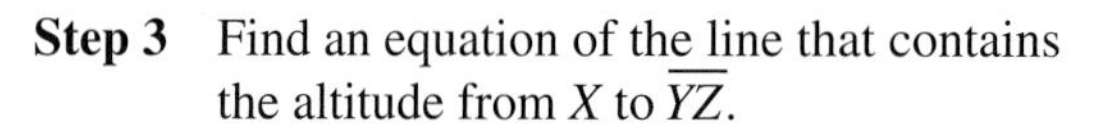

$$\text{slope of } \overleftrightarrow{YZ} = \frac{-1 - 4}{3 - (-2)} = -1$$

Because the product of the slopes of two perpendicular lines is -1, the slope of a line perpendicular to $\overleftrightarrow{YZ}$ is 1. The line passes through $X(-5, -1)$.

$y = mx + b$	Use slope-intercept form.
$-1 = 1(-5) + b$	Substitute -1 for y, 1 for m, and -5 for x.
$4 = b$	Solve for b.

So, the equation of the line is $y = x + 4$.

Step 4 Find the point of intersection of the graphs of the equations $x = -2$ and $y = x + 4$.

Substitute -2 for x in the equation $y = x + 4$. Then solve for y.

$y = x + 4$	Write equation.
$y = -2 + 4$	Substitute -2 for x.
$y = 2$	Solve for y.

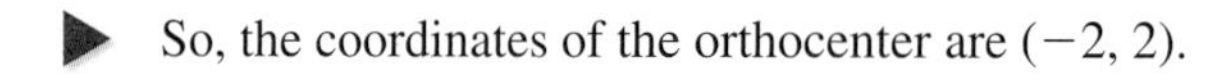

▶ So, the coordinates of the orthocenter are $(-2, 2)$.

Monitoring Progress

Help in English and Spanish at *BigIdeasMath.com*

Tell whether the orthocenter of the triangle with the given vertices is *inside*, *on*, or *outside* the triangle. Then find the coordinates of the orthocenter.

6. $A(0, 3)$, $B(0, -2)$, $C(6, -3)$

7. $J(-3, -4)$, $K(-3, 4)$, $L(5, 4)$

In an isosceles triangle, the perpendicular bisector, angle bisector, median, and altitude from the vertex angle to the base are all the same segment. In an equilateral triangle, this is true for any vertex.

EXAMPLE 4 Proving a Property of Isosceles Triangles

Prove that the median from the vertex angle to the base of an isosceles triangle is an altitude.

SOLUTION

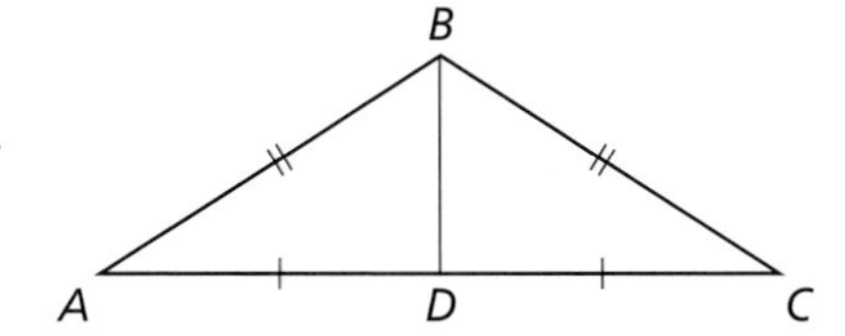

Given $\triangle ABC$ is isosceles, with base $\overline{AC}$.
$\overline{BD}$ is the median to base $\overline{AC}$.

Prove $\overline{BD}$ is an altitude of $\triangle ABC$.

Paragraph Proof Legs $\overline{AB}$ and $\overline{BC}$ of isosceles $\triangle ABC$ are congruent. $\overline{CD} \cong \overline{AD}$ because $\overline{BD}$ is the median to $\overline{AC}$. Also, $\overline{BD} \cong \overline{BD}$ by the Reflexive Property of Congruence (Thm. 2.1). So, $\triangle ABD \cong \triangle CBD$ by the SSS Congruence Theorem (Thm. 5.8). $\angle ADB \cong \angle CDB$ because corresponding parts of congruent triangles are congruent. Also, $\angle ADB$ and $\angle CDB$ are a linear pair. $\overleftrightarrow{BD}$ and $\overleftrightarrow{AC}$ intersect to form a linear pair of congruent angles, so $\overline{BD} \perp \overline{AC}$ and $\overline{BD}$ is an altitude of $\triangle ABC$.

Monitoring Progress

Help in English and Spanish at *BigIdeasMath.com*

8. WHAT IF? In Example 4, you want to show that median $\overline{BD}$ is also an angle bisector. How would your proof be different?

Concept Summary

Segments, Lines, Rays, and Points in Triangles

	Example	Point of Concurrency	Property	Example
perpendicular bisector		circumcenter	The circumcenter P of a triangle is equidistant from the vertices of the triangle.	
angle bisector		incenter	The incenter I of a triangle is equidistant from the sides of the triangle.	
median		centroid	The centroid R of a triangle is two thirds of the distance from each vertex to the midpoint of the opposite side.	
altitude		orthocenter	The lines containing the altitudes of a triangle are concurrent at the orthocenter O.	

6.3 Exercises

Dynamic Solutions available at *BigIdeasMath.com*

Vocabulary and Core Concept Check

1. **VOCABULARY** Name the four types of points of concurrency. Which lines intersect to form each of the points?

2. **COMPLETE THE SENTENCE** The length of a segment from a vertex to the centroid is ____________ the length of the median from that vertex.

Monitoring Progress and Modeling with Mathematics

In Exercises 3–6, point P is the centroid of $\triangle LMN$. Find PN and QP. *(See Example 1.)*

3. $QN = 9$

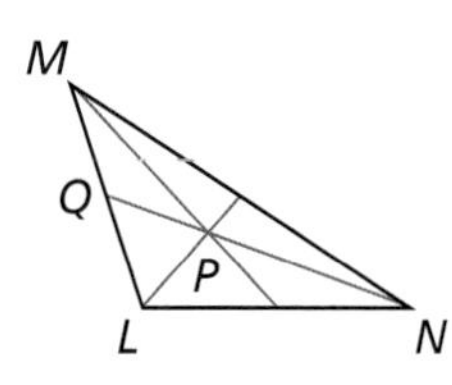

4. $QN = 21$

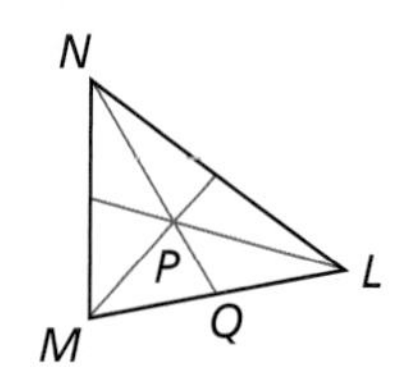

5. $QN = 30$

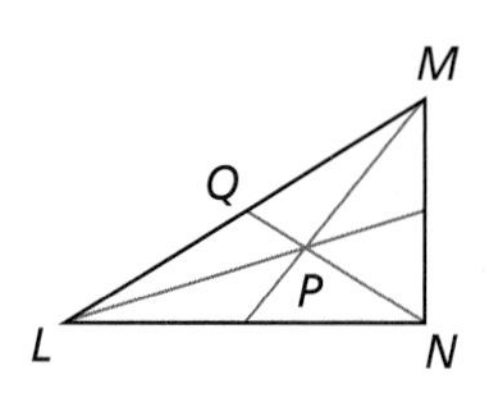

6. $QN = 42$

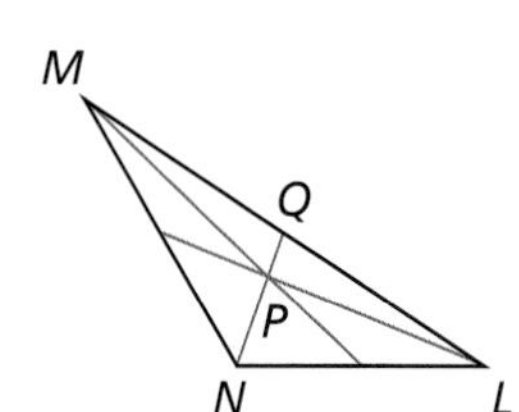

In Exercises 7–10, point D is the centroid of $\triangle ABC$. Find CD and CE.

7. $DE = 5$

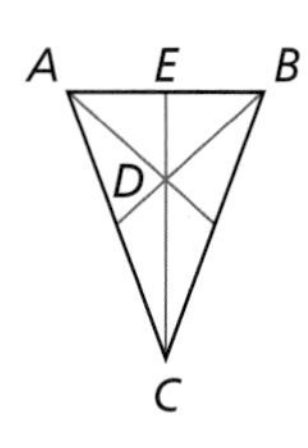

8. $DE = 11$

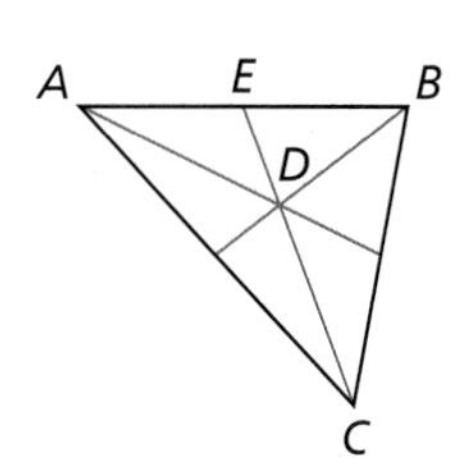

9. $DE = 9$

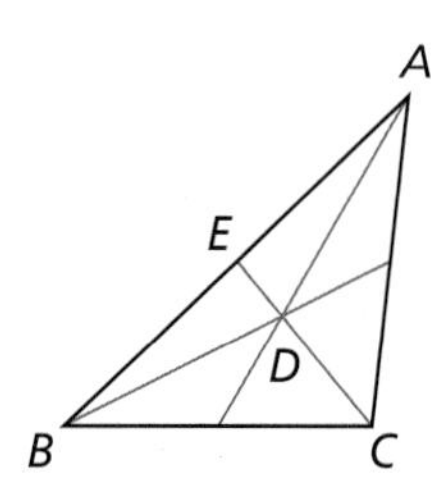

10. $DE = 15$

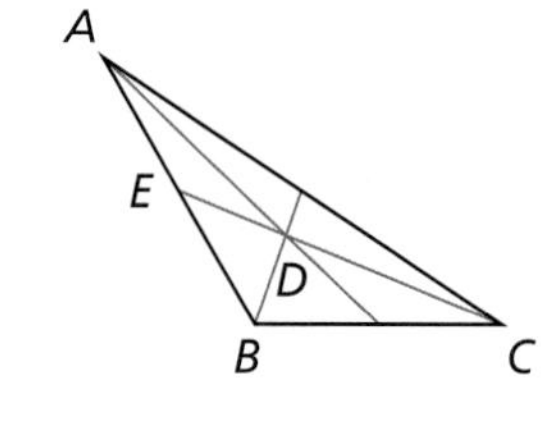

In Exercises 11–14, point G is the centroid of $\triangle ABC$. $BG = 6$, $AF = 12$, and $AE = 15$. Find the length of the segment.

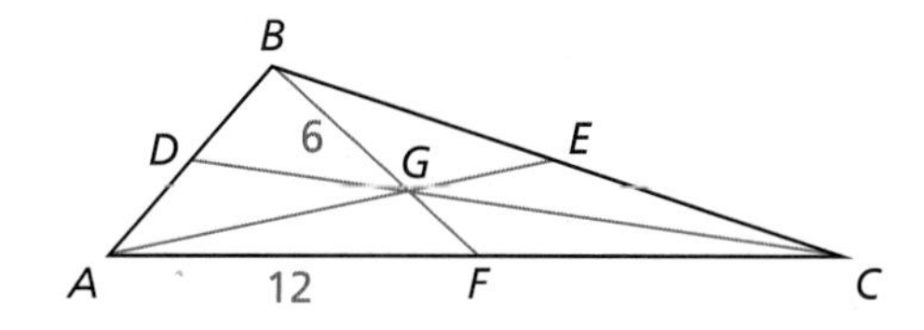

11. $\overline{FC}$

12. $\overline{BF}$

13. $\overline{AG}$

14. $\overline{GE}$

In Exercises 15–18, find the coordinates of the centroid of the triangle with the given vertices. *(See Example 2.)*

15. $A(2, 3)$, $B(8, 1)$, $C(5, 7)$

16. $F(1, 5)$, $G(-2, 7)$, $H(-6, 3)$

17. $S(5, 5)$, $T(11, -3)$, $U(-1, 1)$

18. $X(1, 4)$, $Y(7, 2)$, $Z(2, 3)$

In Exercises 19–22, tell whether the orthocenter is *inside*, *on*, or *outside* the triangle. Then find the coordinates of the orthocenter. *(See Example 3.)*

19. $L(0, 5)$, $M(3, 1)$, $N(8, 1)$

20. $X(-3, 2)$, $Y(5, 2)$, $Z(-3, 6)$

21. $A(-4, 0)$, $B(1, 0)$, $C(-1, 3)$

22. $T(-2, 1)$, $U(2, 1)$, $V(0, 4)$

CONSTRUCTION In Exercises 23–26, draw the indicated triangle and find its centroid and orthocenter.

23. isosceles right triangle

24. obtuse scalene triangle

25. right scalene triangle

26. acute isosceles triangle

ERROR ANALYSIS **In Exercises 27 and 28, describe and correct the error in finding DE. Point D is the centroid of $\triangle ABC$.**

27.

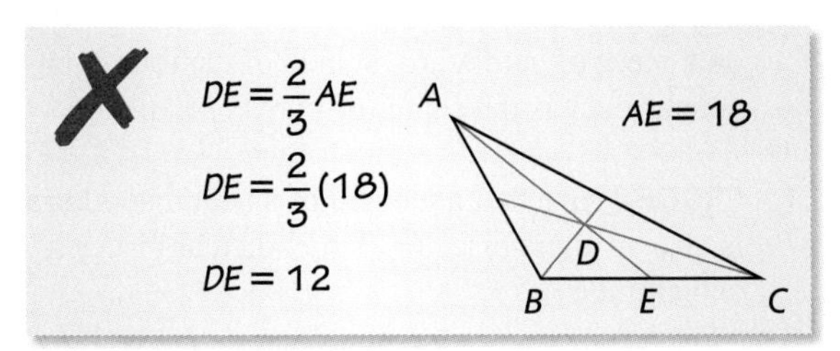

28.

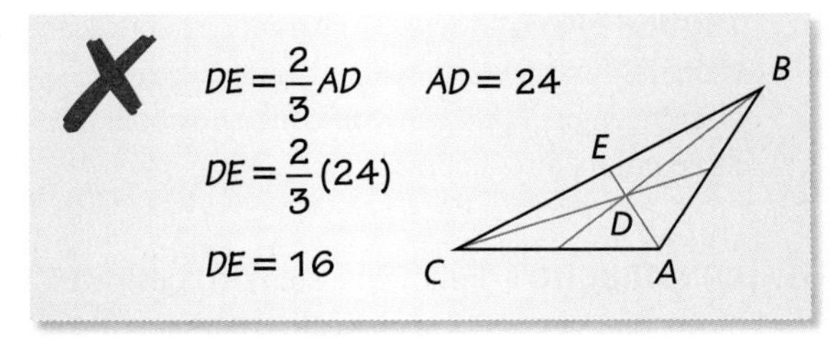

PROOF **In Exercises 29 and 30, write a proof of the statement.** *(See Example 4.)*

29. The angle bisector from the vertex angle to the base of an isosceles triangle is also a median.

30. The altitude from the vertex angle to the base of an isosceles triangle is also a perpendicular bisector.

CRITICAL THINKING **In Exercises 31–36, complete the statement with *always*, *sometimes*, or *never*. Explain your reasoning.**

31. The centroid is __________ on the triangle.

32. The orthocenter is __________ outside the triangle.

33. A median is __________ the same line segment as a perpendicular bisector.

34. An altitude is __________ the same line segment as an angle bisector.

35. The centroid and orthocenter are __________ the same point.

36. The centroid is __________ formed by the intersection of the three medians.

37. **WRITING** Compare an altitude of a triangle with a perpendicular bisector of a triangle.

38. **WRITING** Compare a median, an altitude, and an angle bisector of a triangle.

39. **MODELING WITH MATHEMATICS** Find the area of the triangular part of the paper airplane wing that is outlined in red. Which special segment of the triangle did you use?

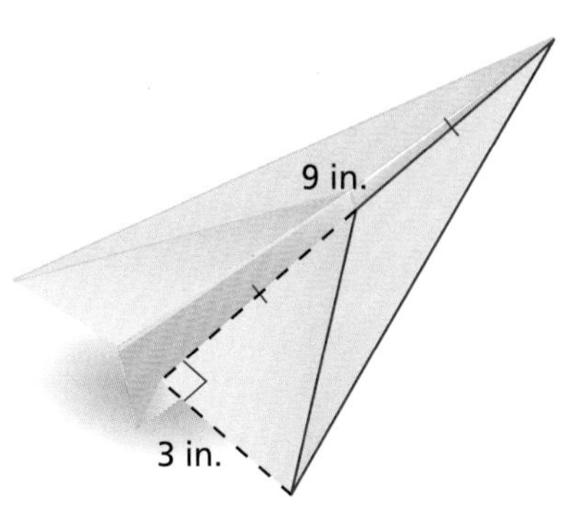

40. **ANALYZING RELATIONSHIPS** Copy and complete the statement for $\triangle DEF$ with centroid K and medians $\overline{DH}$, $\overline{EJ}$, and $\overline{FG}$.

a. $EJ =$ _____ KJ **b.** $DK =$ _____ KH

c. $FG =$ _____ KF **d.** $KG =$ _____ FG

MATHEMATICAL CONNECTIONS **In Exercises 41–44, point D is the centroid of $\triangle ABC$. Use the given information to find the value of x.**

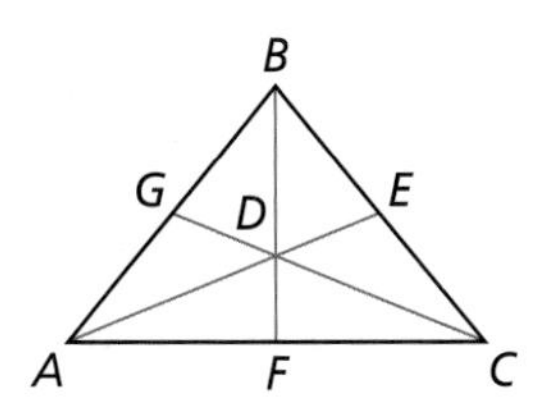

41. $BD = 4x + 5$ and $BF = 9x$

42. $GD = 2x - 8$ and $GC = 3x + 3$

43. $AD = 5x$ and $DE = 3x - 2$

44. $DF = 4x - 1$ and $BD = 6x + 4$

45. **MATHEMATICAL CONNECTIONS** Graph the lines on the same coordinate plane. Find the centroid of the triangle formed by their intersections.

$$y_1 = 3x - 4$$

$$y_2 = \frac{3}{4}x + 5$$

$$y_3 = -\frac{3}{2}x - 4$$

46. **CRITICAL THINKING** In what type(s) of triangles can a vertex be one of the points of concurrency of the triangle? Explain your reasoning.

47. WRITING EQUATIONS Use the numbers and symbols to write three different equations for PE.

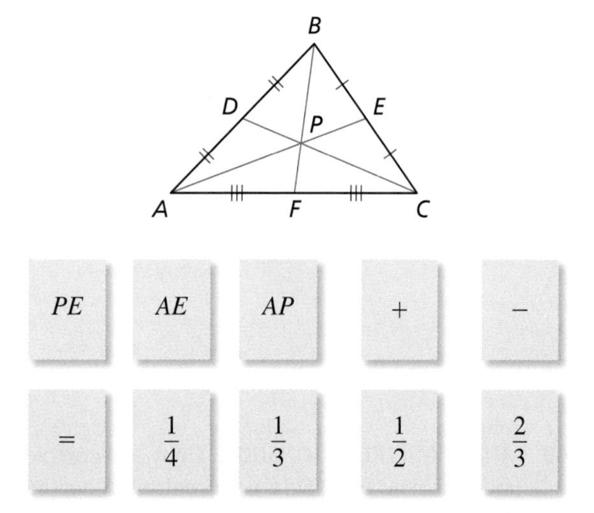

48. HOW DO YOU SEE IT? Use the figure.

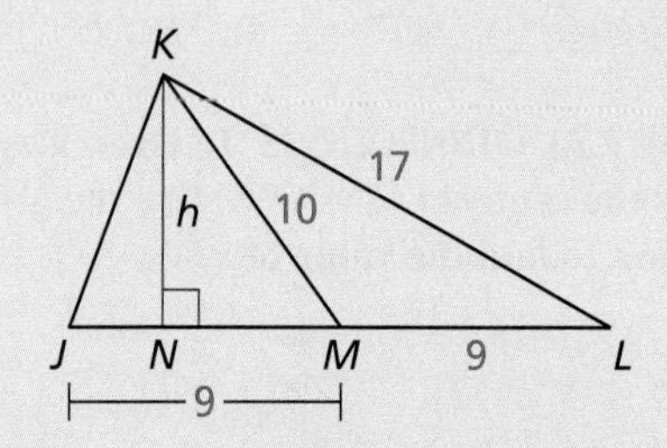

a. What type of segment is $\overline{KM}$? Which point of concurrency lies on $\overline{KM}$?

b. What type of segment is $\overline{KN}$? Which point of concurrency lies on $\overline{KN}$?

c. Compare the areas of $\triangle JKM$ and $\triangle KLM$. Do you think the areas of the triangles formed by the median of any triangle will always compare this way? Explain your reasoning.

49. MAKING AN ARGUMENT Your friend claims that it is possible for the circumcenter, incenter, centroid, and orthocenter to all be the same point. Do you agree? Explain your reasoning.

50. DRAWING CONCLUSIONS The center of gravity of a triangle, the point where a triangle can balance on the tip of a pencil, is one of the four points of concurrency. Draw and cut out a large scalene triangle on a piece of cardboard. Which of the four points of concurrency is the center of gravity? Explain.

51. PROOF Prove that a median of an equilateral triangle is also an angle bisector, perpendicular bisector, and altitude.

52. THOUGHT PROVOKING Construct an acute scalene triangle. Find the orthocenter, centroid, and circumcenter. What can you conclude about the three points of concurrency?

53. CONSTRUCTION Follow the steps to construct a nine-point circle. Why is it called a nine-point circle?

Step 1 Construct a large acute scalene triangle.

Step 2 Find the orthocenter and circumcenter of the triangle.

Step 3 Find the midpoint between the orthocenter and circumcenter.

Step 4 Find the midpoint between each vertex and the orthocenter.

Step 5 Construct a circle. Use the midpoint in Step 3 as the center of the circle, and the distance from the center to the midpoint of a side of the triangle as the radius.

54. PROOF Prove the statements in parts (a)−(c).

Given $\overline{LP}$ and $\overline{MQ}$ are medians of scalene $\triangle LMN$. Point R is on $\overrightarrow{LP}$ such that $\overline{LP} \cong \overline{PR}$. Point S is on $\overrightarrow{MQ}$ such that $\overline{MQ} \cong \overline{QS}$.

Prove **a.** $\overline{NS} \cong \overline{NR}$

b. $\overline{NS}$ and $\overline{NR}$ are both parallel to $\overline{LM}$.

c. R, N, and S are collinear.

Maintaining Mathematical Proficiency

Reviewing what you learned in previous grades and lessons

Determine whether $\overline{AB}$ is parallel to $\overline{CD}$. *(Section 3.5)*

55. $A(5, 6)$, $B(-1, 3)$, $C(-4, 9)$, $D(-16, 3)$

56. $A(-3, 6)$, $B(5, 4)$, $C(-14, -10)$, $D(-2, -7)$

57. $A(6, -3)$, $B(5, 2)$, $C(-4, -4)$, $D(-5, 2)$

58. $A(-5, 6)$, $B(-7, 2)$, $C(7, 1)$, $D(4, -5)$

6.1–6.3 What Did You Learn?

Core Vocabulary

equidistant, *p. 302*
concurrent, *p. 310*
point of concurrency, *p. 310*
circumcenter, *p. 310*
incenter, *p. 313*
median of a triangle, *p. 320*
centroid, *p. 320*
altitude of a triangle, *p. 321*
orthocenter, *p. 321*

Core Concepts

Section 6.1

Theorem 6.1 Perpendicular Bisector Theorem, *p. 302*
Theorem 6.2 Converse of the Perpendicular Bisector Theorem, *p. 302*
Theorem 6.3 Angle Bisector Theorem, *p. 304*
Theorem 6.4 Converse of the Angle Bisector Theorem, *p. 304*

Section 6.2

Theorem 6.5 Circumcenter Theorem, *p. 310*
Theorem 6.6 Incenter Theorem, *p. 313*

Section 6.3

Theorem 6.7 Centroid Theorem, *p. 320*
Orthocenter, *p. 321*
Segments, Lines, Rays, and Points in Triangles, *p. 323*

Mathematical Practices

1. Did you make a plan before completing your proof in Exercise 37 on page 308? Describe your thought process.
2. What tools did you use to complete Exercises 17–20 on page 316? Describe how you could use technological tools to complete these exercises.
3. What conjecture did you make when answering Exercise 46 on page 325? What logical progression led you to determine whether your conjecture was true?

Study Skills

Rework Your Notes

A good way to reinforce concepts and put them into your long-term memory is to rework your notes. When you take notes, leave extra space on the pages. You can go back after class and fill in

- important definitions and rules,
- additional examples, and
- questions you have about the material.

6.1–6.3 Quiz

Find the indicated measure. Explain your reasoning. *(Section 6.1)*

1. UV

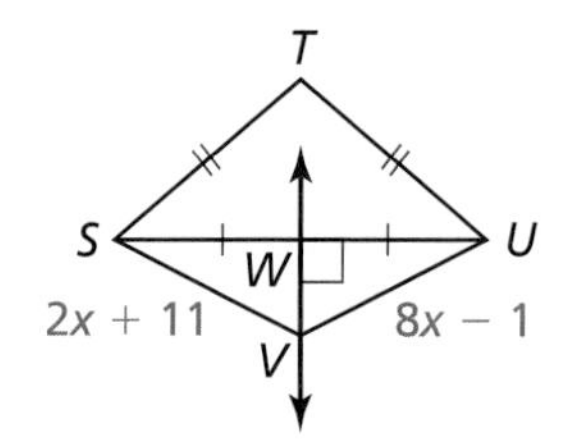

2. QP

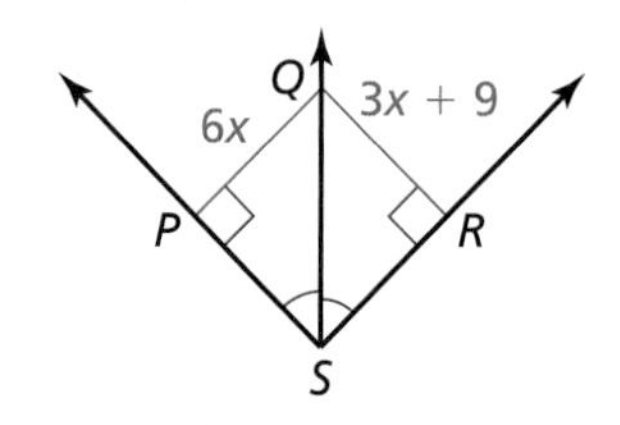

3. $m\angle GJK$

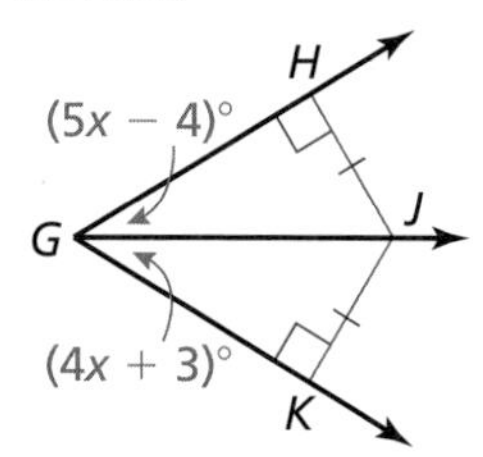

Find the coordinates of the circumcenter of the triangle with the given vertices. *(Section 6.2)*

4. $A(-4, 2)$, $B(-4, -4)$, $C(0, -4)$

5. $D(3, 5)$, $E(7, 9)$, $F(11, 5)$

The incenter of $\triangle ABC$ is point N. Use the given information to find the indicated measure. *(Section 6.2)*

6. $NQ = 2x + 1$, $NR = 4x - 9$
Find NS.

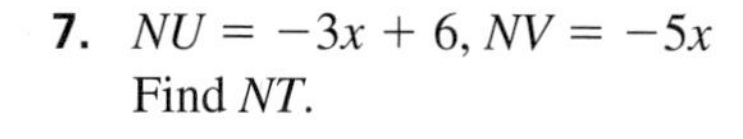

7. $NU = -3x + 6$, $NV = -5x$
Find NT.

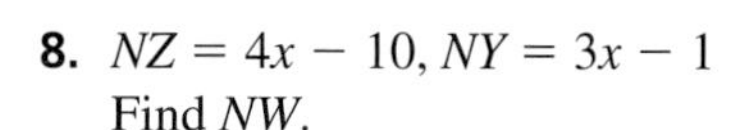

8. $NZ = 4x - 10$, $NY = 3x - 1$
Find NW.

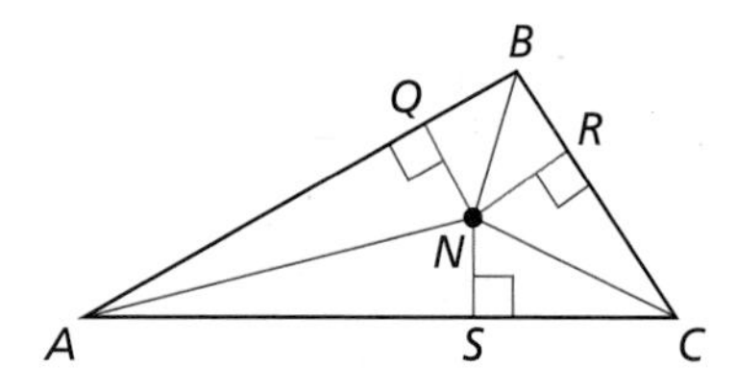

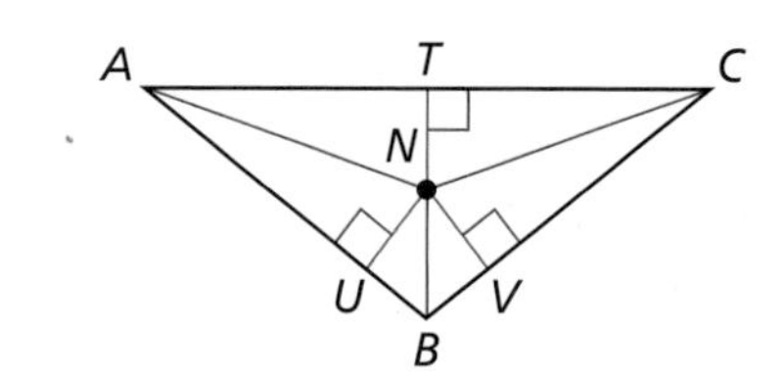

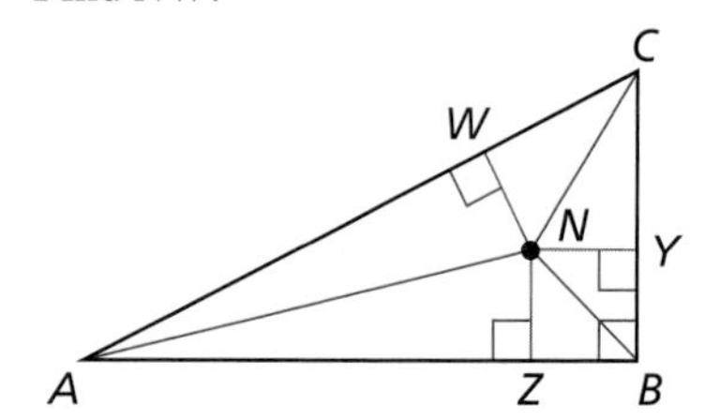

Find the coordinates of the centroid of the triangle with the given vertices. *(Section 6.3)*

9. $J(-1, 2)$, $K(5, 6)$, $L(5, -2)$

10. $M(-8, -6)$, $N(-4, -2)$, $P(0, -4)$

Tell whether the orthocenter is *inside*, *on*, or *outside* the triangle. Then find its coordinates. *(Section 6.3)*

11. $T(-2, 5)$, $U(0, 1)$, $V(2, 5)$

12. $X(-1, -4)$, $Y(7, -4)$, $Z(7, 4)$

13. A woodworker is cutting the largest wheel possible from a triangular scrap of wood. The wheel just touches each side of the triangle, as shown. *(Section 6.2)*

a. Which point of concurrency is the center of the circle? What type of segments are $\overline{BG}$, $\overline{CG}$, and $\overline{AG}$?

b. Which theorem can you use to prove that $\triangle BGF \cong \triangle BGE$?

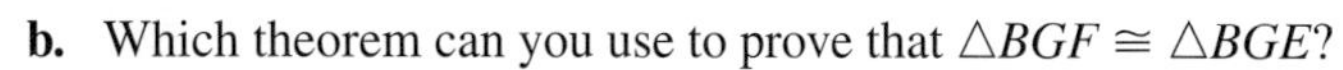

c. Find the radius of the wheel to the nearest tenth of a centimeter. Justify your answer.

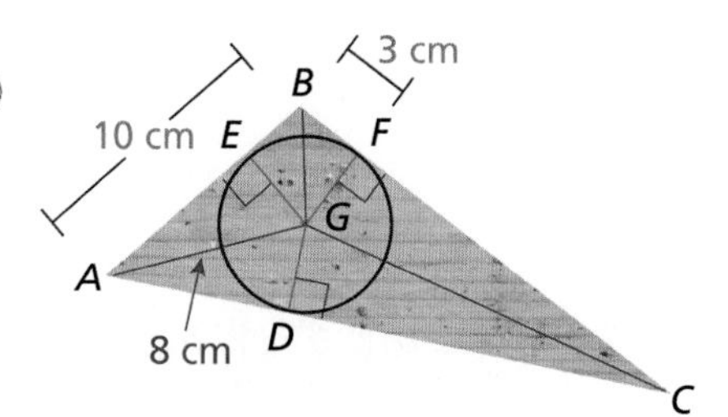

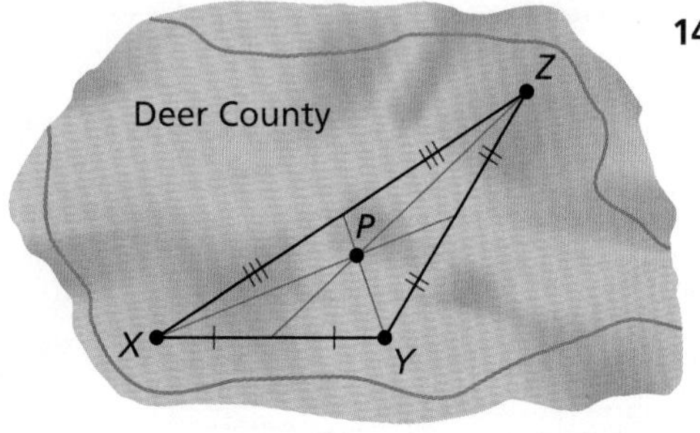

14. The Deer County Parks Committee plans to build a park at point P, equidistant from the three largest cities labeled X, Y, and Z. The map shown was created by the committee. *(Section 6.2 and Section 6.3)*

a. Which point of concurrency did the committee use as the location of the park?

b. Did the committee use the best point of concurrency for the location of the park? If not, which point would be better to use? Explain.

6.4 The Triangle Midsegment Theorem

Essential Question How are the midsegments of a triangle related to the sides of the triangle?

EXPLORATION 1 **Midsegments of a Triangle**

Work with a partner. Use dynamic geometry software. Draw any $\triangle ABC$.

a. Plot midpoint D of $\overline{AB}$ and midpoint E of $\overline{BC}$. Draw $\overline{DE}$, which is a *midsegment* of $\triangle ABC$.

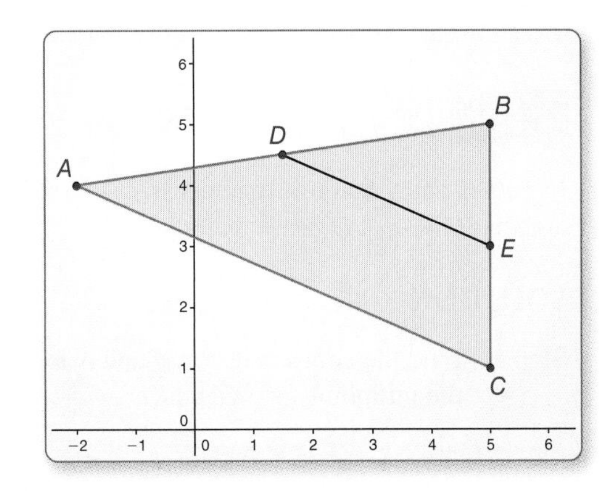

Sample
Points
$A(-2, 4)$
$B(5, 5)$
$C(5, 1)$
$D(1.5, 4.5)$
$E(5, 3)$
Segments
$BC = 4$
$AC = 7.62$
$AB = 7.07$
$DE = ?$

b. Compare the slope and length of $\overline{DE}$ with the slope and length of $\overline{AC}$.

c. Write a conjecture about the relationships between the midsegments and sides of a triangle. Test your conjecture by drawing the other midsegments of $\triangle ABC$, dragging vertices to change $\triangle ABC$, and noting whether the relationships hold.

CONSTRUCTING VIABLE ARGUMENTS

To be proficient in math, you need to make conjectures and build a logical progression of statements to explore the truth of your conjectures.

EXPLORATION 2 **Midsegments of a Triangle**

Work with a partner. Use dynamic geometry software. Draw any $\triangle ABC$.

a. Draw all three midsegments of $\triangle ABC$.

b. Use the drawing to write a conjecture about the triangle formed by the midsegments of the original triangle.

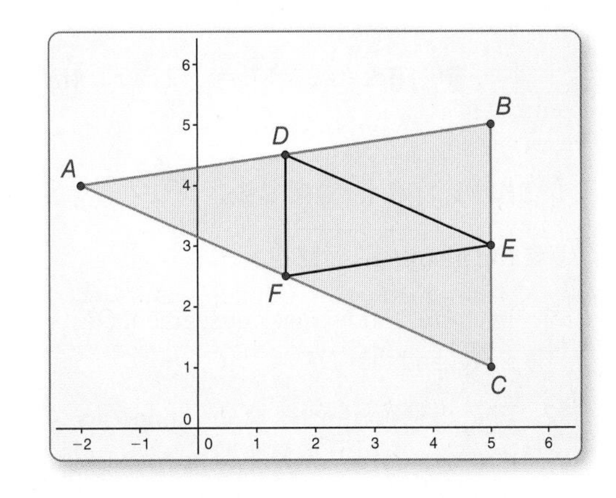

Sample

Points	Segments
$A(-2, 4)$	$BC = 4$
$B(5, 5)$	$AC = 7.62$
$C(5, 1)$	$AB = 7.07$
$D(1.5, 4.5)$	$DE = ?$
$E(5, 3)$	$DF = ?$
	$EF = ?$

Communicate Your Answer

3. How are the midsegments of a triangle related to the sides of the triangle?

4. In $\triangle RST$, $\overline{UV}$ is the midsegment connecting the midpoints of $\overline{RS}$ and $\overline{ST}$. Given $UV = 12$, find RT.

6.4 Lesson

Core Vocabulary

midsegment of a triangle, *p. 330*

Previous
midpoint
parallel
slope
coordinate proof

What You Will Learn

- Use midsegments of triangles in the coordinate plane.
- Use the Triangle Midsegment Theorem to find distances.

Using the Midsegment of a Triangle

A **midsegment of a triangle** is a segment that connects the midpoints of two sides of the triangle. Every triangle has three midsegments, which form the *midsegment triangle*.

The midsegments of $\triangle ABC$ at the right are $\overline{MP}$, $\overline{MN}$, and $\overline{NP}$. The *midsegment triangle* is $\triangle MNP$.

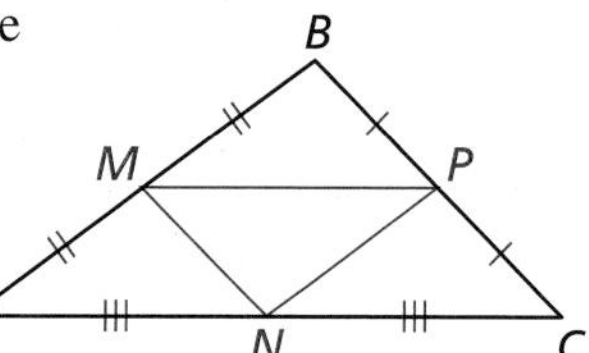

EXAMPLE 1 Using Midsegments in the Coordinate Plane

In $\triangle JKL$, show that midsegment $\overline{MN}$ is parallel to $\overline{JL}$ and that $MN = \frac{1}{2}JL$.

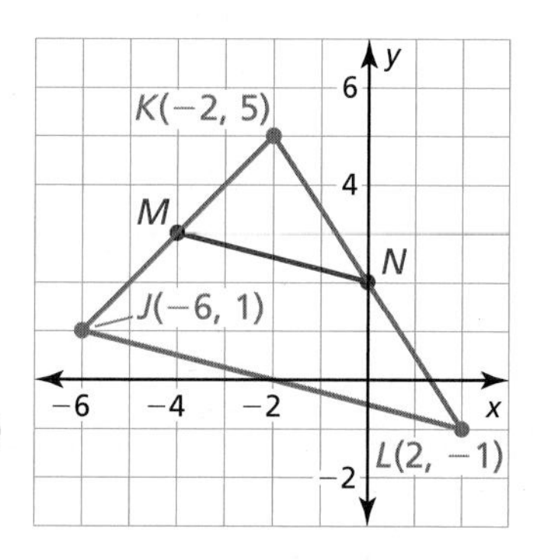

READING

In the figure for Example 1, midsegment $\overline{MN}$ can be called "the midsegment opposite $\overline{JL}$."

SOLUTION

Step 1 Find the coordinates of M and N by finding the midpoints of $\overline{JK}$ and $\overline{KL}$.

$$M\left(\frac{-6+(-2)}{2}, \frac{1+5}{2}\right) = M\left(\frac{-8}{2}, \frac{6}{2}\right) = M(-4, 3)$$

$$N\left(\frac{-2+2}{2}, \frac{5+(-1)}{2}\right) = N\left(\frac{0}{2}, \frac{4}{2}\right) = N(0, 2)$$

Step 2 Find and compare the slopes of $\overline{MN}$ and $\overline{JL}$.

$$\text{slope of } \overline{MN} = \frac{2-3}{0-(-4)} = -\frac{1}{4} \qquad \text{slope of } \overline{JL} = \frac{-1-1}{2-(-6)} = -\frac{2}{8} = -\frac{1}{4}$$

▶ Because the slopes are the same, $\overline{MN}$ is parallel to $\overline{JL}$.

Step 3 Find and compare the lengths of $\overline{MN}$ and $\overline{JL}$.

$$MN = \sqrt{[0-(-4)]^2 + (2-3)^2} = \sqrt{16+1} = \sqrt{17}$$

$$JL = \sqrt{[2-(-6)]^2 + (-1-1)^2} = \sqrt{64+4} = \sqrt{68} = 2\sqrt{17}$$

▶ Because $\sqrt{17} = \frac{1}{2}(2\sqrt{17})$, $MN = \frac{1}{2}JL$.

Monitoring Progress

Help in English and Spanish at *BigIdeasMath.com*

Use the graph of $\triangle ABC$.

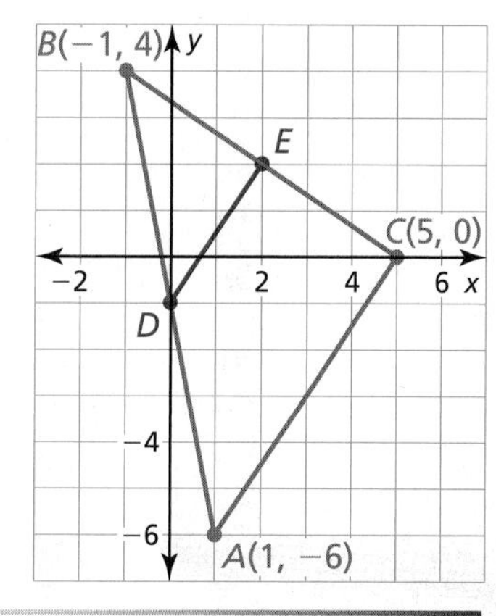

1. In $\triangle ABC$, show that midsegment $\overline{DE}$ is parallel to $\overline{AC}$ and that $DE = \frac{1}{2}AC$.
2. Find the coordinates of the endpoints of midsegment $\overline{EF}$, which is opposite $\overline{AB}$. Show that $\overline{EF}$ is parallel to $\overline{AB}$ and that $EF = \frac{1}{2}AB$.

Using the Triangle Midsegment Theorem

Theorem

Theorem 6.8 Triangle Midsegment Theorem

The segment connecting the midpoints of two sides of a triangle is parallel to the third side and is half as long as that side.

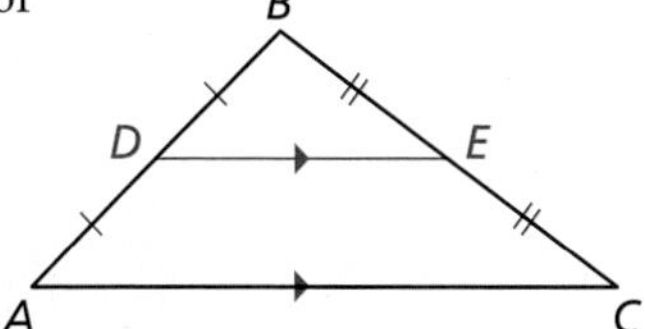

$\overline{DE}$ is a midsegment of $\triangle ABC$, $\overline{DE} \parallel \overline{AC}$, and $DE = \frac{1}{2}AC$.

Proof Example 2, p. 331; Monitoring Progress Question 3, p. 331; Ex. 22, p. 334

EXAMPLE 2 Proving the Triangle Midsegment Theorem

STUDY TIP

When assigning coordinates, try to choose coordinates that make some of the computations easier. In Example 2, you can avoid fractions by using $2p$, $2q$, and $2r$.

Write a coordinate proof of the Triangle Midsegment Theorem for one midsegment.

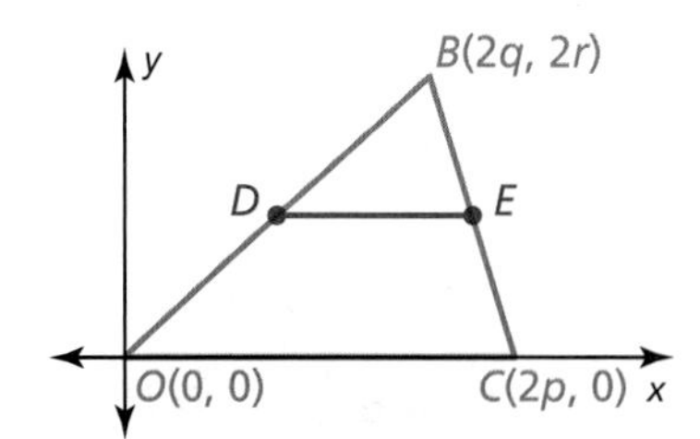

Given $\overline{DE}$ is a midsegment of $\triangle OBC$.

Prove $\overline{DE} \parallel \overline{OC}$ and $DE = \frac{1}{2}OC$

SOLUTION

Step 1 Place $\triangle OBC$ in a coordinate plane and assign coordinates. Because you are finding midpoints, use $2p$, $2q$, and $2r$. Then find the coordinates of D and E.

$$D\left(\frac{2q+0}{2}, \frac{2r+0}{2}\right) = D(q, r) \qquad E\left(\frac{2q+2p}{2}, \frac{2r+0}{2}\right) = E(q+p, r)$$

Step 2 Prove $\overline{DE} \parallel \overline{OC}$. The y-coordinates of D and E are the same, so $\overline{DE}$ has a slope of 0. $\overline{OC}$ is on the x-axis, so its slope is 0.

▶ Because their slopes are the same, $\overline{DE} \parallel \overline{OC}$.

Step 3 Prove $DE = \frac{1}{2}OC$. Use the Ruler Postulate (Post. 1.1) to find DE and OC.

$$DE = |(q+p) - q| = p \qquad OC = |2p - 0| = 2p$$

▶ Because $p = \frac{1}{2}(2p)$, $DE = \frac{1}{2}OC$.

Monitoring Progress

Help in English and Spanish at *BigIdeasMath.com*

3. In Example 2, find the coordinates of F, the midpoint of $\overline{OC}$. Show that $\overline{FE} \parallel \overline{OB}$ and $FE = \frac{1}{2}OB$.

EXAMPLE 3 Using the Triangle Midsegment Theorem

Triangles are used for strength in roof trusses. In the diagram, $\overline{UV}$ and $\overline{VW}$ are midsegments of $\triangle RST$. Find UV and RS.

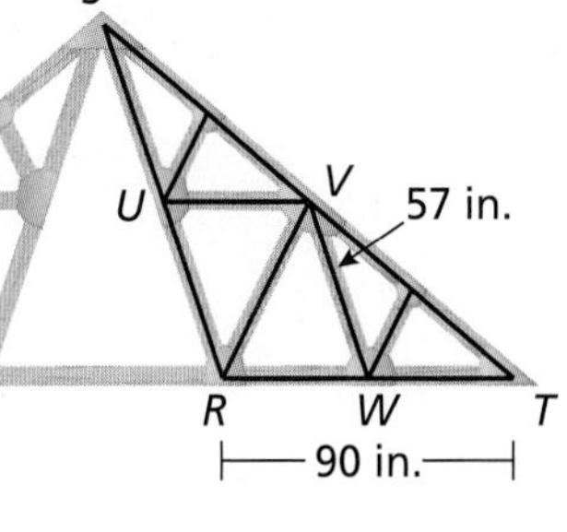

SOLUTION

$UV = \frac{1}{2} \cdot RT = \frac{1}{2}(90 \text{ in.}) = 45 \text{ in.}$

$RS = 2 \cdot VW = 2(57 \text{ in.}) = 114 \text{ in.}$

EXAMPLE 4 Using the Triangle Midsegment Theorem

In the kaleidoscope image, $\overline{AE} \cong \overline{BE}$ and $\overline{AD} \cong \overline{CD}$. Show that $\overline{CB} \parallel \overline{DE}$.

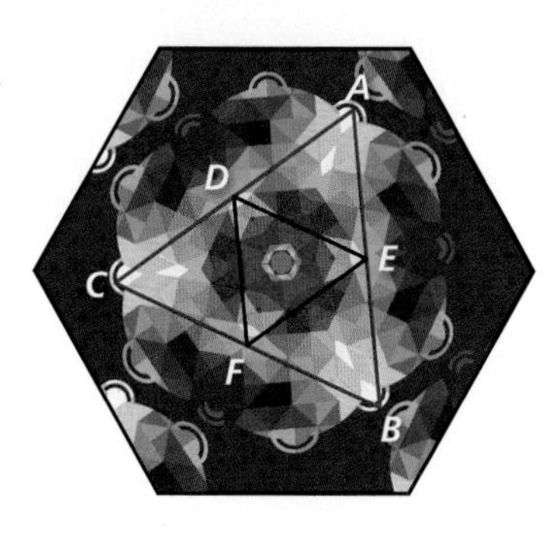

SOLUTION

Because $\overline{AE} \cong \overline{BE}$ and $\overline{AD} \cong \overline{CD}$, E is the midpoint of $\overline{AB}$ and D is the midpoint of $\overline{AC}$ by definition. Then $\overline{DE}$ is a midsegment of $\triangle ABC$ by definition and $\overline{CB} \parallel \overline{DE}$ by the Triangle Midsegment Theorem.

EXAMPLE 5 Modeling with Mathematics

Pear Street intersects Cherry Street and Peach Street at their midpoints. Your home is at point P. You leave your home and jog down Cherry Street to Plum Street, over Plum Street to Peach Street, up Peach Street to Pear Street, over Pear Street to Cherry Street, and then back home up Cherry Street. About how many miles do you jog?

P
1.3 mi
Cherry St.
Pear St.
Peach St.
2.25 mi
Plum St.
1.4 mi

SOLUTION

1. **Understand the Problem** You know the distances from your home to Plum Street along Peach Street, from Peach Street to Cherry Street along Plum Street, and from Pear Street to your home along Cherry Street. You need to find the other distances on your route, then find the total number of miles you jog.
2. **Make a Plan** By definition, you know that Pear Street is a midsegment of the triangle formed by the other three streets. Use the Triangle Midsegment Theorem to find the length of Pear Street and the definition of midsegment to find the length of Cherry Street. Then add the distances along your route.
3. **Solve the Problem**

   length of Pear Street $= \frac{1}{2} \cdot$ (length of Plum St.) $= \frac{1}{2}(1.4 \text{ mi}) = 0.7$ mi

   length of Cherry Street $= 2 \cdot$ (length from P to Pear St.) $= 2(1.3 \text{ mi}) = 2.6$ mi

   distance along your route: $2.6 + 1.4 + \frac{1}{2}(2.25) + 0.7 + 1.3 = 7.125$

   ▶ So, you jog about 7 miles.

4. **Look Back** Use compatible numbers to check that your answer is reasonable.

   total distance:

   $$2.6 + 1.4 + \tfrac{1}{2}(2.25) + 0.7 + 1.3 \approx 2.5 + 1.5 + 1 + 0.5 + 1.5 = 7 \checkmark$$

Monitoring Progress

Help in English and Spanish at *BigIdeasMath.com*

4. Copy the diagram in Example 3. Draw and name the third midsegment. Then find the length of $\overline{VS}$ when the length of the third midsegment is 81 inches.
5. In Example 4, if F is the midpoint of $\overline{CB}$, what do you know about $\overline{DF}$?
6. **WHAT IF?** In Example 5, you jog down Peach Street to Plum Street, over Plum Street to Cherry Street, up Cherry Street to Pear Street, over Pear Street to Peach Street, and then back home up Peach Street. Do you jog more miles in Example 5? Explain.

6.4 Exercises

Dynamic Solutions available at *BigIdeasMath.com*

Vocabulary and Core Concept Check

1. **VOCABULARY** The ___________ of a triangle is a segment that connects the midpoints of two sides of the triangle.

2. **COMPLETE THE SENTENCE** If $\overline{DE}$ is the midsegment opposite $\overline{AC}$ in $\triangle ABC$, then $\overline{DE} \parallel \overline{AC}$ and $DE = $ ___ AC by the Triangle Midsegment Theorem (Theorem 6.8).

Monitoring Progress and Modeling with Mathematics

In Exercises 3–6, use the graph of $\triangle ABC$ with midsegments $\overline{DE}$, $\overline{EF}$, and $\overline{DF}$. *(See Example 1.)*

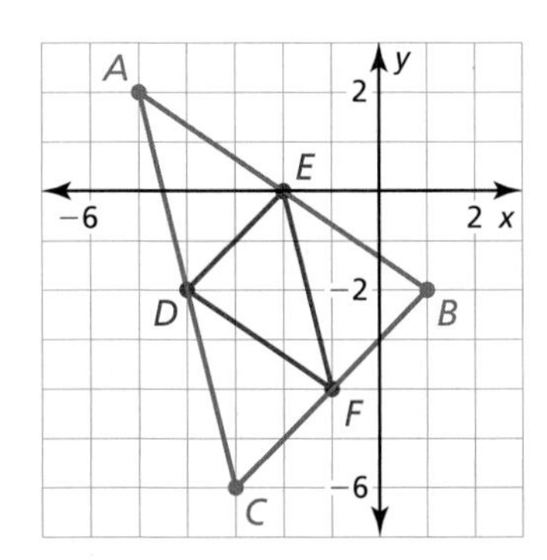

3. Find the coordinates of points *D*, *E*, and *F*.

4. Show that $\overline{DE}$ is parallel to $\overline{CB}$ and that $DE = \frac{1}{2}CB$.

5. Show that $\overline{EF}$ is parallel to $\overline{AC}$ and that $EF = \frac{1}{2}AC$.

6. Show that $\overline{DF}$ is parallel to $\overline{AB}$ and that $DF = \frac{1}{2}AB$.

In Exercises 7–10, $\overline{DE}$ is a midsegment of $\triangle ABC$. Find the value of x. *(See Example 3.)*

7.

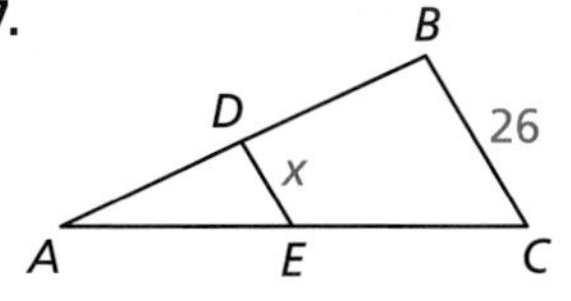

8.

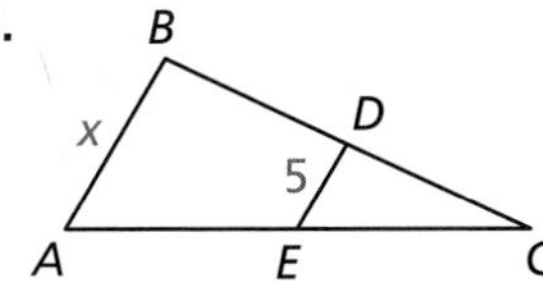

9.

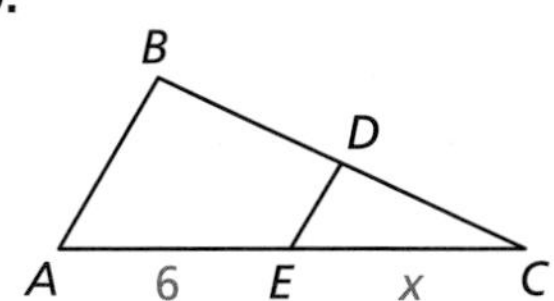

10.

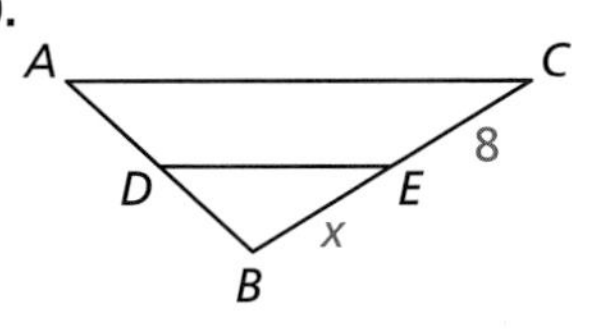

In Exercises 11–16, $\overline{XJ} \cong \overline{JY}$, $\overline{YL} \cong \overline{LZ}$, and $\overline{XK} \cong \overline{KZ}$. Copy and complete the statement. *(See Example 4.)*

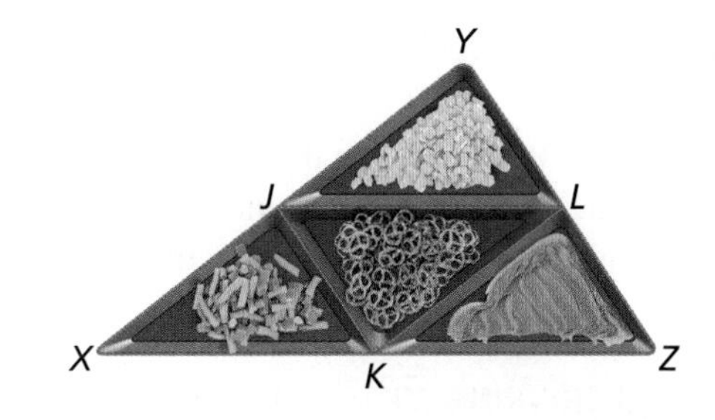

11. $\overline{JK} \parallel$ ___

12. $\overline{JL} \parallel$ ___

13. $\overline{XY} \parallel$ ___

14. $\overline{JY} \cong$ ___ $\cong$ ___

15. $\overline{JL} \cong$ ___ $\cong$ ___

16. $\overline{JK} \cong$ ___ $\cong$ ___

MATHEMATICAL CONNECTIONS In Exercises 17–19, use $\triangle GHJ$, where A, B, and C are midpoints of the sides.

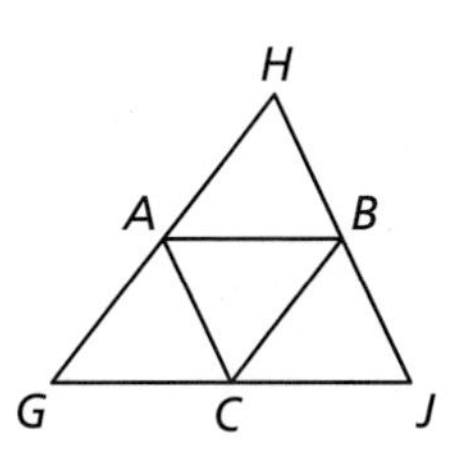

17. When $AB = 3x + 8$ and $GJ = 2x + 24$, what is AB?

18. When $AC = 3y - 5$ and $HJ = 4y + 2$, what is HB?

19. When $GH = 7z - 1$ and $CB = 4z - 3$, what is GA?

20. **ERROR ANALYSIS** Describe and correct the error.

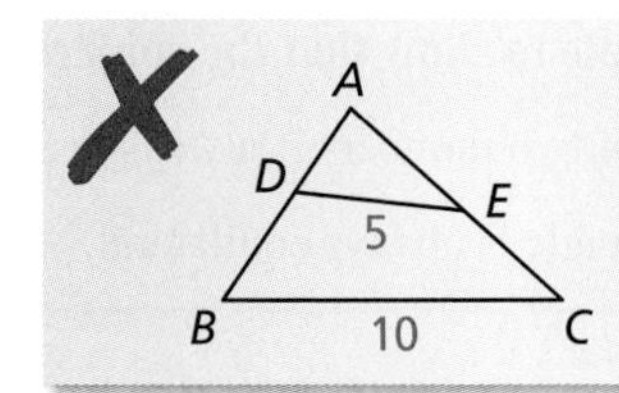

$DE = \frac{1}{2}BC$, so by the Triangle Midsegment Theorem (Thm. 6.8), $\overline{AD} \cong \overline{DB}$ and $\overline{AE} \cong \overline{EC}$.

21. **MODELING WITH MATHEMATICS** The distance between consecutive bases on a baseball field is 90 feet. A second baseman stands halfway between first base and second base, a shortstop stands halfway between second base and third base, and a pitcher stands halfway between first base and third base. Find the distance between the shortstop and the pitcher. *(See Example 5.)*

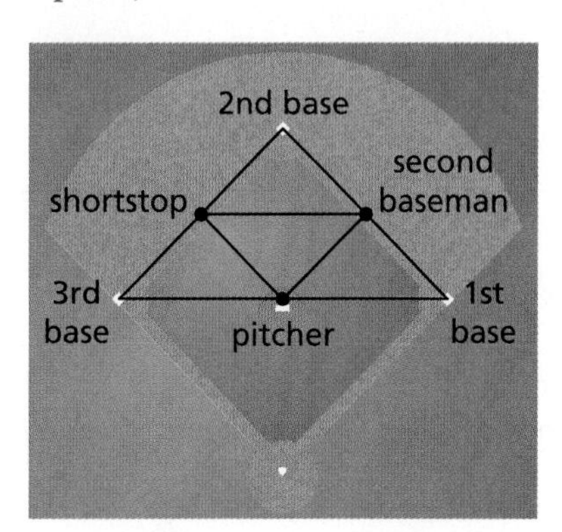

22. **PROVING A THEOREM** Use the figure from Example 2 to prove the Triangle Midsegment Theorem (Theorem 6.8) for midsegment $\overline{DF}$, where F is the midpoint of $\overline{OC}$. *(See Example 2.)*

23. **CRITICAL THINKING** $\overline{XY}$ is a midsegment of $\triangle LMN$. Suppose $\overline{DE}$ is called a "quarter segment" of $\triangle LMN$. What do you think an "eighth segment" would be? Make conjectures about the properties of a quarter segment and an eighth segment. Use variable coordinates to verify your conjectures.

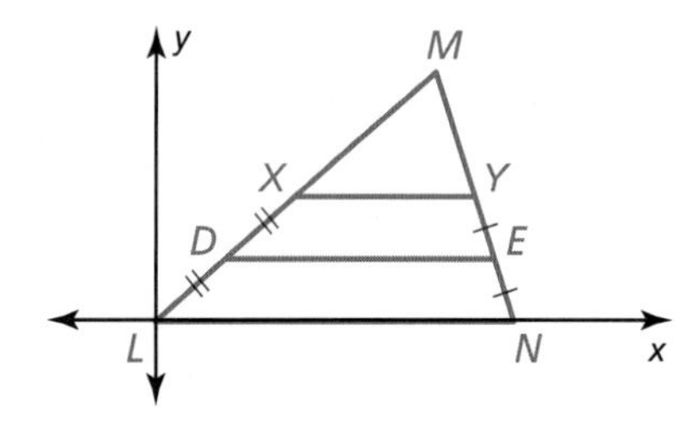

24. **THOUGHT PROVOKING** Find a real-life object that uses midsegments as part of its structure. Print a photograph of the object and identify the midsegments of one of the triangles in the structure.

25. **ABSTRACT REASONING** To create the design shown, shade the triangle formed by the three midsegments of the triangle. Then repeat the process for each unshaded triangle.

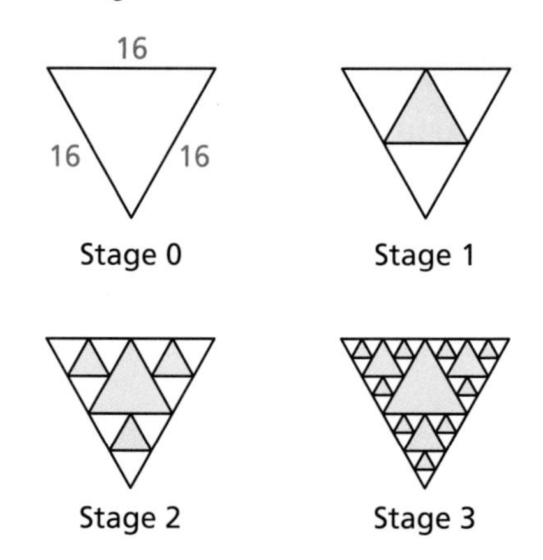

a. What is the perimeter of the shaded triangle in Stage 1?

b. What is the total perimeter of all the shaded triangles in Stage 2?

c. What is the total perimeter of all the shaded triangles in Stage 3?

26. **HOW DO YOU SEE IT?** Explain how you know that the yellow triangle is the midsegment triangle of the red triangle in the pattern of floor tiles shown.

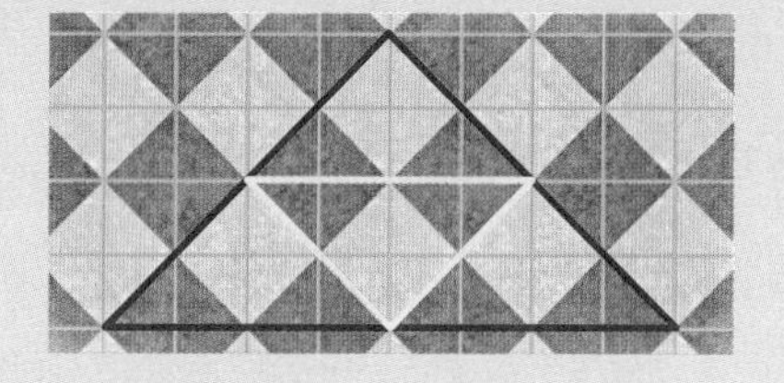

27. **ATTENDING TO PRECISION** The points $P(2, 1)$, $Q(4, 5)$, and $R(7, 4)$ are the midpoints of the sides of a triangle. Graph the three midsegments. Then show how to use your graph and the properties of midsegments to draw the original triangle. Give the coordinates of each vertex.

Maintaining Mathematical Proficiency

Reviewing what you learned in previous grades and lessons

Find a counterexample to show that the conjecture is false. *(Section 2.2)*

28. The difference of two numbers is always less than the greater number.

29. An isosceles triangle is always equilateral.

6.5 Indirect Proof and Inequalities in One Triangle

Essential Question How are the sides related to the angles of a triangle? How are any two sides of a triangle related to the third side?

EXPLORATION 1 **Comparing Angle Measures and Side Lengths**

Work with a partner. Use dynamic geometry software. Draw any scalene $\triangle ABC$.

a. Find the side lengths and angle measures of the triangle.

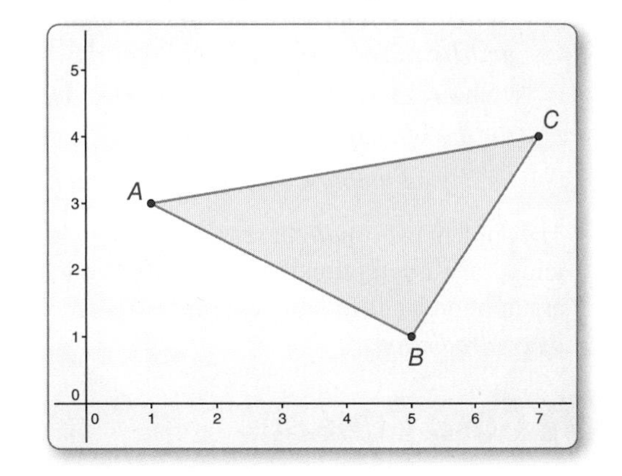

Sample

Points
$A(1, 3)$
$B(5, 1)$
$C(7, 4)$

Segments
$BC = ?$
$AC = ?$
$AB = ?$

Angles
$m\angle A = ?$
$m\angle B = ?$
$m\angle C = ?$

b. Order the side lengths. Order the angle measures. What do you observe?

c. Drag the vertices of $\triangle ABC$ to form new triangles. Record the side lengths and angle measures in a table. Write a conjecture about your findings.

ATTENDING TO PRECISION

To be proficient in math, you need to express numerical answers with a degree of precision appropriate for the content.

EXPLORATION 2 **A Relationship of the Side Lengths of a Triangle**

Work with a partner. Use dynamic geometry software. Draw any $\triangle ABC$.

a. Find the side lengths of the triangle.

b. Compare each side length with the sum of the other two side lengths.

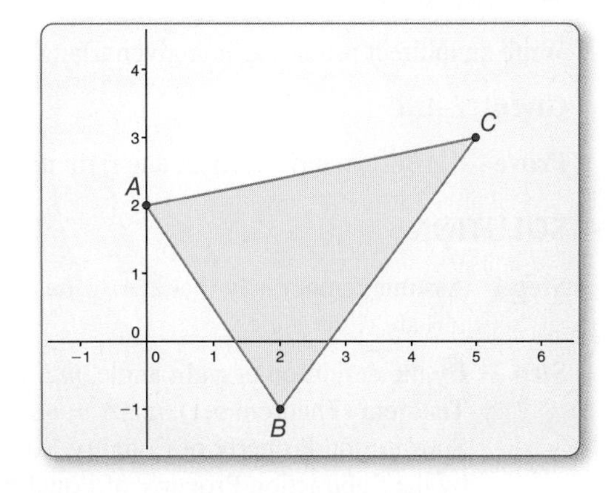

Sample

Points
$A(0, 2)$
$B(2, -1)$
$C(5, 3)$

Segments
$BC = ?$
$AC = ?$
$AB = ?$

c. Drag the vertices of $\triangle ABC$ to form new triangles and repeat parts (a) and (b). Organize your results in a table. Write a conjecture about your findings.

Communicate Your Answer

3. How are the sides related to the angles of a triangle? How are any two sides of a triangle related to the third side?

4. Is it possible for a triangle to have side lengths of 3, 4, and 10? Explain.

6.5 Lesson

Core Vocabulary

indirect proof, *p. 336*

Previous
proof
inequality

What You Will Learn

- Write indirect proofs.
- List sides and angles of a triangle in order by size.
- Use the Triangle Inequality Theorem to find possible side lengths of triangles.

Writing an Indirect Proof

Suppose a student looks around the cafeteria, concludes that hamburgers are not being served, and explains as follows.

> *At first, I assumed that we are having hamburgers because today is Tuesday, and Tuesday is usually hamburger day.*
>
> *There is always ketchup on the table when we have hamburgers, so I looked for the ketchup, but I didn't see any.*
>
> *So, my assumption that we are having hamburgers must be false.*

The student uses *indirect* reasoning. In an **indirect proof**, you start by making the temporary assumption that the desired conclusion is false. By then showing that this assumption leads to a logical impossibility, you prove the original statement true *by contradiction.*

Core Concept

How to Write an Indirect Proof (Proof by Contradiction)

Step 1 Identify the statement you want to prove. Assume temporarily that this statement is false by assuming that its opposite is true.

Step 2 Reason logically until you reach a contradiction.

Step 3 Point out that the desired conclusion must be true because the contradiction proves the temporary assumption false.

EXAMPLE 1 Writing an Indirect Proof

Write an indirect proof that in a given triangle, there can be at most one right angle.

Given $\triangle ABC$

Prove $\triangle ABC$ can have at most one right angle.

SOLUTION

READING

You have reached a *contradiction* when you have two statements that cannot both be true at the same time.

Step 1 Assume temporarily that $\triangle ABC$ has two right angles. Then assume $\angle A$ and $\angle B$ are right angles.

Step 2 By the definition of right angle, $m\angle A = m\angle B = 90°$. By the Triangle Sum Theorem (Theorem 5.1), $m\angle A + m\angle B + m\angle C = 180°$. Using the Substitution Property of Equality, $90° + 90° + m\angle C = 180°$. So, $m\angle C = 0°$ by the Subtraction Property of Equality. A triangle cannot have an angle measure of 0°. So, this contradicts the given information.

Step 3 So, the assumption that $\triangle ABC$ has two right angles must be false, which proves that $\triangle ABC$ can have at most one right angle.

Monitoring Progress

Help in English and Spanish at *BigIdeasMath.com*

1. Write an indirect proof that a scalene triangle cannot have two congruent angles.

Relating Sides and Angles of a Triangle

EXAMPLE 2 Relating Side Length and Angle Measure

Draw an obtuse scalene triangle. Find the largest angle and longest side and mark them in red. Find the smallest angle and shortest side and mark them in blue. What do you notice?

SOLUTION

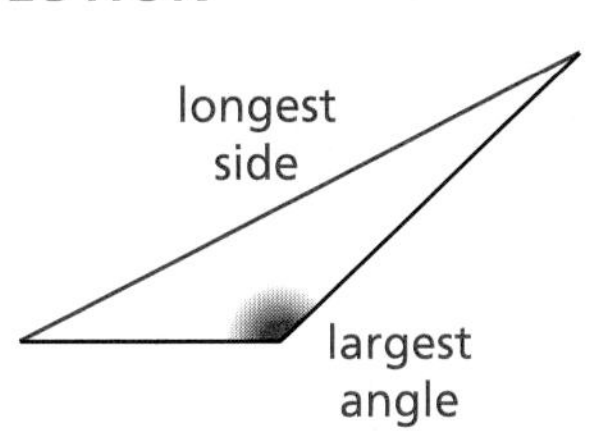

The longest side and largest angle are opposite each other.

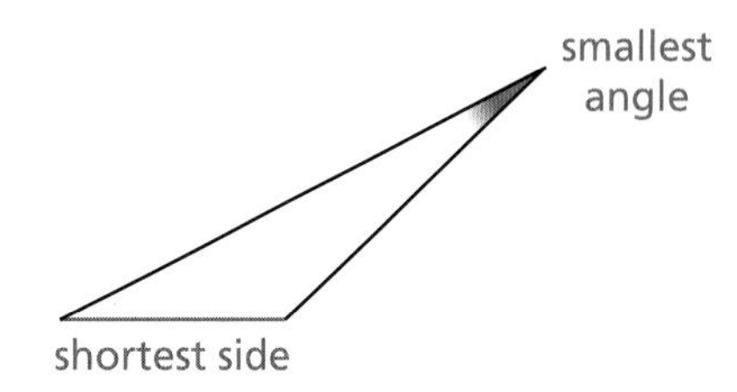

The shortest side and smallest angle are opposite each other.

COMMON ERROR

Be careful not to confuse the symbol ∠ meaning *angle* with the symbol < meaning *is less than.* Notice that the bottom edge of the angle symbol is horizontal.

The relationships in Example 2 are true for all triangles, as stated in the two theorems below. These relationships can help you decide whether a particular arrangement of side lengths and angle measures in a triangle may be possible.

Theorems

Theorem 6.9 Triangle Longer Side Theorem

If one side of a triangle is longer than another side, then the angle opposite the longer side is larger than the angle opposite the shorter side.

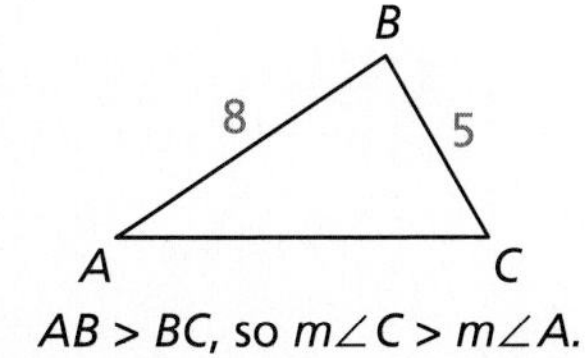

$AB > BC$, so $m\angle C > m\angle A$.

Proof Ex. 43, p. 342

Theorem 6.10 Triangle Larger Angle Theorem

If one angle of a triangle is larger than another angle, then the side opposite the larger angle is longer than the side opposite the smaller angle.

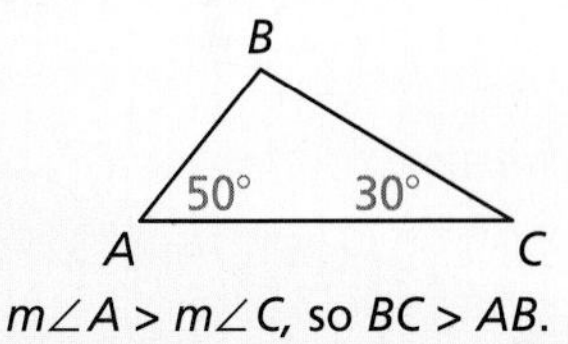

$m\angle A > m\angle C$, so $BC > AB$.

Proof p. 337

PROOF Triangle Larger Angle Theorem

COMMON ERROR

Be sure to consider all cases when assuming the opposite is true.

Given $m\angle A > m\angle C$

Prove $BC > AB$

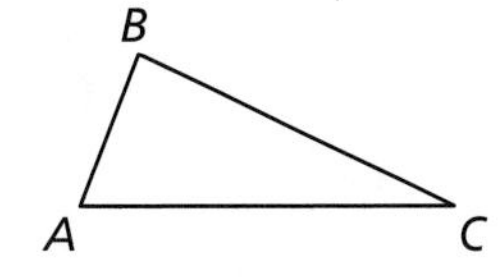

Indirect Proof

Step 1 Assume temporarily that $BC \ngtr AB$. Then it follows that either $BC < AB$ or $BC = AB$.

Step 2 If $BC < AB$, then $m\angle A < m\angle C$ by the Triangle Longer Side Theorem.
If $BC = AB$, then $m\angle A = m\angle C$ by the Base Angles Theorem (Thm. 5.6).

Step 3 Both conclusions contradict the given statement that $m\angle A > m\angle C$. So, the temporary assumption that $BC \ngtr AB$ cannot be true. This proves that $BC > AB$.

EXAMPLE 3 Ordering Angle Measures of a Triangle

You are constructing a stage prop that shows a large triangular mountain. The bottom edge of the mountain is about 32 feet long, the left slope is about 24 feet long, and the right slope is about 26 feet long. List the angles of $\triangle JKL$ in order from smallest to largest.

K

J L

SOLUTION

Draw the triangle that represents the mountain. Label the side lengths.

The sides from shortest to longest are $\overline{JK}$, $\overline{KL}$, and $\overline{JL}$. The angles opposite these sides are $\angle L$, $\angle J$, and $\angle K$, respectively.

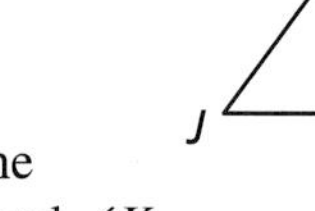

K, 24 ft, 26 ft, J, 32 ft, L

▶ So, by the Triangle Longer Side Theorem, the angles from smallest to largest are $\angle L$, $\angle J$, and $\angle K$.

EXAMPLE 4 Ordering Side Lengths of a Triangle

List the sides of $\triangle DEF$ in order from shortest to longest.

SOLUTION

First, find $m\angle F$ using the Triangle Sum Theorem (Theorem 5.1).

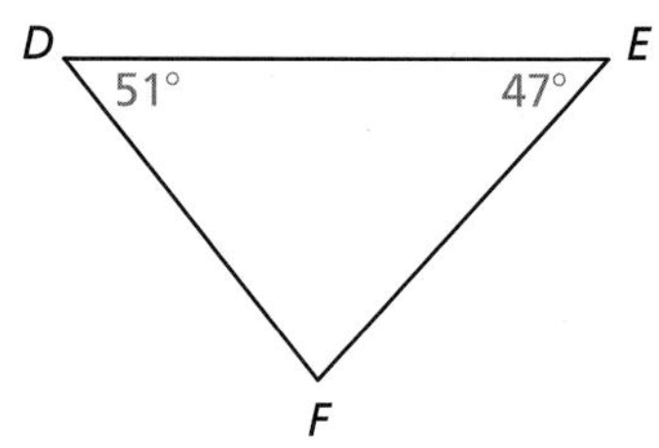

$$m\angle D + m\angle E + m\angle F = 180^\circ$$

$$51^\circ + 47^\circ + m\angle F = 180^\circ$$

$$m\angle F = 82^\circ$$

The angles from smallest to largest are $\angle E$, $\angle D$, and $\angle F$. The sides opposite these angles are $\overline{DF}$, $\overline{EF}$, and $\overline{DE}$, respectively.

▶ So, by the Triangle Larger Angle Theorem, the sides from shortest to longest are $\overline{DF}$, $\overline{EF}$, and $\overline{DE}$.

Monitoring Progress

Help in English and Spanish at *BigIdeasMath.com*

2. List the angles of $\triangle PQR$ in order from smallest to largest.

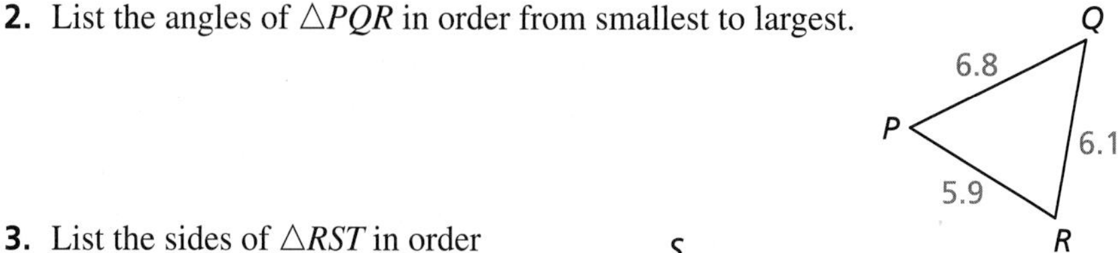

3. List the sides of $\triangle RST$ in order from shortest to longest.

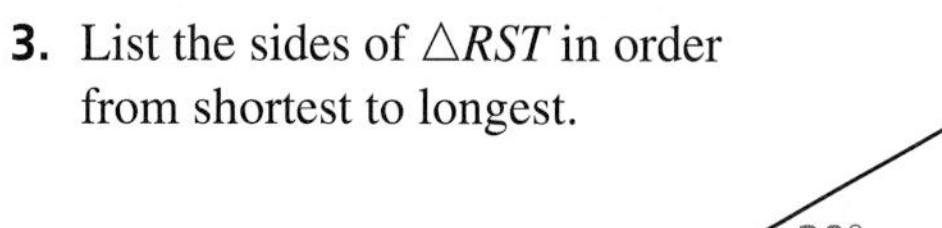

Using the Triangle Inequality Theorem

Not every group of three segments can be used to form a triangle. The lengths of the segments must fit a certain relationship. For example, three attempted triangle constructions using segments with given lengths are shown below. Only the first group of segments forms a triangle.

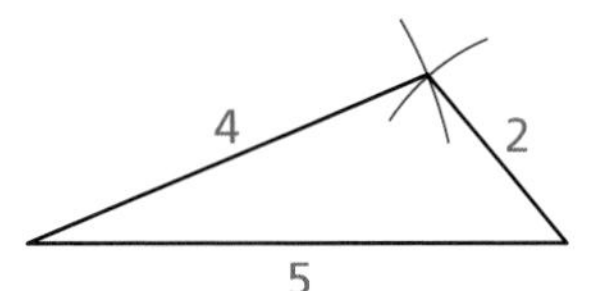

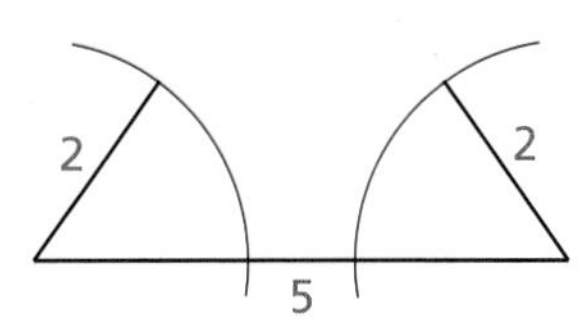

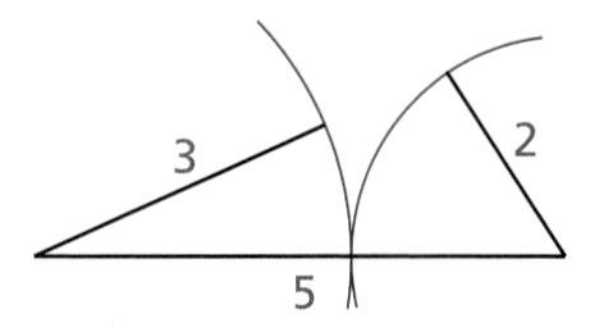

When you start with the longest side and attach the other two sides at its endpoints, you can see that the other two sides are not long enough to form a triangle in the second and third figures. This leads to the *Triangle Inequality Theorem.*

Theorem

Theorem 6.11 Triangle Inequality Theorem

The sum of the lengths of any two sides of a triangle is greater than the length of the third side.

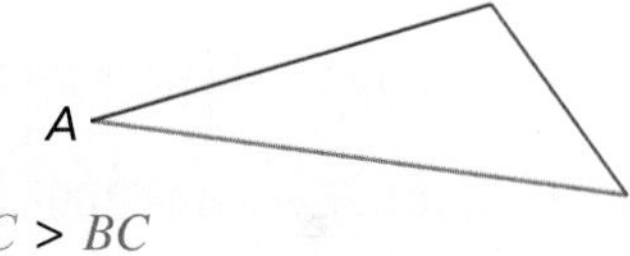

$AB + BC > AC$ $\quad$ $AC + BC > AB$ $\quad$ $AB + AC > BC$

Proof Ex. 47, p. 342

EXAMPLE 5 Finding Possible Side Lengths

A triangle has one side of length 14 and another side of length 9. Describe the possible lengths of the third side.

SOLUTION

Let x represent the length of the third side. Draw diagrams to help visualize the small and large values of x. Then use the Triangle Inequality Theorem to write and solve inequalities.

READING

You can combine the two inequalities, $x > 5$ and $x < 23$, to write the compound inequality $5 < x < 23$. This can be read as *x is between 5 and 23.*

Small values of x

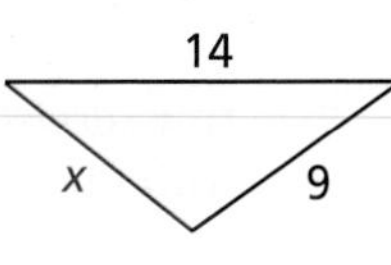

$x + 9 > 14$

$x > 5$

Large values of x

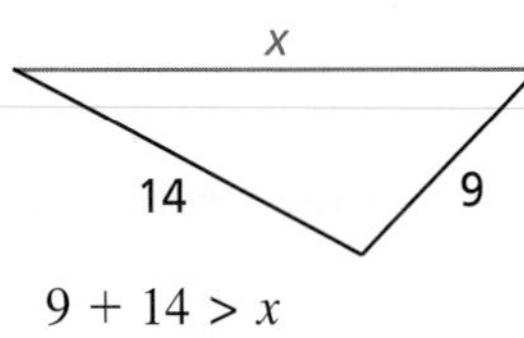

$9 + 14 > x$

$23 > x$, or $x < 23$

▶ The length of the third side must be greater than 5 and less than 23.

Monitoring Progress

Help in English and Spanish at *BigIdeasMath.com*

4. A triangle has one side of length 12 inches and another side of length 20 inches. Describe the possible lengths of the third side.

Decide whether it is possible to construct a triangle with the given side lengths. Explain your reasoning.

5. 4 ft, 9 ft, 10 ft $\quad$ **6.** 8 m, 9 m, 18 m $\quad$ **7.** 5 cm, 7 cm, 12 cm

6.5 Exercises

Dynamic Solutions available at *BigIdeasMath.com*

Vocabulary and Core Concept Check

1. **VOCABULARY** Why is an indirect proof also called a *proof by contradiction*?

2. **WRITING** How can you tell which side of a triangle is the longest from the angle measures of the triangle? How can you tell which side is the shortest?

Monitoring Progress and Modeling with Mathematics

In Exercises 3–6, write the first step in an indirect proof of the statement. *(See Example 1.)*

3. If $WV + VU \neq 12$ inches and $VU = 5$ inches, then $WV \neq 7$ inches.

4. If x and y are odd integers, then xy is odd.

5. In $\triangle ABC$, if $m\angle A = 100°$, then $\angle B$ is not a right angle.

6. In $\triangle JKL$, if M is the midpoint of $\overline{KL}$, then $\overline{JM}$ is a median.

In Exercises 7 and 8, determine which two statements contradict each other. Explain your reasoning.

7. Ⓐ $\triangle LMN$ is a right triangle.
   Ⓑ $\angle L \cong \angle N$
   Ⓒ $\triangle LMN$ is equilateral.

8. Ⓐ Both $\angle X$ and $\angle Y$ have measures greater than 20°.
   Ⓑ Both $\angle X$ and $\angle Y$ have measures less than 30°.
   Ⓒ $m\angle X + m\angle Y = 62°$

In Exercises 9 and 10, use a ruler and protractor to draw the given type of triangle. Mark the largest angle and longest side in red and the smallest angle and shortest side in blue. What do you notice? *(See Example 2.)*

9. acute scalene

10. right scalene

In Exercises 11 and 12, list the angles of the given triangle from smallest to largest. *(See Example 3.)*

11.

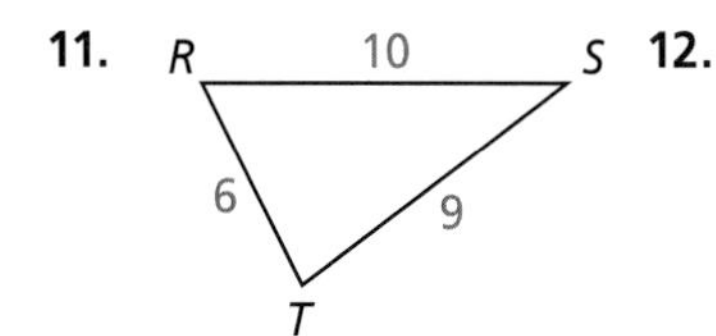

12.

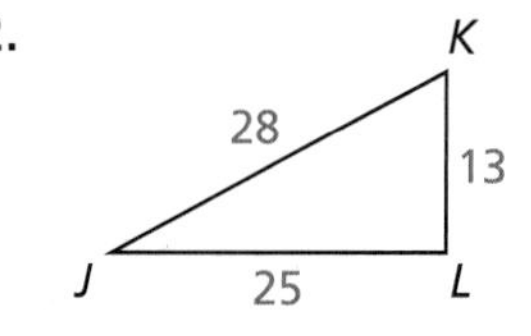

In Exercises 13–16, list the sides of the given triangle from shortest to longest. *(See Example 4.)*

13.

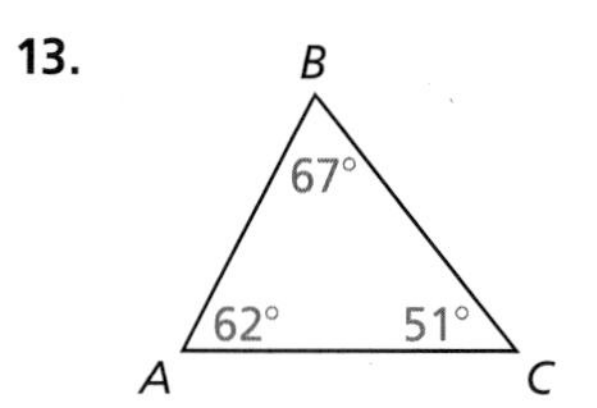

14.

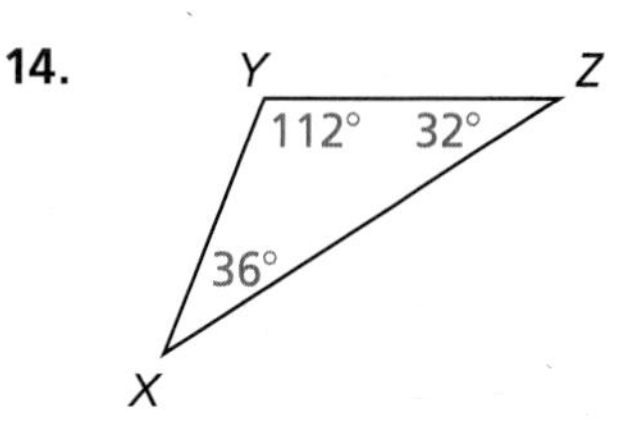

15.

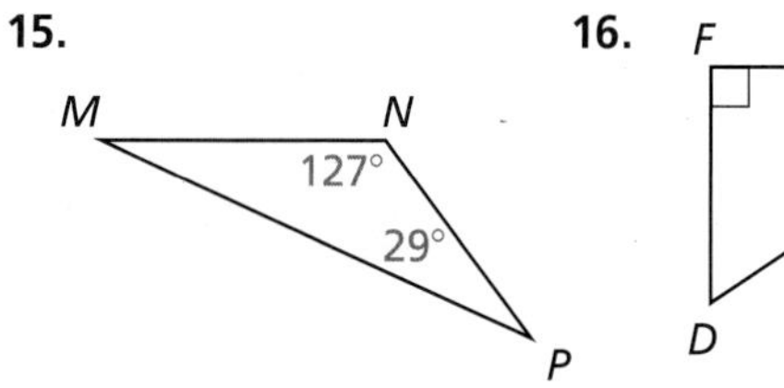

16.

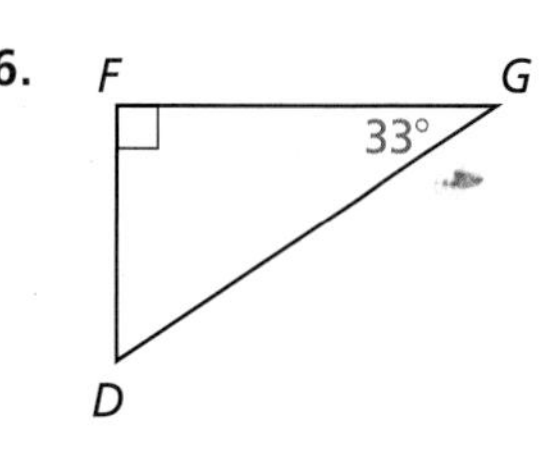

In Exercises 17–20, describe the possible lengths of the third side of the triangle given the lengths of the other two sides. *(See Example 5.)*

17. 5 inches, 12 inches

18. 12 feet, 18 feet

19. 2 feet, 40 inches

20. 25 meters, 25 meters

In Exercises 21–24, is it possible to construct a triangle with the given side lengths? If not, explain why not.

21. 6, 7, 11

22. 3, 6, 9

23. 28, 17, 46

24. 35, 120, 125

25. **ERROR ANALYSIS** Describe and correct the error in writing the first step of an indirect proof.

Show that $\angle A$ is obtuse.

Step 1 Assume temporarily that $\angle A$ is acute.

26. ERROR ANALYSIS Describe and correct the error in labeling the side lengths 1, 2, and $\sqrt{3}$ on the triangle.

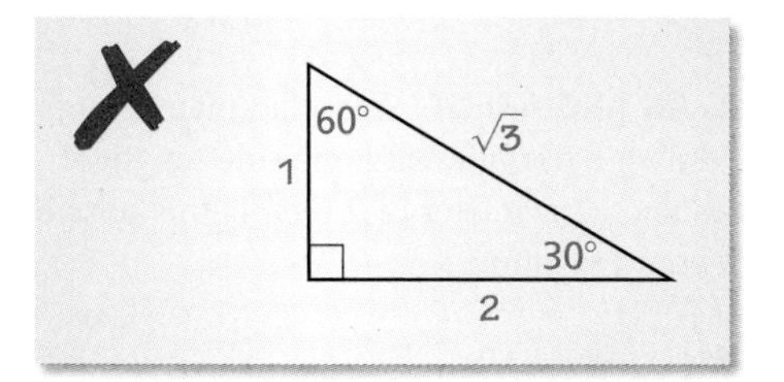

27. REASONING You are a lawyer representing a client who has been accused of a crime. The crime took place in Los Angeles, California. Security footage shows your client in New York at the time of the crime. Explain how to use indirect reasoning to prove your client is innocent.

28. REASONING Your class has fewer than 30 students. The teacher divides your class into two groups. The first group has 15 students. Use indirect reasoning to show that the second group must have fewer than 15 students.

29. PROBLEM SOLVING Which statement about $\triangle TUV$ is false?

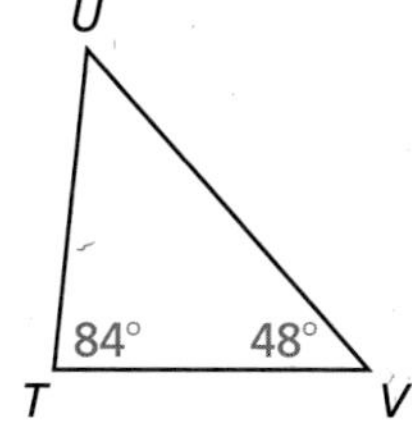

Ⓐ $UV > TU$

Ⓑ $UV + TV > TU$

Ⓒ $UV < TV$

Ⓓ $\triangle TUV$ is isosceles.

30. PROBLEM SOLVING In $\triangle RST$, which is a possible side length for ST? Select all that apply.

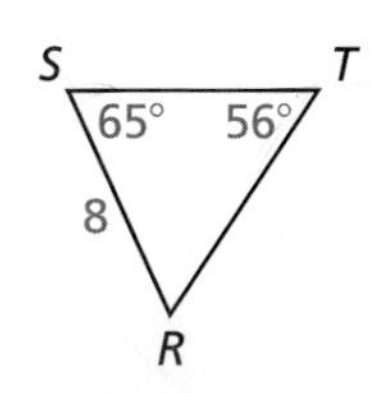

Ⓐ 7

Ⓑ 8

Ⓒ 9

Ⓓ 10

31. PROOF Write an indirect proof that an odd number is not divisible by 4.

32. PROOF Write an indirect proof of the statement "In $\triangle QRS$, if $m\angle Q + m\angle R = 90°$, then $m\angle S = 90°$."

33. WRITING Explain why the hypotenuse of a right triangle must always be longer than either leg.

34. CRITICAL THINKING Is it possible to decide if three side lengths form a triangle without checking all three inequalities shown in the Triangle Inequality Theorem (Theorem 6.11)? Explain your reasoning.

35. MODELING WITH MATHEMATICS You can estimate the width of the river from point A to the tree at point B by measuring the angle to the tree at several locations along the riverbank. The diagram shows the results for locations C and D.

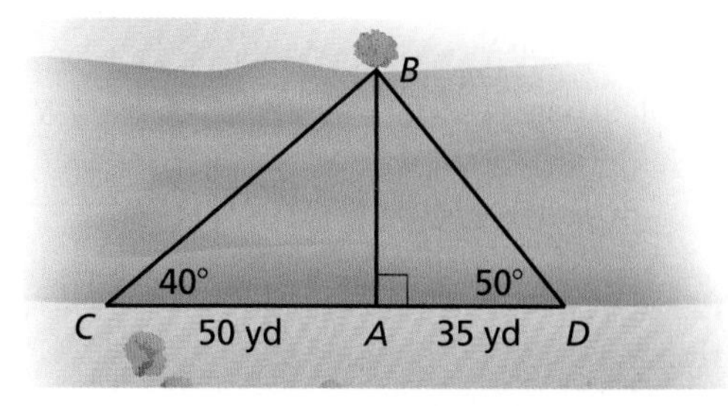

a. Using $\triangle BCA$ and $\triangle BDA$, determine the possible widths of the river. Explain your reasoning.

b. What could you do if you wanted a closer estimate?

36. MODELING WITH MATHEMATICS You travel from Fort Peck Lake to Glacier National Park and from Glacier National Park to Granite Peak.

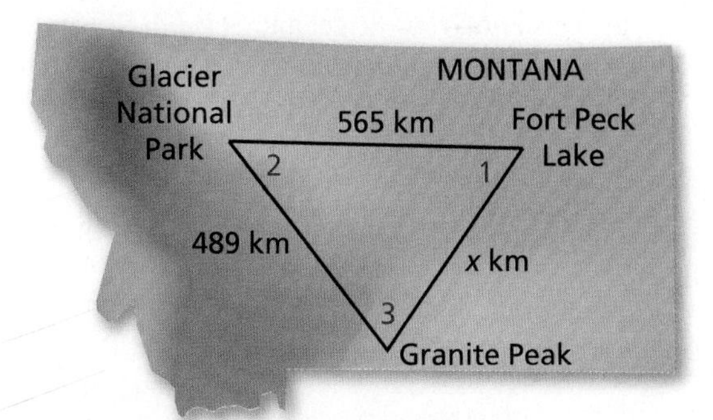

a. Write two inequalities to represent the possible distances from Granite Peak back to Fort Peck Lake.

b. How is your answer to part (a) affected if you know that $m\angle 2 < m\angle 1$ and $m\angle 2 < m\angle 3$?

37. REASONING In the figure, $\overline{XY}$ bisects $\angle WYZ$. List all six angles of $\triangle XYZ$ and $\triangle WXY$ in order from smallest to largest. Explain your reasoning.

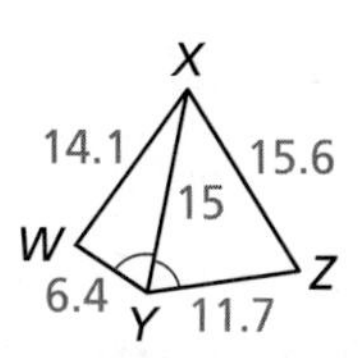

38. MATHEMATICAL CONNECTIONS In $\triangle DEF$, $m\angle D = (x + 25)°$, $m\angle E = (2x - 4)°$, and $m\angle F = 63°$. List the side lengths and angle measures of the triangle in order from least to greatest.

39. ANALYZING RELATIONSHIPS Another triangle inequality relationship is given by the Exterior Angle Inequality Theorem. It states:

The measure of an exterior angle of a triangle is greater than the measure of either of the nonadjacent interior angles.

Explain how you know that $m\angle 1 > m\angle A$ and $m\angle 1 > m\angle B$ in $\triangle ABC$ with exterior angle $\angle 1$.

MATHEMATICAL CONNECTIONS In Exercises 40 and 41, describe the possible values of x.

40.

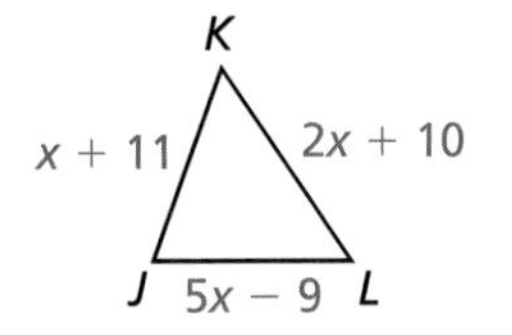

41.

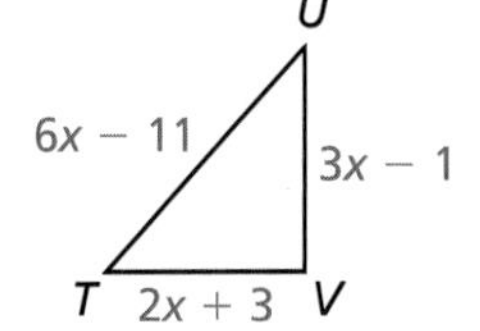

42. HOW DO YOU SEE IT? Your house is on the corner of Hill Street and Eighth Street. The library is on the corner of View Street and Seventh Street. What is the shortest route to get from your house to the library? Explain your reasoning.

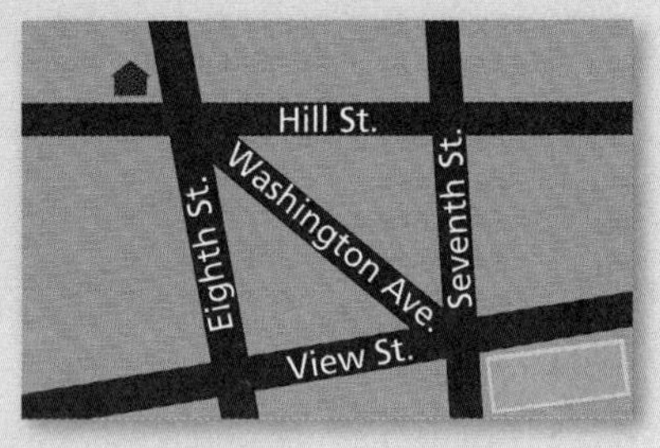

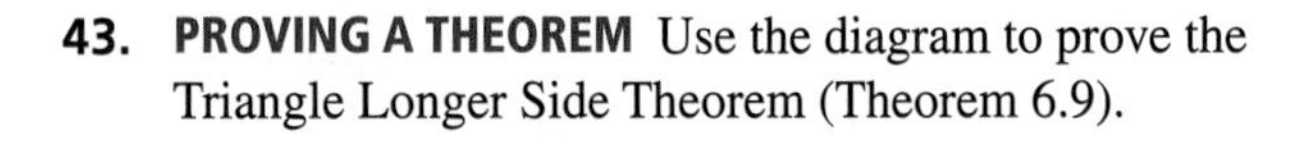

43. PROVING A THEOREM Use the diagram to prove the Triangle Longer Side Theorem (Theorem 6.9).

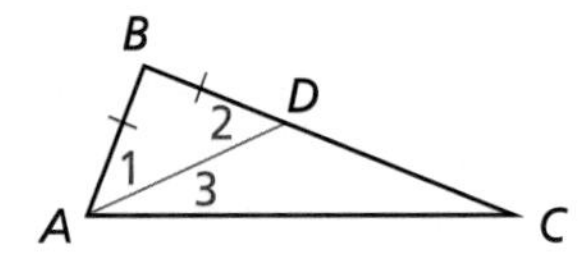

Given $BC > AB$, $BD = BA$

Prove $m\angle BAC > m\angle C$

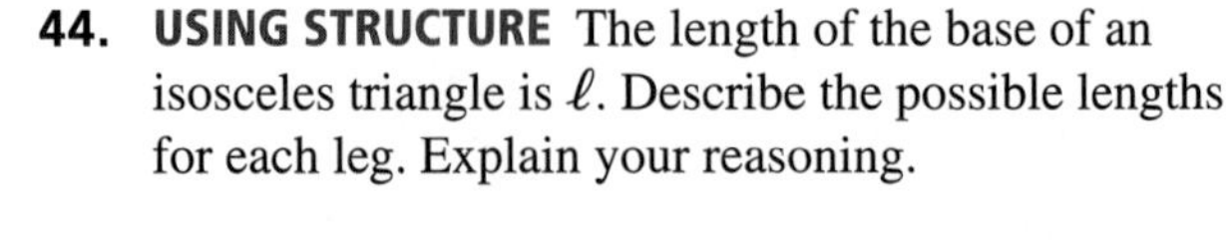

44. USING STRUCTURE The length of the base of an isosceles triangle is ℓ. Describe the possible lengths for each leg. Explain your reasoning.

45. MAKING AN ARGUMENT Your classmate claims to have drawn a triangle with one side length of 13 inches and a perimeter of 2 feet. Is this possible? Explain your reasoning.

46. THOUGHT PROVOKING Cut two pieces of string that are each 24 centimeters long. Construct an isosceles triangle out of one string and a scalene triangle out of the other. Measure and record the side lengths. Then classify each triangle by its angles.

47. PROVING A THEOREM Prove the Triangle Inequality Theorem (Theorem 6.11).

Given $\triangle ABC$

Prove $AB + BC > AC$, $AC + BC > AB$, and $AB + AC > BC$

48. ATTENDING TO PRECISION The perimeter of $\triangle HGF$ must be between what two integers? Explain your reasoning.

49. PROOF Write an indirect proof that a perpendicular segment is the shortest segment from a point to a plane.

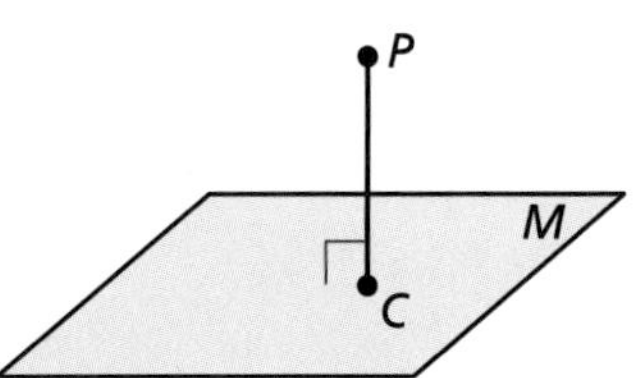

Given $\overline{PC} \perp$ plane M

Prove $\overline{PC}$ is the shortest segment from P to plane M.

Maintaining Mathematical Proficiency

Reviewing what you learned in previous grades and lessons

Name the included angle between the pair of sides given. *(Section 5.3)*

50. $\overline{AE}$ and $\overline{BE}$

51. $\overline{AC}$ and $\overline{DC}$

52. $\overline{AD}$ and $\overline{DC}$

53. $\overline{CE}$ and $\overline{BE}$

6.6 Inequalities in Two Triangles

Essential Question If two sides of one triangle are congruent to two sides of another triangle, what can you say about the third sides of the triangles?

EXPLORATION 1 Comparing Measures in Triangles

Work with a partner. Use dynamic geometry software.

a. Draw $\triangle ABC$, as shown below.

b. Draw the circle with center $C(3, 3)$ through the point $A(1, 3)$.

c. Draw $\triangle DBC$ so that D is a point on the circle.

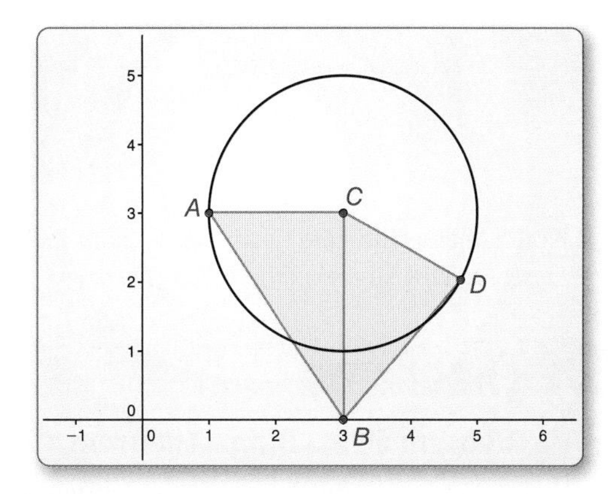

Sample

Points
$A(1, 3)$
$B(3, 0)$
$C(3, 3)$
$D(4.75, 2.03)$

Segments
$BC = 3$
$AC = 2$
$DC = 2$
$AB = 3.61$
$DB = 2.68$

d. Which two sides of $\triangle ABC$ are congruent to two sides of $\triangle DBC$? Justify your answer.

e. Compare the lengths of $\overline{AB}$ and $\overline{DB}$. Then compare the measures of $\angle ACB$ and $\angle DCB$. Are the results what you expected? Explain.

f. Drag point D to several locations on the circle. At each location, repeat part (e). Copy and record your results in the table below.

	D	AC	BC	AB	BD	$m\angle ACB$	$m\angle BCD$
1.	(4.75, 2.03)	2	3				
2.		2	3				
3.		2	3				
4.		2	3				
5.		2	3				

g. Look for a pattern of the measures in your table. Then write a conjecture that summarizes your observations.

CONSTRUCTING VIABLE ARGUMENTS

To be proficient in math, you need to make conjectures and build a logical progression of statements to explore the truth of your conjectures.

Communicate Your Answer

2. If two sides of one triangle are congruent to two sides of another triangle, what can you say about the third sides of the triangles?

3. Explain how you can use the hinge shown at the left to model the concept described in Question 2.

6.6 Lesson

Core Vocabulary

Previous
indirect proof
inequality

What You Will Learn

- Compare measures in triangles.
- Solve real-life problems using the Hinge Theorem.

Comparing Measures in Triangles

Imagine a gate between fence posts A and B that has hinges at A and swings open at B.

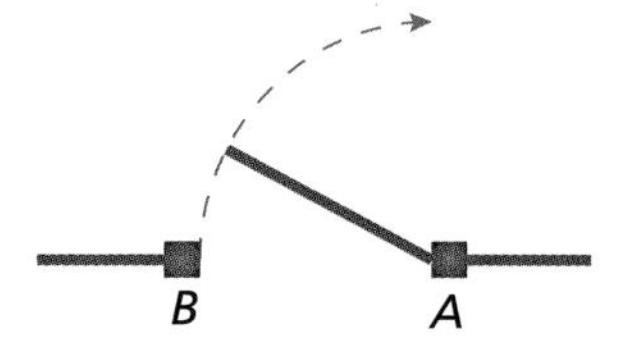

As the gate swings open, you can think of $\triangle ABC$, with side $\overline{AC}$ formed by the gate itself, side $\overline{AB}$ representing the distance between the fence posts, and side $\overline{BC}$ representing the opening between post B and the outer edge of the gate.

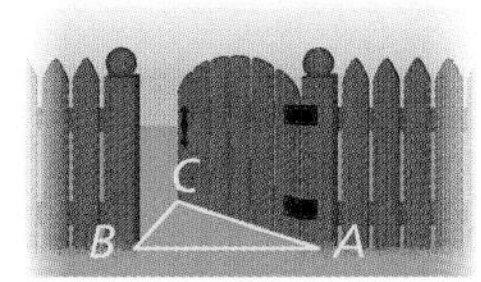

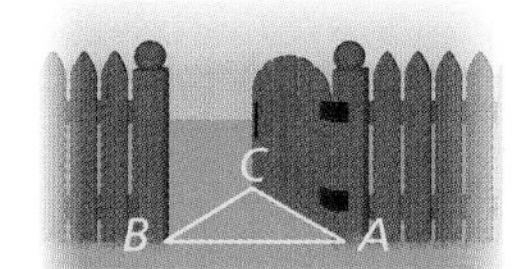

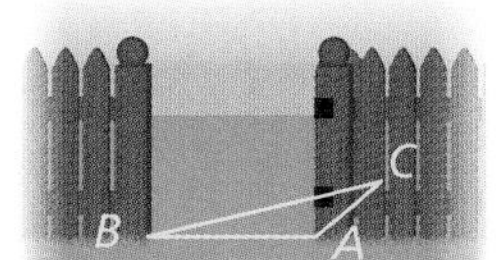

Notice that as the gate opens wider, both the measure of $\angle A$ and the distance BC increase. This suggests the *Hinge Theorem.*

Theorems

Theorem 6.12 Hinge Theorem

If two sides of one triangle are congruent to two sides of another triangle, and the included angle of the first is larger than the included angle of the second, then the third side of the first is longer than the third side of the second.

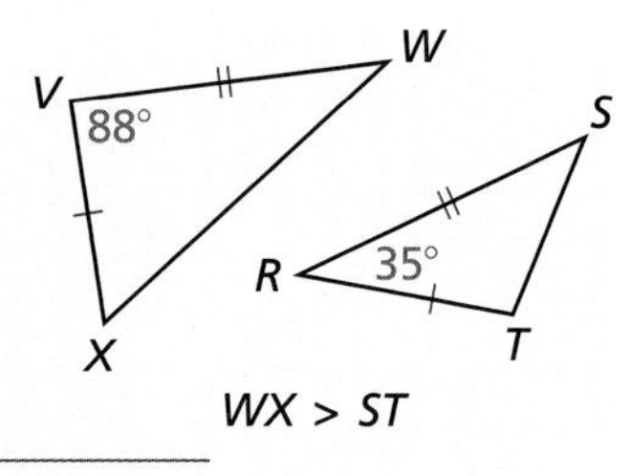

Proof *BigIdeasMath.com*

Theorem 6.13 Converse of the Hinge Theorem

If two sides of one triangle are congruent to two sides of another triangle, and the third side of the first is longer than the third side of the second, then the included angle of the first is larger than the included angle of the second.

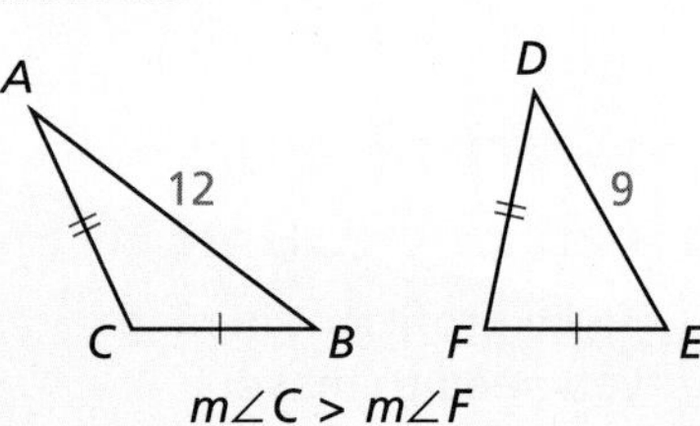

Proof Example 3, p. 345

EXAMPLE 1 Using the Converse of the Hinge Theorem

Given that $\overline{ST} \cong \overline{PR}$, how does $m\angle PST$ compare to $m\angle SPR$?

SOLUTION

You are given that $\overline{ST} \cong \overline{PR}$, and you know that $\overline{PS} \cong \overline{PS}$ by the Reflexive Property of Congruence (Theorem 2.1). Because 24 inches > 23 inches, $PT > SR$. So, two sides of $\triangle STP$ are congruent to two sides of $\triangle PRS$ and the third side of $\triangle STP$ is longer.

- By the Converse of the Hinge Theorem, $m\angle PST > m\angle SPR$.

EXAMPLE 2 Using the Hinge Theorem

Given that $\overline{JK} \cong \overline{LK}$, how does JM compare to LM?

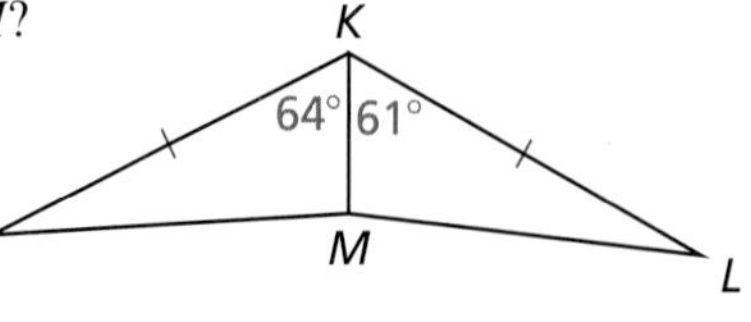

SOLUTION

You are given that $\overline{JK} \cong \overline{LK}$, and you know that $\overline{KM} \cong \overline{KM}$ by the Reflexive Property of Congruence (Theorem 2.1). Because $64° > 61°$, $m\angle JKM > m\angle LKM$. So, two sides of $\triangle JKM$ are congruent to two sides of $\triangle LKM$, and the included angle in $\triangle JKM$ is larger.

▶ By the Hinge Theorem, $JM > LM$.

Monitoring Progress  Help in English and Spanish at *BigIdeasMath.com*

Use the diagram.

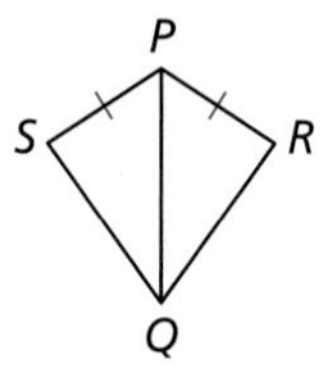

1. If $PR = PS$ and $m\angle QPR > m\angle QPS$, which is longer, $\overline{SQ}$ or $\overline{RQ}$?
2. If $PR = PS$ and $RQ < SQ$, which is larger, $\angle RPQ$ or $\angle SPQ$?

EXAMPLE 3 Proving the Converse of the Hinge Theorem

Write an indirect proof of the Converse of the Hinge Theorem.

Given $\overline{AB} \cong \overline{DE}, \overline{BC} \cong \overline{EF}, AC > DF$

Prove $m\angle B > m\angle E$

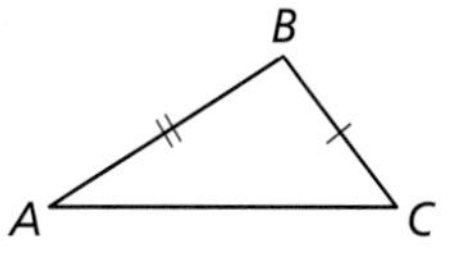

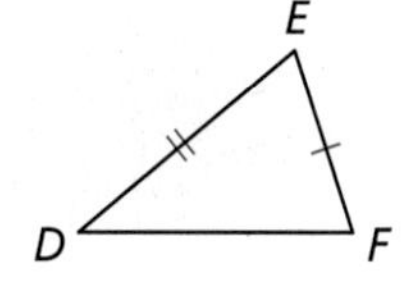

Indirect Proof

Step 1 Assume temporarily that $m\angle B \ngtr m\angle E$. Then it follows that either $m\angle B < m\angle E$ or $m\angle B = m\angle E$.

Step 2 If $m\angle B < m\angle E$, then $AC < DF$ by the Hinge Theorem.

If $m\angle B = m\angle E$, then $\angle B \cong \angle E$. So, $\triangle ABC \cong \triangle DEF$ by the SAS Congruence Theorem (Theorem 5.5) and $AC = DF$.

Step 3 Both conclusions contradict the given statement that $AC > DF$. So, the temporary assumption that $m\angle B \ngtr m\angle E$ cannot be true. This proves that $m\angle B > m\angle E$.

EXAMPLE 4 Proving Triangle Relationships

Write a paragraph proof.

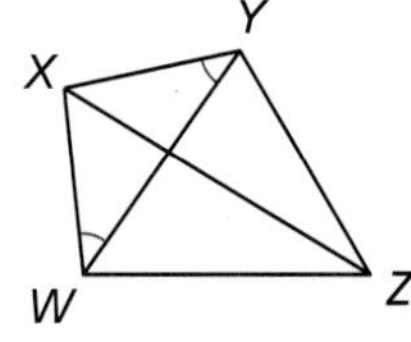

Given $\angle XWY \cong \angle XYW, WZ > YZ$

Prove $m\angle WXZ > m\angle YXZ$

Paragraph Proof Because $\angle XWY \cong \angle XYW$, $\overline{XY} \cong \overline{XW}$ by the Converse of the Base Angles Theorem (Theorem 5.7). By the Reflexive Property of Congruence (Theorem 2.1), $\overline{XZ} \cong \overline{XZ}$. Because $WZ > YZ$, $m\angle WXZ > m\angle YXZ$ by the Converse of the Hinge Theorem.

Monitoring Progress  Help in English and Spanish at *BigIdeasMath.com*

3. Write a temporary assumption you can make to prove the Hinge Theorem indirectly. What two cases does that assumption lead to?

Solving Real-Life Problems

EXAMPLE 5 Solving a Real-Life Problem

Two groups of bikers leave the same camp heading in opposite directions. Each group travels 2 miles, then changes direction and travels 1.2 miles. Group A starts due east and then turns 45° toward north. Group B starts due west and then turns 30° toward south. Which group is farther from camp? Explain your reasoning.

SOLUTION

1. **Understand the Problem** You know the distances and directions that the groups of bikers travel. You need to determine which group is farther from camp. You can interpret a turn of 45° toward north, as shown.

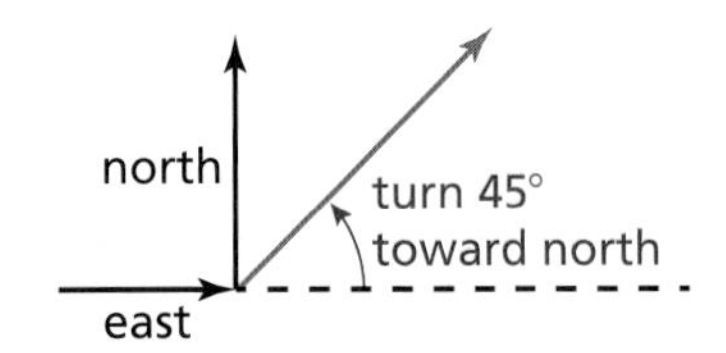

2. **Make a Plan** Draw a diagram that represents the situation and mark the given measures. The distances that the groups bike and the distances back to camp form two triangles. The triangles have two congruent side lengths of 2 miles and 1.2 miles. Include the third side of each triangle in the diagram.

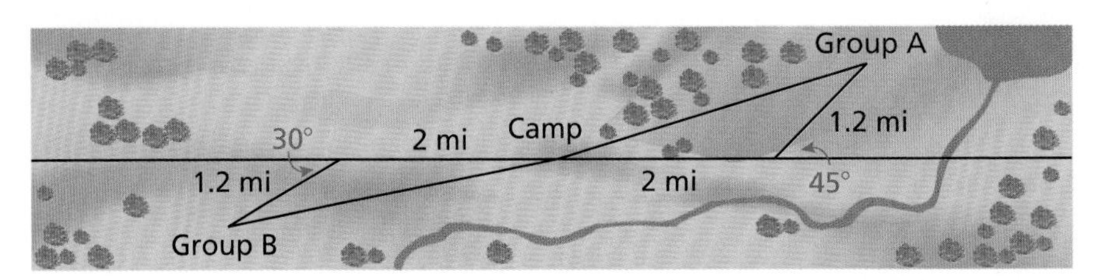

3. **Solve the Problem** Use linear pairs to find the included angles for the paths that the groups take.

Group A: $180° - 45° = 135°$ **Group B:** $180° - 30° = 150°$

The included angles are 135° and 150°.

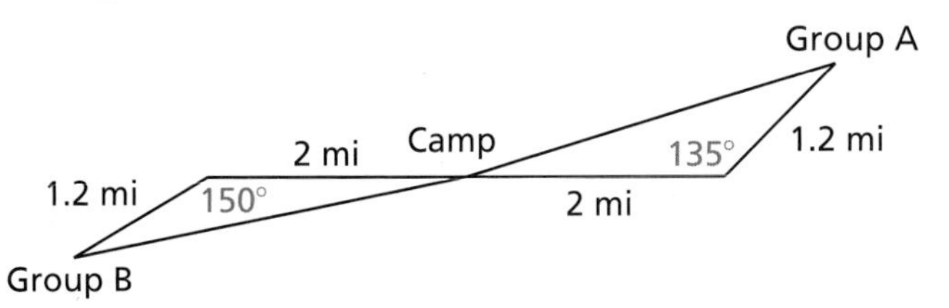

Because $150° > 135°$, the distance Group B is from camp is greater than the distance Group A is from camp by the Hinge Theorem.

▶ So, Group B is farther from camp.

4. **Look Back** Because the included angle for Group A is 15° less than the included angle for Group B, you can reason that Group A would be closer to camp than Group B. So, Group B is farther from camp.

Monitoring Progress

Help in English and Spanish at *BigIdeasMath.com*

4. **WHAT IF?** In Example 5, Group C leaves camp and travels 2 miles due north, then turns 40° toward east and travels 1.2 miles. Compare the distances from camp for all three groups.

6.6 Exercises

Dynamic Solutions available at *BigIdeasMath.com*

Vocabulary and Core Concept Check

1. **WRITING** Explain why Theorem 6.12 is named the "Hinge Theorem."

2. **COMPLETE THE SENTENCE** In $\triangle ABC$ and $\triangle DEF$, $\overline{AB} \cong \overline{DE}$, $\overline{BC} \cong \overline{EF}$, and $AC < DF$. So $m\angle$_____ $> m\angle$_____ by the Converse of the Hinge Theorem (Theorem 6.13).

Monitoring Progress and Modeling with Mathematics

In Exercises 3–6, copy and complete the statement with <, >, or =. Explain your reasoning. *(See Example 1.)*

3. $m\angle 1$ _____ $m\angle 2$

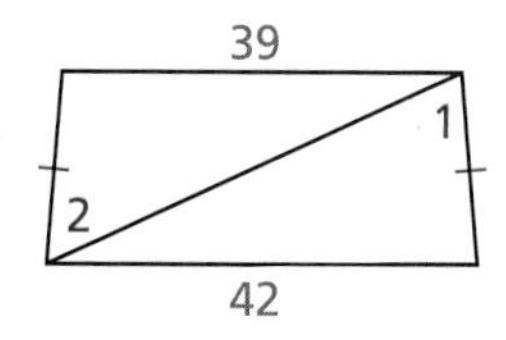

4. $m\angle 1$ _____ $m\angle 2$

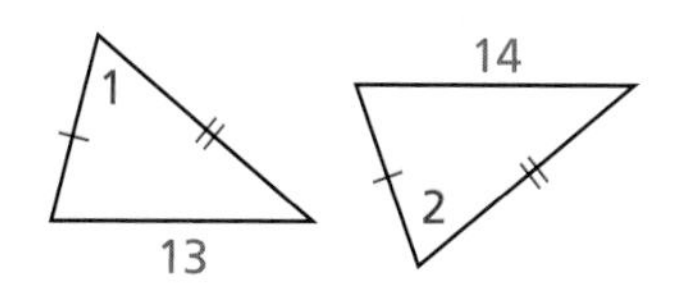

5. $m\angle 1$ _____ $m\angle 2$

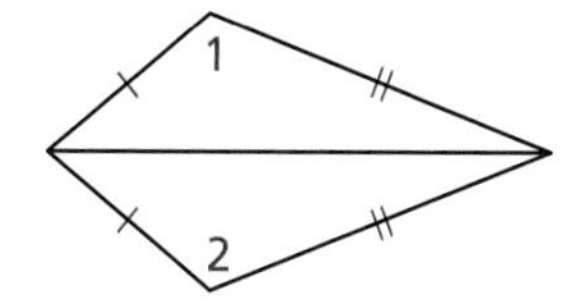

6. $m\angle 1$ _____ $m\angle 2$

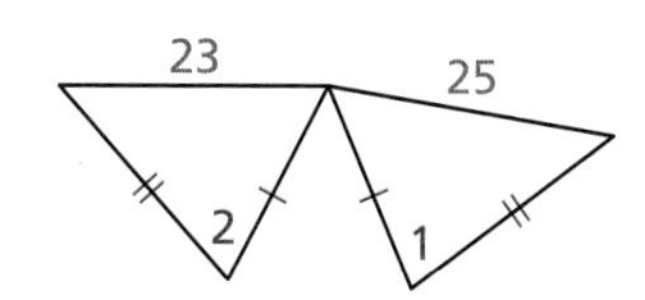

In Exercises 7–10, copy and complete the statement with <, >, or =. Explain your reasoning. *(See Example 2.)*

7. AD _____ CD

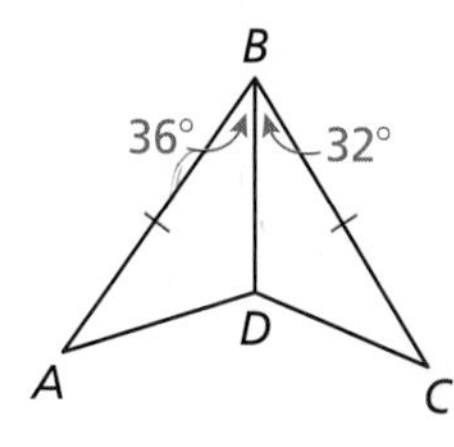

8. MN _____ LK

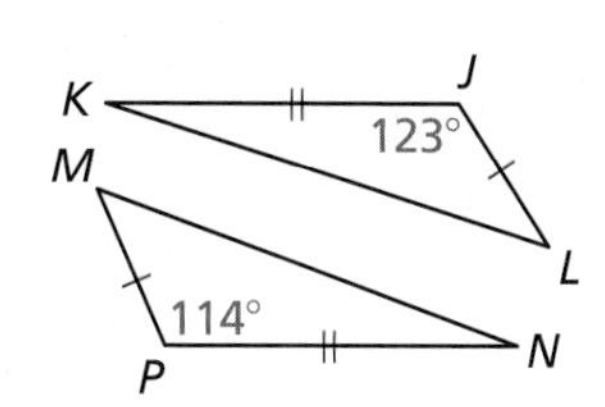

9. TR _____ UR

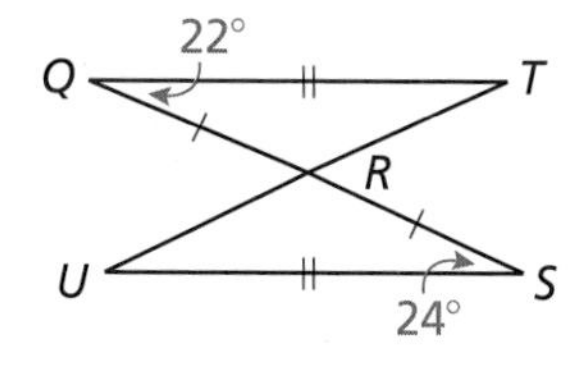

10. AC _____ DC

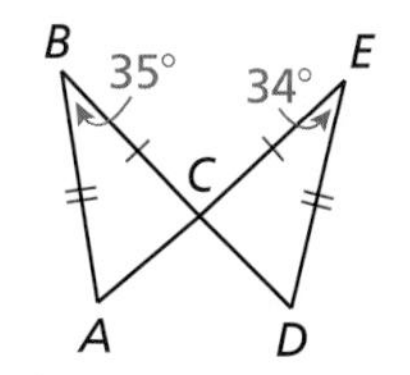

PROOF In Exercises 11 and 12, write a proof. *(See Example 4.)*

11. **Given** $\overline{XY} \cong \overline{YZ}$, $m\angle WYZ > m\angle WYX$

    Prove $WZ > WX$

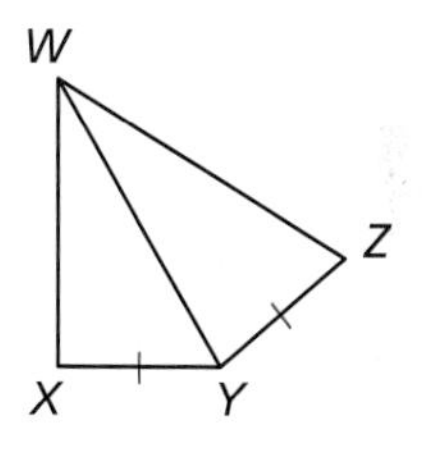

12. **Given** $\overline{BC} \cong \overline{DA}$, $DC < AB$

    Prove $m\angle BCA > m\angle DAC$

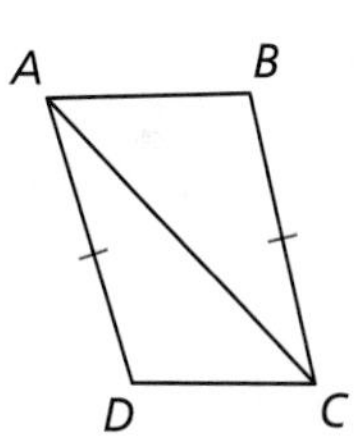

In Exercises 13 and 14, you and your friend leave on different flights from the same airport. Determine which flight is farther from the airport. Explain your reasoning. *(See Example 5.)*

13. **Your flight:** Flies 100 miles due west, then turns 20° toward north and flies 50 miles.

    Friend's flight: Flies 100 miles due north, then turns 30° toward east and flies 50 miles.

14. **Your flight:** Flies 210 miles due south, then turns 70° toward west and flies 80 miles.

    Friend's flight: Flies 80 miles due north, then turns 50° toward east and flies 210 miles.

15. ERROR ANALYSIS Describe and correct the error in using the Hinge Theorem (Theorem 6.12).

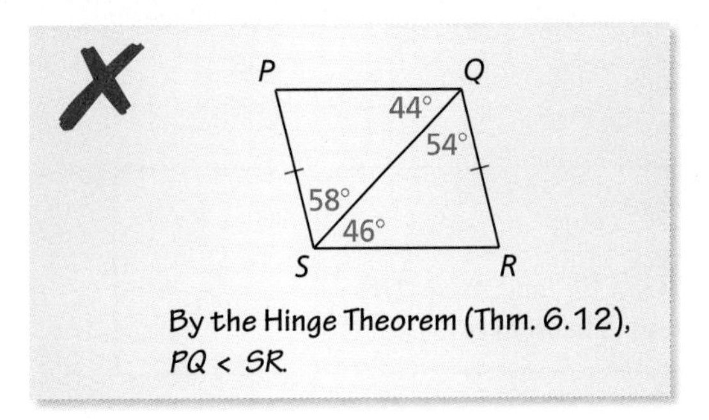

16. REPEATED REASONING Which is a possible measure for $\angle JKM$? Select all that apply.

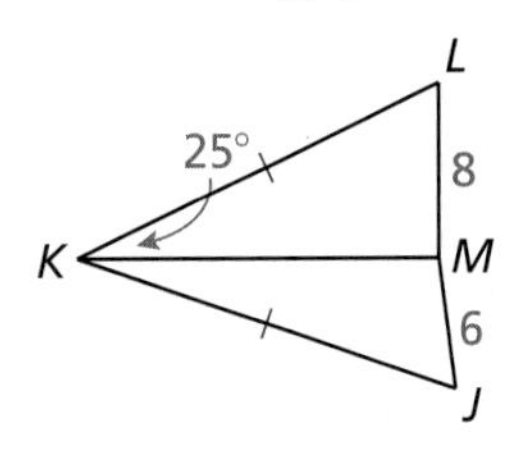

Ⓐ 15°   Ⓑ 22°   Ⓒ 25°   Ⓓ 35°

17. DRAWING CONCLUSIONS The path from E to F is longer than the path from E to D. The path from G to D is the same length as the path from G to F. What can you conclude about the angles of the paths? Explain your reasoning.

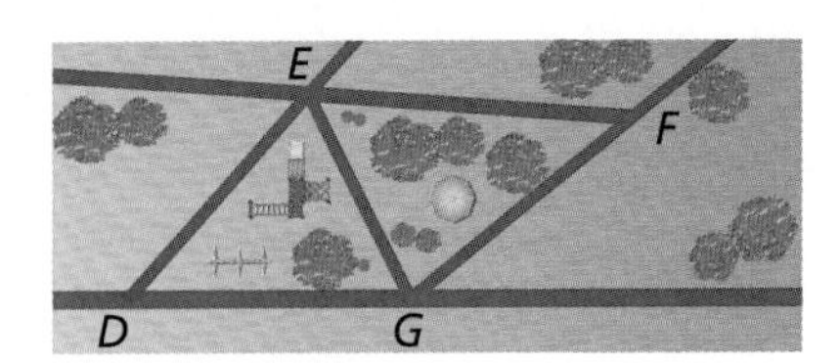

18. ABSTRACT REASONING In $\triangle EFG$, the bisector of $\angle F$ intersects the bisector of $\angle G$ at point H. Explain why $\overline{FG}$ must be longer than $\overline{FH}$ or $\overline{HG}$.

19. ABSTRACT REASONING $\overline{NR}$ is a median of $\triangle NPQ$, and $NQ > NP$. Explain why $\angle NRQ$ is obtuse.

MATHEMATICAL CONNECTIONS In Exercises 20 and 21, write and solve an inequality for the possible values of x.

20.

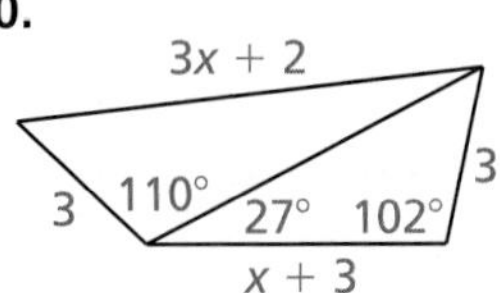

21.

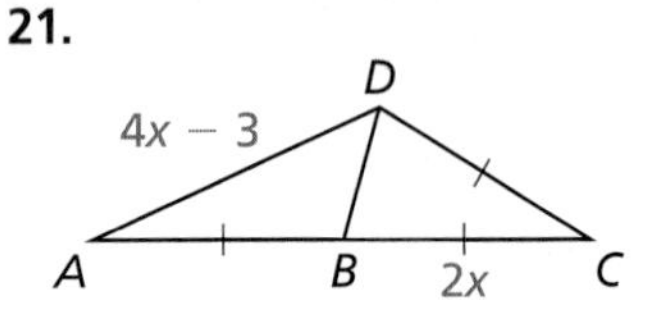

22. HOW DO YOU SEE IT? In the diagram, triangles are formed by the locations of the players on the basketball court. The dashed lines represent the possible paths of the basketball as the players pass. How does $m\angle ACB$ compare with $m\angle ACD$?

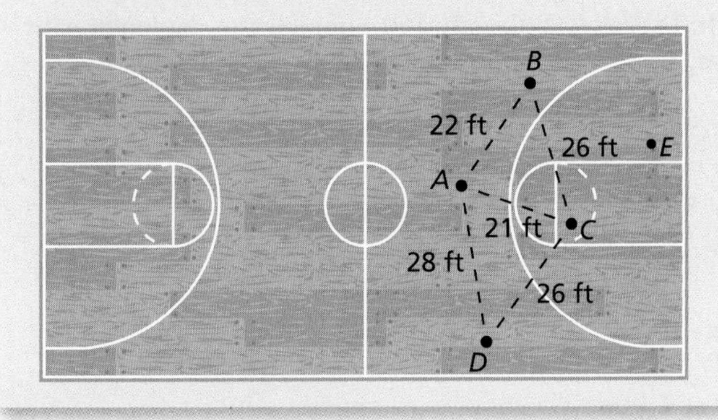

23. CRITICAL THINKING In $\triangle ABC$, the altitudes from B and C meet at point D, and $m\angle BAC > m\angle BDC$. What is true about $\triangle ABC$? Justify your answer.

24. THOUGHT PROVOKING The postulates and theorems in this book represent Euclidean geometry. In spherical geometry, all points are on the surface of a sphere. A line is a circle on the sphere whose diameter is equal to the diameter of the sphere. In spherical geometry, state an inequality involving the sum of the angles of a triangle. Find a formula for the area of a triangle in spherical geometry.

Maintaining Mathematical Proficiency

Reviewing what you learned in previous grades and lessons

Find the value of x. *(Section 5.1 and Section 5.4)*

25.

26.

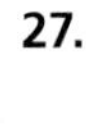

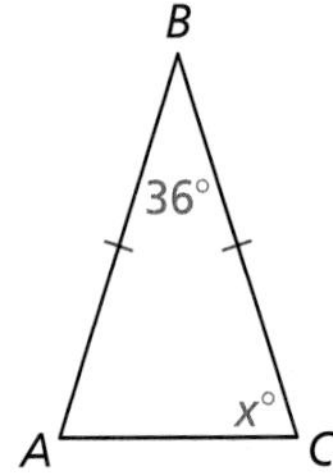

27.

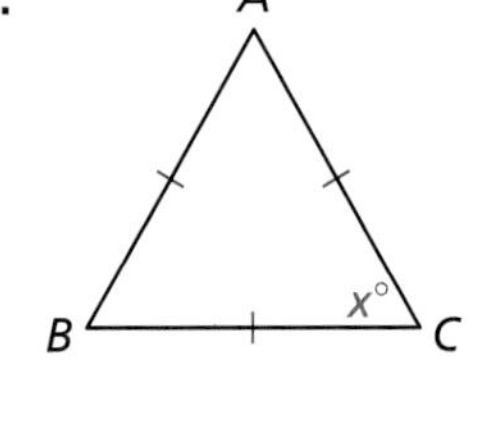

28.

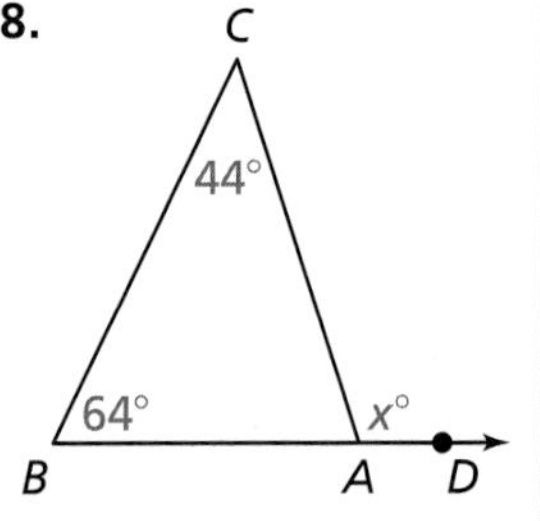

6.4–6.6 What Did You Learn?

Core Vocabulary

midsegment of a triangle, *p. 330*
indirect proof, *p. 336*

Core Concepts

Section 6.4

Using the Midsegment of a Triangle, *p. 330*
Theorem 6.8 Triangle Midsegment Theorem, *p. 331*

Section 6.5

How to Write an Indirect Proof (Proof by Contradiction), *p. 336*
Theorem 6.9 Triangle Longer Side Theorem, *p. 337*
Theorem 6.10 Triangle Larger Angle Theorem, *p. 337*
Theorem 6.11 Triangle Inequality Theorem, *p. 339*

Section 6.6

Theorem 6.12 Hinge Theorem, *p. 344*
Theorem 6.13 Converse of the Hinge Theorem, *p. 344*

Mathematical Practices

1. In Exercise 25 on page 334, analyze the relationship between the stage and the total perimeter of all the shaded triangles at that stage. Then predict the total perimeter of all the shaded triangles in Stage 4.
2. In Exercise 17 on page 340, write all three inequalities using the Triangle Inequality Theorem (Theorem 6.11). Determine the reasonableness of each one. Why do you only need to use two of the three inequalities?
3. In Exercise 23 on page 348, try all three cases of triangles (acute, right, obtuse) to gain insight into the solution.

Performance Task

Bicycle Renting Stations

The city planners for a large town want to add bicycle renting stations around downtown. How will you decide the best locations? Where will you place the rental stations based on the ideas of the city planners?

To explore the answers to these questions and more, go to ***BigIdeasMath.com***.

6 Chapter Review

Dynamic Solutions available at *BigIdeasMath.com*

6.1 Perpendicular and Angle Bisectors (pp. 301–308)

Find *AD*.

From the figure, $\overleftrightarrow{AC}$ is the perpendicular bisector of $\overline{BD}$.

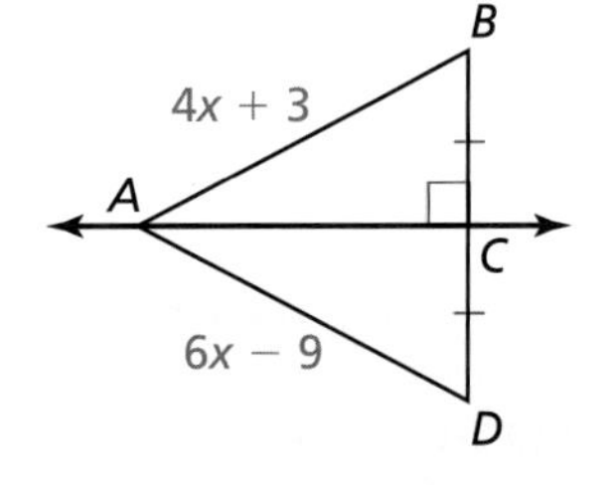

$AB = AD$	Perpendicular Bisector Theorem (Theorem 6.1)
$4x + 3 = 6x - 9$	Substitute.
$x = 6$	Solve for *x*.

▶ So, $AD = 6(6) - 9 = 27$.

Find the indicated measure. Explain your reasoning.

1. *DC*

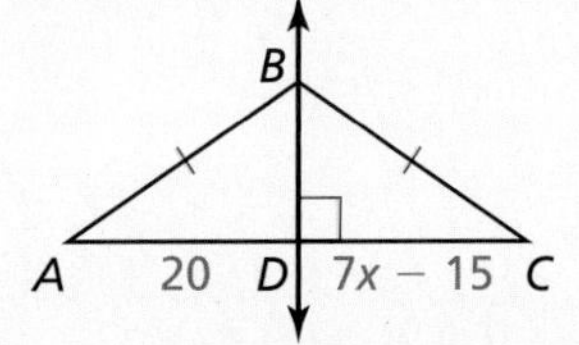

2. *RS*

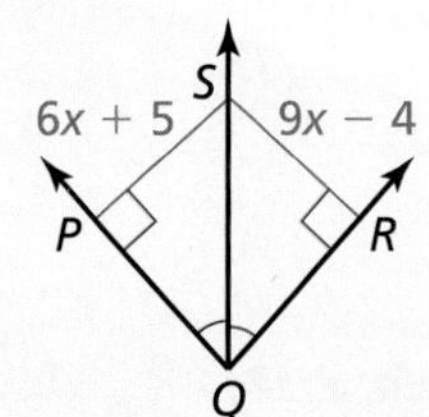

3. $m\angle JFH$

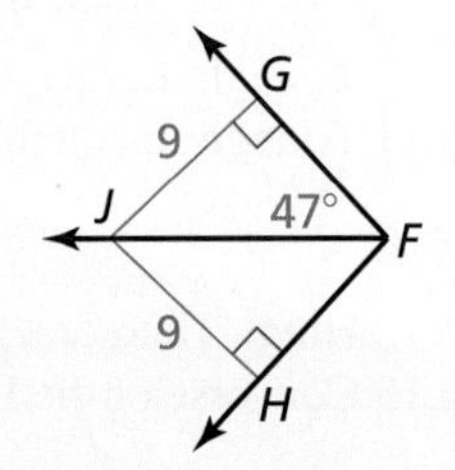

6.2 Bisectors of Triangles (pp. 309–318)

Find the coordinates of the circumcenter of $\triangle QRS$ with vertices $Q(3, 3)$, $R(5, 7)$, and $S(9, 3)$.

Step 1 Graph $\triangle QRS$.

Step 2 Find equations for two perpendicular bisectors.

The midpoint of $\overline{QS}$ is (6, 3). The line through (6, 3) that is perpendicular to $\overline{QS}$ is $x = 6$.

The midpoint of $\overline{QR}$ is (4, 5). The line through (4, 5) that is perpendicular to $\overline{QR}$ is $y = -\frac{1}{2}x + 7$.

Step 3 Find the point where $x = 6$ and $y = -\frac{1}{2}x + 7$ intersect. They intersect at (6, 4).

▶ So, the coordinates of the circumcenter are (6, 4).

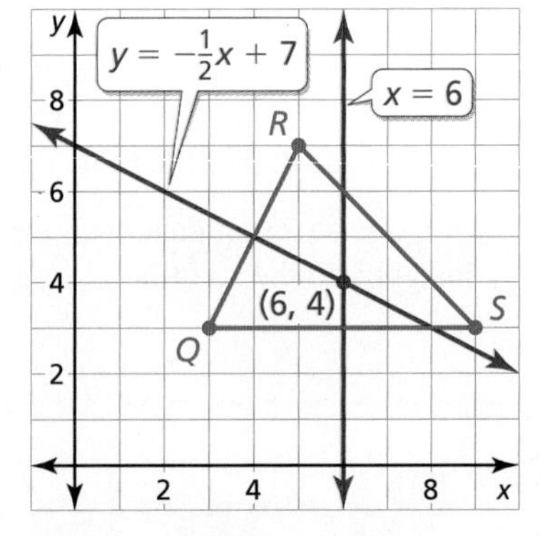

Find the coordinates of the circumcenter of the triangle with the given vertices.

4. $T(-6, -5)$, $U(0, -1)$, $V(0, -5)$

5. $X(-2, 1)$, $Y(2, -3)$, $Z(6, -3)$

6. Point *D* is the incenter of $\triangle LMN$. Find the value of *x*.

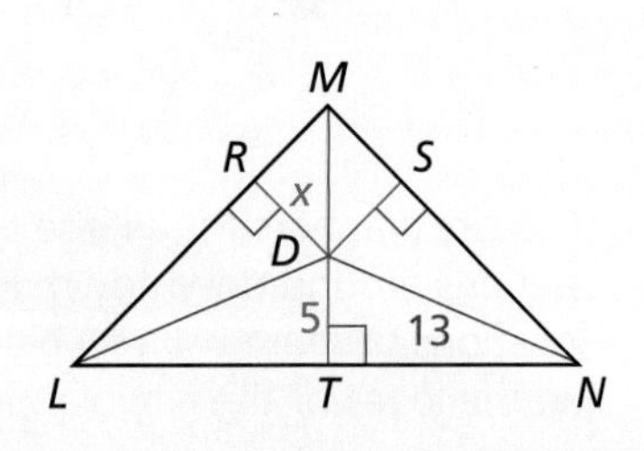

6.3 Medians and Altitudes of Triangles (pp. 319–326)

Find the coordinates of the centroid of $\triangle TUV$ with vertices $T(1, -8)$, $U(4, -1)$, and $V(7, -6)$.

Step 1 Graph $\triangle TUV$.

Step 2 Use the Midpoint Formula to find the midpoint W of $\overline{TV}$. Sketch median $\overline{UW}$.

$$W\left(\frac{1+7}{2}, \frac{-8+(-6)}{2}\right) = (4, -7)$$

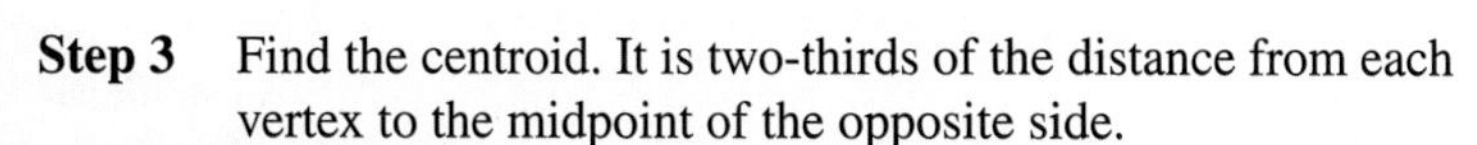

Step 3 Find the centroid. It is two-thirds of the distance from each vertex to the midpoint of the opposite side.

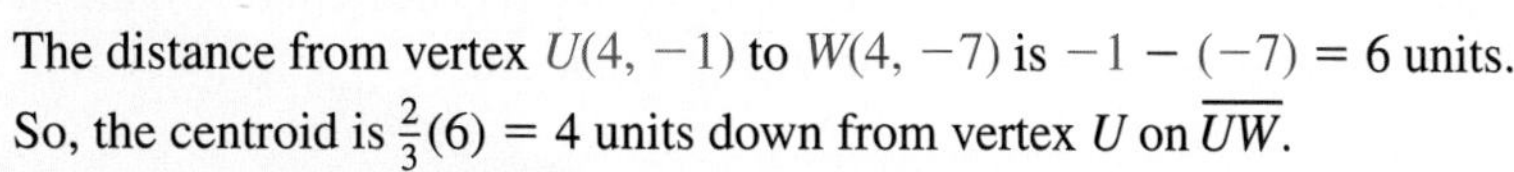

The distance from vertex $U(4, -1)$ to $W(4, -7)$ is $-1 - (-7) = 6$ units.

So, the centroid is $\frac{2}{3}(6) = 4$ units down from vertex U on $\overline{UW}$.

▶ So, the coordinates of the centroid P are $(4, -1 - 4)$, or $(4, -5)$.

Find the coordinates of the centroid of the triangle with the given vertices.

7. $A(-10, 3)$, $B(-4, 5)$, $C(-4, 1)$

8. $D(2, -8)$, $E(2, -2)$, $F(8, -2)$

Tell whether the orthocenter of the triangle with the given vertices is *inside*, *on*, or *outside* the triangle. Then find the coordinates of the orthocenter.

9. $G(1, 6)$, $H(5, 6)$, $J(3, 1)$

10. $K(-8, 5)$, $L(-6, 3)$, $M(0, 5)$

6.4 The Triangle Midsegment Theorem (pp. 329–334)

In $\triangle JKL$, show that midsegment $\overline{MN}$ is parallel to $\overline{JL}$ and that $MN = \frac{1}{2}JL$.

Step 1 Find the coordinates of M and N by finding the midpoints of $\overline{JK}$ and $\overline{KL}$.

$$M\left(\frac{-8+(-4)}{2}, \frac{1+7}{2}\right) = M\left(\frac{-12}{2}, \frac{8}{2}\right) = M(-6, 4)$$

$$N\left(\frac{-4+(-2)}{2}, \frac{7+3}{2}\right) = N\left(\frac{-6}{2}, \frac{10}{2}\right) = N(-3, 5)$$

Step 2 Find and compare the slopes of $\overline{MN}$ and $\overline{JL}$.

$$\text{slope of } \overline{MN} = \frac{5-4}{-3-(-6)} = \frac{1}{3}$$

$$\text{slope of } \overline{JL} = \frac{3-1}{-2-(-8)} = \frac{2}{6} = \frac{1}{3}$$

▶ Because the slopes are the same, $\overline{MN}$ is parallel to $\overline{JL}$.

Step 3 Find and compare the lengths of $\overline{MN}$ and $\overline{JL}$.

$$MN = \sqrt{[-3-(-6)]^2 + (5-4)^2} = \sqrt{9+1} = \sqrt{10}$$

$$JL = \sqrt{[-2-(-8)]^2 + (3-1)^2} = \sqrt{36+4} = \sqrt{40} = 2\sqrt{10}$$

▶ Because $\sqrt{10} = \frac{1}{2}(2\sqrt{10})$, $MN = \frac{1}{2}JL$.

Find the coordinates of the vertices of the midsegment triangle for the triangle with the given vertices.

11. $A(-6, 8)$, $B(-6, 4)$, $C(0, 4)$

12. $D(-3, 1)$, $E(3, 5)$, $F(1, -5)$

6.5 Indirect Proof and Inequalities in One Triangle (pp. 335–342)

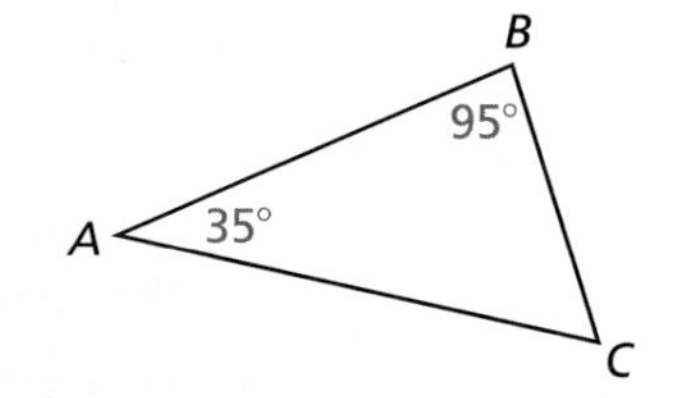

a. List the sides of $\triangle ABC$ in order from shortest to longest.

First, find $m\angle C$ using the Triangle Sum Theorem (Thm. 5.1).

$$m\angle A + m\angle B + m\angle C = 180^\circ$$
$$35^\circ + 95^\circ + m\angle C = 180^\circ$$
$$m\angle C = 50^\circ$$

The angles from smallest to largest are $\angle A$, $\angle C$, and $\angle B$. The sides opposite these angles are $\overline{BC}$, $\overline{AB}$, and $\overline{AC}$, respectively.

▶ So, by the Triangle Larger Angle Theorem (Theorem 6.10), the sides from shortest to longest are $\overline{BC}$, $\overline{AB}$, and $\overline{AC}$.

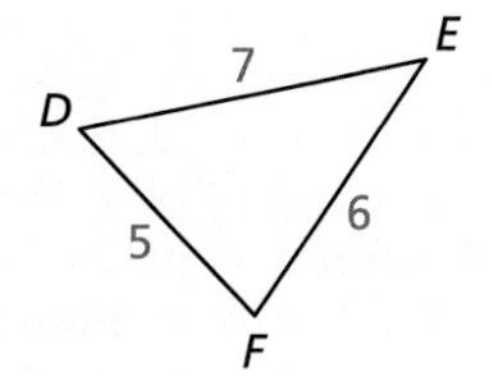

b. List the angles of $\triangle DEF$ in order from smallest to largest.

The sides from shortest to longest are $\overline{DF}$, $\overline{EF}$, and $\overline{DE}$. The angles opposite these sides are $\angle E$, $\angle D$, and $\angle F$, respectively.

▶ So, by the Triangle Longer Side Theorem (Theorem 6.9), the angles from smallest to largest are $\angle E$, $\angle D$, and $\angle F$.

Describe the possible lengths of the third side of the triangle given the lengths of the other two sides.

13. 4 inches, 8 inches **14.** 6 meters, 9 meters **15.** 11 feet, 18 feet

16. Write an indirect proof of the statement "In $\triangle XYZ$, if $XY = 4$ and $XZ = 8$, then $YZ > 4$."

6.6 Inequalities in Two Triangles (pp. 343–348)

Given that $\overline{WZ} \cong \overline{YZ}$, how does XY compare to XW?

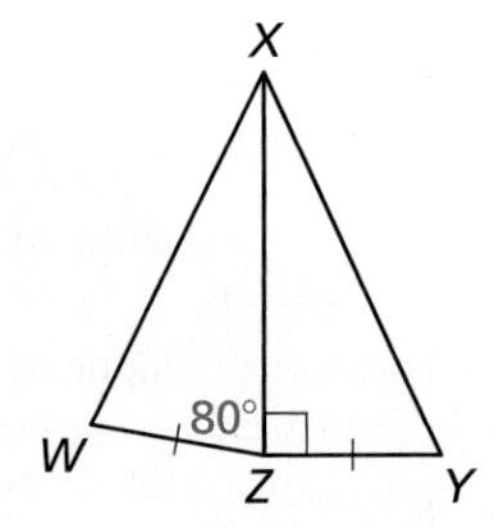

You are given that $\overline{WZ} \cong \overline{YZ}$, and you know that $\overline{XZ} \cong \overline{XZ}$ by the Reflexive Property of Congruence (Theorem 2.1).

Because $90^\circ > 80^\circ$, $m\angle XZY > m\angle XZW$. So, two sides of $\triangle XZY$ are congruent to two sides of $\triangle XZW$ and the included angle in $\triangle XZY$ is larger.

▶ By the Hinge Theorem (Theorem 6.12), $XY > XW$.

Use the diagram.

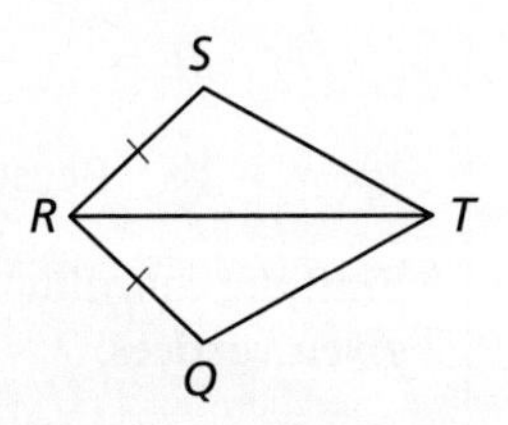

17. If $RQ = RS$ and $m\angle QRT > m\angle SRT$, then how does $\overline{QT}$ compare to $\overline{ST}$?

18. If $RQ = RS$ and $QT > ST$, then how does $\angle QRT$ compare to $\angle SRT$?

6 Chapter Test

In Exercises 1 and 2, $\overline{MN}$ is a midsegment of $\triangle JKL$. Find the value of x.

1.

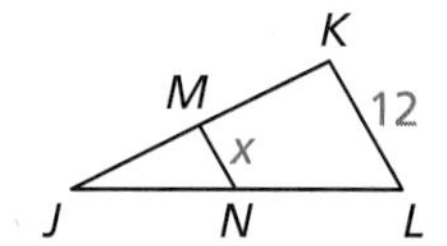

2.

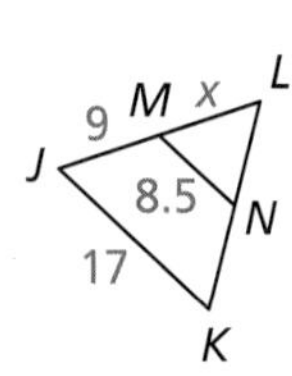

Find the indicated measure. Identify the theorem you use.

3. ST

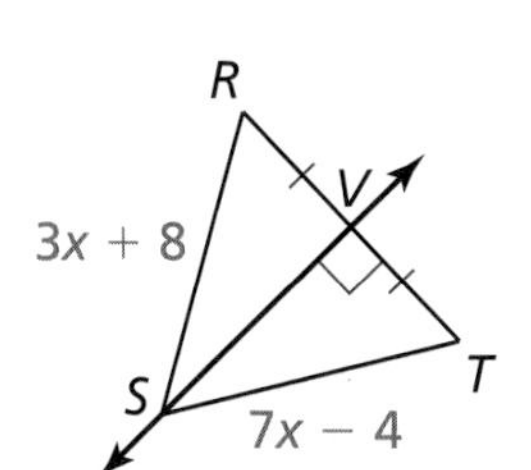

4. WY

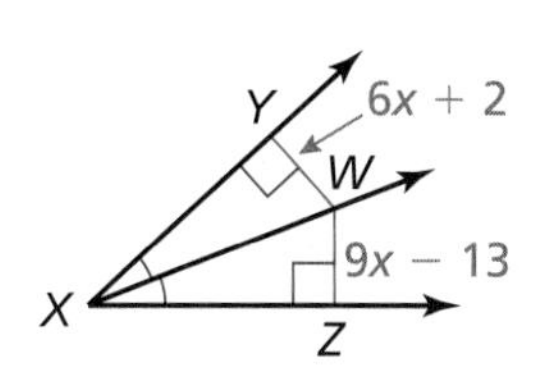

5. BW

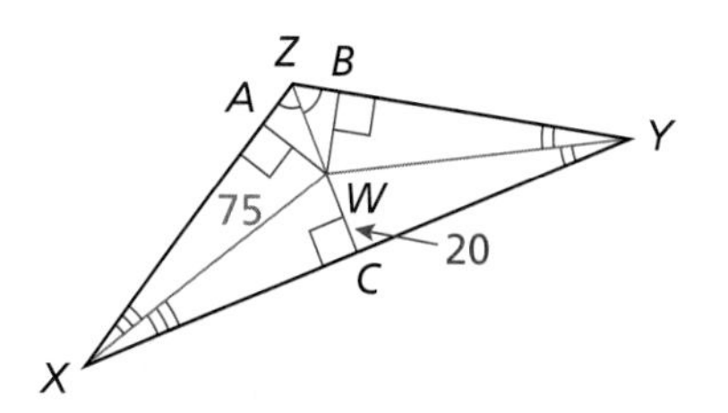

Copy and complete the statement with <, >, or =.

6. AB __ CB

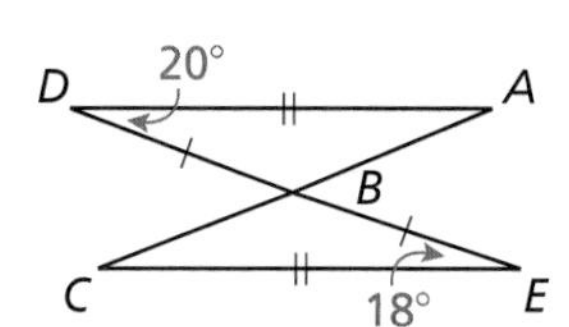

7. $m\angle 1$ ___ $m\angle 2$

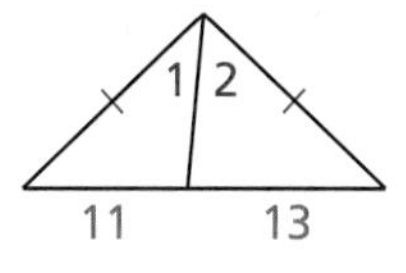

8. $m\angle MNP$ ___ $m\angle NPM$

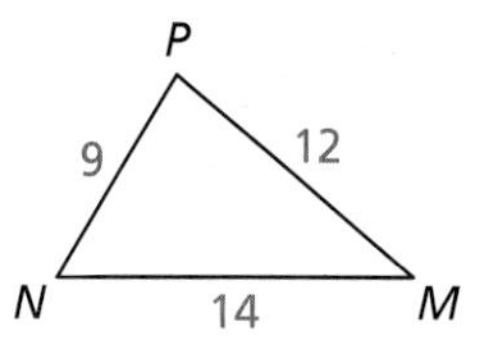

9. Find the coordinates of the circumcenter, orthocenter, and centroid of the triangle with vertices $A(0, -2)$, $B(4, -2)$, and $C(0, 6)$.

10. Write an indirect proof of the Corollary to the Base Angles Theorem (Corollary 5.2): If $\triangle PQR$ is equilateral, then it is equiangular.

11. $\triangle DEF$ is a right triangle with area A. Use the area for $\triangle DEF$ to write an expression for the area of $\triangle GEH$. Justify your answer.

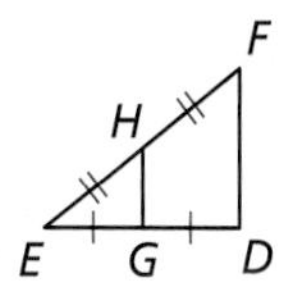

12. Two hikers start at a visitor center. The first hikes 4 miles due west, then turns 40° toward south and hikes 1.8 miles. The second hikes 4 miles due east, then turns 52° toward north and hikes 1.8 miles. Which hiker is farther from the visitor center? Explain how you know.

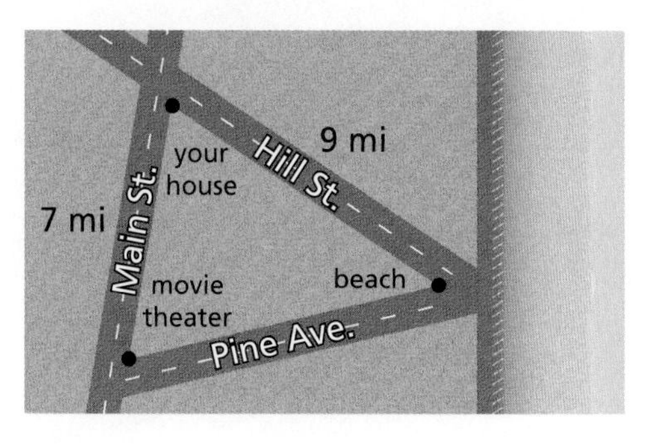

In Exercises 13–15, use the map.

13. Describe the possible lengths of Pine Avenue.

14. You ride your bike along a trail that represents the shortest distance from the beach to Main Street. You end up exactly halfway between your house and the movie theatre. How long is Pine Avenue? Explain.

15. A market is the same distance from your house, the movie theater, and the beach. Copy the map and locate the market.

6 Cumulative Assessment

1. Which definition(s) and/or theorem(s) do you need to use to prove the Converse of the Perpendicular Bisector Theorem (Theorem 6.2)? Select all that apply.

Given $CA = CB$

Prove Point C lies on the perpendicular bisector of $\overline{AB}$.

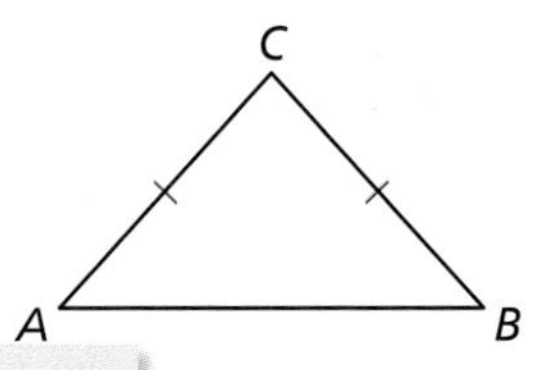

definition of perpendicular bisector	definition of angle bisector
definition of segment congruence	definition of angle congruence
Base Angles Theorem (Theorem 5.6)	Converse of the Base Angles Theorem (Theorem 5.7)
ASA Congruence Theorem (Theorem 5.10)	AAS Congruence Theorem (Theorem 5.11)

2. Use the given information to write a two-column proof.

Given $\overline{YG}$ is the perpendicular bisector of $\overline{DF}$.

Prove $\triangle DEY \cong \triangle FEY$

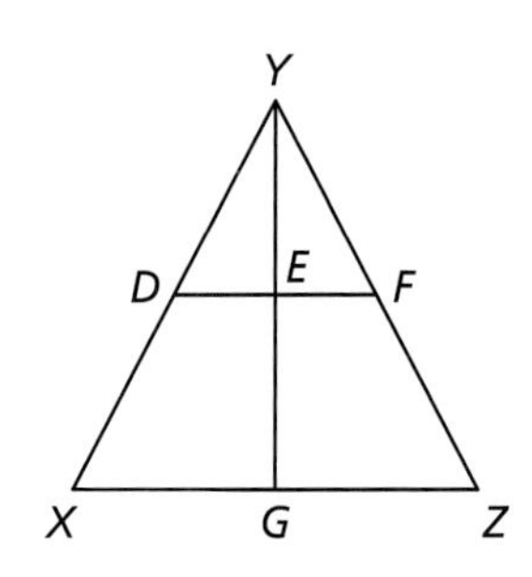

3. What are the coordinates of the centroid of $\triangle LMN$?

Ⓐ (2, 5)

Ⓑ (3, 5)

Ⓒ (4, 5)

Ⓓ (5, 5)

4. Use the steps in the construction to explain how you know that the circle is circumscribed about $\triangle ABC$.

Step 1

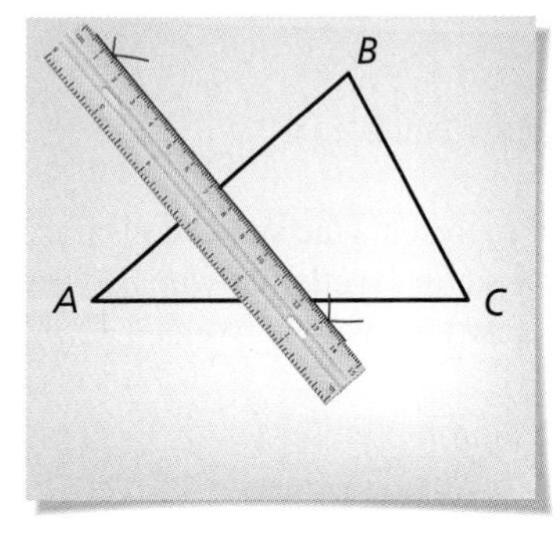

Step 2

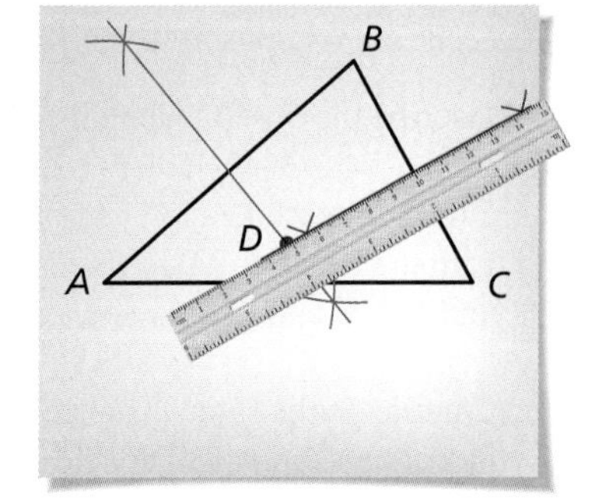

Step 3

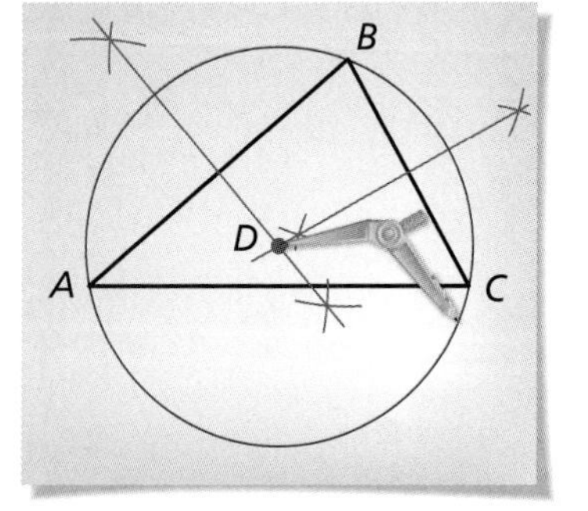

5. Enter the missing reasons in the proof of the Base Angles Theorem (Theorem 5.6).

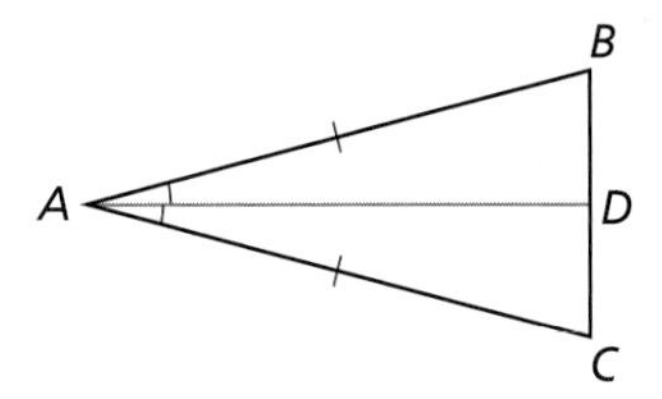

Given $\overline{AB} \cong \overline{AC}$

Prove $\angle B \cong \angle C$

STATEMENTS	REASONS
1. Draw $\overline{AD}$, the angle bisector of $\angle CAB$.	1. Construction of angle bisector
2. $\angle CAD \cong \angle BAD$	2. ________________
3. $\overline{AB} \cong \overline{AC}$	3. ________________
4. $\overline{DA} \cong \overline{DA}$	4. ________________
5. $\triangle ADB \cong \triangle ADC$	5. ________________
6. $\angle B \cong \angle C$	6. ________________

6. Use the graph of $\triangle QRS$.

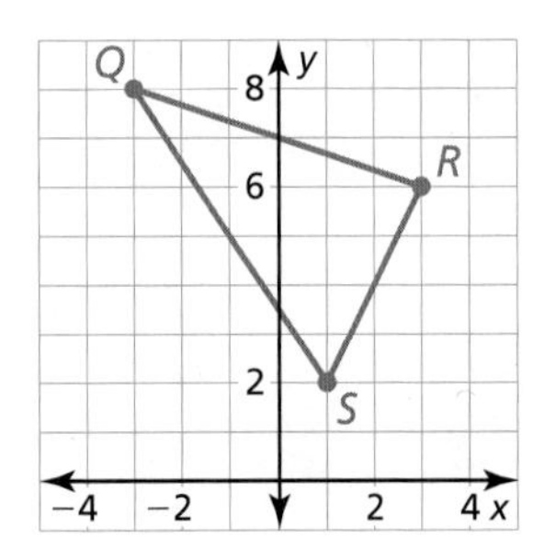

a. Find the coordinates of the vertices of the midsegment triangle. Label the vertices T, U, and V.

b. Show that each midsegment joining the midpoints of two sides is parallel to the third side and is equal to half the length of the third side.

7. A triangle has vertices $X(-2, 2)$, $Y(1, 4)$, and $Z(2, -2)$. Your friend claims that a translation of $(x, y) \rightarrow (x + 2, y - 3)$ and a dilation by a scale factor of 3 will produce a similarity transformation. Do you support your friend's claim? Explain your reasoning.

8. The graph shows a dilation of quadrilateral $ABCD$ by a scale factor of 2. Show that the line containing points B and D is parallel to the line containing points B' and D'.

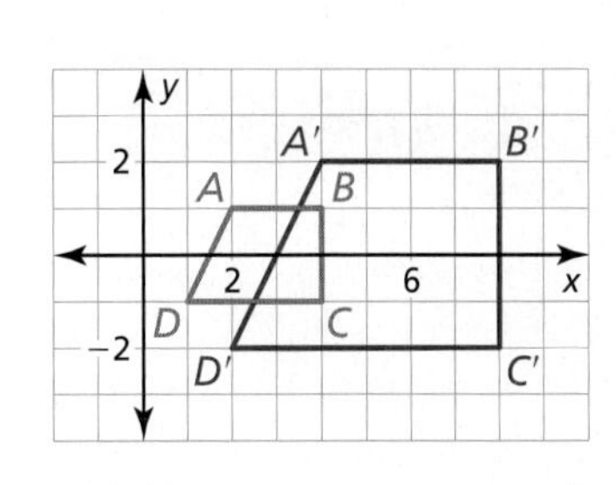

7 Quadrilaterals and Other Polygons

Diamond *(p. 406)*

Window *(p. 395)*

Amusement Park Ride *(p. 377)*

Arrow *(p. 373)*

Gazebo *(p. 365)*

Maintaining Mathematical Proficiency

Using Structure to Solve a Multi-Step Equation

Example 1 **Solve $3(2 + x) = -9$ by interpreting the expression $2 + x$ as a single quantity.**

$3(2 + x) = -9$	Write the equation.
$\frac{3(2 + x)}{3} = \frac{-9}{3}$	Divide each side by 3.
$2 + x = -3$	Simplify.
$\underline{-2} \quad \underline{-2}$	Subtract 2 from each side.
$x = -5$	Simplify.

Solve the equation by interpreting the expression in parentheses as a single quantity.

1. $4(7 - x) = 16$

2. $7(1 - x) + 2 = -19$

3. $3(x - 5) + 8(x - 5) = 22$

Identifying Parallel and Perpendicular Lines

Example 2 **Determine which of the lines are parallel and which are perpendicular.**

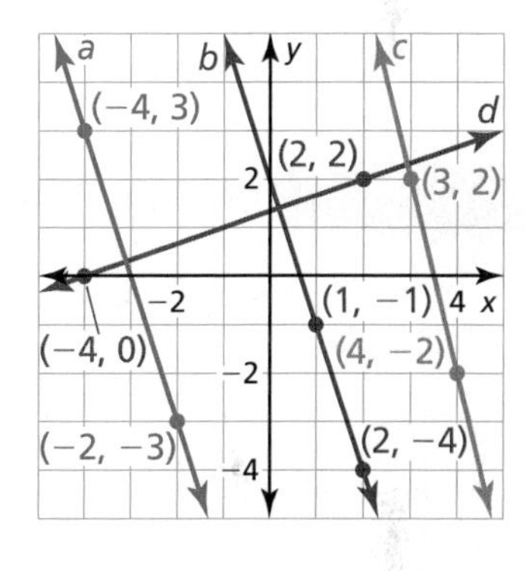

Find the slope of each line.

Line a: $m = \frac{3 - (-3)}{-4 - (-2)} = -3$

Line b: $m = \frac{-1 - (-4)}{1 - 2} = -3$

Line c: $m = \frac{2 - (-2)}{3 - 4} = -4$

Line d: $m = \frac{2 - 0}{2 - (-4)} = \frac{1}{3}$

Because lines a and b have the same slope, lines a and b are parallel. Because $\frac{1}{3}(-3) = -1$, lines a and d are perpendicular and lines b and d are perpendicular.

Determine which lines are parallel and which are perpendicular.

4.

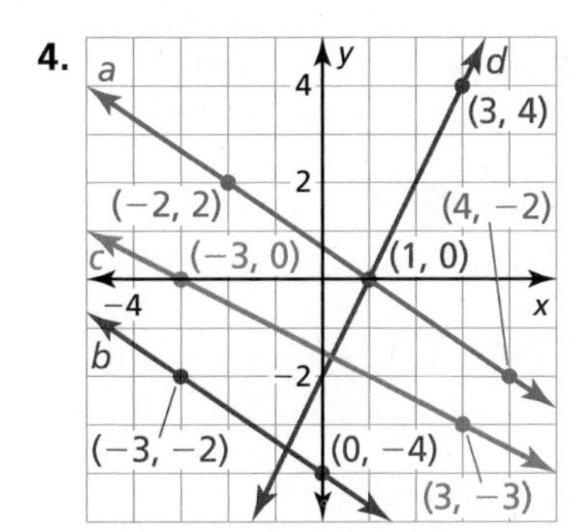

5.

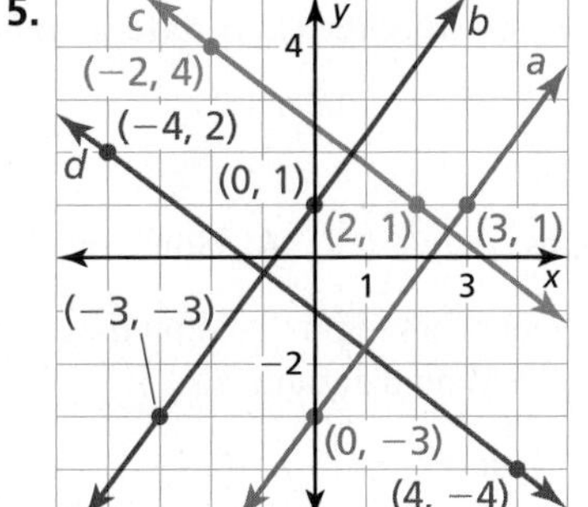

6.

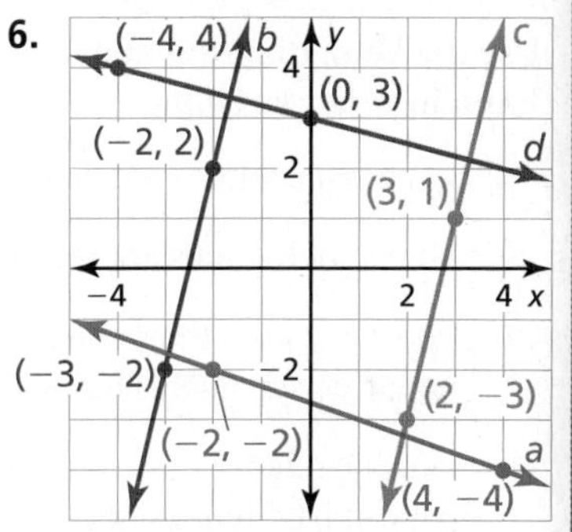

7. **ABSTRACT REASONING** Explain why interpreting an expression as a single quantity does not contradict the order of operations.

Mathematical Practices

Mathematically proficient students use diagrams to show relationships.

Mapping Relationships

Core Concept

Classifications of Quadrilaterals

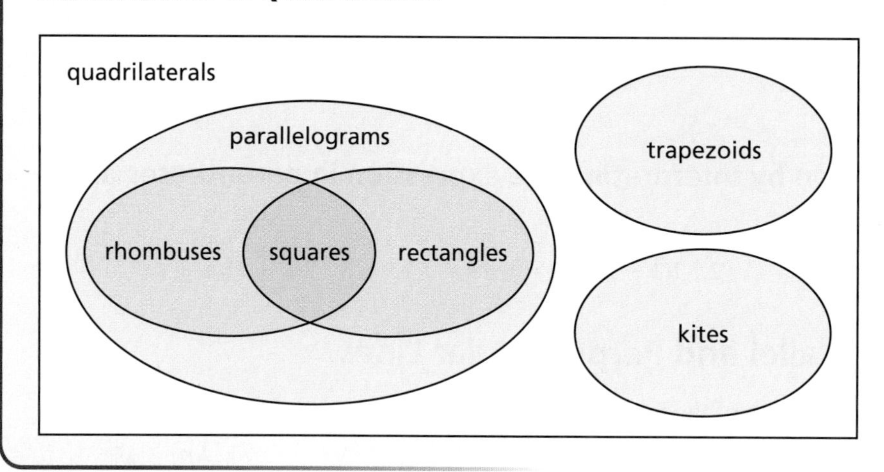

EXAMPLE 1 **Writing Statements about Quadrilaterals**

Use the Venn diagram above to write three true statements about different types of quadrilaterals.

SOLUTION

Here are three true statements that can be made about the relationships shown in the Venn diagram.

- All rhombuses are parallelograms.
- Some rhombuses are rectangles.
- No trapezoids are parallelograms.

Monitoring Progress

Use the Venn diagram above to decide whether each statement is true or false. Explain your reasoning.

1. Some trapezoids are kites.

2. No kites are parallelograms.

3. All parallelograms are rectangles.

4. Some quadrilaterals are squares.

5. Example 1 lists three true statements based on the Venn diagram above. Write six more true statements based on the Venn diagram.

6. A cyclic quadrilateral is a quadrilateral that can be circumscribed by a circle so that the circle touches each vertex. Redraw the Venn diagram so that it includes cyclic quadrilaterals.

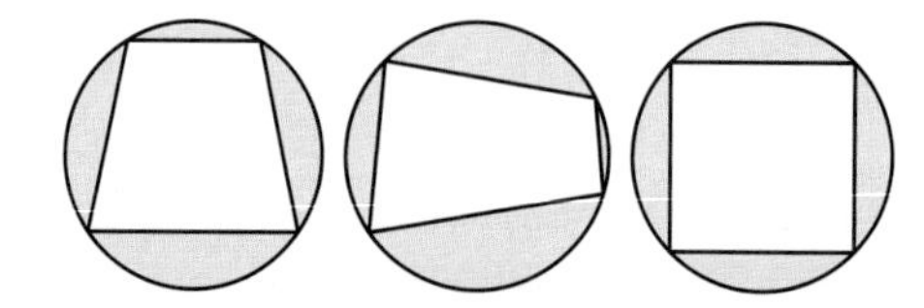

7.1 Angles of Polygons

Essential Question What is the sum of the measures of the interior angles of a polygon?

EXPLORATION 1 The Sum of the Angle Measures of a Polygon

Work with a partner. Use dynamic geometry software.

a. Draw a quadrilateral and a pentagon. Find the sum of the measures of the interior angles of each polygon.

Sample

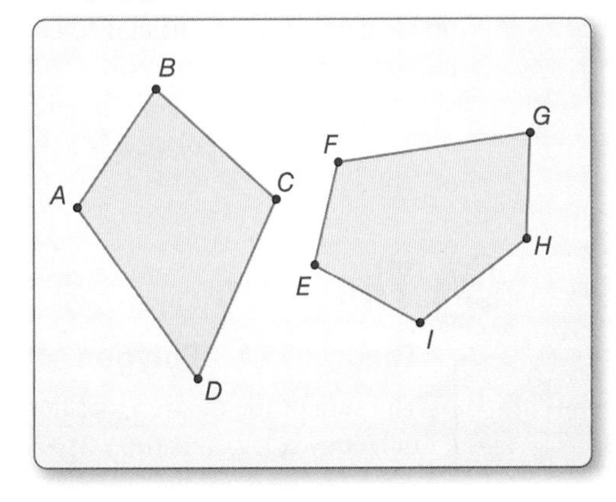

b. Draw other polygons and find the sums of the measures of their interior angles. Record your results in the table below.

Number of sides, n	3	4	5	6	7	8	9
Sum of angle measures, S							

c. Plot the data from your table in a coordinate plane.

d. Write a function that fits the data. Explain what the function represents.

CONSTRUCTING VIABLE ARGUMENTS

To be proficient in math, you need to reason inductively about data.

EXPLORATION 2 Measure of One Angle in a Regular Polygon

Work with a partner.

a. Use the function you found in Exploration 1 to write a new function that gives the measure of one interior angle in a regular polygon with n sides.

b. Use the function in part (a) to find the measure of one interior angle of a regular pentagon. Use dynamic geometry software to check your result by constructing a regular pentagon and finding the measure of one of its interior angles.

c. Copy your table from Exploration 1 and add a row for the measure of one interior angle in a regular polygon with n sides. Complete the table. Use dynamic geometry software to check your results.

Communicate Your Answer

3. What is the sum of the measures of the interior angles of a polygon?

4. Find the measure of one interior angle in a regular dodecagon (a polygon with 12 sides).

7.1 Lesson

Core Vocabulary

diagonal, *p. 360*
equilateral polygon, *p. 361*
equiangular polygon, *p. 361*
regular polygon, *p. 361*

Previous
polygon
convex
interior angles
exterior angles

What You Will Learn

- Use the interior angle measures of polygons.
- Use the exterior angle measures of polygons.

Using Interior Angle Measures of Polygons

In a polygon, two vertices that are endpoints of the same side are called *consecutive vertices*. A **diagonal** of a polygon is a segment that joins two nonconsecutive vertices.

As you can see, the diagonals from one vertex divide a polygon into triangles. Dividing a polygon with n sides into $(n - 2)$ triangles shows that the sum of the measures of the interior angles of a polygon is a multiple of 180°.

Polygon *ABCDE*

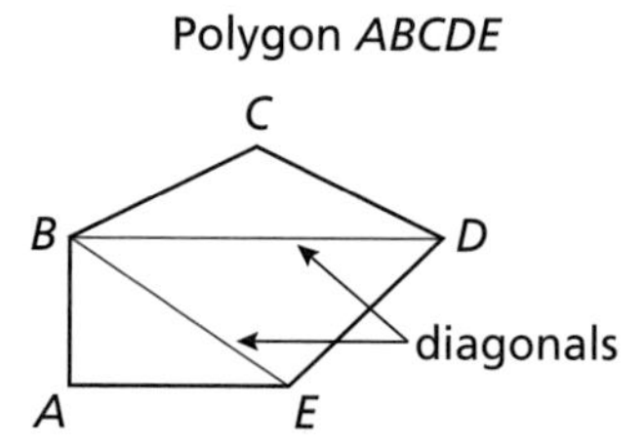

A and *B* are consecutive vertices.
Vertex *B* has two diagonals, $\overline{BD}$ and $\overline{BE}$.

Theorem

Theorem 7.1 Polygon Interior Angles Theorem

The sum of the measures of the interior angles of a convex n-gon is $(n - 2) \cdot 180°$.

$$m\angle 1 + m\angle 2 + \cdots + m\angle n = (n - 2) \cdot 180°$$

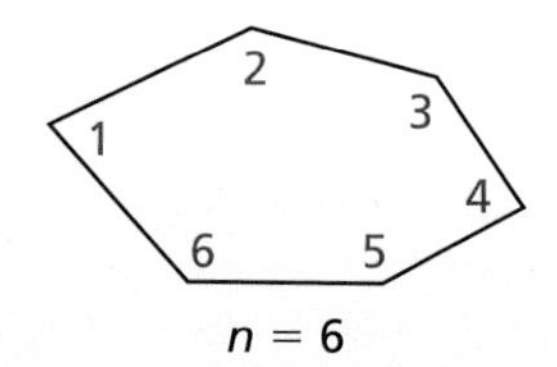

$n = 6$

Proof Ex. 42 (for pentagons), p. 365

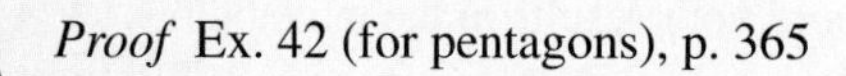

REMEMBER
A polygon is *convex* when no line that contains a side of the polygon contains a point in the interior of the polygon.

EXAMPLE 1 Finding the Sum of Angle Measures in a Polygon

Find the sum of the measures of the interior angles of the figure.

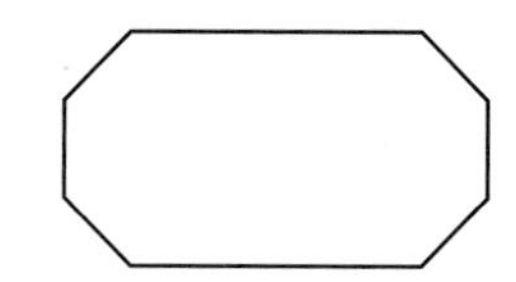

SOLUTION

The figure is a convex octagon. It has 8 sides. Use the Polygon Interior Angles Theorem.

$(n - 2) \cdot 180° = (8 - 2) \cdot 180°$		Substitute 8 for *n*.
$= 6 \cdot 180°$		Subtract.
$= 1080°$		Multiply.

▶ The sum of the measures of the interior angles of the figure is 1080°.

Monitoring Progress

Help in English and Spanish at *BigIdeasMath.com*

1. The coin shown is in the shape of an 11-gon. Find the sum of the measures of the interior angles.

EXAMPLE 2 Finding the Number of Sides of a Polygon

The sum of the measures of the interior angles of a convex polygon is 900°. Classify the polygon by the number of sides.

SOLUTION

Use the Polygon Interior Angles Theorem to write an equation involving the number of sides n. Then solve the equation to find the number of sides.

$(n - 2) \cdot 180° = 900°$	Polygon Interior Angles Theorem
$n - 2 = 5$	Divide each side by 180°.
$n = 7$	Add 2 to each side.

▶ The polygon has 7 sides. It is a heptagon.

Corollary

Corollary 7.1 Corollary to the Polygon Interior Angles Theorem

The sum of the measures of the interior angles of a quadrilateral is 360°.

Proof Ex. 43, p. 366

EXAMPLE 3 Finding an Unknown Interior Angle Measure

Find the value of x in the diagram.

SOLUTION

The polygon is a quadrilateral. Use the Corollary to the Polygon Interior Angles Theorem to write an equation involving x. Then solve the equation.

$x° + 108° + 121° + 59° = 360°$	Corollary to the Polygon Interior Angles Theorem
$x + 288 = 360$	Combine like terms.
$x = 72$	Subtract 288 from each side.

▶ The value of x is 72.

Monitoring Progress  Help in English and Spanish at *BigIdeasMath.com*

2. The sum of the measures of the interior angles of a convex polygon is 1440°. Classify the polygon by the number of sides.

3. The measures of the interior angles of a quadrilateral are $x°$, $3x°$, $5x°$, and $7x°$. Find the measures of all the interior angles.

In an **equilateral polygon**, all sides are congruent.

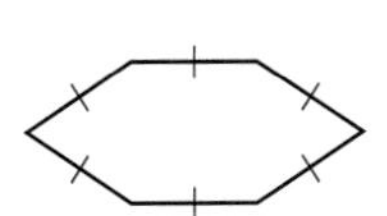

In an **equiangular polygon**, all angles in the interior of the polygon are congruent.

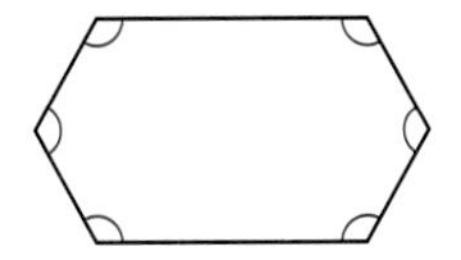

A **regular polygon** is a convex polygon that is both equilateral and equiangular.

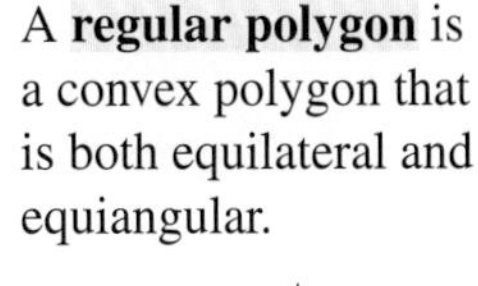

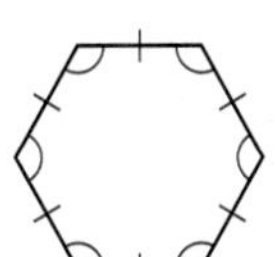

EXAMPLE 4 Finding Angle Measures in Polygons

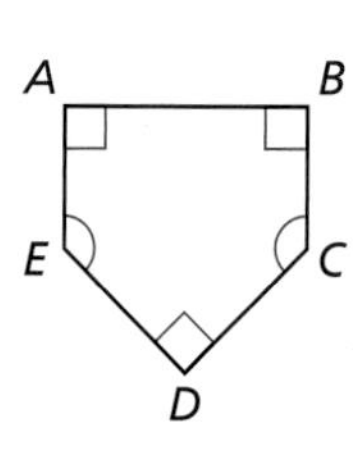

A home plate for a baseball field is shown.

a. Is the polygon regular? Explain your reasoning.

b. Find the measures of $\angle C$ and $\angle E$.

SOLUTION

a. The polygon is not equilateral or equiangular. So, the polygon is not regular.

b. Find the sum of the measures of the interior angles.

$$(n - 2) \cdot 180^\circ = (5 - 2) \cdot 180^\circ = 540^\circ$$ Polygon Interior Angles Theorem

Then write an equation involving x and solve the equation.

$$x^\circ + x^\circ + 90^\circ + 90^\circ + 90^\circ = 540^\circ$$ Write an equation.

$$2x + 270 = 540$$ Combine like terms.

$$x = 135$$ Solve for x.

▶ So, $m\angle C = m\angle E = 135^\circ$.

Monitoring Progress

Help in English and Spanish at *BigIdeasMath.com*

P, Q, R, S, T pentagon: 93°, 156°, 85°

4. Find $m\angle S$ and $m\angle T$ in the diagram.
5. Sketch a pentagon that is equilateral but not equiangular.

Using Exterior Angle Measures of Polygons

Unlike the sum of the interior angle measures of a convex polygon, the sum of the exterior angle measures does *not* depend on the number of sides of the polygon. The diagrams suggest that the sum of the measures of the exterior angles, one angle at each vertex, of a pentagon is 360°. In general, this sum is 360° for any convex polygon.

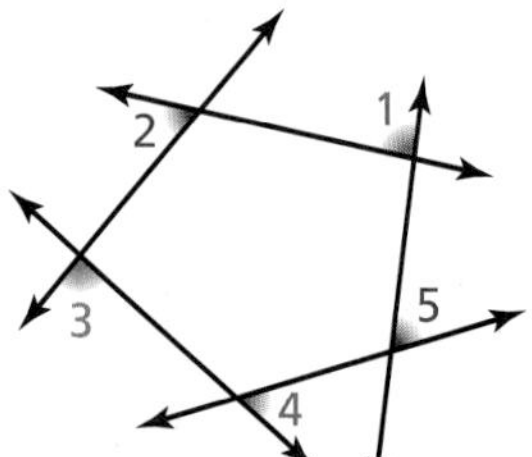

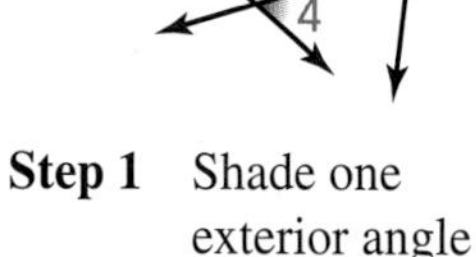

Step 1 Shade one exterior angle at each vertex.

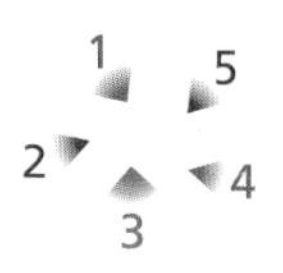

Step 2 Cut out the exterior angles.

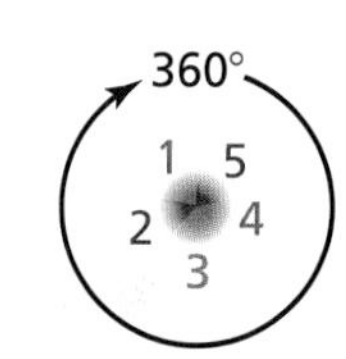

Step 3 Arrange the exterior angles to form 360°.

JUSTIFYING STEPS

To help justify this conclusion, you can visualize a circle containing two straight angles. So, there are 180° + 180°, or 360°, in a circle.

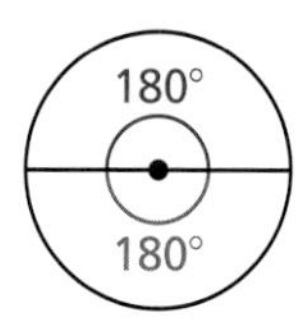

Theorem

Theorem 7.2 Polygon Exterior Angles Theorem

The sum of the measures of the exterior angles of a convex polygon, one angle at each vertex, is 360°.

$$m\angle 1 + m\angle 2 + \cdots + m\angle n = 360^\circ$$

Proof Ex. 51, p. 366

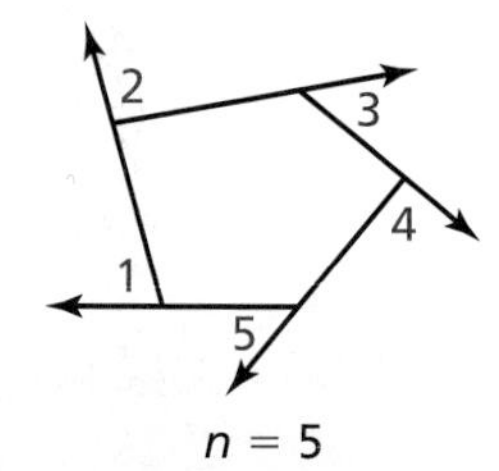

$n = 5$

EXAMPLE 5 Finding an Unknown Exterior Angle Measure

Find the value of x in the diagram.

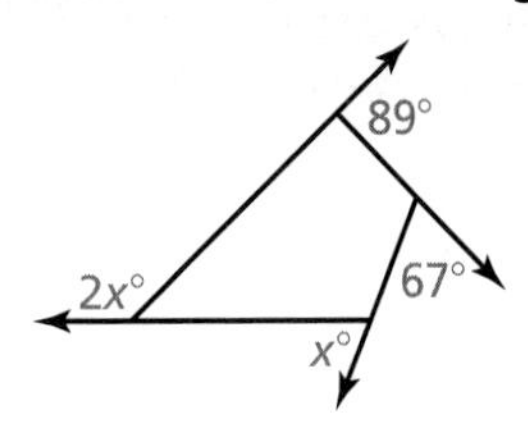

SOLUTION

Use the Polygon Exterior Angles Theorem to write and solve an equation.

$x° + 2x° + 89° + 67° = 360°$	Polygon Exterior Angles Theorem
$3x + 156 = 360$	Combine like terms.
$x = 68$	Solve for x.

▶ The value of x is 68.

REMEMBER

A *dodecagon* is a polygon with 12 sides and 12 vertices.

EXAMPLE 6 Finding Angle Measures in Regular Polygons

The trampoline shown is shaped like a regular dodecagon.

a. Find the measure of each interior angle.

b. Find the measure of each exterior angle.

SOLUTION

a. Use the Polygon Interior Angles Theorem to find the sum of the measures of the interior angles.

$$(n - 2) \cdot 180° = (12 - 2) \cdot 180°$$
$$= 1800°$$

Then find the measure of one interior angle. A regular dodecagon has 12 congruent interior angles. Divide 1800° by 12.

$$\frac{1800°}{12} = 150°$$

▶ The measure of each interior angle in the dodecagon is 150°.

b. By the Polygon Exterior Angles Theorem, the sum of the measures of the exterior angles, one angle at each vertex, is 360°. Divide 360° by 12 to find the measure of one of the 12 congruent exterior angles.

$$\frac{360°}{12} = 30°$$

▶ The measure of each exterior angle in the dodecagon is 30°.

Monitoring Progress  Help in English and Spanish at *BigIdeasMath.com*

6. A convex hexagon has exterior angles with measures 34°, 49°, 58°, 67°, and 75°. What is the measure of an exterior angle at the sixth vertex?

7. An interior angle and an adjacent exterior angle of a polygon form a linear pair. How can you use this fact as another method to find the measure of each exterior angle in Example 6?

7.1 Exercises

Dynamic Solutions available at *BigIdeasMath.com*

Vocabulary and Core Concept Check

1. **VOCABULARY** Why do vertices connected by a diagonal of a polygon have to be nonconsecutive?

2. **WHICH ONE DOESN'T BELONG?** Which sum does *not* belong with the other three? Explain your reasoning.

the sum of the measures of the interior angles of a quadrilateral	the sum of the measures of the exterior angles of a quadrilateral
the sum of the measures of the interior angles of a pentagon	the sum of the measures of the exterior angles of a pentagon

Monitoring Progress and Modeling with Mathematics

In Exercises 3–6, find the sum of the measures of the interior angles of the indicated convex polygon. *(See Example 1.)*

3. nonagon
4. 14-gon
5. 16-gon
6. 20-gon

In Exercises 7–10, the sum of the measures of the interior angles of a convex polygon is given. Classify the polygon by the number of sides. *(See Example 2.)*

7. 720°
8. 1080°
9. 2520°
10. 3240°

In Exercises 11–14, find the value of x. *(See Example 3.)*

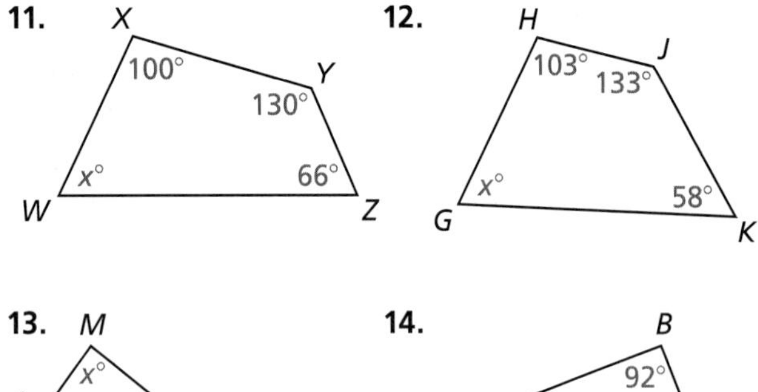

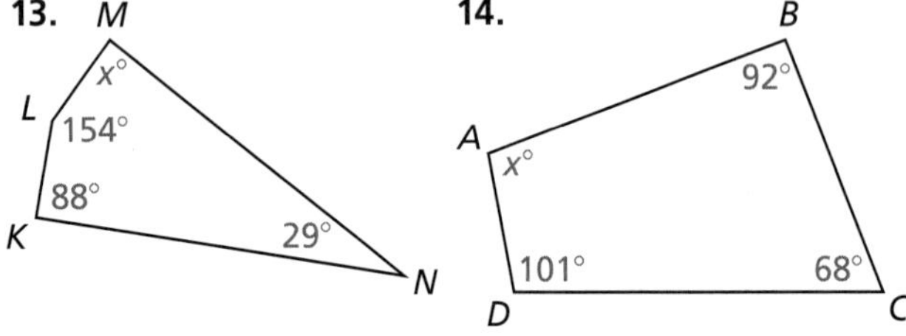

In Exercises 15–18, find the value of x.

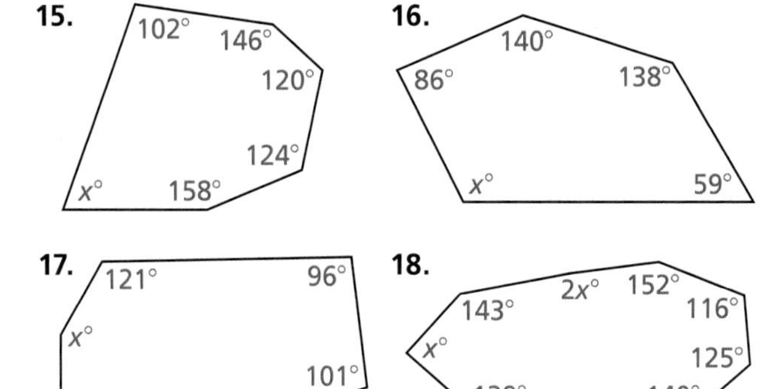

In Exercises 19–22, find the measures of $\angle X$ and $\angle Y$. *(See Example 4.)*

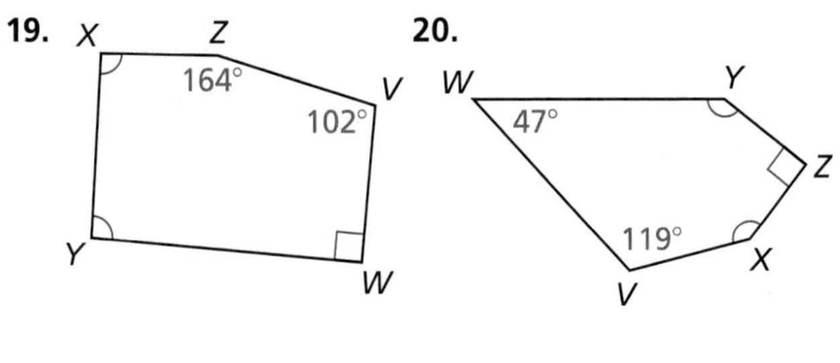

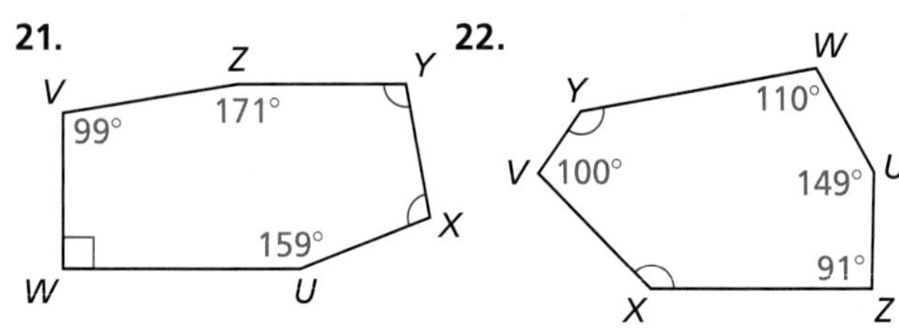

In Exercises 23–26, find the value of x. *(See Example 5.)*

23.

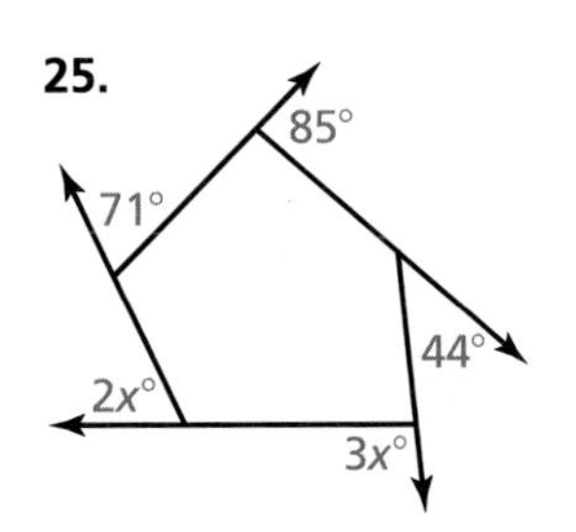

24.

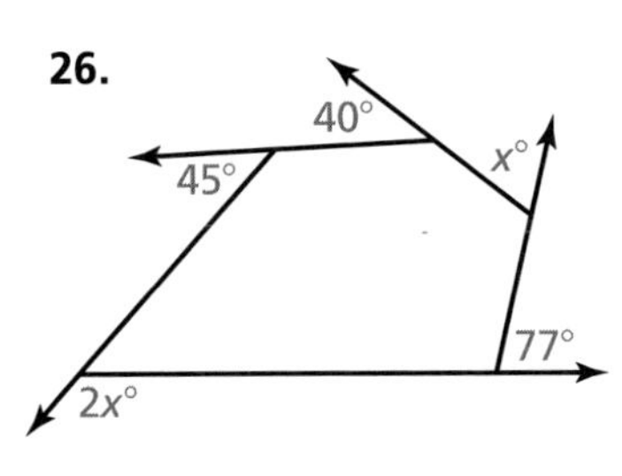

25.

26.

In Exercises 27–30, find the measure of each interior angle and each exterior angle of the indicated regular polygon. *(See Example 6.)*

27. pentagon

28. 18-gon

29. 45-gon

30. 90-gon

ERROR ANALYSIS In Exercises 31 and 32, describe and correct the error in finding the measure of one exterior angle of a regular pentagon.

31.

$(n-2) \cdot 180° = (5-2) \cdot 180°$
$= 3 \cdot 180°$
$= 540°$

The sum of the measures of the angles is 540°. There are five angles, so the measure of one exterior angle is $\frac{540°}{5} = 108°$.

32.

There are a total of 10 exterior angles, two at each vertex, so the measure of one exterior angle is $\frac{360°}{10} = 36°$.

33. MODELING WITH MATHEMATICS The base of a jewelry box is shaped like a regular hexagon. What is the measure of each interior angle of the jewelry box base?

34. MODELING WITH MATHEMATICS The floor of the gazebo shown is shaped like a regular decagon. Find the measure of each interior angle of the regular decagon. Then find the measure of each exterior angle.

35. WRITING A FORMULA Write a formula to find the number of sides n in a regular polygon given that the measure of one interior angle is $x°$.

36. WRITING A FORMULA Write a formula to find the number of sides n in a regular polygon given that the measure of one exterior angle is $x°$.

REASONING In Exercises 37–40, find the number of sides for the regular polygon described.

37. Each interior angle has a measure of 156°.

38. Each interior angle has a measure of 165°.

39. Each exterior angle has a measure of 9°.

40. Each exterior angle has a measure of 6°.

41. DRAWING CONCLUSIONS Which of the following angle measures are possible interior angle measures of a regular polygon? Explain your reasoning. Select all that apply.

Ⓐ 162° Ⓑ 171° Ⓒ 75° Ⓓ 40°

42. PROVING A THEOREM The Polygon Interior Angles Theorem (Theorem 7.1) states that the sum of the measures of the interior angles of a convex n-gon is $(n - 2) \cdot 180°$. Write a paragraph proof of this theorem for the case when $n = 5$.

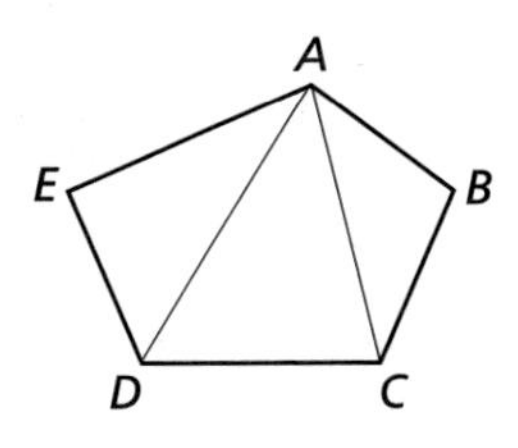

43. PROVING A COROLLARY Write a paragraph proof of the Corollary to the Polygon Interior Angles Theorem (Corollary 7.1).

44. MAKING AN ARGUMENT Your friend claims that to find the interior angle measures of a regular polygon, you do not have to use the Polygon Interior Angles Theorem (Theorem 7.1). You instead can use the Polygon Exterior Angles Theorem (Theorem 7.2) and then the Linear Pair Postulate (Postulate 2.8). Is your friend correct? Explain your reasoning.

45. MATHEMATICAL CONNECTIONS In an equilateral hexagon, four of the exterior angles each have a measure of $x°$. The other two exterior angles each have a measure of twice the sum of x and 48. Find the measure of each exterior angle.

46. THOUGHT PROVOKING For a concave polygon, is it true that at least one of the interior angle measures must be greater than 180°? If not, give an example. If so, explain your reasoning.

47. WRITING EXPRESSIONS Write an expression to find the sum of the measures of the interior angles for a concave polygon. Explain your reasoning.

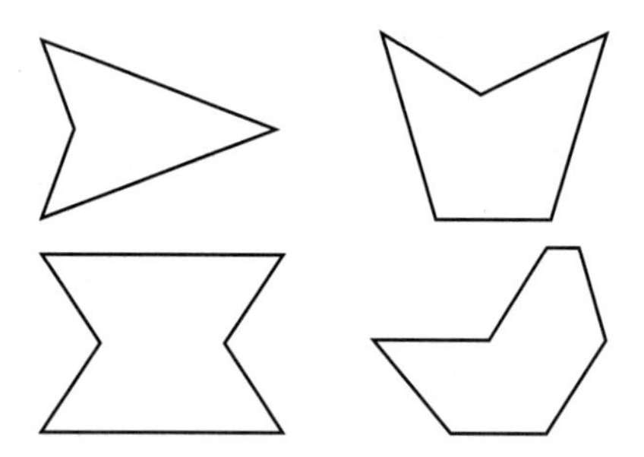

48. ANALYZING RELATIONSHIPS Polygon *ABCDEFGH* is a regular octagon. Suppose sides $\overline{AB}$ and $\overline{CD}$ are extended to meet at a point *P*. Find $m\angle BPC$. Explain your reasoning. Include a diagram with your answer.

49. MULTIPLE REPRESENTATIONS The formula for the measure of each interior angle in a regular polygon can be written in function notation.

a. Write a function $h(n)$, where n is the number of sides in a regular polygon and $h(n)$ is the measure of any interior angle in the regular polygon.

b. Use the function to find $h(9)$.

c. Use the function to find n when $h(n) = 150°$.

d. Plot the points for $n = 3, 4, 5, 6, 7$, and 8. What happens to the value of $h(n)$ as n gets larger?

50. HOW DO YOU SEE IT? Is the hexagon a regular hexagon? Explain your reasoning.

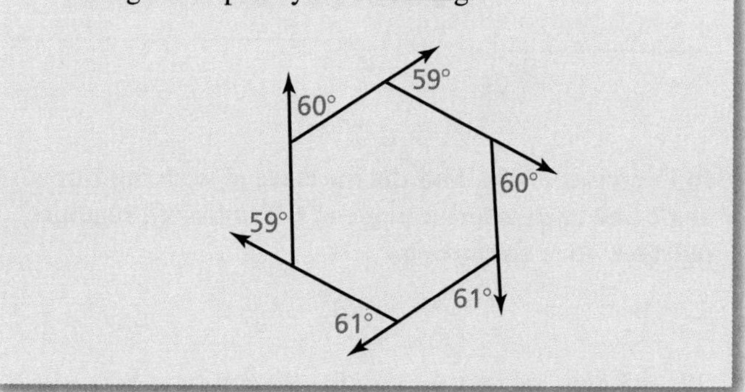

51. PROVING A THEOREM Write a paragraph proof of the Polygon Exterior Angles Theorem (Theorem 7.2). (*Hint:* In a convex n-gon, the sum of the measures of an interior angle and an adjacent exterior angle at any vertex is 180°.)

52. ABSTRACT REASONING You are given a convex polygon. You are asked to draw a new polygon by increasing the sum of the interior angle measures by 540°. How many more sides does your new polygon have? Explain your reasoning.

Maintaining Mathematical Proficiency

Reviewing what you learned in previous grades and lessons

Find the value of *x*. *(Section 3.2)*

53. **54.** **55.** **56.**

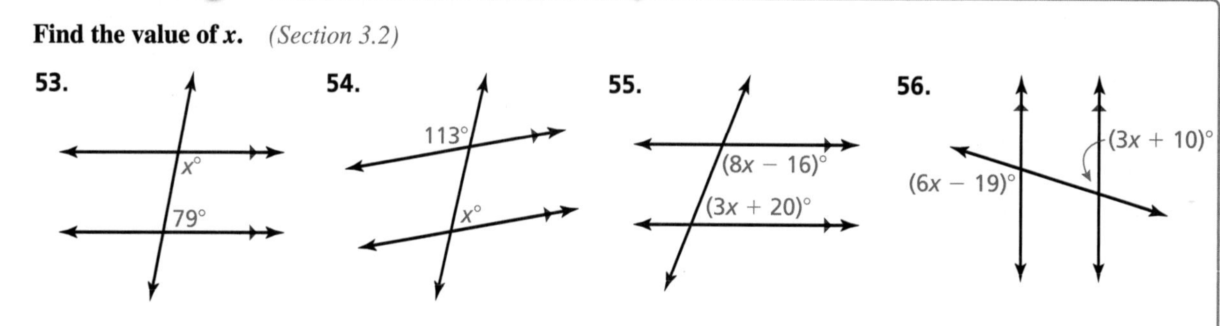

7.2 Properties of Parallelograms

Essential Question What are the properties of parallelograms?

EXPLORATION 1 Discovering Properties of Parallelograms

Work with a partner. Use dynamic geometry software.

a. Construct any parallelogram and label it *ABCD*. Explain your process.

Sample

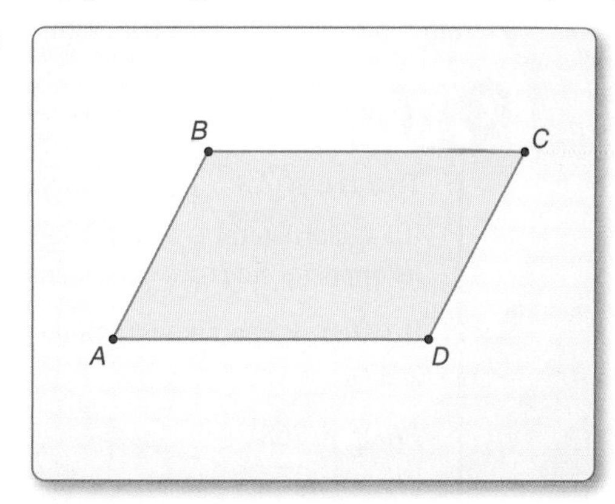

b. Find the angle measures of the parallelogram. What do you observe?

c. Find the side lengths of the parallelogram. What do you observe?

d. Repeat parts (a)–(c) for several other parallelograms. Use your results to write conjectures about the angle measures and side lengths of a parallelogram.

EXPLORATION 2 Discovering a Property of Parallelograms

Work with a partner. Use dynamic geometry software.

a. Construct any parallelogram and label it *ABCD*.

b. Draw the two diagonals of the parallelogram. Label the point of intersection *E*.

Sample

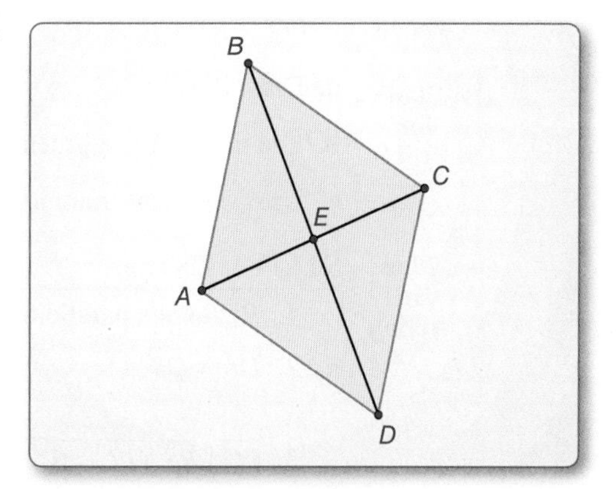

c. Find the segment lengths *AE*, *BE*, *CE*, and *DE*. What do you observe?

d. Repeat parts (a)–(c) for several other parallelograms. Use your results to write a conjecture about the diagonals of a parallelogram.

MAKING SENSE OF PROBLEMS

To be proficient in math, you need to analyze givens, constraints, relationships, and goals.

Communicate Your Answer

3. What are the properties of parallelograms?

7.2 Lesson

Core Vocabulary

parallelogram, *p. 368*

Previous
quadrilateral
diagonal
interior angles
segment bisector

What You Will Learn

- Use properties to find side lengths and angles of parallelograms.
- Use parallelograms in the coordinate plane.

Using Properties of Parallelograms

A **parallelogram** is a quadrilateral with both pairs of opposite sides parallel. In $\square PQRS$, $\overline{PQ} \parallel \overline{RS}$ and $\overline{QR} \parallel \overline{PS}$ by definition. The theorems below describe other properties of parallelograms.

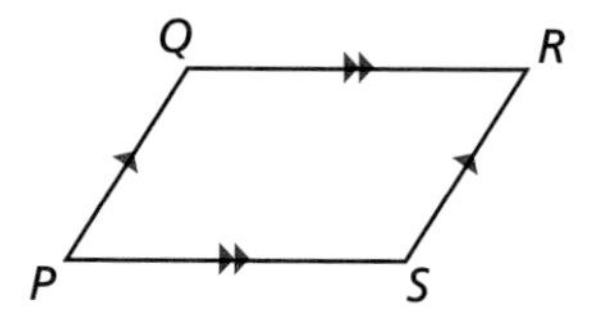

Theorems

Theorem 7.3 Parallelogram Opposite Sides Theorem

If a quadrilateral is a parallelogram, then its opposite sides are congruent.

If $PQRS$ is a parallelogram, then $\overline{PQ} \cong \overline{RS}$ and $\overline{QR} \cong \overline{SP}$.

Proof p. 368

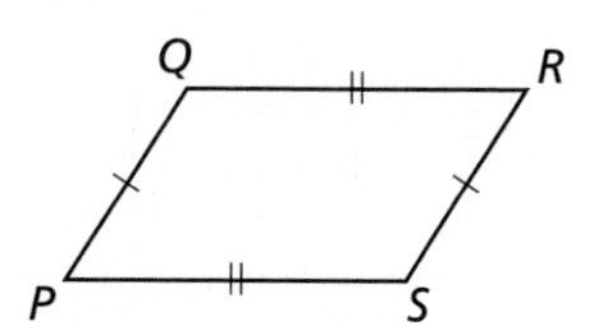

Theorem 7.4 Parallelogram Opposite Angles Theorem

If a quadrilateral is a parallelogram, then its opposite angles are congruent.

If $PQRS$ is a parallelogram, then $\angle P \cong \angle R$ and $\angle Q \cong \angle S$.

Proof Ex. 37, p. 373

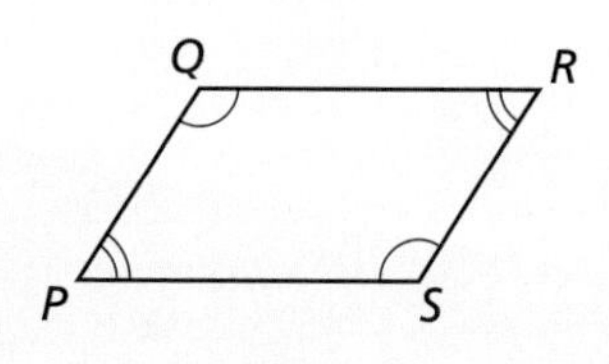

PROOF **Parallelogram Opposite Sides Theorem**

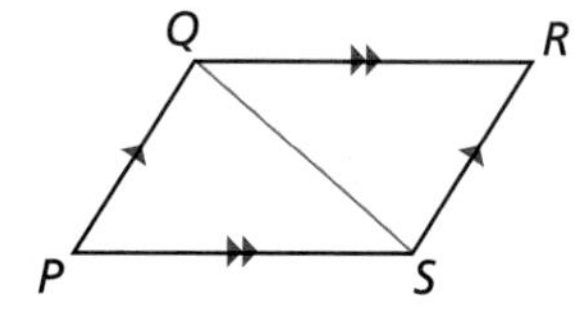

Given $PQRS$ is a parallelogram.

Prove $\overline{PQ} \cong \overline{RS}$, $\overline{QR} \cong \overline{SP}$

Plan for Proof

a. Draw diagonal $\overline{QS}$ to form $\triangle PQS$ and $\triangle RSQ$.

b. Use the ASA Congruence Theorem (Thm. 5.10) to show that $\triangle PQS \cong \triangle RSQ$.

c. Use congruent triangles to show that $\overline{PQ} \cong \overline{RS}$ and $\overline{QR} \cong \overline{SP}$.

Plan in Action

STATEMENTS	REASONS
1. $PQRS$ is a parallelogram.	**1.** Given
a. 2. Draw $\overline{QS}$.	**2.** Through any two points, there exists exactly one line.
3. $\overline{PQ} \parallel \overline{RS}$, $\overline{QR} \parallel \overline{PS}$	**3.** Definition of parallelogram
b. 4. $\angle PQS \cong \angle RSQ$, $\angle PSQ \cong \angle RQS$	**4.** Alternate Interior Angles Theorem (Thm. 3.2)
5. $\overline{QS} \cong \overline{SQ}$	**5.** Reflexive Property of Congruence (Thm. 2.1)
6. $\triangle PQS \cong \triangle RSQ$	**6.** ASA Congruence Theorem (Thm. 5.10)
c. 7. $\overline{PQ} \cong \overline{RS}$, $\overline{QR} \cong \overline{SP}$	**7.** Corresponding parts of congruent triangles are congruent.

EXAMPLE 1 Using Properties of Parallelograms

Find the values of x and y.

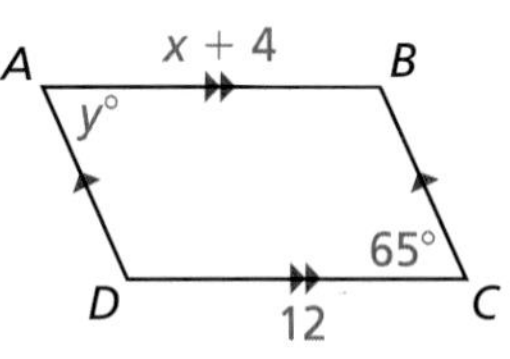

SOLUTION

$ABCD$ is a parallelogram by the definition of a parallelogram. Use the Parallelogram Opposite Sides Theorem to find the value of x.

$AB = CD$	Opposite sides of a parallelogram are congruent.
$x + 4 = 12$	Substitute $x + 4$ for AB and 12 for CD.
$x = 8$	Subtract 4 from each side.

By the Parallelogram Opposite Angles Theorem, $\angle A \cong \angle C$, or $m\angle A = m\angle C$. So, $y° = 65°$.

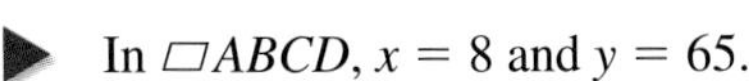

▶ In $\square ABCD$, $x = 8$ and $y = 65$.

Monitoring Progress

Help in English and Spanish at *BigIdeasMath.com*

1. Find FG and $m\angle G$.

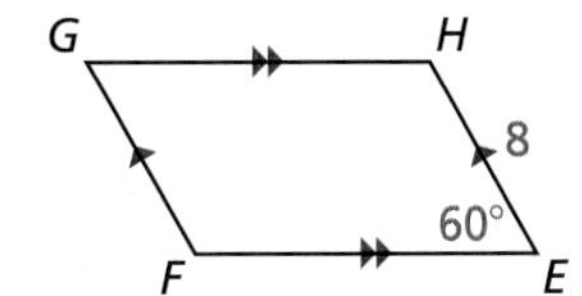

2. Find the values of x and y.

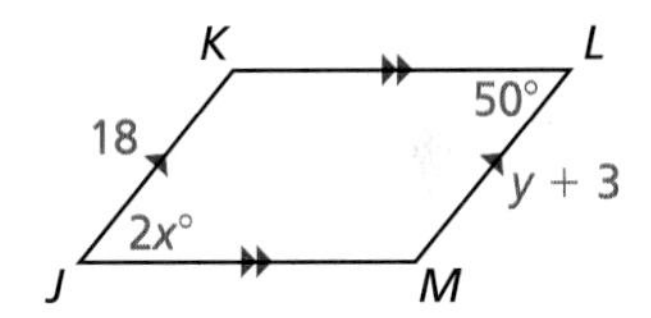

The Consecutive Interior Angles Theorem (Theorem 3.4) states that if two parallel lines are cut by a transversal, then the pairs of consecutive interior angles formed are supplementary.

A pair of consecutive angles in a parallelogram is like a pair of consecutive interior angles between parallel lines. This similarity suggests the Parallelogram Consecutive Angles Theorem.

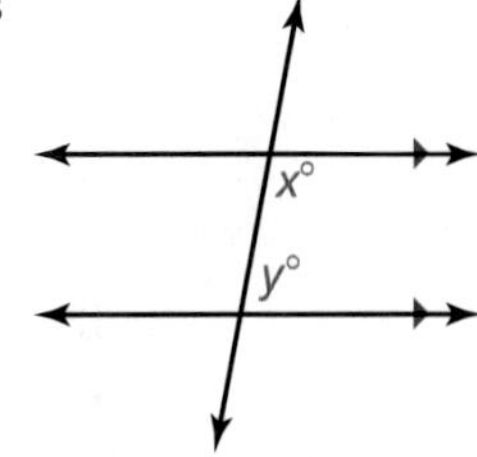

Theorems

Theorem 7.5 Parallelogram Consecutive Angles Theorem

If a quadrilateral is a parallelogram, then its consecutive angles are supplementary.

If $PQRS$ is a parallelogram, then $x° + y° = 180°$.

Proof Ex. 38, p. 373

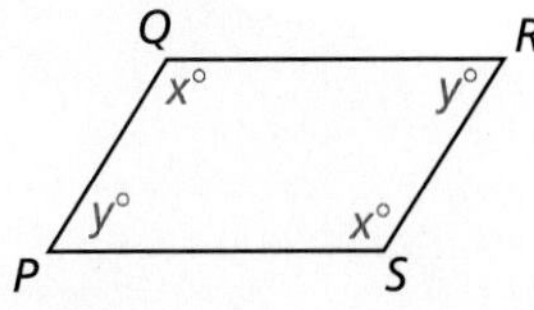

Theorem 7.6 Parallelogram Diagonals Theorem

If a quadrilateral is a parallelogram, then its diagonals bisect each other.

If $PQRS$ is a parallelogram, then $\overline{QM} \cong \overline{SM}$ and $\overline{PM} \cong \overline{RM}$.

Proof p. 370

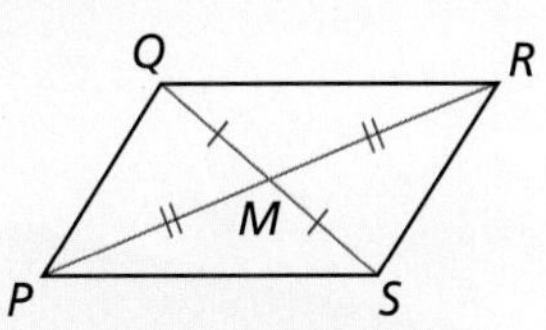

PROOF Parallelogram Diagonals Theorem

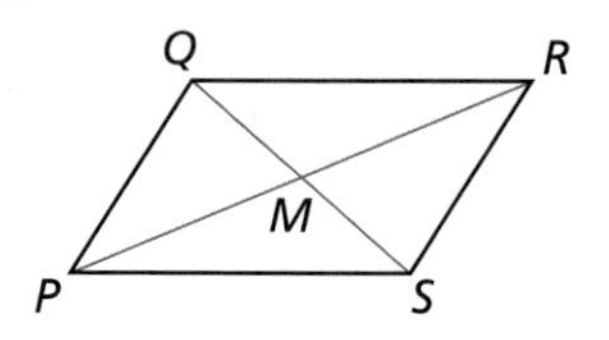

Given $PQRS$ is a parallelogram. Diagonals $\overline{PR}$ and $\overline{QS}$ intersect at point M.

Prove M bisects $\overline{QS}$ and $\overline{PR}$.

STATEMENTS	REASONS
1. $PQRS$ is a parallelogram.	**1.** Given
2. $\overline{PQ} \parallel \overline{RS}$	**2.** Definition of a parallelogram
3. $\angle QPR \cong \angle SRP$, $\angle PQS \cong \angle RSQ$	**3.** Alternate Interior Angles Theorem (Thm. 3.2)
4. $\overline{PQ} \cong \overline{RS}$	**4.** Parallelogram Opposite Sides Theorem
5. $\triangle PMQ \cong \triangle RMS$	**5.** ASA Congruence Theorem (Thm. 5.10)
6. $\overline{QM} \cong \overline{SM}$, $\overline{PM} \cong \overline{RM}$	**6.** Corresponding parts of congruent triangles are congruent.
7. M bisects $\overline{QS}$ and $\overline{PR}$.	**7.** Definition of segment bisector

EXAMPLE 2 Using Properties of a Parallelogram

As shown, part of the extending arm of a desk lamp is a parallelogram. The angles of the parallelogram change as the lamp is raised and lowered. Find $m\angle BCD$ when $m\angle ADC = 110°$.

SOLUTION

By the Parallelogram Consecutive Angles Theorem, the consecutive angle pairs in $\square ABCD$ are supplementary. So, $m\angle ADC + m\angle BCD = 180°$. Because $m\angle ADC = 110°$, $m\angle BCD = 180° - 110° = 70°$.

EXAMPLE 3 Writing a Two-Column Proof

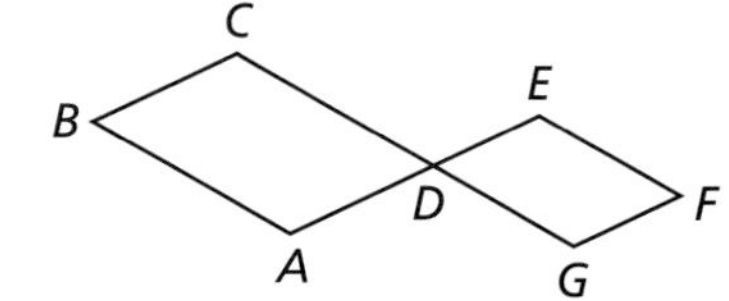

Write a two-column proof.

Given $ABCD$ and $GDEF$ are parallelograms.

Prove $\angle B \cong \angle F$

STATEMENTS	REASONS
1. $ABCD$ and $GDEF$ are parallelograms.	**1.** Given
2. $\angle CDA \cong \angle B$, $\angle EDG \cong \angle F$	**2.** If a quadrilateral is a parallelogram, then its opposite angles are congruent.
3. $\angle CDA \cong \angle EDG$	**3.** Vertical Angles Congruence Theorem (Thm. 2.6)
4. $\angle B \cong \angle F$	**4.** Transitive Property of Congruence (Thm. 2.2)

Monitoring Progress

Help in English and Spanish at *BigIdeasMath.com*

3. WHAT IF? In Example 2, find $m\angle BCD$ when $m\angle ADC$ is twice the measure of $\angle BCD$.

4. Using the figure and the given statement in Example 3, prove that $\angle C$ and $\angle F$ are supplementary angles.

Using Parallelograms in the Coordinate Plane

JUSTIFYING STEPS

In Example 4, you can use either diagonal to find the coordinates of the intersection. Using diagonal $\overline{OM}$ helps simplify the calculation because one endpoint is (0, 0).

EXAMPLE 4 Using Parallelograms in the Coordinate Plane

Find the coordinates of the intersection of the diagonals of ▱*LMNO* with vertices $L(1, 4)$, $M(7, 4)$, $N(6, 0)$, and $O(0, 0)$.

SOLUTION

By the Parallelogram Diagonals Theorem, the diagonals of a parallelogram bisect each other. So, the coordinates of the intersection are the midpoints of diagonals $\overline{LN}$ and $\overline{OM}$.

$$\text{coordinates of midpoint of } \overline{OM} = \left(\frac{7+0}{2}, \frac{4+0}{2}\right) = \left(\frac{7}{2}, 2\right) \quad \text{Midpoint Formula}$$

▶ The coordinates of the intersection of the diagonals are $\left(\frac{7}{2}, 2\right)$. You can check your answer by graphing ▱*LMNO* and drawing the diagonals. The point of intersection appears to be correct.

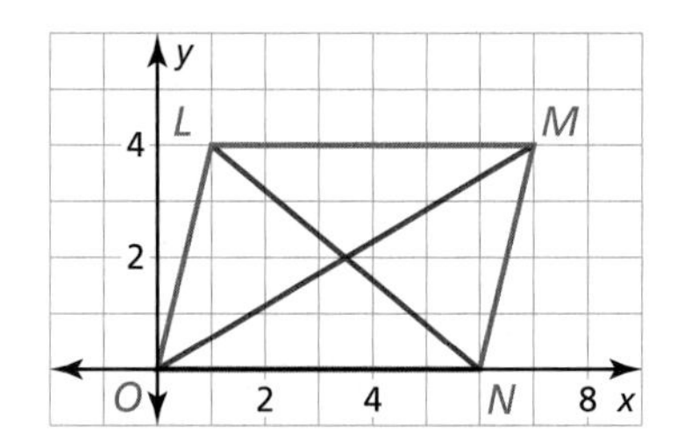

REMEMBER

When graphing a polygon in the coordinate plane, the name of the polygon gives the order of the vertices.

EXAMPLE 5 Using Parallelograms in the Coordinate Plane

Three vertices of ▱*WXYZ* are $W(-1, -3)$, $X(-3, 2)$, and $Z(4, -4)$. Find the coordinates of vertex Y.

SOLUTION

Step 1 Graph the vertices W, X, and Z.

Step 2 Find the slope of $\overline{WX}$.

$$\text{slope of } \overline{WX} = \frac{2-(-3)}{-3-(-1)} = \frac{5}{-2} = -\frac{5}{2}$$

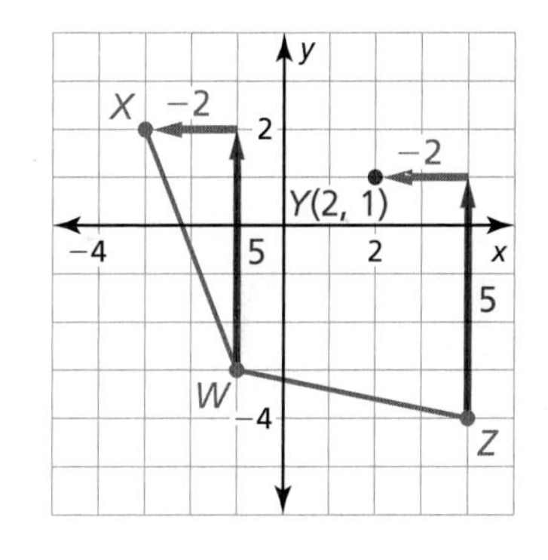

Step 3 Start at $Z(4, -4)$. Use the rise and run from Step 2 to find vertex Y.

A rise of 5 represents a change of 5 units up. A run of -2 represents a change of 2 units left.

So, plot the point that is 5 units up and 2 units left from $Z(4, -4)$. The point is (2, 1). Label it as vertex Y.

Step 4 Find the slopes of $\overline{XY}$ and $\overline{WZ}$ to verify that they are parallel.

$$\text{slope of } \overline{XY} = \frac{1-2}{2-(-3)} = \frac{-1}{5} = -\frac{1}{5} \qquad \text{slope of } \overline{WZ} = \frac{-4-(-3)}{4-(-1)} = \frac{-1}{5} = -\frac{1}{5}$$

▶ So, the coordinates of vertex Y are (2, 1).

Monitoring Progress  Help in English and Spanish at *BigIdeasMath.com*

5. Find the coordinates of the intersection of the diagonals of ▱*STUV* with vertices $S(-2, 3)$, $T(1, 5)$, $U(6, 3)$, and $V(3, 1)$.

6. Three vertices of ▱*ABCD* are $A(2, 4)$, $B(5, 2)$, and $C(3, -1)$. Find the coordinates of vertex D.

7.2 Exercises

Dynamic Solutions available at *BigIdeasMath.com*

Vocabulary and Core Concept Check

1. **VOCABULARY** Why is a parallelogram always a quadrilateral, but a quadrilateral is only sometimes a parallelogram?

2. **WRITING** You are given one angle measure of a parallelogram. Explain how you can find the other angle measures of the parallelogram.

Monitoring Progress and Modeling with Mathematics

In Exercises 3–6, find the value of each variable in the parallelogram. *(See Example 1.)*

3. 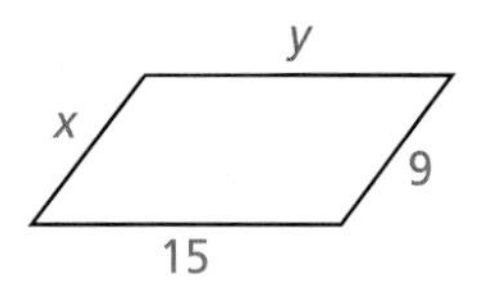

4. 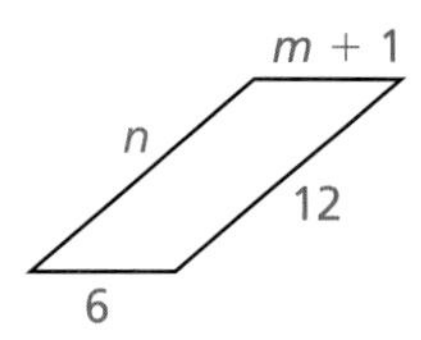

5. 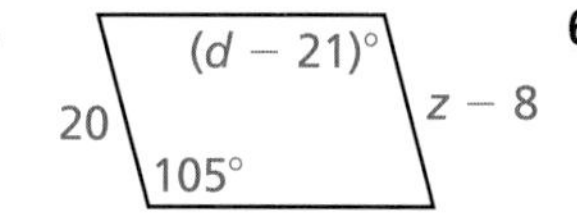

6. 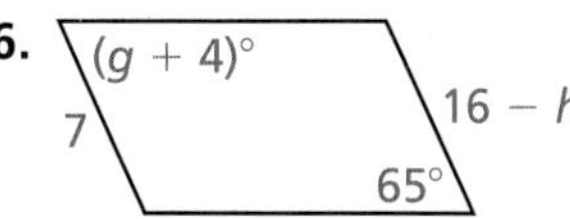

In Exercises 7 and 8, find the measure of the indicated angle in the parallelogram. *(See Example 2.)*

7. Find $m\angle B$.

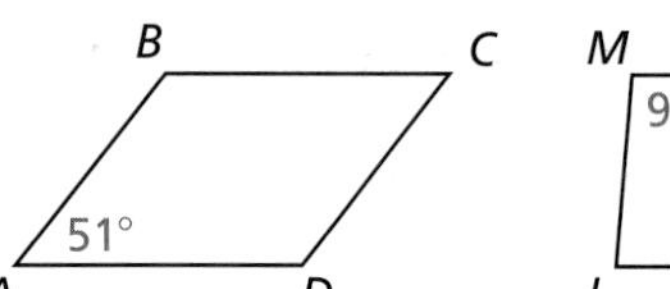

8. Find $m\angle N$.

In Exercises 9–16, find the indicated measure in $\square LMNQ$. Explain your reasoning.

9. LM
10. LP
11. LQ
12. MQ
13. $m\angle LMN$
14. $m\angle NQL$
15. $m\angle MNQ$
16. $m\angle LMQ$

In Exercises 17–20, find the value of each variable in the parallelogram.

17. 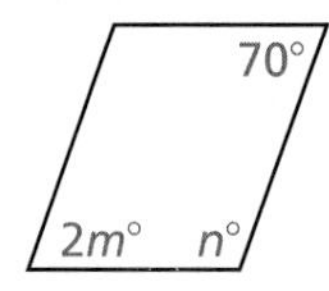

18. 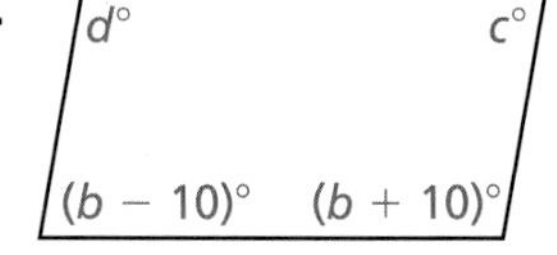

19. 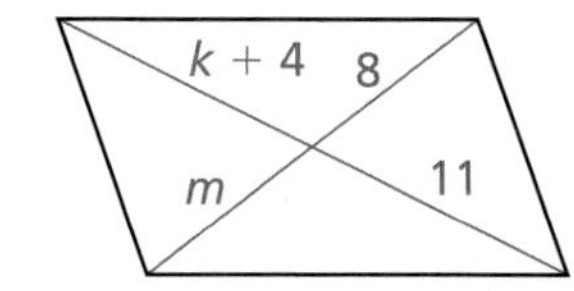

20. 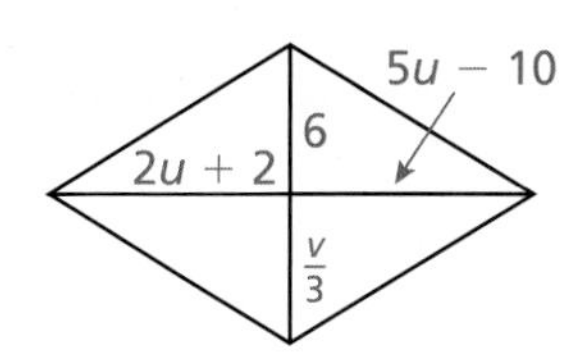

ERROR ANALYSIS In Exercises 21 and 22, describe and correct the error in using properties of parallelograms.

21. 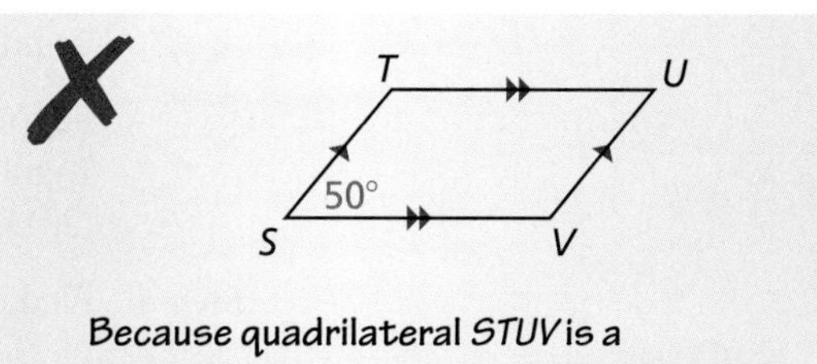

22. 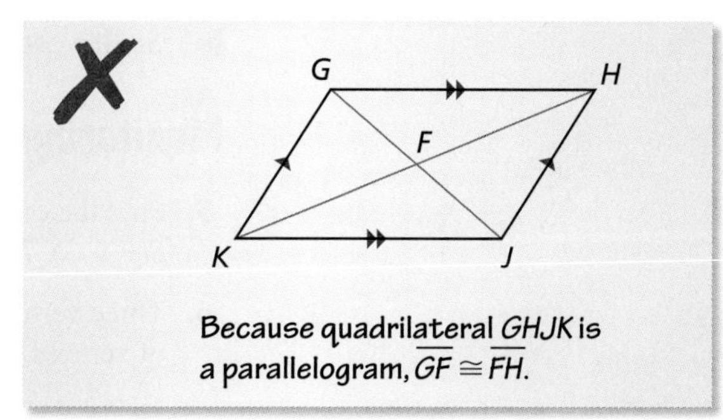

PROOF In Exercises 23 and 24, write a two-column proof. *(See Example 3.)*

23. Given $ABCD$ and $CEFD$ are parallelograms.

Prove $\overline{AB} \cong \overline{FE}$

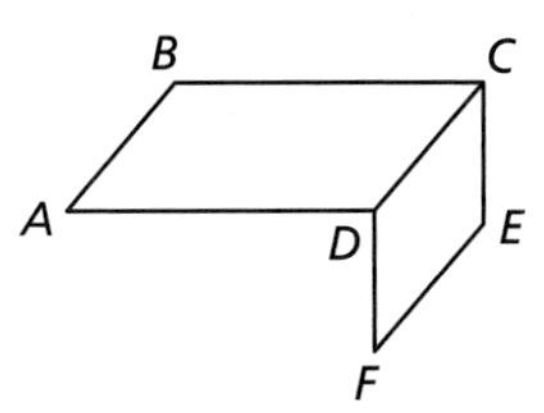

24. Given $ABCD$, $EBGF$, and $HJKD$ are parallelograms.

Prove $\angle 2 \cong \angle 3$

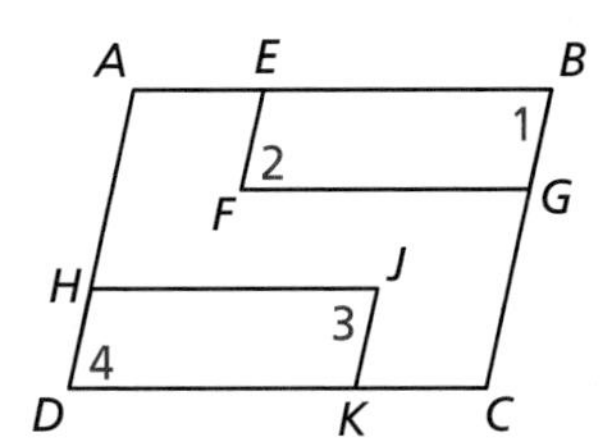

In Exercises 25 and 26, find the coordinates of the intersection of the diagonals of the parallelogram with the given vertices. *(See Example 4.)*

25. $W(-2, 5)$, $X(2, 5)$, $Y(4, 0)$, $Z(0, 0)$

26. $Q(-1, 3)$, $R(5, 2)$, $S(1, -2)$, $T(-5, -1)$

In Exercises 27–30, three vertices of ▱*DEFG* are given. Find the coordinates of the remaining vertex. *(See Example 5.)*

27. $D(0, 2)$, $E(-1, 5)$, $G(4, 0)$

28. $D(-2, -4)$, $F(0, 7)$, $G(1, 0)$

29. $D(-4, -2)$, $E(-3, 1)$, $F(3, 3)$

30. $E(-1, 4)$, $F(5, 6)$, $G(8, 0)$

MATHEMATICAL CONNECTIONS In Exercises 31 and 32, find the measure of each angle.

31. The measure of one interior angle of a parallelogram is 0.25 times the measure of another angle.

32. The measure of one interior angle of a parallelogram is 50 degrees more than 4 times the measure of another angle.

33. MAKING AN ARGUMENT In quadrilateral $ABCD$, $m\angle B = 124°$, $m\angle A = 56°$, and $m\angle C = 124°$. Your friend claims quadrilateral $ABCD$ could be a parallelogram. Is your friend correct? Explain your reasoning.

34. ATTENDING TO PRECISION $\angle J$ and $\angle K$ are consecutive angles in a parallelogram, $m\angle J = (3x + 7)°$, and $m\angle K = (5x - 11)°$. Find the measure of each angle.

35. CONSTRUCTION Construct any parallelogram and label it $ABCD$. Draw diagonals $\overline{AC}$ and $\overline{BD}$. Explain how to use paper folding to verify the Parallelogram Diagonals Theorem (Theorem 7.6) for ▱*ABCD*.

36. MODELING WITH MATHEMATICS The feathers on an arrow form two congruent parallelograms. The parallelograms are reflections of each other over the line that contains their shared side. Show that $m\angle 2 = 2m\angle 1$.

37. PROVING A THEOREM Use the diagram to write a two-column proof of the Parallelogram Opposite Angles Theorem (Theorem 7.4).

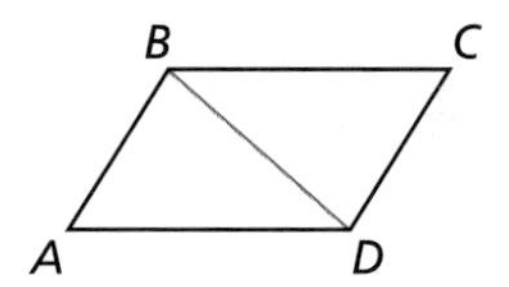

Given $ABCD$ is a parallelogram.

Prove $\angle A \cong \angle C$, $\angle B \cong \angle D$

38. PROVING A THEOREM Use the diagram to write a two-column proof of the Parallelogram Consecutive Angles Theorem (Theorem 7.5).

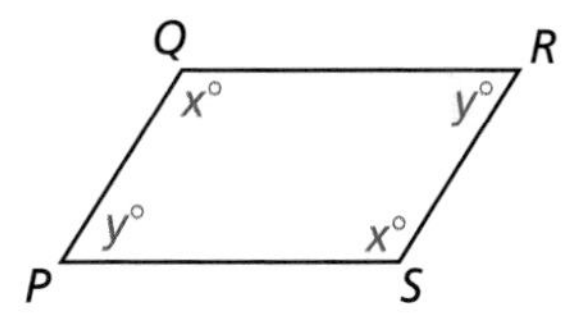

Given $PQRS$ is a parallelogram.

Prove $x° + y° = 180°$

39. PROBLEM SOLVING The sides of ▱*MNPQ* are represented by the expressions below. Sketch ▱*MNPQ* and find its perimeter.

$MQ = -2x + 37$ $\quad$ $QP = y + 14$

$NP = x - 5$ $\quad$ $MN = 4y + 5$

40. PROBLEM SOLVING In ▱*LMNP*, the ratio of LM to MN is 4 : 3. Find LM when the perimeter of ▱*LMNP* is 28.

41. ABSTRACT REASONING Can you prove that two parallelograms are congruent by proving that all their corresponding sides are congruent? Explain your reasoning.

42. HOW DO YOU SEE IT? The mirror shown is attached to the wall by an arm that can extend away from the wall. In the figure, points P, Q, R, and S are the vertices of a parallelogram. This parallelogram is one of several that change shape as the mirror is extended.

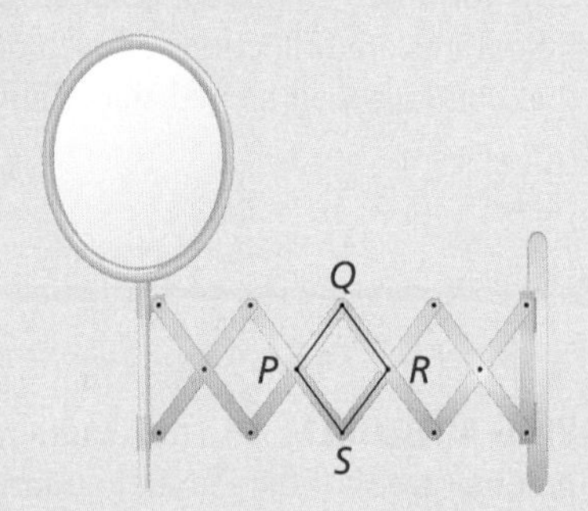

a. What happens to $m\angle P$ as $m\angle Q$ increases? Explain.

b. What happens to QS as $m\angle Q$ decreases? Explain.

c. What happens to the overall distance between the mirror and the wall when $m\angle Q$ decreases? Explain.

43. MATHEMATICAL CONNECTIONS In $\square STUV$, $m\angle TSU = 32°$, $m\angle USV = (x^2)°$, $m\angle TUV = 12x°$, and $\angle TUV$ is an acute angle. Find $m\angle USV$.

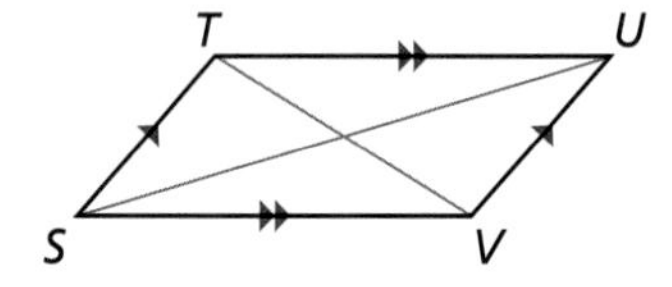

44. THOUGHT PROVOKING Is it possible that any triangle can be partitioned into four congruent triangles that can be rearranged to form a parallelogram? Explain your reasoning.

45. CRITICAL THINKING Points $W(1, 2)$, $X(3, 6)$, and $Y(6, 4)$ are three vertices of a parallelogram. How many parallelograms can be created using these three vertices? Find the coordinates of each point that could be the fourth vertex.

46. PROOF In the diagram, $\overline{EK}$ bisects $\angle FEH$, and $\overline{FJ}$ bisects $\angle EFG$. Prove that $\overline{EK} \perp \overline{FJ}$. (*Hint*: Write equations using the angle measures of the triangles and quadrilaterals formed.)

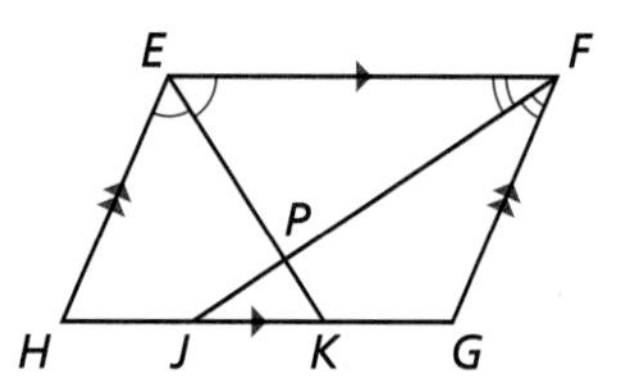

47. PROOF Prove the *Congruent Parts of Parallel Lines Corollary*: If three or more parallel lines cut off congruent segments on one transversal, then they cut off congruent segments on every transversal.

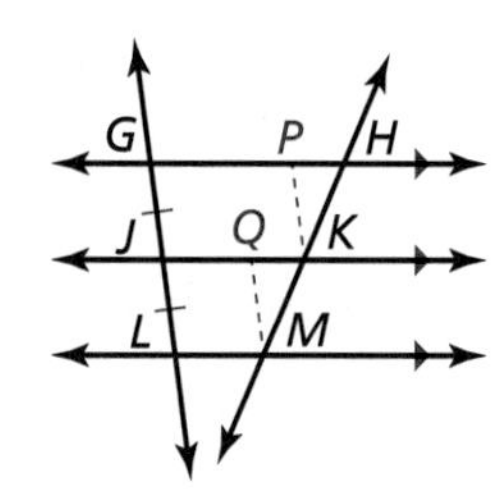

Given $\overleftrightarrow{GH} \parallel \overleftrightarrow{JK} \parallel \overleftrightarrow{LM}$, $\overline{GJ} \cong \overline{JL}$

Prove $\overline{HK} \cong \overline{KM}$

(*Hint:* Draw $\overline{KP}$ and $\overline{MQ}$ such that quadrilateral $GPKJ$ and quadrilateral $JQML$ are parallelograms.)

Maintaining Mathematical Proficiency

Reviewing what you learned in previous grades and lessons

Determine whether lines ℓ and m are parallel. Explain your reasoning. *(Section 3.3)*

48.

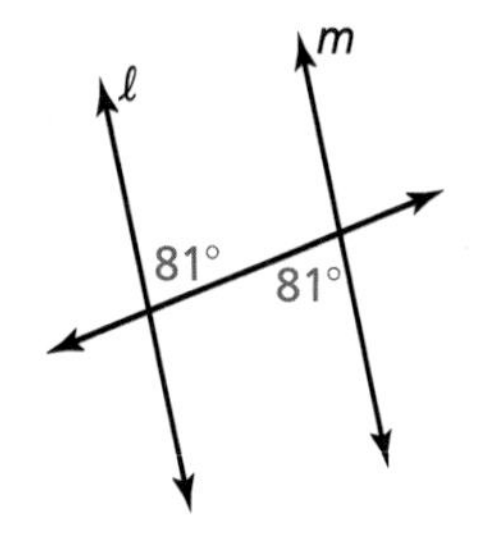

49.

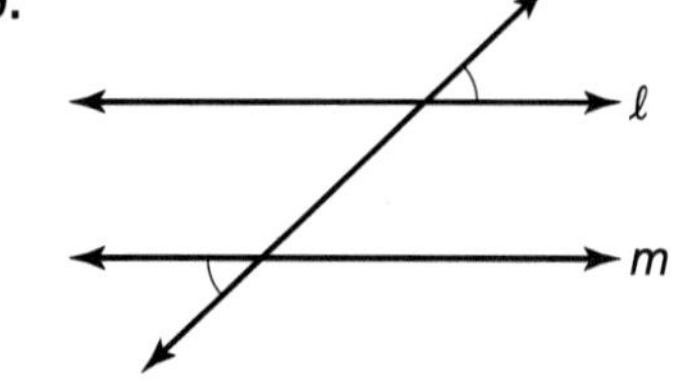

50.

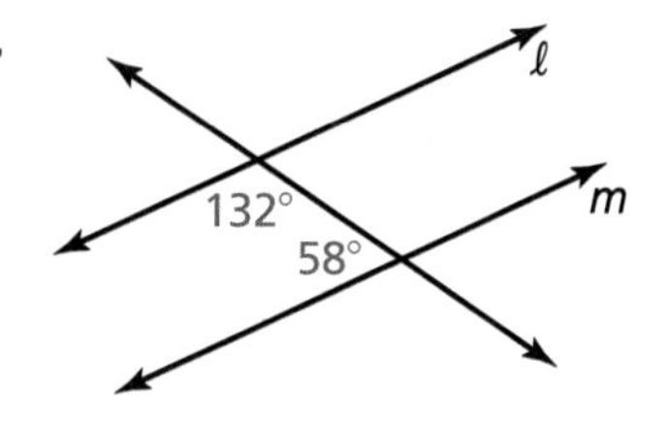

7.3 Proving That a Quadrilateral Is a Parallelogram

Essential Question How can you prove that a quadrilateral is a parallelogram?

EXPLORATION 1 Proving That a Quadrilateral Is a Parallelogram

Work with a partner. Use dynamic geometry software.

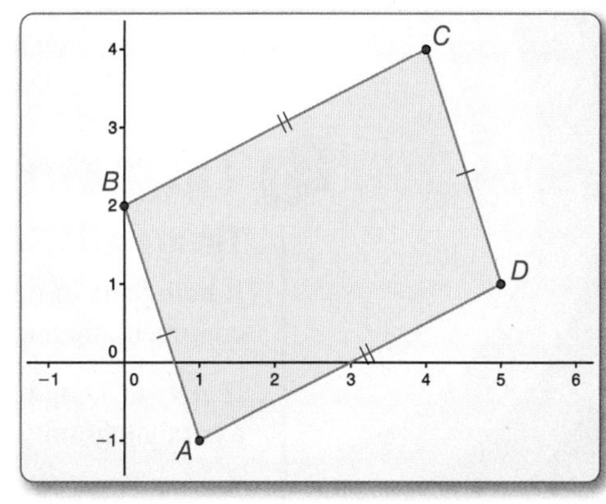

Sample
Points
$A(1, -1)$
$B(0, 2)$
$C(4, 4)$
$D(5, 1)$
Segments
$AB = 3.16$
$BC = 4.47$
$CD = 3.16$
$DA = 4.47$

REASONING ABSTRACTLY

To be proficient in math, you need to know and flexibly use different properties of objects.

a. Construct any quadrilateral $ABCD$ whose opposite sides are congruent.

b. Is the quadrilateral a parallelogram? Justify your answer.

c. Repeat parts (a) and (b) for several other quadrilaterals. Then write a conjecture based on your results.

d. Write the converse of your conjecture. Is the converse true? Explain.

EXPLORATION 2 Proving That a Quadrilateral Is a Parallelogram

Work with a partner. Use dynamic geometry software.

a. Construct any quadrilateral $ABCD$ whose opposite angles are congruent.

b. Is the quadrilateral a parallelogram? Justify your answer.

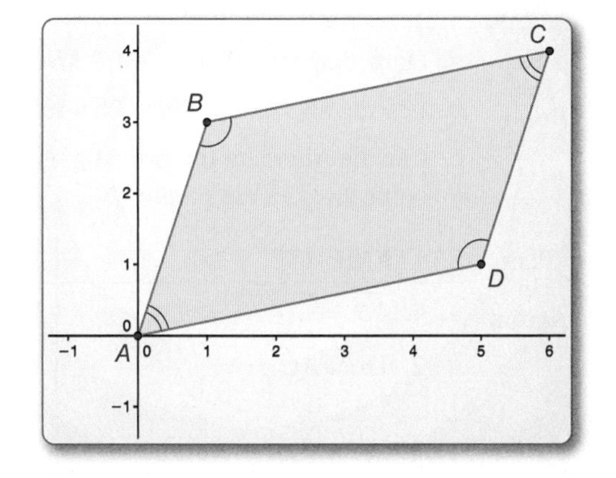

Sample
Points
$A(0, 0)$
$B(1, 3)$
$C(6, 4)$
$D(5, 1)$
Angles
$\angle A = 60.26°$
$\angle B = 119.74°$
$\angle C = 60.26°$
$\angle D = 119.74°$

c. Repeat parts (a) and (b) for several other quadrilaterals. Then write a conjecture based on your results.

d. Write the converse of your conjecture. Is the converse true? Explain.

Communicate Your Answer

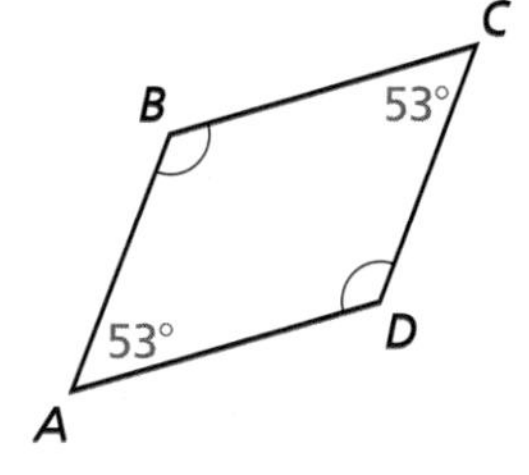

3. How can you prove that a quadrilateral is a parallelogram?

4. Is the quadrilateral at the left a parallelogram? Explain your reasoning.

7.3 Lesson

Core Vocabulary

Previous
diagonal
parallelogram

What You Will Learn

- Identify and verify parallelograms.
- Show that a quadrilateral is a parallelogram in the coordinate plane.

Identifying and Verifying Parallelograms

Given a parallelogram, you can use the Parallelogram Opposite Sides Theorem (Theorem 7.3) and the Parallelogram Opposite Angles Theorem (Theorem 7.4) to prove statements about the sides and angles of the parallelogram. The converses of the theorems are stated below. You can use these and other theorems in this lesson to prove that a quadrilateral with certain properties is a parallelogram.

Theorems

Theorem 7.7 Parallelogram Opposite Sides Converse

If both pairs of opposite sides of a quadrilateral are congruent, then the quadrilateral is a parallelogram.

If $\overline{AB} \cong \overline{CD}$ and $\overline{BC} \cong \overline{DA}$, then $ABCD$ is a parallelogram.

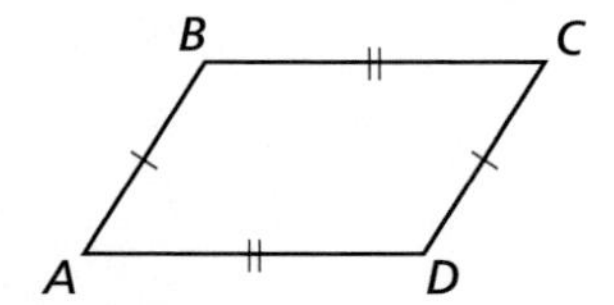

Theorem 7.8 Parallelogram Opposite Angles Converse

If both pairs of opposite angles of a quadrilateral are congruent, then the quadrilateral is a parallelogram.

If $\angle A \cong \angle C$ and $\angle B \cong \angle D$, then $ABCD$ is a parallelogram.

Proof Ex. 39, p. 383

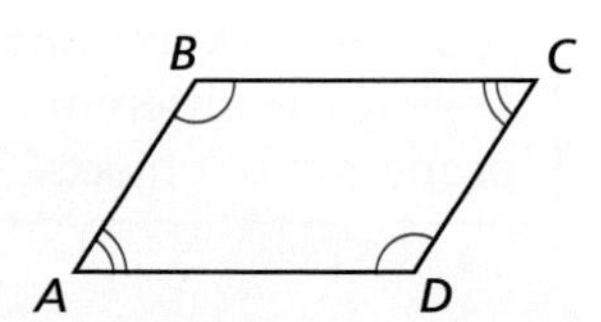

PROOF Parallelogram Opposite Sides Converse

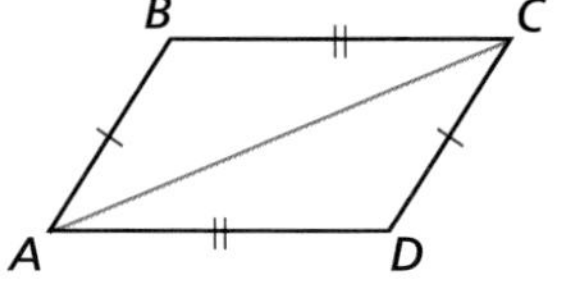

Given $\overline{AB} \cong \overline{CD}, \overline{BC} \cong \overline{DA}$

Prove $ABCD$ is a parallelogram.

Plan for Proof

a. Draw diagonal $\overline{AC}$ to form $\triangle ABC$ and $\triangle CDA$.

b. Use the SSS Congruence Theorem (Thm. 5.8) to show that $\triangle ABC \cong \triangle CDA$.

c. Use the Alternate Interior Angles Converse (Thm. 3.6) to show that opposite sides are parallel.

Plan in Action

	STATEMENTS	REASONS
a.	**1.** $\overline{AB} \cong \overline{CD}, \overline{BC} \cong \overline{DA}$	**1.** Given
	2. Draw $\overline{AC}$.	**2.** Through any two points, there exists exactly one line.
	3. $\overline{AC} \cong \overline{CA}$	**3.** Reflexive Property of Congruence (Thm. 2.1)
b.	**4.** $\triangle ABC \cong \triangle CDA$	**4.** SSS Congruence Theorem (Thm. 5.8)
c.	**5.** $\angle BAC \cong \angle DCA$, $\angle BCA \cong \angle DAC$	**5.** Corresponding parts of congruent triangles are congruent.
	6. $\overline{AB} \parallel \overline{CD}, \overline{BC} \parallel \overline{DA}$	**6.** Alternate Interior Angles Converse (Thm. 3.6)
	7. $ABCD$ is a parallelogram.	**7.** Definition of parallelogram

EXAMPLE 1 Identifying a Parallelogram

An amusement park ride has a moving platform attached to four swinging arms. The platform swings back and forth, higher and higher, until it goes over the top and around in a circular motion. In the diagram below, $\overline{AD}$ and $\overline{BC}$ represent two of the swinging arms, and $\overline{DC}$ is parallel to the ground (line ℓ). Explain why the moving platform $\overline{AB}$ is always parallel to the ground.

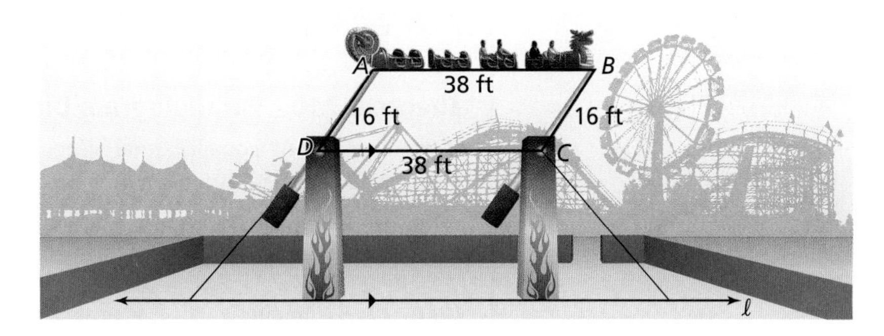

SOLUTION

The shape of quadrilateral *ABCD* changes as the moving platform swings around, but its side lengths do not change. Both pairs of opposite sides are congruent, so *ABCD* is a parallelogram by the Parallelogram Opposite Sides Converse.

By the definition of a parallelogram, $\overline{AB} \parallel \overline{DC}$. Because $\overline{DC}$ is parallel to line ℓ, $\overline{AB}$ is also parallel to line ℓ by the Transitive Property of Parallel Lines (Theorem 3.9). So, the moving platform is parallel to the ground.

Monitoring Progress  Help in English and Spanish at *BigIdeasMath.com*

1. In quadrilateral *WXYZ*, $m\angle W = 42°$, $m\angle X = 138°$, and $m\angle Y = 42°$. Find $m\angle Z$. Is *WXYZ* a parallelogram? Explain your reasoning.

EXAMPLE 2 Finding Side Lengths of a Parallelogram

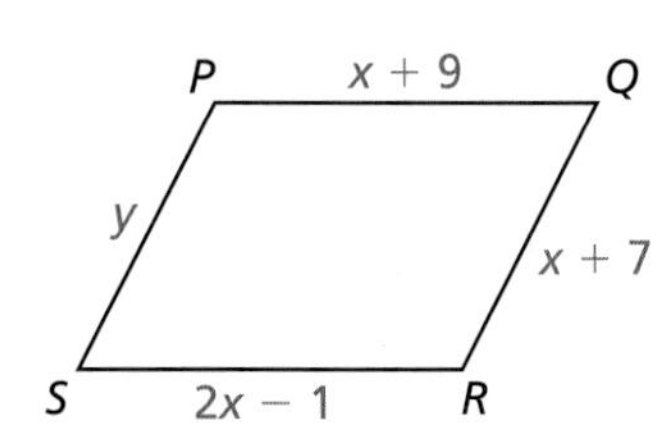

For what values of x and y is quadrilateral *PQRS* a parallelogram?

SOLUTION

By the Parallelogram Opposite Sides Converse, if both pairs of opposite sides of a quadrilateral are congruent, then the quadrilateral is a parallelogram. Find x so that $\overline{PQ} \cong \overline{SR}$.

$PQ = SR$	Set the segment lengths equal.
$x + 9 = 2x - 1$	Substitute $x + 9$ for PQ and $2x - 1$ for SR.
$10 = x$	Solve for x.

When $x = 10$, $PQ = 10 + 9 = 19$ and $SR = 2(10) - 1 = 19$. Find y so that $\overline{PS} \cong \overline{QR}$.

$PS = QR$	Set the segment lengths equal.
$y = x + 7$	Substitute y for PS and $x + 7$ for QR.
$y = 10 + 7$	Substitute 10 for x.
$y = 17$	Add.

When $x = 10$ and $y = 17$, $PS = 17$ and $QR = 10 + 7 = 17$.

▶ Quadrilateral *PQRS* is a parallelogram when $x = 10$ and $y = 17$.

Theorems

Theorem 7.9 Opposite Sides Parallel and Congruent Theorem

If one pair of opposite sides of a quadrilateral are congruent and parallel, then the quadrilateral is a parallelogram.

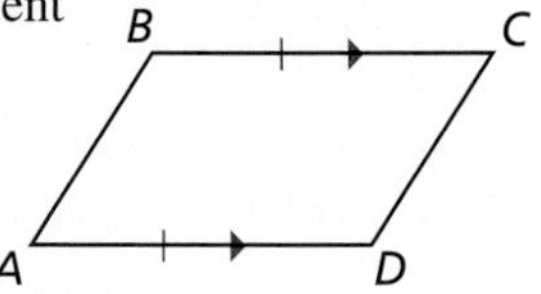

If $\overline{BC} \parallel \overline{AD}$ and $\overline{BC} \cong \overline{AD}$, then $ABCD$ is a parallelogram.

Proof Ex. 40, p. 383

Theorem 7.10 Parallelogram Diagonals Converse

If the diagonals of a quadrilateral bisect each other, then the quadrilateral is a parallelogram.

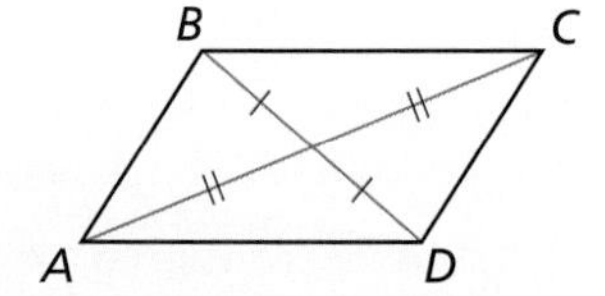

If $\overline{BD}$ and $\overline{AC}$ bisect each other, then $ABCD$ is a parallelogram.

Proof Ex. 41, p. 383

EXAMPLE 3 Identifying a Parallelogram

The doorway shown is part of a building in England. Over time, the building has leaned sideways. Explain how you know that $SV = TU$.

SOLUTION

In the photograph, $\overline{ST} \parallel \overline{UV}$ and $\overline{ST} \cong \overline{UV}$. By the Opposite Sides Parallel and Congruent Theorem, quadrilateral $STUV$ is a parallelogram. By the Parallelogram Opposite Sides Theorem (Theorem 7.3), you know that opposite sides of a parallelogram are congruent. So, $SV = TU$.

EXAMPLE 4 Finding Diagonal Lengths of a Parallelogram

For what value of x is quadrilateral $CDEF$ a parallelogram?

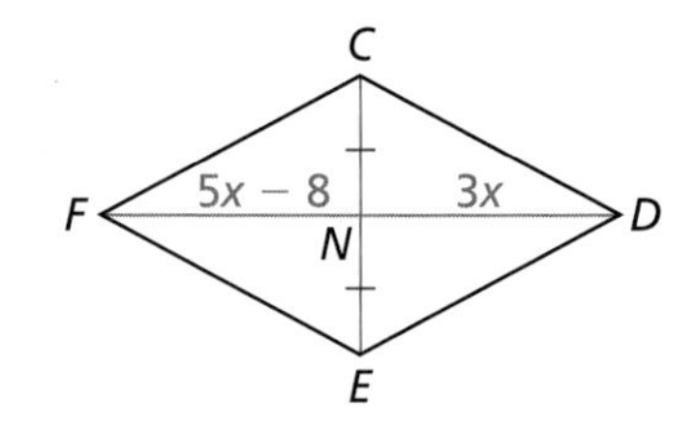

SOLUTION

By the Parallelogram Diagonals Converse, if the diagonals of $CDEF$ bisect each other, then it is a parallelogram. You are given that $\overline{CN} \cong \overline{EN}$. Find x so that $\overline{FN} \cong \overline{DN}$.

$FN = DN$	Set the segment lengths equal.
$5x - 8 = 3x$	Substitute $5x - 8$ for FN and $3x$ for DN.
$2x - 8 = 0$	Subtract $3x$ from each side.
$2x = 8$	Add 8 to each side.
$x = 4$	Divide each side by 2.

When $x = 4$, $FN = 5(4) - 8 = 12$ and $DN = 3(4) = 12$.

▶ Quadrilateral $CDEF$ is a parallelogram when $x = 4$.

Monitoring Progress  Help in English and Spanish at *BigIdeasMath.com*

2. For what values of x and y is quadrilateral $ABCD$ a parallelogram? Explain your reasoning.

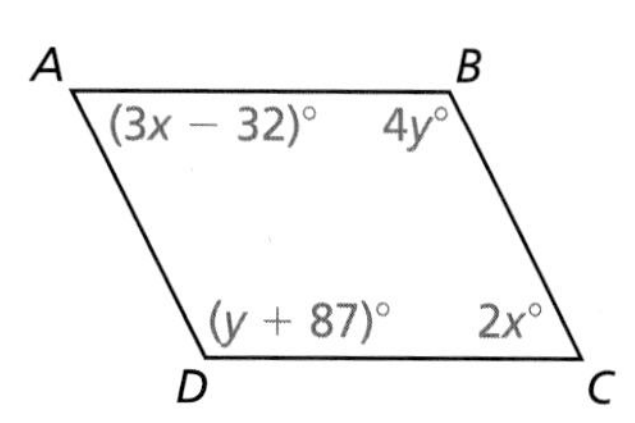

State the theorem you can use to show that the quadrilateral is a parallelogram.

3.

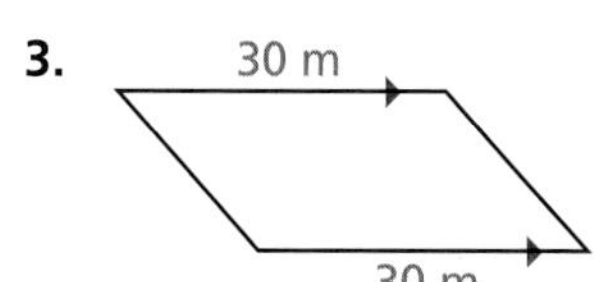

4.

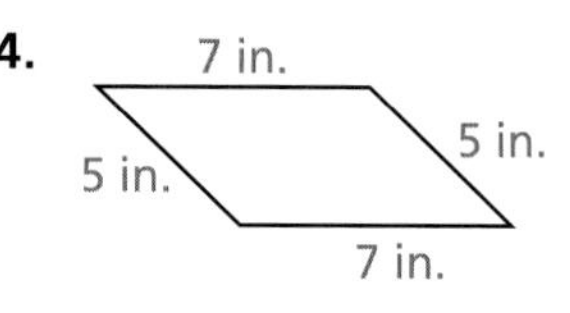

5.

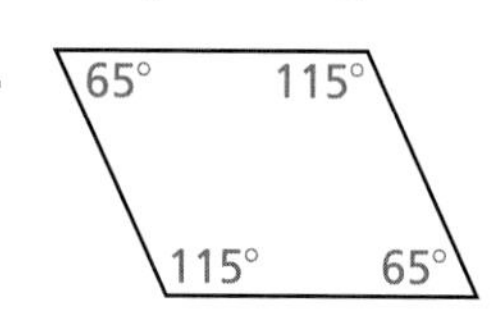

6. For what value of x is quadrilateral $MNPQ$ a parallelogram? Explain your reasoning.

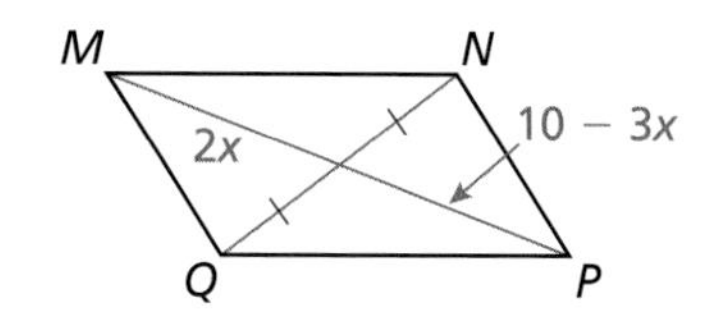

Concept Summary

Ways to Prove a Quadrilateral Is a Parallelogram

1. Show that both pairs of opposite sides are parallel. *(Definition)*	
2. Show that both pairs of opposite sides are congruent. *(Parallelogram Opposite Sides Converse)*	
3. Show that both pairs of opposite angles are congruent. *(Parallelogram Opposite Angles Converse)*	
4. Show that one pair of opposite sides are congruent and parallel. *(Opposite Sides Parallel and Congruent Theorem)*	
5. Show that the diagonals bisect each other. *(Parallelogram Diagonals Converse)*	

Using Coordinate Geometry

EXAMPLE 5 Identifying a Parallelogram in the Coordinate Plane

Show that quadrilateral $ABCD$ is a parallelogram.

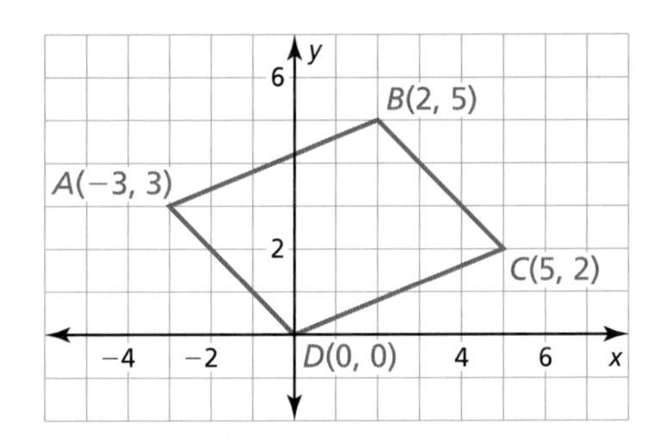

SOLUTION

Method 1 Show that a pair of sides are congruent and parallel. Then apply the Opposite Sides Parallel and Congruent Theorem.

First, use the Distance Formula to show that $\overline{AB}$ and $\overline{CD}$ are congruent.

$$AB = \sqrt{[2 - (-3)]^2 + (5 - 3)^2} = \sqrt{29}$$

$$CD = \sqrt{(5 - 0)^2 + (2 - 0)^2} = \sqrt{29}$$

Because $AB = CD = \sqrt{29}$, $\overline{AB} \cong \overline{CD}$.

Then, use the slope formula to show that $\overline{AB} \parallel \overline{CD}$.

$$\text{slope of } \overline{AB} = \frac{5 - 3}{2 - (-3)} = \frac{2}{5}$$

$$\text{slope of } \overline{CD} = \frac{2 - 0}{5 - 0} = \frac{2}{5}$$

Because $\overline{AB}$ and $\overline{CD}$ have the same slope, they are parallel.

▶ $\overline{AB}$ and $\overline{CD}$ are congruent and parallel. So, $ABCD$ is a parallelogram by the Opposite Sides Parallel and Congruent Theorem.

Method 2 Show that opposite sides are congruent. Then apply the Parallelogram Opposite Sides Converse. In Method 1, you already have shown that because $AB = CD = \sqrt{29}$, $\overline{AB} \cong \overline{CD}$. Now find AD and BC.

$$AD = \sqrt{(-3 - 0)^2 + (3 - 0)^2} = 3\sqrt{2}$$

$$BC = \sqrt{(2 - 5)^2 + (5 - 2)^2} = 3\sqrt{2}$$

Because $AD = BC = 3\sqrt{2}$, $\overline{AD} \cong \overline{BC}$.

▶ $\overline{AB} \cong \overline{CD}$ and $\overline{AD} \cong \overline{BC}$. So, $ABCD$ is a parallelogram by the Parallelogram Opposite Sides Converse.

Monitoring Progress  Help in English and Spanish at *BigIdeasMath.com*

7. Show that quadrilateral $JKLM$ is a parallelogram.
8. Refer to the Concept Summary on page 379. Explain two other methods you can use to show that quadrilateral $ABCD$ in Example 5 is a parallelogram.

7.3 Exercises

Dynamic Solutions available at *BigIdeasMath.com*

Vocabulary and Core Concept Check

1. **WRITING** A quadrilateral has four congruent sides. Is the quadrilateral a parallelogram? Justify your answer.

2. **DIFFERENT WORDS, SAME QUESTION** Which is different? Find "both" answers.

Construct a quadrilateral with opposite sides congruent.	Construct a quadrilateral with one pair of parallel sides.
Construct a quadrilateral with opposite angles congruent.	Construct a quadrilateral with one pair of opposite sides congruent and parallel.

Monitoring Progress and Modeling with Mathematics

In Exercises 3–8, state which theorem you can use to show that the quadrilateral is a parallelogram. *(See Examples 1 and 3.)*

3. 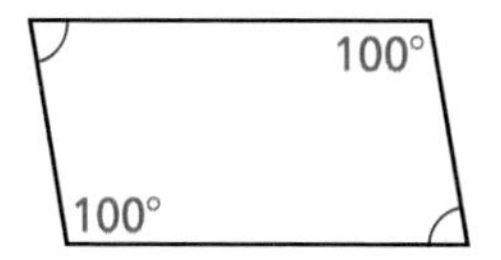

4. 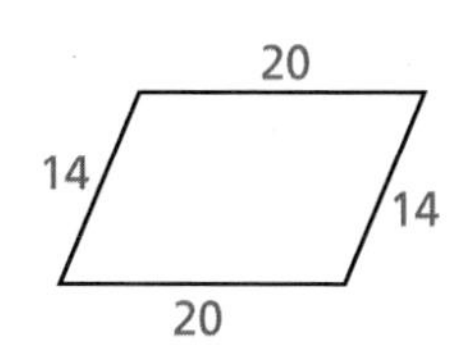

5. 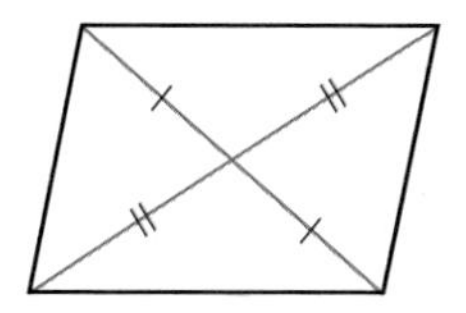

6. 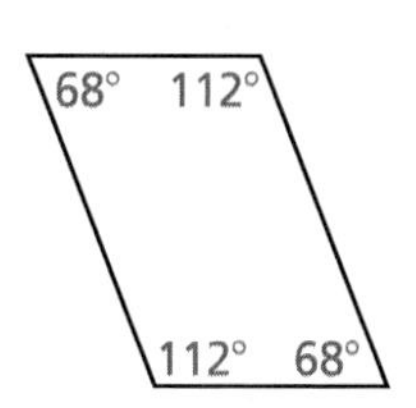

7. 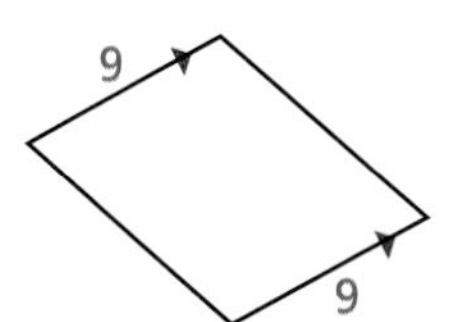

8. 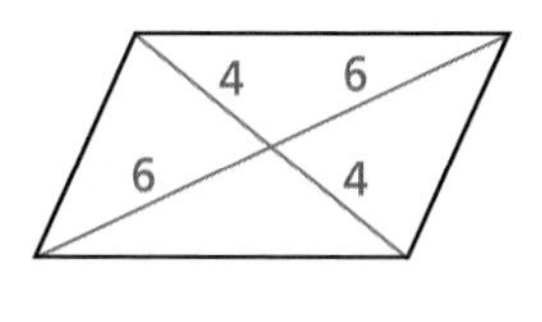

In Exercises 9–12, find the values of x and y that make the quadrilateral a parallelogram. *(See Example 2.)*

9. 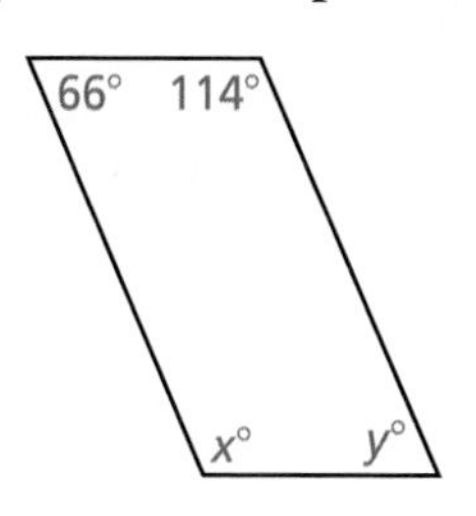

10. 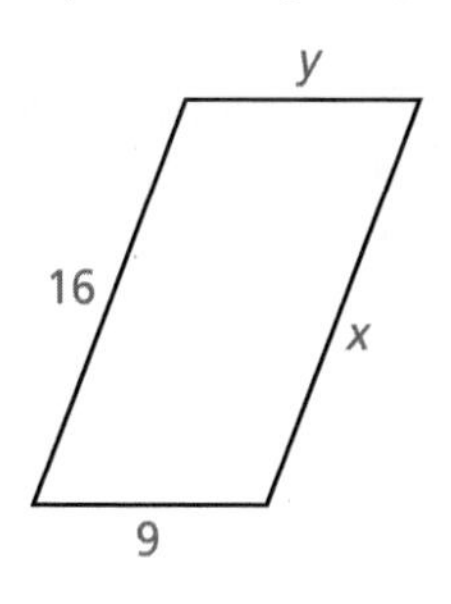

11. 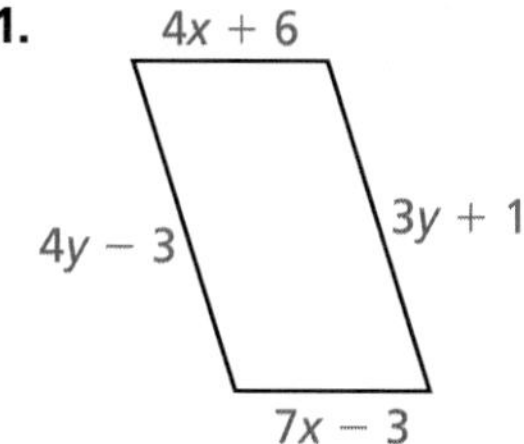

12. 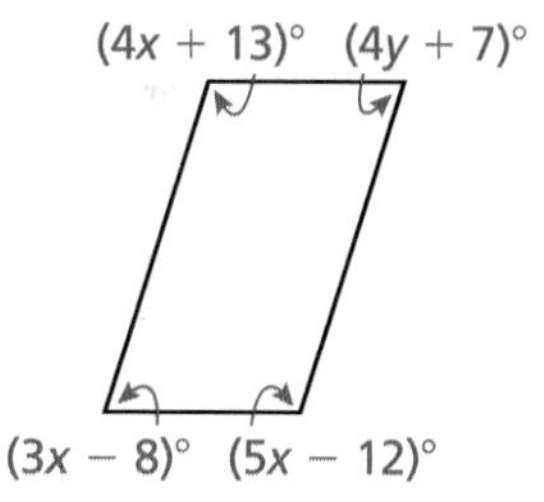

In Exercises 13–16, find the value of x that makes the quadrilateral a parallelogram. *(See Example 4.)*

13. 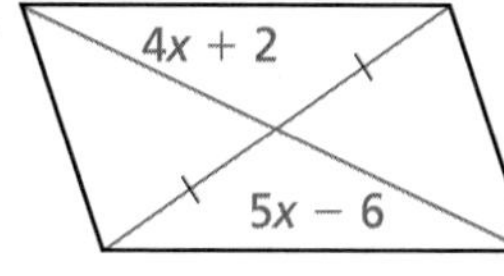

14. 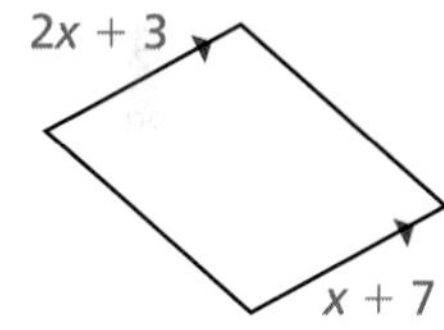

15. 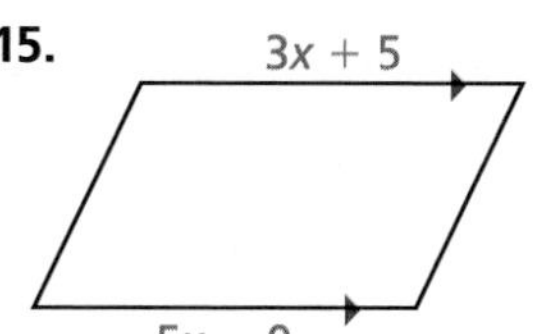

16. 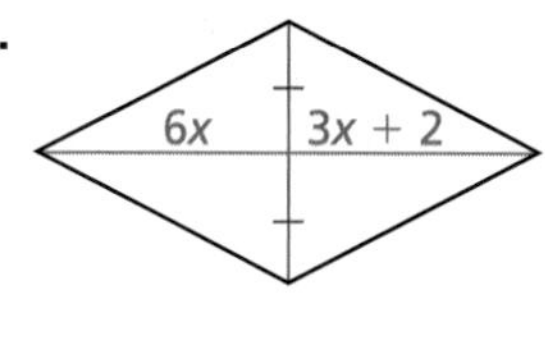

In Exercises 17–20, graph the quadrilateral with the given vertices in a coordinate plane. Then show that the quadrilateral is a parallelogram. *(See Example 5.)*

17. $A(0, 1)$, $B(4, 4)$, $C(12, 4)$, $D(8, 1)$

18. $E(-3, 0)$, $F(-3, 4)$, $G(3, -1)$, $H(3, -5)$

19. $J(-2, 3)$, $K(-5, 7)$, $L(3, 6)$, $M(6, 2)$

20. $N(-5, 0)$, $P(0, 4)$, $Q(3, 0)$, $R(-2, -4)$

ERROR ANALYSIS **In Exercises 21 and 22, describe and correct the error in identifying a parallelogram.**

21.

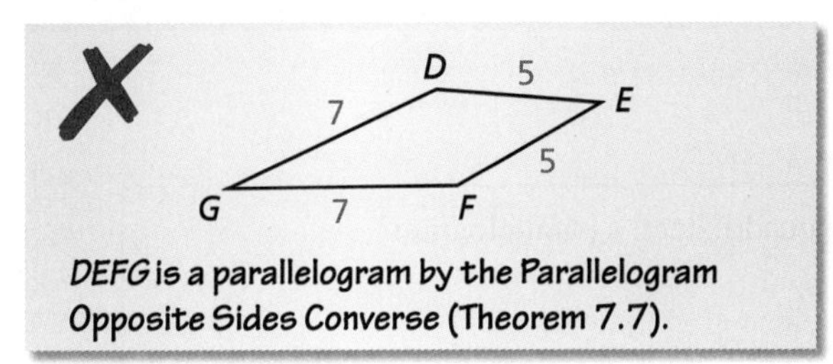

22.

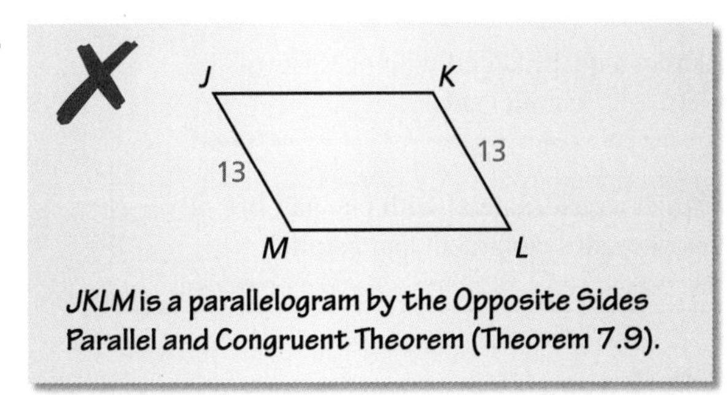

23. MATHEMATICAL CONNECTIONS What value of x makes the quadrilateral a parallelogram? Explain how you found your answer.

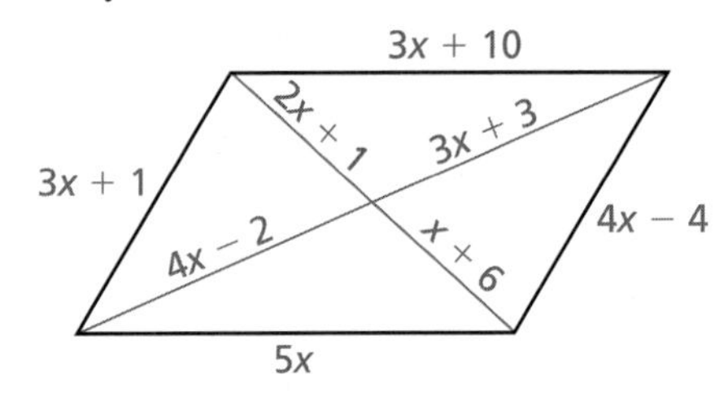

24. MAKING AN ARGUMENT Your friend says you can show that quadrilateral *WXYZ* is a parallelogram by using the Consecutive Interior Angles Converse (Theorem 3.8) and the Opposite Sides Parallel and Congruent Theorem (Theorem 7.9). Is your friend correct? Explain your reasoning.

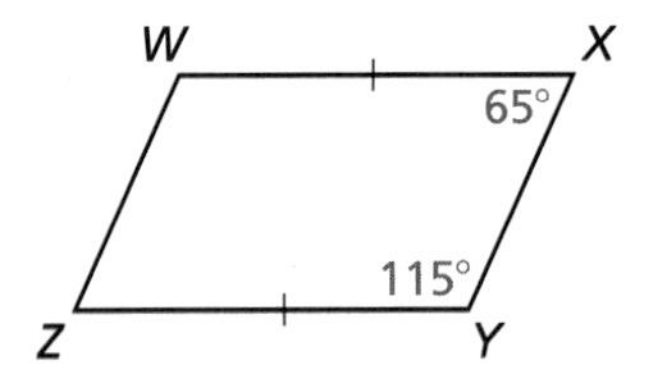

ANALYZING RELATIONSHIPS **In Exercises 25–27, write the indicated theorems as a biconditional statement.**

25. Parallelogram Opposite Sides Theorem (Theorem 7.3) and Parallelogram Opposite Sides Converse (Theorem 7.7)

26. Parallelogram Opposite Angles Theorem (Theorem 7.4) and Parallelogram Opposite Angles Converse (Theorem 7.8)

27. Parallelogram Diagonals Theorem (Theorem 7.6) and Parallelogram Diagonals Converse (Theorem 7.10)

28. CONSTRUCTION Describe a method that uses the Opposite Sides Parallel and Congruent Theorem (Theorem 7.9) to construct a parallelogram. Then construct a parallelogram using your method.

29. REASONING Follow the steps below to construct a parallelogram. Explain why this method works. State a theorem to support your answer.

Step 1 Use a ruler to draw two segments that intersect at their midpoints.

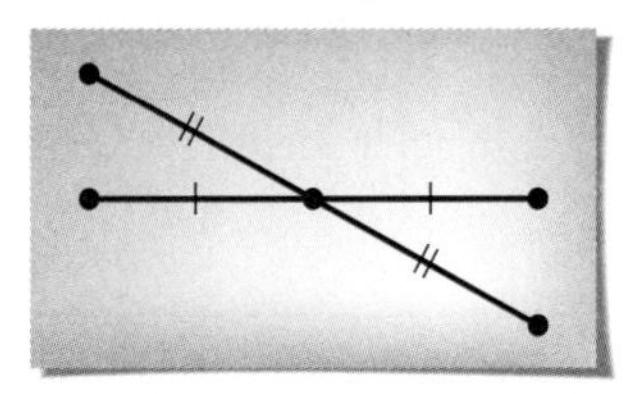

Step 2 Connect the endpoints of the segments to form a parallelogram.

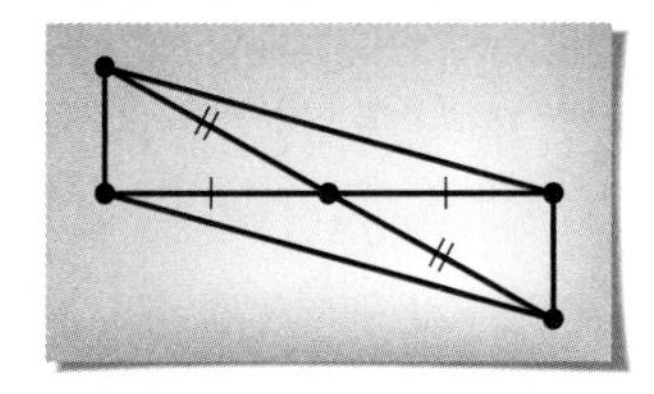

30. MAKING AN ARGUMENT Your brother says to show that quadrilateral *QRST* is a parallelogram, you must show that $\overline{QR} \parallel \overline{TS}$ and $\overline{QT} \parallel \overline{RS}$. Your sister says that you must show that $\overline{QR} \cong \overline{TS}$ and $\overline{QT} \cong \overline{RS}$. Who is correct? Explain your reasoning.

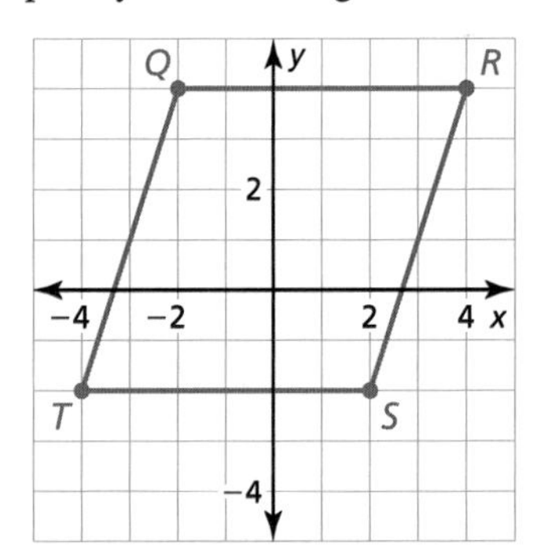

REASONING **In Exercises 31 and 32, your classmate incorrectly claims that the marked information can be used to show that the figure is a parallelogram. Draw a quadrilateral with the same marked properties that is clearly *not* a parallelogram.**

31.

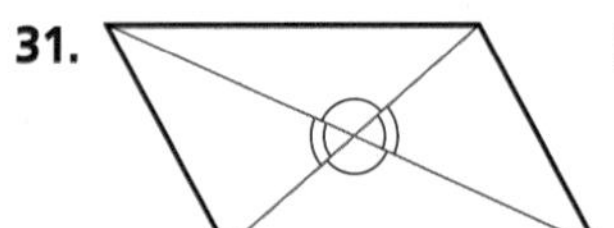

32.

33. **MODELING WITH MATHEMATICS** You shoot a pool ball, and it rolls back to where it started, as shown in the diagram. The ball bounces off each wall at the same angle at which it hits the wall.

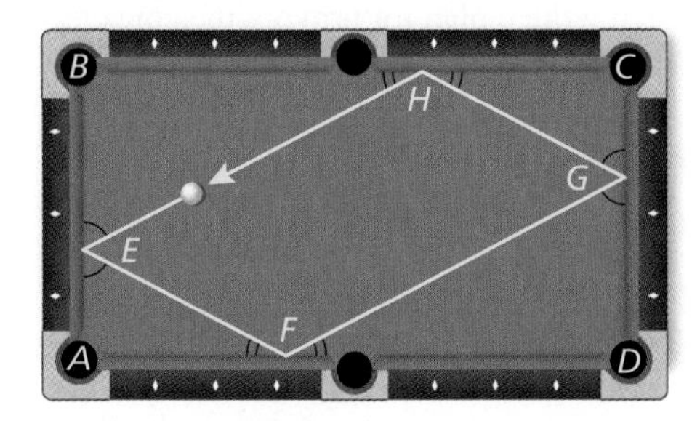

a. The ball hits the first wall at an angle of 63°. So $m\angle AEF = m\angle BEH = 63°$. What is $m\angle AFE$? Explain your reasoning.

b. Explain why $m\angle FGD = 63°$.

c. What is $m\angle GHC$? $m\angle EHB$?

d. Is quadrilateral $EFGH$ a parallelogram? Explain your reasoning.

34. **MODELING WITH MATHEMATICS** In the diagram of the parking lot shown, $m\angle JKL = 60°$, $JK = LM = 21$ feet, and $KL = JM = 9$ feet.

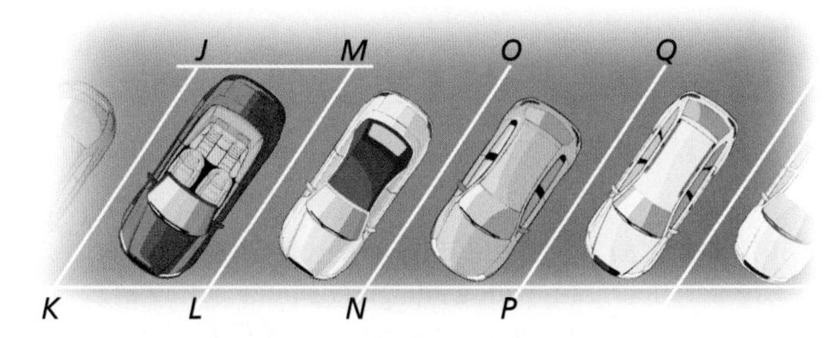

a. Explain how to show that parking space $JKLM$ is a parallelogram.

b. Find $m\angle JML$, $m\angle KJM$, and $m\angle KLM$.

c. $\overline{LM} \parallel \overline{NO}$ and $\overline{NO} \parallel \overline{PQ}$. Which theorem could you use to show that $\overline{JK} \parallel \overline{PQ}$?

REASONING In Exercises 35–37, describe how to prove that *ABCD* is a parallelogram.

35.

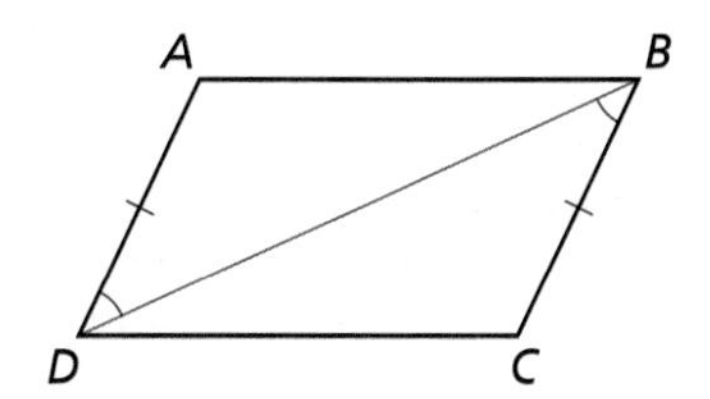

36.

37.

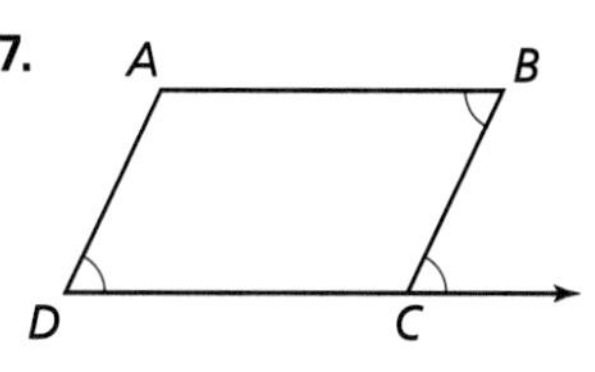

38. **REASONING** Quadrilateral $JKLM$ is a parallelogram. Describe how to prove that $\triangle MGJ \cong \triangle KHL$.

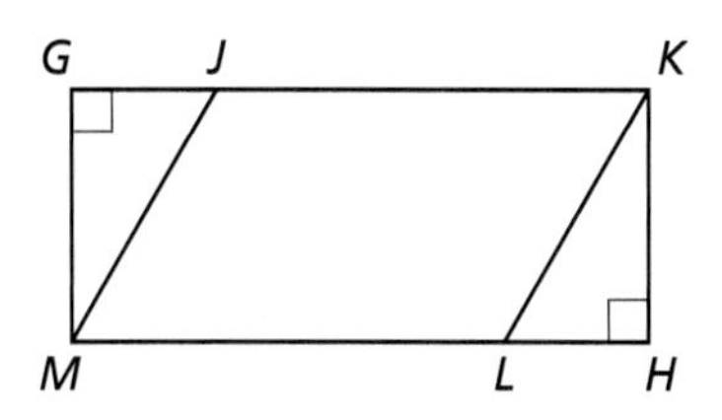

39. **PROVING A THEOREM** Prove the Parallelogram Opposite Angles Converse (Theorem 7.8). (*Hint*: Let $x°$ represent $m\angle A$ and $m\angle C$. Let $y°$ represent $m\angle B$ and $m\angle D$. Write and simplify an equation involving x and y.)

Given $\angle A \cong \angle C$, $\angle B \cong \angle D$

Prove $ABCD$ is a parallelogram.

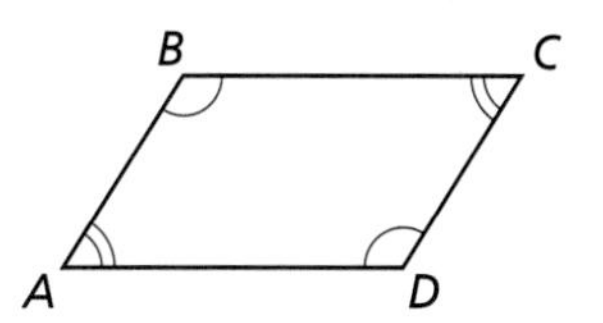

40. **PROVING A THEOREM** Use the diagram of $PQRS$ with the auxiliary line segment drawn to prove the Opposite Sides Parallel and Congruent Theorem (Theorem 7.9).

Given $\overline{QR} \parallel \overline{PS}$, $\overline{QR} \cong \overline{PS}$

Prove $PQRS$ is a parallelogram.

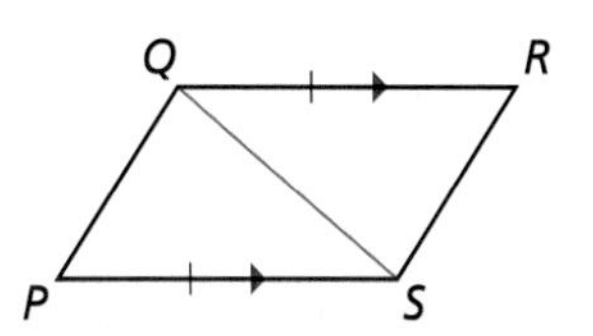

41. **PROVING A THEOREM** Prove the Parallelogram Diagonals Converse (Theorem 7.10).

Given Diagonals $\overline{JL}$ and $\overline{KM}$ bisect each other.

Prove $JKLM$ is a parallelogram.

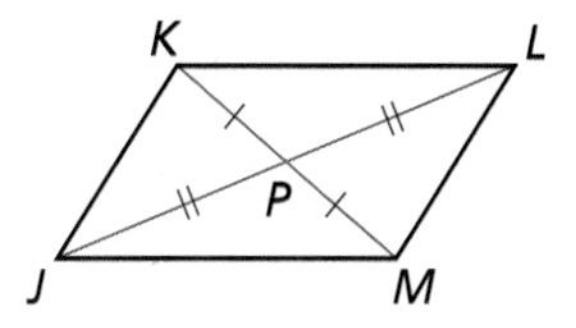

42. PROOF Write a proof.

Given *DEBF* is a parallelogram.
$AE = CF$

Prove *ABCD* is a parallelogram.

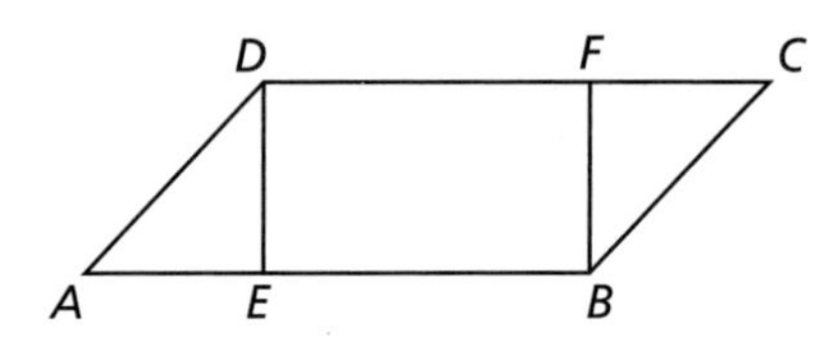

43. REASONING Three interior angle measures of a quadrilateral are 67°, 67°, and 113°. Is this enough information to conclude that the quadrilateral is a parallelogram? Explain your reasoning.

44. HOW DO YOU SEE IT? A music stand can be folded up, as shown. In the diagrams, *AEFD* and *EBCF* are parallelograms. Which labeled segments remain parallel as the stand is folded?

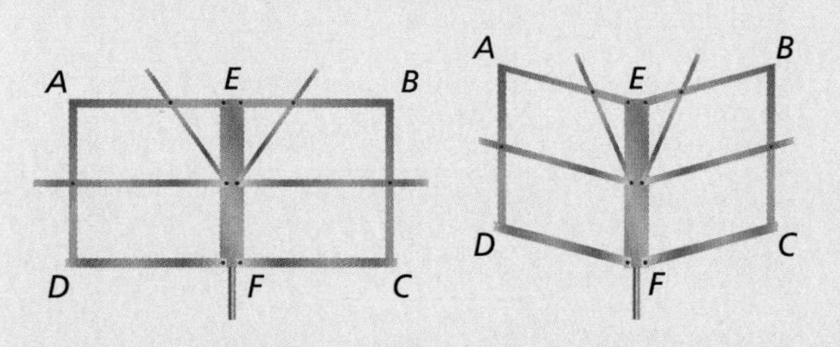

45. CRITICAL THINKING In the diagram, *ABCD* is a parallelogram, $BF = DE = 12$, and $CF = 8$. Find *AE*. Explain your reasoning.

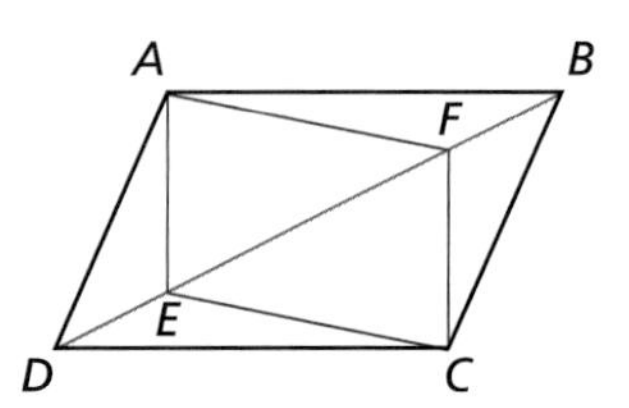

46. THOUGHT PROVOKING Create a regular hexagon using congruent parallelograms.

47. WRITING The Parallelogram Consecutive Angles Theorem (Theorem 7.5) says that if a quadrilateral is a parallelogram, then its consecutive angles are supplementary. Write the converse of this theorem. Then write a plan for proving the converse. Include a diagram.

48. PROOF Write a proof.

Given *ABCD* is a parallelogram.
$\angle A$ is a right angle.

Prove $\angle B$, $\angle C$, and $\angle D$ are right angles.

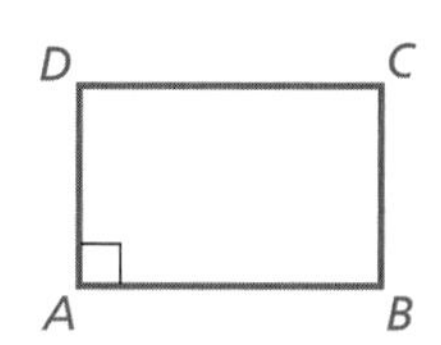

49. ABSTRACT REASONING The midpoints of the sides of a quadrilateral have been joined to form what looks like a parallelogram. Show that a quadrilateral formed by connecting the midpoints of the sides of any quadrilateral is always a parallelogram. (*Hint*: Draw a diagram. Include a diagonal of the larger quadrilateral. Show how two sides of the smaller quadrilateral relate to the diagonal.)

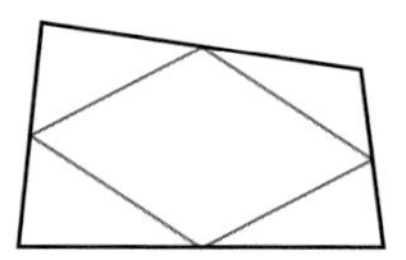

50. CRITICAL THINKING Show that if *ABCD* is a parallelogram with its diagonals intersecting at *E*, then you can connect the midpoints *F*, *G*, *H*, and *J* of $\overline{AE}$, $\overline{BE}$, $\overline{CE}$, and $\overline{DE}$, respectively, to form another parallelogram, *FGHJ*.

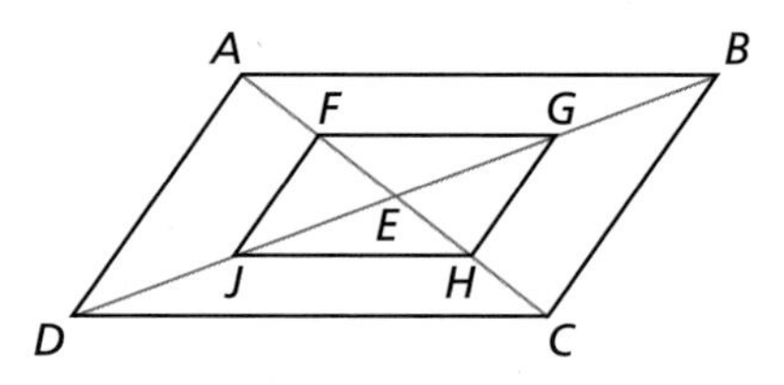

Maintaining Mathematical Proficiency

Reviewing what you learned in previous grades and lessons

Classify the quadrilateral. *(Skills Review Handbook)*

51.

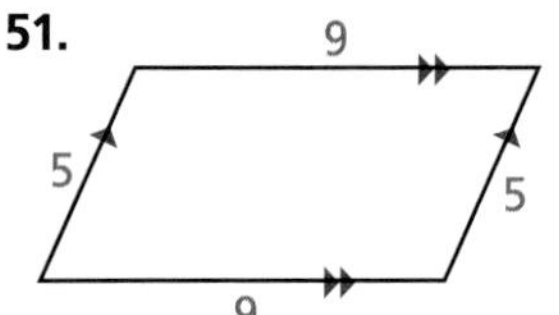

52.

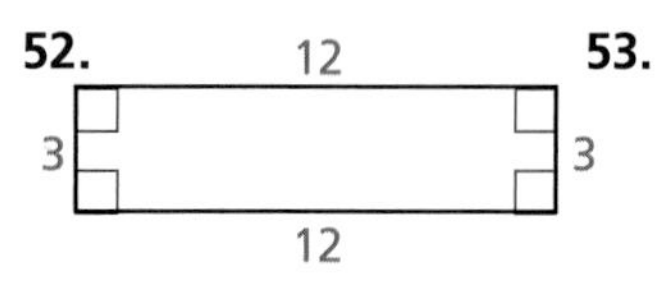

53.

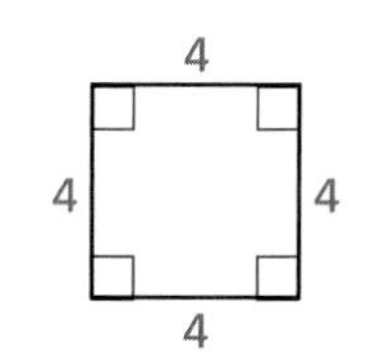

54.

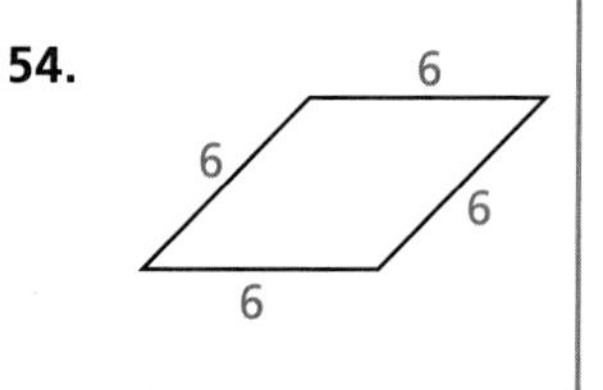

7.1–7.3 What Did You Learn?

Core Vocabulary

diagonal, *p. 360*
equilateral polygon, *p. 361*
equiangular polygon, *p. 361*
regular polygon, *p. 361*
parallelogram, *p. 368*

Core Concepts

Section 7.1

Theorem 7.1 Polygon Interior Angles Theorem, *p. 360*
Corollary 7.1 Corollary to the Polygon Interior Angles Theorem, *p. 361*
Theorem 7.2 Polygon Exterior Angles Theorem, *p. 362*

Section 7.2

Theorem 7.3 Parallelogram Opposite Sides Theorem, *p. 368*
Theorem 7.4 Parallelogram Opposite Angles Theorem, *p. 368*
Theorem 7.5 Parallelogram Consecutive Angles Theorem, *p. 369*
Theorem 7.6 Parallelogram Diagonals Theorem, *p. 369*
Using Parallelograms in the Coordinate Plane, *p. 371*

Section 7.3

Theorem 7.7 Parallelogram Opposite Sides Converse, *p. 376*
Theorem 7.8 Parallelogram Opposite Angles Converse, *p. 376*
Theorem 7.9 Opposite Sides Parallel and Congruent Theorem, *p. 378*
Theorem 7.10 Parallelogram Diagonals Converse, *p. 378*
Ways to Prove a Quadrilateral is a Parallelogram, *p. 379*
Showing That a Quadrilateral Is a Parallelogram in the Coordinate Plane, *p. 380*

Mathematical Practices

1. In Exercise 52 on page 366, what is the relationship between the 540° increase and the answer?
2. Explain why the process you used works every time in Exercise 25 on page 373. Is there another way to do it?
3. In Exercise 23 on page 382, explain how you started the problem. Why did you start that way? Could you have started another way? Explain.

Study Skills

Keeping Your Mind Focused during Class

- When you sit down at your desk, get all other issues out of your mind by reviewing your notes from the last class and focusing on just math.
- Repeat in your mind what you are writing in your notes.
- When the math is particularly difficult, ask your teacher for another example.

7.1–7.3 Quiz

Find the value of *x*. *(Section 7.1)*

1.

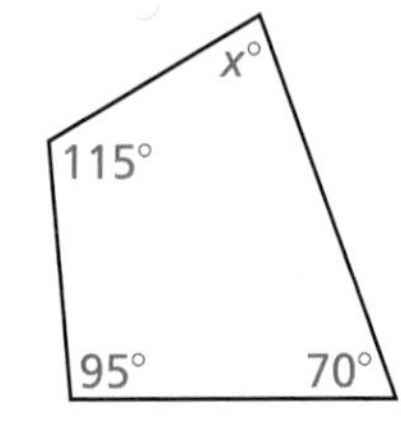

2.

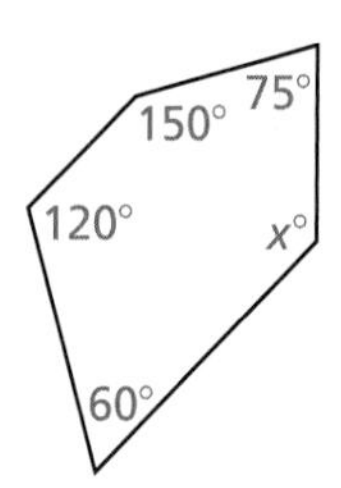

3.

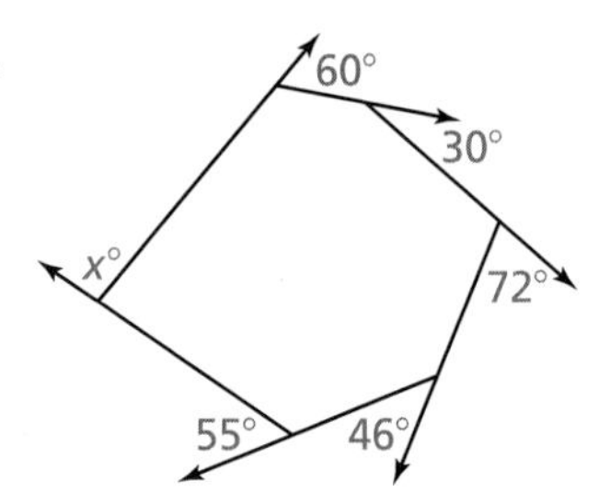

Find the measure of each interior angle and each exterior angle of the indicated regular polygon. *(Section 7.1)*

4. decagon

5. 15-gon

6. 24-gon

7. 60-gon

Find the indicated measure in ▱*ABCD*. Explain your reasoning. *(Section 7.2)*

8. CD

9. AD

10. AE

11. BD

12. $m\angle BCD$

13. $m\angle ABC$

14. $m\angle ADC$

15. $m\angle DBC$

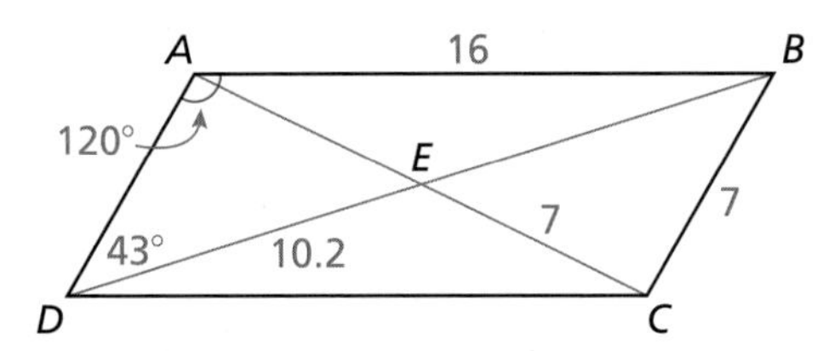

State which theorem you can use to show that the quadrilateral is a parallelogram. *(Section 7.3)*

16.

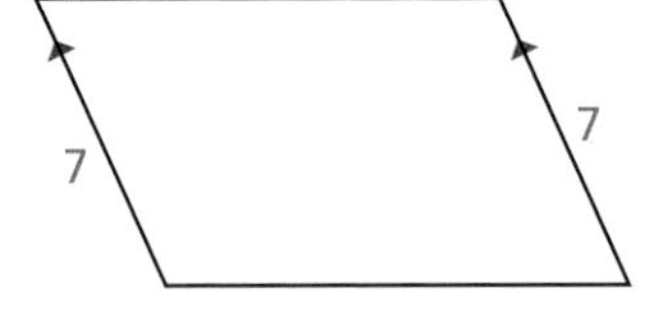

17.

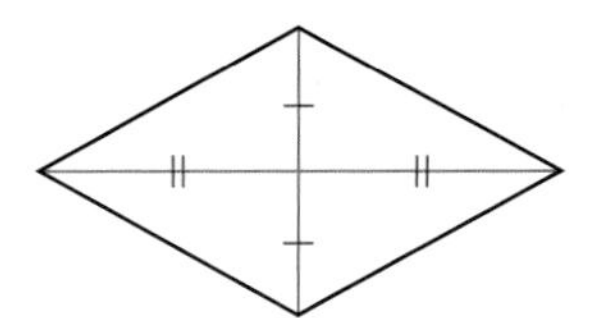

18.

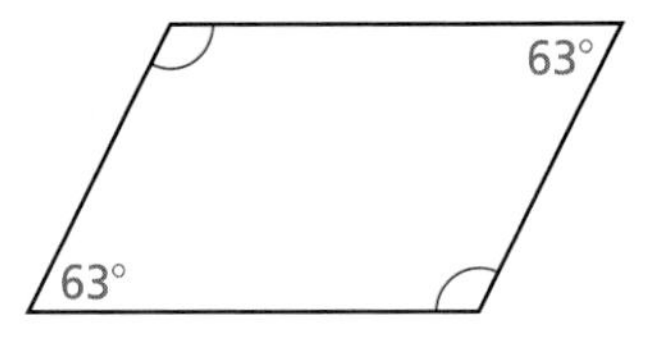

Graph the quadrilateral with the given vertices in a coordinate plane. Then show that the quadrilateral is a parallelogram. *(Section 7.3)*

19. $Q(-5, -2)$, $R(3, -2)$, $S(1, -6)$, $T(-7, -6)$

20. $W(-3, 7)$, $X(3, 3)$, $Y(1, -3)$, $Z(-5, 1)$

21. A stop sign is a regular polygon. *(Section 7.1)*

a. Classify the stop sign by its number of sides.

b. Find the measure of each interior angle and each exterior angle of the stop sign.

22. In the diagram of the staircase shown, $JKLM$ is a parallelogram, $\overline{QT} \parallel \overline{RS}$, $QT = RS = 9$ feet, $QR = 3$ feet, and $m\angle QRS = 123°$. *(Section 7.2 and Section 7.3)*

a. List all congruent sides and angles in ▱*JKLM*. Explain your reasoning.

b. Which theorem could you use to show that $QRST$ is a parallelogram?

c. Find ST, $m\angle QTS$, $m\angle TQR$, and $m\angle TSR$. Explain your reasoning.

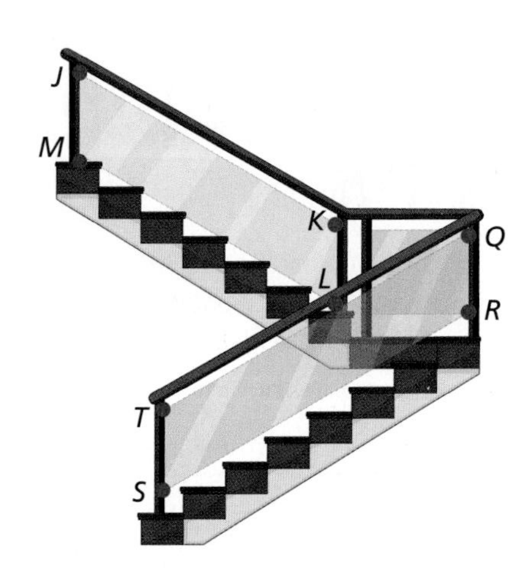

7.4 Properties of Special Parallelograms

Essential Question What are the properties of the diagonals of rectangles, rhombuses, and squares?

Recall the three types of parallelograms shown below.

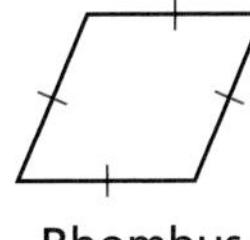

Rhombus

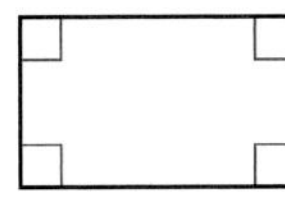

Rectangle

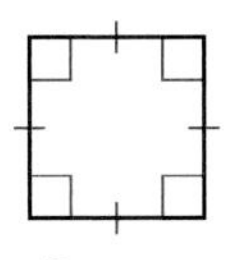

Square

EXPLORATION 1 Identifying Special Quadrilaterals

Work with a partner. Use dynamic geometry software.

a. Draw a circle with center A.

b. Draw two diameters of the circle. Label the endpoints B, C, D, and E.

c. Draw quadrilateral $BDCE$.

d. Is $BDCE$ a parallelogram? rectangle? rhombus? square? Explain your reasoning.

e. Repeat parts (a)–(d) for several other circles. Write a conjecture based on your results.

Sample

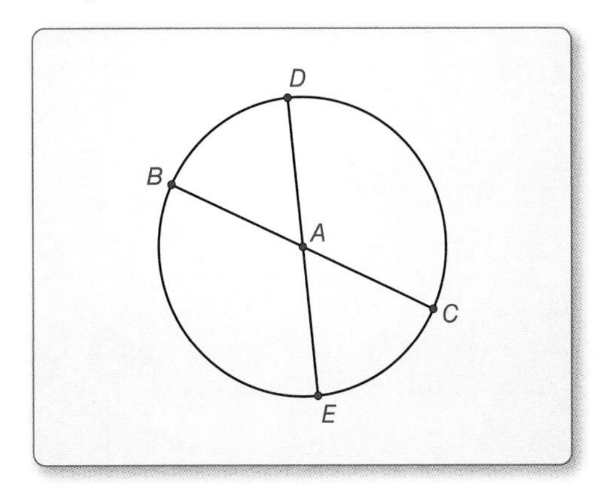

EXPLORATION 2 Identifying Special Quadrilaterals

Work with a partner. Use dynamic geometry software.

a. Construct two segments that are perpendicular bisectors of each other. Label the endpoints A, B, D, and E. Label the intersection C.

b. Draw quadrilateral $AEBD$.

c. Is $AEBD$ a parallelogram? rectangle? rhombus? square? Explain your reasoning.

d. Repeat parts (a)–(c) for several other segments. Write a conjecture based on your results.

Sample

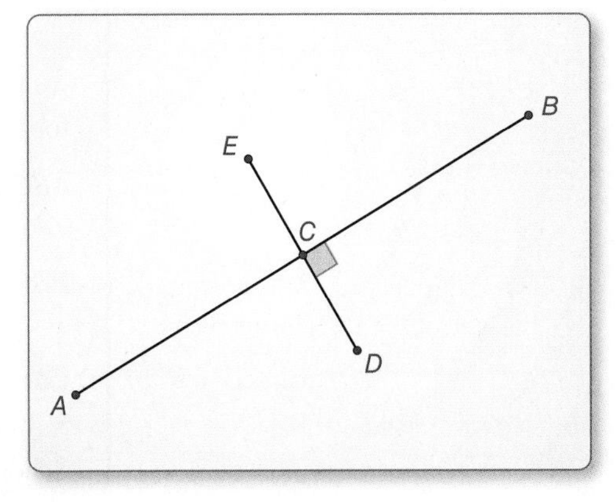

CONSTRUCTING VIABLE ARGUMENTS

To be proficient in math, you need to make conjectures and build a logical progression of statements to explore the truth of your conjectures.

Communicate Your Answer

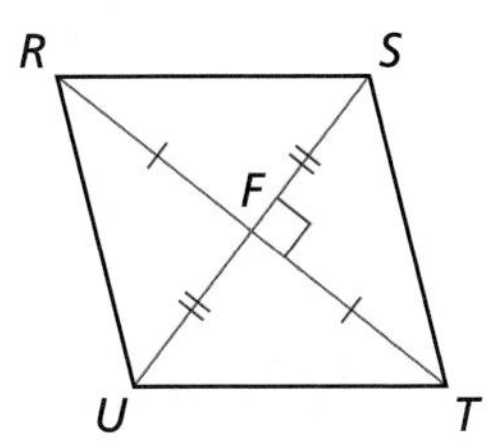

3. What are the properties of the diagonals of rectangles, rhombuses, and squares?

4. Is $RSTU$ a parallelogram? rectangle? rhombus? square? Explain your reasoning.

5. What type of quadrilateral has congruent diagonals that bisect each other?

7.4 Lesson

Core Vocabulary

rhombus, *p. 388*
rectangle, *p. 388*
square, *p. 388*

Previous
quadrilateral
parallelogram
diagonal

What You Will Learn

- Use properties of special parallelograms.
- Use properties of diagonals of special parallelograms.
- Use coordinate geometry to identify special types of parallelograms.

Using Properties of Special Parallelograms

In this lesson, you will learn about three special types of parallelograms: *rhombuses*, *rectangles*, and *squares*.

Core Concept

Rhombuses, Rectangles, and Squares

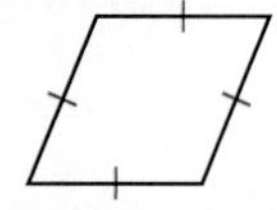

A **rhombus** is a parallelogram with four congruent sides.

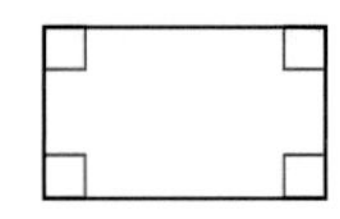

A **rectangle** is a parallelogram with four right angles.

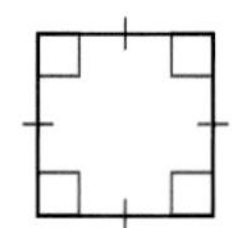

A **square** is a parallelogram with four congruent sides and four right angles.

You can use the corollaries below to prove that a quadrilateral is a rhombus, rectangle, or square, without first proving that the quadrilateral is a parallelogram.

Corollaries

Corollary 7.2 Rhombus Corollary

A quadrilateral is a rhombus if and only if it has four congruent sides.

ABCD is a rhombus if and only if $\overline{AB} \cong \overline{BC} \cong \overline{CD} \cong \overline{AD}$.

Proof Ex. 81, p. 396

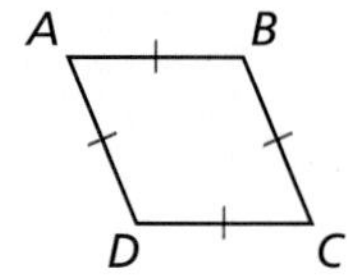

Corollary 7.3 Rectangle Corollary

A quadrilateral is a rectangle if and only if it has four right angles.

ABCD is a rectangle if and only if $\angle A$, $\angle B$, $\angle C$, and $\angle D$ are right angles.

Proof Ex. 82, p. 396

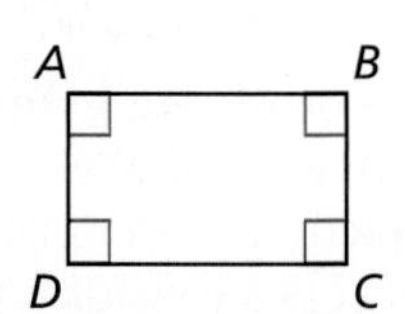

Corollary 7.4 Square Corollary

A quadrilateral is a square if and only if it is a rhombus and a rectangle.

ABCD is a square if and only if $\overline{AB} \cong \overline{BC} \cong \overline{CD} \cong \overline{AD}$ and $\angle A$, $\angle B$, $\angle C$, and $\angle D$ are right angles.

Proof Ex. 83, p. 396

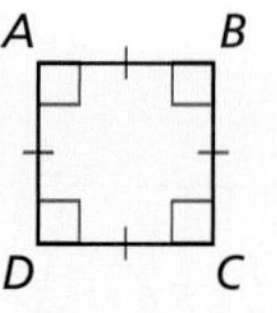

The Venn diagram below illustrates some important relationships among parallelograms, rhombuses, rectangles, and squares. For example, you can see that a square is a rhombus because it is a parallelogram with four congruent sides. Because it has four right angles, a square is also a rectangle.

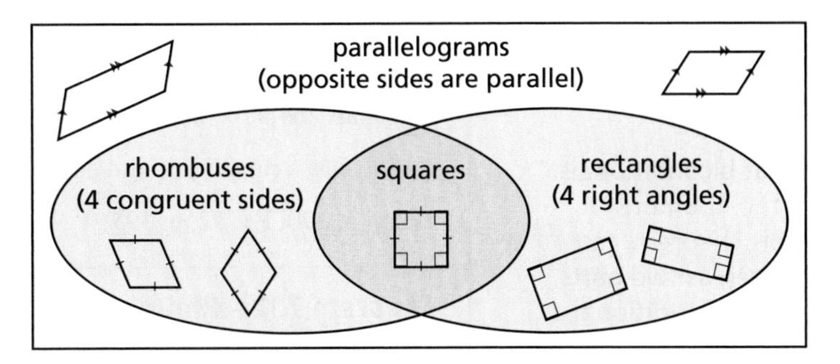

EXAMPLE 1 Using Properties of Special Quadrilaterals

For any rhombus *QRST*, decide whether the statement is *always* or *sometimes* true. Draw a diagram and explain your reasoning.

a. $\angle Q \cong \angle S$ **b.** $\angle Q \cong \angle R$

SOLUTION

a. By definition, a rhombus is a parallelogram with four congruent sides. By the Parallelogram Opposite Angles Theorem (Theorem 7.4), opposite angles of a parallelogram are congruent. So, $\angle Q \cong \angle S$. The statement is *always* true.

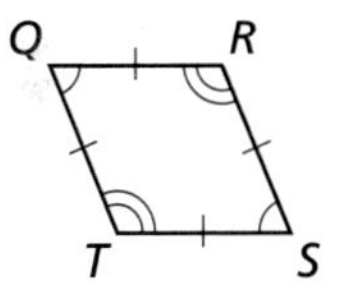

b. If rhombus *QRST* is a square, then all four angles are congruent right angles. So, $\angle Q \cong \angle R$ when *QRST* is a square. Because not all rhombuses are also squares, the statement is *sometimes* true.

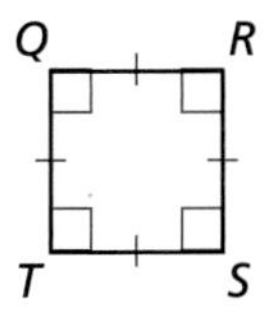

EXAMPLE 2 Classifying Special Quadrilaterals

Classify the special quadrilateral. Explain your reasoning.

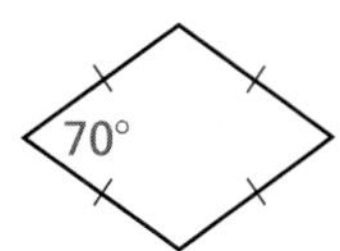

SOLUTION

The quadrilateral has four congruent sides. By the Rhombus Corollary, the quadrilateral is a rhombus. Because one of the angles is not a right angle, the rhombus cannot be a square.

Monitoring Progress

Help in English and Spanish at *BigIdeasMath.com*

1. For any square *JKLM*, is it *always* or *sometimes* true that $\overline{JK} \perp \overline{KL}$? Explain your reasoning.
2. For any rectangle *EFGH*, is it *always* or *sometimes* true that $\overline{FG} \cong \overline{GH}$? Explain your reasoning.
3. A quadrilateral has four congruent sides and four congruent angles. Sketch the quadrilateral and classify it.

Using Properties of Diagonals

Theorems

READING
Recall that biconditionals, such as the Rhombus Diagonals Theorem, can be rewritten as two parts. To prove a biconditional, you must prove both parts.

Theorem 7.11 Rhombus Diagonals Theorem

A parallelogram is a rhombus if and only if its diagonals are perpendicular.

$\square ABCD$ is a rhombus if and only if $\overline{AC} \perp \overline{BD}$.

Proof p. 390; Ex. 72, p. 395

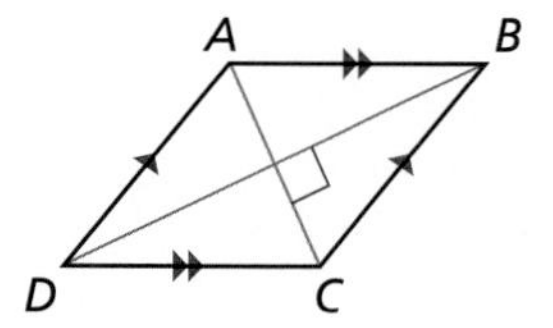

Theorem 7.12 Rhombus Opposite Angles Theorem

A parallelogram is a rhombus if and only if each diagonal bisects a pair of opposite angles.

$\square ABCD$ is a rhombus if and only if $\overline{AC}$ bisects $\angle BCD$ and $\angle BAD$, and $\overline{BD}$ bisects $\angle ABC$ and $\angle ADC$.

Proof Exs. 73 and 74, p. 395

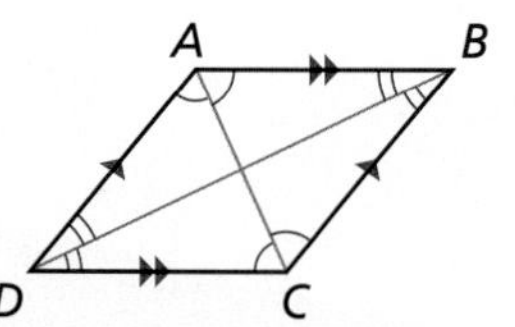

PROOF **Part of Rhombus Diagonals Theorem**

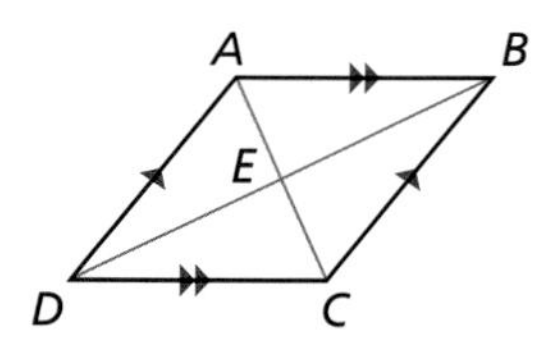

Given $ABCD$ is a rhombus.

Prove $\overline{AC} \perp \overline{BD}$

$ABCD$ is a rhombus. By the definition of a rhombus, $\overline{AB} \cong \overline{BC}$. Because a rhombus is a parallelogram and the diagonals of a parallelogram bisect each other, $\overline{BD}$ bisects $\overline{AC}$ at E. So, $\overline{AE} \cong \overline{EC}$. $\overline{BE} \cong \overline{BE}$ by the Reflexive Property of Congruence (Theorem 2.1). So, $\triangle AEB \cong \triangle CEB$ by the SSS Congruence Theorem (Theorem 5.8). $\angle AEB \cong \angle CEB$ because corresponding parts of congruent triangles are congruent. Then by the Linear Pair Postulate (Postulate 2.8), $\angle AEB$ and $\angle CEB$ are supplementary. Two congruent angles that form a linear pair are right angles, so $m\angle AEB = m\angle CEB = 90°$ by the definition of a right angle. So, $\overline{AC} \perp \overline{BD}$ by the definition of perpendicular lines.

EXAMPLE 3 **Finding Angle Measures in a Rhombus**

Find the measures of the numbered angles in rhombus $ABCD$.

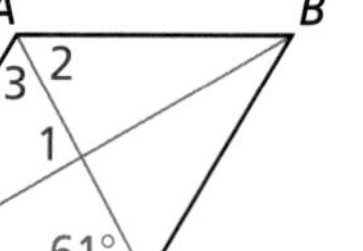

SOLUTION

Use the Rhombus Diagonals Theorem and the Rhombus Opposite Angles Theorem to find the angle measures.

$m\angle 1 = 90°$	The diagonals of a rhombus are perpendicular.
$m\angle 2 = 61°$	Alternate Interior Angles Theorem (Theorem 3.2)
$m\angle 3 = 61°$	Each diagonal of a rhombus bisects a pair of opposite angles, and $m\angle 2 = 61°$.
$m\angle 1 + m\angle 3 + m\angle 4 = 180°$	Triangle Sum Theorem (Theorem 5.1)
$90° + 61° + m\angle 4 = 180°$	Substitute 90° for $m\angle 1$ and 61° for $m\angle 3$.
$m\angle 4 = 29°$	Solve for $m\angle 4$.

▶ So, $m\angle 1 = 90°$, $m\angle 2 = 61°$, $m\angle 3 = 61°$, and $m\angle 4 = 29°$.

Monitoring Progress  Help in English and Spanish at *BigIdeasMath.com*

4. In Example 3, what is $m\angle ADC$ and $m\angle BCD$?

5. Find the measures of the numbered angles in rhombus *DEFG*.

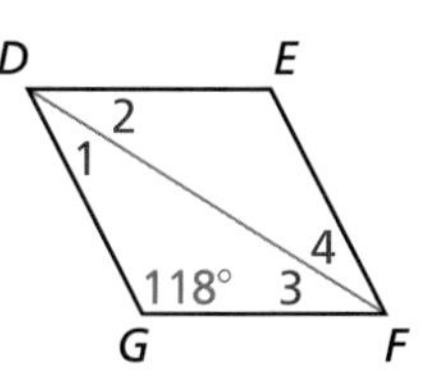

Theorem

Theorem 7.13 Rectangle Diagonals Theorem

A parallelogram is a rectangle if and only if its diagonals are congruent.

$\square ABCD$ is a rectangle if and only if $\overline{AC} \cong \overline{BD}$.

Proof Exs. 87 and 88, p. 396

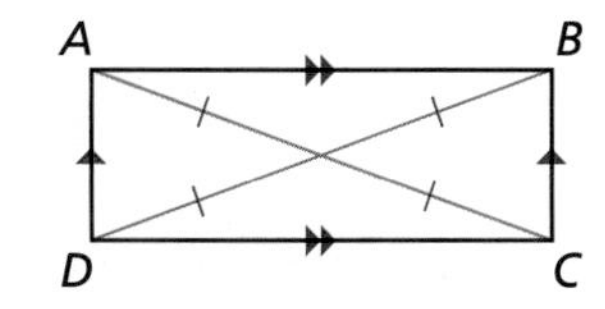

EXAMPLE 4 Identifying a Rectangle

You are building a frame for a window. The window will be installed in the opening shown in the diagram.

a. The opening must be a rectangle. Given the measurements in the diagram, can you assume that it is? Explain.

b. You measure the diagonals of the opening. The diagonals are 54.8 inches and 55.3 inches. What can you conclude about the shape of the opening?

33 in.
44 in.
44 in.
33 in.

SOLUTION

a. No, you cannot. The boards on opposite sides are the same length, so they form a parallelogram. But you do not know whether the angles are right angles.

b. By the Rectangle Diagonals Theorem, the diagonals of a rectangle are congruent. The diagonals of the quadrilateral formed by the boards are not congruent, so the boards do not form a rectangle.

EXAMPLE 5 Finding Diagonal Lengths in a Rectangle

In rectangle *QRST*, $QS = 5x - 31$ and $RT = 2x + 11$. Find the lengths of the diagonals of *QRST*.

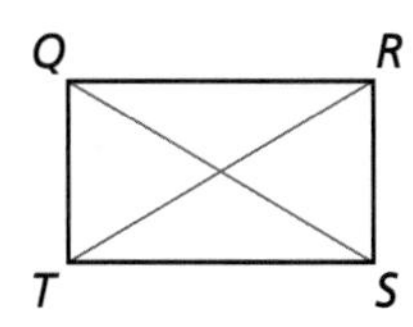

SOLUTION

By the Rectangle Diagonals Theorem, the diagonals of a rectangle are congruent. Find x so that $\overline{QS} \cong \overline{RT}$.

$QS = RT$	Set the diagonal lengths equal.
$5x - 31 = 2x + 11$	Substitute $5x - 31$ for QS and $2x + 11$ for RT.
$3x - 31 = 11$	Subtract $2x$ from each side.
$3x = 42$	Add 31 to each side.
$x = 14$	Divide each side by 3.

When $x = 14$, $QS = 5(14) - 31 = 39$ and $RT = 2(14) + 11 = 39$.

▶ Each diagonal has a length of 39 units.

Monitoring Progress  Help in English and Spanish at *BigIdeasMath.com*

6. Suppose you measure only the diagonals of the window opening in Example 4 and they have the same measure. Can you conclude that the opening is a rectangle? Explain.

7. WHAT IF? In Example 5, $QS = 4x - 15$ and $RT = 3x + 8$. Find the lengths of the diagonals of $QRST$.

Using Coordinate Geometry

EXAMPLE 6 Identifying a Parallelogram in the Coordinate Plane

Decide whether $\square ABCD$ with vertices $A(-2, 6)$, $B(6, 8)$, $C(4, 0)$, and $D(-4, -2)$ is a *rectangle*, a *rhombus*, or a *square*. Give all names that apply.

SOLUTION

1. Understand the Problem You know the vertices of $\square ABCD$. You need to identify the type of parallelogram.

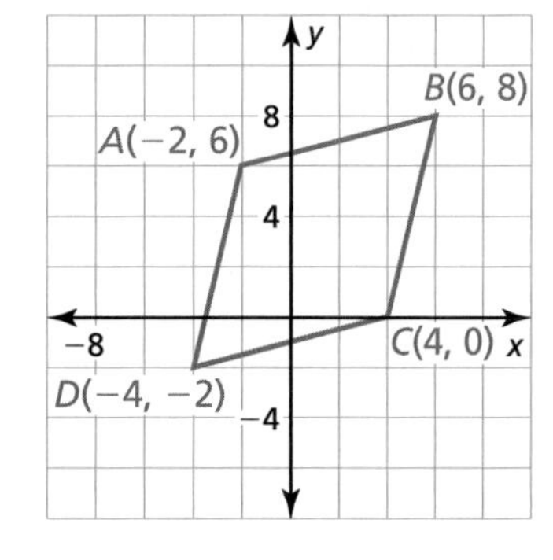

2. Make a Plan Begin by graphing the vertices. From the graph, it appears that all four sides are congruent and there are no right angles.

Check the lengths and slopes of the diagonals of $\square ABCD$. If the diagonals are congruent, then $\square ABCD$ is a rectangle. If the diagonals are perpendicular, then $\square ABCD$ is a rhombus. If they are both congruent and perpendicular, then $\square ABCD$ is a rectangle, a rhombus, and a square.

3. Solve the Problem Use the Distance Formula to find AC and BD.

$$AC = \sqrt{(-2-4)^2 + (6-0)^2} = \sqrt{72} = 6\sqrt{2}$$

$$BD = \sqrt{[6-(-4)]^2 + [8-(-2)]^2} = \sqrt{200} = 10\sqrt{2}$$

Because $6\sqrt{2} \neq 10\sqrt{2}$, the diagonals are not congruent. So, $\square ABCD$ is not a rectangle. Because it is not a rectangle, it also cannot be a square.

Use the slope formula to find the slopes of the diagonals $\overline{AC}$ and $\overline{BD}$.

$$\text{slope of } \overline{AC} = \frac{6-0}{-2-4} = \frac{6}{-6} = -1 \qquad \text{slope of } \overline{BD} = \frac{8-(-2)}{6-(-4)} = \frac{10}{10} = 1$$

Because the product of the slopes of the diagonals is -1, the diagonals are perpendicular.

▶ So, $\square ABCD$ is a rhombus.

4. Look Back Check the side lengths of $\square ABCD$. Each side has a length of $2\sqrt{17}$ units, so $\square ABCD$ is a rhombus. Check the slopes of two consecutive sides.

$$\text{slope of } \overline{AB} = \frac{8-6}{6-(-2)} = \frac{2}{8} = \frac{1}{4} \qquad \text{slope of } \overline{BC} = \frac{8-0}{6-4} = \frac{8}{2} = 4$$

Because the product of these slopes is not -1, $\overline{AB}$ is not perpendicular to $\overline{BC}$.

So, $\angle ABC$ is not a right angle, and $\square ABCD$ cannot be a rectangle or a square. ✓

Monitoring Progress  Help in English and Spanish at *BigIdeasMath.com*

8. Decide whether $\square PQRS$ with vertices $P(-5, 2)$, $Q(0, 4)$, $R(2, -1)$, and $S(-3, -3)$ is a *rectangle*, a *rhombus*, or a *square*. Give all names that apply.

7.4 Exercises

Dynamic Solutions available at *BigIdeasMath.com*

Vocabulary and Core Concept Check

1. **VOCABULARY** What is another name for an equilateral rectangle?

2. **WRITING** What should you look for in a parallelogram to know if the parallelogram is also a rhombus?

Monitoring Progress and Modeling with Mathematics

In Exercises 3–8, for any rhombus *JKLM*, decide whether the statement is *always* or *sometimes* true. Draw a diagram and explain your reasoning. *(See Example 1.)*

3. $\angle L \cong \angle M$

4. $\angle K \cong \angle M$

5. $\overline{JM} \cong \overline{KL}$

6. $\overline{JK} \cong \overline{KL}$

7. $\overline{JL} \cong \overline{KM}$

8. $\angle JKM \cong \angle LKM$

In Exercises 9–12, classify the quadrilateral. Explain your reasoning. *(See Example 2.)*

9. 

10. 

11. 

12. 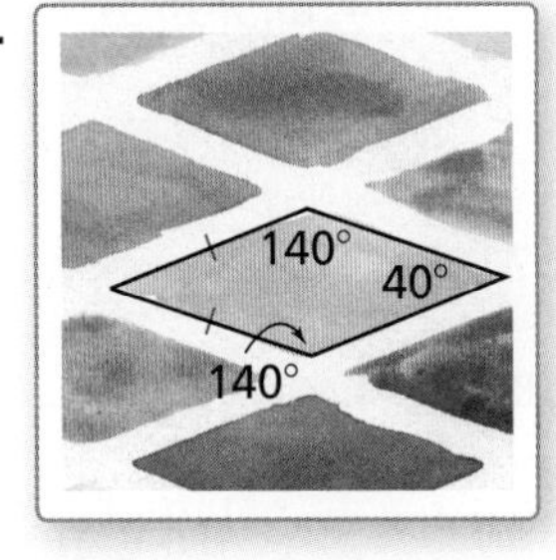

In Exercises 13–16, find the measures of the numbered angles in rhombus *DEFG*. *(See Example 3.)*

13. 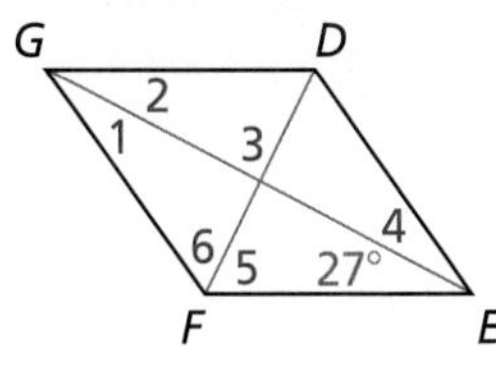

14. 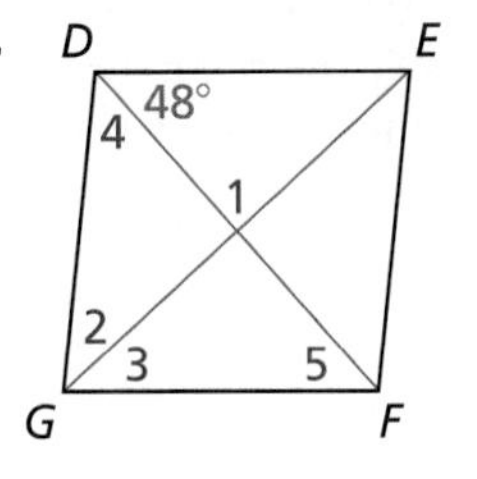

15. 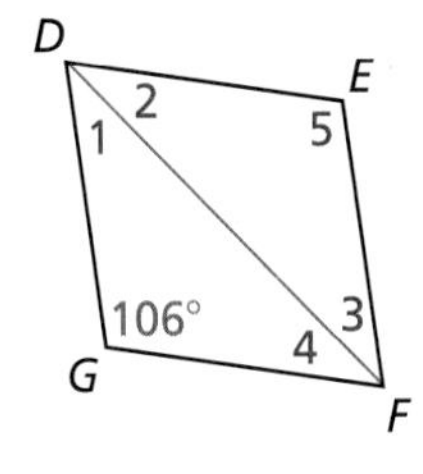

16. 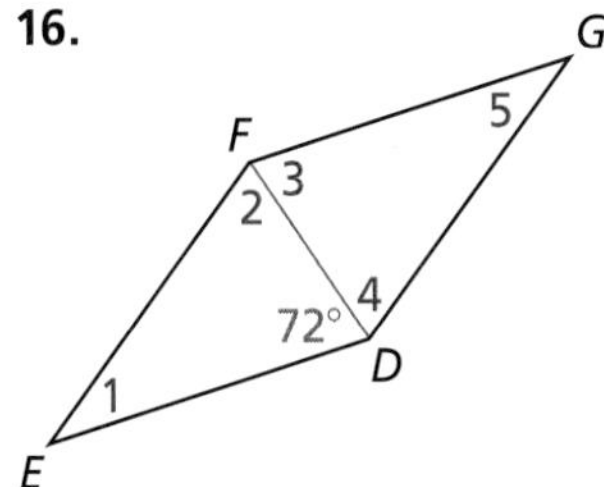

In Exercises 17–22, for any rectangle *WXYZ*, decide whether the statement is *always* or *sometimes* true. Draw a diagram and explain your reasoning.

17. $\angle W \cong \angle X$

18. $\overline{WX} \cong \overline{YZ}$

19. $\overline{WX} \cong \overline{XY}$

20. $\overline{WY} \cong \overline{XZ}$

21. $\overline{WY} \perp \overline{XZ}$

22. $\angle WXZ \cong \angle YXZ$

In Exercises 23 and 24, determine whether the quadrilateral is a rectangle. *(See Example 4.)*

23. 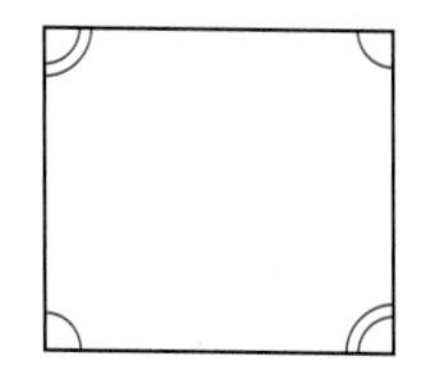

24. 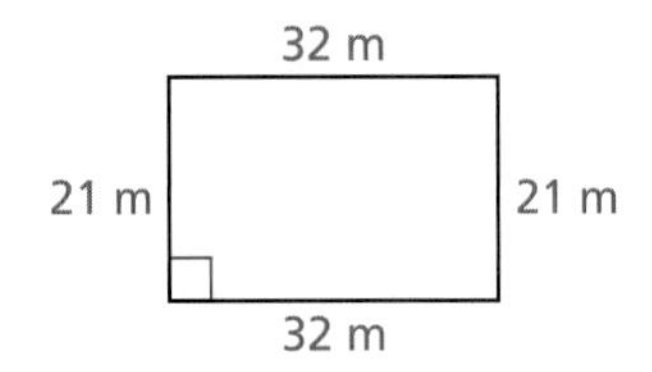

In Exercises 25–28, find the lengths of the diagonals of rectangle *WXYZ*. *(See Example 5.)*

25. $WY = 6x - 7$
$XZ = 3x + 2$

26. $WY = 14x + 10$
$XZ = 11x + 22$

27. $WY = 24x - 8$
$XZ = -18x + 13$

28. $WY = 16x + 2$
$XZ = 36x - 6$

In Exercises 29–34, name each quadrilateral—*parallelogram*, *rectangle*, *rhombus*, or *square*—for which the statement is always true.

29. It is equiangular.

30. It is equiangular and equilateral.

31. The diagonals are perpendicular.

32. Opposite sides are congruent.

33. The diagonals bisect each other.

34. The diagonals bisect opposite angles.

35. **ERROR ANALYSIS** Quadrilateral *PQRS* is a rectangle. Describe and correct the error in finding the value of *x*.

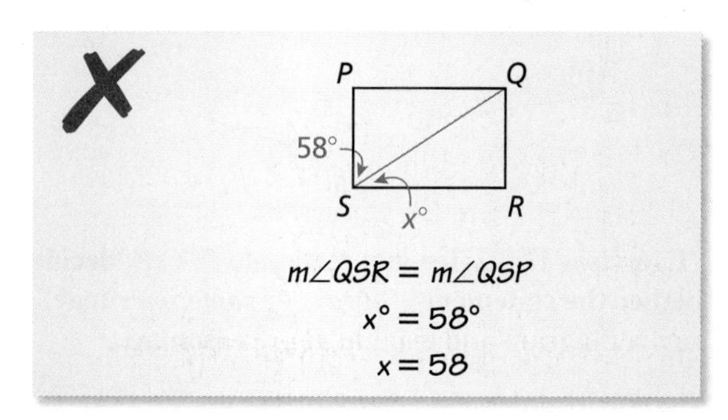

36. **ERROR ANALYSIS** Quadrilateral *PQRS* is a rhombus. Describe and correct the error in finding the value of *x*.

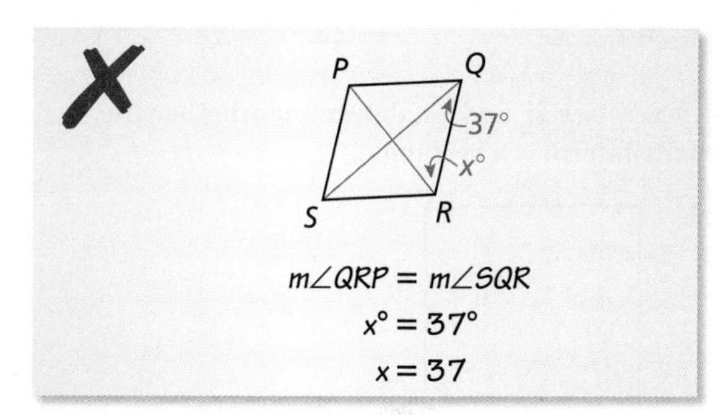

In Exercises 37–42, the diagonals of rhombus *ABCD* intersect at *E*. Given that $m\angle BAC = 53^\circ$, $DE = 8$, and $EC = 6$, find the indicated measure.

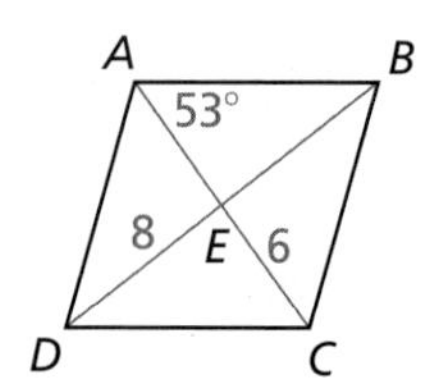

37. $m\angle DAC$

38. $m\angle AED$

39. $m\angle ADC$

40. DB

41. AE

42. AC

In Exercises 43–48, the diagonals of rectangle *QRST* intersect at *P*. Given that $m\angle PTS = 34^\circ$ and $QS = 10$, find the indicated measure.

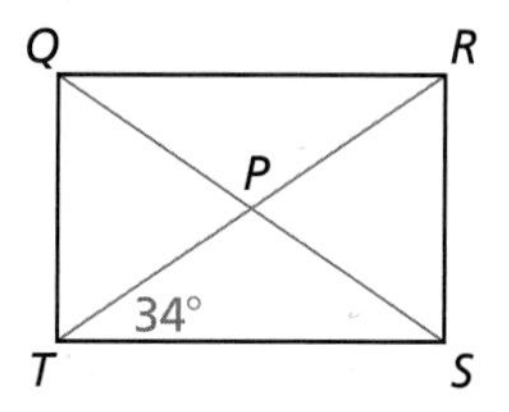

43. $m\angle QTR$

44. $m\angle QRT$

45. $m\angle SRT$

46. QP

47. RT

48. RP

In Exercises 49–54, the diagonals of square *LMNP* intersect at *K*. Given that $LK = 1$, find the indicated measure.

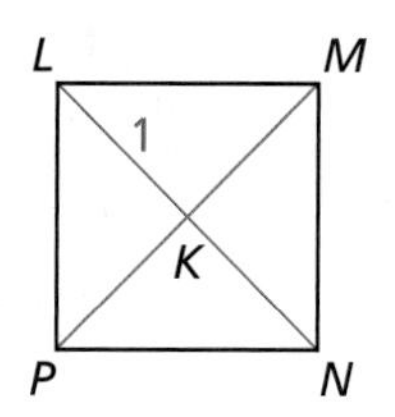

49. $m\angle MKN$

50. $m\angle LMK$

51. $m\angle LPK$

52. KN

53. LN

54. MP

In Exercises 55–60, decide whether $\square JKLM$ is a rectangle, a rhombus, or a square. Give all names that apply. Explain your reasoning. *(See Example 6.)*

55. $J(-4, 2)$, $K(0, 3)$, $L(1, -1)$, $M(-3, -2)$

56. $J(-2, 7)$, $K(7, 2)$, $L(-2, -3)$, $M(-11, 2)$

57. $J(3, 1)$, $K(3, -3)$, $L(-2, -3)$, $M(-2, 1)$

58. $J(-1, 4)$, $K(-3, 2)$, $L(2, -3)$, $M(4, -1)$

59. $J(5, 2)$, $K(1, 9)$, $L(-3, 2)$, $M(1, -5)$

60. $J(5, 2)$, $K(2, 5)$, $L(-1, 2)$, $M(2, -1)$

MATHEMATICAL CONNECTIONS In Exercises 61 and 62, classify the quadrilateral. Explain your reasoning. Then find the values of *x* and *y*.

61.

B
y + 8
3y
104°
A
x°
C
D

62.

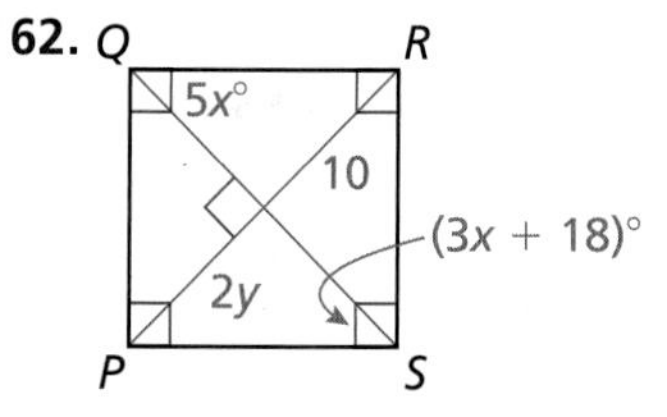

63. DRAWING CONCLUSIONS In the window, $\overline{BD} \cong \overline{DF} \cong \overline{BH} \cong \overline{HF}$. Also, $\angle HAB$, $\angle BCD$, $\angle DEF$, and $\angle FGH$ are right angles.

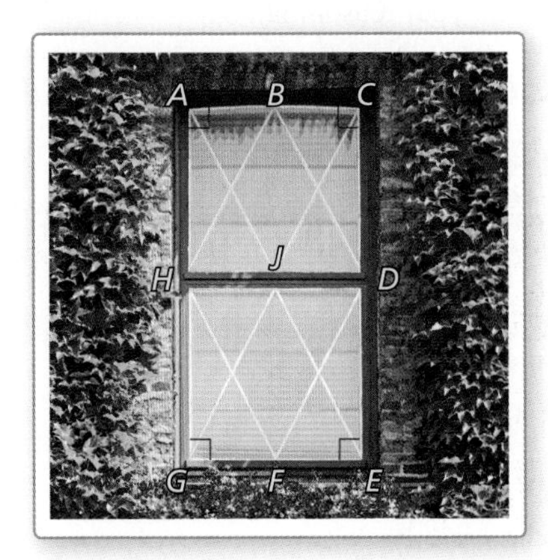

a. Classify *HBDF* and *ACEG*. Explain your reasoning.

b. What can you conclude about the lengths of the diagonals $\overline{AE}$ and $\overline{GC}$? Given that these diagonals intersect at *J*, what can you conclude about the lengths of $\overline{AJ}$, $\overline{JE}$, $\overline{CJ}$, and $\overline{JG}$? Explain.

64. ABSTRACT REASONING Order the terms in a diagram so that each term builds off the previous term(s). Explain why each figure is in the location you chose.

quadrilateral	square
rectangle	rhombus
parallelogram	

CRITICAL THINKING In Exercises 65–70, complete each statement with *always*, *sometimes*, or *never*. Explain your reasoning.

65. A square is ________ a rhombus.

66. A rectangle is ________ a square.

67. A rectangle ________ has congruent diagonals.

68. The diagonals of a square ________ bisect its angles.

69. A rhombus ________ has four congruent angles.

70. A rectangle ________ has perpendicular diagonals.

71. USING TOOLS You want to mark off a square region for a garden at school. You use a tape measure to mark off a quadrilateral on the ground. Each side of the quadrilateral is 2.5 meters long. Explain how you can use the tape measure to make sure that the quadrilateral is a square.

72. PROVING A THEOREM Use the plan for proof below to write a paragraph proof for one part of the Rhombus Diagonals Theorem (Theorem 7.11).

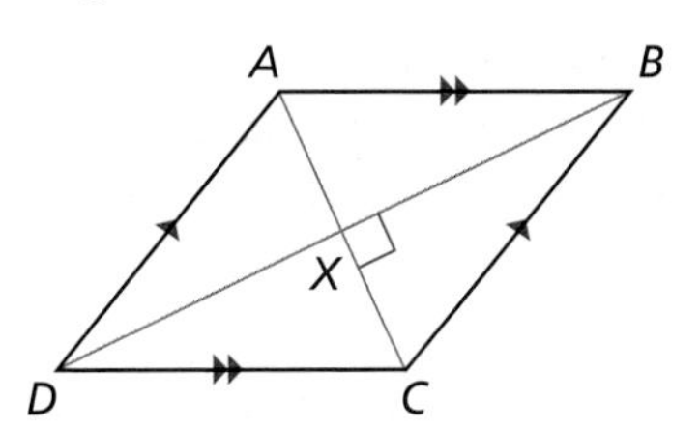

Given *ABCD* is a parallelogram.
$\overline{AC} \perp \overline{BD}$

Prove *ABCD* is a rhombus.

Plan for Proof Because *ABCD* is a parallelogram, its diagonals bisect each other at *X*. Use $\overline{AC} \perp \overline{BD}$ to show that $\triangle BXC \cong \triangle DXC$. Then show that $\overline{BC} \cong \overline{DC}$. Use the properties of a parallelogram to show that *ABCD* is a rhombus.

PROVING A THEOREM In Exercises 73 and 74, write a proof for part of the Rhombus Opposite Angles Theorem (Theorem 7.12).

73. Given *PQRS* is a parallelogram.
$\overline{PR}$ bisects $\angle SPQ$ and $\angle QRS$.
$\overline{SQ}$ bisects $\angle PSR$ and $\angle RQP$.

Prove *PQRS* is a rhombus.

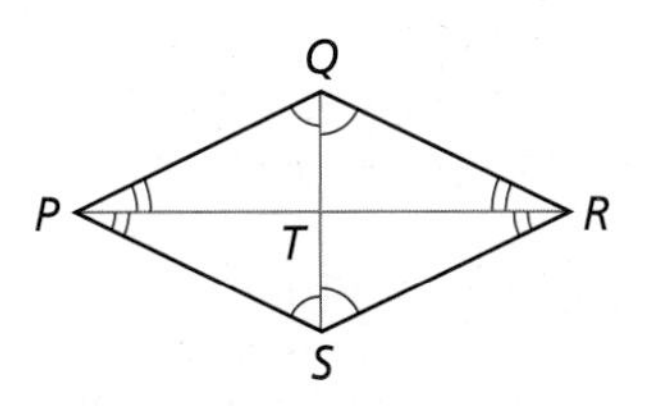

74. Given *WXYZ* is a rhombus.

Prove $\overline{WY}$ bisects $\angle ZWX$ and $\angle XYZ$.
$\overline{ZX}$ bisects $\angle WZY$ and $\angle YXW$.

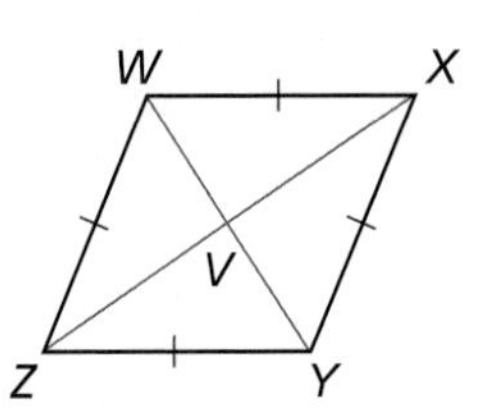

75. ABSTRACT REASONING Will a diagonal of a square ever divide the square into two equilateral triangles? Explain your reasoning.

76. ABSTRACT REASONING Will a diagonal of a rhombus ever divide the rhombus into two equilateral triangles? Explain your reasoning.

77. CRITICAL THINKING Which quadrilateral could be called a regular quadrilateral? Explain your reasoning.

78. HOW DO YOU SEE IT? What other information do you need to determine whether the figure is a rectangle?

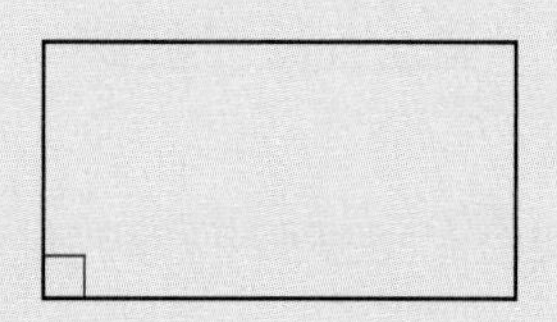

79. REASONING Are all rhombuses similar? Are all squares similar? Explain your reasoning.

80. THOUGHT PROVOKING Use the Rhombus Diagonals Theorem (Theorem 7.11) to explain why every rhombus has at least two lines of symmetry.

PROVING A COROLLARY In Exercises 81–83, write the corollary as a conditional statement and its converse. Then explain why each statement is true.

81. Rhombus Corollary (Corollary 7.2)

82. Rectangle Corollary (Corollary 7.3)

83. Square Corollary (Corollary 7.4)

84. MAKING AN ARGUMENT Your friend claims a rhombus will never have congruent diagonals because it would have to be a rectangle. Is your friend correct? Explain your reasoning.

85. PROOF Write a proof in the style of your choice.

Given $\triangle XYZ \cong \triangle XWZ$, $\angle XYW \cong \angle ZWY$

Prove $WXYZ$ is a rhombus.

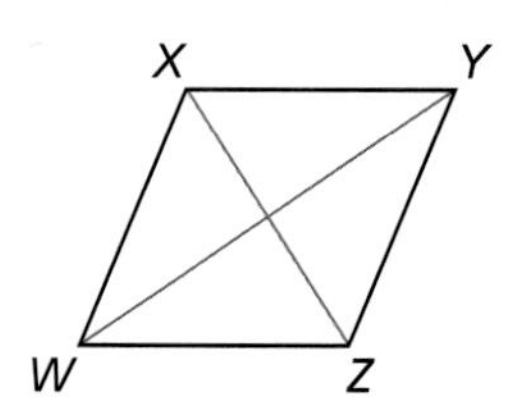

86. PROOF Write a proof in the style of your choice.

Given $\overline{BC} \cong \overline{AD}, \overline{BC} \perp \overline{DC}, \overline{AD} \perp \overline{DC}$

Prove $ABCD$ is a rectangle.

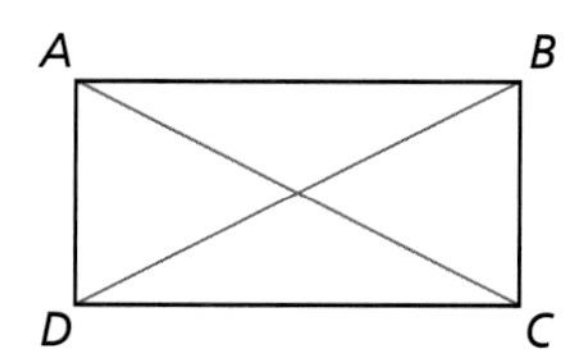

PROVING A THEOREM In Exercises 87 and 88, write a proof for part of the Rectangle Diagonals Theorem (Theorem 7.13).

87. Given $PQRS$ is a rectangle.

Prove $\overline{PR} \cong \overline{SQ}$

88. Given $PQRS$ is a parallelogram.
$\overline{PR} \cong \overline{SQ}$

Prove $PQRS$ is a rectangle.

Maintaining Mathematical Proficiency

Reviewing what you learned in previous grades and lessons

$\overline{DE}$ is a midsegment of $\triangle ABC$. Find the values of x and y. *(Section 6.4)*

89.

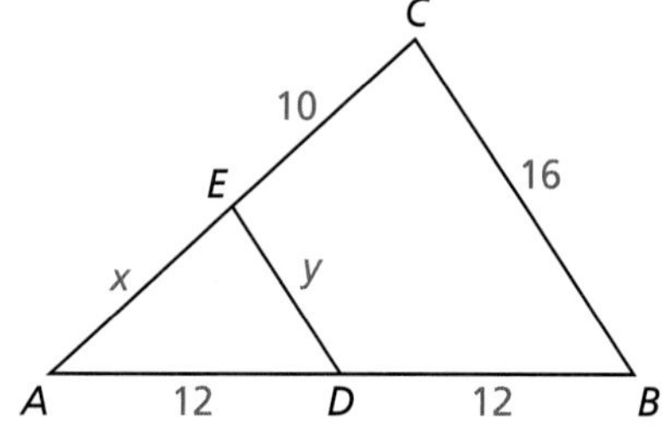

90.

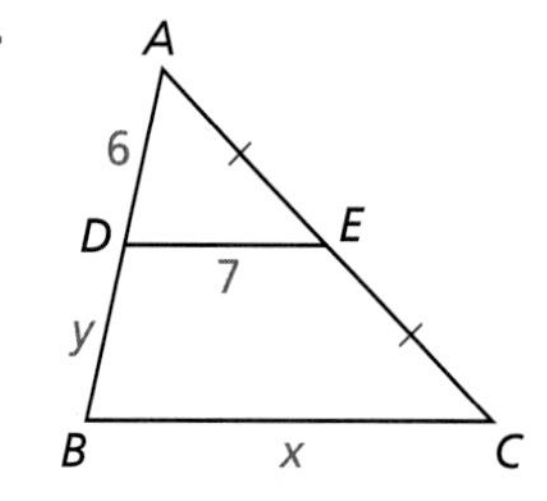

91.

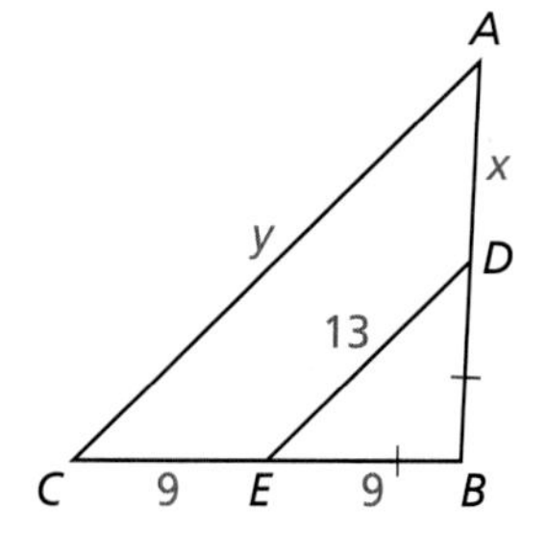

7.5 Properties of Trapezoids and Kites

Essential Question What are some properties of trapezoids and kites?

Recall the types of quadrilaterals shown below.

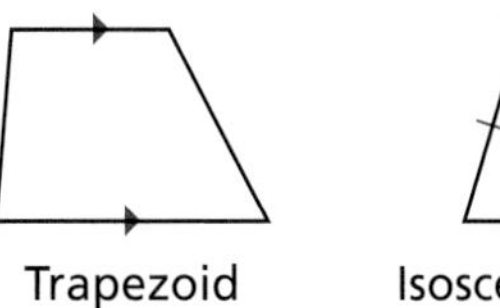

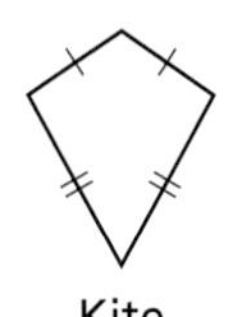

PERSEVERE IN SOLVING PROBLEMS

To be proficient in math, you need to draw diagrams of important features and relationships, and search for regularity or trends.

EXPLORATION 1 Making a Conjecture about Trapezoids

Work with a partner. Use dynamic geometry software.

a. Construct a trapezoid whose base angles are congruent. Explain your process.

b. Is the trapezoid isosceles? Justify your answer.

c. Repeat parts (a) and (b) for several other trapezoids. Write a conjecture based on your results.

Sample

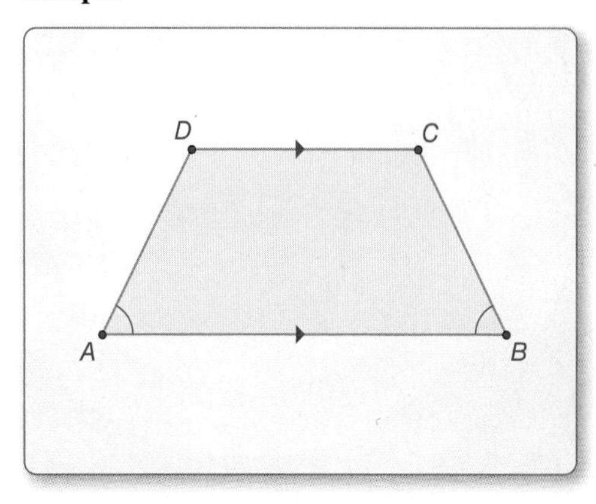

EXPLORATION 2 Discovering a Property of Kites

Work with a partner. Use dynamic geometry software.

a. Construct a kite. Explain your process.

b. Measure the angles of the kite. What do you observe?

c. Repeat parts (a) and (b) for several other kites. Write a conjecture based on your results.

Sample

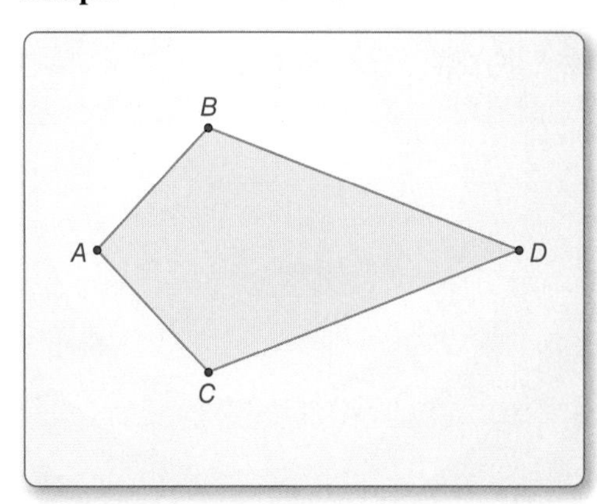

Communicate Your Answer

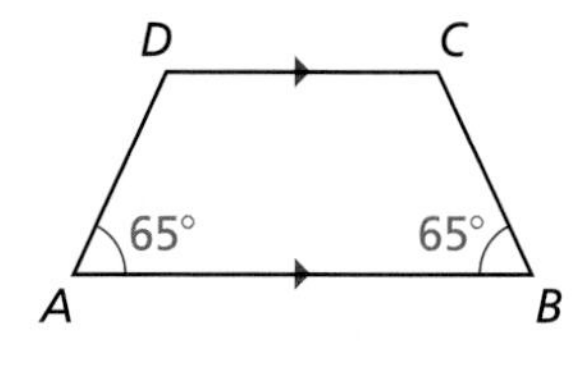

3. What are some properties of trapezoids and kites?

4. Is the trapezoid at the left isosceles? Explain.

5. A quadrilateral has angle measures of 70°, 70°, 110°, and 110°. Is the quadrilateral a kite? Explain.

7.5 Lesson

Core Vocabulary

trapezoid, *p. 398*
bases, *p. 398*
base angles, *p. 398*
legs, *p. 398*
isosceles trapezoid, *p. 398*
midsegment of a trapezoid, *p. 400*
kite, *p. 401*

Previous
diagonal
parallelogram

What You Will Learn

- Use properties of trapezoids.
- Use the Trapezoid Midsegment Theorem to find distances.
- Use properties of kites.
- Identify quadrilaterals.

Using Properties of Trapezoids

A **trapezoid** is a quadrilateral with exactly one pair of parallel sides. The parallel sides are the **bases**.

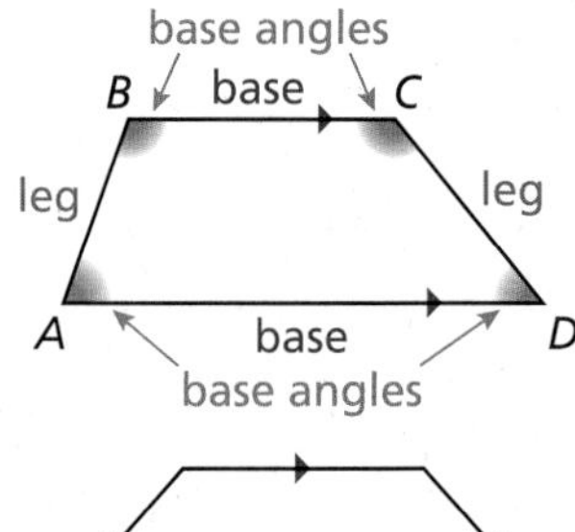

Base angles of a trapezoid are two consecutive angles whose common side is a base. A trapezoid has two pairs of base angles. For example, in trapezoid $ABCD$, $\angle A$ and $\angle D$ are one pair of base angles, and $\angle B$ and $\angle C$ are the second pair. The nonparallel sides are the **legs** of the trapezoid.

If the legs of a trapezoid are congruent, then the trapezoid is an **isosceles trapezoid**.

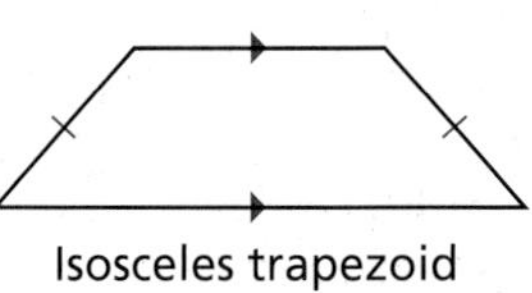
Isosceles trapezoid

EXAMPLE 1 Identifying a Trapezoid in the Coordinate Plane

Show that $ORST$ is a trapezoid. Then decide whether it is isosceles.

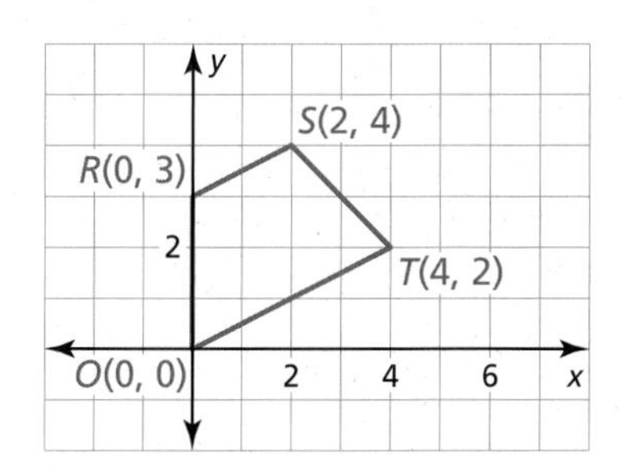

SOLUTION

Step 1 Compare the slopes of opposite sides.

$$\text{slope of } \overline{RS} = \frac{4-3}{2-0} = \frac{1}{2}$$

$$\text{slope of } \overline{OT} = \frac{2-0}{4-0} = \frac{2}{4} = \frac{1}{2}$$

The slopes of $\overline{RS}$ and $\overline{OT}$ are the same, so $\overline{RS} \parallel \overline{OT}$.

$$\text{slope of } \overline{ST} = \frac{2-4}{4-2} = \frac{-2}{2} = -1 \qquad \text{slope of } \overline{RO} = \frac{3-0}{0-0} = \frac{3}{0} \quad \text{Undefined}$$

The slopes of $\overline{ST}$ and $\overline{RO}$ are not the same, so $\overline{ST}$ is not parallel to $\overline{OR}$.

▶ Because $ORST$ has exactly one pair of parallel sides, it is a trapezoid.

Step 2 Compare the lengths of legs $\overline{RO}$ and $\overline{ST}$.

$$RO = |3-0| = 3 \qquad ST = \sqrt{(2-4)^2 + (4-2)^2} = \sqrt{8} = 2\sqrt{2}$$

Because $RO \neq ST$, legs $\overline{RO}$ and $\overline{ST}$ are *not* congruent.

▶ So, $ORST$ is not an isosceles trapezoid.

Monitoring Progress

Help in English and Spanish at *BigIdeasMath.com*

1. The points $A(-5, 6)$, $B(4, 9)$, $C(4, 4)$, and $D(-2, 2)$ form the vertices of a quadrilateral. Show that $ABCD$ is a trapezoid. Then decide whether it is isosceles.

Theorems

Theorem 7.14 Isosceles Trapezoid Base Angles Theorem

If a trapezoid is isosceles, then each pair of base angles is congruent.

If trapezoid $ABCD$ is isosceles, then $\angle A \cong \angle D$ and $\angle B \cong \angle C$.

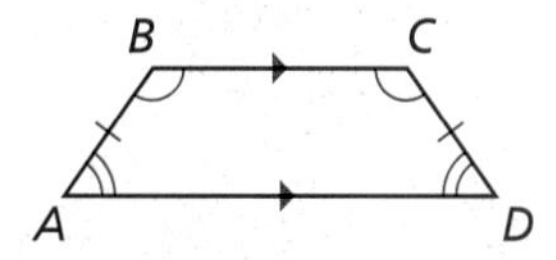

Proof Ex. 39, p. 405

Theorem 7.15 Isosceles Trapezoid Base Angles Converse

If a trapezoid has a pair of congruent base angles, then it is an isosceles trapezoid.

If $\angle A \cong \angle D$ (or if $\angle B \cong \angle C$), then trapezoid $ABCD$ is isosceles.

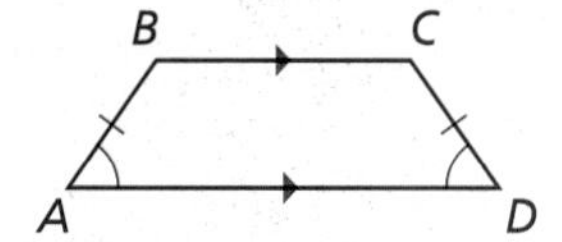

Proof Ex. 40, p. 405

Theorem 7.16 Isosceles Trapezoid Diagonals Theorem

A trapezoid is isosceles if and only if its diagonals are congruent.

Trapezoid $ABCD$ is isosceles if and only if $\overline{AC} \cong \overline{BD}$.

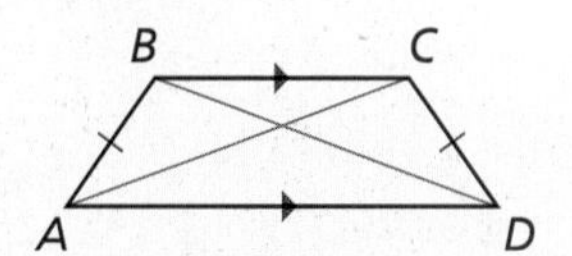

Proof Ex. 51, p. 406

EXAMPLE 2 Using Properties of Isosceles Trapezoids

The stone above the arch in the diagram is an isosceles trapezoid. Find $m\angle K$, $m\angle M$, and $m\angle J$.

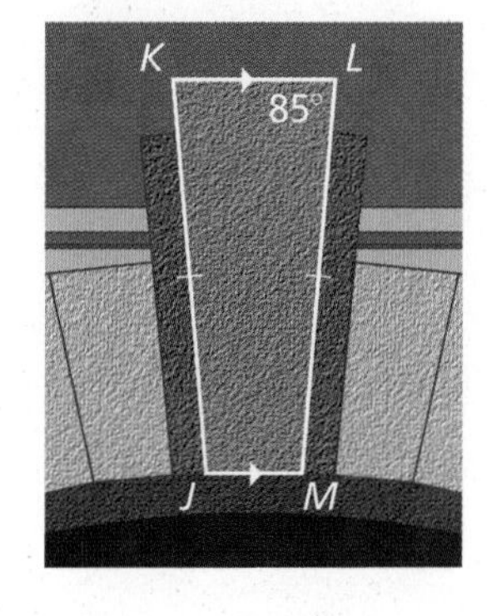

SOLUTION

Step 1 Find $m\angle K$. $JKLM$ is an isosceles trapezoid. So, $\angle K$ and $\angle L$ are congruent base angles, and $m\angle K = m\angle L = 85°$.

Step 2 Find $m\angle M$. Because $\angle L$ and $\angle M$ are consecutive interior angles formed by $\overleftrightarrow{LM}$ intersecting two parallel lines, they are supplementary. So, $m\angle M = 180° - 85° = 95°$.

Step 3 Find $m\angle J$. Because $\angle J$ and $\angle M$ are a pair of base angles, they are congruent, and $m\angle J = m\angle M = 95°$.

▶ So, $m\angle K = 85°$, $m\angle M = 95°$, and $m\angle J = 95°$.

Monitoring Progress  Help in English and Spanish at BigIdeasMath.com

In Exercises 2 and 3, use trapezoid $EFGH$.

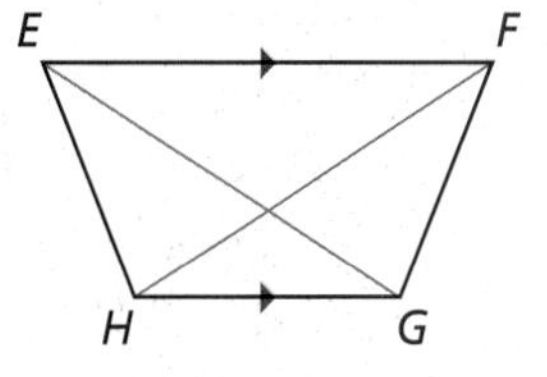

2. If $EG = FH$, is trapezoid $EFGH$ isosceles? Explain.

3. If $m\angle HEF = 70°$ and $m\angle FGH = 110°$, is trapezoid $EFGH$ isosceles? Explain.

Using the Trapezoid Midsegment Theorem

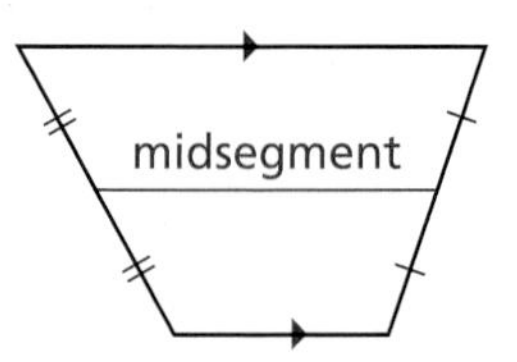

Recall that a midsegment of a triangle is a segment that connects the midpoints of two sides of the triangle. The **midsegment of a trapezoid** is the segment that connects the midpoints of its legs. The theorem below is similar to the Triangle Midsegment Theorem (Thm. 6.8).

READING

The midsegment of a trapezoid is sometimes called the *median* of the trapezoid.

Theorem

Theorem 7.17 Trapezoid Midsegment Theorem

The midsegment of a trapezoid is parallel to each base, and its length is one-half the sum of the lengths of the bases.

If $\overline{MN}$ is the midsegment of trapezoid $ABCD$, then $\overline{MN} \parallel \overline{AB}$, $\overline{MN} \parallel \overline{DC}$, and $MN = \frac{1}{2}(AB + CD)$.

Proof Ex. 49, p. 406

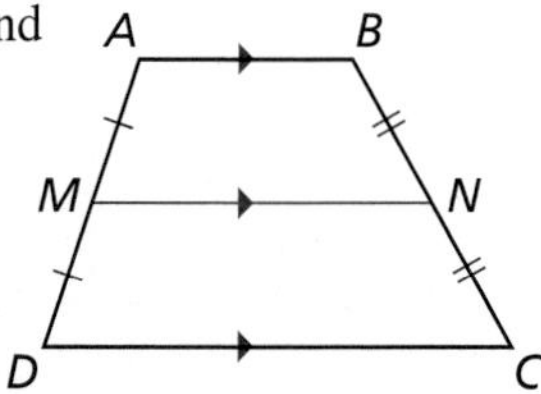

EXAMPLE 3 Using the Midsegment of a Trapezoid

In the diagram, $\overline{MN}$ is the midsegment of trapezoid $PQRS$. Find MN.

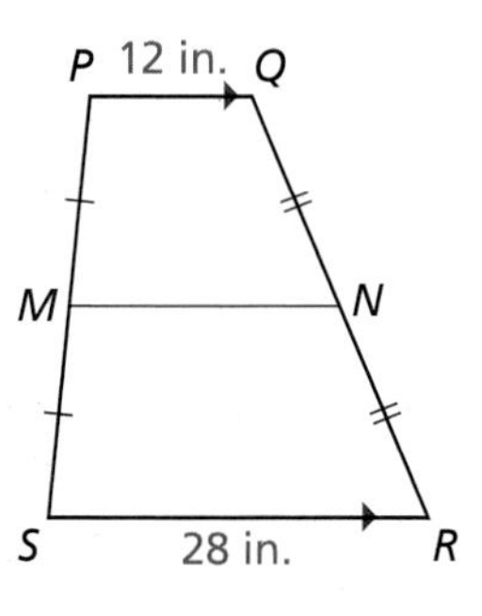

SOLUTION

$MN = \frac{1}{2}(PQ + SR)$	Trapezoid Midsegment Theorem
$= \frac{1}{2}(12 + 28)$	Substitute 12 for PQ and 28 for SR.
$= 20$	Simplify.

▶ The length of $\overline{MN}$ is 20 inches.

EXAMPLE 4 Using a Midsegment in the Coordinate Plane

Find the length of midsegment $\overline{YZ}$ in trapezoid $STUV$.

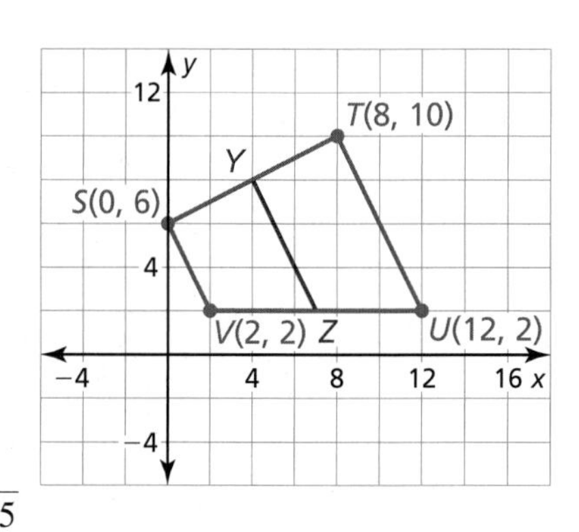

SOLUTION

Step 1 Find the lengths of $\overline{SV}$ and $\overline{TU}$.

$SV = \sqrt{(0 - 2)^2 + (6 - 2)^2} = \sqrt{20} = 2\sqrt{5}$

$TU = \sqrt{(8 - 12)^2 + (10 - 2)^2} = \sqrt{80} = 4\sqrt{5}$

Step 2 Multiply the sum of SV and TU by $\frac{1}{2}$.

$YZ = \frac{1}{2}\left(2\sqrt{5} + 4\sqrt{5}\right) = \frac{1}{2}\left(6\sqrt{5}\right) = 3\sqrt{5}$

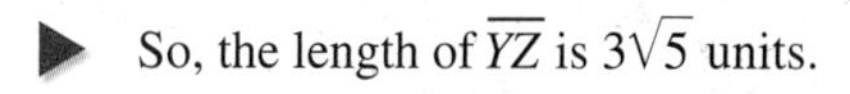

▶ So, the length of $\overline{YZ}$ is $3\sqrt{5}$ units.

Monitoring Progress

Help in English and Spanish at *BigIdeasMath.com*

4. In trapezoid $JKLM$, $\angle J$ and $\angle M$ are right angles, and $JK = 9$ centimeters. The length of midsegment $\overline{NP}$ of trapezoid $JKLM$ is 12 centimeters. Sketch trapezoid $JKLM$ and its midsegment. Find ML. Explain your reasoning.
5. Explain another method you can use to find the length of $\overline{YZ}$ in Example 4.

Using Properties of Kites

A **kite** is a quadrilateral that has two pairs of consecutive congruent sides, but opposite sides are not congruent.

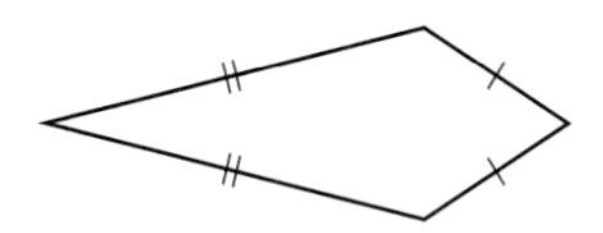

Theorems

Theorem 7.18 Kite Diagonals Theorem

If a quadrilateral is a kite, then its diagonals are perpendicular.

If quadrilateral $ABCD$ is a kite, then $\overline{AC} \perp \overline{BD}$.

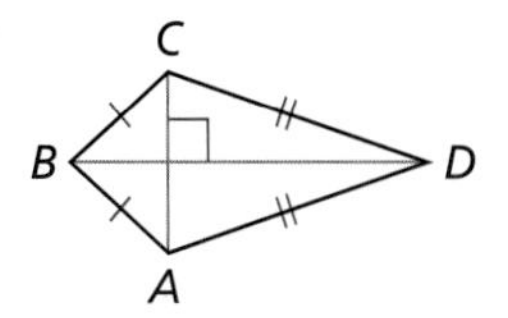

Proof p. 401

Theorem 7.19 Kite Opposite Angles Theorem

If a quadrilateral is a kite, then exactly one pair of opposite angles are congruent.

If quadrilateral $ABCD$ is a kite and $\overline{BC} \cong \overline{BA}$, then $\angle A \cong \angle C$ and $\angle B \not\cong \angle D$.

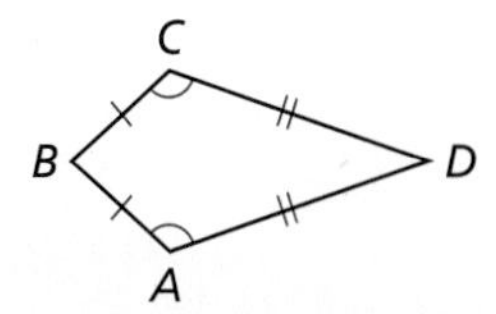

Proof Ex. 47, p. 406

STUDY TIP

The congruent angles of a kite are formed by the noncongruent adjacent sides.

PROOF Kite Diagonals Theorem

Given $ABCD$ is a kite, $\overline{BC} \cong \overline{BA}$, and $\overline{DC} \cong \overline{DA}$.

Prove $\overline{AC} \perp \overline{BD}$

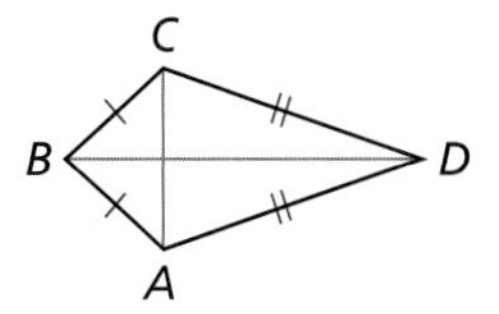

STATEMENTS	REASONS
1. $ABCD$ is a kite with $\overline{BC} \cong \overline{BA}$ and $\overline{DC} \cong \overline{DA}$.	**1.** Given
2. B and D lie on the $\perp$ bisector of $\overline{AC}$.	**2.** Converse of the $\perp$ Bisector Theorem (Theorem 6.2)
3. $\overline{BD}$ is the $\perp$ bisector of $\overline{AC}$.	**3.** Through any two points, there exists exactly one line.
4. $\overline{AC} \perp \overline{BD}$	**4.** Definition of $\perp$ bisector

EXAMPLE 5 Finding Angle Measures in a Kite

Find $m\angle D$ in the kite shown.

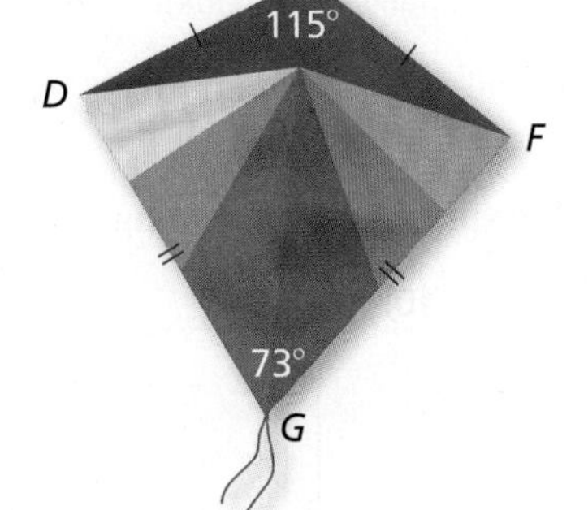

SOLUTION

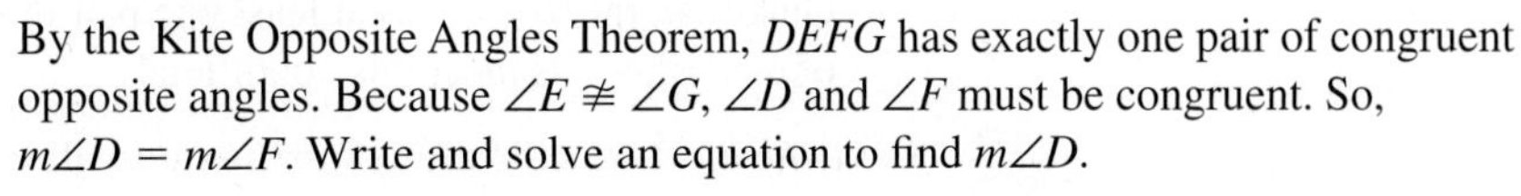

By the Kite Opposite Angles Theorem, $DEFG$ has exactly one pair of congruent opposite angles. Because $\angle E \not\cong \angle G$, $\angle D$ and $\angle F$ must be congruent. So, $m\angle D = m\angle F$. Write and solve an equation to find $m\angle D$.

$m\angle D + m\angle F + 115° + 73° = 360°$	Corollary to the Polygon Interior Angles Theorem (Corollary 7.1)
$m\angle D + m\angle D + 115° + 73° = 360°$	Substitute $m\angle D$ for $m\angle F$.
$2m\angle D + 188° = 360°$	Combine like terms.
$m\angle D = 86°$	Solve for $m\angle D$.

Monitoring Progress  Help in English and Spanish at *BigIdeasMath.com*

6. In a kite, the measures of the angles are $3x^\circ$, 75°, 90°, and 120°. Find the value of x. What are the measures of the angles that are congruent?

Identifying Special Quadrilaterals

The diagram shows relationships among the special quadrilaterals you have studied in this chapter. Each shape in the diagram has the properties of the shapes linked above it. For example, a rhombus has the properties of a parallelogram and a quadrilateral.

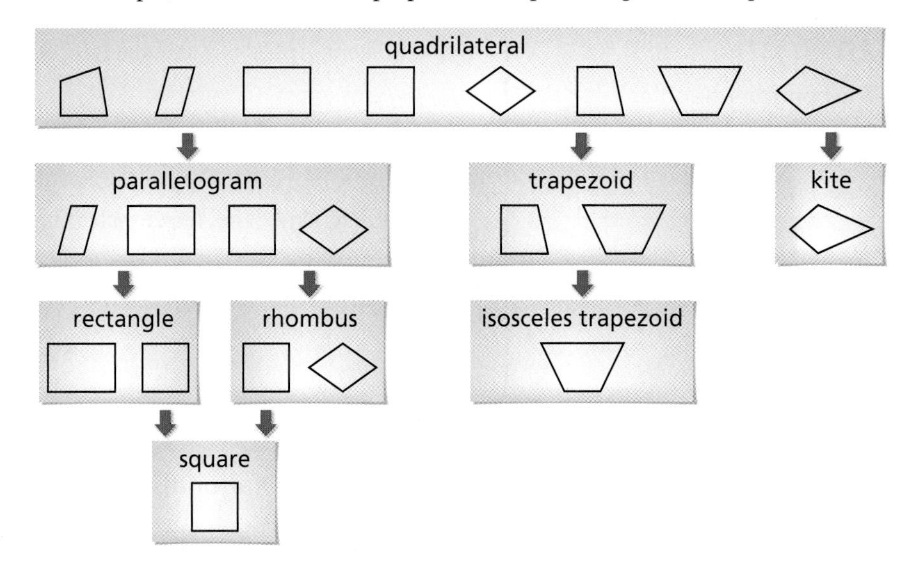

EXAMPLE 6 **Identifying a Quadrilateral**

What is the most specific name for quadrilateral *ABCD*?

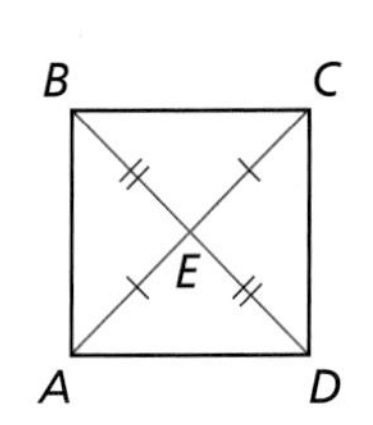

READING DIAGRAMS

In Example 6, *ABCD* looks like a square. But you must rely only on marked information when you interpret a diagram.

SOLUTION

The diagram shows $\overline{AE} \cong \overline{CE}$ and $\overline{BE} \cong \overline{DE}$. So, the diagonals bisect each other. By the Parallelogram Diagonals Converse (Theorem 7.10), *ABCD* is a parallelogram.

Rectangles, rhombuses, and squares are also parallelograms. However, there is no information given about the side lengths or angle measures of *ABCD*. So, you cannot determine whether it is a rectangle, a rhombus, or a square.

▶ So, the most specific name for *ABCD* is a parallelogram.

Monitoring Progress  Help in English and Spanish at *BigIdeasMath.com*

7. Quadrilateral *DEFG* has at least one pair of opposite sides congruent. What types of quadrilaterals meet this condition?

Give the most specific name for the quadrilateral. Explain your reasoning.

8.

S
50 50
R T
51 51
U

9.

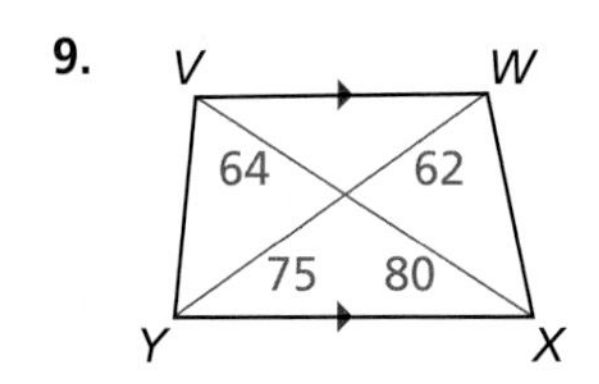

10.

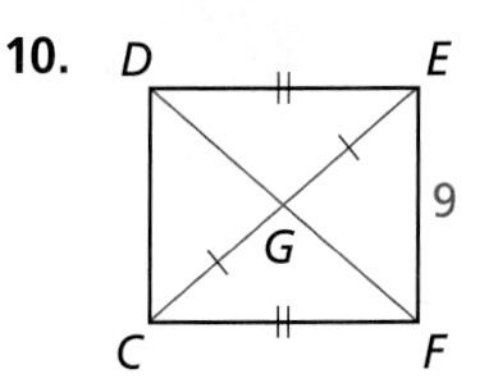

7.5 Exercises

Dynamic Solutions available at *BigIdeasMath.com*

Vocabulary and Core Concept Check

1. **WRITING** Describe the differences between a trapezoid and a kite.

2. **DIFFERENT WORDS, SAME QUESTION** Which is different? Find "both" answers.

Is there enough information to prove that trapezoid $ABCD$ is isosceles?

Is there enough information to prove that $\overline{AB} \cong \overline{DC}$?

Is there enough information to prove that the non-parallel sides of trapezoid $ABCD$ are congruent?

Is there enough information to prove that the legs of trapezoid $ABCD$ are congruent?

Monitoring Progress and Modeling with Mathematics

In Exercises 3–6, show that the quadrilateral with the given vertices is a trapezoid. Then decide whether it is isosceles. *(See Example 1.)*

3. $W(1, 4)$, $X(1, 8)$, $Y(-3, 9)$, $Z(-3, 3)$

4. $D(-3, 3)$, $E(-1, 1)$, $F(1, -4)$, $G(-3, 0)$

5. $M(-2, 0)$, $N(0, 4)$, $P(5, 4)$, $Q(8, 0)$

6. $H(1, 9)$, $J(4, 2)$, $K(5, 2)$, $L(8, 9)$

In Exercises 7 and 8, find the measure of each angle in the isosceles trapezoid. *(See Example 2.)*

7. 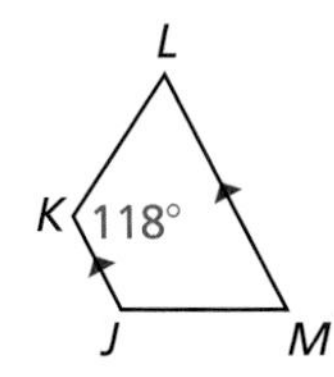

8. 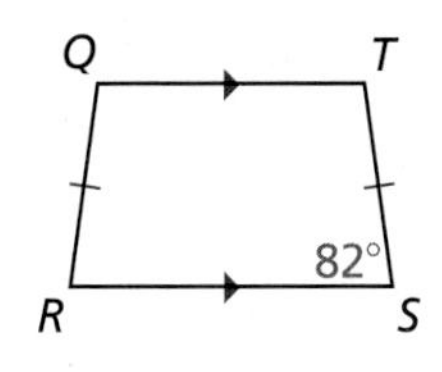

In Exercises 9 and 10, find the length of the midsegment of the trapezoid. *(See Example 3.)*

9. 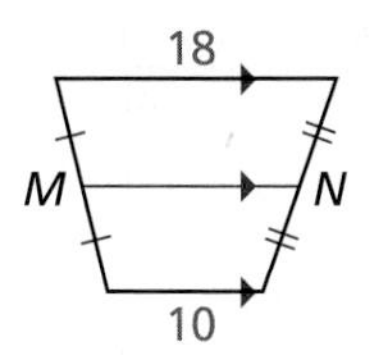

10. 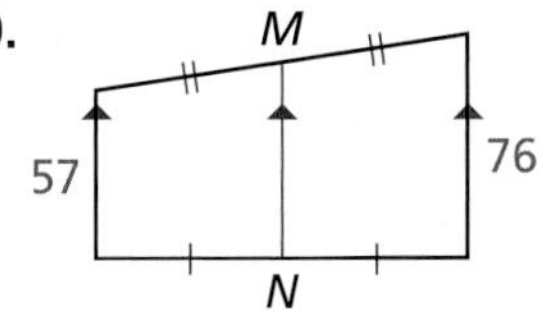

In Exercises 11 and 12, find AB.

11. 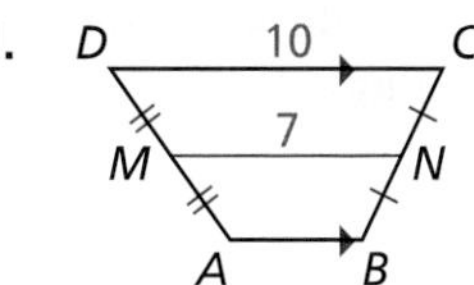

12. 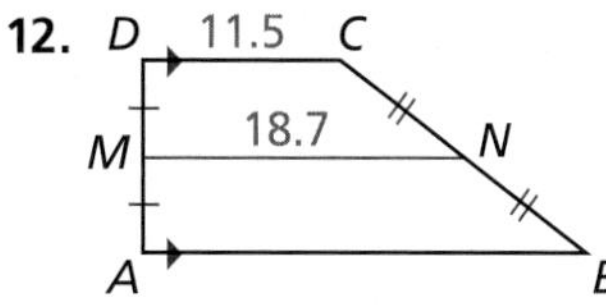

In Exercises 13 and 14, find the length of the midsegment of the trapezoid with the given vertices. *(See Example 4.)*

13. $A(2, 0)$, $B(8, -4)$, $C(12, 2)$, $D(0, 10)$

14. $S(-2, 4)$, $T(-2, -4)$, $U(3, -2)$, $V(13, 10)$

In Exercises 15–18, find $m\angle G$. *(See Example 5.)*

15. 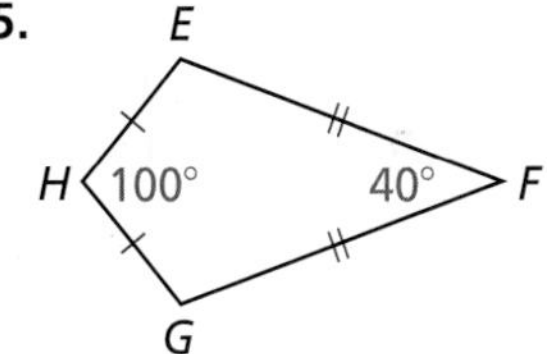

16. 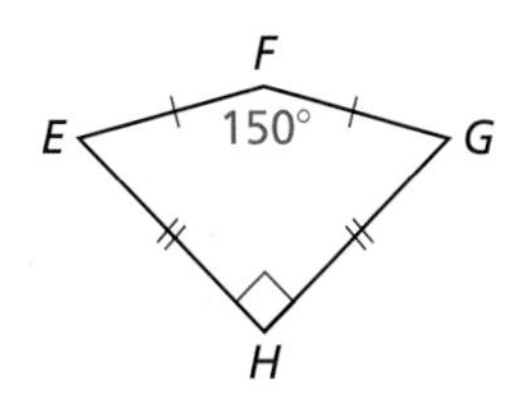

17. 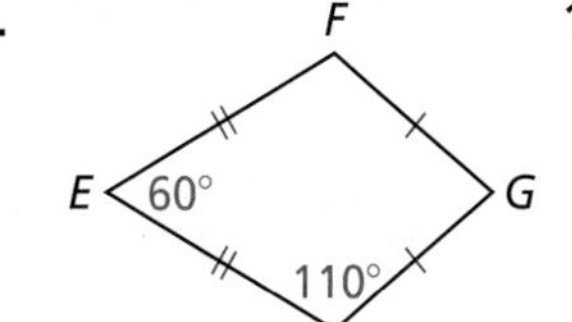

18. 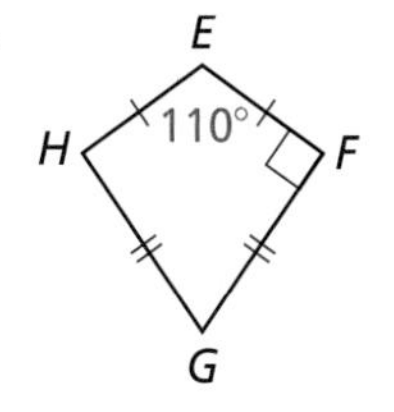

19. ERROR ANALYSIS Describe and correct the error in finding DC.

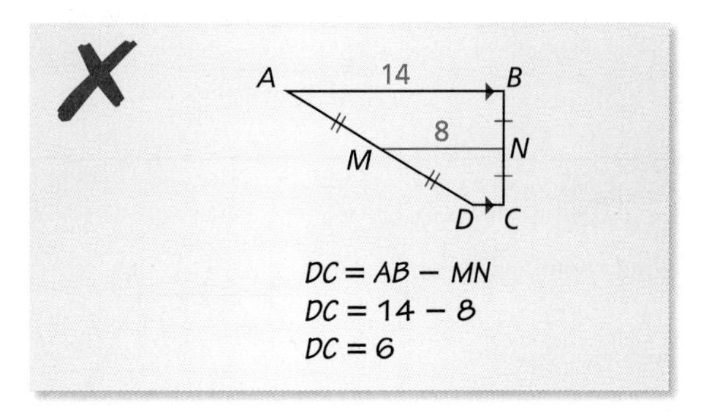

20. ERROR ANALYSIS Describe and correct the error in finding $m\angle A$.

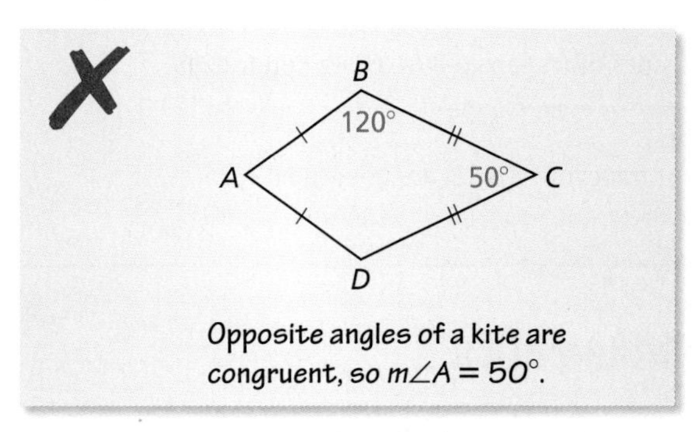

In Exercises 21–24, give the most specific name for the quadrilateral. Explain your reasoning. *(See Example 6.)*

21. J K M L

22.

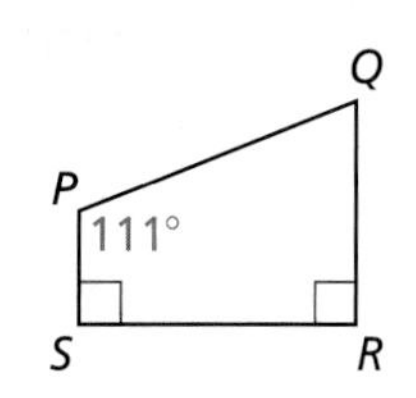

23.

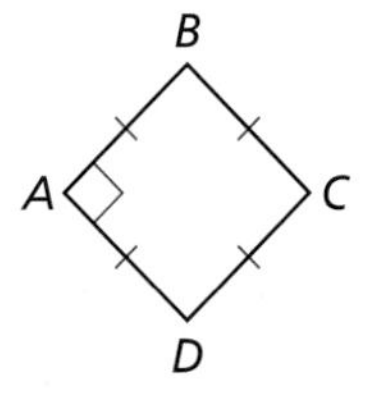

24.

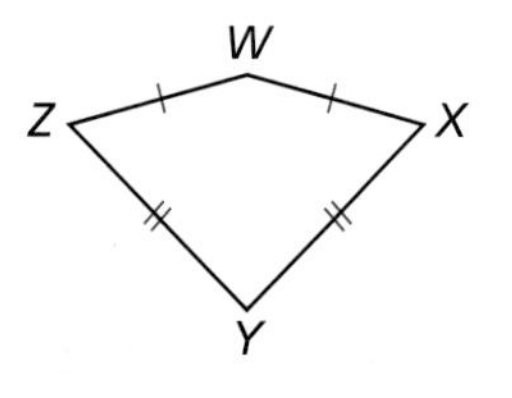

REASONING In Exercises 25 and 26, tell whether enough information is given in the diagram to classify the quadrilateral by the indicated name. Explain.

25. rhombus

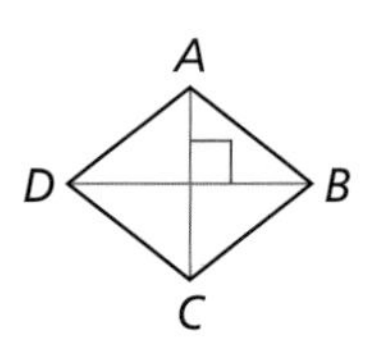

26. square

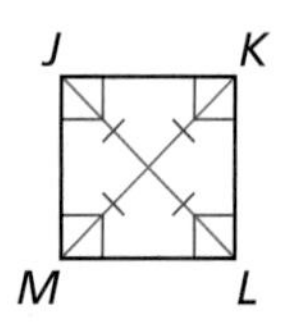

MATHEMATICAL CONNECTIONS In Exercises 27 and 28, find the value of x.

27.

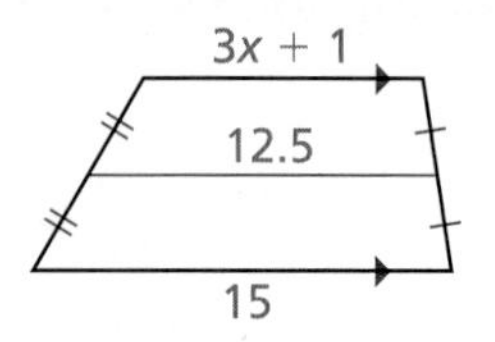

28.

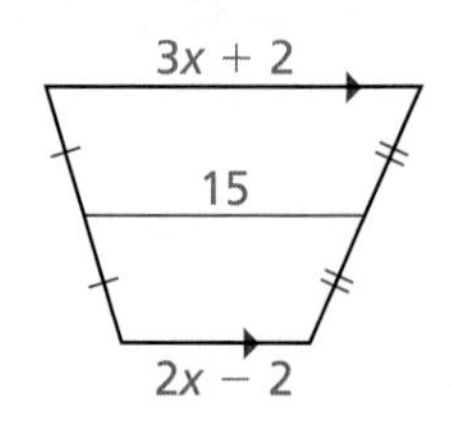

29. MODELING WITH MATHEMATICS In the diagram, $NP = 8$ inches, and $LR = 20$ inches. What is the diameter of the bottom layer of the cake?

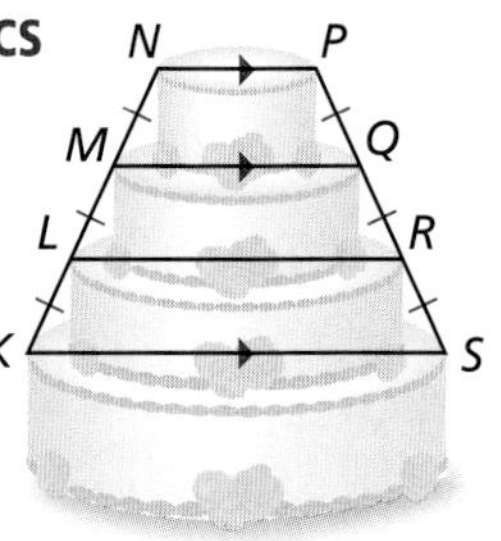

30. PROBLEM SOLVING You and a friend are building a kite. You need a stick to place from X to W and a stick to place from W to Z to finish constructing the frame. You want the kite to have the geometric shape of a kite. How long does each stick need to be? Explain your reasoning.

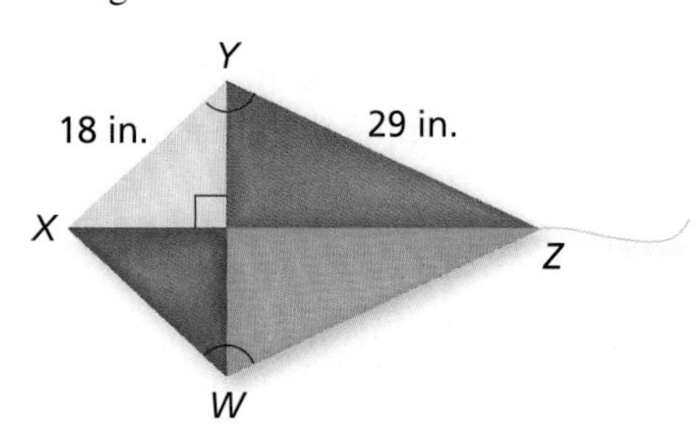

REASONING In Exercises 31–34, determine which pairs of segments or angles must be congruent so that you can prove that $ABCD$ is the indicated quadrilateral. Explain your reasoning. (There may be more than one right answer.)

31. isosceles trapezoid

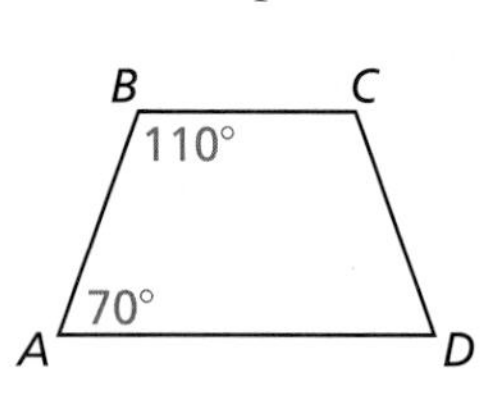

32. kite

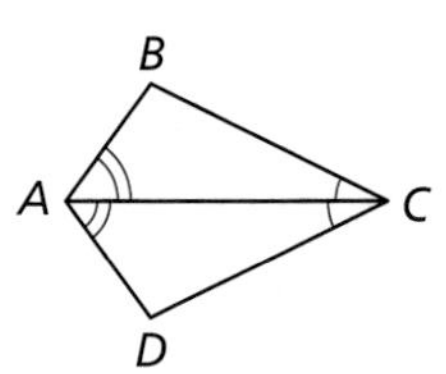

33. parallelogram

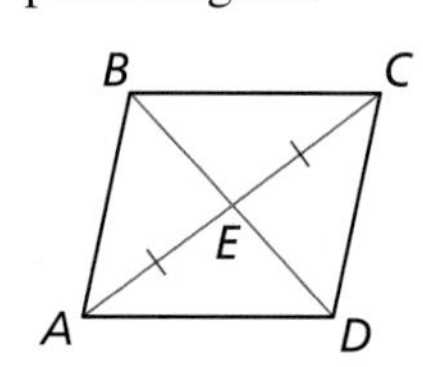

34. square

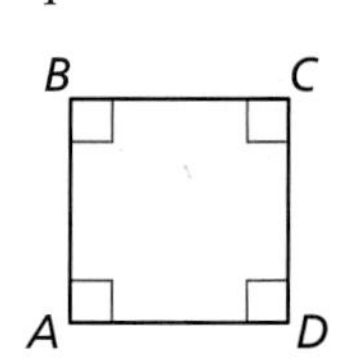

35. PROOF Write a proof.

Given $\overline{JL} \cong \overline{LN}$, $\overline{KM}$ is a midsegment of $\triangle JLN$.

Prove Quadrilateral $JKMN$ is an isosceles trapezoid.

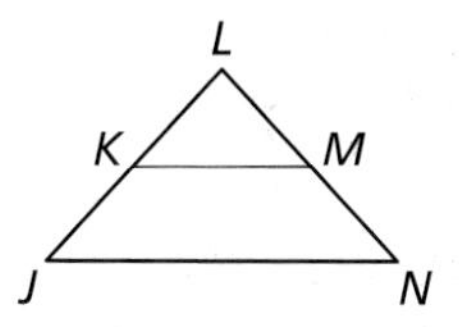

36. PROOF Write a proof.

Given $ABCD$ is a kite.
$\overline{AB} \cong \overline{CB}$, $\overline{AD} \cong \overline{CD}$

Prove $\overline{CE} \cong \overline{AE}$

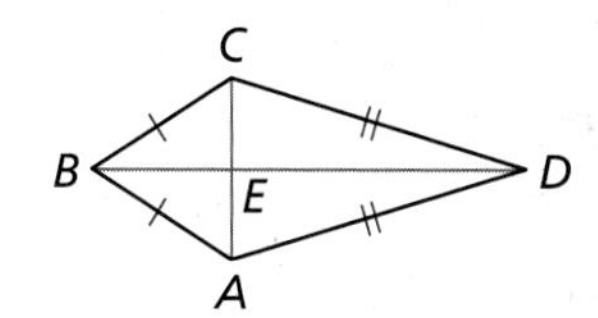

37. ABSTRACT REASONING Point U lies on the perpendicular bisector of $\overline{RT}$. Describe the set of points S for which $RSTU$ is a kite.

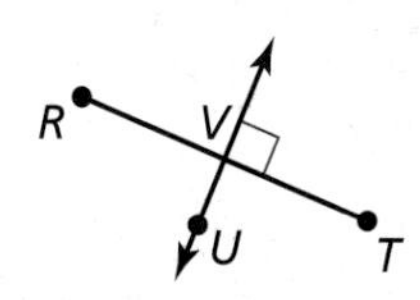

38. REASONING Determine whether the points $A(4, 5)$, $B(-3, 3)$, $C(-6, -13)$, and $D(6, -2)$ are the vertices of a kite. Explain your reasoning,

PROVING A THEOREM In Exercises 39 and 40, use the diagram to prove the given theorem. In the diagram, $\overline{EC}$ is drawn parallel to $\overline{AB}$.

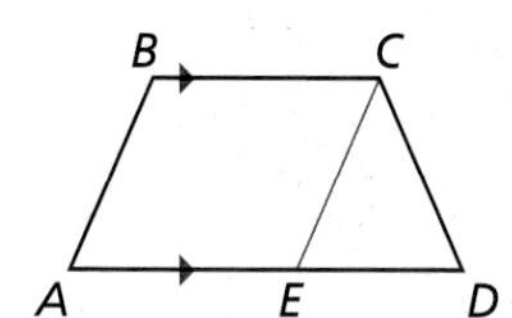

39. Isosceles Trapezoid Base Angles Theorem (Theorem 7.14)

Given $ABCD$ is an isosceles trapezoid.
$\overline{BC} \parallel \overline{AD}$

Prove $\angle A \cong \angle D$, $\angle B \cong \angle BCD$

40. Isosceles Trapezoid Base Angles Converse (Theorem 7.15)

Given $ABCD$ is a trapezoid.
$\angle A \cong \angle D$, $\overline{BC} \parallel \overline{AD}$

Prove $ABCD$ is an isosceles trapezoid.

41. MAKING AN ARGUMENT Your cousin claims there is enough information to prove that $JKLM$ is an isosceles trapezoid. Is your cousin correct? Explain.

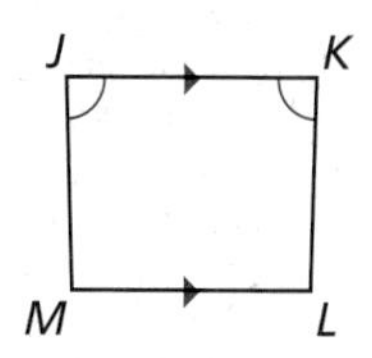

42. MATHEMATICAL CONNECTIONS The bases of a trapezoid lie on the lines $y = 2x + 7$ and $y = 2x - 5$. Write the equation of the line that contains the midsegment of the trapezoid.

43. CONSTRUCTION $\overline{AC}$ and $\overline{BD}$ bisect each other.

a. Construct quadrilateral $ABCD$ so that $\overline{AC}$ and $\overline{BD}$ are congruent, but not perpendicular. Classify the quadrilateral. Justify your answer.

b. Construct quadrilateral $ABCD$ so that $\overline{AC}$ and $\overline{BD}$ are perpendicular, but not congruent. Classify the quadrilateral. Justify your answer.

44. PROOF Write a proof.

Given $QRST$ is an isosceles trapezoid.

Prove $\angle TQS \cong \angle SRT$

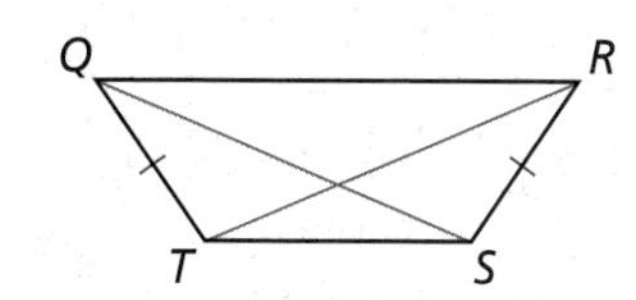

45. MODELING WITH MATHEMATICS A plastic spiderweb is made in the shape of a regular dodecagon (12-sided polygon). $\overline{AB} \parallel \overline{PQ}$, and X is equidistant from the vertices of the dodecagon.

a. Are you given enough information to prove that $ABPQ$ is an isosceles trapezoid?

b. What is the measure of each interior angle of $ABPQ$?

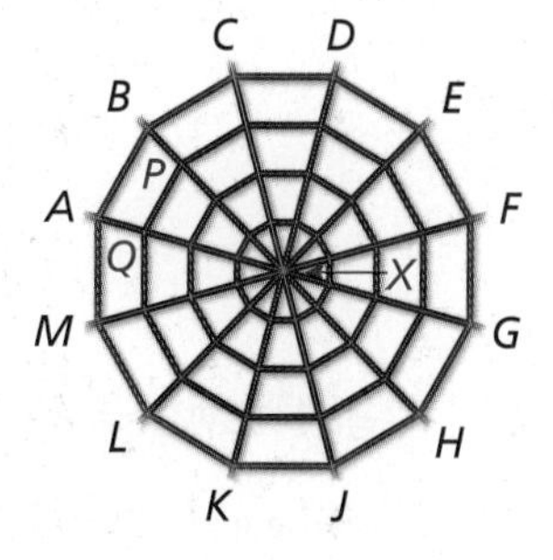

46. ATTENDING TO PRECISION In trapezoid $PQRS$, $\overline{PQ} \parallel \overline{RS}$ and $\overline{MN}$ is the midsegment of $PQRS$. If $RS = 5 \cdot PQ$, what is the ratio of MN to RS?

Ⓐ 3 : 5 Ⓑ 5 : 3

Ⓒ 1 : 2 Ⓓ 3 : 1

47. PROVING A THEOREM Use the plan for proof below to write a paragraph proof of the Kite Opposite Angles Theorem (Theorem 7.19).

Given $EFGH$ is a kite.
$\overline{EF} \cong \overline{FG}, \overline{EH} \cong \overline{GH}$

Prove $\angle E \cong \angle G, \angle F \ncong \angle H$

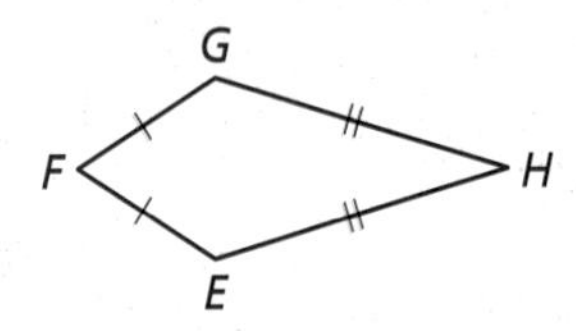

Plan for Proof First show that $\angle E \cong \angle G$. Then use an indirect argument to show that $\angle F \ncong \angle H$.

48. HOW DO YOU SEE IT? One of the earliest shapes used for cut diamonds is called the *table cut*, as shown in the figure. Each face of a cut gem is called a *facet*.

a. $\overline{BC} \parallel \overline{AD}$, and $\overline{AB}$ and $\overline{DC}$ are not parallel. What shape is the facet labeled $ABCD$?

b. $\overline{DE} \parallel \overline{GF}$, and $\overline{DG}$ and $\overline{EF}$ are congruent but not parallel. What shape is the facet labeled $DEFG$?

B C E A D F G

49. PROVING A THEOREM In the diagram below, $\overline{BG}$ is the midsegment of $\triangle ACD$, and $\overline{GE}$ is the midsegment of $\triangle ADF$. Use the diagram to prove the Trapezoid Midsegment Theorem (Theorem 7.17).

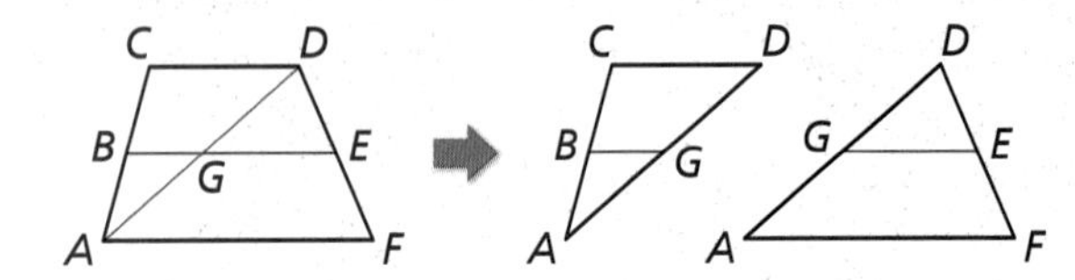

50. THOUGHT PROVOKING Is SSASS a valid congruence theorem for kites? Justify your answer.

51. PROVING A THEOREM To prove the biconditional statement in the Isosceles Trapezoid Diagonals Theorem (Theorem 7.16), you must prove both parts separately.

a. Prove part of the Isosceles Trapezoid Diagonals Theorem (Theorem 7.16).

Given $JKLM$ is an isosceles trapezoid.
$\overline{KL} \parallel \overline{JM}, \overline{JK} \cong \overline{LM}$

Prove $\overline{JL} \cong \overline{KM}$

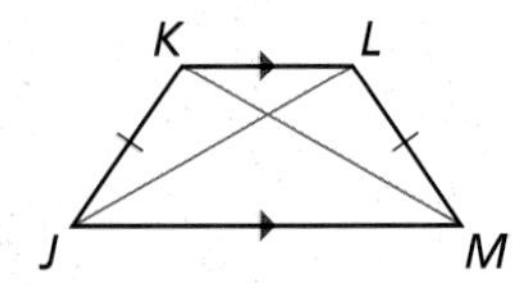

b. Write the other part of the Isosceles Trapezoid Diagonals Theorem (Theorem 7.16) as a conditional. Then prove the statement is true.

52. PROOF What special type of quadrilateral is $EFGH$? Write a proof to show that your answer is correct.

Given In the three-dimensional figure, $\overline{JK} \cong \overline{LM}$. E, F, G, and H are the midpoints of $\overline{JL}$, $\overline{KL}$, $\overline{KM}$, and $\overline{JM}$, respectively.

Prove $EFGH$ is a _______.

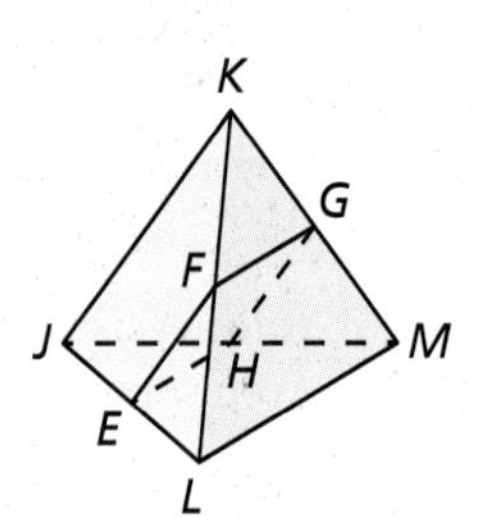

Maintaining Mathematical Proficiency

Reviewing what you learned in previous grades and lessons

Describe a similarity transformation that maps the blue preimage to the green image. *(Section 4.6)*

53.

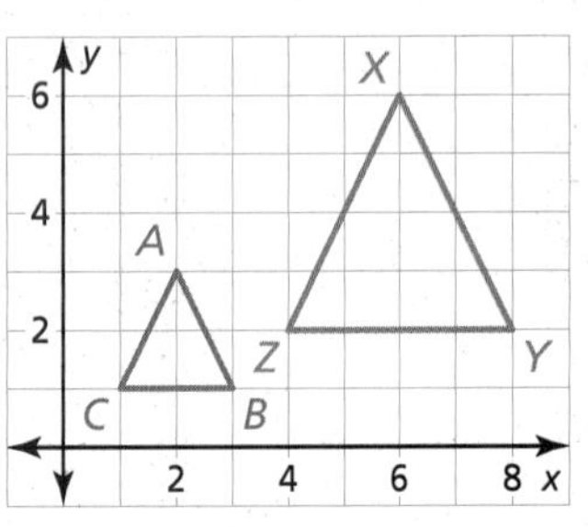

54.

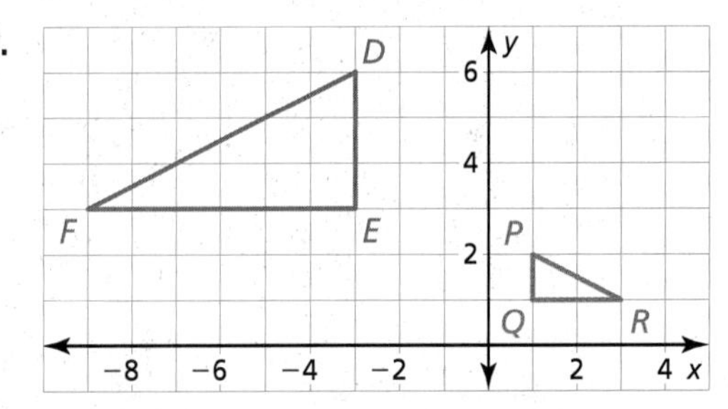

7.4–7.5 What Did You Learn?

Core Vocabulary

rhombus, *p. 388*
rectangle, *p. 388*
square, *p. 388*
trapezoid, *p. 398*
bases (of a trapezoid), *p. 398*
base angles (of a trapezoid), *p. 398*
legs (of a trapezoid), *p. 398*
isosceles trapezoid, *p. 398*
midsegment of a trapezoid, *p. 400*
kite, *p. 401*

Core Concepts

Section 7.4

Corollary 7.2 Rhombus Corollary, *p. 388*
Corollary 7.3 Rectangle Corollary, *p. 388*
Corollary 7.4 Square Corollary, *p. 388*
Relationships between Special Parallelograms, *p. 389*
Theorem 7.11 Rhombus Diagonals Theorem, *p. 390*
Theorem 7.12 Rhombus Opposite Angles Theorem, *p. 390*
Theorem 7.13 Rectangle Diagonals Theorem, *p. 391*
Identifying Special Parallelograms in the Coordinate Plane, *p. 392*

Section 7.5

Showing That a Quadrilateral Is a Trapezoid in the Coordinate Plane, *p. 398*
Theorem 7.14 Isosceles Trapezoid Base Angles Theorem, *p. 399*
Theorem 7.15 Isosceles Trapezoid Base Angles Converse, *p. 399*
Theorem 7.16 Isosceles Trapezoid Diagonals Theorem, *p. 399*
Theorem 7.17 Trapezoid Midsegment Theorem, *p. 400*
Theorem 7.18 Kite Diagonals Theorem, *p. 401*
Theorem 7.19 Kite Opposite Angles Theorem, *p. 401*
Identifying Special Quadrilaterals, *p. 402*

Mathematical Practices

1. In Exercise 14 on page 393, one reason $m\angle 4$, $m\angle 5$, and $m\angle DFE$ are all 48° is because diagonals of a rhombus bisect each other. What is another reason they are equal?
2. Explain how the diagram you created in Exercise 64 on page 395 can help you answer questions like Exercises 65–70.
3. In Exercise 29 on page 404, describe a pattern you can use to find the measure of a base of a trapezoid when given the length of the midsegment and the other base.

Performance Task

Scissor Lifts

A scissor lift is a work platform with an adjustable height that is stable and convenient. The platform is supported by crisscrossing beams that raise and lower the platform. What quadrilaterals do you see in the scissor lift design? What properties of those quadrilaterals play a key role in the successful operation of the lift?

To explore the answers to this question and more, go to ***BigIdeasMath.com***.

7 Chapter Review

Dynamic Solutions available at *BigIdeasMath.com*

7.1 Angles of Polygons *(pp. 359–366)*

Find the sum of the measures of the interior angles of the figure.

The figure is a convex hexagon. It has 6 sides. Use the Polygon Interior Angles Theorem (Theorem 7.1).

$(n - 2) \cdot 180° = (6 - 2) \cdot 180°$	Substitute 6 for *n*.
$= 4 \cdot 180°$	Subtract.
$= 720°$	Multiply.

▶ The sum of the measures of the interior angles of the figure is 720°.

1. Find the sum of the measures of the interior angles of a regular 30-gon. Then find the measure of each interior angle and each exterior angle.

Find the value of *x*.

2.

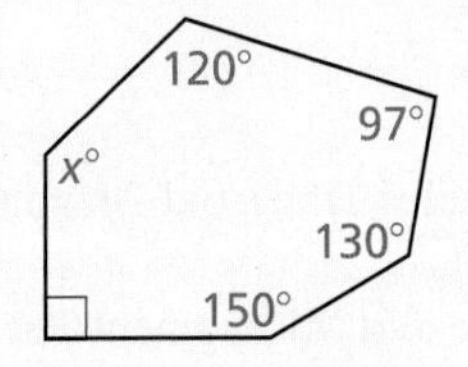

3.

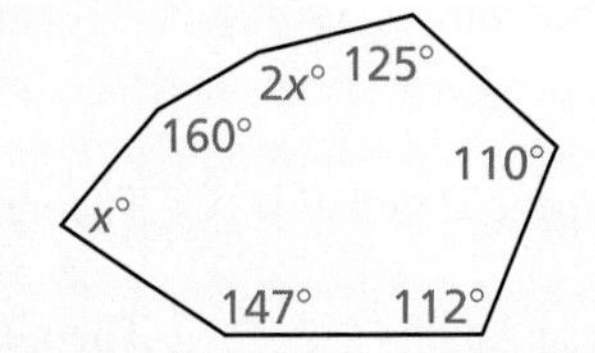

4.

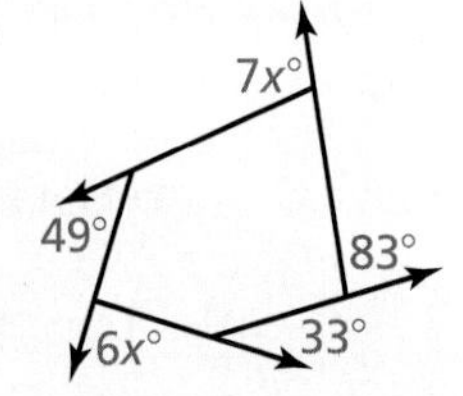

7.2 Properties of Parallelograms *(pp. 367–374)*

Find the values of *x* and *y*.

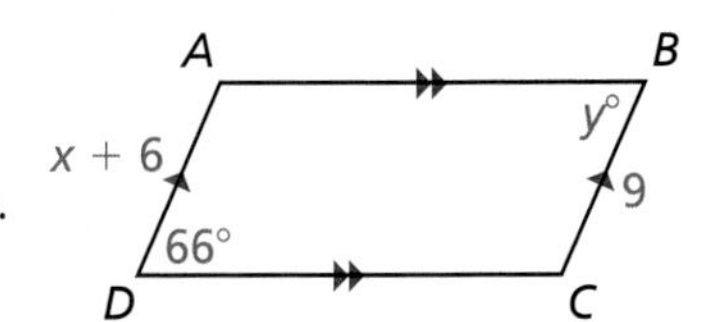

ABCD is a parallelogram by the definition of a parallelogram. Use the Parallelogram Opposite Sides Theorem (Thm. 7.3) to find the value of *x*.

$AD = BC$	Opposite sides of a parallelogram are congruent.
$x + 6 = 9$	Substitute $x + 6$ for *AD* and 9 for *BC*.
$x = 3$	Subtract 6 from each side.

By the Parallelogram Opposite Angles Theorem (Thm. 7.4), $\angle D \cong \angle B$, or $m\angle D = m\angle B$. So, $y° = 66°$.

▶ In ▱*ABCD*, $x = 3$ and $y = 66$.

Find the value of each variable in the parallelogram.

5.

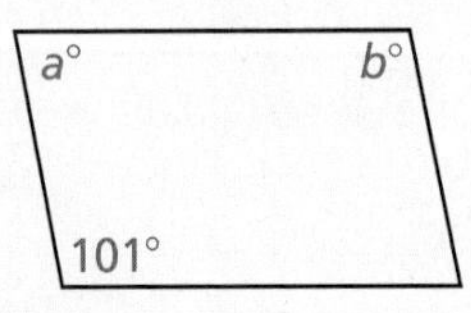

6.

7.

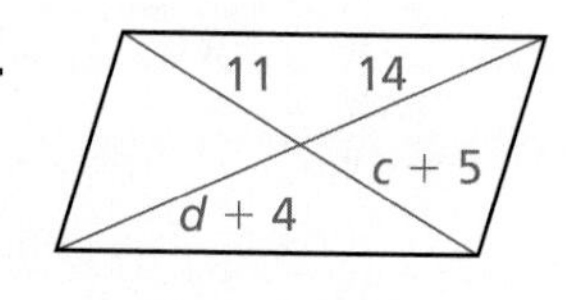

8. Find the coordinates of the intersection of the diagonals of ▱*QRST* with vertices $Q(-8, 1)$, $R(2, 1)$, $S(4, -3)$, and $T(-6, -3)$.

9. Three vertices of ▱*JKLM* are $J(1, 4)$, $K(5, 3)$, and $L(6, -3)$. Find the coordinates of vertex *M*.

7.3 Proving That a Quadrilateral Is a Parallelogram (pp. 375–384)

For what value of x is quadrilateral $DEFG$ a parallelogram?

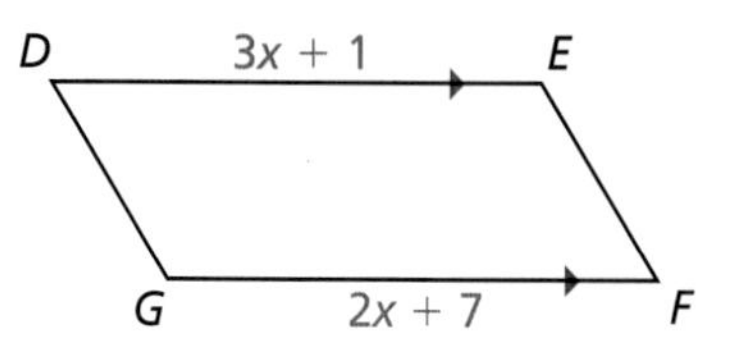

By the Opposite Sides Parallel and Congruent Theorem (Thm. 7.9), if one pair of opposite sides are congruent and parallel, then $DEFG$ is a parallelogram. You are given that $\overline{DE} \parallel \overline{FG}$. Find x so that $\overline{DE} \cong \overline{FG}$.

$DE = FG$	Set the segment lengths equal.
$3x + 1 = 2x + 7$	Substitute $3x + 1$ for DE and $2x + 7$ for FG.
$x + 1 = 7$	Subtract $2x$ from each side.
$x = 6$	Subtract 1 from each side.

When $x = 6$, $DE = 3(6) + 1 = 19$ and $FG = 2(6) + 7 = 19$.

▶ Quadrilateral $DEFG$ is a parallelogram when $x = 6$.

State which theorem you can use to show that the quadrilateral is a parallelogram.

10.

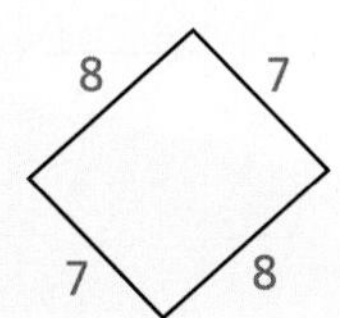

11.

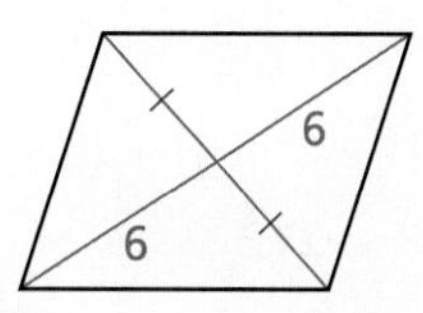

12.

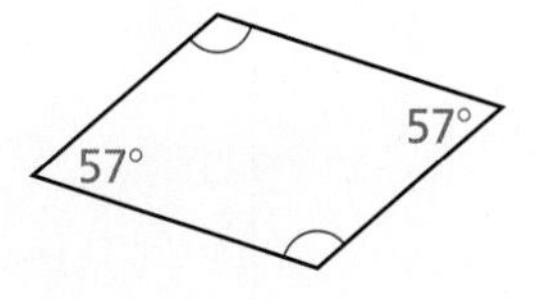

13. Find the values of x and y that make the quadrilateral a parallelogram.

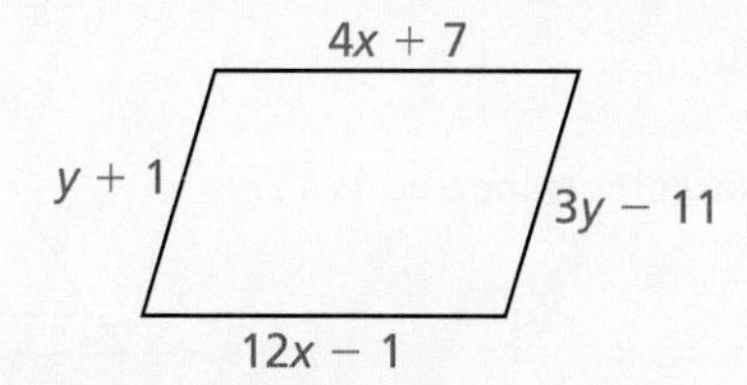

14. Find the value of x that makes the quadrilateral a parallelogram.

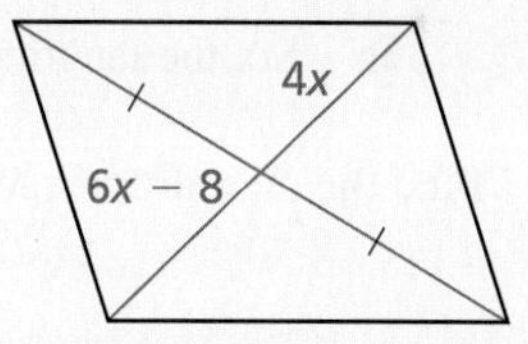

15. Show that quadrilateral $WXYZ$ with vertices $W(-1, 6)$, $X(2, 8)$, $Y(1, 0)$, and $Z(-2, -2)$ is a parallelogram.

7.4 Properties of Special Parallelograms (pp. 387–396)

Classify the special quadrilateral. Explain your reasoning.

The quadrilateral has four right angles. By the Rectangle Corollary (Corollary 7.3), the quadrilateral is a rectangle. Because the four sides are not marked as congruent, you cannot conclude that the rectangle is a square.

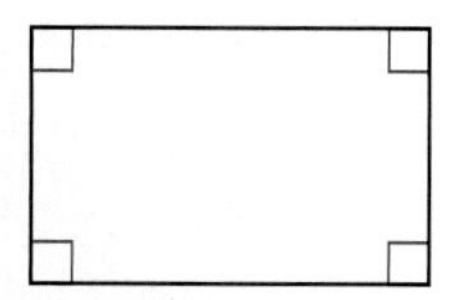

Classify the special quadrilateral. Explain your reasoning.

16.

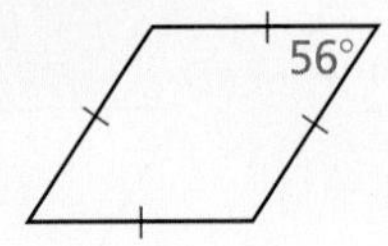

17.

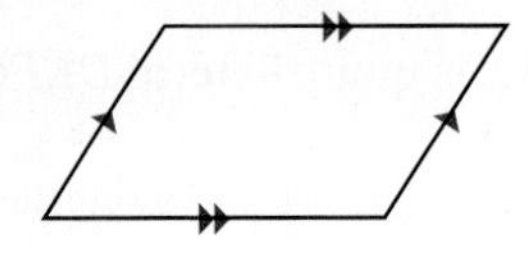

18.

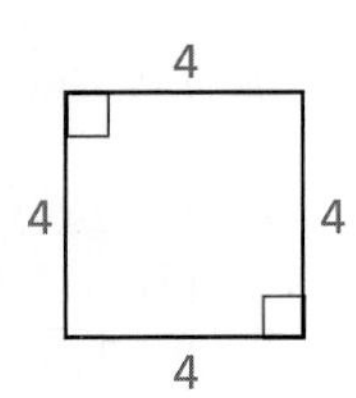

19. Find the lengths of the diagonals of rectangle $WXYZ$ where $WY = -2x + 34$ and $XZ = 3x - 26$.

20. Decide whether $\square JKLM$ with vertices $J(5, 8)$, $K(9, 6)$, $L(7, 2)$, and $M(3, 4)$ is a rectangle, a rhombus, or a square. Give all names that apply. Explain.

7.5 Properties of Trapezoids and Kites (pp. 397–406)

Find the length of midsegment $\overline{EF}$ in trapezoid $ABCD$.

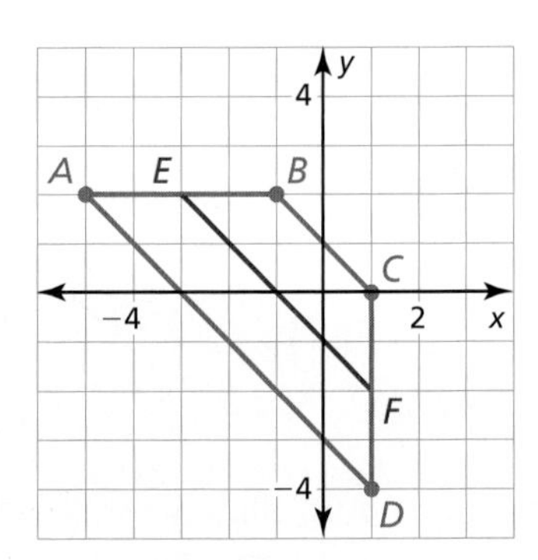

Step 1 Find the lengths of $\overline{AD}$ and $\overline{BC}$.

$$AD = \sqrt{[1 - (-5)]^2 + (-4 - 2)^2}$$
$$= \sqrt{72} = 6\sqrt{2}$$
$$BC = \sqrt{[1 - (-1)]^2 + (0 - 2)^2}$$
$$= \sqrt{8} = 2\sqrt{2}$$

Step 2 Multiply the sum of AD and BC by $\frac{1}{2}$.

$$EF = \tfrac{1}{2}(6\sqrt{2} + 2\sqrt{2}) = \tfrac{1}{2}(8\sqrt{2}) = 4\sqrt{2}$$

▶ So, the length of $\overline{EF}$ is $4\sqrt{2}$ units.

21. Find the measure of each angle in the isosceles trapezoid $WXYZ$.

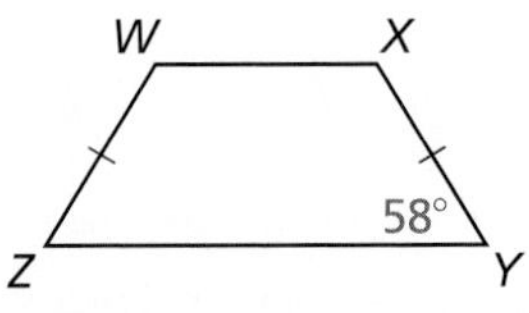

22. Find the length of the midsegment of trapezoid $ABCD$.

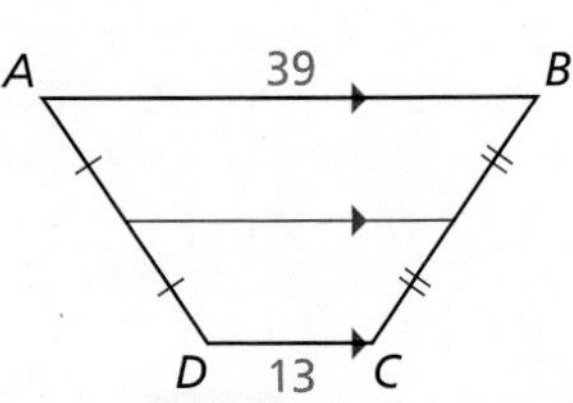

23. Find the length of the midsegment of trapezoid $JKLM$ with vertices $J(6, 10)$, $K(10, 6)$, $L(8, 2)$, and $M(2, 2)$.

24. A kite has angle measures of $7x°$, 65°, 85°, and 105°. Find the value of x. What are the measures of the angles that are congruent?

25. Quadrilateral $WXYZ$ is a trapezoid with one pair of congruent base angles. Is $WXYZ$ an isosceles trapezoid? Explain your reasoning.

Give the most specific name for the quadrilateral. Explain your reasoning.

26.

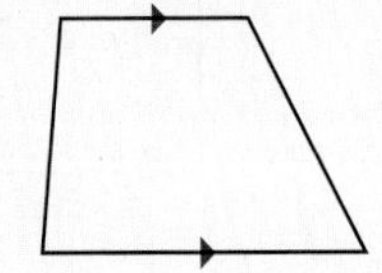

27.

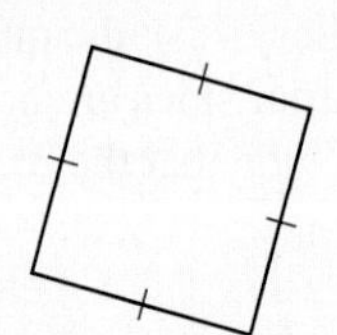

28.

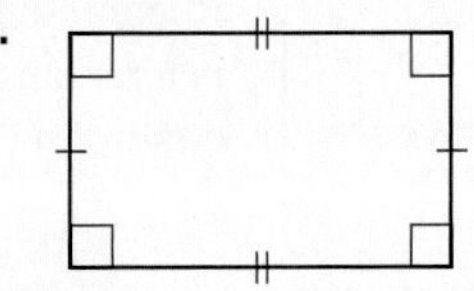

7 Chapter Test

Find the value of each variable in the parallelogram.

1.

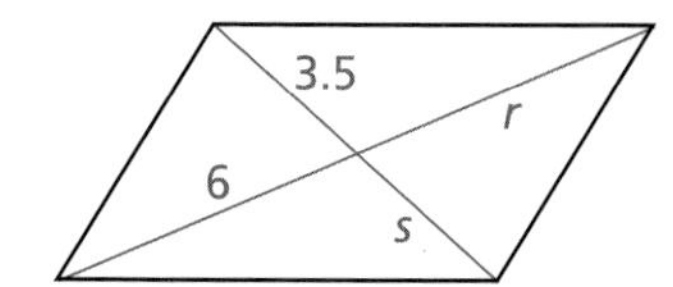

2.

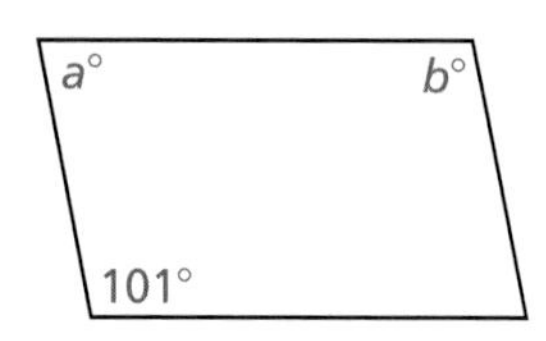

3.

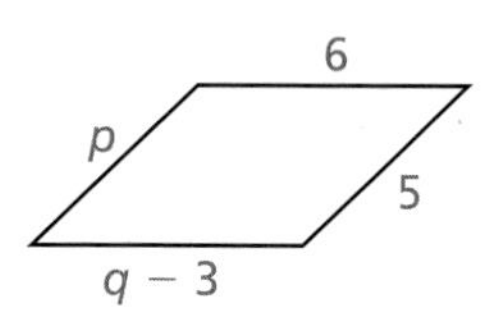

Give the most specific name for the quadrilateral. Explain your reasoning.

4.

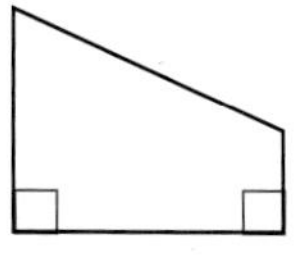

5.

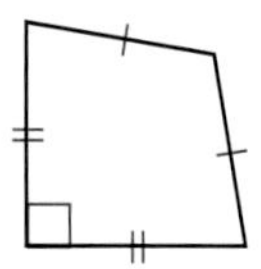

6.

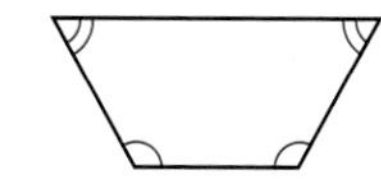

7. In a convex octagon, three of the exterior angles each have a measure of $x°$. The other five exterior angles each have a measure of $(2x + 7)°$. Find the measure of each exterior angle.

8. Quadrilateral $PQRS$ has vertices $P(5, 1)$, $Q(9, 6)$, $R(5, 11)$, and $S(1, 6)$. Classify quadrilateral $PQRS$ using the most specific name.

Determine whether enough information is given to show that the quadrilateral is a parallelogram. Explain your reasoning.

9.

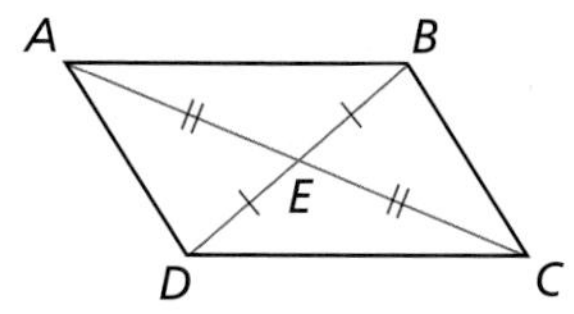

10.

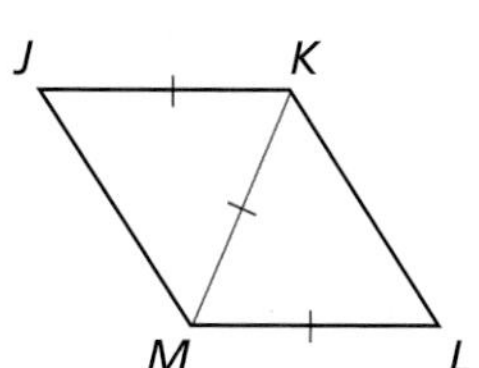

11.

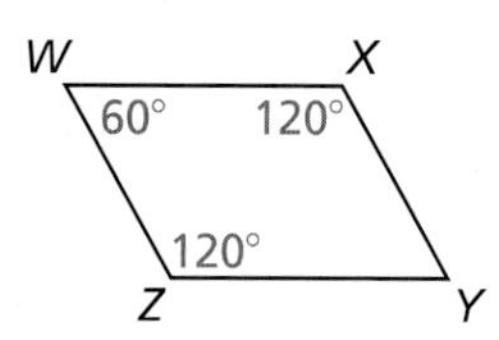

12. Explain why a parallelogram with one right angle must be a rectangle.

13. Summarize the ways you can prove that a quadrilateral is a square.

14. Three vertices of $\square JKLM$ are $J(-2, -1)$, $K(0, 2)$, and $L(4, 3)$.

a. Find the coordinates of vertex M.

b. Find the coordinates of the intersection of the diagonals of $\square JKLM$.

15. You are building a plant stand with three equally-spaced circular shelves. The diagram shows a vertical cross section of the plant stand. What is the diameter of the middle shelf?

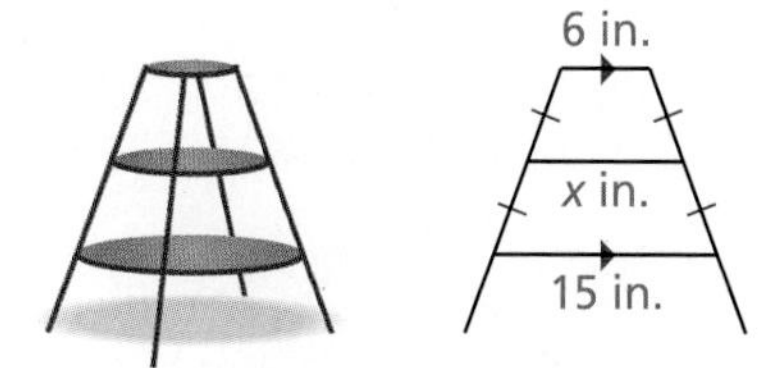

16. The Pentagon in Washington, D.C., is shaped like a regular pentagon. Find the measure of each interior angle.

17. You are designing a binocular mount. If $\overline{BC}$ is always vertical, the binoculars will point in the same direction while they are raised and lowered for different viewers. How can you design the mount so $\overline{BC}$ is always vertical? Justify your answer.

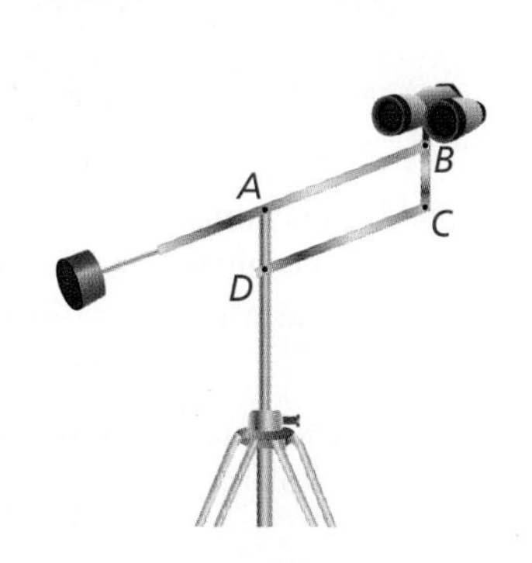

18. The measure of one angle of a kite is 90°. The measure of another angle in the kite is 30°. Sketch a kite that matches this description.

7 Cumulative Assessment

1. Copy and complete the flowchart proof of the Parallelogram Opposite Angles Theorem (Thm. 7.4).

Given *ABCD* is a parallelogram.

Prove $\angle A \cong \angle C$, $\angle B \cong \angle D$

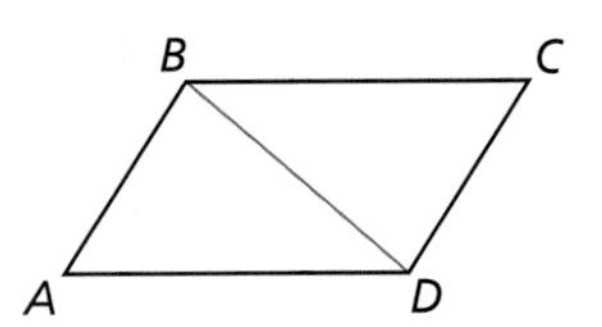

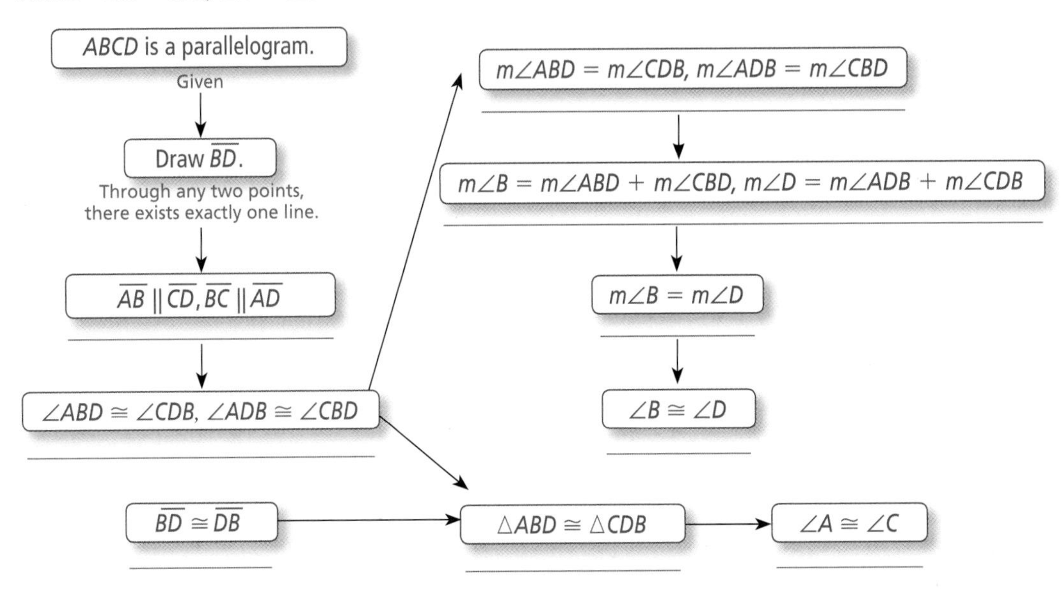

2. Use the steps in the construction to explain how you know that the circle is inscribed within $\triangle ABC$.

Step 1

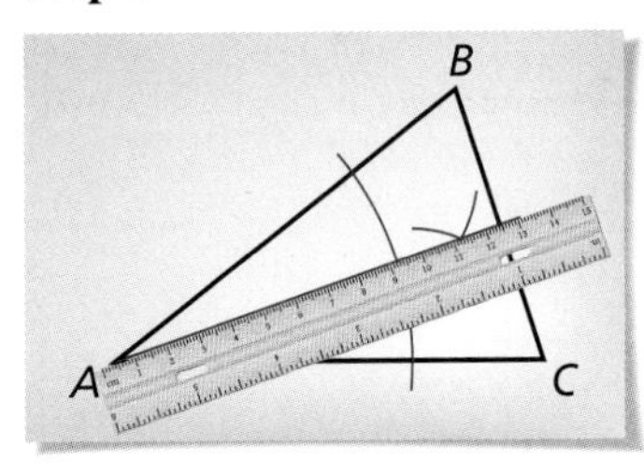

Step 2

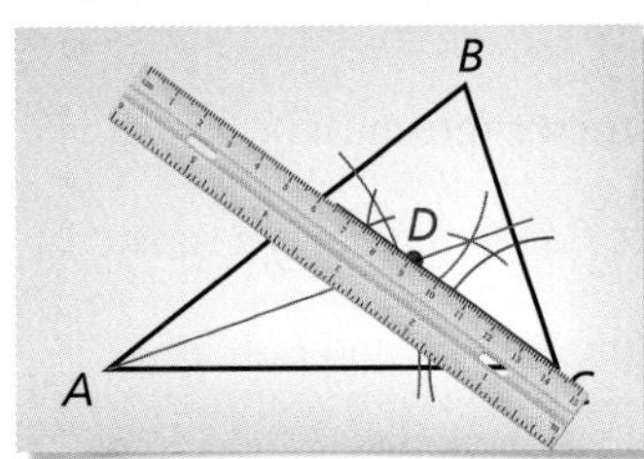

Step 3

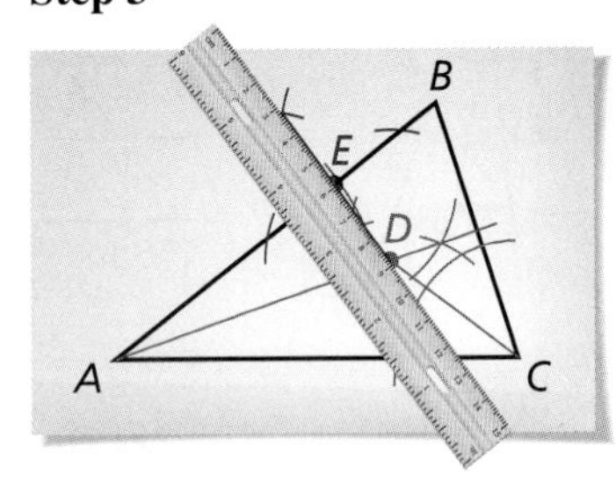

Step 4

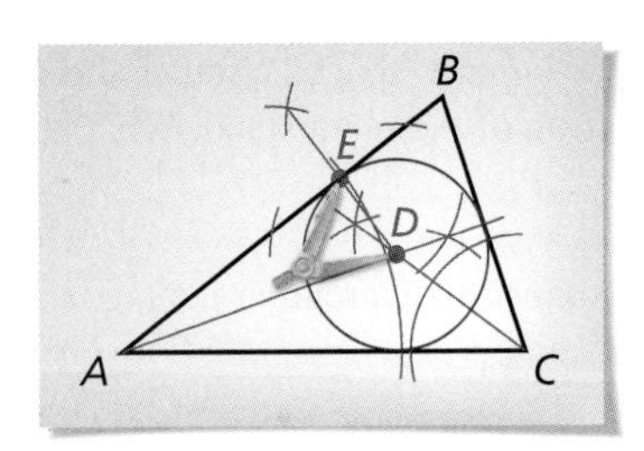

3. Your friend claims that he can prove the Parallelogram Opposite Sides Theorem (Thm. 7.3) using the SSS Congruence Theorem (Thm. 5.8) and the Parallelogram Opposite Sides Theorem (Thm. 7.3). Is your friend correct? Explain your reasoning.

4. Find the perimeter of polygon *QRSTUV*. Is the polygon equilateral? equiangular? regular? Explain your reasoning.

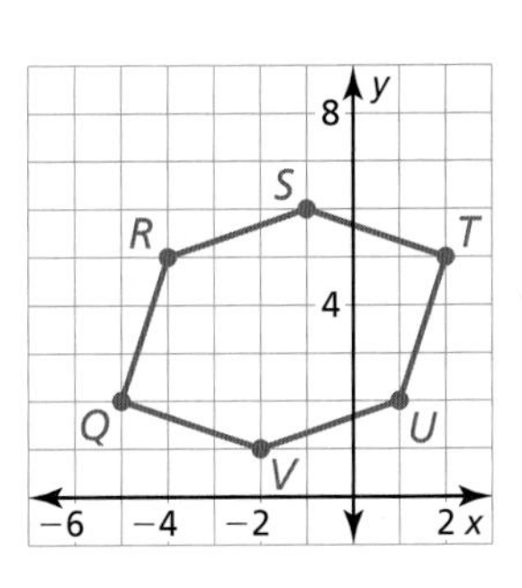

5. Choose the correct symbols to complete the proof of the Converse of the Hinge Theorem (Theorem 6.13).

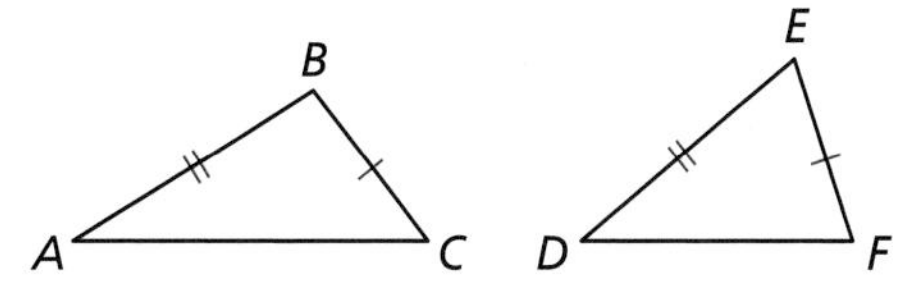

Given $\overline{AB} \cong \overline{DE}, \overline{BC} \cong \overline{EF}, AC > DF$

Prove $m\angle B > m\angle E$

Indirect Proof

Step 1 Assume temporarily that $m\angle B \ngtr m\angle E$. Then it follows that either $m\angle B$ ___ $m\angle E$ or $m\angle B$ ___ $m\angle E$.

Step 2 If $m\angle B$ ___ $m\angle E$, then AC ___ DF by the Hinge Theorem (Theorem 6.12). If $m\angle B$ ___ $m\angle E$, then $\angle B$ ___ $\angle E$. So, $\triangle ABC$ ___ $\triangle DEF$ by the SAS Congruence Theorem (Theorem 5.5) and AC ___ DF.

Step 3 Both conclusions contradict the given statement that AC ___ DF. So, the temporary assumption that $m\angle B \ngtr m\angle E$ cannot be true. This proves that $m\angle B$ ___ $m\angle E$.

6. Use the Isosceles Trapezoid Base Angles Converse (Thm. 7.15) to prove that $ABCD$ is an isosceles trapezoid.

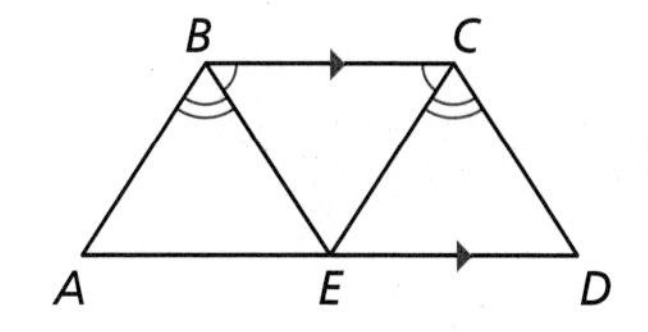

Given $\overline{BC} \parallel \overline{AD}$, $\angle EBC \cong \angle ECB$, $\angle ABE \cong \angle DCE$

Prove $ABCD$ is an isosceles trapezoid.

7. One part of the Rectangle Diagonals Theorem (Thm. 7.13) says, "If the diagonals of a parallelogram are congruent, then it is a rectangle." Using the reasons given, there are multiple ways to prove this part of the theorem. Provide a statement for each reason to form one possible proof of this part of the theorem.

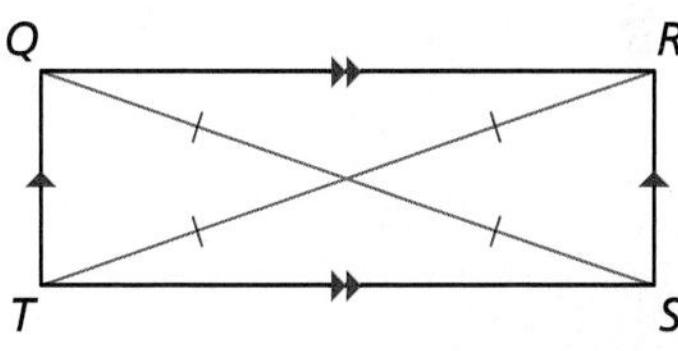

Given $QRST$ is a parallelogram.
$\overline{QS} \cong \overline{RT}$

Prove $QRST$ is a rectangle.

STATEMENTS	REASONS
1. $\overline{QS} \cong \overline{RT}$	**1.** Given
2. ______________	**2.** Parallelogram Opposite Sides Theorem (Thm. 7.3)
3. ______________	**3.** SSS Congruence Theorem (Thm. 5.8)
4. ______________	**4.** Corresponding parts of congruent triangles are congruent.
5. ______________	**5.** Parallelogram Consecutive Angles Theorem (Thm. 7.5)
6. ______________	**6.** Congruent supplementary angles have the same measure.
7. ______________	**7.** Parallelogram Consecutive Angles Theorem (Thm. 7.5)
8. ______________	**8.** Subtraction Property of Equality
9. ______________	**9.** Definition of a right angle
10. ______________	**10.** Definition of a rectangle

Maintaining Mathematical Proficiency

Determining Whether Ratios Form a Proportion

Example 1 **Tell whether $\frac{2}{8}$ and $\frac{3}{12}$ form a proportion.**

Compare the ratios in simplest form.

$$\frac{2}{8} = \frac{2 \div 2}{8 \div 2} = \frac{1}{4}$$

$$\frac{3}{12} = \frac{3 \div 3}{12 \div 3} = \frac{1}{4}$$

The ratios are equivalent.

▶ So, $\frac{2}{8}$ and $\frac{3}{12}$ form a proportion.

Tell whether the ratios form a proportion.

1. $\frac{5}{3}, \frac{35}{21}$

2. $\frac{9}{24}, \frac{24}{64}$

3. $\frac{8}{56}, \frac{6}{28}$

4. $\frac{18}{4}, \frac{27}{9}$

5. $\frac{15}{21}, \frac{55}{77}$

6. $\frac{26}{8}, \frac{39}{12}$

Finding a Scale Factor

Example 2 **Find the scale factor of each dilation.**

a.

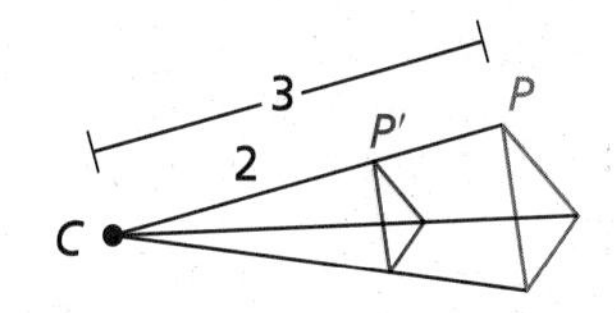

▶ Because $\frac{CP'}{CP} = \frac{2}{3}$, the scale factor is $k = \frac{2}{3}$.

b.

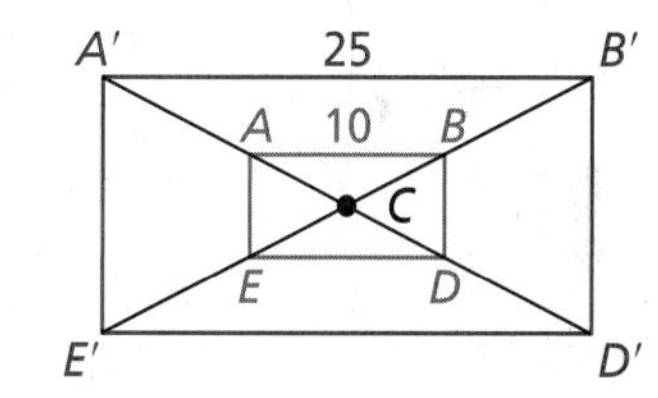

▶ Because $\frac{A'B'}{AB} = \frac{25}{10}$, the scale factor is $k = \frac{25}{10} = \frac{5}{2}$.

Find the scale factor of the dilation.

7.

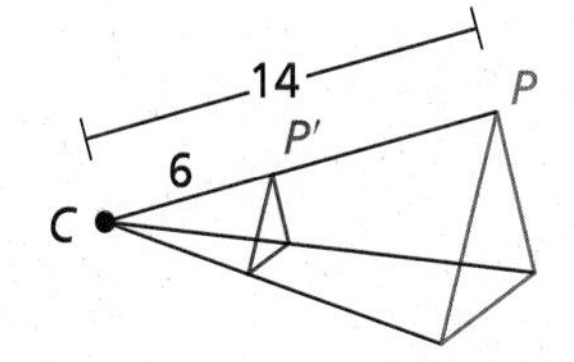

8.

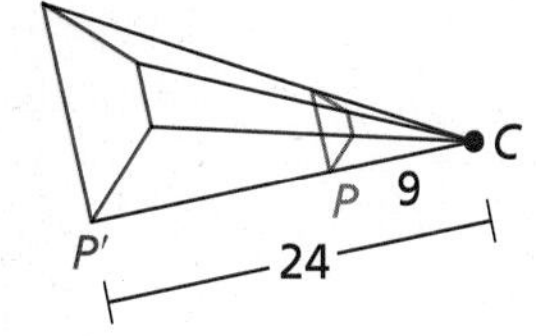

9.

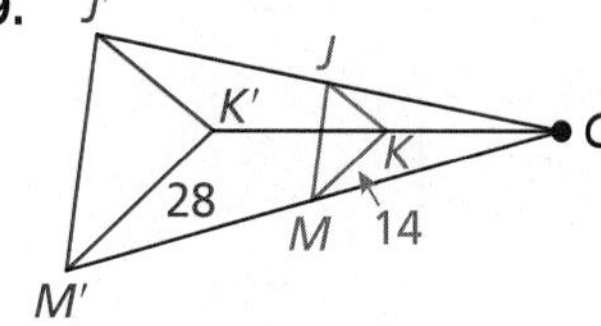

10. ABSTRACT REASONING If ratio X and ratio Y form a proportion and ratio Y and ratio Z form a proportion, do ratio X and ratio Z form a proportion? Explain your reasoning.

Mathematical Practices

Mathematically proficient students look for and make use of a pattern or structure.

Discerning a Pattern or Structure

Core Concept

Dilations, Perimeter, Area, and Volume

Consider a figure that is dilated by a scale factor of k.

1. The perimeter of the image is k times the perimeter of the original figure.
2. The area of the image is k^2 times the area of the original figure.
3. If the original figure is three dimensional, then the volume of the image is k^3 times the volume of the original figure.

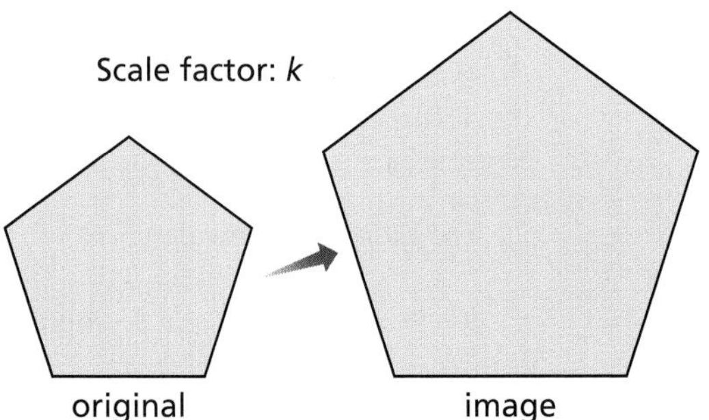

EXAMPLE 1 Finding Perimeter and Area after a Dilation

The triangle shown has side lengths of 3 inches, 4 inches, and 5 inches. Find the perimeter and area of the image when the triangle is dilated by a scale factor of (a) 2, (b) 3, and (c) 4.

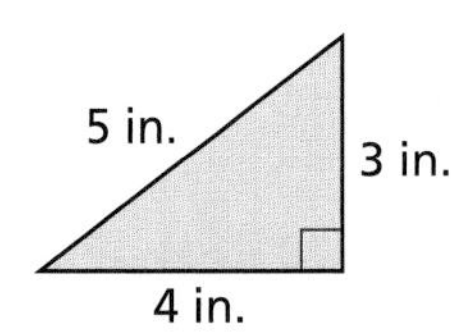

SOLUTION

Perimeter: $P = 5 + 3 + 4 = 12$ in. Area: $A = \frac{1}{2}(4)(3) = 6$ in.^{2}

	Scale factor: k	Perimeter: kP	Area: k^2A
a.	2	$2(12) = 24$ in.	$(2^2)(6) = 24$ in.^{2}
b.	3	$3(12) = 36$ in.	$(3^2)(6) = 54$ in.^{2}
c.	4	$4(12) = 48$ in.	$(4^2)(6) = 96$ in.^{2}

Monitoring Progress

1. Find the perimeter and area of the image when the trapezoid is dilated by a scale factor of (a) 2, (b) 3, and (c) 4.

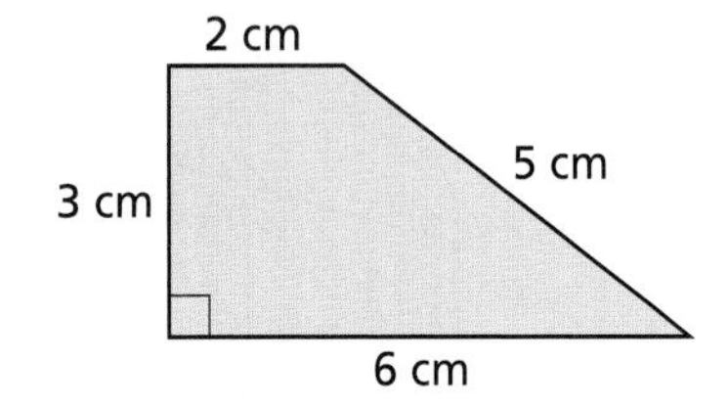

2. Find the perimeter and area of the image when the parallelogram is dilated by a scale factor of (a) 2, (b) 3, and (c) $\frac{1}{2}$.

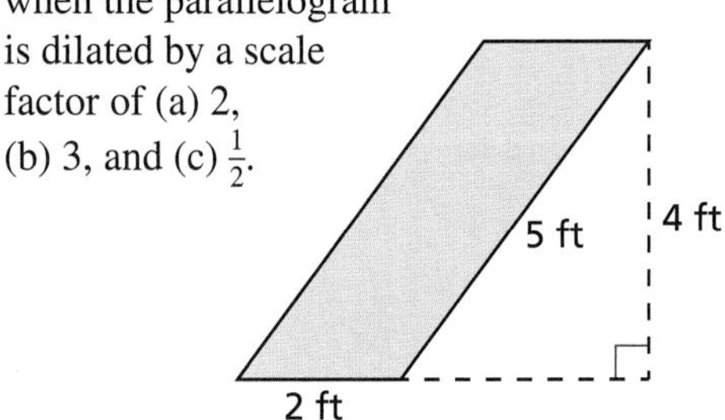

3. A rectangular prism is 3 inches wide, 4 inches long, and 5 inches tall. Find the surface area and volume of the image of the prism when it is dilated by a scale factor of (a) 2, (b) 3, and (c) 4.

8.1 Similar Polygons

Essential Question How are similar polygons related?

EXPLORATION 1 **Comparing Triangles after a Dilation**

Work with a partner. Use dynamic geometry software to draw any $\triangle ABC$. Dilate $\triangle ABC$ to form a similar $\triangle A'B'C'$ using any scale factor k and any center of dilation.

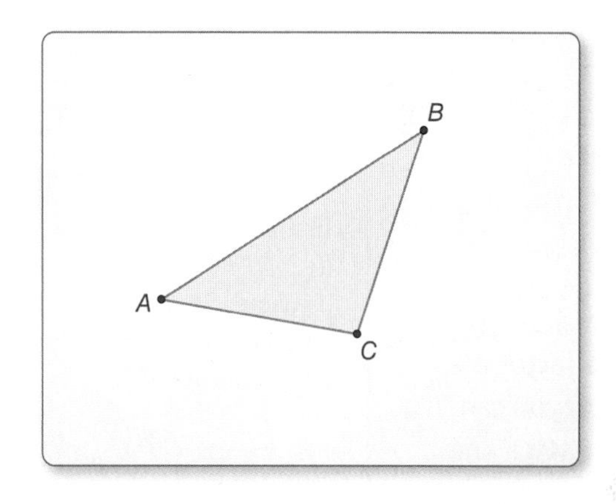

a. Compare the corresponding angles of $\triangle A'B'C'$ and $\triangle ABC$.

b. Find the ratios of the lengths of the sides of $\triangle A'B'C'$ to the lengths of the corresponding sides of $\triangle ABC$. What do you observe?

c. Repeat parts (a) and (b) for several other triangles, scale factors, and centers of dilation. Do you obtain similar results?

EXPLORATION 2 **Comparing Triangles after a Dilation**

Work with a partner. Use dynamic geometry software to draw any $\triangle ABC$. Dilate $\triangle ABC$ to form a similar $\triangle A'B'C'$ using any scale factor k and any center of dilation.

B

A

C

a. Compare the perimeters of $\triangle A'B'C'$ and $\triangle ABC$. What do you observe?

b. Compare the areas of $\triangle A'B'C'$ and $\triangle ABC$. What do you observe?

c. Repeat parts (a) and (b) for several other triangles, scale factors, and centers of dilation. Do you obtain similar results?

LOOKING FOR STRUCTURE

To be proficient in math, you need to look closely to discern a pattern or structure.

Communicate Your Answer

3. How are similar polygons related?

4. A $\triangle RST$ is dilated by a scale factor of 3 to form $\triangle R'S'T'$. The area of $\triangle RST$ is 1 square inch. What is the area of $\triangle R'S'T'$?

8.1 Lesson

What You Will Learn

- Use similarity statements.
- Find corresponding lengths in similar polygons.
- Find perimeters and areas of similar polygons.
- Decide whether polygons are similar.

Core Vocabulary

Previous
similar figures
similarity transformation
corresponding parts

Using Similarity Statements

Recall from Section 4.6 that two geometric figures are similar figures if and only if there is a similarity transformation that maps one figure onto the other.

LOOKING FOR STRUCTURE

Notice that any two congruent figures are also similar. In $\triangle LMN$ and $\triangle WXY$ below, the scale factor is $\frac{5}{5} = \frac{6}{6} = \frac{7}{7} = 1$. So, you can write $\triangle LMN \sim \triangle WXY$ and $\triangle LMN \cong \triangle WXY$.

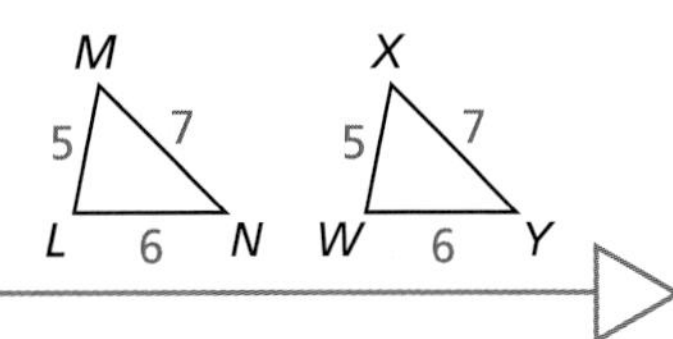

Core Concept

Corresponding Parts of Similar Polygons

In the diagram below, $\triangle ABC$ is similar to $\triangle DEF$. You can write "$\triangle ABC$ is similar to $\triangle DEF$" as $\triangle ABC \sim \triangle DEF$. A similarity transformation preserves angle measure. So, corresponding angles are congruent. A similarity transformation also enlarges or reduces side lengths by a scale factor k. So, corresponding side lengths are proportional.

Corresponding angles

$\angle A \cong \angle D, \angle B \cong \angle E, \angle C \cong \angle F$

Ratios of corresponding side lengths

$\frac{DE}{AB} = \frac{EF}{BC} = \frac{FD}{CA} = k$

EXAMPLE 1 Using Similarity Statements

READING

In a *statement of proportionality*, any pair of ratios forms a true proportion.

In the diagram, $\triangle RST \sim \triangle XYZ$.

a. Find the scale factor from $\triangle RST$ to $\triangle XYZ$.

b. List all pairs of congruent angles.

c. Write the ratios of the corresponding side lengths in a *statement of proportionality*.

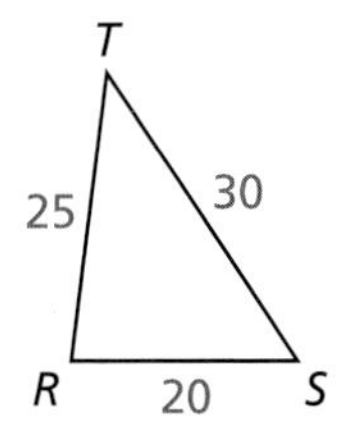

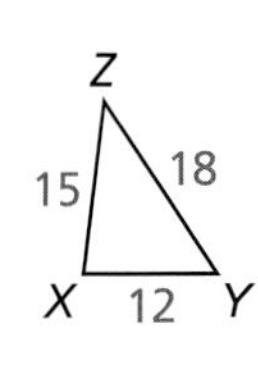

SOLUTION

a. $\frac{XY}{RS} = \frac{12}{20} = \frac{3}{5}$ $\quad \frac{YZ}{ST} = \frac{18}{30} = \frac{3}{5}$ $\quad \frac{ZX}{TR} = \frac{15}{25} = \frac{3}{5}$

So, the scale factor is $\frac{3}{5}$.

b. $\angle R \cong \angle X$, $\angle S \cong \angle Y$, and $\angle T \cong \angle Z$.

c. Because the ratios in part (a) are equal, $\frac{XY}{RS} = \frac{YZ}{ST} = \frac{ZX}{TR}$.

Monitoring Progress  Help in English and Spanish at *BigIdeasMath.com*

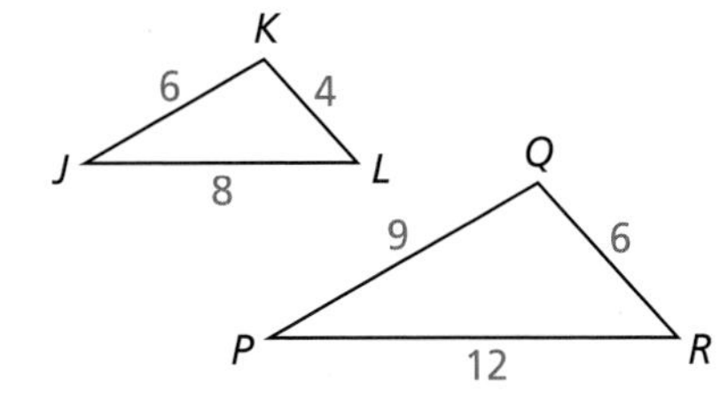

1. In the diagram, $\triangle JKL \sim \triangle PQR$. Find the scale factor from $\triangle JKL$ to $\triangle PQR$. Then list all pairs of congruent angles and write the ratios of the corresponding side lengths in a statement of proportionality.

Finding Corresponding Lengths in Similar Polygons

Core Concept

Corresponding Lengths in Similar Polygons

If two polygons are similar, then the ratio of any two corresponding lengths in the polygons is equal to the scale factor of the similar polygons.

READING

Corresponding lengths in similar triangles include side lengths, altitudes, medians, and midsegments.

EXAMPLE 2 Finding a Corresponding Length

In the diagram, $\triangle DEF \sim \triangle MNP$. Find the value of x.

SOLUTION

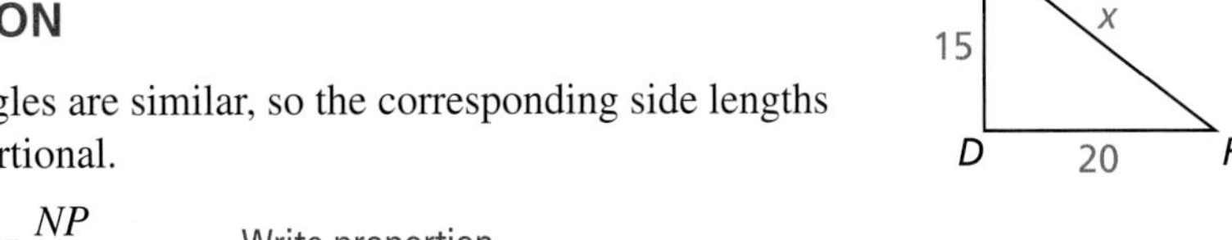

The triangles are similar, so the corresponding side lengths are proportional.

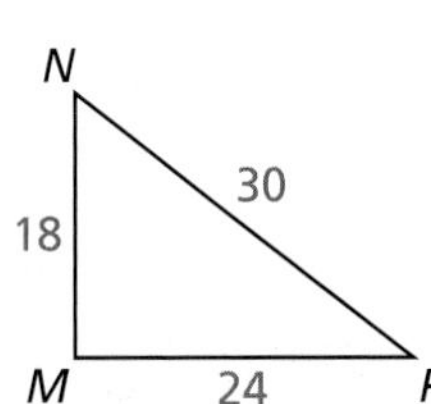

$\frac{MN}{DE} = \frac{NP}{EF}$ Write proportion.

$\frac{18}{15} = \frac{30}{x}$ Substitute.

$18x = 450$ Cross Products Property

$x = 25$ Solve for x.

▶ The value of x is 25.

FINDING AN ENTRY POINT

There are several ways to write the proportion. For example, you could write $\frac{DF}{MP} = \frac{EF}{NP}$.

EXAMPLE 3 Finding a Corresponding Length

In the diagram, $\triangle TPR \sim \triangle XPZ$. Find the length of the altitude $\overline{PS}$.

SOLUTION

First, find the scale factor from $\triangle XPZ$ to $\triangle TPR$.

$$\frac{TR}{XZ} = \frac{6+6}{8+8} = \frac{12}{16} = \frac{3}{4}$$

Because the ratio of the lengths of the altitudes in similar triangles is equal to the scale factor, you can write the following proportion.

$\frac{PS}{PY} = \frac{3}{4}$ Write proportion.

$\frac{PS}{20} = \frac{3}{4}$ Substitute 20 for PY.

$PS = 15$ Multiply each side by 20 and simplify.

▶ The length of the altitude $\overline{PS}$ is 15.

Monitoring Progress

Help in English and Spanish at *BigIdeasMath.com*

2. Find the value of x.

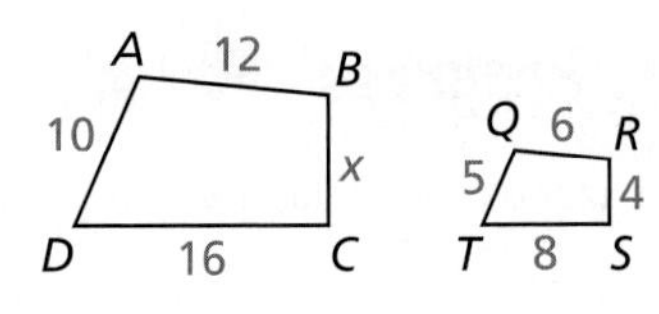

$ABCD \sim QRST$

3. Find KM.

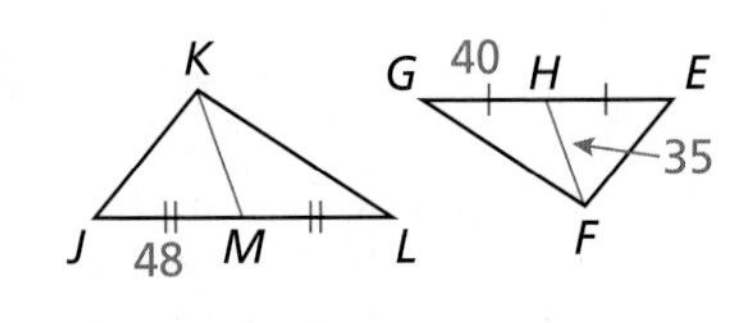

$\triangle JKL \sim \triangle EFG$

Finding Perimeters and Areas of Similar Polygons

Theorem

Theorem 8.1 Perimeters of Similar Polygons

If two polygons are similar, then the ratio of their perimeters is equal to the ratios of their corresponding side lengths.

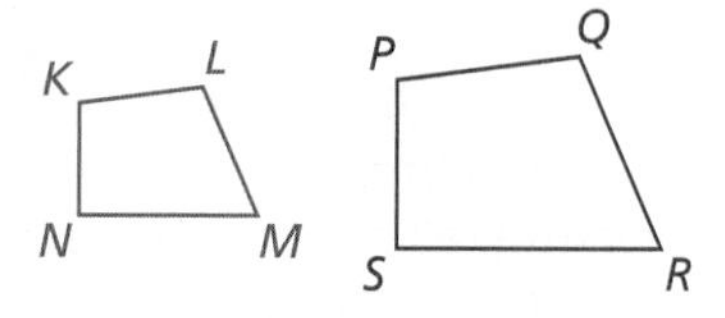

If $KLMN \sim PQRS$, then $\dfrac{PQ + QR + RS + SP}{KL + LM + MN + NK} = \dfrac{PQ}{KL} = \dfrac{QR}{LM} = \dfrac{RS}{MN} = \dfrac{SP}{NK}$.

Proof Ex. 52, p. 426; *BigIdeasMath.com*

ANALYZING RELATIONSHIPS

When two similar polygons have a scale factor of k, the ratio of their perimeters is equal to k.

EXAMPLE 4 Modeling with Mathematics

A town plans to build a new swimming pool. An Olympic pool is rectangular with a length of 50 meters and a width of 25 meters. The new pool will be similar in shape to an Olympic pool but will have a length of 40 meters. Find the perimeters of an Olympic pool and the new pool.

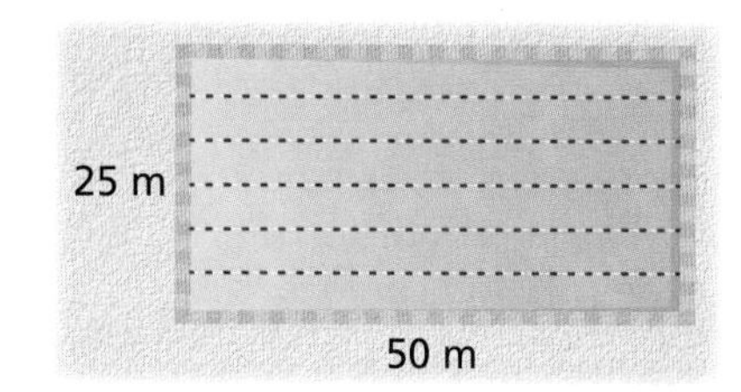

SOLUTION

1. **Understand the Problem** You are given the length and width of a rectangle and the length of a similar rectangle. You need to find the perimeters of both rectangles.

2. **Make a Plan** Find the scale factor of the similar rectangles and find the perimeter of an Olympic pool. Then use the Perimeters of Similar Polygons Theorem to write and solve a proportion to find the perimeter of the new pool.

3. **Solve the Problem** Because the new pool will be similar to an Olympic pool, the scale factor is the ratio of the lengths, $\frac{40}{50} = \frac{4}{5}$. The perimeter of an Olympic pool is $2(50) + 2(25) = 150$ meters. Write and solve a proportion to find the perimeter x of the new pool.

$\dfrac{x}{150} = \dfrac{4}{5}$ Perimeters of Similar Polygons Theorem

$x = 120$ Multiply each side by 150 and simplify.

So, the perimeter of an Olympic pool is 150 meters, and the perimeter of the new pool is 120 meters.

4. **Look Back** Check that the ratio of the perimeters is equal to the scale factor.

$\dfrac{120}{150} = \dfrac{4}{5}$ ✓

STUDY TIP

You can also write the scale factor as a decimal. In Example 4, you can write the scale factor as 0.8 and multiply by 150 to get $x = 0.8(150) = 120$.

Monitoring Progress  Help in English and Spanish at *BigIdeasMath.com*

4. The two gazebos shown are similar pentagons. Find the perimeter of Gazebo A.

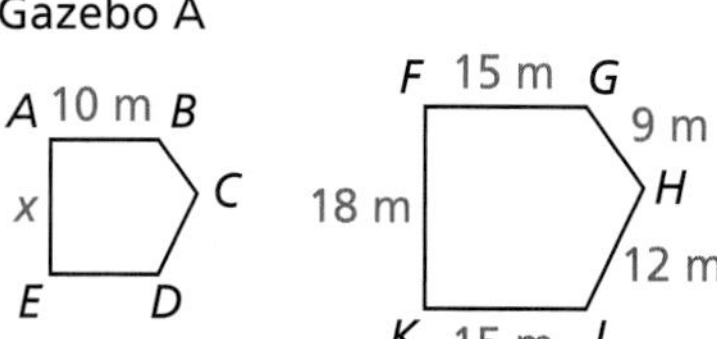

ANALYZING RELATIONSHIPS

When two similar polygons have a scale factor of k, the ratio of their areas is equal to k^2.

Theorem 8.2 Areas of Similar Polygons

If two polygons are similar, then the ratio of their areas is equal to the squares of the ratios of their corresponding side lengths.

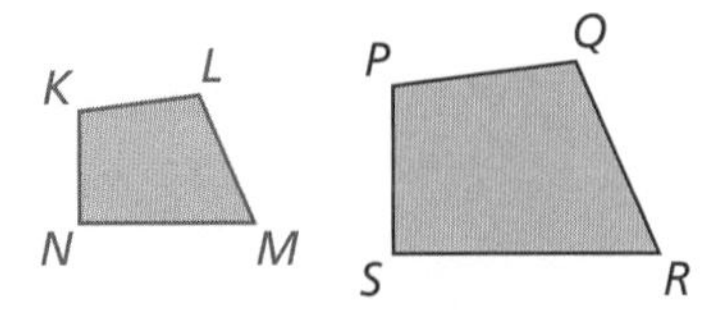

If $KLMN \sim PQRS$, then $\dfrac{\text{Area of } PQRS}{\text{Area of } KLMN} = \left(\dfrac{PQ}{KL}\right)^2 = \left(\dfrac{QR}{LM}\right)^2 = \left(\dfrac{RS}{MN}\right)^2 = \left(\dfrac{SP}{NK}\right)^2$.

Proof Ex. 53, p. 426; *BigIdeasMath.com*

EXAMPLE 5 Finding Areas of Similar Polygons

In the diagram, $\triangle ABC \sim \triangle DEF$. Find the area of $\triangle DEF$.

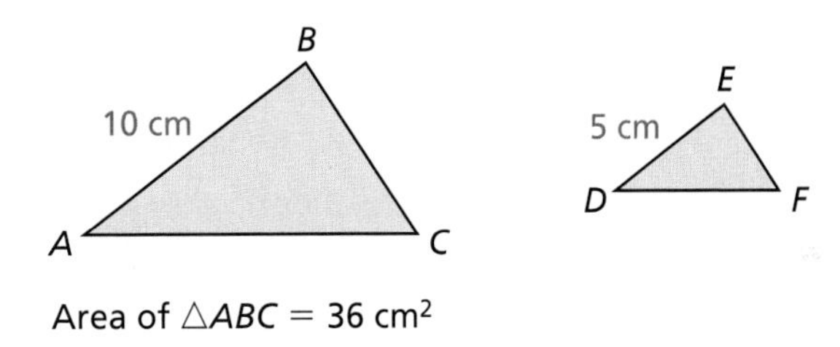

Area of $\triangle ABC = 36$ cm^2

SOLUTION

Because the triangles are similar, the ratio of the area of $\triangle ABC$ to the area of $\triangle DEF$ is equal to the square of the ratio of AB to DE. Write and solve a proportion to find the area of $\triangle DEF$. Let A represent the area of $\triangle DEF$.

$\dfrac{\text{Area of } \triangle ABC}{\text{Area of } \triangle DEF} = \left(\dfrac{AB}{DE}\right)^2$	Areas of Similar Polygons Theorem
$\dfrac{36}{A} = \left(\dfrac{10}{5}\right)^2$	Substitute.
$\dfrac{36}{A} = \dfrac{100}{25}$	Square the right side of the equation.
$36 \cdot 25 = 100 \cdot A$	Cross Products Property
$900 = 100A$	Simplify.
$9 = A$	Solve for A.

▶ The area of $\triangle DEF$ is 9 square centimeters.

Monitoring Progress

Help in English and Spanish at *BigIdeasMath.com*

5. In the diagram, $GHJK \sim LMNP$. Find the area of $LMNP$.

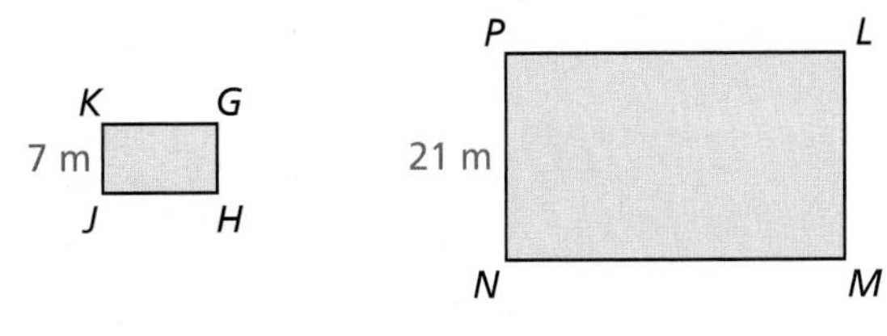

Area of $GHJK = 84$ m^2

Deciding Whether Polygons Are Similar

EXAMPLE 6 Deciding Whether Polygons Are Similar

Decide whether $ABCDE$ and $KLQRP$ are similar. Explain your reasoning.

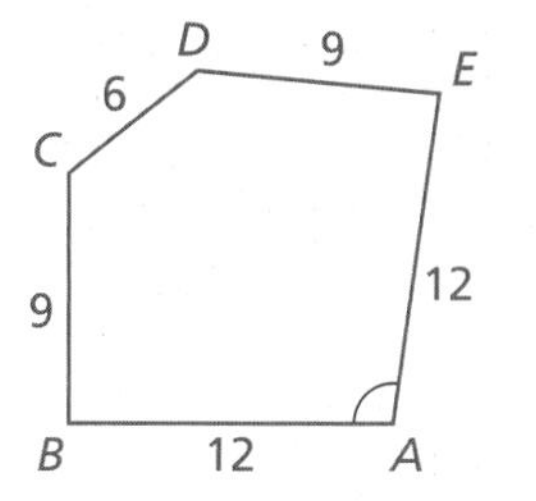

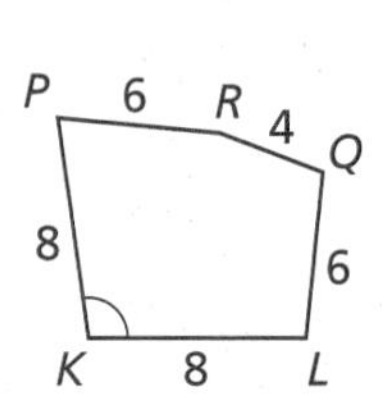

SOLUTION

Corresponding sides of the pentagons are proportional with a scale factor of $\frac{2}{3}$. However, this does not necessarily mean the pentagons are similar. A dilation with center A and scale factor $\frac{2}{3}$ moves $ABCDE$ onto $AFGHJ$. Then a reflection moves $AFGHJ$ onto $KLMNP$.

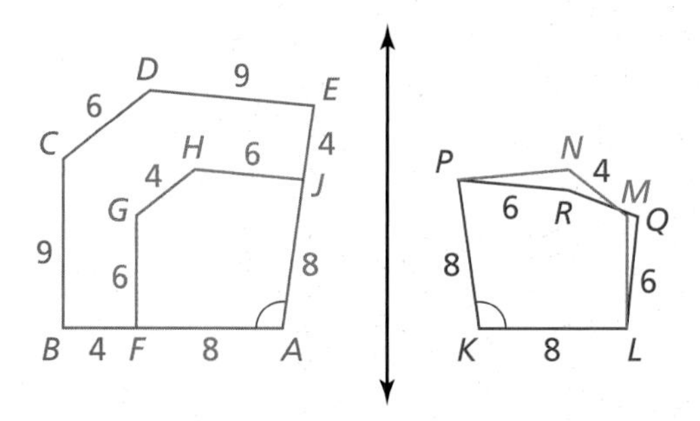

$KLMNP$ does not exactly coincide with $KLQRP$, because not all the corresponding angles are congruent. (Only $\angle A$ and $\angle K$ are congruent.)

▶ Because angle measure is not preserved, the two pentagons are not similar.

Monitoring Progress  Help in English and Spanish at *BigIdeasMath.com*

Refer to the floor tile designs below. In each design, the red shape is a regular hexagon.

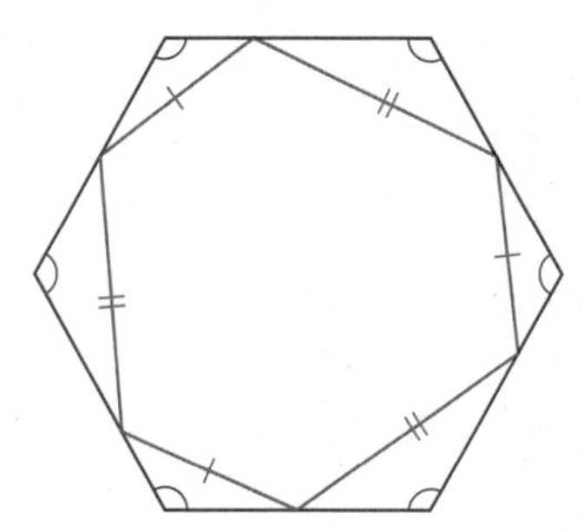

Tile Design 1

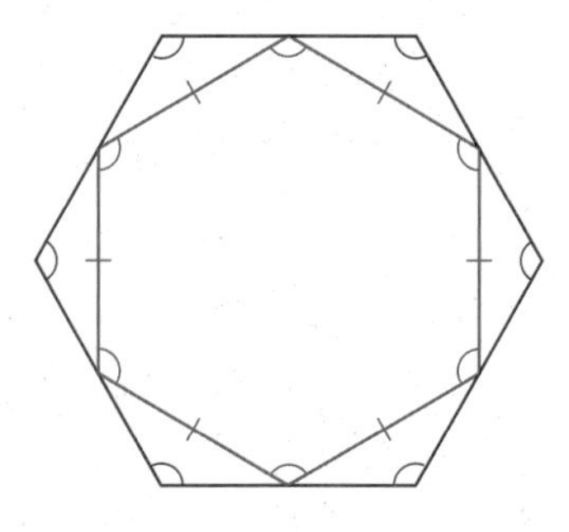

Tile Design 2

6. Decide whether the hexagons in Tile Design 1 are similar. Explain.

7. Decide whether the hexagons in Tile Design 2 are similar. Explain.

8.1 Exercises

Dynamic Solutions available at *BigIdeasMath.com*

Vocabulary and Core Concept Check

1. **COMPLETE THE SENTENCE** For two figures to be similar, the corresponding angles must be ____________, and the corresponding side lengths must be ________________.

2. **DIFFERENT WORDS, SAME QUESTION** Which is different? Find "both" answers.

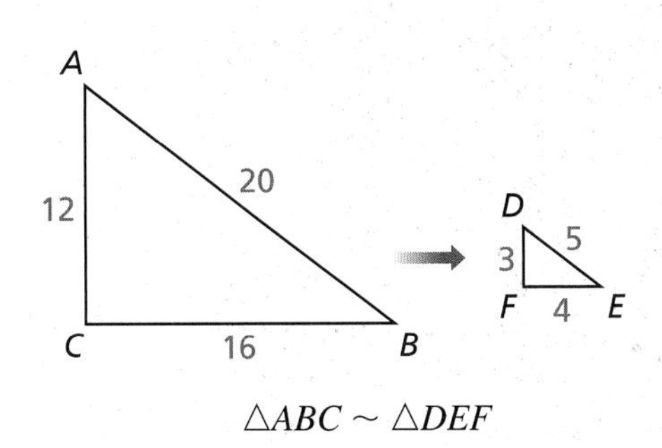

$\triangle ABC \sim \triangle DEF$

What is the scale factor?

What is the ratio of their areas?

What is the ratio of their corresponding side lengths?

What is the ratio of their perimeters?

Monitoring Progress and Modeling with Mathematics

In Exercises 3 and 4, find the scale factor. Then list all pairs of congruent angles and write the ratios of the corresponding side lengths in a statement of proportionality. *(See Example 1.)*

3. $\triangle ABC \sim \triangle LMN$

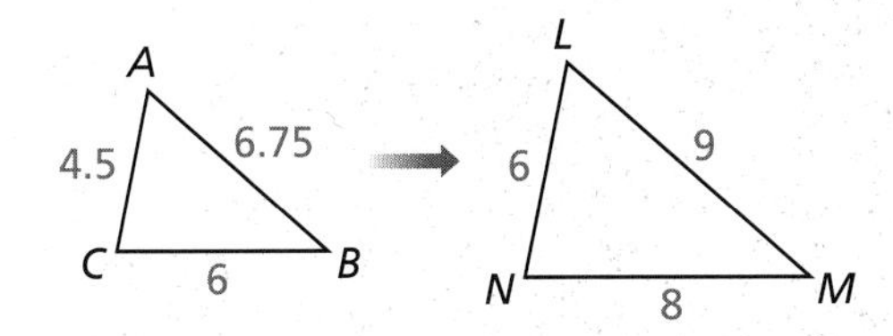

4. $DEFG \sim PQRS$

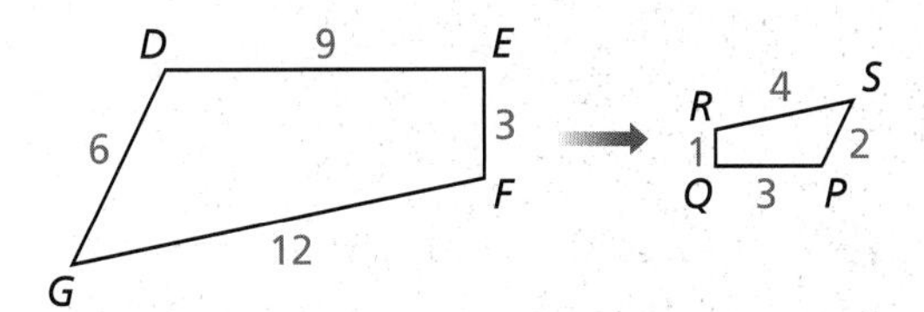

In Exercises 5–8, the polygons are similar. Find the value of x. *(See Example 2.)*

5.

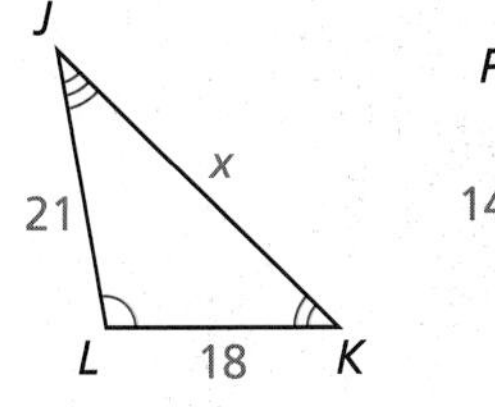

6.

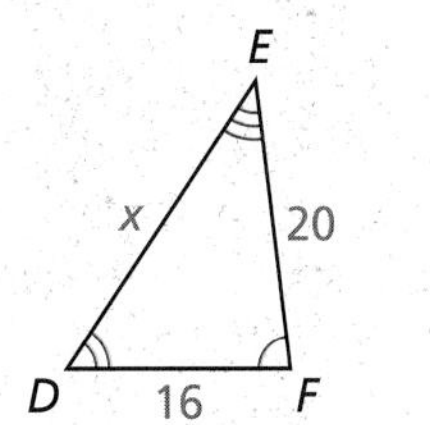

7.

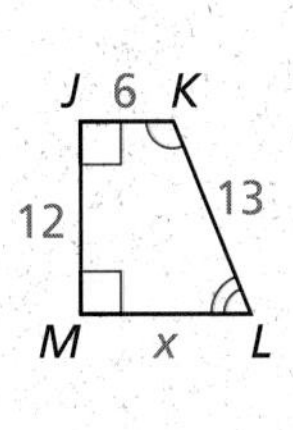

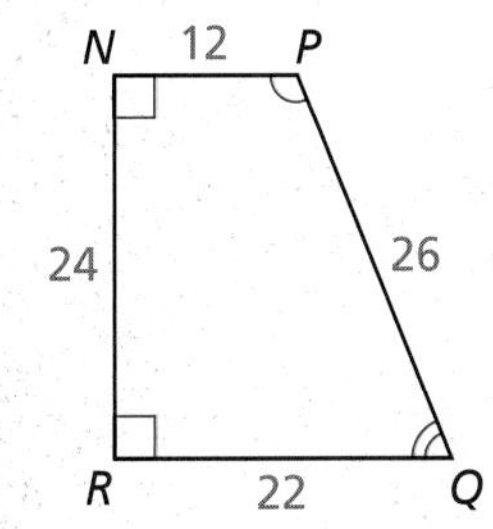

8.

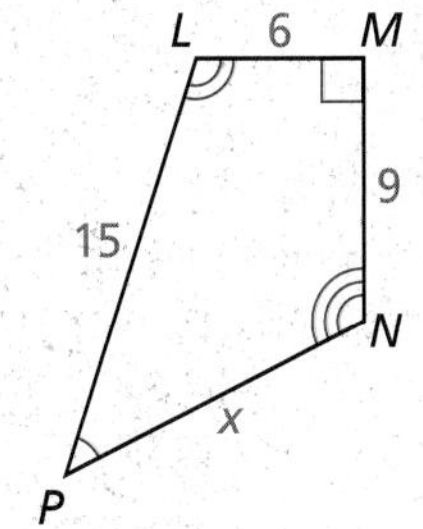

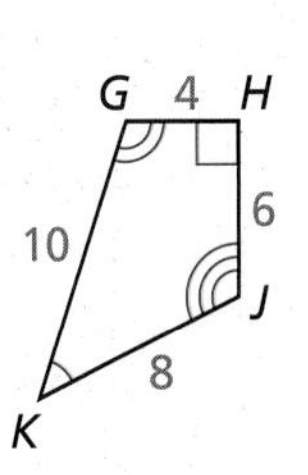

In Exercises 9 and 10, the black triangles are similar. Identify the type of segment shown in blue and find the value of the variable. *(See Example 3.)*

9.

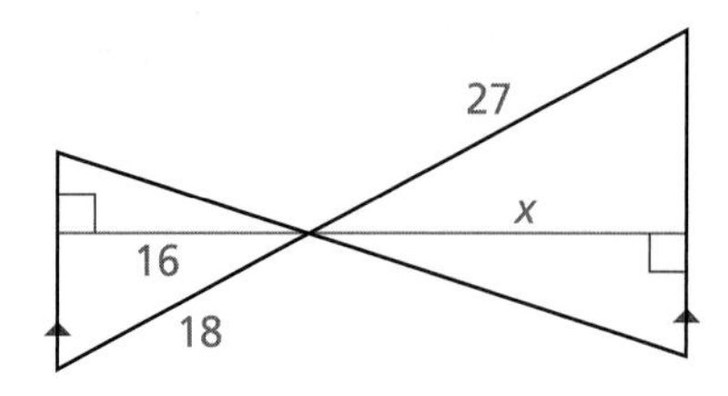

10.

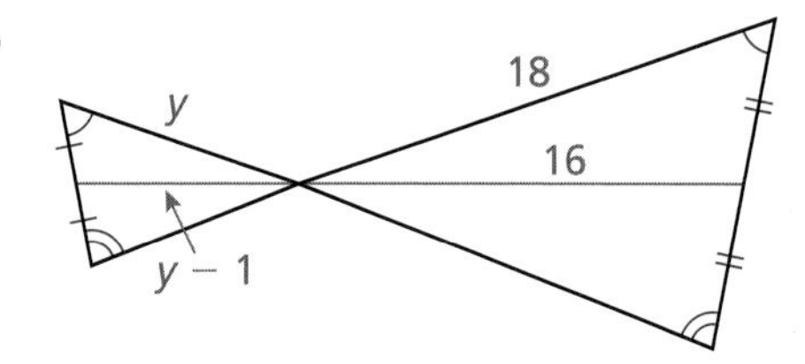

In Exercises 11 and 12, *RSTU* ~ *ABCD*. Find the ratio of their perimeters.

11.

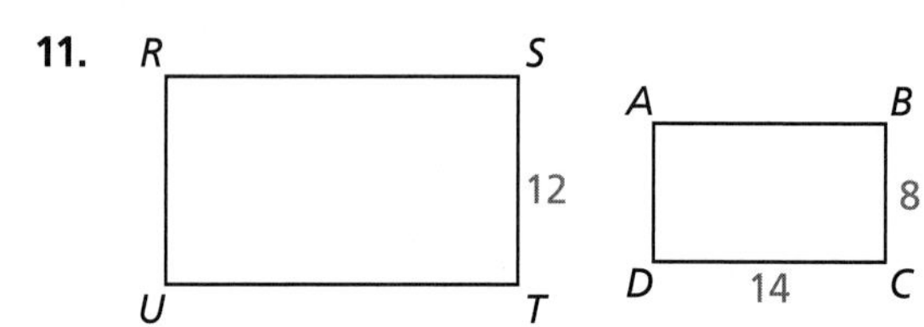

12.

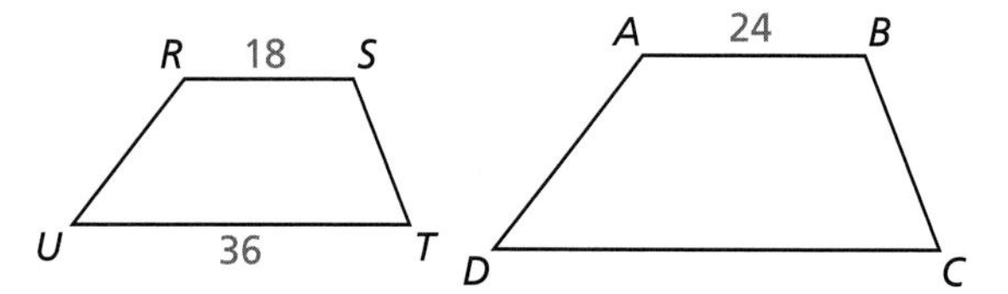

In Exercises 13–16, two polygons are similar. The perimeter of one polygon and the ratio of the corresponding side lengths are given. Find the perimeter of the other polygon.

13. perimeter of smaller polygon: 48 cm; ratio: $\frac{2}{3}$

14. perimeter of smaller polygon: 66 ft; ratio: $\frac{3}{4}$

15. perimeter of larger polygon: 120 yd; ratio: $\frac{1}{6}$

16. perimeter of larger polygon: 85 m; ratio: $\frac{2}{5}$

17. MODELING WITH MATHEMATICS A school gymnasium is being remodeled. The basketball court will be similar to an NCAA basketball court, which has a length of 94 feet and a width of 50 feet. The school plans to make the width of the new court 45 feet. Find the perimeters of an NCAA court and of the new court in the school. *(See Example 4.)*

18. MODELING WITH MATHEMATICS Your family has decided to put a rectangular patio in your backyard, similar to the shape of your backyard. Your backyard has a length of 45 feet and a width of 20 feet. The length of your new patio is 18 feet. Find the perimeters of your backyard and of the patio.

In Exercises 19–22, the polygons are similar. The area of one polygon is given. Find the area of the other polygon. *(See Example 5.)*

19.

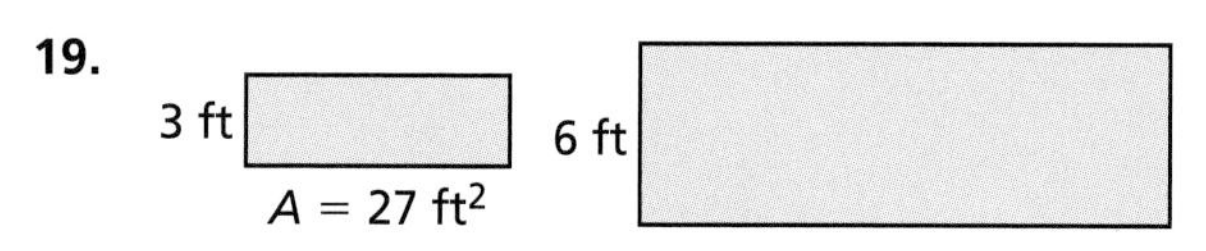

20.

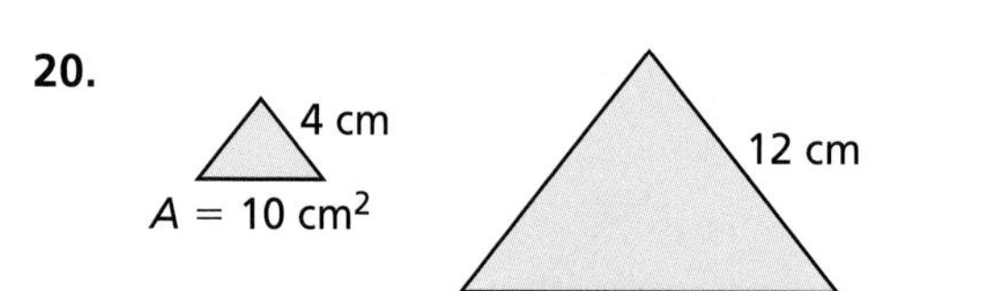

21.

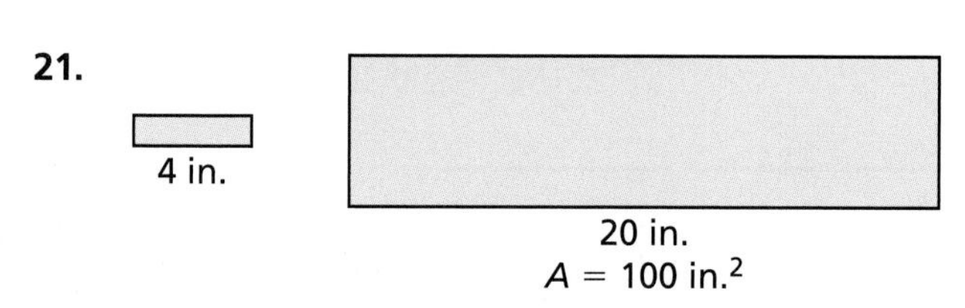

22.

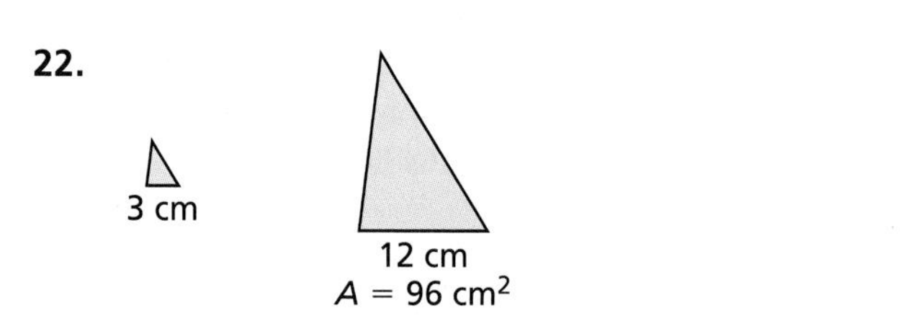

23. ERROR ANALYSIS Describe and correct the error in finding the perimeter of triangle B. The triangles are similar.

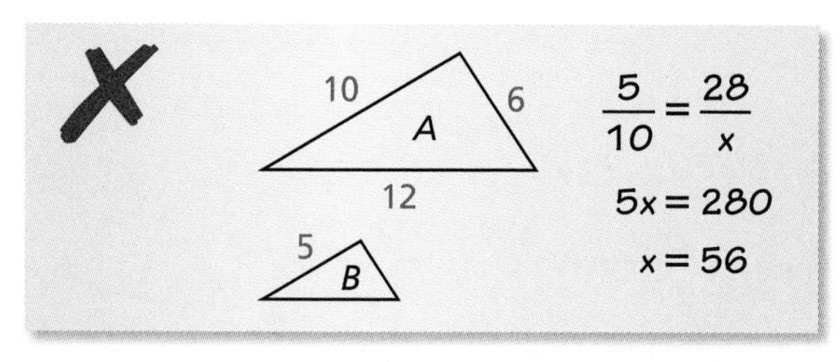

24. ERROR ANALYSIS Describe and correct the error in finding the area of rectangle B. The rectangles are similar.

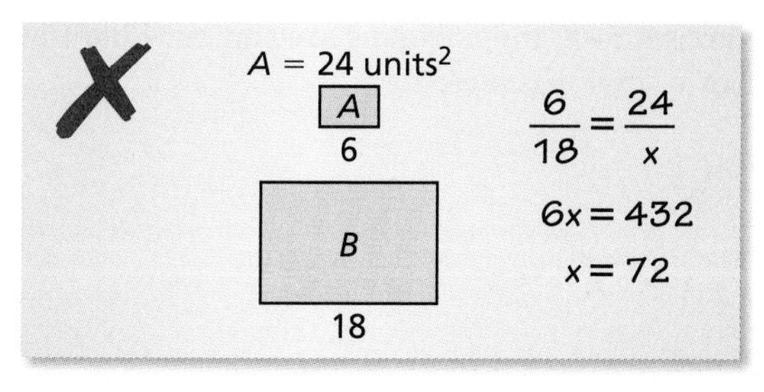

In Exercises 25 and 26, decide whether the red and blue polygons are similar. *(See Example 6.)*

25.

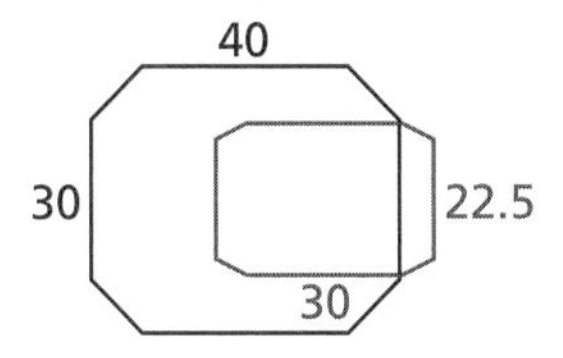

26.

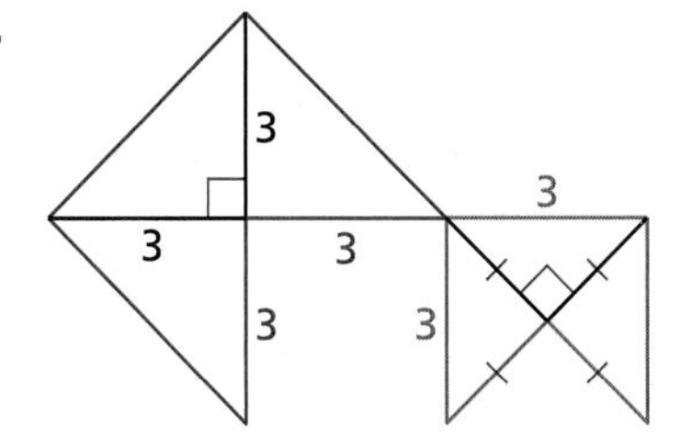

27. REASONING Triangles ABC and DEF are similar. Which statement is correct? Select all that apply.

Ⓐ $\frac{BC}{EF} = \frac{AC}{DF}$ Ⓑ $\frac{AB}{DE} = \frac{CA}{FE}$

Ⓒ $\frac{AB}{EF} = \frac{BC}{DE}$ Ⓓ $\frac{CA}{FD} = \frac{BC}{EF}$

ANALYZING RELATIONSHIPS In Exercises 28–34, *JKLM* ~ *EFGH*.

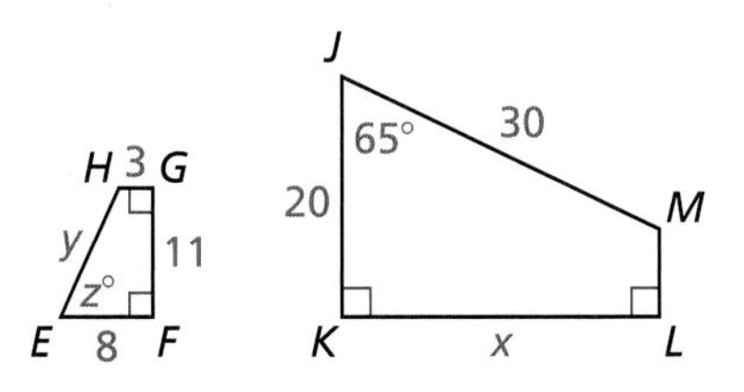

28. Find the scale factor of $JKLM$ to $EFGH$.

29. Find the scale factor of $EFGH$ to $JKLM$.

30. Find the values of x, y, and z.

31. Find the perimeter of each polygon.

32. Find the ratio of the perimeters of $JKLM$ to $EFGH$.

33. Find the area of each polygon.

34. Find the ratio of the areas of $JKLM$ to $EFGH$.

35. USING STRUCTURE Rectangle A is similar to rectangle B. Rectangle A has side lengths of 6 and 12. Rectangle B has a side length of 18. What are the possible values for the length of the other side of rectangle B? Select all that apply.

Ⓐ 6 Ⓑ 9 Ⓒ 24 Ⓓ 36

36. DRAWING CONCLUSIONS In table tennis, the table is a rectangle 9 feet long and 5 feet wide. A tennis court is a rectangle 78 feet long and 36 feet wide. Are the two surfaces similar? Explain. If so, find the scale factor of the tennis court to the table.

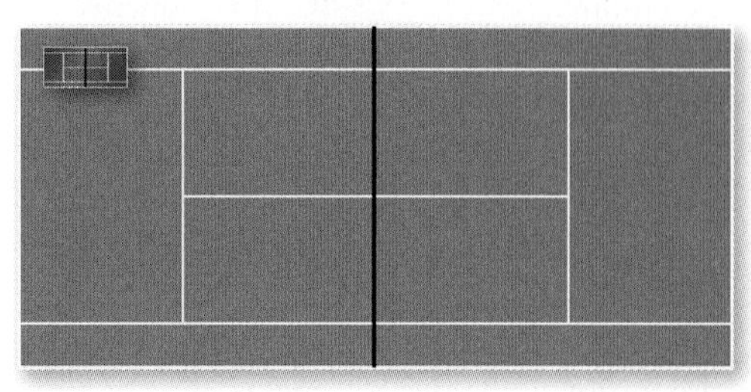

MATHEMATICAL CONNECTIONS In Exercises 37 and 38, the two polygons are similar. Find the values of x and y.

37.

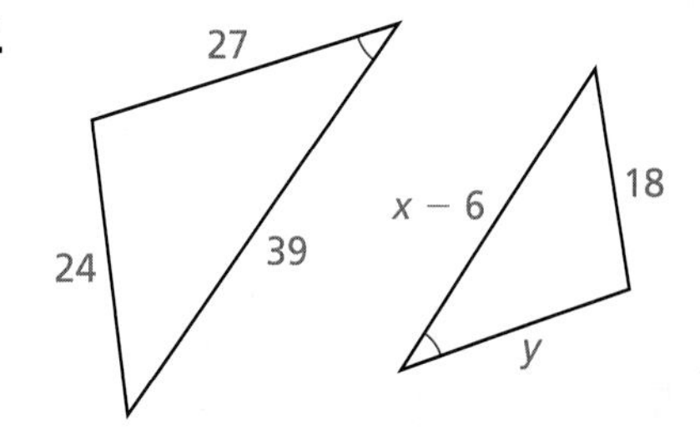

38.

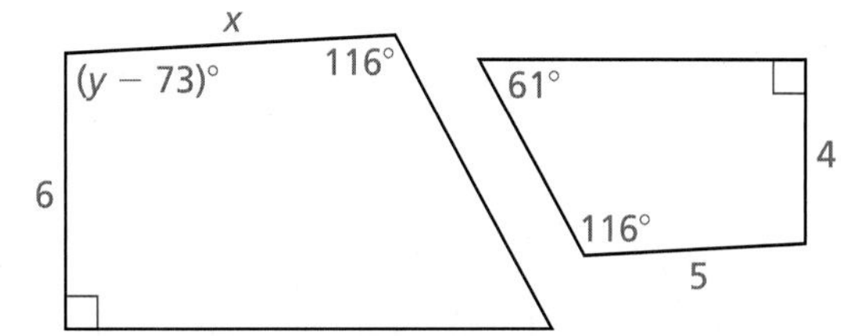

ATTENDING TO PRECISION In Exercises 39–42, the figures are similar. Find the missing corresponding side length.

39. Figure A has a perimeter of 72 meters and one of the side lengths is 18 meters. Figure B has a perimeter of 120 meters.

40. Figure A has a perimeter of 24 inches. Figure B has a perimeter of 36 inches and one of the side lengths is 12 inches.

41. Figure A has an area of 48 square feet and one of the side lengths is 6 feet. Figure B has an area of 75 square feet.

42. Figure A has an area of 18 square feet. Figure B has an area of 98 square feet and one of the side lengths is 14 feet.

CRITICAL THINKING In Exercises 43–48, tell whether the polygons are *always, sometimes,* or *never* similar.

43. two isosceles triangles

44. two isosceles trapezoids

45. two rhombuses

46. two squares

47. two regular polygons

48. a right triangle and an equilateral triangle

49. MAKING AN ARGUMENT Your sister claims that when the side lengths of two rectangles are proportional, the two rectangles must be similar. Is she correct? Explain your reasoning.

50. HOW DO YOU SEE IT? You shine a flashlight directly on an object to project its image onto a parallel screen. Will the object and the image be similar? Explain your reasoning.

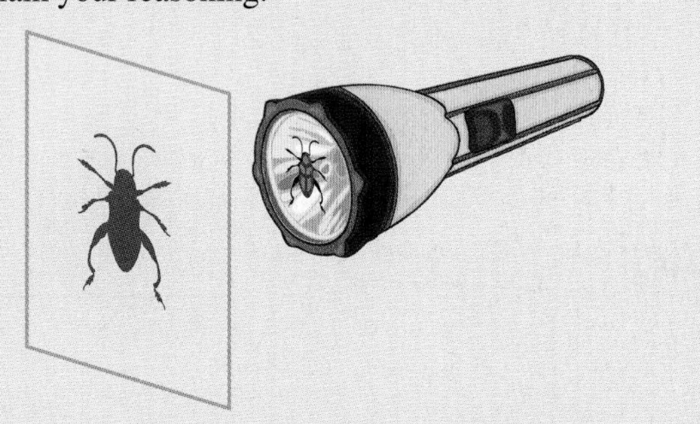

51. MODELING WITH MATHEMATICS During a total eclipse of the Sun, the moon is directly in line with the Sun and blocks the Sun's rays. The distance DA between Earth and the Sun is 93,000,000 miles, the distance DE between Earth and the moon is 240,000 miles, and the radius AB of the Sun is 432,500 miles. Use the diagram and the given measurements to estimate the radius EC of the moon.

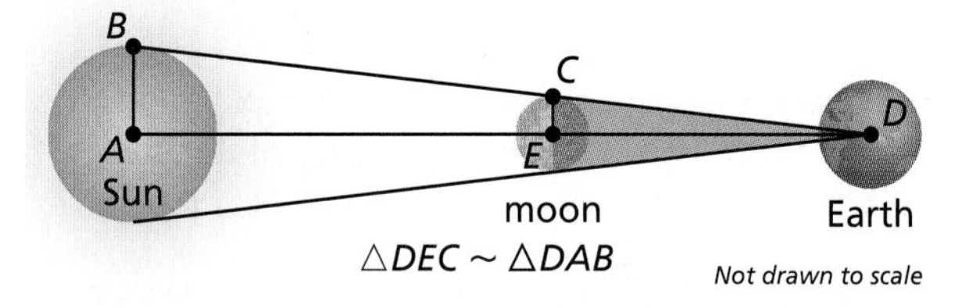

$\triangle DEC \sim \triangle DAB$

Not drawn to scale

52. PROVING A THEOREM Prove the Perimeters of Similar Polygons Theorem (Theorem 8.1) for similar rectangles. Include a diagram in your proof.

53. PROVING A THEOREM Prove the Areas of Similar Polygons Theorem (Theorem 8.2) for similar rectangles. Include a diagram in your proof.

54. THOUGHT PROVOKING The postulates and theorems in this book represent Euclidean geometry. In spherical geometry, all points are points on the surface of a sphere. A line is a circle on the sphere whose diameter is equal to the diameter of the sphere. A plane is the surface of the sphere. In spherical geometry, is it possible that two triangles are similar but not congruent? Explain your reasoning.

55. CRITICAL THINKING In the diagram, $PQRS$ is a square, and $PLMS \sim LMRQ$. Find the exact value of x. This value is called the *golden ratio*. Golden rectangles have their length and width in this ratio. Show that the similar rectangles in the diagram are golden rectangles.

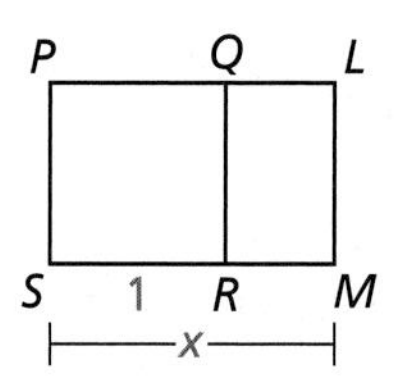

56. MATHEMATICAL CONNECTIONS The equations of the lines shown are $y = \frac{4}{3}x + 4$ and $y = \frac{4}{3}x - 8$. Show that $\triangle AOB \sim \triangle COD$.

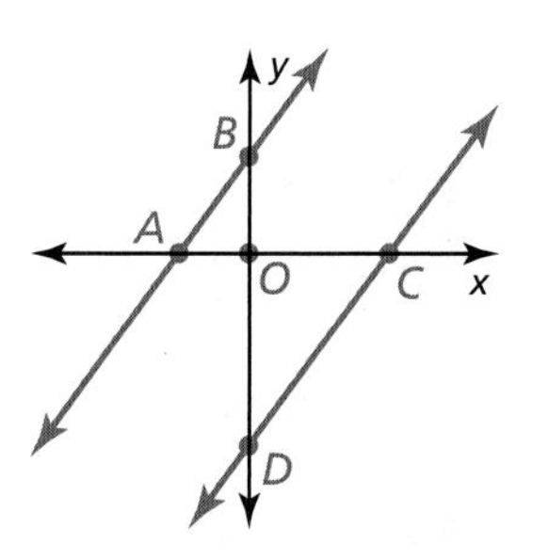

Maintaining Mathematical Proficiency

Reviewing what you learned in previous grades and lessons

Find the value of x. *(Section 5.1)*

57.

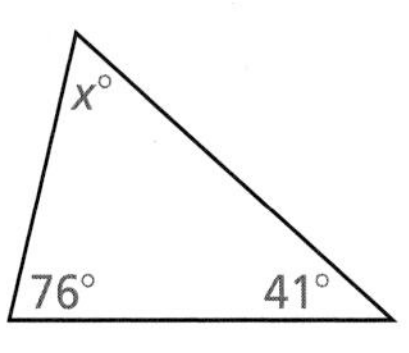

58.

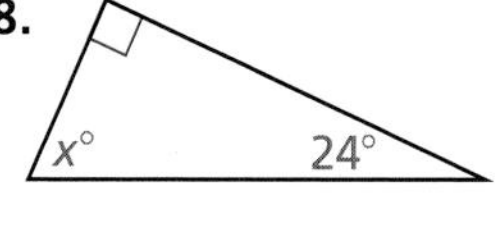

59.

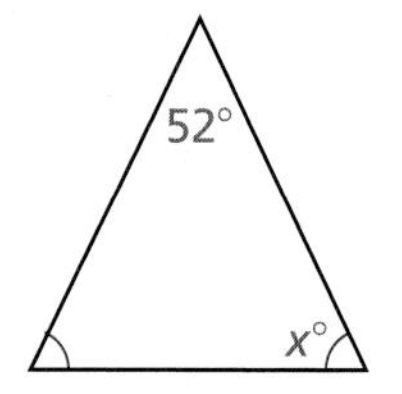

60.

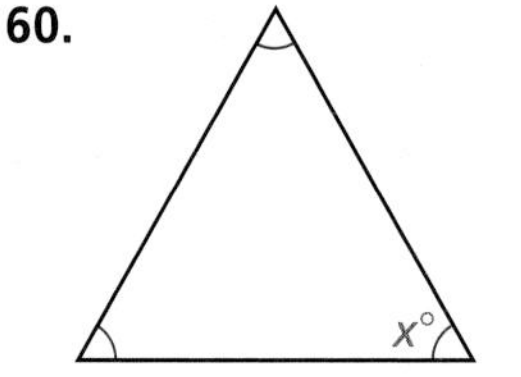

8.2 Proving Triangle Similarity by AA

Essential Question What can you conclude about two triangles when you know that two pairs of corresponding angles are congruent?

EXPLORATION 1 **Comparing Triangles**

Work with a partner. Use dynamic geometry software.

a. Construct $\triangle ABC$ and $\triangle DEF$ so that $m\angle A = m\angle D = 106^\circ$, $m\angle B = m\angle E = 31^\circ$, and $\triangle DEF$ is not congruent to $\triangle ABC$.

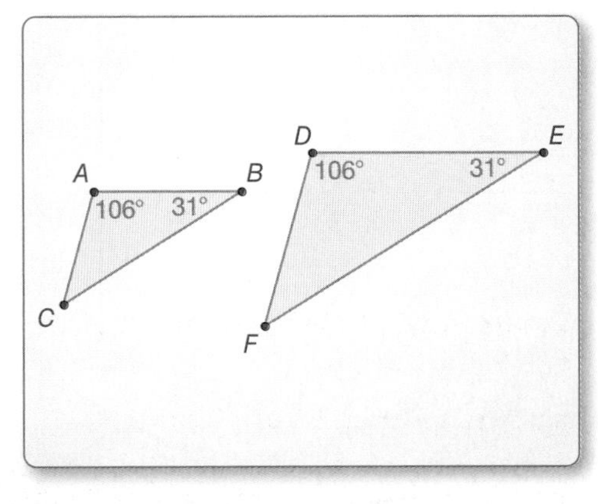

b. Find the third angle measure and the side lengths of each triangle. Copy the table below and record your results in column 1.

	1.	2.	3.	4.	5.	6.
$m\angle A$, $m\angle D$	106°	88°	40°			
$m\angle B$, $m\angle E$	31°	42°	65°			
$m\angle C$						
$m\angle F$						
AB						
DE						
BC						
EF						
AC						
DF						

CONSTRUCTING VIABLE ARGUMENTS

To be proficient in math, you need to understand and use stated assumptions, definitions, and previously established results in constructing arguments.

c. Are the two triangles similar? Explain.

d. Repeat parts (a)–(c) to complete columns 2 and 3 of the table for the given angle measures.

e. Complete each remaining column of the table using your own choice of two pairs of equal corresponding angle measures. Can you construct two triangles in this way that are *not* similar?

f. Make a conjecture about any two triangles with two pairs of congruent corresponding angles.

Communicate Your Answer

2. What can you conclude about two triangles when you know that two pairs of corresponding angles are congruent?

3. Find RS in the figure at the left.

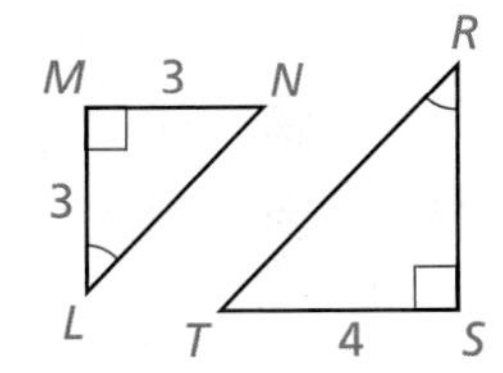

8.2 Lesson

Core Vocabulary

Previous
similar figures
similarity transformation

What You Will Learn

- Use the Angle-Angle Similarity Theorem.
- Solve real-life problems.

Using the Angle-Angle Similarity Theorem

Theorem

Theorem 8.3 Angle-Angle (AA) Similarity Theorem

If two angles of one triangle are congruent to two angles of another triangle, then the two triangles are similar.

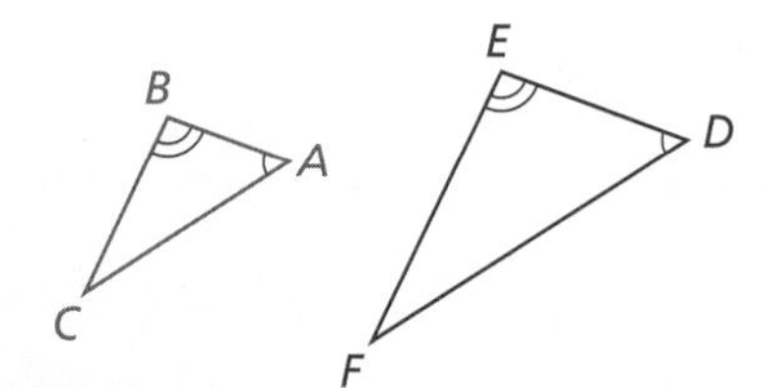

If $\angle A \cong \angle D$ and $\angle B \cong \angle E$, then $\triangle ABC \sim \triangle DEF$.

Proof p. 428

PROOF **Angle-Angle (AA) Similarity Theorem**

Given $\angle A \cong \angle D$, $\angle B \cong \angle E$

Prove $\triangle ABC \sim \triangle DEF$

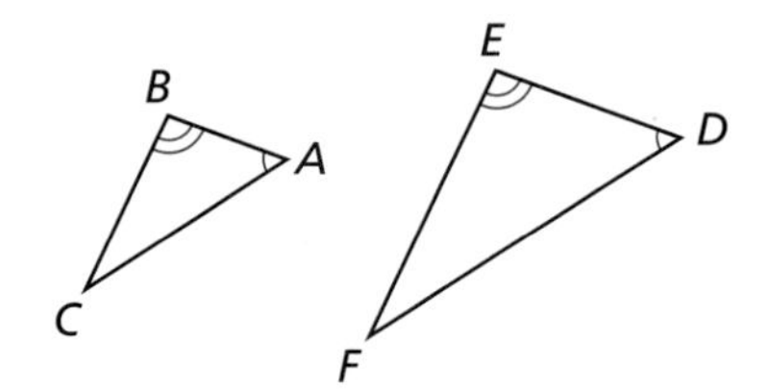

Dilate $\triangle ABC$ using a scale factor of $k = \frac{DE}{AB}$ and center A. The image of $\triangle ABC$ is $\triangle AB'C'$.

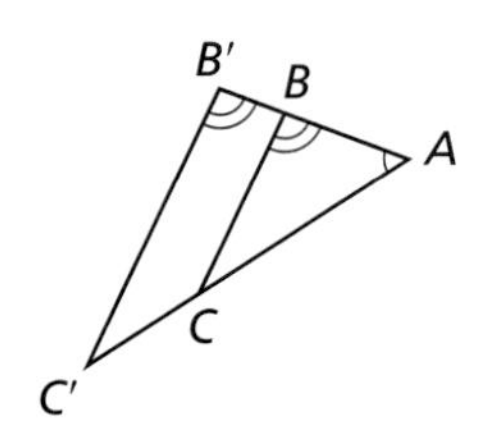

Because a dilation is a similarity transformation, $\triangle ABC \sim \triangle AB'C'$. Because the ratio of corresponding lengths of similar polygons equals the scale factor, $\frac{AB'}{AB} = \frac{DE}{AB}$. Multiplying each side by AB yields $AB' = DE$. By the definition of congruent segments, $\overline{AB'} \cong \overline{DE}$.

By the Reflexive Property of Congruence (Theorem 2.2), $\angle A \cong \angle A$. Because corresponding angles of similar polygons are congruent, $\angle B' \cong \angle B$. Because $\angle B' \cong \angle B$ and $\angle B \cong \angle E$, $\angle B' \cong \angle E$ by the Transitive Property of Congruence (Theorem 2.2).

Because $\angle A \cong \angle D$, $\angle B' \cong \angle E$, and $\overline{AB'} \cong \overline{DE}$, $\triangle AB'C' \cong \triangle DEF$ by the ASA Congruence Theorem (Theorem 5.10). So, a composition of rigid motions maps $\triangle AB'C'$ to $\triangle DEF$.

Because a dilation followed by a composition of rigid motions maps $\triangle ABC$ to $\triangle DEF$, $\triangle ABC \sim \triangle DEF$.

EXAMPLE 1 Using the AA Similarity Theorem

Determine whether the triangles are similar. If they are, write a similarity statement. Explain your reasoning.

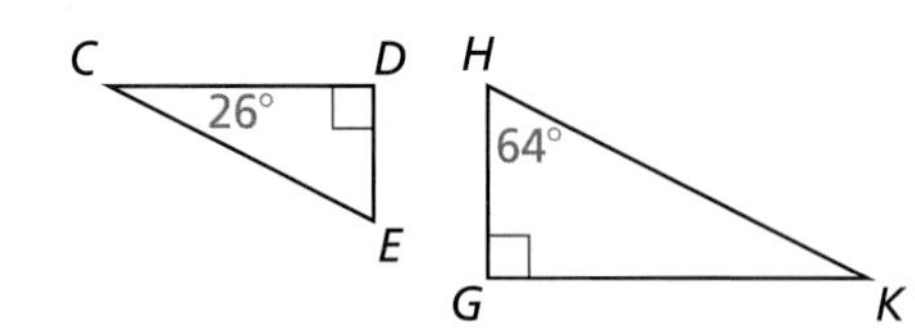

SOLUTION

Because they are both right angles, $\angle D$ and $\angle G$ are congruent.

By the Triangle Sum Theorem (Theorem 5.1), $26° + 90° + m\angle E = 180°$, so $m\angle E = 64°$. So, $\angle E$ and $\angle H$ are congruent.

▶ So, $\triangle CDE \sim \triangle KGH$ by the AA Similarity Theorem.

VISUAL REASONING

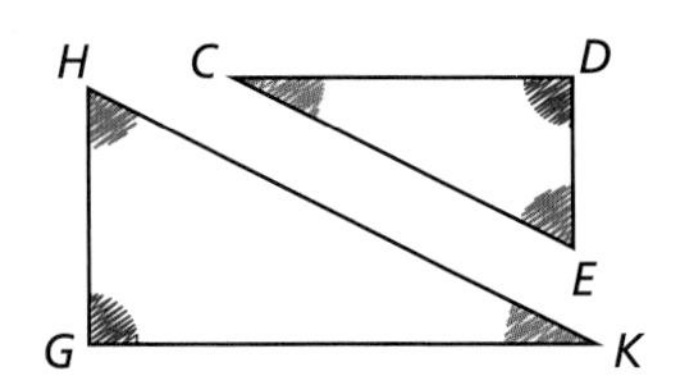

Use colored pencils to show congruent angles. This will help you write similarity statements.

EXAMPLE 2 Using the AA Similarity Theorem

Show that the two triangles are similar.

a. $\triangle ABE \sim \triangle ACD$

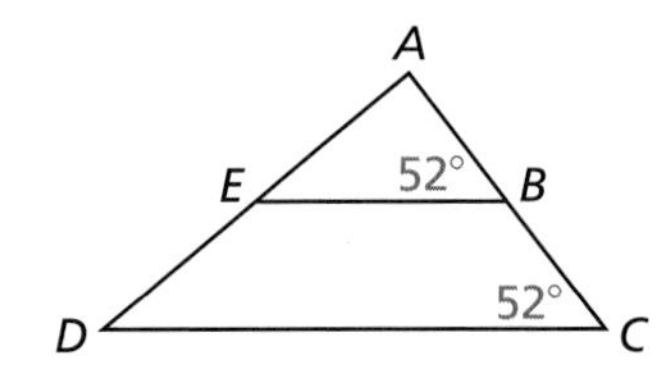

b. $\triangle SVR \sim \triangle UVT$

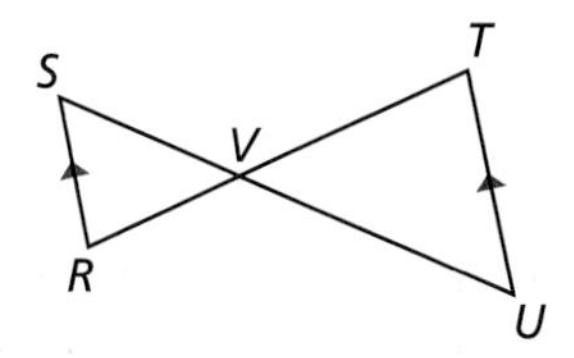

SOLUTION

a. Because $m\angle ABE$ and $m\angle C$ both equal 52°, $\angle ABE \cong \angle C$. By the Reflexive Property of Congruence (Theorem 2.2), $\angle A \cong \angle A$.

▶ So, $\triangle ABE \sim \triangle ACD$ by the AA Similarity Theorem.

b. You know $\angle SVR \cong \angle UVT$ by the Vertical Angles Congruence Theorem (Theorem 2.6). The diagram shows $\overline{RS} \parallel \overline{UT}$, so $\angle S \cong \angle U$ by the Alternate Interior Angles Theorem (Theorem 3.2).

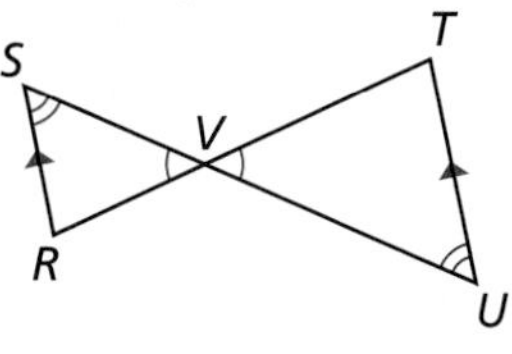

▶ So, $\triangle SVR \sim \triangle UVT$ by the AA Similarity Theorem.

VISUAL REASONING

You may find it helpful to redraw the triangles separately.

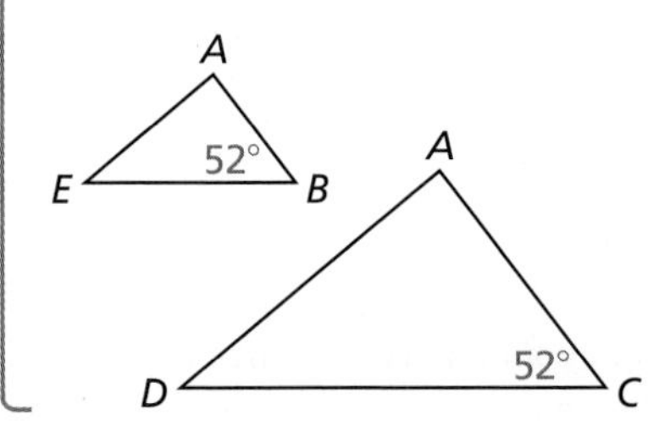

Monitoring Progress

Help in English and Spanish at *BigIdeasMath.com*

Show that the triangles are similar. Write a similarity statement.

1. $\triangle FGH$ and $\triangle RQS$

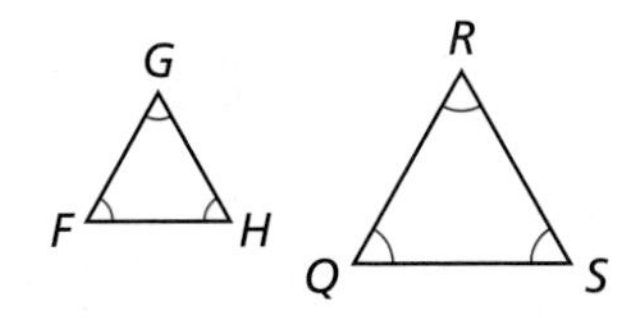

2. $\triangle CDF$ and $\triangle DEF$

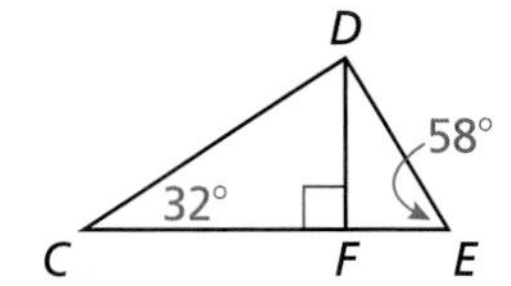

3. **WHAT IF?** Suppose that $\overline{SR} \nparallel \overline{TU}$ in Example 2 part (b). Could the triangles still be similar? Explain.

Solving Real-Life Problems

Previously, you learned a way to use congruent triangles to find measurements indirectly. Another useful way to find measurements indirectly is by using similar triangles.

EXAMPLE 3 Modeling with Mathematics

A flagpole casts a shadow that is 50 feet long. At the same time, a woman standing nearby who is 5 feet 4 inches tall casts a shadow that is 40 inches long. How tall is the flagpole to the nearest foot?

Not drawn to scale

SOLUTION

1. **Understand the Problem** You are given the length of a flagpole's shadow, the height of a woman, and the length of the woman's shadow. You need to find the height of the flagpole.

2. **Make a Plan** Use similar triangles to write a proportion and solve for the height of the flagpole.

3. **Solve the Problem** The flagpole and the woman form sides of two right triangles with the ground. The Sun's rays hit the flagpole and the woman at the same angle. You have two pairs of congruent angles, so the triangles are similar by the AA Similarity Theorem.

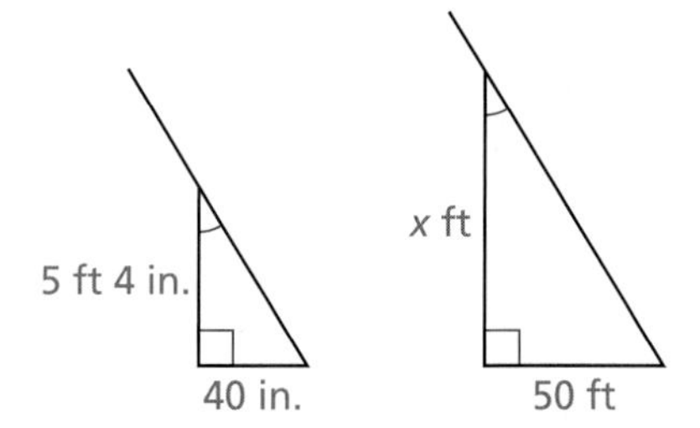

You can use a proportion to find the height x. Write 5 feet 4 inches as 64 inches so that you can form two ratios of feet to inches.

$$\frac{x \text{ ft}}{64 \text{ in.}} = \frac{50 \text{ ft}}{40 \text{ in.}}$$ Write proportion of side lengths.

$$40x = 3200$$ Cross Products Property

$$x = 80$$ Solve for x.

▶ The flagpole is 80 feet tall.

4. **Look Back** Attend to precision by checking that your answer has the correct units. The problem asks for the height of the flagpole to the nearest *foot*. Because your answer is 80 feet, the units match.

   Also, check that your answer is reasonable in the context of the problem. A height of 80 feet makes sense for a flagpole. You can estimate that an eight-story building would be about 8(10 feet) = 80 feet, so it is reasonable that a flagpole could be that tall.

Monitoring Progress  Help in English and Spanish at *BigIdeasMath.com*

4. **WHAT IF?** A child who is 58 inches tall is standing next to the woman in Example 3. How long is the child's shadow?

5. You are standing outside, and you measure the lengths of the shadows cast by both you and a tree. Write a proportion showing how you could find the height of the tree.

8.2 Exercises

Dynamic Solutions available at *BigIdeasMath.com*

Vocabulary and Core Concept Check

1. **COMPLETE THE SENTENCE** If two angles of one triangle are congruent to two angles of another triangle, then the triangles are _______.

2. **WRITING** Can you assume that corresponding sides and corresponding angles of any two similar triangles are congruent? Explain.

Monitoring Progress and Modeling with Mathematics

In Exercises 3–6, determine whether the triangles are similar. If they are, write a similarity statement. Explain your reasoning. *(See Example 1.)*

3.

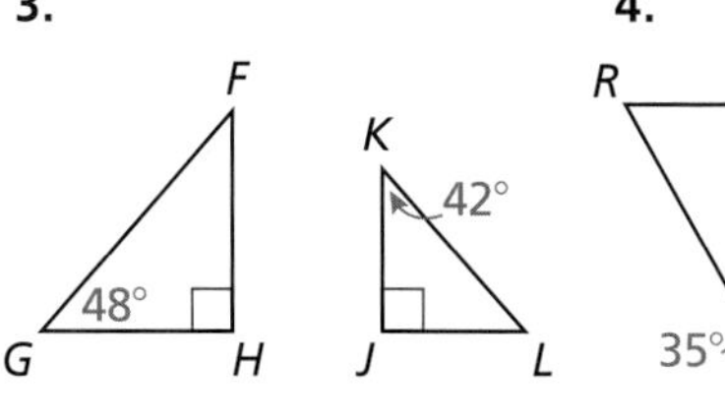

4.

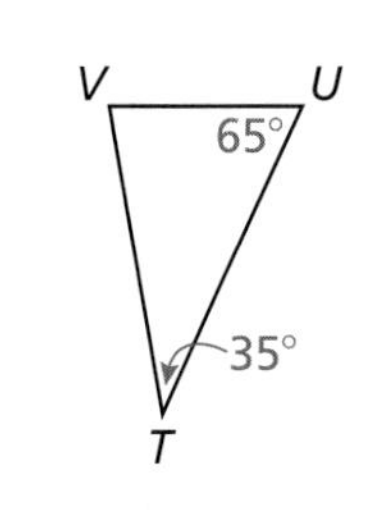

5.

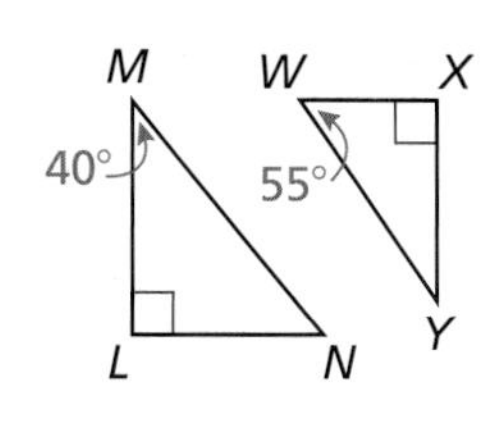

6.

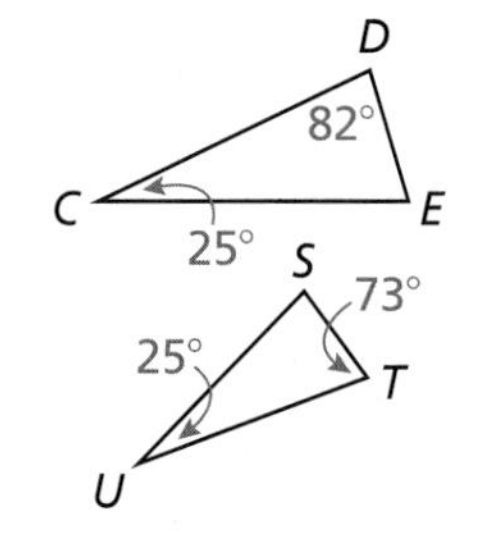

In Exercises 7–10, show that the two triangles are similar. *(See Example 2.)*

7.

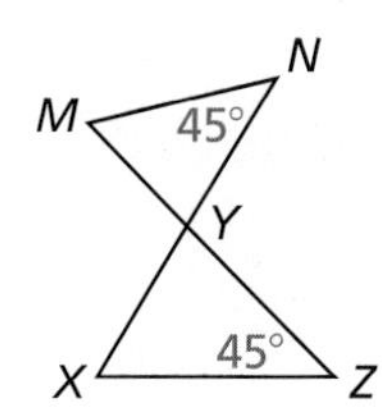

8.

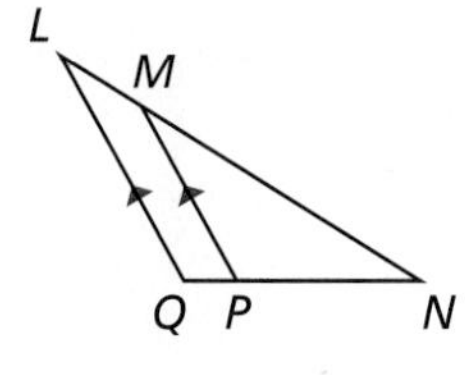

9.

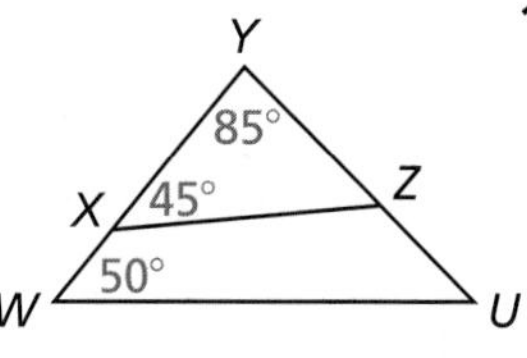

10.

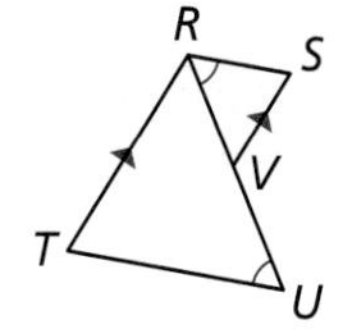

In Exercises 11–18, use the diagram to copy and complete the statement.

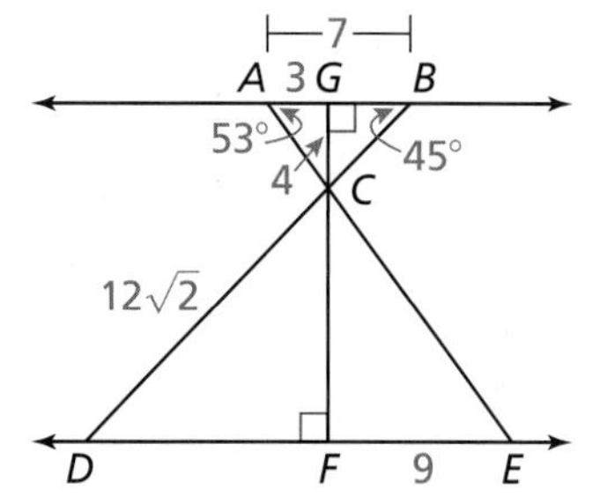

11. $\triangle CAG \sim$ ▒

12. $\triangle DCF \sim$ ▒

13. $\triangle ACB \sim$ ▒

14. $m\angle ECF =$ ▒

15. $m\angle ECD =$ ▒

16. $CF =$ ▒

17. $BC =$ ▒

18. $DE =$ ▒

19. **ERROR ANALYSIS** Describe and correct the error in using the AA Similarity Theorem (Theorem 8.3).

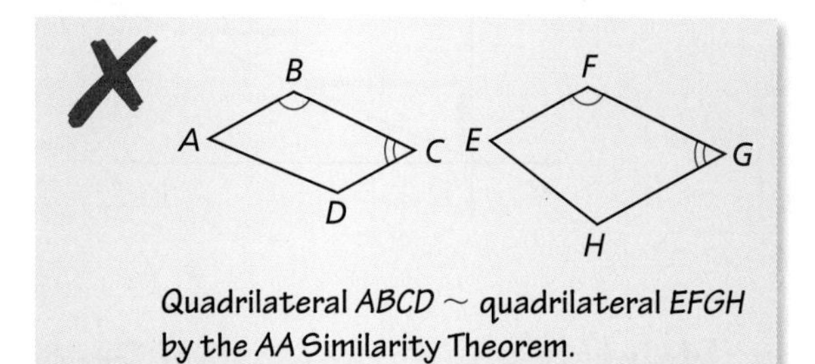

20. **ERROR ANALYSIS** Describe and correct the error in finding the value of x.

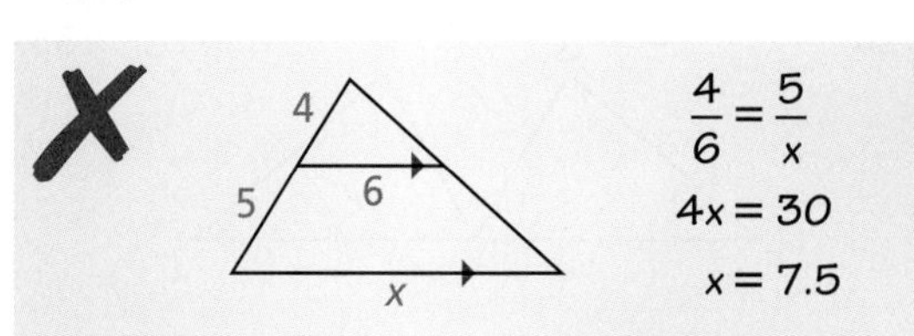

21. **MODELING WITH MATHEMATICS** You can measure the width of the lake using a surveying technique, as shown in the diagram. Find the width of the lake, WX. Justify your answer.

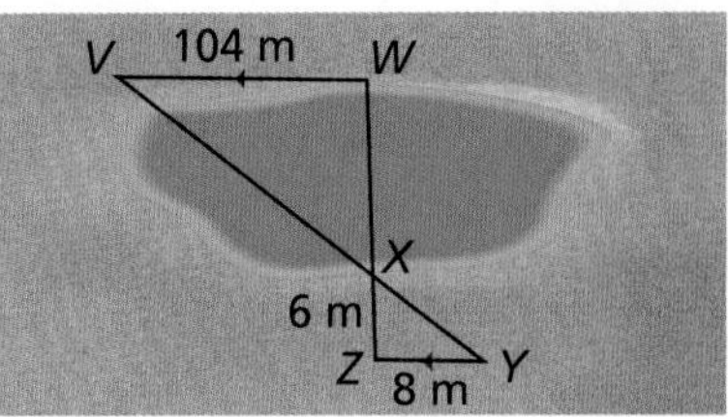

22. **MAKING AN ARGUMENT** You and your cousin are trying to determine the height of a telephone pole. Your cousin tells you to stand in the pole's shadow so that the tip of your shadow coincides with the tip of the pole's shadow. Your cousin claims to be able to use the distance between the tips of the shadows and you, the distance between you and the pole, and your height to estimate the height of the telephone pole. Is this possible? Explain. Include a diagram in your answer.

REASONING In Exercises 23–26, is it possible for $\triangle JKL$ and $\triangle XYZ$ to be similar? Explain your reasoning.

23. $m\angle J = 71^\circ$, $m\angle K = 52^\circ$, $m\angle X = 71^\circ$, and $m\angle Z = 57^\circ$

24. $\triangle JKL$ is a right triangle and $m\angle X + m\angle Y = 150^\circ$.

25. $m\angle L = 87^\circ$ and $m\angle Y = 94^\circ$

26. $m\angle J + m\angle K = 85^\circ$ and $m\angle Y + m\angle Z = 80^\circ$

27. **MATHEMATICAL CONNECTIONS** Explain how you can use similar triangles to show that any two points on a line can be used to find its slope.

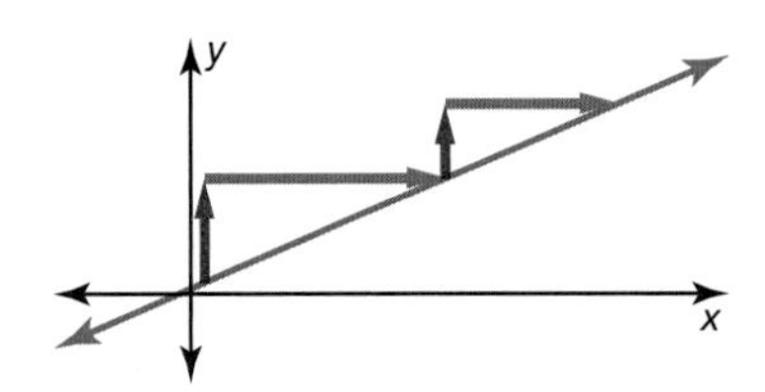

28. **HOW DO YOU SEE IT?** In the diagram, which triangles would you use to find the distance x between the shoreline and the buoy? Explain your reasoning.

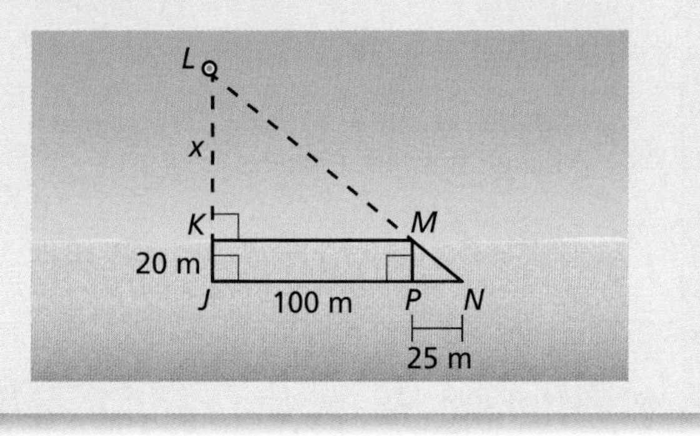

29. **WRITING** Explain why all equilateral triangles are similar.

30. **THOUGHT PROVOKING** Decide whether each is a valid method of showing that two quadrilaterals are similar. Justify your answer.

    a. AAA
    b. AAAA

31. **PROOF** Without using corresponding lengths in similar polygons, prove that the ratio of two corresponding angle bisectors in similar triangles is equal to the scale factor.

32. **PROOF** Prove that if the lengths of two sides of a triangle are a and b, respectively, then the lengths of the corresponding altitudes to those sides are in the ratio $\frac{b}{a}$.

33. **MODELING WITH MATHEMATICS** A portion of an amusement park ride is shown. Find EF. Justify your answer.

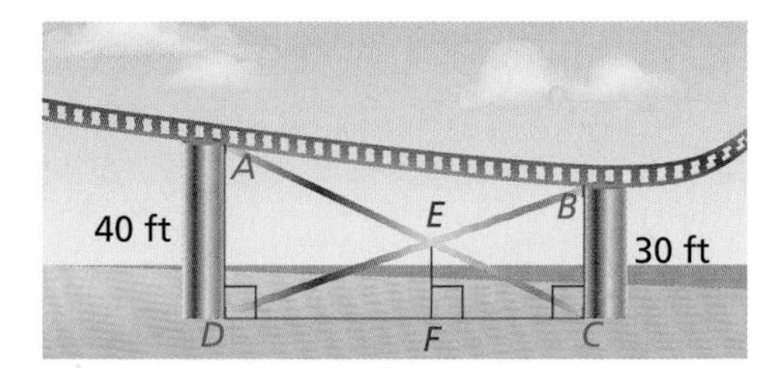

Maintaining Mathematical Proficiency

Reviewing what you learned in previous grades and lessons

Determine whether there is enough information to prove that the triangles are congruent. Explain your reasoning. *(Section 5.3, Section 5.5, and Section 5.6)*

34.

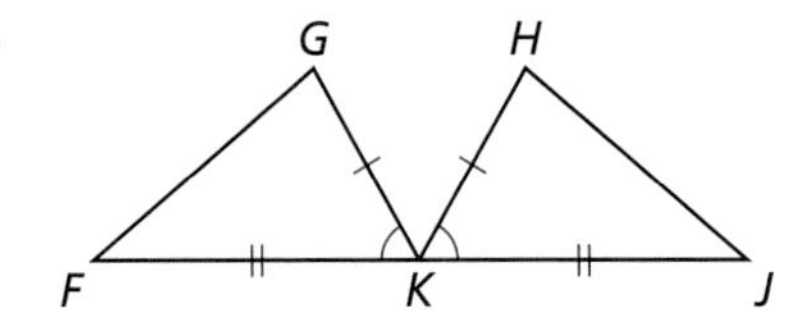

35.

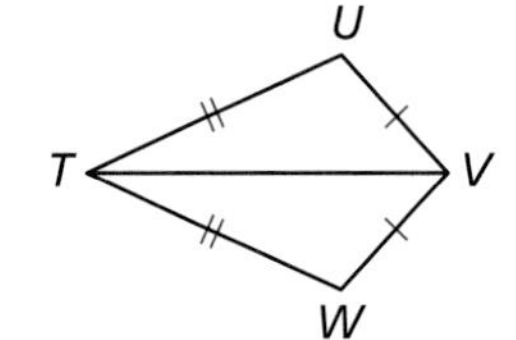

36.

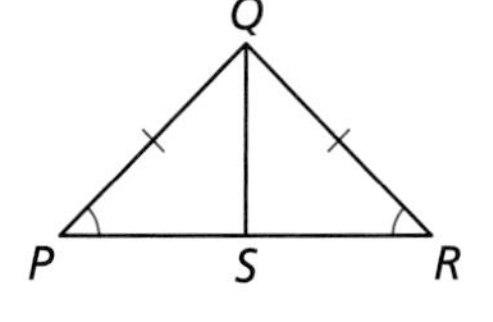

8.1–8.2 What Did You Learn?

Core Concepts

Section 8.1

Corresponding Parts of Similar Polygons, *p. 418*
Corresponding Lengths in Similar Polygons, *p. 419*
Theorem 8.1 Perimeters of Similar Polygons, *p. 420*
Theorem 8.2 Areas of Similar Polygons, *p. 421*

Section 8.2

Theorem 8.3 Angle-Angle (AA) Similarity Theorem, *p. 428*

Mathematical Practices

1. In Exercise 35 on page 425, why is there more than one correct answer for the length of the other side?
2. In Exercise 50 on page 426, how could you find the scale factor of the similar figures? Describe any tools that might be helpful.
3. In Exercise 21 on page 432, explain why the surveyor needs V, X, and Y to be collinear and Z, X, and W to be collinear.

Study Skills

Take Control of Your Class Time

- Sit where you can easily see and hear the teacher, and the teacher can see you. The teacher may be able to tell when you are confused just by the look on your face and may adjust the lesson accordingly. In addition, sitting in this strategic place will keep your mind from wandering.
- Pay attention to what the teacher says about the math, not just what is written on the board. Write problems on the left side of your notes and what the teacher says about the problems on the right side.
- If the teacher is moving through the material too fast, ask a question. Questions help slow the pace for a few minutes and also clarify what is confusing to you.
- Try to memorize new information while learning it. Repeat in your head what you are writing in your notes. That way you are reviewing the information twice.

8.1–8.2 Quiz

List all pairs of congruent angles. Then write the ratios of the corresponding side lengths in a statement of proportionality. *(Section 8.1)*

1. $\triangle BDG \sim \triangle MPQ$

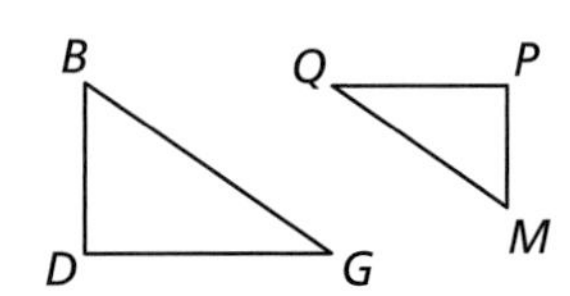

2. $DEFG \sim HJKL$

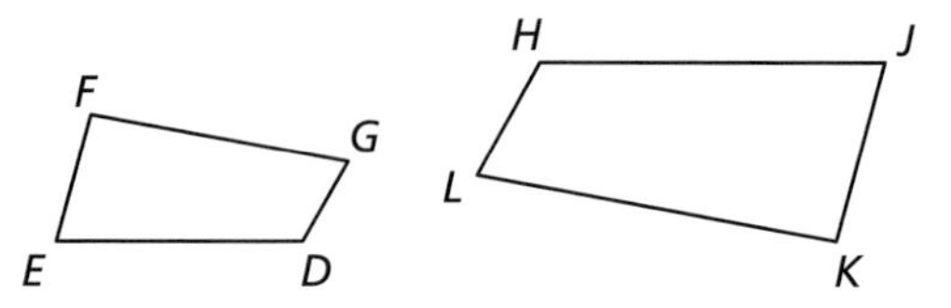

The polygons are similar. Find the value of *x*. *(Section 8.1)*

3.

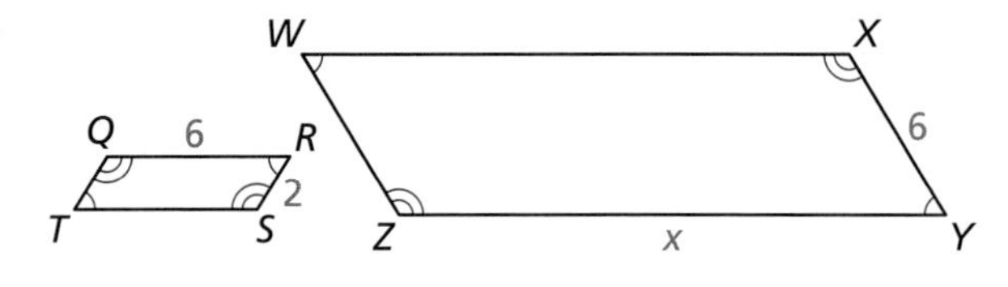

4.

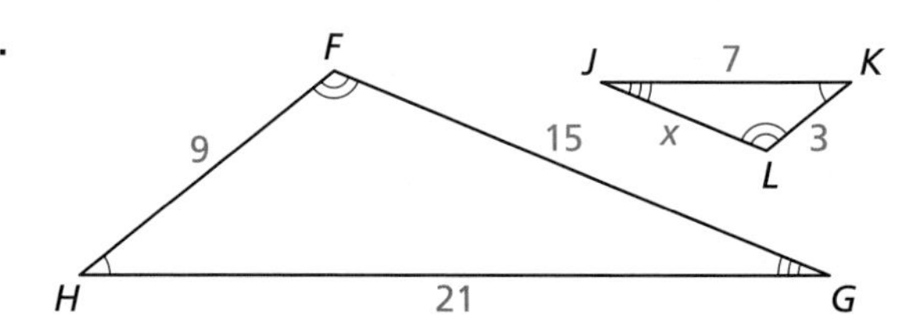

Determine whether the polygons are similar. If they are, write a similarity statement. Explain your reasoning. *(Section 8.1 and Section 8.2)*

5.

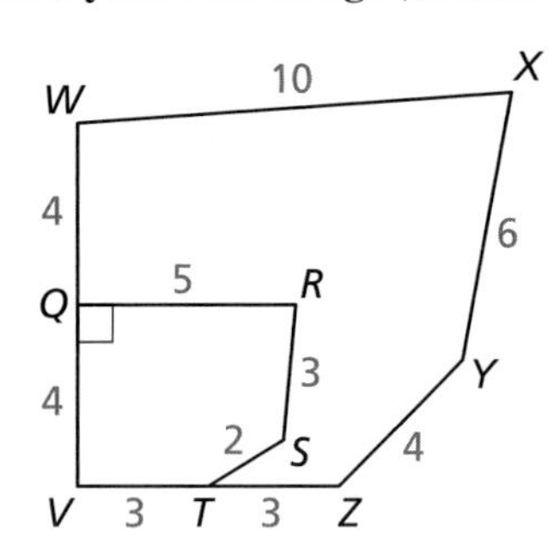

6.

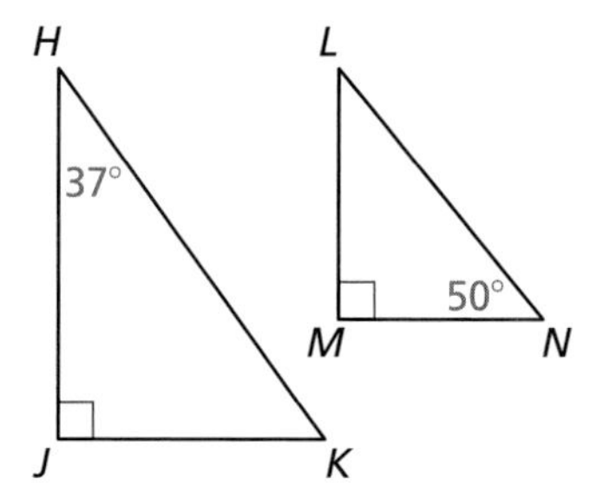

7.

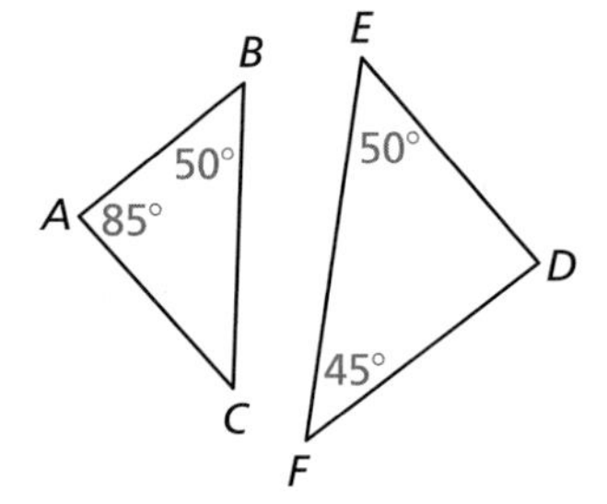

Show that the two triangles are similar. *(Section 8.2)*

8.

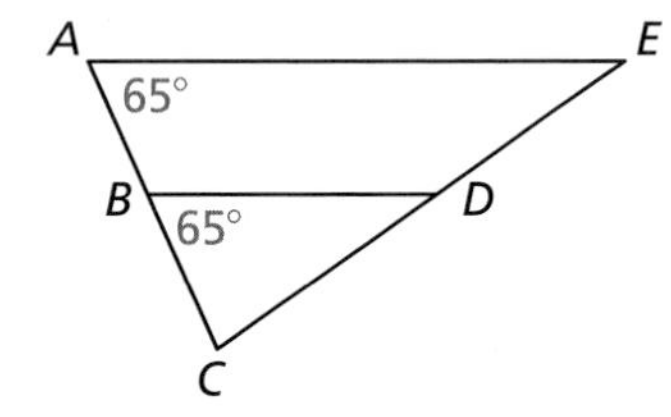

9.

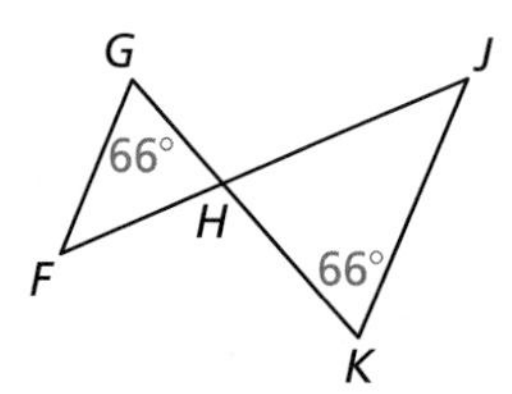

10.

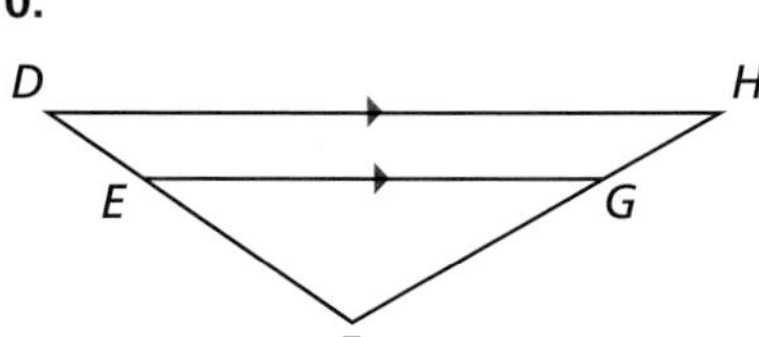

11. The dimensions of an official hockey rink used by the National Hockey League (NHL) are 200 feet by 85 feet. The dimensions of an air hockey table are 96 inches by 40.8 inches. Assume corresponding angles are congruent. *(Section 8.1)*

a. Determine whether the two surfaces are similar.

b. If the surfaces are similar, find the ratio of their perimeters and the ratio of their areas. If not, find the dimensions of an air hockey table that are similar to an NHL hockey rink.

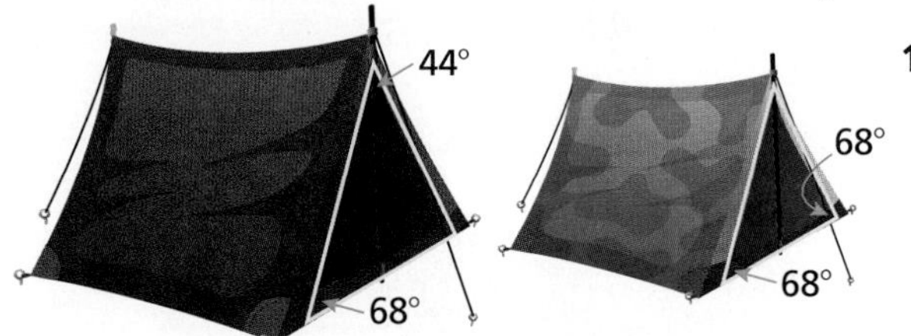

12. You and a friend buy camping tents made by the same company but in different sizes and colors. Use the information given in the diagram to decide whether the triangular faces of the tents are similar. Explain your reasoning. *(Section 8.2)*

8.3 Proving Triangle Similarity by SSS and SAS

Essential Question What are two ways to use corresponding sides of two triangles to determine that the triangles are similar?

EXPLORATION 1 Deciding Whether Triangles Are Similar

Work with a partner. Use dynamic geometry software.

a. Construct $\triangle ABC$ and $\triangle DEF$ with the side lengths given in column 1 of the table below.

	1.	2.	3.	4.	5.	6.	7.
AB	5	5	6	15	9	24	
BC	8	8	8	20	12	18	
AC	10	10	10	10	8	16	
DE	10	15	9	12	12	8	
EF	16	24	12	16	15	6	
DF	20	30	15	8	10	8	
$m\angle A$							
$m\angle B$							
$m\angle C$							
$m\angle D$							
$m\angle E$							
$m\angle F$							

b. Copy the table and complete column 1.

c. Are the triangles similar? Explain your reasoning.

d. Repeat parts (a)–(c) for columns 2–6 in the table.

e. How are the corresponding side lengths related in each pair of triangles that are similar? Is this true for each pair of triangles that are not similar?

f. Make a conjecture about the similarity of two triangles based on their corresponding side lengths.

g. Use your conjecture to write another set of side lengths of two similar triangles. Use the side lengths to complete column 7 of the table.

CONSTRUCTING VIABLE ARGUMENTS

To be proficient in math, you need to analyze situations by breaking them into cases and recognize and use counterexamples.

EXPLORATION 2 Deciding Whether Triangles Are Similar

Work with a partner. Use dynamic geometry software. Construct any $\triangle ABC$.

a. Find AB, AC, and $m\angle A$. Choose any positive rational number k and construct $\triangle DEF$ so that $DE = k \cdot AB$, $DF = k \cdot AC$, and $m\angle D = m\angle A$.

b. Is $\triangle DEF$ similar to $\triangle ABC$? Explain your reasoning.

c. Repeat parts (a) and (b) several times by changing $\triangle ABC$ and k. Describe your results.

Communicate Your Answer

3. What are two ways to use corresponding sides of two triangles to determine that the triangles are similar?

8.3 Lesson

Core Vocabulary

Previous
similar figures
corresponding parts
slope
parallel lines
perpendicular lines

What You Will Learn

- Use the Side-Side-Side Similarity Theorem.
- Use the Side-Angle-Side Similarity Theorem.
- Prove slope criteria using similar triangles.

Using the Side-Side-Side Similarity Theorem

In addition to using congruent corresponding angles to show that two triangles are similar, you can use proportional corresponding side lengths.

Theorem

Theorem 8.4 Side-Side-Side (SSS) Similarity Theorem

If the corresponding side lengths of two triangles are proportional, then the triangles are similar.

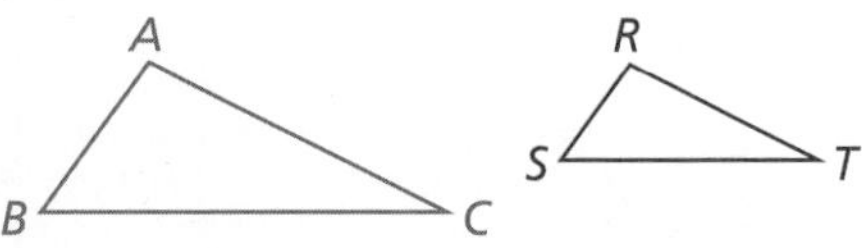

If $\frac{AB}{RS} = \frac{BC}{ST} = \frac{CA}{TR}$, then $\triangle ABC \sim \triangle RST$.

Proof p. 437

EXAMPLE 1 Using the SSS Similarity Theorem

Is either $\triangle DEF$ or $\triangle GHJ$ similar to $\triangle ABC$?

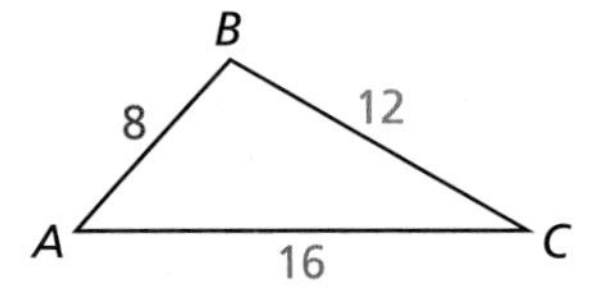

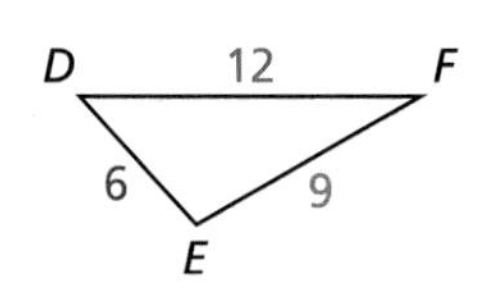

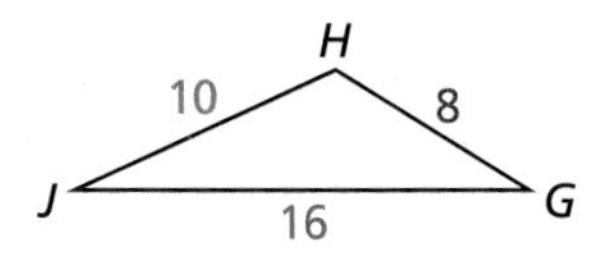

FINDING AN ENTRY POINT

When using the SSS Similarity Theorem, compare the shortest sides, the longest sides, and then the remaining sides.

SOLUTION

Compare $\triangle ABC$ and $\triangle DEF$ by finding ratios of corresponding side lengths.

Shortest sides	Longest sides	Remaining sides
$\frac{AB}{DE} = \frac{8}{6}$	$\frac{CA}{FD} = \frac{16}{12}$	$\frac{BC}{EF} = \frac{12}{9}$
$= \frac{4}{3}$	$= \frac{4}{3}$	$= \frac{4}{3}$

- All the ratios are equal, so $\triangle ABC \sim \triangle DEF$.

Compare $\triangle ABC$ and $\triangle GHJ$ by finding ratios of corresponding side lengths.

Shortest sides	Longest sides	Remaining sides
$\frac{AB}{GH} = \frac{8}{8}$	$\frac{CA}{JG} = \frac{16}{16}$	$\frac{BC}{HJ} = \frac{12}{10}$
$= 1$	$= 1$	$= \frac{6}{5}$

- The ratios are not all equal, so $\triangle ABC$ and $\triangle GHJ$ are not similar.

PROOF SSS Similarity Theorem

Given $\dfrac{RS}{JK} = \dfrac{ST}{KL} = \dfrac{TR}{LJ}$

Prove $\triangle RST \sim \triangle JKL$

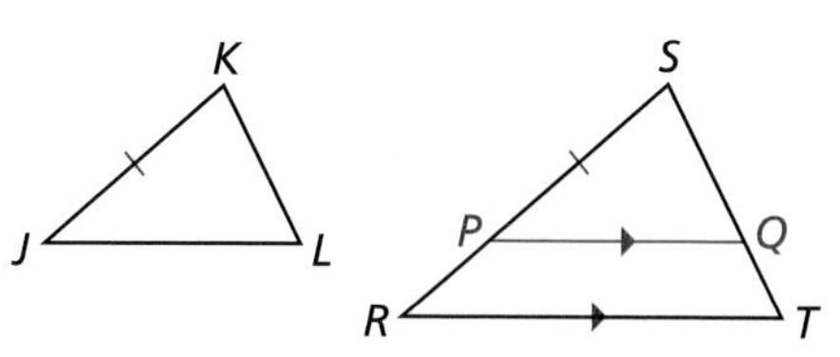

JUSTIFYING STEPS

The Parallel Postulate (Postulate 3.1) allows you to draw an auxiliary line $\overleftrightarrow{PQ}$ in $\triangle RST$. There is only one line through point P parallel to $\overleftrightarrow{RT}$, so you are able to draw it.

Locate P on $\overline{RS}$ so that $PS = JK$. Draw $\overline{PQ}$ so that $\overline{PQ} \parallel \overline{RT}$. Then $\triangle RST \sim \triangle PSQ$ by the AA Similarity Theorem (Theorem 8.3), and $\dfrac{RS}{PS} = \dfrac{ST}{SQ} = \dfrac{TR}{QP}$. You can use the given proportion and the fact that $PS = JK$ to deduce that $SQ = KL$ and $QP = LJ$. By the SSS Congruence Theorem (Theorem 5.8), it follows that $\triangle PSQ \cong \triangle JKL$. Finally, use the definition of congruent triangles and the AA Similarity Theorem (Theorem 8.3) to conclude that $\triangle RST \sim \triangle JKL$.

EXAMPLE 2 Using the SSS Similarity Theorem

Find the value of x that makes $\triangle ABC \sim \triangle DEF$.

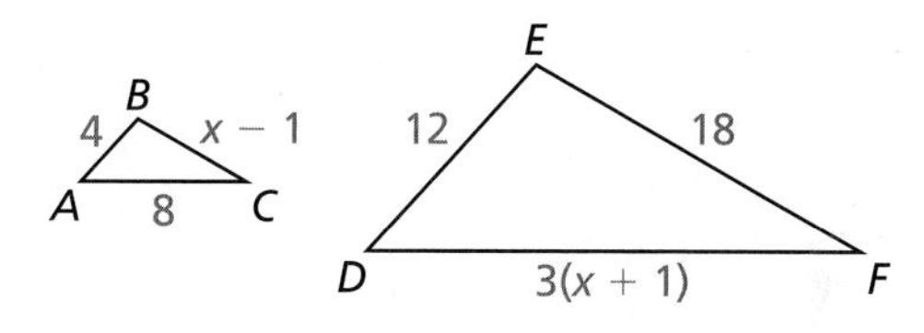

FINDING AN ENTRY POINT

You can use either $\dfrac{AB}{DE} = \dfrac{BC}{EF}$ or $\dfrac{AB}{DE} = \dfrac{AC}{DF}$ in Step 1.

SOLUTION

Step 1 Find the value of x that makes corresponding side lengths proportional.

$\dfrac{AB}{DE} = \dfrac{BC}{EF}$ Write proportion.

$\dfrac{4}{12} = \dfrac{x-1}{18}$ Substitute.

$4 \cdot 18 = 12(x - 1)$ Cross Products Property

$72 = 12x - 12$ Simplify.

$7 = x$ Solve for x.

Step 2 Check that the side lengths are proportional when $x = 7$.

$BC = x - 1 = 6$ $DF = 3(x + 1) = 24$

$\dfrac{AB}{DE} \stackrel{?}{=} \dfrac{BC}{EF} \Rightarrow \dfrac{4}{12} = \dfrac{6}{18}$ ✓ $\dfrac{AB}{DE} \stackrel{?}{=} \dfrac{AC}{DF} \Rightarrow \dfrac{4}{12} = \dfrac{8}{24}$ ✓

▶ When $x = 7$, the triangles are similar by the SSS Similarity Theorem.

Monitoring Progress  Help in English and Spanish at *BigIdeasMath.com*

Use the diagram.

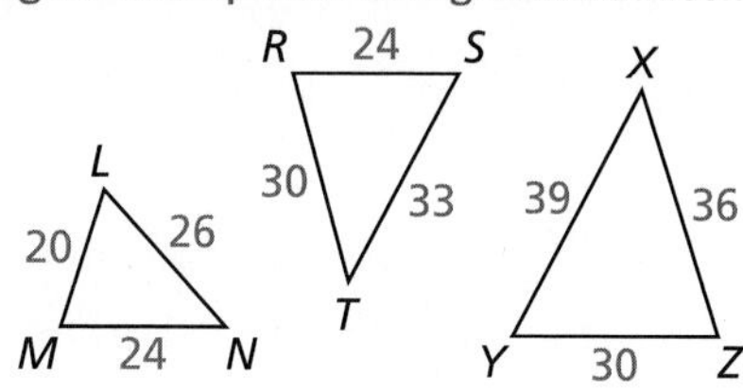

1. Which of the three triangles are similar? Write a similarity statement.
2. The shortest side of a triangle similar to $\triangle RST$ is 12 units long. Find the other side lengths of the triangle.

Using the Side-Angle-Side Similarity Theorem

Theorem 8.5 Side-Angle-Side (SAS) Similarity Theorem

If an angle of one triangle is congruent to an angle of a second triangle and the lengths of the sides including these angles are proportional, then the triangles are similar.

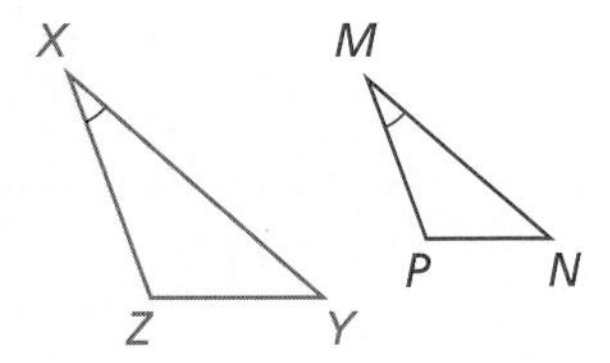

If $\angle X \cong \angle M$ and $\frac{ZX}{PM} = \frac{XY}{MN}$, then $\triangle XYZ \sim \triangle MNP$.

Proof Ex. 33, p. 443

EXAMPLE 3 Using the SAS Similarity Theorem

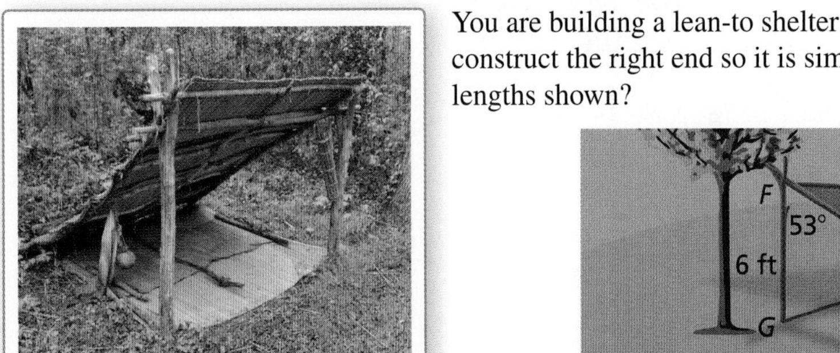

You are building a lean-to shelter starting from a tree branch, as shown. Can you construct the right end so it is similar to the left end using the angle measure and lengths shown?

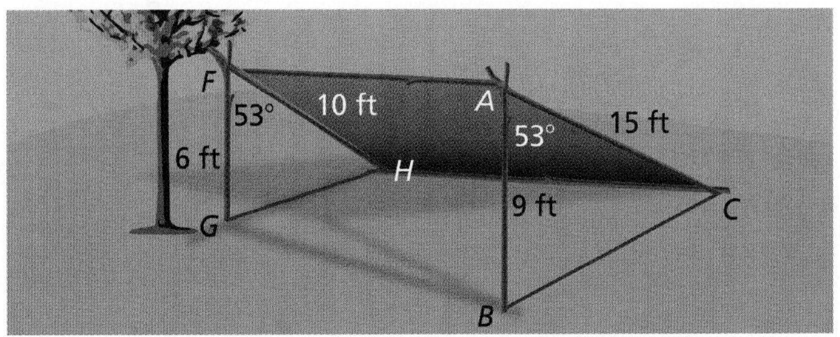

SOLUTION

Both $m\angle A$ and $m\angle F$ equal 53°, so $\angle A \cong \angle F$. Next, compare the ratios of the lengths of the sides that include $\angle A$ and $\angle F$.

Shorter sides

$$\frac{AB}{FG} = \frac{9}{6} = \frac{3}{2}$$

Longer sides

$$\frac{AC}{FH} = \frac{15}{10} = \frac{3}{2}$$

The lengths of the sides that include $\angle A$ and $\angle F$ are proportional. So, by the SAS Similarity Theorem, $\triangle ABC \sim \triangle FGH$.

▶ Yes, you can make the right end similar to the left end of the shelter.

Monitoring Progress

Help in English and Spanish at *BigIdeasMath.com*

Explain how to show that the indicated triangles are similar.

3. $\triangle SRT \sim \triangle PNQ$

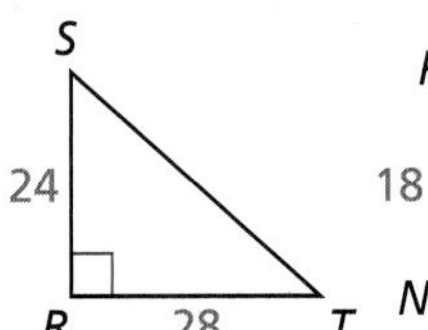

4. $\triangle XZW \sim \triangle YZX$

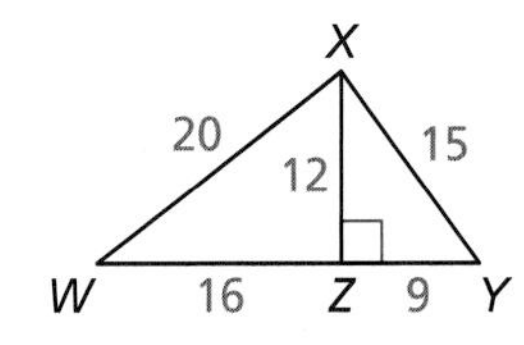

Concept Summary

Triangle Similarity Theorems

AA Similarity Theorem

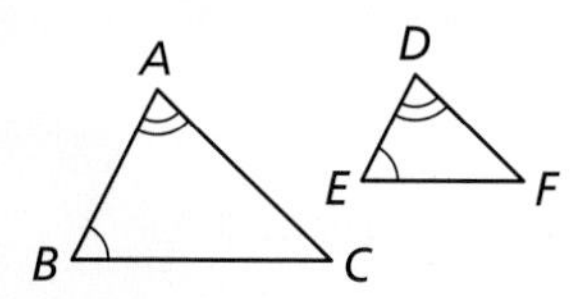

If $\angle A \cong \angle D$ and $\angle B \cong \angle E$, then $\triangle ABC \sim \triangle DEF$.

SSS Similarity Theorem

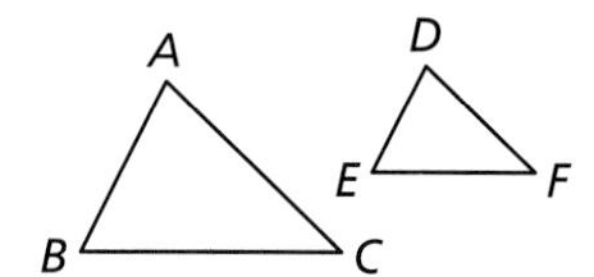

If $\frac{AB}{DE} = \frac{BC}{EF} = \frac{AC}{DF}$, then $\triangle ABC \sim \triangle DEF$.

SAS Similarity Theorem

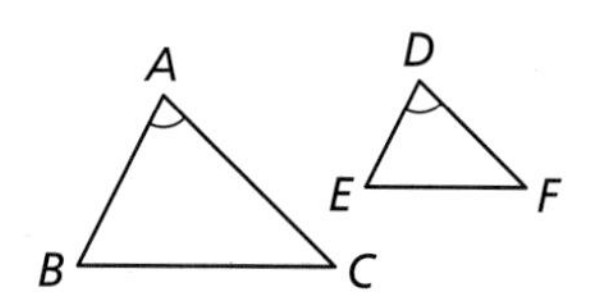

If $\angle A \cong \angle D$ and $\frac{AB}{DE} = \frac{AC}{DF}$, then $\triangle ABC \sim \triangle DEF$.

Proving Slope Criteria Using Similar Triangles

You can use similar triangles to prove the Slopes of Parallel Lines Theorem (Theorem 3.13). Because the theorem is biconditional, you must prove both parts.

1. If two nonvertical lines are parallel, then they have the same slope.
2. If two nonvertical lines have the same slope, then they are parallel.

The first part is proved below. The second part is proved in the exercises.

PROOF **Part of Slopes of Parallel Lines Theorem (Theorem 3.13)**

Given $\ell \parallel n$, ℓ and n are nonvertical.

Prove $m_\ell = m_n$

First, consider the case where ℓ and n are horizontal. Because all horizontal lines are parallel and have a slope of 0, the statement is true for horizontal lines.

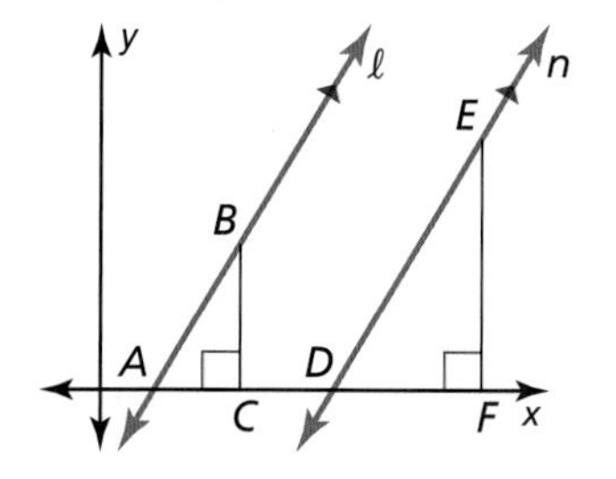

For the case of nonhorizontal, nonvertical lines, draw two such parallel lines, ℓ and n, and label their x-intercepts A and D, respectively. Draw a vertical segment $\overline{BC}$ parallel to the y-axis from point B on line ℓ to point C on the x-axis. Draw a vertical segment $\overline{EF}$ parallel to the y-axis from point E on line n to point F on the x-axis. Because vertical and horizontal lines are perpendicular, $\angle BCA$ and $\angle EFD$ are right angles.

STATEMENTS	REASONS
1. $\ell \parallel n$	**1.** Given
2. $\angle BAC \cong \angle EDF$	**2.** Corresponding Angles Theorem (Thm. 3.1)
3. $\angle BCA \cong \angle EFD$	**3.** Right Angles Congruence Theorem (Thm. 2.3)
4. $\triangle ABC \sim \triangle DEF$	**4.** AA Similarity Theorem (Thm. 8.3)
5. $\frac{BC}{EF} = \frac{AC}{DF}$	**5.** Corresponding sides of similar figures are proportional.
6. $\frac{BC}{AC} = \frac{EF}{DF}$	**6.** Rewrite proportion.
7. $m_\ell = \frac{BC}{AC}$, $m_n = \frac{EF}{DF}$	**7.** Definition of slope
8. $m_n = \frac{BC}{AC}$	**8.** Substitution Property of Equality
9. $m_\ell = m_n$	**9.** Transitive Property of Equality

To prove the Slopes of Perpendicular Lines Theorem (Theorem 3.14), you must prove both parts.

1. If two nonvertical lines are perpendicular, then the product of their slopes is -1.
2. If the product of the slopes of two nonvertical lines is -1, then the lines are perpendicular.

The first part is proved below. The second part is proved in the exercises.

PROOF Part of Slopes of Perpendicular Lines Theorem (Theorem 3.14)

Given $\ell \perp n$, ℓ and n are nonvertical.

Prove $m_\ell m_n = -1$

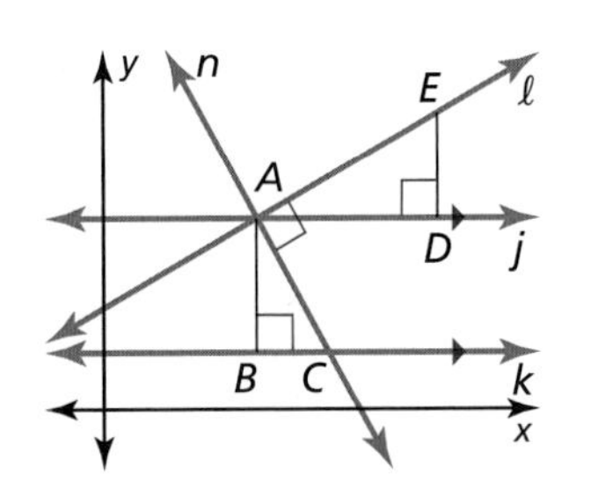

Draw two nonvertical, perpendicular lines, ℓ and n, that intersect at point A. Draw a horizontal line j parallel to the x-axis through point A. Draw a horizontal line k parallel to the x-axis through point C on line n. Because horizontal lines are parallel, $j \parallel k$. Draw a vertical segment $\overline{AB}$ parallel to the y-axis from point A to point B on line k. Draw a vertical segment $\overline{ED}$ parallel to the y-axis from point E on line ℓ to point D on line j. Because horizontal and vertical lines are perpendicular, $\angle ABC$ and $\angle ADE$ are right angles.

STATEMENTS	REASONS
1. $\ell \perp n$	1. Given
2. $m\angle CAE = 90°$	2. $\ell \perp n$
3. $m\angle CAE = m\angle DAE + m\angle CAD$	3. Angle Addition Postulate (Post. 1.4)
4. $m\angle DAE + m\angle CAD = 90°$	4. Transitive Property of Equality
5. $\angle BCA \cong \angle CAD$	5. Alternate Interior Angles Theorem (Thm. 3.2)
6. $m\angle BCA = m\angle CAD$	6. Definition of congruent angles
7. $m\angle DAE + m\angle BCA = 90°$	7. Substitution Property of Equality
8. $m\angle DAE = 90° - m\angle BCA$	8. Solve statement 7 for $m\angle DAE$.
9. $m\angle BCA + m\angle BAC + 90° = 180°$	9. Triangle Sum Theorem (Thm. 5.1)
10. $m\angle BAC = 90° - m\angle BCA$	10. Solve statement 9 for $m\angle BAC$.
11. $m\angle DAE = m\angle BAC$	11. Transitive Property of Equality
12. $\angle DAE \cong \angle BAC$	12. Definition of congruent angles
13. $\angle ABC \cong \angle ADE$	13. Right Angles Congruence Theorem (Thm. 2.3)
14. $\triangle ABC \sim \triangle ADE$	14. AA Similarity Theorem (Thm. 8.3)
15. $\frac{AD}{AB} = \frac{DE}{BC}$	15. Corresponding sides of similar figures are proportional.
16. $\frac{AD}{DE} = \frac{AB}{BC}$	16. Rewrite proportion.
17. $m_\ell = \frac{DE}{AD}, m_n = -\frac{AB}{BC}$	17. Definition of slope
18. $m_\ell m_n = \frac{DE}{AD} \cdot \left(-\frac{AB}{BC}\right)$	18. Substitution Property of Equality
19. $m_\ell m_n = \frac{DE}{AD} \cdot \left(-\frac{AD}{DE}\right)$	19. Substitution Property of Equality
20. $m_\ell m_n = -1$	20. Simplify.

8.3 Exercises

Dynamic Solutions available at *BigIdeasMath.com*

Vocabulary and Core Concept Check

1. **COMPLETE THE SENTENCE** You plan to show that $\triangle QRS$ is similar to $\triangle XYZ$ by the SSS Similarity Theorem (Theorem 8.4). Copy and complete the proportion that you will use: $\frac{QR}{?} = \frac{?}{YZ} = \frac{QS}{?}$.

2. **WHICH ONE DOESN'T BELONG?** Which triangle does *not* belong with the other three? Explain your reasoning.

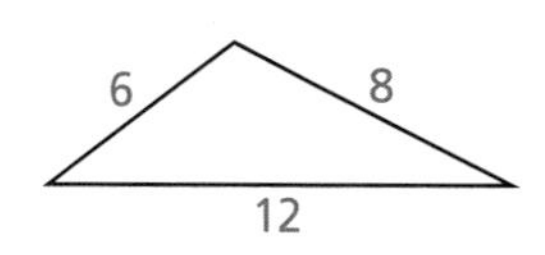

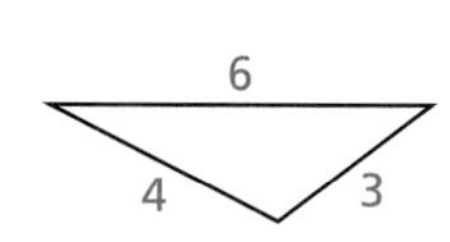

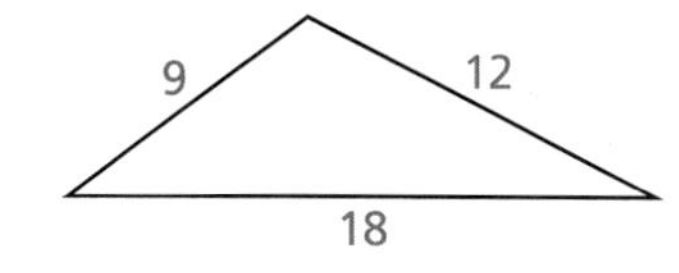

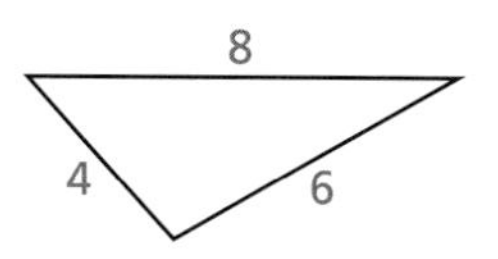

Monitoring Progress and Modeling with Mathematics

In Exercises 3 and 4, determine whether $\triangle JKL$ or $\triangle RST$ is similar to $\triangle ABC$. *(See Example 1.)*

3.

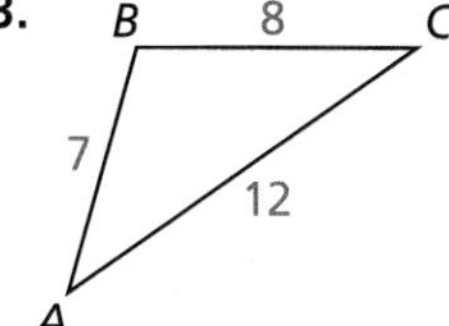

K 7 L, 6, 11, J

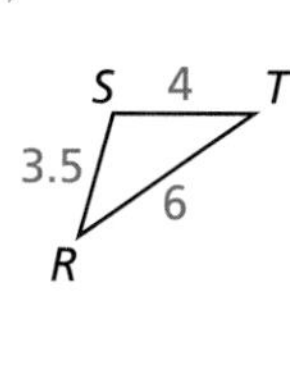

4.

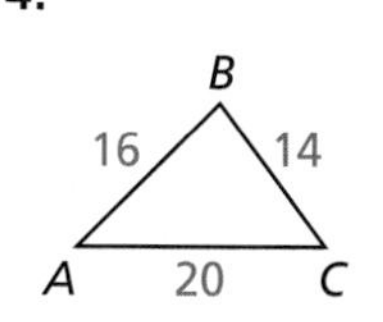

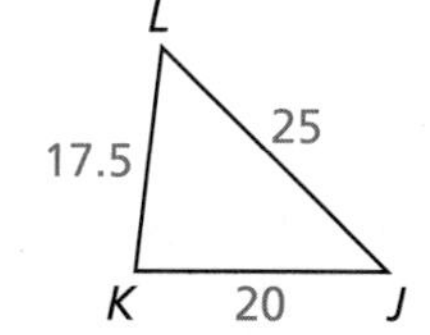

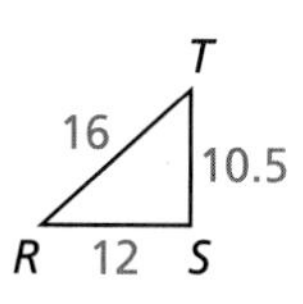

In Exercises 5 and 6, find the value of x that makes $\triangle DEF \sim \triangle XYZ$. *(See Example 2.)*

5.

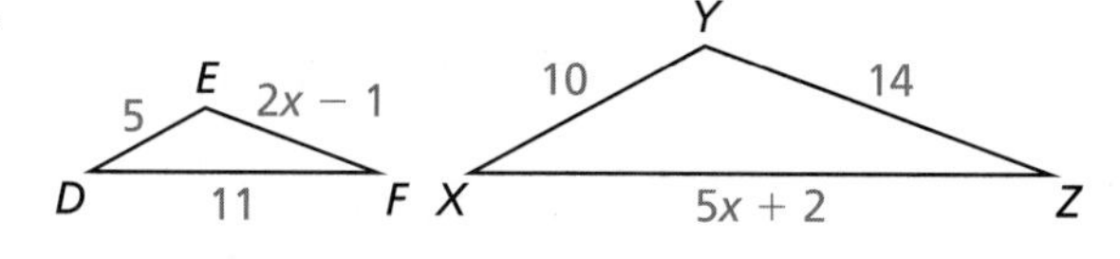

6.

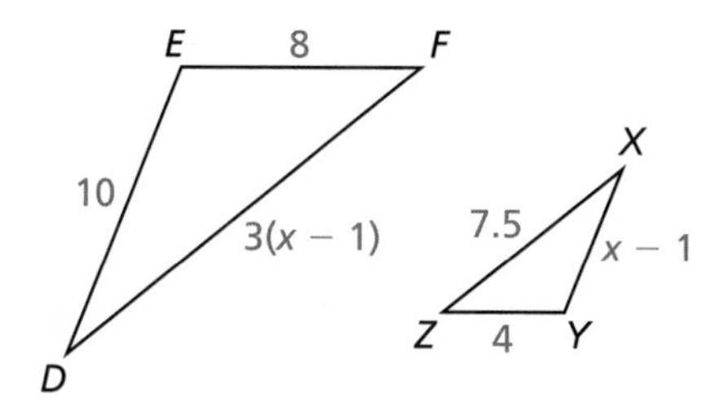

In Exercises 7 and 8, verify that $\triangle ABC \sim \triangle DEF$. Find the scale factor of $\triangle ABC$ to $\triangle DEF$.

7. $\triangle ABC$: $BC = 18$, $AB = 15$, $AC = 12$
   $\triangle DEF$: $EF = 12$, $DE = 10$, $DF = 8$

8. $\triangle ABC$: $AB = 10$, $BC = 16$, $CA = 20$
   $\triangle DEF$: $DE = 25$, $EF = 40$, $FD = 50$

In Exercises 9 and 10, determine whether the two triangles are similar. If they are similar, write a similarity statement and find the scale factor of triangle B to triangle A. *(See Example 3.)*

9.

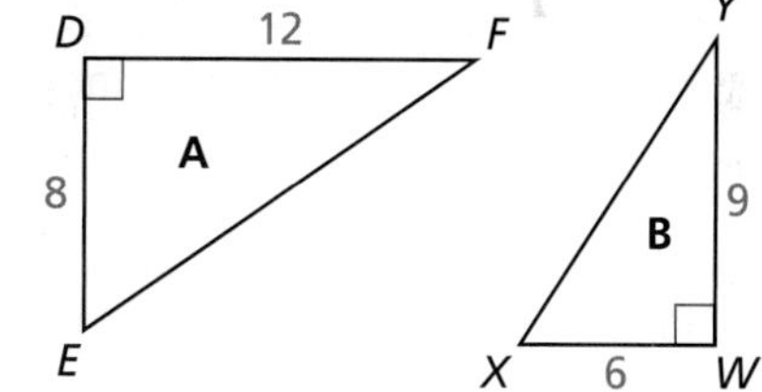

10.

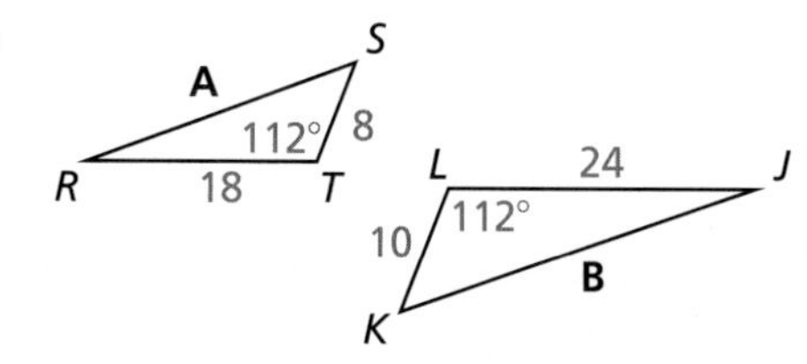

In Exercises 11 and 12, sketch the triangles using the given description. Then determine whether the two triangles can be similar.

11. In $\triangle RST$, $RS = 20$, $ST = 32$, and $m\angle S = 16°$. In $\triangle FGH$, $GH = 30$, $HF = 48$, and $m\angle H = 24°$.

12. The side lengths of $\triangle ABC$ are 24, $8x$, and 48, and the side lengths of $\triangle DEF$ are 15, 25, and $6x$.

In Exercises 13–16, show that the triangles are similar and write a similarity statement. Explain your reasoning.

13.

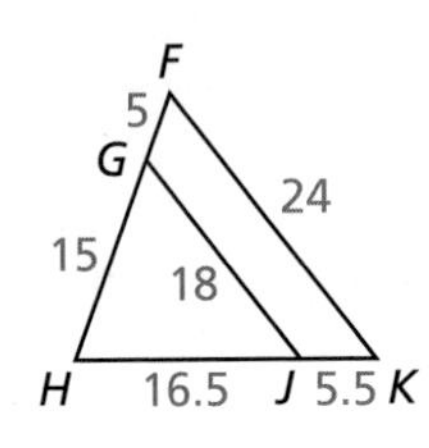

14.

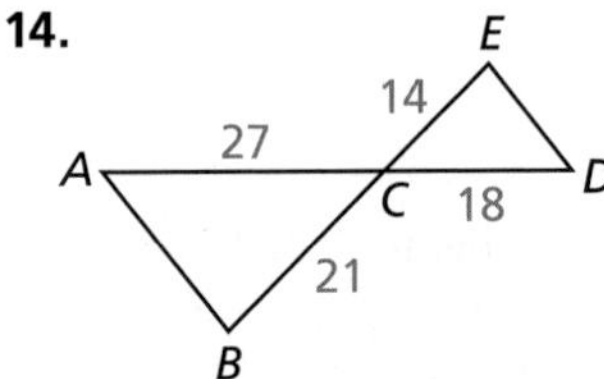

15.

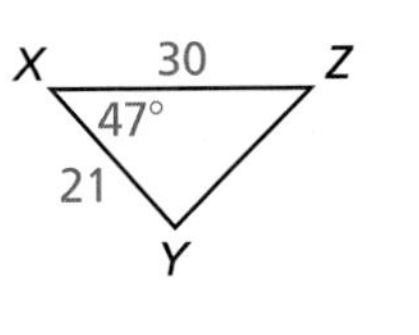

16.

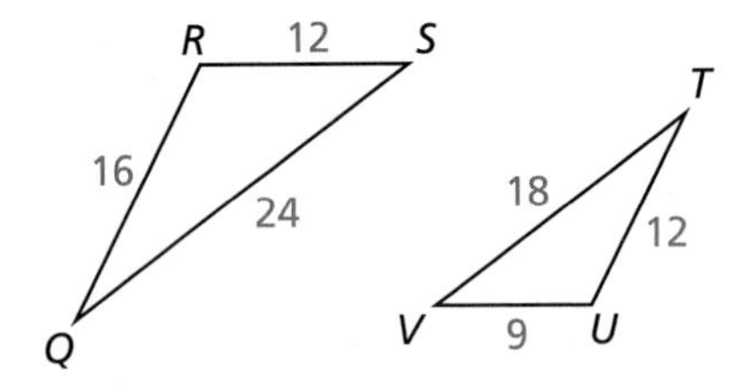

In Exercises 17 and 18, use $\triangle XYZ$.

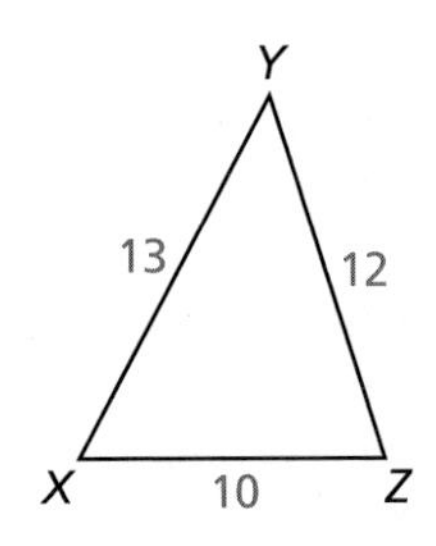

17. The shortest side of a triangle similar to $\triangle XYZ$ is 20 units long. Find the other side lengths of the triangle.

18. The longest side of a triangle similar to $\triangle XYZ$ is 39 units long. Find the other side lengths of the triangle.

19. ERROR ANALYSIS Describe and correct the error in writing a similarity statement.

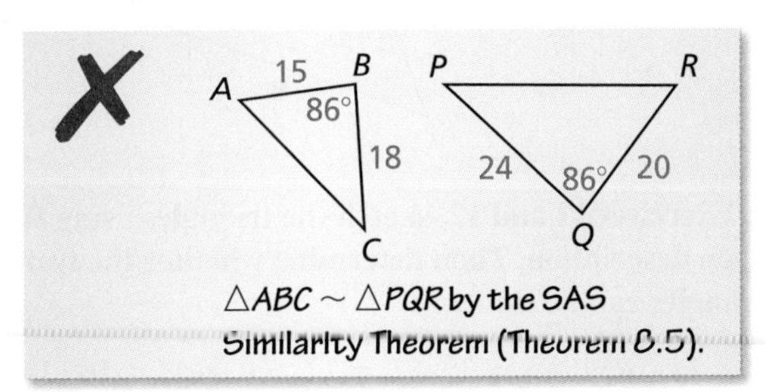

20. MATHEMATICAL CONNECTIONS Find the value of n that makes $\triangle DEF \sim \triangle XYZ$ when $DE = 4$, $EF = 5$, $XY = 4(n + 1)$, $YZ = 7n - 1$, and $\angle E \cong \angle Y$. Include a sketch.

ATTENDING TO PRECISION In Exercises 21–26, use the diagram to copy and complete the statement.

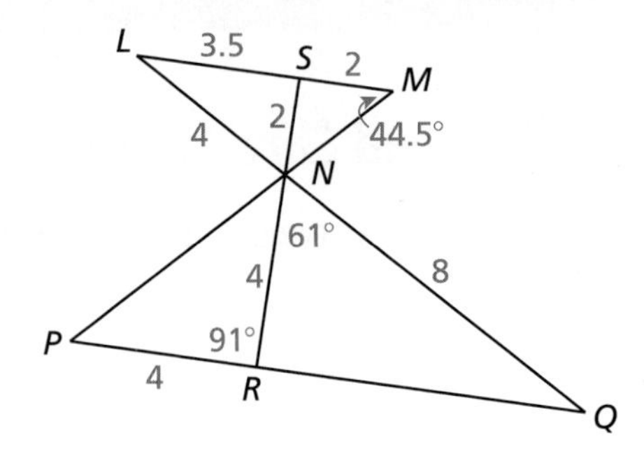

21. $m\angle LNS =$ ▭ **22.** $m\angle NRQ =$ ▭

23. $m\angle NQR =$ ▭ **24.** $RQ =$ ▭

25. $m\angle NSM =$ ▭ **26.** $m\angle NPR =$ ▭

27. MAKING AN ARGUMENT Your friend claims that $\triangle JKL \sim \triangle MNO$ by the SAS Similarity Theorem (Theorem 8.5) when $JK = 18$, $m\angle K = 130°$, $KL = 16$, $MN = 9$, $m\angle N = 65°$, and $NO = 8$. Do you support your friend's claim? Explain your reasoning.

28. ANALYZING RELATIONSHIPS Certain sections of stained glass are sold in triangular, beveled pieces. Which of the three beveled pieces, if any, are similar?

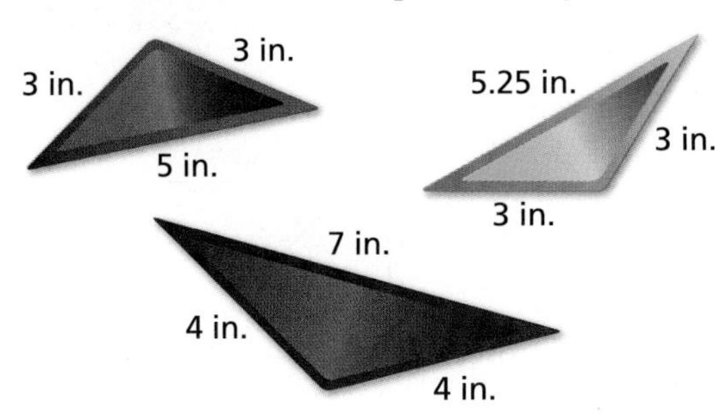

29. ATTENDING TO PRECISION In the diagram, $\frac{MN}{MR} = \frac{MP}{MQ}$. Which of the statements must be true? Select all that apply. Explain your reasoning.

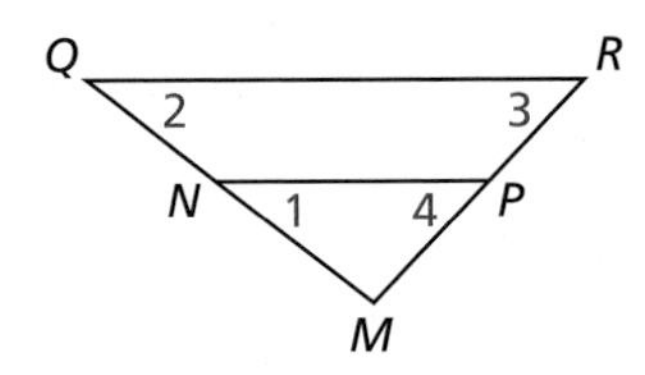

Ⓐ $\angle 1 \cong \angle 2$ Ⓑ $\overline{QR} \parallel \overline{NP}$

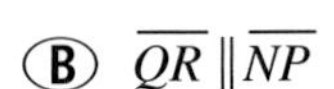

Ⓒ $\angle 1 \cong \angle 4$ Ⓓ $\triangle MNP \sim \triangle MRQ$

30. WRITING Are any two right triangles similar? Explain.

31. **MODELING WITH MATHEMATICS** In the portion of the shuffleboard court shown, $\frac{BC}{AC} = \frac{BD}{AE}$.

a. What additional information do you need to show that $\triangle BCD \sim \triangle ACE$ using the SSS Similarity Theorem (Theorem 8.4)?

b. What additional information do you need to show that $\triangle BCD \sim \triangle ACE$ using the SAS Similarity Theorem (Theorem 8.5)?

32. **PROOF** Given that $\triangle BAC$ is a right triangle and D, E, and F are midpoints, prove that $m\angle DEF = 90°$.

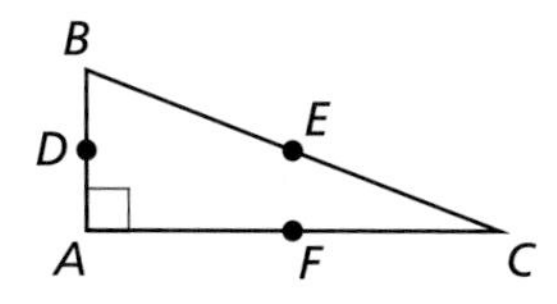

33. **PROVING A THEOREM** Write a two-column proof of the SAS Similarity Theorem (Theorem 8.5).

Given $\angle A \cong \angle D$, $\frac{AB}{DE} = \frac{AC}{DF}$

Prove $\triangle ABC \sim \triangle DEF$

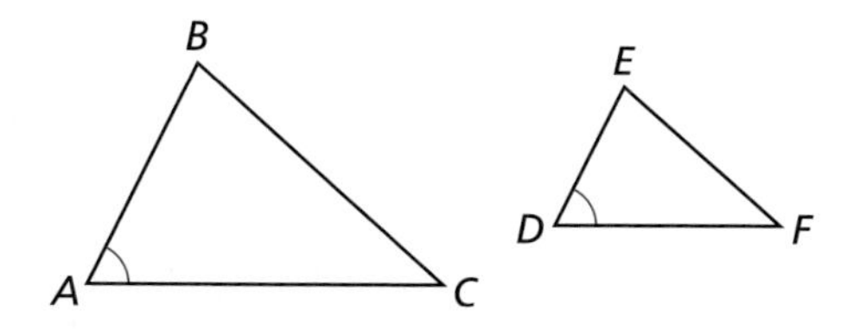

34. **CRITICAL THINKING** You are given two right triangles with one pair of corresponding legs and the pair of hypotenuses having the same length ratios.

a. The lengths of the given pair of corresponding legs are 6 and 18, and the lengths of the hypotenuses are 10 and 30. Use the Pythagorean Theorem to find the lengths of the other pair of corresponding legs. Draw a diagram.

b. Write the ratio of the lengths of the second pair of corresponding legs.

c. Are these triangles similar? Does this suggest a Hypotenuse-Leg Similarity Theorem for right triangles? Explain.

35. **WRITING** Can two triangles have all three ratios of corresponding angle measures equal to a value greater than 1? less than 1? Explain.

36. **HOW DO YOU SEE IT?** Which theorem could you use to show that $\triangle OPQ \sim \triangle OMN$ in the portion of the Ferris wheel shown when $PM = QN = 5$ feet and $MO = NO = 10$ feet?

37. **DRAWING CONCLUSIONS** Explain why it is not necessary to have an Angle-Side-Angle Similarity Theorem.

38. **THOUGHT PROVOKING** Decide whether each is a valid method of showing that two quadrilaterals are similar. Justify your answer.

a. SASA **b.** SASAS **c.** SSSS **d.** SASSS

39. **MULTIPLE REPRESENTATIONS** Use a diagram to show why there is no Side-Side-Angle Similarity Theorem.

40. **MODELING WITH MATHEMATICS** The dimensions of an actual swing set are shown. You want to create a scale model of the swing set for a dollhouse using similar triangles. Sketch a drawing of your swing set and label each side length. Write a similarity statement for each pair of similar triangles. State the scale factor you used to create the scale model.

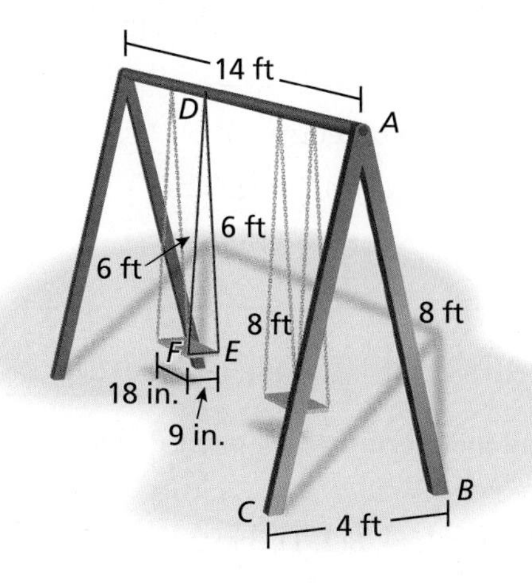

41. PROVING A THEOREM Copy and complete the paragraph proof of the second part of the Slopes of Parallel Lines Theorem (Theorem 3.13) from page 439.

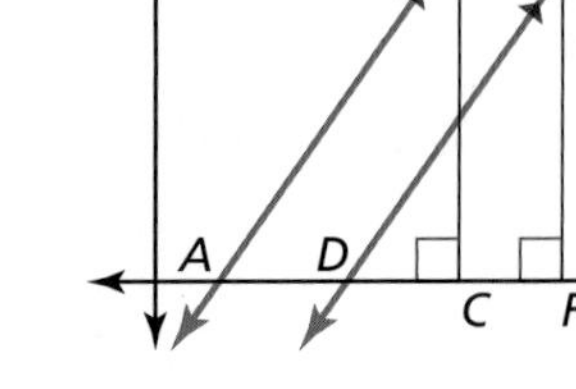

Given $m_\ell = m_n$, ℓ and n are nonvertical.

Prove $\ell \parallel n$

You are given that $m_\ell = m_n$. By the definition of slope, $m_\ell = \frac{BC}{AC}$ and $m_n = \frac{EF}{DF}$. By ____________________, $\frac{BC}{AC} = \frac{EF}{DF}$. Rewriting this proportion yields ____________.

By the Right Angles Congruence Theorem (Thm. 2.3), __________________. So, $\triangle ABC \sim \triangle DEF$ by ____________________. Because corresponding angles of similar triangles are congruent, $\angle BAC \cong \angle EDF$. By ____________________, $\ell \parallel n$.

42. PROVING A THEOREM Copy and complete the two-column proof of the second part of the Slopes of Perpendicular Lines Theorem (Theorem 3.14) from page 440.

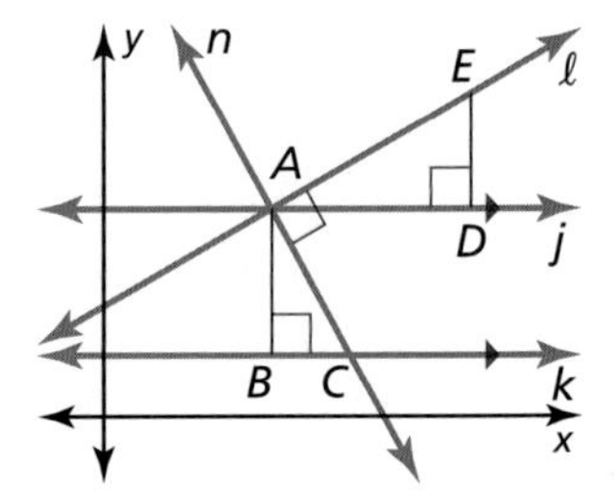

Given $m_\ell m_n = -1$, ℓ and n are nonvertical.

Prove $\ell \perp n$

STATEMENTS	REASONS
1. $m_\ell m_n = -1$	1. Given
2. $m_\ell = \frac{DE}{AD}$, $m_n = -\frac{AB}{BC}$	2. Definition of slope
3. $\frac{DE}{AD} \bullet -\frac{AB}{BC} = -1$	3. ________________
4. $\frac{DE}{AD} = \frac{BC}{AB}$	4. Multiply each side of statement 3 by $-\frac{BC}{AB}$.
5. $\frac{DE}{BC} = $ —	5. Rewrite proportion.
6. ________________	6. Right Angles Congruence Theorem (Thm. 2.3)
7. $\triangle ABC \sim \triangle ADE$	7. ________________
8. $\angle BAC \cong \angle DAE$	8. Corresponding angles of similar figures are congruent.
9. $\angle BCA \cong \angle CAD$	9. Alternate Interior Angles Theorem (Thm. 3.2)
10. $m\angle BAC = m\angle DAE$, $m\angle BCA = m\angle CAD$	10. ________________
11. $m\angle BAC + m\angle BCA + 90° = 180°$	11. ________________
12. ________________	12. Subtraction Property of Equality
13. $m\angle CAD + m\angle DAE = 90°$	13. Substitution Property of Equality
14. $m\angle CAE = m\angle DAE + m\angle CAD$	14. Angle Addition Postulate (Post. 1.4)
15. $m\angle CAE = 90°$	15. ________________
16. ________________	16. Definition of perpendicular lines

Maintaining Mathematical Proficiency

Reviewing what you learned in previous grades and lessons

Find the coordinates of point *P* along the directed line segment *AB* so that *AP* to *PB* is the given ratio. *(Section 3.5)*

43. $A(-3, 6)$, $B(2, 1)$; 3 to 2 **44.** $A(-3, -5)$, $B(9, -1)$; 1 to 3 **45.** $A(1, -2)$, $B(8, 12)$; 4 to 3

8.4 Proportionality Theorems

Essential Question What proportionality relationships exist in a triangle intersected by an angle bisector or by a line parallel to one of the sides?

EXPLORATION 1 **Discovering a Proportionality Relationship**

Work with a partner. Use dynamic geometry software to draw any $\triangle ABC$.

a. Construct $\overline{DE}$ parallel to $\overline{BC}$ with endpoints on $\overline{AB}$ and $\overline{AC}$, respectively.

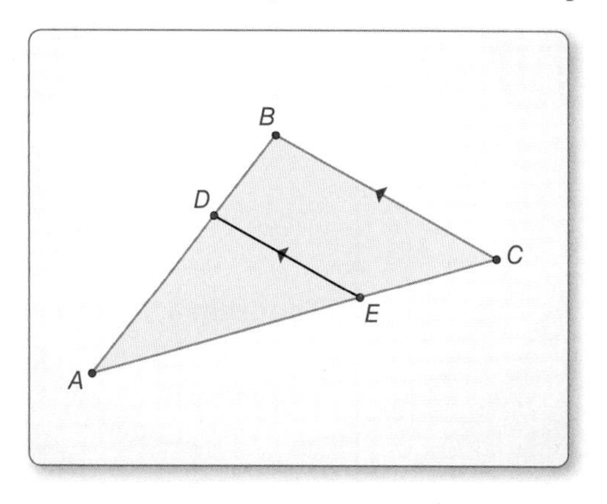

LOOKING FOR STRUCTURE

To be proficient in math, you need to look closely to discern a pattern or structure.

b. Compare the ratios of AD to BD and AE to CE.

c. Move $\overline{DE}$ to other locations parallel to $\overline{BC}$ with endpoints on $\overline{AB}$ and $\overline{AC}$, and repeat part (b).

d. Change $\triangle ABC$ and repeat parts (a)–(c) several times. Write a conjecture that summarizes your results.

EXPLORATION 2 **Discovering a Proportionality Relationship**

Work with a partner. Use dynamic geometry software to draw any $\triangle ABC$.

a. Bisect $\angle B$ and plot point D at the intersection of the angle bisector and $\overline{AC}$.

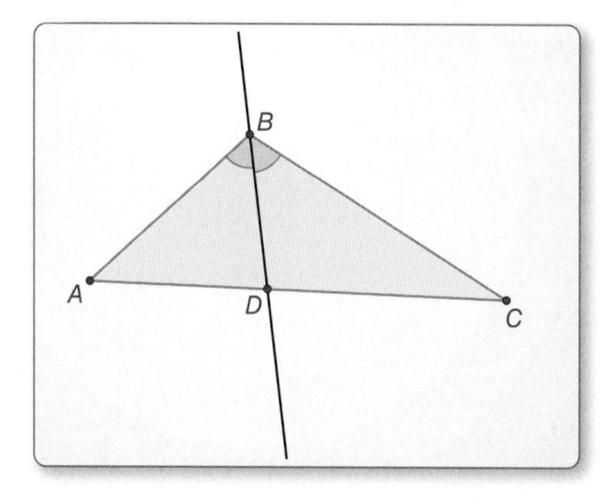

b. Compare the ratios of AD to DC and BA to BC.

c. Change $\triangle ABC$ and repeat parts (a) and (b) several times. Write a conjecture that summarizes your results.

Communicate Your Answer

3. What proportionality relationships exist in a triangle intersected by an angle bisector or by a line parallel to one of the sides?

4. Use the figure at the right to write a proportion.

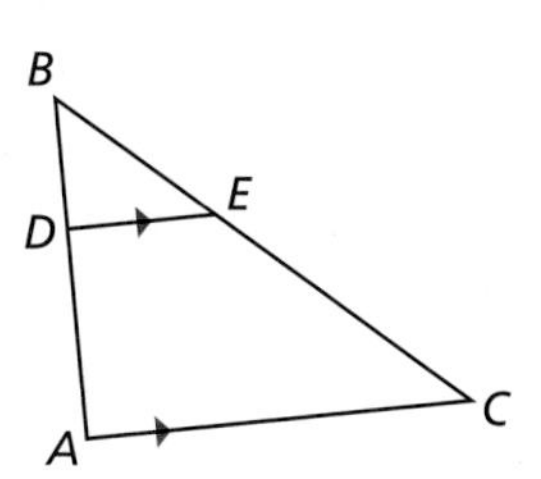

8.4 Lesson

Core Vocabulary

Previous
corresponding angles
ratio
proportion

What You Will Learn

- Use the Triangle Proportionality Theorem and its converse.
- Use other proportionality theorems.

Using the Triangle Proportionality Theorem

Theorems

Theorem 8.6 Triangle Proportionality Theorem

If a line parallel to one side of a triangle intersects the other two sides, then it divides the two sides proportionally.

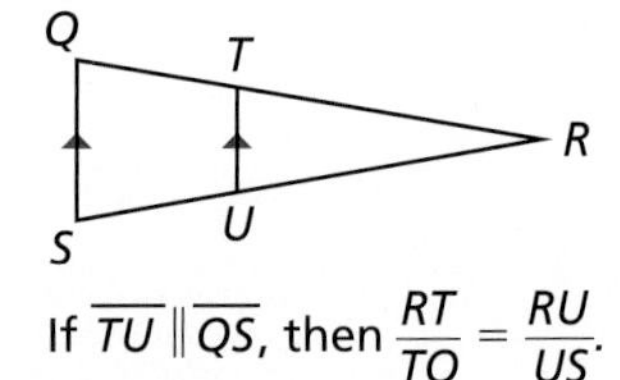

If $\overline{TU} \parallel \overline{QS}$, then $\frac{RT}{TQ} = \frac{RU}{US}$.

Proof Ex. 27, p. 451

Theorem 8.7 Converse of the Triangle Proportionality Theorem

If a line divides two sides of a triangle proportionally, then it is parallel to the third side.

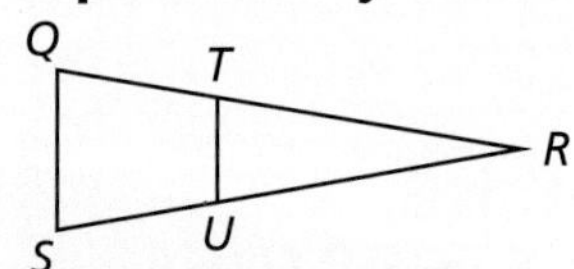

If $\frac{RT}{TQ} = \frac{RU}{US}$, then $\overline{TU} \parallel \overline{QS}$.

Proof Ex. 28, p. 451

EXAMPLE 1 Finding the Length of a Segment

In the diagram, $\overline{QS} \parallel \overline{UT}$, $RS = 4$, $ST = 6$, and $QU = 9$. What is the length of $\overline{RQ}$?

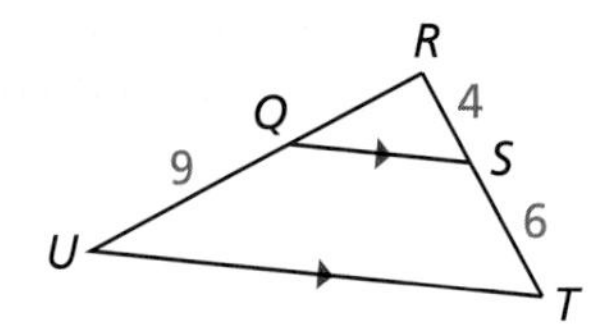

SOLUTION

$\frac{RQ}{QU} = \frac{RS}{ST}$ Triangle Proportionality Theorem

$\frac{RQ}{9} = \frac{4}{6}$ Substitute.

$RQ = 6$ Multiply each side by 9 and simplify.

- The length of $\overline{RQ}$ is 6 units.

Monitoring Progress

Help in English and Spanish at *BigIdeasMath.com*

1. Find the length of $\overline{YZ}$.

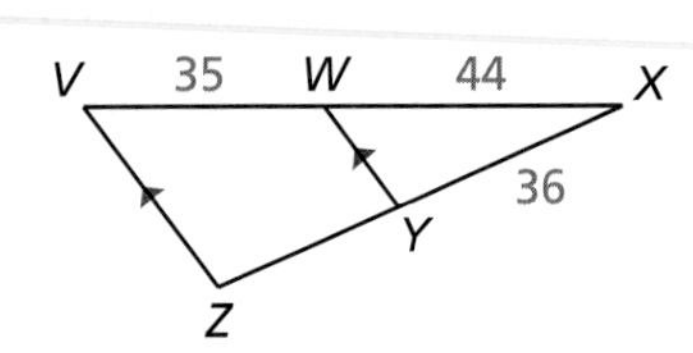

The theorems on the previous page also imply the following:

Contrapositive of the Triangle Proportionality Theorem

If $\frac{RT}{TQ} \neq \frac{RU}{US}$, then $\overline{TU} \nparallel \overline{QS}$.

Inverse of the Triangle Proportionality Theorem

If $\overline{TU} \nparallel \overline{QS}$, then $\frac{RT}{TQ} \neq \frac{RU}{US}$.

EXAMPLE 2 Solving a Real-Life Problem

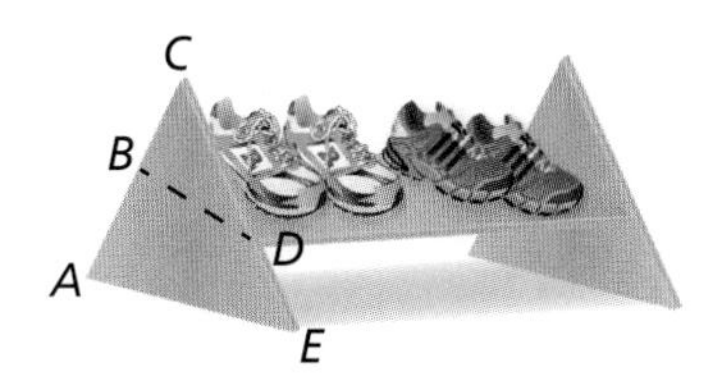

On the shoe rack shown, $BA = 33$ centimeters, $CB = 27$ centimeters, $CD = 44$ centimeters, and $DE = 25$ centimeters. Explain why the shelf is not parallel to the floor.

SOLUTION

Find and simplify the ratios of the lengths.

$$\frac{CD}{DE} = \frac{44}{25} \qquad \frac{CB}{BA} = \frac{27}{33} = \frac{9}{11}$$

▶ Because $\frac{44}{25} \neq \frac{9}{11}$, $\overline{BD}$ is not parallel to $\overline{AE}$. So, the shelf is not parallel to the floor.

Monitoring Progress  Help in English and Spanish at *BigIdeasMath.com*

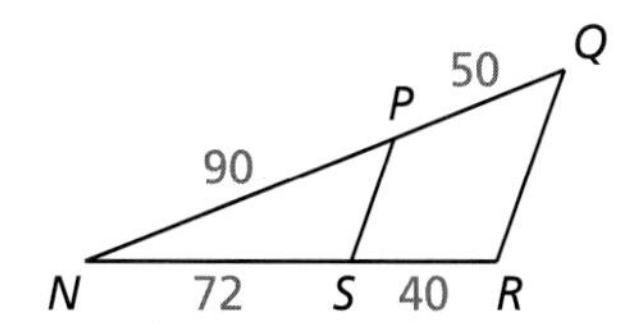

2. Determine whether $\overline{PS} \parallel \overline{QR}$.

Recall that you partitioned a directed line segment in the coordinate plane in Section 3.5. You can apply the Triangle Proportionality Theorem to construct a point along a directed line segment that partitions the segment in a given ratio.

CONSTRUCTION Constructing a Point along a Directed Line Segment

Construct the point L on $\overline{AB}$ so that the ratio of AL to LB is 3 to 1.

SOLUTION

Step 1

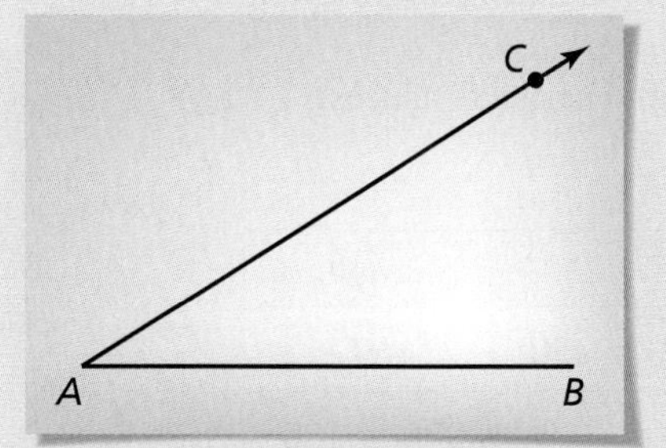

Draw a segment and a ray Draw $\overline{AB}$ of any length. Choose any point C not on $\overleftrightarrow{AB}$. Draw $\overrightarrow{AC}$.

Step 2

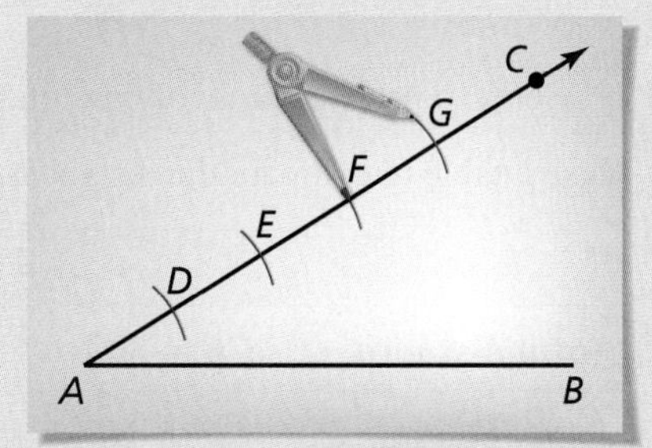

Draw arcs Place the point of a compass at A and make an arc of any radius intersecting $\overrightarrow{AC}$. Label the point of intersection D. Using the same compass setting, make three more arcs on $\overrightarrow{AC}$, as shown. Label the points of intersection E, F, and G and note that $AD = DE = EF = FG$.

Step 3

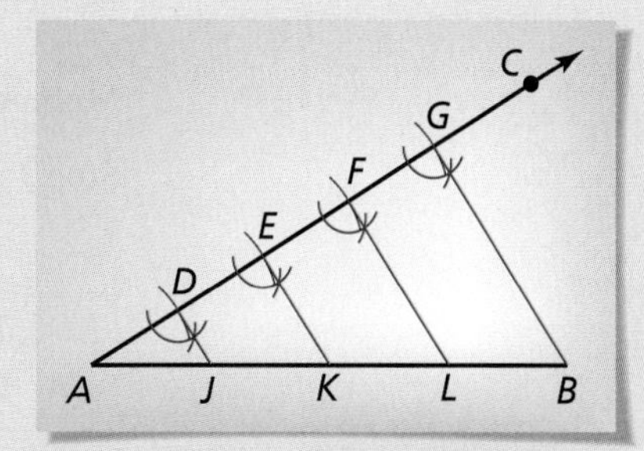

Draw a segment Draw $\overline{GB}$. Copy $\angle AGB$ and construct congruent angles at D, E, and F with sides that intersect $\overline{AB}$ at J, K, and L. Sides $\overline{DJ}$, $\overline{EK}$, and $\overline{FL}$ are all parallel, and they divide $\overline{AB}$ equally. So, $AJ = JK = KL = LB$. Point L divides directed line segment AB in the ratio 3 to 1.

Using Other Proportionality Theorems

Theorem

Theorem 8.8 Three Parallel Lines Theorem

If three parallel lines intersect two transversals, then they divide the transversals proportionally.

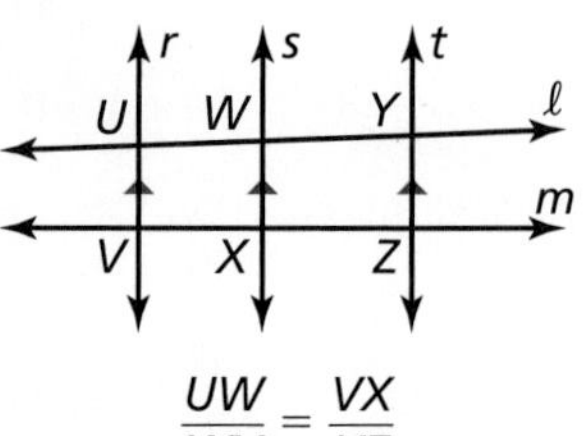

$$\frac{UW}{WY} = \frac{VX}{XZ}$$

Proof Ex. 32, p. 451

EXAMPLE 3 Using the Three Parallel Lines Theorem

In the diagram, $\angle 1$, $\angle 2$, and $\angle 3$ are all congruent, $GF = 120$ yards, $DE = 150$ yards, and $CD = 300$ yards. Find the distance HF between Main Street and South Main Street.

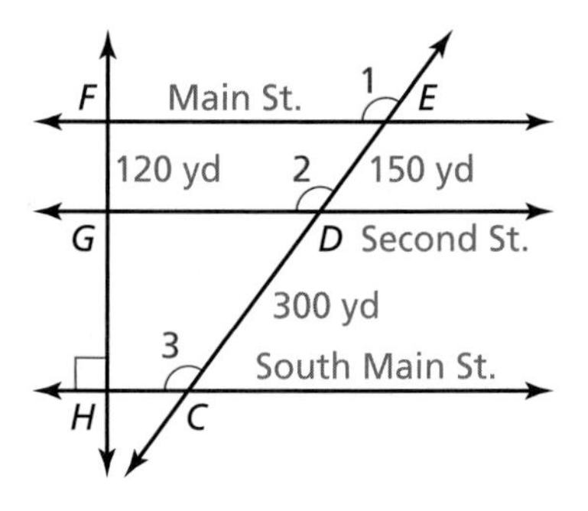

SOLUTION

Corresponding angles are congruent, so $\overleftrightarrow{FE}$, $\overleftrightarrow{GD}$, and $\overleftrightarrow{HC}$ are parallel. There are different ways you can write a proportion to find HG.

Method 1 Use the Three Parallel Lines Theorem to set up a proportion.

$\frac{HG}{GF} = \frac{CD}{DE}$ Three Parallel Lines Theorem

$\frac{HG}{120} = \frac{300}{150}$ Substitute.

$HG = 240$ Multiply each side by 120 and simplify.

By the Segment Addition Postulate (Postulate 1.2), $HF = HG + GF = 240 + 120 = 360$.

▶ The distance between Main Street and South Main Street is 360 yards.

Method 2 Set up a proportion involving total and partial distances.

Step 1 Make a table to compare the distances.

	$\overleftrightarrow{CE}$	$\overleftrightarrow{HF}$
Total distance	$CE = 300 + 150 = 450$	HF
Partial distance	$DE = 150$	$GF = 120$

Step 2 Write and solve a proportion.

$\frac{450}{150} = \frac{HF}{120}$ Write proportion.

$360 = HF$ Multiply each side by 120 and simplify.

▶ The distance between Main Street and South Main Street is 360 yards.

Theorem

Theorem 8.9 Triangle Angle Bisector Theorem

If a ray bisects an angle of a triangle, then it divides the opposite side into segments whose lengths are proportional to the lengths of the other two sides.

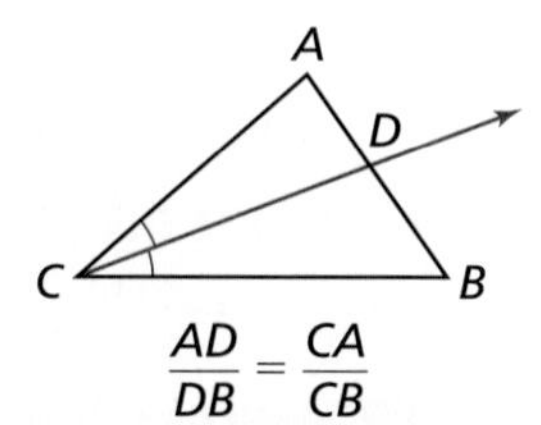

$$\frac{AD}{DB} = \frac{CA}{CB}$$

Proof Ex. 35, p. 452

EXAMPLE 4 Using the Triangle Angle Bisector Theorem

In the diagram, $\angle QPR \cong \angle RPS$. Use the given side lengths to find the length of $\overline{RS}$.

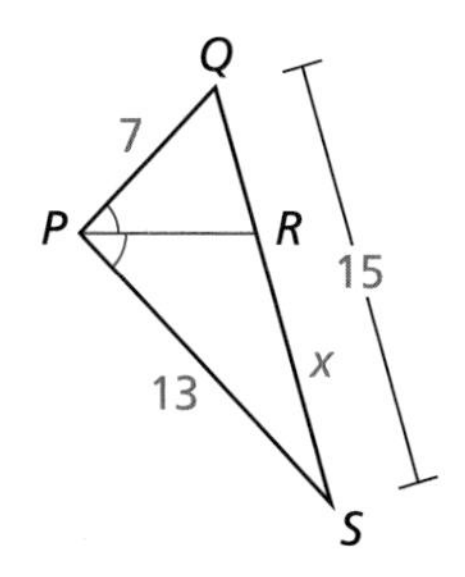

SOLUTION

Because $\overrightarrow{PR}$ is an angle bisector of $\angle QPS$, you can apply the Triangle Angle Bisector Theorem. Let $RS = x$. Then $RQ = 15 - x$.

$\frac{RQ}{RS} = \frac{PQ}{PS}$	Triangle Angle Bisector Theorem
$\frac{15 - x}{x} = \frac{7}{13}$	Substitute.
$195 - 13x = 7x$	Cross Products Property
$9.75 = x$	Solve for *x*.

▶ The length of $\overline{RS}$ is 9.75 units.

Monitoring Progress

Help in English and Spanish at *BigIdeasMath.com*

Find the length of the given line segment.

3. $\overline{BD}$

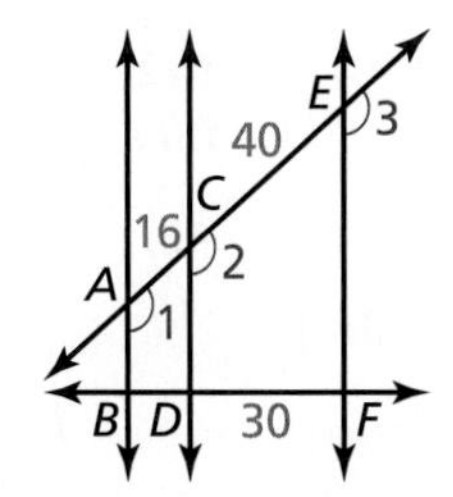

4. $\overline{JM}$

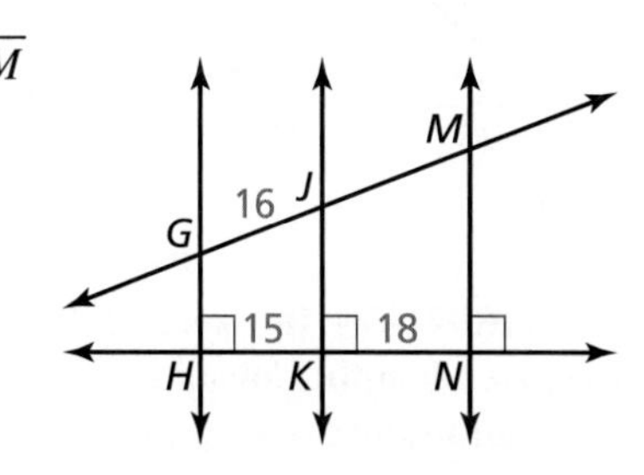

Find the value of the variable.

5.

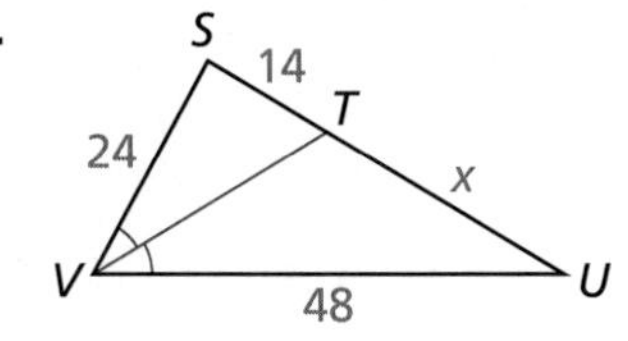

6.

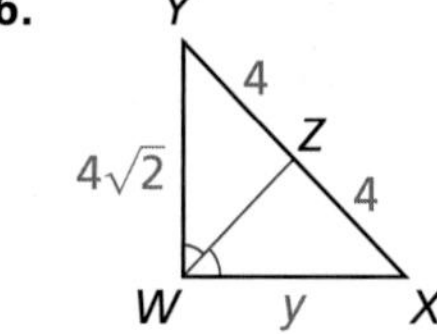

8.4 Exercises

Dynamic Solutions available at *BigIdeasMath.com*

Vocabulary and Core Concept Check

1. **COMPLETE THE STATEMENT** If a line divides two sides of a triangle proportionally, then it is _________ to the third side. This theorem is known as the ___________.

2. **VOCABULARY** In $\triangle ABC$, point R lies on $\overline{BC}$ and $\overrightarrow{AR}$ bisects $\angle CAB$. Write the proportionality statement for the triangle that is based on the Triangle Angle Bisector Theorem (Theorem 8.9).

Monitoring Progress and Modeling with Mathematics

In Exercises 3 and 4, find the length of $\overline{AB}$. ***(See Example 1.)***

3.

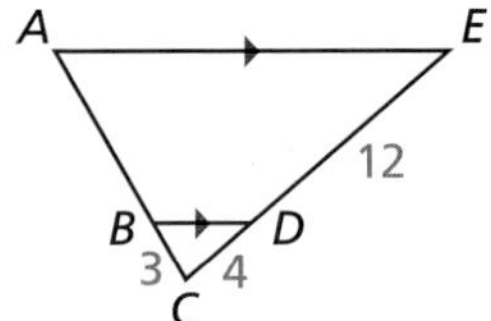

4.

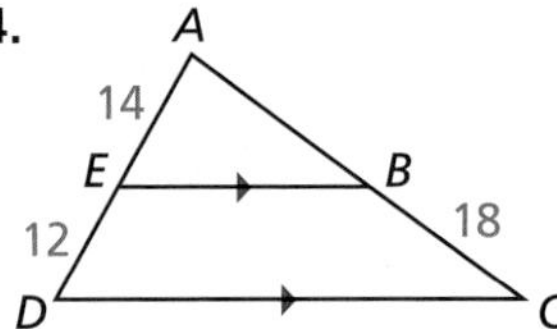

In Exercises 5–8, determine whether $\overline{KM} \parallel \overline{JN}$. ***(See Example 2.)***

5.

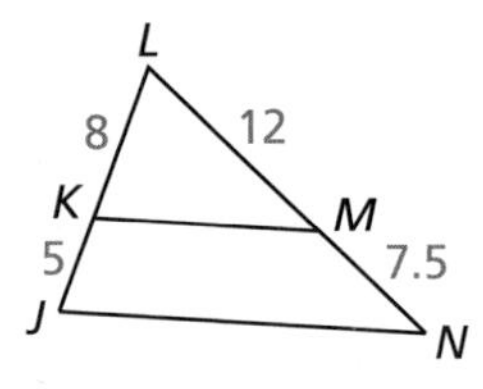

6.

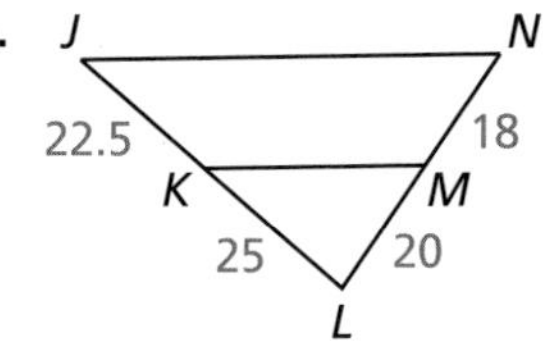

7.

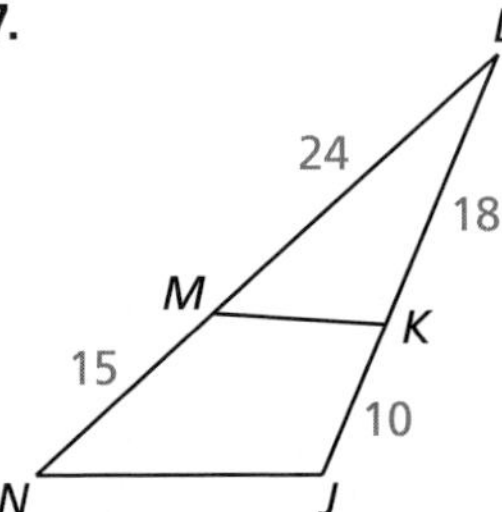

8.

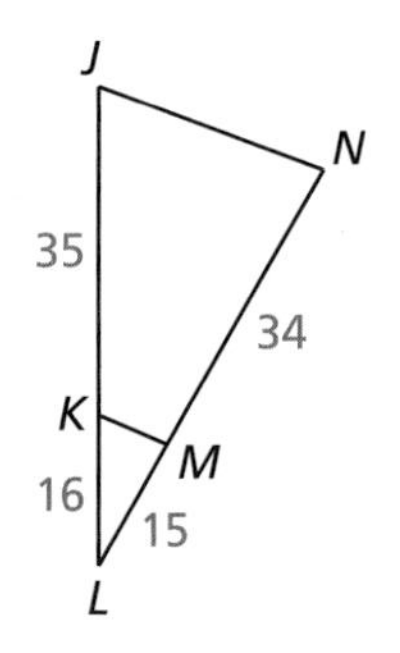

CONSTRUCTION In Exercises 9–12, draw a segment with the given length. Construct the point that divides the segment in the given ratio.

9. 3 in.; 1 to 4

10. 2 in.; 2 to 3

11. 12 cm; 1 to 3

12. 9 cm; 2 to 5

In Exercises 13–16, use the diagram to complete the proportion.

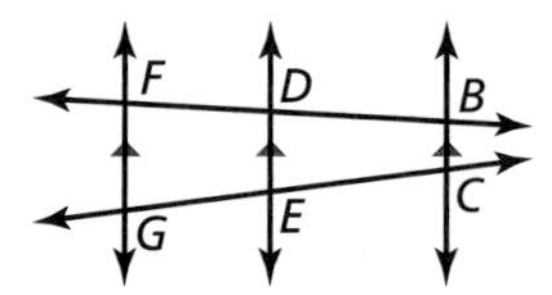

13. $\dfrac{BD}{BF} = \dfrac{\square}{CG}$

14. $\dfrac{CG}{\square} = \dfrac{BF}{DF}$

15. $\dfrac{EG}{CE} = \dfrac{DF}{\square}$

16. $\dfrac{\square}{BD} = \dfrac{CG}{CE}$

In Exercises 17 and 18, find the length of the indicated line segment. ***(See Example 3.)***

17. $\overline{VX}$

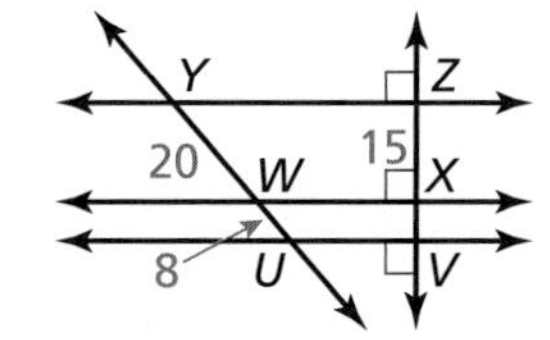

18. $\overline{SU}$

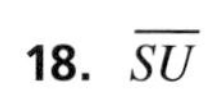

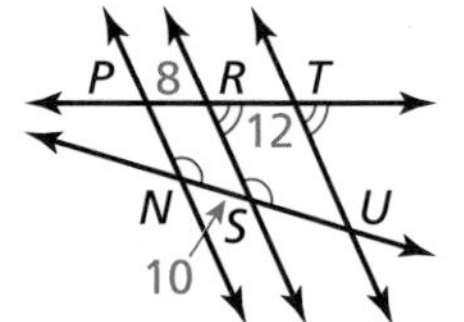

In Exercises 19–22, find the value of the variable. ***(See Example 4.)***

19.

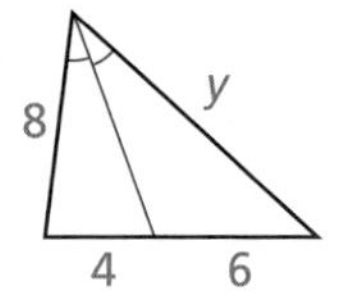

20.

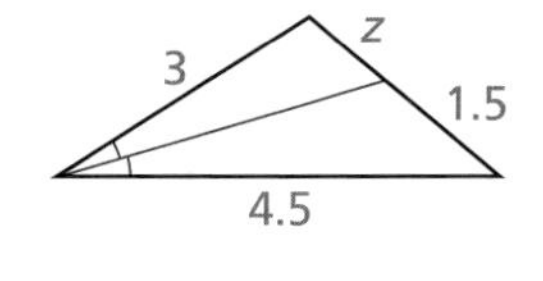

21.

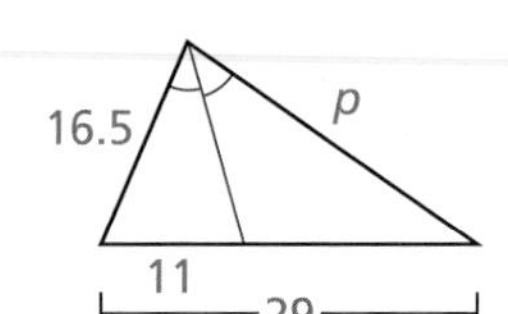

22.

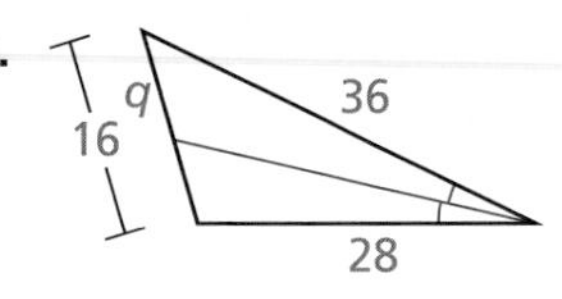

23. ERROR ANALYSIS Describe and correct the error in solving for x.

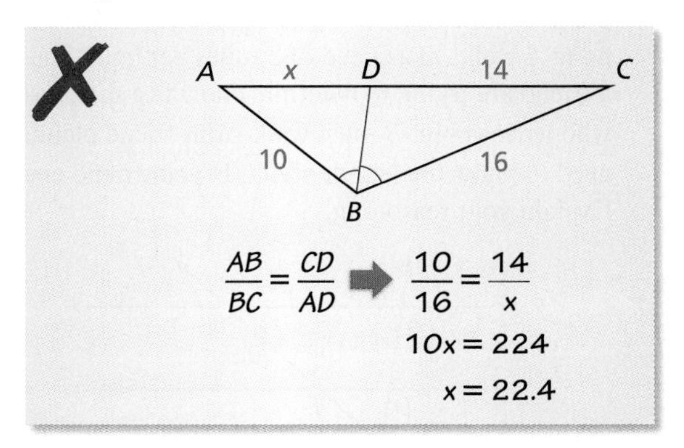

24. ERROR ANALYSIS Describe and correct the error in the student's reasoning.

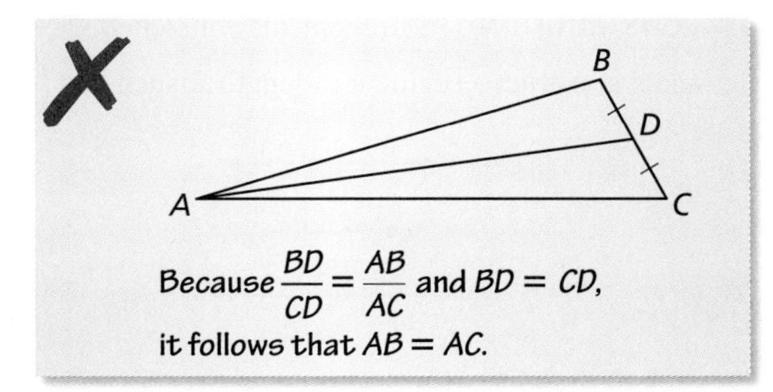

MATHEMATICAL CONNECTIONS In Exercises 25 and 26, find the value of x for which $\overline{PQ} \parallel \overline{RS}$.

25.

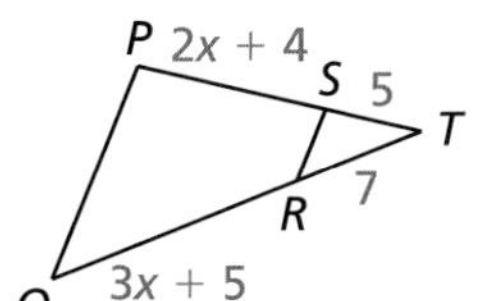

26.

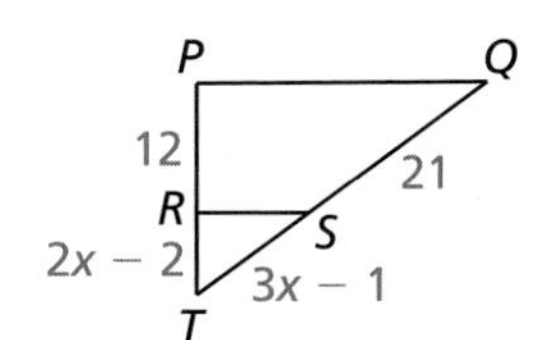

27. PROVING A THEOREM Prove the Triangle Proportionality Theorem (Theorem 8.6).

Given $\overline{QS} \parallel \overline{TU}$

Prove $\frac{QT}{TR} = \frac{SU}{UR}$

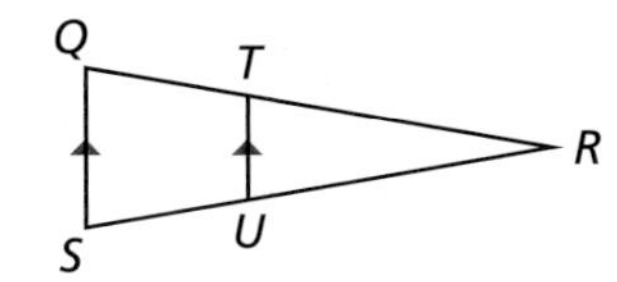

28. PROVING A THEOREM Prove the Converse of the Triangle Proportionality Theorem (Theorem 8.7).

Given $\frac{ZY}{YW} = \frac{ZX}{XV}$

Prove $\overline{YX} \parallel \overline{WV}$

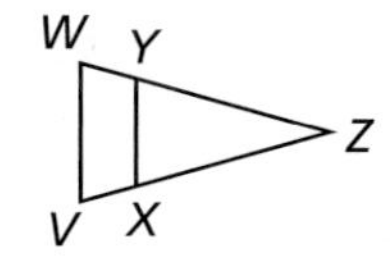

29. MODELING WITH MATHEMATICS The real estate term *lake frontage* refers to the distance along the edge of a piece of property that touches a lake.

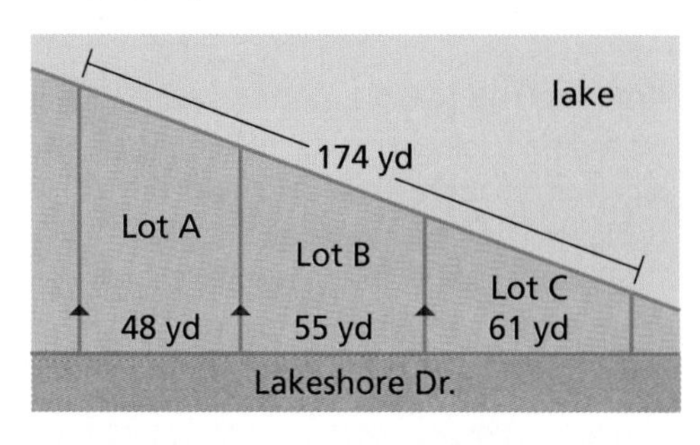

a. Find the lake frontage (to the nearest tenth) of each lot shown.

b. In general, the more lake frontage a lot has, the higher its selling price. Which lot(s) should be listed for the highest price?

c. Suppose that lot prices are in the same ratio as lake frontages. If the least expensive lot is $250,000, what are the prices of the other lots? Explain your reasoning.

30. USING STRUCTURE Use the diagram to find the values of x and y.

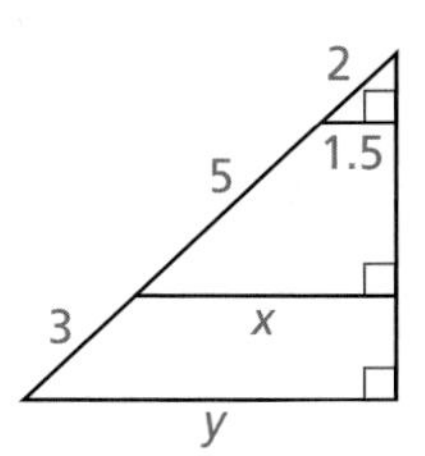

31. REASONING In the construction on page 447, explain why you can apply the Triangle Proportionality Theorem (Theorem 8.6) in Step 3.

32. PROVING A THEOREM Use the diagram with the auxiliary line drawn to write a paragraph proof of the Three Parallel Lines Theorem (Theorem 8.8).

Given $k_1 \parallel k_2 \parallel k_3$

Prove $\frac{CB}{BA} = \frac{DE}{EF}$

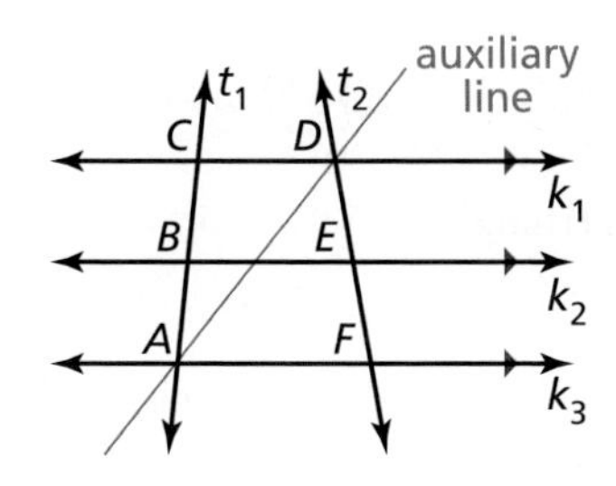

33. CRITICAL THINKING In $\triangle LMN$, the angle bisector of $\angle M$ also bisects $\overline{LN}$. Classify $\triangle LMN$ as specifically as possible. Justify your answer.

34. HOW DO YOU SEE IT? During a football game, the quarterback throws the ball to the receiver. The receiver is between two defensive players, as shown. If Player 1 is closer to the quarterback when the ball is thrown and both defensive players move at the same speed, which player will reach the receiver first? Explain your reasoning.

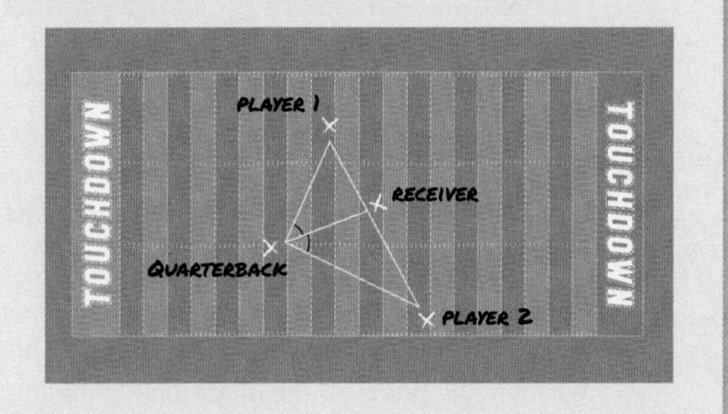

35. PROVING A THEOREM Use the diagram with the auxiliary lines drawn to write a paragraph proof of the Triangle Angle Bisector Theorem (Theorem 8.9).

Given $\angle YXW \cong \angle WXZ$

Prove $\frac{YW}{WZ} = \frac{XY}{XZ}$

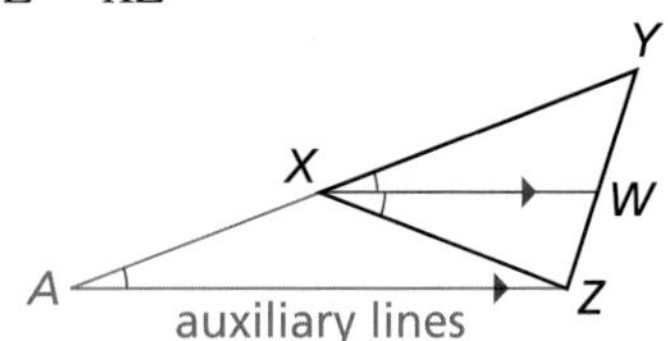

36. THOUGHT PROVOKING Write the converse of the Triangle Angle Bisector Theorem (Theorem 8.9). Is the converse true? Justify your answer.

37. REASONING How is the Triangle Midsegment Theorem (Theorem 6.8) related to the Triangle Proportionality Theorem (Theorem 8.6)? Explain your reasoning.

38. MAKING AN ARGUMENT Two people leave points A and B at the same time. They intend to meet at point C at the same time. The person who leaves point A walks at a speed of 3 miles per hour. You and a friend are trying to determine how fast the person who leaves point B must walk. Your friend claims you need to know the length of $\overline{AC}$. Is your friend correct? Explain your reasoning.

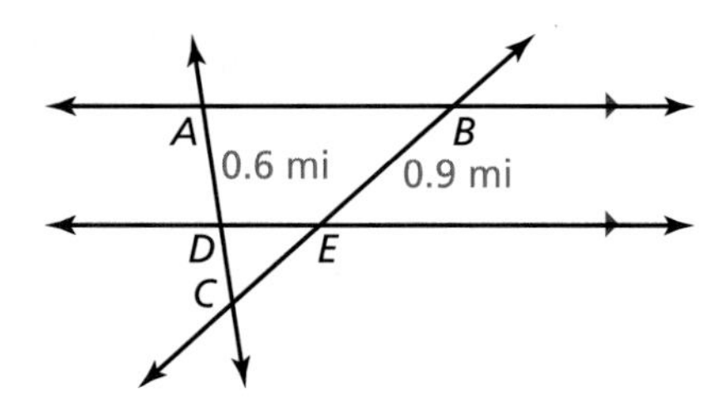

39. CONSTRUCTION Given segments with lengths r, s, and t, construct a segment of length x, such that $\frac{r}{s} = \frac{t}{x}$.

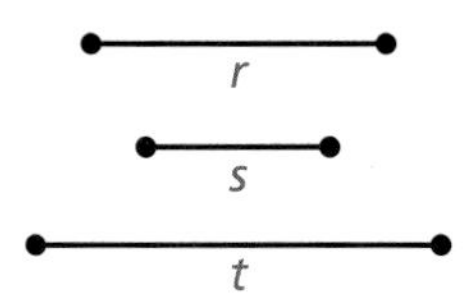

40. PROOF Prove *Ceva's Theorem*: If P is any point inside $\triangle ABC$, then $\frac{AY}{YC} \cdot \frac{CX}{XB} \cdot \frac{BZ}{ZA} = 1$.

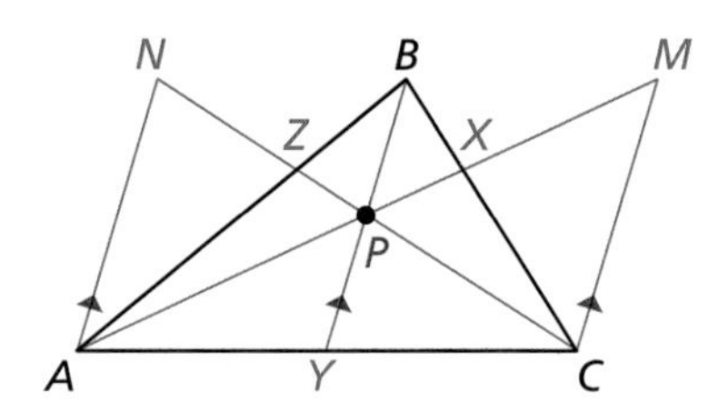

(*Hint*: Draw segments parallel to $\overline{BY}$ through A and C, as shown. Apply the Triangle Proportionality Theorem (Theorem 8.6) to $\triangle ACM$. Show that $\triangle APN \sim \triangle MPC$, $\triangle CXM \sim \triangle BXP$, and $\triangle BZP \sim \triangle AZN$.)

Maintaining Mathematical Proficiency

Reviewing what you learned in previous grades and lessons

Use the triangle. *(Section 5.5)*

41. Which sides are the legs?

42. Which side is the hypotenuse?

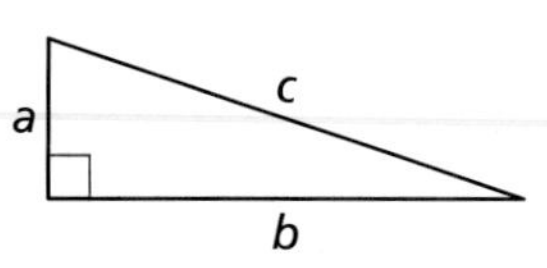

Solve the equation. *(Skills Review Handbook)*

43. $x^2 = 121$

44. $x^2 + 16 = 25$

45. $36 + x^2 = 85$

8.3–8.4 What Did You Learn?

Core Concepts

Section 8.3

Theorem 8.4 Side-Side-Side (SSS) Similarity Theorem, *p. 436*
Theorem 8.5 Side-Angle-Side (SAS) Similarity Theorem, *p. 438*
Proving Slope Criteria Using Similar Triangles, *p. 439*

Section 8.4

Theorem 8.6 Triangle Proportionality Theorem, *p. 446*
Theorem 8.7 Converse of the Triangle Proportionality Theorem, *p. 446*
Theorem 8.8 Three Parallel Lines Theorem, *p. 448*
Theorem 8.9 Triangle Angle Bisector Theorem, *p. 449*

Mathematical Practices

1. In Exercise 17 on page 442, why must you be told which side is 20 units long?
2. In Exercise 42 on page 444, analyze the given statement. Describe the relationship between the slopes of the lines.
3. In Exercise 4 on page 450, is it better to use $\frac{7}{6}$ or 1.17 as your ratio of the lengths when finding the length of $\overline{AB}$? Explain your reasoning.

Performance Task

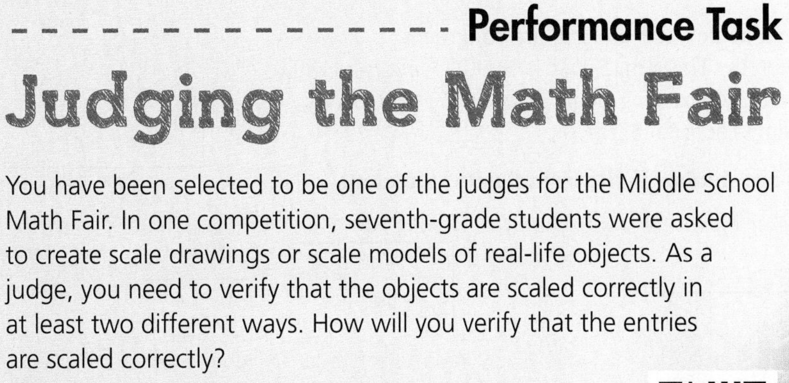

Judging the Math Fair

You have been selected to be one of the judges for the Middle School Math Fair. In one competition, seventh-grade students were asked to create scale drawings or scale models of real-life objects. As a judge, you need to verify that the objects are scaled correctly in at least two different ways. How will you verify that the entries are scaled correctly?

To explore the answers to this question and more, go to ***BigIdeasMath.com***.

8 Chapter Review

Dynamic Solutions available at *BigIdeasMath.com*

8.1 Similar Polygons (pp. 417–426)

In the diagram, $EHGF \sim KLMN$. Find the scale factor from $EHGF$ to $KLMN$. Then list all pairs of congruent angles and write the ratios of the corresponding side lengths in a statement of proportionality.

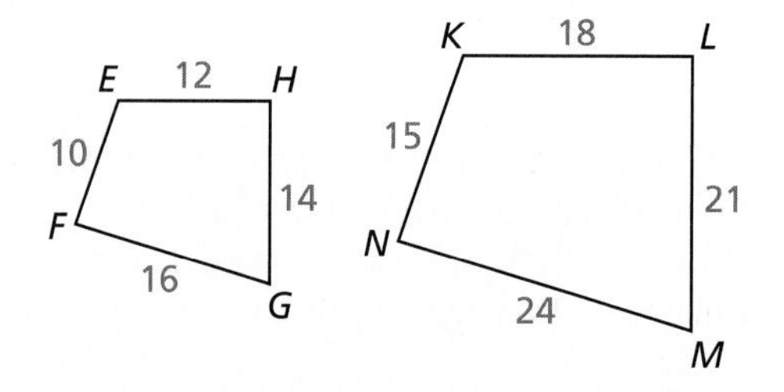

From the diagram, you can see that $\overline{EH}$ and $\overline{KL}$ are corresponding sides. So, the scale factor of $EHGF$ to $KLMN$ is $\frac{KL}{EH} = \frac{18}{12} = \frac{3}{2}$.

$\angle E \cong \angle K$, $\angle H \cong \angle L$, $\angle G \cong \angle M$, and $\angle F \cong \angle N$.

$\frac{KL}{EH} = \frac{LM}{HG} = \frac{MN}{GF} = \frac{NK}{FE}$

Find the scale factor. Then list all pairs of congruent angles and write the ratios of the corresponding side lengths in a statement of proportionality.

1. $ABCD \sim EFGH$

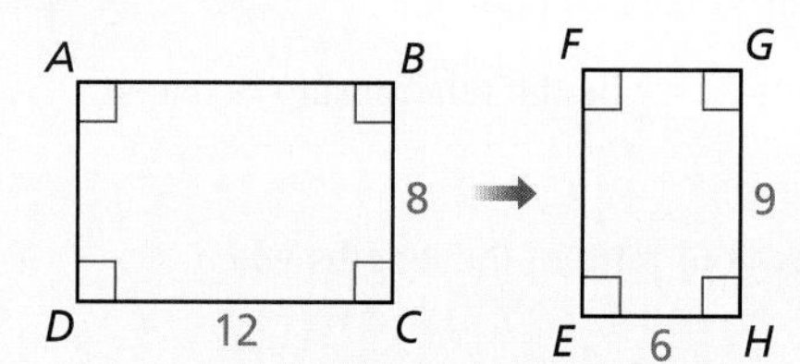

2. $\triangle XYZ \sim \triangle RPQ$

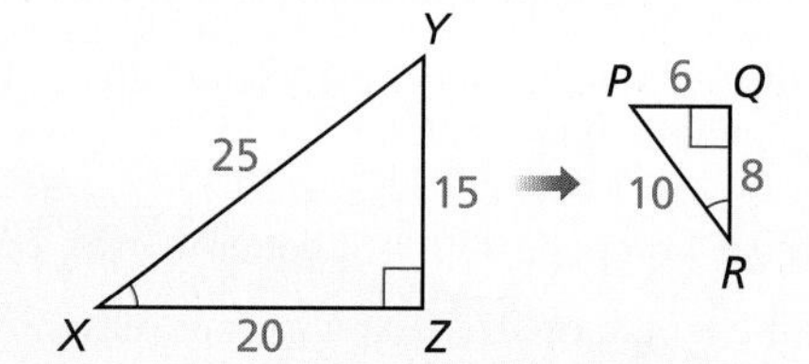

3. Two similar triangles have a scale factor of 3 : 5. The altitude of the larger triangle is 24 inches. What is the altitude of the smaller triangle?

4. Two similar triangles have a pair of corresponding sides of length 12 meters and 8 meters. The larger triangle has a perimeter of 48 meters and an area of 180 square meters. Find the perimeter and area of the smaller triangle.

8.2 Proving Triangle Similarity by AA (pp. 427–432)

Determine whether the triangles are similar. If they are, write a similarity statement. Explain your reasoning.

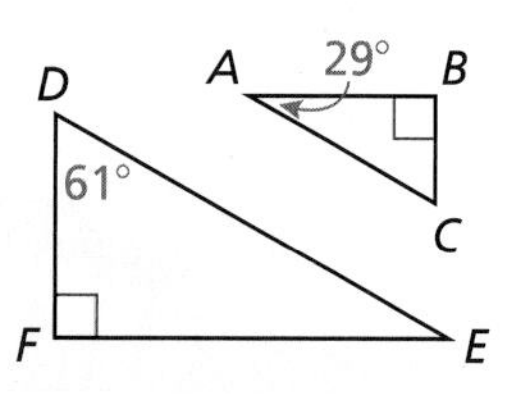

Because they are both right angles, $\angle F$ and $\angle B$ are congruent. By the Triangle Sum Theorem (Theorem 5.1), $61° + 90° + m\angle E = 180°$, so $m\angle E = 29°$. So, $\angle E$ and $\angle A$ are congruent. So, $\triangle DFE \sim \triangle CBA$ by the AA Similarity Theorem (Theorem 8.3).

Show that the triangles are similar. Write a similarity statement.

5.

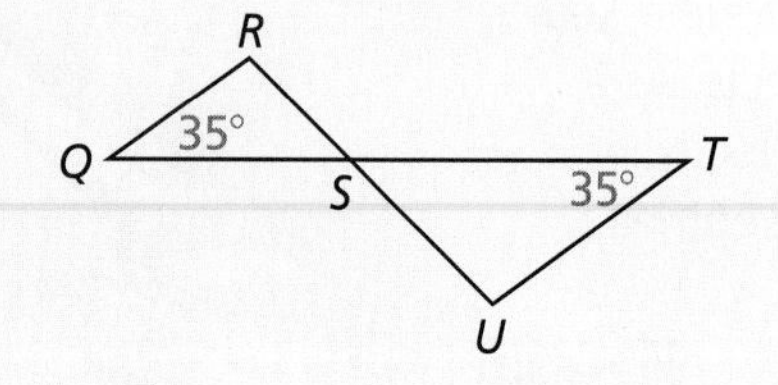

6.

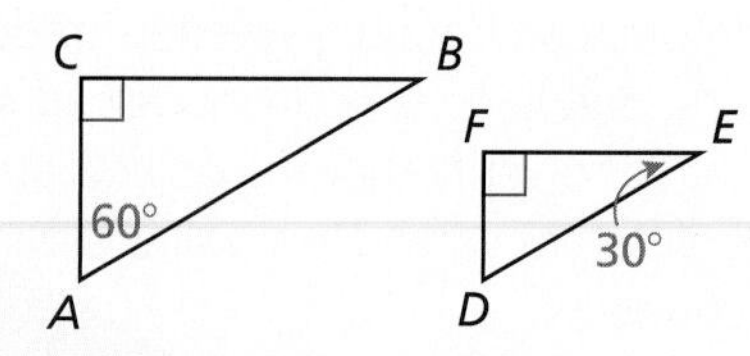

7. A cellular telephone tower casts a shadow that is 72 feet long, while a nearby tree that is 27 feet tall casts a shadow that is 6 feet long. How tall is the tower?

8.3 Proving Triangle Similarity by SSS and SAS (pp. 435–444)

Show that the triangles are similar.

a.

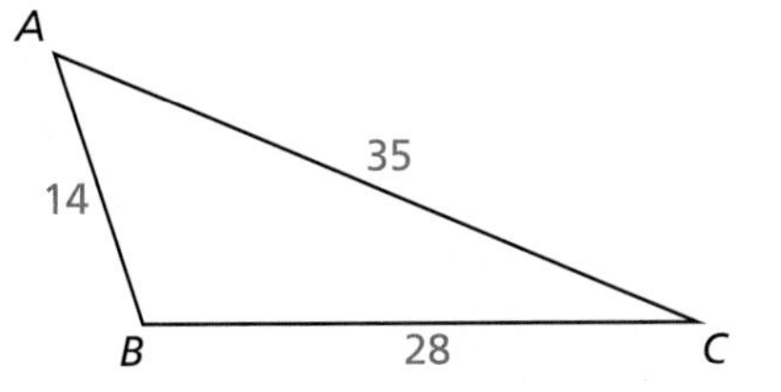

Compare $\triangle ABC$ and $\triangle DEF$ by finding ratios of corresponding side lengths.

Shortest sides	**Longest sides**	**Remaining sides**
$\frac{AB}{DE} = \frac{14}{6} = \frac{7}{3}$	$\frac{AC}{DF} = \frac{35}{15} = \frac{7}{3}$	$\frac{BC}{EF} = \frac{28}{12} = \frac{7}{3}$

All the ratios are equal, so $\triangle ABC \sim \triangle DEF$ by the SSS Similarity Theorem (Theorem 8.4).

b.

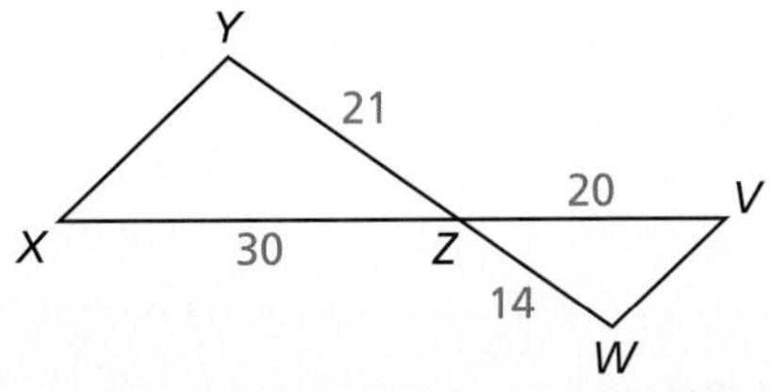

$\angle YZX \cong \angle WZV$ by the Vertical Angles Congruence Theorem (Theorem 2.6). Next, compare the ratios of the corresponding side lengths of $\triangle YZX$ and $\triangle WZV$.

$\frac{WZ}{YZ} = \frac{14}{21} = \frac{2}{3}$ $\frac{VZ}{XZ} = \frac{20}{30} = \frac{2}{3}$

▶ So, by the SAS Similarity Theorem (Theorem 8.5), $\triangle YZX \sim \triangle WZV$.

Use the SSS Similarity Theorem (Theorem 8.4) or the SAS Similarity Theorem (Theorem 8.5) to show that the triangles are similar.

8.

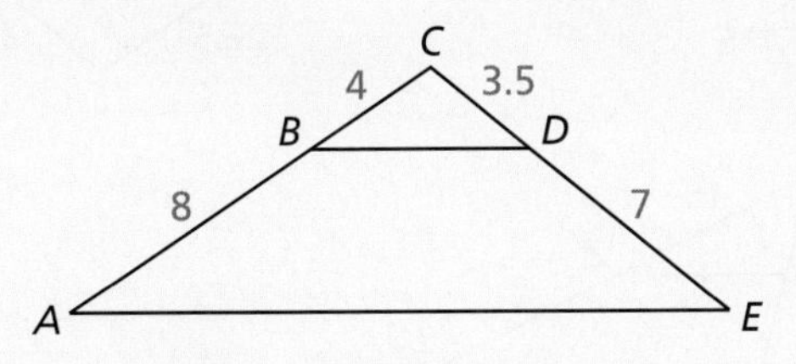

9.

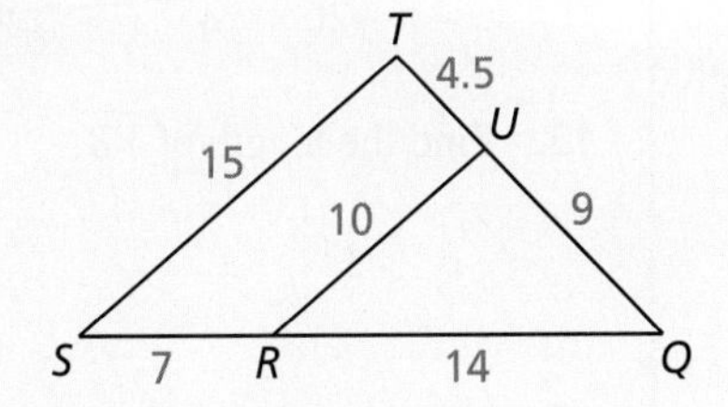

10. Find the value of x that makes $\triangle ABC \sim \triangle DEF$.

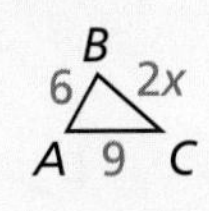

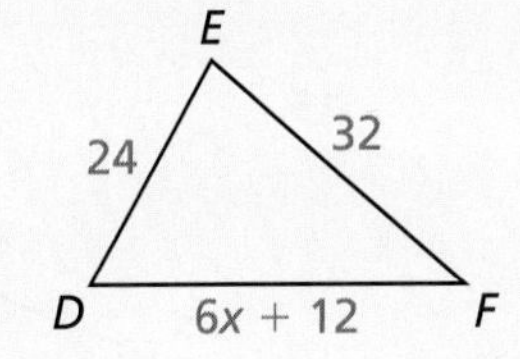

8.4 Proportionality Theorems (pp. 445–452)

a. Determine whether $\overline{MP} \parallel \overline{LQ}$.

Begin by finding and simplifying ratios of lengths determined by $\overline{MP}$.

$$\frac{NM}{ML} = \frac{8}{4} = \frac{2}{1} = 2$$

$$\frac{NP}{PQ} = \frac{24}{12} = \frac{2}{1} = 2$$

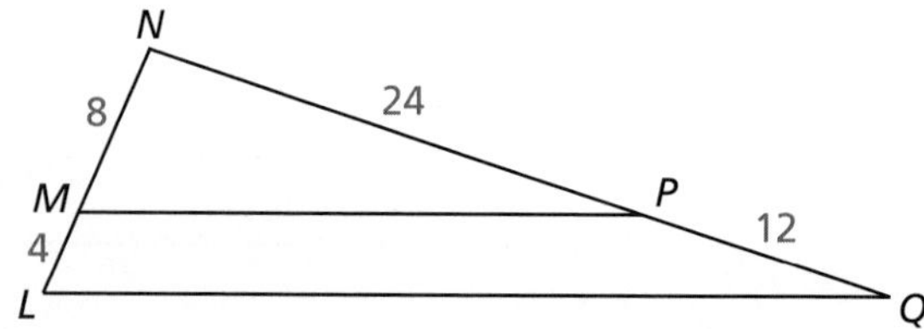

Because $\frac{NM}{ML} = \frac{NP}{PQ}$, $\overline{MP}$ is parallel to $\overline{LQ}$ by the Converse of the Triangle Proportionality Theorem (Theorem 8.7).

b. In the diagram, $\overline{AD}$ bisects $\angle CAB$. Find the length of $\overline{DB}$.

Because $\overline{AD}$ is an angle bisector of $\angle CAB$, you can apply the Triangle Angle Bisector Theorem (Theorem 8.9).

$\frac{DB}{DC} = \frac{AB}{AC}$	Triangle Angle Bisector Theorem
$\frac{x}{5} = \frac{15}{8}$	Substitute.
$8x = 75$	Cross Products Property
$9.375 = x$	Solve for x.

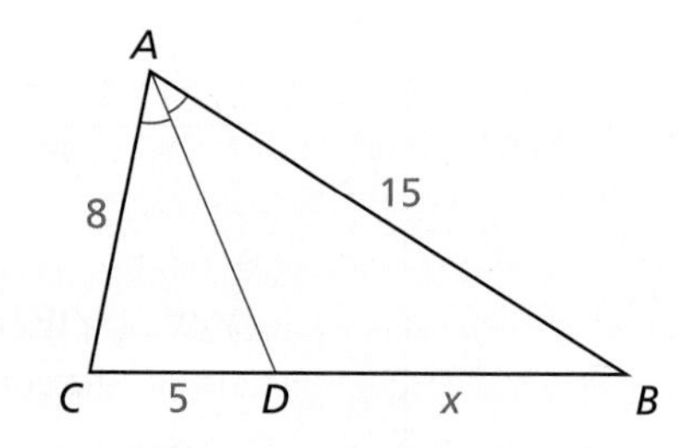

▶ The length of $\overline{DB}$ is 9.375 units.

Determine whether $\overline{AB} \parallel \overline{CD}$.

11.

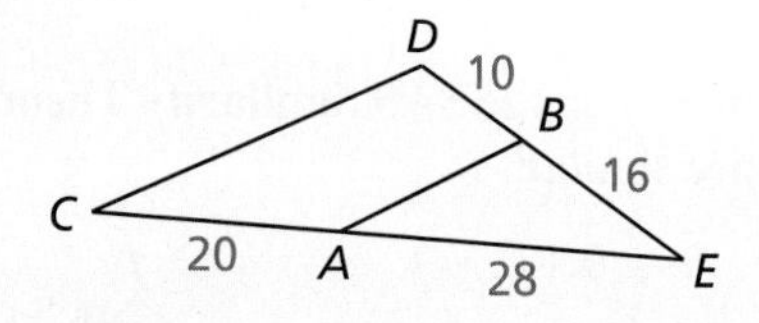

12.

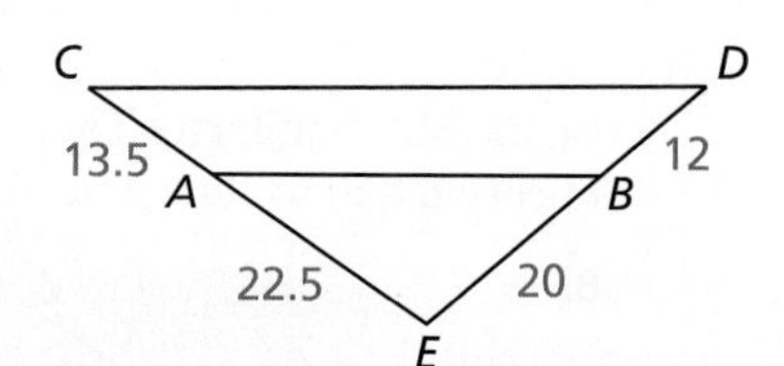

13. Find the length of $\overline{YB}$.

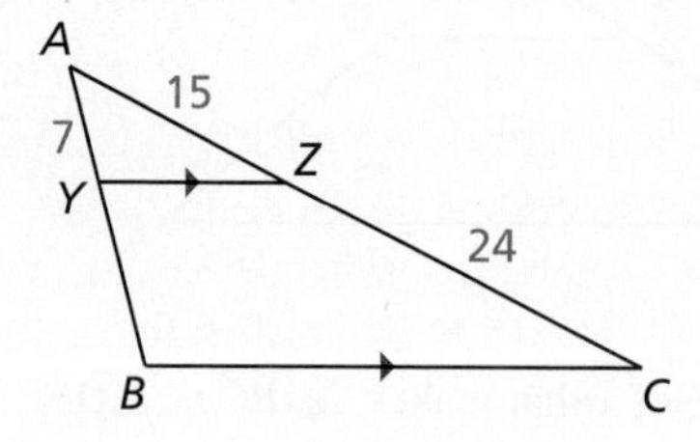

Find the length of $\overline{AB}$.

14.

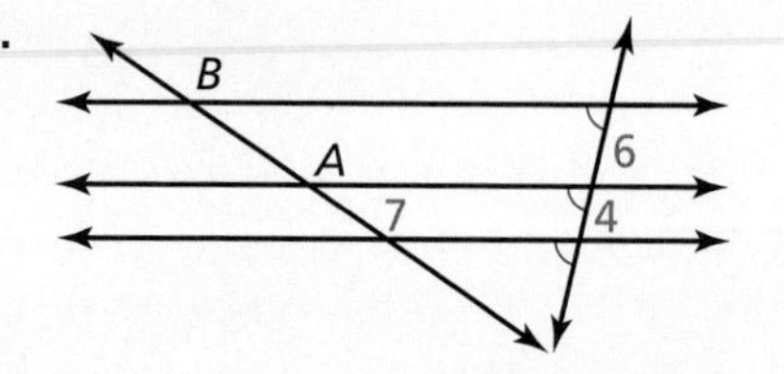

15.

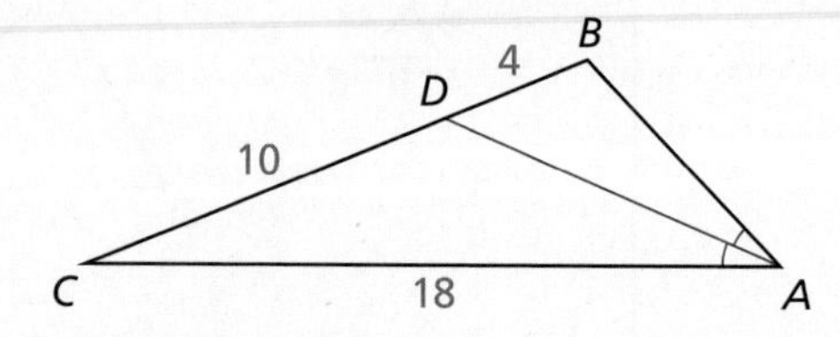

8 Chapter Test

Determine whether the triangles are similar. If they are, write a similarity statement. Explain your reasoning.

1.

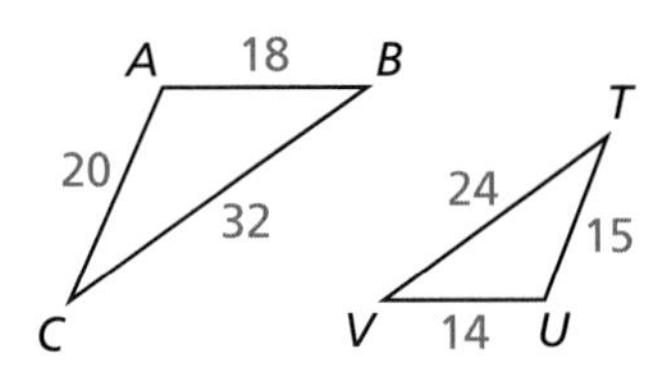

2.

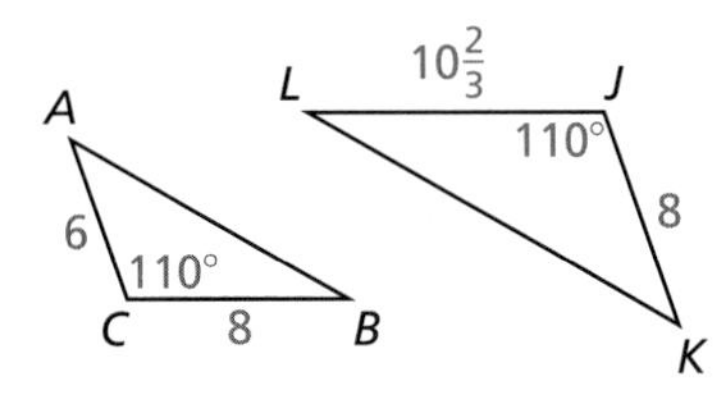

3.

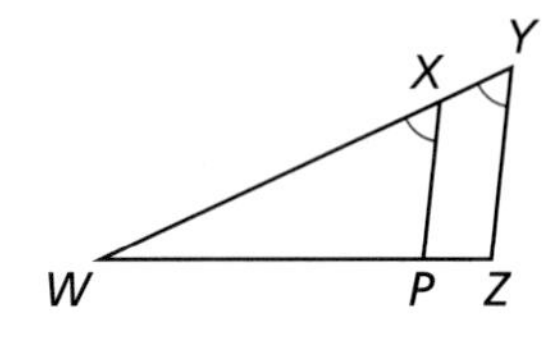

Find the value of the variable.

4.

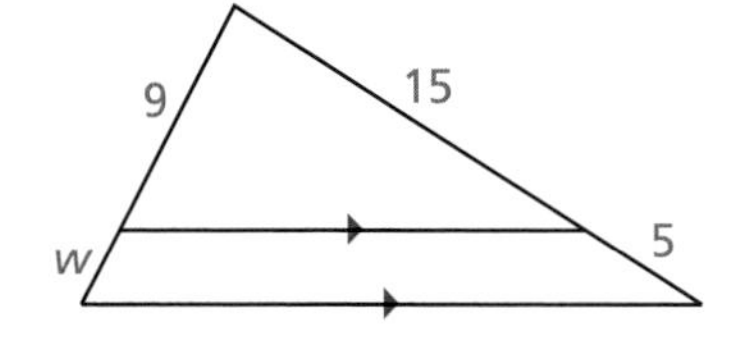

5.

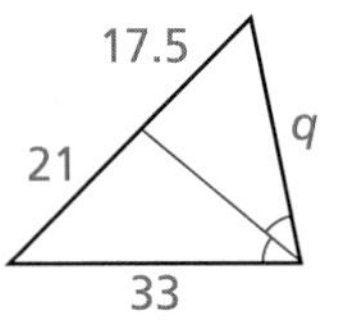

6.

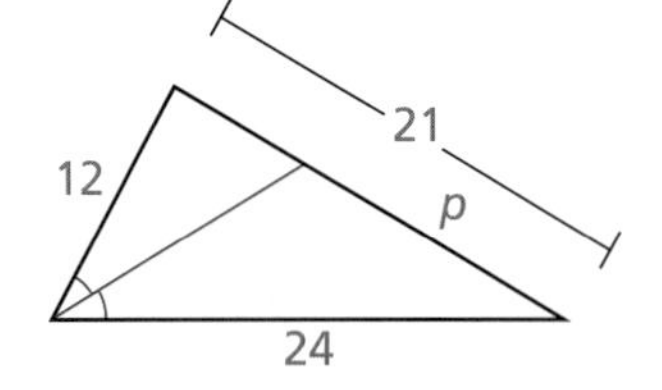

7. Given $\triangle QRS \sim \triangle MNP$, list all pairs of congruent angles. Then write the ratios of the corresponding side lengths in a statement of proportionality.

Use the diagram.

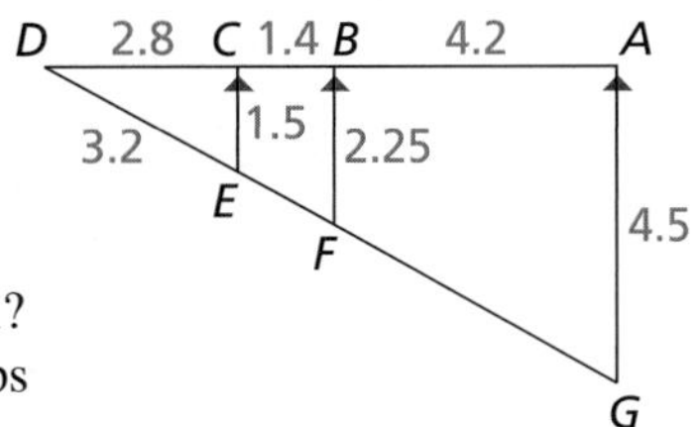

8. Find the length of $\overline{EF}$.

9. Find the length of $\overline{FG}$.

10. Is quadrilateral *FECB* similar to quadrilateral *GFBA*? If so, what is the scale factor of the dilation that maps quadrilateral *FECB* to quadrilateral *GFBA*?

11. You are visiting the Unisphere at Flushing Meadows Corona Park in New York. To estimate the height of the stainless steel model of Earth, you place a mirror on the ground and stand where you can see the top of the model in the mirror. Use the diagram to estimate the height of the model. Explain why this method works.

12. You are making a scale model of a rectangular park for a school project. Your model has a length of 2 feet and a width of 1.4 feet. The actual park is 800 yards long. What are the perimeter and area of the actual park?

13. In a *perspective drawing*, lines that are parallel in real life must meet at a vanishing point on the horizon. To make the train cars in the drawing appear equal in length, they are drawn so that the lines connecting the opposite corners of each car are parallel. Use the dimensions given and the yellow parallel lines to find the length of the bottom edge of the drawing of Car 2.

8 Cumulative Assessment

1. Use the graph of quadrilaterals $ABCD$ and $QRST$.

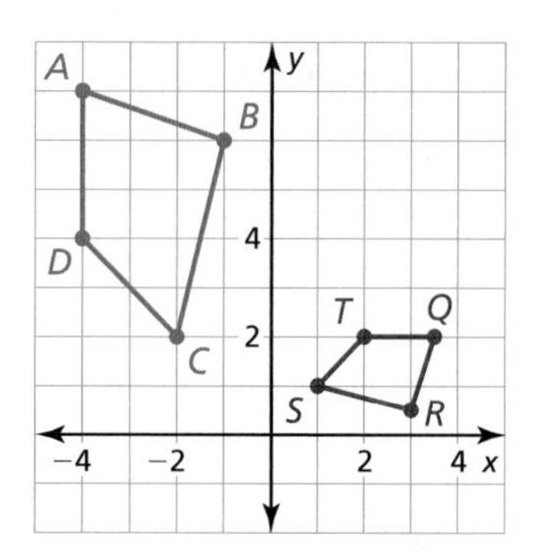

a. Write a composition of transformations that maps quadrilateral $ABCD$ to quadrilateral $QRST$.

b. Are the quadrilaterals similar? Explain your reasoning.

2. In the diagram, $ABCD$ is a parallelogram. Which congruence theorem(s) could you use to show that $\triangle AED \cong \triangle CEB$? Select all that apply.

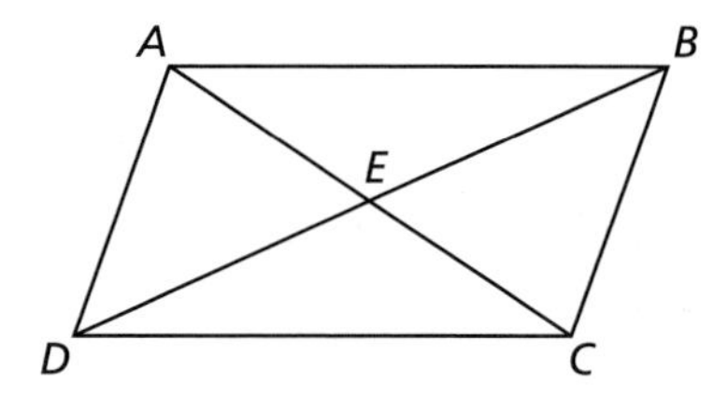

SAS Congruence Theorem (Theorem 5.5)

SSS Congruence Theorem (Theorem 5.8)

HL Congruence Theorem (Theorem 5.9)

ASA Congruence Theorem (Theorem 5.10)

AAS Congruence Theorem (Theorem 5.11)

3. By the Triangle Proportionality Theorem (Theorem 8.6), $\frac{VW}{WY} = \frac{VX}{XZ}$. In the diagram, $VX > VW$ and $XZ > WY$. List three possible values for VX and XZ.

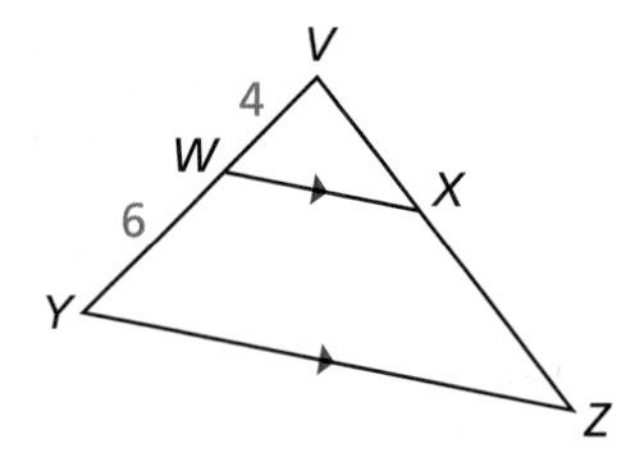

4. The slope of line ℓ is $-\frac{3}{4}$. The slope of line n is $\frac{4}{3}$. What must be true about lines ℓ and n?

Ⓐ Lines ℓ and n are parallel.

Ⓑ Lines ℓ and n are perpendicular.

Ⓒ Lines ℓ and n are skew.

Ⓓ Lines ℓ and n are the same line.

5. Enter a statement or reason in each blank to complete the two-column proof.

Given $\frac{KJ}{KL} = \frac{KH}{KM}$

Prove $\angle LMN \cong \angle JHG$

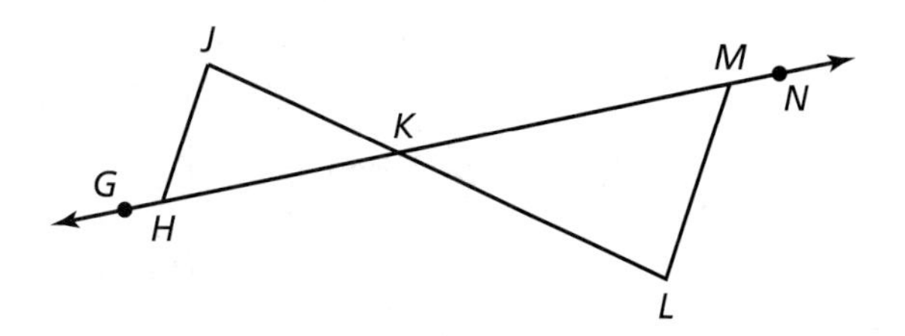

STATEMENTS	REASONS
1. $\frac{KJ}{KL} = \frac{KH}{KM}$	1. Given
2. $\angle JKH \cong \angle LKM$	2. ________
3. $\triangle JKH \sim \triangle LKM$	3. ________
4. $\angle KHJ \cong \angle KML$	4. ________
5. ________	5. Definition of congruent angles
6. $m\angle KHJ + m\angle JHG = 180^\circ$	6. Linear Pair Postulate (Post. 2.8)
7. $m\angle JHG = 180^\circ - m\angle KHJ$	7. ________
8. $m\angle KML + m\angle LMN = 180^\circ$	8. ________
9. ________	9. Subtraction Property of Equality
10. $m\angle LMN = 180^\circ - m\angle KHJ$	10. ________
11. ________	11. Transitive Property of Equality
12. $\angle LMN \cong \angle JHG$	12. ________

6. The coordinates of the vertices of $\triangle DEF$ are $D(-8, 5)$, $E(-5, 8)$, and $F(-1, 4)$. The coordinates of the vertices of $\triangle JKL$ are $J(16, -10)$, $K(10, -16)$, and $L(2, -8)$. $\angle D \cong \angle J$. Can you show that $\triangle DEF \sim \triangle JKL$ by using the AA Similarity Theorem (Theorem 8.3)? If so, do so by listing the congruent corresponding angles and writing a similarity transformation that maps $\triangle DEF$ to $\triangle JKL$. If not, explain why not.

7. Classify the quadrilateral using the most specific name.

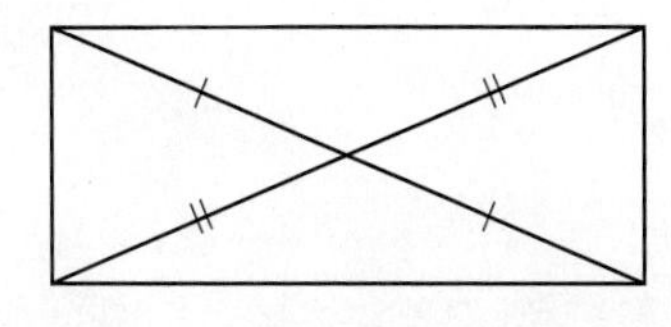

rectangle | square | parallelogram | rhombus

8. Your friend makes the statement "Quadrilateral *PQRS* is similar to quadrilateral *WXYZ*." Describe the relationships between corresponding angles and between corresponding sides that make this statement true.

9 Right Triangles and Trigonometry

Leaning Tower of Pisa *(p. 514)*

Skiing *(p. 497)*

Washington Monument *(p. 491)*

Rock Wall *(p. 481)*

Fire Escape *(p. 469)*

Maintaining Mathematical Proficiency

Using Properties of Radicals

Example 1 **Simplify $\sqrt{128}$.**

$\sqrt{128} = \sqrt{64 \cdot 2}$ — Factor using the greatest perfect square factor.

$= \sqrt{64} \cdot \sqrt{2}$ — Product Property of Square Roots

$= 8\sqrt{2}$ — Simplify.

Example 2 **Simplify $\frac{4}{\sqrt{5}}$.**

$\frac{4}{\sqrt{5}} = \frac{4}{\sqrt{5}} \cdot \frac{\sqrt{5}}{\sqrt{5}}$ — Multiply by $\frac{\sqrt{5}}{\sqrt{5}}$.

$= \frac{4\sqrt{5}}{\sqrt{25}}$ — Product Property of Square Roots

$= \frac{4\sqrt{5}}{5}$ — Simplify.

Simplify the expression.

1. $\sqrt{75}$
2. $\sqrt{270}$
3. $\sqrt{135}$
4. $\frac{2}{\sqrt{7}}$
5. $\frac{5}{\sqrt{2}}$
6. $\frac{12}{\sqrt{6}}$

Solving Proportions

Example 3 **Solve $\frac{x}{10} = \frac{3}{2}$.**

$\frac{x}{10} = \frac{3}{2}$ — Write the proportion.

$x \cdot 2 = 10 \cdot 3$ — Cross Products Property

$2x = 30$ — Multiply.

$\frac{2x}{2} = \frac{30}{2}$ — Divide each side by 2.

$x = 15$ — Simplify.

Solve the proportion.

7. $\frac{x}{12} = \frac{3}{4}$
8. $\frac{x}{3} = \frac{5}{2}$
9. $\frac{4}{x} = \frac{7}{56}$
10. $\frac{10}{23} = \frac{4}{x}$
11. $\frac{x+1}{2} = \frac{21}{14}$
12. $\frac{9}{3x-15} = \frac{3}{12}$
13. **ABSTRACT REASONING** The Product Property of Square Roots allows you to simplify the square root of a product. Are you able to simplify the square root of a sum? of a difference? Explain.

Mathematical Practices

Mathematically proficient students express numerical answers precisely.

Attending to Precision

Core Concept

Standard Position for a Right Triangle

In *unit circle trigonometry*, a right triangle is in **standard position** when:

1. The hypotenuse is a radius of the circle of radius 1 with center at the origin.
2. One leg of the right triangle lies on the x-axis.
3. The other leg of the right triangle is perpendicular to the x-axis.

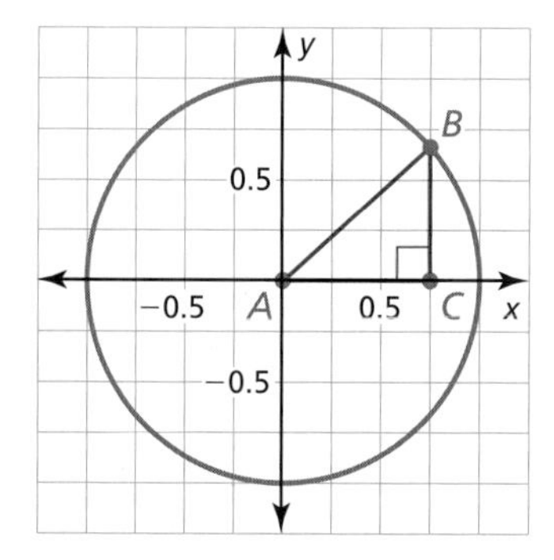

EXAMPLE 1 Drawing an Isosceles Right Triangle in Standard Position

Use dynamic geometry software to construct an isosceles right triangle in standard position. What are the exact coordinates of its vertices?

SOLUTION

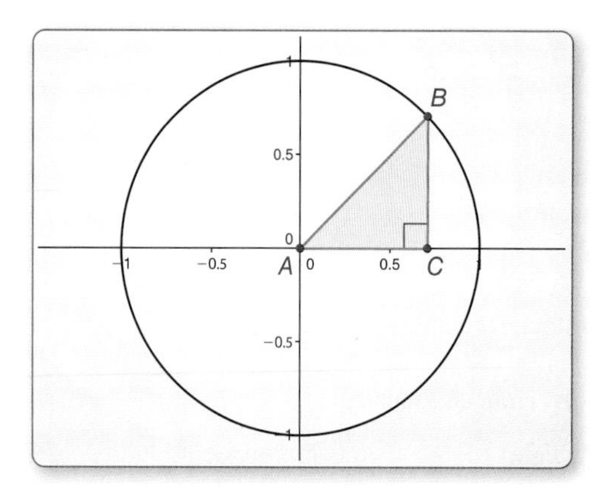

Sample

Points
$A(0, 0)$
$B(0.71, 0.71)$
$C(0.71, 0)$
Segments
$AB = 1$
$BC = 0.71$
$AC = 0.71$
Angle
$m\angle A = 45°$

To determine the exact coordinates of the vertices, label the length of each leg x. By the Pythagorean Theorem, which you will study in Section 9.1, $x^2 + x^2 = 1$. Solving this equation yields

$$x = \frac{1}{\sqrt{2}}, \text{ or } \frac{\sqrt{2}}{2}.$$

▶ So, the exact coordinates of the vertices are $A(0, 0)$, $B\left(\frac{\sqrt{2}}{2}, \frac{\sqrt{2}}{2}\right)$, and $C\left(\frac{\sqrt{2}}{2}, 0\right)$.

Monitoring Progress

1. Use dynamic geometry software to construct a right triangle with acute angle measures of 30° and 60° in standard position. What are the exact coordinates of its vertices?
2. Use dynamic geometry software to construct a right triangle with acute angle measures of 20° and 70° in standard position. What are the approximate coordinates of its vertices?

9.1 The Pythagorean Theorem

Essential Question

How can you prove the Pythagorean Theorem?

EXPLORATION 1 Proving the Pythagorean Theorem without Words

Work with a partner.

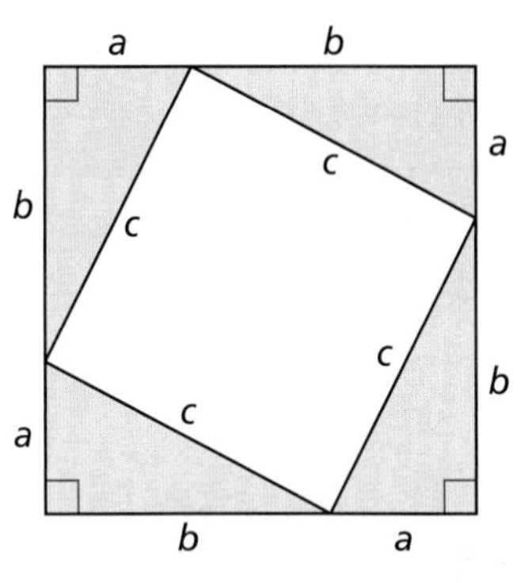

a. Draw and cut out a right triangle with legs a and b, and hypotenuse c.

b. Make three copies of your right triangle. Arrange all four triangles to form a large square, as shown.

c. Find the area of the large square in terms of a, b, and c by summing the areas of the triangles and the small square.

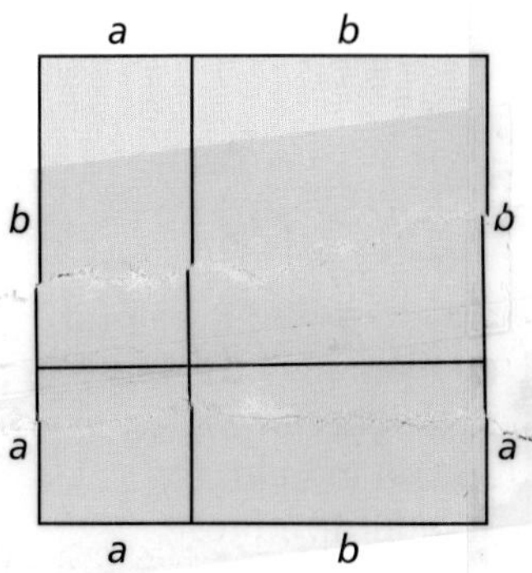

d. Copy the large square. Divide it into two smaller squares and two equally-sized rectangles, as shown.

e. Find the area of the large square in terms of a and b by summing the areas of the rectangles and the smaller squares.

f. Compare your answers to parts (c) and (e). Explain how this proves the Pythagorean Theorem.

EXPLORATION 2 Proving the Pythagorean Theorem

Work with a partner.

a. Draw a right triangle with legs a and b, and hypotenuse c, as shown. Draw the altitude from C to $\overline{AB}$. Label the lengths, as shown.

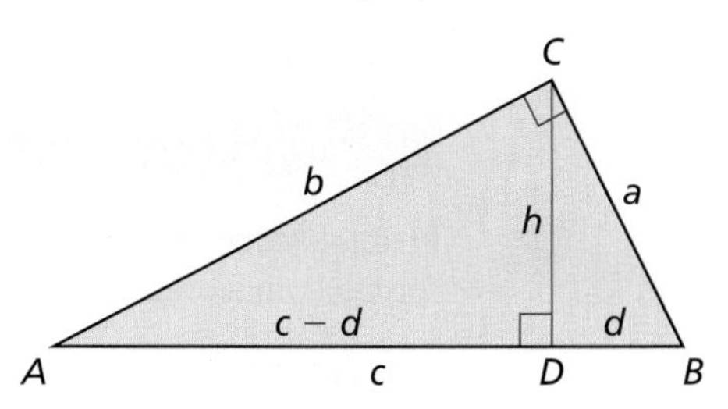

To be proficient in math, you need to know and flexibly use different properties of operations and objects.

b. Explain why $\triangle ABC$, $\triangle ACD$, and $\triangle CBD$ are similar.

c. Write a two-column proof using the similar triangles in part (b) to prove that $a^2 + b^2 = c^2$.

Communicate Your Answer

3. How can you prove the Pythagorean Theorem?

4. Use the Internet or some other resource to find a way to prove the Pythagorean Theorem that is different from Explorations 1 and 2.

EXAMPLE 2 Using the Pythagorean Theorem

Find the value of x. Then tell whether the side lengths form a Pythagorean triple.

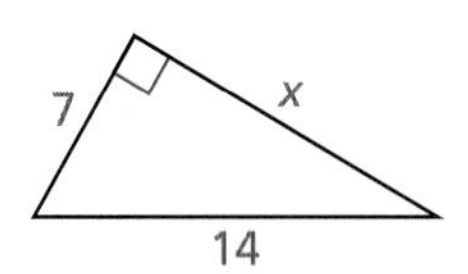

SOLUTION

$c^2 = a^2 + b^2$	Pythagorean Theorem
$14^2 = 7^2 + x^2$	Substitute.
$196 = 49 + x^2$	Multiply.
$147 = x^2$	Subtract 49 from each side.
$\sqrt{147} = x$	Find the positive square root.
$\sqrt{49} \bullet \sqrt{3} = x$	Product Property of Square Roots
$7\sqrt{3} = x$	Simplify.

▶ The value of x is $7\sqrt{3}$. Because $7\sqrt{3}$ is not an integer, the side lengths do not form a Pythagorean triple.

EXAMPLE 3 Solving a Real-Life Problem

The skyscrapers shown are connected by a skywalk with support beams. Use the Pythagorean Theorem to approximate the length of each support beam.

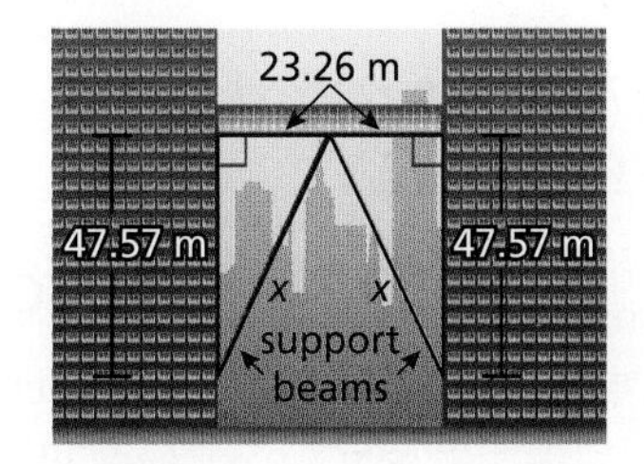

SOLUTION

Each support beam forms the hypotenuse of a right triangle. The right triangles are congruent, so the support beams are the same length.

$x^2 = (23.26)^2 + (47.57)^2$	Pythagorean Theorem
$x = \sqrt{(23.26)^2 + (47.57)^2}$	Find the positive square root.
$x \approx 52.95$	Use a calculator to approximate.

▶ The length of each support beam is about 52.95 meters.

Monitoring Progress

Help in English and Spanish at *BigIdeasMath.com*

Find the value of x. Then tell whether the side lengths form a Pythagorean triple.

1.

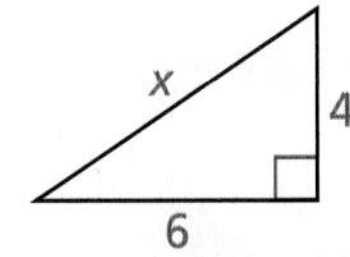

2.

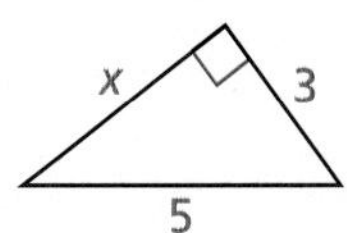

3. An anemometer is a device used to measure wind speed. The anemometer shown is attached to the top of a pole. Support wires are attached to the pole 5 feet above the ground. Each support wire is 6 feet long. How far from the base of the pole is each wire attached to the ground?

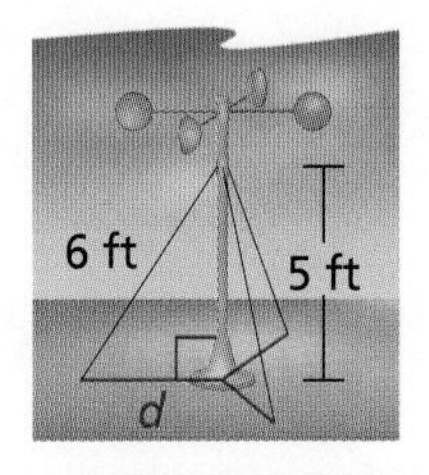

Using the Converse of the Pythagorean Theorem

The converse of the Pythagorean Theorem is also true. You can use it to determine whether a triangle with given side lengths is a right triangle.

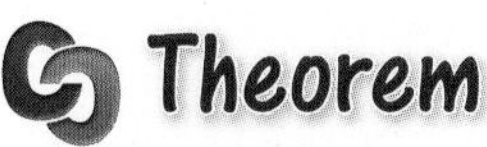

Theorem 9.2 Converse of the Pythagorean Theorem

If the square of the length of the longest side of a triangle is equal to the sum of the squares of the lengths of the other two sides, then the triangle is a right triangle.

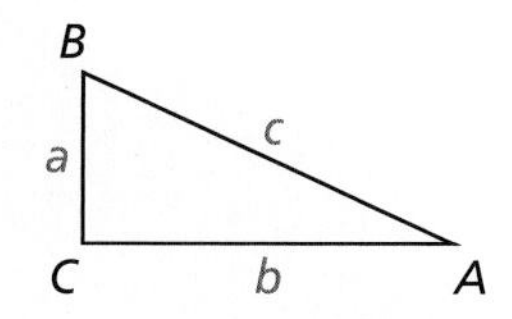

If $c^2 = a^2 + b^2$, then $\triangle ABC$ is a right triangle.

Proof Ex. 39, p. 470

EXAMPLE 4 Verifying Right Triangles

Tell whether each triangle is a right triangle.

a.

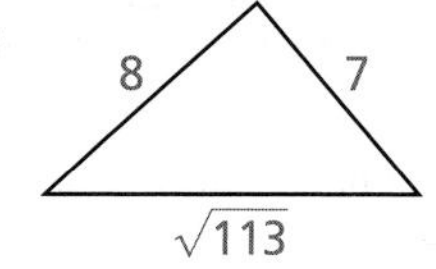

b.

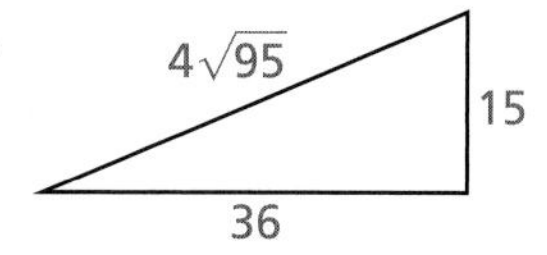

USING TOOLS STRATEGICALLY

Use a calculator to determine that $\sqrt{113} \approx 10.630$ is the length of the longest side in part (a).

SOLUTION

Let c represent the length of the longest side of the triangle. Check to see whether the side lengths satisfy the equation $c^2 = a^2 + b^2$.

a. $(\sqrt{113})^2 \stackrel{?}{=} 7^2 + 8^2$

$113 \stackrel{?}{=} 49 + 64$

$113 = 113$ ✓

▶ The triangle is a right triangle.

b. $(4\sqrt{95})^2 \stackrel{?}{=} 15^2 + 36^2$

$4^2 \cdot (\sqrt{95})^2 \stackrel{?}{=} 15^2 + 36^2$

$16 \cdot 95 \stackrel{?}{=} 225 + 1296$

$1520 \neq 1521$ ✗

▶ The triangle is *not* a right triangle.

Monitoring Progress  Help in English and Spanish at *BigIdeasMath.com*

Tell whether the triangle is a right triangle.

4. 9, $3\sqrt{34}$, 15

5. 22, 14, 26

Classifying Triangles

The Converse of the Pythagorean Theorem is used to determine whether a triangle is a right triangle. You can use the theorem below to determine whether a triangle is acute or obtuse.

Theorem 9.3 Pythagorean Inequalities Theorem

For any $\triangle ABC$, where c is the length of the longest side, the following statements are true.

If $c^2 < a^2 + b^2$, then $\triangle ABC$ is acute.

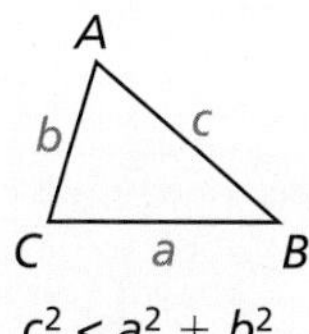

$c^2 < a^2 + b^2$

If $c^2 > a^2 + b^2$, then $\triangle ABC$ is obtuse.

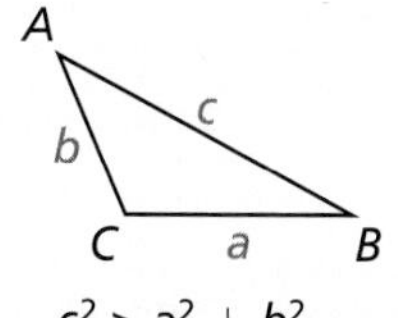

$c^2 > a^2 + b^2$

Proof Exs. 42 and 43, p. 470

REMEMBER

The Triangle Inequality Theorem (Theorem 6.11) on page 339 states that the sum of the lengths of any two sides of a triangle is greater than the length of the third side.

EXAMPLE 5 Classifying Triangles

Verify that segments with lengths of 4.3 feet, 5.2 feet, and 6.1 feet form a triangle. Is the triangle *acute*, *right*, or *obtuse*?

SOLUTION

Step 1 Use the Triangle Inequality Theorem (Theorem 6.11) to verify that the segments form a triangle.

$4.3 + 5.2 \overset{?}{>} 6.1$ $\quad$ $4.3 + 6.1 \overset{?}{>} 5.2$ $\quad$ $5.2 + 6.1 \overset{?}{>} 4.3$

$9.5 > 6.1$ ✓ $\quad$ $10.4 > 5.2$ ✓ $\quad$ $11.3 > 4.3$ ✓

▶ The segments with lengths of 4.3 feet, 5.2 feet, and 6.1 feet form a triangle.

Step 2 Classify the triangle by comparing the square of the length of the longest side with the sum of the squares of the lengths of the other two sides.

c^2		$a^2 + b^2$	Compare c^2 with $a^2 + b^2$.
6.1^2		$4.3^2 + 5.2^2$	Substitute.
37.21		$18.49 + 27.04$	Simplify.
37.21	$<$	45.53	c^2 is less than $a^2 + b^2$.

▶ The segments with lengths of 4.3 feet, 5.2 feet, and 6.1 feet form an acute triangle.

Monitoring Progress  Help in English and Spanish at *BigIdeasMath.com*

6. Verify that segments with lengths of 3, 4, and 6 form a triangle. Is the triangle *acute*, *right*, or *obtuse*?

7. Verify that segments with lengths of 2.1, 2.8, and 3.5 form a triangle. Is the triangle *acute*, *right*, or *obtuse*?

9.1 Exercises

Dynamic Solutions available at *BigIdeasMath.com*

Vocabulary and Core Concept Check

1. **VOCABULARY** What is a Pythagorean triple?

2. **DIFFERENT WORDS, SAME QUESTION** Which is different? Find "both" answers.

Find the length of the longest side.

Find the length of the hypotenuse.

Find the length of the longest leg.

Find the length of the side opposite the right angle.

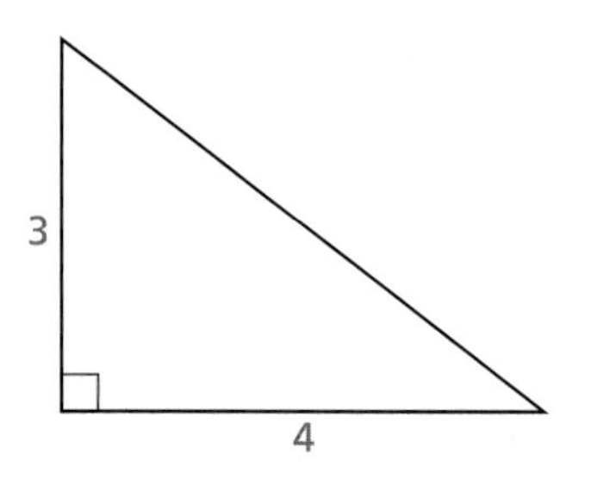

Monitoring Progress and Modeling with Mathematics

In Exercises 3–6, find the value of x. Then tell whether the side lengths form a Pythagorean triple. *(See Example 1.)*

3. 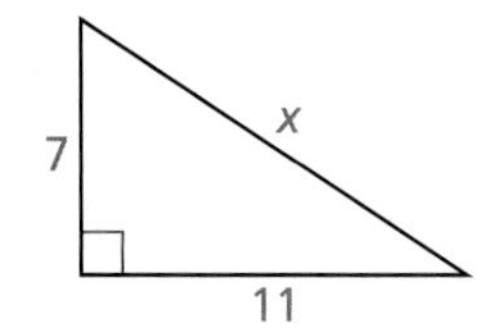

4. 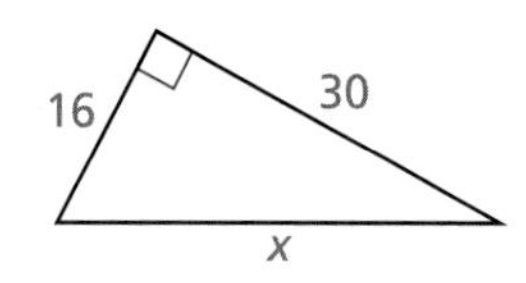

5. 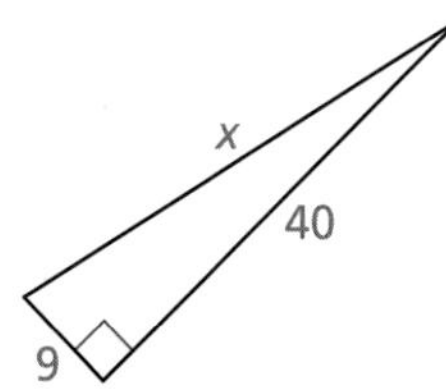

6. 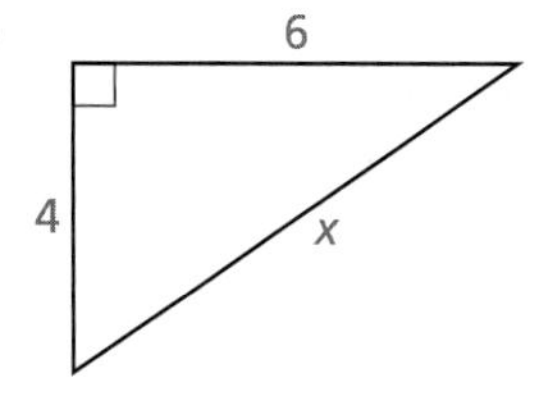

In Exercises 7–10, find the value of x. Then tell whether the side lengths form a Pythagorean triple. *(See Example 2.)*

7. 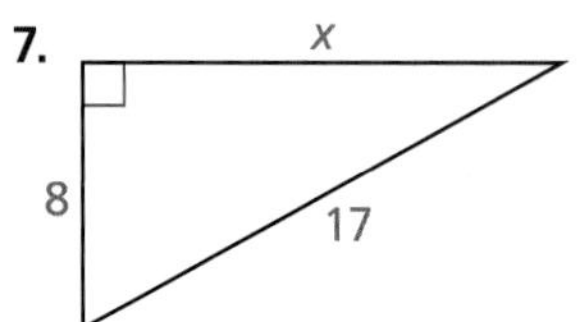

8. 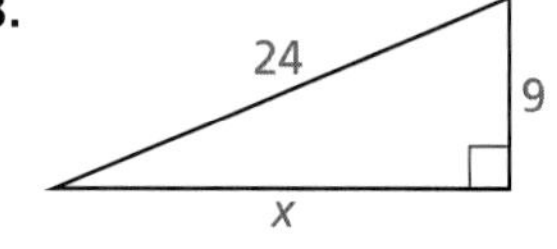

9. 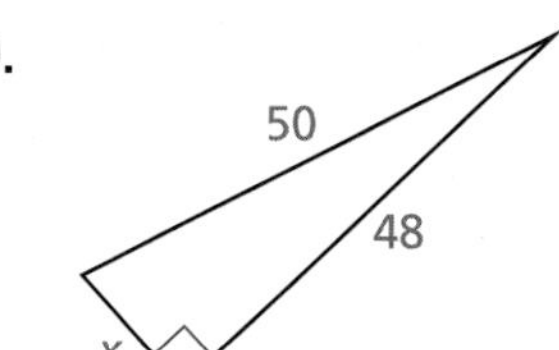

10. 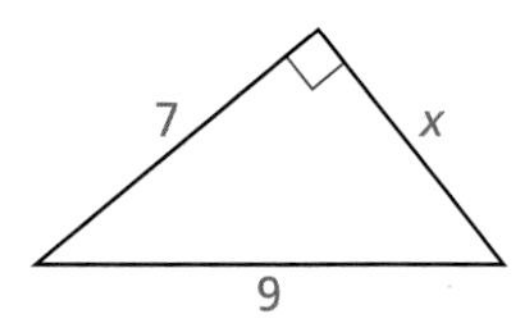

ERROR ANALYSIS **In Exercises 11 and 12, describe and correct the error in using the Pythagorean Theorem (Theorem 9.1).**

11. 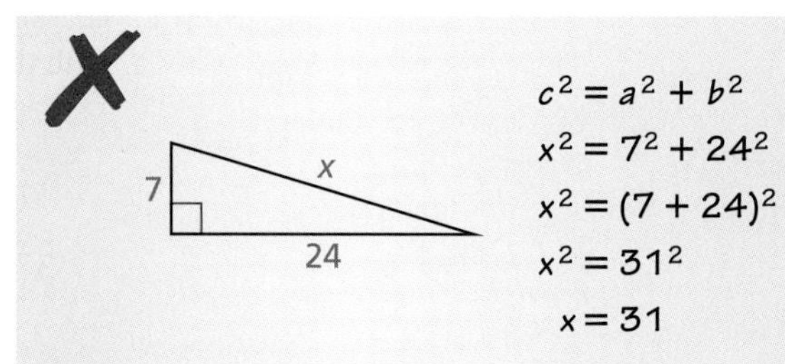

12. 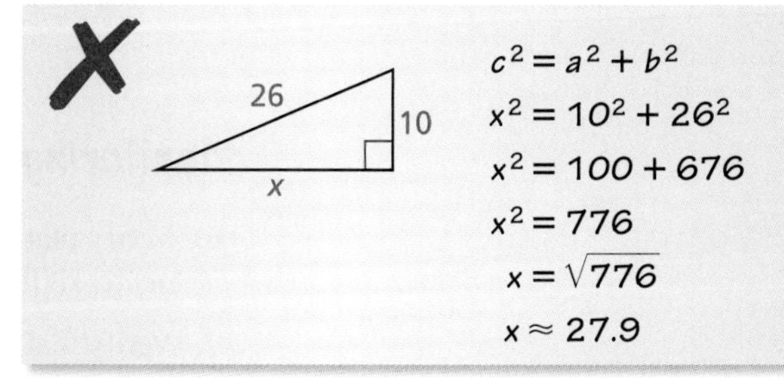

13. **MODELING WITH MATHEMATICS** The fire escape forms a right triangle, as shown. Use the Pythagorean Theorem (Theorem 9.1) to approximate the distance between the two platforms. *(See Example 3.)*

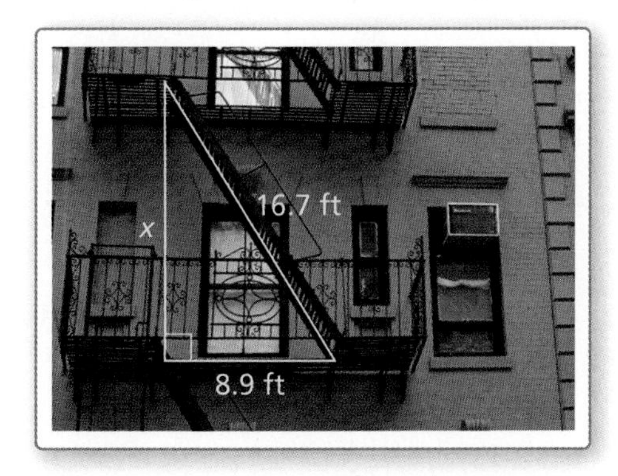

14. **MODELING WITH MATHEMATICS** The backboard of the basketball hoop forms a right triangle with the supporting rods, as shown. Use the Pythagorean Theorem (Theorem 9.1) to approximate the distance between the rods where they meet the backboard.

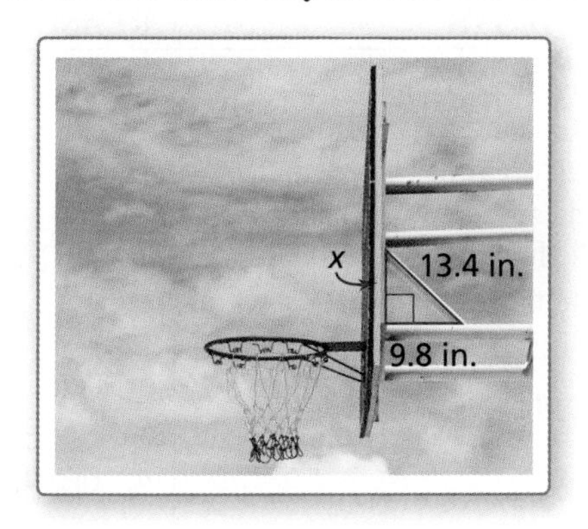

In Exercises 15–20, tell whether the triangle is a right triangle. *(See Example 4.)*

15.

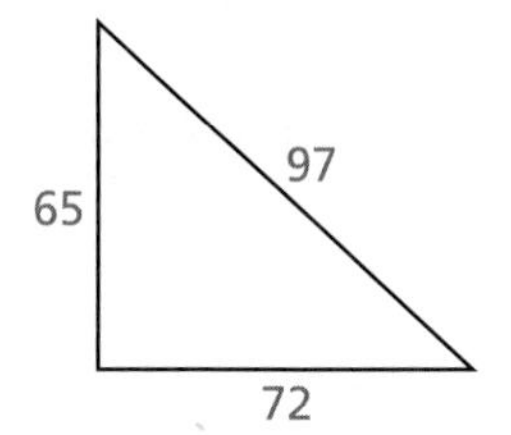

16.

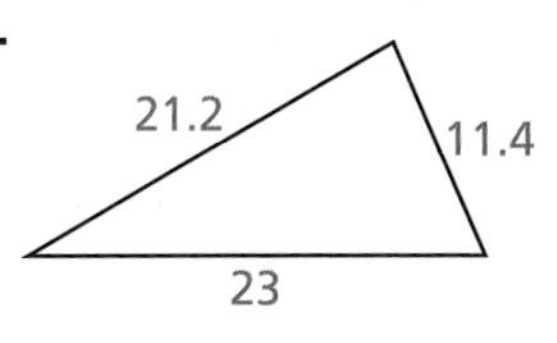

17. 18.

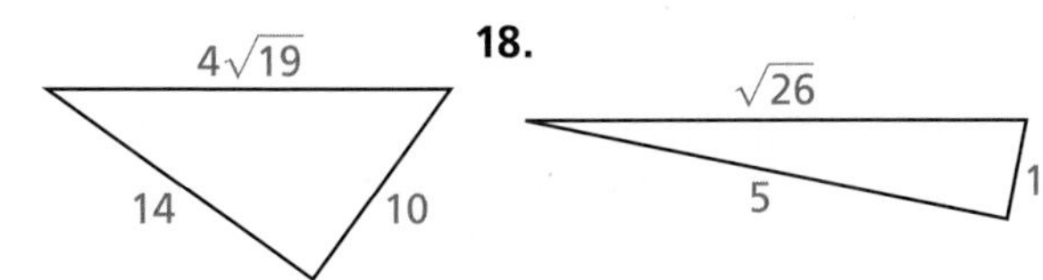

19. 20.

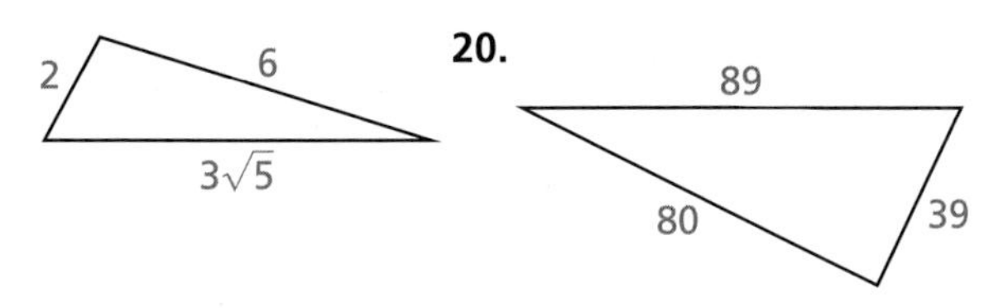

In Exercises 21–28, verify that the segment lengths form a triangle. Is the triangle *acute*, *right*, or *obtuse*? *(See Example 5.)*

21. 10, 11, and 14

22. 6, 8, and 10

23. 12, 16, and 20

24. 15, 20, and 36

25. 5.3, 6.7, and 7.8

26. 4.1, 8.2, and 12.2

27. 24, 30, and $6\sqrt{43}$

28. 10, 15, and $5\sqrt{13}$

29. **MODELING WITH MATHEMATICS** In baseball, the lengths of the paths between consecutive bases are 90 feet, and the paths form right angles. The player on first base tries to steal second base. How far does the ball need to travel from home plate to second base to get the player out?

30. **REASONING** You are making a canvas frame for a painting using stretcher bars. The rectangular painting will be 10 inches long and 8 inches wide. Using a ruler, how can you be certain that the corners of the frame are 90°?

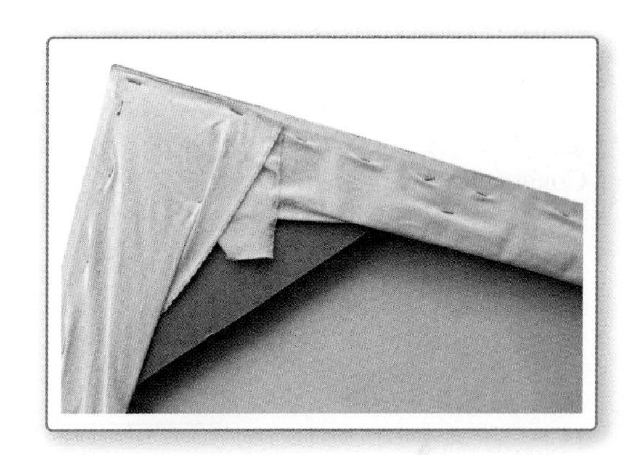

In Exercises 31–34, find the area of the isosceles triangle.

31.

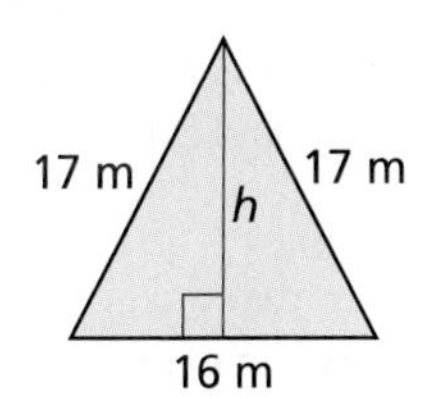

32.

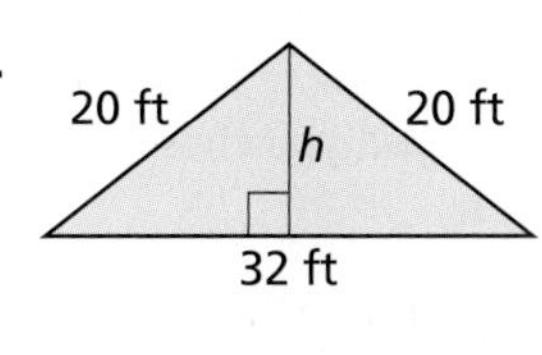

33.

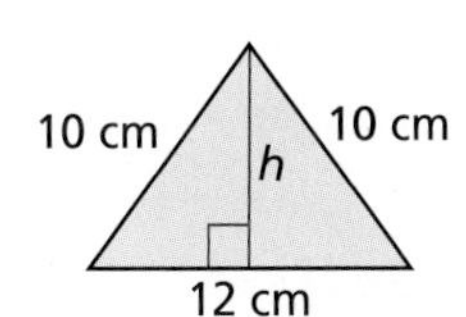

34.

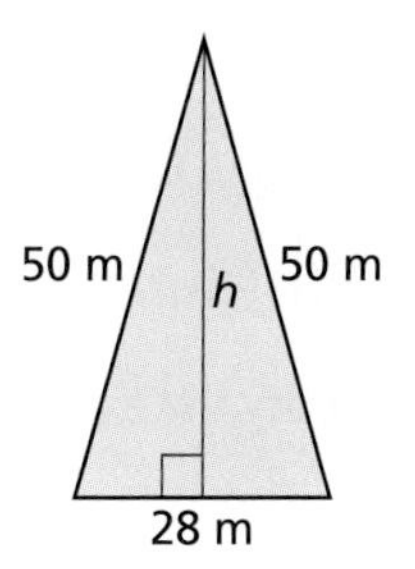

35. ANALYZING RELATIONSHIPS Justify the Distance Formula using the Pythagorean Theorem (Thm. 9.1).

36. HOW DO YOU SEE IT? How do you know $\angle C$ is a right angle without using the Pythagorean Theorem (Theorem 9.1)?

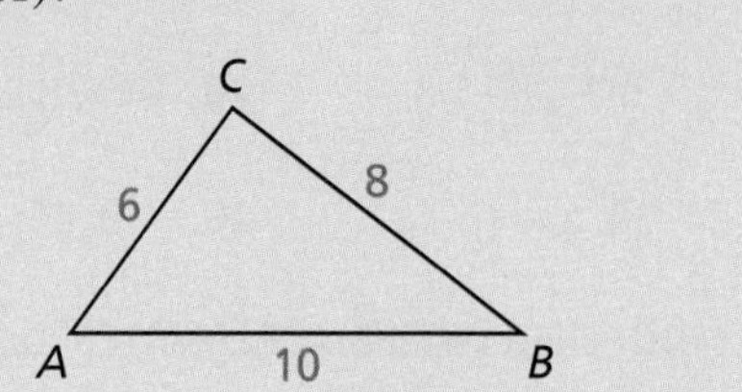

37. PROBLEM SOLVING You are making a kite and need to figure out how much binding to buy. You need the binding for the perimeter of the kite. The binding comes in packages of two yards. How many packages should you buy?

38. PROVING A THEOREM Use the Pythagorean Theorem (Theorem 9.1) to prove the Hypotenuse-Leg (HL) Congruence Theorem (Theorem 5.9).

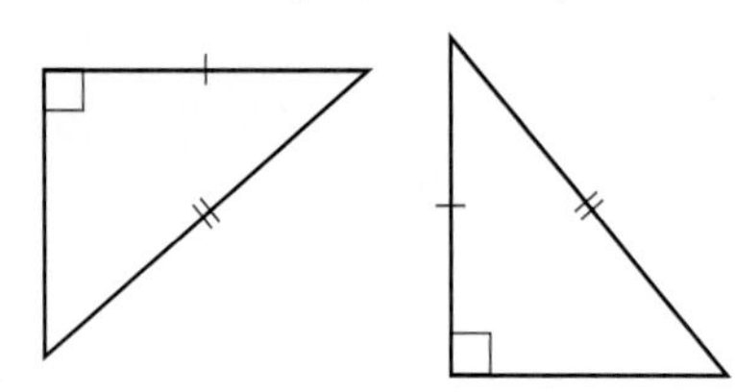

39. PROVING A THEOREM Prove the Converse of the Pythagorean Theorem (Theorem 9.2). (*Hint*: Draw $\triangle ABC$ with side lengths a, b, and c, where c is the length of the longest side. Then draw a right triangle with side lengths a, b, and x, where x is the length of the hypotenuse. Compare lengths c and x.)

40. THOUGHT PROVOKING Consider two integers m and n, where $m > n$. Do the following expressions produce a Pythagorean triple? If yes, prove your answer. If no, give a counterexample.

$$2mn,\ m^2 - n^2,\ m^2 + n^2$$

41. MAKING AN ARGUMENT Your friend claims 72 and 75 cannot be part of a Pythagorean triple because $72^2 + 75^2$ does not equal a positive integer squared. Is your friend correct? Explain your reasoning.

42. PROVING A THEOREM Copy and complete the proof of the Pythagorean Inequalities Theorem (Theorem 9.3) when $c^2 < a^2 + b^2$.

Given In $\triangle ABC$, $c^2 < a^2 + b^2$, where c is the length of the longest side. $\triangle PQR$ has side lengths a, b, and x, where x is the length of the hypotenuse, and $\angle R$ is a right angle.

Prove $\triangle ABC$ is an acute triangle.

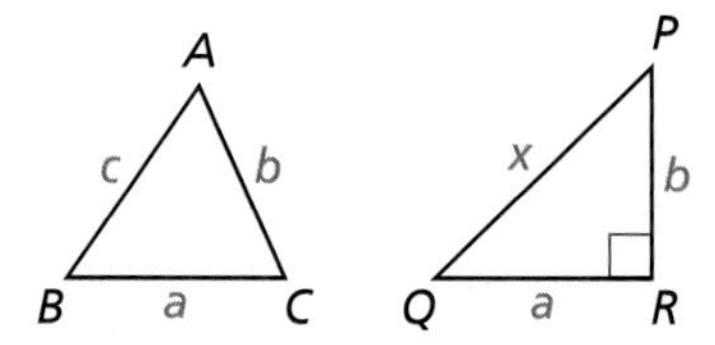

STATEMENTS	REASONS
1. In $\triangle ABC$, $c^2 < a^2 + b^2$, where c is the length of the longest side. $\triangle PQR$ has side lengths a, b, and x, where x is the length of the hypotenuse, and $\angle R$ is a right angle.	**1.** ______
2. $a^2 + b^2 = x^2$	**2.** ______
3. $c^2 < x^2$	**3.** ______
4. $c < x$	**4.** Take the positive square root of each side.
5. $m\angle R = 90°$	**5.** ______
6. $m\angle C < m\angle R$	**6.** Converse of the Hinge Theorem (Theorem 6.13)
7. $m\angle C < 90°$	**7.** ______
8. $\angle C$ is an acute angle.	**8.** ______
9. $\triangle ABC$ is an acute triangle.	**9.** ______

43. PROVING A THEOREM Prove the Pythagorean Inequalities Theorem (Theorem 9.3) when $c^2 > a^2 + b^2$. (*Hint*: Look back at Exercise 42.)

Maintaining Mathematical Proficiency

Reviewing what you learned in previous grades and lessons

Simplify the expression by rationalizing the denominator. *(Skills Review Handbook)*

44. $\frac{7}{\sqrt{2}}$ **45.** $\frac{14}{\sqrt{3}}$ **46.** $\frac{8}{\sqrt{2}}$ **47.** $\frac{12}{\sqrt{3}}$

9.2 Special Right Triangles

Essential Question What is the relationship among the side lengths of 45°- 45°- 90° triangles? 30°- 60°- 90° triangles?

EXPLORATION 1 Side Ratios of an Isosceles Right Triangle

Work with a partner.

a. Use dynamic geometry software to construct an isosceles right triangle with a leg length of 4 units.

b. Find the acute angle measures. Explain why this triangle is called a 45°- 45°- 90° triangle.

> **ATTENDING TO PRECISION**
>
> To be proficient in math, you need to express numerical answers with a degree of precision appropriate for the problem context.

c. Find the exact ratios of the side lengths (using square roots).

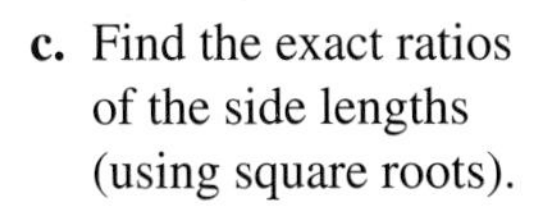

$\frac{AB}{AC} = \square$

$\frac{AB}{BC} = \square$

$\frac{AC}{BC} = \square$

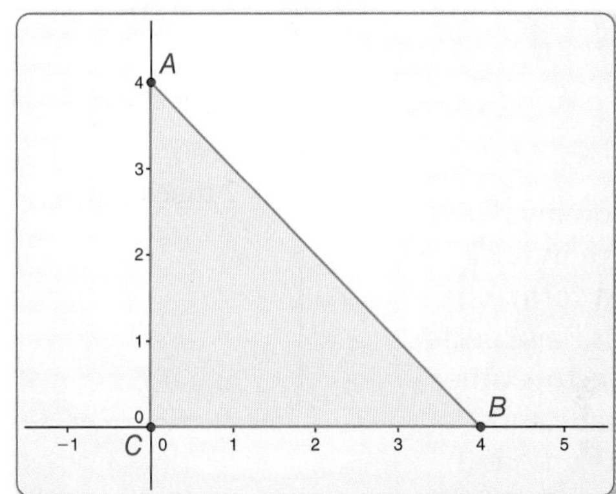

Sample

Points
$A(0, 4)$
$B(4, 0)$
$C(0, 0)$
Segments
$AB = 5.66$
$BC = 4$
$AC = 4$
Angles
$m\angle A = 45°$
$m\angle B = 45°$

d. Repeat parts (a) and (c) for several other isosceles right triangles. Use your results to write a conjecture about the ratios of the side lengths of an isosceles right triangle.

EXPLORATION 2 Side Ratios of a 30°- 60°- 90° Triangle

Work with a partner.

a. Use dynamic geometry software to construct a right triangle with acute angle measures of 30° and 60° (a 30°- 60°- 90° triangle), where the shorter leg length is 3 units.

b. Find the exact ratios of the side lengths (using square roots).

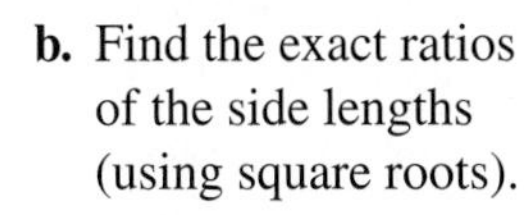

$\frac{AB}{AC} = \square$

$\frac{AB}{BC} = \square$

$\frac{AC}{BC} = \square$

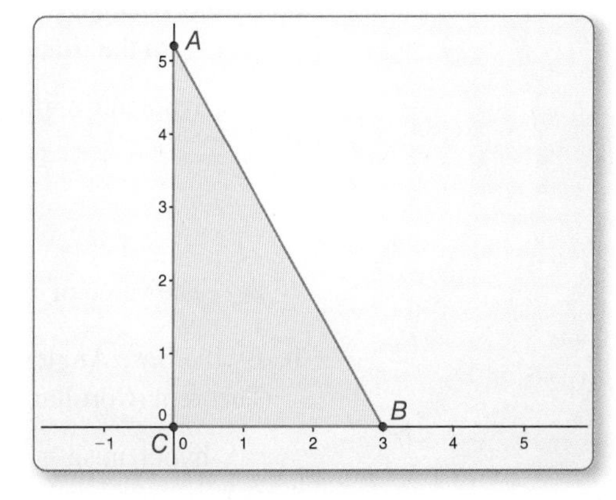

Sample

Points
$A(0, 5.20)$
$B(3, 0)$
$C(0, 0)$
Segments
$AB = 6$
$BC = 3$
$AC = 5.20$
Angles
$m\angle A = 30°$
$m\angle B = 60°$

c. Repeat parts (a) and (b) for several other 30°- 60°- 90° triangles. Use your results to write a conjecture about the ratios of the side lengths of a 30°- 60°- 90° triangle.

Communicate Your Answer

3. What is the relationship among the side lengths of 45°- 45°- 90° triangles? 30°- 60°- 90° triangles?

9.2 Lesson

Core Vocabulary

Previous
isosceles triangle

What You Will Learn

- Find side lengths in special right triangles.
- Solve real-life problems involving special right triangles.

Finding Side Lengths in Special Right Triangles

A 45°- 45°- 90° triangle is an *isosceles right triangle* that can be formed by cutting a square in half diagonally.

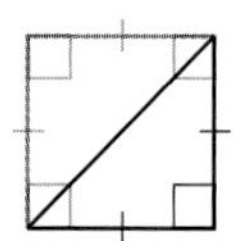

Theorem

Theorem 9.4 45°- 45°- 90° Triangle Theorem

In a 45°- 45°- 90° triangle, the hypotenuse is $\sqrt{2}$ times as long as each leg.

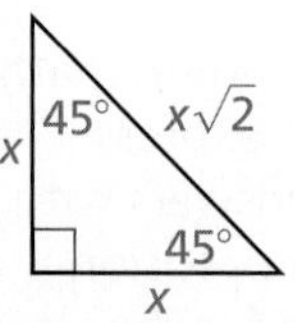

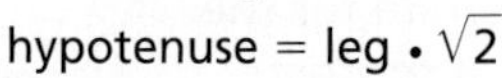

Proof Ex. 19, p. 476

REMEMBER

An expression involving a radical with index 2 is in simplest form when no radicands have perfect squares as factors other than 1, no radicands contain fractions, and no radicals appear in the denominator of a fraction.

EXAMPLE 1 Finding Side Lengths in 45°- 45°- 90° Triangles

Find the value of x. Write your answer in simplest form.

a.

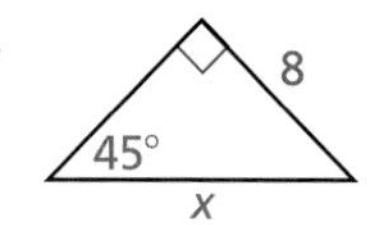

b.

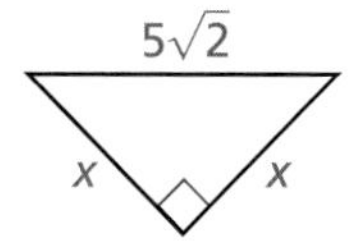

SOLUTION

a. By the Triangle Sum Theorem (Theorem 5.1), the measure of the third angle must be 45°, so the triangle is a 45°- 45°- 90° triangle.

hypotenuse = leg • $\sqrt{2}$	45°- 45°- 90° Triangle Theorem
$x = 8 \cdot \sqrt{2}$	Substitute.
$x = 8\sqrt{2}$	Simplify.

▶ The value of x is $8\sqrt{2}$.

b. By the Base Angles Theorem (Theorem 5.6) and the Corollary to the Triangle Sum Theorem (Corollary 5.1), the triangle is a 45°- 45°- 90° triangle.

hypotenuse = leg • $\sqrt{2}$	45°- 45°- 90° Triangle Theorem
$5\sqrt{2} = x \cdot \sqrt{2}$	Substitute.
$\frac{5\sqrt{2}}{\sqrt{2}} = \frac{x\sqrt{2}}{\sqrt{2}}$	Divide each side by $\sqrt{2}$.
$5 = x$	Simplify.

▶ The value of x is 5.

Theorem

Theorem 9.5 30°-60°-90° Triangle Theorem

In a 30°-60°-90° triangle, the hypotenuse is twice as long as the shorter leg, and the longer leg is $\sqrt{3}$ times as long as the shorter leg.

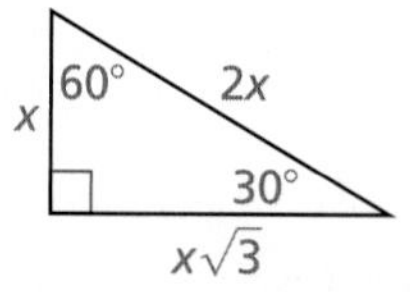

hypotenuse = shorter leg • 2
longer leg = shorter leg • $\sqrt{3}$

Proof Ex. 21, p. 476

REMEMBER

Because the angle opposite 9 is larger than the angle opposite x, the leg with length 9 is longer than the leg with length x by the Triangle Larger Angle Theorem (Theorem 6.10).

EXAMPLE 2 Finding Side Lengths in a 30°-60°-90° Triangle

Find the values of x and y. Write your answer in simplest form.

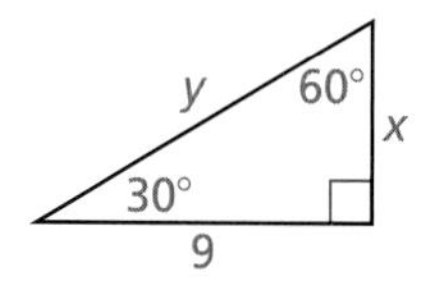

SOLUTION

Step 1 Find the value of x.

longer leg = shorter leg • $\sqrt{3}$	30°-60°-90° Triangle Theorem
$9 = x \cdot \sqrt{3}$	Substitute.
$\frac{9}{\sqrt{3}} = x$	Divide each side by $\sqrt{3}$.
$\frac{9}{\sqrt{3}} \cdot \frac{\sqrt{3}}{\sqrt{3}} = x$	Multiply by $\frac{\sqrt{3}}{\sqrt{3}}$.
$\frac{9\sqrt{3}}{3} = x$	Multiply fractions.
$3\sqrt{3} = x$	Simplify.

▶ The value of x is $3\sqrt{3}$.

Step 2 Find the value of y.

hypotenuse = shorter leg • 2	30°-60°-90° Triangle Theorem
$y = 3\sqrt{3} \cdot 2$	Substitute.
$y = 6\sqrt{3}$	Simplify.

▶ The value of y is $6\sqrt{3}$.

Monitoring Progress

Help in English and Spanish at *BigIdeasMath.com*

Find the value of the variable. Write your answer in simplest form.

1.

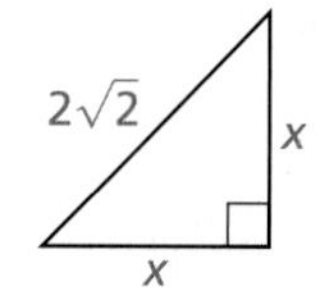

2.

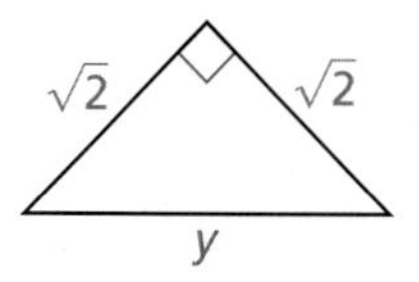

3.

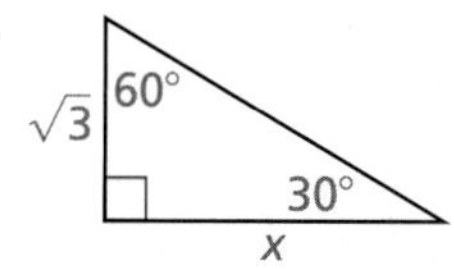

4.

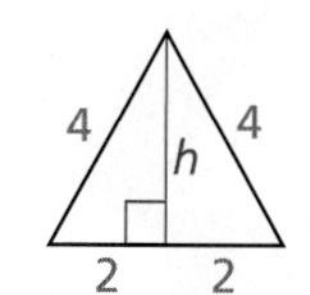

Solving Real-Life Problems

EXAMPLE 3 Modeling with Mathematics

The road sign is shaped like an equilateral triangle. Estimate the area of the sign by finding the area of the equilateral triangle.

SOLUTION

First find the height h of the triangle by dividing it into two 30°-60°-90° triangles. The length of the longer leg of one of these triangles is h. The length of the shorter leg is 18 inches.

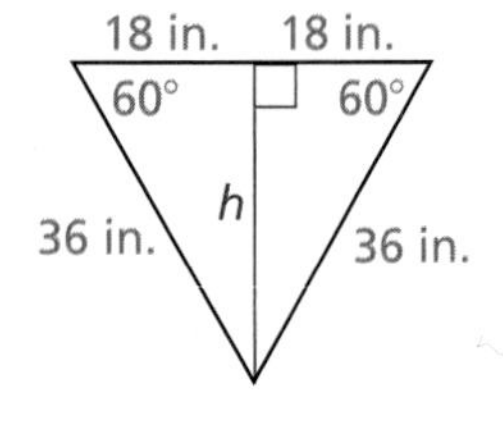

$h = 18 \cdot \sqrt{3} = 18\sqrt{3}$ 30°-60°-90° Triangle Theorem

Use $h = 18\sqrt{3}$ to find the area of the equilateral triangle.

$\text{Area} = \frac{1}{2}bh = \frac{1}{2}(36)\left(18\sqrt{3}\right) \approx 561.18$

▶ The area of the sign is about 561 square inches.

EXAMPLE 4 Finding the Height of a Ramp

A tipping platform is a ramp used to unload trucks. How high is the end of an 80-foot ramp when the tipping angle is 30°? 45°?

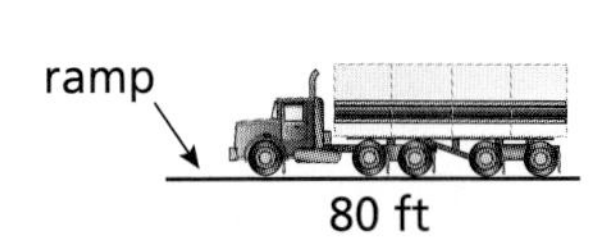

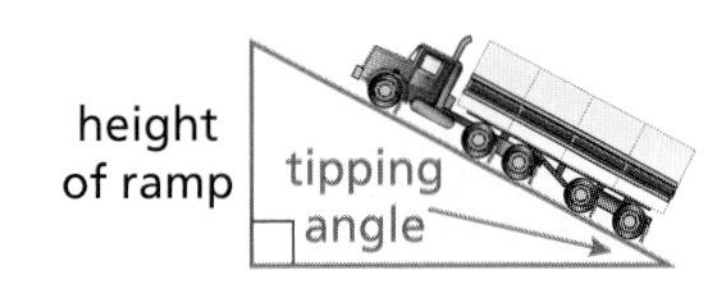

SOLUTION

When the tipping angle is 30°, the height h of the ramp is the length of the shorter leg of a 30°-60°-90° triangle. The length of the hypotenuse is 80 feet.

$80 = 2h$ 30°-60°-90° Triangle Theorem

$40 = h$ Divide each side by 2.

When the tipping angle is 45°, the height h of the ramp is the length of a leg of a 45°-45°-90° triangle. The length of the hypotenuse is 80 feet.

$80 = h \cdot \sqrt{2}$ 45°-45°-90° Triangle Theorem

$\frac{80}{\sqrt{2}} = h$ Divide each side by $\sqrt{2}$.

$56.6 \approx h$ Use a calculator.

▶ When the tipping angle is 30°, the ramp height is 40 feet. When the tipping angle is 45°, the ramp height is about 56 feet 7 inches.

Monitoring Progress

Help in English and Spanish at *BigIdeasMath.com*

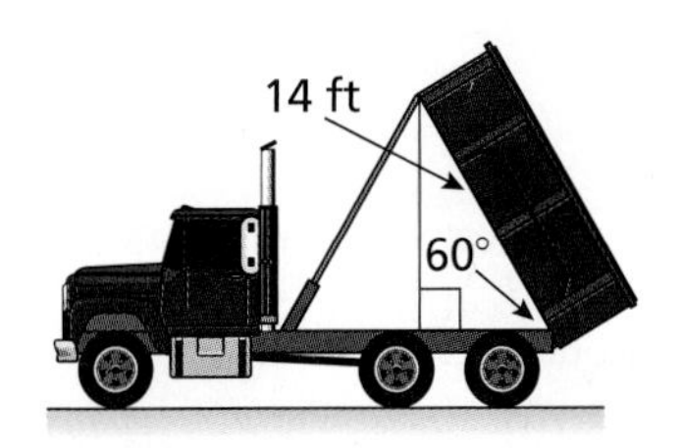

5. The logo on a recycling bin resembles an equilateral triangle with side lengths of 6 centimeters. Approximate the area of the logo.

6. The body of a dump truck is raised to empty a load of sand. How high is the 14-foot-long body from the frame when it is tipped upward by a 60° angle?

9.2 Exercises

Dynamic Solutions available at *BigIdeasMath.com*

Vocabulary and Core Concept Check

1. **VOCABULARY** Name two special right triangles by their angle measures.

2. **WRITING** Explain why the acute angles in an isosceles right triangle always measure 45°.

Monitoring Progress and Modeling with Mathematics

In Exercises 3–6, find the value of x. Write your answer in simplest form. *(See Example 1.)*

3. 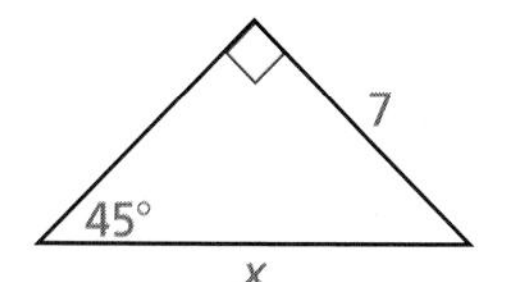

4. 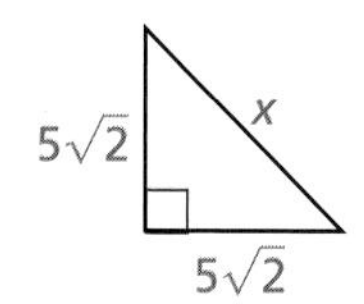

5. 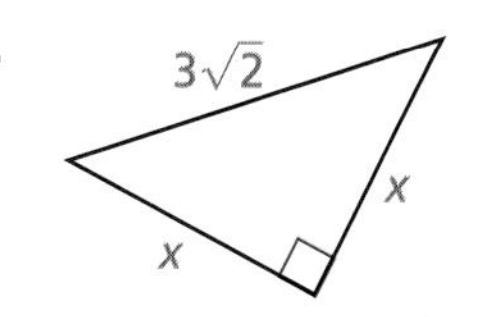

6. 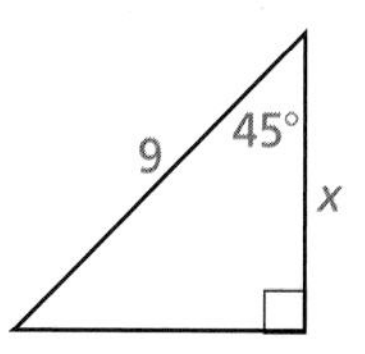

In Exercises 7–10, find the values of x and y. Write your answers in simplest form. *(See Example 2.)*

7. 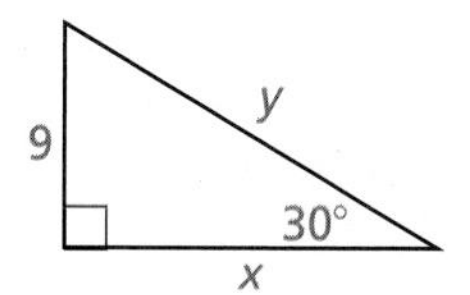

8. 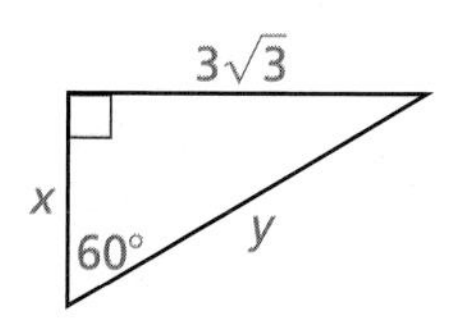

9. 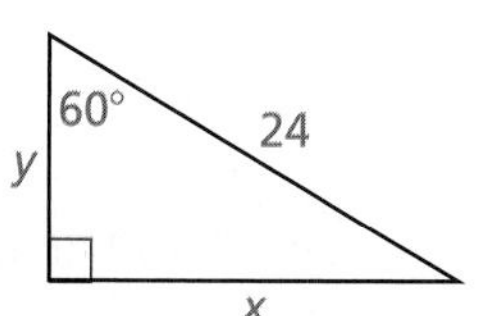

10. 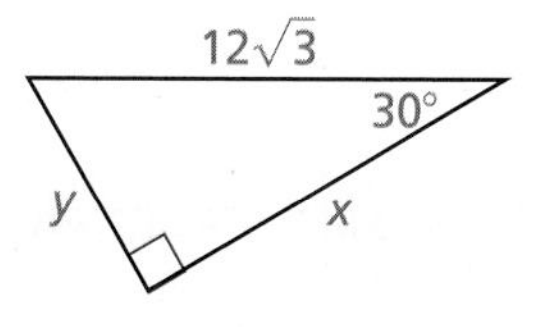

ERROR ANALYSIS In Exercises 11 and 12, describe and correct the error in finding the length of the hypotenuse.

11. ✗ 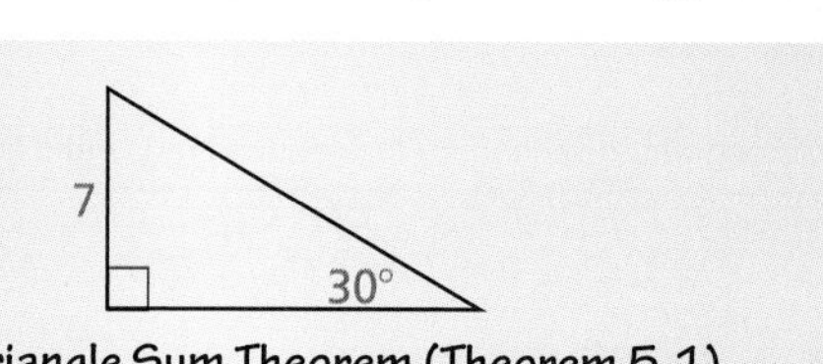

By the Triangle Sum Theorem (Theorem 5.1), the measure of the third angle must be 60°. So, the triangle is a 30°-60°-90° triangle.

hypotenuse = shorter leg • $\sqrt{3} = 7\sqrt{3}$

So, the length of the hypotenuse is $7\sqrt{3}$ units.

12. ✗ 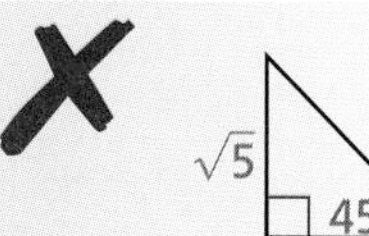

By the Triangle Sum Theorem (Theorem 5.1), the measure of the third angle must be 45°. So, the triangle is a 45°-45°-90° triangle.

hypotenuse = leg • leg • $\sqrt{2} = 5\sqrt{2}$

So, the length of the hypotenuse is $5\sqrt{2}$ units.

In Exercises 13 and 14, sketch the figure that is described. Find the indicated length. Round decimal answers to the nearest tenth.

13. The side length of an equilateral triangle is 5 centimeters. Find the length of an altitude.

14. The perimeter of a square is 36 inches. Find the length of a diagonal.

In Exercises 15 and 16, find the area of the figure. Round decimal answers to the nearest tenth. *(See Example 3.)*

15. 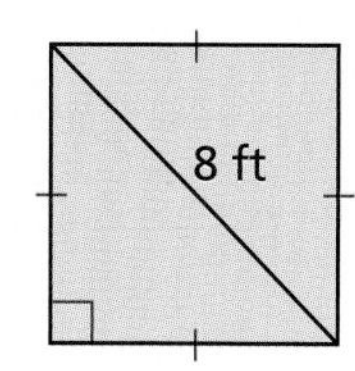

16. 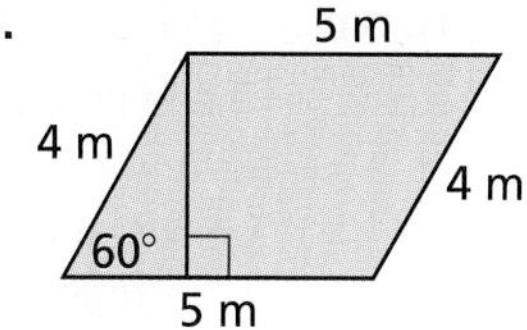

17. **PROBLEM SOLVING** Each half of the drawbridge is about 284 feet long. How high does the drawbridge rise when x is 30°? 45°? 60°? *(See Example 4.)*

18. MODELING WITH MATHEMATICS A nut is shaped like a regular hexagon with side lengths of 1 centimeter. Find the value of x. (*Hint*: A regular hexagon can be divided into six congruent triangles.)

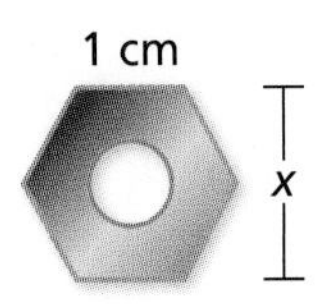

19. PROVING A THEOREM Write a paragraph proof of the 45°-45°-90° Triangle Theorem (Theorem 9.4).

Given $\triangle DEF$ is a 45°-45°-90° triangle.

Prove The hypotenuse is $\sqrt{2}$ times as long as each leg.

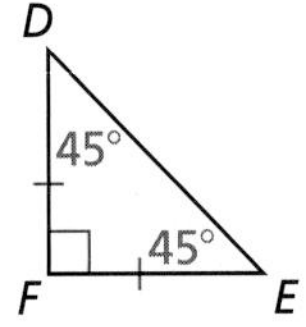

20. HOW DO YOU SEE IT? The diagram shows part of the *Wheel of Theodorus*.

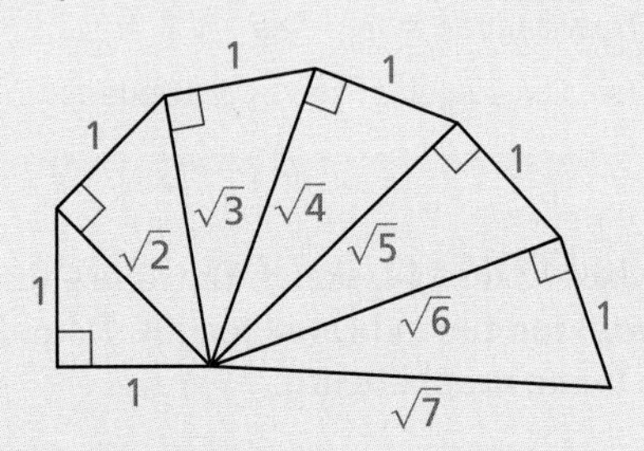

a. Which triangles, if any, are 45°-45°-90° triangles?

b. Which triangles, if any, are 30°-60°-90° triangles?

21. PROVING A THEOREM Write a paragraph proof of the 30°-60°-90° Triangle Theorem (Theorem 9.5). (*Hint*: Construct $\triangle JML$ congruent to $\triangle JKL$.)

Given $\triangle JKL$ is a 30°-60°-90° triangle.

Prove The hypotenuse is twice as long as the shorter leg, and the longer leg is $\sqrt{3}$ times as long as the shorter leg.

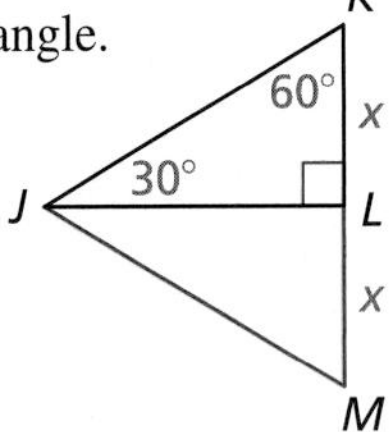

22. THOUGHT PROVOKING A special right triangle is a right triangle that has rational angle measures and each side length contains at most one square root. There are only three special right triangles. The diagram below is called the *Ailles rectangle*. Label the sides and angles in the diagram. Describe all three special right triangles.

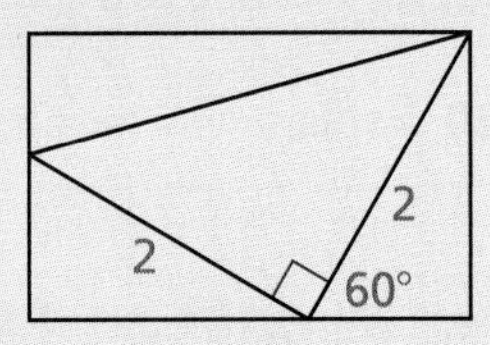

23. WRITING Describe two ways to show that all isosceles right triangles are similar to each other.

24. MAKING AN ARGUMENT Each triangle in the diagram is a 45°-45°-90° triangle. At Stage 0, the legs of the triangle are each 1 unit long. Your brother claims the lengths of the legs of the triangles added are halved at each stage. So, the length of a leg of a triangle added in Stage 8 will be $\frac{1}{256}$ unit. Is your brother correct? Explain your reasoning.

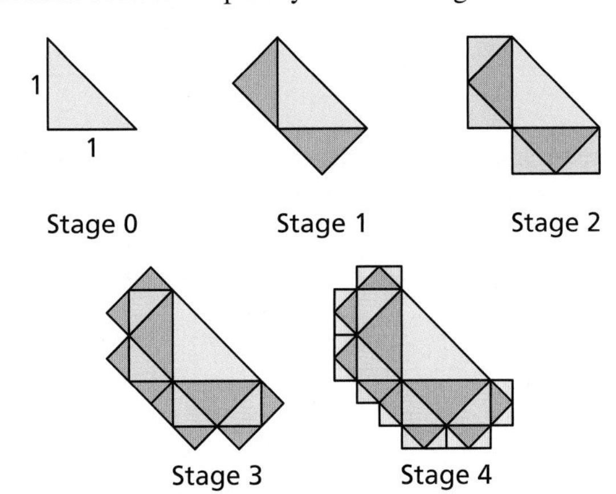

25. USING STRUCTURE $\triangle TUV$ is a 30°-60°-90° triangle, where two vertices are $U(3, -1)$ and $V(-3, -1)$, $\overline{UV}$ is the hypotenuse, and point T is in Quadrant I. Find the coordinates of T.

Maintaining Mathematical Proficiency

Reviewing what you learned in previous grades and lessons

Find the value of x. (*Section 8.1*)

26. $\triangle DEF \sim \triangle LMN$

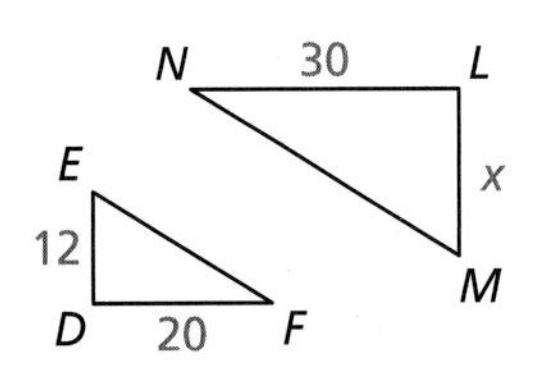

27. $\triangle ABC \sim \triangle QRS$

9.3 Similar Right Triangles

Essential Question How are altitudes and geometric means of right triangles related?

EXPLORATION 1 Writing a Conjecture

Work with a partner.

a. Use dynamic geometry software to construct right $\triangle ABC$, as shown. Draw $\overline{CD}$ so that it is an altitude from the right angle to the hypotenuse of $\triangle ABC$.

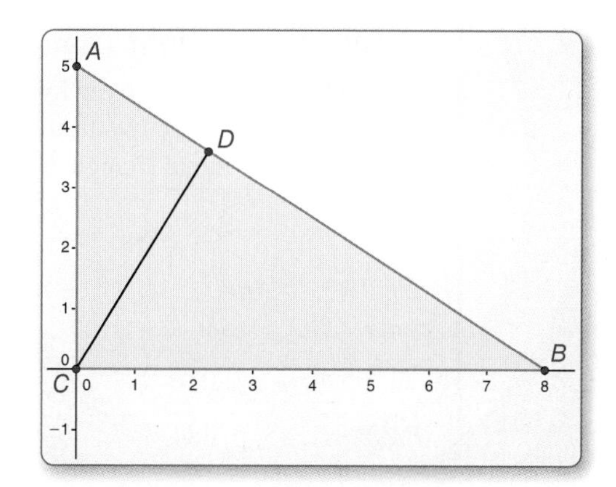

Points
$A(0, 5)$
$B(8, 0)$
$C(0, 0)$
$D(2.25, 3.6)$
Segments
$AB = 9.43$
$BC = 8$
$AC = 5$

b. The **geometric mean** of two positive numbers a and b is the positive number x that satisfies

$$\frac{a}{x} = \frac{x}{b}. \qquad x \text{ is the geometric mean of } a \text{ and } b.$$

Write a proportion involving the side lengths of $\triangle CBD$ and $\triangle ACD$ so that CD is the geometric mean of two of the other side lengths. Use similar triangles to justify your steps.

c. Use the proportion you wrote in part (b) to find CD.

d. Generalize the proportion you wrote in part (b). Then write a conjecture about how the geometric mean is related to the altitude from the right angle to the hypotenuse of a right triangle.

CONSTRUCTING VIABLE ARGUMENTS

To be proficient in math, you need to understand and use stated assumptions, definitions, and previously established results in constructing arguments.

EXPLORATION 2 Comparing Geometric and Arithmetic Means

Work with a partner. Use a spreadsheet to find the arithmetic mean and the geometric mean of several pairs of positive numbers. Compare the two means. What do you notice?

	A	B	C	D
1	a	b	Arithmetic Mean	Geometric Mean
2	3	4	3.5	3.464
3	4	5		
4	6	7		
5	0.5	0.5		
6	0.4	0.8		
7	2	5		
8	1	4		
9	9	16		
10	10	100		
11				

Communicate Your Answer

3. How are altitudes and geometric means of right triangles related?

9.3 Lesson

Core Vocabulary

geometric mean, *p. 480*

Previous
altitude of a triangle
similar figures

What You Will Learn

- Identify similar triangles.
- Solve real-life problems involving similar triangles.
- Use geometric means.

Identifying Similar Triangles

When the altitude is drawn to the hypotenuse of a right triangle, the two smaller triangles are similar to the original triangle and to each other.

Theorem

Theorem 9.6 Right Triangle Similarity Theorem

If the altitude is drawn to the hypotenuse of a right triangle, then the two triangles formed are similar to the original triangle and to each other.

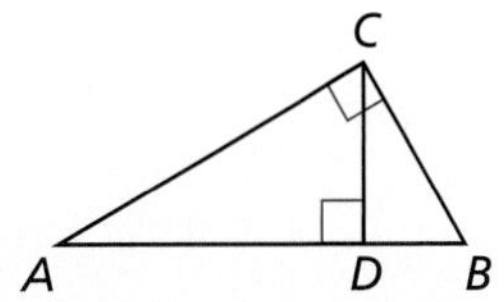

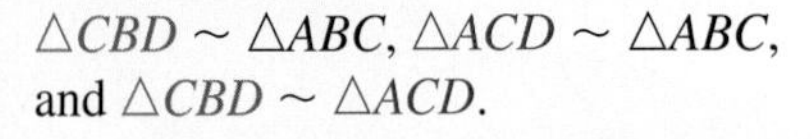
$\triangle CBD \sim \triangle ABC$, $\triangle ACD \sim \triangle ABC$, and $\triangle CBD \sim \triangle ACD$.

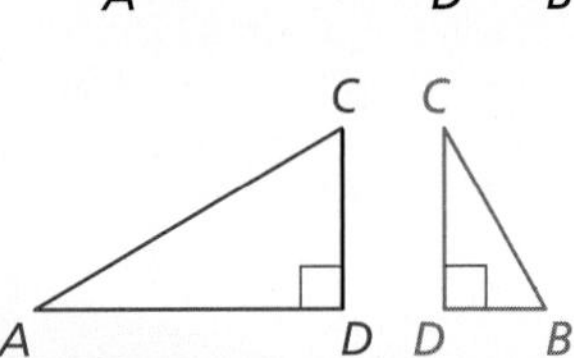

Proof Ex. 45, p. 484

EXAMPLE 1 Identifying Similar Triangles

Identify the similar triangles in the diagram.

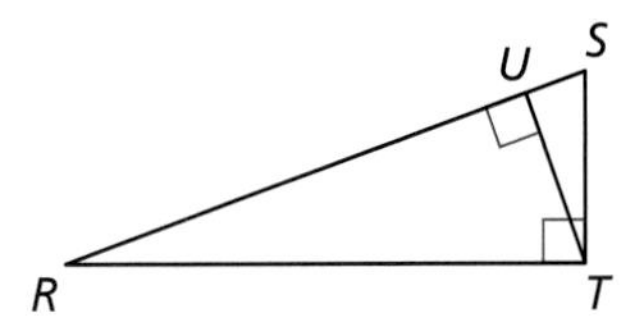

SOLUTION

Sketch the three similar right triangles so that the corresponding angles and sides have the same orientation.

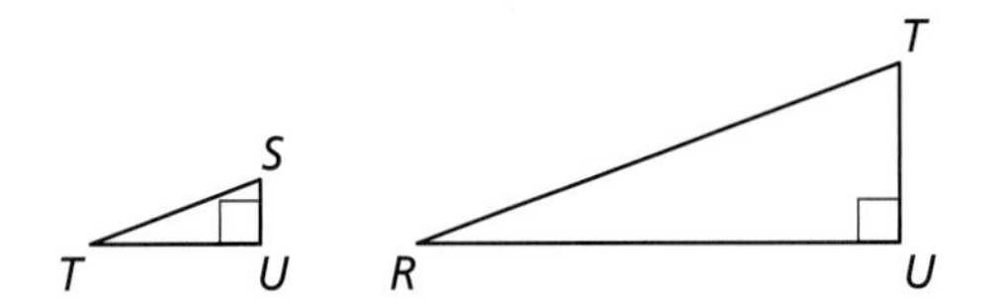

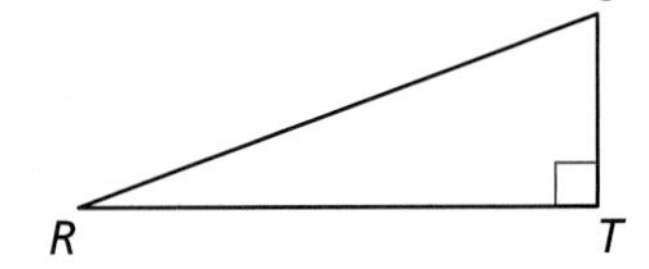

▶ $\triangle TSU \sim \triangle RTU \sim \triangle RST$

Monitoring Progress

Help in English and Spanish at *BigIdeasMath.com*

Identify the similar triangles.

1.

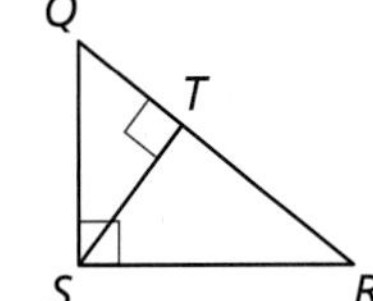

2.

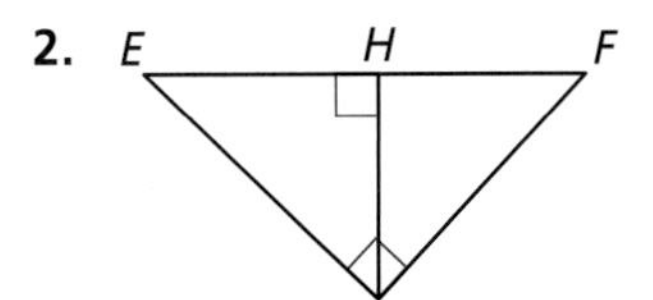

Solving Real-Life Problems

EXAMPLE 2 Modeling with Mathematics

A roof has a cross section that is a right triangle. The diagram shows the approximate dimensions of this cross section. Find the height h of the roof.

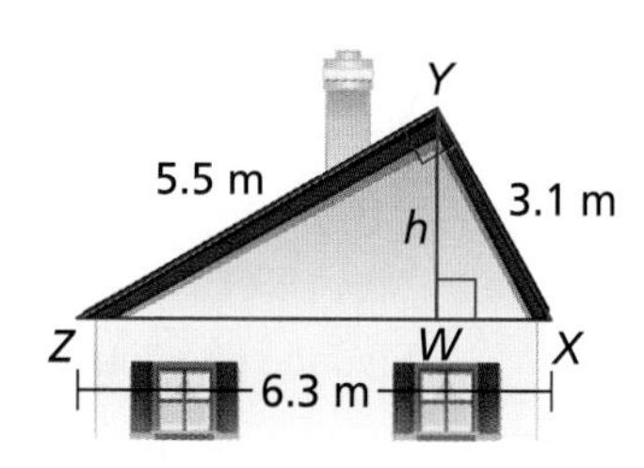

SOLUTION

1. **Understand the Problem** You are given the side lengths of a right triangle. You need to find the height of the roof, which is the altitude drawn to the hypotenuse.
2. **Make a Plan** Identify any similar triangles. Then use the similar triangles to write a proportion involving the height and solve for h.
3. **Solve the Problem** Identify the similar triangles and sketch them.

COMMON ERROR

Notice that if you tried to write a proportion using $\triangle XYW$ and $\triangle YZW$, then there would be two unknowns, so you would not be able to solve for h.

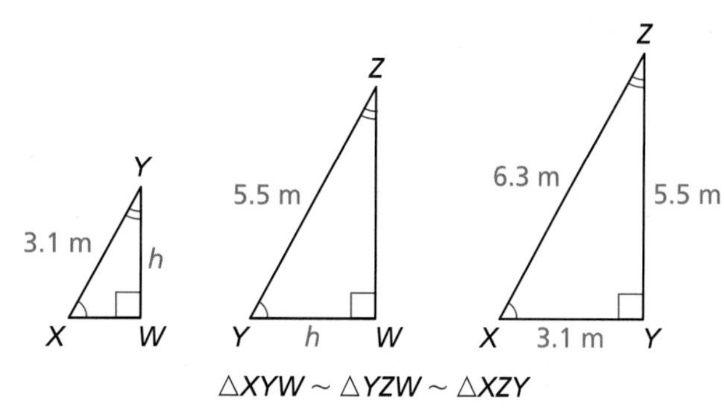

$\triangle XYW \sim \triangle YZW \sim \triangle XZY$

Because $\triangle XYW \sim \triangle XZY$, you can write a proportion.

$\frac{YW}{ZY} = \frac{XY}{XZ}$ Corresponding side lengths of similar triangles are proportional.

$\frac{h}{5.5} = \frac{3.1}{6.3}$ Substitute.

$h \approx 2.7$ Multiply each side by 5.5.

▶ The height of the roof is about 2.7 meters.

4. **Look Back** Because the height of the roof is a leg of right $\triangle YZW$ and right $\triangle XYW$, it should be shorter than each of their hypotenuses. The lengths of the two hypotenuses are $YZ = 5.5$ and $XY = 3.1$. Because $2.7 < 3.1$, the answer seems reasonable.

Monitoring Progress

Help in English and Spanish at *BigIdeasMath.com*

Find the value of x.

3. E, H, G, F; 3, 4, 5, x

4.

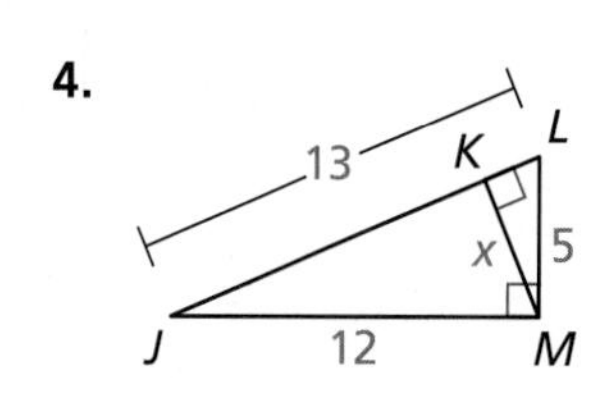

Using a Geometric Mean

Core Concept

Geometric Mean

The **geometric mean** of two positive numbers a and b is the positive number x that satisfies $\frac{a}{x} = \frac{x}{b}$. So, $x^2 = ab$ and $x = \sqrt{ab}$.

EXAMPLE 3 Finding a Geometric Mean

Find the geometric mean of 24 and 48.

SOLUTION

$x^2 = ab$	Definition of geometric mean
$x^2 = 24 \cdot 48$	Substitute 24 for a and 48 for b.
$x = \sqrt{24 \cdot 48}$	Take the positive square root of each side.
$x = \sqrt{24 \cdot 24 \cdot 2}$	Factor.
$x = 24\sqrt{2}$	Simplify.

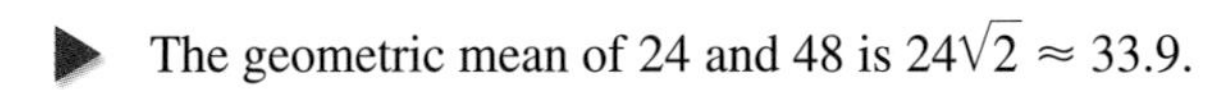

▶ The geometric mean of 24 and 48 is $24\sqrt{2} \approx 33.9$.

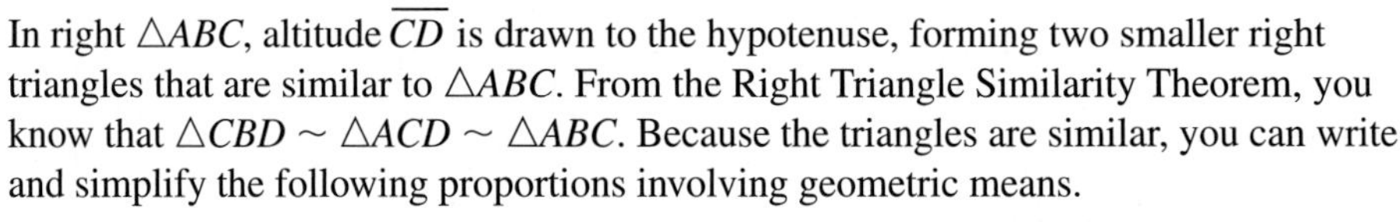

In right $\triangle ABC$, altitude $\overline{CD}$ is drawn to the hypotenuse, forming two smaller right triangles that are similar to $\triangle ABC$. From the Right Triangle Similarity Theorem, you know that $\triangle CBD \sim \triangle ACD \sim \triangle ABC$. Because the triangles are similar, you can write and simplify the following proportions involving geometric means.

$$\frac{CD}{AD} = \frac{BD}{CD} \qquad \frac{CB}{DB} = \frac{AB}{CB} \qquad \frac{AC}{AD} = \frac{AB}{AC}$$

$$CD^2 = AD \cdot BD \qquad CB^2 = DB \cdot AB \qquad AC^2 = AD \cdot AB$$

Theorems

Theorem 9.7 Geometric Mean (Altitude) Theorem

In a right triangle, the altitude from the right angle to the hypotenuse divides the hypotenuse into two segments.

The length of the altitude is the geometric mean of the lengths of the two segments of the hypotenuse.

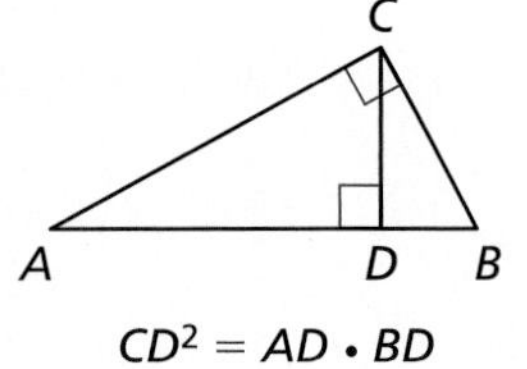

$CD^2 = AD \cdot BD$

Proof Ex. 41, p. 484

Theorem 9.8 Geometric Mean (Leg) Theorem

In a right triangle, the altitude from the right angle to the hypotenuse divides the hypotenuse into two segments.

The length of each leg of the right triangle is the geometric mean of the lengths of the hypotenuse and the segment of the hypotenuse that is adjacent to the leg.

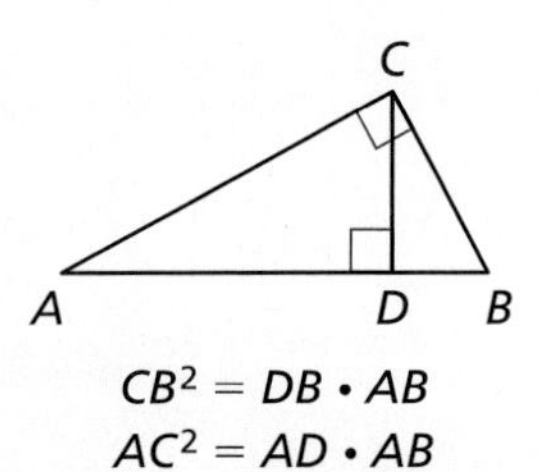

$CB^2 = DB \cdot AB$
$AC^2 = AD \cdot AB$

Proof Ex. 42, p. 484

EXAMPLE 4 Using a Geometric Mean

Find the value of each variable.

a.

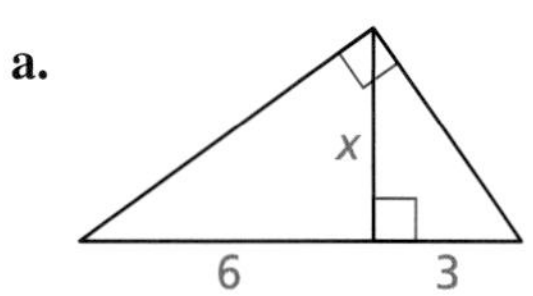

b.

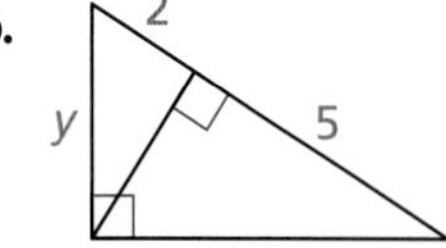

COMMON ERROR

In Example 4(b), the Geometric Mean (Leg) Theorem gives $y^2 = 2 \cdot (5 + 2)$, not $y^2 = 5 \cdot (5 + 2)$, because the side with length y is adjacent to the segment with length 2.

SOLUTION

a. Apply the Geometric Mean (Altitude) Theorem.

$$x^2 = 6 \cdot 3$$
$$x^2 = 18$$
$$x = \sqrt{18}$$
$$x = \sqrt{9} \cdot \sqrt{2}$$
$$x = 3\sqrt{2}$$

▶ The value of x is $3\sqrt{2}$.

b. Apply the Geometric Mean (Leg) Theorem.

$$y^2 = 2 \cdot (5 + 2)$$
$$y^2 = 2 \cdot 7$$
$$y^2 = 14$$
$$y = \sqrt{14}$$

▶ The value of y is $\sqrt{14}$.

EXAMPLE 5 Using Indirect Measurement

To find the cost of installing a rock wall in your school gymnasium, you need to find the height of the gym wall. You use a cardboard square to line up the top and bottom of the gym wall. Your friend measures the vertical distance from the ground to your eye and the horizontal distance from you to the gym wall. Approximate the height of the gym wall.

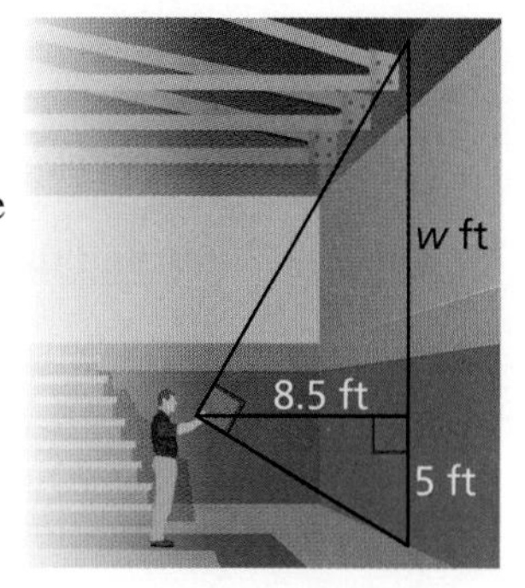

SOLUTION

By the Geometric Mean (Altitude) Theorem, you know that 8.5 is the geometric mean of w and 5.

$8.5^2 = w \cdot 5$	Geometric Mean (Altitude) Theorem
$72.25 = 5w$	Square 8.5.
$14.45 = w$	Divide each side by 5.

▶ The height of the wall is $5 + w = 5 + 14.45 = 19.45$ feet.

Monitoring Progress

Help in English and Spanish at *BigIdeasMath.com*

Find the geometric mean of the two numbers.

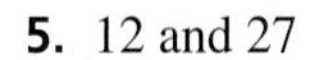

5. 12 and 27 **6.** 18 and 54 **7.** 16 and 18

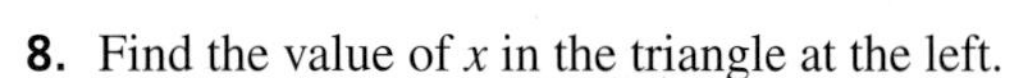

8. Find the value of x in the triangle at the left.

x
4
9

9. WHAT IF? In Example 5, the vertical distance from the ground to your eye is 5.5 feet and the distance from you to the gym wall is 9 feet. Approximate the height of the gym wall.

9.3 Exercises

Dynamic Solutions available at *BigIdeasMath.com*

Vocabulary and Core Concept Check

1. **COMPLETE THE SENTENCE** If the altitude is drawn to the hypotenuse of a right triangle, then the two triangles formed are similar to the original triangle and __________________.

2. **WRITING** In your own words, explain *geometric mean.*

Monitoring Progress and Modeling with Mathematics

In Exercises 3 and 4, identify the similar triangles. *(See Example 1.)*

3. 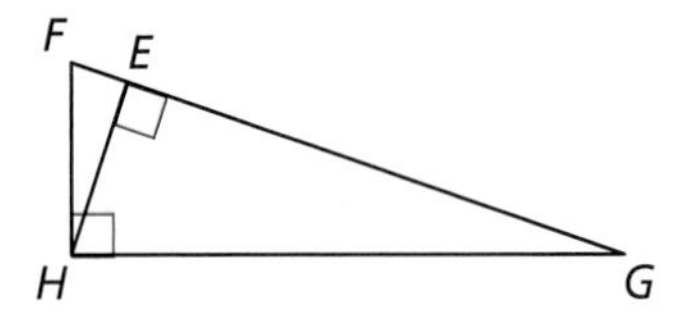

4. 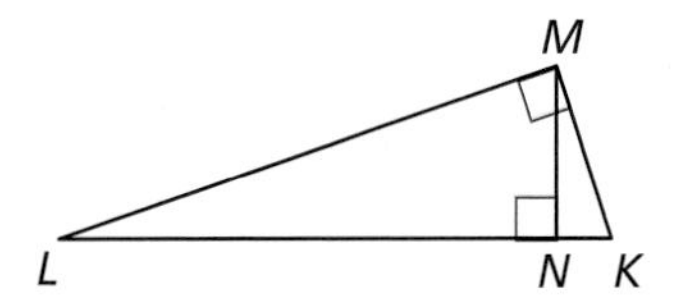

In Exercises 5–10, find the value of x. *(See Example 2.)*

5. 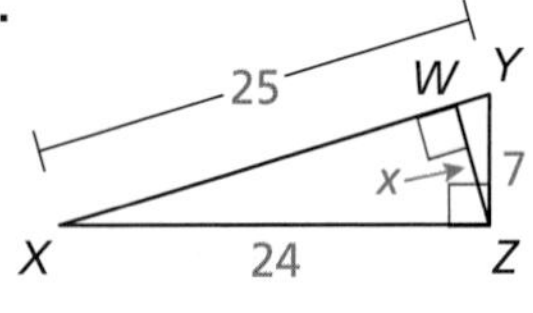

6. 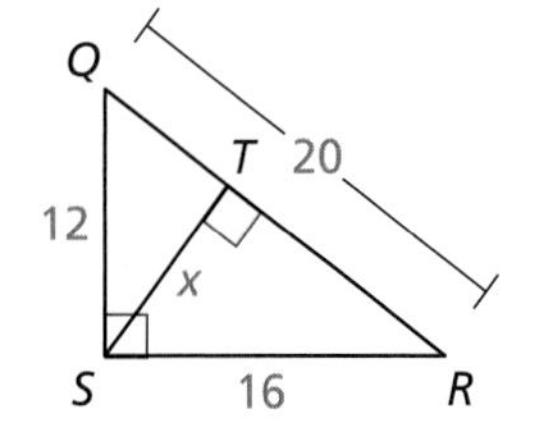

7. 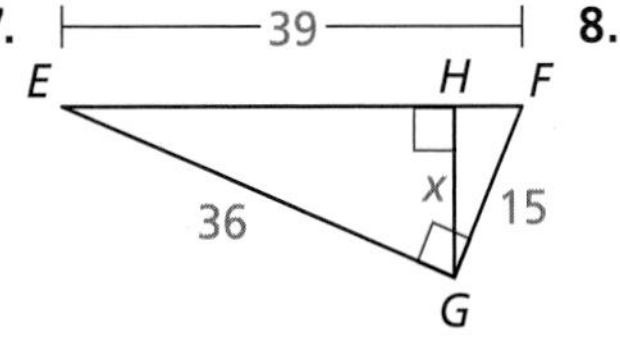

8.

9. 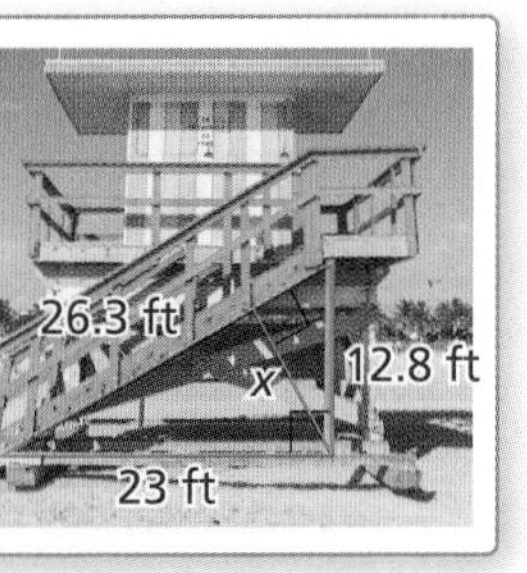

10. 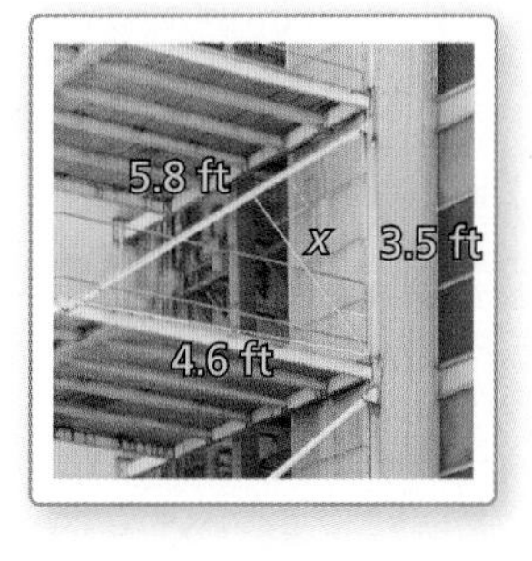

In Exercises 11–18, find the geometric mean of the two numbers. *(See Example 3.)*

11. 8 and 32
12. 9 and 16
13. 14 and 20
14. 25 and 35
15. 16 and 25
16. 8 and 28
17. 17 and 36
18. 24 and 45

In Exercises 19–26, find the value of the variable. *(See Example 4.)*

19. 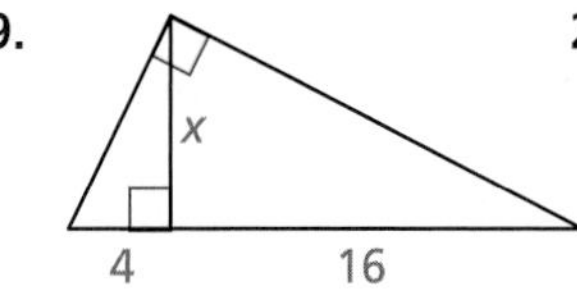

20. 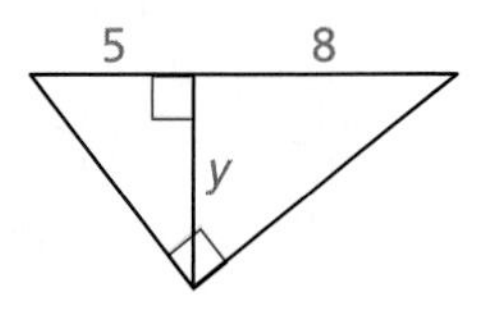

21. 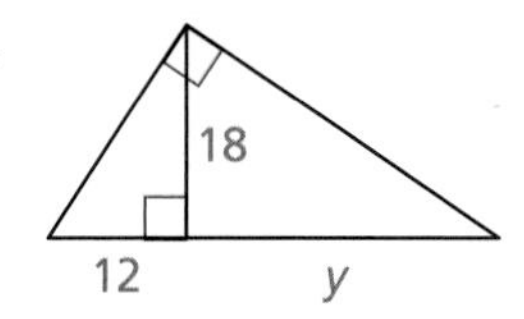

22. 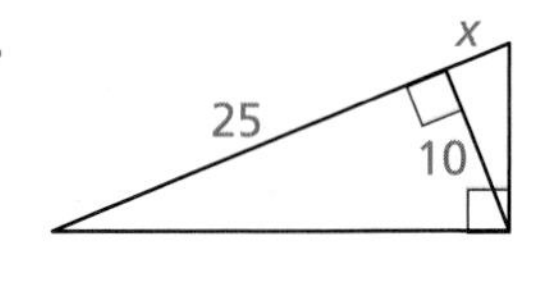

23. 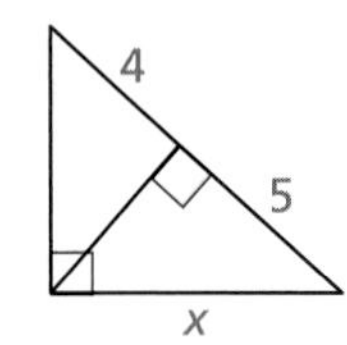

24. 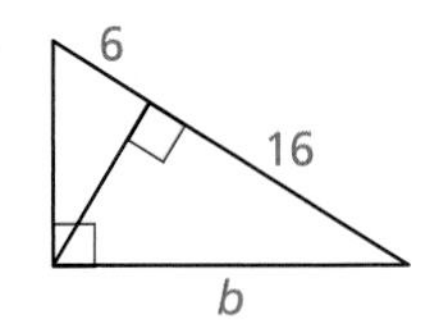

25. 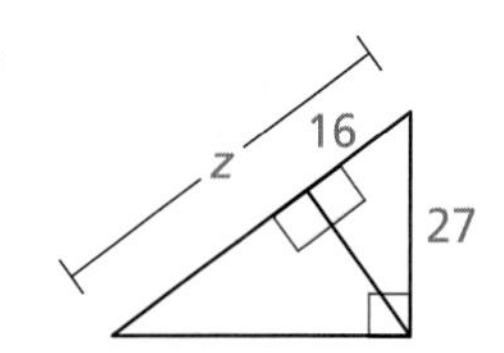

26. 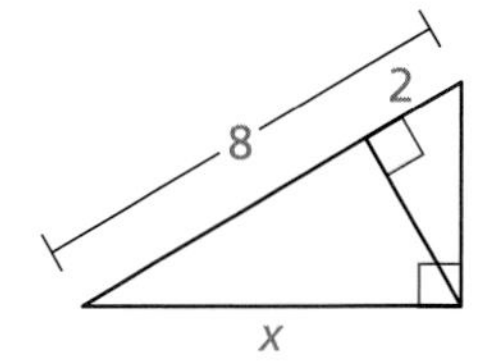

ERROR ANALYSIS In Exercises 27 and 28, describe and correct the error in writing an equation for the given diagram.

27.

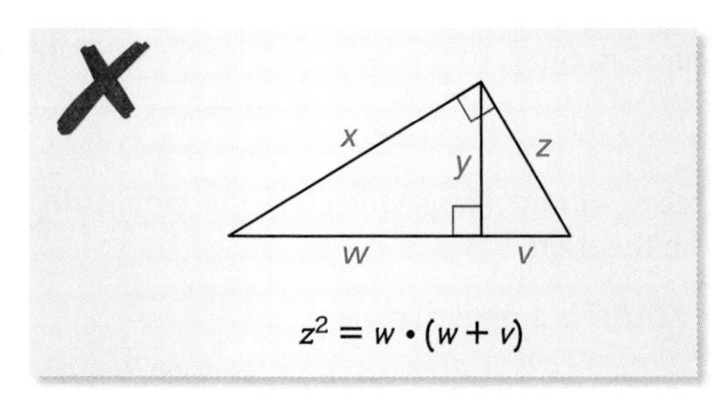

28.

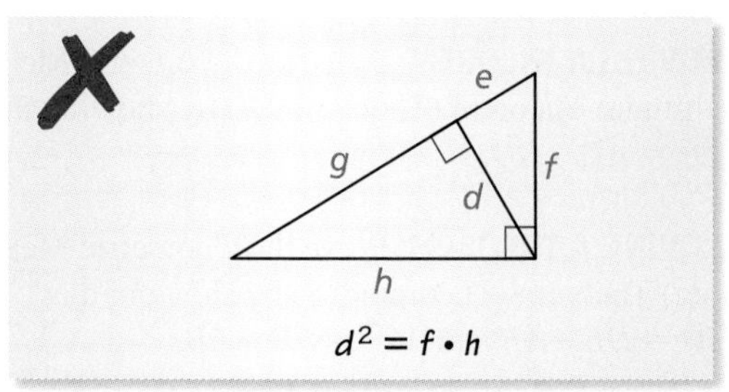

MODELING WITH MATHEMATICS In Exercises 29 and 30, use the diagram. *(See Example 5.)*

29. You want to determine the height of a monument at a local park. You use a cardboard square to line up the top and bottom of the monument, as shown at the above left. Your friend measures the vertical distance from the ground to your eye and the horizontal distance from you to the monument. Approximate the height of the monument.

30. Your classmate is standing on the other side of the monument. She has a piece of rope staked at the base of the monument. She extends the rope to the cardboard square she is holding lined up to the top and bottom of the monument. Use the information in the diagram above to approximate the height of the monument. Do you get the same answer as in Exercise 29? Explain your reasoning.

MATHEMATICAL CONNECTIONS In Exercises 31–34, find the value(s) of the variable(s).

31.

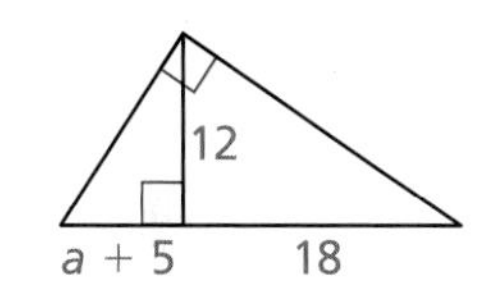

32.

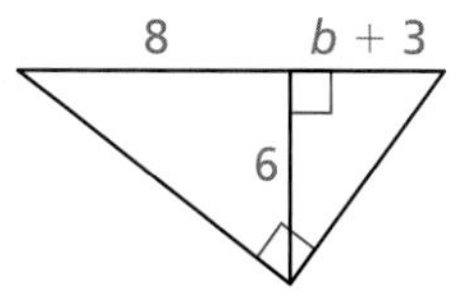

33.

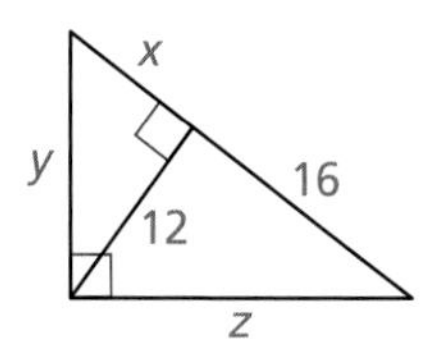

34.

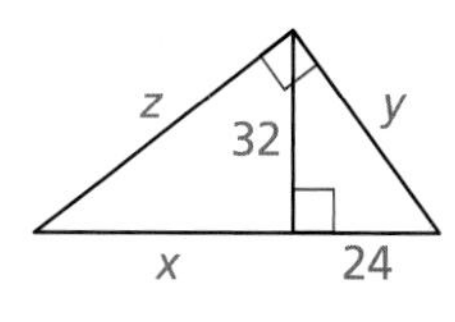

35. REASONING Use the diagram. Decide which proportions are true. Select all that apply.

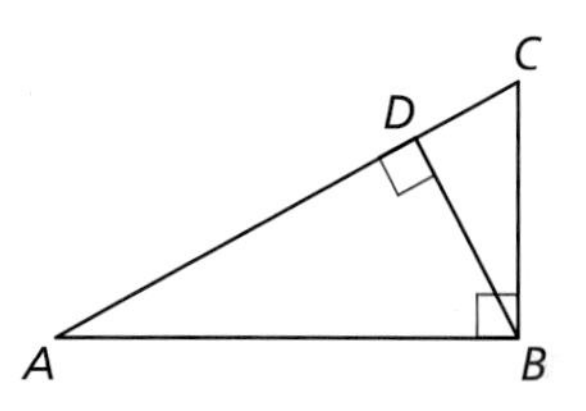

Ⓐ $\frac{DB}{DC} = \frac{DA}{DB}$ Ⓑ $\frac{BA}{CB} = \frac{CB}{BD}$

Ⓒ $\frac{CA}{BA} = \frac{BA}{CA}$ Ⓓ $\frac{DB}{BC} = \frac{DA}{BA}$

36. ANALYZING RELATIONSHIPS You are designing a diamond-shaped kite. You know that $AD = 44.8$ centimeters, $DC = 72$ centimeters, and $AC = 84.8$ centimeters. You want to use a straight crossbar $\overline{BD}$. About how long should it be? Explain your reasoning.

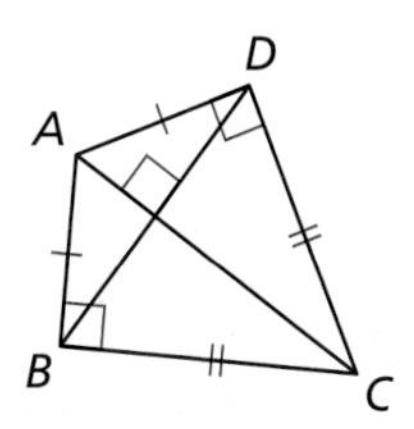

37. ANALYZING RELATIONSHIPS Use the Geometric Mean Theorems (Theorems 9.7 and 9.8) to find AC and BD.

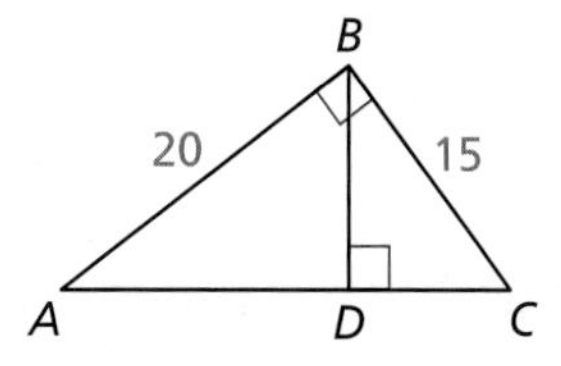

38. **HOW DO YOU SEE IT?** In which of the following triangles does the Geometric Mean (Altitude) Theorem (Theorem 9.7) apply?

Ⓐ Ⓑ

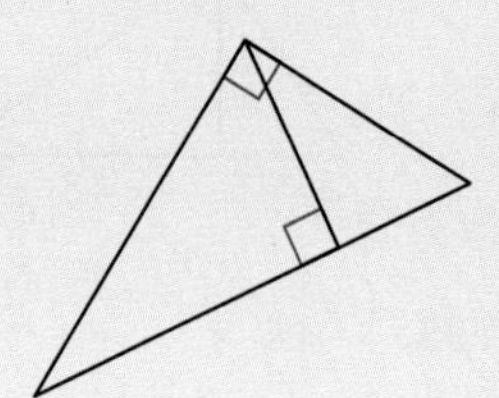

Ⓒ Ⓓ

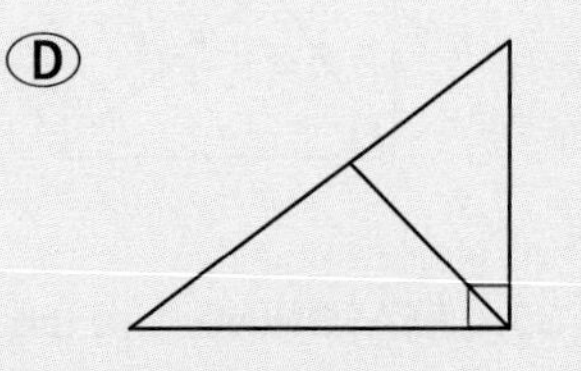

39. **PROVING A THEOREM** Use the diagram of $\triangle ABC$. Copy and complete the proof of the Pythagorean Theorem (Theorem 9.1).

Given In $\triangle ABC$, $\angle BCA$ is a right angle.

Prove $c^2 = a^2 + b^2$

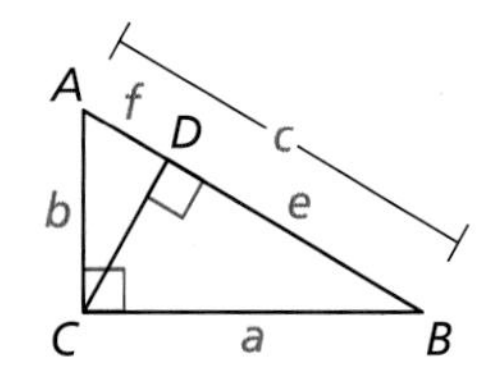

STATEMENTS	REASONS
1. In $\triangle ABC$, $\angle BCA$ is a right angle.	1. ________
2. Draw a perpendicular segment (altitude) from C to $\overline{AB}$.	2. Perpendicular Postulate (Postulate 3.2)
3. $ce = a^2$ and $cf = b^2$	3. ________
4. $ce + b^2 =$ ____ $+ b^2$	4. Addition Property of Equality
5. $ce + cf = a^2 + b^2$	5. ________
6. $c(e + f) = a^2 + b^2$	6. ________
7. $e + f =$ ____	7. Segment Addition Postulate (Postulate 1.2)
8. $c \cdot c = a^2 + b^2$	8. ________
9. $c^2 = a^2 + b^2$	9. Simplify.

40. **MAKING AN ARGUMENT** Your friend claims the geometric mean of 4 and 9 is 6, and then labels the triangle, as shown. Is your friend correct? Explain your reasoning.

In Exercises 41 and 42, use the given statements to prove the theorem.

Given $\triangle ABC$ is a right triangle.
Altitude $\overline{CD}$ is drawn to hypotenuse $\overline{AB}$.

41. **PROVING A THEOREM** Prove the Geometric Mean (Altitude) Theorem (Theorem 9.7) by showing that $CD^2 = AD \cdot BD$.

42. **PROVING A THEOREM** Prove the Geometric Mean (Leg) Theorem (Theorem 9.8) by showing that $CB^2 = DB \cdot AB$ and $AC^2 = AD \cdot AB$.

43. **CRITICAL THINKING** Draw a right isosceles triangle and label the two leg lengths x. Then draw the altitude to the hypotenuse and label its length y. Now, use the Right Triangle Similarity Theorem (Theorem 9.6) to draw the three similar triangles from the image and label any side length that is equal to either x or y. What can you conclude about the relationship between the two smaller triangles? Explain your reasoning.

44. **THOUGHT PROVOKING** The arithmetic mean and geometric mean of two nonnegative numbers x and y are shown.

$$\text{arithmetic mean} = \frac{x + y}{2}$$

$$\text{geometric mean} = \sqrt{xy}$$

Write an inequality that relates these two means. Justify your answer.

45. **PROVING A THEOREM** Prove the Right Triangle Similarity Theorem (Theorem 9.6) by proving three similarity statements.

Given $\triangle ABC$ is a right triangle.
Altitude $\overline{CD}$ is drawn to hypotenuse $\overline{AB}$.

Prove $\triangle CBD \sim \triangle ABC$, $\triangle ACD \sim \triangle ABC$, $\triangle CBD \sim \triangle ACD$

Maintaining Mathematical Proficiency

Reviewing what you learned in previous grades and lessons

Solve the equation for x. *(Skills Review Handbook)*

46. $13 = \frac{x}{5}$

47. $29 = \frac{x}{4}$

48. $9 = \frac{78}{x}$

49. $30 = \frac{115}{x}$

9.1–9.3 What Did You Learn?

Core Vocabulary

Pythagorean triple, *p. 464*
geometric mean, *p. 480*

Core Concepts

Section 9.1

Theorem 9.1 Pythagorean Theorem, *p. 464*
Common Pythagorean Triples and Some of Their Multiples, *p. 464*
Theorem 9.2 Converse of the Pythagorean Theorem, *p. 466*
Theorem 9.3 Pythagorean Inequalities Theorem, *p. 467*

Section 9.2

Theorem 9.4 45°-45°-90° Triangle Theorem, *p. 472*
Theorem 9.5 30°-60°-90° Triangle Theorem, *p. 473*

Section 9.3

Theorem 9.6 Right Triangle Similarity Theorem, *p. 478*
Theorem 9.7 Geometric Mean (Altitude) Theorem, *p. 480*
Theorem 9.8 Geometric Mean (Leg) Theorem, *p. 480*

Mathematical Practices

1. In Exercise 31 on page 469, describe the steps you took to find the area of the triangle.
2. In Exercise 23 on page 476, can one of the ways be used to show that all 30°-60°-90° triangles are similar? Explain.
3. Explain why the Geometric Mean (Altitude) Theorem (Theorem 9.7) does not apply to three of the triangles in Exercise 38 on page 484.

Study Skills

Form a Weekly Study Group, Set Up Rules

Consider using the following rules.

- Members must attend regularly, be on time, and participate.
- The sessions will focus on the key math concepts, not on the needs of one student.
- Students who skip classes will not be allowed to participate in the study group.
- Students who keep the group from being productive will be asked to leave the group.

9.1–9.3 Quiz

Find the value of x. Tell whether the side lengths form a Pythagorean triple. *(Section 9.1)*

1.

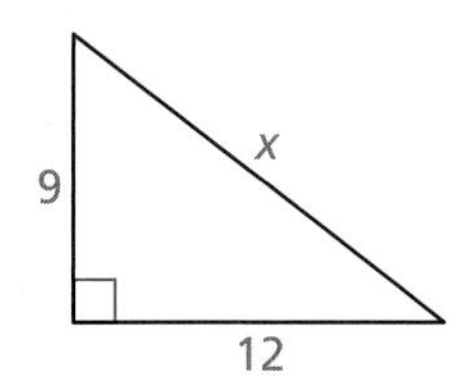

2.

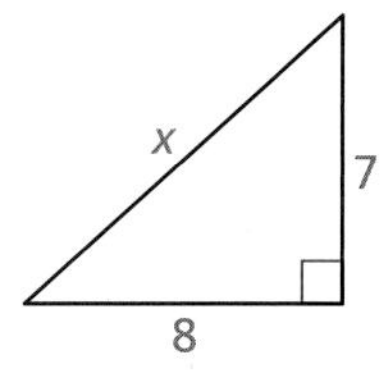

3.

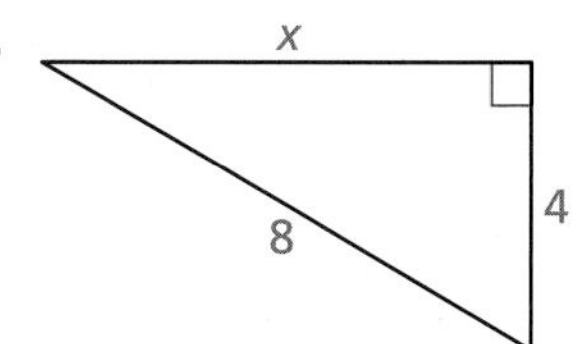

Verify that the segment lengths form a triangle. Is the triangle *acute*, *right*, or *obtuse*? *(Section 9.1)*

4. 24, 32, and 40

5. 7, 9, and 13

6. 12, 15, and $10\sqrt{3}$

Find the values of x and y. Write your answers in simplest form. *(Section 9.2)*

7.

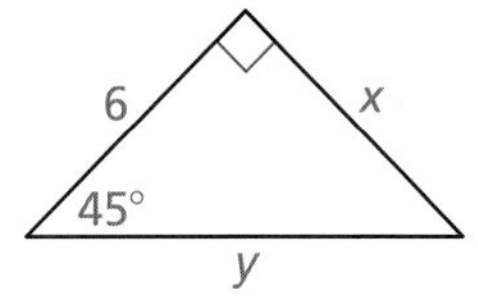

8.

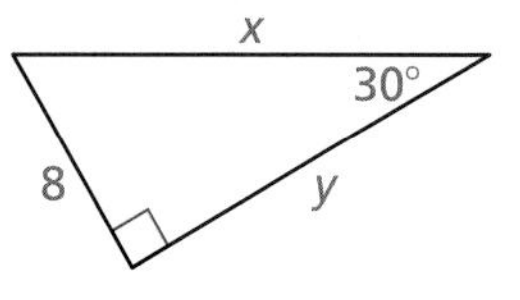

9.

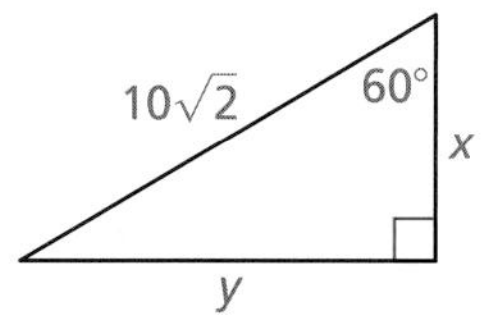

Find the geometric mean of the two numbers. *(Section 9.3)*

10. 6 and 12

11. 15 and 20

12. 18 and 26

Identify the similar right triangles. Then find the value of the variable. *(Section 9.3)*

13.

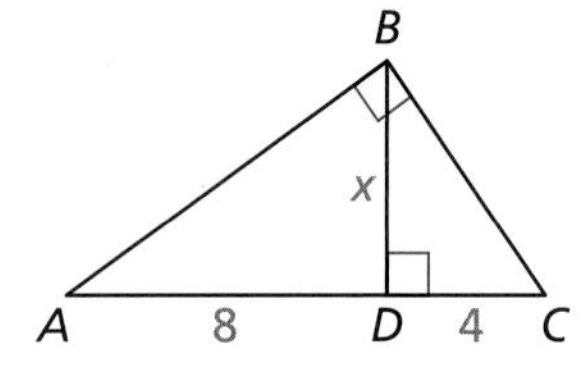

14.

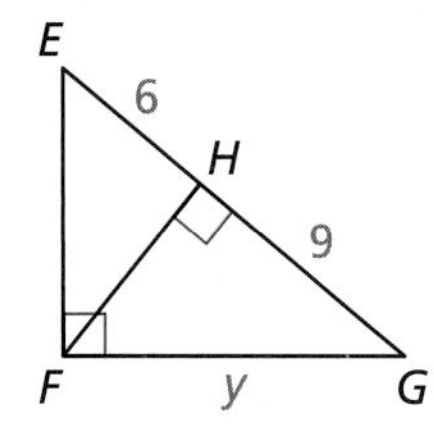

15.

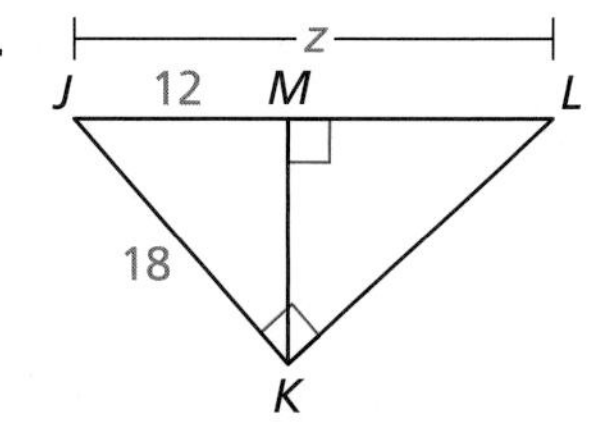

16. Television sizes are measured by the length of their diagonal. You want to purchase a television that is at least 40 inches. Should you purchase the television shown? Explain your reasoning. *(Section 9.1)*

17. Each triangle shown below is a right triangle. *(Sections 9.1–9.3)*

a. Are any of the triangles special right triangles? Explain your reasoning.

b. List all similar triangles, if any.

c. Find the lengths of the altitudes of triangles B and C.

4 A 4

B $3\sqrt{3}$ 3

6 C $6\sqrt{2}$

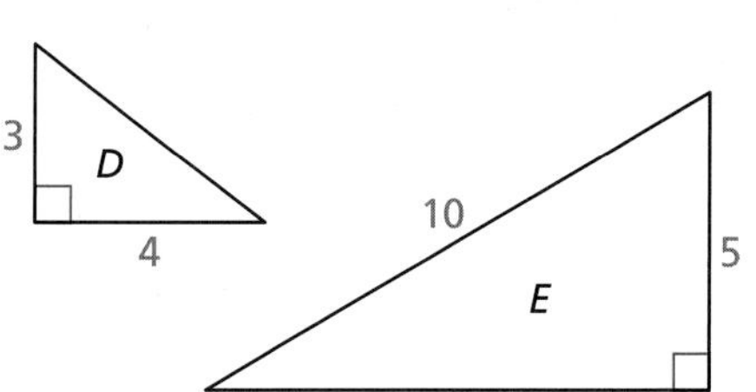

9.4 The Tangent Ratio

Essential Question How is a right triangle used to find the tangent of an acute angle? Is there a unique right triangle that must be used?

Let $\triangle ABC$ be a right triangle with acute $\angle A$. The *tangent* of $\angle A$ (written as tan A) is defined as follows.

$$\tan A = \frac{\text{length of leg opposite } \angle A}{\text{length of leg adjacent to } \angle A} = \frac{BC}{AC}$$

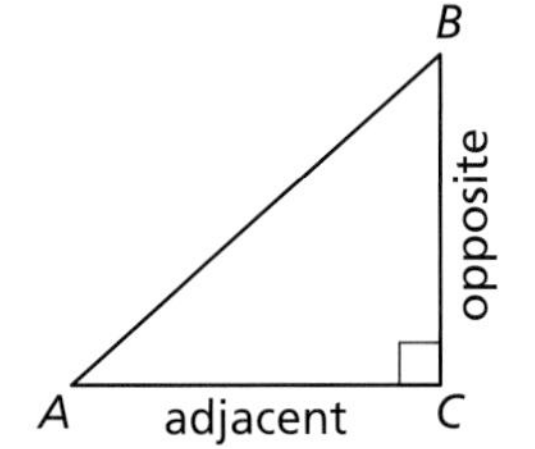

EXPLORATION 1 Calculating a Tangent Ratio

Work with a partner. Use dynamic geometry software.

a. Construct $\triangle ABC$, as shown. Construct segments perpendicular to $\overline{AC}$ to form right triangles that share vertex A and are similar to $\triangle ABC$ with vertices, as shown.

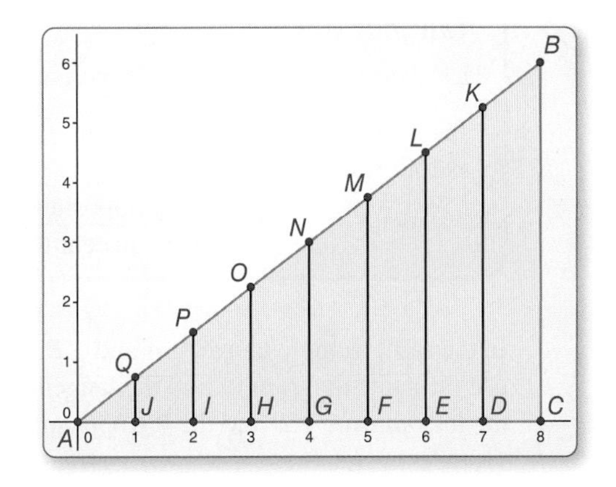

Sample
Points
$A(0, 0)$
$B(8, 6)$
$C(8, 0)$
Angle
$m\angle BAC = 36.87°$

b. Calculate each given ratio to complete the table for the decimal value of tan A for each right triangle. What can you conclude?

Ratio	$\frac{BC}{AC}$	$\frac{KD}{AD}$	$\frac{LE}{AE}$	$\frac{MF}{AF}$	$\frac{NG}{AG}$	$\frac{OH}{AH}$	$\frac{PI}{AI}$	$\frac{QJ}{AJ}$
tan A								

ATTENDING TO PRECISION

To be proficient in math, you need to express numerical answers with a degree of precision appropriate for the problem context.

EXPLORATION 2 Using a Calculator

Work with a partner. Use a calculator that has a tangent key to calculate the tangent of 36.87°. Do you get the same result as in Exploration 1? Explain.

Communicate Your Answer

3. Repeat Exploration 1 for $\triangle ABC$ with vertices $A(0, 0)$, $B(8, 5)$, and $C(8, 0)$. Construct the seven perpendicular segments so that not all of them intersect $\overline{AC}$ at integer values of x. Discuss your results.

4. How is a right triangle used to find the tangent of an acute angle? Is there a unique right triangle that must be used?

9.4 Lesson

Core Vocabulary

trigonometric ratio, *p. 488*
tangent, *p. 488*
angle of elevation, *p. 490*

What You Will Learn

- Use the tangent ratio.
- Solve real-life problems involving the tangent ratio.

Using the Tangent Ratio

A **trigonometric ratio** is a ratio of the lengths of two sides in a right triangle. All right triangles with a given acute angle are similar by the AA Similarity Theorem (Theorem 8.3). So, $\triangle JKL \sim \triangle XYZ$, and you can write $\frac{KL}{YZ} = \frac{JL}{XZ}$. This can be rewritten as $\frac{KL}{JL} = \frac{YZ}{XZ}$, which is a trigonometric ratio. So, trigonometric ratios are constant for a given angle measure.

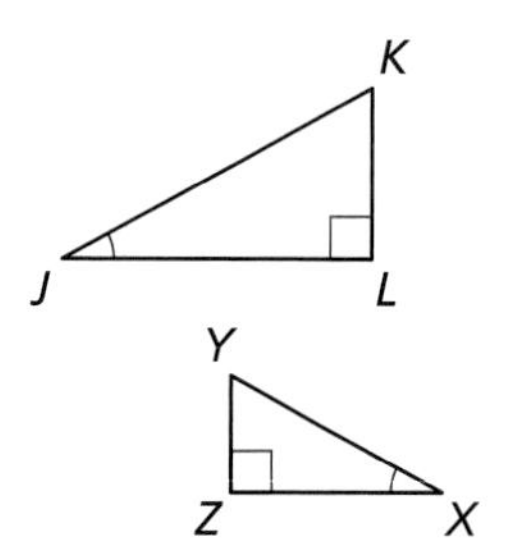

The **tangent** ratio is a trigonometric ratio for acute angles that involves the lengths of the legs of a right triangle.

READING

Remember the following abbreviations.

tangent → tan
opposite → opp.
adjacent → adj.

Core Concept

Tangent Ratio

Let $\triangle ABC$ be a right triangle with acute $\angle A$.

The tangent of $\angle A$ (written as tan A) is defined as follows.

$$\tan A = \frac{\text{length of leg opposite } \angle A}{\text{length of leg adjacent to } \angle A} = \frac{BC}{AC}$$

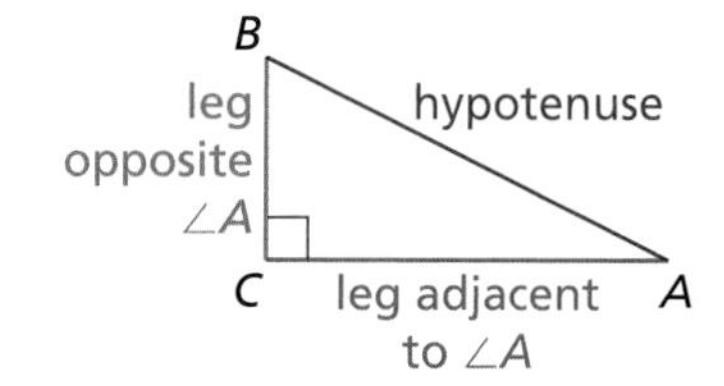

In the right triangle above, $\angle A$ and $\angle B$ are complementary. So, $\angle B$ is acute. You can use the same diagram to find the tangent of $\angle B$. Notice that the leg adjacent to $\angle A$ is the leg *opposite* $\angle B$ and the leg opposite $\angle A$ is the leg *adjacent* to $\angle B$.

ATTENDING TO PRECISION

Unless told otherwise, you should round the values of trigonometric ratios to four decimal places and round lengths to the nearest tenth.

EXAMPLE 1 Finding Tangent Ratios

Find tan S and tan R. Write each answer as a fraction and as a decimal rounded to four places.

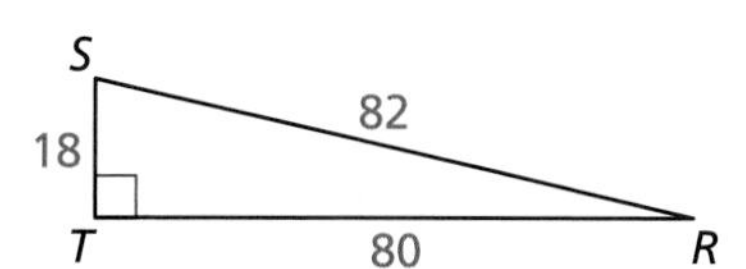

SOLUTION

$$\tan S = \frac{\text{opp. } \angle S}{\text{adj. to } \angle S} = \frac{RT}{ST} = \frac{80}{18} = \frac{40}{9} \approx 4.4444$$

$$\tan R = \frac{\text{opp. } \angle R}{\text{adj. to } \angle R} = \frac{ST}{RT} = \frac{18}{80} = \frac{9}{40} = 0.2250$$

Monitoring Progress

Help in English and Spanish at *BigIdeasMath.com*

Find tan *J* and tan *K*. Write each answer as a fraction and as a decimal rounded to four places.

1.

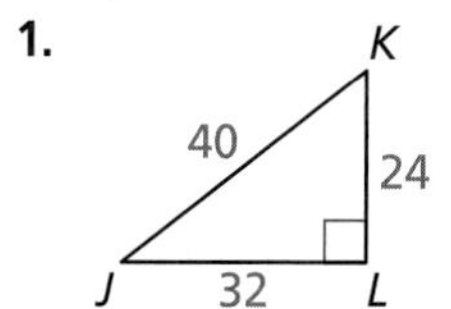

2.

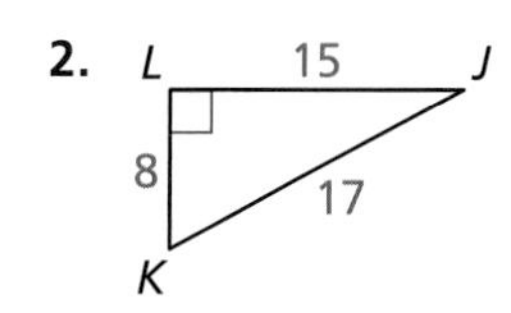

EXAMPLE 2 Finding a Leg Length

Find the value of x. Round your answer to the nearest tenth.

SOLUTION

Use the tangent of an acute angle to find a leg length.

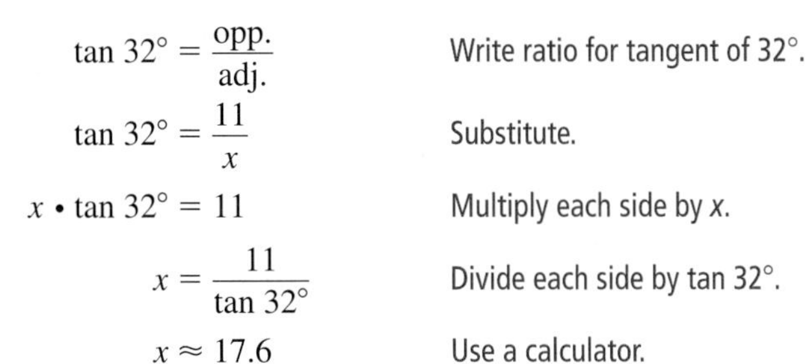

$$\tan 32^\circ = \frac{\text{opp.}}{\text{adj.}}$$ Write ratio for tangent of 32°.

$$\tan 32^\circ = \frac{11}{x}$$ Substitute.

$$x \cdot \tan 32^\circ = 11$$ Multiply each side by x.

$$x = \frac{11}{\tan 32^\circ}$$ Divide each side by tan 32°.

$$x \approx 17.6$$ Use a calculator.

▶ The value of x is about 17.6.

USING TOOLS STRATEGICALLY

You can also use the Table of Trigonometric Ratios available at *BigIdeasMath.com* to find the decimal approximations of trigonometric ratios.

You can find the tangent of an acute angle measuring 30°, 45°, or 60° by applying what you know about special right triangles.

STUDY TIP

The tangents of all 60° angles are the same constant ratio. Any right triangle with a 60° angle can be used to determine this value.

EXAMPLE 3 Using a Special Right Triangle to Find a Tangent

Use a special right triangle to find the tangent of a 60° angle.

SOLUTION

Step 1 Because all 30°-60°-90° triangles are similar, you can simplify your calculations by choosing 1 as the length of the shorter leg. Use the 30°-60°-90° Triangle Theorem (Theorem 9.5) to find the length of the longer leg.

$$\text{longer leg} = \text{shorter leg} \cdot \sqrt{3}$$ 30°-60°-90° Triangle Theorem

$$= 1 \cdot \sqrt{3}$$ Substitute.

$$= \sqrt{3}$$ Simplify.

Step 2 Find tan 60°.

$$\tan 60^\circ = \frac{\text{opp.}}{\text{adj.}}$$ Write ratio for tangent of 60°.

$$\tan 60^\circ = \frac{\sqrt{3}}{1}$$ Substitute.

$$\tan 60^\circ = \sqrt{3}$$ Simplify.

▶ The tangent of any 60° angle is $\sqrt{3} \approx 1.7321$.

Monitoring Progress

Help in English and Spanish at *BigIdeasMath.com*

Find the value of x. Round your answer to the nearest tenth.

5. WHAT IF? In Example 3, the length of the shorter leg is 5 instead of 1. Show that the tangent of 60° is still equal to $\sqrt{3}$.

Solving Real-Life Problems

The angle that an upward line of sight makes with a horizontal line is called the **angle of elevation**.

EXAMPLE 4 Modeling with Mathematics

You are measuring the height of a spruce tree. You stand 45 feet from the base of the tree. You measure the angle of elevation from the ground to the top of the tree to be 59°. Find the height h of the tree to the nearest foot.

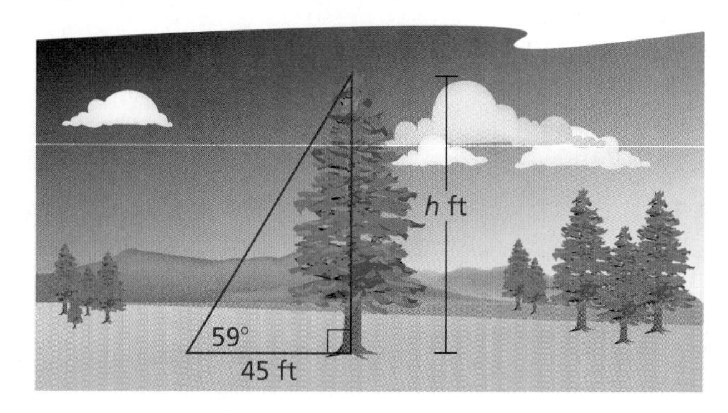

SOLUTION

1. **Understand the Problem** You are given the angle of elevation and the distance from the tree. You need to find the height of the tree to the nearest foot.

2. **Make a Plan** Write a trigonometric ratio for the tangent of the angle of elevation involving the height h. Then solve for h.

3. **Solve the Problem**

$\tan 59^\circ = \frac{\text{opp.}}{\text{adj.}}$	Write ratio for tangent of 59°.
$\tan 59^\circ = \frac{h}{45}$	Substitute.
$45 \cdot \tan 59^\circ = h$	Multiply each side by 45.
$74.9 \approx h$	Use a calculator.

▶ The tree is about 75 feet tall.

4. **Look Back** Check your answer. Because 59° is close to 60°, the value of h should be close to the length of the longer leg of a 30°-60°-90° triangle, where the length of the shorter leg is 45 feet.

$\text{longer leg} = \text{shorter leg} \cdot \sqrt{3}$	30°-60°-90° Triangle Theorem
$= 45 \cdot \sqrt{3}$	Substitute.
≈ 77.9	Use a calculator.

The value of 77.9 feet is close to the value of h. ✓

Monitoring Progress  Help in English and Spanish at *BigIdeasMath.com*

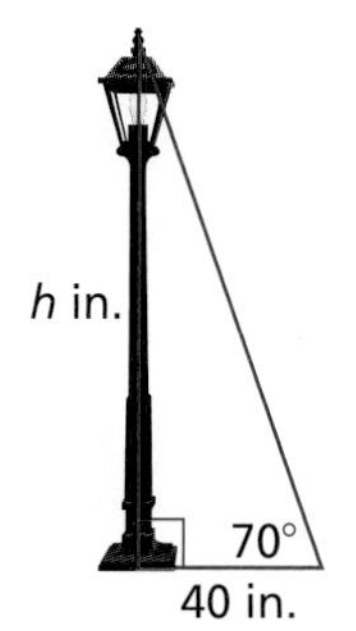

6. You are measuring the height of a lamppost. You stand 40 inches from the base of the lamppost. You measure the angle of elevation from the ground to the top of the lamppost to be 70°. Find the height h of the lamppost to the nearest inch.

9.4 Exercises

Dynamic Solutions available at *BigIdeasMath.com*

Vocabulary and Core Concept Check

1. **COMPLETE THE SENTENCE** The tangent ratio compares the length of _______ to the length of ________.

2. **WRITING** Explain how you know the tangent ratio is constant for a given angle measure.

Monitoring Progress and Modeling with Mathematics

In Exercises 3–6, find the tangents of the acute angles in the right triangle. Write each answer as a fraction and as a decimal rounded to four decimal places. *(See Example 1.)*

3. 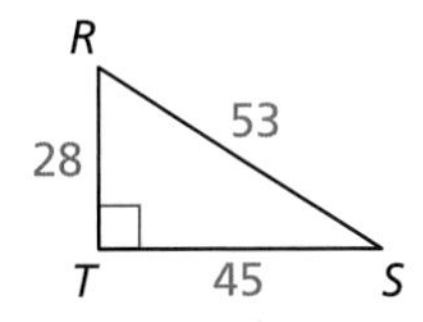

4. 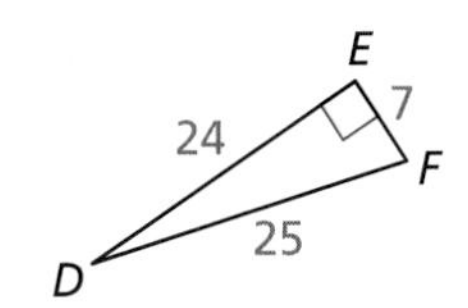

5. 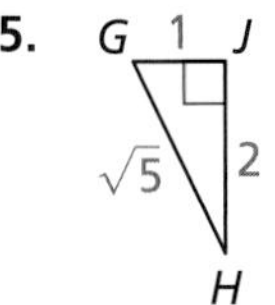

6. 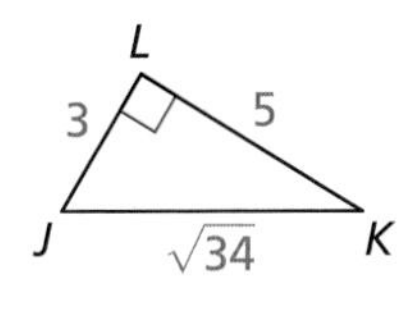

In Exercises 7–10, find the value of x. Round your answer to the nearest tenth. *(See Example 2.)*

7. 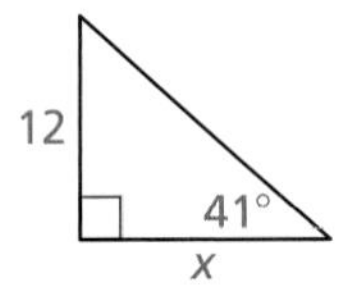

8. 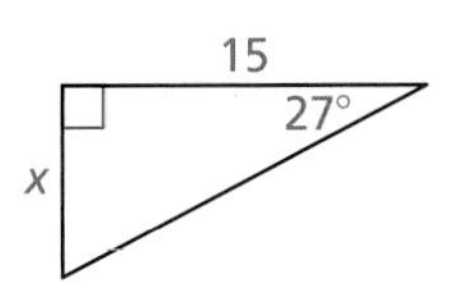

9. 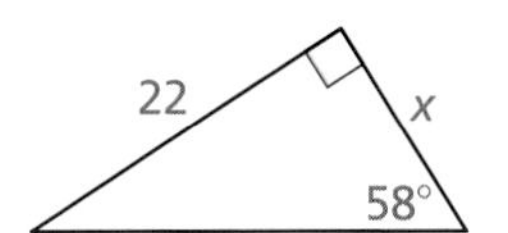

10. 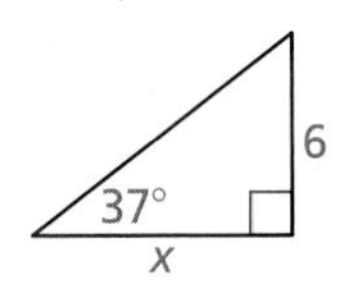

ERROR ANALYSIS In Exercises 11 and 12, describe the error in the statement of the tangent ratio. Correct the error if possible. Otherwise, write not possible.

11. 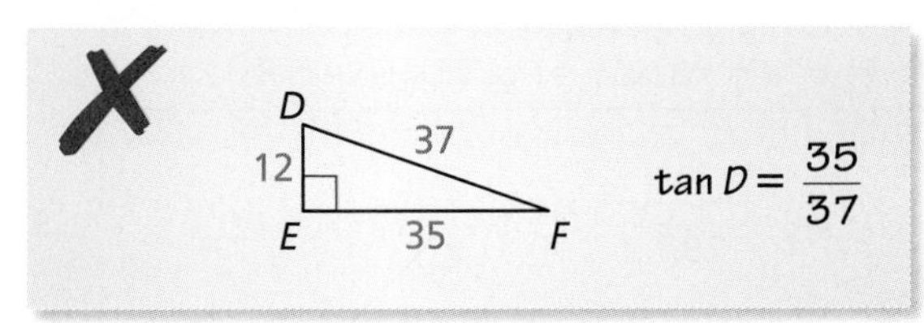

12. 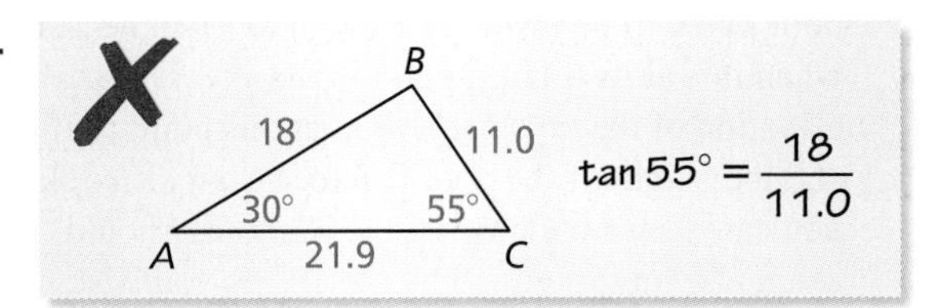

In Exercises 13 and 14, use a special right triangle to find the tangent of the given angle measure. *(See Example 3.)*

13. 45°

14. 30°

15. **MODELING WITH MATHEMATICS** A surveyor is standing 118 feet from the base of the Washington Monument. The surveyor measures the angle of elevation from the ground to the top of the monument to be 78°. Find the height h of the Washington Monument to the nearest foot. *(See Example 4.)*

16. **MODELING WITH MATHEMATICS** Scientists can measure the depths of craters on the moon by looking at photos of shadows. The length of the shadow cast by the edge of a crater is 500 meters. The angle of elevation of the rays of the Sun is 55°. Estimate the depth d of the crater.

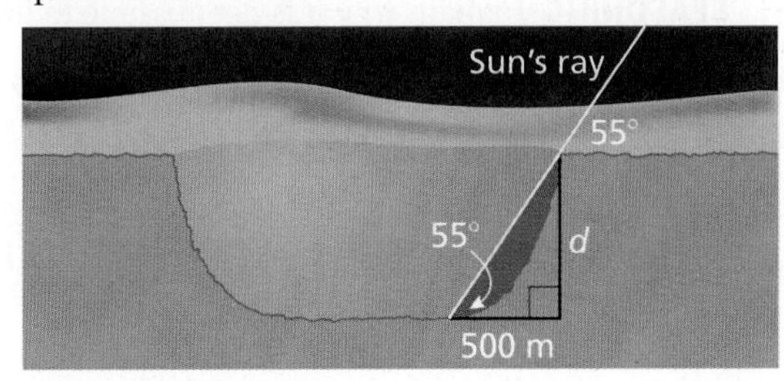

17. **USING STRUCTURE** Find the tangent of the smaller acute angle in a right triangle with side lengths 5, 12, and 13.

18. **USING STRUCTURE** Find the tangent of the larger acute angle in a right triangle with side lengths 3, 4, and 5.

19. **REASONING** How does the tangent of an acute angle in a right triangle change as the angle measure increases? Justify your answer.

20. **CRITICAL THINKING** For what angle measure(s) is the tangent of an acute angle in a right triangle equal to 1? greater than 1? less than 1? Justify your answer.

21. **MAKING AN ARGUMENT** Your family room has a sliding-glass door. You want to buy an awning for the door that will be just long enough to keep the Sun out when it is at its highest point in the sky. The angle of elevation of the rays of the Sun at this point is 70°, and the height of the door is 8 feet. Your sister claims you can determine how far the overhang should extend by multiplying 8 by tan 70°. Is your sister correct? Explain.

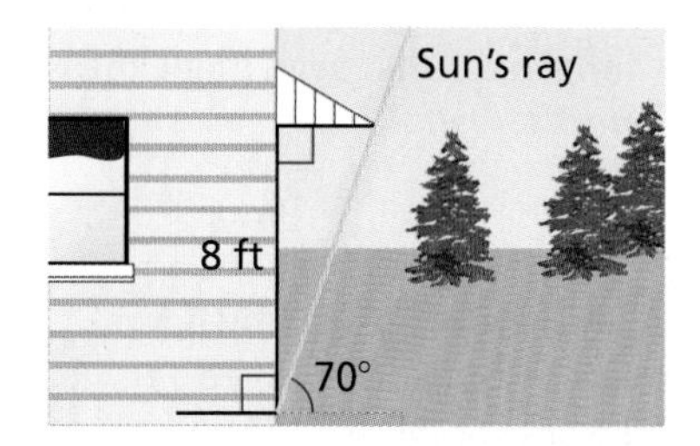

22. **HOW DO YOU SEE IT?** Write expressions for the tangent of each acute angle in the right triangle. Explain how the tangent of one acute angle is related to the tangent of the other acute angle. What kind of angle pair is $\angle A$ and $\angle B$?

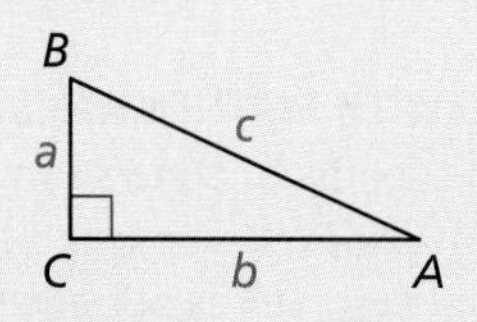

23. **REASONING** Explain why it is not possible to find the tangent of a right angle or an obtuse angle.

24. **THOUGHT PROVOKING** To create the diagram below, you begin with an isosceles right triangle with legs 1 unit long. Then the hypotenuse of the first triangle becomes the leg of a second triangle, whose remaining leg is 1 unit long. Continue the diagram until you have constructed an angle whose tangent is $\frac{1}{\sqrt{6}}$. Approximate the measure of this angle.

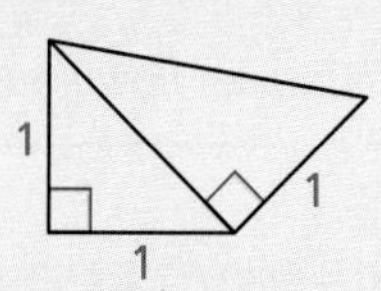

25. **PROBLEM SOLVING** Your class is having a class picture taken on the lawn. The photographer is positioned 14 feet away from the center of the class. The photographer turns 50° to look at either end of the class.

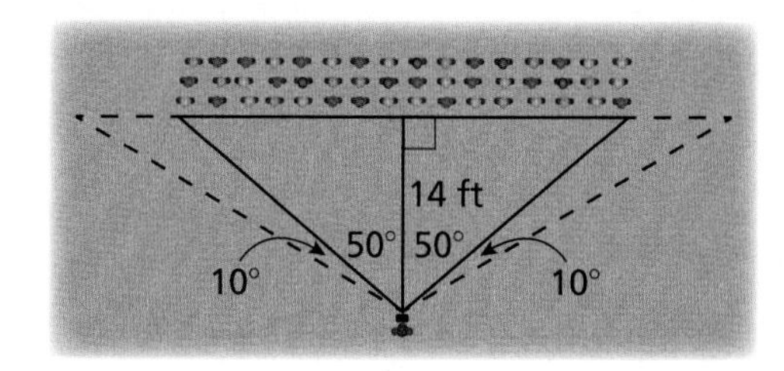

a. What is the distance between the ends of the class?

b. The photographer turns another 10° either way to see the end of the camera range. If each student needs 2 feet of space, about how many more students can fit at the end of each row? Explain.

26. **PROBLEM SOLVING** Find the perimeter of the figure, where $AC = 26$, $AD = BF$, and D is the midpoint of $\overline{AC}$.

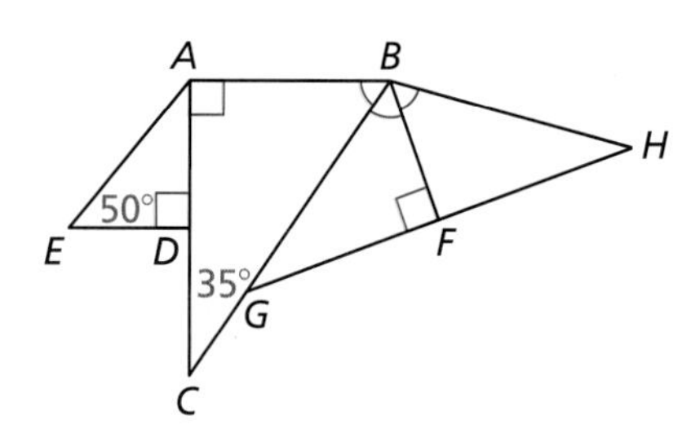

Maintaining Mathematical Proficiency

Reviewing what you learned in previous grades and lessons

Find the value of *x*. *(Section 9.2)*

27.

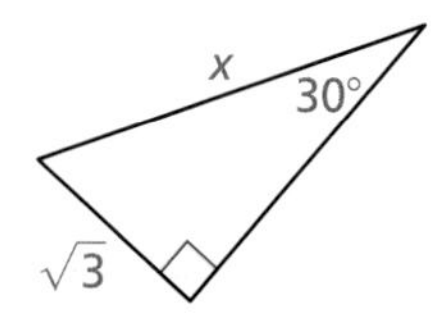

28.

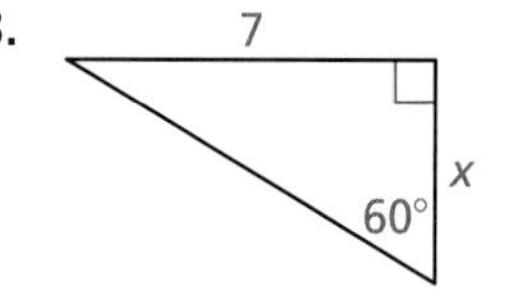

29.

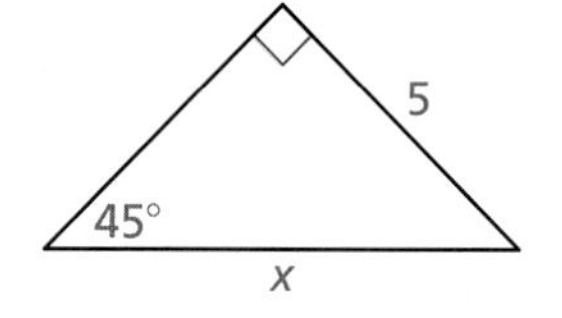

9.5 The Sine and Cosine Ratios

Essential Question

How is a right triangle used to find the sine and cosine of an acute angle? Is there a unique right triangle that must be used?

Let $\triangle ABC$ be a right triangle with acute $\angle A$. The *sine* of $\angle A$ and *cosine* of $\angle A$ (written as sin A and cos A, respectively) are defined as follows.

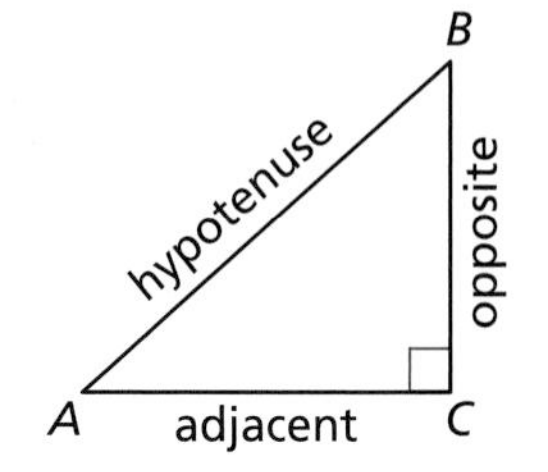

$$\sin A = \frac{\text{length of leg opposite } \angle A}{\text{length of hypotenuse}} = \frac{BC}{AB}$$

$$\cos A = \frac{\text{length of leg adjacent to } \angle A}{\text{length of hypotenuse}} = \frac{AC}{AB}$$

EXPLORATION 1 Calculating Sine and Cosine Ratios

Work with a partner. Use dynamic geometry software.

a. Construct $\triangle ABC$, as shown. Construct segments perpendicular to $\overline{AC}$ to form right triangles that share vertex A and are similar to $\triangle ABC$ with vertices, as shown.

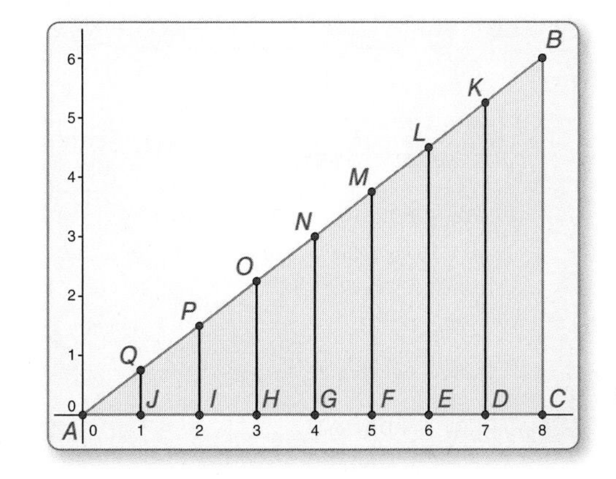

Sample

Points

$A(0, 0)$

$B(8, 6)$

$C(8, 0)$

Angle

$m\angle BAC = 36.87°$

b. Calculate each given ratio to complete the table for the decimal values of sin A and cos A for each right triangle. What can you conclude?

Sine ratio	$\frac{BC}{AB}$	$\frac{KD}{AK}$	$\frac{LE}{AL}$	$\frac{MF}{AM}$	$\frac{NG}{AN}$	$\frac{OH}{AO}$	$\frac{PI}{AP}$	$\frac{QJ}{AQ}$
sin *A*								
Cosine ratio	$\frac{AC}{AB}$	$\frac{AD}{AK}$	$\frac{AE}{AL}$	$\frac{AF}{AM}$	$\frac{AG}{AN}$	$\frac{AH}{AO}$	$\frac{AI}{AP}$	$\frac{AJ}{AQ}$
cos *A*								

LOOKING FOR STRUCTURE

To be proficient in math, you need to look closely to discern a pattern or structure.

Communicate Your Answer

2. How is a right triangle used to find the sine and cosine of an acute angle? Is there a unique right triangle that must be used?

3. In Exploration 1, what is the relationship between $\angle A$ and $\angle B$ in terms of their measures? Find sin B and cos B. How are these two values related to sin A and cos A? Explain why these relationships exist.

9.5 Lesson

Core Vocabulary

sine, *p. 494*
cosine, *p. 494*
angle of depression, *p. 497*

What You Will Learn

- Use the sine and cosine ratios.
- Find the sine and cosine of angle measures in special right triangles.
- Solve real-life problems involving sine and cosine ratios.

Using the Sine and Cosine Ratios

The **sine** and **cosine** ratios are trigonometric ratios for acute angles that involve the lengths of a leg and the hypotenuse of a right triangle.

READING

Remember the following abbreviations.

sine → sin
cosine → cos
hypotenuse → hyp.

Core Concept

Sine and Cosine Ratios

Let $\triangle ABC$ be a right triangle with acute $\angle A$. The sine of $\angle A$ and cosine of $\angle A$ (written as $\sin A$ and $\cos A$) are defined as follows.

$$\sin A = \frac{\text{length of leg opposite } \angle A}{\text{length of hypotenuse}} = \frac{BC}{AB}$$

$$\cos A = \frac{\text{length of leg adjacent to } \angle A}{\text{length of hypotenuse}} = \frac{AC}{AB}$$

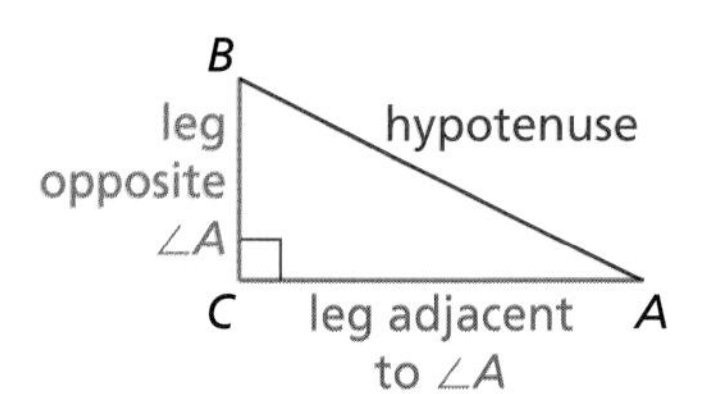

EXAMPLE 1 Finding Sine and Cosine Ratios

Find $\sin S$, $\sin R$, $\cos S$, and $\cos R$. Write each answer as a fraction and as a decimal rounded to four places.

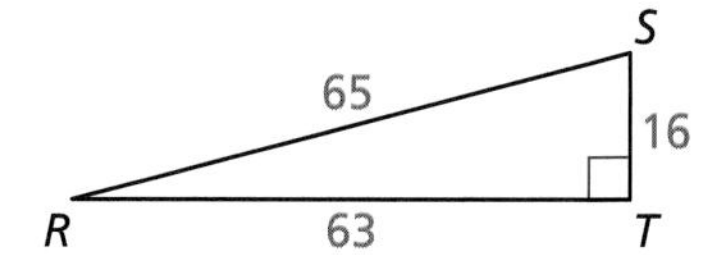

SOLUTION

$$\sin S = \frac{\text{opp. } \angle S}{\text{hyp.}} = \frac{RT}{SR} = \frac{63}{65} \approx 0.9692 \qquad \sin R = \frac{\text{opp. } \angle R}{\text{hyp.}} = \frac{ST}{SR} = \frac{16}{65} \approx 0.2462$$

$$\cos S = \frac{\text{adj. to } \angle S}{\text{hyp.}} = \frac{ST}{SR} = \frac{16}{65} \approx 0.2462 \qquad \cos R = \frac{\text{adj. to } \angle R}{\text{hyp.}} = \frac{RT}{SR} = \frac{63}{65} \approx 0.9692$$

In Example 1, notice that $\sin S = \cos R$ and $\sin R = \cos S$. This is true because the side opposite $\angle S$ is adjacent to $\angle R$ and the side opposite $\angle R$ is adjacent to $\angle S$. The relationship between the sine and cosine of $\angle S$ and $\angle R$ is true for all complementary angles.

Core Concept

Sine and Cosine of Complementary Angles

The sine of an acute angle is equal to the cosine of its complement. The cosine of an acute angle is equal to the sine of its complement.

Let A and B be complementary angles. Then the following statements are true.

$\sin A = \cos(90° - A) = \cos B \qquad \sin B = \cos(90° - B) = \cos A$

$\cos A = \sin(90° - A) = \sin B \qquad \cos B = \sin(90° - B) = \sin A$

EXAMPLE 2 Rewriting Trigonometric Expressions

Write sin 56° in terms of cosine.

SOLUTION

Use the fact that the sine of an acute angle is equal to the cosine of its complement.

$$\sin 56^\circ = \cos(90^\circ - 56^\circ) = \cos 34^\circ$$

▶ The sine of 56° is the same as the cosine of 34°.

You can use the sine and cosine ratios to find unknown measures in right triangles.

EXAMPLE 3 Finding Leg Lengths

Find the values of x and y using sine and cosine. Round your answers to the nearest tenth.

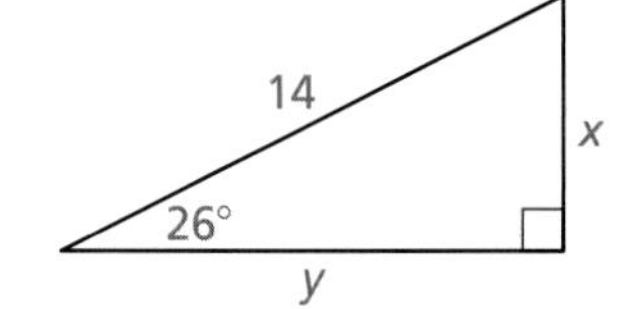

SOLUTION

Step 1 Use a sine ratio to find the value of x.

$\sin 26^\circ = \frac{\text{opp.}}{\text{hyp.}}$	Write ratio for sine of 26°.
$\sin 26^\circ = \frac{x}{14}$	Substitute.
$14 \cdot \sin 26^\circ = x$	Multiply each side by 14.
$6.1 \approx x$	Use a calculator.

▶ The value of x is about 6.1.

Step 2 Use a cosine ratio to find the value of y.

$\cos 26^\circ = \frac{\text{adj.}}{\text{hyp.}}$	Write ratio for cosine of 26°.
$\cos 26^\circ = \frac{y}{14}$	Substitute.
$14 \cdot \cos 26^\circ = y$	Multiply each side by 14.
$12.6 \approx y$	Use a calculator.

▶ The value of y is about 12.6.

Monitoring Progress

Help in English and Spanish at *BigIdeasMath.com*

1. Find sin D, sin F, cos D, and cos F. Write each answer as a fraction and as a decimal rounded to four places.

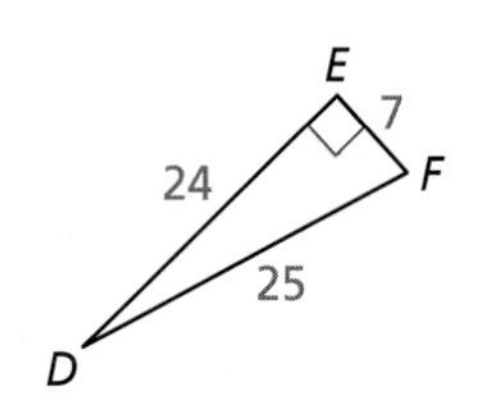

2. Write cos 23° in terms of sine.

3. Find the values of u and t using sine and cosine. Round your answers to the nearest tenth.

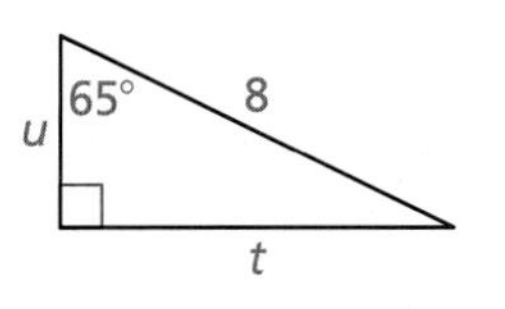

Finding Sine and Cosine in Special Right Triangles

EXAMPLE 4 Finding the Sine and Cosine of 45°

Find the sine and cosine of a 45° angle.

SOLUTION

Begin by sketching a 45°-45°-90° triangle. Because all such triangles are similar, you can simplify your calculations by choosing 1 as the length of each leg. Using the 45°-45°-90° Triangle Theorem (Theorem 9.4), the length of the hypotenuse is $\sqrt{2}$.

Notice that

$\sin 45^\circ = \cos(90 - 45)^\circ$

$= \cos 45^\circ$.

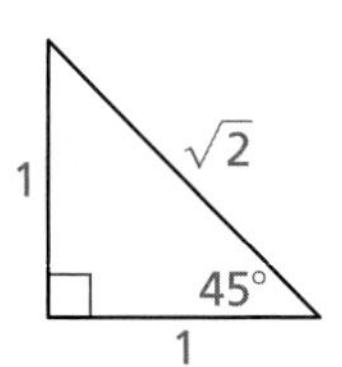

$$\sin 45^\circ = \frac{\text{opp.}}{\text{hyp.}} = \frac{1}{\sqrt{2}} = \frac{\sqrt{2}}{2} \approx 0.7071$$

$$\cos 45^\circ = \frac{\text{adj.}}{\text{hyp.}} = \frac{1}{\sqrt{2}} = \frac{\sqrt{2}}{2} \approx 0.7071$$

EXAMPLE 5 Finding the Sine and Cosine of 30°

Find the sine and cosine of a 30° angle.

SOLUTION

Begin by sketching a 30°-60°-90° triangle. Because all such triangles are similar, you can simplify your calculations by choosing 1 as the length of the shorter leg. Using the 30°-60°-90° Triangle Theorem (Theorem 9.5), the length of the longer leg is $\sqrt{3}$ and the length of the hypotenuse is 2.

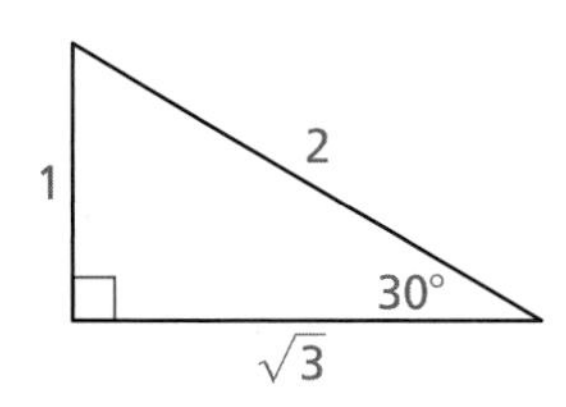

$$\sin 30^\circ = \frac{\text{opp.}}{\text{hyp.}} = \frac{1}{2} = 0.5000$$

$$\cos 30^\circ = \frac{\text{adj.}}{\text{hyp.}} = \frac{\sqrt{3}}{2} \approx 0.8660$$

Monitoring Progress  Help in English and Spanish at *BigIdeasMath.com*

4. Find the sine and cosine of a 60° angle.

Solving Real-Life Problems

Recall from the previous lesson that the angle an upward line of sight makes with a horizontal line is called the *angle of elevation*. The angle that a downward line of sight makes with a horizontal line is called the **angle of depression**.

EXAMPLE 6 Modeling with Mathematics

You are skiing on a mountain with an altitude of 1200 feet. The angle of depression is 21°. Find the distance x you ski down the mountain to the nearest foot.

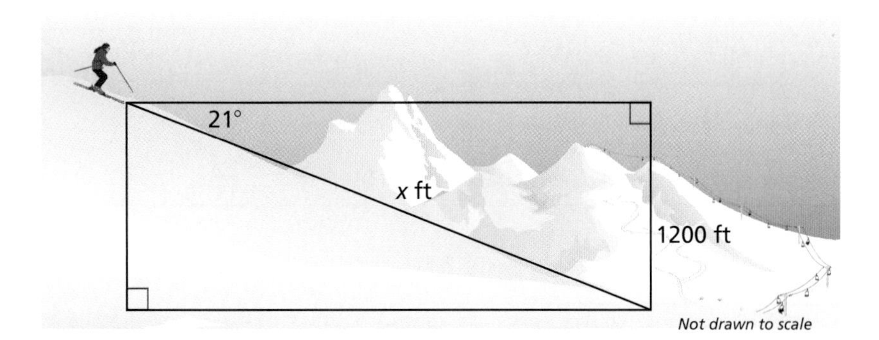

SOLUTION

1. **Understand the Problem** You are given the angle of depression and the altitude of the mountain. You need to find the distance that you ski down the mountain.
2. **Make a Plan** Write a trigonometric ratio for the sine of the angle of depression involving the distance x. Then solve for x.
3. **Solve the Problem**

$$\sin 21^\circ = \frac{\text{opp.}}{\text{hyp.}}$$ Write ratio for sine of 21°.

$$\sin 21^\circ = \frac{1200}{x}$$ Substitute.

$$x \cdot \sin 21^\circ = 1200$$ Multiply each side by x.

$$x = \frac{1200}{\sin 21^\circ}$$ Divide each side by sin 21°.

$$x \approx 3348.5$$ Use a calculator.

▶ You ski about 3349 feet down the mountain.

4. **Look Back** Check your answer. The value of sin 21° is about 0.3584. Substitute for x in the sine ratio and compare the values.

$$\frac{1200}{x} \approx \frac{1200}{3348.5}$$

$$\approx 0.3584$$

This value is approximately the same as the value of sin 21°. ✓

Monitoring Progress

Help in English and Spanish at *BigIdeasMath.com*

5. **WHAT IF?** In Example 6, the angle of depression is 28°. Find the distance x you ski down the mountain to the nearest foot.

9.5 Exercises

Dynamic Solutions available at *BigIdeasMath.com*

Vocabulary and Core Concept Check

1. **VOCABULARY** The sine ratio compares the length of ____________ to the length of ______________.

2. **WHICH ONE DOESN'T BELONG?** Which ratio does *not* belong with the other three? Explain your reasoning.

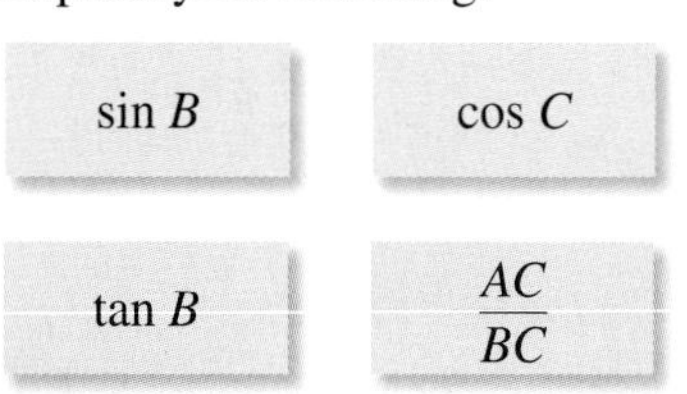

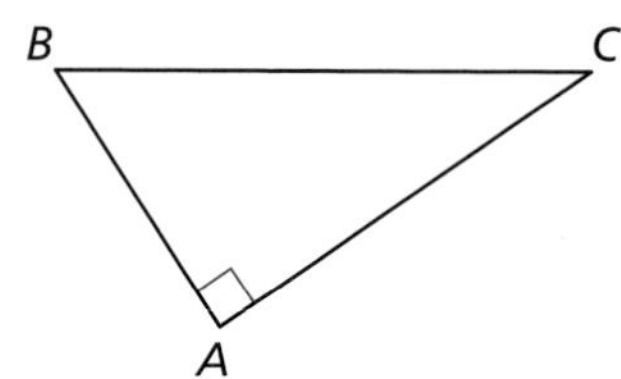

Monitoring Progress and Modeling with Mathematics

In Exercises 3–8, find sin D, sin E, cos D, and cos E. Write each answer as a fraction and as a decimal rounded to four places. *(See Example 1.)*

3. 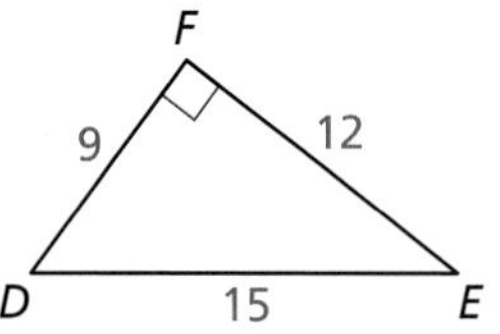

4. 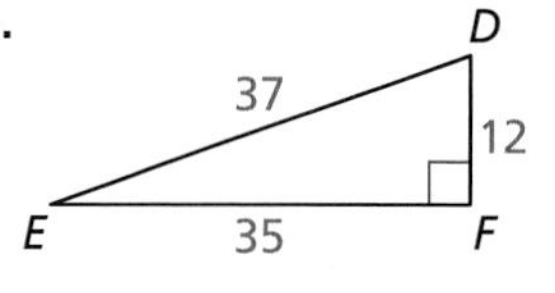

5. 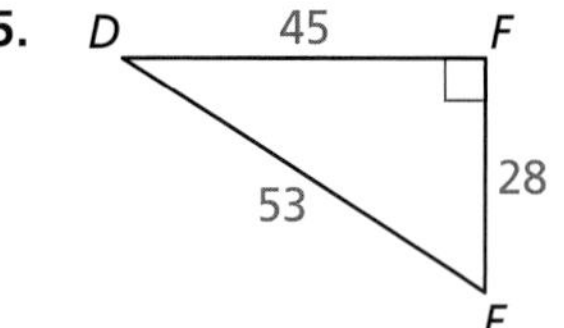

6. 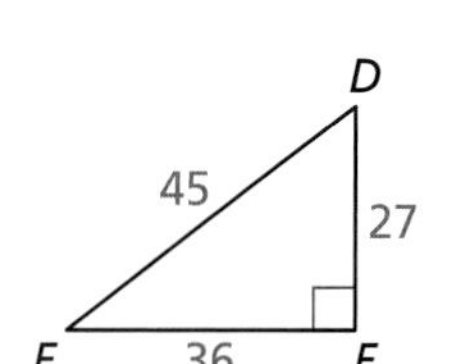

7. 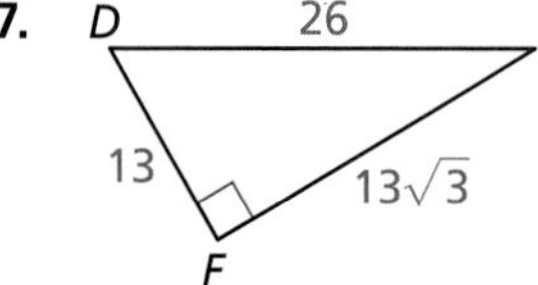

8. 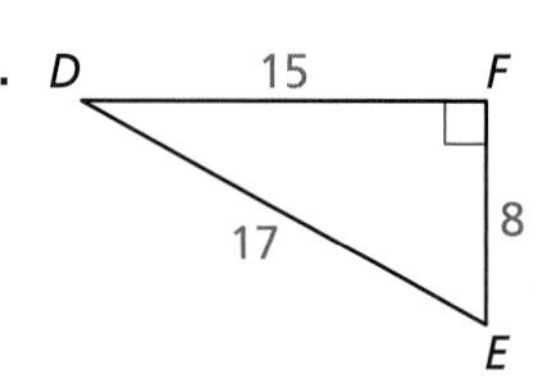

In Exercises 9–12, write the expression in terms of cosine. *(See Example 2.)*

9. sin 37°
10. sin 81°
11. sin 29°
12. sin 64°

In Exercises 13–16, write the expression in terms of sine.

13. cos 59°
14. cos 42°
15. cos 73°
16. cos 18°

In Exercises 17–22, find the value of each variable using sine and cosine. Round your answers to the nearest tenth. *(See Example 3.)*

17. 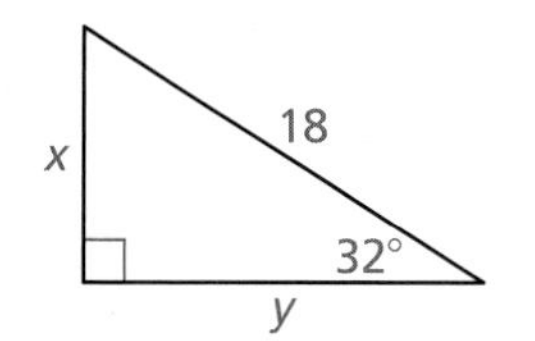

18. 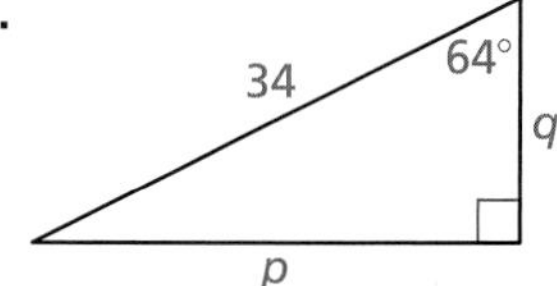

19. 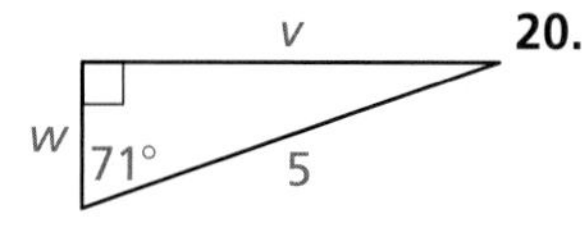

20. 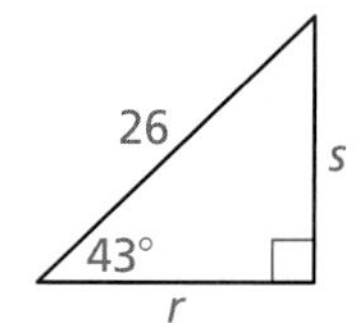

21. 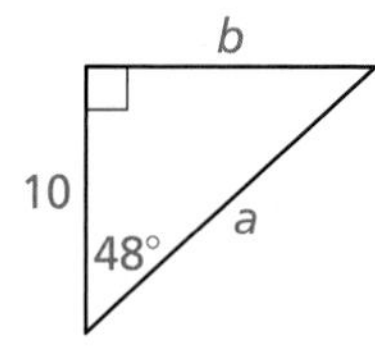

22. 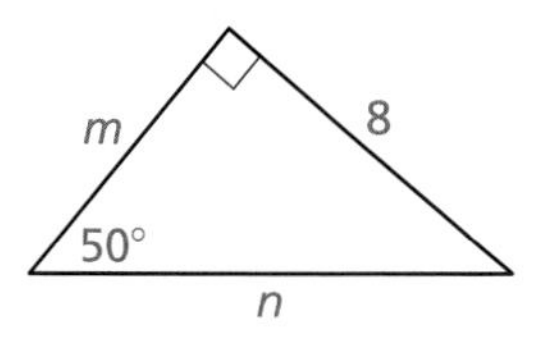

23. **REASONING** Which ratios are equal? Select all that apply. *(See Example 4.)*

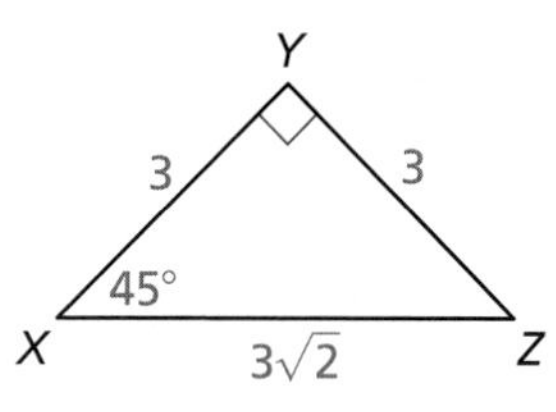

sin X | cos X | sin Z | cos Z

24. REASONING Which ratios are equal to $\frac{1}{2}$? Select all that apply. *(See Example 5.)*

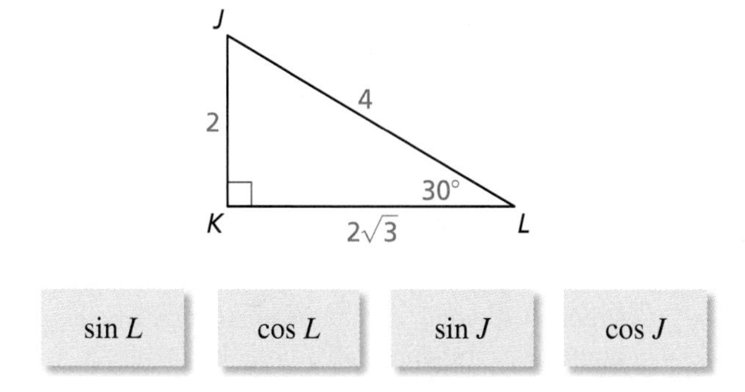

sin L	cos L	sin J	cos J

25. ERROR ANALYSIS Describe and correct the error in finding sin A.

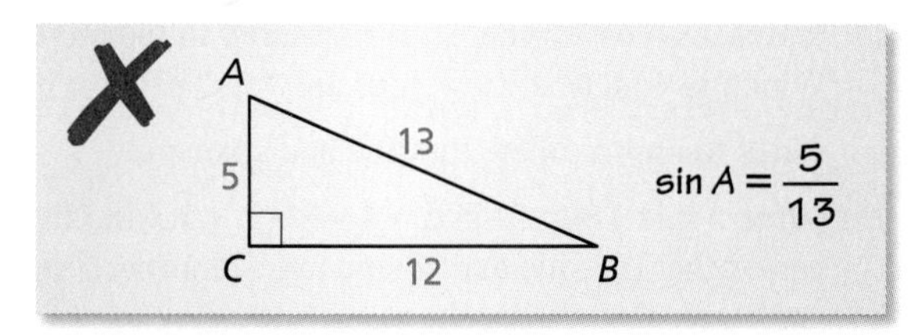

26. WRITING Explain how to tell which side of a right triangle is adjacent to an angle and which side is the hypotenuse.

27. MODELING WITH MATHEMATICS The top of the slide is 12 feet from the ground and has an angle of depression of 53°. What is the length of the slide? *(See Example 6.)*

28. MODELING WITH MATHEMATICS Find the horizontal distance x the escalator covers.

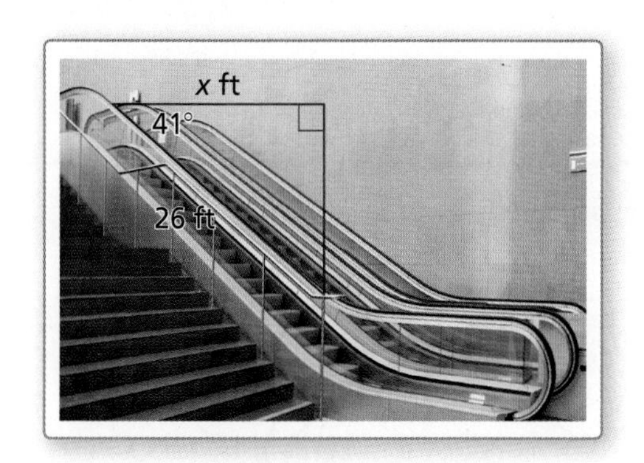

29. PROBLEM SOLVING You are flying a kite with 20 feet of string extended. The angle of elevation from the spool of string to the kite is 67°.

a. Draw and label a diagram that represents the situation.

b. How far off the ground is the kite if you hold the spool 5 feet off the ground? Describe how the height where you hold the spool affects the height of the kite.

30. MODELING WITH MATHEMATICS Planes that fly at high speeds and low elevations have radar systems that can determine the range of an obstacle and the angle of elevation to the top of the obstacle. The radar of a plane flying at an altitude of 20,000 feet detects a tower that is 25,000 feet away, with an angle of elevation of 1°.

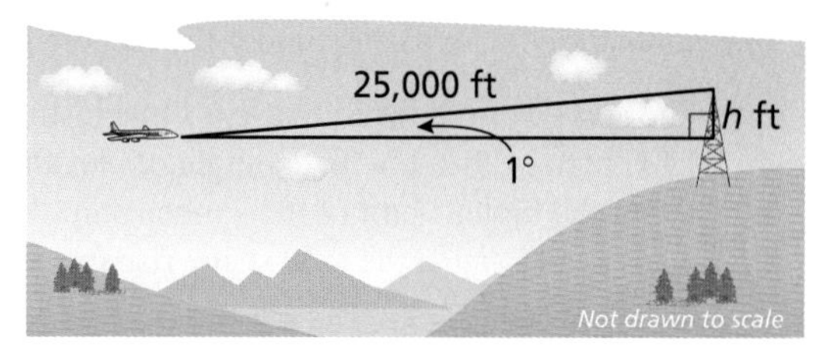

a. How many feet must the plane rise to pass over the tower?

b. Planes cannot come closer than 1000 feet vertically to any object. At what altitude must the plane fly in order to pass over the tower?

31. MAKING AN ARGUMENT Your friend uses the equation $\sin 49° = \frac{x}{16}$ to find BC. Your cousin uses the equation $\cos 41° = \frac{x}{16}$ to find BC. Who is correct? Explain your reasoning.

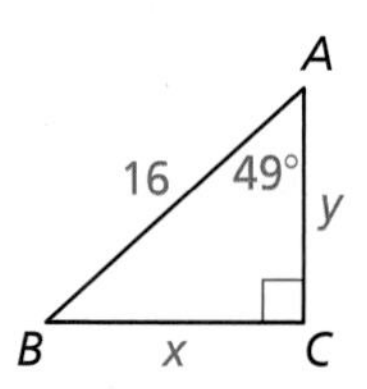

32. WRITING Describe what you must know about a triangle in order to use the sine ratio and what you must know about a triangle in order to use the cosine ratio.

33. MATHEMATICAL CONNECTIONS If $\triangle EQU$ is equilateral and $\triangle RGT$ is a right triangle with $RG = 2$, $RT = 1$, and $m\angle T = 90°$, show that $\sin E = \cos G$.

34. MODELING WITH MATHEMATICS Submarines use sonar systems, which are similar to radar systems, to detect obstacles. Sonar systems use sound to detect objects under water.

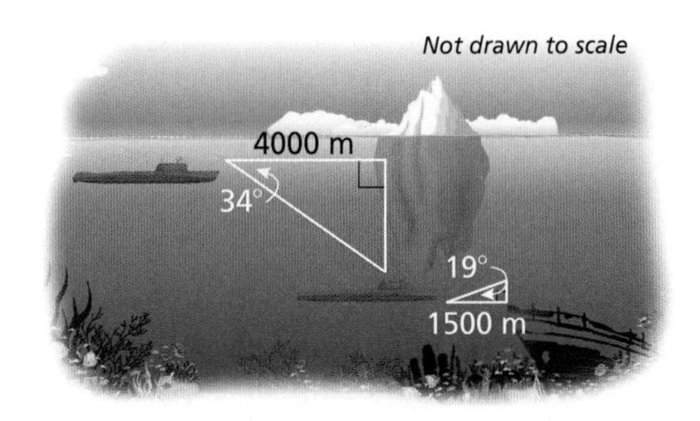

a. You are traveling underwater in a submarine. The sonar system detects an iceberg 4000 meters ahead, with an angle of depression of 34° to the bottom of the iceberg. How many meters must the submarine lower to pass under the iceberg?

b. The sonar system then detects a sunken ship 1500 meters ahead, with an angle of elevation of 19° to the highest part of the sunken ship. How many meters must the submarine rise to pass over the sunken ship?

35. ABSTRACT REASONING Make a conjecture about how you could use trigonometric ratios to find angle measures in a triangle.

36. HOW DO YOU SEE IT? Using only the given information, would you use a sine ratio or a cosine ratio to find the length of the hypotenuse? Explain your reasoning.

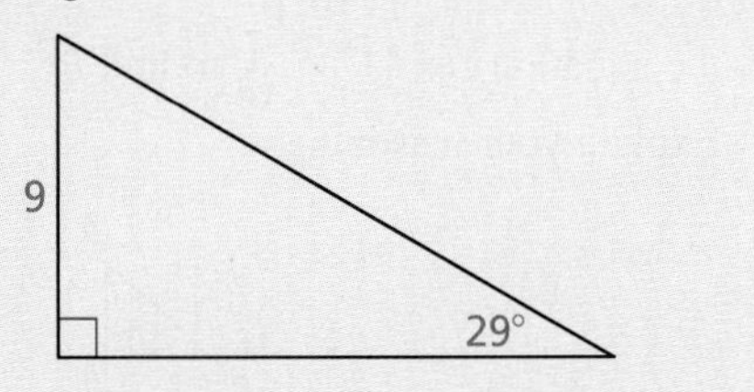

37. MULTIPLE REPRESENTATIONS You are standing on a cliff above an ocean. You see a sailboat from your vantage point 30 feet above the ocean.

a. Draw and label a diagram of the situation.

b. Make a table showing the angle of depression and the length of your line of sight. Use the angles 40°, 50°, 60°, 70°, and 80°.

c. Graph the values you found in part (b), with the angle measures on the x-axis.

d. Predict the length of the line of sight when the angle of depression is 30°.

38. THOUGHT PROVOKING One of the following infinite series represents $\sin x$ and the other one represents $\cos x$ (where x is measured in radians). Which is which? Justify your answer. Then use each series to approximate the sine and cosine of $\frac{\pi}{6}$. (*Hints*: $\pi = 180°$; $5! = 5 \cdot 4 \cdot 3 \cdot 2 \cdot 1$; Find the values that the sine and cosine ratios approach as the angle measure approaches zero.)

a. $x - \frac{x^3}{3!} + \frac{x^5}{5!} - \frac{x^7}{7!} + \cdots$

b. $1 - \frac{x^2}{2!} + \frac{x^4}{4!} - \frac{x^6}{6!} + \cdots$

39. CRITICAL THINKING Let A be any acute angle of a right triangle. Show that (a) $\tan A = \frac{\sin A}{\cos A}$ and (b) $(\sin A)^2 + (\cos A)^2 = 1$.

40. CRITICAL THINKING Explain why the area of $\triangle ABC$ in the diagram can be found using the formula Area $= \frac{1}{2}ab \sin C$. Then calculate the area when $a = 4$, $b = 7$, and $m\angle C = 40°$.

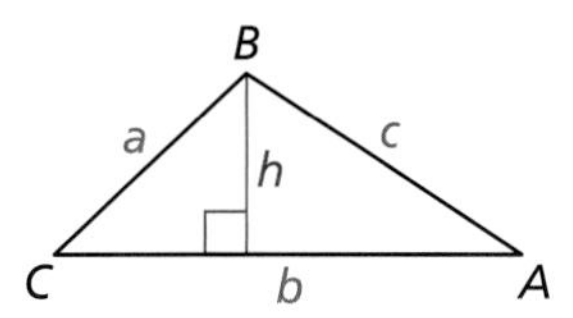

Maintaining Mathematical Proficiency

Reviewing what you learned in previous grades and lessons

Find the value of x. Tell whether the side lengths form a Pythagorean triple. *(Section 9.1)*

41.

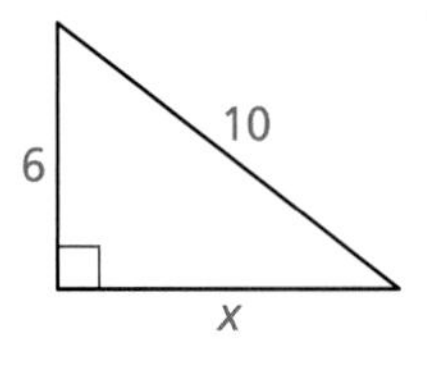

42.

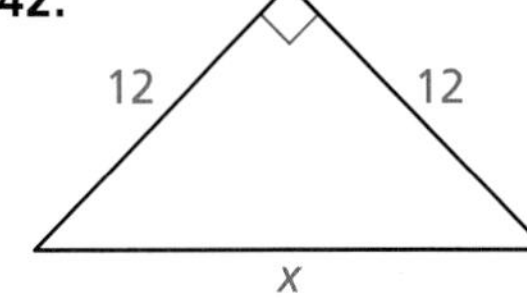

43.

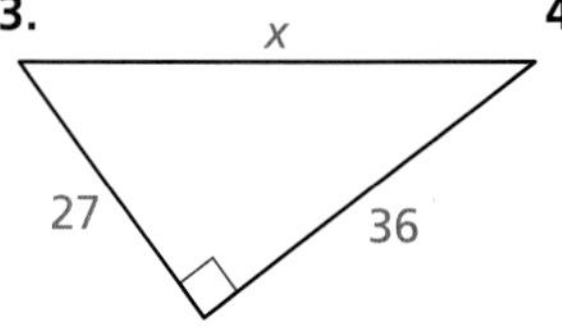

44.

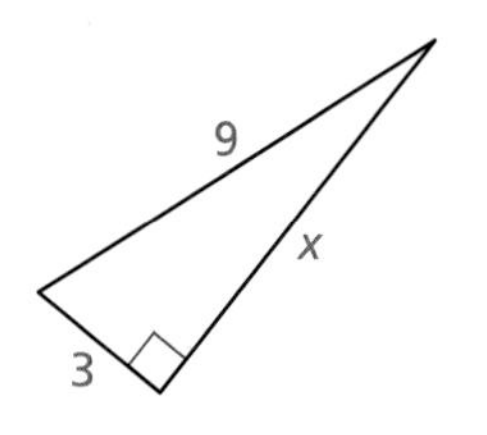

9.6 Solving Right Triangles

Essential Question

When you know the lengths of the sides of a right triangle, how can you find the measures of the two acute angles?

EXPLORATION 1 Solving Special Right Triangles

Work with a partner. Use the figures to find the values of the sine and cosine of $\angle A$ and $\angle B$. Use these values to find the measures of $\angle A$ and $\angle B$. Use dynamic geometry software to verify your answers.

a.

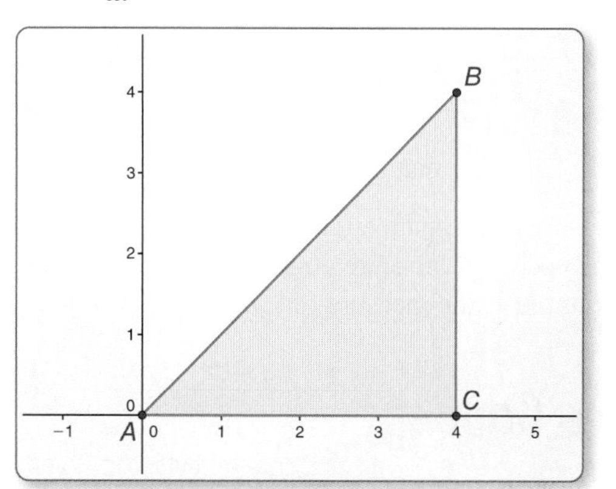

b.

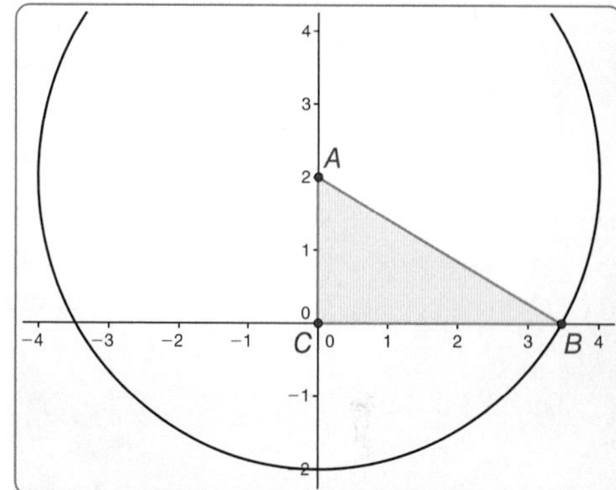

ATTENDING TO PRECISION

To be proficient in math, you need to calculate accurately and efficiently, expressing numerical answers with a degree of precision appropriate for the problem context.

EXPLORATION 2 Solving Right Triangles

Work with a partner. You can use a calculator to find the measure of an angle when you know the value of the sine, cosine, or tangent of the angle. Use the inverse sine, inverse cosine, or inverse tangent feature of your calculator to approximate the measures of $\angle A$ and $\angle B$ to the nearest tenth of a degree. Then use dynamic geometry software to verify your answers.

a.

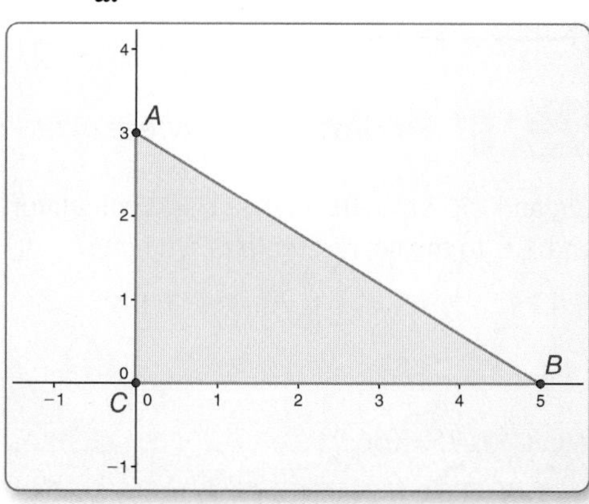

b.

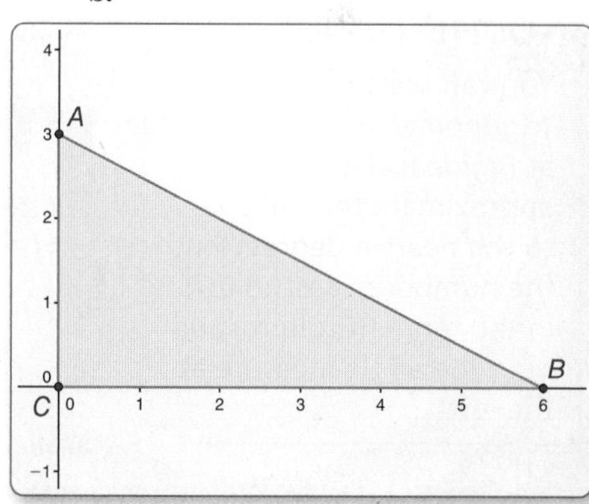

Communicate Your Answer

3. When you know the lengths of the sides of a right triangle, how can you find the measures of the two acute angles?

4. A ladder leaning against a building forms a right triangle with the building and the ground. The legs of the right triangle (in meters) form a 5-12-13 Pythagorean triple. Find the measures of the two acute angles to the nearest tenth of a degree.

9.6 Lesson

Core Vocabulary

inverse tangent, *p. 502*
inverse sine, *p. 502*
inverse cosine, *p.502*
solve a right triangle, *p. 503*

What You Will Learn

- Use inverse trigonometric ratios.
- Solve right triangles.

Using Inverse Trigonometric Ratios

EXAMPLE 1 Identifying Angles from Trigonometric Ratios

Determine which of the two acute angles has a cosine of 0.5.

SOLUTION

Find the cosine of each acute angle.

$$\cos A = \frac{\text{adj. to } \angle A}{\text{hyp.}} = \frac{\sqrt{3}}{2} \approx 0.8660 \qquad \cos B = \frac{\text{adj. to } \angle B}{\text{hyp.}} = \frac{1}{2} = 0.5$$

- The acute angle that has a cosine of 0.5 is $\angle B$.

If the measure of an acute angle is 60°, then its cosine is 0.5. The converse is also true. If the cosine of an acute angle is 0.5, then the measure of the angle is 60°. So, in Example 1, the measure of $\angle B$ must be 60° because its cosine is 0.5.

READING

The expression "$\tan^{-1} x$" is read as "the inverse tangent of x."

Core Concept

Inverse Trigonometric Ratios

Let $\angle A$ be an acute angle.

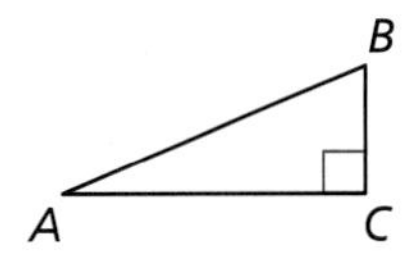

Inverse Tangent If $\tan A = x$, then $\tan^{-1} x = m\angle A$. $\qquad \tan^{-1} \frac{BC}{AC} = m\angle A$

Inverse Sine If $\sin A = y$, then $\sin^{-1} y = m\angle A$. $\qquad \sin^{-1} \frac{BC}{AB} = m\angle A$

Inverse Cosine If $\cos A = z$, then $\cos^{-1} z = m\angle A$. $\qquad \cos^{-1} \frac{AC}{AB} = m\angle A$

ANOTHER WAY

You can use the Table of Trigonometric Ratios available at *BigIdeasMath.com* to approximate $\tan^{-1} 0.75$ to the nearest degree. Find the number closest to 0.75 in the tangent column and read the angle measure at the left.

EXAMPLE 2 Finding Angle Measures

Let $\angle A$, $\angle B$, and $\angle C$ be acute angles. Use a calculator to approximate the measures of $\angle A$, $\angle B$, and $\angle C$ to the nearest tenth of a degree.

a. $\tan A = 0.75$ **b.** $\sin B = 0.87$ **c.** $\cos C = 0.15$

SOLUTION

a. $m\angle A = \tan^{-1} 0.75 \approx 36.9°$

b. $m\angle B = \sin^{-1} 0.87 \approx 60.5°$

c. $m\angle C = \cos^{-1} 0.15 \approx 81.4°$

Monitoring Progress

Help in English and Spanish at *BigIdeasMath.com*

Determine which of the two acute angles has the given trigonometric ratio.

1. The sine of the angle is $\frac{12}{13}$.
2. The tangent of the angle is $\frac{5}{12}$.

Monitoring Progress

Help in English and Spanish at *BigIdeasMath.com*

Let $\angle G$, $\angle H$, and $\angle K$ be acute angles. Use a calculator to approximate the measures of $\angle G$, $\angle H$, and $\angle K$ to the nearest tenth of a degree.

3. $\tan G = 0.43$ **4.** $\sin H = 0.68$ **5.** $\cos K = 0.94$

Solving Right Triangles

Core Concept

Solving a Right Triangle

To **solve a right triangle** means to find all unknown side lengths and angle measures. You can solve a right triangle when you know either of the following.

- two side lengths
- one side length and the measure of one acute angle

EXAMPLE 3 Solving a Right Triangle

Solve the right triangle. Round decimal answers to the nearest tenth.

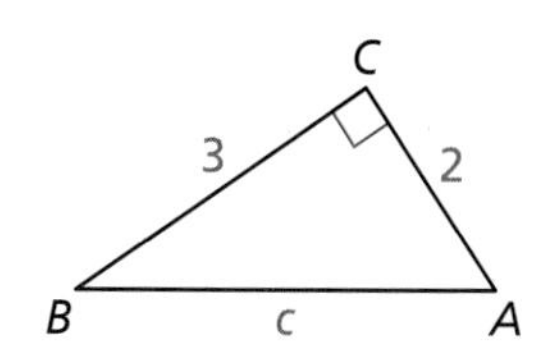

SOLUTION

Step 1 Use the Pythagorean Theorem (Theorem 9.1) to find the length of the hypotenuse.

$c^2 = a^2 + b^2$ Pythagorean Theorem

$c^2 = 3^2 + 2^2$ Substitute.

$c^2 = 13$ Simplify.

$c = \sqrt{13}$ Find the positive square root.

$c \approx 3.6$ Use a calculator.

ANOTHER WAY

You could also have found $m\angle A$ first by finding $\tan^{-1}\frac{3}{2} \approx 56.3°$.

Step 2 Find $m\angle B$.

$m\angle B = \tan^{-1}\frac{2}{3} \approx 33.7°$ Use a calculator.

Step 3 Find $m\angle A$.

Because $\angle A$ and $\angle B$ are complements, you can write

$$m\angle A = 90° - m\angle B$$
$$\approx 90° - 33.7°$$
$$= 56.3°.$$

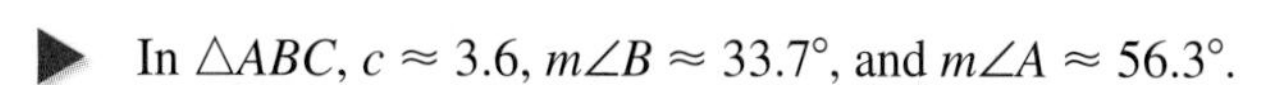

In $\triangle ABC$, $c \approx 3.6$, $m\angle B \approx 33.7°$, and $m\angle A \approx 56.3°$.

Monitoring Progress

Help in English and Spanish at *BigIdeasMath.com*

Solve the right triangle. Round decimal answers to the nearest tenth.

6.

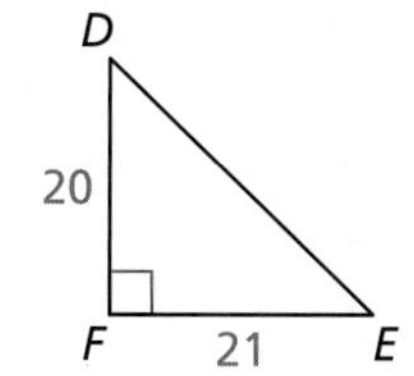

7.

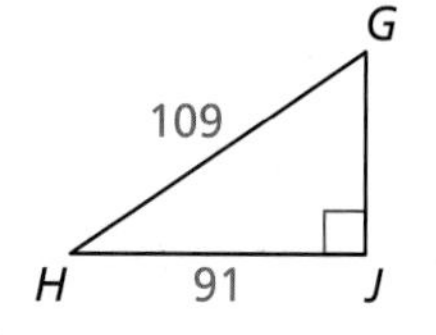

EXAMPLE 4 Solving a Right Triangle

Solve the right triangle. Round decimal answers to the nearest tenth.

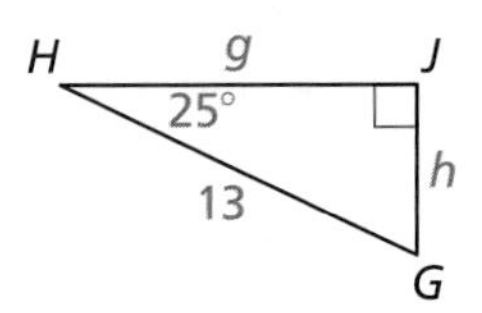

SOLUTION

Use trigonometric ratios to find the values of g and h.

$$\sin H = \frac{\text{opp.}}{\text{hyp.}} \qquad \cos H = \frac{\text{adj.}}{\text{hyp.}}$$

$$\sin 25° = \frac{h}{13} \qquad \cos 25° = \frac{g}{13}$$

$$13 \cdot \sin 25° = h \qquad 13 \cdot \cos 25° = g$$

$$5.5 \approx h \qquad 11.8 \approx g$$

Because $\angle H$ and $\angle G$ are complements, you can write

$$m\angle G = 90° - m\angle H = 90° - 25° = 65°.$$

▶ In $\triangle GHJ$, $h \approx 5.5$, $g \approx 11.8$, and $m\angle G = 65°$.

READING

A *raked stage* slants upward from front to back to give the audience a better view.

EXAMPLE 5 Solving a Real-Life Problem

Your school is building a *raked stage*. The stage will be 30 feet long from front to back, with a total rise of 2 feet. You want the rake (angle of elevation) to be 5° or less for safety. Is the raked stage within your desired range?

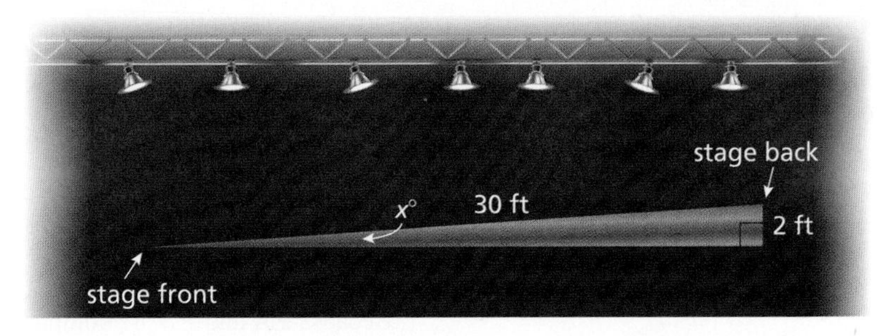

SOLUTION

Use the inverse sine ratio to find the degree measure x of the rake.

$$x \approx \sin^{-1}\frac{2}{30} \approx 3.8$$

▶ The rake is about 3.8°, so it is within your desired range of 5° or less.

Monitoring Progress  Help in English and Spanish at *BigIdeasMath.com*

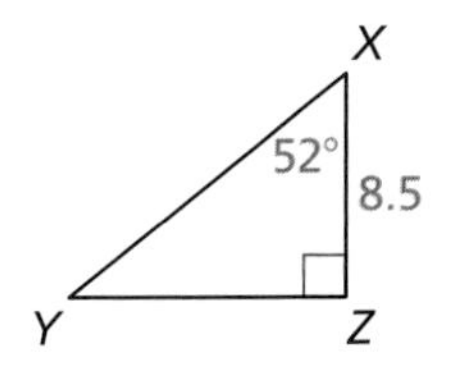

8. Solve the right triangle. Round decimal answers to the nearest tenth.

9. **WHAT IF?** In Example 5, suppose another raked stage is 20 feet long from front to back with a total rise of 2 feet. Is the raked stage within your desired range?

9.6 Exercises

Dynamic Solutions available at *BigIdeasMath.com*

Vocabulary and Core Concept Check

1. **COMPLETE THE SENTENCE** To solve a right triangle means to find the measures of all its ______ and ______.

2. **WRITING** Explain when you can use a trigonometric ratio to find a side length of a right triangle and when you can use the Pythagorean Theorem (Theorem 9.1).

Monitoring Progress and Modeling with Mathematics

In Exercises 3–6, determine which of the two acute angles has the given trigonometric ratio. *(See Example 1.)*

3. The cosine of the angle is $\frac{4}{5}$.

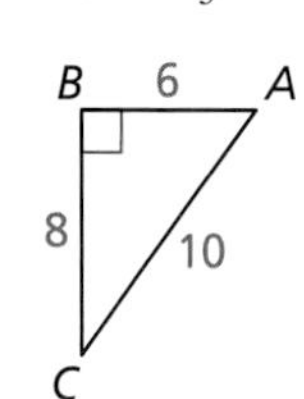

4. The sine of the angle is $\frac{5}{11}$.

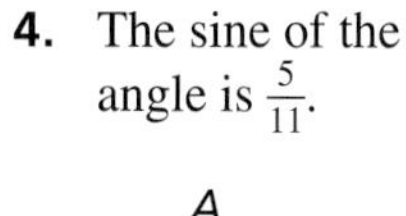
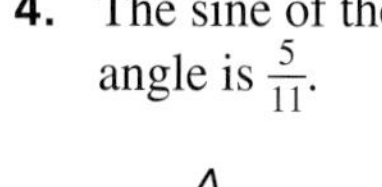
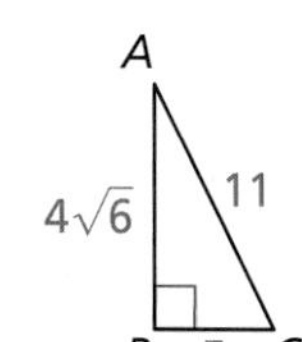

5. The sine of the angle is 0.96.

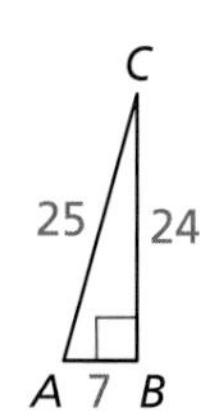

6. The tangent of the angle is 1.5.

C
12
$6\sqrt{13}$
B
18
A

In Exercises 7–12, let $\angle D$ be an acute angle. Use a calculator to approximate the measure of $\angle D$ to the nearest tenth of a degree. *(See Example 2.)*

7. $\sin D = 0.75$
8. $\sin D = 0.19$
9. $\cos D = 0.33$
10. $\cos D = 0.64$
11. $\tan D = 0.28$
12. $\tan D = 0.72$

In Exercises 13–18, solve the right triangle. Round decimal answers to the nearest tenth. *(See Examples 3 and 4.)*

13.

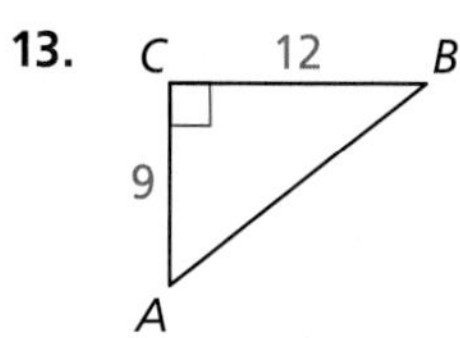

14.

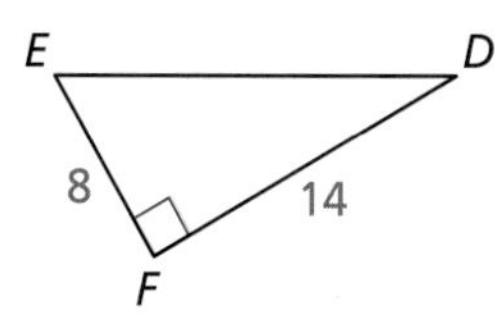

15.

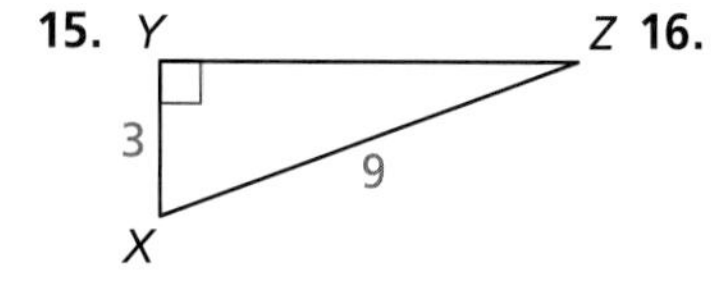

16.

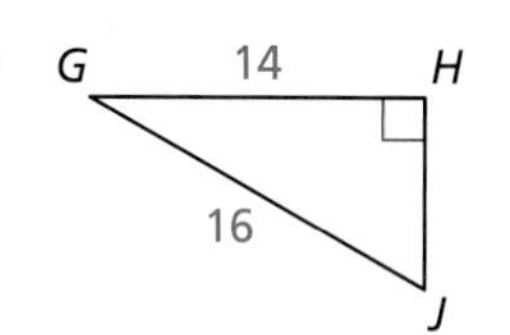

17. 18.

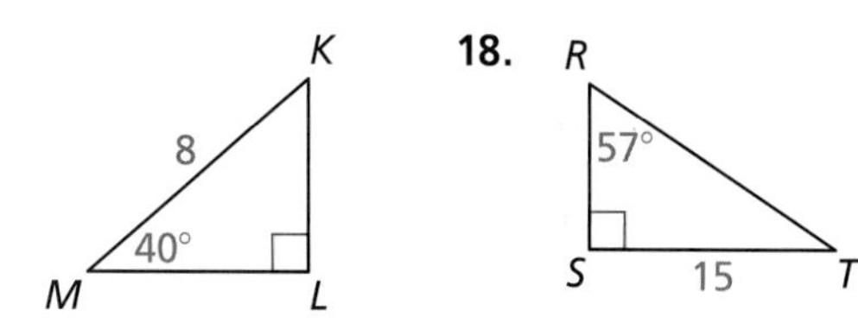

19. **ERROR ANALYSIS** Describe and correct the error in using an inverse trigonometric ratio.

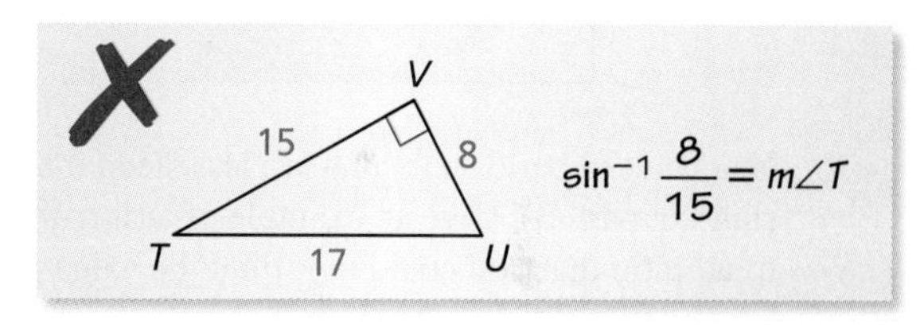

20. **PROBLEM SOLVING** In order to unload clay easily, the body of a dump truck must be elevated to at least 45°. The body of a dump truck that is 14 feet long has been raised 8 feet. Will the clay pour out easily? Explain your reasoning. *(See Example 5.)*

21. **PROBLEM SOLVING** You are standing on a footbridge that is 12 feet above a lake. You look down and see a duck in the water. The duck is 7 feet away from the footbridge. What is the angle of elevation from the duck to you?

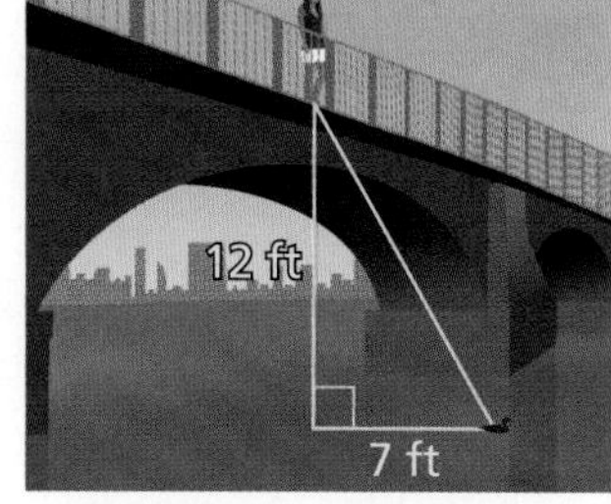

22. HOW DO YOU SEE IT? Write three expressions that can be used to approximate the measure of $\angle A$. Which expression would you choose? Explain your choice.

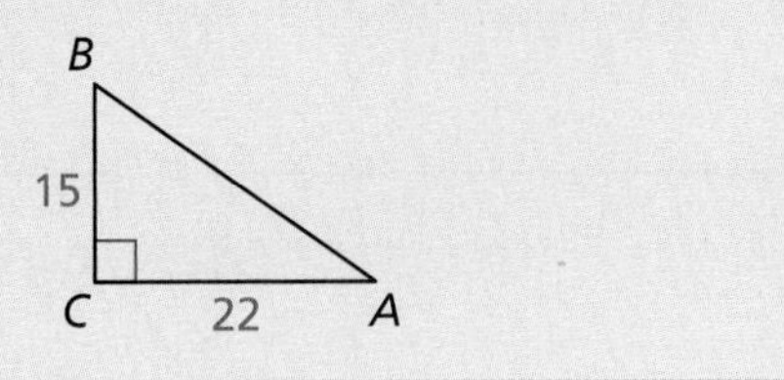

23. MODELING WITH MATHEMATICS The Uniform Federal Accessibility Standards specify that a wheelchair ramp may not have an incline greater than 4.76°. You want to build a ramp with a vertical rise of 8 inches. You want to minimize the horizontal distance taken up by the ramp. Draw a diagram showing the approximate dimensions of your ramp.

24. MODELING WITH MATHEMATICS The horizontal part of a step is called the *tread*. The vertical part is called the *riser*. The recommended riser-to-tread ratio is 7 inches : 11 inches.

a. Find the value of x for stairs built using the recommended riser-to-tread ratio.

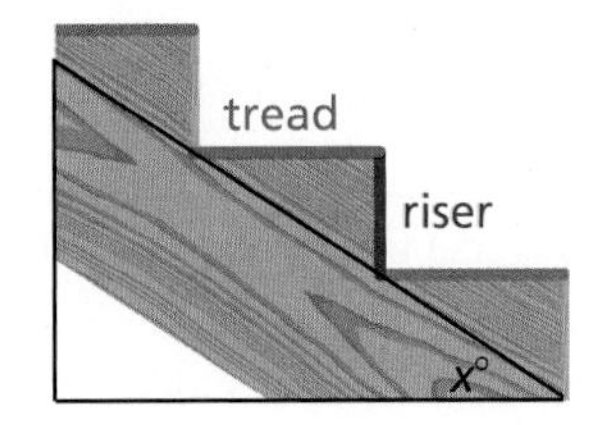

b. You want to build stairs that are less steep than the stairs in part (a). Give an example of a riser-to-tread ratio that you could use. Find the value of x for your stairs.

25. USING TOOLS Find the measure of $\angle R$ without using a protractor. Justify your technique.

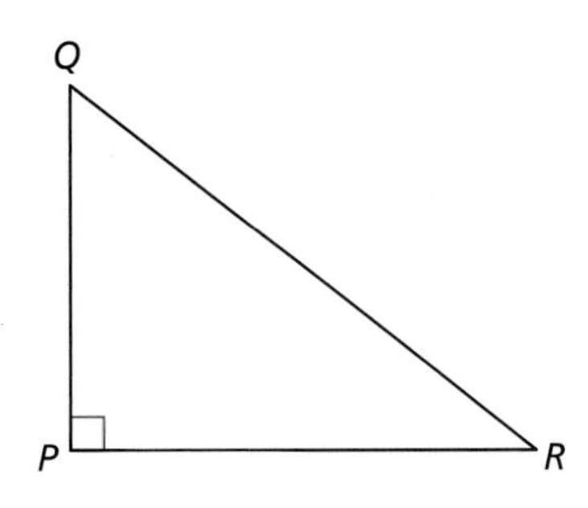

26. MAKING AN ARGUMENT Your friend claims that $\tan^{-1} x = \frac{1}{\tan x}$. Is your friend correct? Explain your reasoning.

USING STRUCTURE In Exercises 27 and 28, solve each triangle.

27. $\triangle JKM$ and $\triangle LKM$

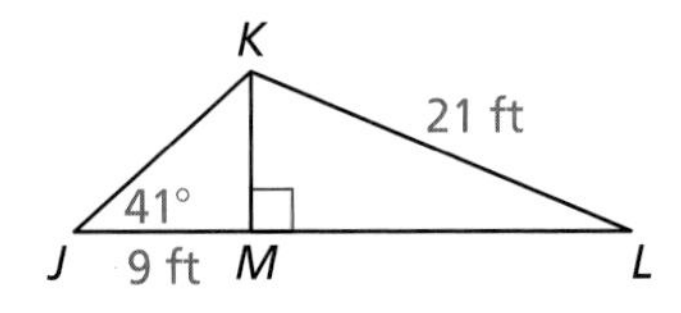

28. $\triangle TUS$ and $\triangle VTW$

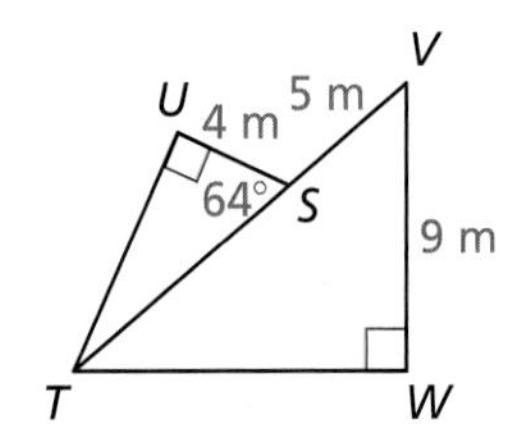

29. MATHEMATICAL CONNECTIONS Write an expression that can be used to find the measure of the acute angle formed by each line and the x-axis. Then approximate the angle measure to the nearest tenth of a degree.

a. $y = 3x$

b. $y = \frac{4}{3}x + 4$

30. THOUGHT PROVOKING Simplify each expression. Justify your answer.

a. $\sin^{-1}(\sin x)$

b. $\tan(\tan^{-1} y)$

c. $\cos(\cos^{-1} z)$

31. REASONING Explain why the expression $\sin^{-1}(1.2)$ does not make sense.

32. USING STRUCTURE The perimeter of rectangle $ABCD$ is 16 centimeters, and the ratio of its width to its length is 1 : 3. Segment BD divides the rectangle into two congruent triangles. Find the side lengths and angle measures of these two triangles.

Maintaining Mathematical Proficiency

Reviewing what you learned in previous grades and lessons

Solve the equation. *(Skills Review Handbook)*

33. $\frac{12}{x} = \frac{3}{2}$

34. $\frac{13}{9} = \frac{x}{18}$

35. $\frac{x}{2.1} = \frac{4.1}{3.5}$

36. $\frac{5.6}{12.7} = \frac{4.9}{x}$

9.7 Law of Sines and Law of Cosines

Essential Question What are the Law of Sines and the Law of Cosines?

EXPLORATION 1 Discovering the Law of Sines

Work with a partner.

a. Copy and complete the table for the triangle shown. What can you conclude?

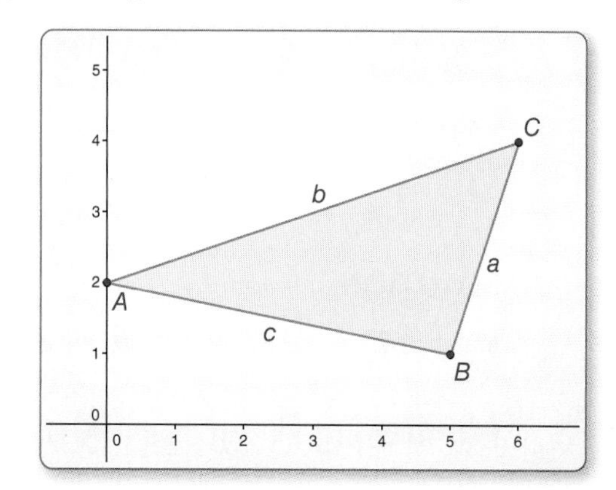

Sample
Segments
$a = 3.16$
$b = 6.32$
$c = 5.10$
Angles
$m\angle A = 29.74°$
$m\angle B = 97.13°$
$m\angle C = 53.13°$

$m\angle A$	a	$\frac{\sin A}{a}$	$m\angle B$	b	$\frac{\sin B}{b}$	$m\angle C$	c	$\frac{\sin C}{c}$

USING TOOLS STRATEGICALLY

To be proficient in math, you need to use technology to compare predictions with data.

b. Use dynamic geometry software to draw two other triangles. Copy and complete the table in part (a) for each triangle. Use your results to write a conjecture about the relationship between the sines of the angles and the lengths of the sides of a triangle.

EXPLORATION 2 Discovering the Law of Cosines

Work with a partner.

a. Copy and complete the table for the triangle in Exploration 1(a). What can you conclude?

c	c^2	a	a^2	b	b^2	$m\angle C$	$a^2 + b^2 - 2ab \cos C$

b. Use dynamic geometry software to draw two other triangles. Copy and complete the table in part (a) for each triangle. Use your results to write a conjecture about what you observe in the completed tables.

Communicate Your Answer

3. What are the Law of Sines and the Law of Cosines?

4. When would you use the Law of Sines to solve a triangle? When would you use the Law of Cosines to solve a triangle?

9.7 Lesson

Core Vocabulary

Law of Sines, *p. 509*
Law of Cosines, *p. 511*

What You Will Learn

- Find areas of triangles.
- Use the Law of Sines to solve triangles.
- Use the Law of Cosines to solve triangles.

Finding Areas of Triangles

So far, you have used trigonometric ratios to solve right triangles. In this lesson, you will learn how to solve any triangle. When the triangle is obtuse, you may need to find a trigonometric ratio for an obtuse angle.

EXAMPLE 1 Finding Trigonometric Ratios for Obtuse Angles

Use a calculator to find each trigonometric ratio. Round your answer to four decimal places.

a. $\tan 150^\circ$ **b.** $\sin 120^\circ$ **c.** $\cos 95^\circ$

SOLUTION

a. $\tan 150^\circ \approx -0.5774$ **b.** $\sin 120^\circ \approx 0.8660$ **c.** $\cos 95^\circ \approx -0.0872$

Monitoring Progress

Help in English and Spanish at *BigIdeasMath.com*

Use a calculator to find the trigonometric ratio. Round your answer to four decimal places.

1. $\tan 110^\circ$ **2.** $\sin 97^\circ$ **3.** $\cos 165^\circ$

Area of a Triangle

The area of any triangle is given by one-half the product of the lengths of two sides times the sine of their included angle. For $\triangle ABC$ shown, there are three ways to calculate the area.

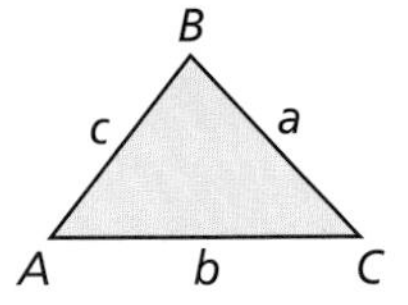

Area $= \frac{1}{2}bc \sin A$ Area $= \frac{1}{2}ac \sin B$ Area $= \frac{1}{2}ab \sin C$

EXAMPLE 2 Finding the Area of a Triangle

Find the area of the triangle. Round your answer to the nearest tenth.

SOLUTION

$$\text{Area} = \frac{1}{2}bc \sin A = \frac{1}{2}(17)(19) \sin 135^\circ \approx 114.2$$

- The area of the triangle is about 114.2 square units.

Monitoring Progress

Help in English and Spanish at *BigIdeasMath.com*

Find the area of $\triangle ABC$ with the given side lengths and included angle. Round your answer to the nearest tenth.

4. $m\angle B = 60^\circ, a = 19, c = 14$ **5.** $m\angle C = 29^\circ, a = 38, b = 31$

Using the Law of Sines

The trigonometric ratios in the previous sections can only be used to solve right triangles. You will learn two laws that can be used to solve any triangle.

You can use the **Law of Sines** to solve triangles when two angles and the length of any side are known (AAS or ASA cases), or when the lengths of two sides and an angle opposite one of the two sides are known (SSA case).

Theorem 9.9 Law of Sines

The Law of Sines can be written in either of the following forms for $\triangle ABC$ with sides of length a, b, and c.

$$\frac{\sin A}{a} = \frac{\sin B}{b} = \frac{\sin C}{c} \qquad \frac{a}{\sin A} = \frac{b}{\sin B} = \frac{c}{\sin C}$$

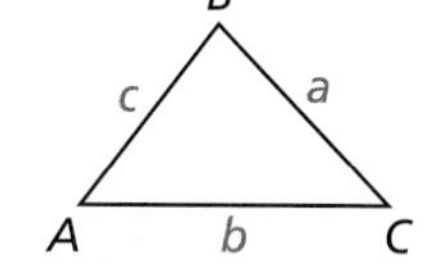

Proof Ex. 51, p. 516

EXAMPLE 3 Using the Law of Sines (SSA Case)

Solve the triangle. Round decimal answers to the nearest tenth.

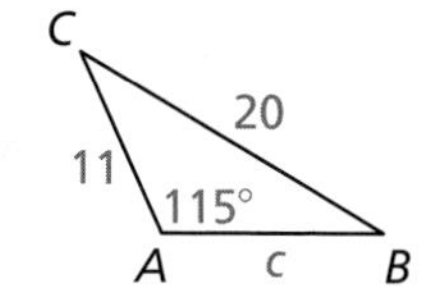

SOLUTION

Use the Law of Sines to find $m\angle B$.

$$\frac{\sin B}{b} = \frac{\sin A}{a} \qquad \text{Law of Sines}$$

$$\frac{\sin B}{11} = \frac{\sin 115^\circ}{20} \qquad \text{Substitute.}$$

$$\sin B = \frac{11 \sin 115^\circ}{20} \qquad \text{Multiply each side by 11.}$$

$$m\angle B \approx 29.9^\circ \qquad \text{Use a calculator.}$$

By the Triangle Sum Theorem (Theorem 5.1), $m\angle C \approx 180^\circ - 115^\circ - 29.9^\circ = 35.1^\circ$.

Use the Law of Sines again to find the remaining side length c of the triangle.

$$\frac{c}{\sin C} = \frac{a}{\sin A} \qquad \text{Law of Sines}$$

$$\frac{c}{\sin 35.1^\circ} = \frac{20}{\sin 115^\circ} \qquad \text{Substitute.}$$

$$c = \frac{20 \sin 35.1^\circ}{\sin 115^\circ} \qquad \text{Multiply each side by sin 35.1°.}$$

$$c \approx 12.7 \qquad \text{Use a calculator.}$$

▶ In $\triangle ABC$, $m\angle B \approx 29.9^\circ$, $m\angle C \approx 35.1^\circ$, and $c \approx 12.7$.

Monitoring Progress  Help in English and Spanish at *BigIdeasMath.com*

Solve the triangle. Round decimal answers to the nearest tenth.

6. A C 51° 17 18 B

7. C 13 16 40° A B

EXAMPLE 4 Using the Law of Sines (AAS Case)

Solve the triangle. Round decimal answers to the nearest tenth.

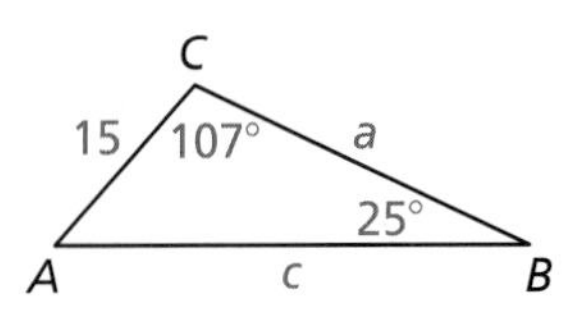

SOLUTION

By the Triangle Sum Theorem (Theorem 5.1), $m\angle A = 180^\circ - 107^\circ - 25^\circ = 48^\circ$.

By the Law of Sines, you can write $\dfrac{a}{\sin 48^\circ} = \dfrac{15}{\sin 25^\circ} = \dfrac{c}{\sin 107^\circ}$.

$\dfrac{a}{\sin 48^\circ} = \dfrac{15}{\sin 25^\circ}$	Write two equations, each with one variable.	$\dfrac{c}{\sin 107^\circ} = \dfrac{15}{\sin 25^\circ}$
$a = \dfrac{15 \sin 48^\circ}{\sin 25^\circ}$	Solve for each variable.	$c = \dfrac{15 \sin 107^\circ}{\sin 25^\circ}$
$a \approx 26.4$	Use a calculator.	$c \approx 33.9$

▶ In $\triangle ABC$, $m\angle A = 48^\circ$, $a \approx 26.4$, and $c \approx 33.9$.

EXAMPLE 5 Using the Law of Sines (ASA Case)

A surveyor makes the measurements shown to determine the length of a bridge to be built across a small lake from the North Picnic Area to the South Picnic Area. Find the length of the bridge.

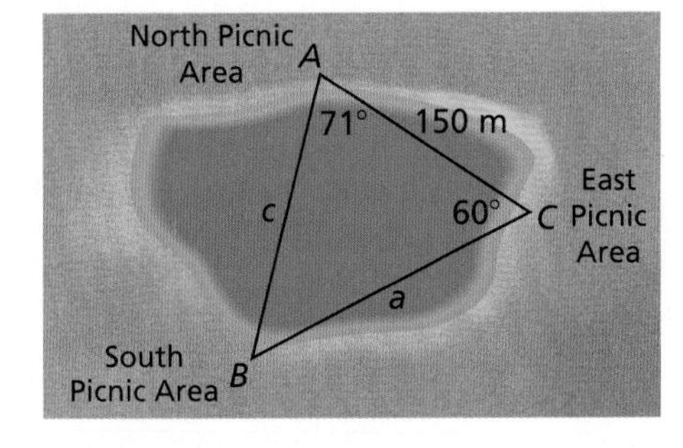

SOLUTION

In the diagram, c represents the distance from the North Picnic Area to the South Picnic Area, so c represents the length of the bridge.

By the Triangle Sum Theorem (Theorem 5.1), $m\angle B = 180^\circ - 71^\circ - 60^\circ = 49^\circ$.

By the Law of Sines, you can write $\dfrac{a}{\sin 71^\circ} = \dfrac{150}{\sin 49^\circ} = \dfrac{c}{\sin 60^\circ}$.

$\dfrac{c}{\sin 60^\circ} = \dfrac{150}{\sin 49^\circ}$	Write an equation involving c.
$c = \dfrac{150 \sin 60^\circ}{\sin 49^\circ}$	Multiply each side by sin 60°.
$c \approx 172.1$	Use a calculator.

▶ The length of the bridge will be about 172.1 meters.

Monitoring Progress

Help in English and Spanish at *BigIdeasMath.com*

Solve the triangle. Round decimal answers to the nearest tenth.

8. **9.**

10. WHAT IF? In Example 5, what would be the length of a bridge from the South Picnic Area to the East Picnic Area?

Using the Law of Cosines

You can use the **Law of Cosines** to solve triangles when two sides and the included angle are known (SAS case), or when all three sides are known (SSS case).

Theorem 9.10 Law of Cosines

If $\triangle ABC$ has sides of length a, b, and c, as shown, then the following are true.

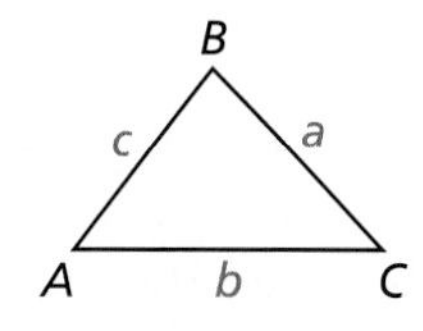

$$a^2 = b^2 + c^2 - 2bc \cos A$$

$$b^2 = a^2 + c^2 - 2ac \cos B$$

$$c^2 = a^2 + b^2 - 2ab \cos C$$

Proof Ex. 52, p. 516

EXAMPLE 6 Using the Law of Cosines (SAS Case)

Solve the triangle. Round decimal answers to the nearest tenth.

B, 14, A, 34°, 11, b, C

SOLUTION

Use the Law of Cosines to find side length b.

$b^2 = a^2 + c^2 - 2ac \cos B$	Law of Cosines
$b^2 = 11^2 + 14^2 - 2(11)(14) \cos 34°$	Substitute.
$b^2 = 317 - 308 \cos 34°$	Simplify.
$b = \sqrt{317 - 308 \cos 34°}$	Find the positive square root.
$b \approx 7.9$	Use a calculator.

ANOTHER WAY

When you know all three sides and one angle, you can use the Law of Cosines or the Law of Sines to find the measure of a second angle.

Use the Law of Sines to find $m\angle A$.

$\dfrac{\sin A}{a} = \dfrac{\sin B}{b}$	Law of Sines
$\dfrac{\sin A}{11} = \dfrac{\sin 34°}{\sqrt{317 - 308 \cos 34°}}$	Substitute.
$\sin A = \dfrac{11 \sin 34°}{\sqrt{317 - 308 \cos 34°}}$	Multiply each side by 11.
$m\angle A \approx 51.6°$	Use a calculator.

By the Triangle Sum Theorem (Theorem 5.1), $m\angle C \approx 180° - 34° - 51.6° = 94.4°$.

▶ In $\triangle ABC$, $b \approx 7.9$, $m\angle A \approx 51.6°$, and $m\angle C \approx 94.4°$.

COMMON ERROR

In Example 6, the smaller remaining angle is found first because the inverse sine feature of a calculator only gives angle measures from 0° to 90°. So, when an angle is obtuse, like $\angle C$ because $14^2 > (7.85)^2 + 11^2$, you will not get the obtuse measure.

Monitoring Progress

Help in English and Spanish at *BigIdeasMath.com*

Solve the triangle. Round decimal answers to the nearest tenth.

11. C, 45, B, 88°, 43, A

12.

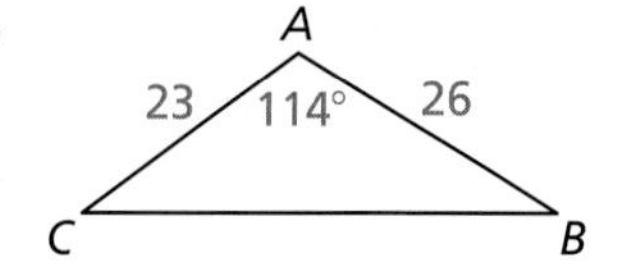

EXAMPLE 7 Using the Law of Cosines (SSS Case)

Solve the triangle. Round decimal answers to the nearest tenth.

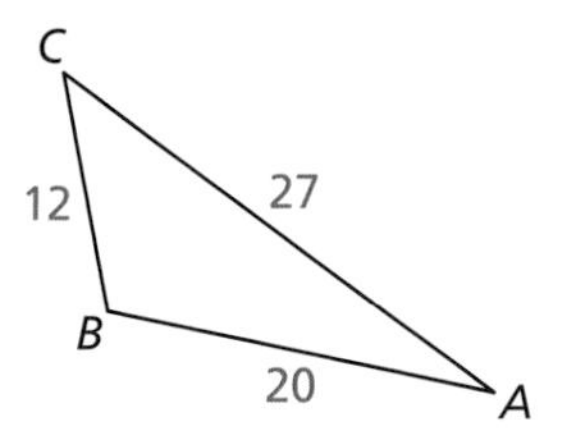

SOLUTION

First, find the angle opposite the longest side, $\overline{AC}$. Use the Law of Cosines to find $m\angle B$.

$b^2 = a^2 + c^2 - 2ac \cos B$	Law of Cosines
$27^2 = 12^2 + 20^2 - 2(12)(20) \cos B$	Substitute.
$\dfrac{27^2 - 12^2 - 20^2}{-2(12)(20)} = \cos B$	Solve for cos *B*.
$m\angle B \approx 112.7°$	Use a calculator.

COMMON ERROR

In Example 7, the largest angle is found first to make sure that the other two angles are acute. This way, when you use the Law of Sines to find another angle measure, you will know that it is between 0° and 90°.

Now, use the Law of Sines to find $m\angle A$.

$\dfrac{\sin A}{a} = \dfrac{\sin B}{b}$	Law of Sines
$\dfrac{\sin A}{12} = \dfrac{\sin 112.7°}{27}$	Substitute for *a*, *b*, and *B*.
$\sin A = \dfrac{12 \sin 112.7°}{27}$	Multiply each side by 12.
$m\angle A \approx 24.2°$	Use a calculator.

By the Triangle Sum Theorem (Theorem 5.1), $m\angle C \approx 180° - 24.2° - 112.7° = 43.1°$.

▶ In $\triangle ABC$, $m\angle A \approx 24.2°$, $m\angle B \approx 112.7°$, and $m\angle C \approx 43.1°$.

EXAMPLE 8 Solving a Real-Life Problem

An organism's step angle is a measure of walking efficiency. The closer the step angle is to 180°, the more efficiently the organism walked. The diagram shows a set of footprints for a dinosaur. Find the step angle *B*.

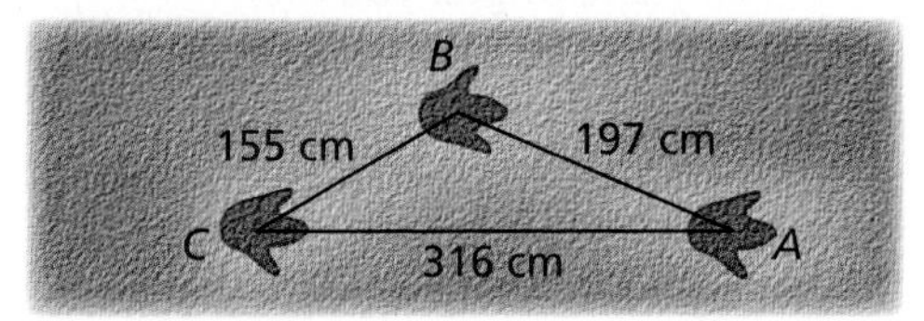

SOLUTION

$b^2 = a^2 + c^2 - 2ac \cos B$	Law of Cosines
$316^2 = 155^2 + 197^2 - 2(155)(197) \cos B$	Substitute.
$\dfrac{316^2 - 155^2 - 197^2}{-2(155)(197)} = \cos B$	Solve for cos *B*.
$127.3° \approx m\angle B$	Use a calculator.

▶ The step angle *B* is about 127.3°.

Monitoring Progress  Help in English and Spanish at *BigIdeasMath.com*

Solve the triangle. Round decimal answers to the nearest tenth.

13. 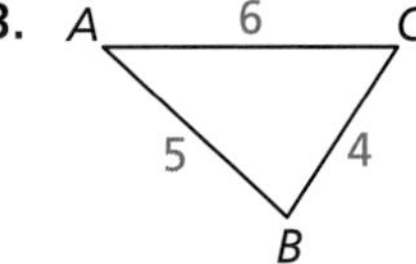

14. 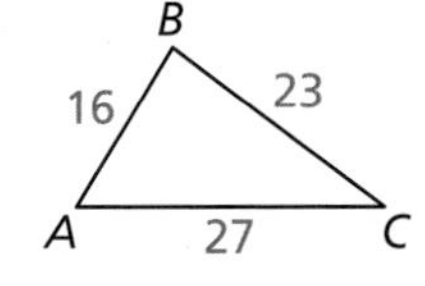

9.7 Exercises

Dynamic Solutions available at *BigIdeasMath.com*

Vocabulary and Core Concept Check

1. **WRITING** What type of triangle would you use the Law of Sines or the Law of Cosines to solve?

2. **VOCABULARY** What information do you need to use the Law of Sines?

Monitoring Progress and Modeling with Mathematics

In Exercises 3–8, use a calculator to find the trigonometric ratio. Round your answer to four decimal places. *(See Example 1.)*

3. sin 127°

4. sin 98°

5. cos 139°

6. cos 108°

7. tan 165°

8. tan 116°

In Exercises 9–12, find the area of the triangle. Round your answer to the nearest tenth. *(See Example 2.)*

9. 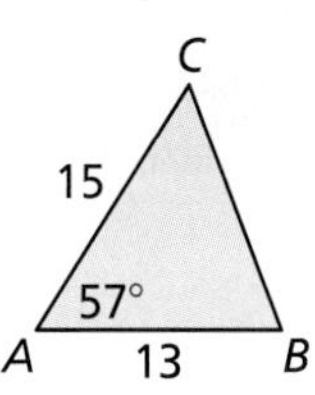

10. 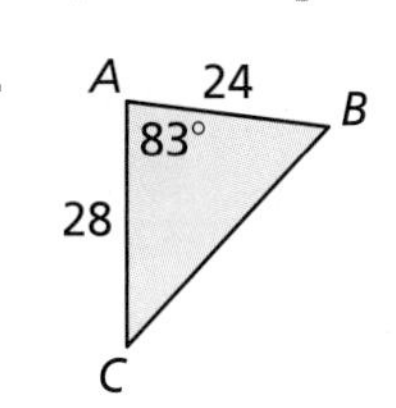

11. 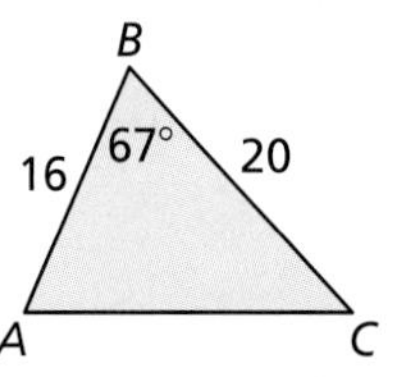

12. 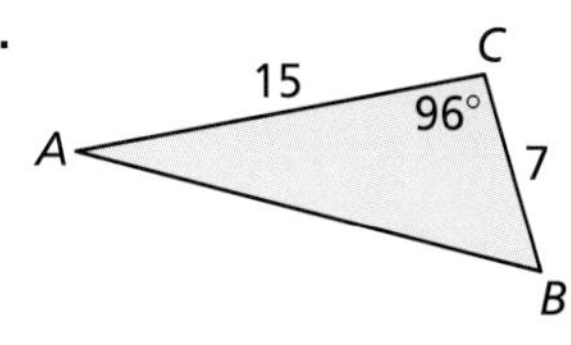

In Exercises 13–18, solve the triangle. Round decimal answers to the nearest tenth. *(See Examples 3, 4, and 5.)*

13. 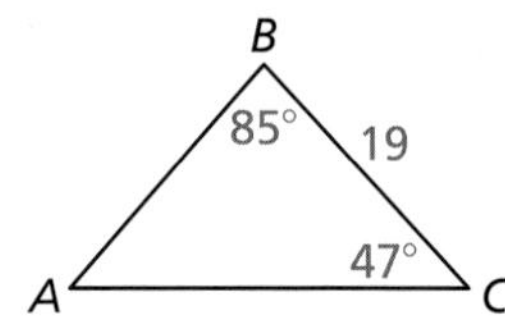

14. 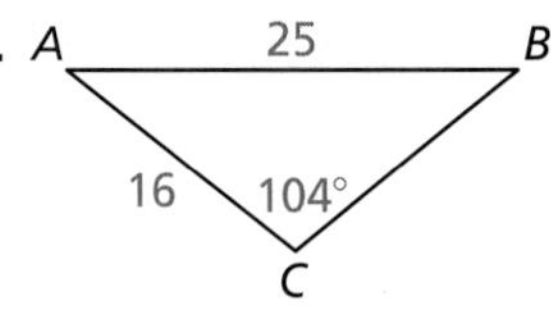

15. 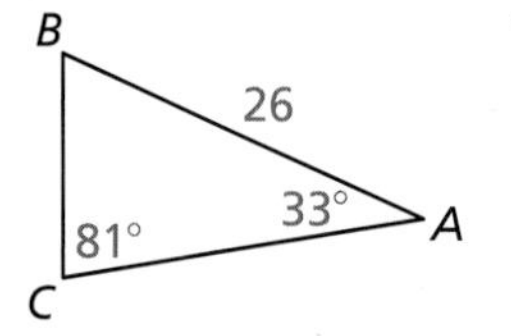

16. 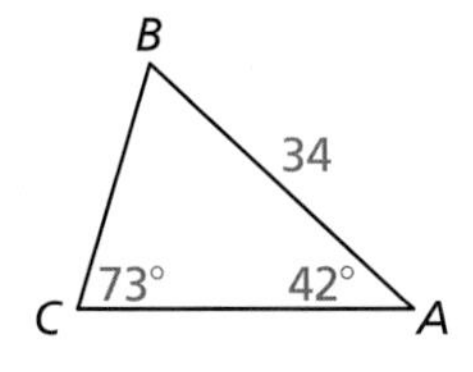

17. 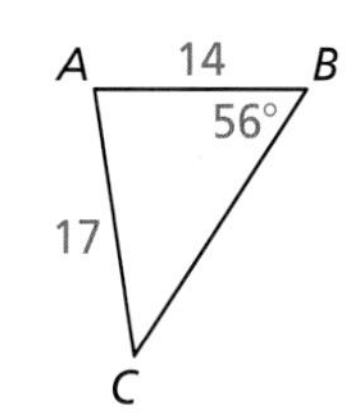

18. 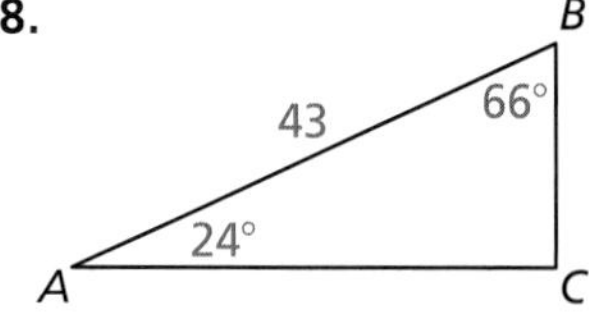

In Exercises 19–24, solve the triangle. Round decimal answers to the nearest tenth. *(See Examples 6 and 7.)*

19. 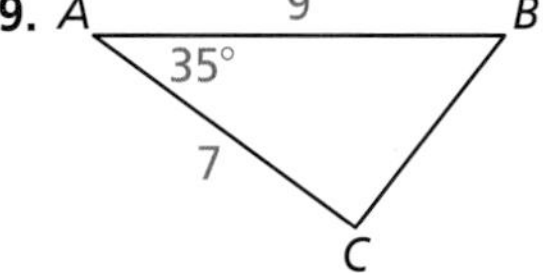

20. 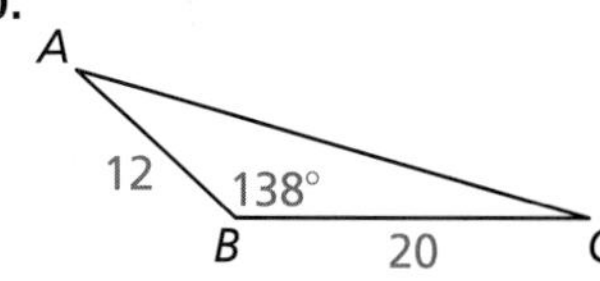

21. 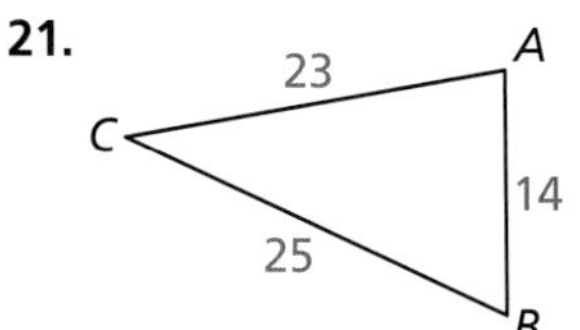

22. 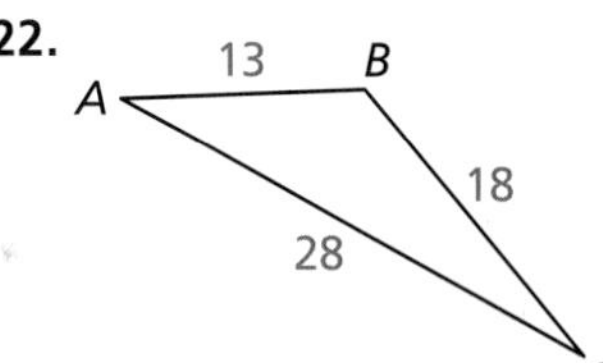

23. 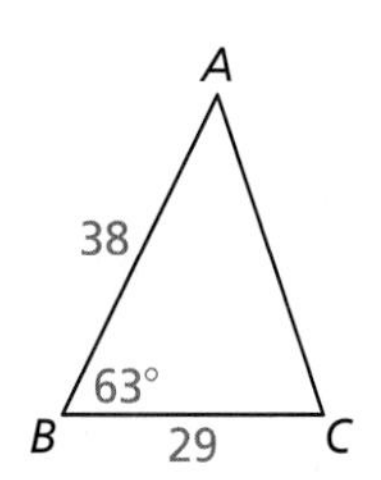

24. 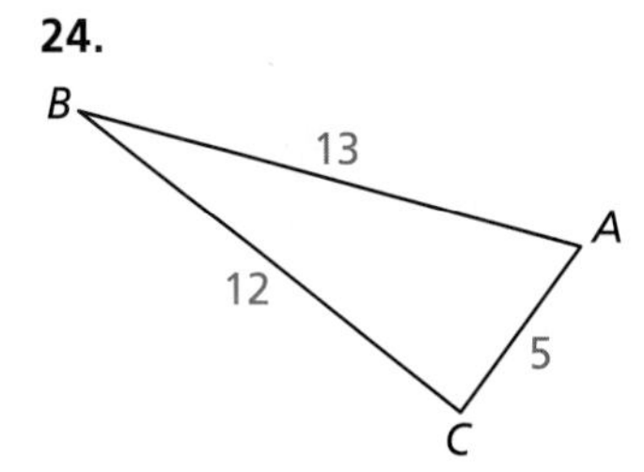

25. **ERROR ANALYSIS** Describe and correct the error in finding $m\angle C$.

A, C, 55°, 5, 6, B

$$\frac{\sin C}{6} = \frac{\sin 55°}{5}$$

$$\sin C = \frac{6 \sin 55°}{5}$$

$$m\angle C \approx 79.4°$$

26. **ERROR ANALYSIS** Describe and correct the error in finding $m\angle A$ in $\triangle ABC$ when $a = 19$, $b = 21$, and $c = 11$.

> ✗
>
> $$\cos A = \frac{19^2 - 21^2 - 11^2}{-2(19)(21)}$$
>
> $m\angle A \approx 75.4°$

COMPARING METHODS In Exercises 27–32, tell whether you would use the Law of Sines, the Law of Cosines, or the Pythagorean Theorem (Theorem 9.1) and trigonometric ratios to solve the triangle with the given information. Explain your reasoning. Then solve the triangle.

27. $m\angle A = 72°, m\angle B = 44°, b = 14$

28. $m\angle B = 98°, m\angle C = 37°, a = 18$

29. $m\angle C = 65°, a = 12, b = 21$

30. $m\angle B = 90°, a = 15, c = 6$

31. $m\angle C = 40°, b = 27, c = 36$

32. $a = 34, b = 19, c = 27$

33. **MODELING WITH MATHEMATICS** You and your friend are standing on the baseline of a basketball court. You bounce a basketball to your friend, as shown in the diagram. What is the distance between you and your friend? *(See Example 8.)*

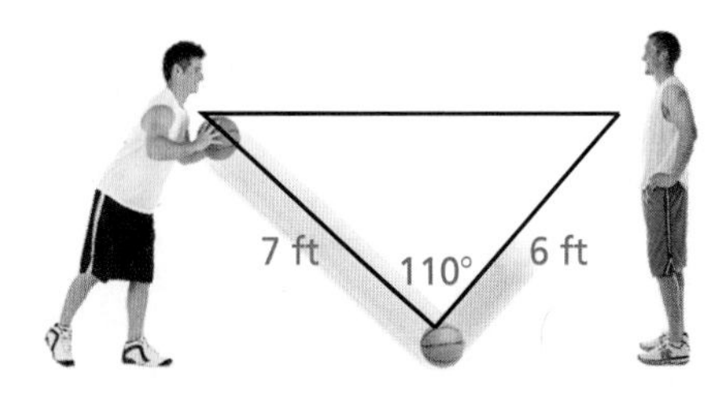

34. **MODELING WITH MATHEMATICS** A zip line is constructed across a valley, as shown in the diagram. What is the width w of the valley?

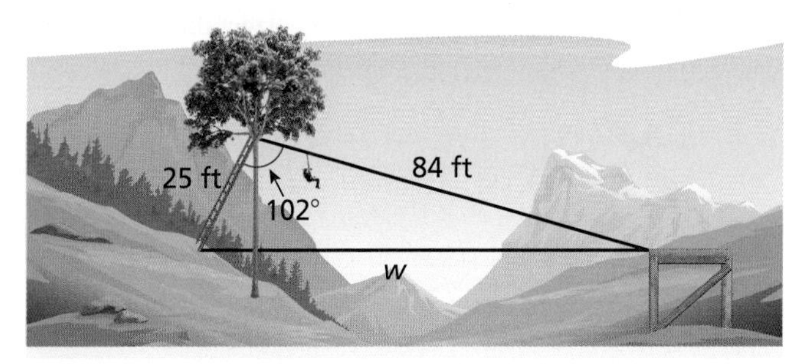

35. **MODELING WITH MATHEMATICS** You are on the observation deck of the Empire State Building looking at the Chrysler Building. When you turn 145° clockwise, you see the Statue of Liberty. You know that the Chrysler Building and the Empire State Building are about 0.6 mile apart and that the Chrysler Building and the Statue of Liberty are about 5.6 miles apart. Estimate the distance between the Empire State Building and the Statue of Liberty.

36. **MODELING WITH MATHEMATICS** The Leaning Tower of Pisa in Italy has a height of 183 feet and is 4° off vertical. Find the horizontal distance d that the top of the tower is off vertical.

37. **MAKING AN ARGUMENT** Your friend says that the Law of Sines can be used to find JK. Your cousin says that the Law of Cosines can be used to find JK. Who is correct? Explain your reasoning.

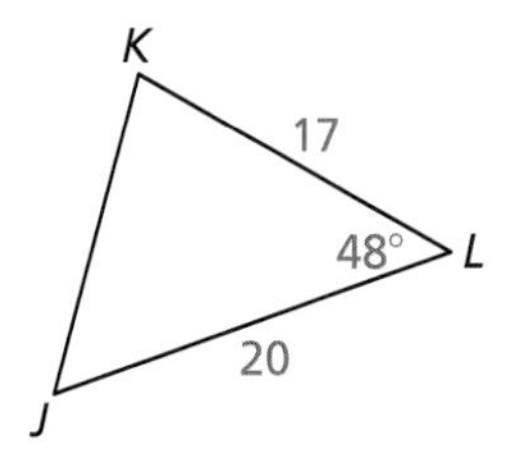

38. **REASONING** Use $\triangle XYZ$.

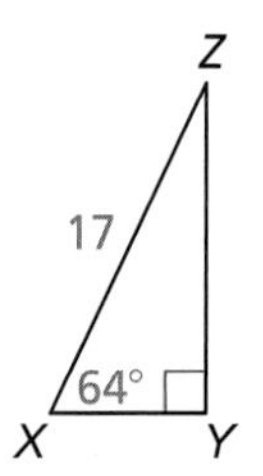

a. Can you use the Law of Sines to solve $\triangle XYZ$? Explain your reasoning.

b. Can you use another method to solve $\triangle XYZ$? Explain your reasoning.

39. MAKING AN ARGUMENT Your friend calculates the area of the triangle using the formula $A = \frac{1}{2}qr \sin S$ and says that the area is approximately 208.6 square units. Is your friend correct? Explain your reasoning.

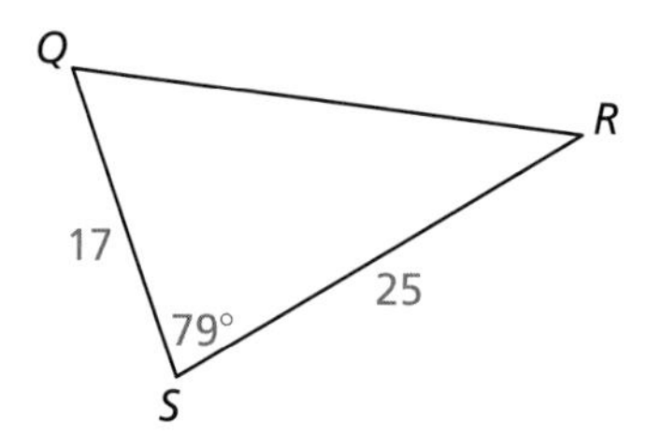

40. MODELING WITH MATHEMATICS You are fertilizing a triangular garden. One side of the garden is 62 feet long, and another side is 54 feet long. The angle opposite the 62-foot side is 58°.

a. Draw a diagram to represent this situation.

b. Use the Law of Sines to solve the triangle from part (a).

c. One bag of fertilizer covers an area of 200 square feet. How many bags of fertilizer will you need to cover the entire garden?

41. MODELING WITH MATHEMATICS A golfer hits a drive 260 yards on a hole that is 400 yards long. The shot is 15° off target.

Not drawn to scale

a. What is the distance x from the golfer's ball to the hole?

b. Assume the golfer is able to hit the ball precisely the distance found in part (a). What is the maximum angle θ (theta) by which the ball can be off target in order to land no more than 10 yards from the hole?

42. COMPARING METHODS A building is constructed on top of a cliff that is 300 meters high. A person standing on level ground below the cliff observes that the angle of elevation to the top of the building is 72° and the angle of elevation to the top of the cliff is 63°.

a. How far away is the person from the base of the cliff?

b. Describe two different methods you can use to find the height of the building. Use one of these methods to find the building's height.

43. MATHEMATICAL CONNECTIONS Find the values of x and y.

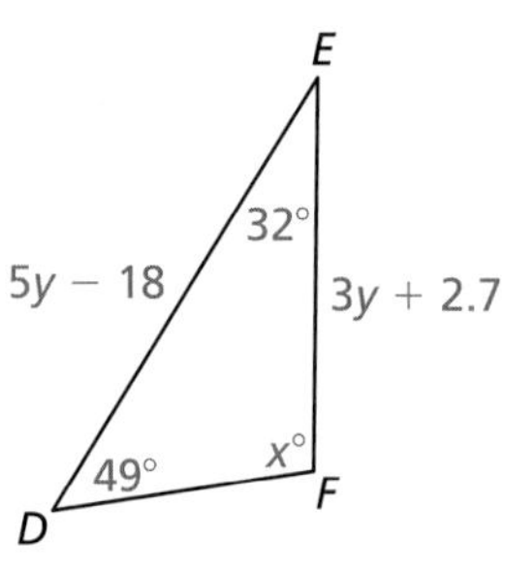

44. HOW DO YOU SEE IT? Would you use the Law of Sines or the Law of Cosines to solve the triangle?

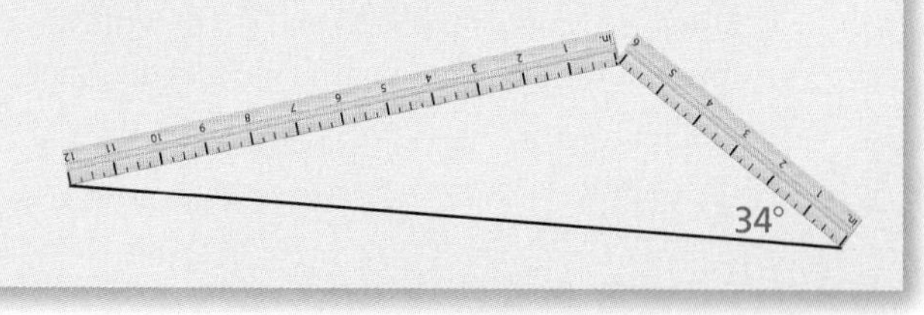

45. REWRITING A FORMULA Simplify the Law of Cosines for when the given angle is a right angle.

46. THOUGHT PROVOKING Consider any triangle with side lengths of a, b, and c. Calculate the value of s, which is half the perimeter of the triangle. What measurement of the triangle is represented by $\sqrt{s(s-a)(s-b)(s-c)}$?

47. ANALYZING RELATIONSHIPS The *ambiguous case* of the Law of Sines occurs when you are given the measure of one acute angle, the length of one adjacent side, and the length of the side opposite that angle, which is less than the length of the adjacent side. This results in two possible triangles. Using the given information, find two possible solutions for $\triangle ABC$. Draw a diagram for each triangle. (*Hint*: The inverse sine function gives only acute angle measures, so consider the acute angle and its supplement for $\angle B$.)

C
b a a
A B B

a. $m\angle A = 40°$, $a = 13$, $b = 16$

b. $m\angle A = 21°$, $a = 17$, $b = 32$

48. ABSTRACT REASONING Use the Law of Cosines to show that the measure of each angle of an equilateral triangle is 60°. Explain your reasoning.

49. CRITICAL THINKING An airplane flies 55° east of north from City A to City B, a distance of 470 miles. Another airplane flies 7° north of east from City A to City C, a distance of 890 miles. What is the distance between Cities B and C?

50. REWRITING A FORMULA Follow the steps to derive the formula for the area of a triangle,

Area $= \frac{1}{2}ab \sin C$.

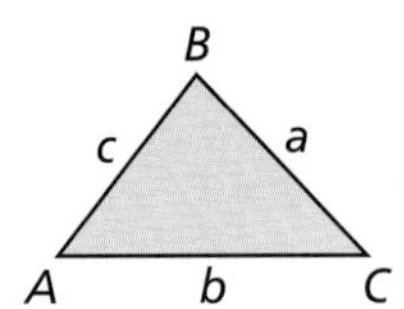

a. Draw the altitude from vertex B to $\overline{AC}$. Label the altitude as h. Write a formula for the area of the triangle using h.

b. Write an equation for $\sin C$.

c. Use the results of parts (a) and (b) to write a formula for the area of a triangle that does not include h.

51. PROVING A THEOREM Follow the steps to use the formula for the area of a triangle to prove the Law of Sines (Theorem 9.9).

a. Use the derivation in Exercise 50 to explain how to derive the three related formulas for the area of a triangle.

Area $= \frac{1}{2}bc \sin A$,

Area $= \frac{1}{2}ac \sin B$,

Area $= \frac{1}{2}ab \sin C$

b. Why can you use the formulas in part (a) to write the following statement?

$$\frac{1}{2}bc \sin A = \frac{1}{2}ac \sin B = \frac{1}{2}ab \sin C$$

c. Show how to rewrite the statement in part (b) to prove the Law of Sines. Justify each step.

52. PROVING A THEOREM Use the given information to complete the two-column proof of the Law of Cosines (Theorem 9.10).

Given $\overline{BD}$ is an altitude of $\triangle ABC$.

Prove $a^2 = b^2 + c^2 - 2bc \cos A$

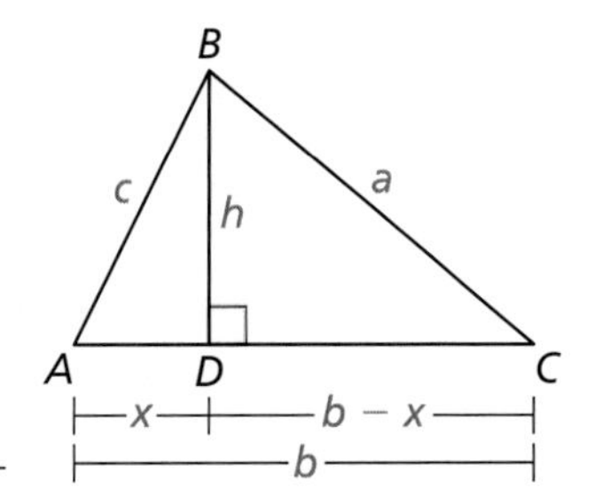

STATEMENTS	REASONS
1. $\overline{BD}$ is an altitude of $\triangle ABC$.	1. Given
2. $\triangle ADB$ and $\triangle CDB$ are right triangles.	2. ________________
3. $a^2 = (b - x)^2 + h^2$	3. ________________
4. ________________	4. Expand binomial.
5. $x^2 + h^2 = c^2$	5. ________________
6. ________________	6. Substitution Property of Equality
7. $\cos A = \frac{x}{c}$	7. ________________
8. $x = c \cos A$	8. ________________
9. $a^2 = b^2 + c^2 - 2bc \cos A$	9. ________________

Maintaining Mathematical Proficiency

Reviewing what you learned in previous grades and lessons

Find the radius and diameter of the circle. *(Skills Review Handbook)*

53.

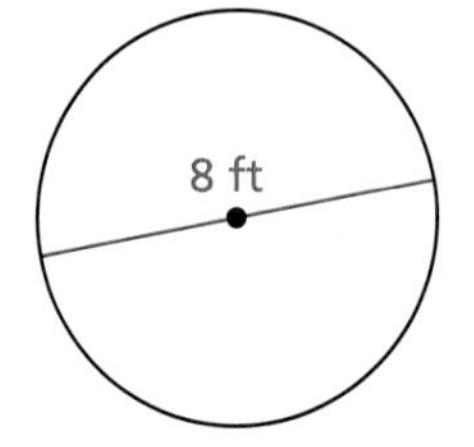

54.

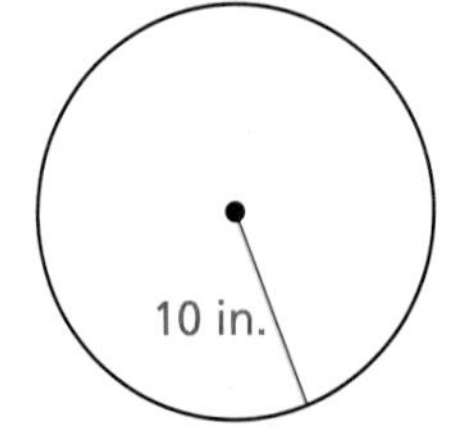

55.

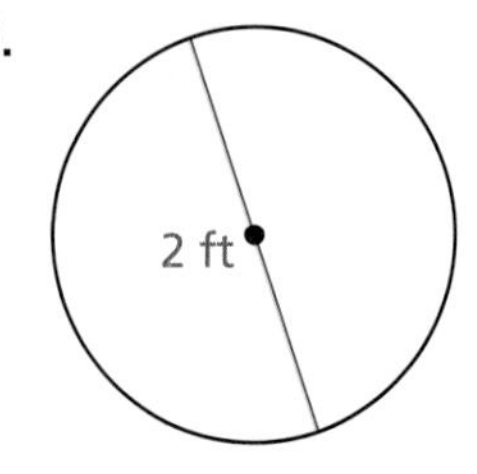

56.

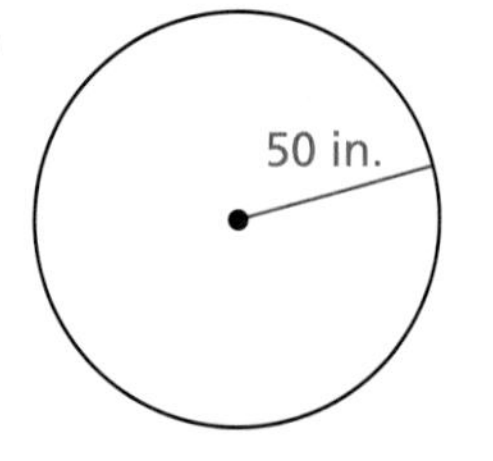

9.4–9.7 What Did You Learn?

Core Vocabulary

trigonometric ratio, *p. 488*
tangent, *p. 488*
angle of elevation, *p. 490*
sine, *p. 494*
cosine, *p. 494*
angle of depression, *p. 497*
inverse tangent, *p. 502*
inverse sine, *p. 502*
inverse cosine, *p. 502*
solve a right triangle, *p. 503*
Law of Sines, *p. 509*
Law of Cosines, *p. 511*

Core Concepts

Section 9.4

Tangent Ratio, *p. 488*

Section 9.5

Sine and Cosine Ratios, *p. 494*
Sine and Cosine of Complementary Angles, *p. 494*

Section 9.6

Inverse Trigonometric Ratios, *p. 502*
Solving a Right Triangle, *p. 503*

Section 9.7

Area of a Triangle, *p. 508*
Theorem 9.9 Law of Sines, *p. 509*
Theorem 9.10 Law of Cosines, *p. 511*

Mathematical Practices

1. In Exercise 21 on page 492, your brother claims that you could determine how far the overhang should extend by dividing 8 by tan 70°. Justify his conclusion and explain why it works.
2. In Exercise 29 on page 499, explain the flaw in the argument that the kite is 18.4 feet high.
3. In Exercise 31 on page 506, for what values does the inverse sine make sense?

Performance Task

Triathlon

There is a big triathlon in town, and you are trying to take pictures of your friends at multiple locations during the event. How far would you need to walk to move between the photography locations?

To explore the answers to this question and more, go to ***BigIdeasMath.com***.

9 Chapter Review

Dynamic Solutions available at *BigIdeasMath.com*

9.1 The Pythagorean Theorem *(pp. 463–470)*

Find the value of x. Then tell whether the side lengths form a Pythagorean triple.

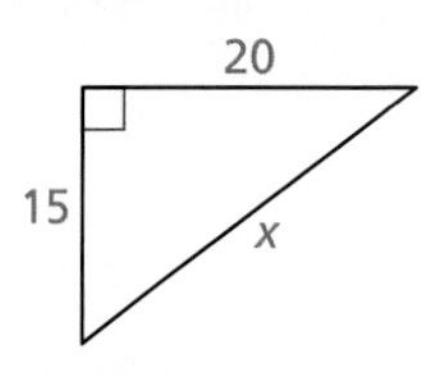

$c^2 = a^2 + b^2$	Pythagorean Theorem (Theorem 9.1)
$x^2 = 15^2 + 20^2$	Substitute.
$x^2 = 225 + 400$	Multiply.
$x^2 = 625$	Add.
$x = 25$	Find the positive square root.

▶ The value of x is 25. Because the side lengths 15, 20, and 25 are integers that satisfy the equation $c^2 = a^2 + b^2$, they form a Pythagorean triple.

Find the value of x. Then tell whether the side lengths form a Pythagorean triple.

1.

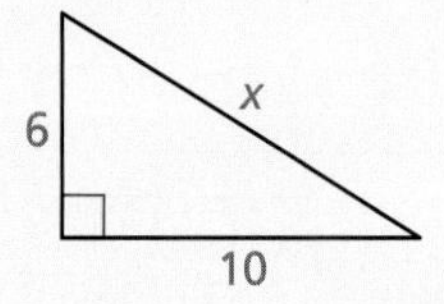

2.

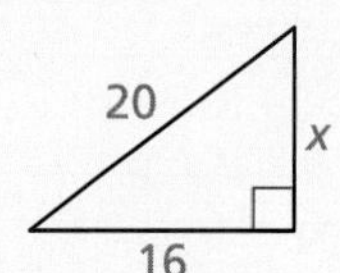

3.

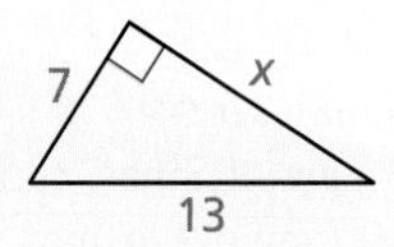

Verify that the segment lengths form a triangle. Is the triangle *acute*, *right*, or *obtuse*?

4. 6, 8, and 9

5. 10, $2\sqrt{2}$, and $6\sqrt{3}$

6. 13, 18, and $3\sqrt{55}$

9.2 Special Right Triangles *(pp. 471–476)*

Find the value of x. Write your answer in simplest form.

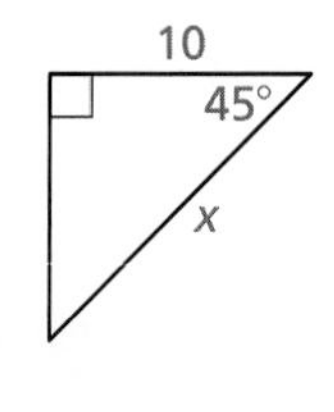

By the Triangle Sum Theorem (Theorem 5.1), the measure of the third angle must be 45°, so the triangle is a 45°-45°-90° triangle.

$\text{hypotenuse} = \text{leg} \cdot \sqrt{2}$	45°-45°-90° Triangle Theorem (Theorem 9.4)
$x = 10 \cdot \sqrt{2}$	Substitute.
$x = 10\sqrt{2}$	Simplify.

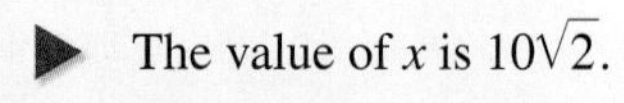

▶ The value of x is $10\sqrt{2}$.

Find the value of x. Write your answer in simplest form.

7.

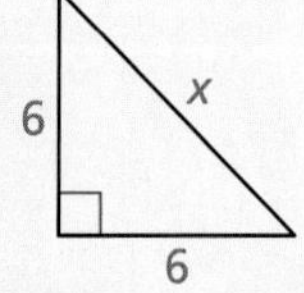

8.

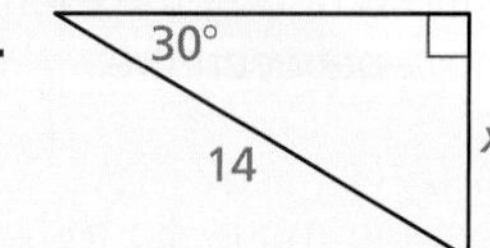

9.

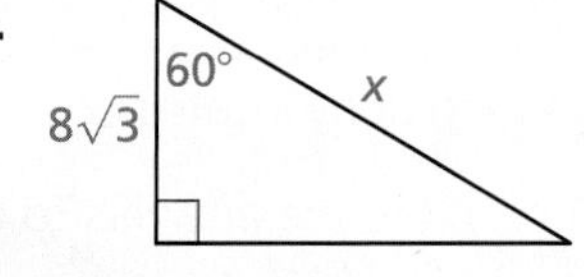

9.3 Similar Right Triangles (pp. 477–484)

Identify the similar triangles. Then find the value of x.

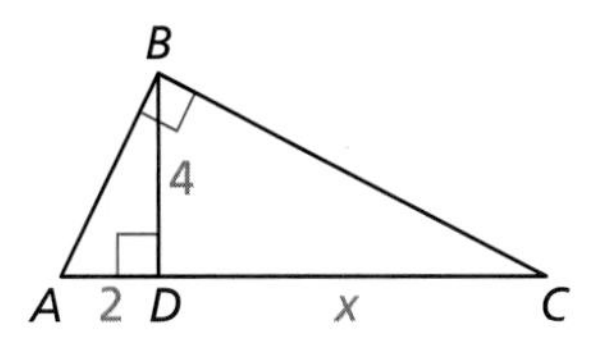

Sketch the three similar right triangles so that the corresponding angles and sides have the same orientation.

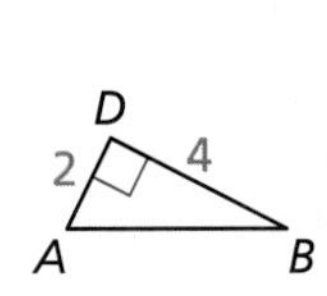

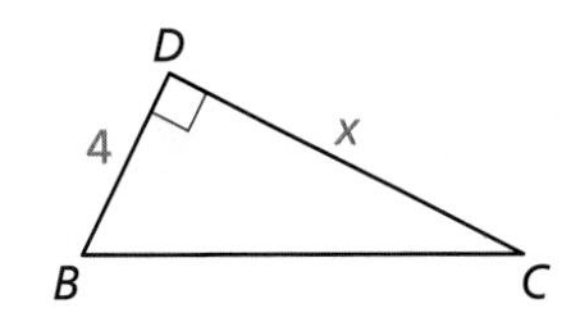

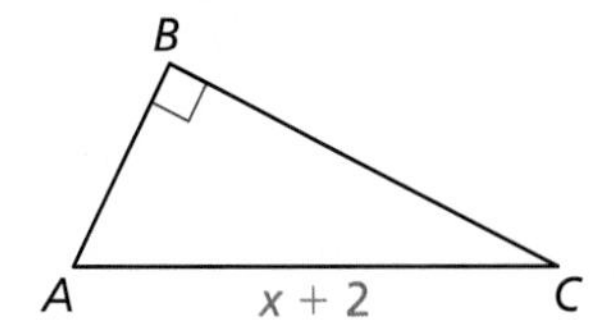

▶ $\triangle DBA \sim \triangle DCB \sim \triangle BCA$

By the Geometric Mean (Altitude) Theorem (Theorem 9.7), you know that 4 is the geometric mean of 2 and x.

$4^2 = 2 \cdot x$	Geometric Mean (Altitude) Theorem
$16 = 2x$	Square 4.
$8 = x$	Divide each side by 2.

▶ The value of x is 8.

Identify the similar triangles. Then find the value of x.

10.

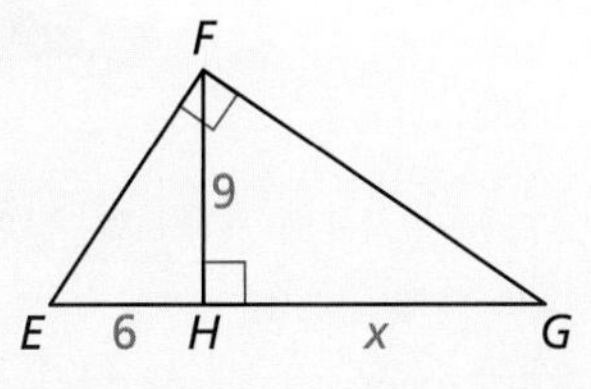

11.

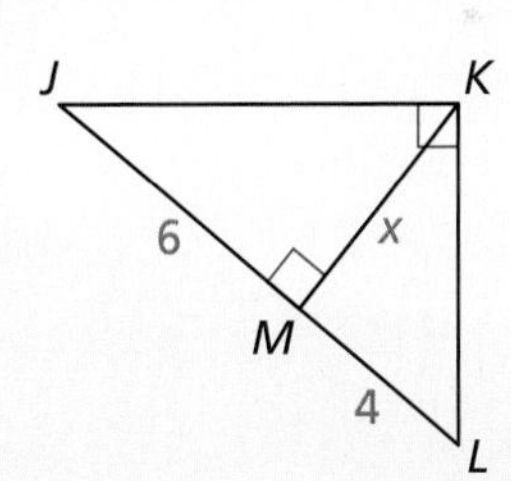

12.

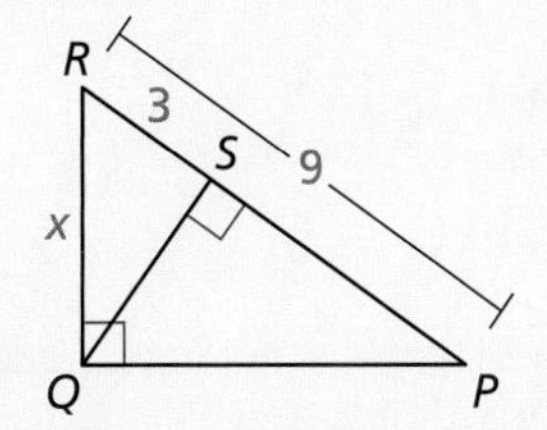

13.

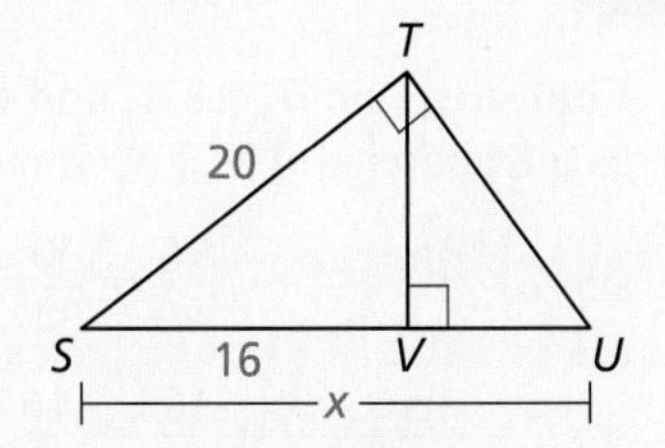

Find the geometric mean of the two numbers.

14. 9 and 25 **15.** 36 and 48 **16.** 12 and 42

9.4 The Tangent Ratio (pp. 487–492)

Find tan M and tan N. Write each answer as a fraction and as a decimal rounded to four places.

$$\tan M = \frac{\text{opp. } \angle M}{\text{adj. to } \angle M} = \frac{LN}{LM} = \frac{6}{8} = \frac{3}{4} = 0.7500$$

$$\tan N = \frac{\text{opp. } \angle N}{\text{adj. to } \angle N} = \frac{LM}{LN} = \frac{8}{6} = \frac{4}{3} \approx 1.3333$$

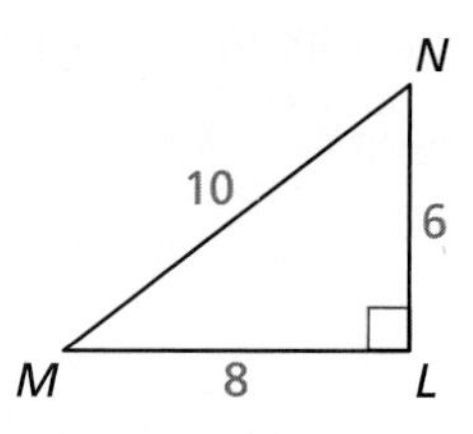

Find the tangents of the acute angles in the right triangle. Write each answer as a fraction and as a decimal rounded to four decimal places.

17.

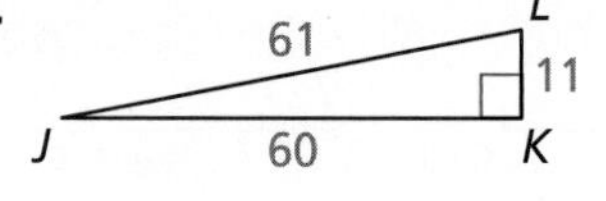

18.

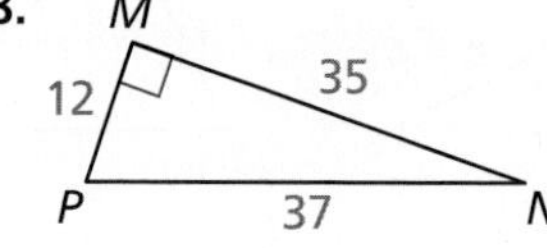

19.

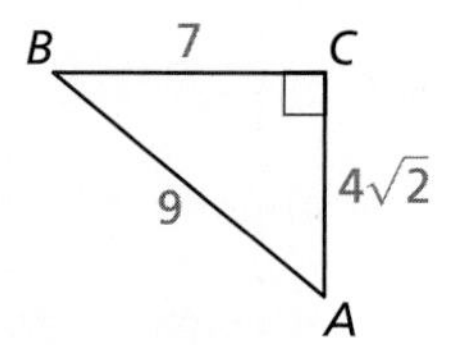

Find the value of x. Round your answer to the nearest tenth.

20.

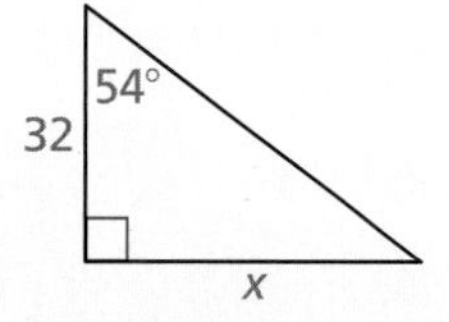

21.

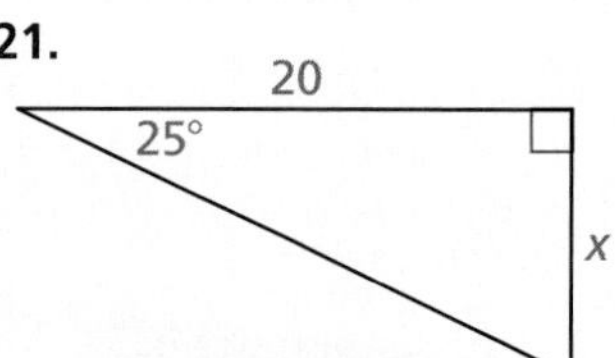

22.

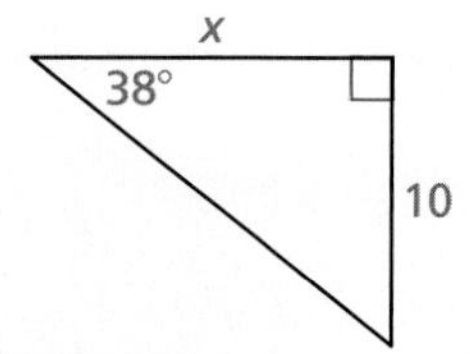

23. The angle between the bottom of a fence and the top of a tree is 75°. The tree is 4 feet from the fence. How tall is the tree? Round your answer to the nearest foot.

9.5 The Sine and Cosine Ratios (pp. 493–500)

Find sin A, sin B, cos A, and cos B. Write each answer as a fraction and as a decimal rounded to four places.

$$\sin A = \frac{\text{opp. } \angle A}{\text{hyp.}} = \frac{BC}{AB} = \frac{30}{34} = \frac{15}{17} \approx 0.8824$$

$$\sin B = \frac{\text{opp. } \angle B}{\text{hyp.}} = \frac{AC}{AB} = \frac{16}{34} = \frac{8}{17} \approx 0.4706$$

$$\cos A = \frac{\text{adj. to } \angle A}{\text{hyp.}} = \frac{AC}{AB} = \frac{16}{34} = \frac{8}{17} \approx 0.4706$$

$$\cos B = \frac{\text{adj. to } \angle B}{\text{hyp.}} = \frac{BC}{AB} = \frac{30}{34} = \frac{15}{17} \approx 0.8824$$

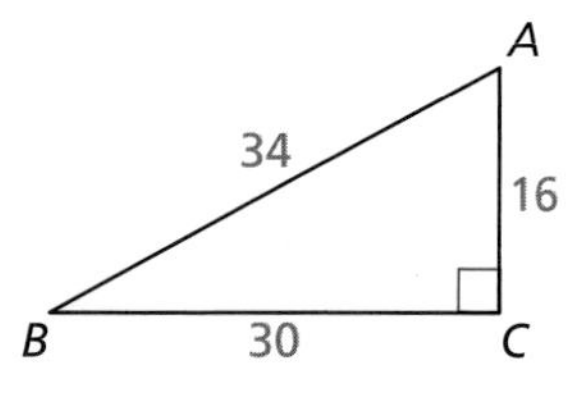

Find sin X, sin Z, cos X, and cos Z. Write each answer as a fraction and as a decimal rounded to four decimal places.

24.

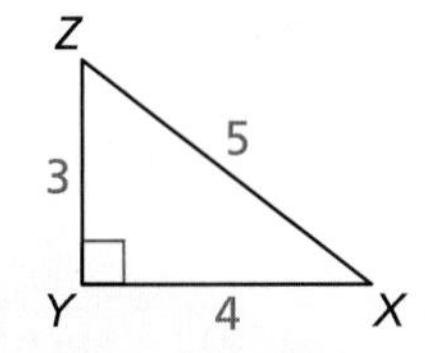

25.

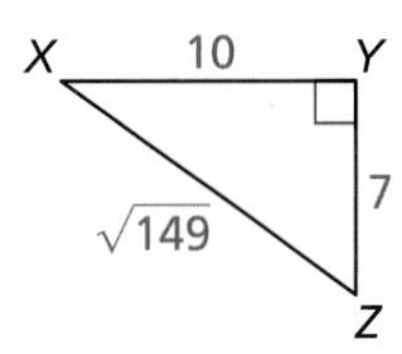

26.

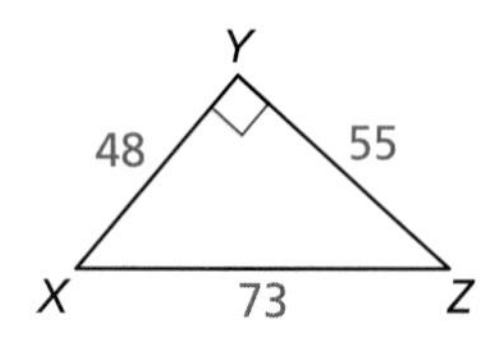

Find the value of each variable using sine and cosine. Round your answers to the nearest tenth.

27.

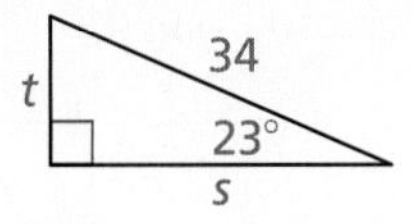

28.

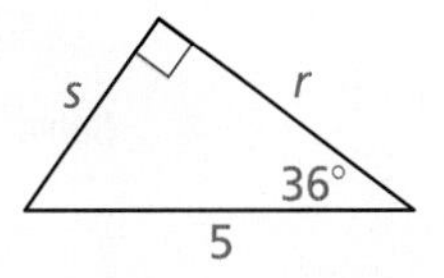

29.

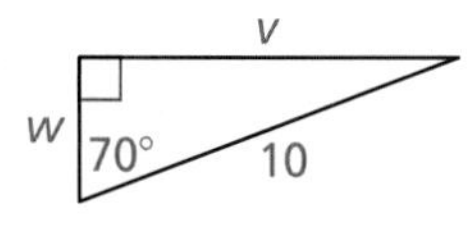

30. Write sin 72° in terms of cosine.

31. Write cos 29° in terms of sine.

9.6 Solving Right Triangles *(pp. 501–506)*

Solve the right triangle. Round decimal answers to the nearest tenth.

A, c, 12, B, 19, C

Step 1 Use the Pythagorean Theorem (Theorem 9.1) to find the length of the hypotenuse.

$c^2 = a^2 + b^2$	Pythagorean Theorem
$c^2 = 19^2 + 12^2$	Substitute.
$c^2 = 505$	Simplify.
$c = \sqrt{505}$	Find the positive square root.
$c \approx 22.5$	Use a calculator.

Step 2 Find $m\angle B$.

$$m\angle B = \tan^{-1}\frac{12}{19} \approx 32.3^\circ \qquad \text{Use a calculator.}$$

Step 3 Find $m\angle A$.

Because $\angle A$ and $\angle B$ are complements, you can write $m\angle A = 90^\circ - m\angle B \approx 90^\circ - 32.3^\circ = 57.7^\circ$.

▶ In $\triangle ABC$, $c \approx 22.5$, $m\angle B \approx 32.3^\circ$, and $m\angle A \approx 57.7^\circ$.

Let $\angle Q$ be an acute angle. Use a calculator to approximate the measure of $\angle Q$ to the nearest tenth of a degree.

32. $\cos Q = 0.32$

33. $\sin Q = 0.91$

34. $\tan Q = 0.04$

Solve the right triangle. Round decimal answers to the nearest tenth.

35.

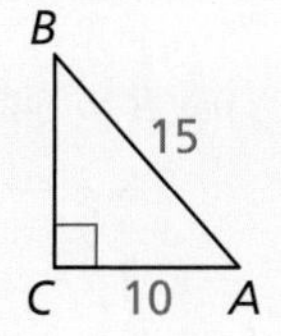

36.

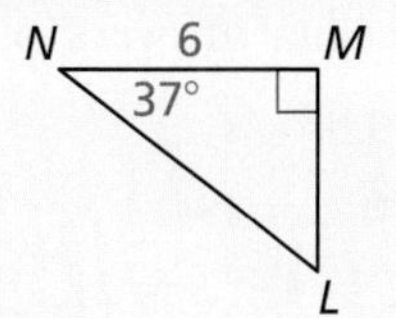

37.

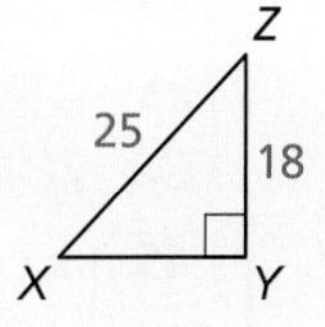

9.7 Law of Sines and Law of Cosines (pp. 507–516)

Solve the triangle. Round decimal answers to the nearest tenth.

a.

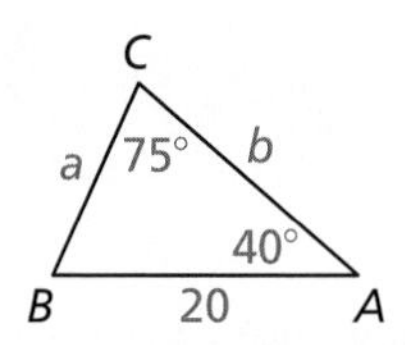

By the Triangle Sum Theorem (Theorem 5.1), $m\angle B = 180° - 40° - 75° = 65°$.

By the Law of Sines, you can write $\dfrac{a}{\sin 40°} = \dfrac{b}{\sin 65°} = \dfrac{20}{\sin 75°}$.

$\dfrac{a}{\sin 40°} = \dfrac{20}{\sin 75°}$	Write two equations, each with one variable.	$\dfrac{b}{\sin 65°} = \dfrac{20}{\sin 75°}$
$a = \dfrac{20 \sin 40°}{\sin 75°}$	Solve for each variable.	$b = \dfrac{20 \sin 65°}{\sin 75°}$
$a \approx 13.3$	Use a calculator.	$b \approx 18.8$

▶ In $\triangle ABC$, $m\angle B = 65°$, $a \approx 13.3$, and $b \approx 18.8$.

b.

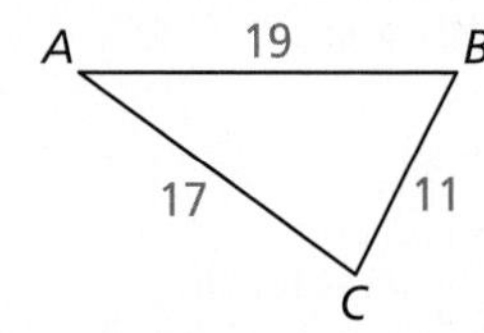

First, find the angle opposite the longest side, $\overline{AB}$. Use the Law of Cosines to find $m\angle C$.

$$19^2 = 11^2 + 17^2 - 2(11)(17) \cos C \qquad \text{Law of Cosines}$$

$$\frac{19^2 - 11^2 - 17^2}{-2(11)(17)} = \cos C \qquad \text{Solve for } \cos C.$$

$$m\angle C \approx 82.5° \qquad \text{Use a calculator.}$$

Now, use the Law of Sines to find $m\angle A$.

$$\frac{\sin A}{a} = \frac{\sin C}{c} \qquad \text{Law of Sines}$$

$$\frac{\sin A}{11} = \frac{\sin 82.5°}{19} \qquad \text{Substitute.}$$

$$\sin A = \frac{11 \sin 82.5°}{19} \qquad \text{Multiply each side by 11.}$$

$$m\angle A \approx 35.0° \qquad \text{Use a calculator.}$$

By the Triangle Sum Theorem (Theorem 5.1), $m\angle B \approx 180° - 35.0° - 82.5° = 62.5°$.

▶ In $\triangle ABC$, $m\angle A \approx 35.0°$, $m\angle B \approx 62.5°$, and $m\angle C \approx 82.5°$.

Find the area of $\triangle ABC$ with the given side lengths and included angle.

38. $m\angle B = 124°, a = 9, c = 11$ **39.** $m\angle A = 68°, b = 13, c = 7$ **40.** $m\angle C = 79°, a = 25, b = 17$

Solve $\triangle ABC$. Round decimal answers to the nearest tenth.

41. $m\angle A = 112°, a = 9, b = 4$

42. $m\angle A = 28°, m\angle B = 64°, c = 55$

43. $m\angle C = 48°, b = 20, c = 28$

44. $m\angle B = 25°, a = 8, c = 3$

45. $m\angle B = 102°, m\angle C = 43°, b = 21$

46. $a = 10, b = 3, c = 12$

9 Chapter Test

Find the value of each variable. Round your answers to the nearest tenth.

1. 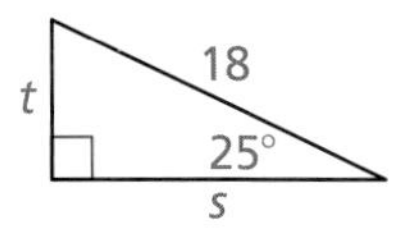

2. 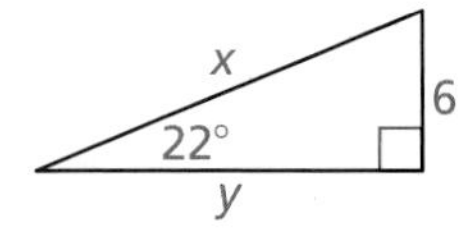

3. 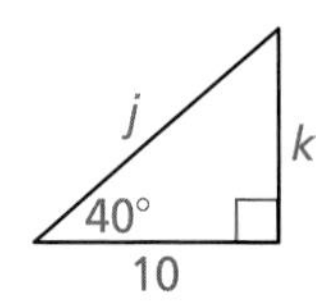

Verify that the segment lengths form a triangle. Is the triangle *acute*, *right*, or *obtuse*?

4. 16, 30, and 34

5. 4, $\sqrt{67}$, and 9

6. $\sqrt{5}$, 5, and 5.5

Solve $\triangle ABC$. Round decimal answers to the nearest tenth.

7. 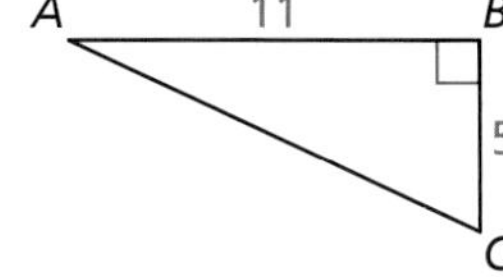

8. 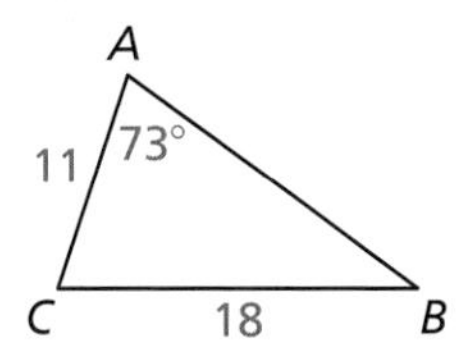

9. 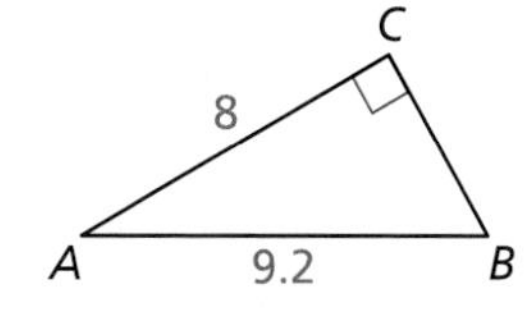

10. $m\angle A = 103°, b = 12, c = 24$

11. $m\angle A = 26°, m\angle C = 35°, b = 13$

12. $a = 38, b = 31, c = 35$

13. Write cos 53° in terms of sine.

Find the value of each variable. Write your answers in simplest form.

14. 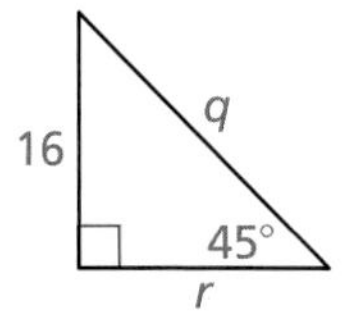

15. 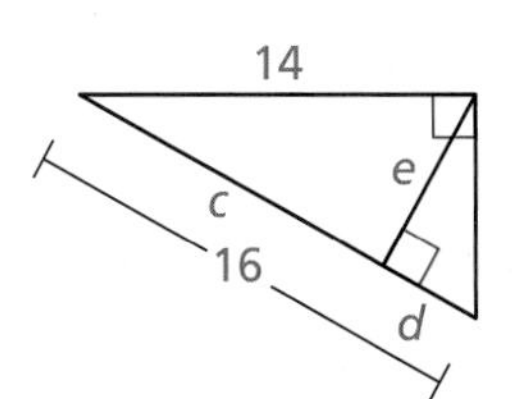

16. 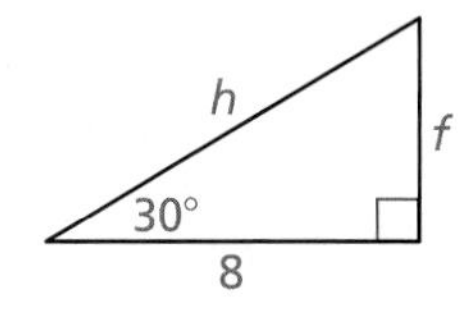

17. In $\triangle QRS$, $m\angle R = 57°$, $q = 9$, and $s = 5$. Find the area of $\triangle QRS$.

18. You are given the measures of both acute angles of a right triangle. Can you determine the side lengths? Explain.

19. You are at a parade looking up at a large balloon floating directly above the street. You are 60 feet from a point on the street directly beneath the balloon. To see the top of the balloon, you look up at an angle of 53°. To see the bottom of the balloon, you look up at an angle of 29°. Estimate the height h of the balloon.

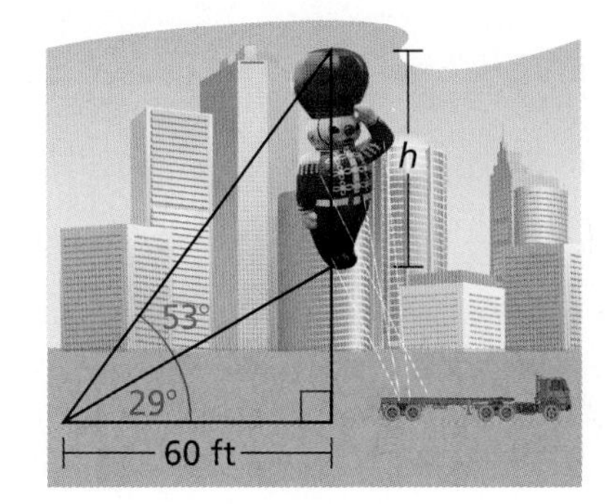

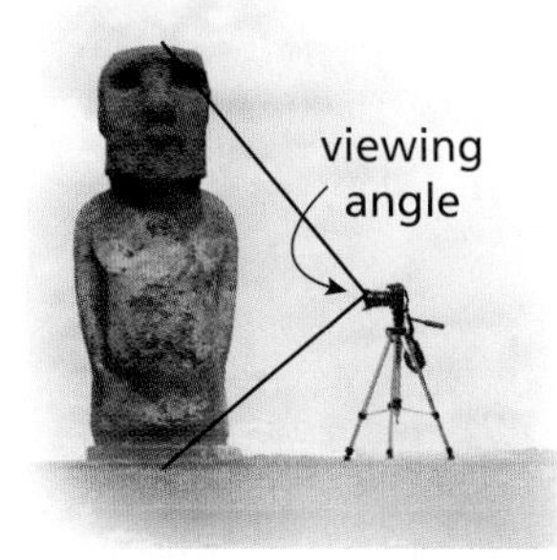

20. You want to take a picture of a statue on Easter Island, called a *moai.* The moai is about 13 feet tall. Your camera is on a tripod that is 5 feet tall. The vertical viewing angle of your camera is set at 90°. How far from the moai should you stand so that the entire height of the moai is perfectly framed in the photo?

9 Cumulative Assessment

1. The size of a laptop screen is measured by the length of its diagonal. You want to purchase a laptop with the largest screen possible. Which laptop should you buy?

Ⓐ
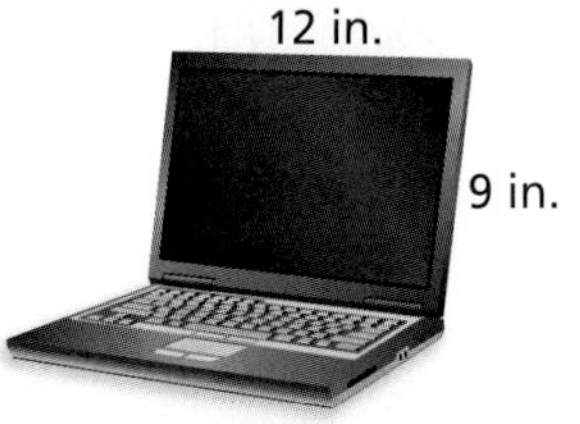

Ⓑ
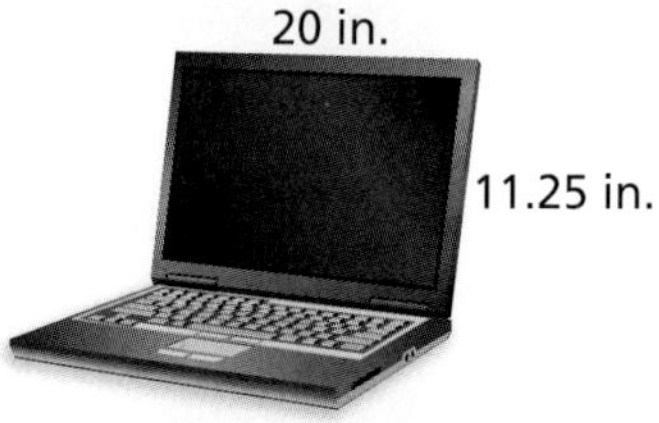

Ⓒ
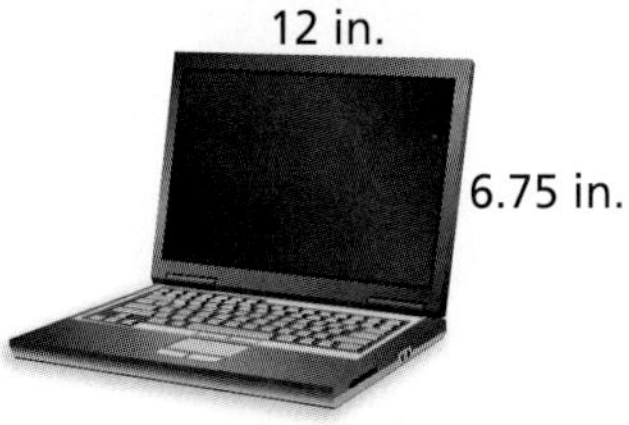

Ⓓ

2. In $\triangle PQR$ and $\triangle SQT$, S is between P and Q, T is between R and Q, and $\frac{QS}{SP} = \frac{QT}{TR}$. What must be true about $\overline{ST}$ and $\overline{PR}$? Select all that apply.

$\overline{ST} \perp \overline{PR}$ | $\overline{ST} \parallel \overline{PR}$ | $ST = PR$ | $ST = \frac{1}{2}PR$

3. In the diagram, $\triangle JKL \sim \triangle QRS$. Choose the symbol that makes each statement true.

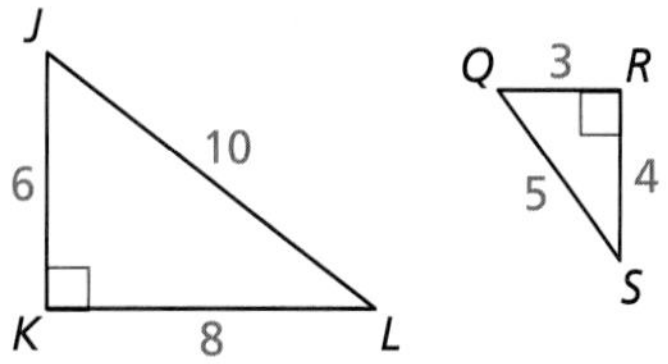

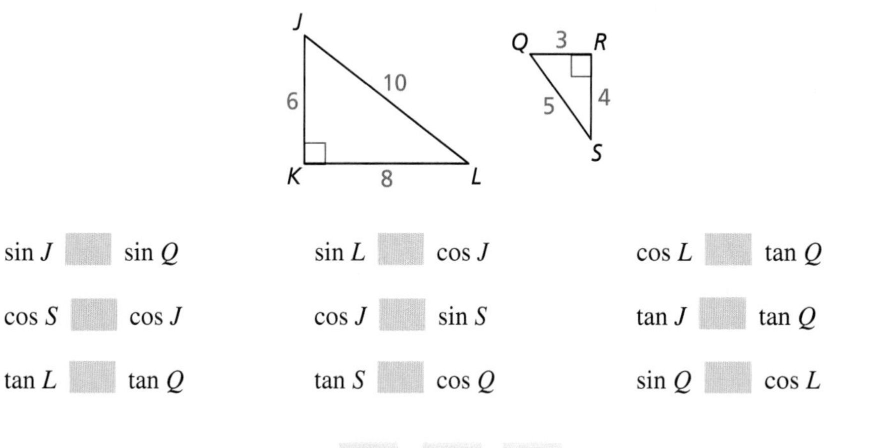

$\sin J$ ☐ $\sin Q$	$\sin L$ ☐ $\cos J$	$\cos L$ ☐ $\tan Q$
$\cos S$ ☐ $\cos J$	$\cos J$ ☐ $\sin S$	$\tan J$ ☐ $\tan Q$
$\tan L$ ☐ $\tan Q$	$\tan S$ ☐ $\cos Q$	$\sin Q$ ☐ $\cos L$

< | = | >

4. A surveyor makes the measurements shown. What is the width of the river?

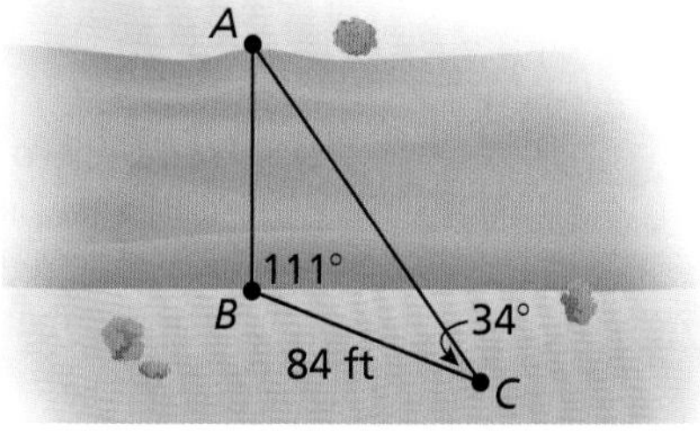

5. Create as many true equations as possible.

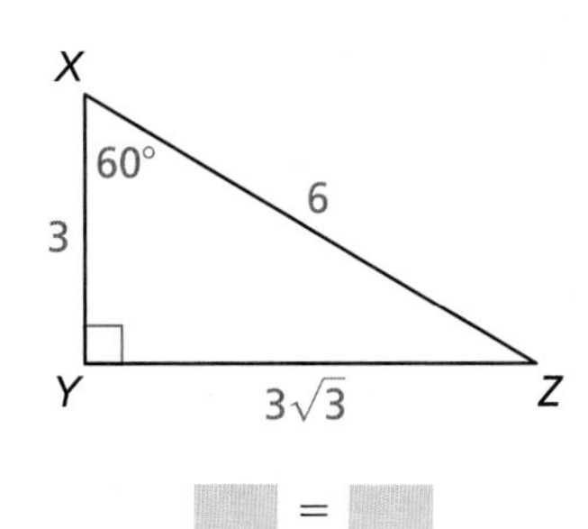

▭ = ▭

$\sin X$	$\cos X$	$\tan X$	$\frac{XY}{XZ}$	$\frac{YZ}{XZ}$
$\sin Z$	$\cos Z$	$\tan Z$	$\frac{XY}{YZ}$	$\frac{YZ}{XY}$

6. Prove that quadrilateral *DEFG* is a kite.

Given $\overline{HE} \cong \overline{HG}$, $\overline{EG} \perp \overline{DF}$

Prove $\overline{FE} \cong \overline{FG}$, $\overline{DE} \cong \overline{DG}$

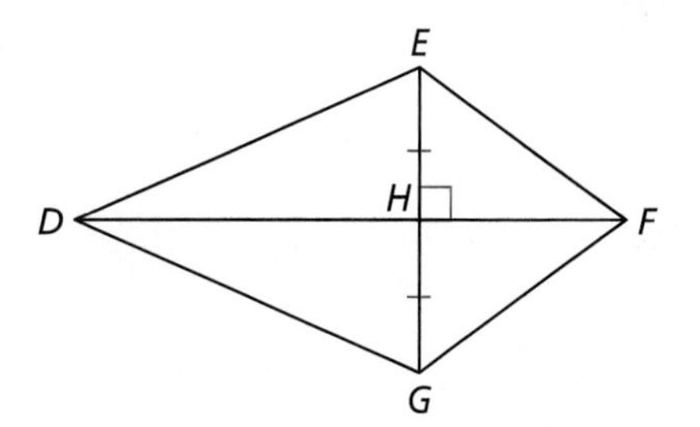

7. What are the coordinates of the vertices of the image of $\triangle QRS$ after the composition of transformations shown?

Translation: $(x, y) \rightarrow (x + 2, y + 3)$
Rotation: 180° about the origin

Ⓐ $Q'(1, 2)$, $R'(5, 4)$, $S'(4, -1)$

Ⓑ $Q'(-1, -2)$, $R'(-5, -4)$, $S'(-4, 1)$

Ⓒ $Q'(3, -2)$, $R'(-1, -4)$, $S'(0, 1)$

Ⓓ $Q'(-2, 1)$, $R'(-4, 5)$, $S'(1, 4)$

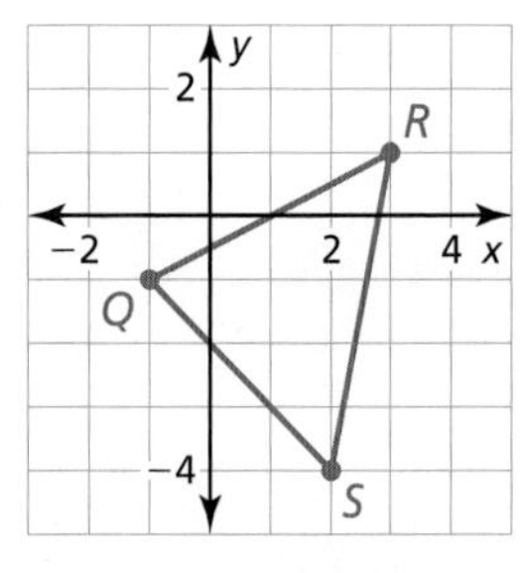

8. The Red Pyramid in Egypt has a square base. Each side of the base measures 722 feet. The height of the pyramid is 343 feet.

a. Use the side length of the base, the height of the pyramid, and the Pythagorean Theorem to find the *slant height*, *AB*, of the pyramid.

b. Find *AC*.

c. Name three possible ways of finding $m\angle 1$. Then, find $m\angle 1$.

10 Circles

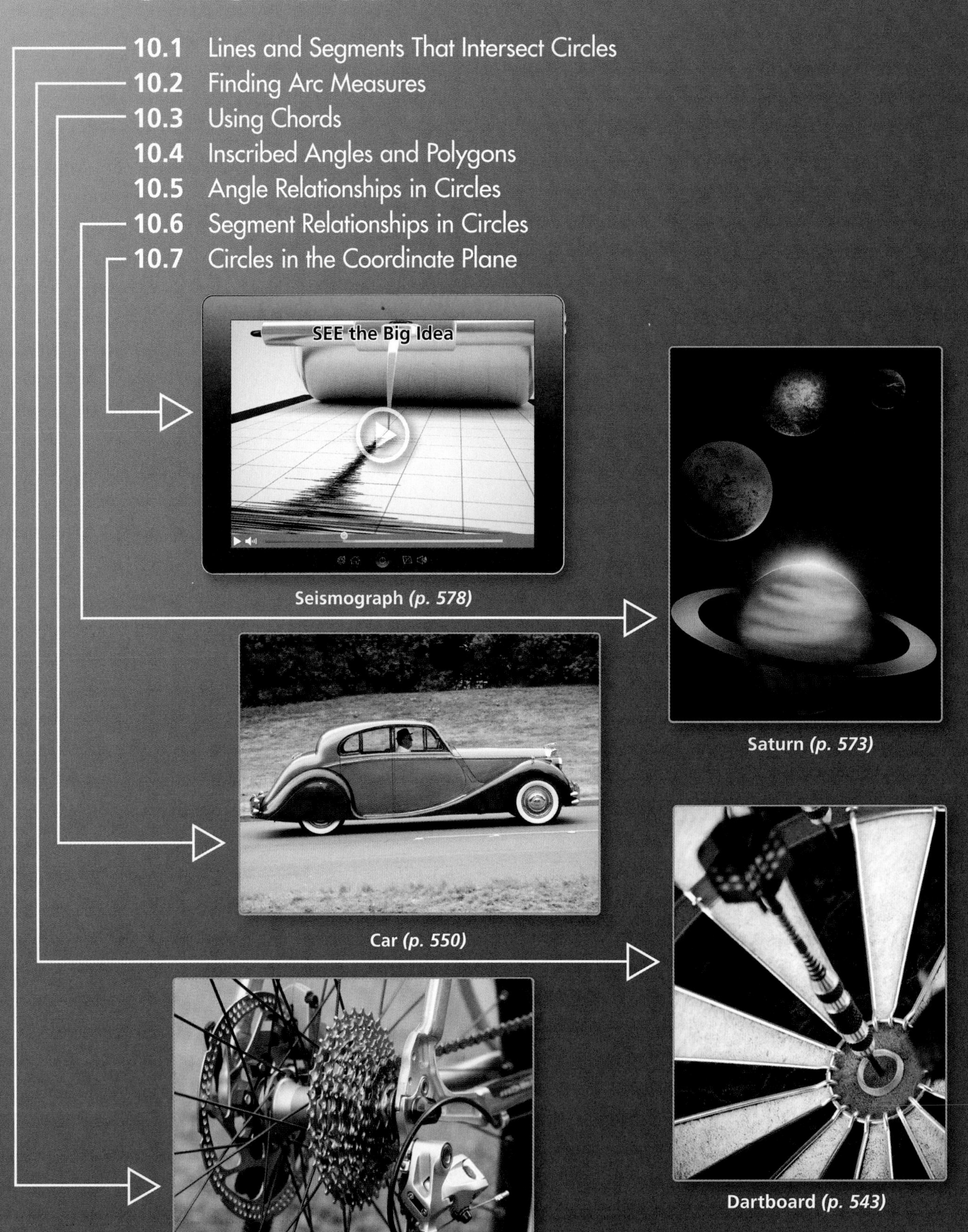

Seismograph (*p. 578*)

Saturn (*p. 573*)

Car (*p. 550*)

Dartboard (*p. 543*)

Bicycle Chain (*p. 535*)

Maintaining Mathematical Proficiency

Multiplying Binomials

Example 1 **Find the product $(x + 3)(2x - 1)$.**

$$\begin{aligned} & \quad\ \text{First} \quad \text{Outer} \quad \text{Inner} \quad \text{Last} \\ (x + 3)(2x - 1) &= x(2x) + x(-1) + 3(2x) + (3)(-1) && \text{FOIL Method} \\ &= 2x^2 + (-x) + 6x + (-3) && \text{Multiply.} \\ &= 2x^2 + 5x - 3 && \text{Simplify.} \end{aligned}$$

▶ The product is $2x^2 + 5x - 3$.

Find the product.

1. $(x + 7)(x + 4)$

2. $(a + 1)(a - 5)$

3. $(q - 9)(3q - 4)$

4. $(2v - 7)(5v + 1)$

5. $(4h + 3)(2 + h)$

6. $(8 - 6b)(5 - 3b)$

Solving Quadratic Equations by Completing the Square

Example 2 **Solve $x^2 + 8x - 3 = 0$ by completing the square.**

$$\begin{aligned} x^2 + 8x - 3 &= 0 && \text{Write original equation.} \\ x^2 + 8x &= 3 && \text{Add 3 to each side.} \\ x^2 + 8x + 4^2 &= 3 + 4^2 && \text{Complete the square by adding } \left(\tfrac{8}{2}\right)^2\text{, or } 4^2\text{, to each side.} \\ (x + 4)^2 &= 19 && \text{Write the left side as a square of a binomial.} \\ x + 4 &= \pm\sqrt{19} && \text{Take the square root of each side.} \\ x &= -4 \pm\sqrt{19} && \text{Subtract 4 from each side.} \end{aligned}$$

▶ The solutions are $x = -4 + \sqrt{19} \approx 0.36$ and $x = -4 - \sqrt{19} \approx -8.36$.

Solve the equation by completing the square. Round your answer to the nearest hundredth, if necessary.

7. $x^2 - 2x = 5$

8. $r^2 + 10r = -7$

9. $w^2 - 8w = 9$

10. $p^2 + 10p - 4 = 0$

11. $k^2 - 4k - 7 = 0$

12. $-z^2 + 2z = 1$

13. **ABSTRACT REASONING** Write an expression that represents the product of two consecutive positive odd integers. Explain your reasoning.

Mathematical Practices

Mathematically proficient students make sense of problems and do not give up when faced with challenges.

Analyzing Relationships of Circles

Core Concept

Circles and Tangent Circles

A **circle** is the set of all points in a plane that are equidistant from a given point called the **center** of the circle. A circle with center D is called "circle D" and can be written as $\odot D$.

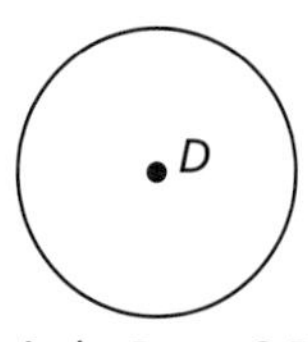

circle D, or $\odot D$

Coplanar circles that intersect in one point are called **tangent circles**.

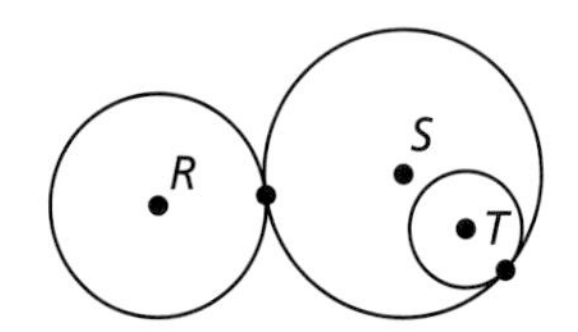

$\odot R$ and $\odot S$ are tangent circles.
$\odot S$ and $\odot T$ are tangent circles.

EXAMPLE 1 Relationships of Circles and Tangent Circles

a. Each circle at the right consists of points that are 3 units from the center. What is the greatest distance from any point on $\odot A$ to any point on $\odot B$?

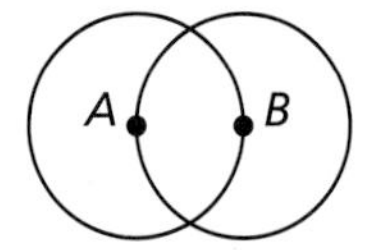

b. Three circles, $\odot C$, $\odot D$, and $\odot E$, consist of points that are 3 units from their centers. The centers C, D, and E of the circles are collinear, $\odot C$ is tangent to $\odot D$, and $\odot D$ is tangent to $\odot E$. What is the distance from $\odot C$ to $\odot E$?

SOLUTION

a. Because the points on each circle are 3 units from the center, the greatest distance from any point on $\odot A$ to any point on $\odot B$ is $3 + 3 + 3 = 9$ units.

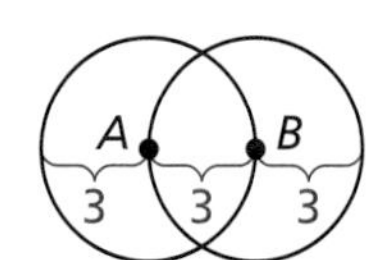

b. Because C, D, and E are collinear, $\odot C$ is tangent to $\odot D$, and $\odot D$ is tangent to $\odot E$, the circles are as shown. So, the distance from $\odot C$ to $\odot E$ is $3 + 3 = 6$ units.

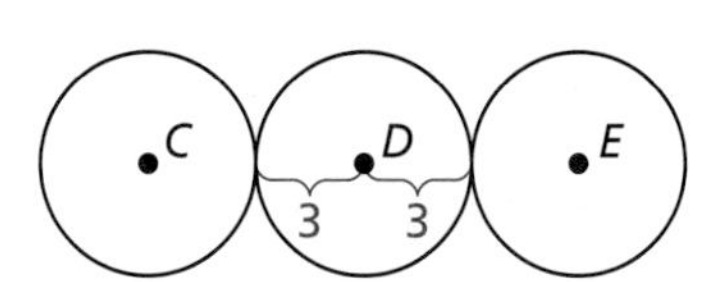

Monitoring Progress

Let $\odot A$, $\odot B$, and $\odot C$ consist of points that are 3 units from the centers.

1. Draw $\odot C$ so that it passes through points A and B in the figure at the right. Explain your reasoning.

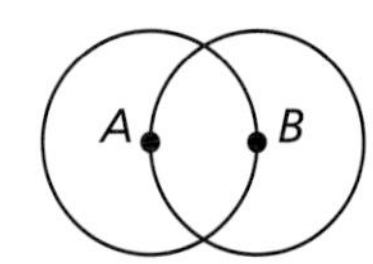

2. Draw $\odot A$, $\odot B$, and $\odot C$ so that each is tangent to the other two. Draw a larger circle, $\odot D$, that is tangent to each of the other three circles. Is the distance from point D to a point on $\odot D$ *less than*, *greater than*, or *equal to* 6? Explain.

10.1 Lines and Segments That Intersect Circles

Essential Question What are the definitions of the lines and segments that intersect a circle?

EXPLORATION 1 **Lines and Line Segments That Intersect Circles**

Work with a partner. The drawing at the right shows five lines or segments that intersect a circle. Use the relationships shown to write a definition for each type of line or segment. Then use the Internet or some other resource to verify your definitions.

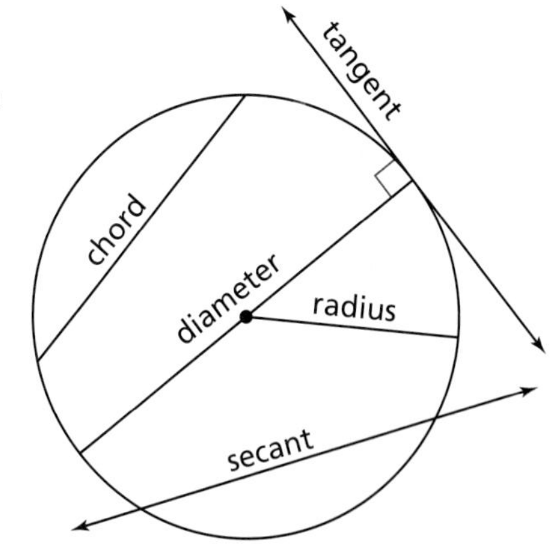

Chord:

Secant:

Tangent:

Radius:

Diameter:

EXPLORATION 2 **Using String to Draw a Circle**

Work with a partner. Use two pencils, a piece of string, and a piece of paper.

a. Tie the two ends of the piece of string loosely around the two pencils.

b. Anchor one pencil on the paper at the center of the circle. Use the other pencil to draw a circle around the anchor point while using slight pressure to keep the string taut. Do not let the string wind around either pencil.

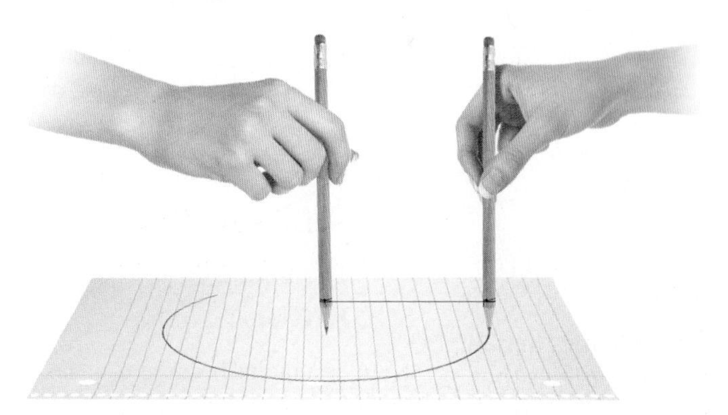

REASONING ABSTRACTLY

To be proficient in math, you need to know and flexibly use different properties of operations and objects.

c. Explain how the distance between the two pencil points as you draw the circle is related to two of the lines or line segments you defined in Exploration 1.

Communicate Your Answer

3. What are the definitions of the lines and segments that intersect a circle?

4. Of the five types of lines and segments in Exploration 1, which one is a subset of another? Explain.

5. Explain how to draw a circle with a diameter of 8 inches.

10.1 Lesson

What You Will Learn

- Identify special segments and lines.
- Draw and identify common tangents.
- Use properties of tangents.

Core Vocabulary

Identifying Special Segments and Lines

A **circle** is the set of all points in a plane that are equidistant from a given point called the **center** of the circle. A circle with center P is called "circle P" and can be written as $\odot P$.

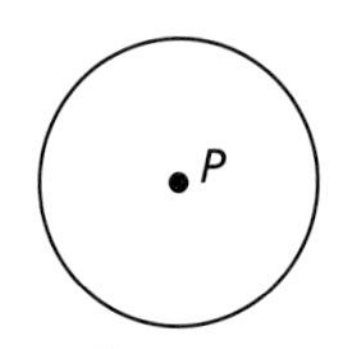

circle P, or $\odot P$

Core Concept

Lines and Segments That Intersect Circles

A segment whose endpoints are the center and any point on a circle is a **radius**.

A **chord** is a segment whose endpoints are on a circle. A **diameter** is a chord that contains the center of the circle.

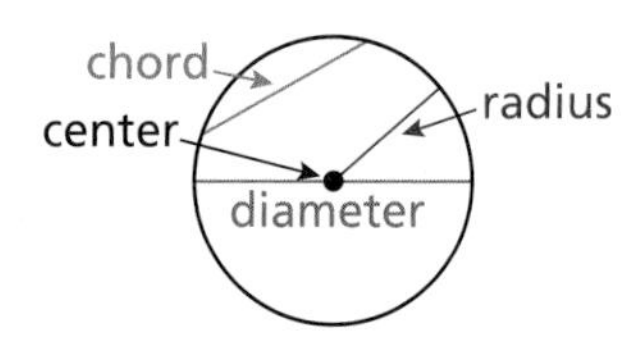

A **secant** is a line that intersects a circle in two points.

A **tangent** is a line in the plane of a circle that intersects the circle in exactly one point, the **point of tangency**. The *tangent ray* $\overrightarrow{AB}$ and the *tangent segment* $\overline{AB}$ are also called tangents.

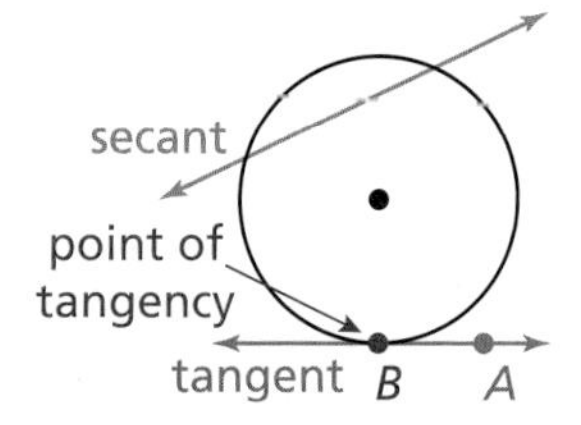

READING

The words "radius" and "diameter" refer to lengths as well as segments. For a given circle, think of *a* radius and *a* diameter as segments and *the* radius and *the* diameter as lengths.

EXAMPLE 1 Identifying Special Segments and Lines

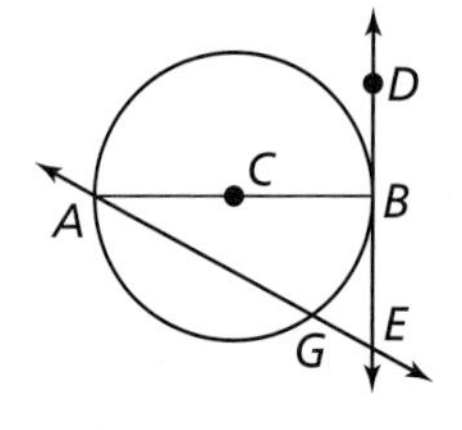

Tell whether the line, ray, or segment is best described as a *radius*, *chord*, *diameter*, *secant*, or *tangent* of $\odot C$.

a. $\overline{AC}$ **b.** $\overline{AB}$

c. $\overrightarrow{DE}$ **d.** $\overleftrightarrow{AE}$

STUDY TIP

In this book, assume that all segments, rays, or lines that appear to be tangent to a circle are tangents.

SOLUTION

a. $\overline{AC}$ is a radius because C is the center and A is a point on the circle.

b. $\overline{AB}$ is a diameter because it is a chord that contains the center C.

c. $\overrightarrow{DE}$ is a tangent ray because it is contained in a line that intersects the circle in exactly one point.

d. $\overleftrightarrow{AE}$ is a secant because it is a line that intersects the circle in two points.

Monitoring Progress

Help in English and Spanish at *BigIdeasMath.com*

1. In Example 1, what word best describes $\overline{AG}$? $\overline{CB}$?

2. In Example 1, name a tangent and a tangent segment.

Drawing and Identifying Common Tangents

Core Concept

Coplanar Circles and Common Tangents

In a plane, two circles can intersect in two points, one point, or no points. Coplanar circles that intersect in one point are called **tangent circles**. Coplanar circles that have a common center are called **concentric circles**.

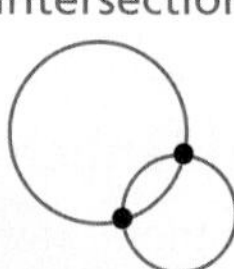

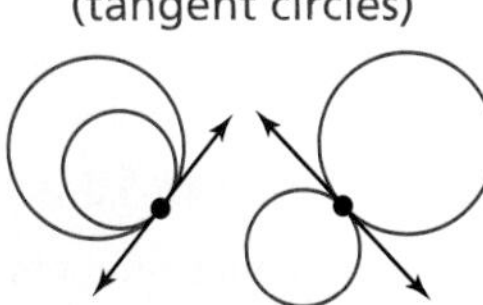

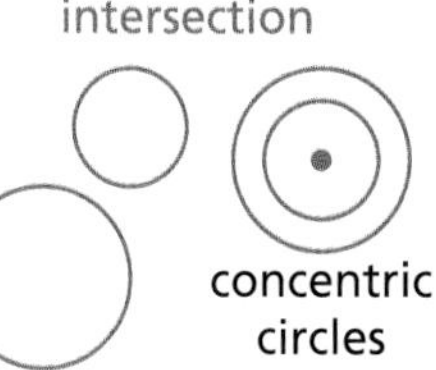

A line or segment that is tangent to two coplanar circles is called a **common tangent**. A *common internal tangent* intersects the segment that joins the centers of the two circles. A *common external tangent* does not intersect the segment that joins the centers of the two circles.

EXAMPLE 2 Drawing and Identifying Common Tangents

Tell how many common tangents the circles have and draw them. Use blue to indicate common external tangents and red to indicate common internal tangents.

a. 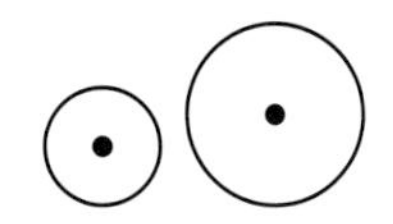

b. 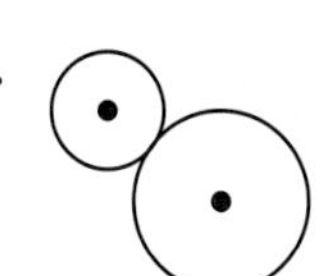

c. 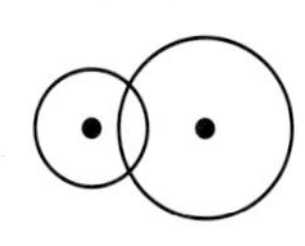

SOLUTION

Draw the segment that joins the centers of the two circles. Then draw the common tangents. Use blue to indicate lines that do not intersect the segment joining the centers and red to indicate lines that intersect the segment joining the centers.

a. 4 common tangents

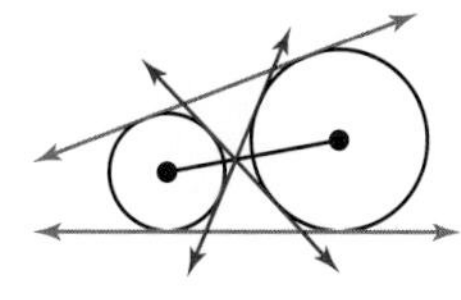

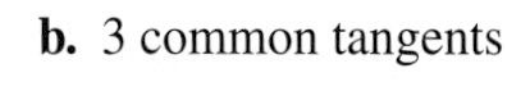

b. 3 common tangents

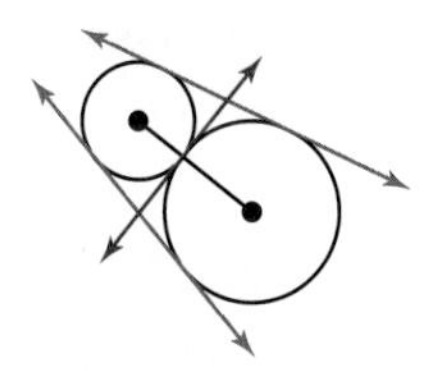

c. 2 common tangents

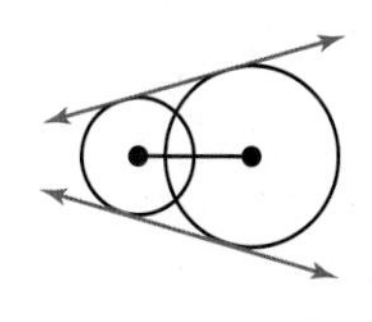

Monitoring Progress  Help in English and Spanish at *BigIdeasMath.com*

Tell how many common tangents the circles have and draw them. State whether the tangents are external tangents or internal tangents.

3. 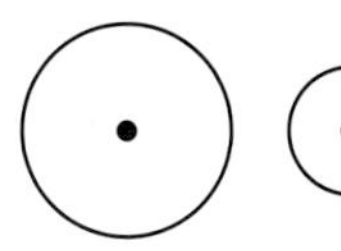

4. 

5. 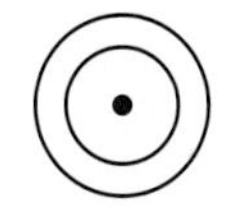

Using Properties of Tangents

Theorems

Theorem 10.1 Tangent Line to Circle Theorem

In a plane, a line is tangent to a circle if and only if the line is perpendicular to a radius of the circle at its endpoint on the circle.

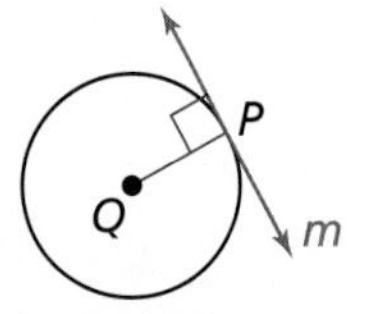

Line m is tangent to $\odot Q$ if and only if $m \perp \overline{QP}$.

Proof Ex. 47, p. 536

Theorem 10.2 External Tangent Congruence Theorem

Tangent segments from a common external point are congruent.

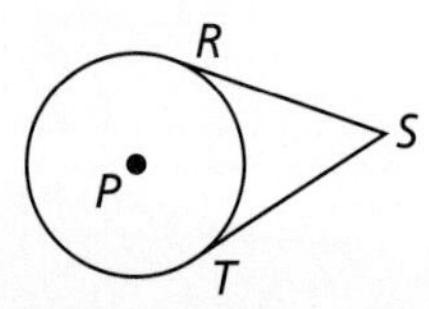

If $\overline{SR}$ and $\overline{ST}$ are tangent segments, then $\overline{SR} \cong \overline{ST}$.

Proof Ex. 46, p. 536

EXAMPLE 3 **Verifying a Tangent to a Circle**

Is $\overline{ST}$ tangent to $\odot P$?

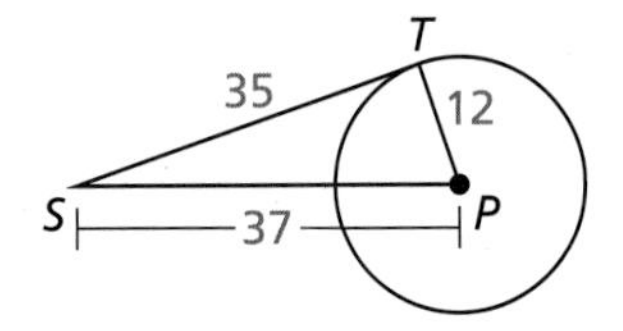

SOLUTION

Use the Converse of the Pythagorean Theorem (Theorem 9.2). Because $12^2 + 35^2 = 37^2$, $\triangle PTS$ is a right triangle and $\overline{ST} \perp \overline{PT}$. So, $\overline{ST}$ is perpendicular to a radius of $\odot P$ at its endpoint on $\odot P$.

▶ By the Tangent Line to Circle Theorem, $\overline{ST}$ is tangent to $\odot P$.

EXAMPLE 4 **Finding the Radius of a Circle**

In the diagram, point B is a point of tangency. Find the radius r of $\odot C$.

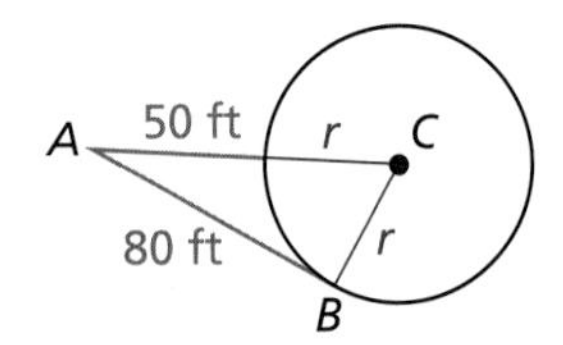

SOLUTION

You know from the Tangent Line to Circle Theorem that $\overline{AB} \perp \overline{BC}$, so $\triangle ABC$ is a right triangle. You can use the Pythagorean Theorem (Theorem 9.1).

$AC^2 = BC^2 + AB^2$	Pythagorean Theorem
$(r + 50)^2 = r^2 + 80^2$	Substitute.
$r^2 + 100r + 2500 = r^2 + 6400$	Multiply.
$100r = 3900$	Subtract r^2 and 2500 from each side.
$r = 39$	Divide each side by 100.

▶ The radius is 39 feet.

CONSTRUCTION Constructing a Tangent to a Circle

Given $\odot C$ and point A, construct a line tangent to $\odot C$ that passes through A. Use a compass and straightedge.

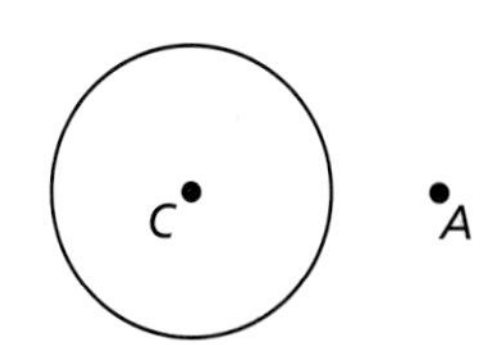

SOLUTION

Step 1

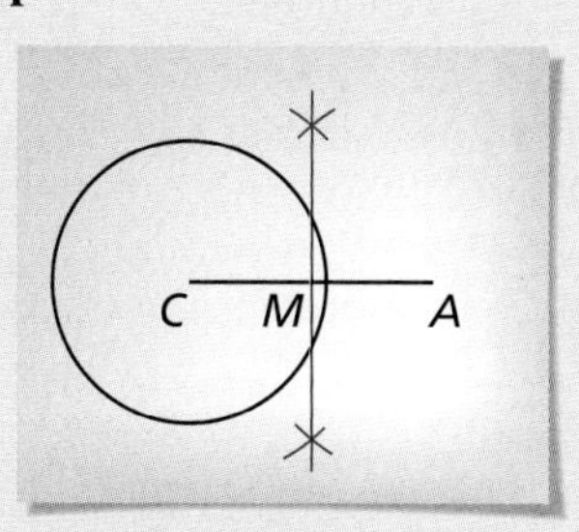

Find a midpoint
Draw $\overline{AC}$. Construct the bisector of the segment and label the midpoint M.

Step 2

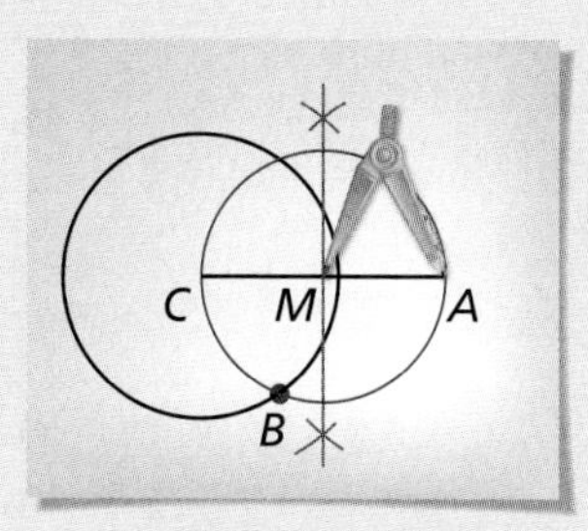

Draw a circle
Construct $\odot M$ with radius MA. Label one of the points where $\odot M$ intersects $\odot C$ as point B.

Step 3

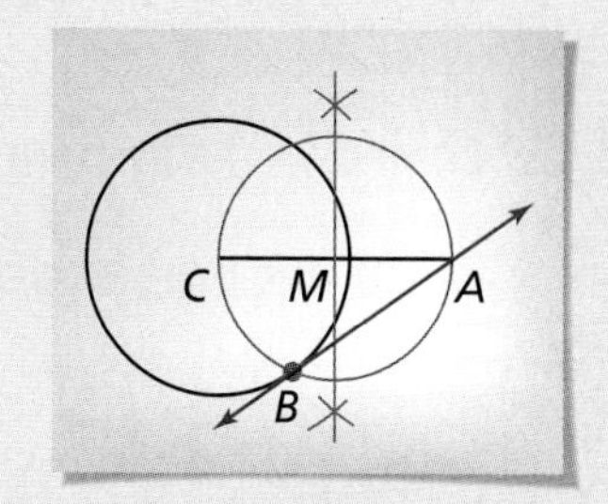

Construct a tangent line
Draw $\overleftrightarrow{AB}$. It is a tangent to $\odot C$ that passes through A.

EXAMPLE 5 Using Properties of Tangents

$\overline{RS}$ is tangent to $\odot C$ at S, and $\overline{RT}$ is tangent to $\odot C$ at T. Find the value of x.

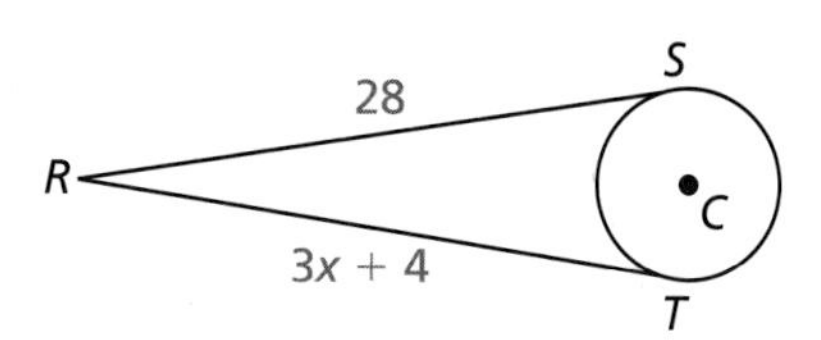

SOLUTION

$RS = RT$	External Tangent Congruence Theorem
$28 = 3x + 4$	Substitute.
$8 = x$	Solve for x.

▶ The value of x is 8.

Monitoring Progress  Help in English and Spanish at *BigIdeasMath.com*

6. Is $\overline{DE}$ tangent to $\odot C$?

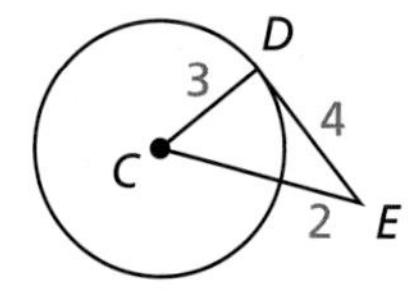

7. $\overline{ST}$ is tangent to $\odot Q$. Find the radius of $\odot Q$.

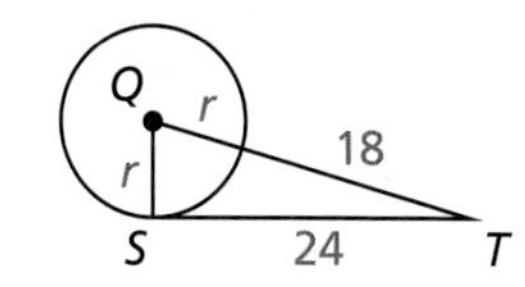

8. Points M and N are points of tangency. Find the value(s) of x.

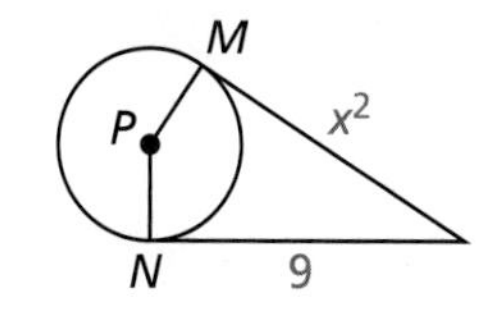

10.1 Exercises

Dynamic Solutions available at *BigIdeasMath.com*

Vocabulary and Core Concept Check

1. **WRITING** How are chords and secants alike? How are they different?

2. **WRITING** Explain how you can determine from the context whether the words *radius* and *diameter* are referring to segments or lengths.

3. **COMPLETE THE SENTENCE** Coplanar circles that have a common center are called ________.

4. **WHICH ONE DOESN'T BELONG?** Which segment does *not* belong with the other three? Explain your reasoning.

chord | radius | tangent | diameter

Monitoring Progress and Modeling with Mathematics

In Exercises 5–10, use the diagram. *(See Example 1.)*

5. Name the circle.

6. Name two radii.

7. Name two chords.

8. Name a diameter.

9. Name a secant.

10. Name a tangent and a point of tangency.

In Exercises 11–14, copy the diagram. Tell how many common tangents the circles have and draw them. *(See Example 2.)*

11. 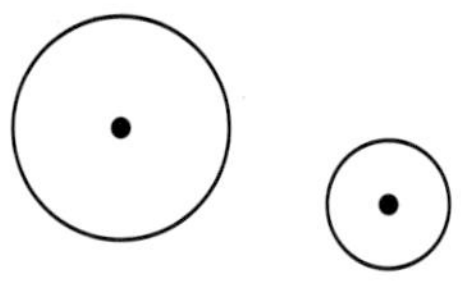

12. 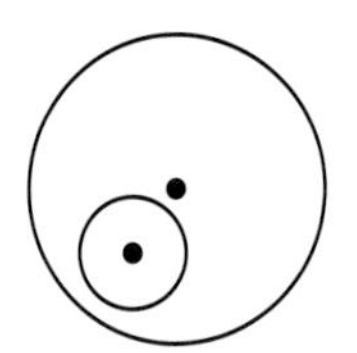

13. 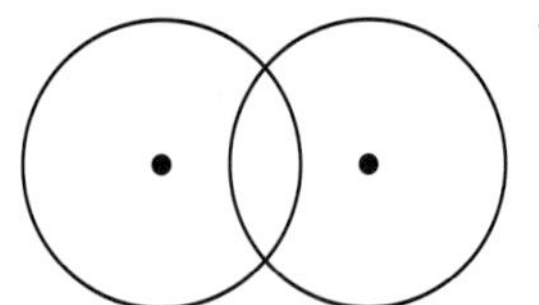

14. 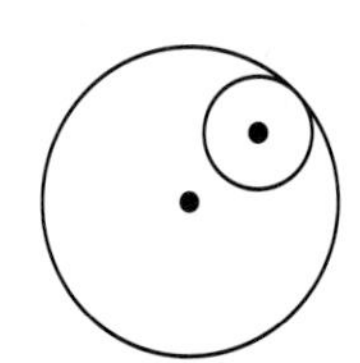

In Exercises 15–18, tell whether the common tangent is *internal* or *external*.

15. 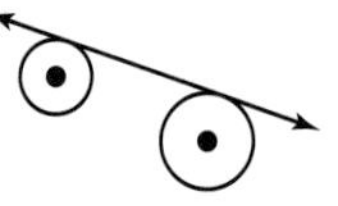

16. 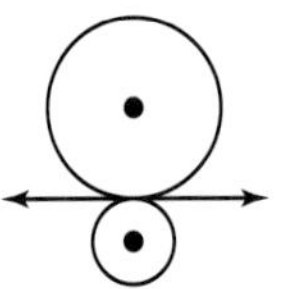

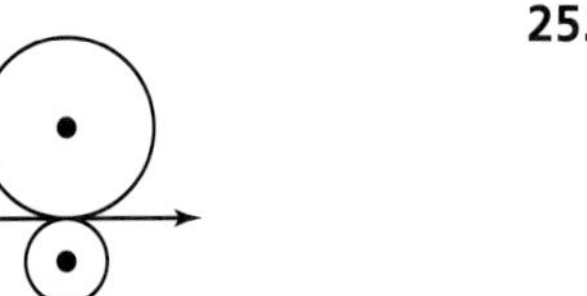

17. 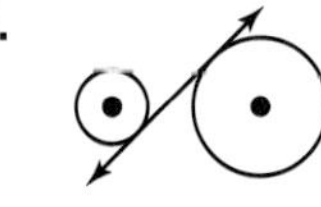

18. 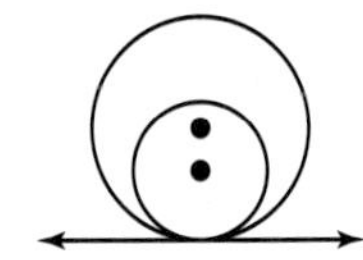

In Exercises 19–22, tell whether $\overline{AB}$ is tangent to $\odot C$. Explain your reasoning. *(See Example 3.)*

19. 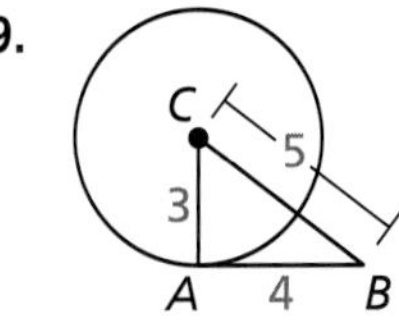

20. 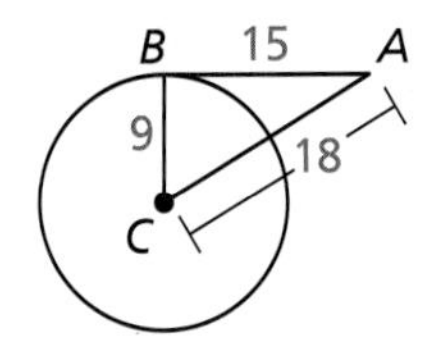

21. 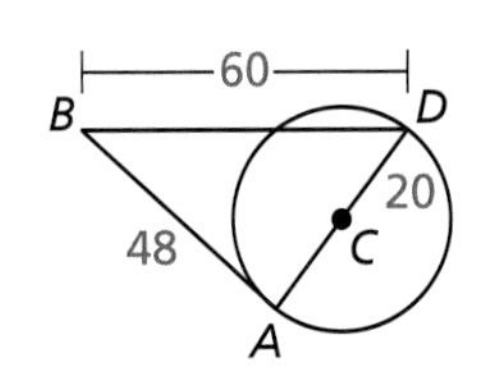

22. 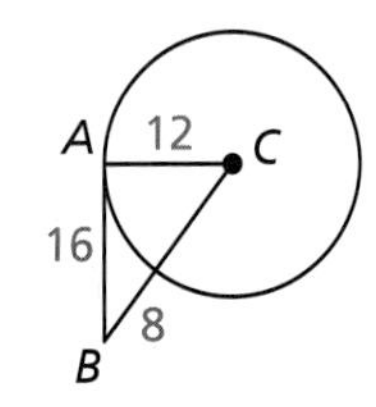

In Exercises 23–26, point B is a point of tangency. Find the radius r of $\odot C$. *(See Example 4.)*

23. 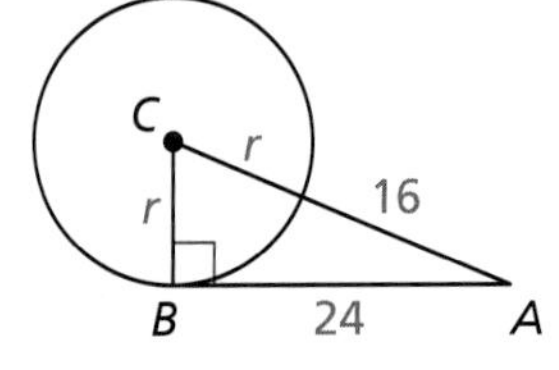

24. 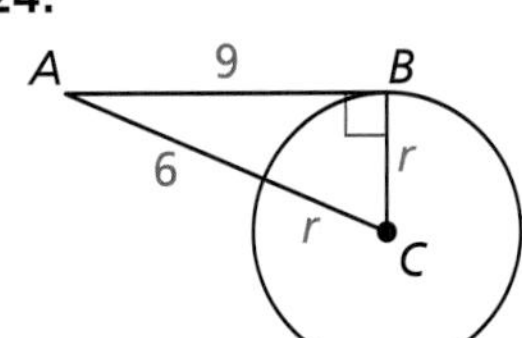

25. 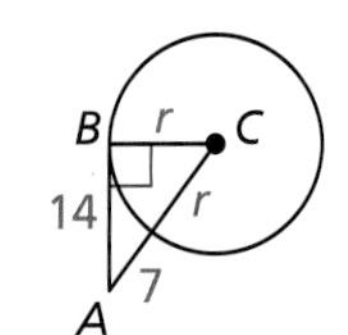

26. 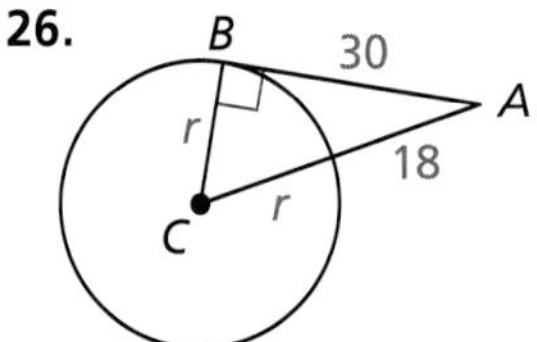

CONSTRUCTION In Exercises 27 and 28, construct $\odot C$ with the given radius and point A outside of $\odot C$. Then construct a line tangent to $\odot C$ that passes through A.

27. $r = 2$ in. **28.** $r = 4.5$ cm

In Exercises 29–32, points B and D are points of tangency. Find the value(s) of x. *(See Example 5.)*

29.

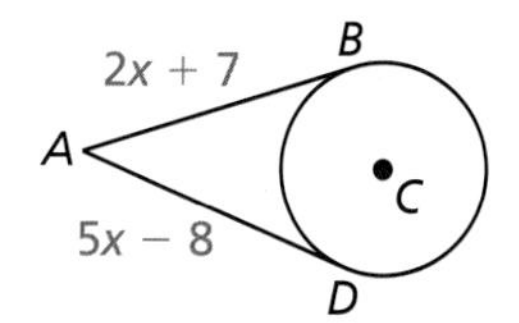

30.

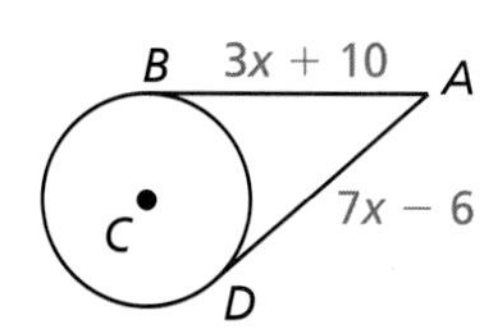

31.

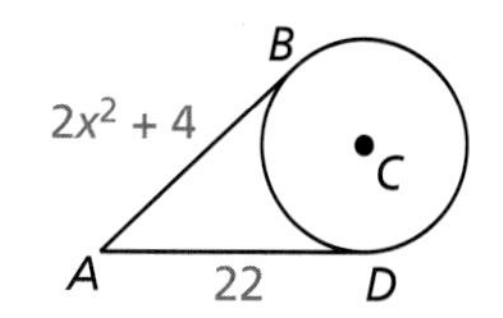

32.

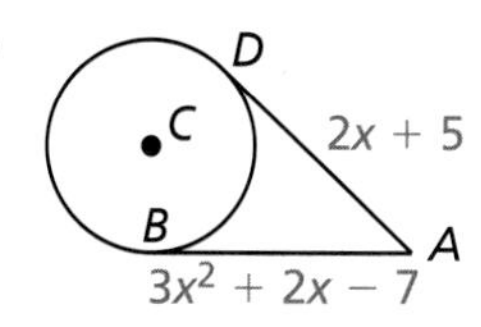

33. ERROR ANALYSIS Describe and correct the error in determining whether $\overline{XY}$ is tangent to $\odot Z$.

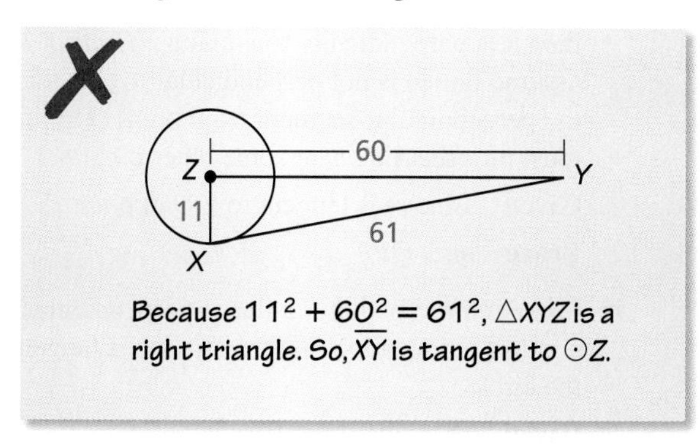

34. ERROR ANALYSIS Describe and correct the error in finding the radius of $\odot T$.

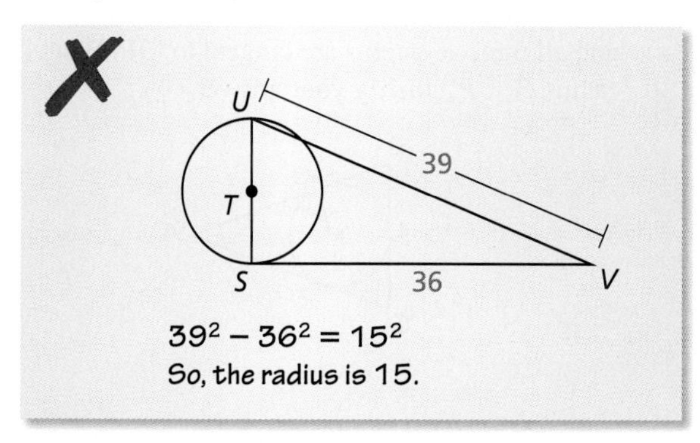

35. ABSTRACT REASONING For a point outside of a circle, how many lines exist tangent to the circle that pass through the point? How many such lines exist for a point on the circle? inside the circle? Explain your reasoning.

36. CRITICAL THINKING When will two lines tangent to the same circle not intersect? Justify your answer.

37. USING STRUCTURE Each side of quadrilateral $TVWX$ is tangent to $\odot Y$. Find the perimeter of the quadrilateral.

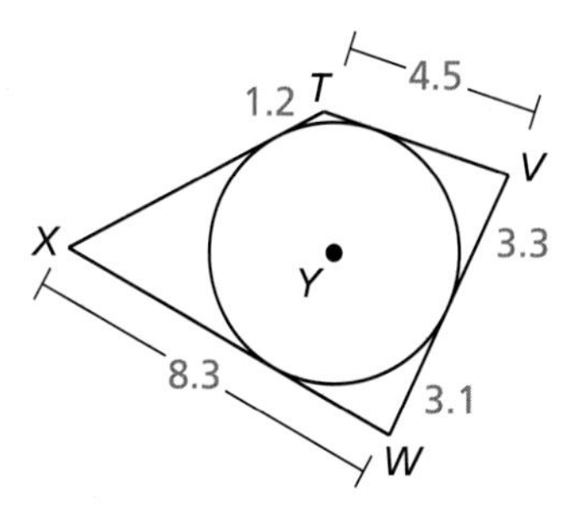

38. LOGIC In $\odot C$, radii $\overline{CA}$ and $\overline{CB}$ are perpendicular. $\overleftrightarrow{BD}$ and $\overleftrightarrow{AD}$ are tangent to $\odot C$.

a. Sketch $\odot C$, $\overline{CA}$, $\overline{CB}$, $\overleftrightarrow{BD}$, and $\overleftrightarrow{AD}$.

b. What type of quadrilateral is $CADB$? Explain your reasoning.

39. MAKING AN ARGUMENT Two bike paths are tangent to an approximately circular pond. Your class is building a nature trail that begins at the intersection B of the bike paths and runs between the bike paths and over a bridge through the center P of the pond. Your classmate uses the Converse of the Angle Bisector Theorem (Theorem 6.4) to conclude that the trail must bisect the angle formed by the bike paths. Is your classmate correct? Explain your reasoning.

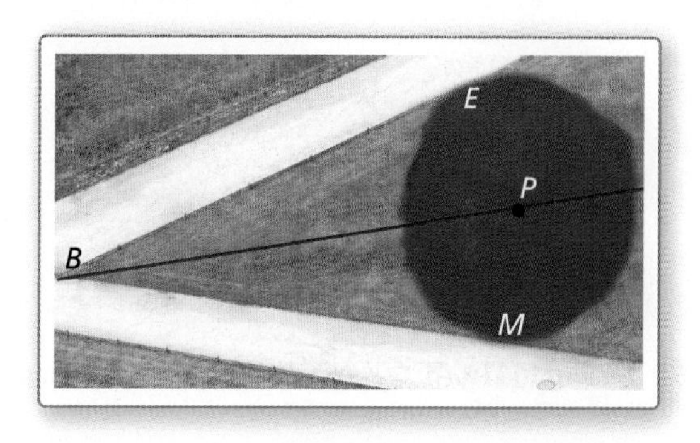

40. MODELING WITH MATHEMATICS A bicycle chain is pulled tightly so that $\overline{MN}$ is a common tangent of the gears. Find the distance between the centers of the gears.

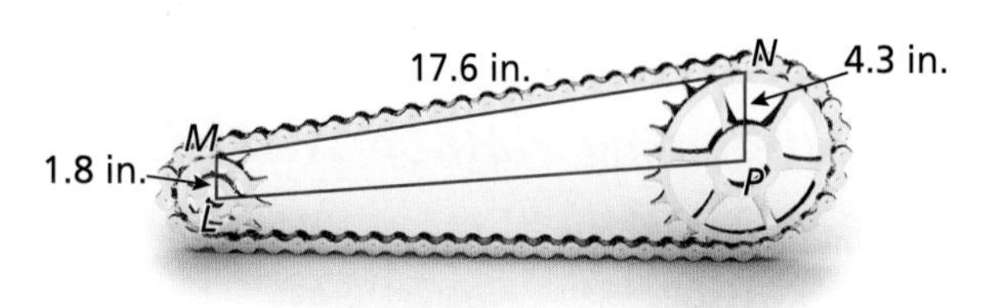

41. WRITING Explain why the diameter of a circle is the longest chord of the circle.

42. HOW DO YOU SEE IT? In the figure, $\overrightarrow{PA}$ is tangent to the dime, $\overrightarrow{PC}$ is tangent to the quarter, and $\overrightarrow{PB}$ is a common internal tangent. How do you know that $\overline{PA} \cong \overline{PB} \cong \overline{PC}$?

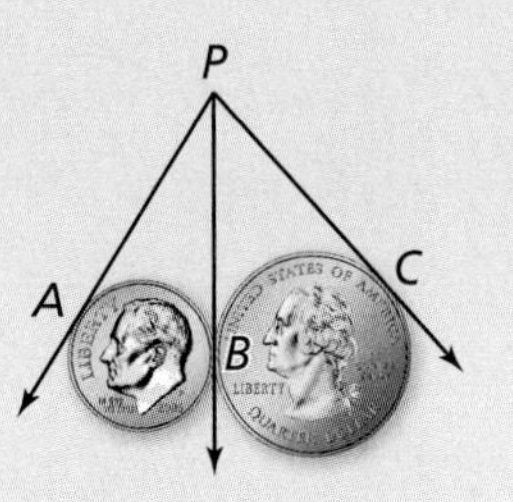

43. PROOF In the diagram, $\overline{RS}$ is a common internal tangent to $\odot A$ and $\odot B$. Prove that $\frac{AC}{BC} = \frac{RC}{SC}$.

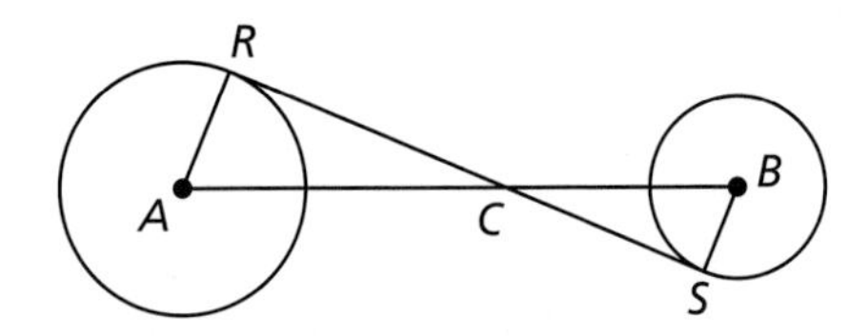

44. THOUGHT PROVOKING A polygon is *circumscribed* about a circle when every side of the polygon is tangent to the circle. In the diagram, quadrilateral $ABCD$ is circumscribed about $\odot Q$. Is it always true that $AB + CD = AD + BC$? Justify your answer.

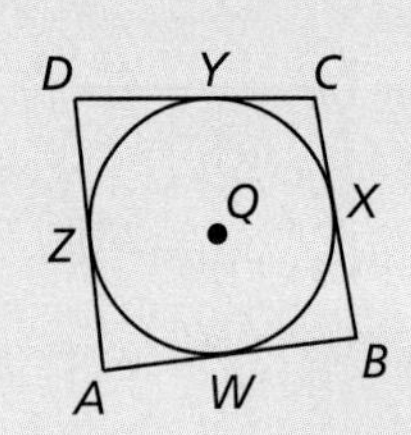

45. MATHEMATICAL CONNECTIONS Find the values of x and y. Justify your answer.

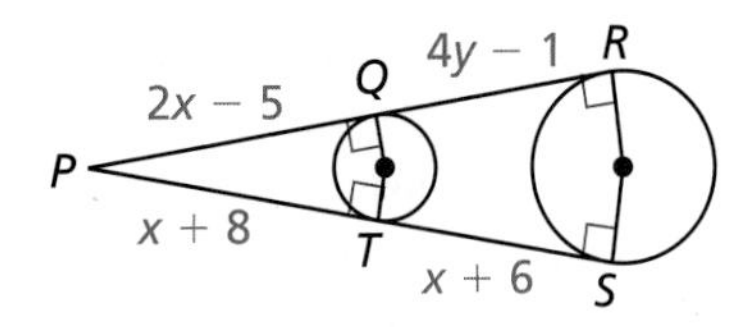

46. PROVING A THEOREM Prove the External Tangent Congruence Theorem (Theorem 10.2).

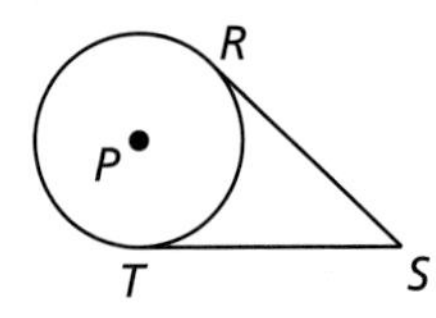

Given $\overline{SR}$ and $\overline{ST}$ are tangent to $\odot P$.

Prove $\overline{SR} \cong \overline{ST}$

47. PROVING A THEOREM Use the diagram to prove each part of the biconditional in the Tangent Line to Circle Theorem (Theorem 10.1).

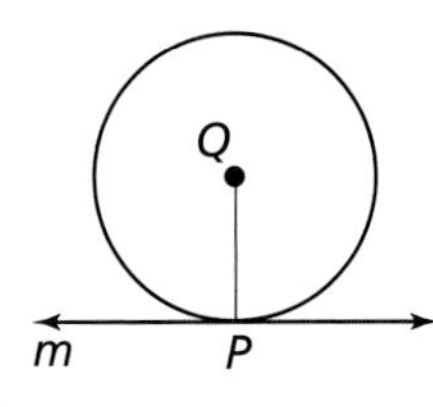

a. Prove indirectly that if a line is tangent to a circle, then it is perpendicular to a radius. (*Hint:* If you assume line m is not perpendicular to $\overline{QP}$, then the perpendicular segment from point Q to line m must intersect line m at some other point R.)

Given Line m is tangent to $\odot Q$ at point P.

Prove $m \perp \overline{QP}$

b. Prove indirectly that if a line is perpendicular to a radius at its endpoint, then the line is tangent to the circle.

Given $m \perp \overline{QP}$

Prove Line m is tangent to $\odot Q$.

48. REASONING In the diagram, $AB = AC = 12$, $BC = 8$, and all three segments are tangent to $\odot P$. What is the radius of $\odot P$? Justify your answer.

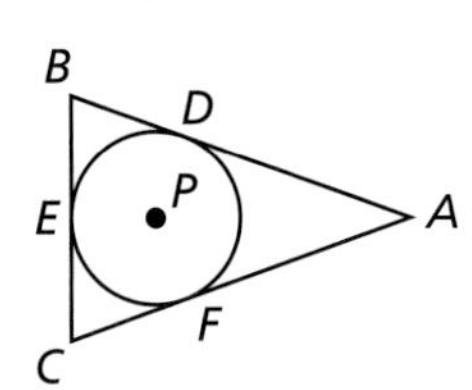

Maintaining Mathematical Proficiency

Reviewing what you learned in previous grades and lessons

Find the indicated measure. *(Section 1.2 and Section 1.5)*

49. $m\angle JKM$

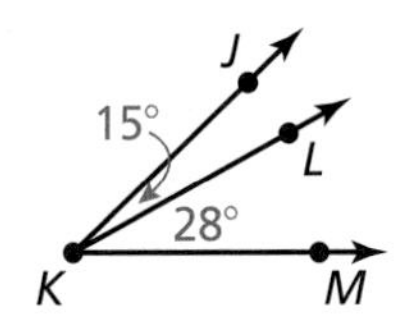

50. AB

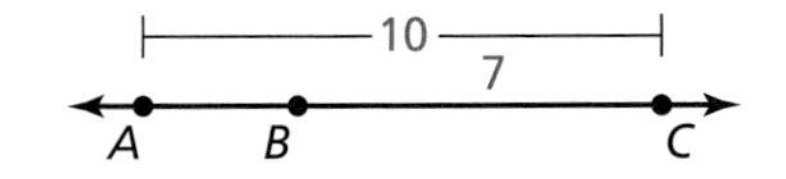

10.2 Finding Arc Measures

Essential Question How are circular arcs measured?

A **central angle** of a circle is an angle whose vertex is the center of the circle. A *circular arc* is a portion of a circle, as shown below. The measure of a circular arc is the measure of its central angle.

If $m\angle AOB < 180^\circ$, then the circular arc is called a **minor arc** and is denoted by $\overparen{AB}$.

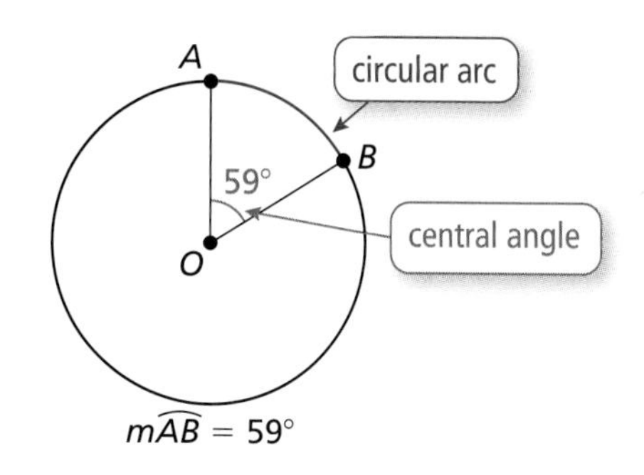

EXPLORATION 1 Measuring Circular Arcs

Work with a partner. Use dynamic geometry software to find the measure of $\overparen{BC}$. Verify your answers using trigonometry.

a.

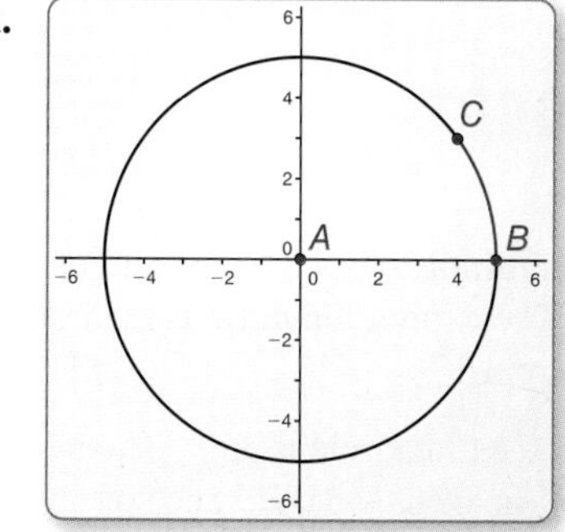
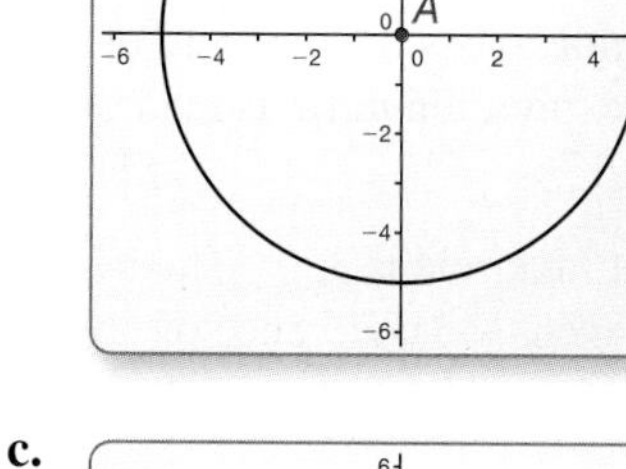

Points
$A(0, 0)$
$B(5, 0)$
$C(4, 3)$

b.

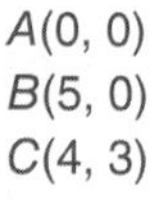

Points
$A(0, 0)$
$B(5, 0)$
$C(3, 4)$

c.

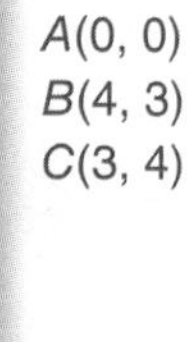

Points
$A(0, 0)$
$B(4, 3)$
$C(3, 4)$

d.

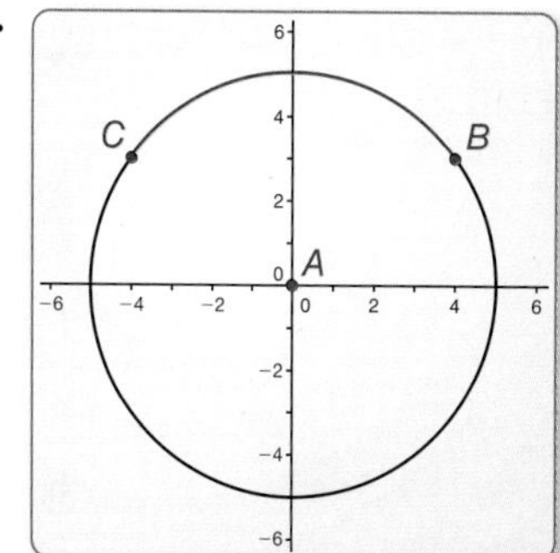

Points
$A(0, 0)$
$B(4, 3)$
$C(-4, 3)$

USING TOOLS STRATEGICALLY

To be proficient in math, you need to use technological tools to explore and deepen your understanding of concepts.

Communicate Your Answer

2. How are circular arcs measured?

3. Use dynamic geometry software to draw a circular arc with the given measure.

a. 30° **b.** 45°
c. 60° **d.** 90°

10.2 Lesson

Core Vocabulary

central angle, *p. 538*
minor arc, *p. 538*
major arc, *p. 538*
semicircle, *p. 538*
measure of a minor arc, *p. 538*
measure of a major arc, *p. 538*
adjacent arcs, *p. 539*
congruent circles, *p. 540*
congruent arcs, *p. 540*
similar arcs, *p. 541*

What You Will Learn

- Find arc measures.
- Identify congruent arcs.
- Prove circles are similar.

Finding Arc Measures

A **central angle** of a circle is an angle whose vertex is the center of the circle. In the diagram, $\angle ACB$ is a central angle of $\odot C$.

If $m\angle ACB$ is less than 180°, then the points on $\odot C$ that lie in the interior of $\angle ACB$ form a **minor arc** with endpoints A and B. The points on $\odot C$ that do not lie on the minor arc AB form a **major arc** with endpoints A and B. A **semicircle** is an arc with endpoints that are the endpoints of a diameter.

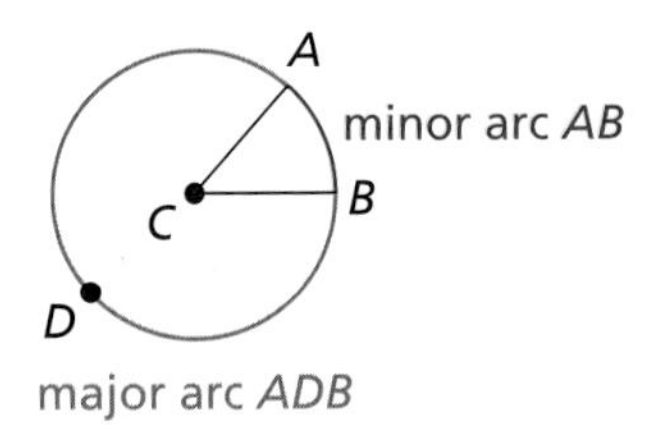

Minor arcs are named by their endpoints. The minor arc associated with $\angle ACB$ is named $\overset{\frown}{AB}$. Major arcs and semicircles are named by their endpoints and a point on the arc. The major arc associated with $\angle ACB$ can be named $\overset{\frown}{ADB}$.

STUDY TIP

The measure of a minor arc is less than 180°. The measure of a major arc is greater than 180°.

Core Concept

Measuring Arcs

The **measure of a minor arc** is the measure of its central angle. The expression $m\overset{\frown}{AB}$ is read as "the measure of arc AB."

The measure of the entire circle is 360°. The **measure of a major arc** is the difference of 360° and the measure of the related minor arc. The measure of a semicircle is 180°.

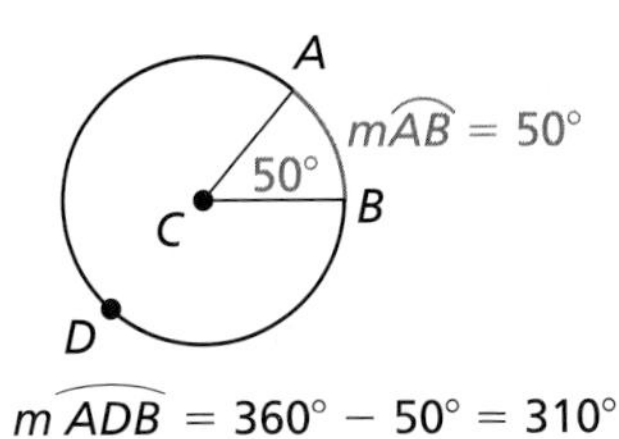

EXAMPLE 1 Finding Measures of Arcs

Find the measure of each arc of $\odot P$, where $\overline{RT}$ is a diameter.

a. $\overset{\frown}{RS}$

b. $\overset{\frown}{RTS}$

c. $\overset{\frown}{RST}$

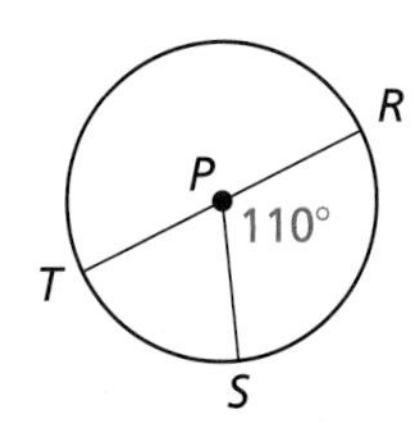

SOLUTION

a. $\overset{\frown}{RS}$ is a minor arc, so $m\overset{\frown}{RS} = m\angle RPS = 110°$.

b. $\overset{\frown}{RTS}$ is a major arc, so $m\overset{\frown}{RTS} = 360° - 110° = 250°$.

c. $\overline{RT}$ is a diameter, so $\overset{\frown}{RST}$ is a semicircle, and $m\overset{\frown}{RST} = 180°$.

Two arcs of the same circle are **adjacent arcs** when they intersect at exactly one point. You can add the measures of two adjacent arcs.

Postulate

Postulate 10.1 Arc Addition Postulate

The measure of an arc formed by two adjacent arcs is the sum of the measures of the two arcs.

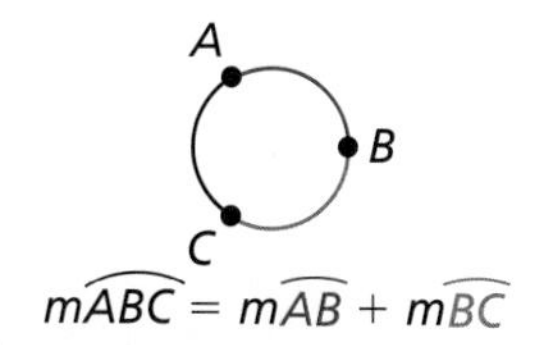

$m\widehat{ABC} = m\widehat{AB} + m\widehat{BC}$

EXAMPLE 2 Using the Arc Addition Postulate

Find the measure of each arc.

a. $\widehat{GE}$ **b.** $\widehat{GEF}$ **c.** $\widehat{GF}$

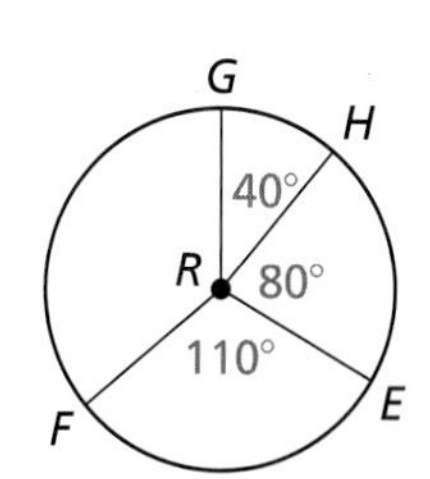

SOLUTION

a. $m\widehat{GE} = m\widehat{GH} + m\widehat{HE} = 40° + 80° = 120°$

b. $m\widehat{GEF} = m\widehat{GE} + m\widehat{EF} = 120° + 110° = 230°$

c. $m\widehat{GF} = 360° - m\widehat{GEF} = 360° - 230° = 130°$

EXAMPLE 3 Finding Measures of Arcs

A recent survey asked teenagers whether they would rather meet a famous musician, athlete, actor, inventor, or other person. The circle graph shows the results. Find the indicated arc measures.

a. $m\widehat{AC}$ **b.** $m\widehat{ACD}$

c. $m\widehat{ADC}$ **d.** $m\widehat{EBD}$

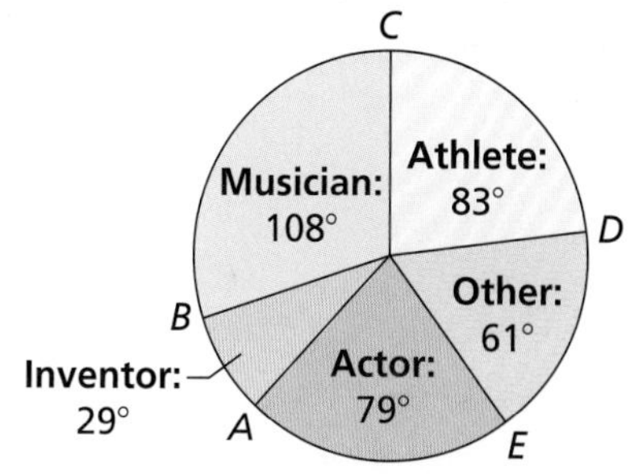

SOLUTION

a. $m\widehat{AC} = m\widehat{AB} + m\widehat{BC}$

$= 29° + 108°$

$= 137°$

b. $m\widehat{ACD} = m\widehat{AC} + m\widehat{CD}$

$= 137° + 83°$

$= 220°$

c. $m\widehat{ADC} = 360° - m\widehat{AC}$

$= 360° - 137°$

$= 223°$

d. $m\widehat{EBD} = 360° - m\widehat{ED}$

$= 360° - 61°$

$= 299°$

Monitoring Progress

Help in English and Spanish at *BigIdeasMath.com*

Identify the given arc as a *major arc*, *minor arc*, or *semicircle*. Then find the measure of the arc.

1. $\widehat{TQ}$ **2.** $\widehat{QRT}$ **3.** $\widehat{TQR}$

4. $\widehat{QS}$ **5.** $\widehat{TS}$ **6.** $\widehat{RST}$

T, Q, R, S; 120°, 80°, 60°

Identifying Congruent Arcs

Two circles are **congruent circles** if and only if a rigid motion or a composition of rigid motions maps one circle onto the other. This statement is equivalent to the Congruent Circles Theorem below.

Theorem 10.3 Congruent Circles Theorem

Two circles are congruent circles if and only if they have the same radius.

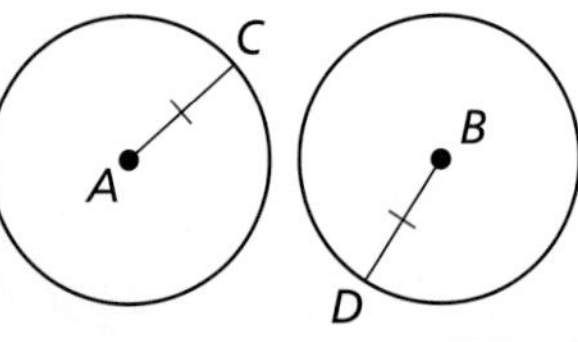

Proof Ex. 35, p. 544

$\odot A \cong \odot B$ if and only if $\overline{AC} \cong \overline{BD}$.

Two arcs are **congruent arcs** if and only if they have the same measure and they are arcs of the same circle or of congruent circles.

Theorem 10.4 Congruent Central Angles Theorem

In the same circle, or in congruent circles, two minor arcs are congruent if and only if their corresponding central angles are congruent.

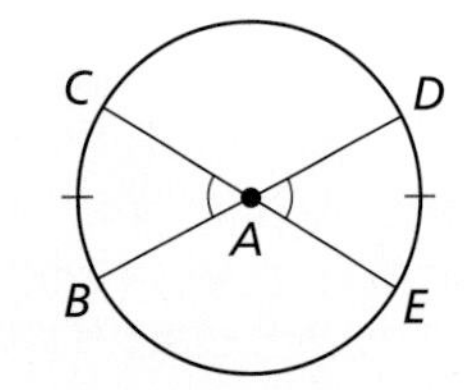

Proof Ex. 37, p. 544

$\widehat{BC} \cong \widehat{DE}$ if and only if $\angle BAC \cong \angle DAE$.

EXAMPLE 4 Identifying Congruent Arcs

Tell whether the red arcs are congruent. Explain why or why not.

a.

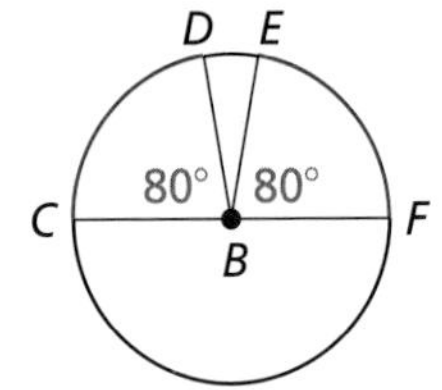

b.

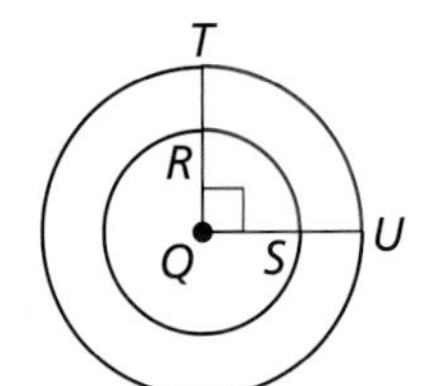

c.

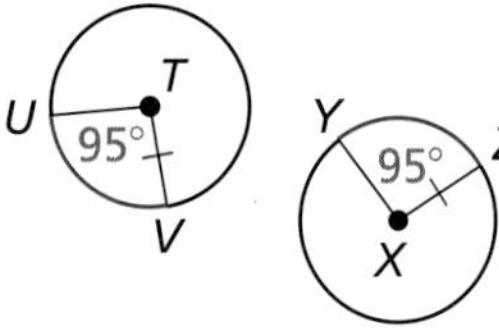

SOLUTION

a. $\widehat{CD} \cong \widehat{EF}$ by the Congruent Central Angles Theorem because they are arcs of the same circle and they have congruent central angles, $\angle CBD \cong \angle FBE$.

b. $\widehat{RS}$ and $\widehat{TU}$ have the same measure, but are not congruent because they are arcs of circles that are not congruent.

c. $\widehat{UV} \cong \widehat{YZ}$ by the Congruent Central Angles Theorem because they are arcs of congruent circles and they have congruent central angles, $\angle UTV \cong \angle YXZ$.

STUDY TIP

The two circles in part (c) are congruent by the Congruent Circles Theorem because they have the same radius.

Monitoring Progress  Help in English and Spanish at *BigIdeasMath.com*

Tell whether the red arcs are congruent. Explain why or why not.

7. 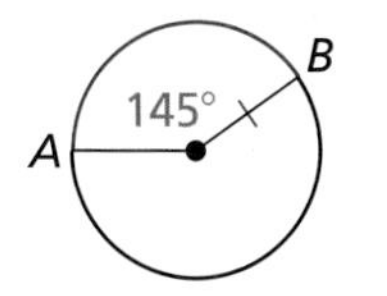

8.

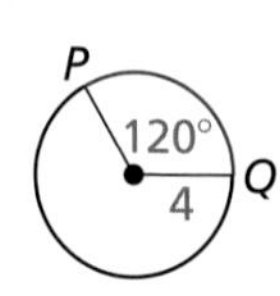

Proving Circles Are Similar

Theorem

Theorem 10.5 Similar Circles Theorem

All circles are similar.

Proof p. 541; Ex. 33, p. 544

PROOF **Similar Circles Theorem**

All circles are similar.

Given $\odot C$ with center C and radius r, $\odot D$ with center D and radius s

Prove $\odot C \sim \odot D$

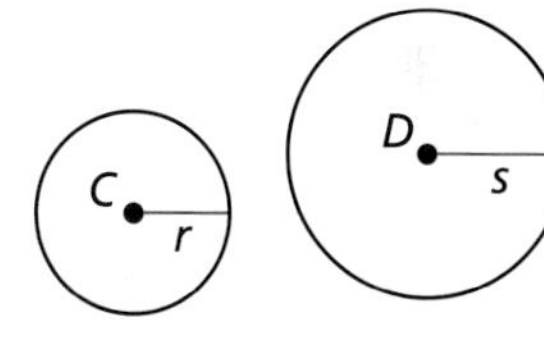

First, translate $\odot C$ so that point C maps to point D. The image of $\odot C$ is $\odot C'$ with center D. So, $\odot C'$ and $\odot D$ are concentric circles.

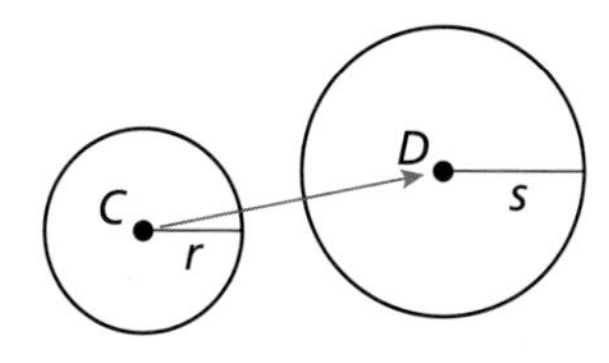

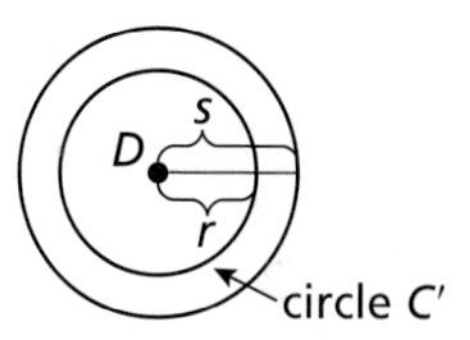

$\odot C'$ is the set of all points that are r units from point D. Dilate $\odot C'$ using center of dilation D and scale factor $\frac{s}{r}$.

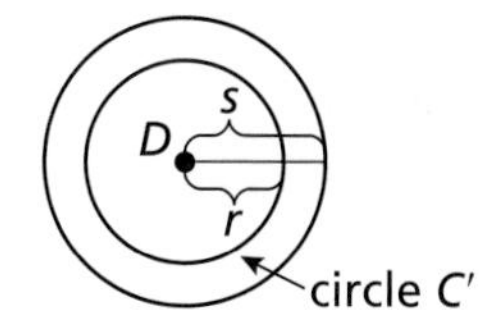

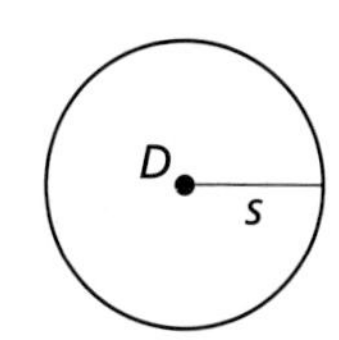

This dilation maps the set of all the points that are r units from point D to the set of all points that are $\frac{s}{r}(r) = s$ units from point D. $\odot D$ is the set of all points that are s units from point D. So, this dilation maps $\odot C'$ to $\odot D$.

Because a similarity transformation maps $\odot C$ to $\odot D$, $\odot C \sim \odot D$.

Two arcs are **similar arcs** if and only if they have the same measure. All congruent arcs are similar, but not all similar arcs are congruent. For instance, in Example 4, the pairs of arcs in parts (a), (b), and (c) are similar but only the pairs of arcs in parts (a) and (c) are congruent.

10.2 Exercises

Dynamic Solutions available at *BigIdeasMath.com*

Vocabulary and Core Concept Check

1. **VOCABULARY** Copy and complete: If $\angle ACB$ and $\angle DCE$ are congruent central angles of $\odot C$, then $\overset{\frown}{AB}$ and $\overset{\frown}{DE}$ are ______________.

2. **WHICH ONE DOESN'T BELONG?** Which circle does *not* belong with the other three? Explain your reasoning.

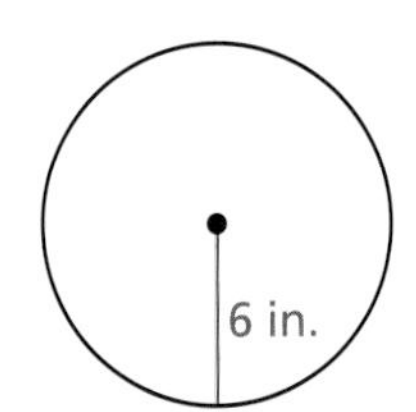

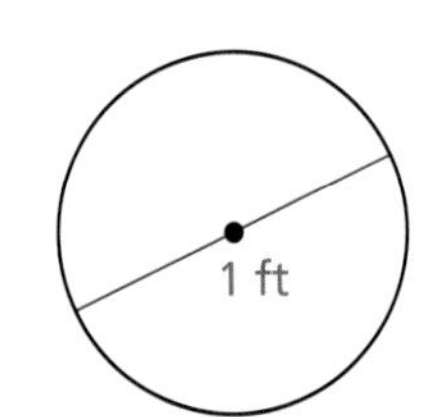

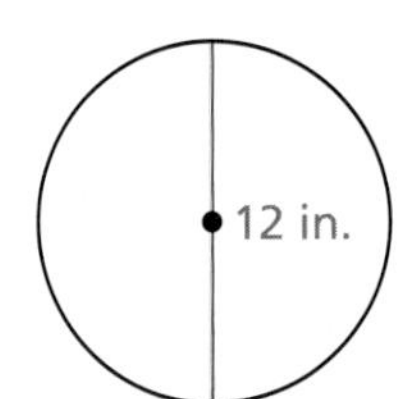

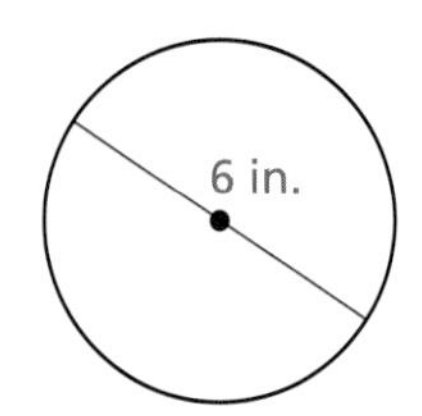

Monitoring Progress and Modeling with Mathematics

In Exercises 3–6, name the red minor arc and find its measure. Then name the blue major arc and find its measure.

3.

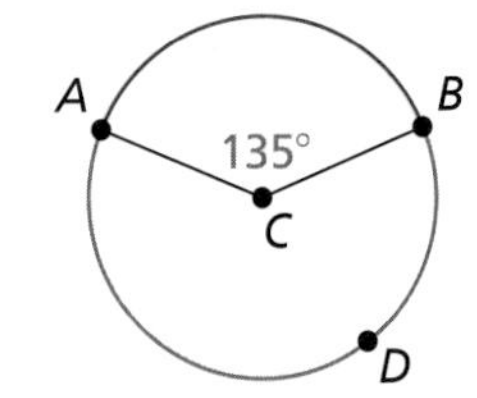

4.

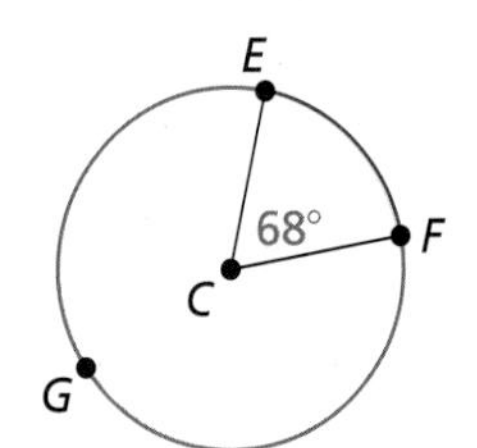

5.

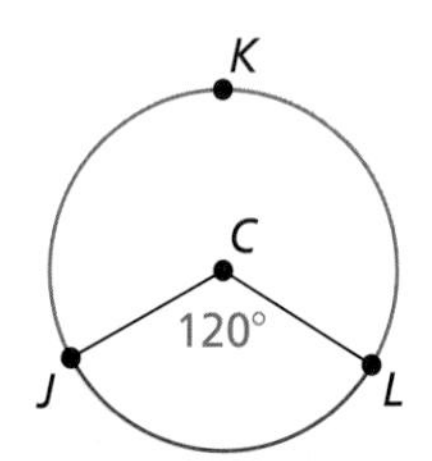

6.

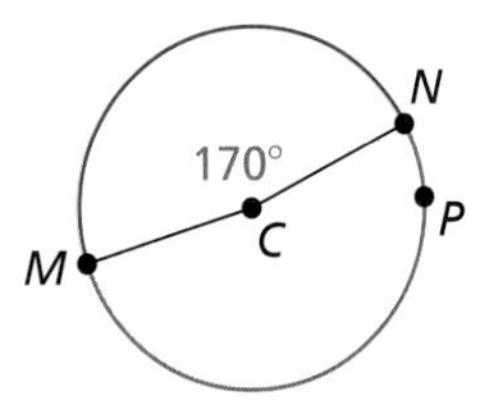

In Exercises 7–14, identify the given arc as a *major arc*, *minor arc*, or *semicircle*. Then find the measure of the arc. *(See Example 1.)*

7. $\overset{\frown}{BC}$

8. $\overset{\frown}{DC}$

9. $\overset{\frown}{ED}$

10. $\overset{\frown}{AE}$

11. $\overset{\frown}{EAB}$

12. $\overset{\frown}{ABC}$

13. $\overset{\frown}{BAC}$

14. $\overset{\frown}{EBD}$

In Exercises 15 and 16, find the measure of each arc. *(See Example 2.)*

15. a. $\overset{\frown}{JL}$
 b. $\overset{\frown}{KM}$
 c. $\overset{\frown}{JLM}$
 d. $\overset{\frown}{JM}$

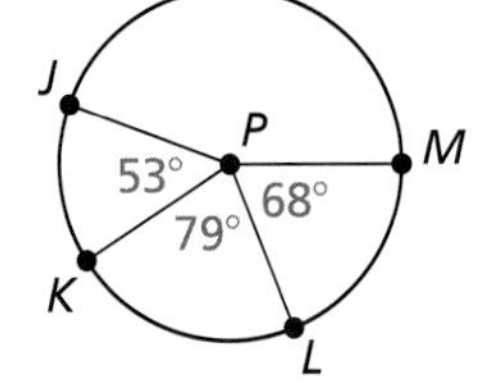

16. a. $\overset{\frown}{RS}$
 b. $\overset{\frown}{QRS}$
 c. $\overset{\frown}{QST}$
 d. $\overset{\frown}{QT}$

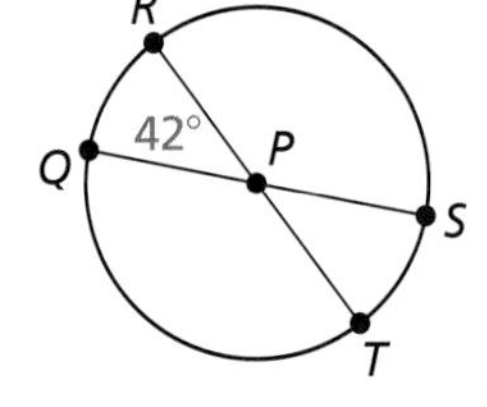

17. **MODELING WITH MATHEMATICS** A recent survey asked high school students their favorite type of music. The results are shown in the circle graph. Find each indicated arc measure. *(See Example 3.)*

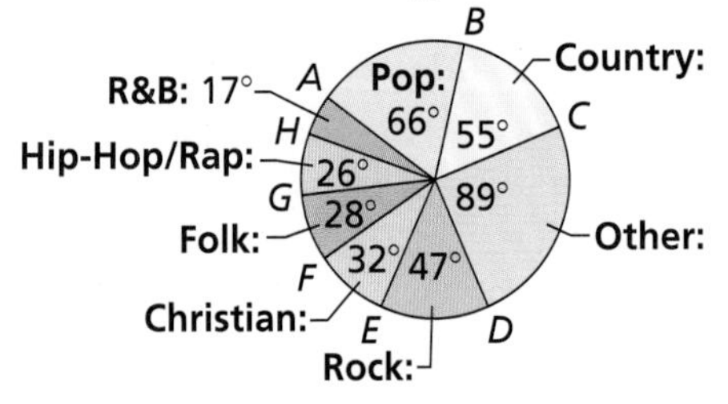

 a. $m\overset{\frown}{AE}$
 b. $m\overset{\frown}{ACE}$
 c. $m\overset{\frown}{GDC}$
 d. $m\overset{\frown}{BHC}$
 e. $m\overset{\frown}{FD}$
 f. $m\overset{\frown}{FBD}$

18. **ABSTRACT REASONING** The circle graph shows the percentages of students enrolled in fall sports at a high school. Is it possible to find the measure of each minor arc? If so, find the measure of the arc for each category shown. If not, explain why it is not possible.

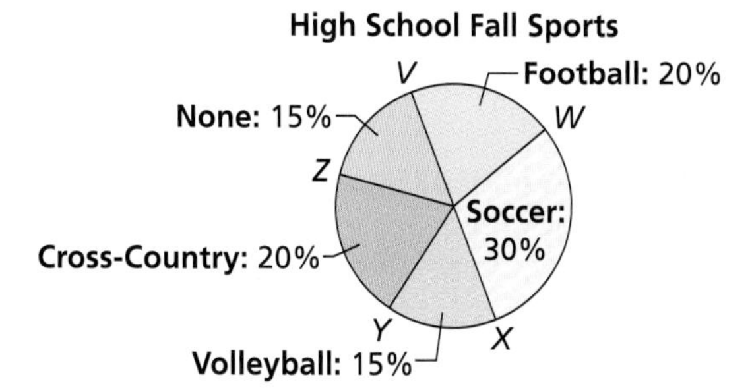

In Exercises 19–22, tell whether the red arcs are congruent. Explain why or why not. *(See Example 4.)*

19.
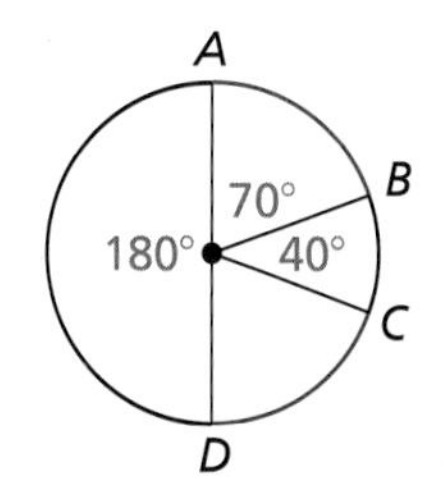

20.
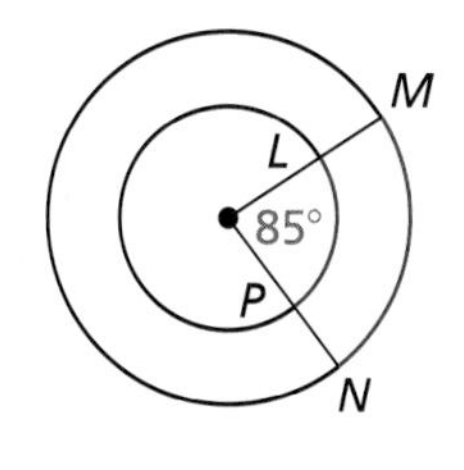

21.

22.
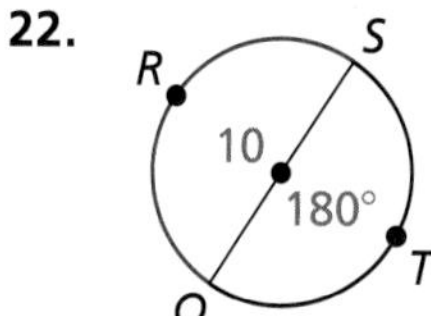

MATHEMATICAL CONNECTIONS In Exercises 23 and 24, find the value of x. Then find the measure of the red arc.

23.
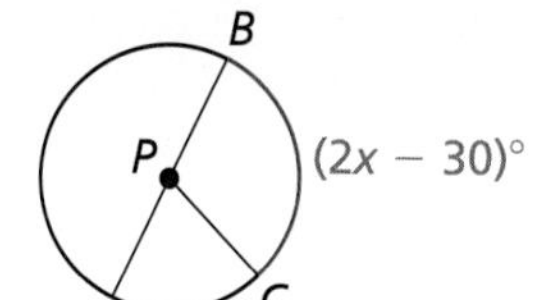

24.
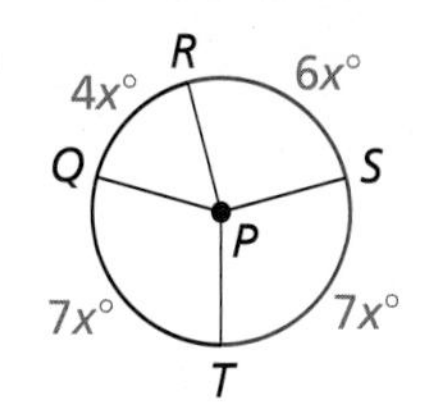

25. **MAKING AN ARGUMENT** Your friend claims that any two arcs with the same measure are similar. Your cousin claims that any two arcs with the same measure are congruent. Who is correct? Explain.

26. **MAKING AN ARGUMENT** Your friend claims that there is not enough information given to find the value of x. Is your friend correct? Explain your reasoning.

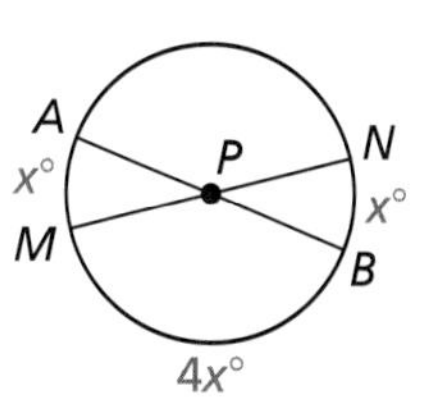

27. **ERROR ANALYSIS** Describe and correct the error in naming the red arc.

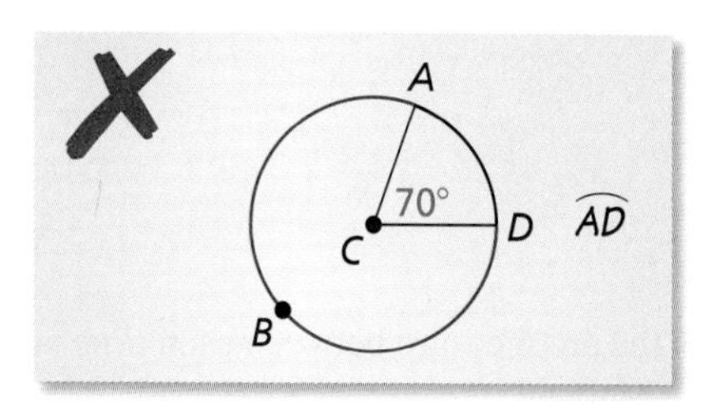

28. **ERROR ANALYSIS** Describe and correct the error in naming congruent arcs.

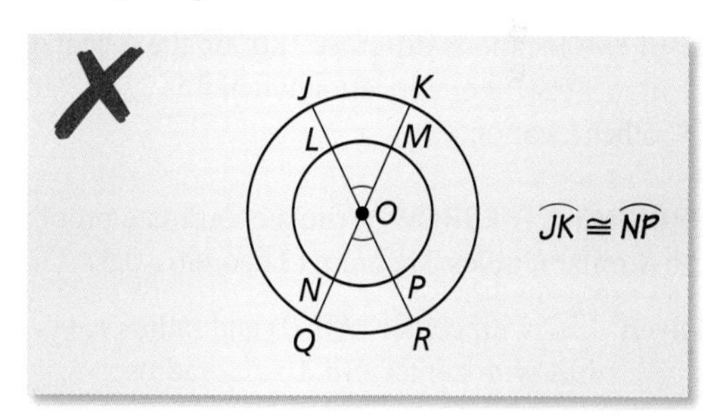

29. **ATTENDING TO PRECISION** Two diameters of $\odot P$ are $\overline{AB}$ and $\overline{CD}$. Find $m\widehat{ACD}$ and $m\widehat{AC}$ when $m\widehat{AD} = 20°$.

30. **REASONING** In $\odot R$, $m\widehat{AB} = 60°$, $m\widehat{BC} = 25°$, $m\widehat{CD} = 70°$, and $m\widehat{DE} = 20°$. Find two possible measures of $\widehat{AE}$.

31. **MODELING WITH MATHEMATICS** On a regulation dartboard, the outermost circle is divided into twenty congruent sections. What is the measure of each arc in this circle?

32. MODELING WITH MATHEMATICS You can use the time zone wheel to find the time in different locations across the world. For example, to find the time in Tokyo when it is 4 P.M. in San Francisco, rotate the small wheel until 4 P.M. and San Francisco line up, as shown. Then look at Tokyo to see that it is 9 A.M. there.

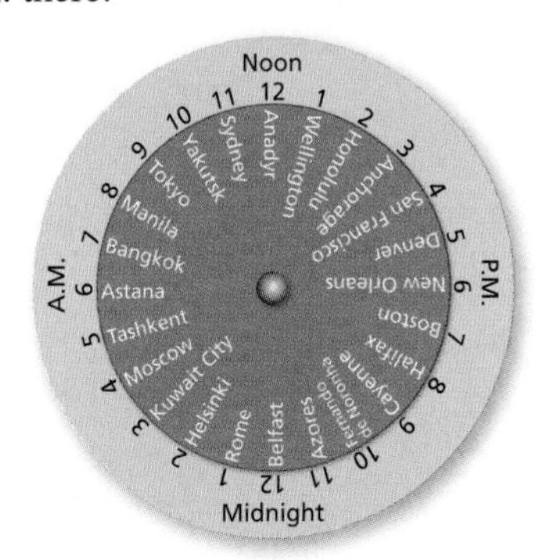

a. What is the arc measure between each time zone on the wheel?

b. What is the measure of the minor arc from the Tokyo zone to the Anchorage zone?

c. If two locations differ by 180° on the wheel, then it is 3 P.M. at one location when it is _____ at the other location.

33. PROVING A THEOREM Write a coordinate proof of the Similar Circles Theorem (Theorem 10.5).

Given $\odot O$ with center $O(0, 0)$ and radius r, $\odot A$ with center $A(a, 0)$ and radius s

Prove $\odot O \sim \odot A$

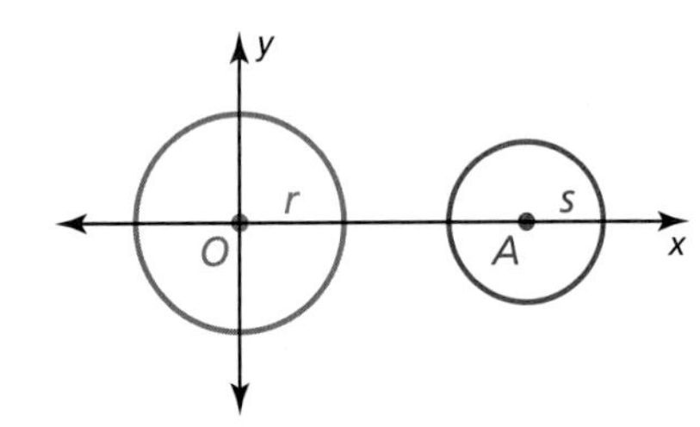

34. ABSTRACT REASONING Is there enough information to tell whether $\odot C \cong \odot D$? Explain your reasoning.

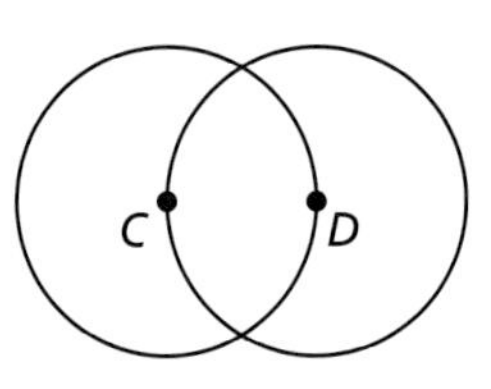

35. PROVING A THEOREM Use the diagram on page 540 to prove each part of the biconditional in the Congruent Circles Theorem (Theorem 10.3).

a. **Given** $\overline{AC} \cong \overline{BD}$
Prove $\odot A \cong \odot B$

b. **Given** $\odot A \cong \odot B$
Prove $\overline{AC} \cong \overline{BD}$

36. HOW DO YOU SEE IT? Are the circles on the target *similar* or *congruent*? Explain your reasoning.

37. PROVING A THEOREM Use the diagram to prove each part of the biconditional in the Congruent Central Angles Theorem (Theorem 10.4).

a. **Given** $\angle BAC \cong \angle DAE$
Prove $\overparen{BC} \cong \overparen{DE}$

b. **Given** $\overparen{BC} \cong \overparen{DE}$
Prove $\angle BAC \cong \angle DAE$

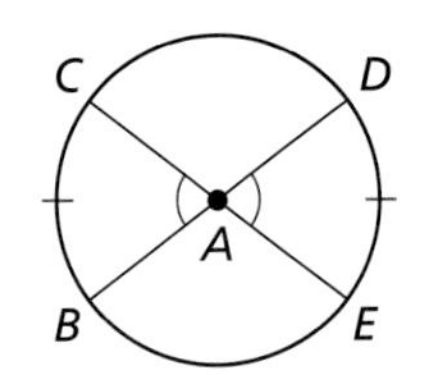

38. THOUGHT PROVOKING Write a formula for the length of a circular arc. Justify your answer.

Maintaining Mathematical Proficiency

Reviewing what you learned in previous grades and lessons

Find the value of x. Tell whether the side lengths form a Pythagorean triple. *(Section 9.1)*

39.

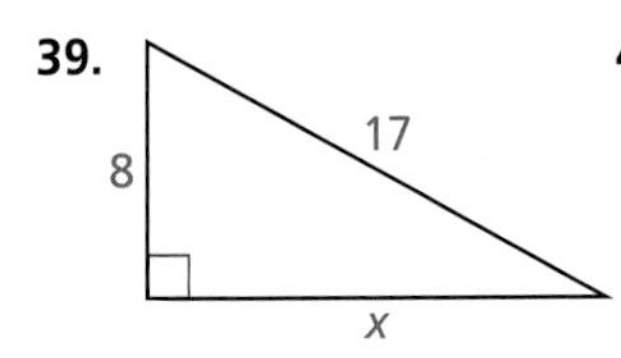

40.

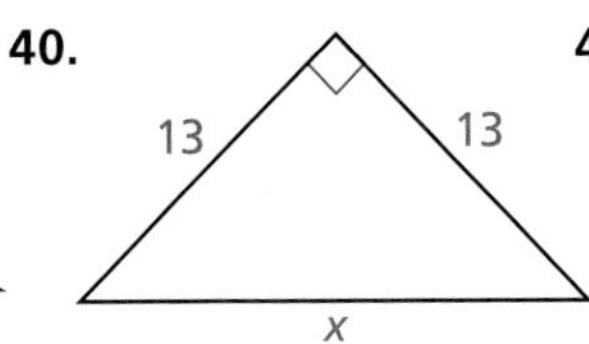

41.

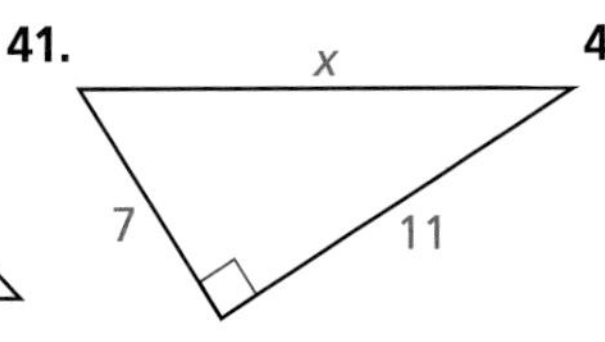

42.

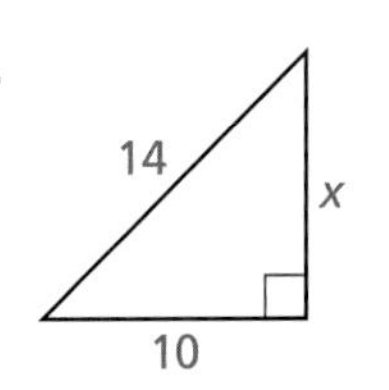

10.3 Using Chords

Essential Question What are two ways to determine when a chord is a diameter of a circle?

EXPLORATION 1 Drawing Diameters

Work with a partner. Use dynamic geometry software to construct a circle of radius 5 with center at the origin. Draw a diameter that has the given point as an endpoint. Explain how you know that the chord you drew is a diameter.

a. $(4, 3)$ **b.** $(0, 5)$ **c.** $(-3, 4)$ **d.** $(-5, 0)$

LOOKING FOR STRUCTURE

To be proficient in math, you need to look closely to discern a pattern or structure.

EXPLORATION 2 Writing a Conjecture about Chords

Work with a partner. Use dynamic geometry software to construct a chord $\overline{BC}$ of a circle A. Construct a chord on the perpendicular bisector of $\overline{BC}$. What do you notice? Change the original chord and the circle several times. Are your results always the same? Use your results to write a conjecture.

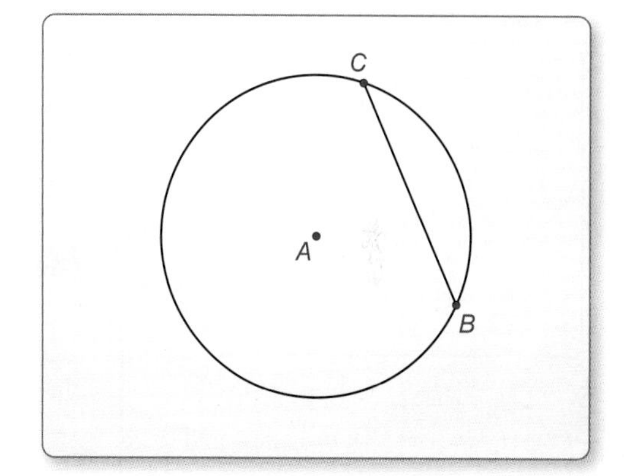

EXPLORATION 3 A Chord Perpendicular to a Diameter

Work with a partner. Use dynamic geometry software to construct a diameter $\overline{BC}$ of a circle A. Then construct a chord $\overline{DE}$ perpendicular to $\overline{BC}$ at point F. Find the lengths DF and EF. What do you notice? Change the chord perpendicular to $\overline{BC}$ and the circle several times. Do you always get the same results? Write a conjecture about a chord that is perpendicular to a diameter of a circle.

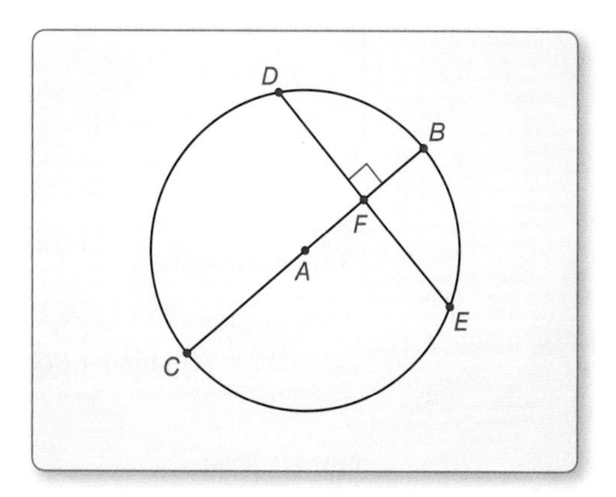

Communicate Your Answer

4. What are two ways to determine when a chord is a diameter of a circle?

10.3 Lesson

Core Vocabulary

Previous
chord
arc
diameter

READING

If $\overset{\frown}{GD} \cong \overset{\frown}{GF}$, then the point G, and any line, segment, or ray that contains G, *bisects* $\overset{\frown}{FD}$.

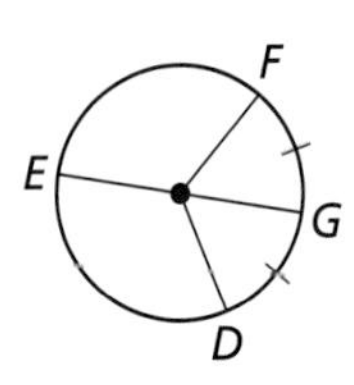

$\overline{EG}$ bisects $\overset{\frown}{FD}$.

What You Will Learn

▶ Use chords of circles to find lengths and arc measures.

Using Chords of Circles

Recall that a *chord* is a segment with endpoints on a circle. Because its endpoints lie on the circle, any chord divides the circle into two arcs. A diameter divides a circle into two semicircles. Any other chord divides a circle into a minor arc and a major arc.

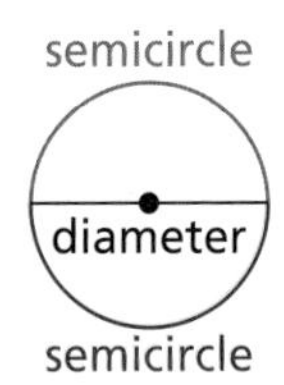

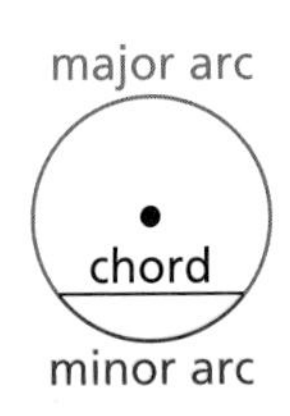

Theorems

Theorem 10.6 Congruent Corresponding Chords Theorem

In the same circle, or in congruent circles, two minor arcs are congruent if and only if their corresponding chords are congruent.

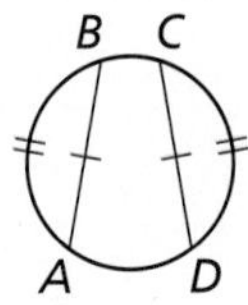

$\overset{\frown}{AB} \cong \overset{\frown}{CD}$ if and only if $\overline{AB} \cong \overline{CD}$.

Proof Ex. 19, p. 550

Theorem 10.7 Perpendicular Chord Bisector Theorem

If a diameter of a circle is perpendicular to a chord, then the diameter bisects the chord and its arc.

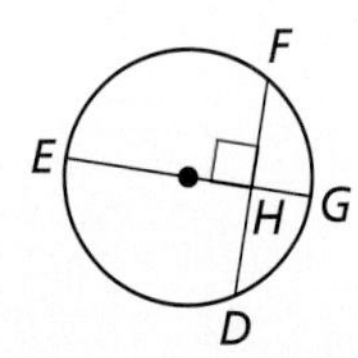

If $\overline{EG}$ is a diameter and $\overline{EG} \perp \overline{DF}$, then $\overline{HD} \cong \overline{HF}$ and $\overset{\frown}{GD} \cong \overset{\frown}{GF}$.

Proof Ex. 22, p. 550

Theorem 10.8 Perpendicular Chord Bisector Converse

If one chord of a circle is a perpendicular bisector of another chord, then the first chord is a diameter.

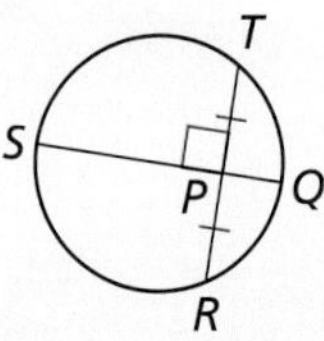

If $\overline{QS}$ is a perpendicular bisector of $\overline{TR}$, then $\overline{QS}$ is a diameter of the circle.

Proof Ex. 23, p. 550

EXAMPLE 1 Using Congruent Chords to Find an Arc Measure

In the diagram, $\odot P \cong \odot Q$, $\overline{FG} \cong \overline{JK}$, and $m\overset{\frown}{JK} = 80°$. Find $m\overset{\frown}{FG}$.

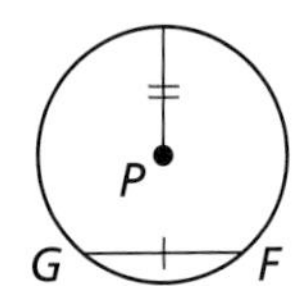

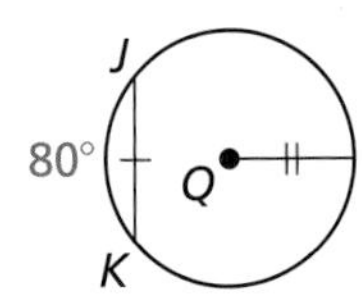

SOLUTION

Because $\overline{FG}$ and $\overline{JK}$ are congruent chords in congruent circles, the corresponding minor arcs $\overset{\frown}{FG}$ and $\overset{\frown}{JK}$ are congruent by the Congruent Corresponding Chords Theorem.

▶ So, $m\overset{\frown}{FG} = m\overset{\frown}{JK} = 80°$.

EXAMPLE 2 Using a Diameter

a. Find HK. **b.** Find $m\widehat{HK}$.

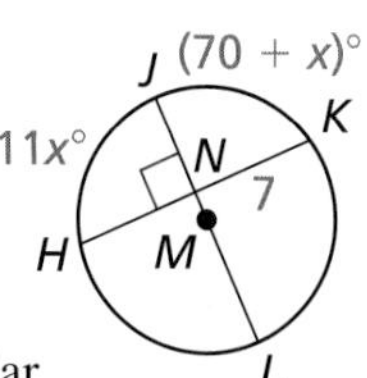

SOLUTION

a. Diameter $\overline{JL}$ is perpendicular to $\overline{HK}$. So, by the Perpendicular Chord Bisector Theorem, $\overline{JL}$ bisects $\overline{HK}$, and $HN = NK$.

▶ So, $HK = 2(NK) = 2(7) = 14$.

b. Diameter $\overline{JL}$ is perpendicular to $\overline{HK}$. So, by the Perpendicular Chord Bisector Theorem, $\overline{JL}$ bisects $\widehat{HK}$, and $m\widehat{HJ} = m\widehat{JK}$.

$m\widehat{HJ} = m\widehat{JK}$	Perpendicular Chord Bisector Theorem
$11x^\circ = (70 + x)^\circ$	Substitute.
$10x = 70$	Subtract x from each side.
$x = 7$	Divide each side by 10.

▶ So, $m\widehat{HJ} = m\widehat{JK} = (70 + x)^\circ = (70 + 7)^\circ = 77^\circ$, and $m\widehat{HK} = 2(m\widehat{HJ}) = 2(77^\circ) = 154^\circ$.

EXAMPLE 3 Using Perpendicular Bisectors

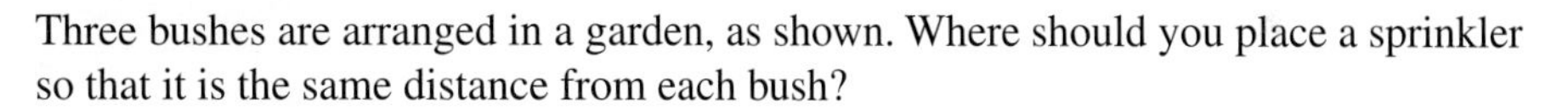

Three bushes are arranged in a garden, as shown. Where should you place a sprinkler so that it is the same distance from each bush?

SOLUTION

Step 1

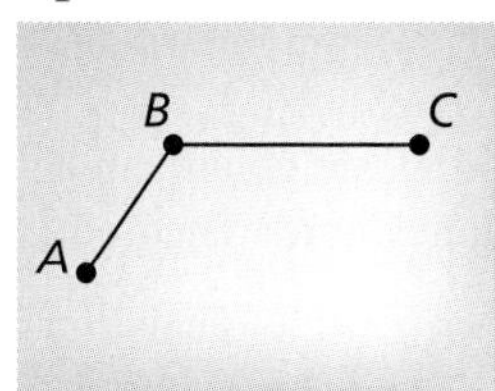

Label the bushes A, B, and C, as shown. Draw segments $\overline{AB}$ and $\overline{BC}$.

Step 2

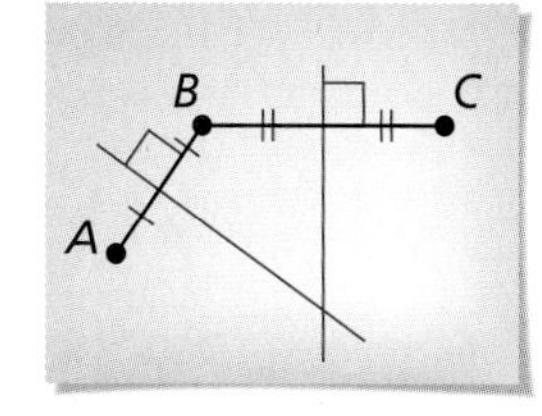

Draw the perpendicular bisectors of $\overline{AB}$ and $\overline{BC}$. By the Perpendicular Chord Bisector Converse, these lie on diameters of the circle containing A, B, and C.

Step 3

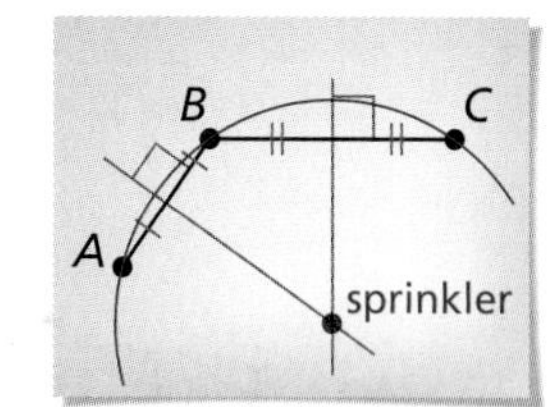

Find the point where the perpendicular bisectors intersect. This is the center of the circle, which is equidistant from points A, B, and C.

Monitoring Progress

Help in English and Spanish at *BigIdeasMath.com*

In Exercises 1 and 2, use the diagram of $\odot D$.

1. If $m\widehat{AB} = 110^\circ$, find $m\widehat{BC}$.

2. If $m\widehat{AC} = 150^\circ$, find $m\widehat{AB}$.

In Exercises 3 and 4, find the indicated length or arc measure.

3. CE

4. $m\widehat{CE}$

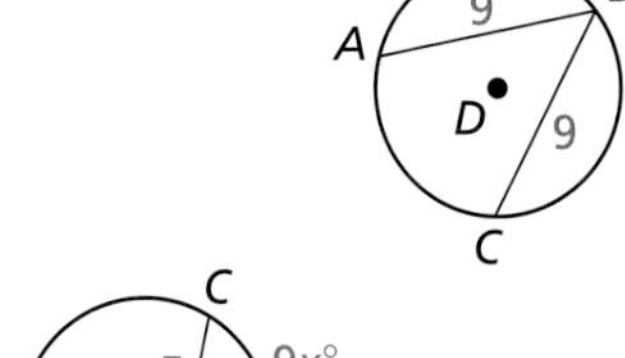

Theorem

Theorem 10.9 Equidistant Chords Theorem

In the same circle, or in congruent circles, two chords are congruent if and only if they are equidistant from the center.

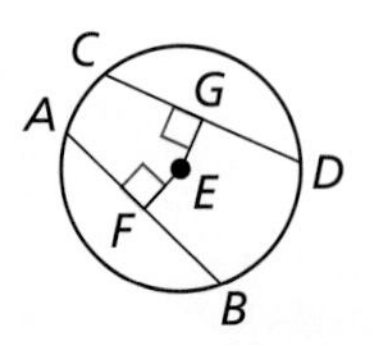

Proof Ex. 25, p. 550

$\overline{AB} \cong \overline{CD}$ if and only if $EF = EG$.

EXAMPLE 4 Using Congruent Chords to Find a Circle's Radius

In the diagram, $QR = ST = 16$, $CU = 2x$, and $CV = 5x - 9$. Find the radius of $\odot C$.

SOLUTION

Because $\overline{CQ}$ is a segment whose endpoints are the center and a point on the circle, it is a radius of $\odot C$. Because $\overline{CU} \perp \overline{QR}$, $\triangle QUC$ is a right triangle. Apply properties of chords to find the lengths of the legs of $\triangle QUC$.

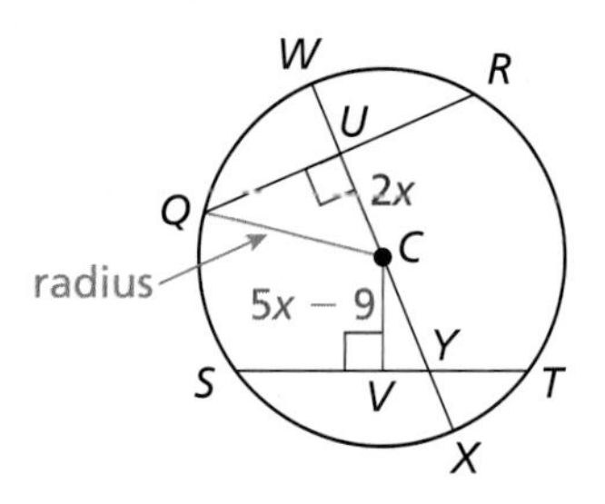

Step 1 Find CU.

Because $\overline{QR}$ and $\overline{ST}$ are congruent chords, $\overline{QR}$ and $\overline{ST}$ are equidistant from C by the Equidistant Chords Theorem. So, $CU = CV$.

$CU = CV$	Equidistant Chords Theorem
$2x = 5x - 9$	Substitute.
$x = 3$	Solve for x.

So, $CU = 2x = 2(3) = 6$.

Step 2 Find QU.

Because diameter $\overline{WX} \perp \overline{QR}$, $\overline{WX}$ bisects $\overline{QR}$ by the Perpendicular Chord Bisector Theorem.

So, $QU = \frac{1}{2}(16) = 8$.

Step 3 Find CQ.

Because the lengths of the legs are $CU = 6$ and $QU = 8$, $\triangle QUC$ is a right triangle with the Pythagorean triple 6, 8, 10. So, $CQ = 10$.

▶ So, the radius of $\odot C$ is 10 units.

Monitoring Progress

Help in English and Spanish at *BigIdeasMath.com*

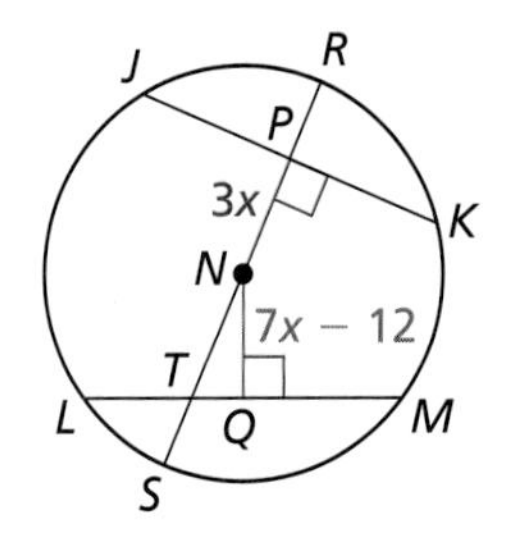

5. In the diagram, $JK = LM = 24$, $NP = 3x$, and $NQ = 7x - 12$. Find the radius of $\odot N$.

10.3 Exercises

Dynamic Solutions available at *BigIdeasMath.com*

Vocabulary and Core Concept Check

1. **WRITING** Describe what it means to bisect a chord.

2. **WRITING** Two chords of a circle are perpendicular and congruent. Does one of them have to be a diameter? Explain your reasoning.

Monitoring Progress and Modeling with Mathematics

In Exercises 3–6, find the measure of the red arc or chord in $\odot C$. *(See Example 1.)*

3. 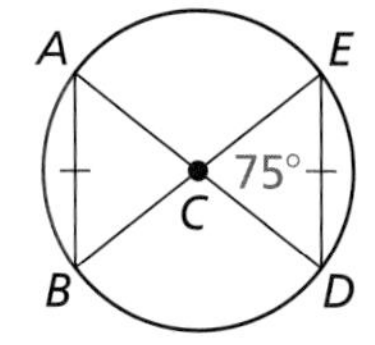

4. 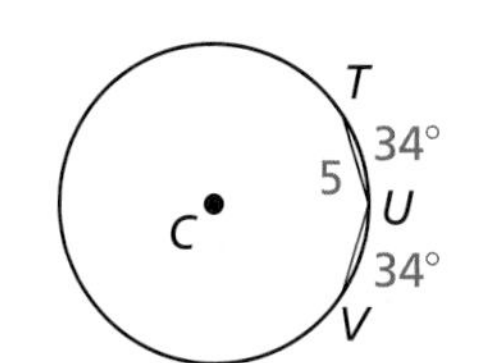

5. 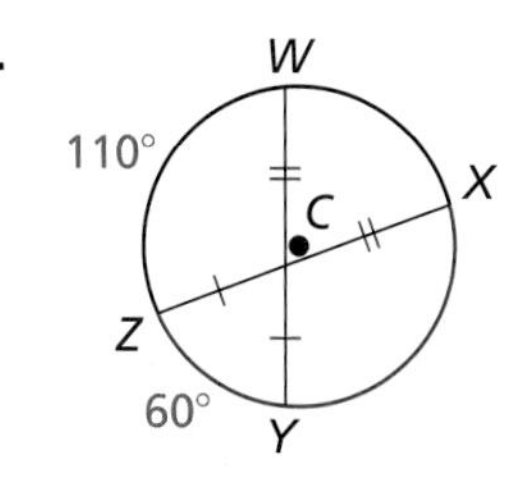

6. 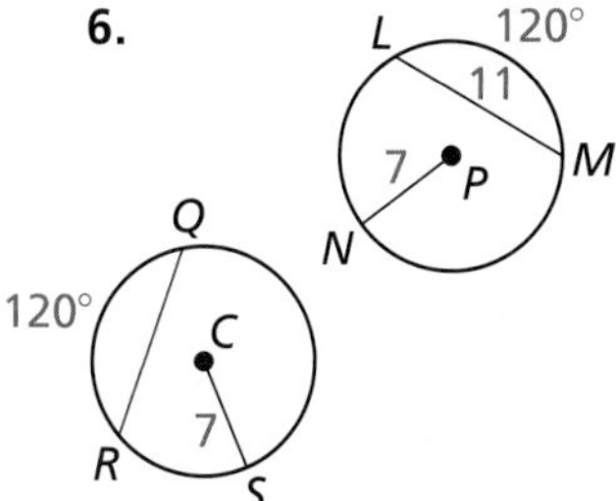

In Exercises 7–10, find the value of x. *(See Example 2.)*

7. 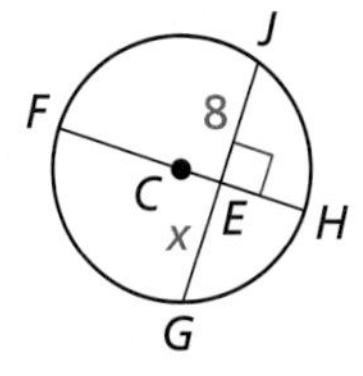

8. 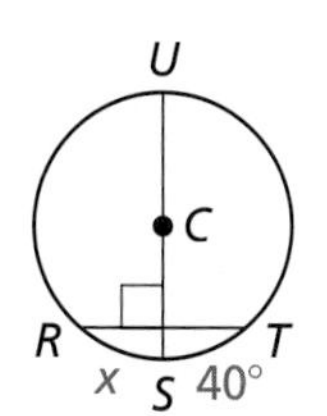

9. 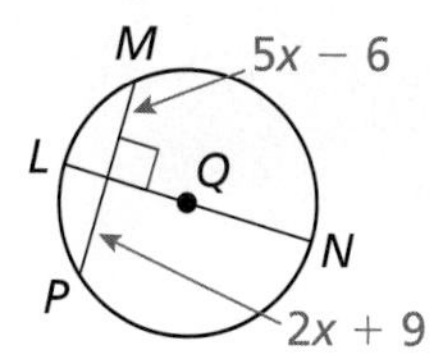

10. 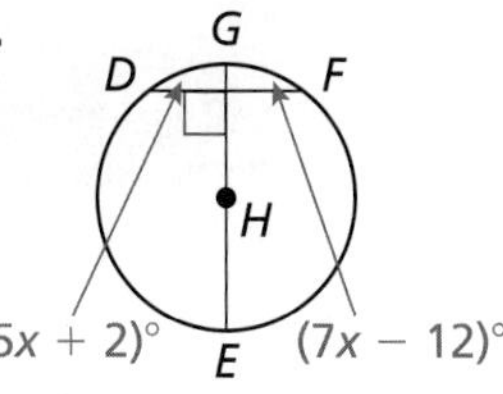

11. **ERROR ANALYSIS** Describe and correct the error in reasoning.

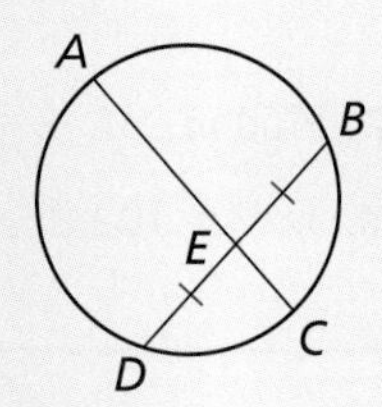

Because $\overline{AC}$ bisects $\overline{DB}$, $\overset{\frown}{BC} \cong \overset{\frown}{CD}$.

12. **PROBLEM SOLVING** In the cross section of the submarine shown, the control panels are parallel and the same length. Describe a method you can use to find the center of the cross section. Justify your method. *(See Example 3.)*

In Exercises 13 and 14, determine whether $\overline{AB}$ is a diameter of the circle. Explain your reasoning.

13. 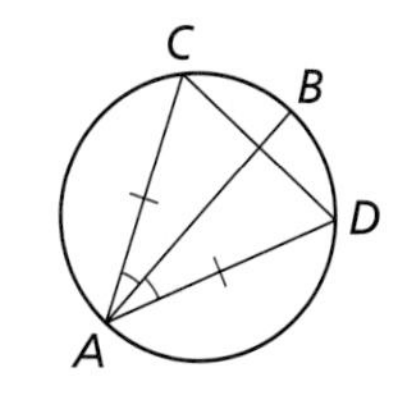

14. 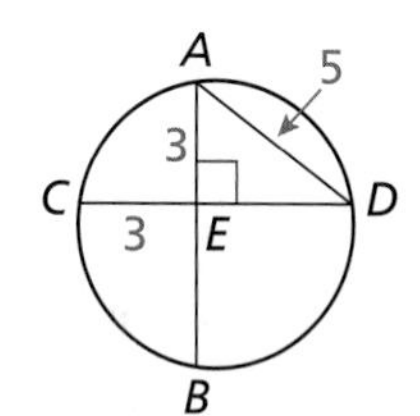

In Exercises 15 and 16, find the radius of $\odot Q$. *(See Example 4.)*

15. 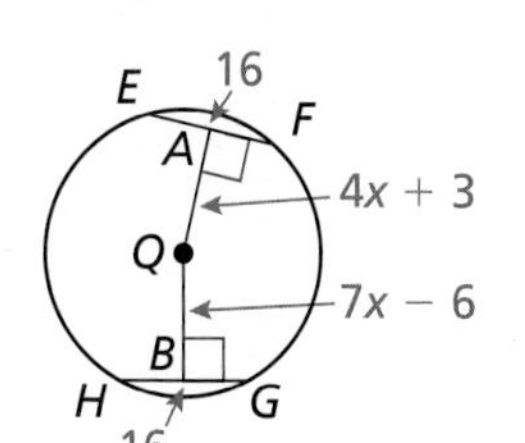

16. 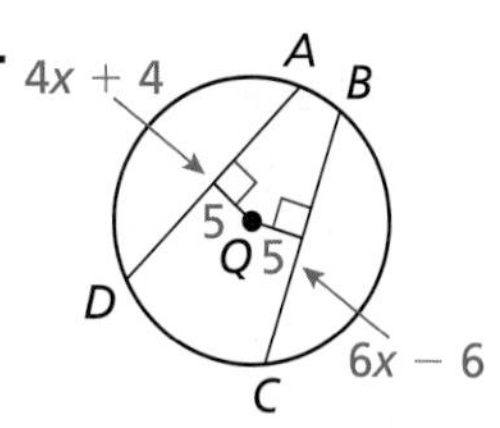

17. **PROBLEM SOLVING** An archaeologist finds part of a circular plate. What was the diameter of the plate to the nearest tenth of an inch? Justify your answer.

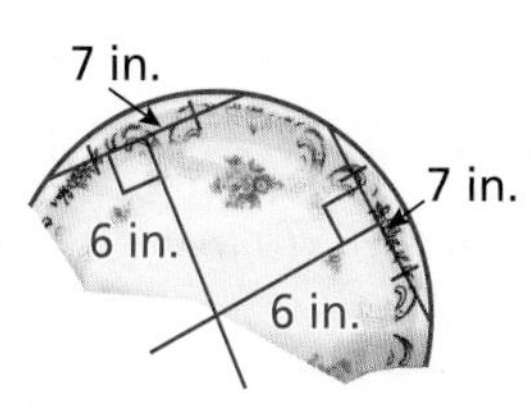

18. HOW DO YOU SEE IT? What can you conclude from each diagram? Name a theorem that justifies your answer.

a.

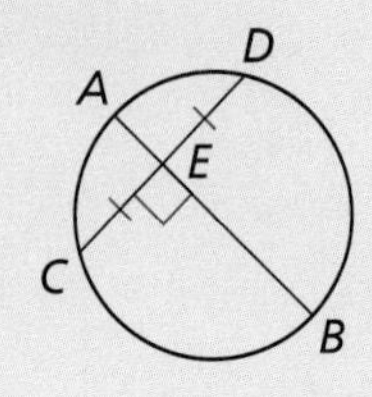

b.

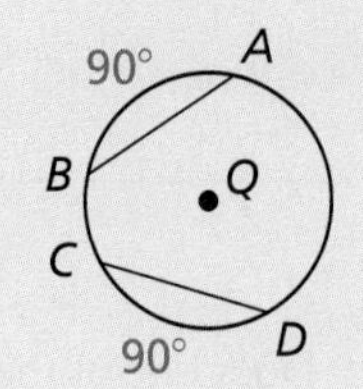

c.

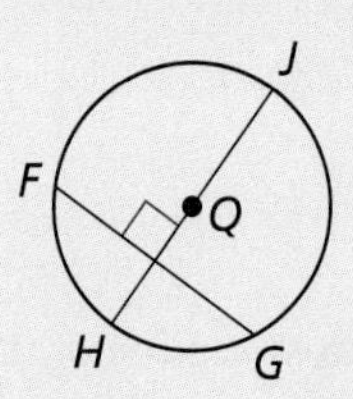

d.

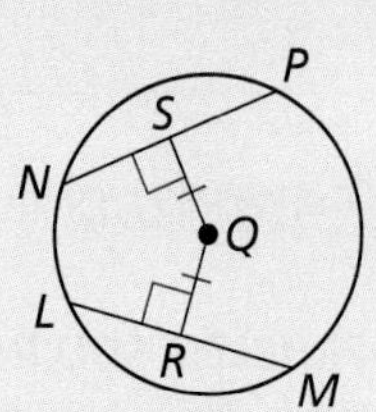

19. PROVING A THEOREM Use the diagram to prove each part of the biconditional in the Congruent Corresponding Chords Theorem (Theorem 10.6).

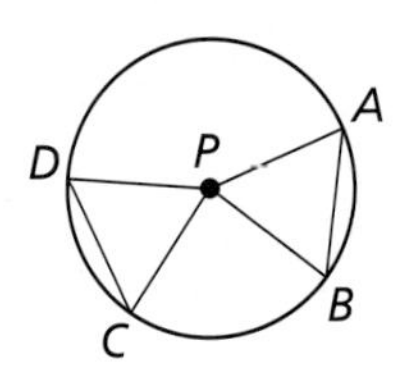

a. Given $\overline{AB}$ and $\overline{CD}$ are congruent chords.
Prove $\widehat{AB} \cong \widehat{CD}$

b. Given $\widehat{AB} \cong \widehat{CD}$
Prove $\overline{AB} \cong \overline{CD}$

20. MATHEMATICAL CONNECTIONS In $\odot P$, all the arcs shown have integer measures. Show that x must be even.

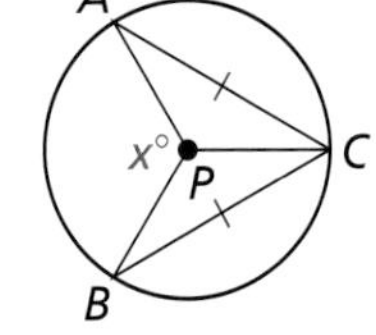

21. REASONING In $\odot P$, the lengths of the parallel chords are 20, 16, and 12. Find $m\widehat{AB}$. Explain your reasoning.

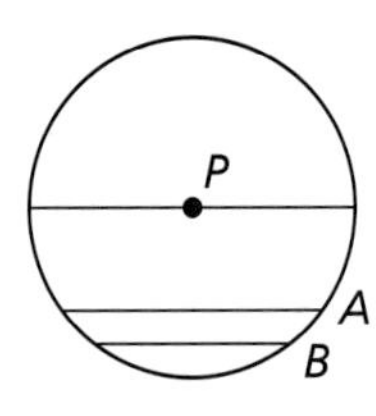

22. PROVING A THEOREM Use congruent triangles to prove the Perpendicular Chord Bisector Theorem (Theorem 10.7).

Given $\overline{EG}$ is a diameter of $\odot L$.
$\overline{EG} \perp \overline{DF}$

Prove $\overline{DC} \cong \overline{FC}$, $\widehat{DG} \cong \widehat{FG}$

23. PROVING A THEOREM Write a proof of the Perpendicular Chord Bisector Converse (Theorem 10.8).

Given $\overline{QS}$ is a perpendicular bisector of $\overline{RT}$.

Prove $\overline{QS}$ is a diameter of the circle L.

(*Hint*: Plot the center L and draw $\triangle LPT$ and $\triangle LPR$.)

24. THOUGHT PROVOKING Consider two chords that intersect at point P. Do you think that $\frac{AP}{BP} = \frac{CP}{DP}$? Justify your answer.

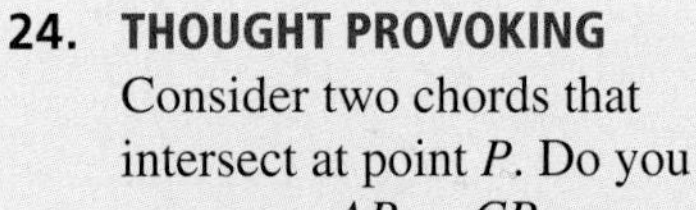

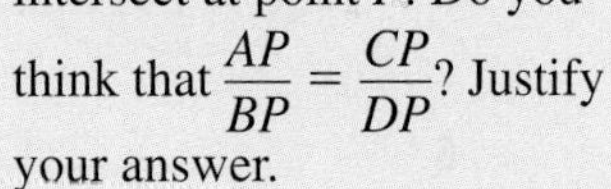

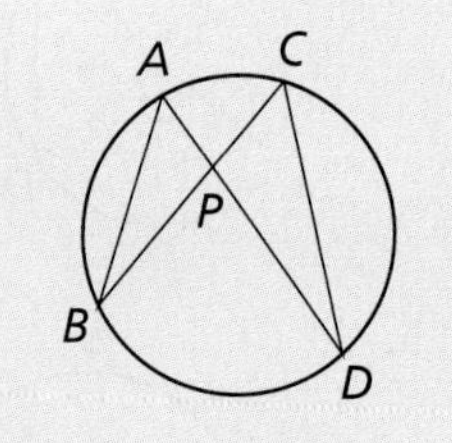

25. PROVING A THEOREM Use the diagram with the Equidistant Chords Theorem (Theorem 10.9) on page 548 to prove both parts of the biconditional of this theorem.

26. MAKING AN ARGUMENT A car is designed so that the rear wheel is only partially visible below the body of the car. The bottom edge of the panel is parallel to the ground. Your friend claims that the point where the tire touches the ground bisects $\widehat{AB}$. Is your friend correct? Explain your reasoning.

Maintaining Mathematical Proficiency

Reviewing what you learned in previous grades and lessons

Find the missing interior angle measure. *(Section 7.1)*

27. Quadrilateral $JKLM$ has angle measures $m\angle J = 32^\circ$, $m\angle K = 25^\circ$, and $m\angle L = 44^\circ$. Find $m\angle M$.

28. Pentagon $PQRST$ has angle measures $m\angle P = 85^\circ$, $m\angle Q = 134^\circ$, $m\angle R = 97^\circ$, and $m\angle S = 102^\circ$. Find $m\angle T$.

10.1–10.3 What Did You Learn?

Core Vocabulary

circle, *p. 530*
center, *p. 530*
radius, *p. 530*
chord, *p. 530*
diameter, *p. 530*
secant, *p. 530*
tangent, *p. 530*
point of tangency, *p. 530*
tangent circles, *p. 531*
concentric circles, *p. 531*
common tangent, *p. 531*
central angle, *p. 538*
minor arc, *p. 538*
major arc, *p. 538*
semicircle, *p. 538*
measure of a minor arc, *p. 538*
measure of a major arc, *p. 538*
adjacent arcs, *p. 539*
congruent circles, *p. 540*
congruent arcs, *p. 540*
similar arcs, *p. 541*

Core Concepts

Section 10.1

Lines and Segments That Intersect Circles, *p. 530*
Coplanar Circles and Common Tangents, *p. 531*
Theorem 10.1 Tangent Line to Circle Theorem, *p. 532*
Theorem 10.2 External Tangent Congruence Theorem, *p. 532*

Section 10.2

Measuring Arcs, *p. 538*
Postulate 10.1 Arc Addition Postulate, *p. 539*
Theorem 10.3 Congruent Circles Theorem, *p. 540*
Theorem 10.4 Congruent Central Angles Theorem, *p. 540*
Theorem 10.5 Similar Circles Theorem, *p. 541*

Section 10.3

Theorem 10.6 Congruent Corresponding Chords Theorem, *p. 546*
Theorem 10.7 Perpendicular Chord Bisector Theorem, *p. 546*
Theorem 10.8 Perpendicular Chord Bisector Converse, *p. 546*
Theorem 10.9 Equidistant Chords Theorem, *p. 548*

Mathematical Practices

1. Explain how separating quadrilateral *TVWX* into several segments helped you solve Exercise 37 on page 535.
2. In Exercise 30 on page 543, what two cases did you consider to reach your answers? Are there any other cases? Explain your reasoning.
3. Explain how you used inductive reasoning to solve Exercise 24 on page 550.

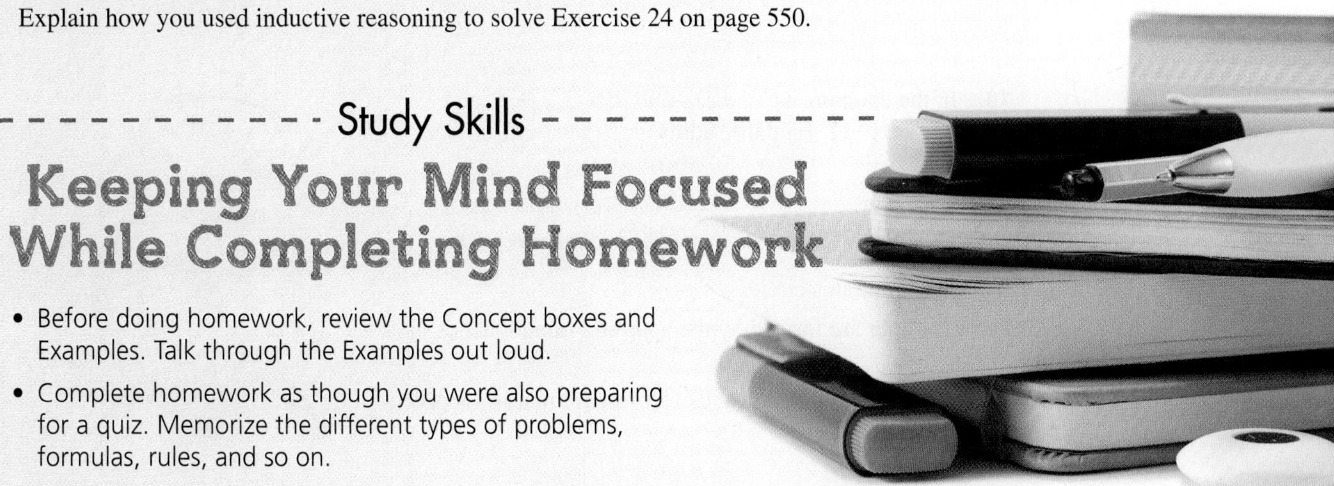

Study Skills

Keeping Your Mind Focused While Completing Homework

- Before doing homework, review the Concept boxes and Examples. Talk through the Examples out loud.
- Complete homework as though you were also preparing for a quiz. Memorize the different types of problems, formulas, rules, and so on.

10.1–10.3 Quiz

In Exercises 1–6, use the diagram. *(Section 10.1)*

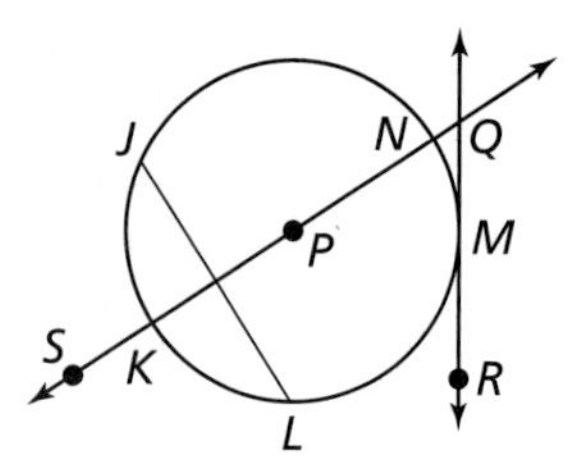

1. Name the circle.

2. Name a radius.

3. Name a diameter.

4. Name a chord.

5. Name a secant.

6. Name a tangent.

Find the value of x. *(Section 10.1)*

7.

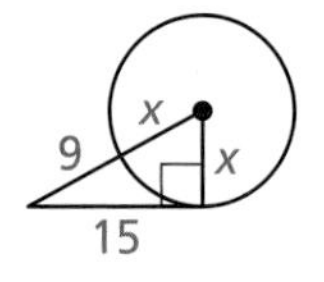

8.

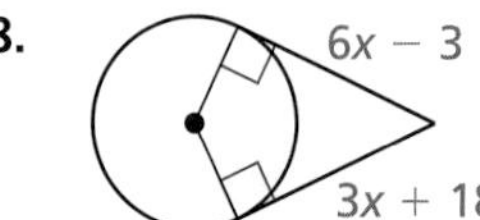

Identify the given arc as a *major arc*, *minor arc*, or *semicircle*. Then find the measure of the arc. *(Section 10.2)*

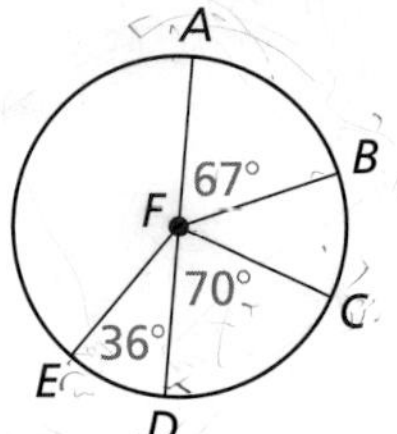

9. $\widehat{AE}$

10. $\widehat{BC}$

11. $\widehat{AC}$

12. $\widehat{ACD}$

13. $\widehat{ACE}$

14. $\widehat{BEC}$

Tell whether the red arcs are congruent. Explain why or why not. *(Section 10.2)*

15.

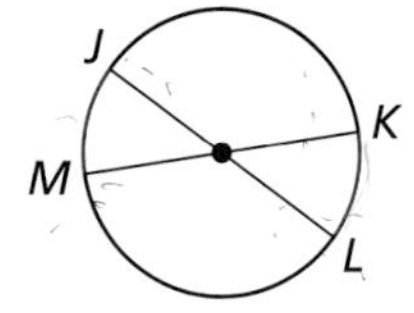

16.

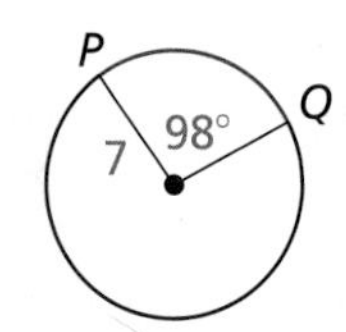

S T 98° 15 R

17. Find the measure of the red arc in $\odot Q$. *(Section 10.3)*

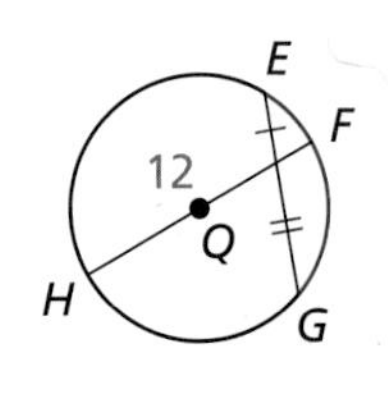

E F 12 Q H G

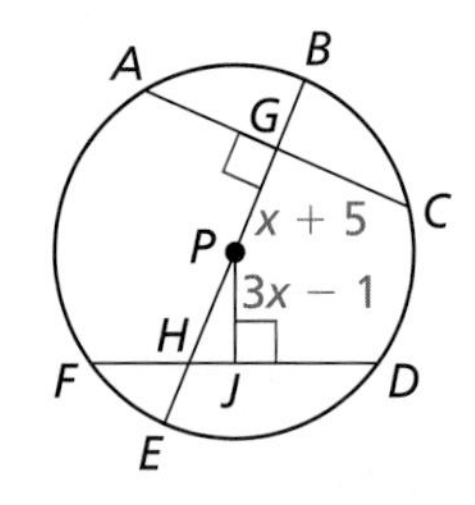

18. In the diagram, $AC = FD = 30$, $PG = x + 5$, and $PJ = 3x - 1$. Find the radius of $\odot P$. *(Section 10.3)*

19. A circular clock can be divided into 12 congruent sections. *(Section 10.2)*

a. Find the measure of each arc in this circle.

b. Find the measure of the minor arc formed by the hour and minute hands when the time is 7:00.

c. Find a time at which the hour and minute hands form an arc that is congruent to the arc in part (b).

10.4 Inscribed Angles and Polygons

Essential Question

How are inscribed angles related to their intercepted arcs? How are the angles of an inscribed quadrilateral related to each other?

An **inscribed angle** is an angle whose vertex is on a circle and whose sides contain chords of the circle. An arc that lies between two lines, rays, or segments is called an **intercepted arc**. A polygon is an **inscribed polygon** when all its vertices lie on a circle.

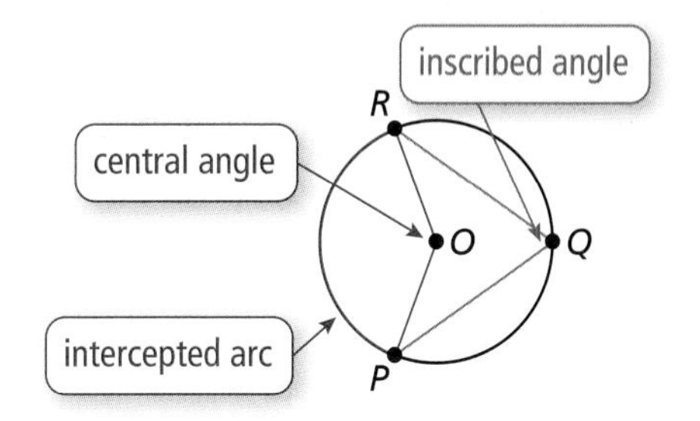

EXPLORATION 1 Inscribed Angles and Central Angles

Work with a partner. Use dynamic geometry software.

a. Construct an inscribed angle in a circle. Then construct the corresponding central angle.

b. Measure both angles. How is the inscribed angle related to its intercepted arc?

c. Repeat parts (a) and (b) several times. Record your results in a table. Write a conjecture about how an inscribed angle is related to its intercepted arc.

Sample

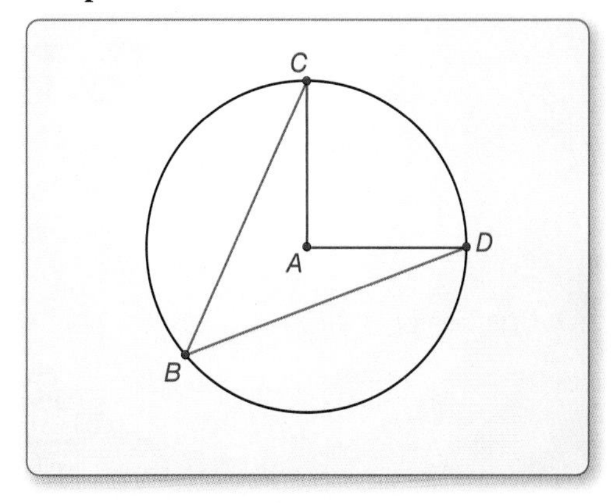

ATTENDING TO PRECISION

To be proficient in math, you need to communicate precisely with others.

EXPLORATION 2 A Quadrilateral with Inscribed Angles

Work with a partner. Use dynamic geometry software.

a. Construct a quadrilateral with each vertex on a circle.

b. Measure all four angles. What relationships do you notice?

c. Repeat parts (a) and (b) several times. Record your results in a table. Then write a conjecture that summarizes the data.

Sample

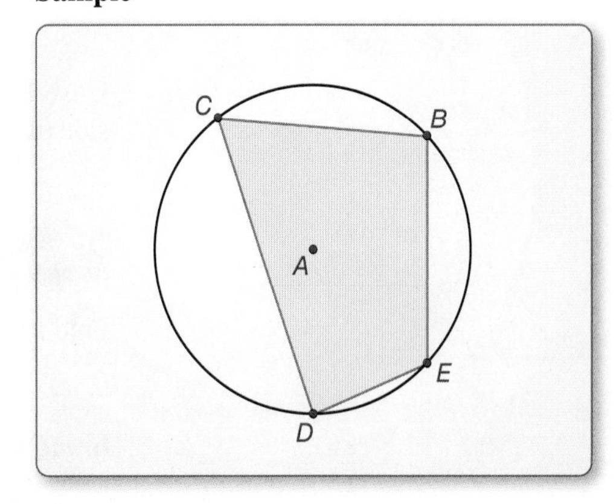

Communicate Your Answer

3. How are inscribed angles related to their intercepted arcs? How are the angles of an inscribed quadrilateral related to each other?

4. Quadrilateral $EFGH$ is inscribed in $\odot C$, and $m\angle E = 80°$. What is $m\angle G$? Explain.

10.4 Lesson

Core Vocabulary

inscribed angle, *p. 554*
intercepted arc, *p. 554*
subtend, *p. 554*
inscribed polygon, *p. 556*
circumscribed circle, *p. 556*

What You Will Learn

- Use inscribed angles.
- Use inscribed polygons.

Using Inscribed Angles

Core Concept

Inscribed Angle and Intercepted Arc

An **inscribed angle** is an angle whose vertex is on a circle and whose sides contain chords of the circle. An arc that lies between two lines, rays, or segments is called an **intercepted arc**. If the endpoints of a chord or arc lie on the sides of an inscribed angle, then the chord or arc is said to **subtend** the angle.

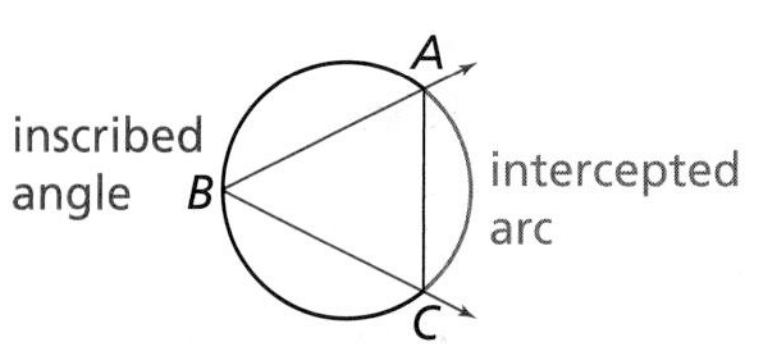

$\angle B$ intercepts $\widehat{AC}$.
$\widehat{AC}$ subtends $\angle B$.
$\overline{AC}$ subtends $\angle B$.

Theorem

Theorem 10.10 Measure of an Inscribed Angle Theorem

The measure of an inscribed angle is one-half the measure of its intercepted arc.

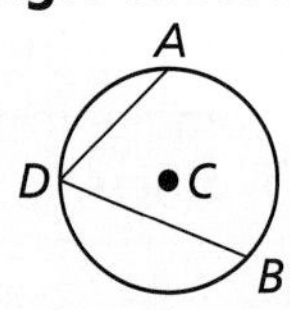

$m\angle ADB = \frac{1}{2}m\widehat{AB}$

Proof Ex. 37, p. 560

The proof of the Measure of an Inscribed Angle Theorem involves three cases.

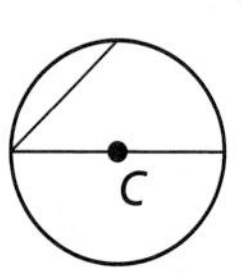

Case 1 Center C is on a side of the inscribed angle.

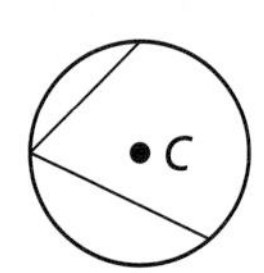

Case 2 Center C is inside the inscribed angle.

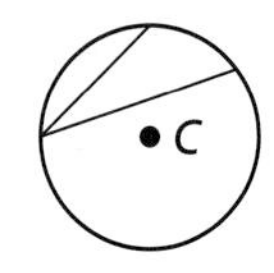

Case 3 Center C is outside the inscribed angle.

EXAMPLE 1 Using Inscribed Angles

Find the indicated measure.

a. $m\angle T$

b. $m\widehat{QR}$

SOLUTION

a. $m\angle T = \frac{1}{2}m\widehat{RS} = \frac{1}{2}(48°) = 24°$

b. $m\widehat{TQ} = 2m\angle R = 2 \cdot 50° = 100°$
Because $\widehat{TQR}$ is a semicircle, $m\widehat{QR} = 180° - m\widehat{TQ} = 180° - 100° = 80°$.

EXAMPLE 2 Finding the Measure of an Intercepted Arc

Find $m\overset{\frown}{RS}$ and $m\angle STR$. What do you notice about $\angle STR$ and $\angle RUS$?

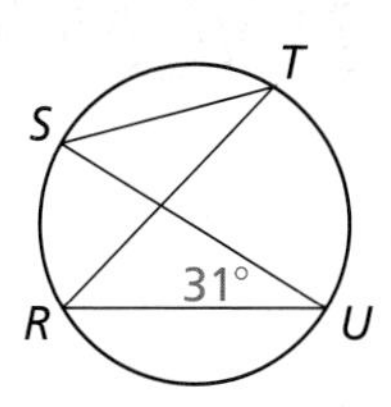

SOLUTION

From the Measure of an Inscribed Angle Theorem, you know that $m\overset{\frown}{RS} = 2m\angle RUS = 2(31^\circ) = 62^\circ$.

Also, $m\angle STR = \frac{1}{2}m\overset{\frown}{RS} = \frac{1}{2}(62^\circ) = 31^\circ$.

▶ So, $\angle STR \cong \angle RUS$.

Example 2 suggests the Inscribed Angles of a Circle Theorem.

Theorem

Theorem 10.11 Inscribed Angles of a Circle Theorem

If two inscribed angles of a circle intercept the same arc, then the angles are congruent.

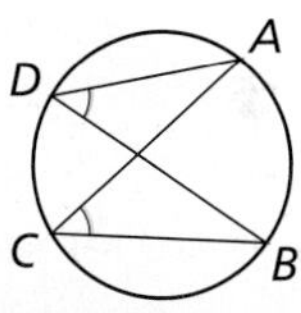

Proof Ex. 38, p. 560

$\angle ADB \cong \angle ACB$

EXAMPLE 3 Finding the Measure of an Angle

Given $m\angle E = 75^\circ$, find $m\angle F$.

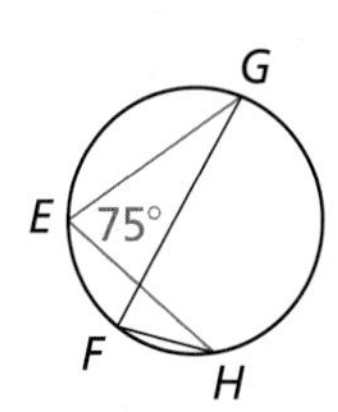

SOLUTION

Both $\angle E$ and $\angle F$ intercept $\overset{\frown}{GH}$. So, $\angle E \cong \angle F$ by the Inscribed Angles of a Circle Theorem.

▶ So, $m\angle F = m\angle E = 75^\circ$.

Monitoring Progress  Help in English and Spanish at *BigIdeasMath.com*

Find the measure of the red arc or angle.

1.

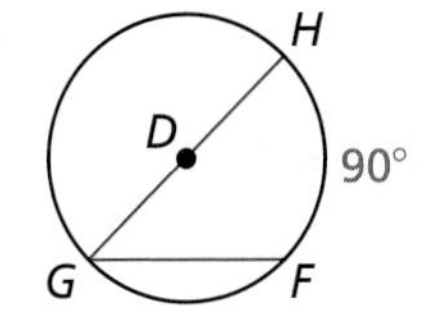

2.

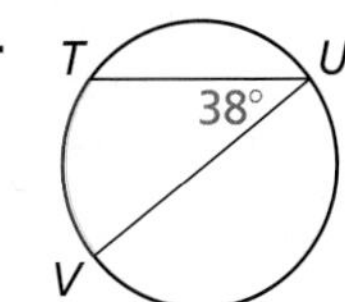

3. Y X 72° W Z

Using Inscribed Polygons

Core Concept

Inscribed Polygon

A polygon is an **inscribed polygon** when all its vertices lie on a circle. The circle that contains the vertices is a **circumscribed circle**.

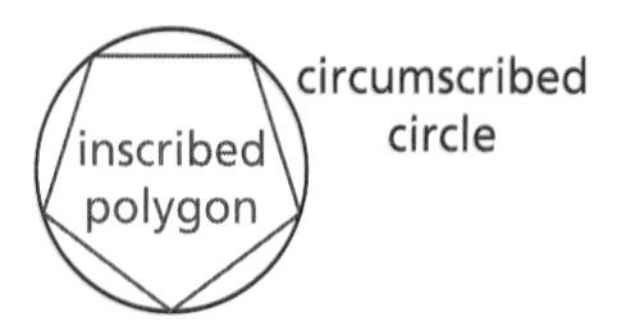

Theorems

Theorem 10.12 Inscribed Right Triangle Theorem

If a right triangle is inscribed in a circle, then the hypotenuse is a diameter of the circle. Conversely, if one side of an inscribed triangle is a diameter of the circle, then the triangle is a right triangle and the angle opposite the diameter is the right angle.

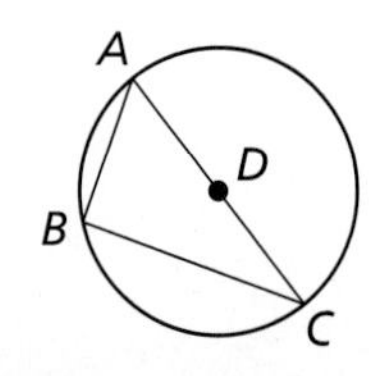

$m\angle ABC = 90°$ if and only if $\overline{AC}$ is a diameter of the circle.

Proof Ex. 39, p. 560

Theorem 10.13 Inscribed Quadrilateral Theorem

A quadrilateral can be inscribed in a circle if and only if its opposite angles are supplementary.

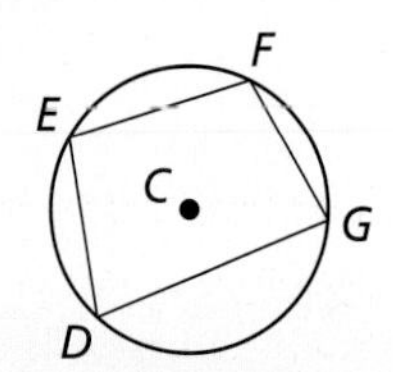

D, E, F, and G lie on $\odot C$ if and only if $m\angle D + m\angle F = m\angle E + m\angle G = 180°$.

Proof Ex. 40, p. 560; *BigIdeasMath.com*

EXAMPLE 4 Using Inscribed Polygons

Find the value of each variable.

a.

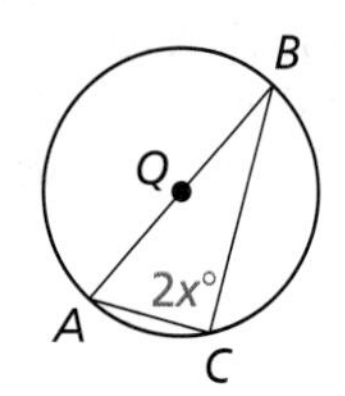

b.

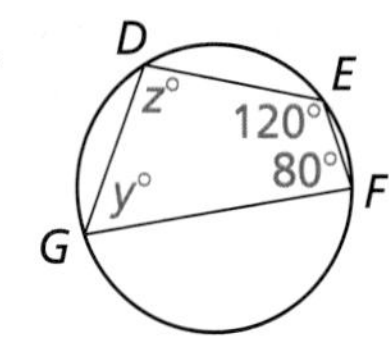

SOLUTION

a. $\overline{AB}$ is a diameter. So, $\angle C$ is a right angle, and $m\angle C = 90°$ by the Inscribed Right Triangle Theorem.

$$2x° = 90°$$

$$x = 45$$

▶ The value of x is 45.

b. *DEFG* is inscribed in a circle, so opposite angles are supplementary by the Inscribed Quadrilateral Theorem.

$$m\angle D + m\angle F = 180° \qquad m\angle E + m\angle G = 180°$$

$$z + 80 = 180 \qquad 120 + y = 180$$

$$z = 100 \qquad y = 60$$

▶ The value of z is 100 and the value of y is 60.

CONSTRUCTION Constructing a Square Inscribed in a Circle

Given $\odot C$, construct a square inscribed in a circle.

SOLUTION

Step 1

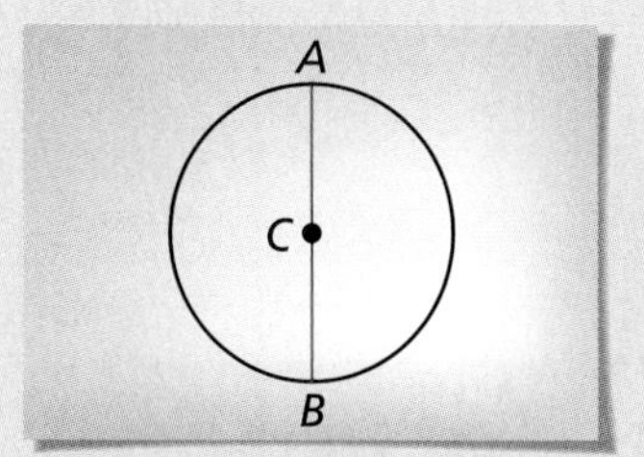

Draw a diameter
Draw any diameter. Label the endpoints A and B.

Step 2

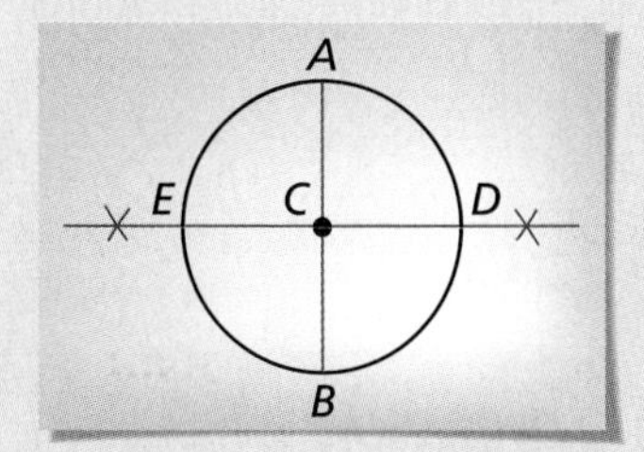

Construct a perpendicular bisector
Construct the perpendicular bisector of the diameter. Label the points where it intersects $\odot C$ as points D and E.

Step 3

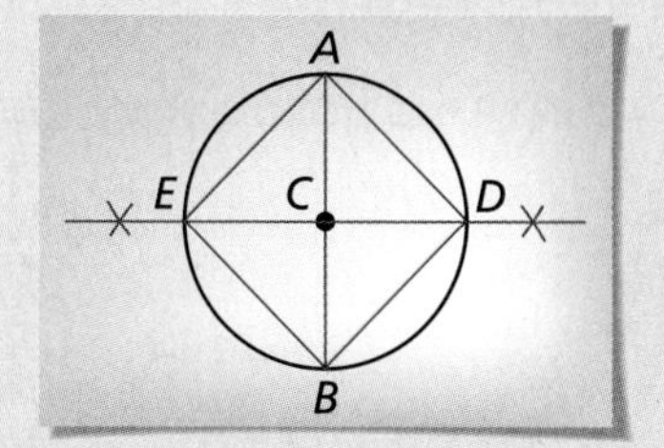

Form a square
Connect points A, D, B, and E to form a square.

EXAMPLE 5 Using a Circumscribed Circle

Your camera has a 90° field of vision, and you want to photograph the front of a statue. You stand at a location in which the front of the statue is all that appears in your camera's field of vision, as shown. You want to change your location. Where else can you stand so that the front of the statue is all that appears in your camera's field of vision?

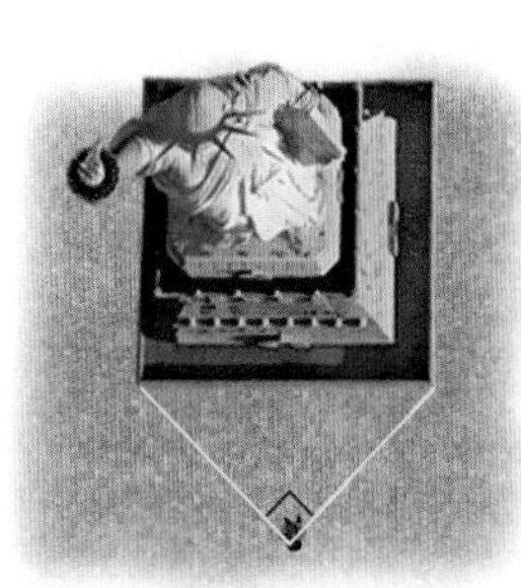

SOLUTION

From the Inscribed Right Triangle Theorem, you know that if a right triangle is inscribed in a circle, then the hypotenuse of the triangle is a diameter of the circle. So, draw the circle that has the front of the statue as a diameter.

▶ The statue fits perfectly within your camera's 90° field of vision from any point on the semicircle in front of the statue.

Monitoring Progress

Help in English and Spanish at *BigIdeasMath.com*

Find the value of each variable.

4.

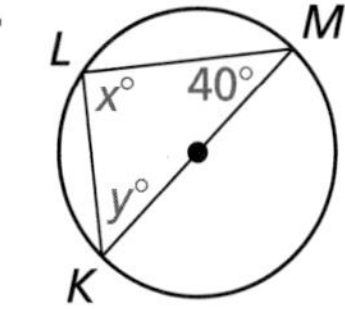

5.

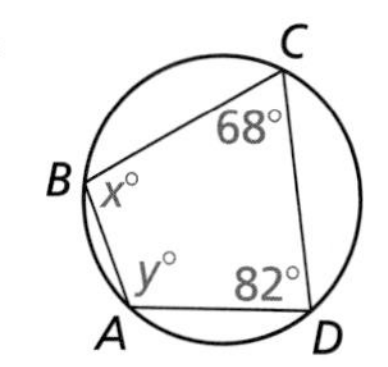

6.

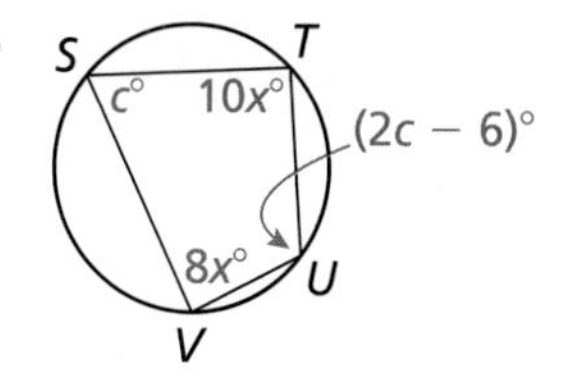

7. In Example 5, explain how to find locations where the left side of the statue is all that appears in your camera's field of vision.

10.4 Exercises

Dynamic Solutions available at *BigIdeasMath.com*

Vocabulary and Core Concept Check

1. **VOCABULARY** If a circle is circumscribed about a polygon, then the polygon is an ______________.

2. **DIFFERENT WORDS, SAME QUESTION** Which is different? Find "both" answers.

Find $m\angle ABC$.	Find $m\angle AGC$.
Find $m\angle AEC$.	Find $m\angle ADC$.

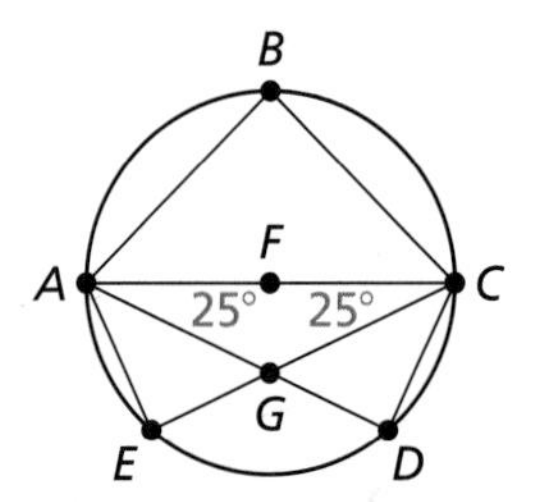

Monitoring Progress and Modeling with Mathematics

In Exercises 3–8, find the indicated measure. *(See Examples 1 and 2.)*

3. $m\angle A$

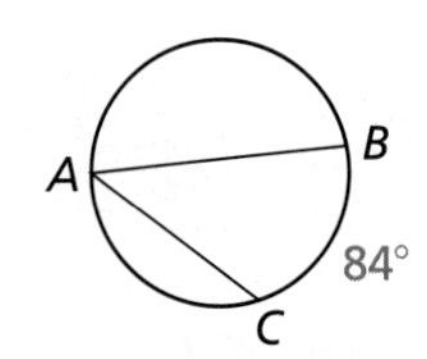

4. $m\angle G$

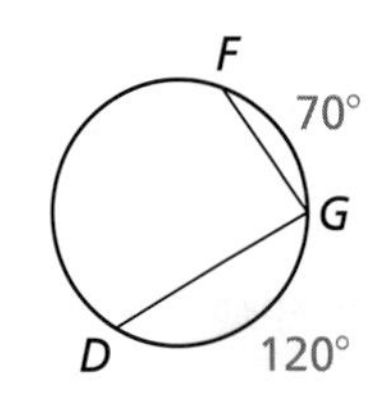

5. $m\angle N$

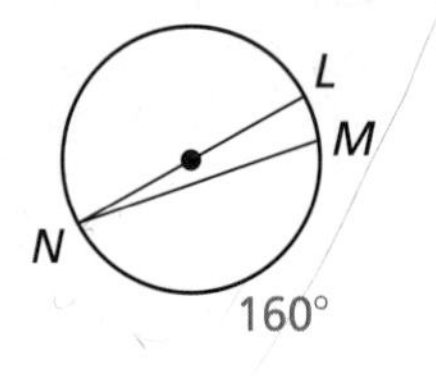

6. $m\widehat{RS}$

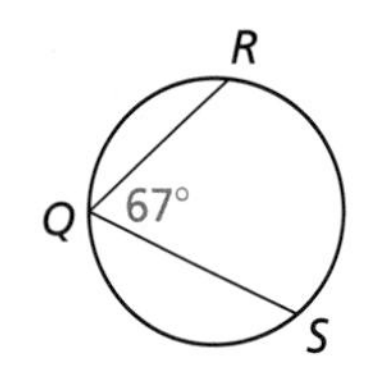

7. $m\widehat{VU}$

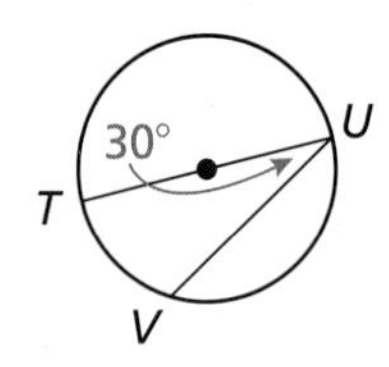

8. $m\widehat{WX}$

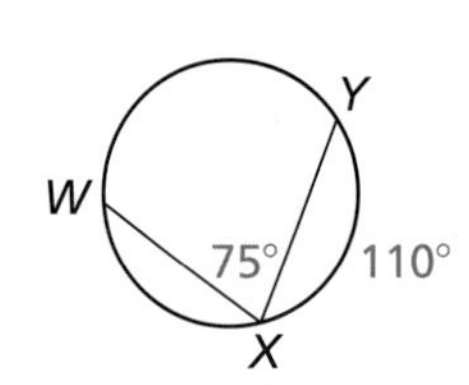

In Exercises 9 and 10, name two pairs of congruent angles.

9.

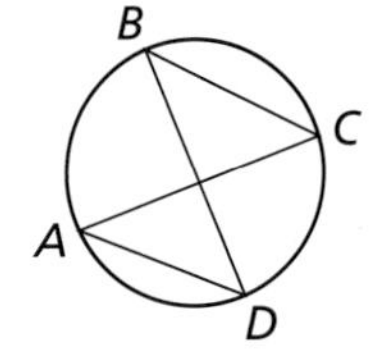

10.

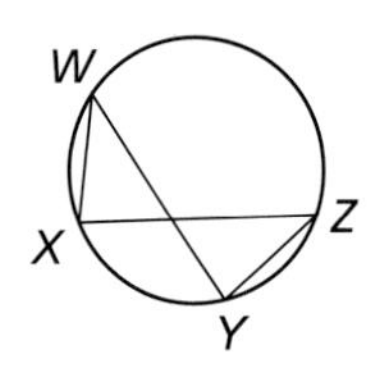

In Exercises 11 and 12, find the measure of the red arc or angle. *(See Example 3.)*

11.

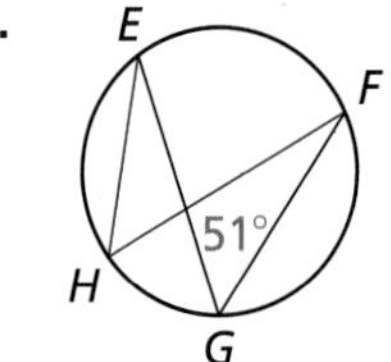

12.

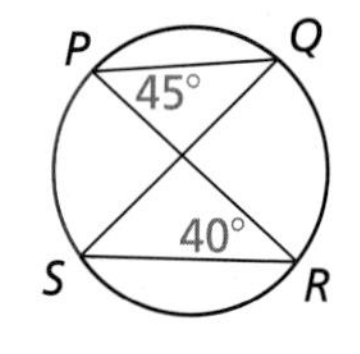

In Exercises 13–16, find the value of each variable. *(See Example 4.)*

13.

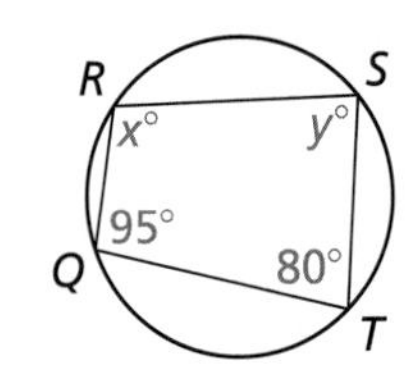

14.

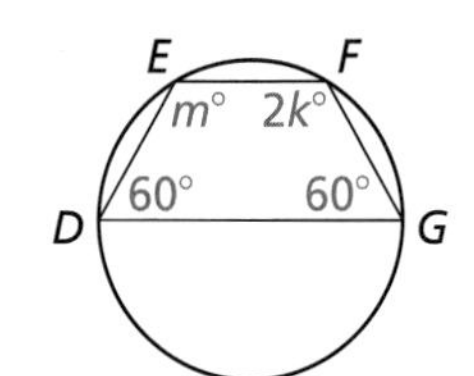

15.

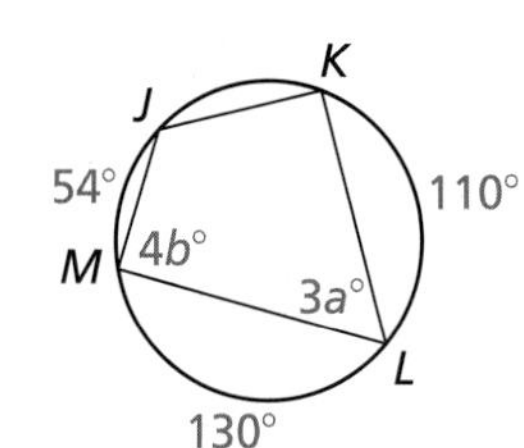

16.

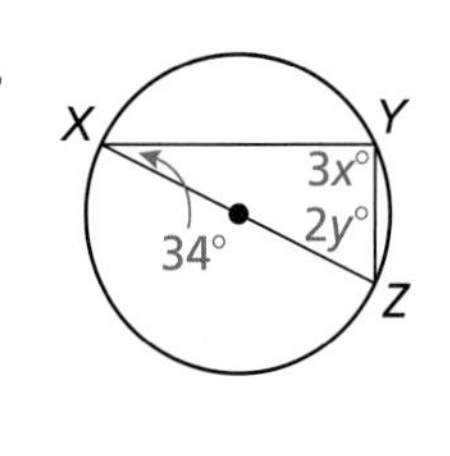

17. **ERROR ANALYSIS** Describe and correct the error in finding $m\widehat{BC}$.

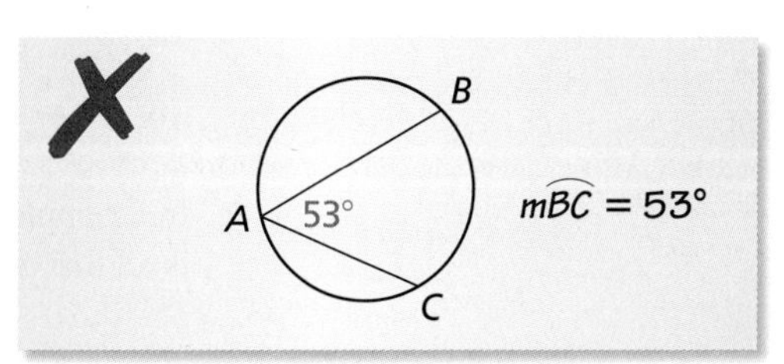

18. MODELING WITH MATHEMATICS A *carpenter's square* is an L-shaped tool used to draw right angles. You need to cut a circular piece of wood into two semicircles. How can you use the carpenter's square to draw a diameter on the circular piece of wood? *(See Example 5.)*

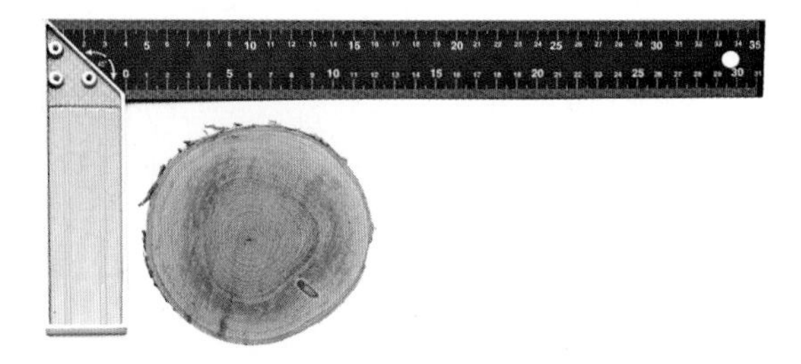

MATHEMATICAL CONNECTIONS In Exercises 19–21, find the values of x and y. Then find the measures of the interior angles of the polygon.

19.

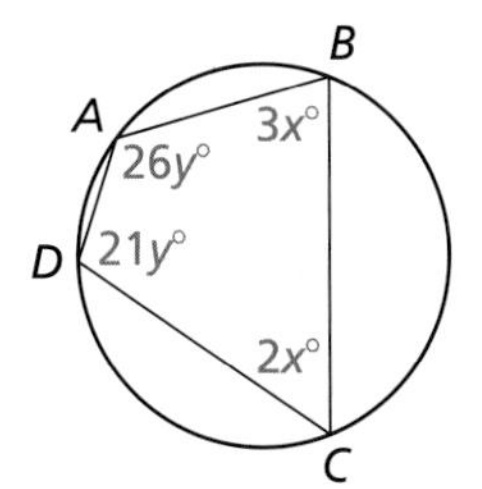

20.

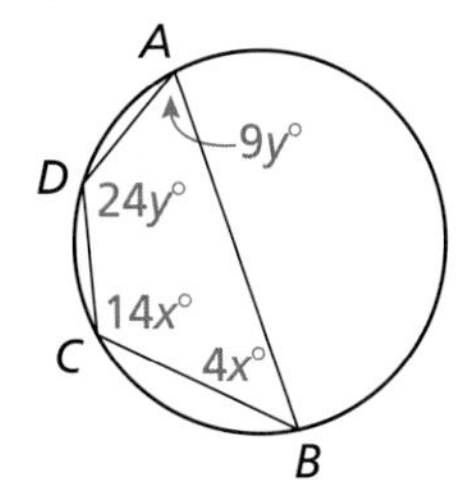

21.

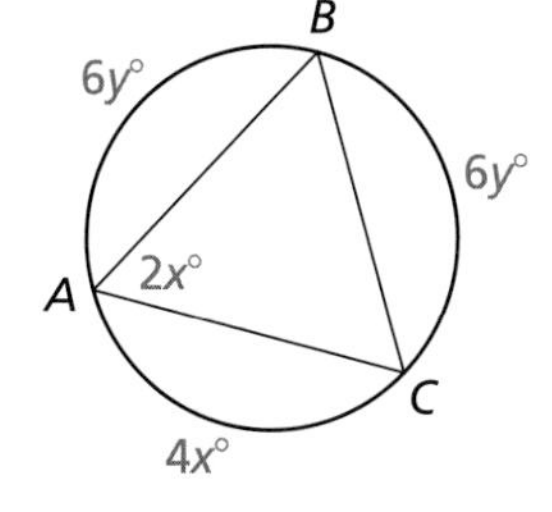

22. MAKING AN ARGUMENT Your friend claims that $\angle PTQ \cong \angle PSQ \cong \angle PRQ$. Is your friend correct? Explain your reasoning.

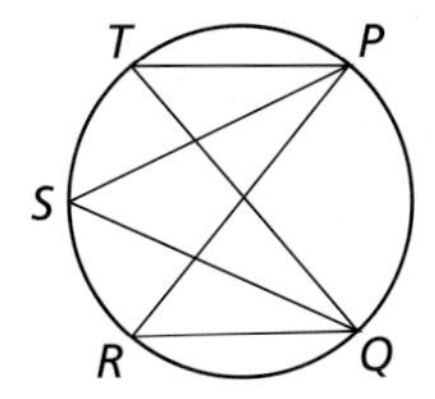

23. CONSTRUCTION Construct an equilateral triangle inscribed in a circle.

24. CONSTRUCTION The side length of an inscribed regular hexagon is equal to the radius of the circumscribed circle. Use this fact to construct a regular hexagon inscribed in a circle.

REASONING In Exercises 25–30, determine whether a quadrilateral of the given type can always be inscribed inside a circle. Explain your reasoning.

25. square

26. rectangle

27. parallelogram

28. kite

29. rhombus

30. isosceles trapezoid

31. MODELING WITH MATHEMATICS Three moons, A, B, and C, are in the same circular orbit 100,000 kilometers above the surface of a planet. The planet is 20,000 kilometers in diameter and $m\angle ABC = 90°$. Draw a diagram of the situation. How far is moon A from moon C?

32. MODELING WITH MATHEMATICS At the movie theater, you want to choose a seat that has the best *viewing angle*, so that you can be close to the screen and still see the whole screen without moving your eyes. You previously decided that seat F7 has the best viewing angle, but this time someone else is already sitting there. Where else can you sit so that your seat has the same viewing angle as seat F7? Explain.

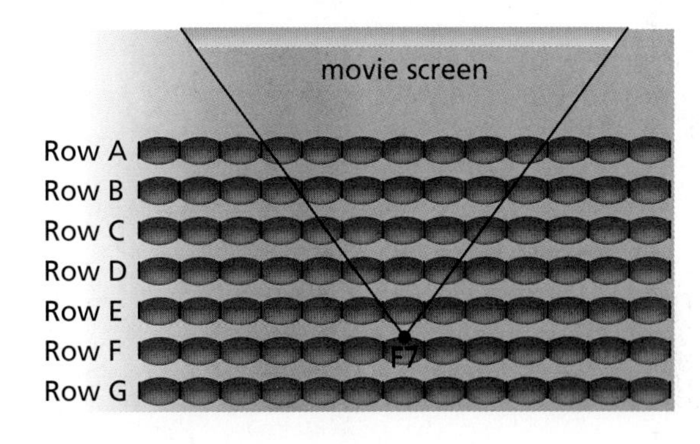

33. WRITING A right triangle is inscribed in a circle, and the radius of the circle is given. Explain how to find the length of the hypotenuse.

34. HOW DO YOU SEE IT? Let point Y represent your location on the soccer field below. What type of angle is $\angle AYB$ if you stand anywhere on the circle except at point A or point B?

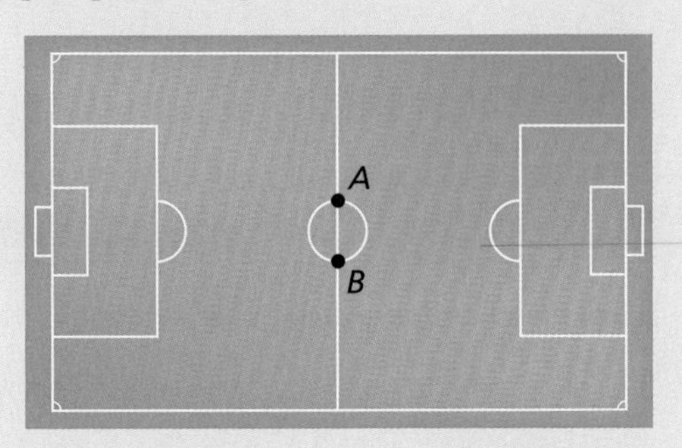

35. WRITING Explain why the diagonals of a rectangle inscribed in a circle are diameters of the circle.

36. THOUGHT PROVOKING The figure shows a circle that is circumscribed about $\triangle ABC$. Is it possible to circumscribe a circle about any triangle? Justify your answer.

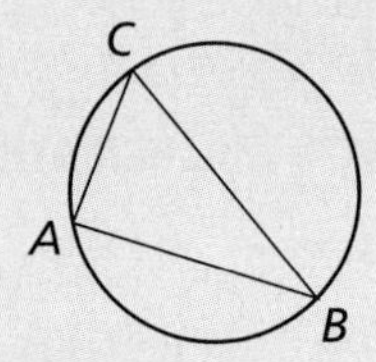

37. PROVING A THEOREM If an angle is inscribed in $\odot Q$, the center Q can be on a side of the inscribed angle, inside the inscribed angle, or outside the inscribed angle. Prove each case of the Measure of an Inscribed Angle Theorem (Theorem 10.10).

a. Case 1

Given $\angle ABC$ is inscribed in $\odot Q$.
Let $m\angle B = x°$.
Center Q lies on $\overline{BC}$.

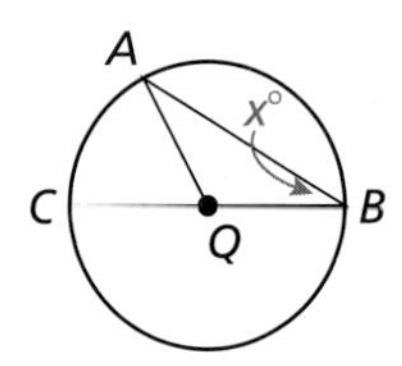

Prove $m\angle ABC = \frac{1}{2} m\overset{\frown}{AC}$

(*Hint*: Show that $\triangle AQB$ is isosceles. Then write $m\overset{\frown}{AC}$ in terms of x.)

b. Case 2 Use the diagram and auxiliary line to write **Given** and **Prove** statements for Case 2. Then write a proof.

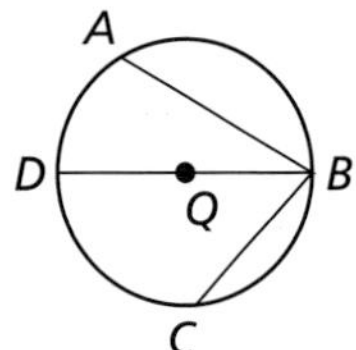

c. Case 3 Use the diagram and auxiliary line to write **Given** and **Prove** statements for Case 3. Then write a proof.

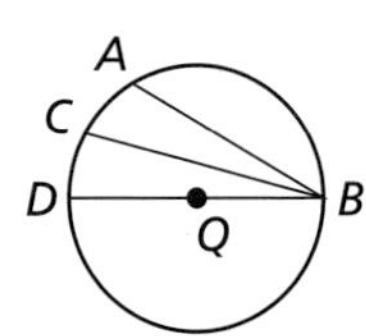

38. PROVING A THEOREM Write a paragraph proof of the Inscribed Angles of a Circle Theorem (Theorem 10.11). First, draw a diagram and write **Given** and **Prove** statements.

39. PROVING A THEOREM The Inscribed Right Triangle Theorem (Theorem 10.12) is written as a conditional statement and its converse. Write a plan for proof for each statement.

40. PROVING A THEOREM Copy and complete the paragraph proof for one part of the Inscribed Quadrilateral Theorem (Theorem 10.13).

Given $\odot C$ with inscribed quadrilateral $DEFG$

Prove $m\angle D + m\angle F = 180°$, $m\angle E + m\angle G = 180°$

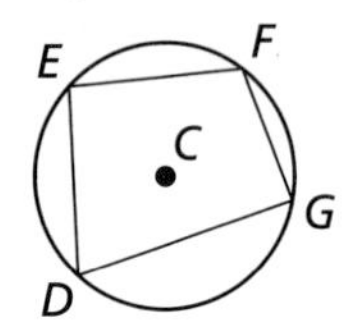

By the Arc Addition Postulate (Postulate 10.1), $m\overset{\frown}{EFG}$ + ____ = 360° and $m\overset{\frown}{FGD} + m\overset{\frown}{DEF} = 360°$. Using the ________ Theorem, $m\overset{\frown}{EDG} = 2m\angle F$, $m\overset{\frown}{EFG} = 2m\angle D$, $m\overset{\frown}{DEF} = 2m\angle G$, and $m\overset{\frown}{FGD} = 2m\angle E$. By the Substitution Property of Equality, $2m\angle D$ + ___ = 360°, so ___. Similarly, ___.

41. CRITICAL THINKING In the diagram, $\angle C$ is a right angle. If you draw the smallest possible circle through C tangent to $\overline{AB}$, the circle will intersect $\overline{AC}$ at J and $\overline{BC}$ at K. Find the exact length of $\overline{JK}$.

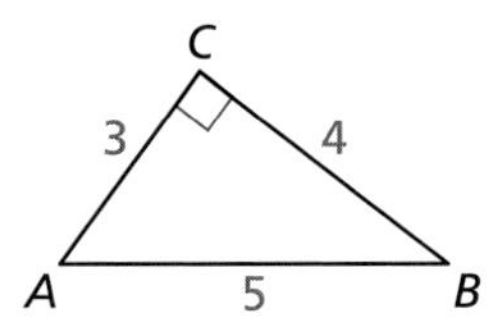

42. CRITICAL THINKING You are making a circular cutting board. To begin, you glue eight 1-inch boards together, as shown. Then you draw and cut a circle with an 8-inch diameter from the boards.

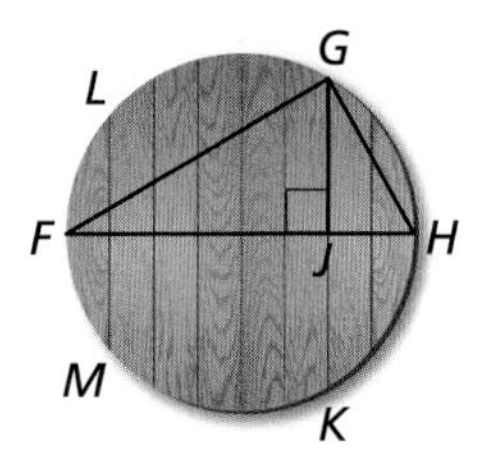

a. $\overline{FH}$ is a diameter of the circular cutting board. Write a proportion relating GJ and JH. State a theorem to justify your answer.

b. Find FJ, JH, and GJ. What is the length of the cutting board seam labeled $\overline{GK}$?

Maintaining Mathematical Proficiency

Reviewing what you learned in previous grades and lessons

Solve the equation. Check your solution. *(Skills Review Handbook)*

43. $3x = 145$

44. $\frac{1}{2}x = 63$

45. $240 = 2x$

46. $75 = \frac{1}{2}(x - 30)$

10.5 Angle Relationships in Circles

Essential Question When a chord intersects a tangent line or another chord, what relationships exist among the angles and arcs formed?

EXPLORATION 1 Angles Formed by a Chord and Tangent Line

Work with a partner. Use dynamic geometry software.

a. Construct a chord in a circle. At one of the endpoints of the chord, construct a tangent line to the circle.

b. Find the measures of the two angles formed by the chord and the tangent line.

c. Find the measures of the two circular arcs determined by the chord.

d. Repeat parts (a)–(c) several times. Record your results in a table. Then write a conjecture that summarizes the data.

Sample

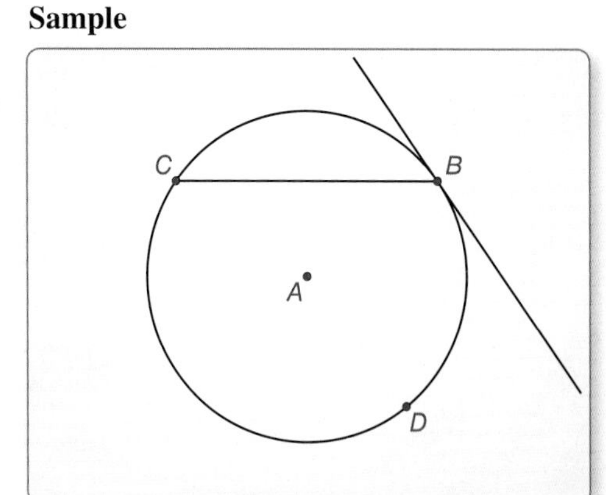

EXPLORATION 2 Angles Formed by Intersecting Chords

Work with a partner. Use dynamic geometry software.

a. Construct two chords that intersect inside a circle.

b. Find the measure of one of the angles formed by the intersecting chords.

c. Find the measures of the arcs intercepted by the angle in part (b) and its vertical angle. What do you observe?

d. Repeat parts (a)–(c) several times. Record your results in a table. Then write a conjecture that summarizes the data.

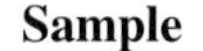
Sample

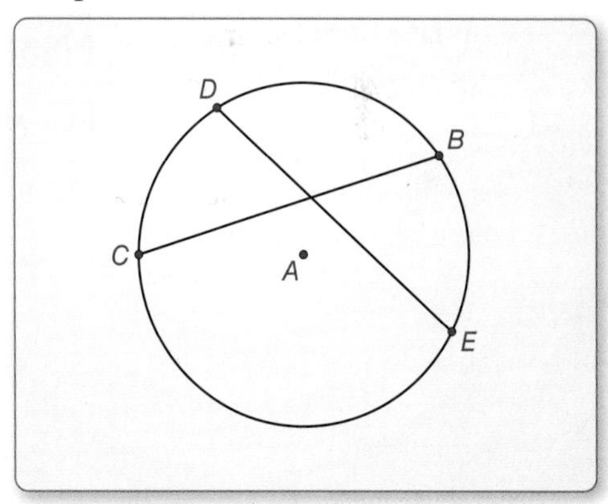

CONSTRUCTING VIABLE ARGUMENTS

To be proficient in math, you need to understand and use stated assumptions, definitions, and previously established results.

Communicate Your Answer

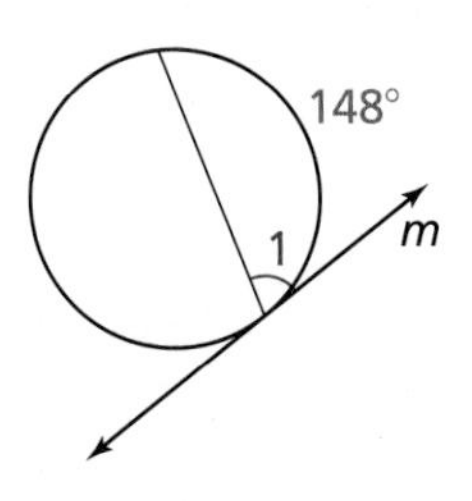

3. When a chord intersects a tangent line or another chord, what relationships exist among the angles and arcs formed?

4. Line m is tangent to the circle in the figure at the left. Find the measure of $\angle 1$.

5. Two chords intersect inside a circle to form a pair of vertical angles with measures of 55°. Find the sum of the measures of the arcs intercepted by the two angles.

10.5 Lesson

Core Vocabulary

circumscribed angle, *p. 564*

Previous
tangent
chord
secant

What You Will Learn

- Find angle and arc measures.
- Use circumscribed angles.

Finding Angle and Arc Measures

Theorem

Theorem 10.14 Tangent and Intersected Chord Theorem

If a tangent and a chord intersect at a point on a circle, then the measure of each angle formed is one-half the measure of its intercepted arc.

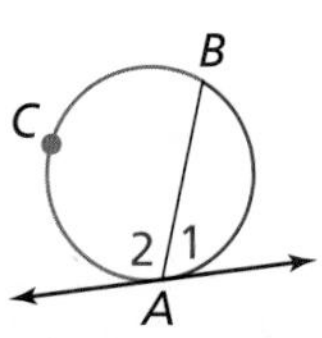

Proof Ex. 33, p. 568

$m\angle 1 = \frac{1}{2}m\overparen{AB}$ $m\angle 2 = \frac{1}{2}m\overparen{BCA}$

EXAMPLE 1 Finding Angle and Arc Measures

Line m is tangent to the circle. Find the measure of the red angle or arc.

a.

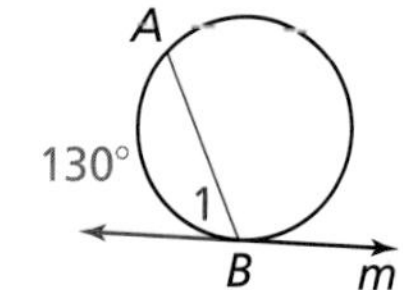

b.

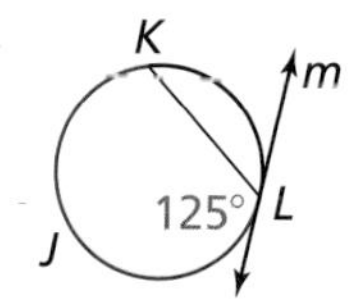

SOLUTION

a. $m\angle 1 = \frac{1}{2}(130^\circ) = 65^\circ$

b. $m\overparen{KJL} = 2(125^\circ) = 250^\circ$

Monitoring Progress

Help in English and Spanish at *BigIdeasMath.com*

Line m is tangent to the circle. Find the indicated measure.

1. $m\angle 1$

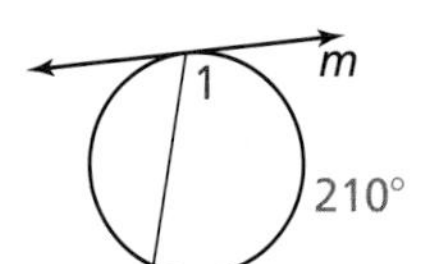

2. $m\overparen{RST}$

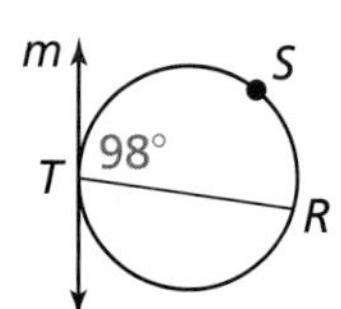

3. $m\overparen{XY}$

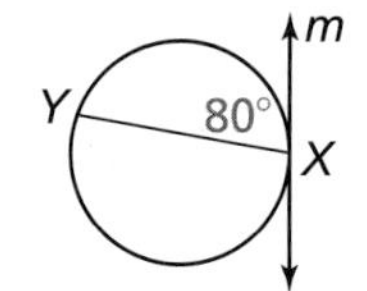

Core Concept

Intersecting Lines and Circles

If two nonparallel lines intersect a circle, there are three places where the lines can intersect.

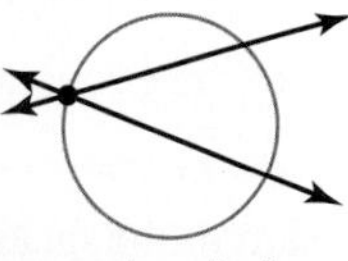

on the circle

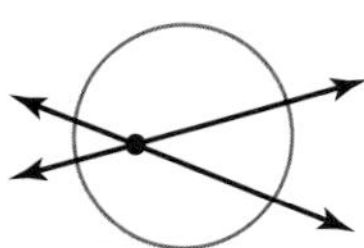

inside the circle

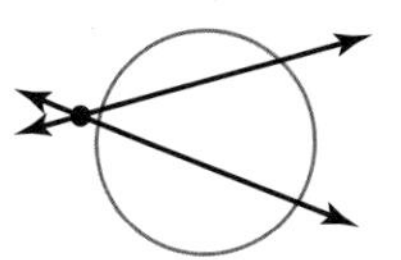

outside the circle

Theorems

Theorem 10.15 Angles Inside the Circle Theorem

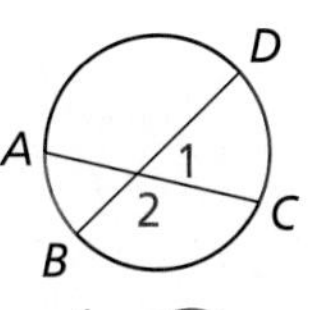

If two chords intersect *inside* a circle, then the measure of each angle is one-half the *sum* of the measures of the arcs intercepted by the angle and its vertical angle.

$m\angle 1 = \frac{1}{2}(m\overset{\frown}{DC} + m\overset{\frown}{AB})$,

$m\angle 2 = \frac{1}{2}(m\overset{\frown}{AD} + m\overset{\frown}{BC})$

Proof Ex. 35, p. 568

Theorem 10.16 Angles Outside the Circle Theorem

If a tangent and a secant, two tangents, or two secants intersect *outside* a circle, then the measure of the angle formed is one-half the *difference* of the measures of the intercepted arcs.

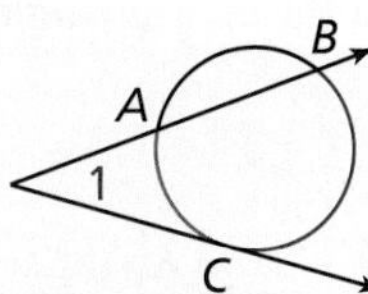

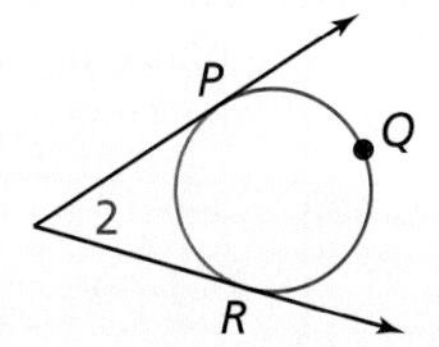

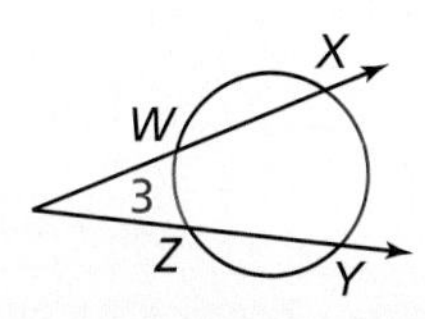

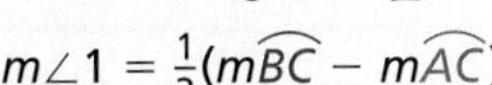

$m\angle 1 = \frac{1}{2}(m\overset{\frown}{BC} - m\overset{\frown}{AC})$ $m\angle 2 = \frac{1}{2}(m\overset{\frown}{PQR} - m\overset{\frown}{PR})$ $m\angle 3 = \frac{1}{2}(m\overset{\frown}{XY} - m\overset{\frown}{WZ})$

Proof Ex. 37, p. 568

EXAMPLE 2 Finding an Angle Measure

Find the value of x.

a.

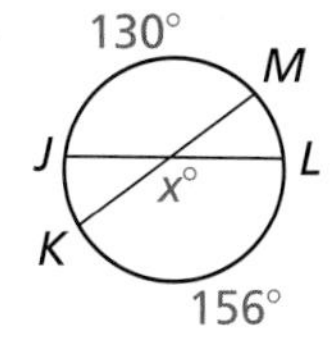

b.

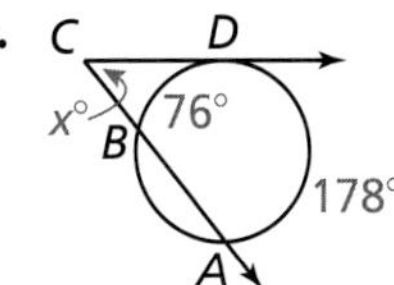

SOLUTION

a. The chords $\overline{JL}$ and $\overline{KM}$ intersect inside the circle. Use the Angles Inside the Circle Theorem.

$$x° = \frac{1}{2}(m\overset{\frown}{JM} + m\overset{\frown}{LK})$$

$$x° = \frac{1}{2}(130° + 156°)$$

$$x = 143$$

▶ So, the value of x is 143.

b. The tangent $\overrightarrow{CD}$ and the secant $\overrightarrow{CB}$ intersect outside the circle. Use the Angles Outside the Circle Theorem.

$$m\angle BCD = \frac{1}{2}(m\overset{\frown}{AD} - m\overset{\frown}{BD})$$

$$x° = \frac{1}{2}(178° - 76°)$$

$$x = 51$$

▶ So, the value of x is 51.

Monitoring Progress

Help in English and Spanish at *BigIdeasMath.com*

Find the value of the variable.

4.

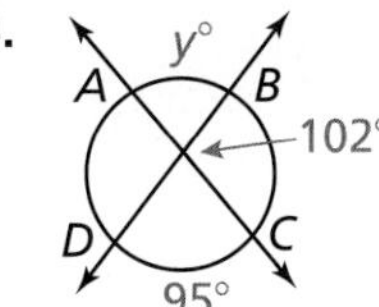

5.

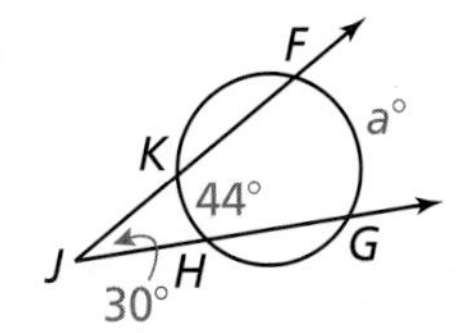

Using Circumscribed Angles

Core Concept

Circumscribed Angle

A **circumscribed angle** is an angle whose sides are tangent to a circle.

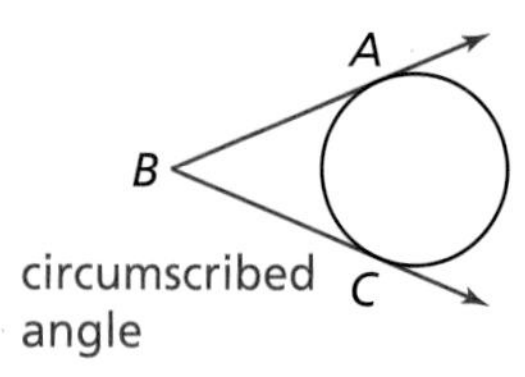

circumscribed angle

Theorem

Theorem 10.17 Circumscribed Angle Theorem

The measure of a circumscribed angle is equal to 180° minus the measure of the central angle that intercepts the same arc.

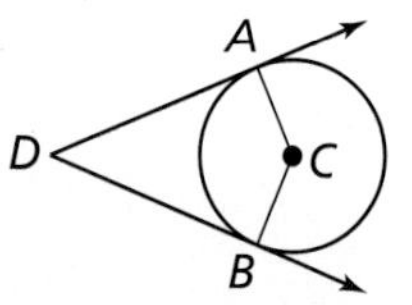

$m\angle ADB = 180° - m\angle ACB$

Proof Ex. 38, p. 568

EXAMPLE 3 Finding Angle Measures

Find the value of x.

a.

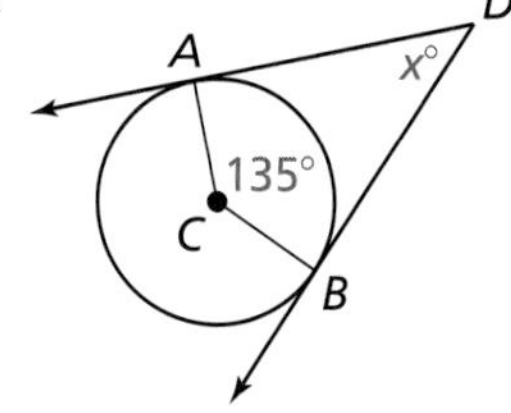

b.

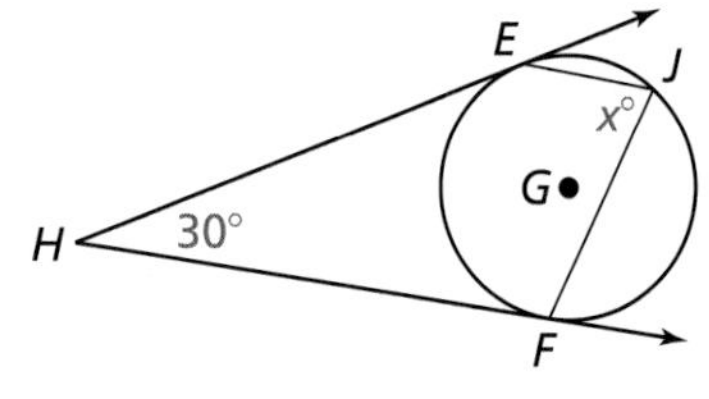

SOLUTION

a. Use the Circumscribed Angle Theorem to find $m\angle ADB$.

$m\angle ADB = 180° - m\angle ACB$	Circumscribed Angle Theorem
$x° = 180° - 135°$	Substitute.
$x = 45$	Subtract.

▶ So, the value of x is 45.

b. Use the Measure of an Inscribed Angle Theorem (Theorem 10.10) and the Circumscribed Angle Theorem to find $m\angle EJF$.

$m\angle EJF = \frac{1}{2}m\widehat{EF}$	Measure of an Inscribed Angle Theorem
$m\angle EJF = \frac{1}{2}m\angle EGF$	Definition of minor arc
$m\angle EJF = \frac{1}{2}(180° - m\angle EHF)$	Circumscribed Angle Theorem
$m\angle EJF = \frac{1}{2}(180° - 30°)$	Substitute.
$x = \frac{1}{2}(180 - 30)$	Substitute.
$x = 75$	Simplify.

▶ So, the value of x is 75.

EXAMPLE 4 Modeling with Mathematics

The northern lights are bright flashes of colored light between 50 and 200 miles above Earth. A flash occurs 150 miles above Earth at point C. What is the measure of $\overset{\frown}{BD}$, the portion of Earth from which the flash is visible? (Earth's radius is approximately 4000 miles.)

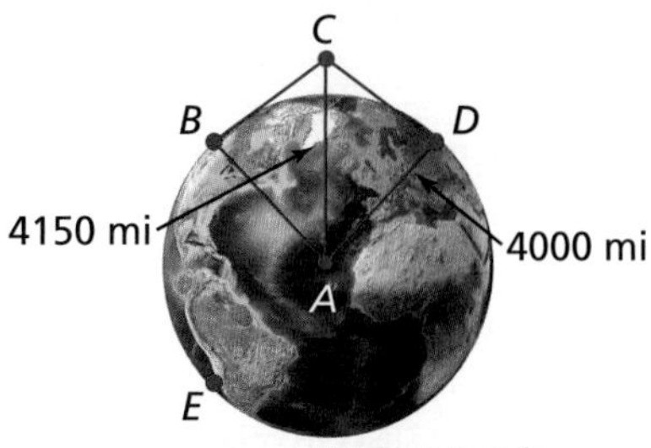

Not drawn to scale

SOLUTION

1. **Understand the Problem** You are given the approximate radius of Earth and the distance above Earth that the flash occurs. You need to find the measure of the arc that represents the portion of Earth from which the flash is visible.

2. **Make a Plan** Use properties of tangents, triangle congruence, and angles outside a circle to find the arc measure.

3. **Solve the Problem** Because $\overline{CB}$ and $\overline{CD}$ are tangents, $\overline{CB} \perp \overline{AB}$ and $\overline{CD} \perp \overline{AD}$ by the Tangent Line to Circle Theorem (Theorem 10.1). Also, $\overline{BC} \cong \overline{DC}$ by the External Tangent Congruence Theorem (Theorem 10.2), and $\overline{CA} \cong \overline{CA}$ by the Reflexive Property of Congruence (Theorem 2.1). So, $\triangle ABC \cong \triangle ADC$ by the Hypotenuse-Leg Congruence Theorem (Theorem 5.9). Because corresponding parts of congruent triangles are congruent, $\angle BCA \cong \angle DCA$. Solve right $\triangle CBA$ to find that $m\angle BCA \approx 74.5°$. So, $m\angle BCD \approx 2(74.5°) = 149°$.

$m\angle BCD = 180° - m\angle BAD$	Circumscribed Angle Theorem
$m\angle BCD = 180° - m\overset{\frown}{BD}$	Definition of minor arc
$149° \approx 180° - m\overset{\frown}{BD}$	Substitute.
$31° \approx m\overset{\frown}{BD}$	Solve for $m\overset{\frown}{BD}$.

▶ The measure of the arc from which the flash is visible is about 31°.

4. **Look Back** You can use inverse trigonometric ratios to find $m\angle BAC$ and $m\angle DAC$.

$$m\angle BAC = \cos^{-1}\left(\frac{4000}{4150}\right) \approx 15.5°$$

$$m\angle DAC = \cos^{-1}\left(\frac{4000}{4150}\right) \approx 15.5°$$

So, $m\angle BAD \approx 15.5° + 15.5° = 31°$, and therefore $m\overset{\frown}{BD} \approx 31°$.

COMMON ERROR

Because the value for $m\angle BCD$ is an approximation, use the symbol $\approx$ instead of $=$.

Monitoring Progress  Help in English and Spanish at *BigIdeasMath.com*

Find the value of x.

6.

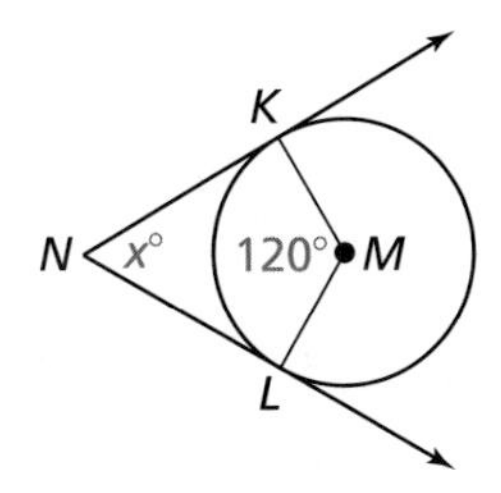

7.

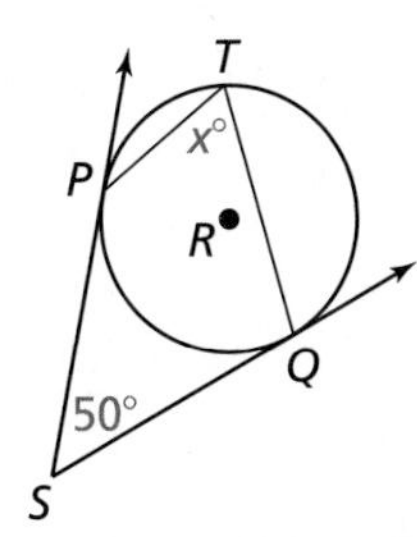

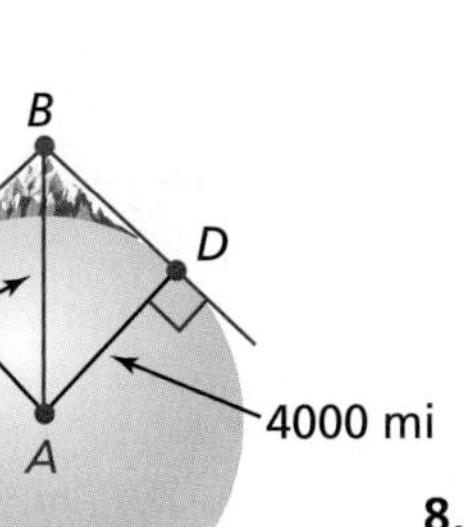

Not drawn to scale

8. You are on top of Mount Rainier on a clear day. You are about 2.73 miles above sea level at point B. Find $m\overset{\frown}{CD}$, which represents the part of Earth that you can see.

10.5 Exercises

Dynamic Solutions available at *BigIdeasMath.com*

Vocabulary and Core Concept Check

1. **COMPLETE THE SENTENCE** Points A, B, C, and D are on a circle, and $\overleftrightarrow{AB}$ intersects $\overleftrightarrow{CD}$ at point P. If $m\angle APC = \frac{1}{2}(m\widehat{BD} - m\widehat{AC})$, then point P is ______ the circle.

2. **WRITING** Explain how to find the measure of a circumscribed angle.

Monitoring Progress and Modeling with Mathematics

In Exercises 3–6, line t is tangent to the circle. Find the indicated measure. *(See Example 1.)*

3. $m\widehat{AB}$

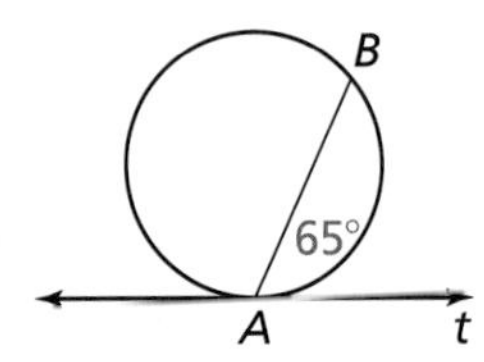

4. $m\widehat{DEF}$

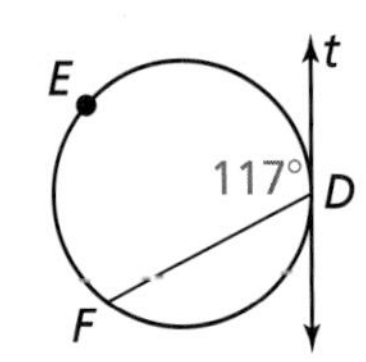

5. $m\angle 1$

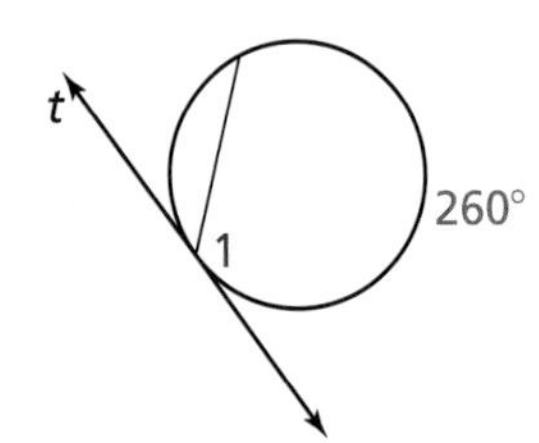

6. $m\angle 3$

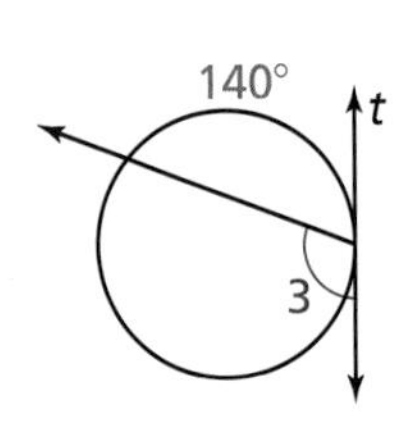

In Exercises 7–14, find the value of x. *(See Examples 2 and 3.)*

7.

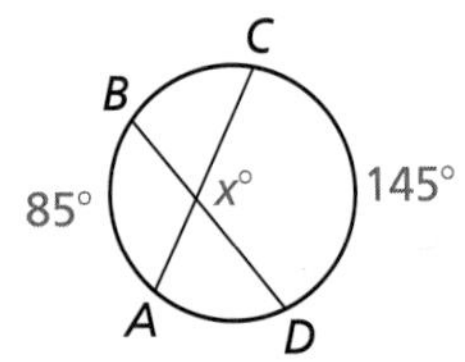

8.

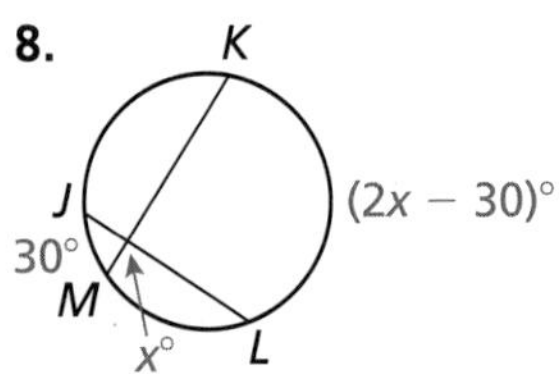

9.

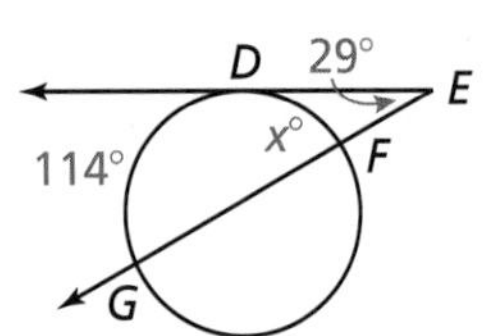

10.

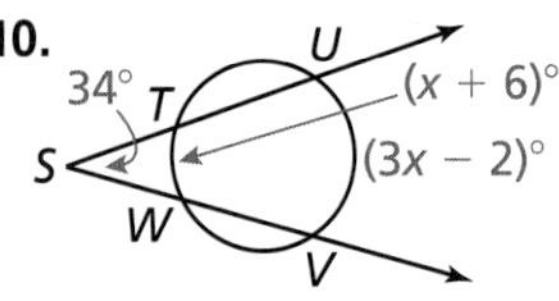

11.

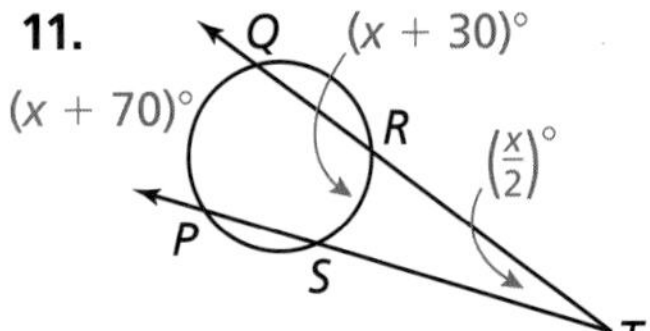

12.

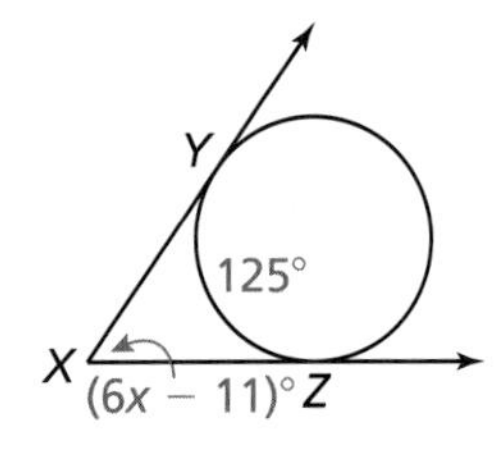

13.

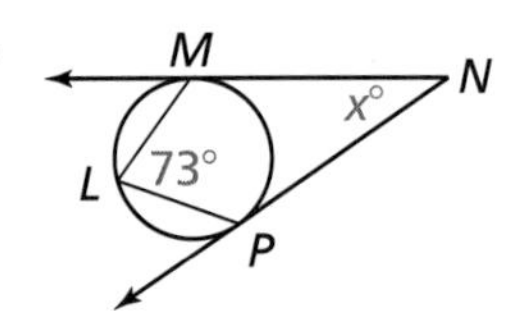

14.

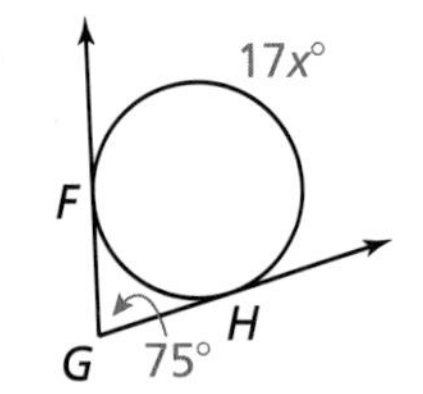

ERROR ANALYSIS **In Exercises 15 and 16, describe and correct the error in finding the angle measure.**

15.

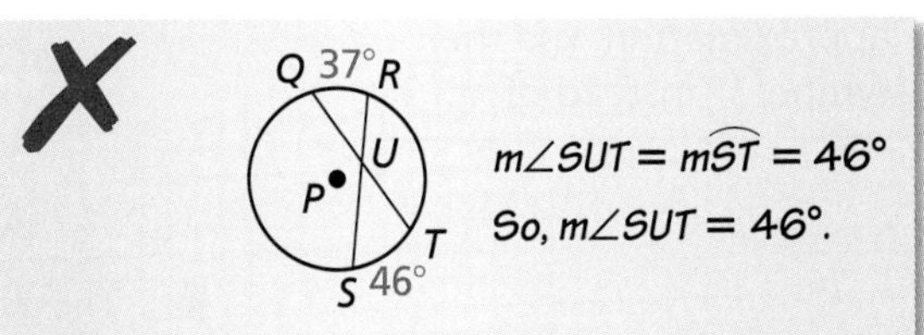

16.

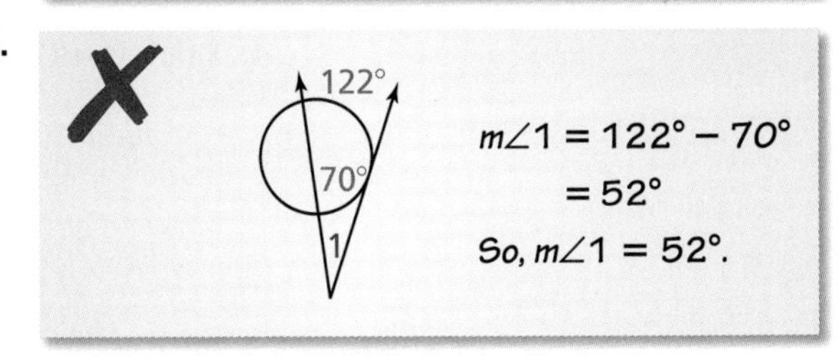

In Exercises 17–22, find the indicated angle measure. Justify your answer.

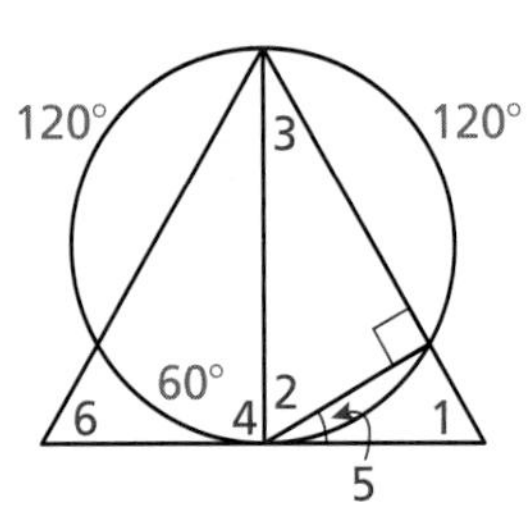

17. $m\angle 1$

18. $m\angle 2$

19. $m\angle 3$

20. $m\angle 4$

21. $m\angle 5$

22. $m\angle 6$

23. PROBLEM SOLVING You are flying in a hot air balloon about 1.2 miles above the ground. Find the measure of the arc that represents the part of Earth you can see. The radius of Earth is about 4000 miles. *(See Example 4.)*

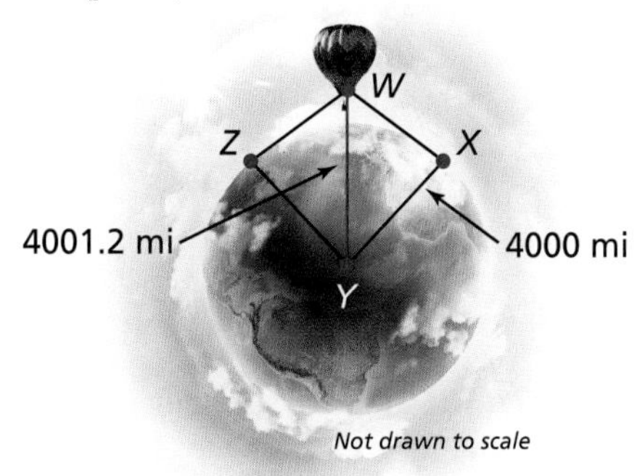

24. PROBLEM SOLVING You are watching fireworks over San Diego Bay S as you sail away in a boat. The highest point the fireworks reach F is about 0.2 mile above the bay. Your eyes E are about 0.01 mile above the water. At point B you can no longer see the fireworks because of the curvature of Earth. The radius of Earth is about 4000 miles, and $\overline{FE}$ is tangent to Earth at point T. Find $m\widehat{SB}$. Round your answer to the nearest tenth.

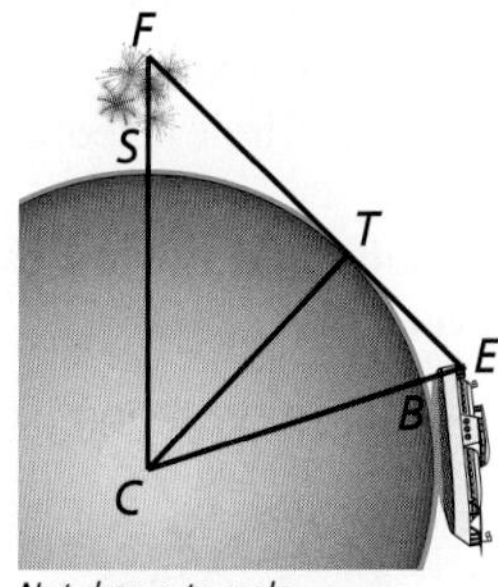

25. MATHEMATICAL CONNECTIONS In the diagram, $\overrightarrow{BA}$ is tangent to $\odot E$. Write an algebraic expression for $m\widehat{CD}$ in terms of x. Then find $m\widehat{CD}$.

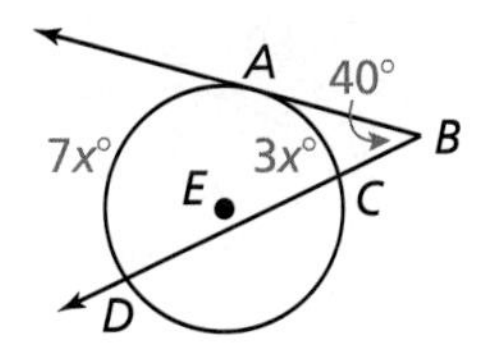

26. MATHEMATICAL CONNECTIONS The circles in the diagram are concentric. Write an algebraic expression for c in terms of a and b.

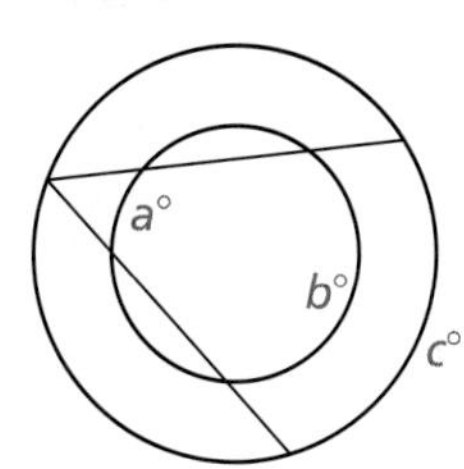

27. ABSTRACT REASONING In the diagram, $\overrightarrow{PL}$ is tangent to the circle, and $\overline{KJ}$ is a diameter. What is the range of possible angle measures of $\angle LPJ$? Explain your reasoning.

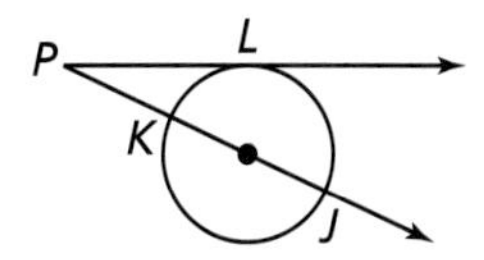

28. ABSTRACT REASONING In the diagram, $\overline{AB}$ is any chord that is not a diameter of the circle. Line m is tangent to the circle at point A. What is the range of possible values of x? Explain your reasoning. (The diagram is not drawn to scale.)

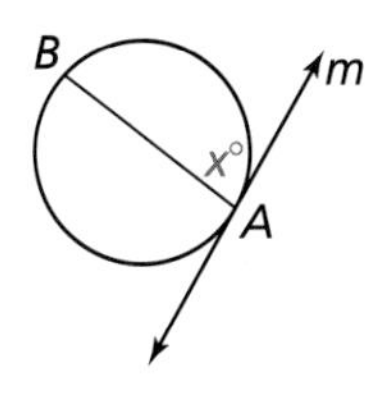

29. PROOF In the diagram, $\overleftrightarrow{JL}$ and $\overleftrightarrow{NL}$ are secant lines that intersect at point L. Prove that $m\angle JPN > m\angle JLN$.

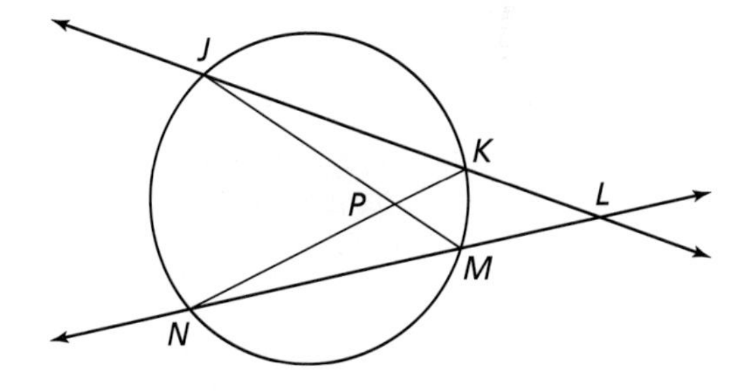

30. MAKING AN ARGUMENT Your friend claims that it is possible for a circumscribed angle to have the same measure as its intercepted arc. Is your friend correct? Explain your reasoning.

31. REASONING Points A and B are on a circle, and t is a tangent line containing A and another point C.

a. Draw two diagrams that illustrate this situation.

b. Write an equation for $m\widehat{AB}$ in terms of $m\angle BAC$ for each diagram.

c. For what measure of $\angle BAC$ can you use either equation to find $m\widehat{AB}$? Explain.

32. REASONING $\triangle XYZ$ is an equilateral triangle inscribed in $\odot P$. $\overline{AB}$ is tangent to $\odot P$ at point X, $\overline{BC}$ is tangent to $\odot P$ at point Y, and $\overline{AC}$ is tangent to $\odot P$ at point Z. Draw a diagram that illustrates this situation. Then classify $\triangle ABC$ by its angles and sides. Justify your answer.

33. PROVING A THEOREM To prove the Tangent and Intersected Chord Theorem (Theorem 10.14), you must prove three cases.

a. The diagram shows the case where $\overline{AB}$ contains the center of the circle. Use the Tangent Line to Circle Theorem (Theorem 10.1) to write a paragraph proof for this case.

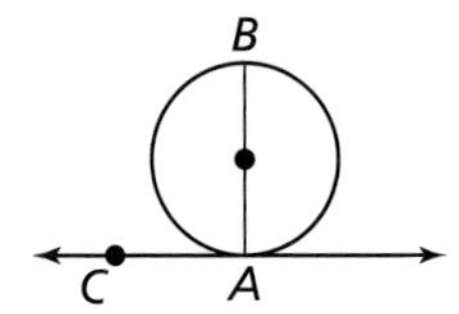

b. Draw a diagram and write a proof for the case where the center of the circle is in the interior of $\angle CAB$.

c. Draw a diagram and write a proof for the case where the center of the circle is in the exterior of $\angle CAB$.

34. HOW DO YOU SEE IT? In the diagram, television cameras are positioned at A and B to record what happens on stage. The stage is an arc of $\odot A$. You would like the camera at B to have a 30° view of the stage. Should you move the camera closer or farther away? Explain your reasoning.

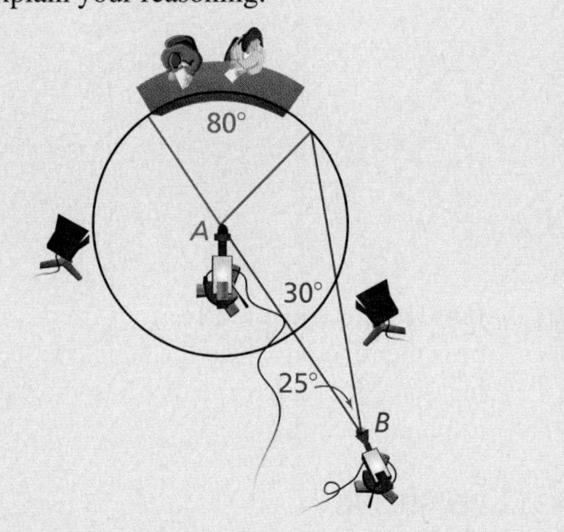

35. PROVING A THEOREM Write a proof of the Angles Inside the Circle Theorem (Theorem 10.15).

Given Chords $\overline{AC}$ and $\overline{BD}$ intersect inside a circle.

Prove $m\angle 1 = \frac{1}{2}(m\widehat{DC} + m\widehat{AB})$

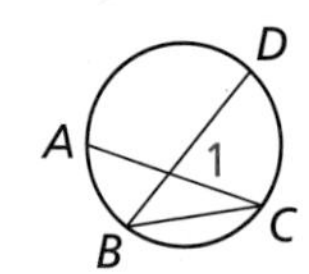

36. THOUGHT PROVOKING In the figure, $\overleftrightarrow{BP}$ and $\overleftrightarrow{CP}$ are tangent to the circle. Point A is any point on the major arc formed by the endpoints of the chord $\overline{BC}$. Label all congruent angles in the figure. Justify your reasoning.

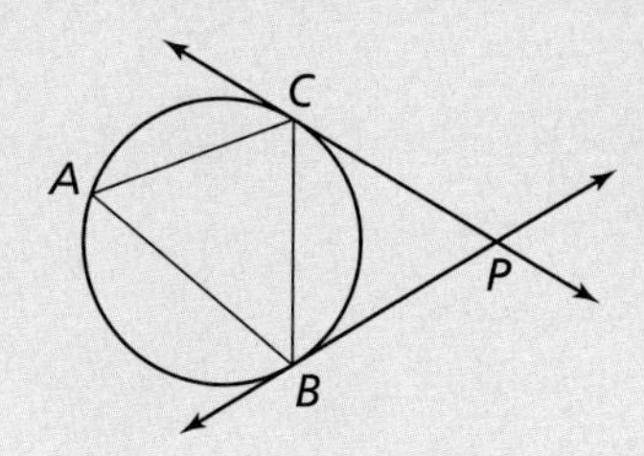

37. PROVING A THEOREM Use the diagram below to prove the Angles Outside the Circle Theorem (Theorem 10.16) for the case of a tangent and a secant. Then copy the diagrams for the other two cases on page 563 and draw appropriate auxiliary segments. Use your diagrams to prove each case.

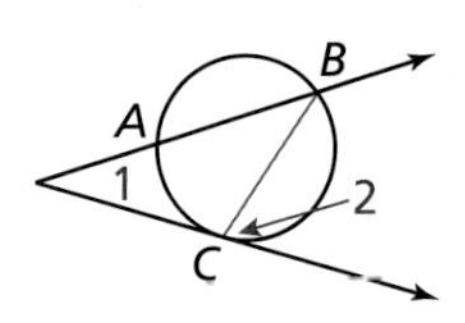

38. PROVING A THEOREM Prove that the Circumscribed Angle Theorem (Theorem 10.17) follows from the Angles Outside the Circle Theorem (Theorem 10.16).

In Exercises 39 and 40, find the indicated measure(s). Justify your answer.

39. Find $m\angle P$ when $m\widehat{WZY} = 200°$.

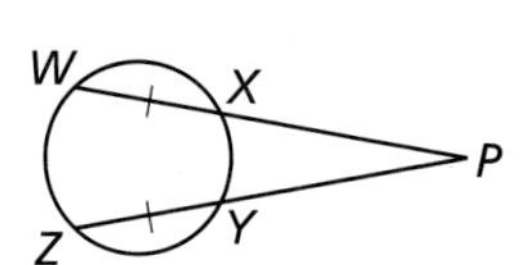

40. Find $m\widehat{AB}$ and $m\widehat{ED}$.

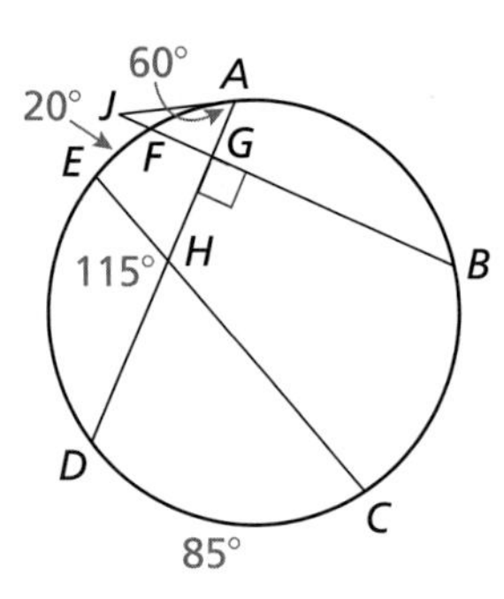

Maintaining Mathematical Proficiency

Reviewing what you learned in previous grades and lessons

Solve the equation. *(Skills Review Handbook)*

41. $x^2 + x = 12$

42. $x^2 = 12x + 35$

43. $-3 = x^2 + 4x$

10.6 Segment Relationships in Circles

Essential Question What relationships exist among the segments formed by two intersecting chords or among segments of two secants that intersect outside a circle?

EXPLORATION 1 Segments Formed by Two Intersecting Chords

Work with a partner. Use dynamic geometry software.

a. Construct two chords $\overline{BC}$ and $\overline{DE}$ that intersect in the interior of a circle at a point F.

b. Find the segment lengths BF, CF, DF, and EF and complete the table. What do you observe?

BF	CF	$BF \cdot CF$
DF	EF	$DF \cdot EF$

Sample

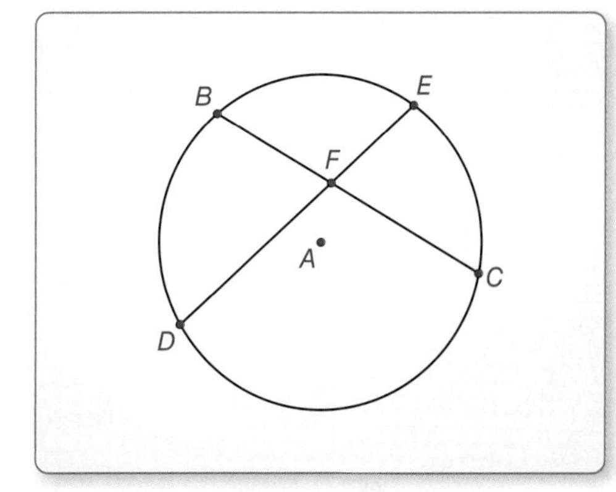

REASONING ABSTRACTLY

To be proficient in math, you need to make sense of quantities and their relationships in problem situations.

c. Repeat parts (a) and (b) several times. Write a conjecture about your results.

EXPLORATION 2 Secants Intersecting Outside a Circle

Work with a partner. Use dynamic geometry software.

a. Construct two secants $\overleftrightarrow{BC}$ and $\overleftrightarrow{BD}$ that intersect at a point B outside a circle, as shown.

b. Find the segment lengths BE, BC, BF, and BD, and complete the table. What do you observe?

BE	BC	$BE \cdot BC$
BF	BD	$BF \cdot BD$

Sample

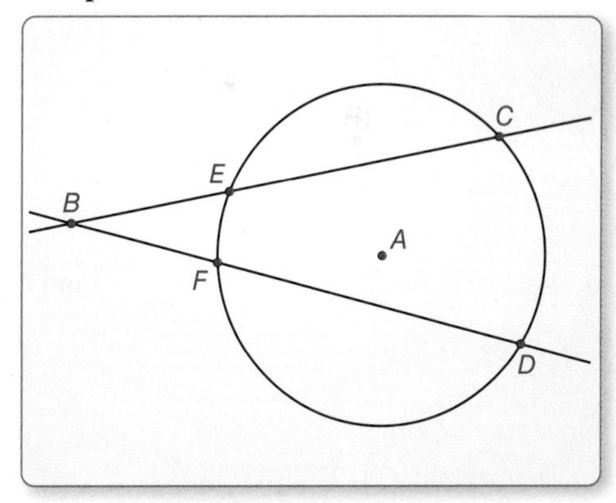

c. Repeat parts (a) and (b) several times. Write a conjecture about your results.

Communicate Your Answer

3. What relationships exist among the segments formed by two intersecting chords or among segments of two secants that intersect outside a circle?

4. Find the segment length AF in the figure at the left.

E, 18, A, F, D, 9, 8, C

10.6 Lesson

Core Vocabulary

segments of a chord, *p. 570*
tangent segment, *p. 571*
secant segment, *p. 571*
external segment, *p. 571*

What You Will Learn

▶ Use segments of chords, tangents, and secants.

Using Segments of Chords, Tangents, and Secants

When two chords intersect in the interior of a circle, each chord is divided into two segments that are called **segments of the chord**.

Theorem

Theorem 10.18 Segments of Chords Theorem

If two chords intersect in the interior of a circle, then the product of the lengths of the segments of one chord is equal to the product of the lengths of the segments of the other chord.

Proof Ex. 19, p. 574

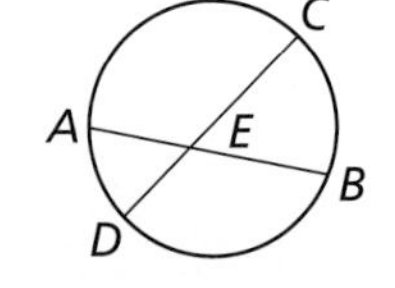

$EA \cdot EB = EC \cdot ED$

EXAMPLE 1 Using Segments of Chords

Find ML and JK.

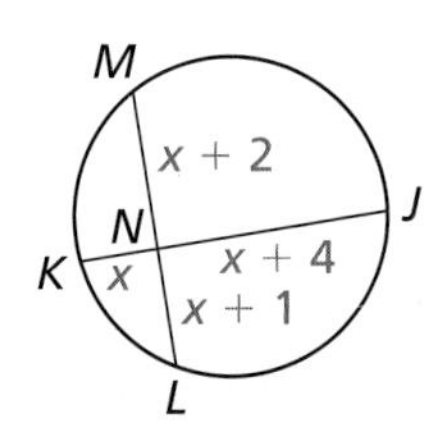

SOLUTION

$NK \cdot NJ = NL \cdot NM$	Segments of Chords Theorem
$x \cdot (x + 4) = (x + 1) \cdot (x + 2)$	Substitute.
$x^2 + 4x = x^2 + 3x + 2$	Simplify.
$4x = 3x + 2$	Subtract x^2 from each side.
$x = 2$	Subtract $3x$ from each side.

Find ML and JK by substitution.

$$ML = (x + 2) + (x + 1) = 2 + 2 + 2 + 1 = 7$$

$$JK = x + (x + 4) = 2 + 2 + 4 = 8$$

▶ So, $ML = 7$ and $JK = 8$.

Monitoring Progress

Help in English and Spanish at *BigIdeasMath.com*

Find the value of x.

1.

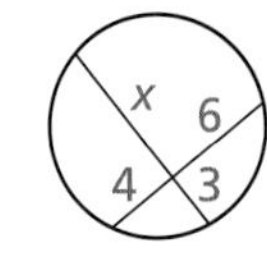

2.

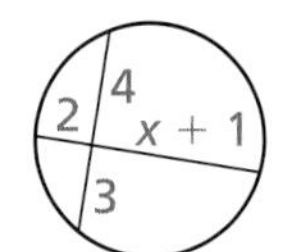

Core Concept

Tangent Segment and Secant Segment

A **tangent segment** is a segment that is tangent to a circle at an endpoint. A **secant segment** is a segment that contains a chord of a circle and has exactly one endpoint outside the circle. The part of a secant segment that is outside the circle is called an **external segment**.

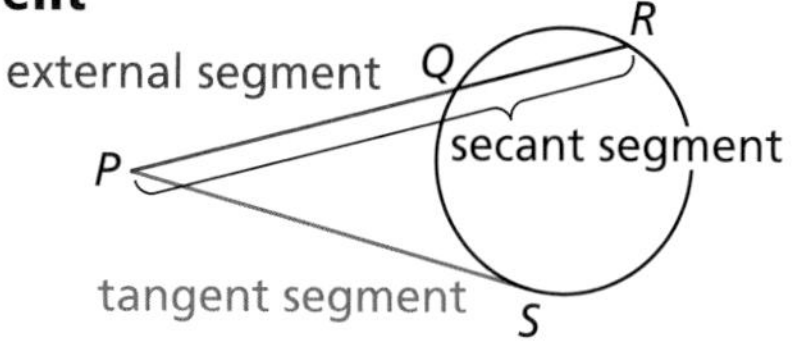

$\overline{PS}$ is a tangent segment.
$\overline{PR}$ is a secant segment.
$\overline{PQ}$ is the external segment of $\overline{PR}$.

Theorem

Theorem 10.19 Segments of Secants Theorem

If two secant segments share the same endpoint outside a circle, then the product of the lengths of one secant segment and its external segment equals the product of the lengths of the other secant segment and its external segment.

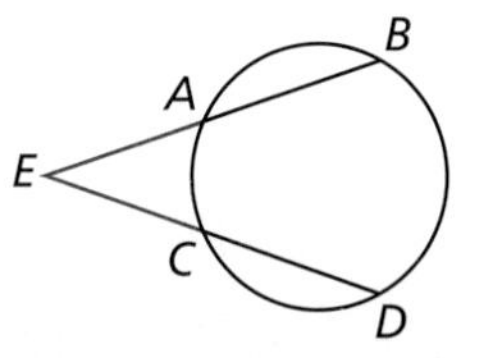

$EA \cdot EB = EC \cdot ED$

Proof Ex. 20, p. 574

EXAMPLE 2 Using Segments of Secants

Find the value of x.

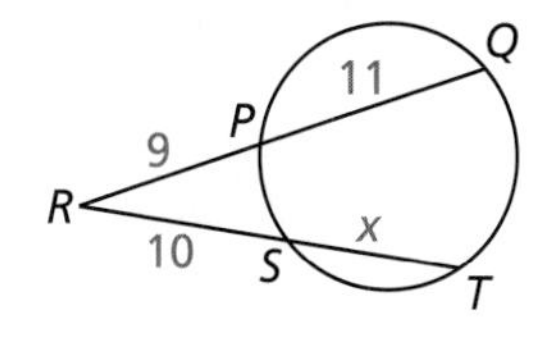

SOLUTION

$RP \cdot RQ = RS \cdot RT$	Segments of Secants Theorem
$9 \cdot (11 + 9) = 10 \cdot (x + 10)$	Substitute.
$180 = 10x + 100$	Simplify.
$80 = 10x$	Subtract 100 from each side.
$8 = x$	Divide each side by 10.

▶ The value of x is 8.

Monitoring Progress

Help in English and Spanish at *BigIdeasMath.com*

Find the value of x.

3.

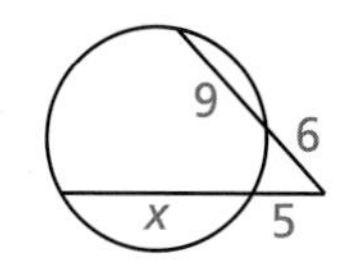

4.

3 $x + 2$

$x + 1$ $x - 1$

Theorem

Theorem 10.20 Segments of Secants and Tangents Theorem

If a secant segment and a tangent segment share an endpoint outside a circle, then the product of the lengths of the secant segment and its external segment equals the square of the length of the tangent segment.

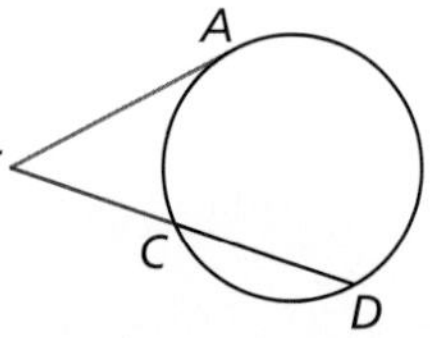

$EA^2 = EC \cdot ED$

Proof Exs. 21 and 22, p. 574

EXAMPLE 3 Using Segments of Secants and Tangents

Find RS.

SOLUTION

$RQ^2 = RS \cdot RT$	Segments of Secants and Tangents Theorem
$16^2 = x \cdot (x + 8)$	Substitute.
$256 = x^2 + 8x$	Simplify.
$0 = x^2 + 8x - 256$	Write in standard form.
$x = \dfrac{-8 \pm \sqrt{8^2 - 4(1)(-256)}}{2(1)}$	Use Quadratic Formula.
$x = -4 \pm 4\sqrt{17}$	Simplify.

Use the positive solution because lengths cannot be negative.

▶ So, $x = -4 + 4\sqrt{17} \approx 12.49$, and $RS \approx 12.49$.

ANOTHER WAY

In Example 3, you can draw segments $\overline{QS}$ and $\overline{QT}$.

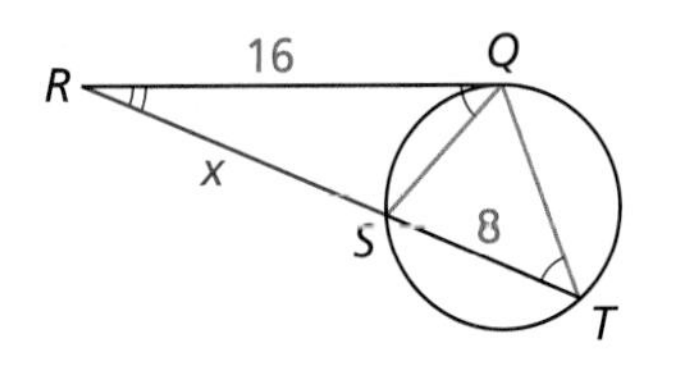

Because $\angle RQS$ and $\angle RTQ$ intercept the same arc, they are congruent. By the Reflexive Property of Congruence (Theorem 2.2), $\angle QRS \cong \angle TRQ$. So, $\triangle RSQ \sim \triangle RQT$ by the AA Similarity Theorem (Theorem 8.3). You can use this fact to write and solve a proportion to find x.

EXAMPLE 4 Finding the Radius of a Circle

Find the radius of the aquarium tank.

SOLUTION

$CB^2 = CE \cdot CD$	Segments of Secants and Tangents Theorem
$20^2 = 8 \cdot (2r + 8)$	Substitute.
$400 = 16r + 64$	Simplify.
$336 = 16r$	Subtract 64 from each side.
$21 = r$	Divide each side by 16.

▶ So, the radius of the tank is 21 feet.

Monitoring Progress

Help in English and Spanish at *BigIdeasMath.com*

Find the value of x.

5.

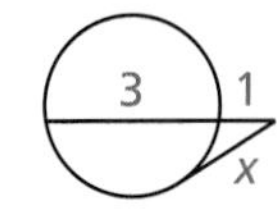

6.

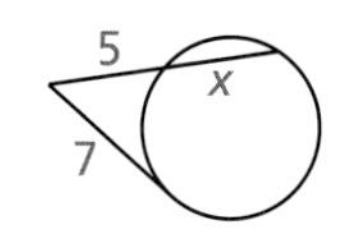

7.

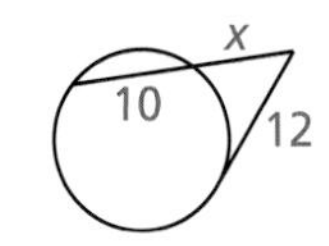

8. WHAT IF? In Example 4, $CB = 35$ feet and $CE = 14$ feet. Find the radius of the tank.

10.6 Exercises

Dynamic Solutions available at *BigIdeasMath.com*

Vocabulary and Core Concept Check

1. **VOCABULARY** The part of the secant segment that is outside the circle is called a(n) ____________.

2. **WRITING** Explain the difference between a tangent segment and a secant segment.

Monitoring Progress and Modeling with Mathematics

In Exercises 3–6, find the value of x. *(See Example 1.)*

3. 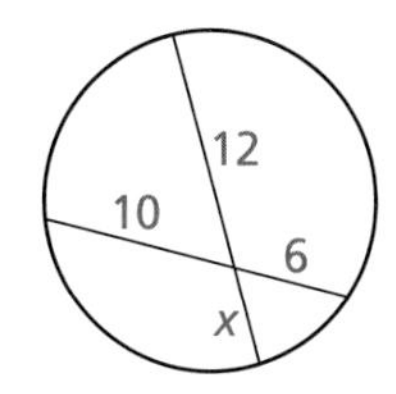

4. 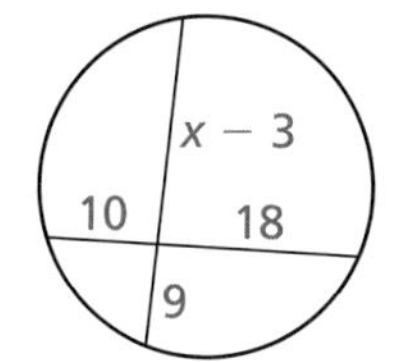

5. 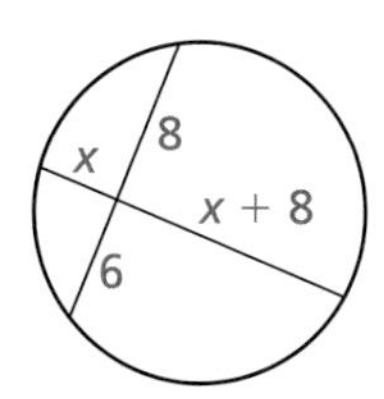

6. 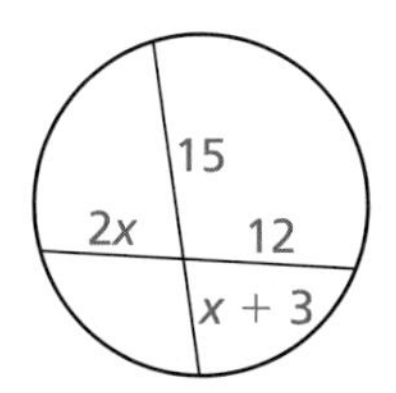

In Exercises 7–10, find the value of x. *(See Example 2.)*

7. 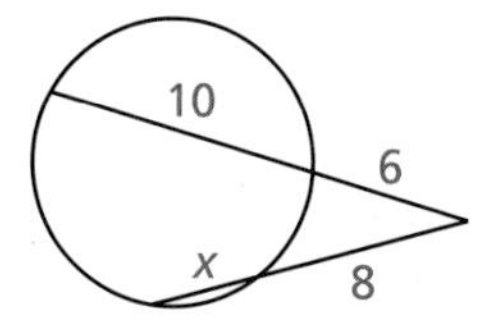

8. 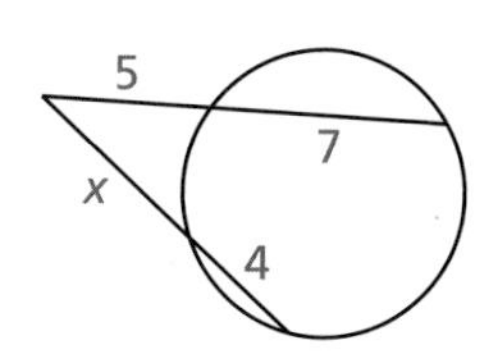

9. 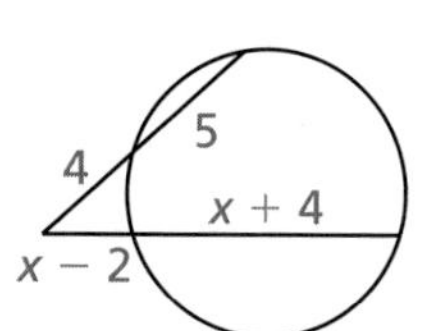

10. 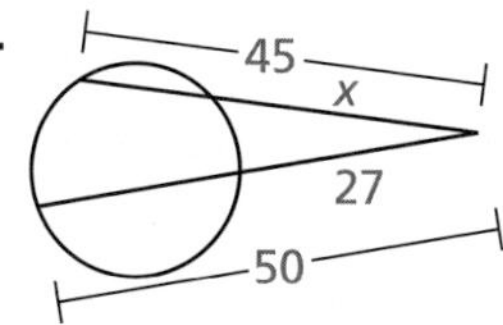

In Exercises 11–14, find the value of x. *(See Example 3.)*

11. 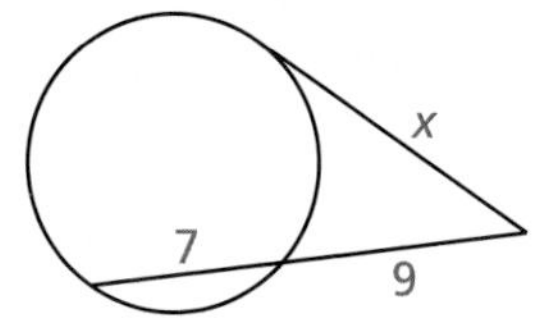

12. 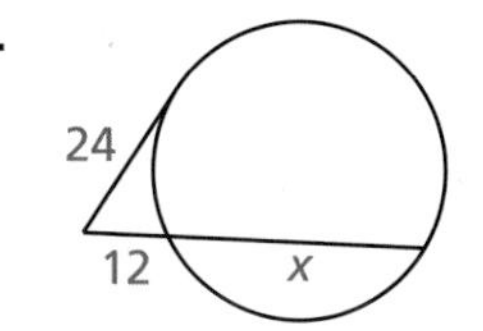

13. 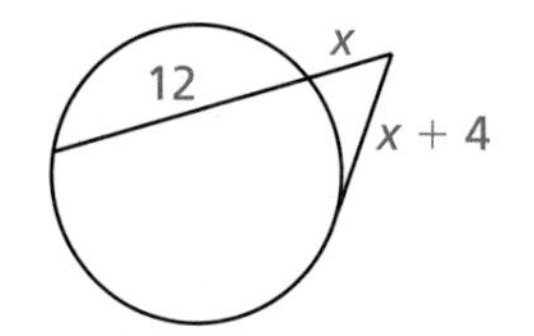

14. 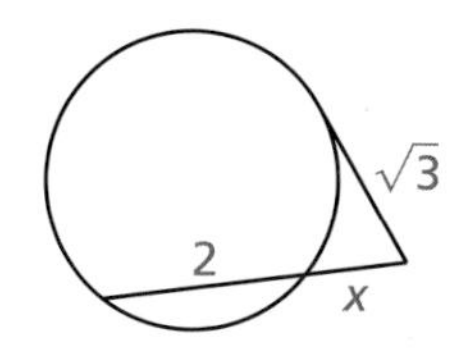

15. **ERROR ANALYSIS** Describe and correct the error in finding CD.

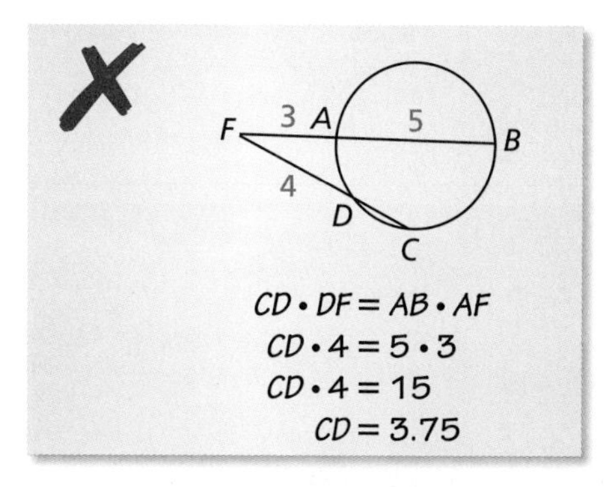

16. **MODELING WITH MATHEMATICS** The Cassini spacecraft is on a mission in orbit around Saturn until September 2017. Three of Saturn's moons, Tethys, Calypso, and Telesto, have nearly circular orbits of radius 295,000 kilometers. The diagram shows the positions of the moons and the spacecraft on one of Cassini's missions. Find the distance DB from Cassini to Tethys when $\overline{AD}$ is tangent to the circular orbit. *(See Example 4.)*

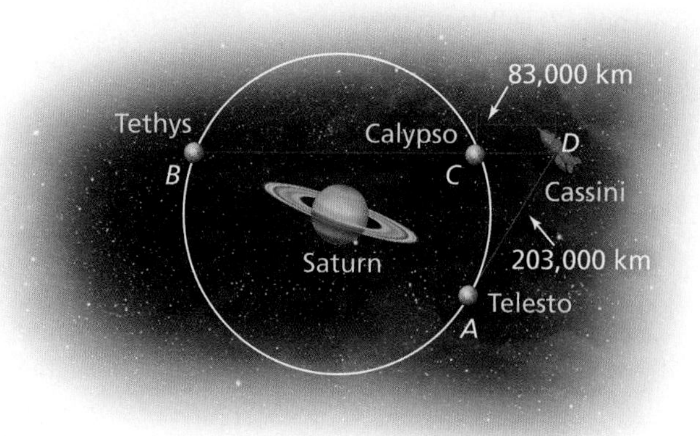

17. MODELING WITH MATHEMATICS The circular stone mound in Ireland called Newgrange has a diameter of 250 feet. A passage 62 feet long leads toward the center of the mound. Find the perpendicular distance x from the end of the passage to either side of the mound.

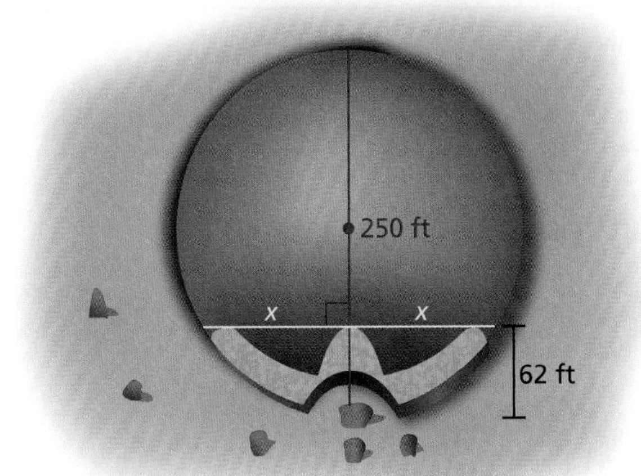

18. MODELING WITH MATHEMATICS You are designing an animated logo for your website. Sparkles leave point C and move to the outer circle along the segments shown so that all of the sparkles reach the outer circle at the same time. Sparkles travel from point C to point D at 2 centimeters per second. How fast should sparkles move from point C to point N? Explain.

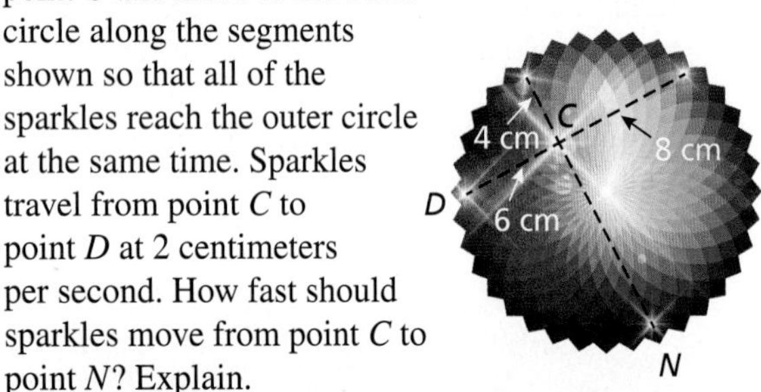

19. PROVING A THEOREM Write a two-column proof of the Segments of Chords Theorem (Theorem 10.18).

Plan for Proof Use the diagram from page 570. Draw $\overline{AC}$ and $\overline{DB}$. Show that $\triangle EAC$ and $\triangle EDB$ are similar. Use the fact that corresponding side lengths in similar triangles are proportional.

20. PROVING A THEOREM Prove the Segments of Secants Theorem (Theorem 10.19). (*Hint*: Draw a diagram and add auxiliary line segments to form similar triangles.)

21. PROVING A THEOREM Use the Tangent Line to Circle Theorem (Theorem 10.1) to prove the Segments of Secants and Tangents Theorem (Theorem 10.20) for the special case when the secant segment contains the center of the circle.

22. PROVING A THEOREM Prove the Segments of Secants and Tangents Theorem (Theorem 10.20). (*Hint*: Draw a diagram and add auxiliary line segments to form similar triangles.)

23. WRITING EQUATIONS In the diagram of the water well, AB, AD, and DE are known. Write an equation for BC using these three measurements.

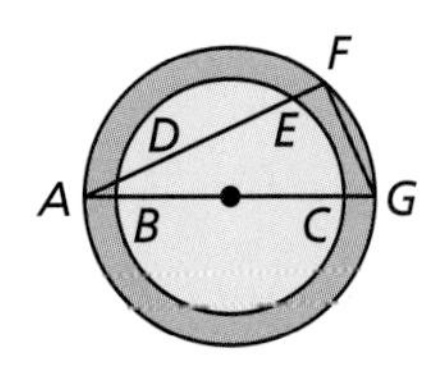

24. HOW DO YOU SEE IT? Which two theorems would you need to use to find PQ? Explain your reasoning.

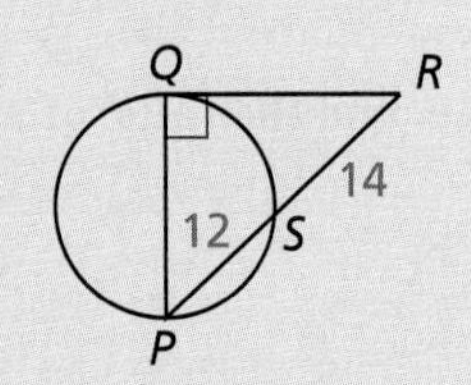

25. CRITICAL THINKING In the figure, $AB = 12$, $BC = 8$, $DE = 6$, $PD = 4$, and A is a point of tangency. Find the radius of $\odot P$.

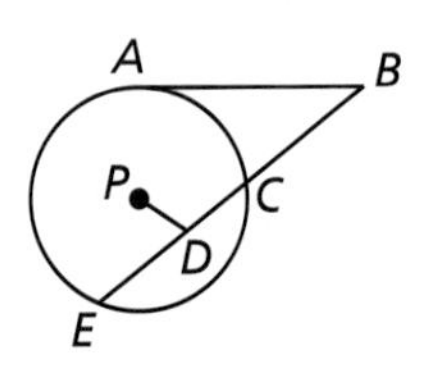

26. THOUGHT PROVOKING Circumscribe a triangle about a circle. Then, using the points of tangency, inscribe a triangle in the circle. Must it be true that the two triangles are similar? Explain your reasoning.

Maintaining Mathematical Proficiency

Reviewing what you learned in previous grades and lessons

Solve the equation by completing the square. *(Skills Review Handbook)*

27. $x^2 + 4x = 45$

28. $x^2 - 2x - 1 = 8$

29. $2x^2 + 12x + 20 = 34$

30. $-4x^2 + 8x + 44 = 16$

10.7 Circles in the Coordinate Plane

Essential Question What is the equation of a circle with center (h, k) and radius r in the coordinate plane?

EXPLORATION 1 The Equation of a Circle with Center at the Origin

Work with a partner. Use dynamic geometry software to construct and determine the equations of circles centered at $(0, 0)$ in the coordinate plane, as described below.

a. Complete the first two rows of the table for circles with the given radii. Complete the other rows for circles with radii of your choice.

b. Write an equation of a circle with center $(0, 0)$ and radius r.

Radius	Equation of circle
1	
2	

EXPLORATION 2 The Equation of a Circle with Center (h, k)

Work with a partner. Use dynamic geometry software to construct and determine the equations of circles of radius 2 in the coordinate plane, as described below.

a. Complete the first two rows of the table for circles with the given centers. Complete the other rows for circles with centers of your choice.

b. Write an equation of a circle with center (h, k) and radius 2.

Center	Equation of circle
(0, 0)	
(2, 0)	

c. Write an equation of a circle with center (h, k) and radius r.

EXPLORATION 3 Deriving the Standard Equation of a Circle

Work with a partner. Consider a circle with radius r and center (h, k).

Write the Distance Formula to represent the distance d between a point (x, y) on the circle and the center (h, k) of the circle. Then square each side of the Distance Formula equation.

How does your result compare with the equation you wrote in part (c) of Exploration 2?

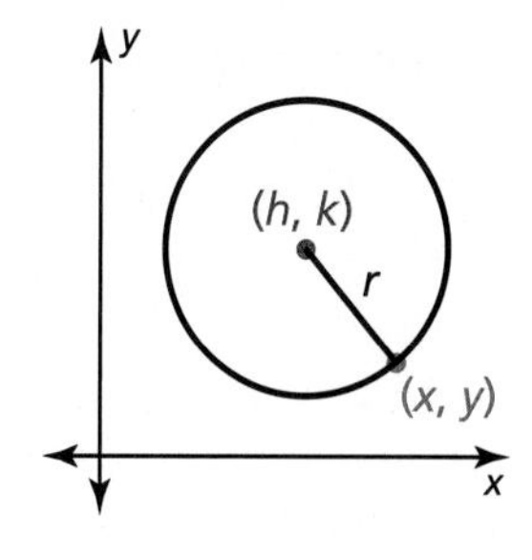

MAKING SENSE OF PROBLEMS

To be proficient in math, you need to explain correspondences between equations and graphs.

Communicate Your Answer

4. What is the equation of a circle with center (h, k) and radius r in the coordinate plane?

5. Write an equation of the circle with center $(4, -1)$ and radius 3.

10.7 Lesson

Core Vocabulary

standard equation of a circle, *p. 576*

Previous

completing the square

What You Will Learn

- Write and graph equations of circles.
- Write coordinate proofs involving circles.
- Solve real-life problems using graphs of circles.

Writing and Graphing Equations of Circles

Let (x, y) represent any point on a circle with center at the origin and radius r. By the Pythagorean Theorem (Theorem 9.1),

$$x^2 + y^2 = r^2.$$

This is the equation of a circle with center at the origin and radius r.

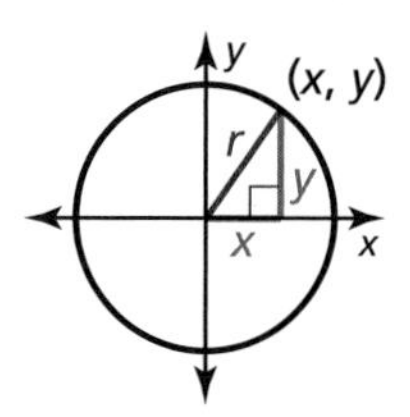

Core Concept

Standard Equation of a Circle

Let (x, y) represent any point on a circle with center (h, k) and radius r. By the Pythagorean Theorem (Theorem 9.1),

$$(x - h)^2 + (y - k)^2 = r^2.$$

This is the **standard equation of a circle** with center (h, k) and radius r.

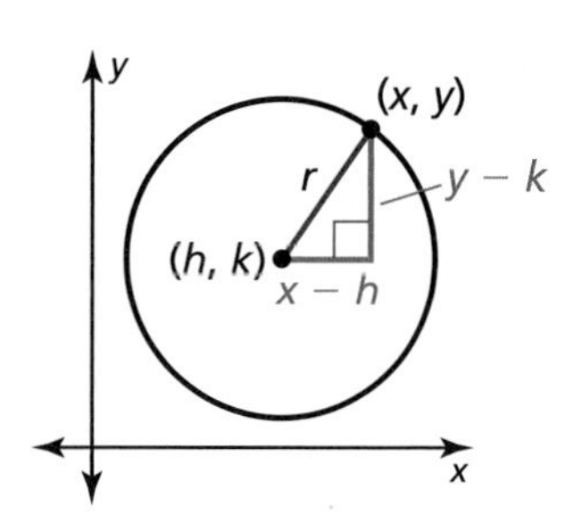

EXAMPLE 1 Writing the Standard Equation of a Circle

Write the standard equation of each circle.

a. the circle shown at the left

b. a circle with center $(0, -9)$ and radius 4.2

SOLUTION

a. The radius is 3, and the center is at the origin.

$(x - h)^2 + (y - k)^2 = r^2$ Standard equation of a circle

$(x - 0)^2 + (y - 0)^2 = 3^2$ Substitute.

$x^2 + y^2 = 9$ Simplify.

▶ The standard equation of the circle is $x^2 + y^2 = 9$.

b. The radius is 4.2, and the center is at $(0, -9)$.

$(x - h)^2 + (y - k)^2 = r^2$

$(x - 0)^2 + [y - (-9)]^2 = 4.2^2$

$x^2 + (y + 9)^2 = 17.64$

▶ The standard equation of the circle is $x^2 + (y + 9)^2 = 17.64$.

Monitoring Progress

Help in English and Spanish at *BigIdeasMath.com*

Write the standard equation of the circle with the given center and radius.

1. center: $(0, 0)$, radius: 2.5

2. center: $(-2, 5)$, radius: 7

EXAMPLE 2 Writing the Standard Equation of a Circle

The point $(-5, 6)$ is on a circle with center $(-1, 3)$. Write the standard equation of the circle.

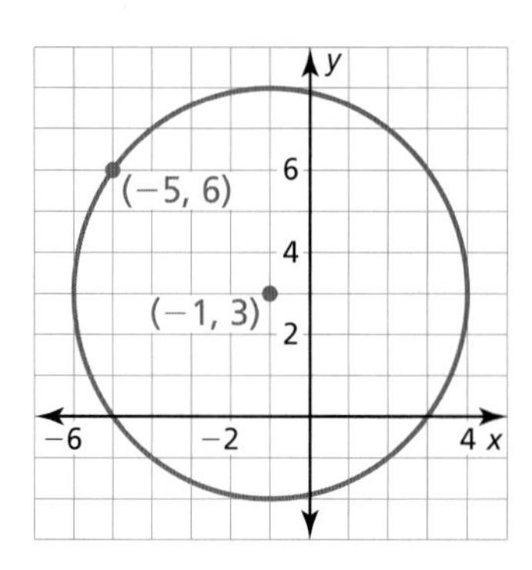

SOLUTION

To write the standard equation, you need to know the values of h, k, and r. To find r, find the distance between the center and the point $(-5, 6)$ on the circle.

$r = \sqrt{[-5 - (-1)]^2 + (6 - 3)^2}$ Distance Formula

$= \sqrt{(-4)^2 + 3^2}$ Simplify.

$= 5$ Simplify.

Substitute the values for the center and the radius into the standard equation of a circle.

$(x - h)^2 + (y - k)^2 = r^2$ Standard equation of a circle

$[x - (-1)]^2 + (y - 3)^2 = 5^2$ Substitute $(h, k) = (-1, 3)$ and $r = 5$.

$(x + 1)^2 + (y - 3)^2 = 25$ Simplify.

▶ The standard equation of the circle is $(x + 1)^2 + (y - 3)^2 = 25$.

REMEMBER

To complete the square for the expression $x^2 + bx$, add the square of half the coefficient of the term bx.

$x^2 + bx + \left(\frac{b}{2}\right)^2 = \left(x + \frac{b}{2}\right)^2$

EXAMPLE 3 Graphing a Circle

The equation of a circle is $x^2 + y^2 - 8x + 4y - 16 = 0$. Find the center and the radius of the circle. Then graph the circle.

SOLUTION

You can write the equation in standard form by completing the square on the x-terms and the y-terms.

$x^2 + y^2 - 8x + 4y - 16 = 0$ Equation of circle

$x^2 - 8x + y^2 + 4y = 16$ Isolate constant. Group terms.

$x^2 - 8x + 16 + y^2 + 4y + 4 = 16 + 16 + 4$ Complete the square twice.

$(x - 4)^2 + (y + 2)^2 = 36$ Factor left side. Simplify right side.

$(x - 4)^2 + [y - (-2)]^2 = 6^2$ Rewrite the equation to find the center and the radius.

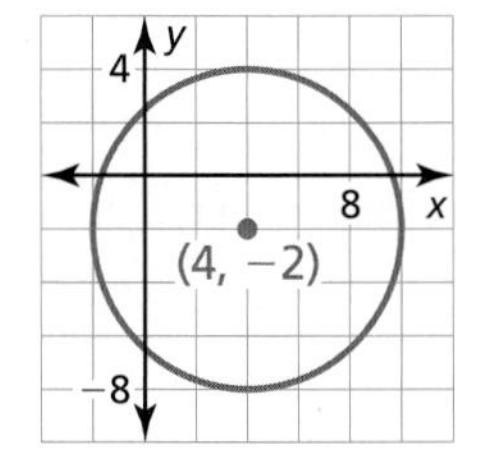

▶ The center is $(4, -2)$, and the radius is 6. Use a compass to graph the circle.

Monitoring Progress  Help in English and Spanish at *BigIdeasMath.com*

3. The point $(3, 4)$ is on a circle with center $(1, 4)$. Write the standard equation of the circle.

4. The equation of a circle is $x^2 + y^2 - 8x + 6y + 9 = 0$. Find the center and the radius of the circle. Then graph the circle.

Writing Coordinate Proofs Involving Circles

EXAMPLE 4 Writing a Coordinate Proof Involving a Circle

Prove or disprove that the point $(\sqrt{2}, \sqrt{2})$ lies on the circle centered at the origin and containing the point (2, 0).

SOLUTION

The circle centered at the origin and containing the point (2, 0) has the following radius.

$$r = \sqrt{(x-h)^2 + (y-k)^2} = \sqrt{(2-0)^2 + (0-0)^2} = 2$$

So, a point lies on the circle if and only if the distance from that point to the origin is 2. The distance from $(\sqrt{2}, \sqrt{2})$ to (0, 0) is

$$d = \sqrt{(\sqrt{2}-0)^2 + (\sqrt{2}-0)^2} = 2.$$

▶ So, the point $(\sqrt{2}, \sqrt{2})$ lies on the circle centered at the origin and containing the point (2, 0).

Monitoring Progress Help in English and Spanish at *BigIdeasMath.com*

5. Prove or disprove that the point $(1, \sqrt{5})$ lies on the circle centered at the origin and containing the point (0, 1).

Solving Real-Life Problems

EXAMPLE 5 Using Graphs of Circles

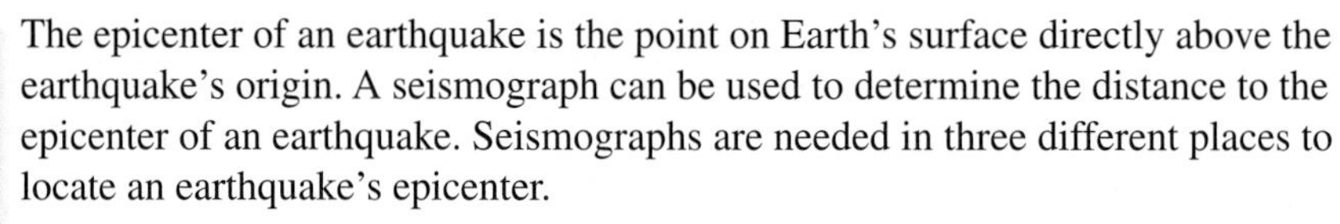

The epicenter of an earthquake is the point on Earth's surface directly above the earthquake's origin. A seismograph can be used to determine the distance to the epicenter of an earthquake. Seismographs are needed in three different places to locate an earthquake's epicenter.

Use the seismograph readings from locations A, B, and C to find the epicenter of an earthquake.

- The epicenter is 7 miles away from $A(-2, 2.5)$.
- The epicenter is 4 miles away from $B(4, 6)$.
- The epicenter is 5 miles away from $C(3, -2.5)$.

SOLUTION

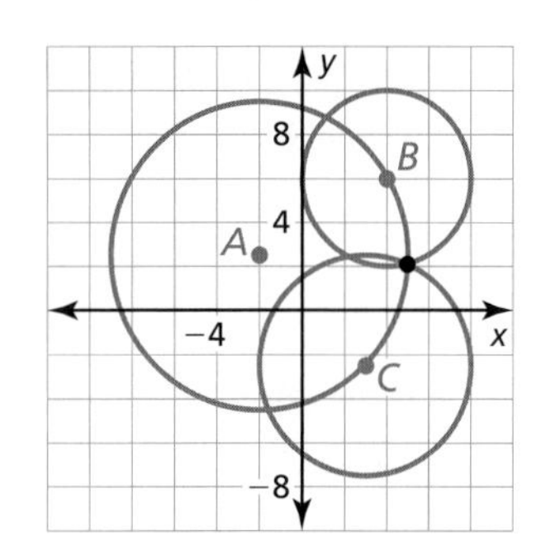

The set of all points equidistant from a given point is a circle, so the epicenter is located on each of the following circles.

$\odot A$ with center $(-2, 2.5)$ and radius 7

$\odot B$ with center $(4, 6)$ and radius 4

$\odot C$ with center $(3, -2.5)$ and radius 5

To find the epicenter, graph the circles on a coordinate plane where each unit corresponds to one mile. Find the point of intersection of the three circles.

▶ The epicenter is at about (5, 2).

Monitoring Progress  Help in English and Spanish at *BigIdeasMath.com*

6. Why are three seismographs needed to locate an earthquake's epicenter?

10.7 Exercises

Dynamic Solutions available at *BigIdeasMath.com*

Vocabulary and Core Concept Check

1. **VOCABULARY** What is the standard equation of a circle?

2. **WRITING** Explain why knowing the location of the center and one point on a circle is enough to graph the circle.

Monitoring Progress and Modeling with Mathematics

In Exercises 3–8, write the standard equation of the circle. *(See Example 1.)*

3. 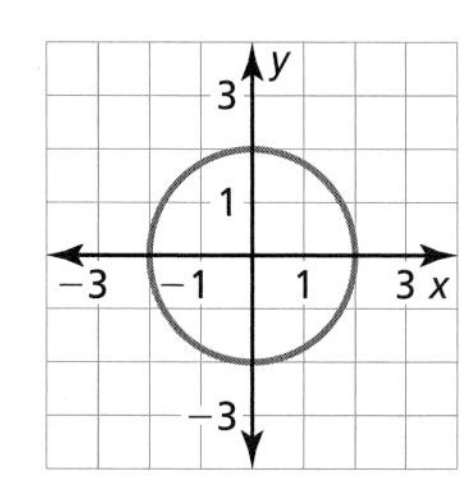

4. 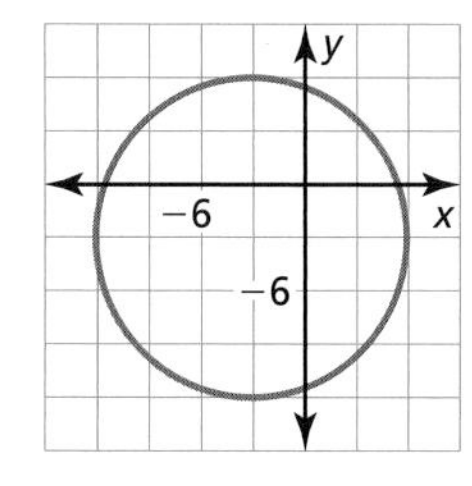

5. a circle with center (0, 0) and radius 7

6. a circle with center (4, 1) and radius 5

7. a circle with center (−3, 4) and radius 1

8. a circle with center (3, −5) and radius 7

In Exercises 9–11, use the given information to write the standard equation of the circle. *(See Example 2.)*

9. The center is (0, 0), and a point on the circle is (0, 6).

10. The center is (1, 2), and a point on the circle is (4, 2).

11. The center is (0, 0), and a point on the circle is (3, −7).

12. **ERROR ANALYSIS** Describe and correct the error in writing the standard equation of a circle.

The standard equation of a circle with center (−3, −5) and radius 3 is $(x - 3)^2 + (y - 5)^2 = 9$.

In Exercises 13–18, find the center and radius of the circle. Then graph the circle. *(See Example 3.)*

13. $x^2 + y^2 = 49$

14. $(x + 5)^2 + (y - 3)^2 = 9$

15. $x^2 + y^2 - 6x = 7$

16. $x^2 + y^2 + 4y = 32$

17. $x^2 + y^2 - 8x - 2y = -16$

18. $x^2 + y^2 + 4x + 12y = -15$

In Exercises 19–22, prove or disprove the statement. *(See Example 4.)*

19. The point (2, 3) lies on the circle centered at the origin with radius 8.

20. The point $(4, \sqrt{5})$ lies on the circle centered at the origin with radius 3.

21. The point $(\sqrt{6}, 2)$ lies on the circle centered at the origin and containing the point (3, −1).

22. The point $(\sqrt{7}, 5)$ lies on the circle centered at the origin and containing the point (5, 2).

23. **MODELING WITH MATHEMATICS** A city's commuter system has three zones. Zone 1 serves people living within 3 miles of the city's center. Zone 2 serves those between 3 and 7 miles from the center. Zone 3 serves those over 7 miles from the center. *(See Example 5.)*

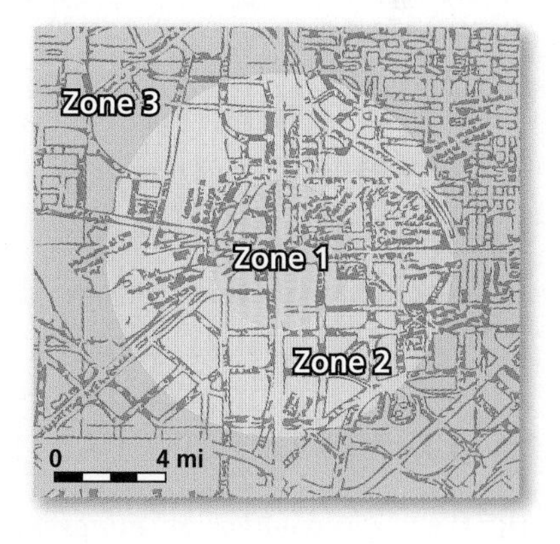

a. Graph this situation on a coordinate plane where each unit corresponds to 1 mile. Locate the city's center at the origin.

b. Determine which zone serves people whose homes are represented by the points (3, 4), (6, 5), (1, 2), (0, 3), and (1, 6).

24. MODELING WITH MATHEMATICS Telecommunication towers can be used to transmit cellular phone calls. A graph with units measured in kilometers shows towers at points (0, 0), (0, 5), and (6, 3). These towers have a range of about 3 kilometers.

a. Sketch a graph and locate the towers. Are there any locations that may receive calls from more than one tower? Explain your reasoning.

b. The center of City A is located at $(-2, 2.5)$, and the center of City B is located at (5, 4). Each city has a radius of 1.5 kilometers. Which city seems to have better cell phone coverage? Explain your reasoning.

25. REASONING Sketch the graph of the circle whose equation is $x^2 + y^2 = 16$. Then sketch the graph of the circle after the translation $(x, y) \rightarrow (x - 2, y - 4)$. What is the equation of the image? Make a conjecture about the equation of the image of a circle centered at the origin after a translation m units to the left and n units down.

26. HOW DO YOU SEE IT? Match each graph with its equation.

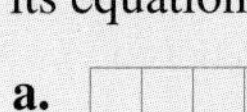

a.

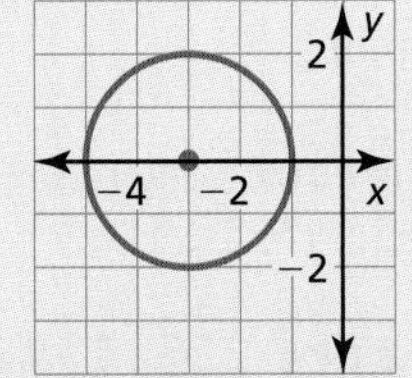

b.

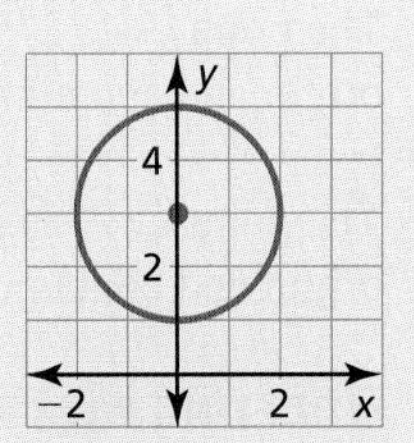

c.

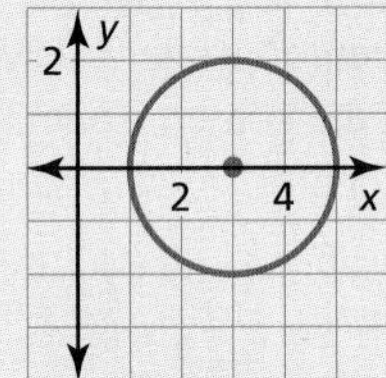

d.

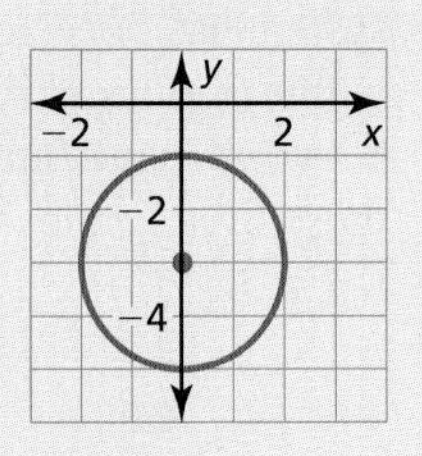

A. $x^2 + (y + 3)^2 = 4$ **B.** $(x - 3)^2 + y^2 = 4$

C. $(x + 3)^2 + y^2 = 4$ **D.** $x^2 + (y - 3)^2 = 4$

27. USING STRUCTURE The vertices of $\triangle XYZ$ are $X(4, 5)$, $Y(4, 13)$, and $Z(8, 9)$. Find the equation of the circle circumscribed about $\triangle XYZ$. Justify your answer.

28. THOUGHT PROVOKING A circle has center (h, k) and contains point (a, b). Write the equation of the line tangent to the circle at point (a, b).

MATHEMATICAL CONNECTIONS In Exercises 29–32, use the equations to determine whether the line is *a tangent*, *a secant*, *a secant that contains the diameter*, or *none of these*. Explain your reasoning.

29. **Circle:** $(x - 4)^2 + (y - 3)^2 = 9$
Line: $y = 6$

30. **Circle:** $(x + 2)^2 + (y - 2)^2 = 16$
Line: $y = 2x - 4$

31. **Circle:** $(x - 5)^2 + (y + 1)^2 = 4$
Line: $y = \frac{1}{5}x - 3$

32. **Circle:** $(x + 3)^2 + (y - 6)^2 = 25$
Line: $y = -\frac{4}{3}x + 2$

33. MAKING AN ARGUMENT Your friend claims that the equation of a circle passing through the points $(-1, 0)$ and $(1, 0)$ is $x^2 - 2yk + y^2 = 1$ with center $(0, k)$. Is your friend correct? Explain your reasoning.

34. REASONING Four tangent circles are centered on the x-axis. The radius of $\odot A$ is twice the radius of $\odot O$. The radius of $\odot B$ is three times the radius of $\odot O$. The radius of $\odot C$ is four times the radius of $\odot O$. All circles have integer radii, and the point (63, 16) is on $\odot C$. What is the equation of $\odot A$? Explain your reasoning.

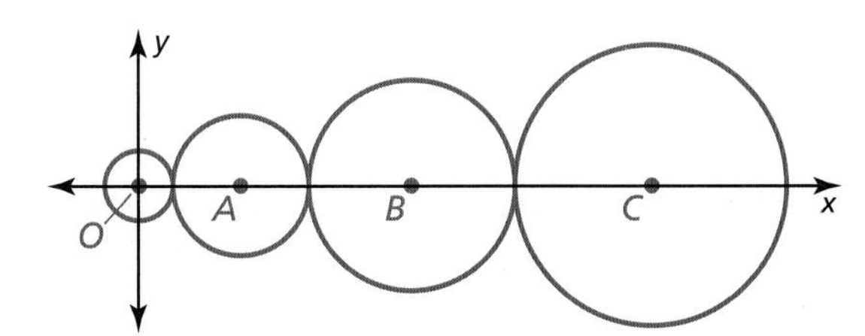

Maintaining Mathematical Proficiency

Reviewing what you learned in previous grades and lessons

Identify the arc as a *major arc*, *minor arc*, or *semicircle*. Then find the measure of the arc. *(Section 10.2)*

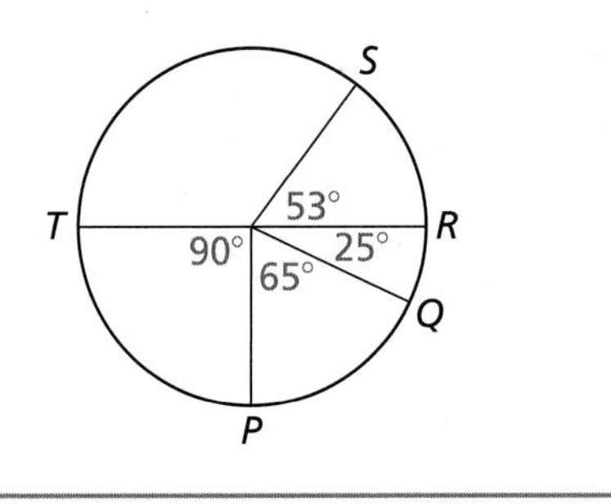

35. $\overset{\frown}{RS}$ **36.** $\overset{\frown}{PR}$

37. $\overset{\frown}{PRT}$ **38.** $\overset{\frown}{ST}$

39. $\overset{\frown}{RST}$ **40.** $\overset{\frown}{QS}$

10.4–10.7 What Did You Learn?

Core Vocabulary

inscribed angle, *p. 554*
intercepted arc, *p. 554*
subtend, *p. 554*
inscribed polygon, *p. 556*
circumscribed circle, *p. 556*
circumscribed angle, *p. 564*
segments of a chord, *p. 570*
tangent segment, *p. 571*
secant segment, *p. 571*
external segment, *p. 571*
standard equation of a circle, *p. 576*

Core Concepts

Section 10.4

Inscribed Angle and Intercepted Arc, *p. 554*
Theorem 10.10 Measure of an Inscribed Angle Theorem, *p. 554*
Theorem 10.11 Inscribed Angles of a Circle Theorem, *p. 555*
Inscribed Polygon, *p. 556*
Theorem 10.12 Inscribed Right Triangle Theorem, *p. 556*
Theorem 10.13 Inscribed Quadrilateral Theorem, *p. 556*

Section 10.5

Theorem 10.14 Tangent and Intersected Chord Theorem, *p. 562*
Intersecting Lines and Circles, *p. 562*
Theorem 10.15 Angles Inside the Circle Theorem, *p. 563*
Theorem 10.16 Angles Outside the Circle Theorem, *p. 563*
Circumscribed Angle, *p. 564*
Theorem 10.17 Circumscribed Angle Theorem, *p. 564*

Section 10.6

Theorem 10.18 Segments of Chords Theorem, *p. 570*
Tangent Segment and Secant Segment, *p. 571*
Theorem 10.19 Segments of Secants Theorem, *p. 571*
Theorem 10.20 Segments of Secants and Tangents Theorem, *p. 572*

Section 10.7

Standard Equation of a Circle, *p. 576*
Writing Coordinate Proofs Involving Circles, *p. 578*

Mathematical Practices

1. What other tools could you use to complete the task in Exercise 18 on page 559?
2. You have a classmate who is confused about why two diagrams are needed in part (a) of Exercise 31 on page 567. Explain to your classmate why two diagrams are needed.

Performance Task

Circular Motion

What do the properties of tangents tell us about the forces acting on a satellite orbiting around Earth? How would the path of the satellite change if the force of gravity were removed?

To explore the answers to this question and more, go to ***BigIdeasMath.com***.

10 Chapter Review

Dynamic Solutions available at *BigIdeasMath.com*

10.1 Lines and Segments That Intersect Circles (pp. 529–536)

In the diagram, $\overline{AB}$ is tangent to $\odot C$ at B and $\overline{AD}$ is tangent to $\odot C$ at D. Find the value of x.

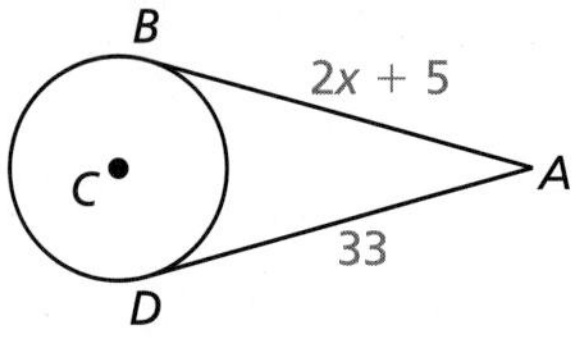

B 2x + 5 C A 33 D

$AB = AD$	External Tangent Congruence Theorem (Theorem 10.2)
$2x + 5 = 33$	Substitute.
$x = 14$	Solve for x.

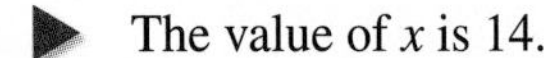

The value of x is 14.

Tell whether the line, ray, or segment is best described as a *radius*, *chord*, *diameter*, *secant*, or *tangent* of $\odot P$.

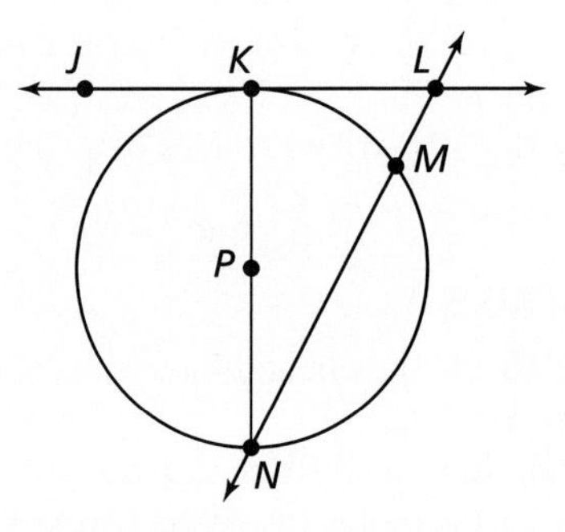

1. $\overline{PK}$

2. $\overline{NM}$

3. $\overrightarrow{JL}$

4. $\overline{KN}$

5. $\overleftrightarrow{NL}$

6. $\overline{PN}$

Tell whether the common tangent is *internal* or *external*.

7.

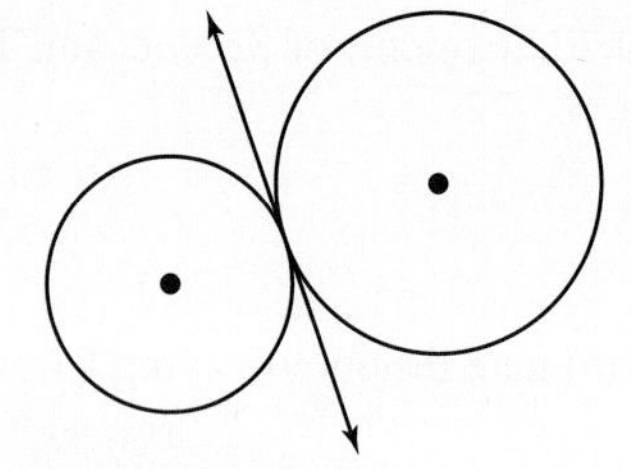

8.

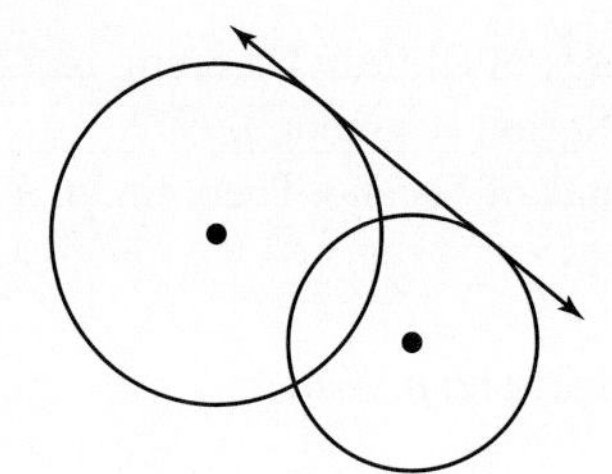

Points Y and Z are points of tangency. Find the value of the variable.

9.

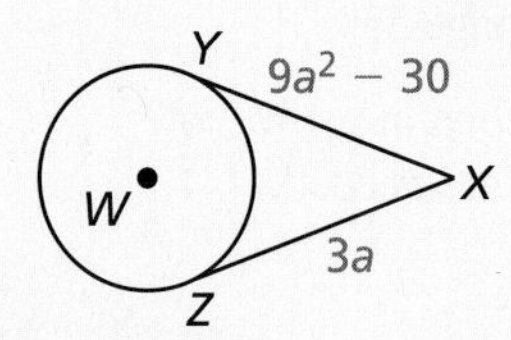

10.

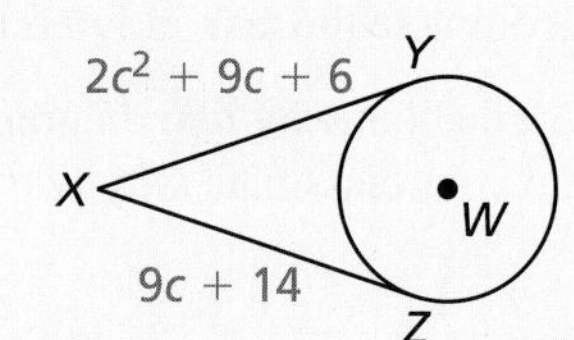

11.

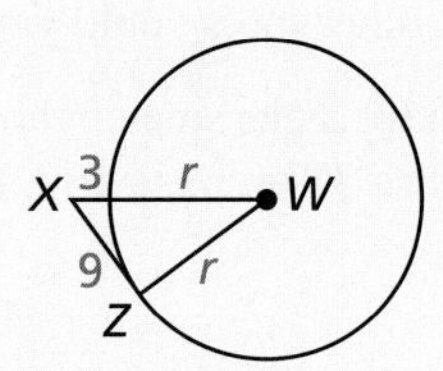

12. Tell whether $\overline{AB}$ is tangent to $\odot C$. Explain.

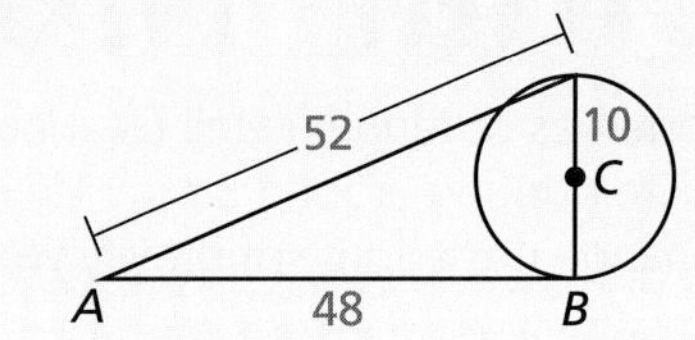

10.2 Finding Arc Measures (pp. 537–544)

Find the measure of each arc of $\odot P$, where $\overline{LN}$ is a diameter.

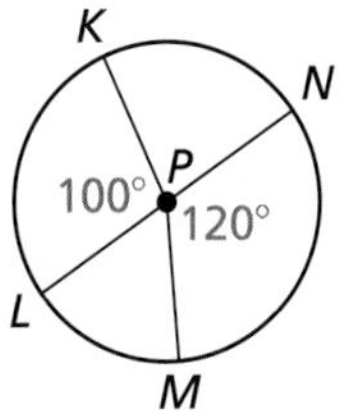

a. $\overparen{MN}$

▶ $\overparen{MN}$ is a minor arc, so $m\overparen{MN} = m\angle MPN = 120°$.

b. $\overparen{NLM}$

▶ $\overparen{NLM}$ is a major arc, so $m\overparen{NLM} = 360° - 120° = 240°$.

c. $\overparen{NML}$

▶ $\overline{NL}$ is a diameter, so $\overparen{NML}$ is a semicircle, and $m\overparen{NML} = 180°$.

Use the diagram above to find the measure of the indicated arc.

13. $\overparen{KL}$ **14.** $\overparen{LM}$ **15.** $\overparen{KM}$ **16.** $\overparen{KN}$

Tell whether the red arcs are congruent. Explain why or why not.

17.

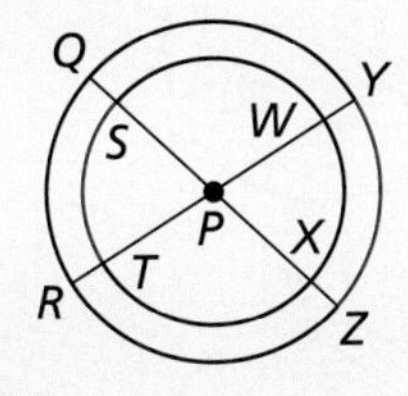

18.

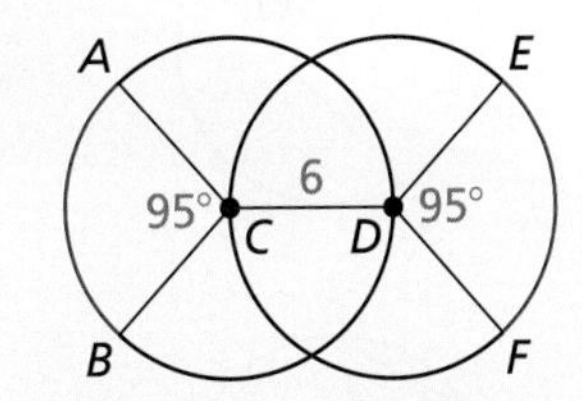

10.3 Using Chords (pp. 545–550)

In the diagram, $\odot A \cong \odot B$, $\overline{CD} \cong \overline{FE}$, and $m\overparen{FE} = 75°$. Find $m\overparen{CD}$.

Because $\overline{CD}$ and $\overline{FE}$ are congruent chords in congruent circles, the corresponding minor arcs $\overparen{CD}$ and $\overparen{FE}$ are congruent by the Congruent Corresponding Chords Theorem (Theorem 10.6).

▶ So, $m\overparen{CD} = m\overparen{FE} = 75°$.

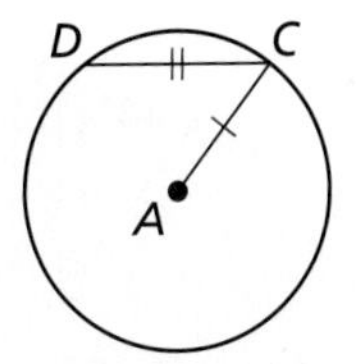

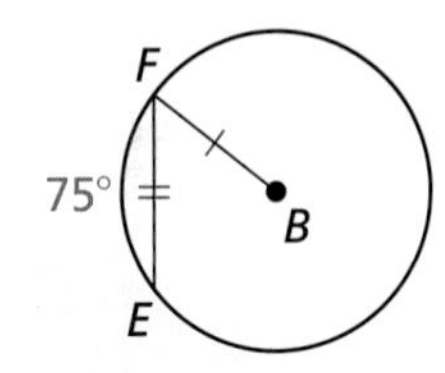

Find the measure of $\overparen{AB}$.

19.

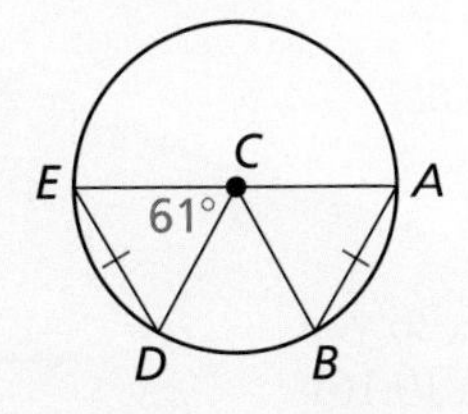

20.

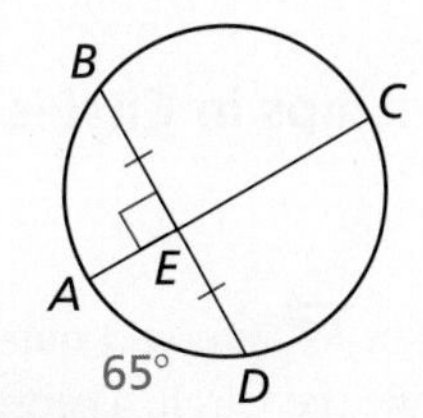

21.

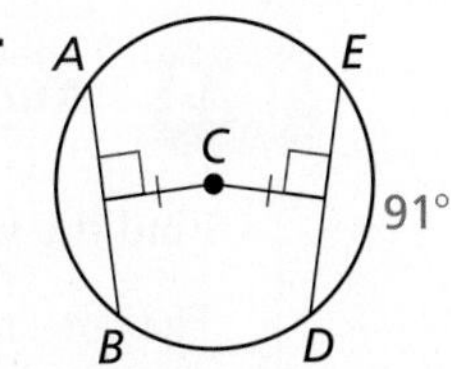

22. In the diagram, $QN = QP = 10$, $JK = 4x$, and $LM = 6x - 24$. Find the radius of $\odot Q$.

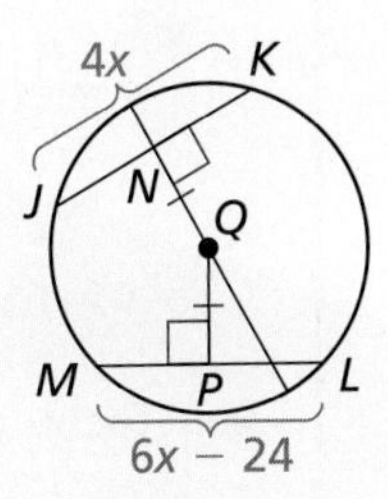

10.4 Inscribed Angles and Polygons (pp. 553–560)

Find the value of each variable.

LMNP is inscribed in a circle, so opposite angles are supplementary by the Inscribed Quadrilateral Theorem (Theorem 10.13).

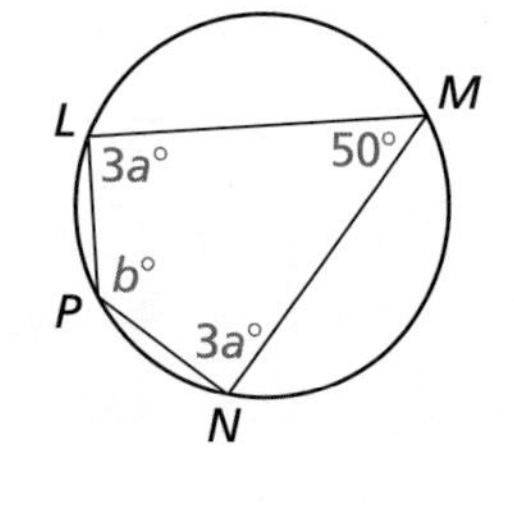

$m\angle L + m\angle N = 180°$ $\qquad$ $m\angle P + m\angle M = 180°$

$3a° + 3a° = 180°$ $\qquad$ $b° + 50° = 180°$

$6a = 180$ $\qquad$ $b = 130$

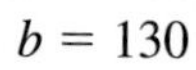

$a = 30$

▶ The value of a is 30, and the value of b is 130.

Find the value(s) of the variable(s).

23.

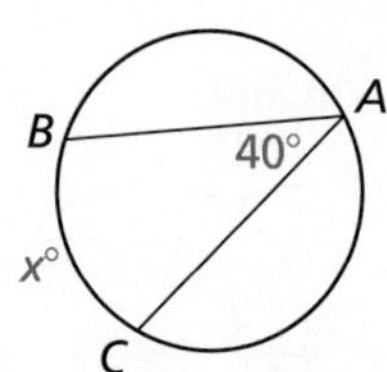

24.

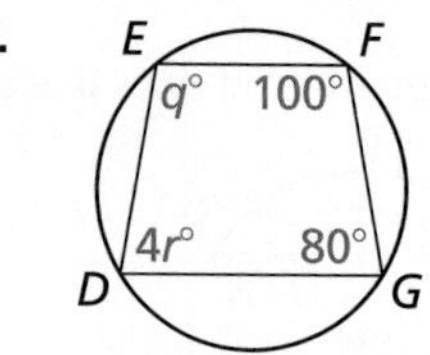

25.

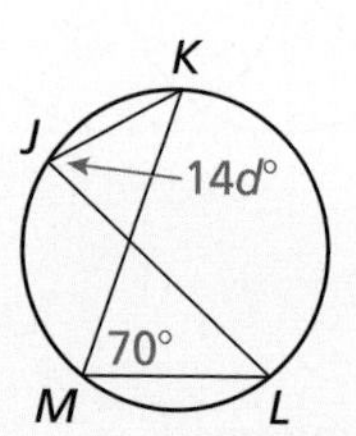

26.

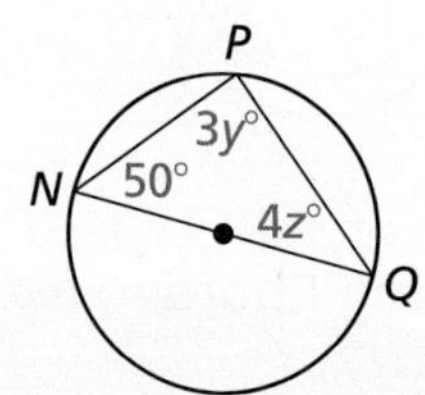

27.

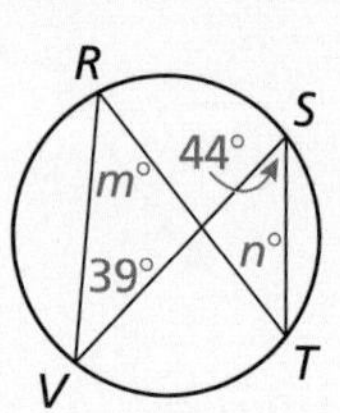

28.

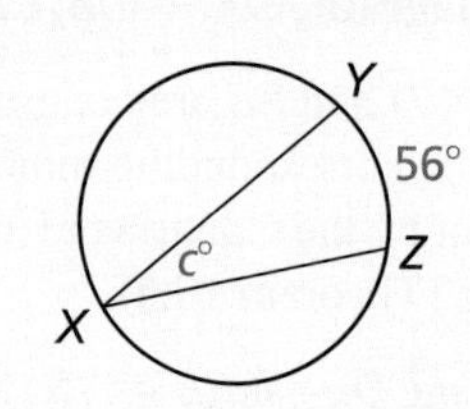

10.5 Angle Relationships in Circles (pp. 561–568)

Find the value of *y*.

The tangent $\overrightarrow{RQ}$ and secant $\overrightarrow{RT}$ intersect outside the circle, so you can use the Angles Outside the Circle Theorem (Theorem 10.16).

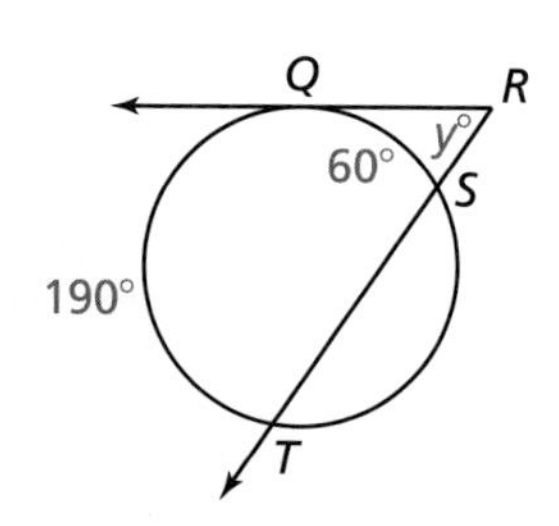

$y° = \frac{1}{2}(m\widehat{QT} - m\widehat{SQ})$ $\qquad$ Angles Outside the Circle Theorem

$y° = \frac{1}{2}(190° - 60°)$ $\qquad$ Substitute.

$y = 65$ $\qquad$ Simplify.

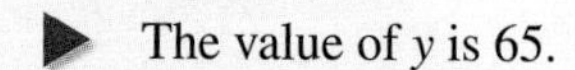

▶ The value of y is 65.

Find the value of x.

29.

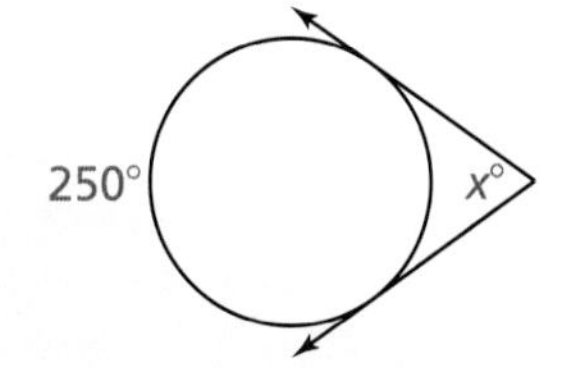

30.

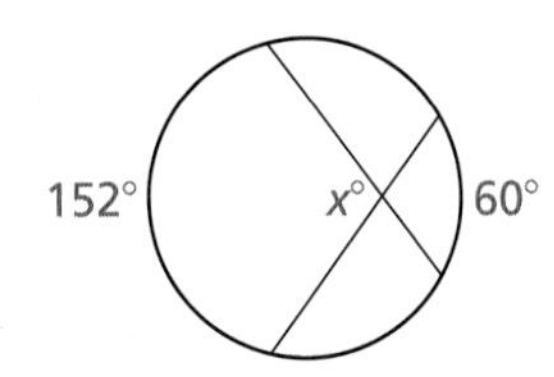

31.

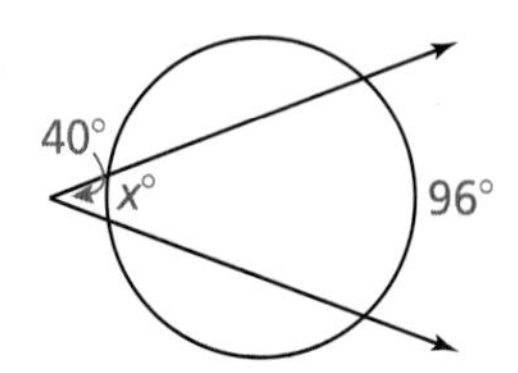

32. Line ℓ is tangent to the circle. Find $m\widehat{XYZ}$.

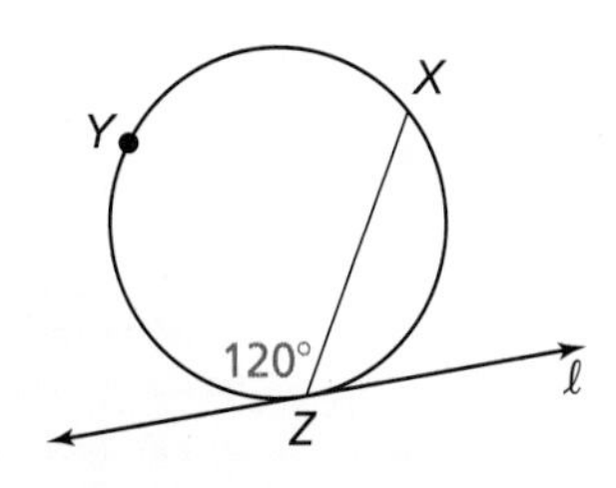

10.6 Segment Relationships in Circles (pp. 569–574)

Find the value of x.

The chords $\overline{EG}$ and $\overline{FH}$ intersect inside the circle, so you can use the Segments of Chords Theorem (Theorem 10.18).

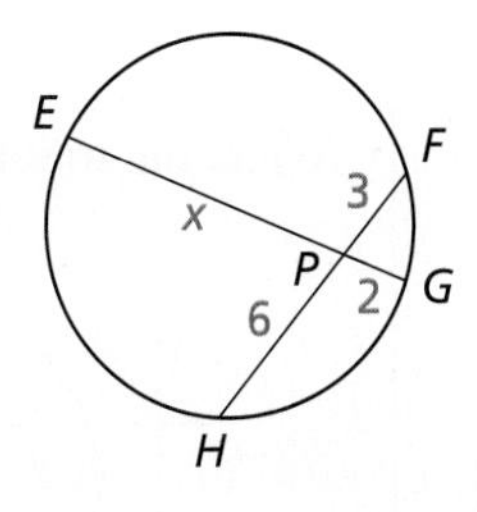

$EP \cdot PG = FP \cdot PH$	Segments of Chords Theorem
$x \cdot 2 = 3 \cdot 6$	Substitute.
$x = 9$	Simplify.

▶ The value of x is 9.

Find the value of x.

33.

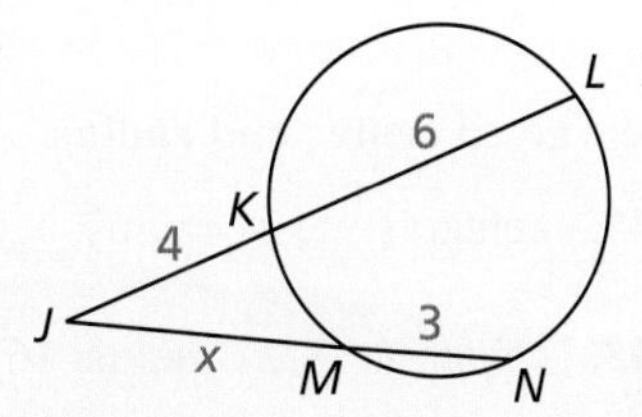

34.

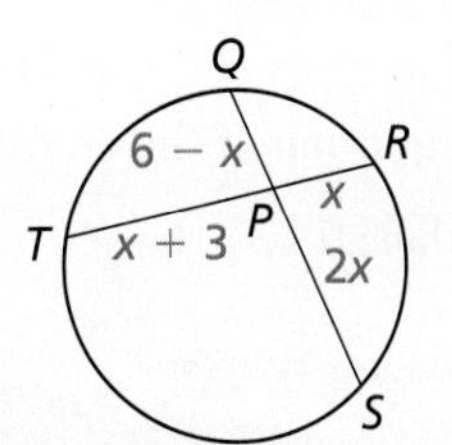

35.

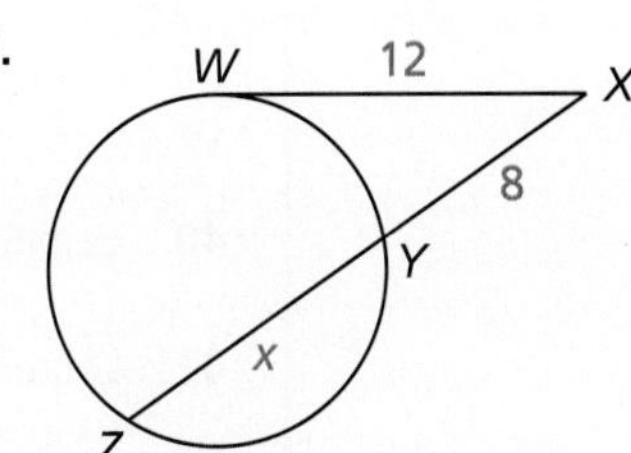

36. A local park has a circular ice skating rink. You are standing at point A, about 12 feet from the edge of the rink. The distance from you to a point of tangency on the rink is about 20 feet. Estimate the radius of the rink.

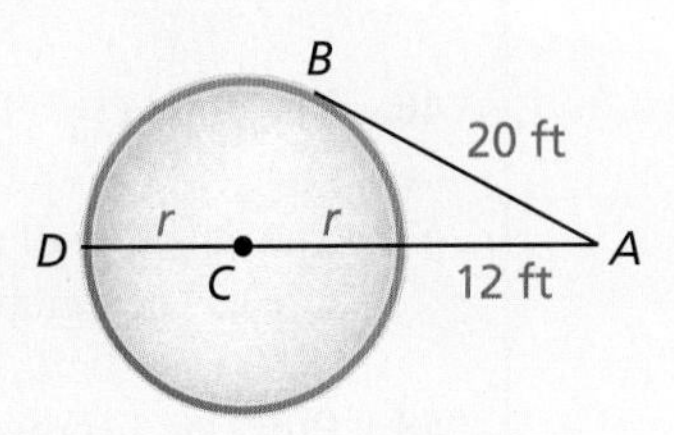

10.7 Circles in the Coordinate Plane (pp. 575–580)

Write the standard equation of the circle shown.

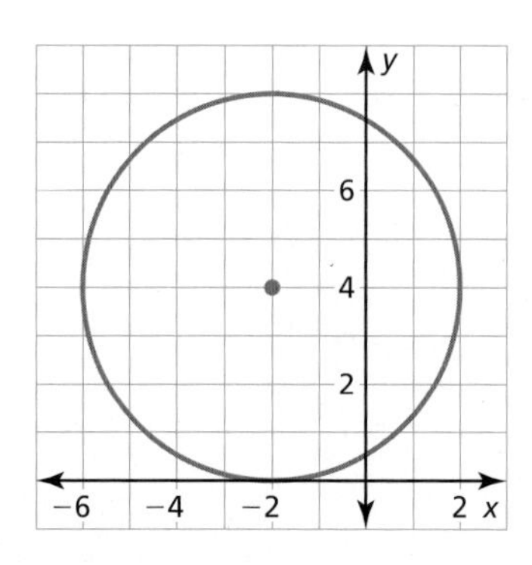

The radius is 4, and the center is $(-2, 4)$.

$(x - h)^2 + (y - k)^2 = r^2$	Standard equation of a circle
$[x - (-2)]^2 + (y - 4)^2 = 4^2$	Substitute.
$(x + 2)^2 + (y - 4)^2 = 16$	Simplify.

▶ The standard equation of the circle is $(x + 2)^2 + (y - 4)^2 = 16$.

Write the standard equation of the circle shown.

37.

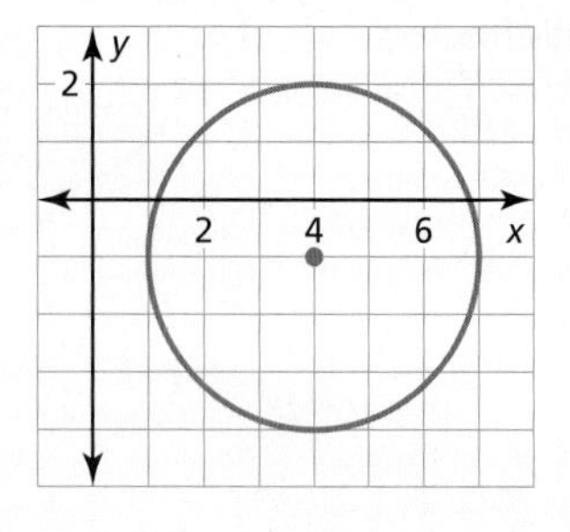

38.

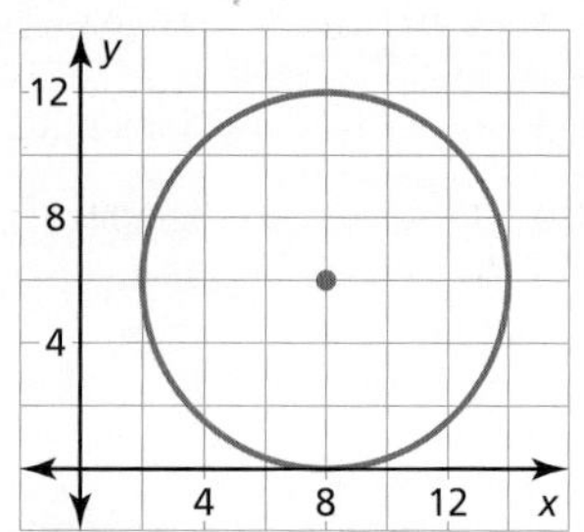

39.

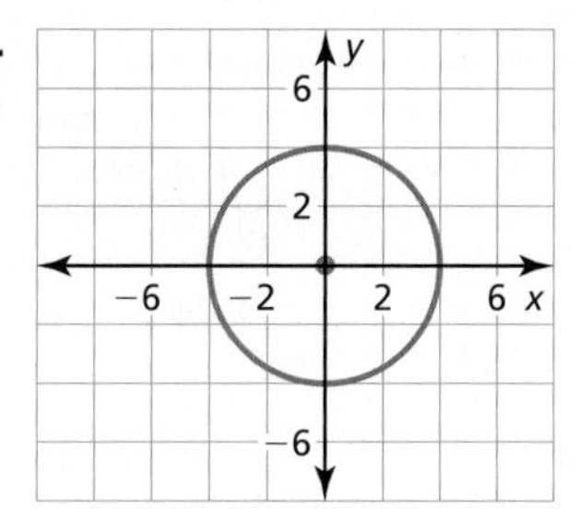

Write the standard equation of the circle with the given center and radius.

40. center: $(0, 0)$, radius: 9

41. center: $(-5, 2)$, radius: 1.3

42. center: $(6, 21)$, radius: 4

43. center: $(-3, 2)$, radius: 16

44. center: $(10, 7)$, radius: 3.5

45. center: $(0, 0)$, radius: 5.2

46. The point $(-7, 1)$ is on a circle with center $(-7, 6)$. Write the standard equation of the circle.

47. The equation of a circle is $x^2 + y^2 - 12x + 8y + 48 = 0$. Find the center and the radius of the circle. Then graph the circle.

48. Prove or disprove that the point $(4, -3)$ lies on the circle centered at the origin and containing the point $(-5, 0)$.

10 Chapter Test

Find the measure of each numbered angle in $\odot P$. Justify your answer.

1.

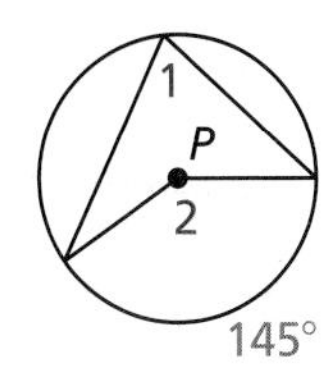

2.

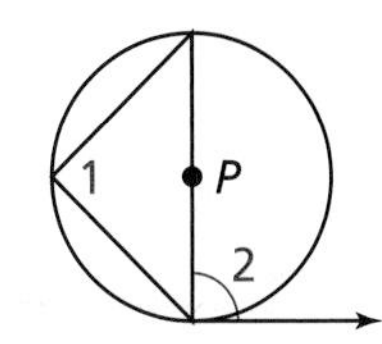

3.

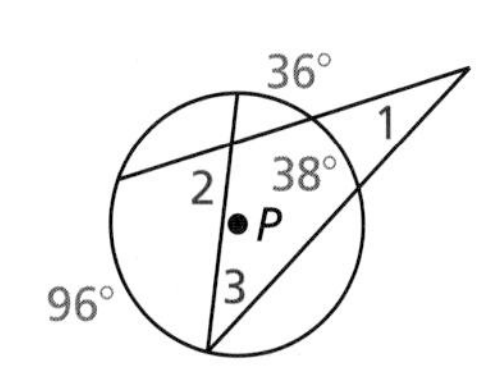

4.

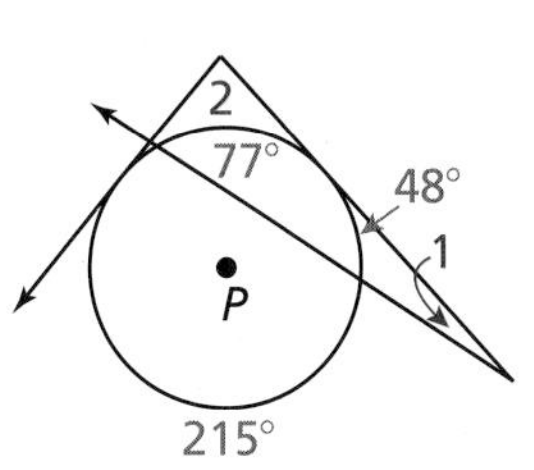

Use the diagram.

5. $AG = 2$, $GD = 9$, and $BG = 3$. Find GF.

6. $CF = 12$, $CB = 3$, and $CD = 9$. Find CE.

7. $BF = 9$ and $CB = 3$. Find CA.

8. Sketch a pentagon inscribed in a circle. Label the pentagon $ABCDE$. Describe the relationship between each pair of angles. Explain your reasoning.

a. $\angle CDE$ and $\angle CAE$ **b.** $\angle CBE$ and $\angle CAE$

Find the value of the variable. Justify your answer.

9.

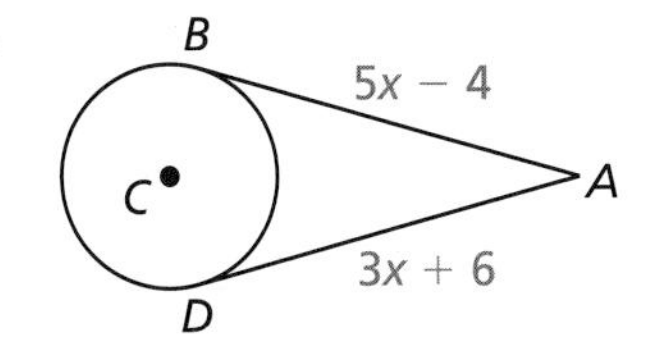

10.

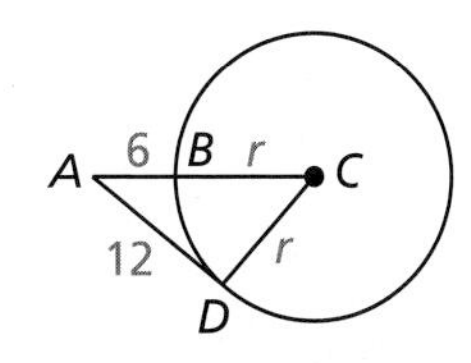

11. Prove or disprove that the point $\left(2\sqrt{2}, -1\right)$ lies on the circle centered at (0, 2) and containing the point (−1, 4).

Prove the given statement.

12. $\widehat{ST} \cong \widehat{RQ}$

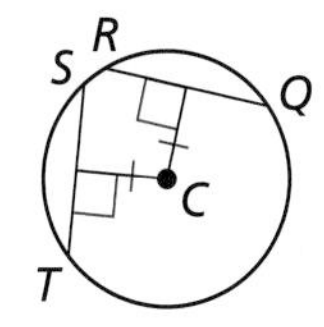

13. $\widehat{JM} \cong \widehat{LM}$

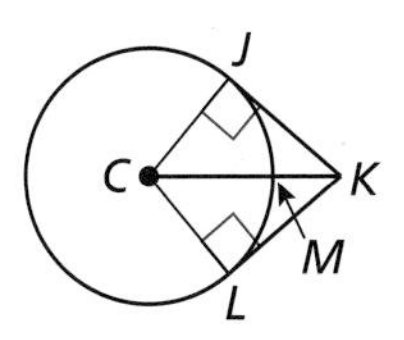

14. $\widehat{DG} \cong \widehat{FG}$

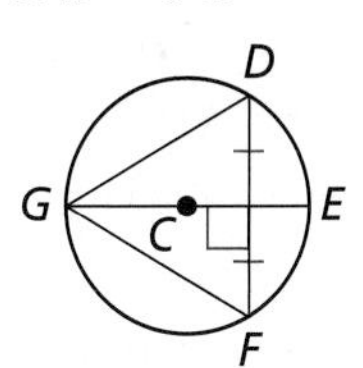

15. A bank of lighting hangs over a stage. Each light illuminates a circular region on the stage. A coordinate plane is used to arrange the lights, using a corner of the stage as the origin. The equation $(x - 13)^2 + (y - 4)^2 = 16$ represents the boundary of the region illuminated by one of the lights. Three actors stand at the points $A(11, 4)$, $B(8, 5)$, and $C(15, 5)$. Graph the given equation. Then determine which actors are illuminated by the light.

16. If a car goes around a turn too quickly, it can leave tracks that form an arc of a circle. By finding the radius of the circle, accident investigators can estimate the speed of the car.

a. To find the radius, accident investigators choose points A and B on the tire marks. Then the investigators find the midpoint C of $\overline{AB}$. Use the diagram to find the radius r of the circle. Explain why this method works.

b. The formula $S = 3.87\sqrt{fr}$ can be used to estimate a car's speed in miles per hour, where f is the *coefficient of friction* and r is the radius of the circle in feet. If $f = 0.7$, estimate the car's speed in part (a).

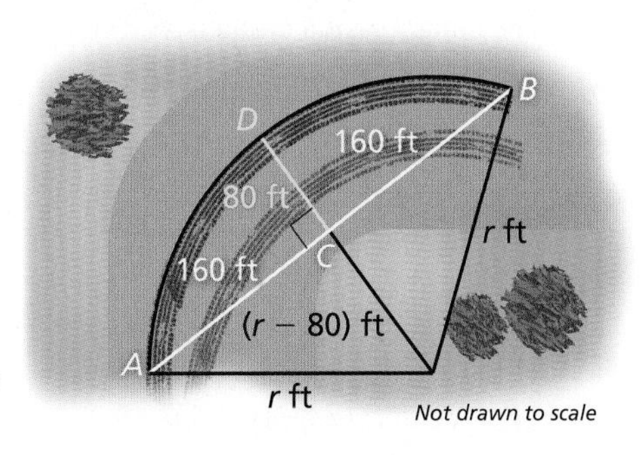

10 Cumulative Assessment

1. Classify each segment as specifically as possible.

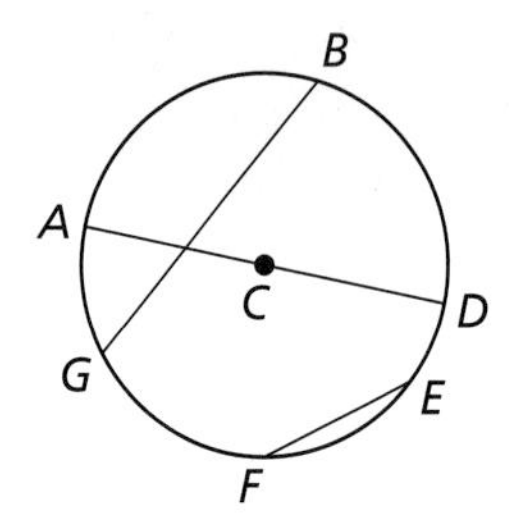

a. $\overline{BG}$ **b.** $\overline{CD}$ **c.** $\overline{AD}$ **d.** $\overline{FE}$

2. Copy and complete the paragraph proof.

Given Circle C with center (2, 1) and radius 1, Circle D with center (0, 3) and radius 4

Prove Circle C is similar to Circle D.

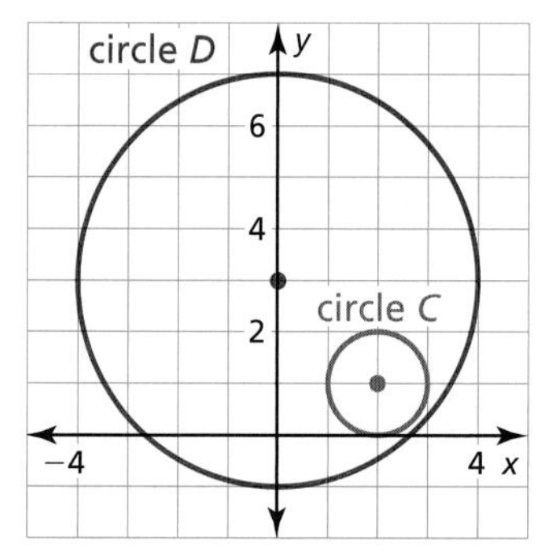

Map Circle C to Circle C' by using the ___________ $(x, y) \rightarrow$ ________ so that Circle C' and Circle D have the same center at (__, __). Dilate Circle C' using a center of dilation (__, __) and a scale factor of ___. Because there is a ________ transformation that maps Circle C to Circle D, Circle C is _________ Circle D.

3. Use the diagram to write a proof.

Given $\triangle JPL \cong \triangle NPL$
$\overline{PK}$ is an altitude of $\triangle JPL$.
$\overline{PM}$ is an altitude of $\triangle NPL$.

Prove $\triangle PKL \sim \triangle NMP$

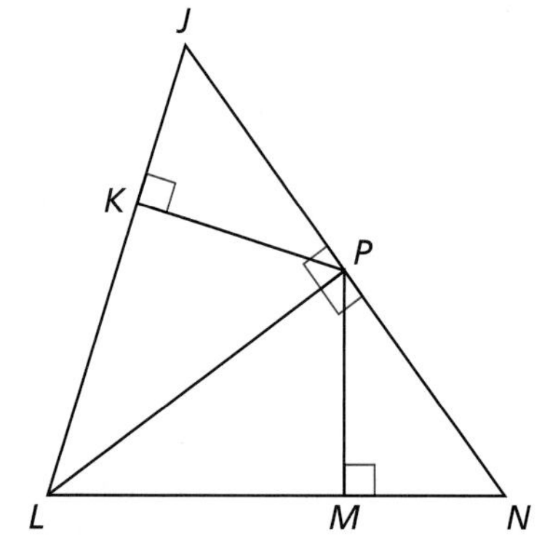

4. The equation of a circle is $x^2 + y^2 + 14x - 16y + 77 = 0$. What are the center and radius of the circle?

Ⓐ center: $(14, -16)$, radius: 8.8

Ⓑ center: $(-7, 8)$, radius: 6

Ⓒ center: $(-14, 16)$, radius: 8.8

Ⓓ center: $(7, -8)$, radius: 5.2

5. The coordinates of the vertices of a quadrilateral are $W(-7, -6)$, $X(1, -2)$, $Y(3, -6)$, and $Z(-5, -10)$. Prove that quadrilateral $WXYZ$ is a rectangle.

6. Which angles have the same measure as $\angle ACB$? Select all that apply.

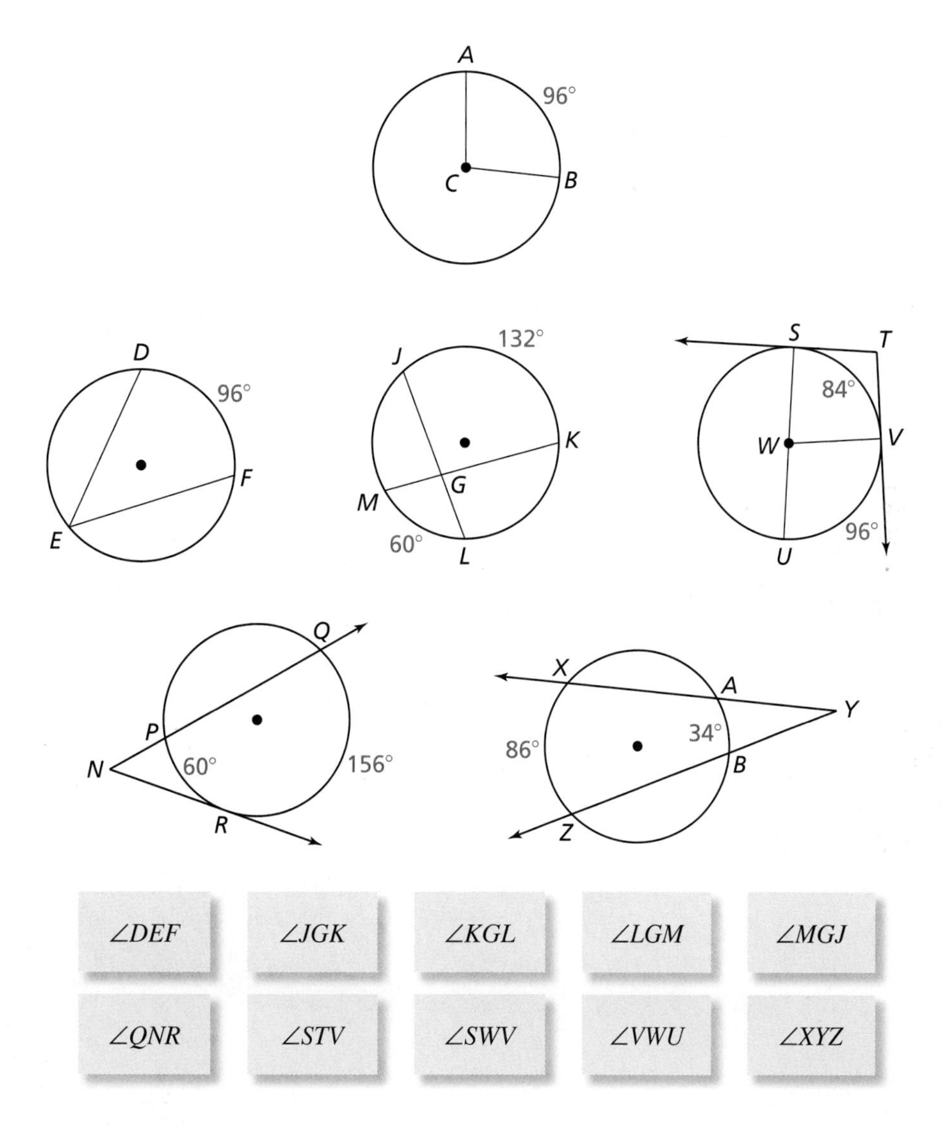

7. Classify each related conditional statement based on the conditional statement "If you are a soccer player, then you are an athlete."

a. If you are not a soccer player, then you are not an athlete.

b. If you are an athlete, then you are a soccer player.

c. You are a soccer player if and only if you are an athlete.

d. If you are not an athlete, then you are not a soccer player.

8. Your friend claims that the quadrilateral shown can be inscribed in a circle. Is your friend correct? Explain your reasoning.

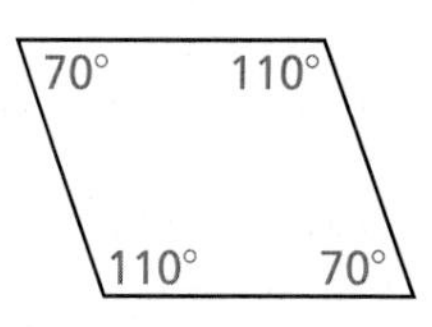

11 Circumference, Area, and Volume

Khafre's Pyramid *(p. 637)*

Gold Density *(p. 628)*

Basaltic Columns *(p. 615)*

Population Density *(p. 603)*

London Eye *(p. 599)*

Maintaining Mathematical Proficiency

Finding Surface Area

Example 1 **Find the surface area of the prism.**

$S = 2\ell w + 2\ell h + 2wh$	Write formula for surface area of a rectangular prism.
$= 2(2)(4) + 2(2)(6) + 2(4)(6)$	Substitute 2 for ℓ, 4 for w, and 6 for h.
$= 16 + 24 + 48$	Multiply.
$= 88$	Add.

▶ The surface area is 88 square inches.

Find the surface area of the prism.

1.

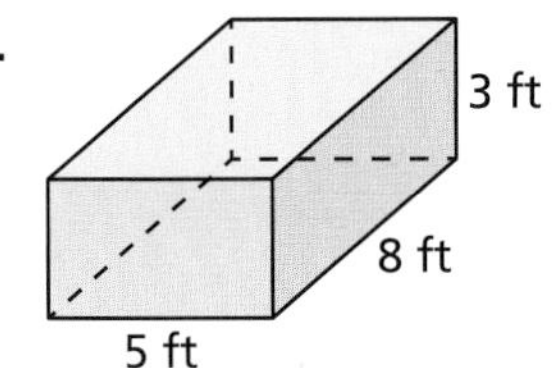

2.

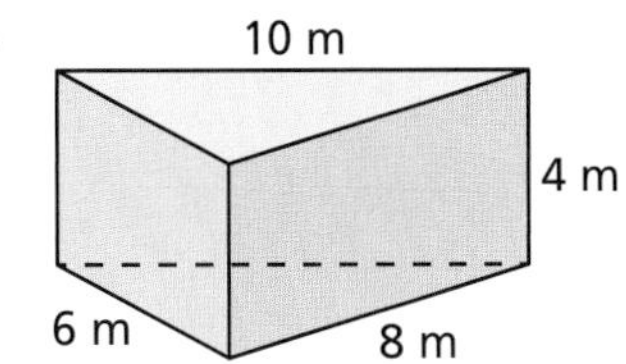

3.

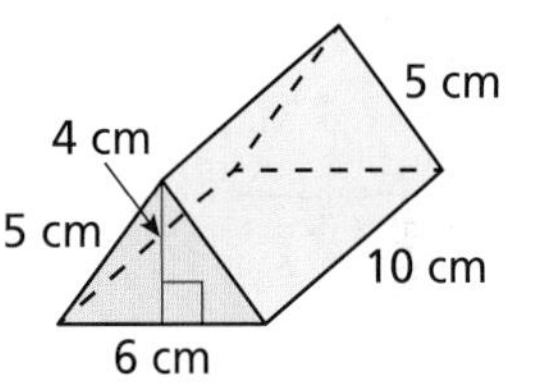

Finding a Missing Dimension

Example 2 **A rectangle has a perimeter of 10 meters and a length of 3 meters. What is the width of the rectangle?**

$P = 2\ell + 2w$	Write formula for perimeter of a rectangle.
$10 = 2(3) + 2w$	Substitute 10 for P and 3 for ℓ.
$10 = 6 + 2w$	Multiply 2 and 3.
$4 = 2w$	Subtract 6 from each side.
$2 = w$	Divide each side by 2.

▶ The width is 2 meters.

Find the missing dimension.

4. A rectangle has a perimeter of 28 inches and a width of 5 inches. What is the length of the rectangle?

5. A triangle has an area of 12 square centimeters and a height of 12 centimeters. What is the base of the triangle?

6. A rectangle has an area of 84 square feet and a width of 7 feet. What is the length of the rectangle?

7. **ABSTRACT REASONING** Write an equation for the surface area of a prism with a length, width, and height of x inches. What solid figure does the prism represent?

Mathematical Practices

Mathematically proficient students create valid representations of problems.

Creating a Valid Representation

Core Concept

Nets for Three-Dimensional Figures

A **net** for a three-dimensional figure is a two-dimensional pattern that can be folded to form the three-dimensional figure.

EXAMPLE 1 Drawing a Net for a Pyramid

Draw a net of the pyramid.

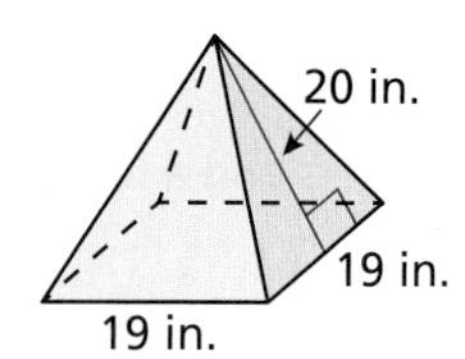

SOLUTION

The pyramid has a square base. Its four lateral faces are congruent isosceles triangles.

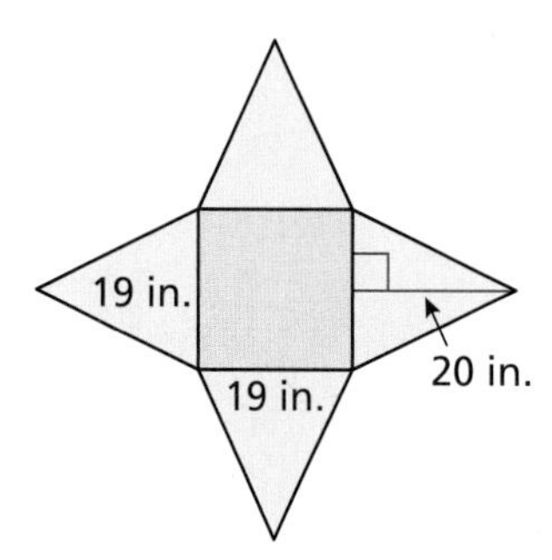

Monitoring Progress

Draw a net of the three-dimensional figure. Label the dimensions.

1.

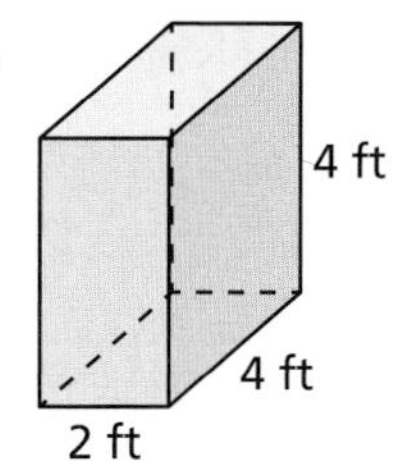

2.

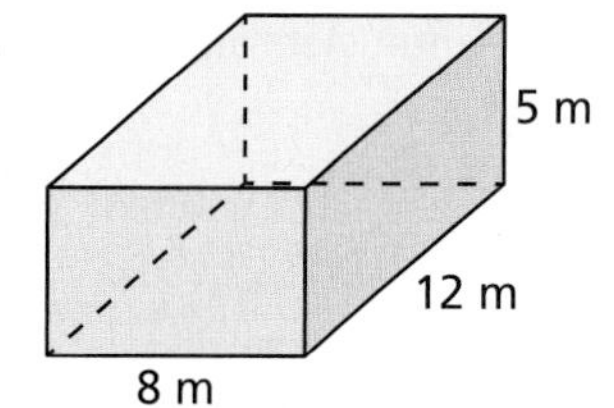

3.

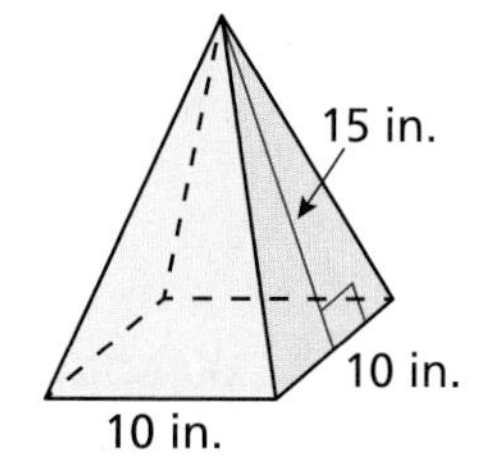

11.1 Circumference and Arc Length

Essential Question

How can you find the length of a circular arc?

EXPLORATION 1 Finding the Length of a Circular Arc

Work with a partner. Find the length of each red circular arc.

a. entire circle

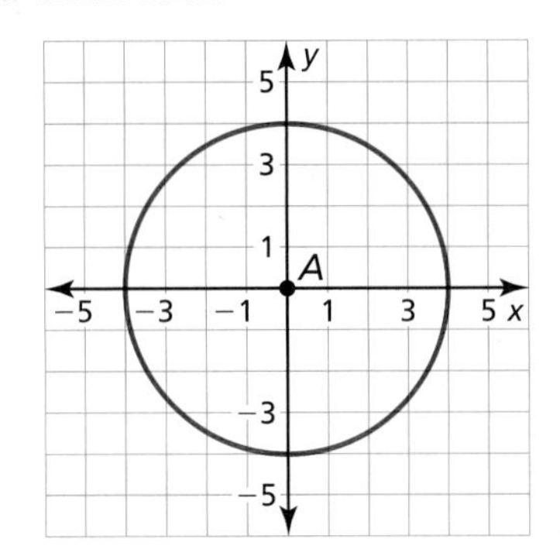

b. one-fourth of a circle

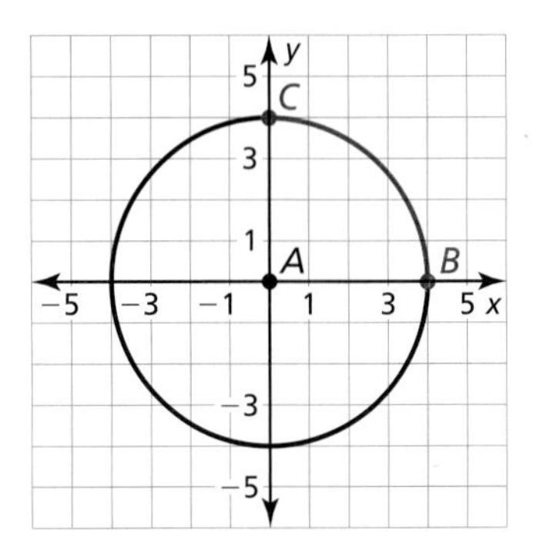

c. one-third of a circle

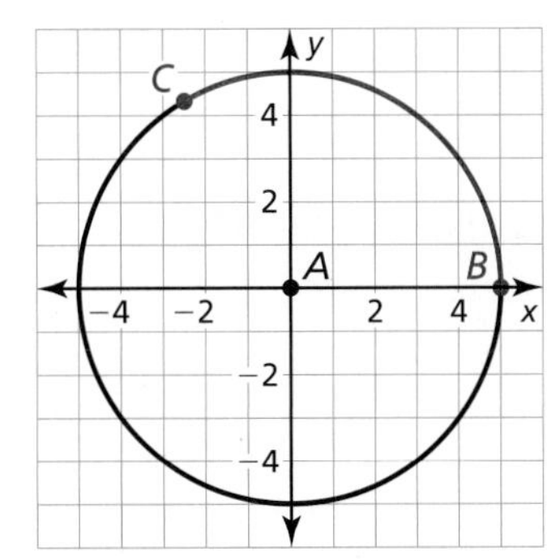

d. five-eighths of a circle

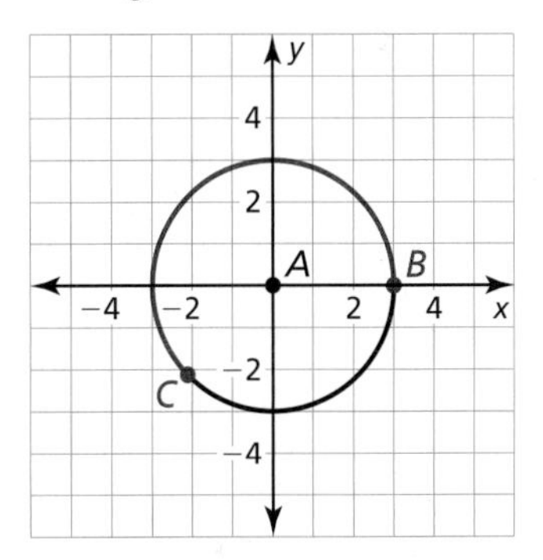

EXPLORATION 2 Using Arc Length

Work with a partner. The rider is attempting to stop with the front tire of the motorcycle in the painted rectangular box for a skills test. The front tire makes exactly one-half additional revolution before stopping. The diameter of the tire is 25 inches. Is the front tire still in contact with the painted box? Explain.

3 ft

LOOKING FOR REGULARITY IN REPEATED REASONING

To be proficient in math, you need to notice if calculations are repeated and look both for general methods and for shortcuts.

Communicate Your Answer

3. How can you find the length of a circular arc?

4. A motorcycle tire has a diameter of 24 inches. Approximately how many inches does the motorcycle travel when its front tire makes three-fourths of a revolution?

11.1 Lesson

Core Vocabulary

circumference, *p. 594*
arc length, *p. 595*
radian, *p. 597*

Previous
circle
diameter
radius

What You Will Learn

- Use the formula for circumference.
- Use arc lengths to find measures.
- Solve real-life problems.
- Measure angles in radians.

Using the Formula for Circumference

The **circumference** of a circle is the distance around the circle. Consider a regular polygon inscribed in a circle. As the number of sides increases, the polygon approximates the circle and the ratio of the perimeter of the polygon to the diameter of the circle approaches $\pi \approx 3.14159\ldots$

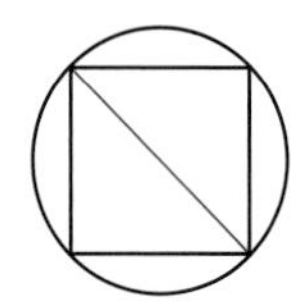

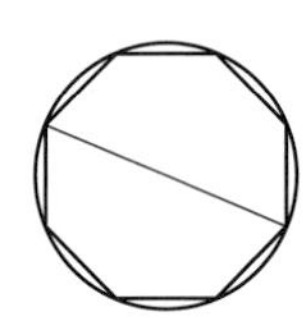

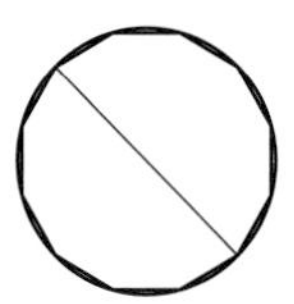

For all circles, the ratio of the circumference C to the diameter d is the same. This ratio is $\frac{C}{d} = \pi$. Solving for C yields the formula for the circumference of a circle, $C = \pi d$. Because $d = 2r$, you can also write the formula as $C = \pi(2r) = 2\pi r$.

Circumference of a Circle

The circumference C of a circle is $C = \pi d$ or $C = 2\pi r$, where d is the diameter of the circle and r is the radius of the circle.

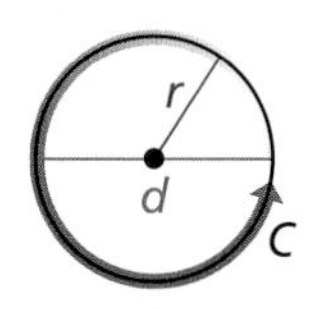

EXAMPLE 1 Using the Formula for Circumference

Find each indicated measure.

a. circumference of a circle with a radius of 9 centimeters

b. radius of a circle with a circumference of 26 meters

SOLUTION

a.
$$C = 2\pi r$$
$$= 2 \cdot \pi \cdot 9$$
$$= 18\pi$$
$$\approx 56.55$$

▶ The circumference is about 56.55 centimeters.

b.
$$C = 2\pi r$$
$$26 = 2\pi r$$
$$\frac{26}{2\pi} = r$$
$$4.14 \approx r$$

▶ The radius is about 4.14 meters.

ATTENDING TO PRECISION

You have sometimes used 3.14 to approximate the value of π. Throughout this chapter, you should use the π key on a calculator, then round to the hundredths place unless instructed otherwise.

Monitoring Progress

Help in English and Spanish at *BigIdeasMath.com*

1. Find the circumference of a circle with a diameter of 5 inches.
2. Find the diameter of a circle with a circumference of 17 feet.

Using Arc Lengths to Find Measures

An **arc length** is a portion of the circumference of a circle. You can use the measure of the arc (in degrees) to find its length (in linear units).

Arc Length

In a circle, the ratio of the length of a given arc to the circumference is equal to the ratio of the measure of the arc to 360°.

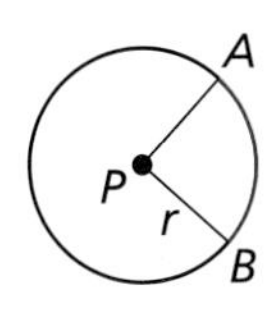

$$\frac{\text{Arc length of } \widehat{AB}}{2\pi r} = \frac{m\widehat{AB}}{360^\circ}, \text{ or}$$

$$\text{Arc length of } \widehat{AB} = \frac{m\widehat{AB}}{360^\circ} \cdot 2\pi r$$

EXAMPLE 2 **Using Arc Lengths to Find Measures**

Find each indicated measure.

a. arc length of $\widehat{AB}$ **b.** circumference of $\odot Z$ **c.** $m\widehat{RS}$

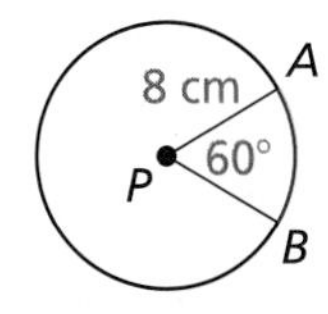

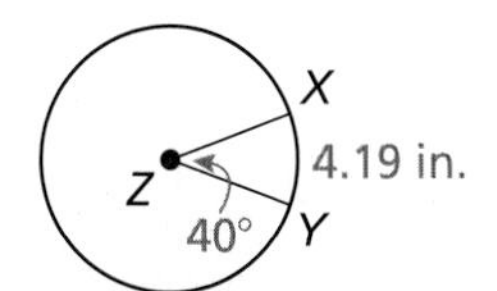

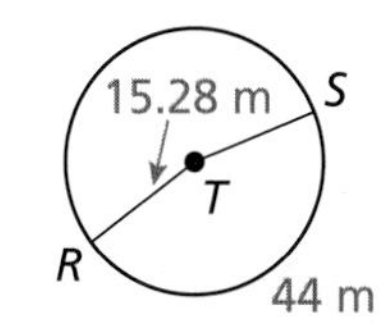

SOLUTION

a. $\text{Arc length of } \widehat{AB} = \frac{60^\circ}{360^\circ} \cdot 2\pi(8)$

$\approx 8.38 \text{ cm}$

b.

$$\frac{\text{Arc length of } \widehat{XY}}{C} = \frac{m\widehat{XY}}{360^\circ}$$

$$\frac{4.19}{C} = \frac{40^\circ}{360^\circ}$$

$$\frac{4.19}{C} = \frac{1}{9}$$

$$37.71 \text{ in.} = C$$

c.

$$\frac{\text{Arc length of } \widehat{RS}}{2\pi r} = \frac{m\widehat{RS}}{360^\circ}$$

$$\frac{44}{2\pi(15.28)} = \frac{m\widehat{RS}}{360^\circ}$$

$$360^\circ \cdot \frac{44}{2\pi(15.28)} = m\widehat{RS}$$

$$165^\circ \approx m\widehat{RS}$$

Monitoring Progress  Help in English and Spanish at *BigIdeasMath.com*

Find the indicated measure.

3. arc length of $\widehat{PQ}$ **4.** circumference of $\odot N$ **5.** radius of $\odot G$

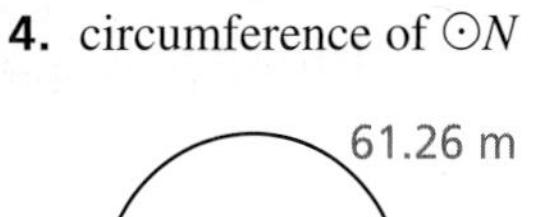

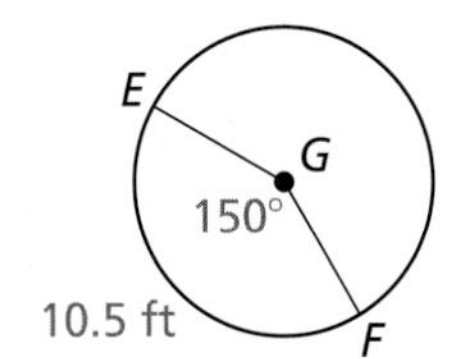

Solving Real-Life Problems

EXAMPLE 3 Using Circumference to Find Distance Traveled

The dimensions of a car tire are shown. To the nearest foot, how far does the tire travel when it makes 15 revolutions?

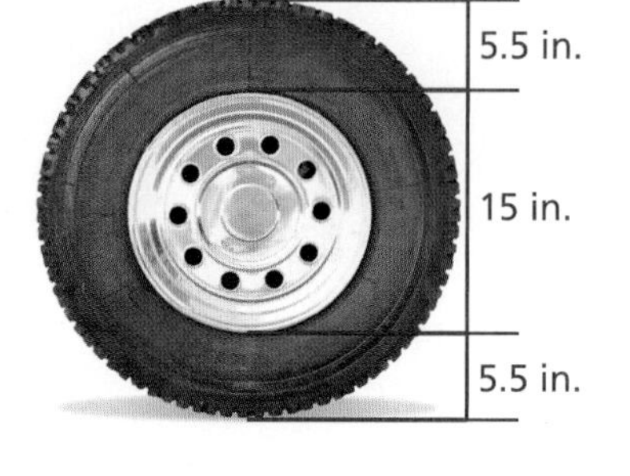

SOLUTION

Step 1 Find the diameter of the tire.

$$d = 15 + 2(5.5) = 26 \text{ in.}$$

Step 2 Find the circumference of the tire.

$$C = \pi d = \pi \cdot 26 = 26\pi \text{ in.}$$

Step 3 Find the distance the tire travels in 15 revolutions. In one revolution, the tire travels a distance equal to its circumference. In 15 revolutions, the tire travels a distance equal to 15 times its circumference.

Distance traveled	=	Number of revolutions	•	Circumference

$$= 15 \cdot 26\pi \approx 1225.2 \text{ in.}$$

COMMON ERROR

Always pay attention to units. In Example 3, you need to convert units to get a correct answer.

Step 4 Use unit analysis. Change 1225.2 inches to feet.

$$1225.2 \text{ in.} \cdot \frac{1 \text{ ft}}{12 \text{ in.}} = 102.1 \text{ ft}$$

▶ The tire travels approximately 102 feet.

EXAMPLE 4 Using Arc Length to Find Distances

The curves at the ends of the track shown are 180° arcs of circles. The radius of the arc for a runner on the red path shown is 36.8 meters. About how far does this runner travel to go once around the track? Round to the nearest tenth of a meter.

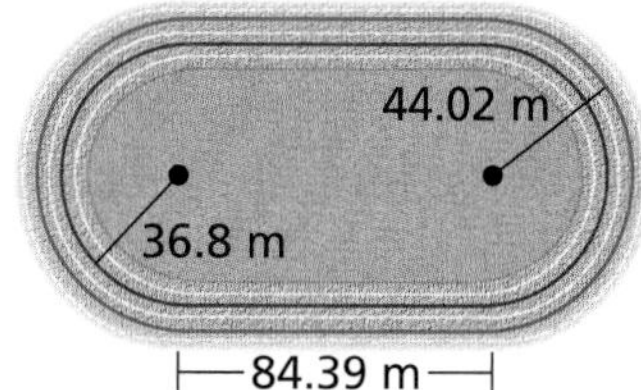

SOLUTION

The path of the runner on the red path is made of two straight sections and two semicircles. To find the total distance, find the sum of the lengths of each part.

Distance	=	2 • Length of each straight section	+	2 • Length of each semicircle

$$= 2(84.39) + 2\left(\frac{1}{2} \cdot 2\pi \cdot 36.8\right)$$

$$\approx 400.0$$

▶ The runner on the red path travels about 400.0 meters.

Monitoring Progress  Help in English and Spanish at *BigIdeasMath.com*

6. A car tire has a diameter of 28 inches. How many revolutions does the tire make while traveling 500 feet?

7. In Example 4, the radius of the arc for a runner on the blue path is 44.02 meters, as shown in the diagram. About how far does this runner travel to go once around the track? Round to the nearest tenth of a meter.

Measuring Angles in Radians

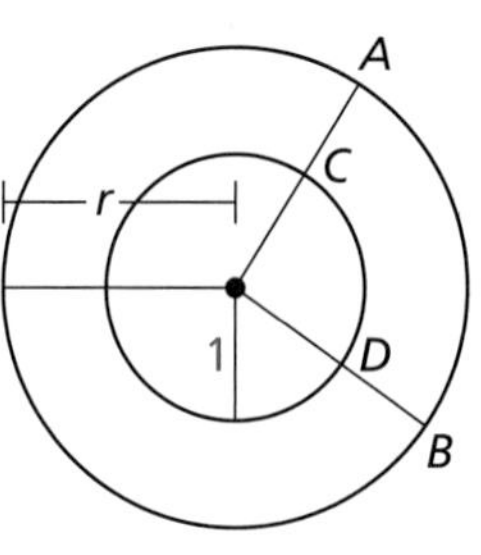

Recall that in a circle, the ratio of the length of a given arc to the circumference is equal to the ratio of the measure of the arc to 360°. To see why, consider the diagram.

A circle of radius 1 has circumference 2π, so the arc length of $\overparen{CD}$ is $\frac{m\overparen{CD}}{360°} \cdot 2\pi$.

Recall that all circles are similar and corresponding lengths of similar figures are proportional. Because $m\overparen{AB} = m\overparen{CD}$, $\overparen{AB}$ and $\overparen{CD}$ are corresponding arcs. So, you can write the following proportion.

$$\frac{\text{Arc length of } \overparen{AB}}{\text{Arc length of } \overparen{CD}} = \frac{r}{1}$$

$$\text{Arc length of } \overparen{AB} = r \cdot \text{Arc length of } \overparen{CD}$$

$$\text{Arc length of } \overparen{AB} = r \cdot \frac{m\overparen{CD}}{360°} \cdot 2\pi$$

This form of the equation shows that the arc length associated with a central angle is *proportional to the radius* of the circle. The constant of proportionality, $\frac{m\overparen{CD}}{360°} \cdot 2\pi$, is defined to be the **radian** measure of the central angle associated with the arc.

In a circle of radius 1, the radian measure of a given central angle can be thought of as the length of the arc associated with the angle. The radian measure of a complete circle (360°) is exactly 2π radians, because the circumference of a circle of radius 1 is exactly 2π. You can use this fact to convert from degree measure to radian measure and vice versa.

Core Concept

Converting between Degrees and Radians

Degrees to radians

Multiply degree measure by

$$\frac{2\pi \text{ radians}}{360°}, \text{ or } \frac{\pi \text{ radians}}{180°}.$$

Radians to degrees

Multiply radian measure by

$$\frac{360°}{2\pi \text{ radians}}, \text{ or } \frac{180°}{\pi \text{ radians}}.$$

EXAMPLE 5 Converting between Degree and Radian Measure

a. Convert 45° to radians.

b. Convert $\frac{3\pi}{2}$ radians to degrees.

SOLUTION

a. $45° \cdot \frac{\pi \text{ radians}}{180°} = \frac{\pi}{4}$ radian

▶ So, $45° = \frac{\pi}{4}$ radian.

b. $\frac{3\pi}{2}$ radians $\cdot \frac{180°}{\pi \text{ radians}} = 270°$

▶ So, $\frac{3\pi}{2}$ radians $= 270°$.

Monitoring Progress

Help in English and Spanish at *BigIdeasMath.com*

8. Convert 15° to radians.

9. Convert $\frac{4\pi}{3}$ radians to degrees.

11.1 Exercises

Dynamic Solutions available at *BigIdeasMath.com*

Vocabulary and Core Concept Check

1. **COMPLETE THE SENTENCE** The circumference of a circle with diameter d is $C =$ ______.

2. **WRITING** Describe the difference between an arc measure and an arc length.

Monitoring Progress and Modeling with Mathematics

In Exercises 3–10, find the indicated measure. *(See Examples 1 and 2.)*

3. circumference of a circle with a radius of 6 inches

4. diameter of a circle with a circumference of 63 feet

5. radius of a circle with a circumference of 28π

6. exact circumference of a circle with a diameter of 5 inches

7. arc length of $\widehat{AB}$

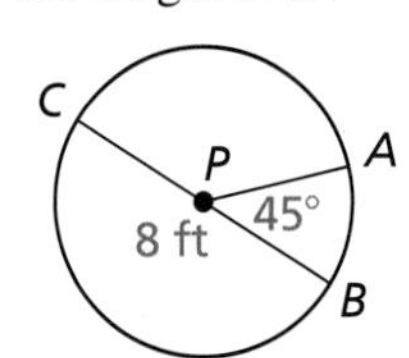

8. $m\widehat{DE}$

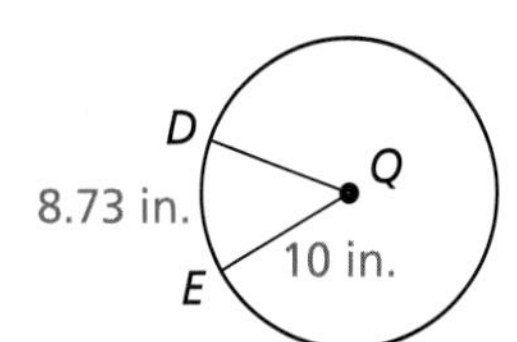

9. circumference of $\odot C$

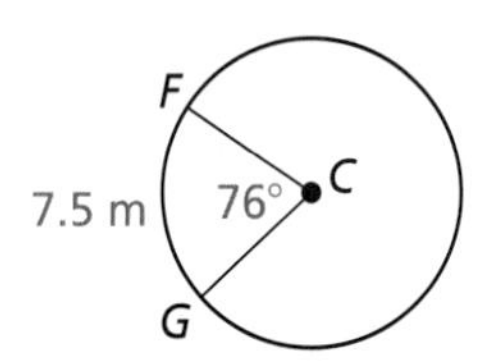

10. radius of $\odot R$

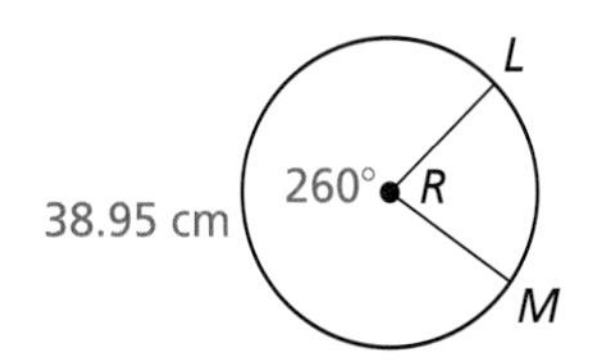

11. **ERROR ANALYSIS** Describe and correct the error in finding the circumference of $\odot C$.

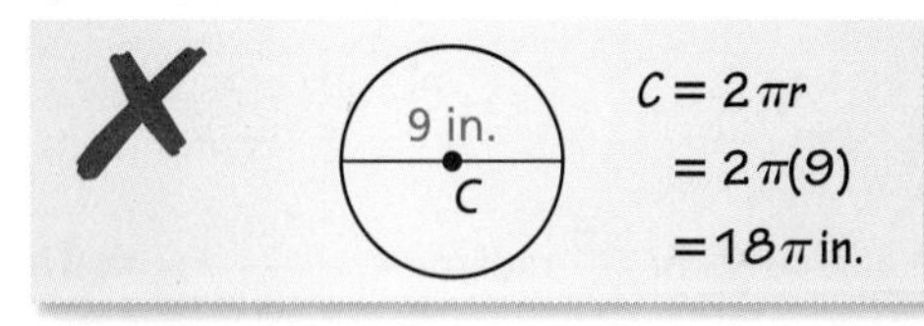

12. **ERROR ANALYSIS** Describe and correct the error in finding the length of $\widehat{GH}$.

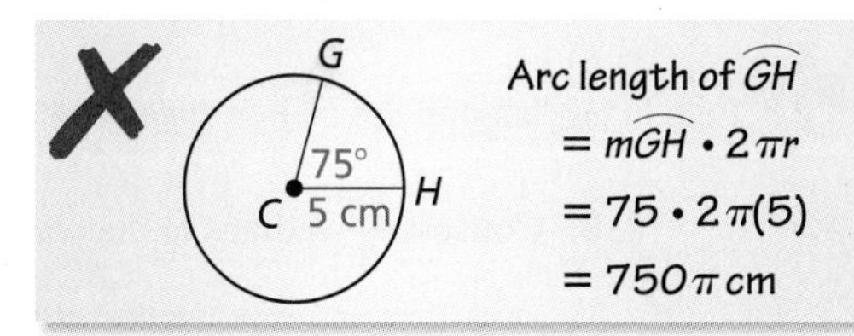

13. **PROBLEM SOLVING** A measuring wheel is used to calculate the length of a path. The diameter of the wheel is 8 inches. The wheel makes 87 complete revolutions along the length of the path. To the nearest foot, how long is the path? *(See Example 3.)*

14. **PROBLEM SOLVING** You ride your bicycle 40 meters. How many complete revolutions does the front wheel make?

In Exercises 15–18, find the perimeter of the shaded region. *(See Example 4.)*

15.

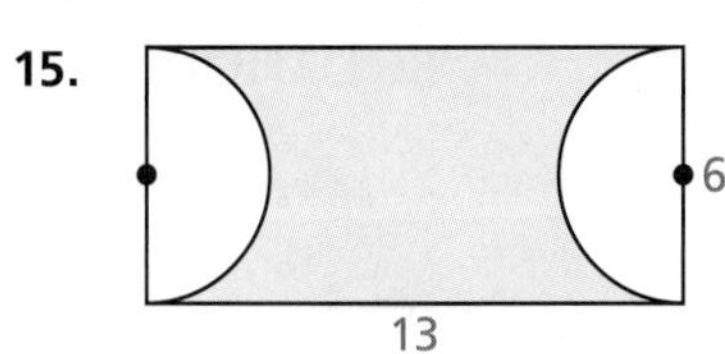

16.

6 3 3 6

17.

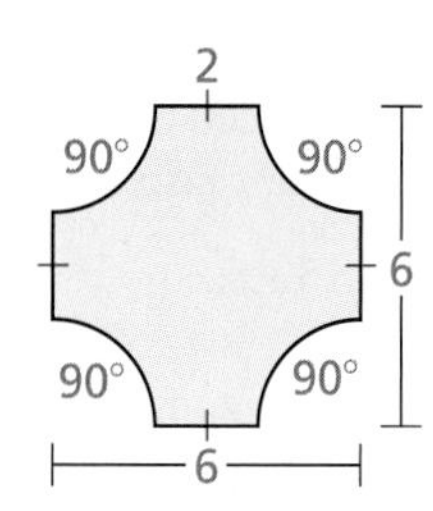

18.

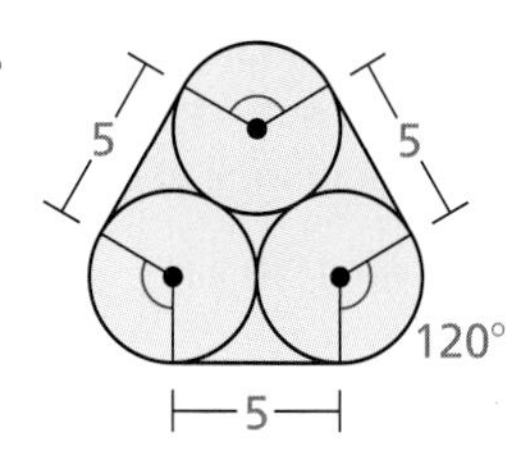

In Exercises 19–22, convert the angle measure. *(See Example 5.)*

19. Convert 70° to radians.

20. Convert 300° to radians.

21. Convert $\frac{11\pi}{12}$ radians to degrees.

22. Convert $\frac{\pi}{8}$ radian to degrees.

23. PROBLEM SOLVING The London Eye is a Ferris wheel in London, England, that travels at a speed of 0.26 meter per second. How many minutes does it take the London Eye to complete one full revolution?

24. PROBLEM SOLVING You are planning to plant a circular garden adjacent to one of the corners of a building, as shown. You can use up to 38 feet of fence to make a border around the garden. What radius (in feet) can the garden have? Choose all that apply. Explain your reasoning.

Ⓐ 7 Ⓑ 8 Ⓒ 9 Ⓓ 10

In Exercises 25 and 26, find the circumference of the circle with the given equation. Write the circumference in terms of π.

25. $x^2 + y^2 = 16$

26. $(x + 2)^2 + (y - 3)^2 = 9$

27. USING STRUCTURE A semicircle has endpoints $(-2, 5)$ and $(2, 8)$. Find the arc length of the semicircle.

28. REASONING $\widehat{EF}$ is an arc on a circle with radius r. Let $x°$ be the measure of $\widehat{EF}$. Describe the effect on the length of $\widehat{EF}$ if you (a) double the radius of the circle, and (b) double the measure of $\widehat{EF}$.

29. MAKING AN ARGUMENT Your friend claims that it is possible for two arcs with the same measure to have different arc lengths. Is your friend correct? Explain your reasoning.

30. PROBLEM SOLVING Over 2000 years ago, the Greek scholar Eratosthenes estimated Earth's circumference by assuming that the Sun's rays were parallel. He chose a day when the Sun shone straight down into a well in the city of Syene. At noon, he measured the angle the Sun's rays made with a vertical stick in the city of Alexandria. Eratosthenes assumed that the distance from Syene to Alexandria was equal to about 575 miles. Explain how Eratosthenes was able to use this information to estimate Earth's circumference. Then estimate Earth's circumference.

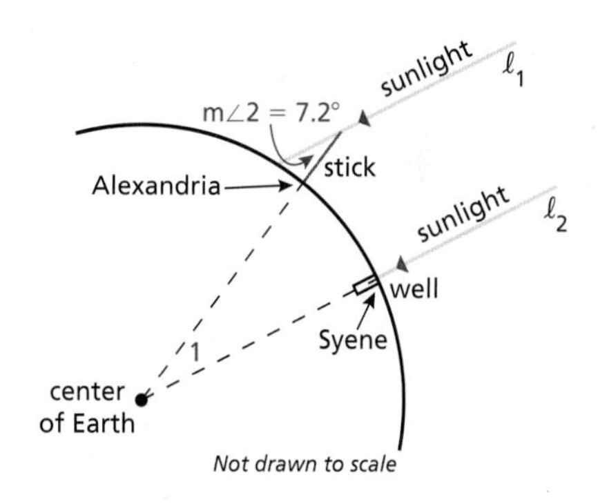

31. ANALYZING RELATIONSHIPS In $\odot C$, the ratio of the length of $\widehat{PQ}$ to the length of $\widehat{RS}$ is 2 to 1. What is the ratio of $m\angle PCQ$ to $m\angle RCS$?

Ⓐ 4 to 1 Ⓑ 2 to 1

Ⓒ 1 to 4 Ⓓ 1 to 2

32. ANALYZING RELATIONSHIPS A 45° arc in $\odot C$ and a 30° arc in $\odot P$ have the same length. What is the ratio of the radius r_1 of $\odot C$ to the radius r_2 of $\odot P$? Explain your reasoning.

33. PROBLEM SOLVING How many revolutions does the smaller gear complete during a single revolution of the larger gear?

34. USING STRUCTURE Find the circumference of each circle.

a. a circle circumscribed about a right triangle whose legs are 12 inches and 16 inches long

b. a circle circumscribed about a square with a side length of 6 centimeters

c. a circle inscribed in an equilateral triangle with a side length of 9 inches

35. REWRITING A FORMULA Write a formula in terms of the measure θ (theta) of the central angle (in radians) that can be used to find the length of an arc of a circle. Then use this formula to find the length of an arc of a circle with a radius of 4 inches and a central angle of $\frac{3\pi}{4}$ radians.

36. HOW DO YOU SEE IT?
Compare the circumference of $\odot P$ to the length of $\widehat{DE}$. Explain your reasoning.

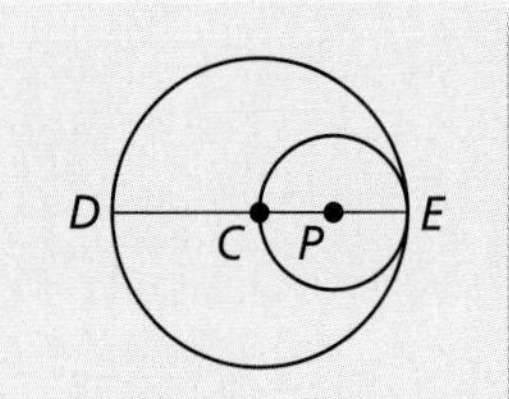

37. MAKING AN ARGUMENT In the diagram, the measure of the red shaded angle is 30°. The arc length a is 2. Your classmate claims that it is possible to find the circumference of the blue circle without finding the radius of either circle. Is your classmate correct? Explain your reasoning.

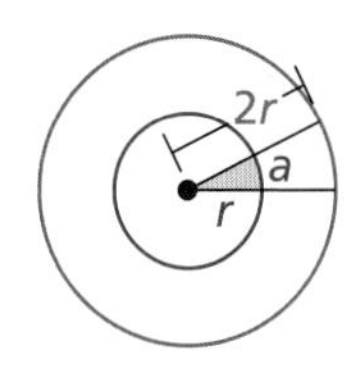

38. MODELING WITH MATHEMATICS What is the measure (in radians) of the angle formed by the hands of a clock at each time? Explain your reasoning.

a. 1:30 P.M.   b. 3:15 P.M.

39. MATHEMATICAL CONNECTIONS The sum of the circumferences of circles A, B, and C is 63π. Find AC.

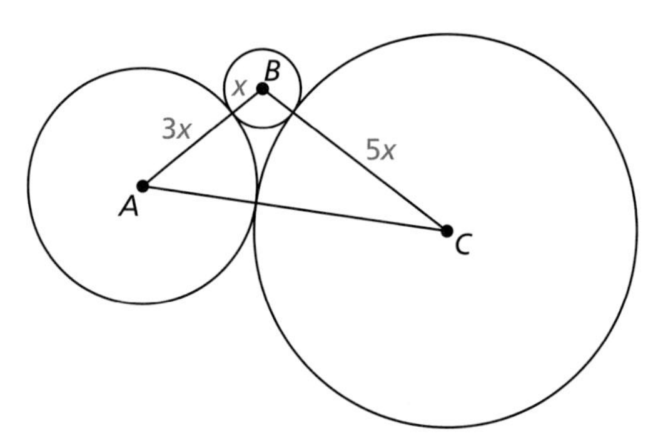

40. THOUGHT PROVOKING Is π a rational number? Compare the rational number $\frac{355}{113}$ to π. Find a different rational number that is even closer to π.

41. PROOF The circles in the diagram are concentric and $\overline{FG} \cong \overline{GH}$. Prove that $\widehat{JK}$ and $\widehat{NG}$ have the same length.

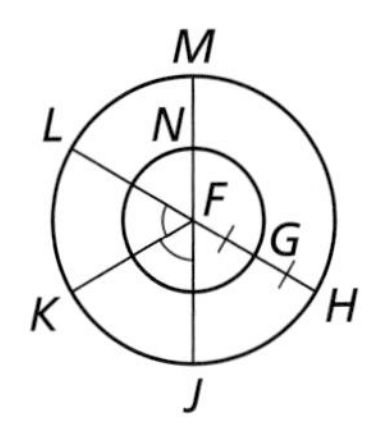

42. REPEATED REASONING $\overline{AB}$ is divided into four congruent segments, and semicircles with radius r are drawn.

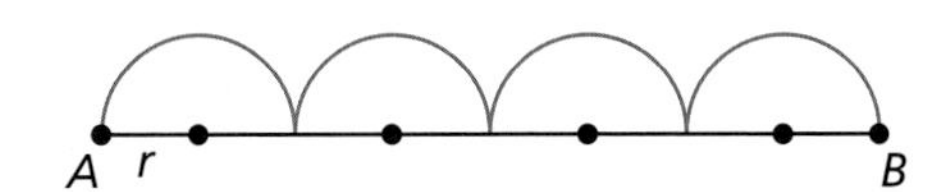

a. What is the sum of the four arc lengths?

b. What would the sum of the arc lengths be if $\overline{AB}$ was divided into 8 congruent segments? 16 congruent segments? n congruent segments? Explain your reasoning.

Maintaining Mathematical Proficiency

Reviewing what you learned in previous grades and lessons

Find the area of the polygon with the given vertices. *(Section 1.4)*

43. $X(2, 4)$, $Y(8, -1)$, $Z(2, -1)$

44. $L(-3, 1)$, $M(4, 1)$, $N(4, -5)$, $P(-3, -5)$

11.2 Areas of Circles and Sectors

Essential Question How can you find the area of a sector of a circle?

EXPLORATION 1 Finding the Area of a Sector of a Circle

Work with a partner. A **sector of a circle** is the region bounded by two radii of the circle and their intercepted arc. Find the area of each shaded circle or sector of a circle.

a. entire circle

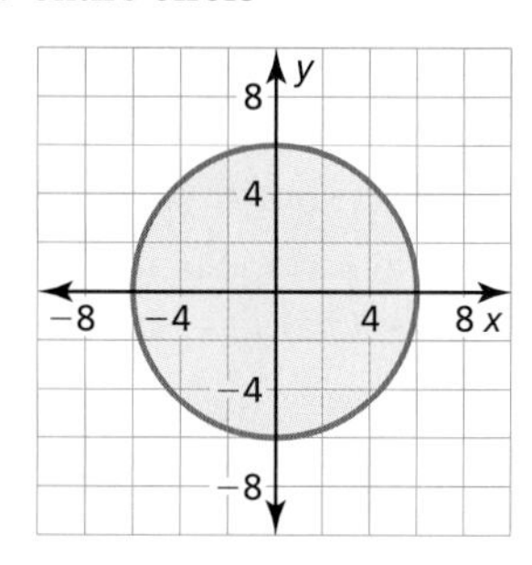

b. one-fourth of a circle

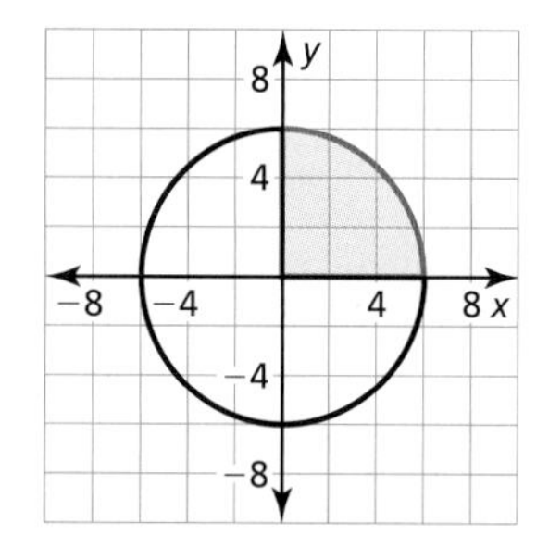

c. seven-eighths of a circle

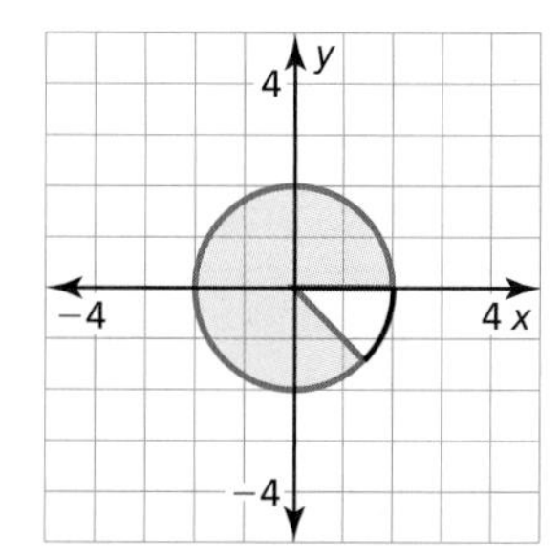

d. two-thirds of a circle

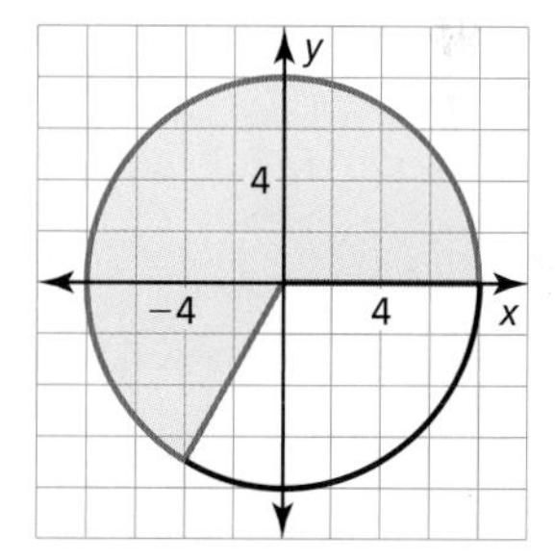

REASONING ABSTRACTLY

To be proficient in math, you need to explain to yourself the meaning of a problem and look for entry points to its solution.

EXPLORATION 2 Finding the Area of a Circular Sector

Work with a partner. A center pivot irrigation system consists of 400 meters of sprinkler equipment that rotates around a central pivot point at a rate of once every 3 days to irrigate a circular region with a diameter of 800 meters. Find the area of the sector that is irrigated by this system in one day.

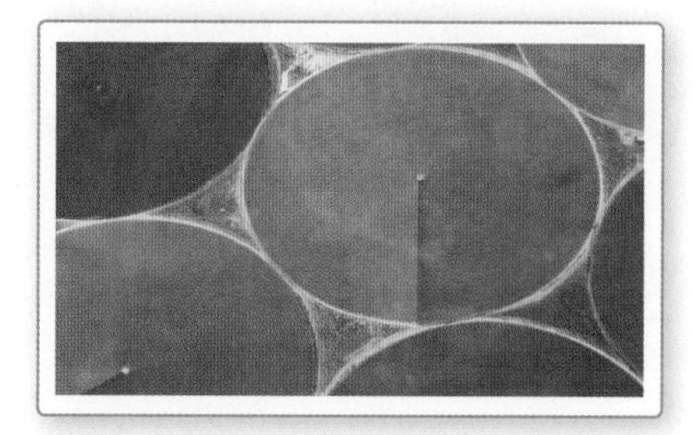

Communicate Your Answer

3. How can you find the area of a sector of a circle?

4. In Exploration 2, find the area of the sector that is irrigated in 2 hours.

11.2 Lesson

Core Vocabulary

population density, *p. 603*
sector of a circle, *p. 604*

Previous
circle
radius
diameter
intercepted arc

What You Will Learn

- Use the formula for the area of a circle.
- Use the formula for population density.
- Find areas of sectors.
- Use areas of sectors.

Using the Formula for the Area of a Circle

You can divide a circle into congruent sections and rearrange the sections to form a figure that approximates a parallelogram. Increasing the number of congruent sections increases the figure's resemblance to a parallelogram.

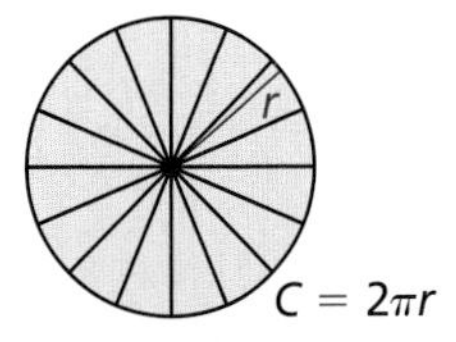

The base of the parallelogram that the figure approaches is half of the circumference, so $b = \frac{1}{2}C = \frac{1}{2}(2\pi r) = \pi r$. The height is the radius, so $h = r$. So, the area of the parallelogram is $A = bh = (\pi r)(r) = \pi r^2$.

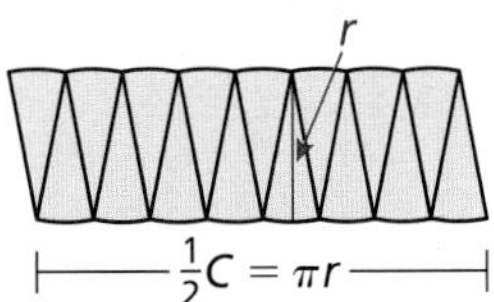

Core Concept

Area of a Circle

The area of a circle is

$$A = \pi r^2$$

where r is the radius of the circle.

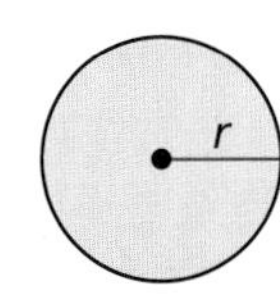

EXAMPLE 1 **Using the Formula for the Area of a Circle**

Find each indicated measure.

a. area of a circle with a radius of 2.5 centimeters

b. diameter of a circle with an area of 113.1 square centimeters

SOLUTION

a.

$A = \pi r^2$	Formula for area of a circle
$= \pi \cdot (2.5)^2$	Substitute 2.5 for r.
$= 6.25\pi$	Simplify.
≈ 19.63	Use a calculator.

▶ The area of the circle is about 19.63 square centimeters.

b.

$A = \pi r^2$	Formula for area of a circle
$113.1 = \pi r^2$	Substitute 113.1 for A.
$\frac{113.1}{\pi} = r^2$	Divide each side by π.
$6 \approx r$	Find the positive square root of each side.

▶ The radius is about 6 centimeters, so the diameter is about 12 centimeters.

Monitoring Progress  Help in English and Spanish at *BigIdeasMath.com*

1. Find the area of a circle with a radius of 4.5 meters.

2. Find the radius of a circle with an area of 176.7 square feet.

Using the Formula for Population Density

The **population density** of a city, county, or state is a measure of how many people live within a given area.

$$\text{Population density} = \frac{\text{number of people}}{\text{area of land}}$$

Population density is usually given in terms of square miles but can be expressed using other units, such as city blocks.

EXAMPLE 2 Using the Formula for Population Density

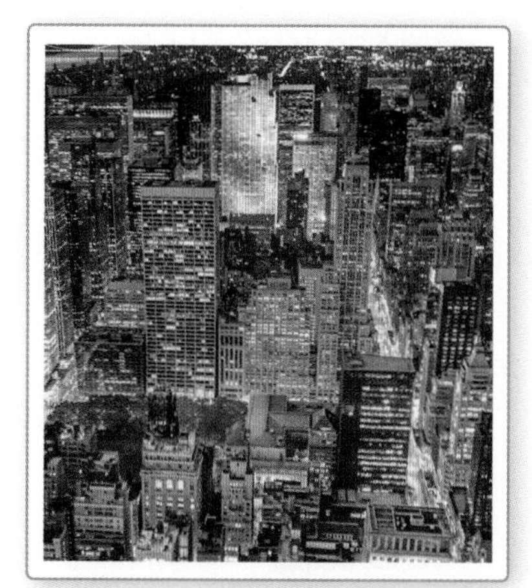

a. About 430,000 people live in a 5-mile radius of a city's town hall. Find the population density in people per square mile.

b. A region with a 3-mile radius has a population density of about 6195 people per square mile. Find the number of people who live in the region.

SOLUTION

a. Step 1 Find the area of the region.

$$A = \pi r^2 = \pi \cdot 5^2 = 25\pi$$

The area of the region is $25\pi \approx 78.54$ square miles.

Step 2 Find the population density.

$$\text{Population density} = \frac{\text{number of people}}{\text{area of land}} \quad \text{Formula for population density}$$

$$= \frac{430{,}000}{25\pi} \quad \text{Substitute.}$$

$$\approx 5475 \quad \text{Use a calculator.}$$

▶ The population density is about 5475 people per square mile.

b. Step 1 Find the area of the region.

$$A = \pi r^2 = \pi \cdot 3^2 = 9\pi$$

The area of the region is $9\pi \approx 28.27$ square miles.

Step 2 Let x represent the number of people who live in the region. Find the value of x.

$$\text{Population density} = \frac{\text{number of people}}{\text{area of land}} \quad \text{Formula for population density}$$

$$6195 \approx \frac{x}{9\pi} \quad \text{Substitute.}$$

$$175{,}159 \approx x \quad \text{Multiply and use a calculator.}$$

▶ The number of people who live in the region is about 175,159.

Monitoring Progress  Help in English and Spanish at *BigIdeasMath.com*

3. About 58,000 people live in a region with a 2-mile radius. Find the population density in people per square mile.

4. A region with a 3-mile radius has a population density of about 1000 people per square mile. Find the number of people who live in the region.

Finding Areas of Sectors

A **sector of a circle** is the region bounded by two radii of the circle and their intercepted arc. In the diagram below, sector APB is bounded by $\overline{AP}$, $\overline{BP}$, and $\widehat{AB}$.

Area of a Sector

The ratio of the area of a sector of a circle to the area of the whole circle (πr^2) is equal to the ratio of the measure of the intercepted arc to 360°.

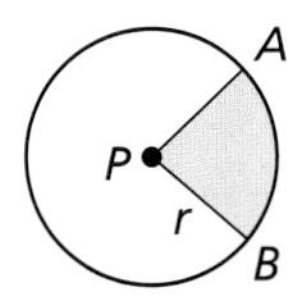

$$\frac{\text{Area of sector } APB}{\pi r^2} = \frac{m\widehat{AB}}{360^\circ}, \text{ or}$$

$$\text{Area of sector } APB = \frac{m\widehat{AB}}{360^\circ} \cdot \pi r^2$$

ANALYZING RELATIONSHIPS

The area of a sector is a fractional part of the area of a circle. The area of a sector formed by a 45° arc is $\frac{45^\circ}{360^\circ}$, or $\frac{1}{8}$ of the area of the circle.

EXAMPLE 3 Finding Areas of Sectors

Find the areas of the sectors formed by $\angle UTV$.

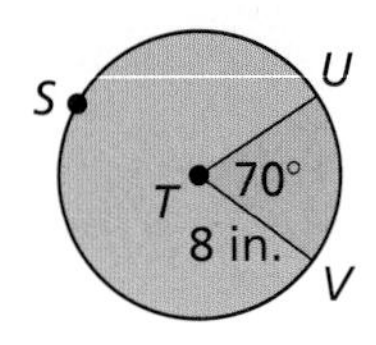

SOLUTION

Step 1 Find the measures of the minor and major arcs.

Because $m\angle UTV = 70^\circ$, $m\widehat{UV} = 70^\circ$ and $m\widehat{USV} = 360^\circ - 70^\circ = 290^\circ$.

Step 2 Find the areas of the small and large sectors.

$\text{Area of small sector} = \frac{m\widehat{UV}}{360^\circ} \cdot \pi r^2$	Formula for area of a sector
$= \frac{70^\circ}{360^\circ} \cdot \pi \cdot 8^2$	Substitute.
≈ 39.10	Use a calculator.
$\text{Area of large sector} = \frac{m\widehat{USV}}{360^\circ} \cdot \pi r^2$	Formula for area of a sector
$= \frac{290^\circ}{360^\circ} \cdot \pi \cdot 8^2$	Substitute.
≈ 161.97	Use a calculator.

▶ The areas of the small and large sectors are about 39.10 square inches and about 161.97 square inches, respectively.

Monitoring Progress  Help in English and Spanish at *BigIdeasMath.com*

Find the indicated measure.

5. area of red sector
6. area of blue sector

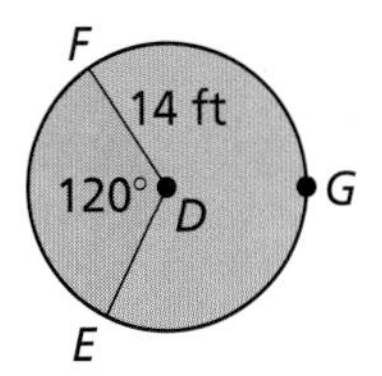

Using Areas of Sectors

EXAMPLE 4 **Using the Area of a Sector**

Find the area of $\odot V$.

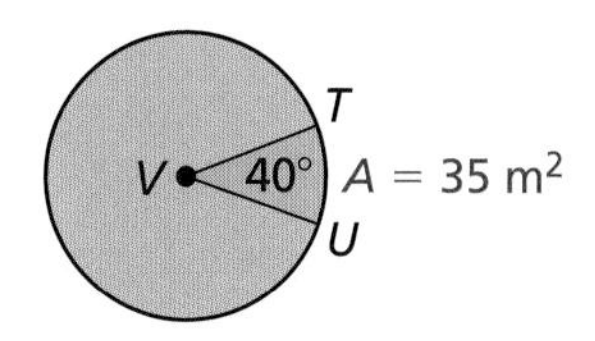

SOLUTION

Area of sector $TVU = \dfrac{m\widehat{TU}}{360^\circ} \cdot \text{Area of } \odot V$		Formula for area of a sector
$35 = \dfrac{40^\circ}{360^\circ} \cdot \text{Area of } \odot V$		Substitute.
$315 = \text{Area of } \odot V$		Solve for area of $\odot V$.

▶ The area of $\odot V$ is 315 square meters.

EXAMPLE 5 **Finding the Area of a Region**

A rectangular wall has an entrance cut into it. You want to paint the wall. To the nearest square foot, what is the area of the region you need to paint?

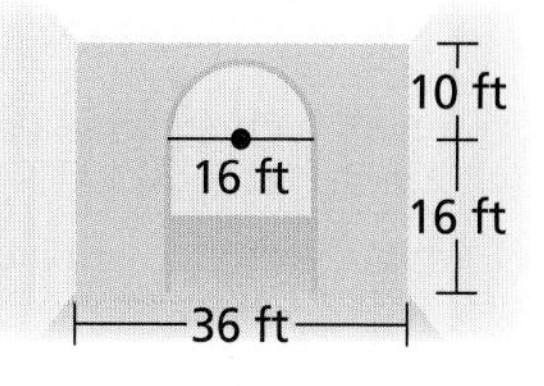

SOLUTION

COMMON ERROR

Use the radius (8 feet), not the diameter (16 feet), when you calculate the area of the semicircle.

The area you need to paint is the area of the rectangle minus the area of the entrance. The entrance can be divided into a semicircle and a square.

Area of wall = Area of rectangle − (Area of semicircle + Area of square)

$$= 36(26) - \left[\frac{180^\circ}{360^\circ} \cdot (\pi \cdot 8^2) + 16^2\right]$$

$$= 936 - (32\pi + 256)$$

$$\approx 579.47$$

▶ The area you need to paint is about 579 square feet.

Monitoring Progress  Help in English and Spanish at *BigIdeasMath.com*

7. Find the area of $\odot H$.

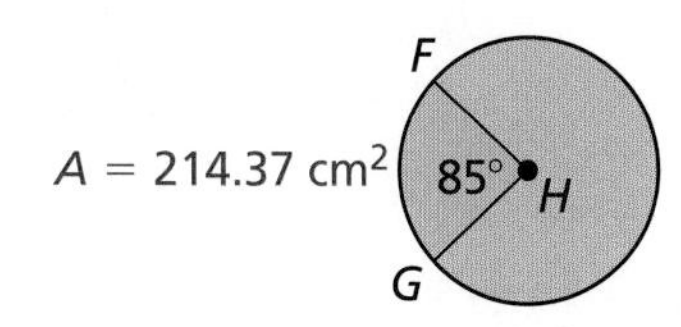

8. Find the area of the figure.

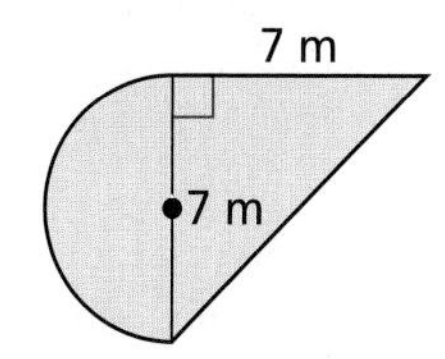

9. If you know the area and radius of a sector of a circle, can you find the measure of the intercepted arc? Explain.

11.2 Exercises

Dynamic Solutions available at *BigIdeasMath.com*

Vocabulary and Core Concept Check

1. **VOCABULARY** A(n) __________ of a circle is the region bounded by two radii of the circle and their intercepted arc.

2. **WRITING** The arc measure of a sector in a given circle is doubled. Will the area of the sector also be doubled? Explain your reasoning.

Monitoring Progress and Modeling with Mathematics

In Exercises 3–10, find the indicated measure. *(See Example 1.)*

3. area of $\odot C$

C
0.4 cm
A

4. area of $\odot C$

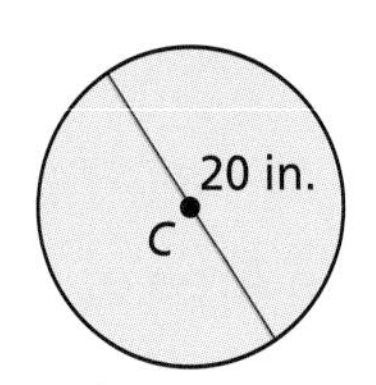

5. area of a circle with a radius of 5 inches

6. area of a circle with a diameter of 16 feet

7. radius of a circle with an area of 89 square feet

8. radius of a circle with an area of 380 square inches

9. diameter of a circle with an area of 12.6 square inches

10. diameter of a circle with an area of 676π square centimeters

In Exercises 11–14, find the indicated measure. *(See Example 2.)*

11. About 210,000 people live in a region with a 12-mile radius. Find the population density in people per square mile.

12. About 650,000 people live in a region with a 6-mile radius. Find the population density in people per square mile.

13. A region with a 4-mile radius has a population density of about 6366 people per square mile. Find the number of people who live in the region.

14. About 79,000 people live in a circular region with a population density of about 513 people per square mile. Find the radius of the region.

In Exercises 15–18, find the areas of the sectors formed by $\angle DFE$. *(See Example 3.)*

15.
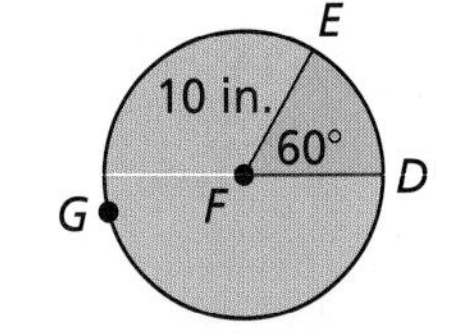

16.
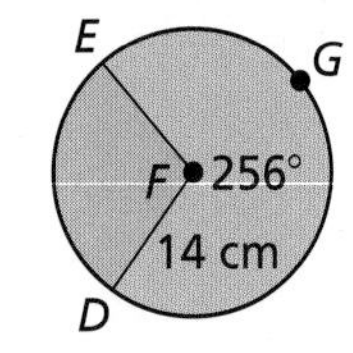

17.
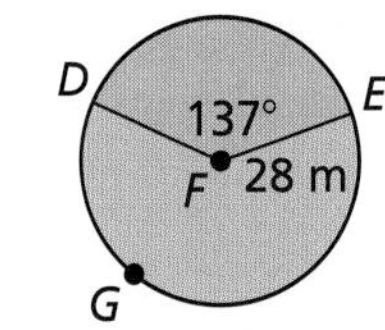

18.
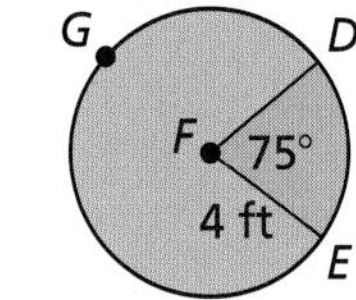

19. **ERROR ANALYSIS** Describe and correct the error in finding the area of the circle.

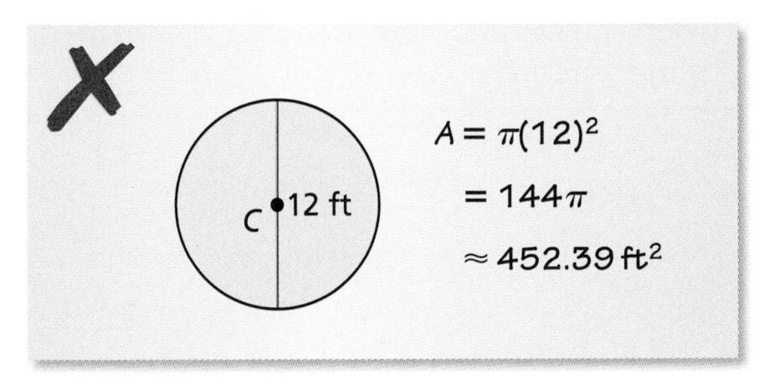

20. **ERROR ANALYSIS** Describe and correct the error in finding the area of sector XZY when the area of $\odot Z$ is 255 square feet.

W
X
Z
115°
Y

Let n be the area of sector XZY.

$$\frac{n}{360} = \frac{115}{255}$$

$$n \approx 162.35 \text{ ft}^2$$

In Exercises 21 and 22, the area of the shaded sector is shown. Find the indicated measure. *(See Example 4.)*

21. area of $\odot M$

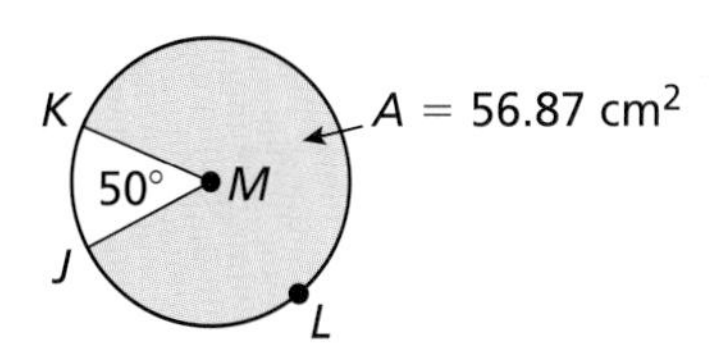

22. radius of $\odot M$

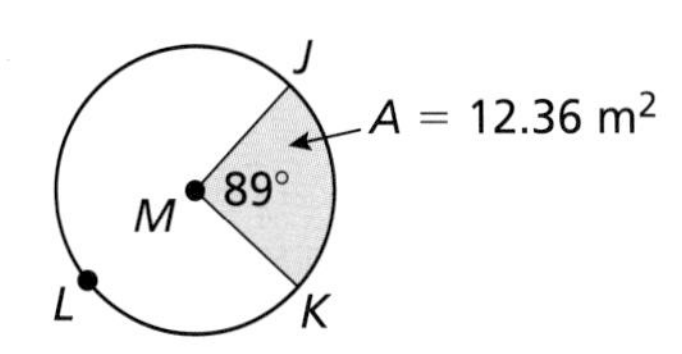

In Exercises 23–28, find the area of the shaded region. *(See Example 5.)*

23.

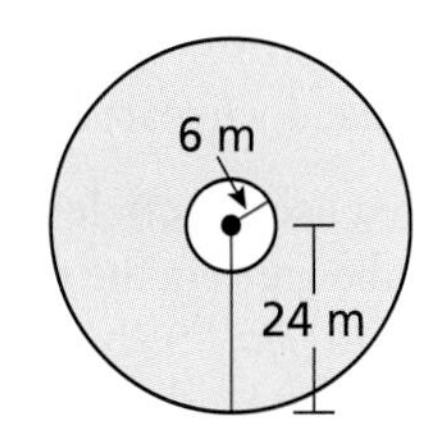

24.

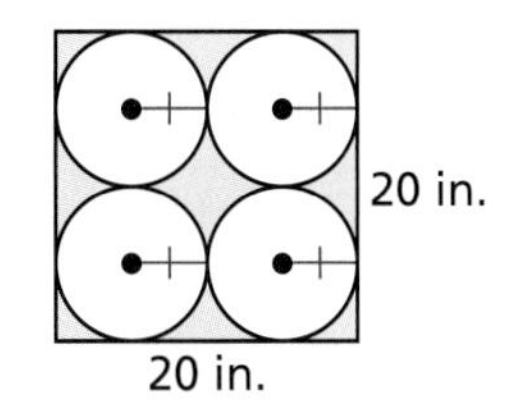

25.

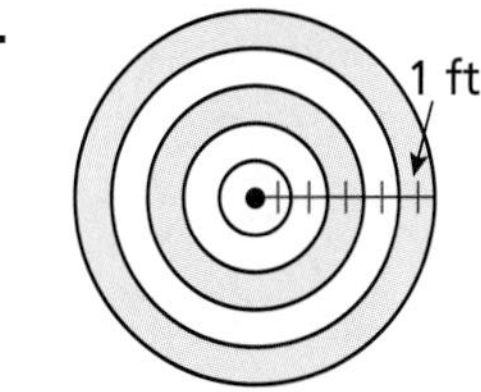

26.

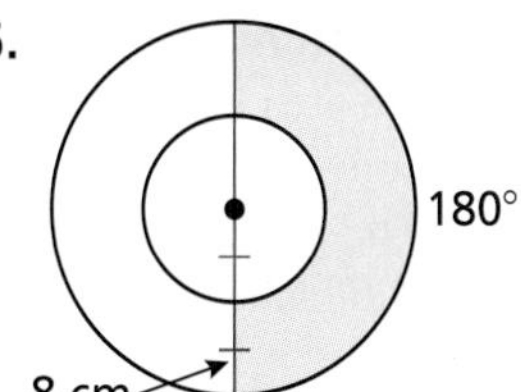

27.

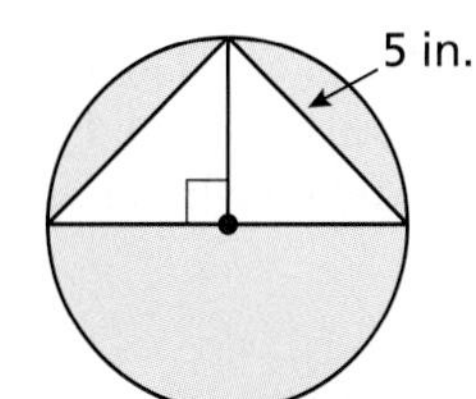

28.

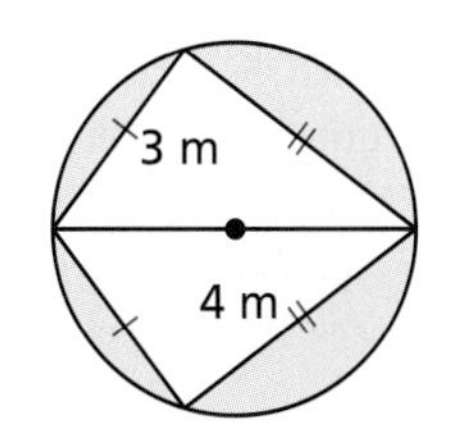

29. PROBLEM SOLVING The diagram shows the shape of a putting green at a miniature golf course. One part of the green is a sector of a circle. Find the area of the putting green.

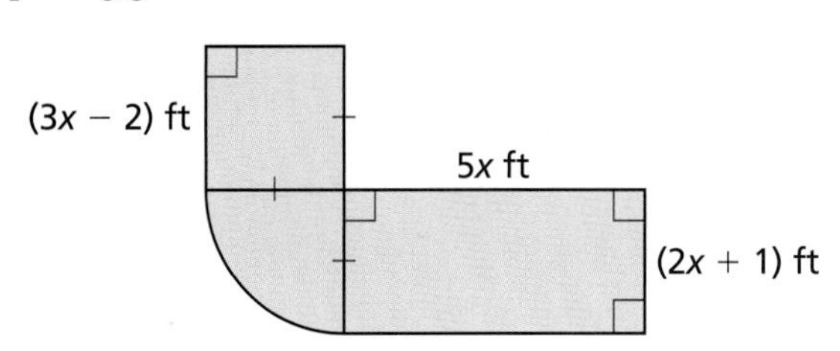

30. MAKING AN ARGUMENT Your friend claims that if the radius of a circle is doubled, then its area doubles. Is your friend correct? Explain your reasoning.

31. MODELING WITH MATHEMATICS The diagram shows the area of a lawn covered by a water sprinkler.

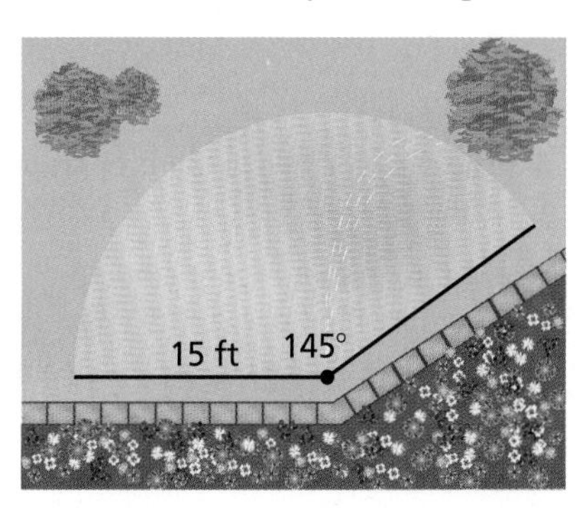

a. What is the area of the lawn that is covered by the sprinkler?

b. The water pressure is weakened so that the radius is 12 feet. What is the area of the lawn that will be covered?

32. MODELING WITH MATHEMATICS The diagram shows a projected beam of light from a lighthouse.

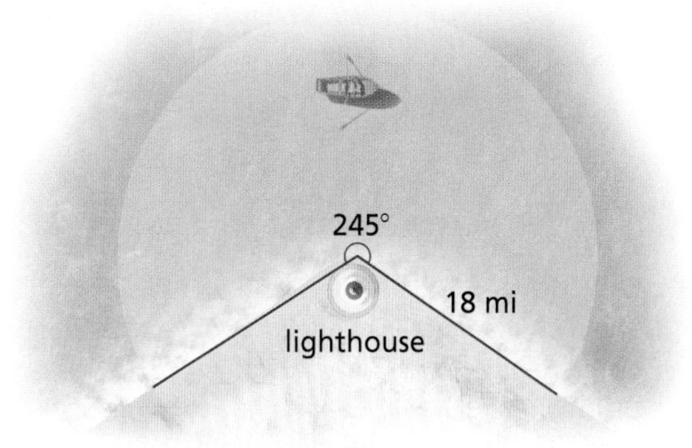

a. What is the area of water that can be covered by the light from the lighthouse?

b. What is the area of land that can be covered by the light from the lighthouse?

33. ANALYZING RELATIONSHIPS Look back at the Perimeters of Similar Polygons Theorem (Theorem 8.1) and the Areas of Similar Polygons Theorem (Theorem 8.2) in Section 8.1. How would you rewrite these theorems to apply to circles? Explain your reasoning.

34. ANALYZING RELATIONSHIPS A square is inscribed in a circle. The same square is also circumscribed about a smaller circle. Draw a diagram that represents this situation. Then find the ratio of the area of the larger circle to the area of the smaller circle.

35. CONSTRUCTION The table shows how students get to school.

Method	Percent of students
bus	65%
walk	25%
other	10%

a. Explain why a circle graph is appropriate for the data.

b. You will represent each method by a sector of a circle graph. Find the central angle to use for each sector. Then construct the graph using a radius of 2 inches.

c. Find the area of each sector in your graph.

36. HOW DO YOU SEE IT? The outermost edges of the pattern shown form a square. If you know the dimensions of the outer square, is it possible to compute the total colored area? Explain.

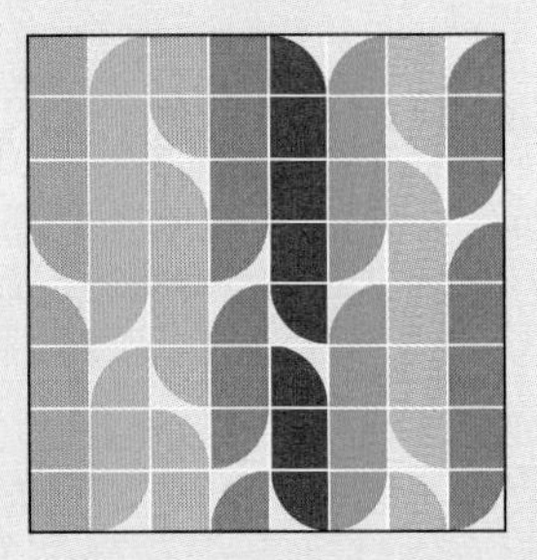

37. ABSTRACT REASONING A circular pizza with a 12-inch diameter is enough for you and 2 friends. You want to buy pizzas for yourself and 7 friends. A 10-inch diameter pizza with one topping costs $6.99 and a 14-inch diameter pizza with one topping costs $12.99. How many 10-inch and 14-inch pizzas should you buy in each situation? Explain.

a. You want to spend as little money as possible.

b. You want to have three pizzas, each with a different topping, and spend as little money as possible.

c. You want to have as much of the thick outer crust as possible.

38. THOUGHT PROVOKING You know that the area of a circle is πr^2. Find the formula for the area of an *ellipse*, shown below.

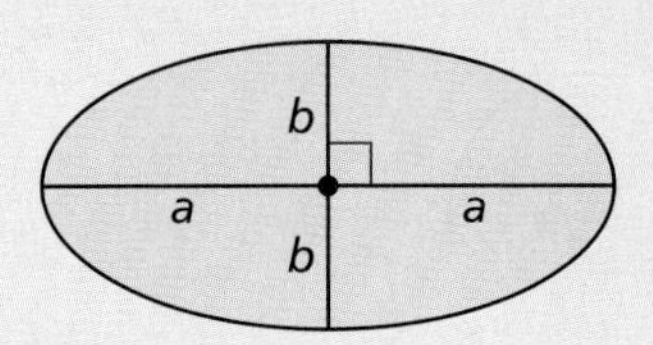

39. MULTIPLE REPRESENTATIONS Consider a circle with a radius of 3 inches.

a. Complete the table, where x is the measure of the arc and y is the area of the corresponding sector. Round your answers to the nearest tenth.

x	30°	60°	90°	120°	150°	180°
y						

b. Graph the data in the table.

c. Is the relationship between x and y linear? Explain.

d. If parts (a)–(c) were repeated using a circle with a radius of 5 inches, would the areas in the table change? Would your answer to part (c) change? Explain your reasoning.

40. CRITICAL THINKING Find the area between the three congruent tangent circles. The radius of each circle is 6 inches.

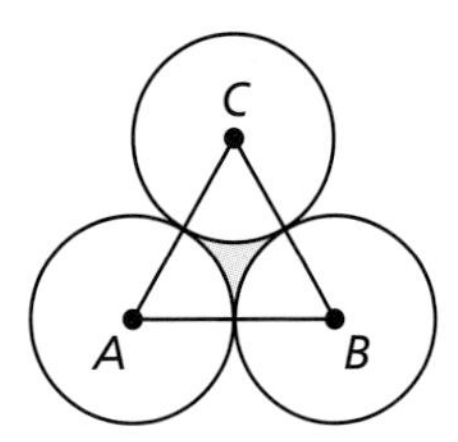

41. PROOF Semicircles with diameters equal to three sides of a right triangle are drawn, as shown. Prove that the sum of the areas of the two shaded crescents equals the area of the triangle.

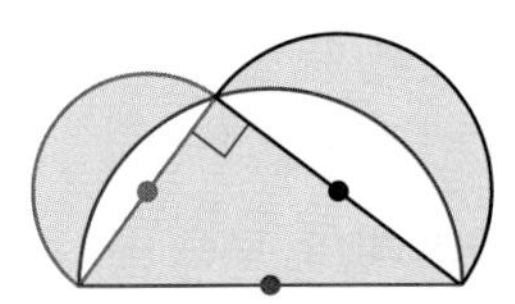

Maintaining Mathematical Proficiency

Reviewing what you learned in previous grades and lessons

Find the area of the figure. *(Skills Review Handbook)*

42.

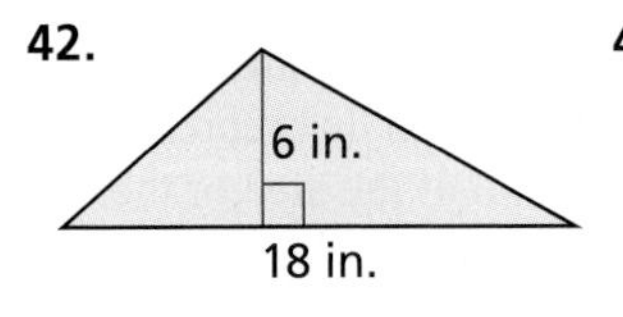

43.

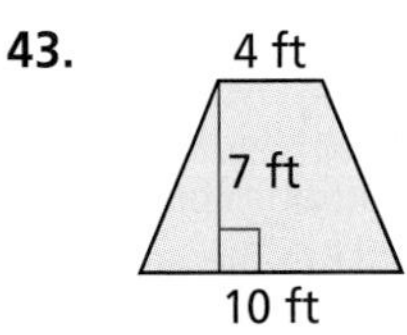

44.

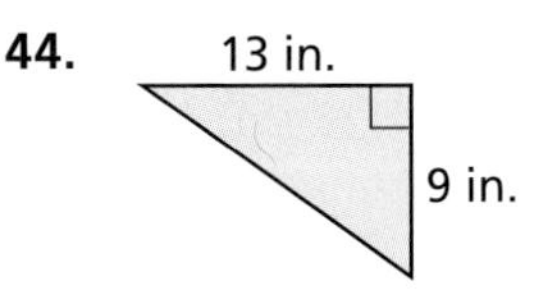

45.

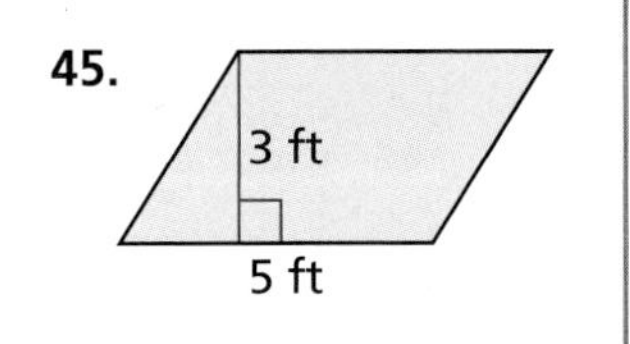

11.3 Areas of Polygons

Essential Question How can you find the area of a regular polygon?

The **center of a regular polygon** is the center of its circumscribed circle.

The distance from the center to any side of a regular polygon is called the **apothem of a regular polygon**.

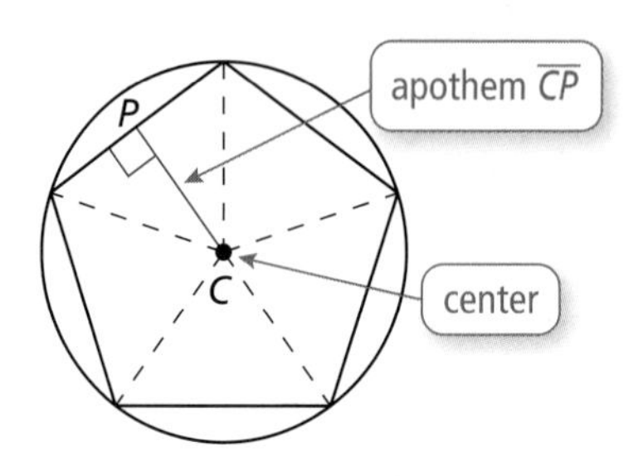

EXPLORATION 1 Finding the Area of a Regular Polygon

Work with a partner. Use dynamic geometry software to construct each regular polygon with side lengths of 4, as shown. Find the apothem and use it to find the area of the polygon. Describe the steps that you used.

a.

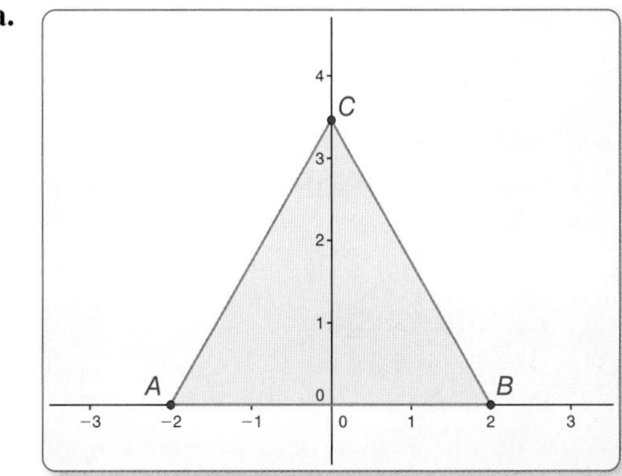

b.

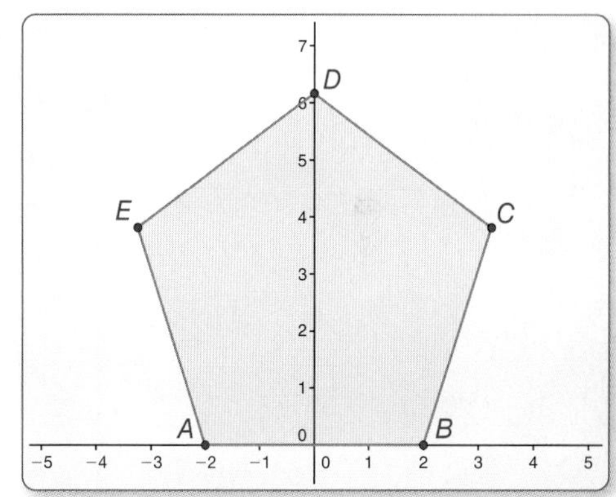

c.

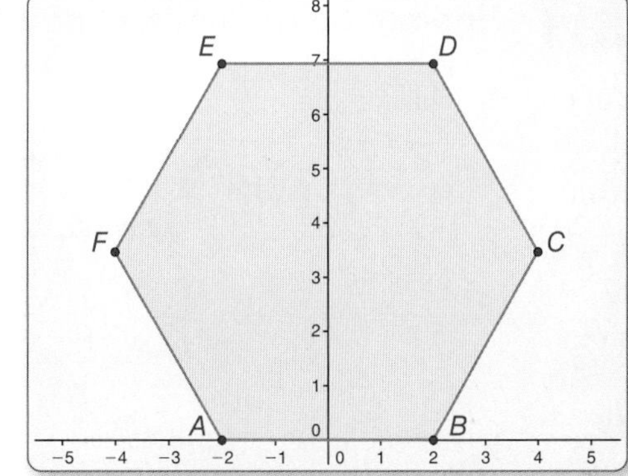

d.

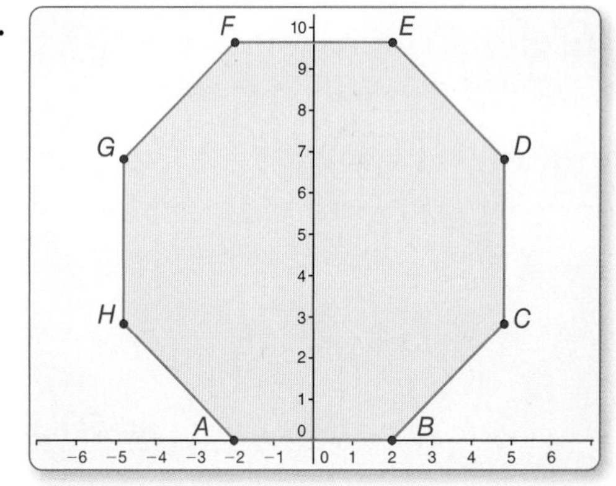

EXPLORATION 2 Writing a Formula for Area

Work with a partner. Generalize the steps you used in Exploration 1 to develop a formula for the area of a regular polygon.

REASONING ABSTRACTLY

To be proficient in math, you need to know and flexibly use different properties of operations and objects.

Communicate Your Answer

3. How can you find the area of a regular polygon?

4. Regular pentagon *ABCDE* has side lengths of 6 meters and an apothem of approximately 4.13 meters. Find the area of *ABCDE*.

11.3 Lesson

Core Vocabulary

center of a regular polygon, p. 611
radius of a regular polygon, p. 611
apothem of a regular polygon, p. 611
central angle of a regular polygon, p. 611

Previous
rhombus
kite

What You Will Learn

- Find areas of rhombuses and kites.
- Find angle measures in regular polygons.
- Find areas of regular polygons.

Finding Areas of Rhombuses and Kites

You can divide a rhombus or kite with diagonals d_1 and d_2 into two congruent triangles with base d_1, height $\frac{1}{2}d_2$, and area $\frac{1}{2}d_1\left(\frac{1}{2}d_2\right) = \frac{1}{4}d_1d_2$. So, the area of a rhombus or kite is $2\left(\frac{1}{4}d_1d_2\right) = \frac{1}{2}d_1d_2$.

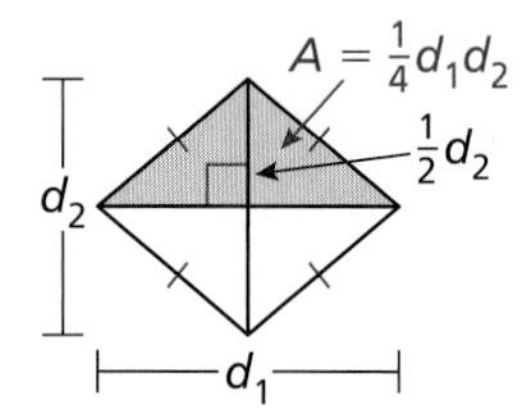

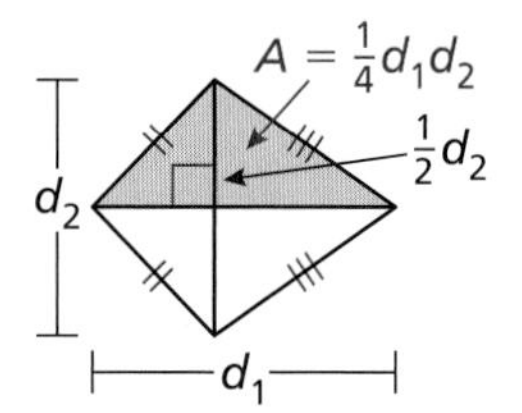

Core Concept

Area of a Rhombus or Kite

The area of a rhombus or kite with diagonals d_1 and d_2 is $\frac{1}{2}d_1d_2$.

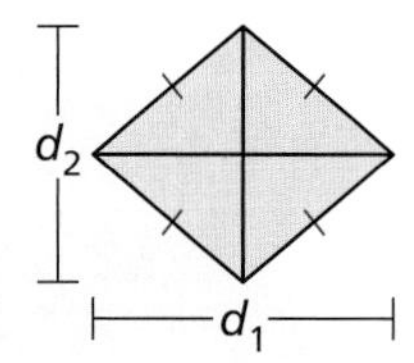

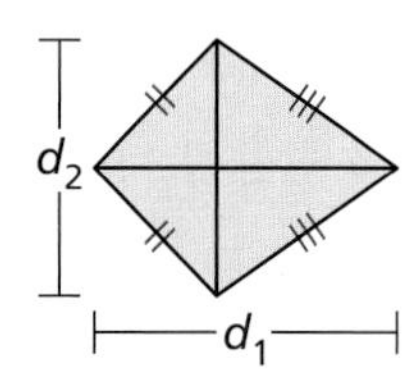

EXAMPLE 1 Finding the Area of a Rhombus or Kite

Find the area of each rhombus or kite.

a. 8 m, 6 m

b.

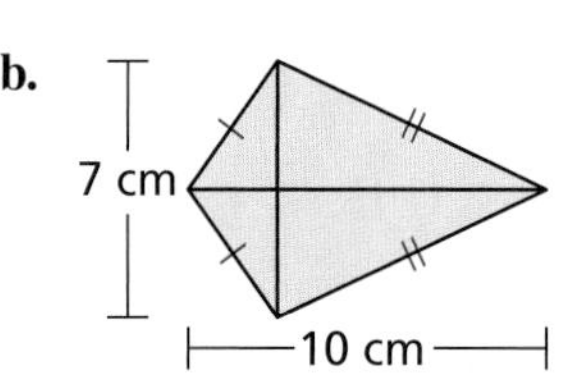

SOLUTION

a. $A = \frac{1}{2}d_1d_2$

$= \frac{1}{2}(6)(8)$

$= 24$

▶ So, the area is 24 square meters.

b. $A = \frac{1}{2}d_1d_2$

$= \frac{1}{2}(10)(7)$

$= 35$

▶ So, the area is 35 square centimeters.

Monitoring Progress

Help in English and Spanish at *BigIdeasMath.com*

1. Find the area of a rhombus with diagonals $d_1 = 4$ feet and $d_2 = 5$ feet.

2. Find the area of a kite with diagonals $d_1 = 12$ inches and $d_2 = 9$ inches.

Finding Angle Measures in Regular Polygons

The diagram shows a regular polygon inscribed in a circle. The **center of a regular polygon** and the **radius of a regular polygon** are the center and the radius of its circumscribed circle.

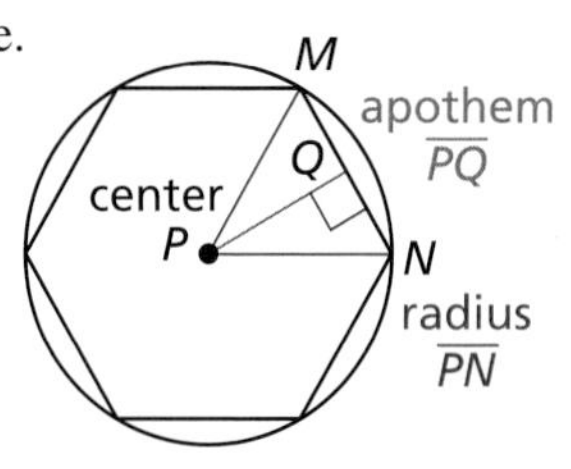

$\angle MPN$ is a central angle.

The distance from the center to any side of a regular polygon is called the **apothem of a regular polygon**. The apothem is the height to the base of an isosceles triangle that has two radii as legs. The word "apothem" refers to a segment as well as a length. For a given regular polygon, think of *an* apothem as a segment and *the* apothem as a length.

A **central angle of a regular polygon** is an angle formed by two radii drawn to consecutive vertices of the polygon. To find the measure of each central angle, divide 360° by the number of sides.

EXAMPLE 2 Finding Angle Measures in a Regular Polygon

In the diagram, $ABCDE$ is a regular pentagon inscribed in $\odot F$. Find each angle measure.

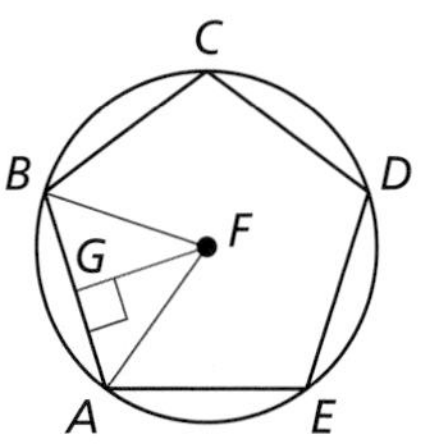

a. $m\angle AFB$ **b.** $m\angle AFG$ **c.** $m\angle GAF$

ANALYZING RELATIONSHIPS

$\overline{FG}$ is an altitude of an isosceles triangle, so it is also a median and angle bisector of the isosceles triangle.

SOLUTION

a. $\angle AFB$ is a central angle, so $m\angle AFB = \frac{360°}{5} = 72°$.

b. $\overline{FG}$ is an apothem, which makes it an altitude of isosceles $\triangle AFB$.

So, $\overline{FG}$ bisects $\angle AFB$ and $m\angle AFG = \frac{1}{2}m\angle AFB = 36°$.

c. By the Triangle Sum Theorem (Theorem 5.1), the sum of the angle measures of right $\triangle GAF$ is 180°.

$$m\angle GAF = 180° - 90° - 36°$$
$$= 54°$$

So, $m\angle GAF = 54°$.

Monitoring Progress

Help in English and Spanish at *BigIdeasMath.com*

In the diagram, $WXYZ$ is a square inscribed in $\odot P$.

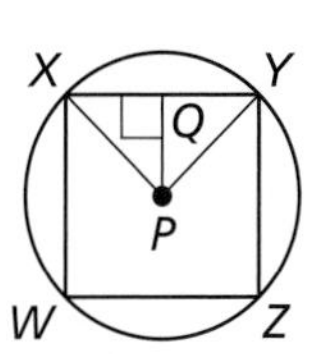

3. Identify the center, a radius, an apothem, and a central angle of the polygon.

4. Find $m\angle XPY$, $m\angle XPQ$, and $m\angle PXQ$.

Finding Areas of Regular Polygons

You can find the area of any regular n-gon by dividing it into congruent triangles.

READING DIAGRAMS

In this book, a point shown inside a regular polygon marks the center of the circle that can be circumscribed about the polygon.

$A =$ Area of one triangle • Number of triangles

$= \left(\frac{1}{2} \cdot s \cdot a\right) \cdot n$ — Base of triangle is s and height of triangle is a. Number of triangles is n.

$= \frac{1}{2} \cdot a \cdot (n \cdot s)$ — Commutative and Associative Properties of Multiplication

$= \frac{1}{2}a \cdot P$ — There are n congruent sides of length s, so perimeter P is $n \cdot s$.

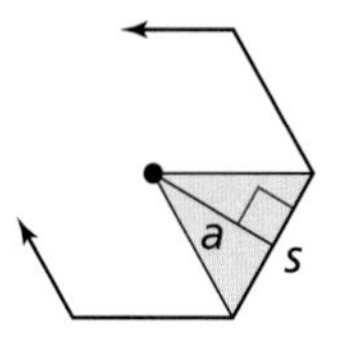

Core Concept

Area of a Regular Polygon

The area of a regular n-gon with side length s is one-half the product of the apothem a and the perimeter P.

$$A = \frac{1}{2}aP, \text{ or } A = \frac{1}{2}a \cdot ns$$

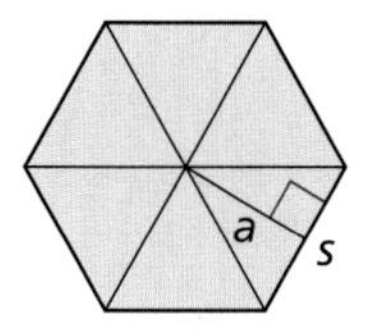

EXAMPLE 3 Finding the Area of a Regular Polygon

A regular nonagon is inscribed in a circle with a radius of 4 units. Find the area of the nonagon.

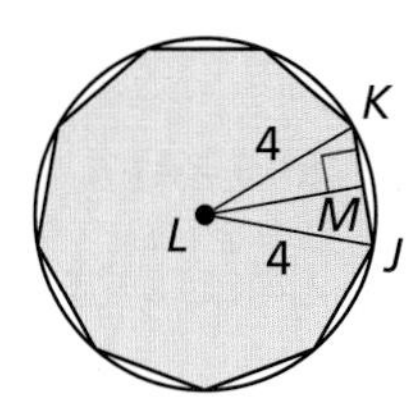

SOLUTION

The measure of central $\angle JLK$ is $\frac{360°}{9}$, or 40°. Apothem $\overline{LM}$ bisects the central angle, so $m\angle KLM$ is 20°. To find the lengths of the legs, use trigonometric ratios for right $\triangle KLM$.

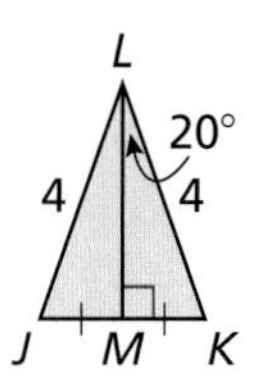

$$\sin 20° = \frac{MK}{LK} \qquad \cos 20° = \frac{LM}{LK}$$

$$\sin 20° = \frac{MK}{4} \qquad \cos 20° = \frac{LM}{4}$$

$$4 \sin 20° = MK \qquad 4 \cos 20° = LM$$

The regular nonagon has side length $s = 2(MK) = 2(4 \sin 20°) = 8 \sin 20°$, and apothem $a = LM = 4 \cos 20°$.

▶ So, the area is $A = \frac{1}{2}a \cdot ns = \frac{1}{2}(4 \cos 20°) \cdot (9)(8 \sin 20°) \approx 46.3$ square units.

EXAMPLE 4 Finding the Area of a Regular Polygon

You are decorating the top of a table by covering it with small ceramic tiles. The tabletop is a regular octagon with 15-inch sides and a radius of about 19.6 inches. What is the area you are covering?

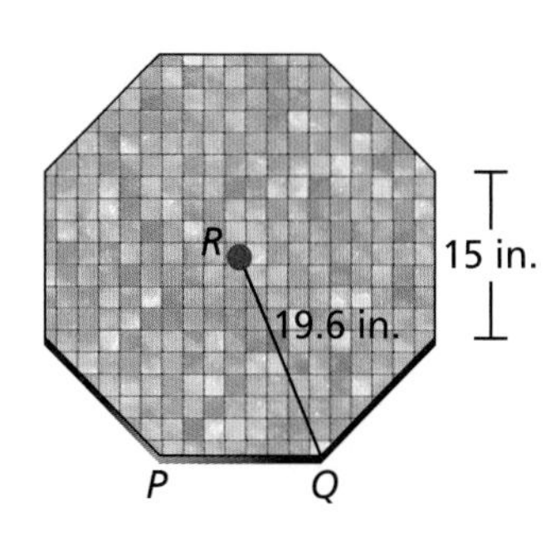

SOLUTION

Step 1 Find the perimeter P of the tabletop.
An octagon has 8 sides, so $P = 8(15) = 120$ inches.

Step 2 Find the apothem a. The apothem is height RS of $\triangle PQR$.

Because $\triangle PQR$ is isosceles, altitude $\overline{RS}$ bisects $\overline{QP}$.

So, $QS = \frac{1}{2}(QP) = \frac{1}{2}(15) = 7.5$ inches.

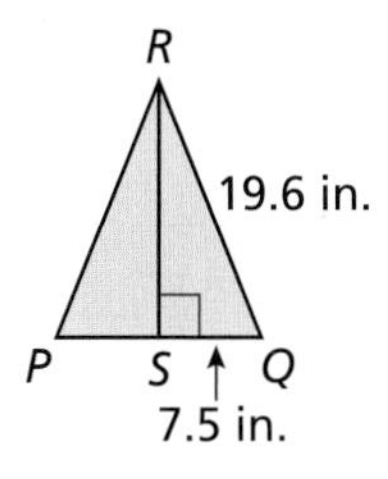

To find RS, use the Pythagorean Theorem (Theorem 9.1) for $\triangle RQS$.

$$a = RS = \sqrt{19.6^2 - 7.5^2} = \sqrt{327.91} \approx 18.108$$

Step 3 Find the area A of the tabletop.

$A = \frac{1}{2}aP$	Formula for area of a regular polygon
$= \frac{1}{2}(\sqrt{327.91})(120)$	Substitute.
≈ 1086.5	Simplify.

▶ The area you are covering with tiles is about 1086.5 square inches.

Monitoring Progress

Help in English and Spanish at *BigIdeasMath.com*

Find the area of the regular polygon.

5.

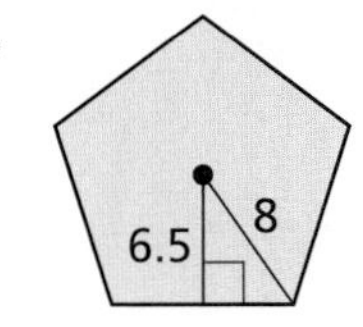

6.

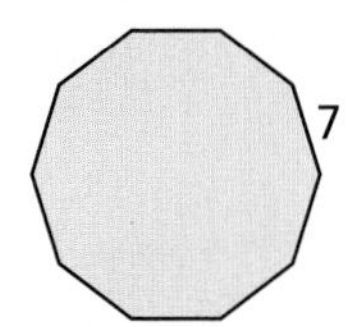

11.3 Exercises

Dynamic Solutions available at *BigIdeasMath.com*

Vocabulary and Core Concept Check

1. **WRITING** Explain how to find the measure of a central angle of a regular polygon.

2. **DIFFERENT WORDS, SAME QUESTION** Which is different? Find "both" answers.

Find the radius of $\odot F$.	Find the apothem of polygon *ABCDE.*
Find *AF.*	Find the radius of polygon *ABCDE.*

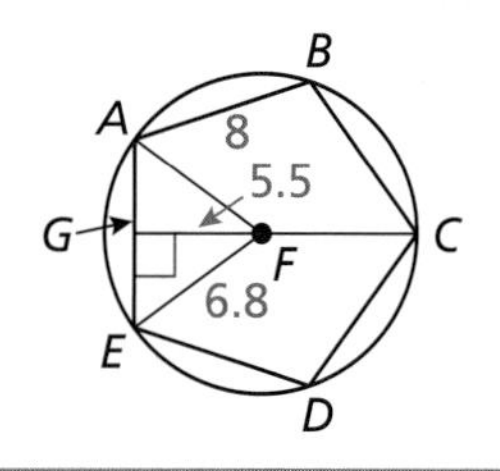

Monitoring Progress and Modeling with Mathematics

In Exercises 3–6, find the area of the kite or rhombus. *(See Example 1.)*

3.

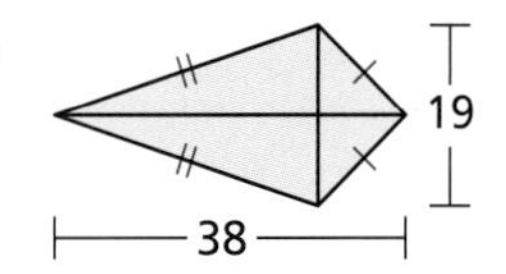

4.

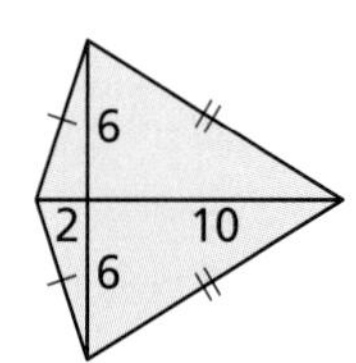

5.

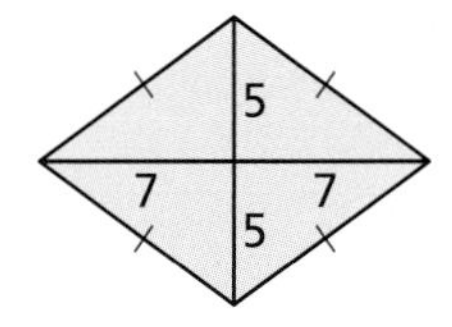

6.

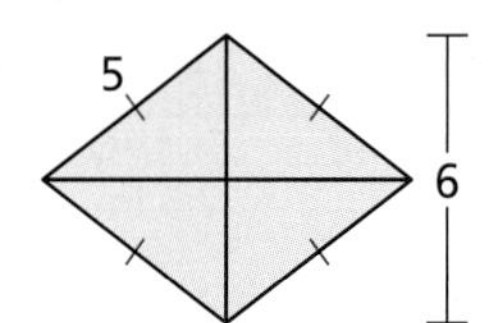

In Exercises 7–10, use the diagram.

7. Identify the center of polygon *JKLMN.*

8. Identify a central angle of polygon *JKLMN.*

9. What is the radius of polygon *JKLMN?*

10. What is the apothem of polygon *JKLMN*?

In Exercises 11–14, find the measure of a central angle of a regular polygon with the given number of sides. Round answers to the nearest tenth of a degree, if necessary.

11. 10 sides

12. 18 sides

13. 24 sides

14. 7 sides

In Exercises 15–18, find the given angle measure for regular octagon *ABCDEFGH.* *(See Example 2.)*

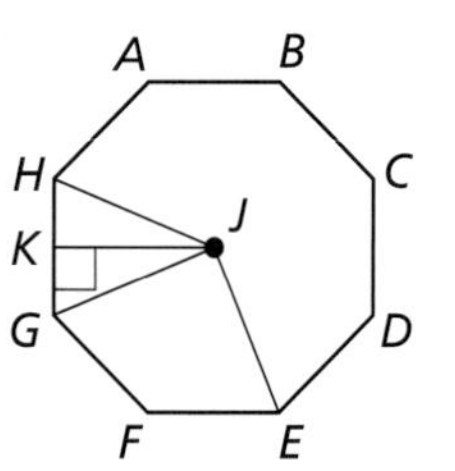

15. $m\angle GJH$

16. $m\angle GJK$

17. $m\angle KGJ$

18. $m\angle EJH$

In Exercises 19–24, find the area of the regular polygon. *(See Examples 3 and 4.)*

19.

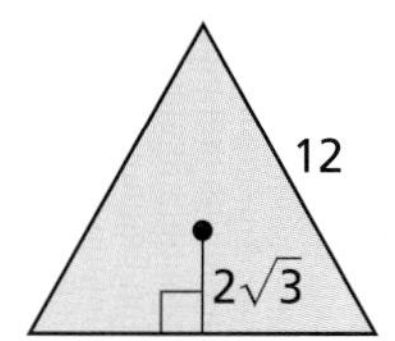

20.

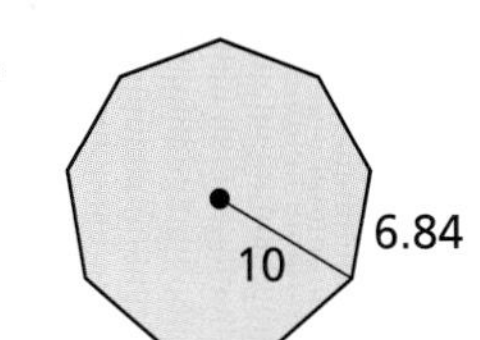

21.

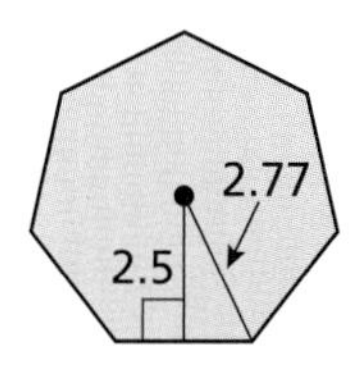

22.

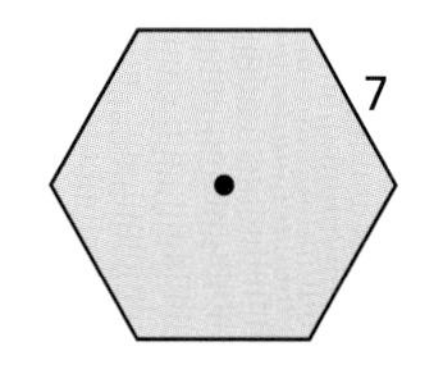

23. an octagon with a radius of 11 units

24. a pentagon with an apothem of 5 units

25. **ERROR ANALYSIS** Describe and correct the error in finding the area of the kite.

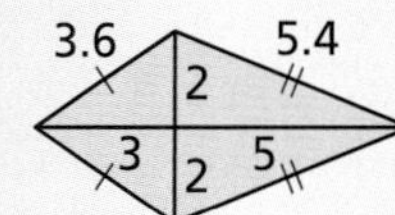

$A = \frac{1}{2}(3.6)(5.4)$

$= 9.72$

So, the area of the kite is 9.72 square units.

26. **ERROR ANALYSIS** Describe and correct the error in finding the area of the regular hexagon.

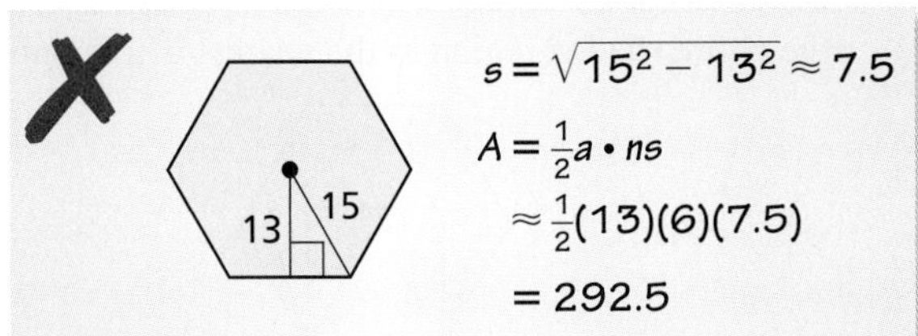

So, the area of the hexagon is about 292.5 square units.

In Exercises 27–30, find the area of the shaded region.

27.

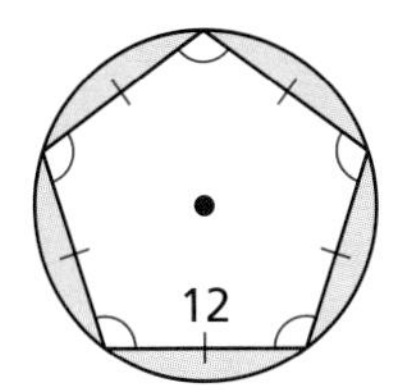

28.

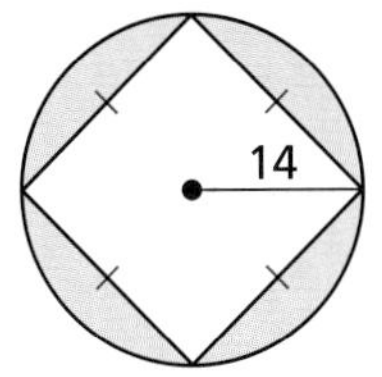

29.

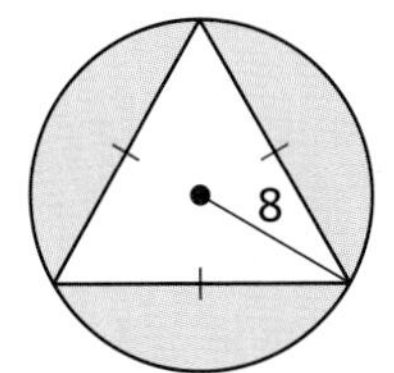

30.

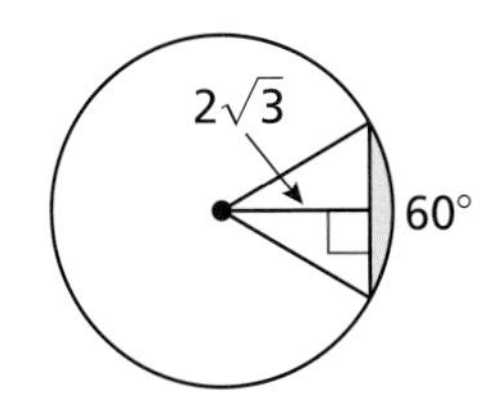

31. **MODELING WITH MATHEMATICS** Basaltic columns are geological formations that result from rapidly cooling lava. Giant's Causeway in Ireland contains many hexagonal basaltic columns. Suppose the top of one of the columns is in the shape of a regular hexagon with a radius of 8 inches. Find the area of the top of the column to the nearest square inch.

32. **MODELING WITH MATHEMATICS** A watch has a circular surface on a background that is a regular octagon. Find the area of the octagon. Then find the area of the silver border around the circular face.

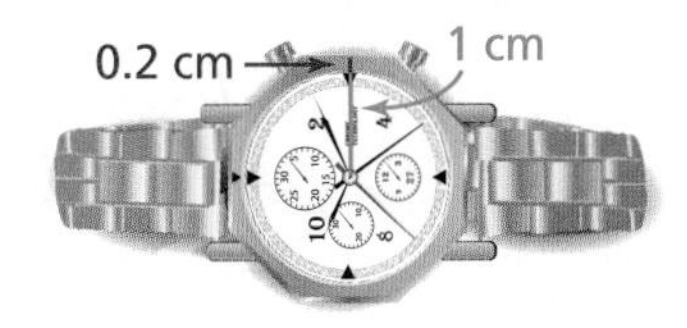

CRITICAL THINKING In Exercises 33–35, tell whether the statement is *true* or *false*. Explain your reasoning.

33. The area of a regular n-gon of a fixed radius r increases as n increases.

34. The apothem of a regular polygon is always less than the radius.

35. The radius of a regular polygon is always less than the side length.

36. **REASONING** Predict which figure has the greatest area and which has the least area. Explain your reasoning. Check by finding the area of each figure.

Ⓐ

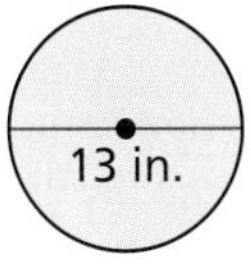

Ⓑ

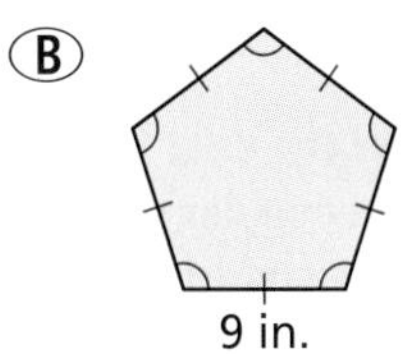

Ⓒ

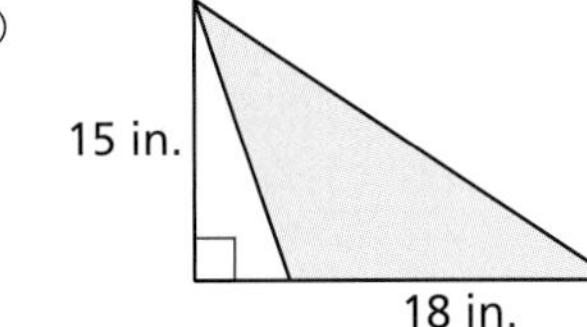

37. **USING EQUATIONS** Find the area of a regular pentagon inscribed in a circle whose equation is given by $(x - 4)^2 + (y + 2)^2 = 25$.

38. **REASONING** What happens to the area of a kite if you double the length of one of the diagonals? if you double the length of both diagonals? Justify your answer.

MATHEMATICAL CONNECTIONS In Exercises 39 and 40, write and solve an equation to find the indicated lengths. Round decimal answers to the nearest tenth.

39. The area of a kite is 324 square inches. One diagonal is twice as long as the other diagonal. Find the length of each diagonal.

40. One diagonal of a rhombus is four times the length of the other diagonal. The area of the rhombus is 98 square feet. Find the length of each diagonal.

41. **REASONING** The perimeter of a regular nonagon, or 9-gon, is 18 inches. Is this enough information to find the area? If so, find the area and explain your reasoning. If not, explain why not.

42. MAKING AN ARGUMENT Your friend claims that it is possible to find the area of any rhombus if you only know the perimeter of the rhombus. Is your friend correct? Explain your reasoning.

43. PROOF Prove that the area of any quadrilateral with perpendicular diagonals is $A = \frac{1}{2}d_1d_2$, where d_1 and d_2 are the lengths of the diagonals.

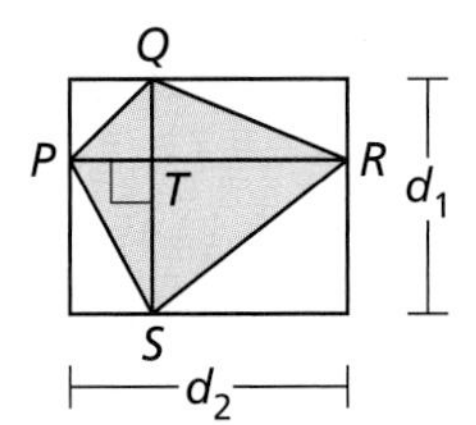

44. HOW DO YOU SEE IT? Explain how to find the area of the regular hexagon by dividing the hexagon into equilateral triangles.

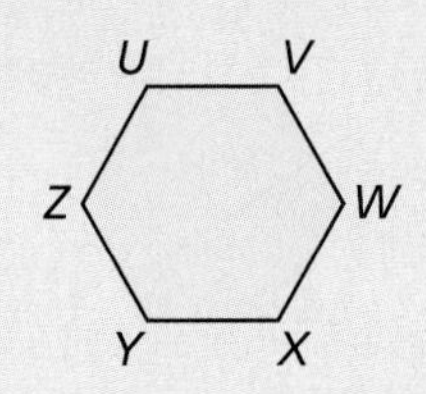

45. REWRITING A FORMULA Rewrite the formula for the area of a rhombus for the special case of a square with side length s. Show that this is the same as the formula for the area of a square, $A = s^2$.

46. REWRITING A FORMULA Use the formula for the area of a regular polygon to show that the area of an equilateral triangle can be found by using the formula $A = \frac{1}{4}s^2\sqrt{3}$, where s is the side length.

47. CRITICAL THINKING The area of a regular pentagon is 72 square centimeters. Find the length of one side.

48. CRITICAL THINKING The area of a dodecagon, or 12-gon, is 140 square inches. Find the apothem of the polygon.

49. USING STRUCTURE In the figure, an equilateral triangle lies inside a square inside a regular pentagon inside a regular hexagon. Find the approximate area of the entire shaded region to the nearest whole number.

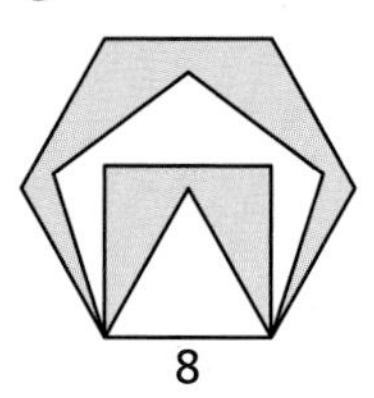

50. THOUGHT PROVOKING The area of a regular n-gon is given by $A = \frac{1}{2}aP$. As n approaches infinity, what does the n-gon approach? What does P approach? What does a approach? What can you conclude from your three answers? Explain your reasoning.

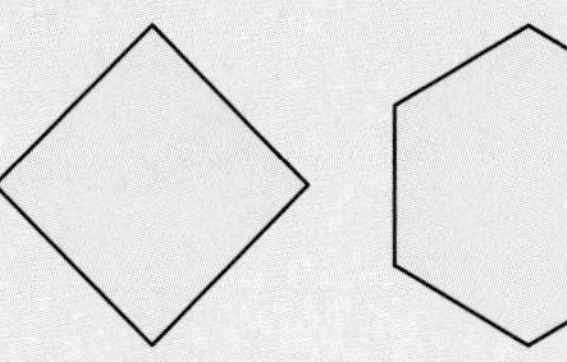

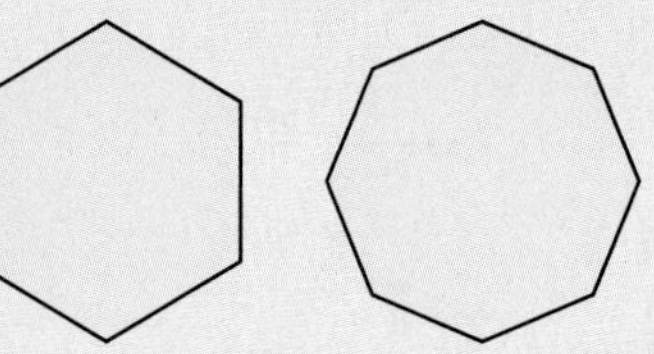

51. COMPARING METHODS Find the area of regular pentagon $ABCDE$ by using the formula $A = \frac{1}{2}aP$, or $A = \frac{1}{2}a \bullet ns$. Then find the area by adding the areas of smaller polygons. Check that both methods yield the same area. Which method do you prefer? Explain your reasoning.

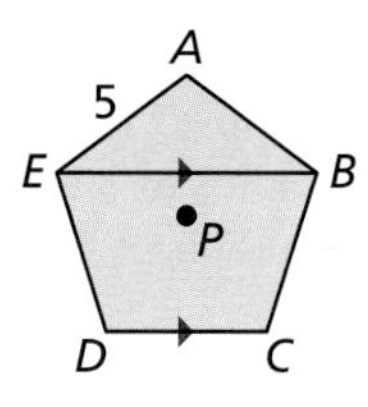

52. USING STRUCTURE Two regular polygons both have n sides. One of the polygons is inscribed in, and the other is circumscribed about, a circle of radius r. Find the area between the two polygons in terms of n and r.

Maintaining Mathematical Proficiency

Reviewing what you learned in previous grades and lessons

Determine whether the figure has *line symmetry*, *rotational symmetry*, *both*, or *neither*. If the figure has line symmetry, determine the number of lines of symmetry. If the figure has rotational symmetry, describe any rotations that map the figure onto itself. *(Section 4.2 and Section 4.3)*

53.

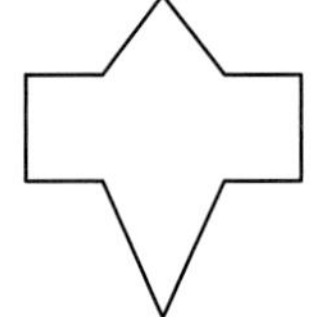

54.

55.

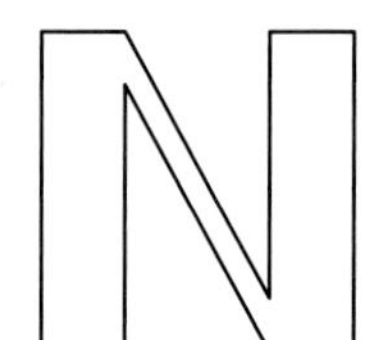

56.

11.4 Three-Dimensional Figures

Essential Question

What is the relationship between the numbers of vertices V, edges E, and faces F of a polyhedron?

A **polyhedron** is a solid that is bounded by polygons, called **faces**.

- Each *vertex* is a point.
- Each *edge* is a segment of a line.
- Each *face* is a portion of a plane.

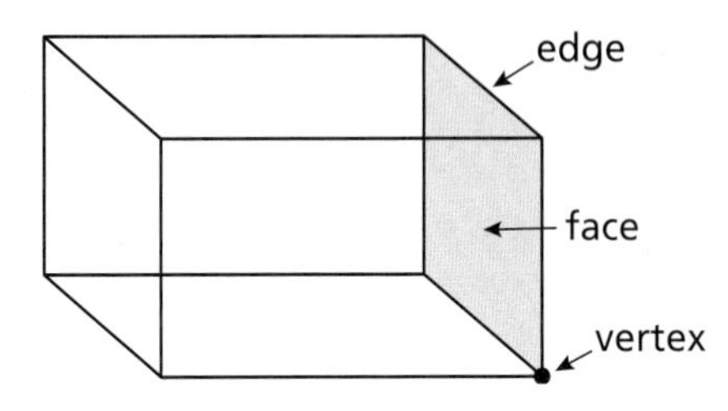

EXPLORATION 1 Analyzing a Property of Polyhedra

Work with a partner. The five *Platonic solids* are shown below. Each of these solids has congruent regular polygons as faces. Complete the table by listing the numbers of vertices, edges, and faces of each Platonic solid.

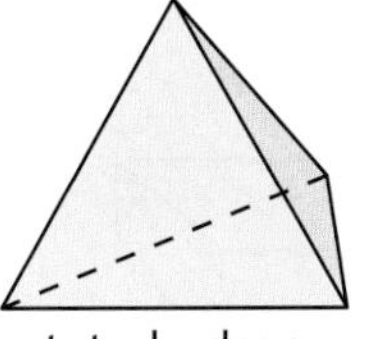
tetrahedron

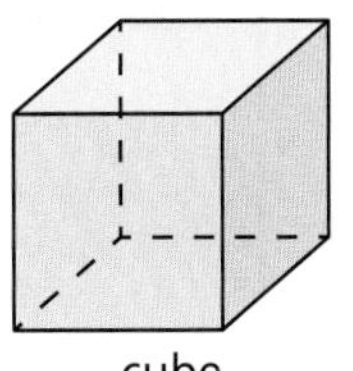
cube

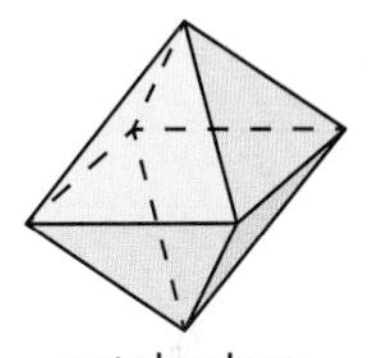
octahedron

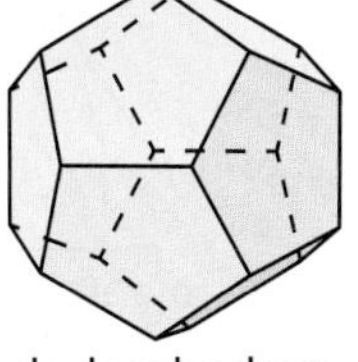
dodecahedron

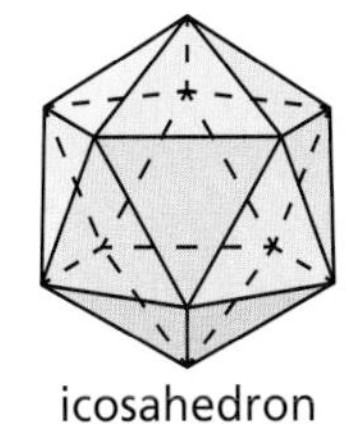
icosahedron

Solid	Vertices, V	Edges, E	Faces, F
tetrahedron			
cube			
octahedron			
dodecahedron			
icosahedron			

CONSTRUCTING VIABLE ARGUMENTS

To be proficient in math, you need to reason inductively about data.

Communicate Your Answer

2. What is the relationship between the numbers of vertices V, edges E, and faces F of a polyhedron? (*Note*: Swiss mathematician Leonhard Euler (1707–1783) discovered a formula that relates these quantities.)

3. Draw three polyhedra that are different from the Platonic solids given in Exploration 1. Count the numbers of vertices, edges, and faces of each polyhedron. Then verify that the relationship you found in Question 2 is valid for each polyhedron.

11.4 Lesson

Core Vocabulary

polyhedron, *p. 618*
face, *p. 618*
edge, *p. 618*
vertex, *p. 618*
cross section, *p. 619*
solid of revolution, *p. 620*
axis of revolution, *p. 620*

Previous
solid
prism
pyramid
cylinder
cone
sphere
base

What You Will Learn

- Classify solids.
- Describe cross sections.
- Sketch and describe solids of revolution.

Classifying Solids

A three-dimensional figure, or solid, is bounded by flat or curved surfaces that enclose a single region of space. A **polyhedron** is a solid that is bounded by polygons, called **faces**. An **edge** of a polyhedron is a line segment formed by the intersection of two faces. A **vertex** of a polyhedron is a point where three or more edges meet. The plural of polyhedron is *polyhedra* or *polyhedrons*.

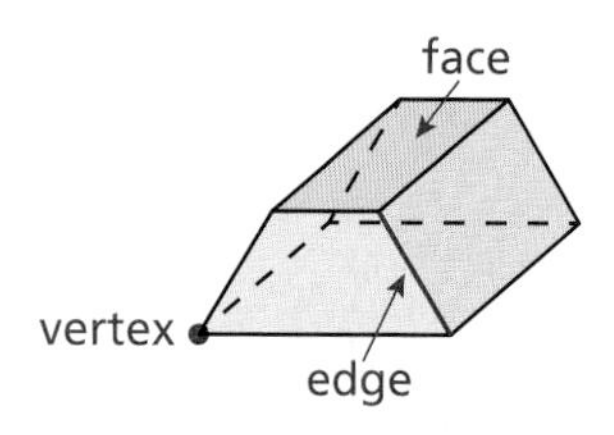

Types of Solids

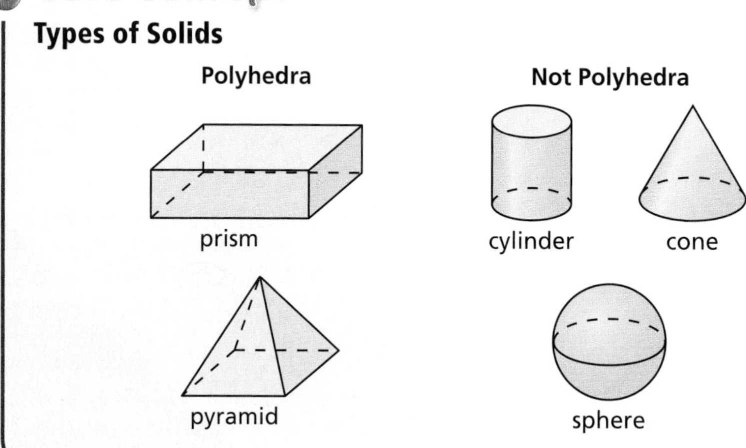

Pentagonal prism

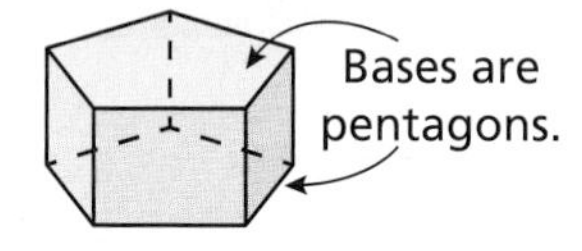

Triangular pyramid

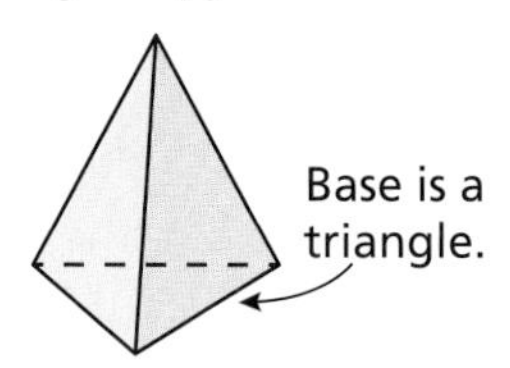

To name a prism or a pyramid, use the shape of the *base*. The two bases of a prism are congruent polygons in parallel planes. For example, the bases of a pentagonal prism are pentagons. The base of a pyramid is a polygon. For example, the base of a triangular pyramid is a triangle.

EXAMPLE 1 Classifying Solids

Tell whether each solid is a polyhedron. If it is, name the polyhedron.

a. 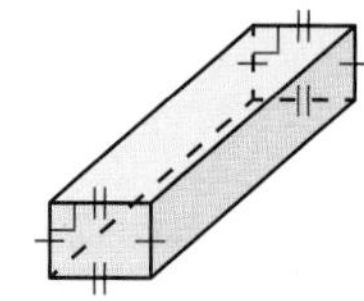

b. 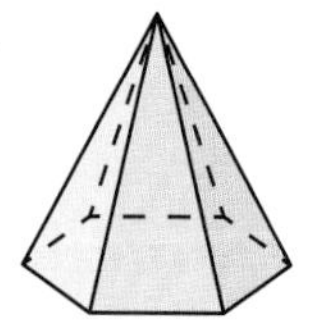

c. 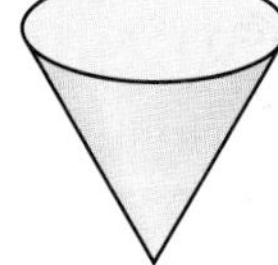

SOLUTION

a. The solid is formed by polygons, so it is a polyhedron. The two bases are congruent rectangles, so it is a rectangular prism.

b. The solid is formed by polygons, so it is a polyhedron. The base is a hexagon, so it is a hexagonal pyramid.

c. The cone has a curved surface, so it is not a polyhedron.

Monitoring Progress  Help in English and Spanish at *BigIdeasMath.com*

Tell whether the solid is a polyhedron. If it is, name the polyhedron.

1. 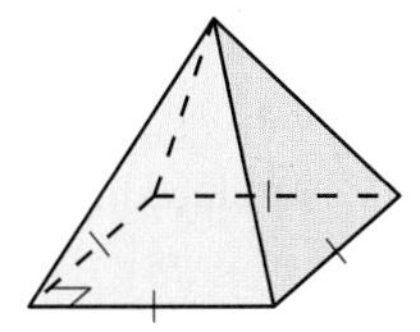

2. 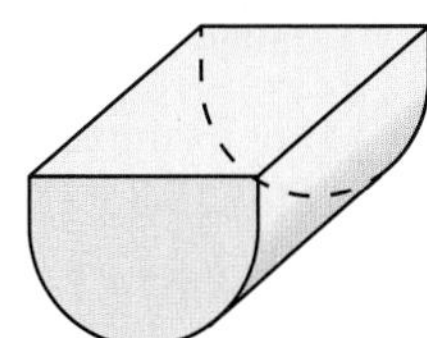

3. 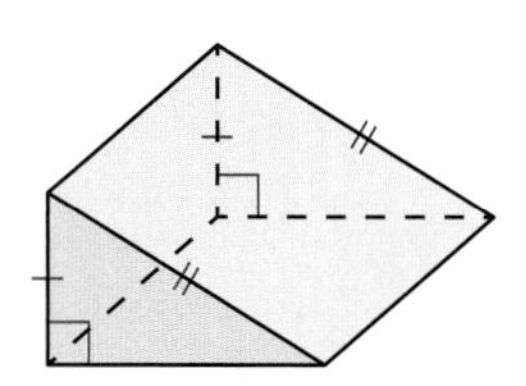

Describing Cross Sections

Imagine a plane slicing through a solid. The intersection of the plane and the solid is called a **cross section**. For example, three different cross sections of a cube are shown below.

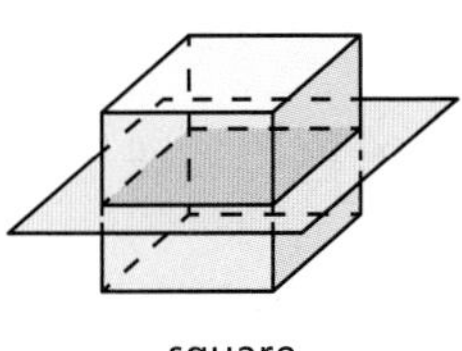

square

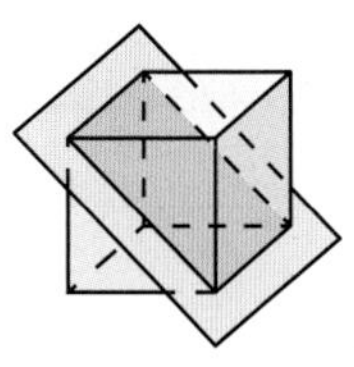

rectangle

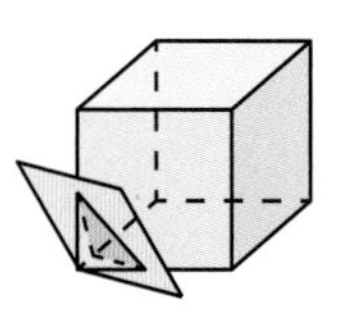

triangle

EXAMPLE 2 Describing Cross Sections

Describe the shape formed by the intersection of the plane and the solid.

a. 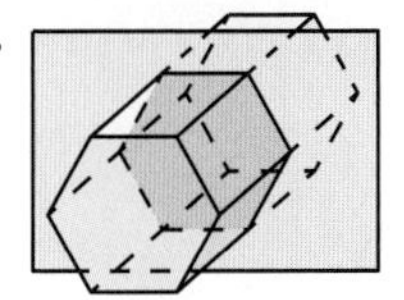

b. 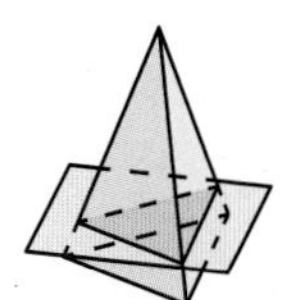

c. 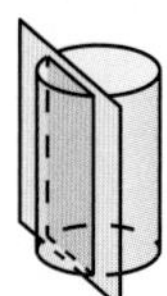

d. 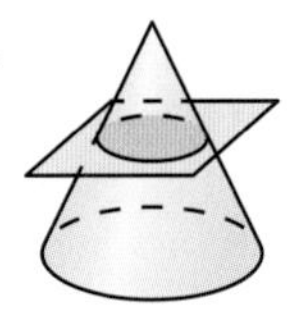

e. 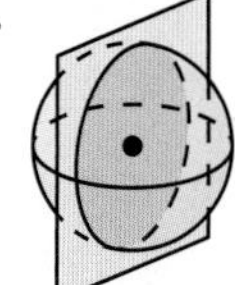

f. 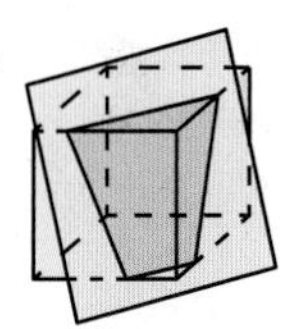

SOLUTION

a. The cross section is a hexagon.

b. The cross section is a triangle.

c. The cross section is a rectangle.

d. The cross section is a circle.

e. The cross section is a circle.

f. The cross section is a trapezoid.

Monitoring Progress  Help in English and Spanish at *BigIdeasMath.com*

Describe the shape formed by the intersection of the plane and the solid.

4. 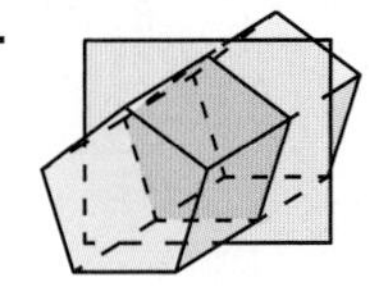

5. 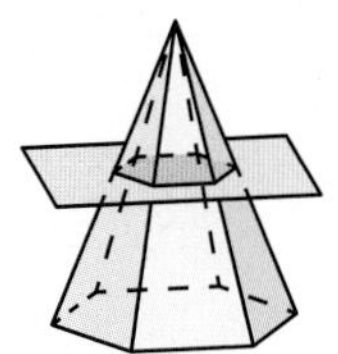

6. 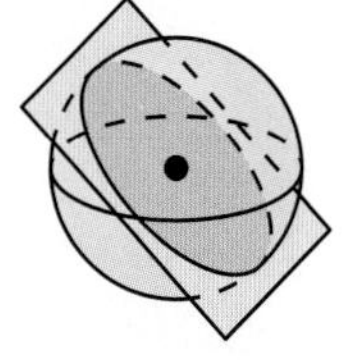

Sketching and Describing Solids of Revolution

A **solid of revolution** is a three-dimensional figure that is formed by rotating a two-dimensional shape around an axis. The line around which the shape is rotated is called the **axis of revolution**.

For example, when you rotate a rectangle around a line that contains one of its sides, the solid of revolution that is produced is a cylinder.

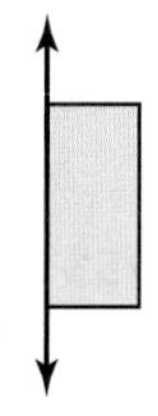
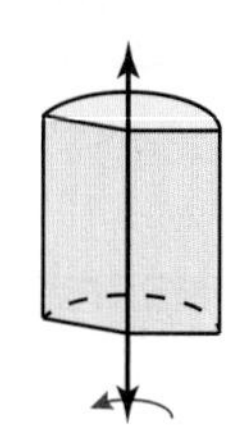
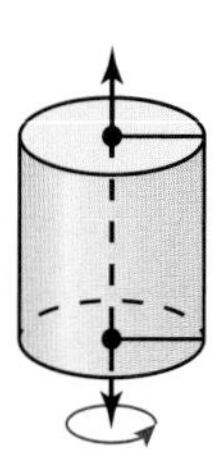

EXAMPLE 3 Sketching and Describing Solids of Revolution

Sketch the solid produced by rotating the figure around the given axis. Then identify and describe the solid.

a.

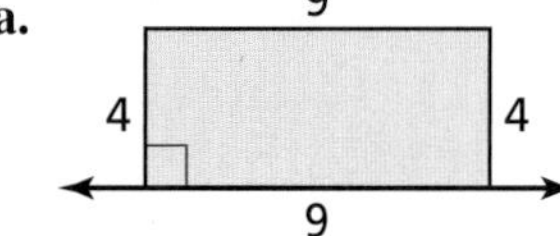

b.

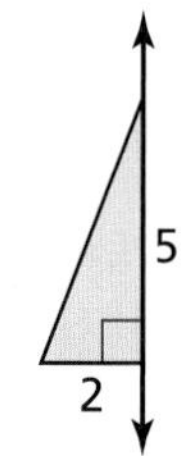

SOLUTION

a.

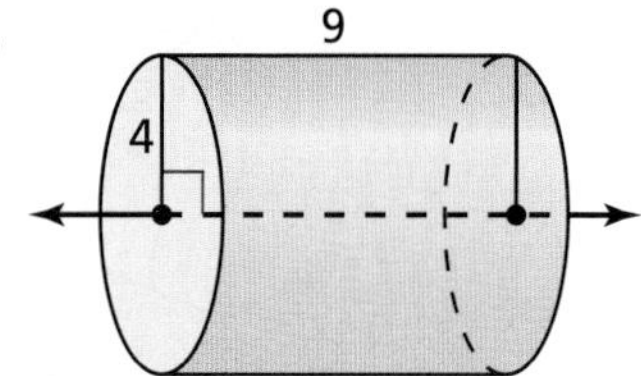

▶ The solid is a cylinder with a height of 9 and a base radius of 4.

b.

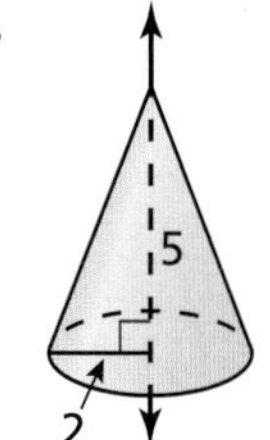

▶ The solid is a cone with a height of 5 and a base radius of 2.

Monitoring Progress  Help in English and Spanish at *BigIdeasMath.com*

Sketch the solid produced by rotating the figure around the given axis. Then identify and describe the solid.

7.

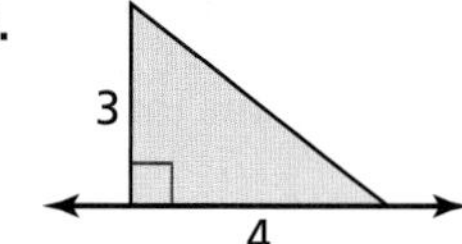

8.

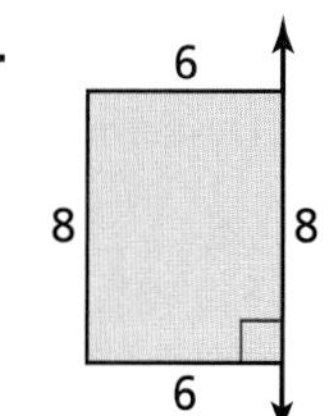

9.

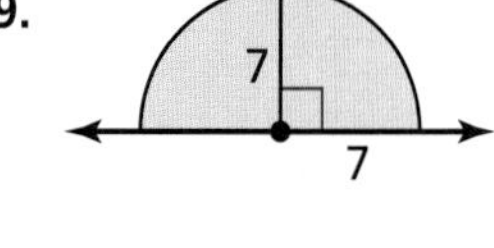

11.4 Exercises

Dynamic Solutions available at *BigIdeasMath.com*

Vocabulary and Core Concept Check

1. **VOCABULARY** A(n) __________ is a solid that is bounded by polygons.

2. **WHICH ONE DOESN'T BELONG?** Which solid does *not* belong with the other three? Explain your reasoning.

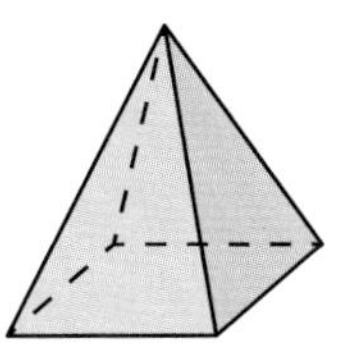
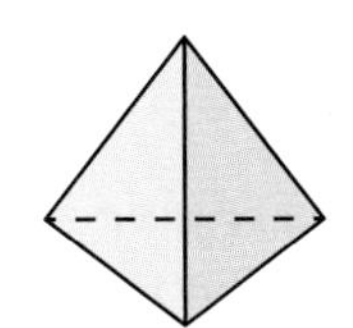
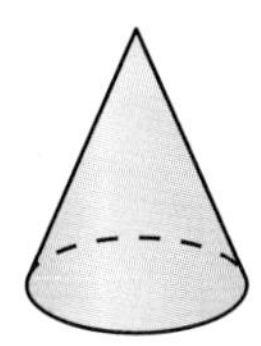
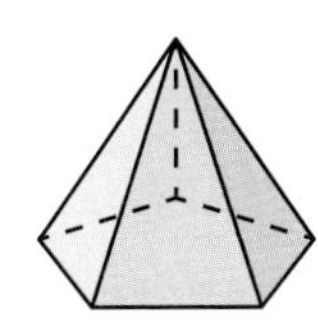

Monitoring Progress and Modeling with Mathematics

In Exercises 3–6, match the polyhedron with its name.

3. 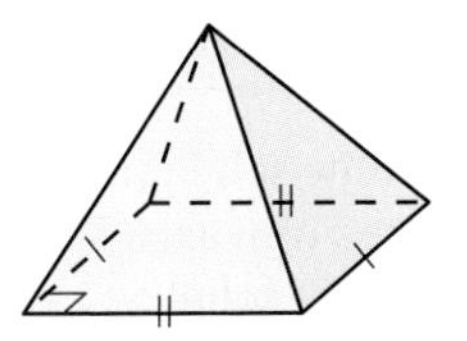

4. 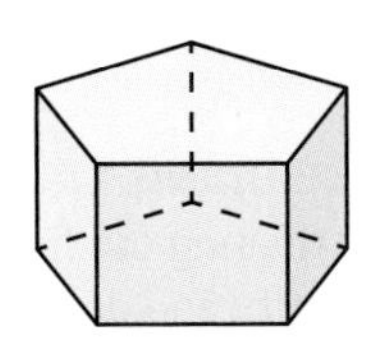

5. 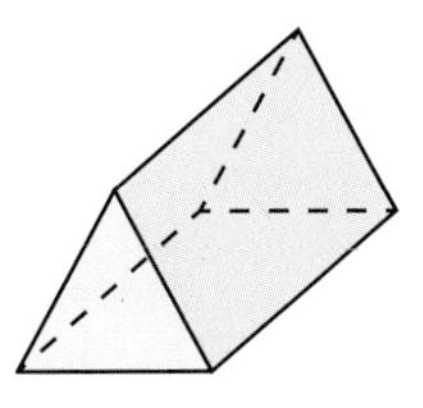

6. 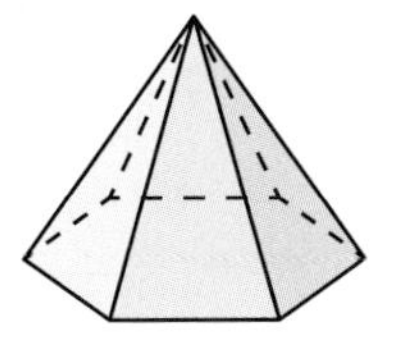

A. triangular prism

B. rectangular pyramid

C. hexagonal pyramid

D. pentagonal prism

In Exercises 7–10, tell whether the solid is a polyhedron. If it is, name the polyhedron. *(See Example 1.)*

7. 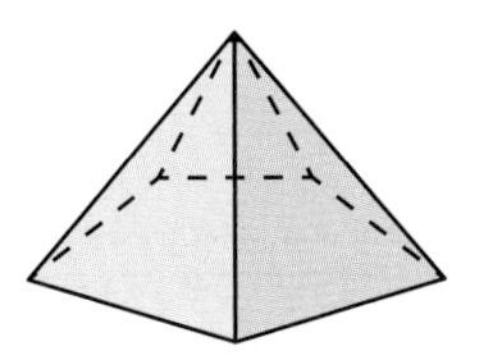

8. 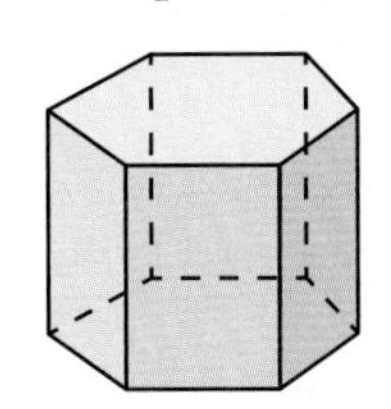

9. 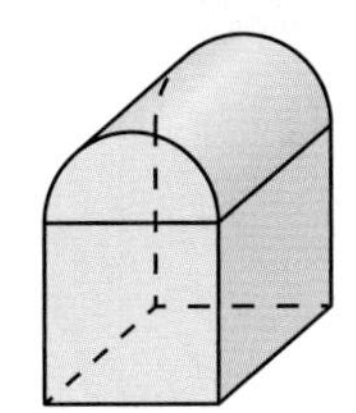

10. 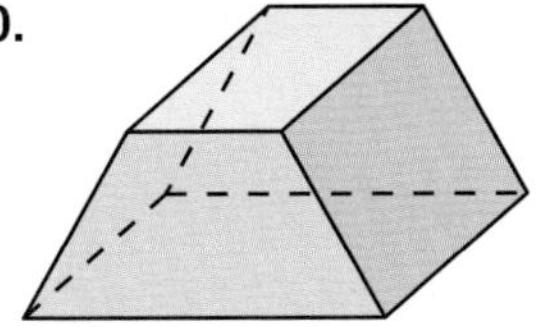

In Exercises 11–14, describe the cross section formed by the intersection of the plane and the solid. *(See Example 2.)*

11. 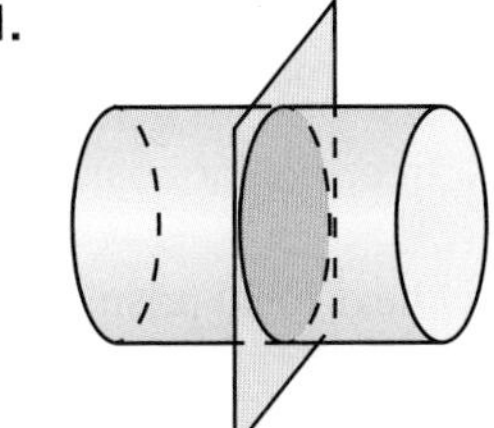

12. 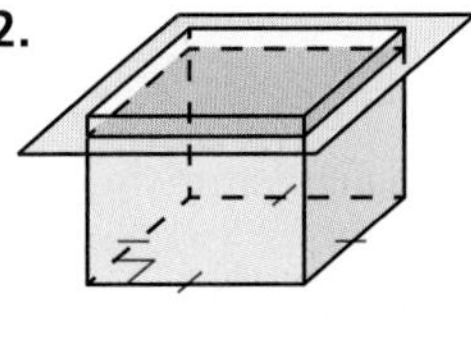

13. 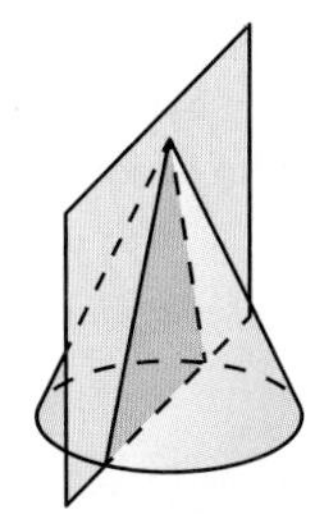

14. 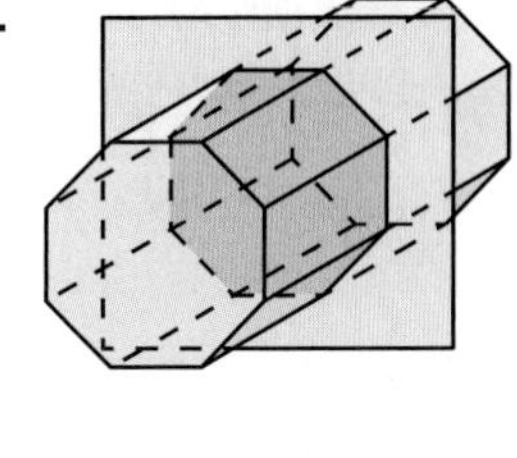

In Exercises 15–18, sketch the solid produced by rotating the figure around the given axis. Then identify and describe the solid. *(See Example 3.)*

15. 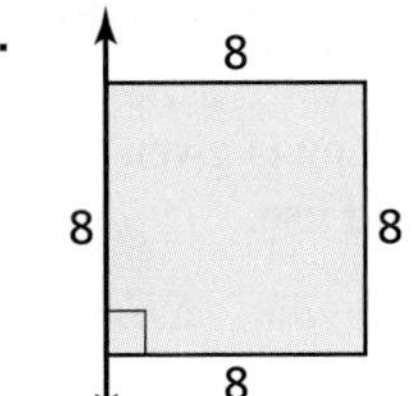

16. 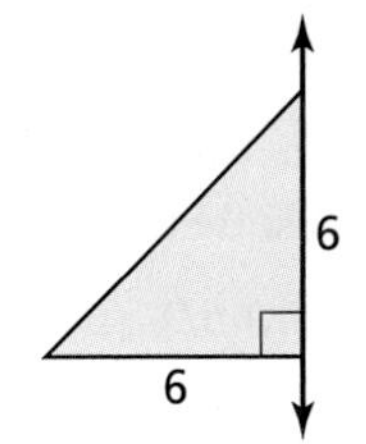

17. 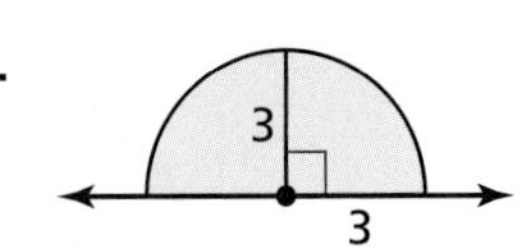

18. 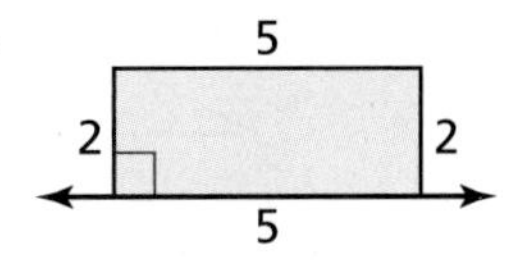

19. **ERROR ANALYSIS** Describe and correct the error in identifying the solid.

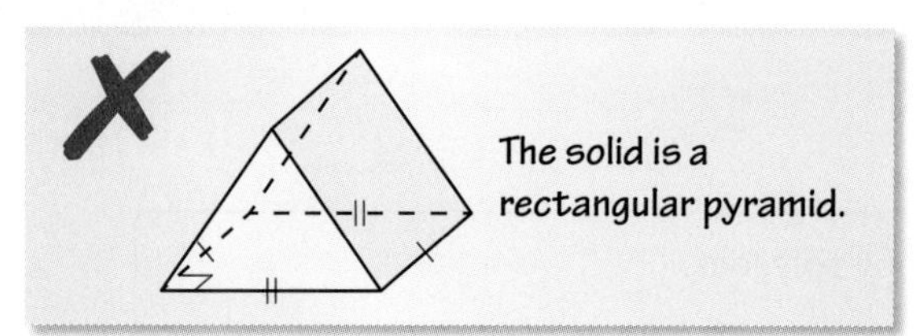

20. **HOW DO YOU SEE IT?** Is the swimming pool shown a polyhedron? If it is, name the polyhedron. If not, explain why not.

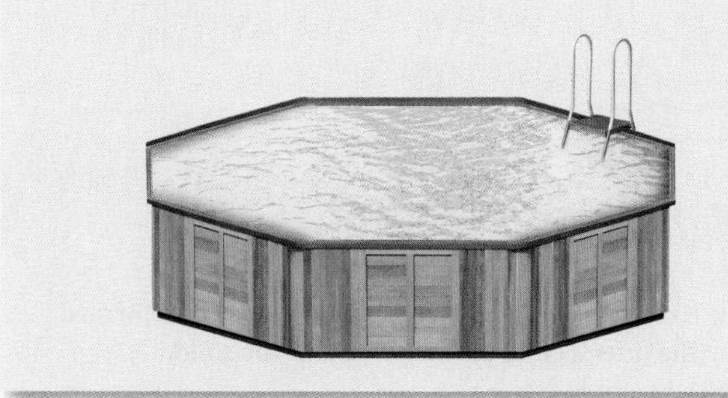

In Exercises 21–26, sketch the polyhedron.

21. triangular prism
22. rectangular prism
23. pentagonal prism
24. hexagonal prism
25. square pyramid
26. pentagonal pyramid

27. **MAKING AN ARGUMENT** Your friend says that the polyhedron shown is a triangular prism. Your cousin says that it is a triangular pyramid. Who is correct? Explain your reasoning.

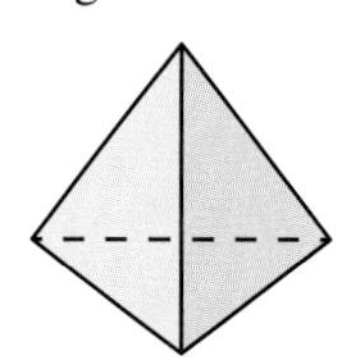

28. **ATTENDING TO PRECISION** The figure shows a plane intersecting a cube through four of its vertices. The edge length of the cube is 6 inches.

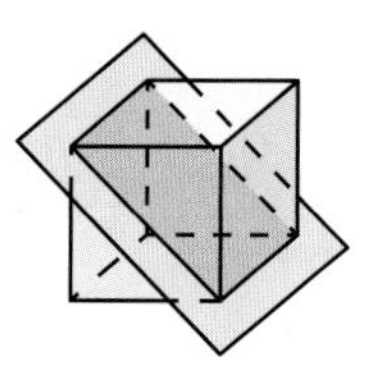

a. Describe the shape of the cross section.

b. What is the perimeter of the cross section?

c. What is the area of the cross section?

REASONING In Exercises 29–34, tell whether it is possible for a cross section of a cube to have the given shape. If it is, describe or sketch how the plane could intersect the cube.

29. circle
30. pentagon
31. rhombus
32. isosceles triangle
33. hexagon
34. scalene triangle

35. **REASONING** Sketch the composite solid produced by rotating the figure around the given axis. Then identify and describe the composite solid.

a.

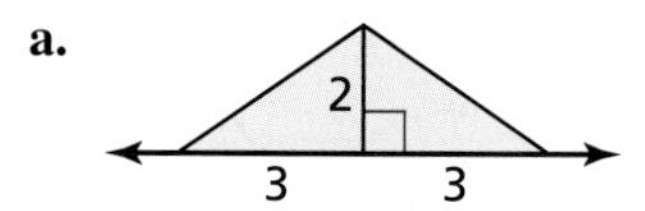

b.

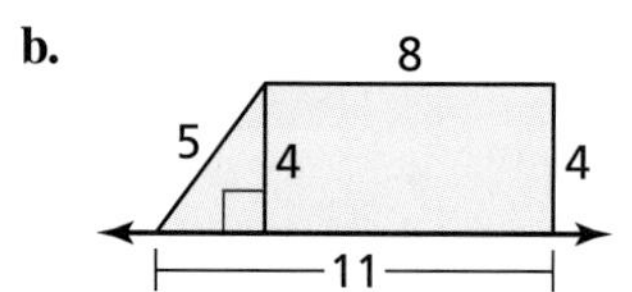

36. **THOUGHT PROVOKING** Describe how Plato might have argued that there are precisely five *Platonic Solids* (see page 617). (*Hint*: Consider the angles that meet at a vertex.)

Maintaining Mathematical Proficiency

Reviewing what you learned in previous grades and lessons

Decide whether enough information is given to prove that the triangles are congruent. If so, state the theorem you would use. *(Sections 5.3, 5.5, and 5.6)*

37. $\triangle ABD$, $\triangle CDB$

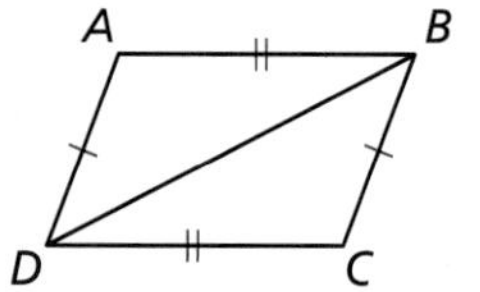

38. $\triangle JLK$, $\triangle JLM$

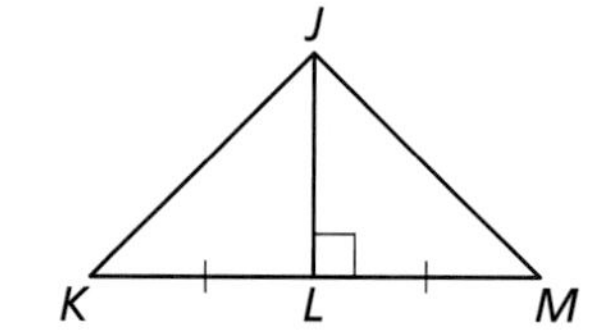

39. $\triangle RQP$, $\triangle RTS$

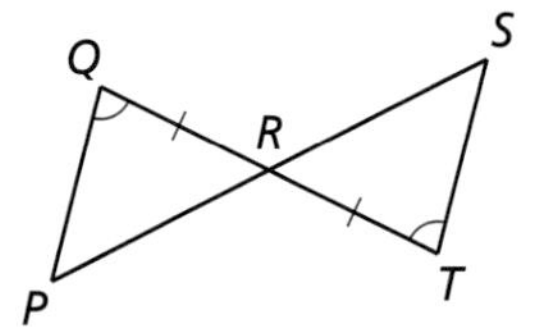

11.1–11.4 What Did You Learn?

Core Vocabulary

circumference, *p. 594*
arc length, *p. 595*
radian, *p. 597*
population density, *p. 603*
sector of a circle, *p. 604*
center of a regular polygon, *p. 611*
radius of a regular polygon, *p. 611*
apothem of a regular polygon, *p. 611*
central angle of a regular polygon, *p. 611*
polyhedron, *p. 618*
face, *p. 618*
edge, *p. 618*
vertex, *p. 618*
cross section, *p. 619*
solid of revolution, *p. 620*
axis of revolution, *p. 620*

Core Concepts

Section 11.1

Circumference of a Circle, *p. 594*
Arc Length, *p. 595*
Converting between Degrees and Radians, *p. 597*

Section 11.2

Area of a Circle, *p. 602*
Population Density, *p. 603*
Area of a Sector, *p. 604*

Section 11.3

Area of a Rhombus or Kite, *p. 610*
Area of a Regular Polygon, *p. 612*

Section 11.4

Types of Solids, *p. 618*
Cross Section of a Solid, *p. 619*
Solids of Revolution, *p. 620*

Mathematical Practices

1. In Exercise 13 on page 598, why does it matter how many revolutions the wheel makes?
2. Your friend is confused with Exercise 19 on page 606. What question(s) could you ask your friend to help them figure it out?
3. In Exercise 38 on page 615, write a proof to support your answer.

Study Skills

Kinesthetic Learners

Incorporate physical activity.

- Act out a word problem as much as possible. Use props when you can.
- Solve a word problem on a large whiteboard. The physical action of writing is more kinesthetic when the writing is larger and you can move around while doing it.
- Make a review card.

11.1–11.4 Quiz

Find the indicated measure. *(Section 11.1)*

1. $m\widehat{EF}$

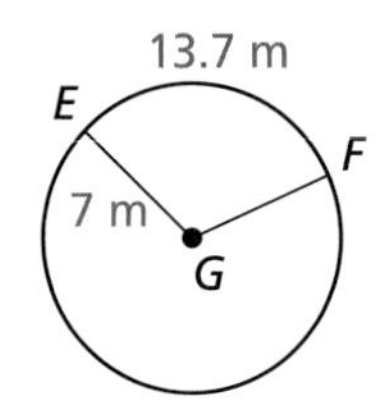

2. arc length of $\widehat{QS}$

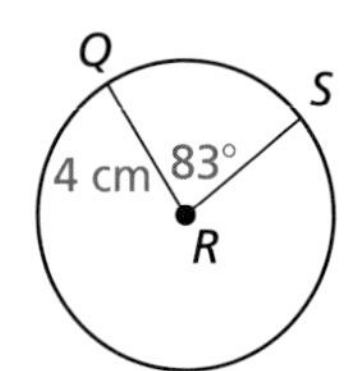

3. circumference of $\odot N$

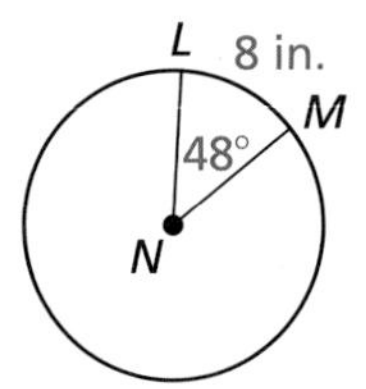

4. Convert 26° to radians and $\frac{5\pi}{9}$ radians to degrees. *(Section 11.1)*

Use the figure to find the indicated measure. *(Section 11.2)*

5. area of red sector

6. area of blue sector

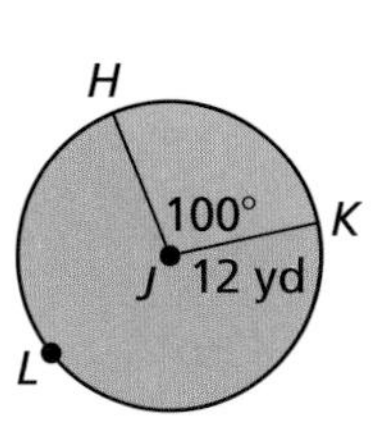

In the diagram, *RSTUVWXY* is a regular octagon inscribed in $\odot C$. *(Section 11.3)*

7. Identify the center, a radius, an apothem, and a central angle of the polygon.

8. Find $m\angle RCY$, $m\angle RCZ$, and $m\angle ZRC$.

9. The radius of the circle is 8 units. Find the area of the octagon.

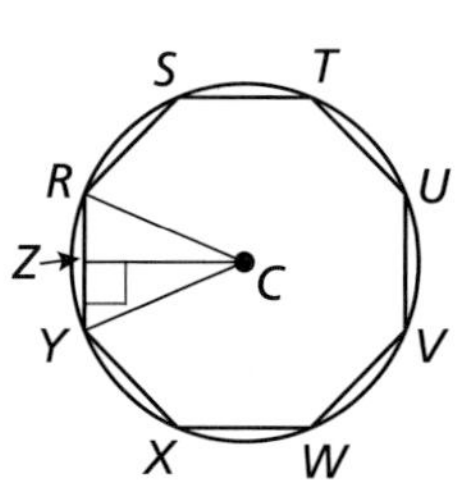

Tell whether the solid is a polyhedron. If it is, name the polyhedron. *(Section 11.4)*

10.

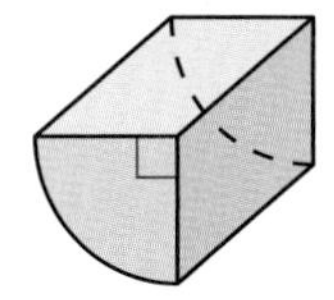

11.

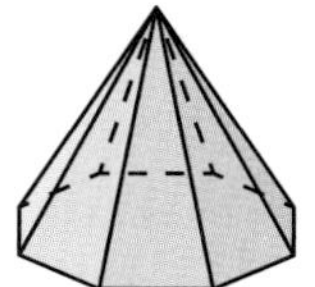

12.

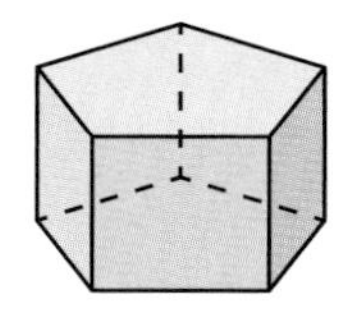

13. Sketch the composite solid produced by rotating the figure around the given axis. Then identify and describe the composite solid. *(Section 11.4)*

14. The two white congruent circles just fit into the blue circle. What is the area of the blue region? *(Section 11.2)*

15. Find the area of each rhombus tile. Then find the area of the pattern. *(Section 11.3)*

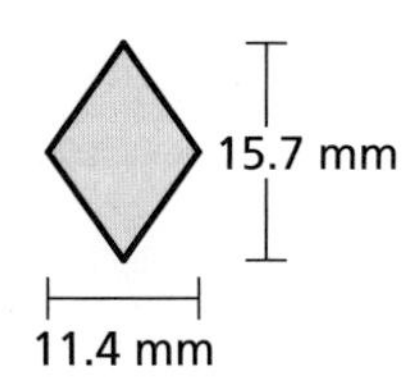

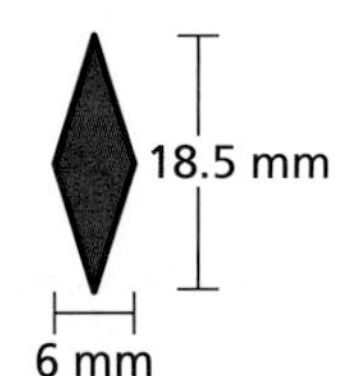

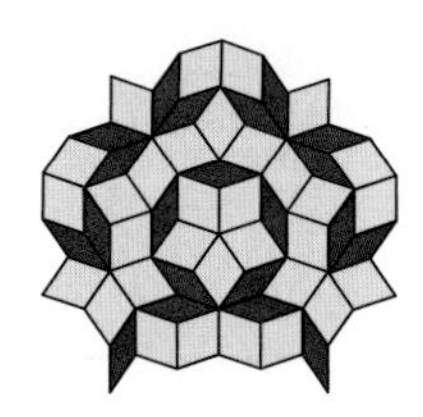

11.5 Volumes of Prisms and Cylinders

Essential Question How can you find the volume of a prism or cylinder that is not a right prism or right cylinder?

Recall that the volume V of a right prism or a right cylinder is equal to the product of the area of a base B and the height h.

$$V = Bh$$

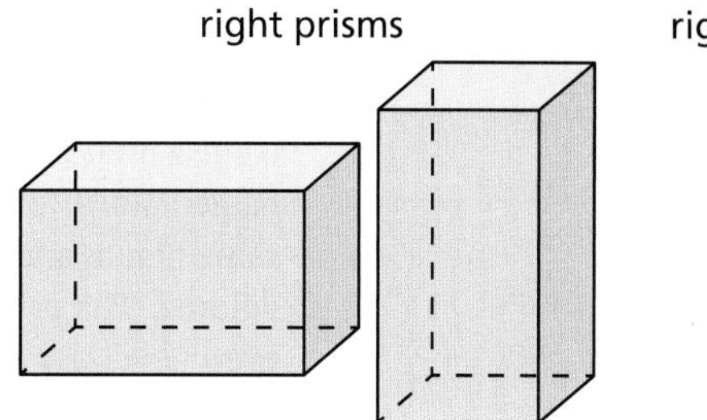

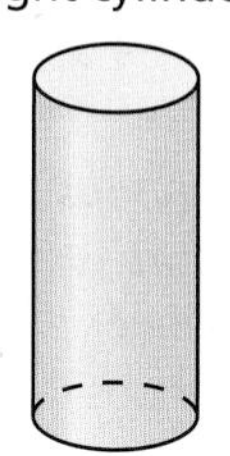

EXPLORATION 1 **Finding Volume**

Work with a partner. Consider a stack of square papers that is in the form of a right prism.

a. What is the volume of the prism?

b. When you twist the stack of papers, as shown at the right, do you change the volume? Explain your reasoning.

c. Write a carefully worded conjecture that describes the conclusion you reached in part (b).

d. Use your conjecture to find the volume of the twisted stack of papers.

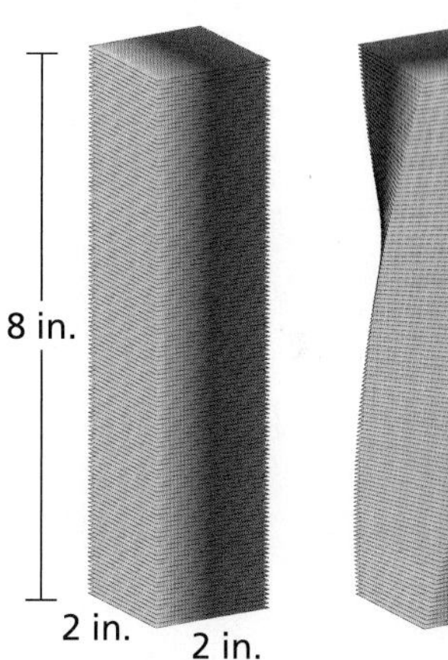

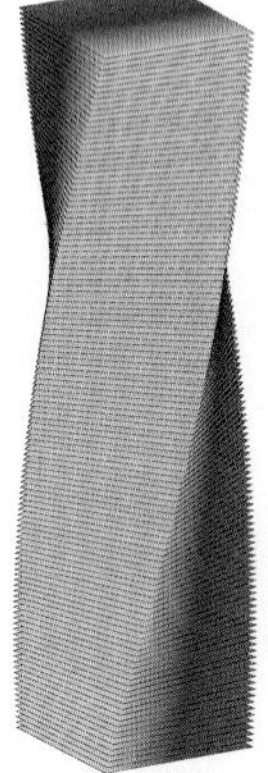

ATTENDING TO PRECISION

To be proficient in math, you need to communicate precisely to others.

EXPLORATION 2 **Finding Volume**

Work with a partner. Use the conjecture you wrote in Exploration 1 to find the volume of the cylinder.

a.

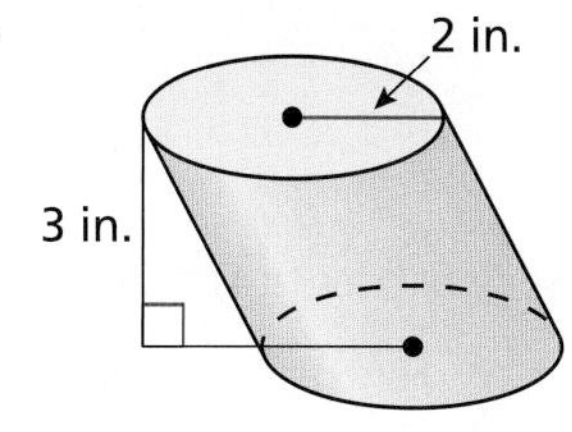

b.

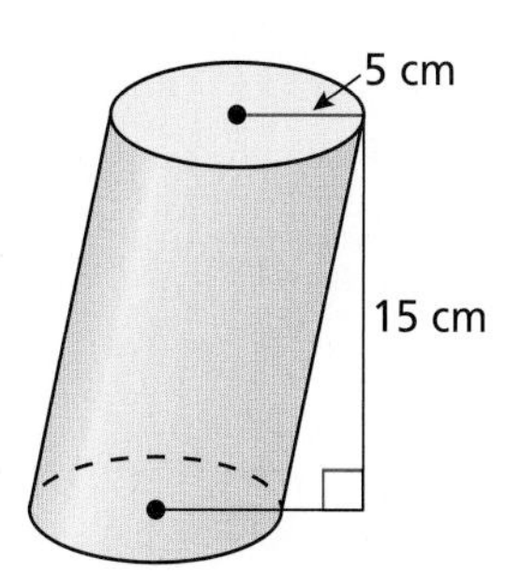

Communicate Your Answer

3. How can you find the volume of a prism or cylinder that is not a right prism or right cylinder?

4. In Exploration 1, would the conjecture you wrote change if the papers in each stack were not squares? Explain your reasoning.

11.5 Lesson

Core Vocabulary

volume, *p. 626*
Cavalieri's Principle, *p. 626*
density, *p. 628*
similar solids, *p. 630*

Previous
prism
cylinder
composite solid

What You Will Learn

- Find volumes of prisms and cylinders.
- Use the formula for density.
- Use volumes of prisms and cylinders.

Finding Volumes of Prisms and Cylinders

The **volume** of a solid is the number of cubic units contained in its interior. Volume is measured in cubic units, such as cubic centimeters (cm^3). **Cavalieri's Principle**, named after Bonaventura Cavalieri (1598–1647), states that if two solids have the same height and the same cross-sectional area at every level, then they have the same volume. The prisms below have equal heights h and equal cross-sectional areas B at every level. By Cavalieri's Principle, the prisms have the same volume.

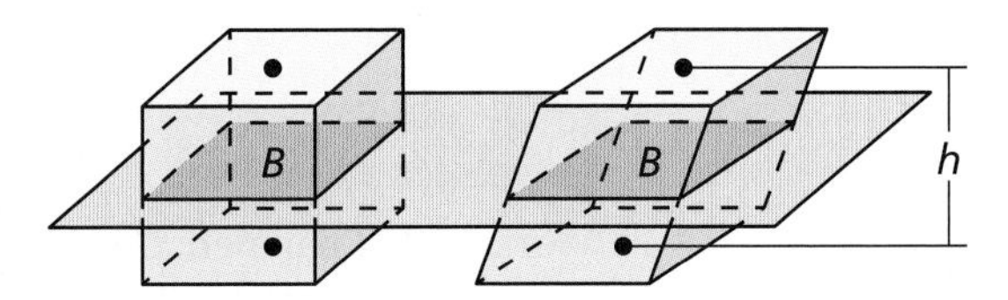

Core Concept

Volume of a Prism

The volume V of a prism is

$$V = Bh$$

where B is the area of a base and h is the height.

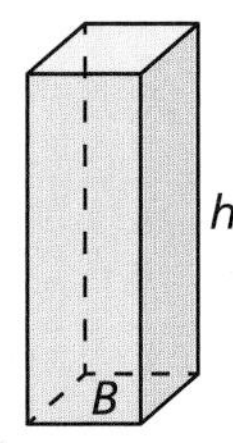

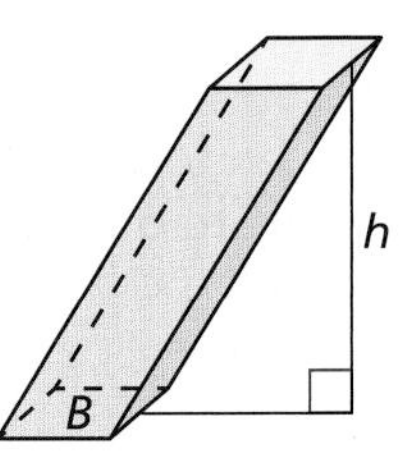

EXAMPLE 1 Finding Volumes of Prisms

Find the volume of each prism.

a.

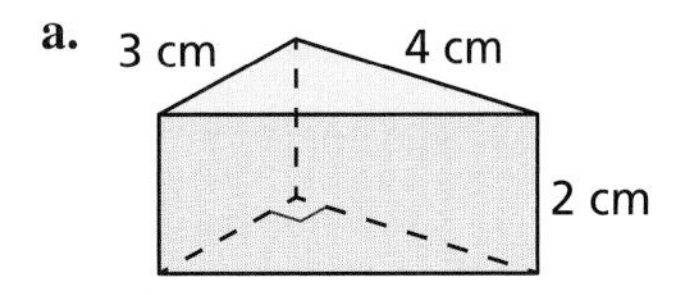

b.

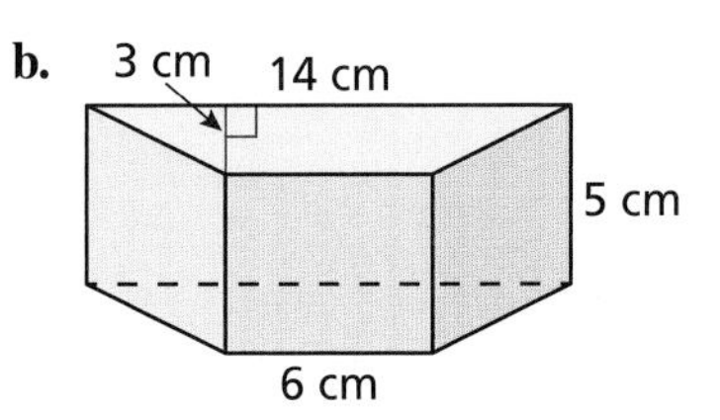

SOLUTION

a. The area of a base is $B = \frac{1}{2}(3)(4) = 6\ \text{cm}^2$ and the height is $h = 2$ cm.

$V = Bh$	Formula for volume of a prism
$= 6(2)$	Substitute.
$= 12$	Simplify.

▶ The volume is 12 cubic centimeters.

b. The area of a base is $B = \frac{1}{2}(3)(6 + 14) = 30\ \text{cm}^2$ and the height is $h = 5$ cm.

$V = Bh$	Formula for volume of a prism
$= 30(5)$	Substitute.
$= 150$	Simplify.

▶ The volume is 150 cubic centimeters.

Consider a cylinder with height h and base radius r and a rectangular prism with the same height that has a square base with sides of length $r\sqrt{\pi}$.

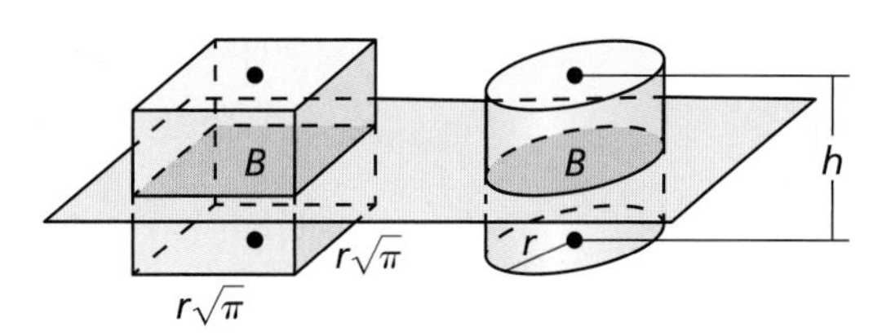

The cylinder and the prism have the same cross-sectional area, πr^2, at every level and the same height. By Cavalieri's Principle, the prism and the cylinder have the same volume. The volume of the prism is $V = Bh = \pi r^2 h$, so the volume of the cylinder is also $V = Bh = \pi r^2 h$.

Volume of a Cylinder

The volume V of a cylinder is

$$V = Bh = \pi r^2 h$$

where B is the area of a base, h is the height, and r is the radius of a base.

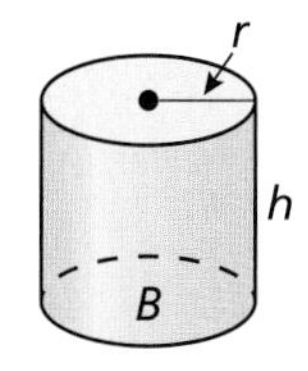

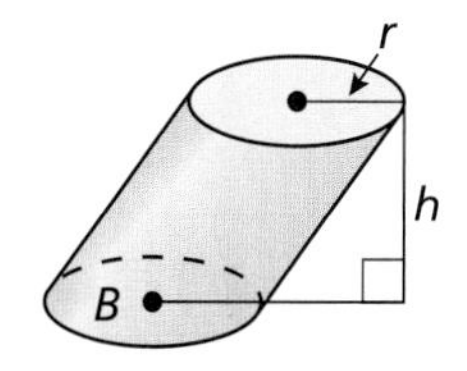

EXAMPLE 2 Finding Volumes of Cylinders

Find the volume of each cylinder.

a.

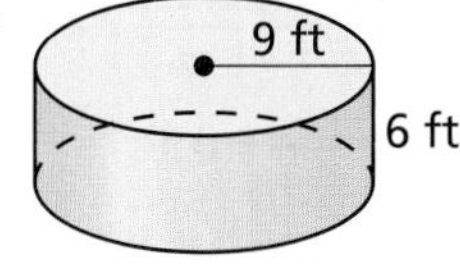

b.

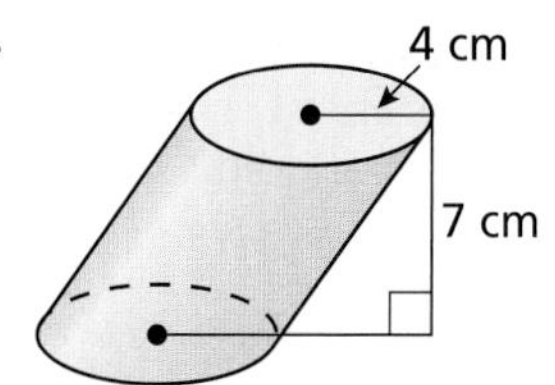

SOLUTION

a. The dimensions of the cylinder are $r = 9$ ft and $h = 6$ ft.

$$V = \pi r^2 h = \pi(9)^2(6) = 486\pi \approx 1526.81$$

▶ The volume is 486π, or about 1526.81 cubic feet.

b. The dimensions of the cylinder are $r = 4$ cm and $h = 7$ cm.

$$V = \pi r^2 h = \pi(4)^2(7) = 112\pi \approx 351.86$$

▶ The volume is 112π, or about 351.86 cubic centimeters.

Monitoring Progress

Help in English and Spanish at *BigIdeasMath.com*

Find the volume of the solid.

1.

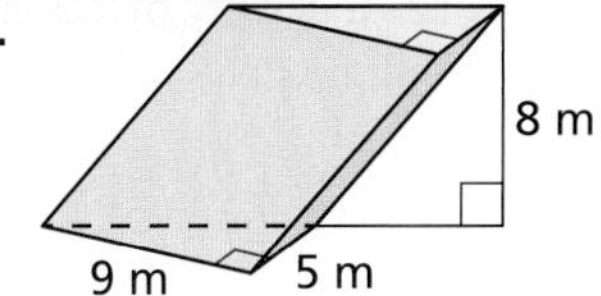

2.

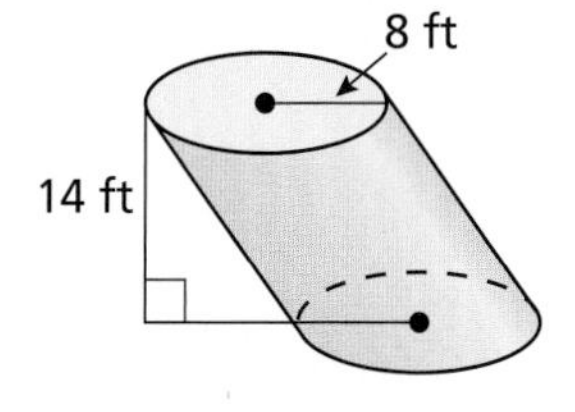

Using the Formula for Density

Density is the amount of matter that an object has in a given unit of volume. The density of an object is calculated by dividing its mass by its volume.

$$\text{Density} = \frac{\text{Mass}}{\text{Volume}}$$

Different materials have different densities, so density can be used to distinguish between materials that look similar. For example, table salt and sugar look alike. However, table salt has a density of 2.16 grams per cubic centimeter, while sugar has a density of 1.58 grams per cubic centimeter.

EXAMPLE 3 Using the Formula for Density

The diagram shows the dimensions of a standard gold bar at Fort Knox. Gold has a density of 19.3 grams per cubic centimeter. Find the mass of a standard gold bar to the nearest gram.

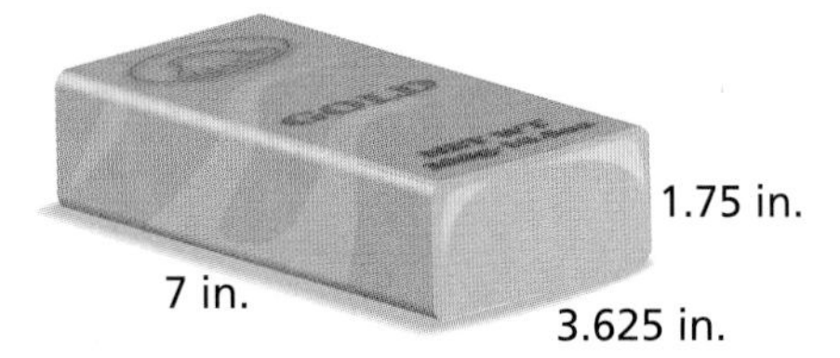

According to the U.S. Mint, Fort Knox houses about 9.2 million pounds of gold.

SOLUTION

Step 1 Convert the dimensions to centimeters using 1 inch = 2.54 centimeters.

$$\textbf{Length}\quad 7\,\cancel{\text{in.}} \cdot \frac{2.54\text{ cm}}{1\,\cancel{\text{in.}}} = 17.78\text{ cm}$$

$$\textbf{Width}\quad 3.625\,\cancel{\text{in.}} \cdot \frac{2.54\text{ cm}}{1\,\cancel{\text{in.}}} = 9.2075\text{ cm}$$

$$\textbf{Height}\quad 1.75\,\cancel{\text{in.}} \cdot \frac{2.54\text{ cm}}{1\,\cancel{\text{in.}}} = 4.445\text{ cm}$$

Step 2 Find the volume.

The area of a base is $B = 17.78(9.2075) = 163.70935\text{ cm}^2$ and the height is $h = 4.445$ cm.

$$V = Bh = 163.70935(4.445) \approx 727.69\text{ cm}^3$$

Step 3 Let x represent the mass in grams. Substitute the values for the volume and the density in the formula for density and solve for x.

$\text{Density} = \frac{\text{Mass}}{\text{Volume}}$	Formula for density
$19.3 \approx \frac{x}{727.69}$	Substitute.
$14{,}044 \approx x$	Multiply each side by 727.69.

▶ The mass of a standard gold bar is about 14,044 grams.

Monitoring Progress  Help in English and Spanish at *BigIdeasMath.com*

3. The diagram shows the dimensions of a concrete cylinder. Concrete has a density of 2.3 grams per cubic centimeter. Find the mass of the concrete cylinder to the nearest gram.

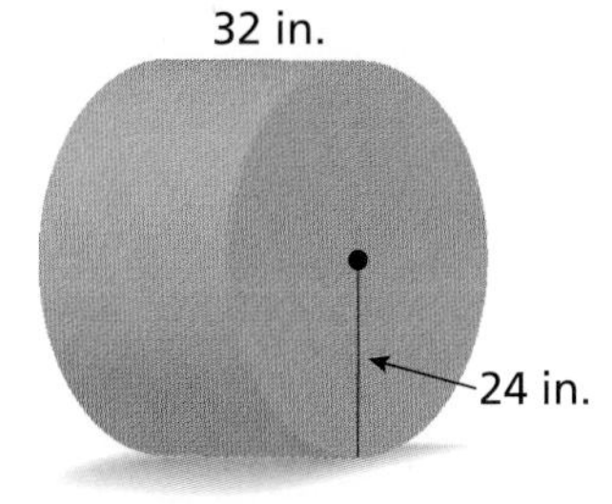

Using Volumes of Prisms and Cylinders

EXAMPLE 4 Modeling with Mathematics

You are building a rectangular chest. You want the length to be 6 feet, the width to be 4 feet, and the volume to be 72 cubic feet. What should the height be?

$V = 72\text{ ft}^3$

h

6 ft

4 ft

SOLUTION

1. **Understand the Problem** You know the dimensions of the base of a rectangular prism and the volume. You are asked to find the height.
2. **Make a Plan** Write the formula for the volume of a rectangular prism, substitute known values, and solve for the height h.
3. **Solve the Problem** The area of a base is $B = 6(4) = 24\text{ ft}^2$ and the volume is $V = 72\text{ ft}^3$.

$V = Bh$	Formula for volume of a prism
$72 = 24h$	Substitute.
$3 = h$	Divide each side by 24.

▶ The height of the chest should be 3 feet.

4. **Look Back** Check your answer.

$V = Bh = 24(3) = 72$ ✓

EXAMPLE 5 Solving a Real-Life Problem

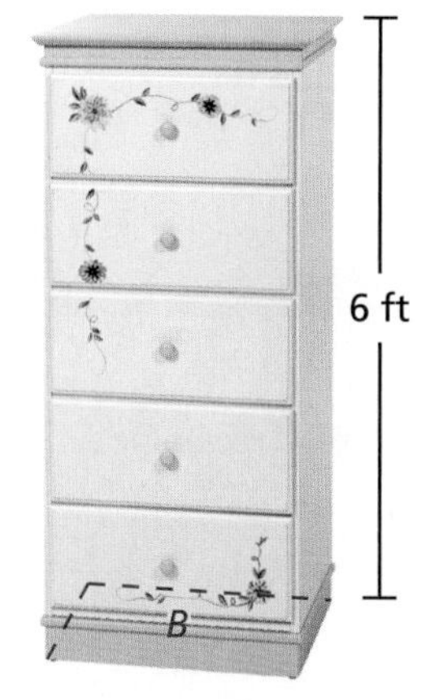

You are building a 6-foot-tall dresser. You want the volume to be 36 cubic feet. What should the area of the base be? Give a possible length and width.

SOLUTION

$V = Bh$	Formula for volume of a prism
$36 = B \cdot 6$	Substitute.
$6 = B$	Divide each side by 6.

▶ The area of the base should be 6 square feet. The length could be 3 feet and the width could be 2 feet.

Monitoring Progress  Help in English and Spanish at *BigIdeasMath.com*

4. **WHAT IF?** In Example 4, you want the length to be 5 meters, the width to be 3 meters, and the volume to be 60 cubic meters. What should the height be?
5. **WHAT IF?** In Example 5, you want the height to be 5 meters and the volume to be 75 cubic meters. What should the area of the base be? Give a possible length and width.

Core Concept

Similar Solids

Two solids of the same type with equal ratios of corresponding linear measures, such as heights or radii, are called **similar solids**. The ratio of the corresponding linear measures of two similar solids is called the *scale factor*. If two similar solids have a scale factor of k, then the ratio of their volumes is equal to k^3.

EXAMPLE 6 Finding the Volume of a Similar Solid

Cylinder A and cylinder B are similar. Find the volume of cylinder B.

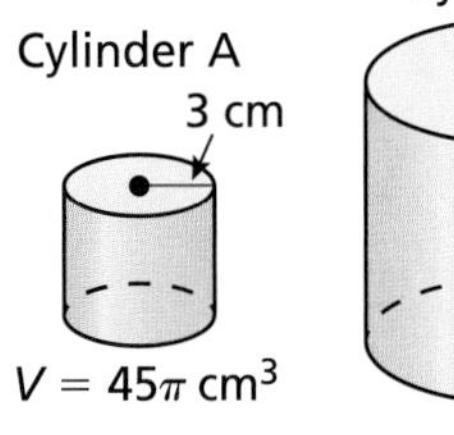

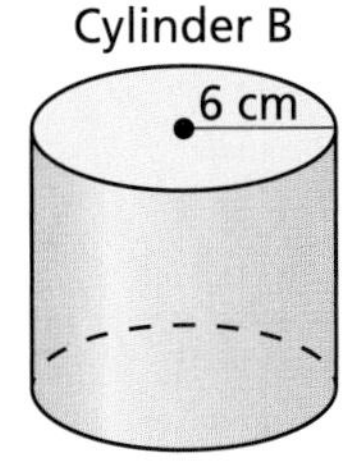

SOLUTION

The scale factor is $k = \dfrac{\text{Radius of cylinder B}}{\text{Radius of cylinder A}}$

$= \dfrac{6}{3} = 2.$

Use the scale factor to find the volume of cylinder B.

$\dfrac{\text{Volume of cylinder B}}{\text{Volume of cylinder A}} = k^3$ — The ratio of the volumes is k^3.

$\dfrac{\text{Volume of cylinder B}}{45\pi} = 2^3$ — Substitute.

$\text{Volume of cylinder B} = 360\pi$ — Solve for volume of cylinder B.

▶ The volume of cylinder B is 360π cubic centimeters.

COMMON ERROR

Be sure to write the ratio of the volumes in the same order you wrote the ratio of the radii.

Prism C

12 m

$V = 1536\ \text{m}^3$

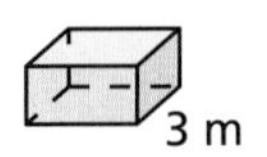

Monitoring Progress

Help in English and Spanish at *BigIdeasMath.com*

6. Prism C and prism D are similar. Find the volume of prism D.

EXAMPLE 7 Finding the Volume of a Composite Solid

Find the volume of the concrete block.

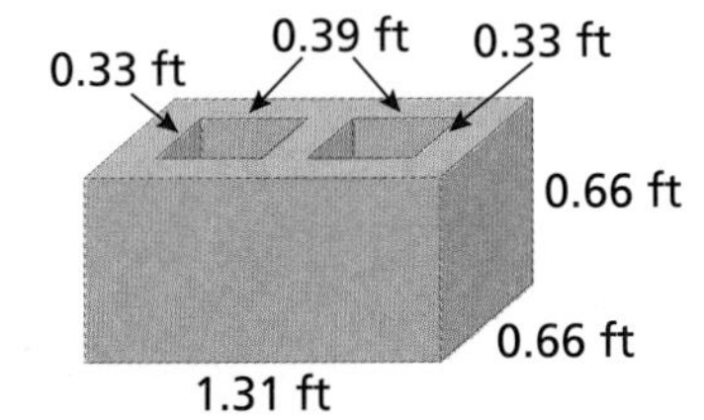

SOLUTION

To find the area of the base, subtract two times the area of the small rectangle from the large rectangle.

$B = \boxed{\text{Area of large rectangle}} - 2 \cdot \boxed{\text{Area of small rectangle}}$

$= 1.31(0.66) - 2(0.33)(0.39)$

$= 0.6072$

Using the formula for the volume of a prism, the volume is

$V = Bh = 0.6072(0.66) \approx 0.40.$

▶ The volume is about 0.40 cubic foot.

Monitoring Progress

Help in English and Spanish at *BigIdeasMath.com*

7. Find the volume of the composite solid.

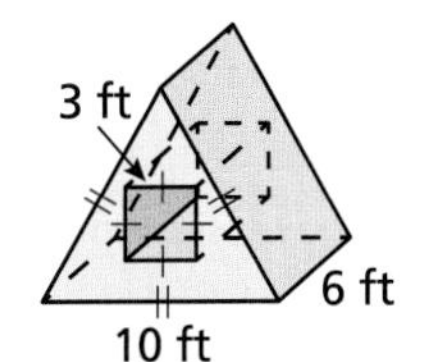

11.5 Exercises

Dynamic Solutions available at *BigIdeasMath.com*

Vocabulary and Core Concept Check

1. **VOCABULARY** In what type of units is the volume of a solid measured?

2. **COMPLETE THE SENTENCE** Density is the amount of ______ that an object has in a given unit of _________.

Monitoring Progress and Modeling with Mathematics

In Exercises 3–6, find the volume of the prism. *(See Example 1.)*

3. 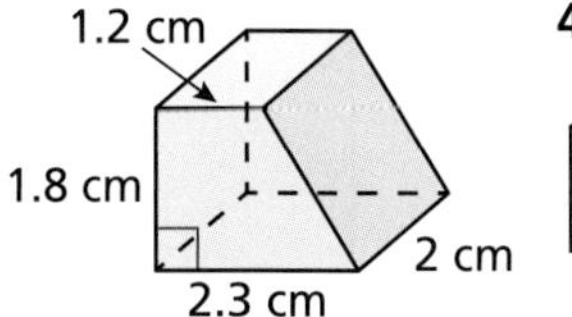

4. 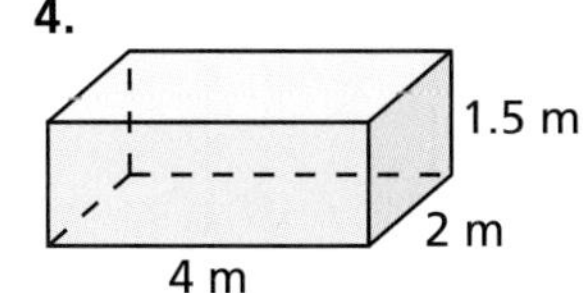

5. 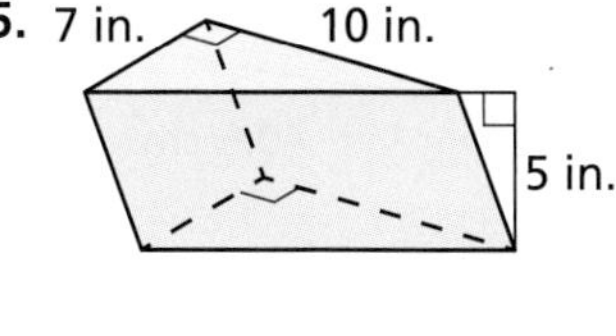

6. 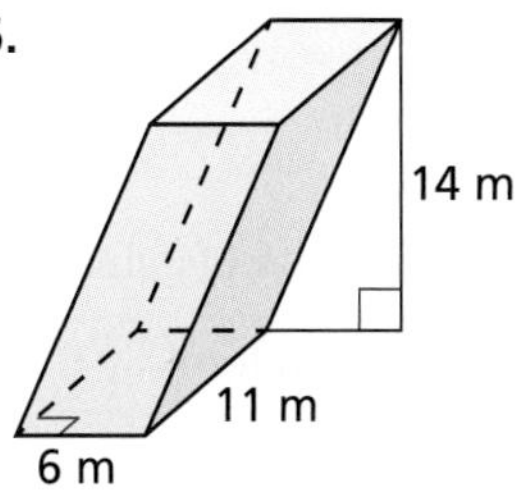

In Exercises 7–10, find the volume of the cylinder. *(See Example 2.)*

7. 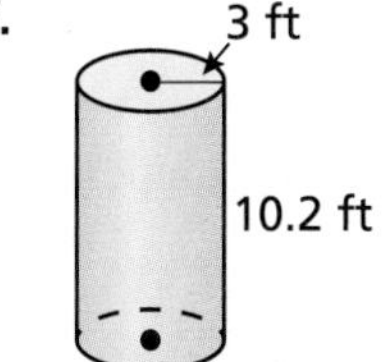

8. 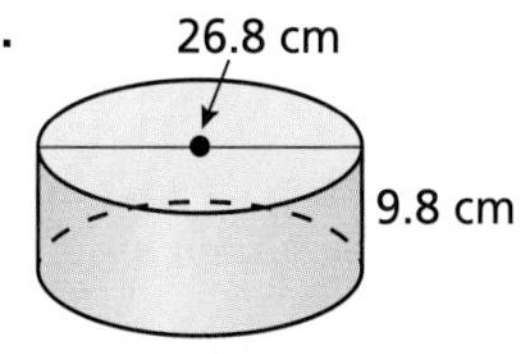

9. 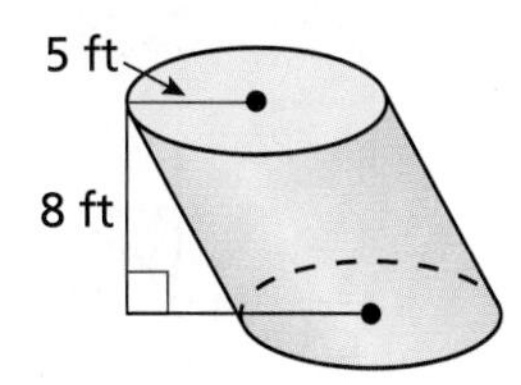

10. 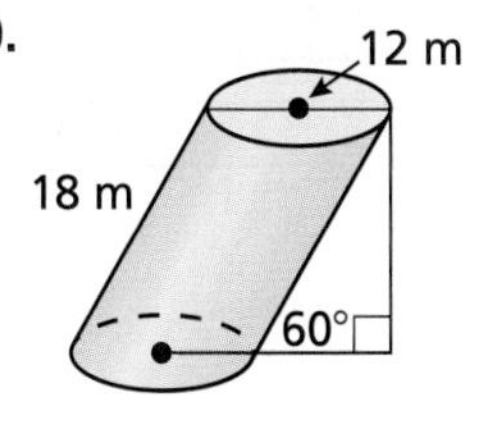

In Exercises 11 and 12, make a sketch of the solid and find its volume. Round your answer to the nearest hundredth.

11. A prism has a height of 11.2 centimeters and an equilateral triangle for a base, where each base edge is 8 centimeters.

12. A pentagonal prism has a height of 9 feet and each base edge is 3 feet.

13. **PROBLEM SOLVING** A piece of copper with a volume of 8.25 cubic centimeters has a mass of 73.92 grams. A piece of iron with a volume of 5 cubic centimeters has a mass of 39.35 grams. Which metal has the greater density?

14. **PROBLEM SOLVING** The United States has minted one-dollar silver coins called the American Eagle Silver Bullion Coin since 1986. Each coin has a diameter of 40.6 millimeters and is 2.98 millimeters thick. The density of silver is 10.5 grams per cubic centimeter. What is the mass of an American Eagle Silver Bullion Coin to the nearest gram? *(See Example 3.)*

15. **ERROR ANALYSIS** Describe and correct the error in finding the volume of the cylinder.

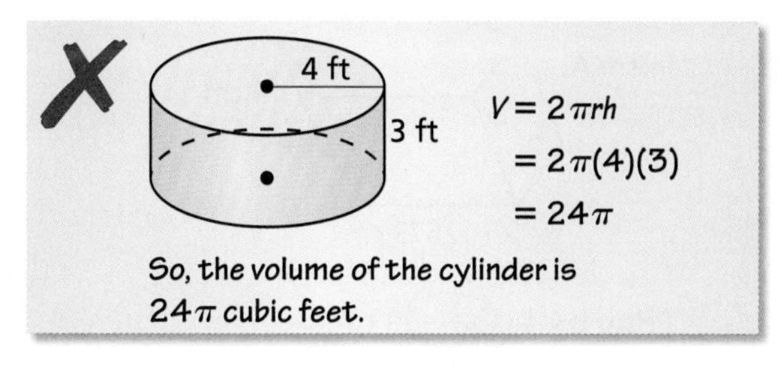

16. ERROR ANALYSIS Describe and correct the error in finding the density of an object that has a mass of 24 grams and a volume of 28.3 cubic centimeters.

$\text{density} = \frac{28.3}{24} \approx 1.18$

So, the density is about 1.18 cubic centimeters per gram.

In Exercises 17–22, find the missing dimension of the prism or cylinder. *(See Example 4.)*

17. Volume = 560 ft^3

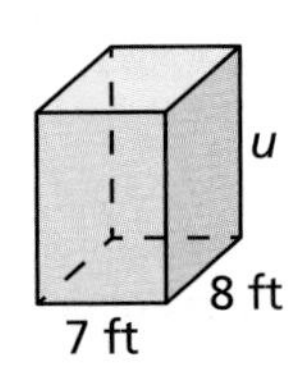

18. Volume = 2700 yd^3

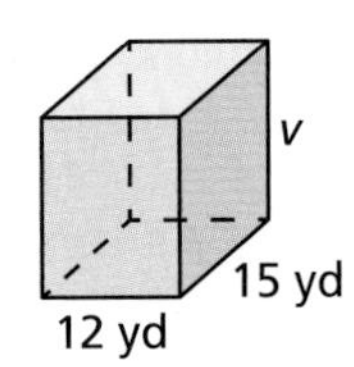

19. Volume = 80 cm^3

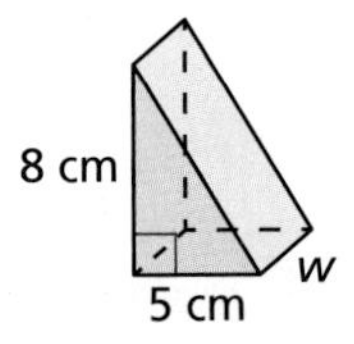

20. Volume = 72.66 in.^{3}

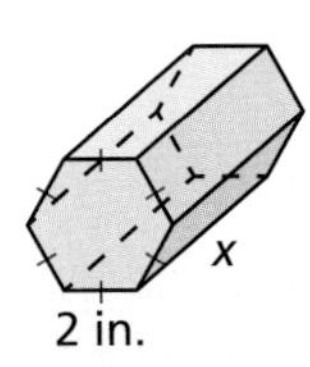

21. Volume = 3000 ft^3

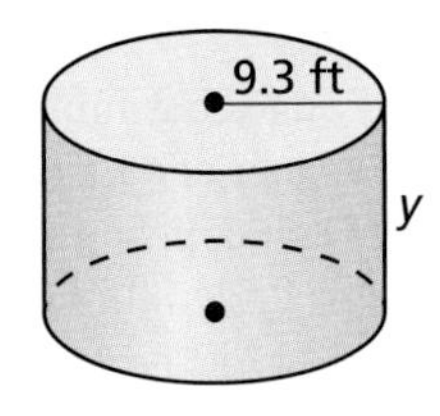

22. Volume = 1696.5 m^3

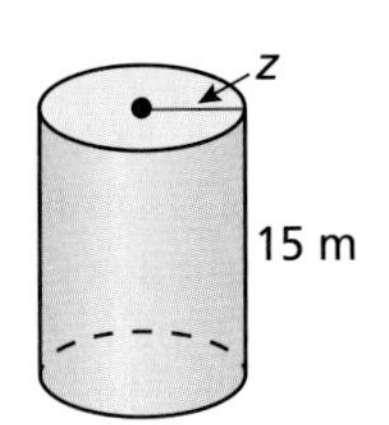

In Exercises 23 and 24, find the area of the base of the rectangular prism with the given volume and height. Then give a possible length and width. *(See Example 5.)*

23. $V = 154$ in.^{3}, $h = 11$ in.

24. $V = 27$ m^3, $h = 3$ m

In Exercises 25 and 26, the solids are similar. Find the volume of solid B. *(See Example 6.)*

25.

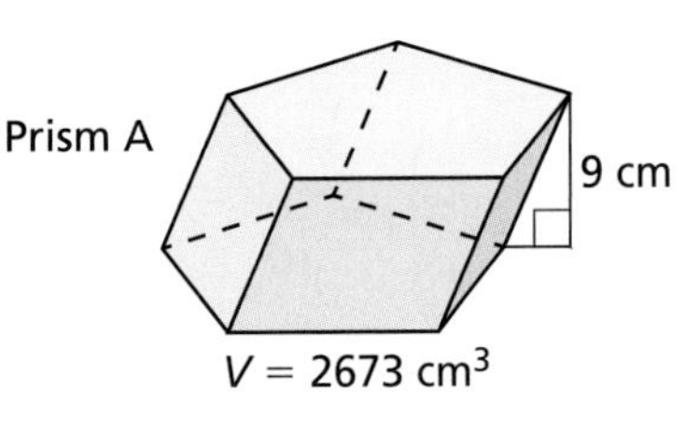

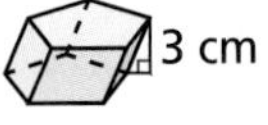

26.

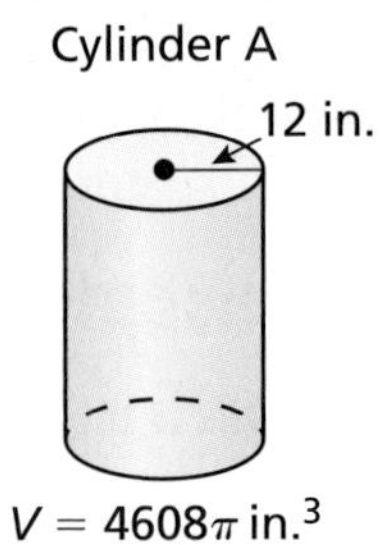

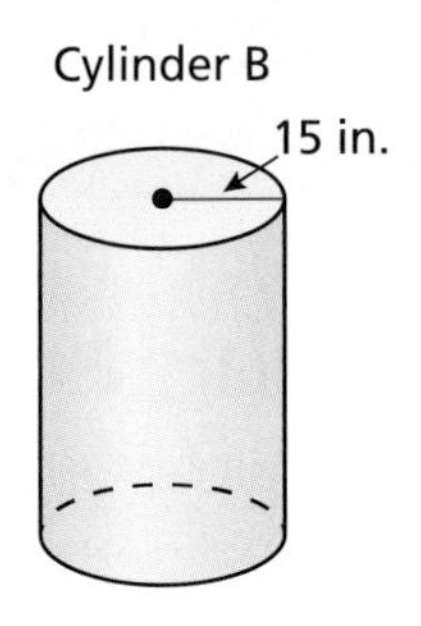

In Exercises 27 and 28, the solids are similar. Find the indicated measure.

27. height x of the base of prism A

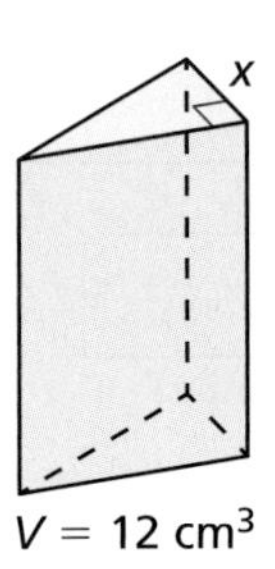

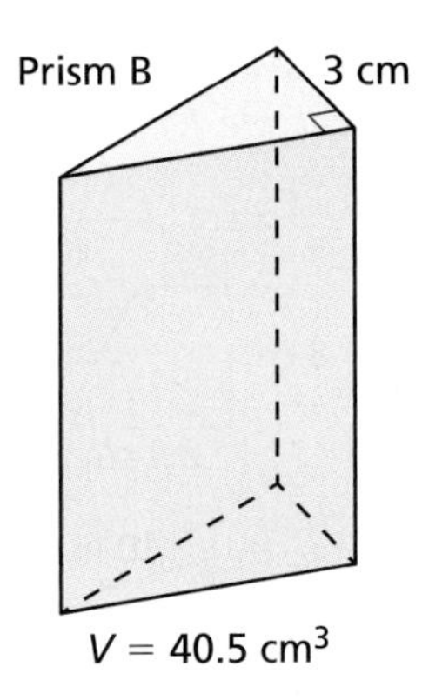

28. height h of cylinder B

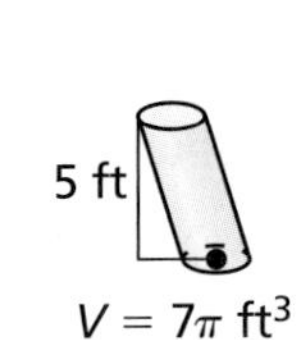

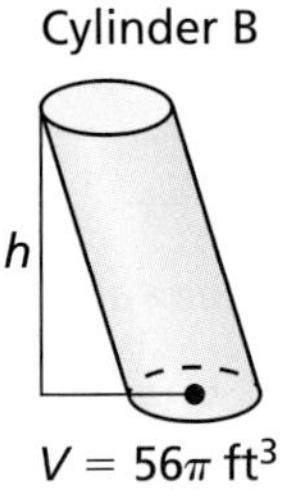

In Exercises 29–32, find the volume of the composite solid. *(See Example 7.)*

29.

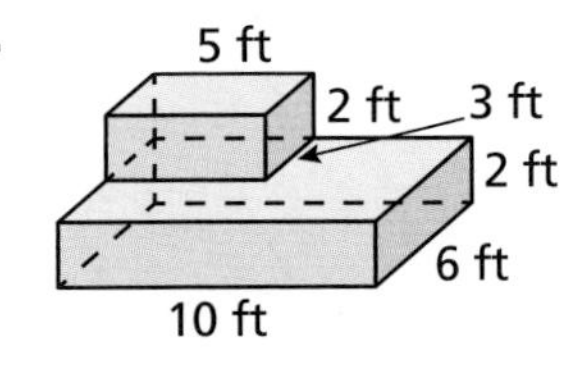

30.

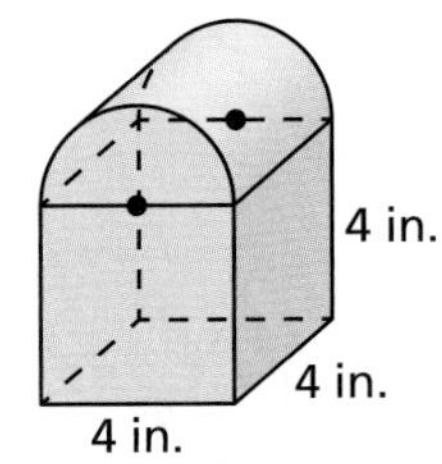

31.

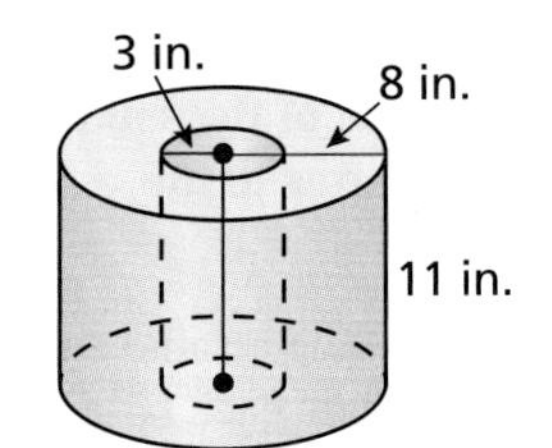

32.

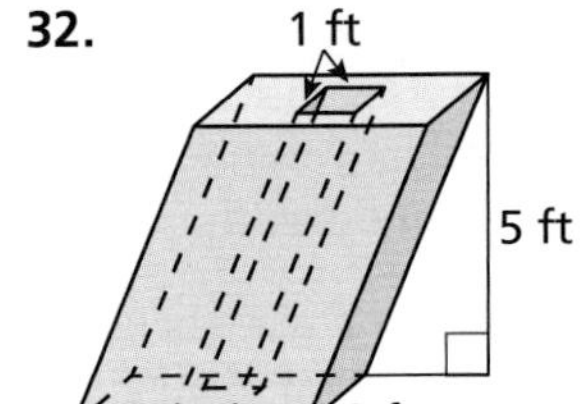

33. MODELING WITH MATHEMATICS The Great Blue Hole is a cylindrical trench located off the coast of Belize. It is approximately 1000 feet wide and 400 feet deep. About how many gallons of water does the Great Blue Hole contain? ($1 \text{ ft}^3 \approx 7.48$ gallons)

34. COMPARING METHODS The *Volume Addition Postulate* states that the volume of a solid is the sum of the volumes of all its nonoverlapping parts. Use this postulate to find the volume of the block of concrete in Example 7 by subtracting the volume of each hole from the volume of the large rectangular prism. Which method do you prefer? Explain your reasoning.

REASONING In Exercises 35 and 36, you are melting a rectangular block of wax to make candles. How many candles of the given shape can be made using a block that measures 10 centimeters by 9 centimeters by 20 centimeters?

35. **36.**

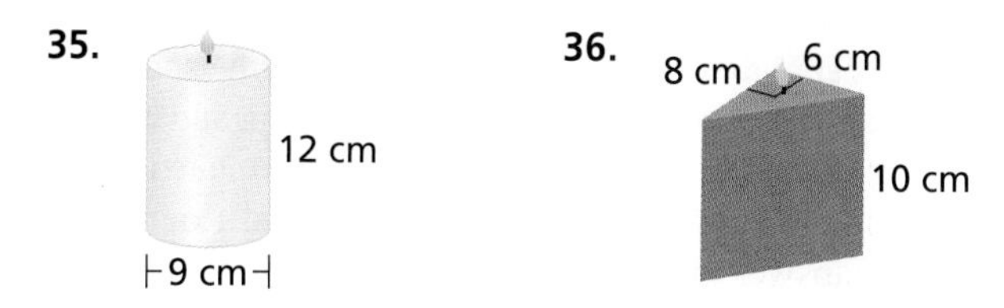

37. PROBLEM SOLVING An aquarium shaped like a rectangular prism has a length of 30 inches, a width of 10 inches, and a height of 20 inches. You fill the aquarium $\frac{3}{4}$ full with water. When you submerge a rock in the aquarium, the water level rises 0.25 inch.

a. Find the volume of the rock.

b. How many rocks of this size can you place in the aquarium before water spills out?

38. PROBLEM SOLVING You drop an irregular piece of metal into a container partially filled with water and measure that the water level rises 4.8 centimeters. The square base of the container has a side length of 8 centimeters. You measure the mass of the metal to be 450 grams. What is the density of the metal?

39. WRITING Both of the figures shown are made up of the same number of congruent rectangles. Explain how Cavalieri's Principle can be adapted to compare the areas of these figures.

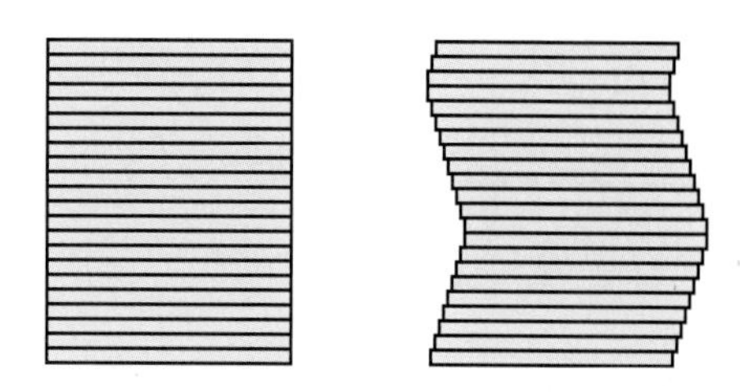

40. HOW DO YOU SEE IT? Each stack of memo papers contains 500 equally-sized sheets of paper. Compare their volumes. Explain your reasoning.

41. USING STRUCTURE Sketch the solid formed by the net. Then find the volume of the solid.

4

2.5

1.5

42. USING STRUCTURE Sketch the solid with the given views. Then find the volume of the solid.

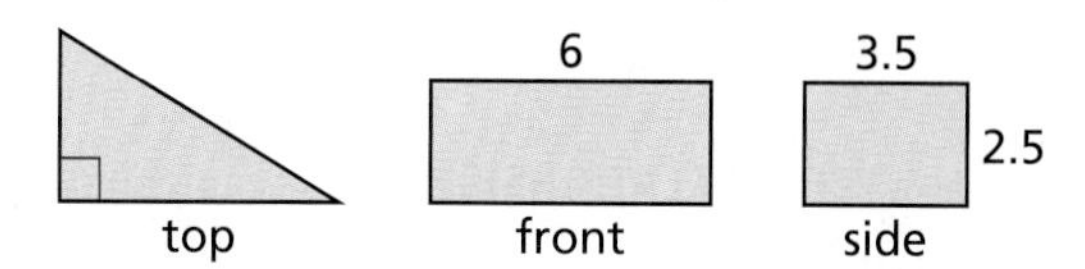

43. OPEN-ENDED Sketch two rectangular prisms that have volumes of 100 cubic inches but different surface areas. Include dimensions in your sketches.

44. MODELING WITH MATHEMATICS Which box gives you more cereal for your money? Explain.

45. CRITICAL THINKING A 3-inch by 5-inch index card is rotated around a horizontal line and a vertical line to produce two different solids. Which solid has a greater volume? Explain your reasoning.

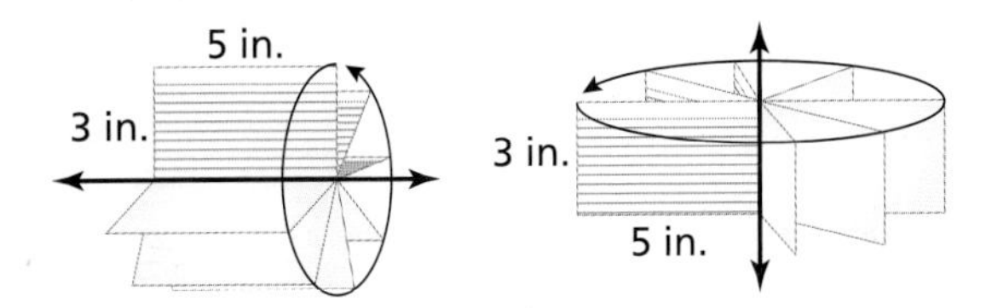

46. CRITICAL THINKING The height of cylinder X is twice the height of cylinder Y. The radius of cylinder X is half the radius of cylinder Y. Compare the volumes of cylinder X and cylinder Y. Justify your answer.

47. USING STRUCTURE Find the volume of the solid shown. The bases of the solid are sectors of circles.

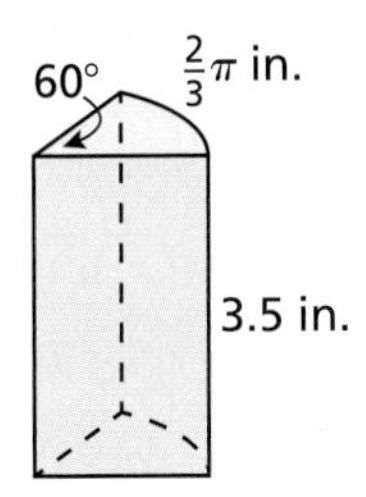

48. MATHEMATICAL CONNECTIONS You drill a circular hole of radius r through the base of a cylinder of radius R. Assume the hole is drilled completely through to the other base. You want the volume of the hole to be half the volume of the cylinder. Express r as a function of R.

49. ANALYZING RELATIONSHIPS How can you change the height of a cylinder so that the volume is increased by 25% but the radius remains the same?

50. ANALYZING RELATIONSHIPS How can you change the edge length of a cube so that the volume is reduced by 40%?

51. MAKING AN ARGUMENT You have two objects of equal volume. Your friend says you can compare the densities of the objects by comparing their mass, because the heavier object will have a greater density. Is your friend correct? Explain your reasoning.

52. THOUGHT PROVOKING Cavalieri's Principle states that the two solids shown below have the same volume. Do they also have the same surface area? Explain your reasoning.

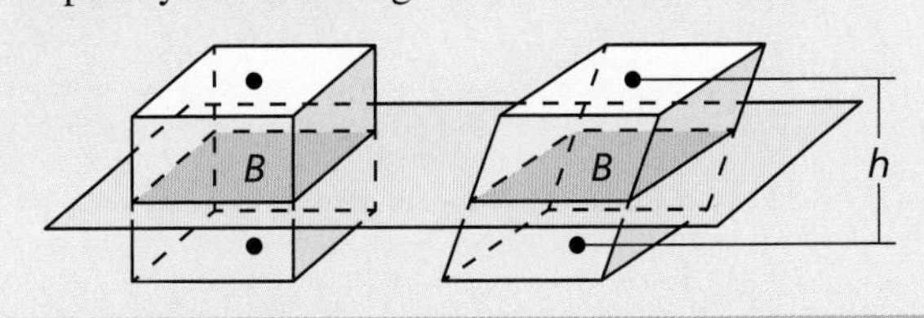

53. PROBLEM SOLVING A barn is in the shape of a pentagonal prism with the dimensions shown. The volume of the barn is 9072 cubic feet. Find the dimensions of each half of the roof.

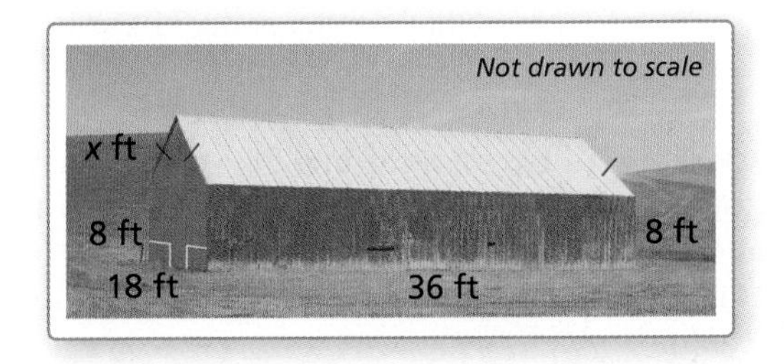

54. PROBLEM SOLVING A wooden box is in the shape of a regular pentagonal prism. The sides, top, and bottom of the box are 1 centimeter thick. Approximate the volume of wood used to construct the box. Round your answer to the nearest tenth.

Maintaining Mathematical Proficiency

Reviewing what you learned in previous grades and lessons

Find the surface area of the regular pyramid. *(Skills Review Handbook)*

55.

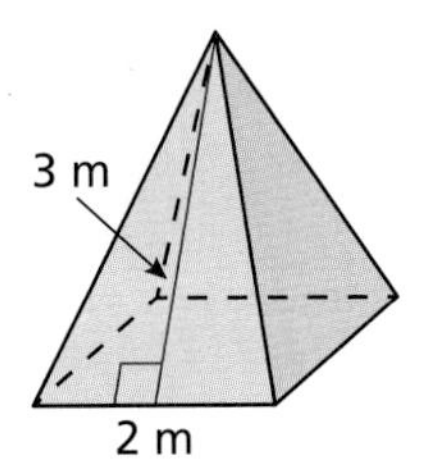

56.

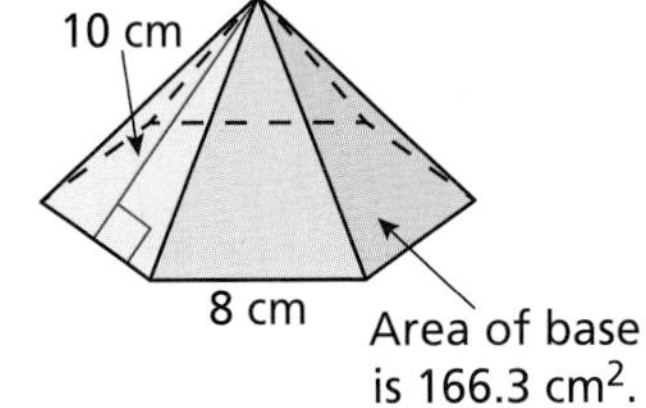

57.

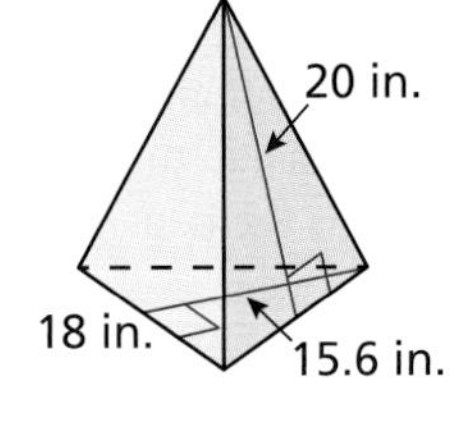

11.6 Volumes of Pyramids

Essential Question How can you find the volume of a pyramid?

EXPLORATION 1 Finding the Volume of a Pyramid

Work with a partner. The pyramid and the prism have the same height and the same square base.

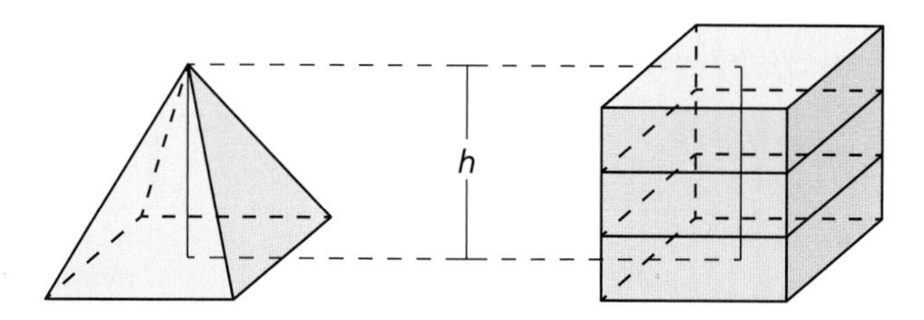

When the pyramid is filled with sand and poured into the prism, it takes three pyramids to fill the prism.

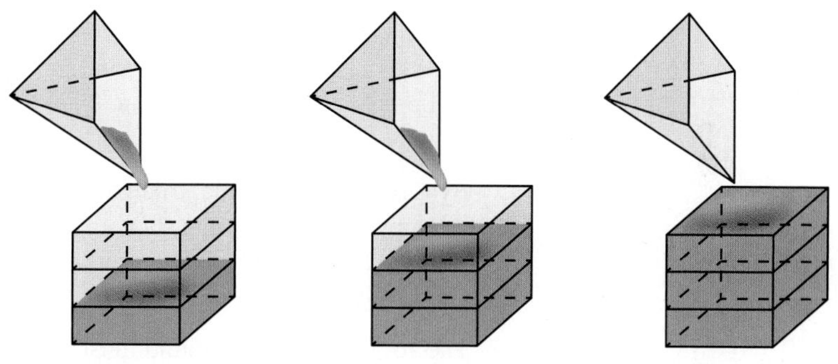

LOOKING FOR STRUCTURE

To be proficient in math, you need to look closely to discern a pattern or structure.

Use this information to write a formula for the volume V of a pyramid.

EXPLORATION 2 Finding the Volume of a Pyramid

Work with a partner. Use the formula you wrote in Exploration 1 to find the volume of the hexagonal pyramid.

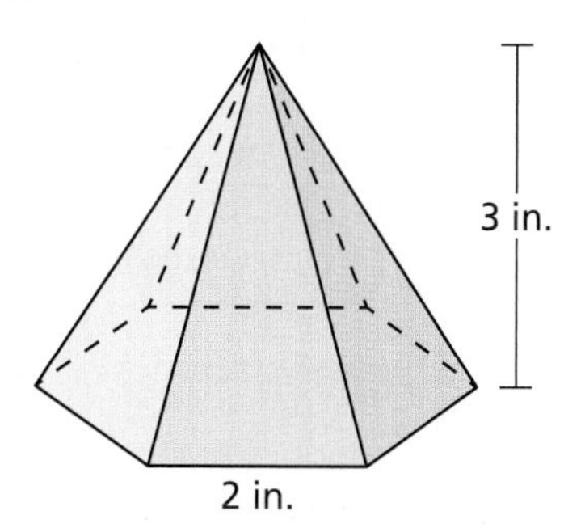

Communicate Your Answer

3. How can you find the volume of a pyramid?

4. In Section 11.7, you will study volumes of cones. How do you think you could use a method similar to the one presented in Exploration 1 to write a formula for the volume of a cone? Explain your reasoning.

11.6 Lesson

Core Vocabulary

Previous
pyramid
composite solid

What You Will Learn

- Find volumes of pyramids.
- Use volumes of pyramids.

Finding Volumes of Pyramids

Consider a triangular prism with parallel, congruent bases $\triangle JKL$ and $\triangle MNP$. You can divide this triangular prism into three triangular pyramids.

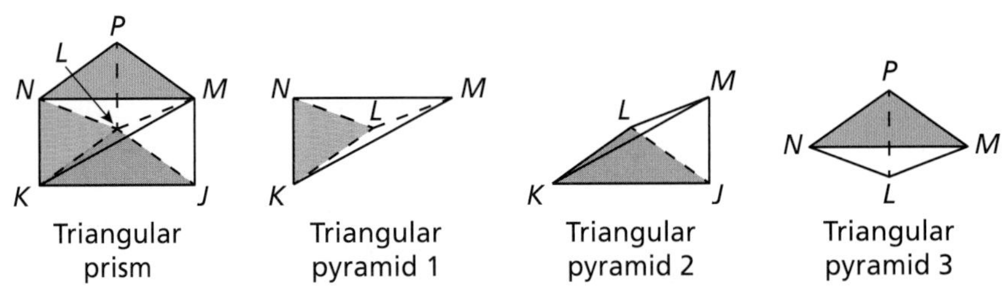

Triangular prism — Triangular pyramid 1 — Triangular pyramid 2 — Triangular pyramid 3

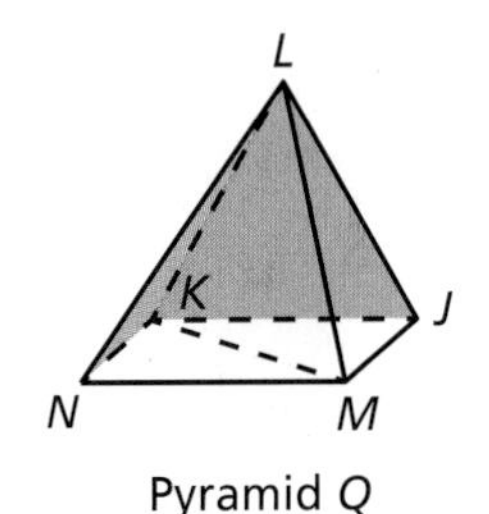

Pyramid Q

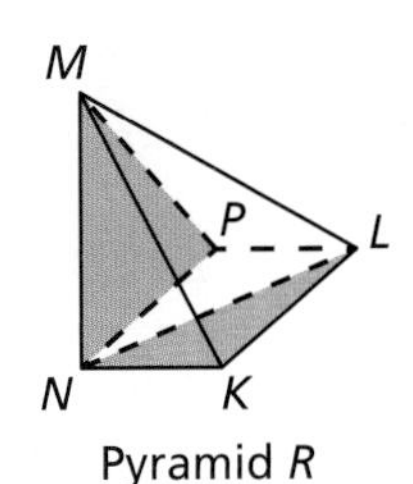

Pyramid R

You can combine triangular pyramids 1 and 2 to form a pyramid with a base that is a parallelogram, as shown at the left. Name this pyramid Q. Similarly, you can combine triangular pyramids 1 and 3 to form pyramid R with a base that is a parallelogram.

In pyramid Q, diagonal $\overline{KM}$ divides $\square JKNM$ into two congruent triangles, so the bases of triangular pyramids 1 and 2 are congruent. Similarly, you can divide any cross section parallel to $\square JKNM$ into two congruent triangles that are the cross sections of triangular pyramids 1 and 2.

By Cavalieri's Principle, triangular pyramids 1 and 2 have the same volume. Similarly, using pyramid R, you can show that triangular pyramids 1 and 3 have the same volume. By the Transitive Property of Equality, triangular pyramids 2 and 3 have the same volume.

The volume of each pyramid must be one-third the volume of the prism, or $V = \frac{1}{3}Bh$. You can generalize this formula to say that the volume of any pyramid with any base is equal to $\frac{1}{3}$ the volume of a prism with the same base and height because you can divide any polygon into triangles and any pyramid into triangular pyramids.

Core Concept

Volume of a Pyramid

The volume V of a pyramid is

$$V = \frac{1}{3}Bh$$

where B is the area of the base and h is the height.

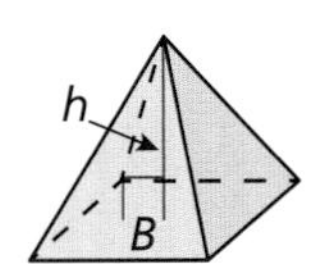

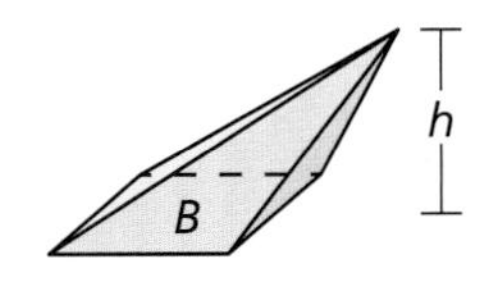

EXAMPLE 1 Finding the Volume of a Pyramid

Find the volume of the pyramid.

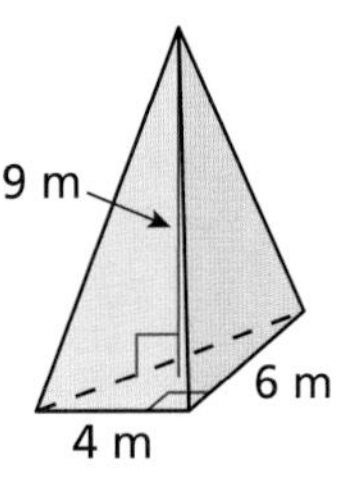

SOLUTION

$V = \frac{1}{3}Bh$	Formula for volume of a pyramid
$= \frac{1}{3}\left(\frac{1}{2} \cdot 4 \cdot 6\right)(9)$	Substitute.
$= 36$	Simplify.

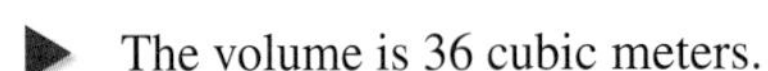

▶ The volume is 36 cubic meters.

Monitoring Progress  Help in English and Spanish at *BigIdeasMath.com*

Find the volume of the pyramid.

1. 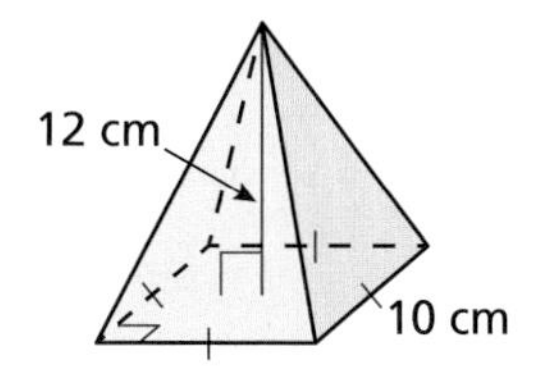

2. 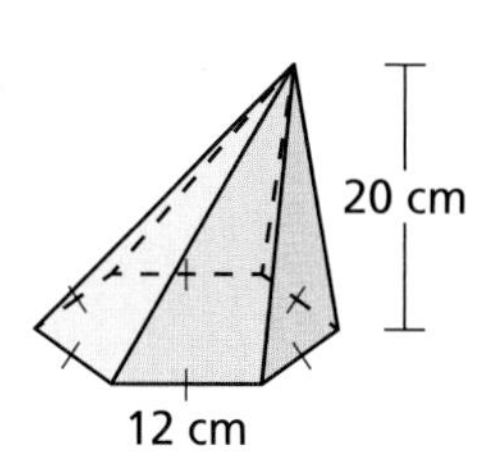

Using Volumes of Pyramids

EXAMPLE 2 Using the Volume of a Pyramid

Khafre's Pyramid, Egypt

Originally, Khafre's Pyramid had a height of about 144 meters and a volume of about 2,218,800 cubic meters. Find the side length of the square base.

SOLUTION

$V = \frac{1}{3}Bh$	Formula for volume of a pyramid
$2{,}218{,}800 \approx \frac{1}{3}x^2(144)$	Substitute.
$6{,}656{,}400 \approx 144x^2$	Multiply each side by 3.
$46{,}225 \approx x^2$	Divide each side by 144.
$215 \approx x$	Find the positive square root.

▶ Originally, the side length of the square base was about 215 meters.

EXAMPLE 3 Using the Volume of a Pyramid

Find the height of the triangular pyramid.

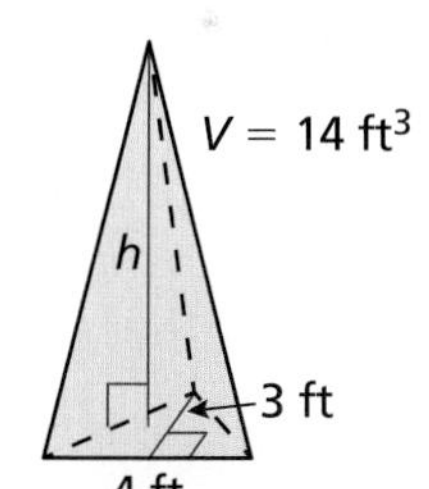

SOLUTION

The area of the base is $B = \frac{1}{2}(3)(4) = 6$ ft^2 and the volume is $V = 14$ ft^3.

$V = \frac{1}{3}Bh$	Formula for volume of a pyramid
$14 = \frac{1}{3}(6)h$	Substitute.
$7 = h$	Solve for h.

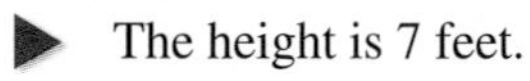
▶ The height is 7 feet.

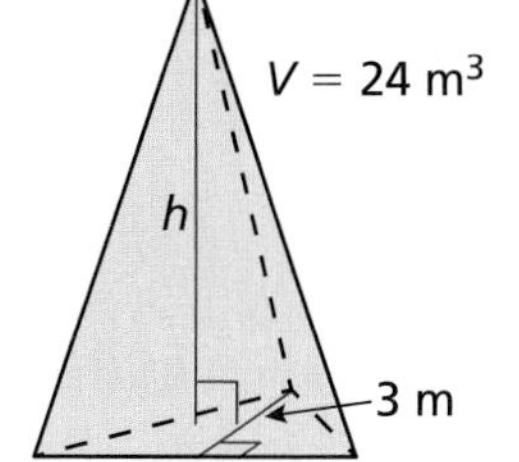

Monitoring Progress  Help in English and Spanish at *BigIdeasMath.com*

3. The volume of a square pyramid is 75 cubic meters and the height is 9 meters. Find the side length of the square base.

4. Find the height of the triangular pyramid at the left.

EXAMPLE 4 Finding the Volume of a Similar Solid

Pyramid A and pyramid B are similar. Find the volume of pyramid B.

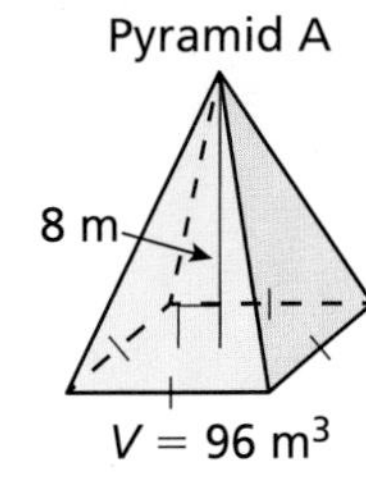

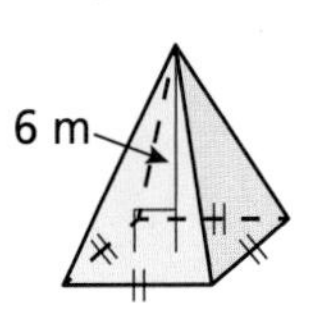

SOLUTION

The scale factor is $k = \dfrac{\text{Height of pyramid B}}{\text{Height of pyramid A}} = \dfrac{6}{8} = \dfrac{3}{4}$.

Use the scale factor to find the volume of pyramid B.

$\dfrac{\text{Volume of pyramid B}}{\text{Volume of pyramid A}} = k^3$	The ratio of the volumes is k^3.
$\dfrac{\text{Volume of pyramid B}}{96} = \left(\dfrac{3}{4}\right)^3$	Substitute.
Volume of pyramid B $= 40.5$	Solve for volume of pyramid B.

▶ The volume of pyramid B is 40.5 cubic meters.

Monitoring Progress

Help in English and Spanish at *BigIdeasMath.com*

5. Pyramid C and pyramid D are similar. Find the volume of pyramid D.

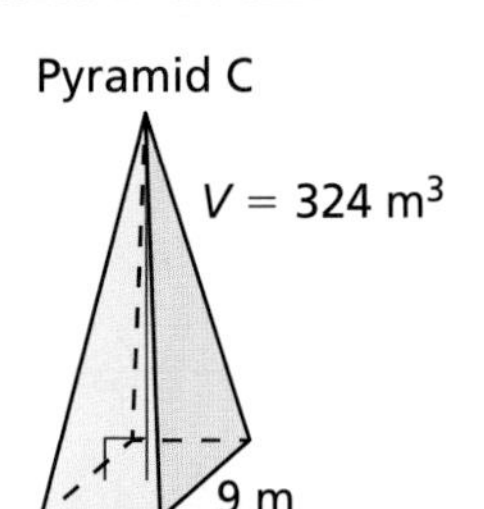

EXAMPLE 5 Finding the Volume of a Composite Solid

Find the volume of the composite solid.

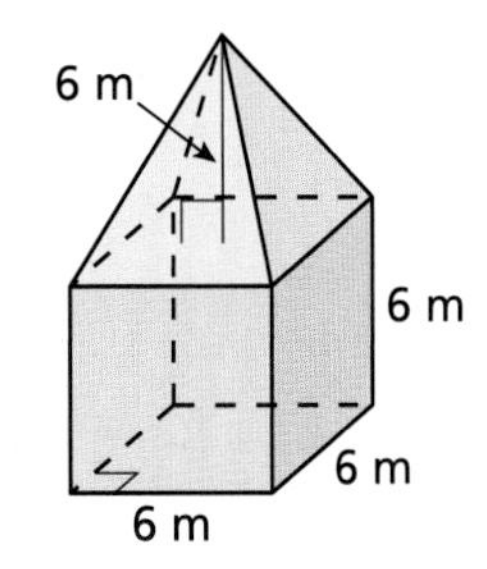

SOLUTION

Volume of solid = Volume of cube + Volume of pyramid

$= s^3 + \frac{1}{3}Bh$	Write formulas.
$= 6^3 + \frac{1}{3}(6)^2 \cdot 6$	Substitute.
$= 216 + 72$	Simplify.
$= 288$	Add.

▶ The volume is 288 cubic meters.

Monitoring Progress

Help in English and Spanish at *BigIdeasMath.com*

6. Find the volume of the composite solid.

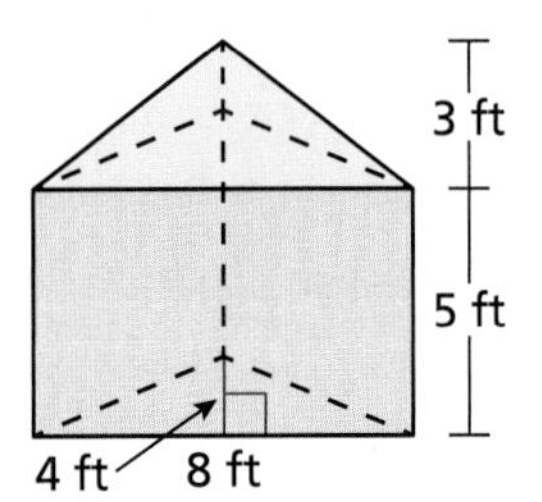

11.6 Exercises

Dynamic Solutions available at *BigIdeasMath.com*

Vocabulary and Core Concept Check

1. **VOCABULARY** Explain the difference between a triangular prism and a triangular pyramid.

2. **REASONING** A square pyramid and a cube have the same base and height. Compare the volume of the square pyramid to the volume of the cube.

Monitoring Progress and Modeling with Mathematics

In Exercises 3 and 4, find the volume of the pyramid. *(See Example 1.)*

3. 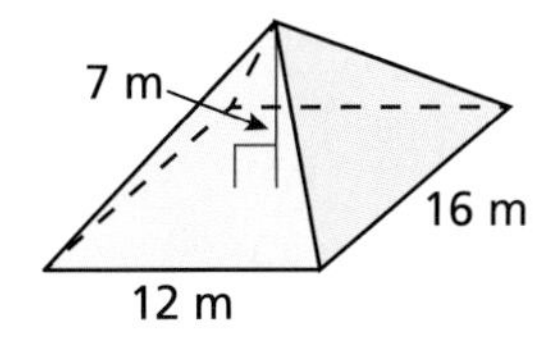

4. 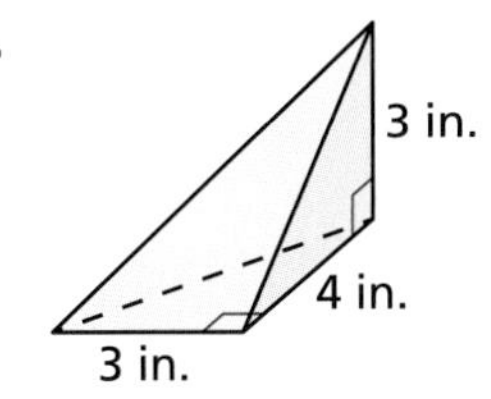

In Exercises 5–8, find the indicated measure. *(See Example 2.)*

5. A pyramid with a square base has a volume of 120 cubic meters and a height of 10 meters. Find the side length of the square base.

6. A pyramid with a square base has a volume of 912 cubic feet and a height of 19 feet. Find the side length of the square base.

7. A pyramid with a rectangular base has a volume of 480 cubic inches and a height of 10 inches. The width of the rectangular base is 9 inches. Find the length of the rectangular base.

8. A pyramid with a rectangular base has a volume of 105 cubic centimeters and a height of 15 centimeters. The length of the rectangular base is 7 centimeters. Find the width of the rectangular base.

9. **ERROR ANALYSIS** Describe and correct the error in finding the volume of the pyramid.

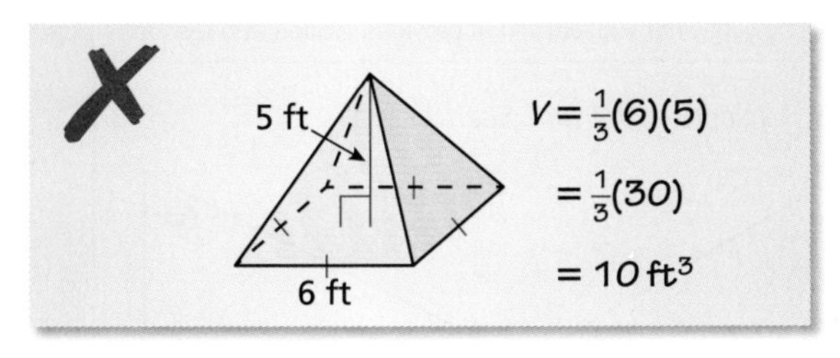

10. **OPEN-ENDED** Give an example of a pyramid and a prism that have the same base and the same volume. Explain your reasoning.

In Exercises 11–14, find the height of the pyramid. *(See Example 3.)*

11. Volume = 15 ft^3

3 ft
3 ft

12. Volume = 224 in.^{3}

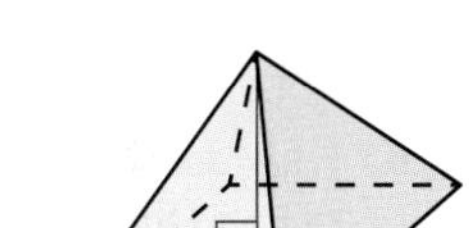
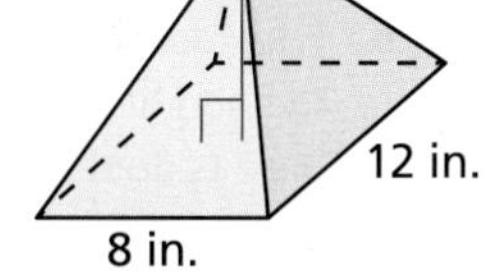

13. Volume = 198 yd^3

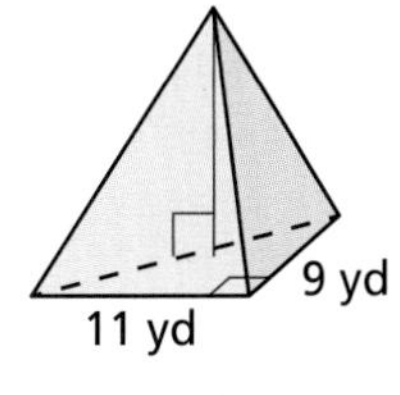

14. Volume = 392 cm^3

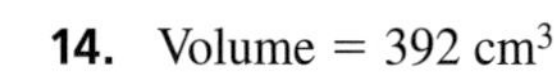
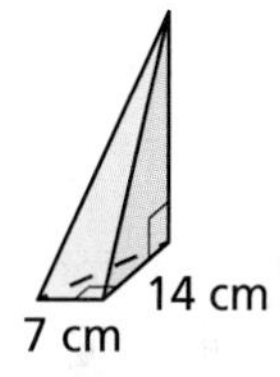

In Exercises 15 and 16, the pyramids are similar. Find the volume of pyramid B. *(See Example 4.)*

15. Pyramid A

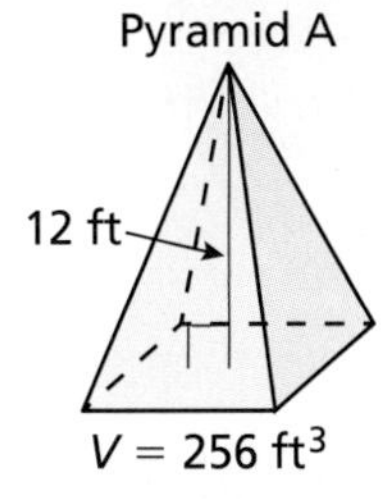

$V = 256$ ft^3

Pyramid B

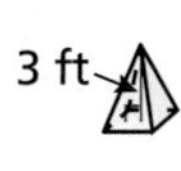

16. Pyramid A

3 in.

$V = 10$ in.^{3}

Pyramid B

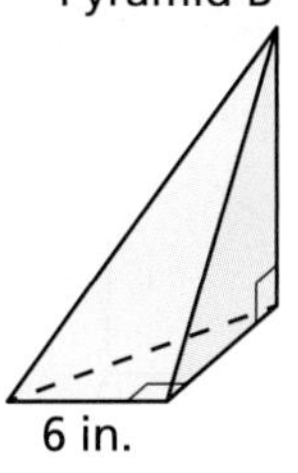

In Exercises 17–20, find the volume of the composite solid. *(See Example 5.)*

17.

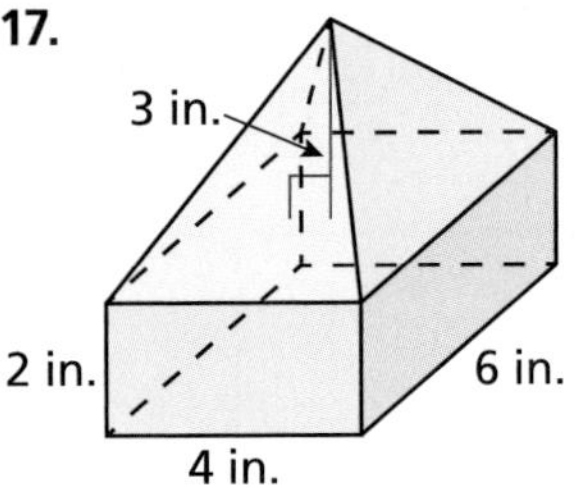

18.

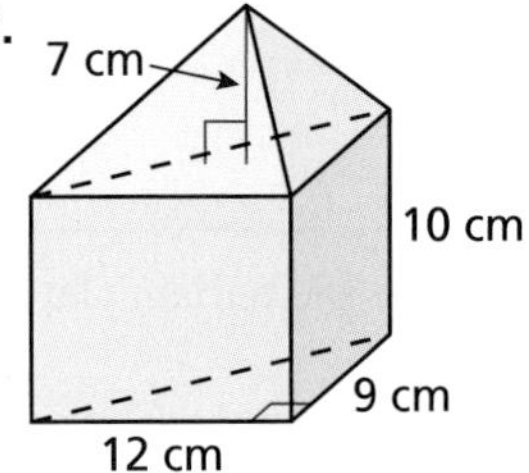

19.

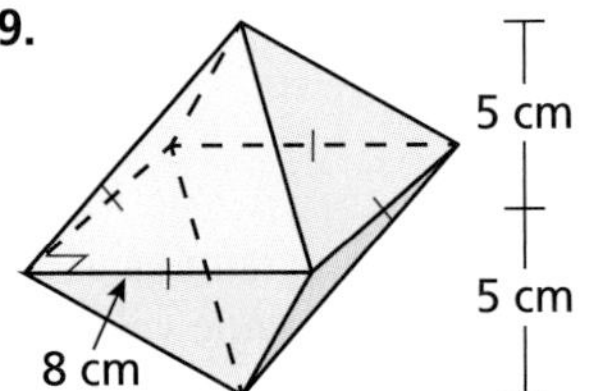

20.

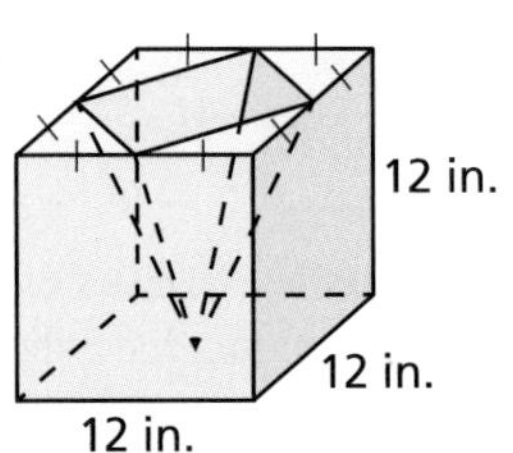

21. **ABSTRACT REASONING** A pyramid has a height of 8 feet and a square base with a side length of 6 feet.

  a. How does the volume of the pyramid change when the base stays the same and the height is doubled?

  b. How does the volume of the pyramid change when the height stays the same and the side length of the base is doubled?

  c. Are your answers to parts (a) and (b) true for any square pyramid? Explain your reasoning.

22. **HOW DO YOU SEE IT?** The cube shown is formed by three pyramids, each with the same square base and the same height. How could you use this to verify the formula for the volume of a pyramid?

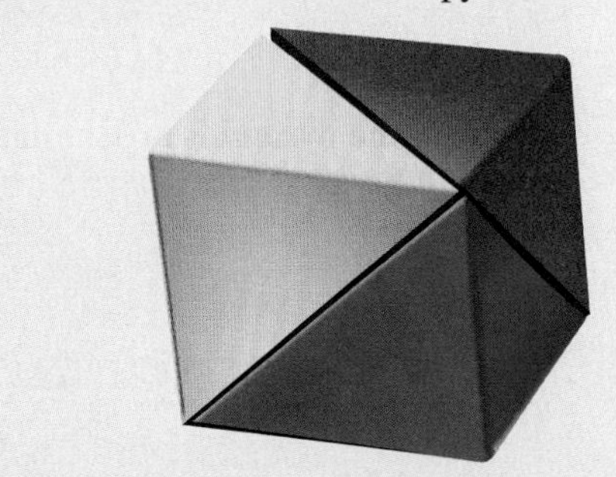

23. **CRITICAL THINKING** Find the volume of the regular pentagonal pyramid. Round your answer to the nearest hundredth. In the diagram, $m\angle ABC = 35°$.

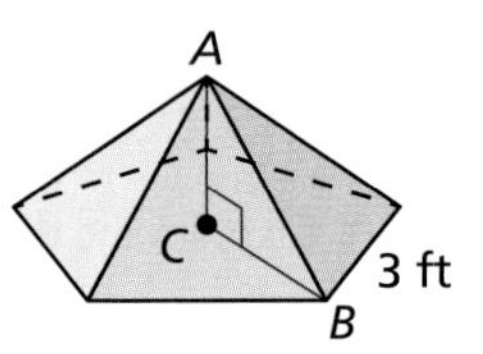

24. **THOUGHT PROVOKING** A *frustum* of a pyramid is the part of the pyramid that lies between the base and a plane parallel to the base, as shown. Write a formula for the volume of the frustum of a square pyramid in terms of a, b, and h. (*Hint*: Consider the "missing" top of the pyramid and use similar triangles.)

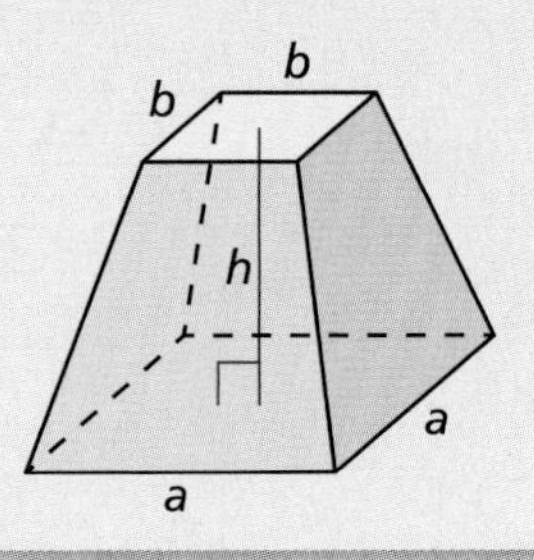

25. **MODELING WITH MATHEMATICS** Nautical deck prisms were used as a safe way to illuminate decks on ships. The deck prism shown here is composed of the following three solids: a regular hexagonal prism with an edge length of 3.5 inches and a height of 1.5 inches, a regular hexagonal prism with an edge length of 3.25 inches and a height of 0.25 inch, and a regular hexagonal pyramid with an edge length of 3 inches and a height of 3 inches. Find the volume of the deck prism.

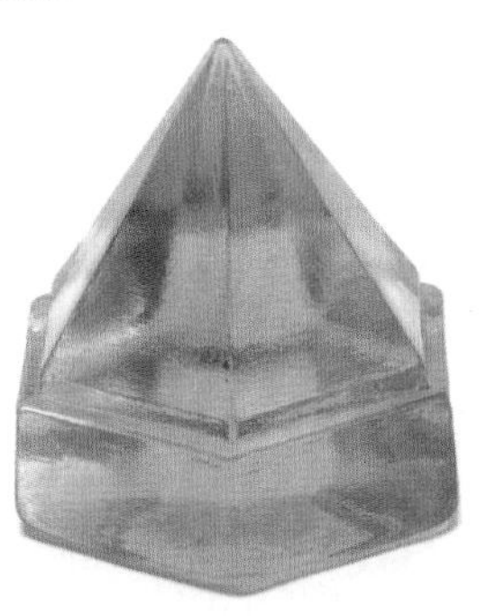

Maintaining Mathematical Proficiency

Reviewing what you learned in previous grades and lessons

Find the value of x. Round your answer to the nearest tenth. *(Section 9.4 and Section 9.5)*

26.

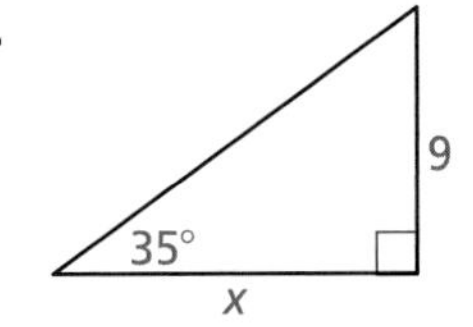

27. 57°, 15, x

28. x, 30°, 10

29. x, 64°, 7

11.7 Surface Areas and Volumes of Cones

Essential Question

How can you find the surface area and the volume of a cone?

EXPLORATION 1 Finding the Surface Area of a Cone

Work with a partner. Construct a circle with a radius of 3 inches. Mark the circumference of the circle into six equal parts, and label the length of each part. Then cut out one sector of the circle and make a cone.

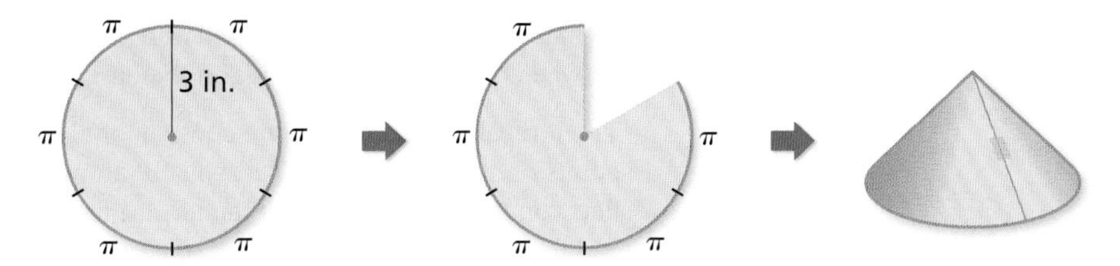

a. Explain why the base of the cone is a circle. What are the circumference and radius of the base?

b. What is the area of the original circle? What is the area with one sector missing?

c. Describe the surface area of the cone, including the base. Use your description to find the surface area.

EXPLORATION 2 Finding the Volume of a Cone

Work with a partner. The cone and the cylinder have the same height and the same circular base.

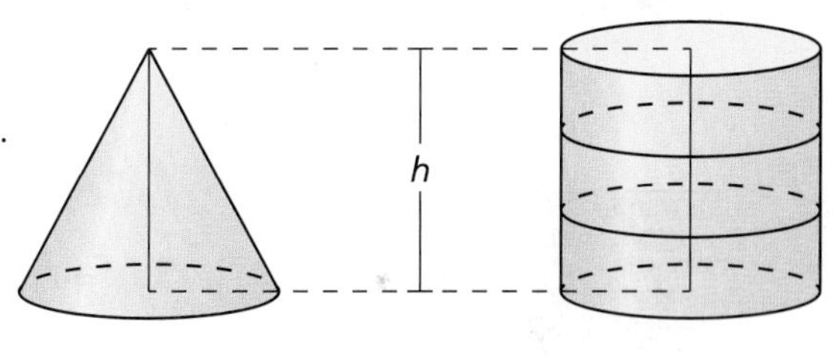

When the cone is filled with sand and poured into the cylinder, it takes three cones to fill the cylinder.

CONSTRUCTING VIABLE ARGUMENTS

To be proficient in math, you need to understand and use stated assumptions, definitions, and previously established results in constructing arguments.

Use this information to write a formula for the volume V of a cone.

Communicate Your Answer

3. How can you find the surface area and the volume of a cone?

4. In Exploration 1, cut another sector from the circle and make a cone. Find the radius of the base and the surface area of the cone. Repeat this three times, recording your results in a table. Describe the pattern.

11.7 Lesson

Core Vocabulary

lateral surface of a cone, *p. 642*

Previous
cone
net
composite solid

What You Will Learn

- Find surface areas of right cones.
- Find volumes of cones.
- Use volumes of cones.

Finding Surface Areas of Right Cones

Recall that a *circular cone*, or *cone*, has a circular *base* and a *vertex* that is not in the same plane as the base. The *altitude*, or *height*, is the perpendicular distance between the vertex and the base. In a *right cone*, the height meets the base at its center and the *slant height* is the distance between the vertex and a point on the base edge.

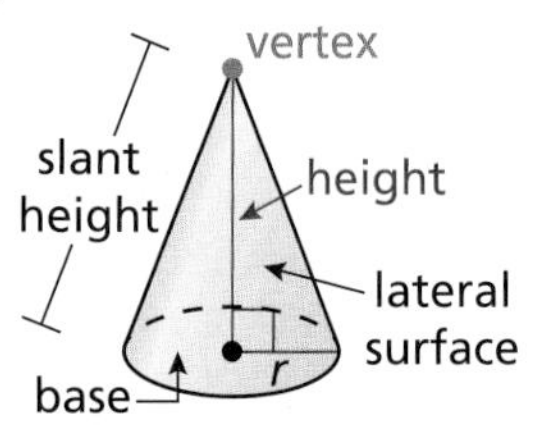

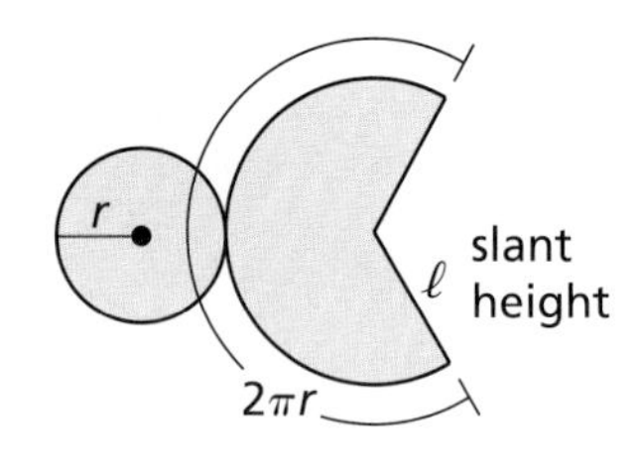

The **lateral surface of a cone** consists of all segments that connect the vertex with points on the base edge. When you cut along the slant height and lay the right cone flat, you get the net shown at the left. In the net, the circular base has an area of πr^2 and the lateral surface is a sector of a circle. You can find the area of this sector by using a proportion, as shown below.

$$\frac{\text{Area of sector}}{\text{Area of circle}} = \frac{\text{Arc length}}{\text{Circumference of circle}}$$ Set up proportion.

$$\frac{\text{Area of sector}}{\pi \ell^2} = \frac{2\pi r}{2\pi \ell}$$ Substitute.

$$\text{Area of sector} = \pi \ell^2 \cdot \frac{2\pi r}{2\pi \ell}$$ Multiply each side by $\pi \ell^2$.

$$\text{Area of sector} = \pi r \ell$$ Simplify.

The surface area of a right cone is the sum of the base area and the lateral area, $\pi r \ell$.

Core Concept

Surface Area of a Right Cone

The surface area S of a right cone is

$$S = \pi r^2 + \pi r \ell$$

where r is the radius of the base and ℓ is the slant height.

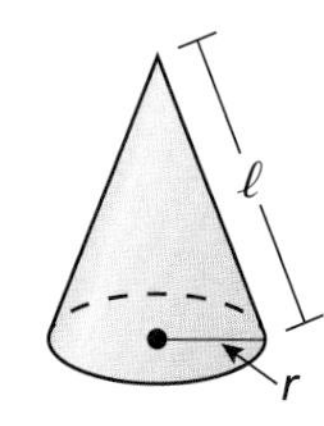

EXAMPLE 1 Finding Surface Areas of Right Cones

Find the surface area of the right cone.

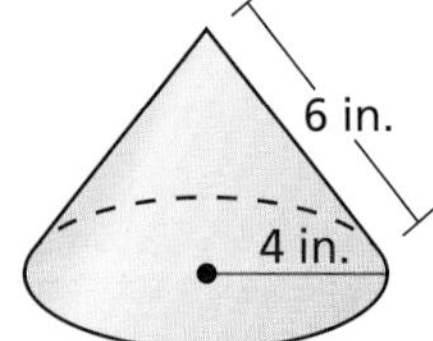

SOLUTION

$$S = \pi r^2 + \pi r \ell = \pi \cdot 4^2 + \pi(4)(6) = 40\pi \approx 125.66$$

▶ The surface area is 40π, or about 125.66 square inches.

Monitoring Progress  Help in English and Spanish at *BigIdeasMath.com*

1. Find the surface area of the right cone.

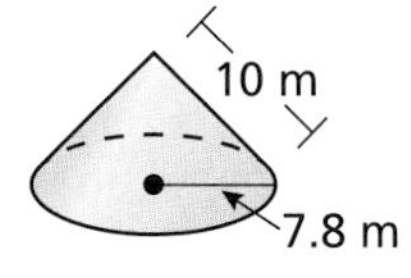

Finding Volumes of Cones

Consider a cone with a regular polygon inscribed in the base. The pyramid with the same vertex as the cone has volume $V = \frac{1}{3}Bh$. As you increase the number of sides of the polygon, it approaches the base of the cone and the pyramid approaches the cone. The volume approaches $\frac{1}{3}\pi r^2 h$ as the base area B approaches πr^2.

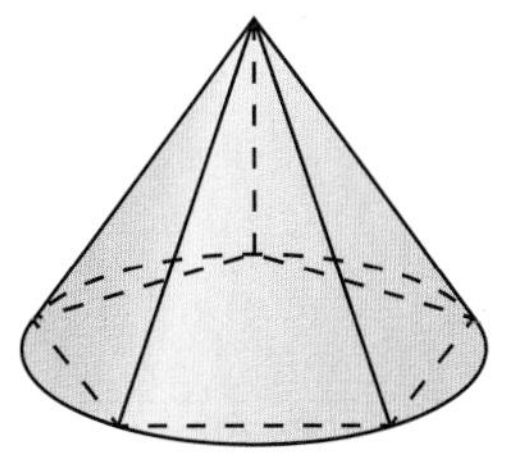
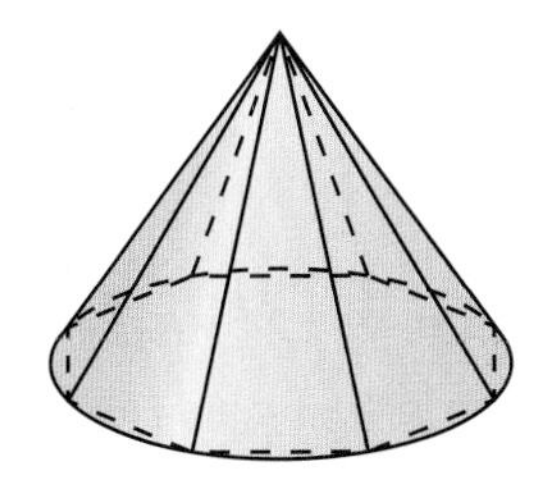
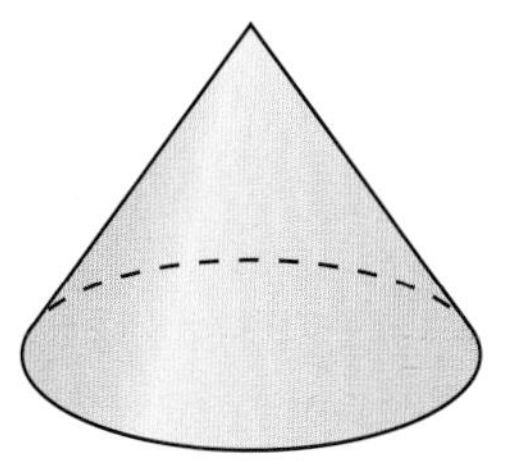

Core Concept

Volume of a Cone

The volume V of a cone is

$$V = \frac{1}{3}Bh = \frac{1}{3}\pi r^2 h$$

where B is the area of the base, h is the height, and r is the radius of the base.

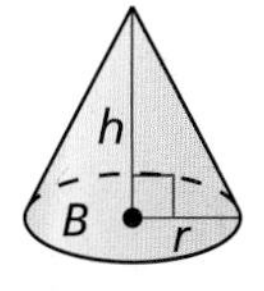

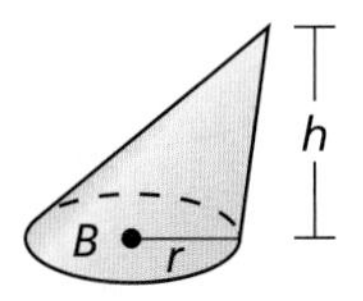

EXAMPLE 2 Finding the Volume of a Cone

Find the volume of the cone.

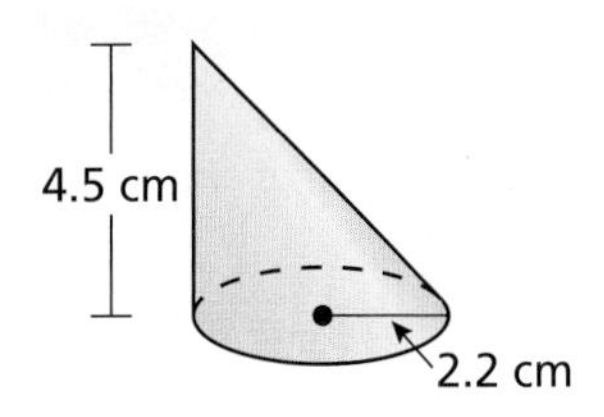

SOLUTION

$V = \frac{1}{3}\pi r^2 h$	Formula for volume of a cone
$= \frac{1}{3}\pi \cdot (2.2)^2 \cdot 4.5$	Substitute.
$= 7.26\pi$	Simplify.
≈ 22.81	Use a calculator.

▶ The volume is 7.26π, or about 22.81 cubic centimeters.

Monitoring Progress  Help in English and Spanish at *BigIdeasMath.com*

Find the volume of the cone.

2.

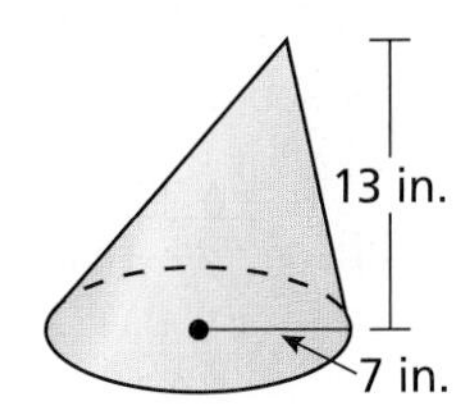

3.

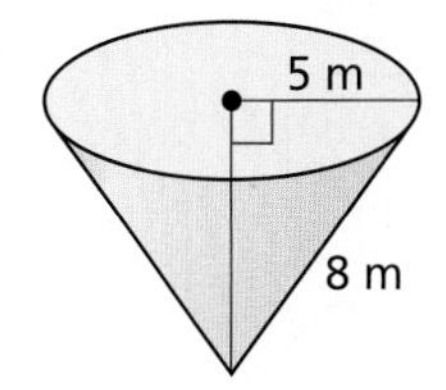

Using Volumes of Cones

EXAMPLE 3 Finding the Volume of a Similar Solid

Cone A and cone B are similar. Find the volume of cone B.

Cone A

$V = 15\pi$ ft³

Cone B

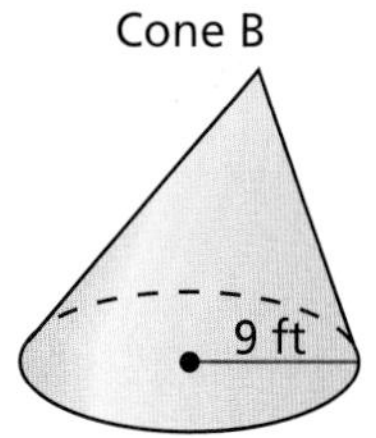

SOLUTION

The scale factor is $k = \dfrac{\text{Radius of cone B}}{\text{Radius of cone A}} = \dfrac{9}{3} = 3$.

Use the scale factor to find the volume of cone B.

$\dfrac{\text{Volume of cone B}}{\text{Volume of cone A}} = k^3$	The ratio of the volumes is k^3.
$\dfrac{\text{Volume of cone B}}{15\pi} = 3^3$	Substitute.
Volume of cone B $= 405\pi$	Solve for volume of cone B.

▶ The volume of cone B is 405π cubic feet.

Cone C

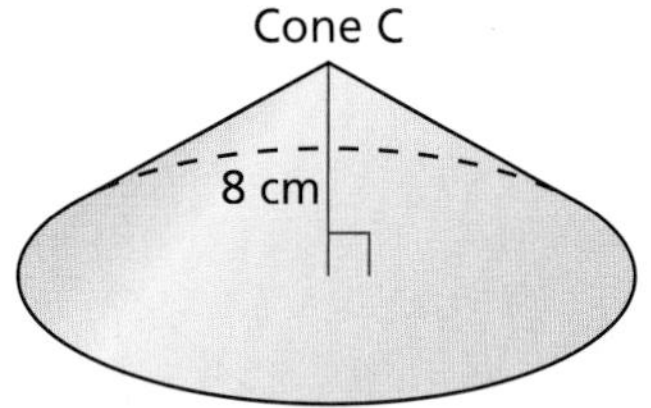

$V = 384\pi$ cm³

Cone D

Monitoring Progress Help in English and Spanish at *BigIdeasMath.com*

4. Cone C and cone D are similar. Find the volume of cone D.

EXAMPLE 4 Finding the Volume of a Composite Solid

Find the volume of the composite solid.

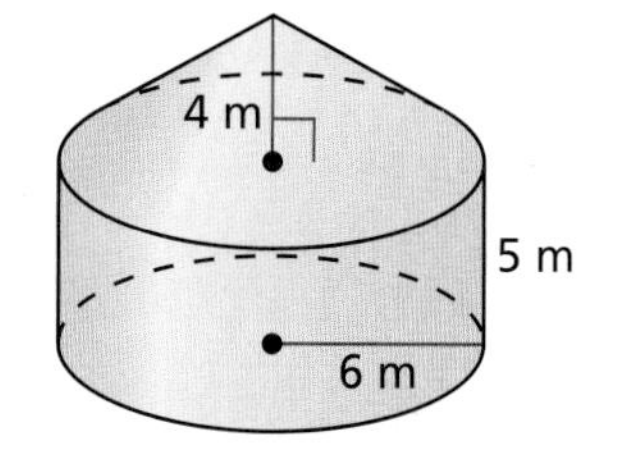

SOLUTION

Let h_1 be the height of the cylinder and let h_2 be the height of the cone.

Volume of solid = Volume of cylinder + Volume of cone

$= \pi r^2 h_1 + \frac{1}{3}\pi r^2 h_2$	Write formulas.
$= \pi \cdot 6^2 \cdot 5 + \frac{1}{3}\pi \cdot 6^2 \cdot 4$	Substitute.
$= 180\pi + 48\pi$	Simplify.
$= 228\pi$	Add.
≈ 716.28	Use a calculator.

▶ The volume is 228π, or about 716.28 cubic meters.

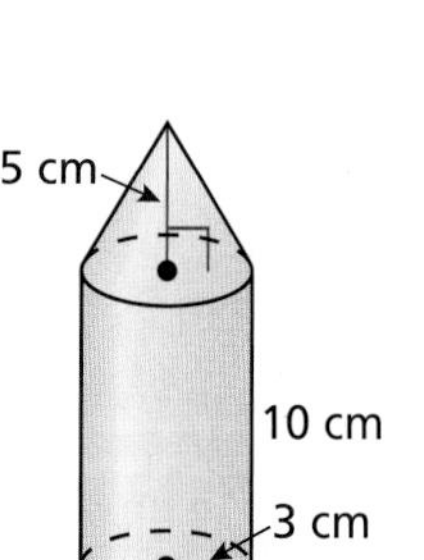

Monitoring Progress  Help in English and Spanish at *BigIdeasMath.com*

5. Find the volume of the composite solid.

11.7 Exercises

Dynamic Solutions available at *BigIdeasMath.com*

Vocabulary and Core Concept Check

1. **WRITING** Describe the differences between pyramids and cones. Describe their similarities.

2. **COMPLETE THE SENTENCE** The volume of a cone with radius r and height h is $\frac{1}{3}$ the volume of a(n) ________ with radius r and height h.

Monitoring Progress and Modeling with Mathematics

In Exercises 3–6, find the surface area of the right cone. *(See Example 1.)*

3. 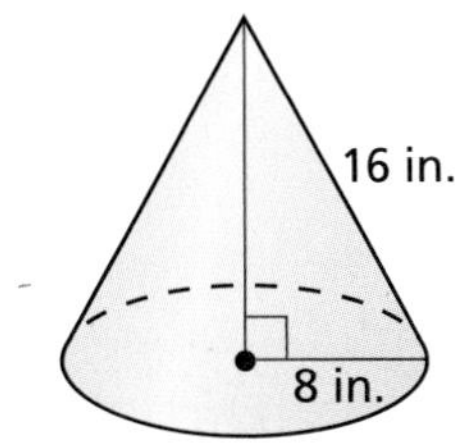

4. 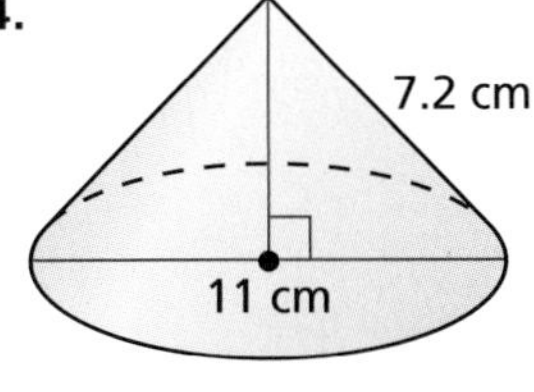

5. A right cone has a radius of 9 inches and a height of 12 inches.

6. A right cone has a diameter of 11.2 feet and a height of 9.2 feet.

In Exercises 7–10, find the volume of the cone. *(See Example 2.)*

7. 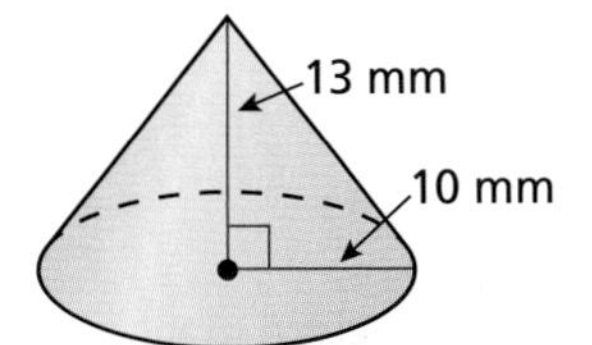

8. 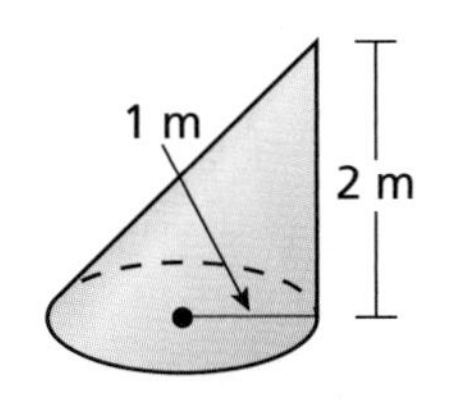

9. A cone has a diameter of 11.5 inches and a height of 15.2 inches.

10. A right cone has a radius of 3 feet and a slant height of 6 feet.

In Exercises 11 and 12, find the missing dimension(s).

11. Surface area = 75.4 cm^2

h
ℓ
3 cm

12. Volume = $216\pi\text{ in.}^3$

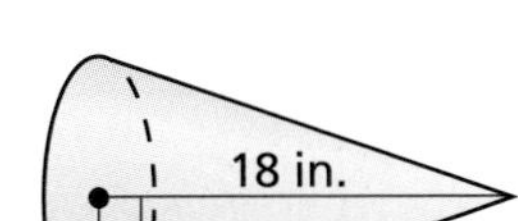
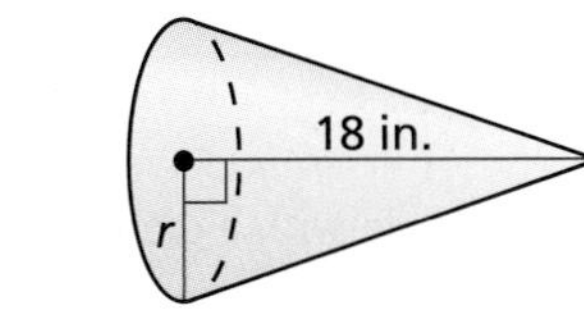

In Exercises 13 and 14, the cones are similar. Find the volume of cone B. *(See Example 3.)*

13. Cone A

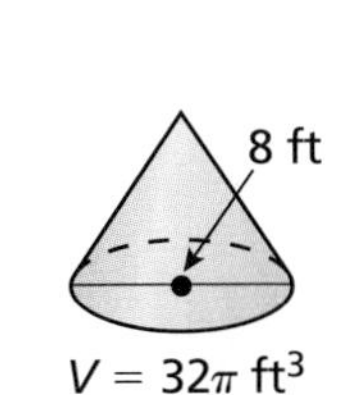

$V = 32\pi\text{ ft}^3$

Cone B

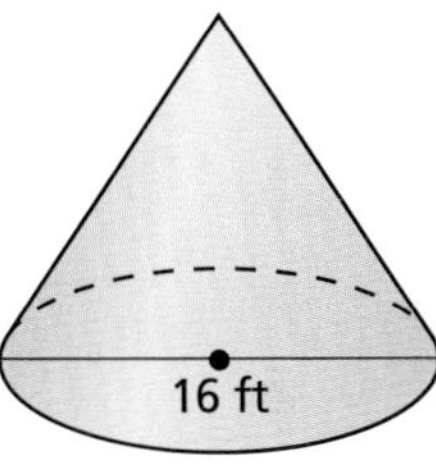

14. Cone A

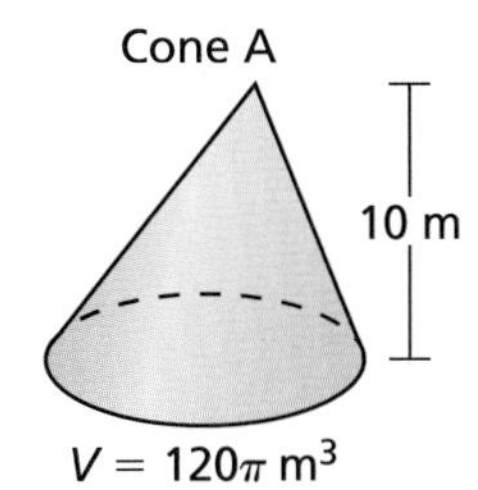

$V = 120\pi\text{ m}^3$

Cone B

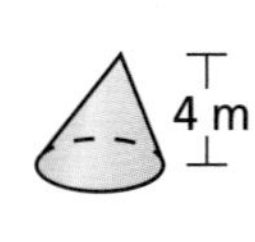

In Exercises 15 and 16, find the volume of the composite solid. *(See Example 4.)*

15. 16.

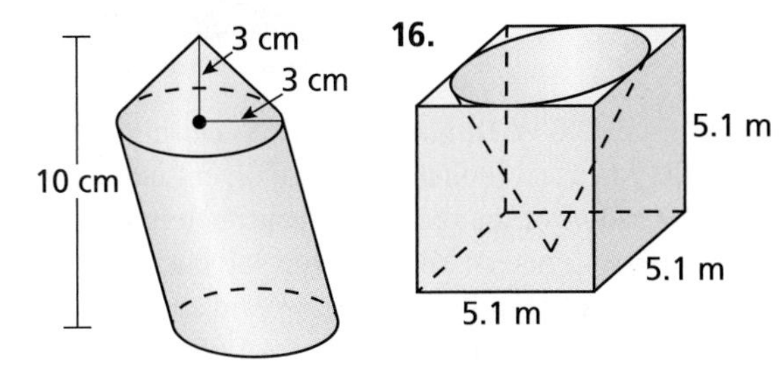

17. **ANALYZING RELATIONSHIPS** A cone has height h and a base with radius r. You want to change the cone so its volume is doubled. What is the new height if you change only the height? What is the new radius if you change only the radius? Explain.

18. **HOW DO YOU SEE IT** A snack stand serves a small order of popcorn in a cone-shaped container and a large order of popcorn in a cylindrical container. Do not perform any calculations.

a. How many small containers of popcorn do you have to buy to equal the amount of popcorn in a large container? Explain.

b. Which container gives you more popcorn for your money? Explain.

In Exercises 19 and 20, find the volume of the right cone.

19. 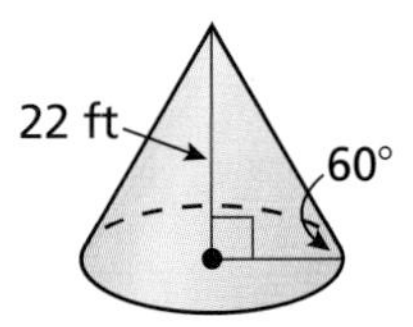

20. 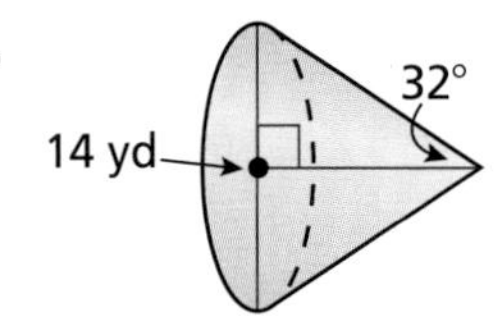

21. **MODELING WITH MATHEMATICS** A cat eats half a cup of food, twice per day. Will the automatic pet feeder hold enough food for 10 days? Explain your reasoning. (1 cup $\approx$ 14.4 in.^{3})

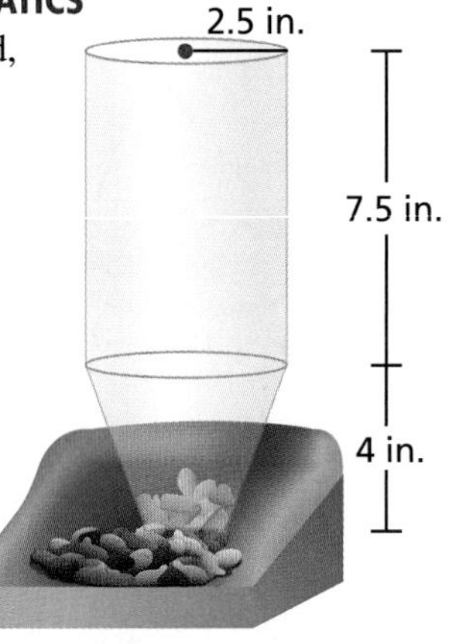

22. **MODELING WITH MATHEMATICS** During a chemistry lab, you use a funnel to pour a solvent into a flask. The radius of the funnel is 5 centimeters and its height is 10 centimeters. You pour the solvent into the funnel at a rate of 80 milliliters per second and the solvent flows out of the funnel at a rate of 65 milliliters per second. How long will it be before the funnel overflows? (1 mL = 1 cm^3)

23. **REASONING** To make a paper drinking cup, start with a circular piece of paper that has a 3-inch radius, then follow the given steps. How does the surface area of the cup compare to the original paper circle? Find $m\angle ABC$.

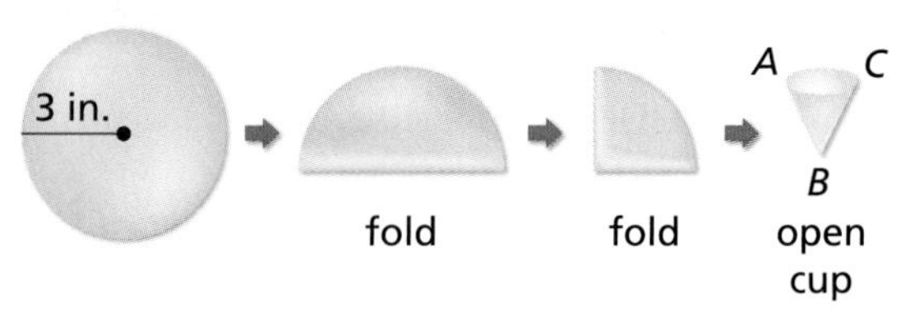

24. **THOUGHT PROVOKING** A *frustum* of a cone is the part of the cone that lies between the base and a plane parallel to the base, as shown. Write a formula for the volume of the frustum of a cone in terms of a, b, and h. (*Hint*: Consider the "missing" top of the cone and use similar triangles.)

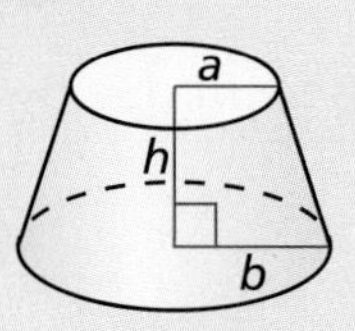

25. **MAKING AN ARGUMENT** In the figure, the two cylinders are congruent. The combined height of the two smaller cones equals the height of the larger cone. Your friend claims that this means the total volume of the two smaller cones is equal to the volume of the larger cone. Is your friend correct? Justify your answer.

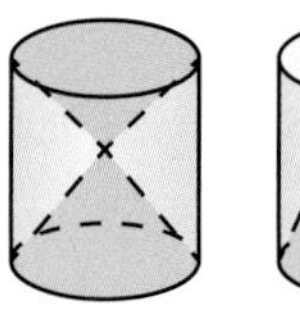 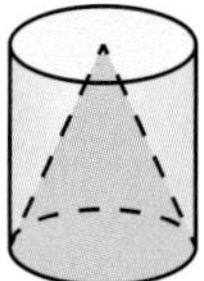

26. **CRITICAL THINKING** When the given triangle is rotated around each of its sides, solids of revolution are formed. Describe the three solids and find their volumes. Give your answers in terms of π.

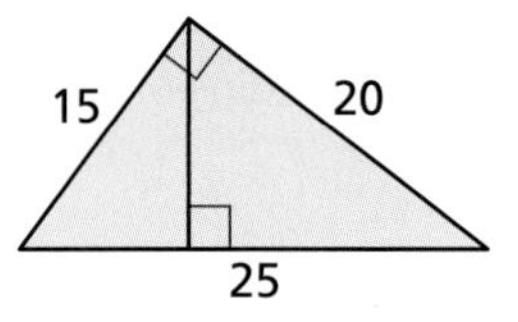

Maintaining Mathematical Proficiency

Reviewing what you learned in previous grades and lessons

Find the indicated measure. *(Section 11.2)*

27. area of a circle with a radius of 7 feet

28. area of a circle with a diameter of 22 centimeters

29. diameter of a circle with an area of 256π square meters

30. radius of a circle with an area of 529π square inches

11.8 Surface Areas and Volumes of Spheres

Essential Question How can you find the surface area and the volume of a sphere?

EXPLORATION 1 Finding the Surface Area of a Sphere

Work with a partner. Remove the covering from a baseball or softball.

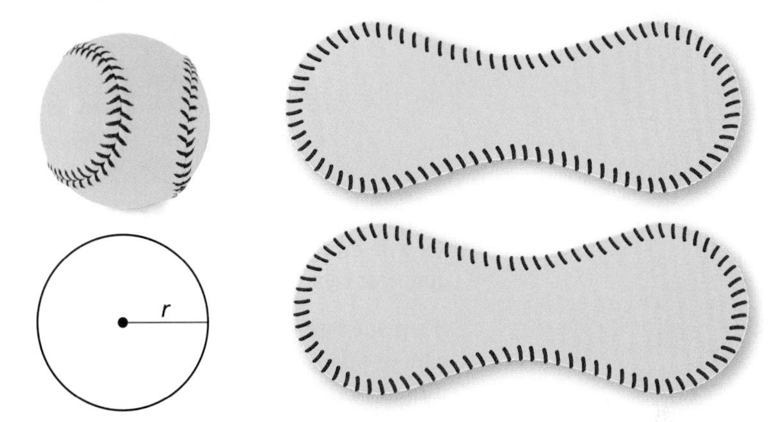

USING TOOLS STRATEGICALLY

To be proficient in math, you need to identify relevant external mathematical resources, such as content located on a website.

You will end up with two "figure 8" pieces of material, as shown above. From the amount of material it takes to cover the ball, what would you estimate the surface area S of the ball to be? Express your answer in terms of the radius r of the ball.

$S =$ ________ Surface area of a sphere

Use the Internet or some other resource to confirm that the formula you wrote for the surface area of a sphere is correct.

EXPLORATION 2 Finding the Volume of a Sphere

Work with a partner. A cylinder is circumscribed about a sphere, as shown. Write a formula for the volume V of the cylinder in terms of the radius r.

$V =$ ________ Volume of cylinder

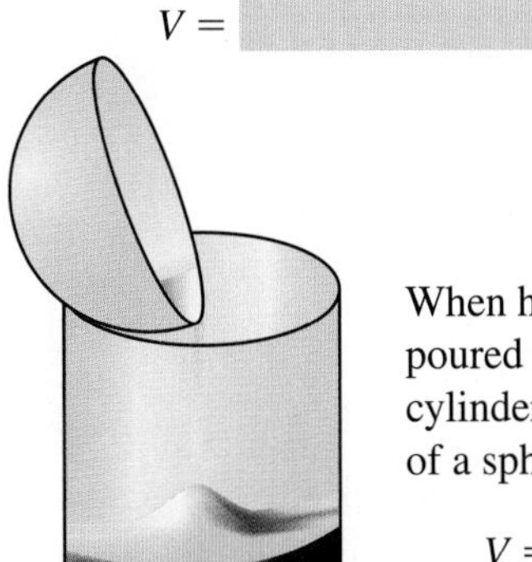

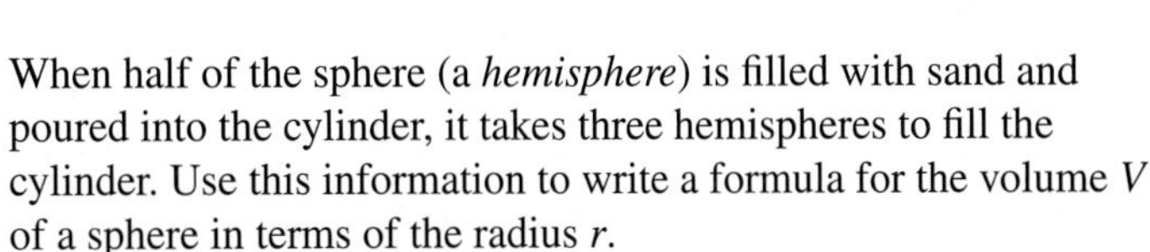

When half of the sphere (a *hemisphere*) is filled with sand and poured into the cylinder, it takes three hemispheres to fill the cylinder. Use this information to write a formula for the volume V of a sphere in terms of the radius r.

$V =$ ________ Volume of a sphere

Communicate Your Answer

3. How can you find the surface area and the volume of a sphere?

4. Use the results of Explorations 1 and 2 to find the surface area and the volume of a sphere with a radius of (a) 3 inches and (b) 2 centimeters.

11.8 Lesson

Core Vocabulary

chord of a sphere, *p. 648*
great circle, *p. 648*

Previous
sphere
center of a sphere
radius of a sphere
diameter of a sphere
hemisphere

What You Will Learn

- Find surface areas of spheres.
- Find volumes of spheres.

Finding Surface Areas of Spheres

A *sphere* is the set of all points in space equidistant from a given point. This point is called the *center* of the sphere. A *radius* of a sphere is a segment from the center to a point on the sphere. A **chord of a sphere** is a segment whose endpoints are on the sphere. A *diameter* of a sphere is a chord that contains the center.

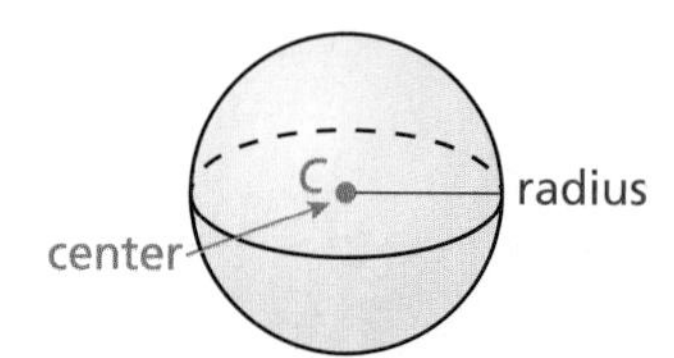

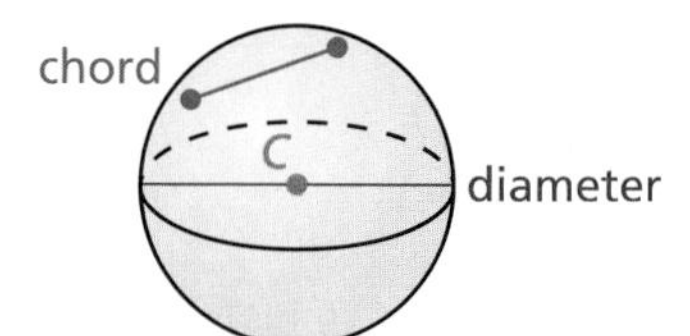

As with circles, the terms radius and diameter also represent distances, and the diameter is twice the radius.

If a plane intersects a sphere, then the intersection is either a single point or a circle. If the plane contains the center of the sphere, then the intersection is a **great circle** of the sphere. The circumference of a great circle is the circumference of the sphere. Every great circle of a sphere separates the sphere into two congruent halves called *hemispheres*.

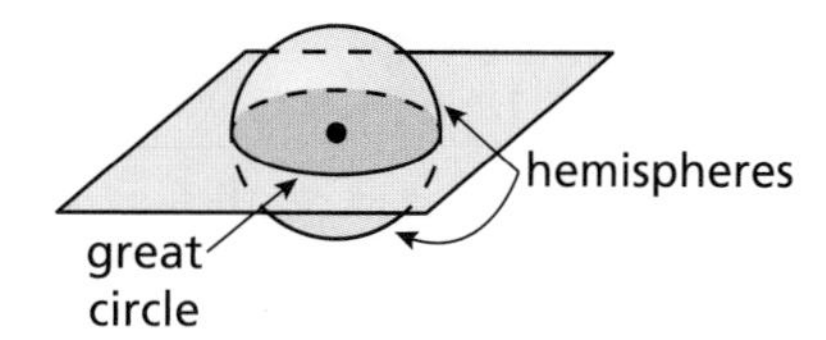

Core Concept

Surface Area of a Sphere

The surface area S of a sphere is

$$S = 4\pi r^2$$

where r is the radius of the sphere.

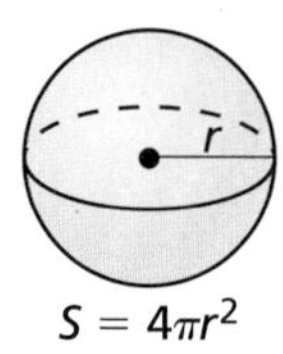

To understand the formula for the surface area of a sphere, think of a baseball. The surface area of a baseball is sewn from two congruent shapes, each of which resembles two joined circles.

So, the entire covering of the baseball consists of four circles, each with radius r. The area A of a circle with radius r is $A = \pi r^2$. So, the area of the covering can be approximated by $4\pi r^2$. This is the formula for the surface area of a sphere.

EXAMPLE 1 Finding the Surface Areas of Spheres

Find the surface area of each sphere.

a.

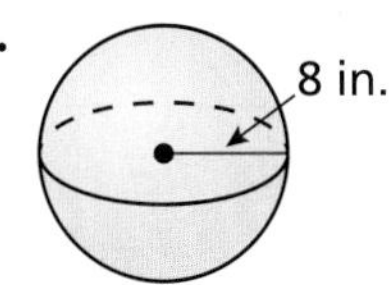

b.

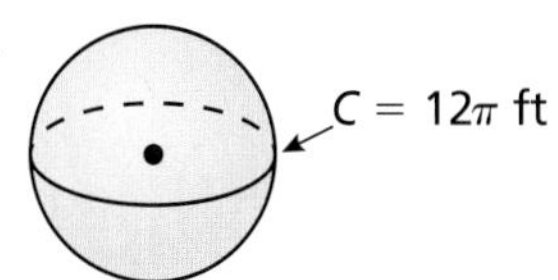

SOLUTION

a. $S = 4\pi r^2$ — Formula for surface area of a sphere

$= 4\pi(8)^2$ — Substitute 8 for r.

$= 256\pi$ — Simplify.

≈ 804.25 — Use a calculator.

▶ The surface area is 256π, or about 804.25 square inches.

b. The circumference of the sphere is 12π, so the radius of the sphere is $\frac{12\pi}{2\pi} = 6$ feet.

$S = 4\pi r^2$ — Formula for surface area of a sphere

$= 4\pi(6)^2$ — Substitute 6 for r.

$= 144\pi$ — Simplify.

≈ 452.39 — Use a calculator.

▶ The surface area is 144π, or about 452.39 square feet.

EXAMPLE 2 Finding the Diameter of a Sphere

Find the diameter of the sphere.

$S = 20.25\pi \text{ cm}^2$

SOLUTION

$S = 4\pi r^2$ — Formula for surface area of a sphere

$20.25\pi = 4\pi r^2$ — Substitute 20.25π for S.

$5.0625 = r^2$ — Divide each side by 4π.

$2.25 = r$ — Find the positive square root.

▶ The diameter is $2r = 2 \cdot 2.25 = 4.5$ centimeters.

COMMON ERROR

Be sure to multiply the value of r by 2 to find the diameter.

Monitoring Progress

Help in English and Spanish at *BigIdeasMath.com*

Find the surface area of the sphere.

1.

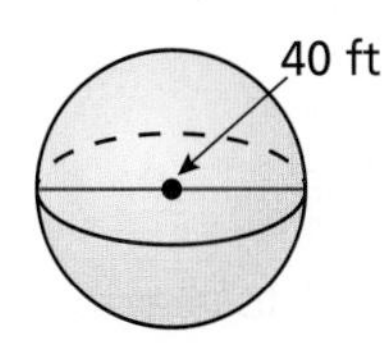

2.

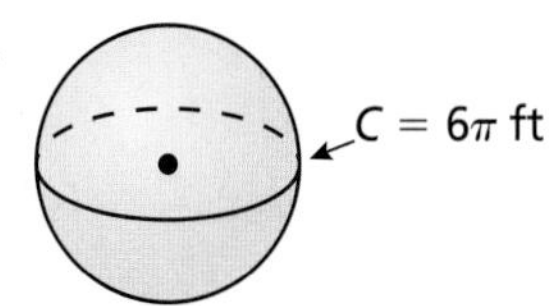

3. Find the radius of the sphere.

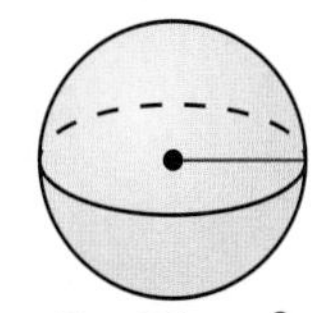

Finding Volumes of Spheres

The figure shows a hemisphere and a cylinder with a cone removed. A plane parallel to their bases intersects the solids z units above their bases.

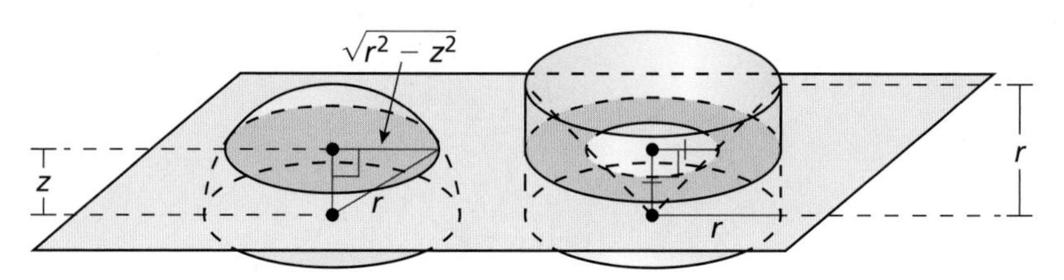

Using the AA Similarity Theorem (Theorem 8.3), you can show that the radius of the cross section of the cone at height z is z. The area of the cross section formed by the plane is $\pi(r^2 - z^2)$ for both solids. Because the solids have the same height and the same cross-sectional area at every level, they have the same volume by Cavalieri's Principle.

$$V_{\text{hemisphere}} = V_{\text{cylinder}} - V_{\text{cone}}$$

$$= \pi r^2(r) - \tfrac{1}{3}\pi r^2(r)$$

$$= \tfrac{2}{3}\pi r^3$$

So, the volume of a sphere of radius r is

$$2 \cdot V_{\text{hemisphere}} = 2 \cdot \tfrac{2}{3}\pi r^3 = \tfrac{4}{3}\pi r^3.$$

Core Concept

Volume of a Sphere

The volume V of a sphere is

$$V = \frac{4}{3}\pi r^3$$

where r is the radius of the sphere.

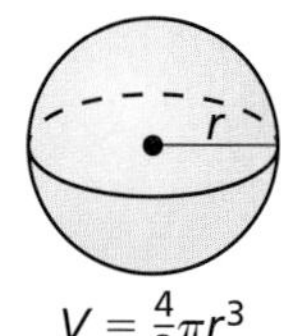

$V = \frac{4}{3}\pi r^3$

EXAMPLE 3 Finding the Volume of a Sphere

Find the volume of the soccer ball.

4.5 in.

SOLUTION

$V = \frac{4}{3}\pi r^3$	Formula for volume of a sphere
$= \frac{4}{3}\pi(4.5)^3$	Substitute 4.5 for r.
$= 121.5\pi$	Simplify.
≈ 381.70	Use a calculator.

▶ The volume of the soccer ball is 121.5π, or about 381.70 cubic inches.

EXAMPLE 4 Finding the Volume of a Sphere

The surface area of a sphere is 324π square centimeters. Find the volume of the sphere.

SOLUTION

Step 1 Use the surface area to find the radius.

$S = 4\pi r^2$	Formula for surface area of a sphere
$324\pi = 4\pi r^2$	Substitute 324π for S.
$81 = r^2$	Divide each side by 4π.
$9 = r$	Find the positive square root.

The radius is 9 centimeters.

Step 2 Use the radius to find the volume.

$V = \frac{4}{3}\pi r^3$	Formula for volume of a sphere
$= \frac{4}{3}\pi(9)^3$	Substitute 9 for r.
$= 972\pi$	Simplify.
≈ 3053.63	Use a calculator.

▶ The volume is 972π, or about 3053.63 cubic centimeters.

EXAMPLE 5 Finding the Volume of a Composite Solid

Find the volume of the composite solid.

2 in.
2 in.

SOLUTION

Volume of solid = Volume of cylinder − Volume of hemisphere

$= \pi r^2 h - \frac{1}{2}\left(\frac{4}{3}\pi r^3\right)$	Write formulas.
$= \pi(2)^2(2) - \frac{2}{3}\pi(2)^3$	Substitute.
$= 8\pi - \frac{16}{3}\pi$	Multiply.
$= \frac{24}{3}\pi - \frac{16}{3}\pi$	Rewrite fractions using least common denominator.
$= \frac{8}{3}\pi$	Subtract.
≈ 8.38	Use a calculator.

▶ The volume is $\frac{8}{3}\pi$, or about 8.38 cubic inches.

Monitoring Progress

Help in English and Spanish at *BigIdeasMath.com*

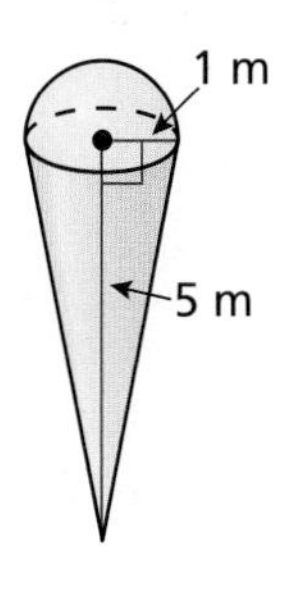

4. The radius of a sphere is 5 yards. Find the volume of the sphere.
5. The diameter of a sphere is 36 inches. Find the volume of the sphere.
6. The surface area of a sphere is 576π square centimeters. Find the volume of the sphere.
7. Find the volume of the composite solid at the left.

11.8 Exercises

Dynamic Solutions available at *BigIdeasMath.com*

Vocabulary and Core Concept Check

1. **VOCABULARY** When a plane intersects a sphere, what must be true for the intersection to be a great circle?

2. **WRITING** Explain the difference between a sphere and a hemisphere.

Monitoring Progress and Modeling with Mathematics

In Exercises 3–6, find the surface area of the sphere. *(See Example 1.)*

3. 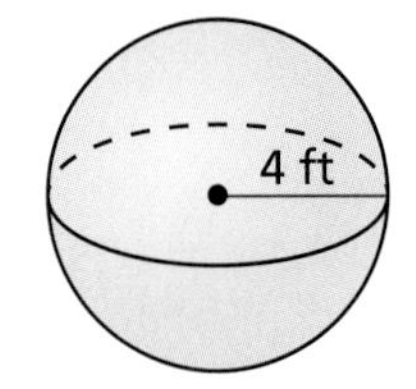

4. 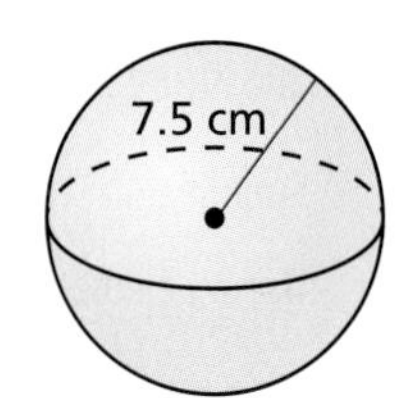

5. 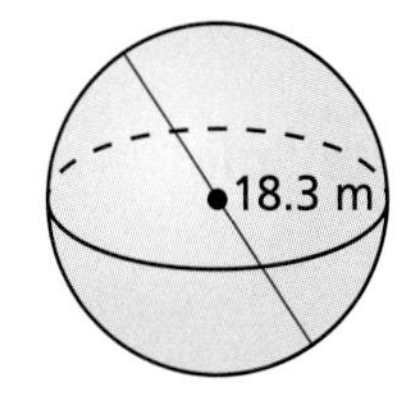

6. 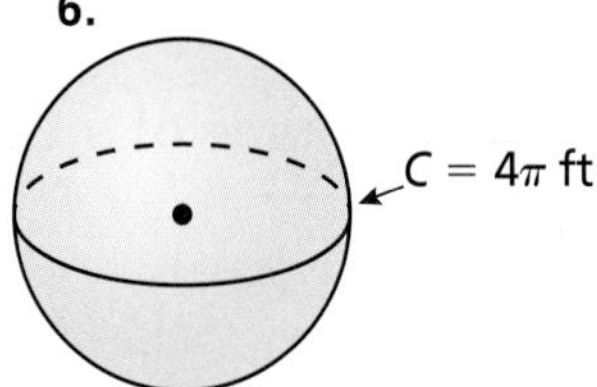

In Exercises 7–10, find the indicated measure. *(See Example 2.)*

7. Find the radius of a sphere with a surface area of 4π square feet.

8. Find the radius of a sphere with a surface area of 1024π square inches.

9. Find the diameter of a sphere with a surface area of 900π square meters.

10. Find the diameter of a sphere with a surface area of 196π square centimeters.

In Exercises 11 and 12, find the surface area of the hemisphere.

11. 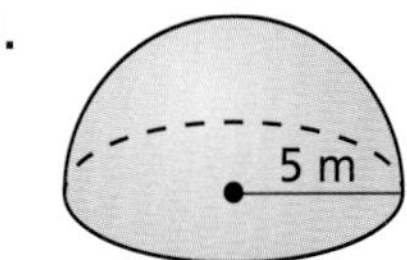

12. 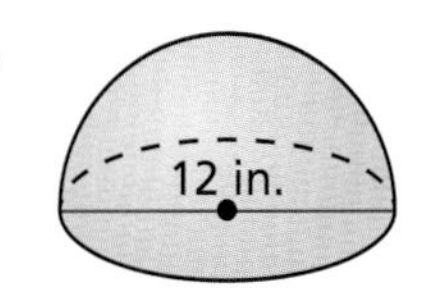

In Exercises 13–18, find the volume of the sphere. *(See Example 3.)*

13. 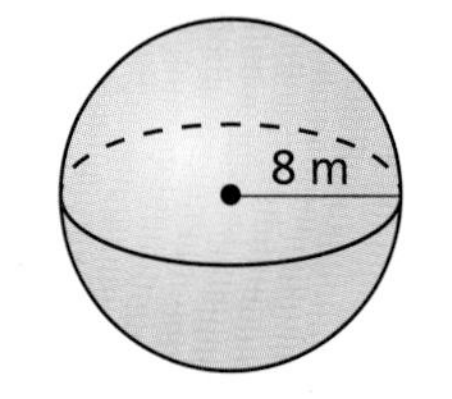

14. 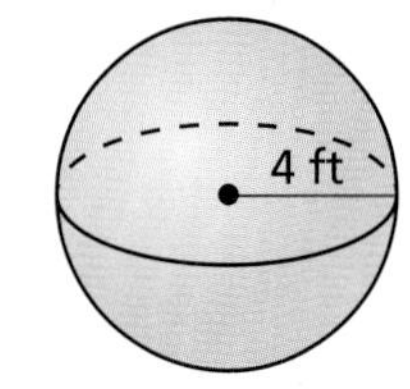

15. 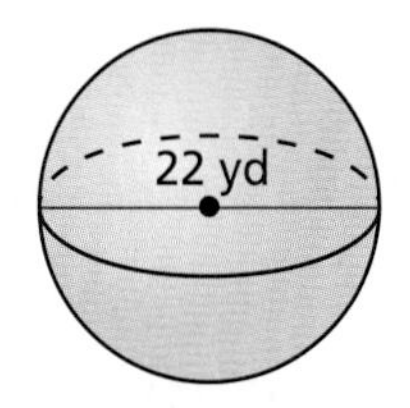

16. 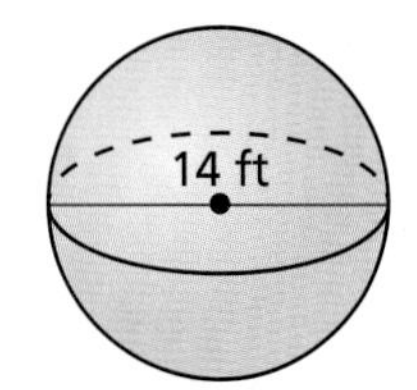

17. 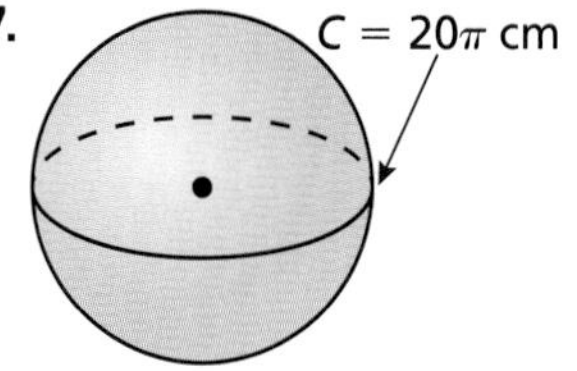

18. 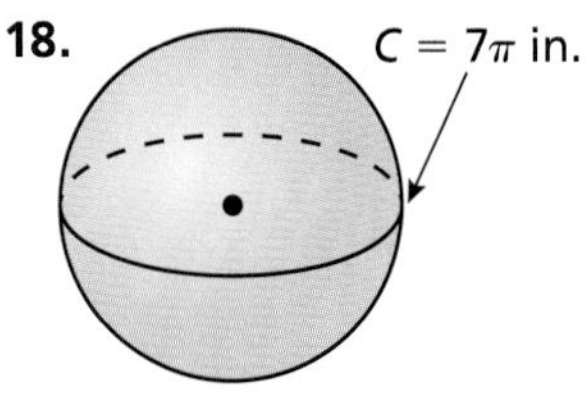

In Exercises 19 and 20, find the volume of the sphere with the given surface area. *(See Example 4.)*

19. Surface area $= 16\pi \text{ ft}^2$

20. Surface area $= 484\pi \text{ cm}^2$

21. **ERROR ANALYSIS** Describe and correct the error in finding the volume of the sphere.

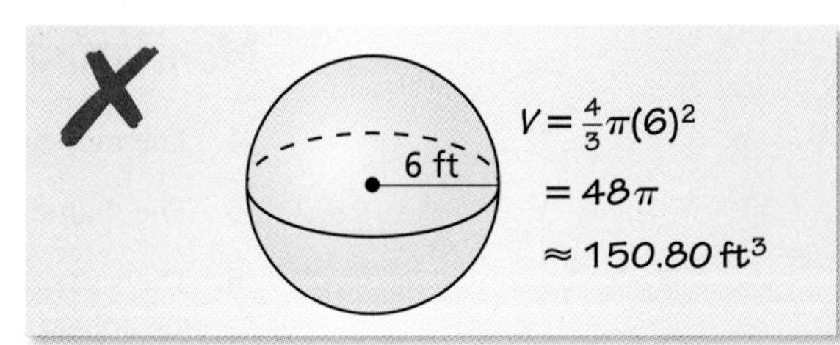

22. ERROR ANALYSIS Describe and correct the error in finding the volume of the sphere.

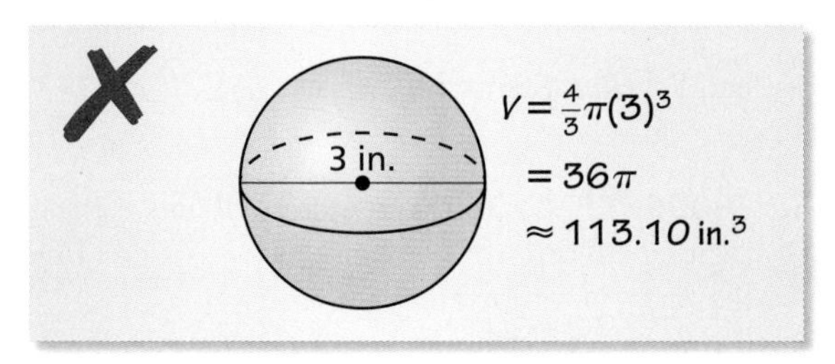

In Exercises 23–26, find the volume of the composite solid. *(See Example 5.)*

23.

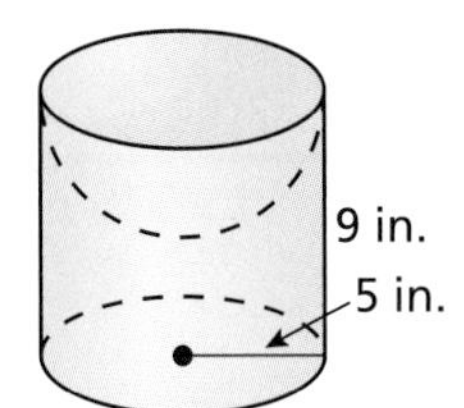

24.

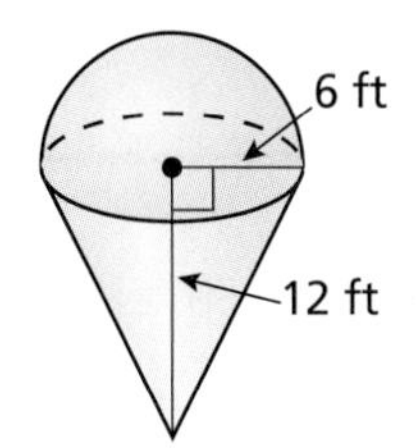

25.

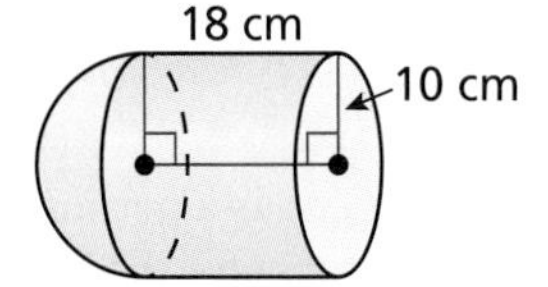

26.

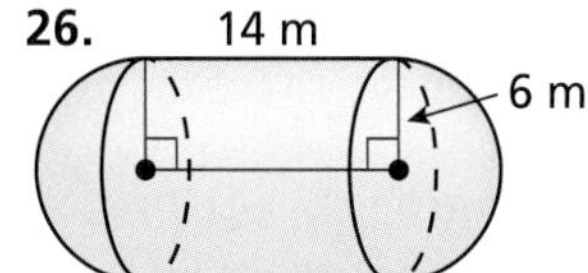

In Exercises 27–32, find the surface area and volume of the ball.

27. bowling ball

$d = 8.5$ in.

28. basketball

$C = 29.5$ in.

29. softball

$C = 12$ in.

30. golf ball

$d = 1.7$ in.

31. volleyball

$C = 26$ in.

32. baseball

$C = 9$ in.

33. MAKING AN ARGUMENT You friend claims that if the radius of a sphere is doubled, then the surface area of the sphere will also be doubled. Is your friend correct? Explain your reasoning.

34. REASONING A semicircle with a diameter of 18 inches is rotated about its diameter. Find the surface area and the volume of the solid formed.

35. MODELING WITH MATHEMATICS A silo has the dimensions shown. The top of the silo is a hemispherical shape. Find the volume of the silo.

36. MODELING WITH MATHEMATICS Three tennis balls are stored in a cylindrical container with a height of 8 inches and a radius of 1.43 inches. The circumference of a tennis ball is 8 inches.

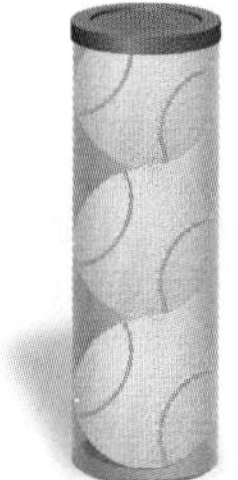

a. Find the volume of a tennis ball.

b. Find the amount of space within the cylinder not taken up by the tennis balls.

37. ANALYZING RELATIONSHIPS Use the table shown for a sphere.

Radius	Surface area	Volume
3 in.	36π in.^{2}	36π in.^{3}
6 in.		
9 in.		
12 in.		

a. Copy and complete the table. Leave your answers in terms of π.

b. What happens to the surface area of the sphere when the radius is doubled? tripled? quadrupled?

c. What happens to the volume of the sphere when the radius is doubled? tripled? quadrupled?

38. MATHEMATICAL CONNECTIONS A sphere has a diameter of $4(x + 3)$ centimeters and a surface area of 784π square centimeters. Find the value of x.

39. MODELING WITH MATHEMATICS The radius of Earth is about 3960 miles. The radius of the moon is about 1080 miles.

a. Find the surface area of Earth and the moon.

b. Compare the surface areas of Earth and the moon.

c. About 70% of the surface of Earth is water. How many square miles of water are on Earth's surface?

40. MODELING WITH MATHEMATICS The Torrid Zone on Earth is the area between the Tropic of Cancer and the Tropic of Capricorn. The distance between these two tropics is about 3250 miles. You can estimate the distance as the height of a cylindrical belt around the Earth at the equator.

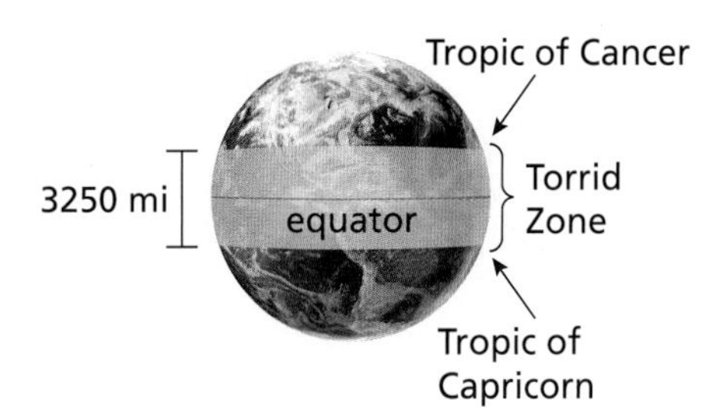

a. Estimate the surface area of the Torrid Zone. (The radius of Earth is about 3960 miles.)

b. A meteorite is equally likely to hit anywhere on Earth. Estimate the probability that a meteorite will land in the Torrid Zone.

41. ABSTRACT REASONING A sphere is inscribed in a cube with a volume of 64 cubic inches. What is the surface area of the sphere? Explain your reasoning.

42. HOW DO YOU SEE IT? The formula for the volume of a hemisphere and a cone are shown. If each solid has the same radius and $r = h$, which solid will have a greater volume? Explain your reasoning.

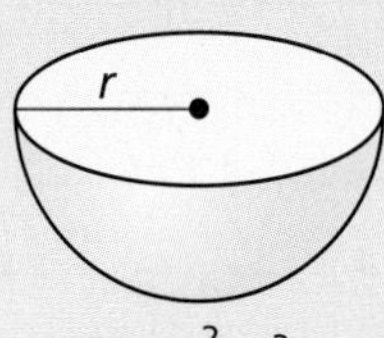

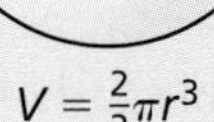

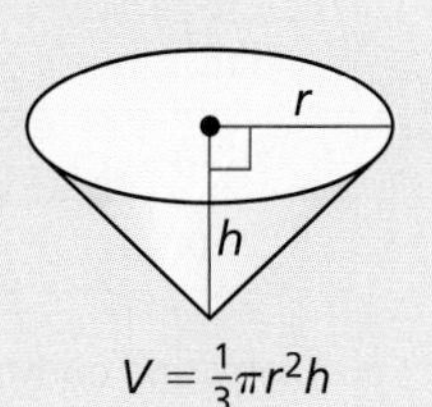

43. CRITICAL THINKING Let V be the volume of a sphere, S be the surface area of the sphere, and r be the radius of the sphere. Write an equation for V in terms of r and S. $\left(\textit{Hint}: \text{Start with the ratio } \frac{V}{S}.\right)$

44. THOUGHT PROVOKING A *spherical lune* is the region between two great circles of a sphere. Find the formula for the area of a lune.

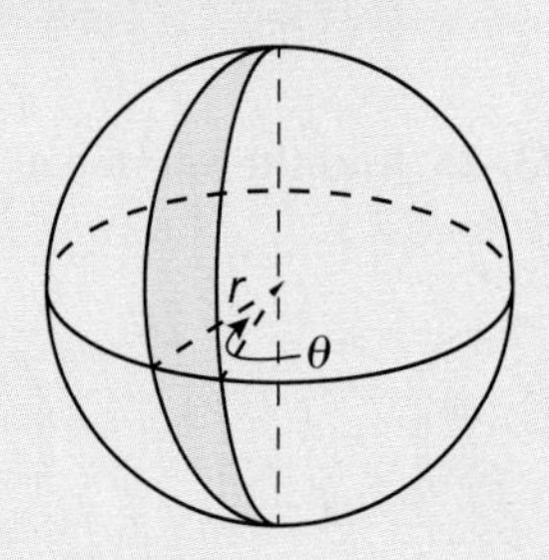

45. CRITICAL THINKING The volume of a right cylinder is the same as the volume of a sphere. The radius of the sphere is 1 inch. Give three possibilities for the dimensions of the cylinder.

46. PROBLEM SOLVING A *spherical cap* is a portion of a sphere cut off by a plane. The formula for the volume of a spherical cap is $V = \frac{\pi h}{6}(3a^2 + h^2)$, where a is the radius of the base of the cap and h is the height of the cap. Use the diagram and given information to find the volume of each spherical cap.

a. $r = 5$ ft, $a = 4$ ft

b. $r = 34$ cm, $a = 30$ cm

c. $r = 13$ m, $h = 8$ m

d. $r = 75$ in., $h = 54$ in.

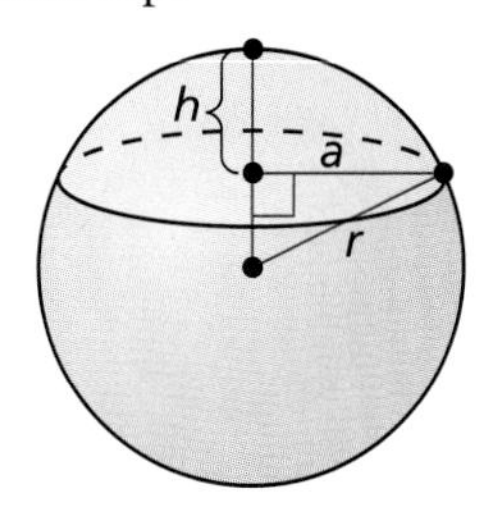

47. CRITICAL THINKING A sphere with a radius of 2 inches is inscribed in a right cone with a height of 6 inches. Find the surface area and the volume of the cone.

Maintaining Mathematical Proficiency

Reviewing what you learned in previous grades and lessons

Solve the triangle. Round decimal answers to the nearest tenth. *(Section 9.7)*

48. $A = 26°, C = 35°, b = 13$

49. $B = 102°, C = 43°, b = 21$

50. $a = 23, b = 24, c = 20$

51. $A = 103°, b = 15, c = 24$

11.5–11.8 What Did You Learn?

Core Vocabulary

Core Concepts

Section 11.5

Section 11.6

Section 11.7

Section 11.8

Mathematical Practices

1. Search online for advertisements for products that come in different sizes. Then compare the unit prices, as done in Exercise 44 on page 633. Do you get results similar to Exercise 44? Explain.
2. In Exercise 15 on page 639, explain why the volume changed by a factor of $\frac{1}{64}$.
3. In Exercise 38 on page 653, explain the steps you used to find the value of x.

Performance Task

Water Park Renovation

The city council will consider reopening the closed water park if your team can come up with a cost analysis for painting some of the structures, filling the pool water reservoirs, and resurfacing some of the surfaces. What is your plan to convince the city council to open the water park?

To explore the answers to these questions and more, go to ***BigIdeasMath.com***.

11.1 Circumference and Arc Length *(pp. 593–600)*

The arc length of $\widehat{QR}$ is 6.54 feet. Find the radius of $\odot P$.

$\dfrac{\text{Arc length of } \widehat{QR}}{2\pi r} = \dfrac{m\widehat{QR}}{360^\circ}$	Formula for arc length
$\dfrac{6.54}{2\pi r} = \dfrac{75^\circ}{360^\circ}$	Substitute.
$6.54(360) = 75(2\pi r)$	Cross Products Property
$5.00 \approx r$	Solve for r.

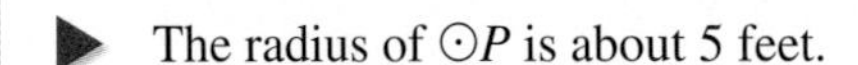

▶ The radius of $\odot P$ is about 5 feet.

Find the indicated measure.

1. diameter of $\odot P$

2. circumference of $\odot F$

3. arc length of $\widehat{AB}$

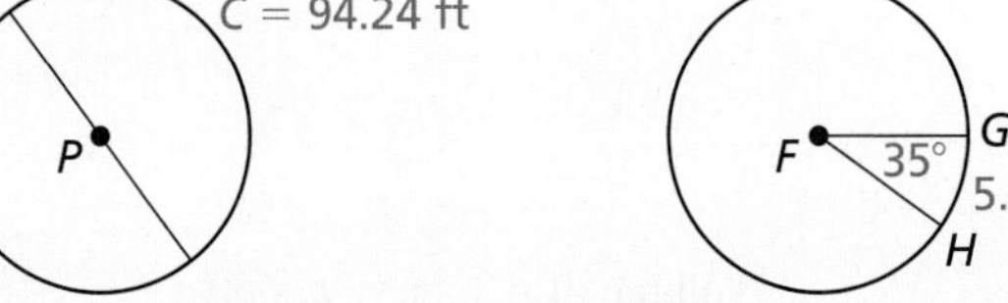

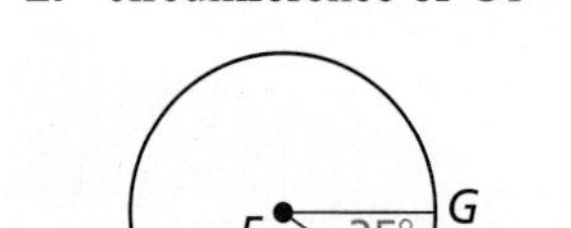

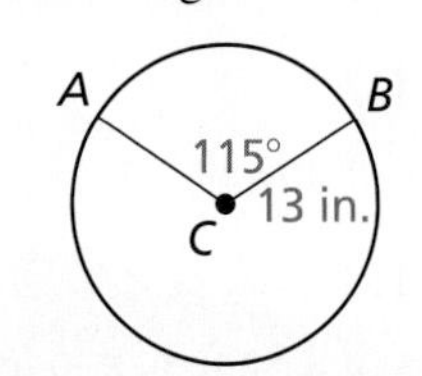

4. A mountain bike tire has a diameter of 26 inches. To the nearest foot, how far does the tire travel when it makes 32 revolutions?

11.2 Areas of Circles and Sectors *(pp. 601–608)*

Find the area of sector *ADB*.

Area of sector $ADB = \dfrac{m\widehat{AB}}{360^\circ} \cdot \pi r^2$	Formula for area of a sector
$= \dfrac{80^\circ}{360^\circ} \cdot \pi \cdot 10^2$	Substitute.
≈ 69.81	Use a calculator.

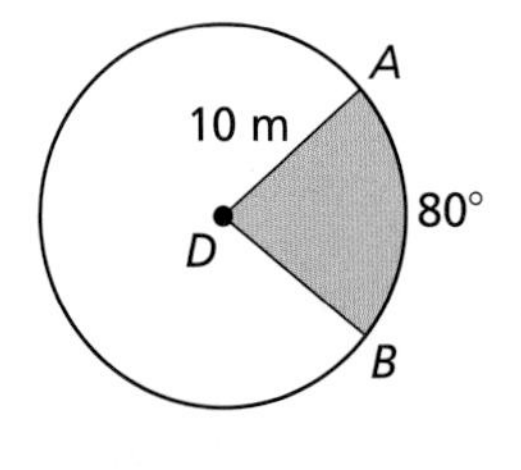

▶ The area of sector *ADB* is about 69.81 square meters.

Find the area of the blue shaded region.

5.

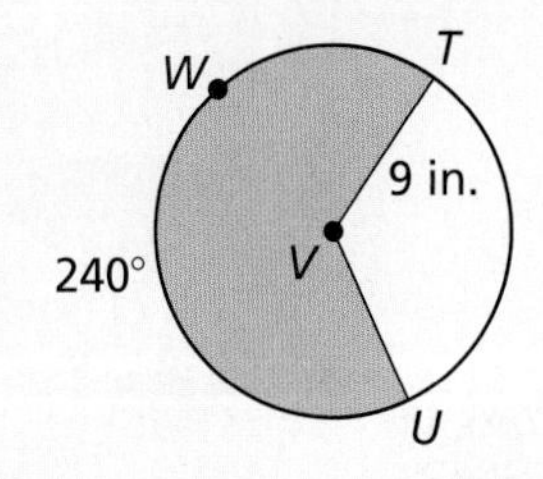

6.

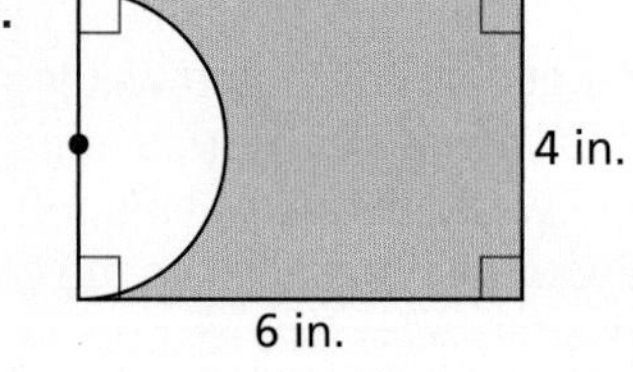

7.

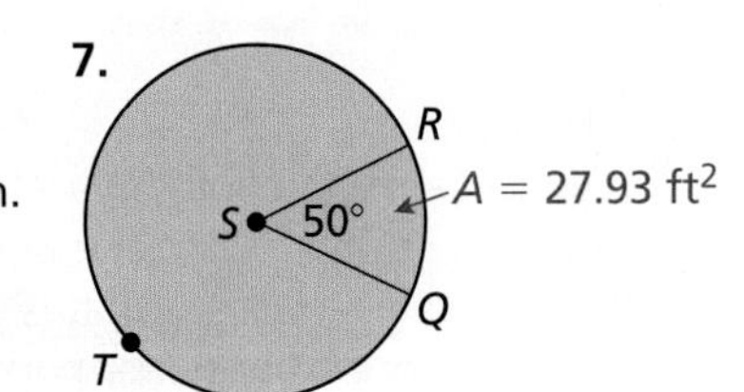

11.3 Areas of Polygons (pp. 609–616)

A regular hexagon is inscribed in $\odot H$. Find (a) $m\angle EHG$, and (b) the area of the hexagon.

a. $\angle FHE$ is a central angle, so $m\angle FHE = \frac{360^\circ}{6} = 60^\circ$.

Apothem $\overline{GH}$ bisects $\angle FHE$.

▶ So, $m\angle EHG = 30^\circ$.

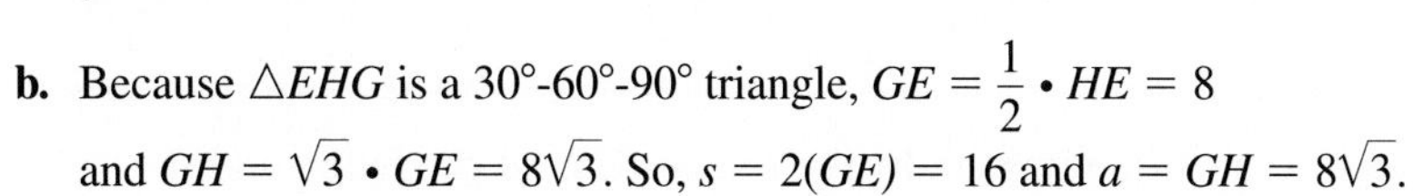

b. Because $\triangle EHG$ is a 30°-60°-90° triangle, $GE = \frac{1}{2} \cdot HE = 8$ and $GH = \sqrt{3} \cdot GE = 8\sqrt{3}$. So, $s = 2(GE) = 16$ and $a = GH = 8\sqrt{3}$.

▶ The area is $A = \frac{1}{2}a \cdot ns = \frac{1}{2}(8\sqrt{3})(6)(16) \approx 665.1$ square units.

Find the area of the kite or rhombus.

8.

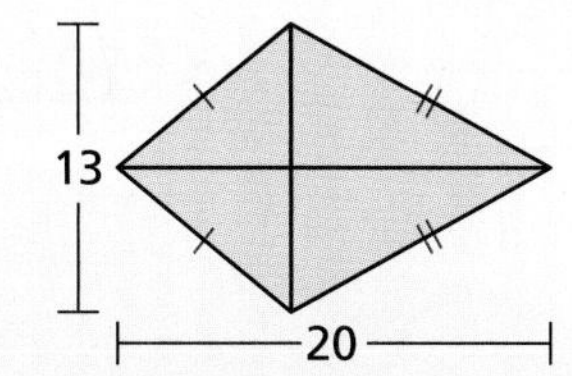

9.

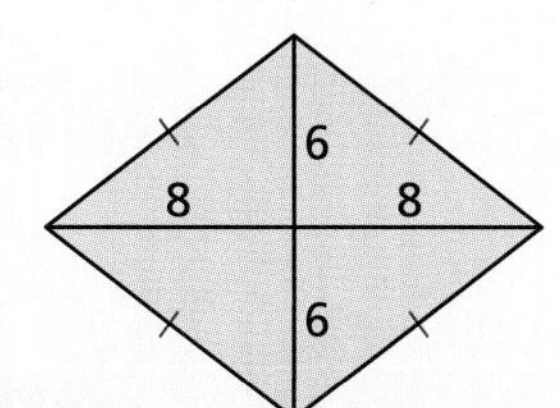

10.

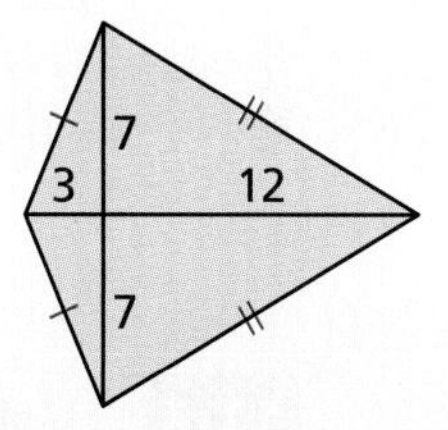

Find the area of the regular polygon.

11.

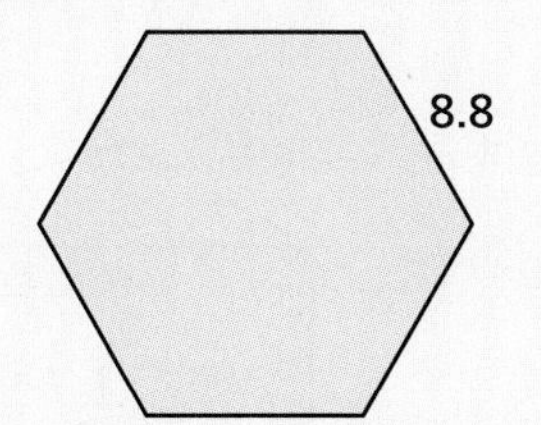

12.

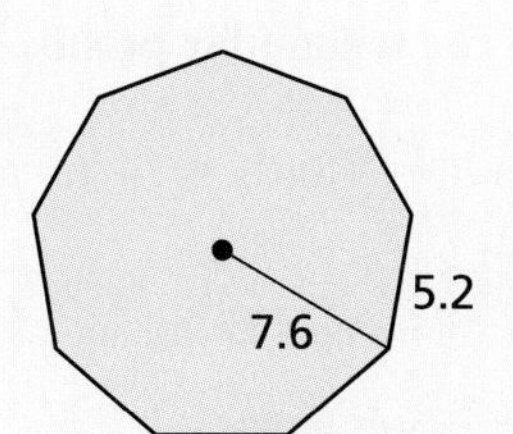

13.

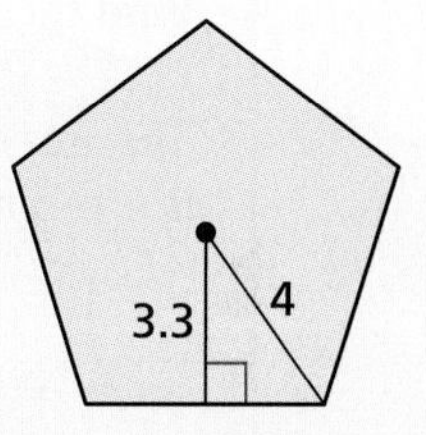

14. A platter is in the shape of a regular octagon with an apothem of 6 inches. Find the area of the platter.

11.4 Three-Dimensional Figures (pp. 617–622)

Sketch the solid produced by rotating the figure around the given axis. Then identify and describe the solid.

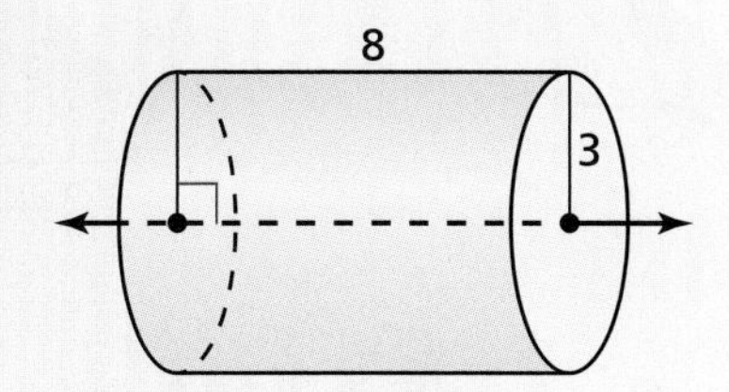

▶ The solid is a cylinder with a height of 8 and a radius of 3.

Sketch the solid produced by rotating the figure around the given axis. Then identify and describe the solid.

15.

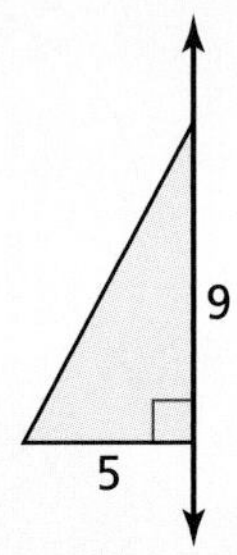

16.

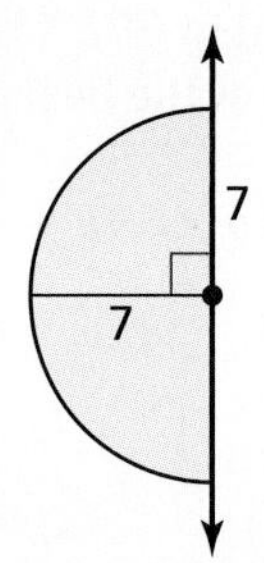

17.

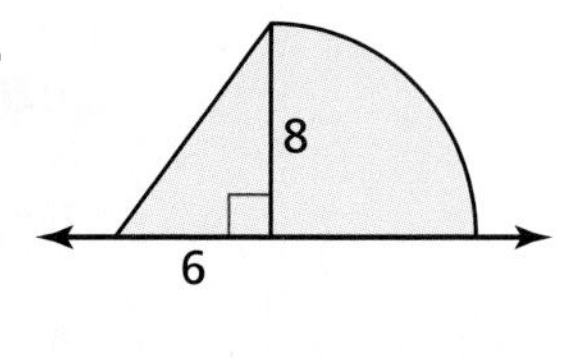

Describe the cross section formed by the intersection of the plane and the solid.

18.

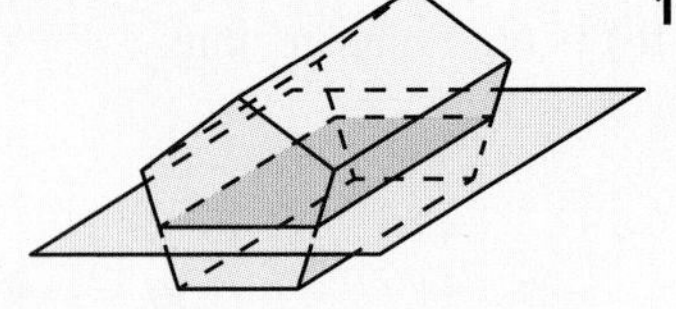

19.

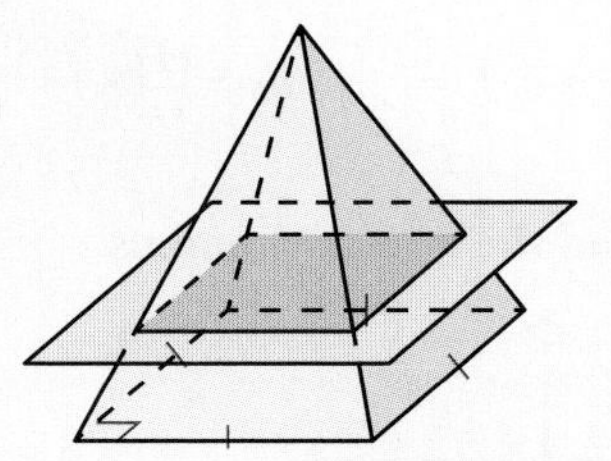

20.

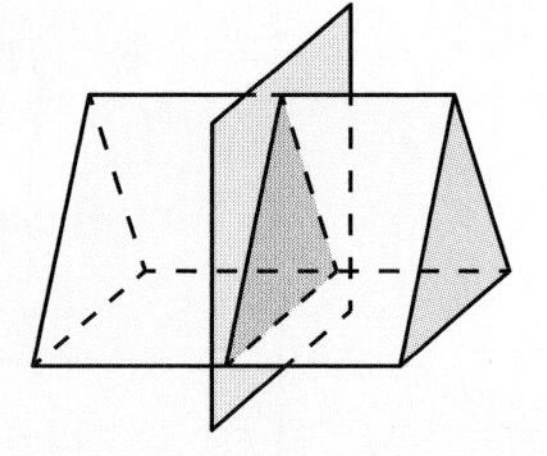

11.5 Volumes of Prisms and Cylinders *(pp. 625–634)*

Find the volume of the triangular prism.

The area of a base is $B = \frac{1}{2}(6)(8) = 24$ in.^{2} and the height is $h = 5$ in.

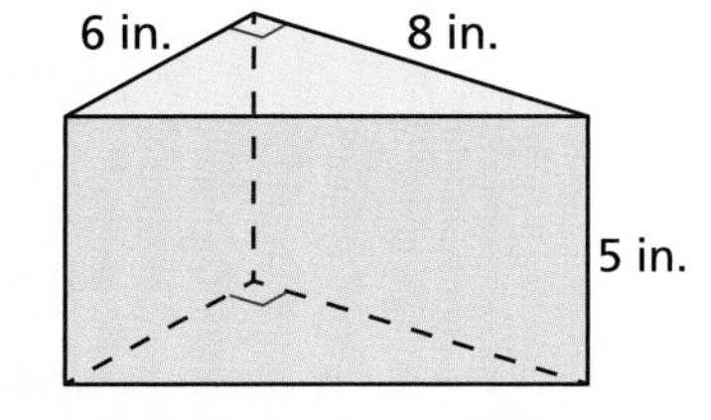

$V = Bh$	Formula for volume of a prism
$= 24(5)$	Substitute.
$= 120$	Simplify.

▶ The volume is 120 cubic inches.

Find the volume of the solid.

21.

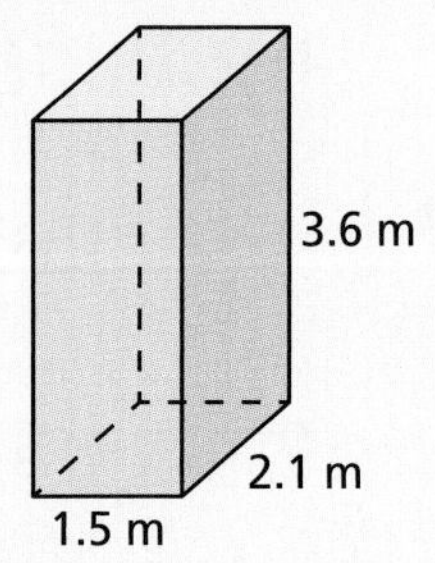

22.

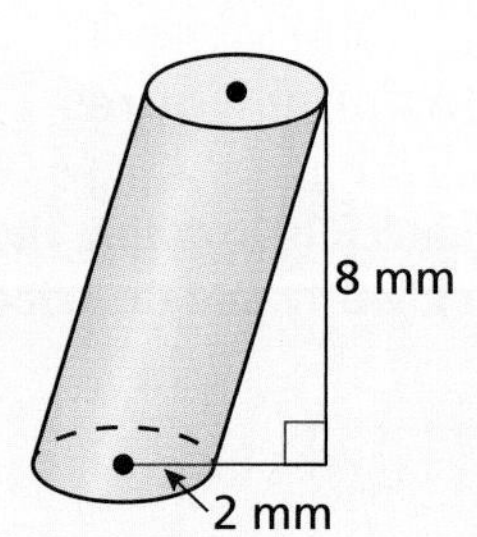

23.

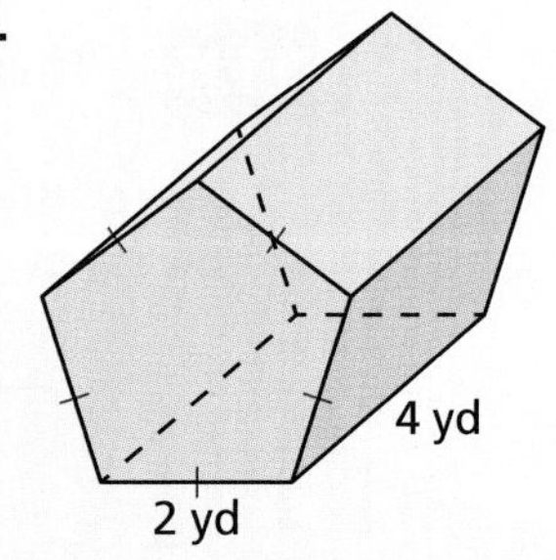

11.6 Volumes of Pyramids (pp. 635–640)

Find the volume of the pyramid.

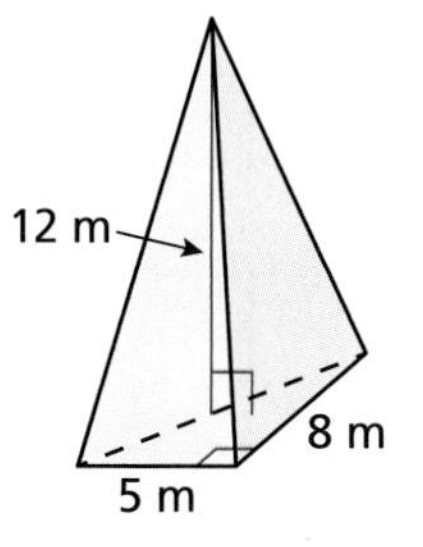

$V = \frac{1}{3}Bh$ Formula for volume of a pyramid

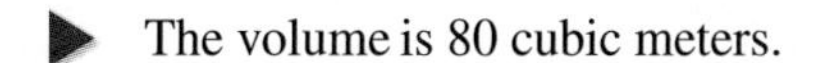

$= \frac{1}{3}\left(\frac{1}{2} \cdot 5 \cdot 8\right)(12)$ Substitute.

$= 80$ Simplify.

▶ The volume is 80 cubic meters.

Find the volume of the pyramid.

24.

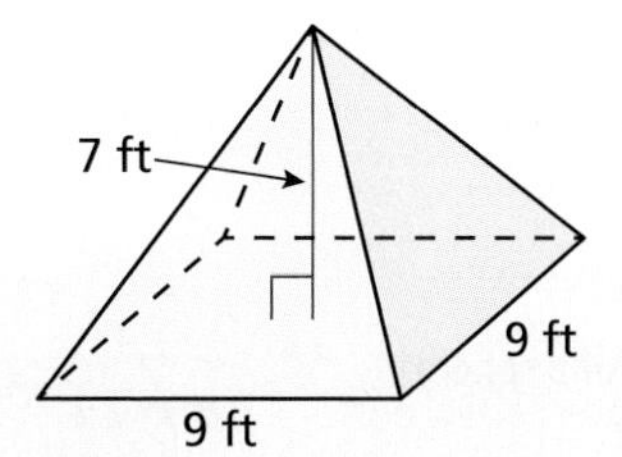

25.

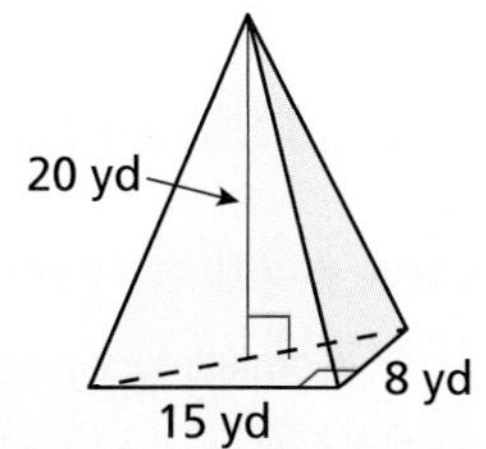

26.

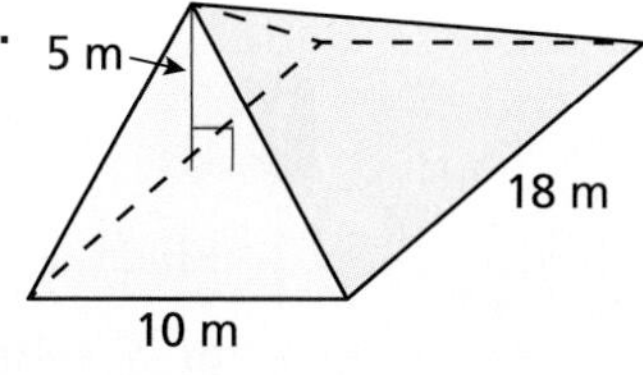

27. The volume of a square pyramid is 60 cubic inches and the height is 15 inches. Find the side length of the square base.

28. The volume of a square pyramid is 1024 cubic inches. The base has a side length of 16 inches. Find the height of the pyramid.

11.7 Surface Areas and Volumes of Cones (pp. 641–646)

Find the (a) surface area and (b) volume of the cone.

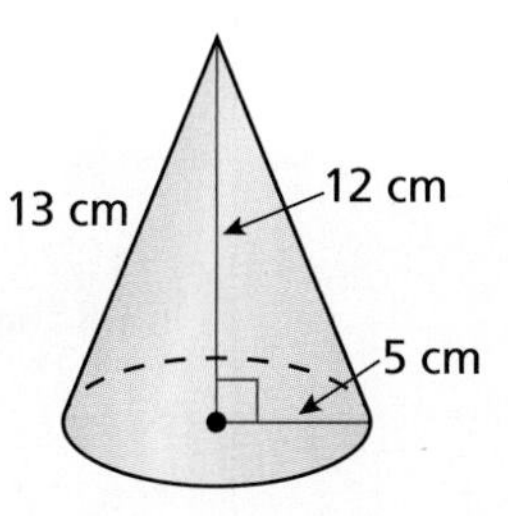

a. $S = \pi r^2 + \pi r\ell$ Formula for surface area of a cone

$= \pi \cdot 5^2 + \pi(5)(13)$ Substitute.

$= 90\pi$ Simplify.

≈ 282.74 Use a calculator.

▶ The surface area is 90π, or about 282.74 square centimeters.

b. $V = \frac{1}{3}\pi r^2 h$ Formula for volume of a cone

$= \frac{1}{3}\pi \cdot 5^2 \cdot 12$ Substitute.

$= 100\pi$ Simplify.

≈ 314.16 Use a calculator.

▶ The volume is 100π, or about 314.16 cubic centimeters.

Find the surface area and the volume of the cone.

29.

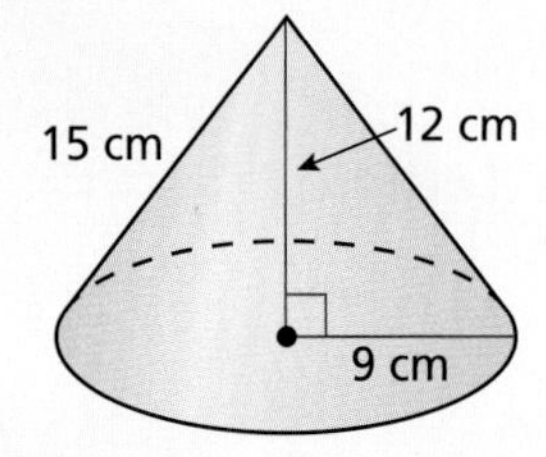

30.

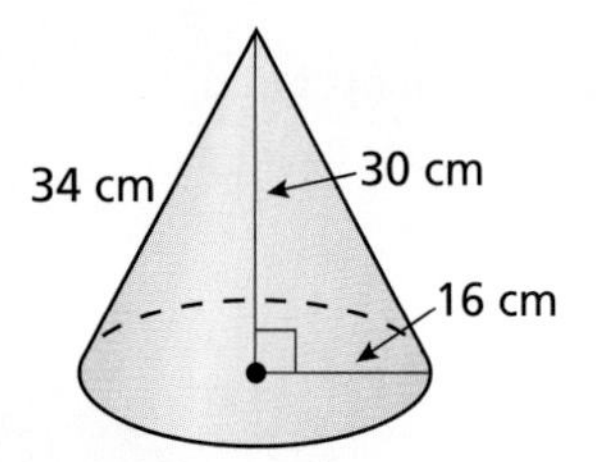

31.

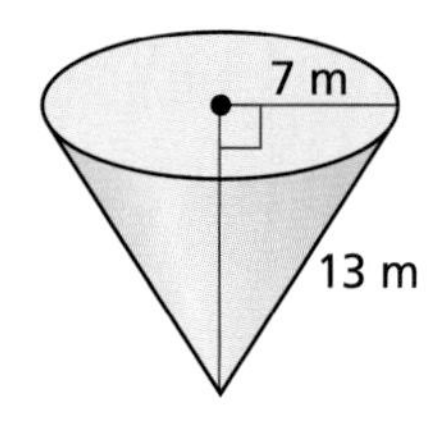

32. A cone with a diameter of 16 centimeters has a volume of 320π cubic centimeters. Find the height of the cone.

11.8 Surface Areas and Volumes of Spheres *(pp. 647–654)*

Find the (a) surface area and (b) volume of the sphere.

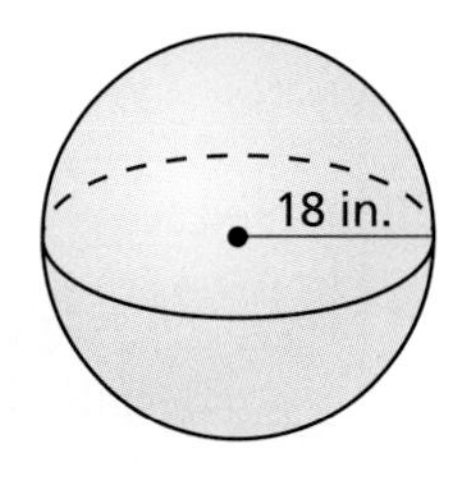

a. $S = 4\pi r^2$ Formula for surface area of a sphere

$= 4\pi(18)^2$ Substitute 18 for *r*.

$= 1296\pi$ Simplify.

≈ 4071.50 Use a calculator.

▶ The surface area is 1296π, or about 4071.50 square inches.

b. $V = \frac{4}{3}\pi r^3$ Formula for volume of a sphere

$= \frac{4}{3}\pi(18)^3$ Substitute 18 for *r*.

$= 7776\pi$ Simplify.

$\approx 24{,}429.02$ Use a calculator.

▶ The volume is 7776π, or about 24,429.02 cubic inches.

Find the surface area and the volume of the sphere.

33.

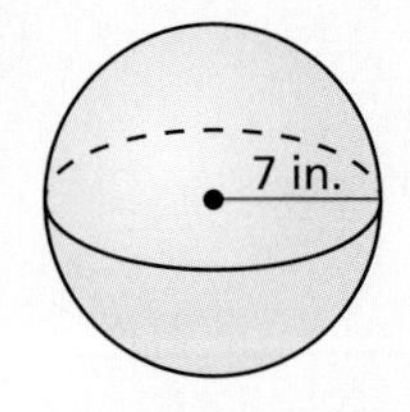

34.

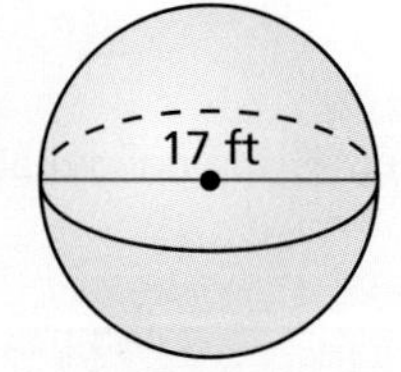

35.

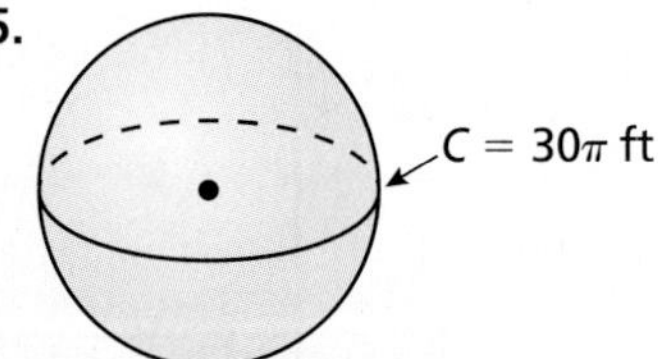

36. The shape of Mercury can be approximated by a sphere with a diameter of 4880 kilometers. Find the surface area and the volume of Mercury.

37. A solid is composed of a cube with a side length of 6 meters and a hemisphere with a diameter of 6 meters. Find the volume of the composite solid.

11 Chapter Test

Find the volume of the solid.

1. 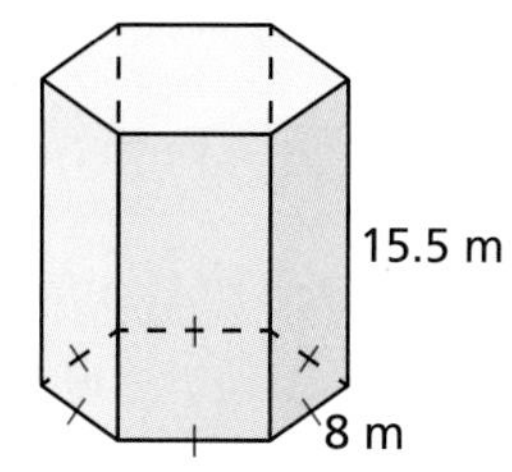

2. 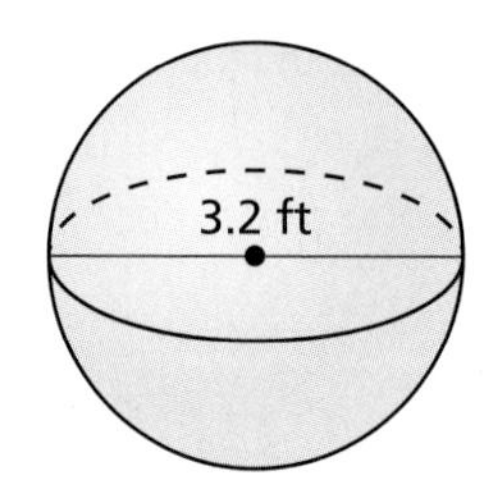

3. 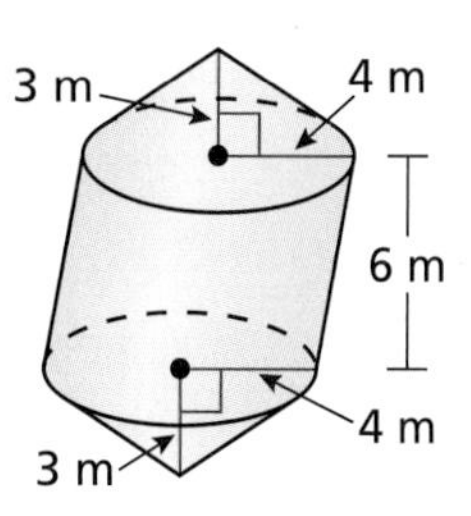

4. 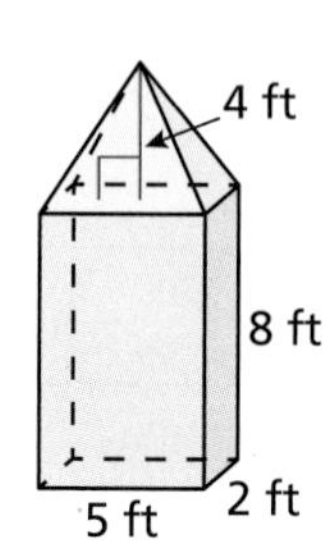

Find the indicated measure.

5. circumference of $\odot F$

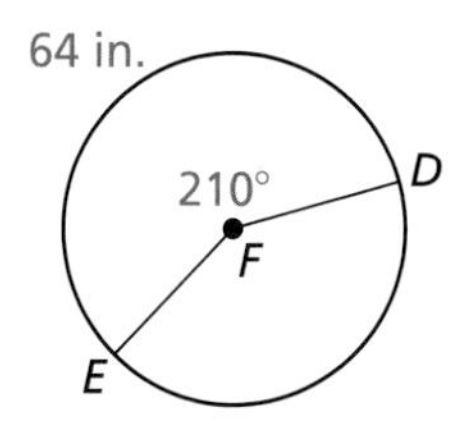

6. $m\widehat{GH}$

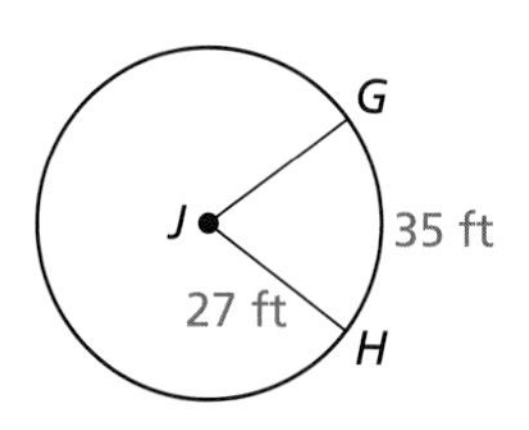

7. area of shaded sector

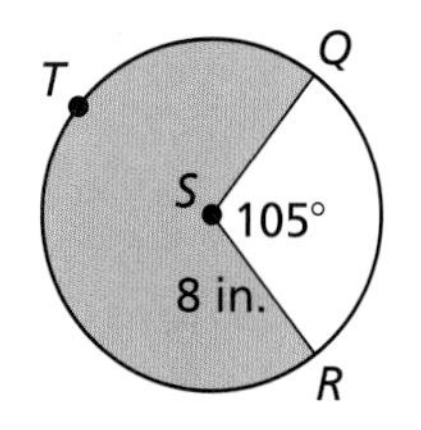

8. Sketch the composite solid produced by rotating the figure around the given axis. Then identify and describe the composite solid.

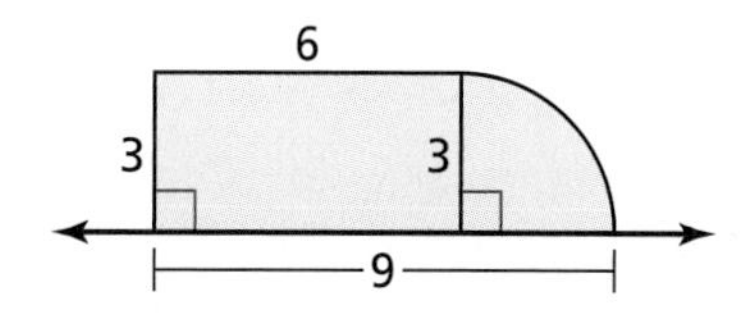

9. Find the surface area of a right cone with a diameter of 10 feet and a height of 12 feet.

10. You have a funnel with the dimensions shown.

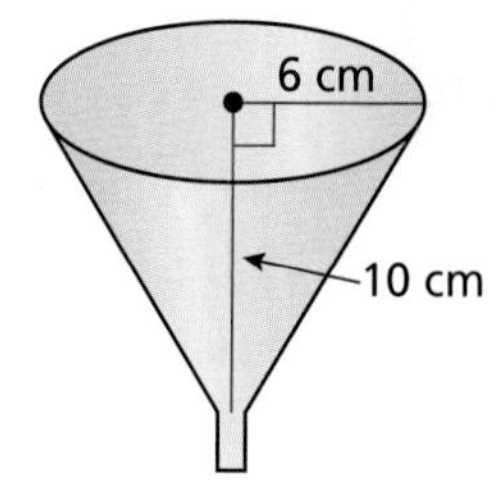

   a. Find the approximate volume of the funnel.

   b. You use the funnel to put oil in a car. Oil flows out of the funnel at a rate of 45 milliliters per second. How long will it take to empty the funnel when it is full of oil? ($1 \text{ mL} = 1 \text{ cm}^3$)

   c. How long would it take to empty a funnel with a radius of 10 centimeters and a height of 6 centimeters if oil flows out of the funnel at a rate of 45 milliliters per second?

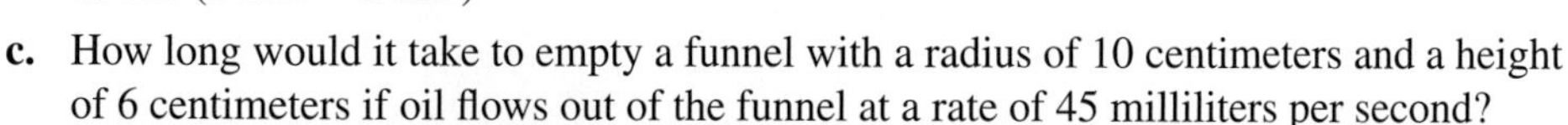

   d. Explain why you can claim that the time calculated in part (c) is greater than the time calculated in part (b) without doing any calculations.

11. A water bottle in the shape of a cylinder has a volume of 500 cubic centimeters. The diameter of a base is 7.5 centimeters. What is the height of the bottle? Justify your answer.

12. Find the area of a dodecagon (12 sides) with a side length of 9 inches.

13. In general, a cardboard fan with a greater area does a better job of moving air and cooling you. The fan shown is a sector of a cardboard circle. Another fan has a radius of 6 centimeters and an intercepted arc of 150°. Which fan does a better job of cooling you?

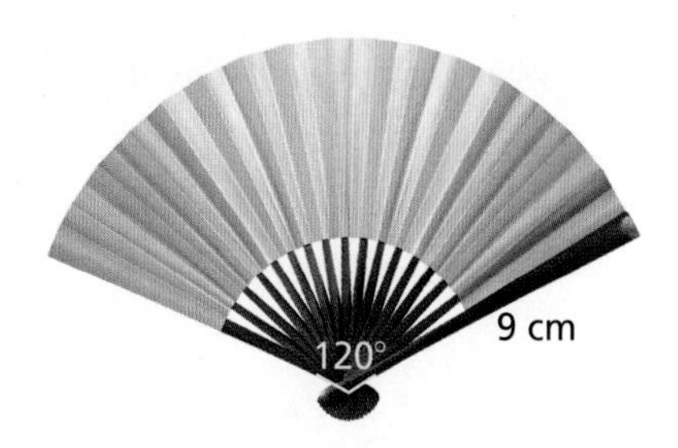

11 Cumulative Assessment

1. Identify the shape of the cross section formed by the intersection of the plane and the solid.

a.

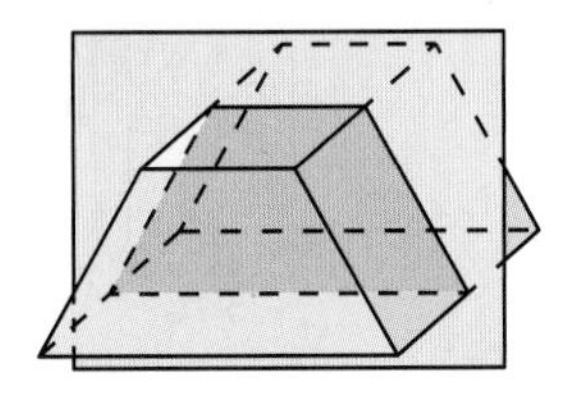

b.

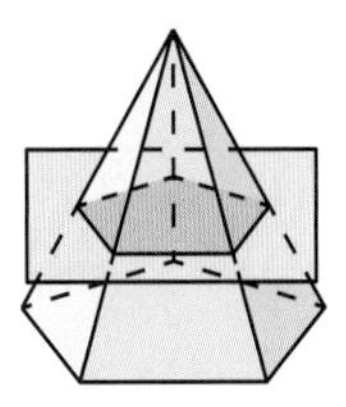

c.

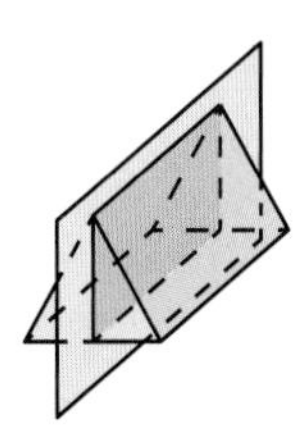

2. In the diagram, $\overleftrightarrow{RS}$ is tangent to $\odot P$ at Q and $\overline{PQ}$ is a radius of $\odot P$. What must be true about $\overleftrightarrow{RS}$ and $\overline{PQ}$? Select all that apply.

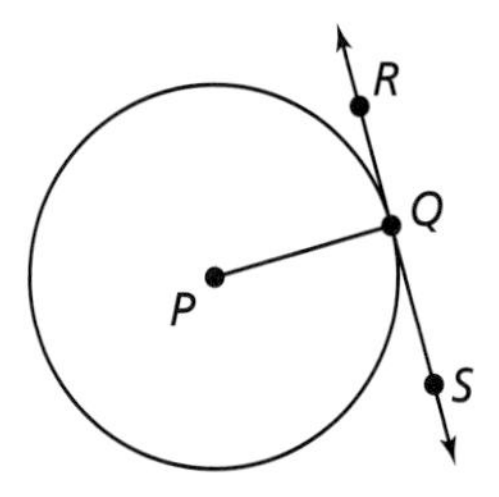

$PQ = \frac{1}{2}RS$	$PQ = RS$	$\overline{PQ}$ is tangent to $\odot P$.	$\overline{PQ} \perp \overleftrightarrow{RS}$

3. A crayon can be approximated by a composite solid made from a cylinder and a cone. A crayon box is a rectangular prism. The dimensions of a crayon and a crayon box containing 24 crayons are shown.

   a. Find the volume of a crayon.

   b. Find the amount of space within the crayon box not taken up by the crayons.

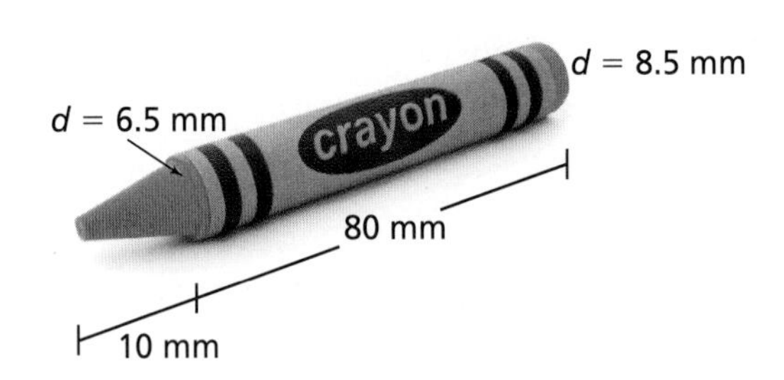

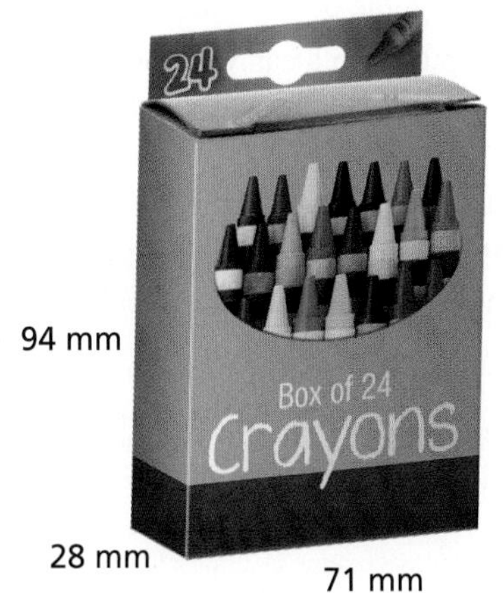

4. What is the equation of the line passing through the point (2, 5) that is parallel to the line $x + \frac{1}{2}y = -1$?

Ⓐ $y = -2x + 9$

Ⓑ $y = 2x + 1$

Ⓒ $y = \frac{1}{2}x + 4$

Ⓓ $y = -\frac{1}{2}x + 6$

5. The top of the Washington Monument in Washington, D.C., is a square pyramid, called a *pyramidion*. What is the volume of the pyramidion?

Ⓐ 22,019.63 ft^3

Ⓑ 172,006.91 ft^3

Ⓒ 66,058.88 ft^3

Ⓓ 207,530.08 ft^3

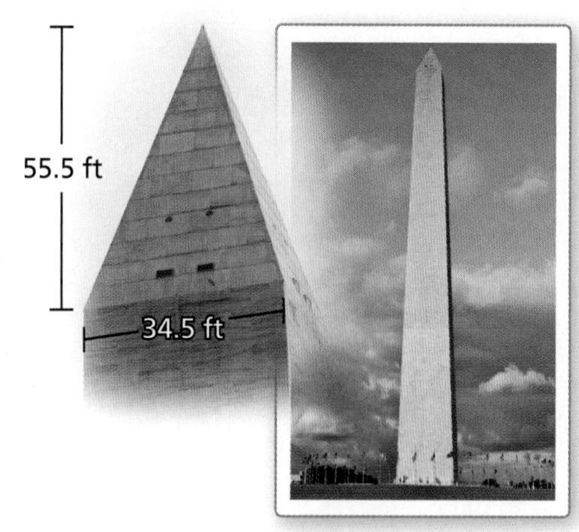

6. Prove or disprove that the point $(1, \sqrt{3})$ lies on the circle centered at the origin and containing the point (0, 2).

7. Your friend claims that the house shown can be described as a composite solid made from a rectangular prism and a triangular prism. Do you support your friend's claim? Explain your reasoning.

8. The diagram shows a square pyramid and a cone. Both solids have the same height, h, and the base of the cone has radius r. According to Cavalieri's Principle, the solids will have the same volume if the square base has sides of length ____.

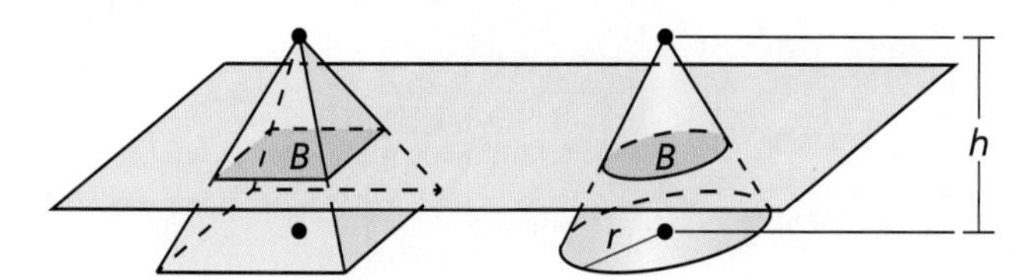

9. About 19,400 people live in a region with a 5-mile radius. Find the population density in people per square mile.

12 Probability

Class Ring *(p. 711)*

Horse Racing *(p. 701)*

Tree Growth *(p. 698)*

Jogging *(p. 687)*

Coaching *(p. 682)*

Maintaining Mathematical Proficiency

Finding a Percent

Example 1 **What percent of 12 is 9?**

$$\frac{a}{w} = \frac{p}{100}$$ Write the percent proportion.

$$\frac{9}{12} = \frac{p}{100}$$ Substitute 9 for *a* and 12 for *w*.

$$100 \cdot \frac{9}{12} = 100 \cdot \frac{p}{100}$$ Multiplication Property of Equality.

$$75 = p$$ Simplify.

▶ So, 9 is 75% of 12.

Write and solve a proportion to answer the question.

1. What percent of 30 is 6? **2.** What number is 68% of 25? **3.** 34.4 is what percent of 86?

Making a Histogram

Example 2 **The frequency table shows the ages of people at a gym. Display the data in a histogram.**

Age	Frequency
10–19	7
20–29	12
30–39	6
40–49	4
50–59	0
60–69	3

Step 1 Draw and label the axes.

Step 2 Draw a bar to represent the frequency of each interval.

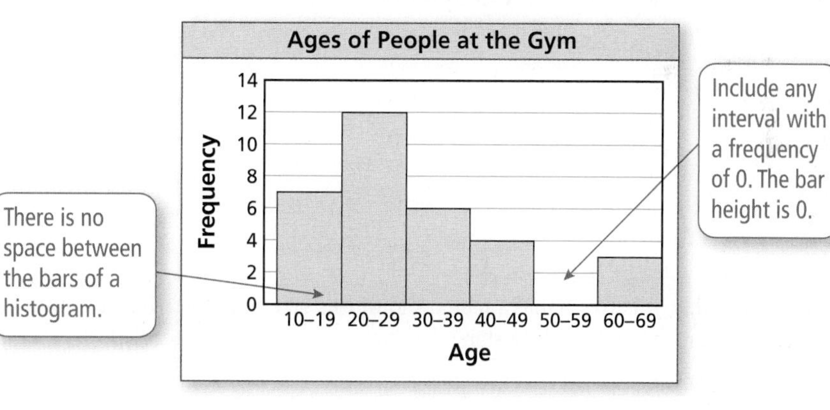

Display the data in a histogram.

4.

	Movies Watched per Week		
Movies	0–1	2–3	4–5
Frequency	35	11	6

5. **ABSTRACT REASONING** You want to purchase either a sofa or an arm chair at a furniture store. Each item has the same retail price. The sofa is 20% off. The arm chair is 10% off, and you have a coupon to get an additional 10% off the discounted price of the chair. Are the items equally priced after the discounts are applied? Explain.

Mathematical Practices

Mathematically proficient students apply the mathematics they know to solve real-life problems.

Modeling with Mathematics

Core Concept

Likelihoods and Probabilities

The **probability of an event** is a measure of the likelihood that the event will occur. Probability is a number from 0 to 1, including 0 and 1. The diagram relates *likelihoods* (described in words) and probabilities.

Words	Impossible	Unlikely	Equally likely to happen or not happen	Likely	Certain
Fraction	0	$\frac{1}{4}$	$\frac{1}{2}$	$\frac{3}{4}$	1
Decimal	0	0.25	0.5	0.75	1
Percent	0%	25%	50%	75%	100%

EXAMPLE 1 **Describing Likelihoods**

Describe the likelihood of each event.

Probability of an Asteroid or a Meteoroid Hitting Earth			
Name	Diameter	Probability of impact	Flyby date
a. Meteoroid	6 in.	0.75	Any day
b. Apophis	886 ft	0	2029
c. 2000 SG344	121 ft	$\frac{1}{435}$	2068–2110

SOLUTION

a. On any given day, it is *likely* that a meteoroid of this size will enter Earth's atmosphere. If you have ever seen a "shooting star," then you have seen a meteoroid.

b. A probability of 0 means this event is *impossible.*

c. With a probability of $\frac{1}{435} \approx 0.23\%$, this event is very *unlikely*. Of 435 identical asteroids, you would expect only one of them to hit Earth.

Monitoring Progress

In Exercises 1 and 2, describe the event as unlikely, equally likely to happen or not happen, or likely. Explain your reasoning.

1. The oldest child in a family is a girl.
2. The two oldest children in a family with three children are girls.
3. Give an example of an event that is certain to occur.

12.1 Sample Spaces and Probability

Essential Question How can you list the possible outcomes in the sample space of an experiment?

The **sample space** of an experiment is the set of all possible outcomes for that experiment.

EXPLORATION 1 Finding the Sample Space of an Experiment

Work with a partner. In an experiment, three coins are flipped. List the possible outcomes in the sample space of the experiment.

EXPLORATION 2 Finding the Sample Space of an Experiment

Work with a partner. List the possible outcomes in the sample space of the experiment.

a. One six-sided die is rolled.

b. Two six-sided dice are rolled.

EXPLORATION 3 Finding the Sample Space of an Experiment

Work with a partner. In an experiment, a spinner is spun.

a. How many ways can you spin a 1? 2? 3? 4? 5?

b. List the sample space.

c. What is the total number of outcomes?

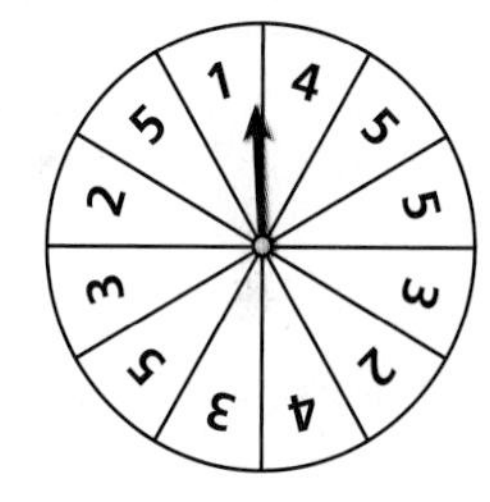

EXPLORATION 4 Finding the Sample Space of an Experiment

Work with a partner. In an experiment, a bag contains 2 blue marbles and 5 red marbles. Two marbles are drawn from the bag.

a. How many ways can you choose two blue? a red then blue? a blue then red? two red?

b. List the sample space.

c. What is the total number of outcomes?

LOOKING FOR A PATTERN

To be proficient in math, you need to look closely to discern a pattern or structure.

Communicate Your Answer

5. How can you list the possible outcomes in the sample space of an experiment?

6. For Exploration 3, find the ratio of the number of each possible outcome to the total number of outcomes. Then find the sum of these ratios. Repeat for Exploration 4. What do you observe?

12.1 Lesson

Core Vocabulary

probability experiment, *p. 668*
outcome, *p. 668*
event, *p. 668*
sample space, *p. 668*
probability of an event, *p. 668*
theoretical probability, *p. 669*
geometric probability, *p. 670*
experimental probability, *p. 671*

Previous
tree diagram

What You Will Learn

- Find sample spaces.
- Find theoretical probabilities.
- Find experimental probabilities.

Sample Spaces

A **probability experiment** is an action, or trial, that has varying results. The possible results of a probability experiment are **outcomes**. For instance, when you roll a six-sided die, there are 6 possible outcomes: 1, 2, 3, 4, 5, or 6. A collection of one or more outcomes is an **event**, such as rolling an odd number. The set of all possible outcomes is called a **sample space**.

EXAMPLE 1 Finding a Sample Space

You flip a coin and roll a six-sided die. How many possible outcomes are in the sample space? List the possible outcomes.

SOLUTION

Use a tree diagram to find the outcomes in the sample space.

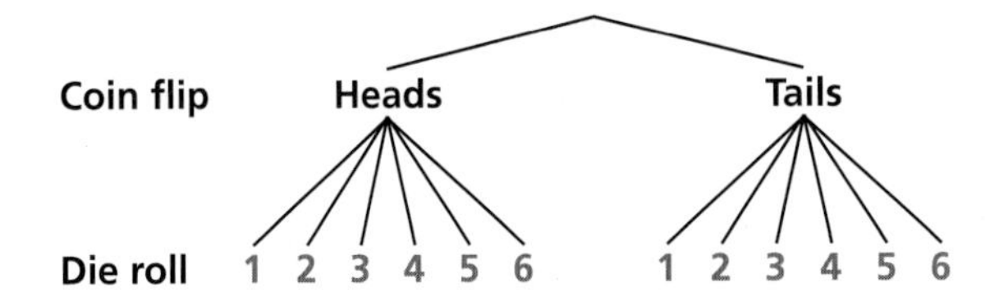

ANOTHER WAY

Using H for "heads" and T for "tails," you can list the outcomes as shown below.

H1	H2	H3	H4	H5	H6
T1	T2	T3	T4	T5	T6

- The sample space has 12 possible outcomes. They are listed below.

Heads, 1	Heads, 2	Heads, 3	Heads, 4	Heads, 5	Heads, 6
Tails, 1	Tails, 2	Tails, 3	Tails, 4	Tails, 5	Tails, 6

Monitoring Progress  Help in English and Spanish at *BigIdeasMath.com*

Find the number of possible outcomes in the sample space. Then list the possible outcomes.

1. You flip two coins.
2. You flip two coins and roll a six-sided die.

Theoretical Probabilities

The **probability of an event** is a measure of the likelihood, or chance, that the event will occur. Probability is a number from 0 to 1, including 0 and 1, and can be expressed as a decimal, fraction, or percent.

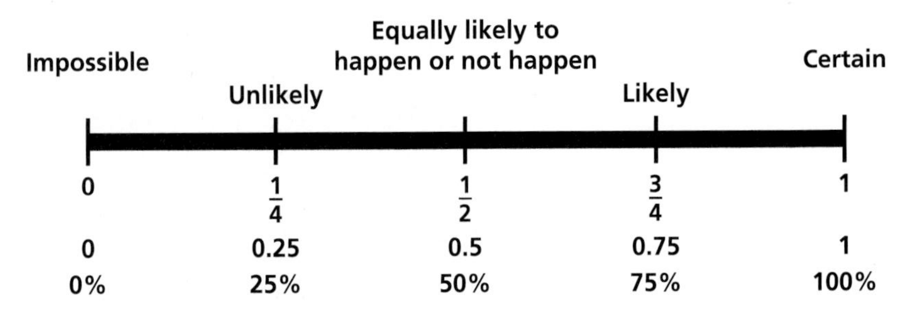

The outcomes for a specified event are called *favorable outcomes*. When all outcomes are equally likely, the **theoretical probability** of the event can be found using the following.

$$\text{Theoretical probability} = \frac{\text{Number of favorable outcomes}}{\text{Total number of outcomes}}$$

The probability of event A is written as $P(A)$.

ATTENDING TO PRECISION

Notice that the question uses the phrase "exactly two answers." This phrase is more precise than saying "two answers," which may be interpreted as "at least two" or as "exactly two."

EXAMPLE 2 Finding a Theoretical Probability

A student taking a quiz randomly guesses the answers to four true-false questions. What is the probability of the student guessing exactly two correct answers?

SOLUTION

Step 1 Find the outcomes in the sample space. Let C represent a correct answer and I represent an incorrect answer. The possible outcomes are:

Number correct	Outcome
0	IIII
1	CIII ICII IICI IIIC
2	IICC ICIC ICCI CIIC CICI CCII
3	ICCC CICC CCIC CCCI
4	CCCC

exactly two correct → 2

Step 2 Identify the number of favorable outcomes and the total number of outcomes. There are 6 favorable outcomes with exactly two correct answers and the total number of outcomes is 16.

Step 3 Find the probability of the student guessing exactly two correct answers. Because the student is randomly guessing, the outcomes should be equally likely. So, use the theoretical probability formula.

$$P(\text{exactly two correct answers}) = \frac{\text{Number of favorable outcomes}}{\text{Total number of outcomes}}$$

$$= \frac{6}{16}$$

$$= \frac{3}{8}$$

▶ The probability of the student guessing exactly two correct answers is $\frac{3}{8}$, or 37.5%.

The sum of the probabilities of all outcomes in a sample space is 1. So, when you know the probability of event A, you can find the probability of the *complement* of event A. The *complement* of event A consists of all outcomes that are not in A and is denoted by $\overline{A}$. The notation $\overline{A}$ is read as "A bar." You can use the following formula to find $P(\overline{A})$.

Core Concept

Probability of the Complement of an Event

The probability of the complement of event A is

$$P(\overline{A}) = 1 - P(A).$$

EXAMPLE 3 Finding Probabilities of Complements

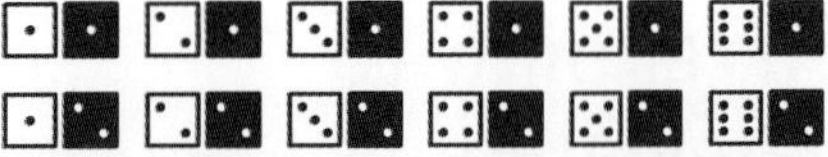

When two six-sided dice are rolled, there are 36 possible outcomes, as shown. Find the probability of each event.

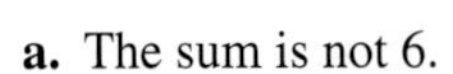

a. The sum is not 6.

b. The sum is less than or equal to 9.

SOLUTION

a. $P(\text{sum is not 6}) = 1 - P(\text{sum is 6}) = 1 - \frac{5}{36} = \frac{31}{36} \approx 0.861$

b. $P(\text{sum} \leq 9) = 1 - P(\text{sum} > 9) = 1 - \frac{6}{36} = \frac{30}{36} = \frac{5}{6} \approx 0.833$

Some probabilities are found by calculating a ratio of two lengths, areas, or volumes. Such probabilities are called **geometric probabilities**.

EXAMPLE 4 Using Area to Find Probability

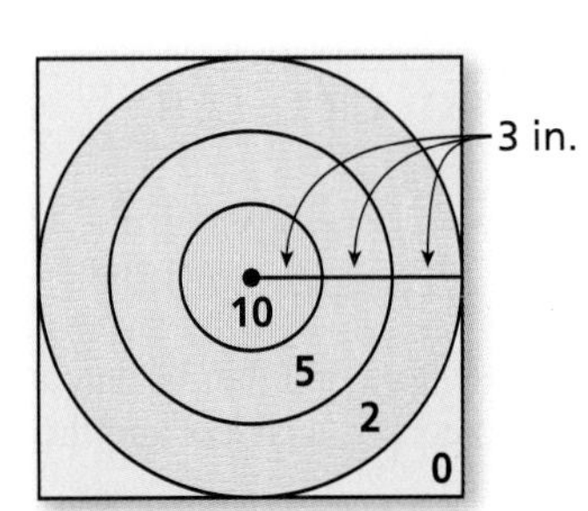

You throw a dart at the board shown. Your dart is equally likely to hit any point inside the square board. Are you more likely to get 10 points or 0 points?

SOLUTION

The probability of getting 10 points is

$$P(\text{10 points}) = \frac{\text{Area of smallest circle}}{\text{Area of entire board}} = \frac{\pi \cdot 3^2}{18^2} = \frac{9\pi}{324} = \frac{\pi}{36} \approx 0.0873.$$

The probability of getting 0 points is

$$\begin{aligned} P(\text{0 points}) &= \frac{\text{Area outside largest circle}}{\text{Area of entire board}} \\ &= \frac{18^2 - (\pi \cdot 9^2)}{18^2} \\ &= \frac{324 - 81\pi}{324} \\ &= \frac{4 - \pi}{4} \\ &\approx 0.215. \end{aligned}$$

▶ You are more likely to get 0 points.

Monitoring Progress  Help in English and Spanish at *BigIdeasMath.com*

3. You flip a coin and roll a six-sided die. What is the probability that the coin shows tails and the die shows 4?

Find $P(\overline{A})$.

4. $P(A) = 0.45$

5. $P(A) = \frac{1}{4}$

6. $P(A) = 1$

7. $P(A) = 0.03$

8. In Example 4, are you more likely to get 10 points or 5 points?

9. In Example 4, are you more likely to score points (10, 5, or 2) or get 0 points?

Experimental Probabilities

An **experimental probability** is based on repeated *trials* of a probability experiment. The number of trials is the number of times the probability experiment is performed. Each trial in which a favorable outcome occurs is called a *success*. The experimental probability can be found using the following.

$$\text{Experimental probability} = \frac{\text{Number of successes}}{\text{Number of trials}}$$

EXAMPLE 5 Finding an Experimental Probability

Spinner Results			
red	**green**	**blue**	**yellow**
5	9	3	3

Each section of the spinner shown has the same area. The spinner was spun 20 times. The table shows the results. For which color is the experimental probability of stopping on the color the same as the theoretical probability?

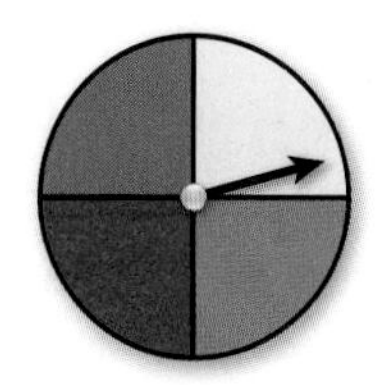

SOLUTION

The theoretical probability of stopping on each of the four colors is $\frac{1}{4}$. Use the outcomes in the table to find the experimental probabilities.

$P(\text{red}) = \frac{5}{20} = \frac{1}{4}$ $\qquad P(\text{green}) = \frac{9}{20}$

$P(\text{blue}) = \frac{3}{20}$ $\qquad P(\text{yellow}) = \frac{3}{20}$

▶ The experimental probability of stopping on red is the same as the theoretical probability.

EXAMPLE 6 Solving a Real-Life Problem

In the United States, a survey of 2184 adults ages 18 and over found that 1328 of them have at least one pet. The types of pets these adults have are shown in the figure. What is the probability that a pet-owning adult chosen at random has a dog?

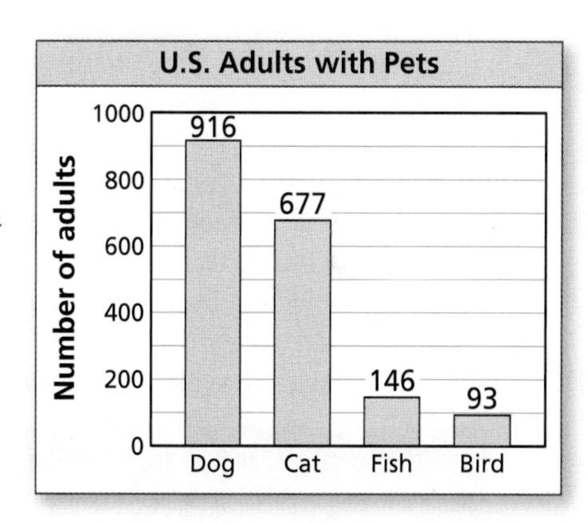

SOLUTION

The number of trials is the number of pet-owning adults, 1328. A success is a pet-owning adult who has a dog. From the graph, there are 916 adults who said that they have a dog.

$$P(\text{pet-owning adult has a dog}) = \frac{916}{1328} = \frac{229}{332} \approx 0.690$$

▶ The probability that a pet-owning adult chosen at random has a dog is about 69%.

Monitoring Progress  Help in English and Spanish at *BigIdeasMath.com*

10. In Example 5, for which color is the experimental probability of stopping on the color greater than the theoretical probability?

11. In Example 6, what is the probability that a pet-owning adult chosen at random owns a fish?

 Dynamic Solutions available at *BigIdeasMath.com*

Vocabulary and Core Concept Check

1. **COMPLETE THE SENTENCE** A number that describes the likelihood of an event is the __________ of the event.

2. **WRITING** Describe the difference between theoretical probability and experimental probability.

Monitoring Progress and Modeling with Mathematics

In Exercises 3–6, find the number of possible outcomes in the sample space. Then list the possible outcomes. *(See Example 1.)*

3. You roll a die and flip three coins.

4. You flip a coin and draw a marble at random from a bag containing two purple marbles and one white marble.

5. A bag contains four red cards numbered 1 through 4, four white cards numbered 1 through 4, and four black cards numbered 1 through 4. You choose a card at random.

6. You draw two marbles without replacement from a bag containing three green marbles and four black marbles.

7. **PROBLEM SOLVING** A game show airs on television five days per week. Each day, a prize is randomly placed behind one of two doors. The contestant wins the prize by selecting the correct door. What is the probability that exactly two of the five contestants win a prize during a week? *(See Example 2.)*

8. **PROBLEM SOLVING** Your friend has two standard decks of 52 playing cards and asks you to randomly draw one card from each deck. What is the probability that you will draw two spades?

9. **PROBLEM SOLVING** When two six-sided dice are rolled, there are 36 possible outcomes. Find the probability that (a) the sum is not 4 and (b) the sum is greater than 5. *(See Example 3.)*

10. **PROBLEM SOLVING** The age distribution of a population is shown. Find the probability of each event.

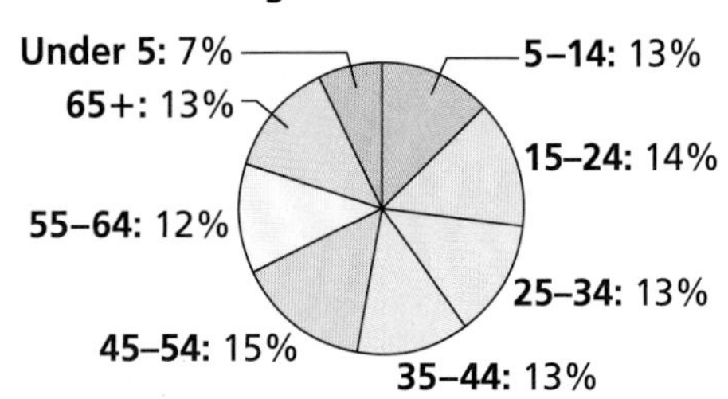

a. A person chosen at random is at least 15 years old.

b. A person chosen at random is from 25 to 44 years old.

11. **ERROR ANALYSIS** A student randomly guesses the answers to two true-false questions. Describe and correct the error in finding the probability of the student guessing both answers correctly.

The student can either guess two incorrect answers, two correct answers, or one of each. So the probability of guessing both answers correctly is $\frac{1}{3}$.

12. **ERROR ANALYSIS** A student randomly draws a number between 1 and 30. Describe and correct the error in finding the probability that the number drawn is greater than 4.

The probability that the number is less than 4 is $\frac{3}{30}$, or $\frac{1}{10}$. So, the probability that the number is greater than 4 is $1 - \frac{1}{10}$, or $\frac{9}{10}$.

13. **MATHEMATICAL CONNECTIONS** You throw a dart at the board shown. Your dart is equally likely to hit any point inside the square board. What is the probability your dart lands in the yellow region? *(See Example 4.)*

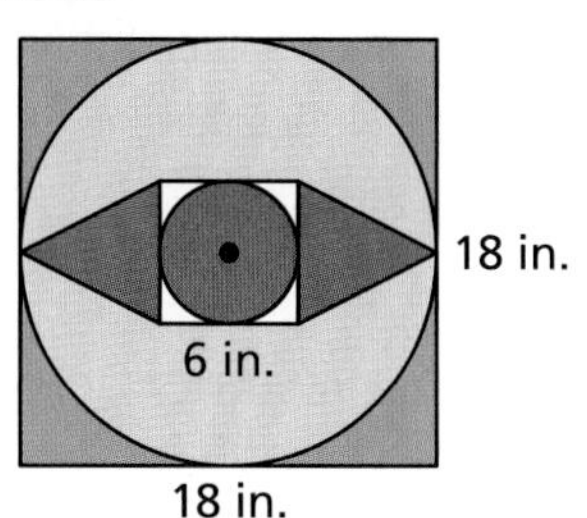

14. **MATHEMATICAL CONNECTIONS** The map shows the length (in miles) of shoreline along the Gulf of Mexico for each state that borders the body of water. What is the probability that a ship coming ashore at a random point in the Gulf of Mexico lands in the given state?

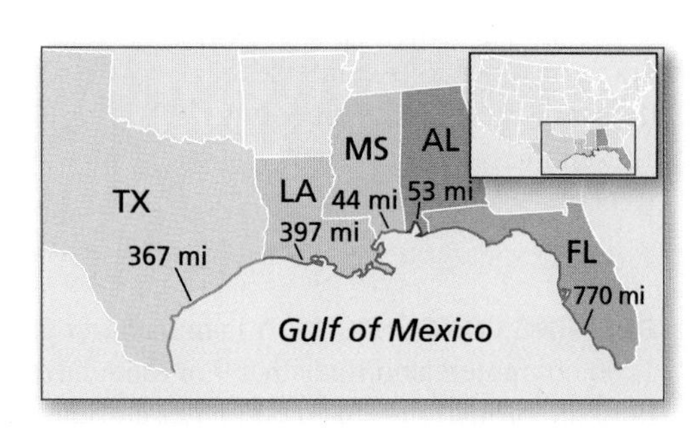

a. Texas
b. Alabama
c. Florida
d. Louisiana

15. **DRAWING CONCLUSIONS** You roll a six-sided die 60 times. The table shows the results. For which number is the experimental probability of rolling the number the same as the theoretical probability? *(See Example 5.)*

Six-sided Die Results					
⚀	⚁	⚂	⚃	⚄	⚅
11	14	7	10	6	12

16. **DRAWING CONCLUSIONS** A bag contains 5 marbles that are each a different color. A marble is drawn, its color is recorded, and then the marble is placed back in the bag. This process is repeated until 30 marbles have been drawn. The table shows the results. For which marble is the experimental probability of drawing the marble the same as the theoretical probability?

Drawing Results				
white	**black**	**red**	**green**	**blue**
5	6	8	2	9

17. **REASONING** Refer to the spinner shown. The spinner is divided into sections with the same area.

a. What is the theoretical probability that the spinner stops on a multiple of 3?

b. You spin the spinner 30 times. It stops on a multiple of 3 twenty times. What is the experimental probability of stopping on a multiple of 3?

c. Explain why the probability you found in part (b) is different than the probability you found in part (a).

18. **OPEN-ENDED** Describe a real-life event that has a probability of 0. Then describe a real-life event that has a probability of 1.

19. **DRAWING CONCLUSIONS** A survey of 2237 adults ages 18 and over asked which sport is their favorite. The results are shown in the figure. What is the probability that an adult chosen at random prefers auto racing? *(See Example 6.)*

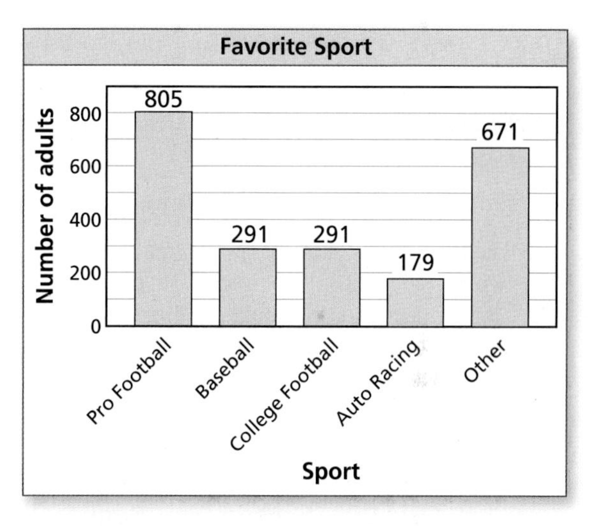

20. **DRAWING CONCLUSIONS** A survey of 2392 adults ages 18 and over asked what type of food they would be most likely to choose at a restaurant. The results are shown in the figure. What is the probability that an adult chosen at random prefers Italian food?

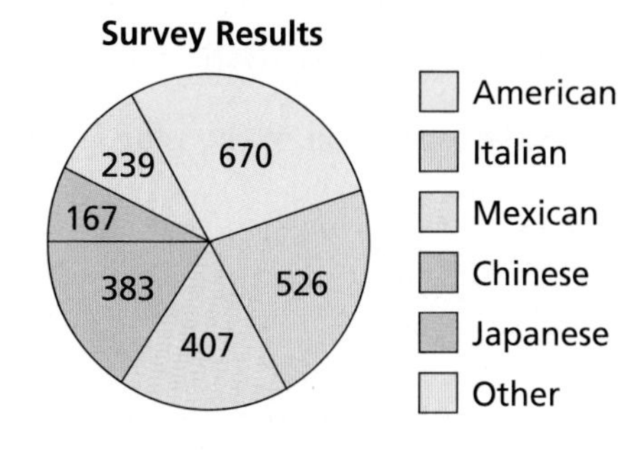

21. **ANALYZING RELATIONSHIPS** Refer to the board in Exercise 13. Order the likelihoods that the dart lands in the given region from least likely to most likely.

A. green **B.** not blue

C. red **D.** not yellow

22. **ANALYZING RELATIONSHIPS** Refer to the chart below. Order the following events from least likely to most likely.

Four-Day Forecast			
Friday	Saturday	Sunday	Monday
Chance of Rain **5%**	Chance of Rain **30%**	Chance of Rain **80%**	Chance of Rain **90%**

A. It rains on Sunday.

B. It does not rain on Saturday.

C. It rains on Monday.

D. It does not rain on Friday.

23. **USING TOOLS** Use the figure in Example 3 to answer each question.

a. List the possible sums that result from rolling two six-sided dice.

b. Find the theoretical probability of rolling each sum.

c. The table below shows a simulation of rolling two six-sided dice three times. Use a random number generator to simulate rolling two six-sided dice 50 times. Compare the experimental probabilities of rolling each sum with the theoretical probabilities.

	A	B	C
1	**First Die**	**Second Die**	**Sum**
2	4	6	10
3	3	5	8
4	1	6	7
5			

24. **MAKING AN ARGUMENT** You flip a coin three times. It lands on heads twice and on tails once. Your friend concludes that the theoretical probability of the coin landing heads up is $P(\text{heads up}) = \frac{2}{3}$. Is your friend correct? Explain your reasoning.

25. **MATHEMATICAL CONNECTIONS** A sphere fits inside a cube so that it touches each side, as shown. What is the probability a point chosen at random inside the cube is also inside the sphere?

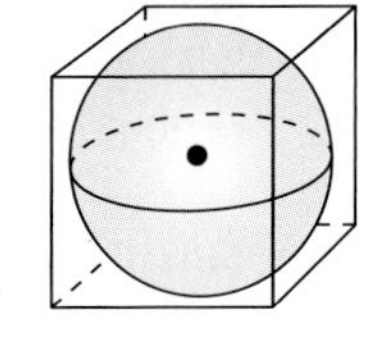

26. **HOW DO YOU SEE IT?** Consider the graph of f shown. What is the probability that the graph of $y = f(x) + c$ intersects the x-axis when c is a randomly chosen integer from 1 to 6? Explain.

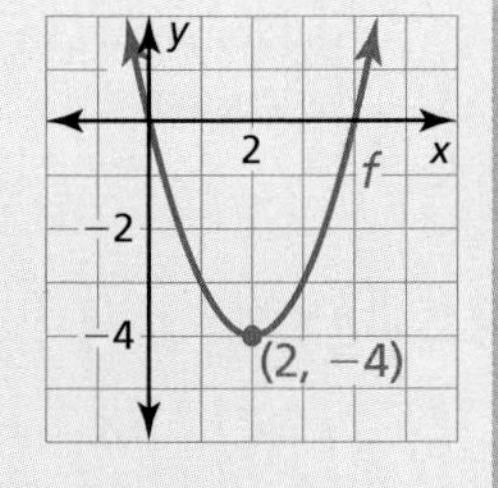

27. **DRAWING CONCLUSIONS** A manufacturer tests 1200 computers and finds that 9 of them have defects. Find the probability that a computer chosen at random has a defect. Predict the number of computers with defects in a shipment of 15,000 computers. Explain your reasoning.

28. **THOUGHT PROVOKING** The tree diagram shows a sample space. Write a probability problem that can be represented by the sample space. Then write the answer(s) to the problem.

Box A	Box B	Outcomes	Sum	Product
1	1	(1, 1)	2	1
	2	(1, 2)	3	2
2	1	(2, 1)	3	2
	2	(2, 2)	4	4
3	1	(3, 1)	4	3
	2	(3, 2)	5	6

Maintaining Mathematical Proficiency

Reviewing what you learned in previous grades and lessons

Simplify the expression. Write your answer using only positive exponents. *(Skills Review Handbook)*

29. $\frac{2x^3}{x^2}$

30. $\frac{2xy}{8y^2}$

31. $\frac{4x^9y}{3x^3y}$

32. $\frac{6y^0}{3x^{-6}}$

33. $(3pq)^4$

34. $\left(\frac{y^2}{x}\right)^{-2}$

12.2 Independent and Dependent Events

Essential Question

How can you determine whether two events are independent or dependent?

Two events are **independent events** when the occurrence of one event does not affect the occurrence of the other event. Two events are **dependent events** when the occurrence of one event *does* affect the occurrence of the other event.

EXPLORATION 1 Identifying Independent and Dependent Events

REASONING ABSTRACTLY

To be proficient in math, you need to make sense of quantities and their relationships in problem situations.

Work with a partner. Determine whether the events are independent or dependent. Explain your reasoning.

a. Two six-sided dice are rolled.

b. Six pieces of paper, numbered 1 through 6, are in a bag. Two pieces of paper are selected one at a time without replacement.

EXPLORATION 2 Finding Experimental Probabilities

Work with a partner.

a. In Exploration 1(a), experimentally estimate the probability that the sum of the two numbers rolled is 7. Describe your experiment.

b. In Exploration 1(b), experimentally estimate the probability that the sum of the two numbers selected is 7. Describe your experiment.

EXPLORATION 3 Finding Theoretical Probabilities

Work with a partner.

a. In Exploration 1(a), find the theoretical probability that the sum of the two numbers rolled is 7. Then compare your answer with the experimental probability you found in Exploration 2(a).

b. In Exploration 1(b), find the theoretical probability that the sum of the two numbers selected is 7. Then compare your answer with the experimental probability you found in Exploration 2(b).

c. Compare the probabilities you obtained in parts (a) and (b).

Communicate Your Answer

4. How can you determine whether two events are independent or dependent?

5. Determine whether the events are independent or dependent. Explain your reasoning.

a. You roll a 4 on a six-sided die and spin red on a spinner.

b. Your teacher chooses a student to lead a group, chooses another student to lead a second group, and chooses a third student to lead a third group.

12.2 Lesson

Core Vocabulary

independent events, *p. 676*
dependent events, *p. 677*
conditional probability, *p. 677*

Previous
probability
sample space

What You Will Learn

- Determine whether events are independent events.
- Find probabilities of independent and dependent events.
- Find conditional probabilities.

Determining Whether Events Are Independent

Two events are **independent events** when the occurrence of one event does not affect the occurrence of the other event.

Core Concept

Probability of Independent Events

Words Two events A and B are independent events if and only if the probability that both events occur is the product of the probabilities of the events.

Symbols $P(A \text{ and } B) = P(A) \cdot P(B)$

EXAMPLE 1 Determining Whether Events Are Independent

A student taking a quiz randomly guesses the answers to four true-false questions. Use a sample space to determine whether guessing Question 1 correctly and guessing Question 2 correctly are independent events.

SOLUTION

Using the sample space in Example 2 on page 669:

$P(\text{correct on Question 1}) = \frac{8}{16} = \frac{1}{2}$ $\qquad P(\text{correct on Question 2}) = \frac{8}{16} = \frac{1}{2}$

$P(\text{correct on Question 1 and correct on Question 2}) = \frac{4}{16} = \frac{1}{4}$

▶ Because $\frac{1}{2} \cdot \frac{1}{2} = \frac{1}{4}$, the events are independent.

EXAMPLE 2 Determining Whether Events Are Independent

A group of four students includes one boy and three girls. The teacher randomly selects one of the students to be the speaker and a different student to be the recorder. Use a sample space to determine whether randomly selecting a girl first and randomly selecting a girl second are independent events.

SOLUTION

Let B represent the boy. Let G_1, G_2, and G_3 represent the three girls. Use a table to list the outcomes in the sample space.

Number of girls	Outcome	
1	G_1B	BG_1
1	G_2B	BG_2
1	G_3B	BG_3
2	G_1G_2	G_2G_1
2	G_1G_3	G_3G_1
2	G_2G_3	G_3G_2

Using the sample space:

$P(\text{girl first}) = \frac{9}{12} = \frac{3}{4}$ $\qquad P(\text{girl second}) = \frac{9}{12} = \frac{3}{4}$

$P(\text{girl first and girl second}) = \frac{6}{12} = \frac{1}{2}$

▶ Because $\frac{3}{4} \cdot \frac{3}{4} \neq \frac{1}{2}$, the events are not independent.

Monitoring Progress  Help in English and Spanish at *BigIdeasMath.com*

1. In Example 1, determine whether guessing Question 1 incorrectly and guessing Question 2 correctly are independent events.
2. In Example 2, determine whether randomly selecting a girl first and randomly selecting a boy second are independent events.

Finding Probabilities of Events

In Example 1, it makes sense that the events are independent because the second guess should not be affected by the first guess. In Example 2, however, the selection of the second person *depends* on the selection of the first person because the same person cannot be selected twice. These events are *dependent*. Two events are **dependent events** when the occurrence of one event *does* affect the occurrence of the other event.

The probability that event B occurs given that event A has occurred is called the **conditional probability** of B given A and is written as $P(B|A)$.

MAKING SENSE OF PROBLEMS

One way that you can find $P(\text{girl second} \mid \text{girl first})$ is to list the 9 outcomes in which a girl is chosen first and then find the fraction of these outcomes in which a girl is chosen second:

$G_1B \quad G_2B \quad G_3B$

$G_1G_2 \quad G_2G_1 \quad G_3G_1$

$G_1G_3 \quad G_2G_3 \quad G_3G_2$

Core Concept

Probability of Dependent Events

Words If two events A and B are dependent events, then the probability that both events occur is the product of the probability of the first event and the conditional probability of the second event given the first event.

Symbols $P(A \text{ and } B) = P(A) \cdot P(B|A)$

Example Using the information in Example 2:

$$P(\text{girl first and girl second}) = P(\text{girl first}) \cdot P(\text{girl second} \mid \text{girl first})$$

$$= \frac{9}{12} \cdot \frac{6}{9} = \frac{1}{2}$$

EXAMPLE 3 Finding the Probability of Independent Events

As part of a board game, you need to spin the spinner, which is divided into equal parts. Find the probability that you get a 5 on your first spin and a number greater than 3 on your second spin.

SOLUTION

Let event A be "5 on first spin" and let event B be "greater than 3 on second spin."

The events are independent because the outcome of your second spin is not affected by the outcome of your first spin. Find the probability of each event and then multiply the probabilities.

$P(A) = \frac{1}{8}$ 1 of the 8 sections is a "5."

$P(B) = \frac{5}{8}$ 5 of the 8 sections (4, 5, 6, 7, 8) are greater than 3.

$$P(A \text{ and } B) = P(A) \cdot P(B) = \frac{1}{8} \cdot \frac{5}{8} = \frac{5}{64} \approx 0.078$$

▶ So, the probability that you get a 5 on your first spin and a number greater than 3 on your second spin is about 7.8%.

EXAMPLE 4 Finding the Probability of Dependent Events

A bag contains twenty \$1 bills and five \$100 bills. You randomly draw a bill from the bag, set it aside, and then randomly draw another bill from the bag. Find the probability that both events A and B will occur.

Event A: The first bill is \$100. **Event B:** The second bill is \$100.

SOLUTION

The events are dependent because there is one less bill in the bag on your second draw than on your first draw. Find $P(A)$ and $P(B|A)$. Then multiply the probabilities.

$P(A) = \frac{5}{25}$ 5 of the 25 bills are \$100 bills.

$P(B|A) = \frac{4}{24}$ 4 of the remaining 24 bills are \$100 bills.

$$P(A \text{ and } B) = P(A) \cdot P(B|A) = \frac{5}{25} \cdot \frac{4}{24} = \frac{1}{5} \cdot \frac{1}{6} = \frac{1}{30} \approx 0.033.$$

▶ So, the probability that you draw two \$100 bills is about 3.3%.

EXAMPLE 5 Comparing Independent and Dependent Events

You randomly select 3 cards from a standard deck of 52 playing cards. What is the probability that all 3 cards are hearts when (a) you replace each card before selecting the next card, and (b) you do not replace each card before selecting the next card? Compare the probabilities.

STUDY TIP

The formulas for finding probabilities of independent and dependent events can be extended to three or more events.

SOLUTION

Let event A be "first card is a heart," event B be "second card is a heart," and event C be "third card is a heart."

a. Because you replace each card before you select the next card, the events are independent. So, the probability is

$$P(A \text{ and } B \text{ and } C) = P(A) \cdot P(B) \cdot P(C) = \frac{13}{52} \cdot \frac{13}{52} \cdot \frac{13}{52} = \frac{1}{64} \approx 0.016.$$

b. Because you do not replace each card before you select the next card, the events are dependent. So, the probability is

$$P(A \text{ and } B \text{ and } C) = P(A) \cdot P(B|A) \cdot P(C|A \text{ and } B)$$

$$= \frac{13}{52} \cdot \frac{12}{51} \cdot \frac{11}{50} = \frac{11}{850} \approx 0.013.$$

▶ So, you are $\frac{1}{64} \div \frac{11}{850} \approx 1.2$ times more likely to select 3 hearts when you replace each card before you select the next card.

Monitoring Progress  Help in English and Spanish at *BigIdeasMath.com*

3. In Example 3, what is the probability that you spin an even number and then an odd number?

4. In Example 4, what is the probability that both bills are \$1 bills?

5. In Example 5, what is the probability that none of the cards drawn are hearts when (a) you replace each card, and (b) you do not replace each card? Compare the probabilities.

Finding Conditional Probabilities

EXAMPLE 6 Using a Table to Find Conditional Probabilities

	Pass	Fail
Defective	3	36
Non-defective	450	11

A quality-control inspector checks for defective parts. The table shows the results of the inspector's work. Find (a) the probability that a defective part "passes," and (b) the probability that a non-defective part "fails."

SOLUTION

a. $P(\text{pass} \mid \text{defective}) = \dfrac{\text{Number of defective parts "passed"}}{\text{Total number of defective parts}}$

$$= \frac{3}{3+36} = \frac{3}{39} = \frac{1}{13} \approx 0.077, \text{ or about } 7.7\%$$

b. $P(\text{fail} \mid \text{non-defective}) = \dfrac{\text{Number of non-defective parts "failed"}}{\text{Total number of non-defective parts}}$

$$= \frac{11}{450+11} = \frac{11}{461} \approx 0.024, \text{ or about } 2.4\%$$

STUDY TIP

Note that when A and B are independent, this rule still applies because $P(B) = P(B \mid A)$.

You can rewrite the formula for the probability of dependent events to write a rule for finding conditional probabilities.

$P(A) \cdot P(B \mid A) = P(A \text{ and } B)$ Write formula.

$P(B \mid A) = \dfrac{P(A \text{ and } B)}{P(A)}$ Divide each side by $P(A)$.

EXAMPLE 7 Finding a Conditional Probability

At a school, 60% of students buy a school lunch. Only 10% of students buy lunch and dessert. What is the probability that a student who buys lunch also buys dessert?

SOLUTION

Let event A be "buys lunch" and let event B be "buys dessert." You are given $P(A) = 0.6$ and $P(A \text{ and } B) = 0.1$. Use the formula to find $P(B \mid A)$.

$P(B \mid A) = \dfrac{P(A \text{ and } B)}{P(A)}$ Write formula for conditional probability.

$= \dfrac{0.1}{0.6}$ Substitute 0.1 for $P(A \text{ and } B)$ and 0.6 for $P(A)$.

$= \dfrac{1}{6} \approx 0.167$ Simplify.

▶ So, the probability that a student who buys lunch also buys dessert is about 16.7%.

Monitoring Progress

Help in English and Spanish at *BigIdeasMath.com*

6. In Example 6, find (a) the probability that a non-defective part "passes," and (b) the probability that a defective part "fails."
7. At a coffee shop, 80% of customers order coffee. Only 15% of customers order coffee and a bagel. What is the probability that a customer who orders coffee also orders a bagel?

12.2 Exercises

Dynamic Solutions available at *BigIdeasMath.com*

Vocabulary and Core Concept Check

1. **WRITING** Explain the difference between dependent events and independent events, and give an example of each.

2. **COMPLETE THE SENTENCE** The probability that event B will occur given that event A has occurred is called the __________ of B given A and is written as _________.

Monitoring Progress and Modeling with Mathematics

In Exercises 3–6, tell whether the events are independent or dependent. Explain your reasoning.

3. A box of granola bars contains an assortment of flavors. You randomly choose a granola bar and eat it. Then you randomly choose another bar.

   Event A: You choose a coconut almond bar first.

   Event B: You choose a cranberry almond bar second.

4. You roll a six-sided die and flip a coin.

   Event A: You get a 4 when rolling the die.

   Event B: You get tails when flipping the coin.

5. Your MP3 player contains hip-hop and rock songs. You randomly choose a song. Then you randomly choose another song without repeating song choices.

   Event A: You choose a hip-hop song first.

   Event B: You choose a rock song second.

6. There are 22 novels of various genres on a shelf. You randomly choose a novel and put it back. Then you randomly choose another novel.

   Event A: You choose a mystery novel.

   Event B: You choose a science fiction novel.

In Exercises 7–10, determine whether the events are independent. *(See Examples 1 and 2.)*

7. You play a game that involves spinning a wheel. Each section of the wheel shown has the same area. Use a sample space to determine whether randomly spinning blue and then green are independent events.

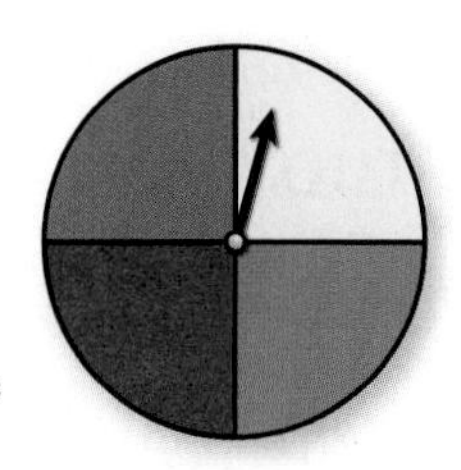

8. You have one red apple and three green apples in a bowl. You randomly select one apple to eat now and another apple for your lunch. Use a sample space to determine whether randomly selecting a green apple first and randomly selecting a green apple second are independent events.

9. A student is taking a multiple-choice test where each question has four choices. The student randomly guesses the answers to the five-question test. Use a sample space to determine whether guessing Question 1 correctly and Question 2 correctly are independent events.

10. A vase contains four white roses and one red rose. You randomly select two roses to take home. Use a sample space to determine whether randomly selecting a white rose first and randomly selecting a white rose second are independent events.

11. **PROBLEM SOLVING** You play a game that involves spinning the money wheel shown. You spin the wheel twice. Find the probability that you get more than \$500 on your first spin and then go bankrupt on your second spin. *(See Example 3.)*

12. **PROBLEM SOLVING** You play a game that involves drawing two numbers from a hat. There are 25 pieces of paper numbered from 1 to 25 in the hat. Each number is replaced after it is drawn. Find the probability that you will draw the 3 on your first draw and a number greater than 10 on your second draw.

13. **PROBLEM SOLVING** A drawer contains 12 white socks and 8 black socks. You randomly choose 1 sock and do not replace it. Then you randomly choose another sock. Find the probability that both events A and B will occur. *(See Example 4.)*

    Event A: The first sock is white.

    Event B: The second sock is white.

14. **PROBLEM SOLVING** A word game has 100 tiles, 98 of which are letters and 2 of which are blank. The numbers of tiles of each letter are shown. You randomly draw 1 tile, set it aside, and then randomly draw another tile. Find the probability that both events A and B will occur.

    Event A: The first tile is a consonant.

    Event B: The second tile is a vowel.

A - 9	H - 2	O - 8	V - 2
B - 2	I - 9	P - 2	W - 2
C - 2	J - 1	Q - 1	X - 1
D - 4	K - 1	R - 6	Y - 2
E - 12	L - 4	S - 4	Z - 1
F - 2	M - 2	T - 6	- 2 Blank
G - 3	N - 6	U - 4	

15. **ERROR ANALYSIS** Events A and B are independent. Describe and correct the error in finding $P(A \text{ and } B)$.

    ✗ $P(A) = 0.6 \quad P(B) = 0.2$

    $P(A \text{ and } B) = 0.6 + 0.2 = 0.8$

16. **ERROR ANALYSIS** A shelf contains 3 fashion magazines and 4 health magazines. You randomly choose one to read, set it aside, and randomly choose another for your friend to read. Describe and correct the error in finding the probability that both events A and B occur.

    Event A: The first magazine is fashion.

    Event B: The second magazine is health.

    ✗ $P(A) = \frac{3}{7} \quad P(B|A) = \frac{4}{7}$

    $P(A \text{ and } B) = \frac{3}{7} \cdot \frac{4}{7} = \frac{12}{49} \approx 0.245$

17. **NUMBER SENSE** Events A and B are independent. Suppose $P(B) = 0.4$ and $P(A \text{ and } B) = 0.13$. Find $P(A)$.

18. **NUMBER SENSE** Events A and B are dependent. Suppose $P(B|A) = 0.6$ and $P(A \text{ and } B) = 0.15$. Find $P(A)$.

19. **ANALYZING RELATIONSHIPS** You randomly select three cards from a standard deck of 52 playing cards. What is the probability that all three cards are face cards when (a) you replace each card before selecting the next card, and (b) you do not replace each card before selecting the next card? Compare the probabilities. *(See Example 5.)*

20. **ANALYZING RELATIONSHIPS** A bag contains 9 red marbles, 4 blue marbles, and 7 yellow marbles. You randomly select three marbles from the bag. What is the probability that all three marbles are red when (a) you replace each marble before selecting the next marble, and (b) you do not replace each marble before selecting the next marble? Compare the probabilities.

21. **ATTEND TO PRECISION** The table shows the number of species in the United States listed as endangered and threatened. Find (a) the probability that a randomly selected endangered species is a bird, and (b) the probability that a randomly selected mammal is endangered. *(See Example 6.)*

	Endangered	Threatened
Mammals	70	16
Birds	80	16
Other	318	142

22. **ATTEND TO PRECISION** The table shows the number of tropical cyclones that formed during the hurricane seasons over a 12-year period. Find (a) the probability to predict whether a future tropical cyclone in the Northern Hemisphere is a hurricane, and (b) the probability to predict whether a hurricane is in the Southern Hemisphere.

Type of Tropical Cyclone	Northern Hemisphere	Southern Hemisphere
tropical depression	100	107
tropical storm	342	487
hurricane	379	525

23. **PROBLEM SOLVING** At a school, 43% of students attend the homecoming football game. Only 23% of students go to the game and the homecoming dance. What is the probability that a student who attends the football game also attends the dance? *(See Example 7.)*

24. PROBLEM SOLVING At a gas station, 84% of customers buy gasoline. Only 5% of customers buy gasoline and a beverage. What is the probability that a customer who buys gasoline also buys a beverage?

25. PROBLEM SOLVING You and 19 other students volunteer to present the "Best Teacher" award at a school banquet. One student volunteer will be chosen to present the award. Each student worked at least 1 hour in preparation for the banquet. You worked for 4 hours, and the group worked a combined total of 45 hours. For each situation, describe a process that gives you a "fair" chance to be chosen, and find the probability that you are chosen.

a. "Fair" means equally likely.

b. "Fair" means proportional to the number of hours each student worked in preparation.

26. HOW DO YOU SEE IT? A bag contains one red marble and one blue marble. The diagrams show the possible outcomes of randomly choosing two marbles using different methods. For each method, determine whether the marbles were selected with or without replacement.

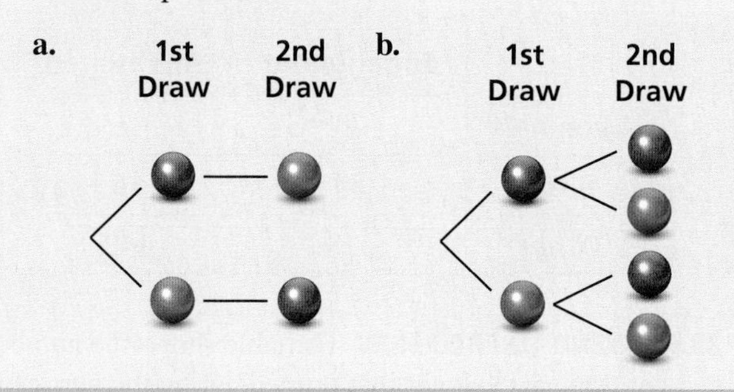

27. MAKING AN ARGUMENT A meteorologist claims that there is a 70% chance of rain. When it rains, there is a 75% chance that your softball game will be rescheduled. Your friend believes the game is more likely to be rescheduled than played. Is your friend correct? Explain your reasoning.

28. THOUGHT PROVOKING Two six-sided dice are rolled once. Events A and B are represented by the diagram. Describe each event. Are the two events dependent or independent? Justify your reasoning.

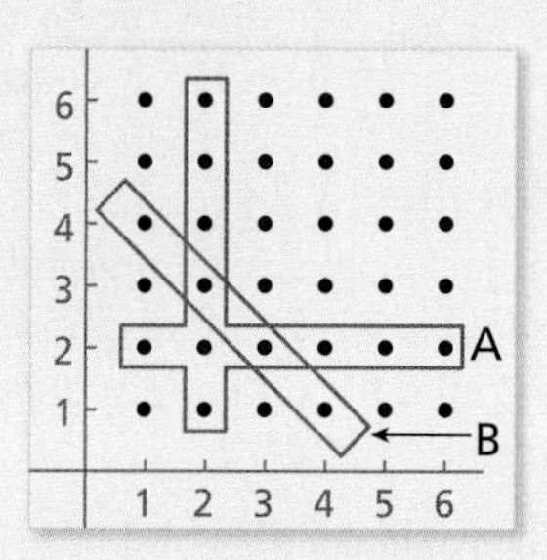

29. MODELING WITH MATHEMATICS A football team is losing by 14 points near the end of a game. The team scores two touchdowns (worth 6 points each) before the end of the game. After each touchdown, the coach must decide whether to go for 1 point with a kick (which is successful 99% of the time) or 2 points with a run or pass (which is successful 45% of the time).

a. If the team goes for 1 point after each touchdown, what is the probability that the team wins? loses? ties?

b. If the team goes for 2 points after each touchdown, what is the probability that the team wins? loses? ties?

c. Can you develop a strategy so that the coach's team has a probability of winning the game that is greater than the probability of losing? If so, explain your strategy and calculate the probabilities of winning and losing the game.

30. ABSTRACT REASONING Assume that A and B are independent events.

a. Explain why $P(B) = P(B \mid A)$ and $P(A) = P(A \mid B)$.

b. Can $P(A \text{ and } B)$ also be defined as $P(B) \cdot P(A \mid B)$? Justify your reasoning.

Maintaining Mathematical Proficiency

Reviewing what you learned in previous grades and lessons

Solve the equation. Check your solution. *(Skills Review Handbook)*

31. $\frac{9}{10}x = 0.18$

32. $\frac{1}{4}x + 0.5x = 1.5$

33. $0.3x - \frac{3}{5}x + 1.6 = 1.555$

12.3 Two-Way Tables and Probability

Essential Question How can you construct and interpret a two-way table?

EXPLORATION 1 Completing and Using a Two-Way Table

Work with a partner. A *two-way table* displays the same information as a Venn diagram. In a two-way table, one category is represented by the rows and the other category is represented by the columns.

Survey of 80 Students

Play an instrument: 25; Both: 16; Speak a foreign language: 30; Neither: 9

The Venn diagram shows the results of a survey in which 80 students were asked whether they play a musical instrument and whether they speak a foreign language. Use the Venn diagram to complete the two-way table. Then use the two-way table to answer each question.

	Play an Instrument	Do Not Play an Instrument	Total
Speak a Foreign Language			
Do Not Speak a Foreign Language			
Total			

a. How many students play an instrument?

b. How many students speak a foreign language?

c. How many students play an instrument and speak a foreign language?

d. How many students do not play an instrument and do not speak a foreign language?

e. How many students play an instrument and do not speak a foreign language?

EXPLORATION 2 Two-Way Tables and Probability

Work with a partner. In Exploration 1, one student is selected at random from the 80 students who took the survey. Find the probability that the student

a. plays an instrument.

b. speaks a foreign language.

c. plays an instrument and speaks a foreign language.

d. does not play an instrument and does not speak a foreign language.

e. plays an instrument and does not speak a foreign language.

MODELING WITH MATHEMATICS

To be proficient in math, you need to identify important quantities in a practical situation and map their relationships using such tools as diagrams and two-way tables.

EXPLORATION 3 Conducting a Survey

Work with your class. Conduct a survey of the students in your class. Choose two categories that are different from those given in Explorations 1 and 2. Then summarize the results in both a Venn diagram and a two-way table. Discuss the results.

Communicate Your Answer

4. How can you construct and interpret a two-way table?

5. How can you use a two-way table to determine probabilities?

12.3 Lesson

Core Vocabulary

two-way table, *p. 684*
joint frequency, *p. 684*
marginal frequency, *p. 684*
joint relative frequency, *p. 685*
marginal relative frequency, *p. 685*
conditional relative frequency, *p. 685*

Previous
conditional probability

READING

A two-way table is also called a *contingency table*, or a *two-way frequency table*.

What You Will Learn

- Make two-way tables.
- Find relative and conditional relative frequencies.
- Use conditional relative frequencies to find conditional probabilities.

Making Two-Way Tables

A **two-way table** is a frequency table that displays data collected from one source that belong to two different categories. One category of data is represented by rows and the other is represented by columns. Suppose you randomly survey freshmen and sophomores about whether they are attending a school concert. A two-way table is one way to organize your results.

Each entry in the table is called a **joint frequency**. The sums of the rows and columns are called **marginal frequencies**, which you will find in Example 1.

		Attendance	
		Attending	Not Attending
Class	Freshman	25	44
	Sophomore	80	32

joint frequency

EXAMPLE 1 Making a Two-Way Table

In another survey similar to the one above, 106 juniors and 114 seniors respond. Of those, 42 juniors and 77 seniors plan on attending. Organize these results in a two-way table. Then find and interpret the marginal frequencies.

SOLUTION

Step 1 Find the joint frequencies. Because 42 of the 106 juniors are attending, $106 - 42 = 64$ juniors are not attending. Because 77 of the 114 seniors are attending, $114 - 77 = 37$ seniors are not attending. Place each joint frequency in its corresponding cell.

Step 2 Find the marginal frequencies. Create a new column and row for the sums. Then add the entries and interpret the results.

		Attendance		
		Attending	Not Attending	Total
Class	Junior	42	64	106
	Senior	77	37	114
	Total	119	101	220

106 juniors responded.
114 seniors responded.
220 students were surveyed.
119 students are attending.
101 students are not attending.

Step 3 Find the sums of the marginal frequencies. Notice the sums $106 + 114 = 220$ and $119 + 101 = 220$ are equal. Place this value at the bottom right.

Monitoring Progress

Help in English and Spanish at *BigIdeasMath.com*

1. You randomly survey students about whether they are in favor of planting a community garden at school. Of 96 boys surveyed, 61 are in favor. Of 88 girls surveyed, 17 are against. Organize the results in a two-way table. Then find and interpret the marginal frequencies.

Finding Relative and Conditional Relative Frequencies

You can display values in a two-way table as frequency counts (as in Example 1) or as *relative frequencies*.

Core Concept

Relative and Conditional Relative Frequencies

A **joint relative frequency** is the ratio of a frequency that is not in the total row or the total column to the total number of values or observations.

A **marginal relative frequency** is the sum of the joint relative frequencies in a row or a column.

A **conditional relative frequency** is the ratio of a joint relative frequency to the marginal relative frequency. You can find a conditional relative frequency using a row total or a column total of a two-way table.

STUDY TIP

Two-way tables can display relative frequencies based on the total number of observations, the row totals, or the column totals.

EXAMPLE 2 Finding Joint and Marginal Relative Frequencies

Use the survey results in Example 1 to make a two-way table that shows the joint and marginal relative frequencies.

SOLUTION

To find the joint relative frequencies, divide each frequency by the total number of students in the survey. Then find the sum of each row and each column to find the marginal relative frequencies.

Class	Attendance: Attending	Attendance: Not Attending	Total
Junior	$\frac{42}{220} \approx 0.191$	$\frac{64}{220} \approx 0.291$	0.482
Senior	$\frac{77}{220} = 0.35$	$\frac{37}{220} \approx 0.168$	0.518
Total	0.541	0.459	1

About 29.1% of the students in the survey are juniors and are *not* attending the concert.

About 51.8% of the students in the survey are seniors.

INTERPRETING MATHEMATICAL RESULTS

Relative frequencies can be interpreted as probabilities. The probability that a randomly selected student is a junior and is *not* attending the concert is 29.1%.

EXAMPLE 3 Finding Conditional Relative Frequencies

Use the survey results in Example 1 to make a two-way table that shows the conditional relative frequencies based on the row totals.

SOLUTION

Use the marginal relative frequency of each *row* to calculate the conditional relative frequencies.

Class	Attendance: Attending	Attendance: Not Attending
Junior	$\frac{0.191}{0.482} \approx 0.396$	$\frac{0.291}{0.482} \approx 0.604$
Senior	$\frac{0.35}{0.518} \approx 0.676$	$\frac{0.168}{0.518} \approx 0.324$

Given that a student is a senior, the conditional relative frequency that he or she is *not* attending the concert is about 32.4%.

Monitoring Progress  Help in English and Spanish at *BigIdeasMath.com*

2. Use the survey results in Monitoring Progress Question 1 to make a two-way table that shows the joint and marginal relative frequencies.

3. Use the survey results in Example 1 to make a two-way table that shows the conditional relative frequencies based on the column totals. Interpret the conditional relative frequencies in the context of the problem.

4. Use the survey results in Monitoring Progress Question 1 to make a two-way table that shows the conditional relative frequencies based on the row totals. Interpret the conditional relative frequencies in the context of the problem.

Finding Conditional Probabilities

You can use conditional relative frequencies to find conditional probabilities.

EXAMPLE 4 Finding Conditional Probabilities

A satellite TV provider surveys customers in three cities. The survey asks whether they would recommend the TV provider to a friend. The results, given as joint relative frequencies, are shown in the two-way table.

		Location		
		Glendale	**Santa Monica**	**Long Beach**
Response	**Yes**	0.29	0.27	0.32
	No	0.05	0.03	0.04

a. What is the probability that a randomly selected customer who is located in Glendale will recommend the provider?

b. What is the probability that a randomly selected customer who will not recommend the provider is located in Long Beach?

c. Determine whether recommending the provider to a friend and living in Long Beach are independent events.

SOLUTION

a. $P(\text{yes} \mid \text{Glendale}) = \dfrac{P(\text{Glendale and yes})}{P(\text{Glendale})} = \dfrac{0.29}{0.29 + 0.05} \approx 0.853$

▶ So, the probability that a customer who is located in Glendale will recommend the provider is about 85.3%.

INTERPRETING MATHEMATICAL RESULTS

The probability 0.853 is a conditional relative frequency based on a column total. The condition is that the customer lives in Glendale.

b. $P(\text{Long Beach} \mid \text{no}) = \dfrac{P(\text{no and Long Beach})}{P(\text{no})} = \dfrac{0.04}{0.05 + 0.03 + 0.04} \approx 0.333$

▶ So, the probability that a customer who will not recommend the provider is located in Long Beach is about 33.3%.

c. Use the formula $P(B) = P(B \mid A)$ and compare $P(\text{Long Beach})$ and $P(\text{Long Beach} \mid \text{yes})$.

$P(\text{Long Beach}) = 0.32 + 0.04 = 0.36$

$P(\text{Long Beach} \mid \text{yes}) = \dfrac{P(\text{Yes and Long Beach})}{P(\text{yes})} = \dfrac{0.32}{0.29 + 0.27 + 0.32} \approx 0.36$

▶ Because $P(\text{Long Beach}) \approx P(\text{Long Beach} \mid \text{yes})$, the two events are independent.

Monitoring Progress  Help in English and Spanish at *BigIdeasMath.com*

5. In Example 4, what is the probability that a randomly selected customer who is located in Santa Monica will not recommend the provider to a friend?

6. In Example 4, determine whether recommending the provider to a friend and living in Santa Monica are independent events. Explain your reasoning.

EXAMPLE 5 Comparing Conditional Probabilities

A jogger wants to burn a certain number of calories during his workout. He maps out three possible jogging routes. Before each workout, he randomly selects a route, and then determines the number of calories he burns and whether he reaches his goal. The table shows his findings. Which route should he use?

	Reaches Goal	Does Not Reach Goal
Route A	~~IIII~~ ~~IIII~~ I	~~IIII~~ I
Route B	~~IIII~~ ~~IIII~~ I	IIII
Route C	~~IIII~~ ~~IIII~~ II	~~IIII~~ I

SOLUTION

Step 1 Use the findings to make a two-way table that shows the joint and marginal relative frequencies. There are a total of 50 observations in the table.

		Result		
		Reaches Goal	Does Not Reach Goal	Total
Route	A	0.22	0.12	0.34
	B	0.22	0.08	0.30
	C	0.24	0.12	0.36
	Total	0.68	0.32	1

Step 2 Find the conditional probabilities by dividing each joint relative frequency in the "Reaches Goal" column by the marginal relative frequency in its corresponding row.

$$P(\text{reaches goal} \mid \text{Route A}) = \frac{P(\text{Route A and reaches goal})}{P(\text{Route A})} = \frac{0.22}{0.34} \approx 0.647$$

$$P(\text{reaches goal} \mid \text{Route B}) = \frac{P(\text{Route B and reaches goal})}{P(\text{Route B})} = \frac{0.22}{0.30} \approx 0.733$$

$$P(\text{reaches goal} \mid \text{Route C}) = \frac{P(\text{Route C and reaches goal})}{P(\text{Route C})} = \frac{0.24}{0.36} \approx 0.667$$

▶ Based on the sample, the probability that he reaches his goal is greatest when he uses Route B. So, he should use Route B.

Monitoring Progress  Help in English and Spanish at *BigIdeasMath.com*

7. A manager is assessing three employees in order to offer one of them a promotion. Over a period of time, the manager records whether the employees meet or exceed expectations on their assigned tasks. The table shows the manager's results. Which employee should be offered the promotion? Explain.

	Exceed Expectations	Meet Expectations
Joy	~~IIII~~ IIII	~~IIII~~ I
Elena	~~IIII~~ ~~IIII~~ II	~~IIII~~ III
Sam	~~IIII~~ ~~IIII~~ I	~~IIII~~ II

12.3 Exercises

Dynamic Solutions available at *BigIdeasMath.com*

Vocabulary and Core Concept Check

1. **COMPLETE THE SENTENCE** A(n) ____________ displays data collected from the same source that belongs to two different categories.

2. **WRITING** Compare the definitions of joint relative frequency, marginal relative frequency, and conditional relative frequency.

Monitoring Progress and Modeling with Mathematics

In Exercises 3 and 4, complete the two-way table.

3.

		Preparation		
		Studied	Did Not Study	Total
Grade	Pass		6	
	Fail			10
	Total	38		50

4.

		Response		
		Yes	No	Total
Role	Student	56		
	Teacher		7	10
	Total		49	

5. **MODELING WITH MATHEMATICS** You survey 171 males and 180 females at Grand Central Station in New York City. Of those, 132 males and 151 females wash their hands after using the public rest rooms. Organize these results in a two-way table. Then find and interpret the marginal frequencies. *(See Example 1.)*

6. **MODELING WITH MATHEMATICS** A survey asks 60 teachers and 48 parents whether school uniforms reduce distractions in school. Of those, 49 teachers and 18 parents say uniforms reduce distractions in school. Organize these results in a two-way table. Then find and interpret the marginal frequencies.

USING STRUCTURE In Exercises 7 and 8, use the two-way table to create a two-way table that shows the joint and marginal relative frequencies.

7.

		Dominant Hand		
		Left	Right	Total
Gender	Female	11	104	115
	Male	24	92	116
	Total	35	196	231

8.

		Gender		
		Male	Female	Total
Experience	Expert	62	6	68
	Average	275	24	299
	Novice	40	3	43
	Total	377	33	410

9. **MODELING WITH MATHEMATICS** Use the survey results from Exercise 5 to make a two-way table that shows the joint and marginal relative frequencies. *(See Example 2.)*

10. **MODELING WITH MATHEMATICS** In a survey, 49 people received a flu vaccine before the flu season and 63 people did not receive the vaccine. Of those who receive the flu vaccine, 16 people got the flu. Of those who did not receive the vaccine, 17 got the flu. Make a two-way table that shows the joint and marginal relative frequencies.

11. **MODELING WITH MATHEMATICS** A survey finds that 110 people ate breakfast and 30 people skipped breakfast. Of those who ate breakfast, 10 people felt tired. Of those who skipped breakfast, 10 people felt tired. Make a two-way table that shows the conditional relative frequencies based on the breakfast totals. *(See Example 3.)*

12. **MODELING WITH MATHEMATICS** Use the survey results from Exercise 10 to make a two-way table that shows the conditional relative frequencies based on the flu vaccine totals.

13. **PROBLEM SOLVING** Three different local hospitals in New York surveyed their patients. The survey asked whether the patient's physician communicated efficiently. The results, given as joint relative frequencies, are shown in the two-way table. *(See Example 4.)*

		Location		
		Glens Falls	Saratoga	Albany
Response	Yes	0.123	0.288	0.338
	No	0.042	0.077	0.131

a. What is the probability that a randomly selected patient located in Saratoga was satisfied with the communication of the physician?

b. What is the probability that a randomly selected patient who was not satisfied with the physician's communication is located in Glens Falls?

c. Determine whether being satisfied with the communication of the physician and living in Saratoga are independent events.

14. **PROBLEM SOLVING** A researcher surveys a random sample of high school students in seven states. The survey asks whether students plan to stay in their home state after graduation. The results, given as joint relative frequencies, are shown in the two-way table.

		Location		
		Nebraska	North Carolina	Other States
Response	Yes	0.044	0.051	0.056
	No	0.400	0.193	0.256

a. What is the probability that a randomly selected student who lives in Nebraska plans to stay in his or her home state after graduation?

b. What is the probability that a randomly selected student who does not plan to stay in his or her home state after graduation lives in North Carolina?

c. Determine whether planning to stay in their home state and living in Nebraska are independent events.

ERROR ANALYSIS In Exercises 15 and 16, describe and correct the error in finding the given conditional probability.

		City			
		Tokyo	London	Washington, D.C.	Total
Response	Yes	0.049	0.136	0.171	0.356
	No	0.341	0.112	0.191	0.644
	Total	0.39	0.248	0.362	1

15. $P(\text{yes} \mid \text{Tokyo})$

✗

$$P(\text{yes} \mid \text{Tokyo}) = \frac{P(\text{Tokyo and yes})}{P(\text{Tokyo})} = \frac{0.049}{0.356} \approx 0.138$$

16. $P(\text{London} \mid \text{no})$

✗

$$P(\text{London} \mid \text{no}) = \frac{P(\text{no and London})}{P(\text{London})} = \frac{0.112}{0.248} \approx 0.452$$

17. **PROBLEM SOLVING** You want to find the quickest route to school. You map out three routes. Before school, you randomly select a route and record whether you are late or on time. The table shows your findings. Assuming you leave at the same time each morning, which route should you use? Explain. *(See Example 5.)*

	On Time	Late
Route A	𝍸 II	IIII
Route B	𝍸 𝍸 I	III
Route C	𝍸 𝍸 II	IIII

18. **PROBLEM SOLVING** A teacher is assessing three groups of students in order to offer one group a prize. Over a period of time, the teacher records whether the groups meet or exceed expectations on their assigned tasks. The table shows the teacher's results. Which group should be awarded the prize? Explain.

	Exceed Expectations	Meet Expectations
Group 1	𝍸 𝍸 II	IIII
Group 2	𝍸 III	𝍸
Group 3	𝍸 IIII	𝍸 I

19. OPEN-ENDED Create and conduct a survey in your class. Organize the results in a two-way table. Then create a two-way table that shows the joint and marginal frequencies.

20. HOW DO YOU SEE IT? A research group surveys parents and coaches of high school students about whether competitive sports are important in school. The two-way table shows the results of the survey.

		Role		
		Parent	Coach	Total
Important	Yes	880	456	1336
	No	120	45	165
	Total	1000	501	1501

a. What does 120 represent?

b. What does 1336 represent?

c. What does 1501 represent?

21. MAKING AN ARGUMENT Your friend uses the table below to determine which workout routine is the best. Your friend decides that Routine B is the best option because it has the fewest tally marks in the "Does Not Reach Goal" column. Is your friend correct? Explain your reasoning.

	Reached Goal	Does Not Reach Goal
Routine A	~~IIII~~	III
Routine B	IIII	II
Routine C	~~IIII~~ II	IIII

22. MODELING WITH MATHEMATICS A survey asks students whether they prefer math class or science class. Of the 150 male students surveyed, 62% prefer math class over science class. Of the female students surveyed, 74% prefer math. Construct a two-way table to show the number of students in each category if 350 students were surveyed.

23. MULTIPLE REPRESENTATIONS Use the Venn diagram to construct a two-way table. Then use your table to answer the questions.

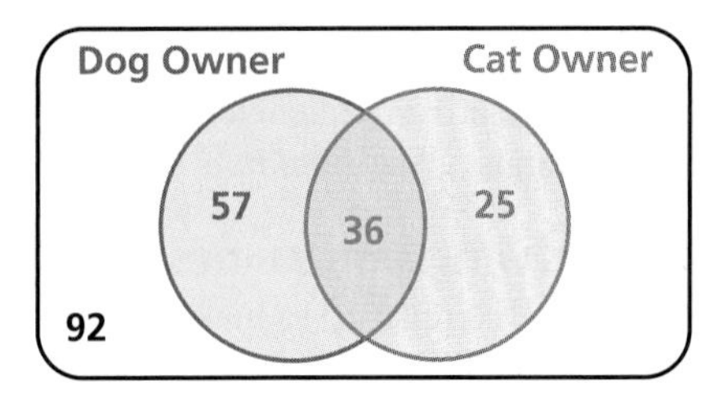

a. What is the probability that a randomly selected person does not own either pet?

b. What is the probability that a randomly selected person who owns a dog also owns a cat?

24. WRITING Compare two-way tables and Venn diagrams. Then describe the advantages and disadvantages of each.

25. PROBLEM SOLVING A company creates a new snack, N, and tests it against its current leader, L. The table shows the results.

	Prefer L	Prefer N
Current L Consumer	72	46
Not Current L Consumer	52	114

The company is deciding whether it should try to improve the snack before marketing it, and to whom the snack should be marketed. Use probability to explain the decisions the company should make when the total size of the snack's market is expected to (a) change very little, and (b) expand very rapidly.

26. THOUGHT PROVOKING Bayes' Theorem is given by

$$P(A|B) = \frac{P(B|A) \cdot P(A)}{P(B)}.$$

Use a two-way table to write an example of Bayes' Theorem.

Maintaining Mathematical Proficiency

Reviewing what you learned in previous grades and lessons

Draw a Venn diagram of the sets described. *(Skills Review Handbook)*

27. Of the positive integers less than 15, set A consists of the factors of 15 and set B consists of all odd numbers.

28. Of the positive integers less than 14, set A consists of all prime numbers and set B consists of all even numbers.

29. Of the positive integers less than 24, set A consists of the multiples of 2 and set B consists of all the multiples of 3.

12.1–12.3 What Did You Learn?

Core Vocabulary

probability experiment, *p. 668*
outcome, *p. 668*
event, *p. 668*
sample space, *p. 668*
probability of an event, *p. 668*
theoretical probability, *p. 669*
geometric probability, *p. 670*
experimental probability, *p. 671*
independent events, *p. 676*
dependent events, *p. 677*
conditional probability, *p. 677*
two-way table, *p. 684*
joint frequency, *p. 684*
marginal frequency, *p. 684*
joint relative frequency, *p. 685*
marginal relative frequency, *p. 685*
conditional relative frequency, *p. 685*

Core Concepts

Section 12.1

Theoretical Probabilities, *p. 668*
Probability of the Complement of an Event, *p. 669*
Experimental Probabilities, *p. 671*

Section 12.2

Probability of Independent Events, *p. 676*
Probability of Dependent Events, *p. 677*
Finding Conditional Probabilities, *p. 679*

Section 12.3

Making Two-Way Tables, *p. 684*
Relative and Conditional Relative Frequencies, *p. 685*

Mathematical Practices

1. How can you use a number line to analyze the error in Exercise 12 on page 672?
2. Explain how you used probability to correct the flawed logic of your friend in Exercise 21 on page 690.

Study Skills

Making a Mental Cheat Sheet

- Write down important information on note cards.
- Memorize the information on the note cards, placing the ones containing information you know in one stack and the ones containing information you do not know in another stack. Keep working on the information you do not know.

12.1–12.3 Quiz

1. You randomly draw a marble out of a bag containing 8 green marbles, 4 blue marbles, 12 yellow marbles, and 10 red marbles. Find the probability of drawing a marble that is not yellow. *(Section 12.1)*

Find $P(\overline{A})$. *(Section 12.1)*

2. $P(A) = 0.32$
3. $P(A) = \frac{8}{9}$
4. $P(A) = 0.01$

5. You roll a six-sided die 30 times. A 5 is rolled 8 times. What is the theoretical probability of rolling a 5? What is the experimental probability of rolling a 5? *(Section 12.1)*

6. Events A and B are independent. Find the missing probability. *(Section 12.2)*

   $P(A) = 0.25$

   $P(B) =$ _____

   $P(A \text{ and } B) = 0.05$

7. Events A and B are dependent. Find the missing probability. *(Section 12.2)*

   $P(A) = 0.6$

   $P(B|A) = 0.2$

   $P(A \text{ and } B) =$ _____

8. Find the probability that a dart thrown at the circular target shown will hit the given region. Assume the dart is equally likely to hit any point inside the target. *(Section 12.1)*

   a. the center circle

   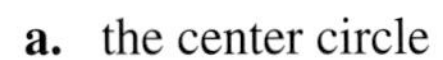

   b. outside the square

   c. inside the square but outside the center circle

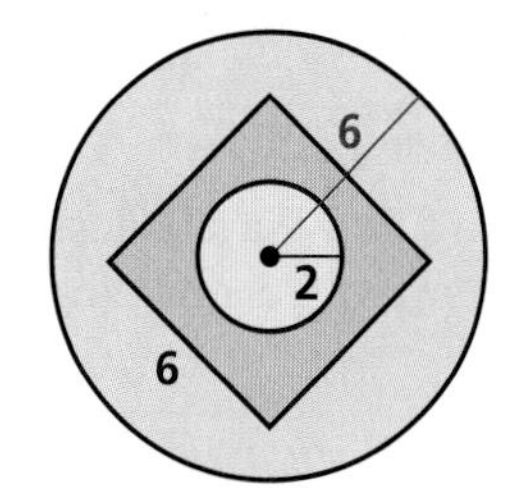

9. A survey asks 13-year-old and 15-year-old students about their eating habits. Four hundred students are surveyed, 100 male students and 100 female students from each age group. The bar graph shows the number of students who said they eat fruit every day. *(Section 12.2)*

   a. Find the probability that a female student, chosen at random from the students surveyed, eats fruit every day.

   b. Find the probability that a 15-year-old student, chosen at random from the students surveyed, eats fruit every day.

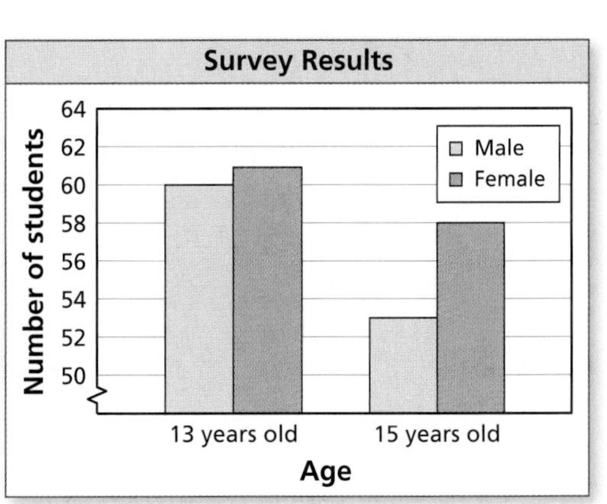

10. There are 14 boys and 18 girls in a class. The teacher allows the students to vote whether they want to take a test on Friday or on Monday. A total of 6 boys and 10 girls vote to take the test on Friday. Organize the information in a two-way table. Then find and interpret the marginal frequencies. *(Section 12.3)*

11. Three schools compete in a cross country invitational. Of the 15 athletes on your team, 9 achieve their goal times. Of the 20 athletes on the home team, 6 achieve their goal times. On your rival's team, 8 of the 13 athletes achieve their goal times. Organize the information in a two-way table. Then determine the probability that a randomly selected runner who achieves his or her goal time is from your school. *(Section 12.3)*

12.4 Probability of Disjoint and Overlapping Events

Essential Question How can you find probabilities of disjoint and overlapping events?

Two events are **disjoint**, or **mutually exclusive**, when they have no outcomes in common. Two events are **overlapping** when they have one or more outcomes in common.

MODELING WITH MATHEMATICS

To be proficient in math, you need to map the relationships between important quantities in a practical situation using such tools as diagrams.

EXPLORATION 1 Disjoint Events and Overlapping Events

Work with a partner. A six-sided die is rolled. Draw a Venn diagram that relates the two events. Then decide whether the events are disjoint or overlapping.

a. Event *A*: The result is an even number.
Event *B*: The result is a prime number.

b. Event *A*: The result is 2 or 4.
Event *B*: The result is an odd number.

EXPLORATION 2 Finding the Probability that Two Events Occur

Work with a partner. A six-sided die is rolled. For each pair of events, find (a) $P(A)$, (b) $P(B)$, (c) $P(A \text{ and } B)$, and (d) $P(A \text{ or } B)$.

a. Event *A*: The result is an even number.
Event *B*: The result is a prime number.

b. Event *A*: The result is 2 or 4.
Event *B*: The result is an odd number.

EXPLORATION 3 Discovering Probability Formulas

Work with a partner.

a. In general, if event *A* and event *B* are disjoint, then what is the probability that event *A* or event *B* will occur? Use a Venn diagram to justify your conclusion.

b. In general, if event *A* and event *B* are overlapping, then what is the probability that event *A* or event *B* will occur? Use a Venn diagram to justify your conclusion.

c. Conduct an experiment using a six-sided die. Roll the die 50 times and record the results. Then use the results to find the probabilities described in Exploration 2. How closely do your experimental probabilities compare to the theoretical probabilities you found in Exploration 2?

Communicate Your Answer

4. How can you find probabilities of disjoint and overlapping events?

5. Give examples of disjoint events and overlapping events that do not involve dice.

12.4 Lesson

Core Vocabulary

compound event, *p. 694*
overlapping events, *p. 694*
disjoint or mutually exclusive events, *p. 694*

Previous
Venn diagram

What You Will Learn

- Find probabilities of compound events.
- Use more than one probability rule to solve real-life problems.

Compound Events

When you consider all the outcomes for either of two events A and B, you form the *union* of A and B, as shown in the first diagram. When you consider only the outcomes shared by both A and B, you form the *intersection* of A and B, as shown in the second diagram. The union or intersection of two events is called a **compound event**.

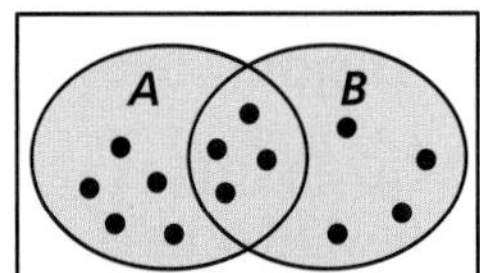

Union of A and B

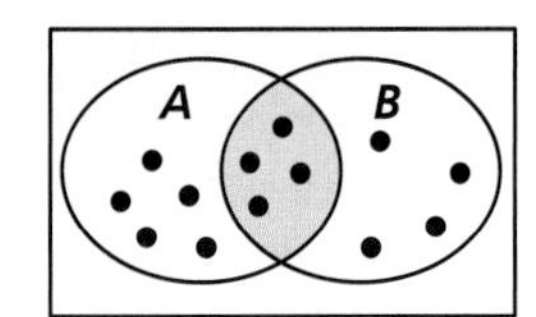

Intersection of A and B

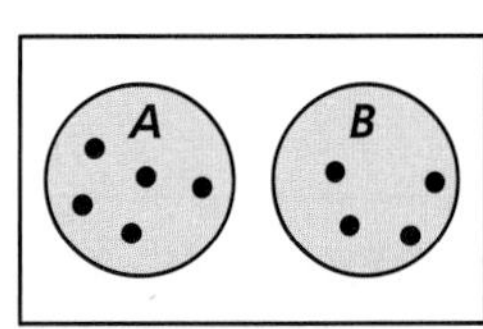

Intersection of A and B is empty.

To find $P(A \text{ or } B)$ you must consider what outcomes, if any, are in the intersection of A and B. Two events are **overlapping** when they have one or more outcomes in common, as shown in the first two diagrams. Two events are **disjoint**, or **mutually exclusive**, when they have no outcomes in common, as shown in the third diagram.

STUDY TIP

If two events A and B are overlapping, then the outcomes in the intersection of A and B are counted *twice* when $P(A)$ and $P(B)$ are added. So, $P(A \text{ and } B)$ must be subtracted from the sum.

Core Concept

Probability of Compound Events

If A and B are any two events, then the probability of A or B is

$$P(A \text{ or } B) = P(A) + P(B) - P(A \text{ and } B).$$

If A and B are disjoint events, then the probability of A or B is

$$P(A \text{ or } B) = P(A) + P(B).$$

EXAMPLE 1 Finding the Probability of Disjoint Events

A card is randomly selected from a standard deck of 52 playing cards. What is the probability that it is a 10 *or* a face card?

SOLUTION

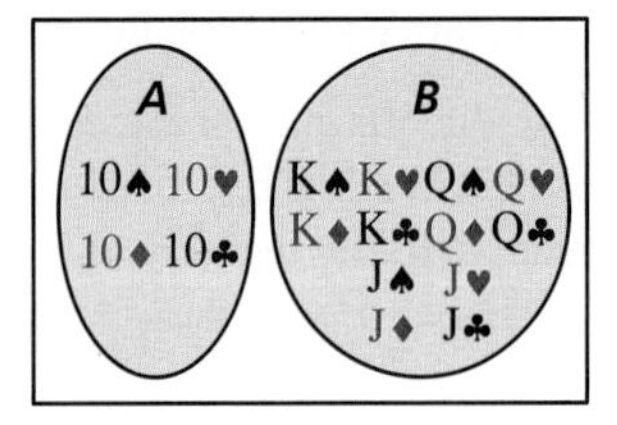

Let event A be selecting a 10 and event B be selecting a face card. From the diagram, A has 4 outcomes and B has 12 outcomes. Because A and B are disjoint, the probability is

$$P(A \text{ or } B) = P(A) + P(B)$$ Write disjoint probability formula.

$$= \frac{4}{52} + \frac{12}{52}$$ Substitute known probabilities.

$$= \frac{16}{52}$$ Add.

$$= \frac{4}{13}$$ Simplify.

$$\approx 0.308.$$ Use a calculator.

COMMON ERROR

When two events A and B overlap, as in Example 2, $P(A \text{ or } B)$ does not equal $P(A) + P(B)$.

EXAMPLE 2 Finding the Probability of Overlapping Events

A card is randomly selected from a standard deck of 52 playing cards. What is the probability that it is a face card *or* a spade?

SOLUTION

Let event A be selecting a face card and event B be selecting a spade. From the diagram, A has 12 outcomes and B has 13 outcomes. Of these, 3 outcomes are common to A and B. So, the probability of selecting a face card or a spade is

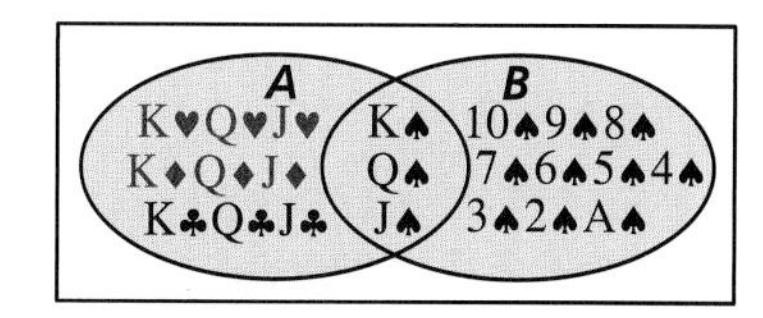

$P(A \text{ or } B) = P(A) + P(B) - P(A \text{ and } B)$ — Write general formula.

$= \frac{12}{52} + \frac{13}{52} - \frac{3}{52}$ — Substitute known probabilities.

$= \frac{22}{52}$ — Add.

$= \frac{11}{26}$ — Simplify.

$\approx 0.423.$ — Use a calculator.

EXAMPLE 3 Using a Formula to Find $P(A \text{ and } B)$

Out of 200 students in a senior class, 113 students are either varsity athletes or on the honor roll. There are 74 seniors who are varsity athletes and 51 seniors who are on the honor roll. What is the probability that a randomly selected senior is both a varsity athlete *and* on the honor roll?

SOLUTION

Let event A be selecting a senior who is a varsity athlete and event B be selecting a senior on the honor roll. From the given information, you know that $P(A) = \frac{74}{200}$, $P(B) = \frac{51}{200}$, and $P(A \text{ or } B) = \frac{113}{200}$. The probability that a randomly selected senior is both a varsity athlete *and* on the honor roll is $P(A \text{ and } B)$.

$P(A \text{ or } B) = P(A) + P(B) - P(A \text{ and } B)$ — Write general formula.

$\frac{113}{200} = \frac{74}{200} + \frac{51}{200} - P(A \text{ and } B)$ — Substitute known probabilities.

$P(A \text{ and } B) = \frac{74}{200} + \frac{51}{200} - \frac{113}{200}$ — Solve for $P(A \text{ and } B)$.

$P(A \text{ and } B) = \frac{12}{200}$ — Simplify.

$P(A \text{ and } B) = \frac{3}{50}, \text{ or } 0.06$ — Simplify.

Monitoring Progress

Help in English and Spanish at *BigIdeasMath.com*

A card is randomly selected from a standard deck of 52 playing cards. Find the probability of the event.

1. selecting an ace *or* an 8
2. selecting a 10 *or* a diamond
3. **WHAT IF?** In Example 3, suppose 32 seniors are in the band and 64 seniors are in the band or on the honor roll. What is the probability that a randomly selected senior is both in the band and on the honor roll?

Using More Than One Probability Rule

In the first four sections of this chapter, you have learned several probability rules. The solution to some real-life problems may require the use of two or more of these probability rules, as shown in the next example.

EXAMPLE 4 Solving a Real-Life Problem

The American Diabetes Association estimates that 8.3% of people in the United States have diabetes. Suppose that a medical lab has developed a simple diagnostic test for diabetes that is 98% accurate for people who have the disease and 95% accurate for people who do not have it. The medical lab gives the test to a randomly selected person. What is the probability that the diagnosis is correct?

SOLUTION

Let event A be "person has diabetes" and event B be "correct diagnosis." Notice that the probability of B depends on the occurrence of A, so the events are dependent. When A occurs, $P(B) = 0.98$. When A does not occur, $P(B) = 0.95$.

A probability tree diagram, where the probabilities are given along the branches, can help you see the different ways to obtain a correct diagnosis. Use the complements of events A and B to complete the diagram, where $\overline{A}$ is "person does not have diabetes" and $\overline{B}$ is "incorrect diagnosis." Notice that the probabilities for all branches from the same point must sum to 1.

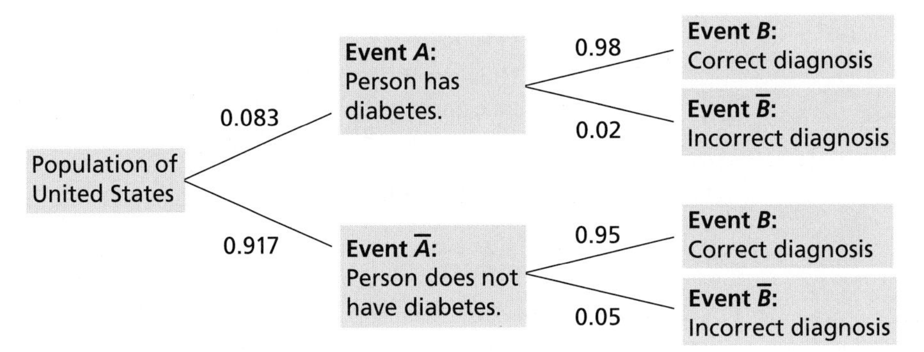

To find the probability that the diagnosis is correct, follow the branches leading to event B.

$P(B) = P(A \text{ and } B) + P(\overline{A} \text{ and } B)$ — Use tree diagram.

$= P(A) \cdot P(B|A) + P(\overline{A}) \cdot P(B|\overline{A})$ — Probability of dependent events

$= (0.083)(0.98) + (0.917)(0.95)$ — Substitute.

≈ 0.952 — Use a calculator.

▶ The probability that the diagnosis is correct is about 0.952, or 95.2%.

Monitoring Progress  Help in English and Spanish at *BigIdeasMath.com*

4. In Example 4, what is the probability that the diagnosis is *incorrect*?

5. A high school basketball team leads at halftime in 60% of the games in a season. The team wins 80% of the time when they have the halftime lead, but only 10% of the time when they do not. What is the probability that the team wins a particular game during the season?

12.4 Exercises

Dynamic Solutions available at *BigIdeasMath.com*

Vocabulary and Core Concept Check

1. **WRITING** Are the events A and $\overline{A}$ disjoint? Explain. Then give an example of a real-life event and its complement.

2. **DIFFERENT WORDS, SAME QUESTION** Which is different? Find "both" answers.

How many outcomes are in the intersection of A and B?

How many outcomes are shared by both A and B?

How many outcomes are in the union of A and B?

How many outcomes in B are also in A?

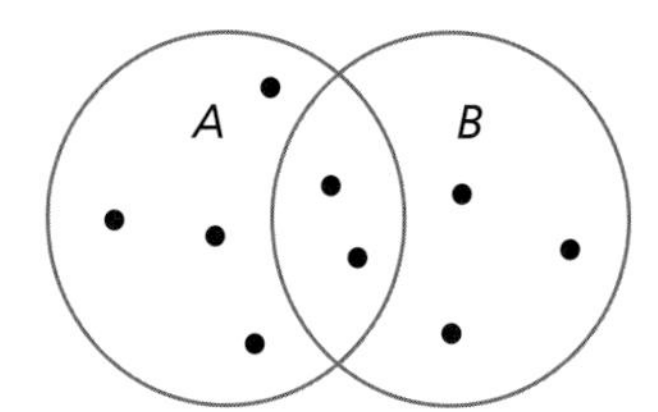

Monitoring Progress and Modeling with Mathematics

In Exercises 3–6, events A and B are disjoint. Find $P(A \text{ or } B)$.

3. $P(A) = 0.3$, $P(B) = 0.1$
4. $P(A) = 0.55$, $P(B) = 0.2$
5. $P(A) = \frac{1}{3}$, $P(B) = \frac{1}{4}$
6. $P(A) = \frac{2}{3}$, $P(B) = \frac{1}{5}$

7. **PROBLEM SOLVING** Your dart is equally likely to hit any point inside the board shown. You throw a dart and pop a balloon. What is the probability that the balloon is red or blue? *(See Example 1.)*

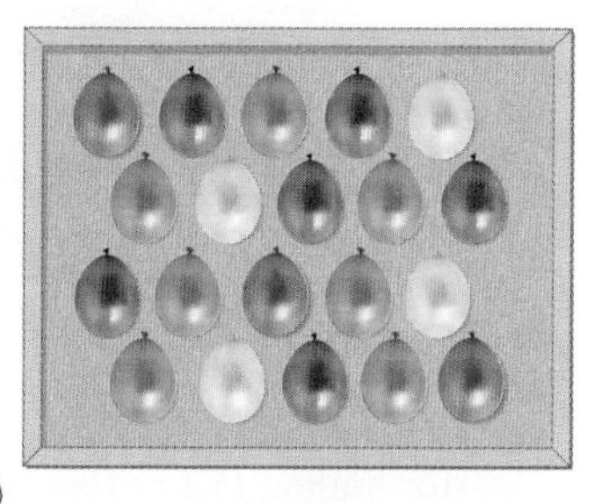

8. **PROBLEM SOLVING** You and your friend are among several candidates running for class president. You estimate that there is a 45% chance you will win and a 25% chance your friend will win. What is the probability that you or your friend win the election?

9. **PROBLEM SOLVING** You are performing an experiment to determine how well plants grow under different light sources. Of the 30 plants in the experiment, 12 receive visible light, 15 receive ultraviolet light, and 6 receive both visible and ultraviolet light. What is the probability that a plant in the experiment receives visible or ultraviolet light? *(See Example 2.)*

10. **PROBLEM SOLVING** Of 162 students honored at an academic awards banquet, 48 won awards for mathematics and 78 won awards for English. There are 14 students who won awards for both mathematics and English. A newspaper chooses a student at random for an interview. What is the probability that the student interviewed won an award for English or mathematics?

ERROR ANALYSIS In Exercises 11 and 12, describe and correct the error in finding the probability of randomly drawing the given card from a standard deck of 52 playing cards.

11. ✗

$P(\text{heart or face card})$
$= P(\text{heart}) + P(\text{face card})$
$= \frac{13}{52} + \frac{12}{52} = \frac{25}{52}$

12. ✗

$P(\text{club or } 9)$
$= P(\text{club}) + P(9) + P(\text{club and } 9)$
$= \frac{13}{52} + \frac{4}{52} + \frac{1}{52} = \frac{9}{26}$

In Exercises 13 and 14, you roll a six-sided die. Find $P(A \text{ or } B)$.

13. Event A: Roll a 6.
Event B: Roll a prime number.

14. Event A: Roll an odd number.
Event B: Roll a number less than 5.

15. DRAWING CONCLUSIONS A group of 40 trees in a forest are not growing properly. A botanist determines that 34 of the trees have a disease or are being damaged by insects, with 18 trees having a disease and 20 being damaged by insects. What is the probability that a randomly selected tree has both a disease and is being damaged by insects? *(See Example 3.)*

16. DRAWING CONCLUSIONS A company paid overtime wages or hired temporary help during 9 months of the year. Overtime wages were paid during 7 months, and temporary help was hired during 4 months. At the end of the year, an auditor examines the accounting records and randomly selects one month to check the payroll. What is the probability that the auditor will select a month in which the company paid overtime wages and hired temporary help?

17. DRAWING CONCLUSIONS A company is focus testing a new type of fruit drink. The focus group is 47% male. Of the responses, 40% of the males and 54% of the females said they would buy the fruit drink. What is the probability that a randomly selected person would buy the fruit drink? *(See Example 4.)*

18. DRAWING CONCLUSIONS The Redbirds trail the Bluebirds by one goal with 1 minute left in the hockey game. The Redbirds' coach must decide whether to remove the goalie and add a frontline player. The probabilities of each team scoring are shown in the table.

	Goalie	No goalie
Redbirds score	0.1	0.3
Bluebirds score	0.1	0.6

a. Find the probability that the Redbirds score and the Bluebirds do not score when the coach leaves the goalie in.

b. Find the probability that the Redbirds score and the Bluebirds do not score when the coach takes the goalie out.

c. Based on parts (a) and (b), what should the coach do?

19. PROBLEM SOLVING You can win concert tickets from a radio station if you are the first person to call when the song of the day is played, or if you are the first person to correctly answer the trivia question. The song of the day is announced at a random time between 7:00 and 7:30 A.M. The trivia question is asked at a random time between 7:15 and 7:45 A.M. You begin listening to the radio station at 7:20. Find the probability that you miss the announcement of the song of the day or the trivia question.

20. HOW DO YOU SEE IT? Are events A and B disjoint events? Explain your reasoning.

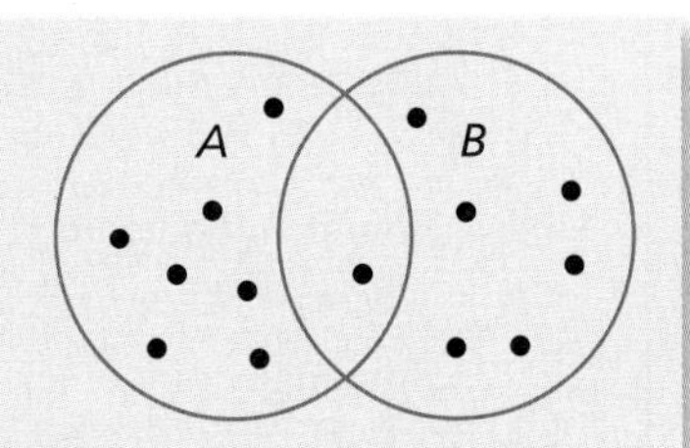

21. PROBLEM SOLVING You take a bus from your neighborhood to your school. The express bus arrives at your neighborhood at a random time between 7:30 and 7:36 A.M. The local bus arrives at your neighborhood at a random time between 7:30 and 7:40 A.M. You arrive at the bus stop at 7:33 A.M. Find the probability that you missed both the express bus and the local bus.

22. THOUGHT PROVOKING Write a general rule for finding $P(A \text{ or } B \text{ or } C)$ for (a) disjoint and (b) overlapping events A, B, and C.

23. MAKING AN ARGUMENT A bag contains 40 cards numbered 1 through 40 that are either red or blue. A card is drawn at random and placed back in the bag. This is done four times. Two red cards are drawn, numbered 31 and 19, and two blue cards are drawn, numbered 22 and 7. Your friend concludes that red cards and even numbers must be mutually exclusive. Is your friend correct? Explain.

Maintaining Mathematical Proficiency

Reviewing what you learned in previous grades and lessons

Find the product. *(Skills Review Handbook)*

24. $(n - 12)^2$ **25.** $(2x + 9)^2$ **26.** $(-5z + 6)^2$ **27.** $(3a - 7b)^2$

12.5 Permutations and Combinations

Essential Question How can a tree diagram help you visualize the number of ways in which two or more events can occur?

EXPLORATION 1 Reading a Tree Diagram

Work with a partner. Two coins are flipped and the spinner is spun. The tree diagram shows the possible outcomes.

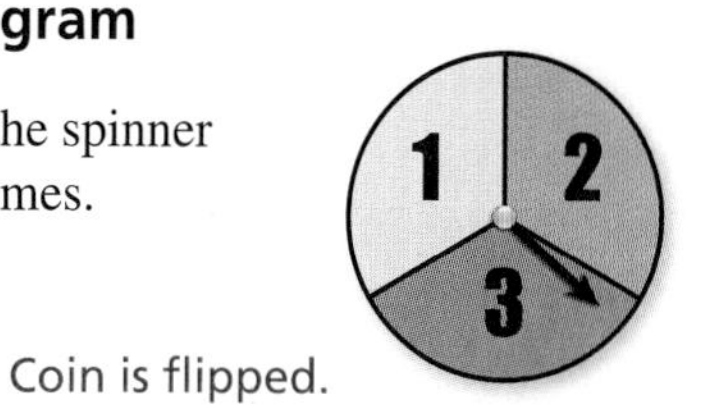

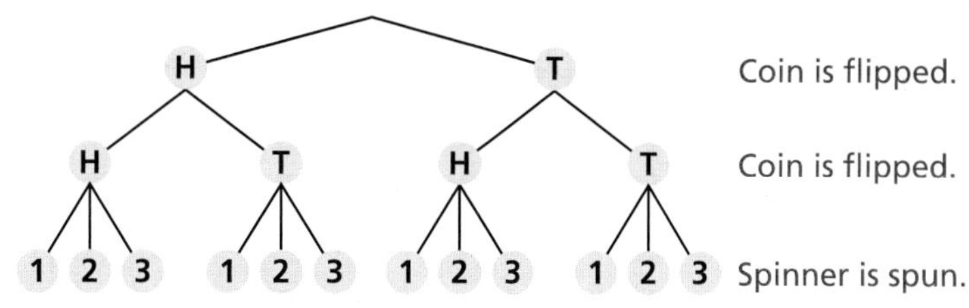

a. How many outcomes are possible?

b. List the possible outcomes.

EXPLORATION 2 Reading a Tree Diagram

Work with a partner. Consider the tree diagram below.

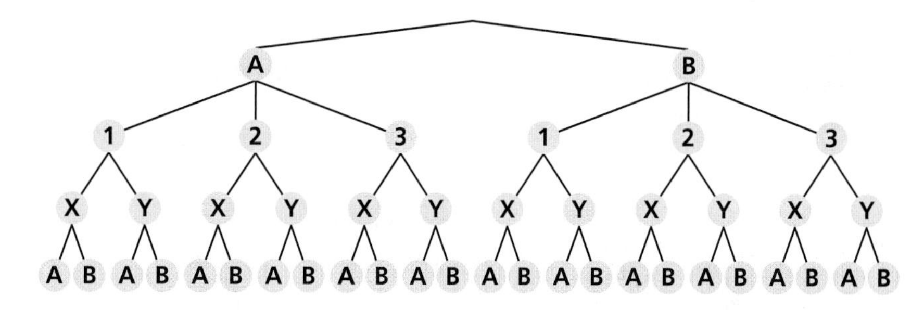

a. How many events are shown?

b. What outcomes are possible for each event?

c. How many outcomes are possible?

d. List the possible outcomes.

CONSTRUCTING VIABLE ARGUMENTS

To be proficient in math, you need to make conjectures and build a logical progression of statements to explore the truth of your conjectures.

EXPLORATION 3 Writing a Conjecture

Work with a partner.

a. Consider the following general problem: Event 1 can occur in m ways and event 2 can occur in n ways. Write a conjecture about the number of ways the two events can occur. Explain your reasoning.

b. Use the conjecture you wrote in part (a) to write a conjecture about the number of ways *more than* two events can occur. Explain your reasoning.

c. Use the results of Explorations 1(a) and 2(c) to verify your conjectures.

Communicate Your Answer

4. How can a tree diagram help you visualize the number of ways in which two or more events can occur?

5. In Exploration 1, the spinner is spun a second time. How many outcomes are possible?

12.5 Lesson

Core Vocabulary

permutation, *p. 700*
n factorial, *p. 700*
combination, *p. 702*

Previous
Fundamental Counting Principle

What You Will Learn

- Use the formula for the number of permutations.
- Use the formula for the number of combinations.

Permutations

A **permutation** is an arrangement of objects in which order is important. For instance, the 6 possible permutations of the letters A, B, and C are shown.

ABC ACB BAC BCA CAB CBA

EXAMPLE 1 Counting Permutations

Consider the number of permutations of the letters in the word JULY. In how many ways can you arrange (a) all of the letters and (b) 2 of the letters?

SOLUTION

REMEMBER

Fundamental Counting Principle: If one event can occur in *m* ways and another event can occur in *n* ways, then the number of ways that both events can occur is $m \cdot n$. The Fundamental Counting Principle can be extended to three or more events.

a. Use the Fundamental Counting Principle to find the number of permutations of the letters in the word JULY.

$$\text{Number of permutations} = \left(\begin{array}{l}\text{Choices for}\\ \text{1st letter}\end{array}\right)\left(\begin{array}{l}\text{Choices for}\\ \text{2nd letter}\end{array}\right)\left(\begin{array}{l}\text{Choices for}\\ \text{3rd letter}\end{array}\right)\left(\begin{array}{l}\text{Choices for}\\ \text{4th letter}\end{array}\right)$$

$$= 4 \cdot 3 \cdot 2 \cdot 1$$

$$= 24$$

▶ There are 24 ways you can arrange all of the letters in the word JULY.

b. When arranging 2 letters of the word JULY, you have 4 choices for the first letter and 3 choices for the second letter.

$$\text{Number of permutations} = \left(\begin{array}{l}\text{Choices for}\\ \text{1st letter}\end{array}\right)\left(\begin{array}{l}\text{Choices for}\\ \text{2nd letter}\end{array}\right)$$

$$= 4 \cdot 3$$

$$= 12$$

▶ There are 12 ways you can arrange 2 of the letters in the word JULY.

Monitoring Progress  Help in English and Spanish at *BigIdeasMath.com*

1. In how many ways can you arrange the letters in the word HOUSE?

2. In how many ways can you arrange 3 of the letters in the word MARCH?

In Example 1(a), you evaluated the expression $4 \cdot 3 \cdot 2 \cdot 1$. This expression can be written as 4! and is read "4 *factorial*." For any positive integer *n*, the product of the integers from 1 to *n* is called ***n* factorial** and is written as

$$n! = n \cdot (n-1) \cdot (n-2) \cdot \cdots \cdot 3 \cdot 2 \cdot 1.$$

As a special case, the value of 0! is defined to be 1.

In Example 1(b), you found the permutations of 4 objects taken 2 at a time. You can find the number of permutations using the formulas on the next page.

Core Concept

Permutations

Formulas	Examples
The number of permutations of n objects is given by $${}_nP_n = n!.$$	The number of permutations of 4 objects is $${}_4P_4 = 4! = 4 \cdot 3 \cdot 2 \cdot 1 = 24.$$
The number of permutations of n objects taken r at a time, where $r \le n$, is given by $${}_nP_r = \frac{n!}{(n-r)!}.$$	The number of permutations of 4 objects taken 2 at a time is $${}_4P_2 = \frac{4!}{(4-2)!} = \frac{4 \cdot 3 \cdot \cancel{2!}}{\cancel{2!}} = 12.$$

USING A GRAPHING CALCULATOR

Most graphing calculators can calculate permutations.

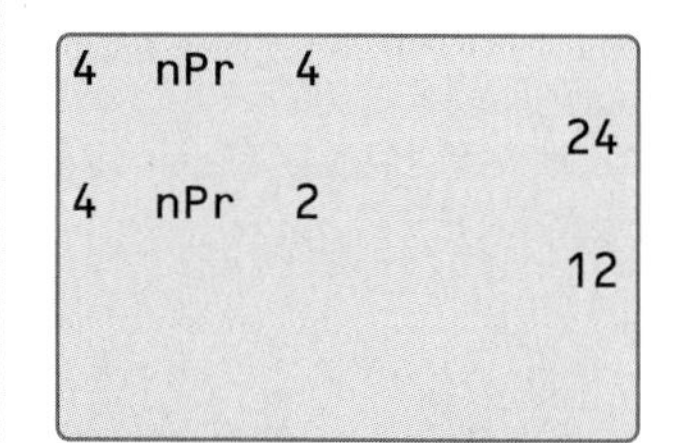

EXAMPLE 2 Using a Permutations Formula

Ten horses are running in a race. In how many different ways can the horses finish first, second, and third? (Assume there are no ties.)

SOLUTION

To find the number of permutations of 3 horses chosen from 10, find ${}_{10}P_3$.

$${}_{10}P_3 = \frac{10!}{(10-3)!} \quad \text{Permutations formula}$$

$$= \frac{10!}{7!} \quad \text{Subtract.}$$

$$= \frac{10 \cdot 9 \cdot 8 \cdot \cancel{7!}}{\cancel{7!}} \quad \text{Expand factorial. Divide out common factor, 7!.}$$

$$= 720 \quad \text{Simplify.}$$

▶ There are 720 ways for the horses to finish first, second, and third.

STUDY TIP

When you divide out common factors, remember that 7! is a factor of 10!.

EXAMPLE 3 Finding a Probability Using Permutations

For a town parade, you will ride on a float with your soccer team. There are 12 floats in the parade, and their order is chosen at random. Find the probability that your float is first and the float with the school chorus is second.

SOLUTION

Step 1 Write the number of possible outcomes as the number of permutations of the 12 floats in the parade. This is ${}_{12}P_{12} = 12!$.

Step 2 Write the number of favorable outcomes as the number of permutations of the other floats, given that the soccer team is first and the chorus is second. This is ${}_{10}P_{10} = 10!$.

Step 3 Find the probability.

$$P(\text{soccer team is 1st, chorus is 2nd}) = \frac{10!}{12!} \quad \text{Form a ratio of favorable to possible outcomes.}$$

$$= \frac{\cancel{10!}}{12 \cdot 11 \cdot \cancel{10!}} \quad \text{Expand factorial. Divide out common factor, 10!.}$$

$$= \frac{1}{132} \quad \text{Simplify.}$$

Monitoring Progress  Help in English and Spanish at *BigIdeasMath.com*

3. **WHAT IF?** In Example 2, suppose there are 8 horses in the race. In how many different ways can the horses finish first, second, and third? (Assume there are no ties.)

4. **WHAT IF?** In Example 3, suppose there are 14 floats in the parade. Find the probability that the soccer team is first and the chorus is second.

Combinations

A **combination** is a selection of objects in which order is *not* important. For instance, in a drawing for 3 identical prizes, you would use combinations, because the order of the winners would not matter. If the prizes were different, then you would use permutations, because the order would matter.

EXAMPLE 4 Counting Combinations

Count the possible combinations of 2 letters chosen from the list A, B, C, D.

SOLUTION

List all of the permutations of 2 letters from the list A, B, C, D. Because order is not important in a combination, cross out any duplicate pairs.

AB	AC	AD	~~BA~~	BC	BD
~~CA~~	~~CB~~	CD	~~DA~~	~~DB~~	~~DC~~

BD and DB are the same pair.

▶ There are 6 possible combinations of 2 letters from the list A, B, C, D.

Monitoring Progress  Help in English and Spanish at *BigIdeasMath.com*

5. Count the possible combinations of 3 letters chosen from the list A, B, C, D, E.

In Example 4, you found the number of combinations of objects by making an organized list. You can also find the number of combinations using the following formula.

USING A GRAPHING CALCULATOR

Most graphing calculators can calculate combinations.

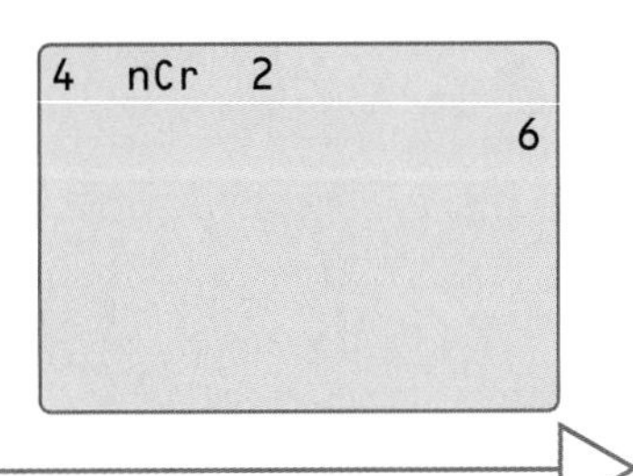

Core Concept

Combinations

Formula The number of combinations of n objects taken r at a time, where $r \leq n$, is given by

$$_nC_r = \frac{n!}{(n-r)! \cdot r!}.$$

Example The number of combinations of 4 objects taken 2 at a time is

$$_4C_2 = \frac{4!}{(4-2)! \cdot 2!} = \frac{4 \cdot 3 \cdot \cancel{2!}}{\cancel{2!} \cdot (2 \cdot 1)} = 6.$$

EXAMPLE 5 Using the Combinations Formula

You order a sandwich at a restaurant. You can choose 2 side dishes from a list of 8. How many combinations of side dishes are possible?

SOLUTION

The order in which you choose the side dishes is not important. So, to find the number of combinations of 8 side dishes taken 2 at a time, find ${}_8C_2$.

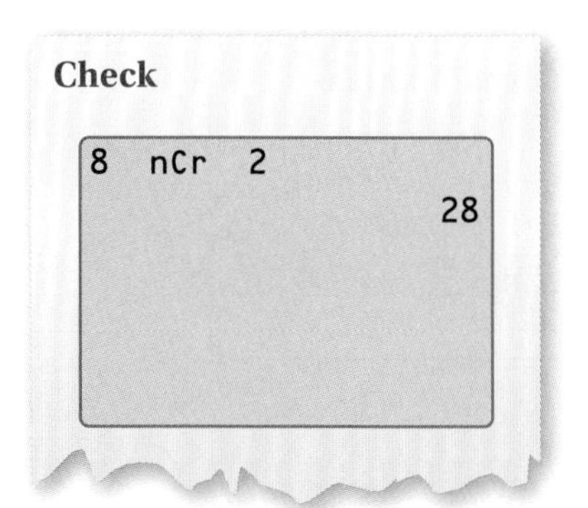

$${}_8C_2 = \frac{8!}{(8-2)! \cdot 2!} \qquad \text{Combinations formula}$$

$$= \frac{8!}{6! \cdot 2!} \qquad \text{Subtract.}$$

$$= \frac{8 \cdot 7 \cdot \cancel{6!}}{\cancel{6!} \cdot (2 \cdot 1)} \qquad \text{Expand factorials. Divide out common factor, 6!.}$$

$$= 28 \qquad \text{Multiply.}$$

▶ There are 28 different combinations of side dishes you can order.

EXAMPLE 6 Finding a Probability Using Combinations

A yearbook editor has selected 14 photos, including one of you and one of your friend, to use in a collage for the yearbook. The photos are placed at random. There is room for 2 photos at the top of the page. What is the probability that your photo and your friend's photo are the 2 placed at the top of the page?

SOLUTION

Step 1 Write the number of possible outcomes as the number of combinations of 14 photos taken 2 at a time, or ${}_{14}C_2$, because the order in which the photos are chosen is not important.

$${}_{14}C_2 = \frac{14!}{(14-2)! \cdot 2!} \qquad \text{Combinations formula}$$

$$= \frac{14!}{12! \cdot 2!} \qquad \text{Subtract.}$$

$$= \frac{14 \cdot 13 \cdot \cancel{12!}}{\cancel{12!} \cdot (2 \cdot 1)} \qquad \text{Expand factorials. Divide out common factor, 12!.}$$

$$= 91 \qquad \text{Multiply.}$$

Step 2 Find the number of favorable outcomes. Only one of the possible combinations includes your photo and your friend's photo.

Step 3 Find the probability.

$$P(\text{your photo and your friend's photos are chosen}) = \frac{1}{91}$$

Monitoring Progress

Help in English and Spanish at *BigIdeasMath.com*

6. **WHAT IF?** In Example 5, suppose you can choose 3 side dishes out of the list of 8 side dishes. How many combinations are possible?

7. **WHAT IF?** In Example 6, suppose there are 20 photos in the collage. Find the probability that your photo and your friend's photo are the 2 placed at the top of the page.

12.5 Exercises

Dynamic Solutions available at *BigIdeasMath.com*

Vocabulary and Core Concept Check

1. **COMPLETE THE SENTENCE** An arrangement of objects in which order is important is called a(n) __________.

2. **WHICH ONE DOESN'T BELONG?** Which expression does *not* belong with the other three? Explain your reasoning.

$\frac{7!}{2! \cdot 5!}$ ${}_7C_5$ ${}_7C_2$ $\frac{7!}{(7-2)!}$

Monitoring Progress and Modeling with Mathematics

In Exercises 3–8, find the number of ways you can arrange (a) all of the letters and (b) 2 of the letters in the given word. *(See Example 1.)*

3. AT
4. TRY
5. ROCK
6. WATER
7. FAMILY
8. FLOWERS

In Exercises 9–16, evaluate the expression.

9. ${}_5P_2$
10. ${}_7P_3$
11. ${}_9P_1$
12. ${}_6P_5$
13. ${}_8P_6$
14. ${}_{12}P_0$
15. ${}_{30}P_2$
16. ${}_{25}P_5$

17. **PROBLEM SOLVING** Eleven students are competing in an art contest. In how many different ways can the students finish first, second, and third? *(See Example 2.)*

18. **PROBLEM SOLVING** Six friends go to a movie theater. In how many different ways can they sit together in a row of 6 empty seats?

19. **PROBLEM SOLVING** You and your friend are 2 of 8 servers working a shift in a restaurant. At the beginning of the shift, the manager randomly assigns one section to each server. Find the probability that you are assigned Section 1 and your friend is assigned Section 2. *(See Example 3.)*

20. **PROBLEM SOLVING** You make 6 posters to hold up at a basketball game. Each poster has a letter of the word TIGERS. You and 5 friends sit next to each other in a row. The posters are distributed at random. Find the probability that TIGERS is spelled correctly when you hold up the posters.

In Exercises 21–24, count the possible combinations of *r* letters chosen from the given list. *(See Example 4.)*

21. A, B, C, D; $r = 3$
22. L, M, N, O; $r = 2$
23. U, V, W, X, Y, Z; $r = 3$
24. D, E, F, G, H; $r = 4$

In Exercises 25–32, evaluate the expression.

25. ${}_5C_1$
26. ${}_8C_5$
27. ${}_9C_9$
28. ${}_8C_6$
29. ${}_{12}C_3$
30. ${}_{11}C_4$
31. ${}_{15}C_8$
32. ${}_{20}C_5$

33. **PROBLEM SOLVING** Each year, 64 golfers participate in a golf tournament. The golfers play in groups of 4. How many groups of 4 golfers are possible? *(See Example 5.)*

34. PROBLEM SOLVING You want to purchase vegetable dip for a party. A grocery store sells 7 different flavors of vegetable dip. You have enough money to purchase 2 flavors. How many combinations of 2 flavors of vegetable dip are possible?

ERROR ANALYSIS In Exercises 35 and 36, describe and correct the error in evaluating the expression.

35. ✗ $_{11}P_7 = \dfrac{11!}{(11-7)} = \dfrac{11!}{4} = 9{,}979{,}200$

36. ✗ $_9C_4 = \dfrac{9!}{(9-4)!} = \dfrac{9!}{5!} = 3024$

REASONING In Exercises 37–40, tell whether the question can be answered using *permutations* or *combinations*. Explain your reasoning. Then answer the question.

37. To complete an exam, you must answer 8 questions from a list of 10 questions. In how many ways can you complete the exam?

38. Ten students are auditioning for 3 different roles in a play. In how many ways can the 3 roles be filled?

39. Fifty-two athletes are competing in a bicycle race. In how many orders can the bicyclists finish first, second, and third? (Assume there are no ties.)

40. An employee at a pet store needs to catch 5 tetras in an aquarium containing 27 tetras. In how many groupings can the employee capture 5 tetras?

41. CRITICAL THINKING Compare the quantities $_{50}C_9$ and $_{50}C_{41}$ without performing any calculations. Explain your reasoning.

42. CRITICAL THINKING Show that each identity is true for any whole numbers r and n, where $0 \le r \le n$.

a. $_nC_n = 1$ **b.** $_nC_r = {}_nC_{n-r}$

c. $_{n+1}C_r = {}_nC_r + {}_nC_{r-1}$

43. REASONING Complete the table for each given value of r. Then write an inequality relating $_nP_r$ and $_nC_r$. Explain your reasoning.

	$r = 0$	$r = 1$	$r = 2$	$r = 3$
$_3P_r$				
$_3C_r$				

44. REASONING Write an equation that relates $_nP_r$ and $_nC_r$. Then use your equation to find and interpret the value of $\dfrac{_{182}P_4}{_{182}C_4}$.

45. PROBLEM SOLVING You and your friend are in the studio audience on a television game show. From an audience of 300 people, 2 people are randomly selected as contestants. What is the probability that you and your friend are chosen? *(See Example 6.)*

46. PROBLEM SOLVING You work 5 evenings each week at a bookstore. Your supervisor assigns you 5 evenings at random from the 7 possibilities. What is the probability that your schedule does not include working on the weekend?

REASONING In Exercises 47 and 48, find the probability of winning a lottery using the given rules. Assume that lottery numbers are selected at random.

47. You must correctly select 6 numbers, each an integer from 0 to 49. The order is not important.

48. You must correctly select 4 numbers, each an integer from 0 to 9. The order is important.

49. MATHEMATICAL CONNECTIONS A polygon is convex when no line that contains a side of the polygon contains a point in the interior of the polygon. Consider a convex polygon with n sides.

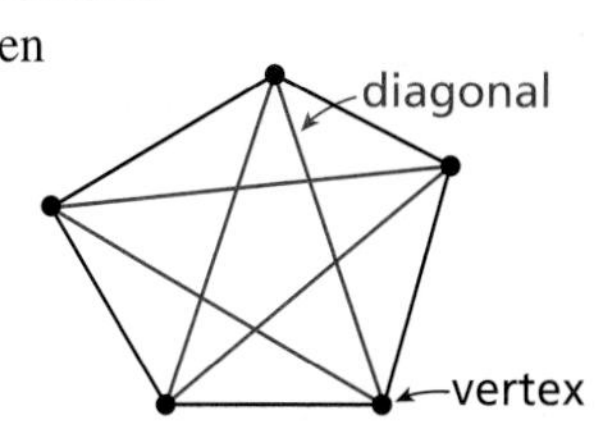

a. Use the combinations formula to write an expression for the number of diagonals in an n-sided polygon.

b. Use your result from part (a) to write a formula for the number of diagonals of an n-sided convex polygon.

50. PROBLEM SOLVING You are ordering a burrito with 2 main ingredients and 3 toppings. The menu below shows the possible choices. How many different burritos are possible?

51. PROBLEM SOLVING You want to purchase 2 different types of contemporary music CDs and 1 classical music CD from the music collection shown. How many different sets of music types can you choose for your purchase?

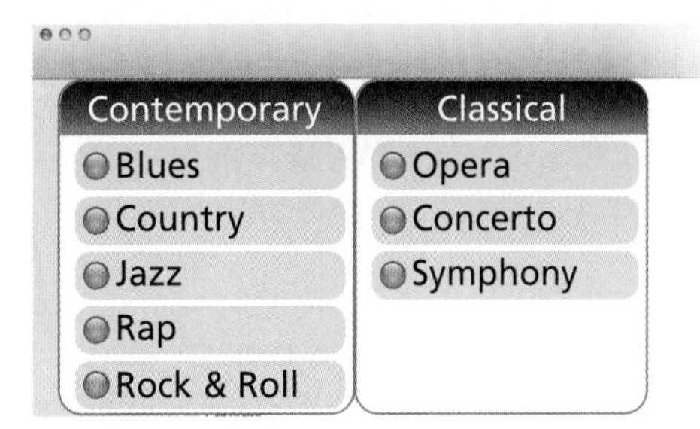

52. HOW DO YOU SEE IT? A bag contains one green marble, one red marble, and one blue marble. The diagram shows the possible outcomes of randomly drawing three marbles from the bag without replacement.

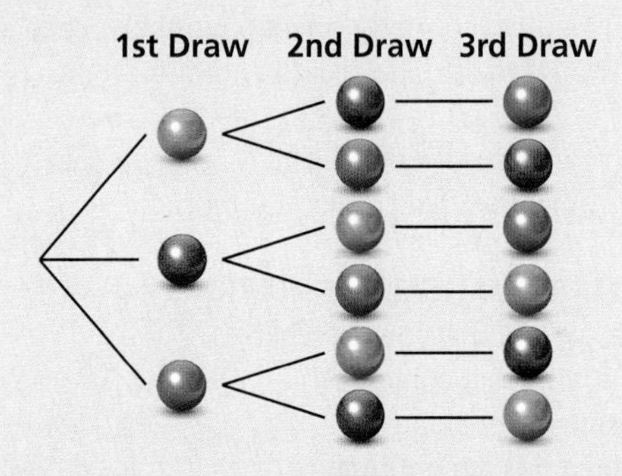

a. How many combinations of three marbles can be drawn from the bag? Explain.

b. How many permutations of three marbles can be drawn from the bag? Explain.

53. PROBLEM SOLVING Every student in your history class is required to present a project in front of the class. Each day, 4 students make their presentations in an order chosen at random by the teacher. You make your presentation on the first day.

a. What is the probability that you are chosen to be the first or second presenter on the first day?

b. What is the probability that you are chosen to be the second or third presenter on the first day? Compare your answer with that in part (a).

54. PROBLEM SOLVING The organizer of a cast party for a drama club asks each of the 6 cast members to bring 1 food item from a list of 10 items. Assuming each member randomly chooses a food item to bring, what is the probability that at least 2 of the 6 cast members bring the same item?

55. PROBLEM SOLVING You are one of 10 students performing in a school talent show. The order of the performances is determined at random. The first 5 performers go on stage before the intermission.

a. What is the probability that you are the last performer before the intermission and your rival performs immediately before you?

b. What is the probability that you are *not* the first performer?

56. THOUGHT PROVOKING How many integers, greater than 999 but not greater than 4000, can be formed with the digits 0, 1, 2, 3, and 4? Repetition of digits is allowed.

57. PROBLEM SOLVING There are 30 students in your class. Your science teacher chooses 5 students at random to complete a group project. Find the probability that you and your 2 best friends in the science class are chosen to work in the group. Explain how you found your answer.

58. PROBLEM SOLVING Follow the steps below to explore a famous probability problem called the *birthday problem*. (Assume there are 365 equally likely birthdays possible.)

a. What is the probability that at least 2 people share the same birthday in a group of 6 randomly chosen people? in a group of 10 randomly chosen people?

b. Generalize the results from part (a) by writing a formula for the probability $P(n)$ that at least 2 people in a group of n people share the same birthday. (*Hint*: Use ${}_nP_r$ notation in your formula.)

c. Enter the formula from part (b) into a graphing calculator. Use the *table* feature to make a table of values. For what group size does the probability that at least 2 people share the same birthday first exceed 50%?

Maintaining Mathematical Proficiency

Reviewing what you learned in previous grades and lessons

59. A bag contains 12 white marbles and 3 black marbles. You pick 1 marble at random. What is the probability that you pick a black marble? *(Section 12.1)*

60. The table shows the result of flipping two coins 12 times. For what outcome is the experimental probability the same as the theoretical probability? *(Section 12.1)*

HH	HT	TH	TT
2	6	3	1

12.6 Binomial Distributions

Essential Question How can you determine the frequency of each outcome of an event?

EXPLORATION 1 Analyzing Histograms

Work with a partner. The histograms show the results when n coins are flipped.

STUDY TIP

When 4 coins are flipped ($n = 4$), the possible outcomes are

TTTT TTTH TTHT TTHH

THTT THTH THHT THHH

HTTT HTTH HTHT HTHH

HHTT HHTH HHHT HHHH.

The histogram shows the numbers of outcomes having 0, 1, 2, 3, and 4 heads.

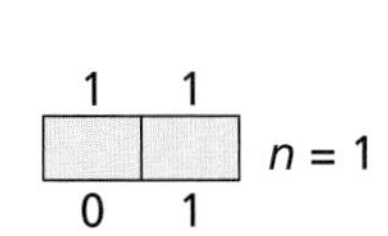

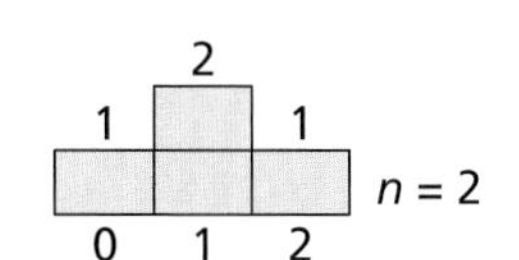

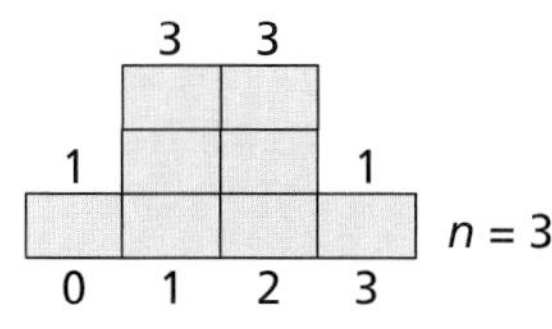

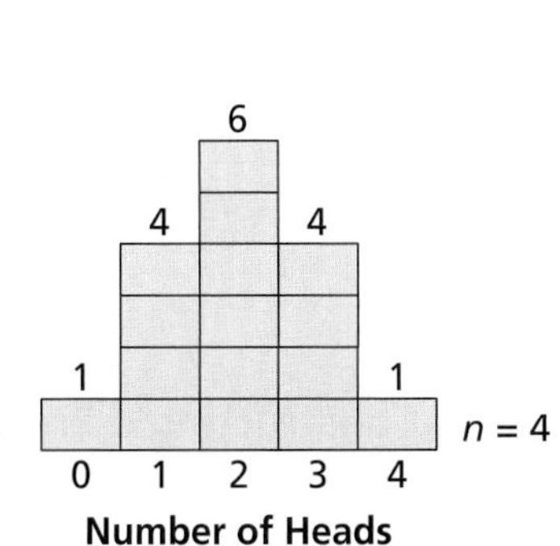

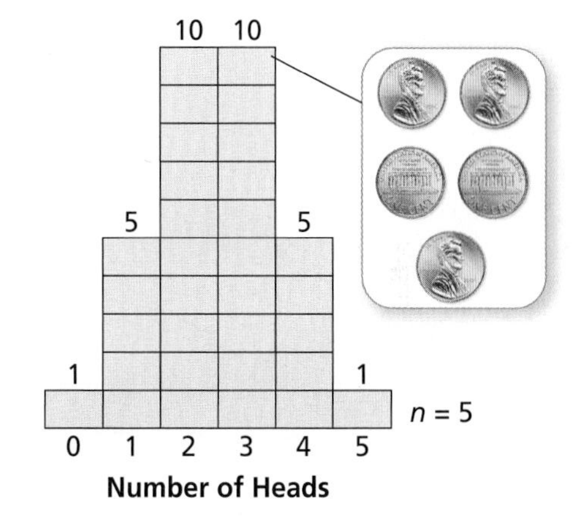

a. In how many ways can 3 heads occur when 5 coins are flipped?

b. Draw a histogram that shows the numbers of heads that can occur when 6 coins are flipped.

c. In how many ways can 3 heads occur when 6 coins are flipped?

EXPLORATION 2 Determining the Number of Occurrences

Work with a partner.

LOOKING FOR A PATTERN

To be proficient in math, you need to look closely to discern a pattern or structure.

a. Complete the table showing the numbers of ways in which 2 heads can occur when n coins are flipped.

n	3	4	5	6	7
Occurrences of 2 heads					

b. Determine the pattern shown in the table. Use your result to find the number of ways in which 2 heads can occur when 8 coins are flipped.

Communicate Your Answer

3. How can you determine the frequency of each outcome of an event?

4. How can you use a histogram to find the probability of an event?

12.6 Lesson

Core Vocabulary

random variable, *p. 708*
probability distribution, *p. 708*
binomial distribution, *p. 709*
binomial experiment, *p. 709*

Previous
histogram

What You Will Learn

- Construct and interpret probability distributions.
- Construct and interpret binomial distributions.

Probability Distributions

A **random variable** is a variable whose value is determined by the outcomes of a probability experiment. For example, when you roll a six-sided die, you can define a random variable x that represents the number showing on the die. So, the possible values of x are 1, 2, 3, 4, 5, and 6. For every random variable, a *probability distribution* can be defined.

Core Concept

Probability Distributions

A **probability distribution** is a function that gives the probability of each possible value of a random variable. The sum of all the probabilities in a probability distribution must equal 1.

Probability Distribution for Rolling a Six-Sided Die						
x	1	2	3	4	5	6
$P(x)$	$\frac{1}{6}$	$\frac{1}{6}$	$\frac{1}{6}$	$\frac{1}{6}$	$\frac{1}{6}$	$\frac{1}{6}$

EXAMPLE 1 Constructing a Probability Distribution

Let x be a random variable that represents the sum when two six-sided dice are rolled. Make a table and draw a histogram showing the probability distribution for x.

SOLUTION

Step 1 Make a table. The possible values of x are the integers from 2 to 12. The table shows how many outcomes of rolling two dice produce each value of x. Divide the number of outcomes for x by 36 to find $P(x)$.

x (sum)	2	3	4	5	6	7	8	9	10	11	12
Outcomes	1	2	3	4	5	6	5	4	3	2	1
$P(x)$	$\frac{1}{36}$	$\frac{1}{18}$	$\frac{1}{12}$	$\frac{1}{9}$	$\frac{5}{36}$	$\frac{1}{6}$	$\frac{5}{36}$	$\frac{1}{9}$	$\frac{1}{12}$	$\frac{1}{18}$	$\frac{1}{36}$

STUDY TIP

Recall that there are 36 possible outcomes when rolling two six-sided dice. These are listed in Example 3 on page 670.

Step 2 Draw a histogram where the intervals are given by x and the frequencies are given by $P(x)$.

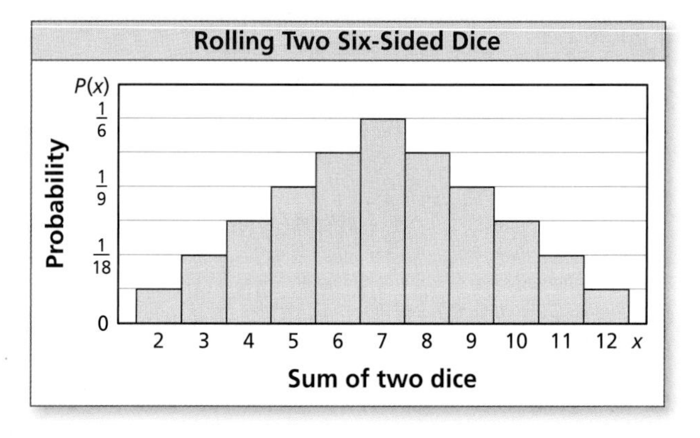

EXAMPLE 2 Interpreting a Probability Distribution

Use the probability distribution in Example 1 to answer each question.

a. What is the most likely sum when rolling two six-sided dice?

b. What is the probability that the sum of the two dice is at least 10?

SOLUTION

a. The most likely sum when rolling two six-sided dice is the value of x for which $P(x)$ is greatest. This probability is greatest for $x = 7$. So, when rolling the two dice, the most likely sum is 7.

b. The probability that the sum of the two dice is at least 10 is

$$\begin{aligned} P(x \geq 10) &= P(x = 10) + P(x = 11) + P(x = 12) \\ &= \frac{3}{36} + \frac{2}{36} + \frac{1}{36} \\ &= \frac{6}{36} \\ &= \frac{1}{6} \\ &\approx 0.167. \end{aligned}$$

▶ The probability is about 16.7%.

Monitoring Progress

Help in English and Spanish at *BigIdeasMath.com*

An octahedral die has eight sides numbered 1 through 8. Let x be a random variable that represents the sum when two such dice are rolled.

1. Make a table and draw a histogram showing the probability distribution for x.
2. What is the most likely sum when rolling the two dice?
3. What is the probability that the sum of the two dice is at most 3?

Binomial Distributions

One type of probability distribution is a **binomial distribution**. A binomial distribution shows the probabilities of the outcomes of a *binomial experiment*.

Core Concept

Binomial Experiments

A **binomial experiment** meets the following conditions.

- There are n independent trials.
- Each trial has only two possible outcomes: success and failure.
- The probability of success is the same for each trial. This probability is denoted by p. The probability of failure is $1 - p$.

For a binomial experiment, the probability of exactly k successes in n trials is

$$P(k \text{ successes}) = {}_nC_k p^k (1 - p)^{n-k}.$$

EXAMPLE 3 Constructing a Binomial Distribution

According to a survey, about 33% of people ages 16 and older in the U.S. own an electronic book reading device, or e-reader. You ask 6 randomly chosen people (ages 16 and older) whether they own an e-reader. Draw a histogram of the binomial distribution for your survey.

ATTENDING TO PRECISION

When probabilities are rounded, the sum of the probabilities may differ slightly from 1.

SOLUTION

The probability that a randomly selected person has an e-reader is $p = 0.33$. Because you survey 6 people, $n = 6$.

$P(k = 0) = {}_6C_0(0.33)^0(0.67)^6 \approx 0.090$

$P(k = 1) = {}_6C_1(0.33)^1(0.67)^5 \approx 0.267$

$P(k = 2) = {}_6C_2(0.33)^2(0.67)^4 \approx 0.329$

$P(k = 3) = {}_6C_3(0.33)^3(0.67)^3 \approx 0.216$

$P(k = 4) = {}_6C_4(0.33)^4(0.67)^2 \approx 0.080$

$P(k = 5) = {}_6C_5(0.33)^5(0.67)^1 \approx 0.016$

$P(k = 6) = {}_6C_6(0.33)^6(0.67)^0 \approx 0.001$

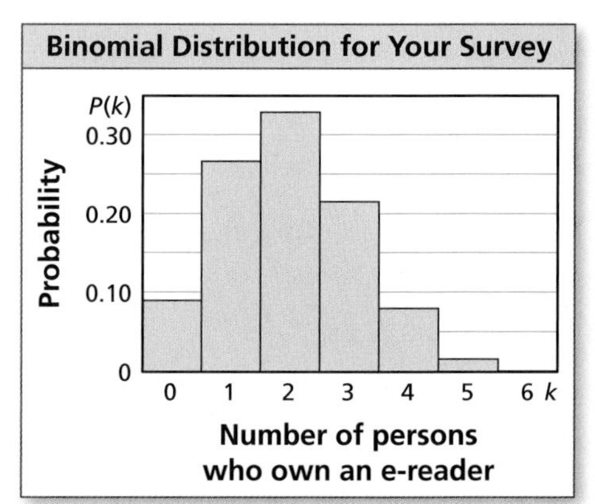

A histogram of the distribution is shown.

EXAMPLE 4 Interpreting a Binomial Distribution

Use the binomial distribution in Example 3 to answer each question.

a. What is the most likely outcome of the survey?

b. What is the probability that at most 2 people have an e-reader?

COMMON ERROR

Because a person may not have an e-reader, be sure you include $P(k = 0)$ when finding the probability that at most 2 people have an e-reader.

SOLUTION

a. The most likely outcome of the survey is the value of k for which $P(k)$ is greatest. This probability is greatest for $k = 2$. The most likely outcome is that 2 of the 6 people own an e-reader.

b. The probability that at most 2 people have an e-reader is

$$P(k \leq 2) = P(k = 0) + P(k = 1) + P(k = 2)$$
$$\approx 0.090 + 0.267 + 0.329$$
$$\approx 0.686.$$

▶ The probability is about 68.6%.

Monitoring Progress

Help in English and Spanish at *BigIdeasMath.com*

According to a survey, about 85% of people ages 18 and older in the U.S. use the Internet or e-mail. You ask 4 randomly chosen people (ages 18 and older) whether they use the Internet or e-mail.

4. Draw a histogram of the binomial distribution for your survey.

5. What is the most likely outcome of your survey?

6. What is the probability that at most 2 people you survey use the Internet or e-mail?

12.6 Exercises

Dynamic Solutions available at *BigIdeasMath.com*

Vocabulary and Core Concept Check

1. **VOCABULARY** What is a random variable?

2. **WRITING** Give an example of a binomial experiment and describe how it meets the conditions of a binomial experiment.

Monitoring Progress and Modeling with Mathematics

In Exercises 3–6, make a table and draw a histogram showing the probability distribution for the random variable. *(See Example 1.)*

3. x = the number on a table tennis ball randomly chosen from a bag that contains 5 balls labeled "1," 3 balls labeled "2," and 2 balls labeled "3."

4. $c = 1$ when a randomly chosen card out of a standard deck of 52 playing cards is a heart and $c = 2$ otherwise.

5. $w = 1$ when a randomly chosen letter from the English alphabet is a vowel and $w = 2$ otherwise.

6. n = the number of digits in a random integer from 0 through 999.

In Exercises 7 and 8, use the probability distribution to determine (a) the number that is most likely to be spun on a spinner, and (b) the probability of spinning an even number. *(See Example 2.)*

7.

Spinner Results
P(x)
Probability
1/2
1/4
0
1 2 3 4 x
Number on spinner

8.

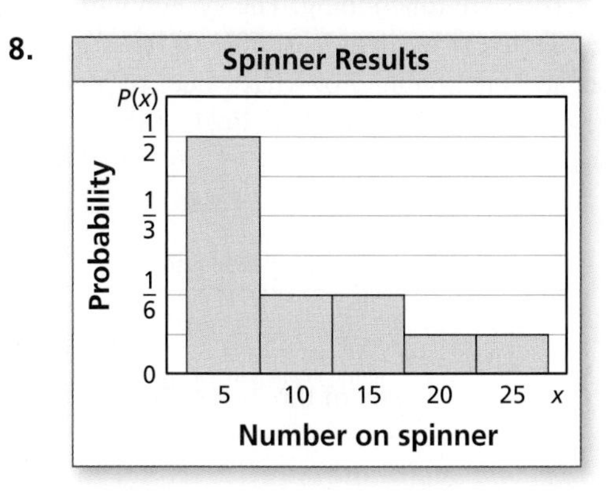

USING EQUATIONS In Exercises 9–12, calculate the probability of flipping a coin 20 times and getting the given number of heads.

9. 1

10. 4

11. 18

12. 20

13. **MODELING WITH MATHEMATICS** According to a survey, 27% of high school students in the United States buy a class ring. You ask 6 randomly chosen high school students whether they own a class ring. *(See Examples 3 and 4.)*

   a. Draw a histogram of the binomial distribution for your survey.

   b. What is the most likely outcome of your survey?

   c. What is the probability that at most 2 people have a class ring?

14. **MODELING WITH MATHEMATICS** According to a survey, 48% of adults in the United States believe that Unidentified Flying Objects (UFOs) are observing our planet. You ask 8 randomly chosen adults whether they believe UFOs are watching Earth.

   a. Draw a histogram of the binomial distribution for your survey.

   b. What is the most likely outcome of your survey?

   c. What is the probability that at most 3 people believe UFOs are watching Earth?

ERROR ANALYSIS In Exercises 15 and 16, describe and correct the error in calculating the probability of rolling a 1 exactly 3 times in 5 rolls of a six-sided die.

15. ✗ $P(k = 3) = {}_5C_3\left(\frac{1}{6}\right)^{5-3}\left(\frac{5}{6}\right)^3$

≈ 0.161

16. ✗ $P(k = 3) = \left(\frac{1}{6}\right)^3\left(\frac{5}{6}\right)^{5-3}$

≈ 0.003

17. MATHEMATICAL CONNECTIONS At most 7 gopher holes appear each week on the farm shown. Let x represent how many of the gopher holes appear in the carrot patch. Assume that a gopher hole has an equal chance of appearing at any point on the farm.

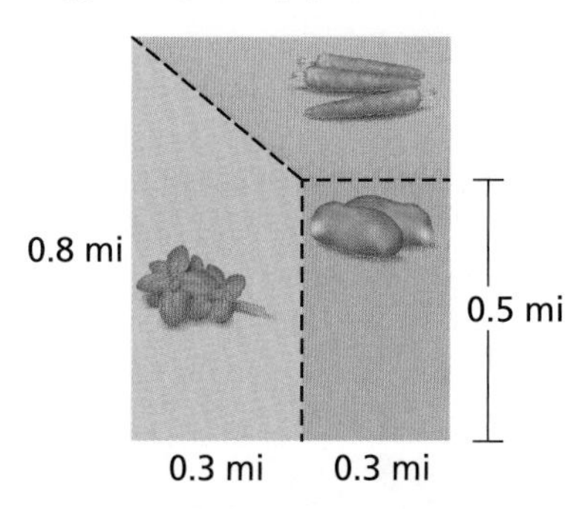

a. Find $P(x)$ for $x = 0, 1, 2, \ldots, 7$.

b. Make a table showing the probability distribution for x.

c. Make a histogram showing the probability distribution for x.

18. HOW DO YOU SEE IT? Complete the probability distribution for the random variable x. What is the probability the value of x is greater than 2?

x	1	2	3	4
$P(x)$	0.1	0.3	0.4	

19. MAKING AN ARGUMENT The binomial distribution shows the results of a binomial experiment. Your friend claims that the probability p of a success must be greater than the probability $1 - p$ of a failure. Is your friend correct? Explain your reasoning.

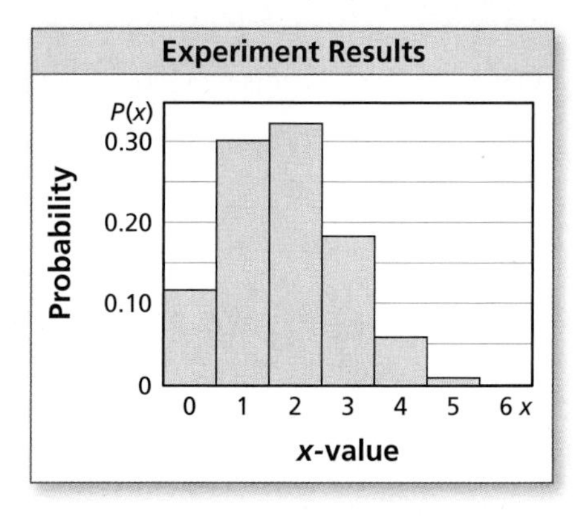

20. THOUGHT PROVOKING There are 100 coins in a bag. Only one of them has a date of 2010. You choose a coin at random, check the date, and then put the coin back in the bag. You repeat this 100 times. Are you certain of choosing the 2010 coin at least once? Explain your reasoning.

21. MODELING WITH MATHEMATICS Assume that having a male and having a female child are independent events, and that the probability of each is 0.5.

a. A couple has 4 male children. Evaluate the validity of this statement: "The first 4 kids were all boys, so the next one will probably be a girl."

b. What is the probability of having 4 male children and then a female child?

c. Let x be a random variable that represents the number of children a couple already has when they have their first female child. Draw a histogram of the distribution of $P(x)$ for $0 \le x \le 10$. Describe the shape of the histogram.

22. CRITICAL THINKING An entertainment system has n speakers. Each speaker will function properly with probability p, independent of whether the other speakers are functioning. The system will operate effectively when at least 50% of its speakers are functioning. For what values of p is a 5-speaker system more likely to operate than a 3-speaker system?

Maintaining Mathematical Proficiency

Reviewing what you learned in previous grades and lessons

List the possible outcomes for the situation. *(Section 12.1)*

23. guessing the gender of three children

24. picking one of two doors and one of three curtains

12.4–12.6 What Did You Learn?

Core Vocabulary

compound event, *p. 694*
overlapping events, *p. 694*
disjoint events, *p. 694*
mutually exclusive events, *p. 694*
permutation, *p. 700*
n factorial, *p. 700*
combination, *p. 702*
random variable, *p. 708*
probability distribution, *p. 708*
binomial distribution, *p. 709*
binomial experiment, *p. 709*

Core Concepts

Section 12.4

Probability of Compound Events, *p. 694*

Section 12.5

Permutations, *p. 701*
Combinations, *p. 702*

Section 12.6

Probability Distributions, *p. 708*
Binomial Experiments, *p. 709*

Mathematical Practices

1. How can you use diagrams to understand the situation in Exercise 22 on page 698?
2. Describe a relationship between the results in part (a) and part (b) in Exercise 52 on page 706.
3. Explain how you were able to break the situation into cases to evaluate the validity of the statement in part (a) of Exercise 21 on page 712.

Performance Task

A New Dartboard

You are a graphic artist working for a company on a new design for the board in the game of darts. You are eager to begin the project, but the team cannot decide on the terms of the game. Everyone agrees that the board should have four colors. But some want the probabilities of hitting each color to be equal, while others want them to be different. You offer to design two boards, one for each group. How do you get started? How creative can you be with your designs?

To explore the answers to these questions and more, go to ***BigIdeasMath.com***.

12 Chapter Review

Dynamic Solutions available at *BigIdeasMath.com*

12.1 Sample Spaces and Probability *(pp. 667–674)*

Each section of the spinner shown has the same area. The spinner was spun 30 times. The table shows the results. For which color is the experimental probability of stopping on the color the same as the theoretical probability?

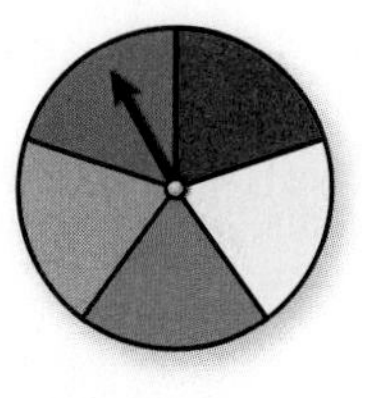

Spinner Results	
green	4
orange	6
red	9
blue	8
yellow	3

SOLUTION

The theoretical probability of stopping on each of the five colors is $\frac{1}{5}$. Use the outcomes in the table to find the experimental probabilities.

$P(\text{green}) = \frac{4}{30} = \frac{2}{15}$ $P(\text{orange}) = \frac{6}{30} = \frac{1}{5}$ $P(\text{red}) = \frac{9}{30} = \frac{3}{10}$ $P(\text{blue}) = \frac{8}{30} = \frac{4}{15}$ $P(\text{yellow}) = \frac{3}{30} = \frac{1}{10}$

▶ The experimental probability of stopping on orange is the same as the theoretical probability.

1. A bag contains 9 tiles, one for each letter in the word HAPPINESS. You choose a tile at random. What is the probability that you choose a tile with the letter S? What is the probability that you choose a tile with a letter other than P?
2. You throw a dart at the board shown. Your dart is equally likely to hit any point inside the square board. Are you most likely to get 5 points, 10 points, or 20 points?

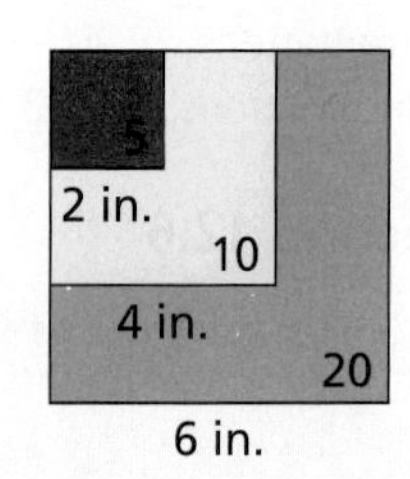

12.2 Independent and Dependent Events *(pp. 675–682)*

You randomly select 2 cards from a standard deck of 52 playing cards. What is the probability that both cards are jacks when (a) you replace the first card before selecting the second, and (b) you do not replace the first card. Compare the probabilities.

SOLUTION

Let event A be "first card is a jack" and event B be "second card is a jack."

a. Because you replace the first card before you select the second card, the events are independent. So, the probability is

$$P(A \text{ and } B) = P(A) \cdot P(B) = \frac{4}{52} \cdot \frac{4}{52} = \frac{16}{2704} = \frac{1}{169} \approx 0.006.$$

b. Because you do not replace the first card before you select the second card, the events are dependent. So, the probability is

$$P(A \text{ and } B) = P(A) \cdot P(B \mid A) = \frac{4}{52} \cdot \frac{3}{51} = \frac{12}{2652} = \frac{1}{221} \approx 0.005.$$

▶ So, you are $\frac{1}{169} \div \frac{1}{221} \approx 1.3$ times more likely to select 2 jacks when you replace the first card before you select the second card.

Find the probability of randomly selecting the given marbles from a bag of 5 red, 8 green, and 3 blue marbles when (a) you replace the first marble before drawing the second, and (b) you do not replace the first marble. Compare the probabilities.

3. red, then green
4. blue, then red
5. green, then green

12.3 Two-Way Tables and Probability (pp. 683–690)

A survey asks residents of the east and west sides of a city whether they support the construction of a bridge. The results, given as joint relative frequencies, are shown in the two-way table. What is the probability that a randomly selected resident from the east side will support the project?

		Location	
		East Side	West Side
Response	Yes	0.47	0.36
	No	0.08	0.09

SOLUTION

Find the joint and marginal relative frequencies. Then use these values to find the conditional probability.

$$P(\text{yes}\,|\,\text{east side}) = \frac{P(\text{east side and yes})}{P(\text{east side})} = \frac{0.47}{0.47 + 0.08} \approx 0.855$$

▶ So, the probability that a resident of the east side of the city will support the project is about 85.5%.

6. What is the probability that a randomly selected resident who does not support the project in the example above is from the west side?

7. After a conference, 220 men and 270 women respond to a survey. Of those, 200 men and 230 women say the conference was impactful. Organize these results in a two-way table. Then find and interpret the marginal frequencies.

12.4 Probability of Disjoint and Overlapping Events (pp. 693–698)

Let A and B be events such that $P(A) = \frac{2}{3}$, $P(B) = \frac{1}{2}$, and $P(A \text{ and } B) = \frac{1}{3}$. Find $P(A \text{ or } B)$.

SOLUTION

$P(A \text{ or } B) = P(A) + P(B) - P(A \text{ and } B)$ — Write general formula.

$= \frac{2}{3} + \frac{1}{2} - \frac{1}{3}$ — Substitute known probabilities.

$= \frac{5}{6}$ — Simplify.

≈ 0.833 — Use a calculator.

8. Let A and B be events such that $P(A) = 0.32$, $P(B) = 0.48$, and $P(A \text{ and } B) = 0.12$. Find $P(A \text{ or } B)$.

9. Out of 100 employees at a company, 92 employees either work part time or work 5 days each week. There are 14 employees who work part time and 80 employees who work 5 days each week. What is the probability that a randomly selected employee works both part time and 5 days each week?

12.5 Permutations and Combinations (pp. 699–706)

A 5-digit code consists of 5 different integers from 0 to 9. How many different codes are possible?

SOLUTION

To find the number of permutations of 5 integers chosen from 10, find ${}_{10}P_5$.

$${}_{10}P_5 = \frac{10!}{(10-5)!} \qquad \text{Permutations formula}$$

$$= \frac{10!}{5!} \qquad \text{Subtract.}$$

$$= \frac{10 \cdot 9 \cdot 8 \cdot 7 \cdot 6 \cdot \cancel{5!}}{\cancel{5!}} \qquad \text{Expand factorials. Divide out common factor, 5!.}$$

$$= 30{,}240 \qquad \text{Simplify.}$$

▶ There are 30,240 possible codes.

Evaluate the expression.

10. ${}_7P_6$ **11.** ${}_{13}P_{10}$ **12.** ${}_6C_2$ **13.** ${}_8C_4$

14. Eight sprinters are competing in a race. How many different ways can they finish the race? (Assume there are no ties.)

15. A random drawing will determine which 3 people in a group of 9 will win concert tickets. What is the probability that you and your 2 friends will win the tickets?

12.6 Binomial Distributions (pp. 707–712)

According to a survey, about 21% of adults in the U.S. visited an art museum last year. You ask 4 randomly chosen adults whether they visited an art museum last year. Draw a histogram of the binomial distribution for your survey.

SOLUTION

The probability that a randomly selected person visited an art museum is $p = 0.21$. Because you survey 4 people, $n = 4$.

$$P(k = 0) = {}_4C_0(0.21)^0(0.79)^4 \approx 0.390$$

$$P(k = 1) = {}_4C_1(0.21)^1(0.79)^3 \approx 0.414$$

$$P(k = 2) = {}_4C_2(0.21)^2(0.79)^2 \approx 0.165$$

$$P(k = 3) = {}_4C_3(0.21)^3(0.79)^1 \approx 0.029$$

$$P(k = 4) = {}_4C_4(0.21)^4(0.79)^0 \approx 0.002$$

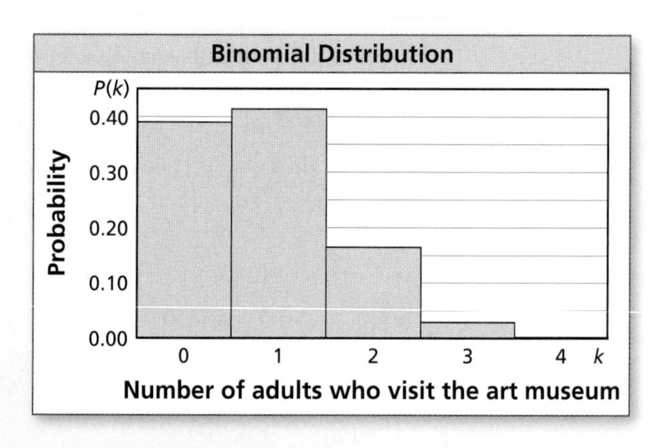

16. Find the probability of flipping a coin 12 times and getting exactly 4 heads.

17. A basketball player makes a free throw 82.6% of the time. The player attempts 5 free throws. Draw a histogram of the binomial distribution of the number of successful free throws. What is the most likely outcome?

12 Chapter Test

You roll a six-sided die. Find the probability of the event described. Explain your reasoning.

1. You roll a number less than 5.

2. You roll a multiple of 3.

Evaluate the expression.

3. ${}_7P_2$

4. ${}_8P_3$

5. ${}_6C_3$

6. ${}_{12}C_7$

7. In the word PYRAMID, how many ways can you arrange (a) all of the letters and (b) 5 of the letters?

8. You find the probability $P(A \text{ or } B)$ by using the equation $P(A \text{ or } B) = P(A) + P(B) - P(A \text{ and } B)$. Describe why it is necessary to subtract $P(A \text{ and } B)$ when the events A and B are overlapping. Then describe why it is *not* necessary to subtract $P(A \text{ and } B)$ when the events A and B are disjoint.

9. Is it possible to use the formula $P(A \text{ and } B) = P(A) \bullet P(B \mid A)$ when events A and B are independent? Explain your reasoning.

10. According to a survey, about 58% of families sit down for a family dinner at least four times per week. You ask 5 randomly chosen families whether they have a family dinner at least four times per week.

a. Draw a histogram of the binomial distribution for the survey.

b. What is the most likely outcome of the survey?

c. What is the probability that at least 3 families have a family dinner four times per week?

11. You are choosing a cell phone company to sign with for the next 2 years. The three plans you consider are equally priced. You ask several of your neighbors whether they are satisfied with their current cell phone company. The table shows the results. According to this survey, which company should you choose?

	Satisfied	Not Satisfied
Company A	IIII	II
Company B	IIII	III
Company C	𝍸 I	𝍸

12. The surface area of Earth is about 196.9 million square miles. The land area is about 57.5 million square miles, and the rest is water. What is the probability that a meteorite that reaches the surface of Earth will hit land? What is the probability that it will hit water?

13. Consider a bag that contains all the chess pieces in a set, as shown in the diagram.

	King	Queen	Bishop	Rook	Knight	Pawn	
Black	1	1	2	2	2	8	
White	1	1	2	2	2	8	

a. You choose one piece at random. Find the probability that you choose a black piece or a queen.

b. You choose one piece at random, do not replace it, then choose a second piece at random. Find the probability that you choose a king, then a pawn.

14. Three volunteers are chosen at random from a group of 12 to help at a summer camp.

a. What is the probability that you, your brother, and your friend are chosen?

b. The first person chosen will be a counselor, the second will be a lifeguard, and the third will be a cook. What is the probability that you are the cook, your brother is the lifeguard, and your friend is the counselor?

12 Cumulative Assessment

1. According to a survey, 63% of Americans consider themselves sports fans. You randomly select 14 Americans to survey.

a. Draw a histogram of the binomial distribution of your survey.

b. What is the most likely number of Americans who consider themselves sports fans?

c. What is the probability at least 7 Americans consider themselves sports fans?

2. What is the arc length of $\widehat{AB}$?

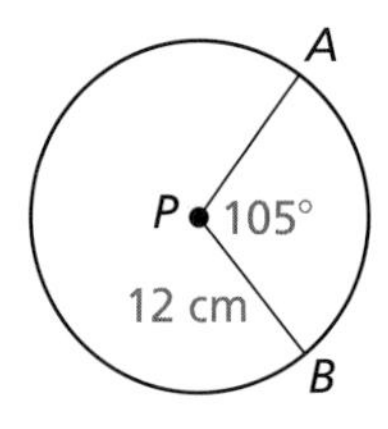

Ⓐ 3.5π cm

Ⓑ 7π cm

Ⓒ 21π cm

Ⓓ 42π cm

3. You order a fruit smoothie made with 2 liquid ingredients and 3 fruit ingredients from the menu shown. How many different fruit smoothies can you order?

4. The point (4, 3) is on a circle with center $(-2, -5)$. What is the standard equation of the circle?

5. Find the length of each line segment with the given endpoints. Then order the line segments from shortest to longest.

a. $A(1, -5)$, $B(4, 0)$

b. $C(-4, 2)$, $D(1, 4)$

c. $E(-1, 1)$, $F(-2, 7)$

d. $G(-1.5, 0)$, $H(4.5, 0)$

e. $J(-7, -8)$, $K(-3, -5)$

f. $L(10, -2)$, $M(9, 6)$

6. Use the diagram to explain why the equation is true.

$$P(A) + P(B) = P(A \text{ or } B) + P(A \text{ and } B)$$

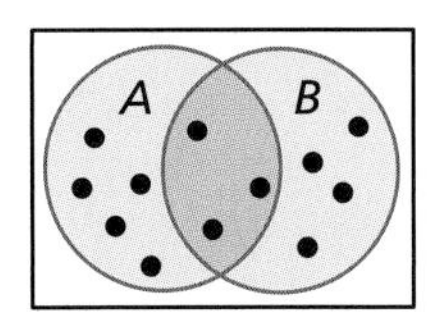

7. A plane intersects a cylinder. Which of the following cross sections cannot be formed by this intersection?

Ⓐ line

Ⓑ triangle

Ⓒ rectangle

Ⓓ circle

8. A survey asked male and female students about whether they prefer to take gym class or choir. The table shows the results of the survey.

		Class		
		Gym	Choir	Total
Gender	Male			50
	Female	23		
	Total		49	106

a. Complete the two-way table.

b. What is the probability that a randomly selected student is female and prefers choir?

c. What is the probability that a randomly selected male student prefers gym class?

9. The owner of a lawn-mowing business has three mowers. As long as one of the mowers is working, the owner can stay productive. One of the mowers is unusable 10% of the time, one is unusable 8% of the time, and one is unusable 18% of the time.

a. Find the probability that all three mowers are unusable on a given day.

b. Find the probability that at least one of the mowers is unusable on a given day.

c. Suppose the least-reliable mower stops working completely. How does this affect the probability that the lawn-mowing business can be productive on a given day?

10. You throw a dart at the board shown. Your dart is equally likely to hit any point inside the square board. What is the probability your dart lands in the yellow region?

Ⓐ $\frac{\pi}{36}$

Ⓑ $\frac{\pi}{12}$

Ⓒ $\frac{\pi}{9}$

Ⓓ $\frac{\pi}{4}$

2 in. 2 in. 2 in.

Additional Topic

Focus of a Parabola

Essential Question What is the focus of a parabola?

EXPLORATION 1 Analyzing Satellite Dishes

Work with a partner. Vertical rays enter a satellite dish whose cross section is a parabola. When the rays hit the parabola, they reflect at the same angle at which they entered. (See Ray 1 in the figure.)

a. Draw the reflected rays so that they intersect the y-axis.

b. What do the reflected rays have in common?

c. The optimal location for the receiver of the satellite dish is at a point called the *focus* of the parabola. Determine the location of the focus. Explain why this makes sense in this situation.

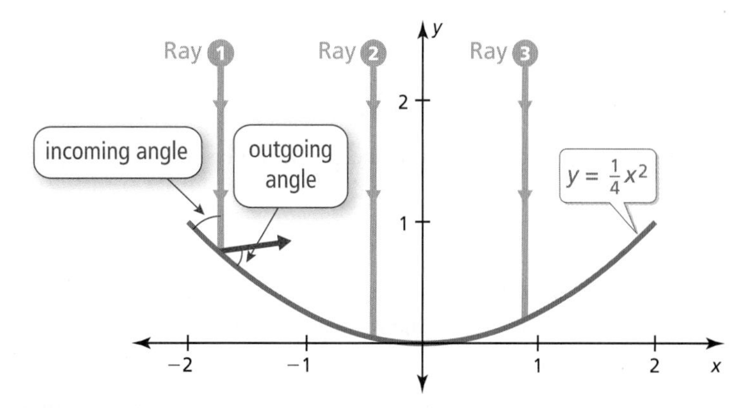

CONSTRUCTING VIABLE ARGUMENTS

To be proficient in math, you need to make conjectures and build logical progressions of statements to explore the truth of your conjectures.

EXPLORATION 2 Analyzing Spotlights

Work with a partner. Beams of light are coming from the bulb in a spotlight, located at the focus of the parabola. When the beams hit the parabola, they reflect at the same angle at which they hit. (See Beam 1 in the figure.) Draw the reflected beams. What do they have in common? Would you consider this to be the optimal result? Explain.

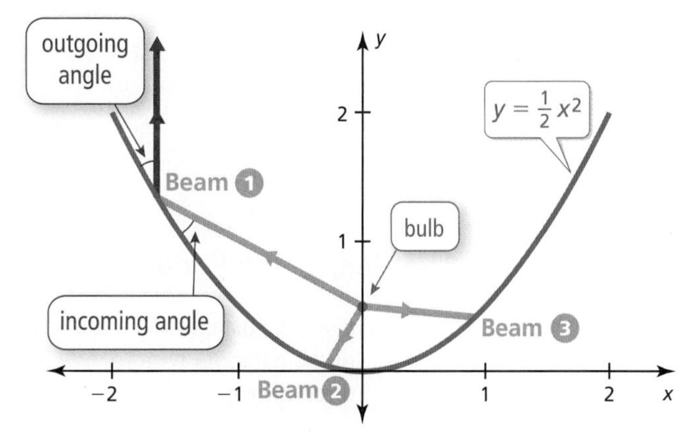

Communicate Your Answer

3. What is the focus of a parabola?

4. Describe some of the properties of the focus of a parabola.

Lesson

Core Vocabulary

focus, *p. 722*
directrix, *p. 722*

Previous
perpendicular
Distance Formula
congruent

What You Will Learn

- Explore the focus and the directrix of a parabola.
- Write equations of parabolas.
- Solve real-life problems.

Exploring the Focus and Directrix

Previously, you learned that the graph of a quadratic function is a parabola that opens up or down. A parabola can also be defined as the set of all points (x, y) in a plane that are equidistant from a fixed point called the **focus** and a fixed line called the **directrix**.

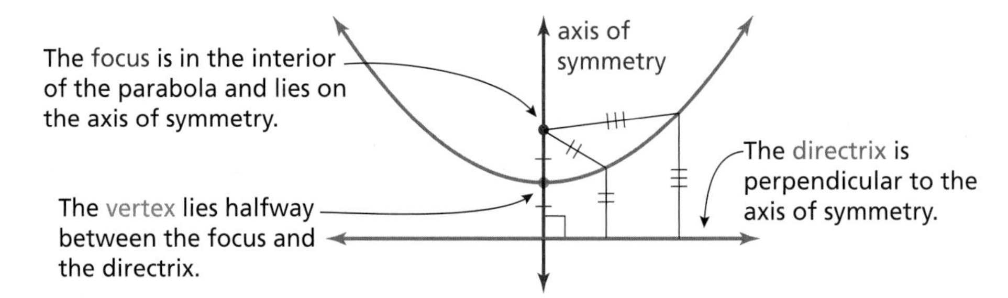

REMEMBER

The distance from a point to a line is defined as the length of the perpendicular segment from the point to the line.

EXAMPLE 1 Using the Distance Formula to Write an Equation

Use the Distance Formula to write an equation of the parabola with focus $F(0, 2)$ and directrix $y = -2$.

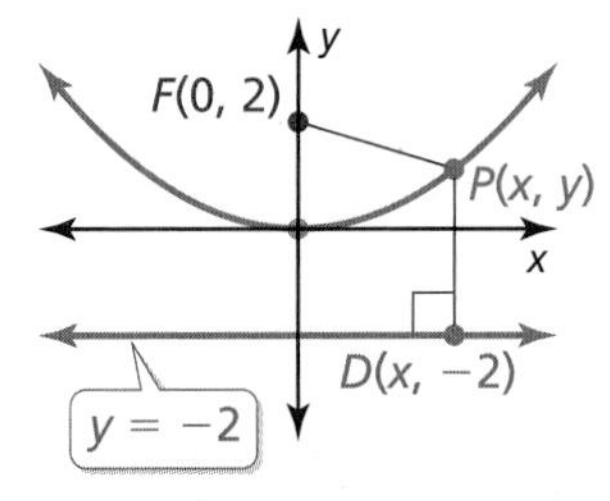

SOLUTION

Notice the line segments drawn from point F to point P and from point P to point D. By the definition of a parabola, these line segments must be congruent.

$PD = PF$	Definition of a parabola
$\sqrt{(x - x_1)^2 + (y - y_1)^2} = \sqrt{(x - x_2)^2 + (y - y_2)^2}$	Distance Formula
$\sqrt{(x - x)^2 + (y - (-2))^2} = \sqrt{(x - 0)^2 + (y - 2)^2}$	Substitute for x_1, y_1, x_2, and y_2.
$\sqrt{(y + 2)^2} = \sqrt{x^2 + (y - 2)^2}$	Simplify.
$(y + 2)^2 = x^2 + (y - 2)^2$	Square each side.
$y^2 + 4y + 4 = x^2 + y^2 - 4y + 4$	Expand.
$8y = x^2$	Combine like terms.
$y = \frac{1}{8}x^2$	Divide each side by 8.

Monitoring Progress  Help in English and Spanish at *BigIdeasMath.com*

1. Use the Distance Formula to write an equation of the parabola with focus $F(0, -3)$ and directrix $y = 3$.

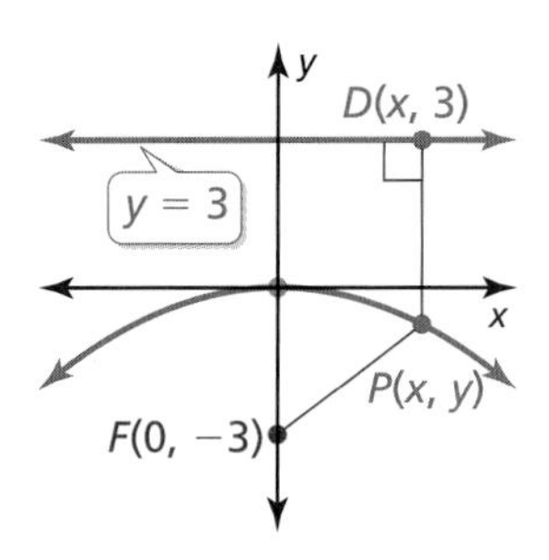

You can derive the equation of a parabola that opens up or down with vertex (0, 0), focus $(0, p)$, and directrix $y = -p$ using the procedure in Example 1.

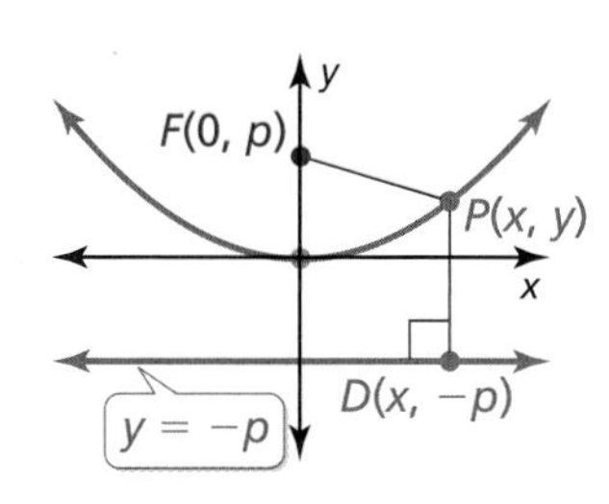

$$\sqrt{(x - x)^2 + (y - (-p))^2} = \sqrt{(x - 0)^2 + (y - p)^2}$$

$$(y + p)^2 = x^2 + (y - p)^2$$

$$y^2 + 2py + p^2 = x^2 + y^2 - 2py + p^2$$

$$4py = x^2$$

$$y = \frac{1}{4p}x^2$$

The focus and directrix each lie $|p|$ units from the vertex. Parabolas can also open left or right, in which case the equation has the form $x = \frac{1}{4p}y^2$ when the vertex is (0, 0).

LOOKING FOR STRUCTURE

Notice that $y = \frac{1}{4p}x^2$ is of the form $y = ax^2$. So, changing the value of p vertically stretches or shrinks the parabola.

Core Concept

Standard Equations of a Parabola with Vertex at the Origin

Vertical axis of symmetry ($x = 0$)

Equation: $y = \frac{1}{4p}x^2$

Focus: $(0, p)$

Directrix: $y = -p$

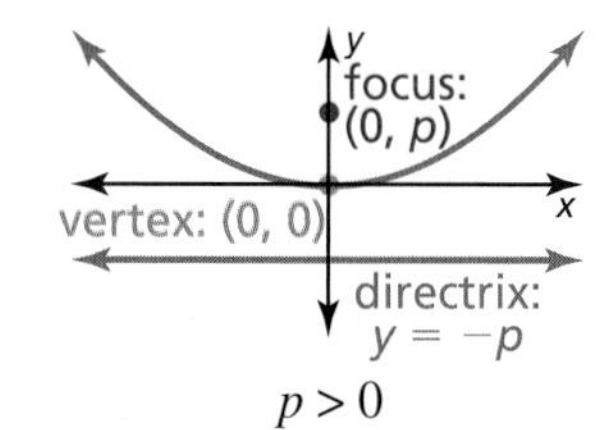

$p > 0$

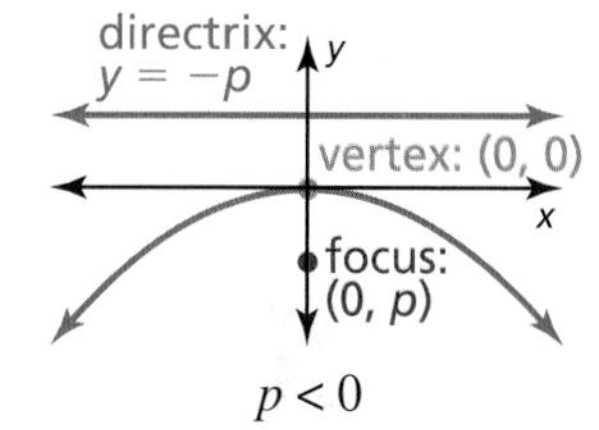

$p < 0$

Horizontal axis of symmetry ($y = 0$)

Equation: $x = \frac{1}{4p}y^2$

Focus: $(p, 0)$

Directrix: $x = -p$

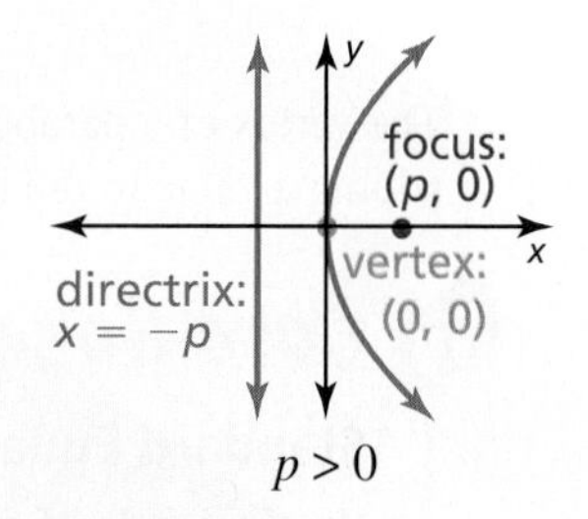

$p > 0$

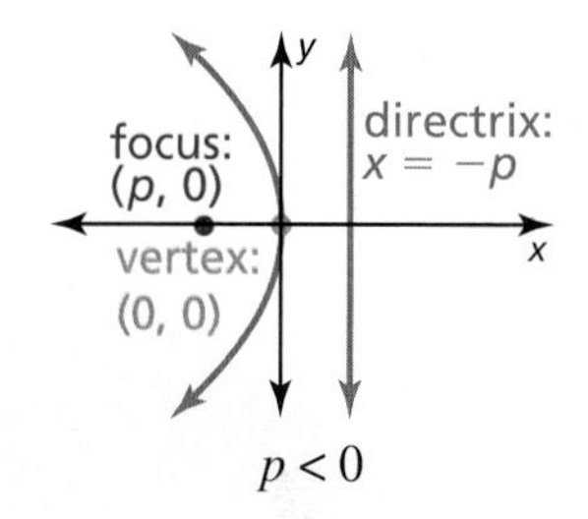

$p < 0$

STUDY TIP

Notice that parabolas opening left or right do *not* represent functions.

EXAMPLE 2 Graphing an Equation of a Parabola

Identify the focus, directrix, and axis of symmetry of $-4x = y^2$. Graph the equation.

SOLUTION

Step 1 Rewrite the equation in standard form.

$-4x = y^2$ Write the original equation.

$x = -\frac{1}{4}y^2$ Divide each side by −4.

Step 2 Identify the focus, directrix, and axis of symmetry. The equation has the form $x = \frac{1}{4p}y^2$, where $p = -1$. The focus is $(p, 0)$, or $(-1, 0)$. The directrix is $x = -p$, or $x = 1$. Because y is squared, the axis of symmetry is the x-axis.

Step 3 Use a table of values to graph the equation. Notice that it is easier to substitute y-values and solve for x. Opposite y-values result in the same x-value.

y	0	±1	±2	±3	±4
x	0	−0.25	−1	−2.25	−4

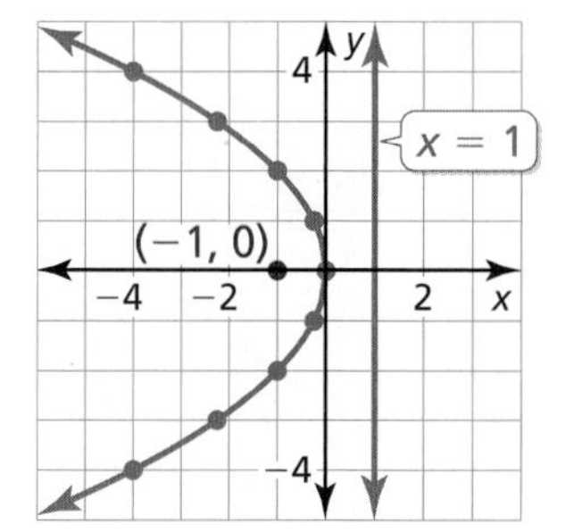

Writing Equations of Parabolas

EXAMPLE 3 Writing an Equation of a Parabola

Write an equation of the parabola shown.

SOLUTION

Because the vertex is at the origin and the axis of symmetry is vertical, the equation has the form $y = \frac{1}{4p}x^2$. The directrix is $y = -p = 3$, so $p = -3$. Substitute -3 for p to write an equation of the parabola.

$$y = \frac{1}{4(-3)}x^2 = -\frac{1}{12}x^2$$

▶ So, an equation of the parabola is $y = -\frac{1}{12}x^2$.

Monitoring Progress  Help in English and Spanish at *BigIdeasMath.com*

Identify the focus, directrix, and axis of symmetry of the parabola. Then graph the equation.

2. $y = 0.5x^2$ **3.** $-y = x^2$ **4.** $y^2 = 6x$

Write an equation of the parabola with vertex at (0, 0) and the given directrix or focus.

5. directrix: $x = -3$ **6.** focus: $(-2, 0)$ **7.** focus: $\left(0, \frac{3}{2}\right)$

The vertex of a parabola is not always at the origin. As in previous transformations, adding a value to the input or output of a function translates its graph.

Core Concept

Standard Equations of a Parabola with Vertex at (*h*, *k*)

Vertical axis of symmetry ($x = h$)

Equation: $y = \frac{1}{4p}(x - h)^2 + k$

Focus: $(h, k + p)$

Directrix: $y = k - p$

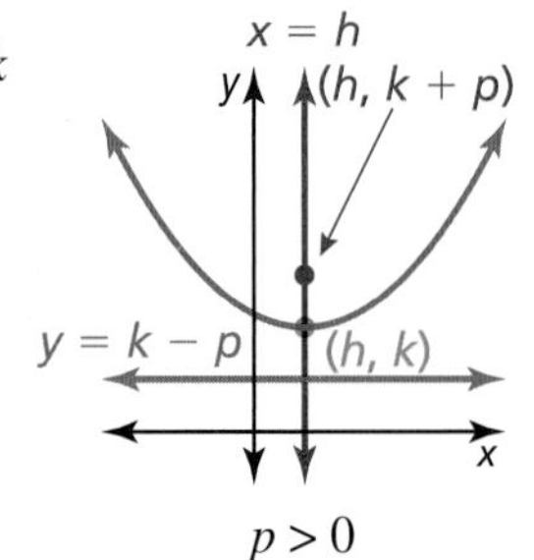

$p > 0$

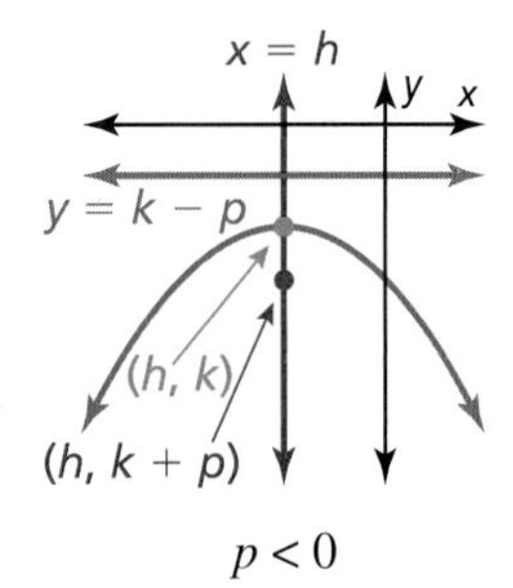

$p < 0$

Horizontal axis of symmetry ($y = k$)

Equation: $x = \frac{1}{4p}(y - k)^2 + h$

Focus: $(h + p, k)$

Directrix: $x = h - p$

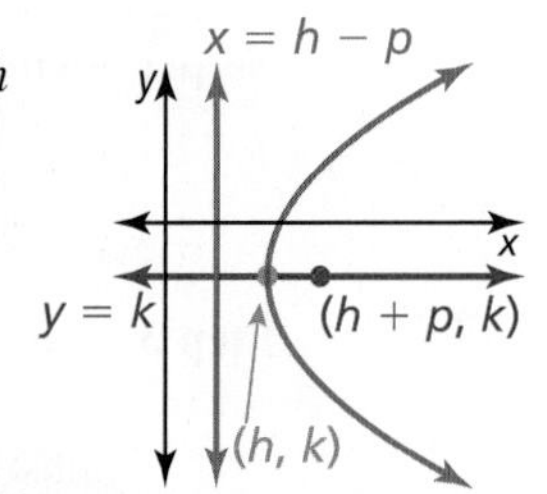

$p > 0$

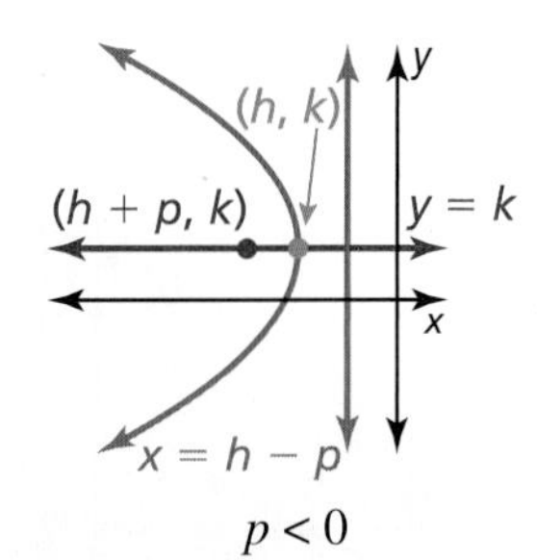

$p < 0$

STUDY TIP

The standard form for a vertical axis of symmetry looks like vertex form. To remember the standard form for a horizontal axis of symmetry, switch *x* and *y*, and *h* and *k*.

EXAMPLE 4 Writing an Equation of a Translated Parabola

Write an equation of the parabola shown.

SOLUTION

Because the vertex is not at the origin and the axis of symmetry is horizontal, the equation has the form $x = \frac{1}{4p}(y - k)^2 + h$. The vertex (h, k) is $(6, 2)$ and the focus $(h + p, k)$ is $(10, 2)$, so $h = 6$, $k = 2$, and $p = 4$. Substitute these values to write an equation of the parabola.

$$x = \frac{1}{4(4)}(y - 2)^2 + 6 = \frac{1}{16}(y - 2)^2 + 6$$

▶ So, an equation of the parabola is $x = \frac{1}{16}(y - 2)^2 + 6$.

Solving Real-Life Problems

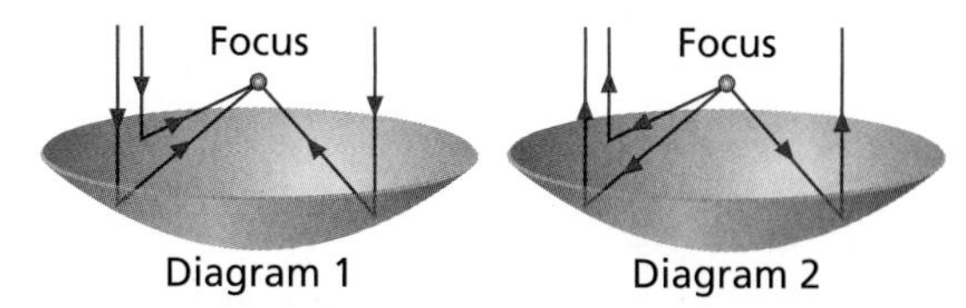

Parabolic reflectors have cross sections that are parabolas. Incoming sound, light, or other energy that arrives at a parabolic reflector parallel to the axis of symmetry is directed to the focus (Diagram 1). Similarly, energy that is emitted from the focus of a parabolic reflector and then strikes the reflector is directed parallel to the axis of symmetry (Diagram 2).

EXAMPLE 5 Solving a Real-Life Problem

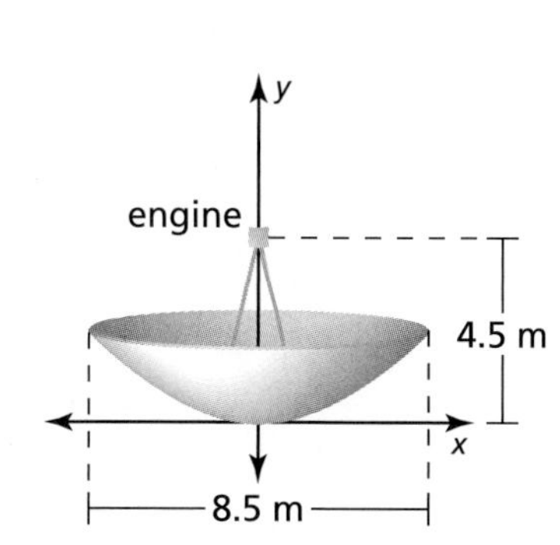

An electricity-generating dish uses a parabolic reflector to concentrate sunlight onto a high-frequency engine located at the focus of the reflector. The sunlight heats helium to 650°C to power the engine. Write an equation that represents the cross section of the dish shown with its vertex at (0, 0). What is the depth of the dish?

SOLUTION

Because the vertex is at the origin, and the axis of symmetry is vertical, the equation has the form $y = \frac{1}{4p}x^2$. The engine is at the focus, which is 4.5 meters above the vertex. So, $p = 4.5$. Substitute 4.5 for p to write the equation.

$$y = \frac{1}{4(4.5)}x^2 = \frac{1}{18}x^2$$

The depth of the dish is the y-value at the dish's outside edge. The dish extends $\frac{8.5}{2} = 4.25$ meters to either side of the vertex (0, 0), so find y when $x = 4.25$.

$$y = \frac{1}{18}(4.25)^2 \approx 1$$

▶ The depth of the dish is about 1 meter.

Monitoring Progress  Help in English and Spanish at *BigIdeasMath.com*

8. Write an equation of a parabola with vertex $(-1, 4)$ and focus $(-1, 2)$.

9. A parabolic microwave antenna is 16 feet in diameter. Write an equation that represents the cross section of the antenna with its vertex at (0, 0) and its focus 10 feet to the right of the vertex. What is the depth of the antenna?

Exercises

Dynamic Solutions available at *BigIdeasMath.com*

Vocabulary and Core Concept Check

1. **COMPLETE THE SENTENCE** A parabola is the set of all points in a plane equidistant from a fixed point called the ______ and a fixed line called the __________ .

2. **WRITING** Explain how to find the coordinates of the focus of a parabola with vertex $(0, 0)$ and directrix $y = 5$.

Monitoring Progress and Modeling with Mathematics

In Exercises 3–10, use the Distance Formula to write an equation of the parabola. *(See Example 1.)*

3.

y
F(0, 1)
P(x, y)
x
D(x, −1)
y = −1

4.

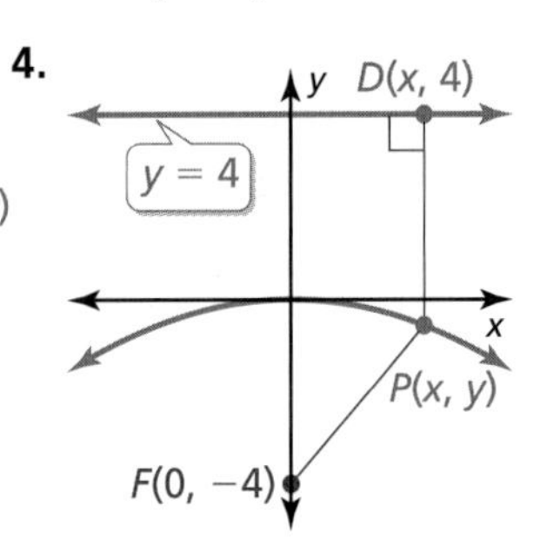

5. focus: $(0, -2)$
directrix: $y = 2$

6. directrix: $y = 7$
focus: $(0, -7)$

7. vertex: $(0, 0)$
directrix: $y = -6$

8. vertex: $(0, 0)$
focus: $(0, 5)$

9. vertex: $(0, 0)$
focus: $(0, -10)$

10. vertex: $(0, 0)$
directrix: $y = -9$

11. **ANALYZING RELATIONSHIPS** Which of the given characteristics describe parabolas that open down? Explain your reasoning.

Ⓐ focus: $(0, -6)$
directrix: $y = 6$

Ⓑ focus: $(0, -2)$
directrix: $y = 2$

Ⓒ focus: $(0, 6)$
directrix: $y = -6$

Ⓓ focus: $(0, -1)$
directrix: $y = 1$

12. **REASONING** Which of the following are possible coordinates of the point P in the graph shown? Explain.

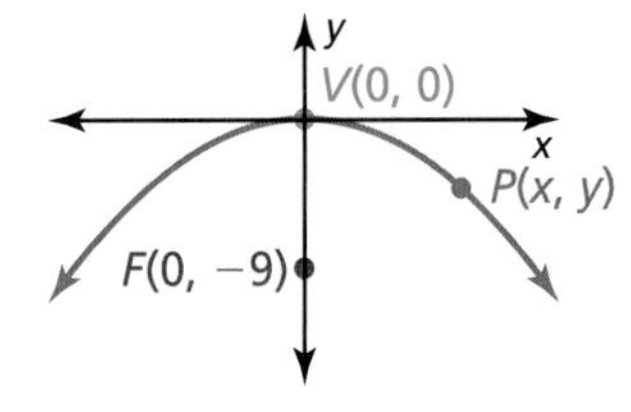

Ⓐ $(-6, -1)$ Ⓑ $\left(3, -\frac{1}{4}\right)$ Ⓒ $\left(4, -\frac{4}{9}\right)$

Ⓓ $\left(1, \frac{1}{36}\right)$ Ⓔ $(6, -1)$ Ⓕ $\left(2, -\frac{1}{18}\right)$

In Exercises 13–20, identify the focus, directrix, and axis of symmetry of the parabola. Graph the equation. *(See Example 2.)*

13. $y = \frac{1}{8}x^2$

14. $y = -\frac{1}{12}x^2$

15. $x = -\frac{1}{20}y^2$

16. $x = \frac{1}{24}y^2$

17. $y^2 = 16x$

18. $-x^2 = 48y$

19. $6x^2 + 3y = 0$

20. $8x^2 - y = 0$

ERROR ANALYSIS In Exercises 21 and 22, describe and correct the error in graphing the parabola.

21.

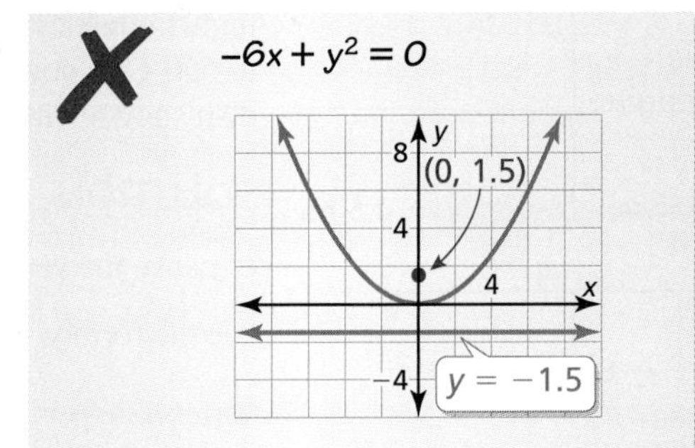

22.

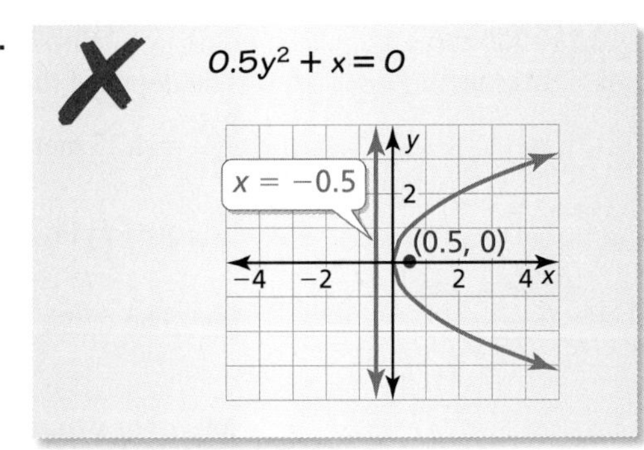

23. **ANALYZING EQUATIONS** The cross section (with units in inches) of a parabolic satellite dish can be modeled by the equation $y = \frac{1}{38}x^2$. How far is the receiver from the vertex of the cross section? Explain.

24. ANALYZING EQUATIONS The cross section (with units in inches) of a parabolic spotlight can be modeled by the equation $x = \frac{1}{20}y^2$. How far is the bulb from the vertex of the cross section? Explain.

In Exercises 25–28, write an equation of the parabola shown. *(See Example 3.)*

25. **26.**

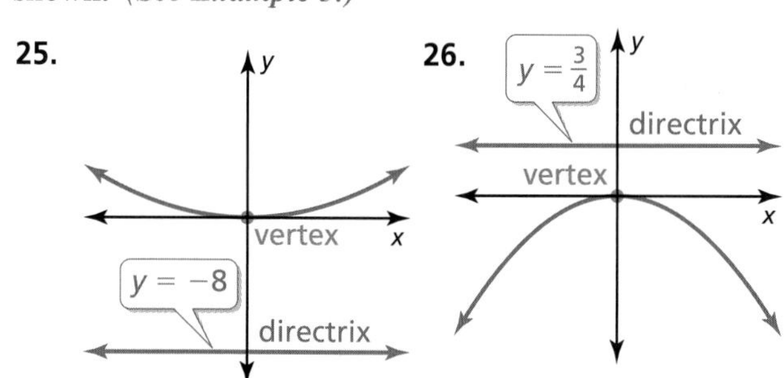

27. **28.**

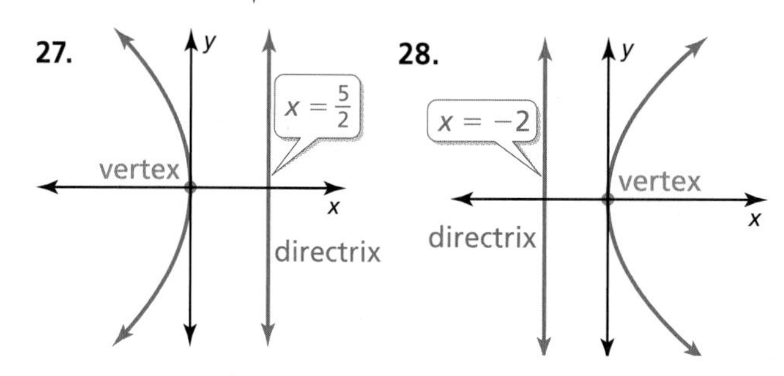

In Exercises 29–36, write an equation of the parabola with the given characteristics.

29. focus: $(3, 0)$
directrix: $x = -3$

30. focus: $\left(\frac{2}{3}, 0\right)$
directrix: $x = -\frac{2}{3}$

31. directrix: $x = -10$
vertex: $(0, 0)$

32. directrix: $y = \frac{8}{3}$
vertex: $(0, 0)$

33. focus: $\left(0, -\frac{5}{3}\right)$
directrix: $y = \frac{5}{3}$

34. focus: $\left(0, \frac{5}{4}\right)$
directrix: $y = -\frac{5}{4}$

35. focus: $\left(0, \frac{6}{7}\right)$
vertex: $(0, 0)$

36. focus: $\left(-\frac{4}{5}, 0\right)$
vertex: $(0, 0)$

In Exercises 37–40, write an equation of the parabola shown. *(See Example 4.)*

37.

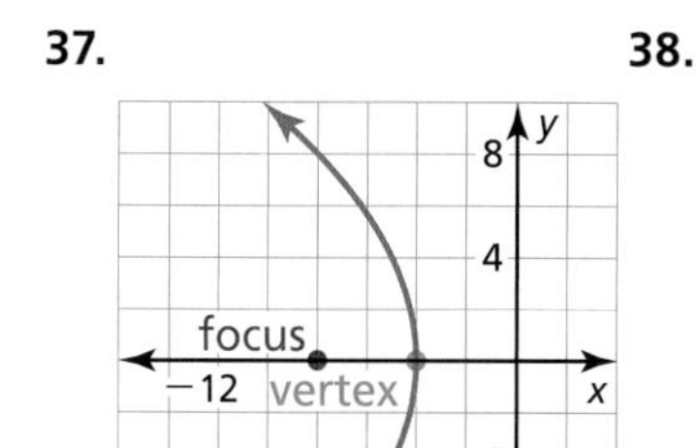

38.

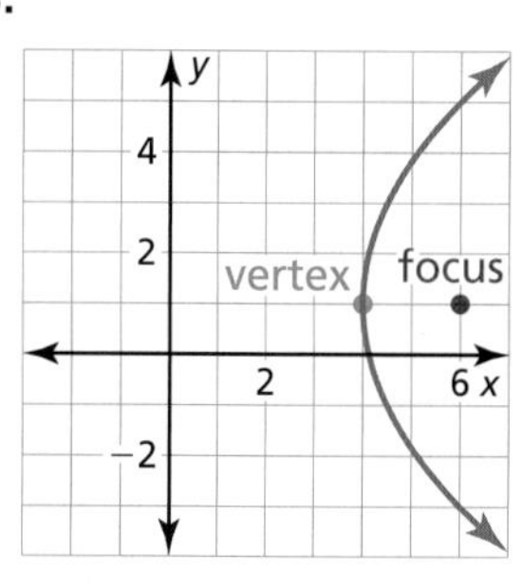

39.

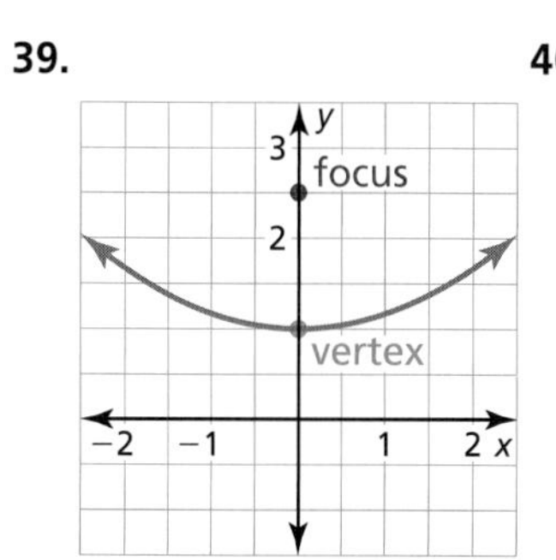

40.

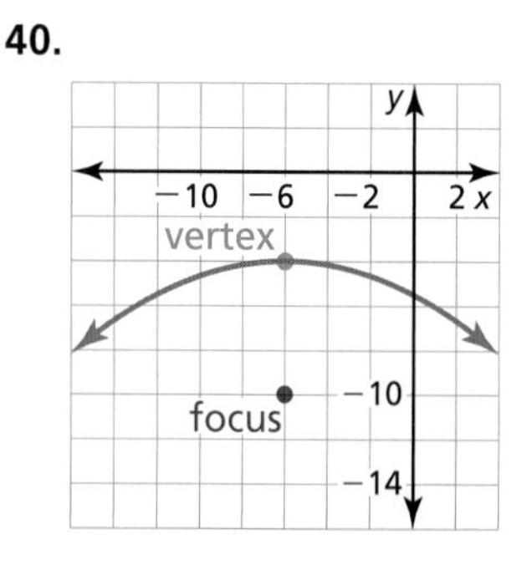

In Exercises 41–46, identify the vertex, focus, directrix, and axis of symmetry of the parabola. Describe the transformations of the graph of the standard equation with $p = 1$ and vertex $(0, 0)$.

41. $y = \frac{1}{8}(x - 3)^2 + 2$

42. $y = -\frac{1}{4}(x + 2)^2 + 1$

43. $x = \frac{1}{16}(y - 3)^2 + 1$

44. $y = (x + 3)^2 - 5$

45. $x = -3(y + 4)^2 + 2$

46. $x = 4(y + 5)^2 - 1$

47. MODELING WITH MATHEMATICS Scientists studying dolphin echolocation simulate the projection of a bottlenose dolphin's clicking sounds using computer models. The models originate the sounds at the focus of a parabolic reflector. The parabola in the graph shows the cross section of the reflector with focal length of 1.3 inches and aperture width of 8 inches. Write an equation to represent the cross section of the reflector. What is the depth of the reflector? *(See Example 5.)*

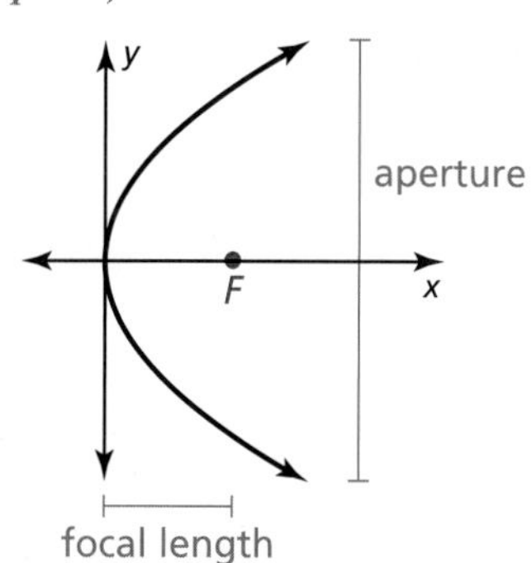

48. MODELING WITH MATHEMATICS Solar energy can be concentrated using long troughs that have a parabolic cross section as shown in the figure. Write an equation to represent the cross section of the trough. What are the domain and range in this situation? What do they represent?

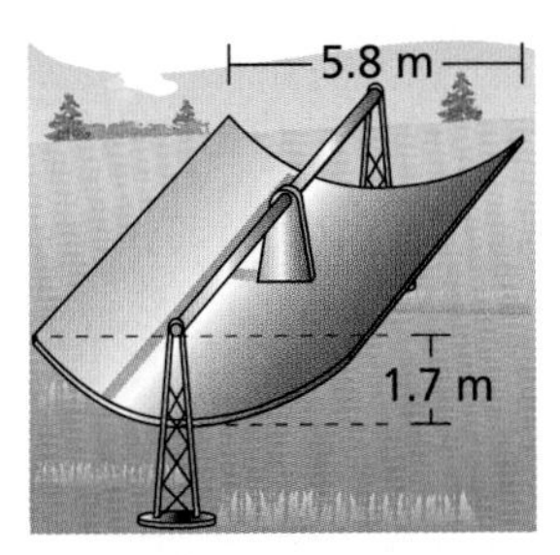

49. ABSTRACT REASONING As $|p|$ increases, how does the width of the graph of the equation $y = \frac{1}{4p}x^2$ change? Explain your reasoning.

50. HOW DO YOU SEE IT? The graph shows the path of a volleyball served from an initial height of 6 feet as it travels over a net.

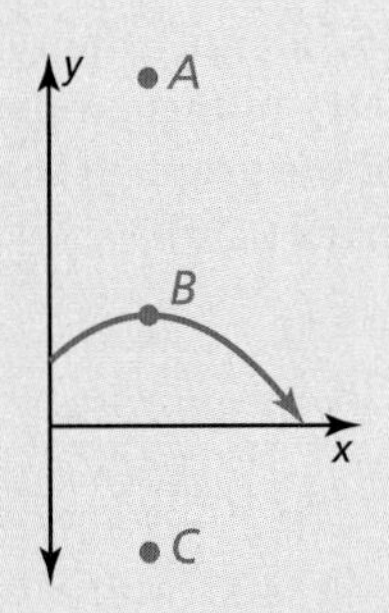

a. Label the vertex, focus, and a point on the directrix.

b. An underhand serve follows the same parabolic path but is hit from a height of 3 feet. How does this affect the focus? the directrix?

51. CRITICAL THINKING The distance from point P to the directrix is 2 units. Write an equation of the parabola.

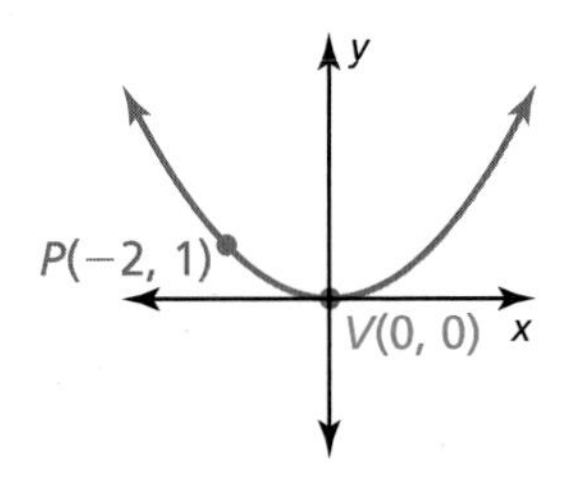

52. THOUGHT PROVOKING Two parabolas have the same focus (a, b) and focal length of 2 units. Write an equation of each parabola. Identify the directrix of each parabola.

53. REPEATED REASONING Use the Distance Formula to derive the equation of a parabola that opens to the right with vertex $(0, 0)$, focus $(p, 0)$, and directrix $x = -p$.

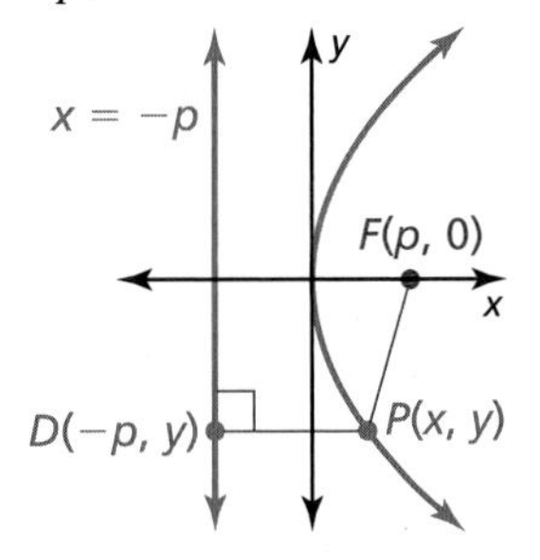

54. PROBLEM SOLVING The *latus rectum* of a parabola is the line segment that is parallel to the directrix, passes through the focus, and has endpoints that lie on the parabola. Find the length of the latus rectum of the parabola shown.

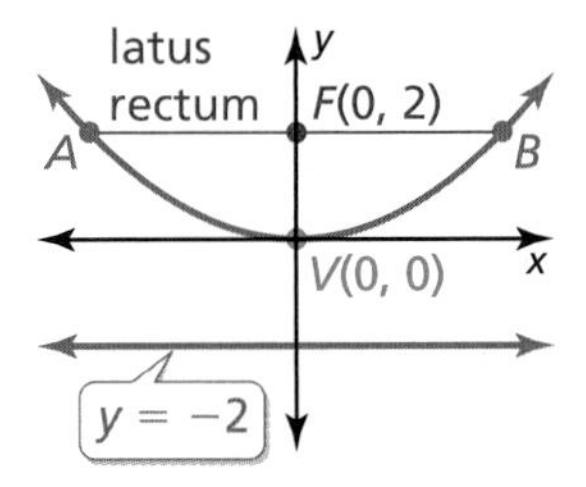

Selected Answers

Chapter 1

Chapter 1 Maintaining Mathematical Proficiency *(p. 1)*

1. 4 **2.** 11 **3.** 5 **4.** 9 **5.** 8 **6.** 6
7. 1 **8.** 5 **9.** 17 **10.** 154 m^2 **11.** 84 yd^2
12. 200 $in.^2$
13. x and y can be any real number, $x \neq y$; $x = y$; no; Absolute value is never negative.

1.1 Vocabulary and Core Concept Check *(p. 8)*

1. Collinear points lie on the same line. Coplanar points lie on the same plane.

1.1 Monitoring Progress and Modeling with Mathematics *(pp. 8–10)*

3. *Sample answer: A, B, D, E* **5.** plane *S*
7. $\overleftrightarrow{QW}$, line *g* **9.** *R, Q, S; Sample answer: T*
11. $\overline{DB}$ **13.** $\overrightarrow{AC}$ **15.** $\overrightarrow{EB}$ and $\overrightarrow{ED}$, $\overrightarrow{EA}$ and $\overrightarrow{EC}$
17. *Sample answer:*

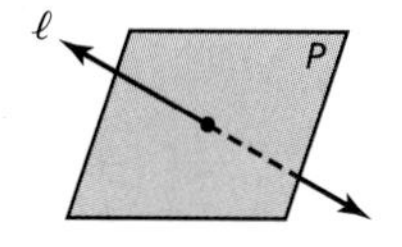

19. *Sample answer:*

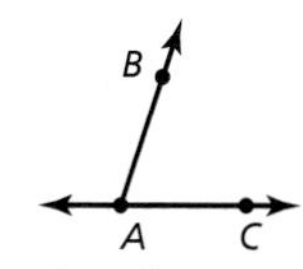

21. *Sample answer:*

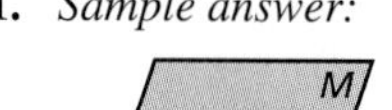

23. *Sample answer:*

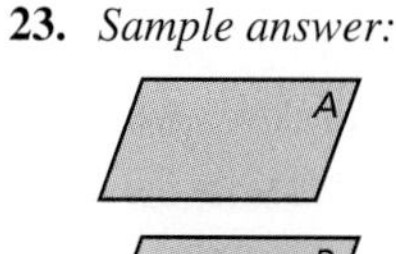

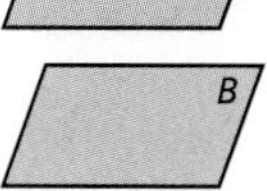

25. $\overrightarrow{AD}$ and $\overrightarrow{AC}$ are not opposite rays because *A, C,* and *D* are not collinear; $\overrightarrow{AD}$ and $\overrightarrow{AB}$ are opposite rays because *A, B,* and *D* are collinear, and *A* is between *B* and *D*.
27. *J* **29.** *Sample answer: D* **31.** *Sample answer: C*
33. $\overleftrightarrow{AE}$ **35.** point **37.** segment **39.** *P, Q, R, S*
41. *K, L, M, N* **43.** *L, M, Q, R*
45. yes; Use the point not on the line and two points on the line to draw the plane.
47. Three legs of the chair will meet on the floor to define a plane, but the point at the bottom of the fourth leg may not be in the same plane. When the chair tips so that this leg is on the floor, the plane defined by this leg and the two legs closest to it now lies in the plane of the floor; no; Three points define a plane, so the legs of the three-legged chair will always meet in the flat plane of the floor.
49. 6; The first two lines intersect at one point. The third line could intersect each of the first two lines. The fourth line can be drawn to intersect each of the first 3 lines. Then the total is $1 + 2 + 3 = 6$.
51. −1 0 1 2 3 4 5
ray
53. −4 −3 −2 −1 0 1 2 3 4 5 6 7
rays
55. **a.** *K, N* **b.** *Sample answer:* plane *JKL*, plane *JQN*
c. *J, K, L, M, N, P, Q*
57. sometimes; The point may be on the line.
59. sometimes; The planes may not intersect.
61. sometimes; The points may be collinear.
63. sometimes; Lines in parallel planes do not intersect, and may not be parallel.

1.1 Maintaining Mathematical Proficiency *(p. 10)*

65. 8 **67.** 10 **69.** $x = 25$ **71.** $x = 22$

1.2 Vocabulary and Core Concept Check *(p. 16)*

1. $\overline{XY}$ represents the segment *XY*, while *XY* represents the distance between points *X* and *Y* (the length of $\overline{XY}$).

1.2 Monitoring Progress and Modeling with Mathematics *(pp. 16–18)*

3. 3.5 cm **5.** 4.5 cm
7.
9.

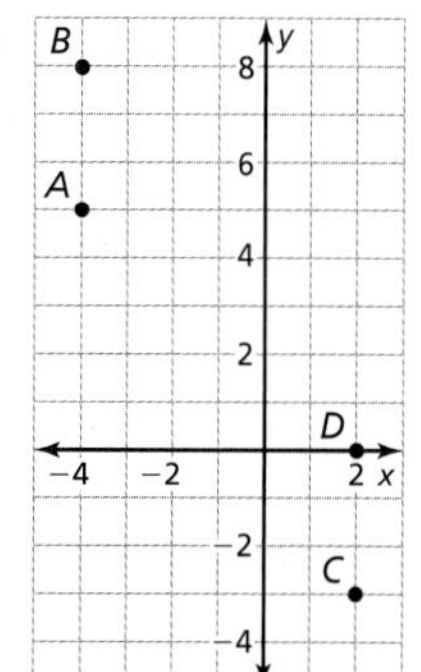

yes

11.

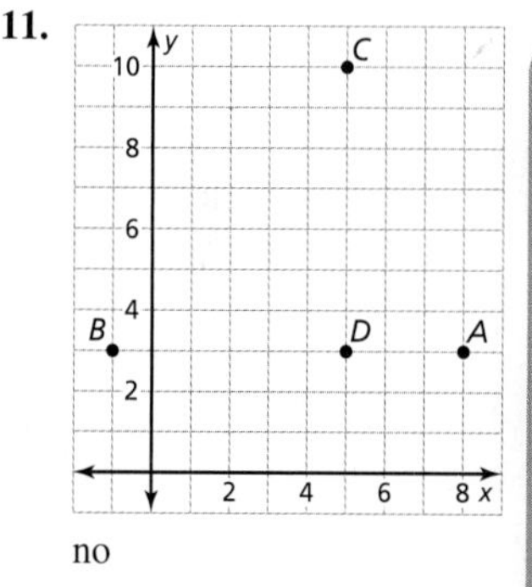

no

13.

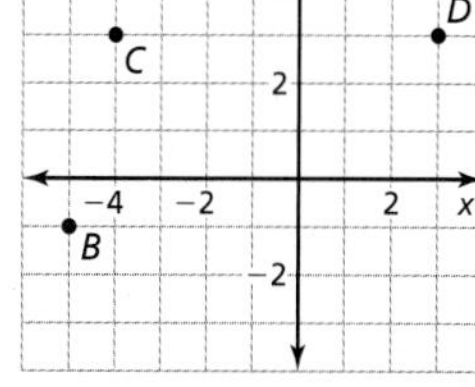

yes

15. 22 **17.** 23 **19.** 24 **21.** 20
23. The absolute value should have been taken;
$AB = |1 - 4.5| = 3.5$
25. $2\frac{1}{4}$ in., $1\frac{3}{4}$ in.; $\frac{1}{2}$ in.; $1\frac{2}{7}$
27. **a.** true; *B* is on $\overleftrightarrow{AC}$ between *A* and *C*.
b. false; *B, C,* and *E* are not collinear.
c. true; *D* is on $\overleftrightarrow{AH}$ between *A* and *H*.
d. false; *C, E,* and *F* are not collinear.
29. **a.** $3x + 6 = 21$; $x = 5$; $RS = 20$; $ST = 1$; $RT = 21$
b. $7x - 24 = 60$; $x = 12$; $RS = 20$; $ST = 40$; $RT = 60$
c. $2x + 3 = x + 10$; $x = 7$; $RS = 6$; $ST = 11$; $RT = 17$
d. $4x + 10 = 8x - 14$; $x = 6$; $RS = 15$; $ST = 19$; $RT = 34$

31. **a.** 64 ft **b.** about 0.24 min
c. You might walk slower if other people are in the hall.

33. 296.5 mi; If the round-trip distance is 647 miles, then the one-way distance is 323.5 miles. $323.5 - 27 = 296.5$

35. $|a - c| = |e - f|$; b and d are not used because when the x-values are the same, you subtract the y-values to find the length of the segment, and vice versa.

37. yes, no; $FC + CB = FB$, so $FB > CB$. $\overline{AC}$ and $\overline{DB}$ overlap but do not share an endpoint.

1.2 Maintaining Mathematical Proficiency *(p. 18)*

39. 5 41. $\frac{13}{2}$, or 6.5 43. $y = 9$ 45. $x = 2$

1.3 Vocabulary and Core Concept Check *(p. 24)*

1. It bisects the segment.

1.3 Monitoring Progress and Modeling with Mathematics *(pp. 24–26)*

3. line k; 34 5. M; 44 7. M; 40 9. $\overrightarrow{MN}$; 32

11. 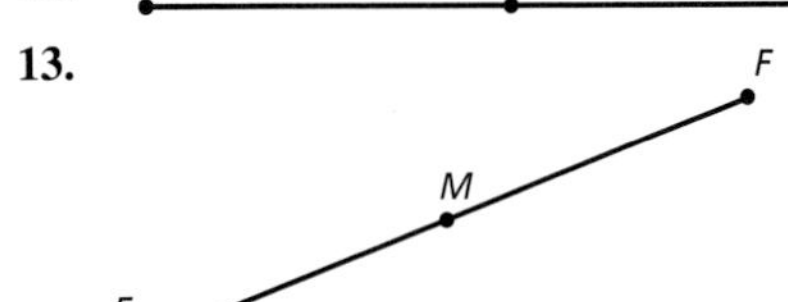

13.

15. (5, 2) 17. $\left(1, \frac{9}{2}\right)$ 19. (3, 12) 21. (18, −9)

23. 10 25. $\sqrt{13}$, or about 3.6 27. $\sqrt{97}$, or about 9.8

29. 6.5

31. The square root should have been taken. $\sqrt{61} \approx 7.8$

33. about 6.7, about 6.3; no; $AB > CD$

35. **a.** To find the x-coordinate of the midpoint, add the x-coordinates of the endpoints, and divide by 2. To find the y-coordinate of the midpoint, add the y-coordinates of the endpoints, and divide by 2.
b. To find the x-coordinate of the other endpoint, multiply the x-coordinate of the midpoint by 2, and subtract the x-coordinate of the given endpoint. To find the y-coordinate of the other endpoint, multiply the y-coordinate of the midpoint by 2, and subtract the y-coordinate of the given endpoint.

37. **a.** about 10.4 m; about 9.2 m **b.** about 18.9 m

39. **a.** about 191 yd **b.** about 40 yd
c. about 1.5 min; $MR \approx 40$ yd, total distance $\approx 40 + 40 + 40 + 70 + 40 = 230$ yd, $\frac{230}{150} \approx 1.5$ min

41. $\left(\frac{a+b}{2}, c\right)$, $|b - a|$

43. location D for lunch; The total distance traveled if you return home is $AM + AM + AB + AB$. The total distance traveled if you go to location D for lunch is $AB + DB + DB + AB$. Because $DB < AM$, the second option involves less traveling.

45. 13 cm

1.3 Maintaining Mathematical Proficiency *(p. 26)*

47. 26 ft, 30 ft^2 49. 36 yd, 60 yd^2

51. $y \geq 13$

53. $z \leq 48$

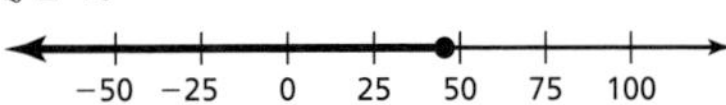

1.4 Vocabulary and Core Concept Check *(p. 34)*

1. $4s$

1.4 Monitoring Progress and Modeling with Mathematics *(pp. 34–36)*

3. quadrilateral; concave 5. pentagon; convex

7. 22 units 9. about 22.43 units 11. about 16.93 units

13. 7.5 square units 15. 9 square units

17. about 9.66 units 19. about 12.17 units

21. 4 square units 23. 6 square units

25. The length should be 5 units; $P = 2\ell + 2w = 2(5) + 2(3) = 16$; The perimeter is 16 units.

27. B

29. **a.** 4 square units; 16 square units; It is quadrupled.
b. yes; If you double the perimeter of the square, which is the same as doubling the side length, then the new area will be $2^2 = 4$ times as big.

31. **a.** 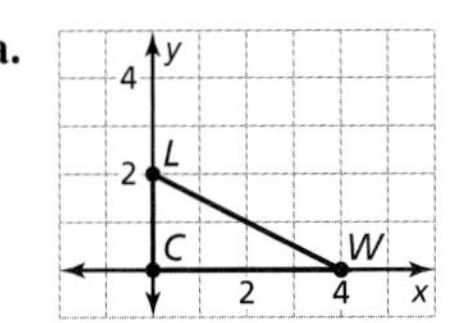

b. about 10.47 mi **c.** about 17.42 mi

33. **a.** y_1 and y_3 **b.** (0, 4), (4, 2), (2, −2)
c. about 15.27 units, 10 square units

35. **a.** 16 units, 16 square units
b. yes; The sides are all the same length because each one is the hypotenuse of a right triangle with legs that are each 2 units long. Because the slopes of the lines of each side are either 1 or −1, they are perpendicular.
c. about 11.31 units, 8 square units; It is half of the area of the larger square.

37. $x = 2$

1.4 Maintaining Mathematical Proficiency *(p. 36)*

39. $x = -1$ 41. $x = 14$ 43. $x = 1$

1.5 Vocabulary and Core Concept Check *(p. 43)*

1. congruent

1.5 Monitoring Progress and Modeling with Mathematics *(pp. 43–46)*

3. $\angle B$, $\angle ABC$, $\angle CBA$ 5. $\angle 1$, $\angle K$, $\angle JKL$ (or $\angle LKJ$)

7. $\angle HMK$, $\angle KMN$, $\angle HMN$ 9. 30°; acute

11. 85°; acute

13. The outer scale was used, but the inner scale should have been used because $\overrightarrow{OB}$ passes through 0° on the inner scale; 150°

15. 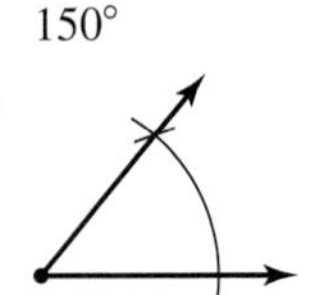

17. $\angle ADE$, $\angle BDC$, $\angle BCD$ 19. 34°

21. 58° 23. 42° 25. 37°, 58° 27. 77°, 103°

29. 32°, 58°

31.

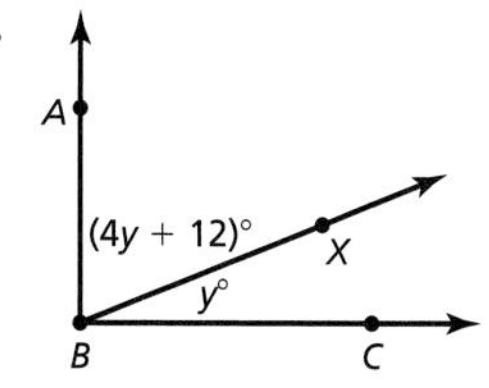

33. 63°, 126° **35.** 62°, 62° **37.** 44°, 44°, 88°

39. 65°, 65°, 130°

41. Subtract $m\angle CBD$ from $m\angle ABC$ to find $m\angle ABD$.

43. 40° **45.** 90°, 90°

47. **a.**

A, (4y + 12)°, X, y°, B, C

b. $4y + 12 + y = 92$, 76°, 16°

49. **a.** acute **b.** acute **c.** acute **d.** right

51. **a.** *Sample answer:* (1, 2) **b.** *Sample answer:* (0, 2)
c. *Sample answer:* (−2, 2) **d.** *Sample answer:* (−2, 0)

53. acute, right, or obtuse; The sum of the angles could be less than 90° (example: 30° + 20° = 50°), equal to 90° (example: 60° + 30° = 90°), or greater than 90° (example: 55° + 45° = 100°).

55. *Sample answer:* You draw a segment, ray, or line in the interior of an angle so that the two angles created are congruent to each other; Angle bisectors and segment bisectors can be segments, rays, or lines, but only a segment bisector can be a point. The two angles/segments created are congruent to each other, and their measures are each half the measure of the original angle/segment.

57. acute; It is likely that the angle with the horizontal is very small because levels are typically used when something appears to be horizontal but still needs to be checked.

1.5 Maintaining Mathematical Proficiency *(p. 46)*

59. $x = 32$ **61.** $x = 71$ **63.** $x = 12$ **65.** $x = 10$

1.6 Vocabulary and Core Concept Check *(p. 52)*

1. Adjacent angles share a common ray, and are next to each other. Vertical angles form two pairs of opposite rays, and are across from each other.

1.6 Monitoring Progress and Maintaining Mathematical Proficiency *(pp. 52–54)*

3. $\angle LJM$, $\angle MJN$ **5.** $\angle EGF$, $\angle NJP$ **7.** 67°

9. 102° **11.** $m\angle QRT = 47°$, $m\angle TRS = 133°$

13. $m\angle UVW = 12°$, $m\angle XYZ = 78°$ **15.** $\angle 1$ and $\angle 5$

17. yes; The sides form two pairs of opposite rays.

19. 60°, 120° **21.** 9°, 81°

23. They do not share a common side, so they are not adjacent; $\angle 1$ and $\angle 2$ are adjacent.

25. 122° **27.** C

29.

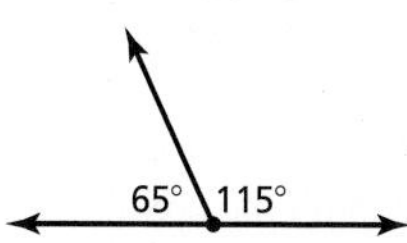

31. B **33.** $x + (2x + 12) = 90$; 26° and 64°

35. $x + \left(\frac{2}{3}x - 15\right) = 180$; 117° and 63°

37. always; A linear pair forms a straight angle, which is 180°.

39. sometimes; This is possible if the lines are perpendicular.

41. always; 45 + 45 = 90

43. The measure of an obtuse angle is greater than 90°. So, you cannot add it to the measure of another angle and get 90°.

45. **a.** 50°, 40°, 140°
b. $\frac{1}{3}$; Because all 4 angles have supplements, the first paper can be any angle. Then there is a 1 in 3 chance of drawing its supplement.

47. yes; Because $m\angle KJL + x° = 90°$ and $m\angle MJN + x° = 90°$, it must be that $m\angle KJL + x° = m\angle MJN + x°$. Subtracting $x°$ from each side of the equation results in the measures being equal. So, the angles are congruent.

49. **a.** $y°$, $(180 - y)°$, $(180 - y)°$
b. They are always congruent; They are both supplementary to the same angle. So, their measures must be equal.

51. 37°, 53°; If two angles are complementary, then their sum is 90°. If x is one of the angles, then $(90 - x)$ is the complement. Write and solve the equation $90 = (x - (90 - x)) + 74$. The solution is $x = 53$.

1.6 Maintaining Mathematical Proficiency *(p. 54)*

53. never; Integers are positive or negative whole numbers. Irrational numbers are decimals that never terminate and never repeat.

55. never; The whole numbers are positive or zero.

57. always; The set of integers includes all natural numbers and their opposites (and zero).

59. sometimes; Irrational numbers can be positive or negative.

Chapter 1 Review *(pp. 56–58)*

1. *Sample answer:* plane XYN **2.** *Sample answer:* line g

3. *Sample answer:* line h **4.** *Sample answer:* $\overrightarrow{XZ}$, $\overrightarrow{YP}$

5. $\overrightarrow{YX}$ and $\overrightarrow{YZ}$ **6.** P **7.** 41 **8.** 11

9.

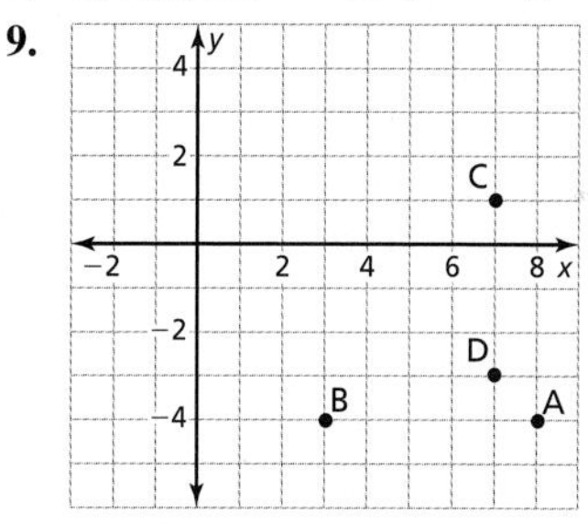

no

10. $\left(\frac{1}{2}, \frac{13}{2}\right)$; about 7.1 **11.** $\left(\frac{13}{2}, -\frac{5}{2}\right)$; about 1.4

12. (−2, −3) **13.** 40 **14.** 20 units, 21 square units

15. about 23.9 units, 24.5 square units **16.** 49°, 28°

17. 88°, 23° **18.** 127° **19.** 78° **20.** 7°

21. 64° **22.** 124°

Chapter 2

Chapter 2 Maintaining Mathematical Proficiency *(p. 63)*

1. $a_n = 6n - 3$; $a_{50} = 297$ **2.** $a_n = 17n - 46$; $a_{50} = 804$

3. $a_n = 0.6n + 2.2$; $a_{50} = 32.2$

4. $a_n = \frac{1}{6}n + \frac{1}{6}$; $a_{50} = \frac{17}{2}$, or $8\frac{1}{2}$

5. $a_n = -4n + 30$; $a_{50} = -170$

6. $a_n = -6n + 14$; $a_{50} = -286$

7. $x = y - 5$ 8. $x = -4y + 3$ 9. $x = y - 3$

10. $x = \frac{y}{7}$ 11. $x = \frac{y - 6}{z + 4}$ 12. $x = \frac{z}{6y + 2}$

13. no; The sequence does not have a common difference.

2.1 Vocabulary and Core Concept Check *(p. 71)*

1. a conditional statement and its contrapositive, as well as the converse and inverse of a conditional statement

2.1 Monitoring Progress and Modeling with Mathematics *(pp. 71–74)*

3. If a polygon is a pentagon, then (it has five sides.)
5. If you run, then (you are fast.)
7. If $x = 2$, then $9x + 5 = 23$.
9. If you are in a band, then you play the drums.
11. If you are registered, then you are allowed to vote.
13. The sky is not blue. 15. The ball is pink.
17. conditional: If two angles are supplementary, then the measures of the angles sum to 180°; true

    converse: If the measures of two angles sum to 180°, then they are supplementary; true

    inverse: If the two angles are not supplementary, then their measures do not sum to 180°; true

    contrapositive: If the measures of two angles do not sum to 180°, then they are not supplementary; true
19. conditional: If you do your math homework, then you will do well on the test; false

    converse: If you do well on the test, then you did your math homework; false

    inverse: If you do not do your math homework, then you will not do well on the test; false

    contrapositive: If you do not do well on the test, then you did not do your math homework; false
21. conditional: If it does not snow, then I will run outside; false

    converse: If I run outside, then it is not snowing; true

    inverse: If it snows, then I will not run outside; true

    contrapositive: If I do not run outside, then it is snowing; false
23. conditional: If $3x - 7 = 20$, then $x = 9$; true

    converse: If $x = 9$, then $3x - 7 = 20$; true

    inverse: If $3x - 7 \neq 20$, then $x \neq 9$; true

    contrapositive: If $x \neq 9$, then $3x - 7 \neq 20$; true
25. true; By definition of right angle, the measure of the right angle shown is 90°.
27. true; If angles form a linear pair, then the sum of the measures of their angles is 180°.
29. A point is the midpoint of a segment, if and only if it is the point that divides the segment into two congruent segments.
31. Two angles are adjacent angles if and only if they share a common vertex and side, but have no common interior points.
33. A polygon has three sides if and only if it is a triangle.
35. An angle is a right angle if and only if it measures 90°.
37. Taking four English courses is a requirement regardless of how many courses the student takes total, and the courses do not have to be taken simultaneously; If students are in high school, then they will take four English courses before they graduate.

39.

p	q	$\sim p$	$\sim p \rightarrow q$
T	T	F	T
T	F	F	T
F	T	T	T
F	F	T	F

41.

p	q	$\sim p$	$\sim q$	$\sim p \rightarrow \sim q$	$\sim(\sim p \rightarrow \sim q)$
T	T	F	F	T	F
T	F	F	T	T	F
F	T	T	F	F	T
F	F	T	T	T	F

43.

p	q	$\sim p$	$q \rightarrow \sim p$
T	T	F	F
T	F	F	T
F	T	T	T
F	F	T	T

45. **a.** If a rock is igneous, then it is formed from the cooling of molten rock; If a rock is sedimentary, then it is formed from pieces of other rocks; If a rock is metamorphic, then it is formed by changing temperature, pressure, or chemistry.

    b. If a rock is formed from the cooling of molten rock, then it is igneous; true; All rocks formed from cooling molten rock are called igneous.

    If a rock is formed from pieces of other rocks, then it is sedimentary; true; All rocks formed from pieces or other rocks are called sedimentary.

    If a rock is formed by changing temperature, pressure, or chemistry, then it is metamorphic; true; All rocks formed by changing temperature, pressure, or chemistry are called metamorphic.

    c. *Sample answer:* If a rock is not sedimentary, then it was not formed from pieces of other rocks; This is the inverse of one of the conditional statements in part (a). So, the converse of this statement will be the contrapositive of the conditional statement. Because the contrapositive is equivalent to the conditional statement and the conditional statement was true, the contrapositive will also be true.
47. no; The contrapositive is equivalent to the original conditional statement. In order to write a conditional statement as a true biconditional statement, you must know that the converse (or inverse) is true.
49. If you tell the truth, then (you don't have to remember anything.)
51. If one is lucky, then (a solitary fantasy can totally transform one million realities.)
53. no; "If $x^2 - 10 = x + 2$, then $x = 4$" is a false statement because $x = -3$ is also possible. The converse, however, of the original conditional statement is true. In order for a biconditional statement to be true, both the conditional statement and its converse must be true.
55. A
57. If today is February 28, then tomorrow is March 1.

59. a. 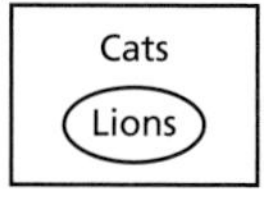

If you see a cat, then you went to the zoo to see a lion; The original statement is true, because a lion is a type of cat, but the converse is false, because you could see a cat without going to the zoo.

b. 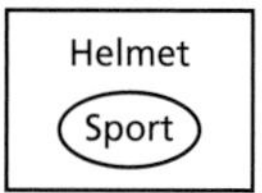

If you wear a helmet, then you play a sport; Both the original statement and the converse are false, because not all sports require helmets and sometimes helmets are worn for activities that are not considered a sport, such as construction work.

c. 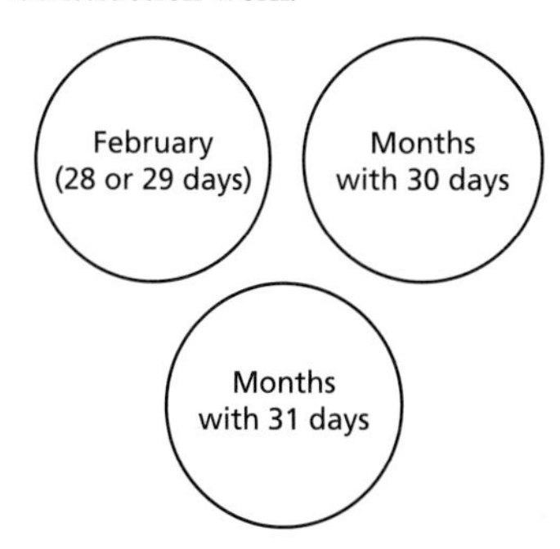

If this month is not February, then it has 31 days; The original statement is true, because February never has 31 days, but the converse is false, because a month that is not February could have 30 days.

61. *Sample answer:* If they are vegetarians, then they do not eat hamburgers.

63. *Sample answer:* Slogan: "This treadmill is a fat-burning machine!" Conditional statement: If you use this treadmill, then you will burn fat quickly.

2.1 Maintaining Mathematical Proficiency *(p. 74)*

65. add a square that connects the midpoints of the previously added square;

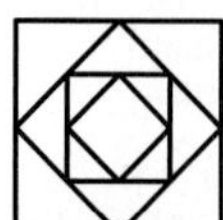

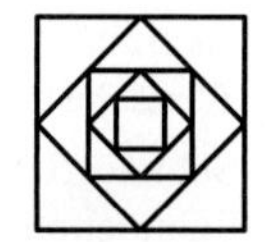

67. add 11; 56, 67 **69.** $1^2, 2^2, 3^2, \ldots; 25, 36$

2.2 Vocabulary and Core Concept Check *(p. 80)*

1. The prefix "Counter-" means "opposing." So, a counterexample opposes the truth of the statement.

2.2 Monitoring Progress and Modeling with Mathematics *(pp. 80–82)*

3. The absolute value of each number in the list is 1 greater than the absolute value of the previous number in the list, and the signs alternate from positive to negative; $-6, 7$

5. The list items are letters in backward alphabetical order; U, T

7. This is a sequence of regular polygons, each polygon having one more side than the previous polygon.

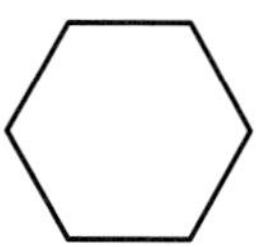

9. The product of any two even integers is an even integer. *Sample answer:* $-2(4) = -8, 6(12) = 72, 8(10) = 80$

11. The quotient of a number and its reciprocal is the square of that number. *Sample answer:* $9 \div \frac{1}{9} = 9 \cdot 9 = 9^2$, $\frac{2}{3} \div \frac{3}{2} = \frac{2}{3} \cdot \frac{2}{3} = \left(\frac{2}{3}\right)^2, \frac{1}{7} \div 7 = \frac{1}{7} \cdot \frac{1}{7} = \left(\frac{1}{7}\right)^2$

13. *Sample answer:* $1 \cdot 5 = 5, 5 \not> 5$

15. They could both be right angles. Then, neither are acute.

17. You passed the class. **19.** not possible

21. not possible

23. If a figure is a rhombus, then the figure has two pairs of opposite sides that are parallel.

25. Law of Syllogism **27.** Law of Detachment

29. The sum of two odd integers is an even integer; Let m and n be integers. Then $(2m + 1)$ and $(2n + 1)$ are odd integers. $(2m + 1) + (2n + 1) = 2m + 2n + 2 = 2(m + n + 1)$; $2(m + n + 1)$ is divisible by 2 and is therefore an even integer.

31. inductive reasoning; The conjecture is based on the assumption that a pattern, observed in specific cases, will continue.

33. deductive reasoning; Laws of nature and the Law of Syllogism were used to draw the conclusion.

35. The Law of Detachment cannot be used because the hypothesis is not true; *Sample answer:* Using the Law of Detachment, because a square is a rectangle, you can conclude that a square has four sides.

37. Using inductive reasoning, we can make a conjecture that male tigers weigh more than female tigers because this was true in all of the specific cases listed in the table.

39. $n(n + 1) =$ the sum of first n positive even integers

41. Argument 2; This argument uses the Law of Detachment to say that when the hypothesis is met, the conclusion is true.

43. The value of y is 2 more than three times the value of x; $y = 3x + 2$; *Sample answer:* If $x = 10$, then $y = 3(10) + 2 = 32$; If $x = 72$, then $y = 3(72) + 2 = 218$.

45. a. true; Based on the Law of Syllogism, if you went camping at Yellowstone, and Yellowstone is in Wyoming, then you went camping in Wyoming.

b. false; When you go camping, you go canoeing, but even though your friend always goes camping when you do, he or she may not choose to go canoeing with you.

c. true; We know that if you go on a hike, your friend goes with you, and we know that you went on a hike. So, based on the Law of Detachment, your friend went on a hike.

d. false; We know that you and your friend went on a hike, but we do not know where. We just know that there is a 3-mile-long trail near where you are camping.

2.2 Maintaining Mathematical Proficiency *(p. 82)*

47. Segment Addition Postulate (Post. 1.2)

49. Ruler Postulate (Post. 1.1)

2.3 Vocabulary and Core Concept Check *(p. 87)*

1. three

2.3 Monitoring Progress and Modeling with Mathematics *(pp. 87–88)*

3. Two Point Postulate (Post. 2.1)

5. *Sample answer:* Line q contains points J and K.

7. *Sample answer:* Through points K, H, and L, there is exactly one plane, which is plane M.

9. 

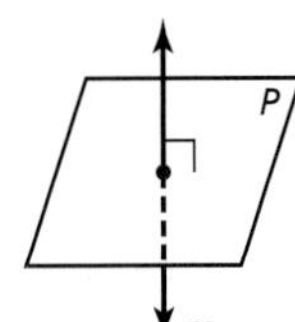

11.

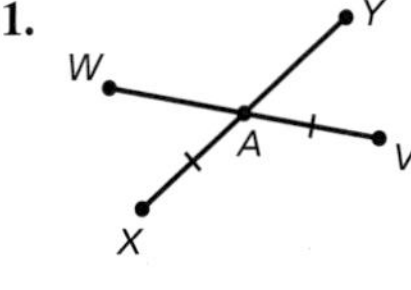

13. yes **15.** no **17.** yes **19.** yes

21. In order to determine that M is the midpoint of $\overline{AC}$ or $\overline{BD}$, the segments that would have to be marked as congruent are $\overline{AM}$ and $\overline{MC}$ or $\overline{DM}$ and $\overline{MB}$, respectively; Based on the diagram and markings, you can assume $\overline{AC}$ and $\overline{DB}$ intersect at point M, such that $\overline{AM} \cong \overline{MB}$ and $\overline{DM} \cong \overline{MC}$.

23. C, D, F, H **25.** Two-Point Postulate (Post. 2.1)

27. a. If there are two points, then there exists exactly one line that passes through them.

b. converse: If there exists exactly one line that passes through a given point or points, then there are two points; false; inverse: If there are not two points, then there is not exactly one line that passes through them; false; contrapositive: If there is not exactly one line that passes through a given point or points, then there are not two points; true

29. $<$

31. yes; For example, the ceiling and two walls of many rooms intersect in a point in the corner of the room.

33. Points E, F, and G must be collinear. They must be on the line that intersects plane P and plane Q; Points E, F, and G can be either collinear or not collinear.

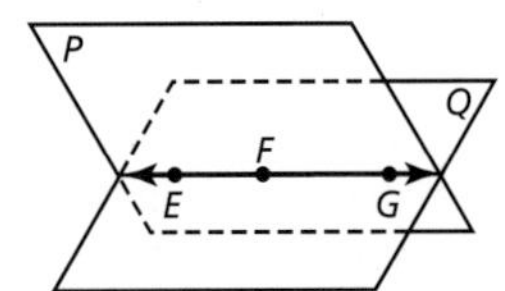

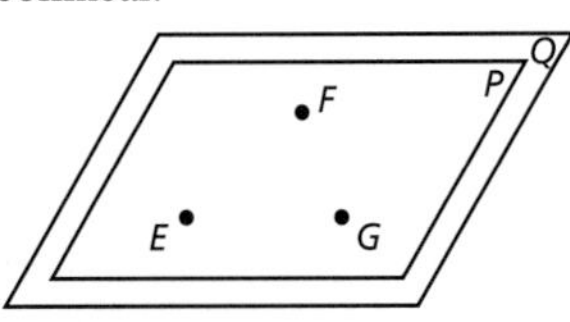

2.3 Maintaining Mathematical Proficiency *(p. 88)*

35. $t = 2$; Addition Property of Equality

37. $x = 4$; Subtraction Property of Equality

2.4 Vocabulary and Core Concept Check *(p. 96)*

1. Reflexive Property of Equality

2.4 Monitoring Progress and Modeling with Mathematics *(p. 96–98)*

3. Subtraction Property of Equality; Addition Property of Equality; Division Property of Equality

5.

Equation	Explanation and Reason
$5x - 10 = -40$	Write the equation; Given
$5x = -30$	Add 10 to each side; Addition Property of Equality
$x = -6$	Divide each side by 5; Division Property of Equality

7.

Equation	Explanation and Reason
$2x - 8 = 6x - 20$	Write the equation; Given
$-4x - 8 = -20$	Subtract $6x$ from each side; Subtraction Property of Equality
$-4x = -12$	Add 8 to each side; Addition Property of Equality
$x = 3$	Divide each side by -4; Division Property of Equality

9.

Equation	Explanation and Reason
$5(3x - 20) = -10$	Write the equation; Given
$15x - 100 = -10$	Multiply; Distributive Property
$15x = 90$	Add 100 to each side; Addition Property of Equality
$x = 6$	Divide each side by 15; Division Property of Equality

11.

Equation	Explanation and Reason
$2(-x - 5) = 12$	Write the equation; Given
$-2x - 10 = 12$	Multiply; Distributive Property
$-2x = 22$	Add 10 to each side; Addition Property of Equality
$x = -11$	Divide each side by -2; Division Property of Equality

13.

Equation	Explanation and Reason
$4(5x - 9) = -2(x + 7)$	Write the equation; Given
$20x - 36 = -2x - 14$	Multiply on each side; Distributive Property
$22x - 36 = -14$	Add $2x$ to each side; Addition Property of Equality
$22x = 22$	Add 36 to each side; Addition Property of Equality
$x = 1$	Divide each side by 22; Division Property of Equality

15.

Equation	Explanation and Reason
$5x + y = 18$	Write the equation; Given
$y = -5x + 18$	Subtract $5x$ from each side; Subtraction Property of Equality

17.

Equation	Explanation and Reason
$2y + 0.5x = 16$	Write the equation; Given
$2y = -0.5x + 16$	Subtract $0.5x$ from each side; Subtraction Property of Equality
$y = -0.25x + 8$	Divide each side by 2; Division Property of Equality

19.

Equation	Explanation and Reason
$12 - 3y = 30x + 6$	Write the equation; Given
$-3y = 30x - 6$	Subtract 12 from each side; Subtraction Property of Equality
$y = -10x + 2$	Divide each side by -3; Division Property of Equality

21.

Equation	Explanation and Reason
$C = 2\pi r$	Write the equation; Given
$\frac{C}{2\pi} = r$	Divide each side by 2π; Division Property of Equality
$r = \frac{C}{2\pi}$	Rewrite the equation; Symmetric Property of Equality

23.

Equation	Explanation and Reason
$S = 180(n - 2)$	Write the equation; Given
$\frac{S}{180} = n - 2$	Divide each side by 180; Division Property of Equality
$\frac{S}{180} + 2 = n$	Add 2 to each side; Addition Property of Equality
$n = \frac{S}{180} + 2$	Rewrite the equation; Symmetric Property of Equality

25. Multiplication Property of Equality

27. Reflexive Property of Equality

29. Reflexive Property of Equality

31. Symmetric Property of Equality

33. $20 + CD$ **35.** $CD + EF$ **37.** $XY - GH$

39. $m\angle 1 = m\angle 3$

41. The Subtraction Property of Equality should be used to subtract x from each side of the equation in order to get the second step.

$7x = x + 24$	Given
$6x = 24$	Subtraction Property of Equality
$x = 4$	Division Property of Equality

43.

Equation	Explanation and Reason
$P = 2\ell + 2w$	Write the equation; Given
$P - 2w = 2\ell$	Subtract $2w$ from each side; Subtraction Property of Equality
$\frac{P - 2w}{2} = \ell$	Divide each side by 2; Division Property of Equality
$\ell = \frac{P - 2w}{2}$	Rewrite the equation; Symmetric Property of Equality

$\ell = 11$ m

45.

Equation	Explanation and Reason
$m\angle ABD = m\angle CBE$	Write the equation; Given
$m\angle ABD = m\angle 1 + m\angle 2$	Add measures of adjacent angles; Angle Addition Postulate (Post. 1.4)
$m\angle CBE = m\angle 2 + m\angle 3$	Add measures of adjacent angles; Angle Addition Postulate (Post. 1.4)
$m\angle ABD = m\angle 2 + m\angle 3$	Substitute $m\angle ABD$ for $m\angle CBE$; Substitution Property of Equality
$m\angle 1 + m\angle 2 = m\angle 2 + m\angle 3$	Substitute $m\angle 1 + m\angle 2$ for $m\angle ABD$; Substitution Property of Equality
$m\angle 1 = m\angle 3$	Subtract $m\angle 2$ from each side; Subtraction Property of Equality

47. Transitive Property of Equality; Angle Addition Postulate (Post. 1.4); Transitive Property of Equality; $m\angle 1 + m\angle 2 = m\angle 3 + m\angle 1$; Subtraction Property of Equality

49.

Equation	Explanation and Reason
$DC = BC, AD = AB$	Marked in diagram; Given
$AC = AC$	AC is equal to itself; Reflexive Property of Equality
$AC + AB + BC = AC + AB + BC$	Add $AB + BC$ to each side of $AC = AC$; Addition Property of Equality
$AC + AB + BC = AC + AD + DC$	Substitute AD for AB and DC for BC; Substitution Property of Equality

51. $ZY = XW = 9$ **53.** A, B, F

55. a.

Equation	Explanation and Reason
$C = \frac{5}{9}(F - 32)$	Write the equation; Given
$\frac{9}{5}C = F - 32$	Multiply each side by $\frac{9}{5}$; Multiplication Property of Equality
$\frac{9}{5}C + 32 = F$	Add 32 to each side; Addition Property of Equality
$F = \frac{9}{5}C + 32$	Rewrite the equation; Symmetric Property of Equality

b.

Degrees Celsius (°C)	Degrees Fahrenheit (°F)
0	32
20	68
32	89.6
41	105.8

c.

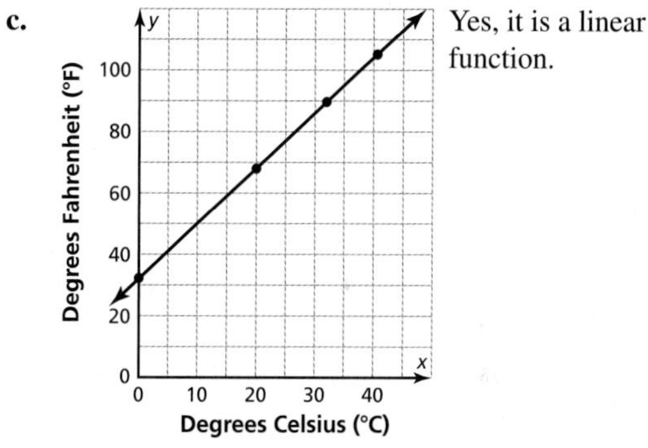

Yes, it is a linear function.

2.4 Maintaining Mathematical Proficiency *(p. 98)*

57. Segment Addition Postulate (Post. 1.2) **59.** Midpoint

2.5 Vocabulary and Core Concept Check *(p. 103)*

1. A postulate is a rule that is accepted to be true without proof, but a theorem is a statement that can be proven.

2.5 Monitoring Progress and Modeling with Mathematics *(pp. 103–104)*

3. Given; Addition Property of Equality; $PQ + QR = PR$; Transitive Property of Equality

5. Transitive Property of Segment Congruence (Thm. 2.1)

7. Symmetric Property of Angle Congruence (Thm. 2.2)

9. Symmetric Property of Segment Congruence (Thm. 2.1)

11.

STATEMENTS	REASONS
1. A segment exists with endpoints A and B.	**1.** Given
2. AB equals the length of the segment with endpoints A and B.	**2.** Ruler Postulate (Post. 1.1)
3. $AB = AB$	**3.** Reflexive Property of Equality
4. $\overline{AB} \cong \overline{AB}$	**4.** Definition of congruent segments

13.

STATEMENTS	REASONS
1. $\angle GFH \cong \angle GHF$	**1.** Given
2. $m\angle GFH = m\angle GHF$	**2.** Definition of congruent angles
3. $\angle EFG$ and $\angle GFH$ form a linear pair.	**3.** Given (diagram)
4. $\angle EFG$ and $\angle GFH$ are supplementary.	**4.** Definition of linear pair
5. $m\angle EFG + m\angle GFH = 180°$	**5.** Definition of supplementary angles
6. $m\angle EFG + m\angle GHF = 180°$	**6.** Substitution Property of Equality
7. $\angle EFG$ and $\angle GHF$ are supplementary.	**7.** Definition of supplementary angles

15. The Transitive Property of Segment Congruence (Thm. 2.1) should have been used; Because if $\overline{MN} \cong \overline{LQ}$ and $\overline{LQ} \cong \overline{PN}$, then $\overline{MN} \cong \overline{PN}$ by the Transitive Property of Segment Congruence (Thm. 2.1).

17. equiangular; By the Transitive Property of Angle Congruence (Thm. 2.2), because $\angle 1 \cong \angle 2$ and $\angle 2 \cong \angle 3$, we know that $\angle 1 \cong \angle 3$. Because all three angles are congruent, the triangle is equiangular. (It is also equilateral and acute.)

19. The purpose of a proof is to ensure the truth of a statement with such certainty that the theorem or rule proved could be used as a justification in proving another statement or theorem. Because inductive reasoning relies on observations about patterns in specific cases, the pattern may not continue or may change. So, the ideas cannot be used to prove ideas for the general case.

21. a. It is a right angle.

b.

STATEMENTS	REASONS
1. $m\angle 1 + m\angle 1 + m\angle 2 + m\angle 2 = 180°$	**1.** Angle Addition Postulate (Post. 1.4)
2. $2(m\angle 1 + m\angle 2) = 180°$	**2.** Distributive Property
3. $m\angle 1 + m\angle 2 = 90°$	**3.** Division Property of Equality

23.

STATEMENTS	REASONS
1. $\overline{QR} \cong \overline{PQ}, \overline{RS} \cong \overline{PQ}$, $QR = 2x + 5, RS = 10 - 3x$	**1.** Given
2. $QR = PQ, RS = PQ$	**2.** Definition of congruent segments
3. $QR = RS$	**3.** Transitive Property of Equality
4. $2x + 5 = 10 - 3x$	**4.** Substitution Property of Equality
5. $5x + 5 = 10$	**5.** Addition Property of Equality
6. $5x = 5$	**6.** Subtraction Property of Equality
7. $x = 1$	**7.** Division Property of Equality

2.5 Maintaining Mathematical Proficiency *(p. 104)*

25. 33°

2.6 Vocabulary and Core Concept Check *(p. 111)*

1. All right angles have the same measure, 90°, and angles with the same measure are congruent.

2.6 Monitoring Progress and Modeling with Mathematics *(pp. 111–114)*

3. $\angle MSN \cong \angle PSQ$ by definition because they have the same measure; $\angle MSP \cong \angle PSR$ by the Right Angles Congruence Theorem (Thm. 2.3). They form a linear pair, which means they are supplementary by the Linear Pair Postulate (Post. 2.8), and because one is a right angle, so is the other by the Subtraction Property of Equality; $\angle NSP \cong \angle QSR$ by the Congruent Complements Theorem (Thm. 2.5) because they are complementary to congruent angles.

5. $\angle GML \cong \angle HMJ$ and $\angle GMH \cong \angle LMJ$ by the Vertical Angles Congruence Theorem (Thm. 2.6); $\angle GMK \cong \angle JMK$ by the Right Angles Congruence Theorem (Thm. 2.3). They form a linear pair, which means they are supplementary by the Linear Pair Postulate (Post. 2.8), and because one is a right angle, so is the other by the Subtraction Property of Equality.

7. $m\angle 2 = 37°$; $m\angle 3 = 143°$; $m\angle 4 = 37°$

9. $m\angle 1 = 146°$; $m\angle 3 = 146°$; $m\angle 4 = 34°$

11. $x = 11$; $y = 17$ **13.** $x = 4$; $y = 9$

15. The expressions should have been set equal to each other because they come from vertical angles;

$$(13x + 45)° = (19x + 3)°$$
$$-6x + 45 = 3$$
$$-6x = -42$$
$$x = 7$$

17. Transitive Property of Angle Congruence (Thm. 2.2); Transitive Property of Angle Congruence (Thm. 2.2)

STATEMENTS	REASONS
1. $\angle 1 \cong \angle 3$	1. Given
2. $\angle 1 \cong \angle 2$, $\angle 3 \cong \angle 4$	2. Vertical Angles Congruence Theorem (Thm. 2.6)
3. $\angle 2 \cong \angle 3$	3. Transitive Property of Angle Congruence (Thm. 2.2)
4. $\angle 2 \cong \angle 4$	4. Transitive Property of Angle Congruence (Thm. 2.2)

19. complementary; $m\angle 1 + m\angle 3$; Transitive Property of Equality; $m\angle 2 = m\angle 3$; congruent angles

STATEMENTS	REASONS
1. $\angle 1$ and $\angle 2$ are complementary. $\angle 1$ and $\angle 3$ are complementary.	1. Given
2. $m\angle 1 + m\angle 2 = 90°$, $m\angle 1 + m\angle 3 = 90°$	2. Definition of complementary angles
3. $m\angle 1 + m\angle 2 = m\angle 1 + m\angle 3$	3. Transitive Property of Equality
4. $m\angle 2 = m\angle 3$	4. Subtraction Property of Equality
5. $\angle 2 \cong \angle 3$	5. Definition of congruent angles

21. Because $\angle QRS$ and $\angle PSR$ are supplementary, $m\angle QRS + m\angle PSR = 180°$ by the definition of supplementary angles. $\angle QRL$ and $\angle QRS$ form a linear pair and by definition are supplementary, which means that $m\angle QRL + m\angle QRS = 180°$. So, by the Transitive Property of Equality, $m\angle QRS + m\angle PSR = m\angle QRL + m\angle QRS$, and by the Subtraction Property of Equality, $m\angle PSR = m\angle QRL$. So, by definition of congruent angles, $\angle PSR \cong \angle QRL$, and by the Symmetric Property of Angle Congruence (Thm. 2.2), $\angle QRL \cong \angle PSR$.

23.

STATEMENTS	REASONS
1. $\angle AEB \cong \angle DEC$	1. Given
2. $m\angle AEB = m\angle DEC$	2. Definition of congruent angles
3. $m\angle DEB = m\angle DEC + m\angle BEC$	3. Angle Addition Postulate (Post. 1.4)
4. $m\angle DEB = m\angle AEB + m\angle BEC$	4. Substitution Property of Equality
5. $m\angle AEC = m\angle AEB + m\angle BEC$	5. Angle Addition Postulate (Post. 1.4)
6. $m\angle AEC = m\angle DEB$	6. Transitive Property of Equality
7. $\angle AEC \cong \angle DEB$	7. Definition of congruent angles

25. Your friend is correct; $\angle 1$ and $\angle 4$ are not vertical angles because they do not form two pairs of opposite rays. So, the Vertical Angles Congruence Theorem (Thm. 2.6) does not apply.

27. no; The converse would be: "If two angles are supplementary, then they are a linear pair." This is false because angles can be supplementary without being adjacent.

29. 50°; 130°; 50°; 130°

2.6 Maintaining Mathematical Proficiency *(p. 114)*

31. *Sample answer:* B, I, and C

33. *Sample answer:* plane ABC and plane BCG

35. *Sample answer:* A, B, and C

Chapter 2 Review *(pp. 116–118)*

1. conditional: If two lines intersect, then their intersection is a point.
   converse: If two lines intersect in a point, then they are intersecting lines.
   inverse: If two lines do not intersect, then they do not intersect in a point.
   contrapositive: If two lines do not intersect in a point, then they are not intersecting lines.
   biconditional: Two lines intersect if and only if their intersection is a point.
2. conditional: If $4x + 9 = 21$, then $x = 3$.
   converse: If $x = 3$, then $4x + 9 = 21$.
   inverse: If $4x + 9 \neq 21$, then $x \neq 3$.
   contrapositive: If $x \neq 3$, then $4x + 9 \neq 21$.
   biconditional: $4x + 9 = 21$ if and only if $x = 3$.
3. conditional: If angles are supplementary, then they sum to 180°.
   converse: If angles sum to 180°, then they are supplementary.
   inverse: If angles are not supplementary, then they do not sum to 180°.
   iontrapositive: If angles do not sum to 180°, then they are not supplementary.
   conditional: Angles are supplementary if and only if they sum to 180°.
4. conditional: If an angle is a right angle, then it measures 90°.
   converse: If an angle measures 90°, then it is a right angle.
   inverse: If an angle is not a right angle, then it does not measure 90°.
   contrapositive: If an angle does not measure 90°, then it is not a right angle.
   biconditional: An angle is a right angle if and only if it measures 90°.
5. The difference of any two odd integers is an even integer.
6. The product of an even and an odd integer is an even integer.
7. $m\angle B = 90°$ **8.** If $4x = 12$, then $2x = 6$. **9.** yes
10. yes **11.** no **12.** no

Selected Answers

13. *Sample answer:*

14. *Sample answer:*

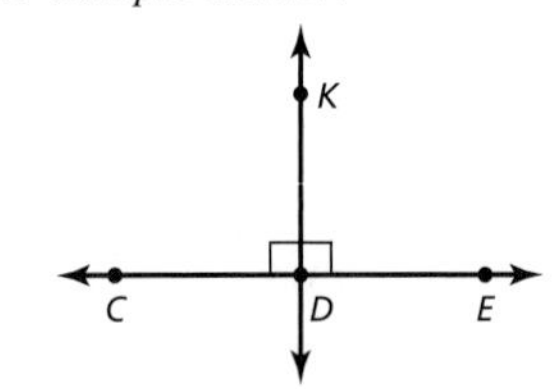

15. *Sample answer:*

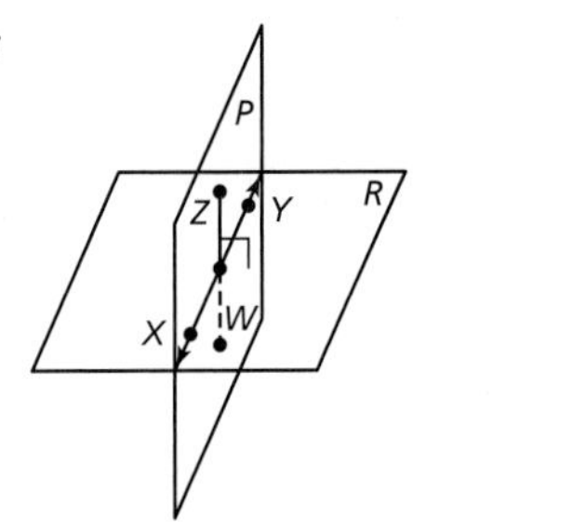

16.

Equation	Explanation and Reason
$-9x - 21 = -20x - 87$	Write the equation; Given
$11x - 21 = -87$	Add $20x$ to each side; Addition Property of Equality
$11x = -66$	Add 21 to each side; Addition Property of Equality
$x = -6$	Divide each side by 11; Division Property of Equality

17.

Equation	Explanation and Reason
$15x + 22 = 7x + 62$	Write the equation; Given
$8x + 22 = 62$	Subtract $7x$ from each side; Subtraction Property of Equality
$8x = 40$	Subtract 22 from each side; Subtraction Property of Equality
$x = 5$	Divide each side by 8; Division Property of Equality

18.

Equation	Explanation and Reason
$3(2x + 9) = 30$	Write the equation; Given
$6x + 27 = 30$	Multiply; Distributive Property
$6x = 3$	Subtract 27 from each side; Subtraction Property of Equality
$x = \frac{1}{2}$	Divide each side by 6; Division Property of Equality

19.

Equation	Explanation and Reason
$5x + 2(2x - 23) = -154$	Write the equation; Given
$5x + 4x - 46 = -154$	Multiply; Distributive Property
$9x - 46 = -154$	Combine like terms; Simplify.
$9x = -108$	Add 46 to each side; Addition Property of Equality
$x = -12$	Divide each side by 9; Division Property of Equality

20. Transitive Property of Equality

21. Reflexive Property of Equality

22. Symmetric Property of Angle Congruence (Thm. 2.2)

23. Reflexive Property of Angle Congruence (Thm. 2.2)

24. Transitive Property of Equality

25.

STATEMENTS	REASONS
1. An angle with vertex A exists.	**1.** Given
2. $m\angle A$ equals the measure of the angle with vertex A.	**2.** Protractor Postulate (Post. 1.3)
3. $m\angle A = m\angle A$	**3.** Reflexive Property of Equality
4. $\angle A \cong \angle A$	**4.** Definition of congruent angles

26.

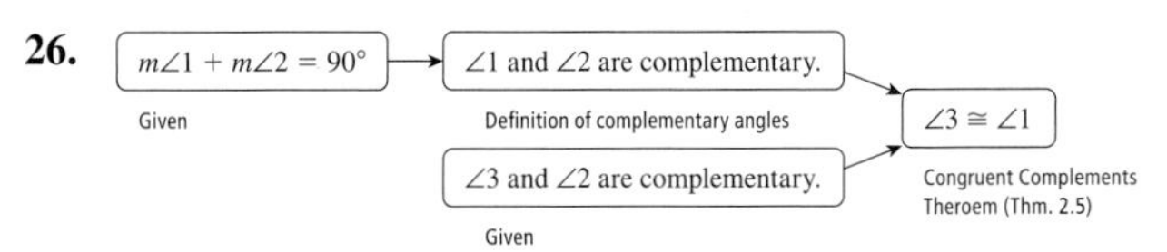

Chapter 3

Chapter 3 Maintaining Mathematical Proficiency *(p. 123)*

1. $m = -\frac{3}{4}$ **2.** $m = 3$ **3.** $m = 0$

4. $y = -3x + 19$ **5.** $y = -2x + 2$ **6.** $y = 4x + 9$

7. $y = \frac{1}{2}x - 5$ **8.** $y = -\frac{1}{4}x - 7$ **9.** $y = \frac{2}{3}x + 9$

10. When calculating the slope of a horizontal line, the vertical change is zero. This is the numerator of the fraction, and zero divided by any number is zero. When calculating the slope of a vertical line, the horizontal change is zero. This is the denominator of the fraction, and any number divided by zero is undefined.

3.1 Vocabulary and Core Concept Check *(p. 129)*

1. skew

3.1 Monitoring Progress and Modeling with Mathematics *(pp. 129–130)*

3. $\overleftrightarrow{AB}$ **5.** $\overleftrightarrow{BF}$ **7.** $\overleftrightarrow{MK}$ and $\overleftrightarrow{LS}$

9. no; They are intersecting lines.

11. $\angle 1$ and $\angle 5$; $\angle 2$ and $\angle 6$; $\angle 3$ and $\angle 7$; $\angle 4$ and $\angle 8$

13. $\angle 1$ and $\angle 8$; $\angle 2$ and $\angle 7$ **15.** corresponding

17. consecutive interior

19. Lines that do not intersect could also be skew; If two coplanar lines do not intersect, then they are parallel.

21. a. true; The floor is level with the horizontal just like the ground.

b. false; The lines intersect the plane of the ground, so they intersect certain lines of that plane.

c. true; The balusters appear to be vertical, and the floor of the tree house is horizontal. So, they are perpendicular.

23. yes; If the original two lines are parallel, and the transversal is perpendicular to both lines, then all eight angles are right angles.

25. $\angle HJG$, $\angle CFJ$ **27.** $\angle CFD$, $\angle HJC$

29. no; They can both be in a plane that is slanted with respect to the horizontal.

3.1 Maintaining Mathematical Proficiency *(p. 130)*

31. $m\angle 1 = 21°$, $m\angle 3 = 21°$, $m\angle 4 = 159°$

3.2 Vocabulary and Core Concept Check *(p. 135)*

1. Both theorems refer to two pairs of congruent angles that are formed when two parallel lines are cut by a transversal, and the angles that are congruent are on opposite sides of the transversal. However with the Alternate Interior Angles Theorem (Thm. 3.2), the congruent angles lie between the parallel lines, and with the Alternate Exterior Angles Theorem (Thm. 3.3), the congruent angles lie outside the parallel lines.

3.2 Monitoring Progress and Modeling with Mathematics *(pp. 135–136)*

3. $m\angle 1 = 117°$ by Vertical Angles Congruence Theorem (Thm. 2.6); $m\angle 2 = 117°$ by Alternate Exterior Angles Theorem (Thm. 3.3)

5. $m\angle 1 = 122°$ by Alternate Interior Angles Theorem (Thm. 3.2); $m\angle 2 = 58°$ by Consecutive Interior Angles Theorem (Thm. 3.4)

7. 64; $2x° = 128°$
$$x = 64$$

9. 12; $m\angle 5 = 65°$
$$65° + (11x - 17)° = 180°$$
$$11x + 48 = 180$$
$$11x = 132$$
$$x = 12$$

11. $m\angle 1 = 100°$, $m\angle 2 = 80°$, $m\angle 3 = 100°$; Because the 80° angle is a consecutive interior angle with both $\angle 1$ and $\angle 3$, they are supplementary by the Consecutive Interior Angles Theorem (Thm. 3.4). Because $\angle 1$ and $\angle 2$ are consecutive interior angles, they are supplementary by the Consecutive Interior Angles Theorem (Thm. 3.4).

13. In order to use the Corresponding Angles Theorem (Thm. 3.1), the angles need to be formed by two parallel lines cut by a transversal, but none of the lines in this diagram appear to be parallel; $\angle 9$ and $\angle 10$ are corresponding angles.

15.

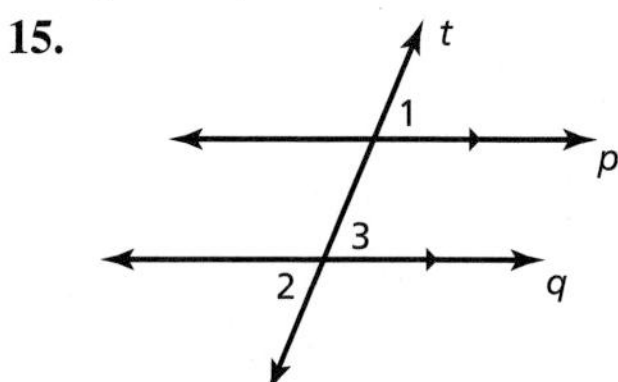

STATEMENTS	REASONS
1. $p \parallel q$	**1.** Given
2. $\angle 1 \cong \angle 3$	**2.** Corresponding Angles Theorem (Thm. 3.1)
3. $\angle 3 \cong \angle 2$	**3.** Vertical Angles Congruence Theorem (Thm. 2.6)
4. $\angle 1 \cong \angle 2$	**4.** Transitive Property of Congruence

17. $m\angle 2 = 104°$; Because the trees form parallel lines, and the rope is a transversal, the 76° angle and $\angle 2$ are consecutive interior angles. So, they are supplementary by the Consecutive Interior Angles Theorem (Thm. 3.4).

19. yes; If two parallel lines are cut by a perpendicular transversal, then the consecutive interior angles will both be right angles.

21. $$19x - 10 = 180$$
$$14x + 2y - 10 = 180; \quad x = 10, y = 25$$

23. no; In order to make the shot, you must hit the cue ball so that $m\angle 1 = 65°$. The angle that is complementary to $\angle 1$ must have a measure of 25° because this angle is alternate interior angles with the angle formed by the path of the cue ball and the vertical line drawn.

3.2 Maintaining Mathematical Proficiency *(p. 136)*

25. If two angles are congruent, then they are vertical angles; false

27. If two angles are supplementary, then they form a linear pair; false

3.3 Vocabulary and Core Concept Check *(p. 142)*

1. corresponding, alternate interior, alternate exterior

3.3 Monitoring Progress and Modeling with Mathematics *(pp. 142–144)*

3. $x = 40$; Lines m and n are parallel when the marked corresponding angles are congruent.
$$3x° = 120°$$
$$x = 40$$

5. $x = 15$; Lines m and n are parallel when the marked consecutive interior angles are supplementary.
$$(3x - 15)° + 150° = 180°$$
$$3x + 135 = 180$$
$$3x = 45$$
$$x = 15$$

7. $x = 60$; Lines m and n are parallel when the marked consecutive interior angles are supplementary.
$$2x° + x° = 180°$$
$$3x = 180$$
$$x = 60$$

9.

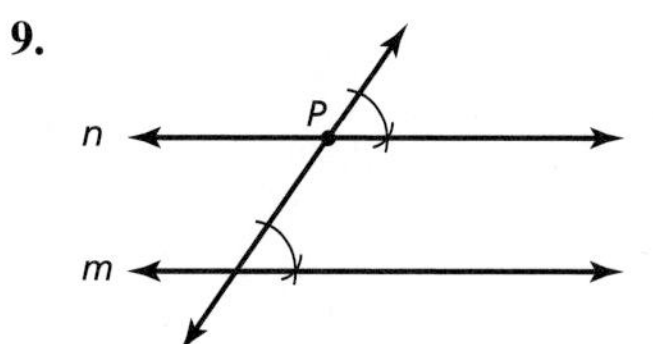

11.

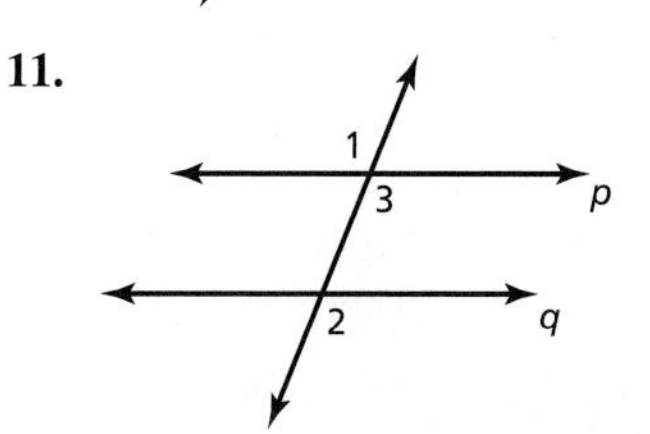

It is given that $\angle 1 \cong \angle 2$. By the Vertical Angles Congruence Theorem (Thm. 2.6), $\angle 1 \cong \angle 3$. Then by the Transitive Property of Congruence (Thm. 2.2), $\angle 2 \cong \angle 3$. So, by the Corresponding Angles Converse (Thm. 3.5), $p \parallel q$.

13. yes; Alternate Interior Angles Converse (Thm. 3.6)

15. no **17.** no

19. This diagram shows that vertical angles are always congruent. Lines a and b are not parallel unless $x = y$, and we cannot assume that they are equal.

21. yes; $m\angle DEB = 180° - 123° = 57°$ by the Linear Pair Postulate (Post. 2.8). So, by definition, a pair of corresponding angles are congruent, which means that $\overleftrightarrow{AC} \parallel \overleftrightarrow{DF}$ by the Corresponding Angles Converse (Thm. 3.5).

23. cannot be determined; The marked angles are vertical angles. You do not know anything about the angles formed by the intersection of $\overleftrightarrow{DF}$ and $\overleftrightarrow{BE}$.

25. yes; E. 20th Ave. is parallel to E. 19th Ave. by the Corresponding Angles Converse (Thm. 3.5). E. 19th Ave. is parallel to E. 18th Ave. by the Alternate Exterior Angles Converse (Thm. 3.7). E. 18th Ave. is parallel to E. 17th Ave. by the Alternate Interior Angles Converse (Thm. 3.6). So, they are all parallel to each other by the Transitive Property of Parallel Lines (Thm. 3.9).

27. The two angles marked as 108° are corresponding angles. Because they have the same measure, they are congruent to each other. So, $m \parallel n$ by the Corresponding Angles Converse (Thm. 3.5).

29. A, B, C, D; The Corresponding Angles Converse (Thm. 3.5) can be used because the angle marked at the intersection of line m and the transversal is vertical angles with, and therefore congruent to, an angle that is corresponding with the other marked angle. The Alternate Interior Angles Converse (Thm. 3.6) can be used because the angles that are marked as congruent are alternate interior angles. The Alternate Exterior Angles Converse (Thm. 3.7) can be used because the angles that are vertical with, and therefore congruent to, the marked angles are alternate exterior angles. The Consecutive Interior Angles Converse (Thm. 3.8) can be used because each of the marked angles forms a linear pair with, and is therefore supplementary to, an angle that is a consecutive interior angles with the other marked angle.

31. two; *Sample answer:* $\angle 1 \cong \angle 5$, $\angle 2 \cong \angle 7$, $\angle 3 \cong \angle 6$, $\angle 4$ and $\angle 7$ are supplementary.

33.

STATEMENTS	REASONS
1. $m\angle 1 = 115°$, $m\angle 2 = 65°$	**1.** Given
2. $m\angle 1 + m\angle 2 = m\angle 1 + m\angle 2$	**2.** Reflexive Property of Equality
3. $m\angle 1 + m\angle 2 = 115° + 65°$	**3.** Substitution Property of Equality
4. $m\angle 1 + m\angle 2 = 180°$	**4.** Simplify.
5. $\angle 1$ and $\angle 2$ are supplementary.	**5.** Definition of supplementary angles
6. $m \parallel n$	**6.** Consecutive Interior Angles Converse (Thm 3.8)

35.

STATEMENTS	REASONS
1. $\angle 1 \cong \angle 2$, $\angle 3 \cong \angle 4$	**1.** Given
2. $\angle 2 \cong \angle 3$	**2.** Vertical Angles Congruence Theorem (Thm. 2.6)
3. $\angle 1 \cong \angle 3$	**3.** Transitive Property of Congruence
4. $\angle 1 \cong \angle 4$	**4.** Transitive Property of Congruence
5. $\overline{AB} \parallel \overline{CD}$	**5.** Alternate Interior Angles Converse (Thm. 3.6)

37. no; Based on the diagram $\overleftrightarrow{AB} \parallel \overleftrightarrow{CD}$ by the Alternate Interior Angles Converse (Thm. 3.6), but you cannot be sure that $\overleftrightarrow{AD} \parallel \overleftrightarrow{BC}$.

39. a.

b. Given: $p \parallel q$, $q \parallel r$
Prove: $p \parallel r$

c.

STATEMENTS	REASONS
1. $p \parallel q$, $q \parallel r$	**1.** Given
2. $\angle 1 \cong \angle 2$, $\angle 2 \cong \angle 3$	**2.** Corresponding Angles Theorem (Thm. 3.1)
3. $\angle 1 \cong \angle 3$	**3.** Transitive Property of Congruence
4. $p \parallel r$	**4.** Corresponding Angles Converse (Thm. 3.5)

3.3 Maintaining Mathematical Proficiency *(p. 144)*

41. about 6.71 **43.** 13

3.4 Vocabulary and Core Concept Check *(p. 152)*

1. midpoint, right

3.4 Monitoring Progress and Modeling with Mathematics *(pp. 152–154)*

3. about 3.2 units

5.

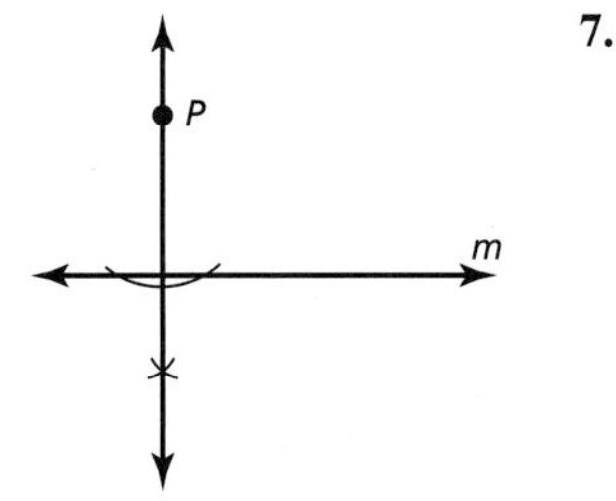

7.

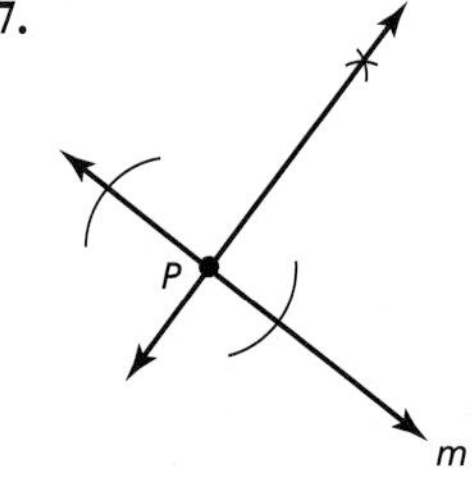

9.

11. In order to claim parallel lines by the Lines Perpendicular to a Transversal Theorem (Thm. 3.12), *both* lines must be marked as perpendicular to the transversal; Lines x and z are perpendicular.

13.

Because $\angle 1 \cong \angle 2$ by definition, $m\angle 1 = m\angle 2$. Also, by the Linear Pair Postulate (Post. 2.8), $m\angle 1 + m\angle 2 = 180°$. Then, by the Substitution Property of Equality, $m\angle 1 + m\angle 1 = 180°$, and $2(m\angle 1) = 180°$ by the Distributive Property. So, by the Division Property of Equality, $m\angle 1 = 90°$. Finally, $g \perp h$ by the definition of perpendicular lines.

15.

STATEMENTS	REASONS
1. $a \perp b$	**1.** Given
2. $\angle 1$ is a right angle.	**2.** Definition of perpendicular lines
3. $\angle 1 \cong \angle 4$	**3.** Vertical Angles Congruence Theorem (Thm. 2.6)
4. $m\angle 1 = 90°$	**4.** Definition of right angle
5. $m\angle 4 = 90°$	**5.** Transitive Property of Equality
6. $\angle 1$ and $\angle 2$ are a linear pair.	**6.** Definition of linear pair
7. $\angle 1$ and $\angle 2$ are supplementary.	**7.** Linear Pair Postulate (Post. 2.8)
8. $m\angle 1 + m\angle 2 = 180°$	**8.** Definition of supplementary angles
9. $90° + m\angle 2 = 180°$	**9.** Substitution Property of Equality
10. $m\angle 2 = 90°$	**10.** Subtraction Property of Equality
11. $\angle 2 \cong \angle 3$	**11.** Vertical Angles Congruence Theorem (Thm. 2.6)
12. $m\angle 3 = 90°$	**12.** Transitive Property of Equality
13. $\angle 1$, $\angle 2$, $\angle 3$, and $\angle 4$ are right angles.	**13.** Definition of right angle

17. none; The only thing that can be concluded in this diagram is that $v \perp y$. In order to say that lines are parallel, you need to know something about both of the intersections between the transversal and the two lines.

19. $m \parallel n$, Because $m \perp q$ and $n \perp q$, lines m and n are parallel by the Lines Perpendicular to a Transversal Theorem (Thm. 3.12). The other lines may or may not be parallel.

21. $n \parallel p$; Because $k \perp n$ and $k \perp p$, lines n and p are parallel by the Lines Perpendicular to a Transversal Theorem (Thm. 3.12).

23. $m\angle 1 = 90°$, $m\angle 2 = 60°$, $m\angle 3 = 30°$, $m\angle 4 = 20°$, $m\angle 5 = 90°$; $m\angle 1 = 90°$, because it is marked as a right angle.
$m\angle 2 = 90° - 30° = 60°$, because it is complementary to the 30° angle.
$m\angle 3 = 30°$, because it is vertical angles with, and therefore congruent to, the 30° angle.
$m\angle 4 = 90° - (30° + 40°) = 20°$, because it forms a right angle together with $\angle 3$ and the 40° angle.
$m\angle 5 = 90°$, because it is vertical angles with, and therefore congruent to, $\angle 1$.

25. $x = 8$ **27.** A, C, D, E

29.

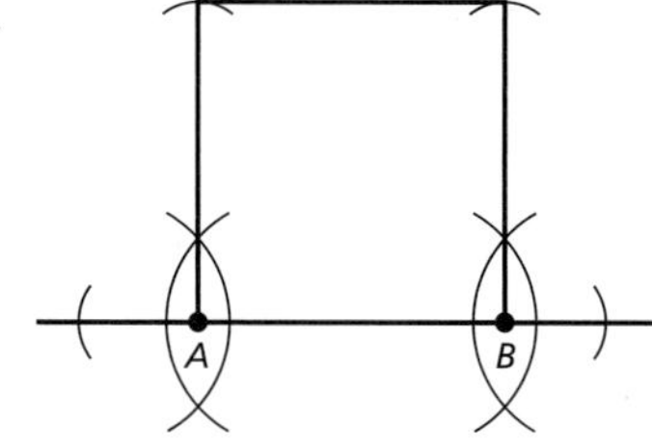

31. rectangle

33. Find the length of the segment that is perpendicular to the plane and that has one endpoint on the given point and one endpoint on the plane; You can find the distance from a line to a plane only if the line is parallel to the plane. Then you can pick any point on the line and find the distance from that point to the plane. If a line is not parallel to a plane, then the distance from the line to the plane is not defined because it would be different for each point on the line.

3.4 Maintaining Mathematical Proficiency *(p. 154)*

35. $-\frac{2}{3}$ **37.** 3 **39.** $m = -\frac{1}{2}$; $b = 7$

41. $m = -8$; $b = -6$

3.5 Vocabulary and Core Concept Check *(p. 160)*

1. directed

3.5 Monitoring Progress and Modeling with Mathematics *(pp. 160–162)*

3. $P(7, -0.4)$ **5.** $P(-1.5, -1.5)$ **7.** $a \parallel c$, $b \perp d$

9. perpendicular; Because $m_1 \cdot m_2 = \left(\frac{2}{3}\right)\left(-\frac{3}{2}\right) = -1$, lines 1 and 2 are perpendicular by the Slopes of Perpendicular Lines Theorem (Thm. 3.14).

11. perpendicular; Because $m_1 \cdot m_2 = 1(-1) = -1$, lines 1 and 2 are perpendicular by the Slopes of Perpendicular Lines Theorem (Thm. 3.14).

13. $y = -2x - 1$

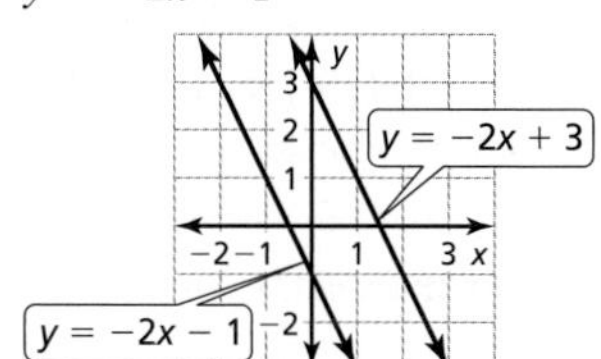

Selected Answers

15. $x = -2$

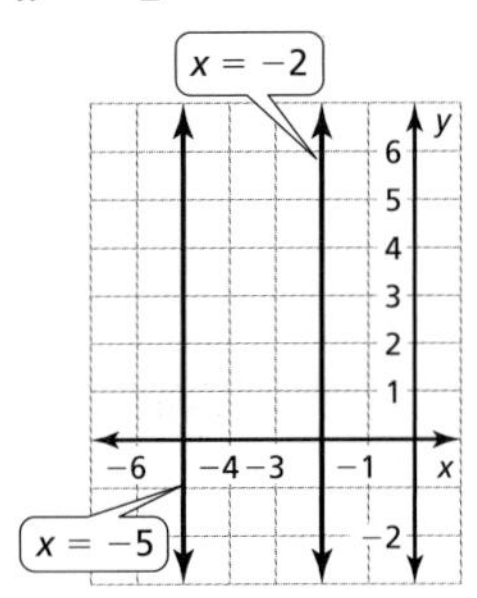

17. $y = \frac{1}{9}x$

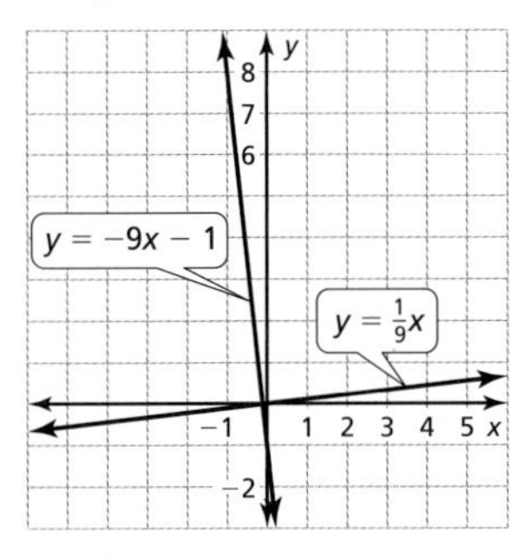

19. $y = \frac{1}{2}x + 2$

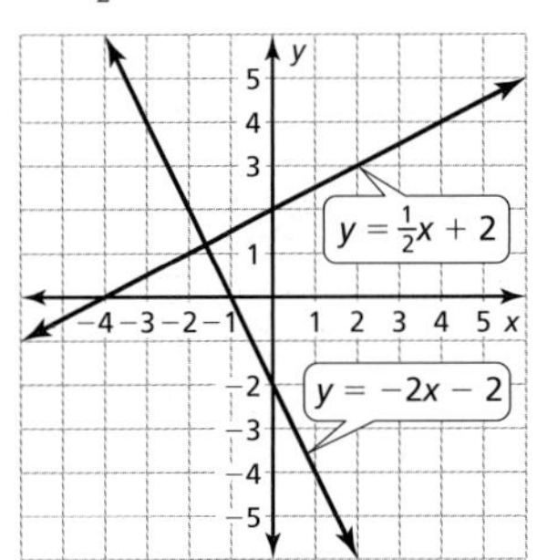

21. about 3.2 units **23.** about 5.4 units

25. Because the slopes are opposites but not reciprocals, their product does not equal -1. Lines 1 and 2 are neither parallel nor perpendicular.

27. $(0, 1)$; $y = 2x + 1$ **29.** $(3, 0)$; $y = \frac{3}{2}x - \frac{9}{2}$

31. $\left(-\frac{11}{5}, -\frac{6}{5}\right)$

33. no; $m_{\overline{LM}} = \frac{2}{5}$, $m_{\overline{LN}} = -\frac{7}{4}$, and $m_{\overline{MN}} = 9$. None of these can pair up to make a product of -1, so none of the segments are perpendicular.

35. $y = \frac{3}{2}x - 1$

37. $m < -1$; The slope of a line perpendicular to ℓ must be the opposite reciprocal of the slope of line ℓ. So, it must be negative, and have an absolute value greater than 1.

39. It will be the same point.

41. **a.** no solution; The lines do not intersect, so they are parallel.

b. $(7, -4)$; The lines intersect in one point.

c. infinitely many solutions; The lines are the same line.

43. $k = 4$

45. Using points $A(3, 2)$ and $B(6, 8)$ find the coordinates of point P that lies beyond point B along $\overrightarrow{AB}$ so that the ratio of AB to BP is 3 to 2. In order to keep the ratio, $\frac{AB}{BP} = \frac{3}{2}$, solve this ratio for BP to get $BP = \frac{2}{3}AB$. Next, find the rise and run from point A to point B. Leave the slope in terms of rise and run and do not simplify. $m_{\overline{AB}} = \frac{8-2}{6-3} = \frac{6}{3} = \frac{\text{rise}}{\text{run}}$. Add $\frac{2}{3}$ of the *run* to the x-coordinate of B, which is $\frac{2}{3} \cdot 3 + 6 = 8$. Add $\frac{2}{3}$ of the *rise* to the y-coordinate of B, which is $\frac{2}{3} \cdot 6 + 8 = 12$. So, the coordinates of P are $(8, 12)$.

47. If lines x and y are perpendicular to line z, then by the Slopes of Perpendicular Lines Theorem (Thm. 3.14), $m_x \cdot m_z = -1$ and $m_y \cdot m_z = -1$. By the Transitive Property of Equality, $m_x \cdot m_z = m_y \cdot m_z$, and by the Division Property of Equality $m_x = m_y$. Therefore, by the Slopes of Parallel Lines Theorem (Thm. 3.13), $x \parallel y$.

49. If lines x and y are vertical lines and they are cut by any horizontal transversal, z, then $x \perp z$ and $y \perp z$ by Theorem 3.14. Therefore, $x \parallel y$ by the Lines Perpendicular to a Transversal Theorem (Thm. 3.12).

51. By definition, the x-axis is perpendicular to the y-axis. Let m be a horizontal line, and let n be a vertical line. Because any two horizontal lines are parallel, m is parallel to the x-axis. Because any two vertical lines are parallel, n is parallel to the y-axis. By the Perpendicular Transversal Theorem (Thm. 3.11), n is perpendicular to the x-axis. Then, by the Perpendicular Transversal Theorem (Thm. 3.11), n is perpendicular to m.

3.5 Maintaining Mathematical Proficiency *(p. 162)*

53. $B(0, -4)$

55. $D(-1, -2)$

57.

x	-2	-1	0	1	2
$y = x - \frac{3}{4}$	$-\frac{11}{4}$	$-\frac{7}{4}$	$-\frac{3}{4}$	$\frac{1}{4}$	$\frac{5}{4}$

Chapter 3 Review *(pp. 164–166)*

1. $\overleftrightarrow{NR}, \overleftrightarrow{MR}, \overleftrightarrow{LQ}, \overleftrightarrow{PQ}$ **2.** $\overleftrightarrow{LM}, \overleftrightarrow{JK}, \overleftrightarrow{NP}$

3. $\overleftrightarrow{JM}, \overleftrightarrow{KL}, \overleftrightarrow{KP}, \overleftrightarrow{JN}$ **4.** plane JKP **5.** $x = 145, y = 35$

6. $x = 13, y = 132$ **7.** $x = 61, y = 29$

8. $x = 14, y = 17$ **9.** $x = 107$ **10.** $x = 133$

11. $x = 32$ **12.** $x = 23$

13. $x \parallel y$; Because $x \perp z$ and $y \perp z$, lines x and y are parallel by the Lines Perpendicular to a Transversal Theorem (Thm. 3.12).

14. none; The only thing that can be concluded in this diagram is that $x \perp z$ and $w \perp y$. In order to say that lines are parallel, you need to know something about *both* of the intersections between the two lines and a transversal.

15. $\ell \parallel m \parallel n$, $a \parallel b$; Because $a \perp n$ and $b \perp n$, lines a and b are parallel by the Lines Perpendicular to a Transversal Theorem (Thm. 3.12). Because $m \perp a$ and $n \perp a$, lines m and n are parallel by the Lines Perpendicular to a Transversal Theorem (Thm. 3.12). Because $\ell \perp b$ and $n \perp b$, lines ℓ and n are parallel by the Lines Perpendicular to a Transversal Theorem (Thm. 3.12). Because $\ell \parallel n$ and $m \parallel n$, lines ℓ and m are parallel by the Transitive Property of Parallel Lines (Thm. 3.9).

16. $a \parallel b$; Because $a \perp n$ and $b \perp n$, lines a and b are parallel by the Lines Perpendicular to a Transversal Theorem (Thm. 3.12).

17. $y = -x - 1$ 18. $y = \frac{1}{2}x + 8$ 19. $y = 3x - 6$

20. $y = \frac{1}{3}x - 2$ 21. $y = \frac{1}{2}x - 4$ 22. $y = 2x + 3$

23. $y = -\frac{1}{4}x + 4$ 24. $y = -7x - 2$

25. about 2.1 units 26. about 2.7 units

Chapter 4

Chapter 4 Maintaining Mathematical Proficiency *(p. 171)*

1. reflection 2. rotation 3. dilation
4. translation
5. no; $\frac{12}{14} = \frac{6}{7} \neq \frac{5}{7}$, The sides are not proportional.
6. yes; The corresponding angles are congruent and the corresponding side lengths are proportional.
7. yes; The corresponding angles are congruent and the corresponding side lengths are proportional.
8. no; Squares have four right angles, so the corresponding angles are always congruent. Because all four sides are congruent, the corresponding sides will always be proportional.

4.1 Vocabulary and Core Concept Check *(p. 178)*

1. $\triangle ABC$ is the preimage, and $\triangle A'B'C'$ is the image.

4.1 Monitoring Progress and Modeling with Mathematics *(pp. 178–180)*

3. $\overrightarrow{CD}$, $\langle 7, -3 \rangle$

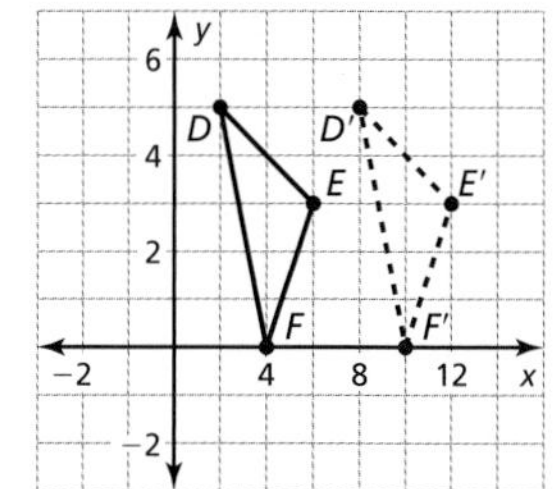

7.

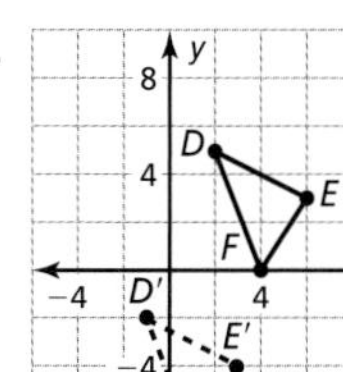

9. $\langle 3, -5 \rangle$ 11. $(x, y) \rightarrow (x - 5, y + 2)$

13. $A'(-6, 10)$ 15. $C(5, -14)$

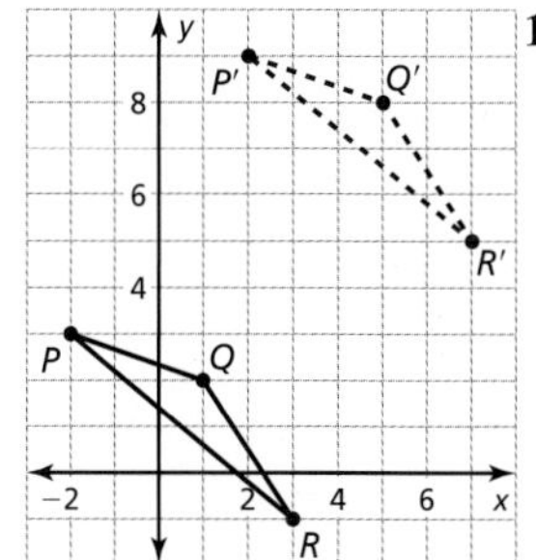

19.

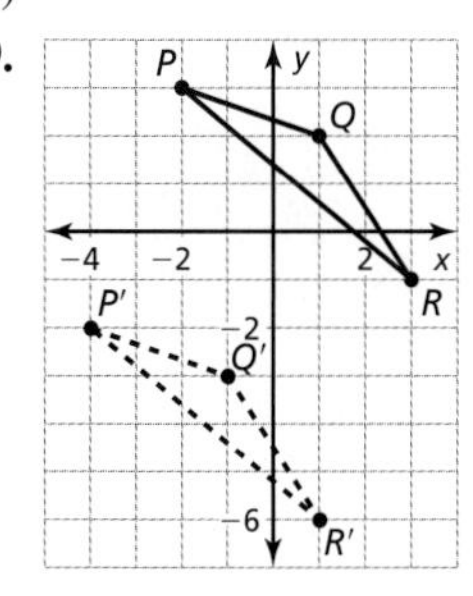

21.

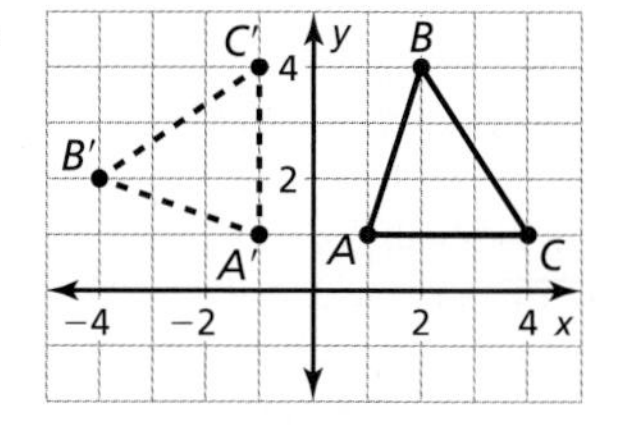

23. translation: $(x, y) \rightarrow (x + 5, y + 1)$, translation: $(x, y) \rightarrow (x - 5, y - 5)$

25. The quadrilateral should have been translated left and down;

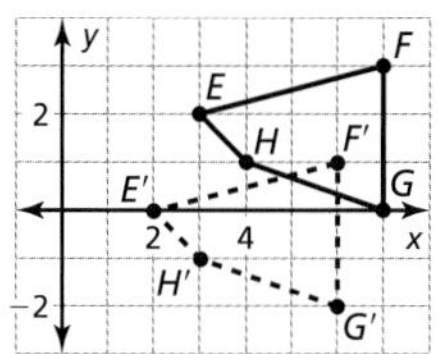

27. **a.** The amoeba moves right 5 squares and down 4 squares.
   b. about 12.8 mm **c.** about 0.52 mm/sec

29. $r = 100, s = 8, t = 5, w = 54$

31. $E'(-3, -4), F'(-2, -5), G'(0, -1)$

33. $(x, y) \rightarrow (x - m, y - n)$; You must go back the same number of units in the opposite direction.

35. If a rigid motion is used to transform figure A to figure A', then by definition of rigid motion, every part of figure A is congruent to its corresponding part of figure A'. If another rigid motion is used to transform figure A' to figure A'', then by definition of rigid motion, every part of figure A' is congruent to its corresponding part of figure A''. So, by the Transitive Property of Congruence, every part of figure A is congruent to its corresponding part of figure A''. So by definition of rigid motion, the composition of two (or more) rigid motions is a rigid motion.

37. Draw a rectangle. Then draw a translation of the rectangle. Next, connect each vertex of the preimage with the corresponding vertex in the image. Finally, make the hidden lines dashed.

39. yes; According to the definition of translation, the segments connecting corresponding vertices will be congruent and parallel. Also, because a translation is a rigid motion, $\overline{GH} \cong \overline{G'H'}$. So, the resulting figure is a parallelogram.

41. no; Because the value of y changes, you are not adding the same amount to each x-value.

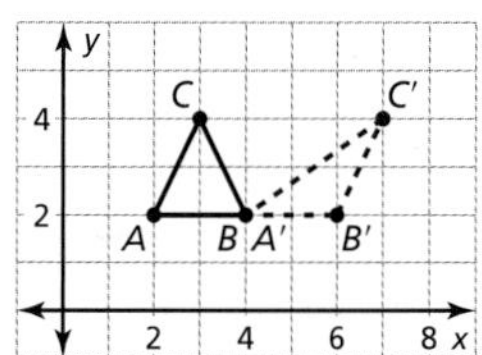

4.1 Maintaining Mathematical Proficiency *(p. 180)*

43. yes 45. no 47. x 49. $6x - 12$

4.2 Vocabulary and Core Concept Check *(p. 186)*

1. translation and reflection

4.2 Monitoring Progress and Modeling with Mathematics *(pp. 186–188)*

3. y-axis 5. neither

7.

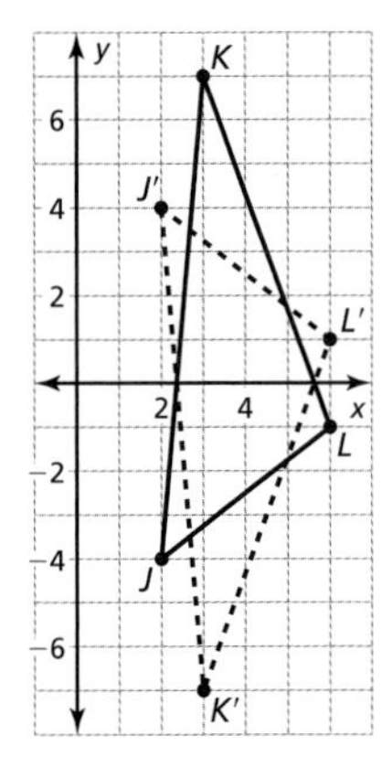

9.

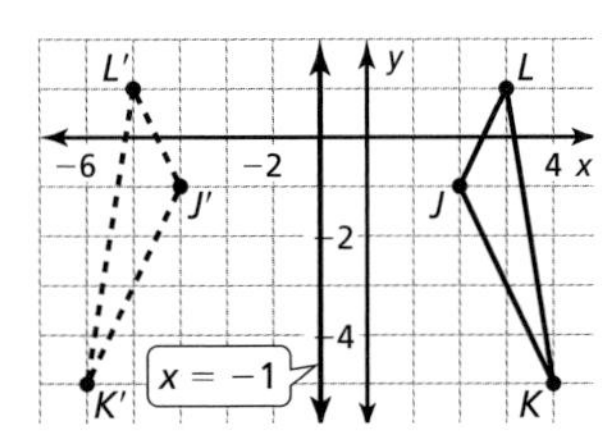

11.

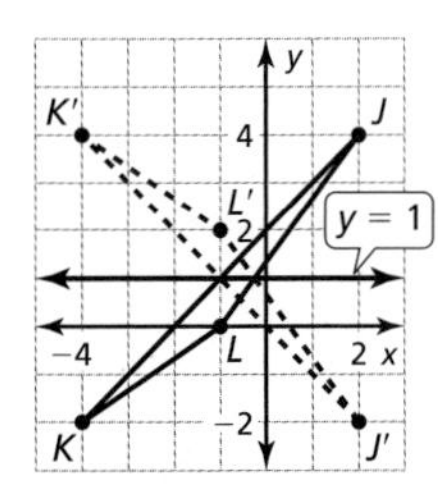

13.

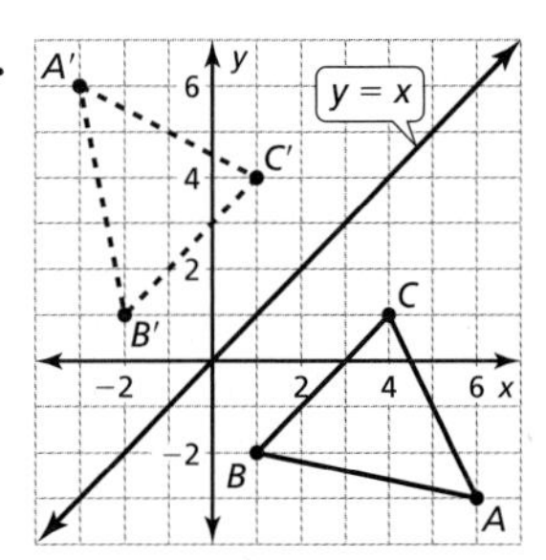

15.

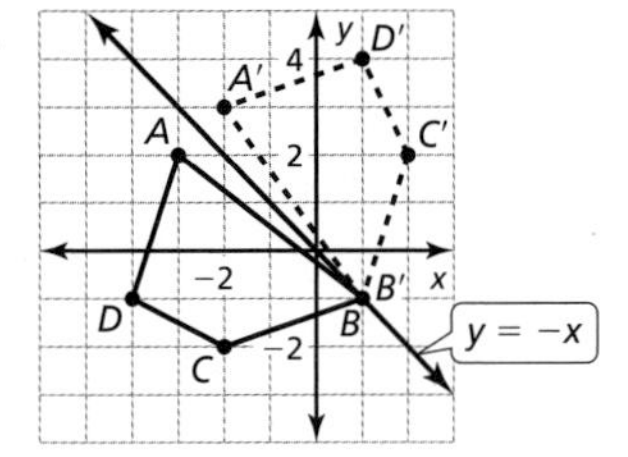

17.

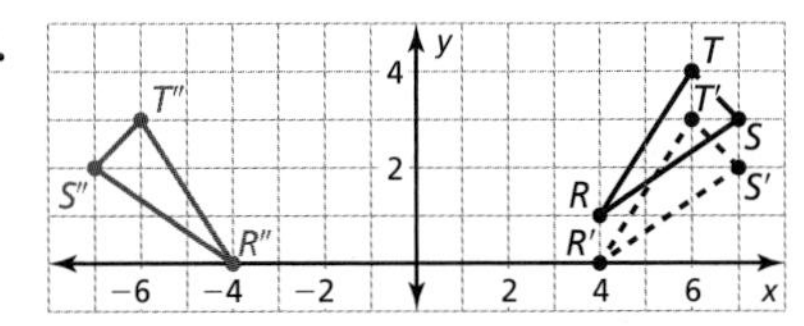

19.

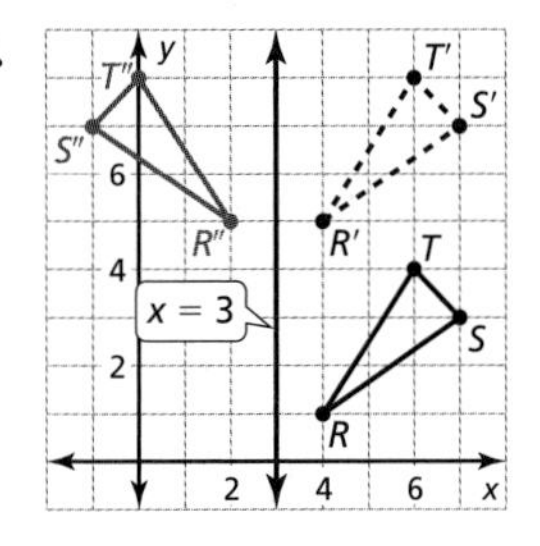

21. 1 **23.** 0

25. **a.** none **b.** MOM **c.** OX **d.** none

27. Reflect H in line n to obtain H'. Then draw $\overline{JH'}$. Label the intersection of JH' and n as K. Because JH' is the shortest distance between J and H' and $HK = H'K$, park at point K.

29. $C(5, 0)$ **31.** $C(-4, 0)$ **33.** $y = -3x - 4$

35.

37.

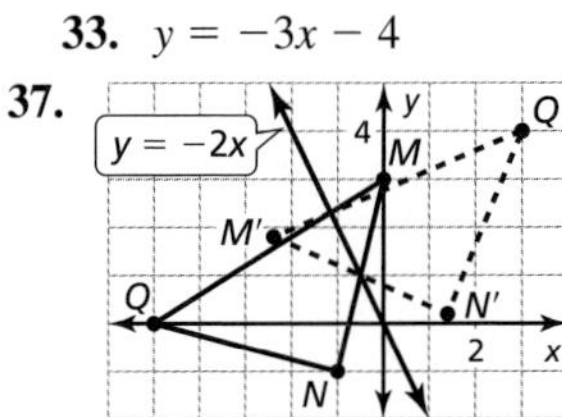

39. $y = x + 1$

4.2 Maintaining Mathematical Proficiency *(p. 188)*

41. 130° **43.** 160° **45.** 30° **47.** 180° **49.** 50°

4.3 Vocabulary and Core Concept Check *(p. 194)*

1. 270°

4.3 Monitoring Progress and Modeling with Mathematics *(pp. 194–196)*

3.

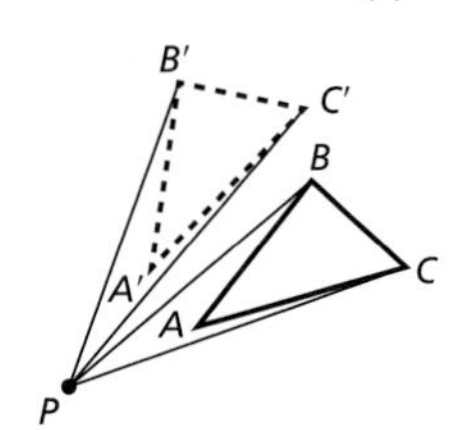

5.

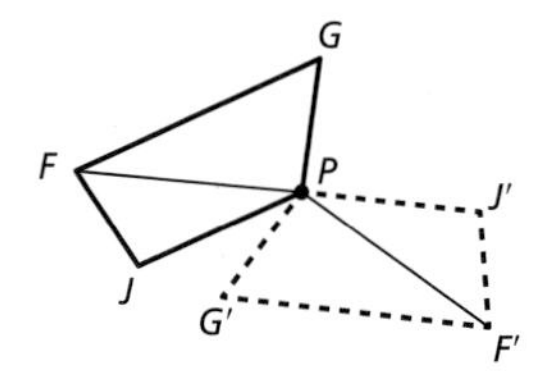

7.

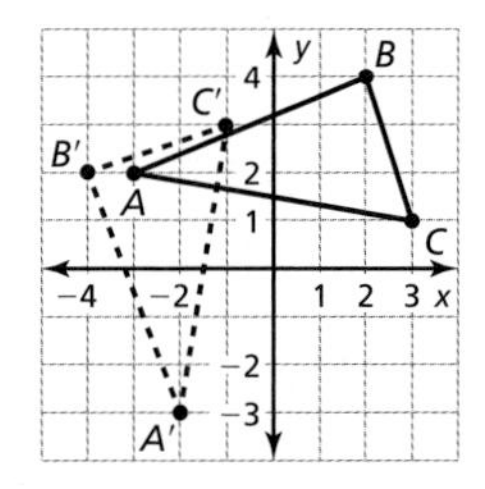

9.

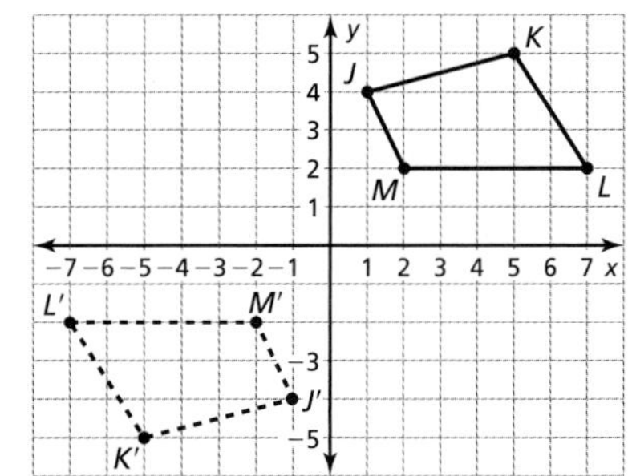

11.

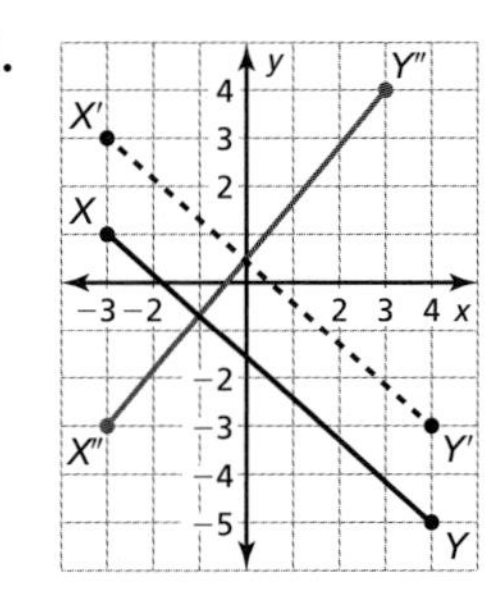

13.

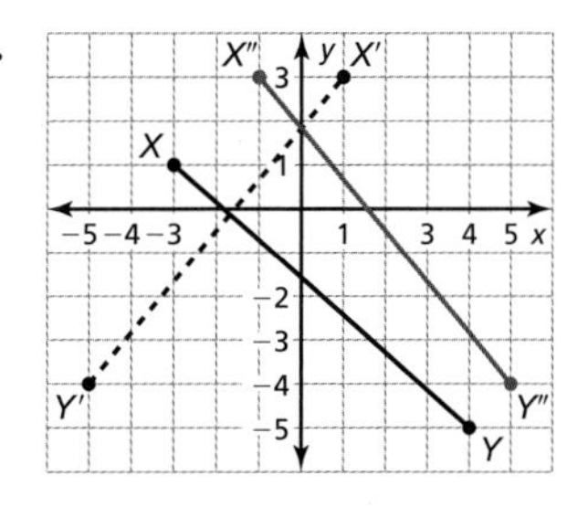

15.

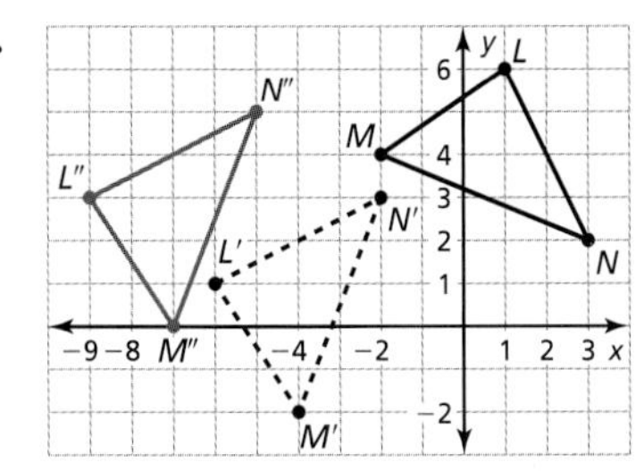

17. yes; Rotations of 90° and 180° about the center map the figure onto itself.

19. yes; Rotations of 45°, 90°, 135°, and 180° about the center map the figure onto itself.

21. F **23.** D, G

25. The rule for a 270° rotation, $(x, y) \to (y, -x)$, should have been used instead of the rule for a reflection in the x-axis; $C(-1, 1) \to C'(1, 1)$, $D(2, 3) \to D'(3, -2)$

27.

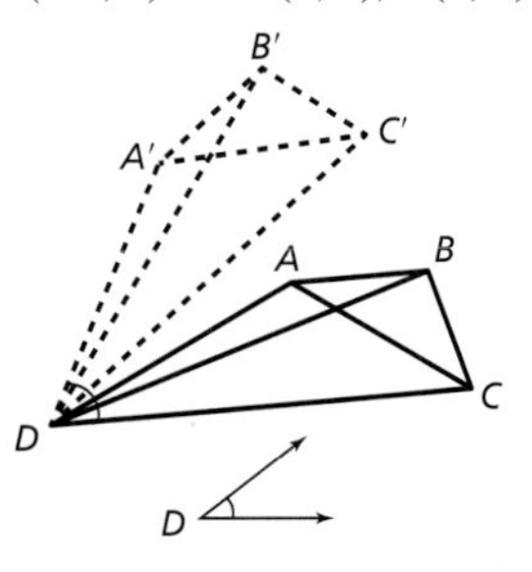

29. **a.** 90°: $y = -\frac{1}{2}x + \frac{3}{2}$, 180°: $y = 2x + 3$, 270°: $y = -\frac{1}{2}x - \frac{3}{2}$, 360°; $y = 2x - 3$; The slope of the line rotated 90° is the opposite reciprocal of the slope of the preimage, and the y-intercept is equal to the x-intercept of the preimage. The slope of the line rotated 180° is equal to the slope of the preimage, and the y-intercepts of the image and preimage are opposites. The slope of the line rotated 270° is the opposite reciprocal of the slope of the preimage, and the y-intercept is the opposite of the x-intercept of the preimage. The equation of the line rotated 360° is the same as the equation of the preimage.

b. yes; Because the coordinates of every point change in the same way with each rotation, the relationships described will be true for an equation with any slope and y-intercept.

31. twice

33. yes; *Sample answer:* A rectangle (that is not a square) is one example of a figure that has 180° rotational symmetry, but not 90° rotational symmetry.

35. **a.** 15°, $n = 12$ **b.** 30°, $n = 6$

37.

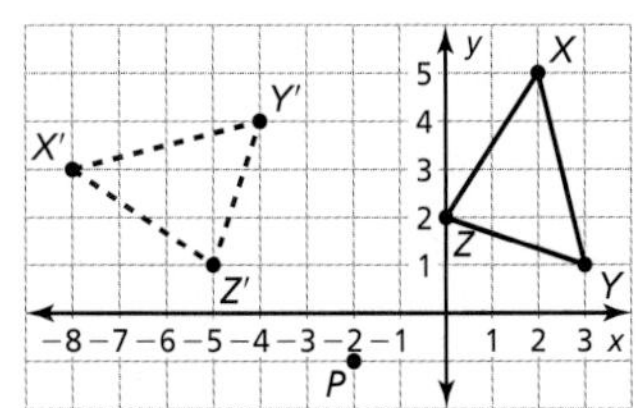

39. (2, 120°); (2, 210°); (2, 300°); The radius remains the same. The angle increases in conjunction with the rotation.

4.3 Maintaining Mathematical Proficiency *(p. 196)*

41. $\angle A$ and $\angle J$, $\angle B$ and $\angle K$, $\angle C$ and $\angle L$, $\angle D$ and $\angle M$; $\overline{AB}$ and $\overline{JK}$, $\overline{BC}$ and $\overline{KL}$, $\overline{CD}$ and $\overline{LM}$, $\overline{DA}$ and $\overline{MJ}$

4.4 Vocabulary and Core Concept Check *(p. 204)*

1. congruent

4.4 Monitoring Progress and Modeling with Mathematics *(pp. 204–206)*

3. $\triangle HJK \cong \triangle QRS$, $\square DEFG \cong \square LMNP$; $\triangle HJK$ is a 90° rotation of $\triangle QRS$. $\square DEFG$ is a translation 7 units right and 3 units down of $\square LMNP$.

5. *Sample answer:* 180° rotation about the origin followed by a translation 5 units left and 1 unit down

7. yes; $\triangle TUV$ is a translation 4 units right of $\triangle QRS$. So, $\triangle TUV \cong \triangle QRS$.

9. no; M and N are translated 2 units right of their corresponding vertices, L and K, but P is translated only 1 unit right of its corresponding vertex, J. So, this is not a rigid motion.

11. $\triangle A''B''C''$ **13.** 5.2 in. **15.** 110°

17. A translation 5 units right and a reflection in the x-axis should have been used; $\triangle ABC$ is mapped to $\triangle A'B'C'$ by a translation 5 units right, followed by a reflection in the x-axis.

19. 42° **21.** 90°

23. Reflect the figure in two parallel lines instead of translating the figure; The third line of reflection is perpendicular to the parallel lines.

25. never; Congruence transformations are rigid motions.

27. sometimes; Reflecting in $y = x$ then $y = x$ is not a rotation. Reflecting in the y-axis then x-axis is a rotation of 180°.

29. no; The image on the screen is larger.

31.

STATEMENTS	REASONS
1. A reflection in line ℓ maps $\overline{JK}$ to $\overline{J'K'}$, a reflection in line m maps $\overline{J'K'}$ to $\overline{J''K''}$, and $\ell \parallel m$.	**1.** Given
2. If $\overline{KK''}$ intersects line ℓ at L and line m at M, then L is the perpendicular bisector of $\overline{KK'}$, and M is the perpendicular bisector of $\overline{K'K''}$.	**2.** Definition of reflection
3. $\overline{KK'}$ is perpendicular to ℓ and m, and $KL = LK'$ and $K'M = MK''$.	**3.** Definition of perpendicular bisector
4. If d is the distance between ℓ and m, then $d = LM$.	**4.** Ruler Postulate (Post. 1.1)
5. $LM = LK' + K'M$ and $KK'' = KL + LK' + K'M + MK''$	**5.** Segment Addition Postulate (Post. 1.2)
6. $KK'' = LK' + LK' + K'M + K'M$	**6.** Substitution Property of Equality
7. $KK'' = 2(LK' + K'M)$	**7.** Distributive Property
8. $KK'' = 2(LM)$	**8.** Substitution Property of Equality
9. $KK'' = 2d$	**9.** Transitive Property of Equality

33. the second classmate that says it is a 180° rotation;
reflections: $P(1, 3) \rightarrow P'(-1, 3) \rightarrow P''(-1, -3)$ and $Q(3, 2) \rightarrow Q'(-3, 2) \rightarrow Q''(-3, -2)$
translation: $P(1, 3) \rightarrow (1 - 4, 3 - 5) \rightarrow (-3, -2)$ and $Q(3, 2) \rightarrow (3 - 4, 2 - 5) \rightarrow Q''(-1, -3)$
180° rotation $P(1, 3) \rightarrow (-1, -3)$ and $Q(3, 2) \rightarrow (-3, -2)$

35.

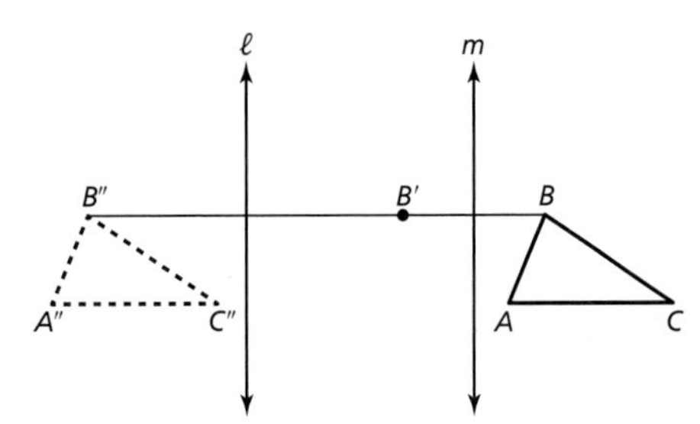

4.4 Maintaining Mathematical Proficiency *(p. 206)*

37. $x = -2$ **39.** $b = 6$ **41.** $n = -7.7$ **43.** 25%

4.5 Vocabulary and Core Concept Check *(p. 212)*

1. $P'(kx, ky)$

4.5 Monitoring Progress and Modeling with Mathematics *(pp. 212–214)*

3. $\frac{3}{7}$; reduction **5.** $\frac{3}{5}$; reduction

7.

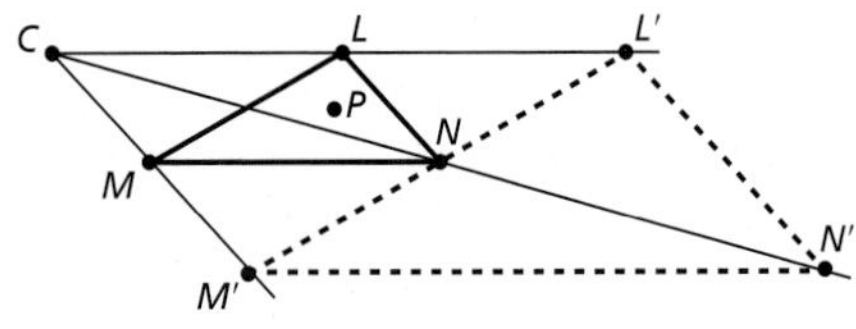

Not drawn to scale.

9.

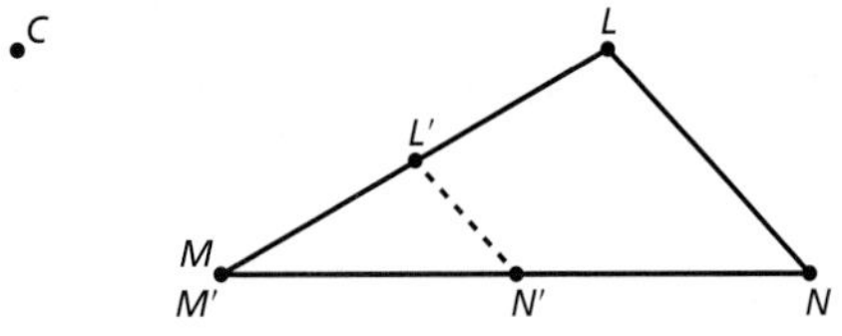

Not drawn to scale.

11.

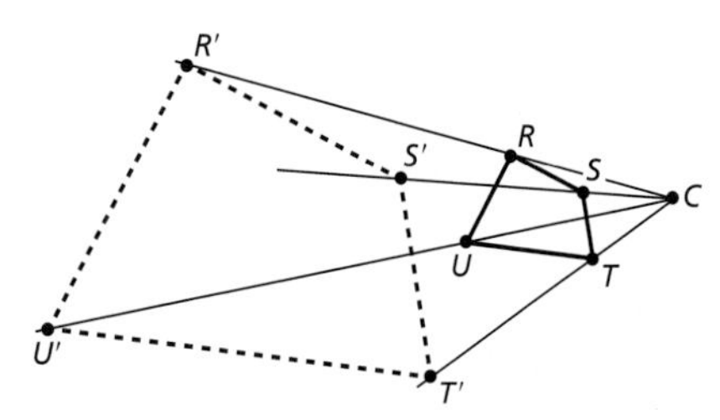

Not drawn to scale.

13.

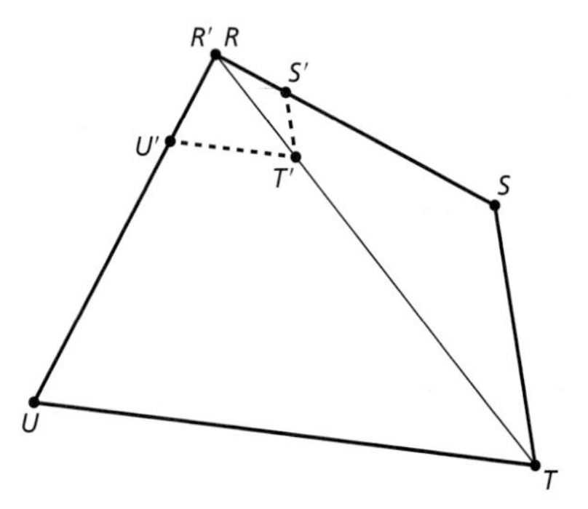

Not drawn to scale.

15.

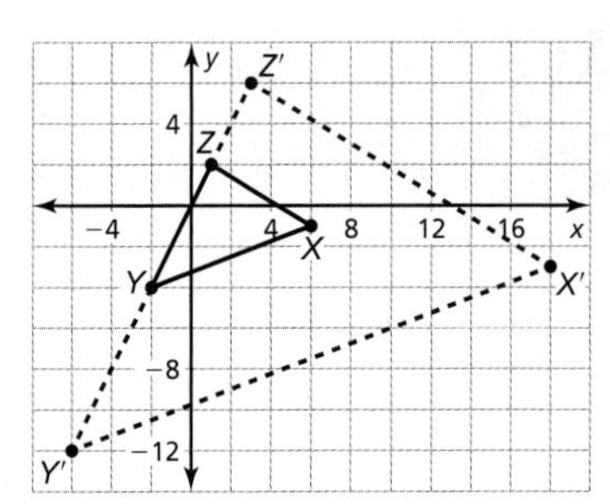

17.

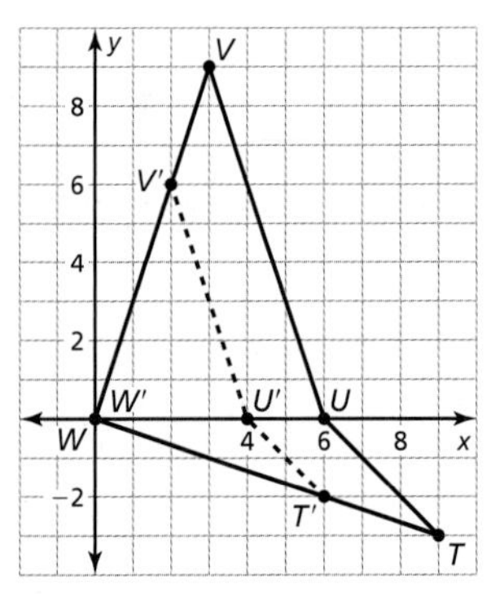

19.

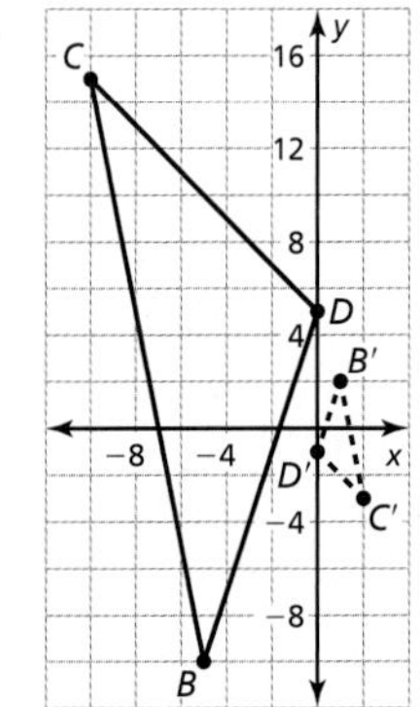

21.

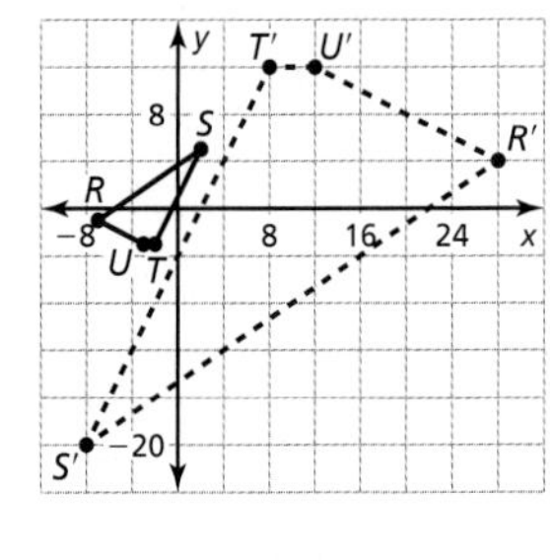

23. The scale factor should be calculated by finding $\frac{CP'}{CP}$, not $\frac{CP}{CP'}$; $k = \frac{3}{12} = \frac{1}{4}$

25. $k = \frac{5}{3}$; $x = 21$ **27.** $k = \frac{2}{3}$; $y = 3$ **29.** $k = 2$

31. 300 mm **33.** 940 mm

35. grasshopper, honey bee, and monarch butterfly; The scale factor for these three is $k = \frac{15}{2}$. The scale factor for the black beetle is $k = 7$.

37. no; The scale factor for the shorter sides is $\frac{8}{4} = 2$, but the scale factor for the longer sides is $\frac{10}{6} = \frac{5}{3}$. The scale factor for both sides has to be the same or the picture will be distorted.

39. $x = 5$, $y = 25$ **41.** original **43.** original

45.

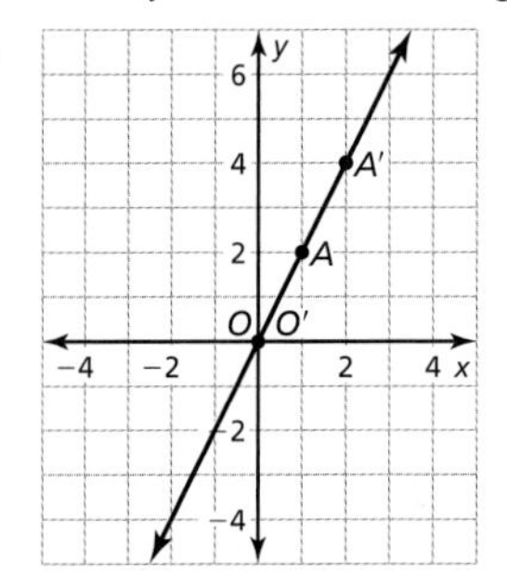

a. $O'A' = 2(OA)$ **b.** $\overleftrightarrow{O'A'}$ coincides with $\overleftrightarrow{OA}$.

47. $k = \frac{1}{16}$

49. **a.** $P = 24$ units, $A = 32$ square units

b. $P = 72$ units, $A = 288$ square units; The perimeter of the dilated rectangle is three times the perimeter of the original rectangle. The area of the dilated rectangle is nine times the area of the original rectangle.

c.

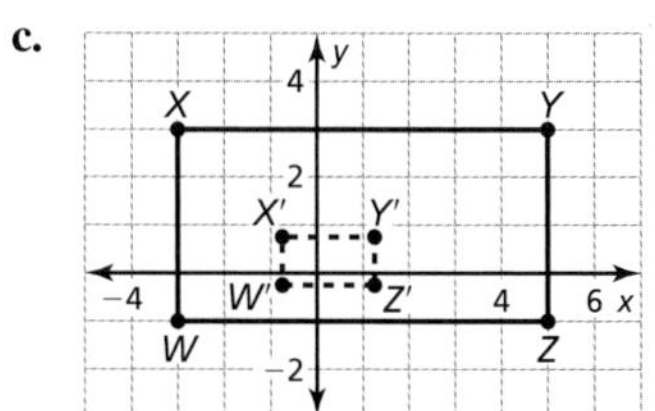

$P = 6$ units, $A = 2$ square units; The perimeter of the dilated rectangle is $\frac{1}{4}$ the perimeter of the original rectangle. The area of the dilated rectangle is $\frac{1}{16}$ the area of the original rectangle.

d. The perimeter changes by a factor of k. The area changes by a factor of k^2.

51. $A'(4, 4)$, $B'(4, 12)$, $C'(10, 4)$

4.5 Maintaining Mathematical Proficiency *(p. 214)*

53. $A'(1, 2)$, $B'(-1, 7)$, $C'(-4, 8)$

55. $A'(0, -1)$, $B'(-2, 4)$, $C'(-5, 5)$

57. $A'(-1, 0)$, $B'(-3, 5)$, $C'(-6, 6)$

4.6 Vocabulary and Core Concept Check *(p. 219)*

1. Congruent figures have the same size and shape. Similar figures have the same shape, but not necessarily the same size.

4.6 Monitoring Progress and Modeling with Mathematics *(pp. 219–220)*

3.

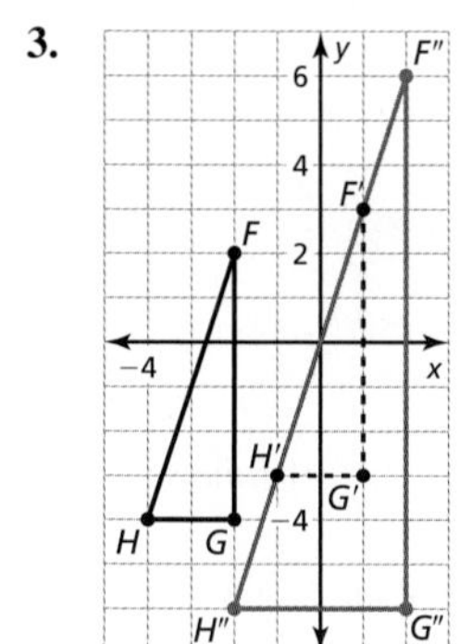

5.

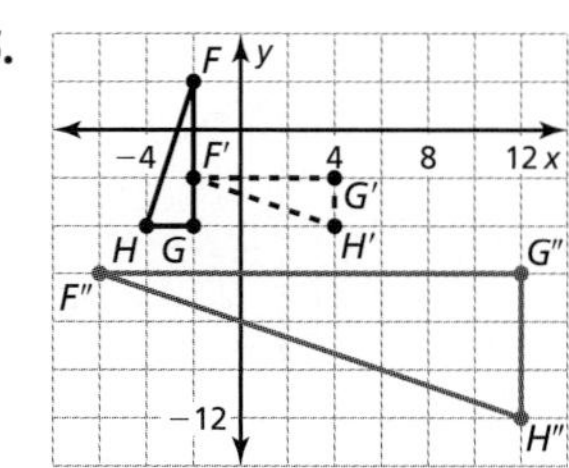

7. *Sample answer:* translation 1 unit down and 1 unit right followed by a dilation with center at $E(2, -3)$ and a scale factor of 2

9. yes; $\triangle ABC$ can be mapped to $\triangle DEF$ by a dilation with center at the origin and a scale factor of $\frac{1}{3}$ followed by a translation of 2 units left and 3 units up.

11. no; The scale factor from $\overline{HI}$ to $\overline{JL}$ is $\frac{2}{3}$, but the scale factor from $\overline{GH}$ to $\overline{KL}$ is $\frac{5}{6}$.

13. Reflect $\triangle ABC$ in $\overleftrightarrow{AB}$. Because reflections preserve side lengths and angle measures, the image of $\triangle ABC$, $\triangle ABC'$, is a right isosceles triangle with leg length j. Also because $\overleftrightarrow{AC} \perp \overleftrightarrow{BA}$, point C' is on $\overleftrightarrow{AC}$. So, $\overline{AC'}$ is parallel to $\overline{RT}$.

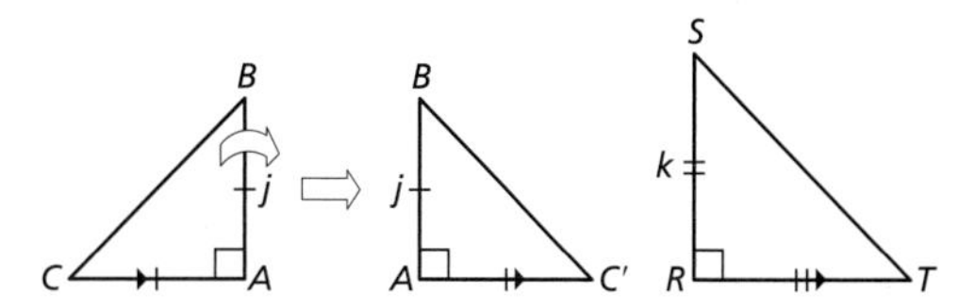

Then translate $\triangle ABC'$ so that point A maps to point R. Because translations map segments to parallel segments and $\overline{AC'} \parallel \overline{RT}$, the image of $\overline{AC'}$ lies on $\overline{RT}$.

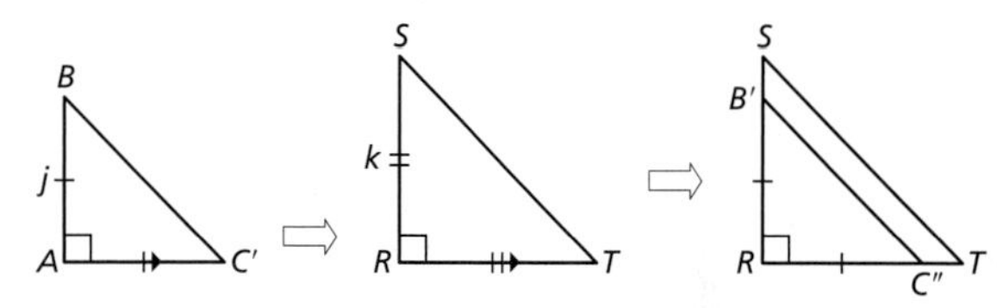

Because translations preserve side lengths and angle measures, the image of $\triangle ABC'$, $\triangle RB'C''$, is a right isosceles triangle with leg length j. Because $\angle B'RC''$ and $\angle SRT$ are right angles, they are congruent. When $\overrightarrow{RC''}$ coincides with $\overrightarrow{RT}$, $\overrightarrow{RB'}$ coincides with $\overrightarrow{RS}$. So, $\overline{RB'}$ lies on $\overrightarrow{RS}$. Next, dilate $\triangle RB'C''$ using center of dilation R. Choose the scale factor to be the ratio of the side lengths of $\triangle RST$ and $\triangle RB'C''$, which is $\frac{k}{j}$.

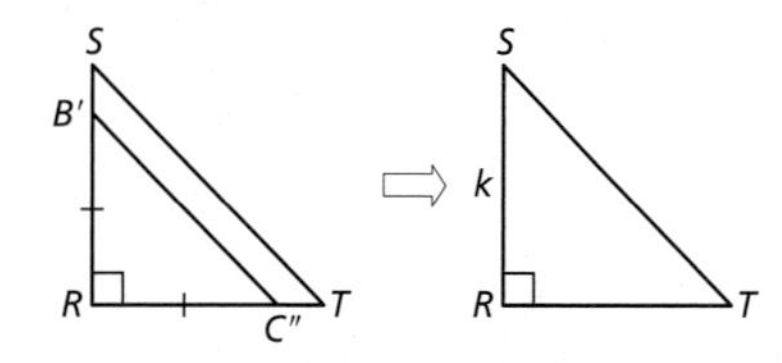

The dilation maps $\overline{RC''}$ to $\overline{RT}$ and $\overline{RB'}$ to $\overline{RS}$ because the images of $\overline{RC''}$ and $\overline{RB'}$ have side length $\frac{k}{j}(j) = k$ and the segments $\overline{RC''}$ and $\overline{RB'}$ lie on lines passing through the center of dilation. So, the dilation maps C'' to T and B' to S. A similarity transformation maps $\triangle ABC$ to $\triangle RST$. So, $\triangle ABC$ is similar to $\triangle RST$.

15. yes; The stop sign sticker can be mapped to the regular-sized stop sign by translating the sticker to the left until the centers match, and then dilating the sticker with a scale factor of 3.15. Because there is a similarity transformation that maps one stop sign to the other, the sticker is similar to the regular-sized stop sign.

17. no; The scale factor is 6 for both dimensions. So, the enlarged banner is proportional to the smaller one.

19. *Sample answer:*

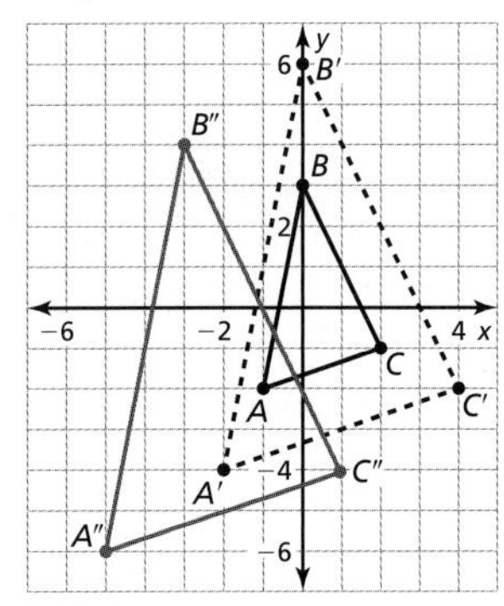

$\triangle A''B''C''$ can be mapped to $\triangle ABC$ by a translation 3 units right and 2 units up, followed by a dilation with center at the origin and a scale factor of $\frac{1}{2}$.

21. $J(-8, 0)$, $K(-8, 12)$, $L(-4, 12)$, $M(-4, 0)$; $J''(-9, -4)$, $K''(-9, 14)$, $L''(-3, 14)$, $M''(-3, -4)$; yes; A similarity transformation mapped quadrilateral $JKLM$ to quadrilateral $J''K''L''M''$.

4.6 Maintaining Mathematical Proficiency *(p. 220)*

23. obtuse **25.** acute

Chapter 4 Review *(pp. 222–224)*

1.

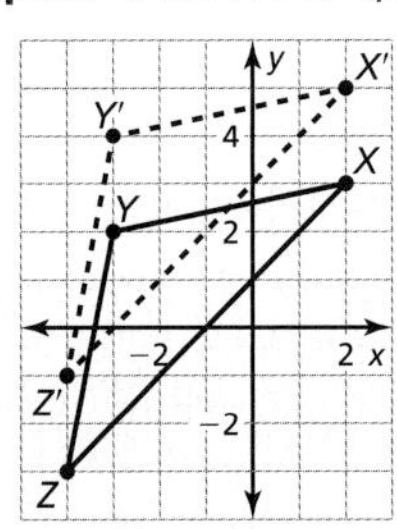

2.

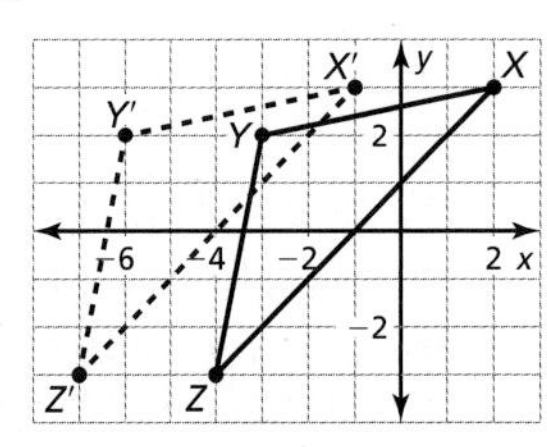

3.

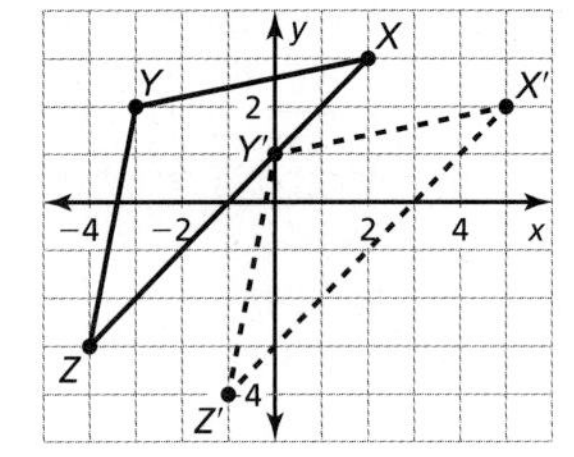

4.

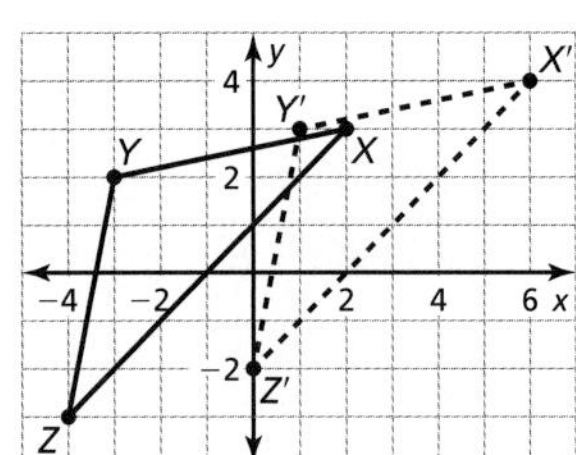

5.

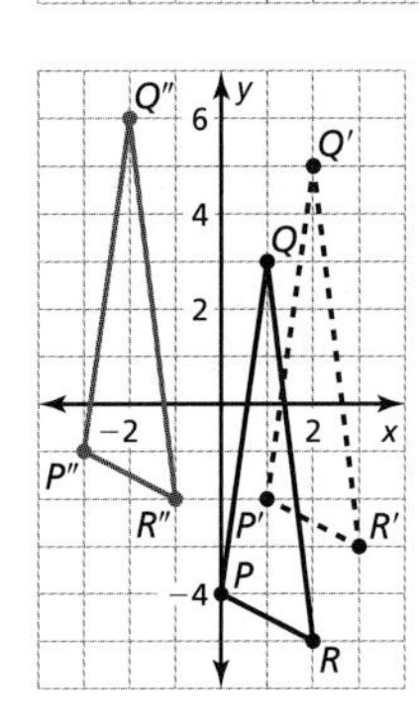

6.

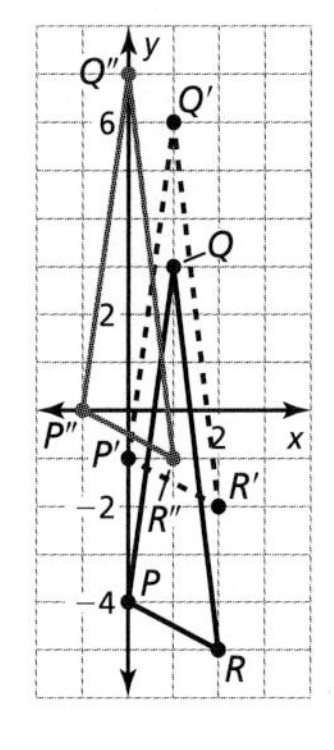

7.

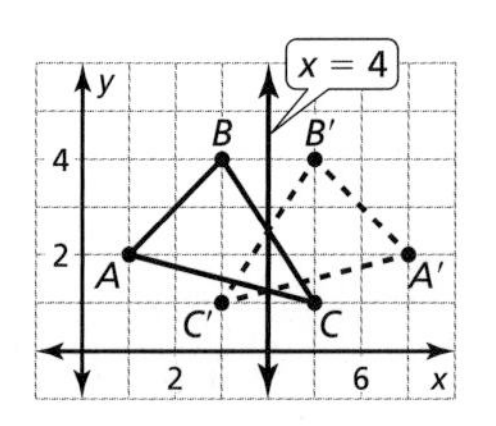

8.

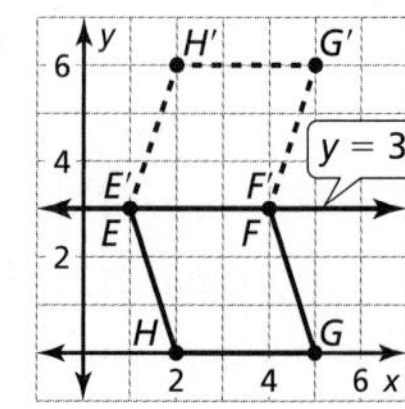

9. 2

10.

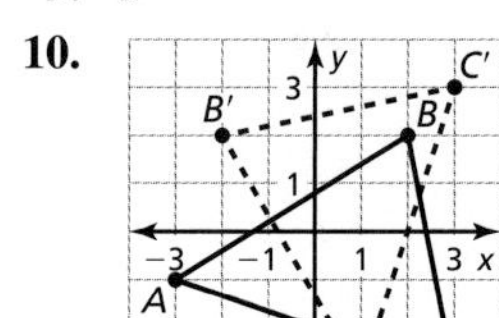

11.

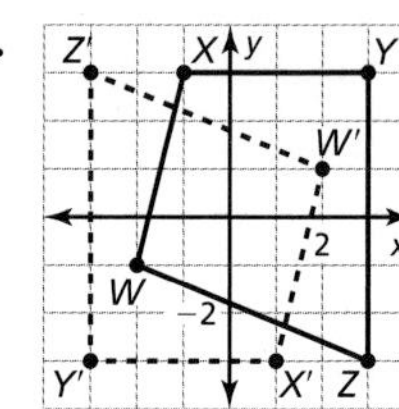

12.

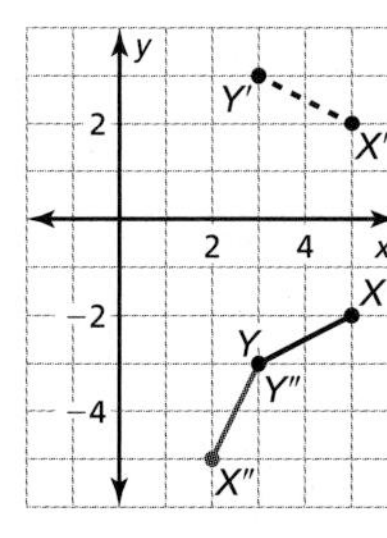

13. yes; Rotations of 60°, 120°, and 180° about the center map the figure onto itself.

14. yes; Rotations of 72° and 144° about the center map the figure onto itself.

15. *Sample answer:* reflection in the y-axis followed by a translation 3 units down

16. *Sample answer:* 180° rotation about the origin followed by a reflection in the line $x = 2$

17. translation; rotation

18.

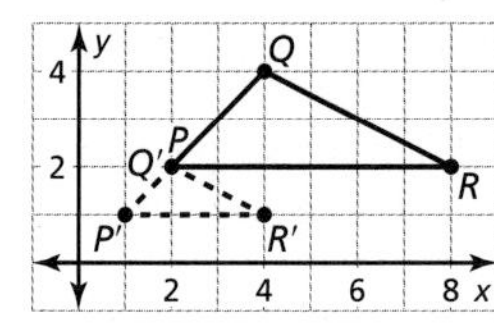

19.

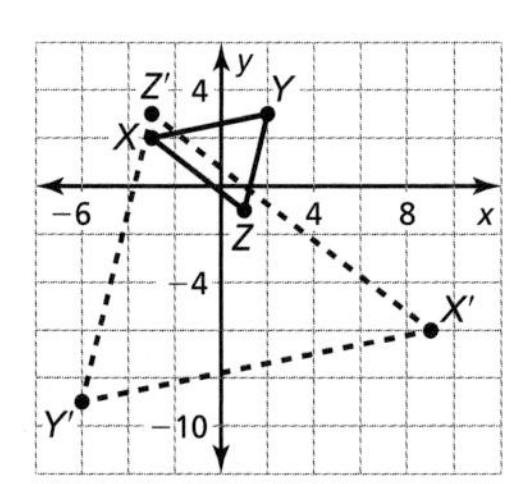

20. 1.9 cm

21. *Sample answer:* reflection in the line $x = -1$ followed by a dilation with center $(-3, 0)$ and $k = 3$

22. *Sample answer:* dilation with center at the origin and $k = \frac{1}{2}$, followed by a reflection in the line $y = x$

23. *Sample answer:* 270° rotation about the origin followed by a dilation with center at the origin and $k = 2$

Chapter 5

Chapter 5 Maintaining Mathematical Proficiency (p. 229)

1. $M(-2, 4)$; about 7.2 units **2.** $M(6, 2)$; 10 units
3. $M\left(\frac{7}{2}, -1\right)$; about 9.2 units **4.** $x = -3$ **5.** $t = 2$
6. $p = 3$ **7.** $w = 2$ **8.** $x = \frac{1}{3}$ **9.** $z = -\frac{3}{4}$
10. yes; The length can be found using the Pythagorean Theorem.

5.1 Vocabulary and Core Concept Check (p. 236)

1. no; By the Corollary to the Triangle Sum Theorem (Cor. 5.1), the acute angles of a right triangle are complementary. Because their measures have to add up to 90°, neither angle could have a measure greater than 90°.

5.1 Monitoring Progress and Modeling with Mathematics (pp. 236–238)

3. right isosceles **5.** obtuse scalene **7.** isosceles; right
9. scalene; not right **11.** 71°; acute **13.** 52°; right
15. 139° **17.** 114° **19.** 36°, 54° **21.** 37°, 53°
23. 15°, 75° **25.** 16.5°, 73.5°
27. The sum of the measures of the angles should be 180°;

$$115° + 39° + m\angle 1 = 180°$$
$$154° + m\angle 1 = 180°$$
$$m\angle 1 = 26°$$

29. 50° **31.** 50° **33.** 40° **35.** 90°
37. acute scalene
39. You could make another bend 6 inches from the first bend and leave the last side 8 inches long, or you could make another bend 7 inches from the first bend and then the last side will also be 7 inches long.

41.

STATEMENTS	REASONS
1. $\triangle ABC$ is a right triangle.	1. Given
2. $\angle C$ is a right angle.	2. Given (marked in diagram)
3. $m\angle C = 90°$	3. Definition of a right angle
4. $m\angle A + m\angle B + m\angle C = 180°$	4. Triangle Sum Theorem (Thm. 5.1)
5. $m\angle A + m\angle B + 90° = 180°$	5. Substitution Property of Equality
6. $m\angle A + m\angle B = 90°$	6. Subtraction Property of Equality
7. $\angle A$ and $\angle B$ are complementary.	7. Definition of complementary angles

43. yes; no

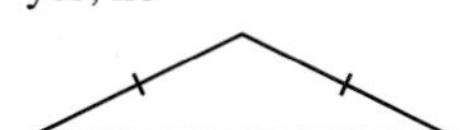

An obtuse equilateral triangle is not possible, because when two sides form an obtuse angle the third side that connects them must be longer than the other two.

45. a. $x = 8, x = 9$ **b.** one ($x = 4$)
47. A, B, F **49.** $x = 43, y = 32$ **51.** $x = 85, y = 65$

53.

STATEMENTS	REASONS
1. $\overleftrightarrow{AB} \parallel \overleftrightarrow{CD}$	1. Given (marked in diagram)
2. $\angle ACD$ and $\angle 5$ form a linear pair.	2. Definition of linear pair
3. $m\angle ACD + m\angle 5 = 180°$	3. Linear Pair Postulate (Post. 2.8)
4. $m\angle 3 + m\angle 4 = m\angle ACD$	4. Angle Addition Postulate (Post. 1.4)
5. $m\angle 3 + m\angle 4 + m\angle 5 = 180°$	5. Substitution Property of Equality
6. $\angle 1 \cong \angle 5$	6. Corresponding Angles Theorem (Thm. 3.1)
7. $\angle 2 \cong \angle 4$	7. Alternate Interior Angles Theorem (Thm. 3.2)
8. $m\angle 1 = m\angle 5$, $m\angle 2 = m\angle 4$	8. Definition of congruent angles
9. $m\angle 3 + m\angle 2 + m\angle 1 = 180°$	9. Substitution Property of Equality

5.1 Maintaining Mathematical Proficiency (p. 238)

55. 86° **57.** 15

5.2 Vocabulary and Core Concept Check (p. 243)

1. To show that two triangles are congruent, you need to show that all corresponding parts are congruent. If two triangles have the same side lengths and angle measures, then they must be the same size and shape.

5.2 Monitoring Progress and Modeling with Mathematics (pp. 243–244)

3. corresponding angles: $\angle A \cong \angle D, \angle B \cong \angle E, \angle C \cong \angle F$; corresponding sides: $\overline{AB} \cong \overline{DE}, \overline{BC} \cong \overline{EF}, \overline{AC} \cong \overline{DF}$; *Sample answer:* $\triangle BCA \cong \triangle EFD$
5. 124° **7.** 23° **9.** $x = 7, y = 8$
11. From the diagram, $\overline{WX} \cong \overline{LM}$, $\overline{XY} \cong \overline{MN}$, $\overline{YZ} \cong \overline{NJ}$, $\overline{VZ} \cong \overline{KJ}$, and $\overline{WV} \cong \overline{LK}$. Also from the diagram, $\angle V \cong \angle K$, $\angle W \cong \angle L$, $\angle X \cong \angle M$, $\angle Y \cong \angle N$, and $\angle Z \cong \angle J$. Because all corresponding parts are congruent, $VWXYZ \cong KLMNJ$.
13. 20°

15.

STATEMENTS	REASONS
1. $\overline{AB} \parallel \overline{DC}$, $\overline{AB} \cong \overline{DC}$, E is the midpoint of $\overline{AC}$ and $\overline{BD}$.	1. Given
2. $\angle AEB \cong \angle CED$	2. Vertical Angles Congruence Theorem (Thm. 2.6)
3. $\angle BAE \cong \angle DCE$, $\angle ABE \cong \angle CDE$	3. Alternate Interior Angles Theorem (Thm. 3.2)
4. $\overline{AE} \cong \overline{CE}$, $\overline{BE} \cong \overline{DE}$	4. Definition of midpoint
5. $\triangle AEB \cong \triangle CED$	5. All corresponding parts are congruent.

Selected Answers

17. The congruence statement should be used to ensure that corresponding parts are matched up correctly; $\angle S \cong \angle Y$; $m\angle S = m\angle Y$; $m\angle S = 90° - 42° = 48°$

19.

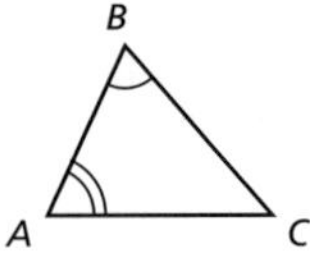

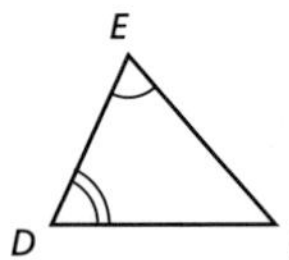

STATEMENTS	REASONS
1. $\angle A \cong \angle D$, $\angle B \cong \angle E$	**1.** Given
2. $m\angle A = m\angle D$, $m\angle B = m\angle E$	**2.** Definition of congruent angles
3. $m\angle A + m\angle B + m\angle C = 180°$, $m\angle D + m\angle E + m\angle F = 180°$	**3.** Triangle Sum Theorem (Thm. 5.1)
4. $m\angle A + m\angle B + m\angle C = m\angle D + m\angle E + m\angle F$	**4.** Transitive Property of Equality
5. $m\angle A + m\angle B + m\angle C = m\angle A + m\angle B + m\angle F$	**5.** Substitution Property of Equality
6. $m\angle C = m\angle F$	**6.** Subtraction Property of Equality
7. $\angle C \cong \angle F$	**7.** Definition of congruent angles

21. corresponding angles: $\angle J \cong \angle X, \angle K \cong \angle Y, \angle L \cong \angle Z$
corresponding sides: $\overline{JK} \cong \overline{XY}, \overline{KL} \cong \overline{YZ}, \overline{JL} \cong \overline{XZ}$

23. $\begin{cases} 17x - y = 40 \\ 2x + 4y = 50 \end{cases}$
$x = 3, y = 11$

25. A rigid motion maps each part of a figure to a corresponding part of its image. Because rigid motions preserve length and angle measure, corresponding parts of congruent figures are congruent, which means that the corresponding sides and corresponding angles are congruent.

5.2 Maintaining Mathematical Proficiency *(p. 244)*

27. $\overline{PQ} \cong \overline{RS}, \angle N \cong \angle T$

29. $\overline{DE} \cong \overline{HI}, \angle D \cong \angle H, \overline{DF} \parallel \overline{HG}, \angle DFE \cong \angle HGI$

5.3 Vocabulary and Core Concept Check *(p. 249)*

1. an angle formed by two sides

5.3 Monitoring Progress and Modeling with Mathematics *(pp. 249–250)*

3. $\angle JKL$ **5.** $\angle KLP$ **7.** $\angle JLK$

9. no; The congruent angles are not the included angles.

11. no; One of the congruent angles is not the included angle.

13. yes; Two pairs of sides and the included angles are congruent.

15.

STATEMENTS	REASONS
1. $\overline{SP} \cong \overline{TP}$, $\overline{PQ}$ bisects $\angle SPT$.	**1.** Given
2. $\overline{PQ} \cong \overline{PQ}$	**2.** Reflexive Property of Congruence (Thm. 2.1)
3. $\angle SPQ \cong \angle TPQ$	**3.** Definition of angle bisector
4. $\triangle SPQ \cong \triangle TPQ$	**4.** SAS Congruence Theorem (Thm. 5.5)

17.

STATEMENTS	REASONS
1. C is the midpoint of $\overline{AE}$ and $\overline{BD}$.	**1.** Given
2. $\angle ACB \cong \angle ECD$	**2.** Vertical Angles Congruence Theorem (Thm. 2.6)
3. $\overline{AC} \cong \overline{EC}, \overline{BC} \cong \overline{DC}$	**3.** Definition of midpoint
4. $\triangle ABC \cong \triangle EDC$	**4.** SAS Congruence Theorem (Thm. 5.5)

19. $\triangle SRT \cong \triangle URT$; $\overline{RT} \cong \overline{RT}$ by the Reflexive Property of Congruence (Thm. 2.1). Also, because all points on a circle are the same distance from the center, $\overline{RS} \cong \overline{RU}$. It is given that $\angle SRT \cong \angle URT$. So, $\triangle SRT$ and $\triangle URT$ are congruent by the SAS Congruence Theorem (Thm. 5.5).

21. $\triangle STU \cong \triangle UVR$; Because the sides of the pentagon are congruent, $\overline{ST} \cong \overline{UV}$ and $\overline{TU} \cong \overline{VR}$. Also, because the angles of the pentagon are congruent, $\angle T \cong \angle V$. So, $\triangle STU$ and $\triangle UVR$ are congruent by the SAS Congruence Theorem (Thm. 5.5)

23.

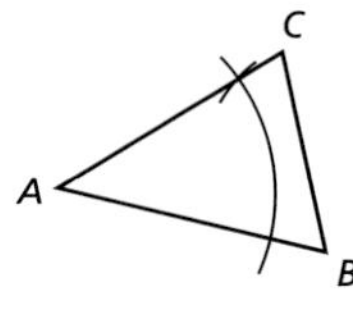

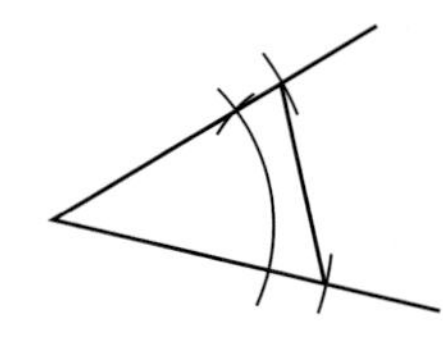

25. $\triangle XYZ$ and $\triangle WYZ$ are congruent so either the expressions for $\overline{XZ}$ and $\overline{WZ}$ or the expressions for $\overline{XY}$ and $\overline{WY}$ should be set equal to each other because they are corresponding sides.

$5x - 5 = 3x + 9$
$2x - 5 = 9$
$2x = 14$
$x = 7$

27. Because $\triangle ABC$, $\triangle BCD$, and $\triangle CDE$ are isosceles triangles, you know that $\overline{AB} \cong \overline{BC}, \overline{BC} \cong \overline{CD}$, and $\overline{CD} \cong \overline{DE}$. So, by the Transitive Property of Congruence (Thm. 2.1), $\overline{AB} \cong \overline{CD}$ and $\overline{BC} \cong \overline{DE}$. It is given that $\angle B \cong \angle D$, so $\triangle ABC \cong \triangle CDE$ by the SAS Congruence Theorem (Thm. 5.5).

29.

STATEMENTS	REASONS
1. $\overline{AC} \cong \overline{DC}, \overline{BC} \cong \overline{EC}$	**1.** Given
2. $\angle ACB \cong \angle DCE$	**2.** Vertical Angles Congruence Theorem (Thm. 2.6)
3. $\triangle ABC \cong \triangle DEC$	**3.** SAS Congruence Theorem (Thm. 5.5)

$x = 4, y = 5$

31.

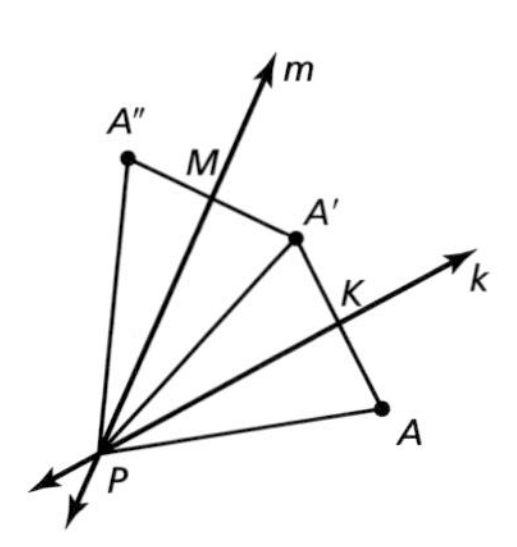

STATEMENTS	REASONS
1. A reflection in line k maps point A to A', a reflection in line m maps A' to A'', and $m\angle MPK = x°$.	**1.** Given
2. Line k is the perpendicular bisector of $\overline{AA'}$, and line m is the perpendicular bisector of $\overline{A'A''}$.	**2.** Definition of reflection
3. $\overline{AK} \cong \overline{KA'}$, $\angle AKP$ and $\angle A'KP$ are right angles, $\overline{A'M} \cong \overline{MA''}$, and $\angle A'MP$ and $\angle A''MP$ are right angles.	**3.** Definition of perpendicular bisector
4. $\angle AKP \cong \angle A'KP$, $\angle A'MP \cong \angle A''MP$	**4.** Right Angles Congruence Theorem (Thm. 2.3)
5. $\overline{KP} \cong \overline{KP}$	**5.** Reflexive Property of Congruence (Thm. 2.1)
6. $\triangle AKP \cong \triangle A'KP$, $\triangle A'MP \cong \triangle A''MP$	**6.** SAS Congruence Theorem (Thm. 5.5)
7. $\overline{AP} \cong \overline{A'P}$, $\overline{A'P} \cong \overline{A''P}$, $\angle APK \cong \angle A'PK$, $\angle A'PM \cong \angle A''PM$	**7.** Corresponding parts of congruent triangles are congruent.
8. $\overline{AP} \cong \overline{A''P}$	**8.** Transitive Property of Congruence (Thm. 2.1)
9. $m\angle APK = m\angle A'PK$, $m\angle A'PM = m\angle A''PM$	**9.** Definition of congruent angles
10. $m\angle MPK = m\angle A'PK + m\angle A'PM$, $m\angle APA'' = m\angle APK + m\angle A'PK + m\angle A'PM + m\angle A''PM$	**10.** Angle Addition Postulate (Post. 1.4)
11. $m\angle APA'' = m\angle A'PK + m\angle A'PK + m\angle A'PM + m\angle A'PM$	**11.** Substitution Property of Equality
12. $m\angle APA'' = 2(m\angle A'PK + m\angle A'PM)$	**12.** Distributive Property
13. $m\angle APA'' = 2(m\angle MPK)$	**13.** Substitution Property of Equality
14. $m\angle APA'' = 2(x°) = 2x°$	**14.** Substitution Property of Equality
15. A rotation about point P maps A to A'', and the angle of rotation is $2x°$.	**15.** Definition of rotation

5.3 Maintaining Mathematical Proficiency *(p. 250)*

33. obtuse isosceles **35.** obtuse scalene

5.4 Vocabulary and Core Concept Check *(p. 256)*

1. The vertex angle is the angle formed by the congruent sides, or legs, of an isosceles triangle.

5.4 Monitoring Progress and Modeling with Mathematics *(pp. 256–258)*

3. *A, D;* Base Angles Theorem (Thm. 5.6)

5. $\overline{CD}, \overline{CE}$; Converse of the Base Angles Theorem (Thm. 5.7)

7. $x = 12$ **9.** $x = 60$ **11.** $x = 79, y = 22$

13. $x = 60, y = 60$ **15.** $x = 30, y = 5$

17.

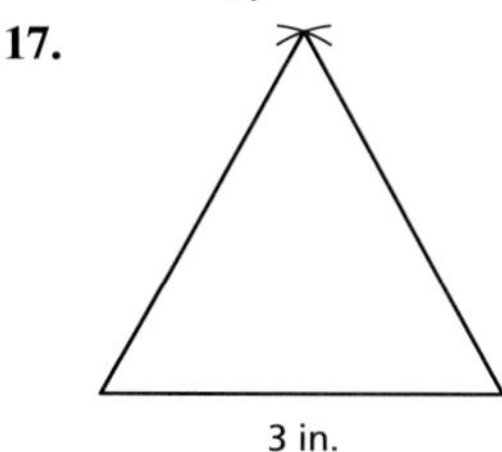

19. When two angles of a triangle are congruent, the sides opposite the angles are congruent; Because $\angle A \cong \angle C$, $\overline{AB} \cong \overline{BC}$. So, $BC = 5$.

21. a. Each edge is made out of the same number of sides of the original equilateral triangle.

b. 1 square unit, 4 square units, 9 square units, 16 square units

c. Triangle 1 has an area of $1^2 = 1$, Triangle 2 has an area of $2^2 = 4$, Triangle 3 has an area of $3^2 = 9$, and so on. So, by inductive reasoning, you can predict that Triangle n has an area of n^2; 49 square units; $n^2 = 7^2 = 49$

23. 17 in.

25. By the Reflexive Property of Congruence (Thm. 2.1), the yellow triangle and the yellow-orange triangle share a congruent side. Because the triangles are all isosceles, by the Transitive Property of Congruence (Thm. 2.1), the yellow-orange triangle and the orange triangle share a side that is congruent to the one shared by the yellow triangle and the yellow-orange triangle. This reasoning can be continued around the wheel, so the legs of the isosceles triangles are all congruent. Because you are given that the vertex angles are all congruent, you can conclude that the yellow triangle is congruent to the purple triangle by the SAS Congruence Theorem (Thm. 5.5).

27.

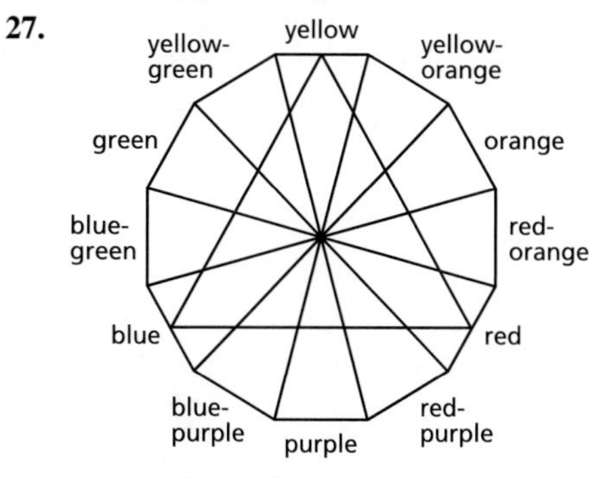

equiangular equilateral

29. no; The two sides that are congruent can form an obtuse angle or a right angle.

Selected Answers

31. 6, 8, 10; If $3t = 5t - 12$, then $t = 6$. If $5t - 12 = t + 20$, then $t = 8$. If $3t = t + 20$, then $t = 10$.

33. If the base angles are $x°$, then the vertex angle is $(180 - 2x)°$, or $[2(90 - x)]°$. Because $2(90 - x)$ is divisible by 2, the vertex angle is even when the angles are whole numbers.

35. **a.** 2.1 mi; By the Exterior Angle Theorem (Thm. 5.2), $m\angle L = 70° - 35° = 35°$. Because $m\angle SRL = 35° = m\angle RLS$, by definition of congruent angles, $\angle SRL \cong \angle RLS$. So, by the Converse of the Base Angles Theorem (Thm. 5.7), $\overline{RS} \cong \overline{SL}$. So, $SL = RS = 2.1$ miles.

b. Find the point on the shore line that has an angle of 45° from the boat. Then, measure the distance that the boat travels until the angle is 90°. That distance is the same as the distance between the boat and the shore line because the triangle formed is an isosceles right triangle.

37.

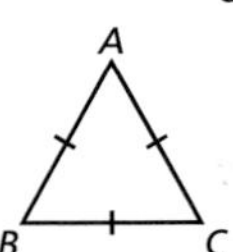

STATEMENTS	REASONS
1. $\triangle ABC$ is equilateral.	1. Given
2. $\overline{AB} \cong \overline{AC}$, $\overline{AB} \cong \overline{BC}$, $\overline{AC} \cong \overline{BC}$	2. Definition of equilateral triangle
3. $\angle B \cong \angle C$, $\angle A \cong \angle C$, $\angle A \cong \angle B$	3. Base Angles Theorem (Thm. 5.6)
4. $\triangle ABC$ is equiangular.	4. Definition of equiangular triangle

39.

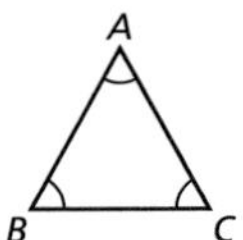

STATEMENTS	REASONS
1. $\triangle ABC$ is equiangular.	1. Given
2. $\angle B \cong \angle C$, $\angle A \cong \angle C$, $\angle A \cong \angle B$	2. Definition of equiangular triangle
3. $\overline{AB} \cong \overline{AC}$, $\overline{AB} \cong \overline{BC}$, $\overline{AC} \cong \overline{BC}$	3. Converse of the Base Angles Theorem (Thm. 5.7)
4. $\triangle ABC$ is equilateral.	4. Definition of equilateral triangle

41.

STATEMENTS	REASONS
1. $\triangle ABC$ is equilateral, $\angle CAD \cong \angle ABE \cong \angle BCF$	1. Given
2. $\triangle ABC$ is equiangular.	2. Corollary to the Base Angles Theorem (Cor. 5.2)
3. $\angle ABC \cong \angle BCA \cong \angle BAC$	3. Definition of equiangular triangle
4. $m\angle CAD = m\angle ABE = m\angle BCF$, $m\angle ABC = m\angle BCA = m\angle BAC$	4. Definition of congruent angles
5. $m\angle ABC = m\angle ABE + m\angle EBC$, $m\angle BCA = m\angle BCF + m\angle ACF$, $m\angle BAC = m\angle CAD + m\angle BAD$	5. Angle Addition Postulate (Post. 1.4)
6. $m\angle ABE + m\angle EBC = m\angle BCF + m\angle ACF = m\angle CAD + m\angle BAD$	6. Substitution Property of Equality
7. $m\angle ABE + m\angle EBC = m\angle ABE + m\angle ACF = m\angle ABE + m\angle BAD$	7. Substitution Property of Equality
8. $m\angle EBC = m\angle ACF = m\angle BAD$	8. Subtraction Property of Equality
9. $\angle EBC \cong \angle ACF \cong \angle BAD$	9. Definition of congruent angles
10. $\angle FEB \cong \angle DFC \cong \angle EDA$	10. Third Angles Theorem (Thm. 5.4)
11. $\angle FEB$ and $\angle FED$ are supplementary, $\angle DFC$ and $\angle EFD$ are supplementary, and $\angle EDA$ and $\angle FDE$ are supplementary.	11. Linear Pair Postulate (Post. 2.8)
12. $\angle FED \cong \angle EFD \cong \angle FDE$	12. Congruent Supplements Theorem (Thm. 2.4)
13. $\triangle DEF$ is equiangular.	13. Definition of equiangular triangle
14. $\triangle DEF$ is equilateral.	14. Corollary to the Converse of the Base Angles Theorem (Cor. 5.3)

5.4 Maintaining Mathematical Proficiency *(p. 258)*

43. $\overline{JK}, \overline{RS}$

5.5 Vocabulary and Core Concept Check *(p. 266)*

1. hypotenuse

5.5 Monitoring Progress and Modeling with Mathematics *(pp. 266–268)*

3. yes; $\overline{AB} \cong \overline{DB}, \overline{BC} \cong \overline{BE}, \overline{AC} \cong \overline{DE}$

5. yes; $\angle B$ and $\angle E$ are right angles, $\overline{AB} \cong \overline{FE}$, $\overline{AC} \cong \overline{FD}$

7. no; You are given that $\overline{RS} \cong \overline{PQ}$, $\overline{ST} \cong \overline{QT}$, and $\overline{RT} \cong \overline{PT}$. So, it should say $\triangle RST \cong \triangle PQT$ by the SSS Congruence Theorem (Thm. 5.8).

9. yes; You are given that $\overline{EF} \cong \overline{GF}$ and $\overline{DE} \cong \overline{DG}$. Also, $\overline{DF} \cong \overline{DF}$ by the Reflexive Property of Congruence (Thm. 2.1). So, $\triangle DEF \cong \triangle DGF$ by the SSS Congruence Theorem (Thm. 5.8).

11. yes; The diagonal supports in this figure form triangles with fixed side lengths. By the SSS Congruence Theorem (Thm. 5.8), these triangles cannot change shape, so the figure is stable.

13. 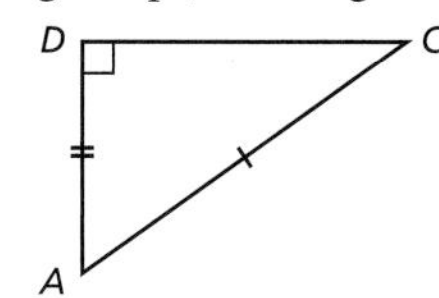

STATEMENTS	REASONS
1. $\overline{AC} \cong \overline{DB}$, $\overline{AB} \perp \overline{AD}$, $\overline{CD} \perp \overline{AD}$	**1.** Given
2. $\overline{AD} \cong \overline{AD}$	**2.** Reflexive Property of Congruence (Thm. 2.1)
3. $\angle BAD$ and $\angle CDA$ are right angles.	**3.** Definition of perpendicular lines
4. $\triangle BAD$ and $\triangle CDA$ are right triangles.	**4.** Definition of a right triangle
5. $\triangle BAD \cong \triangle CDA$	**5.** HL Congruence Theorem (Thm. 5.9)

15.

STATEMENTS	REASONS
1. $\overline{LM} \cong \overline{JK}$, $\overline{MJ} \cong \overline{KL}$	**1.** Given
2. $\overline{JL} \cong \overline{JL}$	**2.** Reflexive Property of Congruence (Thm. 2.1)
3. $\triangle LMJ \cong \triangle JKL$	**3.** SSS Congruence Theorem (Thm. 5.8)

17.

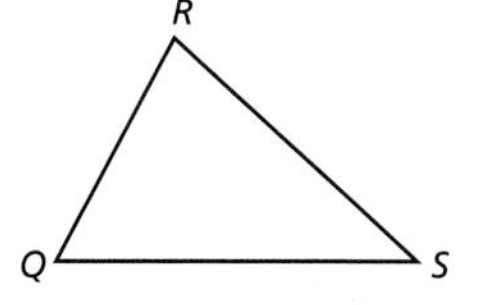

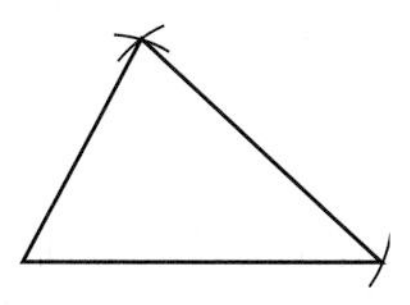

19. The order of the points in the congruence statement should reflect the corresponding sides and angles; $\triangle TUV \cong \triangle ZYX$ by the SSS Congruence Theorem (Thm. 5.8).

21. no; The sides of a triangle do not have to be congruent to each other, but each side of one triangle must be congruent to the corresponding side of the other triangle.

23. **a.** You need to know that the hypotenuses are congruent: $\overline{JL} \cong \overline{ML}$.

b. SAS Congruence Theorem (Thm. 5.5); By definition of midpoint, $\overline{JK} \cong \overline{MK}$. Also, $\overline{LK} \cong \overline{LK}$, by the Reflexive Property of Congruence (Thm. 2.1), and $\angle JKL \cong \angle MKL$ by the Right Angles Congruence Theorem (Thm. 2.3).

25. congruent **27.** congruent

29. yes; Use the string to compare the lengths of the corresponding sides of the two triangles to determine whether SSS Congruence Theorem (Thm. 5.8) applies.

31. both; $\overline{JL} \cong \overline{JL}$ by the Reflexive Property of Congruence (Thm. 2.1), and the other two pairs of sides are marked as congruent. So, the SSS Congruence Theorem (Thm. 5.8) can be used. Also, because $\angle M$ and $\angle K$ are right angles, they are both right triangles, and the legs and hypotenuses are congruent. So, the HL Congruence Theorem (Thm. 5.9) can be used.

33. *Sample answer:*

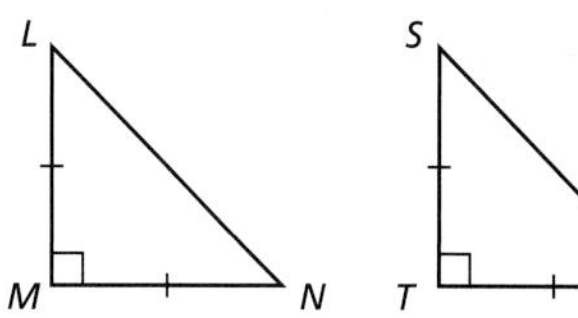

35. **a.** $\overline{BD} \cong \overline{BD}$ by the Reflexive Property of Congruence (Thm. 2.1). It is given that $\overline{AB} \cong \overline{CB}$ and that $\angle ADB$ and $\angle CDB$ are right angles. So, $\triangle ABC$ and $\triangle CBD$ are right triangles and are congruent by the HL Congruence Theorem (Thm. 5.9).

b. yes; Because $\overline{AB} \cong \overline{CB} \cong \overline{CE} \cong \overline{FE}$, $\overline{BD} \cong \overline{EG}$, and they are all right triangles, it can be shown that $\triangle ABD \cong \triangle CBD \cong \triangle CEG \cong \triangle FEG$ by the HL Congruence Theorem (Thm. 5.9).

5.5 Maintaining Mathematical Proficiency *(p. 268)*

37. $\overline{DF}$ **39.** $\angle E$

5.6 Vocabulary and Core Concept Check *(p. 274)*

1. Both theorems are used to prove that two triangles are congruent, and both require two pairs of corresponding angles to be congruent. In order to use the AAS Congruence Theorem (Thm. 5.11), one pair of corresponding nonincluded sides must also be congruent. In order to use the ASA Congruence Theorem (Thm. 5.10), the pair of corresponding included sides must be congruent.

5.6 Monitoring Progress and Modeling with Mathematics *(pp. 274–276)*

3. yes; AAS Congruence Theorem (Thm. 5.11) **5.** no

7. $\angle F$; $\angle L$

9. yes; $\triangle ABC \cong \triangle DEF$ by the ASA Congruence Theorem (Thm. 5.10)

11. no; $\overline{AC}$ and $\overline{DE}$ do not correspond.

13.

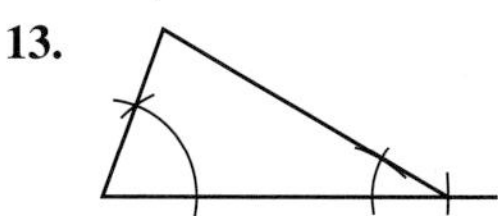

15. In the congruence statement, the vertices should be in corresponding order; $\triangle JKL \cong \triangle FGH$ by the ASA Congruence Theorem (Thm. 5.10).

17.

STATEMENTS	REASONS
1. M is the midpoint of $\overline{NL}$, $\overline{NL} \perp \overline{NQ}$, $\overline{NL} \perp \overline{MP}$, $\overline{QM} \parallel \overline{PL}$	**1.** Given
2. $\angle QNM$ and $\angle PML$ are right angles.	**2.** Definition of perpendicular lines
3. $\angle QNM \cong \angle PML$	**3.** Right Angles Congruence Theorem (Thm. 2.3)
4. $\angle QMN \cong \angle PLM$	**4.** Corresponding Angles Theorem (Thm. 3.1)
5. $\overline{NM} \cong \overline{ML}$	**5.** Definition of midpoint
6. $\triangle NQM \cong \triangle MPL$	**6.** ASA Congruence Theorem (Thm. 5.10)

19.

STATEMENTS	REASONS
1. $\overline{VW} \cong \overline{UW}$, $\angle X \cong \angle Z$	**1.** Given
2. $\angle W \cong \angle W$	**2.** Reflexive Property of Congruence (Thm. 2.2)
3. $\triangle XWV \cong \triangle ZWU$	**3.** AAS Congruence Theorem (Thm. 5.11)

21. You are given two right triangles, so the triangles have congruent right angles by the Right Angles Congruence Theorem (Thm. 2.3). Because another pair of angles and a pair of corresponding nonincluded sides (the hypotenuses) are congruent, the triangles are congruent by the AAS Congruence Theorem (Thm. 5.11).

23. You are given two right triangles, so the triangles have congruent right angles by the Right Angles Congruence Theorem (Thm. 2.3). There is also another pair of congruent corresponding angles and a pair of congruent corresponding sides. If the pair of congruent sides is the included side, then the triangles are congruent by the ASA Congruence Theorem (Thm. 5.10). If the pair of congruent sides is a nonincluded pair, then the triangles are congruent by the AAS Congruence Theorem (Thm. 5.11).

25. yes; When $x = 14$ and $y = 26$, $m\angle ABC = m\angle DBC = m\angle BCA = m\angle BCD = 80°$ and $m\angle CAB = m\angle CDB = 20°$. This satisfies the Triangle Sum Theorem (Thm. 5.1) for both triangles. Because $\overline{CB} \cong \overline{CB}$ by the Reflexive Property of Congruence (Thm. 2.1), you can conclude that $\triangle ABC \cong \triangle DBC$ by the ASA Congruence Theorem (Thm. 5.10) or the AAS Congruence Theorem (Thm. 5.11).

27.

STATEMENTS	REASONS
1. Draw $\overline{AD}$, the angle bisector of $\angle ABC$.	**1.** Construction of angle bisector
2. $\angle CAD \cong \angle BAD$	**2.** Definition of angle bisector
3. $\angle B \cong \angle C$	**3.** Given
4. $\overline{AD} \cong \overline{AD}$	**4.** Reflexive Property of Congruence (Thm. 2.1)
5. $\triangle ABD \cong \triangle ACD$	**5.** AAS Congruence Theorem (Thm. 5.11)
6. $\overline{AB} \cong \overline{AC}$	**6.** Corresponding parts of congruent triangles are congruent.

29. a.

STATEMENTS	REASONS
1. $\angle CDB \cong \angle ADB$, $\overline{DB} \perp \overline{AC}$	**1.** Given
2. $\angle ABD$ and $\angle CBD$ are right angles.	**2.** Definition of perpendicular lines
3. $\angle ABD \cong \angle CBD$	**3.** Right Angles Congruence Theorem (Thm. 2.3)
4. $\overline{BD} \cong \overline{BD}$	**4.** Reflexive Property of Congruence (Thm. 2.1)
5. $\triangle ABD \cong \triangle CBD$	**5.** ASA Congruence Theorem (Thm. 5.10)

b. Because $\triangle ABD \cong \triangle CBD$ and corresponding parts of congruent triangles are congruent, you can conclude that $\overline{AD} \cong \overline{CD}$,which means that $\triangle ACD$ is isosceles by definition.

c. no; For instance, because $\triangle ACD$ is isosceles, the girl sees her toes at the bottom of the mirror. This remains true as she moves backward, because $\triangle ACD$ remains isosceles.

31. *Sample answer:*

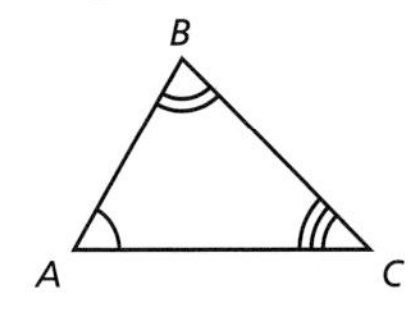

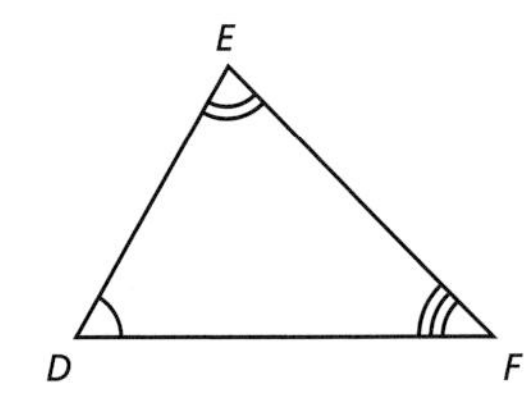

33. a. $\overline{TU} \cong \overline{XY}, \overline{UV} \cong \overline{YZ}, \overline{TV} \cong \overline{XZ}$;
$\overline{TU} \cong \overline{XY}, \angle U \cong \angle Y, \overline{UV} \cong \overline{YZ}$;
$\overline{UV} \cong \overline{YZ}, \angle V \cong \angle Z, \overline{TV} \cong \overline{XZ}$;
$\overline{TV} \cong \overline{XZ}, \angle T \cong \angle X, \overline{TU} \cong \overline{XY}$;
$\angle T \cong \angle X, \overline{TU} \cong \overline{XY}, \angle U \cong \angle Y$;
$\angle U \cong \angle Y, \overline{UV} \cong \overline{YZ}, \angle V \cong \angle Z$;
$\angle V \cong \angle Z, \overline{TV} \cong \overline{XZ}, \angle T \cong \angle X$;
$\angle T \cong \angle X, \angle U \cong \angle Y, \overline{UV} \cong \overline{YZ}$;
$\angle T \cong \angle X, \angle U \cong \angle Y, \overline{TV} \cong \overline{XZ}$;
$\angle U \cong \angle Y, \angle V \cong \angle Z, \overline{TV} \cong \overline{XZ}$;
$\angle U \cong \angle Y, \angle V \cong \angle Z, \overline{TU} \cong \overline{XY}$;
$\angle V \cong \angle Z, \angle T \cong \angle X, \overline{TU} \cong \overline{XY}$;
$\angle V \cong \angle Z, \angle T \cong \angle X, \overline{UV} \cong \overline{YZ}$

b. $\frac{13}{20}$, or 65%

5.6 Maintaining Mathematical Proficiency *(p. 276)*

35. (1, 1)

37.

5.7 Vocabulary and Core Concept Check *(p. 281)*

1. Corresponding

5.7 Monitoring Progress and Modeling with Mathematics *(pp. 281–282)*

3. All three pairs of sides are congruent. So, by the SSS Congruence Theorem (Thm. 5.8), $\triangle ABC \cong \triangle DBC$. Because corresponding parts of congruent triangles are congruent, $\angle A \cong \angle D$.

5. The hypotenuses and one pair of legs of two right triangles are congruent. So, by the HL Congruence Theorem (Thm. 5.9), $\triangle JMK \cong \triangle LMK$. Because corresponding parts of congruent triangles are congruent, $\overline{JM} \cong \overline{LM}$.

7. From the diagram, $\angle JHN \cong \angle KGL$, $\angle N \cong \angle L$, and $\overline{JN} \cong \overline{KL}$. So, by the AAS Congruence Theorem (Thm. 5.11), $\triangle JNH \cong \triangle KLG$. Because corresponding parts of congruent triangles are congruent, $\overline{GK} \cong \overline{HJ}$.

9. Use the AAS Congruence Theorem (Thm. 5.11) to prove that $\triangle FHG \cong \triangle GKF$. Then, state that $\angle FGK \cong \angle GFH$. Use the Congruent Complements Theorem (Thm. 2.5) to prove that $\angle 1 \cong \angle 2$.

11. Use the ASA Congruence Theorem (Thm. 5.10) to prove that $\triangle STR \cong \triangle QTP$. Then, state that $\overline{PT} \cong \overline{RT}$ because corresponding parts of congruent triangles are congruent. Use the SAS Congruence Theorem (Thm. 5.5) to prove that $\triangle STP \cong \triangle QTR$. So, $\angle 1 \cong \angle 2$.

13.

STATEMENTS	REASONS
1. $\overline{AP} \cong \overline{BP}$, $\overline{AQ} \cong \overline{BQ}$	1. Given
2. $\overline{PQ} \cong \overline{PQ}$	2. Reflexive Property of Congruence (Thm. 2.1)
3. $\triangle APQ \cong \triangle BPQ$	3. SSS Congruence Theorem (Thm. 5.8)
4. $\angle APQ \cong \angle BPQ$	4. Corresponding parts of congruent triangles are congruent.
5. $\overline{PM} \cong \overline{PM}$	5. Reflexive Property of Congruence (Thm. 2.1)
6. $\triangle APM \cong \triangle BPM$	6. SAS Congruence Theorem (Thm. 5.5)
7. $\angle AMP \cong \angle BMP$	7. Corresponding parts of congruent triangles are congruent.
8. $\angle AMP$ and $\angle BMP$ form a linear pair.	8. Definition of a linear pair
9. $\overline{MP} \perp \overline{AB}$	9. Linear Pair Perpendicular Theorem (Thm. 3.10)
10. $\angle AMP$ and $\angle BMP$ are right angles.	10. Definition of perpendicular lines

15.

STATEMENTS	REASONS
1. $\overline{FG} \cong \overline{GJ} \cong \overline{HG} \cong \overline{GK}$, $\overline{JM} \cong \overline{LM} \cong \overline{KM} \cong \overline{NM}$	1. Given
2. $\angle FGJ \cong \angle HGK$, $\angle JML \cong \angle KMN$	2. Vertical Angles Congruence Theorem (Thm. 2.6)
3. $\triangle FGJ \cong \triangle HGK$, $\triangle JML \cong \triangle KMN$	3. SAS Congruence Theorem (Thm. 5.5)
4. $\angle F \cong \angle H$, $\angle L \cong \angle N$	4. Corresponding parts of congruent triangles are congruent.
5. $FG = GJ = HG = GK$	5. Definition of congruent segments
6. $HJ = HG + GJ$, $FK = FG + GK$	6. Segment Addition Postulate (Post. 1.2)
7. $FK = HG + GJ$	7. Substitution Property of Equality
8. $FK = HJ$	8. Transitive Property of Equality
9. $\overline{FK} \cong \overline{HJ}$	9. Definition of congruent segments
10. $\triangle HJN \cong \triangle FKL$	10. AAS Congruence Theorem (Thm. 5.11)
11. $\overline{FL} \cong \overline{HN}$	11. Corresponding parts of congruent triangles are congruent.

17. Because $\overline{AC} \perp \overline{BC}$ and $\overline{ED} \perp \overline{BD}$, $\angle ACB$ and $\angle EDB$ are congruent right angles. Because B is the midpoint of $\overline{CD}$, $\overline{BC} \cong \overline{BD}$. The vertical angles $\angle ABC$ and $\angle EBD$ are congruent. So, $\triangle ABC \cong \triangle EBD$ by the ASA Congruence Theorem (Thm. 5.10). Then, because corresponding parts of congruent triangles are congruent, $\overline{AC} \cong \overline{ED}$. So, you can find the distance AC across the canyon by measuring ED.

19.

STATEMENTS	REASONS
1. $\overline{AD} \parallel \overline{BC}$, E is the midpoint of $\overline{AC}$.	1. Given
2. $\overline{AE} \cong \overline{CE}$	2. Definition of midpoint
3. $\angle AEB \cong \angle CED$, $\angle AED \cong \angle BEC$	3. Vertical Angles Congruence Theorem (Thm. 2.6)
4. $\angle DAE \cong \angle BCE$	4. Alternate Interior Angles Theorem (Thm. 3.2)
5. $\triangle DAE \cong \triangle BCE$	5. ASA Congruence Theorem (Thm. 5.10)
6. $\overline{DE} \cong \overline{BE}$	6. Corresponding parts of congruent triangles are congruent.
7. $\triangle AEB \cong \triangle CED$	7. SAS Congruence Theorem (Thm. 5.5)

21. yes; You can show that $WXYZ$ is a rectangle. This means that the opposite sides are congruent. Because $\triangle WZY$ and $\triangle YXW$ share a hypotenuse, the two triangles have congruent hypotenuses and corresponding legs, which allows you to use the HL Congruence Theorem (Thm. 5.9) to prove that the triangles are congruent.

23. $\triangle GHJ$, $\triangle NPQ$

5.7 Maintaining Mathematical Proficiency *(p. 282)*

25. about 17.5 units

5.8 Vocabulary and Core Concept Check *(p. 287)*

1. In a coordinate proof, you have to assign coordinates to vertices and write expressions for the side lengths and the slopes of segments in order to show how sides are related; As with other types of proofs, you still have to use deductive reasoning and justify every conclusion with theorems, proofs, and properties of mathematics.

5.8 Monitoring Progress and Modeling with Mathematics *(pp. 287–288)*

3. *Sample answer:*

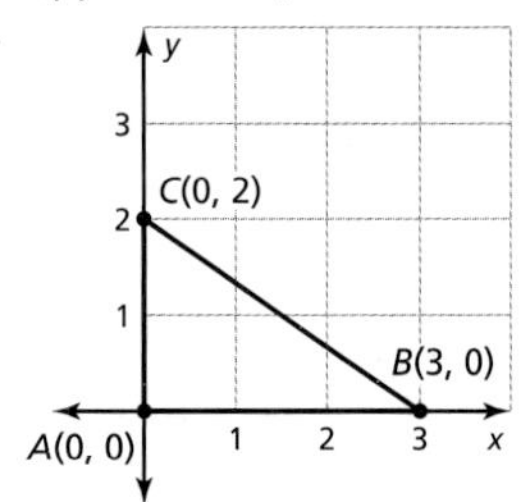

It is easy to find the lengths of horizontal and vertical segments and distances from the origin.

5. *Sample answer:*

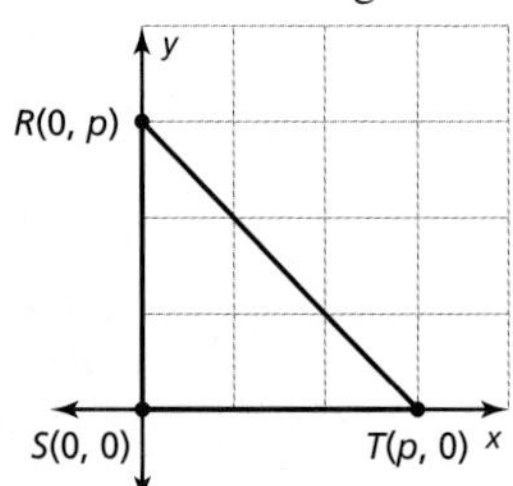

It is easy to find the lengths of horizontal and vertical segments and distances from the origin.

7. Find the lengths of $\overline{OP}$, $\overline{PM}$, $\overline{MN}$, and $\overline{NO}$ to show that $\overline{OP} \cong \overline{PM}$ and $\overline{MN} \cong \overline{NO}$.

9.

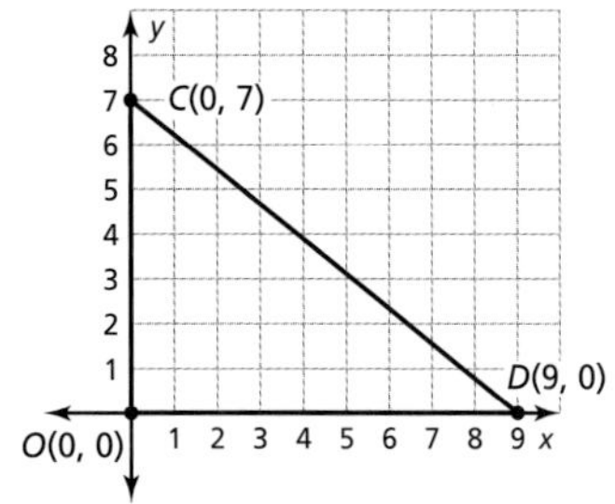

about 11.4 units

11.

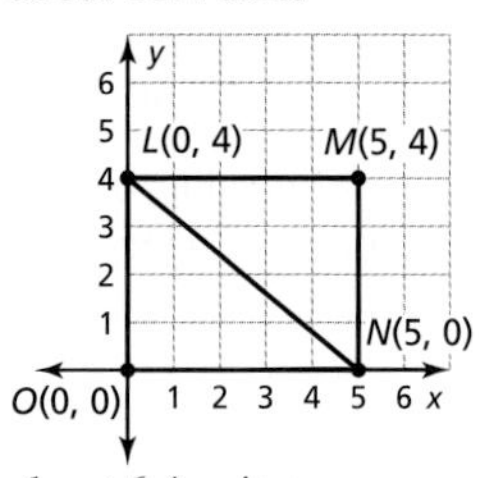

about 6.4 units

13.

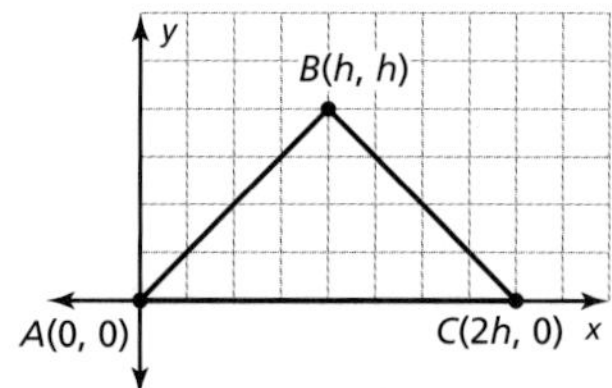

$AB = h\sqrt{2}$, $m_{\overline{AB}} = 1$, $M_{\overline{AB}}\left(\frac{h}{2}, \frac{h}{2}\right)$, $BC = h\sqrt{2}$, $m_{\overline{BC}} = -1$, $M_{\overline{BC}}\left(\frac{3h}{2}, \frac{h}{2}\right)$, $AC = 2h$, $m_{\overline{AC}} = 0$, $M_{\overline{AC}}(h, 0)$; yes; yes; Because $m_{\overline{AB}} \cdot m_{\overline{BC}} = -1$, $\overline{AB} \perp \overline{BC}$ by the Slopes of Perpendicular Lines Theorem (Thm. 3.14). So $\angle ABC$ is a right angle. $\overline{AB} \cong \overline{BC}$ because $AB = BC$. So, $\triangle ABC$ is a right isosceles triangle.

15. $N(h, k)$; $ON = \sqrt{h^2 + k^2}$, $MN = \sqrt{h^2 + k^2}$

17. $DC = k$, $BC = k$, $DE = h$, $OB = h$, $EC = \sqrt{h^2 + k^2}$, $OC = \sqrt{h^2 + k^2}$
So, $\overline{DC} \cong \overline{BC}$, $\overline{DE} \cong \overline{OB}$, and $\overline{EC} \cong \overline{OC}$. By the SSS Congruence Theorem (Thm. 5.8), $\triangle DEC \cong \triangle BOC$.

19.

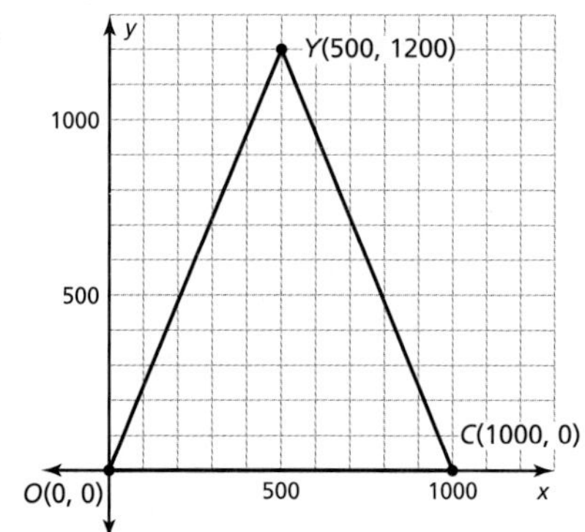

Using the Distance Formula, $OY = 1300$, and $CY = 1300$. Because $\overline{OY} \cong \overline{CY}$, $\triangle OYC$ is isosceles.

21. *Sample answer:* $(-k, -m)$ and (k, m)

23. A

25. $(0, 0)$, $(5d, 0)$, $(0, 5d)$

27. **a.**

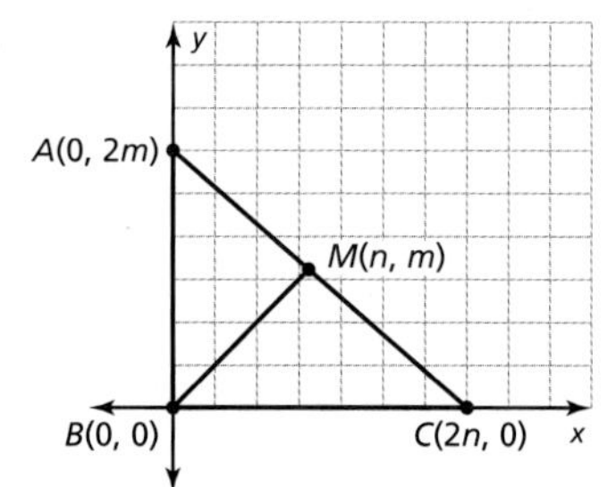

Because M is the midpoint of $\overline{AC}$, the coordinates of M are $M(n, m)$. Using the Distance Formula, $AM = \sqrt{n^2 + m^2}$, $BM = \sqrt{n^2 + m^2}$, and $CM = \sqrt{n^2 + m^2}$. So, the midpoint of the hypotenuse of a right triangle is the same distance from each vertex of the triangle.

b.

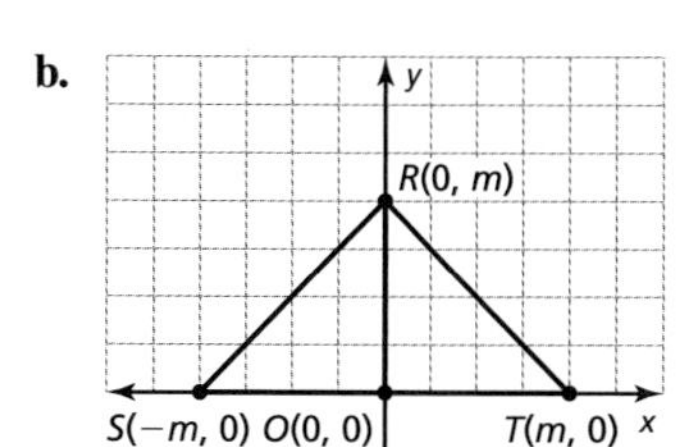

When any two congruent right isosceles triangles are positioned with the vertex opposite the hypotenuse on the origin and their legs on the axes as shown in the diagram, a triangle is formed and the hypotenuses of the original triangles make up two sides of the new triangle. $SR = m\sqrt{2}$ and $TR = m\sqrt{2}$ so these two sides are the same length. So, by definition, $\triangle SRT$ is isosceles.

5.8 Maintaining Mathematical Proficiency *(p. 288)*

29. 34°

Chapter 5 Review *(pp. 290–294)*

1. acute isosceles **2.** 132° **3.** 90° **4.** 42°, 48°

5. 35°, 55°

6. corresponding sides: $\overline{GH} \cong \overline{LM}$, $\overline{HJ} \cong \overline{MN}$, $\overline{JK} \cong \overline{NP}$, and $\overline{GK} \cong \overline{LP}$; corresponding angles: $\angle G \cong \angle L$, $\angle H \cong \angle M$, $\angle J \cong \angle N$, and $\angle K \cong \angle P$; *Sample answer:* $JHGK \cong NMLP$

7. 16°

8. no; There are two pairs of congruent sides and one pair of congruent angles, but the angles are not the included angles.

9. yes;

STATEMENTS	REASONS
1. $\overline{WX} \cong \overline{YZ}$, $\overline{WZ} \parallel \overline{YX}$	**1.** Given
2. $\overline{XZ} \cong \overline{XZ}$	**2.** Reflexive Property of Congruence (Thm. 2.1)
3. $\angle WXZ \cong \angle YZX$	**3.** Alternate Interior Angles Theorem (Thm. 3.2)
4. $\triangle WXZ \cong \triangle YZX$	**4.** SAS Congruence Theorem (Thm. 5.5)

10. P; PRQ **11.** $\overline{TR}$; $\overline{TV}$ **12.** RQS; RSQ

13. $\overline{SR}$; $\overline{SV}$ **14.** $x = 15, y = 5$

15. no; There is only enough information to conclude that two pairs of sides are congruent.

16. yes;

STATEMENTS	REASONS
1. $\overline{WX} \cong \overline{YZ}$, $\angle XWZ$ and $\angle ZYX$ are right angles.	**1.** Given
2. $\overline{XZ} \cong \overline{XZ}$	**2.** Reflexive Property of Congruence (Thm. 2.1)
3. $\triangle WXZ$ and $\triangle YZX$ are right triangles.	**3.** Definition of a right triangle
4. $\triangle WXZ \cong \triangle YZX$	**4.** HL Congruence Theorem (Thm. 5.9)

17. yes;

STATEMENTS	REASONS
1. $\angle E \cong \angle H$, $\angle F \cong \angle J$, $\overline{FG} \cong \overline{JK}$	**1.** Given
2. $\triangle EFG \cong \triangle HJK$	**2.** AAS Congruence Theorem (Thm. 5.11)

18. no; There is only enough information to conclude that one pair of angles and one pair of sides are congruent.

19. yes;

STATEMENTS	REASONS
1. $\angle PLN \cong \angle MLN$, $\angle PNL \cong \angle MNL$	**1.** Given
2. $\overline{LN} \cong \overline{LN}$	**2.** Reflexive Property of Congruence (Thm. 2.1)
3. $\triangle LPN \cong \triangle LMN$	**3.** ASA Congruence Theorem (Thm. 5.10)

20. no; There is only enough information to conclude that one pair of angles and one pair of sides are congruent.

21. By the SAS Congruence Theorem (Thm. 5.5), $\triangle HJK \cong \triangle LMN$. Because corresponding parts of congruent triangles are congruent, $\angle K \cong \angle N$.

22. First, state that $\overline{QV} \cong \overline{QV}$. Then, use the SSS Congruence Theorem (Thm. 5.8) to prove that $\triangle QSV \cong \triangle QTV$. Because corresponding parts of congruent triangles are congruent, $\angle QSV \cong \angle QTV$. $\angle QSV \cong \angle 1$ and $\angle QTV \cong \angle 2$ by the Vertical Angles Congruence Theorem (Thm. 2.6). So, by the Transitive Property of Congruence (Thm. 2.2), $\angle 1 \cong \angle 2$.

23. Using the Distance Formula, $OP = \sqrt{h^2 + k^2}$, $QR = \sqrt{h^2 + k^2}$, $OR = j$, and $QP = j$. So, $\overline{OP} \cong \overline{QR}$ and $\overline{OR} \cong \overline{QP}$. Also, by the Reflexive Property of Congruence (Thm. 2.1), $\overline{QO} \cong \overline{QO}$. So, you can apply the SSS Congruence Theorem (Thm. 5.8) to conclude that $\triangle OPQ \cong \triangle QRO$.

24.

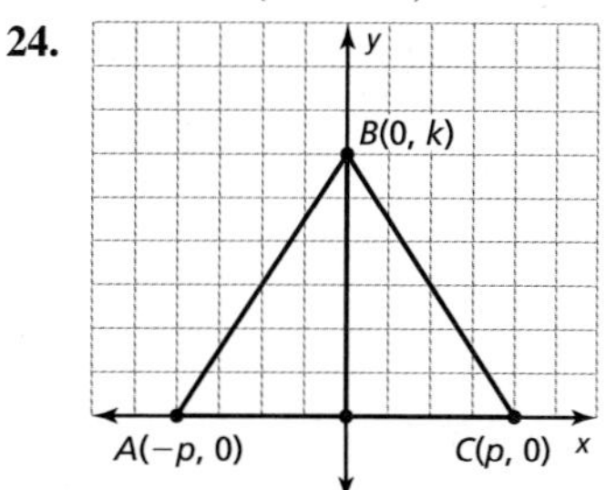

25. $(2k, k)$

Chapter 6

Chapter 6 Maintaining Mathematical Proficiency *(p. 299)*

1. $y = -3x + 10$ **2.** $y = x - 7$ **3.** $y = \frac{1}{4}x - \frac{7}{4}$

4. $-3 \leq w \leq 8$ **5.** $0 < m < 11$ **6.** $s \leq 5$ *or* $s > 2$

7. $d < 12$ *or* $d \geq -7$

8. yes; As with Exercises 6 and 7, if the graphs of the two inequalities overlap going in opposite directions and the variable only has to make one or the other true, then every number on the number line makes the compound inequality true.

6.1 Vocabulary and Core Concept Check *(p. 306)*

1. bisector

6.1 Monitoring Progress and Modeling with Mathematics *(pp. 306–308)*

3. 4.6; Because $GK = KJ$ and $\overleftrightarrow{HK} \perp \overleftrightarrow{GJ}$, point H is on the perpendicular bisector of $\overline{GJ}$. So, by the Perpendicular Bisector Theorem (Thm. 6.1), $GH = HJ = 4.6$.

5. 15; Because $\overleftrightarrow{DB} \perp \overleftrightarrow{AC}$ and point D is equidistant from A and C, point D is on the perpendicular bisector of $\overline{AC}$ by the Converse of the Perpendicular Bisector Theorem (Thm. 6.2). By definition of segment bisector, $AB = BC$. So, $5x = 4x + 3$, and the solution is $x = 3$. So, $AB = 5x = 5(3) = 15$.

7. yes; Because point N is equidistant from L and M, point N is on the perpendicular bisector of $\overline{LM}$ by the Converse of the Perpendicular Bisector Theorem (Thm. 6.2). Because only one line can be perpendicular to $\overline{LM}$ at point K, $\overleftrightarrow{NK}$ must be the perpendicular bisector of $\overline{LM}$, and P is on $\overleftrightarrow{NK}$.

9. no; You would need to know that $\overleftrightarrow{PN} \perp \overleftrightarrow{ML}$.

11. 20°; Because D is equidistant from $\overrightarrow{BC}$ and $\overrightarrow{BA}$, $\overrightarrow{BD}$ bisects $\angle ABC$ by the Converse of the Angle Bisector Theorem (Thm. 6.4). So, $m\angle ABD = m\angle CBD = 20°$.

13. 28°; Because L is equidistant from $\overrightarrow{JK}$ and $\overrightarrow{JM}$, $\overrightarrow{JL}$ bisects $\angle KJM$ by the Angle Bisector Theorem (Thm. 6.3). This means that $7x = 3x + 16$, and the solution is $x = 4$. So, $m\angle KJL = 7x = 7(4) = 28°$.

15. yes; Because H is equidistant from $\overrightarrow{EF}$ and $\overrightarrow{EG}$, $\overrightarrow{EH}$ bisects $\angle FEG$ by the Angle Bisector Theorem (Thm. 6.3).

17. no; Because neither $\overline{BD}$ nor $\overline{DC}$ are marked as perpendicular to $\overrightarrow{AB}$ or $\overrightarrow{AC}$ respectively, you cannot conclude that $DB = DC$.

19. $y = x - 2$ **21.** $y = -3x + 15$

23. Because $\overline{DC}$ is not necessarily congruent to $\overline{EC}$, $\overleftrightarrow{AB}$ will not necessarily pass through point C; Because $AD = AE$, and $\overleftrightarrow{AB} \perp \overline{DE}$, $\overleftrightarrow{AB}$ is the perpendicular bisector of $\overline{DE}$.

25. Perpendicular Bisector Theorem (Thm. 6.1)

27.

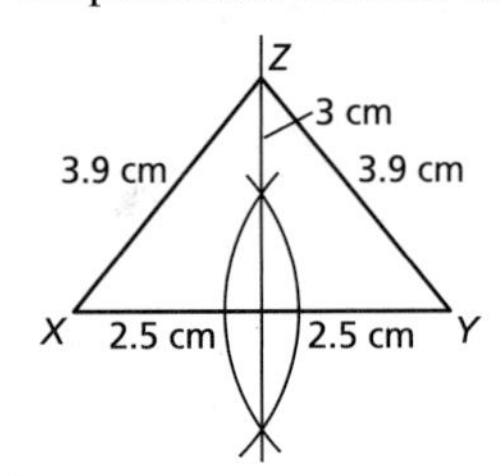

Perpendicular Bisector Theorem (Thm. 6.1)

29. B

31. no; If the triangle is an isosceles triangle, then the angle bisector of the vertex angle will also be the perpendicular bisector of the base.

33. a.

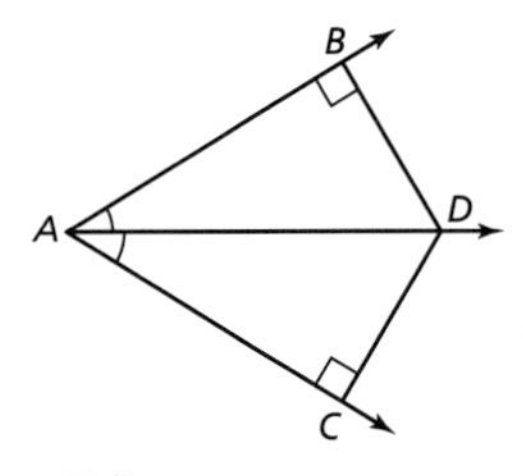

If $\overrightarrow{AD}$ bisects $\angle BAC$, then by definition of angle bisector, $\angle BAD \cong \angle CAD$. Also, because $\overline{DB} \perp \overrightarrow{AB}$ and $\overline{DC} \perp \overrightarrow{AC}$, by definition of perpendicular lines, $\angle ABD$ and $\angle ACD$ are right angles, and congruent to each other by the Right Angles Congruence Theorem (Thm. 2.3). Also, $\overline{AD} \cong \overline{AD}$ by the Reflexive Property of Congruence (Thm. 2.1). So, by the AAS Congruence Theorem (Thm. 5.11), $\triangle ADB \cong \triangle ADC$. Because corresponding parts of congruent triangles are congruent, $DB = DC$. This means that point D is equidistant from each side of $\angle BAC$.

b.

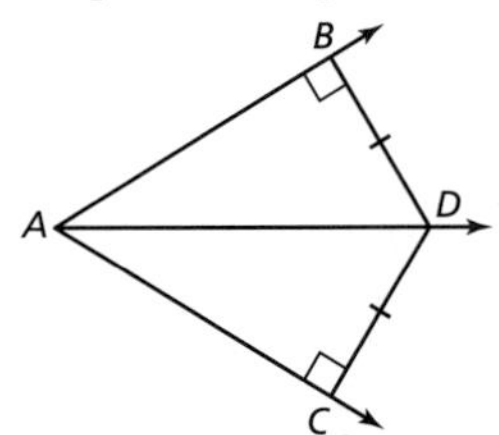

STATEMENTS	REASONS
1. $\overline{DC} \perp \overrightarrow{AC}$, $\overline{DB} \perp \overrightarrow{AB}$, $BD = CD$	**1.** Given
2. $\angle ABD$ and $\angle ACD$ are right angles.	**2.** Definition of perpendicular lines
3. $\triangle ABD$ and $\triangle ACD$ are right triangles.	**3.** Definition of a right triangle
4. $\overline{BD} \cong \overline{CD}$	**4.** Definition of congruent segments
5. $\overline{AD} \cong \overline{AD}$	**5.** Reflexive Property of Congruence (Thm. 2.1)
6. $\triangle ABD \cong \triangle ACD$	**6.** HL Congruence Theorem (Thm. 5.9)
7. $\angle BAD \cong \angle CAD$	**7.** Corresponding parts of congruent triangles are congruent.
8. $\overrightarrow{AD}$ bisects $\angle BAC$.	**8.** Definition of angle bisector

35. a. $y = x$ **b.** $y = -x$ **c.** $y = |x|$

37. Because $\overline{AD} \cong \overline{CD}$ and $\overline{AE} \cong \overline{CE}$, by the Converse of the Perpendicular Bisector Theorem (Thm. 6.2), both points D and E are on the perpendicular bisector of $\overline{AC}$. So, $\overleftrightarrow{DE}$ is the perpendicular bisector of $\overline{AC}$. So, if $\overline{AB} \cong \overline{CB}$, then by the Converse of the Perpendicular Bisector Theorem (Thm. 6.2), point B is also on $\overleftrightarrow{DE}$. So, points D, E, and B are collinear. Conversely, if points D, E, and B are collinear, then by the Perpendicular Bisector Theorem (Thm. 6.2), point B is also on the perpendicular bisector of AC. So, $\overline{AB} \cong \overline{CB}$.

6.1 Maintaining Mathematical Proficiency *(p. 308)*

39. isosceles **41.** equilateral **43.** right

6.2 Vocabulary and Core Concept Check *(p. 315)*

1. concurrent

6.2 Monitoring Progress and Modeling with Mathematics *(pp. 315–318)*

3. 9 **5.** 9 **7.** (5, 8) **9.** (−4, 9) **11.** 16

13. 6 **15.** 32

17. *Sample answer:*

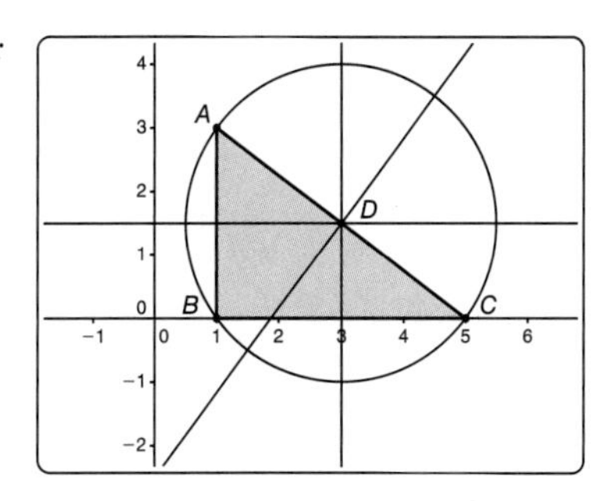

19. *Sample answer:*

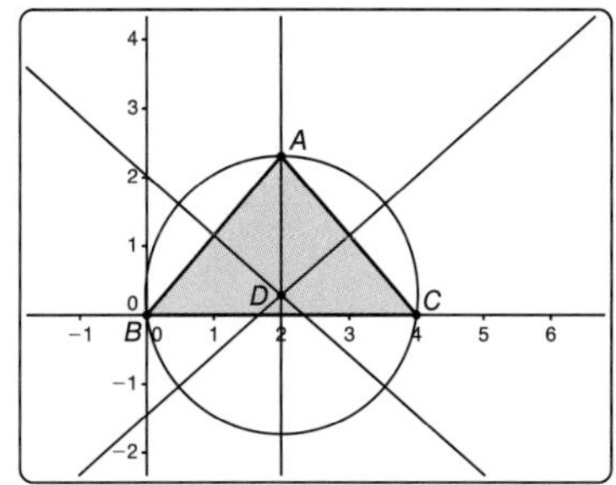

21. *Sample answer:*

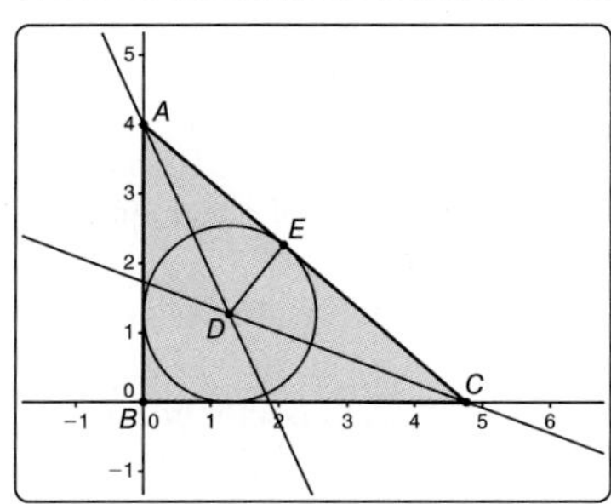

23. *Sample answer:*

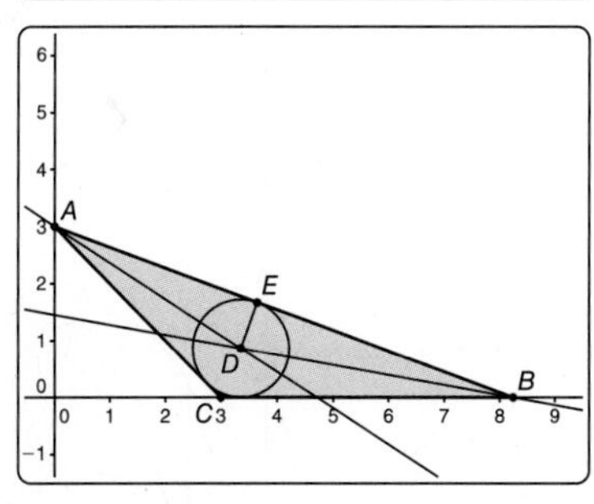

25. Because point G is the intersection of the angle bisectors, it is the incenter. But, because $\overline{GD}$ and $\overline{GF}$ are not necessarily perpendicular to a side of the triangle, there is not sufficient evidence to conclude that $\overline{GD}$ and $\overline{GF}$ are congruent; Point G is equidistant from the sides of the triangle.

27. You could copy the positions of the three houses, and connect the points to draw a triangle. Then draw the three perpendicular bisectors of the triangle. The point where the perpendicular bisectors meet, the circumcenter, should be the location of the meeting place.

29. sometimes; If the scalene triangle is obtuse or right, then the circumcenter is outside or on the triangle, respectively. However, if the scalene triangle is acute, then the circumcenter is inside the triangle.

31. sometimes; This only happens when the triangle is equilateral.

33. $\left(\frac{35}{6}, -\frac{11}{6}\right)$ **35.** $x = 6$

37. The circumcenter of any right triangle is located at the midpoint of the hypotenuse of the triangle.

Let $A(0, 2b)$, $B(0, 0)$, and $C(2a, 0)$ represent the vertices of a right triangle where $\angle B$ is the right angle. The midpoint of $\overline{AB}$ is $M_{\overline{AB}}(0, b)$. The midpoint of $\overline{BC}$ is $M_{\overline{BC}}(a, 0)$. The midpoint of $\overline{AC}$ is $M_{\overline{AC}}(a, b)$. Because $\overline{AB}$ is vertical, its perpendicular b isector is horizontal. So, the equation of the horizontal line passing through $M_{\overline{AB}}(0, b)$ is $y = b$. Because $\overline{BC}$ is horizontal, its perpendicular bisector is vertical. So, the equation of the vertical line passing through $M_{\overline{BC}}(a, 0)$ is $x = a$. The circumcenter of $\triangle ABC$ is the intersection of perpendicular bisectors, $y = b$ and $x = a$, which is (a, b). This point is also the midpoint of $\overline{AC}$.

39. The circumcenter is the point of intersection of the perpendicular bisectors of the sides of a triangle, and it is equidistant from the vertices of the triangle. In contrast, the incenter is the point of intersection of the angle bisectors of a triangle, and it is equidistant from the sides of the triangle.

41. a.

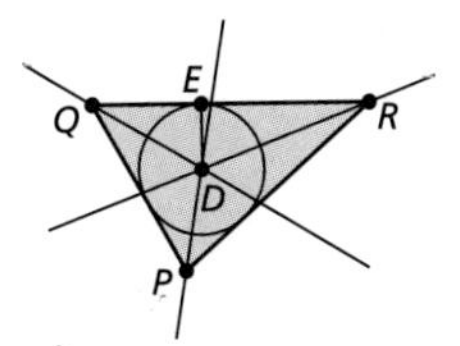

Because this circle is inscribed in the triangle, it is the largest circle that fits inside the triangle without extending into the boundaries.

b. yes; You would keep the center of the pool as the incenter of the triangle, but you would make the radius of the pool at least 1 foot shorter.

43. B

45. yes; In an equilateral triangle, each perpendicular bisector passes through the opposite vertex and divides the triangle into two congruent triangles. So, it is also an angle bisector.

47. a. equilateral; 3; In an equilateral triangle, each perpendicular bisector also bisects the opposite angle.

b. scalene; 6; In a scalene triangle, none of the perpendicular bisectors will also bisect an angle.

49. angle bisectors; about 2.83 in.

51. $x = \dfrac{AB + AC - BC}{2}$ or $x = \dfrac{AB \cdot AC}{AB + AC + BC}$

6.2 Maintaining Mathematical Proficiency *(p. 318)*

53. $M(6, 3)$; $AB \approx 11.3$ **55.** $M(-1, 7)$; $AB \approx 12.6$

57. $x = 6$

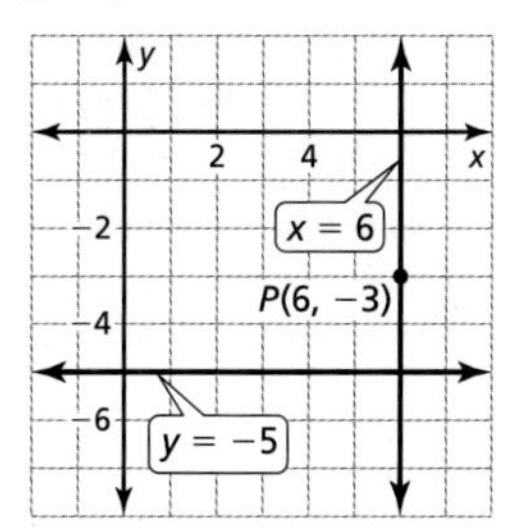

59. $y = \frac{1}{4}x + 2$

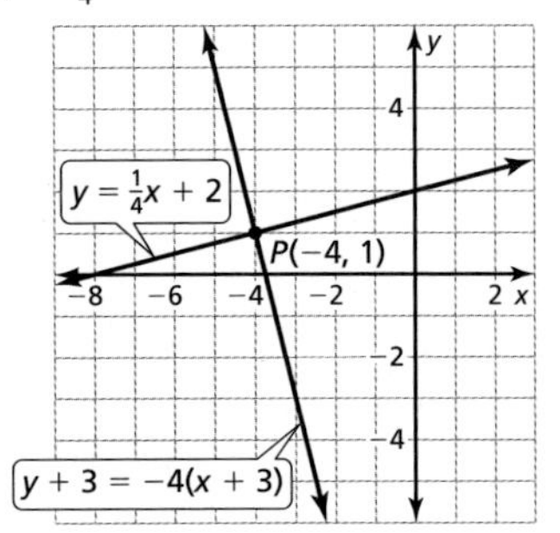

6.3 Vocabulary and Core Concept Check *(p. 324)*

1. circumcenter, incenter, centroid, orthocenter; perpendicular bisectors form the circumcenter, angle bisectors form the incenter, medians form the centroid, altitudes form the orthocenter

6.3 Monitoring Progress and Modeling with Mathematics *(pp. 324–326)*

3. 6, 3 **5.** 20, 10 **7.** 10, 15 **9.** 18, 27 **11.** 12 **13.** 10 **15.** $\left(5, \frac{11}{3}\right)$ **17.** (5, 1) **19.** outside; (0, −5) **21.** inside; (−1, 2)

23.

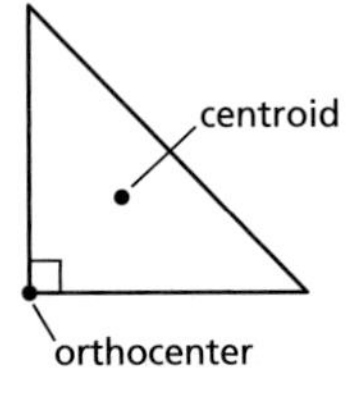

25. *Sample answer:*

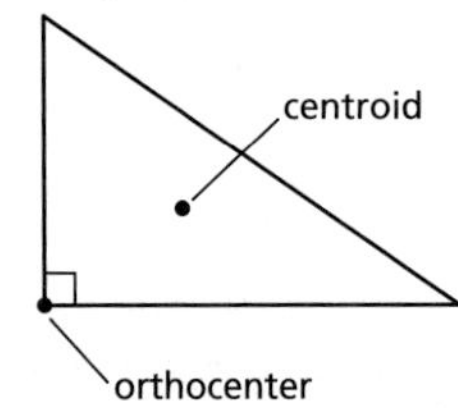

27. The length of $\overline{DE}$ should be $\frac{1}{3}$ of the length of $\overline{AE}$ because it is the shorter segment from the centroid to the side;

$DE = \frac{1}{3}AE$

$DE = \frac{1}{3}(18)$

$DE = 6$

29.

Legs $\overline{AB}$ and $\overline{BC}$ of isosceles $\triangle ABC$ are congruent. $\angle ABD \cong \angle CBD$ because $\overline{BD}$ is an angle bisector of vertex angle ABC. Also, $\overline{BD} \cong \overline{BD}$ by the Reflexive Property ofCongruence (Thm. 2.1). So, $\triangle ABD \cong \triangle CBD$ by the SAS Congruence Theorem (Thm. 5.5). $\overline{AD} \cong \overline{CD}$ because corresponding parts of congruent triangles are congruent. So, $\overline{BD}$ is a median.

31. never; Because medians are always inside a triangle, and the centroid is the point of concurrency of the medians, it will always be inside the triangle.

33. sometimes; A median is the same line segment as the perpendicular bisector if the triangle is equilateral or if the segment is connecting the vertex angle to the base of an isosceles triangle. Otherwise, the median and the perpendicular bisectors are not the same segment.

35. sometimes; The centroid and the orthocenter are not the same point unless the triangle is equilateral.

37. Both segments are perpendicular to a side of a triangle, and their point of intersection can fall either inside, on, or outside of the triangle. However, the altitude does not necessarily bisect the side, but the perpendicular bisector does. Also, the perpendicular bisector does not necessarily pass through the opposite vertex, but the altitude does.

39. 6.75 in.^{2}; altitude **41.** $x = 2.5$ **43.** $x = 4$

45.

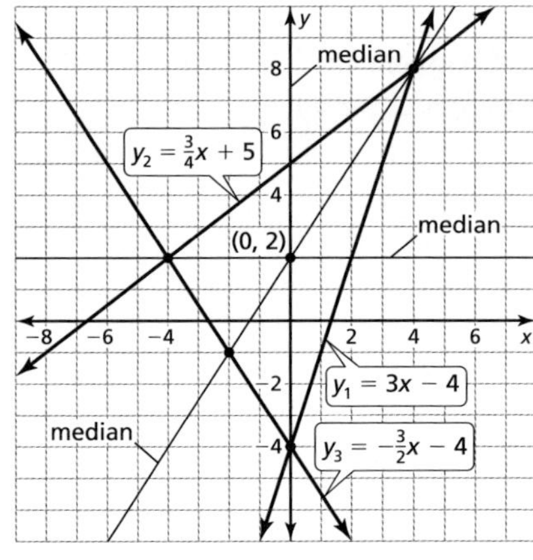

(0, 2)

47. $PE = \frac{1}{3}AE$, $PE = \frac{1}{2}AP$, $PE = AE - AP$

49. yes; If the triangle is equilateral, then the perpendicular bisectors, angle bisectors, medians, and altitudes will all be the same three segments.

51.

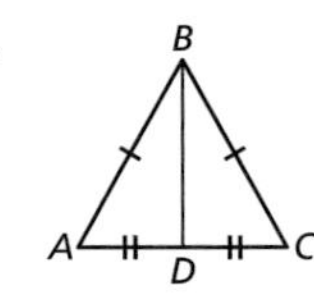

Sides $\overline{AB}$ and $\overline{BC}$ of equilateral $\triangle ABC$ are congruent. $\overline{AD} \cong \overline{CD}$ because $\overline{BD}$ is the median to $\overline{AC}$. Also, $\overline{BD} \cong \overline{BD}$ by the Reflexive Property of Congruence (Thm. 2.1). So, $\triangle ABD \cong \triangle CBD$ by the SSS Congruence Theorem (Thm. 5.8). $\angle ADB \cong \angle CDB$ and $\angle ABD \cong \angle CBD$ because corresponding parts of congruent triangles are congruent. Also, $\angle ADB$ and $\angle CDB$ are a linear pair. Because $\overline{BD}$ and $\overline{AC}$ intersect to form a linear pair of congruent angles, $\overline{BD} \perp \overline{AC}$. So, median $\overline{BD}$ is also an angle bisector, altitude, and perpendicular bisector of $\triangle ABC$.

53. *Sample answer:*

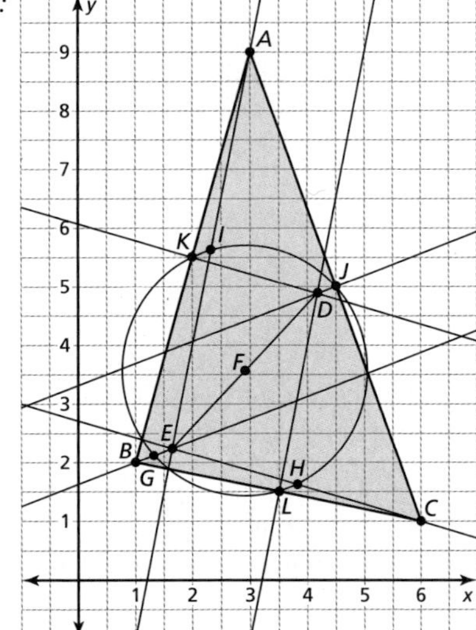

The circle passes through nine significant points of the triangle. They are the midpoints of the sides, the midpoints between each vertex and the orthocenter, and the points of intersection between the sides and the altitudes.

6.3 Maintaining Mathematical Proficiency *(p. 326)*

55. yes **57.** no

6.4 Vocabulary and Core Concept Check *(p. 333)*

1. midsegment

6.4 Monitoring Progress and Modeling with Mathematics *(pp. 333–334)*

3. $D(-4, -2)$, $E(-2, 0)$, $F(-1, -4)$

5. Because the slopes of $\overline{EF}$ and $\overline{AC}$ are the same (-4), $\overline{EF} \parallel \overline{AC}$. $EF = \sqrt{17}$ and $AC = 2\sqrt{17}$. Because $\sqrt{17} = \frac{1}{2}(2\sqrt{17})$, $EF = \frac{1}{2}AC$.

7. $x = 13$ 9. $x = 6$ 11. $\overline{JK} \parallel \overline{YZ}$ 13. $\overline{XY} \parallel \overline{KL}$

15. $\overline{JL} \cong \overline{XK} \cong \overline{KZ}$ 17. 14 19. 17 21. 45 ft

23. An eighth segment, $\overline{FG}$,would connect the midpoints of $\overline{DL}$ and $\overline{EN}$; $\overline{DE} \parallel \overline{LN} \parallel \overline{FG}$, $DE = \frac{3}{4}LN$, and $FG = \frac{7}{8}LN$; Because you are finding quarter segments and eighth segments, use $8p$, $8q$, and $8r$: $L(0, 0)$, $M(8q, 8r)$, and $N(8p, 0)$. Find the coordinates of X, Y, D, E, F, and G. $X(4q, 4r)$, $Y(4q + 4p, 4r)$, $D(2q, 2r)$, $E(2q + 6p, 2r)$, $F(q, r)$, and $G(q + 7p, r)$.
 The y-coordinates of D and E are the same, so $\overline{DE}$ has a slope of 0. The y-coordinates of F and G are also the same, so $\overline{FG}$ also has a slope of 0. $\overline{LM}$ is on the x-axis, so its slope is 0. Because their slopes are the same, $\overline{DE} \parallel \overline{LM} \parallel \overline{FG}$.
 Use the Ruler Postulate (Post. 1.1) to find DE, FG, and LM.
 $DE = 6p$, $FG = 7p$, and $LN = 8p$.
 Because $6p = \frac{3}{4}(8p)$, $DE = \frac{3}{4}LN$. Because $7p = \frac{7}{8}(8p)$, $FG = \frac{7}{8}LN$.

25. a. 24 units b. 60 units c. 114 units

27. After graphing the midsegments, find the slope of each segment. Graph the line parallel to each midsegment passing through the opposite vertex. The intersections of these three lines will be the vertices of the original triangle: $(-1, 2)$, $(9, 8)$, and $(5, 0)$.

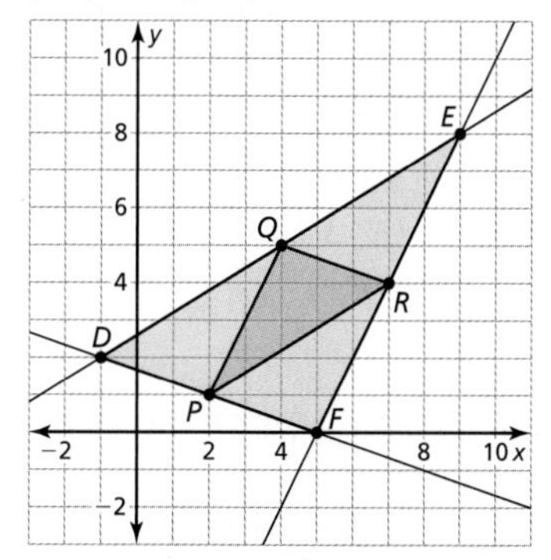

6.4 Maintaining Mathematical Proficiency *(p. 334)*

29. *Sample answer:* An isosceles triangle whose sides are 5 centimeters, 5 centimeters, and 3 centimeters is not equilateral.

6.5 Vocabulary and Core Concept Check *(p. 340)*

1. In an indirect proof, rather than proving a statement directly, you show that when the statement is false, it leads to a contradiction.

6.5 Monitoring Progress and Modeling with Mathematics *(pp. 340–342)*

3. Assume temporarily that $WV = 7$ inches.

5. Assume temporarily that $\angle B$ is a right angle.

7. A and C; The angles of an equilateral triangle are always 60°. So, an equilateral triangle cannot have a 90° angle, and cannot be a right triangle.

9.

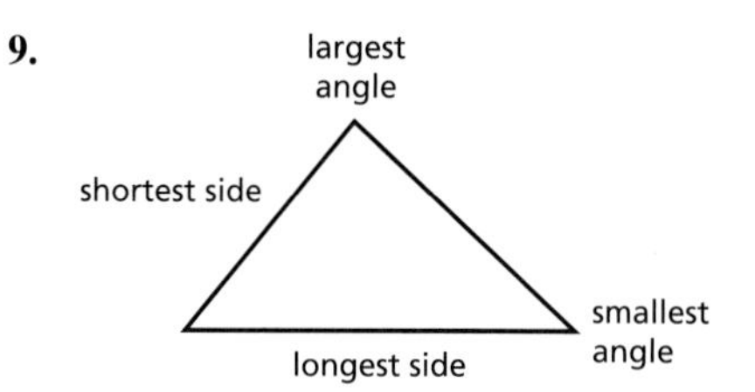

 The longest side is across from the largest angle, and the shortest side is across from the smallest angle.

11. $\angle S$, $\angle R$, $\angle T$ 13. $\overline{AB}$, $\overline{BC}$, $\overline{AC}$ 15. $\overline{NP}$, $\overline{MN}$, $\overline{MP}$

17. 7 in. $< x <$ 17 in. 19. 16 in. $< x <$ 64 in. 21. yes

23. no; $28 + 17 \not> 46$

25. An angle that is not obtuse could be acute or right; Assume temporarily that $\angle A$ is not obtuse.

27. Assume temporarily that the client is guilty. Then the client would have been in Los Angeles, California at the time of the crime. Because the client was in New York at the time of the crime, the assumption must be false, and the client must be innocent.

29. C

31. Assume temporarily that an odd number is divisible by 4. Let the odd number be represented by $2y + 1$ where y is a positive integer. Then, there must be a positive integer x such that $4x = 2y + 1$. However, when you divide each side of the equation by 4, you get $x = \frac{1}{2}y + \frac{1}{4}$, which is not an integer. So, the assumption must be false, and an odd number is not divisible by 4.

33. The right angle of a right triangle must always be the largest angle because the other two will have a sum of 90°. So, according to the Triangle Longer Angle Theorem (Thm. 6.10), because the right angle is larger than either of the other angles, the side opposite the right angle, which is the hypotenuse, will always have to be longer than either of the legs.

35. a. The width of the river must be greater than 35 yards and less than 50 yards. In $\triangle BCA$, the width of the river, $\overline{BA}$, must be less than the length of $\overline{CA}$, which is 50 yards, because the measure of the angle opposite $\overline{BA}$ is less than the measure of the angle opposite $\overline{CA}$, which must be 50°. In $\triangle BDA$, the width of the river, $\overline{BA}$, must be greater than the length of $\overline{DA}$, which is 35 yards, because the measure of the angle opposite $\overline{BA}$ is greater than the measure of the angle opposite $\overline{DA}$, which must be 40°.
 b. You could measure from distances that are closer together. In order to do this, you would have to use angle measures that are closer to 45°.

37. $\angle WXY$, $\angle Z$, $\angle YXZ$, $\angle WYX$ and $\angle XYZ$, $\angle W$; In $\triangle WXY$, because $WY < WX < YX$, by the Triangle Longer Side Theorem (Thm. 6.9), $m\angle WXY < m\angle WYX < m\angle W$. Similarly, in $\triangle XYZ$, because $XY < YZ < XZ$, by the Triangle Longer Side Theorem (Thm. 6.9), $m\angle Z < m\angle YXZ < m\angle XYZ$. Because $m\angle WYX = m\angle XYZ$ and $\angle W$ is the only angle greater than either of them, we know that $\angle W$ is the largest angle. Because $\triangle WXY$ has the largest angle and one of the congruent angles, the remaining angle, $\angle WXY$, is the smallest.

39. By the Exterior Angle Theorem (Thm. 5.2), $m\angle 1 = m\angle A + m\angle B$. Then by the Subtraction Property of Equality, $m\angle 1 - m\angle B = m\angle A$. If you assume temporarily that $m\angle 1 \leq m\angle B$, then $m\angle A \leq 0$. Because the measure of any angle in a triangle must be a positive number, the assumption must be false. So, $m\angle 1 > m\angle B$. Similarly, by the Subtraction Property of Equality, $m\angle 1 - m\angle A = m\angle B$. If you assume temporarily that $m\angle 1 \leq m\angle A$, then $m\angle B \leq 0$. Because the measure of any angle in a triangle must be a positive number, the assumption must be false. So, $m\angle 1 > m\angle A$.

41. $2\frac{1}{7} < x < 13$

43. It is given that $BC > AB$ and $BD = BA$. By the Base Angles Theorem (Thm. 5.6), $m\angle 1 = m\angle 2$. By the Angle Addition Postulate (Post. 1.4), $m\angle BAC = m\angle 1 + m\angle 3$. So, $m\angle BAC > m\angle 1$. Substituting $m\angle 2$ for $m\angle 1$ produces $m\angle BAC > m\angle 2$. By the Exterior Angle Theorem (Thm. 5.2), $m\angle 2 = m\angle 3 + m\angle C$. So, $m\angle 2 > m\angle C$. Finally, because $m\angle BAC > m\angle 2$ and $m\angle 2 > m\angle C$, you can conclude that $m\angle BAC > m\angle C$.

45. no; The sum of the other two sides would be 11 inches, which is less than 13 inches.

47.

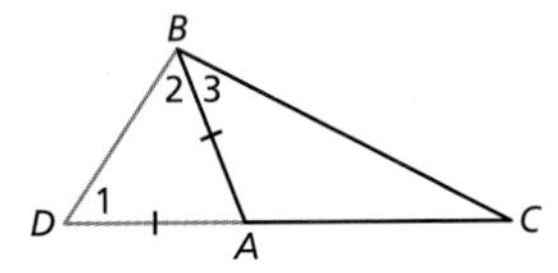

Assume $\overline{BC}$ is longer than or the same length as each of the other sides, $\overline{AB}$ and $\overline{AC}$. Then, $AB + BC > AC$ and $AC + BC > AB$. The proof for $AB + AC > BC$ follows.

STATEMENTS	REASONS
1. $\triangle ABC$	1. Given
2. Extend $\overline{AC}$ to D so that $\overline{AB} \cong \overline{AD}$.	2. Ruler Postulate (Post. 1.1)
3. $AB = AD$	3. Definition of segment congruence
4. $AD + AC = DC$	4. Segment Addition Postulate (Post. 1.2)
5. $\angle 1 \cong \angle 2$	5. Base Angles Theorem (Thm. 5.6)
6. $m\angle 1 = m\angle 2$	6. Definition of angle congruence
7. $m\angle DBC > m\angle 2$	7. Protractor Postulate (Post. 1.3)
8. $m\angle DBC > m\angle 1$	8. Substitution Property
9. $DC > BC$	9. Triangle Larger Angle Theorem (Thm. 6.10)
10. $AD + AC > BC$	10. Substitution Property
11. $AB + AC > BC$	11. Substitution Property

49. Assume temporarily that another segment, $\overline{PA}$, where A is on plane M, is the shortest segment from P to plane M. By definition of the distance between a point and a plane, $\overline{PA} \perp$ plane M. This contradicts the given statement because there cannot be two different segments that share an endpoint and are both perpendicular to the same plane. So, the assumption is false, and because no other segment exists that is the shortest segment from P to plane M, it must be $\overline{PC}$ that is the shortest segment from P to plane M.

6.5 Maintaining Mathematical Proficiency *(p. 342)*

51. $\angle ACD$ **53.** $\angle CEB$

6.6 Vocabulary and Core Concept Check *(p. 347)*

1. Theorem 6.12 refers to two angles with two pairs of sides that have the same measure, just like two hinges whose sides are the same length. Then, the angle whose measure is greater is opposite a longer side, just like the ends of a hinge are farther apart when the hinge is open wider.

6.6 Monitoring Progress and Modeling with Mathematics *(pp. 347–348)*

3. $m\angle 1 > m\angle 2$; By the Converse of the Hinge Theorem (Thm. 6.13), because $\angle 1$ is the included angle in the triangle with the longer third side, its measure is greater than that of $\angle 2$.

5. $m\angle 1 = m\angle 2$; The triangles are congruent by the SSS Congruence Theorem (Thm. 5.8). So, $\angle 1 \cong \angle 2$ because corresponding parts of congruent triangles are congruent.

7. $AD > CD$; By the Hinge Theorem (Thm. 6.12), because $\overline{AD}$ is the third side of the triangle with the larger included angle, it is longer than $\overline{CD}$.

9. $TR < UR$; By the Hinge Theorem (Thm. 6.12), because $\overline{TR}$ is the third side of the triangle with the smaller included angle, it is shorter than $\overline{UR}$.

11. $\overline{XY} \cong \overline{YZ}$ and $m\angle WYZ > m\angle WYX$ are given. By the Reflexive Property of Congruence (Thm. 2.1), $\overline{WY} \cong \overline{WY}$. So, by the Hinge Theorem (Thm. 6.12), $WZ > WX$.

13. your flight; Because $160° > 150°$, the distance you flew is a greater distance than the distance your friend flew by the Hinge Theorem (Thm. 6.12).

15. The measure of the included angle in $\triangle PSQ$ is greater than the measure of the included angle in $\triangle SQR$; By the Hinge Theorem (Thm. 6.12), $PQ > SR$.

17. $m\angle EGF > m\angle DGE$ by the Hinge Theorem (Thm. 6.12).

19. Because $\overline{NR}$ is a median, $\overline{PR} \cong \overline{QR}$. $\overline{NR} \cong \overline{NR}$ by the Reflexive Property of Congruence (Thm. 2.1). So, by the Converse of the Hinge Theorem (Thm. 6.13), $\angle NRQ > \angle NRP$. Because $\angle NRQ$ and $\angle NRP$ form a linear pair, they are supplementary. So, $\angle NRQ$ must be obtuse and $\angle NRP$ must be acute.

21. $x > \frac{3}{2}$

23. $\triangle ABC$ is an obtuse triangle; If the altitudes intersect inside the triangle, then $m\angle BAC$ will always be less than $m\angle BDC$ because they both intercept the same segment, $\overline{CD}$. However, because $m\angle BAC > m\angle BDC$, $\angle A$ must be obtuse, and the altitudes must intersect outside of the triangle.

6.6 Maintaining Mathematical Proficiency *(p. 348)*

25. $x = 38$ **27.** $x = 60$

Chapter 6 Review *(pp. 350–352)*

1. 20; Point B is equidistant from A and C, and $\overleftrightarrow{BD} \perp \overline{AC}$. So, by the Converse of the Perpendicular Bisector Theorem (Thm. 6.2), $DC = AD = 20$.
2. 23; $\angle PQS \cong \angle RQS$, $\overline{SR} \perp \overrightarrow{QR}$, and $\overline{SP} \perp \overrightarrow{QP}$. So, by the Angle Bisector Theorem (Thm. 6.3), $SR = SP$. This means that $6x + 5 = 9x - 4$, and the solution is $x = 3$. So, $RS = 9(3) - 4 = 23$.
3. 47°; Point J is equidistant from $\overrightarrow{FG}$ and $\overrightarrow{FH}$. So, by the Converse of the Angle Bisector Theorem (Thm. 6.4), $m\angle JFH = m\angle JFG = 47°$.
4. $(-3, -3)$ **5.** $(4, 3)$ **6.** $x = 5$ **7.** $(-6, 3)$
8. $(4, -4)$ **9.** inside; $(3, 5.2)$ **10.** outside; $(-6, -1)$
11. $(-6, 6), (-3, 6), (-3, 4)$ **12.** $(0, 3), (2, 0), (-1, -2)$
13. $4 \text{ in.} < x < 12 \text{ in.}$ **14.** $3 \text{ m} < x < 15 \text{ m}$
15. $7 \text{ ft} < x < 29 \text{ ft}$
16. Assume temporarily that $YZ \not> 4$. Then, it follows that either $YZ < 4$ or $YZ = 4$. If $YZ < 4$, then $XY + YZ < XZ$ because $4 + YZ < 8$ when $YZ < 4$. If $YZ = 4$, then $XY + YZ = XZ$ because $4 + 4 = 8$. Both conclusions contradict the Triangle Inequality Theorem (Thm. 6.11), which says that $XY + YZ > XZ$. So, the temporary assumption that $YZ \not> 4$ cannot be true. This proves that in $\triangle XYZ$, if $XY = 4$ and $XZ = 8$, then $YZ > 4$.
17. $QT > ST$ **18.** $m\angle QRT > m\angle SRT$

Chapter 7

Chapter 7 Maintaining Mathematical Proficiency *(p. 357)*

1. $x = 3$ **2.** $x = 4$ **3.** $x = 7$ **4.** $a \parallel b, c \perp d$
5. $a \parallel b, c \parallel d, a \perp c, a \perp d, b \perp c, b \perp d$ **6.** $b \parallel c, b \perp d, c \perp d$
7. You can follow the order of operations with all of the other operations in the equation and treat the operations in the expression separately.

7.1 Vocabulary and Core Concept Check *(p. 364)*

1. A segment connecting consecutive vertices is a side of the polygon, not a diagonal.

7.1 Monitoring Progress and Modeling with Mathematics *(pp. 364–366)*

3. 1260° **5.** 2520° **7.** hexagon **9.** 16-gon
11. $x = 64$ **13.** $x = 89$ **15.** $x = 70$ **17.** $x = 150$
19. $m\angle X = m\angle Y = 92°$ **21.** $m\angle X = m\angle Y = 100.5°$
23. $x = 111$ **25.** $x = 32$ **27.** 108°, 72°
29. 172°, 8°
31. The measure of one interior angle of a regular pentagon was found, but the exterior angle should be found by dividing 360° by the number of angles; $\frac{360°}{5} = 72°$
33. 120° **35.** $n = \dfrac{360}{180 - x}$ **37.** 15 **39.** 40
41. A, B; Solving the equation found in Exercise 35 for n yields a positive integer greater than or equal to 3 for A and B, but not for C and D.
43. In a quadrilateral, when all the diagonals from one vertex are drawn, the polygon is divided into two triangles. Because the sum of the measures of the interior angles of each triangle is 180°, the sum of the measures of the interior angles of the quadrilateral is $2 \cdot 180° = 360°$.
45. 21°, 21°, 21°, 21°, 138°, 138°
47. $(n - 2) \cdot 180°$; When diagonals are drawn from the vertex of the concave angle as shown, the polygon is divided into $n - 2$ triangles whose interior angle measures have the same total as the sum of the interior angle measures of the original polygon.

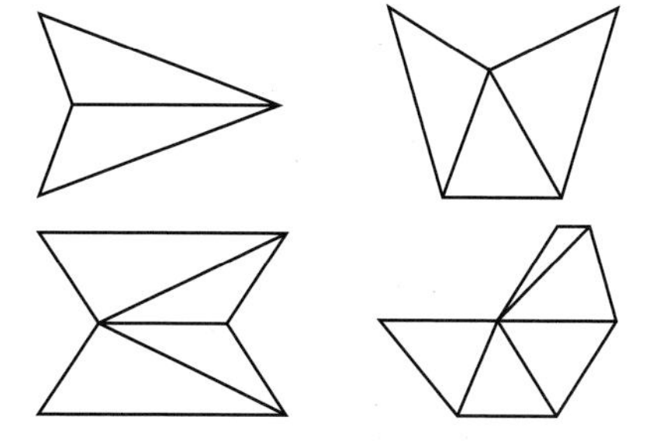

49. **a.** $h(n) = \dfrac{(n - 2) \cdot 180°}{n}$ **b.** $h(9) = 140°$ **c.** $n = 12$

d.

h(n)
(8, 135)
(7, 128.6)
(6, 120)
(5, 108)
(4, 90)
(3, 60)
140 120 100 80 60 40 20
1 2 3 4 5 6 7 8 n

The value of $h(n)$ increases on a curve that gets less steep as n increases.

51. In a convex n-gon, the sum of the measures of the n interior angles is $(n - 2) \cdot 180°$ using the Polygon Interior Angles Theorem (Thm. 7.1). Because each of the n interior angles forms a linear pair with its corresponding exterior angle, you know that the sum of the measures of the n interior and exterior angles is $180n°$. Subtracting the sum of the interior angle measures from the sum of the measures of the linear pairs gives you $180n° - [(n - 2) \cdot 180°] = 360°$.

7.1 Maintaining Mathematical Proficiency *(p. 366)*

53. $x = 101$ **55.** $x = 16$

7.2 Vocabulary and Core Concept Check *(p. 372)*

1. In order to be a quadrilateral, a polygon must have 4 sides, and parallelograms always have 4 sides. In order to be a parallelogram, a polygon must have 4 sides with opposite sides parallel. Quadrilaterals always have 4 sides, but do not always have opposite sides parallel.

7.2 Monitoring Progress and Modeling with Mathematics *(pp. 372–374)*

3. $x = 9, y = 15$ **5.** $d = 126, z = 28$ **7.** 129°
9. 13; By the Parallelogram Opposite Sides Theorem (Thm. 7.3), $LM = QN$.
11. 8; By the Parallelogram Opposite Sides Theorem (Thm. 7.3), $LQ = MN$.
13. 80°; By the Parallelogram Consecutive Angles Theorem (Thm. 7.5), $\angle QLM$ and $\angle LMN$ are supplementary. So, $m\angle LMN = 180° - 100°$.
15. 100°; By the Parallelogram Opposite Angles Theorem (Thm. 7.4), $m\angle QLM = m\angle MNQ$.
17. $m = 35, n = 110$ **19.** $k = 7, m = 8$

21. In a parallelogram, consecutive angles are supplementary; Because quadrilateral *STUV* is a parallelogram, $\angle S$ and $\angle V$ are supplementary. So, $m\angle V = 180° - 50° = 130°$.

23.

STATEMENTS	REASONS
1. *ABCD* and *CEFD* are parallelograms.	**1.** Given
2. $\overline{AB} \cong \overline{DC}$, $\overline{DC} \cong \overline{FE}$	**2.** Parallelogram Opposite Sides Theorem (Thm. 7.3)
3. $\overline{AB} \cong \overline{FE}$	**3.** Transitive Property of Congruence (Thm. 2.1)

25. (1, 2.5) **27.** *F*(3, 3) **29.** *G*(2, 0) **31.** 36°, 144°

33. no; *Sample answer:* $\angle A$ and $\angle C$ are opposite angles, but $m\angle A \neq m\angle C$.

35. *Sample Answer:*

When you fold the parallelogram so that vertex *A* is on vertex *C*, the fold will pass through the point where the diagonals intersect, which demonstrates that this point of intersection is also the midpoint of $\overline{AC}$. Similarly, when you fold the parallelogram so that vertex *B* is on vertex *D*, the fold will pass through the point where the diagonals intersect, which demonstrates that this point of intersection is also the midpoint of $\overline{BD}$.

37.

STATEMENTS	REASONS
1. *ABCD* is a parallelogram.	**1.** Given
2. $\overline{AB} \parallel \overline{DC}$, $\overline{BC} \parallel \overline{AD}$	**2.** Definition of parallelogram
3. $\angle BDA \cong \angle DBC$, $\angle DBA \cong \angle BDC$	**3.** Alternate Interior Angles Theorem (Thm. 3.2)
4. $\overline{BD} \cong \overline{BD}$	**4.** Reflexive Property of Congruence (Thm. 2.1)
5. $\triangle ABD \cong \triangle CDB$	**5.** ASA Congruence Theorem (Thm. 5.10)
6. $\angle A \cong \angle C$, $\angle B \cong \angle D$	**6.** Corresponding parts of congruent triangles are congruent.

39.

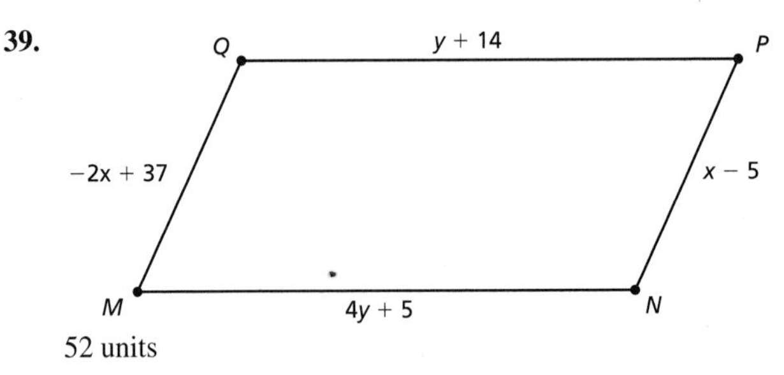

52 units

41. no; Two parallelograms with congruent corresponding sides may or may not have congruent corresponding angles.

43. 16° **45.** 3; (4, 0), (−2, 4), (8, 8)

47.

STATEMENTS	REASONS
1. $\overleftrightarrow{GH} \parallel \overleftrightarrow{JK} \parallel \overleftrightarrow{LM}$, $\overline{GJ} \cong \overline{JL}$	**1.** Given
2. Construct $\overline{PK}$ and $\overline{QM}$ such that $\overline{PK} \parallel \overleftrightarrow{GL} \parallel \overline{QM}$.	**2.** Construction
3. *GPKJ* and *JQML* are parallelograms.	**3.** Definition of parallelogram
4. $\angle GHK \cong \angle JKM$, $\angle PKQ \cong \angle QML$	**4.** Corresponding Angles Theorem (Thm. 3.1)
5. $\overline{GJ} \cong \overline{PK}$, $\overline{JL} \cong \overline{QM}$	**5.** Parallelogram Opposite Sides Theorem (Thm. 7.3)
6. $\overline{PK} \cong \overline{QM}$	**6.** Transitive Property of Congruence (Thm. 2.1)
7. $\angle HPK \cong \angle PKQ$, $\angle KQM \cong \angle QML$	**7.** Alternate Interior Angles Theorem (Thm. 3.2)
8. $\angle HPK \cong \angle QML$	**8.** Transitive Property of Congruence (Thm. 2.2)
9. $\angle HPK \cong \angle KQM$	**9.** Transitive Property of Congruence (Thm. 2.2)
10. $\triangle PHK \cong \triangle QKM$	**10.** AAS Congruence Theorem (Thm. 5.11)
11. $\overline{HK} \cong \overline{KM}$	**11.** Corresponding sides of congruent triangles are congruent.

7.2 Maintaining Mathematical Proficiency *(p. 374)*

49. yes; Alternate Exterior Angles Converse (Thm. 3.7)

7.3 Vocabulary and Core Concept Check *(p. 381)*

1. yes; If all four sides are congruent, then both pairs of opposite sides are congruent. So, the quadrilateral is a parallelogram by the Parallelogram Opposite Sides Converse (Thm. 7.7).

7.3 Monitoring Progress and Modeling with Mathematics *(pp. 381–384)*

3. Parallelogram Opposite Angles Converse (Thm. 7.8)

5. Parallelogram Diagonals Converse (Thm. 7.10)

7. Opposite Sides Parallel and Congruent Theorem (Thm. 7.9)

9. $x = 114, y = 66$ **11.** $x = 3, y = 4$ **13.** $x = 8$

15. $x = 7$

17.

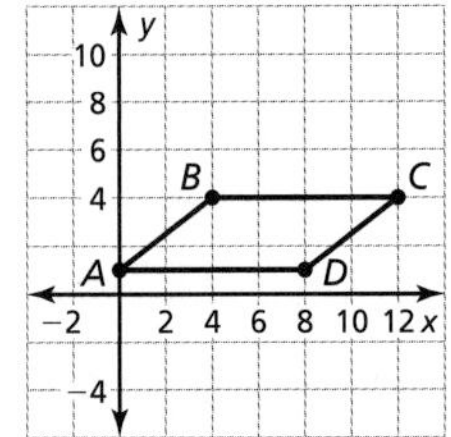

Because $BC = AD = 8$, $\overline{BC} \cong \overline{AD}$. Because both $\overline{BC}$ and $\overline{AD}$ are horizontal lines, their slope is 0, and they are parallel. $\overline{BC}$ and $\overline{AD}$ are opposite sides that are both congruent and parallel. So, *ABCD* is a parallelogram by the Opposite Sides Parallel and Congruent Theorem (Thm. 7.9).

19. 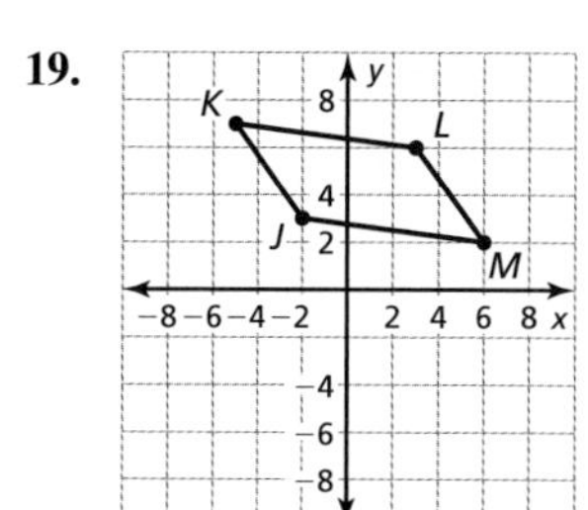

Because $JK = LM = 5$ and $KL = JM = \sqrt{65}$, $\overline{JK} \cong \overline{LM}$ and $\overline{KL} \cong \overline{JM}$. Because both pairs of opposite sides are congruent, quadrilateral *JKLM* is a parallelogram by the Parallelogram Opposite Sides Converse (Thm. 7.7).

21. In order to be a parallelogram, the quadrilateral must have two pairs of opposite sides that are congruent, not consecutive sides; *DEFG* is not a parallelogram.

23. $x = 5$; The diagonals must bisect each other so you could solve for x using either $2x + 1 = x + 6$ or $4x - 2 = 3x + 3$. Also, the opposite sides must be congruent, so you could solve for x using either $3x + 1 = 4x - 4$ or $3x + 10 = 5x$.

25. A quadrilateral is a parallelogram if and only if both pairs of opposite sides are congruent.

27. A quadrilateral is a parallelogram if and only if the diagonals bisect each other.

29. Check students' work; Because the diagonals bisect each other, this quadrilateral is a parallelogram by the Parallelogram Diagonals Converse (Thm. 7.10).

31. *Sample answer:*

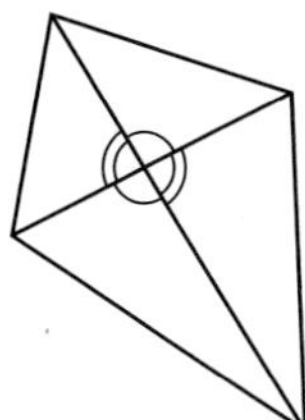

33. a. 27°; Because $\angle EAF$ is a right angle, the other two angles of $\triangle EAF$ must be complementary. So, $m\angle AFE = 90° - 63° = 27°$.

b. Because $\angle GDF$ is a right angle, the other two angles of $\triangle GDF$ must be complementary. So, $m\angle FGD = 90° - 27° = 63°$.

c. 27°; 27°

d. yes; $\angle HEF \cong \angle HGF$ because they both are adjacent to two congruent angles that together add up to 180°, and $\angle EHG \cong \angle GFE$ for the same reason. So, *EFGH* is a parallelogram by the Parallelogram Opposite Angles Converse (Thm. 7.8).

35. You can use the Alternate Interior Angles Converse (Thm. 3.6) to show that $\overline{AD} \parallel \overline{BC}$. Then, $\overline{AD}$ and $\overline{BC}$ are both congruent and parallel. So, *ABCD* is a parallelogram by the Opposite Sides Parallel and Congruent Theorem (Thm 7.9).

37. First, you can use the Linear Pair Postulate (Post. 2.8) and the Congruent Supplements Theorem (Thm. 2.4) to show that $\angle ABC$ and $\angle DCB$ are supplementary. Then, you can use the Consecutive Interior Angles Converse (Thm. 3.8) to show that $\overline{AB} \parallel \overline{DC}$ and $\overline{AD} \parallel \overline{BC}$. So, *ABCD* is a parallelogram by definition.

39.

STATEMENTS	REASONS
1. $\angle A \cong \angle C$, $\angle B \cong \angle D$	1. Given
2. Let $m\angle A = m\angle C = x°$ and $m\angle B = m\angle D = y°$	2. Definition of congruent angles
3. $m\angle A + m\angle B + m\angle C + m\angle D = x° + y° + x° + y° = 360°$	3. Corollary to the Polygon Interior Angles Theorem (Cor. 7.1)
4. $2(x°) + 2(y°) = 360°$	4. Simplify
5. $2(x° + y°) = 360°$	5. Distributive Property
6. $x° + y° = 180°$	6. Division Property of Equality
7. $m\angle A + m\angle B = 180°$, $m\angle A + m\angle D = 180°$	7. Substitution Property of Equality
8. $\angle A$ and $\angle B$ are supplementary. $\angle A$ and $\angle D$ are supplementary.	8. Definition of supplementary angles
9. $\overline{BC} \parallel \overline{AD}$, $\overline{AB} \parallel \overline{DC}$	9. Consecutive Interior Angles Converse (Thm. 3.8)
10. *ABCD* is a parallelogram.	10. Definition of parallelogram

41.

STATEMENTS	REASONS
1. Diagonals $\overline{JL}$ and $\overline{KM}$ bisect each other.	1. Given
2. $\overline{KP} \cong \overline{MP}$, $\overline{JP} \cong \overline{LP}$	2. Definition of segment bisector
3. $\angle KPL \cong \angle MPJ$	3. Reflexive Property of Congruence (Thm. 2.2)
4. $\triangle KPL \cong \triangle MPJ$	4. SAS Congruence Theorem (Thm. 5.5)
5. $\angle MKL \cong \angle KMJ$, $\overline{KL} \cong \overline{MJ}$	5. Corresponding parts of congruent triangles are congruent.
6. $\overline{KL} \parallel \overline{MJ}$	6. Alternate Interior Angles Converse (Thm. 3.6)
7. *JKLM* is a parallelogram.	7. Opposite Sides Parallel and Congruent Theorem (Thm. 7.9)

43. no; The fourth angle will be 113° because of the Corollary to the Polygon Interior Angles Theorem (Cor. 7.1), but these could also be the angle measures of an isosceles trapezoid with base angles that are each 67°.

45. 8; By the Parallelogram Opposite Sides Theorem (Thm. 7.3), $\overline{AB} \cong \overline{CD}$. Also, $\angle ABE$ and $\angle CDF$ are congruent alternate interior angles of parallel segments $\overline{AB}$ and $\overline{CD}$. Then, you can use the Segment Addition Postulate (Post. 1.2), the Substitution Property of Equality, and the Reflexive Property of Congruence (Thm. 2.1) to show that $\overline{DF} \cong \overline{BE}$. So, $\triangle ABE \cong \triangle CDF$ by the SAS Congruence Theorem (Thm. 5.5), which means that $AE = CF = 8$ because corresponding parts of congruent triangles are congruent.

47. If every pair of consecutive angles of a quadrilateral is supplementary, then the quadrilateral is a parallelogram; In $ABCD$, you are given that $\angle A$ and $\angle B$ are supplementary, and $\angle B$ and $\angle C$ are supplementary. So, $m\angle A = m\angle C$. Also, $\angle B$ and $\angle C$ are supplementary, and $\angle C$ and $\angle D$ are supplementary. So, $m\angle B = m\angle D$. So, $ABCD$ is a parallelogram by the Parallelogram Opposite Angles Converse (Thm. 7.8).

49. Given quadrilateral $ABCD$ with midpoints E, F, G, and H that are joined to form a quadrilateral, you can construct diagonal $\overline{BD}$. Then $\overline{FG}$ is a midsegment of $\triangle BCD$, and $\overline{EH}$ is a midsegment of $\triangle DAB$. So, by the Triangle Midsegment Theorem (Thm. 6.8), $\overline{FG} \parallel \overline{BD}$, $FG = \frac{1}{2}BD$, $\overline{EH} \parallel \overline{BD}$, and $EH = \frac{1}{2}BD$. So, by the Transitive Property of Parallel Lines (Thm. 3.9), $\overline{EH} \parallel \overline{FG}$ and by the Transitive Property of Equality, $EH = FG$. Because one pair of opposite sides is both congruent and parallel, $EFGH$ is a parallelogram by the Opposite Sides Parallel and Congruent Theorem (Thm. 7.9).

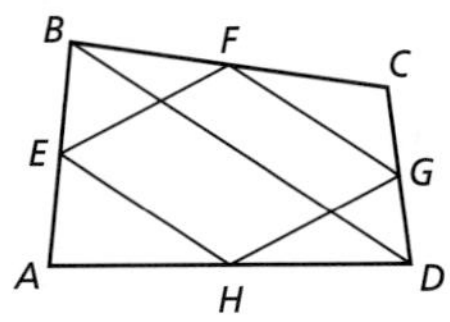

7.3 Maintaining Mathematical Proficiency *(p. 384)*

51. parallelogram **53.** square

7.4 Vocabulary and Core Concept Check *(p. 393)*

1. square

7.4 Monitoring Progress and Modeling with Mathematics *(pp. 393–396)*

3. sometimes; Some rhombuses are squares.

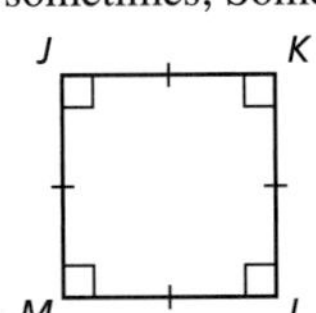

5. always; By definition, a rhombus is a parallelogram, and opposite sides of a parallelogram are congruent.

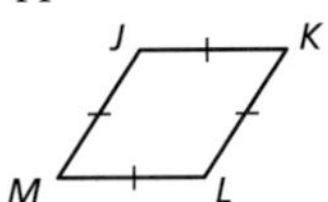

7. sometimes; Some rhombuses are squares.

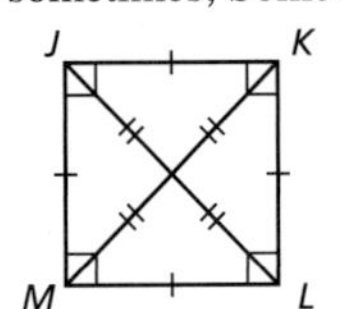

9. square; All of the sides are congruent, and all of the angles are congruent.

11. rectangle; Opposite sides are parallel and the angles are 90°.

13. $m\angle 1 = m\angle 2 = m\angle 4 = 27°$, $m\angle 3 = 90°$; $m\angle 5 = m\angle 6 = 63°$

15. $m\angle 1 = m\angle 2 = m\angle 3 = m\angle 4 = 37°$; $m\angle 5 = 106°$

17. always; All angles of a rectangle are congruent.

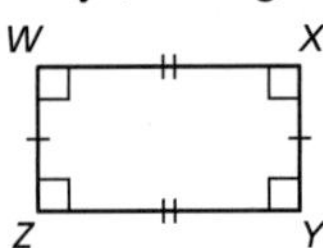

19. sometimes; Some rectangles are squares.

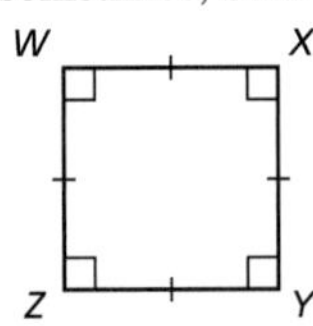

21. sometimes; Some rectangles are squares.

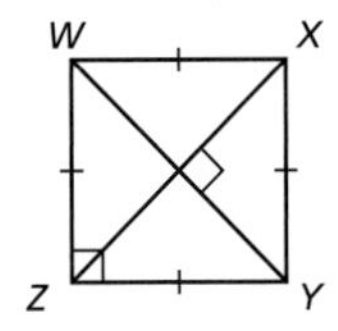

23. no; All four angles are not congruent. **25.** 11 **27.** 4

29. rectangle, square **31.** rhombus, square

33. parallelogram, rectangle, rhombus, square

35. Diagonals do not necessarily bisect opposite angles of a rectangle;

$$m\angle QSR = 90° - m\angle QSP$$
$$x = 32$$

37. 53° **39.** 74° **41.** 6 **43.** 56° **45.** 56°

47. 10 **49.** 90° **51.** 45° **53.** 2

55. rectangle, rhombus, square; The diagonals are congruent and perpendicular.

57. rectangle; The sides are perpendicular and not congruent.

59. rhombus; The diagonals are perpendicular and not congruent.

61. rhombus; The sides are congruent; $x = 76$; $y = 4$

63. **a.** rhombus; rectangle; $HBDF$ has four congruent sides; $ACEG$ has four right angles.

b. $AE = GC$; $AJ = JE = CJ = JG$; The diagonals of a rectangle are congruent and bisect each other.

65. always; By the Square Corollary (Cor. 7.4), a square is a rhombus.

67. always; The diagonals of a rectangle are congruent by the Rectangle Diagonals Theorem (Thm. 7.13).

69. sometimes; Some rhombuses are squares.

71. Measure the diagonals to see if they are congruent.

73.

STATEMENTS	REASONS
1. $PQRS$ is a parallelogram. $\overline{PR}$ bisects $\angle SPQ$ and $\angle QRS$. $\overline{SQ}$ bisects $\angle PSR$ and $\angle RQP$.	1. Given
2. $\angle SRT \cong \angle QRT$, $\angle RQT \cong \angle RST$	2. Definition of angle bisector
3. $\overline{TR} \cong \overline{TR}$	3. Reflexive Property of Congruence (Thm. 2.1)
4. $\triangle QRT \cong \triangle SRT$	4. AAS Congruence Theorem (Thm. 5.11)
5. $\overline{QR} \cong \overline{SR}$	5. Corresponding parts of congruent triangles are congruent.
6. $\overline{QR} \cong \overline{PS}$, $\overline{PQ} \cong \overline{SR}$	6. Parallelogram Opposite Sides Theorem (Thm. 7.3)
7. $\overline{PS} \cong \overline{QR} \cong \overline{SR} \cong \overline{PQ}$	7. Transitive Property of Congruence (Thm. 2.1)
8. $PQRS$ is a rhombus.	8. Definition of rhombus

75. no; The diagonals of a square always create two right triangles.

77. square; A square has four congruent sides and four congruent angles.

79. no; yes; Corresponding angles of two rhombuses might not be congruent; Corresponding angles of two squares are congruent.

81. If a quadrilateral is a rhombus, then it has four congruent sides; If a quadrilateral has four congruent sides, then it is a rhombus; The conditional statement is true by the definition of rhombus. The converse is true because if a quadrilateral has four congruent sides, then both pairs of opposite sides are congruent. So, by the Parallelogram Opposite Sides Converse (Thm. 7.7), it is a parallelogram with four congruent sides, which is the definition of a rhombus.

83. If a quadrilateral is a square, then it is a rhombus and a rectangle; If a quadrilateral is a rhombus and a rectangle, then it is a square; The conditional statement is true because if a quadrilateral is a square, then by definition of a square, it has four congruent sides, which makes it a rhombus by the Rhombus Corollary (Cor. 7.2), and it has four right angles, which makes it a rectangle by the Rectangle Corollary (Cor. 7.3); The converse is true because if a quadrilateral is a rhombus and a rectangle, then by the Rhombus Corollary (Cor. 7.2), it has four congruent sides, and by the Rectangle Corollary (Cor. 7.3), it has four right angles. So, by the definition, it is a square.

85.

STATEMENTS	REASONS
1. $\triangle XYZ \cong \triangle XWZ$, $\angle XYW \cong \angle ZWY$	1. Given
2. $\angle YXZ \cong \angle WXZ$, $\angle YZX \cong \angle WZX$, $\overline{XY} \cong \overline{XW}$, $\overline{YZ} \cong \overline{WZ}$	2. Corresponding parts of congruent triangles are congruent.
3. $\overline{XZ}$ bisects $\angle WXY$ and $\angle WZY$.	3. Definition of angle bisector
4. $\angle XWY \cong \angle XYW$, $\angle WYZ \cong \angle ZWY$	4. Base Angles Theorem (Thm. 5.6)
5. $\angle XYW \cong \angle WYZ$, $\angle XWY \cong \angle ZWY$	5. Transitive Property of Congruence (Thm. 2.2)
6. $\overline{WY}$ bisects $\angle XWZ$ and $\angle XYZ$.	6. Definition of angle bisector
7. $WXYZ$ is a rhombus.	7. Rhombus Opposite Angles Theorem (Thm. 7.12)

87.

STATEMENTS	REASONS
1. $PQRS$ is a rectangle.	1. Given
2. $PQRS$ is a parallelogram.	2. Def. of rectangle
3. $\overline{PS} \cong \overline{QR}$	3. Parallelogram Opposite Sides Thm. (Thm. 7.3)
4. $\angle PQR$ and $\angle QPS$ are right angles.	4. Def. of rectangle
5. $\angle PQR \cong \angle QPS$	5. Right Angles Congruence Thm. (Thm. 2.3)
6. $\overline{PQ} \cong \overline{PQ}$	6. Reflexive Property of Congruence (Thm. 2.1)
7. $\triangle PQR \cong \triangle QPS$	7. SAS Congruence Theorem (Thm. 5.5)
8. $\overline{PR} \cong \overline{SQ}$	8. Corresponding parts of congruent triangles are congruent.

7.4 Maintaining Mathematical Proficiency *(p. 396)*

89. $x = 10, y = 8$ **91.** $x = 9, y = 26$

7.5 Vocabulary and Core Concept Check *(p. 403)*

1. A trapezoid has exactly one pair of parallel sides, and a kite has two pairs of consecutive congruent sides.

7.5 Monitoring Progress and Modeling with Mathematics *(pp. 403–406)*

3. slope of $\overline{YZ}$ = slope of $\overline{XW}$ and slope of $\overline{XY} \neq$ slope of $\overline{WZ}$; $XY = WZ$, so $WXYZ$ is isosceles.

5. slope of $\overline{MQ}$ = slope of $\overline{NP}$ and slope of $\overline{MN} \neq$ slope of $\overline{PQ}$; $MN \neq PQ$, so $MNPQ$ is not isosceles.

7. $m\angle L = m\angle M = 62°$, $m\angle K = m\angle J = 118°$ **9.** 14

11. 4 **13.** $3\sqrt{13}$ **15.** 110° **17.** 80°

19. Because $MN = \frac{1}{2}(AB + DC)$, when you solve for DC, you should get $DC = 2(MN) - AB$; $DC = 2(8) - 14 = 2$.

21. rectangle; $JKLM$ is a quadrilateral with 4 right angles.

23. square; All four sides are congruent and the angles are 90°.

25. no; It could be a kite. **27.** 3 **29.** 26 in.

31. $\angle A \cong \angle D$, or $\angle B \cong \angle C$; $\overline{AB} \parallel \overline{CD}$, so base angles need to be congruent.

33. *Sample answer:* $\overline{BE} \cong \overline{DE}$; Then the diagonals bisect each other.

35.

STATEMENTS	REASONS
1. $\overline{JL} \cong \overline{LN}$, $\overline{KM}$ is a midsegment of $\triangle JLN$.	1. Given
2. $\overline{KM} \parallel \overline{JN}$	2. Triangle Midsegment Theorem (Thm. 6.8)
3. $JKMN$ is a trapezoid.	3. Definition of trapezoid
4. $\angle LJN \cong \angle LNJ$	4. Base Angles Theorem (Thm. 5.6)
5. $JKMN$ is an isosceles trapezoid.	5. Isosceles Trapezoid Base Angles Converse (Thm. 7.15)

37. any point on $\overleftrightarrow{UV}$ such that $UV \neq SV$

39. Given isosceles trapezoid $ABCD$ with $\overline{BC} \parallel \overline{AD}$, construct $\overline{CE}$ parallel to $\overline{BA}$. Then, $ABCE$ is a parallelogram by definition, so $\overline{AB} \cong \overline{EC}$. Because $\overline{AB} \cong \overline{CD}$ by the definition of an isosceles trapezoid, $\overline{CE} \cong \overline{CD}$ by the Transitive Property of Congruence (Thm. 2.1). So, $\angle CED \cong \angle D$ by the Base Angles Theorem (Thm. 5.6) and $\angle A \cong \angle CED$ by the Corresponding Angles Theorem (Thm. 3.1). So, $\angle A \cong \angle D$ by the Transitive Property of Congruence (Thm. 2.2). Next, by the Consecutive Interior Angles Theorem (Thm. 3.4), $\angle B$ and $\angle A$ are supplementary and so are $\angle BCD$ and $\angle D$. So, $\angle B \cong \angle BCD$ by the Congruent Supplements Theorem (Thm. 2.4).

41. no; It could be a square.

43. **a.** rectangle; The diagonals are congruent, but not perpendicular.

b.

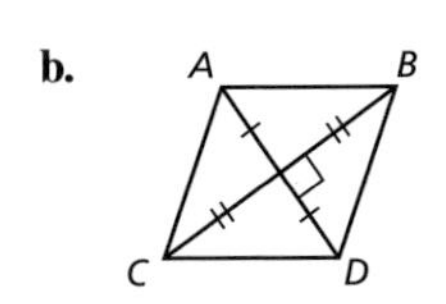

rhombus; The diagonals are perpendicular, but not congruent.

45. **a.** yes **b.** 75°, 75°, 105°, 105°

47. Given kite $EFGH$ with $\overline{EF} \cong \overline{FG}$ and $\overline{EH} \cong \overline{GH}$, construct diagonal $\overline{FH}$, which is congruent to itself by the Reflexive Property of Congruence (Thm. 2.1). So, $\triangle FGH \cong \triangle FEH$ by the SSS Congruence Theorem (Thm. 5.8), and $\angle E \cong \angle G$ because corresponding parts of congruent triangles are congruent. Next, assume temporarily that $\angle F \cong \angle H$. Then $EFGH$ is a parallelogram by the Parallelogram Opposite Angles Converse (Thm. 7.8), and opposite sides are congruent. However, this contradicts the definition of a kite, which says that opposite sides cannot be congruent. So, the assumption cannot be true and $\angle F$ is not congruent to $\angle H$.

49. By the Triangle Midsegment Theorem (Thm. 6.8), $\overline{BG} \parallel \overline{CD}$, $BG = \frac{1}{2}CD$, $\overline{GE} \parallel \overline{AF}$ and $GE = \frac{1}{2}AF$. By the Transitive Property of Parallel Lines (Thm. 3.9), $\overline{CD} \parallel \overline{BE} \parallel \overline{AF}$. Also, by the Segment Addition Postulate (Post. 1.2), $BE = BG + GE$. So, by the Substitution Property of Equality, $BE = \frac{1}{2}CD + \frac{1}{2}AF = \frac{1}{2}(CD + AF)$.

51. **a.**

STATEMENTS	REASONS
1. $JKLM$ is an isosceles trapezoid, $\overline{KL} \parallel \overline{JM}$, $\overline{JK} \cong \overline{LM}$	1. Given
2. $\angle JKL \cong \angle MLK$	2. Isosceles Trapezoid Base Angles Theorem (Thm. 7.14)
3. $\overline{KL} \cong \overline{KL}$	3. Reflexive Property of Congruence (Thm. 2.1)
4. $\triangle JKL \cong \triangle MLK$	4. SAS Congruence Theorem (Thm. 5.5)
5. $\overline{JL} \cong \overline{KM}$	5. Corresponding parts of congruent triangles are congruent.

b. If the diagonals of a trapezoid are congruent, then the trapezoid is isosceles. Let $JKLM$ be a trapezoid, $\overline{KL} \parallel \overline{JM}$ and $\overline{JL} \cong \overline{KM}$. Construct line segments through K and L perpendicular to $\overline{JM}$ as shown. Because $\overline{KL} \parallel \overline{JM}$, $\angle AKL$ and $\angle KLB$ are right angles, so $KLBA$ is a rectangle and $\overline{AK} \cong \overline{BL}$. Then $\triangle JLB \cong \triangle MKA$ by the HL Congruence Theorem (Thm. 5.9). So, $\angle LJB \cong \angle KMA$. $\overline{JM} \cong \overline{JM}$ by the Reflexive Property of Congruence (Thm. 2.1). So, $\triangle KJM \cong \triangle LMJ$ by the SAS Congruence Theorem (Thm. 5.5). Then $\angle KJM \cong \angle LMJ$, and the trapezoid is isosceles by the Isosceles Trapezoid Base Angles Converse (Thm. 7.15).

7.5 Maintaining Mathematical Proficiency *(p. 406)*

53. *Sample answer:* translation 1 unit right followed by a dilation with a scale factor of 2

Chapter 7 Review *(pp. 408–410)*

1. 5040°; 168°; 12° **2.** 133 **3.** 82 **4.** 15

5. $a = 79, b = 101$ **6.** $a = 28, b = 87$

7. $c = 6, d = 10$ **8.** $(-2, -1)$ **9.** $M(2, -2)$

10. Parallelogram Opposite Sides Converse (Thm. 7.7)

11. Parallelogram Diagonals Converse (Thm. 7.10)

12. Parallelogram Opposite Angles Converse (Thm. 7.8)

13. $x = 1, y = 6$ **14.** 4

15. Because $WX = YZ = \sqrt{13}$, $\overline{WX} \cong \overline{YZ}$. Because the slopes of $\overline{WX}$ and $\overline{YZ}$ are both $\frac{2}{3}$, they are parallel. $\overline{WX}$ and $\overline{YZ}$ are opposite sides that are both congruent and parallel. So, $WXYZ$ is a parallelogram by the Opposite Sides Parallel and Congruent Theorem (Thm. 7.9).

16. rhombus; There are four congruent sides.

17. parallelogram; There are two pairs of parallel sides.

18. square; There are four congruent sides and the angles are 90°.

19. 10

20. rectangle, rhombus, square; The diagonals are congruent and perpendicular.

21. $m\angle Z = m\angle Y = 58°$, $m\angle W = m\angle X = 122°$ **22.** 26

23. $3\sqrt{5}$ **24.** $x = 15$; 105°

25. yes; Use the Isosceles Trapezoid Base Angles Converse (Thm. 7.15).

26. trapezoid;There is one pair of parallel sides.

27. rhombus; There are four congruent sides.

28. rectangle; There are four right angles.

Chapter 8

Chapter 8 Maintaining Mathematical Proficiency *(p. 415)*

1. yes **2.** yes **3.** no **4.** no **5.** yes

6. yes **7.** $k = \frac{3}{7}$ **8.** $k = \frac{8}{3}$ **9.** $k = 2$

10. yes; All of the ratios are equivalent by the Transitive Property of Equality.

8.1 Vocabulary and Core Concept Check *(p. 423)*

1. congruent; proportional

8.1 Monitoring Progress and Modeling with Mathematics *(pp. 423–426)*

3. $\frac{4}{3}$; $\angle A \cong \angle L, \angle B \cong \angle M, \angle C \cong \angle N$; $\frac{LM}{AB} = \frac{MN}{BC} = \frac{NL}{CA}$

5. $x = 30$ **7.** $x = 11$ **9.** altitude; 24 **11.** 2 : 3

13. 72 cm **15.** 20 yd **17.** 288 ft, 259.2 ft

19. 108 ft^2 **21.** 4 in.^{2}

23. Because the first ratio has a side length of B over a side length of A, the second ratio should have the perimeter of B over the perimeter of A;

$$\frac{5}{10} = \frac{x}{28}$$

$$x = 14$$

25. no; Corresponding angles are not congruent. **27.** A, D

29. $\frac{5}{2}$ **31.** 34, 85 **33.** 60.5, 378.125 **35.** B, D

37. $x = 35.25$, $y = 20.25$ **39.** 30 m **41.** 7.5 ft

43. sometimes **45.** sometimes **47.** sometimes

49. yes; All four angles of each rectangle will always be congruent right angles.

51. about 1116 mi

53.

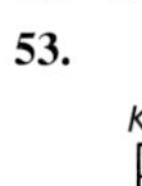
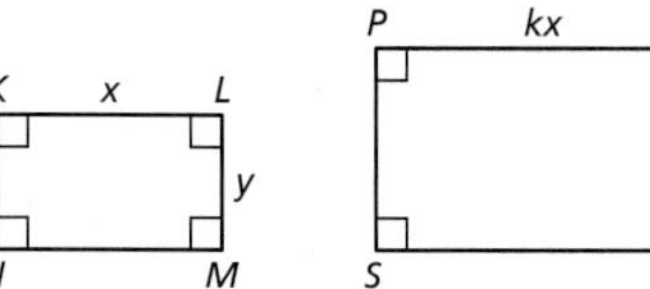

Let $KLMN$ and $PQRS$ be similar rectangles as shown. The ratio of corresponding side lengths is $\frac{KL}{PQ} = \frac{x}{kx} = \frac{1}{k}$. The area of $KLMN$ is xy and the area of $PQRS$ is $(kx)(ky) = k^2xy$. So, the ratio of the areas is $\frac{xy}{k^2xy} = \frac{1}{k^2} = \left(\frac{1}{k}\right)^2$. Because the ratio of corresponding side lengths is $\frac{1}{k}$, any pair of corresponding side lengths can be substituted for $\frac{1}{k}$. So,

$$\frac{\text{Area of } KLMN}{\text{Area of } PQRS} = \left(\frac{KL}{PQ}\right)^2 = \left(\frac{LM}{QR}\right)^2 = \left(\frac{MN}{RS}\right)^2 = \left(\frac{NK}{SP}\right)^2.$$

55. $x = \frac{1 + \sqrt{5}}{2}$; $x = \frac{1 + \sqrt{5}}{2}$ satisfies the proportion $\frac{1}{x} = \frac{x - 1}{1}$.

8.1 Maintaining Mathematical Proficiency *(p. 426)*

57. $x = 63$ **59.** $x = 64$

8.2 Vocabulary and Core Concept Check *(p. 431)*

1. similar

8.2 Monitoring Progress and Modeling with Mathematics *(pp. 431–432)*

3. yes; $\angle H \cong \angle J$ and $\angle F \cong \angle K$, so $\triangle FGH \sim \triangle KLJ$.

5. no; $m\angle N = 50°$

7. $\angle N \cong \angle Z$ and $\angle MYN \cong \angle XYZ$, so $\triangle MYN \sim \triangle XYZ$.

9. $\angle Y \cong \angle Y$ and $\angle YZX \cong \angle W$, so $\triangle XYZ \sim \triangle UYW$.

11. $\triangle CAG \sim \triangle CEF$ **13.** $\triangle ACB \sim \triangle ECD$

15. $m\angle ECD = m\angle ACB$ **17.** $BC = 4\sqrt{2}$

19. The AA Similarity Theorem (Thm. 8.3) does not apply to quadrilaterals. There is not enough information to determine whether or not quadrilaterals $ABCD$ and $EFGH$ are similar.

21. 78 m; Corresponding angles are congruent, so the triangles are similar.

23. yes; Corresponding angles are congruent.

25. no; $94° + 87° > 180°$

27. *Sample answer:* Because the triangles are similar, the ratios of the vertical sides to the horizontal sides are equal.

29. The angle measures are 60°.

31.

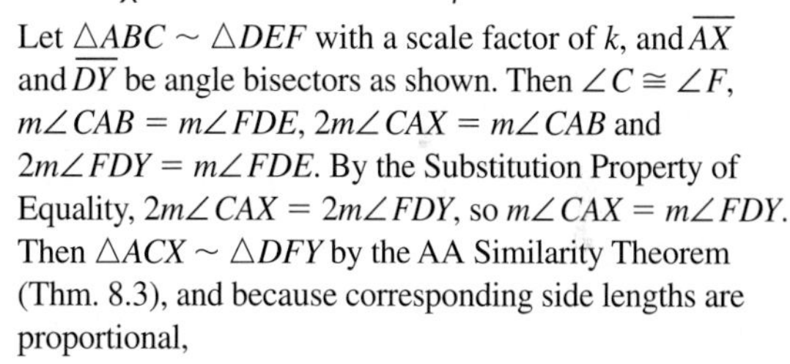

Let $\triangle ABC \sim \triangle DEF$ with a scale factor of k, and $\overline{AX}$ and $\overline{DY}$ be angle bisectors as shown. Then $\angle C \cong \angle F$, $m\angle CAB = m\angle FDE$, $2m\angle CAX = m\angle CAB$ and $2m\angle FDY = m\angle FDE$. By the Substitution Property of Equality, $2m\angle CAX = 2m\angle FDY$, so $m\angle CAX = m\angle FDY$. Then $\triangle ACX \sim \triangle DFY$ by the AA Similarity Theorem (Thm. 8.3), and because corresponding side lengths are proportional,

$$\frac{AX}{DY} = \frac{AC}{DF} = k.$$

33. about 17.1 ft; $\triangle AED \sim \triangle CEB$, so $\frac{DE}{BE} = \frac{4}{3}$. $\triangle DEF \sim \triangle DBC$, so $\frac{EF}{30} = \frac{DE}{DB} = \frac{4}{7}$ and $EF = \frac{120}{7}$.

8.2 Maintaining Mathematical Proficiency *(p. 432)*

35. yes; Use the SSS Congruence Theorem (Thm. 5.8).

8.3 Vocabulary and Core Concept Check *(p. 441)*

1. $\frac{QR}{XY} = \frac{RS}{YZ} = \frac{QS}{XZ}$

8.3 Monitoring Progress and Modeling with Mathematics *(pp. 441–444)*

3. $\triangle RST$ **5.** $x = 4$ **7.** $\frac{12}{18} = \frac{10}{15} = \frac{8}{12} = \frac{2}{3}$

9. similar; $\triangle DEF \sim \triangle WXY$; $\frac{4}{3}$

11. no

16° 20 R S T 32 F 30 24° G H 48

13. $\frac{HG}{HF} = \frac{HJ}{HK} = \frac{GJ}{FK}$, so $\triangle GHJ \sim \triangle FHK$.

15. $\angle X \cong \angle D$ and $\frac{XY}{DJ} = \frac{XZ}{DG}$, so $\triangle XYZ \sim \triangle DJG$.

17. 24, 26

19. Because $\overline{AB}$ corresponds to $\overline{RQ}$ and $\overline{BC}$ corresponds to $\overline{QP}$, the proportionality statement should be $\triangle ABC \sim \triangle RQP$.

21. 61° **23.** 30° **25.** 91°

27. no; The included angles are not congruent.

29. D; $\angle M \cong \angle M$ so, $\triangle MNP \sim \triangle MRQ$ by the SAS Similarity Theorem (Thm. 8.5).

31. **a.** $\frac{CD}{CE} = \frac{BC}{AC}$ **b.** $\angle CBD \cong \angle CAE$

33.

STATEMENTS	REASONS
1. $\angle A \cong \angle D$, $\frac{AB}{DE} = \frac{AC}{DF}$	1. Given
2. Draw $\overline{PQ}$ so that P is on $\overline{AB}$, Q is on $\overline{AC}$, $\overline{PQ} \parallel \overline{BC}$, and $AP = DE$.	2. Parallel Postulate (Post. 3.1)
3. $\angle APQ \cong \angle ABC$	3. Corresponding Angles Theorem (Thm. 3.1)
4. $\angle A \cong \angle A$	4. Reflexive Property of Congruence (Thm. 2.2)
5. $\triangle APQ \sim \triangle ABC$	5. AA Similarity Theorem (Thm. 8.3)
6. $\frac{AB}{AP} = \frac{AC}{AQ} = \frac{BC}{PQ}$	6. Corresponding sides of similar figures are proportional.
7. $\frac{AB}{DE} = \frac{AC}{AQ}$	7. Substitution Property of Equality
8. $AQ \cdot \frac{AB}{DE} = AC$, $DF \cdot \frac{AB}{DE} = AC$	8. Multiplication Property of Equality
9. $AQ = AC \cdot \frac{DE}{AB}$, $DF = AC \cdot \frac{DE}{AB}$	9. Multiplication Property of Equality
10. $AQ = DF$	10. Transitive Property of Equality
11. $\overline{AQ} \cong \overline{DF}$, $\overline{AP} \cong \overline{DE}$	11. Definition of congruent segments
12. $\triangle APQ \cong \triangle DEF$	12. SAS Congruence Theorem (Thm. 5.5)
13. $\overline{PQ} \cong \overline{EF}$	13. Corresponding parts of congruent triangles are congruent.
14. $PQ = EF$	14. Definition of congruent segments
15. $\frac{AB}{DE} = \frac{AC}{DF} = \frac{BC}{EF}$	15. Substitution Property of Equality
16. $\triangle ABC \sim \triangle DEF$	16. SSS Similarity Theorem (Thm. 8.4)

35. no; no; The sum of the angle measures would not be 180°.

37. If two angles are congruent, then the triangles are similar by the AA Similarity Theorem (Thm. 8.3).

39. *Sample answer:*

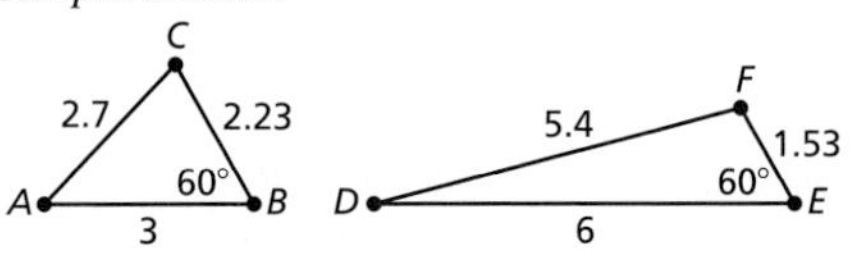

41. the Substitution Property of Equaltiy; $\frac{BC}{EF} = \frac{AC}{DF}$; $\angle ACB \cong \angle DFE$; SAS Similarity Theorem (Thm. 8.5); Corresponding Angles Converse (Thm. 3.5)

8.3 Maintaining Mathematical Proficiency *(p. 444)*

43. $P(0, 3)$ **45.** $P(5, 6)$

8.4 Vocabulary and Core Concept Check *(p. 450)*

1. parallel, Converse of the Triangle Proportionality Theorem (Thm. 8.7)

8.4 Monitoring Progress and Modeling with Mathematics *(pp. 450–452)*

3. 9 **5.** yes **7.** no

9.

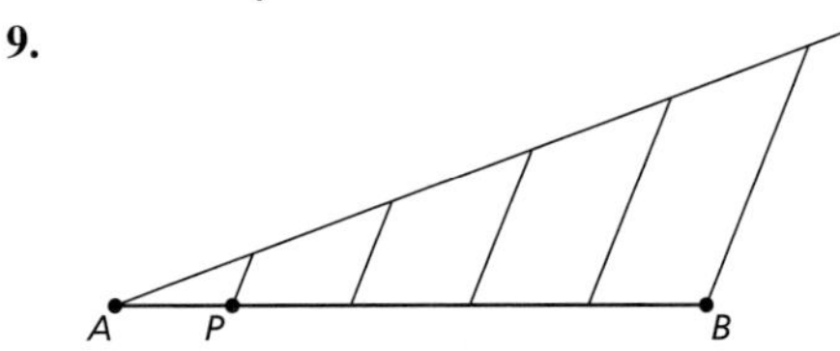

11.

13. CE **15.** BD **17.** 6 **19.** 12 **21.** 27

23. The proportion should show that AD corresponds with DC and BA corresponds with BC;

$$\frac{AD}{DC} = \frac{BA}{BC}$$

$$\frac{x}{14} = \frac{10}{16}$$

$$x = 8.75$$

25. $x = 3$

27.

STATEMENTS	REASONS
1. $\overline{QS} \parallel \overline{TU}$	1. Given
2. $\angle RQS \cong \angle RTU$, $\angle RSQ \cong \angle RUT$	2. Corresponding Angles Theorem (Thm. 3.1)
3. $\triangle RQS \sim \triangle RTU$	3. AA Similarity Theorem (Thm. 8.3)
4. $\frac{QR}{TR} = \frac{SR}{UR}$	4. Corresponding side lengths of similar figures are proportional.
5. $QR = QT + TR$, $SR = SU + UR$	5. Segment Addition Postulate (Post. 1.2)
6. $\frac{QT + TR}{TR} = \frac{SU + UR}{UR}$	6. Substitution Property of Equality
7. $\frac{QT}{TR} + \frac{TR}{TR} = \frac{SU}{UR} + \frac{UR}{UR}$	7. Rewrite the proportion.
8. $\frac{QT}{TR} + 1 = \frac{SU}{UR} + 1$	8. Simplify.
9. $\frac{QT}{TR} = \frac{SU}{UR}$	9. Subtraction Property of Equality

29. a. about 50.9 yd, about 58.4 yd, about 64.7 yd b. Lot C
c. about \$287,000, about \$318,000; $\frac{50.9}{250{,}000} \approx \frac{58.4}{287{,}000}$ and $\frac{50.9}{250{,}000} \approx \frac{64.7}{318{,}000}$

31. Because $\overline{DJ}, \overline{EK}, \overline{FL}$, and $\overline{GB}$ are cut by a transversal $\overrightarrow{AC}$, and $\angle ADJ \cong \angle DEK \cong \angle EFL \cong \angle FGB$ by construction, $\overline{DJ} \parallel \overline{EK} \parallel \overline{FL} \parallel \overline{GB}$ by the Corresponding Angles Converse (Thm. 3.5).

33. isosceles; By the Triangle Angle Bisector Theorem (Thm. 8.9), the ratio of the lengths of the segments of $\overline{LN}$ equals the ratio of the other two side lengths. Because $\overline{LN}$ is bisected, the ratio is 1, and $ML = MN$.

35. Because $\overline{WX} \parallel \overline{ZA}$, $\angle XAZ \cong \angle YXW$ by the Corresponding Angles Theorem (Thm. 3.1) and $\angle WXZ \cong \angle XZA$ by the Alternate Interior Angles Theorem (Thm. 3.2). So, by the Transitive Property of Congruence (Thm. 2.2), $\angle XAZ \cong \angle XZA$. Then $\overline{XA} \cong \overline{XZ}$ by the Converse of the Base Angles Theorem (Thm. 5.7), and by the Triangle Proportionality Theorem (Thm. 8.6), $\frac{YW}{WZ} = \frac{XY}{XA}$. Because $XA = XZ$, $\frac{YW}{WZ} = \frac{XY}{XZ}$.

37. The Triangle Midsegment Theorem (Thm. 6.8) is a specific case of the Triangle Proportionality Theorem (Thm. 8.6) when the segment parallel to one side of a triangle that connects the other two sides also happens to pass through the midpoints of those two sides.

39. (segment labeled x)

8.4 Maintaining Mathematical Proficiency *(p. 452)*

41. a, b 43. $x = \pm 11$ 45. $x = \pm 7$

Chapter 8 Review *(pp 454–456)*

1. $\frac{3}{4}$; $\angle A \cong \angle E$, $\angle B \cong \angle F$, $\angle C \cong \angle G$, $\angle D \cong \angle H$; $\frac{EF}{AB} = \frac{FG}{BC} = \frac{GH}{CD} = \frac{EH}{AD}$

2. $\frac{2}{5}$; $\angle X \cong \angle R$, $\angle Y \cong \angle P$, $\angle Z \cong \angle Q$; $\frac{RP}{XY} = \frac{PQ}{YZ} = \frac{RQ}{XZ}$

3. 14.4 in. 4. $P = 32$ m; $A = 80$ m^2

5. $\angle Q \cong \angle T$ and $\angle RSQ \cong \angle UST$, so $\triangle RSQ \sim \triangle UST$.

6. $\angle C \cong \angle F$ and $\angle B \cong \angle E$, so $\triangle ABC \sim \triangle DEF$.

7. 324 ft

8. $\angle C \cong \angle C$ and $\frac{CD}{CE} = \frac{CB}{CA}$, so $\triangle CBD \sim \triangle CAE$.

9. $\frac{QU}{QT} = \frac{QR}{QS} = \frac{UR}{TS}$, so $\triangle QUR \sim \triangle QTS$. 10. $x = 4$

11. no 12. yes 13. 11.2 14. 10.5 15. 7.2

Chapter 9

Chapter 9 Maintaining Mathematical Proficiency *(p. 461)*

1. $5\sqrt{3}$ 2. $3\sqrt{30}$ 3. $3\sqrt{15}$ 4. $\frac{2\sqrt{7}}{7}$ 5. $\frac{5\sqrt{2}}{2}$

6. $2\sqrt{6}$ 7. $x = 9$ 8. $x = 7.5$ 9. $x = 32$

10. $x = 9.2$ 11. $x = 2$ 12. $x = 17$

13. no; no; Because square roots have to do with factors, the rule allows you to simplify with products, not sums and differences.

9.1 Vocabulary and Core Concept Check *(p. 468)*

1. A Pythagorean triple is a set of three positive integers a, b, and c that satisfy the equation $c^2 = a^2 + b^2$.

9.1 Monitoring Progress and Modeling with Mathematics *(pp. 468–470)*

3. $x = \sqrt{170} \approx 13.0$; no 5. $x = 41$; yes

7. $x = 15$; yes 9. $x = 14$; yes

11. Exponents cannot be distributed as shown in the third line; $c^2 = a^2 + b^2$; $x^2 = 7^2 + 24^2$; $x^2 = 49 + 576$; $x^2 = 625$; $x = 25$

13. about 14.1 ft 15. yes 17. no 19. no

21. yes; acute 23. yes; right 25. yes; acute

27. yes; obtuse 29. about 127.3 ft 31. 120 m^2

33. 48 cm^2

35. The horizontal distance between any two points is given by $(x_2 - x_1)$, and the vertical distance is given by $(y_2 - y_1)$. The horizontal and vertical segments that represent these distances form a right angle, with the segment between the two points being the hypotenuse. So, you can use the Pythagorean Theorem (Thm. 9.1) to say $d^2 = (x_2 - x_1)^2 + (y_2 - y_1)^2$, and when you solve for d, you get the distance formula: $d = \sqrt{(x_2 - x_1)^2 + (y_2 - y_1)^2}$.

37. 2 packages

39.

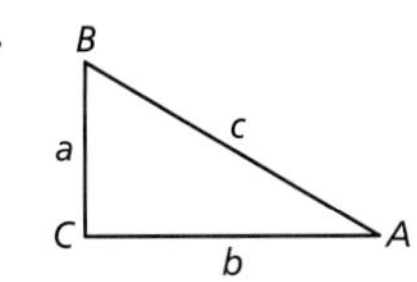

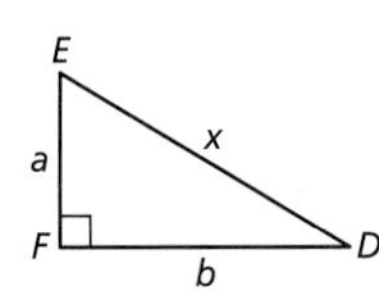

Let $\triangle ABC$ be any triangle so that the square of the length, c, of the longest side of the triangle is equal to the sum of the squares of the lengths, a and b, of the other two sides: $c^2 = a^2 + b^2$. Let $\triangle DEF$ be any right triangle with leg lengths of a and b. Let x represent the length of its hypotenuse. Because $\triangle DEF$ is a right triangle, by the Pythagorean Theorem (Thm. 9.1), $a^2 + b^2 = x^2$. So, by the Transitive Property, $c^2 = x^2$. By taking the positive square root of each side, you get $c = x$. So, $\triangle ABC \cong \triangle DEF$ by the SSS Congruence Theorem (Thm. 5.8).

41. no; They can be part of a Pythagorean triple if 75 is the hypotenuse: $21^2 + 72^2 = 75^2$

43.

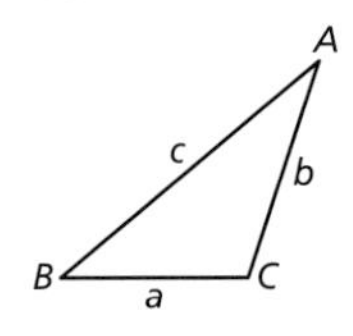

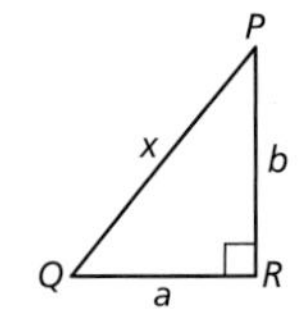

STATEMENTS	REASONS
1. In $\triangle ABC$, $c^2 > a^2 + b^2$, where c is the length of the longest side. $\triangle PQR$ has side lengths a, b, and x, where x is the length of the hypotenuse and $\angle R$ is a right angle.	**1.** Given
2. $a^2 + b^2 = x^2$	**2.** Pythagorean Theorem (Thm. 9.1)
3. $c^2 > x^2$	**3.** Substitution Property
4. $c > x$	**4.** Take the positive square root of each side.
5. $m\angle R = 90°$	**5.** Definition of a right angle
6. $m\angle C > m\angle R$	**6.** Converse of the Hinge Theorem (Thm. 6.13)
7. $m\angle C > 90°$	**7.** Substitution Property
8. $\angle C$ is an obtuse angle.	**8.** Definition of obtuse angle
9. $\triangle ABC$ is an obtuse triangle.	**9.** Definition of obtuse triangle

9.1 Maintaining Mathematical Proficiency *(p. 470)*

45. $\frac{14\sqrt{3}}{3}$ **47.** $4\sqrt{3}$

9.2 Vocabulary and Core Concept Check *(p. 475)*

1. 45°-45°-90°, 30°-60°-90°

9.2 Monitoring Progress and Modeling with Mathematics *(pp. 475–476)*

3. $x = 7\sqrt{2}$ **5.** $x = 3$ **7.** $x = 9\sqrt{3}, y = 18$

9. $x = 12\sqrt{3}, y = 12$

11. The hypotenuse of a 30°-60°-90° triangle is equal to the shorter leg times 2; hypotenuse = shorter leg • 2 = 7 • 2 = 14; So, the length of the hypotenuse is 14 units.

13.

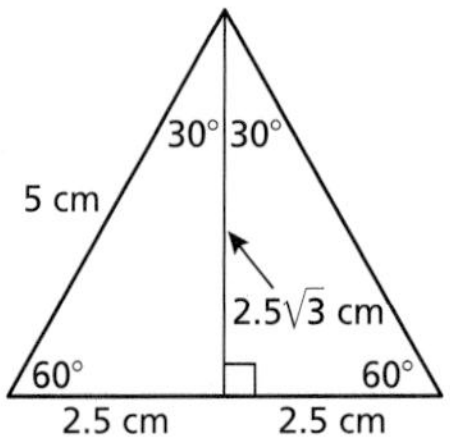

about 4.3 cm

15. 32 ft² **17.** 142 ft; about 200.82 ft; about 245.95 ft

19. Because $\triangle DEF$ is a 45°-45°-90° triangle, by the Converse of the Base Angles Theorem (Thm. 5.7), $\overline{DF} \cong \overline{FE}$. So, let $x = DF = FE$. By the Pythagorean Theorem (Thm. 9.1), $x^2 + x^2 = c^2$, where c is the length of the hypotenuse. So, $2x^2 = c^2$ by the Distributive Property. Take the positive square root of each side to get $x\sqrt{2} = c$. So, the hypotenuse is $\sqrt{2}$ times as long as each leg.

21. Given $\triangle JKL$, which is a 30°-60°-90° triangle, whose shorter leg, $\overline{KL}$, has length x, construct $\triangle JML$, which is congruent and adjacent to $\triangle JKL$. Because corresponding parts of congruent triangles are congruent, $LM = KL = x$, $m\angle M = m\angle K = 60°$, $m\angle MJL = m\angle KJL = 30°$, and $JM = JK$. Also, by the Angle Addition Postulate (Post. 1.4), $m\angle KJM = m\angle KJL + m\angle MJL$, and by substituting, $m\angle KJM = 30° + 30° = 60°$. So, $\triangle JKM$ has three 60° angles, which means that it is equiangular by definition, and by the Corollary to the Converse of the Base Angles Theorem (Cor. 5.3), it is also equilateral. By the Segment Addition Postulate (Post. 1.2), $KM = KL + LM$, and by substituting, $KM = x + x = 2x$. So, by the definition of an equilateral triangle, $JM = JK = KM = 2x$. By the Pythagorean Theorem (Thm. 9.1), $(JL)^2 + (KL)^2 = (JK)^2$. By substituting, we get $(JL)^2 + x^2 = (2x)^2$, which is equivalent to $(JL)^2 + x^2 = 4x^2$, when simplified. When the Subtraction Property of Equality is applied, we get $(JL)^2 = 4x^2 - x^2$, which is equivalent to $(JL)^2 = 3x^2$. By taking the positive square root of each side, $JL = x\sqrt{3}$. So, the hypotenuse of the 30°-60°-90° triangle, $\triangle JKL$, is twice as long as the shorter leg, and the longer leg is $\sqrt{3}$ times as long as the shorter leg.

23. *Sample answer:* Because all isosceles right triangles are 45°-45°-90° triangles, they are similar by the AA Similarity Theorem (Thm. 8.3). Because both legs of an isosceles right triangle are congruent, the legs will always be proportional. So, 45°-45°-90° triangles are all similar by the SAS Similarity Theorem (Thm. 8.5) also.

25. $T(1.5, 1.6)$

9.2 Maintaining Mathematical Proficiency *(p. 476)*

27. $x = 2$

9.3 Vocabulary and Core Concept Check *(p. 482)*

1. each other

9.3 Monitoring Progress and Modeling with Mathematics *(pp. 482–484)*

3. $\triangle HFE \sim \triangle GHE \sim \triangle GFH$ **5.** $x = \frac{168}{25} = 6.72$

7. $x = \frac{180}{13} \approx 13.8$ **9.** About 11.2 ft **11.** 16

13. $2\sqrt{70} \approx 16.7$ **15.** 20 **17.** $6\sqrt{17} \approx 24.7$

19. $x = 8$ **21.** $y = 27$

23. $x = 3\sqrt{5} \approx 6.7$ **25.** $z = \frac{729}{16} \approx 45.6$

27. The length of leg z should be the geometric mean of the length of the hypotenuse, $(w + v)$, and the segment of the hypotenuse that is adjacent to z, which is v, not w; $z^2 = v \cdot (w + v)$

29. about 14.9 ft **31.** $a = 3$ **33.** $x = 9, y = 15, z = 20$

35. A, D **37.** $AC = 25, BD = 12$

39. given; Geometric Mean (Leg) Theorem (Thm. 9.8); a^2; Substitution Property of Equality; Distributive Property; c; Substitution Property of Equality

41.

STATEMENTS	REASONS
1. Draw $\triangle ABC$, $\angle BCA$ is a right angle.	1. Given
2. Draw a perpendicular segment (altitude) from C to $\overline{AB}$, and label the new point on $\overline{AB}$ as D.	2. Perpendicular Postulate (Post. 3.2)
3. $\triangle ADC \sim \triangle CDB$	3. Right Triangle Similarity Theorem (Thm. 9.6)
4. $\frac{BD}{CD} = \frac{CD}{AD}$	4. Corresponding sides of similar figures are proportional.
5. $CD^2 = AD \cdot BD$	5. Cross Products Property

43.

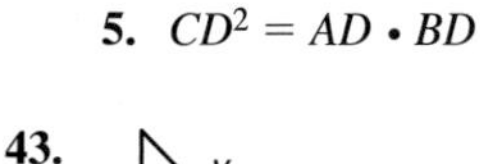

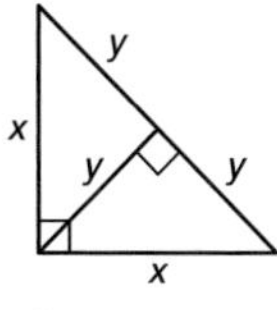

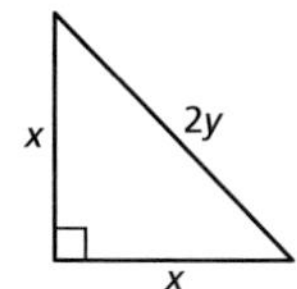

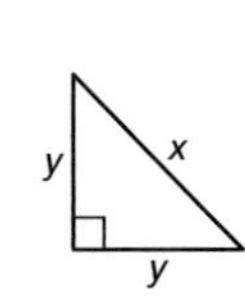

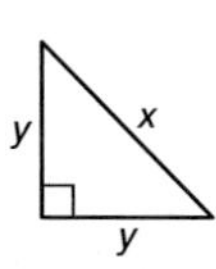

The two smaller triangles are congruent; Their corresponding sides lengths are represented by the same variables. So, they are congruent by the SSS Congruence Theorem (Thm. 5.8).

45.

STATEMENTS	REASONS
1. $\triangle ABC$ is a right triangle. Altitude $\overline{CD}$ is drawn to hypotenuse $\overline{AB}$.	1. Given
2. $\angle BCA$ is a right angle.	2. Definition of right triangle
3. $\angle ADC$ and $\angle BDC$ are right angles.	3. Definition of perpendicular lines
4. $\angle BCA \cong \angle ADC \cong \angle BDC$	4. Right Angles Congruence Theorem (Thm. 2.3)
5. $\angle A$ and $\angle ACD$ are complementary. $\angle B$ and $\angle BCD$ are complementary.	5. Corollary to the Triangle Sum Theorem (Cor. 5.1)
6. $\angle ACD$ and $\angle BCD$ are complementary.	6. Definition of complementary angles
7. $\angle A \cong \angle BCD$, $\angle B \cong \angle ACD$	7. Congruent Complements Theorem (Thm. 2.5)
8. $\triangle CBD \sim \triangle ABC$, $\triangle ACD \sim \triangle ABC$, $\triangle CBD \sim \triangle ACD$	8. AA Similarity Theorem (Thm. 8.3)

9.3 Maintaining Mathematical Proficiency *(p. 484)*

47. $x = 116$ **49.** $x = \frac{23}{6} \approx 3.8$

9.4 Vocabulary and Core Concept Check *(p. 491)*

1. the opposite leg, the adjacent leg

9.4 Monitoring Progress and Modeling with Mathematics *(pp. 491–492)*

3. $\tan R = \frac{45}{28} \approx 1.6071$, $\tan S = \frac{28}{45} \approx 0.6222$

5. $\tan G = \frac{2}{1} = 2.0000$, $\tan H = \frac{1}{2} = 0.5000$

7. $x \approx 13.8$ **9.** $x \approx 13.7$

11. The tangent ratio should be the length of the leg opposite $\angle D$ to the length of the leg adjacent to $\angle D$, not the length of the hypotenuse; $\tan D = \frac{35}{12}$

13. 1 **15.** about 555 ft **17.** $\frac{5}{12} \approx 0.4167$

19. it increases; The opposite side gets longer.

21. no; The Sun's rays form a right triangle with the length of the awning and the height of the door. The tangent of the angle of elevation equals the height of the door divided by the length of the awning, so the length of the awning equals the quotient of the height of the door, 8 feet, and the tangent of the angle of elevation, 70°: $x = \frac{8}{\tan 70°} \approx 6.5$ ft

23. You cannot find the tangent of a right angle, because each right angle has two adjacent legs, and the opposite side is the hypotenuse. So, you do not have an opposite leg and an adjacent leg. If a triangle has an obtuse angle, then it cannot be a right triangle, and the tangent ratio only works for right triangles.

25. **a.** about 33.4 ft

b. 3 students at each end; The triangle formed by the $60°$ angle has an opposite leg that is about 7.5 feet longer than the opposite leg of the triangle formed by the $50°$ angle. Because each student needs 2 feet of space, 3 more students can fit on each end with about 1.5 feet of space left over.

9.4 Maintaining Mathematical Proficiency *(p. 492)*

27. $x = 2\sqrt{3} \approx 3.5$ **29.** $x = 5\sqrt{2} \approx 7.1$

9.5 Vocabulary and Core Concept Check *(p. 498)*

1. the opposite leg, the hypotenuse

9.5 Monitoring Progress and Modeling with Mathematics *(pp. 498–500)*

3. $\sin D = \frac{4}{5} = 0.8000$, $\sin E = \frac{3}{5} = 0.6000$,

$\cos D = \frac{3}{5} = 0.6000$, $\cos E = \frac{4}{5} = 0.8000$

5. $\sin D = \frac{28}{53} \approx 0.5283$, $\sin E = \frac{45}{53} \approx 0.8491$,

$\cos D = \frac{45}{53} \approx 0.8491$, $\cos E = \frac{28}{53} \approx 0.5283$

7. $\sin D = \frac{\sqrt{3}}{2} \approx 0.8660$, $\sin E = \frac{1}{2} = 0.5000$,

$\cos D = \frac{1}{2} = 0.5000$, $\cos E = \frac{\sqrt{3}}{2} \approx 0.8660$

9. $\cos 53°$ **11.** $\cos 61°$ **13.** $\sin 31°$ **15.** $\sin 17°$

17. $x \approx 9.5$, $y \approx 15.3$ **19.** $v \approx 4.7$, $w \approx 1.6$

21. $a \approx 14.9$, $b \approx 11.1$ **23.** $\sin X = \cos X = \sin Z = \cos Z$

25. The sine of $\angle A$ should be equal to the ratio of the length of the leg opposite the angle, to the length of the hypotenuse; $\sin A = \frac{12}{13}$

27. about 15 ft

29. a.

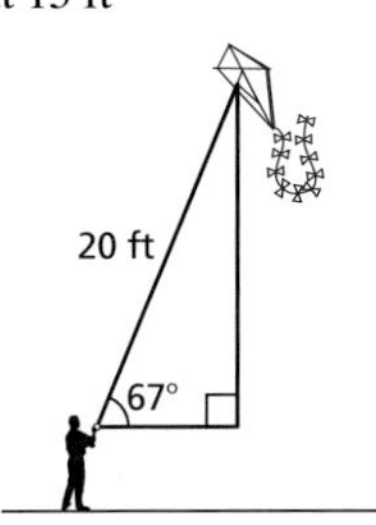

b. about 23.4 ft; The higher you hold the spool, the farther the kite is from the ground.

31. both; The sine of an acute angle is equal to the cosine of its complement, so these two equations are equivalent.

33.

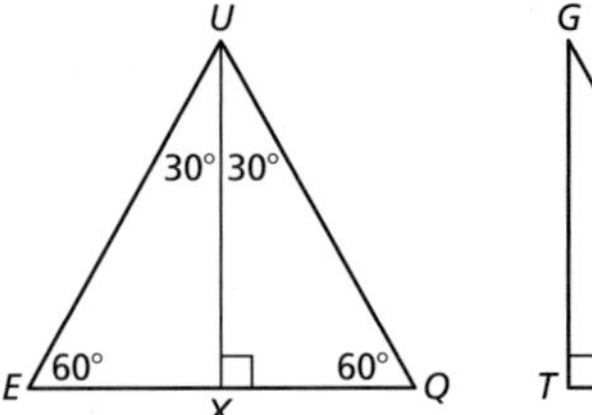

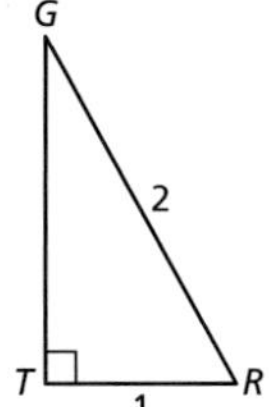

Because $\triangle EQU$ is an equilateral triangle, all three angles have a measure of $60°$. When an altitude, $\overline{UX}$, is drawn from U to $\overline{EQ}$ as shown, two congruent $30°$-$60°$-$90°$ triangles are formed, where $m\angle E = 60°$. So, $\sin E = \sin 60° = \frac{\sqrt{3}}{2}$. Also, in $\triangle RGT$, because the hypotenuse is twice as long as one of the legs, it is also a $30°$-$60°$-$90°$ triangle. Because $\angle G$ is across from the shorter leg, it must have a measure of $30°$, which means that $\cos G = \cos 30° = \frac{\sqrt{3}}{2}$. So, $\sin E = \cos G$.

35. If you knew how to take the inverse of the trigonometric ratios, you could first find the respective ratio of sides and then take the inverse of the trigonometric ratio to find the measure of the angle.

37. a.

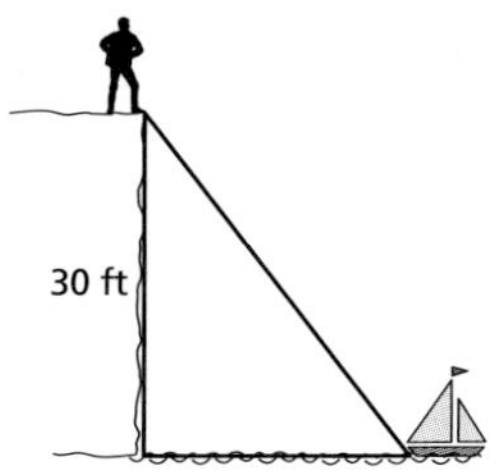

b.

Angle of depression	40°	50°	60°	70°	80°
Approximate length of line of sight (feet)	46.7	39.2	34.6	31.9	30.5

c.

View of Sailboat from Cliff

Approximate Length of Line of Sight (feet)

Angle of Depression (degrees)

d. 60 ft

39. a. $$\frac{\sin A}{\cos A} = \frac{\frac{\text{length of side opposite } A}{\text{length of hypotenuse}}}{\frac{\text{length of side adjacent to } A}{\text{length of hypotenuse}}} \cdot \frac{\text{length of hypotenuse}}{\text{length of hypotenuse}}$$

$$= \frac{\text{length of side opposite } A}{\text{length of side adjacent to } A}$$

$$= \tan A$$

b. $(\sin A)^2 + (\cos A)^2$

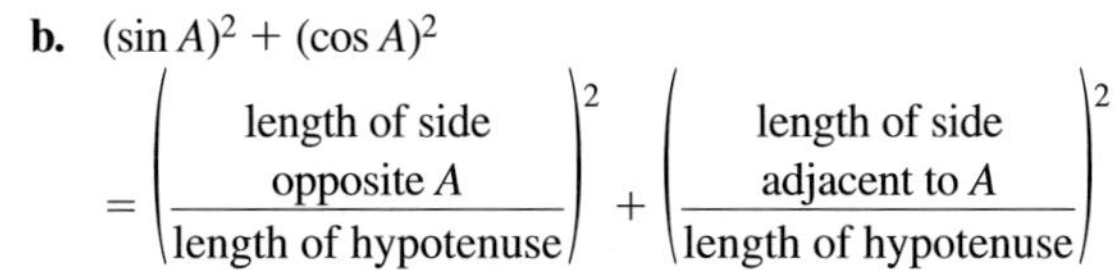

$$= \left(\frac{\text{length of side opposite } A}{\text{length of hypotenuse}}\right)^2 + \left(\frac{\text{length of side adjacent to } A}{\text{length of hypotenuse}}\right)^2$$

$$= \frac{(\text{length of side opposite } A)^2 + (\text{length of side adjacent to } A)^2}{(\text{length of hypotenuse})^2}.$$

By the Pythagorean Theorem (Thm. 9.1), $(\text{length of side opposite } A)^2 + (\text{length of side adjacent to } A)^2 = (\text{length of hypotenuse})^2$.

So, $(\sin A)^2 + (\cos A)^2 = \frac{(\text{length of hypotenuse})^2}{(\text{length of hypotenuse})^2} = 1$.

9.5 Maintaining Mathematical Proficiency *(p. 500)*

41. $x = 8$; yes **43.** $x = 45$; yes

9.6 Vocabulary and Core Concept Check *(p. 505)*

1. sides, angles

9.6 Monitoring Progress and Modeling with Mathematics *(pp. 505–506)*

3. $\angle C$ **5.** $\angle A$ **7.** about $48.6°$ **9.** about $70.7°$

11. about $15.6°$ **13.** $AB = 15, m\angle A \approx 53.1°, m\angle B \approx 36.9°$

15. $YZ \approx 8.5, m\angle X \approx 70.5°, m\angle Z \approx 19.5°$

17. $KL \approx 5.1, ML \approx 6.1, m\angle K = 50°$

19. The sine ratio should be the length of the opposite side to the length of the hypotenuse, not the adjacent side; $\sin^{-1} \frac{8}{17} = m\angle T$

21. about $59.7°$

23.

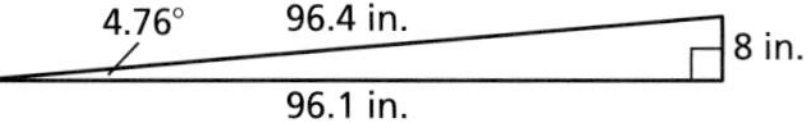

25. about $36.9°$; $PQ = 3$ centimeters and $PR = 4$ centimeters, so $m\angle R = \tan^{-1}\left(\frac{3}{4}\right) \approx 36.9°$.

27. $KM \approx 7.8$ ft, $JK \approx 11.9$ ft, $m\angle JKM = 49°$; $ML \approx 19.5$ ft, $m\angle MKL \approx 68.2°, m\angle L \approx 21.8°$

29. a. *Sample answer:* $\tan^{-1} \frac{3}{1}$; about $71.6°$

b. *Sample answer:* $\tan^{-1} \frac{4}{3}$; about $53.1°$

31. Because the sine is the ratio of the length of a leg to the length of the hypotenuse, and the hypotenuse is always longer than either of the legs, the sine cannot have a value greater than 1.

9.6 Maintaining Mathematical Proficiency *(p. 506)*

33. $x = 8$ **35.** $x = 2.46$

9.7 Vocabulary and Core Concept Check *(p. 513)*

1. Both the Law of Sines (Thm. 9.9) and the Law of Cosines (Thm. 9.10) can be used to solve any triangle.

9.7 Monitoring Progress and Modeling with Mathematics *(pp. 513–516)*

3. about 0.7986 **5.** about -0.7547 **7.** about -0.2679

9. about 81.8 square units **11.** about 147.3 square units

13. $m\angle A = 48°, b \approx 25.5, c \approx 18.7$

15. $m\angle B = 66°, a \approx 14.3, b \approx 24.0$

17. $m\angle A \approx 80.9°, m\angle C \approx 43.1°, a \approx 20.2$

19. $a \approx 5.2, m\angle B \approx 50.5°, m\angle C \approx 94.5°$

21. $m\angle A \approx 81.1°, m\angle B \approx 65.3°, m\angle C \approx 33.6°$

23. $b \approx 35.8, m\angle A \approx 46.2°, m\angle C \approx 70.8°$

25. According to the Law of Sines (Thm. 9.9), the ratio of the sine of an angle's measure to the length of its opposite side should be equal to the ratio of the sine of another angle measure to the length of its opposite side; $\frac{\sin C}{5} = \frac{\sin 55°}{6}$, $\sin C = \frac{5 \sin 55°}{6}$, $m\angle C \approx 43.0°$

27. Law of Sines (Thm. 9.9); given two angle measures and the length of a side; $m\angle C = 64°, a \approx 19.2, c \approx 18.1$

29. Law of Cosines (Thm. 9.10); given the lengths of two sides and the measure of the included angle; $c \approx 19.3$, $m\angle A \approx 34.3°, m\angle B \approx 80.7°$

31. Law of Sines (Thm. 9.9); given the lengths of two sides and the measure of a nonincluded angle; $m\angle A \approx 111.2°$, $m\angle B \approx 28.8°, a \approx 52.2$

33. about 10.7 ft **35.** about 5.1 mi

37. cousin; You are given the lengths of two sides and the measure of their included angle.

39. yes; The area of any triangle is given by one-half the product of the lengths of two sides times the sine of their included angle. For $\triangle QRS$, $A = \frac{1}{2}qr \sin S = \frac{1}{2}(25)(17)\sin 79° \approx 208.6$ square units.

41. a. about 163.4 yd **b.** about $3.5°$

43. $x = 99, y \approx 20.1$ **45.** $c^2 = a^2 + b^2$

47. a. $m\angle B \approx 52.3°, m\angle C \approx 87.7°, c \approx 20.2$; $m\angle B \approx 127.7°, m\angle C \approx 12.3°, c \approx 4.3$

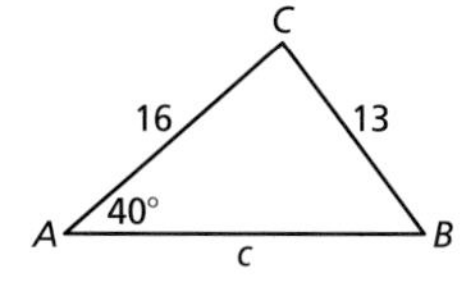

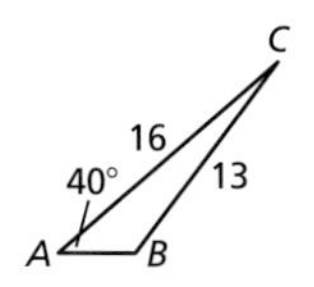

b. $m\angle B \approx 42.4°, m\angle C \approx 116.6°, c \approx 42.4$; $m\angle B \approx 137.6°, m\angle C \approx 21.4°, c \approx 17.3$

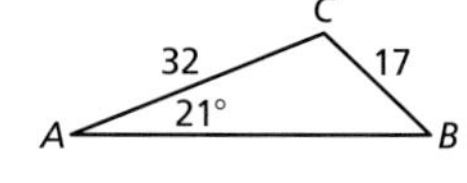

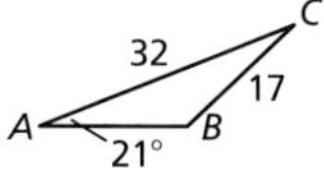

49. about 523.8 mi

51. a.

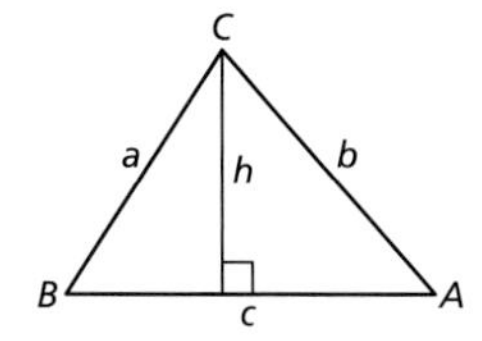

The formula for the area of $\triangle ABC$ with altitude h drawn from C to $\overline{AB}$ as shown is $A = \frac{1}{2}ch$. Because $\sin A = \frac{h}{b}$, $h = b \sin A$. By substituting, you get $A = \frac{1}{2}c(b \sin A) = \frac{1}{2}bc \sin a$.

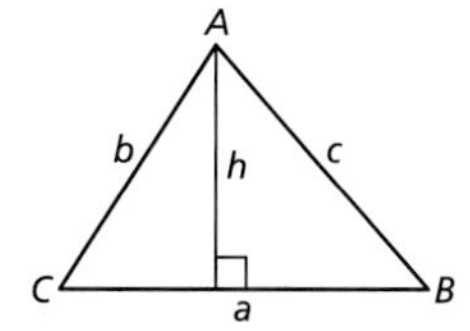

The formula for the area of $\triangle ABC$ with altitude h drawn from A to $\overline{BC}$ as shown is $A = \frac{1}{2}ah$. Because $\sin B = \frac{h}{c}$, $h = c \sin B$. By substituting, you get $A = \frac{1}{2}a(c \sin B) = \frac{1}{2}ac \sin B$. See Exercise 50 for $A = \frac{1}{2}ab \sin C$.

b. They are all expressions for the area of the same triangle, so they are all equal to each other by the Transitive Property.

c. By the Multiplication Property of Equality, multiply all three expressions by 2 to get $bc \sin A = ac \sin B = ab \sin C$. By the Division Property of Equality, divide all three expressions by abc to get $\frac{\sin A}{a} = \frac{\sin B}{b} = \frac{\sin C}{c}$.

9.7 Maintaining Mathematical Proficiency *(p. 516)*

53. $r = 4$ ft, $d = 8$ ft **55.** $r = 1$ ft, $d = 2$ ft

Chapter 9 Review *(pp. 518–522)*

1. $x = 2\sqrt{34} \approx 11.7$; no **2.** $x = 12$; yes

3. $x = 2\sqrt{30} \approx 11.0$; no **4.** yes; acute **5.** yes; right

6. yes; obtuse **7.** $x = 6\sqrt{2}$ **8.** $x = 7$

9. $x = 16\sqrt{3}$ **10.** $\triangle GFH \sim \triangle FEH \sim \triangle GEF$; $x = 13.5$

11. $\triangle KLM \sim \triangle JKM \sim \triangle JLK$; $x = 2\sqrt{6} \approx 4.9$

12. $\triangle QRS \sim \triangle PQS \sim \triangle PRQ$; $x = 3\sqrt{3} \approx 5.2$

13. $\triangle TUV \sim \triangle STV \sim \triangle SUT$; $x = 25$ **14.** 15

15. $24\sqrt{3} \approx 41.6$ **16.** $6\sqrt{14} \approx 22.4$

17. $\tan J = \frac{11}{60} \approx 0.1833$, $\tan L = \frac{60}{11} \approx 5.4545$

18. $\tan N = \frac{12}{35} \approx 0.3429$, $\tan P = \frac{35}{12} \approx 2.9167$

19. $\tan A = \dfrac{7\sqrt{2}}{8} \approx 1.2374$, $\tan B = \dfrac{4\sqrt{2}}{7} \approx 0.8081$

20. $x \approx 44.0$ **21.** $x \approx 9.3$ **22.** $x \approx 12.8$

23. about 15 ft

24. $\sin X = \frac{3}{5} = 0.600$, $\sin Z = \frac{4}{5} = 0.8000$, $\cos X = \frac{4}{5} = 0.8000$, $\cos Z = \frac{3}{5} = 0.6000$

25. $\sin X = \dfrac{7\sqrt{149}}{149} \approx 0.5735$, $\sin Z = \dfrac{10\sqrt{149}}{149} \approx 0.8192$, $\cos X = \dfrac{10\sqrt{149}}{149} \approx 0.8192$, $\cos Z = \dfrac{7\sqrt{149}}{149} \approx 0.5735$

26. $\sin X = \frac{55}{73} \approx 0.7534$, $\sin Z = \frac{48}{73} \approx 0.6575$, $\cos X = \frac{48}{73} \approx 0.6575$, $\cos Z = \frac{55}{73} \approx 0.7534$

27. $s \approx 31.3$, $t \approx 13.3$ **28.** $r \approx 4.0$, $s \approx 2.9$

29. $v \approx 9.4$, $w \approx 3.4$ **30.** $\cos 18°$ **31.** $\sin 61°$

32. $m\angle Q \approx 71.3°$ **33.** $m\angle Q \approx 65.5°$ **34.** $m\angle Q \approx 2.3°$

35. $m\angle A \approx 48.2°$, $m\angle B \approx 41.8°$, $BC \approx 11.2$

36. $m\angle L = 53°$, $ML \approx 4.5$, $NL \approx 7.5$

37. $m\angle X \approx 46.1°$, $m\angle Z \approx 43.9°$, $XY \approx 17.3$

38. about 41.0 square units **39.** about 42.2 square units

40. about 208.6 square units

41. $m\angle B \approx 24.3°$, $m\angle C \approx 43.7°$, $c \approx 6.7$

42. $m\angle C = 88°$, $a \approx 25.8$, $b \approx 49.5$

43. $m\angle A \approx 99.9°$, $m\angle B \approx 32.1°$, $a \approx 37.1$

44. $b \approx 5.4$, $m\angle A \approx 141.4°$, $m\angle C \approx 13.6°$

45. $m\angle A = 35°$, $a \approx 12.3$, $c \approx 14.6$

46. $m\angle A \approx 42.6°$, $m\angle B \approx 11.7°$, $m\angle C \approx 125.7°$

Chapter 10

Chapter 10 Maintaining Mathematical Proficiency *(p. 527)*

1. $x^2 + 11x + 28$ **2.** $a^2 - 4a - 5$ **3.** $3q^2 - 31q + 36$

4. $10v^2 - 33v - 7$ **5.** $4h^2 + 11h + 6$

6. $18b^2 - 54b + 40$ **7.** $x \approx -1.45$; $x \approx 3.45$

8. $r \approx -9.24$; $r \approx -0.76$ **9.** $w = -1$, $w = 9$

10. $p \approx -10.39$; $p \approx 0.39$ **11.** $k \approx -1.32$; $k \approx 5.32$

12. $z = 1$

13. *Sample answer:* $(2n + 1)(2n + 3)$; $2n + 1$ is positive and odd when n is a nonnegative integer. The next positive, odd integer is $2n + 3$.

10.1 Vocabulary and Core Concept Check *(p. 534)*

1. They both intersect the circle in two points; Chords are segments and secants are lines.

3. concentric circles

10.1 Monitoring Progress and Modeling with Mathematics *(pp. 534–536)*

5. $\odot C$ **7.** $\overline{BH}, \overline{AD}$ **9.** $\overleftrightarrow{KG}$

11. 4

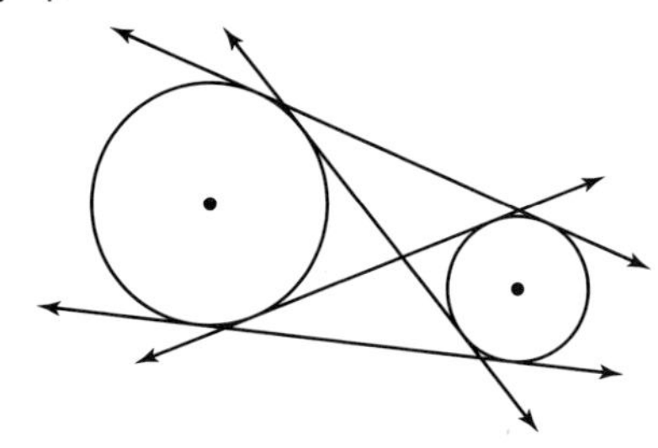

13. 2

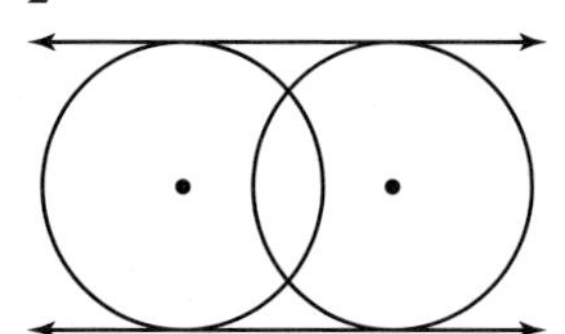

15. external **17.** internal

19. yes; $\triangle ABC$ is a right triangle.

21. no; $\triangle ABD$ is not a right triangle. **23.** 10 **25.** 10.5

27. *Sample answer:*

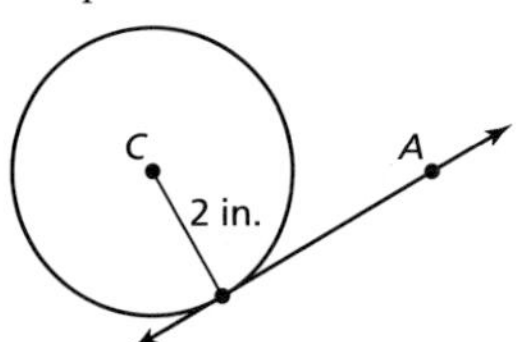

29. 5 **31.** ± 3

33. $\angle Z$ is a right angle, not $\angle YXZ$; $\overline{XY}$ is not tangent to $\odot Z$.

35. 2; 1; 0; *Sample answer:* There are two possible points of tangency from a point outside the circle, one from a point on the circle, and none from a point inside the circle.

37. 25.6 units **39.** yes; $\overline{PE}$ and $\overline{PM}$ are radii, so $\overline{PE} \cong \overline{PM}$.

41. *Sample answer:* Every point is the same distance from the center, so the farthest two points can be from each other is opposite sides of the center.

43. $\angle ARC \cong \angle BSC$ and $\angle ACR \cong \angle BCS$, so $\triangle ARC \sim \triangle BSC$ by the AA Similarity Theorem (Thm. 8.3). Because corresponding sides of similar figures are proportional, $\dfrac{AC}{BC} = \dfrac{RC}{SC}$.

45. $x = 13$, $y = 5$; $2x - 5 = x + 8$ and $2x + 4y - 6 = 2x + 14$.

47. a. Assume m is not perpendicular to $\overline{QP}$. The perpendicular segment from Q to m intersects m at some other point R. Then $QR < QP$, so R must be inside $\odot Q$, and m must be a secant line. This is a contradiction, so m must be perpendicular to $\overline{QP}$.

b. Assume m is not tangent to $\odot Q$. Then m must intersect $\odot Q$ at a second point R. $\overline{QP}$ and $\overline{QR}$ are both radii of $\odot Q$, so $\overline{QP} \cong \overline{QR}$. Because $m \perp \overline{QP}$, $QP < QR$. This is a contradiction, so m must be tangent to $\odot Q$.

10.1 Maintaining Mathematical Proficiency *(p. 536)*

49. 43°

10.2 Vocabulary and Core Concept Check *(p. 542)*

1. congruent arcs

10.2 Monitoring Progress and Modeling with Mathematics *(pp. 542–544)*

3. $\overset{\frown}{AB}$, 135°; $\overset{\frown}{ADB}$, 225° **5.** $\overset{\frown}{JL}$, 120°; $\overset{\frown}{JKL}$, 240°

7. minor arc; 70° **9.** minor arc; 45°

11. semicircle; 180° **13.** major arc; 290°

15. **a.** 132° **b.** 147° **c.** 200° **d.** 160°

17. **a.** 103° **b.** 257° **c.** 196° **d.** 305°
e. 79° **f.** 281°

19. congruent; They are in the same circle and $m\overset{\frown}{AB} = m\overset{\frown}{CD}$.

21. congruent; The circles are congruent and $m\overset{\frown}{VW} = m\overset{\frown}{XY}$.

23. 70; 110°

25. your friend; The arcs must be in the same circle or congruent circles.

27. $\overset{\frown}{AD}$ is the minor arc; $\overset{\frown}{ABD}$ **29.** 340°; 160° **31.** 18°

33. Translate $\odot A$ left a units so that point A maps to point O. The image of $\odot A$ is $\odot A'$ with center O, so $\odot A'$ and $\odot O$ are concentric circles. Dilate $\odot A'$ using center of dilation O and scale factor $\frac{r}{s}$, which maps the points s units from point O to the points $\frac{r}{s}(s) = r$ units from point O.
So, this dilation maps $\odot A'$ to $\odot O$. Because a similarity transformation maps $\odot A$ to $\odot O$, $\odot O \sim \odot A$.

35. **a.** Translate $\odot B$ so that point B maps to point A. The image of $\odot B$ is $\odot B'$ with center A. Because $\overline{AC} \cong \overline{BD}$, this translation maps $\odot B'$ to $\odot A$. A rigid motion maps $\odot B$ to $\odot A$, so $\odot A \cong \odot B$.
b. Because $\odot A \cong \odot B$, the distance from the center of the circle to a point on the circle is the same for each circle. So, $\overline{AC} \cong \overline{BD}$.

37. **a.** $m\overset{\frown}{BC} = m\angle BAC$, $m\overset{\frown}{DE} = m\angle DAE$ and $m\angle BAC = m\angle DAE$, so $m\overset{\frown}{BC} = m\overset{\frown}{DE}$. Because $\overset{\frown}{BC}$ and $\overset{\frown}{DE}$ are in the same circle, $\overset{\frown}{BC} \cong \overset{\frown}{DE}$.
b. $m\overset{\frown}{BC} = m\angle BAC$ and $m\overset{\frown}{DE} = m\angle DAE$. Because $\overset{\frown}{BC} \cong \overset{\frown}{DE}$, $\angle BAC \cong \angle DAE$.

10.2 Maintaining Mathematical Proficiency *(p. 544)*

39. 15; yes **41.** about 13.04; no

10.3 Vocabulary and Core Concept Check *(p. 549)*

1. Split the chord into two segments of equal length.

10.3 Monitoring Progress and Modeling with Mathematics *(pp. 549–550)*

3. 75° **5.** 170° **7.** 8 **9.** 5

11. $\overline{AC}$ and $\overline{DB}$ are not perpendicular; $\overset{\frown}{BC}$ is not congruent to $\overset{\frown}{CD}$.

13. yes; The triangles are congruent, so $\overline{AB}$ is a perpendicular bisector of $\overline{CD}$.

15. 17

17. about 13.9 in.; The perpendicular bisectors intersect at the center, so the right triangle with legs of 6 inches and 3.5 inches have a hypotenuse equal to the length of the radius.

19. **a.** Because $PA = PB = PC = PD$, $\triangle PDC \cong \triangle PAB$ by the SSS Congruence Theorem (Thm. 5.8). So, $\angle DPC \cong \angle APB$ and $\overset{\frown}{AB} \cong \overset{\frown}{CD}$.
b. $PA = PB = PC = PD$, and because $\overset{\frown}{AB} \cong \overset{\frown}{CD}$, $\angle DPC \cong \angle APB$. By the SAS Congruence Theorem (Thm. 5.5), $\triangle PDC \cong \triangle PAB$, so $\overline{AB} \cong \overline{CD}$.

21. about 16.26°; *Sample answer:* $AB = 2\sqrt{2}$ and $PA = PB = 10$, so $m\angle APB \approx 16.26°$ by the Law of Cosines (Thm. 9.10).

23. $\overline{TP} \cong \overline{PR}$, $\overline{LP} \cong \overline{LP}$, and $\overline{LT} \cong \overline{LR}$, so $\triangle LPR \cong \triangle LPT$ by the SSS Congruence Theorem (Thm. 5.8). Then $\angle LPT \cong \angle LPR$, so $m\angle LPT = m\angle LPR = 90°$. By definition, $\overline{LP}$ is a perpendicular bisector of $\overline{RT}$, so L lies on $\overline{QS}$. Because $\overline{QS}$ contains the center, $\overline{QS}$ is a diameter of $\odot L$.

25. If $\overline{AB} \cong \overline{CD}$, then $\overline{GC} \cong \overline{FA}$. Because $\overline{EC} \cong \overline{EA}$, $\triangle ECG \cong \triangle EAF$ by the HL Congruence Theorem (Thm. 5.9), so $\overline{EF} \cong \overline{EG}$ and $EF = EG$.
If $EF = EG$, then because $\overline{EC} \cong \overline{ED} \cong \overline{EA} \cong \overline{EB}$, $\triangle AEF \cong \triangle BEF \cong \triangle DEG \cong \triangle CEG$ by the HL Congruence Theorem (Thm. 5.9). Then $\overline{AF} \cong \overline{BF} \cong \overline{DG} \cong \overline{CG}$, so $\overline{AB} \cong \overline{CD}$.

10.3 Maintaining Mathematical Proficiency *(p. 550)*

27. 259°

10.4 Vocabulary and Core Concept Check *(p. 558)*

1. inscribed polygon

10.4 Monitoring Progress and Modeling with Mathematics *(pp. 558–560)*

3. 42° **5.** 10° **7.** 120°

9. $\angle ACB \cong \angle ADB$, $\angle DAC \cong \angle DBC$ **11.** 51°

13. $x = 100$, $y = 85$ **15.** $a = 20$, $b = 22$

17. The inscribed angle was not doubled; $m\angle BAC = 2(53°) = 106°$

19. $x = 25$, $y = 5$; 130°, 75°, 50°, 105°

21. $x = 30$, $y = 20$; 60°, 60°, 60°

23.

25. yes; Opposite angles are always supplementary.

27. no; Opposite angles are not always supplementary.

29. no; Opposite angles are not always supplementary.

31.

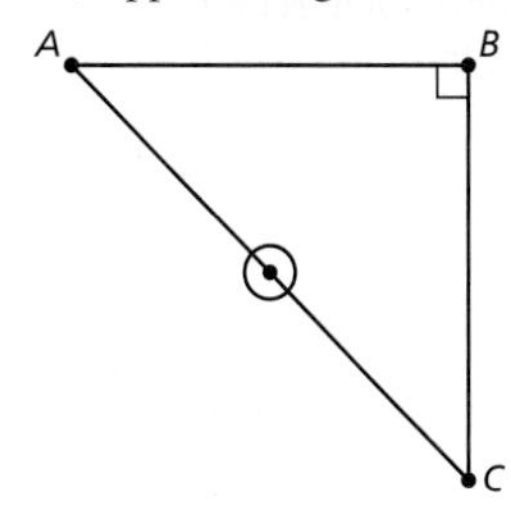

220,000 km

33. double the radius

35. Each diagonal splits the rectangle into two right triangles.

Selected Answers

37. **a.** $\overline{QB} \cong \overline{QA}$, so $\triangle ABC$ is isosceles. By the Base Angles Theorem (Thm. 5.6), $\angle QBA \cong \angle QAB$, so $m\angle BAQ = x°$. By the Exterior Angles Theorem (Thm. 5.2), $m\angle AQC = 2x°$. Then $m\widehat{AC} = 2x°$, so $m\angle ABC = x° = \frac{1}{2}(2x)° = \frac{1}{2}m\widehat{AC}$.

b. Given: $\angle ABC$ is inscribed in $\odot Q$. $\overline{DB}$ is a diameter; Prove: $m\angle ABC = \frac{1}{2}m\widehat{AC}$; By Case 1, proved in part (a), $m\angle ABD = \frac{1}{2}m\widehat{AD}$ and $m\angle CBD = \frac{1}{2}m\widehat{CD}$. By the Arc Addition Postulate (Post. 10.1), $m\widehat{AD} + m\widehat{CD} = m\widehat{AC}$. By the Angle Addition Postulate (Post. 1.4), $m\angle ABD + m\angle CBD = m\angle ABC$.

Then $m\angle ABC = \frac{1}{2}m\widehat{AD} + \frac{1}{2}m\widehat{CD}$
$= \frac{1}{2}\left(m\widehat{AD} + m\widehat{CD}\right)$
$= \frac{1}{2}m\widehat{AC}$.

c. Given: $\angle ABC$ is inscribed in $\odot Q$. $\overline{DB}$ is a diameter; Prove: $m\angle ABC = \frac{1}{2}m\widehat{AC}$; By Case 1, proved in part (a), $m\angle DBA = \frac{1}{2}m\widehat{AD}$ and $m\angle DBC = \frac{1}{2}m\widehat{CD}$. By the Arc Addition Postulate (Post. 10.1), $m\widehat{AC} + m\widehat{CD} = m\widehat{AD}$, so $m\widehat{AC} = m\widehat{AD} - m\widehat{CD}$. By the Angle Addition Postulate (Post. 1.4), $m\angle DBC + m\angle ABC = m\angle DBA$, so $m\angle ABC = m\angle DBA - m\angle DBC$. Then

$m\angle ABC = \frac{1}{2}m\widehat{AD} - \frac{1}{2}m\widehat{CD}$
$= \frac{1}{2}\left(m\widehat{AD} - m\widehat{CD}\right)$
$= \frac{1}{2}m\widehat{AC}$.

39. To prove the conditional, find the measure of the intercepted arc of the right angle and the definition of a semicircle to show the hypotenuse of the right triangle must be the diameter of the circle. To prove the converse, use the definition of a semicircle to find the measure of the angle opposite the diameter.

41. 2.4 units

10.4 Maintaining Mathematical Proficiency *(p. 560)*

43. $x = \frac{145}{3}$ **45.** $x = 120$

10.5 Vocabulary and Core Concept Check *(p. 566)*

1. outside

10.5 Monitoring Progress and Modeling with Mathematics *(pp. 566–568)*

3. 130° **5.** 130° **7.** 115 **9.** 56

11. 40 **13.** 34

15. $\angle SUT$ is not a central angle; $m\angle SUT = \frac{1}{2}\left(m\widehat{QR} + m\widehat{ST}\right) = 41.5°$

17. 60°; Because the sum of the angles of a triangle always equals 180°, solve the equation $90 + 30 + x = 180$.

19. 30°; Because the sum of the angles of a triangle always equals 180°, solve the equation $60 + 90 + x = 180$.

21. 30°; This angle is complementary to $\angle 2$, which is 60°.

23. about 2.8° **25.** $360 - 10x$; 160°

27. $m\angle LPJ < 90$; The difference of $m\widehat{JL}$ and $m\widehat{LK}$ must be less than 180°, so $m\angle LPJ < 90$.

29. By the Angles Inside a Circle Theorem (Thm. 10.15), $m\angle JPN = \frac{1}{2}\left(m\widehat{JN} + m\widehat{KM}\right)$. By the Angles Outside the Circle Theorem (Thm. 10.16), $m\angle JLN = \frac{1}{2}\left(m\widehat{JN} - m\widehat{KM}\right)$. Because the angle measures are positive, $\frac{1}{2}\left(m\widehat{JN} + m\widehat{KM}\right) > \frac{1}{2}m\widehat{JN} > \frac{1}{2}\left(m\widehat{JN} - m\widehat{KM}\right)$, so, $m\angle JPN > m\angle JLN$.

31. **a.**

b. $m\widehat{AB} = 2m\angle BAC$, $m\widehat{AB} = 360° - 2m\angle BAC$

c. 90°; $2m\angle BAC = 360° - 2m\angle BAC$ when $m\angle BAC = 90°$.

33. **a.** By the Tangent Line to Circle Theorem (Thm. 10.1), $m\angle BAC$ is 90°, which is half the measure of the semicircular arc.

b.

By the Tangent Line to Circle Theorem (Thm. 10.1), $m\angle CAD = 90°$. $m\angle DAB = \frac{1}{2}m\widehat{DB}$ and by part (a), $m\angle CAD = \frac{1}{2}m\widehat{AD}$. By the Angle Addition Postulate (Post. 1.4), $m\angle BAC = m\angle BAD + m\angle CAD$. So, $m\angle BAC = \frac{1}{2}m\widehat{DB} + \frac{1}{2}m\widehat{AD} = \frac{1}{2}\left(m\widehat{DB} + m\widehat{AD}\right)$. By the Arc Addition Postulate (Post. 10.1), $m\widehat{DB} + m\widehat{AD} = m\widehat{ADB}$, so $m\angle BAC = \frac{1}{2}\left(m\widehat{ADB}\right)$.

c.

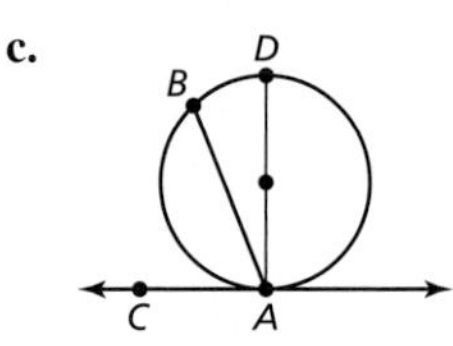

By the Tangent Line to Circle Theorem (Thm. 10.1), $m\angle CAD = 90°$. $m\angle DAB = \frac{1}{2}m\widehat{DB}$ and by part (a), $m\angle DAC = \frac{1}{2}m\widehat{ABD}$. By the Angle Addition Postulate (Post. 1.4), $m\angle BAC = m\angle DAC - m\angle DAB$. So, $m\angle BAC = \frac{1}{2}m\widehat{ABD} - \frac{1}{2}m\widehat{DB} = \frac{1}{2}\left(m\widehat{ABD} - m\widehat{DB}\right)$. By the Arc Addition Postulate (Post. 10.1), $m\widehat{ABD} - m\widehat{DB} = m\widehat{AB}$, so $m\angle BAC = \frac{1}{2}\left(m\widehat{AB}\right)$.

35.

STATEMENTS	REASONS
1. Chords $\overline{AC}$ and $\overline{BD}$ intersect.	1. Given
2. $m\angle ACB = \frac{1}{2}m\widehat{AB}$ and $m\angle DBC = \frac{1}{2}m\widehat{DC}$	2. Measure of an Inscribed Angle Theorem (Thm. 10.10)
3. $m\angle 1 = m\angle DBC + m\angle ACB$	3. Exterior Angle Theorem (Thm. 5.2)
4. $m\angle 1 = \frac{1}{2}m\widehat{DC} + \frac{1}{2}m\widehat{AB}$	4. Substitution Property of Equality
5. $m\angle 1 = \frac{1}{2}\left(m\widehat{DC} + m\widehat{AB}\right)$	5. Distributive Property

37. By the Exterior Angle Theorem (Thm. 5.2), $m\angle 2 = m\angle 1 + m\angle ABC$, so $m\angle 1 = m\angle 2 - m\angle ABC$. By the Tangent and Intersected Chord Theorem (Thm. 10.14), $m\angle 2 = \frac{1}{2}m\widehat{BC}$ and by the Measure of an Inscribed Angle Theorem (Thm. 10.10), $m\angle ABC = \frac{1}{2}m\widehat{AC}$. By the Substitution Property, $m\angle 1 = \frac{1}{2}m\widehat{BC} - \frac{1}{2}m\widehat{AC} = \frac{1}{2}\left(m\widehat{BC} - m\widehat{AC}\right)$;

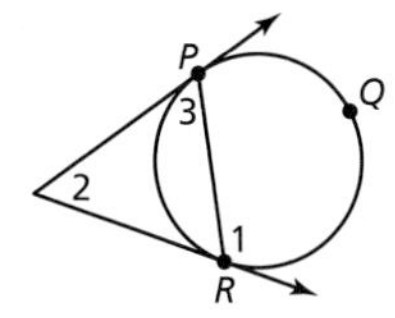

By the Exterior Angle Theorem (Thm. 5.2), $m\angle 1 = m\angle 2 + m\angle 3$, so $m\angle 2 = m\angle 1 - m\angle 3$. By the Tangent and Intersected Chord Theorem (Thm. 10.14), $m\angle 1 = \frac{1}{2}m\widehat{PQR}$ and $m\angle 3 = \frac{1}{2}m\widehat{PR}$. By the Substitution Property, $m\angle 2 = \frac{1}{2}m\widehat{PQR} - \frac{1}{2}m\widehat{PR} = \frac{1}{2}\left(m\widehat{PQR} - m\widehat{PR}\right)$;

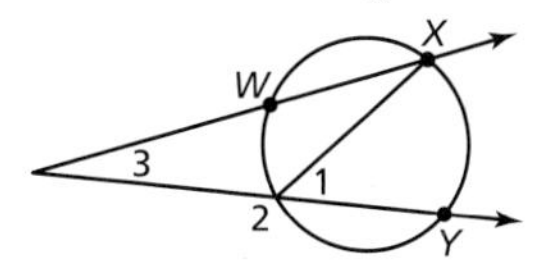

By the Exterior Angle Theorem (Thm. 5.2), $m\angle 1 = m\angle 3 + m\angle WXZ$, so $m\angle 3 = m\angle 1 - m\angle WXZ$. By the Measure of an Inscribed Angle Theorem (Thm. 10.10), $m\angle 1 = \frac{1}{2}m\widehat{XY}$ and $m\angle WXZ = \frac{1}{2}m\widehat{WZ}$. By the Substitution Property, $m\angle 3 = \frac{1}{2}m\widehat{XY} - \frac{1}{2}m\widehat{WZ} = \frac{1}{2}\left(m\widehat{XY} - m\widehat{WZ}\right)$.

39. 20°; *Sample answer:* $m\widehat{WY} = 160°$ and $m\widehat{WX} = m\widehat{ZY}$, so

$$m\angle P = \tfrac{1}{2}\left(m\widehat{WZ} - m\widehat{XY}\right)$$
$$= \tfrac{1}{2}\left(\left(200° - m\widehat{ZY}\right) - \left(160° - m\widehat{WX}\right)\right)$$
$$= \tfrac{1}{2}(40°).$$

10.5 Maintaining Mathematical Proficiency *(p. 568)*

41. $x = -4, x = 3$ **43.** $x = -3, x = -1$

10.6 Vocabulary and Core Concept Check *(p. 573)*

1. external segment

10.6 Monitoring Progress and Modeling with Mathematics *(pp. 573–574)*

3. 5 **5.** 4 **7.** 4 **9.** 5 **11.** 12 **13.** 4

15. The chords were used instead of the secant segments; $CF \cdot DF = BF \cdot AF$; $CD = 2$

17. about 124.5 ft

19.

STATEMENTS	REASONS
1. $\overline{AB}$ and $\overline{CD}$ are chords intersecting in the interior of the circle.	1. Given
2. $\angle AEC \cong \angle DEB$	2. Vertical Angles Congruence Theorem (Thm. 2.6)
3. $\angle ACD \cong \angle ABD$	3. Inscribed Angles of a Circle Theorem (Thm. 10.11)
4. $\triangle AEC \sim \triangle DEB$	4. AA Similarity Theorem (Thm. 8.3)
5. $\dfrac{EA}{ED} = \dfrac{EC}{EB}$	5. Corresponding side lengths of similar triangles are proportional.
6. $EB \cdot EA = EC \cdot ED$	6. Cross Products Property

21.

By the Tangent Line to Circle Theorem (Thm. 10.1), $\angle EAO$ is a right angle, which makes $\triangle AEO$ a right triangle. By the Pythagorean Theorem (Thm. 9.1), $(r + y)^2 = r^2 + x^2$. So, $r^2 + 2yr + y^2 = r^2 + x^2$. By the Subtraction Property of Equality, $2yr + y^2 = x^2$. Then $y(2r + y) = x^2$, so $EC \cdot ED = EA^2$.

23. $BC = \dfrac{AD^2 + (AD)(DE) - AB^2}{AB}$ **25.** $2\sqrt{10}$

10.6 Maintaining Mathematical Proficiency *(p. 574)*

27. $x = -9, x = 5$ **29.** $x = -7, x = 1$

10.7 Vocabulary and Core Concept Check *(p. 579)*

1. $(x - h)^2 + (y - k)^2 = r^2$

10.7 Monitoring Progress and Modeling with Mathematics *(pp. 579–580)*

3. $x^2 + y^2 = 4$ **5.** $x^2 + y^2 = 49$

7. $(x + 3)^2 + (y - 4)^2 = 1$ **9.** $x^2 + y^2 = 36$

11. $x^2 + y^2 = 58$

13. center: (0, 0), radius: 7 **15.** center: (3, 0), radius: 4

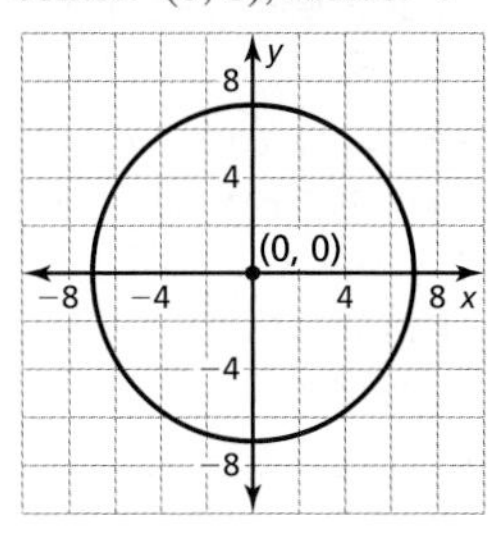

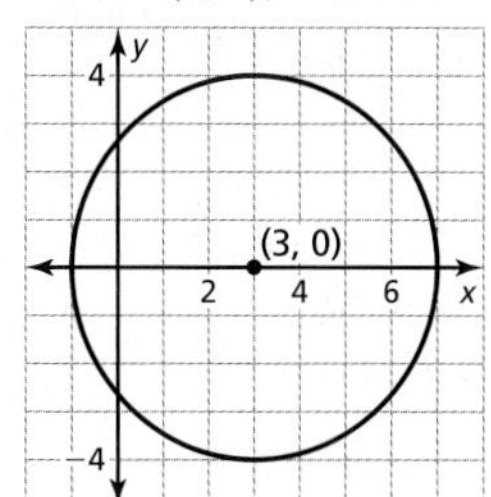

17. center: (4, 1), radius: 1

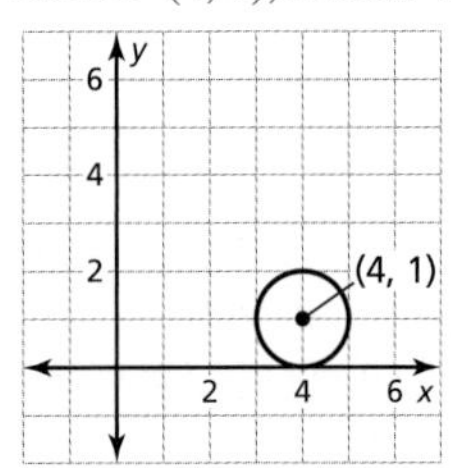

19. The radius of the circle is 8. $\sqrt{(2-0)^2+(3-0)^2}=\sqrt{13}$, so (2, 3) does not lie on the circle.

21. The radius of the circle is $\sqrt{10}$. $\sqrt{(\sqrt{6}-0)^2+(2-0)^2}=\sqrt{10}$, so $(\sqrt{6}, 2)$ does lie on the circle.

23. a.

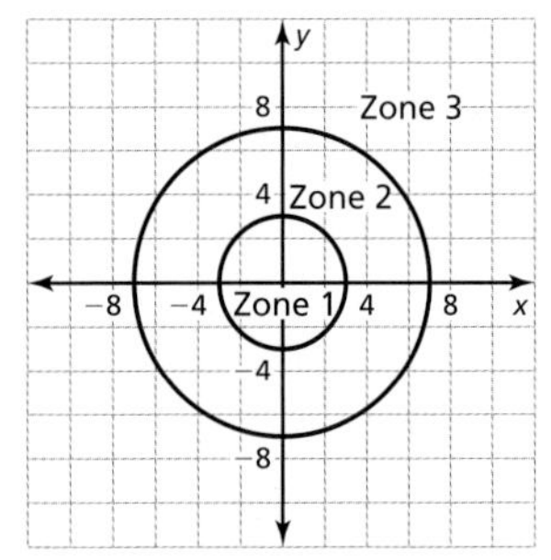

b. zone 2, zone 3, zone 1, zone 1, zone 2

25.

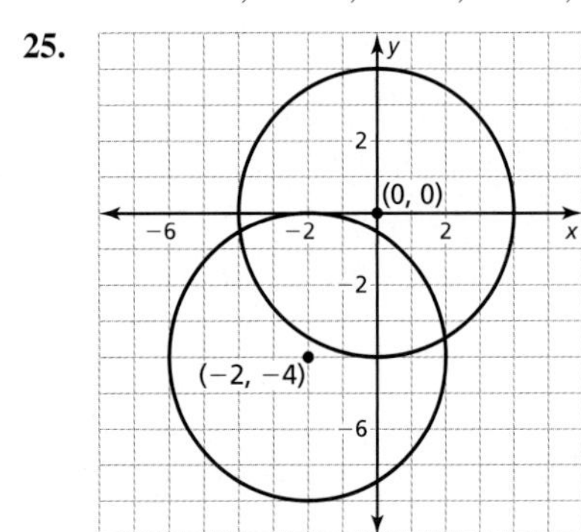

The equation of the image is $(x+2)^2+(y+4)^2=16$; The equation of the image of a circle after a translation m units to the left and n units down is $(x+m)^2+(y+n)^2=r^2$.

27. $(x-4)^2+(y-9)^2=16$; $m\angle Z=90°$, so $\overline{XY}$ is a diameter.

29. tangent; The system has one solution.

31. secant; The system has two solutions, and (5, −1) is not on the line.

33. yes; The diameter perpendicularly bisects the chord from (−1, 0) to (1, 0), so the center is on the y-axis at (0, k) and the radius is k^2+1.

10.7 Maintaining Mathematical Proficiency *(p. 580)*

35. minor arc; 53° **37.** major arc; 270°

39. semicircle; 180°

Chapter 10 Review *(pp. 582–586)*

1. radius **2.** chord **3.** tangent **4.** diameter

5. secant **6.** radius **7.** internal **8.** external

9. 2 **10.** 2 **11.** 12 **12.** tangent; $20^2+48^2=52^2$

13. 100° **14.** 60° **15.** 160° **16.** 80°

17. not congruent; The circles are not congruent.

18. congruent; The circles are congruent and $m\widehat{AB}=m\widehat{EF}$.

19. 61° **20.** 65° **21.** 91° **22.** 26 **23.** 80

24. $q=100, r=20$ **25.** 5 **26.** $y=30, z=10$

27. $m=44, n=39$ **28.** 28 **29.** 70 **30.** 106

31. 16 **32.** 240° **33.** 5 **34.** 3 **35.** 10

36. about 10.7 ft **37.** $(x-4)^2+(y+1)^2=9$

38. $(x-8)^2+(y-6)^2=36$ **39.** $x^2+y^2=16$

40. $x^2+y^2=81$ **41.** $(x+5)^2+(y-2)^2=1.69$

42. $(x-6)^2+(y-21)^2=16$

43. $(x+3)^2+(y-2)^2=256$

44. $(x-10)^2+(y-7)^2=12.25$ **45.** $x^2+y^2=27.04$

46. $(x+7)^2+(y-6)^2=25$

47. center: (6, −4), radius: 2

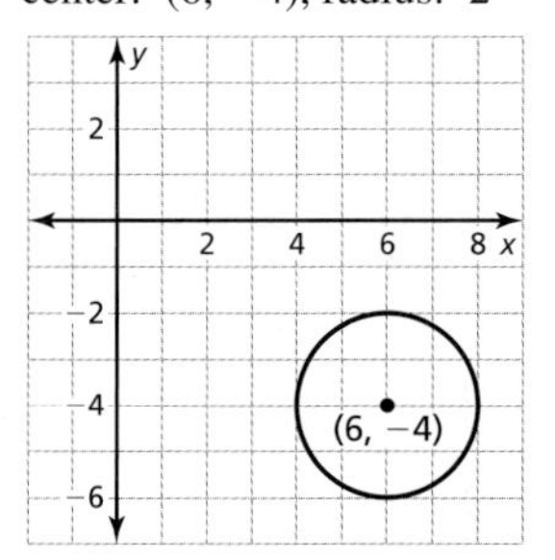

48. The radius of the circle is 5. $d=\sqrt{(0-4)^2+(0+3)^2}=5$, so (4, −3) is on the circle.

Chapter 11

Chapter 11 Maintaining Mathematical Proficiency *(p. 591)*

1. 158 ft^2 **2.** 144 m^2 **3.** 184 cm^2 **4.** 9 in.

5. 2 cm **6.** 12 ft **7.** $S=6x^2$; cube

11.1 Vocabulary and Core Concept Check *(p. 598)*

1. πd

11.1 Monitoring Progress and Modeling with Mathematics *(pp. 598–600)*

3. about 37.70 in. **5.** 14 units **7.** about 3.14 ft

9. about 35.53 m

11. The diameter was used as the radius; $C=\pi d=9\pi$ in.

13. 182 ft **15.** about 44.85 units **17.** about 20.57 units

19. $\frac{7\pi}{18}$ rad **21.** 165° **23.** about 27.19 min

25. 8π units **27.** about 7.85 units

29. yes; *Sample answer:* The arc length also depends on the radius.

31. B **33.** $2\frac{1}{3}$

35. arc length of $\widehat{AB}=r\theta$; about 9.42 in.

37. yes; *Sample answer:* The circumference of the red circle can be found using $2=\frac{30°}{360°}C$. The circumference of the blue circle is double the circumference of the red circle.

39. 28 units

41. *Sample answer:*

STATEMENTS	REASONS
1. $\overline{FG} \cong \overline{GH}$, $\angle JFK \cong \angle KLF$	**1.** Given
2. $FG = GH$	**2.** Definition of congruent segments
3. $FH = FG + GH$	**3.** Segment Addition Postulate (Post. 1.2)
4. $FH = 2FG$	**4.** Substitution Property of Equality
5. $m\angle JFK = m\angle KFL$	**5.** Definition of congruent angles
6. $m\angle JFL = m\angle JFK + m\angle KFL$	**6.** Angle Addition Postulate (Post. 1.4)
7. $m\angle JFL = 2m\angle JFK$	**7.** Substitution Property of Equality
8. $\angle NFG \cong \angle JFL$	**8.** Vertical Angles Congruence Theorem (Thm. 2.6)
9. $m\angle NFG = m\angle JFL$	**9.** Definition of congruent angles
10. $m\angle NFG = 2m\angle JFK$	**10.** Substitution Property of Equality
11. arc length of $\widehat{JK} = \frac{m\angle JFK}{360°} \cdot 2\pi FH$, arc length of $\widehat{NG} = \frac{m\angle NFG}{360°} \cdot 2\pi FG$	**11.** Formula for arc length
12. arc length of $\widehat{JK} = \frac{m\angle JFK}{360°} \cdot 2\pi(2FG)$, arc length of $\widehat{NG} = \frac{2m\angle JFK}{360°} \cdot 2\pi FG$	**12.** Substitution Property of Equality
13. arc length of $\widehat{NG}$ = arc length of $\widehat{JK}$	**13.** Transitive Property of Equality

11.1 Maintaining Mathematical Proficiency *(p. 600)*

43. 15 square units

11.2 Vocabulary and Core Concept Check *(p. 606)*

1. sector

11.2 Monitoring Progress and Modeling with Mathematics *(pp. 606–608)*

3. about 0.50 cm^2 **5.** about 78.54 in.^{2}

7. about 5.32 ft **9.** about 4.00 in.

11. about 464 people per mi^2 **13.** about 319,990 people

15. about 52.36 in.^{2}; about 261.80 in.^{2}

17. about 937.31 m^2; about 1525.70 m^2

19. The diameter was substituted in the formula for area as the radius; $A = \pi(6)^2 \approx 113.10$ ft^2

21. about 66.04 cm^2 **23.** about 1696.46 m^2

25. about 43.98 ft^2 **27.** about 26.77 in.^{2}

29. about 192.48 ft^2

31. **a.** about 285 ft^2 **b.** about 182 ft^2

33. *Sample answer:* change side lengths to radii and perimeter to circumference; Different terms need to be used because a circle is not a polygon.

35. **a.** *Sample answer:* The total is 100%.

b. bus 234°; walk 90°; other 36°

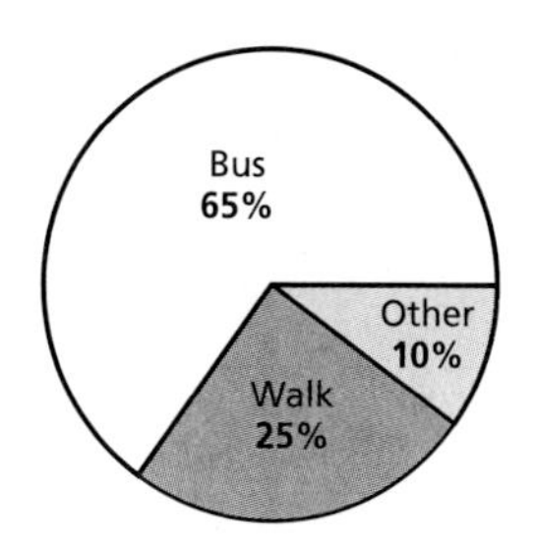

c. bus: about 8.17 in.^{2}; walk: about 3.14 in.^{2}; other: about 1.26 in.^{2}

37. **a.** You should buy two 14-inch pizzas; *Sample answer:* The area is 98π square inches and the cost is $25.98.

b. You should buy two 10-inch pizzas and one 14-inch pizza; *Sample answer:* Buying three 10-inch pizzas is the only cheaper option, and it would not be enough pizza.

c. You should buy four 10-inch pizzas; *Sample answer:* The total circumference is 20π inches.

39. **a.** 2.4 in.^{2}; 4.7 in.^{2}; 7.1 in.^{2}; 9.4 in.^{2}; 11.8 in.^{2}; 14.1 in.^{2}

b.

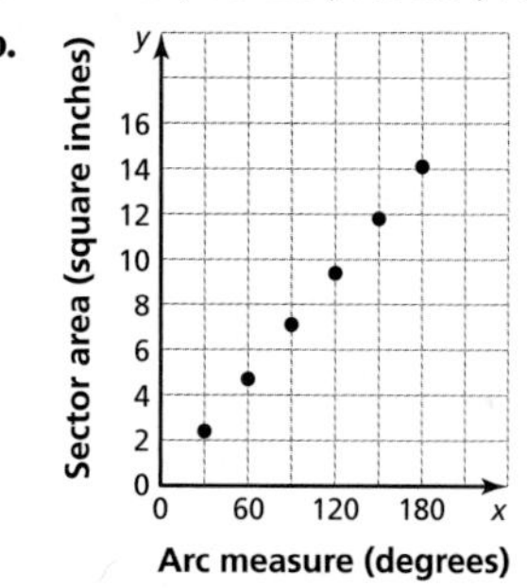

c. yes; *Sample answer:* The rate of change is constant.

d. yes; no; *Sample answer:* The rate of change will still be constant.

41. *Sample answer:* Let $2a$ and $2b$ represent the lengths of the legs of the triangle. The areas of the semicircles are $\frac{1}{2}\pi a^2$, $\frac{1}{2}\pi b^2$, and $\frac{1}{2}\pi(a^2 + b^2)$. $\frac{1}{2}\pi a^2 + \frac{1}{2}\pi b^2 = \frac{1}{2}\pi(a^2 + b^2)$, and subtracting the areas of the unshaded regions from both sides leaves the area of the crescents on the left and the area of the triangle on the right.

11.2 Maintaining Mathematical Proficiency *(p. 608)*

43. 49 ft^2 **45.** 15 ft^2

11.3 Vocabulary and Core Concept Check *(p. 614)*

1. Divide 360° by the number of sides.

11.3 Monitoring Progress and Modeling with Mathematics *(pp. 614–616)*

3. 361 square units **5.** 70 square units **7.** P

9. 5 units **11.** 36° **13.** 15° **15.** 45°

17. 67.5° **19.** about 62.35 square units

21. about 20.87 square units **23.** about 342.24 square units

25. The side lengths were used instead of the diagonals; $A = \frac{1}{2}(8)(4) = 16$

27. about 79.60 square units **29.** about 117.92 square units

31. about 166 in.^{2}

33. true; *Sample answer:* As the number of sides increases, the polygon fills more of the circle.

35. false; *Sample answer:* The radius can be less than or greater than the side length.

37. about 59.44 square units **39.** $x^2 = 324$; 18 in.; 36 in.

41. yes; about 24.73 in.^{2}; *Sample answer:* Each side length is 2 inches, and the central angle is 40°.

43. *Sample answer:* Let $QT = x$ and $TS = y$. The area of $PQRS$ is $\frac{1}{2}d_2x + \frac{1}{2}d_2y = \frac{1}{2}d_2(x + y) = \frac{1}{2}d_2d_1$.

45. $A = \frac{1}{2}d^2$; $A = \frac{1}{2}d^2 = \frac{1}{2}(s^2 + s^2) = \frac{1}{2}(2s^2) = s^2$

47. about 6.47 cm **49.** about 52

51. about 43 square units; *Sample answer:* $A = \frac{1}{2}aP$; There are fewer calculations.

11.3 Maintaining Mathematical Proficiency *(p. 616)*

53. line symmetry; 1 **55.** rotational symmetry; 180°

11.4 Vocabulary and Core Concept Check *(p. 621)*

1. polyhedron

11.4 Monitoring Progress and Modeling with Mathematics *(pp. 621–622)*

3. B **5.** A **7.** yes; pentagonal pyramid **9.** no

11. circle **13.** triangle

15.

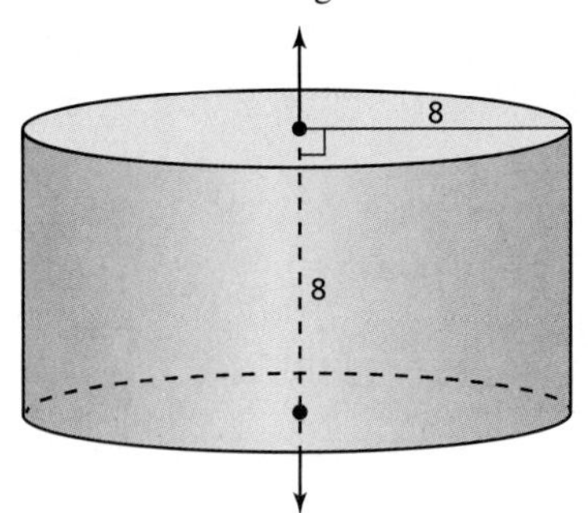

cylinder with height 8 and base radius 8

17.

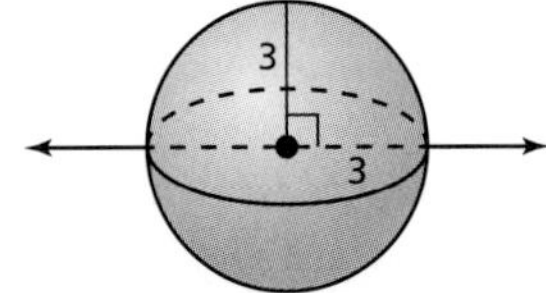

sphere with radius 3

19. There are two parallel, congruent bases, so it is a prism, not a pyramid; The solid is a triangular prism.

21.

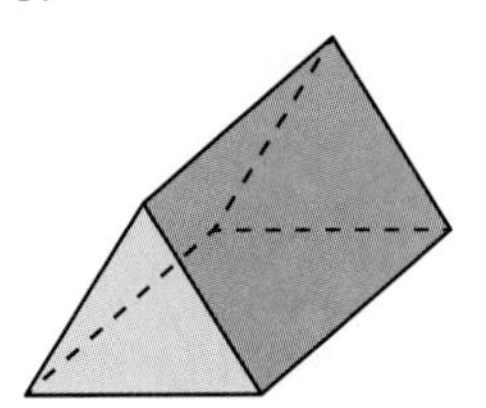

23.

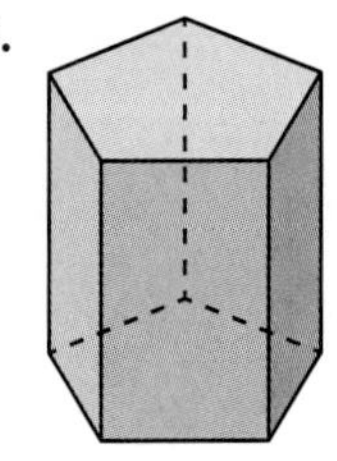

25.

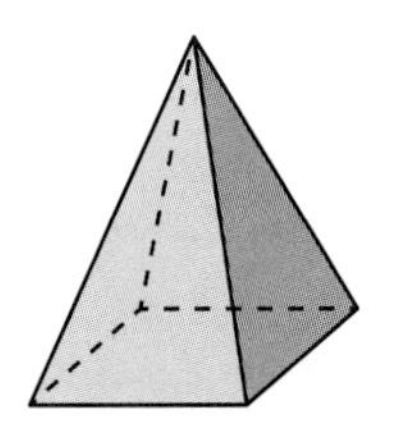

27. your cousin; The sides come together at a point.

29. no

31. yes; *Sample answer:* The plane is parallel to a face.

33. yes; *Sample answer:* The plane passes through six faces.

35. a.

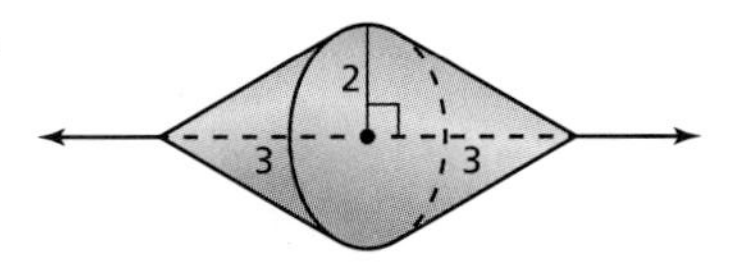

two cones with heights 3 and base radii 2

b.

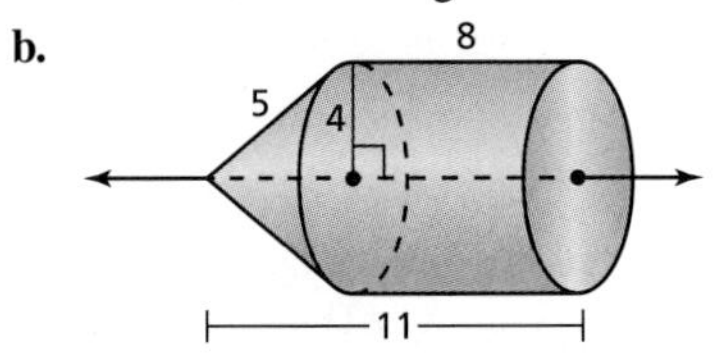

cone with height 3 and base radius 4 and cylinder with height 8 and base radius 4

11.4 Maintaining Mathematical Proficiency *(p. 622)*

37. yes; SSS Congruence Theorem (Thm. 5.8)

39. yes; ASA Congruence Theorem (Thm. 5.10)

11.5 Vocabulary and Core Concept Check *(p. 631)*

1. cubic units

11.5 Monitoring Progress and Modeling with Mathematics *(pp. 631–634)*

3. 6.3 cm^3 **5.** 175 in.^{3} **7.** about 288.40 ft^3

9. about 628.32 ft^3

11.

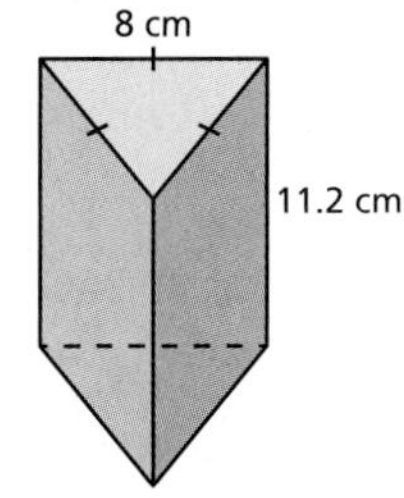

310.38 cm^3

13. copper

15. The base circumference was used instead of the base area; $V = \pi r^2h = 48\pi$ ft^3

17. 10 ft **19.** 4 cm **21.** about 11.04 ft

23. 14 in.^{2}; *Sample answer:* length: 7 in., width: 2 in.

25. 99 cm^3 **27.** 2 cm **29.** 150 ft^3

31. about 1900.66 in.^{3} **33.** about 2,350,000,000 gal

35. 2 **37. a.** 75 in.^{3} **b.** 20

39. *Sample answer:* The stacks have the same height and the rectangles have the same lengths, so the stacks have the same area.

41. 1.5

2.5

4

15 cubic units

43. *Sample answer:*

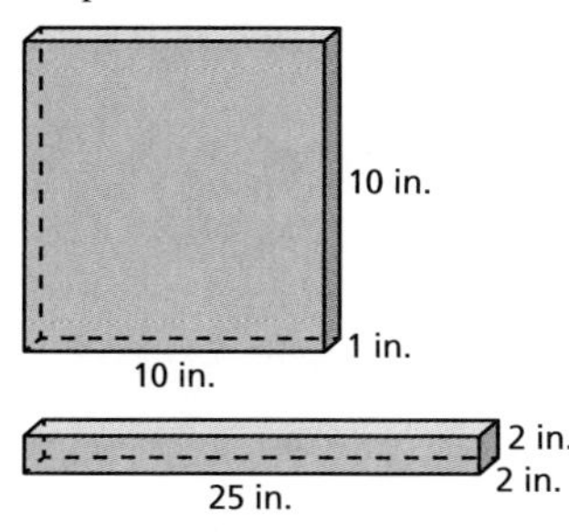

45. the solid produced by rotating around the vertical line; *Sample answer:* The solid produced by rotating around the horizontal line has a volume of 45π cubic inches and the solid produced by rotating around the vertical line has a volume of 75π cubic inches.

47. about 7.33 in.^{3} **49.** Increase the height by 25%.

51. yes; *Sample answer:* Density is proportional to mass when the volume is constant.

53. 36 ft, 15 ft

11.5 Maintaining Mathematical Proficiency *(p. 634)*

55. 16 m^2 **57.** 680.4 in.^{2}

11.6 Vocabulary and Core Concept Check *(p. 639)*

1. *Sample answer:* A triangular prism has two parallel bases that are triangles. A triangular pyramid has one base that is a triangle, and the other faces all intersect at a single point.

11.6 Monitoring Progress and Modeling with Mathematics *(pp. 639–640)*

3. 448 m^3 **5.** 6 m **7.** 16 in.

9. One side length was used in the formula as the base area; $V = \frac{1}{3}(6^2)(5) = 60 \text{ ft}^3$

11. 5 ft **13.** 12 yd **15.** 4 ft^3 **17.** 72 in.^{3}

19. about 213.33 cm^3

21. **a.** The volume doubles. **b.** The volume is 4 times greater.

c. yes; *Sample answer:*

Square Pyramid: $V = \frac{1}{3}s^2h$

Double height: $V = \frac{1}{3}s^2(2h) = 2\left(\frac{1}{3}s^2h\right)$

Double side length of base: $V = \frac{1}{3}(2s)^2h = 4\left(\frac{1}{3}s^2h\right)$

23. about 9.22 ft^3 **25.** about 78 in.^{3}

11.6 Maintaining Mathematical Proficiency *(p. 640)*

27. 12.6 **29.** 16.0

11.7 Vocabulary and Core Concept Check *(p. 645)*

1. *Sample answer:* pyramids have a polygonal base, cones have a circular base; They both have sides that meet at a single vertex.

11.7 Monitoring Progress and Modeling with Mathematics *(pp. 645–646)*

3. about 603.19 in.^{2} **5.** about 678.58 in.^{2}

7. about 1361.36 mm^3 **9.** about 526.27 in.^{3}

11. $\ell \approx 5.00$ cm; $h \approx 4.00$ cm **13.** 256π ft^3

15. about 226.19 cm^3

17. $2h$; $r\sqrt{2}$; *Sample answer:* The original volume is $V = \frac{1}{3}\pi r^2h$ and the new volume is $V = \frac{2}{3}\pi r^2h$.

19. about 3716.85 ft^3

21. yes; *Sample answer:* The automatic pet feeder holds about 12 cups of food.

23. It is half; about 60°

25. yes; *Sample answer:* The base areas are the same and the total heights are the same.

11.7 Maintaining Mathematical Proficiency *(p. 646)*

27. about 153.94 ft^2 **29.** 32 m

11.8 Vocabulary and Core Concept Check *(p. 652)*

1. The plane must contain the center of the sphere.

11.8 Monitoring Progress and Modeling with Mathematics *(pp. 652–654)*

3. about 201.06 ft^2 **5.** about 1052.09 m^2 **7.** 1 ft

9. 30 m **11.** about 157.08 m^2 **13.** about 2144.66 m^3

15. about 5575.28 yd^3 **17.** about 4188.79 cm^3

19. about 33.51 ft^3

21. The radius was squared instead of cubed; $V = \frac{4}{3}\pi(6)^3 \approx 904.78 \text{ ft}^3$

23. about 445.06 in.^{3} **25.** about 7749.26 cm^3

27. $S \approx 226.98$ in.^{2}; $V \approx 321.56$ in.^{3}

29. $S \approx 45.84$ in.^{2}; $V \approx 29.18$ in.^{3}

31. $S \approx 215.18$ in.^{2}; $V \approx 296.80$ in.^{3}

33. no; The surface area is quadrupled.

35. about 20,944 ft^3

37. **a.** 144π in.^{2}, 288π in.^{3}; 324π in.^{2}, 972π in.^{3}; 576π in.^{2}, 2304π in.^{3}

b. It is multiplied by 4; It is multiplied by 9; It is multiplied by 16.

c. It is multiplied by 8; It is multiplied by 27; It is multiplied by 64.

39. **a.** Earth: about 197.1 million mi^2; moon: about 14.7 million mi^2

b. The surface area of the Earth is about 13.4 times greater than the surface area of the moon.

c. about 137.9 million mi^2

41. about 50.27 in.^{2}; *Sample answer:* The side length of the cube is the diameter of the sphere.

43. $V = \frac{1}{3}rS$

45. *Sample answer:* radius 1 in. and height $\frac{4}{3}$ in.; radius $\frac{1}{3}$ in. and height 12 in.; radius 2 in. and height $\frac{1}{3}$ in.

47. $S \approx 113.10$ in.^{2}, $V \approx 75.40$ in.^{3}

11.8 Maintaining Mathematical Proficiency *(p. 654)*

49. $A = 35°$, $a \approx 12.3$, $c \approx 14.6$

51. $a \approx 31.0$, $B \approx 28.1°$, $C \approx 48.9°$

Chapter 11 Review *(pp. 656–660)*

1. about 30.00 ft **2.** about 56.57 cm

3. about 26.09 in. **4.** 218 ft **5.** about 169.65 in.^{2}

6. about 17.72 in.^{2} **7.** 173.166 ft^2

8. 130 square units **9.** 96 square units

10. 105 square units **11.** about 201.20 square units

12. about 167.11 square units

13. about 37.30 square units **14.** about 119.29 in.^{2}

15.

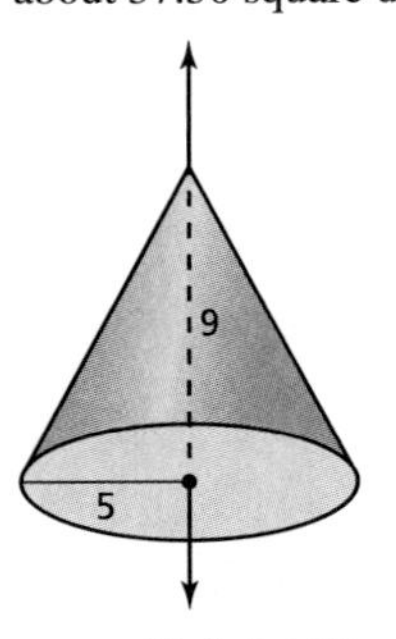

cone with height 9 and base radius 5

16.

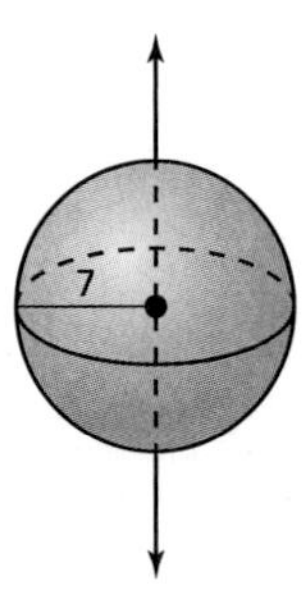

sphere with radius 7

17.

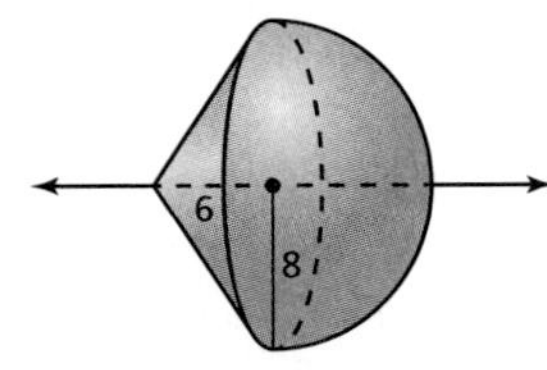

cone with height 6 and base radius 8 and hemisphere with radius 8

18. rectangle **19.** square **20.** triangle

21. 11.34 m^3 **22.** about 100.53 mm^3

23. about 27.53 yd^3 **24.** 189 ft^3 **25.** 400 yd^3

26. 300 m^3 **27.** about 3.46 in. **28.** 12 in.

29. $S \approx 678.58$ cm^2; $V \approx 1017.88$ cm^3

30. $S \approx 2513.27$ cm^2; $V \approx 8042.48$ cm^3

31. $S \approx 439.82$ m^2; $V \approx 562.10$ m^3 **32.** 15 cm

33. $S \approx 615.75$ in.^{2}; $V \approx 1436.76$ in.^{3}

34. $S \approx 907.92$ ft^2; $V \approx 2572.44$ ft^3

35. $S \approx 2827.43$ ft^2; $V \approx 14{,}137.17$ ft^3

36. $S \approx 74.8$ million km^2; $V \approx 60.8$ billion km^3

37. about 272.55 m^3

Chapter 12

Chapter 12 Maintaining Mathematical Proficiency *(p. 665)*

1. $\frac{6}{30} = \frac{p}{100}$, 20% **2.** $\frac{a}{25} = \frac{68}{100}$, 17

3. $\frac{34.4}{86} = \frac{p}{100}$, 40%

4.

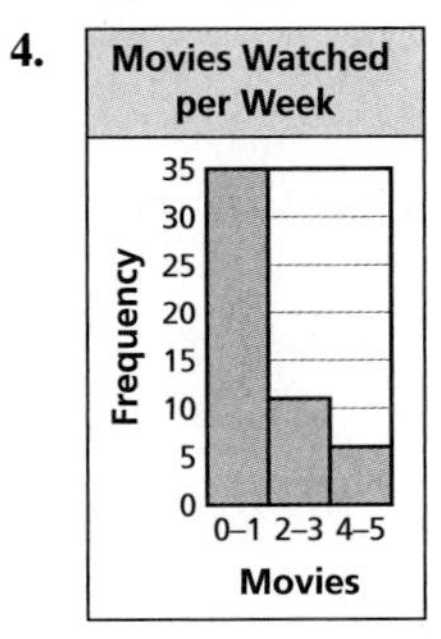

5. no; The sofa will cost 80% of the retail price and the arm chair will cost 81% of the retail price.

12.1 Vocabulary and Core Concept Check *(p. 672)*

1. probability

12.1 Monitoring Progress and Modeling with Mathematics *(pp. 672–674)*

3. 48; 1HHH, 1HHT, 1HTH, 1THH, 1HTT, 1THT, 1TTH, 1TTT, 2HHH, 2HHT, 2HTH, 2THH, 2HTT, 2THT, 2TTH, 2TTT, 3HHH, 3HHT, 3HTH, 3THH, 3HTT, 3THT, 3TTH, 3TTT, 4HHH, 4HHT, 4HTH, 4THH, 4HTT, 4THT, 4TTH, 4TTT, 5HHH, 5HHT, 5HTH, 5THH, 5HTT, 5THT, 5TTH, 5TTT, 6HHH, 6HHT, 6HTH, 6THH, 6HTT, 6THT, 6TTH, 6TTT

5. 12; R1, R2, R3, R4, W1, W2, W3, W4, B1, B2, B3, B4

7. $\frac{5}{16}$, or about 31.25%

9. a. $\frac{11}{12}$, or about 92% **b.** $\frac{13}{18}$, or about 72%

11. There are 4 outcomes, not 3; The probability is $\frac{1}{4}$.

13. about 0.56, or about 56% **15.** 4

17. a. $\frac{9}{10}$, or 90% **b.** $\frac{2}{3}$, or about 67%

c. The probability in part (b) is based on trials, not possible outcomes.

19. about 0.08, or about 8% **21.** C, A, D, B

23. a. 2, 3, 4, 5, 6, 7, 8, 9, 10, 11, 12

b. 2: $\frac{1}{36}$, 3: $\frac{1}{18}$, 4: $\frac{1}{12}$, 5: $\frac{1}{9}$, 6: $\frac{5}{36}$, 7: $\frac{1}{6}$, 8: $\frac{5}{36}$, 9: $\frac{1}{9}$, 10: $\frac{1}{12}$, 11: $\frac{1}{18}$, 12: $\frac{1}{36}$

c. *Sample answer:* The probabilities are similar.

25. $\frac{\pi}{6}$, or about 52%

27. $\frac{3}{400}$, or 0.75%; about 113; $(0.0075)15{,}000 = 112.5$

12.1 Maintaining Mathematical Proficiency *(p. 674)*

29. $2x$ **31.** $\frac{4x^6}{3}$ **33.** $81p^4q^4$

12.2 Vocabulary and Core Concept Check *(p. 680)*

1. When two events are dependent, the occurrence of one event affects the other. When two events are independent, the occurrence of one event does not affect the other. *Sample answer:* choosing two marbles from a bag without replacement; rolling two dice

12.2 Monitoring Progress and Modeling with Mathematics *(pp. 680–682)*

3. dependent; The occurrence of event A affects the occurrence of event B.

5. dependent; The occurrence of event A affects the occurrence of event B.

7. yes **9.** yes **11.** about 2.8% **13.** about 34.7%

15. The probabilities were added instead of multiplied; $P(A \text{ and } B) = (0.6)(0.2) = 0.12$

17. 0.325

19. a. about 1.2% **b.** about 1.0%

You are about 1.2 times more likely to select 3 face cards when you replace each card before you select the next card.

21. a. about 17.1% **b.** about 81.4% **23.** about 53.5%

25. a. *Sample answer:* Put 20 pieces of paper with each of the 20 students' names in a hat and pick one; 5%

b. *Sample answer:* Put 45 pieces of paper in a hat with each student's name appearing once for each hour the student worked. Pick one piece; about 8.9%

27. yes; The chance that it will be rescheduled is $(0.7)(0.75) = 0.525$, which is a greater than a 50% chance.

29. **a.** wins: 0%; loses: 1.99%; ties: 98.01%

b. wins: 20.25%; loses: 30.25%; ties: 49.5%

c. yes; Go for 2 points after the first touchdown, and then go for 1 point if they were successful the first time or 2 points if they were unsuccessful the first time; winning: 44.55%; losing: 30.25%

12.2 Maintaining Mathematical Proficiency *(p. 682)*

31. $x = 0.2$ **33.** $x = 0.15$

12.3 Vocabulary and Core Concept Check *(p. 688)*

1. two-way table

12.3 Monitoring Progress and Modeling with Mathematics *(p. 688–690)*

3. 34; 40; 4; 6; 12

5.

		Gender		
		Male	Female	Total
Response	Yes	132	151	283
	No	39	29	68
	Total	171	180	351

351 people were surveyed, 171 males were surveyed, 180 females were surveyed, 283 people said yes, 68 people said no.

7.

		Dominant Hand		
		Left	Right	Total
Gender	Female	0.048	0.450	0.498
	Male	0.104	0.398	0.502
	Total	0.152	0.848	1

9.

		Gender		
		Male	Female	Total
Response	Yes	0.376	0.430	0.806
	No	0.111	0.083	0.194
	Total	0.487	0.513	1

11.

		Breakfast	
		Ate	Did Not Eat
Feeling	Tired	0.091	0.333
	Not Tired	0.909	0.667

13. **a.** about 0.789 **b.** 0.168

c. The events are independent.

15. The value for $P(\text{yes})$ was used in the denominator instead of the value for $P(\text{Tokyo})$;

$\frac{0.049}{0.39} \approx 0.126$

17. Route B; It has the best probability of getting to school on time.

19. *Sample answer:*

		Transportation to School			
		Rides Bus	Walks	Car	Total
Gender	Male	6	9	4	19
	Female	5	2	4	11
	Total	11	11	8	30

		Transportation to School			
		Rides Bus	Walks	Car	Total
Gender	Male	0.2	0.3	0.133	0.633
	Female	0.167	0.067	0.133	0.367
	Total	0.367	0.367	0.266	1

21. Routine B is the best option, but your friend's reasoning of why is incorrect; Routine B is the best choice because there is a 66.7% chance of reaching the goal, which is higher than the chances of Routine A (62.5%) and Routine C (63.6%).

23. **a.** about 0.438 **b.** about 0.387

25. **a.** More of the current consumers prefer the leader, so they should improve the new snack before marketing it.

b. More of the new consumers prefer the new snack than the leading snack, so there is no need to improve the snack.

12.3 Maintaining Mathematical Proficiency *(p. 690)*

27.

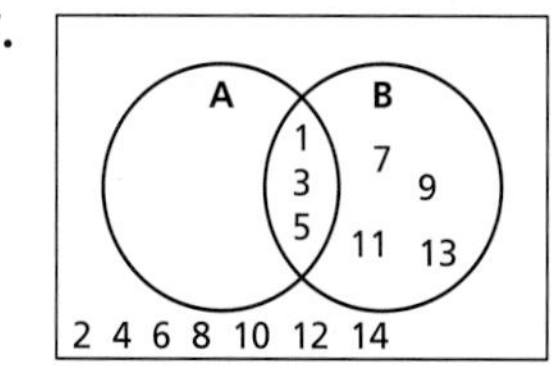

29.

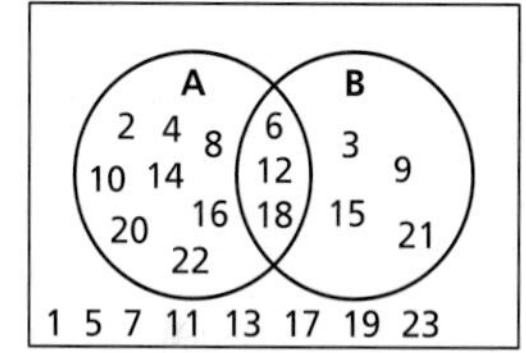

12.4 Vocabulary and Core Concept Check *(p. 697)*

1. yes; $\overline{A}$ is everything not in A; *Sample answer:* event A: you win the game, event $\overline{A}$: you do not win the game

12.4 Monitoring Progress and Modeling with Mathematics *(p. 697–698)*

3. 0.4 **5.** $\frac{7}{12}$, or about 0.58 **7.** $\frac{9}{20}$, or 0.45

9. $\frac{7}{10}$, or 0.7

11. forgot to subtract $P(\text{heart and face card})$;
$P(\text{heart}) + P(\text{face card}) - P(\text{heart and face card}) = \frac{11}{26}$

13. $\frac{2}{3}$ **15.** 10% **17.** 0.4742, or 47.42% **19.** $\frac{13}{18}$

21. $\frac{3}{20}$

23. no; Until all cards, numbers, and colors are known, the conclusion cannot be made.

12.4 Maintaining Mathematical Proficiency *(p. 698)*

25. $4x^2 + 36x + 81$ **27.** $9a^2 - 42ab + 49b^2$

12.5 Vocabulary and Core Concept Check *(p. 704)*

1. permutation

12.5 Monitoring Progress and Modeling with Mathematics *(p. 704–706)*

3. **a.** 2 **b.** 2 **5.** **a.** 24 **b.** 12

7. **a.** 720 **b.** 30 **9.** 20 **11.** 9 **13.** 20,160

15. 870 **17.** 990 **19.** $\frac{1}{56}$ **21.** 4 **23.** 20

25. 5 **27.** 1 **29.** 220 **31.** 6435 **33.** 635,376

35. The factorial in the denominator was left out;

$$_{11}P_7 = \frac{11!}{(11-7)!} = 1{,}663{,}200$$

37. combinations; The order is not important; 45

39. permutations; The order is important; 132,600

41. $_{50}C_9 = {}_{50}C_{41}$; For each combination of 9 objects, there is a corresponding combination of the 41 remaining objects.

43.

	$r = 0$	$r = 1$	$r = 2$	$r = 3$
$_3P_r$	1	3	6	6
$_3C_r$	1	3	3	1

$_nP_r \geq {}_nC_r$; Because $_nP_r = \frac{n!}{(n-r)!}$ and $_nC_r = \frac{n!}{(n-r)! \bullet r!}$, $_nP_r > {}_nC_r$ when $r > 1$ and $_nP_r = {}_nC_r$ when $r = 0$ or $r = 1$.

45. $\frac{1}{44{,}850}$ **47.** $\frac{1}{15{,}890{,}700}$

49. **a.** $_nC_{n-2} - n$ **b.** $\frac{n(n-3)}{2}$

51. 30

53. **a.** $\frac{1}{2}$ **b.** $\frac{1}{2}$; The probabilities are the same.

55. **a.** $\frac{1}{90}$ **b.** $\frac{9}{10}$

57. $\frac{1}{406}$; There are $_{30}C_5$ possible groups. The number of groups that will have you and your two best friends is $_{27}C_2$.

12.5 Maintaining Mathematical Proficiency *(p. 706)*

59. $\frac{1}{5}$

12.6 Vocabulary and Core Concept Check *(p. 711)*

1. a variable whose value is determined by the outcomes of a probability experiment

12.6 Monitoring Progress and Modeling with Mathematics *(pp. 711–712)*

3.

x (value)	1	2	3
Outcomes	5	3	2
P(*x*)	$\frac{1}{2}$	$\frac{3}{10}$	$\frac{1}{5}$

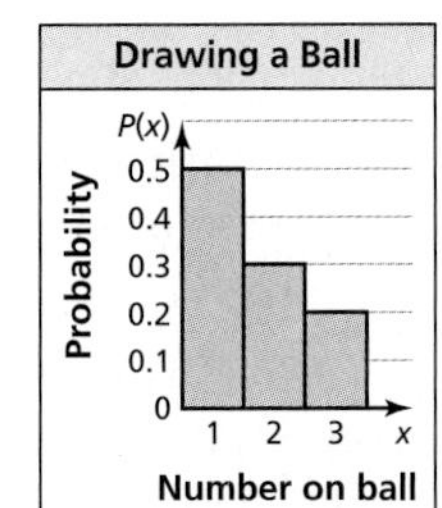

5.

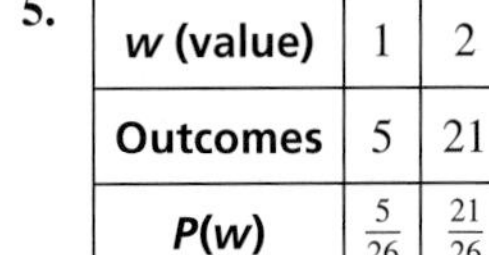

w (value)	1	2
Outcomes	5	21
P(*w*)	$\frac{5}{26}$	$\frac{21}{26}$

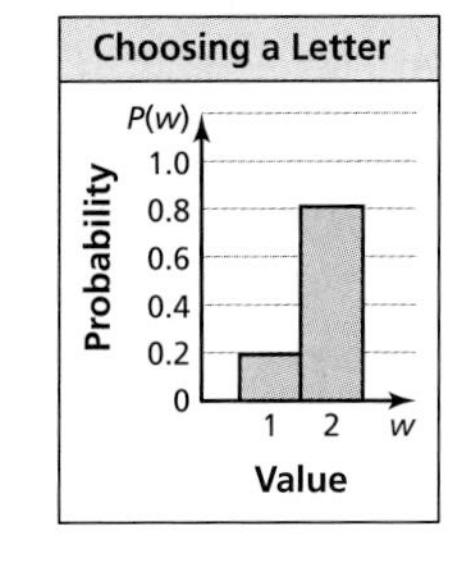

7. **a.** 2 **b.** $\frac{5}{8}$ **9.** about 0.00002 **11.** about 0.00018

13. **a.**

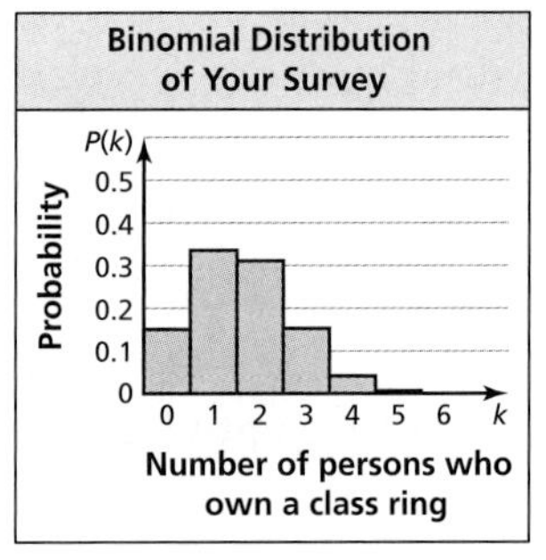

b. The most likely outcome is that 1 of the 6 students owns a ring.

c. about 0.798

15. The exponents are switched;

$P(k = 3) = {}_5C_3\left(\frac{1}{6}\right)^3\left(\frac{5}{6}\right)^{5-3} \approx 0.032$

17. **a.** $P(0) \approx 0.099$, $P(1) \approx 0.271$, $P(2) \approx 0.319$, $P(3) \approx 0.208$, $P(4) \approx 0.081$, $P(5) \approx 0.019$, $P(6) \approx 0.0025$, $P(7) \approx 0.00014$

b.

x	0	1	2	3	4
P(*x*)	0.099	0.271	0.319	0.208	0.081

x	5	6	7
P(*x*)	0.019	0.0025	0.00014

c.

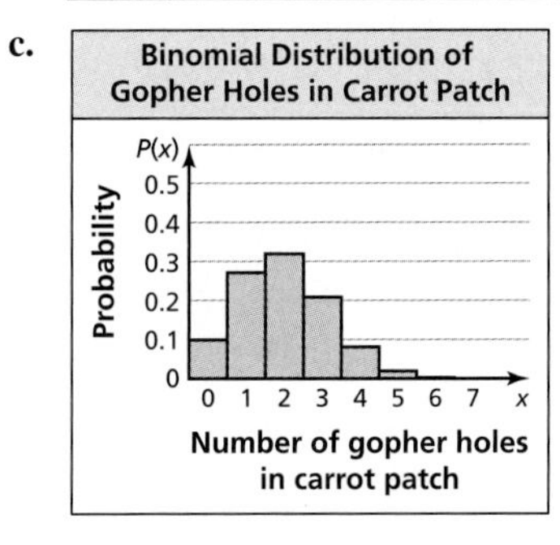

19. no; The data is skewed right, so the probability of failure is greater.

21. **a.** The statement is not valid, because having a male and having a female are independent events.

b. 0.03125

c.

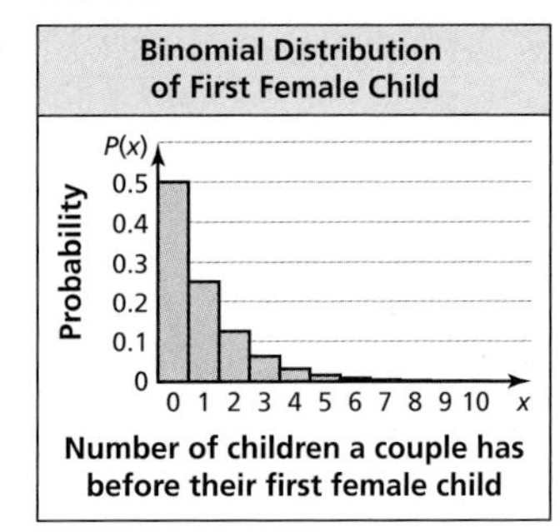

skewed right

12.6 Maintaining Mathematical Proficiency *(p. 712)*

23. FFF, FFM FMF, FMM, MMM, MMF, MFM, MFF

Chapter 12 Review *(pp. 714–716)*

1. $\frac{2}{9}$; $\frac{7}{9}$ **2.** 20 points

3. **a.** 0.15625 **b.** about 0.1667

You are about 1.07 times more likely to pick a red then a green if you do not replace the first marble.

4. **a.** about 0.0586 **b.** 0.0625

You are about 1.07 times more likely to pick a blue then a red if you do not replace the first marble.

5. **a.** 0.25 **b.** about 0.2333

You are about 1.07 times more likely to pick a green and then another green if you replace the first marble.

6. about 0.529

7.

Response		Gender: Men	Women	Total
	Yes	200	230	430
	No	20	40	60
	Total	220	270	490

About 44.9% of responders were men, about 55.1% of responders were women, about 87.8% of responders thought it was impactful, about 12.2% of responders thought it was not impactful.

8. 0.68 **9.** 0.02 **10.** 5040 **11.** 1,037,836,800

12. 15 **13.** 70 **14.** 40,320

15. $\frac{1}{84}$ **16.** about 0.12

17.

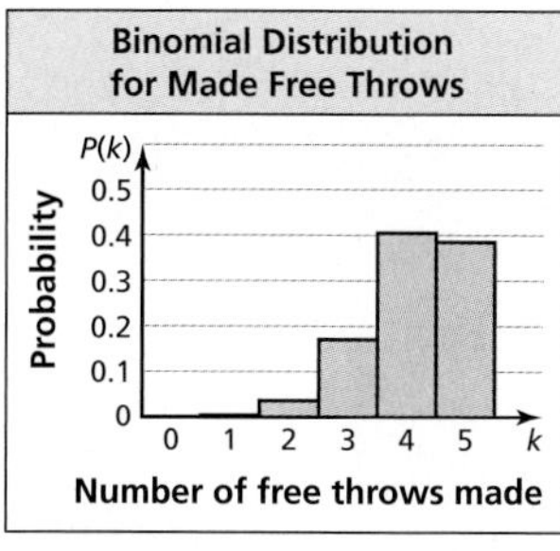

The most likely outcome is that 4 of the 5 free throw shots will be made.

Additional Topic

Vocabulary and Core Concept Check *(p. 726)*

1. focus; directrix

Monitoring Progress and Modeling with Mathematics *(pp. 726–728)*

3. $y = \frac{1}{4}x^2$ **5.** $y = -\frac{1}{8}x^2$ **7.** $y = \frac{1}{24}x^2$

9. $y = -\frac{1}{40}x^2$

11. A, B and D; Each has a value for p that is negative. Substituting in a negative value for p in $y = \frac{1}{4p}x^2$ results in a parabola that has been reflected across the x-axis.

13. The focus is (0, 2). The directrix is $y = -2$. The axis of symmetry is the y-axis.

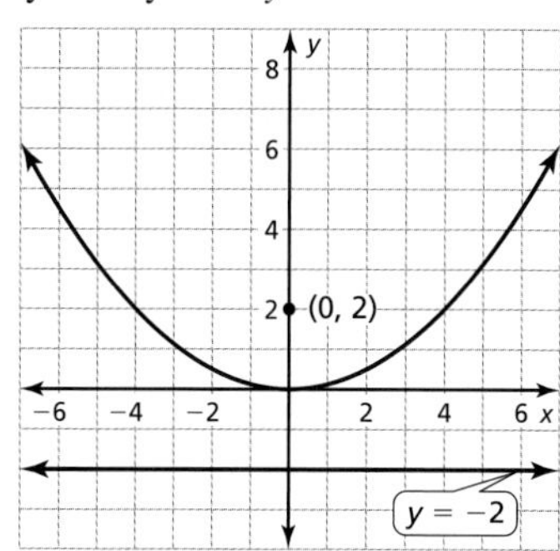

15. The focus is $(-5, 0)$. The directrix is $x = 5$. The axis of symmetry is the x-axis.

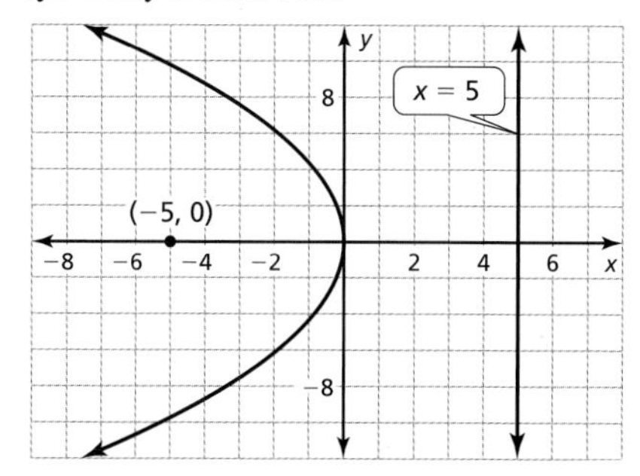

17. The focus is (4, 0). The directrix is $x = -4$. The axis of symmetry is the x-axis.

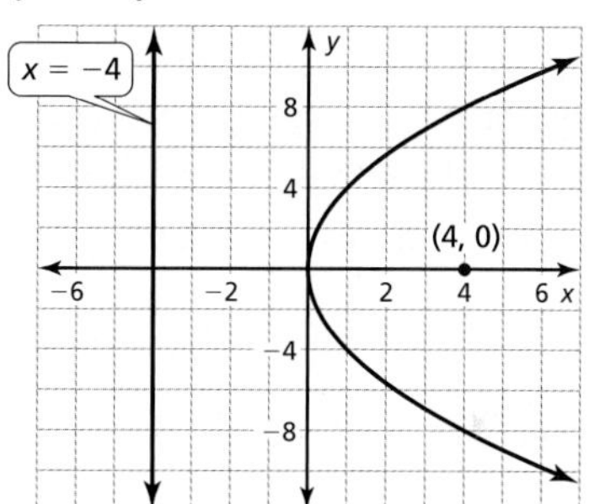

19. The focus is $\left(0, -\frac{1}{8}\right)$. The directrix is $y = \frac{1}{8}$. The axis of symmetry is the y-axis.

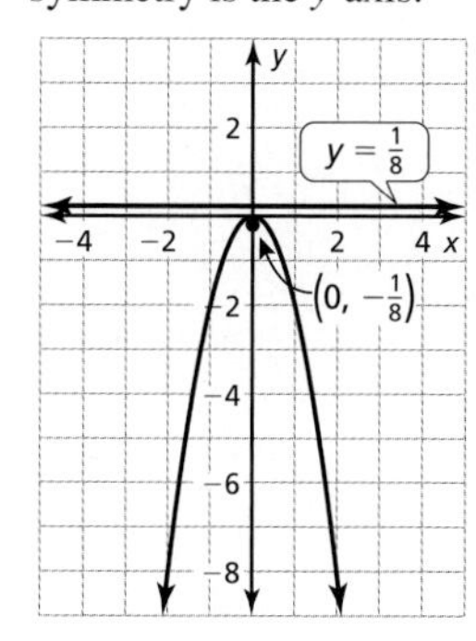

21. Instead of a vertical axis of symmetry, the graph should have a horizontal axis of symmetry.

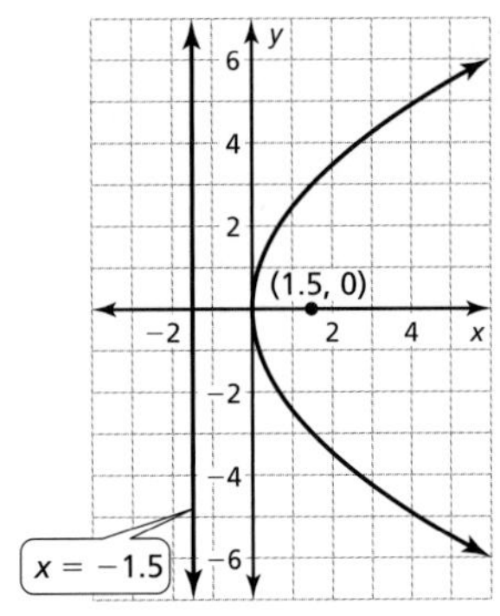

23. 9.5 in.; The receiver should be placed at the focus. The distance from the vertex to the focus is $p = \frac{38}{4} = 9.5$ in.

25. $y = \frac{1}{32}x^2$ **27.** $x = -\frac{1}{10}y^2$ **29.** $x = \frac{1}{12}y^2$

31. $x = \frac{1}{40}y^2$ **33.** $y = -\frac{3}{20}x^2$ **35.** $y = \frac{7}{24}x^2$

37. $x = -\frac{1}{16}y^2 - 4$ **39.** $y = \frac{1}{6}x^2 + 1$

41. The vertex is (3, 2). The focus is (3, 4). The directrix is $y = 0$. The axis of symmetry is $x = 3$. The graph is a vertical shrink by a factor of $\frac{1}{2}$ followed by a translation 3 units right and 2 units up.

43. The vertex is (1, 3). The focus is (5, 3). The directrix is $x = -3$. The axis of symmetry is $y = 3$. The graph is a horizontal shrink by a factor of $\frac{1}{4}$ followed by a translation 1 unit right and 3 units up.

45. The vertex is (2, −4). The focus is $\left(\frac{23}{12}, -4\right)$. The directrix is $x = \frac{25}{12}$. The axis of symmetry is $y = -4$. The graph is a horizontal stretch by a factor of 12 followed by a reflection in the y-axis and a translation 2 units right and 4 units down.

47. $x = \frac{1}{5.2}y^2$; about 3.08 in.

49. As $|p|$ increases, the graph gets wider; As $|p|$ increases, the constant in the function gets smaller which results in a vertical shrink, making the graph wider.

51. $y = \frac{1}{4}x^2$ **53.** $x = \frac{1}{4p}y^2$

English-Spanish Glossary

English

acute angle *(p. 39)* An angle that has a measure greater than 0° and less than 90°

adjacent angles *(p. 48)* Two angles that share a common vertex and side, but have no common interior points

adjacent arcs *(p. 539)* Arcs of a circle that have exactly one point in common

alternate exterior angles *(p. 128)* Two angles that are formed by two lines and a transversal that are outside the two lines and on opposite sides of the transversal

alternate interior angles *(p. 128)* Two angles that are formed by two lines and a transversal that are between the two lines and on opposite sides of the transversal

altitude of a triangle *(p. 321)* The perpendicular segment from a vertex of a triangle to the opposite side or to the line that contains the opposite side

angle *(p. 38)* A set of points consisting of two different rays that have the same endpoint

angle bisector *(p. 42)* A ray that divides an angle into two angles that are congruent

angle of depression *(p. 497)* The angle that a downward line of sight makes with a horizontal line

angle of elevation *(p. 490)* The angle that an upward line of sight makes with a horizontal line

angle of rotation *(p. 190)* The angle that is formed by rays drawn from the center of rotation to a point and its image

apothem of a regular polygon *(p. 611)* The distance from the center to any side of a regular polygon

arc length *(p. 595)* A portion of the circumference of a circle

axiom *(p. 12)* A rule that is accepted without proof

Spanish

ángulo agudo *(p. 39)* Un ángulo que tiene una medida mayor que 0° y menor que 90°

ángulos adyacentes *(p. 48)* Dos ángulos que comparten un vértice y lado en común, pero que no tienen puntos interiores en común

arcos adyacentes *(p. 539)* Arcos de un círculo que tienen exactamente un punto en común

ángulos exteriores alternos *(p. 128)* Dos ángulos que son formados por dos rectas y una transversal que están fuera de las dos rectas y en lados opuestos de la transversal

ángulos interiores alternos *(p. 128)* Dos ángulos que son formados por dos rectas y una transversal que están entre las dos rectas y en lados opuestos de la transversal

altitud de un triángulo *(p. 321)* El segmento perpendicular desde el vértice de un triángulo al lado opuesto o a la recta que contiene el lado opuesto

ángulo *(p. 38)* Un conjunto de puntos que consiste en dos rayos distintos que tienen el mismo punto extremo

bisectriz de un ángulo *(p. 42)* Un rayo que divide un ángulo en dos ángulos congruentes

ángulo de depresión *(p. 497)* El ángulo formado entre una recta de vista descendente y una recta horizontal

ángulo de elevación *(p. 490)* El ángulo formado entre una recta de vista ascendente y una recta horizontal

ángulo de rotación *(p. 190)* El ángulo que está formado por rayos dibujados desde el centro de rotación hacia un punto y su imagen

apotema de un polígono regular *(p. 611)* La distancia desde el centro a cualquier lado de un polígono regular

longitud de arco *(p. 595)* Una porción de la circunferencia de un círculo

axioma *(p. 12)* Una regla que es aceptada sin demostración

axis of revolution *(p. 620)* The line around which a two-dimensional shape is rotated to form a three-dimensional figure

eje de revolución *(p. 620)* La recta alrededor de la cual una forma bidimensional rota para formar una figura tridimensional

B

base angles of an isosceles triangle *(p. 252)* The two angles adjacent to the base of an isosceles triangle

base angles of a trapezoid *(p. 398)* Either pair of consecutive angles whose common side is a base of a trapezoid

base of an isosceles triangle *(p. 252)* The side of an isosceles triangle that is not one of the legs

bases of a trapezoid *(p. 398)* The parallel sides of a trapezoid

between *(p. 14)* When three points are collinear, one point is between the other two.

biconditional statement *(p. 69)* A statement that contains the phrase "if and only if"

binomial distribution *(p. 709)* A type of probability distribution that shows the probabilities of the outcomes of a binomial experiment

binomial experiment *(p. 709)* An experiment in which there are a fixed number of independent trials, exactly two possible outcomes for each trial, and the probability of success is the same for each trial.

ángulos de la base de un triángulo isósceles *(p. 252)* Los dos ángulos adyacentes a la base de un triángulo isósceles

ángulos de la base de un trapecio *(p. 398)* Cualquier par de ángulos consecutivos cuyo lado común es la base de un trapezoide

base de un triángulo isósceles *(p. 252)* El lado de un triángulo isósceles que no es uno de los catetos

bases de un trapecio *(p. 398)* Los lados paralelos de un trapezoide

entre *(p. 14)* Cuando tres puntos son colineales, un punto está entre los otros dos.

enunciado bicondicional *(p. 69)* Un enunciado que contiene la frase "si y sólo si"

distribución del binomio *(p. 709)* Un tipo de distribución de probabilidades que muestra las probabilidades de los resultados posibles de un experimento del binomio

experimento del binomio *(p. 709)* Un experimento en el que hay un número fijo de pruebas independientes, exactamente dos resultados posibles para cada prueba, y la probabilidad de éxito es la misma para cada prueba

Cavalieri's Principle *(p. 626)* If two solids have the same height and the same cross-sectional area at every level, then they have the same volume.

center of a circle *(p. 530)* The point from which all points on a circle are equidistant

center of dilation *(p. 208)* The fixed point in a dilation

center of a regular polygon *(p. 611)* The center of a polygon's circumscribed circle

center of rotation *(p. 190)* The fixed point in a rotation

center of symmetry *(p. 193)* The center of rotation in a figure that has rotational symmetry

central angle of a circle *(p. 538)* An angle whose vertex is the center of a circle

Principio de Cavalieri *(p. 626)* Si dos sólidos tienen la misma altura y la misma área transversal en todo nivel, entonces tienen el mismo volumen.

centro de un círculo *(p. 530)* El punto desde donde todos los puntos en un círculo son equidistantes

centro de dilatación *(p. 208)* El punto fijo en una dilatación

centro de un polígono regular *(p. 611)* El centro del círculo circunscrito de un polígono

centro de rotación *(p. 190)* El punto fijo en una rotación

centro de simetría *(p. 193)* El centro de rotación en una figura que tiene simetría rotacional

ángulo central de un círculo *(p. 538)* Un ángulo cuyo vértice es el centro de un círculo

central angle of a regular polygon *(p. 611)* An angle formed by two radii drawn to consecutive vertices of a polygon

ángulo central de un polígono regular *(p. 611)* Un ángulo formado por dos radios extendidos a vértices consecutivos de un polígono

centroid *(p. 320)* The point of concurrency of the three medians of a triangle

centroide *(p. 320)* El punto de concurrencia de las tres medianas de un triángulo

chord of a circle *(p. 530)* A segment whose endpoints are on a circle

cuerda de un círculo *(p. 530)* Un segmento cuyos puntos extremos están en un círculo

chord of a sphere *(p. 648)* A segment whose endpoints are on a sphere

cuerda de una esfera *(p. 648)* Un segmento cuyos puntos extremos están en una esfera

circle *(p. 530)* The set of all points in a plane that are equidistant from a given point

círculo *(p. 530)* El conjunto de todos los puntos en un plano que son equidistantes de un punto dado

circumcenter *(p. 310)* The point of concurrency of the three perpendicular bisectors of a triangle

circuncentro *(p. 310)* El punto de concurrencia de las tres bisectrices perpendiculares de un triángulo

circumference *(p. 594)* The distance around a circle

circunferencia *(p. 594)* La distancia alrededor de un círculo

circumscribed angle *(p. 564)* An angle whose sides are tangent to a circle

ángulo circunscrito *(p. 564)* Un ángulo cuyos lados son tangentes a un círculo

circumscribed circle *(p. 556)* A circle that contains all the vertices of an inscribed polygon

círculo circunscrito *(p. 556)* Un círculo que contiene todos los vértices de un polígono inscrito

collinear points *(p. 4)* Points that lie on the same line

puntos colineales *(p. 4)* Puntos que descansan en la misma recta

combination *(p. 702)* A selection of objects in which order is not important

combinación *(p. 702)* Una selección de objetos en la que el orden no es importante

common tangent *(p. 531)* A line or segment that is tangent to two coplanar circles

tangente común *(p. 531)* Una recta o segmento que es tangente a dos círculos coplanarios

complementary angles *(p. 48)* Two angles whose measures have a sum of 90°

ángulos complementarios *(p. 48)* Dos ángulos cuyas medidas suman 90°

component form *(p. 174)* A form of a vector that combines the horizontal and vertical components

forma componente *(p. 174)* Una forma de un vector que combina los componentes horizontales y verticales

composition of transformations *(p. 176)* The combination of two or more transformations to form a single transformation

composición de transformaciones *(p. 176)* La combinación de dos o más transformaciones para formar una transformación única

compound event *(p. 694)* The union or intersection of two events

evento compuesto *(p. 694)* la unión o intersección de dos eventos

concentric circles *(p. 531)* Coplanar circles that have a common center

círculos concéntricos *(p. 531)* Círculos coplanarios que tienen un centro en común

conclusion *(p. 66)* The "then" part of a conditional statement written in if-then form

conclusión *(p. 66)* La parte después de "entonces" en un enunciado condicional escrito de la forma "si..., entonces..."

concurrent *(p. 310)* Three or more lines, rays, or segments that intersect in the same point

concurrente *(p. 310)* Tres o más rectas, rayos o segmentos que se intersectan en el mismo punto

conditional probability *(p. 677)* The probability that event B occurs given that event A has occurred, written as $P(B|A)$

probabilidad condicional *(p. 677)* La probabilidad de que el evento B ocurra dado que el evento A ha ocurrido, escrito como $P(B|A)$

conditional relative frequency *(p. 685)* The ratio of a joint relative frequency to the marginal relative frequency in a two-way table

frecuencia relativa condicional *(p. 685)* La razón de una frecuencia relativa conjunta a la frecuencia relativa marginal en una tabla de doble entrada

conditional statement *(p. 66)* A logical statement that has a hypothesis and a conclusion

enunciado condicional *(p. 66)* Un enunciado lógico que tiene una hipótesis y una conclusión

congruence transformation *(p. 201)* A transformation that preserves length and angle measure

See rigid motion.

transformación de congruencia *(p. 201)* Una transformación que preserva la longitud y medida del ángulo

Ver movimiento rígida.

congruent angles *(p. 40)* Two angles that have the same measure

ángulos congruentes *(p. 40)* Dos ángulos que tienen la misma medida

congruent arcs *(p. 540)* Arcs that have the same measure and are of the same circle or of congruent circles

arcos congruentes *(p. 540)* Arcos que tienen la misma medida y que son del mismo círculo o de círculos congruentes

congruent circles *(p. 540)* Circles that can be mapped onto each other by a rigid motion or a composition of rigid motions

círculos congruentes *(p. 540)* Círculos que pueden superponerse sobre sí mismos mediante un movimiento rígido o una composición de movimientos rígidos

congruent figures *(p. 200)* Geometric figures that have the same size and shape

figuras congruentes *(p. 200)* Figuras geométricas que tienen el mismo tamaño y forma

congruent segments *(p. 13)* Line segments that have the same length

segmentos congruentes *(p. 13)* Segmentos de rectas que tienen la misma longitud

conjecture *(p. 76)* An unproven statement that is based on observations

conjetura *(p. 76)* Una afirmación no comprobada que se basa en observaciones

consecutive interior angles *(p. 128)* Two angles that are formed by two lines and a transversal that lie between the two lines and on the same side of the transversal

ángulos interiores consecutivos *(p. 128)* Dos ángulos que son formados por dos rectas y una transversal que descansan entre las dos rectas y en el mismo lado de la transversal

construction *(p. 13)* A geometric drawing that uses a limited set of tools, usually a compass and a straightedge

construcción *(p. 13)* Un dibujo geométrico que usa un conjunto limitado de herramientas, generalmente una regla y compás

contrapositive *(p. 67)* The statement formed by negating both the hypothesis and conclusion of the converse of a conditional statement

contrapositivo *(p. 67)* El enunciado formado por la negación de la hipótesis y conclusión del converso de un enunciado condicional

converse *(p. 67)* The statement formed by exchanging the hypothesis and conclusion of a conditional statement

converso *(p. 67)* El enunciado formado por el intercambio de la hipótesis y conclusión de un enunciado condicional

coordinate *(p. 12)* A real number that corresponds to a point on a line

coordenada *(p. 12)* Un número real que corresponde a un punto en una línea

coordinate proof *(p. 284)* A style of proof that involves placing geometric figures in a coordinate plane

prueba de coordenadas *(p. 284)* Un estilo de prueba que implica colocar figuras geométricas en un plano coordenado

coplanar points *(p. 4)* Points that lie in the same plane

puntos coplanarios *(p. 4)* Puntos que descansan en el mismo plano

corollary to a theorem *(p. 235)* A statement that can be proved easily using the theorem

corolario de un teorema *(p. 235)* Un enunciado que puede comprobarse fácilmente usando el teorema

corresponding angles *(p. 128)* Two angles that are formed by two lines and a transversal that are in corresponding positions

ángulos correspondientes *(p. 128)* Dos ángulos que están formados por dos líneas y una transversal que están en las posiciones correspondientes

corresponding parts *(p. 240)* A pair of sides or angles that have the same relative position in two congruent figures

partes correspondientes *(p. 240)* Un par de lados o ángulos que tienen la misma posición relativa en dos figuras congruentes

cosine *(p. 494)* For an acute angle of a right triangle, the ratio of the length of the leg adjacent to the acute angle to the length of the hypotenuse

coseno *(p. 494)* Para un ángulo agudo de un triángulo rectángulo, la razón de la longitud del cateto adyacente al ángulo agudo a la longitud de la hipotenusa

counterexample *(p. 77)* A specific case for which a conjecture is false

contraejemplo *(p. 77)* Un caso específico para el que una conjetura es falsa

cross section *(p. 619)* The intersection of a plane and a solid

sección transversal *(p. 619)* La intersección de un plano y un sólido

D

deductive reasoning *(p. 78)* A process that uses facts, definitions, accepted properties, and the laws of logic to form a logical argument

razonamiento deductivo *(p. 78)* Un proceso que usa hechos, definiciones, propiedades aceptadas y las leyes de la lógica para formar un argumento lógico

defined terms *(p. 5)* Terms that can be described using known words, such as *point* or *line*

términos definidos *(p. 5)* Términos que pueden describirse usando palabras conocidas, como *punto* o *línea*

density *(p. 628)* The amount of matter that an object has in a given unit of volume

densidad *(p. 628)* La cantidad de materia que tiene un objeto en una unidad de volumen dada

dependent events *(p. 677)* Two events in which the occurrence of one event does affect the occurrence of the other event

eventos dependientes *(p. 677)* Dos eventos en los que la ocurrencia de un evento afecta la ocurrencia del otro evento

diagonal *(p. 360)* A segment that joins two nonconsecutive vertices of a polygon

diagonal *(p. 360)* Un segmento que une dos vértices no consecutivos de un polígono

diameter *(p. 530)* A chord that contains the center of a circle

diámetro *(p. 530)* Una cuerda que contiene el centro de un círculo

dilation *(p. 208)* A transformation in which a figure is enlarged or reduced with respect to a fixed point

dilatación *(p. 208)* Una transformación en la cual una figura se agranda o reduce con respecto a un punto fijo

directed line segment *(p. 156)* A segment that represents moving from point A to point B is called the directed line segment AB.

segmento de línea dirigido *(p. 156)* Un segmento que representa el moverse del punto A al punto B se llama el segmento de línea dirigido AB.

directrix *(p. 722)* A fixed line perpendicular to the axis of symmetry, such that the set of all points (x, y) of the parabola are equidistant from the focus and the directrix

directriz *(p. 722)* Una recta fija perpendicular al eje de simetría de modo tal, que el conjunto de todos los puntos (x, y) de la parábola sean equidistantes del foco y la directriz

disjoint events *(p. 694)* Two events that have no outcomes in common

eventos disjunto *(p. 694)* Dos eventos que no tienen resultados en común

distance *(p. 12)* The absolute value of the difference of two coordinates on a line

distancia *(p. 12)* El valor absoluto de la diferencia de dos coordenadas en una línea

distance from a point to a line *(p. 148)* The length of the perpendicular segment from the point to the line

distancia desde un punto a una línea *(p. 148)* La longitud del segmento perpendicular desde el punto a la línea

E

edge *(p. 618)* A line segment formed by the intersection of two faces of a polyhedron

borde *(p. 618)* Un segmento de línea formado por la intersección de dos caras de un poliedro

endpoints *(p. 5)* Points that represent the ends of a line segment or ray

puntos extremos *(p. 5)* Punto que representan los extremos de un rayo o segmento de línea

enlargement *(p. 208)* A dilation in which the scale factor is greater than 1

agrandamiento *(p. 208)* Una dilatación en donde el factor de escala es mayor que 1

equiangular polygon *(p. 361)* A polygon in which all angles are congruent

polígono equiangular *(p. 361)* Un polígono en donde todos los ángulos son congruentes

equidistant *(p. 302)* A point is equidistant from two figures when it is the same distance from each figure.

equidistante *(p. 302)* Un punto es equidistante desde dos figuras cuando está a la misma distancia de cada figura.

equilateral polygon *(p. 361)* A polygon in which all sides are congruent

polígono equilátero *(p. 361)* Un polígono en donde todos los lados son congruentes

equivalent statements *(p. 67)* Two related conditional statements that are both true or both false

enunciados equivalentes *(p. 67)* Dos enunciados condicionales relacionados que son ambos verdaderos, o ambos falsos

event *(p. 668)* A collection of one or more outcomes in a probability experiment.

evento *(p. 668)* Una colección de uno o más resultados en un experimento de probabilidades

experimental probability *(p. 671)* The ratio of the number of successes, or favorable outcomes, to the number of trials in a probability experiment

probabilidad experimental *(p. 671)* La razón del número de éxitos, o resultados favorables, con respecto al número de pruebas en un experimento de probabilidades

exterior of an angle *(p. 38)* The region that contains all the points outside of an angle

exterior de un ángulo *(p. 38)* La región que contiene todos los puntos fuera de un ángulo

exterior angles *(p. 233)* Angles that form linear pairs with the interior angles of a polygon

ángulos exteriores *(p. 233)* Ángulos que forman pares lineales con los ángulos interiores de un polígono

external segment *(p. 571)* The part of a secant segment that is outside the circle

segmento externo *(p. 571)* La parte de un segmento secante que está fuera del círculo

face *(p. 618)* A flat surface of a polyhedron

cara *(p. 618)* Una superficie plana de un poliedro

flowchart proof (flow proof) *(p. 106)* A type of proof that uses boxes and arrows to show the flow of a logical argument

prueba de organigrama (prueba de flujo) *(p. 106)* Un tipo de prueba que usa casillas y flechas para mostrar el flujo de un argumento lógico

focus *(p. 722)* A fixed point in the interior of a parabola, such that the set of all points (x, y) of the parabola are equidistant from the focus and the directrix

foco *(p. 722)* Un punto fijo en el interior de una parábola, de tal forma que el conjunto de todos los puntos (x, y) de la parábola sean equidistantes del foco y la directriz

G

geometric mean *(p. 480)* The positive number x that satisfies $\frac{a}{x} = \frac{x}{b}$

So, $x^2 = ab$ and $x = \sqrt{ab}$.

media geométrica *(p. 480)* El número positivo x que satisface $\frac{a}{x} = \frac{x}{b}$

Entonces, $x^2 = ab$ and $x = \sqrt{ab}$.

geometric probability *(p. 670)* A probability found by calculating a ratio of two lengths, areas, or volumes

probabilidad geométrica *(p. 670)* Una probabilidad hallada al calcular la razón de dos longitudes, áreas o volúmenes

glide reflection *(p. 184)* A transformation involving a translation followed by a reflection

reflexión por deslizamiento *(p. 184)* Una transformación que implica una traslación seguida de una reflexión

great circle *(p. 648)* The intersection of a plane and a sphere such that the plane contains the center of the sphere

gran círculo *(p. 648)* La intersección de un plano y una esfera, de tal forma que el plano contiene el centro de la esfera

H

horizontal component *(p. 174)* The horizontal change from the starting point of a vector to the ending point

componente horizontal *(p. 174)* El cambio horizontal desde el punto de inicio de un vector hasta el punto final

hypotenuse *(p. 264)* The side opposite the right angle of a right triangle

hipotenusa *(p. 264)* El lado opuesto al ángulo recto de un triángulo recto

hypothesis *(p. 66)* The "if" part of a conditional statement written in if-then form

hipótesis *(p. 66)* La parte después de "si" en un enunciado condicional escrito de la forma "si..., entonces..."

I

if-then form *(p. 66)* A conditional statement in the form "if p, then q"

forma "si..., entonces..." *(p. 66)* Un enunciado condicional en la forma de "si p, entonces q"

image *(p. 174)* A figure that results from the transformation of a geometric figure

imagen *(p. 174)* Una figura que resulta de la transformación de una figura geométrica

incenter *(p. 313)* The point of concurrency of the angle bisectors of a triangle

incentro *(p. 313)* El punto de concurrencia de las bisectrices de los ángulos de un triángulo

independent events *(p. 676)* Two events in which the occurrence of one event does not affect the occurrence of another event

eventos independientes *(p. 676)* Dos eventos en los que la ocurrencia de un evento no afecta la ocurrencia de otro evento

indirect proof *(p. 336)* A style of proof in which you temporarily assume that the desired conclusion is false, then reason logically to a contradiction

This proves that the original statement is true.

prueba indirecta *(p. 336)* Un estilo de prueba en donde uno asume temporalmente que la conclusión deseada es falsa, luego se razona de forma lógica hasta llegar a una contradicción

Esto prueba que el enunciado original es verdadero.

inductive reasoning *(p. 76)* A process that includes looking for patterns and making conjectures

razonamiento inductivo *(p. 76)* Un proceso que incluye buscar patrones y hacer conjeturas

initial point *(p. 174)* The starting point of a vector

punto inicial *(p. 174)* El punto de inicio de un vector

inscribed angle *(p. 554)* An angle whose vertex is on a circle and whose sides contain chords of the circle

ángulo inscrito *(p. 554)* Un ángulo cuyo vértice está en un círculo y cuyos lados contienen cuerdas del círculo

inscribed polygon *(p. 556)* A polygon in which all the vertices lie on a circle

polígono inscrito *(p. 556)* Un polígono en donde todos los vértices descansan sobre un círculo

intercepted arc *(p. 554)* An arc that lies between two lines, rays, or segments

arco interceptado *(p. 554)* Un arco que descansa entre dos rectas, rayos o segmentos

interior of an angle *(p. 38)* The region that contains all the points between the sides of an angle

interior de un ángulo *(p. 38)* La región que contiene todos los puntos entre los lados de un ángulo

interior angles *(p. 233)* Angles of a polygon

ángulos interiores *(p. 233)* Los ángulos de un polígono

intersection *(p. 6)* The set of points two or more geometric figures have in common

intersección *(p. 6)* El conjunto de puntos que dos o más figuras geométricas tienen en común

inverse *(p. 67)* The statement formed by negating both the hypothesis and conclusion of a conditional statement

inverso *(p. 67)* El enunciado formado por la negación de la hipótesis y conclusión de un enunciado condicional

inverse cosine *(p. 502)* An inverse trigonometric ratio, abbreviated as $\cos^{-1}$

For acute angle A, if $\cos A = z$, then $\cos^{-1} z = m\angle A$.

coseno inverso *(p. 502)* Una razón trigonométrica inversa, abreviada como $\cos^{-1}$

Para un ángulo agudo A, si $\cos A = z$, entonces $\cos^{-1} z = m\angle A$.

inverse sine *(p. 502)* An inverse trigonometric ratio, abbreviated as $\sin^{-1}$

For acute angle A, if $\sin A = y$, then $\sin^{-1} y = m\angle A$.

seno inverso *(p. 502)* Una razón trigonométrica inversa, abreviada como $\sin^{-1}$

Para un ángulo agudo A, si $\sin A = y$, entonces $\sin^{-1} y = m\angle A$.

inverse tangent *(p. 502)* An inverse trigonometric ratio, abbreviated as $\tan^{-1}$

For acute angle A, if $\tan A = x$, then $\tan^{-1} x = m\angle A$.

tangente inversa *(p. 502)* Una razón trigonométrica inversa, abreviada como $\tan^{-1}$

Para un ángulo agudo A, si $\tan A = x$, entonces $\tan^{-1} x = m\angle A$.

isosceles trapezoid *(p. 398)* A trapezoid with congruent legs

trapecio isósceles *(p. 398)* Un trapecio con catetos congruentes

joint frequency *(p. 684)* Each entry in a two-way table

frecuencia conjunta *(p. 684)* Cada valor en una tabla de doble entrada

joint relative frequency *(p. 685)* The ratio of a frequency that is not in the total row or the total column to the total number of values or observations in a two-way table

frecuencia relativa conjunta *(p. 685)* La razón de una frecuencia que no está en la hilera total o columna total del número total de valores u observaciones en una tabla de doble entrada

kite *(p. 401)* A quadrilateral that has two pairs of consecutive congruent sides, but opposite sides are not congruent

papalote *(p. 401)* Un cuadrilátero que tiene dos pares de lados congruentes consecutivos, pero los lados opuestos no son congruentes

L

lateral surface of a cone *(p. 642)* Consists of all segments that connect the vertex with points on the base edge of a cone

superficie lateral de un cono *(p. 642)* Consiste en todos los segmentos que conectan el vértice con puntos en el borde base de un cono

Law of Cosines *(p. 511)* For $\triangle ABC$ with side lengths of a, b, and c,

$$a^2 = b^2 + c^2 - 2bc \cos A,$$
$$b^2 = a^2 + c^2 - 2ac \cos B, \text{ and}$$
$$c^2 = a^2 + b^2 - 2ab \cos C.$$

Ley de cosenos *(p. 511)* Para $\triangle ABC$ con longitudes de lados de a, b, y c,

$$a^2 = b^2 + c^2 - 2bc \cos A,$$
$$b^2 = a^2 + c^2 - 2ac \cos B, \text{ y}$$
$$c^2 = a^2 + b^2 - 2ab \cos C.$$

Law of Sines *(p. 509)* For $\triangle ABC$ with side lengths of a, b, and c,

$$\frac{\sin A}{a} = \frac{\sin B}{b} = \frac{\sin C}{c} \text{ and}$$
$$\frac{a}{\sin A} = \frac{b}{\sin B} = \frac{c}{\sin C}.$$

Ley de senos *(p. 509)* Para $\triangle ABC$ con longitudes de lados de a, b, y c,

$$\frac{\sin A}{a} = \frac{\sin B}{b} = \frac{\sin C}{c} \text{ y}$$
$$\frac{a}{\sin A} = \frac{b}{\sin B} = \frac{c}{\sin C}.$$

legs of an isosceles triangle *(p. 252)* The two congruent sides of an isosceles triangle

catetos de un triángulo isósceles *(p. 252)* Los dos lados congruentes de un triángulo isósceles

legs of a right triangle *(p. 264)* The sides adjacent to the right angle of a right triangle

catetos de un triángulo recto *(p. 264)* Los lados adyacentes al ángulo recto de un triángulo recto

legs of a trapezoid *(p. 398)* The nonparallel sides of a trapezoid

catetos de un trapecio *(p. 398)* Los lados no paralelos de un trapezoide

line *(p. 4)* A line has one dimension. It is represented by a line with two arrowheads, but it extends without end.

recta *(p. 4)* Una recta tiene una dimensión. Se representa por una línea con dos flechas, pero se extiende sin fin.

line perpendicular to a plane *(p. 86)* A line that intersects the plane in a point and is perpendicular to every line in the plane that intersects it at that point

recta perpendicular a un plano *(p. 86)* Una recta que intersecta el plano en un punto y es perpendicular a cada recta en el plano que la intersecta en ese punto

line of reflection *(p. 182)* A line that acts as a mirror for a reflection

recta de reflexión *(p. 182)* Una recta que actúa como un espejo para una reflexión

line segment *(p. 5)* Consists of two endpoints and all the points between them

See segment.

segmento de recta *(p. 5)* Consiste en dos puntos extremos y todos los puntos entre ellos

Ver segmento.

line symmetry *(p. 185)* A figure in the plane has line symmetry when the figure can be mapped onto itself by a reflection in a line.

simetría de recta *(p. 185)* Una figura en el plano tiene simetría de recta cuando la figura puede superponerse sobre sí misma por una reflexión en una recta.

line of symmetry *(p. 185)* A line of reflection that maps a figure onto itself

recta de simetría *(p. 185)* Una recta de reflexión que superpone una figura sobre sí misma

linear pair *(p. 50)* Two adjacent angles whose noncommon sides are opposite rays

par lineal *(p. 50)* Dos ángulos adyacentes cuyos lados no comunes son rayos opuestos

M

major arc *(p. 538)* An arc with a measure greater than 180°

arco mayor *(p. 538)* Un arco con una medida mayor de 180°

marginal frequency *(p. 684)* The sums of the rows and columns in a two-way table

frecuencia marginal *(p. 684)* Las sumas de las hileras y columnas en una tabla de doble entrada

marginal relative frequency *(p. 685)* The sum of the joint relative frequencies in a row or a column in a two-way table

frecuencia relativa marginal *(p. 685)* La suma de las frecuencias relativas conjuntas en una hilera o columna en una tabla de doble entrada

measure of an angle *(p. 39)* The absolute value of the difference between the real numbers matched with the two rays that form the angle on a protractor

medida de un ángulo *(p. 39)* El valor absoluto de la diferencia entre los números reales asociados con los dos rayos que forman el ángulo en un transportador

measure of a major arc *(p. 538)* The measure of a major arc's central angle

medida de arco mayor *(p. 538)* La medida del ángulo central de un arco mayor

measure of a minor arc *(p. 538)* The measure of a minor arc's central angle

medida de arco menor *(p. 538)* La medida del ángulo central de un arco menor

median of a triangle *(p. 320)* A segment from a vertex of a triangle to the midpoint of the opposite side

mediana de un triángulo *(p. 320)* Un segmento desde el vértice de un triángulo hasta el punto medio del lado opuesto

midpoint *(p. 20)* The point that divides a segment into two congruent segments

punto medio *(p. 20)* El punto que divide un segmento en dos segmentos congruentes

midsegment of a trapezoid *(p. 400)* The segment that connects the midpoints of the legs of a trapezoid

segmento medio de un trapezoide *(p. 400)* El segmento que conecta los puntos medios de los catetos de un trapezoide

midsegment of a triangle *(p. 330)* A segment that connects the midpoints of two sides of a triangle

segmento medio de un triángulo *(p. 330)* Un segmento que conecta los puntos medios de dos lados de un triángulo

minor arc *(p. 538)* An arc with a measure less than 180°

arco menor *(p. 538)* Un arco con una medida menor de 180°

mutually exclusive events *(p. 694)* Two events that have no outcomes in common

eventos mutuamente exclusivos *(p. 694)* Dos eventos que no tienen resultados en común

N

***n* factorial** *(p. 700)* The product of the integers from 1 to n, for any positive integer n

factorial de *n* *(p. 700)* El producto de los números enteros de 1 a n, para cualquier número entero positivo n

negation *(p. 66)* The opposite of a statement

If a statement is p, then the negation is "not p," written $\sim p$.

negación *(p. 66)* Lo opuesto de un enunciado o afirmación

Si un enunciado es p, entonces la negación es "no p," y se escribe $\sim p$.

net *(p. 592)* A two-dimensional pattern than can be folded to form a three-dimensional figure

desarrollo de poliedros *(p. 592)* Un patrón bidimensional que puede doblarse para formar una figura tridimensional

O

obtuse angle *(p. 39)* An angle that has a measure greater than 90° and less than 180°

ángulo obtuso *(p. 39)* Un ángulo que tiene una medida mayor que 90° y menor que 180°

opposite rays *(p. 5)* If point C lies on $\overleftrightarrow{AB}$ between A and B, then $\overrightarrow{CA}$ and $\overrightarrow{CB}$ are opposite rays.

rayos opuestos *(p. 5)* Si el punto C descansa en $\overleftrightarrow{AB}$ entre A y B, entonces $\overrightarrow{CA}$ y $\overrightarrow{CB}$ son rayos opuestos.

orthocenter *(p. 321)* The point of concurrency of the lines containing the altitudes of a triangle

ortocentro *(p. 321)* El punto de concurrencia de las líneas que contienen las alturas de un triángulo

outcome *(p. 668)* The possible result of a probability experiment

resultado *(p. 668)* El resultado posible de un experimento de probabilidad

overlapping events *(p. 694)* Two events that have one or more outcomes in common

eventos superpuestos *(p. 694)* Dos eventos que tienen uno o más resultados en común

P

paragraph proof *(p. 108)* A style of proof that presents the statements and reasons as sentences in a paragraph, using words to explain the logical flow of an argument

prueba en forma de párrafo *(p. 108)* Un estilo de prueba que presenta los enunciados y motivos como oraciones en un párrafo, usando palabras para explicar el flujo lógico de un argumento

parallel lines *(p. 126)* Coplanar lines that do not intersect

rectas paralelas *(p. 126)* Rectas coplanarias que no se intersectan

parallel planes *(p. 126)* Planes that do not intersect

planos paralelos *(p. 126)* Planos que no se intersectan

parallelogram *(p. 368)* A quadrilateral with both pairs of opposite sides parallel

paralelogramo *(p. 368)* Un cuadrilátero con ambos pares de lados opuestos paralelos

permutation *(p. 700)* An arrangement of objects in which order is important

permutación *(p. 700)* Una disposición de objectos en la que el orden es importante

perpendicular bisector *(p. 149)* A line that is perpendicular to a segment at its midpoint

bisectriz perpendicular *(p. 149)* Una recta que es perpendicular a un segmento en su punto medio

perpendicular lines *(p. 68)* Two lines that intersect to form a right angle

rectas perpendiculares *(p. 68)* Dos líneas que se intersectan para formar un ángulo recto

plane *(p. 4)* A flat surface made up of points that has two dimensions and extends without end and is represented by a shape that looks like a floor or wall

plano *(p. 4)* Una superficie plana formada por puntos que tiene dos dimensiones y se extiende sin fin y que está representada por una forma que parece un piso o una pared

point *(p. 4)* A location in space that is represented by a dot and has no dimension

punto *(p. 4)* Un lugar en el espacio que está representado por un punto y no tiene dimensión

point of concurrency *(p. 310)* The point of intersection of concurrent lines, rays, or segments

punto de concurrencia *(p. 310)* El punto de intersección de rectas, rayos o segmentos concurrentes

point of tangency *(p. 530)* The point at which a tangent line intersects a circle

punto de tangencia *(p. 530)* El punto en donde una recta tangente intersecta a un círculo

polyhedron *(p. 618)* A solid that is bounded by polygons

poliedro *(p. 618)* Un sólido que está encerrado por polígonos

population density *(p. 603)* A measure of how many people live within a given area

densidad de población *(p. 603)* Medición de la cantidad de personas que habitan un área dada

postulate *(p. 12)* A rule that is accepted without proof

postulado *(p. 12)* Una regla que es aceptada sin demostración

preimage *(p. 174)* The original figure before a transformation

preimagen *(p. 174)* La figura original antes de una transformación

probability distribution *(p. 708)* A function that gives the probability of each possible value of a random variable

distribución de probabilidad *(p. 708)* Una función que da la probabilidad de cada valor posible de una variable aleatoria

probability of an event *(p. 668)* A measure of the likelihood, or chance, that an event will occur

probabilidad de un evento *(p. 668)* Una medida de la probabilidad o posibilidad de que ocurrirá un evento

probability experiment *(p. 668)* An action, or trial, that has varying results

experimento de probabilidad *(p. 668)* Una acción o prueba que tiene resultados variables

proof *(p. 100)* A logical argument that uses deductive reasoning to show that a statement is true

prueba *(p. 100)* Un argumento lógico que usa el razonamiento deductivo para mostrar que un enunciado es verdadero

Pythagorean triple *(p. 464)* A set of three positive integers a, b, and c that satisfy the equation $c^2 = a^2 + b^2$

triple pitagórico *(p. 464)* Un conjunto de tres números enteros positivos a, b, y c que satisfacen la ecuación $c^2 = a^2 + b^2$

R

radian *(p. 597)* A unit of measurement for angles

radián *(p. 597)* Una unidad de medida para ángulos

radius of a circle *(p. 530)* A segment whose endpoints are the center and any point on a circle

radio de un círculo *(p. 530)* Un segmento cuyos puntos extremos son el centro y cualquier punto en un círculo

radius of a regular polygon *(p. 611)* The radius of a polygon's circumscribed circle

radio de un polígono regular *(p. 611)* El radio del círculo circunscrito de un polígono

random variable *(p. 708)* A variable whose value is determined by the outcomes of a probability experiment

variable aleatoria *(p. 708)* Una variable cuyo valor está determinado por los resultados de un experimento de probabilidad

ray *(p. 5)* $\overrightarrow{AB}$ is a ray if it consists of the endpoint A and all points on $\overleftrightarrow{AB}$ that lie on the same side of A as B.

rayo *(p. 5)* $\overrightarrow{AB}$ es un rayo, si consiste del punto extremo A y todos los puntos en $\overleftrightarrow{AB}$ que descansan en el mismo lado de A como B.

rectangle *(p. 388)* A parallelogram with four right angles

rectángulo *(p. 388)* Un paralelogramo con cuatro ángulos rectos

reduction *(p. 208)* A dilation in which the scale factor is greater than 0 and less than 1

reducción *(p. 208)* Una dilatación en donde el factor de escala es mayor que 0 y menor que 1

reflection *(p. 182)* A transformation that uses a line like a mirror to reflect a figure

reflexión *(p. 182)* Una transformación que usa una recta como un espejo para reflejar una figura

regular polygon *(p. 361)* A convex polygon that is both equilateral and equiangular

polígono regular *(p. 361)* Un polígono convexo que es tanto equilátero como equiángulo

rhombus *(p. 388)* A parallelogram with four congruent sides

rombo *(p. 388)* Un paralelogramo con cuatro lados congruentes

right angle *(p. 39)* An angle that has a measure of 90°

ángulo recto *(p. 39)* Un ángulo que tiene una medida de 90°

rigid motion *(p. 176)* A transformation that preserves length and angle measure
See congruence transformation.

movimiento rígido *(p. 176)* Una transformación que preserva la longitud y medida del ángulo
Ver transformación de congruencia.

rotation *(p. 190)* A transformation in which a figure is turned about a fixed point

rotación *(p. 190)* Una transformación en la cual una figura gira sobre un punto fijo

rotational symmetry *(p. 193)* A figure has rotational symmetry when the figure can be mapped onto itself by a rotation of 180° or less about the center of the figure.

simetría de rotación *(p. 193)* Una figura tiene simetría de rotación cuando la figura puede superponerse sobre sí misma mediante una rotación de 180° o menos en el centro de la figura.

S

sample space *(p. 668)* The set of all possible outcomes for an experiment

espacio de muestra *(p. 668)* El conjunto de todos los resultados posibles de un experimento

scale factor *(p. 208)* The ratio of the lengths of the corresponding sides of the image and the preimage of a dilation

factor de escala *(p. 208)* La razón de las longitudes de los lados correspondientes de la imagen y la preimagen de una dilatación

secant *(p. 530)* A line that intersects a circle in two points

secante *(p. 530)* Una recta que intersecta a un círculo en dos puntos

secant segment *(p. 571)* A segment that contains a chord of a circle and has exactly one endpoint outside the circle

segmento de secante *(p. 571)* Un segmento que contiene una cuerda de un círculo y que tiene exactamente un punto extremo fuera del círculo

sector of a circle *(p. 604)* The region bounded by two radii of the circle and their intercepted arc

sector de un círculo *(p. 604)* La región encerrada por dos radios del círculo y su arco interceptado

segment *(p. 5)* Consists of two endpoints and all the points between them
See line segment.

segmento *(p. 5)* Consiste en dos puntos extremos y todos los puntos entre ellos
Ver segmento de recta.

segment bisector *(p. 20)* A point, ray, line, line segment, or plane that intersects the segment at its midpoint

bisectriz de segmento *(p. 20)* Un punto, rayo, recta, segmento de recta o plano que intersecta el segmento en su punto medio

segments of a chord *(p. 570)* The segments formed from two chords that intersect in the interior of a circle

segmentos de una cuerda *(p. 570)* Los segmentos formados a partir de dos cuerdas que se intersectan en el interior de un círculo

semicircle *(p. 538)* An arc with endpoints that are the endpoints of a diameter

semicírculo *(p. 538)* Un arco con puntos extremos que son los puntos extremos de un diámetro

sides of an angle *(p. 38)* The rays of an angle

lados de un ángulo *(p. 38)* Los rayos de un ángulo

similar arcs *(p. 541)* Arcs that have the same measure

arcos similares *(p. 541)* Arcos que tienen la misma medida

similar figures *(p. 216)* Geometric figures that have the same shape but not necessarily the same size

figuras similares *(p. 216)* Figuras geométricas que tienen la misma forma pero no necesariamente el mismo tamaño

similar solids *(p. 630)* Two solids of the same type with equal ratios of corresponding linear measures

sólidos similares *(p. 630)* Dos sólidos del mismo tipo con razones iguales de medidas lineales correspondientes

similarity transformation *(p. 216)* A dilation or a composition of rigid motions and dilations

transformación de similitud *(p. 216)* Una dilatación o composición de movimientos rígidos y dilataciones

sine *(p. 494)* For an acute angle of a right triangle, the ratio of the length of the leg opposite the acute angle to the length of the hypotenuse

seno *(p. 494)* Para un ángulo agudo de un triángulo rectángulo, la razón de la longitud del cateto enfrente del ángulo agudo a la longitud de la hipotenusa

skew lines *(p. 126)* Lines that do not intersect and are not coplanar

rectas sesgadas *(p. 126)* Rectas que no se intersectan y que no son coplanarias

solid of revolution *(p. 620)* A three-dimensional figure that is formed by rotating a two-dimensional shape around an axis

sólido de revolución *(p. 620)* Una figura tridimensional que se forma por la rotación de una forma bidimensional alrededor de un eje

solve a right triangle *(p. 503)* To find all unknown side lengths and angle measures of a right triangle

resolver un triángulo recto *(p. 503)* Para encontrar todas las longitudes de los lados y las medidas de los ángulos desconocidas de un triángulo recto

square *(p. 388)* A parallelogram with four congruent sides and four right angles

cuadrado *(p. 388)* Un paralelogramo con cuatro lados congruentes y cuatro ángulos rectos

standard equation of a circle *(p. 576)* $(x - h)^2 + (y - k)^2 = r^2$, where r is the radius and (h, k) is the center

ecuación estándar de un círculo *(p. 576)* $(x - h)^2 + (y - k)^2 = r^2$, donde r es el radio y (h, k) es el centro

standard position *(p. 462)* A right triangle is in standard position when the hypotenuse is a radius of the circle of radius 1 with center at the origin, one leg lies on the x-axis, and the other leg is perpendicular to the x-axis.

posición estándar *(p. 462)* Un triángulo recto se encuentra en posición estándar cuando la hipotenusa es un radio del círculo de radio 1 con centro en el origen, un cateto descansa en el eje x y el otro cateto es perpendicular al eje x.

straight angle *(p. 39)* An angle that has a measure of 180°

ángulo llano *(p. 39)* Un ángulo que tiene una medida de 180°

subtend *(p. 554)* If the endpoints of a chord or arc lie on the sides of an inscribed angle, the chord or arc is said to subtend the angle.

subtender *(p. 554)* Si los puntos extremos de una cuerda o arco descansan en los lados de un ángulo inscrito, se dice que la cuerda o arco subtiende el ángulo.

supplementary angles *(p. 48)* Two angles whose measures have a sum of 180°

ángulos suplementarios *(p. 48)* Dos ángulos cuyas medidas suman 180°

tangent *(p. 488)* For an acute angle of a right triangle, the ratio of the length of the leg opposite the acute angle to the length of the leg adjacent to the acute angle

tangente *(p. 488)* Para un ángulo agudo de un triángulo rectángulo, la razón de la longitud del cateto enfrente del ángulo agudo a la longitud del cateto adyacente al ángulo agudo

tangent of a circle *(p. 530)* A line in the plane of a circle that intersects the circle at exactly one point

tangente de un círculo *(p. 530)* Una recta en el plano de un círculo que intersecta el círculo en exactamente un punto

tangent circles *(p. 531)* Coplanar circles that intersect in one point

círculos tangentes *(p. 531)* Círculos coplanarios que se intersectan en un punto

tangent segment *(p. 571)* A segment that is tangent to a circle at an endpoint

segmento de tangente *(p. 571)* Un segmento que es tangente a un círculo en un punto extremo

terminal point *(p. 174)* The ending point of a vector

punto terminal *(p. 174)* El punto final de un vector

theorem *(p. 101)* A statement that can be proven

teorema *(p. 101)* Un enunciado que puede comprobarse

theoretical probability *(p. 669)* The ratio of the number of favorable outcomes to the total number of outcomes when all outcomes are equally likely

probabilidad teórica *(p. 669)* La razón del número de resultados favorables con respecto al número total de resultados cuando todos los resultados son igualmente probables

transformation *(p. 174)* A function that moves or changes a figure in some way to produce a new figure

transformación *(p. 174)* Una función que mueve o cambia una figura de cierta manera para producir una nueva figura

translation *(p. 174)* A transformation that moves every point of a figure the same distance in the same direction

traslación *(p. 174)* Una transformación que mueve cada punto de una figura la misma distancia en la misma dirección

transversal *(p. 128)* A line that intersects two or more coplanar lines at different points

transversal *(p. 128)* Una recta que intersecta dos o más rectas coplanarias en puntos distintos

trapezoid *(p. 398)* A quadrilateral with exactly one pair of parallel sides

trapecio *(p. 398)* Un cuadrilátero con exactamente un par de lados paralelos

trigonometric ratio *(p. 488)* A ratio of the lengths of two sides in a right triangle

razón trigonométrica *(p. 488)* Una razón de las longitudes de dos lados en un triángulo recto

truth table *(p. 70)* A table that shows the truth values for a hypothesis, conclusion, and conditional statement

tabla de verdad *(p. 70)* Una tabla que muestra los verdaderos valores para una hipótesis, conclusión y enunciado condicional

truth value *(p. 70)* True (T) or false (F)

valor de verdad *(p. 70)* Verdadero (V) o falso (F)

two-column proof *(p. 100)* A type of proof that has numbered statements and corresponding reasons that show an argument in a logical order

prueba de dos columnas *(p. 100)* Un tipo de prueba que tiene enunciados numerados y motivos correspondientes que muestran un argumento en un orden lógico

two-way table *(p. 684)* A frequency table that displays data collected from one source that belong to two different categories

tabla de doble entrada *(p. 684)* Una tabla de frecuencia que muestra los datos recogidos de una fuente que pertenece a dos categorías distintas

U

undefined terms *(p. 4)* Words that do not have formal definitions, but there is agreement about what they mean

In geometry, the words *point*, *line*, and *plane* are undefined terms.

términos no definidos *(p. 4)* Palabras que no tienen definiciones formales, pero hay un consenso acerca de lo que significan

En geometría, las palabras *punto*, *línea* y *plano* son términos no definidos.

vector *(p. 174)* A quantity that has both direction and magnitude and is represented in the coordinate plane by an arrow drawn from one point to another

vector *(p. 174)* Una cantidad que tiene tanto dirección como magnitud y que está representada en el plano coordenado por una flecha dibujada de un punto a otro

vertex angle *(p. 252)* The angle formed by the legs of an isosceles triangle

ángulo del vértice *(p. 252)* El ángulo formado por los catetos de un triángulo isósceles

vertex of an angle *(p. 38)* The common endpoint of two rays

vértice de un ángulo *(p. 38)* El punto extremo que dos rayos tienen en común

vertex of a polyhedron *(p. 618)* A point of a polyhedron where three or more edges meet

vértice de un poliedro *(p. 618)* Un punto de un poliedro donde se encuentran tres o más bordes

vertical angles *(p. 50)* Two angles whose sides form two pairs of opposite rays

ángulos verticales *(p. 50)* Dos ángulos cuyos lados forman dos pares de rayos opuestos

vertical component *(p. 174)* The vertical change from the starting point of a vector to the ending point

componente vertical *(p. 174)* El cambio vertical desde el punto de inicio de un vector hasta el punto final

volume *(p. 626)* The number of cubic units contained in the interior of a solid

volumen *(p. 626)* El número de unidades cúbicas contenidas en el interior de u sólido

Index

Index

D

Q

R

S

Postulates

1.1 Ruler Postulate

The points on a line can be matched one to one with the real numbers. The real number that corresponds to a point is the coordinate of the point. The distance between points A and B, written as AB, is the absolute value of the difference of the coordinates of A and B.

1.2 Segment Addition Postulate

If B is between A and C, then $AB + BC = AC$.
If $AB + BC = AC$, then B is between A and C.

1.3 Protractor Postulate

Consider $\overleftrightarrow{OB}$ and a point A on one side of $\overleftrightarrow{OB}$. The rays of the form $\overrightarrow{OA}$ can be matched one to one with the real numbers from 0 to 180. The measure of $\angle AOB$, which can be written as $m\angle AOB$, is equal to the absolute value of the difference between the real numbers matched with $\overrightarrow{OA}$ and $\overrightarrow{OB}$ on a protractor.

1.4 Angle Addition Postulate

If P is in the interior of $\angle RST$, then the measure of $\angle RST$ is equal to the sum of the measures of $\angle RSP$ and $\angle PST$.

2.1 Two Point Postulate

Through any two points, there exists exactly one line.

2.2 Line-Point Postulate

A line contains at least two points.

2.3 Line Intersection Postulate

If two lines intersect, then their intersection is exactly one point.

2.4 Three Point Postulate

Through any three noncollinear points, there exists exactly one plane.

2.5 Plane-Point Postulate

A plane contains at least three noncollinear points.

2.6 Plane-Line Postulate

If two points lie in a plane, then the line containing them lies in the plane.

2.7 Plane Intersection Postulate

If two planes intersect, then their intersection is a line.

2.8 Linear Pair Postulate

If two angles form a linear pair, then they are supplementary.

3.1 Parallel Postulate

If there is a line and a point not on the line, then there is exactly one line through the point parallel to the given line.

3.2 Perpendicular Postulate

If there is a line and a point not on the line, then there is exactly one line through the point perpendicular to the given line.

4.1 Translation Postulate

A translation is a rigid motion.

4.2 Reflection Postulate

A reflection is a rigid motion.

4.3 Rotation Postulate

A rotation is a rigid motion.

10.1 Arc Addition Postulate

The measure of an arc formed by two adjacent arcs is the sum of the measures of the two arcs.

Theorems

2.1 Properties of Segment Congruence

Segment congruence is reflexive, symmetric, and transitive.

Reflexive For any segment AB, $\overline{AB} \cong \overline{AB}$.

Symmetric If $\overline{AB} \cong \overline{CD}$, then $\overline{CD} \cong \overline{AB}$.

Transitive If $\overline{AB} \cong \overline{CD}$ and $\overline{CD} \cong \overline{EF}$, then $\overline{AB} \cong \overline{EF}$.

2.2 Properties of Angle Congruence

Angle congruence is reflexive, symmetric, and transitive.

Reflexive For any angle A, $\angle A \cong \angle A$.

Symmetric If $\angle A \cong \angle B$, then $\angle B \cong \angle A$.

Transitive If $\angle A \cong \angle B$ and $\angle B \cong \angle C$, then $\angle A \cong \angle C$.

2.3 Right Angles Congruence Theorem

All right angles are congruent.

2.4 Congruent Supplements Theorem

If two angles are supplementary to the same angle (or to congruent angles), then they are congruent.

2.5 Congruent Complements Theorem

If two angles are complementary to the same angle (or to congruent angles), then they are congruent.

2.6 Vertical Angles Congruence Theorem

Vertical angles are congruent.

3.1 Corresponding Angles Theorem

If two parallel lines are cut by a transversal, then the pairs of corresponding angles are congruent.

3.2 Alternate Interior Angles Theorem

If two parallel lines are cut by a transversal, then the pairs of alternate interior angles are congruent.

3.3 Alternate Exterior Angles Theorem

If two parallel lines are cut by a transversal, then the pairs of alternate exterior angles are congruent.

3.4 Consecutive Interior Angles Theorem

If two parallel lines are cut by a transversal, then the pairs of consecutive interior angles are supplementary.

3.5 Corresponding Angles Converse

If two lines are cut by a transversal so the corresponding angles are congruent, then the lines are parallel.

3.6 Alternate Interior Angles Converse

If two lines are cut by a transversal so the alternate interior angles are congruent, then the lines are parallel.

3.7 Alternate Exterior Angles Converse

If two lines are cut by a transversal so the alternate exterior angles are congruent, then the lines are parallel.

3.8 Consecutive Interior Angles Converse

If two lines are cut by a transversal so the consecutive interior angles are supplementary, then the lines are parallel.

3.9 Transitive Property of Parallel Lines

If two lines are parallel to the same line, then they are parallel to each other.

3.10 Linear Pair Perpendicular Theorem

If two lines intersect to form a linear pair of congruent angles, then the lines are perpendicular.

3.11 Perpendicular Transversal Theorem

In a plane, if a transversal is perpendicular to one of two parallel lines, then it is perpendicular to the other line.

3.12 Lines Perpendicular to a Transversal Theorem

In a plane, if two lines are perpendicular to the same line, then they are parallel to each other.

3.13 Slopes of Parallel Lines

In a coordinate plane, two distinct nonvertical lines are parallel if and only if they have the same slope. Any two vertical lines are parallel.

3.14 Slopes of Perpendicular Lines

In a coordinate plane, two nonvertical lines are perpendicular if and only if the product of their slopes is -1. Horizontal lines are perpendicular to vertical lines.

4.1 Composition Theorem

The composition of two (or more) rigid motions is a rigid motion.

4.2 Reflections in Parallel Lines Theorem

If lines k and m are parallel, then a reflection in line k followed by a reflection in line m is the same as a translation. If A'' is the image of A, then

1. $\overline{AA''}$ is perpendicular to k and m, and
2. $AA'' = 2d$, where d is the distance between k and m.

4.3 Reflections in Intersecting Lines Theorem

If lines k and m intersect at point P, then a reflection in line k followed by a reflection in line m is the same as a rotation about point P. The angle of rotation is $2x^\circ$, where x° is the measure of the acute or right angle formed by lines k and m.

5.1 Triangle Sum Theorem

The sum of the measures of the interior angles of a triangle is 180°.

5.2 Exterior Angle Theorem

The measure of an exterior angle of a triangle is equal to the sum of the measures of the two nonadjacent interior angles

Corollary 5.1 Corollary to the Triangle Sum Theorem

The acute angles of a right triangle are complementary.

5.3 Properties of Triangle Congruence

Triangle congruence is reflexive, symmetric, and transitive.

Reflexive For any triangle $\triangle ABC$, $\triangle ABC \cong \triangle ABC$.

Symmetric If $\triangle ABC \cong \triangle DEF$, then $\triangle DEF \cong \triangle ABC$.

Transitive If $\triangle ABC \cong \triangle DEF$ and $\triangle DEF \cong \triangle JKL$, then $\triangle ABC \cong \triangle JKL$.

5.4 Third Angles Theorem

If two angles of one triangle are congruent to two angles of another triangle, then the third angles are also congruent.

5.5 Side-Angle-Side (SAS) Congruence Theorem

If two sides and the included angle of one triangle are congruent to two sides and the included angle of a second triangle, then the two triangles are congruent.

5.6 Base Angles Theorem

If two sides of a triangle are congruent, then the angles opposite them are congruent.

5.7 Converse of the Base Angles Theorem

If two angles of a triangle are congruent, then the sides opposite them are congruent.

Corollary 5.2 Corollary to the Base Angles Theorem

If a triangle is equilateral, then it is equiangular.

Corollary 5.3 Corollary to the Converse of the Base Angles Theorem

If a triangle is equiangular, then it is equilateral.

5.8 Side-Side-Side (SSS) Congruence Theorem

If three sides of one triangle are congruent to three sides of a second triangle, then the two triangles are congruent.

5.9 Hypotenuse-Leg (HL) Congruence Theorem

If the hypotenuse and a leg of a right triangle are congruent to the hypotenuse and a leg of a second right triangle, then the two triangles are congruent.

5.10 Angle-Side-Angle (ASA) Congruence Theorem

If two angles and the included side of one triangle are congruent to two angles and the included side of a second triangle, then the two triangles are congruent.

5.11 Angle-Angle-Side (AAS) Congruence Theorem

If two angles and a non-included side of one triangle are congruent to two angles and the corresponding non-included side of a second triangle, then the two triangles are congruent.

6.1 Perpendicular Bisector Theorem

In a plane, if a point lies on the perpendicular bisector of a segment, then it is equidistant from the endpoints of the segment.

6.2 Converse of the Perpendicular Bisector Theorem

In a plane, if a point is equidistant from the endpoints of a segment, then it lies on the perpendicular bisector of the segment.

6.3 Angle Bisector Theorem

If a point lies on the bisector of an angle, then it is equidistant from the two sides of the angle.

6.4 Converse of the Angle Bisector Theorem

If a point is in the interior of an angle and is equidistant from the two sides of the angle, then it lies on the bisector of the angle.

6.5 Circumcenter Theorem

The circumcenter of a triangle is equidistant from the vertices of the triangle.

6.6 Incenter Theorem

The incenter of a triangle is equidistant from the sides of the triangle.

6.7 Centroid Theorem

The centroid of a triangle is two-thirds of the distance from each vertex to the midpoint of the opposite side.

6.8 Triangle Midsegment Theorem

The segment connecting the midpoints of two sides of a triangle is parallel to the third side and is half as long as that side.

6.9 Triangle Longer Side Theorem

If one side of a triangle is longer than another side, then the angle opposite the longer side is larger than the angle opposite the shorter side.

6.10 Triangle Larger Angle Theorem

If one angle of a triangle is larger than another angle, then the side opposite the larger angle is longer than the side opposite the smaller angle.

6.11 Triangle Inequality Theorem

The sum of the lengths of any two sides of a triangle is greater than the length of the third side.

6.12 Hinge Theorem

If two sides of one triangle are congruent to two sides of another triangle, and the included angle of the first is larger than the included angle of the second, then the third side of the first is longer than the third side of the second.

6.13 Converse of the Hinge Theorem

If two sides of one triangle are congruent to two sides of another triangle, and the third side of the first is longer than the third side of the second, then the included angle of the first is larger than the included angle of the second.

7.1 Polygon Interior Angles Theorem

The sum of the measures of the interior angles of a convex n-gon is $(n - 2) \cdot 180°$.

Corollary 7.1 Corollary to the Polygon Interior Angles Theorem

The sum of the measures of the interior angles of a quadrilateral is 360°.

7.2 Polygon Exterior Angles Theorem

The sum of the measures of the exterior angles of a convex polygon, one angle at each vertex, is 360°.

7.3 Parallelogram Opposite Sides Theorem

If a quadrilateral is a parallelogram, then its opposite sides are congruent.

7.4 Parallelogram Opposite Angles Theorem

If a quadrilateral is a parallelogram, then its opposite angles are congruent.

7.5 Parallelogram Consecutive Angles Theorem

If a quadrilateral is a parallelogram, then its consecutive angles are supplementary.

7.6 Parallelogram Diagonals Theorem

If a quadrilateral is a parallelogram, then its diagonals bisect each other.

7.7 Parallelogram Opposite Sides Converse

If both pairs of opposite sides of a quadrilateral are congruent, then the quadrilateral is a parallelogram.

7.8 Parallelogram Opposite Angles Converse

If both pairs of opposite angles of a quadrilateral are congruent, then the quadrilateral is a parallelogram.

7.9 Opposite Sides Parallel and Congruent Theorem

If one pair of opposite sides of a quadrilateral are congruent and parallel, then the quadrilateral is a parallelogram.

7.10 Parallelogram Diagonals Converse

If the diagonals of a quadrilateral bisect each other, then the quadrilateral is a parallelogram.

Corollary 7.2 Rhombus Corollary

A quadrilateral is a rhombus if and only if it has four congruent sides.

Corollary 7.3 Rectangle Corollary

A quadrilateral is a rectangle if and only if it has four right angles.

Corollary 7.4 Square Corollary

A quadrilateral is a square if and only if it is a rhombus and a rectangle.

7.11 Rhombus Diagonals Theorem

A parallelogram is a rhombus if and only if its diagonals are perpendicular.

7.12 Rhombus Opposite Angles Theorem

A parallelogram is a rhombus if and only if each diagonal bisects a pair of opposite angles.

7.13 Rectangle Diagonals Theorem

A parallelogram is a rectangle if and only if its diagonals are congruent.

7.14 Isosceles Trapezoid Base Angles Theorem

If a trapezoid is isosceles, then each pair of base angles is congruent.

7.15 Isosceles Trapezoid Base Angles Converse

If a trapezoid has a pair of congruent base angles, then it is an isosceles trapezoid.

7.16 Isosceles Trapezoid Diagonals Theorem

A trapezoid is isosceles if and only if its diagonals are congruent.

7.17 Trapezoid Midsegment Theorem

The midsegment of a trapezoid is parallel to each base, and its length is one-half the sum of the lengths of the bases.

7.18 Kite Diagonals Theorem

If a quadrilateral is a kite, then its diagonals are perpendicular.

7.19 Kite Opposite Angles Theorem

If a quadrilateral is a kite, then exactly one pair of opposite angles are congruent.

8.1 Perimeters of Similar Polygons

If two polygons are similar, then the ratio of their perimeters is equal to the ratios of their corresponding side lengths.

8.2 Areas of Similar Polygons

If two polygons are similar, then the ratio of their areas is equal to the squares of the ratios of their corresponding side lengths.

8.3 Angle-Angle (AA) Similarity Theorem

If two angles of one triangle are congruent to two angles of another triangle, then the two triangles are similar.

8.4 Side-Side-Side (SSS) Similarity Theorem

If the corresponding side lengths of two triangles are proportional, then the triangles are similar.

8.5 Side-Angle-Side (SAS) Similarity Theorem

If an angle of one triangle is congruent to an angle of a second triangle and the lengths of the sides including these angles are proportional, then the triangles are similar.

8.6 Triangle Proportionality Theorem

If a line parallel to one side of a triangle intersects the other two sides, then it divides the two sides proportionally.

8.7 Converse of the Triangle Proportionality Theorem

If a line divides two sides of a triangle proportionally, then it is parallel to the third side.

8.8 Three Parallel Lines Theorem

If three parallel lines intersect two transversals, then they divide the transversals proportionally.

8.9 Triangle Angle Bisector Theorem

If a ray bisects an angle of a triangle, then it divides the opposite side into segments whose lengths are proportional to the lengths of the other two sides.

9.1 Pythagorean Theorem

In a right triangle, the square of the length of the hypotenuse is equal to the sum of the squares of the lengths of the legs.

9.2 Converse of the Pythagorean Theorem

If the square of the length of the longest side of a triangle is equal to the sum of the squares of the lengths of the other two sides, then the triangle is a right triangle.

9.3 Pythagorean Inequalities Theorem

For any $\triangle ABC$, where c is the length of the longest side, the following statements are true.
If $c^2 < a^2 + b^2$, then $\triangle ABC$ is acute.
If $c^2 > a^2 + b^2$, then $\triangle ABC$ is obtuse.

9.4 45°-45°-90° Triangle Theorem

In a 45°-45°-90° triangle, the hypotenuse is $\sqrt{2}$ times as long as each leg.

9.5 30°-60°-90° Triangle Theorem

In a 30°-60°-90° triangle, the hypotenuse is twice as long as the shorter leg, and the longer leg is $\sqrt{3}$ times as long as the shorter leg.

9.6 Right Triangle Similarity Theorem

If the altitude is drawn to the hypotenuse of a right triangle, then the two triangles formed are similar to the original triangle and to each other.

9.7 Geometric Mean (Altitude) Theorem

In a right triangle, the altitude from the right angle to the hypotenuse divides the hypotenuse into two segments. The length of the altitude is the geometric mean of the lengths of the two segments of the hypotenuse.

9.8 Geometric Mean (Leg) Theorem

In a right triangle, the altitude from the right angle to the hypotenuse divides the hypotenuse into two segments. The length of each leg of the right triangle is the geometric mean of the lengths of the hypotenuse and the segment of the hypotenuse that is adjacent to the leg.

9.9 Law of Sines

The Law of Sines can be written in either of the following forms for $\triangle ABC$ with sides of length a, b, and c.

$$\frac{\sin A}{a} = \frac{\sin B}{b} = \frac{\sin C}{c}$$

$$\frac{a}{\sin A} = \frac{b}{\sin B} = \frac{c}{\sin C}$$

9.10 Law of Cosines

If $\triangle ABC$ has sides of length a, b, and c, then the following are true.

$a^2 = b^2 + c^2 - 2bc \cos A$

$b^2 = a^2 + c^2 - 2ac \cos B$

$c^2 = a^2 + b^2 - 2ab \cos C$

10.1 Tangent Line to Circle Theorem

In a plane, a line is tangent to a circle if and only if the line is perpendicular to a radius of the circle at its endpoint on the circle.

10.2 External Tangent Congruence Theorem

Tangent segments from a common external point are congruent.

10.3 Congruent Circles Theorem

Two circles are congruent circles if and only if they have the same radius.

10.4 Congruent Central Angles Theorem

In the same circle, or in congruent circles, two minor arcs are congruent if and only if their corresponding central angles are congruent.

10.5 Similar Circles Theorem

All circles are similar.

10.6 Congruent Corresponding Chords Theorem

In the same circle, or in congruent circles, two minor arcs are congruent if and only if their corresponding chords are congruent.

10.7 Perpendicular Chord Bisector Theorem

If a diameter of a circle is perpendicular to a chord, then the diameter bisects the chord and its arc.

10.8 Perpendicular Chord Bisector Converse

If one chord of a circle is a perpendicular bisector of another chord, then the first chord is a diameter.

10.9 Equidistant Chords Theorem

In the same circle, or in congruent circles, two chords are congruent if and only if they are equidistant from the center.

10.10 Measure of an Inscribed Angle Theorem

The measure of an inscribed angle is one-half the measure of its intercepted arc.

10.11 Inscribed Angles of a Circle Theorem

If two inscribed angles of a circle intercept the same arc, then the angles are congruent.

10.12 Inscribed Right Triangle Theorem

If a right triangle is inscribed in a circle, then the hypotenuse is a diameter of the circle. Conversely, if one side of an inscribed triangle is a diameter of the circle, then the triangle is a right triangle and the angle opposite the diameter is the right angle.

10.13 Inscribed Quadrilateral Theorem

A quadrilateral can be inscribed in a circle if and only if its opposite angles are supplementary.

10.14 Tangent and Intersected Chord Theorem

If a tangent and a chord intersect at a point on a circle, then the measure of each angle formed is one-half the measure of its intercepted arc.

10.15 Angles Inside the Circle Theorem

If two chords intersect inside a circle, then the measure of each angle is one-half the sum of the measures of the arcs intercepted by the angle and its vertical angle.

10.16 Angles Outside the Circle Theorem

If a tangent and a secant, two tangents, or two secants intersect outside a circle, then the measure of the angle formed is one-half the difference of the measures of the intercepted arcs.

10.17 Circumscribed Angle Theorem

The measure of a circumscribed angle is equal to 180° minus the measure of the central angle that intercepts the same arc.

10.18 Segments of Chords Theorem

If two chords intersect in the interior of a circle, then the product of the lengths of the segments of one chord is equal to the product of the lengths of the segments of the other chord.

10.19 Segments of Secants Theorem

If two secant segments share the same endpoint outside a circle, then the product of the lengths of one secant segment and its external segment equals the product of the lengths of the other secant segment and its external segment.

10.20 Segments of Secants and Tangents Theorem

If a secant segment and a tangent segment share an endpoint outside a circle, then the product of the lengths of the secant segment and its external segment equals the square of the length of the tangent segment.

Reference

Properties

Properties of Equality

Addition Property of Equality
If $a = b$, then $a + c = b + c$.

Subtraction Property of Equality
If $a = b$, then $a - c = b - c$.

Multiplication Property of Equality
If $a = b$, then $a \cdot c = b \cdot c, c \neq 0$.

Division Property of Equality
If $a = b$, then $\frac{a}{c} = \frac{b}{c}, c \neq 0$.

Reflexive Property of Equality
$a = a$

Symmetric Property of Equality
If $a = b$, then $b = a$.

Transitive Property of Equality
If $a = b$ and $b = c$, then $a = c$.

Substitution Property of Equality
If $a = b$, then a can be substituted for b (or b for a) in any equation or expression.

Properties of Segment and Angle Congruence

Reflexive Property of Congruence
For any segment AB, $\overline{AB} \cong \overline{AB}$.

For any angle A, $\angle A \cong \angle A$.

Symmetric Property of Congruence
If $\overline{AB} \cong \overline{CD}$, then $\overline{CD} \cong \overline{AB}$.

If $\angle A \cong \angle B$, then $\angle B \cong \angle A$.

Transitive Property of Congruence
If $\overline{AB} \cong \overline{CD}$ and $\overline{CD} \cong \overline{EF}$, then $\overline{AB} \cong \overline{EF}$.

If $\angle A \cong \angle B$ and $\angle B \cong \angle C$, then $\angle A \cong \angle C$.

Other Properties

Transitive Property of Parallel Lines
If $p \parallel q$ and $q \parallel r$, then $p \parallel r$.

Distributive Property
Sum
$a(b + c) = ab + ac$
Difference
$a(b - c) = ab - ac$

Triangle Inequalities

Triangle Inequality Theorem

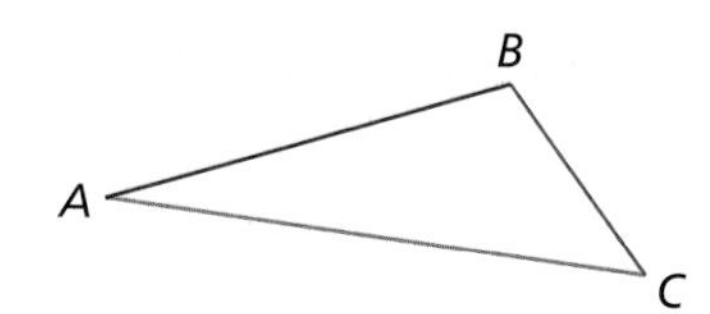

$AB + BC > AC$
$AC + BC > AB$
$AB + AC > BC$

Pythagorean Inequalities Theorem

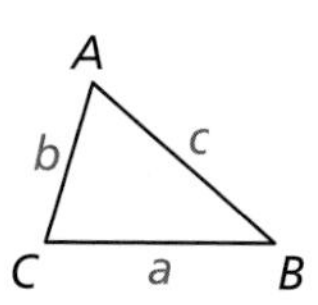

If $c^2 < a^2 + b^2$, then $\triangle ABC$ is acute.

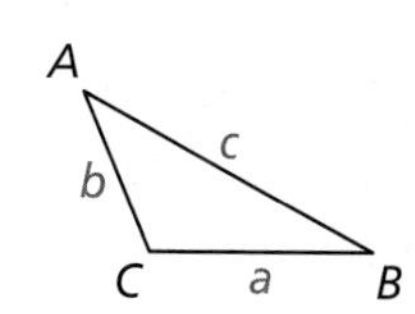

If $c^2 > a^2 + b^2$, then $\triangle ABC$ is obtuse.

Formulas

Coordinate Geometry

Slope

$m = \dfrac{y_2 - y_1}{x_2 - x_1}$

Slope-intercept form

$y = mx + b$

Point-slope form

$y - y_1 = m(x - x_1)$

Standard form of a linear equation

$Ax + By = C$

Midpoint Formula

$\left(\dfrac{x_1 + x_2}{2}, \dfrac{y_1 + y_2}{2}\right)$

Standard equation of a circle

$(x - h)^2 + (y - k)^2 = r^2$, with center (h, k) and radius r

Distance Formula

$d = \sqrt{(x_2 - x_1)^2 + (y_2 - y_1)^2}$

Polygons

Triangle Sum Theorem

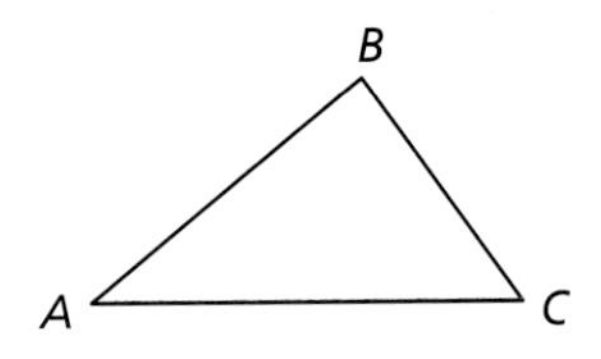

$m\angle A + m\angle B + m\angle C = 180°$

Exterior Angle Theorem

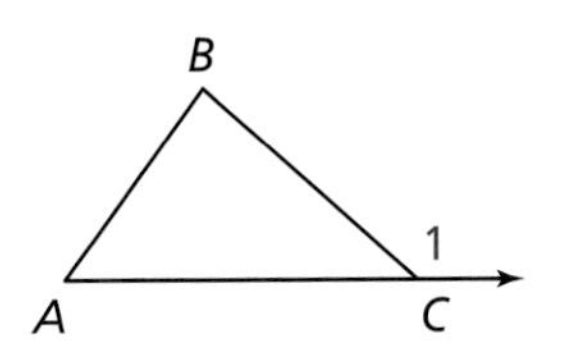

$m\angle 1 = m\angle A + m\angle B$

Triangle Midsegment Theorem

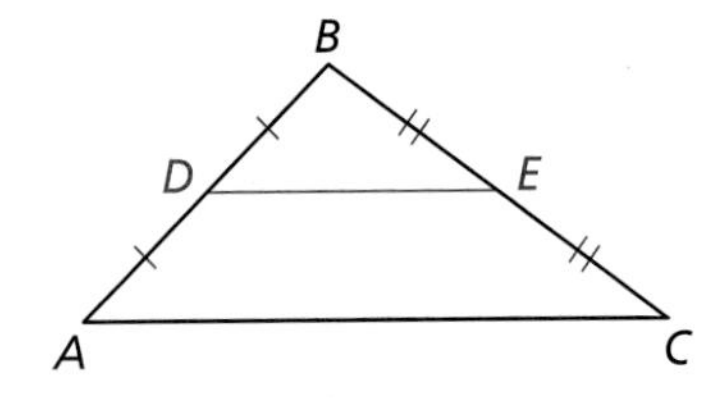

$\overline{DE} \parallel \overline{AC}$, $DE = \frac{1}{2}AC$

Trapezoid Midsegment Theorem

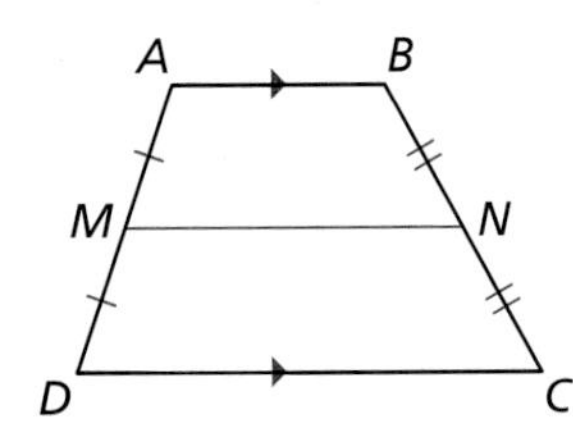

$\overline{MN} \parallel \overline{AB}, \overline{MN} \parallel \overline{DC}$, $MN = \frac{1}{2}(AB + CD)$

Polygon Interior Angles Theorem

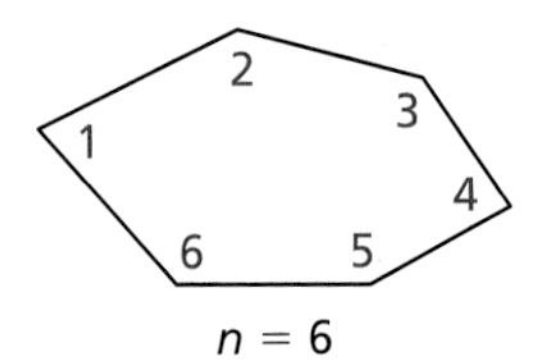

$m\angle 1 + m\angle 2 + \cdots + m\angle n = (n - 2) \cdot 180°$

Polygon Exterior Angles Theorem

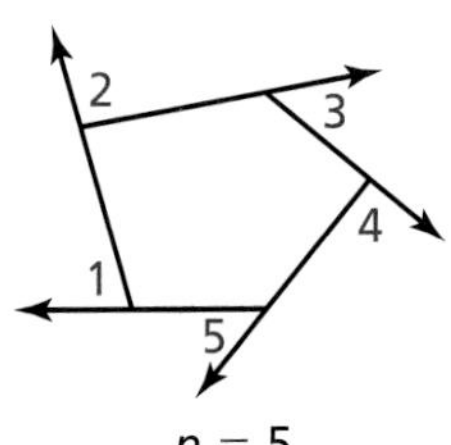

$m\angle 1 + m\angle 2 + \cdots + m\angle n = 360°$

Geometric Mean (Altitude) Theorem

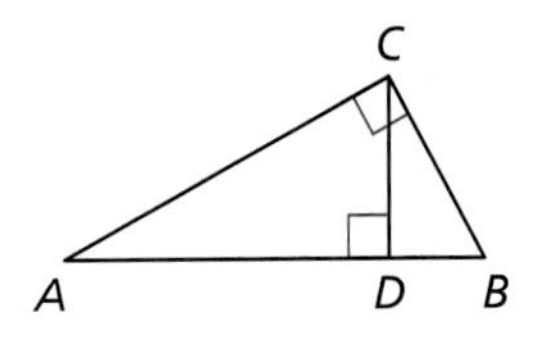

$CD^2 = AD \cdot BD$

Geometric Mean (Leg) Theorem

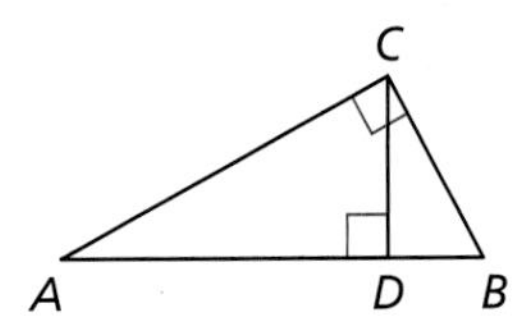

$CB^2 = DB \cdot AB$ $AC^2 = AD \cdot AB$

Right Triangles

Pythagorean Theorem

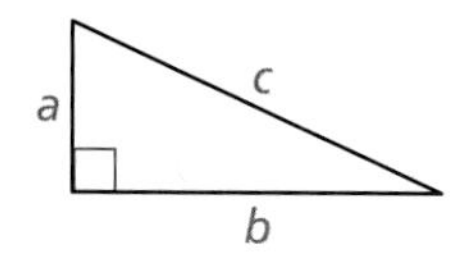

$a^2 + b^2 = c^2$

45°-45°-90° Triangles

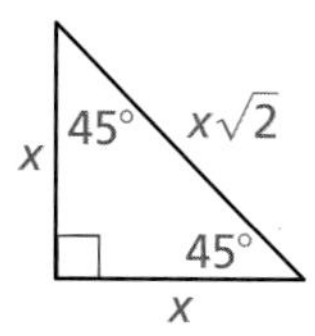

hypotenuse = leg • $\sqrt{2}$

30°-60°-90° Triangles

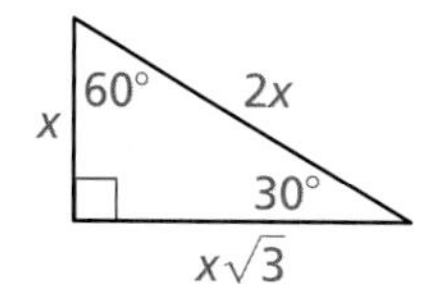

hypotenuse = shorter leg • 2
longer leg = shorter leg • $\sqrt{3}$

Trigonometry

Ratios

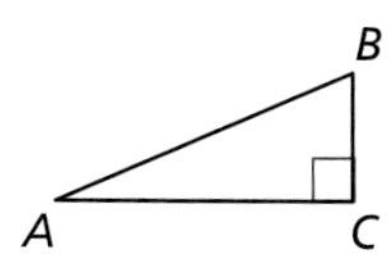

$\sin A = \frac{BC}{AB}$ $\cos A = \frac{AC}{AB}$ $\tan A = \frac{BC}{AC}$

$\sin^{-1}\frac{BC}{AB} = m\angle A$ $\cos^{-1}\frac{AC}{AB} = m\angle A$ $\tan^{-1}\frac{BC}{AC} = m\angle A$

Conversion between degrees and radians
$180° = \pi$ radians

Sine and cosine of complementary angles
Let A and B be complementary angles. Then the following statements are true.

$\sin A = \cos(90° - A) = \cos B$ $\sin B = \cos(90° - B) = \cos A$

$\cos A = \sin(90° - A) = \sin B$ $\cos B = \sin(90° - B) = \sin A$

Any Triangle

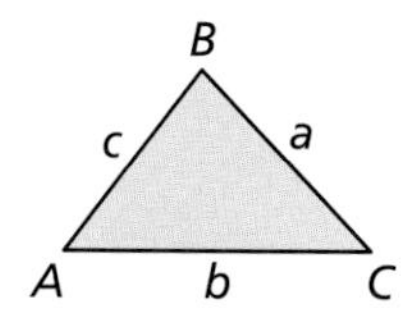

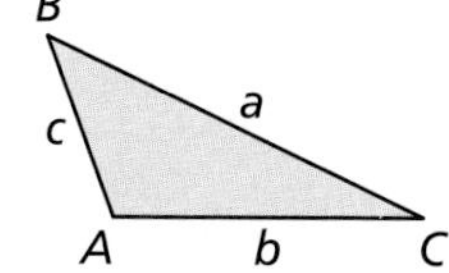

Area

$\text{Area} = \frac{1}{2}bc \sin A$

$\text{Area} = \frac{1}{2}ac \sin B$

$\text{Area} = \frac{1}{2}ab \sin C$

Law of Sines

$\frac{\sin A}{a} = \frac{\sin B}{b} = \frac{\sin C}{c}$

$\frac{a}{\sin A} = \frac{b}{\sin B} = \frac{c}{\sin C}$

Law of Cosines

$a^2 = b^2 + c^2 - 2bc \cos A$
$b^2 = a^2 + c^2 - 2ac \cos B$
$c^2 = a^2 + b^2 - 2ab \cos C$

Circles

Arc length

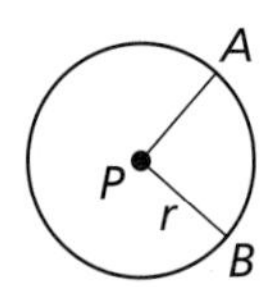

Arc length of $\widehat{AB} = \dfrac{m\widehat{AB}}{360°} \cdot 2\pi r$

Area of a sector

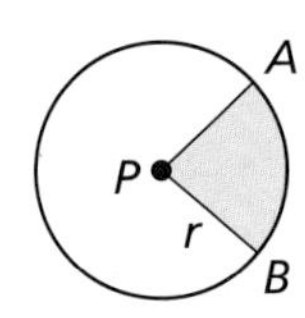

Area of sector $APB = \dfrac{m\widehat{AB}}{360°} \cdot \pi r^2$

Central angles

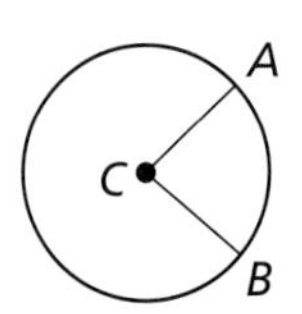

$m\angle ACB = m\widehat{AB}$

Inscribed angles

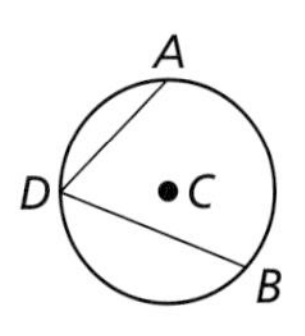

$m\angle ADB = \frac{1}{2}m\widehat{AB}$

Tangent and intersected chord

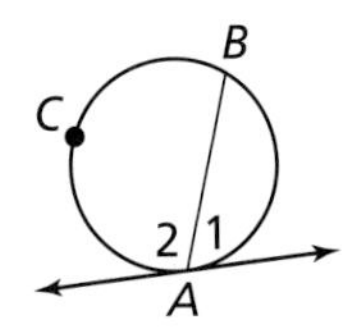

$m\angle 1 = \frac{1}{2}m\widehat{AB}$

$m\angle 2 = \frac{1}{2}m\widehat{BCA}$

Angles and Segments of Circles

Two chords

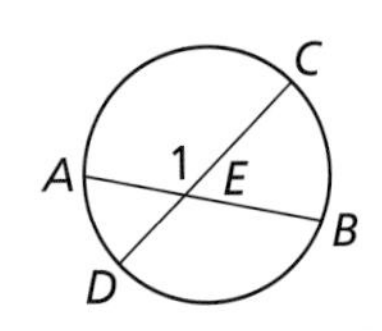

$m\angle 1 = \frac{1}{2}\left(m\widehat{AC} + m\widehat{DB}\right)$

$EA \cdot EB = EC \cdot ED$

Two secants

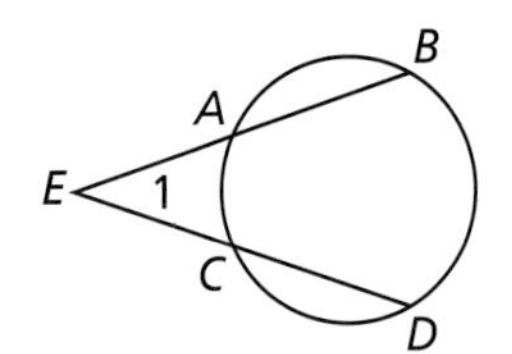

$m\angle 1 = \frac{1}{2}\left(m\widehat{BD} - m\widehat{AC}\right)$

$EA \cdot EB = EC \cdot ED$

Tangent and secant

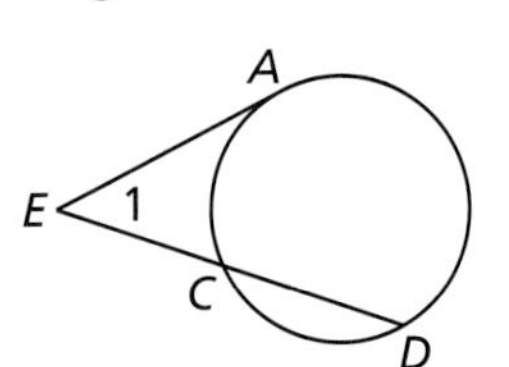

$m\angle 1 = \frac{1}{2}\left(m\widehat{AD} - m\widehat{AC}\right)$

$EA^2 = EC \cdot ED$

Two tangents

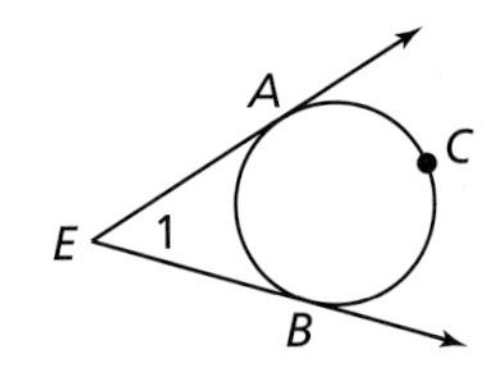

$m\angle 1 = \frac{1}{2}\left(m\widehat{ACB} - m\widehat{AB}\right)$

$EA = EB$

Probability and Combinatorics

Theoretical Probability $= \dfrac{\text{Number of favorable outcomes}}{\text{Total number of outcomes}}$

Experimental Probability $= \dfrac{\text{Number of successes}}{\text{Number of trials}}$

Probability of the complement of an event
$P(\overline{A}) = 1 - P(A)$

Probability of independent events
$P(A \text{ and } B) = P(A) \cdot P(B)$

Probability of dependent events
$P(A \text{ and } B) = P(A) \cdot P(B \mid A)$

Probability of compound events
$P(A \text{ or } B) = P(A) + P(B) - P(A \text{ and } B)$

Permutations

$${}_nP_r = \frac{n!}{(n-r)!}$$

Combinations

$${}_nC_r = \frac{n!}{(n-r)! \cdot r!}$$

Binomial experiments

$$P(k \text{ successes}) = {}_nC_k p^k (1-p)^{n-k}$$

Perimeter, Area, and Volume Formulas

Square

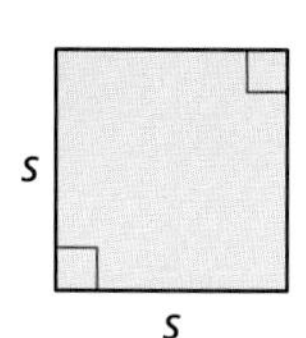

$P = 4s$
$A = s^2$

Rectangle

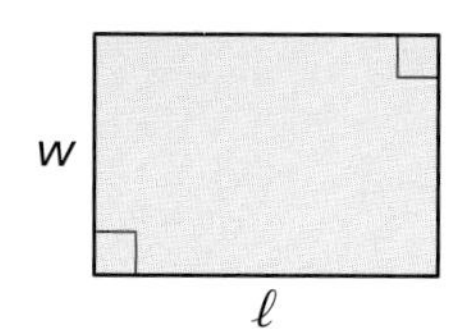

$P = 2\ell + 2w$
$A = \ell w$

Triangle

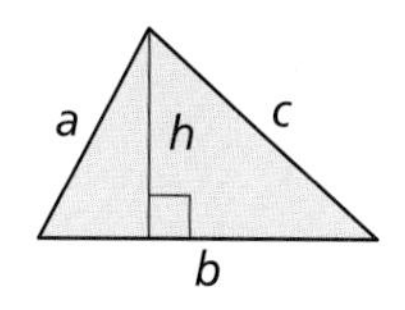

$P = a + b + c$
$A = \frac{1}{2}bh$

Circle

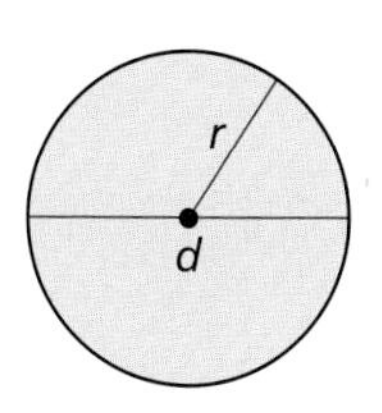

$C = \pi d$ or $C = 2\pi r$
$A = \pi r^2$

Parallelogram

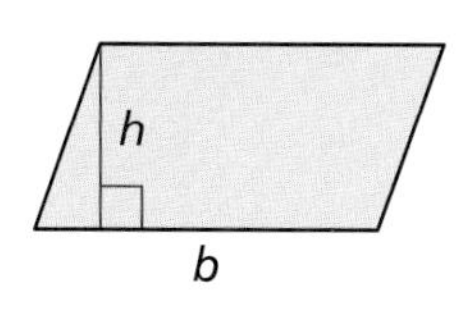

$A = bh$

Trapezoid

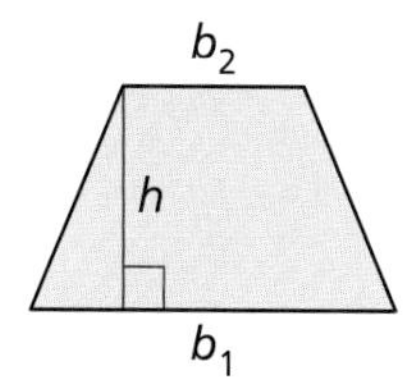

$A = \frac{1}{2}h(b_1 + b_2)$

Rhombus/Kite

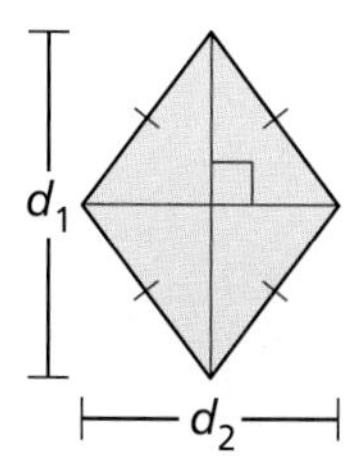

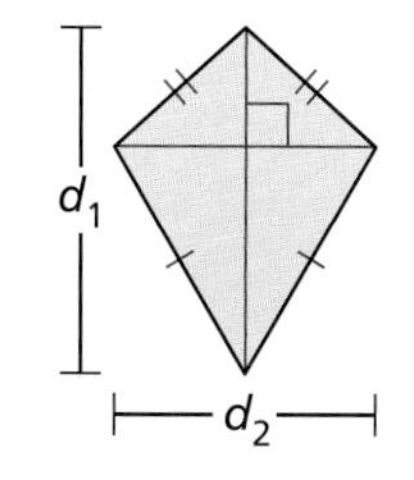

$A = \frac{1}{2}d_1d_2$

Regular *n*-gon

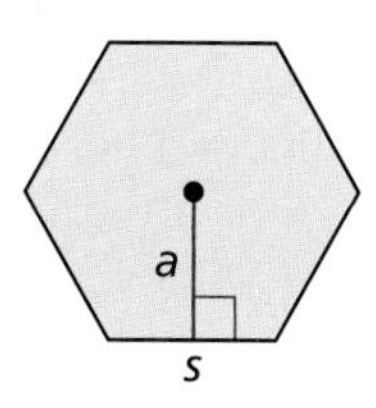

$A = \frac{1}{2}aP$ or $A = \frac{1}{2}a \cdot ns$

Prism

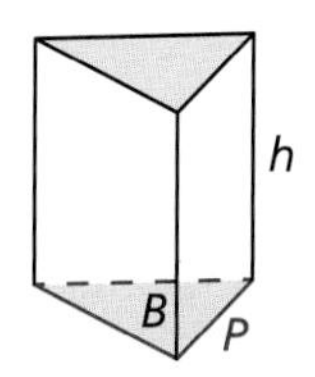

$L = Ph$
$S = 2B + Ph$
$V = Bh$

Cylinder

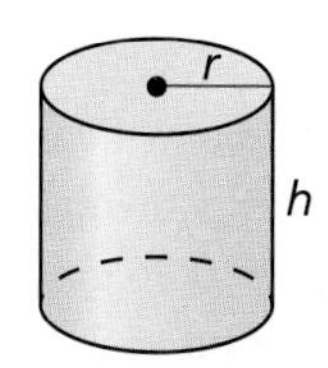

$L = 2\pi rh$
$S = 2\pi r^2 + 2\pi rh$
$V = \pi r^2h$

Pyramid

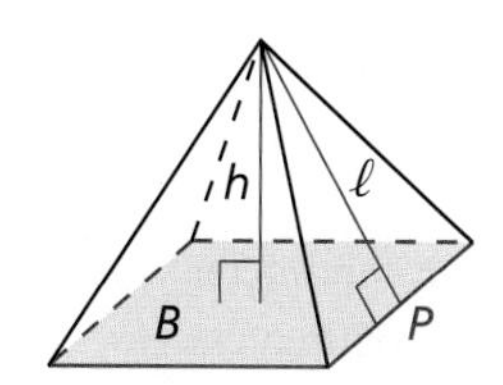

$L = \frac{1}{2}P\ell$
$S = B + \frac{1}{2}P\ell$
$V = \frac{1}{3}Bh$

Cone

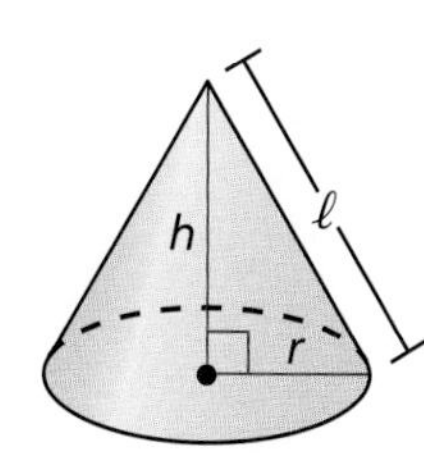

$L = \pi r\ell$

$S = \pi r^2 + \pi r\ell$

$V = \frac{1}{3}\pi r^2h$

Sphere

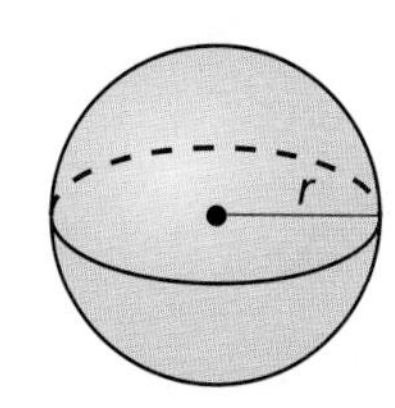

$S = 4\pi r^2$

$V = \frac{4}{3}\pi r^3$

Reference

Other Formulas

Geometric mean

$x = \sqrt{a \cdot b}$

Quadratic Formula

$x = \dfrac{-b \pm \sqrt{b^2 - 4ac}}{2a}$,

where $a \neq 0$ and $b^2 - 4ac \geq 0$

Density

$\text{Density} = \dfrac{\text{Mass}}{\text{Volume}}$

Similar polygons or similar solids with scale factor $a : b$

Ratio of perimeters $= a : b$

Ratio of areas $= a^2 : b^2$

Ratio of volumes $= a^3 : b^3$

Conversions

U.S. Customary

1 foot = 12 inches
1 yard = 3 feet
1 mile = 5280 feet
1 mile = 1760 yards
1 acre = 43,560 square feet
1 cup = 8 fluid ounces
1 pint = 2 cups
1 quart = 2 pints
1 gallon = 4 quarts
1 gallon = 231 cubic inches
1 pound = 16 ounces
1 ton = 2000 pounds

U.S. Customary to Metric

1 inch = 2.54 centimeters
1 foot ≈ 0.3 meter
1 mile ≈ 1.61 kilometers
1 quart ≈ 0.95 liter
1 gallon ≈ 3.79 liters
1 cup ≈ 237 milliliters
1 pound ≈ 0.45 kilogram
1 ounce ≈ 28.3 grams
1 gallon ≈ 3785 cubic centimeters

Time

1 minute = 60 seconds
1 hour = 60 minutes
1 hour = 3600 seconds
1 year = 52 weeks

Temperature

$C = \frac{5}{9}(F - 32)$

$F = \frac{9}{5}C + 32$

Metric

1 centimeter = 10 millimeters
1 meter = 100 centimeters
1 kilometer = 1000 meters
1 liter = 1000 milliliters
1 kiloliter = 1000 liters
1 milliliter = 1 cubic centimeter
1 liter = 1000 cubic centimeters
1 cubic millimeter = 0.001 milliliter
1 gram = 1000 milligrams
1 kilogram = 1000 grams

Metric to U.S. Customary

1 centimeter ≈ 0.39 inch
1 meter ≈ 3.28 feet
1 meter ≈ 39.37 inches
1 kilometer ≈ 0.62 mile
1 liter ≈ 1.06 quarts
1 liter ≈ 0.26 gallon
1 kilogram ≈ 2.2 pounds
1 gram ≈ 0.035 ounce
1 cubic meter ≈ 264 gallons

Credits

Front Matter

viii © Francisco Javier Alcerreca Gomez | Dreamstime.com; **ix** Colette3/Shutterstock.com; **x** © Kaehub | Dreamstime.com; **xi** ©iStockphoto.com/bradwieland; **xii** Edimur/Shutterstock.com; **xiii** sculpies/Shutterstock.com; **xiv** FreshPaint/Shutterstock.com; **xv** koh sze kiat/Shutterstock.com; **xvi** ©iStockphoto.com/Pamela Moore; **xvii** ©iStockphoto.com/-hakusan-; **xviii** CristinaMuraca/Shutterstock.com; **xix** Monkey Business Images/Shutterstock.com

Chapter 1

0 *top left* © Francisco Javier Alcerreca Gomez | Dreamstime.com; *top right* CHEN WS/Shutterstock.com; *center left* Dimitar Kunev/Shutterstock.com; *bottom right* Maxim Blinkov/Shutterstock.com; *bottom left* Sukpaiboonwat/Shutterstock.com; **7** *top right* Benjah-bmm27; *top left* Sukpaiboonwat/Shutterstock.com; *bottom right* Benjah-bmm27; **9** *Exercise 35* koosen/Shutterstock.com; *Exercise 36* Vladru/Shutterstock.com, Jeff Schultes/Shutterstock.com; *Exercise 37* Winston Link/Shutterstock.com; *Exercise 38* Nyord/Shutterstock.com; **17** paulrommer/Shutterstock.com; **20** Maxim Blinkov/Shutterstock.com; **25** Mark Herreid/Shutterstock.com; **27** Aleksander Erin/Shutterstock.com; **33** Dimitar Kunev/Shutterstock.com; **40** malamalama/Shutterstock.com; **45** Lewis Calvert Cooper (*gonzoshots.com*); **46** Ruslan Gi/Shutterstock.com; **49** CHEN WS/Shutterstock.com; **53** *center left* © Francisco Javier Alcerreca Gomez | Dreamstime.com; *bottom left* Shane Trotter/Shutterstock.com; **55** pirita/Shutterstock.com

Chapter 2

62 *top left* Josef Hanus/Shutterstock.com; *top right* 60 Degrees, sculpture by Kevin O'Dwyer, Sculpture in the Parklands, Ireland; *center left* spirit of america/Shutterstock.com; *bottom right* Colette3/Shutterstock.com; *bottom left* sint/Shutterstock.com; **67** sint/Shutterstock.com; **72** *Exercise 45 top* mikeledray/Shutterstock.com; *Exercise 45 center and bottom* Tyler Boyes/Shutterstock.com; **73** Darren J. Bradley/Shutterstock.com; **81** BMCL/Shutterstock.com; **82** Siim Sepp/Shutterstock.com, © Vevesoran | Dreamstime.com; **89** Aleksander Erin/Shutterstock.com; **90** Kolopach/Shutterstock.com; **95** spirit of america/Shutterstock.com; **104** *top left* keko-ka/Shutterstock.com, tele52/Shutterstock.com; *bottom left* 60 Degrees, sculpture by Kevin O'Dwyer, Sculpture in the Parklands, Ireland; **108** Josef Hanus/Shutterstock.com; **113** kastianz/Shutterstock.com; **115** TsuneoMP/Shutterstock.com; **119** Natykach Nataliia/Shutterstock.com; **121** JeniFoto/Shutterstock.com

Chapter 3

122 *top left* gyn9037/Shutterstock.com; *top right* CC BY-NC-SA/ElizabethGarbee; *center left* EpicStockMedia/Shutterstock.com; *bottom right* © Kaehub | Dreamstime.com; *bottom left* ©iStockphoto.com/Brian McEntire; **129** Iraidak/Shutterstock.com; **130** *center left* © Kaehub | Dreamstime.com; *bottom right* ©iStockphoto.com/skynesher; **141** bikeriderlondon/Shutterstock.com; **143** *center left* Aleksangel/Shutterstock.com; *bottom left* EpicStockMedia/Shutterstock.com; **145** Aleksander Erin/Shutterstock.com; **146** © Bingram | Dreamstime.com; **151** *top* ©iStockphoto.com/Renphoto; *bottom* ©iStockphoto.com/kokouu; **153** *Exercise 21* Palto/Shutterstock.com; *Exercise 22* Mircea Maties/Shutterstock.com; **161** gyn9037/Shutterstock.com; **163** Irmairma/Shutterstock.com; **167** © Chris Eriksson | Dreamstime.com

Chapter 4

170 *top left* Roman Gorielov/Shutterstock.com; *top right* Odua Images/Shutterstock.com; *center left* Sybille Yates/Shutterstock.com; *bottom right* ©iStockphoto.com/bradwieland; *bottom left* Dima Sobko/Shutterstock.com; **179** Tribalium/Shutterstock.com; **181** Big Ideas Learning, LLC; **185** *top left* Butterfly Hunter/Shutterstock.com; *bottom right* Melinda Fawver/Shutterstock.com; **195** ©iStockphoto.com/bradwieland; **196** *Exercise 35a* Laura Mountainspring/Shutterstock.com; *Exercise 35b* Sybille Yates/Shutterstock.com; *top right* nito/Shutterstock.com; **197** Aleksander Erin/Shutterstock.com; **205** Dario Sabljak/Shutterstock.com; **211** *top right* Odua Images/Shutterstock.com; *center right and bottom left* Henrik Larsson/Shutterstock.com; **213** *Exercise 31* sarra22/Shutterstock.com; *Exercise 32* irin-k/Shutterstock.com; *Exercise 33* paulrommer/Shutterstock.com; *Exercise 34 and Exercise 35 top and bottom left* Evgeniy Ayupov/Shutterstock.com; *Exercise 35 top right* irin-k/Shutterstock.com; *Exercise 35 bottom right* Ambient Ideas/Shutterstock.com; **221** ©iStockphoto.com/OlgaChertova; **225** *Exercise 7* serg_dibrova/Shutterstock.com; *Exercise 8* Africa Studio/Shutterstock.com; *bottom right* greggsphoto/Shutterstock.com

Chapter 5

228 *top left* © Edimur/Shutterstock.com; *top right* Andrey Smirnov/Shutterstock.com; *center left* MaxyM/Shutterstock.com; *bottom right* racorn/Shutterstock.com; *bottom left* clivewa/Shutterstock.com; **232** © Ralf Broskvar | Dreamstime.com; **235** clivewa/Shutterstock.com; **241** racorn/Shutterstock.com; **248** MaxyM/Shutterstock.com; **255** Andrey Smirnov/Shutterstock.com; **257** *top left* Jeff Whyte/Shutterstock.com; *Exercise 23* Lucie Lang/Shutterstock.com; *Exercise 24* Yezepchyk Oleksandr/Shutterstock.com; **259** Aleksander Erin/Shutterstock.com; **268** *center left* Victoria Kalinina/Shutterstock.com; *center right* Koksharov Dmitry/Shutterstock.com; **276** szefei/Shutterstock.com; **278** Edimur/Shutterstock.com; **279** Kert/Shutterstock.com; **282** *top left* pisaphotography/Shutterstock.com; *top right* MarcelClemens/Shutterstock.com; **288** Mikael Damkier/Shutterstock.com; **289** wanpatsorn/Shutterstock.com; **295** Viacheslav Lopatin/Shutterstock.com

Chapter 6

298 *top left* Warren Goldswain/Shutterstock.com; *top right* puttsk/Shutterstock.com; *center left* sculpies/Shutterstock.com; *bottom right* Olena Mykhaylova/Shutterstock.com; *bottom left* Nikonaft/Shutterstock.com; **303** Nikonaft/Shutterstock.com; **305** sababa66/Shutterstock.com; **307** Tobias W/Shutterstock.com; **311** filip robert/Shutterstock.com, John T Takai/Shutterstock.com; **316** *top right* Matthew Cole/Shutterstock.com; *center right* artkamalov/Shutterstock.com, anton_novik/Shutterstock.com; **318** *center left* Olena Mykhaylova/Shutterstock.com; *top right* Lonely/Shutterstock.com; **325** Laschon Maximilian/Shutterstock.com; **327** Aleksander Erin/Shutterstock.com; **331** sculpies/Shutterstock.com; **332** Shumo4ka/Shutterstock.com; **333** Alexandra Lande/Shutterstock.com, Dan Kosmayer/Shutterstock.com; **334** clearviewstock/Shutterstock.com; **338** Dzm1try/Shutterstock.com; **344** ©iStockphoto.com/enchanted_glass; **346** Warren Goldswain/Shutterstock.com; **349** Portokalis/Shutterstock.com

Chapter 7

356 *top left* FreshPaint/Shutterstock.com; *top right* © Swinnerrr | Dreamstime.com; *center left* CC-PD by Ppntori; *bottom right* nacroba/ Shutterstock.com; *bottom left* Reprinted with permission of The Photo News.; **360** © Peter Spirer | Dreamstime.com; **365** Reprinted with permission of The Photo News.; **370** Marco Prati/Shutterstock.com; **377** ©iStockphoto.com/A-Digit; **378** Ann Paterson; **383** kokandr/ Shutterstock.com; **385** Aleksander Erin/Shutterstock.com; **386** Matthew Cole/Shutterstock.com; **393** *Exercise 9* Ksenia Palimski/Shutterstock.com; *Exercise 10* Sundraw Photography/Shutterstock.com; *Exercise 11* EvenEzer/Shutterstock.com; *Exercise 12* donatas1205/Shutterstock.com; **395** © Swinnerrr | Dreamstime.com; **407** © Dmitry Kalinovsky | Dreamstime.com

Chapter 8

414 *top left* © Luis Santos | Dreamstime.com; *top right* Sfocato/ Shutterstock.com; *center left* Vicky SP/Shutterstock.com; *bottom right* © Paul Maguire | Dreamstime.com; *bottom left* koh sze kiat/ Shutterstock.com; **420** koh sze kiat/Shutterstock.com; **426** pilgrim.artworks/ Shutterstock.com; **430** Rigucci/Shutterstock.com; **433** Aleksander Erin/ Shutterstock.com; **434** Sarawut Padungkwan/Shutterstock.com; **438** bkp/Shutterstock.com; **443** *top left* © Luis Santos | Dreamstime.com; *top right* Sfocato/Shutterstock.com; **453** Felix Mizioznikov/ Shutterstock.com; **457** *center right* Stuart Monk/Shutterstock.com, Viorel Sima/Shutterstock.com; *bottom right* ssguy/Shutterstock.com

Chapter 9

460 *top left* Samot/Shutterstock.com; *top right* ©iStockphoto.com/ Sportstock; *center left* © Vacclav | Dreamstime.com; *bottom right* ©iStockphoto.com/Pamela Moore; *bottom left* robert paul van beets/ Shutterstock.com; **469** *top left* robert paul van beets/Shutterstock.com; *center left* Naypong/Shutterstock.com; *center right* ©iStockphoto.com/ susandaniels; **475** AND Inc/Shutterstock.com; **481**©iStockphoto.com/ Pamela Moore; **482** *Exercise 9* © Jorg Hackemann | Dreamstime.com; *Exercise 10* © Det-anan Sunonethong | Dreamstime.com; **483** © Alex Zarubin | Dreamstime.com, Kamenetskiy Konstantin/ Shutterstock.com; **485** Aleksander Erin/Shutterstock.com; **491** JASON TENCH/Shutterstock.com; **497** *top left* ©iStockphoto.com/ Sportstock; *top right* IPFL/Shutterstock.com; **499** ©iStockphoto.com/ollo; **500** robin2/Shutterstock.com, Vladimir Zadvinskii/Shutterstock.com, Natali Snailcat/Shutterstock.com; **514** *center left* ©iStockphoto.com/4x6; *bottom left* NEILRAS/Shutterstock.com, Mariya Ermolaeva/ Shutterstock.com; *center right* Christopher Salerno/Shutterstock.com; **515** FloridaStock/Shutterstock.com; **517** Martin Good/Shutterstock.com; **523** *bottom right* Andrew McDonough/Shutterstock.com, zzveillust/ Shutterstock.com; *bottom right* Alberto Loyo/Shutterestock.com, Bairachnyi Dmitry/Shutterstock.com; **525** Nestor Noci/Shutterstock.com

Chapter 10

526 *top left* ©iStockphoto.com/-hakusan-; *top right* © Aaron Rutten | Dreamstime.com; *center left* ©iStockphoto.com/falun; *bottom right* FikMik/ Shutterstock.com; *bottom left* ©iStockphoto.com/martin_k; **529** ©iStockphoto.com/Rebirth3d; **535** kbgen/Shutterstock.com; **543** Markus Gann/Shutterstock.com; **549** *top right* TsuneoMP/ Shutterstock.com, Kjpargeter/Shutterstock.com; *bottom right* M. Unal Ozmen/Shutterstock.com; **550** ©iStockphoto.com/falun; **551** Aleksander Erin/Shutterstock.com; **552** tale/Shutterstock.com; **559** © Vadim Yerofeyev | Dreamstime.com, Andrey Eremin/ Shutterstock.com; **565** leonello calvetti/Shutterstock.com; **567** Sergey Nivens/Shutterstock.com, sonya etchison/Shutterstock.com; **573** OHishiapply/Shutterstock.com, Paul B. Moore/Shutterstock.com, BlueRingMedia/Shutterstock.com; **574** *center left* © Stbernarstudio | Dreamstime.com; *bottom left* Alhovik/Shutterstock.com; **578** ©iStockphoto.com/-hakusan-; **581** lexaarts/Shutterstock.com

Chapter 11

590 *top left* Pius Lee/Shutterstock.com; *top right* Kotomiti Okuma/ Shutterstock.com; *center left* NCG/Shutterstock.com; *bottom right* CristinaMuraca/Shutterstock.com; *bottom left* maigi/Shutterstock.com; **593** Mauro Rodrigues/Shutterstock.com; **596** Gwoeii/Shutterstock.com; **598** *top right* ©iStockphoto.com/Prill Mediendesign & Fotografie; *center right* Gena73/Shutterstock.com; **599** Mikio Oba/Shutterstock.com; **600** ©iStockphoto.com/Charles Mann; **601** *bottom left* Elena Elisseeva/ Shutterstock.com*; bottom right* © Sam Beebe / CC-BY-2.0; **603** CristinaMuraca/Shutterstock.com; **608** *top left* Victoria Kalinina/ Shutterstock.com; **613** *top left* Baloncici/Shutterstock.com; *top right* ©iStockphoto.com/Olgertas, keellla/Shutterstock.com; **615** *center left* NCG/ Shutterstock.com; *bottom left* Laschon Maximilian/Shutterstock.com; **622** LesPalenik/Shutterstock.com; **623** Aleksander Erin/Shutterstock.com; **628** Krasowit/Shutterstock.com; **629** *top righ* © Bombaert | Dreamstime.com; *bottom left* Photobac/Shutterstock.com; **631** *Exercise 13 left* Zelenskaya/Shutterstock.com; *Exercise 13 right* Kompaniets Taras/Shutterstock.com; *bottom right* Patryk Kosmider/ Shutterstock.com; **633** *top left* Wata51/Shutterstock.com; *bottom right* ronstik/Shutterstock.com, YuriyZhuravov/Shutterstock.com, Petrenko Andriy/Shutterstock.com, Rafa Irusta/Shutterstock.com; **634** © Fallsview | Dreamstime.com; **637** Pius Lee/Shutterstock.com; **640** © Richard Lowthian | Dreamstime.com; **647** Mark Herreid/ Shutterstock.com; **648** Dan Thornberg/Shutterstock.com; **653** *Exercise 27* Nomad_Soul/Shutterstock.com; *Exercise 28* Lightspring/ Shutterstock.com; *Exercise 29* Mark Herreid/Shutterstock.com; *Exercise 30* Keattikorn/Shutterstock.com; *Exercise 31* Lightspring/Shutterstock.com; *Exercise 32* Dan Thornberg/Shutterstock.com; *top right* MaxyM/ Shutterstock.com; **654** leonello calvetti/Shutterstock.com; **655** haveseen/ Shutterstock.com; **661** viritphon/Shutterstock.com; **662** *bottom left* Lucie Lang/Shutterstock.com; *bottom right* Stephanie Frey/Shutterstock.com; **663** *top right* Orhan Cam/Shutterstock.com, pmphoto/Shutterstock.com; *center* Pavel L Photo and Video/Shutterstock.com; *bottom right* neelsky/ Shutterstock.com

Chapter 12

664 *top left* Margaret M Stewart/Shutterstock.com; *top right* Mikhail Pogosov/Shutterstock.com; *center left* Nadiia Gerbish/ Shutterstock.com; *bottom right* Tyler Olson/Shutterstock.com; *bottom left* bikeriderlondon/Shutterstock.com; **667** *top* Sascha Burkard/ Shutterstock.com; *center* Ruslan Gi/Shutterstock.com; *bottom right* cobalt88/Shutterstock.com, Sanchai Khudpin/Shutterstock.com; **674** ©iStockphoto.com/hugolacasse; **675** *top right* cTermit/ Shutterstock.com; *top center* Picsfive/Shutterstock.com; **678** ©iStockphoto.com/andriikoval; **680** *center left* Marques/ Shutterstock.com, Laurin Rinder/Shutterstock.com; *bottom left* oliveromg/ Shutterstock.com; **682** *center left* cobalt88/Shutterstock.com; *center right* bikeriderlondon/Shutterstock.com; **687** Tyler Olson/Shutterstock.com; **688** *bottom left* Andrey Bayda/Shutterstock.com; *bottom right* Image Point Fr/Shutterstock.com; **691** Aleksander Erin/Shutterstock.com; **693** CrackerClips Stock Media/Shutterstock.com; **697** ©iStockphoto.com/ Andy Cook; **698** *top left* Nadiia Gerbish/Shutterstock.com; *center right* bikeriderlondon/Shutterstock.com; **701** Mikhail Pogosov/Shutterstock.com; **703** *center top* Alan Bailey/Shutterstock.com; *center bottom* ©iStockphoto.com/zorani; **705** Tony Oshlick/Shutterstock.com; **706** Telnov Oleksii/Shutterstock.com; **707** Sascha Burkard/ Shutterstock.com; **709** claudiofichera/Shutterstock.com; **711** Margaret M Stewart/Shutterstock.com; **712** ecco/Shutterstock.com; **713** Dmitry Melnikov/Shutterstock.com; **717** Christos Georghiou/ Shutterstock.com

Additional Topic

727 inxti/Shutterstock.com